Alpena Co. Library
211 N. First Ave.
Alpena, MI 49707

C0-DYB-453

SECOND EDITION

Quagga *and* Zebra Mussels

Biology Control

CRC Press is a
Taylor & Francis business

The opinions expressed in this book are solely those of the authors and do not reflect the policy or official opinions of NOAA, USGS, Department of Commerce or any other agency of the Federal Government.

Cover Image Credit: Chris Houghton, Rob Paddock, and John Janssen, University of Wisconsin-Milwaukee, School of Freshwater Sciences. This image of quagga mussels (profunda/deepwater morph) presents a composite of photos taken on the southern slope of Northeast Reef in Lake Michigan (43° 45.11′N, 87° 34.70′W) on April 25, 2012. Image photos were taken at a depth of 52 m with a deepwater ROV and clearly show the elongated siphons of the profunda morph.

CRC Press
Taylor & Francis Group
6000 Broken Sound Parkway NW, Suite 300
Boca Raton, FL 33487-2742

© 2014 by Taylor & Francis Group, LLC
CRC Press is an imprint of Taylor & Francis Group, an Informa business

No claim to original U.S. Government works

Printed on acid-free paper
Version Date: 20130912

International Standard Book Number-13: 978-1-4398-5436-5 (Hardback)

This book contains information obtained from authentic and highly regarded sources. Reasonable efforts have been made to publish reliable data and information, but the author and publisher cannot assume responsibility for the validity of all materials or the consequences of their use. The authors and publishers have attempted to trace the copyright holders of all material reproduced in this publication and apologize to copyright holders if permission to publish in this form has not been obtained. If any copyright material has not been acknowledged please write and let us know so we may rectify in any future reprint.

Except as permitted under U.S. Copyright Law, no part of this book may be reprinted, reproduced, transmitted, or utilized in any form by any electronic, mechanical, or other means, now known or hereafter invented, including photocopying, microfilming, and recording, or in any information storage or retrieval system, without written permission from the publishers.

For permission to photocopy or use material electronically from this work, please access www.copyright.com (http://www.copyright.com/) or contact the Copyright Clearance Center, Inc. (CCC), 222 Rosewood Drive, Danvers, MA 01923, 978-750-8400. CCC is a not-for-profit organization that provides licenses and registration for a variety of users. For organizations that have been granted a photocopy license by the CCC, a separate system of payment has been arranged.

Trademark Notice: Product or corporate names may be trademarks or registered trademarks, and are used only for identification and explanation without intent to infringe.

Library of Congress Cataloging-in-Publication Data

Zebra mussels.
  Quagga and zebra mussels : biology, impacts, and control / editors, Thomas F. Nalepa and Don Schloesser. -- Second edition.
       pages cm
  "A CRC title, part of the Taylor & Francis imprint, a member of the Taylor & Francis Group, the academic division of T&F Informa plc."
       Includes bibliographical references and index.
  ISBN 978-1-4398-5436-5 (hardcover : acid-free paper) 1. Zebra mussel. 2. Quagga mussel. 3. Zebra mussel--Control. 4. Quagga mussel--Control. 5. Zebra mussel--Environmental aspects. 6. Quagga mussel--Environmental aspects. I. Nalepa, T. F. II. Schloesser, Donald W. III. Title.

  QL430.7.D8Z45 2013
  594'.4--dc23   2013017601

Visit the Taylor & Francis Web site at
http://www.taylorandfrancis.com

and the CRC Press Web site at
http://www.crcpress.com

We may infer what havoc the introduction of any new beast of prey might cause in a country before the instincts of the indigenous inhabitants become adapted to the strangers' craft and power.

Charles Darwin (1834)

Contents

Foreword xi
Preface.. xiii
Acknowledgments... xv
Editors.. xvii
Contributors xix

**Part I
Prelude**

Chapter
My Story on Finding the First Zebra Mussel (*Dreissena polymorpha*) in North America....................... 3

Sonya Gutschi Santavy

In Recognition: John Glenn Sparks a Giant Leap for Environmental Protection 5

Allegra Cangelosi

**Part II
Distribution and Spread**

Chapter 1
Chronological History of Zebra and Quagga Mussels (Dreissenidae) in North America, 1988–2010 9

Amy J. Benson

Chapter 2
Influence of Environmental Factors on Zebra Mussel Population Expansion in Lake Champlain, 1994–2010 33

J. Ellen Marsden, Pete Stangel, and Angela D. Shambaugh

Chapter 3
Replacement of Zebra Mussels by Quagga Mussels in the Erie Canal, New York, USA 55

Kenton M. Stewart

Chapter 4
Invasion of Quagga Mussels (*Dreissena rostriformis bugensis*) to the Mid-Lake Reef Complex
in Lake Michigan: A Photographic Montage 65

Jeffrey S. Houghton, Robert Paddock, and John Janssen

Chapter 5
Long-Term Change in the Hudson River's Bivalve Populations: A History of Multiple Invasions (and Recovery?)............71

David L. Strayer and Heather M. Malcom

Chapter 6
Spread of the Quagga Mussel (*Dreissena rostriformis bugensis*) in Western Europe 83

Abraham bij de Vaate, Gerard van der Velde, Rob S.E.W. Leuven, and Katharina C.M. Heiler

v

Chapter 7
Origin and Spread of Quagga Mussels (*Dreissena rostriformis bugensis*) in Eastern Europe with Notes on Size Structure of Populations 93

Marina I. Orlova

Chapter 8
Summary of Zebra Mussel *(Dreissena polymorpha)* in Polish Lakes over the Past 50 Years with Emphasis on the Masurian Lakes (Northeastern Poland) 103

Krzysztof Lewandowski and Anna Stańczykowska

Part III
Response, Management, and Mitigation

Chapter 9
One Reporter's Perspective on the Invasion of Dreissenid Mussels in North America: Reflections on the News Media ... 117

Steve Pollick

Chapter 10
Early Responses to Zebra Mussels in the Great Lakes: A Journey from Information Vacuum to Policy and Regulation 135

Ronald W. Griffiths, Don W. Schloesser, and William P. Kovalak

Chapter 11
Catalyst for Change: "The Little Dreissenid That Did" (Change National Policy on Aquatic Invasive Species) 177

David F. Reid and Dean Wilkinson

Chapter 12
Invading Dreissenid Mussels Transform the 100-Year-Old International Joint Commission 187

Mark J. Burrows

Chapter 13
Eradication of Zebra Mussels *(Dreissena polymorpha)* from Millbrook Quarry, Virginia: Rapid Response in the Real World 195

Raymond T. Fernald and Brian T. Watson

Chapter 14
Management and Control of Dreissenid Mussels in Water Infrastructure Facilities of the Southwestern United States 215

Rajat K. Chakraborti, Sharook Madon, Jagjit Kaur, and Dale Gabel

Chapter 15
Impact of Dreissenid Mussels on the Infrastructure of Dams and Hydroelectric Power Plants 243

Thomas H. Prescott, Renata Claudi, and Katherine L. Prescott

Chapter 16
Managing Expansion of Dreissenids within Traditional Parameters: The Story of Quagga Mussels in Lake Mead National Recreation Area 259

Valerie Hickey

Chapter 17
Dreissenid Mussels as Sentinel Biomonitors for Human and Zoonotic Pathogens 265

David Bruce Conn, Frances E. Lucy, and Thaddeus K. Graczyk

Chapter 18
Contaminant Concentrations in Dreissenid Mussels from the Laurentian Great Lakes: A Summary of Trends from the Mussel Watch Program 273

Kimani L. Kimbrough, W. Edward Johnson, Annie P. Jacob, and Gunnar G. Lauenstein

Part IV
Morphology, Physiology, and Behavior

Chapter 19
Morphological Variability of *Dreissena polymorpha* and *Dreissena rostriformis bugensis* (Mollusca: Bivalvia) 287

Vera Pavlova and Yuri Izyumov

Chapter 20
Variation in the Quagga Mussel (*Dreissena rostriformis bugensis*) with Emphasis on the Deepwater Morphotype in Lake Michigan 315

Thomas F. Nalepa, Vera Pavlova, Wai H. Wong, John Janssen, Jeffrey S. Houghton, and Kerrin Mabrey

Chapter 21
Behavior of Juvenile and Adult Zebra Mussels (*Dreissena polymorpha*) 331

Jarosław Kobak

Chapter 22
Antipredator Strategy of Zebra Mussels (*Dreissena polymorpha*): From Behavior to Life History 345

Marcin Czarnoleski and Tomasz Müller

Chapter 23
Variation in Predator–Prey Interactions between Round Gobies and Dreissenid Mussels 359

Christopher J. Houghton and John Janssen

Chapter 24
Density, Growth, and Reproduction of Zebra Mussels (*Dreissena polymorpha*) in Two Oklahoma Reservoirs 369

Chad J. Boeckman and Joseph R. Bidwell

Chapter 25
Limiting Environmental Factors and Competitive Interactions between Zebra and Quagga Mussels in North America 383

David W. Garton, Robert McMahon, and Ann M. Stoeckmann

Chapter 26
Evolutionary, Biogeographic, and Population Genetic Relationships of Dreissenid Mussels, with Revision of Component Taxa 403

Carol A. Stepien, Igor A. Grigorovich, Meredith A. Gray, Timothy J. Sullivan, Shane Yerga-Woolwine, and Gokhan Kalayci

Chapter 27
Effects of Algal Composition, Seston Stoichiometry, and Feeding Rate on Zebra Mussel (*Dreissena polymorpha*) Nutrient Excretion in Two Laurentian Great Lakes... 445

Thomas H. Johengen, Henry A. Vanderploeg, and James R. Liebig

Chapter 28
Chemical Regulation of Dreissenid Reproduction 461

Donna R. Kashian and Jeffrey L. Ram

Chapter 29
Role of Fluid Dynamics in Dreissenid Mussel Biology..471

Josef Daniel Ackerman

Part V
Impacts

Chapter 30
Meta-Analysis of Dreissenid Effects on Freshwater Ecosystems.. 487

Scott N. Higgins

Chapter 31
Effects of Invasive Quagga Mussels (*Dreissena rostriformis bugensis*) on Chlorophyll and Water Clarity in Lakes Mead and Havasu of the Lower Colorado River Basin, 2007–2009 495

Wai H. Wong, G. Chris Holdren, Todd Tietjen, Shawn Gerstenberger, Bryan Moore, Kent Turner, and Doyle C. Wilson

Chapter 32
Role of Selective Grazing by Dreissenid Mussels in Promoting Toxic *Microcystis* Blooms and Other Changes in Phytoplankton Composition in the Great Lakes.. 509

Henry A. Vanderploeg, Alan E. Wilson, Thomas H. Johengen, Julianne Dyble Bressie, Orlando Sarnelle, James R. Liebig, Sander D. Robinson, and Geoffrey P. Horst

Chapter 33
Trends in Phytoplankton, Zooplankton, and Macroinvertebrates in Saginaw Bay Relative to Zebra Mussel (*Dreissena polymorpha*) Colonization: A Generalized Linear Model Approach 525

Sara Adlerstein, Thomas F. Nalepa, Henry A. Vanderploeg, and Gary L. Fahnenstiel

Chapter 34
Lake Michigan after Dreissenid Mussel Invasion: Water Quality and Food Web Changes during the Late Winter/Spring Isothermal Period... 545

Steven A. Pothoven and Gary L. Fahnenstiel

Chapter 35
Nutrient Cycling by Dreissenid Mussels: Controlling Factors and Ecosystem Response.................................555

Harvey A. Bootsma and Qian Liao

CONTENTS ix

Chapter 36
Benthification of Freshwater Lakes: Exotic Mussels Turning Ecosystems Upside Down 575

Christine M. Mayer, Lyubov E. Burlakova, Peter Eklöv, Dean Fitzgerald, Alexander Y. Karatayev, Stuart A. Ludsin, Scott Millard, Edward L. Mills, A. P. Ostapenya, Lars G. Rudstam, Bin Zhu, and Tataina V. Zhukova

Chapter 37
Variability of Zebra Mussel (*Dreissena polymorpha*) Impacts in the Shannon River System, Ireland 587

Dan Minchin and Anastasija Zaiko

Chapter 38
Impacts of *Dreissena* on Benthic Macroinvertebrate Communities: Predictable Patterns Revealed by Invasion History...... 599

Jessica M. Ward and Anthony Ricciardi

Chapter 39
Interactions between Exotic Ecosystem Engineers (*Dreissena* spp.) and Native Burrowing Mayflies (*Hexagenia* spp.) in Soft Sediments of Western Lake Erie..611

Kristen M. DeVanna, Don W. Schloesser, Jonathan M. Bossenbroek, and Christine M. Mayer

Chapter 40
Zebra Mussel Impacts on Unionids: A Synthesis of Trends in North America and Europe.. 623

Frances E. Lucy, Lyubov E. Burlakova, Alexander Y. Karatayev, Sergey E. Mastitsky, and David T. Zanatta

Chapter 41
Impacts of Dreissenid Mussels on the Distribution and Abundance of Diving Ducks on Lake St. Clair.......................... 647

David R. Luukkonen, Ernest N. Kafcas, Brendan T. Shirkey, and Scott R. Winterstein

Chapter 42
Context-Dependent Changes in Lake Whitefish Populations Associated with Dreissenid Invasion 661

Michael D. Rennie

Chapter 43
Effects of Dreissenids on Monitoring and Management of Fisheries in Western Lake Erie 681

Martin A. Stapanian and Patrick M. Kocovsky

**Part VI
General**

Chapter 44
General Overview of Zebra and Quagga Mussels: What We Do and Do Not Know.. 695

Alexander Y. Karatayev, Lyubov E. Burlakova, and Dianna K. Padilla

Chapter 45
Comparative Role of Dreissenids and Other Benthic Invertebrates as Links for Type-E Botulism Transmission in the Great Lakes... 705

Alicia Pérez-Fuentetaja, Mark D. Clapsadl, and W. Theodore Lee

Chapter 46
A Comparison of Consumptive Demand of *Diporeia* spp. and *Dreissena* in Lake Michigan Based on Bioenergetics Models 713

Daniel J. Ryan, Thomas F. Nalepa, Lori N. Ivan, Maria S. Sepúlveda, and Tomas O. Höök

Chapter 47
Variation in Length–Frequency Distributions of Zebra Mussels (*Dreissena polymorpha*) within and between Three Baltic Sea Subregions: Szczecin Lagoon, Curonian Lagoon, and Gulf of Finland 725

Christiane Fenske, Anastasija Zaiko, Adam Woźniczka, Sven Dahlke, and Marina I. Orlova

Chapter 48
A Note on Dreissenid Mussels and Historic Shipwrecks 741

Russ Green

Part VII
Appendix: Narratives for Video Clips

Video Clip 1: A Visual Documentation of the Eradication of Zebra Mussels (*Dreissena polymorpha*) from Millbrook Quarry, Virginia 747

Raymond T. Fernald and Brian T. Watson

Video Clip 2: Invasion of Quagga Mussels (*Dreissena rostriformis bugensis*) to the Midlake Reef Complex in Lake Michigan: A Video Montage 749

Jeffrey S. Houghton, Robert Paddock, and John Janssen

Video Clip 3: Close-Up View of Inhalant Siphons of Quagga Mussels (*Dreissena rostriformis bugensis*, Deepwater Morph) on the Midlake Reef Complex in Lake Michigan 751

Thomas F. Nalepa, Jeffrey S. Houghton, Robert Paddock, and John Janssen

Video Clip 4: Visual Documentation of Quagga Mussels (*Dreissena rostriformis bugensis*) at Two Depths in Southeastern Lake Michigan 753

Russ Miller, Nathan Hawley, and Steven A. Ruberg

Video Clip 5: Zebra Mussel Movements on the Bottom of Lake Michigan 755

Barry M. Lesht and Nathan Hawley

Video Clip 6: Behavior of Zebra Mussels Exposed to *Microcystis* Colonies from Natural Seston and Laboratory Cultures 757

Henry A. Vanderploeg and J. Rudi Strickler

Video Clip 7: Visual Evidence a Native Mussel Population (Unionidae: Bivalvia) in the St. Clair River (Laurentian Great Lakes) Has Survived Despite the Presence of *Dreissena* 759

Greg Lashbrook and Kathy Johnson

Postlude–Synopsis 761

Thomas F. Nalepa and Don W. Schloesser

Index 763

Foreword

The discovery of the zebra mussel, *Dreissena polymorpha*, in Lake St. Clair in 1988 signaled what was to become a major threat to the St. Lawrence Great Lakes. It became a matter of great concern when it combined with frazil ice to clog the water intake pipes to Monroe, Michigan, in 1989 and also when its rapid spread and exponential population growth threatened cooling water supplies to electric power plants, especially nuclear power plants, and industries around the Great Lakes. The zebra mussel also brought attention to aquatic organisms as invasive species. *D. polymorpha* and the related species, *D. rostriformis bugensis* (quagga mussel), would adversely impact other aquatic organisms and bring about changes in all the Great Lakes, except Lake Superior, by altering the food web, water clarity, water chemistry, and even the sediments and beaches. Furthermore, it appears that these changes may have altered the environment to favor the establishment of other invasive species.

Terrestrial invasive species had been of concern to agriculture for many years, but as pointed out in Chapter 11 by David F. Reid and Dean Wilkinson, the importance of aquatic nonindigenous species as potential invasive species had largely been ignored or not recognized as a threat. Despite the number of nonindigenous species that had made their way or had been introduced to the Great Lakes—and the collapse of the lake trout fishery attributed to predation by the sea lamprey, *Petromyzon marinus*—there had been no major call to deal with invasive species in the Great Lakes. The fact that the zebra mussel was adversely impacting our water supplies only a year after their presence in the Great Lakes was discovered was especially alarming and brought to light the need to deal with aquatic invasive species.

It is to the credit of the Great Lakes research community that they responded quickly to undertake studies of the zebra mussel. We had little knowledge about the zebra mussel in 1988–1989. Little was known about its life history, means of distribution, native habitat, or origin of the invasive population. Representatives from the Great Lakes Environmental Research Laboratory (GLERL) of the National Oceanic and Atmospheric Administration, the National Fisheries Research Center—Great Lakes of the U.S. Fish and Wildlife Service (USFWS), Cooperative Institute for Limnology and Ecosystems Research (CILER), the U.S. Environmental Protection Agency, and representatives of each of the six Great Lakes Regional Sea Grant Programs met at the Detroit Metropolitan Airport on June 6, 1990, to discuss the need for research on the zebra mussel and possible development of a cooperative, coordinated research plan. This group was joined by representatives of the Great Lakes Commission, the Great Lakes Fishery Commission, the U.S. Coast Guard, and the U.S. Army Corps of Engineers to form the U.S. Great Lakes Non-Indigenous Species Research Coordinating Committee. This committee subsequently became the Great Lakes Panel on Exotic Species of the Aquatic Nuisance Species Task Force established by the Nonindigenous Aquatic Nuisance Prevention and Control Act of 1990.

The Great Lakes Task Force of the Northeast–Midwest Congressional Coalition also reacted quickly to the presence of the zebra mussel, recognizing that it was an invasive species problem. Diane Blagman, of Congressman Bob Carr's staff, and Allegra Cangelosi, of the Northeast–Midwest Congressional Coalition, worked to get legislation introduced. The congressional delegations from the Great Lake states provided strong support for such legislation, and the Nonindigenous Aquatic Nuisance Species Act was introduced in the House (HR 4214) and the Senate (S2244) in March 1990. Because of the great concern about the zebra mussel, the House Appropriation Committee earmarked research funds before action on these two bills. I joined other scientists in a hearing on the zebra mussel problem before the Subcommittee on Oceanography and Great Lakes; Subcommittee on Fisheries, Wildlife and Conservation of the Environment; and the Subcommittee on Coast Guard and Navigation on June 14, 1990. Closely similar testimony was presented at a hearing on S2244 held by the Senate Subcommittee on Environmental Protection on June 19, 1990. These two original bills were replaced by another bill, HR 5390, which was introduced in August 1990 and later passed by the House and adopted by the Senate (in lieu of S2244) to become the Nonindigenous Aquatic Nuisance Prevention and Control Act of 1990 (Public Law 101-646).

A subgroup of the U.S. Great Lakes Non-Indigenous Species Research Coordinating Committee, John Gannon (USFWS), Don Schloesser (USFWS), Russ Moll (CILER), Tom Nalepa (GLERL), and I (GLERL) met at GLERL on July 27, 1990, to develop a framework document for what was to become the Great Lakes Non-Indigenous Species Research Framework. It had six major topics for further research: (1) biology and life history, (2) effects on ecosystems, (3) socioeconomic analysis of costs and benefits, (4) control and mitigation, (5) preventing new introductions, and (6) reducing the spread. The framework was adopted for the Great Lakes by the U.S. research communities, as well as Canadian biologists, to focus their research on nonindigenous species. Subsequently, the plan was adopted in other regions of the United States and Canada.

The Great Lakes research community met the original challenge of the zebra mussel and quickly came together with a coordinated and cooperative research program. The first edition of this book, *Zebra Mussel: Biology, Impacts, and Control*, published in 1993, provided a valuable service by pulling together and documenting the initial response and early results of research on the impacts of the North American (e.g., Great Lakes) zebra mussel invasion and also included valuable insights from decades of research conducted in other countries. This edition continues this important service by providing an update on research on zebra and quagga mussels and a summary of new findings from almost 20 years of North American research since the first book was published.

Alfred M. Beeton
Ann Arbor, Michigan

Preface

When our first edited book on zebra mussels (*Dreissena polymorpha*) was published (Nalepa and Schloesser 1993), it was only a few years after this species had been reported in North America, and our intention was to document initial impacts, summarize control responses, and provide basic information on the biology of this new invader. Even though zebra mussels had colonized Eurasia over a century ago, and many published articles were available, we believed it was important to compile information on this species into one reference that would be useful to ecologists, industrial and municipal water users, resource managers, and the general public. Contributions focused to a great extent on developments in the Laurentian Great Lakes where the first mussels were found, but other contributions provided information on European populations as well.

Much has happened in the 20 years since the first book was published. While zebra mussels have continued to spread through North America, quagga mussels (*Dreissena rostriformis bugensis*), which were newly reported in North America when the first book was being compiled, have proven to be a greater threat than zebra mussels. For instance, in the Great Lakes, quagga mussels have mostly displaced zebra mussels, have spread to deep, offshore regions where zebra mussels were never found, and have caused wide-scale ecological changes far beyond those caused by zebra mussels. Further, this species is now well-established in the southwestern United States, where it has presented new control challenges for facilities managers. Hence, we felt that an updated compilation that included recent events would be beneficial to a wide audience.

This Second Edition builds on the First Edition in several different ways. First, it has the advantage of additional hindsight. A number of contributions provide details of how agencies, legislatures, policy-makers, and researchers have responded to dreissenid-related events. Other contributions provide retrospectives/summaries of topics such as ecological impacts and dreissenid biology. As both dreissenids continue to spread, some contributions provide new insights on how both species respond and adapt to different environmental conditions. Second, we include video clips on the companian CD that provide visual documentation of covered topics and are meant to enrich chapter text. With dreissenids, in some cases a video provides an understanding not easily captured by the written word or a figure.

The amount of information on zebra and quagga mussels has increased almost exponentially since these species were first discovered in North America in the late 1980s. Undoubtedly, large amounts of information will continue to be generated as control strategies evolve and as dreissenids adapt to old environments, spread to new environments, and induce and respond to shifts in ecosystems. We hope that this book provides a good sense of where we have been, where we are, and where we need to go along this time–information continuum.

Thomas F. Nalepa
Don W. Schloesser

Acknowledgments

We thank all the authors for their willingness to contribute to this project and for their cooperation and prompt responses during the review process. It was a pleasure to work with all of you. We especially thank all those authors who also contributed to the First Edition: J. Ackerman, A. bij de Vaate, R. Claudi, D. Garton, R. Griffiths, W. Kovalak, R. McMahon, K. Lewandowski, J. Ram, A. Stańczykowski, and D. Strayer. We thank Marie Colton and John Bratton of the Great Lakes Environmental Research Laboratory, NOAA, Don Scavia of the Graham Environmental Sustainability Institute, University of Michigan, and Russell Strach of the Great Lakes Science Center, USGS, for their support. The assistance and advice of Jill Jurgensen, Joette Lynch, and John Sulzycki of CRC Press and attentiveness of Arunkumar Aranganathan of SPi Global are greatly appreciated. Finally, we are deeply grateful to our wives, Cheri and Cindy, for their understanding, support, and encouragement during the long road to publication.

This is Contribution Number 1667 of the Great Lakes Environmental Research Laboratory, NOAA, and Contribution Number 1761 of the Great Lakes Science Center, USGS.

Editors

Circa 1993

Circa 2013

Thomas F. Nalepa (on right in both images) was a research biologist with the Great Lakes Environmental Research Laboratory, National Oceanic and Atmospheric Administration (NOAA), Ann Arbor, Michigan, for 37 years before retiring from federal service in 2011. He is now a part-time research scientist with the Graham Environmental Sustainability Institute, University of Michigan, and also maintains an emeritus position with NOAA.

Don W. Schloesser is a fisheries scientist with the Great Lakes Science Center, U.S. Geological Survey, Ann Arbor, Michigan. He has been at the center since 1977. Tom and Don shared a mutual interest in benthic communities long before the first zebra mussel was reported in the Great Lakes in 1988, but it was the discovery of this organism that initiated joint collaborations and research projects. Both attended the first organized meeting on zebra mussels in North America in 1989, and in 1993 co-edited the first edition of this book. Since then, they have continued to share ideas, assess research needs, and pool resources. Joint projects on dreissenids have focused on long-term trends and spread, population dynamics, biology, and impacts on the ecosystem, particularly impacts on other components of the benthic community.

Contributors

Josef Daniel Ackerman
Department of Integrative Biology
University of Guelph
Guelph, Ontario, Canada

Sara Adlerstein
School of Natural Resources and Environment
University of Michigan
Ann Arbor, Michigan

Amy J. Benson
U.S. Geological Survey
Southeast Ecological Science Center
Gainesville, Florida

Joseph R. Bidwell
School of Environmental and Life Sciences
The University of Newcastle
Callaghan, New South Wales, Australia

and

Ecotoxicology and Water Quality Research Laboratory
Department of Zoology
Oklahoma State University
Stillwater, Oklahoma

Abraham bij de Vaate
Waterfauna Hydrobiological Consultancy
Lelystad, the Netherlands

Chad J. Boeckman
Pioneer Hi-Bred
DuPont Agricultural Biotechnology
Ankeny, Iowa

and

Ecotoxicology and Water Quality Research Laboratory
Department of Zoology
Oklahoma State University
Stillwater, Oklahoma

Harvey A. Bootsma
School of Freshwater Sciences
University of Wisconsin-Milwaukee
Milwaukee, Wisconsin

Jonathan M. Bossenbroek
The Lake Erie Center
and
Department of Environmental Sciences
The University of Toledo
Oregon, Ohio

Julianne Dyble Bressie
Great Lakes Environmental Research Laboratory
National Oceanic and Atmospheric Administration
Ann Arbor, Michigan

Lyubov E. Burlakova
Great Lakes Center
and
The Research Foundation
SUNY Buffalo State
The State University of New York
Buffalo, New York

Mark J. Burrows
International Joint Commission's
Great Lakes Regional Office
Windsor, Ontario, Canada

Allegra Cangelosi
Northeast–Midwest Institute
and
Great Ships Initiative
Washington, District of Columbia

Rajat K. Chakraborti
Water Engineer
CH2M HILL
Thousand Oaks, California

Mark D. Clapsadl
Great Lakes Center
SUNY Buffalo State
Buffalo, New York

Renata Claudi
RNT Consulting Inc.
Picton, Ontario, Canada

David Bruce Conn
One Health Center for Zoonotic and Emerging Diseases
School of Mathematical and Natural Sciences
Berry College
Mount Berry, Georgia

and

Department of Invertebrate Zoology
Museum of Comparative Zoology
Harvard University
Cambridge, Massachusetts

Marcin Czarnoleski
Institute of Environmental Sciences
Jagiellonian University
Kraków, Poland

Sven Dahlke
Biological Station
University of Greifswald
Hiddensee, Germany

Kristen M. DeVanna
Aquatic Ecology Laboratory
Department of Evolution, Ecology, and Organismal Biology
The Ohio State University
Columbus, Ohio

Peter Eklöv
Department of Limnology
Evolutionary Biology Centre
Uppsala University
Uppsala, Sweden

Gary L. Fahnenstiel
Great Lakes Research Center
and
Michigan Tech Research Institute
Michigan Technological University
Houghton, Michigan

Christiane Fenske
Zoological Institute and Museum
University of Greifswald
Greifswald, Germany

Raymond T. Fernald
Virginia Department of Game and Inland Fisheries
Bureau of Wildlife Resources
Richmond, Virginia

Dean Fitzgerald
Department of Natural Resources
Cornell Biological Field Station
Bridgeport, New York

and

Ecological Services for Ontario
Hamilton, Ontario, Canada

Dale Gabel
CH2M HILL
Englewood, Colorado

David W. Garton
School of Biology
Georgia Institute of Technology
Atlanta, Georgia

Shawn Gerstenberger
Department of Environmental and Occupational Health
University of Nevada, Las Vegas
Las Vegas, Nevada

Thaddeus K. Graczyk
Centre for Biomolecular Environmental and Public Health Research
Institute of Technology
School of Science
Sligo, Ireland

and

Department of Biology and Environmental Sciences
Northern Arizona University
Yuma, Arizona

Meredith A. Gray
The Lake Erie Center
The University of Toledo
Toledo, Ohio

Russ Green
Thunder Bay National Marine Sanctuary
National Oceanic and Atmospheric Administration
Alpena, Michigan

Ronald W. Griffiths
Oregon Hatchery Research Center
Oregon State University
Corvallis, Oregon

Igor A. Grigorovich
Canada North Environmental Services
Saskatoon, Saskatchewan, Canada

Nathan Hawley
Great Lakes Environmental Research Laboratory
National Oceanic and Atmospheric Administration
Ann Arbor, Michigan

Katharina C.M. Heiler
Department of Animal Ecology and Systematics
Justus Liebig University
Giessen, Germany

Valerie Hickey
The World Bank
Washington, District of Columbia

Scott N. Higgins
Center for Limnology
University of Wisconsin-Madison
Madison, Wisconsin

and

Department of Fisheries and Oceans Canada
Winnipeg, Manitoba, Canada

G. Chris Holdren
Bureau of Reclamation
U.S. Department of the Interior
Denver, Colorado

Tomas O. Höök
Department of Forestry and Natural Resources
Purdue University
West Lafayette, Indiana

Geoffrey P. Horst
Department of Fisheries and Wildlife
Michigan State University
East Lansing, Michigan

Christopher J. Houghton
Department of Biological Sciences
School of Freshwater Sciences
University of Wisconsin-Milwaukee
Milwaukee, Wisconsin

Jeffrey S. Houghton
School of Freshwater Sciences
University of Wisconsin-Milwaukee
Milwaukee, Wisconsin

Lori N. Ivan
Cooperative Institute for Limnology and Ecosystems Research
University of Michigan
Ann Arbor, Michigan

and

Department of Forestry and Natural Resources
Purdue University
West Lafayette, Indiana

Yuri Izyumov
Papanin Institute for Biology of Inland Waters
Russian Academy of Sciences
Borok, Russia

Annie P. Jacob
Consolidated Safety Services
Fairfax, Virginia

John Janssen
Department of Biological Sciences
and
Great Lakes Water Institute
and
School of Freshwater Sciences
University of Wisconsin-Milwaukee
Milwaukee, Wisconsin

CONTRIBUTORS

Thomas H. Johengen
Cooperative Institute for Limnology
  and Ecosystems Research
University of Michigan
Ann Arbor, Michigan

Kathy Johnson
Gregory A.D.
Lakeport, Michigan

W. Edward Johnson
National Centers for Coastal Ocean
  Science
National Oceanic and Atmospheric
  Administration
Silver Spring, Maryland

Ernest N. Kafcas
Michigan Department of Natural
  Resources
Mt. Clemens Fisheries Station
St. Clair, Michigan

Gokhan Kalayci
Great Lakes Genetics/Genomics
The Lake Erie Center
The University of Toledo
Toledo, Ohio

and

Faculty of Fisheries
Molecular Biology and Genetic Lab
Department of Aquaculture
Rize University
Rize, Turkey

Alexander Y. Karatayev
Great Lakes Center
SUNY Buffalo State
Buffalo, New York

Donna R. Kashian
Department of Biological Sciences
Wayne State University
Detroit, Michigan

Jagjit Kaur
CH2M HILL
Thousand Oaks, California

Kimani L. Kimbrough
National Centers for Coastal Ocean
  Science
National Oceanic and Atmospheric
  Administration
Silver Spring, Maryland

Jarosław Kobak
Faculty of Biology
  and Environmental
  Protection
Department of Invertebrate Zoology
Nicolaus Copernicus University
Toruń, Poland

Patrick M. Kocovsky
Lake Erie Biological Station
Great Lakes Science Center
U.S. Geological Survey
Sandusky, Ohio

William P. Kovalak
Detroit Edison Company
Detroit, Michigan

Greg Lashbrook
Gregory A.D.
Lakeport, Michigan

Gunnar G. Lauenstein (Retired)
(Formerly) National Centers for
  Coastal Ocean Science
National Oceanic and Atmospheric
  Administration
Silver Spring, Maryland

W. Theodore Lee
Department of Biology
State University of New York at
  Fredonia
Fredonia, New York

Barry M. Lesht
CSC, Inc.
and
Earth and Environmental Sciences
  Department
University of Illinois at Chicago
Chicago, Illinois

Rob S.E.W. Leuven
Department of Environmental Science
Institute for Water and Wetland
  Research
Radboud University Nijmegen
Nijmegen, the Netherlands

Krzysztof Lewandowski
Department of Biological Education
  and Protection of Nature
Siedlce University of Natural
  Sciences and Humanities
Siedlce, Poland

Qian Liao
Department of Civil Engineering and
  Mechanics
University of Wisconsin-Milwaukee
Milwaukee, Wisconsin

James R. Liebig
Great Lakes Environmental Research
  Laboratory
National Oceanic and Atmospheric
  Administration
Ann Arbor, Michigan

Frances E. Lucy
Department of Environmental Science
and
Centre for Environmental Research,
  Innovation and Sustainability
Institute of Technology
Sligo, Ireland

and

Environmental Services Ireland
Leitrim, Ireland

Stuart A. Ludsin
Aquatic Ecology Laboratory
Department of Evolution,
  Ecology, and Organismal
  Biology
The Ohio State University
Columbus, Ohio

David R. Luukkonen
Rose Lake Wildlife Research Center
Michigan Department of Natural
  Resources
East Lansing, Michigan

Kerrin Mabrey
Cooperative Institute Limnology
 and Ecosystems Research
University of Michigan
Ann Arbor, Michigan

Sharook Madon
Water Resources and Ecosystems
 Management
CH2M HILL
San Diego, California

Heather M. Malcom
Cary Institute of Ecosystem Studies
Millbrook, New York

J. Ellen Marsden
The Rubenstein School of
 Environment and Natural Resources
The University of Vermont
Burlington, Vermont

Sergey E. Mastitsky
RNT Consulting Inc.
Picton, Ontario, Canada

Christine M. Mayer
The Lake Erie Center
and
Department of Environmental
 Sciences
The University of Toledo
Oregon, Ohio

Robert McMahon
Department of Biology
The University of Texas at Arlington
Arlington, Texas

Scott Millard
Great Lakes Laboratory for
 Fisheries and Aquatic Science
Department of Fisheries and Oceans
Burlington, Ontario, Canada

Russ Miller
Cooperative Institute for Limnology
 and Ecosystems Research
University of Michigan
Ann Arbor, Michigan

Edward L. Mills (Retired)
(Formerly) Department of Natural
 Resources
Cornell Biological Field Station
Bridgeport, New York

Dan Minchin
Marine Organism Investigations
Ballina, Killaloe Ireland

and

Lough Derg Science Group
Castlelough
Nenagh, Ireland

and

Coastal Research and Planning
 Institute
University of Klaipeda
Klaipeda, Lithuania

Bryan Moore
Lake Mead National Recreational
 Area
National Park Service
Boulder City, Nevada

Tomasz Müller
Institute of Nature Conservation
Polish Academy of Sciences
Kraków, Poland

Thomas F. Nalepa
Graham Environmental Sustainability
 Institute
University of Michigan
and
Great Lakes Environmental Research
 Laboratory
National Oceanic and Atmospheric
 Administration
Ann Arbor, Michigan

Marina I. Orlova
Zoological Institute
Russian Academy of Science
St. Petersburg, Russia

A.P. Ostapenya
Laboratory of Hydroecology
College of Biology
Belarussian State University
Minsk, Belarus

Robert Paddock
School of Freshwater Sciences
University of Wisconsin-Milwaukee
Milwaukee, Wisconsin

Dianna K. Padilla
Department of Ecology
 and Evolution
Stony Brook University
Stony Brook, New York

Vera Pavlova
Papanin Institute for Biology of
 Inland Waters
Russian Academy of Sciences
Borok, Russia

Alicia Pérez-Fuentetaja
Department of Biology
and
Great Lakes Center
SUNY Buffalo State
Buffalo, New York

Steve Pollick
Conservation Editor
Fremont, Ohio

Steven A. Pothoven
Great Lakes Environmental
 Research Laboratory
National Oceanic and Atmospheric
 Administration
Muskegon, Michigan

Katherine L. Prescott
RNT Consulting Inc.
Picton, Ontario, Canada

Thomas H. Prescott
RNT Consulting Inc.
Picton, Ontario, Canada

Jeffrey L. Ram
Department of Physiology
Wayne State University
Detroit, Michigan

David F. Reid (Retired)
(Formerly) Great
 Lakes Environmental
 Research Laboratory
National Oceanic and Atmospheric
 Administration
Ann Arbor, Michigan

CONTRIBUTORS

Michael D. Rennie
Environmental and Life Sciences
  Program
Trent University
Peterborough, Ontario, Canada

and

Fisheries and Oceans Canada
Freshwater Institute
Winnipeg, Manitoba, Canada

Anthony Ricciardi
Redpath Museum
McGill University
Montreal, Quebec, Canada

Sander D. Robinson
Cooperative Institute for Limnology
  and Ecosystems Research
University of Michigan
Ann Arbor, Michigan

Steven A. Ruberg
Great Lakes Environmental
  Research Laboratory
National Oceanic and Atmospheric
  Administration
Ann Arbor, Michigan

Lars G. Rudstam
Department of Natural Resources
Cornell Biological Field Station
Bridgeport, New York

Daniel J. Ryan
Division of Environmental Service
Pennsylvania Fish and Boat Commission
Bellefonte, Pennsylvania

and

Department of Forestry and Natural
  Resources
Purdue University
West Lafayette, Indiana

Sonya Gutschi Santavy
Department of Fisheries and Oceans
Transport Canada
Sarnia, Ontario, Canada

Orlando Sarnelle
Department of Fisheries and Wildlife
Michigan State University
East Lansing, Michigan

Don W. Schloesser
Great Lakes Science Center
U.S. Geological Survey
Ann Arbor, Michigan

Maria S. Sepúlveda
Department of Forestry and Natural
  Resources
Purdue University
West Lafayette, Indiana

Angela D. Shambaugh
Watershed Management Division
Vermont Department of
  Environmental Conservation
Montpelier, Vermont

Brendan T. Shirkey
Department of Fisheries and Wildlife
Michigan State University
East Lansing, Michigan

Anna Stańczykowska
Department of Ecology and
  Environment Protection
Siedlce University of Natural
  Sciences and Humanities
Siedlce, Poland

Pete Stangel
Watershed Management Division
Vermont Department of
  Environmental Conservation
Montpelier, Vermont

Martin A. Stapanian
Lake Erie Biological Station
Great Lakes Science Center
U.S. Geological Survey
Sandusky, Ohio

Carol A. Stepien
Great Lakes Genetics/Genomics
  Laboratory
The Lake Erie Center
and
Department of Environmental
  Sciences
The University of Toledo
Toledo, Ohio

Kenton M. Stewart
Department of Biological Science
State University of New York
Buffalo, New York

Ann M. Stoeckmann
Department of Biology
Francis Marion University
Florence, South Carolina

David L. Strayer
Cary Institute of Ecosystem Studies
Millbrook, New York

J. Rudi Strickler
School of Freshwater Sciences
University of Wisconsin-Milwaukee
Milwaukee, Wisconsin

Timothy J. Sullivan
Great Lakes Genetics/Genomics
  Laboratory
The Lake Erie Center
and
Department Environmental Sciences
The University Toledo
Toledo, Ohio

Todd Tietjen
Southern Nevada Water Authority
Henderson, Nevada

Kent Turner
Lake Mead National Recreational
  Area
National Park Service
Boulder City, Nevada

Henry A. Vanderploeg
Great Lakes Environmental Research
  Laboratory
National Oceanic and Atmospheric
  Administration
Ann Arbor, Michigan

Gerard van der Velde
Department of Animal Ecology
  and Ecophysiology
Institute for Water and Wetland
  Research
Radboud University Nijmegen
Nijmegen, the Netherlands

and

Naturalis Biodiversity Center
Leiden, the Netherlands

Jessica M. Ward
Department of Biology
McGill University
Montreal, Quebec, Canada

and

AECOM
Markham, Ontario, Canada

Brian T. Watson
Virginia Department of Game and
  Inland Fisheries
Bureau of Wildlife Resources
Forest, Virginia

Dean Wilkinson (Retired)
(Formerly) National Marine Fisheries
  Service
National Oceanic and Atmospheric
  Administration
Silver Spring, Maryland

Alan E. Wilson
Department of Fisheries and Allied
  Aquacultures
Auburn University
Auburn, Alabama

Doyle C. Wilson
Lake Havasu City Operations
  Department
Lake Havasu, Arizona

Scott R. Winterstein
Department of Fisheries
  and Wildlife
Michigan State University
East Lansing, Michigan

Wai H. Wong
Department of Environmental
  and Occupational Health
University of Nevada, Las Vegas
Las Vegas Nevada

and

Department of Biology
State University of New York
Oneonta, New York

Adam Woźniczka
Research Station
National Marine Fisheries Research
  Institute
Świnoujście, Poland

Shane Yerga-Woolwine
Great Lakes Genetics/Genomics
  Laboratory
The Lake Erie Center
The University Toledo
Toledo, Ohio

Anastasija Zaiko
Coastal Research and Planning
  Institute
University of Klaipeda
Klaipeda, Lithuania

David T. Zanatta
Department of Biology
Institute of Great Lakes Research
Central Michigan University
Mount Pleasant, Michigan

Bin Zhu
Department Biology
University Hartford
West Hartford, Connecticut

Tataina V. Zhukova
Naroch Biological Station
Belarussian State University
Minsk, Belarus

PART I

Prelude

My Story on Finding the First Zebra Mussel (*Dreissena polymorpha*) in North America

Sonya Gutschi Santavy

In May 1988, I was working as a research assistant for Dr. Paul Hebert at the Great Lakes Institute at the University of Windsor, in Windsor, Ontario. I had just graduated from the University of Guelph and was hired for a 1-year position funded by the Environmental Youth Corps Program of the Ontario Ministry of Environment. My initial task was to establish some research sites for Dr. Lawrence Weider who was not yet at the university, but would be arriving later on in June or July. His research would focus on the genetics of benthic communities with an emphasis on tubificid worms.

On June 1, 1988, we were sampling in Lake St. Clair in a small 5 m research vessel (the Monarch) that was owned by The Great Lakes Institute (GLI), which is now called the Great Lakes Institute for Environmental Research (GLIER). I used a sampling grid of Lake St. Clair established earlier by Pugsley et al. (1985). The site grid provided us with information on the bottom substrate, so I focused on a few sites with soft muddy substrate. I accompanied a graduate student, Bernie Muncaster, who was doing experimental research on native unionids in the lake, and Ronald Allison, an engineering student also contracted through the Youth Corps Program who agreed to assist me.

I sampled several sites using a Ponar grab (see Figure Prelude 1) before Bernie headed to his traps to check on his unionids. While at this site, I asked if we could spend a little time taking a few extra samples, even though I knew that the bottom substrate would not be suitable for the tubificid worms that I was trying to obtain. The rocky bottom was more suited to Bernie's unionids. As expected, the Ponar sample came up filled with stones and gravel, and as we went through the sample picking out benthos and looking for worms, I found one stone that actually looked like two stones stuck together. I tried to pull them apart, but they were firmly stuck together. I called out to the others, holding the "stones" out for them to see. Bernie said the smaller stone looked like a mussel, but one he didn't recognize. We decided to save the stone and the mussel to show to others.

We did not find any other mussels on the pebbles and rocks collected.

Upon returning to GLI, the lone specimen was shown to Dr. Paul Hebert before Bernie sent it to Dr. Mackie at the University of Guelph for positive identification. After its identification, a postdoctoral fellow, Dr. Neil Billington, commented to me "I could have told them what it was, we have these in England."

Shortly after Dr. Mackie's identification, Hebert, Muncaster, and Mackie began to research the population of zebra mussels in the lake. Divers retrieved zebra mussels at various sites across the lake in July and August, sampling at some of the same sites we had used in May. Shell measurements determined the ages of the mussels, while body tissues were used to determine reproductive status and genetic variation of establishing populations. I did the shell measurements and aided with other aspects of the study. I didn't think I wanted to ever see (or smell!) another zebra mussel again. By September the following year, their paper was published (see Hebert et al. 1989). I continued at the University as Dr. Weider's assistant, doing genetic research, until 1990. I was interviewed by Michigan Natural Resources Magazine (West 1990) and Allison and I were also mentioned in Maclean's (Steacy 1989) and Sports Illustrated (Boyle 1990).

Meanwhile, there were discussions with the Ministry of Natural Resources and the Ministry of Environment. The Ministry of Natural Resources was established as the lead agency on the zebra mussel (an invasive species) issue because of the threat to native species. However, soon thereafter, the Ministry of Environment became the lead agency when the mussels began to clog water intake pipes and temporarily shut down some plants in the area.

In spring 1990, I was approached by Ron Griffiths, a biologist at the Ministry of Environment, to help establish a zebra mussel monitoring program for the southwestern region of Ontario. For the next 2 years, I made presentations and designed zebra mussel displays. I counted and tracked

Figure Prelude 1 Sonya Gutschi Santavy, the first person to collect a zebra mussel, *Dreissena polymorpha*, in North America. She is holding a Ponar grab, which is the sampler used to collect the specimen. (Courtesy of David Kenyon, Michigan Department of Natural Resources; photo taken Circa 1990).

zebra mussels in harbors, lakes, rivers, and water pipelines in southwestern Ontario (more shell measurements!). I remember discovering how sharp those shells were after I foolishly used a harbor wall covered with zebra mussels to help hoist myself out of the cold waters. I was cold from the water and didn't notice the cuts on my feet until I warmed up and someone noticed the bloody footprints!

Because of firsthand experience, I looked suspiciously at every freighter I saw as mass transporters of invasive species. I continued to make zebra mussel/invasive species presentations at schools as technical officer for Environment Canada's Great Lakes Pollution Prevention Program. The topic (of zebra mussels) continued to make headlines in the papers for a few years, and it didn't seem long before zebra mussels were an established species throughout the Great Lakes. Children at the schools I visited had never seen the beaches without zebra mussel shells scattered about.

Eighteen years after I found the first zebra mussel in the Great Lakes, I returned to the invasive species issue and am now with Transport Canada, tracking all ballast water management processes of freighters arriving at Canadian ports. Although invasive species still threaten ecosystems globally, I am doing my small part so the next threat to the Great Lakes won't be carried in ballast water.

REFERENCES

Boyle, R. H. October 15, 1990. *In Shock over Shells*. Sports Illustrated. pp. 88–95. New York: Time Inc. Magazine Co.

Hebert, P. D. N., B. W. Muncaster, and G. L. Mackie. 1989. Ecological and genetic studies on *Dreissena polymorpha*: A new mollusc in the Great Lakes. *Can. J. Fish. Aquat. Sci.* 46: 1581–1591.

Pugsley, C. W., P. D. N. Hebert, G. H. Wood, G. Brotea, and T. W. Obal. 1985. Distribution of contaminants in clams and sediments from the Huron-Erie corridor. I-PCBs and octachlorostyrene. *J. Great Lakes Res.* 11: 275–289.

Steacy, A. November 6, 1989. *An Alien Invasion. Foreign Mussels are Disrupting the Lakes*, Macleans 97.

West, J. July/August 1990. Mussel menace. *Michigan Natural Resource Magazine* 4–11.

In Recognition
John Glenn Sparks a Giant Leap for Environmental Protection

Allegra Cangelosi

Many know that astronaut John Glenn was the first human to orbit the Earth, leading space exploration to its eventual "Giant Leap." Few are aware, however, of Senator John Glenn's later role in catalyzing a Giant Leap in environmental protection—prevention and control of aquatic invasive species.

The introduction, spread, and disruptive nature of invasive species—now recognized as a top threat to aquatic ecosystems globally—was not widely recognized in the late twentieth century, except in specific places like Senator Glenn's home region, the Laurentian Great Lakes. In 1990, Senator Glenn looked ahead and, in his role as Co-chair of the Senate Great Lakes Task Force of the Northeast-Midwest Coalition, led a bipartisan national effort to enact pioneering federal policy and regulations to prevent and control the introduction and spread of invasive aquatic species. The outcome was the passage of the Nonindigenous Aquatic Nuisance Prevention and Control Act of 1990 (reauthorized as the National Invasive Species Act of 1996), for which he received an Aquatic Nuisance Species Task Force Award (Figure Prelude 2).

Senator Glenn's original impetus was to protect the Great Lakes from nuisance species like the zebra mussel, but the multi-taxonomic and global magnitude of the invasive species problem became readily apparent to him. He therefore focused his legislation on interrupting vectors of new species introductions, like ballast discharge from transoceanic commercial vessels, while controlling the spread of problem species already established in U.S. waters. He took pains to involve in his legislation the entire array of federal and state agencies that would need to collaborate so that prevention and control efforts would be effective and efficient. This interjurisdictional scope required hours of additional negotiations between congressional committees and chambers to achieve a consensus approach.

The time and effort was well spent. The U.S. federal laws crafted by Senator Glenn immediately invigorated and strengthened the International Maritime Organization's efforts to establish international standards for management of ships' ballast and continue to provide a solid statutory foundation for U.S. federal actions in this issue area, including promulgation in 2012 of numeric federal standards limiting live organisms in ballast discharge to U.S. waters. Clearly, much more work is needed, but thanks to Senator Glenn's leadership, the federal policy and research framework for that work at the federal level are firmly in place.

I worked in association with Senator Glenn on the issue of aquatic invasive species for many years, as staff assistant and Great Lakes Task Force Director from 1989 to 1992 and as Great Lakes Washington Program Director at the Northeast-Midwest Institute through 1999, when Senator Glenn took his well-earned retirement. I know firsthand that these achievements would not have been possible without Senator Glenn's unwavering commitment to aquatic ecosystem protection and abiding respect for the entire range of interest groups involved in meeting that objective. Perhaps, it was the unique vantage of the Earth's magnificent but finite aquatic and human resources that astronaut John Glenn took in so many years ago that spurred this commitment and respect. No matter what the cause, Earth's diverse aquatic ecosystems, and all those people they enrich with sustenance, livelihood, and inspiration, now benefit from Senator Glenn's efforts.

It was an honor to work with Senator Glenn on this and other issues; I wish for more statesmen like him in the halls of Congress now and in the years ahead. His legacy shows how leadership, vision, and respect can culminate in a competent national response to even the most difficult of issues.

Figure Prelude 2 Senator John Glenn with an award presented to him by the Aquatic Nuisance Species Task Force for his efforts and leadership in addressing the problem of invasive species.

PART II

Distribution and Spread

CHAPTER 1

Chronological History of Zebra and Quagga Mussels (Dreissenidae) in North America, 1988–2010

Amy J. Benson

CONTENTS

Abstract 9
Introduction 10
Distribution History 10
    1988 10
    1989 10
    1990 11
    1991 12
    1992 14
    1993 14
    1994 17
    1995 18
    1996 19
    1997 20
    1998–2001 20
    2002–2004 21
    2005–2006 22
    2007–2010 23
Discussion 25
Appendix 29
Acknowledgments 29
References 29

ABSTRACT

An unprecedented invasion began in North America in the mid-/late-1980s when two Eurasian mussel species, *Dreissena polymorpha* (zebra mussel) and *Dreissena rostriformis bugensis* (quagga mussel), became established in the Laurentian Great Lakes. It is believed that Lake Erie was the initial location of establishment for both species, and within 3 years, zebra mussels had been found in all the Great Lakes. Since 1986, the combined distribution of two dreissenids has expanded throughout the Great Lakes region and the St. Lawrence River in Canada and also in the United States from the Great Lakes to the Mississippi Basin including the Arkansas, Cumberland, Illinois, Missouri, Ohio, and Tennessee river basins. The distribution of dreissenid mussels in the Atlantic drainage has been limited to the Hudson and Susquehanna rivers. In the western United States, the quagga mussel established a large population in the lower Colorado River and spread to reservoirs in Arizona, California, Colorado, Nevada, and Utah. Overall, dreissenid species have been documented in 131 river systems and 772 inland lakes, reservoirs, and impoundments in the United

States and Canada. The rapid colonization of North America by these two species was likely aided by human activity.

INTRODUCTION

A native of Eastern Europe, the zebra mussel (*Dreissena polymorpha*) began appearing in Western Europe in the late 1700s and early 1800s, rapidly spreading after construction of major canal systems (Mackie et al. 1989, Counts and Handwerker 1990). Canals would also play an early role in the transport and distribution of zebra mussel veligers between major drainage basins in the United States. Even before zebra mussels and quagga mussels (*Dreissena rostriformis bugensis*) were discovered in North America, Sinclair and Isom (1963) compared the dispersal of zebra mussels in Europe to the spread of *Corbicula*, a small bivalve from Asia already known to be a biofouling organism in the United States. They concluded that zebra mussels could likely be the next invader from Europe and would cause similar fouling problems if they colonized North America. Strayer (1991) predicted that, once established, zebra mussels would probably have a wide distribution in rivers and lakes of North America based on the size, water hardness, and range of temperatures tolerated in European waters.

Soon after establishment in the Great Lakes in the late-1980s, zebra mussels fell under public scrutiny when a city in southeastern Michigan was crippled by the loss of a significant portion of its water pumping capability and suffered several water outages in 1989 and 1990 (LePage 1993). Industrial water users quickly agreed that this species posed a serious fouling threat to any facility drawing water from infested lakes. As a result of concern by industry, the scientific community, and the general public, the National Aquatic Nuisance Species Prevention and Control Act was passed in 1990.

In response to this federal legislation, efforts to compile information about the distribution of zebra and quagga mussels in North America were initiated as part of an existing U.S. Department of Interior database created to track occurrences of nonindigenous fishes in the United States (Williams and Jennings 1991). In the present chapter, we provide a summary of compiled data from 1988 to 2010. Occurrence data for both species were gathered through an informal network of biologists in federal, state, and local governments and also of nongovernmental contributors such as biologists and environmental specialists from universities, biological consulting companies, public utilities, and volunteer monitoring programs. A majority of mussel reports were incidental observations made by scientists conducting non-dreissenid studies and from dreissenid research and local monitoring programs. Each report represented a collected specimen or an observation of adults, juveniles, or veligers. Locations of many specific collections given herein have not yet been published; therefore, personal communications from many sources are listed in Appendix. Scientific literature was reviewed to further track the spatial and temporal distribution of the two dreissenid species. Because of the large number of zebra mussel collections made that document its distribution and the fact that museums can only accept a limited number of voucher specimens per species, most records in the database do not have voucher specimens. All reports were spatially referenced in a geographic information system as part of the U.S. Geological Survey's Nonindigenous Aquatic Species database. Because of the "it is everywhere" syndrome, which is the decrease in reports once mussels are widely known to occur in an area, it is most likely that not all known locations of zebra and quagga mussels were reported to this tracking effort.

DISTRIBUTION HISTORY

1988

The first reported zebra mussel was collected on June 1, 1988 from Lake St. Clair (Hebert et al. 1989), and a month later in July, the first adults were seen in the western basin of Lake Erie (Leach 1993). Surveys at several sites in Lake St. Clair later that same year revealed densities of 1–200/m^2 and a distribution that extended over most of the southern portion of the lake (Hebert et al. 1989). It was hypothesized that the initial colonization probably occurred in 1985–1986, most likely as the result of ballast water discharge (Hebert et al. 1989, 1991, Griffiths et al. 1991). Carlton (2008) provided evidence that zebra mussels actually occurred in Lake Erie on natural gas wellheads in Canadian waters between Point Pelee and Erieau and between Long Point and Port Colborne in 1986. Therefore, based on descriptions of these infested wellheads, zebra mussels may have arrived slightly earlier than previously concluded. Leach (1993) reported the presence of veligers in the western basin of Lake Erie in June 1988 during a study of phytoplankton, which at the time was before adult zebra mussels had been actually reported in the lake. By the end of 1988, mussel populations were documented in the western two-thirds of Lake Erie, as far east as Long Point, Ontario (Leach 1993; Figure 1.1).

1989

Hebert et al. (1991) surveyed Lake St. Clair in 1989 and found dramatic increases in zebra mussel distribution and densities at many sites previously sampled in 1988. The greatest density increase at any one site was from 0.5 to 4500/m^2 (Hebert et al. 1991). The highest density of zebra mussels on one of several fish spawning reefs in western Lake Erie was 233,000/m^2 (Leach 1993). Peak mean larval abundance in the extreme western basin of Lake Erie increased 20/L in 1988 to 116/L in 1989 (Leach 1993).

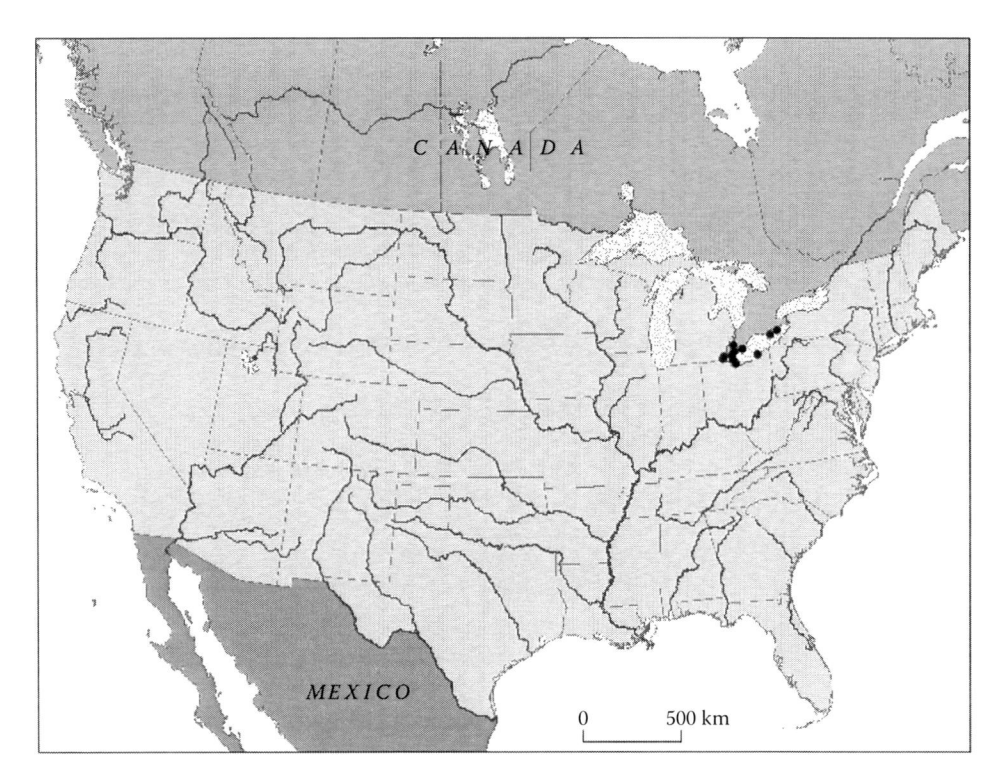

Figure 1.1 Locations of new discoveries of dreissenid mussels in the United States and Canada from 1986 to 1988 (black points).

Impacts of zebra mussels came to the attention of the Great Lakes community in 1989 (Nalepa and Schloesser 1993). During a routine inspection of an ozonation system on January 29, 1989, a large quantity of zebra mussels was found inside a water treatment plant in Monroe, Michigan, that drew its water from the western basin of Lake Erie (LePage 1993). Mussels in the water treatment plant grew quickly, as the previous inspection in April 1988 revealed no mussels. By September 1, 1989, the first pumping outage of water attributed to zebra mussels occurred at this plant (LePage 1993). Zebra mussels affected more than just water treatment plants. Kovalak et al. (1993) reported a zebra mussel density of 700,000–800,000/m^2 in an intake canal of an electric generating plant also along the western basin in Monroe, Michigan. Three of nine other power plants in the vicinity north and south of Lake St. Clair reported zebra mussels, but at lower densities, including a nuclear power plant that reported a density of 20,000–30,000/m^2 (Kovalak et al. 1993).

By late summer 1989, the downstream flow of water transported zebra mussel veligers from the western to the eastern basin of Lake Erie (Riessen et al. 1993) and to Port Colborne, Ontario, near the entrance of the Welland Canal (Mills et al. 1993). In September, zebra mussels were found downstream in the western basin of Lake Ontario at St. Catharines, Ontario (Griffiths et al. 1991). At the same time, a few small individuals were found attached to two buoys in the St. Lawrence River much farther east near Massena, New York (Conn et al. 1992a). Conn et al. (1992b) suggested this finding was the result of a separate ballast water introduction because no zebra mussels were found in a 160 km stretch of the St. Lawrence River upstream of Massena. In 1989, mussels were also found at the farthest point west in the Great Lakes, Duluth Harbor on Lake Superior. These mussels may have been inadvertently transported to the harbor by a research vessel (D. Pratt, pers. comm.). In addition, the southernmost portion of Lake Michigan was colonized by mussels at a mean density of 4000/m^2 off Whiting, Indiana (M. Schurder, pers. comm.).

Two discoveries of dreissenid mussels in 1989 would later prove to be very significant. First, zebra mussels were found in the Chicago Sanitary and Ship Canal (I. Polls, pers. comm. in Marsden et al. 1991). This was important because the canal links Lake Michigan to the Illinois River that carries water away from the lake, thus allowing zebra mussels to exit from the Great Lakes Basin into the Mississippi River Basin. Second, the first quagga mussel (*D. rostriformis bugensis*), a small, single specimen first identified as a zebra mussel, was collected from Lake Erie near Port Colborne, Ontario, in September 1989 (Mills et al. 1993). By late 1989, zebra mussels were found at the westernmost portion of the Great Lakes east to the St. Lawrence River (Figure 1.2).

1990

Two years after the finding of zebra mussels in Lake St. Clair, mussel densities in the southern and eastern portions of the lake reached 10,000/m^2 (Griffiths 1993). Veligers

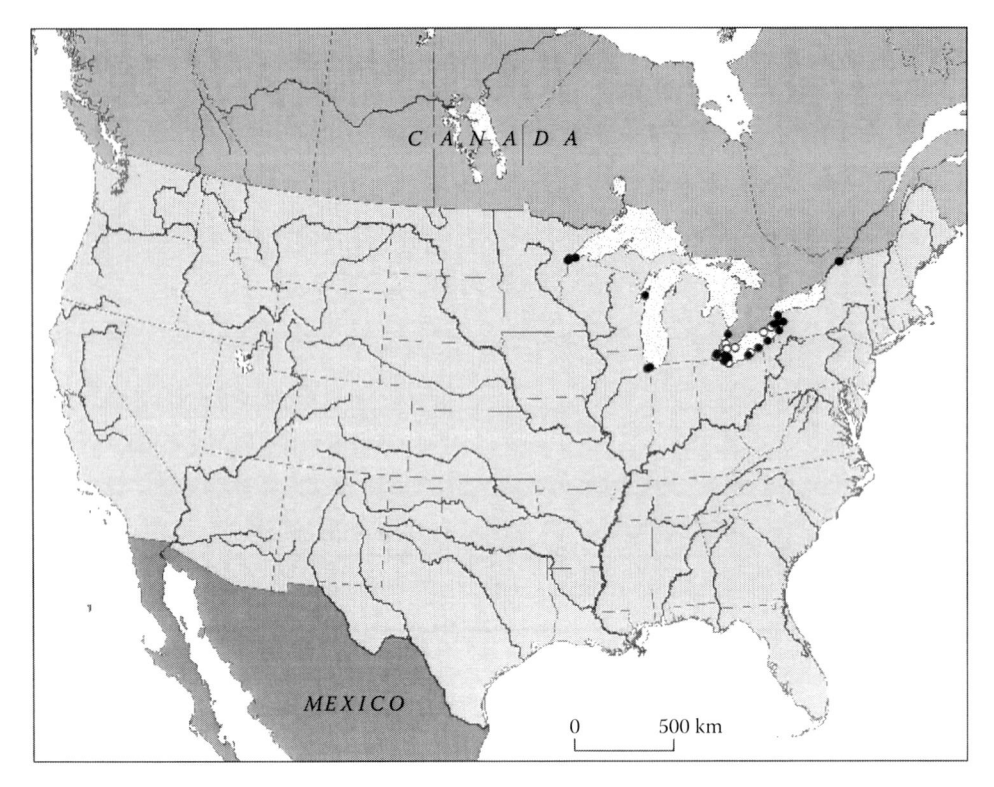

Figure 1.2 Locations of new discoveries of dreissenid mussels in the United States and Canada in 1989 (black points). Also given are locations where mussels were present in previous years (white points).

continued to increase in the western basin of Lake Erie; mean numbers of larvae increased 220% between 1989 and 1990 to reach a maximum of 1225/L in August 1990 (Leach 1993). Maximum mean density of mussels on one fish spawning reef was 342,000/m², an increase of 148% over the previous year (Leach 1993). Surprisingly, at this early stage in the invasion, Hebert et al. (1991) found more than 10,000 zebra mussels attached to a single unionid mussel in Lake St. Clair. Overall, zebra mussels expanded their distribution along nearshore waters of Lake Erie and western and southern Lake Ontario (Griffiths et al. 1991).

In addition to the canal in Chicago, the Erie Canal in upstate New York was found to be a significant vector for dispersal of zebra mussels as it connects Lake Erie to the Mohawk and Hudson rivers in the Atlantic drainage (Mills et al. 2000). An inland monitoring effort documented zebra mussel veligers and juveniles in the Erie Canal west of Rochester at Albion and Lockport, New York, in July (C. Lange, pers. comm.). Two months later, they were detected in the canal at Palmyra, 25 miles east of Rochester and the Genesee River (E. Mills, pers. comm.).

As zebra mussels continued to spread along the north shore of Lake Ontario east to the Bay of Quinte region in Ontario, Canada, mussels were found for the first time in Lake Huron near Goderich, Port Elgin, and Southampton, Ontario. Further, mussels were found in the St. Mary's River near Sault Sainte Marie and as far west as Lake Superior near Thunder Bay, Ontario (Ontario Ministry of Natural Resources, pers. comm.). Power generating plants continued to struggle with zebra mussel fouling. The plant in Monroe, Michigan, had an estimated 500 m³ of zebra mussels deposited in the intake canal and 300 m³ impinged on the traveling screens in 1 year's fouling (Kovalak et al. 1993).

In December 1990, navigation buoys in the upper St. Lawrence River were decommissioned for the year and inspected for zebra mussels. Twelve of 115 buoys had zebra mussels, and all 12 were located in the vicinity of the locks near Massena, New York (Conn et al. 1992a). In 1990, these buoys extended 12 km into Lake Ontario from 180 km in the upper St. Lawrence River downstream to Cornwall, New York. Although the extent of mussel occurrence did not change in 1990 compared to 1989, more localized populations were being found in all the Great Lakes except for Lake Superior (Figure 1.3).

1991

Expansion was significant in several major drainage basins in the state of New York. A population was found in the Hudson River near the town of Catskill in May, perhaps a result of the eastward flow of mussels from Lake Erie via the Erie Canal (Strayer et al. 1996). However, this may have been a separate introduction by vessel transport to the far eastern end of the canal in 1991 (Strayer et al. 1993). By the end of 1991, mussels had expanded their

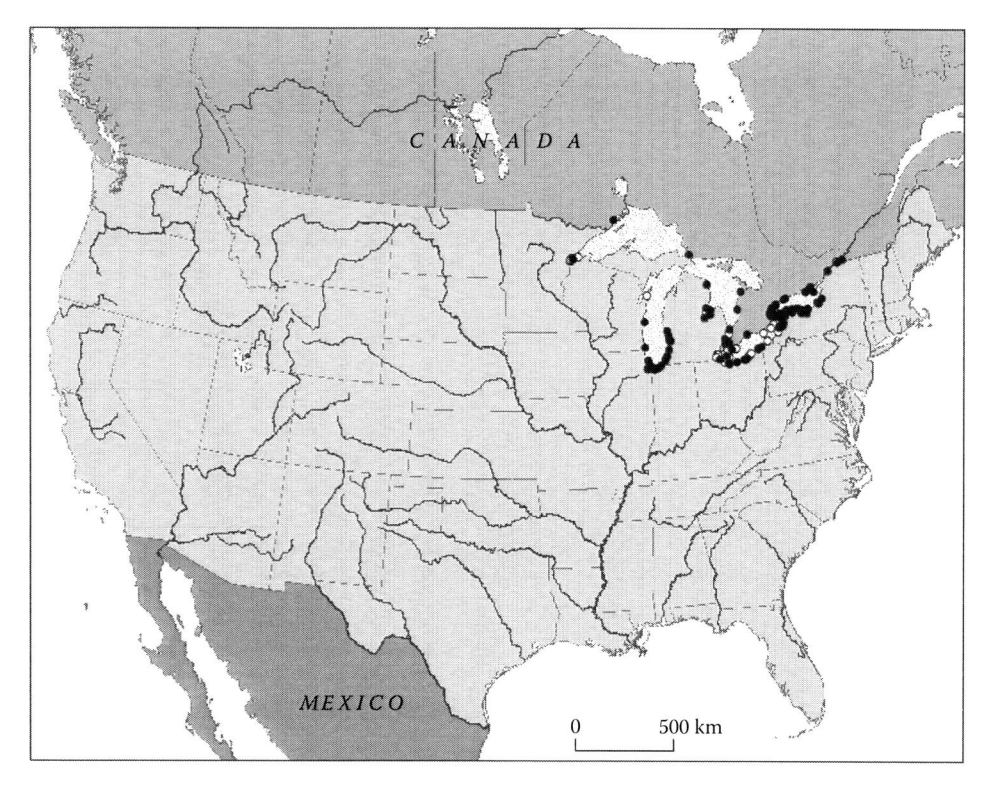

Figure 1.3 Locations of new discoveries of dreissenid mussels in the United States and Canada in 1990 (black points). Also given are locations where mussels were present in previous years (white points).

range nearly 40 miles downstream in the Hudson River as far as Newburgh where densities varied from 5 to 50/m² (Strayer et al. 1996). The Susquehanna drainage was also colonized in 1991 as veligers were identified in the upper Susquehanna River at Johnson City, New York. However, no adult population was located as the source of those veligers (K. Young, pers. comm.). Mussels also spread into the Finger Lakes region of central New York as adults were found in Cayuga, Oneida, Onondaga, and Seneca lakes (C. O'Neill, pers. comm.).

The first report of zebra mussels from the Illinois River occurred in June 1991 in a side channel near Bath, Illinois, which is located approximately 322 km downstream of Chicago, and mussels were also reported in September near the river mouth (Marsden et al. 1991). The earliest reports of zebra mussels in the Mississippi River also occurred in 1991 at Melvin Price Locks and Dam, approximately 26 km below the Illinois River mouth near Alton, Illinois (T. Keevin, pers. comm.); mussels were also reported near La Crosse, Wisconsin (Cope et al. 1997). From the Mississippi River, zebra mussels spread into the Ohio River and its tributaries in Tennessee, the Tennessee River as far upriver as Chattanooga, and the Cumberland River as far upriver as Nashville (B. Kerley, pers. comm.).

In August, a single mussel similar to *D. polymorpha* but with a slightly different shell morphology was collected in the Erie Canal near Palmyra, New York. More of these different-looking zebra mussels were collected from Lake Ontario near Rochester in November. They were genetically identified as another dreissenid species and given the name "quagga" mussel (May and Marsden 1992). It was later identified as *D. rostriformis bugensis*, a native of the Black Sea drainage in Ukraine (Spidle et al. 1994). After this announcement, researchers reexamined previously collected samples of zebra mussels from Lake Ontario and discovered that quagga mussels were present as early as July 1990 in Lake Ontario near Niagara-on-the-Lake, Ontario (Marsden et al. 1992).

In Ontario, Canada, zebra mussels were beginning to appear in small inland lakes including Balsam, Big Bald, Muskoka, Rice, Rosseau, and Simcoe (B. McKay, pers. comm.). As in 1990, buoys in the St. Lawrence River were again examined in 1991. This time 78 of the 105 examined buoys had zebra and quagga mussels present (Conn et al. 1992a). In northeastern Lake Erie, the density of zebra mussels reached 21,000/m² (Dermott et al. 1993). By the end of 1991, zebra mussels had dispersed throughout most of the nearshore waters of lakes St. Clair, Erie, Ontario, Michigan (notably the southern portion and Green Bay), and Huron (notably Saginaw Bay). Quagga mussels were now being accurately identified; its distribution included Lake Erie, Lake Ontario, St. Lawrence River, and Erie Canal (May and Marsden 1992, Lei and Miller 1994). By the end of 1991, zebra mussels were present in all five of the Great Lakes, in the Mississippi River including two of its major tributaries (Marsden et al. 1991, Ludyanskiy et al. 1993, O'Neill and Dextrase 1994, Miller and Payne 1997), and the Hudson

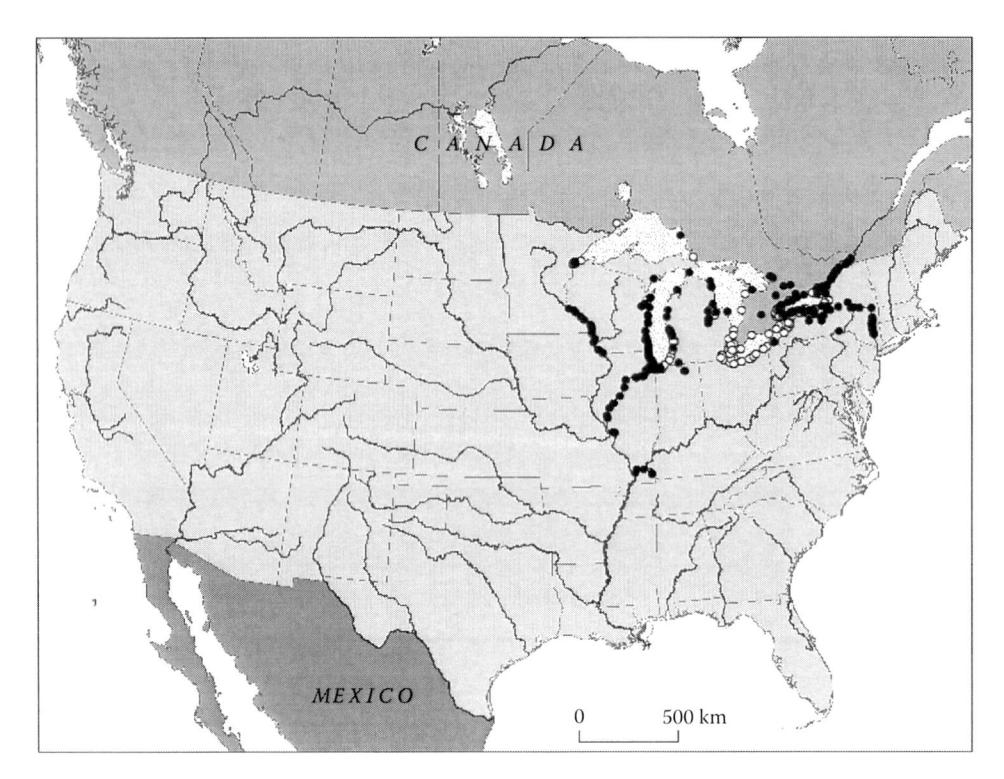

Figure 1.4 Locations of new discoveries of dreissenid mussels in the United States and Canada in 1991 (black points). Also given are locations where mussels were present in previous years (white points).

River (Strayer et al. 1993; Figure 1.4). Zebra mussels continued to spread outside the Great Lakes proper, with populations being reported in 18 small lakes within the Great Lakes basin. Figure 1.5 shows the cumulative total of lakes, reservoirs, and other water bodies infested with dreissenids over time for each U.S. state and Canadian province.

1992

In 1992, zebra mussels spread upstream in the Ohio River as far as Willow Island Lock and Dam, which is a distance of 1318 river kilometers from its confluence with the Mississippi River and only 261 river kilometers from its origin in Pittsburgh, Pennsylvania. In addition, mussels were discovered at all three locks and dams on the Kanawha River, an Ohio River tributary in West Virginia (B. Cremeans, pers. comm.). In the lower Mississippi River, adult zebra mussels and veligers were found as far south as Vicksburg, Mississippi (Forester et al. 1993). The first population discovered west of the Mississippi River occurred in Lake Dardanelle on the Arkansas River, a southern tributary of the Mississippi River (R. Muia, pers. comm.).

Strayer et al. (1993) found zebra mussels at all visited sites in the Hudson River, from Troy to Haverstraw, New York, while none were found below Haverstraw where the salinity of the river reaches 3–6 parts per thousand (ppt). Densities across sampled sites suggested increases of 100–1000 times from 1991 to 1992 (Strayer et al. 1993). By the end of 1992, the number of mussels in the Hudson River was estimated at 550 billion (Strayer et al. 1996).

At the end of 1992, the range of zebra mussels in North America included all the Great Lakes (to a lesser extent in Lake Superior where living conditions appeared less than optimal), portions of all navigable rivers of the Mississippi Basin, east of the Mississippi River, and Arkansas River (Figure 1.6). The cumulative number of small lakes with zebra mussel populations increased from 18 to 24 between 1991 and 1992 (Figure 1.5).

1993

Zebra mussels continued to spread down the Mississippi River to Louisiana and westward into Arkansas and Oklahoma (Figure 1.7). In January, mussels were discovered in Oklahoma at three locks and dams on the Arkansas River, including W. D. Mayo Lock and Dam No. 14, R. S. Kerr Lock and Dam No. 15, and Webbers Fall Lock and Dam No. 16 (E. Laney, pers. comm.). In June, mussels were also found at the Chouteau Lock and Dam No. 17 on the Verdigris River, a tributary of the Arkansas River (E. Laney, pers. comm.). In the Arkansas River in Arkansas, mussels were found at locks and dams 3, 4, 5, 6, and 7 (J. Harris, pers. comm.). In the lower Mississippi River, new populations were found near Memphis, Tennessee (D. Hubbs, pers. comm.), and in Louisiana near Lettsworth, New Orleans, Berwick, and Bayou Teche at the confluence with the Atchafalaya River (J. Forester, pers. comm.).

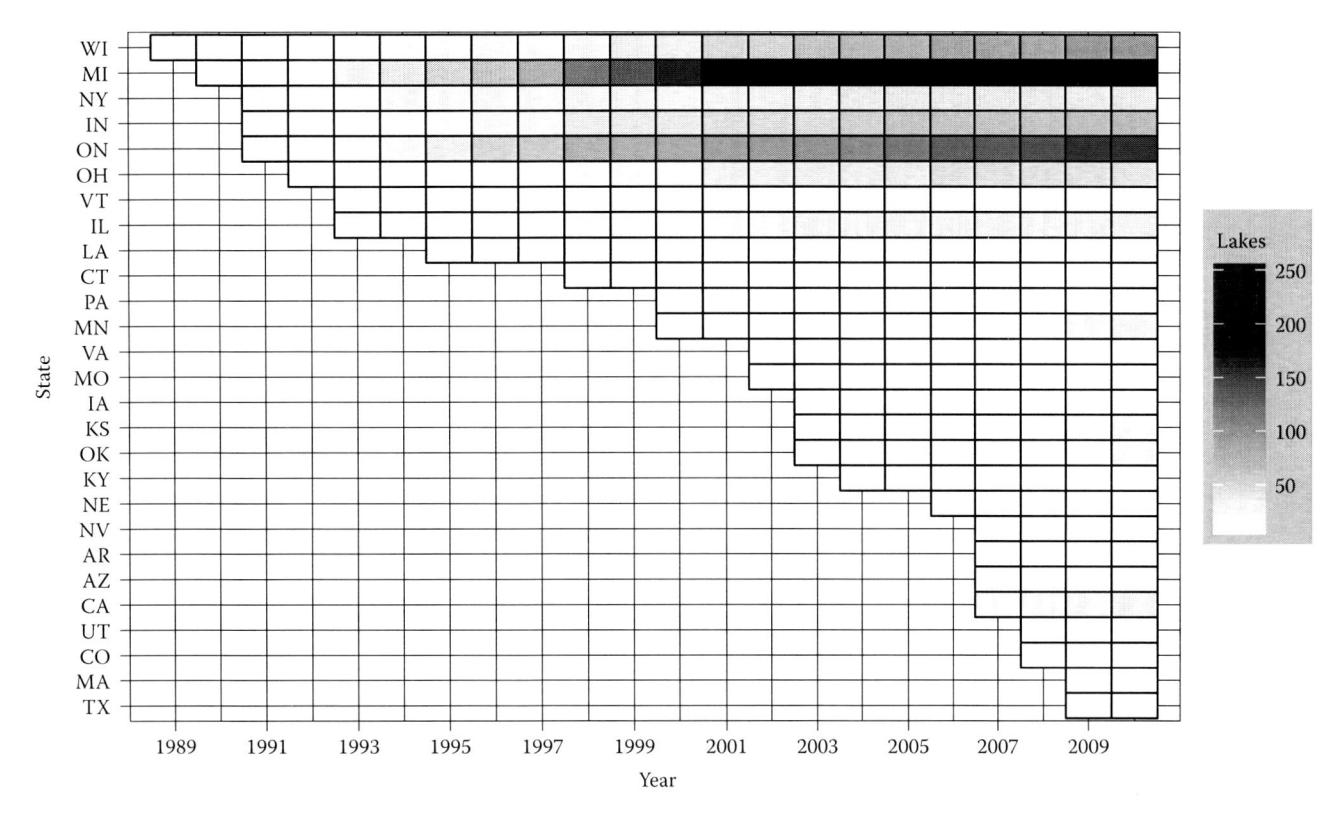

Figure 1.5 Number of natural lakes, small reservoirs/impoundments, and quarries outside the five Laurentian Great Lakes and Lake St. Clair where dreissenids were present by state/province and year (1989–2010). Light gray shade given for contrast only.

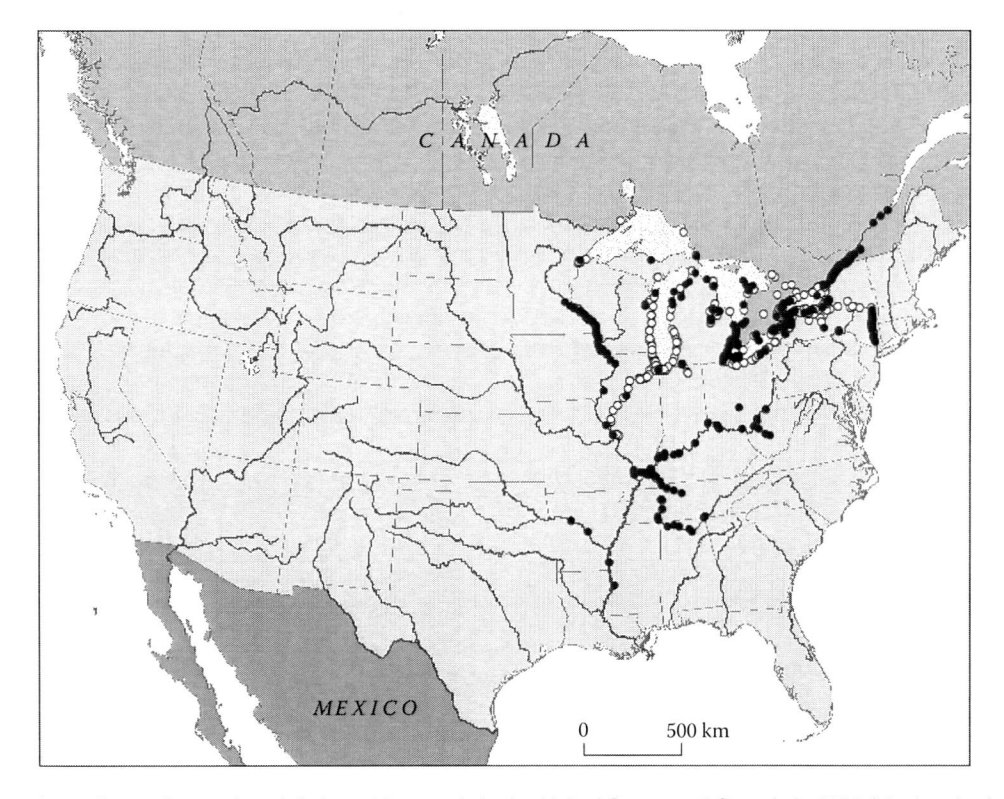

Figure 1.6 Locations of new discoveries of dreissenid mussels in the United States and Canada in 1992 (black points). Also given are locations where mussels were present in previous years (white points).

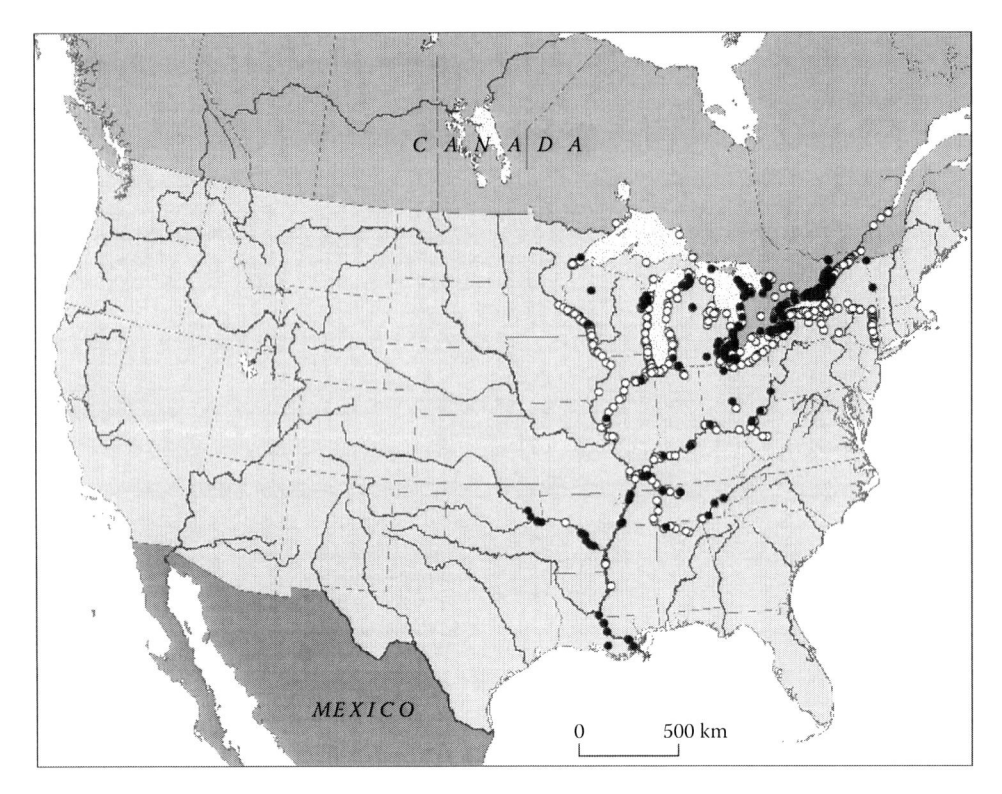

Figure 1.7 Locations of new discoveries of dreissenid mussels in the United States and Canada in 1993 (black points). Also given are locations where mussels were present in previous years (white points).

As previously mentioned, canals have proved to be an important secondary vector for dispersal of dreissenid mussels. The Champlain Canal, which connects Lake Champlain to the Erie Canal and the Hudson River, most likely played a role in the dispersal of mussels to Lake Champlain where, in June, zebra mussels were reported for the first time in the southern end of the lake near Benson, Vermont (J. Kellogg, pers. comm.). In August, zebra mussels were also found in the upper portion of the lake as well near Grand Isle (J. Kellogg, pers. comm.). Nearby, zebra mussels were found in the St. Lawrence River from Kingston, Ontario, north to the city of Quebec, Quebec (Wormington et al. 1993), in the Snye and Rideau rivers, and in the Rideau Canal in eastern Ontario, Canada (Ontario Federation of Anglers and Hunters, unpublished data).

In July, zebra mussels were discovered in the Tennessee River above the city of Chattanooga and soon thereafter found at Chickamauga, Watts Bar, and Fort Loudon locks and dams in Tennessee and in Alabama at Wilson Lock and Dam (S. Ahlstedt, pers. comm.). New discoveries of mussels in the Cumberland River included Old Hickory Reservoir, Old Hickory Navigation Lock, and the Cumberland City Steam Plant (S. Ahlstedt, pers. comm.). Zebra mussels continued to move into the upper reaches of the Ohio River as far as Pike Island Lock and Dam, located above Wheeling, West Virginia, and just 135 river kilometers from Pittsburgh (M. Fowles, pers. comm.). Also in the region, veligers were collected in July in Griggs Reservoir on the Scioto River, a tributary of the Ohio River in central Ohio (P. Orndorff, pers. comm.). A quagga mussel was also collected from the Ohio River in October (Kraft 1994).

As in previous years and locations, mussel densities were quite high. For example, in the lower Illinois River in early 1993, the density of newly settled mussels reached 100,000/m^2 (Whitney et al. 1996). However, by October, densities decreased by as much as 41% at sites where densities were greatest (Whitney et al. 1996). The density decrease may have been due to degraded water quality, such as low dissolved oxygen levels, high turbidity, or high temperatures in the river (Whitney et al. 1996).

With the knowledge that mussels could be transported overland attached to boat hulls, state agencies became concerned early in the zebra mussel invasion that mussels could be spread over great distances. Their concerns were realized when dead zebra mussels were found on a boat trailer at the Needles agricultural inspection station, California, in November 1993. The trailer had been transported from Lake Erie (J. Janik, pers. comm.). A second overland dispersal was reported in 1993 from Smith Mountain Lake, Virginia, when a marina operator spotted mussels on a boat hull just before it was to enter the lake. Further inspection revealed a small number of live mussels among many dead mussels attached to the hull (R. Eades, pers. comm.).

The cumulative number of small lakes with mussels in the Great Lakes region increased from 24 to 46 (Figure 1.5).

1994

By 1994, zebra mussels reached the upper navigable reaches of nearly all major rivers in the Mississippi Basin in the eastern United States with exception of the Missouri River (Figure 1.8). Densities of zebra mussels in the upper Mississippi River at Lock and Dam 6 averaged 21/m^2 in February, and densities at Lock and Dam 7 averaged 829/m^2 in December. Densities at Black Rock Lock on the Niagara River in New York averaged approximately 10,000/m^2 in January after 5 years of colonization (Miller and Payne 1997). Mussel populations in the Illinois River declined as much as 99% from 1993, possibly due to water level fluctuations making conditions in the river unfavorable (Whitney et al. 1996). Elsewhere in the Illinois River near Havana, the number of veligers drifting past a sampling site during the summer months was estimated to be as high as 70 million/s (Stoeckel et al. 1997). Additional mussel populations were found in the Arkansas River at Locks and Dams 2, 8, 9, 10, 12, and 13 and also for the first time in the White River in Arkansas (G. Bartelt, pers. comm.). Mussels were discovered at the Newt Graham Lock and Dam No. 18 on the Verdigris River in eastern Oklahoma (E. Laney, pers. comm.). Electric power stations and waterworks plants drawing water from the Mississippi River in and near New Orleans reported zebra mussels at their facilities. The southernmost report of zebra mussels was in the Mississippi River near Venice, Louisiana, close the river mouth (J. Forester, pers. comm.).

Zebra mussels were found in the headwaters of the Ohio River and beyond in Pennsylvania at Emsworth Locks and Dam, just west of Pittsburgh, and in the Allegheny River at Locks 4 and 7, near Natrona and Kittanning, Pennsylvania, respectively (M. Fowles, pers. comm.). In Vermont, zebra mussels continued to spread in Lake Champlain, with mussels inhabiting approximately the lower third of the lake, while several sites in the middle portion of the lake were becoming newly infested (N. Kamman, pers. comm.).

Quagga mussels were reported from Lake Erie near the city of Erie, Pennsylvania (E. Mastellar, pers. comm.). Within a few years, a local beach near the city had to be closed because of the large volume of shells that washed ashore, of which approximately one-third were identified as quagga mussels (E. Obert, pers. comm.). In Ontario, Canada, zebra mussels were discovered in the Ausable, Sauble, and Trent rivers (Ontario Federation of Anglers and Hunters, unpublished data).

Since the initial finding of mussels on a boat at a California agricultural inspection station in 1993, five additional boats with zebra mussels attached to their hulls were detected at three different inspection stations. The second finding of zebra mussels occurred in California in November 1994. Forty specimens were removed from a boat on a flatbed trailer being shipped from Toledo, Ohio, to San Diego, California. Their condition was listed as "live" at the time of collection according to the California Department of Water Resources.

By the end of 1994, a total of 77 small lakes within the Great Lakes Basin in Michigan, New York, Ohio, Wisconsin, and Ontario were reported to have zebra mussels present (Figure 1.5).

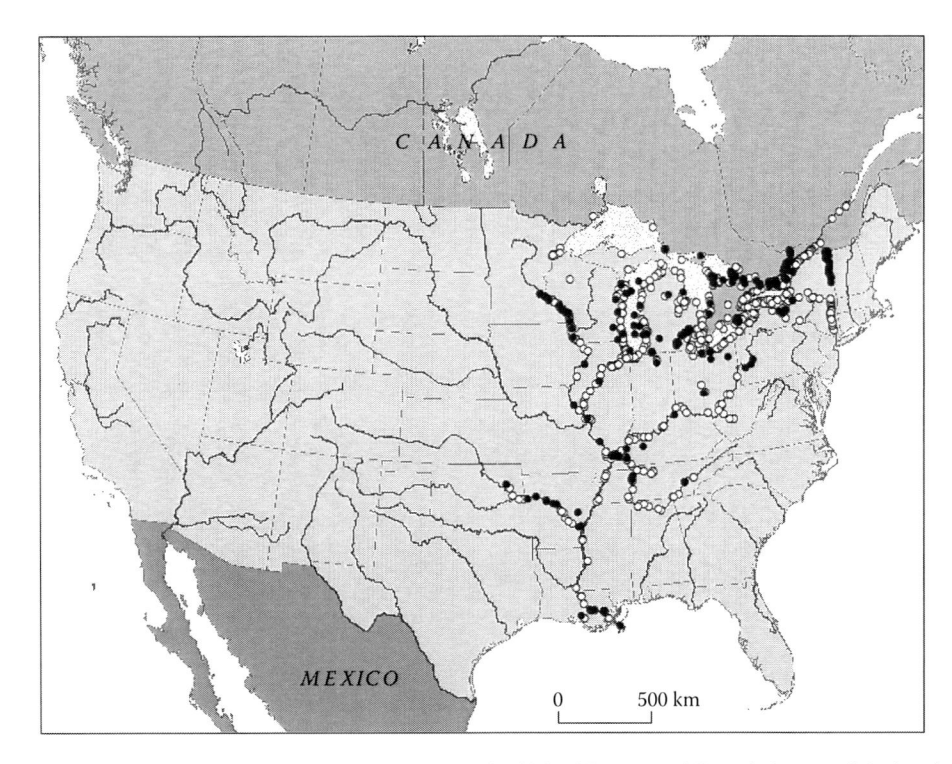

Figure 1.8 Locations of new discoveries of dreissenid mussels in the United States and Canada in 1994 (black points). Also given are locations where mussels were present in previous years (white points).

1995

Zebra mussels infested all locks and dams and some power generation plants in the upper Mississippi River by 1995 (Figure 1.9). Densities of mussels at the northernmost dams remained relatively low, ranging from $1/m^2$ at Lock and Dam 1 in St. Paul, Minnesota, to $11,432/m^2$ at Lock and Dam 13 in Fulton, Illinois (Cope et al. 1997). After a massive die-off in 1994, populations in the Illinois River apparently recovered and densities exceeded $1000/m^2$ at three sites in the upper, middle, and lower sections of the river (Whitney et al. 1996). As in 1994, the number of veligers drifting past the Havana site was estimated at nearly 70 million/s (Stoeckel et al. 1997). However, the successful recruitment in summer 1995 did not lead to sustained high abundances in the Illinois River and mussels had all but disappeared by fall 1995; it was suggested that only under flood events (i.e., improved water quality) is the water in the Illinois River suitable for zebra mussel recruitment and survival (Whitney et al. 1996). Several successful recruitment events between 1993 and 1995 corresponded with elevated water levels (Whitney et al. 1996). The first quagga mussels reported from the Mississippi River were found along with zebra mussels near St. Louis, Missouri (Kraft 1995).

In 1995, zebra mussels were found for the first time in the Monongahela River of the upper Ohio River drainage in Pennsylvania; locations included Lock and Dam No. 2 (River Mile 11.2) near Braddock and Maxwell Lock and Dam (River Mile 61.2) near East Millsboro (M. Fowles, pers. comm.). In July, several zebra mussels were reported from Chautauqua Lake in southwestern New York (S. Mooradian, pers. comm.). Chautauqua Lake is located in the headwaters of the Allegheny River and established populations may serve as a source of zebra mussel veligers in both the Allegheny and Ohio rivers. Also in the Ohio drainage, a large monitoring effort began in 1995 to study the effects zebra mussels may have on native mussels in the Ohio River. Native mussel beds in the river were monitored for densities of zebra mussels, which ranged from $4/m^2$ at an upriver site to $3540/m^2$ at a downriver site (P. Morrison, unpublished data). In the Kentucky River, an Ohio River tributary, zebra mussels were found in Pools 1 through 6 of the river in August (B. Higgins, pers. comm.).

Zebra mussels continued to expand in southern rivers. Veligers were found at all monitoring stations downstream from River Mile 260 in the Tennessee River, and moderate numbers of adults were found at all nine Tennessee River locks upstream to River Mile 602, near Knoxville (B. Kerley, pers. comm.). Densities in the lower portion of the Cumberland River were slightly higher than in 1994 (B. Kerley, pers. comm.) and, in the lower Mississippi River, mussel density was nearly $50,000/m^2$ at the Jefferson Parish Waterworks in New Orleans (B. Grant, pers. comm.).

In the St. Croix River, a Mississippi River tributary in Wisconsin, several boats pulled from the river in 1994 had zebra mussels attached to their hulls even though the river was not considered to be infested at the time. However, zebra mussels began to occur on monitoring plates in 1995 (D. Wedan, pers. comm.).

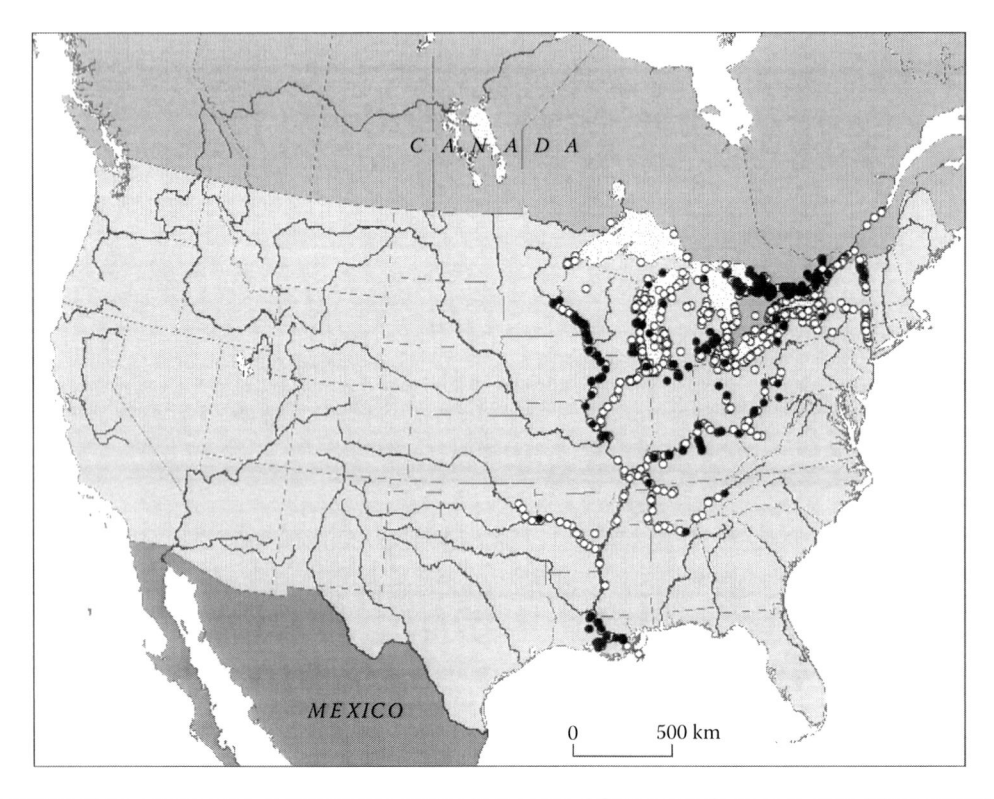

Figure 1.9 Locations of new discoveries of dreissenid mussels in the United States and Canada in 1995 (black points). Also given are locations where mussels were present in previous years (white points).

Mussels continued to be found attached to boats at various California agricultural inspection stations; live mussels were found attached to a yacht from Michigan that was destined for San Diego (J. Janik, pers. comm.).

Zebra mussels continued to spread in Ontario. First appearances were recorded in four rivers, the Cataraqui, Otonabee, Ottawa, and Severn, and also in 12 additional lakes (Ontario Federation of Anglers and Hunters, unpublished data). The most notable change in the distribution of the zebra mussel was the infestation of more small lakes in close proximity to the Great Lakes. In 1995, adults or veligers were found at 41 additional lakes in the Great Lakes region and one in the State of Louisiana bringing the cumulative total to 119 (Figure 1.5).

1996

In 1996, seven power plants along the lower Mississippi River reported the presence of veligers in January. However, it was suggested that the warm summer temperatures in the river were too high for successful reproduction and long-term survival. It was suggested that recruitment likely came from reproduction upriver or reintroduction by barges (Y. Allen, pers. comm.). Densities of zebra mussels increased throughout the Arkansas River in Arkansas and mussels were found for the first time in the White River near Clarendon (B. Wagner, pers. comm.).

Densities of adult zebra mussels in Lake Champlain were generally greater in 1996 than in 1995. Density of new recruits at one site in the southern portion of the lake was 3,676,000/m^2 in midsummer, and veligers were found for the first time in the northeast arm of the lake (Vermont Agency of Natural Resources, unpublished data). In contrast to an expanding population in Lake Champlain, a strong 1992 year class of zebra mussels continued to dominate in the Hudson River, but biomass of this year class was at its lowest since 1992 (D. Strayer, pers. comm.). Also in eastern New York, zebra mussels (veligers only) were observed for the first time in Saratoga Lake (C. O'Neill, pers. comm.) and in Lake George (S. Nierzwicki-Bauer, pers. comm.).

In the Ohio River, zebra mussel densities within native mussel beds increased 10- to 30-fold over the previous year (P. Morrison, per. comm.), and a small number of adults were collected from Lock 4 on the Monongahela River (Mile 41.5) near Monessen, Pennsylvania (M. Fowles, pers. comm.). Zebra mussel densities increased slightly from the previous year in the Tennessee River, and reproduction was still evident at all power plants located along the river (B. Kerley, pers. comm.).

The California Department of Water Resources reported four boats entering the state with zebra mussels attached; three of the boats had dead mussels and one had live mussels. The number of boats with live or dead mussels totaled thirteen since 1993. (J. Janik, pers. comm.).

Zebra mussels continued to spread in Ontario. Mussels were newly discovered in the Severn River and Trent Canal (Figure 1.10; Ontario Federation of Anglers and Hunters, unpublished data). By the end of 1996, zebra mussels had been reported from a total of 159 small lakes and reservoirs in the United States and Canada (Figure 1.5).

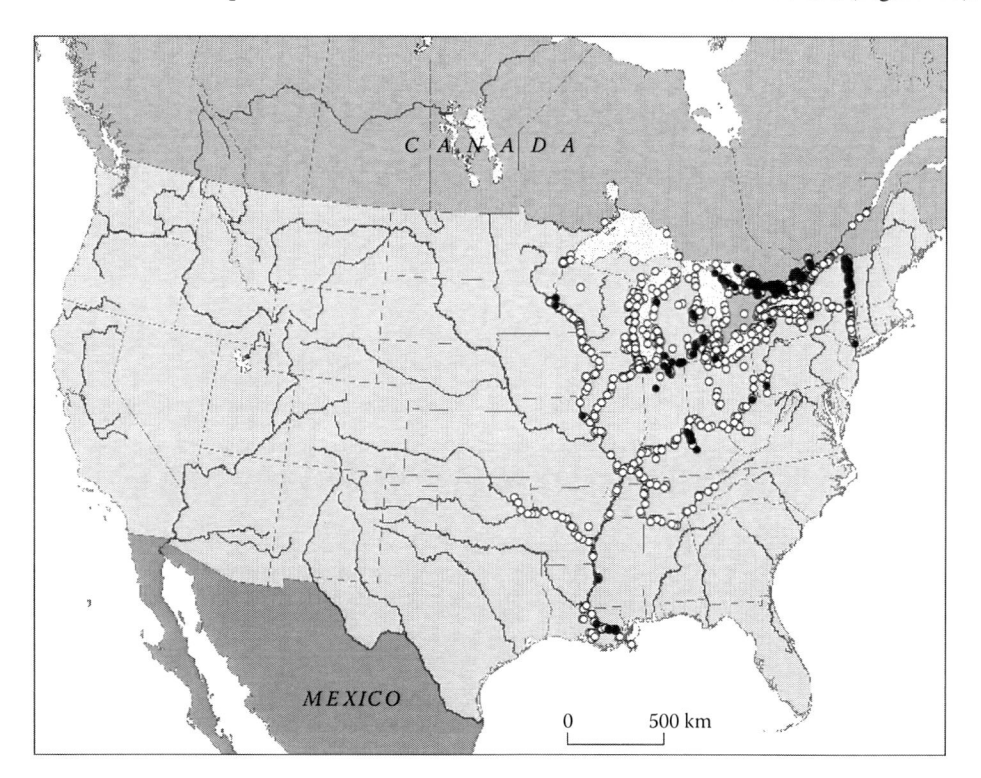

Figure 1.10 Locations of new discoveries of dreissenid mussels in the United States and Canada in 1996 (black points). Also given are locations where mussels were present in previous years (white points).

1997

Monitoring efforts in 1997 showed increased densities at many locations and continued range expansion of both zebra and quagga mussels. The first quagga mussels in Lake Michigan were found just west of the Straits of Mackinac at a maximum density of 14/m^2 in 1997 (Nalepa et al. 2001). Two additional locations with zebra mussels were discovered in the upper Ohio River drainage in Pennsylvania at Dashields Lock (Ohio River Mile 13.3), near Pittsburgh, and at Lock 3 on the Monongahela River (Mile 23.8) near Elizabeth (M. Fowles, pers. comm.). Continued monitoring in Lake Champlain showed peak veliger densities increased each year since sampling began in 1994, and by 1997, densities reached nearly 500,000/m^3 in the southern part of the lake and over 100,000/m^3 in the recently colonized northern part (Vermont DEC, unpublished data). Mussels were now found throughout Lake Champlain and also the freshwater portion of the Hudson River. Density of adults in the river was approximately 40,000/m^2, with the 1996 year class to be the largest to date (D. Strayer, pers. comm.).

A record spawning of mussels took place in Lake Dardanelle on the Arkansas River in spring 1997. In May, the settling rate of new recruits was 2.6 million/m^2 near a power plant intake, which resulted in adult densities of over 77,000/m^2 in mid-July (C. Gagen, pers. comm.). A higher density of 12.6 million/m^2 of new recruits was recorded at another lake location in June (C. Gagen, pers. comm.). By the end of July, populations at various sampling sites suffered large die-offs of 80%–100% as water temperatures reached 32°C for nearly a week (C. Gagen, pers. comm.). In the Illinois River, the number of adult zebra mussels remained lower than in previous years; however, veliger numbers were higher than in 1996 after low numbers in 1994–1995 (D. Blodgett, pers. comm.).

Adult zebra mussel densities in the Tennessee River were slightly higher in 1997 when compared to 1996 (B. Kerley, pers. comm.). Densities increased twofold from 1996 to 60/m^2 in the upper portion of the river, while densities in the lower portions of the river remained low at 7/m^2 except just above Kentucky Dam nearest the river mouth where the density was 45/m^2 (B. Kerley, unpublished data). As in 1996, reproduction was evident at all power plants along the river.

In the eight states and one province surrounding the Great Lakes, zebra mussels continued to expand their range into many small lakes (Figure 1.11). By summer 1997, the cumulative number of lakes where zebra mussels had been reported was as follows: Michigan (78), Ontario (54), Indiana (33), New York (17), Wisconsin (11), and Ohio (8) for an overall total of 190 lakes (Figure 1.5).

1998–2001

The spread of zebra mussels continued in the years 1998–2001 with reports of first occurrences at several new locations (Figure 1.12). In Connecticut, a small number of adult mussels were discovered in East Twin Lake in 1998 when

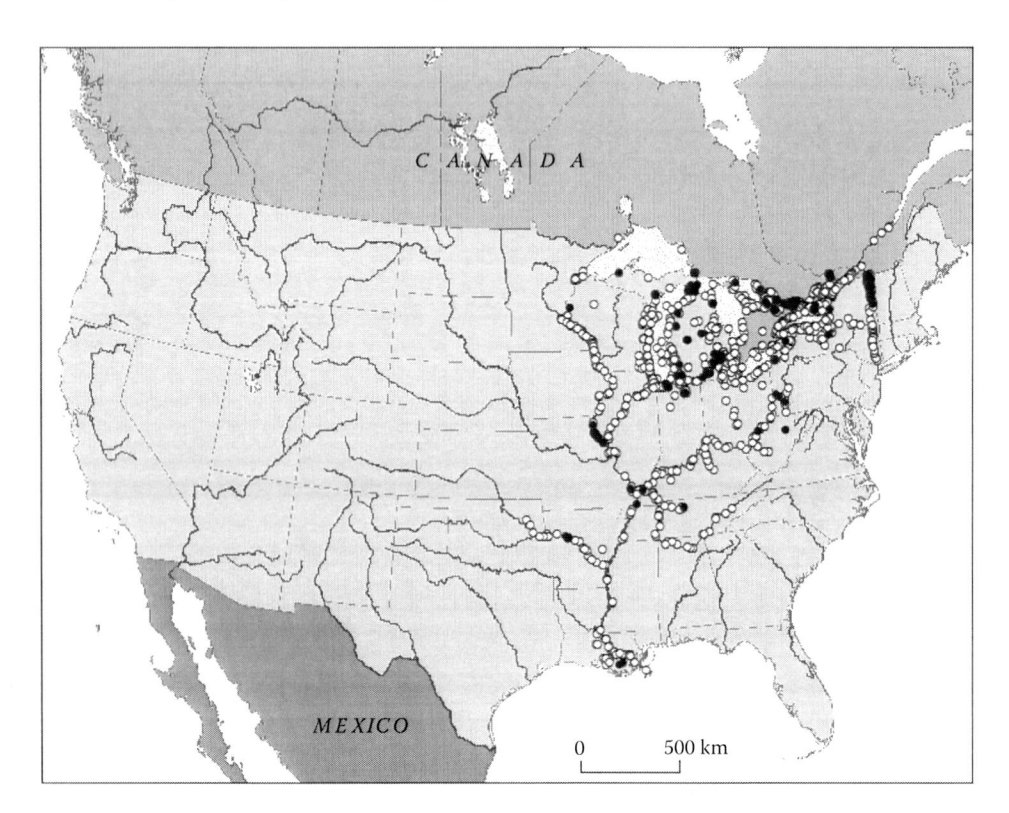

Figure 1.11 Locations of new discoveries of dreissenid mussels in the United States and Canada in 1997 (black points). Also given are locations where mussels were present in previous years (white points).

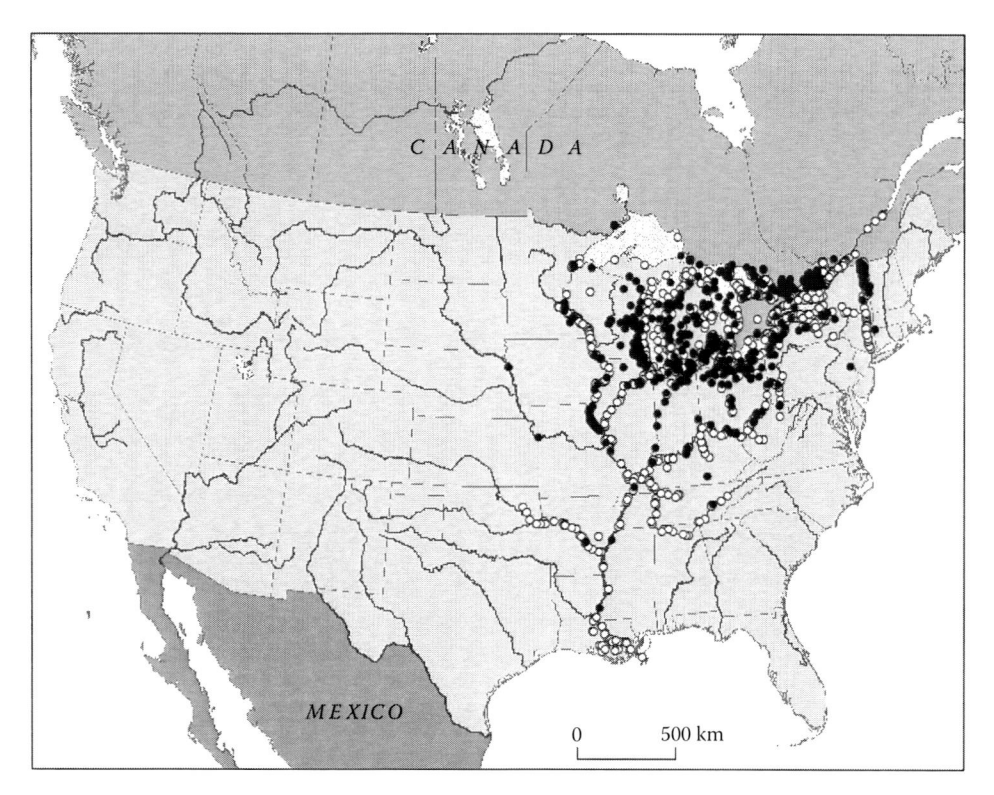

Figure 1.12 Locations of new discoveries of dreissenid mussels in the United States and Canada in 1998–2001 (black points). Also given are locations where mussels were present in previous years (white points).

biologists were conducting an aquatic weed removal project (D. Strayer, pers. comm.) and in adjacent West Twin Lake in 2001 (N. Balcom, pers. comm.). Until this time period, mussels were not reported from the Missouri River despite a steady flow of barge traffic from the infested Mississippi River. Speculation of why mussels were not present in the Missouri River focused on the river's high flow (velocity) and high turbidity. However, a single zebra mussel was eventually identified from the Missouri River in April 1999 at a power company located 24 km south of Sioux City, Iowa. The specimen was found on the boot of the intake structure's traveling screen (G. Weise, pers. comm.). In 2001, mussels were found at a river intake of a power station near Kansas City, Kansas, indicating that the population in the Missouri River had expanded (S. McMurray, pers. comm.).

Large numbers of zebra mussels continued to be a problem in the upper Ohio River where over 3600 kg were removed from a single flood gate at Pike Island Lock and Dam facility in 1999 (M. Fowles, pers. comm.). Zebra mussel densities near native mussel beds in the upper portion of the river reached approximately 6000/m² in 1998; densities at sites in the lower portion of the river decreased from the previous year, possibly due to summer floods (P. Morrison, pers. comm.). In a 48 km stretch of the Seneca River from Cross Lake to Phoenix, New York, the mussel population was estimated to be 4.5 billion in 1999 (C. Lange, pers. comm.).

In 2000, zebra mussels were discovered in Edinboro Lake, a 101 hectare reservoir in Pennsylvania located 23 km east of Lake Erie. The population was well established by the time of discovery and estimated to be over 37 million individuals (T. Shaw, pers. comm.). In December 2000, a water drawdown of 1.5 m was conducted on the reservoir to expose the zebra mussels to freezing air temperatures. The drawdown eliminated over 40% of the population when surveyed the following spring. Eleven months after the first drawdown, a second one was carried out that further reduced the population by 98% of the original population. Unfortunately, the population rebounded to densities of over 40,000/m² in 2009 (J. Grazio, pers. comm.).

Zebra mussels were found in two more rivers in Ontario, the South Nation and the Mississippi, along with 36 additional lakes (Ontario Federation of Anglers and Hunters, unpublished data). The number of small inland lakes reported with infestations increased from 190 at the beginning of 1998 to 432 by the end of 2001. Range expansion was particularly evident in Michigan where 107 lakes were newly infested between 1998 and 2001 (Figure 1.5).

2002–2004

In 2002, several reports of zebra mussels in waters of Midwestern states indicated a significant range expansion (Figure 1.13). Mussels were found in Lake of the Ozarks, an impoundment on the Osage River, which is a tributary of the Missouri River in central Missouri (S. McMurray, pers. comm.). Adult mussels were also found in El Dorado and

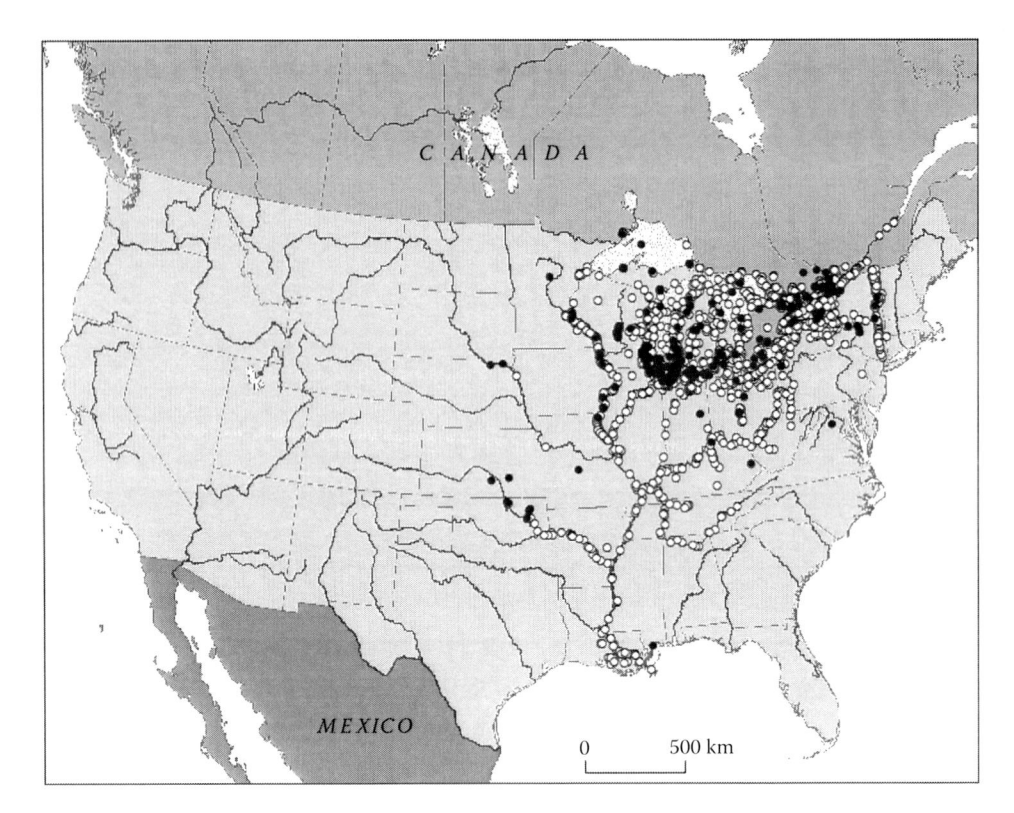

Figure 1.13 Locations of new discoveries of dreissenid mussels in the United States and Canada in 2002–2004 (black points). Also given are locations where mussels were present in previous years (white points).

Cheney reservoirs in 2003 and 2004, respectively, in south-central Kansas (T. Mosher, pers. comm.). Veligers were collected from the Missouri River several kilometers above and below Lewis and Clark Lake along the Nebraska/South Dakota border (J. Shearer, pers. comm.). The source of those veligers was never discovered (A. Burgess, pers. comm.).

In summer 2002, zebra mussels were discovered in a 5-hectare quarry in northern Virginia, the first found in open waters of the state (L. Kozlowsky, pers. comm.). The source was suspected to be a contaminated diving gear as the quarry is used heavily by recreational divers. Because of the quarry's close proximity to drinking water supply of the city of Manassas, this population was viewed as a threat. Therefore, in 2006, the Virginia Department of Game and Inland Fisheries successfully eradicated the population of zebra mussels in the quarry, treating it with 1.0 ppm potassium chloride (Fernald and Watson 2013).

A few zebra mussels were found in French Creek near the town of Venango, Pennsylvania, in 2002 (D. Crabtree, pers. comm.). French Creek is a tributary of the Allegheny River in western Pennsylvania, known for its diverse native mussel populations. It is located downstream from Edinboro Lake where a large mussel population was discovered in 2000. No reproduction in the creek was evident in 2002 and 2003, suggesting Edinboro Lake may serve as source for occasional settlement of juveniles in the creek (D. Crabtree, pers. comm.). In eastern Pennsylvania, both quagga and zebra mussels were collected from a private quarry near Easton (D. Strayer, pers. comm.).

Quagga mussels had not been found in many locations outside the Great Lakes until Grigorovich et al. (2008) reported finding low densities ($1/m^2$) of quagga mussels in the middle Ohio River in 2004; this was the first report of quagga mussels being present in the river. In the same study, they found quagga mussels in low densities ($5/m^2$) in the upper Mississippi River near Dresbach, Minnesota. In Ontario, Canada, quagga mussels were reported in Lake Simcoe in 2004 (Ontario Federation of Anglers and Hunters, unpublished data).

New locations of zebra mussels included Oologah Lake on the Verdigris River in northeastern Oklahoma (E. Laney, pers. comm.), Clear Fork Reservoir in Ohio (R. Sanders, pers. comm.), Dewey Lake in eastern Kentucky (S. Foster, pers. comm.), Ossawinnamakee Lake in central Minnesota (G. Montz, pers. comm.), and several locations in the upper Susquehanna River drainage in south-central New York (T. Horvath, pers. comm.).

The number of small inland lakes with zebra mussels increased from 432 in 2002 to 532 in 2004, and states/provinces with the most infested lakes were Michigan (218), Ontario (106), Wisconsin (67), and Indiana (46) (Figure 1.5).

2005–2006

In 2005 and 2006, Grigorovich et al. (2008) found additional locations with quagga mussels in both the Ohio and upper Mississippi rivers (Figure 1.14). They reported a

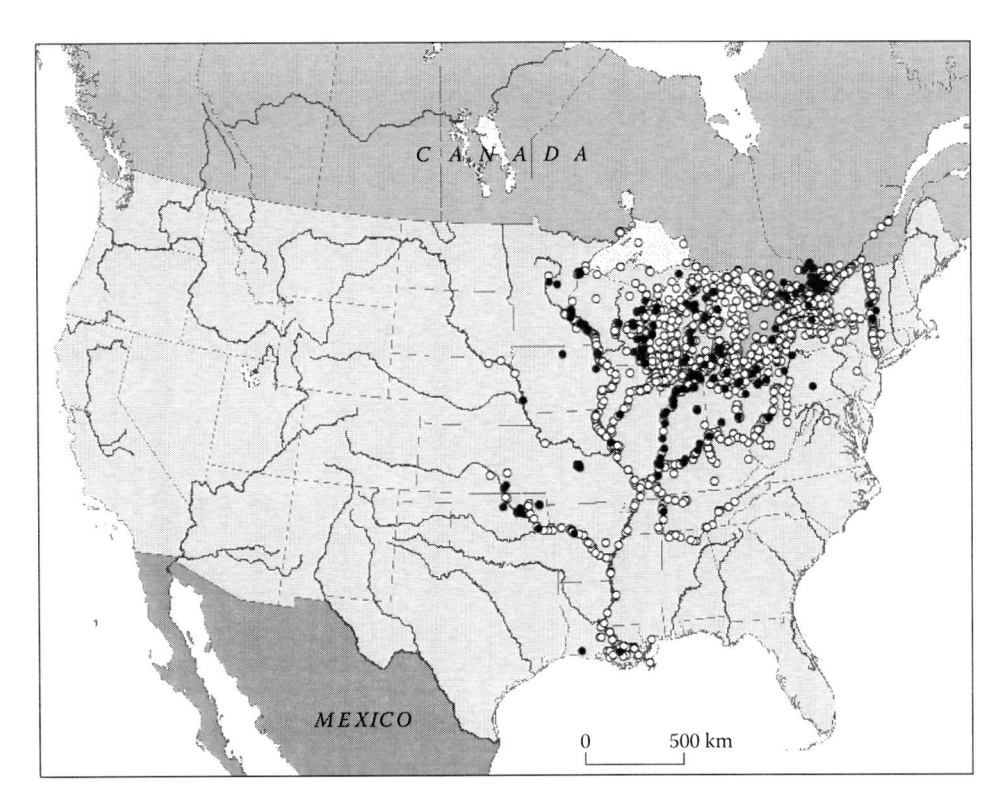

Figure 1.14 Locations of new discoveries of dreissenid mussels in the United States and Canada in 2005–2006 (black points). Also given are locations where mussels were present in previous years (white points).

maximum density of 118/m² in the Mississippi River near Reads Landing, Minnesota. In the Susquehanna River drainage of south-central Pennsylvania, quagga mussels were discovered in Clover Creek Quarry (N. Polato, pers. comm.). With the closest population of quagga mussels located over 241 km away in Lake Erie, this introduction was likely a result of recreational diving activities as suspected in previous quarry infestations. From 1993 to 2006, 10 quarries in the United States became infested with dreissenid mussels.

A large population of zebra mussels was discovered in a 47 hectare man-made lake, located on Offutt Air Force Base in eastern Nebraska (Figure 1.14). The lake is adjacent to the Missouri River and was the first water body entirely within Nebraska to be colonized. Because the lake was seen as a constant source of veligers to the Missouri River where conditions appear suboptimal for reproduction, the decision was made to eradicate the lake population. The eradication effort consisted of treatments with 1.0 ppm copper sulfate in September 2008 and April 2009 (U.S. Air Force 2009). Continued monitoring of the lake found no live mussels or veligers 12 months after the final treatment, but in late 2010, mussels were once again found in the lake (S. Schainost, pers. comm.).

The total number of small inland lakes with zebra mussels grew to 631 by the end of 2006. Ninety-nine newly infested lakes and reservoirs were documented in this 2 year period, of which 27 occurred in Michigan alone. This large increase in Michigan may have been a function of an increase in monitoring and not an increase in the rate of dispersal. Other states/provinces, some outside the Great Lakes region, with newly infested lakes included Iowa, Indiana, Kansas, Minnesota, Missouri, Nebraska, New York, Oklahoma, Pennsylvania, and Wisconsin (Figure 1.5).

2007–2010

The most significant range expansion of dreissenids since they entered the Mississippi River drainage was discovered on January 6, 2007 when quagga mussels were found on a mussel-monitoring substrate in Lake Mead, Nevada (W. Baldwin, pers. comm.; Figure 1.15). Lake Mead is a large impoundment on the Colorado River in Nevada and Arizona covering nearly 64,750 hectares. It supplies water to much of the southwestern United States and is also widely used by recreational boaters. Although it would be impossible to determine the specific vector that introduced quagga mussels to Lake Mead, most likely, it was via a fouled boat arriving from infested waters. Genetic evidence suggests that the source population was from Lake Ontario (Stepien et al. 2013). Within a week of the initial discovery, quagga mussels were found downriver from Lake Mead in two additional impoundments, Lakes Mohave and Havasu (S. Ellis, pers. comm.). Diver surveys of Lake Mead in May 2007 showed that the distribution of quagga mussels was nearly lake-wide and that the mussels had probably been there for several years undetected (B. Moore, pers. comm.).

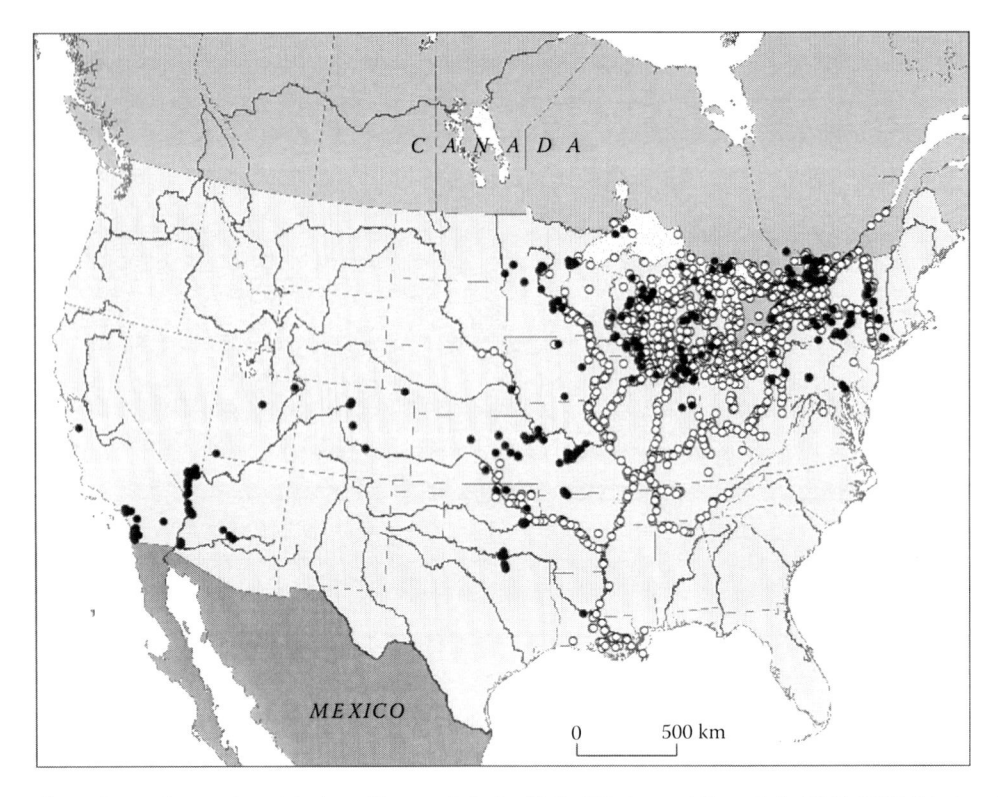

Figure 1.15 Locations of new discoveries of dreissenid mussels in the United States and Canada in 2007–2010 (black points). Also given are locations where mussels were present in previous years (white points).

The presence of quagga mussels in the Colorado River system led to the inadvertent transport of veligers, pumped by way of the Colorado River Aqueduct, into 10 water supply reservoirs in southern California. By late 2009, the state of California reported finding quagga mussels in 19 reservoirs in the southern part of the state. In a separate introduction in northern California, zebra mussels were collected from San Justo Reservoir in January 2008 (D. Norton, pers. comm.). A single quagga mussel was also found in the Central Arizona Project Canal near Scottsdale, Arizona, in August 2007 (L. Riley, pers. comm.). This canal system also draws water from the Colorado River. One year later, mussels were found in the Central Arizona Project–Salt River Project canal connection at the Granite Reef Diversion Dam located 35 km east of Phoenix (L. Swanson, pers. comm.).

Reports of zebra and quagga mussels in new locations indicated continued expansion in western states. Dreissenids were first found in the state of Colorado in January 2008 at Pueblo Reservoir on the upper Arkansas River (V. Milano, pers. comm.). By October, six additional reservoirs in Colorado were found to have quagga mussel veligers, two of which also had zebra mussel veligers; however, no adults or veligers were found in any Colorado waters in 2010 (E. Brown, pers. comm.). In Utah, a water sample collected from Electric Lake in November 2008 was found to contain zebra mussel DNA, and, although no adults were found, this was the first indication of possible presence of mussels in the state (L. Dalton, pers. comm.). In February 2009, quagga mussel veligers were positively identified in Red Fleet Reservoir in northeastern Utah and a lone live adult was collected from Sand Hollow Reservoir in May 2010 (L. Dalton, pers. comm.). Sampling efforts were unable to find any evidence of veligers or additional adult mussels in any Utah reservoir in 2010 (Utah Division of Wildlife Resources 2011). In April 2009, a single zebra mussel specimen was collected in Lake Texoma, Texas, which is an impoundment on the Red River. A search of the lake in the following months revealed large numbers of mussels at multiple locations, including a smaller number of individuals in an adjacent creek north of Dallas (B. Hysmith, pers. comm.). An attempt to eradicate mussels from the Dallas-area creek with potassium chloride was made in September 2010.

Monitoring of zebra mussels within the upper Susquehanna River Basin found small numbers at 13 new sites in northeast Pennsylvania and south-central New York in 2007 (A. Faulds, pers. comm.). Although veligers were collected from this area previously in 1991, no adult population was found to exist until 2004 (T. Horvath, pers. comm.). In the lower Susquehanna River Basin, several dead adult zebra mussels were discovered in November 2008 after a drawdown at Muddy Run Reservoir, a pumped storage facility off the lower Susquehanna River in southeast Pennsylvania. Mussels were also found at Conowingo Dam, which is approximately 19 km downriver from Muddy Run Reservoir. This find was the first report of live mussels in the state of Maryland (R. Klauda, pers. comm.). Elsewhere, an established zebra mussel population was discovered for the first time in the state of Massachusetts at Laurel Lake in 2009. Shortly thereafter, more mussels were found downstream from the lake in the Housatonic River (T. Flannery, pers. comm.). A few individual zebra mussels were reported from two impoundments on the Housatonic River in southwestern Connecticut in October 2010. In New York, quagga mussels had not been found in a small lake since 1994, but from 2008 to 2010, four more small lakes were added to the list of infested lakes, bringing the total to seven in the state. Zebra mussels were discovered in a second Nebraska reservoir in late November 2010 (J. LaRandeau, pers. comm.). The reservoir was promptly drained in an attempt to eradicate the mussels.

By the end of 2010, either one or both dreissenid species had been detected in 772 small natural lakes, reservoirs, impoundments, and quarries in the United States and Canada outside the initial Great Lakes invasion (Figure 1.5). In the United States, mussels have been found in 611 lakes in 26 states, of which 568 had zebra mussel adults or veligers only, 33 had quagga mussels only, and 10 had both species. In Canada, only zebra mussels had been documented in 161 small water bodies, all within the province of Ontario. In addition to lakes, dreissenids had been found in 131 rivers, of which 110 were in the United States and 21 in Canada (Table 1.1). A final observation on the distributions of both species is that quagga mussels are replacing the zebra mussels in Lake Ontario (Mills et al. 1999, Wilson et al. 2006), Lake Michigan (Nalepa et al. 2009), and Lake Erie (E. Obert, pers. comm.). This is not unique to North America as Ludyanskiy (1991) reviewed the literature and found that *D. r. bugensis* replaced *D. polymorpha* in the Dnieper River system in Ukraine.

DISCUSSION

The initial rate of spread of dreissenids in North America was so rapid that, besides passive dispersal, human intervention must have played a significant role, especially to upstream locations and disconnected waterways. Carlton (1993) listed 20 examples of such activities or mechanisms. At the same time, it cannot be assumed that passive, downstream range expansion of dreissenids was not aided by human activities in the Great Lakes, not only because there are large gaps in their distribution patterns but also because of the rapid appearance of dreissenids in Lake Ontario. In just 2 years after their discovery, zebra mussels were distributed across the entire Great Lakes drainage from Lake Superior to the St. Lawrence River (Johnson and Carlton 1996).

Small vessels and larger ship traffic would likely be a primary mechanism of mussel dispersion. Microscopic mussel larvae are capable of being transported in ship ballast, and adults can adhere to vessel hulls. However, distribution by vessels cannot be totally substantiated due to the lack of

Table 1.1 Country, State and Province, Water Body, and Year Dreissenid Mussels Were First Discovered in North America

Country	State/Province	Water Body	Year
CA	ONT	Lake Erie	1986
CA	ONT	Lake St. Clair	1988
US	MI	St. Clair River	1989
CA	ONT	Welland Canal	1989
CA	ONT	Lake Ontario	1989
US	WI	Lake Michigan	1989
US	MN	Lake Superior	1989
US	MI	River Raisin	1989
US	NY	St. Lawrence River	1989
US	WI	St. Louis River	1989
US	IL	Chicago Sanitary & Ship Canal	1989
US	MI	Lake Huron	1990
US	IL	Calumet River	1990
US	NY	Genesee River	1990
US	NY	Oswego River	1990
CA	ONT	St. Clair River	1990
CA	ONT	St. Lawrence River	1990
CA	ONT	St. Mary's River	1990
US	IL	Chicago River	1991
US	IL	Des Plaines River	1991
US	NY	Erie Canal	1991
US	NY	Hudson River	1991
US	IL	Illinois River	1991
US	IL	Kankakee River	1991
US	IL	Mississippi River	1991
US	WI	Mississippi River	1991
US	NY	Mohawk River	1991
CA	ONT	Niagara River	1991
US	IL	Ohio River	1991
US	NY	Susquehanna River	1991
US	KY	Tennessee River	1991
US	AR	Arkansas River	1992
US	MI	Au Gres River	1992
US	MI	Belle River	1992
US	MI	Black River	1992
US	WI	Black River	1992
US	MI	Clinton River	1992
US	KY	Cumberland River	1992
US	TN	Cumberland River	1992
CA	ONT	Detroit River	1992
US	MI	Detroit River	1992
US	WV	Kanawha River	1992
US	AR	Mississippi River	1992
US	IA	Mississippi River	1992
US	LA	Mississippi River	1992
US	MN	Mississippi River	1992
US	MS	Mississippi River	1992
US	NY	Niagara River	1992

(*continued*)

Table 1.1 (continued) Country, State and Province, Water Body, and Year Dreissenid Mussels Were First Discovered in North America

Country	State/Province	Water Body	Year
US	IN	Ohio River	1992
US	KY	Ohio River	1992
US	OH	Ohio River	1992
US	WV	Ohio River	1992
US	MI	Pigeon River	1992
US	MI	Pine River	1992
US	MI	Saginaw River	1992
CA	ONT	Soulanges Canal	1992
US	AL	Tennessee River	1992
US	TN	Tennessee River	1992
US	MI	Thunder Bay River	1992
US	OK	Arkansas River	1993
US	LA	Bayou Teche	1993
US	WI	Flambeau River	1993
US	MI	Menominee River	1993
US	TN	Mississippi River	1993
CA	ONT	Rideau Canal	1993
CA	ONT	Rideau River	1993
US	OH	Scioto River	1993
CA	ONT	Snye River	1993
US	OK	Verdigris River	1993
US	PA	Allegheny River	1994
US	MI	Au Sable River	1994
CA	ONT	Ausable River	1994
US	WI	Fox River	1994
US	KY	Green River	1994
US	MO	Mississippi River	1994
CA	ONT	Murray Canal	1994
CA	ONT	Sauble River	1994
CA	ONT	Trent River	1994
US	AR	White River	1994
US	LA	Atchafalaya River	1995
CA	ONT	Cataraqui River	1995
US	TN	French Broad River	1995
US	KY	Kentucky River	1995
US	PA	Monongahela River	1995
CA	ONT	Otonabee River	1995
CA	ONT	Ottawa River	1995
US	WI	Sheboygan River	1995
US	MN	St. Croix River	1995
US	MI	Huron River	1996
CA	ONT	Severn River	1996
US	IN	St. Joseph River	1996
US	IN	Tippecanoe River	1996
CA	ONT	Trent Canal	1996
US	WV	Buckhannon River	1997
US	OH	Hocking River	1997
CA	ONT	Napanee River	1997
US	PA	Ohio River	1997

Table 1.1 (continued) Country, State and Province, Water Body, and Year Dreissenid Mussels Were First Discovered in North America

Country	State/Province	Water Body	Year
US	MI	Ontonagon River	1997
US	IN	Deep River	1998
US	MI	Flint River	1998
US	IN	Wabash River	1998
US	WI	Bark River	1999
US	IN	Fawn River	1999
US	OH	Little Miami River	1999
US	WI	Menominee River	1999
US	MO	Meramec River	1999
US	IA	Missouri River	1999
US	WV	Monongahela River	1999
US	WI	Oconomowoc River	1999
US	WI	Oconto River	1999
US	WI	Pewaukee River	1999
US	IL	Rock River	1999
US	NY	Seneca River	1999
US	IL	Wabash River	1999
US	WI	Wisconsin River	1999
US	WI	Wolf River	1999
US	OH	Auglaize River	2000
US	IL	Fox River	2000
US	IL	Mazon River	2000
US	OH	Sandusky River	2000
US	IL	Skokie River	2000
CA	ONT	South Nation River	2000
US	MN	Zumbro River	2000
US	OH	Grand River	2001
US	OH	Maumee River	2001
CA	ONT	Mississippi River	2001
US	KY	Mississippi River	2001
US	KS	Missouri River	2001
US	WI	Muskego Canal	2001
US	OH	Muskingum River	2001
US	WI	White River	2001
US	PA	French Creek	2002
US	WI	Mukwonago River	2002
US	MO	Osage River	2002
US	NE	Missouri River	2003
US	KS	Walnut River	2003
US	KS	North Fork Ninnescah River	2004
US	OH	Olentangy River	2004
CA	ONT	Lynn River	2005
CA	ONT	Muskrat River	2005
US	OK	Neosho River	2005
US	OH	Chagrin River	2006
US	OK	Hominy River	2006
US	IA	Maquoketa River	2006
US	OH	Rocky River	2006

Table 1.1 (continued) Country, State and Province, Water Body, and Year Dreissenid Mussels Were First Discovered in North America

Country	State/Province	Water Body	Year
US	WI	Amnicon River	2007
US	AZ	Colorado River	2007
US	CA	Colorado River	2007
US	NV	Colorado River	2007
US	KS	Delaware River	2007
CA	ONT	Madawaska River	2007
US	MN	Pine River	2007
US	PA	Susquehanna River	2007
US	NV	Virgin River	2007
US	NY	West Branch Tioughnioga River	2007
US	MO	White River	2007
US	CO	Arkansas River	2008
US	IA	Cedar River	2008
US	CO	Colorado River	2008
US	KS	North Cottonwood River	2008
US	AZ	Salt River CAP-SRP Canal	2008
US	MD	Susquehanna River	2008
US	IA	Chariton River	2009
US	MA	Housatonic River	2009
US	KS	Kansas River	2009
US	MO	Missouri River	2009
US	TX	Red River	2009
US	KS	Republican River	2009
US	KS	Saline River	2009
US	MN	St. Louis River	2009
US	OK	Washita River	2009
US	OK	Canadian River	2010
US	KS	Cottonwood River	2010
US	CT	Housatonic River	2010
US	MO	Little Platte River	2010
US	KS	Neosho River	2010
US	OK	North Canadian River	2010
US	LA	Red River	2010
US	ND	Red River	2010

The 1986 record was derived from Carlton (2008); US, United States; CA, Canada.

sampling in areas with and without ports at the time (Johnson and Padilla 1996). Nonetheless, as early as fall 1989, live mussels were discovered on buoys decommissioned from Lake Superior at Duluth Harbor, and a small number of mussels were found on buoys located in the harbor the following April (D. Pratt, pers. comm.). In 1990, commercial vessels at several locations in the upper Great Lakes including Superior Harbor (Lake Superior), Milwaukee Harbor and Sturgeon Bay (Lake Michigan), and Saginaw Bay (Lake Huron) were found to have mussels on their hulls, and all had spent time in either Lake St. Clair or Lake Erie (C. Kraft, pers. comm.).

Prior to dispersal of mussels into drainages outside the Great Lakes, Griffiths et al. (1991) mentioned that both the Erie Canal and the Chicago Sanitary and Ship Canal were pathways by which mussels could disperse. Most likely, a combination of two man-mediated vectors, canals and ships, expedited the spread of zebra mussels outside the Great Lakes early in their North American invasion. Griffiths et al. (1991) also noted recreational boating and bait buckets as possible methods of spreading zebra mussels overland between drainages. Ricciardi et al. (1995) showed that both species of dreissenids could withstand periods of desiccation for 3–5 days in cool, moist conditions. With these conditions, such a time period would allow the transport of mussels from the Great Lakes to any part of North America.

Either by passive drift or by active transport through vessel traffic, zebra mussels dispersed out of the Great Lakes (specifically from Lake Michigan) down the Illinois River and into the Mississippi River. A single mussel was found in September 1991 in the Mississippi River near river mile marker 695 at La Crosse, Wisconsin, which was located hundreds of kilometers away from known populations, suggesting it most likely became dislodged from a barge hull (Cope et al. 1997). A similar example of active transport was the finding of a single zebra mussel on an intake screen at a power plant on the Missouri River in 1999; barges pass within 9 m of the intake screen (G. Weise, pers. comm.). The power plant was located 1152 km upriver from the nearest known dreissenid population. Yet another example occurred in April 1992 when a barge dry-docked for repairs on the Mississippi River, just below the mouth of the Illinois River at Hartford, Illinois. The barge was found to have over 1000 zebra mussels attached to its hull (Keevin et al. 1992). The barge's logbook showed that it had traveled 20,563 km up and down the Mississippi River from Minnesota to Louisiana before dry-docking (Keevin et al. 1992). These incidences, along with known natural dispersal patterns, give credibility to the assumption that barge traffic has been the primary cause of dispersal in navigable waters of major rivers.

Overland movement on recreational boats became of great concern once mussels were found in waters not directly connected to the Great Lakes. Numerous incidences were documented in which highway inspection stations and marina operators found zebra mussels attached to boats and underwater equipment that were hauled overland by trailers. As early as 1993, the state of California began to document mussels on boat hulls stopped at agricultural inspection stations. Approximately 34 boats were detected with zebra mussels between 1993 and 2004 (J. Janik and B. Sandige, pers. comm.). After the discovery of mussels in Lake Mead, boat inspections were increased, and from January 2009 through March 2010, an astonishing 361,370 boats were inspected, of which 27,699 were cleaned and 561 were either contaminated or suspected of being contaminated with mussels (California Department of Food and Agriculture, unpublished data). The inspection program in the state of Washington reported intercepting and decontaminating 17 boats with attached zebra mussels from 2007 to 2009 (A. Pleus, pers. comm.). In Idaho, two boats with mussels attached were intercepted in 2009, and eight boats were intercepted in 2010 at newly deployed inspection stations (A. Ferriter, pers. comm.). In addition to their ability to attach to hard substrates, zebra mussels have been known to colonize submerged aquatic vegetation (Garton and Haag 1993), thus providing a less obvious pathway of overland dispersal when boat trailers are fouled with aquatic plants. During a 2002 seasonal boat launch survey at the Allegheny Reservoir in northwestern Pennsylvania, zebra mussels were found attached to aquatic vegetation on 14 boat trailers (Hillyard 2003).

The reproductive strategy of zebra mussels does not appear to favor small isolated introductions because external fertilization can dilute gametes to where fertilization is greatly reduced (Johnson and Padilla 1996). Yet many isolated water bodies became infested with mussels. Bossenbroek et al. (2007) predicted that Lake Mead would be very vulnerable to mussel invasion, but it is not known how many times adults or veligers of quagga mussels were introduced into Lake Mead before the population became established. Similarly, the population of zebra mussels in Lake Texoma, Texas, was discovered many kilometers upriver from an existing population. Over several years, five boats at the lake were documented with mussels attached to the hulls prior to the discovery of the population in 2009. Because the finding of these infested boats was not part of an organized inspection program, the possibility exists that many infested boats may have been launched into the reservoir before mussels became established (B. Hysmith, pers. comm.). These instances, along with infested boats coming into California, indicate overland transport of zebra mussels on or in boats, and boat trailers from thousands of kilometers away pose a real threat to uninfested waters. The discovery of mussels in Lake Mead and Lake Texoma represents examples of populations becoming established in disjunct locations, and there can be little doubt that mussels arrived by overland transport on a boat coming from infested waters. In the western United States, where water is in short supply and efficient delivery is crucial, inspection and monitoring are essential to early detection and rapid response, and the first step in management or control.

In view of the western expansion, a question to consider is how impoundments may actually enhance the distribution of some invasive species. Johnson et al. (2008) viewed impoundments as possible "stepping stones" for invasive species to expand their distributions. After examining boating activity, accessibility, and physicochemical characteristics of 5281 water bodies in Wisconsin, Johnson et al. (2008) concluded that impoundments were significantly more likely to be invaded than natural lakes. This being the case, river systems with impoundments that are commonly found in the western United States would be particularly vulnerable. Another logical question is whether or not the quagga mussel would be able to survive conditions in the Colorado River if it were not for Lake Mead. A risk assessment after mussels were discovered in Lake Mead indicated that the probability of mussels reaching high densities in the main portion of the Colorado River was very low (Kennedy 2007). It was determined that high water current velocities and high suspended loads in the river would severely limit populations. However, quagga mussels are rapidly expanding in Lake Mead, even in deeper regions where sediments consist of silt and other fine materials (Wong et al. 2013).

In addition to unintentional introductions by boats and ships, several unique pathways have been reported. In 2000, zebra mussels were found attached to the roots of approximately 2500 flowers harvested from canals in The Netherlands and imported to the state of Michigan (L. Drees, pers. comm.). The flower shipment was accompanied by a Netherlands plant quarantine certificate stating that it was "free from quarantine pests and practically free from other injurious pests." Another plant that provided a unique pathway was a green alga, *Aegagropila linnaei*, referred to as the "Marimo ball" or "Japanese moss ball" in the aquarium trade. Juvenile quagga mussels were found inside one of these moss balls in a large shipment from Singapore bound for sale in the United States (K. Rondeau, pers. comm.). The shipment was intercepted at the Miami International Airport by U.S. Department of Agriculture inspectors. These unusual and unforeseen pathways for introduction could potentially distribute zebra and quagga mussels to any water body in North America.

Fortunately, few intentional introductions have been documented. However, in 1998, a Florida conservation officer confiscated zebra mussels from a bait shop that were brought illegally into the state from Lake Champlain in Vermont (G. Warren, pers. comm.). The shop owner wanted to see if they would survive in Florida, most likely to be used to increase the visibility of water for diving activities. Zebra mussels have turned up in several quarries known for recreational diving probably for the same reason, increased water clarity. Another example of an intentional introduction occurred in Ontario, Canada, where a home aquarium release of mussels into a nearby water body was reported. Fortunately, a rapid response to the site resulted in removal of the mussels before they became established (C. O'Neill, pers. comm.).

It is often stated that when an invasive species becomes common over a large area, the prevailing assumption is that it is "everywhere" and thus, prevention of future introductions becomes complacent. The facts indicate that zebra and quagga mussels are not everywhere, and whether intentional or unintentional, it is important that new pathways are recognized and exposed to help prevent further spread.

APPENDIX

List of Personal Communications

Ahlstedt, Steve, Tennessee Valley Authority
Allen, Yvonne, Louisiana State University
Balcom, Nancy, Connecticut Sea Grant
Baldwin, Wen, Lake Mead Boat Owner's Association
Bartelt, Gary, U.S. Army Corps of Engineers
Blodgett, Douglas, Illinois Natural History Survey
Brown, Elizabeth, Colorado Division of Wildlife
Burgess, Andy, South Dakota Game Fish and Parks
Crabtree, Darrin, The Nature Conservancy
Cremeans, Bill, U.S. Army Corps of Engineers
Dalton, Larry, Utah Division of Wildlife Resources
Drees, Linda, U.S. Fish and Wildlife Service
Eades, Rick, Virginia Department of Game and Inland Fisheries
Ellis, Susan, California Department of Fish and Game
Faulds, Ann, Pennsylvania Sea Grant
Ferriter, Amy, Idaho Fish and Game
Flannery, Tom, Massachusetts Department of Conservation
Forester, John, U.S. Fish and Wildlife Service
Foster, Steven, U.S. Army Corps of Engineers
Fowles, Mike, U.S. Army Corps of Engineers
Gagen, Charles, Arkansas Tech University
Grant, Bob, Jefferson Parrish Waterworks
Harris, John, Arkansas Highway Department
Janik, Jeff, California Department of Water Resources
Higgins, Brian, Kentucky River Authority
Horvath, Thomas, SUNY College at Oneonta
Hubbs, Don, Tennessee Wildlife Resources Agency
Hysmith, Bruce, Texas Department of Parks and Wildlife
Kamman, Neil, Vermont Department of Environmental Conservation
Keevin, Tom, U.S. Army Corps of Engineers
Kellogg, Jim, Vermont Department of Environmental Conservation
Kerley, Bennie, Tennessee Valley Authority
Klauda, Ron, Maryland Department of Natural Resources
Kozlowsky, Louis, National Oceanic and Atmospheric Administration
Kraft, Clifford, Wisconsin Sea Grant Institute
Laney, Everett, U.S. Army Corps of Engineers
Lange, Cameron, Acres International, Inc.
LaRandeau, John, U.S. Army Corps of Engineers
Mastellar, Ed, The Pennsylvania State University
McKay, Beth, Ontario Ministry of Natural Resources
McMurray, Steve, Missouri Department of Conservation
Milano, Vicky, Colorado Division of Wildlife
Mills, Ed, Cornell University
Mooradian, Steve, New York Department of Environmental Conservation
Montz, Gary, Minnesota Department of Natural Resources
Morrison, Patti, U.S. Fish and Wildlife Service
Mosher, Tom, Kansas Department of Wildlife and Parks
Muia, Ray, Calgon Corporation
Nierzwicki-Bauer, Sandra, Rensselaer Polytechnic Institute
Obert, Eric, The Pennsylvania State University
O'Neill, Charles, Jr., New York Sea Grant
Orndorff, Perry, Dublin Road Water Treatment Plant
Pleus, Alan, Washington Department of Fish and Wildlife
Polato, Nick, The Pennsylvania State University
Pratt, Dennis, Wisconsin Department of Natural Resources
Riley, Larry, Arizona Game and Fish Department
Rondeau, Kristian, U.S. Department of Agriculture
Sanders, Randy, Ohio Department of Natural Resources
Sandige, William, California Department of Water Resources
Schainost, Steve, Nebraska Game and Parks Commission
Schurder, Marla, Illinois-Indiana Sea Grant
Shaw, Tony, Pennsylvania Department of Environmental Protection
Shearer, Jeff, South Dakota Game Fish and Parks
Strayer, David, Cary Institute of Ecosystem Studies
Swanson, Lesly, Salt River Project Environmental Services
Warren, Gary, Florida Fish and Wildlife Conservation Commission
Wedan, Dave, U.S. Fish and Wildlife Service
Weise, Gary, MidAmerican Energy Company
Young, Kevin, Acres International, Inc.

ACKNOWLEDGMENTS

I thank the many individuals too numerous to list who reported hundreds of observations and collections to this effort. Special thanks goes to the Ontario Federation of Anglers and Hunters for much of the Canadian observations. Other significant contributors include NOAA Sea Grant College Programs, NOAA Great Lakes Environmental Research Laboratory, U.S. Fish and Wildlife Service, U.S. Environmental Protection Agency, U.S. Army Corps of Engineers, National Park Service, Ontario Ministry of Natural Resources, Fisheries and Oceans Canada, Illinois Natural History Survey, Iowa Department of Natural Resources, Tennessee Valley Authority, California Department of Water Resources, California Department of Fish and Game, Vermont Department of Environmental Conservation, Indiana Department of Natural Resources, Missouri Department of Conservation, Kansas Department of Wildlife and Parks, Minnesota Department of Natural Resources, Ohio Department of Natural Resources, Wisconsin Department of Natural Resources, Colorado Division of Wildlife, Pennsylvania Department of Environmental Protection, Utah Division of Wildlife Resources, Kentucky River Authority, Texas Parks and Wildlife, Arizona Game and Fish Department, Arkansas Tech University, Cornell University, Louisiana State University, Cary Institute of Ecosystem Studies, and Rensselaer Polytechnic Institute. A thank you also goes to this book's editors for the opportunity, to the U.S. Geological Survey's Invasive Species Program, and to two anonymous reviewers.

REFERENCES

Bossenbroek, J. M., L. E. Johnson, B. Peters, and D. M. Lodge. 2007. Forecasting the expansion of zebra mussels in the United States. *Conserv. Biol.* 21: 800–810.

Carlton, J. T. 1993. Dispersal mechanisms of the zebra mussel (*Dreissena polymorpha*), In *Zebra Mussels: Biology, Impacts, and Control*, T. F. Nalepa and D. W. Schloesser, eds., pp. 677–698. Boca Raton, FL: CRC Press.

Carlton, J. T. 2008. The zebra mussel *Dreissena polymorpha* found in North America in 1986 and 1987. *J. Great Lakes Res.* 34: 770–773.

Conn, D. B., S. R. LaPan, and G. C. LeTendre. 1992a. Range expansion of zebra mussels (*Dreissena polymorpha*) and quagga mussels (*Dreissena* sp.) in the St. Lawrence River and the eastern basin of Lake Ontario. *Dreissena polymorpha Inform. Rev.* 3: 1.

Conn, D. B., K. A. Shoen, and S. Lee. 1992b. The spread of the zebra mussel, *Dreissena polymorpha*, in the St. Lawrence River, and its potential interactions with native benthic biota. *J. Shellfish Res.* 11: 222.

Cope, W. G., M. R. Bartsch, and R. R. Hayden. 1997. Longitudinal patterns in abundance of the zebra mussel (*Dreissena polymorpha*) in the Upper Mississippi River. *J. Freshwat. Ecol.* 12: 235–238.

Counts, C. L., III and T. S. Handwerker. 1990. A review of the biology of *Dreissena polymorpha* (Pallas, 1771) with considerations for research and control. Princess Anne, MD: University of Maryland Eastern Shore.

Dermott, R., J. Mitchell, I. Murray, and E. Fear. 1993. Biomass and production of zebra mussels (*Dreissena polymorpha*) in shallow waters of northeastern Lake Erie, In *Zebra Mussels: Biology, Impacts, and Control*, T. F. Nalepa and D. W. Schloesser, eds., pp. 399–413. Boca Raton, FL: CRC Press.

Fernald, R. T. and B. T. Watson. 2013. Eradication of zebra mussels (*Dreissena polymorpha*) from Milbrook Quarry, Virginia: Rapid response in a real world. In *Quagga and Zebra Mussels: Biology, Impacts, and Control*, T. F. Nalepa and D. W. Schloesser, eds., pp. 195–215. Boca Raton, FL: CRC Press.

Forester, J. S., R. H. Kay, R. L. Walter, and D. J. Fruge. 1993. Zebra mussel monitoring in the lower Mississippi River. *Dreissena polymorpha Inform. Rev.* 4: 3–5.

Garton, D. W. and W. R. Haag. 1993. Seasonal reproductive cycles and settlement patterns of *Dreissena polymorpha* in western Lake Erie, In *Zebra Mussels Biology, Impacts, and Control*, T. F. Nalepa and D. W. Schloesser, eds., pp. 111–128. Boca Raton, FL: CRC Press.

Griffiths, R. W. 1993. Effects of zebra mussels (*Dreissena polymorpha*) on the benthic fauna of Lake St. Clair, In *Zebra Mussels Biology, Impacts, and Control*, T. F. Nalepa and D. W. Schloesser, eds., pp. 415–437. Boca Raton, FL: CRC Press.

Griffiths, R. W., D. W. Schloesser, J. H. Leach, and W. P. Kovalak. 1991. Distribution and dispersal of the zebra mussel (*Dreissena polymorpha*) in the Great Lakes region. *Can. J. Fish. Aquat. Sci.* 48: 1381–1388.

Grigorovich, I A., T. R. Angradi, and C. A. Stepien. 2008. Occurrence of the quagga mussel (*Dreissena bugensis*) and the zebra mussel (*Dreissena polymorpha*) in the Upper Mississippi River System. *J. Freshwat. Ecol.* 23: 429–435.

Hebert, P. D. N., B. W. Muncaster, and G. L. Mackie. 1989. Ecological and genetic studies of *Dreissena polymorpha* (Pallas): A new mollusc in the Great Lakes. *Can. J. Fish. Aquat. Sci.* 46: 1587–1591.

Hebert, P. D. N., C. C. Wilson, M. H. Murdoch, and R. Lazar. 1991. Demography and ecological impacts of the invading mollusc *Dreissena polymorpha*. *Can. J. Zool.* 69: 405–409.

Hillyard, A. 2003. Zebra mussel program 2003. U.S. Forest Service, Allegheny National Forest, Bradford Ranger District, Bradford, Pennsylvania.

Johnson, L. E. and J. T. Carlton. 1996. Post-establishment spread in large-scale invasions: Dispersal mechanisms of the zebra mussel *Dreissena polymorpha*. *Ecology* 77: 1686–1690.

Johnson, P. T. J., J. D. Olden, and M. J. Vander Zanden. 2008. Dam invaders: Impoundments facilitate biological invasions into freshwaters. *Front. Ecol. Environ.* 6: 357–363.

Johnson, L. E. and D. K. Padilla. 1996. Geographic spread of exotic species: Ecological lessons and opportunities from the invasion of the zebra mussel *Dreissena polymorpha*. *Biol. Conserv.* 78: 23–33.

Keevin, T., R. Yarborough, A. C. Miller, and E. A. Theriot. 1992. Inadvertent transport of live zebra mussels on barges—Experiences in the St. Louis District, Spring 1992, Zebra Mussel Technical Notes ZMR-1-07. U.S. Army Corps of Engineers, Waterways Experiment Station, Vicksburg, MS.

Kennedy, T. A. 2007. A Dreissena risk assessment for the Colorado River ecosystem. U.S. Geological Survey Open-File Report 2007-1085, http://pubs.usgs.gov/of/2007/1085/of2007-1085.pdf.

Kovalak, W. P., G. D. Longton, and R. D. Smithee. 1993. Infestation of power plant water systems by the zebra mussel (*Dreissena polymorpha* Pallas), In *Zebra Mussels: Biology, Impacts, and Control*, T. F. Nalepa and D. W. Schloesser, eds., pp. 359–380. Boca Raton, FL: CRC Press.

Kraft, C. 1994. Quagga mussel found in the Ohio River. Zebra mussel update #21. Wisconsin Sea Grant Institute, Madison, WI, 6pp.

Kraft, C. 1995. New sightings. Zebra mussel update #24. Wisconsin Sea Grant Institute, Madison, WI, 6pp.

Leach, J. H. 1993. Impacts of the zebra mussel (*Dreissena polymorpha*) on water quality and fish spawning reefs in western Lake Erie, In *Zebra Mussels: Biology, Impacts, and Control*, T. F. Nalepa and D. W. Schloesser, eds., pp. 381–397. Boca Raton, FL: CRC Press.

Lei, J. and A. C. Miller. 1994. Shell shape differences in *Dreissena* spp. Zebra mussel Research Technical Note ZMR-1-21, U.S. Army Corps of Engineer Waterways Experiment Station, Vicksburg, Mississippi.

LePage, W. L. 1993. The impact of *Dreissena polymorpha* on waterworks operations at Monroe, Michigan: A case history, In *Zebra mussels: Biology, Impacts, and Control*, T. F. Nalepa and D. W. Schloesser, eds., pp. 333–358. Boca Raton, FL: CRC Press.

Ludyanskiy, M. L. 1991. Are *Dreissena polymorpha* and *Dreissena bugensis* synonymous? *Dreissena polymorpha Inform. Rev.* 2: 2–3.

Ludyanskiy, M. L., D. McDonald, and D. MacNeill. 1993. Impact of the zebra mussel, a bivalve invader. *BioScience* 43: 533–544.

Mackie, G. L., W. N. Gibbons, B. W. Muncaster, and I. M. Gray. 1989. *The Zebra Mussel, Dreissena polymorpha: A Synthesis of European Experiences and a Preview for North America*. Queen's Printer for Ontario, Ontario, Canada.

Marsden, J. E., E. Mills, A. Spidle, and B. May. 1992. Quagga mussel update. *Dreissena polymorpha Inform. Rev.* 3: 1–2.

Marsden, J. E., R. E. Sparks, and K. D. Blodgett. 1991. Overview of the zebra mussel invasion: Biology, impacts, and projected spread, In *Proceedings of the 1991 Governor's Conference on the Management of the Illinois River System*, pp. 88–95. Urbana, IL: University of Illinois.

May, B. and J. E. Marsden. 1992. Genetic identification and implications of another invasive species of Dreissenid mussel in the Great Lakes. *Can. J. Fish. Aquat. Sci.* 49: 1501–1506.

Miller, A. C. and B. S. Payne. 1997. Density and size demography of newly established populations of *Dreissena polymorpha* in the U.S. inland waterway system, In *Zebra Mussels and Aquatic Nuisance Species*, F. M. D'Itri, ed., pp. 99–116. Chelsea, MI: Ann Arbor Press.

Mills, E. L., J. R. Chrisman, B. Baldwin, R. W. Owens, R. O'Gorman, T. Howells, E. F. Roseman, and M. K. Raths. 1999. Changes in the Dreissenid community in the lower Great Lakes with emphasis on southern Lake Ontario. *J. Great Lakes Res.* 25: 187–197.

Mills, E. L., J. R. Chrisman, and K. R. Holeck. 2000. The role of canals in the spread on nonindigenous species in North America, In *Nonindigenous Freshwater Organisms*, R. Claudi and J. H. Leach, eds., pp. 347–379. Boca Raton, FL: CRC Press.

Mills, E. L., R. M. Dermott, E. F. Roseman, D. Dustin, E. Mellina, D. B. Conn, and A. P. Spidle. 1993. Colonization, ecology, and population structure of the "quagga" mussel (Bivalvia: Dreissenidae) in the Lower Great Lakes. *Can. J. Fish. Aquat. Sci.* 50: 2305–2314.

Nalepa, T. F., D. L. Fanslow, and G. A. Lang. 2009. Transformation of the offshore benthic community in Lake Michigan: Recent shift from the native amphipod *Diporeia* spp. to the invasive mussel *Dreissena rostriformis bugensis. Freshwat. Biol.* 54: 466–479.

Nalepa, T. F. and D. W. Schloesser. 1993. *Zebra Mussels: Biology, Impacts, and Control*. Boca Raton, FL: CRC Press.

Nalepa, T. F., D. W. Schloesser, S. A. Pothoven, D. W. Hondorp, D. L. Fanslow, M. L. Tuchman, and G. W. Fleischer. 2001. First finding of the amphipod *Echinogammarus ischnus* and the mussel *Dreissena bugensis* in Lake Michigan. *J. Great Lakes Res.* 27: 384–391.

O'Neill, C. R., Jr. and A. Dextrase. 1994. The introduction and spread of the zebra mussel in North America. In *Proceedings of the 4th International Zebra Mussel Conference*, March 7–10, 1994, Madison, WI.

Ricciardi, A., R. Serrouya, and F. G. Whorisky. 1995. Aerial exposure tolerance of zebra and quagga mussels (Bivalvia: Dreissenidae): Implications for overland dispersal. *Can. J. Fish. Aquat. Sci.* 52: 470–477.

Riessen, H. P., T. A. Ferro, and R. A. Kamman. 1993. Distribution of zebra mussel (*Dreissena polymorpha*) veligers in eastern Lake Erie during the first year of colonization, In *Zebra Mussels: Biology, Impacts, and Control*, T. F. Nalepa and D. W. Schloesser, eds., pp. 143–152. Boca Raton, FL: CRC Press.

Sinclair, R. M. and B. G. Isom. 1963. Further studies on the introduced Asiatic clam *Corbicula* in Tennessee. Tennessee Stream Pollution Control Board, Tennessee Department of Public Health, Nashville, TN.

Spidle, A. P., J. E. Marsden, and B. May. 1994. Identification of the Great Lakes quagga mussel as *Dreissena bugensis* from the Dnieper River, Ukraine, on the basis of allozyme variation. *Can. J. Fish. Aquat. Sci.* 51: 1485–1489.

Stoeckel, J. A., D. W. Schneider, L. A. Soeken, K. D. Blodgett, and R. E. Sparks. 1997. Larval dynamics of a riverine metapopulation: Implications for zebra mussel recruitment, dispersal, and control in a large-river system. *J. N. Am. Benthol. Soc.* 16: 586–601.

Strayer, D. L. 1991. Projected distribution of the zebra mussel, *Dreissena polymorpha*, in North America. *Can. J. Fish. Aquat. Sci.* 48: 1389–1395.

Strayer, D. L., J. Powell, P. Ambrose, L. C. Smith, M. L. Pace, and D. T. Fischer. 1996. Arrival, spread, and early dynamics of a zebra mussel (*Dreissena polymorpha*) population in the Hudson River estuary. *Can. J. Fish. Aquat. Sci.* 53: 1143–1149.

Strayer, D., J. Powell, B. Walton, and E. Mellina. 1993. Spread of zebra mussels in the Hudson River estuary in 1992. *Dreissena polymorpha Inform. Rev.* 4: 5.

Stepien, C. A., I. A. Grigorovich, M. A. Gray, T. J. Sullivan, S. Verga-Woolwine, and G. Kalayei. 2003. Evolutionary, biogeographic, and population genetic relationships of dreissenid mussels, with revision of component taxa. In *Quagga and Zebra Mussels: Biology, Impacts, and Control*, 2nd Edn., T.F. Nalepa and D.W. Schloesser, eds., pp. 403–444. Boca Raton, FL: CRC Press.

U.S. Air Force. 2009. Final summary report, zebra mussel eradication project, Lake Offutt, Offutt Air Force Base, Nebraska. U.S. Air Force, Offutt Air Force Base, NB.

Utah Division of Wildlife Resources. 2011. Attack against the invasion of the quagga & zebra mussels (2010 Boating Season Summary). http://wildlife.utah.gov/mussels/PDF/ais_summary_annual_2010.pdf (accessed April 8, 2011).

Whitney, S. D., K. D. Blodgett, and R. E. Sparks. 1996. Where have all the zebra mussels gone? *Dreissena* 7: 8.

Williams, J. D. and D. P. Jennings. 1991. Computerized data base for exotic fishes: The western United States. *Calif. Fish Game* 77: 86–93.

Wilson, K. A., E. T. Howell, and D. A. Jackson. 2006. Replacement of zebra mussels by quagga mussels in the Canadian nearshore of Lake Ontario: The importance of substrate, round goby abundance, and upwelling frequency. *J. Great Lakes Res.* 32: 11–28.

Wong, W. H., G. C. Holden, T. Tietjen, S. L. Gertenberger, B. Moore, K. Turner, and D. Wilson. 2013 Effects of invasive quagga mussels (*Dreissena rostriformis bugensis*) on chlorophyll and water clarity in Lakes Mead and Havasu of the lower Colorado River basin, 2007–2009. In *Quagga and Zebra Mussels: Biology, Impacts, and Control*, 2nd Edn., T. F. Nalepa and D. W. Schloesser, eds., pp. 495–509. Boca Raton, FL: CRC Press.

Wormington, A., C. A. Timmins, and R. M. Dermott. 1993. Distribution of zebra mussels on Canadian navigation buoys on the Great Lakes and upper St. Lawrence River, December 1992. Canadian Manuscript Report of Fisheries and Aquatic Sciences No. 2186. Fisheries and Oceans, Burlington, Ontario, Canada.

CHAPTER 2

Influence of Environmental Factors on Zebra Mussel Population Expansion in Lake Champlain, 1994–2010

J. Ellen Marsden, Pete Stangel, and Angela D. Shambaugh

CONTENTS

Abstract 33
Introduction 34
   Lake Morphometry and Spread of Zebra Mussels 34
Methods 35
   Zebra Mussel Sampling: Field Methods 35
   Zebra Mussel Sampling: Laboratory Methods 38
   Water Chemistry and Plankton Sampling 38
   Data Analysis 39
Results 39
   Thermal, Chemical, and Biological Patterns in Lake Champlain 39
   Veliger Densities 41
   Settled Larvae Densities 41
   Adult Densities and Sizes 46
   Changes in Water Quality 46
Discussion 49
   Factors Affecting Zebra Mussel Colonization and Growth 49
   Effects of Zebra Mussels in Lake Champlain 50
Acknowledgments 51
References 51

ABSTRACT

The spread of zebra mussels is facilitated by currents and human traffic and may be limited by temperature, calcium availability, and productivity. Lake Champlain is an interesting system for examining factors that affect spread of zebra mussels, as the lake has a complex structure consisting of bays separated by islands and causeways, water flows from south to north, temperature and calcium concentrations decrease from south to north, and productivity is highest in southern areas of the lake and in regions of the Northeast Arm. Zebra mussels were first discovered in Lake Champlain in 1993 and spread to all regions of the lake by 1996. Monitoring was conducted at 24 stations lake-wide between 1994 and 2010 to track the spread of zebra mussels and monitor water-quality parameters and plankton densities. Spatial expansion of zebra mussels in the lake was largely mediated by currents and barriers to flow. Veliger production, density, and recruitment of juveniles appear to be most strongly associated with productivity. Correlation of zebra mussel colonization rate and densities with calcium concentrations is confounded by currents and by temperature and productivity gradients. Colonization has been slowest, and remains at lowest densities in areas furthest from the initial invasion that have the lowest calcium concentrations, lowest annual mean temperatures, and lowest productivity measures. Local areas of high productivity within the low-density basins have high densities of mussels. Water clarity has increased in some areas of the lake, abundance of some benthic invertebrates has increased locally, and unionid mussels have declined, but other changes in the lake that might be attributable to zebra mussels, such as changes in productivity, have not occurred.

INTRODUCTION

Information on invasion dynamics (e.g., the rate of spread and factors that limit spread) of exotic organisms is key to predicting their effects on ecosystems and probability of invasion. Zebra mussels (*Dreissena polymorpha*) spread rapidly throughout North America since they were first discovered in the late 1980s, and predicted range limits based on temperature and calcium requirements were rapidly exceeded (e.g., Strayer 1991, Ramcharan et al. 1992). Downstream expansion of mussel populations was largely mediated by passive drift of veligers and wind-generated currents that transported mussels attached to debris. Upstream and overland expansion occurred largely as a consequence of human activities, particularly associated with boat traffic (Johnson and Carlton 1996, Allen and Ramcharan 2001). Establishment and persistence of populations in flowing water where there are no upstream populations depends on availability of refugia from currents that would prevent settlement of veligers near parental populations (Allen and Ramcharan 2001). Thus, population expansion and growth of zebra mussels involves an integration of probability of colonization (dispersal vectors), suitable conditions for settlement of new colonists (substrate, refugia from currents, and possibly local ion concentrations), and suitable conditions for growth and reproduction (temperature, calcium, and food supply). Predators may influence population density, but predation has not been shown to substantially affect ability of dreissenid populations to become established.

Lake Champlain is an interesting system to study factors that affect the spread and population growth of zebra mussels: the lake is moderately large (1127 km^2), divided by natural and artificial barriers into several segments with various trophic and calcium characteristics, and has complex currents. Calcium concentrations decline on a gradient from south to north; temperature generally declines northward but is high in small-volume bays and basins; and areas of high productivity occur in the southern and northeastern ends of the lake as a consequence of local high nutrient inputs to the lake. Within a year of discovery of zebra mussels in the lake, a standardized, long-term plan was established to monitor zebra mussels at 21 stations throughout the lake. Sampling was conducted in concert with lake-wide water quality monitoring, and additional stations were added later. The purpose of this chapter is to (1) document the spread and growth of the zebra mussel population in the lake, (2) identify factors that may have affected the spread and population growth, and (3) examine changes that have occurred in the lake since zebra mussels became established.

Lake Morphometry and Spread of Zebra Mussels

To provide a context for purposes of discussion, an understanding of lake morphometry is required. Lake Champlain lies between New York and Vermont, with its northernmost bay extending into Quebec (Figure 2.1). The lake is relatively deep (average depth = 19.5 m, maximum depth = 122 m), narrow (19 km), long (193 km), and has a volume of 25.8 km^3. The lake is divided by islands and causeways into four well-defined basins (Marsden and Langdon 2012). The two largest basins are the deep (maximum = 122 m), largely oligotrophic Main Lake and the mesotrophic Northeast Arm, locally known as the Inland Sea (maximum depth = 50 m). The Northeast Arm is a basin naturally defined by three large islands to the west; construction of five causeways between the islands further isolated this basin from the Main Lake and created Carry Bay and the Gut, two small eutrophic bays between these islands. To the south of the Northeast Arm is Malletts Bay (maximum depth = 32 m), also mesotrophic, which is cut off from the Northeast Arm to the north and from the Main Lake to the south by two causeways with narrow openings. A broad channel, obstructed by a causeway with a 168 m passage, links the Northeast Arm at its northern end to Missisquoi Bay, a shallow (maximum depth = 5 m), highly eutrophic bay. The passage through the causeway was widened to 268 m in 2007. The three eastern bays together are

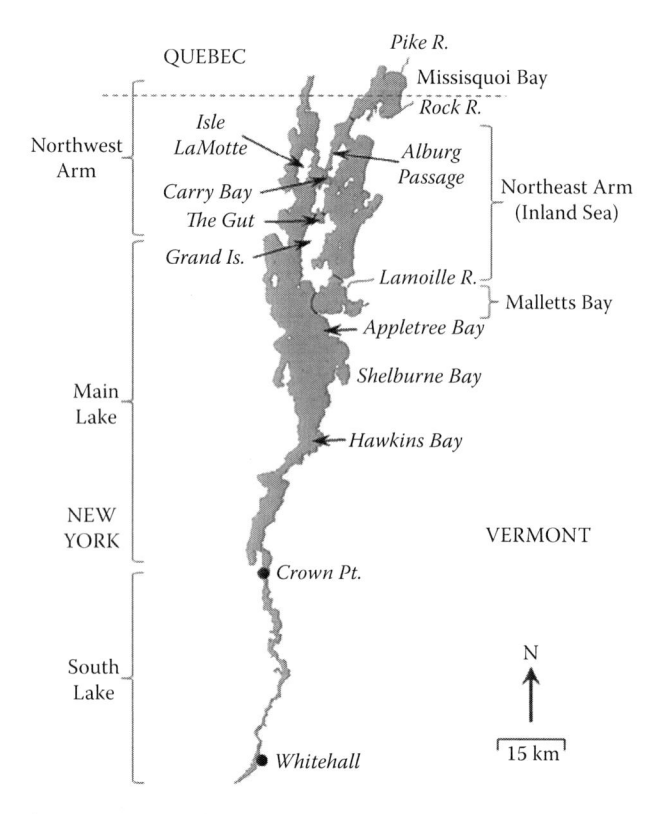

Figure 2.1 Lake Champlain, showing major lake basins, bays, and rivers mentioned in the text.

referred to as the northeast section of the lake. To the west of the islands is the Northwest Arm, an extension of the Main Lake. To the south, the lake becomes narrower and highly eutrophic, in part due to elevated inputs of phosphorus and sediments from deforestation in the 1800s and from agriculture beginning in the 1930s. Though contiguous with the Main Lake, this region is known as South Lake and is generally defined as the segment south of Crown Point.

Zebra mussels were first found in Lake Champlain in 1993 at Benson Landing in South Lake; the initial invasion is presumed to have been via the Champlain Canal that enters the lake at Whitehall (Marsden and Hauser 2009). The Champlain Canal is connected to the Erie Canal at Troy, New York, and mussels were established in the latter canal by 1991 (May and Marsden 1992). Passive flow through the Erie Canal could be sufficient to spread veligers as far as Troy, but north of Troy, the canal extends uphill to a height of land at Fort Ann, New York. Zebra mussels attached to floating vegetation could be passively transported through stretches of the canal by wind-generated currents, but this adventitious route would likely be slow. Transportation of mussels through the canal could also occur on boat hulls or vegetation entrained on boat propellers or anchors.

After zebra mussels entered Lake Champlain, water currents were probably the primary vector of distribution. Overall, mass transport of water in Lake Champlain is from south to north and, in the northern portion of the lake, from east to west. However, complex water flow regimes occur in the Main Lake, and lake-wide flow is constrained and affected by lake morphometry. Based on current flow patterns, predicted expansion of mussels from Benson Landing would be rapid and northward, because currents in the Main Lake can exceed 16 cm/s (Myer and Gruendling 1979). However, currents in surface waters, where the primary movement of veligers is likely to occur, are generally ≤0.5 cm/s.

Transport of zebra mussels between lake basins is hypothesized to be limited by water flow and connectivity. Malletts Bay receives flow from the Lamoille River (average = 37 m^3/s) and, in turn, drains through one channel in the northern causeway (10–20 m wide, depending on lake level) and two channels in the western causeway (25 and 54 m wide), with a retention time estimated from 0.41 to 0.71 years (Myer and Gruendling 1979). Percent flow through the northern channel is estimated to be 56% southward and 44% northward and is largely dictated by wind direction. In calm winds, 71% of the water flows through the western causeway channels into the Main Lake; in a north wind, flow is 99% out of the bay; and in a south wind, this is reversed (Henson and Gruendling 1977, Myer and Gruendling 1979). Consequently, zebra mussels could enter Malletts Bay from the Main Lake during periods of inward flow due to south winds. The Northeast Arm receives water from Missisquoi and Malletts bays and drains westward through the Gut and Carry Bay; in a south wind,

there is limited flow (16%–18% of total volume) into the arm through these bays (Myer and Gruendling 1979). Therefore, the Northeast Arm could receive zebra mussels from Malletts Bay and the Main Lake, but at a considerably reduced rate relative to transport in the Main Lake. Missisquoi Bay receives water from the Pike River (average = 8 m^3/s) at the north and the Rock and Missisquoi rivers (average 51 m^3/s) at the south. About 86% of water volume flows into the Main Lake through the Alburg Passage and Carry Bay, but some water flows into the Northeast Arm. This outward drainage pattern would tend to prevent zebra mussel invasion into Missisquoi Bay by water currents, particularly prior to the removal of 100 m of the causeway at the entrance to the bay in 2007. Thus, based on water flow alone, we predict that zebra mussels would expand from south to north along the western side and later enter Malletts Bay, then the Northeast Arm, and finally Missisquoi Bay. Cul-de-sac bays, such as Shelburne, Willsboro, and Cumberland Bays, would likely have had later colonization times but higher ultimate population densities as flow into the bays is slow and veligers from settled populations in the bays would not be moved away by currents.

METHODS

Within a year of the first discovery of zebra mussels in Lake Champlain in 1993, annual zebra mussel monitoring began at 21 stations throughout the lake (Figure 2.2, Table 2.1) and was integrated with the Lake Champlain Long-Term Water Quality and Biological Monitoring Program that began in 1992. Stations were added in 1997 (Willsboro Bay), 1998 (Basin Harbor), and 2006 (a second station in Missisquoi Bay). Sampling at eight shoreline stations was discontinued after 2003 and discontinued at all Main Lake stations after 2005 because zebra mussels were well established. Monitoring focused on semiquantitative veliger sampling to track population spread, and settlement plates to track density and seasonal growth of settled individuals. Documentation of adult zebra mussel presence and density on natural bottom substrates was sporadic; thus, much of the data on adult densities were acquired from other studies, or were anecdotal (e.g., Beekey et al. 2004, Ellrott and Marsden 2004, unpublished data).

Zebra Mussel Sampling: Field Methods

Zebra mussel veligers in open water were sampled at 12 lake stations, with a new station added in Missisquoi Bay in 2006 (Figure 2.2, Table 2.1). Samples were collected twice each month from May to October using a 13 cm aperture Wisconsin-style plankton net with 63 µm mesh, towed vertically from a depth of 10 m to the surface at a rate of 0.5 m/s (Marsden 1992). When station depth was less than 10 m, the net was towed from 1 m off the bottom to the

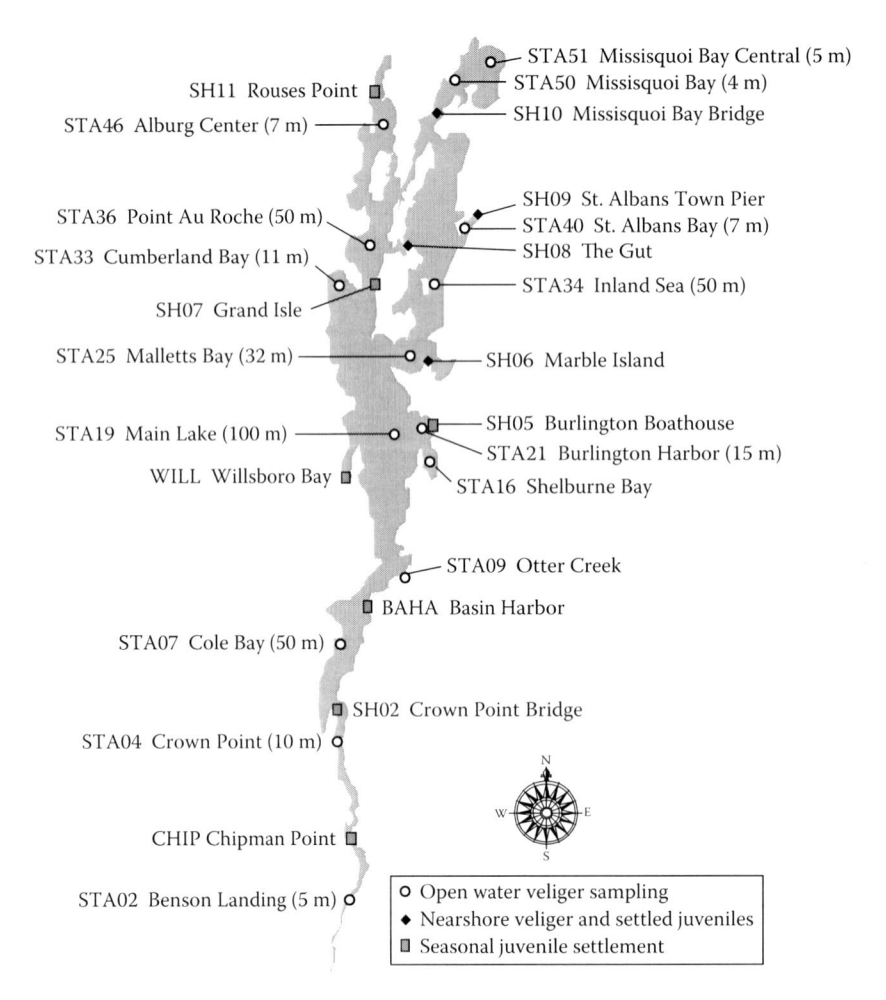

Figure 2.2 Location of stations in Lake Champlain at which samples were collected between 1994 and 2010.

surface. Five vertical tows of equal depth were combined to make one sample per site. A net efficiency of 100% was assumed and the water volumes filtered were estimated based on length of tow and net aperture. Volumes of water filtered for each sample ranged from 0.13 to 0.66 m³. Net contents were concentrated and preserved in a 50% ethanol solution. Surface water temperature and Secchi-depth transparency were also measured at each station. In 2003, veliger densities estimated from samples collected twice monthly, June to September, by vertical net tows at four stations were compared to densities estimated from samples collected at the same stations by use of a peristaltic pump that sampled a known volume of water. These comparisons indicated that plankton net efficiency was highly variable, likely due to net clogging. Consequently, veliger densities reported here are specific to our Lake Champlain sampling and are not necessarily comparable with densities from other monitoring programs using other techniques.

Occurrence and density of veligers at four nearshore stations located in shallow water near marinas and bays in the Northeast arm were determined bimonthly from May to October (Table 2.1) located in shallow water near marinas and in bays in the Northeast Arm. Samples of veligers were collected using horizontal plankton net tows. The net was thrown from shore and slowly towed horizontally below the surface at a rate of 0.5 m/s (VTDEC 2006, method 4.2.2). An individual sample was a composite of five tows of equal length. Surface water temperature and Secchi-disk transparency were recorded at each station. Sampling protocols and sample preservation were the same as for open-water veliger samples.

Occurrence and density of settled juveniles were determined at four nearshore stations (Table 2.1) by deploying an array of three 15 × 15 cm gray polyvinyl chloride (PVC) settling plates in mid-May (Figure 2.3; Marsden 1992). The plates were arranged horizontally along a stainless steel threaded eyebolt and separated by approximately 3 cm. Plate arrays were suspended in the water column from a dock or bridge abutment, with the top plate 2–3 m below the water surface. The bottom of each plate array was attached to a rope with a weight resting on the lake bottom. Top plates remained in the water for the entire summer to estimate seasonal recruitment. Middle and bottom plates were collected and replaced alternately every 2 weeks. Thus, plates were available for settlement by juvenile mussels for a total of 4 weeks. Settlement over an entire season was determined by placing plate arrays at seven nearshore stations (Table 2.1) in areas of the lake where

Table 2.1 Location and Depth of Sites Sampled as Part of the Long-Term Zebra Mussel Monitoring Program in Lake Champlain, 1994–2010. "STA" Sites Were Offshore; "SH" Sites Were Adjacent to Shore. Calcium, Total Phosphorus, Chlorophyll, Phytoplankton, Zooplankton, and Veligers Were Measured at All "STA" Sites; Veligers and Juvenile Settlement Was Measured at Sites SH06, SH08, SH09, SH10; Veliger Densities (Ending in 2003) and Seasonal Juvenile Settlement Were Measured at Sites SH02, SH05, SH07, SH11, BAHA, CHIP, and WILL. Sampling at STA51 Was Added in 2006 and Was Discontinued at SH07 in 2006.

Location	Description	Latitude and Longitude	Depth (m)
STA02	Benson's Landing, VT	N 43°42.89′ W 73°22.98′	4
STA04	Crown Point, NY	N 43°57.10′ W 73°24.47′	12
STA07	Cole Bay, NY	N 44°07.56′ W 73°24.77′	50
STA09	Otter Creek	N 44°14.53′ W 73°19.75′	97
STA16	Shelburne Bay	N 44°22.55′ W 73°13.92′	21
STA19	Main Lake, near Burlington, VT	N 44°28.26′ W 73°17.95′	100
STA21	Burlington Harbor, VT	N 44°28.49′ W 73°13.90′	16
STA25	Outer Malletts Bay, VT	N 44°34.92′ W 73°16.87′	32
STA33	Cumberland Bay, NY	N 44°42.07′ W 73°25.09′	11
STA34	Inland Sea, VT	N 44°42.49′ W 73°13.61′	50
STA36	Point Au Roche, NY	N 44°45.37′ W 73°21.30′	50
STA40	St. Albans Bay, VT	N 44°47.12′ W 73°09.73′	7
STA46	Alburg Center, NY	N 44°56.90′ W 73°20.40′	7
STA50	Missisquoi Bay, VT	N 45°00.80′ W 73°10.43′	4
STA51	Missisquoi Bay Central	N 45°00.00′ W 73°10.00′	5
SH02	Crown Point Bridge, NY, at old ferry dock at campground	N 44°01.80′ W 73°25.33′	NA
SH05	Burlington Boathouse, VT, at dock	N 44°28.57′ W 73°13.39′	NA
SH06	Marble Island Club, Colchester, VT, at dock	N 44°34.24′ W 73°13.83′	NA
SH07	Grand Isle Ferry Dock, VT	N 44°41.31′ W 73°20.98′	NA
SH08	The Gut, Tudhope Sailing Center, Grand Isle, VT, in the Gut at dock	N 44°45.98′ W 73°17.50′	NA
SH09	St. Albans Bay, VT, town pier	N 44°48.39′ W 73°08.45′	NA
SH10	Missisquoi Bay Bridge, VT, in bay	N 44°57.85′ W 73°13.23′	NA
SH11	Lighthouse Point Marina, near Rouses Point, NY, Lighthouse Point Marina, at dock	N 44°49.95′ W 73°21.00′	NA
BAHA	Basin Harbor, VT, at dock	N 44°11.48′ W 73°21.53′	NA
CHIP	Chipman Point Marina, VT, at dock	N 43°48.01′ W 72°22.35′	NA
WILL	Willsboro Bay Marina, Willsboro, NY, at dock	N 44°24.30′ W 73°23.30′	NA

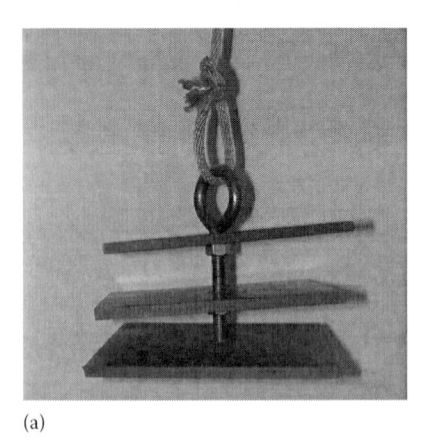

Figure 2.3 (See color insert.) Settling plate array used to monitor seasonal recruitment of juvenile zebra mussels. (a) Plate prior to deployment and (b) plate after 24 weeks deployment at station CHIP in 2003.

zebra mussels had been established for many years. Plates were positioned in the water in May and retrieved in October. After retrieval, individual plates were stored in airtight plastic containers and preserved with 95% ethanol for transport to the laboratory where they were stored in a refrigerator at 4°C.

Adult zebra mussels were sampled in 2005 along two shoreline transects, one north of Isle La Motte (near STA46) and one west of Grand Isle near Point Au Roche (near STA36). At each site, three 50 m long transects were positioned perpendicular to shore in 3–5 m water depth. Divers sampled five 0.20 m² quadrats along each transect (Marsden 1992). Each quadrat sample was taken in a different bottom substrate (e.g., soft sediment, macrophytes, cobble, and bedrock/boulder) as visually determined by the divers. Mussels within each quadrat were bagged, rinsed in a 1 mm sieve, and placed on ice.

Zebra Mussel Sampling: Laboratory Methods

Analytical procedures used in the laboratory followed methods detailed in Marsden (1992). Veligers were identified using a dissecting stereomicroscope (30× magnification) under cross-polarized light (Johnson 1995). For samples that contained few veligers (approximately 100 per sample), all individuals were counted. If veligers were too abundant to count (>100 per sample), samples were diluted and three 1 mL subsamples were extracted into 1 mL Sedgewick–Rafter cells. Densities were extrapolated to an entire sample and reported as number of veligers/m³.

Settling plates were examined under a dissecting stereomicroscope at 30× magnification and new recruits on the underside of each plate were counted. If the number of individuals was too high to count accurately, a 1 cm² square was randomly placed on each plate and individuals in five separate replicates were counted. For plates with extremely dense infestations and uniform distribution of individuals, 25% of each plate was counted. Plate density was reported as number of individuals/m² (method modified from Marsden 1992).

All mussels in the quadrat samples collected by divers were counted unless the number was greater than 200 and then the sample was subdivided. A maximum of 200 mussels from each quadrat was evaluated (e.g., live, dead, bleached) and measured, and total volume was estimated by water displacement (Marsden 1992).

Water Chemistry and Plankton Sampling

Water clarity, chlorophyll, phosphorus, calcium, and plankton data were collected at 12 offshore stations from 1994 to 2010 (Table 2.1). Visual transparency was measured using a Secchi disk. Samples for chlorophyll-a analysis were collected using a vertically integrated hose sampler between the lake surface and a depth twice the Secchi depth (VTDEC 2006). Water samples (100 mL) were filtered in the field through 47 mm diameter GF/A glass fiber filter and wrapped in a 90 mm No. 3 glass fiber filter. Filters were placed in a dark container on ice for transport to the laboratory and then frozen until analyzed.

Beginning in 2006, vertically integrated phytoplankton samples were collected using a 63 μm mesh plankton net with a 13 cm opening, towed upward at a rate of 0.5 m/s from a depth of twice the Secchi-disk depth (VTDEC 2006). The net was rinsed with lake water, and collected material was placed in a 50 mL centrifuge tube and preserved with Lugol's solution. In the laboratory, 1 mL subsamples were analyzed using an inverted microscope and Sedgewick–Rafter cells. Natural algal units (e.g., unicells, colonies, and filaments) with at least one dimension greater than 50 μm were identified to the lowest feasible taxonomic level. Cell counts were recorded for each.

Vertically integrated zooplankton samples were collected using a 153 μm mesh plankton net with a 30 cm opening, towed upward a rate of 0.5 m/s from a depth just above the sediments. The net was rinsed with lake water. The collected material was placed in a 125 mL bottle and narcotized with cold club soda or antacid tablets. Buffered 10%

formalin–sucrose–rose bengal solution was added to achieve a final concentration of 5% formalin. In the laboratory, 1 mL subsamples were analyzed using an inverted microscope and Sedgewick–Rafter cells. Individuals were identified to the lowest feasible taxon.

Water was sampled for total phosphorus using Kemmerer or Van Dorn water bottles in the epilimnion. Total phosphorus samples were immediately placed without filtration or preservation into 75 mL borosilicate glass test tubes. Samples were analyzed using acid–persulfate digestion followed by colorimetric analysis using the ascorbic acid method.

Data Analysis

Comparisons of veliger and settled juvenile densities among lake stations and years were based on seasonal-weighted mean estimates. Simpson's integral was used to calculate area of density versus time plots for each year, and areas were divided by the duration of the sampling season. Seasonal-weighted mean density estimates were based on a standard season length of 150 days defined by zero density values at the start and end of the sampling seasons. This analysis provided a more appropriate index of overall larval and juvenile production at each site than mean density, due to extreme within-season variation of densities.

Secchi-depth, chlorophyll, oxygen, and phosphorus data from 1992 to 2010 (STA2, STA4, STA21, STA33, STA34, STA36, STA40, STA46, and STA50) were analyzed using linear regression to detect trends in these variables after the appearance of zebra mussels in 1993. Regressions were calculated for the entire monitoring season (May–October) and for three shorter periods to approximate periods of increasing (May–June), stable (July–August), and decreasing (September–October) temperatures and productivity.

RESULTS

All data from the Lake Champlain long-term zebra mussel monitoring program can be accessed online at http://www.anr.state.vt.us/dec/waterq/lakes/htm/lp_lczebramon.htm; data from the long-term water quality and biological monitoring project can be accessed at http://www.anr.state.vt.us/dec/waterq/lakes/htm/lp_longterm.htm.

Thermal, Chemical, and Biological Patterns in Lake Champlain

Summer temperatures throughout the lake were in the range for optimal zebra mussel growth, with summer maxima between 23°C and 28°C. Maximum summer temperatures occurred in the small and shallow bays (e.g., Malletts Bay, 27.8°C; Missisquoi Bay, 26.8°C; and the Gut, 26.8°C) and the South Lake (28°C), and minimum temperatures occurred in the area of highest volume, the Main Lake (among-site average 24.9°C; Figure 2.4). One exception was Willsboro Bay, which consistently had the coldest summer temperatures and in 2000 had the coldest summer maximum, 20.1°C. Similarly, the growing season for zebra mussels (number of days when water temperatures were above 12°C) was longest in the South Lake (159–165 days), moderate in St. Albans Bay and Missisquoi Bay (145–147 days), and shortest in the Main Lake (120 days) (Table 2.2). These estimates of growing-season length provide only a general comparison among areas,

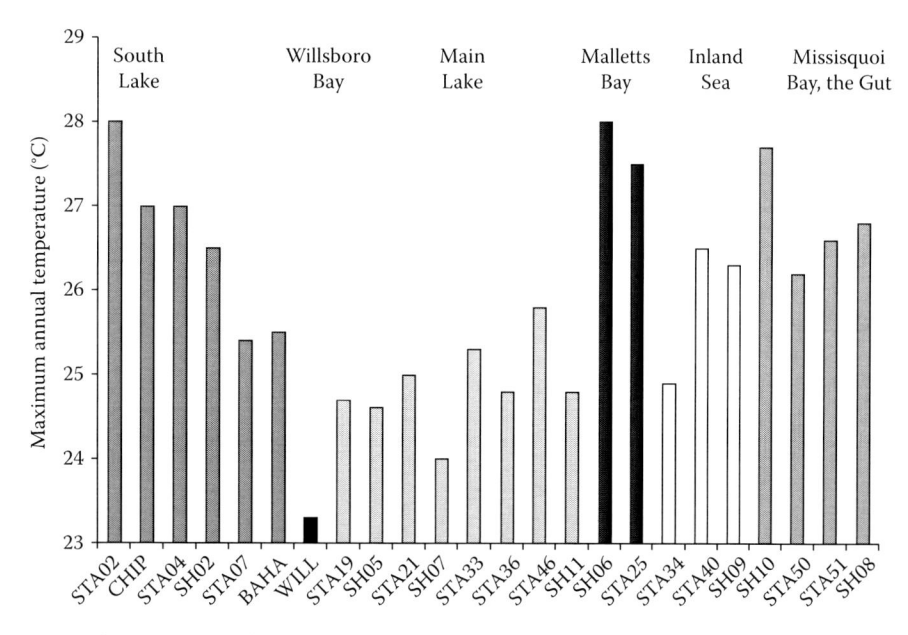

Figure 2.4 Maximum annual temperatures (°C) measured at each of the sites sampled in Lake Champlain between 1992 and 2010. Different bar shadings indicate groupings of sites by lake basin, with lake basins identified above the bars.

Table 2.2 Estimated Growing Season (Days) and Calcium Concentrations (mg/L) at 12 Sites in Lake Champlain. Data Are Averages of 19 Years of Monitoring (1992–2010). Zebra Mussel Growing Season Was Estimated as the Number of Days above 12°C (See Text)

Lake Segment	Station	Estimated Growing Season	Mean Ca±±
South Lake	STA2	165	28.3
	STA4	159	22.2
	STA7	138	17.5
Main Lake	STA19	120	16.7
	STA21	123	17.1
	STA33	131	14.6
Northwest Arm	STA36	128	16.5
	STA46	139	17.0
Malletts Bay	STA25	129	12.3
Northeast Arm	STA34	140	16.4
	STA40	145	17.4
Missisquoi Bay	STA50	147	13.5

as in some years, temperature was above 12°C when sampling began, and data only reflect temperatures above the thermocline. The lake was totally or partially ice covered in most winters, with bays beginning to freeze in December; the average date of ice closure is February 15 in Lake Champlain.

Calcium concentrations decline from the south portions of the lake (28.3 mg/L at Benson Landing) to the north (16.6 mg/L at Point Au Roche), with lowest concentrations in Malletts Bay (12.3 mg/L) and Missisquoi Bay (13.5 mg/L) (Table 2.2, Figure 2.5). Overall, Secchi-depth transparency was highest in the Main Lake (average 5.3 m) and lowest in South Lake (0.9 m), St. Albans Bay (2.8 m), and Missisquoi Bay (1.7 m) (Figure 2.6).

Overall, chlorophyll and zooplankton densities were highest in Missisquoi Bay, St. Albans Bay, and South Lake and lowest at the Main Lake stations (Figures 2.7 and 2.8). Zooplankton were dominated by rotifers, copepods, and cladocerans (Mihuc et al. 2012). Phytoplankton densities were also highest at Missisquoi

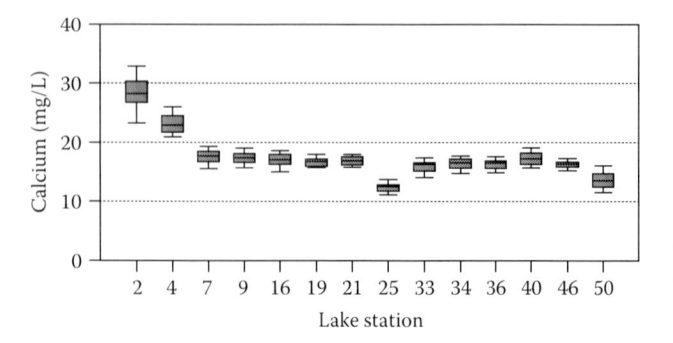

Figure 2.5 Mean calcium concentrations (mg/L) measured from 1996 to 2010 at 14 sites in Lake Champlain. Boxes represent 25th and 75th percentiles; whiskers represent 10th and 90th percentiles.

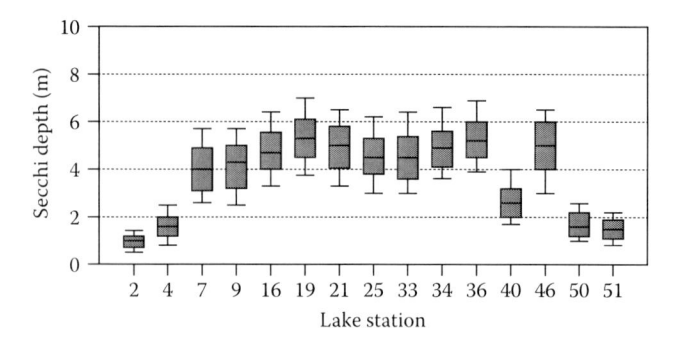

Figure 2.6 Mean Secchi-depth transparency (m) measured from 1994 to 2010 at 15 stations in Lake Champlain. Boxes represent 25th and 75th percentiles; whiskers represent 10th and 90th percentiles.

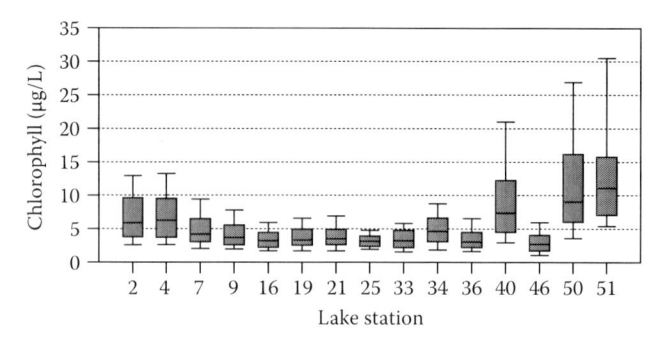

Figure 2.7 Mean chlorophyll concentration (μg/L) measured from 1994 to 2010 at 15 stations in Lake Champlain. Boxes represent 25th and 75th percentiles; whiskers represent 10th and 90th percentiles.

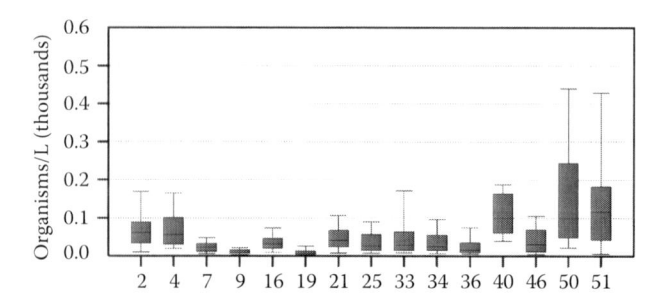

Figure 2.8 Mean zooplankton densities (no./L × 10³) measured from 1994 to 2010 at 15 stations in Lake Champlain. Boxes represent 25th and 75th percentiles; whiskers represent 10th and 90th percentiles.

Bay and St. Albans Bay but were low in South Lake (Figure 2.9). Phytoplankton were dominated by diatoms throughout the sampling period and throughout the lake, except for a shift toward cyanobacteria in the northeastern lake (Missisquoi Bay and Northeast Arm, particularly St. Albans Bay; Smeltzer et al. 2012). The spatial pattern of productivity from north to south may be dictated in part by phosphorus availability; total dissolved phosphorus was high in Missisquoi Bay (18–20.5 μg/L) and moderate in St. Albans Bay (10 μg/L) compared with

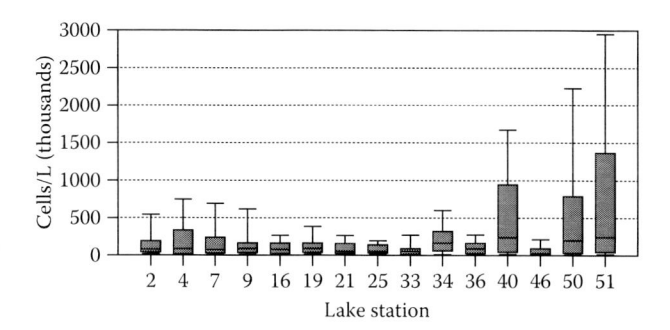

Figure 2.9 Mean phytoplankton densities (no. cells/L × 10³) measured from 1994 to 2010 at 15 stations in Lake Champlain. Boxes represent 25th and 75th percentiles; whiskers represent 10th and 90th percentiles.

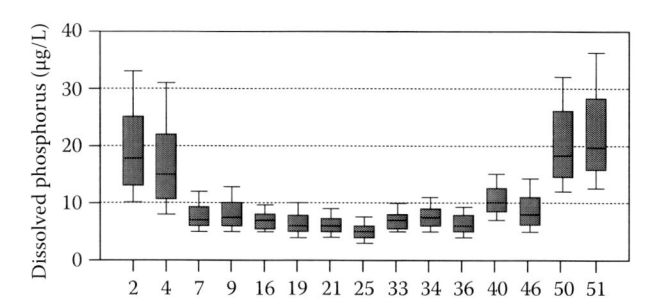

Figure 2.10 Mean phosphorus concentration (μg/L) measured from 1994 to 2010 at 15 stations in Lake Champlain. Boxes represent 25th and 75th percentiles; whiskers represent 10th and 90th percentiles.

the rest of the lake, where phosphorus concentrations did not exceed 8 μg/L (Figure 2.10). However, high phosphorus concentrations in South Lake (15–16 μg/L) were not reflected in high productivity.

If zebra mussel filtering controlled the growth of phytoplankton and zooplankton in South Lake, we would expect a general trend of declining plankton densities as zebra mussel populations rapidly increased; however, no temporal trends within or among stations were evident. Peak zooplankton densities that were two to four times higher than the long-term average occurred at several sites during the 17 years of monitoring, but there was no consistency in which year the peak occurred (Figure 2.11). Phytoplankton densities were more consistent over time and also did not show any lake-wide trends in abundance (Figure 2.12). However, total zooplankton density data masked a community shift that occurred beginning in 1996, when rotifers declined lake-wide; average densities declined at shallow sites from 140 to 20/L and at deep sites from 30 to 5/L (Mihuc et al. 2012). Rotifer abundance began to rebound after 2005, but with a species shift from *Polyarthra*, *Kellicottia*, and *Keratella* to a preponderance of *Conochilus*, a colonial rotifer that may be less vulnerable to dreissenid filtration than single-celled rotifers (Mihuc et al. 2012).

Veliger Densities

In 1994, 1 year after zebra mussels were found in Lake Champlain, veligers were present throughout the lake except at stations at Grand Isle, Cumberland Bay, Malletts Bay, and the Northeast Arm (Figures 2.13a,b and 2.14). Densities were less than 10 veligers/m³, except in South Lake, where densities reached 545/m³. By 1996, veligers were found at all stations. Densities peaked in South Lake in 1999 and 2000, with a peak density of 83,567/m³ at the Crown Point Bridge. Densities in the Main Lake and Northwest Arm peaked in 2000 and 2005; in both years, veliger abundance in the Northwest Arm was approximately double that in the Main Lake. Densities in the Northeast Arm did not rise above 300/m³ until 2004 and remained low in Malletts Bay (≤205/m³), in the Northeast Arm (≤765/m³ except for St. Albans Bay), and in Missisquoi Bay (≤541/m³). In 2005, densities rose abruptly at most stations. Maximum densities were observed at Benson Landing (75,699/m³ in 1999), at the Crown Point Bridge (83,567/m³ in 2000), and in St. Albans Bay (52,056/m³ in 2008).

Veligers were first detected each year in South Lake in mid-May, and then 1–4 weeks later in the Main Lake, Northwest Arm, Northeast Arm, Malletts Bay, and Missisquoi Bay. Densities peaked in late June to mid-July in South Lake and as late as early August in Missisquoi Bay. Peak veliger densities were frequently followed by a second, smaller peak, usually 6 weeks later but sometimes up to 10 weeks later, so that the second peak occurred as late as mid- to late September. Water temperatures at first appearance of veligers varied from 11°C to 18°C.

Settled Larvae Densities

Spread of the zebra mussel population occurred sequentially from south to north as defined by the first appearance of juveniles on settlement plates. Settled juveniles were present at the South and Main Lake stations in 1994, at Grand Isle in 1995, and the Northwest Arm by 1997 (Figure 2.15). Colonization of plates in the northeast lake was progressively later; juveniles were first found in the Gut in 1997, St. Albans in 1999, the Missisquoi Bay bridge in 2003, Missisquoi Bay in 2008, and Malletts Bay in 2009 (Figure 2.16). The peak of settled juvenile densities generally occurred in late July or early August, either simultaneously or 2 weeks after the peak of veliger densities. Settled juvenile densities were high, often exceeding 100,000/m², at the South Lake, Main Lake, and Northeast Arm sites, whereas densities remained low (<2500/m²) in Malletts Bay, the Northeast Arm, and Missisquoi Bay, with the exception of 2008 in St. Albans Bay (12,400/m²). Highest average densities were found at Chipman Point (498,000/m²) in 1995. Juvenile densities rarely exceeded 750,000/m³, and the highest density observed was 3,676,000/m² at Benson

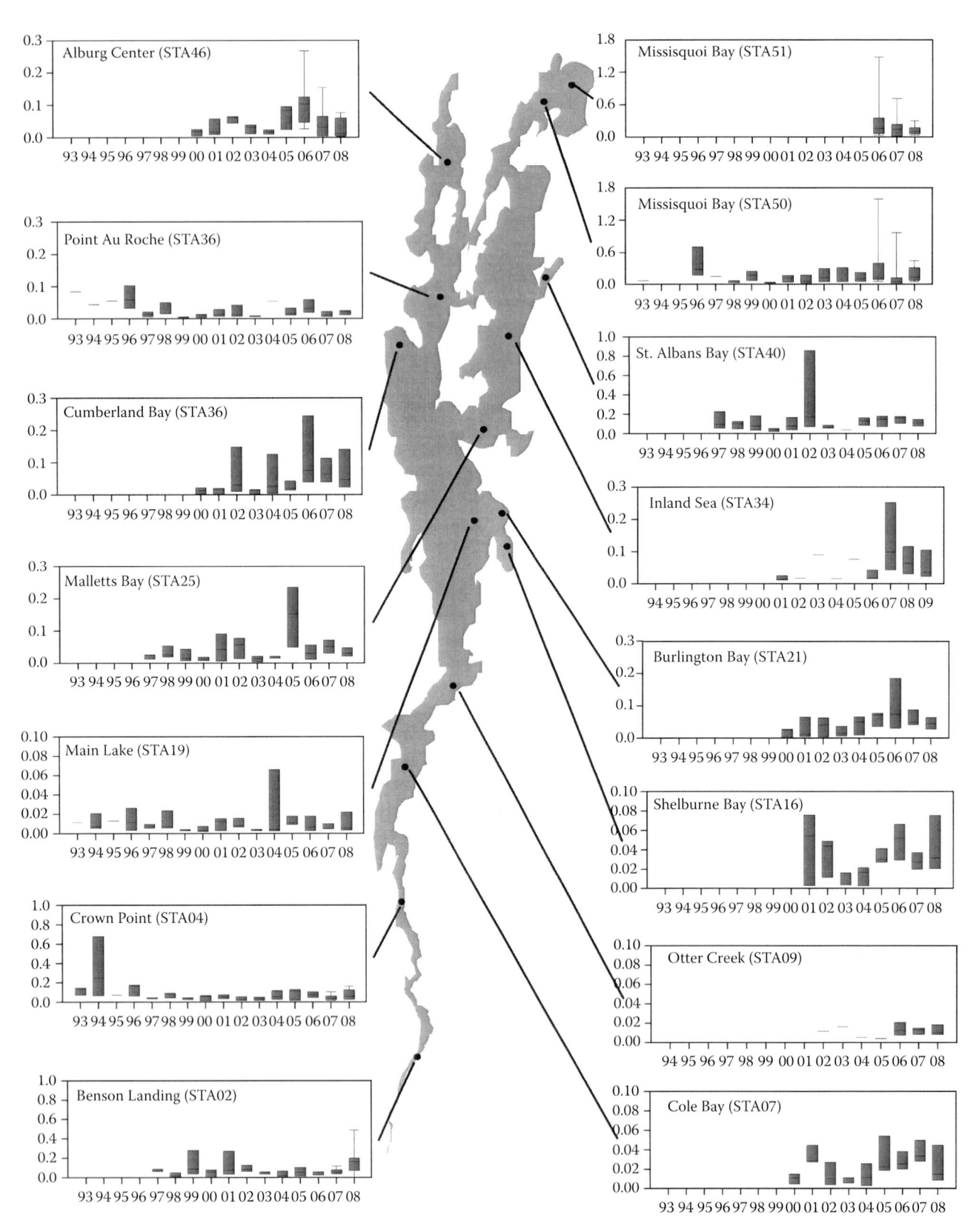

Figure 2.11 Annual weighted mean zooplankton densities (no./L) from 1994 to 2010 at 15 stations in Lake Champlain.

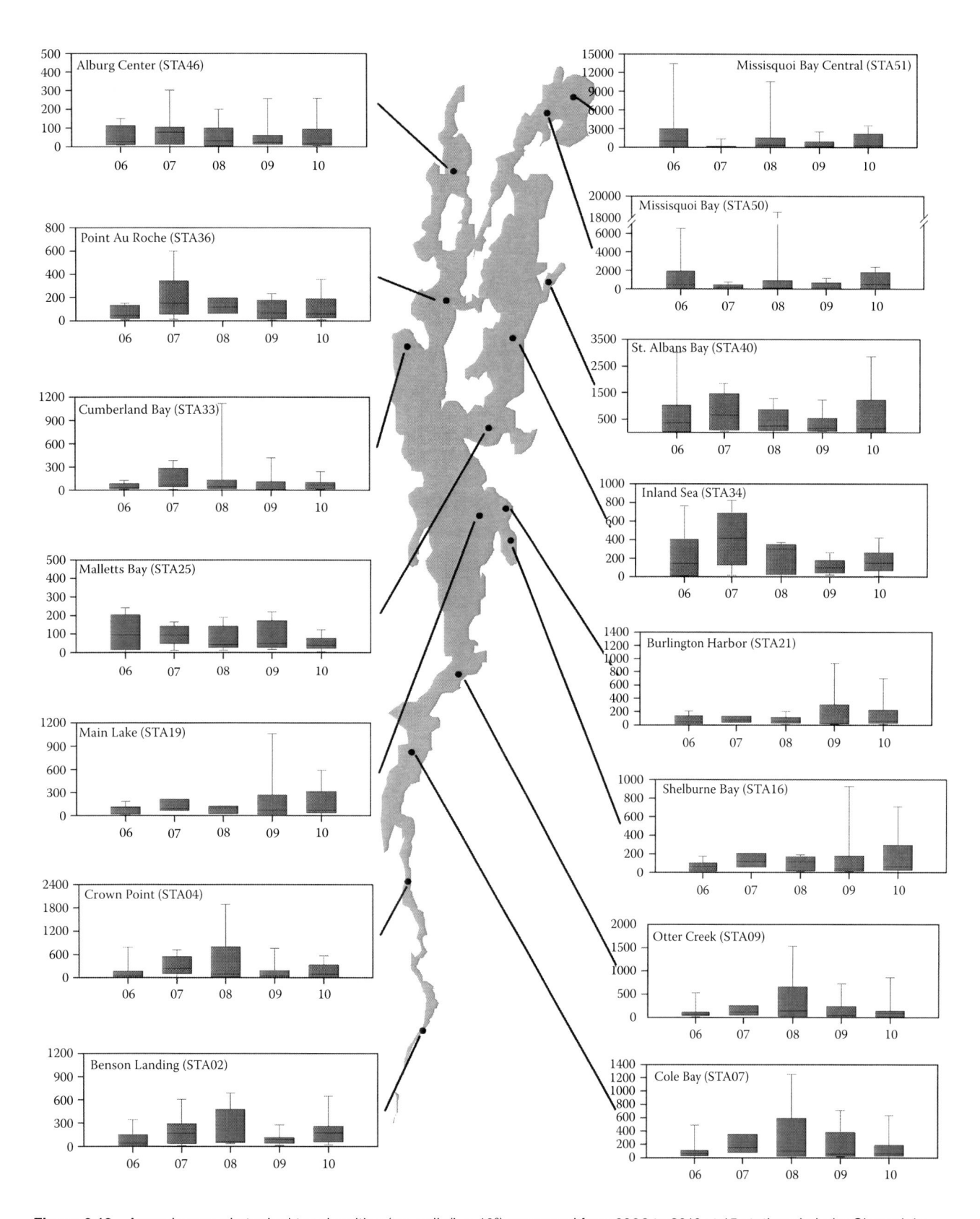

Figure 2.12 Annual mean phytoplankton densities (no. cells/L × 10³) measured from 2006 to 2010 at 15 stations in Lake Champlain.

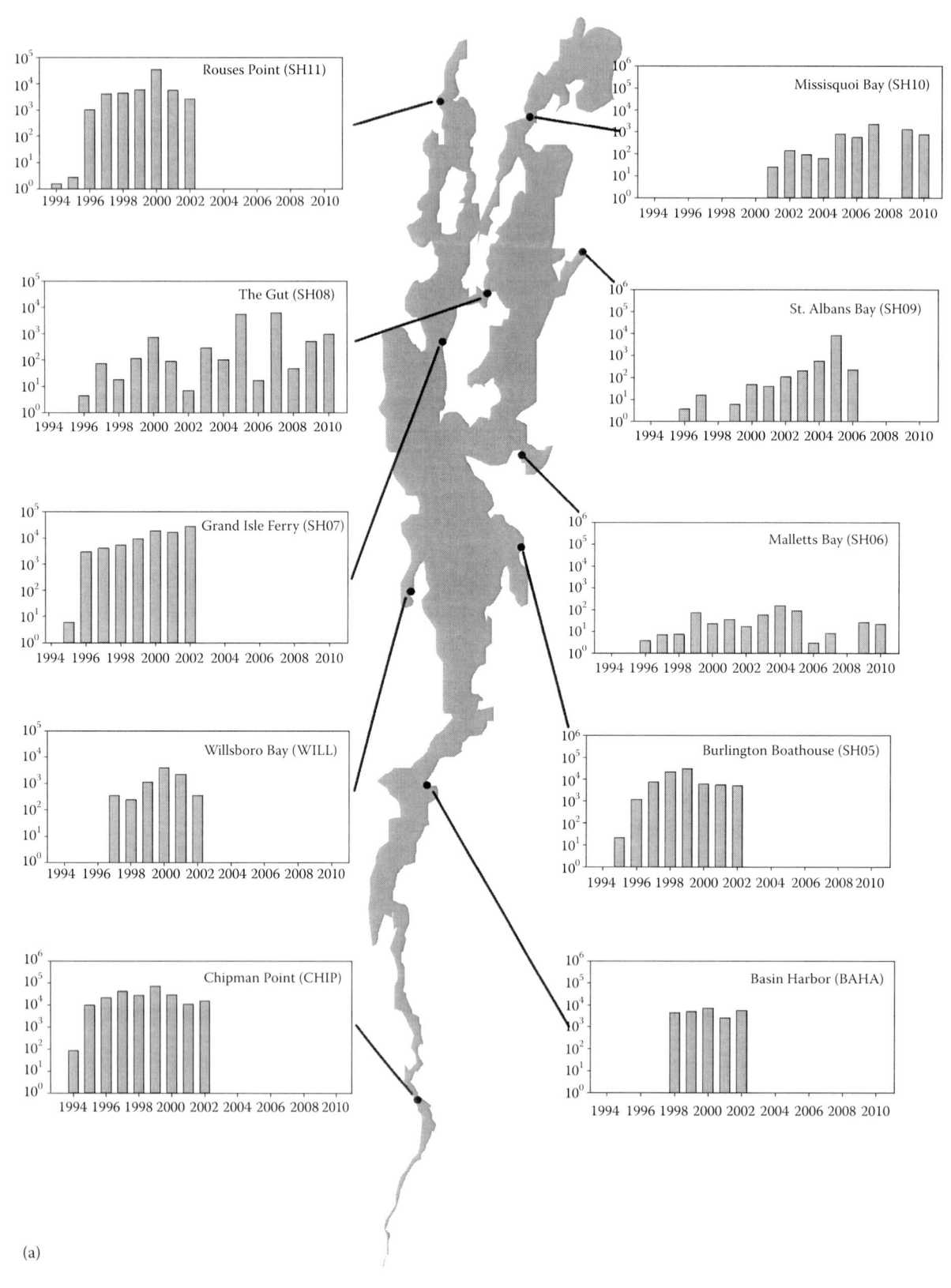

Figure 2.13 Mean veliger densities (no./L) between 1992 and 2010 at (a) 10 nearshore sites.

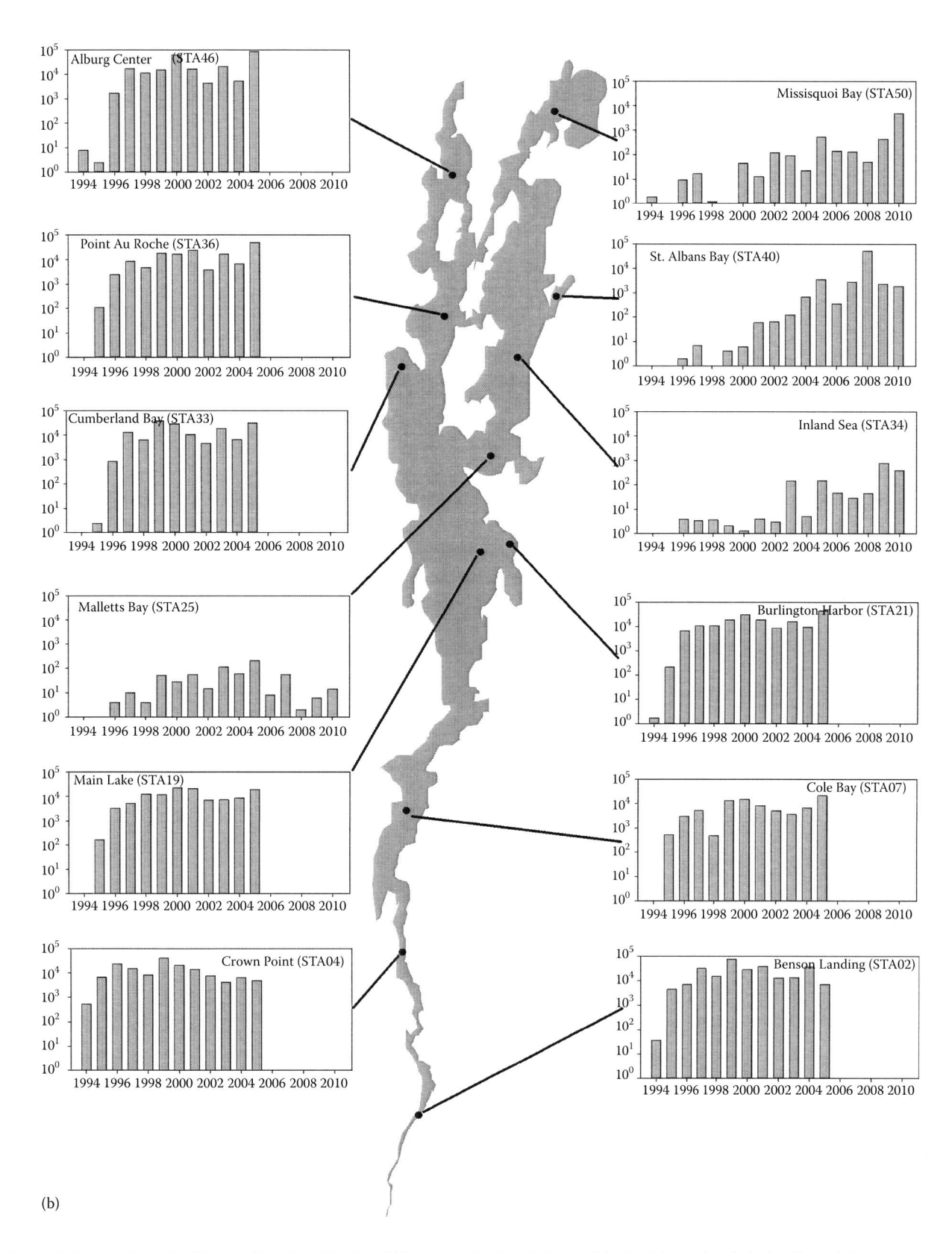

Figure 2.13 (continued) Mean veliger densities (no./L) between 1992 and 2010 at (b) 12 offshore sites in Lake Champlain.

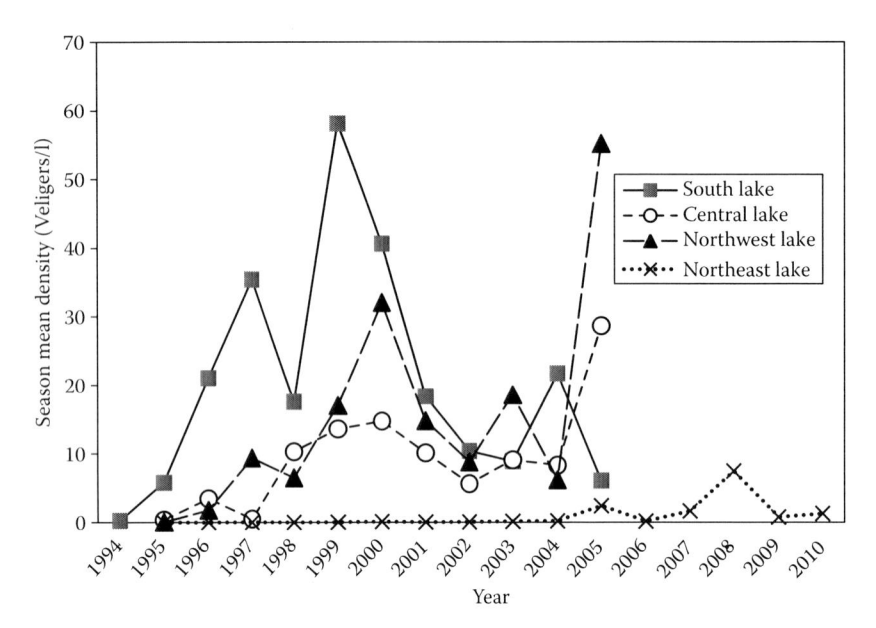

Figure 2.14 Annual weighted mean veliger densities (no./L) in Lake Champlain, grouped by lake basin, from 1994 to 2010.

Harbor in 1996. Growth of juveniles was also highest at Chipman Point, where shell length of settled mussels measured 13.5–16.5 mm at the end of their first year. In the Main Lake, the largest mussels on settlement plates were 4–8 mm, and mussels did not exceed 5 mm at Rouses Point, Grand Isle, or Missisquoi Bay. Settled mussels at St. Albans Bay in the Northeast Arm ranged in shell length from 7 to 14.5 mm. At the end of the year, plates at Chipman Point in 1997–1999 and 2004 had two size classes of juveniles, 4–6 and 14.5–16 mm; this appears to represent spawning by the first cohort, that is, a second generation within the year.

Adult Densities and Sizes

In quadrat samples collected by divers in 2005, mean (±SD) densities of zebra mussels were 1610 ± 1620/m² at Grand Isle and 4395 ± 905/m² at Isle La Motte. Two modal lengths of zebra mussels were present at Isle La Motte, with peaks at 16 and 28 mm; a single modal peak at 28–30 mm was present at Grand Isle (Figure 2.17).

Adult zebra mussels heavily colonized native unionid shells at sites throughout the southern half of the Main Lake in 1997 (Hallac and Marsden 2000, Marsden, unpublished data). In 2000, silty sediments at the base of piers at the Crown Point Bridge were observed to be densely carpeted with zebra mussel shells, with a layer of zebra mussel shells several centimeters thick under the surface layer (Marsden, pers. obs.). Colonization of soft sediments was well established by 2001, when contiguous carpets of zebra mussels were seen on sand substrates in Appletree Bay (31,312/m²) and silt–mud substrates in Hawkins Bay (38,173/m²; Beekey et al. 2004). Adult zebra mussels were first seen at the Missisquoi Bay Bridge in 2003 and in Missisquoi Bay in 2004 during a snorkel survey. The individual found in the bay was 33.5 mm long, indicating it was at least 2 and probably 3 years old or older. In 2005, an extensive walleye seining operation (21 hauls with a 180 m seine) incidentally sampled 6835 native mussels and 250 zebra mussels; the zebra mussels, which were mostly attached to unionids, ranged from 13 to 30 mm with a mean length of 20.1 mm (Pientka 2005).

Changes in Water Quality

Over the 18 year monitoring period, Secchi-depth transparency increased significantly ($P < 0.05$) by up to a meter at 4 of 12 sites: STA2 and STA4 (South Lake), STA33 (Cumberland Bay), and STA46 (Isle LaMotte) (Table 2.3). No pre-zebra mussel data were available for phytoplankton; therefore, we used chlorophyll as a surrogate to evaluate changes in primary productivity. Of the 9 sites in which chlorophyll was measured over the 18 years, concentrations increased significantly ($P < 0.05$) at one site (STA2) and decreased significantly at two sites (STA21 and STA33) (Table 2.3). However, the power of these analyses was low, due to wide variation in measurements and low r^2 values. Partitioning the data into 2 month seasonal periods reduced variation and increased power only slightly; a significant ($P < 0.05$) increase in chlorophyll occurred between July and August at STA46 and September to October at STA33 and STA36. Total phosphorus concentrations were significantly ($P < 0.05$) greater at 10 of 12 sites. The two sites where concentrations did not change significantly at were STA02 (South Lake) and STA50 (Missisquoi Bay) (Table 2.3).

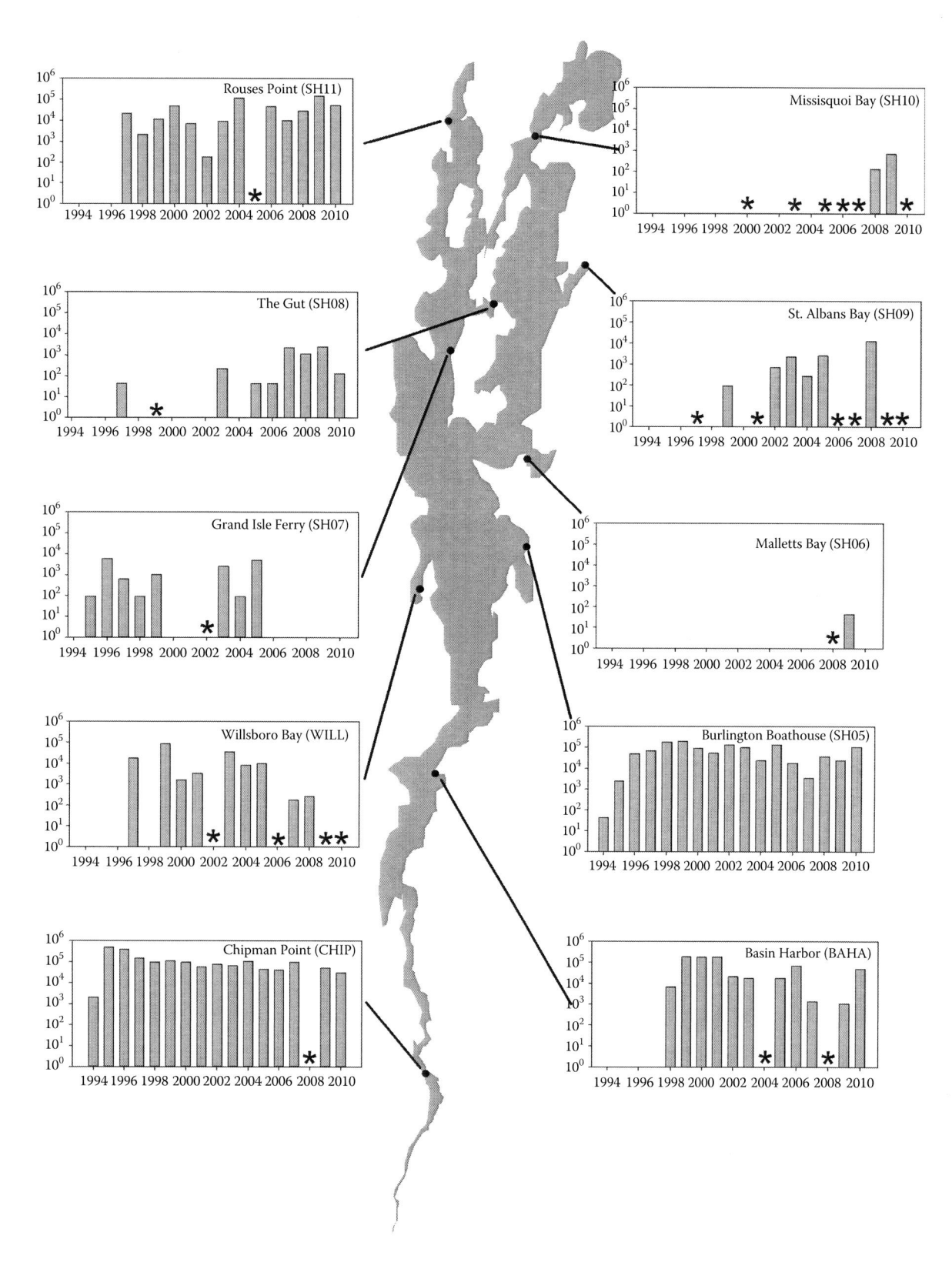

INFLUENCE OF ENVIRONMENTAL FACTORS ON ZEBRA MUSSEL POPULATION EXPANSION

Figure 2.15 Annual cumulative juvenile zebra mussel densities (no./m^2) from season plates at 10 sites in Lake Champlain from 1994 to 2010. Asterisks denote lack of data due to lost or damaged settlement plates.

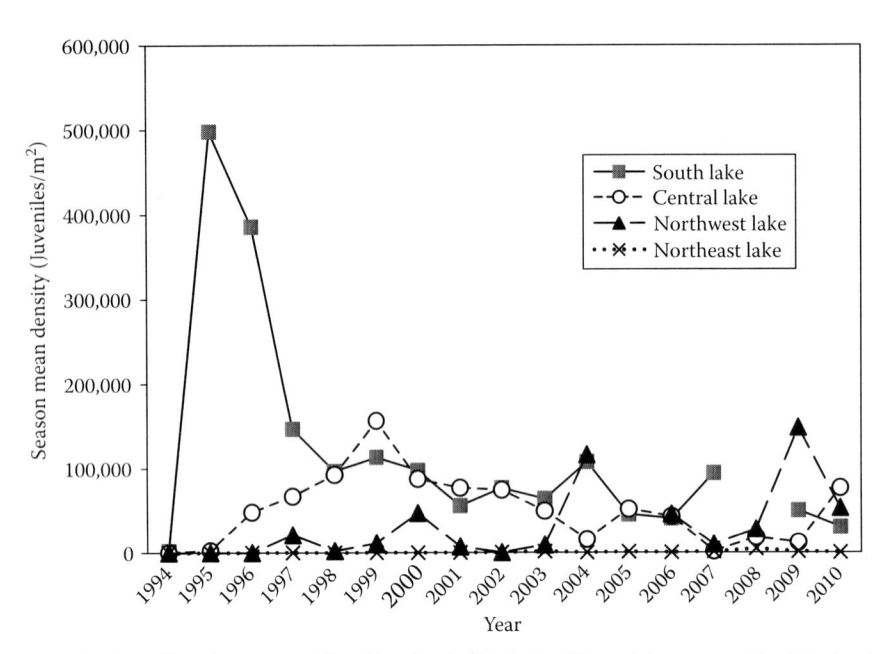

Figure 2.16 Annual cumulative juvenile zebra mussel densities (no./m^2) in Lake Champlain, grouped by lake basin, from 1994 to 2010.

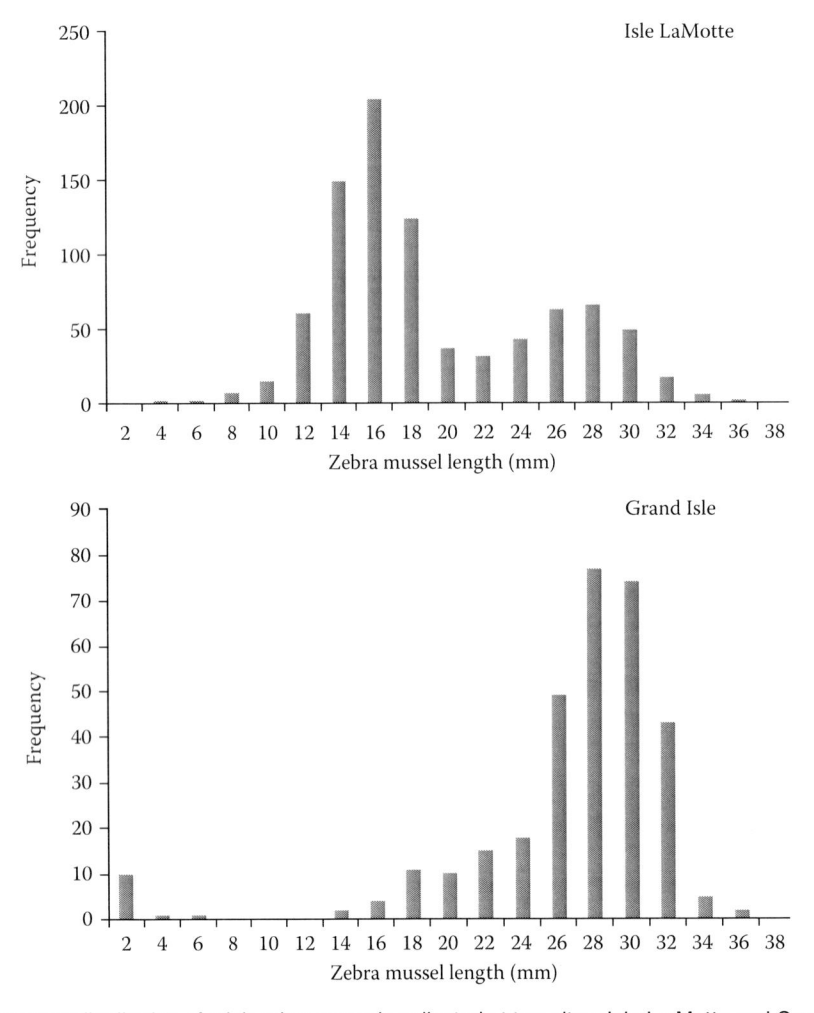

Figure 2.17 Length–frequency distribution of adult zebra mussels collected at two sites, Isle La Motte and Grand Isle, Lake Champlain, in 2005.

Table 2.3 Regression Analysis of Changes in Secchi-Depth Transparency, Total Phosphorus, and Chlorophyll in Lake Champlain between 1992 and 2010, Measured from May to September Each Year. When Regressions Were Significant (Indicated by Asterisks), Secchi-Depth Transparency and Total Phosphorus Increased and Chlorophyll Decreased over Time

Variable	Station	N	R^2	F	P
Chlorophyll	STA2	268	0.021	5.754	0.017*
	STA4	270	0.008	2.275	0.133
	STA21	251	0.018	4.625	0.032*
	STA33	252	0.051	13.286	<0.001*
	STA34	265	0.002	0.602	0.439
	STA36	252	0.004	0.907	0.342
	STA40	263	0	0.064	0.794
	STA46	256	0.005	1.358	0.245
	STA50	258	0	0.062	0.803
Secchi depth	STA2	273	0.045	12.93	<0.001*
	STA4	275	0.2	68.20	<0.001*
	STA7	242	0.013	3.157	0.077
	STA19	247	0	0.0003	0.986
	STA21	244	0.005	1.228	0.269
	STA25	261	0.004	1.151	0.284
	STA33	252	0.02	5.196	0.023*
	STA34	261	0	0.143	0.705
	STA36	248	0.009	2.387	0.124
	STA40	261	0.005	1.552	0.214
	STA46	260	0.105	30.366	<0.001*
	STA50	261	0.014	3.636	0.058
Total phosphorus	STA2	311	0.001	0.404	0.526
	STA4	314	0.039	12.615	<0.001*
	STA7	274	0.151	48.434	<0.001*
	STA19	280	0.021	5.972	0.015*
	STA21	285	0.021	6.009	0.015*
	STA25	294	0.062	19.189	<0.001*
	STA33	289	0.018	5.104	0.025*
	STA34	298	0.106	35.013	<0.001*
	STA36	285	0.034	10.092	0.002*
	STA40	301	0.030	9.129	0.003*
	STA46	303	0.016	4.921	0.027*
	STA50	295	0.006	1.623	0.204

DISCUSSION

Zebra mussels appeared in southern Lake Champlain in 1993, and veligers spread rapidly throughout the lake; however, recruitment of settled juveniles and establishment of adult populations were slower. Colonization rates, population expansion, and growth of zebra mussels appeared to have been influenced by calcium concentrations, overall productivity, and water currents. As found in other water bodies, the introduction of zebra mussels in Lake Champlain led to increased water clarity in local areas, a decline in unionid mussels, and increased abundance of some benthic invertebrates; however, impacts on zooplankton and phytoplankton productivity were not apparent.

Factors Affecting Zebra Mussel Colonization and Growth

Spatial expansion of zebra mussels in Lake Champlain was largely mediated by currents and barriers to flow, whereas veliger production, density, and recruitment of juveniles appeared to be most strongly associated with productivity. Correlation of colonization rates and densities of zebra mussels with calcium concentrations was confounded by currents and by temperature and productivity gradients. Colonization was slowest, and remains at lowest densities, in areas furthest from the initial invasion that have the lowest calcium concentrations, lowest annual mean temperatures, and lowest productivity measures. However, within basins of lowest overall densities, local areas of high productivity (St. Albans Bay, the Gut) had high densities of mussels.

Based on overall productivity, we would predict that zebra mussel growth would be highest in the more productive bays (Missisquoi Bay, Carry Bay, the Gut) and South Lake, moderate in the Northeast Arm, and lowest in the Main Lake; high summer temperatures in the bays would allow rapid growth, although rapid cooling of these low-volume areas in fall shortens the growing season relative to the high-volume Main Lake. In contrast, low calcium concentrations would be expected to limit mussel survival and growth in Malletts and Missisquoi bays.

Calcium was initially thought to be a major factor limiting distribution of zebra mussels in North America, with lower limits variously estimated between 12 and 28 mg/L (Ramcharan et al. 1992, Cohen and Weinstein 2001). In Lake Champlain, zebra mussels became established and reproduce in areas with calcium concentrations as low as 12.3 mg/L. At present, zebra mussels are established in only one other water body with lower calcium concentrations (11 mg/L) and that is Lake George, NY, which is within the Lake Champlain drainage. Calcium is required by mussels for shell growth, sperm motility, tissue ion balance, larval survival, and juvenile growth. However, data to support the hypothesis that calcium is a vital limiting ion are mixed. Molluscs may, but do not always, incorporate calcium into their shells in proportion to ambient calcium concentrations (Russel-Hunter et al. 1967, Mackie and Flippance 1983, Hinch et al. 1989). Molluscs from areas with low calcium concentrations may increase calcium utilization efficiency (Pynnonen 1991). A study in Lake Champlain indicated that there was, in fact, a significant negative relationship between calcium content of zebra mussel shells and ambient calcium levels. For example, calcium content of zebra mussel shells from Lake Champlain was higher than in shells from Lake Michigan, even though calcium content in Lake Michigan is higher than in Lake Champlain (Eliopoulos and Stangel 1998). As of 2005, adult zebra mussels were established in Missisquoi Bay, which is one of two areas with the lowest calcium concentrations; 250 zebra mussels, ranging in size from 13 to 30 mm, were found attached to native unionid mussels during a seining operation (Pientka 2005). Thus, calcium is not the only factor that limits expansion of zebra mussels in Lake Champlain, although mussels do show symptoms likely associated with low calcium availability. Shells tend to lose their periostracum layer, as evidenced by shells that are mostly white and chalky. Without the protection of this proteinaceous layer, shells appear to be more fragile, as they are easily crushed, and live mussels are often found with holes worn through their shells. If the periostracum serves as a barrier to loss of calcium ions to the environment, low ambient calcium may lead to loss of shell calcium in Lake Champlain. In contrast, native mussels in the lake (of which the most common species are *Elliptio complanata*, *Lampsilis radiata*, *L. cardium*, *Pyganodon grandis*, *Leptodea fragilis*, and *Potamilus alatus*) do not show unusual wear or shell fragility. The exception is the shells of *E. complanata*, which frequently show deep erosions around the umbo.

Low calcium levels may also affect recruitment dynamics. The rapid dispersion of zebra mussels throughout Europe and subsequently throughout North America is a consequence of the presence of a planktonic veliger, coupled with post-metamorphic drift and human activities, particularly commercial and recreational boating (e.g., Griffiths et al. 1991, Martel 1993, Minchin et al. 2002). A comparison of models of successful and failed zebra mussel invasions based on water quality and vector (connectivity and boat traffic) data concluded that factors that best predicted distribution of zebra mussels were presence of an upstream source of veligers and suitable ionic concentrations—primarily magnesium, calcium, and total hardness or conductivity (Allen and Ramcharan 2001). In areas examined by Allen and Ramcharan (2001) with invasion success, calcium concentrations were higher than 22 mg/L in all cases but one, in which the calcium concentration was 16 mg/L. The spread of mussels in Lake Champlain appears to be a consequence of dispersion from "upstream" areas, with spread inhibited in places by barriers and/or limited flow. However, successful colonization does not appear to be correlated to low calcium levels. Distribution of zebra mussels in Lake Champlain may be a result of source–sink dynamics in which mussels in low-calcium areas are supplemented from high-calcium upstream areas. Rapid appearance of veligers at northern sites, within 1 year of the first discovery of zebra mussels in Lake Champlain, suggests that these veligers were likely drifters rather than progeny of local mussels in the north. In addition, population growth in low-calcium areas (Malletts Bay, Missisquoi Bay) has been slow, as might be expected given the limited water exchange from the Main Lake. In support of this hypothesis, De Lafontaine and Cusson (1997) concluded that zebra mussel veligers in the Richelieu River largely originated from Lake Champlain and local production was very low. Finally, densities of settled larvae did not increase in proportion to increases in densities of veligers, which suggests that veliger survival is compromised in areas of low calcium. If this is the case, then populations will increase slowly or not at all in northern areas of Lake Champlain with low calcium concentrations (e.g., Horvath et al. 1996).

In South Lake, earlier warming of the water, higher overall summer temperatures, and greater productivity appear to support higher fecundity and more rapid growth of zebra mussels compared to more northern areas of the lake. Reproduction in South Lake began earlier in the spring each year and was more likely to occur in two distinct peaks as compared to more northern stations. A second cohort of settled juveniles, which appeared to represent reproduction by the first cohort, was only seen in South Lake. Interestingly, seasonal onset of reproduction, indicated by appearance of veligers in the plankton samples, was closely tied to date but not to temperature, which varied widely at the time when veligers first appeared.

Zebra mussel populations in Lake Champlain spread into five tributaries within 6 years. Adult zebra mussels were found up to 3.5 km upstream in the LaPlatte River in 1997, and in Lewis and Otter Creek in 1998; veligers were found in Little Otter Creek and the Winooski River in 1999. Boat traffic in these rivers is limited to small recreational paddlers and fishermen who may have transported mussels to ramps after excursions into the lake. The rivers have little elevation until the fall line, 10–20 km inland, so the spread of mussels upstream may have been mediated by wind-generated drifting on floating material or surges of lake water during onshore winds and periods of high lake level.

Effects of Zebra Mussels in Lake Champlain

Zebra mussel colonization in the Great Lakes and inland waters has resulted in increased water clarity (Barbiero and Tuchman 2004), decreased pelagic productivity (Fahnenstiel et al. 2010a,b), blooms of cyanobacteria (Makarewicz et al. 1999, Vanderploeg et al. 2001, Raikow et al. 2004), and *Cladophora* (Ozersky et al. 2009) as a result of increased bioavailability of phosphorus and light penetration, changes in benthic invertebrate density and richness (Haynes et al. 1999), and high mortality of native mussels (Nalepa 1994). Indirect effects include widespread decreases in burrowing amphipod *Diporeia* (Watkins et al. 2007), shifts in diets and condition factor of lake whitefish (*Coregonus clupeaformis*) (Pothoven et al. 2001, Hoyle et al. 2008), and changes in depth distribution of walleye as a consequence of increased light penetration (Fitzsimons et al. 1995). Of these effects, the most apparent in Lake Champlain was an increase in water clarity in South Lake and Northwest Arm during the 1990s. Smeltzer et al. (2012) noted increased Secchi-depth transparency in the Main Lake, Northwest Arm, and Cumberland Bay before the arrival of zebra mussels in the mid-1990s, but their data indicated that the most rapid and substantial increase in water clarity occurred in South Lake, likely as a consequence of zebra mussel filtration.

Phosphorus has increased in most areas of Lake Champlain since the early 1990s. Based on studies elsewhere, we expected a decrease in pelagic phosphorus as nutrients were redirected to the benthos (Fahnenstiel et al. 1995, 2010a,b). However, changes in phosphorus in Lake Champlain were confounded by effects of historic anthropogenic phosphorus inputs throughout much of the lake, particularly Missisquoi Bay and St. Albans Bay. In these bays, increased phosphorus was attributed to resuspension of phosphorus-loaded sediments during storm events and increased benthic bioavailable phosphorus due to deposition of fecal and pseudofecal material (Ozersky et al. 2009).

Despite an increase in phosphorus, chlorophyll levels have not increased except in Missisquoi Bay (this study, Smeltzer et al. 2012). Cyanobacteria blooms dominated by *Microcystis* have occurred with some regularity in Missisquoi Bay during the 2000s, and blooms of *Anabaena*, *Aphanizomenon*, and *Microcystis* occurred in the St. Albans Bay during the same period. Dreissenids selectively reject cyanobacteria, leading to increases in light-tolerant taxa of cyanobacteria such as *Microcystis* as they subsequently have a competitive advantage for nutrients relative to phytoplankton species that are filtered by mussels (Vanderploeg et al. 2001, Fishman et al. 2010). However, the density of adult zebra mussels in Missisquoi Bay and St. Albans Bay are still too low to expect detectable effects due to phytoplankton filtering; therefore, the increase in frequency of cyanobacteria blooms appears to be unrelated to the expansion of zebra mussels (Smeltzer et al. 2012).

Total zooplankton abundance did not show any consistent trends over time at the sites sampled in this study. However, zooplankton are not equally vulnerable to zebra mussel filtration; dreissenid impacts on zooplankton seen elsewhere have been most pronounced in microzooplankton such as rotifers (e.g., David et al. 2009). Using long-term data collected at five sites by state agencies in Vermont and New York beginning in 1992, Carling et al. (2004) and Mihuc et al. (2012) found substantial decreases in rotifer densities at locations throughout Lake Champlain after 1996, followed by an increase in abundance of the colonial rotifer *Conochilus* since 2005. These zooplankton community shifts are likely a direct consequence of the rapid increase in population abundance of zebra mussels.

Other studies in Lake Champlain indicate that, similar to the Great Lakes, abundance and species richness of benthic invertebrates increased in association with expansion of zebra mussel colonies (Beekey et al. 2004). *Diporeia* is a notable exception; this burrowing amphipod was highly abundant in Great Lakes sediments but decreased precipitously after the expansion of zebra mussels (Watkins et al. 2007). *Diporeia* was historically scarce in Lake Champlain (Myer and Gruendling 1979), so any changes in their abundance would have a relatively minor effect on the lake food web. Heavy fouling of unionid mussel shells led to substantial mortality and the need to add six species to the threatened and endangered species list (Hallac and Marsden 2001). In contrast to the Great Lakes, lake whitefish in Lake Champlain have not altered their diet to consume zebra mussels and their condition has remained high compared to Great Lakes populations in the post-dreissenid period (Herbst et al. 2011). Walleye abundance has steadily declined in Lake Champlain for several decades, but no recent changes in their abundance and distribution have been noted. Overall, effects of zebra mussels in Lake Champlain have not been as great as those observed in the Great Lakes, in part due to the minimum food web role of vulnerable species such as *Diporeia* and in part due to effects such as phosphorus load that may have masked effects of zebra mussels on productivity.

Another marked contrast with the Great Lakes is the absence of quagga mussels in Lake Champlain. Quagga mussels were present in the Erie Canal by 1991, but have not been observed in the Champlain Canal that connects the Erie Canal to Lake Champlain. The slow expansion of quagga mussels relative to zebra mussels may be due to their slow production of byssal attachments, so they are not as readily transported on boats or floating debris. It seems likely that quagga mussels will eventually become established in Lake Champlain. Subsequently, deep, offshore areas that are currently sparsely occupied by zebra mussels may become heavily colonized, which will exacerbate effects caused by zebra mussels.

ACKNOWLEDGMENTS

We thank Cathi Eliopoulos, Neil Kamman, and Michaela Stickney who conducted zebra mussel monitoring during the early years of the study. Funding for zebra mussel monitoring was provided by the Lake Champlain Basin Program.

REFERENCES

Allen, Y. C. and C. W. Ramcharan. 2001. *Dreissena* distribution in commercial waterways of the U.S.: Using failed invasions to identify limiting factors. *Can. J. Fish. Aquat. Sci.* 58:898–907.

Barbiero, R. P. and M. L. Tuchman. 2004. Long-term dreissenid impacts on water clarity in Lake Erie. *J. Great Lakes Res.* 30:557–565.

Beekey, M. A., D. J. McCabe, and J. E. Marsden. 2004. Soft sediment colonization by zebra mussels facilitates invertebrate communities. *Freshwat. Biol.* 49:1–11.

Carling, K., T. B. Mihuc, C. Siegfried, F. Dunlap, and R. Bonham. 2004. Where have all the rotifers gone? Zooplankton community patterns in Lake Champlain from 1992–2001. In *Lake Champlain: Partnerships and Research in the New Millennium*, T. Manley, P. Manley, and T. B. Mihuc, eds., pp. 259–270. Dordrecht, The Netherlands: Kluwer Academic Press.

Cohen, A. N. and A. Weinstein. 2001. *Zebra Mussel's Calcium Threshold and Implications for Its Potential Distribution in North America*. Richmond, CA: San Francisco Estuary Institute.

David, K. A., B. M. Davis, and R. D. Hunter. 2009. Lake St. Clair zooplankton: Evidence for post-Dreissena changes. *J. Freshwat. Ecol.* 24:199–209.

De Lafontaine, Y. and B. Cusson. 1997. Veligers of zebra mussels in the Richelieu River: An intrusion from Lake Champlain? In *Proceedings of the Second Northeast Conference on Nonindigenous Aquatic Nuisance Species*, Burlington, VT, April 18–19, 1997, ed. N. C. Balcom. Connecticut Sea Grant College Program, University of Connecticut, Groton, CT, pp. 30–40, Publ. No. CTSG-97-02.

Eliopoulos, C. and P. Stangel. 1998. Lake Champlain 1997 zebra mussel monitoring program. Final report. Waterbury, VT: Vermont Department of Environmental Conservation.

Ellrott, B. E. and J. E. Marsden. 2004. Lake trout restoration in Lake Champlain. *Trans. Am. Fish. Soc.* 133:252–264.

Fahnenstiel, G. L., T. B. Bridgeman, G. A. Lang, M. J. McCormick, and T. F. Nalepa. 1995. Phytoplankton productivity in Saginaw Bay, Lake Huron: Effects of zebra mussel *(Dreissena polymorpha)* colonization. *J. Great Lakes Res.* 21:465–475.

Fahnenstiel, G., T. Nalepa, S. Pothoven, H. Carrick, and D. Scavia. 2010a. Lake Michigan lower food web: Long-term observations and *Dreissena* impact. *J. Great Lakes Res.* 36(Suppl. 3):1–4.

Fahnenstiel, G., S. Pothoven, H. Vanderploeg, D. Klarer, T. Nalepa, and D. Scavia. 2010b. Recent changes in primary production and phytoplankton in the offshore region of southeastern Lake Michigan. *J. Great Lakes Res.* 36:20–20.

Fishman, D. B., S. A. Adlerstein, H. A. Vanderploeg, G. L. Fahnenstiel, and D. Scavia. 2010. Phytoplankton community composition of Saginaw Bay, Lake Huron, during the zebra mussel *(Dreissena polymorpha)* invasion: A multivariate analysis. *J. Great Lakes Res.* 36:9–19.

Fitzsimons, J. D., J. H. Leach, S. J. Nepszy, and V. W. Cairns. 1995. Impacts of zebra mussel on walleye *(Stizostedion vitreum)* reproduction in western Lake Erie. *Can. J. Fish. Aquat. Sci.* 52:578–586.

Griffiths, R. W., D. W. Schloesser, J. H. Leach, and W. P. Kovalak. 1991. Distribution and dispersal of the zebra mussel *(Dreissena polymorpha)* in the Great Lakes region. *Can. J. Fish. Aquat. Sci.* 48:1381–1388.

Hallac, D. E. and J. E. Marsden. 2000. Tolerance to and recovery from zebra mussel *(Dreissena polymorpha)* fouling in *Elliptio complanata* and *Lampsilis radiata*. *Can. J. Zool.* 78:161–166.

Hallac, D. E. and J. E. Marsden. 2001. A comparison of conservation strategies between *Elliptio complanata* and *Lampsilis radiata* threatened by zebra mussels *(Dreissena polymorpha)*: Periodic cleaning vs quarantine and translocation. *J. N. Am. Benthol. Soc.* 20:200–210.

Haynes, J. M., T. W. Stewart, and G. E. Cook. 1999. Benthic macroinvertebrate communities in south western Lake Ontario following invasion of Dreissena: Continuing change. *J. Great Lakes Res.* 25:828–838.

Henson, E. B. and G. K. Gruendling. 1977. The trophic status and phosphorus loadings of Lake Champlain. Ecological Research Series EPA 600/3-77-106. Boston, MA: U.S. Environmental Protection Agency.

Herbst, S. J., J. E. Marsden, and S. J. Smith. 2011. Lake whitefish in Lake Champlain after commercial fishery closure and ecosystem changes. *N. Am. J. Fish. Manag.* 31:1106–1115.

Hinch, S. G., L. J. Kelly, and R. H. Green. 1989. Morphological variation of *Elliptio complanata* (Bivalvia: Unionidae) in differing sediments of soft-water lakes exposed to acid deposition. *Can. J. Zool.* 67:1895–1899.

Horvath, T. G., G. A. Lamberti, D. M. Lodge, and W. L. Perry. 1996. Zebra mussel dispersal in lake-stream systems: Source-sink dynamics? *J. N. Am. Benthol. Soc.* 15:564–575.

Hoyle, J. A., J. N. Bowlby, and B. J. Morrison. 2008. Lake whitefish and walleye population responses to dreissenid mussel invasion in Eastern Lake Ontario. *Aquat. Ecosyst. Health Manage.* 11:403–411.

Johnson, L. E. 1995. Enhanced early detection and enumeration of zebra mussel *(Dreissena* spp.) veligers using cross-polarized light microscopy. *Hydrobiologia* 312:139–146.

Johnson, L. E. and J. T. Carlton. 1996. Post-establishment spread in large-scale invasions: Dispersal mechanisms of the zebra mussel, *Dreissena polymorpha*. *Ecology* 77:1686.

Mackie, G. L. and L. A. Flippance. 1983. Intra- and interspecific variations in calcium content of freshwater mollusca in relation to calcium content of the water. *J. Mollusc. Stud.* 46:204–212.

Makarewicz, J. C., T. W. Lewis, and P. Bertram. 1999. Phytoplankton composition and biomass in the offshore waters of Lake Erie: Pre and post-dreissena introduction (1983–1993). *J. Great Lakes Res.* 25:135–148.

Marsden, J. E. 1992. Standard protocols for monitoring and sampling zebra mussels. Illinois Natural History Survey Biological Notes 138. Champaign, IL.

Marsden, J. E. and M. Hauser. 2009. Exotic species in Lake Champlain. *J. Great Lakes Res.* 35:250–265.

Marsden, J. E. and R. W. Langdon. 2012. The history and future of Lake Champlain's fish and fisheries. *J. Great Lakes Res.* 38(Suppl. 1):19–34.

Martel, A. 1993. Dispersal and recruitment of zebra mussel *(Dreissena polymorpha)* in a nearshore area in west-central Lake Erie—The significance of postmetamorphic drifting. *Can. J. Fish. Aquat. Sci.* 50:3–12.

May, B. and J. E. Marsden. 1992. Genetic identification and implications of a second invasive species of dreissenid mussel in the Great Lakes. *Can. J. Fish. Aquat. Sci.* 49:1501–1506.

Mihuc, T. B., F. Dunlap, C. Binggeli, L. Myers, C. Pershyn, A. Groves, and A. Waring. 2012. Long-term patterns in Lake Champlain's zooplankton. *J. Great Lakes Res.* 38(Suppl. 1):49–57.

Minchin, D., F. Lucy, and M. Sullivan. 2002. Zebra mussel: Impacts and spread. In *Invasive Aquatic Species of Europe: Distribution, Impacts and Management*, E. Leppäkoski, S. Gollasch, and S. Olenin, eds., pp. 135–146. Dordrecht, The Netherlands: Kluwer Academic Press.

Myer, G. E. and G. K. Gruendling. 1979. *Limnology of Lake Champlain*. Burlington, VT: Lake Champlain Basin Study.

Nalepa, T. F. 1994. Decline of native unionid bivalves in Lake St Clair after infestation by the zebra mussel, *Dreissena polymorpha*. *Can. J. Fish. Aquat. Sci.* 51:2227–2233.

Ozersky, T., S. Y. Malkin, D. R. Barton, and R. E. Hecky. 2009. Dreissenid phosphorus excretion can sustain *C. glomerata* growth along a portion of Lake Ontario shoreline. *J. Great Lakes Res.* 35:321–328.

Pientka, B. 2005. Summary of walleye seining in 2005 at Sandy Point in Missisquoi Bay of Lake Champlain with a focus on addressing the specific conditions of the threatened and endangered species permit. Essex Junction, VT: Vermont Department of Fish and Wildlife.

Pothoven, S. A., T. F. Nalepa, P. J. Schneeberger, and S. B. Brandt. 2001. Changes in diet and body condition of lake whitefish in southern Lake Michigan associated with changes in benthos. *N. Am. J. Fish. Manage.* 21:876–883.

Pynnonen, K. 1991. Accumulation of ^{45}Ca in the freshwater Unionids *Anodonta anatina* and *Unio tumidus*, as influenced by water hardness, protons and aluminum. *J. Exp. Zool.* 260:18–27.

Raikow, D. F., O. Sarnelle, W. E. Wilson, and S. K. Hamilton. 2004. Dominance of the noxious cyanobacterium *Microcystis aeruginosa* in low-nutrient lakes is associated with exotic zebra mussels. *Limnol. Oceanogr.* 49:482–487.

Ramcharan, C. W., D. K. Padilla, and S. I. Dodson. 1992. Models to predict potential occurrence and density of the zebra mussel, *Dreissena polymorpha*. *Can. J. Fish. Aquat. Sci.* 49:2611–2620.

Russel-Hunter, W., M. L. Apley, A. J. Burky, and R. T. Meadows. 1967. Interpopulation variations in calcium metabolism in the stream limpet *Ferrissia rivularis* (Say). *Science* 155:338–340.

Smeltzer, E. A., D. Shambaugh, and P. Stangel. 2012. Environmental change in Lake Champlain revealed by long-term monitoring. *J. Great Lakes Res.* 38(Suppl. 1):6–18.

Strayer, D. L. 1991. Projected distribution of the zebra mussel, *Dreissena polymorpha*, in North America. *Can. J. Fish. Aquat. Sci.* 48:1389–1395.

Vanderploeg, H. A., J. R. Liebig, W. W. Carmichael, M. A. Agy, T. H. Johengen, G. L. Fahnenstiel, and T. F. Nalepa. 2001. Zebra mussel (*Dreissena polymorpha*) selective filtration promoted toxic *Microcystis* blooms in Saginaw Bay (Lake Huron) and Lake Erie. *Can. J. Fish. Aquat. Sci.* 58:1208–1221.

VTDEC (Vermont Department of Environmental Conservation). 2006. *Field Methods Manual*. Waterbury, VT: Vermont Department of Environmental Conservation.

Watkins, J. M., R. Dermott, S. J. Lozano, E. L. Mills, L. G. Rudstam, and J. V. Scharold. 2007. Evidence for remote effects of dreissenid mussels on the amphipod *Diporeia*: Analysis of Lake Ontario benthic surveys, 1972–2003. *J. Great Lakes Res.* 33:642–657.

CHAPTER 3

Replacement of Zebra Mussels by Quagga Mussels in the Erie Canal, New York, USA

Kenton M. Stewart

CONTENTS

Abstract 55
Introduction 55
Study Area 56
Methods 56
Results 58
Discussion 59
Annotation 62
Acknowledgments 62
References 62

ABSTRACT

Dreissenid mussels were sampled along a 300 km section of the Erie Canal between Buffalo and Syracuse, New York, over the period 1991–2009. Samples were collected at 2 sites in summer 1991, 20 sites in 1993, and unevenly (temporally) at up to 21 sites through winter 2009. Zebra mussels (*Dreissena polymorpha*) were found at all sites and dominated the population early in the study period. However, over time, quagga mussels (*Dreissena rostriformis bugensis*) made up an increasing proportion of the dreissenid population. Quagga mussels expanded their range from west to east and, by 2009, were found at all sites. Also, quagga mussels displaced zebra mussels as the dominant dreissenid mussel (near 100%) at many sites, particularly those sites in the western portion of the canal that were colonized by quagga mussels early in the study period. The percentage of quagga mussels varied widely between sites by 2009, but it was not apparent if this was a result of the west to east expansion or a difference in rate of increases at various sites. Regardless, the increasing dominance of quagga mussels provided evidence that quagga mussels can outcompete zebra mussels under the low-flow and shallow-warm conditions found in the canal.

INTRODUCTION

The first report of dreissenid mussels, specifically *Dreissena polymorpha* (zebra mussel), in North America was that of Hebert et al. (1989). Zebra mussels were first found in Lake St. Clair, a relatively shallow lake between Lakes Huron and Erie of the Laurentian Great Lakes, and may have been in Lake St. Clair since 1986 (Griffiths 1993). Carlton (2008) also suggested 1986 as the year when zebra mussels became established in North America. Leech (1993, p. 382) noted that four adult zebra mussels were "observed attached to the hull of a commercial fishing tug dry docked at Kingsville, Ontario in December 1987." Kingsville is a town located on the northern shore of western Lake Erie. Mussels can be distributed by boat movement, but their most common dispersal mechanism is probably when their microscopic "veliger" larvae are transported by currents. Veligers were observed in eastern Lake Erie in late summer 1989 and, by summer 1990, could be routinely collected in the water column of the eastern basin of Lake Erie (Riessen et al. 1993). Dermott and Munawar (1993) and Dermott and Kerec (1997) provided additional evidence of zebra mussel expansion when they found mussels in offshore sediments and deepwater benthos of eastern Lake Erie.

The first sightings of another species of dreissenid mussel in the Great Lakes occurred when a slightly different mussel was found (September 1989) near Port Colborne, Ontario (Mills et al. 1993a). Port Colborne is a small town located on the north shore of eastern Lake Erie and is near the Lake Erie entrance of the Welland Canal that allows ships to pass between Lakes Erie and Ontario. Both May and Marsden (1992) and Spidle et al. (1994) confirmed the identity of the new dreissenid mussel as *D. rostriformis bugensis* or the "quagga mussel." Interest and concern about the quagga mussel expanded rapidly and prompted an extensive review of its biology and ecology (Mills et al. 1996).

Through a chance boat trip to the City of Buffalo water intake building on June 5, 1991, I was able to scrape hundreds of live dreissenids (most less than 10 mm long in size) off the inside walls of the intake structure. The round and historic building is a stable site for dreissenid colonization because the walls were built extra thick (~6 m of reinforced concrete) to prevent damage from occasional ice shoves during winter. The building is located about 1.8 km off the extreme eastern end of Lake Erie and is positioned in about 7 m of water. On the day of that initial collection, no notice was made of subtle differences in shell morphology among the hundreds of animals collected. Indeed, it was not until several weeks later, after the mussels had been well preserved, washed, and dried, that after careful examination the June 1991 collection also included some quagga mussels.

The finding of two species of dreissenids initiated the present study to sample at multiple sites and times along the Erie Canal. The objective was to monitor (with time) the easterly movement and changes in population structure of both species of dreissenids.

STUDY AREA

Although very small in comparison to the large outflow of the Niagara River from Lake Erie, some outflow may enter the western end of the Erie Canal (also the mouth of Tonawanda Creek near North Tonawanda, NY). The slope of Tonawanda Creek is minimal at North Tonawanda and it does not take much of a hydraulic head difference between the Niagara River and Tonawanda Creek to have some Lake Erie water move upstream. Observations from the Main St. Bridge in North Tonawanda indicate water may flow downstream and sometimes upstream. Although episodic upstream movement may aid the initial easterly transport of dreissenid veligers, most of the easterly flow in the canal, from Lockport to Rochester, and again from Rochester to, New York, is a consequence of decreases in elevation toward Syracuse. Other than at times of low flow, unusually warm air temperatures, and little to no wind, stratification in the relatively shallow canal would be expected to range from very transient to nonexistent.

METHODS

Dreissenid mussels were usually collected in winter (mostly in February) between 1991 and 2009 at up to 21 locations along the Buffalo–Syracuse portion of the Erie Canal (Figure 3.1, and see Table 3.1 for specific sampling sites). The Buffalo–Syracuse portion constitutes the westerly half of the Erie Canal and is roughly half the total distance (Buffalo, NY, to Albany/Troy, NY) of the canal. The sites were labeled numerically from 1 to 21, with Site 1 the most westerly site located at the offshore Water Intake Building (of the City of Buffalo) and Site 21 the most easterly site near Syracuse, New York. Dreissenids collected at Site 1 were regarded as being reflective of the eastern end of Lake Erie. As mentioned, some tiny fraction of the huge Niagara River outflow from Lake Erie occasionally entered the western end of the Erie Canal; the easterly end of Lake Erie had the potential to serve as a source of veliger larvae for some westerly portions of the canal.

The Buffalo–Syracuse portion is entirely frozen over some winters and partially covered with ice in others. Figure 3.2 depicts conditions at 6 of the 21 sites where dreissenids were collected. These sites illustrate some of the variability in ice cover that may be encountered during one winter (1999). The first four sites, Middleport (Site 6), Albion (Site 8), Brockport (Site 9), and Rochester Route 31 (Site 11), were within a section of the Buffalo–Syracuse part of the canal where the water level was lowered each winter. In contrast, water levels in winter at Palmyra (Site 15) and Clyde (Site 19) were maintained at roughly the normal summer level. In specific days of February 1999, there was no ice cover at Rochester Route 31 site; some ice and open water at the Middleport, Albion, and Brockport sites; and a full ice cover of variable thickness at the Palmyra and Clyde sites.

Sampling required several days each winter to collect dreissenid mussels at all sites along the Buffalo–Syracuse stretch of the Erie Canal. Depending on the extent of ice cover on the frozen canal, mussels were collected by cutting holes in the ice and then scrapping live animals off bridge abutments or rocks on canal sides/bottom with a small-mesh wire basket on end of a long pole. If sites had little or no ice, mussels were scrapped off multiple nearshore rocks. On the day of collection, all mussels were preserved in ~5% formalin or high-grade ethanol. After about a month in the laboratory, the preservative was drained off, and mussels were rinsed well with tap water, spread out in a low-sided box, and then dried in a ventilated hood for ≥1 month. Subsets of mussels were periodically checked during the drying period and no significant change in length occurred. However, weight of the wet preserved animals initially dropped rapidly during the drying period but after about a month, there was little further weight loss. After drying, all mussels were manually sized (shell length in 1 mm increments) and weighed (total dreissenids in every size class, to nearest 0.001 g) (see annotation).

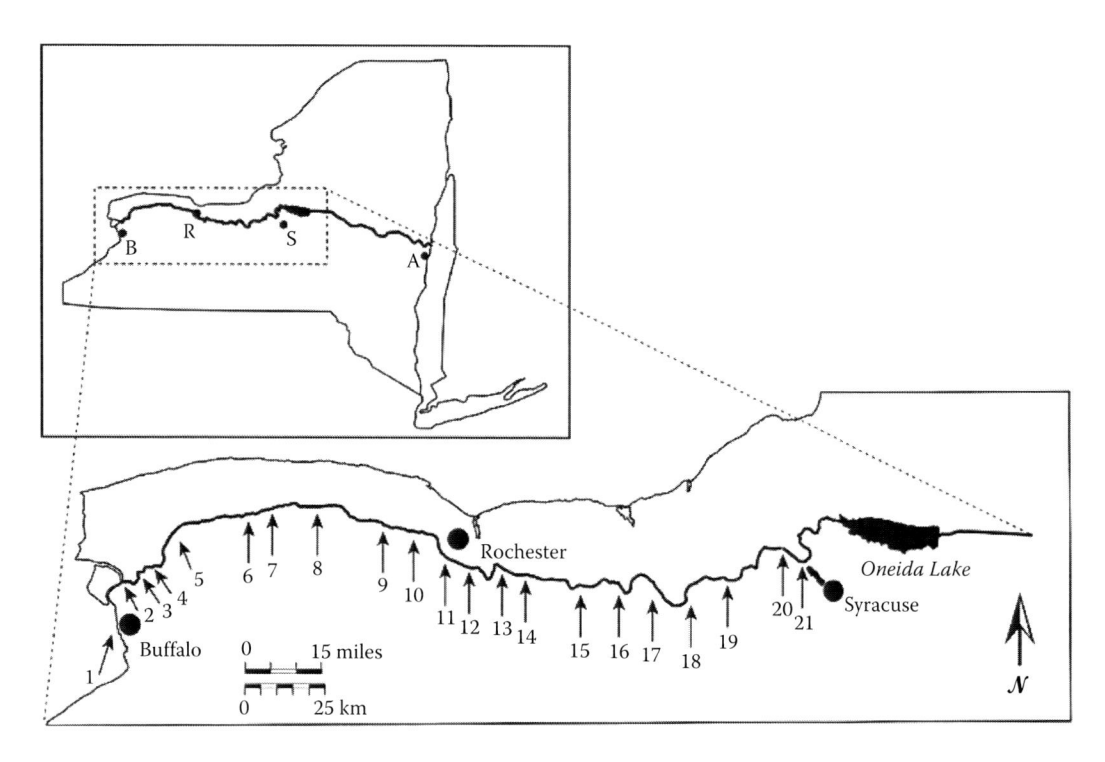

REPLACEMENT OF ZEBRA MUSSELS BY QUAGGA MUSSELS IN THE ERIE CANAL, NEW YORK, USA 57

Figure 3.1 Locations of 21 sites sampled for dreissenid mussels along the west half of the Erie Canal (Buffalo to Syracuse, NY) between 1991 and 2009. Inset: Major cities along the entire Erie Canal are B, Buffalo; R, Rochester; S, Syracuse; and A, Albany/Troy.

Table 3.1 Locations and Substrate Types of 21 Sites Where Dreissenids Were Collected Along the West Half of the Erie Canal, New York. during Winter, the Water Level in the Erie Canal from Just before Site 5 to Just after Site 14 is Lowered from an Average Depth of ~3.6 to ≤0.5 m. However, Some Easterly Flow Is Always Present

Site Number	Location and Substrate Type
1	City of Buffalo Water Intake Bldg., ~1.8 km offshore in east end of Lake Erie
2	Beneath the Main St. Bridge, North Tonawanda, NY
3	Beneath the east Robinson St. Bridge, near N.F. Blvd., North Tonawanda, NY
4	Beneath the Campbell Blvd. (Hwy 270) Bridge., North Tonawanda, NY
5	Beneath the broadest bridge and immediately above locks, Lockport, NY
6	Off many rocks on canal bottom, Middleport, NY
7	Off hard substrates at Medina, NY
8	Off rocks and other hard substrates at Albion, NY
9	Off rocks and other hard substrates at Brockport, NY
10	Off rocks and other hard substrates at Spencerport, NY
11	Off hard substrates below Hwy 31 Bridge
12	Off rocks and hard substrates by locks near University of Rochester housing, Rochester, NY
13	Off rocks and other hard substrates at Pittsford, NY
14	Off rocks and other hard substrates near Hwy 250, Fairport, NY
15	Off rocks below Hwy 21 Bridge, Palmyra, NY
16	Off rocks below Hwy 88 Bridge, Newark, NY
17	Off rocks and hard substrates, Clyde, NY
18	Off rocks and other hard substrates, Montezuma, NY
19	Off rocks and Hwy 34 Bridge abutments, Weedsport, NY
20	Off rocks and other hard substrates just below locks, Baldwinsville, NY
21	Off rocks and other hard substrates, Onondaga Lake outlet to canal

Figure 3.2 Photographs of 6 of 21 sites that show variability of habitat conditions and ice cover in winter 1999 along the Erie Canal in New York. Middleport (Site 6), Albion (Site 8), Brockport (Site 9), Rochester Route 31 (Site 11), Palymra (Site 15), and Clyde (Site 17).

RESULTS

An important goal of this research was to track the changing percentage of quagga mussels over time in the Erie Canal in New York State. However, because total counts were made routinely of both dreissenid species from every collection site, results also tracked the percent change of zebra mussels at each site. Initially, the dreissenid population at most sites sampled along the Erie Canal was dominated by zebra mussels. With time, variable percentages of both zebra and quagga mussels were found at most sites, and then percentages of quagga mussels increased. Table 3.2 shows that over time, there was a decline in percent of zebra mussels and a corresponding increase in percent quagga mussels, from westerly sites near Buffalo, NY (e.g., Sites 1–3), to easterly sites near Syracuse, NY (e.g., Sites 19–21).

A graphic representation of the shift from zebra mussels to quagga mussels is given in Figure 3.3. At Site 1, during the first two sampling years in 1991 and 1993, the percentage of quagga mussels was <20% and ≤40%, respectively. However, by 1994, the percentage of quagga mussels increased to greater than 80% and, by winter 1996, the percentage of quagga mussels at Site 1 was nearly 100%. Percent quaggas also increased rapidly at other proximate sites. Figure 3.3 also shows that, by winter 2002, quagga mussels accounted for 12%–100% of the dreissenid population at Sites 1–15, which were located in the first (west to east) 200 km of the canal. Some variability existed, but by

Table 3.2 Percent of Zebra Mussels in the Total Dreissenid Population at Each Site in Each Sampling Area. All Samples Were Collected in Winter Except Those with an Asterisk (*) That Were Sampled in Summer. Blank, Not Sampled

Site	1991	1993	1994	1995	1996	1998	1999	2000	2002	2009
1	84*	59		8	1	0	0	0	0	0
2		68		38	34	6	3	0	0	0
3	99*	97		64	39	50	4	2	1	5
4		99		72	45	21	12	4	3	3
5		98		81	79	38	34	12	9	3
6		100		93	89	62	74	28	21	13
7		100	100	92*	88	98	78	54	15	
8		100		99	98	98	96	94	92	9
9		100	100	100	98	99	98	96	94	10
10		100	100	100	99	100	100	96	98	15
11		100	100	100	100		100	98	96	
12		100	100	100	100	100	100	99	98	79
13		100	100		100	100	100	99	98	98
14		100	100		100	100	100	99	94	22
15		100	100	100	100	100	100	99	88	64
16		100	100	100	100	100	100	100	100	59
17		100	100	100	100	100	100	100	100	94
18		100	100	100		100	100	100	100	76
19		100	100	100	100	100	100	100	100	80
20			100*	100*		100*	100	100	100	58*
21		100		100*	100	100*	100	100	100	7*

winter 2009, quagga mussels accounted for more than 50% of the dreissenid population at over half of the 21 sampled sites along the Buffalo–Syracuse portion of the Erie Canal.

Representative length–frequency histograms of all dreissenids (zebra and quagga) at six different sites along the Erie Canal indicate substantial variation in life-history structure (Figure 3.4). Histograms for populations at Middleport (Site 6), Medina (Site 7), and Albion (Site 8) suggest a bimodal or two-cohort age structure. The population histogram at Clyde (Site 17) was unusual because it appeared to show three cohorts, which was uncommon along the canal. The length–frequency histogram for the population at Montezuma (Site 18) showed only one cohort. Sites with one cohort was more common than sites with three cohorts, but interestingly, the Montezuma site with one cohort was only a few kilometers down the canal from the Clyde site where there were three cohorts. The dreissenid population at Baldwinsville (Site 20) had two cohorts.

DISCUSSION

Large numbers of exotic plant and animal species have been introduced into the Great Lakes over the past century (Mills et al. 1993b) and into a variety of lake–stream systems generally (Horvath et al. 1996, Stoeckel et al. 1997, Pimentel 2005). The finding that zebra mussels were replaced by quagga mussels in the Erie Canal was consistent with studies in other locations. For example, a shift in dominance from zebra mussels to quagga mussels was noted in the Soulanges Canal of the upper St. Lawrence River in Canada (Ricciardi and Whoriskey 2004). Indeed, such a shift from one dreissenid species to the other has been reported in areas of all of the Laurentian Great Lakes, for example, Lake Ontario (Mills et al. 1999, Wilson et al. 2006), Lake Erie (Stoeckmann 2003), Lake Huron (Nalepa et al. 2007), Lake Michigan (Nalepa et al. 2009, 2010), and Lake Superior (Grigorovich et al. 2008). The present study was unique in that it showed temporal and spatial trends in the progressive replacement of zebra mussels by quagga mussels at multiple sites over many years.

Although not dramatic, there tends to be a slow west-to-east flow of water, along the Buffalo–Syracuse portion of the Erie Canal. As seems to be the case for larval dispersal in riverine populations (Stoeckel et al. 1997), portions of the Erie Canal may serve as a *Dreissena* metapopulation from which larvae are dispersed "downstream" or "down canal." An overview of the 10 histograms in Figure 3.4 suggested an easterly transport of quagga mussels (most likely as veliger larvae) along the Erie Canal over time. Obviously, the Buffalo–Syracuse portion of the Erie Canal is not a closed tube and there are some places along the Canal where boaters launch and retrieve their boats. These activities may have introduced quagga mussels at these

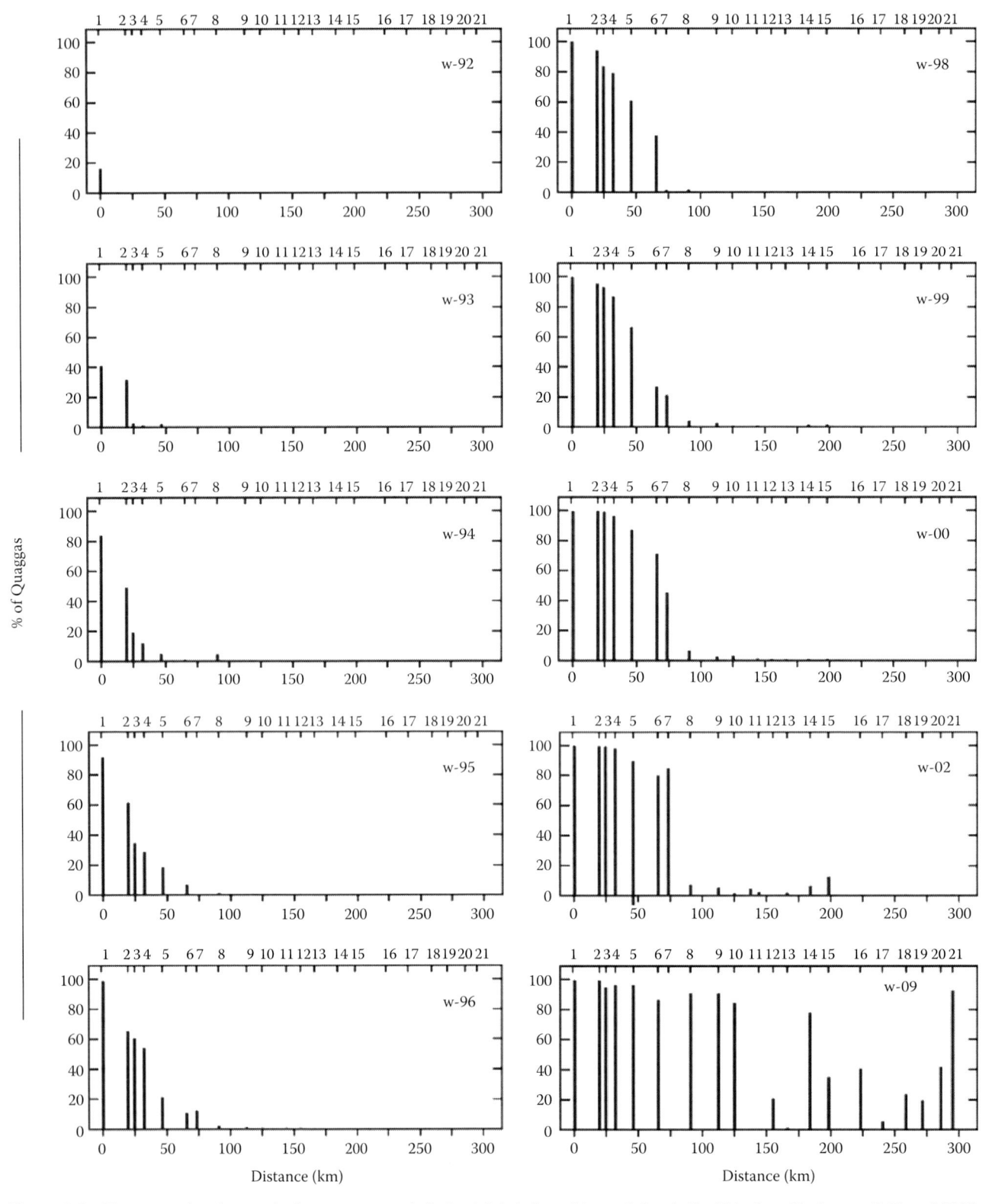

Figure 3.3 The proportion (percent) of quagga mussels in the total dreissenid population in the Erie Canal between 1991 and 2009. Each panel represents a different year (W-92 = winter 1992, W-00 = winter 2000, W-09 = winter 2009, etc.). Numbers over the top of each panel refer to sampling sites along the Canal as given in Figure 3.1. The X-axis shows approximate distance (km) from the most easterly site (Site 1).

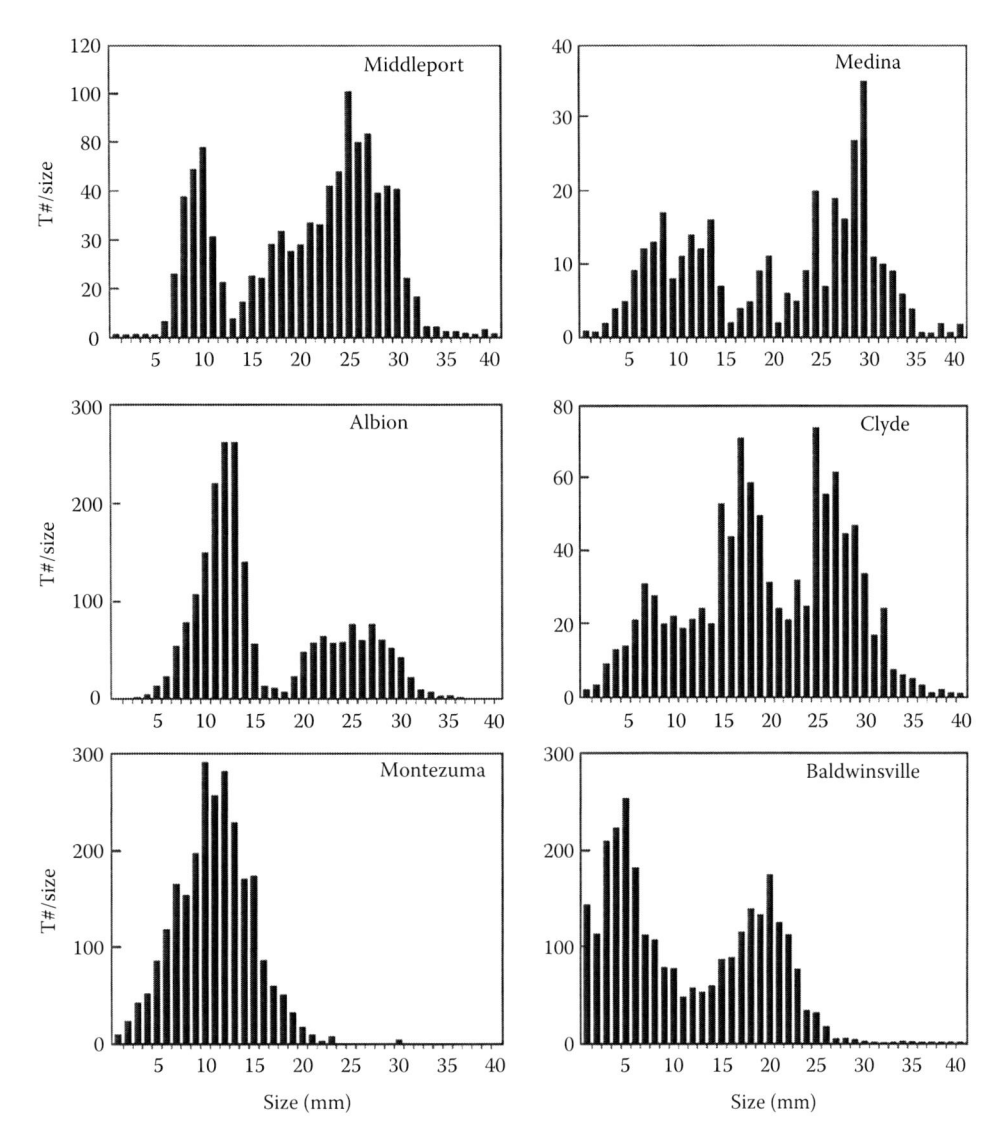

Figure 3.4 Representative length–frequency histograms of the dreissenid population (both species combined) at six sites along the Erie Canal in February 2000. T#/size = total number of mussels measured in each 1 mm size category. Middleport (Site 6), n = 1217; Medina (Site 7), n = 316; Albion (Site 8), n = 2092; Clyde (Site 17), n = 1002; Montezuma (Site 18), n = 2512; Baldwinsville (Site 20), n = 2804, n = total number of mussels sized at each site.

launch sites. Further, there are water inflows from the Genesee River (near Rochester, NY) and the Seneca River (NW of Auburn, NY) and minor water outflows at a few locations along the canal. Additionally, the proximate connection of the Seneca River (near the Onondaga Site 21) to the Oswego River (connecting to Lake Ontario) allowed boaters from Lake Ontario (where quagga populations have expanded rapidly, Mills et al. 1999) to possibly be an inadvertent vector for quaggas to enter the study area of the canal from Lake Ontario.

When both species of dreissenid mussels are present in a water body, the precise reasons for a "selective competitive edge" where one species becomes dominant over the other are not clear. In some extensive beds of macrophytes, there may be a spatial partitioning of mussel species that favors zebra mussels (Diggins et al. 2004). However, any possible advantage of quaggas over zebra mussels in deeper (and cooler) habitats (Dermott and Munawar 1993, Spidle et al. 1994, Dermott and Kerec 1997, Roe and MacIsaac 1997, Wilson et al. 2006, Nalepa et al. 2007, 2010) does not seem to hold true in habitats that are relatively shallow and thermally homogeneous (Ricciardi and Whoriskey 2004, Jones and Ricciardi 2005). Indeed, Mitchell et al. (1996) suggested that there should not be a generalization that quagga mussels dominate in deep-cold water and zebra mussels dominate in shallow-warm waters. It may be that some reproductive, physiological (Stoeckmann 2003), or subtle physiochemical factors

(Spidle et al. 1995, Jones and Ricciardi 2005) are more important in the eventual dominance of quagga mussels over zebra mussels than generally realized. Whatever the single or combined factors, the present study showed that, over time, quagga mussels clearly became the dominant dreissenid mussel in the Erie Canal between 1991 and 2009.

ANNOTATION

If mussels are allowed to die in the absence of preservative, shells may gap and some flesh may fall out causing errors in the dried weight measurements. If just routine length–frequency and weight–frequency determinations are made, then a weak formalin or alcohol solution works quite well for preservation and shells stay closed for ease in handling for the appropriate measurements. If mussels are collected for genetic analysis (May and Marsden 1992, Claxton and Boulding 1998, Brown and Stepien 2010), then common aspirin can be added to a container of water and freshly collected mussels. Mussels tend to open or gap their shells in response to the aspirin and, at that stage, a high-grade ethanol can then be added to preserve the inside flesh while the shells are gapped. Otherwise, mussels may close their shells before the inside flesh is adequately preserved, some decay may take place, and good molecular/genetic characterization would be difficult (pers. comm., J. Brown, Lake Erie Center, May 2009).

ACKNOWLEDGMENTS

I greatly acknowledge the Buffalo Bureau of Water for letting me ride along on their historic icebreaker boat, the "Cotter," to collect dreissenid mussels (mostly during February) from the Buffalo Water Intake Building. Mike Weimer (USFWS), Deana Amico, Mark Graham, and Juliette Smith (the latter three were students at SUNY/Buffalo at the time), and David Zimmer, each helped collect mussels from selected sites along the Erie Canal on one day in the 1990s. James Stamos provided great help with the figures. No funds were received from any agency to support this research over the years.

REFERENCES

Brown, J.E. and C.A. Stepien. 2010. Population genetic history of the dreissenid mussel invasions: Expansion patterns across North America. *Biol. Invasions* 12: 3687–3710.

Carlton, J.T. 2008. The zebra mussel *Dreissena polymorpha* found in North America in 1986 and 1987. *J. Great Lakes Res.* 34: 770–773.

Claxton, W.T. and E.G. Boulding. 1998. A molecular technique for identifying field collections of zebra mussel (*Dreissena polymorpha*) and quagga mussel (*Dreissena bugensis*) veliger larvae applied to eastern Lake Erie, Lake Ontario, and Lake Simcoe. *Can. J. Zool.* 76: 194–198.

Dermott, R. and D. Kerec. 1997. Changes to the deepwater benthos of eastern Lake Erie since the invasion of *Dreissena*: 1979–1993. *Can. J. Fish. Aquat. Sci.* 54: 922–930.

Dermott, R. and M. Munawar. 1993. The invasion of the Lake Erie offshore sediments by *Dreissena* and its ecological implications. *Can. J. Fish Aquat. Sci.* 50: 2298–2304.

Diggins, T.P., M. Weimer, K.M. Stewart et al. 2004. Epiphytic refugium: Are two species of invading freshwater bivalves partitioning spatial resources. *Biol. Invasions* 6: 83–88.

Griffiths, R.W. 1993. Effects of zebra mussels (*Dreissena polymorpha*) on the benthic fauna of Lake St. Clair, In *Zebra Mussels, Biology, Impacts, and Control*, T.F. Nalepa and D.W. Schloesser, Eds., pp. 415–437. Boca Raton, FL: CRC Press.

Grigorovich, I.A., J.R. Kelly, J.A. Darling, and C.W. West. 2008. The quagga mussel invades the Lake Superior Basin. *J. Great Lakes Res.* 34: 342–350.

Hebert, P.D.N., B.W. Muncaster, and G.L. Mackie. 1989. Ecological and genetic studies on *Dreissena polymorpha* (Pallas): A new mollusk in the Great Lakes. *Can. J. Fish. Aquat. Sci.* 46: 1587–1591.

Horvath, T.G., G.A. Lamberti, D.M. Lodge, and W.L. Perry. 1996. Zebra mussel dispersal in lake-stream systems: Source-sink dynamics? *J. N. Am. Benthol. Soc.* 15: 564–575.

Jones, L.A. and A. Ricciardi. 2005. Influence of physiochemical factors on the distribution and biomass of invasive mussels (*Dreissena polymorpha* and *Dreissena bugensis*) in the St. Lawrence River. *Can. J. Fish. Aquat. Sci.* 62: 1953–1962.

Leech, J.H. 1993. Impacts of the zebra mussel (*Dreissena polymorpha*) on water quality and fish spawning reefs in western Lake Erie, In *Zebra Mussels: Biology, Impacts, and Control*, T.F. Nalepa and D.W. Schloesser, Eds., pp. 381–397. Boca Raton, FL: CRC Press.

May, B. and J.E. Marsden. 1992. Genetic identification and implications of another invasive species of dreissenid mussel in the Great Lakes. *Can. J. Fish. Aquat. Sci.* 49: 1501–1506.

Mills, E.L., J.R. Chrisman, B. Baldwin et al. 1999. Changes in the dreissenid community in the lower Great Lakes with emphasis on southern Lake Ontario. *J. Great Lakes. Res.* 25: 187–197.

Mills, E.L., R.M. Dermott, E.F. Roseman et al. 1993a. Colonization, ecology and population structure of the "quagga" mussel (Bivalvia: Dreissenidae) in the lower Great Lakes. *Can. J. Fish. Aquat. Sci.* 50: 2305–2314.

Mills, E.L., J.H. Leach, J.T. Carlton, and C.L. Secor. 1993b. Exotic species in the Great Lakes: A history of biotic crises and anthropogenic introductions. *J. Great Lakes Res.* 19: 1–54.

Mills, E.L., G. Rosenberg, A.P. Spidle, M. Ludyanskiy, Y. Pligin, and B. May. 1996. A review of the biology and ecology of the quagga mussel (*Dreissena bugensis*), a second species of freshwater dreissenid introduced to North America. *Am. Zool.* 36: 271–286.

Mitchell, J.S., R.C. Bailey, and R.W. Knapton. 1996. Abundance of *Dreissena polymorpha* and *Dreissena bugensis* in a warm water plume: Effects of depth and temperature. *Can. J. Fish. Aquat. Sci.* 53: 1705–1712.

Nalepa, T.F., D.L. Fanslow, and G.A. Lang. 2009. Transformation of the offshore benthic community in Lake Michigan: Recent shift from the native amphipod *Diporeia* spp. to the invasive mussel *Dreissena rostriformis bugensis. Freshwat. Biol.* 543: 466–479.

Nalepa, T.F., D.L. Fanslow, S.A. Pathogen, A.F. Foley, III, and G.A. Lang. 2007. Long-term trends in benthic macroinvertebrate populations in Lake Huron over the past four decades. *J. Great Lakes Res.* 33: 41–436.

Nalepa, T.F., D.L. Fanslow, and S.A. Pothoven. 2010. Recent changes in density, biomass, recruitment, size structure, and nutritional state of *Dreissena* populations in southern Lake Michigan. *J. Great Lakes Res.* 36: 5–19.

Pimentel, D. 2005. Aquatic nuisance species in the New York State Canal and Hudson River systems and the Great Lakes Basin: An economic and environmental assessment. *Environ. Manage.* 35: 692–702.

Ricciardi, A. and F.G. Whoriskey. 2004. Exotic species replacement: Shifting dominance of dreissenid mussels in the Soulanges Canal, upper St. Lawrence River, Canada. *J. N. Am. Benthol. Soc.* 23: 507–514.

Riessen, H.P., T.A. Ferro, and R.A. Kamman. 1993. Distribution of zebra mussel (*Dreissena polymorpha*) veligers in Eastern Lake Erie during the first year of colonization, In *Zebra Mussels, Biology, Impacts, and Control*, T.F. Nalepa and D.W. Schloesser, Eds., pp. 415–437. Boca Raton, FL: CRC Press.

Roe, S.L. and H.J. MacIsaac. 1997. Deepwater population structure and reproductive state of quagga mussels (*Dreissena bugensis*) in Lake Erie. *Can. J. Fish. Aquat. Sci.* 54: 2428–2433.

Spidle, A.P., J.E. Marsden, and B. May. 1994. Identification of the Great Lakes quagga mussel as *Dreissena bugensis* from the Dnieper River, Ukraine, on the basis of allozyme variation. *Can. J. Fish. Aquat. Sci.* 51: 1485–1489.

Spidle, A., E.L. Mills, and B. May. 1995. Limits to tolerance of temperature and salinity in the quagga mussel (*Dreissena bugensis*). *Can. J. Fish. Aquat. Sci.* 57: 2108–2119.

Stoeckel, J.A., D.W. Schneider, L.A. Soeken, K.D. Blodgett, and R.E. Sparks. 1997. Larval dynamics of a riverine metapopulation: Implications for zebra mussel recruitment, dispersal, and control in large river systems. *J. N. Am. Benthol. Soc.* 16: 586–601.

Stoeckmann, A. 2003. Physiological energetics of Lake Erie dreissenid mussels. A basis for the displacement of *Dreissena polymorpha* by *Dreissena bugensis*. *Can. J. Fish. Aquat. Sci.* 60: 126–134.

Wilson, K.A., E.T. Howell, and D.A. Jackson. 2006. Replacement of zebra mussels in the Canadian nearshore of Lake Ontario: The importance of substrate, round goby abundance, and upwelling frequency. *J. Great Lakes Res.* 32: 11–28.

CHAPTER 4

Invasion of Quagga Mussels (*Dreissena rostriformis bugensis*) to the Mid-Lake Reef Complex in Lake Michigan
A Photographic Montage

Jeffrey S. Houghton, Robert Paddock, and John Janssen

CONTENTS

Abstract 65
Introduction 65
Methods 66
Results and Discussion 66
Summary 70
Acknowledgments 70
References 70

ABSTRACT

We provide photodocumentation of the invasion of quagga mussels (*Dreissena rostriformis bugensis*) to the Mid-Lake Reef Complex (MLRC) in Lake Michigan. The MLRC is an ensemble of reefs between the northern and southern basins of the lake. The reefs have summits that range from a depth of about 40–50 m and lie entirely deeper than the euphotic zone. We assembled individual frames from 2002, 2003, 2005, 2006, and 2009 from video taken via a remotely operated vehicle (ROV). While the ROV dives and videos targeted lake trout (*Salvelinus namaycush*) reproduction at the MLRC, they nonetheless provided clear visual documentation of the quagga mussel invasion of the reef complex. Quagga mussels were nearly nonexistent in 2002 but completely covered cobble substrate of East Reef and Sheboygan Reef by 2006.

INTRODUCTION

The distribution and abundance trends of dreissenids in Lake Michigan have been well documented in several recent papers (Nalepa et al. 2009, 2010). Based on these latest population surveys, spatial patterns indicate low abundances in the middle of the lake, extending along the lake's north–south axis. Such a distribution is not surprising since this offshore region is deep (>100 m), and dreissenid populations may be limited by low food availability. Yet the middle of the lake is also the location of the mid-lake reef complex (MLRC), which is a system of several Paleozoic carbonate reefs with summits at depths of about 40–50 m and with bases ≥100 m. This reef complex separates Lake Michigan's northern and southern basins (Figure 4.1). The lake-wide surveys of dreissenid populations in Lake Michigan by Nalepa et al. (2009, 2010) were conducted with a Ponar grab, which cannot effectively collect samples on hard substrates such as those found within the MLRC; hence, no sampling sites were located within the MLRC in the dreissenid surveys by Nalepa et al. (2009, 2010). In general, dreissenids prefer hard substrates (Jones and Ricciardi 2005), so abundances in the middle of the lake, as depicted by density contour maps (see Figure 3 in Nalepa et al. 2009), may be misleading if the MLRC is heavily colonized by dreissenids. Hard substrates are a challenge to sample, but the stability of hard substrates also means they can be disproportionately important habitats for dreissenid colonization (Janssen et al. 2005).

In this chapter, we present visual evidence of the importance of the MLRC as a substrate for colonization by the quagga mussel (*Dreissena rostriformis bugensis*) in Lake Michigan.

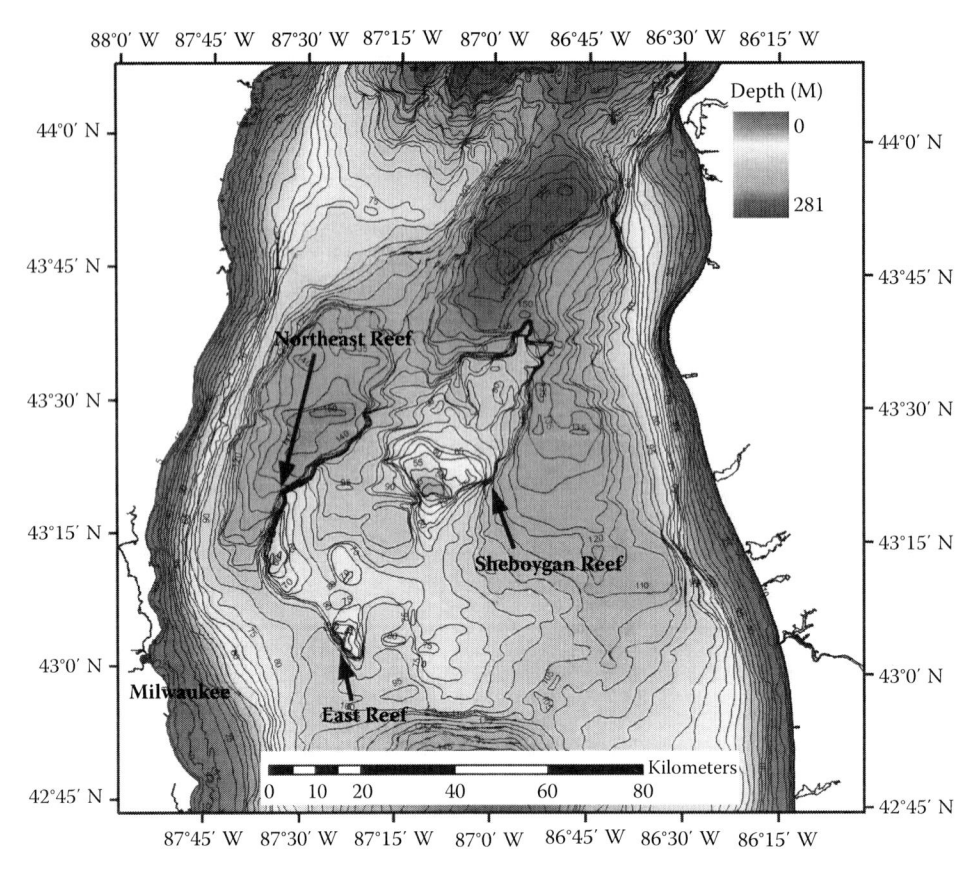

Figure 4.1 (See color insert.) Bathymetry of central Lake Michigan's MLRC indicating positions of the three major summits, East Reef, Northeast Reef, and Sheboygan Reef. (Modified from NOAA: Great Lakes Data Rescue Project—Lake Michigan Bathymetry, Area III, 2010.)

We provide time-series photographs captured from videos that were taken as part of studies to define lake trout spawning on the reef complex (Janssen et al. 2006). These are complemented with other images that serve to illustrate the impact of quagga mussels on the MLRC. Our observations began in 2001 and extended through 2010.

METHODS

Most underwater images are single frames from video taken via a remotely operated vehicle (ROV). Details of ROV configuration, operation, and sampling locations are given in Janssen et al. (2006). Images shown here are mainly from East Reef (about 43°01′ N, 87°21′ W), Sheboygan Reef (about 43°21′ N, 87°09′ W), and Northeast Reef (about 43°15′ N and 87°35′ W) (Figure 4.1). Cameras and recording devices were upgraded during the 9 years of the study. Early video recordings were on Hi-8 video tapes and later recordings (2005 and subsequent) were digital. For video documentation of mussels on these reefs (Houghton et al. 2013). Still photography was done only in 2010, and these images, all from Northeast Reef, were taken with a Wolfvision™ SCB1 XGA (resolution 1024 × 768 pixels). Most individual video frames were chosen because they showed a similar range of rock sizes and light conditions. Lighting was provided by a combination of natural light and lights on the ROV. Several frames show a pair of electrodes pointing away from the camera. These electrodes were 35 cm long and separated by 35 cm. Plexiglas sampling nozzles, visible in every image, were about 50 mm in diameter. At East Reef in 2006–2009, we towed a beam trawl designed to be towed over cobble and boulders (Hudson et al. 1995) to collect lake trout fry, but it also collected a considerable bycatch of quagga mussels. Tows were 5 min long and covered a distance of about 300 m.

RESULTS AND DISCUSSION

The quagga mussel invasion on the MLRC was most completely documented by individual video frames from East Reef in 2002, 2003 (spring and fall), 2005, 2006, and 2009 (Figures 4.2a through f). The East Reef images showed virtually no quagga mussels in 2002 (Figure 4.2a; a few whitish spots are individual quagga mussels). By 2003, quagga mussels primarily occurred in clusters (Figure 4.2b), but by 2005 (Figure 4.2c) and thereafter (Figures 4.2d through f), quagga mussels more or less completely covered all rocky substrates. This time sequence of quagga mussel expansion on the reef coincided well with abundance increases at depths of 30–50 m along the shore in the southern basin of the lake

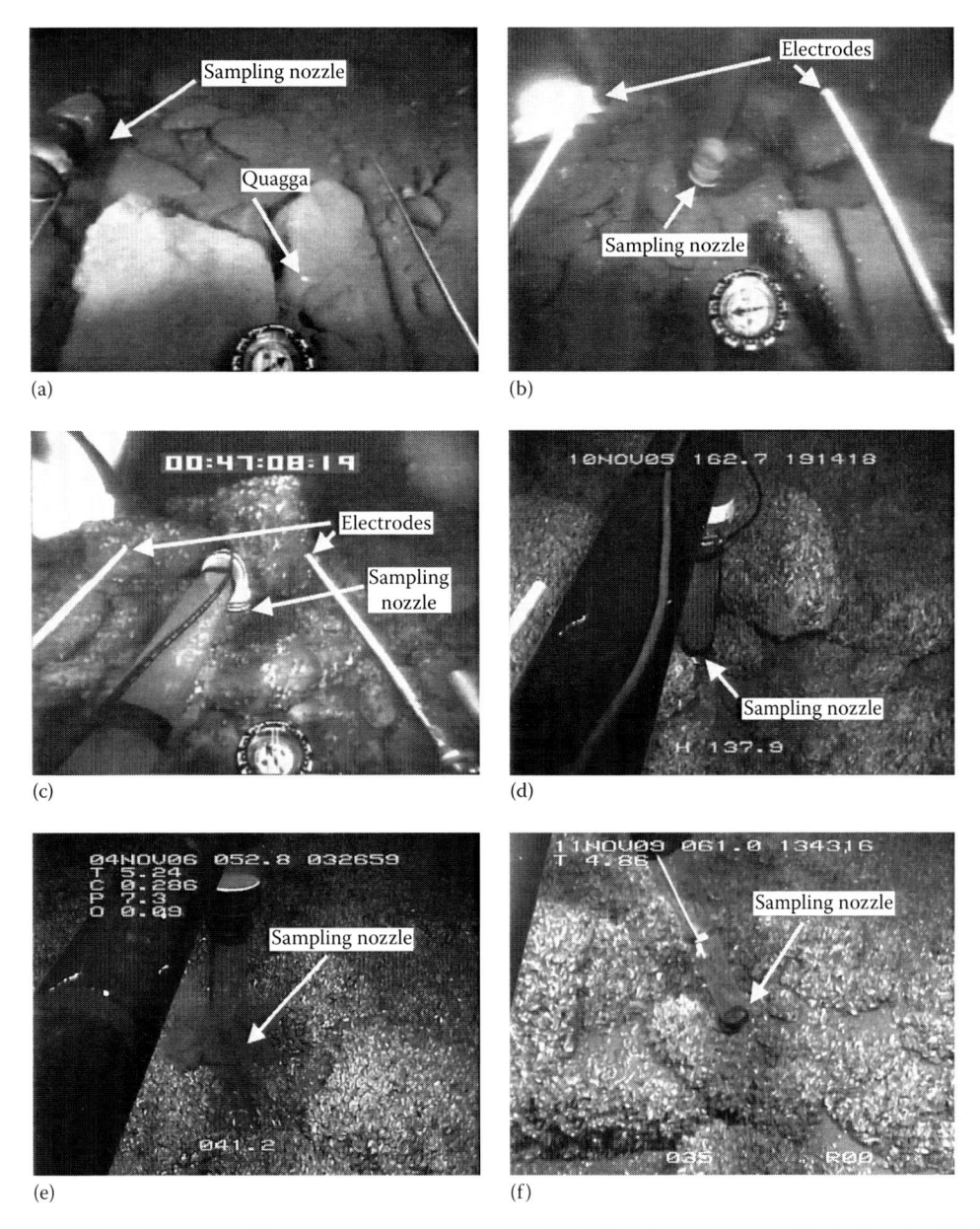

Figure 4.2 (See color insert.) Video–still images of quagga mussels at East Reef (approximately 43°01′ N and 87°21′ W). The nozzle (50 mm diameter Plexiglas) and electrodes (35 cm long and 35 cm separation) are labeled for size reference. (a) April 2002; this was our first dive at East Reef, very few quagga mussels were present, and they appeared as scattered white spots on the image. (b) May 2003; quagga mussels were still uncommon. (c) November 2003; quagga mussels were now common and in clusters. (d) November 2005; quagga mussels now covered most rocks, but there are occasional areas of bare rock. (e) November 2006; all cobble and boulder surfaces were now totally covered with quagga mussels. (f) November 2009; quagga mussels covered all boulders and cobbles, but there was less coverage on bedrock underlying cobble and boulders.

(Nalepa et al. 2010). At these depths, the quagga mussel population began to increase in 2002 and reached a peak in 2006.

We spent less time and collected less video footage at Sheboygan Reef than at East Reef. However, images in Figure 4.3 showed a pattern of invasion similar to East Reef. Sheboygan Reef had very few quagga mussels in November 2001 (Figure 4.3a) but was nearly completely covered by mussels in November 2004 (Figure 4.3b). The extent of mussel colonization in an area of smooth bedrock at Sheboygan Reef (June 2004) is shown in Figure 4.4; this area is just south of the cobble/boulder area of Figure 4.3. We have not revisited this site, but such sites might be useful for monitoring purposes because they lack heterogeneity and the bottom smoothness may permit Ponar grab sampling. East Reef also has flat areas of smooth bedrock, but we have not investigated their extent.

Close-up still photographs (Northeast Reef, November 2010) show in more detail how tightly packed the quagga mussels were (Figure 4.5). In Figure 4.5a, extended siphons of the mussels are readily seen in a somewhat lateral view.

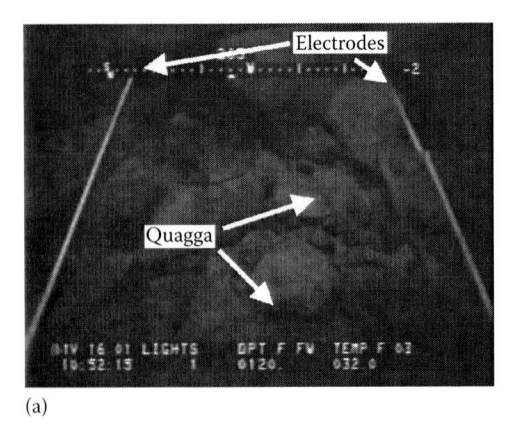

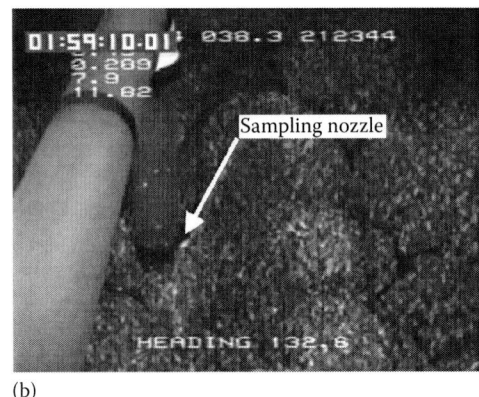

Figure 4.3 **(See color insert.)** Video–still images of quagga mussels at Sheboygan Reef (approximately 43°21′ N and 87°09′ W) at a depth of approximately 40 m. Sheboygan Reef was sampled less frequently than East Reef, but the invasion timing appeared to be similar to that for East Reef. (a) November 2001; only a few quagga mussels were visible. (b) November 2004; quagga mussels covered nearly all rocks.

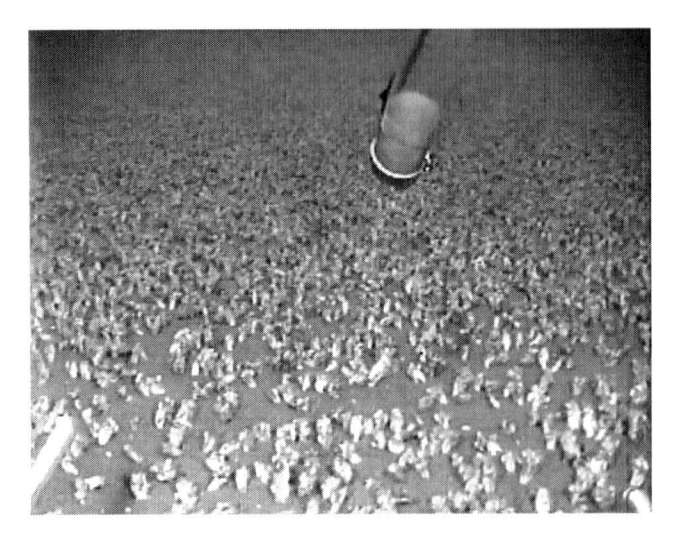

Figure 4.4 **(See color insert.)** Quagga mussels on smooth bedrock immediately south of the sites in Figure 4.3. (Sheboygan Reef, June 2004, approximately 43° 20.5′ N and 87°09′ W.)

While fairly transparent, siphons tended to blur images of the mussels. In Figures 4.2 and 4.3, siphons were extended, but distances and limited resolution of the videos did not resolve the images. Siphons are also extended in Figure 4.5b, but views of mussel valves were less obscured by siphons because the camera angle was more orthogonal to the substrate surface. Siphons were retracted in Figure 4.5c and most of Figure 4.5d because an electroshocker to collect lake trout fry had been activated.

Hydra was seen attached to some quagga mussels in Figure 4.5b. We believe *Hydra* were probably present at the MLRC historically since it is commonly found on hard substrates as deep as 400 m (Nalepa et al. 1987).

Beam trawl sampling targeted rocky habitat for trout fry, but mussel bycatch in the beam trawl was typically very large (Figure 4.6). However, since the trawl only grazes the tops of cobble and boulders (based on post-tow ROV dives), these samples could not be considered quantitative for mussels. Indeed, abundances estimated from trawls are severely underestimated (Nalepa et al. 2010). The trawl sample of mussels in Figure 4.6 was about 100 L and represented about one-half of the total sample. The trawl occasionally encountered coarse, sandy substrate at East Reef, and we often found some collected mussels attached to coarse sand by their byssal threads (Figure 4.7). Because we were exploring for lake trout spawning sites, we do not investigate areas with a sand bottom with the ROV.

Estimates of mussel abundance/biomass or percent coverage at the MLRC could not be easily quantified for at least two reasons. First, there are multiple scales of substrate heterogeneity, and we mapped only areas near summits as they are most likely to be used by lake trout for spawning. Substrate heterogeneity included more or less rounded rocks, flattened (flagstone) rocks often at talus slopes, flat bedrock, clay, and sand. Rocks were typically in clusters of varied size on the reef plateaus and talus slopes. Second, cobble and boulders were heterogeneous in size and shape, and therefore, accurate estimates of surface areas were not possible. Hence, we did not attempt to quantify quagga mussel density or coverage from our images in part because of habitat heterogeneity but also because our images were biased toward areas of cobble believed to be used by lake trout for spawning.

An initial approach to quantify mussel populations might follow the habitat mapping design of Waples et al. (2005), who used a combination of multibeam bathymetry and seabed classification (with ROV ground truthing) to map habitats along the Wisconsin coast of Lake Michigan. The ROV used in the present study was capable of suctioning mussels from cobble and boulder and also returning samples to the surface; thus, it was feasible to collect mussel samples from these surfaces. It may be feasible to perform a regression of photographic/video samples with suction samples and then use the regression as a calibration for photographs for extensive photographic/video surveys. This is potentially complicated because mussel siphons obscure individual mussels (Figure 4.5a). However, it might be possible to quantify mussels from images by counting extended siphons.

The ecological importance and dynamics of the MLRC in Lake Michigan before mussel colonization are not well

INVASION OF QUAGGA MUSSELS (*DREISSENA ROSTRIFORMIS BUGENSIS*) 69

Figure 4.5 **(See color insert.)** Still images of Northeast Reef, November 2010 (approximately 43°15′ N and 87°35′ W). (a) Somewhat lateral view of quagga mussels with their siphons extended. Note that siphons obscured resolution of individual quagga mussels. (b) View of quagga mussels with siphons extended with the camera nearly orthogonal to substrate surface. Note the cluster of *Hydra* (circle), and closer inspection reveals other scattered *Hydra*. (c) Quagga mussels with siphons contracted due to electroshocking. The slimy sculpin (*C. cognatus*) had been stunned by the electroshocker. (d) Quagga mussels with siphons contracted near an electroshocked burbot (*Lota lota*). Away from the electroshocked area, mussel siphons (upper corners) were more extended and the image was hazier.

Figure 4.6 **(See color insert.)** Bycatch of quagga mussels in beam trawl used to collect lake trout fry at East Reef (June 2010). The total catch was about 200 L of live quagga mussels.

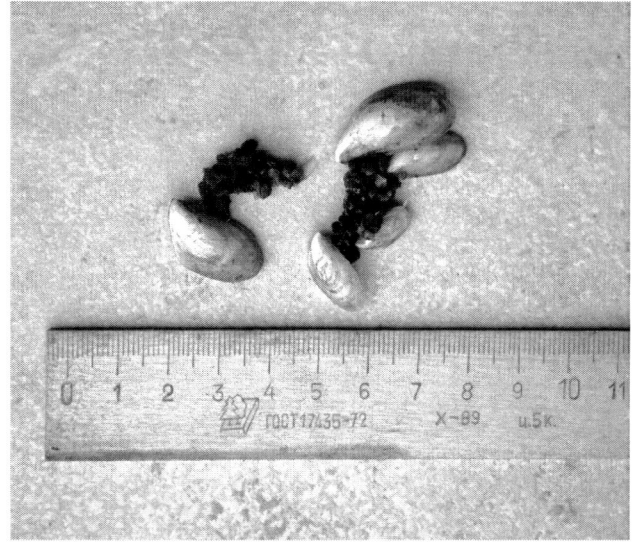

Figure 4.7 **(See color insert.)** Quagga mussels with byssal threads attached to coarse sand from East Reef. These mussels were collected via beam trawl.

understood, but commercial fishers recognized the area as highly productive for native populations of lake trout (*Salvelinus namaycush*) (Coberly and Horrall 1980). Lake trout were likely sustained by the mysid shrimp *Mysis diluviana*, slimy sculpin (*Cottus cognatus*), and bloater (*Coregonus hoyi*). Bloaters were heavily fished by commercial fishers until the area was closed to fishing to protect lake trout restoration efforts (see Janssen et al. 2006, Houghton et al. 2010).

The reefs and their summits lie entirely deeper than the euphotic zone, so Houghton et al. (2010) argued that the likely reason for fish productivity was hydrodynamic processes similar to those found at oceanic sea mounts (see Genin 2004 for a review). They argued that allochthonous transport of *Mysis* onto the reefs during their nocturnal vertical migrations, followed by topographic blockage of the diurnal descent (Genin 2004), concentrated *Mysis* and thus provided a food resource for slimy sculpins. Water currents around sea mounts are complex and create areas of upwelling and downwelling (Genin 2004), which may be the mechanism by which quagga mussels within the MLRC obtain phytoplankton. A common belief was that bottom substrates of the MLRC would not be heavily colonized by quagga mussels because of limited food availability, but our images indicate that large populations of quagga mussels have now colonized this reef area. These events parallel those in the marine environment where many assumed that little life would be found in deep water habitats, but subsequent oceanic exploration proved this assumption wrong (McLain 2010).

SUMMARY

In a span of only a few years, the MLRC in Lake Michigan went from being essentially free of dreissenid mussels in fall 2001 to completely covered by quagga mussels in 2006. We hope these images, while not quantitative, serve as visual evidence of a significant deepwater invasion in an unsampled area. We also suggest that the highly successful invasion of the reefs provide evidence that these reefs function in a manner analogous to oceanic sea mounts as documented by Houghton et al. (2010).

ACKNOWLEDGMENTS

We acknowledge the crew of the R/V Neeskay for their hard work and support, the NOAA Explore and National Underwater Research Programs, Great Lakes Fishery Trust, and Wisconsin Sea Grant.

REFERENCES

Coberly, C.E. and R.M. Horrall. 1980. Fish Spawning grounds in Wisconsin water of the Great Lakes. University of Wisconsin Sea Grant Institute, Madison, WI, Pub. No. Wis-SG-80-235.

Genin, A. 2004. Bio-physical coupling in the formation of zooplankton and fish aggregations over abrupt topographies. *J. Mar. Syst.* 50: 3–20.

Houghton, C.J., C.R. Bronte, R.W. Paddock, and J. Janssen. 2010. Evidence for allochthonous prey delivery to Lake Michigan's Mid-Lake Reef Complex: Are deep reefs analogs to oceanic sea mounts? *J. Great Lakes Res.* 36: 666–673.

Houghton, J.S., R. Paddock, and J. Janssen. 2013. Video Clip No.2: Invasion of quagga mussels (*Dreissena rostriformis bugensis*) to the Mid-lake Reef Complex in Lake Michigan: A Video Montage. In *Quagga and Zebra Mussels: Biology, Impacts, and Control*, 2nd Edn., T. F. Nalepa and D. W. Schloesser, Eds., p. 746. Boca Raton, FL: CRC Press.

Hudson, P.L., J.F. Savino, and C.R. Bronte. 1995. Predator-prey relations and competition for food between age-0 lake trout and slimy sculpins in the Apostle Island region of Lake Superior. *J. Great Lakes Res.* 21: 445–457.

Janssen, J., M. Berg, and S. Lozano. 2005. Submerged terra incognita: The abundant but unknown rocky zones. In *The Lake Michigan Ecosystem: Ecology, Health and Management*, T. Edsall and M. Munawar, Eds., pp. 113–139. Amsterdam, The Netherlands: Academic Publishing.

Janssen, J., D.J. Jude, T.A. Edsall, M. Toneys, and P. McKee. 2006. Evidence of lake trout reproduction at Lake Michigan's Mid-Lake Reef Complex. *J. Great Lakes Res.* 32: 749–763.

Jones, L.A. and A. Ricciardi. 2005. Influence of physicochemical factors on the distribution and biomass of invasive mussels in the St. Lawrence River. *Can. J. Fish. Aquat. Sci.* 62: 1953–1962.

McLain, C. 2010. An empire lacking food. *Am. Sci.* 98: 470–477.

Nalepa, T.F., D.L. Fanslow, and G.A. Lang. 2009. Transformation of the offshore benthic community in Lake Michigan: Recent shift from the native amphipod *Diporeia* spp. to the invasive mussel *Dreissena rostriformis bugensis*. *Freshwat. Biol.* 54: 466–479.

Nalepa, T.F., D.L. Fanslow, and S.A. Pothoven. 2010. Recent changes in density, biomass, recruitment, size structure, and nutritional state of *Dreissena* populations in southern Lake Michigan. *J. Great Lakes Res.* 36: 5–19.

Nalepa, T.F., C.C. Remsen, and J.V. Klump. 1987. Observations of *Hydra* from a submersible at two deepwater sites in Lake Superior. *J. Great Lakes Res.* 13: 84–87.

Waples, J.T., R. Paddock, J. Janssen, D. Lovalvo, B. Schulze, J. Kaster, and J.V. Klump. 2005. High resolution bathymetry and lakebed characterization in the nearshore of western Lake Michigan. *J. Great Lakes Res.* 31: 64–74.

CHAPTER 5

Long-Term Change in the Hudson River's Bivalve Populations
A History of Multiple Invasions (and Recovery?)

David L. Strayer and Heather M. Malcom

CONTENTS

Abstract 71
Introduction 72
Study Area 72
Methods 72
Results and Discussion 73
    Invaders 73
        Zebra Mussel (*Dreissena polymorpha*) 73
        Quagga Mussel (*Dreissena rostriformis bugensis*) 75
        Asian Clam (*Corbicula fluminea*) 75
        Other Invaders 75
    Responses of the Native Bivalves 75
        Pearly Mussels (Unionidae) 75
        Pea and Fingernail Clams (Sphaeriidae) 78
General Discussion 78
    Bivalves and Ecosystem Instability 78
    Notes on Exploitation Competition 79
    Return of the Natives 79
Acknowledgments 80
References 80

ABSTRACT

We studied bivalve populations in the freshwater tidal Hudson River, New York, from 1990 to 2010, a period in which three invasive bivalves (zebra mussels [*Dreissena polymorpha*], quagga mussels [*Dreissena rostriformis bugensis*], and the Asian clam [*Corbicula fluminea*]) appeared. Although the latter two species have not yet become abundant, zebra mussels became very abundant and affected many parts of the ecosystem (plankton biomass, water clarity, and chemistry, much of the upper food web). In response, populations of native unionid mussels (chiefly *Elliptio complanata*, *Anodonta implicata*, and *Leptodea ochracea*) and sphaeriid clams (chiefly *Pisidium* spp.) plummeted through the 1990s and showed reduced recruitment and body condition. Survivorship of zebra mussels dropped steadily from ~50%/year in the early-1990s to <1%/year after 2005, and population density, recruitment, and body condition of native bivalves partially recovered. It appears that exploitation competition, rather than fouling, was the main way by which zebra mussels have harmed native bivalves. The Hudson's bivalve community is likely to remain dynamic over the coming decades, leading to instability in the ecosystem. It is difficult to predict the future, but long-term coexistence of invaders and native bivalves appears to be at least possible.

INTRODUCTION

During the twentieth century, invasions of *Dreissena* spp., *Corbicula* spp., and *Limnoperna fortunei* were occurring around the world at the same time as native unionoid bivalves were declining extensively as a result of habitat change, overharvest, and pollution. As a result of this combination of invasion and loss of native faunas, bivalve communities in many freshwaters in 2011 are very different than they were in 1900, in terms of species composition, ecological traits, biomass, and ultimately ecological function. Because bivalves can play so many trophic and engineering roles (e.g., Strayer et al. 1999, Vaughn and Hakenkamp 2001), these changes in bivalve communities had large, unanticipated consequences on the structure and function of many freshwater ecosystems.

Here, we describe the results of a 21-year study of bivalve populations, both native and nonnative, in the Hudson River, New York. During this dynamic period of study (1990–2010), the river was invaded by three ecologically important nonnative bivalve species, which led to rapid and deep losses in the native bivalve community. Now, it appears that the native species are recovering, perhaps indicating the possibility that the invaders and the natives can coexist over the long term. This chapter describes the dynamics of both the invaders and the natives in detail and extends the findings of previous papers that were based on shorter runs of data (Strayer et al. 1994, 1996, 2011, Strayer and Smith 1996, 2001, Strayer and Malcom 2006, 2007). The broader effects of these dynamics on the Hudson River ecosystem were described in detail elsewhere (e.g., Caraco et al. 1997, 2000, 2006, Pace et al. 1998, 2010, Strayer et al. 1999, 2004, 2008, 2011, Strayer 2009).

STUDY AREA

The study area is the freshwater tidal section of the Hudson River in eastern New York (Figure 5.1), which extends from RKM 100–248 (RKM, river kilometer, as measured from the Battery at the southern end of Manhattan in New York City). The average width and depth of the river in the study area are 900 and 8 m, respectively, and mean annual discharge is ~500 m³/s, depending on location. Approximately 15% of the study area is <3 m deep at low tide, and much of this shallow region supports rooted vegetation (chiefly *Vallisneria americana*). The entire study area is tidal; twice-daily tides have a typical range of 0.8–1.6 m, reverse the direction of water flow except during spates, and keep the water column from stratifying. Nevertheless, sea salt is present only during extended periods of low freshwater flow and only in the most downriver parts of the study area (below RKM 120); even in this lower region, salinity is usually <1 psu. The water is turbid (growing-season Secchi depths usually are 1–2 m), hard (calcium = 25–30 mg/L), and nutrient rich (~0.5 mg/L NO_3–N, ~30 μg/L PO_4–P) (Caraco et al. 1997). Most of the river bottom in the study area is sand or mud, but ~7% of the study area is rocky. The macrozoobenthos is dominated by tubificid oligochaetes, chironomids, amphipods, and various bivalves (Strayer and Smith 2001).

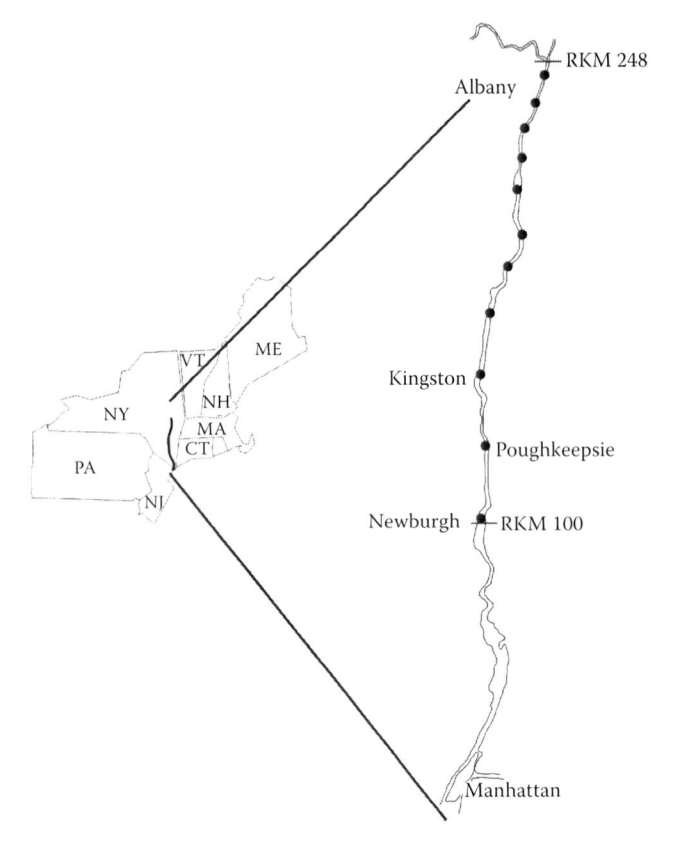

Figure 5.1 Map of the study area in the freshwater tidal reach of the Hudson River. Dots show the location of transects where samples of bivalves living on soft sediments were taken in 1991–2010.

METHODS

The results reported here come from three different sampling programs, which were described in detail by Strayer and Malcom (2006, 2007) and Strayer et al. (2011). Unionid mussels, *Corbicula*, and *Dreissena* living on sand and mud were sampled annually in 1991–2010 along 11 cross-channel transects deployed in a stratified random design. Each transect contained four stations that were initially chosen randomly, then revisited each year in late June to early August. At each station, we took five samples with a standard PONAR (23 × 23 cm) grab. We sieved the samples in the field through a 2.8 mm mesh sieve. Adult unionids were removed from the sample and placed (along with any attached *Dreissena*) into individual Whirl-Pak bags, then they and the remaining sieve residue were placed into a cooler, taken to the laboratory, and frozen. Later, samples were thawed and carefully examined for small unionids, *Corbicula*, and *Dreissena*. We used calipers to measure shell lengths of *Dreissena* and shell length, width, and height of unionids. Soft tissues of unionids were removed, dried at 60°C, and weighed to determine dry mass.

Populations of *Dreissena* living on sediments too rocky to sample with a PONAR were sampled by SCUBA divers at six or seven sites in early August and (in most years) in early June in 1993–2010. At each site, divers haphazardly picked up ten rocks 15–40 cm in maximum dimension. Rocks were put into coolers, taken to the laboratory, and kept in a cold room until processed (usually within 2 days). We removed and counted all *Dreissena* from each rock and traced the outline of the rock onto paper, which we later cut out and weighed to estimate the projected area of each rock. Shell lengths were measured on random subsamples of animals from each site (n = 300/site, if possible) to develop size–frequency distributions. In addition, a subsample of animals (n = 50/site, if possible) spanning a wide range of shell length was chosen to develop shell length–dry body mass regressions.

Population densities of sphaeriid clams and *Rangia cuneata* were determined as part of an annual program to monitor macrozoobenthos (Strayer and Smith 2001, Strayer et al. 2011). We sampled two stations at each of four sites along the study area in late September to early October 1990–1999, 2001–2002, and 2005–2006. Five replicate samples were taken at each station with a petite PONAR grab (15 × 15 cm). Samples were sieved in the field through a 0.5 mm mesh sieve and preserved in buffered formalin. In the laboratory, samples were stained with Rose Bengal, rinsed, and sorted at 6–12× magnification. At least 20% of these replicate samples were double sorted to estimate sorting efficiencies using the removal method (Zippin 1958).

"Body condition" of both zebra mussels and unionids was calculated as follows. We calculated the $\log_{10}$–$\log_{10}$ regression of dry body mass on shell length (or shell length, width, and height for unionids) for all individuals of a given species and defined "condition" of each individual as the residual from the overall regression for each species. Because condition is calculated in $\log_{10}$ units, a difference in condition of 0.3 corresponds to a two-fold difference in body mass at a given shell length.

Unless noted otherwise, we use the term "filtration rate" to refer to the aggregate filtration rate of all zebra mussels in the population, as an areally weighted average over the entire study area. Filtration rate was estimated by applying the regression of Kryger and Riisgard (1988) to the measured population densities and body sizes of Hudson River zebra mussels. Filtration rates estimated from Kryger and Riisgard's regression agreed well with measurements of filtration rates of Hudson River zebra mussels made in the laboratory (Roditi et al. 1996) and the field (Roditi et al. 1997). Estimates of filtration rates assume a water temperature of 20°C. The units of this filtration rate are m/day, which is equivalent to m^3/m^2/day.

RESULTS AND DISCUSSION

Invaders

During the study period (1990–2010), three nonnative bivalves (*D. polymorpha*, *D. bugensis*, and *C. fluminea*) became established in the freshwater tidal Hudson River. Prior to this period, two nonnative species (*Sphaerium corneum* and *Pisidium amnicum*) were already established in the area and two additional nonnative species (*Mytilopsis leucophaeata* and *R. cuneata*) were established in the adjoining brackish waters of the Hudson. Here, we briefly describe the invasion history and population dynamics of each of these species.

Zebra Mussel (Dreissena polymorpha)

Zebra mussels were first found in the Hudson River in May 1991 (Strayer et al. 1996). The population grew very rapidly and by October 1992 accounted for more than 50% of all heterotrophic biomass in the ecosystem (Strayer et al. 1996). Because of its large filtration rates, the zebra mussel population had deep and wide-ranging effects on many parts of the Hudson River ecosystem (e.g., Caraco et al. 1997, 2000, 2006, Pace et al. 1998, 2010, Strayer et al. 1999, 2004, 2008, 2011).

Over its first 20 years, dynamics of the Hudson's zebra mussel population (Figure 5.2) were dominated by two processes: (1) strong cycling apparently related to strong interactions between adult and larval–juvenile zebra mussels (Strayer and Malcom 2006) and (2) increased mortality of adults (Strayer et al. 2011). Recruitment success was very high in many (but not all) years in which the adult population was small. Once a large year-class was established (e.g., in 1992, 1996, 2001, 2006), recruitment and growth of juveniles tended to be poor until that year-class nearly disappeared (Strayer and Malcom 2006). Out-of-phase cycles in population density (Figure 5.2A) and individual body size (Figure 5.2B), especially in the early years of the invasion, tended to stabilize interannual variation in population filtration rate (Figure 5.2C) (and therefore impacts on the ecosystem).

Initially, survival of adult zebra mussels in the river was high (~50%/year) but fell steadily in subsequent years (~20%/year around the year 2000, <1%/year since 2005; Strayer et al. 2011). The effect of this reduced survivorship was to decrease mean body size of zebra mussels (Figure 5.2B) (essentially, all of the animals in the river are now <1 year old) and therefore total filtration rate (Figure 5.2C). Presumably in response to these changes, some parts of the ecosystem recovered toward preinvasion conditions (Strayer and Malcom 2007, Pace et al. 2010, Strayer et al. 2011). The causes of increased mortality on zebra mussels are only partly known. Part of the increased mortality is a result of increased mortality from blue crabs (although the number of crabs in the river hasn't increased), and part results from

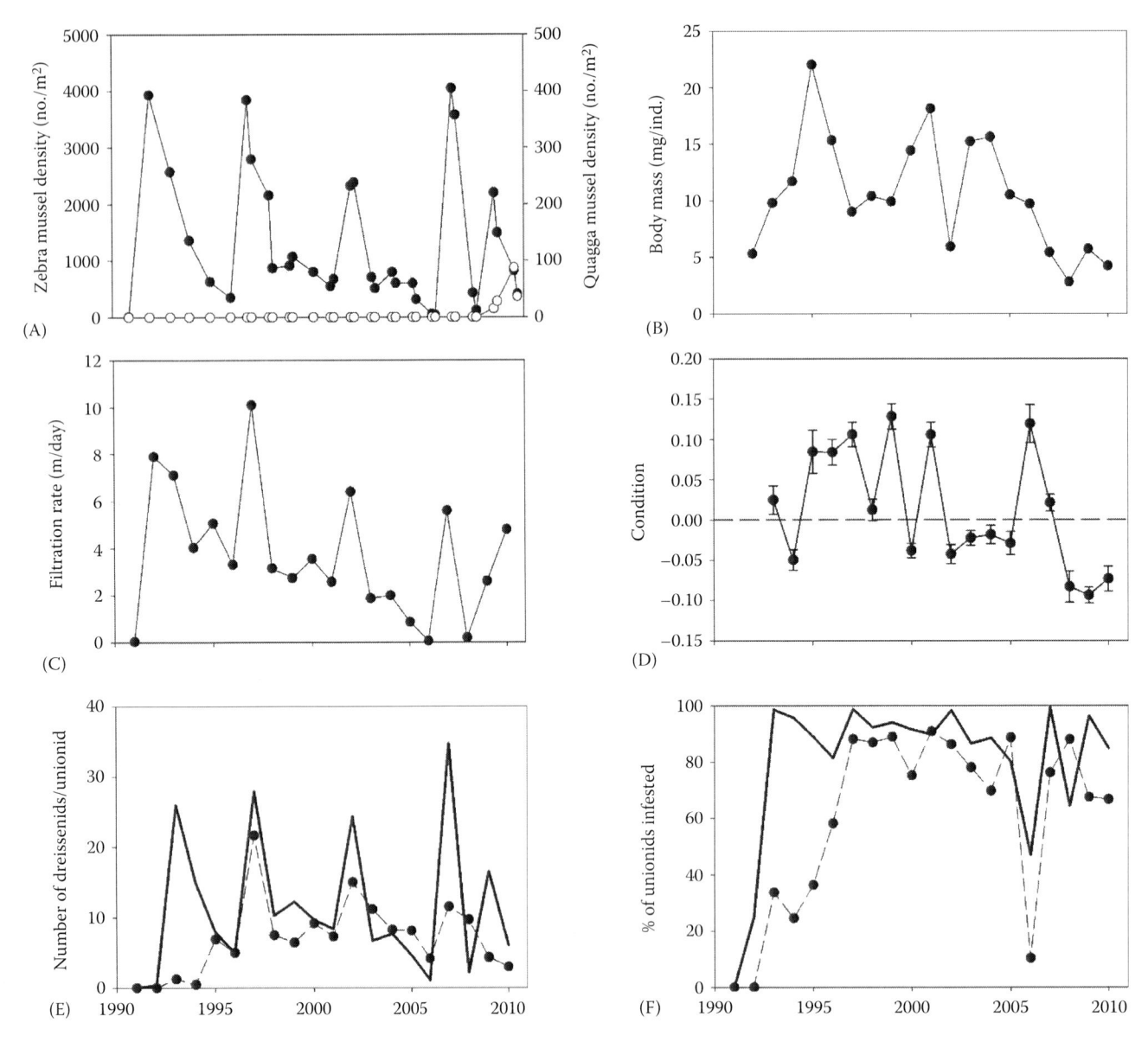

Figure 5.2 Dynamics of *Dreissena* populations in the freshwater Hudson River. (A) Riverwide mean population densities of zebra mussels (closed circles) and quagga mussels (open circles); note the 10-fold difference in scaling of the *y*-axes. (B) Mean body mass of zebra mussels. (C) Estimated riverwide mean filtration rate of the *Dreissena* spp. populations during the annual August sampling. (D) Mean body condition (±1 SE; see text for explanation) of zebra mussels. (E) The number of dreissenids attached to unionids. (F) The percentage of unionids with at least one dreissenid attached. Data in (E) and (F) exclude juvenile unionids (*Elliptio* < 20 mm long, *Anodonta* < 40 mm long, *Leptodea* < 30 mm long), which were rarely fouled (2 of 385 animals). Points and dashed lines in (E) and (F) show data from our study, and solid lines show predictions from models of Ricciardi et al. (1995) based on density estimates of the entire dreissenid population.

increased mortality from some unknown source capable of passing through 2.5 cm mesh (Carlsson et al. 2011).

Three other aspects of zebra mussel population dynamics in the Hudson River are worthy of comment. First, because increased mortality uncoupled the cycles of population density and body size (Figures 5.2A and B), interannual variation in population filtration rate has been increasing (Figure 5.2C). We would therefore expect to see increased interannual variation in ecosystem effects. Second, for unknown reasons, very few zebra mussels settled on native unionid mussels in the first 3–5 years of the invasion (Figures 5.2E and F).

Nevertheless, as will be discussed later, mortality of these nearly unfouled unionids was very high, showing that fouling is not the only (or even the primary) way in which zebra mussels harm unionids. After the mid-1990s, fouling rates on unionids in the Hudson River were similar to what has been seen elsewhere in North America, and unionid mortality did not increase. Third, although body condition of zebra mussels varied considerably between years (Figure 5.2D), this interannual variation was not correlated with phytoplankton biomass ($r^2 = 0.06$, $p = 0.34$), freshwater flow ($r^2 = 0.15$, $p = 0.19$), or total filtration rate of the *Dreissena* population

($r^2 = 0.05$, $p = 0.41$). Nevertheless, it appears that body condition may have declined in recent years, perhaps contributing to the recent increase in zebra mussel mortality.

Quagga Mussel (Dreissena rostriformis bugensis)

Quagga mussels were first found in the Hudson River in June 2008. By 2010, they had been seen throughout the freshwater tidal reach and constituted a small but growing fraction of the *Dreissena* population (Figure 5.2A).

It is difficult to predict how well this species will do in the Hudson River in the future. Although it had been thought that quagga mussels tend to outcompete zebra mussels from bodies of water in which both occur (e.g., Mills et al. 1996, Karatayev et al. 2011), a recent review (Zhulidov et al. 2010) concluded that the quagga mussels do not always displace zebra mussels.

Asian Clam (Corbicula fluminea)

A few empty shells collected at Kingston on June 27, 2008, constitute the first record of *Corbicula fluminea* from the Hudson River. We did not reliably record this species from our samples in 2008, so our observations for that year are incomplete. In 2009, we found scattered empty shells of this species from Newburgh (RKM 100) to South Troy (RKM 235), but no living animals. In 2010, we found a few living animals ($5/m^2$) at Hudson (RKM 185) and empty shells at several other stations. Evidently, a sparse population of *Corbicula* now exists throughout the length of the freshwater tidal reach.

This species often becomes abundant and can have large effects on other parts of the ecosystem (see reviews by Strayer 1999, Vaughn and Hakenkamp 2001, Sousa et al. 2008). It remains to be seen whether this species will establish a large population in the Hudson River. Although it has been suggested that *Corbicula* cannot establish large, persistent populations in cold climates (McMahon and Bogan 2001), large populations already occur in some of the Hudson River's tributaries (Strayer, personal observation), and climatic warming should make the Hudson River more suitable for this species.

Other Invaders

Four other nonnative bivalve species are known from the freshwater tidal region of the Hudson River, although none is abundant or widespread enough to be ecologically important in this region. The false dark mussel *M. leucophaeata* is a dreissenid native to estuaries of the southeastern United States. It first appeared in the Hudson River in the 1930s, probably introduced with ballast water or oysters (Mills et al. 1997). This brackish-water species appeared in low numbers in our samples from Newburgh (RKM 100), especially in dry summers. Although not abundant enough in the freshwater tidal region of the Hudson to be ecologically important, it attained densities >$1000/m^2$ in brackish regions downriver (Walton 1996) and may affect other benthic species in that part of the river.

R. cuneata, a brackish-water clam native to the southeastern United States, appeared in the Hudson River in the 1980s, perhaps brought in with ballast water or oyster reintroductions (Mills et al. 1997). It can develop large populations and affect other parts of estuarine ecosystems (Wong et al. 2010). *Rangia* is very abundant in the brackish parts of the Hudson River downriver of our study area. Llanso et al. (2003) reported mean densities of $609/m^2$ in the oligohaline parts of the river in 2000–2001, where it probably affects both benthic and pelagic communities. *Rangia* does not reproduce in freshwater, though, so it occurs only in low numbers (<$100/m^2$) in the most downriver parts of the study area, only during dry summers, and probably is of little ecological importance in the freshwater tidal reach.

Two Eurasian sphaeriids (*S. corneum* and *P. amnicum*) have been reported from the study area. There are just a few doubtful records of the former species from the river (Bode et al. 1986, Strayer 1987), which apparently occurs in very low densities in the upper part of the estuary. The latter species also occurs chiefly in the upper part of the estuary (RKM 213–248), where local densities may exceed $100/m^2$ (Simpson et al. 1984, Bode et al. 1986, Strayer and Smith 2001). Little is known about the time of arrival, population dynamics, or ecological roles of these two species in the Hudson River. While *P. amnicum* may have some local impacts where it is abundant, it seems unlikely that either species plays a large role in the river's ecology.

Responses of the Native Bivalves

Two families of bivalves, the Unionidae and the Sphaeriidae, are native to the Hudson River. Although many aspects of their biology are very different, both families showed broadly similar responses to recent invasions of the river by nonnative bivalves.

Pearly Mussels (Unionidae)

The Hudson River estuary contained eight species of unionids in the late nineteenth century (Strayer 1987). Two of these, *Lasmigona subviridis* and *Strophitus undulatus*, apparently disappeared by the mid-twentieth century, probably as a result of the large changes in channel morphology and pollution in the upper estuary (Strayer and Smith 2001, Miller et al. 2006), and another species that persisted into the late twentieth century, *Lampsilis cariosa* (Strayer 1987), was too rare ever to appear in our quantitative samples. Of the five species that did appear in our samples, two (*L. radiata* and *Ligumia nasuta*) were represented by single individuals, so we can say nothing about their recent distribution or population dynamics. Based on the number and collection localities of nineteenth-century specimens in museums, *L. cariosa*, *L. radiata*, and *L. nasuta* were once common and widespread in the river.

The three remaining species (*Anodonta implicata*, *Elliptio complanata*, and *Leptodea ochracea*) all were common and widespread in the Hudson River when our sampling program began in 1991 and showed broadly similar distributions and population dynamics. Densities of all three were highest in the upper, more riverine, sections of the estuary, where very dense populations existed (Figure 5.3), and all declined sharply when zebra mussels first appeared in the river (Figure 5.4). Declines in population density were accompanied by sharp drops in recruitment (Figure 5.5) and body condition (Figure 5.6). These downward trends in population densities stopped ~2000, and at least in the case of *E. complanata*, actually may have reversed since then (Figure 5.4).

Recruitment of juvenile *E. complanata* fell to low levels in 1993–2000, then began to rise, and is now at or

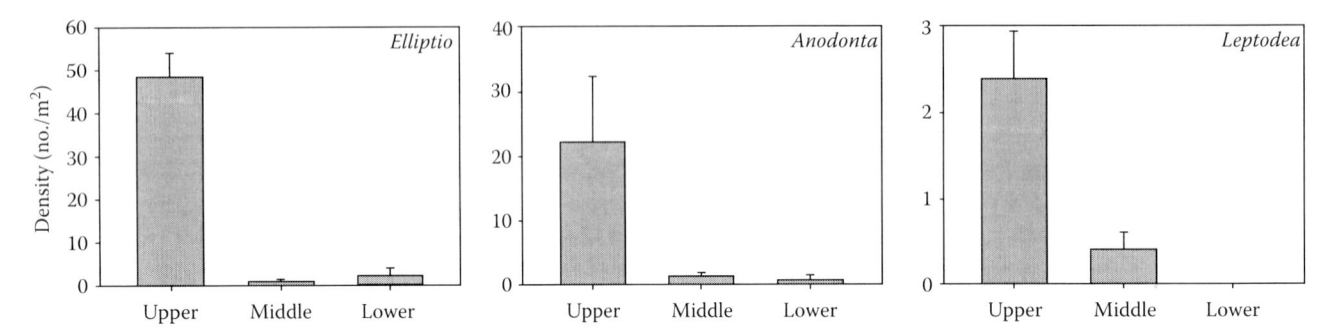

Figure 5.3 Mean densities (+1 SE) of three unionid species (*E. complanata*, *A. implicata*, *L. ochracea*) in different sections of the freshwater tidal Hudson River (upper, RKM 213–248; middle, RKM 151–213; lower, RKM 100–151) in 1991–1992, before the outbreak of zebra mussels.

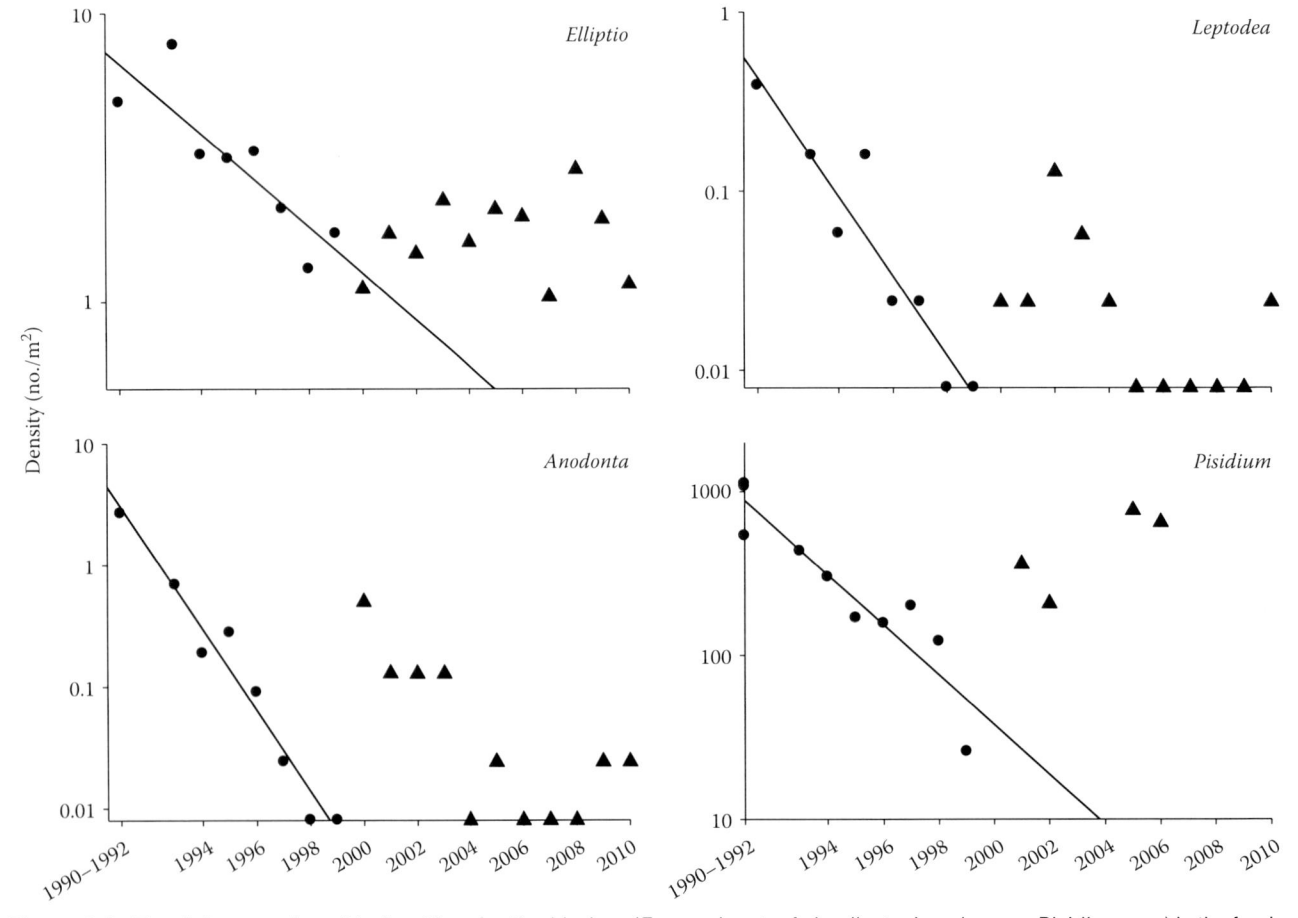

Figure 5.4 Trends in mean riverwide densities of native bivalves (*E. complanata*, *A. implicata*, *L. ochracea*, *Pisidium* spp.) in the freshwater tidal region of the Hudson River, 1990–2010. Data were divided into two phases: decline (1990–1999, solid circles, lines fitted through data points) and "recovery" (2000–2010, solid triangles). Note that the y-axes are logarithmically scaled.

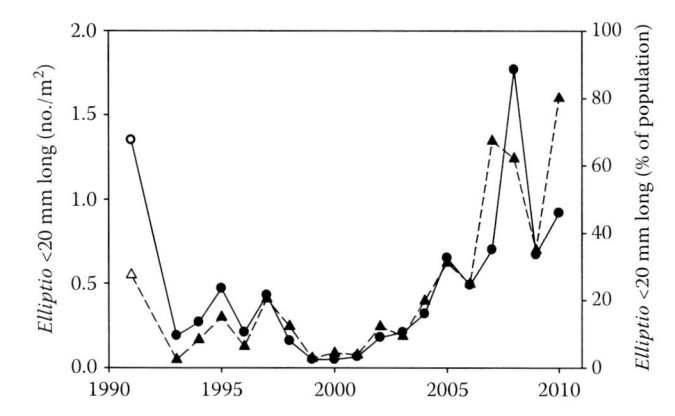

Figure 5.5 Riverwide mean density of juvenile *E. complanata* (<20 mm long) in the freshwater tidal region of the Hudson River, expressed both as absolute densities (circles and solid line) and as percentage of the population (triangles and dashed line). Open symbols show data from before the zebra mussel outbreak (1991 and 1992 data are combined) and filled symbols show data after the zebra mussel outbreak.

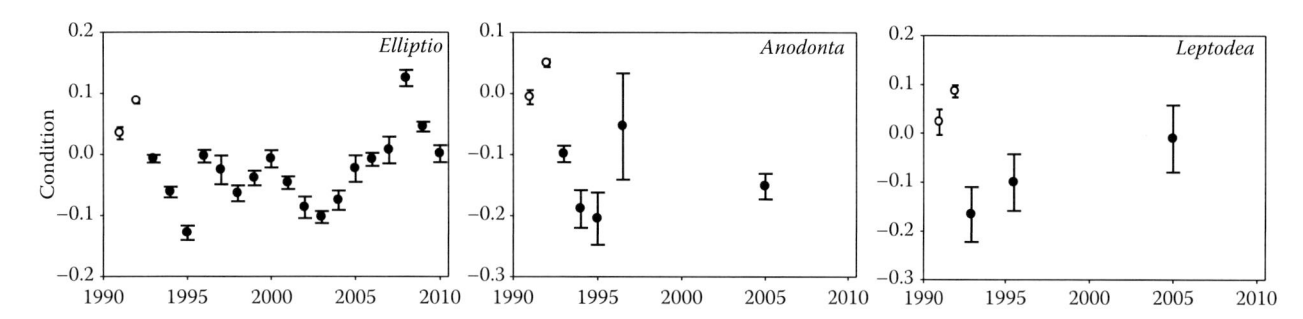

Figure 5.6 Trends in mean body condition (+1 SE, see text for definition) of unionids (*E. complanata*, *A. implicata*, *L. ochracea*) in the freshwater tidal Hudson River, 1991–2010. Open circles show data from before the outbreak of zebra mussels and closed circles show data from after the outbreak. Because of insufficient data, multiple years were combined for *Anodonta* and *Leptodea*; the last two data points for these species represent 1996–1997 and 2000–2010 and 1994–1997 and 2000–2010, respectively.

above preinvasion levels (Figure 5.5). Although the data on *A. implicata* and *L. ochracea* are too sparse to support a formal analysis of temporal trends in recruitment, the number of juveniles has increased since 2000 (data not shown).

Body condition of *E. complanata* fell after zebra mussels became abundant and stayed low until ~2005 (Figure 5.6). In the last few years, though, body condition improved markedly and is now near preinvasion values. Data for the two other unionid species are sparse because of the small number of specimens collected but seem consistent with the pattern shown by *E. complanata*. Although unionid mortality and body condition often have been correlated with fouling intensity in other bodies of water (e.g., Haag et al. 1993, Ricciardi et al. 1995, Baker and Hornbach 2000, Burlakova et al. 2000), body condition of unionids in the Hudson River was much better correlated with the riverwide filtration rate of the *Dreissena* population than with fouling on individual unionids. We found only very weak correlations between fouling and unionid condition (Figure 5.7), which supports the conclusion of an earlier statistical analysis (Strayer and Malcom 2007) that fouling was not of primary importance in determining unionid condition in the Hudson River. Instead, we found that the condition of *E. complanata* could be predicted by the riverwide filtration rate of *Dreissena* (Figure 5.8). This relationship was especially strong when filtration rate of only the largest zebra mussels was considered, suggesting that the strength of exploitation competition between unionids and *Dreissena* may depend on body size of the latter. This parallels the finding of Pace et al. (2010) that the largest zebra mussels had the strongest effect on microzooplankton populations in the river.

Although all three unionid species showed broadly similar changes over time, there were species-specific differences in severity of impacts consistent with what has been seen elsewhere in North America. Declines in population size, recruitment, and body condition were much more severe for *L. ochracea* and especially *A. implicata* than for *E. complanata*. Other researchers have found that anodontines (such as *Anodonta*) and lampsilines (such as *Leptodea*) were more severely affected by the *Dreissena* invasion than amblemines and pleurobemines (such as *Elliptio*) (Haag et al. 1993, Strayer 1999).

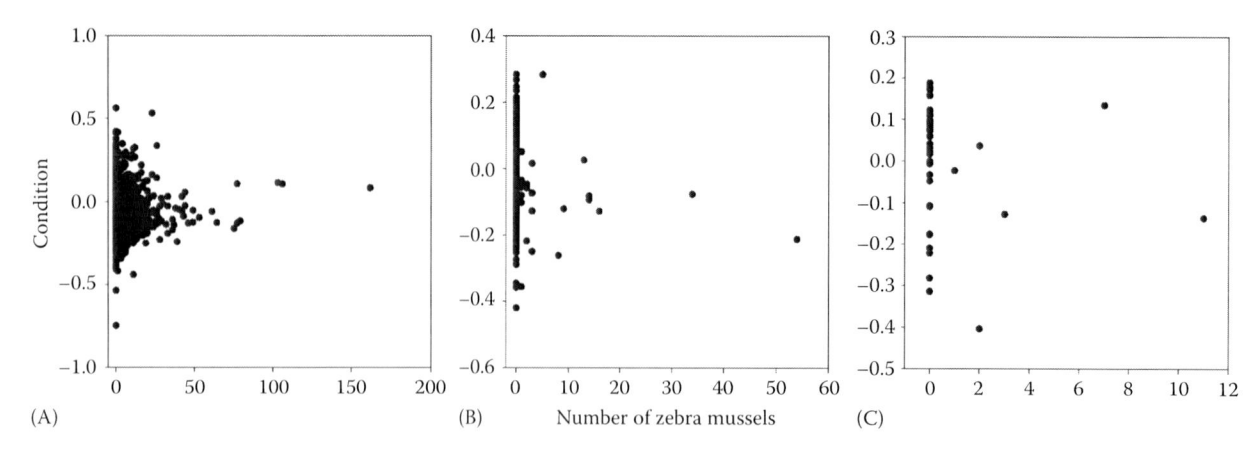

Figure 5.7 Mean body condition (see text for definition) of unionid mussels (*E. complanata, A. implicata, L. ochracea*) as a function of the number of zebra mussels attached to each mussel. Each dot represents an individual unionid (excluding juveniles: *Elliptio* <20 mm, *Anodonta* <40 mm, *Leptodea* <30 mm). For all three species, there is a very weak negative relationship between unionid condition and number of attached zebra mussels (A: $n = 2050$, $r^2 = 0.01$, $p < 0.001$; B: $n = 364$, $r^2 = 0.019$, $p = 0.009$; C: $n = 49$, $r^2 = 0.019$, $p = 0.35$).

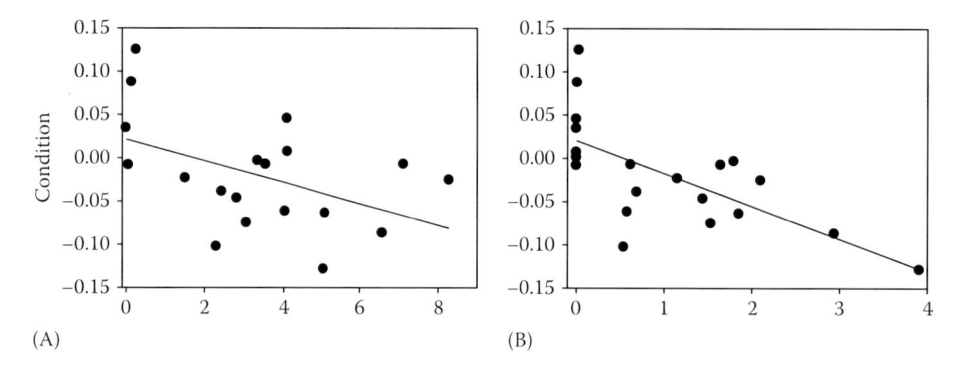

Figure 5.8 Relationship between mean body condition of *E. complanata* (see text for definition) and riverwide mean filtration rate of the *Dreissena* population in the freshwater tidal region of the Hudson River. *Dreissena* filtration rate is expressed either as filtration of all animals (A: Total *Dreissena* filtration rate (m/day) $r^2 = 0.25$, $p = 0.02$) or as filtration rate of just the largest animals (>20 mm shell length) (B: Large *Dreissena* filtration rate (m/day) $r^2 = 0.46$, $p < 0.001$).

Pea and Fingernail Clams (Sphaeriidae)

We have far less information about the native sphaeriids of the Hudson River than the unionids. The only taxon common throughout the study area is a small-bodied species (or multiple species?) of *Pisidium* (as noted earlier, two nonnative sphaeriids live in the upper 25 km of the estuary, along with the native *S. striatinum*—Strayer and Smith 2001). As noted by Strayer et al. (2011), populations of *Pisidium* declined sharply in the initial years after the zebra mussel outbreak (1993–1999), then recovered to near preinvasion levels (Figure 5.4). Thus, it appears that the response of native sphaeriids to the *Dreissena* invasion of the Hudson River was similar to that of the unionids.

GENERAL DISCUSSION

Bivalves and Ecosystem Instability

Bivalve communities in the freshwater tidal Hudson River changed dramatically in the twentieth century, and especially during 1990–2010, after *Dreissena* spp. first appeared. Similar dramatic changes were seen globally in fresh and coastal waters as habitat change, overharvest, and pollution damaged or destroyed native populations (e.g., Lydeard et al. 2004, Strayer 2008, Beck et al. 2011), and aggressive bivalve invaders such as *Dreissena* spp., *Corbicula* spp., *L. fortunei*, *Corbula amurensis*, *Mytilus galloprovincialis*, *Mya arenaria*, and *Musculista senhousia* were carried around the world as a result of careless human activities (e.g., Carlton 1992, 1999, Strayer 1999, Karatayev et al. 2007). Because of these changes, bivalve communities in many freshwaters, estuaries, and coastal waters in 2010 are entirely different than they were just 100 years ago, in terms of species composition, abundance, and interactions with their ecosystems.

During this same time period (1990–2010), it was finally appreciated that bivalves play major roles in aquatic ecosystems, controlling the amount and type of suspended particles, water chemistry and clarity, the upper food web, and the physical character of the sediments (e.g., Alpine and Cloern 1992, Dame 1996, Strayer et al. 1999, Vaughn and Hakenkamp 2001, Newell 2004). Thus, the widespread changes to bivalve communities that occurred over the past century or two probably drove large, unintended, and in most cases, unobserved, changes in the structure and function of aquatic ecosystems around the world. It seems likely that the events that occurred in the Hudson River also occurred in many bodies of water around the world, even if they were not always well documented.

We expect that the Hudson's bivalve communities will continue to be very dynamic over the coming decades and lead to long-lasting instability in the ecosystem, for four reasons. First, it seems likely that new nonnative species will continue to invade the river. The list of possible invaders includes both bivalves (e.g., *L. fortunei*—Karatayev et al. 2007; *A. woodiana*—Watters 1997) and species that may interact strongly with bivalves (e.g., the round goby, the black carp, the New Zealand mudsnail). Second, several invaders arrived in the Hudson River only recently, and their ultimate population size and impacts still are unclear. Species such as the quagga mussel, *C. fluminea*, and the Chinese mitten crab appear to be well established in the river, but only as small populations. We cannot yet predict whether they eventually will form large populations and play important roles in the ecosystem or remain bit players. Third, the demography of the zebra mussel population and its effects on the ecosystem are still unsettled even 20 years after its arrival in the Hudson River. It appears that at least decades are required for such a dominant invader to reach some sort of equilibrium with the ecosystem, if indeed such an equilibrium ever develops. Fourth, the Hudson River will be affected in the near future by forces other than species invasions that are likely to interact with bivalve populations. Rising sea level, changing climate, and changing land use all will affect the Hudson River and many other aquatic ecosystems in the twenty-first century. The slow rates of some of the key processes (e.g., sediment routing) associated with these drivers and the long life span of some bivalves may cause decades-long delays before the ultimate effects of these changes are observed. For all these reasons, bivalve populations and their ecosystem-level effects are likely to continue to be very dynamic over the coming decades in the Hudson River and in aquatic ecosystems around the world.

Notes on Exploitation Competition

It is well-established that overgrowth of native bivalves by *Dreissena* spp. can harm or kill the natives (e.g., Haag et al. 1993, Ricciardi et al. 1995, Baker and Hornbach 2000, Burlakova et al. 2000). Nevertheless, this does not appear to be the primary mechanism by which zebra mussels affected native bivalves in the Hudson River. Instead, evidence suggests that exploitation competition was the primary way by which zebra mussels harmed or killed native bivalves. To begin with, populations of the sphaeriid *Pisidium* declined steeply (Figure 5.4), even though this tiny clam was never fouled by zebra mussels in the Hudson River. Furthermore, unionid population density (Figure 5.4), recruitment (Figure 5.5), and body condition (Figure 5.6) all fell sharply in the early years of the zebra mussel invasion (1993–1995), even though few unionids were fouled at all during this period (Figure 5.2). Finally, statistical models show only the weakest of evidence that fouling affected body condition of unionids (Figure 5.7). Instead, the riverwide filtration rate of the zebra mussel population was clearly related to unionid body condition (Figure 5.8; see also Strayer and Malcom 2007 for a formal statistical analysis of this problem).

There is a tantalizing hint that large zebra mussels may compete more strongly than small zebra mussels with unionids (Figure 5.8). It has been shown that large zebra mussels are capable of capturing different kinds of particles than small mussels (MacIsaac et al. 1995), which may underlie the recent recovery of zooplankton in the Hudson (Pace et al. 2010). The subject of particle capture by bivalves of different sizes clearly deserves further investigation.

The mechanism by which zebra mussels affect native bivalves has important implications. Obviously, mitigation strategies such as cleaning dreissenids from infested unionids (e.g., Schloesser 1996, Hallac and Marsden 2001) will be successful only if fouling, not exploitation competition, is the primary mechanism of interaction. Furthermore, local fouling intensity will not always be well correlated with food depletion by zebra mussels. The latter will depend on characteristics of the ecosystem (e.g., water residence time, intensity of nutrient limitation, morphometry and mixing depth, and amount of non-algal turbidity—Dame 1996, Strayer et al. 1999, Vaughn 2010), as well as the density and activities of the bivalve population, and may be uncoupled from local fouling rates (Figure 5.2). Therefore, the relative importance of fouling and exploitation competition will need to be understood if we are to assess or predict effects of zebra mussels on native bivalves.

Return of the Natives

The most interesting aspect of our study to a conservation biologist is the possibility that populations of native bivalves may be returning to the Hudson River after an initial decade or so of severe losses. If such recovery occurs in the Hudson River and elsewhere, it raises the possibility that short-term treatments such as removing native bivalves to refuges (e.g., Newton et al. 2001) or cleaning them in situ (e.g., Schloesser 1996, Hallac and Marsden 2001) (or even doing nothing, as in the Hudson River) might be enough to allow native species to withstand the zebra mussel invasion.

Although the recovery of the Hudson River's native bivalves is promising from the point of view of conservation (the native unionids aren't all dead, as predicted by models—Ricciardi et al. 1998), the long-term fate of the river's native unionids and sphaeriids still is far from clear. The clearest lesson of our study is that the zebra mussel population and the Hudson River ecosystem have not settled down to any sort of stable state. Furthermore, we don't fully understand the mechanisms that are causing the situation in the Hudson River to be so dynamic, so it would be very risky to make predictions about the future from what we now know. A particular point of uncertainty is whether the large number of juvenile unionids that have appeared in recent years (Figure 5.5) will successfully grow into adults or will die before reaching maturity and simply represent a demographic dead end. Therefore, we think it is too early to declare that the native bivalves of the Hudson River will coexist over the long term with zebra mussels and other invaders, although this now appears to be a real possibility.

ACKNOWLEDGMENTS

We are grateful to our colleagues at the Cary Institute of Ecosystem Studies for many useful discussions and ideas, to two reviewers for helpful comments on the manuscript, and to many assistants for help with sample collection and analysis. This work was supported by grants from the Hudson River Foundation and the National Science Foundation.

REFERENCES

Alpine, A.E. and J.E. Cloern. 1992. Trophic interactions and direct physical effects control phytoplankton biomass and production in an estuary. *Limnol. Oceanogr.* 37: 946–955.

Baker, S.M. and D.J. Hornbach. 2000. Physiological status and biochemical composition of a natural population of unionid mussels (*Amblema plicata*) infested by zebra mussels (*Dreissena polymorpha*). *Am. Midl. Nat.* 143: 443–452.

Beck, M.W., R.D. Brumbaugh, L. Airoldi et al. 2011. Oyster reefs at risk and recommendations for conservation, restoration, and management. *BioScience* 61: 107–116.

Bode, R.W., M.A. Novak, J.P. Fagnani, and D.M. DeNicola. 1986. The benthic macroinvertebrates of the Hudson River from Troy to Albany, New York. Final Report to the Hudson River Foundation, Grant No. 78:84A. Available at http://www.hudsonriver.org/ls/reports/Simpson_017_84A_final_report.pdf (accessed July 20, 2011).

Burlakova, L.E., A.Y. Karatayev, and D.K. Padilla. 2000. The impact of *Dreissena polymorpha* (Pallas) invasion on unionid mussels. *Int. Rev. Hydrobiol.* 85: 529–541.

Caraco, N.F., J.J. Cole, S.E.G. Findlay et al. 2000. Dissolved oxygen declines in the Hudson River associated with the invasion of the zebra mussel (*Dreissena polymorpha*). *Environ. Sci. Technol.* 34: 1204–1210.

Caraco, N.F., J.J. Cole, P.A. Raymond et al. 1997. Zebra mussel invasion in a large, turbid river: Phytoplankton response to increased grazing. *Ecology* 78: 588–602.

Caraco, N.F., J.J. Cole, and D.L. Strayer. 2006. Top down control from the bottom: Regulation of eutrophication in a large river by benthic grazing. *Limnol. Oceanogr.* 51: 664–670.

Carlsson, N.O.L., H. Bustamante, D.L. Strayer, and M.L. Pace. 2011. Biotic resistance on the move: Native predators structure invasive zebra mussel populations. *Freshwat. Biol.* 56: 1630–1637.

Carlton, J.T. 1992. Introduced marine and estuarine mollusks of North America: An end-of-the-20th-century perspective. *J. Shellfish Res.* 11: 489–505.

Carlton, J.T. 1999. Molluscan invasions in marine and estuarine communities. *Malacologia* 41: 439–454.

Dame, R.F. 1996. *Ecology of Marine Bivalves: An Ecosystem Approach*. Boca Raton, FL: CRC Press.

Haag, W.R., D.J. Berg, D.W. Garton, and J.L. Farris. 1993. Reduced survival and fitness in native bivalves in response to fouling by the introduced zebra mussel (*Dreissena polymorpha*) in western Lake Erie. *Can. J. Fish. Aquat. Sci.* 50: 13–19.

Hallac, D.E. and J.E. Marsden. 2001. Comparison of conservation strategies for unionid threatened by zebra mussels (*Dreissena polymorpha*): Periodic cleaning vs. quarantine and translocation. *J. N. Am. Benthol. Soc.* 20: 200–210.

Karatayev, A.Y., D. Boltovskoy, D.K. Padilla, and L.E. Burlakova. 2007. The invasive bivalves *Dreissena polymorpha* and *Limnoperna fortunei*: Parallels, contrasts, potential spread and invasion impacts. *J. Shellfish Res.* 26: 205–213.

Karatayev, A.Y., S.E. Mastitsky, D.K. Padilla, L.E. Burlakova, and M.M. Hajduk. 2011. Differences in growth and survivorship of zebra and quagga mussels: Size matters. *Hydrobiologia* 668: 183–194.

Kryger, J. and H.U. Riisgard. 1988. Filtration rate capacities in 6 species of European freshwater bivalves. *Oecologia* 77: 34–38.

Llanso, R., M. Southerland, J. Vølstad et al. 2003. Hudson River estuary biocriteria final report. Report to the New York State Department of Environmental Conservation, Albany. Available at http://nysl.nysed.gov/Archimages/90611.PDF (accessed July 20, 2011).

Lydeard, C., R.H. Cowie, W.F. Ponder et al. 2004. The global decline of nonmarine mollusks. *BioScience* 54: 321–330.

MacIsaac, H.J., C.J. Lonnee, and J.H. Leach. 1995. Suppression of microzooplankton by zebra mussels: Importance of mussel size. *Freshwat. Biol.* 34: 379–387.

McMahon, R.F. and A.E. Bogan. 2001. Mollusca: Bivalvia. In *Ecology and Classification of North American Freshwater Invertebrates*, 2nd Edn., J.H. Thorp and A.P. Covich, eds., pp. 331–429. San Diego, CA: Academic Press.

Miller, D., J. Ladd, and W.C. Nieder. 2006. Channel morphology in the Hudson River estuary: Historical changes and opportunities for restoration. In *Hudson River Fishes and Their Environment*, J.R. Waldman, K.E. Limburg, and D.L. Strayer, eds., pp. 29–37. Bethesda, MD: American Fisheries Society.

Mills, E.L., J.T. Carlton, M.D. Scheuerell, and D.L. Strayer. 1997. Biological invasions in the Hudson River: An inventory and historical analysis. *N.Y. State Mus. Circ.* 57: 1–51.

Mills, E.L., G. Rosenberg, A.P. Spidle, M. Ludyanskiy, Y. Pligin, and B. May. 1996. A review of the biology and ecology of the quagga mussel (*Dreissena bugensis*), a second species of dreissenid mussel introduced to North America. *Am. Zool.* 36: 271–286.

Newell, R.I.E. 2004. Ecosystem influences of natural and cultivated populations of suspension-feeding bivalve mollusks: A review. *J. Shellfish Res.* 23: 51–61.

Newton, T.J., E.M. Monroe, R. Kenyon, S. Gutreuter, K.I. Welke, and P.A. Thiel. 2001. Evaluation of relocation of unionid mussels into artificial ponds. *J. N. Am. Benthol. Soc.* 20: 468–485.

Pace, M.L., S.E.G. Findlay, and D. Fischer. 1998. Effects of an invasive bivalve on the zooplankton community of the Hudson river. *Freshwat. Biol.* 39: 103–116.

Pace, M.L., D.L. Strayer, D.T. Fischer, and H.M. Malcom. 2010. Increased mortality of zebra mussels associated with recovery of zooplankton in the Hudson River. *Ecosphere* 1(1): art3. Doi: 10.1890/ES10-00002.1.

Ricciardi, A., R.J. Neves, and J.B. Rasmussen. 1998. Impending extinctions of North American freshwater mussels (Unionoida) following the zebra mussel (*Dreissena polymorpha*) invasion. *J. Anim. Ecol.* 67: 613–619.

Ricciardi, A., F.G. Whoriskey, and J.B. Rasmussen. 1995. Predicting the intensity and impact of *Dreissena* infestation on native unionid mussels from *Dreissena* field density. *Can. J. Fish. Aquat. Sci.* 52: 1449–1461.

Roditi, H.A., N.F. Caraco, J.J. Cole, and D.L. Strayer. 1996. Filtration of Hudson River water by the zebra mussel (*Dreissena polymorpha*). *Estuaries* 19: 824–832.

Roditi, H.A., D.L. Strayer, and S. Findlay. 1997. Characteristics of zebra mussel (*Dreissena polymorpha*) biodeposits in a tidal freshwater estuary. *Arch. Hydrobiol.* 140: 207–219.

Schloesser, D.W. 1996. Mitigation of unionid mortality caused by zebra mussel infestation: Cleaning of unionids. *N. Am. J. Fish. Manage.* 16: 942–946.

Simpson, K.W., R.W. Bode, J.P. Fagnani, and D.M. DeNicola. 1984. The freshwater macrobenthos of the main channel, Hudson River. Part B. Biology, taxonomy and distribution of resident macrobenthic species. Final report to the Hudson River Foundation, Grant 8/83A/39. Available at http://www.hudsonriver.org/ls/reports/Simpson_008_83A_final_report.pdf (accessed July 20, 2011).

Sousa, R., C. Antunes, and L. Guilhermino. 2008. Ecology of the invasive Asian clam *Corbicula fluminea* (Müller, 1774) in aquatic ecosystems: An overview. *Ann. Limnol.* 44: 85–94.

Strayer, D. 1987. Ecology and zoogeography of the freshwater mollusks of the Hudson River basin. *Malacol. Rev.* 20: 1–68.

Strayer, D.L. 1999. Effects of alien species on freshwater mollusks in North America. *J. N. Am. Benthol. Soc.* 18: 74–98.

Strayer, D.L. 2008. *Freshwater Mussel Ecology: A Multifactor Approach to Distribution and Abundance*. Berkeley, CA: University of California Press.

Strayer, D.L. 2009. Twenty years of zebra mussels: Lessons from the mollusk that made headlines. *Front. Ecol. Environ.* 7: 135–141.

Strayer, D.L., N.F. Caraco, J.J. Cole, S. Findlay, and M.L. Pace. 1999. Transformation of freshwater ecosystems by bivalves: A case study of zebra mussels in the Hudson River. *BioScience* 49: 19–27.

Strayer, D.L., N. Cid, and H.M. Malcom. 2011. Long-term changes in a population of an invasive bivalve and its effects. *Oecologia* 165: 1063–1072.

Strayer, D.L., K. Hattala, and A. Kahnle. 2004. Effects of an invasive bivalve (*Dreissena polymorpha*) on fish populations in the Hudson River estuary. *Can. J. Fish. Aquat. Sci.* 61: 924–941.

Strayer, D.L., D.C. Hunter, L.C. Smith, and C. Borg. 1994. Distribution, abundance, and role of freshwater clams (Bivalvia: Unionidae) in the freshwater tidal Hudson River. *Freshwat. Biol.* 31: 239–248.

Strayer, D.L. and H.M. Malcom. 2006. Long-term demography of a zebra mussel (*Dreissena polymorpha*) population. *Freshwat. Biol.* 51: 117–130.

Strayer, D.L. and H.M. Malcom. 2007. Effects of zebra mussels (*Dreissena polymorpha*) on native bivalves: The beginning of the end or the end of the beginning? *J. N. Am. Benthol. Soc.* 26: 111–122.

Strayer, D.L., M.L. Pace, N.F. Caraco, J.J. Cole, and S.E.G. Findlay. 2008. Hydrology and grazing jointly control a large-river food web. *Ecology* 89: 12–18.

Strayer, D.L., J. Powell, P. Ambrose, L.C. Smith, M.L. Pace, and D.T. Fischer. 1996. Arrival, spread, and early dynamics of a zebra mussel (*Dreissena polymorpha*) population in the Hudson River estuary. *Can. J. Fish. Aquat. Sci.* 53: 1143–1149.

Strayer, D.L. and L.C. Smith. 1996. Relationships between zebra mussels (*Dreissena polymorpha*) and unionid clams during the early stages of the zebra mussel invasion of the Hudson River. *Freshwat. Biol.* 36: 771–779.

Strayer, D.L. and L.C. Smith. 2001. The zoobenthos of the freshwater tidal Hudson River and its response to the zebra mussel (*Dreissena polymorpha*) invasion. *Arch. Hydrobiol. Suppl.* 139: 1–52.

Vaughn, C.C. 2010. Biodiversity losses and ecosystem function in freshwaters: Emerging conclusions and research directions. *BioScience* 60: 25–35.

Vaughn, C.C. and C.C. Hakenkamp. 2001. The functional role of burrowing bivalves in freshwater ecosystems. *Freshwat. Biol.* 46: 1431–1446.

Walton, W.C. 1996. Occurrence of the zebra mussel (*Dreissena polymorpha*) in the oligohaline Hudson River, New York. *Estuaries* 19: 612–618.

Watters, G.T. 1997. A synthesis and review of the expanding range of the Asian freshwater mussel *Anodonta woodiana* (Lea, 1834) (Bivalvia: Unionidae). *Veliger* 40: 152–156.

Wong, W.H., N.N. Rabelais, and R.E. Turner. 2010. Abundance and ecological significance of the clam *Rangia cuneata* (Sowerby, 1831) in the upper Barataria Estuary (Louisiana, USA). *Hydrobiologia* 651: 305–315.

Zhulidov, A.V., A.V. Kozhara, G.H. Scherbina, T.F. Nalepa, A. Protasov, S.A. Afanasiev, E.G. Pryanichnikova, D.A. Zhulidov, T.Y. Gurtovaya, and D.F. Pavlov. 2010. Invasion history, distribution, and relative abundances of *Dreissena bugensis* in the old world: A synthesis of data. *Biol. Invasions* 12: 1923–1940.

Zippin, C. 1958. The removal method of population estimation. *J. Wildl. Manage.* 22: 82–90.

CHAPTER 6

Spread of the Quagga Mussel (*Dreissena rostriformis bugensis*) in Western Europe

Abraham bij de Vaate, Gerard van der Velde, Rob S.E.W. Leuven, and Katharina C.M. Heiler

CONTENTS

Abstract 83
Introduction 84
Range Expansion of the Zebra Mussel 84
Range Expansion of the Quagga Mussel in Western Europe 86
Source and Vector of Quagga Mussel Introduction into Western Europe 87
Quagga and Zebra Mussel Interactions in Western Europe 88
Future Considerations 88
Acknowledgments 90
References 90

ABSTRACT

The spread of quagga mussels in western Europe is described in this chapter including their impact on densities of zebra mussels. Being a congener with a life cycle similar to zebra mussels, quagga mussels have vectors for range expansion that are also similar to those for zebra mussels. Starting in the second part of the eighteenth century, the zebra mussel was able to extend its range in Europe because of canal construction that connected river systems. Adult and juvenile mussels dispersed through attachment to boat hulls, and drifting larvae dispersed through waters that were interconnected. Through jump dispersal, zebra mussels were able to overcome dispersal barriers such as the Baltic, North, and Irish Seas and mountains of the Alps and Pyreneans. High levels of water pollution formed barriers for dispersal as well and led to the disappearance of established zebra mussel populations in the past. However, due to improved water quality during the past few decades, the role of pollution as a barrier diminished.

Range expansion of the quagga mussel in Europe started in the 1940s from the northern portion of the Black Sea through rivers in Ukraine. In 2004, it was discovered in the Romanian section of the River Danube, 2005 in the Main River and in 2006 was discovered in the Rhine River delta, The Netherlands. We assume that western Europe was colonized by quagga mussels via two separate corridors: one via eastern Europe through upstream dispersal in the River Danube from its initial distribution range and the other via jump dispersal into in western Europe. Phylogeographic relationships were not clear enough to reveal founder lineages of quagga mussels in Germany where the Rhine and Danube River basins are partly situated and the two rivers are connected by the Main–Danube canal. It has been shown that quagga mussels are displacing zebra mussels in the dreissenid community. The introduction of quagga mussels in two large lakes in The Netherlands resulted in an increase in overall dreissenid density.

Quagga mussels have continued their spread in western Europe at a faster rate than the past spread of zebra mussels because of completed networks of canal–river systems that link all major river basins, creation of suitable habitat by river engineering, and improvements of water quality. Shipping is expected to be the primary vector for the spread of quagga mussels in the area; a secondary vector would be veliger transport by drift, as enhanced by canal management. Three major corridors will serve as pathways for mussels to migrate and spread throughout continental Europe. Currently, quagga mussels are established only in water bodies where zebra mussels were already present.

INTRODUCTION

The quagga mussel (*Dreissena rostriformis bugensis*, Andrusov 1897) has recently been reported from western Europe (Molloy et al. 2007, Son 2007, Bij de Vaate and Beisel 2011) and is now one of three dreissenid species to have invaded and colonized this area. The other two species are the Conrad's false mussel or false dark mussel (*Mytilopsis leucophaeata*, Conrad 1831) (Heiler et al. 2010, Van der Velde et al. 2010) and the zebra mussel (*D. polymorpha*, Pallas 1771). Conrad's false mussel is a brackish-water species that originated from the eastern Atlantic coast of North America including the Gulf of Mexico (Marelli and Gray 1983). In western Europe, it occurs in brackish waters in Germany (e.g., Kiel Canal), The Netherlands (e.g., Noordzee canal, Canal of Gent to Terneuzen), Belgium (port of Antwerp), France (e.g., port of Dunkerque and Canal of Caen), and the United Kingdom (e.g., Cardiff docks) (Verween et al. 2010). The zebra mussel originated from the Ponto-Caspian region (Azov, Black, and Caspian Seas), and its spread created the first human-induced mass invasion of a Ponto-Caspian species in Europe. This species provides a good example of how and in which directions a Ponto-Caspian invader is able to extend its range (e.g., Bij de Vaate et al. 2002). Presently, the zebra mussel is widely distributed in Europe. It is found from Sweden and Finland in the north to Italy, Spain, and Montenegro in the south and is still expanding its range (Bidwell 2010, Pollux et al. 2010, Van der Velde et al. 2010, Wilke et al. 2010).

The objectives of this chapter are to: (1) review data to identify possible pathways and vectors available for quagga mussels to colonize and spread through western Europe; (2) document changes in dreissenid densities after the introduction of quagga mussels; and (3) compare the invasion history of the quagga mussel to that of the zebra mussel.

RANGE EXPANSION OF THE ZEBRA MUSSEL

Range expansion of the zebra mussel in Europe occurred by continuous dispersal through invasion corridors from neighboring populations as well as by jump dispersal (Van der Velde et al. 2010; Chapter 7). Starting in the second part of the eighteenth century, it extended its range because of canal construction that connected river systems (Ludyanskiy et al. 1993, Bij de Vaate et al. 2002). Continuous canal construction resulted in the present-day European network of waterways that are presently completely colonized by zebra mussels (Kinzelbach 1992, Leuven et al. 2009, Heiler et al. 2012). Adult and juvenile mussels dispersed through attachment to boat hulls, and drifting larvae dispersed through waters that were interconnected (e.g., Carlton 1993).

Construction of canals in the eighteenth century that connected the Dnieper, Neman, and Vistula Rivers marked the start of the northward and subsequently westward expansion of zebra mussels in Europe (Decksbach 1935, Kinzelbach 1992). In general, this expansion took place through a central and southern corridor via canals that connected large rivers flowing from south to north (Bij de Vaate et al. 2002, Leuven et al. 2009; Figure 6.1). Canal construction within the Volga River basin also allowed dispersal of this species from its native range toward the Baltic Sea from where it was further transported to harbors along the Baltic Sea attached to rafts of tree trunks (Bij de Vaate et al. 2002). From the Baltic Sea, jump dispersal occurred when trunks were further transported by cargo ships. In this way, zebra mussels were introduced in Great Britain in 1824 (Aldridge 2010) and in The Netherlands in 1826 (Pollux et al. 2010). The last important connection in the network of river and canals in western Europe was the construction of the Main–Danube Canal that linked the rivers Danube and Rhine. This canal was opened in 1992, and it provided a pathway for genetic interchange between zebra mussel lineages from both river basins (Müller et al. 2002).

For a long time, European mountain ridges such as the Alps and Pyrenees formed dispersal barriers just like straits in marine waters (i.e., Irish Sea). However, zebra mussels attached to recreational boats transported on trailers bypassed the Alps and colonized one alpine lake after another in the 1970s and also colonized Italian water bodies southward; a process that is still going on (Kinzelbach 1992, Cianfanelli et al. 2010). Ireland was colonized by zebra mussels attached to recreational boats that were transported from the United Kingdom in seagoing vessels, hence bypassing the Irish Sea. All these barriers could be passed when delays in transport were substantially reduced as the European Union expanded and inner borders and abolition of the value added tax (VAT) on secondhand boats were removed (Pollux et al. 2003). Finally, mountains of the Pyrenees were passed, and dispersal of the zebra mussel on the Iberian Peninsula is still going on (Durán et al. 2010). The newly established populations appeared genetically most similar to nearby populations (Pollux et al. 2003, Astanei et al. 2005, Rajagopal et al. 2009).

After the late 1870s, a new vector started to play a role in the dispersal of zebra mussels—the use of ballast water instead of dry ballast by seagoing vessels (Carlton 1985). This development increased the probability that discharged ballast water by these vessels would lead to release of mussel

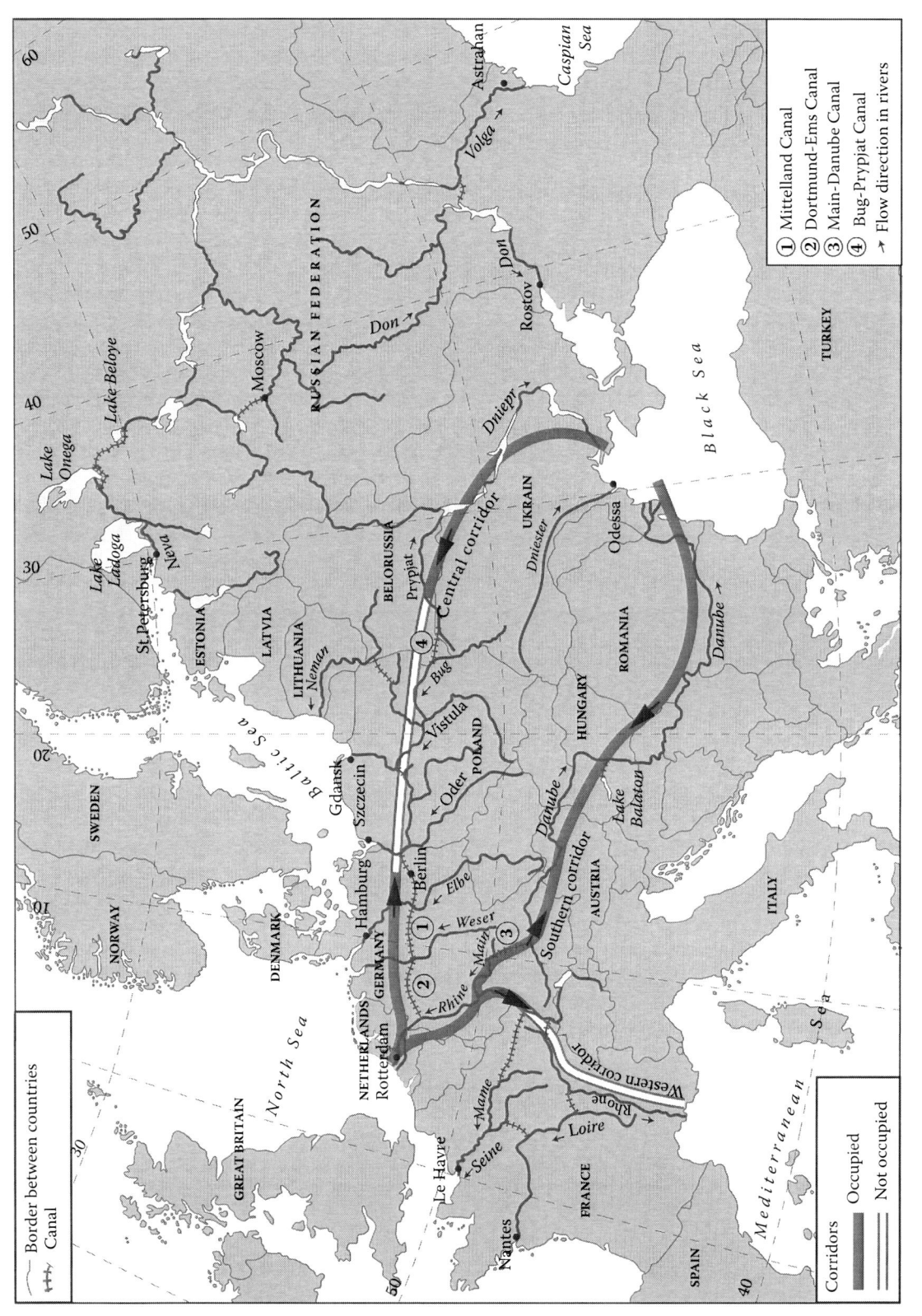

Figure 6.1 Three main corridors by which quagga mussels can spread through western Europe. The gray portion of the corridors indicates the extent of colonization as of July 2011. The white portion indicates the portion that will likely be colonized in the future. (Adapted from bij de Vaate, A. et al., *Can. J. Fish. Aquat. Sci.*, 59, 1159, 2002.)

larvae into the Great Lakes in North America. This was, however, not possible until the St. Lawrence Seaway was constructed in 1959 and allowed ocean vessels from the Atlantic Ocean to travel to the Great Lakes. In the late 1980s, the Great Lakes were colonized by both zebra and quagga mussels, although the latter species was not recognized until 1991 (May and Marsden 1992, Spidle et al. 1994).

Recently, further expansion of both zebra and quagga mussels was observed in Europe (Van der Velde et al. 2010 and literature therein) and in North America (Brown and Stepien 2010, Benson 2013). In Europe, past levels of water pollution formed barriers for dispersal and led to the disappearance of established populations (Van der Velde et al. 2002). However, water quality has greatly improved during the past few decades, diminishing the role of this barrier. Aldridge (2010), who observed zebra mussel reintroductions in Great Britain, hypothesized that in addition to improvements in water quality, improvements in transport (i.e., reduced transit time) also facilitated dispersal.

RANGE EXPANSION OF THE QUAGGA MUSSEL IN WESTERN EUROPE

Being a congener with a life cycle similar to that of zebra mussels, quagga mussels use vectors for range expansion that are similar to those for zebra mussels (Carlton 1993). The start of range expansion by the quagga mussel in Europe occurred about two centuries later than that of the zebra mussel (Orlova 2013). However, once initiated, quagga mussels dispersed at a higher rate than zebra mussels because dispersal pathways and vectors were already in place and opportunities for spread increased. In addition, habitat changes such as improvements in water quality and an increase of solid substrates (e.g., riprap) also facilitated spread.

Range expansion of the quagga mussel started in the 1940s (Orlova et al. 2003, Son 2007). Before the 1940s, the range of this species was restricted to mouths of two rivers that discharged into the Black Sea, the Southern Bug and Dnieper Rivers in Ukraine (Orlova et al. 2003, Son 2007). From the 1940s to the 1980s, quagga mussels expanded only in water bodies within Russia and Ukraine (Orlova et al. 2004, 2005, Zhulidov et al. 2005, Orlova 2013). However, in the 1980s, quagga mussels along with zebra mussels made the jump dispersal into North America. Both species arrived from multiple sources in the Black Sea region, including the mouth of the Volga River (Therriault et al. 2005, Brown and Stepien 2010). The appearance of the quagga mussel in eastern Lake Erie in 1989 (Mills et al. 1999) marked the start of its range expansion in North America (Brown and Stepien 2010). In Europe, the discovery of quagga mussels in the Romanian section of the Danube River in 2004 marked the beginning of a westward expansion (Micu and Telembici 2004). In 2006, quagga mussels were observed in the Hollandsch Diep, a freshwater section of a former Rhine–Meuse estuary, The Netherlands (Bij de Vaate 2006, Bij de Vaate and Jansen 2007, Molloy et al. 2007, Schonenberg and Gittenberger 2008), and soon thereafter more observations in The Netherlands were recorded, including distributaries and connected lakes of the Rhine River such as the rivers Nederrijn, IJssel, and Meuse (Soes 2008, Bij de Vaate and Jansen 2009) and lakes IJsselmeer and Markermeer, Frisian lakes in the northern part of the country, and Bathse spuikanaal in the southwestern part (Zuid-Beveland, Province of Zeeland) (Raad 2010). By 2011, nearly all main water courses and larger lakes in The Netherlands were colonized (Matthews et al. 2012). In 2007, quagga mussels were discovered in the Main River (Germany) (Van der Velde and Platvoet 2007), which is a tributary of the Rhine River, and this was the start of more recorded observations in Germany. Heiler et al. (2011) found individuals of the freshwater mussel *Anodonta anatina* and the snail *Viviparus viviparus* overgrown by both zebra and quagga mussels in the Main River in the vicinity of Hanau-Steinheim, Germany. Martens et al. (2007) discovered quagga mussels in a series of harbors along the upper Rhine River, while Haybach and Christmann (2009) found quagga mussels in the lower Rhine River between Dormagen and Bimmen in 2008. Mayer et al. (2009) found quagga mussels on ship hulls on the slipway of a shipyard at Speyer, along the upper Rhine River. Bij de Vaate and Beisel (2011) found them in the French section of the Moselle River, another tributary of the Rhine River, and this was the first recorded occurrence in France. In 2009, Sablon et al. (2010) and Sablon and Vercauteren (2011) recorded quagga mussels for the first time from Belgium (Albert Canal, in the vicinity of Grobbendonk), while Marescaux et al. (2012) observed upstream dispersal of quagga mussels in the Belgian section of the Meuse River in 2010.

Imo et al. (2010) studied aspects of quagga mussel populations in the Main and Rhine Rivers including current distributions, time of arrival, population structure, and genetics. Population genetic analysis did not reveal any sign of founder effects. Based on noncontinuous distributions, shell size structure, and chronology of the invasion records, two independent introductions occurred in these two rivers. The first introduction occurred in the Main River before 2005 that, according to these authors, was probably a result of jump dispersal. A second introduction occurred later in the upper Rhine River. By modeling the invasion history based on zebra and quagga mussel population dynamics, Heiler et al. (2013) specified that the first introduction into western Europe occured in the Main-Danube Canal in early 2004. According to this reconstruction, the first introduction event in the Rhine Delta happened later, in mid 2005. Heiler et al. (2012) observed that range expansion in Germany occurred in an easterly direction through canals of which the Mittelland Canal is the most important as it connects basins of the Weser, Elbe, and Oder Rivers (Figure 6.1).

The frequent occurrence of quagga mussels in slow-flowing or large standing waters in Europe (e.g., lakes

IJsselmeer and Markermeer, The Netherlands) is in agreement with the observed expansion in rivers via reservoirs in Russia (Orlova et al., 2004, 2005). From all these observations, it is obvious that quagga mussels are rapidly expanding in western Europe. However, the source and direction of dispersal are not clear. To further address these issues, we need to examine distribution data in more detail.

SOURCE AND VECTOR OF QUAGGA MUSSEL INTRODUCTION INTO WESTERN EUROPE

There are four possible scenarios for the introduction of quagga mussels into western Europe: (1) continuous westward dispersal via the Danube River starting in Romania from the Black Sea region; (2) introduction via ballast water in ships from the Black Sea region, the Danube River delta, or the Volga River that was discharged in western Europe (3) introduction from ballast water in ships from North America that was discharged in western Europe and (4) introduction by boats that were transported overland from eastern Europe and launched into western Europe.

Molloy et al. (2007) proposed that range expansion in western Europe started in or before 2004 based on shell sizes of specimens found in the Hollandsch Diep in 2006. The introduction of quagga mussels to the Hollandsch Diep was thought by then to have taken place from the Danube River through the southern corridor (Figure 6.1). This assumption was based on earlier publications on the colonization of the Danube River that indicated range expansion in an upstream direction (Micu and Telembici 2004, Popa and Popa 2006). Moreover, before 2006, several other Ponto-Caspian macroinvertebrate species succeeded to colonize the Rhine basin via the southern corridor (Bij de Vaate et al. 2002, Leuven et al. 2009). However, Van der Velde et al. (2010) doubted that the quagga mussel had taken this route because there was a gap of >2000 km between known records in western Europe and the Danube River. Moreover, Van der Velde and Platvoet (2007) did not observe quagga mussels in the Danube basin section of the Main–Danube Canal and the upper Danube River until about 3 years after their arrival in the Hollandsch Diep. Bij de Vaate (2010) demonstrated that quagga mussels were present in the Rhine basin section of the Main–Danube Canal in 2008, while only a single specimen was found in the Danube basin section of the canal. Hence, Bij de Vaate (2010) suggested that the Rhine basin was colonized from a founder population in the Hollandsch Diep situated in the Rhine delta. In addition, Heiler et al. (2012) still did not find a continuous distribution of quagga mussels even in 2009, but instead found a gap between populations in the upper Danube River near the Main–Danube Canal and populations in the Danube River in Austria.

Phylogeographic relationships were not clear enough to reveal founder lineages of quagga mussels in Germany, and therefore, the origin of founder populations could not be confirmed based on genetic analysis (Imo et al. 2010). This indicates that genetic studies do not support nor reject hypotheses that origins of quagga mussels in western Europe were from North America or from the Ponto-Caspian area. Overland transport from eastern Europe has shown to be theoretically possible since quagga and zebra mussels can survive several days without water depending on weather conditions (temperature, humidity) (Ricciardi et al. 1995). However, this vector does not seem very likely given the long distance gap between populations in west and east Europe.

All known observations of the range expansion of quagga mussels into and in western Europe indicate that this species did not use the Danube River to arrive into the Rhine River basin. Rather, western Europe has been colonized by quagga mussels via two separate events: one from eastern Europe through upstream dispersal in the Danube River from its initial distribution range (see Orlova 2013) and the other via jump dispersal. Unknown is whether the jump dispersal of quagga mussels into western Europe can be considered as a primary (directly from the Ponto-Caspian area) or a secondary introduction (from the mouth of the Danube delta, the Volga River, or from North America).

Range expansion, e.g., in the Rhine River has been facilitated by inland navigation (Mayer et al. 2009). The Rhine River is one of the heaviest navigated rivers in the world. Mussels likely attached to ships and then hitchhiked in an upstream direction (Van der Velde and Platvoet 2007, Haybach and Christmann 2009). Also, upstream movement of mussels in the Danube River most likely depended on inland navigation of vessels. Quagga mussels were first observed in this river in 2004 at Cernavodă (Romania) (Micu and Telembici 2004). This was followed by a report of mussels at Drobeta-Turnu Severin (Romania), which was located just upstream from where mussels were first found (Popa and Popa 2006). In 2008, range expansion had extended upstream at least to Komárom (Hungary) (Szekeres et al. 2008). By 2009, quagga mussels were found further upstream at several locations in the Austrian and German section of the river (Heiler et al. 2012). These observations together indicate an upstream migration rate in the Danube River of 500–600 km/year, which is about the same as estimated for the Rhine River (unpublished data).

Upstream migration of quagga mussels was also observed in the canalized Meuse River, which discharges into the Hollandsch Diep. Surveys in the Dutch section of this river revealed that the upstream-most location where quagga mussels were found was at Sambeek in September 2007, at Linne in September 2008, at Maastricht (all locations in The Netherlands) in July 2010, and at Gives (Belgium) in August 2010. Together, these observations indicate an estimated upstream migration rate of about 50–70 km/year.

As of July 2011, there were further reports of quagga mussels in western Europe from The Netherlands, Germany, Belgium, France, and Austria.

QUAGGA AND ZEBRA MUSSEL INTERACTIONS IN WESTERN EUROPE

In areas outside western Europe, quagga mussels were able to nearly outcompete zebra mussels within a relatively short period (e.g., Mills et al. 1999, Orlova et al. 2004, Wilson et al. 2006, Zhulidov et al. 2010). Heiler et al. (2013) reported that relative abundance of quagga mussels in the total dreissenid population increased 26% per year at several studied locations in western Europe. Changes in dreissenid populations were also studied in the Lake IJsselmeer area, The Netherlands (Bij de Vaate, unpublished). This area was originally an inland sea, called Zuyder Zee. After this inland sea was isolated by a dam in 1932, it became a freshwater lake (named Ijsselmeer) in which zebra mussels soon became abundant (De Jong and Bij de Vaate 1989, Bij de Vaate 1991). Over time, land reclamation reduced the size of the lake and another dam, completed in 1975, divided the lake into a northern part called Lake IJsselmeer and a southern part called Lake Markermeer (Noordhuis et al. 2010). Colonization of Lake IJsselmeer by quagga mussels in 2007 was followed by yearly sampling (in October) of riprap in the littoral zone at a location in the southern part of the lake. In 2007, 1% of dreissenids were quagga mussels, but this percentage increased to 94% in 2011 (Figure 6.2). In the deeper parts of the lake (3.5–5.0 m), dreissenids mainly occur attached to empty shells of the marine bivalve *Mya arenaria*, which was abundant during the Zuyder Zee period. The average percentage of quagga mussels in the dreissenid population at these depths increased from <1% in 2007 to 47% (northern part) and 94% (southern part) in 2010. In 2010, the average quagga mussel contribution to the dreissenid population in Lake Markermeer was 50%. Temporal trends in percentages of quagga mussels in the dreissenid community in Lake IJsselmeer are consistent with those from the Hollandsch Diep (Figure 6.2).

Previous surveys between 1981 and 2007 in Lakes IJsselmeer and Markermeer indicated a decrease in zebra mussel densities (Noordhuis et al. 2010). However, average biovolume (i.e., the amount of water displaced by living mussels as described by Smit and Dudok van Heel [1992]) of dreissenids in Lake IJsselmeer increased from 33 mL/m^2 in 2007 to 152 mL/m^2 in 2010, and in Lake Markermeer increased from 22 mL/m^2 in 2006 to 55 mL/m^2 in 2010. Contribution of quagga mussels to total dreissenid biovolume in 2010 was on average 85% and 60% for lakes IJsselmeer and Markermeer, respectively.

Trends in biomass of zebra and quagga mussels were investigated monthly at one location in both Lake Markermeer and Lake IJsselmeer between June 2009 and July 2011. Since mussels were collected with a trawl net, only relative changes in biomass could be monitored. Relative biomass was calculated separately for the two species based on the relationship between soft body ash-free dry weight (AFDW) and shell length and determination of size frequency distributions. While biomass of quagga and zebra mussels was generally similar when the study was initiated in 2009, biomass of quagga mussels exceeded that of zebra mussels by 2011, particularly in Lake IJsselmeer (Figure 6.3). Both Lake IJsselmeer and Lake Markermeer are heavily utilized by overwintering populations of diving ducks (tufted duck, scaup, pochard, and goldeneye), and quagga mussels are now probably the most important food source for these populations (Noordhuis et al. 2010, Van Eerden and De Leeuw 2010). Filtration activities of zebra mussels are considered to be the primary reason for increased water clarity in eutrophic Dutch lakes (Reeders and Bij de Vaate 1990, Reeders et al. 1993, Dionisio Pires et al. 2010). With quagga mussels becoming dominant, similar if not greater increases in water clarity might be expected (Diggins 2001).

FUTURE CONSIDERATIONS

Quagga mussels will continue to spread in western Europe at a more rapid rate than zebra mussels did in the past because of the completion of a canal–river network linking major river systems and increased river transportation. Ship traffic is expected to be the primary vector for spread of quagga mussels in the area, and veliger transport by drift, as enhanced by canal management, would be a secondary vector. Three major corridors, termed the central, southern, and western (Figure 6.1), serve as pathways for mussels to migrate and spread throughout the region.

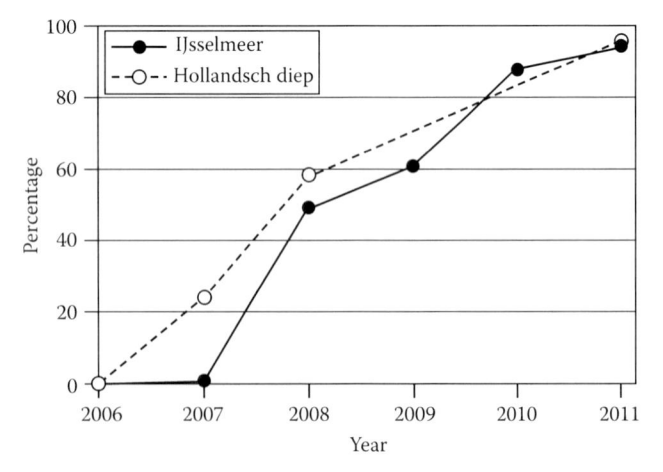

Figure 6.2 Trends in relative abundance (percent) of quagga mussels in the total dreissenid population in the Hollandsch Diep and on riprap in the southern part of Lake IJsselmeer, 2006–2011.

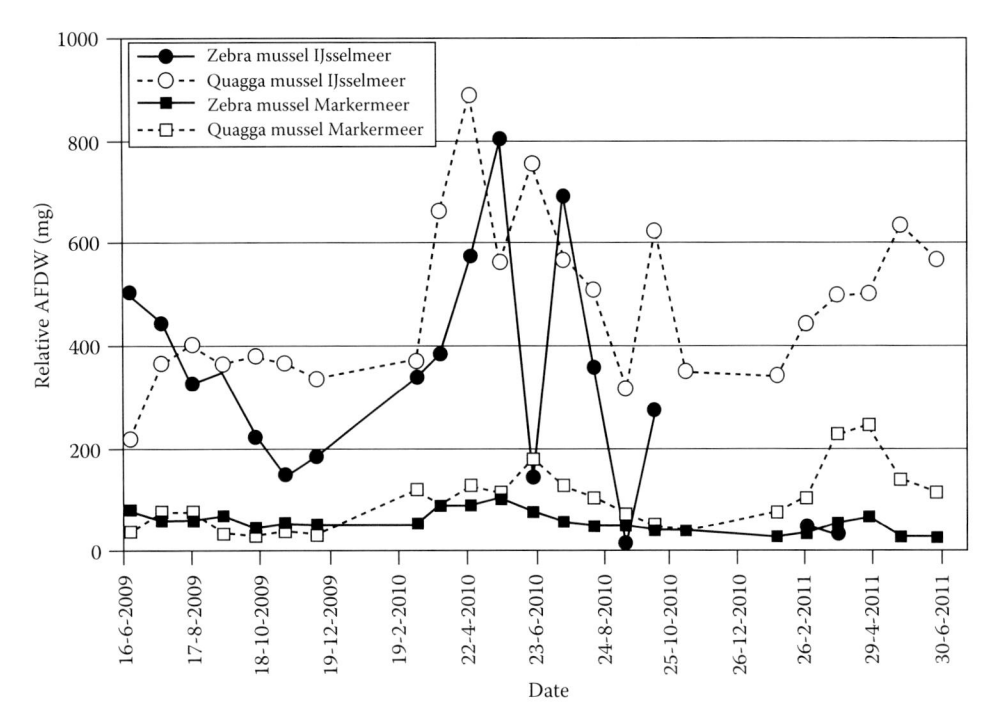

Figure 6.3 Trends in relative biomass (AFDW, ash-free dry weight of the soft tissue) of zebra and quagga mussels at a site in the southern part of Lake IJsselmeer and a site in the western part of Lake Markermeer, 2009–2011. Data for zebra mussels in Lake IJsselmeer in 2010 and 2011 were not calculated due to their relatively low density.

The Alps and Pyrenees mountains were long-time barriers that prevented zebra mussels from spreading into Italy and the Iberian Peninsula, respectively. However, overland transport of mussels attached to recreational boats or with fishing equipment allowed dispersal beyond these mountain barriers (Giusti and Oppi 1972, Rajagopal et al. 2009). This vector of dispersal has also been important in the spread of zebra mussels from continuous waterways to isolated water bodies (Bidwell 2010). If management options fail to prevent spread of dreissenids, quagga mussels can also be expected to spread via overland transport, and mountain barriers and relative isolation of a water body would be no hindrance to dispersal.

Compared with zebra mussels, quagga mussels have a competitive advantage based on physiological characteristics. Quagga mussels prefer still water over flowing water (Ackerman 1999), which is consistent with their relative abundance in harbors and reservoirs in the Rhine basin. They flourish better in turbid waters than zebra mussels due to their higher assimilation efficiency (Baldwin et al. 2002) and lower respiration rate (Stoeckmann 2003). For example, construction of reservoirs in the Dnieper River (Ukraine) led to increased silt concentrations in the seston and a corresponding increase in the relative dominance of quagga mussels in the dreissenid population (Karatayev et al. 1998). Also, this trend to dominate in relatively turbid lakes is apparent in The Netherlands (viz., lakes Markermeer and IJsselmeer).

In conclusion, we hypothesize that quagga mussels will ultimately appear in all inland waters in western Europe that are already colonized by zebra mussels (Pollux et al. 2010) and its spread will be a much faster rate than found for zebra mussels. Given an upstream migration rate of up to 500–600 km/year, we assume that all connected waterways with a suitable habitat will be colonized within a few more years. More time will be needed to colonize water bodies that are more isolated (e.g., lakes and river tributaries without shipping). For isolated areas like Italy, the Iberian Peninsula, Ireland, and the United Kingdom, colonization will depend on the effectiveness of measures taken to prevent dreissenid spread.

Presently, quagga mussels are mostly found where zebra mussels were already present. Whether quagga mussels will dominate dreissenid populations depends on local factors of which stream velocity, food availability, and silt concentrations in the seston are probably the most important. In most freshwater bodies of western Europe, it is expected that quagga mussels will dominate the dreissenid population as has already been observed in eastern Europe. Orlova et al. (2004) observed an increase in the percentage of quagga mussels in the dreissenid population in the Volga Delta (Russia) from 4% in 1994, 24% in 1995, 32% in 1996, to 96% in 2000. In contrast, Zhulidov et al. (2006) found that the dominance of quagga mussels relative to zebra mussels gradually increased but then rapidly declined in the Don River system (Russia). Continued studies will be needed to determine if the presence of quagga mussels will lead to lower densities of zebra mussels in all cases and whether quagga mussels (plus zebra mussels) will produce more total dreissenid biomass than zebra mussels alone. Future impacts on the functioning of aquatic ecosystems in western Europe will likely depend on the ultimate proportion of quagga and zebra mussels and the ultimate dreissenid biomass realized.

ACKNOWLEDGMENTS

Dr. Nándor Oertel provided additional information of quagga mussels in the Danube River; Bert Jansen is acknowledged for preparing the figures. One of the authors (K.C.M. Heiler) was financially supported by the DBU (German Federal Foundation for the Environment).

REFERENCES

Ackerman, J.D. 1999. Effect of velocity on the filter feeding of dreissenid mussels (*Dreissena polymorpha* and *Dreissena bugensis*): Implications for trophic dynamics. *Can. J. Fish. Aquat. Sci.* 56: 1551–1561.

Aldridge, D.C. 2010. *Dreissena polymorpha* in Great Britain: History of spread, impacts and control. In *The Zebra Mussel in Europe*, G. van der Velde, S. Rajagopal, and A. bij de Vaate, eds., pp. 79–91. Leiden, The Netherlands: Backhuys Publishers.

Astanei, J., E. Gosling, J. Wilson, and E. Powell. 2005. Genetic variability and phylogeography of the invasive zebra mussel, *Dreissena polymorpha* (Pallas). *Mol. Ecol.* 14: 1655–1666.

Baldwin, B.S., M.S. Mayer, J. Dayton, N. Pau, J. Mendillo, M. Sullivan, A. Moore, A. Ma, and E.L. Mills. 2002. Comparative growth and feeding in zebra and quagga mussels (*Dreissena polymorpha* and *Dreissena bugensis*): Implications for North American lakes. *Can. J. Fish. Aquat. Sci.* 59: 680–694.

Benson, A. 2013. Chronological history of zebra and quagga mussels (Dreissenidae) in North America, 1988–2010. In *Quagga and Zebra Mussels: Biology, Impacts, and Control*, 2nd Edn., T. F. Nalepa and D. W. Schloesser, eds., pp. 9–31. Boca Raton: CRC Press.

Bidwell, J.R. 2010. Range expansion of *Dreissena polymorpha*: A review of major dispersal vectors in Europe and North America. In *The Zebra Mussel in Europe*, G. van der Velde, S. Rajagopal, and A. bij de Vaate, eds., pp. 69–78. Leiden, The Netherlands: Backhuys Publishers.

Bij de Vaate, A. 1991. Distribution and aspects of population dynamics of the zebra mussel, *Dreissena polymorpha* (Pallas, 1771), in the lake IJsselmeer area (The Netherlands). *Oecologia* 86: 40–50.

Bij de Vaate, A. 2006. De quaggamossel, *Dreissena rostriformis bugensis* (Andrusov 1897), een nieuwe zoetwater mosselsoort voor Nederland. *Spirula Corr. Bl. Ned. Malac. Ver.* 353: 143–144.

Bij de Vaate, A. 2010. Some evidence for ballast water transport being the vector of the quagga mussel (*Dreissena rostriformis bugensis* Andrusov 1897) introduction into Western Europe and subsequent upstream dispersal in the River Rhine. *Aquat. Invasions* 5: 207–209.

Bij de Vaate, A., and J.-N. Beisel. 2011. Range expansion of the quagga mussel (*Dreissena rostriformis bugensis* Andrusov 1897) in Western Europe: First observation from France. *Aquat. Invasions* 6 (Suppl. 1): 71–74.

Bij de Vaate, A. and E.A. Jansen. 2007. Onderscheid tussen de driehoeksmossel en de quaggamossel. *Spirula Corr. Bl. Ned. Malac. Ver.* 356: 78–81.

Bij de Vaate, A. and E.A. Jansen. 2009. De verspreiding van de quaggamossel in de rijkswateren. *Spirula Corr. Bl. Ned. Malac. Ver.* 368: 72–75.

Bij de Vaate, A., K. Jazdzewski, H. Ketelaars, S. Gollasch, and G. van der Velde. 2002. Geographical patterns in range extension of macroinvertebrate Ponto-Caspian species in Europe. *Can. J. Fish. Aquat. Sci.* 59: 1159–1174.

Brown, J.E. and C.A. Stepien. 2010. Population genetic history of the dreissenid mussel invasions: Expansion patterns across North America. *Biol. Invasions* 12: 3687–3710.

Carlton, J.T. 1985. Transoceanic and interoceanic dispersal of coastal marine organisms: The biology of ballast water. *Oceanogr. Mar. Biol. Ann. Rev.* 23: 313–371.

Carlton, J.T. 1993. Dispersal mechanisms of the zebra mussel (*Dreissena polymorpha*). In *Zebra Mussels: Biology, Impact and Control*, T.F. Nalepa and D.W. Schloesser, eds., pp. 677–697. Boca Raton, FL: CRC Press.

Cianfanelli, S., E. Lori, and M. Bodon. 2010. *Dreissena polymorpha*: Current status of knowledge about the distribution in Italy. In *The Zebra Mussel in Europe*, G. van der Velde, S. Rajagopal, and A. bij de Vaate, eds., pp. 93–100. Leiden, The Netherlands: Backhuys Publishers.

Decksbach, N.K. 1935. *Dreissena polymorpha*: Verbreitung im europäischen Teile der UdSSR und die Sie bedingenden Faktoren. *Verh. Int. Verein. Theor. Angew. Limnol.* 7: 432–438.

De Jong, J. and A. bij de Vaate. 1989. Dams and the environment. The Zuiderzee damming. International Commission on Large Dams (ICOLD), Bulletin 66, Paris, France.

Diggins, T.P. 2001. A seasonal comparison of suspended sediment filtration by quagga (*Dreissena bugensis*) and zebra (*D. polymorpha*) mussels. *J. Great Lakes Res.* 27: 457–466.

Dionisio Pires, L.M., B.W. Ibelings, and E. van Donk. 2010. Zebra mussels as a potential tool in the restoration of eutrophic shallow lakes dominated by toxic cyanobacteria. In *The Zebra Mussel in Europe*, G. van der Velde, S. Rajagopal, and A. bij de Vaate, eds., pp. 331–341. Leiden, The Netherlands: Backhuys Publishers.

Durán, C., M. Lanao, A. Anadón, and V. Touyá. 2010. Management strategies for the zebra mussel invasion in the Ebro River basin. *Aquat. Invasions* 5: 309–316.

Giusti, F. and E. Oppi. 1972. *Dreissena polymorpha* (Pallas) nuovamente in Italia. *Mem. Mus. Civ. St. Nat. Verona* 20: 45–49.

Haybach, A. and K.-H. Christmann. 2009. Erster Nachweis der Quaggamuschel *Dreissena rostriformis bugensis* (Andrusov, 1897) (Bivalvia: Dreissenidae) im Niederrhein-Westfalen. *Lauterbornia* 67: 69–72.

Heiler, K.C.M., S. Brandt, C. Albrecht, T. Hauffe, and T. Wilke. 2012. A new approach for dating introduction events of the quagga mussel (*Dreissena rostriformis bugensis*). *Biol. Invasions* 14: 1311–1316.

Heiler, K.C.M., S. Brandt, and P.V. von Oheimb. 2011. Introduction into *Dreissena rostriformis bugensis* and observations of attachment on native molluscs in the Main River (Bivalvia: Veneroidea: Dreissenidae). *Mitt. Dtsch. Malakozool. Ges.* 84: 53–58.

Heiler, K.C.M., Bij de Vaate, A., Ekschmitt, K., Oheimb, P.V. von, Albrecht, C. and T. Wilke. 2013. Reconstruction of the early invasion history of the quagga mussel (*Dreissena rostriformis bugensis*) in Western Europe. *Aquatic Invasions.* 8: 53–57.

Heiler, K.C.M., N. Nahavandi, and C. Albrecht. 2010. A new invasion into an ancient lake—The invasion history of the dreissenid mussel *Mytilopsis leucophaeata* (Conrad, 1831) and its first record in the Caspian Sea. *Malacologia* 53: 185–192.

Imo, M., A. Seitz, and J. Johannesen. 2010. Distribution and invasion genetics of the quagga mussel (*Dreissena rostriformis bugensis*) in German rivers. *Aquat. Ecol.* 44: 731–740.

Karatayev, A.Y., L.E. Burlakova, and D.K. Padilla. 1998. Physical factors that limit the distribution and abundance of *Dreissena polymorpha* (Pall.). *J. Shellfish Res.* 17: 1219–1235.

Kinzelbach, R. 1992. The main features of the phylogeny and dispersal of the zebra mussel *Dreissena polymorpha*. In *The Zebra Mussel Dreissena polymorpha. Ecology, Biological Monitoring and First Applications in the Water Quality Management*, D. Neumann and H.A. Jenner, eds., *Limnol. Aktuell* 4: 5–17. Stuttgart, Germany: Fischer Verlag.

Leuven, R.S.E.W., G. van der Velde, I. Baijens, J. Snijders, C. van der Zwart, H.J.R. Lenders, and A. bij de Vaate. 2009. The river Rhine: A global highway for dispersal of aquatic invasive species. *Biol. Invasions* 11: 1989–2008.

Ludyanskiy, M.L., D. MacDonald, and D. MacNeill. 1993. Impact of the zebra mussel, a bivalve invader. *BioScience* 43: 533–544.

Marelli, D.C. and S. Gray. 1983, Conchological redescriptions of *Mytilopsis sallei* and *Mytilopsis leucophaeta* of the brackish western Atlantic (Bivalvia, Dreissenidae). *Veliger* 25: 185–193.

Marescaux, J., A. bij de Vaate, and K. Van Doninck. 2012. First records of *Dreissena rostriformis bugensis* (Andrusov 1897) in the Meuse River. *BioInvasions Records.* 1: 109–114.

Martens, A., K. Grabow, and G. Schoolmann. 2007. Die Quagga-Muschel *Dreissena rostriformis bugensis* (Andrusov, 1897) am Oberrhein (Bivalvia: Dreissenidae). *Lauterbornia* 61: 145–152.

Matthews, J., G. van der Velde, A. bij de Vaate, and R.S.E.W. Leuven. 2012. Key factors for spread, impacts and management of Quagga mussels in The Netherlands. Final report 24 February 2012. Radboud University Nijmegen, Institute for Water and Wetland Research, Department of Environmental Sciences & Department of Animal Ecology and Ecophysiology & Waterfauna Hydrobiologisch Adviesbureau, Lelystad. Commissioned by Invasive Alien Species Team, Plant Protection Service, Wageningen. Ministry of Agriculture, Nature and Food Quality. Reports Environmental Science No. 404, pp. 1–121.

May, B. and J.E. Marsden. 1992. Genetic identification and implications of another invasive species of dreissenid mussel in the Great Lakes. *Can. J. Fish. Aquat. Sci.* 49: 1501–1506.

Mayer, S., A. Rander, K. Grabow, and A. Martens. 2009. Binnenfrachtschiffe als Vektoren der Quagga-Muschel *Dreissena rostriformis bugensis* (Andrusov) im Rhein (Bivalvia: Dreissenidae). *Lauterbornia* 67: 63–67.

Micu, D. and A. Telembici. 2004. First record of *Dreissena bugensis* (Andrusov 1897) from the Romanian stretch of river Danube. In *Abstracts of the International Symposium of Malacology*, August 19–22, 2004, Sibiu, Romania.

Mills, E.L., R.M. Dermott, E.F. Roseman, D. Dustin, E. Mellina, D.B. Conn, and A.P. Spidle. 1999. Colonization, ecology, and population structure of the "quagga" mussel (Bivalvia: Dreissenidae) in the lower Great Lakes. *Can. J. Fish. Aquat. Sci.* 50: 2305–2314.

Molloy, D.P., A. bij de Vaate, T. Wilke, and L. Giamberini. 2007. Discovery of *Dreissena rostriformis bugensis* (Andrusov 1897) in Western Europe. *Biol. Invasions* 9: 871–874.

Müller, J.C., D. Hidde, and A. Seitz. 2002. Canal construction destroys the barrier between major European lineages of the zebra mussel. *Proc. R. Soc. Lond. Ser. B* 269: 1139–1142.

Noordhuis, R., M.R. van Eerden, and M. Roos. 2010. Crash of zebra mussel, transparency and water bird populations in Lake Markermeer. In *The Zebra Mussel in Europe*, G. van der Velde, S. Rajagopal, and A. bij de Vaate, eds., pp. 265–277. Leiden, The Netherlands: Backhuys Publishers.

Orlova, M.I., P.I. Antonov, G.Kh. Shcherbina, and T.W. Therriault. 2003. *Dreissena bugensis*: Evolutionary underpinning for invasion success based on its range extension in Europe. In *Invasion of Alien Species in Holarctic. Proceedings of U.S.–Russia Invasive Species Workshop*, August 27–31, 2001, Borok, Russia, pp. 452–466.

Orlova, M.I., J.R. Muirhead, P.I. Antonov, G.Kh. Shcherbina, Y.I. Starobogatov, G.I. Biochino, T.W. Therriault, and H.J. MacIsaac. 2004. Range expansion of quagga mussels *Dreissena rostriformis bugensis* in the Volga River and Caspian Sea basin. *Aquat. Ecol.* 38: 561–573.

Orlova, M.I., T.W. Therriault, P.I. Antonov, and G.K. Shcherbina. 2005. Invasion ecology of quagga mussels (*Dreissena rostriformis bugensis*): A review of evolutionary and phylogenetic impacts. *Aquat. Ecol.* 39: 401–418.

Orlova, M. I. 2013. Origin and spread of quagga mussels (*Dreissena rostriformis bugensis*) in eastern Europe with notes on size structure of populations. In *Quagga and Zebra Mussels, Biology, Impacts, and Control*, 2nd Edn., T. F. Nalepa and D. W. Schloesser, eds., pp. 93–102, Boca Raton: CRC Press.

Pollux, B.J.A., D. Minchin, G. van der Velde, T. van Alen, S. Moon-van der Staay, and J.H.P. Hackstein. 2003. Zebra mussels (*Dreissena polymorpha*) in Ireland, AFLP fingerprinting and boat traffic both indicate an origin from Britain. *Freshwat. Biol.* 48: 1127–1139.

Pollux, B.J.A., G. van der Velde, and A. bij de Vaate. 2010. A perspective on global spread of *Dreissena polymorpha*: A review on possibilities and limitations. In *The Zebra Mussel in Europe*, G. van der Velde, S. Rajagopal, and A. bij de Vaate, eds., pp. 45–58. Leiden, The Netherlands: Backhuys Publishers.

Popa, O.P. and L.O. Popa. 2006. The most westward European occurrence point for *Dreissena bugensis* (Andrusov 1897). *Malacol. Bohemoslov.* 5: 3–5.

Raad, H. 2010. Molluskeninventarisatie Bathse Spuikanaal en omgeving (Zuid-Beveland, prov. Zeeland). *Spirula Corr. Bl. Ned. Malac. Ver.* 374: 68–70.

Rajagopal, S., B.J.A. Pollux, J.L. Peters, G. Cremers, S. Yeo Moon-van der Staay, T. van Alen, J. Eygensteyn, A. van Hoek, A. Palau, A. bij de Vaate, and G. van der Velde. 2009. Origin of Spanish invasion by the zebra mussel, *Dreissena polymorpha* (Pallas, 1771) revealed by Amplified Fragment Length Polymorphism (AFLP) fingerprinting. *Biol. Invasions* 11: 2147–2159.

Reeders, H.H. and A. bij de Vaate. 1990. Zebra mussels (*Dreissena polymorpha*): A new perspective for water quality management. *Hydrobiologia* 200/201: 437–450.

Reeders, H.H., A. bij de Vaate, and R. Noordhuis. 1993. Potential of the zebra mussel (*Dreissena polymorpha*) for water quality management. In *Zebra Mussels: Biology, Impact and Control*, T.F. Nalepa and D.W. Schloesser, Eds., pp. 439–451. Boca Raton, FL: CRC Press.

Ricciardi, A., R. Serrouya, and F.G. Whoriskey. 1995. Aerial exposure tolerance of zebra and quagga mussels (Bivalvia: Dreissenidae): Implications for overland dispersal. *Can. J. Fish. Aquat. Sci.* 52: 470–477.

Sablon, R. and T. Vercauteren. 2011. Exotische soorten weekdieren in (Antwerpse) rivieren en stilstaande waters. Evolutie van de voorbije 20 jaar. *ANTenne* 5: 9–19.

Sablon, R., T. Vercauteren, and P. Jacobs. 2010. De quaggamossel (*Dreissena rostriformis bugensis* (Andrusov, 1897)), een recent gevonden invasieve zoetwatermossel in Vlaanderen. *ANTenne* 4: 32–36.

Schonenberg, D.B. and A. Gittenberger. 2008. The invasive quagga mussel *Dreissena rostriformis bugensis* (Andrusov, 1879) (Bivalvia: Dreissenidae) in the Dutch Haringvliet, an enclosed freshwater Rhine-Meuse estuary, the westernmost record for Europe. *Basteria* 72: 345–352.

Smit, H. and E. Dudok van Heel. 1992. Methodical aspects of a simple allometric biomass determination of *Dreissena polymorpha* aggregations. In *The Zebra Mussel Dreissena polymorpha. Ecology, Biological Monitoring and First Applications in the Water Quality Management*, D. Neumann and H.A. Jenner, eds. *Limnol. Aktuell* 4: 79–86. Stuttgart, Germany: Fischer Verlag.

Soes, D.M. 2008. Quagga-mossels bij Wageningen. *Spirula Corr. Bl. Ned. Malac. Ver.* 362: 42–43.

Son, M.O. 2007. Native range of the zebra mussel and quagga mussel and new data on their invasions within the Ponto-Caspian Region. *Aquat. Invasions* 2: 174–184.

Spidle, A.P., J.E. Marsden, and B. May. 1994. Identification of the Great Lakes Quagga Mussel as *Dreissena bugensis* from the Dnieper River, Ukraine, on the basis of allozyme variation. *Can. J. Fish. Aquat. Sci.* 51: 1485–1489.

Stoeckmann, A. 2003. Physiological energetics of Lake Erie dreissenid mussels: A basis for the displacement of *Dreissena polymorpha* by *Dreissena bugensis*. *Can. J. Fish. Aquat. Sci.* 60: 126–134.

Szekeres, J., Z. Szalóky, and K. Bodolai. 2008. Első adat a *Dreissena bugensis* (Andrusov, 1897) (Bivalvia: Dreissenidae) magyarországi megjelenéséről. *Malakológiai Tájékoztató* 26: 33–36.

Therriault, T.W., M.I. Orlova, M.F. Docker, H.J. MacIsaac, and D.D. Health. 2005. Invasion genetics of a freshwater mussel (*Dreissena rostriformis bugensis*) in eastern Europe: High gene flow and multiple introductions. *Heredity* 95: 16–23.

Van Eerden, M.R. and J. de Leeuw. 2010. How *Dreissena* sets the winter scene for water birds: Dynamic interactions between diving ducks and zebra mussels. In *The Zebra Mussel in Europe*, G. van der Velde, S. Rajagopal, and A. bij de Vaate, eds., pp. 251–264. Leiden, The Netherlands: Backhuys Publishers.

Van der Velde, G., I. Nagelkerken, S. Rajagopal, and A. bij de Vaate. 2002. Invasions by alien species in inland freshwater bodies in Western Europe. In: *Invasive Aquatic Species in Europe. Distribution, Impacts and Management*, E. Leppäkoski, S. Gollasch, and S. Olenin, eds., pp. 360–372. Dordrecht, The Netherlands: Kluwer Academic Publishers.

Van der Velde, G. and D. Platvoet. 2007. Quagga mussels *Dreissena rostriformis bugensis* (Andrusov, 1897) in the Main River (Germany). *Aquat. Inv.* 2: 261–264.

Van der Velde, G., S. Rajagopal, and A. bij de Vaate. 2010. From zebra mussel to quagga mussels: An introduction to the Dreissenidae. In *The Zebra Mussel in Europe*, G. van der Velde, S. Rajagopal, and A. bij de Vaate, eds., pp. 1–10. Leiden, The Netherlands: Backhuys Publishers.

Verween, A., M. Vincx, and S. Degraer. 2010. *Mytilopsis leucophaeata*: The brackish water equivalent of *Dreissena polymorpha*? A review. In *The Zebra Mussel in Europe*, G. van der Velde, S. Rajagopal, and A. bij de Vaate, eds., pp. 29–43. Leiden, The Netherlands: Backhuys Publishers.

Wilke, T., R. Schultheiß, C. Albrecht, N. Bornmann, S. Trajanovski, and T. Kevrekidis. 2010. Native *Dreissena* freshwater mussels in the Balkans: In and out of ancient lakes. *Biogeosciences* 7: 3051–3065.

Wilson, K.A., E.T. Howell, and D.A. Jackson. 2006. Replacement of zebra mussels by quagga mussels in the Canadian nearshore of Lake Ontario: The importance of substrate, round goby abundance, and upwelling frequency. *J. Great Lakes Res.* 32: 11–28.

Zhulidov, A.V., A.V. Kozhara, G.H. Scherbina, T.F. Nalepa, A. Protasov, S.A. Afanasiev, E.G. Pryanichnikova, D.A. Zhulidov, T. Yu. Gurtovaya, and D.F. Pavlov. 2010. Invasion history, distribution, and relative abundances of *Dreissena bugensis* in the old world: A synthesis of data. *Biol. Invasions* 12: 1923–1940.

Zhulidov, A.V., T.F. Nalepa, A.V. Kozhara, D.A. Zhulidov, and T.Y. Gurtovaya. 2006. Recent trends in relative abundance of two dreissenid species, *Dreissena polymorpha* and *Dreissena bugensis* in the lower Don River System, Russia. *Arch. Hydrobiol.* 165: 209–220.

Zhulidov, A.V., D.A. Zhulidov, D.F. Pavlov, T.F. Nalepa, and T.Y. Gurtovaya. 2005. Expansion of the invasive bivalve mollusk *Dreissena bugensis* (quagga mussel) in the Don and Volga River Basins: Revisions based on archived specimens. *Ecohydrol. Hydrobiol.* 5: 127–133.

CHAPTER 7

Origin and Spread of Quagga Mussels (*Dreissena rostriformis bugensis*) in Eastern Europe with Notes on Size Structure of Populations

Marina I. Orlova

CONTENTS

Abstract 93
Introduction 93
Origins of Dreissenidae and *D. r. bugensis* 94
General Spread (Dreissenidae) 95
  Eastern Invasion Corridor 97
Quagga Mussels versus Zebra Mussels 97
Comparative Size Structure: Quagga Mussels and Zebra Mussels 99
Summary 100
References 100

ABSTRACT

Multiple lines of evidence indicate the Dreissenidae family, including ancestors of quagga and zebra mussels, appeared in the Jurassic–Cretaceous period (ca. 175 million years ago). The family evolved in the Pontocaspian region that includes the Black, Azov, and Caspian Seas, where the fauna experienced many series of changes in habitat conditions that produced evolutionary divergence of many species in a relatively small area. In recent times (i.e., past 200 years), dreissenid mussels and other Pontocaspian fauna have spread from their endemic range to new areas in Europe by way of three main geographic corridors as primarily mediated by human activity. Of the many dreissenid taxa that evolved in the Pontocaspian region (number varies because of taxonomic changes), about 25%–30% have exhibited "invasive" histories of range expansion. Two of these taxa—*Dreissena rostriformis bugensis* (Andrusov 1897) (quagga mussel) and *Dreissena polymorpha* (Pallas 1771) (zebra mussel)—have wide habitat tolerances and are characterized as "invasive." Over the past 200 years, both of these dreissenid taxa have utilized the same vectors for range expansion. However, there are four differences between the spread of quagga and zebra mussels: (1) Quagga mussels spread slower than zebra mussels, and therefore, the present range of quagga mussels is smaller (in Europe) than that of zebra mussels; (2) first records of quagga mussels are typically in areas already colonized by zebra mussels; (3) quagga mussels "naturalize" to new habitats faster than zebra mussels, which allows quagga mussels to become more abundant than previously established populations of zebra mussels; and (4) quagga mussels colonize deeper water than zebra mussels. These traits have led to the belief that zebra mussels are the first to invade an area where they act as ecosystem "engineers" that create favorable conditions for later invasion by quagga mussels, which in most cases become the dominant "successional" dreissenid taxa.

INTRODUCTION

The man-induced spread of species within the family Dreissenidae in eastern Europe has a long, well-documented history (Andrusov 1897, Marelli and Gray 1985, Nuttall 1990, Starobogatov and Andreeva 1994, Therriault and Orlova 2010). Of the many taxa within this family, perhaps the most successful invaders, and presently the two taxa with the widest distribution, are *Dreissena polymorpha* (Pallas 1771) (zebra mussel) and *Dreissena rostriformis bugensis* (Andrusov 1897) (quagga mussel). Both of these taxa have remarkable, yet very different invasion histories, particularly in eastern Europe. The zebra mussel has an extensive history of spread dating back to the eighteenth century (Nowak 1971, Kinzelbach 1992, bij de Vaate et al. 2002, 2013). Although the spread of zebra mussels through Europe likely occurred naturally since the late-Holocene period. (Starobogatov and Andreevea 1994), the rate of dispersal increased greatly since the establishment of interbasin connections (canals) in Europe over the past 200 years (e.g., Andrusov 1897, Nowak 1971, Kinzelbach 1992, Minchin et al. 2002). In contrast, distributions of the quagga mussel remained relatively restricted until the mid-1990s, despite having access to the same invasion corridors as the zebra mussel.

Understanding reasons for such differences in invasion histories of zebra mussels and quagga mussels, despite overlapping native ranges and similar life habits, requires not only comparisons of biology and responses to environmental conditions (Baldwin et al. 2002, Stoeckmann 2003, Diggins et al. 2004) but also a perspective on phylogenetic and evolutionary relationships. In this chapter, we examine the origins of *D. r. bugensis* and the development of characteristics important for range extension. Also, we closely examine the spread of these taxa in eastern Europe with particular emphasis on the Volga River basin.

ORIGINS OF DREISSENIDAE AND *D. R. BUGENSIS*

Dreissenidae (excluding *Mytilopsis*) is a family of species represented by sessile organisms that has its origins in marine environments but now inhabit fresh to brackish waters mostly in the northern hemisphere (Andrusov 1897, Marelli and Gray 1985, Hebert et al. 1989, Nuttall 1990, Starobogatov and Andreeva 1994, Van der Velde and Platvoet 2007, Therriault and Orlova 2010). Despite morphological and behavioral characteristics that resemble those of the ancient order Mytiloida, this family is within the order Veneroida and possesses all peculiarities of this relatively young evolutionary group of bivalves (Andrusov 1897, Meisenheimer 1901, Younge and Campbell 1968, Morton 1970, Starobogatov 1994). Representatives of fossil dreissenids have been found in Eocene and Oligocene deposits of central and western Europe and also in Miocene deposits along the north Mediterranean basin and in deposits of the Paratethys basin (Nuttall 1990).

Recent molecular phylogenetic analyses (Park and O'Foighil 2000) are consistent with the hypothesis of Starobogatov (1994) that Dreissenidae appeared around the Jurassic–Cretaceous time with common ancestors of Corbiculidae (e.g., invasive genus *Corbicula*) and Sphaeriidae (e.g., noninvasive fingernail clams). Molecular analysis by Gelembuik et al. (2006) determined that a split in the Dreissenidae lineage occurred in the Tethys Sea period and coincided approximately with the Cretaceous–Tertiary mass extinction event 63 million years ago. During the late Miocene–Pliocene period 3–5 million years ago, the family flourished in the Paratethys basin, presently comprised of basins of the Black, Azov, and Caspian Seas, and collectively termed the Pontocaspian region. In this period, the basin was subjected to wide fluctuations in specific salinity (from fresh to mesohaline) as various geologic events led to alternating periods of large-lake isolations and open connections to the Mediterranean basin and the world ocean. This created many ecological niches previously not available to marine and freshwater seston feeders, such as those in the family Dreissenidae. As a result, numerous ancestral forms characterized by high phenotypic (and probably genetic) plasticity evolved into a wide spectrum of ecotypes (Andrusov 1897, Nevesskaya 1971, Baback 1983, Nuttall 1990, Starobogatov 1994, Geary et al. 2001). Hence, present-day dreissenids evolved out of a geographic region that experienced a wide "mix" of evolutionary events, variable water salinities, and increased ecological niches.

There are three genera within the family Dreissenidae—*Mytilopsis*, *Congeria*, and *Dreissena*—and within the genus *Dreissena*, there are two primary lineages as defined by morphological studies and phylogenetic trees based on DNA studies: *Dreissena* and *Pontodreissena* (Andrusov 1897, Therriault et al. 2004, Gelembuik et al. 2006). The former lineage includes *D. polymorpha* and is characterized by species and subspecies that historically inhabited coastal zones, estuaries, rivers, and smaller lakes (Nuttall 1990, Rosenberg and Ludyanskiy 1994, Starobogatov 1994). The latter lineage includes *D. r. bugensis* and is characterized by species and subspecies that inhabited more lacustrine, deeper-water environments of ancient large lakes, and inhabit the modern-day Caspian Sea.

There are several hypotheses regarding the evolutionary origin of *D. r. bugensis*, but all attribute the area of origin to the Black Sea basin. *D. r. bugensis* is a subspecies of *D. rostriformis*, an extant dreissenid species well represented in the fossil records. Fossils of *D. rostriformis* occur in the western part of the Paratethys basin (Pannon Lake) and date back to the late Miocene 3.5 million years ago (Geary et al. 2001). Presently, there are four other recognized subspecies of *D. rostriformis*: *D. r. distincta*, *D. r. pontocaspica*, *D. r. compressa*, and *D. r. gimmi* (Stepien et al. 2013). All four of these subspecies are considered brackish-water taxa. One of these, *D. r. distincta*, invaded the Black Sea basin during the Novoeuxinian epoch (late Pleistocene) where it probably served as an ancestor of *D. r. bugensis*. Despite having different tolerances to salinity, the two subspecies are genetically very closely related (Therriault et al. 2004). The divergence of *D. r. bugensis* from *D. r. distincta* probably occurred during the late Novoeuxinian period (cf. 5000–7500 years ago). There are no fossil records of *D. r. bugensis* before this time, and this was the glacial period when the Black Sea basin was completely formed and regions of low salinity/freshwater were permanently established (Mordukhai-Boltovskoi 1960). These regions in the Black Sea basin (termed limans) resulted from river valley impoundment (natural headwater barricades) and creation of river estuaries. *D. r. distincta* inhabited cold, profundal regions of ancient large lakes and is still found in the profundal zone of the Caspian Sea (Baback 1983, Orlova et al. 2005). It seems *D. r. bugensis* retained this ability to inhabit profundal regions as it evolved from *D. r. distincta* and adapted to freshwater.

D. r. bugensis was first recorded in the Bug River liman in the northern portion of the Black Sea and was described as *D. rostriformis* (Andrusov 1890) but later described as *D. bugensis* (Andrusov 1897). Recently, *D. bugensis* was determined to be genetically similar to *D. rostriformis* and thus not a separate species (Therriault et al. 2004). Yet there are great differences in habit preferences of the two taxa. In particular, *D. bugensis* is found only in freshwater and in waters of low salinity (<3 ppt), and *D. rostriformis* and associated subspecies are only found in waters that are mesohaline (5–13 ppt). Although *D. bugensis* was not genetically separate from *D. rostriformis*, Therriault et al. (2004) considered it a freshwater subspecies of *D. rostriformis* and hence provided the provisional name *D. r. bugensis*. Based on reexamination of collections taken at the end of the nineteenth century, *D. r. bugensis* was also present in the Dnieper River liman in the northern portion of the Black Sea (Son 2007a). Thus, the native range of *D. r. bugensis* is presently defined as the entire Dnieper–Bug estuary system, including the Dnieper River delta and the lower Inguletz River.

GENERAL SPREAD (DREISSENIDAE)

The Pontocaspian region, including the native region of *D. r. bugensis*, is considered a main source of aquatic invaders to inland waters of both Eurasia and North America (Reid and Orlova 2002). The region is located in southeastern Europe in arid and semiarid climatic zones (Figure 7.1). In 1952, the Taganrog Gulf of the Azov and Black Seas were connected to the Caspian basin via the Volga–Don Canal, and thus allowed an exchange of species between these three water bodies/basins. All three water bodies contain a variety of salinity conditions ranging from freshwater to salinity values up to 26 ppt in the Black Sea. Freshwater to water of relatively low salinity occurs in areas where freshwater mixes with marine water near large rivers, such as the Kuban, Danube, and Dniester Rivers in the Black Sea, the Don River in the Azov Sea, and Volga, Ural, and Terek Rivers in the Caspian Sea.

At present, two modern faunas in the Pontocaspian region are of great interest to invasion biologists: liman-relict species and Caspian species. The former group of species has been the most common invasives of inland waters, while the latter group consists of autochthonous Caspian species that are basically confined to the Caspian Sea and to the lower portions of the Volga River. These two groups can be genetically differentiated. For instance, phylogeographic studies of genetic divergences in populations of dreissenids from the Caspian Sea and estuaries and rivers of the Pontocaspian region have shown individuals representing mt DNA haplotypes are characteristic to the Black and Azov Seas and associated estuaries of major rivers (May et al. 2006). Individuals from these areas are evidently highly invasive and have spread to northern Europe and North America. In contrast, mt DNA haplotypes unique to dreissenids in the Caspian Sea are restricted to the Caspian Sea and Volga River upstream to about Samara Luka. The only exception to this distribution is reported by Voroshilova (2009) for *D. polymorpha* in Rybinsk Reservoir, which is attributed to peripheral speciation.

In Europe, there are three invasion corridors that serve as major routes of species spread from the Pontocaspian region (Figure 7.1): (1) an eastern corridor composed of a long route that extends along the Volga River north from the Caspian Sea and a shorter route that extends east then north from the Azov Sea along the Don River; (2) a central corridor composed of the Dnieper River and along the northern shores of the Black Sea; and (3) a southern corridor that originates in the Black Sea and extends west up to the Danube, Main, and Rhine Rivers. Although some differences occur as to which invasion route(s) was used by individual species, the literature clearly indicates that these three corridors were the primary routes that facilitated invasions of Pontocaspian fauna (Son 2007a,b). One particular region, the Dnieper–Bug Liman estuarine system in the northern Black Sea, has a great number of endemic estuarine relict invertebrate species besides *D. r. bugensis* that have become the source of Pontocaspian invasions for inland waters of the Holarctic region (Mordukhai-Boltovskoi 1960). As noted, this region has been subjected to large temporal and spatial variability of conditions (e.g., geologic structures, salinity, temperature, stratification of water masses), which produced evolutionary divergence in a relatively small area.

Natural mechanisms that could have facilitated the spread of *D. r. bugensis* include: (1) movement upstream in rivers as transported by other fauna (i.e., waterfowl); (2) movement laterally along coasts of the Black, Azov, and Caspian Seas as transported by fauna and debris carried in estuarine currents; and (3) movement through interbasin and local canal connections during flood events (Zhuravel 1951, Pligin 1984, Kharchenko 1995, Mills et al. 1996, Son 2007a,b). However, the vast majority of mechanisms for spread of *D. r. bugensis* have been facilitated by human disturbance. For example, construction of the Severokrymsky canal resulted in the

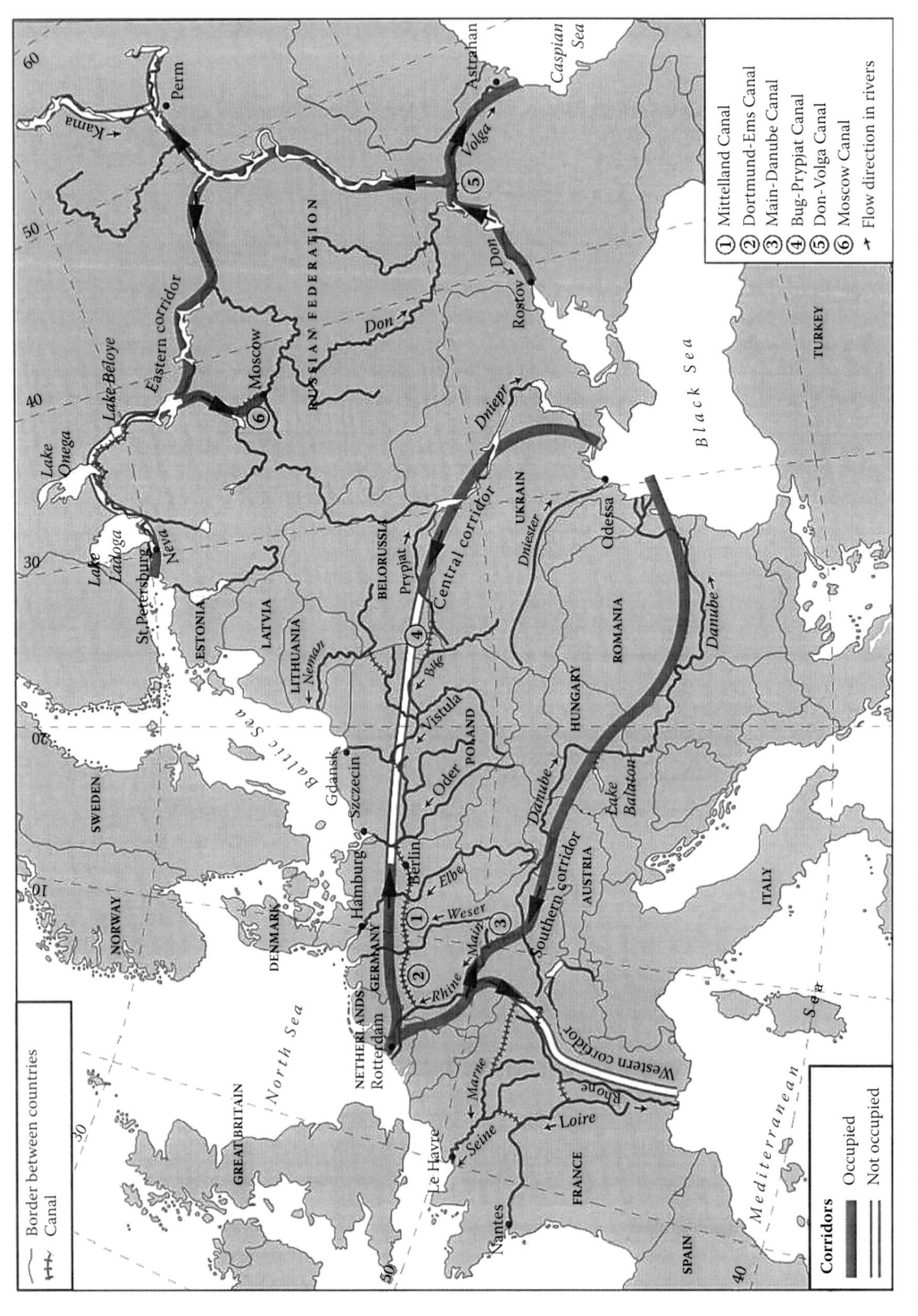

Figure 7.1 Location of three corridors that serve as routes for the spread of quagga mussels (*Dreissena rostriformis bugensis*) and other Pontocaspian taxa. Southern corridor, Danube and Main–Rhine Rivers; central corridor, Dnieper River; and eastern corridor, Don and Volga Rivers. The eastern corridor is discussed in this chapter, while the other corridors are discussed in (bij de Vaate et al. 2013).

spread of *D. r. bugensis* into Dnieper basin, some water bodies of the Crimea peninsula, and into the Dniester Liman and Dniester River delta (Stadnichenko 1979, Son 2007a).

Prior to the 1940s, *D. r. bugensis* was restricted to its native range but thereafter began to spread to the lower part of Dnieper River basin (Zhuravel 1951) and then to the Ponto-Azov area (Figure 7.1). The expansion of *D. r. bugensis* was facilitated by removal of natural geographic barriers by construction of irrigation canals, which resulted in rapid invasion of corridors along coasts of the north Black Sea (Son 2007a). In addition, impoundment of rivers and changes in freshwater and estuarine habitats of reservoirs predisposed many water bodies to colonization and range expansion (Mills et al. 1996, Orlova et al. 2005, Son 2007a,b). Human activities undoubtedly increased rates of dreissenid spread in all three major geographic corridors because each has had some form of human construction that would have contributed to the spread of dreissenid mussels. Examples include the following: (1) formation of a cascade of dams, reservoirs, and river modifications in the most eastern corridor in the 1940s provided a gateway for the spread of dreissenids and other fauna; (2) irrigation canals and impoundments in the Dnieper River in the central corridor allowed dreissenids to rapidly expand soon after these structures were created (Zhuravel 1951, Tzeyeb et al. 1966, Pligin 1984, Mills et al. 1996).

Support for increased rate of spread caused by human activities is found in the speed and intensity by which quagga mussels invaded rivers in these invasion corridors. Mussel spread appears to have occurred quickly by a large number of individuals as shown by the absence of genetic "bottlenecks" of founding populations. Genetic diversity (by polymorphic microsatellite loci) of quagga mussel populations from the lower Danube River reported by Popa et al. (2009) was high compared with those of *D. polymorpha*. Therefore, differentiation between populations was low and no founder effect was detected. Popa et al. (2009) attribute their results to two factors: (1) a large number of individuals in newly established populations that retained a substantial proportion of original genetic diversity of the invading population and (2) multiple colonization events similar to east European populations (Therriault et al. 2005). Large founding numbers of mussels into new habitats through natural mechanisms are unlikely to occur because of the small number of mussels likely to be carried by migrating fauna and drift debris, and there are no known instances where natural flooding between waterways facilitated spread of mussels.

Eastern Invasion Corridor

As noted, the expansion of *D. r. bugensis* in eastern Europe was facilitated by the construction of canals, dams, and reservoirs. In particular, completion of the Don–Volga Canal in 1952 connected the Ponto-Azov basin (Black and Azov Seas) with the Caspian basin (Volga River, Caspian Sea) and provided a pathway for quagga mussels to expand outside its native range. Also, a series of dams and reservoirs were constructed along the Volga River between the 1950s and early 1980s, which provided lacustrine conditions preferred by this species. First reports of *D. r. bugensis* east of its historical range (Ponto-Azov basin) were in the Volga River system in 1992 at multiple sites in the Kuybyshev Reservoir, which is in the middle reaches of the Volga River (Antonov 1993). However, reexamination of archived specimens showed that quagga mussels were actually present in the lower reaches of the river in 1981 near the town of Akhtubinsk (Zhulidov et al. 2004). While it is unclear why it took so long for this species to spread in the Volga River if found as early as 1981, it was likely related to the limited number of surveys conducted over that time period. In the decade after 1992, it spread widely through the system. In the lower reaches of the river, it was found in the Volga Delta in 1994 and in the northern portion of the Caspian Sea in 1996 (Orlova et al. 2004) where salinities varied from freshwater to 3 ppt.

Quagga mussels were first recorded in the upper Volga River in 1997 in Rybinsk Reservoir and by 2000 were found in several other reservoirs downstream, such as Gorkov and Uglich Reservoirs. By 2001, it was found in seven of nine reservoirs in the system, and its range extended 3000 km from Rybinsk Reservoir in the north to the Caspian Sea delta in the south (Orlova and Scherbina 2002, Orlova and Panov 2004, Orlova et al. 2004a, 2005, Therriault et al. 2005, Zhulidov et al. 2004, 2005, 2006, Son 2007b). In 2003, specimens of *D. r. bugensis* were found in the Moskva River in waterways of Moscow (Lvova 2004) and in 2003–2004 in the Izh River, about 30 km upstream from the Nizhnekamskoye Reservoir (Zhulidov et al. 2010). In 2007, quagga mussels were discovered in Volga River tributaries and in Sharony Lake, which is connected to the Volga River by an irrigation canal (Zhulidov et al. 2010).

The rapid invasion of the Volga River by quagga mussels is supported by genetic information from populations in the Volga River and Caspian Sea. This information suggests that the rapid spread was a result of multiple introductions from various sources, which included localities from within the Ponto-Azov basin and within the Volga River system itself (Wilson et al. 1999, Therriault et al. 2005). Also, the spread within the Volga River system was probably facilitated by large spatial and genetic variability of parental and larval populations (Orlova et al. 2004, 2005). These characteristics of multiple colonization events into the Volga River basin and large sizes of inoculated populations agree with studies for both species in the Great Lakes of North America (Wilson et al. 1999).

QUAGGA MUSSELS VERSUS ZEBRA MUSSELS

To date, quagga and zebra mussels appear to utilize the same vectors for range expansion. However, there are four differences in the spread of quagga and zebra mussels in eastern Europe: (1) Quagga mussels spread slower than zebra mussels, and therefore, to date, the range of quagga mussels is less than that of zebra mussels; (2) quagga mussels "naturalize" to new habitats faster than zebra mussels that allow quagga mussels to gain dominance over previously established populations of zebra mussels in some habitats (Tzeyeb et al. 1966, Zhulidov et al. 2006, 2010, and as follows); (3) first records of quagga mussels are always reported from established beds of zebra mussels; and (4) in contrast to zebra mussels, quagga mussels colonize deepwater habitats (Tzeyeb et al. 1966, Orlova et al. 2005). Examples that illustrate these differences in spread characteristics are numerous. For example, in the Dnieper River, zebra mussels were present and thrived before the arrival of quagga mussels and before the river was transformed into a cascade of reservoirs (Zhuravel 1934). This supports the theory that zebra mussels colonize new and unaltered river systems faster than quagga mussels (Mordukhai-Boltovskoi 1960, Mills et al. 1996, Orlova et al. 2005). In addition, once quagga mussels arrived in deeper reservoirs of the Dnieper River (e.g., Dnieprovskoe and Kakhovskoe Reservoirs), quagga mussels quickly replaced zebra mussels as the dominant dreissenid mussel, probably as a result of "naturalization" to local conditions and competition for available foods (Tzeyeb et al. 1966). Instances where quagga mussels "naturalize" to new habitat conditions are quite common, whereas zebra mussels appear to lack this capability, at least in the presence of quagga mussels (Tzeyeb et al. 1966, Zhulidov et al. 2006, 2010). In the deeper reservoirs of the Dnieper River, quagga mussels occupy greater depths than zebra mussels, up to depths of 28 m where temperatures are 10°C lower than the shallower areas throughout the summer (Mills et al. 1996). The time period for quagga mussels to replace zebra range appears to be between 5 and 10 years (Tzeyeb et al. 1966, Mills et al. 1996). This period of time is about the same time as that required for environmental conditions in newly created reservoirs to stabilize (Tzeyeb et al. 1966).

At the same time quagga mussels replaced zebra mussels in older and deeper reservoirs, zebra mussels occurred extensively and quagga mussels occurred sporadically in newly constructed and shallower reservoirs of the Dnieper River system (Tzeyeb et al. 1966).

There are many examples of the expansion of quagga mussels relative to zebra mussels throughout the Volga River system. When quagga mussels became established in the Volga River Delta, Saratov Reservoir (lower Volga River), Kuybyshev Reservoir (middle Volga River), and the Rybinsk Reservoir, it took 3–10 years to become the dominate dreissenid and comprise nearly 100% of all dreissenids found (Figure 7.2).

The basis for observed "naturalization" to habitat conditions by quagga mussels and not by zebra mussels was developed from studies of other taxa. Khalaman (2008) explained successive changes in associations of biofouling invertebrates by emulating adaptive strategies observed for weeds (Grime 2001). Zebra and quagga mussels are slightly divergent in their life and colonization strategies, despite both being predominantly competitors with other invertebrates. Zebra mussels are typically "violent" (i.e., competitive, k-strategist) invertebrates, whereas quagga mussels are more "explerency" (i.e., r-strategist, able to adjust to new habitats and use available foods more efficiently). These traits may explain why zebra mussels spread more successfully, especially to areas rich with available resources (e.g., plankton from eutrophication), while quagga mussels have a greater capacity to tolerate unfavorable trophic conditions, including biological oligotrophy that can occur after zebra mussels become established (Wilson et al. 2006) and a greater flexibility/capacity to colonize a wider variety of habitats.

Another explanation for the "naturalization" advantage of quagga mussels is that populations of this species typically increase after other invaders have become established, as consistent with the invasion "meltdown" theory (Simberloff and Von Holle 1999, Ricciardi 2001).

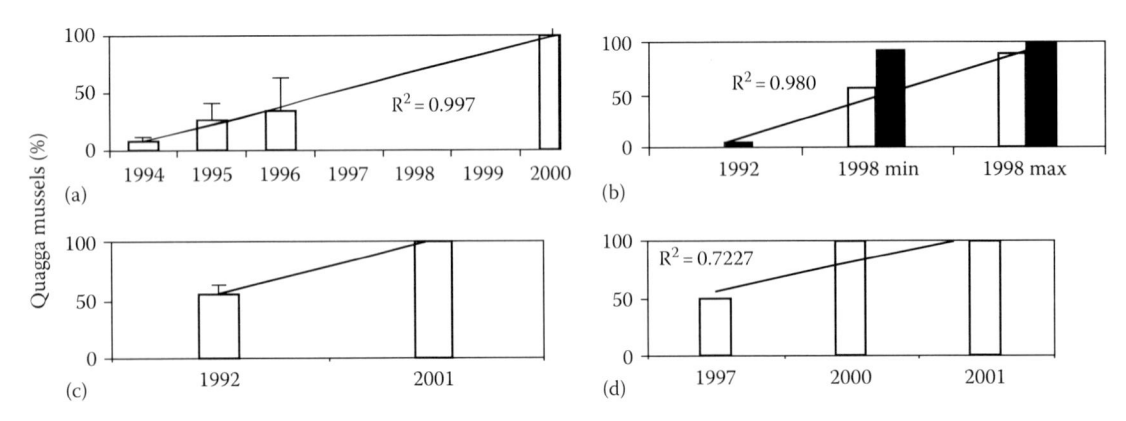

Figure 7.2 Change in relative percentage of quagga mussels (*Dreissene rostriformis bugensis*) in the total dreissenid population at various locations along different reaches of the Volga River. Samples collected in September of each year. (a) Volga River Delta, (b) Saratov Reservoir (black bar) and nearby relict lake (white bar) in the lower Volga River, (c) Kuybyshev Reservoir in the middle Volga River, and (d) Rybinsk Reservoir in the upper Volga River. (From Orlova, M.I. et al., *Aquat. Ecol.*, 38, 561, 2004.)

In this paradigm, an established habitat engineer (e.g., zebra mussel) causes biological oligotrophy in the water column that favors another invasive species (e.g., quagga mussel). In general, quagga mussels are more adapted to oligotrophic-like conditions than zebra mussels, so the observed "oligotrophication" of waters after establishment by zebra mussels provides circumstances that favor quagga mussels (Baldwin et al. 2002, Stoeckmann 2003). Therefore, quagga mussels are likely to be the ultimate successional species as it spreads and replaces zebra mussels.

There are several notable exceptions to these generalizations about the two species. The first is the range expansion of quagga mussels in North America, where it became established and rapidly spread in the southwestern region of the United States despite zebra mussels having a wider range in distribution at the time (Benson 2013). The introduction was a result of mussels being transported long distances overland as associated with a trailered boat. As in this case, when quagga mussels arrive before zebra mussels, they may "naturalize" the habitat that may prevent colonization by zebra mussels. The other exception is in the lower Don River system, Russia (Zhulidov et al. 2006). For unknown reasons, quagga mussels increased relative to zebra mussels for a number of years after introduction but then unexpectedly declined.

COMPARATIVE SIZE STRUCTURE: QUAGGA MUSSELS AND ZEBRA MUSSELS

Length–frequency distributions of quagga and zebra mussels can change rapidly in the early years of an invasion. Both zebra mussels and quagga mussels invaded the lower Volga Delta in the early-1990s, and populations on perennial plants were sampled in 1994, 1995, and 1996 (Figure 7.3). In 1994, sizes of zebra mussels ranged from 1 to 27 mm. Individuals <10 mm occurred in greatest frequency, but the presence of individuals up to 27 mm indicated the species had become established several years earlier. In contrast, only a few small (1.5 mm) quagga mussels were found in 1994, which indicated only recent establishment. By 1996, both species exhibited relatively continuous length distributions, and the size of greatest frequency was 10–11 mm for zebra mussels and 14–15 mm for quagga mussels, which seems to indicate a faster growth rate for quagga mussels in the cohort settled the previous year. When considering only individuals >5 mm, differences in size structure of populations of quagga mussels and zebra mussels are readily apparent in areas where the two species co-occurred over long periods. In the delta area of the Dnieper River along a riverbed slope, the quagga mussel population had a number of distinct cohorts, with peak frequencies at 8–9, 12–13, 17–18, and 24 mm (Figure 7.4). In contrast, the zebra mussel population displayed only one distinct cohort at 11–12 mm (Figure 7.4).

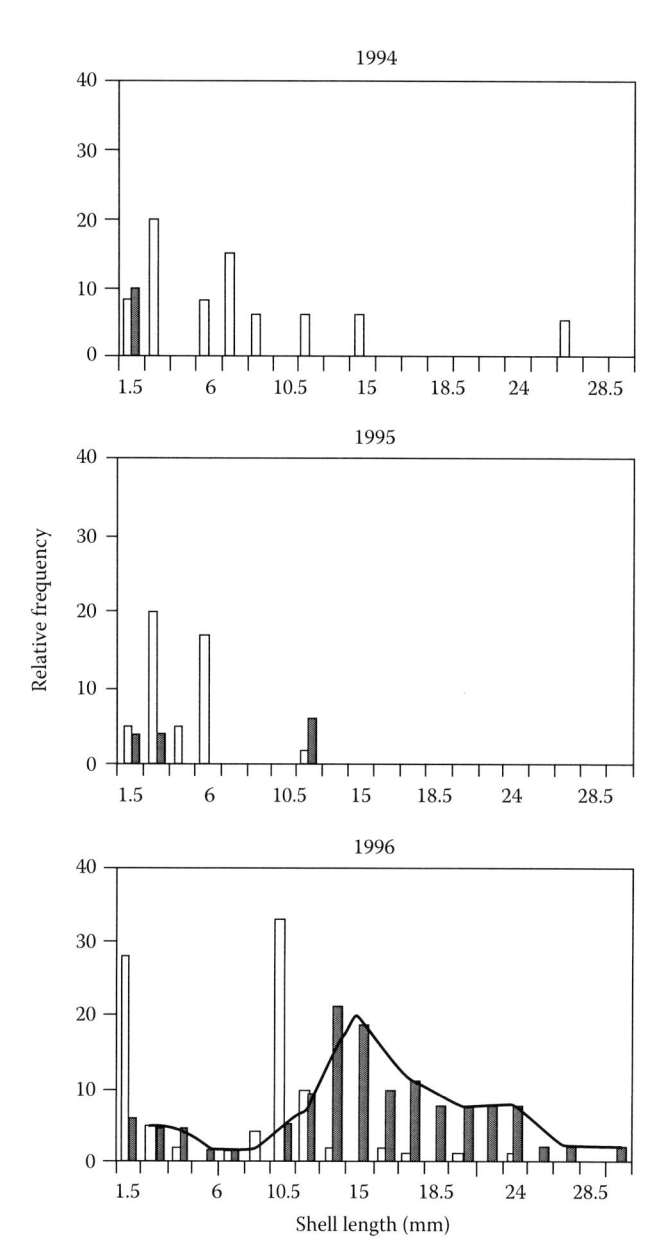

Figure 7.3 Relative size frequency (shell length, mm) of quagga mussels (black bars) and zebra mussels (white bars) in 1994, 1995, and 1996 soon after both species invaded the Volga River Delta.

In the Kuybyshev Reservoir, populations of both species displayed two distinct cohorts, but corresponding cohorts were distinctly larger in shell length for quagga mussels than for zebra mussels. For quagga mussels, sizes of the two cohorts were 16 and 24–25 mm, whereas for zebra mussels, sizes were 12–13 and 22 mm. Assuming these cohorts represent comparable year classes, growth rates of quagga mussels were distinctly greater for quagga mussels than for zebra mussels.

As noted, quagga mussels became established in Rybinsk Reservoir in the upper portion of the Volga River in 1997. By 2000, it had spread to all areas of this reservoir. Yet despite a wide distribution, there were great spatial variations in

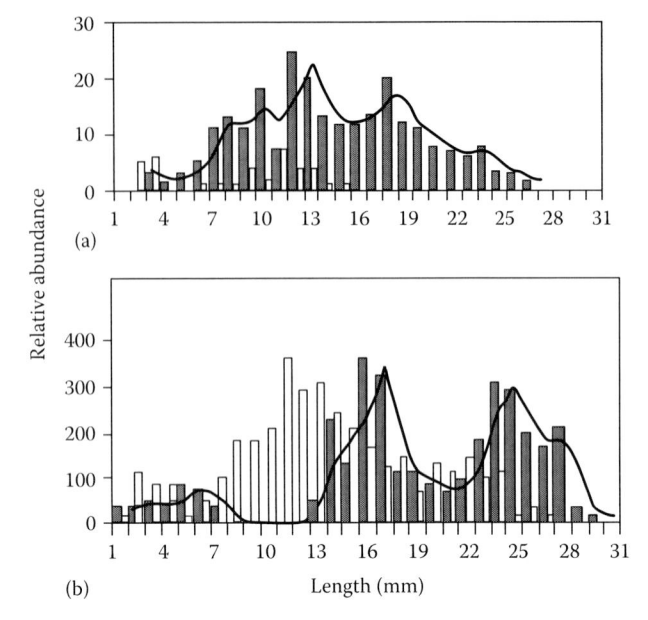

Figure 7.4 Relative size structure (shell length, mm) of quagga mussels (black bars) and zebra mussels (white bars) in the Dnieper River Delta (a) and Kuybyshev Reservoir (b). Samples in both locations were collected on the riverbed slope. Populations in the Dnieper River Delta and Kuybyshev Reservoir typified size structures in native and invasion ranges, respectively.

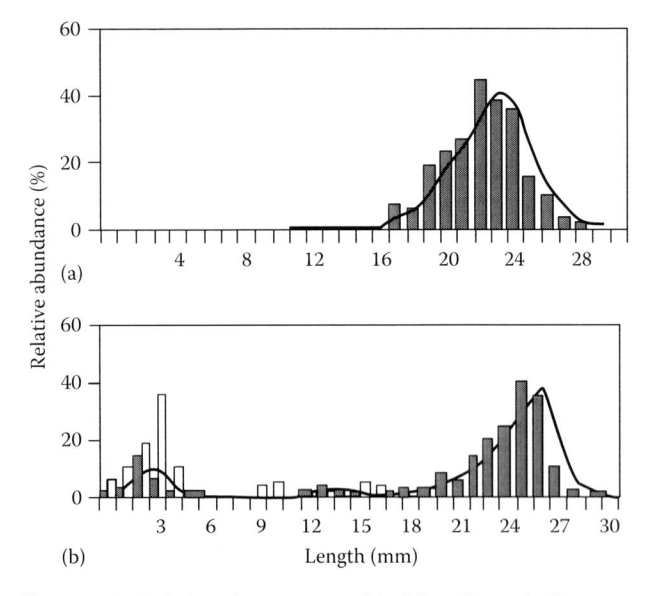

Figure 7.5 Relative size structure (shell length, mm) of quagga mussels (black bars) and zebra mussels (white bars) in two proximate sites 4 years after their colonization in the Rybinsk Reservoir, July 2001. (a) Riverbed slope and (b) shallow submerged tree stumps.

size–frequency distributions. The size structure of quagga mussel populations found at two proximate sites with different substrates in 2001 is given in Figure 7.5. At both sites, there was a peak in individuals at 22–26 mm, which likely represented the first, large recruitment after establishment several years earlier. However, smaller individuals were present at one site but were absent at the other. This would indicate that site-specific recruitment was highly variable, and that veliger drift and settlement within the reservoir was not uniform. The site with no small individuals was located on the slope of the riverbed where the current was likely greater than at the other site, which may have prevented settlement. Yet it does not explain the presence of large individuals that likely settled there several years earlier.

SUMMARY

To date, quagga mussels occupy the eastern European countries of Russia, Ukraine, Moldavia, Romania, Hungary, and Bulgaria. These countries are situated in the Pontocaspian region and are in catchment areas of basins that form the three major invasion corridors utilized by dreissenid mussels to spread in eastern Europe. Mussels occur in most mainstream portions of the invasion corridors that are often regulated by dams (e.g., Dnieper and Volga Rivers) to transform rivers into cascades of reservoirs that provide habitats for mussel colonization. In Europe, quagga mussels have mostly been reported in relatively shallow waters compared to reports in the deeper, colder waters of the Laurentian Great Lakes in North America. Recently, however, the profundal phenotype of the quagga mussel has been found in a reservoir in the middle reaches of the Volga River system (Pavlova 2012). Variations of morphological characteristics of quagga mussels are greater than in co-occurring zebra mussels; that is, the zebra mussel does not possess a deepwater morph as does the quagga mussel (Claxton et al. 1998). Hence the extended distribution of quagga mussels into deeper waters is not unexpected (Bogutskaya et al. 2013 in Press). Ultimately, quagga mussels appear to be a more dominant successional species than zebra mussels, and as such, they are likely to be the most common species found in relatively stable habitat conditions in Europe and North America.

REFERENCES

Andrusov, N. I. 1890. *Dreissena rostriformis* Desh. in the Bug River. Trudy S-Peterburgskago Obschestva Estestvoispitatelei. *Dept. Geol. Mineral.* 6: 1–2. (In Russian.)

Andrusov, N. I. 1897. Fossil and recent Dreissenidae of Eurasia. Trudy Sankt-Peterburgskago Obschestva Estestvoispitatelei. *Dept. Geol. Mineral.* 25: 1–683. (In Russian with German summary.)

Antonov, P. I. 1993. About invasion of bivalve *Dreissena bugensis* (Andr.) into the Volga River Reservoirs. Ecological problems of large rivers basins. In *Book of Abstracts of International Conference*, Togliatti, Russia, September 6–10, 1993. Togliatti, Russia: Publishers of IEVB RAS. pp. 52–53. (In Russian.)

Baback, E. V. 1983. The pliocene and quaternary Dreissenidae of the Euxinian Basin. In *Proceedings of Palaeontological Institute Academy of Sciences of USSR*, Moscow, Russia, pp. 1–204. (In Russian.)

Baldwin, B. S., M. S. Mayer, J. Dayton et al. 2002. Comparative growth and feeding in zebra and quagga mussels (*Dreissena polymorpha* and *Dreissena bugensis*): Implications for North American lakes. *Can. J. Fish. Aqua. Sci.* 59: 680–694.

bij de Vaate, A., K. Jazdzewski, H. A. M. Ketelaars, S. Gollasch, and G. Van der Velde. 2002. Geographical patterns in range extension of Ponto-Caspian macroinvertebrate species in Europe. *Can. J. Fish. Aquat. Sci.* 59: 1159–1174.

bij de Vaate, A., G. Van der Velde, R. S. E. W. Leuven, and K. C. M. Heiler. 2013. Spread of the quagga mussel, *Dreissena rostriformis bugensis*, in western Europe. In *Quagga and Zebra Mussels: Biology, Impacts, and Control*, 2nd edn., T. F. Nalepa and D. W. Schloesser, Eds., pp. 83–92. Boca Raton, FL: CRC Press.

Bogutskaya, N.G., Kiyashko, P.I., Orlova, M.I., and Naseka A.M. 2013. Keys to fish and invertebrates of the Caspian see. Vol. 1. Fishes and molluscs. Moscow: KMK Scientific Publishers (in press) (in Russian).

Claxton, W.T., Wilson, A.B., Mackie, G.L., and Boulding, E.G. 1998. A genetic and morphological comparison of shallow- and deep-water populations of the introduced dreissenid bivalve *Dreissena bugensis*. *Can. J. Zool.* 76: 1269–1276.

Diggins, T. P., M. Weime, K. M. Stewart et al. 2004. Epiphytic refugium: Are two species of invading freshwater bivalves partitioning spatial resources? *Biol. Invasions* 6: 83–88.

Geary, D. H., I. Magyar, and P. Muller. 2001. Ancient Lake Pannon and its endemic molluscan fauna (Central Europe; Mio-Pliocene). *Adv. Ecol. Res.* 31: 463–482.

Gelembuik, G. W., G. E. May, and C. E. Lee. 2006. Phylogeography and systematics of zebra mussels and related species. *Mol. Ecol.* 15: 1033–1050.

Grime, J. P. 2001. *Plant Strategies, Vegetation Processes, and Ecosystem Properties*, 2nd Edn. New York: Wiley & Sons.

Hebert, P. D. N., B. W. Muncaster, and G. L. Mackie. 1989. Ecological and genetic studies on *Dreissena polymorpha* (Pallas): A new mollusc in the Great Lakes. *Can. J. Fish. Aquat. Sci.* 46: 1587–1591.

Khalaman, V. V. 2008. Succession of fouling communities and interspecific interrelationships between fouling organisms in the White Sea. Abstract of DSc thesis. St. Petersburg, Russia, 48p. (In Russian.)

Kharchenko, T. A. 1995. *Dreissena*: Range, ecology, fouling impacts. *Hydrobiol. J.* 31: 3–20. (In Russian.)

Kinzelbach, R. 1992. The main features of the phylogeny and dispersal of the zebra mussel *Dreissena polymorpha*. *Limnol. Aktuell* 4: 5–17.

Lvova, A. A. 2004. On penetration of *Dreissena bugensis* (Bivalvia, Dreissenidae) into the Ucha reservoir (Moscow province) and the Moskva River. *Zool. Zhurn.* 83: 766–768. (In Russian.)

Marelli, D. and S. Gray. 1985. Comments on the status of recent members of the genus *Mytilopsis* (Bivalvia: Dreissenidae). *Malacol. Rev.* 18: 117–122.

May, G., G. Gelembuik, V. Panov, M. Orlova, and C. Lee. 2006. Molecular ecology of zebra mussel invasions. *Mol. Ecol.* 15: 1021–1031.

Meisenheimer, J. 1901. Entwicklungsgeschichte von *Dreissena polymorpha* Pall. *Ztshr. Wiss. Zooll. Bd.* 69: 1–137.

Mills, E. L., G. Rosenberg, A. P. Spidle, M. Ludyanskiy, Y. Pligin, and B. May. 1996. A review of the biology and ecology of the quagga mussel (*Dreissena bugensis*), a second species of freshwater dreissenid introduced to North America. *Am. Zool.* 36: 271–286.

Minchin, D., F. Lucy, and M. Sullivan. 2002. Zebra mussel: Impacts and spread. In *Invasive Aquatic Species of Europe: Distribution, Impacts and Spread*. E. Leppäkoski, S. Gollasch, and S. Olenin, Eds., pp. 135–146. Dordrecht, The Netherlands: Kluwer Press.

Mordukhai-Boltovskoi, F. D. 1960. *Caspian Fauna in Azov and Black Sea Basin*. Moscow, Russia: Publishers of the Academy of Science, 288 pp. (In Russian.)

Morton, B. 1970. The evolution of the heteromyarian condition in the Dreissenacea (Bivalvia). *Palaeontology* 13: 563–572.

Nevesskaya, L. A. 1971. To the classification of ancient closed and semiclosed water basins to base on the character of their fauna. *Tr. Paleont. Inst. Acad. Nauk. SSSR* 130: 258–279. (In Russian.)

Nowak, E. 1971. The range expansion of animals and its cause. *Zeszyty Naukowe* 3: 1–255. Translation by the Smithsonian Institution and National Science Foundation, Washington, DC. Foreign Science Publication Department of the National Center for Science, Technology, and Economic Information, Warsaw, Poland, 1975: pp. 1–163.

Nuttall, C. P. 1990. Review of the Caenozoic heterodont bivalve superfamily Dreissenacea. *Palaeontology (Lond.)* 33: 707–737.

Orlova, M. I., J. R. Muirhead, P. I. Antonov et al. 2004. Range expansion of quagga mussels *Dreissena rostriformis bugensis* in the Volga river and Caspian sea basin. *Aquat. Ecol.* 38: 561–573.

Orlova, M. I. and V. E. Panov. 2004. Establishment of the zebra mussel, *Dreissena polymorpha* (Pallas) in the Neva Estuary (Gulf of Finland, Baltic Sea): Distribution, population structure and possible impact on local unionid bivalves. *Dev. Hydrobiol.* 176: 207–217.

Orlova, M. I. and G. H. Scherbina. 2002. About range expansion of *Dreissena bugensis* Andr. (Dreissenidae, Bivalvia) in Upper Volga Reservoirs. *Zool. J.* 81: 515–520. (In Russian.)

Orlova, M. I., T. W. Therriault, P. I. Antonov, and G. H. Shcherbina. 2005. Invasion ecology of quagga mussels (*Dreissena rostriformis bugensis*): A review of evolutionary and phylogenetic impacts. *Aquat. Ecol.* 39: 401–418.

Park, J. K. and D. O'Foighil. 2000. Sphaeriid and corbiculid clams represent separate heterodont bivalve radiations into freshwater environments. *Mol. Phylogenet. Evol.* 14: 75–88.

Pavlova, V. 2012. First finding of deepwater profunda morph of quagga mussel *Dreissena bugensis* in the European part of its range. *Biol. Invasions* 14: 509–514.

Pligin, Y. 1984. Extension of the distribution of *Dreissena bugensis*. *Malacol. Rev.* 17: 143–144.

Popa, O., T. Trichkova, D. Kozuharov, Z. Hubenov, and L. Popa. 2009. Of *Dreissena bugensis* in the Lower Danube Basin. *Aquatic Biodiversity International Conference*, 2009, Sibiu, Romania, Book of Abstracts.

Reid, D. F. and M. I. Orlova. 2002. Geological and evolutionary underpinnings for the success of Ponto-Caspian species invasion in the Baltic Sea and North American Great Lakes. *Can. J. Fish. Aquat. Sci.* 59: 1144–1158.

Ricciardi, A. 2001. Facilitative interactions among aquatic invaders: Is an "invasional meltdown" occurring in the Great Lakes? *Can. J. Fish. Aquat. Sci.* 58: 2513–2525.

Rosenberg, G. and M. L. Ludyanskiy. 1994. A nomenclatural review of *Dreissena* (Bivalvia: Dreissenidae) with identification of the quagga mussel as *Dreissena bugensis*. *Can. J. Fish. Aquat. Sci.* 51: 1474–1484.

Simberloff, D. and B. Von Holle. 1999. Positive interactions of nonindigenous species: Invasional meltdown? *Biol. Invasions* 1: 21–32.

Son, M. O. 2007a. *Invasive Molluscs in Fresh and Brackish Waters of the Northern Black Sea Region*. Odessa, Ukarine: Druk Press. (In Russian.)

Son, M. O. 2007b. Native range of zebra mussel and quagga mussel and new data on their invasions within the ponto-caspian region. *Aquat. Invasions* 3: 174–184.

Stadnichenko, A. P. 1979. A review of the fauna of freshwater molluscs of Crimea. *Vestnik Zool.* 13: 44–49. (In Russian.)

Starobogatov, Y. I. 1994. Taxonomy and palaeontology. In *Dreissena polymorpha (Pall,.) (Bivalvia, Dreissenidae). Taxonomy, Ecology and Practical Meaning*, Y. I. Starobogatov, Ed., pp. 18–46. Moscow, Russia: Nauka Press. (In Russian.)

Starobogatov, Y. I. and S. I. Andreeva. 1994. Range. In *Dreissena polymorpha (Pall,.) (Bivalvia, Dreissenidae). Taxonomy, Ecology and Practical Meaning*, Y. I. Starobogatov, Ed., pp. 47–53. Moscow, Russia: Nauka Press. (In Russian.)

Stepien, C. A., I. A. Grigorovich, D. J. Murphy et al. 2013. Evolutionary, biogeographic, and population genetic relationships of dreissenid mussels, with revision of component taxa. In *Quagga and Zebra Mussels: Biology, Impacts, and Control*, 2nd Edn., T. F. Nalepa and D. W. Schloesser, eds., pp. 403–444. Boca Raton, FL: CRC Press.

Stoeckmann, A. 2003. Physiological energetics of Lake Erie dreissenid mussels: A basis for the displacement of *Dreissena polymorpha* by *Dreissena bugensis*. *Can. J. Fish. Aquat. Sci.* 60: 126–134.

Therriault, T. W., M. F. Docker, M. I. Orlova, D. D. Heath, and H. J. MacIsaac. 2004. Molecular resolution of Dreissenidae (Mollusca: Bivalvia) including the first report of *Mytilopsis leucophaeata* in the Black Sea basin. *Mol. Phyl. Evol.* 30: 479–489.

Therriault, T. W. and M. I. Orlova. 2010. Invasion success within the family Dreissenidae: Prerequisites, mechanisms and perspectives. In *The Zebra Mussel in Europe*, G. Van der Velde, S. Rajagopal, and A. bij de Vaate, eds., pp. 59–68. Leiden, Holland: Backhuys Publishers.

Therriault, T. W., M. I. Orlova, M. F. Docker, H. J. MacIsaac, and D. D. Heath. 2005. Invasion genetics of a freshwater mussel (*Dreissena rostriformis bugensis*) in eastern Europe: High gene flow and multiple introductions. *Heredity* 95: 16–23.

Tzeyeb, Y. Y., A. M. Almazov, and V. I. Vladimirov. 1966. Regularities in the changes of the hydrological, hydrochemical and hydrobiological regimes of the Dnieper River on flow-off regulation and their effect on fish biology and sanitary state of the reservoirs. *Hydrobiol. J.* 2: 3–18. (In Russian.)

Van der Velde, G. and D. Platvoet. 2007. Quagga mussels *Dreissena rostriformis bugensis* (Andrusov, 1897) in the main river (Germany). *Aquat. Invasions* 3: 261–264.

Voroshilova, I. S. 2009. Origin and population structure of peripheral settlements of *Dreissena polymorpha* (Pallas, 1771) from north-east boundary of species range. Abstract of PhD thesis, Borok, 24p. (In Russian.)

Wilson, K. A., E. T. Howell, and J. A. Jackson. 2006. Replacement of zebra mussels by quagga mussels in the Canadian nearshore of Lake Ontario: The importance of substrate, round goby abundance, and upwelling frequency. *J. Great Lakes Res.* 32: 11–28.

Wilson, A. B., K. A. Naish, and E. G. Boulding. 1999. Multiple dispersal strategies of the invasive quagga mussel (*Dreissena bugensis*) as revealed by microsatellite analysis. *Can. J. Fish. Aquat. Sci.* 56: 2248–2261.

Younge, C. M. and J. I. Campbell. 1968. On the heteromyarian condition in the Bivalvia with special Reference to *Dreissena polymorpha* and Certain Mytilacea. *Trans. R. Soc. Edinburgh* 68: 4–42.

Zhulidov, A. V., A. V. Kozhara, G. H. Scherbina et al. 2010. Invasion history, distribution, and relative abundances of *Dreissena bugensis* in the old world: A synthesis of data. *Biol. Invasions* 12: 1923–1940.

Zhulidov, A. V., T. F. Nalepa, A. V. Kozhara, D. A. Zhulidov, and T. Y. Gurtovaya. 2006. Recent trends in relative abundance of two dreissenid species, *Dreissena polymorpha* and *Dreissena bugensis* in the lower Don river system, Russia. *Arch. Hydrobiol.* 165: 209–220.

Zhulidov, A. V., D. F. Pavlov, T. F. Nalepa, G. H. Scherbina, D. A. Zhulidov, and T. Y. Gurtovaya. 2004. Relative distributions of *Dreissena bugensis* and *Dreissena polymorpha* in the lower Don river system, Russia. *Int. Rev. Hydrobiol.* 89: 326–333.

Zhulidov, A. V., D. A. Zhulidov, D. F. Pavlov, T. F. Nalepa, and T. Y. Gurtovaya. 2005. Expansion of the invasive bivalve mollusc *Dreissena bugensis* (quagga mussel) in the Don and Volga River basins: Revisions based on archived specimens. *Ecohydrol. Hydrobiol.* 5: 127–133.

Zhuravel, P. A. 1934. Some notes on changes in fauna of rapid region of the Dnieper River in connection with Dnieproges. *Priroda* 8: 50–56. (In Russian.)

Zhuravel, P. A. 1951. About *Dreissena bugensis* (Mollusca) from the system of the Dnieper River and about its recent appearance in Dneprovskoye Reservoir. *Zool. J.* 30: 186–188. (In Russian.)

CHAPTER 8

Summary of Zebra Mussel (*Dreissena polymorpha*) in Polish Lakes over the Past 50 Years with Emphasis on the Masurian Lakes (Northeastern Poland)

Krzysztof Lewandowski and Anna Stańczykowska

CONTENTS

Abstract 103
Introduction 103
Masurian Lakeland 104
Suwalskie Lakeland 109
Pomeranian Lakeland 110
Present Distribution in Poland 110
Summary and Conclusions 110
Appendix 111
Acknowledgments 112
References 112

ABSTRACT

The ecology of *Dreissena polymorpha* (zebra mussel) was studied over the past 50 years (1959–2010) in lakes of northeastern Poland, especially lakes in the Masurian Lakeland. This species was very common, being found in 60%–100% of lakes studied during the first 30 years of this period. Densities differed between and within lakes over time. Examination of environmental parameters in different lakes indicated densities were not strongly dependent on any single factor but could be related to trophic status. In general, densities were highest in large and deep lakes of mesotrophic and meso-eutrophic status. *D. polymorpha* was absent or occurred in low densities in shallow, hypereutrophic, and polymictic lakes. In some lakes, populations declined to low levels in the 1970s through the 1990s as a result of increased eutrophication. With recent changes in agricultural practices and construction of sewage treatment plants, some recovery is now being observed. In most Masurian lakes, comparative densities were determined primarily by mortality of planktonic veligers during settlement and in the postveliger stage.

Predation by fish and waterfowl was believed to be relatively minor. Currently, we are comparing ongoing changes in populations of *D. polymorpha* in Masurian lakes to changes in other Polish lakelands to determine the importance of landscape features, climate conditions, and water body characteristics on population dynamics.

INTRODUCTION

The invasive species *Dreissena polymorpha* (Pall.) (zebra mussel) is one of the most common bivalves in Polish waters, particularly in lakes of northern Poland (e.g., Masurian Lakeland). Because this species is so widespread and abundant, its biology and ecology have been the subject of numerous studies in Poland as well as in eastern Europe (e.g., Limanova 1978, Stańczykowska 1997). While *D. polymorpha* has been present in northeastern Poland for 200 years, regular studies on the ecology of this species in Poland only started in the 1950s and 1960s (Wiktor 1969). In particular, studies in the Masurian lakes were initiated in 1959–1960 (Stańczykowska

1961) and have continued for over 50 years (e.g., Stańczykowska 1964, 1977, Stańczykowska and Lewandowski 1980, 1997, Lewandowski 1991, 2001, Stańczykowska et al. 2010).

The results of the first 30 years of studies (1959–1990) in the Masurian Lakeland were summarized in a chapter in the first edition of this book (Nalepa and Schloesser 1993, Stańczykowska and Lewandowski 1993a). The chapter reviewed long-term studies in several dozen lakes and focused mainly on distributions and population dynamics (density, biomass, production, and dynamics of planktonic larvae and postveligers), including how population dynamics were related to habitat conditions such as trophic status, depth, and substrate type. These studies also involved some aspects of *D. polymorpha* ecology such as aggregations, individual growth, age structure, filtration, role in nutrient cycling, and relationships with other bivalves such as Unionidae (Lewandowski 1976, 1982, Stańczykowska et al. 1976, Stańczykowska and Planter 1985, Stańczykowska and Lewandowski 1995). Changes in *D. polymorpha* populations observed between 1959 and 1990 were mainly related to changes in environmental conditions (e.g., Gliwicz et al. 1980, Zdanowski and Hutorowicz 1993), whereas intrinsic population and biotic community factors were less important.

The next 20 years of study (1990–2010) were marked by a continuation of earlier studies but were expanded to address new and more specific issues. More attention was paid to populations at the local level and to biotic community factors such as relationships with other malacofauna, effects of plants on settlement and survival of larvae and postveligers, effects of predators, and roles of *D. polymorpha* in lake ecosystems (Prejs et al. 1990, Lewandowski and Ozimek 1997, Lewandowski 1999). We were particularly interested in roles of large zebra mussel populations in ecosystem function, mainly effects of filtration on seston removal from water, nutrient cycling, accumulation of heavy metals, and importance as food for predators (Stańczykowska and Lewandowski 1993b, Stańczykowska 1994, Lewandowski and Stańczykowska 2000).

This chapter examines if and how *D. polymorpha* populations in the Masurian Lakelands changed over time relative to habitat conditions. The same approach and methods were used throughout the entire 50 year period, and all data were derived mainly from field studies and field experiments. In addition, we examine and compare the occurrence of *D. polymorpha* in two other large lakelands—the Pomeranian Lakeland located in northwestern Poland and the Suwalskie Lakeland located in the most northeastern portion of Poland (Figure 8.1; Kołodziejczyk 1994, 1999, Stańczykowska et al. 1997, Świerczyński 1997). Finally, we compared occurrence of *D. polymorpha* in northern Poland to occurrence in central and southern Poland (different lakes, reservoirs, lagoons, rivers) (Dusoge et al. 1999, Stańczykowska et al. 2010).

MASURIAN LAKELAND

Studies of *D. polymorpha* were carried out between 1959 and 2010 mainly in the central part of the Masurian Lakeland, which is often called the Great Masurian Lakes region. This region includes the two largest lakes in Poland, Lake Śniardwy and a complex of lakes called Lake Mamry; both lakes have an area larger than 100 km^2. There are dozens of other lakes in this region that differ in size, depth, mictic type, and water chemistry (see Appendix). Many are interconnected by rivers and by a system of artificial channels built in the eighteenth and nineteenth centuries. The trophic status of lakes within the Masurian Lakeland is highly variable. Lakes in the north (Lake Mamry complex) are mostly mesoeutrophic, lakes in the middle (Lake Jagodne to Lake Tałty) are highly eutrophic, and lakes in the south (Śniardwy, Nidzkie) are mesoeutrophic.

During 50 years of our studies, overall environmental conditions in the Masurian lakes underwent remarkable changes. Initially, the lakes were relatively oligotrophic, but in the 1970s and 1980s, nutrient inputs increased because of increased surface runoff from agricultural catchments, poorly treated waste waters, and mass tourism (Zdanowski and Hutorowicz 1993). In the 1990s, some measures were undertaken to counteract lake eutrophication such as construction or modification of sewage treatment plants in main towns of the region (Mikołajki, Giżycko). Even so, in the early-2000s, most lakes in the Masurian Lakeland (90%) were considered eutrophic, and only a few were considered mesotrophic.

Changes in populations of *D. polymorpha* paralleled alterations of environmental conditions in most Masurian lakes. In the 1960s, *D. polymorpha* was found in most of the lakes (Table 8.1), and it inhabited littoral zones at depths between 1 and 12 m. Densities in some lakes ranged from several hundred to several thousand individuals per square meter. In the 1970s, *D. polymorpha* declined in some of the lakes—mainly in those lakes that were small, shallow, and highly eutrophic. The decline of *D. polymorpha* expanded in the 1980s and 1990s as large and deep lakes like Lake Ryńskie, Tałty, and Mikołajskie were affected by tourism and domestic sewage input. In the 1990s, *D. polymorpha* was present in only 60% of the studied lakes, down from 81% in the early-1960s (Table 8.1). Densities decreased to only several hundred individuals per square meter, and distributions in nearshore zones became limited as populations occurred no deeper than 6 m. Data on *D. polymorpha* densities and depth distributions relative to bathymetric data allowed estimation of percent lake bottom area colonized (Table 8.2). Results showed large decreases in bottom surface area colonized between 1960 and 1994. Relatively small changes were observed in lakes of the Lake Mamry complex, whereas the largest change occurred in Lake Śniardwy.

By the early-2000s, improvements in lake water quality occurred as a result of changes in agricultural practices, replacement of large state farms, building of tourist centers equipped with sewage treatment facilities, and

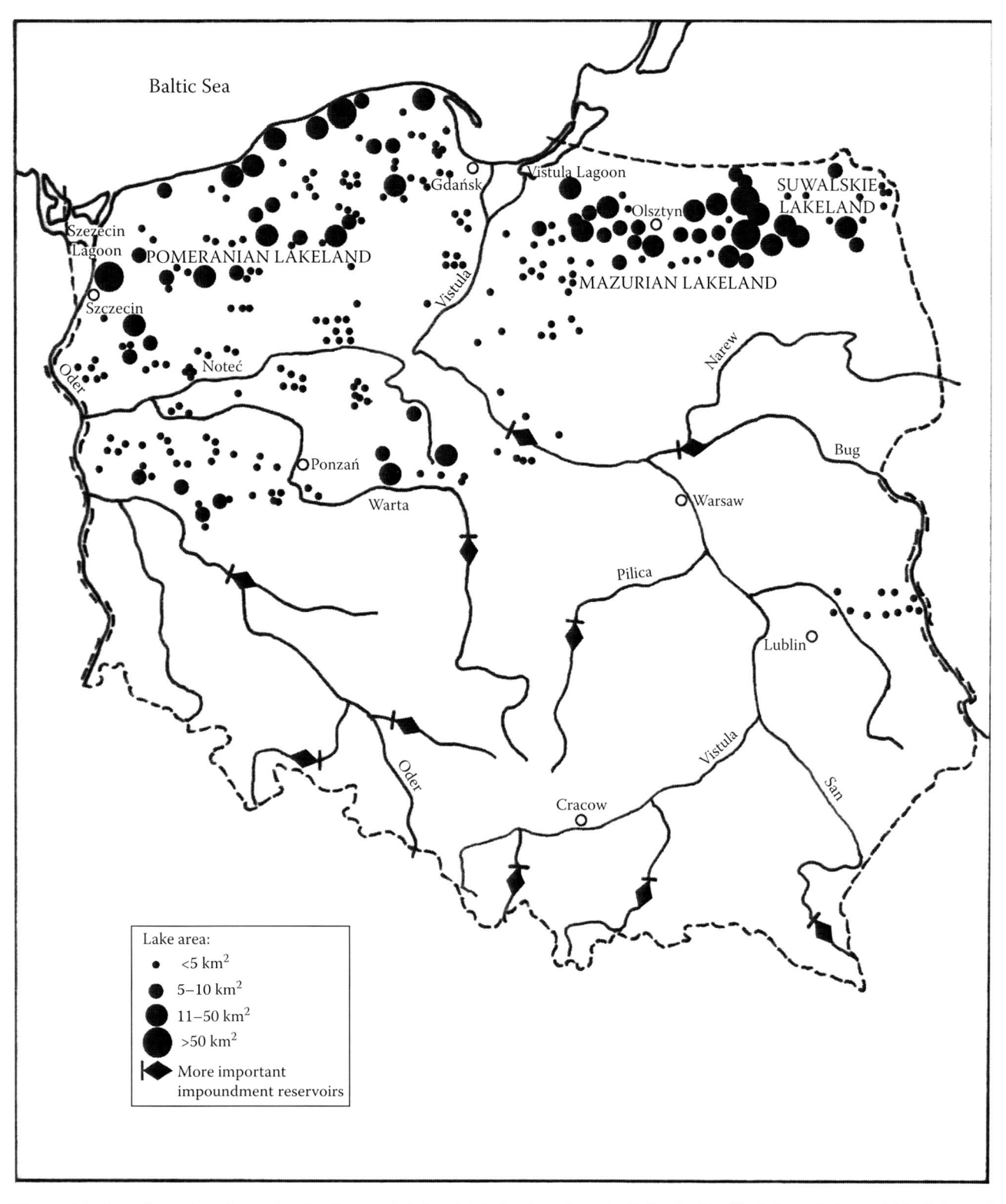

SUMMARY OF *DREISSENA POLYMORPHA* (ZEBRA MUSSEL) IN POLISH LAKES

Figure 8.1 Location of major water bodies and defined lakeland regions in Poland. (Modified from Dobrowolski, K.A. and Lewandowski, K., *The Strategy of Wetland Protection in Poland*, Institute of Ecology PAS Publishing Office, Dziekanów Leśny, Poland, 1998.)

Table 8.1 Frequency of Occurrence of *D. polymorpha* in Lakes in the Masurian Lakeland between 1959 and 2010

Year	No. of Lakes Examined	No. of Lakes with *D. polymorpha*	% of lakes with *D. polymorpha*
1959–1962	36	29	81
1972	36	26	73
1988	36	23	64
1993–1995	20	12	60
2005–2007	22	15	67
2010	23	17	74

Data derived as follows: 1959–1988 (Stańczykowska and Lewandowski 1993a), 1993–1995 (Stańczykowska and Stoczkowski 1997), and 2005–2010 (this study).

construction of municipal sewage treatment plants. Water quality improvements led to increased numbers of lakes in which *D. polymorpha* was found. In 2010, *D. polymorpha* occurred in 74% of lakes of the region (Table 8.1).

In our earlier chapter, trends in population densities of *D. polymorpha* in various lakes were placed into three different categories: stable, unstable or variable, and decreasing (Stańczykowska and Lewandowski 1993a). These categories were still basically applicable when describing more recent trends (1990–2010). However, populations in some lakes that initially were categorized as having decreasing densities have recently shown some recovery. Stable populations with high densities were noted throughout the 50-year period in most northern lakes of the Masurian Lakeland region. Typical examples include Lake Mamry (mean annual densities of 700–1600 ind. m^{-2}), Lake Święcajty (900–1800 ind. m^{-2}), and Lake Kisajno (500–1000 ind. m^{-2}) (Figure 8.2). Lakes with stable populations but with lower mean densities (below 50 ind. m^{-2}) were mostly found in the southern part of the system (e.g., Lake Bełdany). Examples of lakes in which populations were considered to be unstable or variable with no defined trends were Lake Boczne (200–1500 ind. m^{-2}), Lake Szymon (0–1800 ind. m^{-2}), Lake Tałtowisko (0–1500 ind. m^{-2}), and Lake Śniardwy (100–1500 ind. m^{-2} (Figure 8.3). Among the lakes with populations that initially decreased but then recently recovered were Lake Talty, Lake Jagodne, and Lake Niegocin. In Lake Talty and Lake Jagodne, *D. polymorpha* was not found in the 1980s and 1990s but was found at densities of 150–200 ind. m^{-2} in 2005–2010. In Lake Niegocin, mean annual densities were 1000–1300 ind. m^{-2} in 1960–1970, declined to 100–150 ind. m^{-2} in 1980–1990, and then increased to 700–1200 ind. m^{-2} in 2005–2010 (Figure 8.4). Of lakes where populations decreased, total absence seemed permanent only in Lake Ryńskie. In this lake, mean density was 1600 ind. m^{-2} in 1959 and declined to 100 ind. m^{-2} in the early 1980s, and then no individuals were found between 1985 and 2010. As noted, long-term trends in densities of *D. polymorpha* populations in these lakes were temporally correlated with changes in trophic status and water quality. Long-term assessments of water quality performed by the State Inspectorate of Environmental Protection showed a general decrease in water quality followed by water quality improvements after the installation of sewage treatment plants (e.g., Kochańska 2006).

In our studies of lakes in the Masurian Lakeland region, we focused primarily on various aspects of *D. polymorpha* ecology in Lake Mikołajskie, which is situated in the central part of this region and where the hydrobiological station of the Polish Academy of Sciences is located. Sampling of *D. polymorpha* populations was carried out in Lake Mikołajskie with high frequency over the 50-year period. Mean densities varied widely in the early portion of this period (1959–1976) (Figure 8.5), and much of this variation was attributed to intrinsic population dynamics rather than to anthropomorphic impacts (Stańczykowska and Lewandowski 1993a). The population collapsed in the 1970s, and very low densities continued through the 1980s and 1990s. Low densities occurred during a time when water quality in the lake severely declined. Recovery of the population began in 2007 (similar to that observed in Lake Niegocin) and occurred after water quality improved due to the installation of a municipal sewage treatment plant for the city of Mikołajki.

Another area in the Masurian Lakeland region consists of lakes associated with the Krutynia River. The 17 lakes in this river system are quite different in size and mean depth; surface areas vary from 0.2 to 8.4 km^2, and maximum depths vary from 3 to 51 m. These lakes are surrounded by forests, croplands, and villages. Most lakes are moderately eutrophic, but there are mesotrophic and very eutrophic lakes as well. The latter are situated in the upper and middle parts of the river system (Hillbricht-Ilkowska and Wiśniewski 1996). Distribution and abundance of *D. polymorpha* in lakes of the Krutynia River system were studied twice, once in 1989 (Lewandowski 1996) and again in 2010. In 1989, *D. polymorpha* inhabited 15 of the 17 lakes (88%), and mean densities of 700–3500 ind. m^{-2} were found in mesotrophic lakes. *D. polymorpha* was found in low densities (<30 ind. m^{-2}), or was absent, in lakes considered to be highly eutrophic. Mean densities in other lakes varied

Table 8.2 Percent of Bottom Area Colonized by *D. polymorpha* in Lakes in the Masurian Lakeland in 1960, 1980, and 1994

Year	Święcajty	Mamry	Kisajno	Dobskie	Mikołajskie	Nidzkie	Śniardwy
1960	49	55	64	65	19	77	84
1980	21	34	38	34	11	51	45
1994	22	33	30	29	—	50	34

Source: Modified from Stańczykowska, A. and Stoczkowski, R., *Pol. Arch. Hydrobiol.*, 44, 417, 1997.

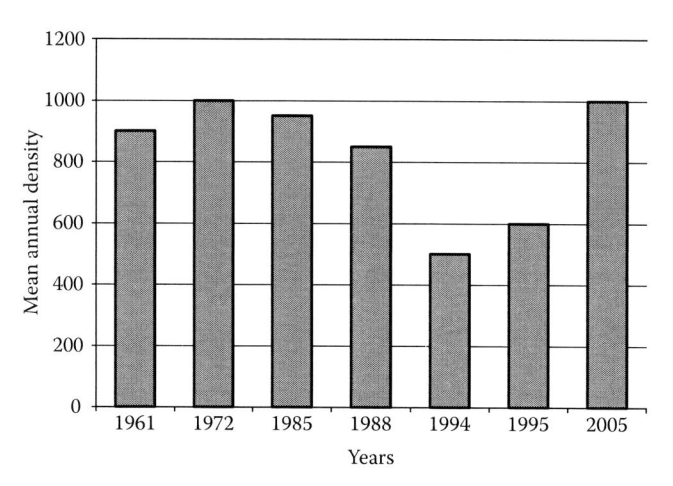

SUMMARY OF *DREISSENA POLYMORPHA* (ZEBRA MUSSEL) IN POLISH LAKES

Figure 8.2 Mean annual densities (ind. m^{-2}) of *D. polymorpha* in Lake Kisajno. The population in this lake was considered "stable" over time.

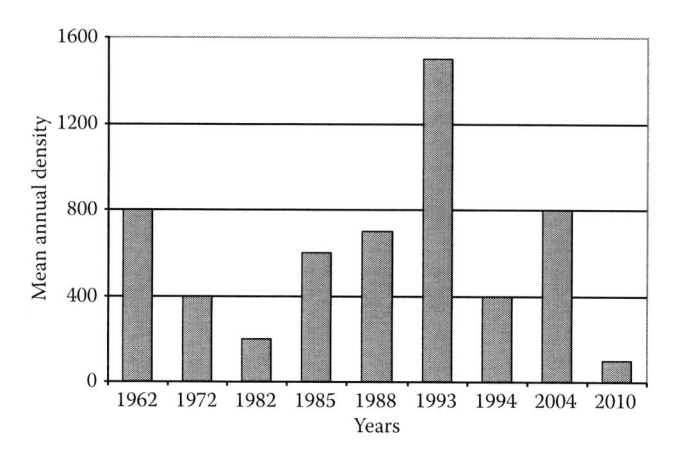

Figure 8.3 Mean annual densities (ind. m^{-2}) of *D. polymorpha* in Lake Śniardwy. The population in this lake was considered "unstable or variable" over time.

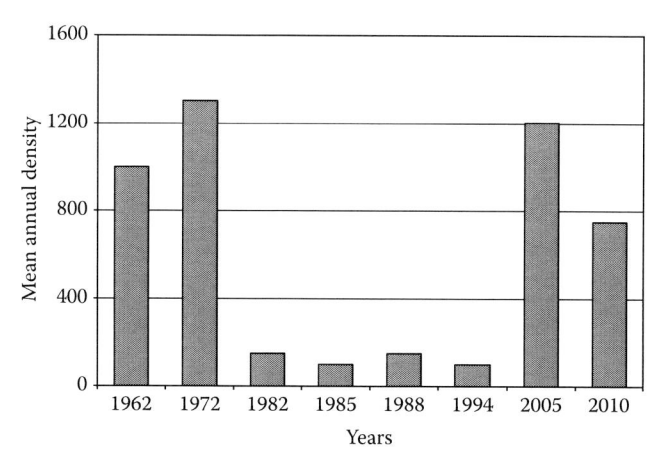

Figure 8.4 Mean annual densities (ind. m^{-2}) of *D. polymorpha* in Lake Niegocin. The population in this lake was considered to be "decreased" but with recent recovery.

from several dozen to several hundred ind. m^{-2} (Lewandowski 1996). In 2010, *D. polymorpha* was found in only 12 of the 17 lakes (69%), and densities were generally lower than found in 1989. *D. polymorpha* was totally absent from highly eutrophic lakes where it had occurred at very low densities in 1989. In mesotrophic lakes, densities ranged from 600 to 3200 ind. m^{-2}, and in moderately eutrophic lakes, densities ranged from 10 to 130 ind. m^{-2} (see Appendix).

Another group of lakes in the Masurian Lakeland is interconnected by the small (12 km long) and shallow Jorka River. This group consists of 5 lakes with surface areas that vary from 0.12 to 1.74 km^2. The uppermost and largest of these is Lake Majcz, which is considered to be mesotrophic, and four other smaller, more eutrophic lakes: Inulec, Głębokie, Zelwążek, and Jorzec. In the 1970s, a cage culture industry for rainbow trout (*Salmo gairdneri*) in Lake Głębokie strongly enriched this lake and the two lakes situated downstream along the Jorka River (Lakes Zelwążek and Jorzec). The industry ceased operations in 1988, but its effects persisted for many years afterward in the form of nutrient-rich bottom sediments. Environmental conditions and community structure of all five lakes were studied in detail (Hillbricht-Ilkowska 1983, 2002, Hillbricht-Ilkowska and Ławacz 1985), and the occurrence of *D. polymorpha* was examined over many years (Lewandowski et al. 1997). Between 1976 and 1997, *D. polymorpha* remained relatively abundant in the two upper lakes of the system that were unaffected by nutrient enrichment activities (Lakes Majcz and Inulec; Table 8.3). However, densities of *D. polymorpha* in Lake Głębokie and the two downstream lakes declined between 1979 and 1994. In 1994, several years after culture operations ceased, the density of *D. polymorpha* in Lake Głębokie was 10 times lower than in 1976, and populations were absent from the two downstream lakes. We believe that drastic density declines and disappearance of *D. polymorpha* from these downstream lakes was a result of accumulation and decomposition of excess food from the culture industry, which negatively impacted water quality in these three lakes. Declines in densities of other mollusk species were attributed to the same cause (Kołodziejczyk et al. 2009).

Subsequent studies of *D. polymorpha* in these five lakes in the late-1990s and 2000s showed some recovery of populations. *D. polymorpha* reappeared and gradually increased in Lake Zelwążek, and densities in Lake Głębokie increased such that by 2007, which was the last year densities were determined, mean density in this lake was highest of all lakes examined in the system. Also, in 2007, mussels reappeared in Lake Jorzec after an absence since 1976. Studies of other malacofauna carried out in the 1990s did not show similar recoveries (Kołodziejczyk et al. 2009).

Our 50-year long studies on various aspects of *D. polymorpha* ecology in the Masurian Lakeland showed that this species dominated density and biomass of the total malacofauna in most lakes. This was particularly true for deep, dimictic lakes. For shallow, polymictic lakes, *D. polymorpha* dominated those that were oligotrophic (low phosphorus

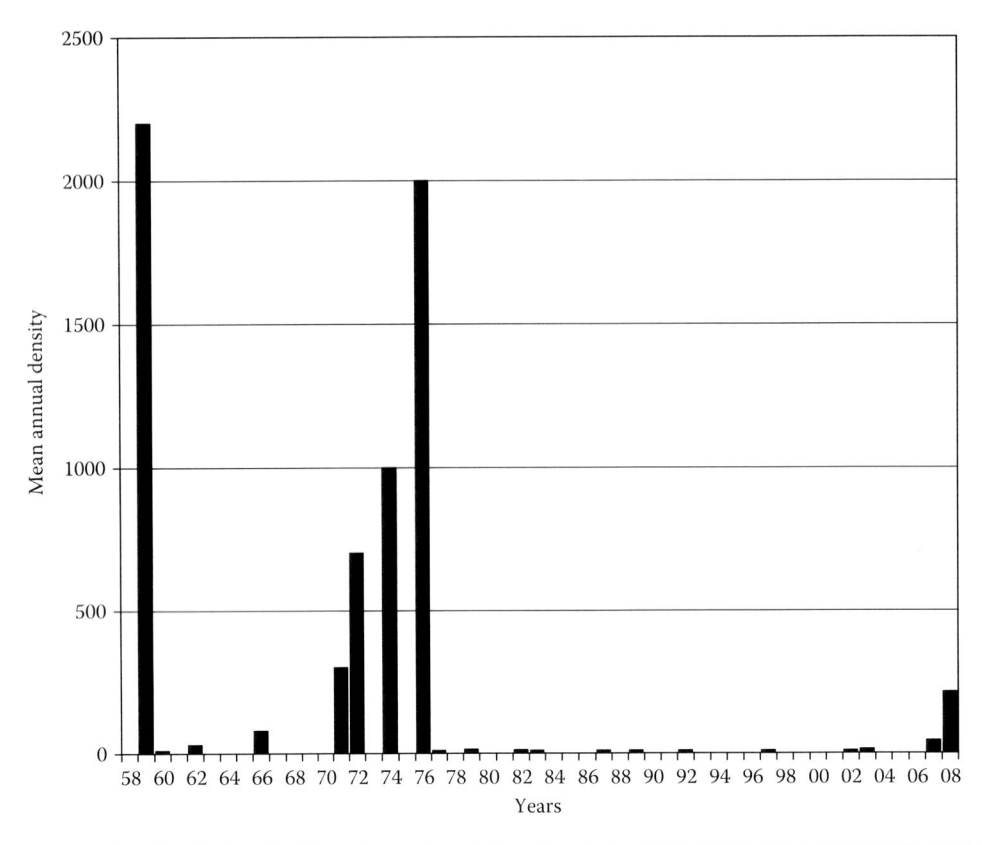

Figure 8.5 Mean annual densities (ind. m^{-2}) of *D. polymorpha* in Lake Mikołajskie, 1959–2008. The population in this lake was unstable in the early portion of this period, decreased to low levels, and then began to recover.

Table 8.3 Mean Annual Densities (ind. m^{-2}) of *D. polymorpha* in Its Zone of Occurrence in Lakes of the Jorka River System between 1976 and 2007

Lake	1976	1994	1997	2005	2007
Majcz	510	360	180	700	410
Inulec	400	1300	860	80	300
Głębokie	280	20	80	280	830
Zelwążek	175	0	5	15	25
Jorzec	130	0	0	0	2

Data derived as follows: 1976 (Stańczykowska et al. 1983b), 1994 (Lewandowski et al. 1997), 1997, 2005, and 2007 (this study).

concentrations) but was totally absent from those that were highly eutrophic (Stańczykowska et al. 1983a).

In addition to water quality and trophic and mictic status, the occurrence of *D. polymorpha* in lakes of the Masurian Lakeland may be related to many other factors. For example, early life stages (e.g., planktonic veligers and settled postveligers) are dependent on appropriate substrate on which to settle and survive. The finding of such substrate is a function of physical features such as wind speed/direction and water currents, and these features play a prominent role in lake distributions. Strong winds may move planktonic larvae deeper in the water column and transport them to windward shores. Far greater numbers of postveligers were found at windward sites than at leeward sites in studies carried out in Lake Inulec, which has a very complex shoreline (Lewandowski 1999).

Impacts of river currents on distributions of *D. polymorpha* were very apparent in lakes of the Krutynia River system. Extremely high densities and biomasses of *D. polymorpha* were noted in transition zones between lakes and the river (e.g., where the river enters and exits lakes). In such zones, the currents are moderate and concentrations of seston are high. Under these conditions, densities were often 11,000 m^{-2} and biomasses were 7.9 kg m^{-2}, which were values far greater than found in other lake zones or in the river proper. In addition, higher densities of postveligers were noted near shores of some lakes that were affected by river currents (Lewandowski 1996).

Densities are also a function of the type of substratum available for postveliger settlement. Settlement and site selection is a passive process associated with mechanical sinking of older veligers due to increased weight of their shells but is also an active process associated with individuals searching for appropriate substrate. An experiment performed in Lake Majcz where settling postveligers were given many choices (stones, colonies of adult *D. polymorpha*, submerged macrophytes, sand, and mud) found that submerged macrophytes were colonized most frequently (Lewandowski 1982).

Table 8.4 Mean Densities (ind. m⁻²) of *D. polymorpha* on Submersed Macrophyte Taxa in Lake Majcz in 1994

Macrophyte Taxa	Mean Density	Macrophyte Relative Abundance (%)
Chara spp.	2892	32.2
C. demersum	1750	32.3
N. obtusa	975	27.1
S. aloides	888	6.8
M. spicatum	196	<0.1
E. canadensis	182	<0.1
Others	0	1.5

Source: Modified from Lewandowski, K. and Ozimek, T., *Pol. Arch. Hydrobiol.*, 44, 457, 1997; Ozimek, T., *Pol. Arch. Hydrobiol.*, 44, 445, 1997.

While substrates in littoral regions of Masurian lakes are diverse, most littoral regions are dominated by submerged macrophytes. For example, over 30% of the surface area of Lake Majcz was covered with vegetation (Ozimek 1997). Different plant species provide different conditions for settling postveligers. Studies of many Masurian lakes showed that the density of postveligers on perennial plants was much higher than on annual plants (Lewandowski and Ozimek 1997). Settlement on perennial plants occurred year round (also under ice cover), whereas settlement on annual plants occurred only in midsummer. Highest mussel densities in Lake Majcz were recorded on four plant taxa: *Chara* spp., *Ceratophyllum demersum*, *Nitellopsis obtusa*, and *Stratiotes aloides*, while far fewer individuals settled on *Myriophyllum spicatum* and *Elodea canadensis* (Lewandowski and Ozimek 1997) (Table 8.4).

Detailed studies showed that the abundance of young mussels on submerged macrophytes depended mostly on the area covered, plant compactness, and species composition. In many Masurian lakes, densities of postveligers in excess of 100,000 per m² are not uncommon on vegetated bottoms (Lewandowski 1982). Individuals that settled on plants usually dominate the population; in most Masurian lakes, these individuals contribute to more than 80% of the entire population. Seasonal die-off of plants (particularly annual plants) is the largest cause of mortality in these individuals (over 90% annually). Some of these individuals may resettle and supplement the young population found on bottom substrates.

Densities of *D. polymorpha* may be affected by predators, mainly benthic-feeding fish and waterfowl. However, many studies of birds (mainly the coot *Fulica atra*) feeding on mussels in various regions of Poland indicated that impacts of bird predation on densities of *D. polymorpha* in Masurian lakes are minimal (Stempniewicz 1974, Stańczykowska et al. 1990, Stoczkowski and Stańczykowska 1994). This is especially true in the 2000s since waterfowl populations in the region are at low levels (R. Halba, personal communication). Reduced densities of waterfowl is likely a result of an increase in densities of the main predator of waterfowl, the American mink (*Mustela vison*) (Bartoszewicz and Zalewski 2003, Bonesi and Palazon 2007). Despite recent declines in waterfowl, however, decreased predation is not likely the reason for increased densities of *D. polymorpha* in many lakes.

Fish predation, especially by roach (*Rutilus rutilus*), may exert some impact on mussel densities in Masurian lakes. Results of field and laboratory studies showed that roach feeding on *D. polymorpha* was highly selective and efficient. Fish within a defined size range selected mussels of optimum (mean) size from mussel colonies with individuals having a wide range of sizes. Very small mussels provide low amounts of energy relative to energy expended by the fish predator, and very large mussels require predators to expend high amounts of energy to tear individuals off the substratum and to crush the tough shell (Prejs et al. 1990).

An important aspect of population status in *D. polymorpha* is the size and weight (i.e., condition) of individuals in relation to environmental conditions (trophic status, temperature, etc.). Studies of mussel populations in lakes of the Masurian Lakeland in the 1960s demonstrated low variability of mussel condition within a lake but great variability between lakes (Stańczykowska 1964). Populations of *D. polymorpha* attained largest body sizes in eutrophic lakes and in through-flow lakes in the Krutynia River system. Smallest body sizes were found in two extreme habitats: highly eutrophic lakes (e.g., Lake Tałty, Niegocin, and Bełdany) and oligotrophic lakes (e.g., Lake Mamry and Majcz) (Sarnowska and Lewandowski 2003). Stańczykowska (1964, 1977) also found that sizes of individuals from lakes with high densities were smaller than individuals from lakes with low densities. Similar results were obtained in field experiments that examined relative growth rates in lakes of various densities; that is, slower growth was observed in lakes with high densities (Stańczykowska and Lewandowski 1995).

SUWALSKIE LAKELAND

The Suwalskie Lakeland region is located in the northeastern part of Poland (Figure 8.1), and lakes in this region are far less enriched than lakes in the Masurian Lakeland. Studies of lakes in this region were initiated in the 1920s and 1930s. Two of the most studied lakes are Lake Hańcza and Lake Wigry. With a maximum depth of >100 m, Lake Hańcza is the deepest lake in Poland and, overall, the deepest lake in the Central European lowlands. Lake Hańcza has the lowest trophic status (oligo- or α-mesotrophy) of any Polish lowland lake. The littoral zone of this lake has a stony bottom along most of the shoreline, which is a feature rarely found in other lowland lakes. *D. polymorpha* was first reported from this lake in the 1920s (Poliński 1922). Historic studies showed that *D. polymorpha* was the most widespread benthic species in the lake, occurred to depths of 12–13 m, and had relatively high densities (sometimes over 6000 ind. m⁻²) that remained relatively stable over time (Stańczykowska 1966, Kołodziejczyk 1994, 1999).

More recent studies (in 2002) found that *D. polymorpha* occurred to a depth of 10 m, but highest densities occurred between 6 and 8 m (Lewandowski, unpublished).

Lake Wigry has a surface area of 22 km² and is the largest lake in the Suwalskie Lakeland. This lake has a maximum depth of 75 m and is considered to be mesotrophic. Similar to Lake Hañcza, *D. polymorpha* was also found in Lake Wigry in the 1920s (Poliñski 1922). More recently, Lewandowski (1992) reported that *D. polymorpha* in Lake Wigry reached a depth of 6 m and had a mean and maximum density of 130 and 1000 ind. m^{-2}, respectively. There are 56 lakes in close proximity to Lake Wigry that vary in size, depth, and trophic status. Kołodziejczyk (1994) found *D. polymorpha* in 24% of all lakes that he investigated, but in all lakes, this species was subdominant. In a comparative study of malacofauna in isolated lakes and lakes interconnected within this region, Kołodziejczyk (1989) found that *D. polymorpha* occurred only in lakes with high river throughput, and it was most abundant in areas where rivers entered lakes. Although the role of streams and rivers in transporting *D. polymorpha* between lakes is insufficiently studied, these studies support the common belief that water movement has a major impact on dispersal and occurrence of *D. polymorpha* in these lakes.

POMERANIAN LAKELAND

The Pomeranian Lakeland is situated in northwestern Poland between the lower stretches of the two largest rivers in Poland—the Vistula and the Odra (Figure 8.1). There are many lakes in this region, but they are more dispersed than those in the Masurian Lakeland and of smaller size. Overall, lakes in this region are less eutrophic than lakes in the Masurian and Suwalskie Lakelands because there is less agriculture and tourism.

In the 1990s, studies of *D. polymorpha* were carried out in lakes in the Pomeranian Lakeland concurrent with studies in Masurian lakes. These studies focused mainly on two large lakes, Lake Iñsko and Lake Miedwie, but studies were also conducted in dozens of smaller water bodies (Świerczyñski 1997). Comparisons of *D. polymorpha* in the Pomeranian and Masurian Lakelands showed remarkable differences in distributions. In the Pomeranian Lakeland, *D. polymorpha* was found in most studied water bodies, and overall densities were higher than in Masurian lakes. In addition, mussel populations extended into much deeper waters in Pomeranian than in Masurian lakes. In Lake Iñsko, for example, *D. polymorpha* occurred to a depth of 38 m, and in Lake Miedwie to 26 m (Świerczyñski 1997). In Lake Wdzydze (surface area 15 km², depth 68 m), *D. polymorpha* was most numerous between 2.5 and 10.0 m depth but was sporadically recorded to 55 m (L. Żmudziñski, personal communication). Differences in densities and distributions of *D. polymorpha* between the two lake regions were attributed to lower pollution, lower trophic status, and more favorable conditions for veliger settlement in Pomeranian than in Masurian lakes (Stañczykowska et al. 1997).

PRESENT DISTRIBUTION IN POLAND

There are marked differences in the occurrence of *D. polymorpha* between the northern, central, and southern regions of Poland. These regions differ in topography, climatic conditions, and water resources. The northern region consists of areas with postglacial and coastal lakes, small rivers flowing directly to the sea, and lagoons of the Vistula and Odra Rivers. As described for the Masurian, Suwalskie, and Pomeranian Lakelands, many water bodies in this portion of Poland are interconnected, which favors rapid and widespread dispersal of this species. However, distributions and abundances vary widely in this region, primarily because of differences in development and degree of nutrient enrichment.

In central Poland, *D. polymorpha* occurs in large reservoirs (i.e., Sulejowski, Zegrzyñski, Włocławski) and in middle sections of large rivers but does not occur in many lakes (Abraszewska-Kowalczyk et al. 1999, Dusoge et al. 1999, Jurkiewicz-Karnkowska 2004). Overall, densities are much lower than densities in water bodies in the northern region.

In southern Poland, *D. polymorpha* is not widely distributed. Studies in mountain rivers, retention reservoirs, and small but unusually deep oligotrophic lakes in the Tatras and Karkonosze Mountains have not reported the presence of this species. In the late-1990s, however, *D. polymorpha* was recorded for the first time in some small post-exploitation reservoirs used for recreation (M. Strzelec and A. Kownacki, personal communications). This seems to indicate that *D. polymorpha* is extending its range in southern Poland (Stañczykowska et al. 2010).

SUMMARY AND CONCLUSIONS

D. polymorpha is an invasive species that has been recorded in Poland for ca. 200 years. Our studies in the Masurian Lakeland and in other lake regions in northern Poland span a time period of about 50 years. These studies focused on distributions, factors that affect distributions, and long-term changes in populations.

D. polymorpha is the dominant species in the benthic malacofauna of many lakes in the Masurian and Pomeranian Lakelands but is not dominant in lakes in the Suwalskie Lakeland (Kołodziejczyk 1989, 1994, 1999). The latter region is less developed and less nutrient-enriched compared to the other two regions. In addition, *D. polymorpha* in the Suwalskie Lakeland seems to be confined to lakes with a high river throughput. Studies in the Masurian lakes beginning in 1959 indicate mean population densities can be defined as stable, unstable with no clear trends, or decreased but with some recent recovery. The latter population pattern occurred in lakes strongly influenced by nutrient enrichment from agriculture, general development, and tourism. Densities in these lakes decreased to low levels in the 1980s and 1990s but tended to increase in the 2000s mainly because of changed agriculture practices and construction of sewage treatment plants. Overall, mussel densities were greater in mesotrophic than in eutrophic lakes and relatively low to zero in highly eutrophic lakes. While trophic status is the most important factor influencing *D. polymorpha* populations in Masurian lakes, physical features also influence mussel densities. Many lakes are components of river systems and therefore interconnected. Densities in these lakes are generally greatest near river inputs and outflows. Other factors also influence densities such as lake mixing, shoreline configuration, and substrate type. Littoral regions of many Masurian lakes are covered with submerged macrophytes, which serve as substrates for newly settled individuals that enhance populations when plants die back and some individuals resettle on bottom substrates. Predation by fish and birds has little influence on population densities and distributions in these lakes.

The distribution of *D. polymorpha* is highly variable across Poland. This species occurs widely in the northern region but occurs less frequently and is less abundant in water bodies in the central and southern regions. Greater numbers and closer proximity of water bodies in the north facilitate spread through interconnected lakes and rivers, while generally mesotrophic conditions lead to high densities. In central and southern Poland, *D. polymorpha* is found in some reservoirs and large rivers but in relatively few lakes. However, *D. polymorpha* continues to spread in these two regions as evidenced by new reports of this species in reservoirs.

APPENDIX

Area, Maximum Depth, Trophic Status, and Range in Annual Mean Densities of *D. polymorpha* in Lakes Discussed in Text. The Density Range Is for the Period between 1959 and 2010 in Most Lakes

Lake	Area (ha)	Maximum Depth (m)	Trophic Status	Densities of *D. polymorpha* (Number m^{-2})
Masurian Lakeland				
Bełdany	780	31.0	Eutrophy	0–50
Boczne	190	15.0	Eutrophy	200–1,800
Dobskie	1,760	21.0	Early eutrophy	400–800
Głębokie	47	34.3	Eutrophy	20–800
Inulec	178	10.1	Eutrophy	80–1,300
Jagodne	936	34.0	Eutrophy	100–1,500
Jorzec	42	16.4	Eutrophy	0–150
Kisajno	2,536	24.0	Early eutrophy	500–1,000
Majcz	163	16.4	Mesotrophy	180–700
Mamry Pn	2,478	40.0	Mesotrophy	700–1,600
Mikołajskie	460	27.8	Eutrophy	0–2,200
Nidzkie	1,724	25.0	Eutrophy	300–700
Niegocin	2,499	40.0	Eutrophy	100–1,300
Ryńskie	620	47.0	Eutrophy	0–1,600
Szymon	204	34.0	Eutrophy	0–1,800
Śniardwy	10,598	25.0	Early eutrophy	100–1,500
Święcajty	814	28.0	Eutrophy	900–1,800
Tałtowisko	323	35.0	Eutrophy	0–1,600
Tałty	1,162	37.5	Eutrophy	0–3,600
Zelwążek	12	7.4	Eutrophy	0–200
Pomeranian Lakeland				
Ińsko	590	42.0	Mesotrophy	0–45,000
Miedwie	3,530	44.0	Mesotrophy	20–1,200
Suwalskie Lakeland				
Hańcza	311	106.1	Mesotrophy	300–6,000
Wigry	2,187	75.0	Mesotrophy	100–1,000

ACKNOWLEDGMENTS

Many thanks to our colleagues for discussions and practical help on the lakes, especially to Dr. A. Kołodziejczyk and M. Sc. M. Świerczyński. We are very thankful to anonymous reviewers and editors for constructive guidelines.

REFERENCES

Abraszewska-Kowalczyk, A., I. Jobczyk, and E. Pazera. 1999. Inwazja racicznicy zmiennej *Dreissena polymorpha* (Pallas, 1771) w Zbiorniku Sulejowskim i w dolnym biegu Pilicy [Invasion of zebra mussel *Dreissena polymorpha* (Pallas, 1771) in Sulejowski Reservoir and the lower Pilica]. 3–5. XV Krajowe Seminarium Malakologiczne, 23–25 IX 1999, Łódź, Polska.

Bartoszewicz, M. and A. Zalewski. 2003. American mink, *Mustela vison* diet and predation on waterfowl in the Słońsk Reserve, western Poland. *Folia Zool.* 52:225–238.

Bonesi, L. and S. Palazon. 2007. The American mink in Europe: Status, impacts and control. *Biol. Conserv.* 134:470–483.

Dobrowolski, K.A. and K. Lewandowski. 1998. *The Strategy of Wetland Protection in Poland*. Dziekanów Leśny, Poland: Institute of Ecology PAS Publishing Office.

Dusoge, K., K. Lewandowski, and A. Stańczykowska. 1999. Benthos of various habitats in the Zegrzyński Reservoir (central Poland). *Acta Hydrobiol.* 41:103–116.

Gliwicz, Z.M., A. Kowalczewski, T. Ozimek, E. Pieczynska, A. Prejs, K. Prejs, and J.I. Rybak. 1980. *An Assessment of the State of Eutrophication of the Great Masurian Lakes*. Warszawa, Poland: Wydawnictwo Instytutu Kształtowania Srodowiska.

Hillbricht-Ilkowska, A. 1983. Biotic structure and processes in the lake system of River Jorka watershed (Masurian Lakeland, Poland). *Ekol. Pol.* 31:535–834.

Hillbricht-Ilkowska, A. 2002. River-lake system in a mosaic landscape: Main results and some implications for theory and practice from studies on the River Jorka system (Masurian Lakeland, Poland). *Pol. J. Ecol.* 50:543–550.

Hillbricht-Ilkowska, A. and W. Ławacz. 1985. Factors affecting nutrient budget in lakes of the r. Jorka watershed (Masurian Lakeland, Poland). *Ekol. Pol.* 33:171–381.

Hillbricht-Ilkowska, A. and R.J. Wiśniewski, eds. 1996. Funkcjonowanie systemów rzeczno-jeziornych w krajobrazie pojeziernym: Rzeka Krutynia (Pojezierze Mazurskie) [The functioning of river-lake system in a lakeland landscape: River Krutynia (Masurian Lakeland, Poland)]. *Zeszyty Naukowe Komitetu Człowiek i Środowisko* 13:1–461.

Jurkiewicz-Karnkowska, E. 2004. Malacocoenoses of large lowland dam reservoirs of Vistula River basin and selected aspects of their function. *Folia Malacol.* 12:1–56.

Kochańska, E., Ed. 2006. *Raport o Stanie Środowiska Województwa Warmińsko-Mazurskiego w 2005 r.* [Report on environmental status in the Marmińsko-Mazurskie Province for the year 2005]. Olsztyn, Poland: Biblioteka Monitoringu Środowiska.

Kołodziejczyk, A. 1989. Malacofauna in isolated and interconnected lakes. *Arch. Hydrobiol.* 114:431–441.

Kołodziejczyk, A. 1994. Mięczaki słodkowodne Suwalskiego Parku Krajobrazowego [The freshwater molluscs of Suwałki Landscape Park]. *Zeszyty Naukowe Komitetu Człowiek i Środowisko* 7:243–265.

Kołodziejczyk, A. 1999. Molluscs on Characeae in an oligotrophic Hańcza Lake (NE Poland). *Folia Malacol.* 7:47–50.

Kołodziejczyk, A., K. Lewandowski, and A. Stańczykowska. 2009. Long-term changes of mollusc assemblages in bottom sediments of small semi-isolated lakes of different trophic state. *Pol. J. Ecol.* 57:331–339.

Lewandowski, K. 1976. Unionidae as a substratum for *Dreissena polymorpha* (Pall.). *Pol. Arch. Hydrobiol.* 23:409–420.

Lewandowski, K. 1982. The role of early developmental stages in the dynamics of *Dreissena polymorpha* (Pall.) (*Bivalvia*) populations in lakes. II. Settling of larvae and the dynamics of numbers of settled individuals. *Ekol. Pol.* 30:223–286.

Lewandowski, K. 1991. The occurrence of *Dreissena polymorpha* (Pall.) in some mesotrophic lakes of the Masurian Lakeland (Poland). *Ekol. Pol.* 39:273–286.

Lewandowski, K. 1992. Występowanie i rozmieszczenie mięczaków, ze szczególnym uwzględnieniem małża *Dreissena polymorpha* (Pall.) w litoralu kilku jezior Wigierskiego Parku Narodowego [The occurrence and distribution of molluscs with particular reference to bivalve *Dreissena polymorpha* (Pall.) in the littoral of some lakes of the Wigry National Park]. *Zeszyty Naukowe Komitetu Człowiek i Środowisko* 3:145–151.

Lewandowski, K. 1996. Występowanie *Dreissena polymorpha* (Pall.) oraz małży z rodziny Unionidae w systemie rzeczno-jeziornym Krutyni (Pojezierze Mazurskie) [The occurrence of *Dreissena polymorpha* (Pall.) and bivalves of the family Unionidae in the Krutynia river-lake system (Masurian Lakeland)]. *Zeszyty Naukowe Komitetu Człowiek i Środowisko* 13:173–185.

Lewandowski, K. 1999. The occurrence of zebra mussel *Dreissena polymorpha* (Pall.) in a lake of diversified shoreline. *Pol. Arch. Hydrobiol.* 46:303–316.

Lewandowski, K. 2001. Development of populations of *Dreissena polymorpha* (Pall.) in lakes. *Folia Malacol.* 9:171–216.

Lewandowski, K and T. Ozimek. 1997. Relationship of *Dreissena polymorpha* (Pall.) to various species of submerged macrophytes. *Pol. Arch. Hydrobiol.* 44:457–466.

Lewandowski, K. and A. Stańczykowska. 2000. Rola małża *Dreissena polymorpha* (Pall.) (racicznica zmienna) w ekosystemach słodkowodnych [The role of *Dreissena polymorpha* (Pall.) in freshwater ecosystems]. *Przegl. Zool.* 43:13–21.

Lewandowski, K., R. Stoczkowski, and A. Stańczykowska. 1997. Distribution of *Dreissena polymorpha* (Pall.) in lakes of the Jorka river watershed. *Pol. Arch. Hydrobiol.* 44:431–443.

Limanova, N.A. 1978. *Drejssena. Bibliografitcheskij ukazatel* [*Dreissena*. Bibliography]. Moskva, Russia: Akademia Nauk SSSR.

Nalepa, T.F. and D.W. Schloesser. 1993. *Zebra Mussels: Biology, Impacts, and Control*. Boca Raton, FL: CRC Press.

Ozimek, T. 1997. Submerged macrophytes as a substrate for *Dreissena polymorpha* (Pall.) in five lakes of the Jorka river watershed. *Pol. Arch. Hydrobiol.* 44:445–455.

Poliński, W. 1922. O faunie mięczaków ziemi suwalskiej [On the molluscan fauna of Suwałki region]. *Sprawozdania Stacji Hydrobiologicznej na Wigrach* 1:37–43.

Prejs, A., K. Lewandowski, and A. Stańczykowska-Piotrowska. 1990. Size-selective predation by roach (*Rutilus rutilus*) on zebra mussel (*Dreissena polymorpha*): Field studies. *Oecologia* 83:378–384.

Sarnowska, K. and K. Lewandowski. 2003. Maksymalne rozmiary małża racicznica zmienna (*Dreissena polymorpha* (Pall.)) w jeziorach polskich [Maximum size of *Dreissena polymorpha* (Pall.), zebra mussels in Polish lakes]. *Przegl. Zool.* 47:105–108.

Stańczykowska, A. 1961. Gwałtowna redukcja liczebności *Dreissensia polymorpha* Pall. w kilku jeziorach mazurskich okolic Mikołajek [A rapid density reduction of *Dreissena polymorpha* Pall. in several Masurian lakes near Mikołajki]. *Ekol. Pol. B* 7:151–153.

Stańczykowska, A. 1964. On the relationship between abundance aggregations and "condition" of *Dreissena polymorpha* (Pall.) in 36 Masurian lakes. *Ekol. Pol. A* 12:653–690.

Stańczykowska, A. 1966. Einige Gesetzmässigkeiten des Vorkommens von *Dreissena polymorpha* Pall. *Verh. Int. Verein Limnol.* 16:1761–1776.

Stańczykowska, A. 1977. Ecology of *Dreissena polymorpha* (Pall.) (*Bivalvia*) in lakes. *Pol. Arch. Hydrobiol.* 24:461–530.

Stańczykowska, A. 1994. Long-term changes in some *Dreissena polymorpha* populations in Poland. *Int. Ver. Theor. Ang. Limnol. Verh.* 25:2352–2354.

Stańczykowska, A. 1997. Review of studies on *Dreissena polymorpha* (Pall.). *Pol. Arch. Hydrobiol.* 44:401–415.

Stańczykowska, A., E. Jurkiewicz-Karnkowska, and K. Lewandowski. 1983a. Ecological characteristics of lakes in north-eastern Poland versus their trophic gradient. X. Occurrence of molluscs in 42 lakes. *Ekol. Pol.* 31:459–475.

Stańczykowska, A., W. Ławacz, J. Mattice, and K. Lewandowski. 1976. Bivalves as a factor effecting circulation of matter in Lake Mikołajskie (Poland). *Limnologica (Berlin)* 10:347–352.

Stańczykowska, A. and K. Lewandowski. 1980. Studies on the ecology of *Dreissena polymorpha* (Pall.) in some lakes. In *Atti IV Congresso S.M.J., Siena, 6–9 Ottobre 1978*, F. Giusti, Ed., pp. 369–374. Siena, Italy: Atti Accademia Fisiocritici.

Stańczykowska, A. and K. Lewandowski. 1993a. Thirty years of studies of *Dreissena polymorpha* ecology in Masurian lakes of northeastern Poland. In *Zebra Mussels: Biology, Impacts, and Control*, T.F. Nalepa and D.W. Schloesser, eds., pp. 3–37. Boca Raton, FL: CRC Press.

Stańczykowska, A. and K. Lewandowski. 1993b. Effect of filtering activity of *Dreissena polymorpha* (Pall.) on the nutrient budget of the littoral of Lake Mikołajskie. *Hydrobiologia* 251:73–79.

Stańczykowska, A. and K. Lewandowski. 1995. Individual growth of the freshwater mussel *Dreissena polymorpha* (Pall.) in Mikołajskie Lake; estimates in situ. *Ekol. Pol.* 43:267–276.

Stańczykowska, A. and K. Lewandowski, eds. 1997. *Dreissena polymorpha* (Pall.) in the lakes of northern Poland. *Pol. Arch. Hydrobiol.* 44:401–520.

Stańczykowska, A., K. Lewandowski, and M. Czarnoleski. 2010. Distribution and densities of *Dreissena polymorpha* in Poland—Past and present. In *The Zebra Mussel in Europe*, G. Van der Velde, S. Rajagopal, and A. Bij de Vaate, eds., pp. 119–126. Leiden, Holland: Backhuys Publishers.

Stańczykowska, A., K. Lewandowski, and J. Ejsmont-Karabin. 1983b. Biotic structure and processes in the lake system of r. Jorka watershed (Masurian Lakeland, Poland). IX. Occurrence and distribution of molluscs with special consideration to *Dreissena polymorpha* (Pall.). *Ekol. Pol.* 31:761–780.

Stańczykowska, A., K. Lewandowski, and M. Świerczyński. 1997. Summary of studies on *Dreissena polymorpha* (Pall.) conducted in the period 1993–95 in the Masurian and Pomeranian lakelands. *Pol. Arch. Hydrobiol.* 44:517–520.

Stańczykowska, A. and M. Planter. 1985. Factors affecting nutrient budget in lakes of the r. Jorka watershed (Masurian Lakeland, Poland) X. Role of the mussel *Dreissena polymorpha* (Pall.) in N and P cycles in a lake ecosystem. *Ekol. Pol.* 33:345–356.

Stańczykowska, A. and R. Stoczkowski. 1997. Are the changes in *Dreissena polymorpha* (Pall.) distribution in the Great Masurian Lakes related to trophic state? *Pol. Arch. Hydrobiol.* 44:417–429.

Stańczykowska, A., P. Zyska, A. Dombrowski, H. Kot, and E. Zyska. 1990. The distribution of waterfowl in relation to mollusc populations in the man-made Lake Zegrzyńskie. *Hydrobiologia* 191:233–240.

Stempniewicz, L. 1974. The effect of feeding of coot (*Fulica atra* L.) on the character of the shoals of *Dreissena polymorpha* Pall. in the Lake Gopło. *Acta Univ. N. Copernici* 34:84–103.

Stoczkowski, R. and A. Stańczykowska. 1994. The diet of the Coot *Fulica atra* in the Zegrzyński Reservoir (Central Poland). *Acta Orn.* 29:171–176.

Świerczyński, M. 1997. Occurrence of *Dreissena polymorpha* (Pall.) in lakes Miedwie and Ińsko. *Pol. Arch. Hydrobiol.* 44:487–503.

Wiktor, J. 1969. Biologia *Dreissena polymorpha* (Pall.) i jej ekologiczne znaczenie w Zalewie Szczecińskim [The biology of *Dreissena polymorpha* (Pall.) and its ecological importance in the Firth of Szczecin]. *Stud. Mat. Morsk. Inst. Ryb. Gdynia, Ser. A* 5:1–88.

Zdanowski, B. and A. Hutorowicz. 1993. Trofia i czystość Wielkich Jezior Mazurskich [Trophy and water quality in the Great Masurian Lakes]. *Komunikaty Rybackie* 6:1–5.

PART III

Response, Management, and Mitigation

CHAPTER 9

One Reporter's Perspective on the Invasion of Dreissenid Mussels in North America
Reflections on the News Media

Steve Pollick

CONTENTS

First Contact 117
Before the Perfect Storm 118
Perfect Storm: 1988–1993 119
    Spreading the Alarm 119
Recognizing the Biological/Ecological Threat..122
Waterworks and Power Plants versus Zebra Mussels 124
    Stemming the Zebra Mussel Tide 125
Legislation and Politics..128
Aftermath of the Perfect Storm, 1994–2011 129
Internet Age 130
One Newspaperman's Take on the Dreissenid Mussel Tragedy 133

FIRST CONTACT

My first hands-on encounter with a zebra mussel occurred in 1989 or 1990. The precise year is a little vague because catching a zebra mussel while on a fishing trip hardly is indelibly memorable. After all, a 1/2 in. zebra mussel is hardly a 10 lb walleye or a 5 lb smallmouth bass or a tandem-hooked "double" of jumbo-size, 12 in. yellow perch. Those kinds of catches would have been memorable. But the mussel catch from Lake Erie—near the best rocky, reef fishing around Bass Islands area of the western basin—left a little to be desired.

This fishing trip occurred on a perfect, sunny morning on Lake Erie, late spring or early summer. Our boat was drifting near a reef and my lure snagged on the rocky bottom. I had "hooked the planet," as my buddies teased. After working my lure free of the bottom, I retrieved my line quickly in anticipation of another cast. But as the baited lure broke the surface, I noticed something—a tiny "clam"—skewered on the needlepoint of the hook. "Hey, it's a zebra mussel,"

I hollered to my companions. Everybody stopped casting briefly to examine the curiosity. This was early days for the zebra mussel in Lake Erie, but we all had heard and read about these little invaders. In fact, I even had written about them and their discovery in Lake St. Clair in June 1988 and subsequently in western Lake Erie in October 1988.

I squeezed the fragile shell in my fingers, perhaps in contempt, and flicked the remains back into the lake. I have an olfactory memory of a "sour-musty, fishy, muddy-water" odor from handling the crushed mussel. I do not exactly recall the ensuing conversation on board, but I know it revolved around the mussels, at least for a few minutes. And I can guarantee the conversation hardly was complimentary of the pesky invader. But none of us fishing together that day—good, seasoned anglers and experienced lake men—had a clue as to the magnitude of what was to come.

We did not know it, but we who lived on western Lake Erie and upstream near Lake St. Clair were sitting at ground zero of an eco-nuclear explosion that would send reverberations and shock waves across North America so quickly that science and political institutions never could catch up, let alone get ahead of the invasion and ensuing ubiquitous infestation, which continues in the western reaches of the country. At the time there were prophets preaching of impending doom to the ecology and economy, but not enough could be done. Indeed, it is a matter of record that political institutions never did catch up with the zebra mussel threat. Some truly determined, valiant efforts were made in those early years, the most notable of which was the enactment of the Nonindigenous Aquatic Nuisance Prevention and Control Act of 1990, informally called the Nonindigenous Species Act. The writing, development, and marshaling of the bill through Congress were spearheaded by former astronaut and U.S. Senator John Glenn (D., Ohio).

As outdoors editor of *The Blade* newspaper in Toledo, Ohio, I witnessed these developments with great interest. I had been writing about Great Lakes issues since 1971, so when first reports of this new invader surfaced, I wrote a 3 day series on invasive species in 1989 that ran on the front page. The series caught the attention of Congresswoman Marcy Kaptur (D., Toledo) who promised to bring the issue to the attention of fellow lawmakers.

Well, fast-forward 20 years from that initial western Lake Erie mussel encounter to the summer of 2010. Six older anglers, including me, were on a western Lake Erie yellow perch fishing trip northwest of West Sister Island, a few miles west of that first mussel encounter. We were bringing up masses of zebra mussels as bycatch of our fishing excursion, and we carried memories of Monroe as the site of an infamous, highly publicized, zebra mussel-caused "meltdown" of the city water supply in December 1989. Its water intake had been clogged to a trickle by millions of mussels.

Several times during the day's fishing, in 26 ft of water, one or another fisherman would wind up not with a perch but with a gob of mussels. It remains somewhat amazing that what once was soft, muddy lake bottom—where 20 years prior, zebra mussels were not thought to become a problem—was now a carpet of mussels. The mussels had created their own substrate, colonizing progressively on the spent shells of previous generations of their own kind.

Nowadays a fisherman might snag a mussel on hook's end just about anywhere on western Lake Erie. It is noteworthy that such an eventuality as snagging a mussel gob hardly stopped conversation about the usual man-things on that yellow perch trip. Sometimes the reeled-in mussels would be scraped off and redeposited into the lake without comment, without even calling any attention to the inconvenience. In short, the zebra mussel had become woven into the fabric of the western Lake Erie sportfishing community within 20 years. All on board fatefully accepted the occurrence of mussels with tough love. After all, the highly prized walleye didn't go away because of the mussels' presence, though their stocks have had ups and downs over the 20 years, including one of the most successful year classes ever in 2003, long after zebra mussels had blanketed the lake floor. Yellow perch and smallmouth bass didn't go away either, despite ups and downs, though both of those species were being affected by zebra mussels in nefariously subtle ways, as we shall see later. Sportfishermen, moreover, can see that Lake Erie has become "cleaner." But those with more than anecdotal knowledge of the reasons behind the vastly improved water clarity quickly had to step in and fill in the obvious blanks in the learning curve, such that "clearer" does not mean "cleaner" or "purer." More about that later as well. A fair reading of how the zebra mussel went from a curiosity on the tip of a fishhook to an accepted—if subtly acknowledged by the public—economic and ecological nuisance can be obtained by surveying hundreds of newspaper articles, columns, and features of educational technical bulletins of public agencies and of newsletters and broadcast media news and documentaries.

Such a review need not be exhaustive to tell the dreissenid story in North America, which still is unfolding in 2011. But an essay of some 925 articles and dozens of videotapes (yes, tapes, it was that long ago) sufficiently paints a picture.

BEFORE THE PERFECT STORM

To begin at a convenient beginning—sort of going back to the Big Bang some 14 billion years ago—go back to 1963. That is the year of publication of a guide called *Pond Life of Europe*, in which the zebra mussel, *Dreissena polymorpha*, is described, and that is the same year the Tennessee Valley Stream Pollution Control Association presciently warned in a newsletter of the potential mussel menace. Then in 1968, several other U.S. and foreign-government papers and documents warned of the potential economic and environmental damage that could visit North America, should *Dreissena* ever be introduced here, however unintentionally. According to an article in the Canadian press in 1990 (*Chronicle-Journal Times News*, March 15, 1990), European scientists had been warning their North American counterparts about zebra mussels as far back as the 1950s. In any case, the mussels' now too-familiar origins in the Caspian Sea region of Russia were on record as a possible source for the invader to come to North America. Zebra mussels had a history of ship travel in barge canals across Europe as documented by their arrival in Great Britain. For better or for worse, however, such distant early warnings were supplanted by more pressing environmental, economical, and political issues of the late 1960s and through the 1970s.

Then in 1981, a consultant's report (Bio-Environmental Services, Georgetown, Ontario, for Environment Canada) loudly sounded an alarm about the dangers of aquatic invasive species entering North America via the St. Lawrence Seaway and Great Lakes in unexchanged ballast water of cargo ships from overseas. The consultants had sampled and analyzed ballast water from 55 overseas ships that entered the seaway in the summer and fall of 1980. They found no less than 56 foreign species in the ships' tanks. The report to Environment Canada was shown to the Canadian and U.S. coast guards and presently was shelved. The warning went unheeded in the high places where it needed to be taken seriously.

PERFECT STORM: 1988–1993

Seven years passed, overseas shipping continued to ply the seaway and its ports, dumping infested ballast water with little oversight. Then on June 1, 1988, biology students from the University of Windsor found samples of some small unknown mussels in Lake St. Clair. They were quickly identified as *D. polymorpha*, the dreaded zebra mussel. Worse, the specimens were aged as being at least 2 years old. So the proverbial cat was long out of the bag, though the full impact of the revelation was yet to be told. A headline in the *Windsor Star* on July 27, 1988, said it all: "New Clams Post Threat to Lakes" (Figure 9.1), and it was the first of many more headlines to come. The article among other things noted: "There are at least 50 exotic species in the Great Lakes."

What followed over the next few years was a veritable flood of news articles, bulletins, pamphlets, newsletters, and broadcasts (Figure 9.2). In the end, no one in the public, government, or business and industry could complain that the word on zebra mussels had not been widely and thoroughly disseminated. Zebra mussels—or at least their free-swimming larvae or veligers—seemingly flowed "downhill" in the seaway, from Lake St. Clair through the short, fast chute of the Detroit River into western Lake Erie, arguably one of the most productive freshwater spawning and nursery grounds for fish in the world. Lake Ontario, just downstream via the Niagara River and Welland Canal, soon would be next. But zebra mussels also were moved "upstream" by ships moving cargoes within the Great Lakes, having been found in Lake Michigan and Lake Superior in 1989. In October 1988, just 4 months after the Lake St. Clair discovery, a researcher at the Ohio State University's Stone Laboratory in the western Lake Erie confirmed that zebra mussels had become established in the lake. Within 2 years, the mussels colonized almost every hard surface in the lake. In March 1989, a stolen car was pulled from the harbor at Wheatley, Ontario. It was coated heavily inside and out with zebra mussels as densely aggregated as 37,000/m^2. A photograph of the "zebra mussel car," widely circulated in both the Canadian and U.S. press, quickly dramatized the mussel invasion and along with a spate of articles thereafter galvanized public and research attention.

Spreading the Alarm

Among examples of the many news reports sounding the zebra mussel alarms was the previously mentioned three-part series on alien invaders, which included extensive attention to the new zebra mussel threat, published in *The Blade* in Toledo in May 1989. On September 30, 1989, *The Buffalo News* reported zebra mussels at Ashtabula, Ohio, near the

Figure 9.1 The first newspaper headline about the establishment of zebra mussels in North America. (From *Windsor Star*, July 27, 1988.)

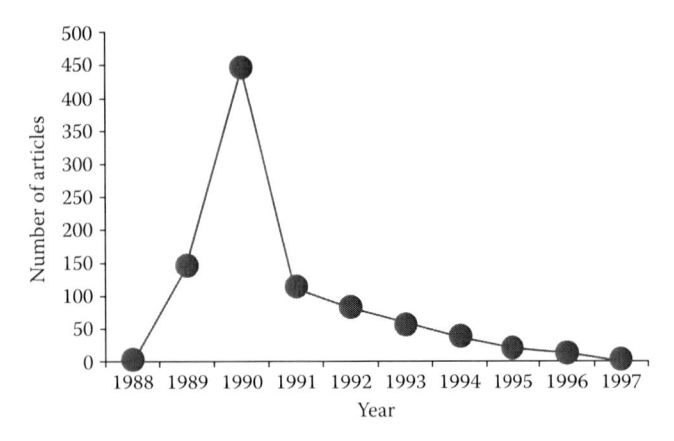

Figure 9.2 Trend in number of newspaper articles, public and industry technical bulletins, newsletters, radio and television broadcasts, and documentaries between 1988 and 1997. (Courtesy of D.W. Schloesser.)

Pennsylvania line, and at Port Maitland, Ontario. The article noted that the zebra mussels could be expected at Buffalo, New York, on Lake Erie's eastern terminus, by autumn of 1989 and in Lake Ontario no later than 1990. Prophetically, another article in southeast Michigan's *Monroe Evening News* on December 18, 1989, quoted an environmental adviser to the Detroit Edison electric power-generating facility in Monroe thusly "It will only be a matter of time before the zebra mussel enters the Mississippi and Missouri rivers and starts appearing down south."

Writing in the *Mariners Weather Log*, scientists charted the way west as being through the man-made Chicago Sanitary and Ship Canal diversions, which provided a water link between the Great Lakes and the Mississippi River watersheds via Lake Michigan and the Illinois River. The Mississippi River and its tributaries cover about two-thirds of the lower 48 states. Scientists also suggested that trailered recreational boats also might be a vector for the spread of the mussels. These forecasts proved too true across the continent over the next 20 years, as we shall see later. In any event, it seemed as though the dreaded zebra mussels were being discovered all across the Great Lakes, seemingly within months, as more and more individuals began searching for a pest that already may have been widespread and well established.

In the winter of 1990, scientists agreed that zebra mussels would be "coming to Lake Michigan." In January 1990, a biologist with Environment Canada also received belated vindication for raising the warning flags in 1981 with the ill-fated ballast-water report to the two coast guards. This came in a lengthy feature detailing the mussel invasion in *Toronto Life* magazine. One suspects that the biologist, who forecast the arrival of mussels in 1981, gladly would have foregone the attention in 1990 if he had been listened to 9 years previously. On January 15, 1990, *The Los Angeles Times* quoted a researcher of the Great Lakes Fishery Commission in Ann Arbor, Michigan, who reinforced the long-term, strategic implications of the mussels' coming: "The zebra mussel will be plaguing a large part of North America in the next 100 years." This statement left little room for misunderstanding the threat. Out in Nebraska in the Corn and Cattle Belt, the *Omaha World-Herald* on February 19, 1990, described how zebra mussels had imperiled power plants in the Great Lakes—at Monroe, Michigan, and Dunkirk, New York—and it went on to tell readers that once the mussels reached Lake Michigan they "almost certainly" would reach the Mississippi basin via the Chicago diversions. The article renewed the specter of fear for infestation of much of North America's surface water. Readers as widely separated as those of the *Kansas City Times* and *Philadelphia Inquirer* were treated to similar articles the same day.

The very next day, February 20, 1990, the *Owen Sound Sun Times* in Ontario told its readers "Zebra Mussels Discovered in Lake Huron." This account also noted that mussels had been confirmed in upper Lake Michigan's Green Bay, citing U.S. Fish and Wildlife Service and Detroit Edison sources. On March 19, the *Pittsburgh Post-Gazette* warned readers of the inland spread of zebra mussels, forecasting their pestiferous presence in the Ohio, Allegheny, and Monongahela rivers—the famous tri-rivers—by 1993. The circulation of warnings and advisories of mussel advances also was being pumped through the internal pipelines of public agencies in 1990. Witness an office memorandum of the Illinois Department of Conservation on March 30 that was spot on in summarizing the situation: "I urge this news release be given high priority on the part of our division, the director's level, and our information and education staff to promptly inform the general public about this destructive little critter," the memo reads. Further, "Public participation is critical in stemming the invasion of the zebra mussel." The next day, the *Buffalo News* carried an article that dovetailed precisely into the state of the public information art: "Zebra Mussels Expected in Chautauqua Lake." And so the beloved Finger Lakes were not to escape notice. By the end of April 1990, southeast Michigan's *Monroe Evening News* was reporting "Zebra Mussels Found in Lake Superior Ports."

The Great Lakes, indeed the entire St. Lawrence Seaway, appeared to be invaded entirely by the end of April of 1990, the same month Wisconsin Sea Grant, in its *Littoral Drift*, announced, "Sea Grant Launches Zebra Mussel Watch." And in May, a telling graphic in New York Sea Grant's *Coastlines* carried a pointed summary caption: "Zebra mussels can now be found in all of Lake St. Clair and Lake Erie, the Niagara River, the southwest corner of Lake Ontario; the St. Lawrence River from slightly downstream of Lake Ontario to the New York state/province of Ontario border, the northeast corner of Lake Ontario; Green Bay of western Lake Michigan, and Duluth Harbor at the far west end of Lake Superior." On May 10, 1990, the *Cincinnati Enquirer* reported that the Ohio River might be infested within 3–5 years; on the 20th, the *Grand Rapids Press* in western lower Michigan carried an article: "Grand Rapids Water

System Braces for Zebra Mussels," and on May 26, even the *Durham Sun* in North Carolina ran a story that headlined "Tiny Mussels Pose Threat to Nuclear Plants." Also on the 26th, nationally syndicated columnist Jack Anderson quoted Ohio Sea Grant as: "It may be [Ohio's] problem now, but before long it will be a problem across the country." Both the *New York Times* and *Toronto Globe and Mail* weighed in with articles in early July 1990, suggesting that the mussels were spreading far and wide to the detriment of the economy and ecology. On July 13, the *Chicago Tribune* noted that mussels had reached the south end of Lake Michigan at Gary, Indiana.

Overall, 1990 was the second consecutive year filled with frenetic press activity about zebra mussels (Figure 9.2). On April 5, 1990, the *Detroit Free Press* played its version of a now-recurring story, "Mussels May Leapfrog Inland by Way of Boats." In a documentary broadcast by Wisconsin Public Television in April 1990, in celebration of the 20th Earth Day, April 22, mussel movement was forecast to reach beyond the Great Lakes into the Mississippi watershed. The broadcast quoted a federal researcher who noted that the spread to date was "only the tip of the iceberg." In *The Blade* in Toledo on June 15, 1990, was a Washington dateline with this: "Federal officials, sounding ever more urgent alarms, warn that zebra mussels would infest all the freshwater east of the Rockies if they aren't stopped." It was a theme also addressed by the esteemed *Atlantic Monthly* at length in July. But in contrast to *The Blade* story earlier, another story in the same edition of the paper by the *Associated Press* stated "Don't jump the gun on zebra mussels, experts tell Congress." Perhaps the respective reporters heard different individuals with opposing opinions testify that day. But such mixed messages did little to clarify and only served to confuse both public and politicians. The year wound down with news of continued expansion of zebra mussels. "Unwelcome Guests" in Wisconsin Magazine of the *Milwaukee Journal* on November 26 detailed the coming of the mussel to Lake Michigan, while 2 days later the *New York Times* announced to the world "Scientists Find Zebra Mussels in Hudson River Ahead of Schedule." One wonders whose schedule that might have been. Clearly the mussels were not paying attention. The *Associated Press* kept the theme of mussel spread alive at year's end with a wire story on December 6 that stated "Zebra Mussels Wending Way to Inland Waters, Experts Say."

The turn of the year in 1991 brought an excellent summary to the public arena about the status of the zebra mussel issue to date from Ohio Sea Grant in a four-page fact sheet entitled, "Zebra Mussels in the Great Lakes, the Invasion and Its Implications." The publication among other points outlined how zebra mussels likely would spread south and west beyond the Great Lakes via the Chicago diversion into the Mississippi River and east via the Erie Canal from Lake Erie to the Mohawk River and thence to the Hudson River. The Illinois Department of Conservation promised a full-court press on monitoring for zebra mussels in 1991. One official interviewed was less worried about Lake Michigan, which was considered too deep and cold for zebras, but remained concerned about their spread to inland lakes and streams unless they were "nipped in the bud." Ironically, while zebras did not become fully established across Lake Michigan, the big, deep lake's bottom eventually was carpeted with its cousin, the quagga mussel (*Dreissena rostriformis bugensis*).

The summer of 1991 proceeded swimmingly for the mussels. The *Cincinnati Enquirer* announced on June 12, "Zebra Mussels in the Great Miami Watershed." The Great Miami River, which comes down to the river city from the northeast, was Ohio's first-designated scenic river. On June 25, in Illinois, the *Peoria Journal Star* told the world, "Multibillion-Dollar Pests Found in River." It referred to the state's beloved Illinois River and sited mussels within 60 miles of the queen of America's streams, the mighty Mississippi River. The Mississippi, as it turned out, had not escaped zebra mussel notice.

In September 1991, *Wildlife Conservation* magazine pointed out that mussels likely would be trailered on boats to all parts of the United States within 20 years. On September 13, the *St. Paul Pioneer Press* reported mussels in the Mississippi River at La Crosse, Wisconsin, far above the confluence of the Illinois and the Mississippi rivers. That raised more questions for subsequent stories to reveal to the reading public. Two weeks after the La Crosse, Wisconsin, discovery and reports, the Corps of Engineers found mussels far downstream on the Mississippi River from La Crosse, just north of St. Louis and at Dubuque, Iowa. In November 1991, *Biome*, the journal of the Canadian Museum of Natural History, well summed up the course of events with a major piece, "Zebra Mussels: North American Environmental and Economic Nightmare of the Nineties." No kidding. And a year-end report in *Soundings* noted, "Mussels Discovered Inland in Indiana." So 1991 came to a close, and awareness increased that zebra mussels were barreling on westward through connected waterways to inland impoundments, likely carried on trailered recreational boats despite an array of cautions and urging from various state and federal conservation, environmental, wildlife, and watercraft agencies.

In March 1992, the journal *Environmental Science & Engineering* detailed "How the Zebra Mussel Is Winning the West." The piece was indicative of the depth of understanding of the zebra mussel threat, which by that point went beyond more superficial treatment in the popular press, probably because the mussel threat indeed rapidly was spreading westward. On May 16, the *Cincinnati Enquirer* reported that mussels were found in the tristate region of Ohio, Indiana, and Kentucky. In the June edition of *Water Quality*, by the U.S. Water News, zebra mussels were reported in the Tennessee and Cumberland rivers. So the advance marched south to Dixie as well as west, following the water highways. On June 30, 1992, the *Detroit Free*

Press carried a report about the speed by which mussels advanced down the Mississippi watershed; the headline of the report read, "Zebra Mussels Spread to Warmer Waters" with a subheadline, "They Weren't Expected to Move South." Mussels had moved 300 river miles in a year, from St. Louis to St. Francisville, Louisiana. Ohio Sea Grant's *Twine Line* in October 1992 reported that zebra mussels had been confirmed in an Ohio inland body for the first time, Hargus Lake near Circleville, Ohio. The mussels were suspected to have hitchhiked on a trailered boat to Hargus Lake from Lake Erie. They appeared to have been in the water body for 3 years.

On November 11, 1992, the *Southwest Missourian* reported that zebra mussels had been confirmed in the Mississippi River at Cape Girardeau, Missouri, and all the way downstream to Vicksburg, Mississippi. On December 8, 1992, Michigan's *Grand Rapids Press* reported, "Zebra Mussels Found at Arkansas Nuclear Plant."

In April 1993, Ohio's *Twine Line* reported that zebra mussels were being transported inland by fishermen and by scuba divers. Griggs O'Shaughnessy Reservoir at Columbus, a fishing venue, and White Star Park quarry at Gibsonburg, in northwest Ohio, both had mussels. The latter location had no connection with streams and allowed no motorboats, so it was suggested that mussels were introduced to the reservoir, cold and deep, by scuba divers who frequently use the water for recreation. On April 22, 1993, the *Baton Rouge Advocate* reported that zebra mussels were being found in Louisiana rivers. It was another indication of the rapid advance of the pests to the southern and western United States. So the mussels had made it from the U.S. "North Coast" of the Great Lakes to its "South Coast" on the Gulf of Mexico.

An article in the May 1993 *Water Farming Journal* noted that zebra mussels had spread all the way to New Orleans. Mussels also expanded in the East, according to a July 27 edition of the *St. Albans Messenger* in Vermont. "They're here," the paper said with "Zebra Mussels Found in Lake Champlain." On September 4, 1993, the *Minneapolis Star-Tribune* noted that zebra mussels also had moved inland into northern Wisconsin's fabled Flambeau River watershed.

Amid this news came some of the initial findings of the invasion of a larger, deepwater cousin of the zebra mussel, the quagga mussel. One early article was entitled "From Tough Ruffe to Quagga," in the July 25, 1992 issue of *Science News*. The article among other things described a mussel similar to the zebra, noticed in the fall of 1990 in deep sample trawls in the southern basin of Lake Ontario. By the summer of 1991, quaggas, 20%–50% larger than zebras, were in the Erie Canal, which connects to Lake Erie "upstream." On August 18, 1992, Michigan's *Grand Rapids Press* carried an *Associated Press* story with the headline "Hungrier, Bigger Mussel Invades the Great Lakes!" Perhaps this reflected on a recent *Science News* report and possibly others that noted the quagga mussel was hardier and better equipped to survive colder waters than the zebra mussel.

Noting its discovery in Lake Ontario in October 1991, the article said that by August 1992, the quagga mussel was confirmed in Lake Erie, the St. Lawrence River, and inland western New York's Onondaga Lake. In December 1992, the *Research Biologist*, the American Institutes of Fishing's journal, added to the public evidence of quagga mussels with "Second Nuisance Mussel in Lake Erie." The article noted quaggas at two sites in the eastern basin of Lake Erie. At this point, the zebra mussel was well entrenched in the public consciousness, as reflected in the gradual decline of reports in popular media after 1993 (Figure 9.2). Zebra and quagga mussels, however, were not reading or listening or watching and instead kept marching toward all points of the U.S. compass, particularly to the west.

The Illinois Department of Conservation promised a full-court press on monitoring for zebra mussels in 1991. The department documented possible damage to engines and drivetrains of recreational boats, cited the potential impacts on commercial and recreational fishing, and warned beachgoers about the stench of accumulated mussel shells on shorelines. The take-home message was that mussels were here to stay and the aim should be to try to slow their spread. One official interviewed was less worried about Lake Michigan, which was considered too deep and cold for zebras, but the official remained concerned about zebra mussels' spread to inland lakes and streams unless they were "nipped in the bud," which, of course, did not happen. Lake Michigan, ironically, would become the eventual domain of the zebra's cousin, the quagga mussel.

RECOGNIZING THE BIOLOGICAL/ECOLOGICAL THREAT

A University of Guelph zoologist who figured prominently in the discovery and study of zebra mussels was emphatic that mussels were a threat with a dramatic statement in *Equinox* magazine in October 1990, in an article entitled, "Alien Onslaught." The zoologist stated: "The explosive growth of *D. polymorpha* is shaping up to be the largest, costliest, and most disruptive biological event in Great Lakes history." The feature went on to discuss the problem with ongoing ballast-water pest introductions and infestations, discussed such permanent solutions as chemical or heat treatment, and even discussed the possible redesign of ballast tanks in commercial ships. But it concluded with a telling statement from the zoologist that "concrete action is still years away."

It is worth quoting at length an article authored by Ohio Sea Grant for *Ontario Fisherman* magazine in April 1990: "The feeding method of zebra mussels points to one of the growing concerns in regard to Lake Erie's food chain. Each mussel is capable of filtering about a liter of water per day. Nearly all particulate matter, including the plankton, is strained from the water. The zebra mussel removes a considerable quantity of plankton from the water. The part that isn't eaten is tied up, unavailable to microscopic crustaceans which feed larval and juvenile fishes and unavailable to the plankton-feeding forage fish that support Lake Erie's sport and commercial fisheries." This was plain-spoken, clear information that fishermen, whether hook-and-line sport anglers or commercial netters, could understand. The article further discussed the theory that much of the western Lake Erie mudflats couldn't be colonized by mussels. It was a theory based on the state of knowledge at the time, but that later would be disproved by the mussels themselves, for it eventually was found that even a stray rock, or a bed of dead mussel shells on a vast muddy bottom, could and would provide an attractive substrate for larval veligers to attach to and develop. The foregoing biological concerns bring up an additional consideration, detailed in the Cleveland Museum of Natural History's *Explorer* quarterly for the spring of 1990. An article therein illustrated how some of the plankton in the water column, filtered out by the mussels but not ingested or assimilated, become bound up in mucus pellets. The pellets, called pseudofeces, along with feces accumulate on the lake bottom, decompose, and contribute to low oxygen that contributes to stress and possible mortality to fish and other aquatic fauna.

In August, the Michigan Department of Natural Resources issued a news release, "PBS Special Examines Threat to Great Lakes," which contained some rather dramatic statements for the popular press to chew on. One spoke of "zebra mussels terrorizing industrial, municipal, and recreational water users in the Great Lakes and will eventually affect anyone who takes water from the lakes." The release went on to restate and simplify the case for a Michigan audience that "zebra mussels remove algae and plankton from the water column, which will deplete the food chain for walleye and [yellow] perch." The release concluded that the mussels pose a greater possibility for disaster than any dramatic oil spill.

Stories circulated in the early 1990s also suggested mussels were a positive food for prey fish species, and there had not yet been any confirmation that mussels posed a danger for the bioaccumulation of contaminants up the food chain en route to the human dinner table. Nor did laymen take into consideration the possible decimation of the base of the ecosystem food webs by the filter-feeding masses of zebra mussels, which individually filter up to a liter of water a day. Such reports raised hopes of sport anglers' wishful thinking that clearer water equaled cleaner water. Subsequent research would disabuse the public of that notion, for those who were listening. As evident time and again throughout the zebra mussel saga, information about consequences readily was available in the public domain. Just as often as not, however, that information was not acted on quickly enough if at all, or simply not recognized and absorbed.

Early in 1991, an excellent summary appeared in the public arena about the status of the zebra mussel issue, this from Ohio Sea Grant in a four-page fact sheet entitled "Zebra Mussels in the Great Lakes, the Invasion and Its Implications." The publication among other things represented the state of the scientific art on the aquatic pests. It noted, among other points, the following: mussels were thriving at depths of 6–45 ft. They generally preferred to colonize hard surfaces, but "surprisingly they can also colonize soft, muddy bottoms; mussels even would form mats on the remains of spent mussel shells." The fact sheet also importantly traced how and where mussels interrupted the traffic on the established food highway in the lakes via the mussels' voluminous filter-feeding. The route started with the smallest of the small, phytoplankton, and continued up the microscopic chain through zooplankton and connecting them as food to larval and juvenile fishes and plankton-feeding forage fishes that support important sport and commercial fishes (e.g., walleye, yellow perch). This was just the sort of information that a vitally interested public and fishermen both need and needed to understand. The fact sheet also went on to introduce the concept of bioaccumulation of contaminants, noting that the mussels could concentrate toxic organic pollutants suspended in the water column by a surprising factor of 300,000 and then redeposit the concentrated contamination in the bottom sediments via expulsion of pseudofeces. "Zebra mussel now are a permanent part of the lake environment" ended the fact sheet with a note of finality. In one sense, the importance of this publication is that it institutionalized the zebra mussel story, carried it beyond mere daily news curiosity, and gave it the depth and impact of permanence. In April 1991, the *Columbus Dispatch* reported in the daily press what Sea Grant had written months before in the fact sheet, headlining "Zebra Mussels Found Laden with Toxic Chemicals," followed up with "Food fish that eat them could pose risk." Thus, the proverbial other shoe had fallen. The dangers of bioaccumulation of toxics in the food chain were beginning to see popular light.

The zebra mussel issue also was being institutionalized on the Canadian side of the lakes, even down to the elementary education level. The Ontario Ministry of Natural Resources, in a Youth Fisheries Education Program booklet, a student manual for fisheries education, complete with lesson plans for teachers, included a well-vetted section on zebra mussels.

In June 1992, Ohio's *Twine Line* reported on another aspect of mussel meanness with "Invader is Out-Musseling Native Mussels and Clams." The report noted that in the summer of 1992, federal researchers could find no viable populations of native mussels in Lake St. Clair. The publication's October issue went on to report that about 30 species of native mussels and clams had been extirpated by zebra mussels in Lake Erie. The article also noted that plans afoot to use black carp against mussels had been declared "too risky."

On September 2, 1993, Michigan's *Ann Arbor News* asked, "Lake Erie Now Being Overfiltered?" The article noted that *Diporeia*, a tiny shrimp and a pillar of the lake's food base for valuable fishes such as lake trout, whitefish, sculpin, and smelt, had virtually disappeared because of mussels.

WATERWORKS AND POWER PLANTS VERSUS ZEBRA MUSSELS

Two of the more dramatic battles that drew national attention in the early days of the zebra mussel war both occurred at Monroe, Michigan, on the west end of Lake Erie. In August 1989, the mammoth Detroit Edison coal-fired power plant began grappling with monstrous mussel problems, as the water intake to the twin-stack facility was found to be choked with mussels to the incredible tune of 700,000/m^2. Also, the city of Monroe declared a water emergency in December because zebra mussels also had clogged its lake water intake and reduced flow to a mere trickle. The front page of *The Monroe Evening News* displayed a photograph of the city's residents at a grocery store, waiting to buy bottled water (Figure 9.3). The cities of Vermilion, Ohio, on the south shore of Lake Erie, and Tilbury, Ontario, on the north shore would later experience similar water woes and similarly would make local headlines as a result. On June 4, 1990, the *Ann Arbor News* displayed a photograph of a "pig," shaped like a World War II bomb, that was used to mechanically bore its way through mussel clusters clogging the mussel-plagued Monroe, Michigan, water intake (Figure 9.4). The mussel take-over of the City of Monroe was cleverly depicted in a series of editorial cartoons (Figure 9.5).

By 1990, the industry was paying very close attention to mussels, especially the power industry. The Electric Power Research Institute produced a 16 min informational video for its members called "Zebra Mussels, the Silent Invaders." It covered the mussel influx and how it was affecting electric power plant operations, how to collect and recognize mussels, how fast and far they spread, and ways to reduce their impacts, including thermal, chemical, mechanical, and antimussel coating regimes. The video, interestingly, still was targeted toward just Great Lakes power plants. An article on August 17, 1990, in the *Grand Rapids Press* in southwest Michigan entitled "Mussels Tracked at Muskegon Plant," illustrated the recurring themes about mussel-related problems and damage to industrial plants, waterworks, and power plants as the zebra mussel invasion expanded over time. In each case, the press responded by duly reporting on the issue.

Even the U.S. Army Corps of Engineers got into the act, deciding to flex its own mussel muscle by enlisting its Waterway Experiment Station in Vicksburg, Mississippi, to do a 4 year, $2 million research study to assess impacts of zebra mussels on waterworks, according to the *Sunday Post* in Vicksburg on February 17, 1991. At this relatively late stage in the game—given the rapid and widespread of the pest and the size of the incoming damage receipts—$2 million seemed a paltry, even token sum in retrospect, and a 4 year time frame seemed like a classic case of too little too late. Some members of the media recognized as much even

Figure 9.3 Front page of the *Monroe Evening News* on December 16, 1989.

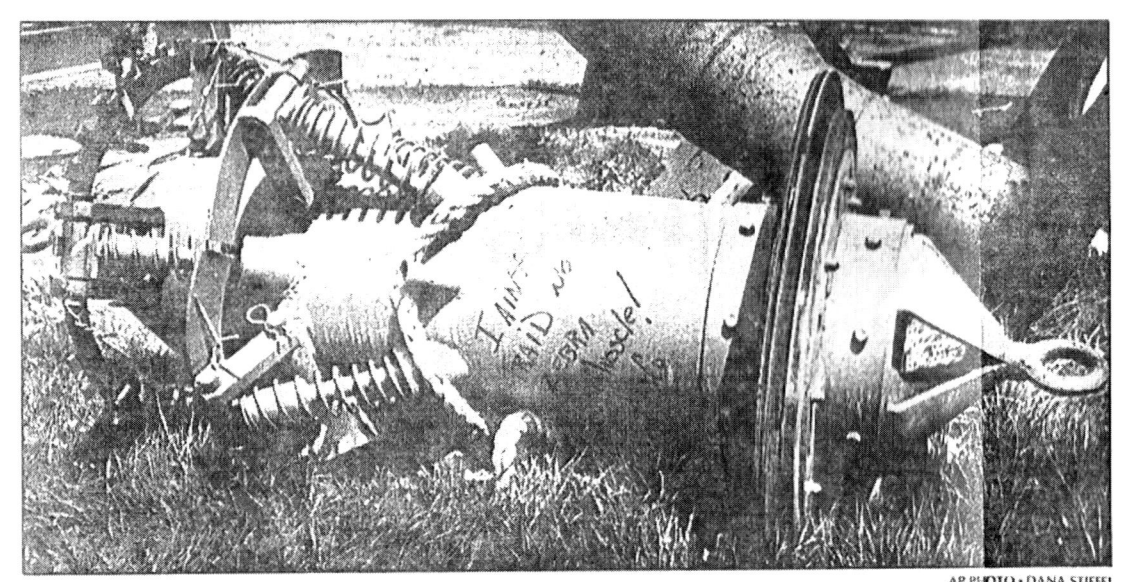

PIG'S AWAY! — With a spirit reminiscent of World War II bomber crews who scrawled messages such as 'A kiss for Adolph' on their bombs, this pipe-scouring 'pig,' as it's called, contains a message for the zebra mussels it cleaned out of Monroe's water intake pipes. Even though the spelling is suspect, the device is being monitored nationally as a solution to the zebra mussel problem.

AP PHOTO • DANA STIEFEL

Figure 9.4 Photo of "pigger" used to remove mussels from the inside of Monroe water plant's intake. Written on the pigger is "I ain't fraid of no muscles!." The photo was in the *Ann Arbor News*, June 4, 1990.

a year sooner than the Corps of Engineers. A column on zebra mussels on February 4, 1990, in *The Blade* in Toledo succinctly summed up the state of the research response that was to be a companion to the political one: "It is happening more quickly than science can react."

Because the zebra mussel explosion occurred in the western Lake Erie–Detroit River and Lake St. Clair region, with print media abuzz almost daily about developments, it is no surprise that television news organizations in the area picked up on the issue. In the spring of 1989 right on into late 1992, Detroit television channels 2, 4, 7, and 50 and channel 9 in Windsor, Ontario, and WGTE-TV channel 30 of the Public Broadcasting System in Toledo all featured news segments on mussels, particularly the more graphic issues, such as the clogged water treatment and power plant intakes at Monroe, or the "mussel car" at Wheatley, Ontario, not too far east on Lake Erie's north shore. Review of videotapes of news programs at the time illustrated the power of television and video images. Among several that come to mind was one showing mussel-cleaning crews filmed inside the huge intakes (about the size of small gas stations) at Detroit Edison's Monroe electric power plant, where the crew scraped and shoveled and power-washed thick layers of mussels from walls, water pumps, and pipe surfaces. A second notable video was of a scuba diver surfacing alongside a workboat with a mass of mussels and telling how walking on the mussel-carpeted lake bottom in the Bass Islands region of western Lake Erie was "like walking on Velcro." Then there was a documentary produced by Wisconsin Public Television, in observance of the 20th Earth Day, called "The Zebra Mussel." This television piece was rebroadcast nationally on the PBS NewsHour MacNeil–Lehrer Report on July 30, 1990, and had been filmed in the fall of 1989. The program began with the now-familiar clog up of water intakes of Monroe, Michigan, and graphically covered the unexpected, uncontrolled population explosion of zebra mussels, traced its invasion route from Europe via ballast water, and ominously noted that it was too late to close the barn door on the zebra. The PBS piece also stressed how mussels were using up the base elements of the food web of the waters they infested. Videos of native crayfish coated with mussels, and even a length of fishing line to which clumps of mussels had attached, were dramatic presentations for viewers. Among telling points made in the commentary were that warm, shallow, western Lake Erie had developed the highest known mussel concentrations in the world at 750,000/yd^2 and that federal researchers were busy searching and hoping for zebra mussel-eating fish. The point about hoping for a fish that would eat zebra mussels points to the saying about "be careful what you wish for," as we shall see.

Stemming the Zebra Mussel Tide

In September 1991 came word of a major "uh-oh." The *Great Lakes Reporter*, a publication of the Center for the Great Lakes, detailed the arrival of two more ballast-water hitchhikers, the tubenose goby and the round goby, traced to the Lake St. Clair River in April 1990. The round goby in particular would have an impact, especially in Lake Erie, that ecologically, if not economically, rivaled that of the

Figure 9.5 Series of editorial cartoons that depict the infestation of dreissenid mussels in the water treatment plant of the city of Monroe, MI. (From The *Monroe Evening News*, December 26–29, 1989.)

zebra mussel. The explosion of round gobies lay just over the horizon. Gobies were shown to feast on zebra mussels, but later research again raised the specter of bioaccumulation of contaminants up the food web. In any event, no fish species, including freshwater drum among others, had been able to even dent mussel proliferation.

On January 21, 1990, the *Orleans Times Herald* in New York state published an article, indicative of others to come for many years, about the supposed benefits of zebra mussels. For one thing, the article quoted authorities as saying that zebra mussels were first-rate fare for diving ducks, sturgeon, and eels of Lake Ontario. In retrospect, it almost is quaint to consider that ducks and fish could eat down the intimidatingly, overwhelming masses of mussels.

In June 1991, the Illinois Department of Conservation, apparently in anticipation of the coming mussel tsunami in the state, produced an 11 min videotape as a "Special Report" on zebras for general public dissemination. Mussel colonies, the narrator said, "wreak havoc on waters they infest." The tape placed the mussels' likely starting point here as 1985 in Lake St. Clair, where the plankton-rich habitat was an ideal breeding ground. As of the date of the videotaping, it was noted that the mussels were in all the Great Lakes and connecting rivers and had reached infestation levels in Lake Erie in particular. Several Sea Grant and other public agency publications at the same time were busy discussing use of special chemicals to treat water intakes and the use of antifouling paint to reduce mussel encrustation of boat hulls.

At the time, the chapter and verse of the mussels' need for hard surfaces still were being projected without reference to their ability to carpet even soft muddy bottoms, often using the shells of their spent brethren as an attachment substrate. Diving ducks and drum, carp, and sturgeon were presented as mussel predators but at least with the proviso that they would not control mussel expansion because of the pests' fecundity. Even as late as May 1993, *Water Farming Journal* reported that "Three U.S. Fish Farms Preparing to Spawn Black Carp for Use Against Zebra Mussels." It again was to be a case on using one invasive species to eat another. The black carp was one of the "Asian carps" that were to be regarded as the most severe ecological threat 15 years later. However, black carp turned out not to be as big a threat as its cousins, the bighead and silver carp.

Biological predators were not the only remedies under consideration in the war against the mussels. Chemical solutions also were being sought, and with them came another suite of concerns. For example, an *Associated Press* wire story on April 19, 1990, raised a new issue on the public radar with a headline "Mussel Chemical Prompts Concern." It discussed Clam-Trol CT-1, a compound being introduced by utilities to combat mussels in water intakes. Like the issue of mussel leapfrogging on trailered boats, Clam-Trol and its use by utilities would become a regional issue around the lakes, as duly reported by various local newspapers large and small.

Along the way in these early mussel years, all sorts of anti-mussel remedies were offered. On September 1, 1990, the *Associated Press* reported that two brothers in Defiance, Ohio, had devised a machine that would dispense chlorine tablets in timely fashion, and the tablets were designed to dissolve and kill mussels. On September 20, the *Toronto Star* carried a report touting "African Plant Found to Kill Zebra Mussels." It was a theme to be recycled regularly for the next month as newspaper after newspaper "discovered" the story and ran with it.

By October 1990, even *Sports Illustrated* magazine had gotten into the act, with an article on the 15th entitled "In Shock Over Mussels." The end of December 1990 brought a spate of articles about a new anti-mussel remedy, potassium phosphate. But the chemical, like most else, was not found to be a magic bullet and had limited applications.

Almost as a side show to all this came the first print-media reports of proffered assistance from researchers in the still-then, Soviet Union. On May 3, 1990, the *Ann Arbor News* reported "Soviets Offer Aid on Mussels." The promise was for a biological control to the creatures, a familiar organism "over there." Soviet interactions were rife with false starts and empty promises and would become a dead-end road in coming months. Southwest Michigan's *Grand Rapids Press* on August 23, 1990, carried a report about the ongoing, purported mussel-control aid proffered by Soviet researchers. As feared, the headline said it all: "Soviets Offer No Help in Halting Zebra Mussels, EPA Official Says." Some notions of control would not go away. For example, on September 27, the *Associated Press* was telling the public, in another iteration, that "Zebra Mussels Have Good Side, Researchers Tell Michigan Group." The story quoted a Soviet scientist as saying that the mussels "are a wonderful filter of water." Such unqualified statements, if seen by a reader who had no other frame of reference through other information or stories, thus might reach a false conclusion about the mussels. At least the same article went on to repeat a telling forecast that "75 percent of the continent has habitat that could support the zebra mussel."

Then there was the lighter side of the mussel story. The *Geauga Times Leader* in northeast Ohio on May 9, 1990, carried a story entitled "Zebra Mussel Jewelry Brings Profit to Shopkeeper." It was the story of Kent Floro, of Oak Harbor, Ohio, a member of a marina family on the Toussaint River on western Lake Erie who struck on the idea of making novelty jewelry—earrings and necklaces to tie tacks—with mussel shells, to which goggle-eyes were glued. It was a breath of comic relief. To further add to the fun, several years later, an article on how blue crabs may control mussels was carried by several newspapers and the article included a "file" photo of a mussel with goggle-eyes.

Along with proposed control schemes came some erroneous notions, including a prevalent one that held that clearer water, thanks to zebra mussel filtering, spelled cleaner water. The media jumped on the contention with both feet. On August 30, 1991, the *Cleveland Plain Dealer* articulated that the public learning curve was rising in a story that said that clearer did not equal cleaner or better when it came to Lake Erie's water quality. The story went on to say that increased water clarity because of mussel filtering allowed sunlight to penetrate to the lake bottom at Put-in-Bay. The clearer water paved the way for renewed growth of bottom-oriented, rooted vegetation and eventual return of various high-value fish species that relate to such habitat, such as largemouth bass, northern pike, and muskellunge. But a year later, the notion persisted in the public mind, as evidenced in a *Grand Rapids Press* editorial on April 6, 1992, "Mussels May Help Purify the Water." Further, on August 13, 1992, a headline in Michigan's Ann Arbor News noted "Notorious Mussels Help Clean Up Water." The common misconception that clearer is cleaner again rose its ugly head. The story, however, had more ominous tales to tell. The story reported on a research article published in the journal *Archives of Environmental Contamination and Toxicology*. In this article, scientists reported that they had found concentrations of cadmium, selenium, and PCBs in zebra mussel bodies and shells. An environmental engineer was quoted as: "There's no doubt Lake Erie is clearer because of the zebra mussel. But I don't know if they're cleaning the water."

On July 27, 1990, a story in the *Port Clinton News Herald* on Lake Erie's southwest shore took note of "beaches piled deeply with empty shells." The article detailed how windrows of shells of dead zebra mussels had washed ashore in the fall of 1989. It was not the first or last news item about the shoreline shell shock. Often such pieces included warnings about not going barefoot on the fragile, razor-sharp shells.

LEGISLATION AND POLITICS

Over time, the issue of dreissenid mussels in North America gained political "muscle." A guest editorial in the news pamphlet *Advisor* (Great Lakes Commission) in February 1990 was authored by the prominent U.S. Senator John Glenn (D., Ohio), who was hard at work generalizing what he hoped would be a significant political response, the Nonindigenous Aquatic Nuisance Prevention and Control Act of 1990. "This problem will not wait," the esteemed senator and former astronaut asserted. "We're already behind. I hope the Great Lakes Commission will endorse and support this legislation." From the vantage point of a 20-year look back, the senator could not have been more on target when it came to being behind the curve in the mussel war. The act he sponsored was indeed passed by late 1990, but succeeding Congresses did not fully appropriate all funds authorized by the Act.

The year of 1992 opened with word in the news wires that the cavalry, in the form of federal aid, would not be coming to the rescue, at least in any numbers anytime soon. The *Port Clinton News Herald* quoted the home-state hero Senator Glenn as being disappointed that President (George H.W.) Bush's budget of $1.52 trillion contained no money to control zebra mussels. Washington still did not take the pests or invasive species' massive economic impact, seriously (Figure 9.6). Even nearly two decades later, in 2010, there was sluggish political response to invasive species, as evidenced by the slow walking of action to halt the Asian carp from entering the Great Lakes—also at the Chicago choke point between the Mississippi watershed and the lakes. Just days earlier, on April 3, 1992, the Great Lakes Commission in a newsletter quoted the anti-mussel champion, Senator Glenn, who gamely was trying to keep the issue on the front burner. The senator stressed "carried overland in the transport of recreational boats from one body to another, the zebra mussel probably will invade two-thirds of North America's freshwater if left uncontrolled." He added that mussels would cost the Great Lakes states alone $500 million a year for 10 years. The senator's efforts even were noted in a national sportfishing magazine, *Bassmaster*, in June 1991, in an article entitled, "The Zebra Mussel Invasion." The *Bassmaster* piece was a thorough examination of the status of the species' pestiferous intrusion and outlining its threat to beyond the Great Lakes. The article also devoted space to the Nonindigenous Species Act of 1990 and discussed how the act was designed to provide $20 million over 5 years toward zebra

Figure 9.6 Editorial cartoon characterizing the minimal amount of federal funds devoted to the zebra mussel problem. (From *Windsor Star*, April 7, 1990.)

mussel research and control. In retrospect, such news illustrated the tremendous credibility gap that exists between passing legislation and implementing it. The latter takes money, and Congress and the White House weren't giving, not nearly enough or quickly enough, to do much good. At this point in the Great Lakes in 1991, the invasive species tally now was up to about 115 species, more than double the total listed by the shelved Environment Canada report a decade earlier. Elsewhere in the article, a federal biologist noted that at that point, "the Mississippi River is our major concern." The *Bassmaster* article also noted that as many as 600 ships steamed into the Great Lakes each year, many of them carrying as much as 1.25 million gallons of ballast water per vessel. The magazine also noted that as of May 1989, 2 years prior, when the invasion was known to be well under way, the U.S. and Canadian Coast Guards still only were asking for voluntary exchange of ballast at sea, this on the assumption that saltwater would kill the potentially nasty invaders in the water. It still had not dawned on regulators and enforcers that even "empty" ballast tanks truly were not empty because of assorted detritus, sediment, and organisms that had settled in the bottom of the tanks. Ships with no ballast water, or with "empty" tanks, were not required to exchange at sea. The situation begged for disinfection treatment.

Some of the first, half-hearted attempts to restrict contaminated ballast-water dumping were reported on April 9, 1993, in the *Buffalo Evening News*, with "New Ballast Water Rules Aim to Slow Spread of Pests." It described U.S. Coast Guard regulations requiring that ocean-going ships to take on seawater as ballast as they crossed the open ocean before entering the St. Lawrence Seaway and the Great Lakes.

The lack of timely political and legislative action did not seem in synch with public attitudes. The April 1992 edition of *Twine Line* included a telling outline of public perception of the zebra mussel problem. It discussed an annual survey of 700 visitors to the Mid-American Boat Show in Cleveland, in which 54.2% of respondents considered mussels a threat to Lake Erie sportfishing and boating industries. However, just 46.7% thought they were harmful to the environment, while 63% thought the environment and economic impacts were not well known. But the August 1992 issue of *Twine Line* updated common perceptions in a public survey of 285 individuals, noting that 76% of them had heard of zebra mussels but only 43% of them had seen one.

Thus, the initial tsunami of the zebra mussel and quagga mussel storm had washed across eastern and central North America. What followed after about 1993 was essentially more of the same, just the geography changed as the mussel waves washed south and west, often hitching rides on trailered boats. So this perhaps reflected reduced broad media and public attention as compared to 1988–1993 (Figure 9.2).

AFTERMATH OF THE PERFECT STORM, 1994–2011

The year 1994 reflected the beginning of a decline in broad media and public interest in the topic of zebra and quagga mussels, as if the public had reached at least a grudging acceptance of the economically and ecologically costly invasion. The principal exceptions were those geographical regions and locales where the pests represented a new and nasty, expensive threat to industry, public waterworks, fisheries, and recreational opportunities. But reporting where and when it happened followed the template of topics stamped out in the prior 5 or so years. Headlines from 1994 gave a sampling of things to come: February 7, *Grand Rapids Press,* "95 U.S. Budget Slashes Funding for Zebra Mussel Research"; March 7, *Ann Arbor News,* "Budget Would Cut Zebra Mussel Research"; August 2, the *Democrat Argos* in Caruthersville, MO, "Zebra Mussel Spreads Around Missouri"; and August 27, *Associated Press* wire story, "Almost 150 Exotics into Great Lakes Now."

The mussel mania continued to slow through 1995, but reporting that did appear was more substantial and significant than former, simpler "the-mussels-are-coming" themes. On January 22, 1995, for example, the *Columbus Dispatch* carried a lengthy interview with the director of a prominent research center on Lake Erie. The headline of the article was telling "Zebra Mussels Changing Life in Lake Erie." The article went on to say: "The zebra mussel is making unbelievable changes in the ecosystem," this in reference to impacts from New York to the Gulf of Mexico to Oklahoma. In a reflection on the subtlety of the larger invasive issue, the article reported that the loss of native clams to the mussel infestations even was costing clamshell collectors millions of dollars in losses annually. Pieces of native clamshell bits are used to high-grade pearls, and zebra mussels were killing clams that no longer could be used for commercial production of oysters. In the same article, a state biologist restated the contention that clearer water is not necessarily cleaner water because zebra mussels could concentrate such contaminants as phosphorus and PCBs.

A headline in the August 22, 1995, edition of *The Blade* in Toledo displayed an almost tongue-in-cheek acceptance of the pest by stating "Go West Young Zebra Mussels: California Battles New Menace." The article detailed how mussels were being found on trailered boat hulls during stops at interstate border inspection stations. The reach of the infamous pests now stretched coast to coast. A decade after the mussels colonized Lake St. Clair, they had been absorbed into the larger suite of environmental problems known variously as aquatic invasive species or aquatic nuisance species (ANS). In July 1996, for example, the Great Lakes Commission's *ANS Bulletin* outlined "Battling Biological Invaders" and rolled zebra mussels into the "pest" group that also included the infamous early invader, the sea lamprey, plus the ruffe, gobies, and purple loosestrife, among others. In all, the bulletin noted that 139 species of invaders were known to infest the Great Lakes.

In August 1996, *Scientific American* still was reporting on zebra mussels with "Mussel Mayhem, Continued." Its detailed article included the menacing news that an estimated 50% of the contamination at ground zero, Lake St. Clair, now was embedded in zebra mussel tissue. On August 22, a *New York Times* wire story in the *Ann Arbor News* raised a chilling thought, noting that Ohio State University researchers were trying to determine whether zebra mussel contamination was reaching the dinner table. At the same time, the National Research Council was reporting "U.S. Urged to Regulate Ballast Water." Clearly, after a decade of damage and hundreds of millions of dollars thrown at the problem since the initial mussel outbreaks, effective ballast-water treatment and control had not been achieved.

In any case, the mold had been set for the issue of public reaction to zebra and quagga mussels as reflected in the media. It is possible to obtain a fair picture of media coverage as it transpired from the late 1990s through the first decade of the twenty-first century by sampling some summary pieces published in 2010 and 2011. In some cases, zebra mussels and quagga mussels still made news, though not of the nearly daily variety of reports 20 years prior. Often they also were not cast as stand-alone pests, but rather they were rolled into the larger arena of invaders, which by now were so common as to wear the acronym in certain circles of ANS. A typical example is the treatment of invasives, which refer to both the mussels, in the December 10 issue of the news magazine *The Week*. The issue's briefing page was devoted to "Nature's Marauders," and it covered the gamut from kudzu vine and long-horned beetles to northern snakehead fish, Asian carp, and the mussels. The zebra mussel appears in a photograph, showing a familiar "gob" of zebras that had attached to a submerged stick. Quagga mussels were in the article's text in a section devoted to their origins in the rivers of the Ukraine and transport overseas in ballast water. Interestingly, federal scientists noted that quagga mussels by 2010 carpeted the entire bottom of Lake Michigan, in densities averaging 5,000/m^2, and that they busily were colonizing lakes and reservoirs in the southwestern United States, "having arrived via boats hauled by trailer from the Midwest."

An excellent summary tracking the path of zebra and quagga mussels to the southwest appeared in the summer/fall quarterly edition of *Eddies*, a fisheries' conservation publication of the U.S. Fish and Wildlife Service. The article, "Conservation in a Quagga-mire," traced the now-familiar history and basics of the biology, ecology, and economy of mussel invasion south and west through the sprawling mid-continent Mississippi River watershed. In particular the article detailed how quagga mussels, largely restricted to the Great Lakes basin until 2007, leap-frogged some 1,800 miles to the lower Colorado River at Lake Mead on the Arizona/Nevada border. "It is likely that quagga mussels came [to Lake Mead] attached to a houseboat brought from the Great Lakes," *Eddies* noted. The quaggas then simply followed the flow of water, including into artificial irrigation and water-supply diversions and associated drainages in southern California and Arizona. Trailered boats transferred quaggas to isolated impoundments in Colorado and Utah. Various graphics with the article display all-too-familiar representations, from mussel-encrusted marine equipment in Lake Oologah, Oklahoma, to a heavily coated penstock gate at Lake Mohave on the Colorado River in Arizona. The western invasion was costing the Metropolitan Water District of Southern California $10 million to $15 million a year for mussel controls. In an aside, the *Eddies* piece noted that between 1993 and 1999, the "combined economic impact to industries, businesses, and communities may have exceeded $5 billion during that six-year span."

In the July/August 2011 issue of *The Wildlife Volunteer*, a publication of the Michigan Wildlife Conservancy revisited a now-familiar theme with "Clear Water Not Necessarily Good." It spoke of the incredible explosion of quagga mussels between 2000 and 2010. Another article in the *Detroit Free Press* on October 2, 2011, headlined "Voracious Invasive Quagga Mussels Gobbling Great Lakes Food Chain." It noted that "quagga numbers are staggering" and that "there are 437 trillion in Lake Michigan alone based on 2010 surveys." Large predatory fish such as salmon and trout were growing more slowly because of fewer forage fish, in turn because of mussel disruptions down the food chain, especially the loss of *Diporeia*. The MWC article noted that quagga mussels had replaced zebra mussels in Lake Michigan and are present in Lake Huron and Lake Ontario as well but that zebra mussel even had made it to remote Isle Royale in northwest Lake Superior, "despite an order prohibiting ships from emptying ballast tanks near Isle Royale." Obviously, the weak-kneed federal response to implementing the spirit and intent of the 1990 Nonindigenous Species Act continues without a successful challenge. Lastly, the MWC piece notes that zebra mussels "are showing up in inland lakes in Michigan."

INTERNET AGE

Unlike the state of the media in the mid-1980s, when news clipping services and handling of "paper" cuttings of news stories and photocopies of reports were the norm, cyberspace and the electronic age and personal computers—collectively the Internet—made it far easier to keep track of mussel activities in the news. One of the prime, go-to websites on the Internet is www.protectyourwaters.net. It is part of the "Stop Aquatic Hitchhikers!" project of the federal Aquatic Nuisance Species Task Force. The website is updated weekly with an array of news reports on all manner of plant and animal invaders, and more often than not, a new report turns up about the mussels' spread. A survey in the spring and summer of 2011 presented a fair sample of reports of how the mussels had progressed toward and into the West in the prior 10–15 years. The reports mainly are from the websites of newspapers and television broadcasting stations, a sign of the times and an indication of where the communication media are headed. Most mussel-related topics in the West now are familiar: trying to stop the spread of zebra mussels and/or quagga mussels via regulations on specific bodies of water still thought to be mussel-free, and the latest round of attempts to control mussel-specific bodies of water and locales.

News services in more than a dozen states were tracking stories of zebra and quagga mussels in the summer of 2011. The following is a summary of those developments, state by state:

Connecticut—a popular recreational body of water, Candlewood Lake, had posting of 20 warning signs at five public launch ramps and several private ones advising boaters about the zebra mussel threat, given the recent discovery of zebra mussels in nearby Lake Lillinonah and Lake Zoar. The advisory extended to kayaks, canoes, and personal watercraft, which were not being targeted 15 years ago. This was according to the July 21 edition of the *Litchfield County Times*. The warnings direct boaters to "clean, drain, and dry" their boats, trailers, and gear. The same article cites other mussel-infested waters in Connecticut and across Massachusetts, New York, Pennsylvania, and Vermont. A similar article in the *Danbury Patch* noted that inspections of incoming, out-of-town watercraft are being done by volunteers in hopes of staving off a mussel invasion at Candlewood Lake.

Minnesota—while zebra mussels were no stranger to this state, they had not infested all waters of this "land of 10,000 lakes." Thus, it is no surprise to see news stories in the spring and summer of 2011 describing grassroots attempts to prevent infestations at individual lakes. The *Minneapolis Star Tribune* on May 22 reported on attempts by the 140-member homeowner association at Christmas Lake to prevent the spread of zebra mussels from nearby Lake Minnetonka, where mussels were discovered in the summer of 2010. Among other things, the association planned to install a code-activated gate on the lake's lone launch ramp. "The idea is to require boaters to go to one of several locations for an invasive species inspection, where they could get a punch code, similar to an automatic car wash, to raise the gate at the ramp," the paper reported. It also said that large numbers of lakeside homeowners did not think that the state department of natural resources was doing enough to prevent the spread of zebra mussels in inland lakes. In July 22, 2011, *Echo Press* in Alexandria, Minnesota, reported that the Douglas County Lakes Association has invited a retired scientist from New York to discuss the possible use of a mussel biopesticide, Zequanox. The scientist and a colleague found that a common bacteria strain, *Pseudomonas fluorescens*, could kill zebra and quagga mussels. The bacteria are the principal active ingredient in Zequanox. The approval for testing this biopesticide in several Minnesota lakes was being sought. Also in July 22, the *Sacramento Business Journal* in California announced the first, full-scale field test of Zequanox in the United States. The test would involve treating for quagga mussels at Davis Dam at Laughlin, Nevada, on the lower Colorado River. The trial was approved by the U.S. Bureau of Reclamation after 2 years of trials in Canada.

Missouri—a man from Independence, Missouri received a maximum fine of $1000 and 6 months' probation for introducing zebra mussels into Smithville Lake in 2010, according to the July 11, 2011, edition of the *Smithville Herald*. It was the first prosecution under a provision in the Missouri Wildlife Code designed to stop the spread of invasive species. The case involved an individual installing a private boat lift moved to Smithville Lake from Lake of the Ozarks, where zebra mussels already were established. A quick action following a swimming inspection of a marina by conservation authorities was undertaken by the U.S. Army Corps of Engineers and Missouri Department of Conservation. The agencies removed the offending boat lift and decontaminated it with a copper-based algae killer to eradicate the mussel infestation in August 2010. No zebra mussels have been detected in the lake since, though monitoring continues, the newspaper reported.

Nebraska—the state's boaters were reminded of the importance of watercraft inspections in stopping the spread of zebra and quagga mussels, according to the *Imperial Republican* newspaper on May 11. In particular, the Nebraska Game and Parks Commission and University of Nebraska teamed up to alert boaters on such popular lakes as Enders Reservoir about the mussel threat. The newspaper reported that the NGPC began educating anglers about the dangers of mussels in 1999–2000. A major concern, the article said, was about out-of-state boaters, many of them from Colorado, which has confirmed zebra mussel invasions of some of its impoundments and on trailered boats. Mussels have been found at Zorinsky Lake in Omaha and at one near Offutt Air Force Base there, the newspaper said. Voluntary inspections of boats in eastern Nebraska, informational handouts on the clean–drain–dry process, and decontamination as needed were under way through the university's invasive species project.

North Dakota—the state's boaters were reminded that immature zebra mussels are present in the southern reaches of the Red River, based on water samples below the Kidder Dam near Wahpeton in June 2010, and again in 2011, according to the July 1 edition of the *Jamestown Sun*. The clean–drain–dry mantra was included in the advisory article, which noted that zebra mussels are "found in areas east and south of North Dakota. Adult mussels have not been found in North Dakota waters."

Texas—zebra mussels were causing huge, potentially very expensive problems in the northern portions of the state in 2011 according to several reports in the spring and summer. *WFAA-TV* in Dallas/Fort Worth noted in a broadcast summary on zebra mussel woes on May 19, 2011, that in 2009, zebra mussels first were found in Lake Texoma, on which about 1.5 million north Texans rely for water. But since this discovery, the North Texas Municipal Water District had not dumped a "single drop of water" to its treatment plant at Lake Devon, which in turn led to conservation measures in cities across the northern tier of the state. In the lake-to-lake pumping shutdown, the water district was trying to prevent the spread of zebra mussels from the Red River valley to the Trinity River valley, *WFAA* said. To safely resume, pumping might first entail a complex filtration system costing $50 million to $100 million. It was acknowledged, however, that a single boat infected with zebra mussels could just as easily transfer the mussels from lake to lake, and the state of Texas had done little to educate boaters about the threat or ways to combat it through clean–drain–dry. A state government spokesman said that the state could not afford an aggressive boater education campaign. The television station noted that a crew traveled to Minnesota to investigate that state's successes in slowing zebra mussel spread and found that Minnesota spends $5 million a year, through boat-license fees, on fighting invasive species. On May 23 in another report, *WFAA-TV* said that "what started as an issue at Lake Texoma is now a problem in a second north Texas lake." Mussels were found on a boat that came to Lake Ray Hubbard (the reservoir that supplies water for Dallas) from Lake Texoma and had sat in a marina slip for 10 days till it was pulled out for repairs. Mussels were discovered on the hull at that time. The infested boat was sprayed with chlorine and quarantined before it was returned to the lake, and several traps were placed in the marina slips to see whether mussels that hitchhiked on its hull have multiplied. On June 27, the Texas Parks and Wildlife Department scheduled a news conference to warn 4th of July weekend boaters about the mussel invasion and how to take preventive precautions, according to the *Dallas Morning News*. The newspaper noted that zebra mussels had infested Lake Texoma and rendered it useless as a local drinking-water source.

New Mexico—the western neighbor state to Texas was having its own zebra mussel problems in mid-2011. KOB-TV in Albuquerque reported on May 27 that Lake Sumner in southeastern New Mexico was to be closed temporarily to boating following discovery of quagga mussel larvae in lake water. Sumner Lake was said to be the first body of water in the state to see quagga mussels. State park authorities were fanning out to spot-check boats for both zebra mussels and quagga mussels in the hope of preventing their spread. On June 4, the *Santa Fe New Mexican* reported that Sumner Lake would remain closed to boating at least through midmonth as authorities continue to try to determine whether a quagga mussel infestation had taken place. The paper said that the first indications of quagga presence occurred in the fall of 2010.

Utah—the Utah Division of Wildlife Resources stopped several boats, found to be carrying quagga mussels, from launching into Lake Powell on the Colorado River system, according to KCPW public radio on May 23. The broadcast said that transporting live mussels within the state or across its borders is illegal, and the state was recommending the clean–drain–dry technique before transporting boats to new waters. The station's website carried an on-line stream of an array of quagga mussel photographs, including a widely circulated stock photograph of a quagga mussel-encrusted outboard propeller. KSTU-TV in Salt Lake City carried a similar story on May 21, discussing the state's first checkpoint to inspect trailered boats for invasives, which is at the I-15 point of entry from Arizona. Of 20 boats inspected initially, two had to be decontaminated, using high-pressure water at a temperature of 140°F, the station said. Next door in Nevada, the *Carson City News* reported its version of the I-15 inspection story, noting that Lake Mead and other reservoirs in the lower Colorado River system are infested with quagga mussels. In May 2010, a single quagga mussel was found in Utah's Sand Hollow Reservoir, southwest of Hurricane, but no additional mussels had been found there or elsewhere in Utah since, the newspaper said. On May 26, the *Salt Lake Tribune* traced the first appearance of a quagga mussel in Lake Powell on the lower Colorado River in the summer 2007, this on the Arizona side of the lake's south end. The article went on to discuss boat inspections and decontamination and presence of mussels at the state's Sand Hollow Reservoir. It also discussed the threat to food webs in Lake Powell because of mussel filtering and how that could impact the lake's striped bass fishery, much as mussels had affected Great Lakes salmon fisheries. The *Tribune* piece also made note of other, familiar mussel issues, such as the threat of shells cutting the feet of swimmers and beachgoers and the clogging of pipes and turbines at the power-generating stations associated with the dams.

Wyoming—Utah's neighbor to the northeast was busy by June 2011, sounding the mussel alarms as the Wyoming Game and Fish Department set up boat inspection stations along routes to major boating destinations in the state, according to a June 10 report in the *Jackson Hole News and Guide*. As with other western states, inspections applied to hand-powered watercraft as well as powerboats. The reports noted that zebra and quagga mussels were present in neighboring states but had not yet been found in Wyoming.

Montana—KPAX-TV in Missoula reported in June 26, 2011, that Montana Fish, Wildlife, and Parks and the Confederated Salish and Kootenai Tribes were conducting boat inspections on U.S. 93, a major north–south route in the northwest corner of the state, in an effort to detect and stop the spread of zebra and quagga mussels and other ANS. All watercraft, including kayaks and canoes, were to be inspected. A month earlier, on May 19, the *Hungry Horse News* in Columbia Falls, Montana, reported that Glacier National Park was stepping up its boat inspection and permit program because of "the rapid westward migration of aquatic invasive species on recreational watercraft." The paper noted that a live quagga mussel was found in February 2011, prior to its launch in nearby Flathead Lake, followed by the discovery of mussels on two boats in neighboring Idaho. To the north, a similar program was required by Parks Canada at Waterton Lake, which is part of the Waterton–Glacier International Peace Park on the Alberta–Montana border. The U.S. National Park Service (NPS) at Waterton–Glacier in the spring of 2011 issued an aquatic invasive species bulletin to alert the public about the threats. "It's not science fiction" the NPS stated, adding that "impacts are already occurring in waters in the Great Lakes, eastern (Canadian) provinces and states, the prairies and plains, and more recently in the southwest United States. Since the 1980s, freshwater zebra and quagga mussels have steadily advanced westward, presumably transported on trailered boats," the NPS summarized. It also referred to the mussels found on the sailboat at Flathead Lake as "just downstream from Glacier."

Washington—a 40 ft long sailboat contaminated with zebra mussels was stopped in Spokane County in late May 2011, according to the *Spokane Spokesmen-Review* on May 25. The Washington State Patrol noticed the adult-size mussels during a routine commercial vehicle inspection at the port of entry on I-90, just west of the Idaho border. The sailboat was being delivered from the Great Lakes to a new owner in Vancouver, British Columbia. The commercial driver was cited and fined $1600 for bringing prohibited aquatic species into the state, the paper said. On June 24, the newspaper also reported the setup of aquatic invasive species check stations in eastern Washington with mandatory stops for any vehicle carrying a watercraft of any type. Zebra and quagga mussels and other invasives were the targets. The article noted that mussels had not yet been found in the state.

Oregon—about the same time the zebra mussel-infested sailboat was being intercepted at Washington's eastern border, a state boat-decontamination crew discovered the first zebra mussels in Oregon at the Ashland port of entry on I-5, this after the boat supposedly passed muster as clean in California, according to a May 20 article in the *Mail Tribune* in Medford, Oregon. More than two dozen mussels were found on an outboard motor of a trailered runabout. The boat's owner, a Washington man, said that he was transporting the craft from Arizona to his home in Everett, Washington, and that he had bought the boat on eBay from a party in Michigan. The newspaper noted that boat inspections in Oregon at the time were voluntary, but a mandatory inspection bill was under *consideration*. The infested boat was hot-wash decontaminated before being released to travel. On April 7 in Hood River, the paper said, quagga mussels were found on a boat entering Oregon from Idaho. It too was decontaminated.

California—quagga mussels first appeared in California waters in 2007 in Lake Havasu on the Arizona border and then went west in the Colorado River Aqueduct, appearing over the next 3 years in dams and drinking-water reservoirs and even in a golf course pond that uses Colorado River water, according to a June 29, 2011, story in the *Press-Enterprise* in Riverside, California. The article went on to discuss mandatory boat inspections, which have prevented mussel outbreaks in at least some southern California reservoirs. In Los Angeles County, authorities agreed to spend $1.8 million on boat inspections to keep mussels out of Castaic and Pyramid lakes, according to the *Santa Clarita Valley Signal* on June 3. On May 20, 2011, five boaters were cited by Lake County authorities for not having current quagga mussel inspection stickers, according to the Lake County *Record-Bee*. The county has one of the strongest anti-mussel laws in the state, with a fine of $1000 and court costs of $1700 for a violation, the paper said. The situation is different for northern California, where authorities are watching and inspecting boats. The famed Lake Tahoe, for example, remained mussel-free, according to the *Sierra Sun* in Truckee, California, in an article on May 23 about mandatory boat inspections. Thus, in northern California, as elsewhere in the edges of the West and Northwest where zebra mussels and quagga mussels were not yet known to be established, the summer of 2011 was a period of watchful waiting, surveillance, inspection, and decontamination, principally of trailered watercraft. Only time will tell if the proverbial "other shoe" will fall in those locales still thought to be mussel-free.

ONE NEWSPAPERMAN'S TAKE ON THE DREISSENID MUSSEL TRAGEDY

The zebra/quagga mussel story, arguably an environmental tragedy of national and international scope, has been part of my outdoors writing career for over 25 years. The presence of mussels as the "poster children" of invasives has made its way into the social fabric of North America. Invasive species are now a common political, social, and satirical component of our culture. From my perch, on the verge of retirement, it appears that mussels are poised to complete the intrusion that began in the late 1980s, reaching into the last untouched regions of the remaining 48 states and southern Canadian provinces. Maybe the more northerly subarctic and arctic regions will prove too inhospitable. However, if these mussels do adapt to new environments, there are simply too many pathways for them to spread. Not every road and trail can have a check station, and not every veliger-infested watercraft can be stopped, inspected, identified, and decontaminated. Mussels simply are too pervasive, too prolific, and too hardy.

At the end of 2011, Congress still ineffectually was waffling and battling with individual Great Lakes states over ballast-water controls on overseas shipping, which so early was fingered as the potential culprit for mussel entry into North America. Yet, truly effective prevention and enforcement against the influx of still more alien pests remain beyond the horizon of political reality. We dither at our peril.

One needs only to look at the devastation of the emerald ash borer, from China, on valuable ash forests and stands in less than a decade to remind us of the urgency in preventing unwanted introductions. Several species of Asian carp are just a few strands of electrical current away from invading the Great Lakes at Chicago via the same water superhighway that contributed to the spread of mussels several decades ago.

The so-called threats to the economics of shipping that would result because of additional ballast-water controls pale in comparison to the billions of dollars in costs already born, and continuing, across the continent by users of raw water. And the toll in ecological disruption is almost incalculable. The mussel invasion was classic, so swift, and so widespread. While the media can report events or offer opinions through editorials or satiric cartoons, it is ultimately the will of the public that pressures preventive actions. So consider this a reminder, a quiet warning, summed in the altruism that if we do not heed history and learn from it, we are destined to repeat it.

CHAPTER 10

Early Responses to Zebra Mussels in the Great Lakes
A Journey from Information Vacuum to Policy and Regulation

Ronald W. Griffiths, Don W. Schloesser, and William P. Kovalak

CONTENTS

Abstract	135
Introduction	136
1988 ... A Curiosity Is Discovered	139
Early 1989 ... Calm before the Storm	140
Late 1989 ... Zebra Mussels Take Over (Jansen 1989)	144
Spreading the Word	146
1990 ... Political Responses Emerge	147
Canadian Governments	147
U.S. Governments	149
Binational Cooperation	151
Zebra Mussels Continue to Spread	151
Spreading the Word	152
1991 ... Implementation	154
Canadian Governments	154
U.S. Governments	156
Zebra Mussels Disperse beyond the Great Lakes Basin	157
Spreading the Word	159
Discussion	160
Just Another Invasive Species	160
Policy Shift: Recognition of Aquatic Nuisance Species	163
Research Funding	164
Role of Science	165
Concluding Remarks	168
Acknowledgments	170
References	170

ABSTRACT

Invasive species such as zebra mussels pose a threat to the economies and environments of coastal and freshwater habitats around the world. Consequently, it is important that government policies and programs be adequate to protect these waters from invaders. This chapter documents key events that took place in the early years (1988–1991) of zebra mussel colonization of the Laurentian Great Lakes and evaluates government responses (policies and programs) to this disruptive, invasive, freshwater species.

Responses to the zebra mussel invasion followed an "event sequence" typical of most unintentionally introduced exotic species to the Great Lakes: initial release, discovery, taxonomic verification, information-search media reports, scientific presentation/publication, expert committee(s) establishment, distribution and containment, outreach and education, abatement and adaptation, ecosystem impact monitoring, acceptance as naturalized; then wait for the next invasive species. However, the "abatement and adaptation" event was the most costly that the Great Lake's community had ever experienced from an aquatic exotic species. Zebra mussel impacts initially occurred in municipal and industrial facilities taking raw water for domestic, cooling, and process use. Companies and government agencies that operated multiple facilities such as Ontario Ministry of the Environment's water treatment plants, Ontario Hydro's electric-generating stations, and Detroit Edison's electric-generating stations had the advantage of deep financial pockets, biological-knowledgeable staff, and engineering expertise that were able to quickly establish monitoring and abatement programs to prevent severe impacts. Consequently, zebra mussels had the largest impacts on smaller, single-facility companies, severely restricting and interrupting water supplies to water treatment facilities in Monroe and Bay City, Michigan; Ford's Engine Plant in Windsor, Ontario; and a food processing plant in Wheatley, Ontario. Biofouling impacts on facilities drawing water from the Great Lakes declined rapidly after the first year, as facility operators put monitoring and abatement programs into place.

Zebra mussels also impacted the Great Lakes ecosystem. Their filtering of seston from the water column for food increased water clarity in many localities, thus prompting aquatic plants to grow in shallow waters, which caused further changes in the composition of invertebrate and fish communities. This trophic cascade impacted recreational, hunting, and fishing activities.

The cost of abatement programs, impacts on aquatic environments, along with the realization that zebra mussels would spread across North America, prompted U.S. political action that resulted in a new national policy. This action took the form of the Nonindigenous Aquatic Nuisance Prevention and Control Act of 1990 that shifted government focus from dealing with individual invasive species as they appeared (i.e., containment) to dealing with the source(s) of invasive species (i.e., prevention). The legislation created a National Task Force to specifically implement programs to control the spread and prevent introduction of invasive species. It also started a 20-year "prevention policy process" that began with voluntary, ballast-water exchange guidelines for transoceanic ships and finally ended with mandatory guidelines for all ships that included legally enforceable biological discharge limits. Ratification of the 2004 Ballast Water Management Convention developed by the International Maritime Organization with its proposed ballast standard limiting "living organisms in ship's ballast water discharge" would greatly help prevent the establishment of invasive species populations in coastal and freshwater systems throughout the world.

INTRODUCTION

Over the past century, at least 162 species have been introduced into the Laurentian Great Lakes (Ricciardi 2001). Some of these introduced species were a direct result of government policy to address identified environmental problems and reestablish (temporary) economic activities (e.g., fisheries). However, most of these species were unintentionally introduced into the Great Lakes via ballast-water discharges, constructed canals, and accidental releases, for example, bait buckets, aquaria, and aquaculture escapees (Mills et al. 1993). Fortunately, only a handful of these unintentionally introduced species, such as the sea lamprey and zebra mussel, have caused considerable economic and ecologic damage. Still, other potentially damaging aquatic species have been identified that are waiting for an opportunity to immigrate to the Great Lakes (Grigorovich et al. 2003).

Invasive species pose a risk to the economy and environments of coastal and freshwater habitats around the world—essentially any waterbody may be at risk. Protection of ecosystems from invasive (i.e., exotic) species depends primarily on government policies and programs authorized by legislation and administered by cooperative interaction among responsible agencies and societal stakeholders (e.g., industry, academia, public). Key elements to prevent, exterminate, and contain invasive-aquatic species include regulation of ballast-water discharges, routine monitoring to detect new introductions, outreach to inform stakeholders, and a permanent network of scientists in government, academia, and private sector collaborating to minimize the time to assemble information, identify control and implement abatement strategies, and obtain administrative approvals for action plans.

The purpose of this chapter is to review how U.S. and Canadian governments and society responded to the initial invasion and dispersal years of zebra mussels in North America. This case study evaluates: economic costs and environmental impacts imposed by this invasive species; roles of science, media, and conferences in policy development; responses by government agencies; and success of agency programs. This assessment will help clarify how governments can respond more rapidly and adequately to invasive species to decrease economic and environmental disruption, thus protecting public goods and services in the future. Materials for this review come from published articles in newspapers, magazines, scientific literature, and conference materials, along with unpublished information from government reports, personal notes, interviews, and video recordings of the authors and others who actively participated in the earliest years of the zebra-mussel "invasion" in North America (see Table 10.1). Unreferenced events and information throughout this chapter are unpublished, firsthand accounts of the authors.

Table 10.1 Presentations at the First Four North American Zebra-Mussel Conferences

Presenter	Affiliation[a]	Title
A. Presentations at the *First North American Zebra Mussel Conference: Fouling of Water Systems by Zebra Mussels* held in Ann Arbor, Michigan, on June 13, 1989		
Marg Dochoda	GLFC	Progress in controlling introductions of exotics in the Great Lakes: ballast-water update
Gerry Mackie	University of Guelph	Biology of zebra mussels
Don Schloesser	USFWS	Potential interactions of the zebra mussel with indigenous species in North America
Ronald Griffiths	OMOE	Fouling of drinking-water systems
William Kovalak	Detroit Edison	Fouling of industrial water intakes
John Oyer	McNamee, Porter, and Seeley	Water treatment plant design overview
Leif Marking	USFWS	Problems involved in development of control agents for nuisance organisms
Wayne Brusate	Commercial Diving, Inc.	Underwater inspections
Larry Lyons	Betz Chemicals	Monitoring and control
David Long and Robert McMahon	Buckman Chemicals and UTA	Chemical controls
Steve Ryan	Taprogge America Corp	Water straining systems
Jon Stanley	USFWS	Open discussion and identification of research needs
B. Presentations at the *Second North American Zebra Mussel Conference: Zebra Mussels in the Great Lakes, in Rochester*, New York, on November 28–29, 1989		
James Carlton	Williams College	Controlling ballast introductions of exotic species: applying the international experience to the Great Lakes
Don Schloesser	USFWS	Update on ballast-water introduction of exotic species into the Great Lakes
Abraham bij de Vaate	IIWM	Occurrence and population dynamics of the zebra mussel in Lake Ijsselmeer, The Netherlands
Paul Hebert	University of Windsor	Discovery of the zebra mussel in the Great Lakes—genetic studies
Gerry Mackie	University of Guelph	Biology of the zebra mussel in the Great Lakes
Tom Nalepa	NOAA	Spread of the zebra mussel in Europe and the Great Lakes
Joe Leach	OMNR	Impacts of the zebra mussel on the ecology of Lake Erie
Ronald Griffiths	OMOE	Zebra mussel impacts on surface water uses
Wilfred LaPage	MMWA	Zebra mussel fouling of water intakes: impacts on water treatment plants
William Kovalak	Detroit Edison	Zebra mussel fouling of water intakes: impacts on industrial water facilities
		Panel discussion: impacts of zebra mussels on the Great Lakes
Leif Marking	USFWS	Development and approval of chemical control agents
Robert McMahon	UTA	Chemical control of zebra mussels
Henk Jenner	NVTKEM	Control of zebra mussels in power plants and industrial settings: the European experience
Phil Hecht	Malcolm Pirnie Environmental Engineers	Physical control devices: mechanical filtration and microstraining
Mohammed Karim	OMOE	Design implications of zebra mussels on water-treatment facilities
Christopher Gross	Long Island Lighting Co.	Use of antibiofouling coatings for mussel control
Duane Nietzle	Battelle Northwest Labs	Biofouling and the nuclear energy industry
Robert Malouf and John Gannon	NYSGI and USFWS	Panel discussion: identification of information and research needs
C. Presentations at the *Third North American Zebra Mussel Conference: Zebra Mussel Symposium, in Cleveland*, Ohio, on December 6, 1989		
George Voinovich	Mayor, City of Cleveland	Opening address
David W. Garton	Ohio State University	Biology of zebra mussels
Joe Leach	OMNR	Impact of zebra mussels on natural resources and water quality
William Kovalak	Detroit Edison	Current conditions at Detroit Edison
Ronald Griffiths	OMOE	Current conditions of industrial raw-water intakes

(continued)

Table 10.1 (continued) Presentations at the First Four North American Zebra-Mussel Conferences

Presenter	Affiliation[a]	Title
John Morrison	Ohio EPA	Regulatory perspective from the Division of Water Pollution and Water Quality
Lorie Chase	Ohio EPA	Regulatory perspective from the Division of Public Water Supply
Mike Auger	OMOE	Current conditions on the northern shore of Lake Erie
Wilfred LaPage	MMWA	Current experience of the Monroe Water System
Ronald Griffiths	OMOE	Current European practices to control of zebra mussels
David Bohley	Bohley's Diving Service	Intake inspection techniques
David Erck	Buckman Laboratories	Chemical control experiments and delivery techniques
		Water utility round table

D. Presentations at the *Fourth North American Zebra Mussel Conference: Zebra Mussels: The Great Lakes Experience*, Guelph, Ontario, Canada, on February 19, 1990

Presenter	Affiliation[a]	Title
Gerald L. Mackie Don Schloesser	University of Guelph* USFWS	The past, present, and future of zebra-mussel research in North America
David Garton and Wendell Haag	Ohio State University	Reproduction and recruitment of *Dreissena* during the first invasion year in western Lake Erie
Joe Leach	OMNR	Potential ecological impacts of the zebra mussel, *D. polymorpha*, in Lake Erie
Gerry Mackie	University of Guelph	Outstanding biological and life history attributes of the zebra mussel, *D. polymorpha*, and the ecological implications to native species of bivalves in the Great Lakes
Don Schloesser William Kovalak	USFWS Detroit Edison	Infestation of native unionids by *D. polymorpha*, in a power-plant canal in Lake Erie
S.J. Nichols	NFRC	Maintenance of zebra mussels (*D. polymorpha*) in captivity
William Kovalak, G.D. Longton, and R.D. Smithee	Detroit Edison	Infestation of power-plant water systems by zebra mussels (*D. polymorpha*)
A.B. Greenburg	City of Cleveland	Zebra mussel infestation conditions at Cleveland, Ohio, and the south shore of Lake Erie
Wilfred LaPage and L.J. Bollyky	MMWA Bollyky Associates Inc.	The impact of *D. polymorpha* on water works operations at Monroe, Michigan
Bruce Kilgour and Gerry Mackie	Mackie and Associates Water Systems Analysts	Colonization of different construction materials by the zebra mussel (*D. polymorpha*)
Don Lewis and J. Sferrazza	Aquatic Sciences Inc.	Control of the zebra mussel, *D. polymorpha*, at two infested metal casting plants with chlorine gas and liquid hypochlorite and the effect of temperature and pH on treatment effectiveness
Ian Martin, M.A. Baker, and Gerry Mackie	University of Guelph	Comparative efficacies of sodium hypochlorite and registered biocides for controlling the zebra mussel, *D. polymorpha*
Renata Claudi	Ontario Hydro	Response of zebra mussels to sodium hypochlorite treatment at the Nanticoke Thermal Generating Station
David Erck	Buckman Laboratories Inc.	Laboratory studies of the efficacy of an ionene polymer for control of *D. polymorpha*, the zebra mussel
L.A. Lyons, J.C. Petrille	Betz Laboratories Inc.	Zebra-mussel fouling control to cooling systems using a nonoxidizing molluscicide
Mohammed Karim Mike Auger	OMOE	Design considerations of intakes at water treatment plants against infestations by zebra mussels
Ronald Griffiths	OMOE	Spatial and temporal distribution of the zebra mussel in the Great Lakes region
Gerry Mackie and C. Wright	MAWSA	Removal of seston, nutrients, and BOD from activated sewage sludge and its biodeposition by the zebra mussel, *D. polymorpha*
Gerry Mackie Don Schloesser	University of Guelph USFWS	Open forum: future research needs

[a] GLFC, Great Lakes Fisheries Commission; IIWM, Institute for Inland Water Management, The Netherlands; MMWA, Monroe Michigan Water Authority; NFRC, National Fisheries Research Center—Great Lakes; NOAA, National Oceanic and Atmospheric Administration, GLERL; NVTKEM, N.V. Tot Keuring van Elektrotechnische Materialen, The Netherlands; NYSGI, New York Sea Grant Institute; OMNR, Ontario Ministry of Natural Resources; OMOE, Ontario Ministry of Environment; USFWS, U.S. Fish and Wildlife Service; UTA, University of Texas at Arlington.

1988 ... A CURIOSITY IS DISCOVERED

The first documented collection of a zebra mussel in North America occurred from the southern portion of Lake St. Clair on June 1, 1988 (Hebert et al. 1989). Zebra mussels were found in bottom samples collected by Sonya Gutschi (see Gutschi Santavy 2013). It is likely that other people had observed zebra mussels during the previous 2 years because mussels had been present in Lake St. Clair and western Lake Erie since 1986 (Griffiths et al. 1991, Carlton 2008), but they did not understand the importance of their find, and thus did not report the sighting to authorities for investigation. In contrast, Gutschi recognized that this "wart on a rock" was sufficiently unusual that it required closer examination back at the University of Windsor, an action that was the spark for the ever-increasing recognition of zebra mussels in North America. Bernie Muncaster at the University of Windsor then took several specimens to his former advisor, Gerry Mackie at the University of Guelph, an expert in freshwater bivalves, who identified the mussels as *Dreissena polymorpha*. Meanwhile, Paul Hebert, director of the Great Lakes Institute at University of Windsor, collected information on the distribution, abundance, and genetics of this new invader, which formed the basis of the first scientific publication on zebra mussels in North America (Hebert et al. 1989).

Also, during early summer 1988, a fisherman brought live specimens of small "stripped clams" attached to the shell of a native mussel from Lake St. Clair to Joe Leach, a scientist with Ontario Ministry of Natural Resources at Wheatley Harbour, located on the western basin of Lake Erie. Leach contacted Gerry Mackie about these stripped clams and learned that he had similar specimens of zebra mussels from Lake St. Clair. Thus, Leach was likely the first government official to know about the occurrence of zebra mussels in North America. After a quick literature search, Leach realized the potential impacts this species could inflict on water uses. He phoned the *Windsor Star* newspaper about the presence of this new exotic species. The first newspaper article on the "zebra mussel invasion" of the Great Lakes was published on July 27, 1988 (vander Doelen and Hornberger 1988; see Figure 9.1, Pollick 2013). Leach then conducted a shoreline survey along the western basin of Lake Erie in October 1988 and found zebra mussels near Leamington, Wheatley, and Kingsville, Ontario. He phoned colleagues along the south shore of western Lake Erie in Ohio, but no one had noticed zebra mussels in this area of the lake. With a detailed description of the free-swimming larval "veliger" stage from his colleagues in Britain, Leach re-examined water samples from central and western Lake Erie and discovered that zebra mussel larvae were present in the summer and autumn of 1988.

One day after starting work with the Southwest Region of the Ontario Ministry of Environment, Ron Griffiths was handed a memo from Dick Brown, manager of Utility Operations Branch for the Southwest Region, on November 1, 1988, stating that zebra mussels were present in western Lake Erie. Griffiths found out later that Joe Leach had written a memo in late October to both the Ontario Ministry of Environment and Ontario Hydro expressing concerns that newly-established zebra mussels may colonize water intake pipes in Lake Erie and seriously impede water flow. Consequently, Griffiths immediately organized a shoreline survey from Pt. Bruce west to Amherstburg on Lake Erie, then north to Sarnia at the head of the St. Clair River near Lake Huron to document the distribution of zebra mussels. Zebra mussel shells were found along several beaches, and live mussels were found attached to hard objects from Pt. Stanley on Lake Erie to Lighthouse Cove along south Lake St. Clair (Griffiths 1989).

A quick review of some European literature (e.g., Clarke 1952, Morton 1969) was sufficient to convince anyone that zebra mussels would likely have drastic impacts on water users (e.g., boaters, fishers, drinking-water facilities, industrial facilities, electric-generating facilities) and the ecology of the Great Lakes. Assessments were provided to Dick Brown, and within 2 months, he organized inspections of nine water-supply facilities located in the known distribution of mussels based on shoreline surveys. Zebra mussels were found in seven water facilities, three of which had "heavy" infestations that resulted in 15%–25% loss of water intake capacity. Within 1 month, water superintendents for these water-supply facilities formed a zebra mussel committee that met monthly to report on impacts of mussels at their facilities and evaluate potential solutions.

In the United States, the first documented collection of zebra mussels was a single large specimen (>20 mm) found by commercial divers in the St. Clair River near St. Clair, Michigan, July 31, 1988. This specimen was identified by Bill Kovalak, a biologist with Detroit Edison Company, who recognized it from illustrations in a report on biofouling mollusks, which anticipated the eventual introduction of zebra mussels into North American waters (Sinclair 1964). Aware of its history fouling raw-water systems in Europe, Detroit Edison began systematic inspections of its cooling water intakes along the Lake Huron–Lake Erie corridor in early August 1988. By September 1988, it was clear that zebra mussels were established at low densities (<100/m^2) in the cooling water intake at Detroit Edison's power plant on the western shore of Lake Erie at Monroe, Michigan. Because of risks to plant operations, in mid-September, Detroit Edison contracted a title–abstract search of European literature for information on zebra-mussel biology and control that was later given to the team headed by Gerry Mackie that prepared the first assessment of the "zebra mussel issue" in 1989.

In October 1988, after learning about zebra mussels in Lake St. Clair, Don Schloesser, a biologist with the U.S. Fish and Wildlife Service, and Tom Nalepa, a biologist with National Oceanic and Atmospheric Administration, conducted a quick survey in the southeast part of Lake St. Clair and surprisingly collected zebra mussels attached to rocks and unionid mussels.

Zebra mussels had not been apparent in a benthic survey of the Lake St. Clair–Detroit River corridor in 1985 (Schloesser et al. 1991, unpublished), nor in a lakewide survey of unionids in Lake St. Clair in 1986 (Nalepa and Gauvin 1988).

By the end of 1988, zebra mussels were perceived by individuals of several government agencies in Canada and the United States as an important "regional issue" confined to western Lake Erie and Lake St. Clair. While a handful of academic and government scientists gathered information on zebra mussels from Europe and staff from water-supply utilities in Ontario and electric-generating industry in the United States (e.g., Detroit Edison Company) monitored zebra mussels at their facilities, few others in industry, government, or the public appeared to realize the potential impact this invader could have on water users and aquatic ecosystems, despite four published newspaper articles that detailed many of the critical issues (vander Doelen and Hornberger 1988, Hornberger 1988, Geigen-Miller 1988a,b).

EARLY 1989 ... CALM BEFORE THE STORM

By early 1989, it was apparent to a number of biologists that zebra mussels could have profound impacts on Great Lakes industries and ecology. Based on documented impacts of zebra mussels in Europe, biologists knew there was an urgent need to compile more information, promote public awareness of issues, and network with other investigators to share the workload and minimize duplication of effort. Thus, Griffiths and Kovalak organized a zebra mussel workshop to share information and outline a coordinated scientific and communication strategy.

Twelve people (Figure 10.1), representing government, academia, and water-supply and electric-generating utilities, accepted an invitation to the *First Zebra Mussel Workshop* in North America held on March 9, 1989, at the Ontario Ministry of Environment office in London, Ontario. The workshop consisted of eight formal presentations followed by a discussion and planning session. The workshop was videotaped for later distribution to participants, thereby providing a record of the proceedings (available from RWG). Formal presentations were made by

1. Gerry Mackie, who provided an overview of the life history and biology of zebra mussels based on European literature and compared traits to a more commonly known bivalve invader (i.e., *Corbicula*). Mackie emphasized the variability of life history and growth rates of these two taxa.
2. Don Schloesser, who provided information on past benthic studies conducted in the St. Clair River–Lake Erie corridor in 1985. He emphasized the absence of zebra mussels in the 1985 survey and reviewed introductions of past invasive species into the Great Lakes. Schloesser noted his discovery of another exotic species (a macrophyte plant, *Nitellopsis obtusa*) believed to have been introduced into the Great Lakes by ballast water discharged from ships. He also identified potential impacts of zebra mussels on indigenous ducks, fish, and unionid mussels.
3. Joe Leach, who provided information on the location of zebra mussels along Canadian shores of Lake Erie and presented abundances of veligers in central and western basins of Lake Erie. Leach showed pictures of zebra mussels attached to buoys, boat hulls, fishing nets, and other solid objects, raised concerns for walleye spawning because zebra mussels had covered fish-spawning reefs, and reported freshwater drum were feeding on zebra mussels.
4. Ron Griffiths, who provided information on distribution of zebra mussels along Canadian shores of Lake Erie, the Detroit River, and Lake St. Clair. He noted differences in the sizes of zebra mussels in Lake St. Clair and Lake Erie, showed mussels attached to macrophytes, noted the absence of zebra mussels in the St. Clair River in 1985, noted mussels could suspend themselves at the surface of the water (drifting) in aquaria, and emphasized that mussels were effective filter feeders and as a result could consume large proportions of suspended resources, thereby increasing water clarity and altering benthic communities. Finally, he emphasized zebra mussels were seen only as a regional issue.
5. Peter Kauss, who provided information on contaminant uptake by native mussels via the water and food chain and worried about potential for zebra mussels to alter contaminant cycling in the Great Lakes.
6. Eileen Leitch, who provided information on the biofouling, marine mussel, *Mytilopsis*, which causes flow restrictions in water pipes and promotes metal corrosion beneath the organisms. She expressed concern about the potential of zebra mussels to cause similar problems for electric-generating plants using freshwater.
7. Rick Turnbull, who provided information on abundances of zebra mussels in pipes and on intake structures at water-treatment facilities along the north shore of western Lake Erie. He noted mussel densities as high as 3500 individuals/m^2, indicated water plants have up to 16 km of various sized pipes inside and outside the plant that could be plugged if an accumulation of mussels occurred at "critical" points, noted mechanical cleaning with pressure sprayers required 5.6 kg/cm^2 to remove mussels, and suggested chemical treatment might include "prechlorination" of water before it entered facilities to kill zebra mussels.
8. Bill Kovalak, who provided information on mussel colonization of an electric-generating facility in Monroe, Michigan. He indicated highest densities were around 2500/m^2, observed mussels would clump together to form golf ball sized colonies (i.e., druses, nodules) that could clog small-diameter water pipes, expressed concern that any disruption to water flow in a facility had the potential to cost hundreds of thousands of dollars a day, showed that mussels were largely attached to intake structures and only a few mussels were found in service-water systems (e.g., systems for fire protection), and observed in the laboratory that unionids suffered high mortality when colonized by zebra mussels.

Figure 10.1 Participants At the *First Zebra Mussel Workshop* in North America held in London, Ontario, Canada, on March 9, 1989. Top row: Gerry Mackie, Ron Griffiths, Joe Leach, and Doug Huber. Middle row: Tom Nalepa, Don Schloesser, Bill Kovalak, and Dick Brown. Bottom row: Peter Kauss, Rick Turnbull, Eileen Leitch, and Paul Wianko.

Workshop participants concurred that zebra mussels:

1. Posed a threat to Great Lakes water users because they could clog raw-water intakes of municipal-water plants, electric-generating facilities, and other industries that require cooling and process waters. In addition, they could foul (attach to) hulls of commercial ships and recreational boats, commercial fishing gear, navigation buoys, and shipwrecks, and interfere with recreation activities along beaches.
2. Posed a threat to aquatic ecosystems in North America because they filter phytoplankton and suspended solids from the water column, potentially altering food webs and energy flow, and attach to any solid object, potentially fouling fish-spawning shoals and smothering native mussels.
3. Would rapidly spread and colonize aquatic ecosystems across the Great Lakes and North America because of their high fecundity, free-swimming larval stage, and wide environmental tolerances.

4. Were essentially unknown to the public and water users in the Great Lakes basin.
5. Showed ballast waters were a source of aquatic invasive species, and this problem needed to be addressed.
6. Did not recognize national boundaries, and as a result, an international response was needed to address zebra mussels and prevent the next "invader."

A discussion and planning session identified several urgent research and communication needs to be addressed: (1) understand the specific life history and growth rates of zebra mussels in the Great Lakes (i.e., which European population does the Great Lakes population mimic); (2) summarize European literature as an initial information base for people in North America; (3) research chemical (e.g., chlorine) and nonchemical control techniques; (4) prevent (minimize risk) further exotic introductions into the Great Lakes (North America) by addressing vectors (e.g., ship ballast-water discharges); (5) establish a central "clearinghouse" to coordinate information and address questions from public, media, agencies, and industries; (6) map mussel distribution, determine dispersal rate, and estimate the potential distribution in North America; (7) determine if zebra mussels accumulate and store organic contaminants; (8) document and report impacts on water users; (9) document and report impacts on ecosystems; and (10) notify water users about potential impacts of zebra mussels and establish monitoring programs to detect veligers and adults (e.g., American Water Works Association, Electrical Power Research Institute, boating groups, fisherman, industrial cooling water users).

Consequently, workshop participants outlined an action plan based on available funds, resources, and agency mandates:

1. Mackie agreed to submit a proposal to the Research Advisory Committee of Ontario Ministry of Environment to investigate the population biology of zebra mussels in Lake St. Clair and Lake Erie. Subsequently, the committee provided $87,000 for this proposal in April 1989 (see Mackie 1991).
2. The Water Resources Branch of Ontario Ministry of Environment agreed to fund (~$10,000) a study to review and summarize European literature on zebra mussels (see Mackie et al. 1989).
3. Ontario Hydro, the largest electric utility in North America at the time, agreed to fund research on chlorine and other chemical (e.g., ozone, peroxide) and nonchemical (e.g., electric pulses, coatings, shock waves) methods to control zebra mussel in pipes. In August 1989, $11 million in funding was made available.
4. All agreed to forward concerns of ballast-water discharge to Marg Dochoda of the Great Lakes Fishery Commission, who had highlighted this critical issue in various political, government, and public forums (GLFC 1988) after the discovery of the European ruffe (*Gymnocephalus cernuus*) in the Great Lakes in 1986. The U.S. Fish and Wildlife Service directly supported this effort through staff on various committees and encouraged support from other environmental and fisheries groups (USFWS 1988). Consequently, Griffiths organized a letter of support for the need to regulate ballast-water discharges that the regional director signed on March 22, 1989, and sent to the Intergovernmental Relations Office.
5. Schloesser and Griffiths agreed to act as "zebra mussel coordinators" for the United States and Ontario, respectively. In this role, they would actively accumulate and dispense information; answer questions from the public, media, agencies, and industries; and exchange information between the two countries at regular intervals. They encouraged all parties to report new sightings so they could track the spread of zebra mussels and notify affected water users.
6. Schloesser, Leach, Kovalak, and Griffiths agreed to sample zebra mussels in their respective local areas to analyze size–frequency distributions in order to identify the initial-invasion location as well as measure the rate of dispersal (see Griffiths et al. 1991).
7. Griffiths agreed to obtain zebra mussels from low-lift wells at several water-treatment plants in Ontario and have mussels tested for chlorinated organic contaminants, e.g., polychlorinated biphynols, heptachlor, DDT. All tested contaminants were below detection limits in young (0+ years) mussels, and thus there was no immediate human health concern related to mussels found in water-treatment plants.
8. Brown agreed to provide monthly updates from water-treatment plants in the zebra-mussel-infested area to the zebra mussel coordinators and have staff present their findings to the media and at local and regional conferences (e.g., American Water Works Association).
9. Kovalak agreed to provide updates to zebra-mussel coordinators on the spread, density, and impacts of zebra mussels on their electric-generating stations and success of various techniques to remove mussels from different sections of the facilities, as well as present this information to the media, local and regional meetings, and conferences.
10. Wianko agreed to implement a monitoring program for the detection of mussels (veligers and adults) at all Ontario Hydro facilities. Data from this monitoring program would be made available to zebra mussel coordinators for distribution. A standard sampling program would be made available for use by other industries. He also agreed to report on any impacts of zebra mussels in facilities when they did occur and the success of methods to abate problems.
11. Schloesser and Kovalak agreed to continue their investigations of zebra mussel impacts on native unionid mussels.

Other action items as a result of the workshop included the following:

1. Nalepa and Schloesser suggested that they would repeat the 1986 native mussel survey in Lake St. Clair to examine potential effects of zebra mussels on native mussels. In concert with this study, Griffiths proposed to examine the effect of zebra mussels on the Lake

St. Clair ecosystem by measurement of water clarity, water and sediment chemistry, aquatic vegetation, and benthic macroinvertebrates.
2. Leach indicated that he would continue to examine effects of zebra mussels on spawning success of walleye at three Lake Erie shoals.
3. Kovalak, Schloesser, and Griffiths agreed that they would start planning another workshop that would be held in the United States.

The next workshop was entitled "Workshop on Fouling of Water Systems by Zebra Mussels" and was held in Ann Arbor, Michigan, on June 13, 1989. This workshop was originally to take place at the U.S. Fish & Wildlife Service office, but the demand to attend the workshop was so great that the venue was changed in late May to the office of McNamee, Porter, and Seeley. This gathering became known as the *First North American Zebra Mussel Conference* (Table 10.1 and Figure 10.2). Total attendance was about 100 people, some from as far away as Texas, Washington, DC, and Mississippi. The morning session discussed concerns about exotic species invasions, zebra mussel biology, and zebra mussel impacts on aquatic species and industries, while the afternoon session addressed industrial control and abatement techniques.

The great interest in the *First North American Zebra Mussel Conference* was probably a direct response to media reports of zebra mussels in April and May 1989. Articles in several local newspapers highlighted the occurrence of zebra mussels and possible impacts to the Great Lakes, utilities, and industries. For example, *The Toledo Blade* newspaper printed a three-part zebra-mussel article of considerable length over a 2-day period (Pollick 1989a–c). Magazine articles (e.g., Lowe 1989) and television news items also began to cover zebra mussels and often projected the subject as a "serious" threat. The recovery of a red Camaro automobile from the waters of Lake Erie that was encrusted with zebra mussels (i.e., the Mussel Car) after being submerged for just 6 months was a big media highlight (Griffiths et al. 1989, Wright 1989). Government agencies and industry newsletters highlighted the zebra mussel issue in the Great Lakes (e.g., Edison Electric Institute and U.S. Fish and Wildlife Service). Two presentations at the *32nd Annual Conference of the International Association for Great Lakes Research* in Wisconsin (one showing a slide of zebra-mussel shells piled about half a meter high along a beach) clearly brought the message of zebra mussels to the Great Lakes research community. Even the introduction of the Great Lakes Exotic Species Prevention Act (H.R. 1497) to Congress on March 16 calling for the U.S. Coast Guard to report back to Congress on options to

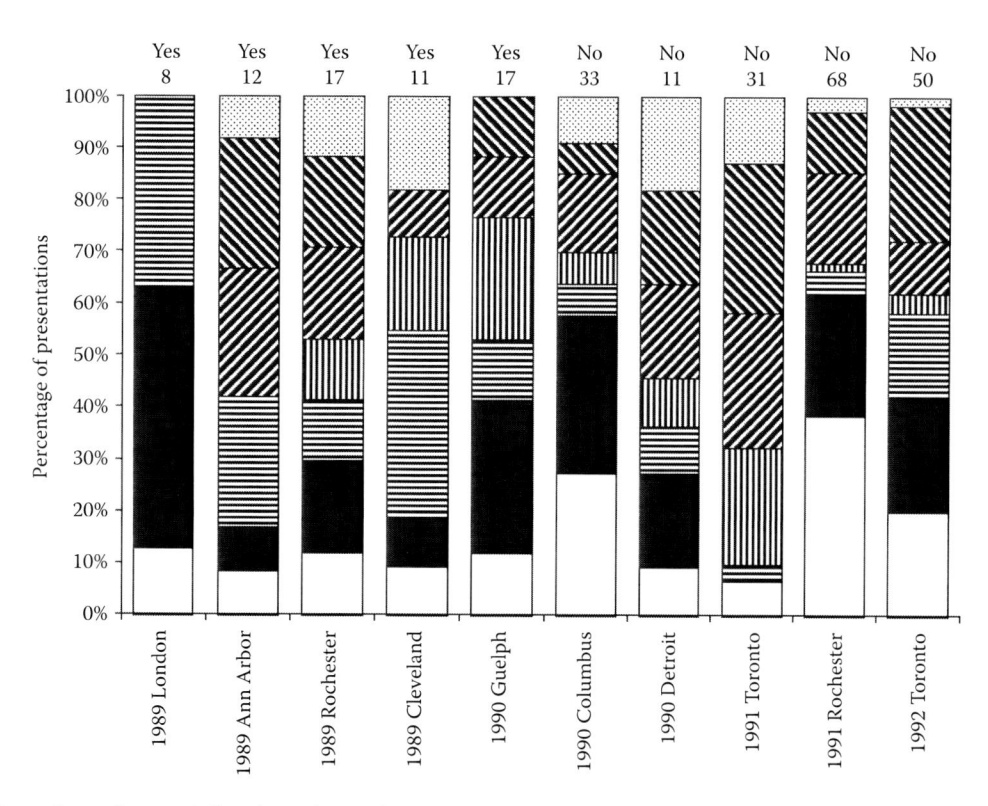

Figure 10.2 Percentage of presentations by topic at zebra mussel conferences, 1989–1991. Conferences provided by year and location. Given above each bar graph is whether the conference included a forum on research priorities (yes or no) and the total number of presentations at the conference (n). Key to topics: biology, open; ecology; solid; industrial impacts, horizontal lines; industrial treatments, vertical lines; chemical controls, right diagonal lines; nonchemical controls, left diagonal lines; others, dots.

control zebra mussel infestations in waters of the United States brought political attention to this issue. This bill was incorporated into the Coast Guard Authorization Act and passed by Congress in November 1989.

The *First North American Zebra Mussel Conference* not only informed water users of potential problems but provided a forum for those already studying zebra mussels to network and exchange information. Examples include Dave Garton from Ohio State University, who discovered zebra mussels around South Bass Island in western Lake Erie during a university class trip in fall 1988, and Robert (Bob) McMahon from the Center for *Corbicula* Research at University of Texas, Arlington, who had zebra mussels shipped from Lake Erie to his lab just a month before the first conference to assess the toxicity of a couple of molluscicides used by industry to control biofouling. Both these researchers became core members of the zebra mussel research effort.

Exhibitors of mussel-control equipment and consulting services were a valuable addition to the first conference. These companies, over the next few months, sent out information on zebra mussels to a large number of utilities and industries in the Great Lakes region to advertise products and services and thus greatly helped spread the word about impacts of zebra mussels to water users.

The afternoon discussion at the first conference emphasized that researchers, water users, and service and supplies companies would all greatly benefit from a central "information exchange office" that supplied up-to-date information on monitoring techniques, research activities, and control options. Until a funded information network was operational, Schlosser and Griffiths agreed to continue to act as leads of a Zebra Mussel Watch Committee. This committee was, in essence, a continuation of the ad hoc responsibilities of the coordinator roles established at the first workshop in London, Ontario, in March 1989. By mid-1989, this new ad hoc Zebra Mussel Watch Committee consisted of four members from the United States, Schloesser (Coordinator), Kovalak, Nalepa, and Garton, and four members from Ontario, Griffiths (Coordinator), Mackie, Leach, and Renata Claudi. Fortunately, it was organized just in time for a massive demand for information that occurred in late 1989.

The afternoon discussion at the first conference also revealed that utility managers were skeptical that zebra mussels could/would impact their facilities. Alan Greenberg, chief of water purification at Cleveland, Ohio, and Wil LaPage, chief of water purification at Monroe, Michigan, Water Authority, were both in attendance and represented two of the larger water utilities in lower Great Lakes that were likely to be affected by zebra mussels. Both were amazed zebra mussels could become a problem. Wil LaPage said that over the decades, he had encountered and solved many problems, but really, "how can such a small thing cause any problems?"

LATE 1989 … ZEBRA MUSSELS TAKE OVER (JANSEN 1989)

Unbeknownst to researchers in early- and mid-1989, summer 1989 provided optimal environment conditions for reproduction, growth, and survival of zebra mussels in western Lake Erie. Abundances of free-swimming veligers increased 700% from 1988 and peaked at 100,000–500,000/m^3, which yielded settling rates of 1200–1900 veligers/m^2/day (Garton and Haag 1993, Leach 1993). Abundance of juvenile and adult mussels increased 300% between 1988 and 1989 to a maximum density of 233,000/m^2 at Hen Island reef (Leach 1993). Zebra mussels had spread throughout Lake Erie by the end of 1989.

Drinking-water-treatment facilities were the first group to widely report impacts from these newly recruited mussels, particularly at a facility in Monroe, Michigan. This facility delivered a summer average of 42,000 m^3/day of water to a population of 45,000 people (LePage 1993). While a few mussels had been discovered in the facility as early as January 1989, rapid colonization of the intake by new recruits in July suddenly restricted the availability of raw water from western Lake Erie to just 35,000 m^3/day (i.e., 80% intake capacity) whenever a west wind occurred. West winds lowered the water depth over the intake crib, thereby reducing the flow of water into the intake pipe. On September 1, 1989, suction was completely lost and delivery of water to the facility stopped for the first time since the plant started operations in 1924. Although water flow was restored the next day, this event prompted immediate installation of a separate "backup" water-supply system to draw water from the Raisin River in October. After the September event, plans were made to chlorinate the 14 km long intake pipe from Lake Erie to clear mussels from the pipe. However, on December 14, 1989, the water flow again stopped for a period of 56 h because frazil ice, which was last observed 28 years prior, plugged the water intake. Apparently, turbulent flow conditions created by the mussels on the intake crib enhanced transformation of supercooled water into shards of frazil ice that subsequently accumulated in the intake pipe. This forced a water-emergency declaration that restricted and closed down operations of industrial and commercial water users, schools, hospitals, bars, and restaurants in Monroe. Over the next 2 years, the Monroe facility spent $300,000 to control zebra mussels and maintain water flow to residents.

Zebra mussels created similar sudden flow restrictions of water at drinking-water-treatment facilities in Ontario. During 1989, zebra mussels spread from 7 to 12 of the 14 public water-supply facilities in the mussel-colonized area of the Great Lakes. In early 1989, only two water-treatment facilities reported a significant reduction in pumping capacity: the West Lorne facility, which drew water from Lake Erie, was reduced to 75% capacity, and the Tilbury facility, which drew water from Lake St. Clair, was reduced to

85% capacity. This changed in November, when the water facility at Blenheim reported a sudden flow reduction to just 45% of the typical water intake capacity. Like the Monroe facility, the Blenheim facility quickly installed a temporary intake to maintain water flow for users. A visual underwater inspection revealed that an 8 cm layer of zebra mussels had accumulated in the bell of the intake, which chocked off the flow of water. It was estimated that 70 metric tons of mussels had colonized the intake pipe since it was inspected earlier in the year. Shortly thereafter, pumping capacity at West Lorne suddenly dropped from 75% to 60%, which was just sufficient to meet winter water demands. Cleaning this concrete intake, however, was questionable because of its age and poor condition. As a result, a new intake, end-of-pipe chlorination system and low-lift well were constructed at a cost $3 million. The Tilbury facility also saw a sudden drop in water capacity from 85% to 70% that required installation of a shore siphon to maintain sufficient water to meet demands. Mussel densities on the intake structure were estimated to be 140,000/m^2 (Griffiths et al. 1991). Temporary interruptions to water flow occurred throughout the winter as frazzle ice accumulated and clogged the intake pipe when water temperature approached 0°C, which is an unusually high temperature for frazzle ice to develop. Although the Kent County water facility had not yet seen a reduction in water capacity, an inspection of its 120 cm diameter, 750 m long intake pipe revealed a mussel population estimated at 127 tonnes.

While zebra mussels were observed at all electric-generating stations along Lake Erie by late-1989, only the operation of the station at Monroe, Michigan, was threatened by a shutdown from zebra mussels (Kovalak et al. 1993). In midsummer, mussel densities in the intake canal at Monroe increased to a maximum just over 800,000/m^2. This was a 100-fold increase over the densities observed in early-1989. The highest maximum density ever recorded worldwide was just over 1,000,000/m^2 in a water intake canal of the Chernobyl nuclear power plant in the former U.S.S.R. (Kaftannikova et al. 1987). At Monroe, mussels had colonized trash bars (screens to exclude debris) at sufficient abundances to block 75% of water flow into power-plant screen houses, where water enters the plant for heat exchangers. Mussels formed a 5–8 cm layer on the intake canals, screenhouse walls, and circulation pumps. Large clusters of mussels (druses) frequently "sloughed" off and became lodged on traveling screens and in tubes of heat exchangers downstream of screen houses. High-pressure water streams were required every 8 h to backwash about 3 m^3 of mussels from traveling screens to maintain water flows. Intake bays in the two screenhouses were dewatered during scheduled outages to mechanically remove mussels from walls, floor, trash bars, and circulation pumps with high-pressure water spray (ca. 3000 psi), while tubes of heat exchangers were manually cleaned with hooked picks. About 10 metric tons of mussels was removed from the Monroe facility in 1989. In addition, by late 1989, removal of mussels at entrances to screen houses was essential because mussels in trash bars created a spillway dam to the flow of water, which could have prevented water flow to the power plant when west winds lowered the water level in the lake. Total blockage of water did not occur, but even an average flow reduction of 15% could cost the facility $500,000 in revenue.

Zebra mussels also interfered with various other companies that withdrew water from the Great Lakes (Griffiths et al. 1989). At Omstead Foods in Wheatley, Ontario, staff was surprised when they turned on three water pumps to wash summer produce and found that low-lift wells were filled with zebra mussels to a depth of 1 m. It appeared mussels had colonized the intake pipe from Lake Erie and were dislodged from the pipe wall when water flow increased when additional pumps were turned on. After removal of about 2000 kg of mussels from the low-lift well on June 25 and July 9, 1989, about 400 kg of mussels per week was removed until the end of the recruitment season for mussels in mid-September. Meanwhile, at the Ford Engine Casting Plant in Windsor, Ontario, staff witnessed an unusual steady increase in temperature of hydraulic fluids and cooling water of one large blast furnace, as well as a reduction in efficiency of air conditioners. When they replaced the heat-exchange units during summer shutdown, they found zebra mussels had infested the complete service-water system and accounted for almost 50% of the internal volume of pipes, including one 46 cm diameter water main. Veliger abundances in the water main were about 200,000/m^3 and almost 1,000,000/m^3 in a reservoir that held water used for fire fighting. Since the blast furnaces could not be shut down, the main water lines were injected with chlorine in early September to deal with this severe biofouling problem to ensure safety of personnel and maintain plant production.

Zebra mussels also interfered with non-industrial activities in and along the shores of the Great Lakes. Mussels attached to hulls of recreation, fishing, and commercial vessels, forming layers up to 3 cm deep, reduced fuel efficiency, pitted metal hauls, and required manual removal. Small juvenile mussels were sucked into cooling systems of inboard and outboard motors where they became lodged and caused motors to overheat. Veligers settled on trap nets in nearshore waters and grew rapidly, adding considerable weight to fishing gear. In addition, fouled nets did not fish efficiently, were difficult to retrieve from the water, and once on shore required extensive work to remove the mussels. At the same time, some navigation buoys and other floats began to sink as the weight of attached zebra mussels continuously increased from new recruits and growth, which led to concerns for ship and boat-navigation safety.

Zebra mussels also were found attached to a variety of live organisms including crayfish, snails, and sturgeon. In western Lake Erie, Schloesser and Kovalak (1991) found the biomass of zebra mussels encrusted on native mussels averaged 74% of host biomass, a mass sufficient to interfere with the mussel's feeding and locomotion.

Mussel populations in some areas of the lakes were sufficiently large that their combined filter-feeding activity was able to remove almost all algae and suspended material from the water column on a daily basis (Roberts 1990, Bunt et al. 1993). This accounted for increased water clarity repeatedly reported by the public at numerous beaches along Lake Erie and Lake St. Clair throughout summer 1989, although many erroneously credited government pollution-abatement programs for the sudden improvement in clarity. Summer Secchi-depth measurements confirmed that water clarity increased by 85% from 1988 to 1989 in western Lake Erie (Leach 1993) and 50% in Lake St. Clair (RWG, unpublished data). Makarewicz et al. (1999) showed that mussel filtration also reduced algal biomass in offshore waters of western Lake Erie, despite little change in water clarity because of storm-induced suspension of sediments and continual high loadings of suspended solids from the Maumee River. Still shipwreck divers commented to the Ontario Minister of Environment about the sudden ease of sighting shipwrecks in western and central Lake Erie and expressed concerns about safeguarding these historic treasures. The clearer waters may have also revealed zebra mussels as a new source of food to diving ducks. Autumn counts of diving ducks on annual northbound migrations showed a significant shift of ducks from hunting refuges (e.g., Anchor Bay on Lake St. Clair) that were essentially free of zebra mussels to southern Lake St. Clair and northern Lake Erie where zebra mussels were plentiful (Wormington and Leach 1993).

Other water users were also directly impacted by zebra mussels in 1989. Swimmers reported cutting their feet on sharp mussel shells when walking in and near the water. At some beaches, abundances of zebra mussels were great enough that authorities posted signs informing swimmers that footwear was required to enter the water. Beach visitors also had to deal with the foul odor of decaying mussels washed up on shore. Shipwreck SCUBA divers could only watch as visible structures of underwater ships were obstructed from view by layers of mussels (SOS 1989). Finally, duck hunters in the southeastern United States had to deal with higher levels of organic contaminants because diving ducks (e.g., lesser scaup, greater scaup) feeding on zebra mussels in the Great Lakes region before flying south to winter (Hamilton et al. 1994, Custer and Custer 1996) bioaccumulated more organic contaminants than those feeding on vegetation (Mazak et al. 1997). Ironically, the ducks were returning some contaminants to the south since they closed the loop on a regional cycle. That is, agricultural pesticides used in the southern United States (Gianessi and Anderson 1995) were volatilized into the air (Harner et al. 2001, Li and Bidleman 2001), blown northward by prevailing winds (Rapaport and Eisenteich 1986), and then fell on the Great Lakes (International Air Quality Advisory Board 1999). These contaminants were filtered from the water by zebra mussels (Bruner et al. 1994a) and then carried back south by mussel-eating ducks.

Spreading the Word

The highly unusual events caused by zebra mussels were eagerly reported by the press (newspapers, magazines, TV, radio; see Pollick 2013), which kept zebra mussels in the eye of the public for more than 6 months as events sequentially unfolded in 1989. As a result, the demand for information exploded in July with each new event fueling a temporary flurry of information requests. Members of the ad hoc Zebra Mussel Watch Committee, established at the *First North American Zebra Mussel Conference*, were central contact points for information, especially for the media, since there was no designated "government office" or funded clearinghouse for zebra mussel and exotic species issues. Griffiths and Schloesser, as coordinators of the Zebra Mussel Watch Committee, received over 1,050 phone calls requesting information and/or comments about zebra mussels in 1989, with 99% of these requests occurring after July 1. Additionally, Mike Auger and Rick Turnbull from Utility Operations Branch of Ontario Ministry of Environment handled over 510 phone calls specifically related to water-supply issues caused by zebra mussels. Information and comments provided by committee members appeared in 100s of newspaper and magazine articles, including some that were syndicated in Canada and United States. In addition, committee members participated in more than 50 television broadcasts and 20 radio interviews.

Awareness of the zebra mussel invasion was directly brought to attention of the wider scientific community with the September 1989 publication of the Hebert et al. (1989) paper. Additionally, Griffiths attended the International Society of Limnology (SIL) conference in Munich, Germany, in August 1989 to draw attention to the zebra mussel invasion in North America among European scientists as well as gain insights from several European zebra mussel researchers about this species. Griffiths also sampled zebra mussels around Europe (e.g., Lake Constance, Lake Zurich) and talked to water-treatment plant operators about control strategies for zebra mussels. A synthesis of this information was presented in November to a group of scientists at the Canadian Centre for Inland Waters in Burlington, Ontario, at the invitation of Jack Vallentyne, senior scientist with Fisheries and Oceans Canada. The lecture included a population dynamics model based on catastrophe theory and European invasion and population data (Figure 10.3). The model showed that typically mussel population abundance increased exponentially for a 3–5 year period following invasion, then fluctuated around a high value for another 3–5 years (potentially limited by food, habitat, weather) before suddenly declining to a lower and more stable value. Comments by scientists at the lecture ranged from "zebra mussels were just a flash in the pan and would quickly decline like other exotics with no lasting impact to the Great Lakes ecosystems" to "a high abundance over a period of 5 years would likely affect forage fish populations since it was equivalent to the time of their life cycle."

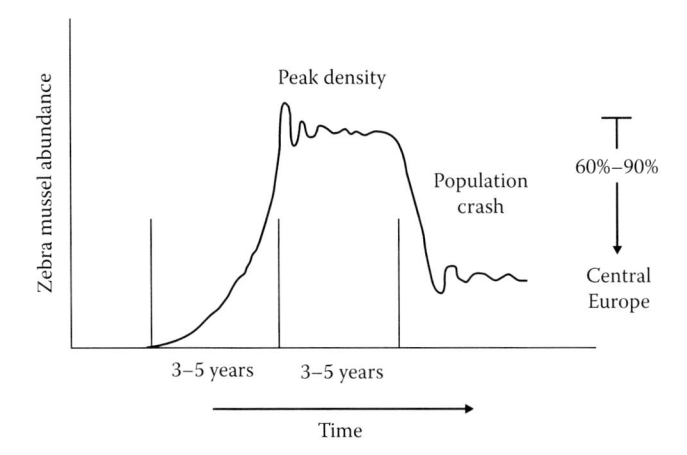

Figure 10.3 Population dynamics of zebra mussels based on central European and Swedish invasion data. The model predicts that after a 3–5 year period of establishment and exponential growth, population cycles fluctuate for about 3–5 years before they decline by 60%–90%.

Meanwhile, awareness of zebra mussels was brought to the attention of managers of utilities and industries throughout the Great Lakes through mailings of professional society bulletins (e.g., American Water Works Association, Electric Power Research Institute [EPRI]) and pamphlets from environmental and engineering consultants. Additionally, Griffiths and Schloesser traveled to Charlotte, North Carolina, in November, and gave a presentation on potential impacts of zebra mussels at the annual meeting of the EPRI. This action was prompted by concerns over electric-power supplies in North America. To many of the attendees, this was the first information they had that electric facilities in North America were vulnerable to fouling by "such a small critter." In addition, this was the first presentation to any water-resource user group outside the Great Lakes region. Contents of the presentation, which relied heavily on the zebra mussel invasion of the electric power plant in Monroe, Michigan, were published in their conference proceedings and distributed to managers of electric-generating facilities across the United States (Griffiths et al. 1989).

Wide-scale reports of zebra mussels helped generate a large attendance (225) at the *Second North American Zebra Mussel Conference* entitled *Zebra Mussels in the Great Lakes* held in Rochester, New York, in November 1989 (Figure 10.2). The conference was unable to accommodate all potential attendees, with many referred to the "next" zebra mussel conference. The *Third North American Zebra Mussel Conference* entitled *Zebra Mussel Symposium* was subsequently held in Cleveland, Ohio, in December 1989. Interest in zebra mussels was so great that the conference had to change its venue to accommodate the larger than expected attendance (350 attendees).

Although there was broad overlap of environmental and biological information provided at the two conferences, the second conference emphasized impacts and controls for electric-power users, while the third conference focused on impacts and controls for water-supply users. Presenters at both conferences reported on direct impacts of zebra mussels on Great Lakes utility operations, methods to prevent and remove zebra mussels from these facilities, pro and cons of the various techniques, and associated costs and lost revenue attributed to zebra mussels. Notable events at the second conference included presentations by two European investigators, Henk Jenner and Abraham bij de Vaate, who provided a European perspective on zebra mussels and various methods of control used in Europe, and a presentation by Paul Hebert that showed zebra mussels in Lake St. Clair were abundant enough to filter the entire water column at least once a day that could account for increased water clarity noted in summer. In addition, Griffiths presented his population dynamics model for zebra mussels and suggested that the most intense impacts to industry and ecosystems of the Great Lakes would occur over a 3–5 year period (maximum abundance of zebra mussels). Finally, in the conference summary, John Gannon of the U.S. Fish and Wildlife Service presented a 10 year economic impact projection of zebra mussels of $655 million cost for water-supply facilities, $870 million cost for electric-generation facilities, $502 million cost for manufacturing industries, $65 million cost for shipping and boating, and $2710 million lost in fisheries revenue, for a total of $480 million/year for 10 years and total cost of $4.8 billion. This economic-cost projection was immediately (Breckenridge 1989, Frischkorn 1989) and widely reported (e.g., Roberts 1990) and likely responsible for political and government action on issues of zebra mussels and other exotic species. After the third conference, Bob Peoples, a legislative development specialist of the U.S. Fish and Wildlife Service, noted in a memo that "This may be the ecological catastrophe many of those concerned about introduced species have anticipated, but hoped would never happen."

1990 ... POLITICAL RESPONSES EMERGE

Canadian Governments

In 1990, politicians in Canada began to organize responses to complaints and concerns reported by multiple-user groups about zebra mussels. In early January 1990, Doug Dodge, director of Southwestern Region of the Ontario Ministry of Environment, asked Griffiths to represent him at a meeting in the legislative building in Toronto. The meeting was chaired by the Minister of Natural Resources and attended by staff from the Ministries of Natural Resources, Municipal Affairs, Tourism and Recreation, and Environment. The fisheries coordinator for Ontario Ministry of Natural Resources gave a cursory presentation on zebra mussels and their impacts. Griffiths then provided detailed information on the current situation and impacts on water users. The difference in perspective between the two ministries was quite evident: the Ministry of Natural Resources viewed zebra mussels as a pest control issue similar to sea lamprey. The spread of mussels had to be managed (i.e., slowed or restricted) so that it did not invade inland waters with important recreational fisheries, such as lake trout. However, zebra mussels had not caused any impacts to fish populations in the Great Lakes and were believed not to be able to survive in the soft-water trout lakes in central and northern Ontario. In contrast, the Ministry of Environment viewed zebra mussels as a biological pollutant (i.e., a deleterious material discharge). Consequently, the central focus of the ministry was to document the extent and spread of this biological spill, provide assistance (information and funding) to abate and prevent damage to infrastructures (e.g., water-treatment and electric-generating plants), and warn water users about this pollutant. Hundreds of thousands of dollars had already been spent to maintain water flows at a dozen water plants alone, while research into controlling and preventing mussels from restricting the flow-through intakes was underway with funding from Ministry of Environment and Ontario Hydro.

In late January, 1990, the Ontario Cabinet decided that the Ministry of Natural Resources would be the lead agency to develop and coordinate a provincial strategy on zebra mussels: "Since the invaders are living animals, and since the Ministry of Natural Resources has the fisheries expertise, the Ministry of Natural Resources was the logical ministry to lead the battle" (OMNR 1990). Jim Bradley, Minister of Environment, told Griffiths at a later date that a large provincial ministry, such as the Ministry of Natural Resources, was required to deal with this issue because the Ontario Cabinet would not support increased funding to address zebra mussels; the Ministry of Environment was simply too small a provincial agency to tackle this issue with its existing funds. Subsequently, the Minister of Natural Resources sent a letter to the Ministers of Environment, Municipal Affairs, Intergovernmental Affairs, Tourism and Recreation, and Treasury and Economics requesting their Parliamentary Assistants be appointed to an Inter-Ministerial Steering Committee to be chaired by Jack Riddell, Parliamentary Assistant to Minister of Natural Resources. This political steering committee was charged with the development and coordination of a Provincial Zebra Mussel Program. The letter also proposed the formation of

1. An Inter-Ministerial Coordinating Committee composed of members of the steering committee and Ontario Hydro and chaired by the Ministry of Natural Resources Provincial Coordinator that would be responsible to (a) coordinate provincial zebra mussel programs (i.e., biological research, control methods, monitoring, impact assessments, mitigation strategies, contaminant cycling), (b) develop government policy on compensation and assistance to water users, (c) develop a water conservation strategy, (d) establish technical task forces to address specific issues, (e) review policy and program proposals with Deputy Ministers, and (f) recommend policy and programs to a steering committee for approval
2. A Zebra Mussel Secretariat in the Ministry of Natural Resources composed of professional and clerical staff that would be responsible to gather and report zebra mussel activities and information to the steering committee

In addition, the letter proposed immediate establishment of an ad hoc Inter-Ministerial Coordinating Committee to be composed of representatives from key ministries and Ontario Hydro and chaired by the Great Lakes Fisheries Program Coordinator of the Ministry of Natural Resources until the Inter-Ministerial Coordinating Committee and the Zebra Mussel Secretariat were established and staffed. This committee would be responsible for (a) refinement of the proposed strategy, (b) preparation of an initial budget for cabinet approval, (c) communication with public and private stakeholders, and (d) review of program needs and agency roles.

In early April 1990, the Ontario government announced to the public that the Ministry of Natural Resources would be the lead government agency for zebra-mussel issues and officially introduced members of the Inter-Ministerial Steering Committee. In addition, the Inter-Ministerial Coordinating Committee held its first meeting at the Sheraton Centre in Toronto. At this first meeting:

1. Representatives of the Ontario Ministry of Environment questioned the need for a secretariat and a coordinator, informed committee members about the range of activities that occurred over the past 2 years, acknowledged the central communication role of the existing Zebra Mussel Watch Committee established at the first zebra-mussel conference in 1989, and emphasized an urgency to fund programs to monitor, combat, and control zebra mussels in municipal-water intakes in the lower Great Lakes.
2. The Ontario Hydro representative detailed plans for installation of chlorination systems to protect service-water systems at electric-generating plants by June at a cost of $6 million and noted an additional $5 million could be used for research into chemical and nonchemical methods to control larval and adult mussels.
3. Representatives of the Ministry of Natural Resources noted their initiatives included a public education program (reduce the spread of zebra mussels via boats, bait), an invasion-monitoring program (monitor zebra mussel distribution and spread), research program to specifically examine effects of zebra mussels on fish (e.g., walleye, whitefish), a provincial coordination program to avoid duplication among agencies, and a central clearinghouse for zebra mussel information.
4. The committee agreed the immediate priority was to produce a budget for submission to Management Board of Cabinet for "enhanced funding"; until then, each Ministry

would redirect its funds to cover zebra mussel initiatives. During 1990, the Southwest Region of the Ministry of Environment redirected internal funds to several zebra-mussel issues, including $35,000 for a zebra mussel assistant, $15,000 for research, $36,000 for expenses in the Technical Assessment Section, $605,000 for design and installation of control facilities at 12 water-treatment plants, and $75,000 for video inspections of water intakes in the Utilities Operations section.

In early-May 1990, the ad hoc Inter-Ministerial Coordinating Committee met again to review and incorporate items into a budget proposal. Funding for Ontario Ministry of Environment initiatives surpassed all others and for the 1990–1991 fiscal year included $2 million for research and monitoring, $15.2 million for zebra mussel control and cleaning of municipal intakes, and $50,000 for public communications and presentations. For the 1991–1992 fiscal year, the committee included $3.5 million for research and monitoring and $36 million for zebra mussel control and cleaning of municipal intakes. The budget was finalized by the end of May and submitted to Management Board of Cabinet for enhanced provincial funding. Media reports (e.g., Coulson 1990) suggested that a budget of $8 million was requested from the provincial government.

In June 1990, the Ministry of Natural Resources sponsored a zebra mussel information session at a meeting in Toronto specifically organized to update government staff about zebra-mussel issues in Ontario. Chaired by Dodge, five morning presentations were followed by a question and answer session in the afternoon. In August 1990, Dodge flew to The Netherlands to confer with European representatives about zebra mussels. His trip report stated that he was surprised to learn that Europeans "did not view zebra mussels as a problem (Dodge, unpublished report)." Soon after this trip, Dodge resigned as zebra-mussel coordinator and the proposed budget to address zebra mussel issues failed to get approval from the Ontario Cabinet. With the defeat of the Liberal government in an election in October, members of the Inter-Ministerial Steering Committee were dismissed, and the ad hoc Zebra Mussels Coordination Committee was disbanded. Soon after, Chris Brousseau, district manager of Moosonee, was announced to be the new zebra mussel coordinator of the Zebra Mussel Coordination Office in the Ministry of Natural Resources.

In March 1990, the Hazardous Contaminants Branch of Agriculture Canada clarified in a memo to the Ontario Ministry of Environment that chlorine had to be registered under the Canadian Pest Control Act if it was to be used to control zebra mussels. As a result, the Ontario Ministry of Environment requested emergency registration of chlorine and hypochlorite as a pesticide to control zebra mussels. However, the request was denied because there was insufficient information on the environmental fate and distribution of these chemicals. Consequently, staff of the Ministry of Environment were instructed to send a memo to clients stating that any "inferred approval" for the use of chlorine to treat and control zebra mussels in water intakes and other pipes was withdrawn. This action placed domestic and fire-water supplies for tens of thousands and electric-power supplies to hundreds of thousands of citizens at risk of disruption.

In late-March 1990, the Ontario Ministry of Environment met with federal representatives of Fisheries and Oceans Canada, Environment Canada, Agriculture Canada, and Health and Welfare Canada to discuss the use of chlorine to control biofouling by zebra mussels. Federal agencies indicated chlorine could be used to control zebra mussels but only as a short-term solution. Therefore, the Ministry of Environment proposed a "Certificate of Approvals" approach, whereby approval of chlorine, as chlorine gas and hypochlorite, for the purpose of discouraging and eliminating attachment of mollusks in facilities requiring raw water would be exempted from the Ontario Pesticides Act and granted under the Ontario Water Resources Act. In June 1990, the Ontario Ministry of Environment filed an amendment to Regulation 751 under the Pesticides Act as Ontario Regulation 358/90 that took immediate effect pending federal approval. In July, a guide for the use of chlorine to control mollusks was published, and it incorporated previous recommendations on the maximum concentrations of chlorine application based on water temperature and veliger occurrence in water. In addition, the guide required that benthic-macroinvertebrate studies be conducted around intakes before and after chlorination to assess possible environmental impacts. In August, the Ministry of the Environment received federal approval for chlorine and hypochlorite as control agents. The first approvals were signed for the Nanticoke electric-generating station of Ontario Hydro and the Amherstburg water-supply facility just before mussel settlement in 1990.

Zebra mussels prompted many changes to the approval process and programs of the Ontario Ministry of Environment in 1990. Changes were so numerous that a 30 min presentation on zebra mussels at the annual surface water workshop in February 1990 to update staff now required a whole-day session in October. Presentations at the session included biology and distribution, impacts on aquatic biota, impacts and controls at electric-generating stations, impacts and controls at water-treatment plants, chlorination approvals, chlorine treatment alternatives, environment changes in the Great Lakes, effectiveness of various control methods to protect water intakes from biofouling, and, most notably, changes to regulatory approvals.

U.S. Governments

In the United States, Senator John Glenn of Ohio introduced Bill S2244 "the Nonindigenous Aquatic Species Act of 1990" to the Senate in early March 1990. This bill was subsequently referred to the Committee on Environment and Public Works. About the same time, Congressman Henry Nowak of New York introduced a companion bill HR4214 to the House, which was referred to the Merchant Marine and Fisheries Committee.

In June, these committees heard testimony from Gilman Veith, of the Environmental Protection Agency; Michael Donahue, executive director of the Great Lakes Commission; Constance Harriman of the Department of the Interior; Alfred Beeton of the National Oceanic and Atmospheric Administration; and Thomas Thompson of the U.S. Coast Guard.

Additionally, representatives from various Sea Grants, U.S. Fish and Wildlife, National Oceanic and Atmospheric Administration, U.S. Environmental Protection Agency, and Cooperative Institute for Limnological and Ecosystems Research met at the Detroit Metro Airport, Michigan, in June 1990, to discuss development of an interagency coordinated research program for nonindigenous species in general and, in particular, zebra mussels. Representatives of these groups agreed to establish an ad hoc U.S. Great Lakes Nonindigenous Species Coordinating Committee with representation from 14 organizations: 6 State Sea Grants, 5 federal agencies, Cooperative Institute for Limnology and Ecosystem Research, Great Lakes Commission, and Great Lakes Fisheries Commission. The resulting structure and the research agenda that was put together influenced the composition of the Aquatic Nuisance Species Task Force and the Great Lakes Panel on Aquatic Nuisance Species developed specifically through NANCPA. The objective of this committee was to establish a focused and coordinated research program that provided the scientific basis for sound policy development and facilitate coordination of research, education, communication, and technology transfer about nonindigenous species in the Great Lakes basin. By August, this committee had developed a research program that grouped specific ideas and studies into six major research themes that had been identified at the *Fourth North American Zebra Mussel Conference* in Guelph, Ontario, in February 1990 (discussed in the succeeding text). The six theme areas were biology/life history, ecosystem effects, socioeconomic analysis, control and mitigation, prevention, and reduction of spread.

In late-September 1990, the U.S. Environmental Protection Agency sponsored a 3-day workshop on "Ecology and Management of Zebra Mussels and other Aquatic Nuisance Species" at Saginaw Valley State University in Saginaw, Michigan. This workshop brought together researchers and resource managers from the United States, Canada, Europe, and Russia. The purpose of the workshop was to review and evaluate information on the ecology and management of introduced aquatic nuisance species, particularly zebra mussels, and make recommendations with respect to research needs, management options, information coordination, policy, and legislation. The workshop was an initial step for the agency to develop a coordinated program of research and management of nonindigenous species in the Great Lakes. After a day of public presentations from various experts, individual workshop participants were placed in one of four groups to specifically evaluate: (a) potential ecological effects of zebra mussels, (b) dispersal rate and final North American distribution of zebra mussels, (c) means to prevent new introductions, and (d) ways to control established introduced species. Each group developed a set of recommendations that included research needs to fill information gaps, public education requirements, information coordination activities, management options, dispersal mechanisms, and biological, chemical, and physical methods of control (U.S. EPA 1991).

Despite the short-time period until congressional adjournment at the end of October, the House passed H.R. 5390 (in lieu of H.R. 4212) by voice vote on October 1 and the Senate passed S.2244 before the end of October. The speed with which these bills passed through Congress reflected the urgency about the zebra-mussel issue in the Great Lakes region and the expected importance of not only zebra mussels but all aquatic nuisance species.

Although there were concerns over the size of government and costs required to manage zebra mussels and other exotic species, the President signed the Nonindigenous Aquatic Nuisance Prevention and Control Act (NANPCA) into law on November 29, 1990. Public Law 101–646 provided clear direction to federal agencies and authorized $40 million over 5 years to (a) undertake research and public education on zebra mussels and other aquatic nuisance species, (b) coordinate federally conducted and funded activities through establishment of an interagency commission, (c) develop and implement safe and sound control methods for aquatic nuisance species, and (d) prevent unintentional introduction and dispersal of nonindigenous species through ballast-water management. Specifically, provisions in NANPCA:

1. Directed the Army Corps of Engineers to develop environmentally sound techniques to prevent and control zebra-mussel infestations at public facilities, including water intakes, navigation locks, gated dams, pumping plants, hydropower plants, dredges, and commercial vessels.
2. Authorized grants for research and public education on zebra mussels and other nonindigenous aquatic nuisance species.
3. Established an interagency Aquatic Nuisance Species Task Force charged with implementation of NANPCA. The task force consisted of representatives from 10 federal agencies and included 12 ex officio members; the task force was co-chaired by U.S. Fish and Wildlife Service and National Oceanography and Atmospheric Administration.
4. Specifically provided a charter for the Great Lakes Panel on Aquatic Nuisance Species (first regional panel under the Aquatic Nuisance Species Task Force).
5. Required the U.S. Coast Guard to issue voluntary guidelines to prevent the introduction and spread of aquatic nuisance species into the Great Lakes though the exchange of ballast water not later than 6 months after enactment and to issue corresponding regulations within 24 months, i.e., mandatory ballast-water exchange regulation by November 29, 1992.
6. Designated zebra mussels as an injurious species under the Lacey Act to restrict intentional transport of this species into and throughout the United States.

During the period when Congress was assessing federal legislation, other local organizations and state agencies met to address zebra mussels. For example, in April, Dave MacNeill and Chuck O'Neill of New York Sea Grant and Ellen Marsden of Cornell University held a workshop in Syracuse, New York, entitled "Ecosystem Impacts and Management of Zebra Mussels in the Great Lakes: A Cooperative Research Planning Workshop" sponsored by the Great Lakes Research Consortium. Its purpose was to bring research scientists and natural resource managers (19 participants) together for an informal work session to explore: (a) impacts of zebra mussels on plankton, water clarity, benthos, and fishes; (b) methods to predict spread of zebra mussels; (c) criteria to standardize sample designs; and (d) management techniques and control agents. Recognition of potential ecosystem impacts by zebra mussels was important because funding priorities of various government agencies were based on recommendations of research-planning workshops that were conducted at the local and regional level in the Great Lakes.

Binational Cooperation

Zebra mussels were sufficiently important to be discussed at bilateral meetings between Canada and the United States. For example, on July 16, 1990, the Canadian Minister of Environment and the Administrator of the U.S. Environmental Protection Agency agreed to coordinate actions to address problems in the Great Lakes caused by zebra mussels. On October 10, the first Canada/U.S. bilateral meeting on zebra mussels was held in Washington, DC. Participants shared information on research monitoring and control programs, committed to continue communications, examined possibilities to establish an information clearing house, identified areas for cooperation and action, and agreed to joint sponsorship of a single annual international zebra-mussel conference starting in 1991 or 1992.

A special report entitled "Exotic Species and the Shipping Industry" was presented to both federal governments from the International Joint Commission and Great Lakes Fishery Commission in September (Dochoda et al. 1990). It summarized the proceedings of a Toronto workshop (February 28–March 2, 1990) that concluded the health and integrity of the Great Lakes ecosystem were jeopardized by colonization of exotic species transported in the ballast water of ocean-crossing ships. This issue was reinforced with the discovery of two more exotic species: the round goby (*Neogobius melanostomus*) and tubenose goby (*Proterorhinus marmoratus*) in the St. Clair River in early 1990 (Jude et al. 1992). Media reports about exotic species and ballast water maintained visibility of this issue throughout 1990 (e.g., White 1990).

Zebra Mussels Continue to Spread

As governments responded, zebra mussels continued to spread, and by late-1990, mussels were found throughout the Great Lakes from Duluth Harbor in Lake Superior to Cornwall on the St. Lawrence River, as well as in the Erie Barge Canal (Griffiths et al. 1991). Documented dispersal of zebra mussels was probably assisted by public awareness and the increased ability to detect veligers with the publication of the first photographic guide to veligers (Hopkins 1990). This guide spurred implementation of industrial and open-water monitoring programs of veligers throughout the Great Lakes. For example, analysis of veligers in water samples between May and November 1990 from intakes of eight water facilities located between south Lake Huron and the St. Lawrence River documented that veligers occurred consistently at only two Lake Erie facilities: the Union Water Treatment Plant in Kingsville between June and late July and the Rosehill Water Treatment Plant in the Niagara Region between July and late August.

The summer of 1990 again provided optimal environmental conditions for zebra-mussel reproduction, growth, and survival. Abundances of free-swimming veligers in 1990 exceeded those in 1989, peaking between 250,000 and 700,000/m^3 across the western basin of Lake Erie (Garton and Haag 1993, Leach 1993). Consequently, the abundance of juvenile and adult mussels exceeded those of 1989 with a maximum density of 270,000/m^2 at Hen Island reef and 340,000/m^2 at Sunken Chicken reef (Leach 1993). Juvenile mussels (2–11 mm) on these reefs were estimated to filter the entire water column every 1.0–2.5 days (Bunt et al. 1993), and the entire population of mussels (2–29 mm) was estimated to filter the entire water column between 4 and 19 times/day (MacIsaac et al. 1992). This filtration activity likely accounted for the low summer chlorophyll-*a* concentrations (rarely exceeded 2 µg/L), which were more than 50% less than measured in 1988 (Leach 1993).

In September 1990, zebra mussels occurred throughout the main basin of Lake St. Clair at densities up to 10,000/m^2, which was 50 times peak densities reported in 1988 (Hebert et al. 1989). Meanwhile, nearshore densities of 350,000/m^2 were reported at the mouth of Thames River (Bailey and Hunter 1991) and 200,000/m^2 along the south shore (Mackie 1991). Since it was estimated that a mussel population of about 6000/m^2 was sufficient to filter the entire water column twice daily (Hebert et al. 1991), zebra mussels alone could account for unusually high water clarity, that is Secchi depth of 1.8–2.8 m (Griffiths 1993), in the lake that was more than double that reported prior to colonization by zebra mussels (Leach 1972, 1980). Increased water clarity probably accounted for the presence of aquatic weeds and filamentous algae on sediments at several locations where none had been previously observed (Schloesser and Manny 1982, RWG, personal observations). Garton and Haag (1993) similarly noted macrophytes for the first time in shallow waters around docks at Stone Laboratory in western Lake Erie in response to clearer waters. Zebra mussels shifted habitat characteristics from a relatively homogeneous environment of silt and sand substrates that were easily disturbed by winds to generate turbid waters to a heterogeneous environment of weed beds, zebra mussel shells, filamentous algae, and silt and sand patches that were more resistant to wind disturbance. These changes likely accounted for the observed 11-fold increase in abundance of amphipods, flatworms, snails, and trichopterans in the benthic environment (Griffiths 1993). Dermott et al. (1993) similarly noted that abundances of benthos, particularly amphipods, chironomids, and mollusks, were about 10 times greater where zebra mussels were present than in similar areas where zebra mussels were absent along northeast shores of Lake Erie. In contrast, zebra mussels had a devastating impact on populations of native unionids. Abundances of unionids in Lake St. Clair declined 25% after the zebra mussel invasion (Nalepa 1994), and unionid populations in western Lake Erie declined even more where intensities of infestation exceeded several thousand zebra mussels per unionid (Haag et al. 1993, Schloesser and Nalepa 1994). An infestation intensity of only 100–200 zebra mussels per unionid appeared sufficient to cause unionid mortality. This was not encouraging news for the unionid population in Presque Isle, Lake Erie, where by late 1990 the average abundance of attached zebra mussels was 121/unionid (Schloesser and Masteller 1999).

Only minor impacts to water supplies at utilities and industries were reported in 1990. This was likely a result of precautions and implementation of cleaning and preventative technologies initiated in both 1989 and, especially, in spring 1990. Overall, chlorination became the principal short-term method to control zebra mussel abundance at municipal and industrial facilities. In spring, the Monroe water plant installed a temporary water intake in the Raisin River and then removed mussels from their 1.6 km long water intake that extended into Lake Erie. The removal resulted in a pile of mussels sufficient to fill "seven to eight large dump trucks" at a cost of about $100,000 (LePage 1993). Subsequently, a polyethylene hose was run down the interior length of the pipe to a crib house that supplied a continuous stream of chlorine to prevent future zebra mussel colonization. Similar chlorination equipment was installed at eight water-treatment plants in Ontario to protect raw-water intakes at a cost of about $200,000 each. In addition, the Tilbury Water Treatment Plant received $1.6 million to build an infiltration gallery to collect water from Lake St. Clair. Industries along Canadian shores of the St. Clair River also installed chlorination equipment to protect raw-water intakes at a combined cost of about $3 million. Ontario Hydro installed chlorination equipment to protect service-water supplies at several electric-generating stations and at the Bruce nuclear station at a combined estimated cost of about $6 million. The Detroit Edison Company removed 20 ton of zebra mussels from screen houses at the Monroe Power Plant in September and installed a chlorination system to protect service-water systems against zebra mussel colonization (Kovalak et al. 1993). Despite the greater volume of live mussels and shell debris removed in 1990, costs of cleaning at the Monroe plant did not increase proportionately because of continuous improvements in removal and disposal techniques.

While chlorine became the standard short-term method of mussel control, applied research on potential long-term control methods that used nonchemical treatments became an immediate priority. In March, Ontario Hydro provided $5.2 million to its Zebra Mussel Coordinating Committee to fund research on alternative control methods. This research program included testing ozone, gamma radiation, ultraviolet (UV) light, acoustics, mechanical filtration, electrofiltration, thermal shock, high water pressure, and nontoxic and controlled-release coatings. The program also examined effects of control methods on metals and materials commonly found in water-withdrawal facilities. Research studies in 1990 were largely "proof of principal" experiments to identify potential strategies and methods for subsequent field trials. In addition, funds from the Electrical Power Research Institute were used to develop a zebra mussel monitoring and control manual to be distributed to its members. In addition, the institute funded a study that tested effectiveness of proprietary molluscicides (e.g., Bulab6002, 6009; Clam-Trol) on zebra mussel veligers. The Niagara Mohawk Power Corporation also funded an evaluation of proprietary molluscicides to treat mussels including an evaluation of environmental effects and fates, along with hydrogen peroxide, ozone, and other oxidants. The Empire State Electrical Energy Research Corporation funded a zebra mussel deterrence study that tested effectiveness of underwater sound and acoustic technology.

The water-supply industry also funded research to control zebra mussels. The American Water Works Association Research Foundation funded a study to explore seasonal occurrence and ecology of veligers and to examine effects of hypochlorite, chlorine dioxide, chloramines, potassium permanganate, unionized ammonia, and hydrogen peroxide with iron on zebra mussel veligers in the western (City of Toledo) and central (City of Cleveland) basins of Lake Erie. The Erie County Water Authority funded examination of possible inactivation of larval and adult zebra mussels by chlorine, ozone, and hydrogen peroxide.

Spreading the Word

Monthly or seasonal meetings were a regular feature among most industrial and municipal water users, such as Ontario Hydro, Detroit Edison, Lambton Industrial Society,

Ontario Water Treatment Plants, and Electrical Power Research Institute. Furthermore, Ontario Hydro sponsored a seminar about zebra mussels in January 1990 in Toronto to specifically inform staff of its research, control, and progress on prevention. Numerous regional conferences were also held to address water-user impacts and control of zebra mussels: Western Ontario Water Works Association (April), Pollution Control Association of Ontario (April), Georgian Bay Waterworks Conference (April), American Power Conference (April), and Ontario Municipal Water Association (May). National conferences were also held such as the International Macrofouling Symposium in Florida sponsored by Electrical Power Research Institute (December).

In 1990, progress was made in establishing offices to provide information on zebra mussels that would set the stage for a funded "zebra mussel clearing house." Throughout 1989 into early 1990, industries, nonprofits, and governments established contact offices and departments to handle the large volume of calls and information requests about zebra mussels in North America. Examples include Ontario Federation of Anglers and Hunters, American Water Works Association, EPRI, and Zebra Mussel Coordinators Committee, Zebra Mussel Watch Committee, and U.S. Great Lakes Nonindigenous Species Coordinating Committee. These groups were "rolled over" to support a central source of information on zebra mussels: the Zebra Mussel Clearinghouse. This clearinghouse was a partnership of private industry and government agencies housed with New York Sea Grant. This contact point facilitated sharing of zebra mussel information among researchers, government agencies, industries, and others throughout North America. It provided access to research, technological and policy information on the biology, spread, impacts, and controls of zebra mussels. Later, the Federal Aquatic Nuisance Specie Task Force and U.S. Army Corps of Engineers would use the clearinghouse to report federal initiatives and activities. The clearinghouse satisfied a need for a central source of information that was suggested at the *First Zebra Mussel Workshop* in March 1989 and the subsequent three *North American Zebra Mussel Conferences* in 1989.

"Newsletters" replaced the single-issued, stand-alone "fact sheets" produced by U.S. Fish and Wildlife Service, Great Lakes Fishery Commission, Great Lakes Sea Grants, and Ontario Ministry Natural Resources as the prime source of printed information about zebra mussels. The Wisconsin Sea Grant published the first issue of "Zebra Mussel Update Bulletin" written by Cliff Kraft at University of Wisconsin in May, while the Zebra Mussel Information Clearinghouse published the first issue of the bimonthly newsletter entitled "*Dreissena polymorpha—Information Review*" in September. These multipage newsletters presented summaries of research, meetings, legislation, and impacts on the environment and industries, along with dates for future meetings and conferences, sighting locations of zebra mussels, and other information to facilitate communications. A wallet-sized identification card with a color photograph of a zebra mussel produced by Wisconsin Sea Grant, however, was the most requested printed item for public distribution by agencies around the Great Lakes. Over a million of these cards were produced by late 1993, including a version in French for distribution in Ontario.

In 1990, an innumerable number of "zebra mussel" public information sessions, industrial seminars, and scientific symposia were held. Some notable events included: (1) a series of well-attended public information sessions sponsored by Ohio Sea Grant that were held in towns throughout Ohio (e.g., Port Clinton, Defiance, Fremont), (2) an evening session on zebra mussels at the Harbourfront Limnology Lecture Series held at York Quay Centre in Toronto attended by about 400 people, and (3) a public session in Grosse Pointe Farms, Michigan, entitled "Environmental and Economic Impacts of Zebra Mussels on the Great Lakes" sponsored by the Great Lakes Forum.

Public media articles in 1990 reported on occurrences of mussels at new locations (e.g., Kehoe 1990), proposed provincial ($8 million) (e.g., Coulson 1990) and U.S. funding ($40 million) (e.g., Edmonds 1990), potential environmental impacts of mussels (e.g., David 1990), and possible control methods for zebra mussels (e.g., Grossi 1990, Mackie 1990, Ruscitti and McKenzie 1990). Even the *Los Angeles Times* in California reported about zebra mussels in the Great Lakes (Green 1990). Zebra mussels were also noted in the "prestigious" journal *Science* with an article written by Leslie Roberts and published in September (Roberts 1990). Similar information was made available to the general public in magazines such as Toronto Life (Lees 1990), Equinox (Banks 1990), Sports Illustrated (Boyle 1990), The Atlantic (Hart 1990), Ontario Fisherman (Snyder 1990), and The Sportsman's Outdoors Annual (Griffiths 1990).

The *Fourth North American Zebra Mussel Conference*, entitled *Zebra Mussels: The Great Lakes Experience*, was held at the University of Guelph in Guelph, Ontario, in February 1990 (Table 10.1, Figure 10.2). This conference was sponsored by the University of Guelph and U.S. Fish and Wildlife Service. It was attended by 250 people and consisted of 18 presentations. In addition to presentations on effects of zebra mussels on native bivalves and lake environments, biofouling at specific power and water facilities was highlighted. About one-third of the presentations provided information on the effectiveness of various control chemicals (e.g., chlorine, hypochlorite, ozone, molluscicides) based on laboratory and full-facility testing. The afternoon session of this conference focused on identification of key research questions to clarify potential environmental problems, understand dispersal mechanisms, and minimize impacts associated with zebra mussels.

The first session at a major scientific conference, entitled *Zebra Mussels in the Great Lakes*, was chaired by Gerry Mackie and Don Schloesser and held at the *38th Annual*

Meeting of the North American Benthological Society at Virginia Polytechnic Institute and State University in May 1990. Seventeen papers were presented in this special session, with 14 of those being updates of presentations given at the *Fourth North American Zebra Mussel Conference*. Two of the added presentations were given by Dutch researchers: Abraham bij de Vaate presented information on growth of zebra mussels in lotic and lentic waters and Harro Reeders presented information on filtration rates and pseudofeces production. The third presentation, given by David Strayer, showed for the first time that the potential distribution of zebra mussels extended across the continental United States. Also in May, two zebra mussel presentations were given at the annual conference of the International Association for Great Lakes Research held at the University of Windsor, Ontario, in a species-invasion session chaired by Joe Leach. In addition, a special session on exotic bivalves was held in June at an annual meeting of the American Society of Limnology and Oceanography at College of William and Mary in Virginia, with 8 of 16 presentations about zebra mussels.

In December 1990, the *Fifth North American Zebra Mussel Conference* entitled *International Zebra Mussel Research Conference* was hosted by Ohio Sea Grant at Ohio State University in Columbus (Figure 10.2). It was attended by about 200 people and consisted of the largest number of presentation to that date (35). Anna Stańczykowska-Piotrowska from Poland, a dedicated, long-time European zebra-mussel researcher, gave the keynote address. Presentations on biology, ecology, control, and prevention were equally represented with remaining presentations dealing with intake fouling, monitoring methods, and socioeconomic issues. A presentation by Ron Griffiths on dispersal mechanisms emphasized the role that fishing boats and trailers could play to spread zebra mussels outside the Great Lakes region; this point was brought home by a later presentation that indicated zebra mussels had been found in an inland lake in New York state and, subsequently, by reports of mussels found on boats and trailers in inland waters (Kraft 1990, 1991a). Ruth Beeton reported an 85% decline in spring diatom populations in Hatchery Bay of western Lake Erie that provided another example of dramatic effects zebra mussels had on phytoplankton resources. Peter Landrum and colleagues showed zebra mussels could rapidly accumulate organic contaminants, such as PAHs and PCBs, and concentrate them to a much greater extent than fish, which implied zebra mussels would likely change nutrient cycling and transport of contaminants in the Great Lakes. Panel discussions on ecosystem impacts, control alternatives, socioeconomics, and monitoring provided a forum to document research needs and clarify issues. Two socioeconomic presentations provided added insights into the zebra mussel issue based on measurements of direct costs and public attitudes. Leroy Hushak reported that port authorities along the south shore of Lake Erie spent no money on zebra mussel control in 1989 and 1990, while commercial shippers spent only $1500, and charter boat captains spent about $200/captain on preventative actions, such as antifouling paints. A handful of private boaters and a few marinas did incur costs for zebra-mussel damages in 1989, but no damages were reported in 1990 after these users implemented preventative actions, primarily antifouling paints in 1989. Industries and municipalities that drew raw water from Lake Erie reported the highest costs, typically to clean and remove zebra mussels from water intakes, but this totaled only $210,000. The study concluded that regular maintenance and preventative procedures were adequate for zebra mussel control along the south shore of Lake Erie and these added actions had just a small cost to users. This was in stark contrast to costs paid by utilities along the western shore of Lake Erie (e.g., Detroit Edison, City of Monroe, City of Toledo) and in the province of Ontario, where Ontario Hydro spent almost $10 million to install chlorination equipment at its facilities and various Ontario municipalities spent almost $4 million to remove zebra mussels from intakes and prevent recolonization by late 1990. Meanwhile, survey results from attendees at the Mid-America Boat Show in 1990 indicated boat owners saw zebra mussels as a threat to boating and sportfishing and that these owners approved use of public money for zebra mussel research (Lichkoppler et al. 1993).

In December 1990, the *Sixth North American Zebra Mussel Conference* entitled *A Symposium on Zebra Mussels: 1990 Developments* was held in Detroit, Michigan, and sponsored by the Detroit Water and Sewage Department and WW Engineering and Science, Inc. (Figure 10.2). About 160 people attended to hear 11 presentations. Eight presentations addressed zebra mussel control and prevention (e.g., chemical, operational, physical) at facilities in the lower Great Lakes. One presentation documented taste and odor problems in waters of the east basin of Lake Erie, an issue that would occur over all the lower lakes in the next few years.

Zebra mussels were honored as the "pest of the year" for its newsworthiness by *Discover* magazine (Reynolds 1991) and labeled the "North American Environmental and Economic Nightmare of the Nineties" by Canadian Museum of Nature (Topping 1991).

1991 ... IMPLEMENTATION

Canadian Governments

The newly elected Ontario government moved quickly to respond to issues related to zebra mussels and other exotic species. In late January, the Legislative Assembly's Standing Committee on Resources Development, chaired by Peter Kormos, heard from several experts including Joe Leach, Gerry Mackie, Jon Stanley, and Dave Garton about the ecological and economic impacts of zebra mussels, possible control strategies, and potential policy initiatives.

The committee released an 80-page report on "Exotic Species in Ontario" in May that included 30 recommendations to the provincial government, with 13 specifically about zebra mussels. Most of these recommendations, however, had already been put in place the previous year by the Ministry of Natural Resources (e.g., signage at all boat ramps identifying measures to reduce risk of spreading zebra mussels), Ministry of the Environment (e.g., use of chlorine as a control method should be strictly monitored and used only as an interim measure), and Ontario Hydro (e.g., support research related to development of physical and mechanical methods of control including nontoxic coatings). A few recommendations left unaddressed included the following: (1) inspect boats to reduce spread of zebra mussels in live wells; (2) develop containment guidelines for mussels used in classrooms, research labs, and other legitimate purposes and initiate licensing procedures to eliminate transport of mussels by all others; (3) establish research priorities to assist in resource allocation; (4) coordinate programs related to zebra-mussel distribution and control; (5) investigate ways (grants, loans, tax incentives) to encourage private sector participation in zebra-mussel research; and (6) develop a geographic information system to distribute data of the known geographic range of zebra mussels and other exotics.

Meanwhile, the Ontario Ministry of Natural Resources held a zebra mussel workshop in Toronto in January 1991 to review programs, update staff, and plan future activities. On the first day of the workshop, the Lake Ontario, Lake Erie, and Inland Fisheries Assessment Units provided updates on zebra mussel initiatives and monitoring results. These updates were followed by presentations on: (a) zebra mussel impacts on water intakes, (b) zebra mussel control efforts, (c) compliance with voluntary ballast-water guidelines, (d) impacts on the Great Lakes, (e) and standardized sampling protocols. On the second day, work groups were convened to: (1) clarify the role of the Ministry of Natural Resources in research, monitoring, information, and communication; (2) establish standard sampling protocols; (3) identify vectors to slow the dispersal of zebra mussels; and (4) formulate a communication plan. The workshop concluded that the Ontario Zebra Mussel Program should only focus on: (a) coordinating scientific research, (b) providing information to slow dispersal, (c) monitoring distribution and dispersal, and (d) investigating impacts on aquatic environments. In addition, Ministry of Natural Resources agreed to use the standardized monitoring protocols of Marsden (1992) for its invasion-monitoring program. Furthermore, an education program was proposed for recreational fishery and boater groups because they were identified as the primary vector of zebra-mussel dispersal. The program included the development of fact sheets and information signs for marinas and boat launches. The "Zebra Mussels Boaters Guide" released in late 1991 provided an excellent array of easy methods for boaters to slow the spread of mussels by cleaning boats with chlorinated water, scrapping adults off surfaces and disposing on land, drying boats in the sun for 3 days, running hot water through engines, and removing aquatic plants. Finally, Mussel Morsels, a single-page (typically) newsletter to be published by Ministry of Natural Resources (approximately monthly in 1991, bimonthly in 1992), was presented as a communication tool to keep staff and others up to date on current zebra mussel activities and information.

In March 1991, the Ontario Cabinet of the provincial government provided enhanced funding for 1991–1992 to the Ontario Zebra Mussel Program. Ontario Ministry of Natural Resources received $2.8 million of $3.5 million requested. Although it was estimated zebra mussels would cost Ontario more than a $1 billion over the next decade (Gorrie 1990), the budget was constrained by a forecasted economic recession, which had begun to reduce government revenues.

The Zebra Mussel Coordination Office received $320,000 and the Communications Office received $200,000. A single Zebra Mussel Coordination Office replaced the previously proposed governance structure, and the new coordination office became functional in May when three staff were hired. The office was essentially a communication hub that received information and answered questions from agencies, media, and citizens and supplied materials including zebra mussel identification cards, a zebra mussel slide presentation, fact sheets, signs, newsletters (regular Mussel Morsels and annual Zebra Mussel Watch), and distribution maps. The office assembled a catalogue of research projects in Ontario that was printed in late 1991 (OMNR 1991). It also funded and helped organize annual zebra-mussel conferences.

The Research and Assessment Project of the Ontario Zebra Mussel Program received $575,000. Seventy-five percent of these funds supported projects conducted by Ministry of Natural Resources staff in Lake Erie ($474,200), Lake St. Clair ($16,000), Lake Ontario ($72,000), and the Rideau River ($8,000), with a general theme that focused on impacts of zebra mussels on fish abundance, growth, diet, and fish composition, especially walleye in Lake Erie and lake whitefish in Lake Ontario (OMNR 1991). The other 25% of the funds ($148,300) supported studies by university researchers who examined zebra mussel dispersal, size variation of mussels in Lake Ontario, effects on benthos, impacts of predation by diving duck in western Lake Erie, grazing impacts on phytoplankton, and contribution to beach material composition, with results published in the primary literature (e.g., MacIsaac et al. 1992, Bunt et al. 1993, Hamilton et al. 1994, Hincks and Mackie 1997, Bailey et al. 1999, Chase and Bailey 1999). These same projects were allocated $386,900 in 1990 by Ministry of Natural Resources (OMNR 1991) with 92% of funds supporting projects conducted by Ministry of Natural Resources staff, primarily in Lake Erie ($310,100).

The invasion-monitoring project in the Ontario Zebra Mussel Program received $1.2 million, essentially to provide new boats, motors, and trailers to participating Fisheries Assessment Units across Ontario. A total of $559,000 was allocated in 1991 to 10 Fisheries Assessment Units (4 Great Lakes and 6 inland water units) that participated in monitoring zebra-mussel veligers and adults at 150 stations across Ontario (OMNR 1991). The Lake Ontario Fisheries Assessment Unit and three inland fisheries units received $177,200 for invasion monitoring in 1990. The remainder of the budget was not allocated to specific projects. Interestingly, monitoring sites were located on a number of soft-water lakes in the Muskoka region of Ontario where zebra mussels were unlikely to establish large populations because of the low calcium concentrations in the water but none in several hard-water lakes in southwest Ontario, such as Chesley Lake, which likely was the first inland lake to be colonized by zebra mussels. Unfortunately, these monitoring results, except from the Lake Ontario Fisheries Unit, were not made available to the Ontario Ministry of Environment and other agencies. This million-dollar monitoring program replaced the low-cost, informal reporting network used by the Ontario Ministry of Environment.

Data from the invasion-monitoring project were used to produce a distribution map of zebra mussels at the end of 1991. This map appeared in the January 1992 Zebra Mussel Watch newsletter (OMNR 1992) and was reproduced in many newspapers (e.g., van Moorsel 1992), magazines, and engineering journals (e.g., Lewis 1992) throughout the Great Lakes basin. Unfortunately, the depicted distribution of mussel-infested waters was underrepresented, which left water users in a number of communities believing mussels were not present when in fact they were. In addition, this map indicated zebra mussels had colonized Lake Winnebago in Wisconsin just west of Lake Michigan. Based on this information, Appleton Common Council was just about to allocate $55,000 to initiate a zebra mussel monitoring and control program when the council was informed that zebra mussels had not been detected in the lake (Kraft 1992a). This error was corrected on the subsequent distribution map of zebra mussels in 1992, the last map produced by Ontario Ministry of Natural Resources.

With no enhanced funding from the Ontario Cabinet, the Ministry of Environment immediately reduced its involvement in zebra mussel activities. Consequently, few staff of the Ministry of Environment were able to attend meetings and conferences. This loss of visibility with the Great Lakes public resulted in a dramatic decline in phone calls reporting possible mussel locations in the latter half of the year, and reduced interaction with colleagues around the Great Lakes basin resulted in the loss of another valuable source of local colonization information. Municipalities were informed that there was no new money specifically for control and prevention projects but that they should continue requesting funds through existing provincial programs.

In April 1991, the Ontario Ministry of Environment organized its own Zebra Mussel Coordinating Committee that would focus on effects of zebra mussels on water quality, research on control and prevention techniques in partnership with Ontario Hydro, and assistance to municipalities whose source waters were infested with zebra mussels. Proposals for zebra-mussel research were directed to the Research and Technology Branch of the ministry, but no public notice of this program was made because of limited funds. Funds were provided to Gerry Mackie to continue work on zebra-mussel life history in Lake St. Clair ($25,100) and examine the distribution of mussels in Lake Ontario ($74,000), as well as to Ralph Mitchell at Harvard University to explore potential biological control agents ($50,100). Internal funds were provided to Bernie Neary to develop a model based on water chemistry to predict dispersal of zebra mussels to inland Ontario lakes, Ken Nicholls to assess impacts of zebra mussels on phytoplankton, and Ron Griffiths to assess the impact of zebra mussels on the fauna of Lake St. Clair. Meanwhile, Ontario Hydro provided $1.35 million for 16 studies, 13 conducted internally, that examined various chemical and physical methods of control and prevention as well as $90,000 for benthic studies to assess any impact of chlorine treatment on the environment.

U.S. Governments

In the United States, funds authorized from the NANPCA of 1990 spurred implementation of zebra-mussel programs by federal agencies in 1991. This act was passed in late 1990, which allowed for implementation before the reproduction season in 1991. The U.S. Coast Guard issued a joint policy statement with the Canadian Coast Guard that established voluntary guidelines for ballast-water discharges from ships in the Great Lakes that took effect on March 15. This policy aimed to reduce the probability of introduction and spread of additional aquatic nuisance species into the Great Lakes. Under the control act, the U.S. Coast Guard was to issue a mandatory ballast-water exchange regulation by November 29, 1992. On April 8, 1993, the Coast Guard published final rules in the Federal Register that required ballast-water management practices for vessels that enter the Great Lakes after operation on waters beyond the exclusive economic zone of the United States (the 200 mile limit for territorial waters). The rules required exchanged water to have a salinity of 30 parts per thousand and records of where the exchange occurred, with violators facing penalties of up to $25,000. Unfortunately, the Canadian government did not issue similar mandatory, ballast-water regulations.

In early 1991, the Great Lakes Sea Grants received $1.4 million to fund zebra-mussel research. They received 58 proposals in early March and provided funds for 12 projects in mid-June. Studies included research into dispersal mechanisms, byssal thread adhesive qualities, genetics, osmoregulatory physiology, control through reproductive intervention, safety aspects of ballast-water exchange, UV B as a control agent, impacts on food-web dynamics, and impacts on carbon and phosphorus dynamics in plankton. An additional $500,000 was allocated for outreach, education, and technology transfer primarily to support the Zebra Mussel Clearinghouse and its efforts to initiate a zebra mussel graphics library, produce literature on control, and create an annual research conference that would rotate among six Sea Grant states.

In July 1991, the Great Lakes Panel on Aquatic Nuisance Species, established under NANPCA, held its first meeting. The panel replaced the U.S. Great Lakes Nonindigenous Species Coordinating Committee established in June 1990. The research program developed by the coordinating committee was partially used by the Great Lakes Panel on Aquatic Nuisance Species and later by various Sea Grants as a basis for funds awarded to projects under NANPCA in early 1991 (Kraft 1991a). This Great Lakes Panel included representation of 14 organizations: 6 Sea Grants, 5 federal agencies, Great Lakes Commission, and Great Lakes Fisheries Commission. The Great Lakes Panel was charged to coordinate exotic species activities throughout the binational Great Lakes system. The panel also represented regional interests on the National Task Force that mobilized federal resources to address the problem of aquatic nuisance species nationwide. Canadian representatives from government and industries were invited as panel observers.

With funding from the Task Force established under NANPCA, the National Fisheries Research Center (NFRC) of the U.S. Fish and Wildlife Service in Gainesville, Florida, announced in mid-1991 that it would incorporate information on the distribution of zebra mussels into an existing geographic information system database for nonindigenous aquatic species. This initiative began to produce maps that showed the North America distribution of zebra mussels in late 1992.

Later in 1991, the U.S. Army Corp of Engineers announced initiation of a 4 year program at its Waterways Experimental Station in Vicksburg, Mississippi, to develop environmentally sound prevention and control strategies for infestations at public facilities, including water intakes, navigation locks, gated dams, pumping plants, hydropower plants, dredges, and commercial vessels. The program stressed development of physical rather than chemical methods to ensure that native biota and potable-water supplies were not adversely affected. Control strategies included early detection, infestation removal, and prevention actions. Laboratory studies were also conducted to evaluate tolerances of zebra mussels to desiccation, high temperature, and hypoxia. The experiment station also monitored the biology, physiology, physical condition, and size demography of natural populations at key sites on major waterways. Finally, impacts on native biota, especially unionid mussels, were also to be evaluated (Ingram and Miller 1992).

In late July 1991, the Congressional Office of Technology Assessment (OTA) convened a panel to address the broader issue of "harmful nonindigenous species in the United States" that included Jim Carleton of Williams College–Mystic Seaport, an expert on ballast-water introductions, and Bill Kovalak of Detroit Edison. While not published until 1993, findings and recommendations in the final report validated actions and results of the largely voluntary international cooperative effort among government agencies, academia, and water users.

Zebra Mussels Disperse beyond the Great Lakes Basin

By late 1991, zebra mussels had spread out of the Great Lakes basin. To the east, mussels had moved down the New York Barge Canal to the Hudson River (Lange and Cap 1991), while to the west, mussels had moved through the Chicago Sanitation Canal and down the Illinois River to the Mississippi River (Kraft 1991b, Tucker et al. 1993). Mussels were also discovered to the south in the Susquehanna River (Lange and Cap 1991), which is not physically connected to Great Lakes, as well as in inland lakes and reservoirs in New York, Indiana, Ohio, and as far south as Kentucky Lake, an impoundment on the Tennessee River (Kraft 1991b, Ludyanskiy et al. 1993). This jump out of the Great Lakes basin prompted state agencies in Florida, Tennessee, and Missouri to initiate monitoring activities.

In 1991, it became apparent that zebra mussels likely reached their carrying capacity in the Lake St. Clair–Lake Erie region (i.e., initial colonized region of the Great Lakes). For the first time, veliger abundances in Lake Erie were not greater than the previous year. Peak abundance of 530,000/m^3 near Monroe, Michigan, in western Lake Erie (Bidwell et al. 1991) and 316,000/m^3 in open waters of central Lake Erie near Cleveland, Ohio (Greenberg et al. 1991), was similar to those in 1990. Meanwhile, average veliger densities declined in the western basin from 96,000/m^3 in 1990 to 85,000/m^3 in 1991 (Leach 1992). Furthermore, zebra-mussel densities on Sunken Chicken Reef declined from 350,000/m^2 in 1990 to 110,000/m^2 in 1991 (Fitzsimons et al. 1995), while those on Hen Island declined from 268,000 to 253,000/m^2 (MacIsaac et al. 1992). Similarly, biomass near Point Pelee in Lake Erie declined from almost 600 to around 30 g/m^2 (Hamilton et al. 1994).

Food limitations may explain decreases in populations in 1991. Zebra mussels were estimated to consume 1/3 to 1/2 of summer phytoplankton production in western Lake Erie (Madenjian 1995, Stoeckmann and Garton 1997). This intense grazing pressure alone likely accounted for the average 65% reduction in chlorophyll *a* and associated 77% increase in water transparency (i.e., 1.2 m) in the western basin of Lake Erie (Leach 1992), the 86% reduction in planktonic diatoms in Hatchery Bay (Holland 1993), and the 90% decline in total phytoplankton measured at water intakes across Lake Erie (Nicholls and Hopkins 1993) after mussel colonization. With summer concentrations of chlorophyll *a* less than 2 μg/L (Leach 1992, MacIsaac et al. 1992, Nicholls and Hopkins 1993), a lack of food sufficient to maintain high abundances was very probable. This would account for the observation by Stoeckmann and Garton (1991) that mussel growth in flow-through, lake-water aquaria at South Bass Island was stimulated when supplemental food was provided. Similarly, the lack of optimum food supplies would account for observed declines in size at maturity from 13 mm in May 1990 to 5 mm in May 1991 in the western basin (Nichols and Kollar 1991) and greater tissue degrowth measured in 1991 over 1990 in Lake St. Clair (Nalepa et al. 1993).

A walleye population localized in Lake St. Clair and connecting channels between Lakes Huron and Erie became the first fish population to be impacted by zebra mussels. This walleye population ascends the Thames River in April to spawn near Komoka, Ontario, where it supported a seasonal recreational fishery of 200 fishers/day (RWG, personal observation). In spring 1991, however, a severe decline in catch rates was widely reported by local fishers (MacLennan 1992), and the riverine fishery essentially disappeared by 1995. At that time, sport-walleye landings in Lake St. Clair declined from 125,000 fish in the late 1980s to just 16,000 fish in the early 1990s (OMNR 1993a). Survivorship of adult walleye, however, was high (S = 0.68), which suggested fish mortality in Lake St. Clair, and the Thames River was not the cause of the population decline (MacLennan 1992). Walleye-egg hatchability studies in the laboratory and field suggested nitrate and other water chemistry concentrations in the river did not cause failed recruitment (Flood and Lee 1995, 1996). Although egg hatchability was lower in spawning reaches of the Thames River than in upstream tributaries to the Thames River, it was sufficient to provide 27 million larval walleye to Lake St. Clair from May to mid-June 1992 (OMNR 1993b). The zebra mussel induced environment shift in Lake St. Clair from a turbid, muddy, vegetation-poor habitat to a clear-water, vegetation-rich habitat likely accounted for the collapse of the walleye population. Catches of young-of-the-year walleye in Lake St. Clair were essentially zero between 1989 and 1991 (MacLennan 1992). Coincidental increases in population and growth of visual feeding predators, such as muskellunge and bass, after colonization by zebra mussels (OMNR 1993c) support this hypothesis. Predation by these fishes on larval and young-of-the-year walleye, along with an aversion of walleye to clear waters (Lester et al. 2002), may have been mechanisms that reduced this walleye population. This was in contrast to western Lake Erie, where zebra mussels were not found to have a direct impact on walleye-egg viability on reefs (Fitzsimons et al. 1995). However, larval and young-of-the-year walleye in Lake Erie did not have to swim through waters populated by bass and muskellunge to find the safety of more turbid habitats in the lake.

The potential of zebra mussels to change contaminant cycling in the Great Lakes became evident when studies showed they accumulated up to 10 times the PCB concentrations found in native unionids in Lake St. Clair (Brieger and Hunter 1993) and western Lake Erie (Kreis et al. 1991). These organic contaminants were available directly to predators, such as ducks and fishes (Mazak et al. 1997), and indirectly through pseudofeces to detritus feeders, such as amphipods and worms (Bruner et al. 1994b). Stewart and Hynes (1994) showed abundances of benthic macroinvertebrates (especially annelids, gastropods, amphipods, and crayfish) were higher in southwestern Lake Ontario in 1991 after zebra mussels were established. This study supported previous observations in Lake St. Clair (Griffiths 1993) and in northeastern Lake Erie (Dermott et al. 1993) that zebra mussels caused shifts in benthic fauna of the Great Lakes.

In July 1990, a benthic survey in Rondeau Bay on the north shore of Lake Erie found large abundances of zebra mussels and the presence of the exotic Asian clam, *Corbicula fluminea* (Griffiths 1992). Until 1977, Rondeau Bay was known as a shallow, clear-water, weedy environment with abundant yellow perch, largemouth bass, and panfish. Extensive rains in spring 1977 caused tons of sediments and associated herbicides and nutrients to flow into the bay that resulted in prolonged turbid-water conditions and loss of a large proportion of submergent vegetation, primarily Eurasian milfoil (*Myriophyllum spicatum*) and water celery (*Vallisneria americana*) (OMOE 1982). Total fish catch from Rondeau Bay declined 78% from 1977 to 1981 (Witzel 1981, Hyatt 1982). Sailboats now were able to ply the weed-free central basin of the bay (2 m depth), and swimmers enjoyed plant-free littoral areas. Repeated attempts to reestablish *Potamogeton pectinatus* populations in the bay from tubers failed. However, after just 2 years, zebra mussels restored the environment of the bay to pre-1977 conditions. Summer clear-water conditions returned, aquatic macrophytes, especially *Vallisneria*, grew throughout the bay, and yellow perch populations increased. Requests for aquatic-herbicide permits returned, sailboats no longer could use the central basin, and swimmers moved to weed-free waters along Lake Erie.

Chlorination of municipal and industrial water intakes and in service-water systems of electric-generating facilities likely minimized impacts of zebra mussels along both Canadian and U.S. shores of Lake Erie in 1991. After excellent treatment results were reported by utilities along Lake Erie, water-treatment plants along Lake Michigan installed chlorine feed lines to discourage settlement and attachment of veligers and juveniles in water intakes, including the Green Bay water utility ($470,000), City of Wilmette ($250,000), and Gary-Hobart Water Corporation, and a private utility in Gary Indiana ($750,000–$1.25 million). Veligers had been found at 20 of 24 intakes drawing water from Lake Michigan in Indiana and Illinois in summer 1991 (Kraft 1992b). Meanwhile, water-withdrawal operators in the Niagara River and Lake Ontario area, including Niagara County Water District, Erie County Water Authority, Regional Municipality of Hamilton-Wentworth, and City of Buffalo, indicated their intention to install chlorination equipment (San Giacomo and Cavalcoli 1991). Similarly, electric-generating facilities in the Lake Michigan basin announced plans to install treatment systems for zebra mussels.

In March 1991, Ontario Hydro submitted a report to the Ministry of Environment entitled "Zebra Mussel Chlorination Benthic Survey 1990," in compliance with its Certificate of Approval. This report showed no impacts on surrounding benthic-macroinvertebrate communities from its chlorination program to control zebra-mussel biofouling at four facilities. Studies of benthic macroinvertebrates surrounding intakes of water-treatment plants and industrial facilities also showed no impacts from chlorination (i.e., no chlorine escaping from intake). These reports provided confidence in the safety of the chlorination treatment process and addressed concerns of the Canadian federal government.

Unfortunately, water-supply utilities were again reminded of the need to chlorinate when zebra mussels reduced a 122 cm diameter intake pipe at Bay City, Michigan, by 15–60 cm along its 6 km length, which restricted intake flow capacity to 75% (Conway 1991). Staff feared the uneven layer of mussels that colonized the intake pipe (estimated at 220 dump trucks in volume) would create turbulent flows in the winter season and cause frazil-ice formation that would shut down the intake completely. Consequently, Bay City chlorinated the intake pipe in November to kill off the zebra-mussel population, using their old 122 cm diameter intake to obtain water while the new intake was treated.

The Cleveland Division of Water warned that if zebra mussels continued to reduce summer turbidity levels in the central basin of Lake Erie, water treatment would have to be switched from alum to a more expensive synthetic coagulating agent because the efficiency of alum to coagulate particles at low turbidity levels was poor (Greenberg et al. 1991). Meanwhile, some Ontario water plants had to add bentonite clay to raw water in order for alum and filtration treatment systems to clear water of turbidity. Fears expressed by water-treatment managers over high trihalomethane levels in treated water as a result of intake chlorination of raw water failed to materialize because mussel filtration had greatly reduced organic matter contents in the lakes.

Paul Rufolo received a U.S. Patent in April 1991 for a preventative maintenance system used in underwater intake pipes, which consisted of placement of a small-diameter, chemical feed line inside a water intake. Rufolo sent letters to industrial and municipal water users around the Great Lakes to discuss license arrangements to allow installation of the patented system. This action delayed installation of some chlorination systems. The patent was rebuked on both sides of the border because this system had been proposed earlier by Wil LePage at the zebra-mussel conference in December 1989 as a means to control zebra mussels in water intakes.

Spreading the Word

A flood of zebra mussel articles appeared in scientific journals in 1991. Articles reported on distribution and dispersal (Griffiths et al. 1991), biology (Garton and Haag 1991, Mackie 1991), ecology (Hebert et al. 1991, MacIsaac et al. 1991), impact on native bivalves (Schloesser and Kovalak 1991), and control (Sblendorio et al. 1991), in addition to an editorial overview of this issue (Cooley 1991). One article predicted that the potential distribution of zebra mussels extended to all Canadian provinces and U.S. states except Florida (Strayer 1991). This was an important prediction because zebra mussels were perceived by many politicians and stakeholders to be a regional issue confined to the Great Lakes.

Presentations about zebra mussels were common at both engineering (e.g., impacts on infrastructure, control, and prevention studies) and scientific (e.g., dispersal rates, population dynamics, impacts on the environment and biota, toxicity test results) society meetings in 1991, with special sessions held at annual meetings of the American Water Works Association, International Association for Great Lakes Research, and American Society of Limnology and Oceanography. In addition, the EPRI hosted a zebra-mussel-control technology conference in October and another zebra-mussel conference for its members in November. Courses for water-treatment plant operators on zebra-mussel-control procedures were offered by the Center of Continuing Studies at Niagara University, Niagara Falls, New York (March), and by New York Sea Grant and New York State Rural Water Association in Syracuse (April).

The *Seventh North American Zebra Mussel Conference* occurred in January 1991. It was the third zebra-mussel conference in a month, reflecting the great demand for information by people around the Great Lakes basin. The conference entitled *Zebra Mussels: Implications for Industries and Municipalities* was held in East Lansing, Michigan, and was sponsored by Michigan Sea Grant and Michigan Department of Natural Resources. The conference was targeted to the regional level and attended by almost 200 people, primarily from Michigan. Industrial operators, such as Bill Kovalak, and water-plant operators, such as Wil LePage, shared their experiences and perspectives. Other presentations mostly dealt with aspects of zebra mussel control, treatment, and prevention (e.g., chemical, operational, nonchemical) based on studies in the laboratory and undertaken at facilities along the lower Great Lakes, along with government approval processes needed for chemical control programs.

The *Eighth North American Zebra Mussel Conference* entitled *Zebra Mussels: Mitigation Options for Industries* was held at the Royal York Hotel in Toronto, Ontario, in February 1991 and was sponsored by Ontario Hydro (Figure 10.2). It was attended by 500 people and consisted of 31 presentations. This conference would later be recognized as the *First*

International Zebra Mussel Conference (Lucy and Muckle-Jeffs 2010). It focused on zebra-mussel treatment and prevention, with sessions on: (a) current mitigation strategies undertaken at facilities (8 presentations), (b) nonchemical mitigation options (11 presentations), and (c) chemical effectiveness and government regulation (12 presentations). None of the presentations dealt with ecological impacts of mussels.

The *Ninth North American Zebra Mussel Conference* entitled the *Second International Zebra Mussel Research Conference* was held at the Genesee Plaza Holiday Inn in Rochester, New York, in November 1991 (Figure 10.2). This conference was sponsored by Great Lakes Sea Grant, Environment Canada, and Fisheries and Oceans Canada and was attended by almost 500 people. The conference had 60 presentations and 9 posters, lasted 4 days (the longest conference to date), and had keynote presentations from well-known, European zebra mussel researchers, including Norbert Walz from Germany, who presented information on the dispersal of zebra mussels throughout Europe, Michael Ludyanskiy from Russia, who presented information on the Soviet experience to control zebra mussels, and Abraham bij de Vaate from The Netherlands, who presented information on the beneficial uses of zebra mussels in water-quality management. Contributed papers at the conference were organized around six themes, which were similar to the original research themes put forward at the *Fourth International Conference on Zebra Mussels* in February 1990: (1) biology, physiology, behavior, and population dynamics; (2) ecosystem impacts, fishing impacts, nutrient/contaminant cycles, and predator/prey interactions; (3) mechanisms and rate of spread, range extension, and monitoring; (4) biological control/mitigation; (5) physical control/mitigation; and (6) chemical control/mitigation (Figure 10.2). Highlights among the first three themes included the following:

1. The discovery of a second species of *Dreissena*, the quagga mussel, in the Erie Barge Canal by Ellen Marsden and Bernie May. This species was later identified as *Dreissena bugensis* (Rosenberg and Ludyanskiy 1994) and was likely introduced via ballast water into Lake Ontario in 1989 (May and Marsden 1992). This announcement again cast doubts about the effectiveness of voluntary ballast-water guidelines.
2. Field evidence that post-metamorphic zebra mussels (shell length 300–900 µm) could drift in large numbers in the water column. This ability of young mussels to behaviorally "set sail" at the water's surface and/or in the water column was important in understanding mussel dispersal and recruitment.
3. Data from voluntary monitoring networks to the west, south, and east of the Great Lakes showed that zebra mussels had dispersed from the region.

Highlights of the last three themes included: (1) systematic testing of various microorganisms as potential control agents for zebra mussels and (2) effectiveness of physical control methods, such as low-voltage electric fields, UV-B radiation, acoustic energy, and antifouling coatings.

The *Tenth North American Zebra Mussel Conference* entitled the *Second International Zebra Mussel Conference* was held at the Westin Harbour Castle in Toronto, Ontario, in February 1992. The conference was sponsored by Ontario Ministry of Natural Resources, Ontario Ministry of Environment, Canadian Department of Fisheries and Oceans, and Ontario Hydro. It was attended by 650 people and consisted of 70 presentations and 35 posters. The theme for an all-day plenary session focused on the roles, activities, and programs of various Canadian and American agencies, industries, and industrial associations. Contributed presentations were split into two concurrent sessions that emphasized: (a) biology and ecology and (b) chemical and physical controls (Figure 10.2). Surprisingly, only about 20% of these 50 contributed presentations had been given at the zebra-mussel conference held just 3 months before, indicating the volume of research conducted on zebra mussels. Highlights in the biology, ecology, and environment session included: (1) the lack of a measurable impact of mussels on walleye spawning, egg viability, and young-of-the-year growth, despite high densities of zebra mussels on spawning shoals in Lake Erie; (2) data of ballast-water compliance of vessels entering the St. Lawrence Seaway (94% in the previous year, that is mid-ocean exchange of ballast waters); and (3) zebra-mussel shells as a new source of beach material along shores of the Great Lakes.

After the conference in February 1992, a single *North American Zebra Mussel Conference* was held annually, sponsored by agencies from both Canada and the United States as agreed by the two federal governments in October 1990. Consequently, the *Third International Zebra Mussel Conference* was held in Toronto, Ontario, in February 1993, while the *Fourth International Zebra Mussel Conference* was held in Madison, Wisconsin, in March 1994. These zebra-mussel conferences eventually "morphed" into annual International Conferences on Aquatic Invasive Species, which have been held in many countries throughout the world since 2004 (Lucy and Muckle-Jeffs 2010).

DISCUSSION

Just Another Invasive Species

Prior to zebra mussels, interest in invasive species rarely extended beyond mention in scientific papers, despite the more than 100 species that had been unintentionally introduced into the Great Lakes (Mills et al. 1993). Consequently, zebra mussels initially followed the same "event sequence" as previous unintentionally introduced species: (1) initial release, (2) discovery, (3) taxonomic verification, (4) information search, (5) media reports, (6) scientific presentations/publications, (7) expert committee(s) establishment, (8) distribution and containment, (9) outreach and education, (10) abatement and adaptation, (11) ecosystem impact monitoring, and (12) acceptance as naturalized.

The unintentional release of zebra mussel veligers (Event 1) probably occurred in summer 1986 in or near the shipping channel through Lake St. Clair (Griffiths et al. 1991). This is the shallowest section of the Great Lakes Seaway, and the discharge of ballast water would assist in a ship's passage. It is not uncommon to see turbid waters of resuspended bottom sediments trailing ships as they pass through Lake St. Clair (RWG, personal observations). These veligers then would be swept by currents into the central basin of Lake St. Clair and down through the Detroit River to the western basin of Lake Erie. Alternatively, Carlton (2008) provided evidence that mussels were already present in western Lake Erie in 1986 and subsequently spread to Lake St. Clair. Regardless, the first mussels in North America were likely introduced in 1986.

From its initial release, 2 years passed before zebra mussels were sufficiently abundant to be discovered in summer of 1988 (Event 2) and taxonomically identified by experts (Event 3). This immediately started an information scramble to understand the biology and ecology of this species (Event 4). Media reports (Event 5) of this newest invader followed in July 1988 (i.e., vander Doelen and Hornberger 1988) along with the submission of a manuscript in October 1988 (i.e., Hebert et al. 1989) to a scientific journal (Events 5 and 6). Staff at government agencies were alerted and were able to easily find mussels in the fall 1988. Events 2 through 6 occurred over just a 6-month period by a handful of scientifically curious people from universities, government agencies, and utilities in both the United States and Canada.

The *First Zebra Mussel Workshop*, in March 1989, brought 12 members from U.S. federal and Ontario provincial agencies, water and electric-generating utilities, and academia (Figure 10.1) together in London, Ontario, to share information. These 12 people previously had little interaction with each other. This group quickly agreed that the potential economic and environmental damage posed by zebra mussels exceeded that of all previous invasive species. By the end of the workshop, this group of individuals had: (a) organized a management structure to oversee a North American response, that is an expert committee (Event 7); (b) formulated a communication and outreach strategy to increase awareness among stakeholders; (c) developed a research and funding priority plan to obtain essential scientific information; and (d) developed important research questions that quickly needed to be addressed to improve understanding of economic and environmental impacts. The zebra mussel coordinators expanded the Zebra Mussel Watch Committee to eight members after the first Zebra Mussel Conference in June. This expert committee functioned as a central research coordinator and clearinghouse of information until about the time the New York Zebra Mussel Information Clearinghouse became active in late 1990. The demand for information from government, media, water users, and the public across the Great Lakes in the first 2 years was so great that regional experts, government agencies, and funded non-profit groups had to take over these functions.

Two essential government-response programs to any invasive species are distribution monitoring and containment (i.e., slowing dispersal). Initial response to the detection of a biological pollutant (i.e., invasive species) thus is essentially the same as that of a chemical spill—where is it, how fast is it spreading, can it be contained (Event 8). Up-to-date information on distributions was a critical element in planning zebra-mussel abatement programs and consequently always in demand from water users. A voluntary stakeholder monitoring program was initiated in early 1989 whereby members of industry, academia, public, and media were encouraged to report new sightings of zebra mussels in Ontario waters to Ron Griffiths at the Ontario Ministry of Environment and in U.S. waters to Don Schloesser at the U.S. Fish and Wildlife Service. Members of the Zebra Mussel Watch Committee then confirmed these sightings. This information was regularly updated on maps (e.g., Griffiths et al. 1991) and shared with the public and water users. The replacement of this low-cost, voluntary monitoring program with a $1.2 million government monitoring program by the Ontario Ministry of Natural Resources actually resulted in underestimation of infested waters and errors in mussel distribution. Consequently, water users in some areas were unaware that mussels were present and were not taking appropriate actions, while in other areas, water users were spending money on monitoring and treatment when no mussels were present. Fortunately, this program ceased by late 1992. Beginning in mid-1992, mussel sightings from voluntary monitoring programs throughout North America were reported to the U.S. Geological Survey, which produced maps showing North American mussel distribution on an annual basis.

Containment of zebra mussels to specific geographic areas (e.g., bays) was not possible. By the time zebra mussels were confirmed in 1988, they already occupied an area from Lake St. Clair through to the central basin of Lake Erie (Griffiths et al. 1991). Their high fecundity (i.e., tens of thousands of eggs per female) and free-floating veliger stage (Mackie 1991) promoted rapid dispersal. Still, attempts to slow the dispersal of zebra mussels were a major initiative of the Ontario Ministry of Natural Resources and state agencies in the United States beginning about mid-1990. Unfortunately, these agencies had no direct influence over commercial shipping in the Great Lakes and thus could not slow dispersal throughout the Great Lakes and its interconnected waterways (e.g., Erie Canal). These programs essentially used an educational approach consisting of posting signs at boat launch sites, bait and tackle shops, and marinas and providing pamphlets and other printed materials primarily to boating and fishing groups. In Ontario, dispersal-reduction programs had little effect as mussels spread to all inland waterways connected to the Great Lakes by early 1991. The rapid spread of mussels to inland (fishing) lakes in the Great Lakes states and then outside the Great Lakes basin similarly suggested that these programs had little effect at reducing dispersal in the United States. Essentially, there was no implementation of roadside inspection programs that could have prevented people with pleasure and fishing boats from moving mussels among water bodies. Later watercraft-interception programs appeared to have slowed the dispersal of mussels in the western United States (Zook and Philips 2009), as evidenced by a substantial delay between when mussels were detected on trailered watercraft and occurrence in nearby lakes.

Frequent and continuous media reporting (TV, radio, print) throughout 1989 and 1990 kept zebra-mussel issues in front of the public and politicians. At least once every week, a member of the Zebra Mussel Watch Committee was interviewed for comments on the latest developments and impacts. Meanwhile, public outreach events held throughout the Great Lakes region provided more detailed information and answered specific concerns (Event 9). However, it was the "zebra-mussel conference" that became the central education and outreach event to increase awareness, update stakeholders, highlight new research, and provide engineering solutions, especially in an era with limited internet services. Each zebra-mussel conference attracted people with unusually diverse backgrounds, for example, policy people, political staff from all levels of government, scientists, resource managers, engineering and environmental consultants, water users, industrial representatives, legal staff, and media (print, TV, radio). These conferences provided a relaxed atmosphere in which these various groups of people could interact, share the latest information (e.g., biological, environmental, industrial impacts, control and prevention technology, legal, engineering, policy), coordinate research, foster cooperation, and become familiar with available engineering and consulting services. Consequently, as information on the biology, ecology, impact, and control aspects of zebra mussels was discovered and documented, it was being rapidly transferred (0.5–1 year) to those who formulated policy and developed treatment programs, as opposed to waiting for information through traditional sources, which took years (journal publication, books, proceedings, etc.). Importance of these conferences was reflected in the larger than expected participation (venues of two of the earliest conferences had to be moved to accommodate demand) and funding support from numerous agencies in the United States and Canada. Attendance grew from 12 people at the *First Zebra Mussel Workshop* in early-1989 to 350 at the zebra mussel conference in Cleveland in late-1989, to 500 at the conference in Toronto in early-1991, and to 650 at the conference in Toronto in early-1992. A survey in 1994 of industrial and municipal water users found that 59% of respondents participated in zebra mussel conferences and listed these conferences as the second highest ranked source for information value (personal contact with specialists was given the highest ranking in information value) (Suvedi and Heinze 1994). The success of these education and outreach events is best shown by the continuous evolution from regional gatherings, to annual binational events, to inclusion of all aquatic invasive species, and finally its expansion into an international event for the past decade (Lucy and Muckle-Jeffs 2010). Networking among participants at this theme-focused conference generated sufficient interest and publications to spawn new journals (*Biological Invasions*; *Aquatic Invasions*) and thereby helped solidify a new scientific field of interest.

Conference presentations reflected the knowledge and priorities at individual periods in time (Figure 10.2). A doubling in the number of presentations occurred by end of 1990 (Columbus) and again at the end of 1991 (Rochester) reflecting the growth in knowledge and investment in research. Biology (e.g., life history, growth, genetics, and spatial distribution) and ecology (e.g., dispersal, population dynamics, feeding, impacts on biota and the environment) of zebra mussels were topics always of great interest at conferences, with information from Europe continuously being replaced with information gathered in North America. Presentations of zebra mussel impacts on water intakes and industrial operations were most common in early conferences and declined in number after mid-1990. Meanwhile, results of treatment programs at specific facilities began in late 1989 and were common until mid-1991. Research into various chemical and nonchemical control methods was another mainstay of all conferences, and presentations on this topic represented about one-half of all presentations in Toronto 1991. Over the period 1989–1991, topics of other presentations progressed from ballast-water issues to estimates of control costs and regulations and to standard sampling protocols. Research priority forums were well-attended sessions held at ends of early conferences to identify research priorities, coordinate research activities, and minimize duplication of efforts among researchers.

Initial urgency surrounding zebra mussels was not driven by environmental concerns but by their potential impacts on public water supply, electric-power generation, and industrial production because they reduced intake capacity for potable, cooling, and process waters. Such impacts prompted immediate abatement measures (Event 10). Detroit Edison in Michigan was the first company to implement a monitoring and abatement (i.e., mechanical removal) program for zebra mussels. These actions were a direct consequence of a staff member, Bill Kovalak, who recognized specimens of zebra mussels collected from their facilities and knew this was a common biofouling species in Europe. This early response likely prevented a complete shutdown of any of their electric-generating facilities along the Great Lakes.

In Ontario, the foresight of Joe Leach to inform both Ontario Hydro and Ministry of Environment by letter in October 1988 that the newly introduced zebra mussel could restrict water intake capacity of facilities along the Great Lakes initiated government-response activities to evaluate these concerns. The structure of Canadian society provided a distinct advantage in initially dealing with zebra mussels.

With the single province of Ontario spanning the Great Lakes, a single agency responsible for assessing water quality, regulating water takings, and operating public water-treatment plants (Ontario Ministry of Environment), and all electric-generating facilities operated by a single crown corporation (Ontario Hydro), it was much easier (i.e., fewer people, direct lines of communication, financial resources) to respond to threats posed by zebra mussels in a coordinated fashion (i.e., undertake monitoring, inform facility managers, and permit abatement activities). Response could be quickly initiated because Ontario agencies had the flexibility to internally redirect staff, resources, and funds as necessary to address an evolving issue. Also, information could move quickly between government scientists, regulators, and facility managers. This accounted for rapid establishment of monitoring activities and abatement programs (i.e., establishment of backup water supplies and mechanical removal of mussel from intake structures) at water and electric-generating facilities throughout the area infested with zebra mussels. This quick, coordinated government response was responsible for the lack of interruptions at water-supply and electric-generating facilities in Ontario.

Unfortunately, the numerous Ontario industries drawing process and cooling waters from the Great Lakes were disadvantaged. With less financial resources, a lack of biological–knowledgeable staff, delays in receiving zebra-mussel information (e.g., distribution and abatement), and unfamiliarity with biofouling, managers at these facilities were less informed, unaware, and unprepared for water capacity impacts, especially those located close to the initial populations of zebra mussels in Lake St. Clair and the western basin of Lake Erie. This situation was similar to that on the U.S. side of the Great Lakes. With eight U.S. states and numerous independently owned water-treatment, electric-generating, and industrial facilities bordering the Great Lakes, a coordinated response to a new threat (e.g., monitoring, abatement, communications) among several government agencies and hundreds of facilities was very difficult to achieve, especially over a short-time period. The complete loss of raw water to the Monroe Water Treatment Plant, severe restrictions in flow of an automobile engine plant in Windsor, Ontario, and problems at a food plant in Wheatley, Ontario, were a direct result of these constraints. Skepticism by operating engineers that zebra mussels could cause severe water-withdrawal problems probably contributed to the severity of impacts because of delayed action by managers.

Once the threat of mussels was shown to be credible by well-reported problems, most water users installed temporary water intakes, monitored zebra-mussel abundance, and mechanically removed mussels from intakes and water intake cribs. With cooperation among state and provincial regulators, municipal and industrial owners, and engineering and environmental consultants, chemical delivery systems were installed inside water intakes of many facilities within a year. This "adaptation" by facilities to zebra mussels accounted for the almost complete absence of water restrictions after the first year of mussel-induced problems. Studies on additional chemical (e.g., chlorine, oxidants, molluscicides), physical (e.g., UV, sieves, acoustics), and biological control methods, applied in various manners (e.g., press, pulse) on different life stages (eggs, veligers, adults), rapidly followed, providing alternatives and improving efficiency of methods to control and prevent biofouling at industrial and municipal facilities (e.g., Claudi and Mackie 1994). Interestingly, while state agencies in the United States had authority to permit numerous chemicals (e.g., chlorine, potassium permanganate, ozone, molluscicides), the Ontario Ministry of Environment actually could not legally permit any chemical for mussel control, since none were registered with the Canadian federal government as a molluscicide. Thus, a creative solution had to be found whereby chlorine was exempted from the Ontario Pesticides Act and regulated under the Ontario Water Resources Act to permit its use to control mollusks inside water intakes. No chemical has yet received federal approval to control zebra mussels in Canada (as of 2012). Costs of these abatement and adaptation responses to mussels were estimated to be $120 million from 1989–1994 (Park and Hushak 1999), more than had been spent on any previous invasive species.

Policy Shift: Recognition of Aquatic Nuisance Species

The introduction of the Nonindigenous Aquatic Nuisance Act (NANCPA) to the U.S. Senate by Senator John Glenn in March 1990, and a companion bill introduced to the House by Representative Henry Nowak, was the "event" that ended all controversy about zebra mussels being "just another invasive species." The senator noted that this bill was necessary "because at present there is nothing in U.S. law to prevent another nuisance species from entering the lakes or any other of our waters the same way the zebra mussel did. We are also concerned that the Great Lakes region and other regions are unprepared to address the demands upon research and control resources created by such nuisance species."

Unlike previous invasive species, zebra mussels prompted a major shift in U.S. national policy with respect to "pest management," which historically had developed in response to "agricultural/terrestrial" species (largely insects and plants) (see Reid and Wilkinson 2013). Instead of taking the usual biological approach focusing on "species containment," this legislation took an ecological approach focusing on the "vectors of aquatic nuisance species" in order to reduce the spread of invaders and prevent further invasions. It created a national regulatory and policy framework, with dedicated funding, that authorized regional and state implementation (i.e., shared responsibilities—states can improve on federal standards and decide implementation structure) and gave legal definition to "nonindigenous species" and "aquatic-nuisance species (ANS)." This shift was likely a result of years of effort to focus on ballast-water discharges as the means to prevent species introductions, and thus protect Great Lakes fisheries (e.g., Dochoda 1989, 1991, Dochoda et al. 1990). Although many details in initial drafts were revised as a result of comments from federal agencies on funding and key roles, recommendations from the U.S. Great Lakes Nonindigenous Species Coordinating Committee, testimony at committee hearings (e.g., Michael Donahue of the Great Lakes Commission to the Senate Subcommittee on Environmental Protection on June 19, 1990), and lobbying from the shipping industry, the essential elements of the bill prevailed in the final version. The fact that the act passed both houses in the late days of Congress indicated that it was supported by a wide section of society. This legislative vision had no equivalent in Canada.

NANPCA not only provided direction and funding to existing federal agencies (e.g., U.S. Fish and Wildlife Service, National Oceanic and Atmospheric Administration, Army Core of Engineers, Coast Guard) but created new institutions to deal specifically with aquatic nuisance species. This legislation established a regional Great Lakes Panel on Aquatic Nuisance Species and a National Task Force to address the issue. The mandate of this panel included: (a) identifying priorities for the Great Lakes region; (b) making recommendations to the national task force; (c) coordinating, where possible, aquatic nuisance species program activities in the Great Lakes; and (d) reporting on prevention, research, and control activities in the Great Lakes region. Interestingly, although the original compact that created the Great Lakes Commission provided full participation by Canadian provinces as though they were states, NANCPA only encouraged the Great Lakes Panel "to invite representatives from the federal, provincial and territorial governments of Canada to participate as observers." The panel consisted of 35 individuals drawn from United States and Canadian federal agencies, the eight Great Lakes states and the province of Ontario, regional agencies, concerned citizen groups, municipal representatives, tribal authorities, commercial interests, and the academic community. The National Invasive Species Act of 1996 reauthorized funding and government agency mandates to continue nuisance species research and initiatives into the next century.

In contrast, the nonlegislative "lead agency" approach of Ontario provided a rapid political response but of short duration. In late 1992, the Zebra Mussel Coordination Office was closed, and funds were redirected to other programs. Production and distribution of newsletters, warning signs, pamphlets, education material, and distribution maps ended. Fortunately, the Ministry of Natural Resources had partnered with the Ontario Federation of Anglers and Hunters in early 1992 to operate an "Invading Species Awareness Program" that allowed for continued activities. This program listed various invasive species and explained how to prevent them from spreading to forests, lakes, and streams in Ontario. An "invading species hotline" established by the Ministry received 565 phone calls about zebra mussels in 1992 (25% from media), which was similar to that received by Ontario Ministry of the Environment in 1989. However, there is no evidence of any government response to this information. Thus, despite continued needs for zebra-mussel information from the public, industry, and media, a specific government department no longer existed. In a survey of 10 stakeholder groups on important policy issues in the Great Lakes, provincial representatives rated the issue of invasive species, especially zebra mussels, lowest among all other stakeholder groups (Hartig et al. 2001); in contrast, other stakeholders such as native, academic, and state representatives continued to recognize invasive species as a substantial issue in the Great Lakes region.

Research Funding

Few invasive species prompted more research funding than zebra mussels. From a few thousand dollars in late 1988 at the University of Windsor to support a single project, funding increased to several million dollars in 1991 (Table 10.2) to support over a hundred projects (OMNR 1991, Coscarelli and Rendall 1996). Research funding was critical to engage academia and consultants (i.e., grease for the wheels) in order to have various ecological, biological, societal, and economic questions addressed quickly.

Table 10.2   Zebra Mussel Research Funding from Ontario Government (Ontario Ministry of Natural Resources, Ontario Ministry of Environment, and Ontario Hydro) and U.S. Sources from 1989 to 1992

	Funding Year			
Funding Source	**1989**	**1990**	**1991**	**1992**
Ontario: (CDN$) Ministry of Environment	128,500	32,000	124,100	245,000
Ontario Hydro	—	837,600	645,000	552,000
Ministry of Natural Resources	—	564,100	1,129,200	638,150
Total	128,500	1,433,700	1,898,300	1,435,150
USA (US$)	—	435,500	5,569,330	5,443,353

Ontario data from OMNR (1991, 1992); USA data from Coscarelli and Rendall (1996).

During the early invasion years, more funding was directed to zebra mussel research than any other invasive species (Coscarelli and Rendall 1996). Ontario Ministry of Environment was the first government agency to fund zebra-mussel research (Table 10.2). In 1989, funds were provided to summarize European knowledge on biology, ecology, and control (Mackie et al. 1989) and to study the life history of zebra mussels and their impacts on native bivalves in Lake St. Clair (Mackie 1991). Ontario Hydro, the largest electric-generating company in North America at that time, partnered early with the Ontario Ministry of Environment to pool environmental research expertise and economic resources. Together, Ontario Ministry of Environment and Ontario Hydro provided $2.56 million to fund research during 1989–1992 (Table 10.2), with 84.2% going to control and mitigation, 10.1% to ecosystem effects, and 5.7% to biology and life history. With enhanced funding from the Ontario government, a shift in research priorities from primarily applied (i.e., control and mitigation) to ecosystem-based studies (e.g., fisheries) occurred in 1991–1992. In total, the Ontario Ministry of Natural Resource provided $2.33 million in research funds with 53.8% going to ecosystem effects, 30.7% to spatial distribution and dispersal, 15.3% to biology and life history, and 0.2% to control and mitigation. No provincial agency supported studies on prevention or socioeconomics.

In the United States, NANCPA authorized millions of dollars in funding for: (a) education and outreach, (b) environmental research, and (c) research into environmentally sound methods (i.e., chemicals, coating, cleaning techniques) to protect public facilities from biofouling through the Army Corps of Engineers. NANPCA sharply increased funding for zebra-mussel research to $5.5 million in 1991, which was more than 10 times the funding available in the United States a year earlier (Table 10.2). Funding support peaked in 1991 as Ontario and U.S. sources contributed $7.4 million to zebra-mussel research. After 1992, U.S. sources accounted for an ever-increasing proportion of the total funding. Over the 5 year period of NANCPA funding, $17.1 million was spent on zebra-mussel research. Most of these funds went to studies of ecosystem effects ($11.1 million—64.9%), with intermediate support for biology and life history ($2.5 million—15%) and control and mitigation ($2.5 million—15%), while studies of prevention and socioeconomics received little support (<$0.2 million each—0.6%) (Coscarelli and Rendall 1996).

Role of Science

Science played an essential role in the early response by water users and governments (e.g., initiatives and policies) by providing sound predictions (information) to address specific questions including potential distributions in North America, expected duration of impacts, and possible ecosystem impacts. Fortunately, a century of European scientific studies provided a solid foundation to understand zebra-mussel biology (e.g., see summary by Mackie et al. 1989) and predict environmental and industrial impacts in North America.

Predictions of the potential distribution of zebra mussels at regional (e.g., Neary and Leach 1992, Murray et al. 1993) and continental scales (Strayer 1991) based on water chemistry and temperature requirements were essential to show that zebra mussels were not only a Great Lakes issue but an issue across the continent. Evidence that zebra mussels were a national problem was crucial in prompting U.S. legislative action.

Model predictions of population dynamics based on catastrophe theory and central European information (e.g., Stańczykowska 1977) provided a temporal basis for management action (Figure 10.3). This model indicated that it would take 3–5 years for a founding population to reach peak abundance, with absolute values of peak abundances probably being dependent on primary production and habitat conditions. Since mussels were typically discovered 1–2 years after invasion, local effects could be expected 1–3 years after they were detected. Peak abundance levels would last about 3–5 years, and this was the time period when mussels were expected to have their greatest impacts on the environment and water users. Mussel abundances then would likely collapse to a lower level (similar to most invading species, Sakai et al. 2001), because of an induced reduction in food supplies (i.e., less phytoplankton), although a change in habitat (e.g., loss if native unionid shells) and/or an increase in predation by ducks and fishes (e.g., gobies) may contribute to decreased abundances in specific localities. At lower abundances, mussels were less likely to impact water users but would still likely be sufficient to maintain altered environmental conditions (e.g., clearer waters, reduced phytoplankton, and increased submerged vegetation).

Prediction of zebra-mussel impacts at the ecosystem level was essential to show resource managers and policymakers that this invasive species was unlike previous invaders. European studies (e.g., Stańczykowska 1977) clearly showed that zebra mussels could induce a trophic cascade whereby their feeding activity reduces seston concentrations such that the resulting increase in water clarity allowed reestablishment of aquatic macrophytes that subsequently induced other biotic changes (e.g., invertebrate and fish composition). This trophic cascade was applied to the Canadian basin of Lake St. Clair to explain observed changes in 1990 and predict future implications (Figure 10.4). Prior to zebra mussels, Lake St. Clair was well documented (UGLCCS 1988) as a shallow, turbid, muddy-bottomed waterbody that had a perch and walleye fishery (Figure 10.4a). With the introduction of zebra mussels (Figure 10.4b), optimal food and environmental conditions allowed the population to rapidly multiply (Figure 10.4c). In 1989, the zebra-mussel population was sufficiently large such that its feeding activity cleared waters of the lake, causing an immediate growth response

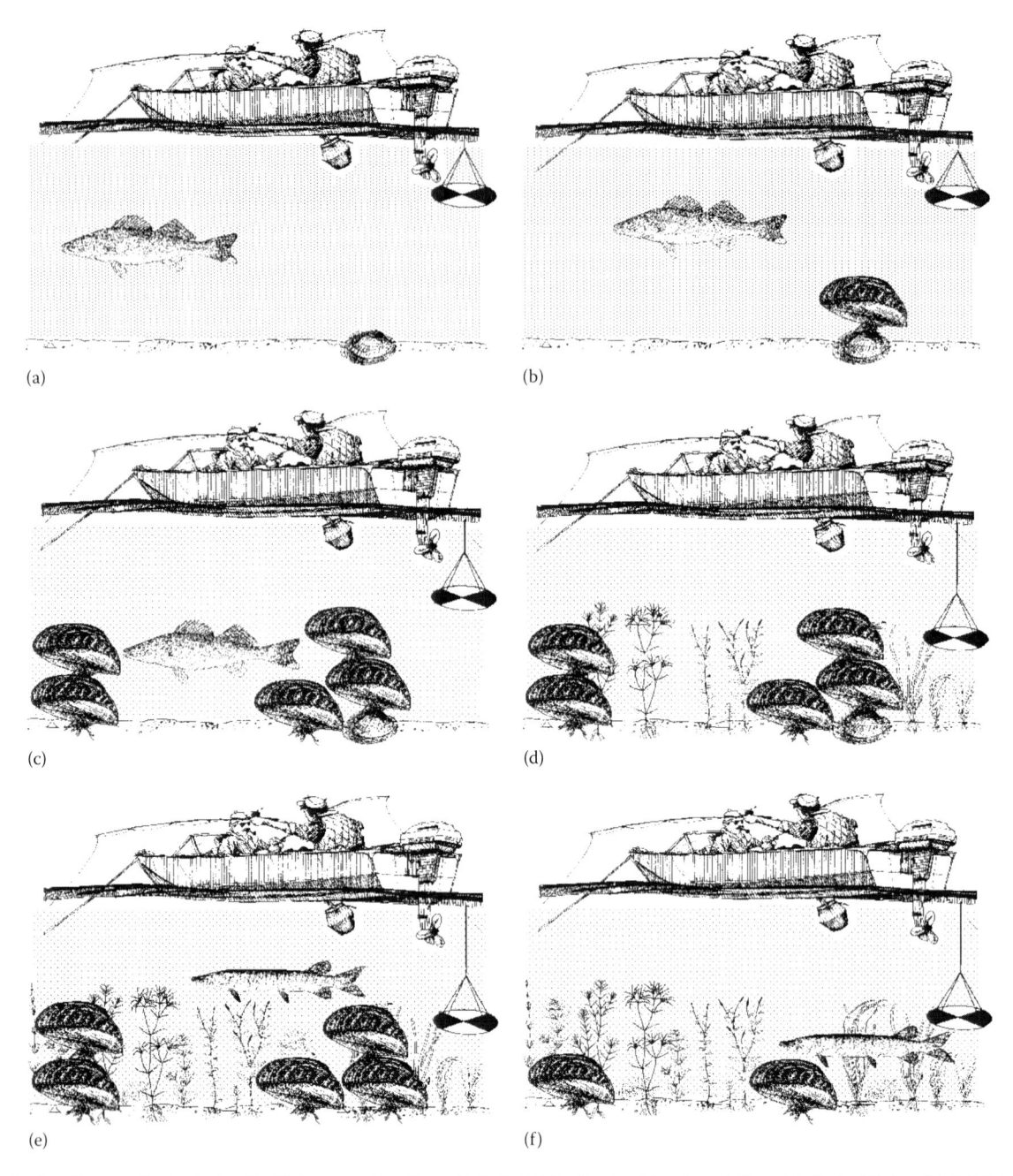

Figure 10.4 Changes in the Lake St. Clair ecosystem from 1980 to 2000. See text for description. (a) Early 1980s, (b) 1986, (c) 1988, (d) 1989, (e) 1990, and (f) after 1994.

by aquatic plants and forced light-sensitive walleye into deeper waters (Figure 10.4d). Native fauna responded to these changes in 1990 with: marked increases in populations of amphipods, snails, and crayfishes; a reduction in unionid mussel populations; and increases in sight-feeding fishes, such as muskellunge and basses (Figure 10.4e). Finally, after 1994, ecosystem structure stabilized when zebra-mussel abundance (biomass) collapsed because of reduced phytoplankton supply (Figure 10.4f). The documented increase in clarity (Hebert et al. 1991, Bellmore 1992, Griffiths 1993), extensive growth of aquatic vegetation (Griffiths 1993), crash in phytoplankton abundance (Bellmore 1992), enrichment of bottom sediments with phosphorus (RWG, unpublished data), shift in benthic composition (Griffiths 1993), loss of unionid mussels (Nalepa 1994), loss of Thames River walleye fishery (RWG, personal observation), and reduced catch of walleye along with increased abundance of clear-water musky and bass in Lake St. Clair (Magner 1993, OMNR 1995) all supported this predictive framework. Similar changes followed in Lake Erie, Rondeau Bay, Saginaw Bay, Oneida Lake, Lake Michigan, and Hudson River as zebra mussels spread across the Great Lakes Basin and beyond (e.g., MacIsaac 1996,

Vanderploeg et al. 2002, Nalepa et al. 2003, Miehls et al. 2009, Strayer 2009, Fahnenstiel et al. 2010).

However, not all scientists agreed that zebra mussels were the cause of observed environmental changes. One scientist stated that it is not necessary to invoke zebra mussels to explain changes in clarity in 1989; it is correlated with the great abundance and mean size of *Daphnia* at those times in the lake (Wu and Culver 1991). In other words, a seasonally abundant, filter-feeding zooplankter was responsible for the increased clarity in western Lake Erie (Wu and Culver 1991). Another U.S. scientist stated that seasonal low-water levels in Lake St. Clair caused the extensive growth of aquatic vegetation in the lake, which subsequently cleared the water (Thomas and Haas 2012). One scientist in Canada suggested zebra mussels did not improve water clarity in Lake Erie as a whole, based on measurements in the offshore, pelagic, and relatively unproductive parts of the central and eastern basins of the lake (Charlton 2001). Furthermore, it was suggested zebra mussels had little to do with the decline in algal abundance even in the western basin of Lake Erie as it was solely related to lower phosphorus loadings. Finally, the same scientist dismissed any fisheries impacts from zebra mussels: "at one time there was speculation that fundamental deleterious effects on fisheries would occur" (Charlton 2001).

Perhaps the greatest failure of science was the inability by fisheries scientists to predict the impacts of zebra mussels on fisheries. This stems from the fact that models of fisheries management in the Great Lakes in the early years of the mussel invasion simply allocated fish abundance to sport and commercial fisheries; that is, fish production is essentially a black box and viewed to be independent of the environment. Despite declines in phytoplankton biomass of 68%–86% (Johannsson and Millard 1998) and primary production of 22%–55% (Madenjian 1995, Stoeckmann and Garton 1997, Millard et al. 1999), the disappearance of *Diporeia* in 1993 (Dermott and Kerec 1997), water clarity improvements in the western (Holland 1993, Leach 1993) and eastern (Barbiero and Tuchman 2004) basins, and increased aquatic vegetation growth in the western basin of Lake Erie (Holland et al. 1995) following the zebra-mussel invasion, fisheries managers were unable to determine any consequences for fisheries. While leading a workshop for fishery biologists in eastern Lake Erie in 1991, Clifford Kraft noted that he "was impressed during our discussions of fishery issues that zebra mussels were seldom mentioned as a current problem … in this mussel-infested lake." When asked for details, biologists expressed long-range concerns about potential zebra-mussel impacts but said that little evidence of such impacts had been observed to date. "Bad weather, overfishing, and overstocking of salmon and trout were discussed as more likely immediate causes" (Kraft 1991c). Ron Griffiths had a similar experience with senior Ontario Ministry of Natural Resource fisheries biologists in late 1997 in Toronto, when he linked impacts of zebra mussels in Lake St. Clair to the loss of the Thames River walleye spawning stock and increased abundance of basses and muskellunge. The fisheries biologists firmly disagreed with the link between zebra mussels and fisheries but were unable to provide an alternate cause for these observations. Recently, there is some recognition by fisheries biologists that dreissenid mussels reduced phytoplankton and zooplankton populations in Lake Erie (Johnson 2009), assisted in the recovery of burrowing mayflies (Edsall et al. 1999), reduced the abundance of forage fish (Johnson 2009) and walleye (LEC 2005), reduced relative abundance of planktivores (Zhu et al. 2008), and even assisted in the recovery of eutrophic-intolerant fish species (e.g., minnows, rock bass, lake whitefish) in the western basin (Ludsin et al. 2001).

Unfortunately, this gap in understanding the linkage between zebra mussels and fisheries led to the greatest uncertainty in the 1989 economic impact statement by Jon Stanley at U.S. Fish and Wildlife Service because his estimate of a $2710 million loss in fisheries revenue over 10 years was made with little support from fisheries science. However, this economic statement is credited by many as being the single most important prediction for gaining political and government action on zebra mussels.

Monitoring of ecosystem impacts of zebra mussels (Event 11) in the Great Lakes began essentially with the discovery of zebra mussels, as environmental changes in Lakes St. Clair (Hebert et al. 1991) and western Lake Erie (Leach 1993) had already been observed and measured. These studies continued to spread across the Great lakes as the mussels dispersed. With millions of dollars in research funds from NANPCA, monitoring of ecosystem impacts greatly increased, covering the complete range where mussels occurred and were predicted to invade (Coscarelli and Rendall 1996).

Continued monitoring of ecosystem impacts showed that original zebra mussel predictions (see earlier section) largely came true. The distribution of dreissenid mussels (i.e., zebra and quagga mussels) approached the range predicted across the Great Lakes basin and North America (Benson 2013). Dramatic declines in abundance (biomass) occurred in Lake Erie (Patterson et al. 2005), Lake St. Clair (Hunter and Simons 2004), and elsewhere after 3–5 years of high abundances, and changes in food-web components (e.g., phytoplankton declines, zooplankton declines, profundal benthos declines, littoral benthos increases, pelagic and turbid-water fish declines, benthic and sight-feeding fish increases) have been linked to dreissenids (e.g., MacIsaac 1996, Vanderploeg et al. 2002, Strayer 2009). As expected, however, secondary factors were also found to alter predictions in specific situations.

Early investigators of zebra mussels recognized quickly that environmental conditions in the Great Lakes were ideal for zebra mussels and thus they would integrate into the ecosystem (Event 12). Once mussels had dispersed throughout the Great Lakes, it was evident to all that this species was now a permanent member of the Great Lakes community. The eventual collapse of population abundances is indicative of the point in time that zebra mussels became a "naturalized" species within local ecosystems.

CONCLUDING REMARKS

Zebra mussels represent yet another example of "policy by tragedy." Nonindigenous aquatic species were rightly perceived as ecosystem wildcards that had to be prevented from colonizing the Great Lakes (Dochoda 1989). However, senior policy advisors were reluctant to enact preventative regulations because so few of the past introductions caused substantial economic costs to water-users or environmental damage to the Great Lakes ecosystem. Pollutants (e.g., nutrients, pesticides, metals, and organic compounds), overfishing, and habitat destruction were regularly cited as causes for collapses of valued fish and other native species, which subsequently allowed exotic fishes (e.g., alewife, smelt) and algae to flourish and thereby affected a range of water users (e.g., Campbell et al. 1969, Hartman 1972, Christie 1974, Ashworth 1986). Exotic fishes like Pacific salmon and brown trout were purposefully stocked by government agencies to try and correct ecosystem problems while creating a valued (temporary) fishery until native stocks were rehabilitated (e.g., Christie 1974, Brandt 1986). Zebra mussels, a well-known and highly documented pest species in Europe (e.g., Clarke 1952), represented the problem that no one wanted in North America and yet was rarely mentioned in the campaign to prevent exotic introductions until they appeared in 1988. Today, zebra mussels are well recognized in North America, but the general issue of aquatic nuisance species is still poorly understood. Yet, Grigorovich et al. (2003) have identified an additional 47 invertebrates that pose an immediate invasion risk through ballast water. One of these species is the golden mussel, *Limnoperna fortunei*, a native of Southeast Asia, that has caused widespread ecological and biofouling impacts in South America since 1991, similar to those documented for zebra mussels in North America (Boltovskoy et al. 2009).

Zebra mussels clearly showed that invasive species have real economic and environmental costs, although these costs are difficult to establish. Abatement costs presented in this chapter show an increase from hundreds of thousands to tens of millions of dollars per year for electric-generating and municipal-water-treatment facilities in only a few years (1989–1991). Connelly et al. (2007) calculated the total cost of abatement programs for zebra mussels by electric-generating and municipal-water-treatment facilities in the Great Lakes at $267 million over the period 1989–2004 (average of $18 million/year). Unfortunately, the bulk of the costs imposed by zebra mussels are related to ecological and environmental impacts (e.g., fish abundance, nuisance algal populations, nuisance vegetation growth, fish kills, bacterial levels, and beach drift) that cause losses to fisheries, recreation, tourism, and property values. Often, these costs are not inconsequential: Rothlisberger et al. (2012) estimated that invasive species introduced by shipping to the Great Lakes conservatively costs water users (fishers, raw-water users, wildlife watchers) $200 million/year; the Commissioner of the Environment and Sustainable Development (2001) estimated that $500 million is spent annually on efforts to control invasive-aquatic species in the Great Lakes; and Pimentel et al. (2005) estimated zebra mussels cost $1 billion/year in damages and control costs in the United States. These costs are essentially permanent since zebra mussels are now a naturalized species in the Great Lakes ecosystem and in many other aquatic ecosystems in North America.

Zebra mussels started both the Canadian and U.S. federal governments along the inevitable path to enacting a "prevention policy" as it was the only way to prevent ecologic and economic impacts of invasive species. Federal governments in both Canada and United States have an exclusive role to regulate international trade and thus are responsible for preventing introductions via ballast-water discharges. Although the Canadian Coast Guard had commissioned a study in 1981 (Bio-Environmental Services Ltd. 1981) that indicated ballast water in ships carried exotic species to the Great Lakes, the uncertainty surrounding the impacts of exotic species on the ecology and economy of the Great Lakes likely deferred specific and strong action. However, once zebra mussels began to impact Great Lakes water users, the Canadian Coast Guard responded quickly by promulgating voluntary ballast-water guidelines in May 1989. This action was possible because no legislative change was required and they already had discussions about this issue with the U.S. Coast Guard, the St. Lawrence Seaway Authority, and the shipping industry in the previous 2 years. Ship safety for ballast-water exchange in the open ocean, and who made the final decision to exchange water, approved secondary exchange areas, and enforced regulations, were major concern considered in the decision process. After authorization provided by NANCPA, the U.S. Coast Guard issued similar voluntary guidelines for ballast discharges in the Great Lakes that took effect March 15, 1991. These were followed by mandatory ballast-water exchange regulations on April 8, 1993, for ships entering the Great Lakes beyond the exclusive economic zone. However, the continued appearance of new species and increased rate of introductions following zebra mussel (Ricciardi 2001) provided firm evidence that voluntary guidelines and mandatory regulations for ballast-water exchange were ineffective. Invasion risks associated with ships reporting no ballast on board (NOBOB) and the 30 ppt salinity standard (Reeves 2000) were finally addressed with amendments to the Canadian Shipping Act in 2006 and U.S. Coast Guard amendments to its regulations on ballast-water management in 2012. Consequently, saltwater flushing is now a standard requirement for all vessels entering the Great Lakes and coastal waters of North America, that is vessels declaring ballast or NOBOB, while the biological standard in regulation D-2 of the Ballast Water Management Convention, adopted by the International Marine Organization (a United Nations agency) in 2004, was accepted as the legal water-quality standard for all ballast discharges. These policy changes greatly reduce opportunities for aquatic invasive species to reach North American waters via ballast water. It should be noted that no new invader has been reported in the Great Lakes via ballast discharges since *Hemimysis anomala* was discovered in 2006 (Pothoven et al. 2007). Ratification of the 2004 Ballast Water Management Convention developed by the International Marine Organization, which includes a water standard in regulation D-2 that imposes an enforceable limit on "living organisms in ship's ballast-water discharge," would greatly reduce the transfer of invasive species throughout the world.

Given that introductions of invasive species are still possible via ballast water (under some circumstances), escapes from the aquarium trade, bait industry, and live food imports, the ability to detect newly-introduced species as soon as possible is imperative to contain and possibly exterminate a new invasive (vander Zanden et al. 2010). Unfortunately, biological surveys and monitoring activities have been greatly reduced in the Great Lakes with loss of equipment and agency personnel beginning in the early 1990s. In Ontario, intensive biological surveys in many areas of western Lake Erie, Lake Huron, and the joining corridor, where many exotic species were first found, were essentially stopped in 1996 (RWG, personal observation). The separation of Ontario Hydro into three companies also resulted in the loss of environmental monitoring and survey capabilities. Furthermore, no effort has been made to establish an effective response protocol at the national and provincial level to deal with invasive species (see references in Kerr et al. 2005). Establishing a response protocol and guidelines for the extermination of newly-established species agreed upon by both federal governments, Ontario, and the Great Lakes states would greatly assist the decision-making process for extermination and minimize response times to act when a new species is found. Fortunately, good news in the fight against invasive species occurred with amendments to the Great Lakes Water Quality Agreement (GLWQA) in 2012. The GLWQA now explicitly recognizes aquatic invasive species as a significant water-quality issue (Annex 6) and provides provisions for, among other initiatives, developing and implementing an early detection and a rapid-response program within 2 years.

The Great Lakes Fisheries Commission was provided with about $22 million in 2006 to protect a Great Lakes fisheries industry estimated at $4 billion annually through control of sea lamprey (GLFC 2006). Clearly, a similar sum of money needs to be available for federal agencies to monitor for invasive species and prevent them from establishing populations in the Great Lakes following the adage that "an ounce of prevention is worth a pound of cure" (Meyerson and Reaser 2003).

Finally, ecosystem effects caused by zebra mussels challenge the scientific concepts and frameworks that government agencies have used to manage water quality and fisheries of the Great Lakes. These filter-feeders have effectively decoupled the algae–phosphorus relationship (Mellina et al. 1995) that was the basis of Great Lakes pollution-abatement programs since the early-1970s. In fact, their intense feeding has caused a "center-outwards" trophic cascade throughout lake ecosystems: their removal of phytoplankton and suspended matter from the water column has increased water clarity that has promoted growth of submerged aquatic vegetation and may have subsequently contributed to mid-summer *Microcystis* blooms and nuisance growths of *Cladophora* in nearshore areas (Vanderploeg et al. 2002, Hecky et al. 2004). Further, dreissenid activities have induced a cascade of changes in the benthic and pelagic food webs, ultimately affecting top fish predators (Riley et al. 2008, Nalepa et al. 2009). Response by the International Joint Commission (IJC) to the re-occurrence of cyanobacterial blooms and nuisance, nearshore *Cladophora* growth was to propose a further reduction in phosphorus loadings to the Great Lakes. However, empirical evidence suggests that any further reductions in phosphorus loadings will reduce economically important sport and commercial fishes such as walleye in Lake Erie (Gopalan et al. 1998, Kershner and Stein 1998). In 1998, the Lake Erie Committee, which reports to the Great Lakes Fishery Commission (GLFC), passed a motion stating that "given the incomplete scientific understanding of the relationship of phosphorus to fish production and fish community structure in Lake Erie, the Lake Erie Committee does not support deviation from the phosphorus targets established with the Great Lakes Water Quality Agreement until a thorough scientific review of the target phosphorus concentration for Lake Erie has been carried out in an ecosystem (rather than control) context". Fortunately, the trophic cascade induced by a single filter-feeding species, tolerant of suspended sediments, nutrient, and insecticide loadings, suggest that water quality management no longer has to rely solely on phosphorus controls to regulate phytoplankton and periphyton abundance; management of littoral food-webs (i.e. bio-manipulation), particularly benthic filter-feeders and grazers, can also contribute to attain the goals of the Great Lakes Water Quality Agreement. Controlling discharges of sediments (i.e. sands and silts) and pesticides from agriculture lands, in particular, will allow for the possible return of the native (historical) filter-feeding insects and their associated biotic community in littoral areas of the Great Lakes. It was this native benthic fauna, characterized by snails and net-spinning caddisflies (e.g., *Hydropsyche-Goniobasis* community) and by burrowing mayflies (e.g. *Hexagenia-Oecetis* community) (Shelford and Boesel 1942), in the western basin of Lake Erie that was decimated by unregulated discharges of nutrients, pesticides, and sediments via soil erosion in the 1950s (Britt 1955, Krieger 1998). This biota historically controlled phytoplankton abundance in the pelagic zone and filamentous green algae in the littoral zone, while supporting the productivity of fisheries (e.g. perch, walleye, basses, pikes, trouts). Consequently, restoration of these elements of the benthic community would reduce phytoplankton and periphyton abundance, while allowing mesotrophic phosphorus conditions for fisheries productivity. This policy action would avoid the conflict between maintaining phosphorus loadings to meet fisheries goals and reducing phosphorus loadings to meet algal water-quality and anoxia-depletion goals of the Great Lakes Water Quality Agreement.

ACKNOWLEDGMENTS

We thank numerous researchers who commented and contributed perspectives of historic events outlined in this chapter. We thank chapter reviewers for comments and suggested improvements that often added length to the text and, we hope, to the documentation aspect of this work. These "memory jogs" often added things once forgotten. This article is Contribution 1722 of the U.S. Geological Survey Great Lakes Science Center, Ann Arbor, Michigan.

REFERENCES

Ashworth, W. 1986. *The Late, Great Lakes: An Environmental History*. Detroit, MI: Wayne State University Press.

Bailey, R.C., L. Grapentine, T.J. Stewart et al. 1999. Dreissenidae in Lake Ontario: Impact assessment at the whole lake and Bay of Quinte spatial scales. *J. Great Lakes Res.* 25: 482–491.

Bailey, J.F. and R.D. Hunter. 1991. Factors influencing *Dreissena* recruitment and biomass accumulation on an artificial substratum. *J. Shellfish Res.* 11: 217–218.

Banks, B. 1990. Alien onslaught. *Equinox* 53: 69–75.

Barbiero, R.P. and M.L. Tuchman, 2004. Long-term dreissenid impacts on water clarity in Lake Frie. *J. Great Lakes Res.* 30: 557–565.

Bellmore, L.A. 1992. The effects of zebra mussels (*Dreissena polymorpha*) on the water clarity of Lake St. Clair. BSc thesis. Waterloo, Ontario, Canada: Wilfrid Laurier University.

Benson, A.J. 2013. Chronological history of zebra and quagga mussels (Dreissenidae) in North America, 1988–2010. In *Quagga and Zebra Mussels: Biology, Impacts, and Control*, 2nd Edn., T.F. Nalepa and D.W. Schloesser, eds., pp. 9–33. Boca Raton, FL: CRC Press.

Bidwell, J.R., L.A. Lyons, D.S. Cherry, and J.C. Petrille. 1991. Surveillance of zebra mussels (*Dreissena polymorpha*), larval densities, settling, and growth at a power plant on western Lake Erie. *J. Shellfish Res.* 11: 218.

Bio-Environmental Services Ltd. 1981. The presence and implication of foreign organisms in ship ballast waters discharged into the Great Lakes. Report prepared for Water Pollution Control Directorate. Ottawa, Ontario, Canada: Environmental Protection Service, Environment Canada.

Boltovskoy, D., A. Karatayev, L. Burlakova et al. 2009. Significant ecosystem-wide effects of the swiftly spreading invasive freshwater bivalve *Limnoperna fortunei*. *Hydrobiologia* 636: 271–284.

Boyle, R.H. 1990. In shock over shells, *Sports Illustrated*, October 15, 88–95.

Brandt, S.B. 1986. Food of trout and salmon in Lake Ontario. *J. Great Lakes Res.* 12: 200–205.

Breckenridge, T. 1989. Mussels may cost region $5 billion. The Plain Dealer, Cleveland, OH, December 7.

Brieger, G. and R.D. Hunter. 1993. Uptake and depuration of PCB 77, PCB 169, and hexachlorobenzene by zebra mussels (*Dreissena polymorpha*). *Ecotox. Environ. Saf.* 26: 153–165.

Britt, N.W. 1955. Stratification in western Lake Erie in summer of 1953: Effects on the *Hexagenia* (Ephemeroptera) population. *Ecology* 36: 239–244.

Bruner, K.A., S.W. Fisher, and P.F. Landrum. 1994a. The role of zebra mussel, *Dreissena polymorpha*, in contaminant cycling: I. The effect of body size and lipid content on the bioconcentration of PCBs and PAHs. *J. Great Lakes Res.* 20: 725–734.

Bruner, K.A., S.W. Fisher, and P.F. Landrum. 1994b. The role of zebra mussel, *Dreissena polymorpha*, in contaminant cycling: II. Zebra mussel contaminant accumulation from algae and suspended particles, and transfer to the benthic invertebrate, *Gammarus fasciatus*. *J. Great Lakes Res.* 20: 735–750.

Bunt, C.M., H.J. MacIsaac, and W.G. Sprules. 1993. Pumping rates and projected filtering impacts of juvenile zebra mussels (*Dreissena polymorpha*) in western Lake Erie. *Can. J. Fish. Aquat. Sci.* 50: 1017–1022.

Campbell, N.J., J.P. Bruce, J.F. Hendrickson et al. 1969. Pollution of Lake Erie, Lake Ontario, and the international section of the St. Lawrence River. Report to the International Lake Erie Water Pollution Board and the International Lake Ontario-St. Lawrence River Water Pollution Board. Vol. I. Windsor, Ontario, Canada: International Joint Commission.

Carlton, J.T. 2008. The zebra mussel (*Dreissena polymorpha*) found in North America in 1986 and 1987. *J. Great Lakes Res.* 34: 770–773.

Charlton, M. 2001. Did zebra mussels clean up Lake Erie? *Great Lakes Res. Rev.* 5: 11–15.

Chase, M.E. and R.C. Bailey. 1999. The ecology of the zebra mussel (*Dreissena polymorpha*) in the lower Great Lakes of North America: I. Population dynamics and growth. *J. Great Lakes Res.* 25: 107–121.

Christie, W.J. 1974. Changes in the fish species composition of the Great Lakes. *J. Fish. Res. Bd. Can.* 31: 827–854.

Clarke, K.B. 1952. The infestation of waterworks by *Dreissena polymorpha*, a freshwater mussel. *J. Inst. Water Eng.* 6: 370–379.

Claudi, R. and G. Mackie. 1994. *Practical Manual for Zebra Mussel Monitoring and Control*. Boca Raton, FL: CRC Press.

Commissioner of the Environment and Sustainable Development. 2001. A legacy worth protecting: Charting a sustainable course in the Great Lakes and St. Lawrence River basin. 2001 report of the Commissioner of the Environment and Sustainable Development. Commissioner of the Environment and Sustainable Development, Ottawa, Ontario, Canada.

Connelly, N.A., C.R. O'Neill, B.A. Knuth, and T.L. Brown. 2007. Economic impacts of zebra mussels on drinking water treatment and electric power generation facilities. *Environ. Manage.* 40: 105–112.

Conway, B. 1991. Mussels blocking water line. *The Bay City Times*, Bay City, MI, November 13.

Cooley, J.M. 1991. Zebra mussels. *J. Great Lakes Res.* 17: 1–2.

Coscarelli, M. and J. Rendall. 1996. Aquatic nuisance species research relevant to the Great Lakes basin: Research guidance and descriptive inventory. Great Lakes Panel on Aquatic Nuisance Species, Ann Arbor, MI.

Coulson, S. 1990. *Ontario Seeks to Stop Spread of Pesky Zebra Mussels*. London, Ontario, Canada: The London Free Press, June 19.

Custer, C.M. and T.W. Custer. 1996. Food habits of diving ducks in the Great Lakes after the zebra mussel invasion. *J. Field Ornithol.* 67: 86–99.

David, F. 1990. Mussel invasion. *Syracuse Herald American*, Syracuse, New York, April 22.

Dermott, R. and D. Kerec. 1997. Changes to the deep-water benthos of eastern Lake Erie since the invasion of *Dreissena*: 1979–1993. *Can. J. Fish. Aquat. Sci.* 54: 922–930.

Dermott, R., J. Mitchell, I. Murray, and E. Fear. 1993. Biomass and production of zebra mussels, *Dreissena polymorpha*, in shallow waters of northeastern Lake Erie. In *Zebra Mussels: Biology, Impacts and Control*, T.F. Nalepa and D.W. Schloesser, eds., pp. 399–413. Boca Raton: CRC Press.

Dochoda, M.A. 1989. Preventing ballast water introductions in the Great Lakes. *FOCUS* 14: 15–17.

Dochoda, M.A. 1991. Meeting the challenge of exotics in the Great Lakes: The role of an International Commission. *Can. J. Fish. Aquat. Sci.* 48(Suppl. 1): 171–176.

Dochoda, M.A., A.L. Hamilton, and B.L. Bandurski. 1990. *Workshop on Exotic Species and the Shipping Industry: Summary and Recommendations*. Windsor, Ontario, Canada: International Joint Commission; Ann Arbor, MI: Great Lakes Fishery Commission.

vander Doelen, V. and R. Hornberger. 1988. New clams pose threat to intakes. *Windsor Star*, Windsor, Ontario, Canada, July 7.

Edmonds, P. 1990. *$40 Million Proposed for Fighting Tiny Mussels*. Detroit, MI: The Detroit Free Press, March 8.

Edsall, T.A., C.P. Madenjian, and B.A. Manny. 1999. Burrowing mayflies in Lake Erie—a review. In *State of Lake Erie: Past, Present and Future*, N. M. Munawar, T. Edsall, and I.F. Munawar, eds., pp. 219–231. Leiden, The Netherlands: Backhuys Publishers.

Fahnenstiel, G., T. Nalepa, S. Pothoven, H. Carrick, and D. Scavia. 2010. Lake Michigan lower food web: Long-term observations and *Dreissena* impact. *J. Great Lakes Res.* 36: 1–4.

Fitzsimons, J.D., J.H. Leach, S.J. Nepszy, and V.W. Cairns. 1995. Impacts of zebra mussel on walleye (*Stizostedion vitreum*) reproduction in western Lake Erie. *Can. J. Fish. Aquat. Sci.* 52: 578–586.

Flood, K. and J. Lee. 1995. Walleye embryo/larval bioassay results from Thames River/nitrate study. Ontario Ministry of Environment, Toronto, Ontario, Canada.

Flood, K. and J. Lee. 1996. Thames River walleye study. Ontario Ministry of Environment, Toronto, Ontario, Canada.

Frischkorn, J.L. 1989. Tiny mussels but big cost: $5 billion. *The News Herald*, Willoughby, OH, December 7.

Garton, D.W. and W.R. Haag. 1991. Heterozygosity, shell length, and metabolism in the European mussel, *Dreissena polymorpha*, from a recently established population in Lake Erie. *Comp. Biochem. Physiol.* 99A: 45–48.

Garton, D.W. and W.R. Haag. 1993. Seasonal reproductive cycles and settlement patterns of *Dreissena polymorpha* in western Lake Erie. In *Zebra Mussels: Biology, Impacts and Control*, T.F. Nalepa and D.W. Schloesser, eds., pp. 111–128. Boca Raton, FL: CRC Press.

Geigen-Miller, P. 1988a. *Losses in Millions Feared from Erie Mussel Menace*. London, Ontario, Canada: The London Free Press, November 25.

Geigen-Miller, P. 1988b. *Ships Pressured to Dump Ballast at Sea*. London, Ontario, Canada: The London Free Press, November 29.

Gianessi, L.P. and J.E. Anderson. 1995. *Pesticide Use in U.S. Crop Production*. Washington, DC: National Data Report.

Gopalan, G., D.A. Culver, L. Wu, and B.K. Trauben. 1998. Effects of recent ecosystem changes on the recruitment of young-of-the-year fish in western Lake Erie. *Can. J. Fish. Aquat. Sci.* 55: 2572–2579.

Gorrie, P. 1990. Exotics push out native species. *Canadian Geographic*, December 1990/January 1991.

Great Lakes Fishery Commission. 1988. Fact sheet on the transport and control of exotic biota introduced into the Great Lakes through ship ballast water. Ann Arbor, MI: Great Lakes Fisheries Commission.

Great Lakes Fishery Commission. 2006. Annual Report 2006. Ann Arbor, MI: Great Lakes Fisheries Commission.

Green, L. 1990. Invasion of zebra mussels threatens U.S. waterways. *Los Angeles Times*, Los Angeles, CA, January 15.

Greenberg, A., G. Matisoff, G. Gubanich, and J. Ciaccia. 1991. Zebra mussel veliger densities and water quality parameters in Lake Erie at the Cleveland water intakes. *J. Shellfish Res.* 11: 227.

Griffiths, R.W. 1989. Introduction of zebra mussels into the Great Lakes: Truth and consequences. In *Monthly Newsletter # 45*. London, Ontario, Canada: Ontario Ministry of Environment. Ontario.

Griffiths, R.W. 1990. The zebra mussel: Just another addition to the Great Lakes? The sportsman's outdoor annual: A special edition of the *Sydenham Sportsman's Newsletter*, pp. 12–13. Owen Sound, Ontario, Canada, July.

Griffiths, R.W. 1992. The Asian Clam: Another exotic species visits Ontario. In *Monthly Newsletter #66*. London, Ontario, Canada: Ontario Ministry of Environment.

Griffiths, R.W. 1993. Effect of zebra mussels (*Dreissena polymorpha*) on the benthic fauna of Lake St. Clair. In *Zebra Mussels: Biology, Impacts and Control*, T.F. Nalepa and D.W. Schloesser, eds., pp. 415–437. Boca Raton, FL: CRC Press.

Griffiths, R.W., W.P. Kovalak, and D.W. Schloesser. 1989. The zebra mussel, *Dreissena polymorpha*, in North America: Impact on raw water users. In *Proceedings of the EPRI Service Water System Reliability Improvement Seminar*, held in Charlotte, N.C., November 6–8, 1989, pp. 11–26. Palo Alto, CA: Electric Power Research Institute.

Griffiths, R.W., D.W. Schloesser, J.H. Leach, and W.P. Kovalak. 1991. Distribution and dispersal of the zebra mussel (*Dreissena polymorpha*) in the Great Lakes. *Can. J. Fish. Aquat. Sci.* 48: 1381–1388.

Grigorovich, I.A., R.I. Colautti, E.L. Mills, K. Holeck, A.G. Ballert, and H.J. MacIsaac. 2003. Ballast-mediated animal introductions in the Laurentian Great Lakes: Retrospective and prospective analyses. *Can. J. Fish. Aquat. Sci.* 60: 740–756.

Grossi, F. 1990. Company wages war on tiny zebra mussels. *The Globe and Mail*, Toronto, Ontario, Canada, July 30.

Gutschi Santavy, S. 2013. My story on finding the first zebra mussel (*Dreissena polymorpha*) in North America. In *Quagga and Zebra Mussels: Biology, Impacts, and Control*, 2nd Edn., T.F. Nalepa and D.W. Schloesser, eds., pp. 3–5. Boca Raton, FL: CRC Press.

Haag, W.R., D.J. Berg, and D.W. Garton. 1993. Reduced survival and fitness in native bivalves in response to fouling by the introduced zebra mussel (*Dreissena polymorpha*) in western Lake Erie. *Can. J. Fish. Aquat. Sci.* 50: 13–19.

Hamilton, D.J., C.D. Ankney, and R.C. Bailey. 1994. Predation of zebra mussels by diving ducks—An enclosure study. *Ecology* 75: 521–531.

Harner, T., T.F. Bidleman, L.M. Jantunen, and D. Mackay. 2001. Soil-air exchange model of persistent pesticides in the United States cotton belt. *Environ. Toxicol. Chem.* 20: 1612–1621.

Hart, M. 1990. Invasion of the zebra mussel. *The Atlantic*, 266: 81–87.

Hartig, J., J.V. DePinto, S. Bocking, J.V. Stone, and P. McIntyre. 2001. Great Lakes science and policy: Strengthening the connection. Report of a survey of Great Lakes policy issues. Prepared for the Joyce Foundation, Chicago, IL.

Hartman, W.L. 1972. Lake Erie: Effects of exploitation, environmental changes, and new species on the fisheries resources. *J. Fish. Res. Bd. Can.* 29: 899–912.

Hebert, P.D.N., B.W. Muncaster, and G.L. Mackie. 1989. Ecological and genetic studies on *Dreissena polymorpha* (Pallas): A new mollusc in the Great Lakes *Can. J. Fish. Aquat. Sci.* 46: 1587–1591.

Hebert, P.D.N., C.C. Wilson, M.H. Murdoch, and R. Lazar. 1991. Demography and ecological impacts of the invading mollusc, *Dreissena polymorpha*. *Can. J. Zool.* 69: 405–409.

Hecky, R.E., R.E.H. Smith, and D.R. Barton et al. 2004. The nearshore phosphorus shunt: A consequence of ecosystem engineering by dreissenids in the Laurentian Great Lakes. *Can. J. Fish. Aquat. Sci.* 61: 1285–1293.

Hincks, S.S. and G.L Mackie. 1997. Effects of pH, calcium, alkalinity, hardness, and chlorophyll on the survival, growth and reproductive success of zebra mussel (*Dreissena polymorpha*) in Ontario lakes. *Can. J. Fish. Aquat. Sci.* 54: 2049–2057.

Holland, R.E. 1993. Changes in planktonic diatoms and water transparency in Hatchery Bay, Bass Island area, western Lake Erie since the establishment of the zebra mussel. *J. Great Lakes Res.* 19: 717–624.

Holland, R.E., T.H. Johengen, and A.M. Beeton. 1995. Trends in nutrient concentrations in Hatchery Bay, western Lake Erie, before and after *Dreissena polymorpha*. *Can. J. Fish. Aquat. Sci.* 52: 1202–1209.

Hopkins, G. 1990. The zebra mussel: A photographic guide to the identification of microscopic veligers. Ontario Ministry of Environment, Toronto, Ontario, Canada.

Hornberger, R. 1988. War declared on intruder clam. *Windsor Star*, Windsor, Ontario, Canada, November 24.

Hunter, R.D. and K.A. Simons. 2004. Dreissenids in Lake St. Clair in 2001: Evidence for population regulation. *J. Great Lakes Res.* 30: 528–537.

Hyatt, R.A. 1982. Summer and winter creel census on Rondeau Bay Lake Erie, 1981. Lake Erie Fisheries Assessment Unit Report 1982-2. Toronto, Ontario, Canada: Ministry of Natural Resources.

Ingram, J. and A.C. Miller. 1992. Hydraulic structures versus zebra mussels. In *Proceedings of the Hydraulic Engineering Session at Water Forum 1992*. Baltimore, MD, August 2–6, 1992, pp. 606–611. American Society of Civil Engineers.

International Air Quality Advisory Board. 1999. Linking Canada and United States sources and source regions of selected persistent toxic substances to deposition in the Great Lakes basin: A progress report, July 1999. International Air Quality Advisory Board 1997–1999 Priorities Report. Report to the International Joint Commission. www.ijc.org/rel/boards/iaqab/pr9799/index.html

Jansen, J. 1989. Zebra mussels take over. *North Essex News*, Essex, U.K., July 26.

Jude, D.J, R.H. Reider, and G.R. Smith. 1992. Establishment of Gobiidae in the Great Lakes Basin. *Can. J. Fish. Aquat. Sci.* 49: 416–421.

Johannsson, O.E. and E.S. Millard. 1998. Impairment assessment of beneficial uses: Degradation of phytoplankton and zooplankton populations. Prepared for the Beneficial Uses Subcommittee of the Lake Erie Lakewide Management Plan (LaMP). Technical Report No. 13.

Johnson, T.B. 2009. Base of the Lake Erie food web. In *The State of Lake Erie, 2004*, Great Lakes Fishery Commission Special Publication 09-02, J.T. Tyson, R.A. Stein, and J.M. Dettmers, eds., pp. 13–22. Ann Arbor, MI: Great Lakes Fishery Commission.

Kaftannikova, O.G., A.A. Protasov, R.A. Kalinichenko, and S. Afans'yev. 1987. Biofouling in the reservoirs and water supply systems of fossil and nuclear power stations. In *Izucheniye Morskogo Obrastaniya i Pazrabotka Metidov Borbys Nim (Marine Biofouling Study and Development of New Control Methods)*, O.A. Skarlato, ed., pp. 56–61. Leningrad, Russia: ZIN Press.

Kehoe, M. 1990. *Pests at Home in St. Clair?* London, Ontario, Canada: The London Free Press, September 28.

Kerr, S.J., C.S. Brousseau, and M. Muschett. 2005. Invasive aquatic species in Ontario: A review and analysis of potential pathways for introduction. *Fisheries* 30: 21–30.

Kershner, M.W. and R.A. Stein. 1998. *Food Web Modelling in the Western and Central Basins of Lake Erie*. Federal Aid in Sport Fish Restoration, Project F-69-P, Final Report. Columbus, OH: Ohio Department of Natural Resources.

Kovalak, W.P., G.D. Longton, and R.D. Smithee. 1993. Infestation of power plant water systems by the zebra mussel (*Dreissena polymorpha* Pallas. In *Zebra Mussels: Biology, Impacts and Control*, T.F. Nalepa and D.W. Schloesser, eds., pp. 359–380. Boca Raton, FL: CRC Press.

Kraft, C. 1990. Zebra mussel update #5. University of Wisconsin Sea Grant Institute, Green Bay, WI.

Kraft, C. 1991a. Zebra mussel update #6. University of Wisconsin Sea Grant Institute, Green Bay, WI.

Kraft, C. 1991b. Zebra mussel update #11. University of Wisconsin Sea Grant Institute, Green Bay, WI.

Kraft, C. 1991c. Zebra mussel update #9. University of Wisconsin Sea Grant Institute, Green Bay, WI.

Kraft, C. 1992a. Zebra mussel update #13. University of Wisconsin Sea Grant Institute, Green Bay, WI, June 5, 4pp.

Kraft, C. 1992b. Zebra mussel update #14. University of Wisconsin Sea Grant Institute, Green Bay, WI.

Kreis, R.G., M.D. Mullin, R. Rossmann, and L.L. Wallace. 1991. Organic contaminant and heavy metal concentrations in zebra mussel tissue from western Lake Erie. *J. Shellfish Res.* 11: 228.

Krieger, K. 1998. *Mayflies and Lake Erie: A Sign of the Times*. FS-069. Columbus, OH: Ohio Sea Grant.

Lake Erie Committee. 2005. *Lake Erie Walleye Management Plan*. Ann Arbor, MI: Great Lakes Fishery Commission.

Lange, C.L. and R.K. Cap. 1991. The range extension of the zebra mussel (*Dreissena polymorpha*) in inland water of New York state. *J. Shellfish Res.* 11: 228.

Leach, J.H. 1972. Distribution of chlorophyll a and related variables in Ontario waters of Lake St. Clair. In *Proceedings of the 16th Conference on Great Lakes Research*, pp. 80–82. Ann Arbor, MI: International Association of Great Lakes Research.

Leach, J.H. 1980. Limnological sampling intensity in Lake St. Clair in relation to distribution of water masses. *J. Great Lakes Res.* 6: 141–145.

Leach, J.H. 1992. Population dynamics of larvae and adult zebra mussels and changes in water quality in western Lake Erie. In *Mussel Morsals* 2: 3. Maple, Ontario, Canada: Ontario Ministry of Natural Resources.

Leach, J.H. 1993. Impacts of the Zebra Mussel (*Dreissena polymorpha*) on water quality and fish spawnings in western Lake Erie. In *Zebra Mussels: Biology, Impacts and Control*, T.F. Nalepa and D.W. Schloesser, eds., pp. 381–397. Boca Raton, FL: CRC Press.

Lees, D. 1990. Musseling in. *Toronto Life*, pp. 17–20.

LePage, W.L. 1993. The impact of *Dreissena polymorpha* on waterworks operations at Monroe, Michigan: A case study. In *Zebra Mussels: Biology, Impacts and Control*, T.F. Nalepa and D.W. Schloesser, eds., pp. 333–358. Boca Raton, FL: CRC Press.

Lester, N.P., P.A. Ryan, R.S. Kushneriuk, A.J. Dextrase, and M.R. Rawson. 2002. The effect of water clarity on walleye (*Stizostedion vitreum*) habitat and yield. Percid Community Synthesis. Ontario Ministry of Natural Resources, Peterborough, Ontario, Canada.

Lewis, D. 1992. How the zebra mussel is winning the west. *Environ. Sci. Eng.* February/March: 68–72.

Li, Y.F. and T.F. Bidleman. 2001. Toxaphene in the United States. 2. Emissions and residues. *J. Geophys. Res.* 106: 17929–17938.

Lichkoppler, F.R., D.O. Kelch, and M.A. Berry. 1993. Attitudes of 1990, 1991 and 1992 mid-America Boat Show and 1991 Fairport Fishing Symposium patrons concerning the zebra mussel, Lake Erie, and Great Lakes pollution. *J. Great Lakes Res.* 19: 129–135.

Lowe, K. 1989. An unwelcome newcomer. *Michigan Outdoors*. April.

Lucy, F.E. and E. Muckle-Jeffs. 2010. History of the zebra mussel/ICAIS conference series. *Aquat. Invasions* 5: 1–3.

Ludsin, S.A., M.W. Kershner, K.A. Blocksom, R.L. Knight, and R.A. Stein. 2001. Life after death in Lake Erie: Nutrient controls drive fish species richness, rehabilitation. *Ecol. Appl.* 11: 731–746.

Ludyanskiy, M.L., D. McDonald, and D. MacNeill. 1993. Impact of the zebra mussel, a bivalve invader. *BioScience* 43: 533–544.

MacIsaac, H.J. 1996. Potential abiotic and biotic impacts of zebra mussels on the inland waters of North America. *Am. Zool.* 36: 287–299.

MacIsaac, H.J., W.G. Sprules, O.E. Johannsson, and J.H. Leach. 1992. Filtering impacts of larval and sessile zebra mussels (*Dreissena polymorpha*) in western Lake Erie. *Oecologia* 92: 30–39.

MacIsaac, H.J., W.G. Sprules, and J.H. Leach. 1991. Ingestion of small-bodied zooplankton by zebra mussels (*Dreissena polymorpha*): Can cannibalism on larvae influence population dynamics? *Can. J. Fish. Aquat. Sci.* 48: 2051–2060.

Mackie, R. 1990. Mussels spreading rapidly. *The Globe and Mail*, Toronto, Ontario, Canada, July 5.

Mackie, G.L. 1991. Biology of the exotic zebra mussel, *Dreissena polymorpha*, in relation to native bivalves and its potential impact in Lake St. Clair. *Hydrobiologia* 219: 251–268.

Mackie, G.L., W.N. Gibbons, B.W. Muncaster, and I.M. Gray. 1989. The zebra mussel, *Dreissena polymorpha*: A synthesis of European experiences and a preview for North America. Report prepared by B.A.R. Environmental for Ministry of Environment, Toronto, Ontario, Canada.

MacLennan, D.S. 1992. *Lake St. Clair Walleye Stock Assessment*. Tilbury, Ontario, Canada: Ontario Ministry of Natural Resources.

Madenjian, C.P. 1995. Removal of algae by the zebra mussel (*Dreissena polymorpha*) population in western Lake Erie: A bioenergetics approach. *Can. J. Fish. Aquat. Sci.* 52: 381–390.

Makarewicz, J.C., T.W. Lewis, and P. Bertram. 1999. Phytoplankton composition and biomass in the offshore waters of Lake Erie: Pre- and post-*Dreissena* introductions (1983–1993). *J. Great Lakes Res.* 25: 135–148.

Magner, M. 1993. Scientists fear mussel pushing game fish out. *Jackson Citizen Patriot*, Jackson, MI, April 25.

Marsden, J.E. 1992. Standard protocols for monitoring and sampling zebra mussels. Biological Notes 138. Illinois Natural History Survey, Champaign, IL.

May, B. and J.E. Marsden. 1992. Genetic identification and implications of a second invasive species of dreissenid mussel in the Great Lakes. *Can. J. Fish. Aquat. Sci.* 49: 1501–1506.

Mazak, E.J., H.J. MacIsaac, M.R. Servos, and R. Hesslein. 1997. Influence of feeding habits on organochlorine contaminant accumulation in waterfowl on the Great Lakes. *Ecol. Appl.* 7: 1133–1143.

Mellina, E., E. Mills, and J.B. Rasmussen. 1995. The impact of zebra mussels (*Dreissena polymorpha*) on the phosphorus budgets and phytoplankton biomass in lakes. *Can. J. Fish. Aquat. Sci.* 52: 2553–2573.

Meyerson, L.A. and J.K. Reaser. 2003. Bioinvasions, bioterrorism and biosecurity. *Front. Ecol. Environ.* 6: 307–314.

Miehls, A.L.J., D.M. Mason, K.A. Frank, A.E. Krause, S.D. Peacor, and W.W. Taylor. 2009. Invasive species impacts on ecosystem structure and function: A comparison of Oneida Lake, New York, USA, before and after zebra mussel invasion. *Ecol. Model.* 220: 3194–3209.

Millard, E.S., E.J. Fee, D.D. Myles, and J.A. Dahl. 1999. Comparison of phytoplankton photosynthesis using 14C-incubator technique and numerical modelling in Lakes Erie, Ontario, the Bay of Quinte, and the Northwest Ontario Lake Size Series (NOLSS). In *State of Lake Erie: Past, Present and Future*, M. Munawar, T. Edsall, and I.F. Munawar, eds., pp. 441–468. Leiden, The Netherlands: Backhuys Publishers.

Mills, E.L., J.H. Leach, J.T. Carleton, and C.L. Secor. 1993. Exotic species in the Great Lakes: A history of biotic crisis and anthropogenic introductions. *J. Great Lakes Res.* 19: 1–54.

van Moorsel, G. 1992. *Control Cost Likely $6 Billion over 10 Years*. London, Ontario, Canada: The London Free Press, February 21.

Morton, B.S. 1969. Studies on the biology of *Dreissena polymorpha* Pall. IV. Habits, habitats, distribution, and control. *Water Treat. Exam.* 18: 233–241.

Murray, T.E., P.H. Rich, and E.H. Jokinen. 1993. Invasion potential of the zebra mussel, *Dreissena polymorpha*, in Connecticut: Predictions from water chemistry data. Special Report No. 36. Connecticut Institute of Water Resources, Storrs-Mansfield, CT.

Nalepa, T.F. 1994. Decline of native bivalves (Unionidae: Bivalvia) in Lake St. Clair after infestation by the zebra mussels *Dreissena polymorpha*. *Can. J. Fish. Aquat. Sci.* 51: 2227–2233.

Nalepa, T.F., J.F. Cavaletto, M. Ford, W.M. Gordon, and M. Wimmer. 1993. Seasonal and annual variation in weight and biochemical content of the zebra mussel, *Dreissena polymorpha*, in Lake St. Clair. *J. Great Lakes Res.* 19: 541–552.

Nalepa, T.F., D.L. Fanslow, M.B. Lansing, and G.A. Lang. 2003. Trends in the benthic macroinvertebrate community of Saginaw Bay, Lake Huron, 1987 to 1996: Response to phosphorus abatement and the zebra mussel, *Dreissena polymorpha*. *J. Great Lakes Res.* 29: 14–33.

Nalepa, T.F., D.L Fanslow, and G.A. Lang. 2009. Transformation of the offshore benthic community in Lake Michigan: Recent shift from the native amphipad *Diporeia* spp. to the invasive mussel *Dreissena rostritormis Bugensis*. *Freshwat. Biol.* 54: 466–479.

Nalepa, T.F. and J.M. Gauvin. 1988. Distribution, abundance, and biomass of freshwater mussels (Bivalvia, Unionidae) in Lake St. Clair. *J. Great Lakes Res.* 14: 411–419.

Neary, B.P. and J.H. Leach. 1992. Mapping the potential spread of the zebra mussel (*Dreissena polymorpha*) in Ontario. *Can. J. Fish. Aquat. Sci.* 49: 406–415.

Nichols, S.J. and B. Kollar. 1991. Reproductive cycle of zebra mussel (*Dreissena polymorpha*) in western Lake Erie at Monroe, Michigan. *J. Shellfish Res.* 11: 235.

Nicholls, K.H. and G.J. Hopkins. 1993. Recent changes in Lake Erie (north shore) phytoplankton: Cumulative impacts of phosphorus loading reductions and the zebra mussel introduction. *J. Great Lakes Res.* 19: 637–647.

OMOE. 1982. Water quality studies of Rondeau Bay and watershed, Kent County. Ontario Ministry of Environment, London, Ontario, Canada.

OMNR. 1990. Zebra mussel informer. Ontario Ministry of Natural Resources, Toronto, Ontario, Canada.

OMNR. 1991. Catalogue of Ontario zebra mussel research projects 1990–91. Ontario Ministry of Natural Resources, Toronto, Ontario, Canada.

OMNR. 1992. Zebra mussel watch. Ontario Ministry of Natural Resources, Toronto, Ontario, Canada.

OMNR. 1993a. The future of walleye. Lake St. Clair/Thames River Fact Sheet, March 1993. Ontario Ministry of Natural Resources, Toronto, Ontario, Canada.

OMNR. 1993b. Walleye facts. Lake St. Clair/Thames River Fact Sheet, March 1993. Ontario Ministry of Natural Resources, Toronto, Ontario, Canada.

OMNR. 1993c. Lake St. Clair fisheries report 1992. Ontario Ministry of Natural Resources, Tilbury, Ontario, Canada.

OMNR. 1995. Lake St. Clair fisheries report 1994. Ontario Ministry of Natural Resources, Tilbury, Ontario, Canada.

Park, J. and L. Hushak. 1999. Zebra mussel control costs in surface water using facilities. Technical Summary Series, Publication OSHU-TS-028. Columbus, OH: Ohio Sea Grant College Program.

Patterson, M.W.R., J.J.H. Ciborowski, and D.R. Barton. 2005. The distribution and abundance of *Dreissena* species in Lake Erie, 2002. *J. Great Lakes Res.* 31(Suppl. 2): 223–237.

Pimentel, D., R. Zuniga, and D. Morrison. 2005. Update on the environmental and economic costs associated with alien-invasive species in the United States. *Ecol. Econ.* 52: 273–288.

Pollick, S. 1989a. Pesky invaders plague fishing in Great Lakes. *The Blade*, Toledo, Spain, May 10.

Pollick, S. 1989b. Explosive reproduction common characteristic of new lake pests. *The Blade*, Toledo, Spain, May 11.

Pollick, S. 1989c. Cleaner water believed boosting pests' numbers. *The Blade*, Toledo, Spain, May 12.

Pollick, S. 2013. One reporter's perspective on the invasion of dreissenid mussels in North America: Reflections on the news media. In *Quagga and Zebra Mussels: Biology, Impacts, and Control*, 2nd Edn., T.F. Nalepa and D.W. Schloesser, eds., pp. 117–135. Boca Raton, FL: CRC Press.

Pothoven, S.A., I.A. Grigorovich, G.L. Fahnenstiel, and M.D. Balcer. 2007. Mysid *Hemimysis anomala* into the Lake Michigan Basin. *J. Great Lakes Res.* 33: 285–292.

Rapaport, R.A. and S.J. Eisenteich. 1986. Atmospheric deposition of toxaphene to eastern North America derived from peat accumulations. *Atmos. Environ.* 20: 2367–2379.

Reeves, E. 2000. Exotic politics: An analysis of the law and politics of exotic invasions of the Great Lakes. *Toledo J. Great Lakes' Law Sci. Pol.* 2: 125–206.

Reid, D.F. and D. Wilkinson. 2013. Catalyst for change—The little dreissenid that did (change national policy on aquatic invasive species). In *Quagga and Zebra Mussels: Biology, Impacts, and Control*, 2nd Edn., T.F. Nalepa and D.W. Schloesser, eds., pp. 177–187. Boca Raton, FL: CRC Press.

Reynolds, C.V. 1991. Invasion of the zebra mussels. *Discover* 12: 44.

Ricciardi, A. 2001. Facilitative interactions among aquatic invaders: Is an "invasional meltdown" occurring in the Great Lakes? *Can. J. Fish. Aquat. Sci.* 58: 2513–2525.

Riley, S.C., E.F. Roseman, S.J. Nichols, T.R. O'Brien, C.S. Kiley, and J.S. Schaeffer. 2008. Deepwater demersal fish community collapse in Lake Huron. *Trans. Amer. Fish. Soc.* 137: 1879–1890.

Roberts, L. 1990. Zebra mussel invasion threatens U.S. water. *Science* 249: 1370–1372.

Rosenberg, G. and M. Ludyanskiy. 1994. A nomenclatural review of *Dreissena* (Bivalvia: Dreissenidae), with identification of the quagga mussel as *Dreissena bugensis*. *Can. J. Fish. Aquat. Sci.* 51: 1474–1484.

Rothlisberger, J., D. Finnoff, R. Cooke, and D. Lodge. 2012. Shipborne nonindigenous species diminish Great Lakes ecosystem services. *Ecosystems* 15: 463–476.

Ruscitti, J. and R. McKenzie. 1990. *Saving Water Intakes Expensive*. London, Ontario, Canada: The London Free Press, March 14.

Sakai, A.K., F.W. Allendorf, J.S. Holt et al. 2001. The population biology of invasive species. *Ann. Rev. Ecol. Syst.* 32: 305–332.

San Giacomo, R. and M. Cavalcoli. 1991. Case studies for the engineering of mussel control facilities in raw water intake systems. *J. Shellfish Res.* 11: 237.

Sblendorio, R.P., J.C. Malinchock, and R. Claudi. 1991. Controlling zebra mussel infestations at hydroelectric plants. *Hydro Rev.* 10: 42–50.

Schloesser, D.W. and W.P. Kovalak. 1991. Infestation on unionids by *Dreissena polymorpha* in a power canal in Lake Erie. *J. Shellfish Res.* 10: 355–359.

Schloesser, D.W. and T.F. Nalepa. 1994. Dramatic decline of unionids bivalves in offshore waters of western Lake Erie after infestation by the zebra mussel, *Dreissena polymorpha. Can. J. Fish. Aquat. Sci.* 51: 2234–2242.

Schloesser, D.W. and B.A. Manny. 1982. Distribution and relative abundance of submerged aquatic macrophytes in the St. Clair-Detroit River ecosystem. U.S.G.S. Great Lakes Science Center Report Number 82-7. U.S. Geologic Survey, Ann Arbor, MI.

Schloesser, D.W. and E.C. Masteller. 1999. Mortality of unionid bivalves associated with dreissenid mussels (*Dreissena polymorpha* and *D. bugensis*) in Presque Isle Bay, Lake Erie. *Northeast. Natural.* 6: 341–352.

Shelford, V. E. and M. W. Boesel. 1942. Bottom animal communities of the island area of Western Lake Erie in the summer of 1937. *Ohio J. Sci.* 42: 179–190.

Sinclair, R.M. 1964. Clam pests in Tennessee water supplies. *J. Am. Wat. Works Assoc.* 56: 592–599.

Snyder, F.L. 1990. Zebra mussels in Lake Erie: The invasion and its implications. *Ontario Fisherman*, April/May.

SOS. 1989. Zebra mussels on the move: Mapping survey undertaken. *Save Ontario Shipwrecks Newsletter*, pp. 20–21. Orleans, Ontario, Canada.

Stańczykowska, A. 1977. Ecology of *Dreissena polymorpha* (Pallas) (Bivalvia) in lakes. *Pol. Arch. Hydrobiol.* 24: 461–530.

Stewart, T.W. and J.M. Hynes. 1994. Benthic macroinvertebrate communities of southwestern Lake Ontario following the invasion of *Dreissena. J. Great Lakes Res.* 20: 479–493.

Stoeckmann, A.M. and D.W. Garton. 1991. Metabolic responses to increased food supply and induced spawning. *J. Shellfish Res.* 11: 239.

Stoeckmann, A.M. and D.W. Garton. 1997. A seasonal energy budget for zebra mussels (*Dreissena polymorpha*) in western Lake Erie. *Can. J. Fish. Aquat. Sci.* 54: 2743–2751.

Strayer, D.L. 1991. Projected distribution of the zebra mussel, *Dreissena polymorpha*, in North America. *Can. J. Fish. Aquat. Sci.* 48: 1389–1395.

Strayer, D.L. 2009. Twenty years of zebra mussels: Lessons from the mollusk that made headlines. *Front. Ecol. Environ.* 7: 135–141.

Suvedi, M. and K. Heinze. 1994. Evaluation of the Great Lakes Sea Grant Network's zebra mussel outreach activities for industrial and municipal water users. Report prepared for: Great Lakes Sea Grant Network. Michigan State University, East Lansing, MI.

Thomas, M.V. and R.C. Haas. 2012. Status of Lake St. Clair submerged plants, fish community, and sport fishery. Fisheries Research Report 2009, p. 90. Lansing: Michigan Department of Natural Resources.

Topping, J. 1991. North American environmental and economic nightmare of the nineties. *Biome* 11: 1–2.

Tucker, J.K., C.H. Theiling, K.D. Blodgett, and P.A. Thiel. 1993. Initial occurrences of zebra mussels on freshwater mussels in the upper Mississippi River system. *J. Freshwat. Ecol.* 8: 245–251.

UGLCCS. 1988. *Upper Great Lakes Connecting Channels Study.* Volume II. Environment Canada and U.S. Environmental Protection Agency.

U.S. Congress, Office of Technology Assessment. 1993. *Harmful Non-Indigenous Species in the United States.* OTA-F-565. Washington, DC: U.S. Government Printing Office.

U.S. EPA. 1991. Ecology and management of the zebra mussel and other introduced aquatic nuisance species. EPA/600/3-91/003. U.S. Environmental Protection Agency, Washington, DC.

USFWS. 1988. Ballast-transported organisms pose a threat to Great Lakes. U.S. Department of the Interior, Fish and Wildlife Service Research Information Bulletin No. 88-45. National Fisheries Research Center-Great Lakes, Ann Arbor, MI.

USGS. 2011. United States Geological Survey Nonindigenous Aquatic Species Database: The Zebra Mussel and Quagga Mussel Information Resource Page. http://nas.er.usgs.gov/taxgroup/mollusks/zebramussel/

Vanderploeg, H.A., T.F. Nalepa, D.J. Jude et al. 2002. Dispersal and emerging ecological impacts of Ponto-Caspian species in the Laurentian Great Lakes. *Can. J. Fish. Aquat. Sci.* 59: 1209–1228.

White, S. 1990. *Biologists Brace for Other Exotic Species.* London, Ontario, Canada: The London Free Press, June 20.

Witzel, L. 1981. Summer and winter creel census on Rondeau Bay, Lake Erie 1980. Ontario Ministry of Natural Resources, Wheatley, Ontario, Canada.

Wormington, A. and J.H. Leach. 1993. Concentrations of migrant diving ducks at Point Pelee National Park, Ontario, in response to invasion of zebra mussels, *Dreissena polymorpha. Can. Field Nat.* 106: 376–380.

Wright, S. "Mussel Car" recovered. Monthly Newsletter # 48. Ontario Ministry of Environment, London, Ontario, Canada.

Wu, L. and D.A. Culver. 1991. Zooplankton grazing and phytoplankton abundance: An assessment before and after invasion of *Dreissena polymorpha. J. Great Lakes Res.* 17: 425–436.

vander Zanden, M.J., G.J.A. Hansen, S.N. Higgins, and M.S. Kornis. 2010. A pound of prevention, plus a pound of cure: Early detection and eradication of invasive species in the Laurentian Great Lakes. *J. Great Lakes Res.* 36: 199–205.

Zhu, X., T.B. Johnson, and J.T. Tyson. 2008. Synergistic changes in the fish community of western Lake Erie as modified by non-indigenous species and environmental fluctuations. In *Checking the Pulse of Lake Erie*, M. Munawar and R. Heath, eds., pp. 439–474. Burlington, VT: Aquatic Ecosystem Health and Management Society (Ecovision).

Zook, W. and S. Philips. 2009. A survey of watercraft interception programs for dreissenid mussels in the western United States: Results of an on-line survey completed in February 2009. Report prepared for Western Regional Panel of the National Aquatic Nuisance Species Task Force. Pacific States Marine Fisheries Commission, Portland, OR.

CHAPTER 11

Catalyst for Change:
"The Little Dreissenid That Did" (Change National Policy on Aquatic Invasive Species)

David F. Reid and Dean Wilkinson

CONTENTS

Abstract 177
Introduction 178
Background and Legislative/Policy History 178
    Terminology 178
    U.S. National Policy 178
    The Zebra Mussel: Catalyst for National Policy Change 180
National Invasive Species Policy Evolves 182
Summary 183
Acknowledgments 184
References 184

ABSTRACT

The earliest and most extensive legislation that established laws for prevention and control of nonindigenous and invasive (nuisance) species in the United States was for protection of crops and livestock. Federal legislation to protect aquatic ecosystems was absent until the late-twentieth century, and federal policies most often took a "benign neglect" approach, sometimes even encouraging introductions of new species in attempts to make natural ecosystems more responsive to human needs. Policies against species introductions were rare and primarily reactive to individual species that became nuisances after they were established. Each new species invasion was treated as a separate event. The discovery of the zebra mussel (*Dreissena polymorpha*) in 1988 led to recognition that aquatic invasions were also a problem. While discoveries of nonindigenous species raised concerns prior to the late-1980s, it was the discovery of the zebra mussel that catalyzed awareness and rapid policy changes, including establishment of new regulations at many levels of government across the Unites States and Canada. In the United States, the centerpiece of legislation was the Nonindigenous Aquatic Nuisance Prevention and Control Act, passed in 1990 and reauthorized and expanded as the National Invasive Species Act in 1996. It included legislative mandates for prevention (emphasizing ballast water), zebra mussel control, and other actions such as ecosystem surveys. It established the Aquatic Nuisance Species Task Force to encourage a coordinated federal approach and provided for establishment of regional panels; ultimately, it provided the impetus for Executive Order 13112, which resulted in the first comprehensive national plan for managing all invasive species in the United States. Although little additional progress was made after 1996 through 2011, the policy changes started in 1990 established aquatic invasive species as recognized national problems with regional and local implications. The legislation of the 1990s led to growth of sustained federal and state programs that continue to respond to new introductions while working toward preventing future introductions.

INTRODUCTION

Laws are rules of conduct or actions prescribed or formally recognized as binding or enforceable by an authority; *laws* are established by legislation, which is the act of making and enacting laws, but "legislation" can also refer to specific acts approved by a legislature. *Policy* is a course or plan of action adopted or pursued by an individual entity under some legal authority. Policy is not a law, although laws may incorporate policies in their text.* In this chapter, we review the history of legislation and government policy related to nonindigenous species (see "Terminology" section) in the United States. We show that legislation and associated government policies associated with nonindigenous biota were first established in the late-1800s, but were primarily for protection of terrestrial crops and livestock. Natural aquatic ecosystems were not considered until the late 1980s after the invasion of the Great Lakes by zebra mussels (*Dreissena polymorpha*).

BACKGROUND AND LEGISLATIVE/ POLICY HISTORY

Terminology

The reader is cautioned that the terminology related to this topic is not consistent and several terms with the same or very similar meaning are used. The term "invasive species" is widely, but mistakenly, used when referring to any species not native to an ecosystem. The term was defined and incorporated into U.S. policy by Executive Order 13112, issued in 1999, as "an alien species whose introduction does or is likely to cause economic or environmental harm or harm to human health." To provide consistency, it also defined "alien species" as "any species, including its seeds, eggs, spores, or other biological material capable of propagating that species, that is not native to (a particular ecosystem)." Thus, an "alien species" can also be called a "nonnative species" yet is not necessarily invasive unless it causes or may cause some form of harm. Other terms for "native (species)" are "indigenous" or "endemic," while "nonindigenous," "nonendemic," "introduced," and "exotic" are also used instead of "nonnative." The terms "injurious" and "harmful" have also been used in place of "invasive." Biologists, on the other hand, may use the term "invasive" to describe any nonindigenous (nonnative/ nonendemic/introduced/exotic) species that spreads rapidly beyond its native range (Jeschke and Strayer 2005), but make no inference as to harm. The reader is referred to ANSTF (1994b), Carlton (2001), Colautti and MacIsaac (2004), and ISAC (2006) for further information.

* http://dictionary.reference.com/

U.S. National Policy

The history of invasive species laws and policies in the United States was one of slow evolution over more than a century. As soon as the Civil War (1861–1865) ended, Congress passed "*An Act to prevent the spread of foreign diseases among cattle of the United States*" (39th Congress, December 1865; Farrington 1909), the first in a series of federal laws spanning the late-1800s aimed specifically at protecting livestock. During the same period, federal policy allowed increasing numbers of immigrants and encouraged the growth of agriculture. In response, the U.S. population expanded westward to take advantage of plentiful farm lands, and transportation improvements made it possible to move crops thousands of miles to their final markets. Trade across regions and with other countries increased, and farmers began to recognize additional threats posed by arrival of various nonindigenous plants, insects, and diseases (Burgess 1909). However, laws to protect agriculture other than livestock were absent at the federal level until the Plant Quarantine Act of 1912 (7 U.S.C. 151 et seq.), followed by the Animal Damage Control Act of 1931 (7 U.S.C. 426), Federal Seed Act of 1940 (7 U.S.C. 1551 et seq.), Organic Act of 1944, Federal Plant Pest Act of 1957 (7 U.S.C. 150aa–150jj), and the Federal Noxious Weed Act of 1974 (U.S.C. 2801 et seq.), among others (USDA 2011). These provided powerful tools to regulate import and control of invasive species, but their primary goal and application were limited to protection of agricultural production.

Federal and state policies for aquatic ecosystems in the nineteenth century and through the first half of the twentieth century can best be characterized as "benign neglect." They provided only limited protection for aquatic ecosystems against species introductions, and sometimes resource management agencies even encouraged introductions of nonindigenous species in attempts to "improve" ecosystems and make them more responsive to human needs (Crossman 1991, Mills et al 1993, U.S. Congress 1993). Little attention was paid to the fact that such introductions could have negative impacts on target ecosystems and their associated native species or that nonindigenous species might become invasive by expanding beyond the ecosystem to which they were introduced.

The Lacey Act of 1900 (16 U.S.C. 3371–3378) was one of the earliest laws with provisions against transportation and unpermitted possession of invasive (injurious) animal species, including fish. However, even though it is over 100 years old, application of the Lacey Act has been limited. More than half of the taxa were already present in the United States when listed, and most of these taxa that were already established in the wild continued to spread even after the listing (Fowler et al. 2007). At the end of 2010, the list of injurious wildlife included only 20 species or families of closely related species* and only six more were under evaluation (USFWS 2011). One of the difficulties in adding such species to the list is how to identify potentially injurious species before they become established in the United States. Even if a species is known to present serious risks, the Lacey Act contains no provisions to allow emergency listing.

At times there have been proposals to change the Lacey Act to use a "clean" list approach, which would only allow importation of species determined in advance to be noninjurious. However, such a change would shift the burden of proof to the importer. Whenever a "clean list" approach has been proposed, substantial stakeholder and political oppositions have stymied the implementation of such changes (Peoples et al. 1992, U.S. Congress 1993). For example, in 1973, the U.S. Fish and Wildlife Service proposed the modification of its injurious wildlife regulations to a "clean list" approach. This created such an overwhelming number of negative responses during the public comment period that it went through two revisions and additional public comment periods over the next 4 years before it was eventually abandoned (ANSTF 1994b, Appendix E).

The first law that focused entirely on aquatic invasive species was the Rivers and Harbors Appropriation Act of 1899 (33 U.S.C. 403; Chapter 425). It gave the U.S. Army Corps of Engineers (USACE) authority to remove aquatic weeds as part of their responsibility for navigation channels and was primarily enacted in response to the establishment and spread of water hyacinth (*Eichhornia crassipes*). This authority was expanded in the River and Harbor Act of 1958 (Public Law 85–500). The 1958 Act specifically mentioned water hyacinth, alligator weed (*Alternanthera philoxeroides*), Eurasian water milfoil (*Myriophyllum spicatum*), and "other obnoxious plant growths" and authorized "a comprehensive project to provide for control and progressive eradication of (the listed species) from the navigable waters, tributary streams, connecting channels, and other allied waters (in several specific states) used for navigation, flood control, drainage, agriculture, fish and wildlife conservation, public health, and related purposes." In addition, it recognized the importance of research to develop effective and economical weed control and established a cost-share program between the USACE and the states to fund such research.

The 1955 Convention on Great Lakes Fisheries was ratified by the governments of the United States and Canada, at least in part, in response to an aquatic invasive species—the sea lamprey (*Petromyzon marinus*). The Great Lakes Fishery Commission (GLFC), a binational commission composed of representatives from both countries, was created by Article II of this Convention. Article IV (d) assigned the GLFC responsibility "to formulate and implement a comprehensive program for the purpose of eradicating or minimizing the sea lamprey populations in the Convention Area." In some ways, this Convention was a precursor to the present binational cooperation on invasive species along both the north and south U.S. borders via the Commission for Environmental Cooperation under the 1994 North American Agreement on Environmental Cooperation.

Passage of the Endangered Species Act of 1973 (ESA, 16 U.S.C. 1531 et seq.) indirectly helped broaden the focus of U.S. invasive species policy from agriculture to natural ecosystems. Passage of ESA was not directed toward invasive species, but the ESA process to designate a species as "endangered" or "threatened" requires publication of a final rule that identifies factors contributing to that designation as well as development of a species recovery plan. ANSTF (1994a) analyzed final rules that included citations of contributory factors in listings for 69 fish species in the United States and found that, after habitat alteration (cited as a factor in 91% of listed species), nonindigenous species were cited as a cause of decline or potential threat in 70% of cases. Wilcove et al. (1998) used data from ESA listings and other data sources to examine and rank the threats to "imperiled" species. Similar to the findings in ANSTF (1994a), they reported that the second most common threat (affecting 49% of all imperiled species and 53% of fish species) after habitat alteration was the spread of nonindigenous species. Because of the requirement to develop a recovery plan for species listed under ESA, managers were confronted with reducing impacts caused by invasive species.

Executive orders are presidential directions to the Executive Branch to use existing legal authority to achieve specific policy goals. However, executive orders may be canceled or ignored by succeeding presidents, and without aggressive implementation, there is no assurance that policy intent will be carried out. Executive Order 11987 issued by President Carter in 1977 was the first official U.S. policy to recognize "exotic species" as a broadly generic and serious problem in both terrestrial and aquatic ecosystems. The order adopted a precautionary policy to restrict the introduction of exotic organisms unless there was a finding† that an introduction would not have an adverse effect on natural ecosystems. However, it was not successful in changing federal policy and, except for agricultural protection, government and societal approaches continued to be predominantly reactive, responding "after-the-fact." New invasions continued to be treated as independent events and broader-scale impacts on ecosystems were generally not considered.

---

* For example, snakeheads (Family Channidae) include 28 species.

† A "finding" in this context is a formal decision by the Department of Agriculture or Department of Interior; Carter Executive Order 11987, Section 2(d): "This Order does not apply to the introduction of any exotic species, or the export of any native species, if the Secretary of Agriculture or the Secretary of the Interior finds that such introduction or exportation will not have an adverse effect on natural ecosystems."

Figure 11.1 Some of the original zebra mussels found in Lake St. Clair in June, 1988, are now archived at the Canadian Museum of Nature, Ottawa, Ontario, Canada. (Photo courtesy of Brent Gibson, Great Lakes United.)

The Zebra Mussel: Catalyst for National Policy Change

By the late-1980s, the increasing number of nonindigenous aquatic species discovered in the Great Lakes raised concern within the region (IJC and GLFC 1990), but it was the discovery of a single species—the zebra mussel (*Dreissena polymorpha*; Hebert et al. 1989, Carlton 2008a; see Figure 11.1)—that catalyzed major changes in national legislation, policy, and the approach of the U.S. federal government toward *aquatic* invasive species.

The GLFC became alarmed in the late-1980s when another potentially harmful species, the Eurasian ruffe (*Gymnocephalus cernuus*), was found in the St. Louis River (Lake Superior) in 1987,* and ballast discharge was mentioned as a possible source (Pratt et al 1988, Dachoda 2009). Details of the discovery and potential impacts were presented to the Lake Superior Committee at the Annual Meeting of the GLFC in March 1988. Dr. James T. Carlton (Williams College, CT), then at University of Oregon, gave a presentation at the June 1988 annual meeting of the GLFC on the diversity of live biota found in ballast tanks. Dr. Carlton's presentation stunned the audience, and the GLFC and the International Joint Commission (IJC), already concerned about nonindigenous species in the Great Lakes, were galvanized into joint action to prevent further introductions via ballast water (Carlton 2008b, Dachoda 2009). In consultation with the shipping industry and various other stakeholders, the two commissions developed a plan to require ballast water exchange for ocean-going vessels before they entered the Great Lakes. Letters from both commissions to the governments of the United States and Canada were in preparation in summer 1988 when discovery of the zebra mussel in Lake St. Clair (Hebert et al. 1989) was reported. The letters were modified to include a last-minute reference to the zebra mussel discovery and were transmitted in August 1988 (IJC and GLFC 1990).

In 1990, slightly more than 2 years after zebra mussels were first reported in the Great Lakes and the commissions sent their letters, Congress passed the Nonindigenous Aquatic Nuisance Prevention and Control Act (NANPCA, 11/29/90, P.L. 101–636). The Great Lakes Task Force of the Northeast–Midwest Congressional Coalition, and especially Senator John Glenn (OH) and Congressmen Dennis Hertel (MI), Robert Davis (MI), and Henry Nowak (NY), introduced the Senate and House versions of the bill and were instrumental in getting it passed (see Foreword to this book by A. Beeton).

Dr. Carlton, who was extensively involved in activities leading up to passage of NANPCA, identified what he believed were the two most significant factors that pushed Congress "over the top" to pass this major legislation (J. T. Carlton 2008b):

> The immediate catalysts for NANPCA were events such as the closing of the water plant at Monroe, Michigan in September and December 1989†, but the real driver—and I believe the major reason NANPCA passed—was Jon Stanley's estimate that the cost of zebra mussels would be about $5B between 1989 and 1999‡. Without that number, NANPCA would not exist. Although he publicly revised his estimate down to $4B at the August 1990 American Fishery Society meeting in Pittsburgh, $5B stuck as *the* number for the next 10–15 years, and still appears today. All else followed—from Senator John Glenn's staffers walking into his office and telling him about zebra mussels (it's said he

---

* Pratt et al.'s (1988) report on the first recognition of ruffe was in 1987 when the Wisconsin Department of Natural Resources captured and identified ruffe in the St. Louis River. Subsequent reexamination of archived samples by other agencies revealed that ruffe specimens collected in 1986 had been misidentified as johnny darters (*Etheostoma nigrum*).

† Zebra mussels had lined the main water intake pipe in layers, reducing its effective cross-sectional diameter of 36 in. to less than 12 in.

‡ Dr. Jon Stanley, then Director of the U.S. Fish and Wildlife Service National Fisheries Research Center—Great Lakes, in Ann Arbor, Michigan (now the Great Lakes Science Center, U.S. Geological Survey), made the cited $5B "off-the-cuff" estimate in November 1989 in response to a reporter's question.

first thought it was a practical joke being played on him), from the seeds planted at the International Maritime Organization in London at virtually the same time by the Australians, who were reacting to the arrival of Japanese dinoflagellates in Tasmania via ballast water… and a dozen other key pivotal moments and steps, which seemed all to be happening at a break-neck speed.

Congress could have passed a species-specific "zebra mussel control act," which would have continued the long-standing U.S. policy to treat and react to each new species invasion as an isolated and independent event. Instead, while emphasizing actions against the zebra mussel, Congress gave specific recognition to the broader problem of aquatic species invasions. NANPCA was a "call to arms" against aquatic invasive species for federal and state governments. It encouraged a comprehensive approach that emphasized ecosystems rather than individual species and aimed at prevention of introductions via analysis and control of pathways and vectors. It established the Aquatic Nuisance Species Task Force (ANSTF) whose members consisted of representatives from five federal agencies (Fish and Wildlife Service [Department of Interior], National Oceanic and Atmospheric Administration [Department of Commerce], Environmental Protection Agency, Coast Guard [then Department of Transportation], U.S. Army Corps of Engineers [Department of Defense]), plus "any other Federal agency" the chairpersons deemed appropriate, and a number of non-Federal ex-officio members. The ANSTF was charged with coordinating NANPCA-related actions among federal agencies and to "develop and implement a program for waters of the United States to prevent introduction and dispersal of aquatic nuisance species." Core elements of the new approach established legislatively by NANPCA included prevention, detection, monitoring, and control of aquatic nuisance species and recognized both research and education as key activities necessary to address the invasive species problem. It also established legislative definitions of "aquatic nuisance species" and "nonindigenous species" (ANSTF 1994b).

After passage of NANPCA in 1990, significant changes were made in the federal policy approach to invasive species. Almost immediately there was increased emphasis on preventive measures. Inclusion of ballast water provisions led to implementation of new ballast water management regulations (U.S. Coast Guard 1993) and new research related to the better understanding of the physical and biological characteristics and delivery of ballast water (Buck 2010). An often overlooked consequence of the zebra mussel invasion and passage of NANPCA was a marked increase in coordination and cooperation among government entities with responsibilities for dealing with invasive species. The formation of the ANSTF under NANPCA fostered a common approach and free exchange of ideas. Among the first actions taken by the task force was to set up interagency groups charged with identifying priorities for zebra mussel research and ballast water management.

While passage of NANPCA provided the framework, cooperation across agencies was sometimes difficult—the process of establishing federal agency programs is normally internal and involves competing programmatic and budget priorities. In addition, a high priority program in one agency may be considered tangential in another, and at times agencies may unknowingly work at cross purposes. For example, as late as 2006, one federal agency funded a project to control the invasive giant reed (*Arundo donax*) in California, while another federal agency considered promoting it as a source of bioenergy in the southeast.

NANPCA included two provisions to encourage participation by states: (1) development of state management plans (Section 1204) and, perhaps because the Great Lakes region was a driving force behind the legislation, (2) establishment of a regional panel (Section 1203) under the auspices of the ANSTF and specific to the Great Lakes, to be convened by the Great Lakes Commission (Great Lakes Panel on Aquatic Nuisance Species, http://www.glc.org/ans/panel.html). This was the only panel specifically mandated in the original 1990 legislation, and the first meeting of the panel was convened in late-1991. In 1996, the original 1990 Act was reauthorized (Nonindigenous Invasive Species Act [NISA], 1996), and language was added to form a Western Regional Panel and allow additional regional panels as well. Membership in all panels was comprised of regional representatives from "Federal, State, and local agencies and from private environmental and commercial interests." In addition, the Great Lakes Panel was encouraged to invite representatives from the federal, provincial, and territorial governments of Canada to participate as observers. Expansion of regional panels under NISA resulted in creation of four additional panels: Gulf and South Atlantic (1999), Northeast (2001), Mississippi River Basin (2002), and Mid-Atlantic (2003). Cooperation on invasive species problems within these regions increased considerably. In 2011, the ANSTF included 13 federal agency representatives plus 12 ex-officio members and six regional panels. Each panel reports to the ANSTF on priorities and program activities in its region and provides advice to public and private interests on regional invasive species issues.

The regional panel concept provided a formalized infrastructure for discussing common cross-boundary problems posed by invasive species and opportunities for coordinated and/or collaborative activities within and across regions as well as nationally and internationally (Canada, Mexico). Depending on regional needs, regional panels have undertaken a variety of actions, such as development of model guidance for state legislation, regulation, and policy (Great Lakes Panel), surveys of coastal waters (Northeast Panel), development of management plans for particular species (e.g., Asian carp [*Hypophthalmichthys* spp.]—Mississippi River Basin Panel; green crab [*Carcinus maenas*]—Western

Panel), online species identification and alert systems (Gulf and South Atlantic Panel), and preparation of rapid response plans for potential future invasions (dreissenids—Western Panel). Another major activity of regional panels has been encouragement and assistance with preparation of individual state management plans. As of winter 2011, there were 33 state and 3 interstate management plans approved by the ANSTF.

NATIONAL INVASIVE SPECIES POLICY EVOLVES

The passage of NISA in 1996 solidified the "sea change" in national policy initiated by the 1990 legislation and also broadened the scope of the original Act (Figure 11.2). NISA expanded the geographic emphasis of the original Act to areas outside the Great Lakes while retaining focus on aquatic ecosystems. Subsequently, there was interest in developing a national program to address all invasive species, both aquatic and terrestrial. In April 1997, a large group of scientists and natural resource managers organized by Don C. Schmitz (then Florida Department of Environmental Protection), Dr. Jim Carlton (Williams College), and Dr. Dan Simberloff (then at Florida State University) co-signed a letter to Vice President Al Gore advocating a comprehensive approach to invasive species. At the direction of Vice President Gore, the Departments of Agriculture, Commerce, and Interior began work on a response, which ultimately led to another executive order.

In February 1999, President Bill Clinton issued Executive Order 13112 to strengthen Executive Branch focus and coordination on invasive species problems. Many elements of this order mirrored activities covered by NANPCA and included similar priorities, such as prevention, ecological surveys and monitoring for impacts, early detection, control, and public education. It also established another interagency coordinating body—the National Invasive Species Council (NISC) consisting of the Secretaries of State, Treasury, Defense, Interior, Agriculture, Commerce, Transportation, and the Administrator of the Environmental Protection Agency. The Council was tasked with providing national leadership for all invasive species.

The Clinton executive order recognized the existing responsibilities and programs assigned to the ANSTF and mandated the NISC to work with the ANSTF and other already-established federal groups, such as the Federal Interagency Committee for the Management of Noxious and Exotic Weeds (established in 1994). The NISC was also tasked with preparing a national management plan with recommendations for detailed performance-oriented

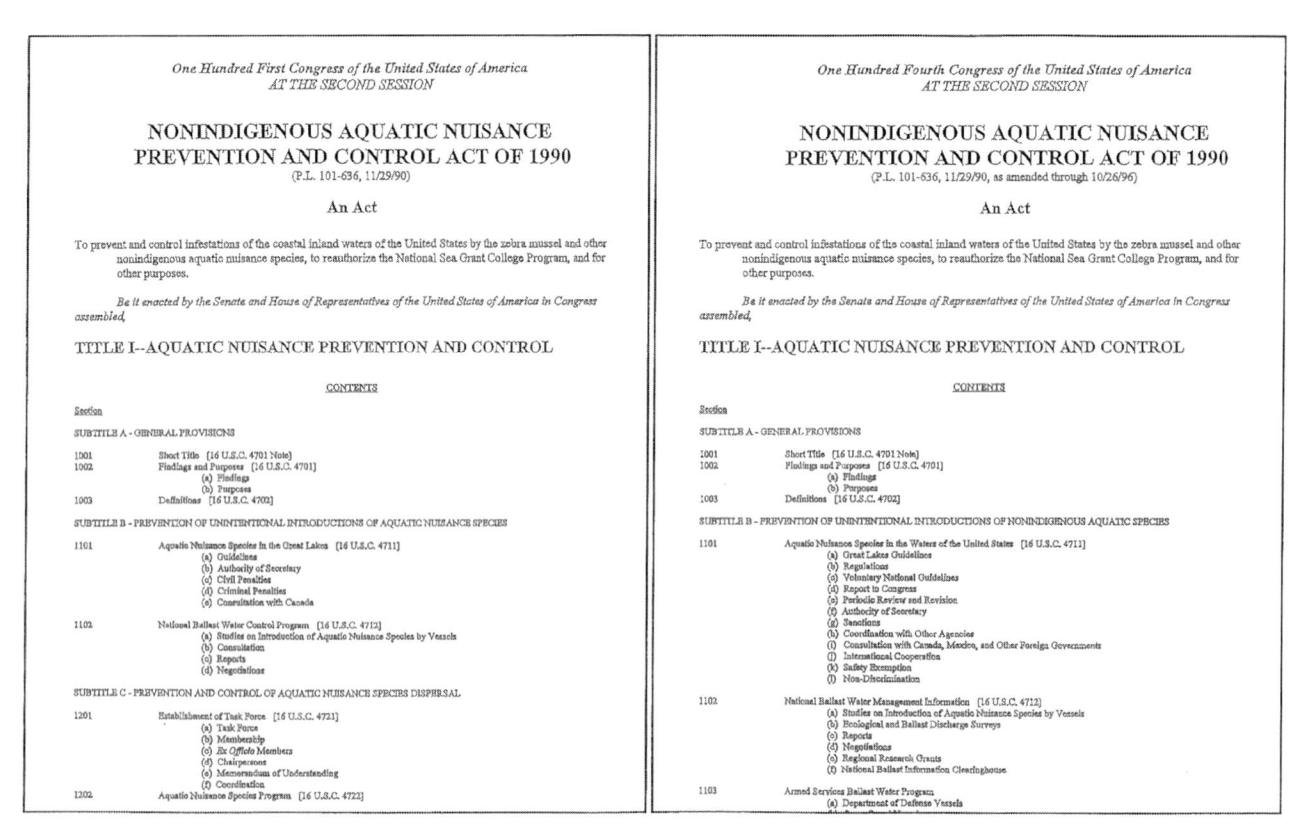

Figure 11.2 The Nonindigenous Aquatic Nuisance Prevention and Control Act of 1990 (left) was reauthorized as the National Invasive Species Act of 1996 (right). The scope of the original 1990 Act was substantially broadened by the 1996 reauthorization, as illustrated by this side-by-side comparison of the first page of each Table of Contents.

goals and measures of success for federal agency efforts concerned with invasive species. Executive Order 13112 fostered a strong link to stakeholder input by creating the Invasive Species Advisory Committee (ISAC), which reports to the council and consists of non-federal members. The order placed new proactive obligations on federal agencies to identify agency actions that might affect the status of invasive species and determine the likelihood that proposed actions would cause or promote introduction or spread of invasive species. It specified that such actions were to proceed only after the responsible agency made a public determination that potential benefits of proposed action would outweigh potential harm. This established an implicit requirement for a formal risk assessment procedure, reflecting work on a generic risk assessment protocol completed in 1996 by the ANSTF (ANSTF 1996). By also including an explicit restriction on funding until risk assessment was completed and publicly vetted, it affected far more than just federal agencies since there are many programs through which funds are passed to other entities that potentially involve invasive species. As an example, a large portion of funds used by states for fish stocking comes through the 1950 Federal Aid in Sport Fish Restoration Act (as amended; 16 U.S.C. 777–777k, 64 Stat. 430). Before Executive Order 13112, these funds were occasionally used to stock nonindigenous species. Under the provisions of the executive order, a more rigorous process to evaluate such stocking actions became necessary. As required by Executive Order 13112, the NISC produced a national invasive species management plan in 2001, which was revised and updated in 2008 (NISC 2001, 2008).

The U.S. Department of Agriculture maintains the National Invasive Species Information Center (http://www.invasivespeciesinfo.gov/index.shtml), which includes information about the content and status of each invasive-species-related bill introduced in Congress since 2001 (http://www.invasivespeciesinfo.gov/laws/bills.shtml, accessed August 30, 2012). Multiple failed attempts by Congress since 2001 to reauthorize and expand NISA, or to pass any other broad new legislation covering invasive species, are documented. In spite of the lack of further legislative progress, federal agencies have continued to act both individually and collaboratively on the provisions and mandates established by the 1990 and 1996 Acts. For example, U.S. Fish and Wildlife Service, National Oceanic and Atmospheric Administration, the State of California, and a private utility company worked together from 2000 to 2006 to successfully eradicate the marine macroalga *Caulerpa taxifolia* from two bays in southern California (CA EPA 2006, Merkel & Associates 2006). The NISC and ANSTF completed a joint pathways analysis and ranking guide (NISC & ANSTF 2005), and both are exploring the potential effects of global climate change on species invasions and invasion processes. ANSTF updated and revised an Aquatic Nuisance Species Research Risk Analysis Protocol (ANSTF 2010) and approved a new 5 year Strategic Plan (ANSTF 2012). The U.S. Coast Guard, after a lengthy deliberative process, issued a final rule for "Standards for Living Organisms in Ships' Ballast Water Discharged in U.S. Waters" (USCG 2012), which established a national standard with numeric limits on the concentration of live organisms allowed in ballast water discharged into U.S. waters. Numerous federal and state agencies continue to maintain up-to-date online nonindigenous species information and mapping databases (see U.S. Department of Agriculture, National Invasive Species Information Center for a list: http://www.invasivespeciesinfo.gov/aquatics/databases.shtml [accessed August 30, 2012]).

Although little federal legislative progress has been made since 1996, the policy changes started by NANPCA and expanded by NISA established aquatic invasive species as recognized national problems with regional and even local implications. The legislation of the 1990s led to growth of sustained federal and state programs that continue to respond to new introductions while working to prevent future introductions.

SUMMARY

The earliest and most extensive legislation in the United States to address invasive species was for protection of crops and livestock. Until the late-twentieth century, government policy often encouraged intentional introduction of nonindigenous species, particularly fish, associated with attempts to "improve" the benefits natural ecosystems could provide to society. Regulations against species introductions were rare, were primarily in response to individual nuisance species after they already became established, and were mostly aimed at terrestrial ecosystem protection. Each nuisance species that became established was treated as a separate and independent problem, and there was little acknowledgement that introduction of nonindigenous species into natural areas could cause serious ecologic damage.

Discovery of the zebra mussel in 1988 catalyzed major changes in the U.S. national policy and for the first time there was specific recognition of aquatic invasive species as a national concern. The Nonindigenous Aquatic Nuisance Prevention and Control Act of 1990 (NANPCA) and its reauthorization as the National Invasive Species Act of 1996 (NISA) became centerpieces for a dramatically different national approach to invasive species, which was supplemented and broadened by Executive Order 13112 in 1999. Without the zebra mussel invasion, it seems unlikely Congress would have acted so quickly or created such a comprehensive approach to this problem.

Unfortunately, as of 2012, Congress had repeatedly failed to reauthorize or amend NISA (1996; i.e., NANPCA [1990] as amended) or pass any other broad new legislation covering invasive species. However, the policy changes of the 1990s, which started in response to the zebra mussel invasion, led to the establishment and growth of sustained federal and state programs that continue responding to new aquatic invasive species introductions while also working to prevent future introductions.

ACKNOWLEDGMENTS

The authors appreciate the constructive comments and suggestions provided by the two editors of this book. We also thank Dr. Jim Carlton and Ms. Marg Dachoda for their valuable insights on events that occurred in the late 1980s leading up to the passage of the Nonindigenous Aquatic Nuisance Prevention and Control Act of 1990.

REFERENCES

ANSTF (Aquatic Nuisance Species Task Force). 1994a. *Report to Congress: Findings, Conclusions, and Recommendations of the Intentional Introductions Policy Review*. Washington, DC: U.S. Fish & Wildlife Service.

ANSTF (Aquatic Nuisance Species Task Force). 1994b. *Aquatic Nuisance Species Program*. Washington, DC: U.S. Fish & Wildlife Service.

ANSTF (Aquatic Nuisance Species Task Force). 1996. *Generic Nonindigenous Aquatic Organisms Risk Analysis Review Process*. Washington, DC: Risk Assessment and Management Committee, Aquatic Nuisance Species Task Force. http://www.anstaskforce.gov/Documents/ANSTF_Risk_Analysis.pdf (accessed November 11, 2011).

ANSTF (Aquatic Nuisance Species Task Force). 2010. *Federal Aquatic Nuisance Species (ANS) Research Risk Analysis Protocol*. Washington, DC: Aquatic Nuisance Species Task Force. http://www.anstaskforce.gov/Revised_ANSTF_Research_Protocol_FINAL.pdf (accessed November 11, 2011).

ANSTF (Aquatic Nuisance Species Task Force). 2012. *Aquatic Nuisance Species Task Force Strategic Plan (2013–2017)*. Washington, DC: Aquatic Nuisance Species Task Force. http://anstaskforce.gov/Documents/ANSTF%20Strategic%20Plan%202013-2017.pdf (accessed August 30, 2012).

Buck, E.H. 2010. Ballast water management to combat invasive species. Report RL32344. Washington, DC: Congressional Research Service—Library of Congress.

Burgess, A.F. 1909. Laws for the control of pests affecting agriculture. In *Cyclopedia of American Agriculture*, ed. L.H. Bailey, pp. 527–533. London, U.K.: The Macmillan Co.

CA EPA (California Environmental Protection Agency, State Water Resources Control Board). 2006. *News Release: Caulerpa Taxifolia Eradication—Officials Proclaim Victory Over "Killer Algae" But Remain Vigilant to New Sightings*. Carlsbad, CA: California Environmental Protection Agency, State Water Resources Control Board. http://www.cal-ipc.org/resources/pdf/CaulerpaEradicated.pdf (accessed February 8, 2012).

Carlton, J.T. 2001. *Introduced Species in U.S. Coastal Waters: Environmental Impacts and Management Priorities*. Arlington, TX: Pew Oceans Commission.

Carlton, J.T. 2008a. The zebra mussel *Dreissena polymorpha* found in North America in 1986 and 1987. *J. Great Lakes Res.* 34: 770–773.

Carlton, J.T. 2008b. "Re: ZM anniversary," Message to D. Reid, March 31, 2008. E-mail.

Colautti, R.I. and H.J. MacIsaac. 2004. A neutral terminology to define "invasive" species. *Divers. Distrib.* 10: 135–141.

Crossman, E.J. 1991. Introduced freshwater fishes: A review of the North American perspective with emphasis on Canada. *Can. J. Fish. Aquat. Sci.* 48(Suppl. 1): 46–57.

Dachoda, M. 2009. "Re: Memories of the past," Message to D. Reid. November 23, 2009. E-mail.

Farrington, A.M. 1909. Quarantine laws and practices. In *Cyclopedia of American Agriculture,* ed. L.H. Bailey, pp. 486–489. London, U.K.: The Macmillan Co.

Fowler, A.J., D.M. Lodge, and J.F. Hsia. 2007. Failure of the Lacey Act to protect U.S. ecosystems against animal invasions. *Front. Ecol. Environ.* 5: 353–359.

Hebert, P.D.N., B.W. Muncaster, and G.L. Mackie. 1989. Ecological and genetic-studies on *Dreissena polymorpha* (Pallas)—A new mollusk in the Great Lakes. *Can. J. Fish. Aquat. Sci.* 46: 1587–1591.

IJC (International Joint Commission) and GLFC (Great Lakes Fishery Commission). 1990. *Exotic Species and the Shipping Industry: The Great Lakes-St. Lawrence Ecosystem at Risk: A Special Report to the Governments of the United States and Canada*. Windsor, Ontario, Canada: International Joint Commission.

ISAC (Invasive Species Advisory Council). 2006. Invasive species definition clarification and guidance white paper. http://www.invasivespeciesinfo.gov/docs/council/isacdef.pdf (accessed October 7, 2011).

Jeschke, J.M. and D.L. Strayer. 2005. Invasion success of vertebrates in Europe and North America. *Proc. Nat. Acad. Sci.* 102: 7198–202.

Merkel & Associates. 2006. Final report on eradication of the invasive seaweed *Caulerpa Taxifolia* from Agua Hedionda Lagoon and Huntington Harbour, California. Report prepared by Rachel Woodfield, Merkel & Associates, Inc. and Keith Merkel, Merkel & Associates, Inc. for Steering Committee of the Southern California *Caulerpa* Action Team. http://www.globalrestorationnetwork.org/uploads/files/CaseStudyAttachments/71_c._taxifolia_eradication.pdf (accessed May 13, 2013).

Mills, E.L., J.H. Leach, J.T. Carlton, and C.L. Secor. 1993. Exotic species in the Great Lakes: A history of biotic crises and anthropogenic introductions. *J. Great Lakes Res.* 19: 1–54.

NISC (National Invasive Species Council). 2001. *Meeting the Invasive Species Challenge: National Invasive Species Management Plan*. Washington, DC: National Invasive Species Council.

NISC (National Invasive Species Council). 2008. *2008–2012 National Invasive Species Management Plan*. Washington, DC: National Invasive Species Council.

NISC (National Invasive Species Council) and ANSTF (Aquatic Nuisance Species Task Force). 2005. *Focus Group Conference Report and Pathways Ranking Guide*. Washington, DC: National Invasive Species Council and Aquatic Nuisance Species Task Force.

Peoples, R.A. Jr., J.A. McCann, and L.B. Starnes. 1992. Introduced organisms: Policies and activities of the U.S. Fish & Wildlife Service. In *Dispersal of Living Organisms into Aquatic Ecosystems,* eds. A. Rosenfield, and R. Mann, pp. 325–352. College Park, MD: Maryland Sea Grant.

U.S. Coast Guard. 1993. CGD 91–066, 58 FR 18334, April 8, 1993, as amended by CGD 94–003, 59 FR 67634, December 30, 1994; USCG–1998–3423, 64 FR 26682, May 17, 1999.

U.S. Coast Guard. 2012. 77 FR 17254, March 23, 2012.

U.S. Congress. 1993. *Harmful Non-Indigenous Species in the United States*. OTA-F-565. Washington, DC: Office of Technology Assessment, U.S. Government Printing Office.

USDA (U.S. Department of Agriculture). 2011. Federal laws and regulations-public laws and acts. Agricultural Research Service. http://www.invasivespeciesinfo.gov/laws/publiclaws.shtml (accessed October 7, 2011).

USFWS (U.S. Fish and Wildlife Service). 2011. Current list of injurious wildlife. http://www.fws.gov/fisheries/ans/Current_Listed_IW.pdf (accessed October 7, 2011).

Wilcove, D.S., D. Rothstein, J. Dubow, A. Phillips, and E. Losos. 1998. Quantifying threats to imperiled species in the United States. *BioScience* 48: 607–615.

CHAPTER 12

Invading Dreissenid Mussels Transform the 100-Year-Old International Joint Commission

Mark J. Burrows*

CONTENTS

Abstract 187
Background 188
Ecosystem or Ecotoxicology Approach? 190
Calls to Action 190
Catalyst for Change..191
Increased Awareness and Action 191
Regulatory Discord...192
Preventing the "Next Zebra Mussel" Invasion...193
Rapid Response..194
References..194

ABSTRACT

The introduction of dreissenid mussels (zebra mussel *Dreissena polymorpha* and quagga mussel *Dreissena rostriformis bugensis*) into the Laurentian Great Lakes was a substantial event in the 100-year-old history of the International Joint Commission (IJC). These mussels caused the IJC to expand its scope of efforts beyond persistent toxic substances to address the complex biological, physical, and chemical changes that were occurring. Since its inception, the IJC maintained a strong focus on water quality issues caused by industrial waste, persistent toxic substances, and eutrophication. Therefore, the IJC did not immediately consider dreissenid mussels and other aquatic invasive species as a high priority because of uncertainty about the extent to which invasive species could impact Great Lakes ecosystems. As early as 1912, an IJC study identified how ballast water could serve as a pathway for dispersing biological as well as chemical pollution. A decade before zebra mussels were discovered in the Great Lakes, the IJC clearly described the harm that the transfer of aquatic invasive species between drainage basins could inflict on native ecosystems. However, for many years, there was no clear consensus among scientists, or political will within governments, to institute mandatory regulations to prevent the influx of aquatic invasive species. Prior to the introduction of dreissenid mussels, the invasive sea lamprey (*Petromyzon marinus*) and alewife (*Alosa pseudoharengus*) caused alarm but were addressed on a species-specific basis. Ultimately, it was the severe consequences of the dreissenid mussel invasion that tipped the scales and caused IJC to more aggressively pursue the problem of aquatic invasive species and integrate an ecosystem approach in all aspects of boundary-waters treaty activities. The IJC became a consistent advocate for action. In addition to calling for binational action to establish consistent and biologically effective regulations and to address pathways of introduction, the IJC drew attention to chronic problems possibly linked to dreissenids, such as altered nutrient cycling and algae blooms within the nearshore areas of the Great Lakes. Over the past 20 years, the IJC has partnered with other organizations to facilitate binational exchange of knowledge about the state of science and technology and to provide a collaborative binational

* Mark J. Burrows is a physical scientist at the International Joint Commission's Great Lakes Regional Office in Windsor, Ontario. The views expressed in this chapter are his alone and do not represent the opinions of the IJC or any of its affiliated offices or advisory groups.

response to new discoveries of aquatic invasive species in the Great Lakes. At present, the IJC has a broad ecological approach and is strongly committed to binational efforts to prevent the "next zebra mussel" invasion.

BACKGROUND

The International Joint Commission (IJC) was formed by the Boundary Waters Treaty of 1909 to resolve disputes over the diversion and use of waterways that criss-cross the border between Canada and the United States (Figure 12.1). The treaty includes the following passage about pollution:

> ... it is further agreed that the waters herein defined as boundary waters and water flowing across the boundary shall not be polluted on either side to the injury of health or property on the other. (Article IV).

Early studies by the IJC investigated the scope of problems in the Great Lakes and recognized the impacts of pollutants and other issues, some of which are still relevant today. These studies were reported in the first half of the 1900s, but there was not enough support to motivate governments to act. For approximately 60 years after the Boundary Waters Treaty was signed, the IJC had a narrow definition of "pollution" limited to oils, chemicals, untreated waste water, and trash. The final report of an IJC reference study in 1912 on the pollution of boundary waters (International Joint Commission 1918) described conditions as

> The waters of Rainy River, St. Marys River, St. Clair River, and of the Detroit and Niagara Rivers, in consequence of the unrestricted discharge of sewage from vessels and towns, are no longer fit for domestic use unless subjected to extensive treatment in water-purification plants. Below the cities of Detroit and Buffalo the waters of the Detroit and Niagara Rivers, respectively, are so intensely polluted that it is highly questionable whether by the aid of any ordinary purification plant they can be made at all suitable for drinking purposes.

The report listed unacceptably high death rates from typhoid and also reported two kinds of pollution from vessels: raw sewage/garbage and ballast water discharges. The report stated:

> This pollution is a serious menace to public health, not alone through the possible contamination of the public water supplies near their intakes, but also by reason of its effect upon the water supplies of other vessels following or crossing the same routes.

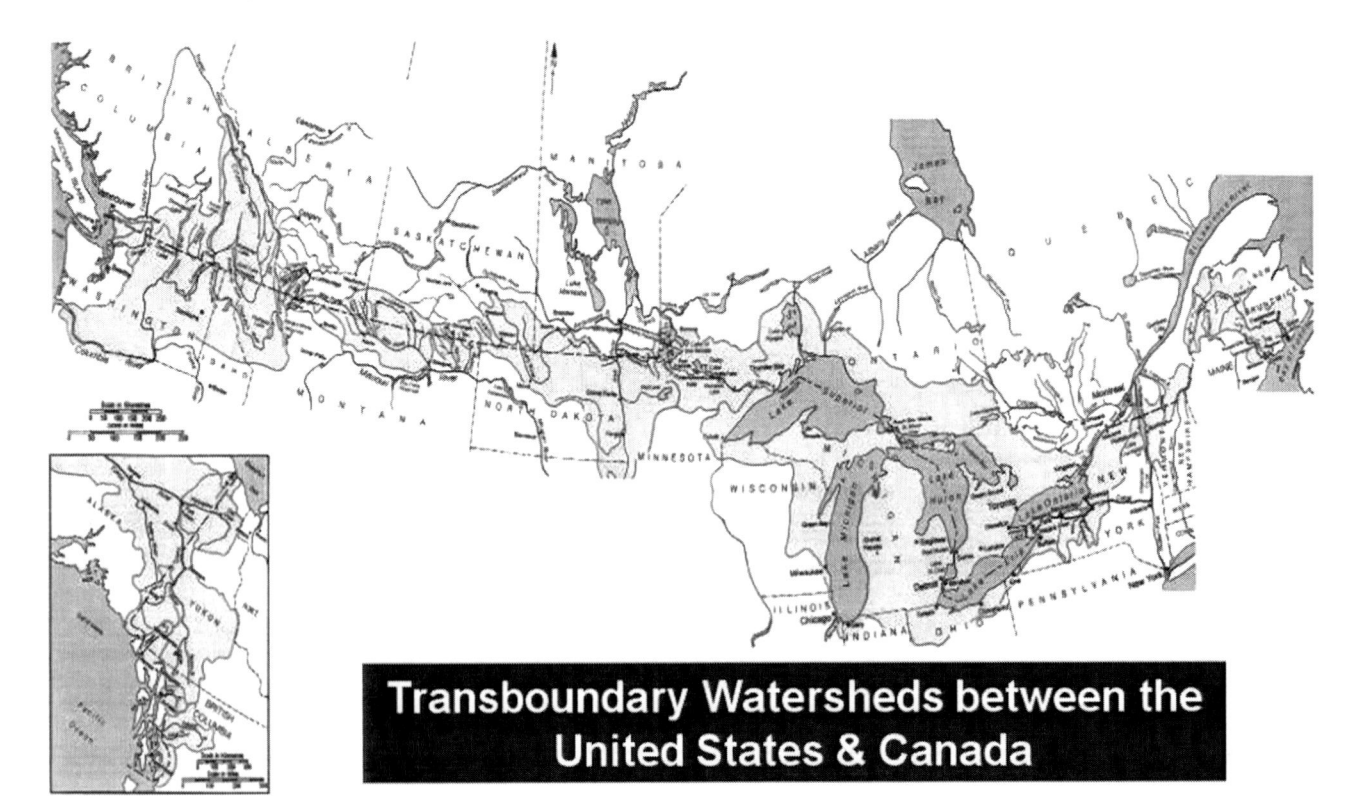

Figure 12.1 Location of boundary waters between the United States and Canada. The International Joint Commission was formed by a treaty in 1909 to resolve water disputes between the two countries. (Modified from International Joint Commission, The International Joint Commission and the boundary waters treaty Canada—United States, Windsor, Ontario, Canada, 24pp., 1990.)

The report goes on to mention the difficulty of treating ballast water:

> Pollution by water ballast constitutes a more difficult problem. There has not yet come to the notice of the commission any feasible means of purifying the rather large quantities of water which vessels while in the polluted areas of inner harbors frequently take on board for purposes of ballast, and which they afterwards discharge upon approaching their destination, often while passing water intakes. It will probably be sufficient for the present at least to control this practice by regulations designed to limit or prevent the discharge of ballast water in the neighborhood of intakes. In the event of failure of such control by regulations, more expensive and time-consuming methods of treatment will have to be developed and prescribed.

The 1918 report inspired a 1930 field study of chlorination of ballast water tanks, and the ensuing publication (Ferguson 1932) concluded that

> In general, there do not seem to be any real physical difficulties in the way of effectively chlorinating ballast water. It has been shown that ballast water tanks are usually filthy, containing accumulations of rust and other sediment; and when to these tanks is added foul harbor water it is at once apparent that pollution of otherwise uncontaminated water, particularly near municipal intakes, is quite within the realm of possibility. Further study was discontinued on this problem as it was felt that sufficient data had been gathered for present purposes and that direct pollution of Great Lakes waters by vessel sewage is a far more serious menace and one which should receive first attention, rather than the lesser menace of vessel ballast water.

No effective action was taken by governments in response to these early reports, and water quality conditions continued to deteriorate. In a 1950 report (International Joint Commission 1950), the IJC stated:

> Industrial wastes, which were of little concern in 1912, are now a major problem. The daily discharge into these boundary waters now averages more than 2 billion U.S. gallons (approximately 1¾ billion Imperial gallons). While much of this is condenser and cooling water which has not been adversely affected by its use for industrial purposes, an appreciable volume of harmful pollutants is discharged daily. These include some 13,000 pounds of phenols, 8,000 pounds of cyanides, 25,000 pounds of ammonium compounds and large quantities of oils and suspended solids of all types.

The report goes on to say

> The wastes from vessel traffic through these international waters also constitute a pollution problem. The sewage from vessels at the height of the navigation season is the equivalent of the wastes contributed from a population of 1000 in the St. Marys River area and a population of 3900 in the St. Clair-Detroit River area. In addition, disposal of garbage, bilge water and water ballast creates problems in ports and congested areas. Such disposal is particularly objectionable near water intakes and bathing beaches.

The 1918 and 1950 reports clearly showed the IJC recognized multiple pathways of potential biological pollution, including ballast water discharges. However, the focus of IJC was on industrial and urban pollution. Broader concerns about ecosystem integrity and biological pollution were not part of the organizational culture before the 1950s, although some aquatic invasive species such as sea lamprey did attract attention.

Sea lampreys (*Petromyzon marinus*) were first discovered in Lake Erie in 1921 and are believed to have arrived via the Welland Canal. From Lake Erie, lampreys quickly colonized the upper Great Lakes and developed large infestations in Lakes Michigan and Huron. The collapse of fisheries in the Great Lakes caused by sea lamprey prompted the Convention of Great Lakes Fisheries between Canada and the United States and the creation of the Great Lakes Fishery Commission (GLFC) in 1955. In addition to control of sea lamprey, the GLFC was authorized to conduct research targeted to maximize sustained productivity of fish stocks and publish scientific information to inform fisheries managers. This was a big step forward; however, aquatic invasive species were still addressed on a species-specific basis rather than addressed by taking a broader view and dealing with general pathways. Many scientists at the time recognized the risk of further introductions of invasive species through canals, ballast water, and other pathways. However, there was no clear consensus among scientists about the overall long-term environmental effects of invasive species; therefore, it was not a priority policy issue for the IJC and its advisory boards. Environmental activism of the 1960s and 1970s brought attention to the proclamation that Lake Erie was "dead" as a result of advanced eutrophication, and the Cuyahoga River was so polluted with surface oil that it caught fire. Events like these led to new environmental legislation and passage of the 1972 Great Lakes Water Quality Agreement. Ironically, these developments served to further institutionalize the focus of the IJC on persistent toxic substances. True to the objectives of the 1972 agreement, the Great Lakes Regional Office of the IJC (created by the agreement) and its advisory boards concentrated efforts on spill prevention, toxic impacts, discharge limits, point source control, and eutrophication. Along with implementing objectives of the 1972 agreement, the U.S. and Canadian governments assigned two reference studies to the IJC. The first study involved assessment of water quality in Lakes Huron and Superior, and the second study involved an investigation of pollution from land use in the Great Lakes. For the latter study, extensive work by the Pollution from Land Use Activities Reference Group (PLUARG) within the

IJC between 1971 and 1980 produced 121 scientific reports and many recommendations regarding environmental stresses associated with urban development and anthropogenic impacts to the ecosystem. Scientists recognized the contribution of runoff from impervious surfaces, soil erosion, habitat loss, loss of biodiversity, and water demands associated with human development and population growth as factors to consider in addition to toxic pollution. The 1978 revision of the Great Lakes Water Quality Agreement reflected this broader ecosystem approach, but did little to address the issue of aquatic invasive species.

ECOSYSTEM OR ECOTOXICOLOGY APPROACH?

Records of advisory board meetings of the IJC in the 1980s documented continued emphasis on long-term ecosystem effects of persistent toxic substances. In addition, there were expressed concerns about the "limited approach to water quality management" and growing support to expand the scope of IJC activities to address aquatic invasive species. By 1981, invasive alewives (*Alosa pseudoharengus*) were thought to comprise 60% of the standing fish biomass in Lake Michigan (D. Haffner, personal correspondence with acting Regional Director, IJC). This drew attention of IJC scientists at the Great Lakes regional office to an unpublished manuscript by Stanford H. Smith, which documented the historical declines of fisheries in Lake Ontario. This manuscript linked ecosystem changes to the aquatic invasive species problem and demonstrated the value of a broad approach to ecosystem management. Although the document was peer reviewed, it remained unpublished until 1995 when it was released by the GLFC as a technical report (Smith 1995). Several years later a strong case was made to adopt an ecosystem management approach in a joint report by the IJC and GLFC (International Joint Commission 1985). This report was commissioned by the IJC through the Science Advisory Board (SAB) work group on Indicators and Ecosystem Quality and highlighted integral effects of a variety of stressors including aquatic invasive species. It discussed general effects of intentional and unintentional species introductions and included sea lamprey, rainbow smelt (*Osmerus mordax*), alewife (*Alosa (Pomolobus) pseudoharengus*), pacific salmonids (*oncorhynchus* spp.), and a copepod (*Eurytemora affinis*). Discussion of the spread of the copepod *E. affinis* described how this brackish water copepod was discovered in Lake Ontario in 1958 and later spread throughout all of the Great Lakes (Anderson and Clayton 1959). The report states that the introduction of the copepod was likely "caused in some fashion by the construction of the St. Lawrence Seaway… (and)… it may have been carried upstream in the bilge or ballast water of the ships using this waterway." The IJC work group report clearly exposed the same pathway that zebra mussels (*Dreissena polymorpha*) and quagga mussels (*Dreissena rostriformis bugensis*) used to invade the Great Lakes; however, no adverse impacts from the copepod were noted. The report stated that, other than sea lamprey, the effect of these particular invaders at the time was either "marginal or unknown and facilitated by other chemical and physical ecosystem stressors." Although the 1985 work group report strongly supported a broad ecosystem approach, it did not serve as a call to action to prevent the introduction of invasive species. Through the leadership of individual advisory board members (e.g., Dr. Jack Valentine, Department of Fisheries and Oceans, Canada), IJC moved toward incorporating the ecosystem approach into the Great Lakes Water Quality Agreement. However, there were doubts about an expanded scope for IJC beyond the issue of persistent toxic substances. A broadly defined ecosystem approach was thought to be too ambiguous and difficult to implement with limited resources. Advocates of an "ecotoxicology" approach would have included aquatic invasive species as one of many routes of exposures of Great Lakes' organisms to persistent toxic substances, but would not expand the discussion beyond the scope of toxicological impacts. Some members, however, recommended a broad ecosystem management approach to address the large number of ecosystem stressors in addition to toxic pollutants. Challenges to adopting a broad ecosystem approach coupled with the uncertainty about the long-term effects of invasive species resulted in IJC reports that addressed the problem from various perspectives. The IJC's overall approach to ecosystem management as well as its reporting about the aquatic invasive species problem would soon change as the invasion of zebra mussels and quagga mussels unfolded.

CALLS TO ACTION

IJC was tasked with responsibilities related to aquatic invasive species as a result of the 1987 Protocol that amended the Great Lakes Water Quality Agreement of 1978. The protocol included a requirement for the U.S. and Canadian Coast Guards to review practices related to pollution from shipping, including studies to determine if live fish or invertebrates in ballast water discharges into the Great Lakes constituted a threat. The two Coast Guards were tasked to submit progress reports to the IJC, which then had a responsibility to address ballast water dangers and report to the two governments. This change to the agreement was prompted by the July 1987 discovery of European river ruffe (*Gymnocephalus cernua*) in Duluth Harbor (St. Louis River) in west Lake Superior. Reports from Europe about ruffe included its invasive nature, low commercial value, and potential adverse effect on lake whitefish (*Coregonus clupeaformis*), an important commercial fish species. This information prompted a debate about the feasibility to conduct a massive eradication of the ruffe in Duluth Harbor. Controversy surrounding the discovery of ruffe led to some sparse language that was inserted into Annex 6 of the Great Lakes Water Quality Agreement just before the Protocol was signed in November 1987. Coast Guards of both countries included the ruffe issue in their meeting agenda in 1988. In response to a request by GLFC staff, the Science Advisory Board and Water Quality Board of the IJC documented serious concerns about the threat of invasive species. IJC advisory boards quickly organized a workshop on "Exotic Species and the Shipping Industry," which resulted in recommendations to both IJC and GLFC to alert the U.S. and Canadian governments to the global threat posed by aquatic invasive species discharged in ballast water (Dachoda et al. 1990).

When zebra mussel populations dramatically increased in the Great Lakes and began to clog water intakes, there was a clear connection between invasive species, high economic costs, and potential harm to humans. It was the increase and rapid spread of zebra mussels that prompted the IJC to convey stronger advice regarding the impacts of invasive species to the two governments. In 1989, IJC expressed the need to "expedite controls which ensure that this source of [invasive species] contamination to the system is curtailed" and "sufficient studies have been conducted to confirm the threat that this pollution source poses to the Great Lakes and that action is required" (International Joint Commission 1989). The GLFC and IJC issued a joint report in 1990 recommending consistent U.S. and Canadian requirements for ballast water exchange, research to improve exchange and treatment technology, and action through national delegations to the International Maritime Organization to "augment and strengthen existing conventions, codes and processes" to promote a consistent approach to solve the global problem of ship-mediated introductions of aquatic invasive species (International Joint Commission and Great Lakes Fishery Commission 1990).

The 5th Biennial Report of IJC in 1990 recognized zebra mussels as a more serious problem than ruffe, described impacts of zebra mussels, and acknowledged the threat posed by untreated ballast. Biennial reports from the IJC are required by the Great Lakes Water Quality Agreement of 1978 to chronicle progress and assess the effectiveness of government programs. The report stated, "the potential for introduction of other exotic species is real and such introduction could have calamitous consequences." Although IJC recognized aquatic invasive species as a serious "biological integrity issue," it returned its attention to persistent toxic substances, zero discharge, and cleanup efforts as major themes in the next four biennial reports. For example, the 9th Biennial Report in 1998 reflected on persistent toxic substances and reaffirmed an intent to focus on water quality objectives. However, by 1999 change was underway; IJC returned to the urgent tone of earlier reports as zebra and quagga mussels were joined by other invasive species including round goby (*Neogobius melanostomus*), New Zealand mudsnail (*Potamopyrgus antipodarum*), fishhook waterflea (*Cercopagis pengoi*), and the virus viral hemorrhagic septicemia (VHS) (*Novirhabdovirus* sp.).

CATALYST FOR CHANGE

In the 10 years after the GLFC and IJC completed their joint report in 1990, there was little progress in policies to reduce the number of aquatic invasive species introduced in ballast water. Failure of the U.S. and Canadian governments to develop improved measures to exchange or treat ballast water as recommended in the 1990 report prompted the IJC to refocus its efforts and draw attention to the need for effective binational action. In 1999, IJC commissioned a white paper (Reeves 1999) and organized a workshop on exotics and public policy held during the Biennial Great Lakes Water Quality Forum in Milwaukee. This event provided the basis for the IJC's 10th Biennial Report in 2000, which contained aggressive recommendations about aquatic invasive species. The IJC highlighted the high cost of damages caused by dreissenid mussels and was critical of the lack of effective legislation, regulation, and inter-jurisdictional consistency. In addition, the 10th Biennial Report recommended stronger actions than ever before as it requested a reference (official task under the treaty funded by the U.S. and Canadian governments) to develop binational standards for ballast water discharges and recommendations to implement those standards.

INCREASED AWARENESS AND ACTION

IJC made a sustained effort in the first decade of the twenty-first century to raise public awareness about aquatic invasive species and build support for government action. After the release of the 10th Biennial Report, the Water Quality Board of the IJC issued a report that called for consistent regional standards for ballast water discharges, technology development, and contingency plans for response to newly discovered populations of aquatic invasive species, including "chemical treatment on a short-term emergency basis" as a stop-gap measure (International Joint Commission 2001). The report was well received by the public and recommendations were incorporated in the 11th and 12th Biennial Reports in 2002 and 2004, respectively. IJC repeated its request to the governments for a reference to assist with development of regional standards. However, both governments did not concur and continued to pursue national programs through their respective administrative processes. Despite this setback, IJC continued to press for action and joined the Great Lakes Panel on Aquatic Nuisance Species, the Great Lakes Commission, GLFC, and nongovernmental organizations to sponsor events to raise public awareness. The IJC sponsored workshops that highlighted different aspects of the issue at each of its Biennial Public Forums in 2003, 2005, 2007, and 2009. These workshops sparked a much publicized debate about a study that analyzed the costs and benefits of ocean vessel shipping into the Great Lakes (Taylor and Roach 2005). In particular, the IJC highlighted this study during a workshop at the June

2005 Biennial Meeting in Kingston, Ontario. The study challenged the assumption that cost benefits of ocean shipping of goods into the Great Lakes exceeded costs to the environment caused by aquatic invasive species introduced in ballast water. It reported that ocean vessels entering the Great Lakes contributed a relatively modest savings in transportation costs for users of the system. An end to ocean shipping into the Great Lakes would result in a transportation cost penalty of $54.9 million per year (US$). This cost penalty would equate to a 5.9% increase in door-to-door transportation costs for goods shipped in the Great Lakes based on 2005 estimates. Comparison of that cost penalty to the billions of dollars in costs attributed to aquatic invasive species made the point that nothing should be "off the table" to stop the introduction of aquatic invasive species into the Great Lakes. Great Lakes United (GLU), a citizen's coalition dedicated to protecting and restoring the Great Lakes, acted to publicize the debate, which resulted in a strong response from the ocean-shipping industry. These actions all served to highlight the need to accelerate government action to implement ballast water discharge regulations that are biologically effective.

While the IJC and other organizations advocated government action for many years, it was the introduction of zebra and quagga mussels and other aquatic invaders publicized in the media that resulted in widespread public support and political action. Between 1986 and 2010, zebra and quagga mussels spread throughout the Great Lakes, the Mississippi River, and as far west as California. As mussels clogged water intakes, media coverage spread the object lesson about the damage caused by aquatic invasive species. In total, the number of nonindigenous aquatic species in the Great Lakes grew to over 180 during this time period, and visible, new threats to the Great Lakes such as silver carp (*Hypophthalmichthys molitrix*), bighead carp (*H. nobilis*), and northern snakehead (*Channa argus*) attracted international media attention.

REGULATORY DISCORD

The public became impatient with implementation schedules for legislative action and Coast Guard ballast water regulations. Increased demands for action motivated passage of legislation at state levels, and states filed legal action to force the U.S. federal government to use the Clean Water Act to control the introduction of aquatic invasive species. Michigan was the first state to independently develop ballast water regulations. It instituted mandatory reporting requirements in 2001 (PA 114) and implemented a technology-based standard in 2005 (Act 451, Section 324.3112). Michigan requirements for ballast water treatment came into effect in 2007 and effectively prohibited discharge of ballast water by oceangoing ships in Michigan waters because no approved treatment systems were installed (Table 12.1). Ships ordinarily discharge ballast water when loading cargo, so a prohibition on ballast water discharge prevents ballasted ships from loading cargo. The economic impact from lost cargo capacity may vary by vessel depending on the trade, loading conditions, stability requirements, and load-line capacity. During a public meeting in January 2012, representatives of the Detroit port authority stated that in the absence of consistent regional standards, state ballast water regulations have become a barrier to exports and have created an economic hardship for Michigan ports.

California and other states successfully sued the United States Environmental Protection Agency (U.S.EPA) to regulate ballast water under provisions of the Clean Water Act. The lawsuit began in 2005, and the decision was upheld on appeal in 2008. As a result, Vessel General Permit regulations were developed and promulgated by the U.S.EPA from 2007 to 2009. These regulations incorporated the Section 401 discharge requirements set by states. Most state requirements used the International Maritime Organization (IMO)

Table 12.1 Government Agency and Corresponding Biological Standards for Ballast Water Discharge into the Great Lakes (as of February 2012)

Government Agency	Ballast Water Standard for New Vessels	Ballast Water Standard for Existing Vessels
Transport Canada	IMO D-1 and D-2	IMO D-1 and D-2
U.S. Coast Guard, U.S. EPA	BWE/BMPs	BWE/BMPs
Michigan	BMPs and discharge prohibited without approved treatment	BMPs and discharge prohibited without approved treatment
Illinois, Minnesota, Pennsylvania	IMO in 2012	IMO by 2016
Ohio	Lake Carriers-IMO in 2016; Oceangoing-IMO in 2012	IMO by 2016
New York	1000 × IMO in 2013	100 × IMO by 2013
Wisconsin	IMO by 2012	IMO by 2014

Source: Stollenwerk, J., Minnesota Control Agency, (2010 updated February, 2012). IMO, International Maritime Organization; BWE, ballast water exchange; SWF, saltwater flush; BMPs, best management practices. For definition of standards depicted by IMO, IMO D-1, and IMO D-2, see: http://globallast.imo.org/index.asp?page=mepc.htm.

Table 12.2 Status (as of January 31, 2012) of Ratification of the International Convention for the Control and Management of Ships Ballast Water and Sediments, 2004. All Countries Ratified by Accession Except for Brazil, Maldives, Spain, and Syrian Arab Republics, Which Ratified by Actual Ratification, and the Netherlands, Which Ratified by Approval. The Convention will be Implemented 12 Months after Being Ratified by 30 Nations That Account for at Least 35% of the World Merchant Shipping Tonnage. To Date, 33 Nations Representing 26.46% of Tonnage Have Ratified the Convention

Country	Date of Instrument Deposit	Country	Date of Instrument Deposit
Albania	January 15, 2009	Mexico	March 18, 2008
Antigua and Barbuda	December 19, 2008	Mongolia	September 28, 2011
Barbados	May 11, 2007	Montenegro	November 29, 2011
Brazil	April 14, 2010	The Netherlands	May 10, 2010
Canada	April 8, 2010	Nigeria	October 13, 2005
Cook Islands	February 2, 2010	Norway	March 29, 2007
Croatia	June 29, 2010	Palau	September 28, 2011
Egypt	May 18, 2007	Republic of Korea	December 10, 2009
France	September 24, 2008	Saint Kitts and Nevis	August 30, 2005
Iran	April 6, 2011	Sierra Leone	November 21, 2007
Kenya	January 14, 2008	South Africa	April 15, 2008
Kiribati	February 5, 2007	Spain	September 14, 2005
Lebanon	December 15, 2011	Sweden	November 24, 2009
Liberia	September 18, 2008	Syrian Arab Republic	September 2, 2005
Malaysia	September 27, 2010	Trinidad and Tobago	January 3, 2012
Maldives	June 22, 2005	Tuvalu	December 2, 2005
Marshall Islands	November 26, 2009		

Source: International Maritime Organization (www.imo.org/About/Conventions/StatusOfConventions/Documents/Status%20-%2013.pdf).

standards as outlined in the 2004 International Convention for the Control and Management of Ships Ballast Water & Sediments (for standards see: http://globallast.imo.org/index.asp?page=mepc.htm) as a basis for regulations, but specified more aggressive implementation schedules and smaller allowable concentrations of organisms (Table 12.1).

On an international basis, the IMO Ballast Water Convention Discharge Standard will enter into force 12 months after it is ratified by 30 nations that account for at least 35% of the world merchant shipping tonnage. As of January 31, 2012, 33 nations representing 26.46% of world merchant shipping tonnage have ratified the Convention, including Canada in April 2010 (Table 12.2). Although the United States did not ratify the Convention, final rules proposed by the U.S. Coast Guard are the same as the convention discharge standard, with provisions to implement stricter standards when practicable. At present, no uniform regional standard has been established, which has resulted in an inconsistent patchwork of state and federal ballast water regulations as of February 2012.

PREVENTING THE "NEXT ZEBRA MUSSEL" INVASION

To prevent the introduction of aquatic invasive species via all potential pathways, including ballast water, hull fouling, recreational boats, pet trade, live food, bait, ornamental plants, and aquaculture, it is important that any binational approach to regulation be consistent. A key step in preventing another dreissenid mussel–like organism from entering the Great Lakes is to harmonize state and federal ballast water regulation schemes so that an effective, uniform standard can be strictly enforced. To this end, the IJC joined the Saint Lawrence Seaway Development Corporation (SLSDC) to facilitate the Great Lakes Ballast Water Collaborative. The idea for the collaborative was conceived by representatives of IJC, SLSDC, Minnesota Sea Grant, and others at the 16th International Conference on Aquatic Invasive Species in 2009. The purpose of the collaborative was to provide a forum where available information on efforts to prevent discharges of aquatic invasive species in ballast water could be shared and evaluated in order to find universal and workable regulatory solutions. The Ballast Water Collaborative served as a clearing house for up-to-date information about ballast water treatment technology, testing, and enforcement. It also provided an important forum for dialogue between ship owners and state, provincial, and federal regulators, scientists, and nongovernment organizations. The IJC and SLSDC cohosted a series of meetings that started with the premise that ballast water treatment systems would be required in the near future, which allowed discussion to concentrate on what could be done in the near term, how to verify effectiveness of treatment, and how to certify compliance with regulatory standards. New treatment technologies and combinations of technologies were examined, as well as risks associated with ballast water transfers between ports within the Great Lakes. Efforts of the Ballast Water Collaborative resulted in an open and inclusive approach where diverse views were examined on the merits of facts and science. Importantly, the collaborative afforded opportunities to examine facts, validate opinions, build trust among parties, and establish a common understanding of key issues. The Ballast Water Collaborative is consistent with responsibilities of the IJC under the Boundary Waters Treaty to act in the best interests of the shared waters of the Great Lakes from a basis of sound science and informed decision making. To date, the collaborative shows great potential to help identify workable solutions and avoid potential conflicts between ballast water discharge requirements in the United States and Canada.

RAPID RESPONSE

With efforts invested to halt introductions of aquatic invasive species in ballast water, what should be done if prevention efforts fail and a new species is introduced? The short answer is to be ready with rapid response! The history of failures to contain dreissenids, ruffe, gobies, and other Great Lakes aquatic invasive species has underscored the need for early detection and rapid response efforts. IJC recognized the need for coordinated planning between the United States and Canada for early detection and rapid response for any new introductions and formed a rapid-response work group in 2007. The work group recommended a policy framework for consistent binational approaches to rapid response that IJC carried forward in its 15th Biennial Report in 2011. IJC then carried this effort further when it directed the work group to assess detection and response capabilities of all jurisdictions and to develop a pilot binational rapid response plan in 2012.

Clearly, problems caused by dreissenid mussels and other aquatic invasive species have proved to be a highly complex problem for Great Lakes policy makers. For example, initiatives to address problems that have reemerged in nearshore areas of the Great Lakes such as nuisance aquatic plants and benthic algae suggest a complex linkage between aquatic invasive species, nutrient cycling, and changes to substrates. This biological, chemical, and physical linkage exacerbates chronic issues like eutrophication and reverses progress made through point source controls of nutrients.

Many factors influenced the IJC to recognize the threat of aquatic invasive species as an urgent priority; however, the impact of dreissenid mussels ultimately prompted it to change. IJC committed major efforts to prevent and eliminate the spread of aquatic invasive species and to understand how invasive species contributed to complex problems in nearshore areas of the Great Lakes. As the IJC expanded its focus beyond a historical perspective, it pressed the U.S. and Canadian governments for action and addressed aspects of biological integrity just as aggressively as persistent chemical contamination of the Great Lakes ecosystem.

REFERENCES

Anderson, D.V. and D. Clayton. 1959. *Plankton in Lake Ontario*. Ontario Department of Lands and Forests, Div. Res., Phy. Sect., Phy. Note No. 1, 7pp.

Dachoda, M., A. Hamilton, and B. Bandurski. 1990. *Exotic Species and the Shipping Industry a Workshop* held February 28–March 2, 1990. Workshop summary and recommendations. International Joint Commission and Great Lakes Fishery Commission. Windsor, Ontario, Canada.

Ferguson, G. 1932. The chlorination of ballast water on Great Lakes vessels. *Public Health Reports* 47: 256–258.

International Joint Commission. 1918. Final report of the International Joint Commission on the pollution of boundary waters reference. Washington, DC.

International Joint Commission. 1950. Report of the International Joint Commission on the pollution of boundary waters. Washington, DC.

International Joint Commission. 1985. A conceptual approach for the application of biological indicators of ecosystem quality in the Great Lakes basin—A joint effort of the International Joint Commission and the Great Lakes Fisheries Commission. Report of Science Advisory Board Work Group on Indicators and Ecosystem Quality. Windsor, Ontario, Canada.

International Joint Commission. 1989. Great Lakes water quality. Report of the Great Lakes Water Quality Board to the International Joint Commission. Washington, DC, 128pp.

International Joint Commission. 1990. The International Joint Commission and the boundary waters treaty Canada—United States 1990. Windsor, Ontario, Canada, 24pp.

International Joint Commission. 2001. Alien invasive species and biological pollution of the Great Lakes basin ecosystem. Report of the Great Lakes Water Quality Board to the International Joint Commission. Windsor, Ontario, Canada, 21pp.

International Joint Commission and Great Lakes Fishery Commission. 1990. Exotic species and the shipping industry: The Great Lakes-St. Lawrence River ecosystem at risk. Report to the U.S. and Canadian governments. Washington, DC, 74pp.

Reeves, E. 1999. Exotic policy, International Joint Commission white paper on policies for the prevention of the invasion of the Great Lakes by exotic organisms. *Prepared for International Joint Commission Workshop on "Exotic Policy"*. Milwaukee, WI.

Smith, S.H. 1995. Responses of fish communities to early ecological changes in the Laurentian Great Lakes, and their relation to the invasion and establishment of the alewife and sea lamprey. GLFC Technical Report 60. Great Lakes Fishery Commission. Ann Arbor, TX.

Stollenwerk, J. 2010. Great lakes ballast water collaborative meeting report, updated February, 2012. Minnesota Pollution Control Agency. See: http://www.greatlakes-seawat.com/en/pdf/Ballast_Collaborative_Report_and_WEReports_Duluth%28Final%29.pdf

Taylor, J. and J. Roach. 2005. Ocean shipping in the Great lakes: Transportation cost increases that would result from a cessation of ocean vessel shipping. Paper presented at the *International Joint Commission Biennial Workshop*, Kingston, Ontario, Canada.

CHAPTER **13**

Eradication of Zebra Mussels (*Dreissena polymorpha*) from Millbrook Quarry, Virginia
Rapid Response in the Real World

Raymond T. Fernald and Brian T. Watson

CONTENTS

Abstract	196
Introduction	196
Background and Pre-Procurement Actions	198
Millbrook Quarry Zebra Mussel Workgroup	198
The Virginia Nonindigenous Aquatic Nuisance Species Act	199
Investigations of Millbrook Quarry	200
Emergency Procurement and Solicitation of Funds	200
Funding	200
Continuing Geochemical and Hydrologic Investigation	201
Alternative Selection and Procurement	201
The Competitive Negotiation Process	201
Millbrook Quarry RFP Evaluation Panel	201
Criteria for Evaluation of Proposals	201
Review of Proposals	201
Muriate of Potash (KCl)	201
Spectrus CT-1300 (Clamtrol)	202
Liquid CO_2	202
Negotiation and Contractor Selection	202
Other Alternatives Considered	202
No Action	203
Chlorine	203
pH Shift	203
Increasing Salinity	203
Dewatering the Quarry	203
Copper Sulfate	203
Environmental Compliance and Public Review	204
National Environmental Policy Act	204
Hydrologic and Geochemical Setting	204
Threatened or Endangered Species	206
Aquatic Wildlife	206
Streams, Rivers, Lakes, and Other Surface Waters	207
Groundwater	207
Other Environmental Considerations	207

Coastal Zone Management Act and Virginia Coastal Resources Management Program 207
Federal Insecticide, Fungicide, and Rodenticide Act (FIFRA) 207
State and Local Environmental Requirements 208
Interagency Review and Public Involvement 208
Eradication and Confirmation.. 208
Discussion 210
Appendix..212
Acknowledgments..212
References..212

ABSTRACT

In August 2002, zebra mussels (*Dreissena polymorpha*) were found in an abandoned quarry in northern Virginia (Millbrook Quarry) and reported to the Virginia Department of Game and Inland Fisheries. The quarry was and remains a popular SCUBA recreational and training dive site in the Washington, DC, metropolitan area. Upon confirmation of the infestation, the Department embarked on a course of "rapid response" to provide the first successful eradication of a large open-water population of this invasive aquatic nuisance species. This chapter narrates the course taken by the Department and its partner agencies; explores the statutory, regulatory, procedural, and technical hurdles encountered; and offers recommendations for enhancing the ability of government agencies to effectively deal with such invasions. In this instance, "rapid response" involved nearly 3½ years of planning, investigation, negotiation, and project design, followed by 3 weeks of project implementation. Based on our experience, we conclude that the following steps could dramatically shorten the time required to respond to future invasions: (1) prepared interagency rapid response plans that address agency responsibilities and authority; (2) clear guidelines, protocols, and decision matrices for project design and implementation; (3) streamlined procurement and environmental review; and (4) stable funding for invasive species management.

INTRODUCTION

In late August 2002, the Virginia Department of Game and Inland Fisheries (VDGIF; see Appendix for a list of acronyms) received an unconfirmed report of a zebra mussel (*Dreissena polymorpha*) infestation in Millbrook Quarry, in western Prince William County, Virginia. Although zebra mussels had been discovered and removed from boats before launch at Smith Mountain Lake in 1993 and in Norfolk in 1997 (VDGIF, unpubl. data), an open-water infestation had never before been documented in Virginia. Mussel specimens were collected from Millbrook Quarry and forwarded to Dr. Richard Neves (Virginia Cooperative Fish and Wildlife Research Unit [VCFWRU] at Virginia Polytechnic Institute and State University [VPISU]) and Mike Pinder (VDGIF) for identification. On September 3, 2002, these specimens were confirmed as zebra mussels, thus documenting the first zebra mussel infestation in the state of Virginia.

Millbrook Quarry (Figure 13.1) is located in Prince William County in the northern portion of Virginia (Figure 13.2) and was opened in 1947 to produce stone for construction of Virginia Highway 55 (VDMME 2003). Gooch et al. (1960) reported that in 1958 the quarry was about 122 m wide, 183 m long, and 30 m deep. In 2003, a survey and bathymetric analysis by VDGIF revealed a maximum depth of 30 m, surface area of approximately 0.049 km^2, and volume of approximately 680,000 m^3 (VDGIF 2003a, unpubl. data) (Figure 13.3). Records indicate the quarry had been inactive since at least February 1963 (VDMME 2003).

The Dive Shop in Fairfax, Virginia, began using the quarry for SCUBA diving in the early 1970s and first leased the quarry as a training and recreational dive site in 1978 (J. Wall, pers. comm.). Through *The Dive Shop*, the quarry was accessed regularly by more than a dozen other dive establishments in the northern Virginia/Washington, DC, metropolitan region. Use of the quarry occurred on weekends only and primarily from April through mid-November (http://www.thediveshop-va.com/aboutus/millbrook/). The quarry had a single gated entrance and offered limited Saturday camping for divers, but there was no running water or electricity on site. No public non-diving recreational use was permitted.

Figure 13.1 View of Millbrook Quarry looking southwest from the northeastern shoreline. The surface buoys mark submerged dive platforms.

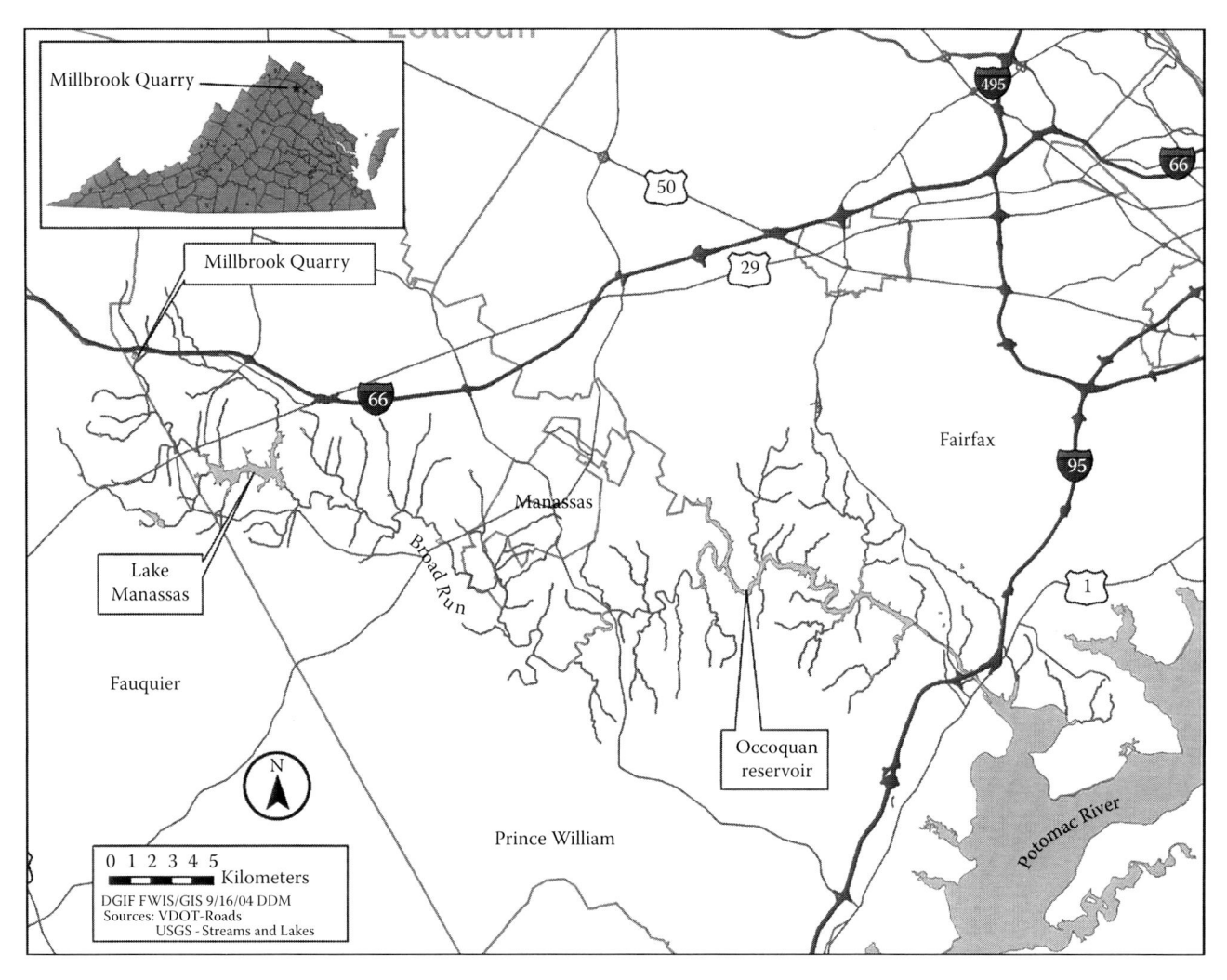

Figure 13.2 Regional location of Millbrook Quarry, Prince William County, in northern Virginia (insert). The quarry is located near tributaries of the Occoquan River, which flows into the Potomac River.

Millbrook Quarry is separated from Broad Run (Figure 13.4), a perennial tributary of the Occoquan River, by a 60–90 m wide berm (Figure 13.5). Broad Run flows into Lake Manassas about 9 km downstream of the quarry, and this reservoir serves as the primary water supply for the City of Manassas and neighboring communities (Figure 13.2). Just downstream of Lake Manassas is the Occoquan Reservoir, a primary water supply for over 1 million people in northern Virginia (Figure 13.2). If zebra mussels were to infest Occoquan Reservoir, Fairfax Water (formerly the Fairfax County Water Authority) estimated that it would incur a $2–$4 million capital outlay for chemical feed facilities and $500,000–$850,000 per year for chemicals and system maintenance (J. B. Hedges, pers. comm.). The City of Manassas would likely have incurred similar expenses to treat zebra mussels at its facility on Lake Manassas. Furthermore, if zebra mussels were to spread throughout Virginia, water intakes and other industrial facilities would have been vulnerable.

In addition, many rare and declining species of freshwater mussels (Unionidae) could suffer significant losses. Impacts to native freshwater mussel communities could be devastating since there are 81 species of mussels found in Virginia, including 39 that are listed as state or federally endangered or threatened and 62 that are identified as Species of Greatest Conservation Need in Virginia's Wildlife Action Plan (VDGIF 2005a, 2010, 2012). In the Virginia reaches of the upper Tennessee River basin alone, there are approximately 55 species of freshwater mussels that occurred historically, of which at least 46 are extant and 33 are state or federally listed as endangered or threatened (VDGIF 2010, 2012).

Unfortunately, given the proximity of Millbrook Quarry to Broad Run, it was very likely that zebra mussels in the quarry would spread to surrounding waters. Broad Run was a likely pathway for escape, since this stream historically flooded the bank separating it from Millbrook Quarry (1972, Hurricane Agnes). In addition, the unintentional transport

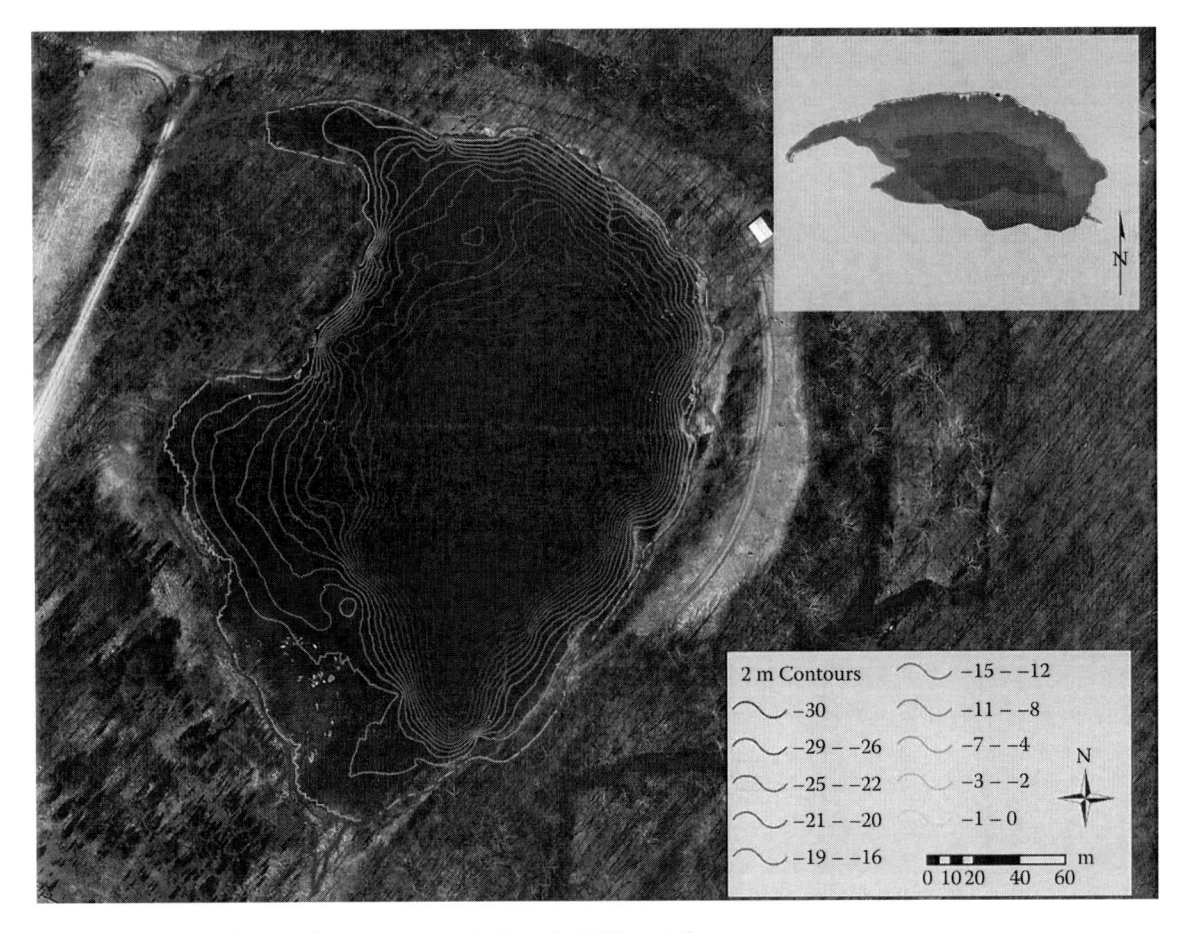

Figure 13.3 Bathymetry and three-dimensional graphic (insert) of Millbrook Quarry.

Figure 13.4 View of Broad Run, a tributary of the Occoquan River located adjacent to Millbrook Quarry.

of veligers or juvenile mussels by divers from the quarry to other state waters was considered likely. Conversely, while a groundwater connection between Millbrook Quarry and Broad Run was suspected, there was no direct surface outflow or inflow, effectively eliminating most natural dispersal modes and rendering this population relatively isolated. Also, since this was the only known zebra mussel infestation in Virginia, its eradication would completely remove this invasive species from the state, a factor of significant sociopolitical, if not geographical, importance.

BACKGROUND AND PRE-PROCUREMENT ACTIONS

Millbrook Quarry Zebra Mussel Workgroup

Upon confirmation of the presence of zebra mussels in the quarry, VDGIF began a preliminary assessment of the threat to native wildlife communities and a review of its legal jurisdiction to intervene. Subsequently, VDGIF organized a meeting of numerous federal, state, and local agencies and organizations, which resulted in the establishment of the Millbrook Quarry Zebra Mussel Workgroup (for a list of participating agencies, see USFWS 2005). Membership in the workgroup was open to all interests, and a number of parties were encouraged to participate, especially local governments and agencies with pertinent technical expertise or with responsibility for wildlife, environmental protection, public health, and potable water.

In October 2002, the workgroup held its first meeting to review the situation and determine potential courses of

Figure 13.5 Aerial view of Millbrook Quarry and surrounding roads, rivers, and streams. Note the land separation (60–90 m wide berm) between the quarry and Broad Run, a tributary of the Occoquan River.

action. At that time, all parties agreed that eradication of the population, if possible, should be the ultimate goal, despite recognition that eradication of a large open-water population of zebra mussels had never been attempted. The only applicable case was the documented eradication of another dreissenid mussel (black-striped mussel *Mytilopsis* sp.) from several double-locked marinas in Darwin Harbour, Australia, through treatment with copper sulfate, sodium hypochlorite, and calcium hypochlorite (Ferguson 2000, Bax et al. 2002).

The workgroup quickly concluded that, before an appropriate eradication plan could be developed, substantial technical information needed to be collected about the quarry and degree of infestation. Information needs were grouped into three categories: (1) water chemistry, (2) hydrogeologic and physical parameters of the quarry, and (3) characteristics of the zebra mussel population. Workgroup agencies (primarily VDGIF, VDMME, and the Occoquan Watershed Monitoring Laboratory [OWML] of VPISU) initially sought to conduct the necessary onsite data collection in mid-November 2002. That effort was unsuccessful because of access constraints imposed by the property and dive shop owners, which stemmed from the lack of statutory authority of government agencies to compel landowner assistance or cooperation to address invasive species issues. The workgroup then met again in late-November 2002 to review potential eradication options that had been developed by VDGIF staff via literature review and consultation with zebra mussel experts around the United States. Control options included adding chlorine, acid (decrease pH), salt/brine (increase salinity), copper sulfate, potassium, or Clamtrol (a commercially available molluscicide). Dewatering the entire quarry also was considered.

The Virginia Nonindigenous Aquatic Nuisance Species Act

Partly in response to the Millbrook Quarry infestation and lack of statutory authority for state agencies to address invasive species on private lands, legislation was introduced and passed by the 2003 Virginia General Assembly authorizing VDGIF to respond to such invasive species incidents. The resulting act (§29.1-571-577) designated zebra mussels and quagga mussels (*Dreissena rostriformis bugensis*) as nonindigenous aquatic nuisance species (NANS) and authorized VDGIF to designate additional NANS by regulation. The act also authorized VDGIF to implement measures and conduct operations to suppress, control, eradicate, prevent, or retard the spread of any such species and to obtain warrants to search private or public property in Virginia and eradicate or seize any NANS found. Furthermore, the act established civil penalties (up to $25,000 per incident) for violating the act or for knowingly obstructing the VDGIF from carrying out its duties under the act and also established liability for costs of investigation, control, and eradication measures incurred by any state or local agency or authority. A prescient provision of the act was to prohibit VDGIF from using traditional sources of income (e.g., hunting and fishing license revenues) for such activities. Immediately upon passage of the act in March 2003, the landowner and dive shop owner granted VDGIF and cooperating agencies access to the property to pursue eradication of the infestation.

Investigations of Millbrook Quarry

In April 2003, upon implementation of decontamination protocols for all fieldwork at Millbrook Quarry (USFWS 2005), staff from key agencies conducted the prescribed fieldwork. Primary tasks undertaken by VDGIF included: (1) construction of maps of the quarry surface and bathymetry and determination of quarry volume; (2) visual (scuba) assessment of quarry walls, fractures, and fissures; (3) confirmation of the distribution of zebra mussels throughout the quarry; (4) assessment of the general density and layering of zebra mussels in the quarry; and (5) surveys of Broad Run and Lake Manassas to confirm that zebra mussels were isolated and present only in the quarry. Specific tasks completed by OWML and VDMME included: (1) determination of surface drainage and watershed of the quarry; (2) assessment of regional groundwater elevations, geochemistry, and flow from existing data and investigations of local wells; (3) review of past geological surveys, quarry records, and studies of the Millbrook Formation and associated faults; (4) survey of surface water elevations in Millbrook Quarry, Broad Run, and Catletts Branch; (5) analysis of base and event-related surface flows and elevations in Millbrook Quarry, Broad Run, and Catletts Branch; and (6) analysis of base and event-related water chemistry of Broad Run, Catletts Branch, Millbrook Quarry, and five local wells.

Data analyses from these investigations were completed in June 2003 (OWML 2003, VDGIF 2003b, VDMME 2003). Analysis confirmed widespread occurrence of zebra mussels of several year classes throughout the quarry, from near the surface to depths exceeding 21 m (see Fernald and Watson 2013). Also, analysis established a baseline hydrologic and geochemical characterization of the quarry, groundwater, and Broad Run. The workgroup met again in July 2003 to review the data and analyses and to reconsider treatment options in light of this information. Treatment with chlorine, acid, or salt/brine, or dewatering the quarry were considered unlikely to succeed or otherwise considered undesirable because of environmental concerns, technical infeasibility, logistics, or expense. Likewise, copper sulfate treatment was disfavored because of potential environmental concerns. Treatments with Clamtrol or potassium, however, were considered to potentially offer acceptable solutions, with caveats that selection of a treatment would require formal evaluation via the Virginia procurement process, that VDGIF probably would be the purchasing agency, and that no funds had been identified to pay for the eradication.

Emergency Procurement and Solicitation of Funds

In late July 2003, VDGIF initiated formal efforts to obtain funding for eradication of the infestation and, at the same time, sought eradication proposals from potential vendors via the state's Emergency Procurement Solicitation process. Unfortunately, confirmed offers of funding were not forthcoming, and, in late October 2003, the emergency solicitation was canceled due to lack of funding to proceed. This action ended review and evaluation of the proposals that had been submitted in response to the Emergency Procurement Solicitation. Furthermore, VDGIF staff was instructed not to pursue further solicitation or evaluation of proposals until funding for the eradication was secured.

Funding

Based on preliminary review of the emergency procurement responses, the eradication effort was anticipated to cost between $150,000 and $800,000. Because such an eradication had never been attempted and a wide range of technologies were under consideration, there were no established methods, treatments, or models by which to accurately forecast project cost. Therefore, $800,000 was established as the amount of funding needed for eradication. A wide variety of funding options was explored, including numerous federal agencies; state, regional, and local governments likely to be adversely impacted by presence of zebra mussels in Virginia's waters; and prominent Virginia industries that would be similarly impacted. The goal of securing $800,000 in funding commitments was reached in September 2004, thereby enabling VDGIF to begin formal development of the Request for Proposals (RFP). Primary funding for the eradication included $300,000 from the U.S. Department of Agriculture (USDA) as a Wildlife Habitat Incentive Program (WHIP) grant, $100,000 from the U.S. Fish and Wildlife Service (USFWS) as a State and Tribal Wildlife Grant (SWG), a $200,000 contribution and an additional pledge of up to $163,000 from Fairfax Water, and $37,000 in contributions from Virginia industries and local governments.

Continuing Geochemical and Hydrologic Investigation

As suggested in the initial report (VDMME 2003), VDGIF focused on raising funds to pursue the eradication project while OWML and VDMME continued investigations of the hydrology and geochemistry of the quarry. These efforts provided further documentation of the relationships between the quarry, regional groundwater, and Broad Run and a greater understanding of the potential environmental implications of pursuing eradication of the infestation (Lassetter et al. 2004, 2005, OWML 2004).

ALTERNATIVE SELECTION AND PROCUREMENT

The Competitive Negotiation Process

Procurement of goods and services by public agencies often entails complex protocols. In Virginia, Competitive Negotiation is a preferred procurement method used for goods and nonprofessional services when it is not practicable or fiscally advantageous to use competitive sealed bids (Code of Virginia, §§ 2.2-4301 & 2.2-4303C). This process has the advantage of flexibility for describing in general terms what is being sought and the factors to be used in evaluating responses. It offers the opportunity, through negotiation, to change the content of an offer and pricing after opening.

This method of procurement requires issuance of an RFP that describes in general terms the requirement, factors that will be used to evaluate the proposal, the Commonwealth General Terms and Conditions, and any special conditions including unique capabilities or qualifications that will be required. Responses to the RFP are held unopened until the date and time specified for their receipt. After the closing date of the RFP, all valid responses are evaluated on the basis of the criteria set forth in the RFP, using the scoring weights previously determined. Pricing constitutes 25% of the evaluation, and the other specified criteria determine 75% of the scoring for each proposal. The score for each proposal is expressed as points received based on a maximum possible score of 100 points.

If appropriate, negotiations are then conducted with each of the responders selected. Negotiation allows modification of proposals, including price. Offers and counteroffers may be made as many times with each responder as is necessary to secure a reasonable contract. After negotiations have been conducted with each of the selected responders, the agency posts an Intent to Award to the responder which, in its opinion, has made the best proposal.

Millbrook Quarry RFP Evaluation Panel

In anticipation of the breadth of expertise required to properly evaluate proposals submitted in response to the RFP, and recognizing the wide range of technical, public health, environmental, and socioeconomic issues to be considered, an interagency RFP Evaluation Panel of eight members representing seven agencies was assembled: VDGIF (two members), VDMME, OWML, VCFWRU, the Virginia Department of Environmental Quality (VDEQ), the Virginia Department of Health (VDH), and Fairfax Water. The inclusion of primary stakeholders, academic researchers, and Virginia regulatory agencies on the panel was intended to ensure broad interagency support during subsequent environmental and permit reviews.

Criteria for Evaluation of Proposals

As described earlier, the RFP evaluation process requires establishment of formal criteria by which each panel member must score each proposal submitted. The panel recognized several primary objectives that must be satisfied by any acceptable proposal: (1) the process must achieve 100% mortality of zebra (and quagga) mussels in Millbrook Quarry; (2) the process must not significantly or unacceptably affect nontarget wildlife, the environment, or human health; and (3) to the greatest possible extent, the process must not involve any adverse off-site environmental impacts. These concerns guided establishment of the formal evaluation criteria as they were presented in the RFP released in late-November 2004 (VDGIF 2004). The deadline for submission of proposals was January 10, 2005.

Review of Proposals

Three proposals were submitted in response to the RFP, and each offered a different treatment. Upon receipt of the proposals, all panel members reviewed each proposal for completeness and clarity and submitted to the panel chairman any questions or issues that required clarification. A list of questions and issues that required clarification was then compiled and sent to each vendor. A teleconference regarding these issues was held with each vendor in March 2005 and, upon receipt of written responses from each vendor, each panel member individually scored the proposals according to the established criteria and forwarded those preliminary scores to the panel chairman. Proposals were then formally reviewed and scored at a panel meeting in April 2005. Control treatments of the three proposals were as follows:

Muriate of Potash (KCl)

The first proposal involved infusing the entire water column of Millbrook Quarry with potassium by pumping 131,000 kg of muriate of potash (potassium chloride—MOP 98%), in aqueous solution, from land-based storage tanks into the quarry. A floating supply line would extend to a small workboat outfitted with a specially designed diffuser assembly. Treatment was to occur within zones determined by depth and by presence of thermoclines within the water column.

Although the exact mode of action was unknown, potassium evidently kills mussels by interfering with the ability to transfer oxygen across gill tissue, resulting in asphyxia (Aquatic Sciences Inc. 1997). To ensure lethal concentrations of potassium throughout the water column and yet minimize likelihood of "hotspots" within the quarry, a "target" potassium concentration of 100 mg/L (or parts per million: ppm) throughout the water column was adopted. A concentration of 50 ppm was assigned as the minimum concentration to initiate bioassays (McMahon et al. 1994), though long-term exposure to 30–40 ppm was deemed sufficient to kill 100% of all zebra mussels of all life stages in the quarry (Aquatic Sciences L.P. 2005). At these concentrations, potassium was considered to pose no risks to human health nor any risks of substantial harm to non-molluscan aquatic wildlife, vegetation, or terrestrial wildlife (Waller et al. 1993, Aquatic Sciences L.P. 2005). Further, the proposal suggested that potassium would provide long-term (estimated at 33 years by the contractor) protection of Millbrook Quarry against future infestation by zebra mussels.

Potassium concentrations would be monitored at various depths along transects established throughout the quarry, both during and after application of the treatment solution. Mortality of zebra mussels would be confirmed by bioassay with zebra mussels imported to the quarry, and by direct and video confirmation of zebra mussel mortality by SCUBA divers.

Very little, if any, land disturbance would be required, as the staging area and setup would occur within the previously disturbed uplands surrounding the quarry. No disturbance of substrate or bottom sediments within the quarry would be necessary and no disturbance of land in or adjacent to Broad Run would occur, though Broad Run would be monitored throughout the project for groundwater infiltration of potassium from the quarry.

Spectrus CT-1300 (Clamtrol)

The second proposal was to treat the entire quarry with the molluscicide Spectrus CT-1300 (Clamtrol). The target concentration of 8–10 ppm would require approximately 7,571 L of molluscicide. The chemical would be applied by a boat-based system consisting of a chemical injection distribution header and hoses. An air diffuser system would be operated in water deeper than 3 m to facilitate complete mixing.

Clamtrol is effective at killing zebra mussels but, at the proposed concentration, the vendor anticipated that most zooplankton, macroinvertebrates, and fishes would also be killed. The evaluation panel expressed concerns that Clamtrol may not be 100% effective in the quarry due to the possibility of inadequate mixing of the water column and the short life span of the chemical. Also, there were concerns about monitoring protocols and the stated warranty (i.e., how it would be determined whether live zebra mussels discovered after treatment had survived the treatment or had been subsequently introduced).

Liquid CO_2

The third proposal entailed injecting 102,000–114,000 kg of liquid CO_2 into the quarry water column to lower the dissolved oxygen concentration to 4.0 ppm or less. A system of pipes would be constructed to remove water from the quarry, inject the CO_2, and return the treated water to the bottom of the quarry through a sparging hose.

The panel was skeptical that the proposed drop in dissolved oxygen concentration to 4 ppm would be sufficient to achieve 100% eradication of the zebra mussel population in the quarry and questioned whether that concentration could even be maintained in an open water system. The panel had other reservations about the proposal, including the lack of a warranty to eradicate the zebra mussel population and that no monitoring of zebra mussel mortality or detailed confirmation of the resultant dissolved oxygen concentration would be provided by the prospective contractor.

Negotiation and Contractor Selection

By unanimous decision, the panel recommended that VDGIF negotiate with the vendors that submitted proposals for treatment with potassium and treatment with Clamtrol. The panel also unanimously recommended that the vendor that proposed treatment with liquid CO_2 be dropped from further consideration because the proposal lacked technical merit, assurances, a reasonable chance of success, and adequate documentation.

Initial negotiations were conducted with the two selected vendors in May 2005, and revised proposals were submitted to VDGIF by each prospective vendor in June 2005. At a series of meetings in June–July 2005, the panel formally rescored the revised proposals, and final scores were submitted to VDGIF's Director of Purchasing in late-July 2005.

Upon review of the revised proposals, the panel unanimously selected the muriate of potash treatment proposal as the preferred alternative, expressing confidence that use of potassium offered the greatest likelihood of successfully eradicating the zebra mussel population in Millbrook Quarry with virtually no significant adverse environmental impacts. Furthermore, the panel believed this treatment would provide long-term protection against reinfestation of the quarry by zebra mussels. A contract ($365,069) was awarded in August 2005, and the detailed chronological procurement documentation, as mandated by Virginia procurement law, was included as Appendix D of the final Environmental Assessment (EA) for the project (USFWS 2005).

Other Alternatives Considered

As described earlier, VDGIF conducted a literature review and interagency consultations to develop a preliminary set of eradication alternatives for review by the workgroup. The following alternatives were considered generally less desirable because of environmental concerns, technical infeasibility, logistics, or expense. No proposals were submitted for any of these potential alternatives.

No Action

As required under the National Environmental Policy Act (NEPA), a "no action" alternative was considered, whereupon zebra mussels would continue to thrive in Millbrook Quarry and pose a significant risk of spread throughout the Occoquan Watershed and the rest of the state. It was highly unlikely that the infestation could be contained forever, especially since the quarry was used by the public as a recreational and training dive site. Even if diving was prohibited or restricted, mandatory decontamination protocols were enforced by the quarry operator, or the quarry was purchased by a public institution for "quarantine" purposes, zebra mussels would likely spread via human trespass, flooding, movement of contaminated or encrusted wildlife from the quarry to Broad Run, or perhaps even groundwater transport of veligers. The long-term impacts anticipated by continued zebra mussel infestation of Millbrook Quarry, and their eventual escape, were deemed to be environmentally and economically unacceptable.

Chlorine

Chlorine, an oxidizing agent, is among the most widely used compounds for zebra mussel control in North America and Europe (Waller et al. 1993, McMahon et al. 1994, Aquatic Sciences L.P. 2005). Chlorine kills zebra mussels through asphyxiation and limited glycolysis over a prolonged period of exposure. In various applications, adult zebra mussels are treated at concentrations of 0.3–2.0 ppm; veligers and post-veliger larvae are more sensitive to chlorine than adults (McMahon et al. 1994). Intermittent or short-term treatment of juveniles and adults at higher concentrations can be less effective because mussels close their valves to avoid exposure of soft tissues (McMahon et al. 1994). Primary concerns with chlorine as a treatment are its toxicity to nontarget organisms and the production of carcinogenic trihalomethanes (Waller et al. 1993, Aquatic Sciences L.P. 2005). Maintaining an adequate chlorine concentration over the required exposure period is problematic: cooler temperatures prevent "boiling off" of chlorine, but also increase the exposure time required.

pH Shift

Zebra mussels typically inhabit water bodies with pH levels of approximately 6.6–8.5 (Benson and Raikow 2011). Addition of acid to the water column could lower the pH of Millbrook Quarry and kill the zebra mussels through shock and ionic loss of calcium, sodium, and potassium. There is evidence, however, that after several days at pH 5.5–6.0, adults can adapt and may succumb only at pH levels below 5.2 (Heath 2003). Concerns that made this treatment undesirable included lack of clarity regarding the pH level required for 100% mortality, exposure period required, effectiveness at varying temperatures, impacts on nontarget organisms, and ability to achieve the required pH reduction at all depths and locations throughout the quarry. High buffering capacity of Millbrook Quarry and associated groundwater and regulations prohibiting acid discharges were also important considerations that led to dismissal of this alternative.

Increasing Salinity

Elevated salinity kills zebra mussels through ionic tissue imbalance. Populations in North America generally tolerate salinity up to 4 ppt (parts per thousand), while European populations tolerate salinities up to about 10 ppt (Benson and Raikow 2011). Increasing salinity is not widely used as a treatment or control agent for zebra mussels. Though it would be effective at all temperatures, this treatment option is relatively expensive, would significantly affect nontarget organisms, and would pose risk of saltwater encroachment into groundwater and wells.

Dewatering the Quarry

Aerial exposure of adult zebra mussels would result in 100% mortality through asphyxiation. The required exposure time, however, could vary from hours to weeks depending on air temperature, humidity, and thickness of the zebra mussel clumps. This treatment option clearly would impact nontarget aquatic organisms. A primary deterrent for this treatment was that it may not have been feasible to dewater the quarry due to groundwater inflow: the quarry likely would simply refill from groundwater as it was being drained. Additionally, the water would have to be pumped away from the quarry without exposing Broad Run to zebra mussel adults or veligers.

Copper Sulfate

Copper, a nonoxidizing agent, is toxic to aquatic organisms including algae, mussels, and other invertebrates. The cupric ion is available in a variety of salts (e.g., copper sulfate—$CuSO_4$) and primarily affects fishes and aquatic invertebrates through rapid binding of copper to gill membranes, which causes damage and interferes with osmoregulatory processes (USEPA 2008). The amount of cupric ion needed to induce mortality in aquatic animals is dependent on a number of water quality parameters including pH, alkalinity, and dissolved organic carbon (USEPA 2008). McMahon et al. (1994) reported 100% mortality of zebra mussels in 24 h at an ionic copper concentration of 5 ppm. Treatment with copper sulfate to achieve 1 ppm copper was used in an attempt to eradicate zebra mussels at Offutt Air

Force Base Lake near Omaha, Nebraska (URS Group Inc. 2009, Schainost 2010). Copper sulfate also was used in conjunction with chlorine in the eradication of black-striped mussels from Darwin Harbor, Australia (Bax et al. 2002). Though copper is relatively benign to most fish species at concentrations needed to kill zebra mussels, a significant fish kill occurred following treatment of Offutt Air Force Base Lake (URS Group Inc. 2009). Impacts on nontarget species and on drinking water quality would be of concern if copper sulfate was applied at Millbrook Quarry.

ENVIRONMENTAL COMPLIANCE AND PUBLIC REVIEW

During preliminary consideration of treatments, it was recognized that any eradication proposal would be subject to extensive environmental review by federal, state, and local agencies and by various non-government interests. Upon confirmation of federal funding for the eradication, it was acknowledged that compliance with NEPA and with Section 7 of the Endangered Species Act (ESA) would be required. The EA prepared for NEPA review also would serve as the Virginia Environmental Impact Assessment (VEIA) submitted to VDEQ for interagency review pursuant to the Virginia Environmental Impact Report (VEIR) law (§10.1–1188). Because Millbrook Quarry was located within Virginia's Coastal Resources Management Zone, compliance with the federal Coastal Zone Management Act (CZMA) and the Virginia Coastal Resources Management Program (VCP) was necessary. Further, because potassium chloride was not a labeled pesticide in Virginia, a permit would be needed from the Virginia Department of Agriculture and Consumer Services (VDACS) for the intended use, and an emergency exemption would be needed from the U.S. Environmental Protection Agency (USEPA) under Section 18 of the Federal Insecticide, Fungicide, and Rodenticide Act (FIFRA). In addition, several other state or local permits, waivers, or agency statements of compliance would be necessary for the intended eradication, including those related to sediment and erosion control, pollutant discharge, handling of hazardous materials, and wetland disturbance. Table 13.1 provides a synopsis of environmental permits, approvals, and reviews conducted for the eradication, and the primary roles of each agency involved.

National Environmental Policy Act

The normal procedure for compliance with NEPA is to prepare a draft EA for federal and public review prior to selection of a treatment option. However, in the case of mussel eradication from Millbrook Quarry, which vendors would submit proposals for consideration, or what treatment options would be submitted as formal proposals, could not be anticipated. Further, the state procurement process mandated confidentiality throughout the evaluation and selection process. Therefore, an EA for this effort could not be developed until the proposals were formally submitted and reviewed, and the procurement process concluded. After consultation with the U.S. Fish and Wildlife Service, which served as the lead federal agency for NEPA compliance, the following statement was included in the RFP to specifically mandate compliance with NEPA:

> The contractor shall eradicate zebra and quagga mussels at the Millbrook Quarry, Prince William County, Virginia, within the contracted time period and shall provide all goods, services and expertise necessary to complete this task…. The treatment process and associated design, construction and monitoring efforts must comply with all NEPA guidelines and with applicable NEPA compliance and reporting requirements. Documentation of such compliance is mandatory as a condition of federal grants providing funds for this effort. All proposals must document how the vendor would comply with this condition.

For purposes of impact analysis, the project area included Millbrook Quarry, the surrounding, previously disturbed uplands, Broad Run, and the groundwater and private wells to the south of the quarry. Because all treatment options considered (apart from the no-action alternative) would entail construction of a shore-based pump station, storage tanks, and appurtenant facilities, the anticipated construction impacts would be similar, albeit inconsequential, for each treatment option. In addition to review of treatment options, discussion of the following topics was required in the NEPA environmental assessment (USFWS 2005).

Hydrologic and Geochemical Setting

Assessments of the hydrologic and geological setting of Millbrook Quarry and Broad Run were of critical importance because virtually any treatment option to eradicate zebra mussels from the quarry would affect and be affected by the hydrology and water chemistry of the quarry. Also, such assessments were critical to safeguard against impacting regional groundwater and wells and surface waters of Broad Run, Lake Manassas, and other water bodies. These assessments were conducted by VDMME and OWML as described earlier. In their presentation at the 2004 Virginia Water Resources Symposium (Lassetter et al. 2004), the following summary statements regarding Millbrook Quarry were provided:

1. Geologic studies indicate that calcareous bedrock in Millbrook Quarry has a very high capacity for acid neutralization.
2. North/northeast-trending faults and fractures and north/northwest-striking bedrock strata in the quarry area are very likely pathways for southward-directed groundwater flow.

Table 13.1 Environmental Permits, Approvals, and Reviews Required for Millbrook Quarry Zebra Mussel Eradication, and Agency Roles

Permit, Approval, or Resource/Issue Reviewed	Federal	State	Local	Permit, Approval, or Concurrence	Confirmation of No-Permit-Needed	Approval of On-site Measures	Review of EA/VEIA	Post-Treatment Report
NEPA Environmental Assessment	USFWS USDA			X			X X	X X
FIFRA Section 18 Special Exemption	USEPA	VDACS		X X[a]				X X
VA Environmental Impact Assessment		VSA VDEQ		X			X X	
Threatened/Endangered Wildlife (Section 7 ESA, VESA)	USFWS	VDGIF		X X[a]			X X	
Threatened/Endangered Plants/Insects (Section 7 ESA)	USFWS	VDACS VDCR		X X[a]			X X X	
Coastal Resources (CZMA/VCP)		VDEQ		X[a]			X	
Chesapeake Bay Preservation Act		VDCR		X			X	
Historic/archeological resources		VDHR		X[a]			X	
NANS Importation Permit		VDGIF		X			X	
Wetlands and Aquatic Resources	USACOE	VDEQ VMRC			X X X		X X	
Water quality (CWA §401 cert.)		VDEQ			X[a]		X	
Water pollutant discharge (NPDES/VPDES)		VDEQ VDCR			X[a] X[a]		X X	
Sediment and erosion control and stormwater management		VDCR	PWC		X	X	X X	
Hazardous materials and solid wastes		VDEQ	COM		X	X	X X	
Drinking water resources		VDH			X		X	
Air quality impacts		VDEQ			X		X	
Emergency response procedures			PWC			X	X	
Natural Heritage Resources		VDCR					X	

[a] Federal regulatory program shared with, implemented by, or delegated to state agency.

3. The north/northeast-trending section of Broad Run just south of the quarry probably reflects the prevailing bedrock fracture pattern and, depending upon fracture interconnectivity, may be conducting water as leakage from the quarry.
4. Analysis of stream base flow conditions in Broad Run during August 2004 indicates an upper bound for groundwater inflow ~0.65 cfs (~0.0184 m^3/s).
5. If a significant hydrologic connection exists, it would likely occur southeast and south of the quarry, where the water elevation in Broad Run is ~5 to 10 ft (~1.5 to 3.0 m) below that of the quarry.
6. The present water level in Millbrook Quarry likely reflects equilibrium conditions with the regional, unconfined groundwater system, that moves southward through the quarry area.

7. Water wells to the north, west, and east of the quarry are located up gradient and should not be affected by changes in water quality that may occur in the quarry; wells to the south and southeast are potentially exposed to impacts related to pumping and/or changes in water quality in the quarry.
8. Water balance calculations indicate an annual rate of subsurface outflow to be ~3.6% of the volume of water in the quarry; ~6.5M gal/year (~24,600 m^3/year).
9. The major ion compositions of groundwater and surface water in the study area reflect the different geologic characteristics of the Culpeper Basin and Blue Ridge rocks.
10. The combined geochemical and hydrologic data suggest that leakage from the quarry undergoes mixing with regional groundwater before emerging in Broad Run.
11. Stable isotope compositions of oxygen and hydrogen (analytical results received in October 2004) may provide the means to confirm and refine the water balance calculations and interactions between groundwater and surface waters.

In essence, these assessments established that water chemistry of Millbrook Quarry largely reflected regional groundwater, that groundwater flowed through the quarry from north to south, and that leakage from the quarry to the groundwater occurred at a rate of about 3%–4% of quarry volume per year. As postulated (see result 11), the stable isotope analysis of October 2004 provided further support of the water balance calculations between Millbrook Quarry, regional groundwater, and Broad Run (Lassetter et al. 2005). Benthic sediments in the quarry were derived from historic quarry operations, surface water runoff from the very small watershed (0.121 km^2, excluding the quarry itself) (VDMME 2003), and biogenic deposition since the quarry's abandonment. Through interagency consultation with each prospective contractor, it was determined that the quarry sediments would have little to no effect on treatment outcome.

Threatened or Endangered Species

As a component of the EA, and to comply with Section 7 of the Endangered Species Act pursuant to grants funding the eradication effort, databases of the Department's Virginia Fish and Wildlife Information System were queried and it was determined that two federally listed species occurred within a 3.2 km radius of Millbrook Quarry: the bald eagle (*Haliaeetus leucocephalus*), and a native mussel, the dwarf wedgemussel (*Alasmidonta heterodon*). The database confirmed that no bald eagle nests were known to occur on property around the quarry and, as presented earlier, no long-term alteration of existing terrestrial habitat would occur. Therefore, none of the treatment options would be expected to impact bald eagles. Similarly, there was no record of occurrence of the dwarf wedgemussel in Broad Run. Because another native mussel, the brook floater (*Alasmidonta varicosa*), was listed as an endangered species under the Virginia Endangered Species Act (VESA—4VAC15-20-130) and was known to occur in Broad Run downstream of Lake Manassas, VDGIF experts conducted a mussel survey of Broad Run from just upstream of Millbrook Quarry to within 0.8 km of the confluence of Broad Run into Lake Manassas (VDGIF 2005b). No dwarf wedgemussels or brook floaters were discovered during the survey, so it was determined that these listed species would not be impacted by the proposed project.

Aquatic Wildlife

Aquatic wildlife in the quarry included largemouth bass (*Micropterus salmoides*) and sunfish, catfish, turtles, mollusks, and crayfish as observed by VDGIF staff during site visits and scuba dives. Because the quarry is a man-made body of water, all aquatic life in the quarry had been introduced (i.e., by humans, other wildlife, or from Broad Run during flood events). The selected treatment option (muriate of potash) was expected to kill only zebra mussels and exotic Asian clams (*Corbicula fluminea*) because toxicity data indicated that the proposed final concentration was not lethal to nontarget organisms other than freshwater mollusks. For example, the threshold effect concentration (TEC) for potassium is 272.6 ppm for the zooplankter *Ceriodaphnia* and 426.7 ppm for fathead minnows (*Pimephales promelas*) (Aquatic Sciences Inc. 1997).

The primary concern regarding off-site impacts to aquatic life was the potential for potassium to infiltrate Broad Run via groundwater and thus offer a potential threat to native mussels in that stream. Below Lake Manassas, VDGIF databases indicated that at least six native mussel species inhabited Broad Run, including the brook floater and the yellow lance (*Elliptio lanceolata*), which were species designated as endangered and of special concern, respectively, by the state. No records, however, were available for the reach upstream of Lake Manassas to the quarry. Therefore, VDGIF staff conducted a mussel survey of Broad Run downstream of Millbrook Quarry in August 2005 (VDGIF 2005b). A significant population of the common eastern elliptio (*Elliptio complanata*) and lesser numbers of lance mussels (*Elliptio* spp.) were documented in the downstream section of the surveyed reach, but numbers of mussels decreased significantly closer to the quarry. No individuals of the brook floater or yellow lance were observed during the survey.

Toxicity data were not available for the mussel species found in Broad Run, but elevated potassium levels in the range of 10–15 ppm have been reported as lethal to other freshwater mussel species over a several-week period (Imley 1973). Based on flow and isotope data gathered during the pre-procurement studies, VDMME estimated that Millbrook Quarry contributed approximately 25% of the groundwater inflow to Broad Run below the quarry and that groundwater typically constituted approximately 10% of the surface flow of Broad Run (Lassetter et al. 2005). Therefore, if target concentrations in Millbrook Quarry were 100 ppm of potassium, under normal flow conditions the potassium addition from Millbrook Quarry to Broad Run would be on the order of 2–3 ppm (the background potassium concentration in Broad Run is approximately 1 ppm). Under extreme low flow conditions (i.e., no flow into Broad Run except for groundwater contributions), the highest potassium concentration that could occur in Broad Run would be approximately 25 ppm if the concentration of potassium in Millbrook Quarry was 100 ppm. Few mussels were documented to occur within the first kilometer of Broad Run downstream of the quarry, and potassium concentrations would decrease through dilution further downstream; therefore, insignificant impacts to native freshwater mussels in Broad Run were anticipated. Available toxicity data indicated that potassium concentrations anticipated in Broad Run would have no adverse impact on other aquatic taxa (Aquatic Sciences Inc. 1997).

Streams, Rivers, Lakes, and Other Surface Waters

All treatment options were designed to alter the water chemistry of Millbrook Quarry and, if successful, would result in the total mortality and ultimate decomposition of the entire organic mass of the zebra mussel population. The impact of the decomposition process, though unmeasured, was anticipated to be minimal when compared to the environmental risks posed by the zebra mussels.

Via the selected treatment, elevated levels of potassium were projected to remain in the quarry for over 30 years, effectively protecting the quarry from reinfestation with zebra mussels for that period. As discussed earlier, however, the anticipated post-treatment potassium level in Broad Run under normal flow conditions would be approximately 3–4 ppm, and under extreme low flow conditions, the highest potassium concentrations that should temporarily occur in Broad Run would be approximately 25 ppm. Given the distance from Millbrook Quarry to Lake Manassas (Figure 13.2) and the size of Lake Manassas, no adverse impacts to Lake Manassas or other downstream waters were anticipated.

Groundwater

Hydrological studies by VDMME found that groundwater entered Millbrook Quarry from the north, east, and west and exited the quarry to the south. Therefore, any chemical treatment would primarily affect groundwater to the south of the quarry, but impacts were anticipated to be negligible and benign. Migration into potable-water wells was not anticipated to be a problem for any of the treatment options because there were very few wells in the vicinity of Millbrook Quarry to the south, and none of the proposed treatment options were known to pose any human health impacts at the anticipated treatment concentrations.

The national secondary (non-mandatory) drinking water standard for chlorides is 250 ppm (USEPA 2012), and the VDEQ has adopted this as a state standard for public water supplies. By comparison, the final chloride concentration in Millbrook Quarry after treatment was anticipated to be approximately 90 ppm, well below this standard. There is no federal or state water quality standard for potassium, but potassium chloride is widely used as the primary alternative to sodium chloride in home water softeners, and many health benefits are attributed to diets rich in potassium. The 2010 Dietary Guidelines for Americans issued by the U.S. Department of Agriculture and the U.S. Department of Health and Human Services (2010) recommend an adult daily potassium intake of at least 4,700 mg.

Other Environmental Considerations

As required under NEPA and VEIR, numerous other potentially harmful impacts of treatment options were considered, including impacts on terrestrial wildlife, wetlands, natural areas and unique or important vegetation, cultural or historic resources, and recreational or socioeconomic resources. As documented in the EA (USFWS 2005), none of the treatment options were considered to pose significant risks to these resources.

Coastal Zone Management Act and Virginia Coastal Resources Management Program

Pursuant to the federal CZMA and applicable Virginia Coastal Resources Management Program regulations, VDEQ was required to certify that the proposed eradication complied with the VCP enforceable policies. Furthermore, Prince William County had designated the entire county as a Chesapeake Bay Preservation Area under the Chesapeake Bay Preservation Act; thus, documentation was needed to confirm the treatment was in compliance with the general performance criteria pursuant to that statute and implementing regulations. The Virginia Department of Conservation and Recreation (VDCR) Division of Chesapeake Bay Local Assistance concurred that the treatment was consistent with the Chesapeake Bay Preservation Act and associated regulations.

Federal Insecticide, Fungicide, and Rodenticide Act (FIFRA)

Because the selected treatment of infusing the water column with potassium chloride was not a registered pesticide use, an application was needed for an emergency exemption (quarantine) under Section 18 of FIFRA. The formal application explained the risk posed by zebra mussels, detailed the proposed use of potassium chloride and anticipated results, and examined the need to use an unregistered pesticide in lieu of existing registered molluscicides or alternative control practices. The application required approval by the Virginia Department of Agriculture and Consumer Services (VDACS) and then approval by the USEPA. The most difficult issue was convincing USEPA of the need to use potassium, an unregistered product, rather than a registered molluscicide. Approval of our FIFRA exemption, though not as complex or lengthy a process as compliance with NEPA, nonetheless required nearly 3 months to achieve.

State and Local Environmental Requirements

In addition to compliance with state regulatory programs discussed earlier, VDEQ reviewed the draft EA and determined that no permits under the Virginia Pollution Discharge Elimination System, Virginia Water Protection Permit program, or other regulatory programs administered by VDEQ were required. VDEQ also stipulated that the proposed eradication project was in compliance with water program procedures, regulations, and laws under its administration.

The Virginia Department of Health was also consulted, and it concurred that the proposed concentration of potassium chloride posed no human health risks, that impacts on water quality in Lake Manassas would be minimal, and that impacts on the quality of raw water entering the Lake Manassas Water Treatment Plant would be insignificant. VDH further concurred that no impact on the quality of raw water entering other public water supplies was anticipated.

VDGIF and VDEQ together reviewed appropriate state and federal databases to confirm that there were no solid or hazardous waste sites likely to impact or be impacted by the proposed project and stipulated that any soil suspected of contamination or wastes generated by the project would be tested and disposed of in accordance with applicable federal, state, and local laws and regulations.

Interagency Review and Public Involvement

The draft EA was submitted to VDEQ for coordination of interagency and public review and for determination of consistency with the Virginia Coastal Resources Management Program. Notice of availability of the draft EA and of a 30 day public comment period was posted on VDEQ and VDGIF websites, a press release was issued by VDGIF to announce the availability of the draft EA and the scheduled public information meeting, and a notice of availability of the draft EA was published in the Washington Post. The draft EA was available for viewing at VDEQ headquarters in Richmond and available for viewing or downloading on VDGIF's website.

In addition, all landowners of parcels adjacent to Millbrook Quarry and Broad Run from the northern limits of the quarry southward to the convergence of Broad Run into Lake Manassas were identified from the Prince William County and Fauquier County internet sites depicting tax records and ownership of land parcels in the project area. Each landowner was mailed a letter notifying them of the project, of the availability of the draft EA, and of available methods for submission of questions or comments regarding the project to VDEQ. Each owner of a potable-water well that could be potentially affected was visited at their home, both to discuss the anticipated project and to secure permission and access for pre- and post-eradication testing of their well water.

Though several members of the press indicated plans to attend the public meeting, no members of the public or media attended, and no opposition or concerns regarding the project were expressed. Relatively few public or agency comments, and no significant objections, were submitted in response to the public notice. Primary concerns expressed by commenting agencies related to hazardous material spill containment, sediment and erosion control, hazardous waste management, and air quality.

Upon completion of the VDEQ's public and interagency review of the draft EA and certification of consistency with Virginia's Coastal Resources Management Program, responses were provided to all submitted questions and concerns, after which approvals of the final EA and other environmental documents were secured from USFWS, USEPA, VDACS, VDCR, VDEQ, and Virginia's Secretary of Administration (VSA). The VSA approved the project on December 1, 2005; a Finding of No Significant Impact (FONSI) was issued by the U.S. Fish and Wildlife Service on December 2, 2005; and the FIFRA special quarantine exemption was approved on January 20, 2006. Thus, after nearly 3½ years of site evaluation, fundraising, review of treatment options, and pre-eradication environmental compliance, final approval was received to proceed with the eradication (Figure 13.6).

ERADICATION AND CONFIRMATION

To kill the zebra mussels through exposure to potassium, the entire quarry was injected with approximately 659 m^3 of a 12% potassium solution (131,000 kg MOP 98%) over a 3 week period from January 31 to February 17, 2006. The solution was mixed offsite at a subcontractor's facility, delivered each Monday through Friday to the site by tanker truck, and pumped into land-based storage tanks (Figure 13.7). Each day's supply of solution (approximately 43.9 m^3) was then pumped through a floating supply line to a workboat outfitted with a specially designed diffuser manifold on its bow (Figure 13.8). The diffuser manifold fed solution through a set of interchangeable hoses ranging from 1.5 to 15.2 m in length. During the first week of treatment, the solution was discharged via 3 m hoses and sprayed on the water surface, though we planned to use longer discharge hoses during the second and third weeks of treatment.

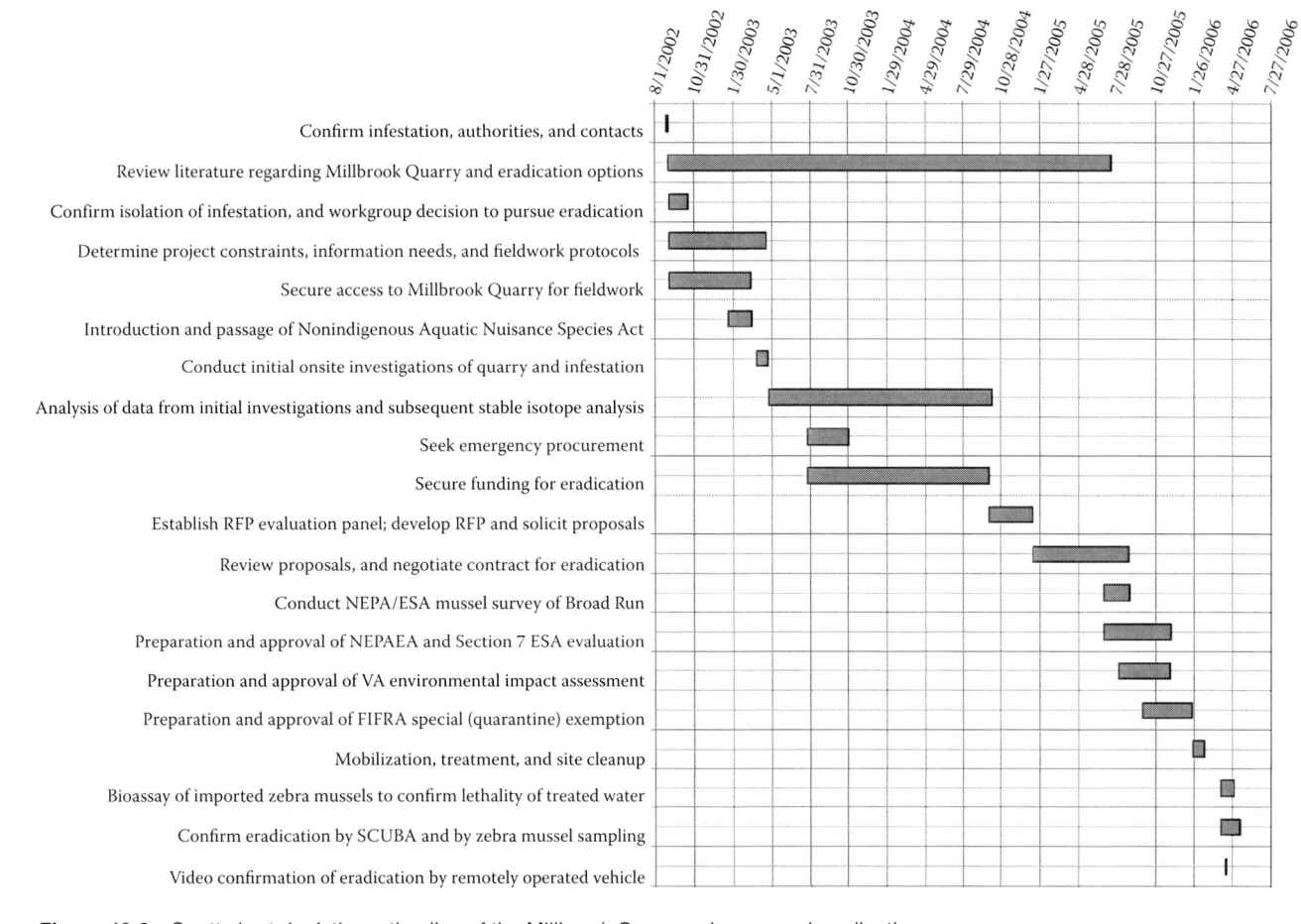

ERADICATION OF ZEBRA MUSSELS (*DREISSENA POLYMORPHA*) FROM MILLBROOK QUARRY, VIRGINIA

Figure 13.6 Gantt chart depicting a timeline of the Millbrook Quarry zebra mussel eradication.

Figure 13.7 Storage tanks and delivery tanker (insert) used to transport and store a 12% solution of potassium used to eradicate zebra mussels in Millbrook Quarry, Virginia, over a 3 week period from January 31 to February 17, 2006.

Figure 13.8 Workboat outfitted with a diffuser manifold and floating supply line (line not visible in photograph) used to distribute a 12% solution of potassium to eradicate zebra mussels in Millbrook Quarry, Virginia, over a 3 week period from January 31 to February 17, 2006.

After each 5 days (Monday through Friday) of treatment, the contractor measured potassium concentrations at 59–67 points along seven horizontal transects across the quarry, with the fixed sampling depth for each point ranging from 0.5 to 29.0 m. Potassium concentrations were also measured in Broad Run. The target concentration for the quarry was 100 ppm; far below the level that would invoke environmental or human health concerns, but more than twice the minimum concentration needed to kill zebra mussels. To our surprise, after the first week of treatment, all by spray and discharge into the top 3 m of the water column, measurements indicated virtually complete mixing of the potassium throughout the water column. Hence, the solution was applied the same way in the second and third weeks. After 3 weeks of treatment, further measurements revealed that potassium concentrations ranged from 98 to 115 ppm throughout the water column, with no apparent potassium leakage from the quarry into adjacent waters (Aquatic Sciences L.P. 2006, Fernald and Watson 2013).

Several weeks after treatment was completed, four separate methods were used to confirm that the treatment was successful. First, over a thousand zebra mussels were scraped from shallow (<1 m) subsurface rocks at numerous sites around the quarry during informal assessments, and no live mussels were found. Second, VDGIF SCUBA divers who had documented the extent of the infestation during pre-eradication assessments conducted another visual inspection of the quarry and observed no live mussels. Third, the contractor conducted an extensive video survey with a robotic camera and documented that all observed mussels were dead (Fernald and Watson 2013). Fourth, 80 mesh bags containing 100 live zebra mussels each (imported for this purpose) were placed at various locations and depths throughout the quarry, thus exposing the mussels to the treated quarry water. Mortality among the 100-mussel samples averaged 60% after 17 days of exposure to the treated quarry water, and mortality was 100% after 31 days of exposure (Aquatic Sciences L.P. 2006). In contrast, none of the 100 zebra mussels held in untreated water drawn from Broad Run died after 17 days of on-site monitoring. Other aquatic wildlife including turtles, fishes, aquatic insects, and snails appeared unaffected and continued to thrive in the quarry.

Follow-up assessments of Millbrook Quarry in August 2006 and August 2007 revealed average potassium concentrations of 109.5 and 76.3 ppm on the two sampling dates, respectively (Aquatic Sciences L.P. 2007). Monthly monitoring by OWML from February 2006 through January 2008 indicated a gradual decline in potassium concentration from about 93 to 70 ppm (OWML 2008). Though concentrations measured by the two contractors differed, both studies confirmed that, while remaining lethal to mussels, potassium concentrations in Millbrook Quarry declined somewhat during the 2 years after treatment, possibly due to changes in water chemistry and biological uptake in the quarry (Aquatic Sciences L.P. 2007). OWML (2008) also reported increased potassium concentrations (maximum of 4.15 ppm) in Broad Run during drought conditions experienced in summer and fall 2007, though they reported that potassium concentrations in Broad Run and potable wells generally remained low and near the detection limit (1.0 ppm) throughout their study. The highest post-treatment potassium concentration measured in a potable well downstream of the quarry was 1.72 ppm (OWML 2008). It was suggested that monitoring of potassium concentrations in the quarry be continued so that risks of reinfestation by zebra mussels can be evaluated over the long term (Aquatic Sciences L.P. 2007).

DISCUSSION

In many respects, the successful eradication of zebra mussels from Millbrook Quarry, Virginia, reflected the nature of the infestation and the setting of the quarry itself. The quarry was essentially isolated from other surface waters, and groundwater connections were judged insufficient to significantly dilute the potassium below the desired concentration or to result in downstream contamination. Further, there was no uncontrolled public use or access to the site; that is, all fieldwork including treatment and post-treatment assessment could occur on weekdays when there was no public use or occur during the winter when no diving was scheduled. Finally, there were no sensitive or imperiled native species that precluded use of the selected method of eradication. Even so, the eradication effort revealed strengths and weaknesses in the 3½ year approval process that may prove useful in future efforts to eradicate invasive species.

Once zebra mussels were confirmed in the quarry, a number of important decisions were made that led to a successful outcome. The first decision proved critical to the success of the eradication effort: rather than accept the presence of zebra mussels in Virginia as inevitable and insurmountable, the decision was made to evaluate all reasonable measures for eradication. The second decision proved equally important: recognition that the eradication effort needed input from a variety of sources. This led to formation of a technical workgroup comprised of experts from various state and federal agencies and local interests. The workgroup served three major objectives: (1) to enlist technical expertise in public health, water chemistry, and geochemical hydrology; (2) to garner project support from landowners, non-governmental conservation interests, potentially affected corporations, and from local, state, and federal agencies, many of whom would ultimately provide critical public support for the eradication;

and (3) to enlist the support of non-government activists, which led to passage of the Virginia Nonindigenous Aquatic Nuisance Species Act. The third major decision that led to eradication success again evolved around the recognized need for external partners. The evaluation panel for the RFP consisted of representatives from different agencies and of experts within specific fields. The evaluation panel served to: (1) ensure that all technical, public health, and environmental issues were fully considered; (2) maintain focus and engagement of essential project partners through a sometimes arduous and frustrating path to project selection and approval; and (3) integrate the primary political and regulatory agencies that ultimately would review the project proposal into the decision process, thereby ensuring (to the extent possible) that the proposal would be acceptable to those agencies during their regulatory review. Another important factor that led to eradication success was the use of the competitive negotiation process. The process was not just a procedure to simply attain the most convenient or lowest-cost contract for the eradication attempt, but the process itself was instrumental in developing a final project that maximized chances of success.

Even so, significant obstacles were encountered during the 3½ year project, a few of which may be instructional for dealing with similar situations in the future. First and foremost, and as recognized in many existing invasive species management or rapid response plans, it is imperative that responsibility and legal authority to address invasive species issues be clearly established in appropriate legislation and in agency strategic/operational plans. Second, to the extent possible, interagency "rapid response" plans should establish guidelines and step-by-step procedures for decision making and agency actions to address such environmental issues, thereby avoiding the "design as you build" approach used in the eradication of zebra mussels from Millbrook Quarry. Third, efficient and streamlined procurement and environmental review procedures (state and federal) for responding to invasive species emergencies could greatly reduce the time required for responding to future invasions or discoveries. Finally, availability of staff and funds to address invasive species issues is critical to program or project implementation and ultimately will determine success or failure.

Despite advanced planning, however, compliance with environmental laws and the unique nature of any eradication effort will require project coordination and planning that likely will be time consuming. For example, planning for the 2008 attempt to eradicate zebra mussels from Offutt Air Force Base Lake via treatment with copper sulfate took about 2½ years (URS Group Inc. 2009, Schainost 2010). The planning effort entailed establishment of a project-specific zebra mussel working group, preliminary field investigations and consideration of alternatives, preparation and review of an EA, and securing of a USEPA Section 24(c) Special Local Need Label (URS Group Inc. 2009). Similarly, in November 2010, zebra mussels were discovered in Lake Zorinsky, a 1.03 km^2 lake constructed by the U.S. Army Corps of Engineers (USACOE, Corps) and managed by the City of Omaha, Nebraska. The 10 m deep lake was partially drawn down (−6 m) in winter to kill zebra mussels by exposure to desiccation and freezing temperatures. That eradication project, developed as a cooperative effort of the USACOE, the City of Omaha, and the Nebraska Game and Parks Commission, also entailed development of an EA (USACOE et al. 2010). The Corps was able to develop the EA and issue their FONSI in less than 2 months, however, and the drawdown was implemented from December 2010 through July 2011 (K. Decker, Nebraska Invasive Species Project; pers. comm.). Both of these case studies illustrate that site-specific considerations must be addressed in designing projects to eradicate invasive species.

In remarkable contrast to these three (including Millbrook Quarry) eradication projects in the United States, the eradication of black-striped mussels in a water body in Australia was conducted in less than 1 month from discovery to completion (Bax et al. 2002). In that case, the Northern Territory Government and Australian national agencies adopted a "whole-of-government" approach that facilitated immediate mobilization of numerous agencies, passage of enabling regulations in 1 day, and initiation of chemical treatment only 4 days after discovery of the infestation (Bax et al. 2002). Though such impressively "rapid response" may not be feasible in the United States under most circumstances, development of rapid response protocols in pursuit of such goals is encouraged.

In summary, the successful eradication of zebra mussels from Millbrook Quarry, Virginia, reflected a fortuitous and nearly ideal setting for treatment success, a firm resolve to attempt what many considered unlikely or impossible, a decision to include as many technical and political experts as was warranted, and a commitment to meaningful and productive public involvement throughout the planning process. "Rapid response" in this real-world effort involved nearly 3½ years of planning, investigation, negotiation, and project design, followed by 3 weeks of project implementation. The following could dramatically shorten the time required to respond to future invasions: preparation of interagency rapid response plans that address agency responsibilities and authority; guidelines, protocols, and decision matrices for project design and implementation; streamlined procurement and environmental review; and enhanced funding for invasive species management.

APPENDIX

LIST OF ACRONYMS

COM	City of Manassas
CWA	Clean Water Act
CZMA	Coastal Zone Management Act of 1972
EA	Environmental Assessment
ESA	Endangered Species Act
FIFRA	Federal Insecticide, Fungicide, and Rodenticide Act
FONSI	Finding of No Significant Impact
MOP	Muriate of Potash
NANS	Nonindigenous Aquatic Nuisance Species
NEPA	National Environmental Policy Act
NPDES	National Pollutant Discharge Elimination System
OWML	Occoquan Watershed Monitoring Laboratory of VPISU
PWC	Prince William County
RFP	Request for Proposals
SWG	State and Tribal Wildlife Grant
TEC	Threshold Effect Concentration
USACOE	United States Army Corps of Engineers
USDA	United States Department of Agriculture
USEPA	United States Environmental Protection Agency
USFWS	United States Fish and Wildlife Service
VAC	Virginia Administrative Code
VCFWRU	Virginia Cooperative Fish and Wildlife Research Unit at VPISU
VCP	Virginia Coastal Resources Management Program
VDACS	Virginia Department of Agriculture and Consumer Services
VDCR	Virginia Department of Conservation and Recreation
VDEQ	Virginia Department of Environmental Quality
VDGIF	Virginia Department of Game and Inland Fisheries
VDH	Virginia Department of Health
VDHR	Virginia Department of Historic Resources
VDMME	Virginia Department of Mines, Minerals, and Energy
VEIA	Virginia Environmental Impact Assessment
VEIR	Virginia Environmental Impact Report
VESA	Virginia Endangered Species Act
VMRC	Virginia Marine Resources Commission
VPDES	Virginia Pollutant Discharge Elimination System
VPISU	Virginia Polytechnic Institute and State University (Virginia Tech)
VSA	Virginia Secretary of Administration
WHIP	Wildlife Habitat Incentive Program

ACKNOWLEDGMENTS

First and foremost, we acknowledge the Millbrook Quarry Zebra Mussel Workgroup, whose members independently and collectively generated public and institutional support, provided technical and political expertise, spurred legislative action, and ultimately secured funding for the eradication. Likewise, the Millbrook Quarry RFP Evaluation Panel, including (in addition to the authors) Mr. Rick Browder of VDEQ, Ms. Jamie Bain Hedges of Fairfax Water, Mr. William Lassetter, Jr., of VDMME, Dr. Richard Neves of VCFWRU, Mr. Harold Post of OWML, and Dr. Khizar Wasti of VDH, contributed their expertise during review of the formal eradication proposals and collectively guided our treatment selection. We consulted several noted experts on zebra mussels and their eradication during preliminary consideration of treatment options: Dr. Robert McMahon, Ms. Susan Jerrine Nichols, Mr. Charles O'Neill, Jr., and Mr. Don Schloesser graciously provided their time and expertise to advise us in our quest. Mr. Marshall Trammel of VDACS was instrumental in securing our FIFRA exemption from USEPA. Staffs from VDMME and OWML conducted the hydrologic, geochemical, and water chemistry investigations throughout the project, including post-treatment monitoring. Many persons from VDGIF conducted or assisted with various phases of the project including our Department administrators, GIS/FWIS section, fisheries and wildlife biologists, Purchasing Office, Resource Education Division, and Law Enforcement Division including the Dive Team and SLAP Team. Mr. Mark Miller, Ms. Karee Miller, Mr. John Wall, and Mr. Pete Swinzow provided access to Millbrook Quarry to conduct the eradication and monitoring. This project could never have been implemented without the financial support of USDA, USFWS, Fairfax Water, Prince William County, the City of Manassas, and Dominion Resources, Inc. Finally, we acknowledge and greatly appreciate the efficiency and expertise of Mr. Dan Butts and his crew from Aquatic Sciences, L.P., who conducted the eradication, bioassay, robotic video confirmation of success, and subsequent water chemistry monitoring.

REFERENCES

Aquatic Sciences, Inc. 1997. Ontario Hydro baseline toxicity testing of potash using standard acute and chronic methods: ASI Project E9015. In *Eradication of zebra mussels at Millbrook Quarry, Prince William County, Virginia. Proposal M20065 submitted to the Virginia Department of Game and Inland Fisheries in response to RFP 00375-352.* Orchard Park, NY: Aquatic Sciences L.P.

Aquatic Sciences L.P. 2005. Eradication of zebra mussels at Millbrook Quarry, Prince William County, Virginia. Proposal M20065 submitted to the Virginia Department of Game and Inland Fisheries in response to RFP 00375-352. Orchard Park, NY: Aquatic Sciences L.P.

Aquatic Sciences L.P. 2006. Eradication of zebra and quagga mussels at Millbrook Quarry, Prince William County, Virginia. Final report submitted to the Virginia Department of Game and Inland Fisheries. Orchard Park, NY: Aquatic Sciences L.P.

Aquatic Sciences L.P. 2007. August 2007 water quality and chemistry results: follow up to the 2006 zebra mussel eradication program, Millbrook Quarry, Haymarket, Virginia. Report submitted to the Virginia Department of Game and Inland Fisheries. Orchard Park, NY: Aquatic Sciences L.P.

Bax, N., K. Hayes, A. Marshall, D. Parry, and R. Thresher. 2002. Man-made marinas as sheltered islands for alien marine organisms; establishment and eradication of an alien invasive marine species. In *Turning the Tide: The Eradication of Invasive Species*, eds. C. R. Veitch, M. N. Clout, pp. 26–39. Cambridge, U.K.: IUCN SSC.

Benson, A. J. and D. Raikow. 2011. *Dreissena polymorpha*. USGS Nonindigenous Aquatic Species Database, Gainesville, FL. http://nas.er.usgs.gov/queries/FactSheet.aspx?speciesID = 5.

Ferguson, R. 2000. *The Effectiveness of Australia's Response to the Black-Striped Mussel Incursion in Darwin, Australia*. Report of the Marine Pest Incursion Management Workshop, Carlton Hotel, pp. 27–28 August 1999, Darwin, Australia: Department of Environmental and Heritage.

Fernald, R. T., and B. T. Watson. 2013. Video Clip 1: A visual documentation of the eradication of zebra mussels (*Dreissena polymorpha*) from Millbrook Quarry, Virginia. In *Quagga and Zebra Mussels: Biology, Impacts, and Control*. 2nd Edn., T. F. Nalepa and D. W. Schloesser, eds., p. 747. Boca Raton, FL: CRC Press.

Gooch, E. O., R. S. Wood, and W.T. Parrott. 1960. Sources of aggregate used in Virginia highway construction. Virginia Division of Mineral Resources Report 1.

Heath, R .T. 1993. Zebra mussel migration to lakes and reservoirs. Report FS-058. Kent, OH: Ohio Sea Grant College Program.

Imley, M. M. 1973. Effects of potassium on survival and distribution of freshwater mussels. *Malacologia* 12: 97–113.

Lassetter, W. Jr., R. Sobeck, Jr., and B. T. Watson. 2004. *Geochemical and Hydrologic Investigations of Ground Water and Surface Water Interactions at Millbrook Quarry, Prince William County, Virginia: Implications for Zebra Mussel Control*. Presentation at the Virginia Water Research Symposium on October 6, 2004. Blacksburg, VA.

Lassetter, W. Jr., R. Sobeck, Jr., and B. T. Watson. 2005. *Update to: Geochemical and Hydrologic Investigations of Ground Water and Surface Water Interactions at Millbrook Quarry, Prince William County, Virginia: Implications for Zebra Mussel Control*. Presentation to the Millbrook Quarry RFP Evaluation Panel on June 28, 2005. Richmond, VA.

McMahon, R., T. A. Ussery, and M. Clarke. 1994. Review of zebra mussel control methods. Technical Note ZMR-2-14. Vicksburg, MS: U.S. Army Engineer Waterways Experiment Station.

Occoquan Watershed Monitoring Laboratory. 2003. Chemistry data from Millbrook Quarry, nearby streams, and wells. Presentation and report to the Virginia Department of Game and Inland Fisheries and to the Millbrook Quarry Zebra Mussel Workgroup on July 22, 2003. Fairfax, VA.

Occoquan Watershed Monitoring Laboratory. 2004. Chemistry data from Millbrook Quarry, nearby streams, and wells. Unpublished data submitted to the Virginia Department of Game and Inland Fisheries, Richmond, VA.

Occoquan Watershed Monitoring Laboratory. 2008. Millbrook Quarry final report. Report submitted to the Virginia Department of Game and Inland Fisheries, Richmond, VA.

Schainost, S. 2010. Zebra mussels in a Nebraska lake. PowerPoint Presentation. Nebraska Game and Parks Commission. http://snr.unl.edu/invasives/documents/NebraskasZebraMusselInfestationandEradicationatLakeOffutt.pdf (accessed March 9, 2012).

U.S. Army Corps of Engineers, the City of Omaha, and the Nebraska Game and Parks Commission. 2010. Finding of no significant impact and environmental assessment, zebra mussel control project, Lake Zorinsky, Omaha, Douglas County, Nebraska. http://www.nwo.usace.army.mil/FONSI.LakeZorinskyZebraMussel.Dec2010.pdf

U.S. Department of Agriculture and U.S. Department of Health and Human Services. 2010. *Dietary Guidelines for Americans*, 7th Edn. Washington, DC: U.S. Government Printing Office. http://health.gov/dietaryguidelines/dga2010/DietaryGuidelines2010.pdf

U.S. Environmental Protection Agency. 2008. Coppers Facts. EPA 738-F-06-014. http://www.epa.gov/oppsrrd1/REDs/factsheets/copper_red_fs.pdf (accessed March 9, 2012).

U.S. Environmental Protection Agency. 2012. Secondary drinking water regulations: Guidance for nuisance chemicals. http://water.epa.gov/drink/contaminants/secondarystandards.cfm (accessed March 9, 2012).

U.S. Fish and Wildlife Service. 2005. Final environmental assessment, Millbrook Quarry zebra mussel and quagga mussel eradication. Virginia Department of Game and Inland Fisheries, Richmond, VA. http://www.dgif.virginia.gov/wildlife/final_zm_ea.pdf

URS Group Inc. 2009. Final summary report: Zebra mussel eradication project, Lake Offutt, Offutt Air force Base, Nebraska. http://www.aquaticnuisance.org/wordpress/wp-content/uploads/2009/01/OAFB-ZM-Final-Summary-Report.pdf

Virginia Department of Game and Inland Fisheries. 2003a. Volume estimation and three dimensional modeling of a Virginia quarry. Unpublished poster. Richmond, VA.

Virginia Department of Game and Inland Fisheries. 2003b. Millbrook Quarry update and field analysis. Presentation and unpublished data submitted to the Millbrook Quarry Zebra Mussel Workgroup on July 22, 2003. Fairfax, VA.

Virginia Department of Game and Inland Fisheries. 2004. Request for Proposals 00375–352: Eradication of zebra and quagga mussels at Millbrook Quarry, Prince William County, Virginia. Richmond, VA.

Virginia Department of Game and Inland Fisheries. 2005a. Virginia's comprehensive wildlife conservation strategy. Richmond, VA. http://www.bewildvirginia.org/wildlifeplan/

Virginia Department of Game and Inland Fisheries. 2005b. Freshwater mussel survey of Broad Run: Millbrook Quarry to Lake Manassas. Unpublished report. Richmond, VA.

Virginia Department of Game and Inland Fisheries. 2010. List of native and naturalized fauna of Virginia, March 2010. Richmond, VA. http://www.dgif.virginia.gov/wildlife/virginianativenaturalizedspecies.pdf

Virginia Department of Game and Inland Fisheries. 2012. Special legal status faunal species in Virginia. Richmond, VA. http://www.dgif.virginia.gov/wildlife/virginiatescspecies.pdf

Virginia Department of Mines, Minerals and Energy. 2003. Hydrologic studies in support of VDGIF's investigation of zebra mussels at Millbrook Quarry, Prince William County, Virginia. PowerPoint presentation and unpublished data submitted to the Virginia Department of Game and Inland Fisheries and to the Millbrook Quarry Zebra Mussel Workgroup on July 22, 2003. Fairfax, VA.

Waller, D. L., J. J. Rach, W. G. Cope, L. L. Marking, S. W. Fisher, and H. Dabrowska. 1993. Toxicity of candidate molluscicides to zebra mussels (*Dreissena polymorpha*) and selected non-target organisms. *Journal of Great Lakes and Research* 19: 695–702.

CHAPTER 14

Management and Control of Dreissenid Mussels in Water Infrastructure Facilities of the Southwestern United States

Rajat K. Chakraborti, Sharook Madon, Jagjit Kaur, and Dale Gabel

CONTENTS

Abstract 216
Introduction 216
Impacts of Dreissenid Mussels on Ecology and Water Resources 218
    Ecological Impacts 218
    Water Resources Impacts 218
Blue Green Algae in Source Water and Water Treatment Processes 219
Environmental Requirements of Dreissenid Mussels Relative to Control Approaches 219
Monitoring and Early Detection 220
Dreissenid Mussel Control in Water Infrastructure Facilities 221
Specific Control Approaches 222
    Desiccation 222
    Reservoir Drawdown 222
    Chemical Treatment 222
    Materials of Construction and Coatings 223
    Oxygen Deprivation 224
    Copper Ions 224
    Biological Control 224
Key Elements of a Dreissenid Mussel Control Plan 224
Conceptual Framework for Developing a Mussel Control Strategy for Water Treatment Plant Protection 225
    Control in Source Water 226
    Control at Water Intake Structure 226
    Control in Water Treatment Plant 227
    Selection of Appropriate Control Method 228
Example Applications of Mussel Control Strategies 228
Example 1: Protection of a Large Intake Structure 228
    Background 228
    Chlorine Treatment at Intake No. 3 230
    Final Design Criteria 231
    Materials Analysis 232
    Evolution of a Mussel Control Strategy 232
    Summary of Protection: A Large Intake Structure 233
Example 2: Protection of a Reservoir Dam Outlet Structure and Transmission Pipeline 233
    Background 233
    Facility 234

Design of a Dreissenid Mussel Control System 235
Summary of Protection: A Reservoir Dam Outlet Structure and Transmission Pipeline 236
Example 3: Pre-Selection of a Multi-Barrier Approach to Protect a Raw Water Intake and Transmission System 236
Background 236
Multi-Barrier Control Approach 236
Antifouling Coating 236
Chemical Treatment Plan 237
Operational Actions 237
Summary of Protection: Multi-Barrier Approach for Raw Water Intake and Transmission System 238
Example 4: Strategy to Protect a Water Conveyance System 238
Background 238
Treatment Options 238
Summary of Protection: A Water Conveyance System 239
Acknowledgments 240
References 240

ABSTRACT

The discovery and spread of quagga mussels (*Dreissena rostriformis bugensis*) in source waters in the southwestern United States have presented an economic and operational threat to water conveyance facilities in this region. Infestation of lakes and reservoirs by quagga and zebra mussels (*Dreissena polymorpha*) can negatively impact water quality, food webs, and recreational fisheries. In addition, increased maintenance is required to dislodge mussels and remove dead shells from tanks, basins, and water treatment processes. Among various impacts of dreissenid mussels, clogging of pipelines and water conveyance systems including intake structures poses severe problems to water facilities. Planning and implementation of dreissenid mussel control are intense tasks for water utilities. Often various chemicals have been used to control dreissenid mussels in water infrastructure, particularly oxidizing chlorine-based chemicals. While these chemicals are effective in controlling mussel populations, they can in certain cases adversely impact the quality of product water. The formation of disinfection by-products is one of several drawbacks of using oxidizing chemicals such as chlorine. Due to chemical properties and physicochemical conditions of particular water systems, suitability of each chemical application is system-specific. Detailed investigations of available control options under specific mussel control strategies require optimization of the control process based on the life stages of dreissenid mussels, degree of infestation, aquatic species present in the water body, water chemistry, existing treatment processes, and regulatory compliance requirements.

In this chapter, background information is provided on factors that should be considered when planning dreissenid mussel control and mitigation in water infrastructure facilities, including dreissenid mussel biology, ecological and economic impacts, and environmental requirements of mussels. In addition, dreissenid control strategies are defined, and guidance is provided on the implementation of specific mussel management and control measures for protecting water infrastructure facilities. A conceptualized strategy is provided that includes a road map that water utility managers could implement for mussel management and control strategy. Implementation strategies should minimize impacts to resources and recreation while facilitating normal operations of water supply and conveyance systems.

A "toolbox" of mussel control approaches is presented with specific examples on how to take aggressive and concerted actions to eradicate, contain, and monitor dreissenid mussels in facility distribution systems and to prevent the spread of mussels to uninfested water bodies. Case studies present strategies for effective control of mussels at a water intake structure, at a dam outlet structure, and in various pipelines. Guidance is provided to water utility managers regarding (1) the use of chemicals with measures from a toolbox of proven mussel control strategies and technologies to deliver effective and operationally feasible mussel control and management options, ensuring environmental compliance, and (2) potential modifications in operations and maintenance required in water treatment plants and in conveyance systems for water utilities to achieve the goal of supplying uninterrupted, good-quality potable water.

INTRODUCTION

The discovery of quagga mussels (*Dreissena rostriformis bugensis*) in the Colorado River and their subsequent spread to surface waters in the southwestern United States presents an economic and water quality threat to numerous water treatment facilities that depend on these source waters. In the southwestern United States, quagga mussels were first discovered in Lake Mead in January 2007, and they spread rapidly downstream to Lake Havasu and the Colorado River Aqueduct and west through the aqueduct to reservoirs near San Diego, California. Mussel infestations of rivers, lakes, and reservoirs have been reported in various locations in Nevada, Arizona, Colorado, Utah, Oklahoma, Texas, New Mexico, and southern California (USGS 2011). There have also been recent reports of quagga mussels in Lahontan and Rye Patch Reservoirs in Nevada in May 2011 (USGS 2011). In January 2008, zebra mussels (*Dreissena polymorpha*) were found in San Justo Reservoir near Hollister, California, and were also observed a year later in the Hollister Conduit, which is used to convey Central Valley Project water to San Justo Reservoir and for distribution to water users. Zebra mussels were also recently confirmed in Lake Ray Roberts, which is part of the Trinity River Basin in Texas (TPWD 2012).

These two species of dreissenid mussels (zebra mussels and quagga mussels) are prolific breeders and can rapidly overwhelm source-water ecosystems and water utility facilities that depend on these source waters. Large numbers of mussels can cause hydraulic problems including clogging in pumps, pipes, screens, intakes, outfalls, and other structures associated with water distribution and treatment facilities. In the eastern United States and Europe, water supply, power generation, navigation, shipping, commercial and sport fishing, and a variety of other activities have all been heavily impacted by dreissenid infestations. Dreissenids will generally settle, attach to, and grow on surfaces where flows are approximately <2 m/s (Boelman et al. 1997). Both fixed and moveable infrastructure components, ranging from fish screens, trash racks, and pump stations to pipes, aqueducts, valves, gates, cables, chains, and filters, are vulnerable to mussel infestations. Economic costs associated with the control and management of mussel infestations in these facilities can be considerable. U.S. congressional researchers have estimated that the zebra mussel infestation in the Great Lakes area cost the power industry $3.1 billion between 1993 and 1999, with an economic impact to industries, businesses, and communities of more than $5 billion (De Leon 2008).

Approximately $0.5 million is spent annually on zebra mussel control at individual water systems in the Great Lakes region. Direct economic costs are approximately $100 million annually in eastern North America; unquantified secondary and environmental costs could be substantially larger. Potential economic and environmental impacts in the western United States could be greater than those in the eastern United States (California Science Advisory Panel 2007). Although no specific estimates are available for amounts spent directly on mussel control activities (e.g., control and prevention) or amounts indirectly lost (e.g., potential fishery decline) in the western United States, Turner et al. (2011) provided a list of expenditures for four federal agencies to deal with quagga mussels. In 2008, the National Park Service spent $5.0 million for inspection, the U.S. Fish and Wildlife Service spent $1.8 million for its aquatic invasive species program in the western United States, and the U.S. Geological Survey spent $0.2 million for support to deal with dreissenids. During 2008 through 2010, the U.S. Bureau of Reclamation (USBR) spent $12.6 million for research, prevention and control, early detection, and education (Turner et al. 2011). The potential cost for upgrades to 13 hydropower facilities in the Colorado River Basin alone has been estimated to be $23.6 million, with chemical costs estimated at another $1.3 million per year. It is estimated that the Metropolitan Water District of Southern California (MWD) alone will spend $10 to 15 million annually in operations and maintenance costs to address the quagga mussel infestation in its Colorado River Aqueduct and terminal reservoirs. Of this, about $7.2 million will be spent for design and installation of infrastructure to control mussels (De Leon 2008).

Most of the southwestern region of the United States is extremely vulnerable to mussel infestation due to favorable environmental factors such as water temperature, available food sources, and sufficient calcium content in surface waters (Wong and Gerstenberger 2011). Therefore, this region is on high alert, and proper planning is critical to prevent the spread of mussels to uninfested waters and to control mussel infestations in water infrastructure facilities. Many water bodies in the southwest are navigable and support a large recreational industry that could serve as a potential vector for mussel introductions to uninfested waters. In addition, large aqueducts and conveyance systems could be impacted by dreissenids. Approximately 1200 water treatment plants operate in the western United States and rely on lakes and rivers for their raw water supply. Each one could potentially incur additional costs for facility improvements and annual maintenance to control dreissenids (De Leon 2008).

The goal of this chapter is to outline and define control strategies and to provide guidance on the implementation of specific management options and measures to deal with mussel infestations. Strategies are based on experiences in managing and controlling mussel infestations in water infrastructure facilities throughout the United States, and particularly in facilities located in the southwest. Specific objectives are to (1) provide an overview of the potential impacts of dreissenid mussels on water infrastructure facilities and source-water ecosystems; (2) outline environmental factors that contribute to dreissenid mussel settlement and growth; (3) explore potential options available for the control and management of dreissenid mussel infestations; (4) conceptualize mussel control approaches for protecting water infrastructure facilities; and (5) provide specific examples on mussel control in water infrastructure facilities with an emphasis on water treatment plant operations and drinking water distribution systems in the southwestern United States.

IMPACTS OF DREISSENID MUSSELS ON ECOLOGY AND WATER RESOURCES

The arid southwestern United States relies heavily on complex networks of interconnected water storage and supply systems consisting of lakes, reservoirs, rivers, pump stations, pipelines, and canals. This extensive water infrastructure is vulnerable to mussel infestations and is one reason why quagga mussel populations, once established in Lake Mead, spread relatively quickly to downstream areas through the Colorado River Aqueduct and to several lakes and reservoirs in southern California. The continuous flow of water into pipes and intake structures of water treatment plants serves as a source of food and oxygen for the mussels (O'Neill 1993). Dreissenids can not only impact water infrastructure facilities but can also have profound effects on water quality and ecology of raw water sources used for treatment plant operations.

Ecological Impacts

The dreissenid invasion in the Great Lakes offers valuable insights into potential impacts that mussels may have on lake and reservoir ecosystems in the western United States. In less than two decades, the composition and function of Great Lakes' ecosystems have been dramatically altered by dreissenids. Mussels can affect ecosystem dynamics by altering or disrupting the food web, degrading water quality, and creating conditions conducive for blue-green algal blooms (e.g., *Microcystis*). *Microsystis* is a non-nitrogen-fixing alga that produces toxins that impact aquatic organisms, wildlife, domestic animals, and humans that drink or ingest algae in the water (Carmichael 1996). A good example of dreissenid impacts on water quality and *Microsystis* blooms can be derived from studies of Saginaw Bay, Lake Huron (Nalepa and Fahnenstiel 1996a). Prior to mussel introduction in 1991, primary consequences of cultural eutrophication in Saginaw Bay in the 1980s were adverse taste and odor problems and algae-induced, filter-clogging problems experienced by municipal water treatment plants. Raw water odor was found to be strongly correlated with blue-green algae in the intake at Whitestone Point (Bierman et al. 1984). Between 1972 and 1988, over $500 million was spent on municipal wastewater treatment plant improvements in the Saginaw Bay watershed (Great Lakes Water Quality Board 1989). Combined with a ban by the State of Michigan on the use of high-phosphate detergents, these improvements resulted in substantial decreases in phosphorus loadings from these sources over the 16 year period, and blooms of *Microcystis* were all but eliminated. However, after zebra mussels became established, blooms of *Microcystis* once again became apparent (Lavrentyev et al. 1995, Vanderploeg et al. 2001, 2002, Bierman et al. 2005). Zebra mussels (and quagga mussels) are known to actively reject some of the toxic species of algae (e.g., toxic blue-green algae) that they filter out of the water column (Vanderploeg and Strickler 2013). Mussels can sort unpalatable algae from the remainder of the food particles before they enter the digestive tract, bind the rejected algae in mucus (pseudofeces), and return it to the water where these blue-green algae can continue to grow. Bierman et al. (2005) described such episodes of blue-green algal (*Microcystis*) blooms throughout the Great Lakes that were likely caused by selective filtration activities of zebra mussels.

Other impacts of zebra mussels in Saginaw Bay included declines in chlorophyll *a* concentrations (Fahnenstiel et al. 1995b, Nalepa et al. 1999), increases in water clarity and enhanced growth of benthic algae and submerged vegetation (Fahnenstiel et al. 1995a, Lowe and Pillsbury 1995, Skubinna et al. 1995), alterations of nutrient levels (Fahnenstiel et al. 1995a, Heath et al. 1995, Johengen et al. 1995), and shifts in primary production from pelagic to benthic compartments (Lowe and Pillsbury 1995, Kaur et al. 2000, Pillsbury et al. 2002).

Field studies in various other systems have also reported changes in phytoplankton community structure after zebra mussel invasions (e.g., Makarewicz 1993, Nicholls and Hopkins 1993, Horgan and Mills 1997, Vanderploeg et al. 2001). Dreissenids can affect (1) phytoplankton biomass, productivity, and community structure by nonselective filtration (MacIsaac 1996, James et al. 1997); (2) selective rejection of certain types of ingested algae as pseudofeces (Vanderploeg et al. 2001); (3) filtration of herbivorous zooplankton (Shevtsova et al. 1986, MacIsaac et al. 1991); (4) removal of suspended solids from the water column, which impacts water clarity and light penetration (Reeders et al. 1989, MacIsaac et al. 1992, Holland 1993, Fahnenstiel et al. 1995b); and (5) excretion of soluble forms of nutrients (Nalepa et al. 1991, Quigley et al. 1993, Heath et al. 1995, Mellina et al. 1995, Effler et al. 1997, James et al. 1997, Canale and Chapra 2002).

While impacts of dreissenids on phytoplankton communities in water bodies in the southwestern United States are still in the early stages of evaluation, similar changes may be expected. For example, after quagga mussels became established in the Boulder Basin of Lake Mead, water clarity increased 13% and chlorophyll concentrations declined 45% (Wong et al. 2013).

Water Resources Impacts

Dreissenids can cause changes in water quality that have strong implications to water supply systems. For example, an increase in total organic carbon (TOC) and toxic algal blooms in source water may contribute to an increase in disinfectant byproducts (DBP) in the treated water and increase potential complications in treatment processes to eliminate this toxicity in the water. DBP formation depends on TOC levels, water temperature, chlorine dose, pH, bromide level, and contact time. The range of these parameters in the water system will impact treatment system designs and operations. Increased TOC and algal toxins may require changing the water treatment processes in order to meet state and federal regulatory limits.

Dreissenids alter effective use of water resources for recreational activities. Mussels are known to foul many types of surfaces, including docks, boats, outboard motors, fishing gear, and beaches, all of which can adversely affect recreational activities such as fishing, boating, swimming, and active and passive recreation. To protect non-infested source waters from mussel infestation, five large boat-wash facilities were installed at marinas on Lake Mead, Nevada–Arizona, so that all boats slipped and moored could be inspected and washed with hot water (Wong et al. 2011). In addition, a boat cleaning training course was offered to all marina workers, and quagga mussel disinfection workshops were conducted for all staff. Use of the "Clean, Drain and Dry" public message campaign was adopted to encourage boaters to prevent the spread of quagga mussels from Lake Mead (Turner et al. 2011). Another concern with mussel infestations on recreational activities is that mussels may cause health hazards including cuts from the shells piled at the beaches (O'Neill 1993).

BLUE GREEN ALGAE IN SOURCE WATER AND WATER TREATMENT PROCESSES

The quality of source water and compliance of finished water quality within regulatory requirements are key factors driving the selection of water treatment processes. As described previously, dreissenid mussels can cause water quality changes and blooms of toxic blue-green algae in source waters from lakes and reservoirs. Such toxic blooms (e.g., *Microcystis*), as found in the eastern United States, potentially pose public health risks (Carmichael 1996), and one of the more common toxins, *Microcystin-LR,* is a hepatotoxin that affects the human liver (Hitzfeld et al. 2000). An increase of algal toxins in source water may require changes to the potable water treatment processes in order to achieve health standard goals for finished water.

Alvarez et al. (2010) evaluated the effectiveness of various water treatment processes under elevated concentrations of algal toxins (*Microcystin-LR*) in source waters and compared toxin removal results by treatment methods such as ultraviolet/hydrogen peroxide (UV/H_2O_2) oxidation, ozone oxidation, powdered activated carbon (PAC) adsorption, granulated activated carbon (GAC) adsorption, biological degradation, nanofiltration (NF), and reverse osmosis (RO) membrane filtration. Major results included the following:

- Treatments involving both UV and H_2O_2 were effective at removal of *Microcystin-LR*, but the level of removal was dictated by H_2O_2 concentrations and availability for treatment. UV alone was not effective at removing *Microcystin-LR*, providing only about 10% removal at doses normally used for disinfection. An UV dose of 120 mJ/cm^2 is needed for 50% *Microcystin-LR* removal. Most current UV systems use a low-pressure or medium-pressure mercury vapor lamp and expose water to UV by pumping the water around a sleeve within which the UV lamp is supported. Typical system designs deliver UV dosages of 25–35 mJ/cm^2 and are adequate to deactivate bacteria and viruses (Wolfe 1990).
- GAC was effective at removing *Microcystin-LR* when the GAC was replaced frequently and TOC concentrations were low. It was found that nearly 100% removal was achieved in new GAC filters with 5 min of contact time; generally, 50% removal of *Microcystin-LR* is achieved using GAC.
- The ozone/advanced oxidation potential (AOP) combination was effective for removal of *Microcystin-LR* if the pH was below 7. Doses as low as 0.4 mg/L achieved *Microcystin-LR* removals greater than 97%.
- PAC removed *Microcystin-LR* at a dose of 10 mg/L and a contact time of 30 min.
- Biological degradation alone provided 35% removal of *Microcystin-LR*. It is not a preferred single treatment for removing *Microcystin-LR*, but it could be used as a polishing step in conjunction with other treatment methods, such as UV or ozone oxidation or PAC and GAC adsorption.
- Filtration through RO and NF membranes removed *Microcystin-LR* efficiently; a minimum of 95% of the toxin was removed under most circumstances.

Based on these results, the ozone/AOP and/or RO/NF membrane treatments are the most effective in removing algal toxins such as *Microcystin-LR* from source water. This information is helpful to design/adjust water treatment plant operations if source water becomes polluted with toxic, blue-green algae caused by dreissenid infestation.

ENVIRONMENTAL REQUIREMENTS OF DREISSENID MUSSELS RELATIVE TO CONTROL APPROACHES

An understanding of the environmental requirements of dreissenids and their responses to specific environmental factors is particularly useful for developing and implementing mussel control and management strategies. As such, mussel control approaches (e.g., chemical treatments) can be targeted during times when dreissenids are physiologically most vulnerable to specific environmental factors or combinations of environmental factors. A range of environmental conditions are known to affect the ability of quagga and zebra mussels to reproduce, grow, and complete their life cycle in natural systems, and these are comprehensively reviewed by Mackie and Claudi (2010). Environmental factors that can mediate the physiological responses and tolerances of dreissenids include water temperature, dissolved oxygen, pH, calcium, total alkalinity, total hardness, salinity, total dissolved solids, conductivity, nutrients (and phytoplankton levels), turbidity and total suspended solids (TSS), flow velocity, and substrate quality. For example, water temperatures between 16°C and 26°C are optimal for growth and reproduction of zebra mussels. Outside of this optimal temperature range, growth and reproduction rates can slow down considerably with a resulting decline in mussel production. Water temperatures in excess of 30°C–32°C can be lethal (Cohen 2007, Mackie and Claudi 2010).

The fact that dreissenids are physiologically stressed at certain water temperatures can potentially be exploited in the implementation of control strategies. Chemical treatments that use oxidizing chemicals such as chlorine will be more effective at warmer temperatures close to the upper tolerance limits where adult mussels are stressed. Water temperature appears to have a strong effect on chlorine-related mortality in adult zebra mussels (Van Benschoten et al. 1995), and similar temperature effects are also expected to occur in quagga mussels due to physiological similarities to zebra mussels (Garton et al. 2013). For example, chlorine concentrations of 0.5 mg/L resulted in 100% mortality of adult mussels in 9 days at water temperatures of 18°C–22°C, whereas 100% mortality was not achieved even after 25 days of continuous chlorination at 9°C–15°C. The time to achieve 100% mortality can be further reduced at warm temperatures by increasing chlorine concentrations (Van Benschoten et al. 1993a,b, 1995).

In reservoirs that stratify, low dissolved oxygen levels in the hypolimnion during summer can be exploited as a natural mechanism for mussel control and can be combined with other strategies such as manipulation of reservoir water levels (CDFG 2011). Susceptibility of zebra mussels to hypoxia is also significantly more acute at warmer temperatures (Johnson and McMahon 1998), and treatment with hypoxic waters, where feasible, will be more effective at controlling mussels at warmer water temperatures. Besides using hypolimnetic waters, conditions of hypoxia/anoxia can be induced either chemically or physically (sealing off water to mussels). Filtration and metabolic rates of dreissenids are also higher at warmer temperatures (up to a limit) and effectiveness of targeted mussel toxins/treatment chemicals will thus be greater at warmer temperatures.

MONITORING AND EARLY DETECTION

Effective programs for the prevention and control of veligers and adult mussels need to be focused on the most vulnerable components of both the raw water source and the water treatment/distribution system. Transport of veligers and adults between water bodies can occur passively with the downstream flow of water (Figure 14.1, top panel) and can occur actively by transport on trailers or boats (Figure 14.1, bottom panel). Boats that have been in the water for more than 1 or 2 days may have mussels attached to their hulls, anchors, chains, trailers, equipment, and engine drive units. Boat inspections can be an important element of a monitoring program and are typically supported by use of trained boat inspectors and construction of boat cleaning stations. Conducting boat inspections requires coordination with relevant authorities and state agencies.

If mussels become established in a raw water supply system, monthly or more frequent monitoring is recommended so that mussel population dynamics can be documented. Monitoring veliger abundance provides valuable information about the seasonal timing of reproduction. An understanding of temporal and spatial patterns in mussel growth and reproduction will aid in refining control strategies. For example, use of an overlaying capability in a geographical information system (GIS) will define dreissenid growth relative to environmental factors and may help focus on key areas in a water body for monitoring populations (Ramcharan et al. 1992, Miller and Ignacio 1994, Chakraborti et al. 2002). In an effort to identify "hotspot" areas for mussels in Saginaw Bay, Lake Huron in the Great Lakes, Chakraborti et al. (2002) developed a statistical model using GIS that provided insights on spatial relationships between limnological variables and zebra mussel growth, as defined by changes in biomass over a specified time. In addition to substrate quality, temperature, phytoplankton biomass (measured as chlorophyll *a*), and TSS were considered to be the most important variables affecting growth of zebra mussels. Three layers of attributes were developed from these variables and were overlaid in a GIS environment according to their respective weighting factors, which were calculated in statistical analysis of spatially matched data. The model showed that, among the factors studied, chlorophyll *a* contributed the

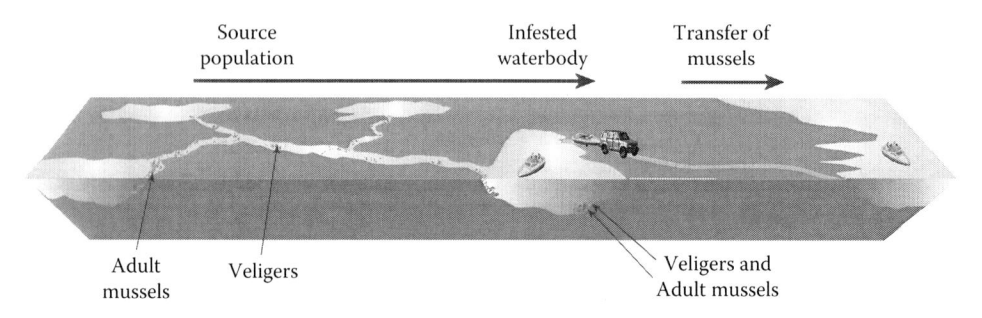

Figure 14.1 **(See color insert.)** Pathways by which dreissenid mussels can be transported between water bodies.

most to mussel growth. Also, shallow portions of the bay and areas in close proximity to shorelines were found to be the most suitable growth regions. These model predictions were confirmed by field data. The model provided a rapid, objective, reliable, and cost-effective approach to define areas of greatest mussel growth potential and can be used to prioritize locations for monitoring activities.

Early detection of dreissenids is critical for implementing effective control strategies. Methods such as cross-polarized microscopy and DNA analysis using polymerase chain reaction (PCR) can be helpful for early detection of mussel presence. Cross-polarized microscopy, the most commonly used method to confirm veliger presence in water samples, is highly accurate and reliable, but it can be labor intensive (USBR 2011). The PCR method confirms the presence of mussel DNA in samples where veligers have been detected using microscopy. The discovery of mussels in the very early stage of infestation affords the opportunity to track changes in the water body as the infestation progresses (Hosler 2011, USBR 2011). Results of DNA tests indicate that veliger numbers must reach greater thresholds compared to microscopic methods for consistency of detection (Hosler 2011). Early-detection activities may also include placement of artificial substrates at strategic locations. In Lake Mead, concrete-backed boards were placed at marinas to detect settled juveniles and track growth rates (Turner et al. 2011).

DREISSENID MUSSEL CONTROL IN WATER INFRASTRUCTURE FACILITIES

Control strategies for dreissenids are typically separated into the following three categories: preventative, proactive, and reactive. Preventive controls are designed to prevent or delay the establishment of mussels through public education, regulation, and inspections. Certainly, the best outcome is to prevent the establishment of mussels into the waters of concern; however, if prevention is unattainable, delaying the eventual infestation will provide a water treatment facility with added financial and functional benefits. Also, preventative measures can benefit the management of water systems by reducing eventual costs of actively controlling dreissenids with proactive and reactive measures (Thomas et al. 2010).

Public education lies at the foundation of all preventative measures. The first step of an effective educational program is to develop an attitude of concern. In order for any control strategy to be successful, the public, particularly potential users of reservoirs and other portions of water conveyance systems, must be aware of the problem. The second step is acceptance of the fact that user activities are a potential means of dreissenid transport. The public must understand that they have a stake in prevention and that their activities are the most likely transport mechanism. The third step is that users must adhere to recommendations put forth by resource managers. The public must be kept informed on what is required of them and that their response is of utmost importance. An informed public is the foundation of prevention, and effective strategies include newspaper articles, television coverage, presentations made to user groups, mailings, installation of signs on boat ramps, handing out fact sheets, and handing out pocket cards with prevention strategies to boat owners. It is expected that the more effort that is put toward education, the more likely a recreational user will take steps to assist in the prevention process. Coordinating educational efforts with the relevant groups and agencies should provide synergistic benefits and opportunities to share costs.

Although important, public education alone is not sufficient to control mussel infestations; educational efforts need to be supplemented with regulatory measures. For instance, recreational boating has been identified as a major mechanism for mussel spread (Figure 14.1); therefore, restriction of boats can be an effective way to prevent mussel establishment in an uninfested water body. Restrictions and regulations may include use of the following: (1) prohibit use of bait from outside the watershed; (2) require mandatory boat inspections prior to off-loading; (3) install washing stations for decontaminating boats; and (4) implement fines for traveling on highways with debris on boat trailers.

The primary focus of proactive treatment is to inhibit attachment of veligers or the translocation of adult mussels to infrastructure components (e.g., water intakes or pipes, equipment, water infrastructure) where functionality would be adversely impacted. Some proactive control techniques include (1) chemical treatment using a variety of oxidizing or non-oxidizing chemicals; (2) use of antifouling or foul-release paints and coatings on infrastructure surfaces; (3) use of mechanical in-line strainers and filters, infiltration galleries, or sand filters; (4) UV radiation; and (5) maintenance of flows above approximately 2 m/s to minimize veliger settlement (Boelman et al. 1997, Mackie and Claudi 2010). Other non-chemical proactive strategies that have been suggested include the use of electric currents, acoustics, high-speed agitators, and magnetic fields, but tests of these approaches either yielded mixed results or were deemed impractical for application in industrial settings (Mackie and Claudi 2010).

Reactive control approaches are implemented when dreissenids have become abundant and widely established in a water system. Reactive controls may involve (1) chemical treatments using oxidants or non-oxidant chemicals; (2) physical removal of adult mussels using mechanical means such as power washing, scraping, and pigging; (3) periodic shock treatment involving thermal shock or freezing; (4) desiccation, including reservoir drawdown and draining and drying of system components; and (5) oxygen deprivation via inducing hypoxic or anoxic conditions (Mackie and Claudi 2010). Specific examples of some of these control measures to protect various components of water infrastructure facilities are presented later in this chapter.

SPECIFIC CONTROL APPROACHES

Desiccation

Drying and periodic cleaning of components that could serve as settlement and attachment sites for veligers is an effective control option. Vulnerability to desiccation provides an environmentally sound control technique that may be useful. Operational controls, such as diverting water during the period of cleaning a facility, are frequently required and should be incorporated into the design of new water infrastructure. Humidity, air temperature, duration of treatment, and mussel size determine the effectiveness of desiccation (Boelman et al. 1997).

Reservoir Drawdown

Reservoir drawdown can be used in some cases for inducing mortality by exposing dreissenids to hot (>33°C), dry, or extreme cold (<0°C) air temperatures. This method usually involves only partial reservoir drawdown due to operational constraints and therefore will not completely eliminate mussels from the water body. However, it will help reduce population densities in the shallow areas most often used by the public. Drawdown could be followed by mechanical removal and proper disposal (e.g., in a landfill) of the dead mussels.

Chemical Treatment

The ultimate selection of a chemical control method will depend on consideration of site-specific opportunities and constraints involving the desired chemical treatment. Factors influencing selection of chemicals are costs, proven effectiveness of control methods, and resulting impacts on water quality. Table 14.1 presents a selected suite of oxidizing and non-oxidizing chemicals (along with other non-chemical control methods) that could be considered in planning for a potential dreissenid infestation at water infrastructure facilities. Advantages and potential environmental and water quality impacts of each control method are also given. While not an exhaustive list of all potential chemical treatment options are available for mussel control, Table 14.1 provides some common control methods based on site-specific opportunities and constraints dictated by a facility's use.

Application of chlorine as an oxidizing chemical is the most widespread chemical treatment employed for control of dreissenids (Belanger et al. 1991, Van Benschoten et al. 1993a,b). The widespread use of chlorine is based on three factors: it is an effective molluscicide; it often already exists at water infrastructure facilities; and its use and side effects are well understood by operators and regulators. Over the last decade, discharge of any type of chlorine compound into receiving waters has been increasingly

Table 14.1 Advantages and Disadvantages of Various Chemical Methods for Control of Dreissenid Mussels at a Water Infrastructure Facility

Compound	Advantages	Disadvantages
Chlorine (gas, sodium hypochlorite, calcium hypochlorite)	• Inexpensive • Compatible with raw water • Applied at low dosages	• THMs/HAA formed • Potential water quality violations • Safety • Leaves residual that is toxic to fish, especially for control of adult mussels
Permanganate (potassium or sodium)	• THMs/HAAs not produced • Mitigates taste and odor	• Long contact time • Pink water if overdosed • Effective for veligers
Chlorine dioxide	• Effective at low dosages • THMs/HAAs not produced • Independent of pH	• On-site generation • High-oxidant demand • Chlorite formation
Ozone	• Disinfection credit • Increased filterability • Does not accelerate corrosion	• Bromate formation • Power cost • Maintain residual in feed • Due to its short half-life, relatively large dosages and long contact times are required for control of adult mussels. Expensive and difficult to use in a subsurface pipeline/intake application
Chloramines	• THM/HAA not produced	• Requires chlorine and ammonia facilities • Safety considerations • May not be compatible with current disinfection systems, and interconnected conveyance system
Cationic	• THM/HAA not produced	• Jar testing required
Polymers (non-oxidant)	• Effective on both veligers and adult mussels • Potentially causes release of byssal threads	• Storage must protect the polymer from bacteria and fungus • Chemical costs

HAA, haloacetic acid; THM, trihalomethanes.

restricted by regulators due to chlorine's impact on aquatic life. For this reason, chlorine applications at the intake mouth should be fed from a solution line inside the intake pipe, such that any leakage in the solution line will not result in chlorine entering the environment. Additionally, the chlorine feed can only be operated during the active intake of raw water. A major concern with the use of chlorine compounds is byproduct generation, usually referred to as DBPs, including trihalomethanes (THM) and haloacetic acids (HAA), which are carcinogenic or suspected to be carcinogenic if present in the drinking water at high levels. The application of chlorine will form DBPs, depending on the organic matter present in water (i.e., TOC levels averaging about 3 mg/L or more), chlorine dosage, specific chlorine compound used, temperature, pH, alkalinity, hardness, and the time of contact of chlorine with the water. As noted, dreissenids can cause toxic algal blooms and an increase of TOC in source waters, and hence the problem of DBP formation becomes of greater importance within water treatment plant operations. DBPs need to be considered as part of the water plant's overall treatment strategy, and bench-scale testing is recommended to provide the required information on the potential formation of DBPs with existing physical–chemical conditions of the source water and chemical dose. Free chlorine is the most prevalent molluscicide and is often used in a liquid form as sodium hypochlorite (NaOCl) or in the gaseous form (Cl$_2$). Chlorine dioxide and chloramines are also used, but to a lesser extent since their handling and preparation are more complex (DeGirolamo et al. 1991). Evaluating chlorine as a treatment option (or any chemical option) also includes consideration of applicable permits, environmental regulations, chemical delivery and storage, feed systems, economics, and existing facility systems. Chlorine can cause interactions with plant equipment as well. Chlorine attacks rubber gaskets and can impact pump and pipe integrity over long-term use, depending on the materials of construction.

In addition to chlorine compounds, a number of non-oxidizing molluscicides (cationic polymers) have been developed (Table 14.1). These compounds have advantages over chlorine in that they are noncorrosive and do not form toxic byproducts (McMahon et al. 1993). In addition, application times may be shorter than needed for chlorine since bivalves generally do not detect these chemicals and so do not exhibit avoidance through valve closure (Jenner 1990). Disadvantages include (1) impacts on non-target organisms, which may preclude these molluscicides from being released into certain receiving systems (Bidwell 1993), and (2) impacts on water treatment unit processes or water conveyance infrastructure facilities comprising of water intake structure, water transmission pipeline and valves, pump station, water storage basin, and water distribution system.

Materials of Construction and Coatings

Toxic and nontoxic coating materials are available that can either prevent dreissenid settlement or cause weak attachment of byssal threads. Materials such as copper or brass are toxic, and mussels find surfaces made of these materials inhospitable and avoid attachment. Infrastructure components could be coated with a toxic coating material to slow down colonization and the spread of mussels; however, many of these will oxidize over time, thus reducing their effectiveness. Additionally, unintended consequences of coating toxicity can be a matter of concern from a water supply or ecological perspective, and coatings may require certifications if they are used in potable water conveyance systems. Coatings based on foul-release mechanisms are effective and limit initial settlement and attachment, but can be mechanically weak and subject to failure due to detachment and abrasion. Wells and Sytsma (2009) reported that heavy metal–based coatings are both effective and durable, but they work by releasing biocides such as copper into the surrounding water, which may impact non-target aquatic fauna. Therefore, leach rates of heavy metal–based coatings should be investigated to determine if rates meet regulatory requirements.

Wells and Sytsma (2009) estimated that costs of silicone-based, foul-release coatings over a 5 year period were $127/m^2 and that their maximum, effective life span was 6 years. If advances in silicone-based coatings result in an effective life span greater than 10 years, and leaching from the surface is not environmentally harmful, such coatings may prove to be cost effective for preventing mussel attachment on water treatment plant components such as trash racks, intake bays, intake tunnels, piping, and pump wells (Wells and Sytsma 2009). Recently, the U.S. Bureau of Reclamation (Reclamation 2012) studied the effectiveness of various antifouling coating materials for mussel control over a 3 year period at Parker Dam, California. This study showed that many products are effective at keeping mussels off surfaces, but that (1) conventional epoxies, ASTM A788 steel, and 304 stainless steel were not suitable antifouling material and (2) copper, bronze, and brass-based coatings were effective, but their toxicity to other aquatic organisms limits their use. Silicone-based, foul-release products were effective for mussel control on infrastructure components that do not have significant debris (wear) or gouges. U.S. Bureau of Reclamation is also testing the strength of mussel attachment to various antifouling materials so that the relative ease of mechanical cleaning can be compared (Allen Skaja, personal communications).

The development and certification of silicone-based, foul-release coatings is continuing. In November 2011, one product (FujiFilm Smart Surface, LLC Duplex silicone) received NSF 61 certification from Underwriters Laboratories for potable water use under the following conditions (Charles Fisher, personal communication): tank size $\geq$7570 L, pipe diameter ≥1.37 m, curing time before 7 days at 26.7°C, and surface area (cm^2) to volume (L) ratio = 29.03. Because such silicone-based coatings presently have a limited life span, particularly on infrastructure components subject to wear, surfaces would require periodic recoating and thus the design of facility components should enable access for recoating.

Oxygen Deprivation

Oxygen deprivation (suffocation), accomplished by sealing pipes long enough for the water to lose its dissolved oxygen, can be used as a mussel control method. As noted, mussel demand for oxygen is greatest in warm water; therefore, oxygen deprivation tends to work best in summer. Hypoxic (less than 0.8 mg/L dissolved oxygen) conditions can cause mortality in juveniles after 2–3 days of exposure and cause mortality in adults after more than 2 weeks of exposure (Johnson and McMahon 1998). Further, time to death for adult zebra mussels (13.4–33.2 mm shell length) is 12 days at 5% oxygen saturation at 25°C. Time to death at the same oxygen concentration increases to 70 days at 15°C. Hence, there is a significant temperature dependence of hypoxia effects on mortality of adult mussels. These results also suggest that dreissenid control by oxygen deprivation would require relatively long exposure times, but could be effective, particularly during late summer months when the species are thermally stressed. One of the drawbacks and precautions for use of oxygen deprivation as a viable control method is the potential for increased corrosion and pitting of surfaces. This may occur in either of two forms: (1) lack of oxygen may increase the presence of sulfate-reducing bacteria, a class of bacteria that induces corrosion (Claudi and Mackie 1994), and (2) release of sodium-meta-bisulfite or hydrogen sulfide gas into the water under oxygen deprived conditions creates acidity, which in turn increases corrosion (Claudi and Mackie 1994).

Copper Ions

The use of copper-ion generators to protect facilities from mussel settlement has been identified as a potential control mechanism. Copper ionizes in water, and these copper ions apparently prevent dreissenid attachment (O'Neill 1993). Copper ions at a concentration of 0.01 mg/L are most toxic to mussels including dreissenids during their early developmental stages, from fertilization to the veliger stage (Yaroslavtseva and Sergeeva 2007, Mackie and Claudi 2010). The use of copper ions as a control is dependent on water hardness. An example of the application of copper ion generators for control of dreissenids in a lake intake is provided later in this chapter.

Biological Control

A toxin from naturally occurring soil bacteria, *Pseudomonas fluorescens*, has been identified that specifically targets dreissenids (Malloy and Mayer 2007). *P. fluorescens* has a worldwide distribution and is present in soils and in some water bodies. In nature, it is a harmless species that is a common component of the root–soil matrix near the root zone of plants and is known to protect plants from diseases. Mussels that consume sufficient amounts of these bacteria as a food source ingest a toxin within the bacterial cells that destroys the digestive system and results in death (USBR 2011). The toxin-based mode of action underscores an advantage of this potential control technology; even dead microbes contain active toxin and can be used for mussel control. Trials indicate that it is easier to use, less labor intensive, and requires shorter treatment time than chlorine. This control measure is currently in the development stage. However, in March 2012, dead cells of *P. fluorescens*, patented as Zequanox, did receive approval from the U.S. Environmental Protection Agency (U.S. EPA) for use as a molluscicide (Zequanox 2012). State-specific approvals may still be necessary. This product may be used as an alternative or as a complement to chemical treatments for controlling mussels in a water body.

KEY ELEMENTS OF A DREISSENID MUSSEL CONTROL PLAN

The development and successful implementation of a multifaceted control plan for dreissenids requires the navigation of key challenges (Figure 14.2). Important factors that need to be considered include the following:

1. Exploration of chemical and non-chemical treatment options as proactive or reactive control measures based on the status of the mussel infestation (such as low or high degree of infestation) and life stage of mussels targeted for control (adults vs. veligers).
2. Impacts on water quality and aquatic species in source water or receiving water ecosystems may limit the spectrum of chemical and non-chemical control options.
3. Impacts on hydraulic flow caused by mussel infestations such as reduced capacity of pipes due to clogging, and impacts from changes in infrastructure associated with retrofitting for mussel control.
4. Impacts on water treatment plants due to treatment of source water may require adjustment of treatment processes. For example, chlorine treatment in source water may require adjustment in unit processes to keep DBPs and residual chlorine within acceptable limits. In addition, algal toxins in source water may require adjustment in treatment methods for toxicity and taste and odor control.
5. Impacts on wholesale customers due to increased capital and maintenance costs of water treatment and delivery.
6. Storage and safety of control chemicals in addition to chemicals required for regular treatment may require adjustment of control options.
7. Regulatory compliance requirements of chemicals are a key component of selecting the most suitable chemical control option.

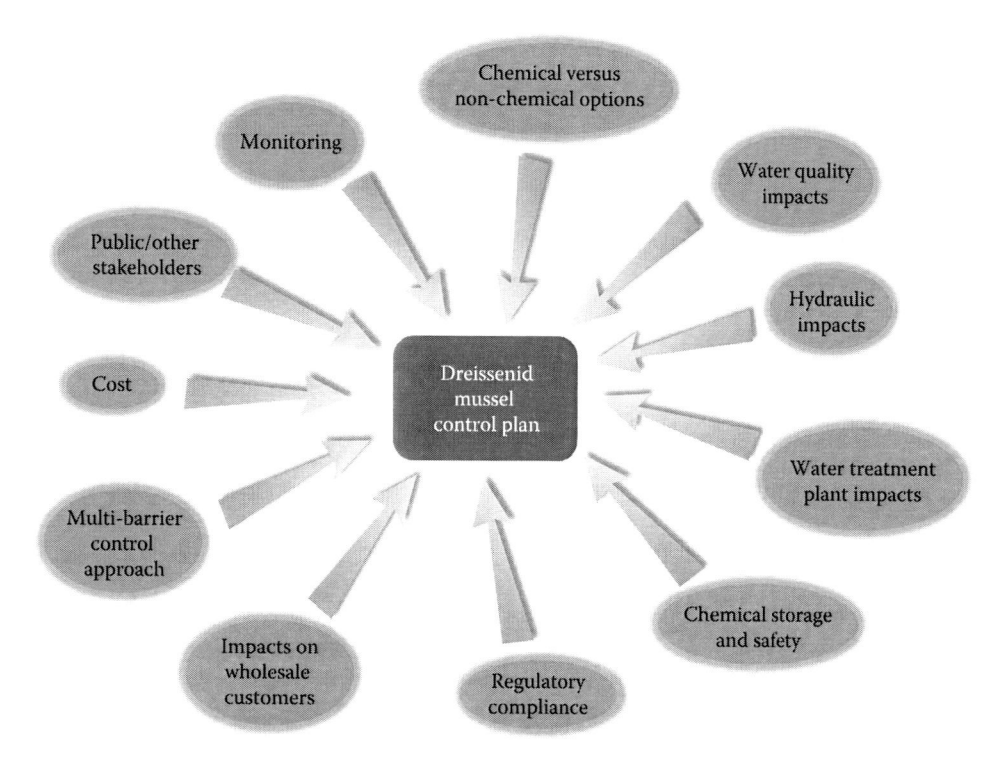

Figure 14.2 Key elements and considerations of a control plan for dreissenid mussels.

8. The involvement of public and other stakeholders in management of the water body providing source waters to water infrastructure facilities may strongly influence control selections.
9. Implementation of a multi-barrier approach for controlling mussels in diverse areas (water intake, conveyance pipelines, storage tanks, pump stations, etc.) may be necessary.
10. Cost for implementation of controls will often drive selection of the most suitable multi-barrier approach including primary and secondary controls from the suit of options.
11. Monitoring is the key to a successful control plan because monitoring will help assess effectiveness of the plan and therefore assist in evaluating any needed adjustments.

CONCEPTUAL FRAMEWORK FOR DEVELOPING A MUSSEL CONTROL STRATEGY FOR WATER TREATMENT PLANT PROTECTION

In preparation for an impending mussel infestation or upon first detection of mussels in a raw water source, key responsibilities of water facility managers include (1) conducting an infestation vulnerability assessment of their systems/components based on environmental conditions for mussel growth and reproduction; (2) maintaining infrastructure functionality with no or minimal disruptions to operations; (3) protecting the ecosystem of the raw water source or receiving waters; (4) minimizing impacts on end users; and (5) maintaining recreational opportunities to the extent possible. Water managers may need to consider several criteria before selecting mussel control options that are best suited for their system. Important factors to consider (in no particular order) include the following:

- *Effectiveness of control measures*, which may be used for first screening of potential control options out of a suite of physical, chemical, and mechanical methods for a particular system based on a literature study, monitoring results, and laboratory tests.
- *Environmental impacts*, which may imply that preferred treatments are those that do not harm the environment, ecosystem health, and non-target species.
- *Ease of operation*, which may dictate preferred measures if they require less operation and maintenance requirements and less operator training for handling, preparation, and application.
- *Operational flexibility*, which may be an important selection factor if the preferred control method is similar to methods already in place for water treatment.
- *Regulatory requirements*, which may favor one control approach over another if there are no or fewer regulatory constraints (such as Clean Water Act [CWA], National Pollutant Discharge Elimination System [NPDES], and U.S. EPA regulations) on using the preferred approach.
- *Cost*, which obviously impacts the decision-making process when selecting a preferred mussel control solution.
- *Record of performance*, which may favor one method if the control is tested in similar systems and provided better results than other methods.

- *Health and safety concerns*, which may favor a particular control method with the least health and safety concerns.
- *Feasibility of application to the specific infrastructure and needs*, which may be a factor in decision making if application of one control method is more feasible in a particular system while still providing best possible results.

Other specific selection criteria could be used based on the specific needs of the facility, and it is recommended that the control measures be evaluated against these criteria.

Figure 14.3 presents a conceptualization of the process and tools available for protecting various components of a water treatment system infested with dreissenids. Various control options are conceptualized from the influent of a treatment system, including source water and intake structure, to the effluent of the system. As is typically the case, a multi-barrier approach is recommended for effective control of mussels in water systems since a single control method is often not optimal or may not provide the required redundancy for protecting complex water facilities. For example, a multi-barrier approach for controlling mussels will be different for a system in which the infestation is very localized and detected early than for a heavily infested system with both veligers and adult mussels. Also, it is important to design an approach such that the infestation can ideally be isolated from the rest of the system and treated with one control measure. A design for mussel control can be different for (1) a flowing water system versus a stagnant system; (2) a system populated by many aquatic species versus a system with only a few species; and (3) an infestation located near the intake structure of a drinking water treatment plant versus an infestation where water is not primarily used for drinking purposes (i.e., power plant). Any control strategy that is adopted will require some level of coordination among water infrastructure agencies and regulatory authorities and should be based on a combination of the proactive and reactive treatment options discussed earlier.

Control in Source Water

The primary goal of various control options in water bodies with water intakes is to limit the growth of adult mussels. Control Option 1 (Figure 14.3) describes an approach if the source water (including reservoirs, lakes, and rivers) is infested. The most suitable control strategy of various physical and chemical methods described earlier is the core of this control option, that is, to keep the population of mussels under control if complete eradication is not possible. A combination of chemical and physical methods meeting regulatory requirements would be an appropriate alternative for controlling dreissenids in source water. Chemical control methods that may not cause any adverse environmental effects could be applied to source water in a controlled area isolated from water bodies used for drinking water purpose or used by public for recreation.

Although containment or eradication of dreissenids may be possible in small water bodies (Heimowitz and Phillips 2006), eradication from large river systems and reservoirs is likely impossible. Biologists in the Great Lakes region have explored options for mussel eradication for years without much success. A reported successful eradication of zebra mussels in a 16 acre quarry lake (Millbrook Quarry, Virginia) used 174,000 gal (659 m^3) of potassium chloride (Fernald and Watson 2013). Therefore, application of this technique, while successful, is likely only practical in small, closed, and controlled systems.

Control at Water Intake Structure

The water intake structure is the most vulnerable area for mussel infestation if mussels are established in source waters. Components of the intake structure at risk include log boom and piles, fish screens, solid panels, diffuser panels, forebay structures including piping, chain and scrapers, pumps, pipes and valves, surge tanks, settling pond systems,

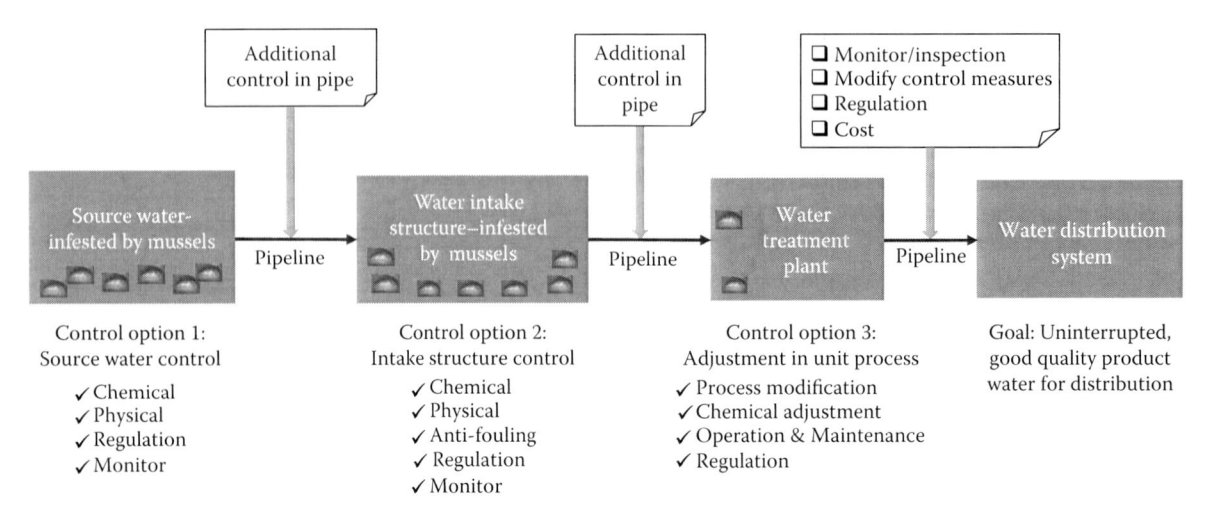

Figure 14.3 Conceptual toolbox for developing a control strategy for dreissenid mussels in a water treatment plant.

control structures, adjustable weirs, and instrumentation. Mussels within intake pipes cause loss of flow initially due to increased coefficient of friction and later due to volume restrictions. Control Option 2 (Figure 14.3) uses combination of physical and chemical methods, including antifouling materials, to protect parts of the intake structure from mussel impacts. Physical control measures such as scraping, oxygen deprivation, and desiccation could be effective control measures for protecting intake structures (Claudi and Mackie 2000). Mussel-infested pipes connected to the intake structure can also be protected using physical measures such as pigging. Pipeline pigging involves forcing a flexible plug through pipelines to remove debris and mussels that accumulate on interior pipe surfaces. It applies a differential pressure across a short length flexible plug (typically through pumping) to force it through the pipe. Manual cleaning, including hydroblasting, abrasive blasting, and wire brush cleaning, can also be used to remove mussel infestations particularly from small-diameter outfall diffuser nozzles. Other potential strategies to control mussels at the intake and pump stations include installation of screens, chemical treatment, and heated water.

The installation of screens made of copper, brass, or nickel are effective because veligers avoid attaching to these surfaces (Marsden and Lansky 2000, Wells and Sytsma 2009). Regardless of materials, screens should be cleaned periodically as entrapped debris may serve as attachment sites for mussels and should be designed for easy replacement and cleaning. The small size of veligers (approximately 70–300 µm) requires appropriately designed screening systems to be effective. Sand filtration systems or intakes buried in infiltration beds are effective at excluding most mussel veligers from water intakes (Texas Parks and Wildlife Department 2010), as are online filters of 40 µm mesh (Mackie and Claudi 2010).

Pumps and intake structures beyond the screens can be protected by using a chemical control system. Any chemical treatment that prevents veligers from attaching to and growing on structures (e.g., pipes, pumps, and valves) during the reproductive season can be used. Continued monitoring of veligers will indicate if the reproductive season is year round or only during warmer months. Chemical treatment can be conducted as a low-dose, continuous feed targeting the veligers during the reproductive season or intermittently at higher doses to target the settled juveniles. The chemical of choice depends on availability, cost, need for taste and odor control, and need to avoid the production of DBPs. Some potential choices include (1) ozone in gaseous form, which is attractive from the perspective of effectiveness, simplicity, and treatment chemistry; (2) potassium permanganate, which could be effective against dreissenids, especially veligers; and (3) chlorine compounds. Chlorine compounds are an effective control method, but free chlorine application may cause undesirable levels of DBPs. Research on transmission loss will be required to confirm costs and practicality. The location of the feed facilities should be selected appropriately. Regardless of chemical method, changes in flow rate through the intake and connecting piping should pace chemical application to achieve required effectiveness of treatment for controlling mussels. Gaseous-source chemical feeds, such as chlorine and ozone, are normally constant-flow, variable-concentration designs. Changing the amount of chemical applied at the intake will therefore have a response lag time equal to the travel time from the chemical source to the intake. This may be undesirable for plant operations. Liquid-source chemicals (e.g., hypochlorite) can be more easily applied at constant concentrations in variable flow systems since response lag time will be minimal. Also, plant operations and regulations may require that a separate, sample-point upstream of the chemical application point at the intake be provided in order to analyze for a "true" raw water sample. An example presenting chemical lag time estimates for the design of chemical control option for a water utility is presented later in this chapter.

An effective and environmentally sound method of controlling dreissenids in intake pipes is systematic, periodic flushing with heated water. However, local resource management agencies should be consulted to determine whether the amount of hot water discharged into the environment would require a discharge permit. Under a controlled application, water temperatures must exceed 37°C for about 1 h to ensure 100% mortality of mussels (McMahon et al. 1995). After a thermal flush, mussels may be removed by pumping water through the pipe at a high rate of flow. Moving intakes to greater depths for cooler source water may not be a feasible solution for protecting intake structures because mussels can survive near-freezing water temperatures. In Lake Michigan, zebra mussels are generally limited to depths of <50 m, but quagga mussels are very abundant at depths of >90 m (Nalepa et al. 2010). Examples of effective, mussel-control methods for protecting transmission pipes are provided later in this chapter.

Control in Water Treatment Plant

Once dreissenids are found in the treatment plant, adjustments in plant processes may be required (Figure 14.3, Control Option 3). These may include changes in coagulant dose at the coagulation–flocculation process, adjustments in the dose of disinfectants such as chlorine and/or ozone, and changes in the filtration process in a conventional water treatment plant operation. Process modifications are needed relative to the density and life stage of mussels entering the treatment plant. As described previously, adjustments in unit processes of water treatment (such as PAC/GAC vs. NF/UV/RO) may be necessary if toxic algal blooms are present in the source water. Obviously, any adjustments in unit processes and chemicals used will likely increase operation and maintenance costs of a water infrastructure facility. Meeting regulatory requirements to achieve drinking water standards after adjustments of unit processes and chemical doses will be required for water treatment plant operations.

Selection of Appropriate Control Method

A comparative analysis can be used to screen all control alternatives and to subsequently select a few for further definition and development of life-cycle cost estimates. Cost estimates and comparative analysis will enable facility managers to select the control strategy that best fits their needs based on consideration of internal and external constraints and financial limitations. Further, evaluations of any strategy must include the potential for unintended consequences within the system and the ecosystem of the source water. Hence, it is important that managers select appropriate sites for monitoring. As previously described, unintended consequences of mussel control measures include impacts on non-target organisms, public health, and water quality.

To avoid the spread of mussels, facilities need to develop a multi-pronged approach including public education and outreach, monitoring, assessment, and consideration of various control, containment, and management strategies. The goal of developing a control plan is to ensure that operation of the water infrastructure facility continues with minimal disruptions. To avoid further spread of mussel infestation and to control and reduce the size of infestations, the plan should document efforts and measures to: (1) protect water supply and delivery systems and (2) prevent spread of mussels to uninfested water bodies (e.g., California Fish and Game Act 2301).

In summary, a facility should consider the following recommendations for developing a mussel monitoring and control plan:

1. As a first key step, a facility with water bodies uninfested by dreissenids needs to conduct a vulnerability assessment if its source water has potential for infestation based on physical and chemical analyses (e.g., calcium, pH, dissolved oxygen, turbidity, temperature, and alkalinity). Common water quality parameters that correlate well with mussel survival and density are typically used to assess mussel infestation risk potential in water bodies (CH2M HILL 2007, CH2M HILL 2009, Mackie and Claudi 2010). This type of analysis provides insights into colonization potential or potential spread of populations based on a range of environmental parameters as they affect population growth. Accordingly, suitable proactive, preventative, and reactive measures should be selected from a toolbox of approved technologies as outlined previously. It is important to select appropriate control strategies for various components at various locations with an understanding of the potential impacts to non-target species. For a vulnerable but yet uninfested water body, a rapid response plan should be developed that takes a multi-pronged approach, including public education and outreach and monitoring.
2. Develop and implement a systematic monitoring program for measuring changes in conditions after mussel infestation.
3. Collaborate with public, local, and state agencies to ensure that appropriate measures are taken for protecting infrastructure and systems.
4. Prepare and implement a boat inspection program, and develop a decontamination plan for equipment and boats that have had contact with infested waters.
5. Conduct evaluations of the control options based on the efficacy, costs, ability to meet regulatory requirements, adaptability, and safety for use.

EXAMPLE APPLICATIONS OF MUSSEL CONTROL STRATEGIES

This section provides a few examples of mussel mitigation programs that have been developed for various water facilities in the southwestern United States since dreissenid mussels were first discovered in Lake Mead in January 2007. These examples illustrate management responses to infestations using the toolbox of options described earlier and include (1) protection of a large intake structure; (2) protection of a large outlet structure and transmission pipeline; (3) pre-selection of a multi-barrier approach to protect a raw water intake and transmission system; and (4) a strategy to protect a water conveyance system.

EXAMPLE 1: PROTECTION OF A LARGE INTAKE STRUCTURE

Background

The Southern Nevada Water Authority (SNWA) is a cooperative agency formed in 1991 to address southern Nevada's unique water needs on a regional basis. SNWA officials are charged with managing the region's water resources and providing for present and future water needs of residents and businesses in the Las Vegas Valley. The SNWA currently operates two large diameter raw water intakes in Lake Mead (Intake No. 1 and Intake No. 2). Each intake is capable of withdrawing approximately 2.27 million cubic meters of water per day (m^3/day) (600 mgd). The intakes feed two separate water treatment plants, one located on the shore of Lake Mead and the other approximately 16 (km) away within the Las Vegas Valley. Each intake has a coarse bar screen to help keep large debris from entering. A raw water intake pumping station is also associated with each intake to convey the water to separate treatment plants.

Due to persistent drought conditions in the Colorado River Watershed, low lake levels were close to impacting SNWA's ability to withdraw water from the original Intake No. 1, which is located approximately 15 m higher than the newer Intake No. 2. To provide long-term protection for the Las Vegas Valley water supply, the SNWA authority has embarked on constructing a third intake offshore in deeper water. This new Intake No. 3 will be about 42 m lower than Intake No. 2. Water will be conveyed from Intake No. 3 through approximately 4.8 km of 6.1 m diameter tunnel. To provide system flexibility, this tunnel will tie into the Intake No. 2 system, with the future ability to either be tied into the Intake No. 1 system or be tied to a third pump station facility located in the near vicinity.

Since its inception in 1991, the SNWA has worked proactively to ensure water needs of the area are met. This includes managing all of its intake, treatment, and conveyance systems. In 1997, part of the design of Intake No. 2 included a treatment system for dreissenids. Although at the time dreissenids were not yet present in Lake Mead, such a system would be very difficult to construct later into an operating and flooded tunnel system. SNWA therefore took advantage of the tunnel construction to install a chemical system and chemical feed lines to the intake. The chemical system included a dry feed silo for granular potassium permanganate and a day tank for mixing a permanganate solution for delivery through the chemical feed lines located within the tunnel to a dosing ring around the intake. The small treatment system was tested, but then put into storage until a possible need arose.

In 2007, quagga mussels were found in Lake Mead and have since spread throughout the lake. Between 2007 and 2010, mussels did not cause a significant impediment to water delivery and treatment, likely in part due to the large size of the intake facilities. Mussels have been found on coarse bar screens, and divers have been employed several times to examine the intakes and scrape off any mussels that were found. In December 2009, the Intake No. 2 system (tunnel and pumping station forebay) were shut down and dewatered to construct a connection for the future Intake No. 3. Some of the original components of the intake were replaced, and the 488 m long tunnel was inspected and cleaned. Figure 14.4 shows a typical infestation found on

Figure 14.4 **(See color insert.)** Quagga mussels on a carbon-steel debris rack of the water treatment plant in Lake Mead. Pen included for scale.

Figure 14.5 **(See color insert.)** Quagga mussels on the stainless-steel, flare end of the water intake in Lake Mead.

the coarse bar screen, and Figure 14.5 shows the infestation found on the flared end of the intake. Similar infestations were found throughout the tunnel, and approximately 6.0–7.5 m^3 of mussel debris (shells and other waste) was found in the bottom of the pumping station forebay.

Since the original infestation, SNWA has met with several agencies around the country to discuss their findings and experiences with mussel control. It has also partnered with several agencies such as the USBR and the Metropolitan Water District of Southern California (MWD) in performing in-lake studies to understand growth rates, growth cycles, and depths of the water column where veligers exist. Various materials and coatings have been studied to deter mussel attachment, and various treatment options have been devised and tested to kill mussels within the water conveyance system.

Prior to 2007, SNWA was injecting chlorine at Intake Pumping Station No. 1 for other treatment purposes. It was found that this chlorine treatment was also discouraging mussel growth on downstream portions of infrastructure components. In 2007, chlorine injection facilities were also added at Intake Pumping Station No. 2 with similar results. Neither of these facilities as constructed, however, protected the intake and tunnel system. SNWA considered starting up the potassium permanganate treatment system at Intake No. 2; however, preliminary pilot testing indicated that the treatment may impart a slight discoloration to the water, which would be undesirable. With planned construction for Intake No. 3, SNWA decided to move forward with chlorine treatment for mussel control at the intake itself. The decision was based on several factors, including previously noted successes in the system and pilot testing. Another important factor was the chlorination infrastructure that was already in place at the nearby treatment plant, which had sufficient capacity to also supply the mussel-treatment system. These existing facilities would provide a large savings over other options that would require additional infrastructure to be constructed.

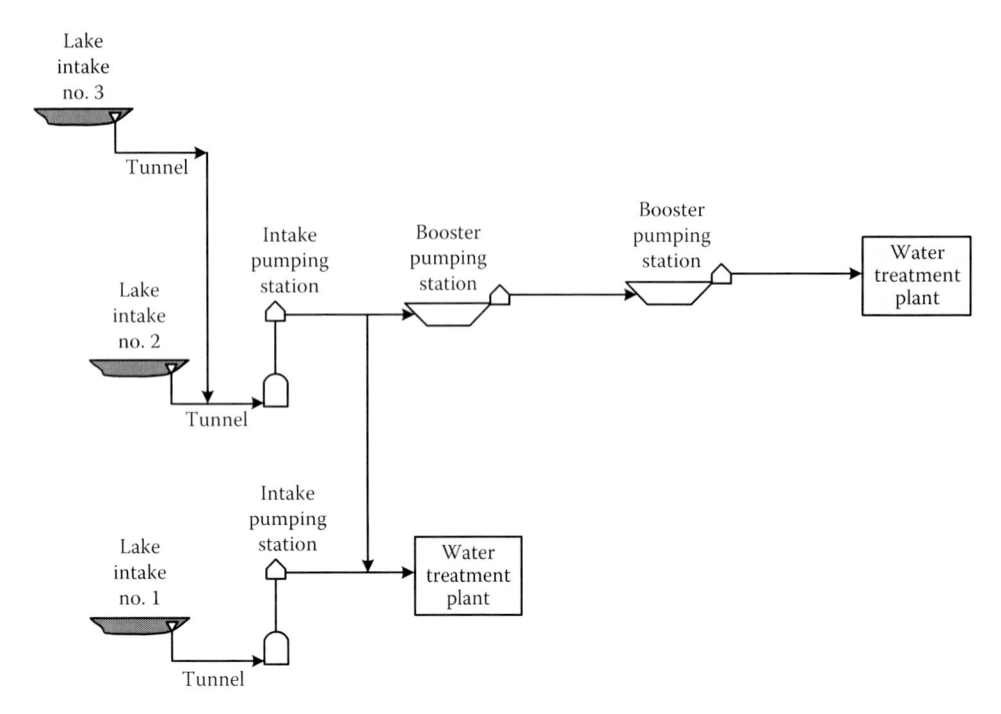

Figure 14.6 System schematic of the water intake and pipeline infrastructure of the water treatment plant in Lake Mead. The plant is operated and maintained by the Southern Nevada Water Authority.

Figure 14.6 provides a general sketch of the overall system. Intake No. 1 supplies raw water to the nearby water treatment plant on the shore of the lake, and Intake No. 2 supplies a separate water treatment plant located in the Las Vegas Valley. The third intake will connect into the Intake No. 2 tunnel system, with a possible future connection to the tunnel for Intake No. 1. An emergency bypass pipeline also allows raw water delivery from the Intake Pumping Station No. 2 to the treatment plant at the lake. Intake No. 2 is the only facility currently set up for chemical addition at the mouth of the intake. Intake No. 3, currently under construction, will also have this capability. Additional chemical application points are available throughout the raw water system, including locations at two booster pumping stations.

Chlorine Treatment at Intake No. 3

Chlorine treatment at the intake structure was planned using existing facilities located at the nearby water treatment plant. The chlorination system consists of bulk tanker storage for gaseous chlorine, evaporators, chlorinators, and injectors for production of a chlorine/water solution (CLS). The CLS would then be conveyed through chemical feed pipes installed within the tunnel system to the mouth of the intakes. For the long tunnel at Intake No. 3, there were two main challenges identified with this strategy. The first challenge was travel time through the long distance between the chlorinators and the application point at the intake (approximately 5.6 km). The second challenge was finding materials that would have maximum design life in the 4.8 km long tunnel. The design life criterion for the tunnel was 100 years.

Systems using gaseous chlorine typically have injectors to set up a vacuum that draws chlorine gas through chlorinators. The gas is mixed with the water flowing through the injectors and then introduced at the point of application. Injectors inherently have a fairly narrow flow range; therefore, chemical concentrations are usually modified by varying the rate of gas flow through the chlorinators at a constant injector flow. Pumping rates from Intake No. 3 can vary significantly over a 24 h period, ranging from 0 m^3/day at a minimum to about 3.48×10^6 m^3/day (920 mgd) at a maximum. The latter rate is based on existing capacity of potable water storage in the overall system and the desire to replenish storage using off-peak power for both pumping and water treatment facilities. When the chlorine solution is being conveyed 5.6 km away, travel time becomes an important factor, and chemical concentrations need to be predicted hours before the actual change in pumping rate. This increases chances for over- or underdosing due to uncertainty about when the chemical concentration change delivered in the chemical feed piping actually reaches the intake. The response to this challenge was to limit one of the variables by setting up a constant strength chlorine water solution that will be delivered through the 5 km of piping that makes up the chemical feed system to the intake. Dosing rates can then be immediately paced with raw water flow rates by varying pumping rates of the chemical solution. This removes travel time and variable CLS concentrations from the equation.

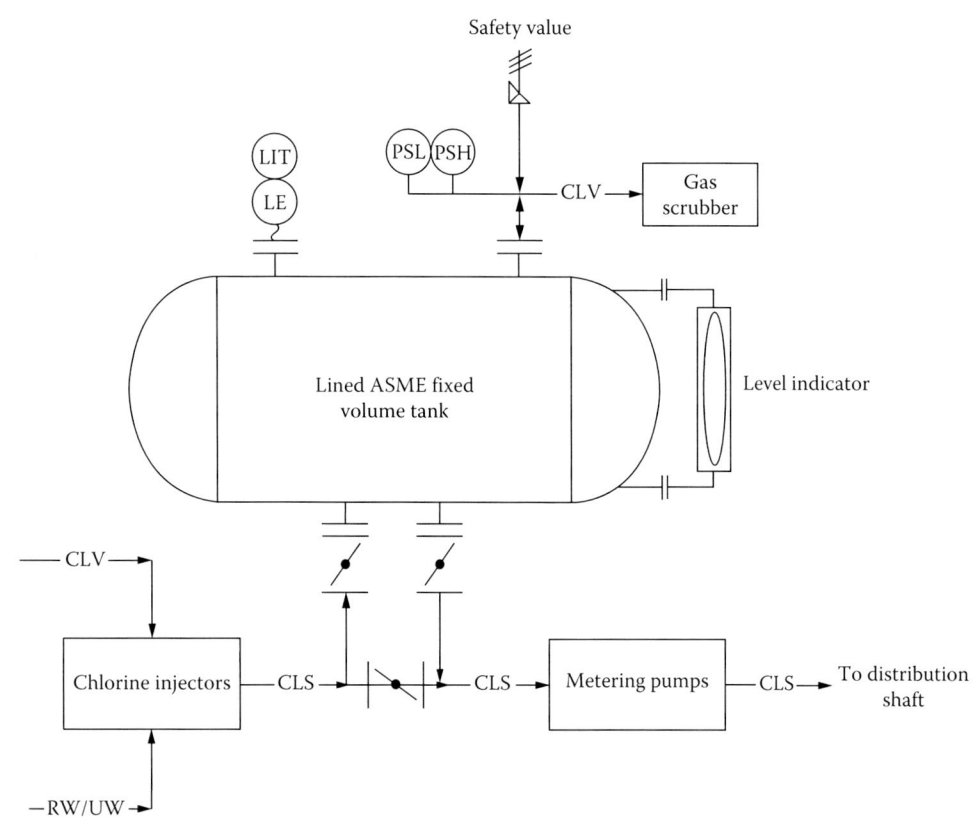

Figure 14.7 System schematic of the chemical preparation and storage facilities of the water treatment plant in Lake Mead. The system allows flexibility and immediate response to operation changes in raw water flows. ASME, American Society of Mechanical Engineers; LIT, level indicating transmitter; LE, level element; CLV, chlorine vacuum; PSH, pressure switch high; PSL, pressure switch low; RW/UW, raw water/utility water; CLS, chlorine/water solution.

To maintain a constant concentration of chlorine solution through injectors and allow immediate adjustments in dosing flow, both injector flow and chlorinator flow needed to be varied. Since injectors have a limited flow range, several injectors would need to be connected in parallel to cover the needed range of raw water pumping flows. This complicated control and operation of the overall system. To solve this issue, an alternative approach was developed of preparing a stock constant concentration of CLS that would be stored in a day tank and delivered to the intake through chemical metering pumps. On the upstream end, the day tank allowed a more constant flow through the injectors with the ability to store excess CLS. A constant concentration of CLS could then be metered to the intake to match the diurnal swings in flow. This allowed a more precise and immediate pacing of chemical dosing with raw water pumping rate. A schematic of the system is shown in Figure 14.7.

Final Design Criteria

The final design of the system allowed flexibility to provide both a shock dosage of chlorine and a lower concentration maintenance dose. The maintenance dose was for targeting and killing mussel veligers, and the shock dosage was established to kill any established adult mussels. Based on these two conditions, the following conservative design parameters were used:

Minimum raw water intake flow	6.4×10^5 m³/day (170 mgd)
Maximum raw water intake flow	5.0×10^6 m³/day (1320 mgd)
Maintenance dose rate	0.5 mg/L
Shock dose rate	2.0 mg/L
Tunnel component design life	100 years

Based on the raw water flow rate and dosing criteria, the final design included a 15 cm diameter chemical feed pipe extending from a new chemical control building, through the shaft and tunnel system, and to the deep water intake. The new building housed the CLS day tank and associated chemical metering pumps.

Special care was required at the intake structure to strike a balance between controlling mussel attachment and growth, preventing chemical releases into the lake, and preventing corrosion of bar screens. In order to confirm hydraulics of the system and mechanics of the chemical mixing, a computational fluid dynamics model was prepared of the intake and the chemical application points. The full range of raw water flows and chemical injection flows was modeled. Based on these results, the final design included a 15 cm diameter dosing ring located just inside of the circular bar screen at the top of the intake. Sixteen small 2.5 cm diameter chemical injection ports were distributed around the ring to help promote a quick and equal distribution of the chemical into the raw water.

Materials Analysis

A full materials analysis was performed to determine the best materials for conveying the acidic CLS to the intake. The analysis studied plastics such as high-density polyethylene (HPDE), polyvinyl chloride (PVC), chlorinated polyvinyl chloride (CPVC), and fiberglass. Teflon-lined pipes and tubing were also analyzed. Ultimately, the only material determined suitable for maximum life (approaching 100 years) was titanium, and therefore titanium was selected for the chemical feed piping installed in the tunnel and the dosing ring at the intake. PVC has also been used for similar solution applications; however, manufacturers were reluctant to associate more than a 20 year design life for the material. To save expense, PVC was used for portions of the system outside the tunnel that could more easily be replaced.

Evolution of a Mussel Control Strategy

Chemical feed piping and dosing rings for Intake No. 2 and Intake No. 3 were included in three separate contracts covering the construction of the new tunnel systems. Construction of the above-grade chlorine systems were planned for a later date following tunnel construction. In the interim, SNWA continued to study treatment approaches, discussed experiences of other system owners, and ran pilot plant trials. One of the concerns with chlorine application was the possible additional formation of undesirable DBPs.

One of the alternative treatments that showed promise for the SNWA system was the use of chloramines. Chloramines would be formed by introducing both a chlorinated water solution and ammonia at each intake. As with the gaseous chlorine system, SNWA also had sufficient aqueous ammonia storage and handling facilities at the nearby treatment plant to supply such a system at the intakes.

At Intake No. 2, most of the needed infrastructure was already available for running a full-scale test of treating mussels with chloramines. This infrastructure included a newly installed titanium feed pipe for the CLS and existing stainless steel pipes that were installed during the original construction as a feed lines for potassium permanganate. Late in 2011, SNWA started up a full-scale test of chloramine application at Intake No. 2. The test included delivering a 4:1 ratio of CLS to aqueous ammonia solution to the intake. The target application was 0.2 ppm (mg/L) chlorine and 0.05 ppm (mg/L) of ammonia. Ammonia was added about 1.2 m below the CLS in the vertical intake shaft. After about 3 weeks of dosing, the Intake No. 2 system was shut down, and divers inspected the 18.3 m tall vertical riser section. It was found that there was still some minor infestation between the CLS and ammonia application points. Below the zone of mixing, however, no mussels were found. Other benefits of chloramine treatment over straight CLS included (1) a smaller amount of total chemicals used; (2) minimal (0.1 ppm) degradation of chloramine concentrations through approximately 16 km of piping and three small balancing reservoirs; and (3) no detectable DBP formation. On the downside, there was significant scale formation of calcium hydroxide in the ammonia piping when raw water was used to dilute the aqueous ammonia solution. Further tests are planned, using softened water for dilution.

The aforementioned example focused on the use of chlorine as a chemical control, but a widely used chemical option in large treatment facilities is potassium permanganate. As discussed earlier, potassium permanganate is often used to assist in manganese removal and more commonly for taste and odor control. An oxidant, potassium permanganate has not been associated with the formation of DBPs and is not subject to the stringent environmental discharge restrictions of chlorine. Potassium permanganate in both powder or liquid solution was used as an example to evaluate feed rate. For potential use of potassium permanganate ($KMnO_4$) to treat quagga mussels in the SNWA system, feed rates based on dosage, solution strength, and tunnel flow rate parameters were calculated. Table 14.2 presents an example calculation of $KMnO_4$ solutions varying from 0.5% to 5.0% for doses of 0.15 and 0.30 mg/L of $KMnO_4$ under tunnel

Table 14.2 Feed Rates of Potassium Permanganate ($KMnO_4$) under Different Flow Rates at the Intake of the SNWA at Different Doses, % Solutions, and Solution Strengths. Min., Minimum; Max., Maximum

$KMnO_4$ Dose (mg/L)	$KMnO_4$ % Solution	$KMnO_4$ Solution Strength (kg/L)	$KMnO_4$ Feed Rate (m³/day) Raw Water Flow Rate of 379,000 m³/day	$KMnO_4$ Feed Rate (m³/day) Raw Water Flow Rate of 3,785,000 m³/day
0.15 (Min)	0.5 (Min)	0.005	11.3	113.6
	3.0 (Average)	0.031	1.9	18.9
	5.0 (Max)	0.051	1.1	11.3
0.30 (Max)	0.5 (Min)	0.005	22.7	227.1
	3.0 (Average)	0.031	3.8	37.9
	5.0 (Max)	0.051	2.3	22.7

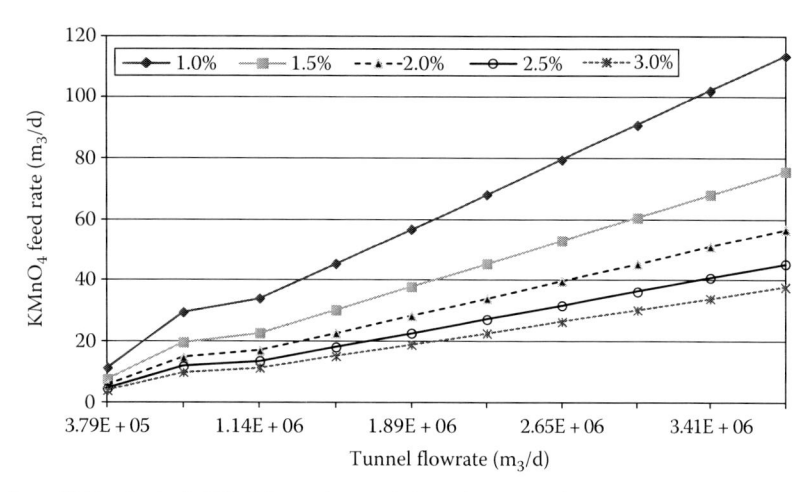

Figure 14.8 (See color insert.) Feed rate (m³/day) for various solutions (percent) of potassium permanganate (KMnO$_4$) relative to water flow rate (m³/day) in the intake tunnel at the water treatment plant in Lake Mead. Rates were calculated based on a KMnO$_4$ dose of 0.3 mg/L.

flow rates of 379,000 and 3,785,000 m³/day. In general, KMnO$_4$ feed rates increase with the increase in the raw water flow rate. A plot of feed rate of KMnO$_4$ for various solutions (percent) based on a KMnO$_4$ dose of 0.3 mg/L relative to various raw water flow rates in the intake tunnel is presented in Figure 14.8. The rate of change of KMnO$_4$ feed rate for 1% solution with the increase in tunnel flow rate is significantly higher than 3% solutions. Calculations were performed based on the following example parameters for the permanganate system:

Range of raw water flow	3.79×10^5 – 3.79×10^6 m³/day (100–1000 mgd)
Maximum dosing rate	0.30 mg/L
Minimum dosing rate	0.15 mg/L

KMnO$_4$ solution percentages range between 0.5% and 5.0%, with 3.0% being typical. Solution strengths are based on a potassium permanganate solution density of 1.02 kg/L (8.34 lbs/gal). Feed rates were calculated using the following equation:

$$\text{Feed rate} = \frac{[\text{Dosing rate}][\text{Tunnel raw water flow rate}]}{[\text{KMnO}_4 \text{ solution density}]\,[\text{KMnO}_4 \text{ solution strength}]}$$

Based on the flow rates and dosage parameters for the permanganate system, a minimum of 56.7 kg/day (125 lb/day) of solid KMnO$_4$ and a maximum of 1135 kg/day (2500 lb/day) of solid KMnO$_4$ are estimated. At 3% KMnO$_4$ solution, this equates to 1.9 m³/day (500 gal/day) at minimum to 38 m³/day (10,000 gal/day) at maximum.

Summary of Protection: A Large Intake Structure

The experiences of a large public water system, in this case SNWA drawing water from Lake Mead, demonstrates the challenges, considerations, and careful planning that are required to match a treatment approach for dreissenids with the goals and objectives of a facility to best serve their customers. Research, testing, design, and construction based on the experience of SNWA reveal the following considerations for similar systems: (1) treatment objectives and alternatives are dynamic and can change over time; (2) material and coating research is ongoing with somewhat mixed results to date—these items may be part of an overall scheme, but probably not the single answer for mussel control; (3) the best approach is somewhat dependent on individual system size and characteristics; (4) criteria for life-time designs should be evaluated on how easy it is to replace a component in the future, and requirements for long design life can greatly impact facility cost (i.e., use of titanium components); (5) careful selection of system materials provides the flexibility for treating with a variety of chemicals if strategies change over time; (6) chemicals already in use can save costs in infrastructure and chemical delivery; (7) consider how mussel treatment chemicals may impact other treatment processes or what byproducts may be formed from chemical use; (8) understand the hydraulics of the raw water system that may affect treatment configuration (i.e., range of flows, diurnal flow variations, chemical travel time, and chemical demand); and (9) consider impacts of treatment to finished water (i.e., odor, taste, or color).

EXAMPLE 2: PROTECTION OF A RESERVOIR DAM OUTLET STRUCTURE AND TRANSMISSION PIPELINE

Background

Pueblo Dam is located on the Arkansas River about 10 km upstream and west of the city of Pueblo, Colorado. The dam and associated reservoir is a major water storage facility owned and

Figure 14.9 (See color insert.) Pueblo Dam and Reservoir on the Arkansas River near Pueblo, Colorado. See also Prescott et al. (2013).

operated by USBR as a component of the Fryingpan-Arkansas (Fry-Ark) project. Fry-Ark is owned and operated by the Reclamation. The Southeastern Colorado Water Conservancy District (SECWCD) distributes water diverted by the project from Colorado's western slope. Fry-Ark was constructed in the early 1970s and has a total storage capacity of approximately 4.4×10^8 m^3 (358,000 acre ft) for multiple benefits including agricultural irrigation, municipal and industrial water supply, flood control, and recreation (Figure 14.9).

The dam has both earth fill and concrete gravity sections. The central concrete buttress-type gravity section is approximately 533 m long and has a maximum structural height of 76 m. This central section includes a 167 m wide overflow spillway designed for a maximum spill discharge of 5,423 m^3/s (191,500 cfs). The concrete buttress section includes several outlet works in addition to the river outlet. The presence of dreissenid (veligers) in Pueblo Reservoir was confirmed in 2008. Veligers were detected again in a single sample in May 2011. However, adult mussels have not been identified during ongoing reservoir or facility sampling activities.

As part of Colorado Springs Utilities' Southern Delivery System (SDS), a connection to the dam was made at the River Outlet Works (known as the Pueblo Dam Connection) for source water. SDS is a comprehensive water supply project that will convey raw water from the Pueblo Reservoir to end users in Pueblo West, Colorado Springs, Fountain, and Security, Colorado. Elements of the system will include a connection to the Pueblo Reservoir, three raw water pump stations, over 80 km of raw water conveyance pipeline, a terminal storage reservoir, a water treatment plant (WTP), a water pump station, and water transmission pipelines from the WTP to connection points within the Colorado Springs treated water distribution system (Figure 14.10). The project is hydraulically sized to convey up to 295,300 m^3/day (78 mgd) from Pueblo Reservoir to the WTP and 113,600 m^3/day (30 mgd) to the Pueblo West Pump Station via the raw water conveyance system. The WTP will be sized with the ultimate capacity to deliver up to 681,400 m^3/day (180 mgd) to end users on a peak-day basis once the new terminal storage reservoir is constructed. During the design of the Pueblo Dam Connection, the design team developed control approaches for dreissenids and management strategies that included pipeline coatings, chemical control, and construction materials to minimize mussel impacts.

Facility

The Pueblo Dam Connection begins with a connection to an existing cast iron flange on the square outlet gate of the River Outlet Works inside Buttress No. 16 of Pueblo Dam. The design accommodates a maximum velocity of 21 m/s (70 fps) within the 1.2 m outlet tunnel. Flow through the river outlet will be dependent on reservoir water levels as well as passage through a new fixed cone valve prior to discharge; however, water velocity should be sufficient to prevent veliger settlement. A tapered rectangular stainless steel liner will be installed into the existing concrete discharge tunnel and will transition to a circular pipe section. The circular pipe transition will taper up to a 2.3 m pipe over the length of the existing dam tunnel. After the 2.3 m pipe exits the outlet tunnel, it bifurcates into two segments. The primary pipeline segment continues following the centerline of Buttress No. 16 to the fixed cone valve facility and discharge to the Arkansas River. The secondary pipeline segment is the 2.3 m turnout to the SDS/joint users pipeline, which is exposed welded steel pipeline (approximately 23 m) as it

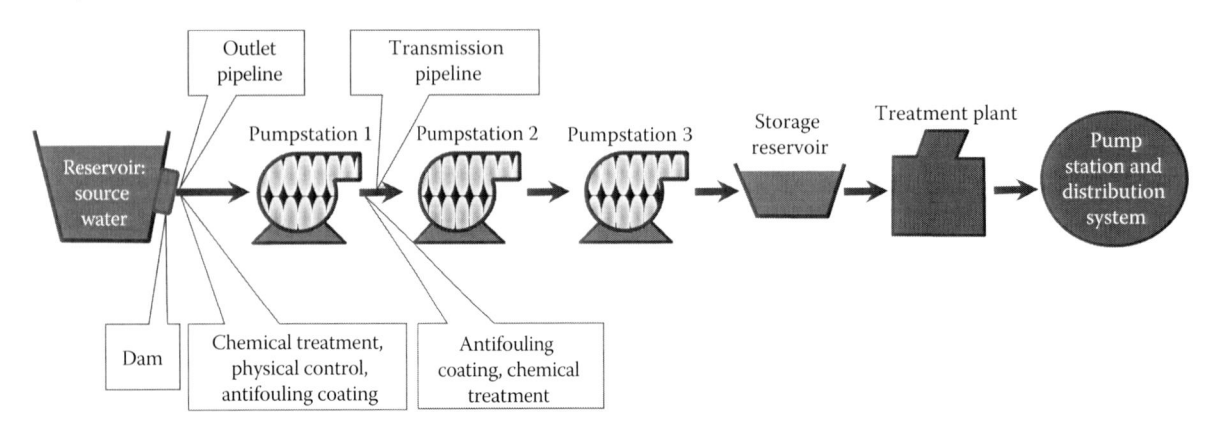

Figure 14.10 System components and chemical injection points for the control of dreissenid mussels at Pueblo Dam and Reservoir.

spans the Arkansas River and becomes buried on the north bank of the river. The turnout pipeline continues along the north bank of the Arkansas River to a meter vault and then east and north until it reaches the first of the SDS raw water pump stations. The pipe from the bifurcation to meter vault located on the north river bank will be coated with mussel control protective coatings. Chemical injection capabilities will be provided at the meter vault for future chemical injection. The pipeline downstream of the meter vault will be protected with this future chemical injection system, which will be fed from the SDS raw water pump station.

Design of a Dreissenid Mussel Control System

As noted, veligers (from mussel DNA tests) have been detected in Pueblo Reservoir, but adults have yet to be found. If mussel abundances increase in the reservoir, it is likely that surfaces associated with existing dam facilities and the new Pueblo Dam Connection may become infested with mussels. Additionally, if dense colonies of mussels were to establish in the reservoir, the organic matter and metabolic wastes produced in the form of pseudofeces, together with the decomposition of dead mussels, could potentially impact the quality of the raw water by increasing TOC and taste and odor compounds. Potential impacts of mussel infestation to the dam outlet can be grouped into the following three categories: (1) hydraulic impacts, which reduce hydraulic capacities of trash racks and internal walls of the outlet piping; (2) functional impacts, which affect trash racks, sluice gate guides and stems, and pipeline valves; and (3) maintenance impacts, which increase maintenance efforts as related to mechanical scraping or line pigging and collection and removal of dead-mussel shells. Primary mussel control in the piping to the raw water pump station will be accomplished through chemical treatment. Chemical feed systems will be installed in a 6.4 cm diameter chemical feed line routed in a secondary containment pipe and installed parallel to the raw water pipeline. The pipe will convey a chemical molluscicide, such as potassium permanganate, to the butterfly valve vault that is used to control the flow from the river outlet pipeline connection. A diffuser ring was installed in the raw water pipeline at the vault to inject the chemical (potassium permanganate) into the raw water supply at a dosage ranging from 0.5 to 5 mg/L whenever water is withdrawn from the reservoir.

The dose is approximately 0.5 to 1 mg/L greater than the oxidant demand of the raw water. Sensors (i.e., to measure oxidation reduction potential) are installed in the pipe downstream of the diffuser to ensure a residual concentration is maintained. The location of the diffuser ring, approximately 152 m downstream of the reservoir, prevents the chemical from diffusing back into the reservoir or outlet tunnel. Locating the diffuser ring in the raw-water pipeline and downstream of the connection to the river-outlet works minimizes accidental release of the chemical into the river. Small-diameter pipings located upstream of the chemical diffuser that cannot be effectively coated are constructed of copper tubing.

Protective coatings are also being planned between the bifurcation structure and the valve vault where the chemical diffuser ring is located. These coatings are recommended for use in a continuous water flow at a minimum velocity to keep mussels from pipe attachment. One tested coating (Sher-Release Coating System) is a four-coat epoxy/silicone technology that recommends a minimum velocity of 0.15 m/s (0.5 fps) in fresh water, which is less than the expected velocity in the pipeline. Therefore, normal use of the pipeline should provide adequate velocities. Under stagnant water conditions, mussels can attach, but byssal adhesion should be weak; per the manufacturer, mussels can be flushed out when velocities are subsequently increased to greater than 4.3 m/s (14 fps). The coating can withstand velocities up to 21.9 m/s (72 fps) per manufacturer testing on boat hulls. Therefore, scheduled flushing can be part of operational procedures. Even in stagnant water conditions where flushing cannot be performed, coatings ease the cleaning process. Coatings are anticipated to last 5 to 10 years, although this time frame can vary significantly.

As noted previously, only one manufacturer of foul-release coatings has obtained NSF 61 approval (Charles Fisher, personal communication). Protective coatings may not be necessary for some of the deeper intake trash racks at the dam facility because low oxygen concentrations may inhibit mussel colonization. For instance, Table 14.3 provides a summary of dissolved oxygen concentrations at various depth intervals in spring and summer in 2006 and 2008. Some intakes of the facility are located at 15 m and, at this depth, DO concentrations are <4 mg/L on a consistent basis during the late summer months. Hence, mussel colonization potential is expected to be very low.

Table 14.3 Dissolved Oxygen Concentrations (mg/L) at Various Depth Intervals near the Intake Structure of a Dam Facility in the Southwestern United States. The Intake Is Located 15 m from the Water Surface. SD, Standard Deviation; Max., Maximum; Min., Minimum

Depth		All Seasons				Spring			Summer		
From (m)	To (m)	Mean	Max	Min	SD	Mean	Max	Min	Mean	Max	Min
—	<8	6.72	8.40	2.40	1.31	7.24	8.40	6.50	5.59	7.60	2.80
8.1	14	5.38	8.10	0.90	1.99	6.31	8.20	4.50	3.94	6.20	0.90
14.1	19	4.43	8.10	0.70	2.31	5.53	8.10	4.20	2.94	5.50	0.70
19.1	25	3.93	7.70	0.00	2.53	5.17	7.70	3.50	1.99	3.90	0.00
>25	—	3.60	7.50	0.00	2.51	4.38	7.50	1.40	0.62	3.00	0.00

Summary of Protection: A Reservoir Dam Outlet Structure and Transmission Pipeline

The water transport system associated with Pueblo Dam provides an example of protecting a water infrastructure facility from mussel infestation at a dam intake structure, transmission pipeline, and water treatment plant as outlined in Figure 14.10. A series of control measures were implemented including velocity control, foul-release coating, and chemical control. Potassium permanganate was identified as a potential chemical control for mussels that could also address potential taste and odor issues. Coordinating with various regulatory agencies to receive approval for treatment and applying coating material are the core control measures proposed for this system.

EXAMPLE 3: PRE-SELECTION OF A MULTI-BARRIER APPROACH TO PROTECT A RAW WATER INTAKE AND TRANSMISSION SYSTEM

Background

This example of a multi-barrier approach features a water plant on a lake in the southwestern United States that receives source water from a segment of a large river that is not yet infested with dreissenid mussels, but has habitat conditions to potentially support future mussel infestations. The development of the multi-barrier mussel control strategy includes preventative approaches, non-chemical operational control measures, and proactive and reactive chemical control methods to minimize impacts from any future mussel infestations. A screened intake of the water treatment plant draws up to 757,000 m³/day (200 mgd) of water from the river. The intake includes a pump station that discharges raw water into the overall system. The full system also includes pipelines, a water treatment plant, flow control structures, and a canal discharge structure (Figure 14.11). The flow discharged to the canal eventually reaches water supply reservoirs further downstream in the system. A key challenge of any program to control and eradicate mussels is the need to protect various components of the overall system while minimizing negative impacts of control measures. A multiple barrier approach was determined to be the most applicable mussel control strategy for the protection of this facility.

Multi-Barrier Control Approach

The multiple barrier control approach planned for this system includes a combination of various methods as outlined previously, including desiccation, velocity control, antifouling coatings, operational controls, and chemical treatment of waters containing veligers. Complete drying of system components, when possible, was recommended as the preferred method to kill settled veligers and adults, but, in reality, desiccation was a viable approach for mussel control only on the intake screens and panels. Under a velocity control strategy, the system was designed to keep a minimum 6 fps (1.8 m/s) velocity in the pipelines to prevent veliger settlement (Boelman et al. 1997). Velocity control provided the following maximum flow rates in various sections of the distribution system:

Between intake and bifurcation	564,000 m³/day (149 mgd)
Between bifurcation and canal	416,000 m³/day (110 mgd)
Between bifurcation and WTP	348,000 m³/day (92 mgd)

These flow rates are important for the consideration of chemical control and available chemical contact time as discussed later. Note that effectiveness of chemical control measures is a function of chemical contact time and dose strategy. Infrequent high doses can be used for shock treatment of adults or continuous low doses can be used for veliger treatment as based on monitoring results of mussel reproductive periods.

Antifouling Coating

From the previously mentioned studies performed by USBR on coating materials, it was found that antifouling materials provided reasonable performance, especially when

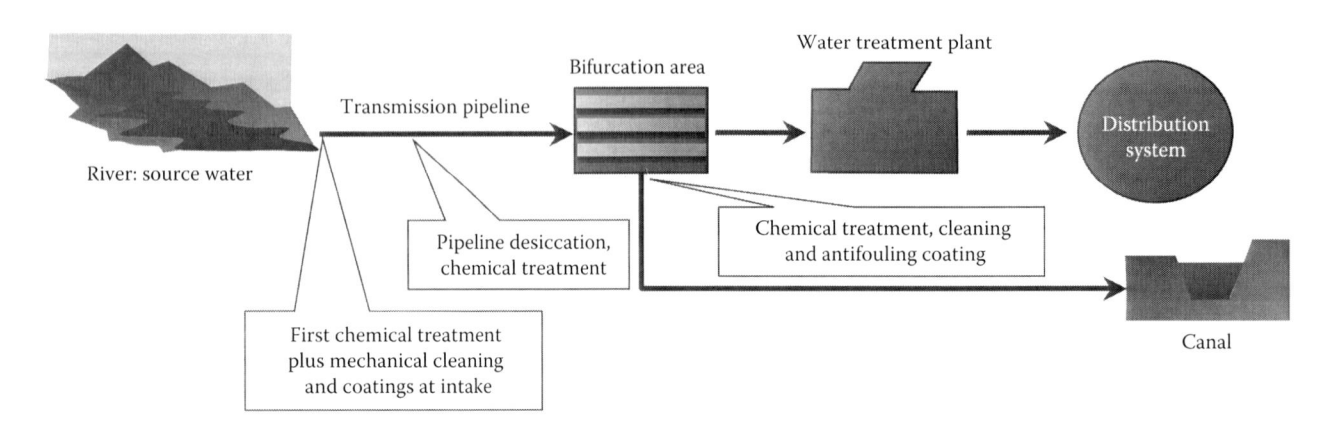

Figure 14.11 Infrastructure of a water treatment plant showing potential locations for chemical injection to control dreissenid mussels.

combined with increased velocity (Allen Skaja, Personal communication). Antifouling coatings were found to reduce cleaning effort in most cases, particularly for areas with low water velocity (USBR 2012). As mentioned earlier, effective life of coating materials is reported to be about 6 years (Wells and Sytsma 2009). It was determined that coating materials could be applied on diffuser panels, on exposed piping/hardware, and on some appurtenant structure facilities. It was recommended that copper piping be considered in fish screens and fasteners. After recent NSF approval of the Sher-Release coating (as described previously), it was determined that this coating could be applied on the internal surface of pipes larger than 1.37 m in diameter.

Chemical Treatment Plan

The strategy for implementation of a chemical treatment approach for the system included investigations of commonly implemented control approaches with a proven record of performance in similar systems. The chemical control plan was based on monitoring results and the degree of infestation (adult and/or veliger stages). Both infrequent high doses and seasonally adjusted low doses were planned. Treatment options depended on the status of infestation at the time of application.

Chlorine was selected as the primary chemical for controlling mussels. Flow in various sections of the system was an important parameter in assessing chemical options. As shown in Figure 14.12, detention time for contact with chemicals was higher at lower flow rates, which allows chemicals to react with mussels for longer periods of time. The transmission pipeline between the intake structure and the bifurcation area had large variations in flow and detention times that varied from about 2 to 22 h. By the time water reached the water treatment plant, the chlorine had sufficient time to react with the mussels even with higher flow rates (about 5 h at 379,000 m^3/day [100 mgd]). Higher flow rates also allowed sufficient mixing. Obviously, lower flow rates provided relatively longer detention times for the water treated at the intake structure and allowed longer reaction time before reaching the treatment plant. In general, at a design flow rate of about 700,000 m^3/day (185 mgd) for this particular water infrastructure facility, a chlorine dose of 1.6 mg/L was determined to have a contact time of about 4 h, and this dose was deemed sufficient to control dreissenids (Brady et al. 1996).

Sodium bisulfite or sulfur dioxide was planned for dechlorination or quenching if the residual chlorine concentration became higher than the acceptable limit in water discharged to the canal. The dosage of the dechlorination chemical depended on residual chlorine concentration requirements in the treated water.

Operational Actions

Various operational actions were planned to evaluate the effectiveness of control measures, and such actions included monitoring of vulnerable areas. Effective monitoring is a function of the frequency of measurements, which in turn depends on the intensity of mussel infestation in the system and location of sampling. Monitoring also helps in assessing if further modification to the treatment method is

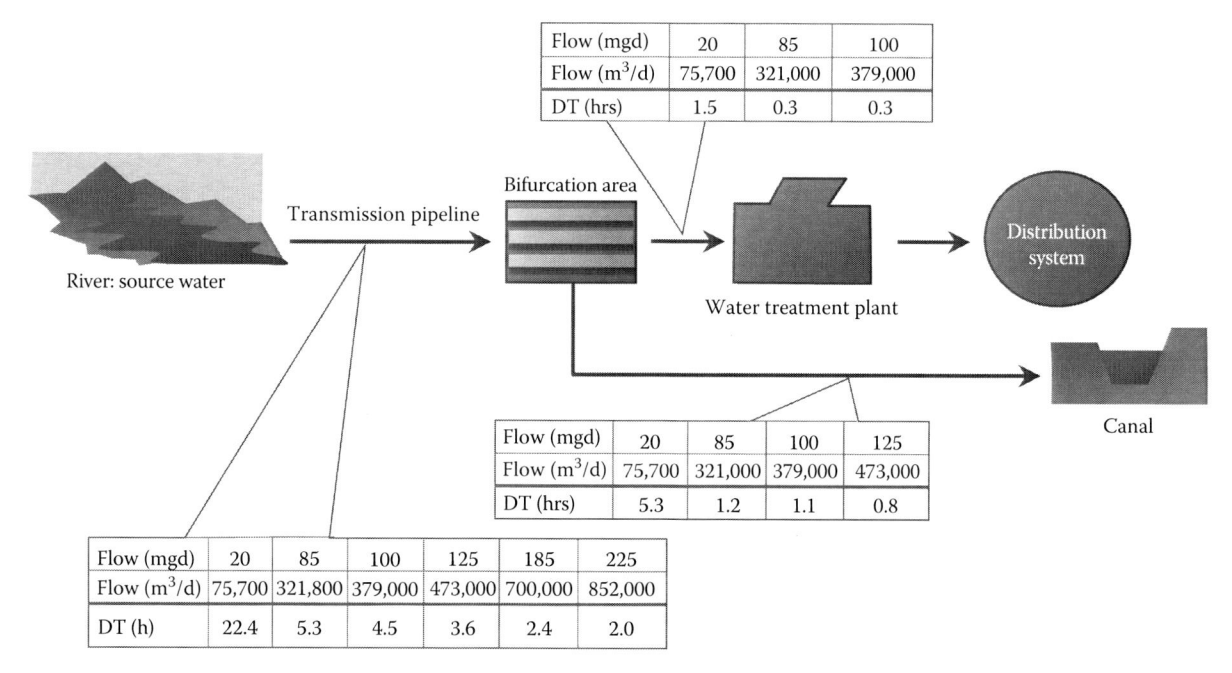

Flow (mgd)	20	85	100
Flow (m^3/d)	75,700	321,000	379,000
DT (hrs)	1.5	0.3	0.3

Flow (mgd)	20	85	100	125
Flow (m^3/d)	75,700	321,000	379,000	473,000
DT (hrs)	5.3	1.2	1.1	0.8

Flow (mgd)	20	85	100	125	185	225
Flow (m^3/d)	75,700	321,800	379,000	473,000	700,000	852,000
DT (h)	22.4	5.3	4.5	3.6	2.4	2.0

Figure 14.12 Direction of water flow, water flow rate (mgd = million gallons per day, m^3/day = cubic meters per day) and detention time (DT in hours) of chemicals in various infrastructure components of a water treatment plant in California. In this particular example, the chemical is chlorine (dose = 1.6 mg/L). Million gallons per day (mgd) is equivalent to 3,785 m^3/day.

required. Routine activities and actions were also planned to evaluate the effectiveness of chemical control actions. For instance, underwater inspections of wet wells were planned to understand sensitivity of physical/chemical conditions on growth of the mussels and to maintain function of the wet wells.

Operational actions are a critical part of assessing the multiple barrier approach. Such actions included cleaning and scraping the log boom annually with possible application of a foul-release coating, managing the fish screen by physical cleaning annually or quarterly based on the status of infestation, rotating spare panels for desiccation, and acquiring spare parts when one screen is down for maintenance. Key operational actions for maintenance of the pumps included drying pumps periodically, cycling pumps using a chemical flush to remove any mussels sheltered within components of the pump, and minimizing the pump idle period so that dreissenids do not have enough time to settle within the pipes and pump components. Continuous operation of chains and scrapers was planned in order to keep moving parts free of mussel settlement, and continuous recirculation of water in tanks and wet wells was deemed an efficient way to control mussels. Continuous chemical flush, physical cleaning/scraping, and periodic dry-out of tanks were also used to control mussels in elevated or underground tanks.

Finally, as a performance measure, monitoring of the system was planned using information from the Supervisory Control and Data Acquisition (SCADA) system. Information on water flow and water quality in tanks and pipes would indicate any mussel growth and infestation.

Summary of Protection: Multi-Barrier Approach for Raw Water Intake and Transmission System

Based on status of the facility and potential for mussel impacts, a cost-effective, permanent control strategy with a multi-barrier approach for an intake and transmission system in the southwestern United States included mechanical (such as scraping), chemical (chlorine), desiccation, and antifouling coatings (foul-release coating). Areas for control measures included the water intake, transmission pipeline, and bifurcation area.

The plan included continuous monitoring of the intake structure to document status of mussels, installing systems to clean pipes and facilities periodically, providing environmentally safe, effective coatings on intake structures, acquiring regulatory approvals before applying any control option, and coordinating with state and federal regulatory agencies along with partnerships with research agencies for effective mussel control planning. A key component of the plan was to implement automatic control for injection of chemicals and online continuous monitoring so that proper actions could be taken at any stage of the control plan.

EXAMPLE 4: STRATEGY TO PROTECT A WATER CONVEYANCE SYSTEM

Background

A study was conducted to evaluate various treatment options for control of dreissenids and slime for a water conveyance system that supplies drinking water to a water supply district. There was concern that mussels would enter the source-water lake and then eventually enter the holding reservoir. Raw water is screened and chlorinated at the intake and then flows 160 km by gravity to a holding reservoir. A water treatment plant draws water from the reservoir and treats it with dissolved air floatation (DAF), ozonation, deep bed biological activated carbon (BAC), free chlorine, and UV before discharge to the distribution system. In order to prevent mussels from entering the reservoir, and to control biofilm and mussels in the water transmission and distribution pipeline, a vulnerability assessment was conducted in the lake and several control and chemical treatment options at the intake were evaluated.

Treatment Options

Table 14.4 provides a list of water quality conditions judged suitable for high, moderate, and low mussel growth potential compared to conditions in the lake that provided the source water to the water treatment plant. It was found that the lake was vulnerable for mussel settlement and growth due to water quality conditions suitable for high to moderate mussel growth (highlighted in bold type in the table). Table 14.5 provides a list of various chemical options for controlling slime and mussels at the intake structure, conveyance system, and in the reservoir. As shown, chlorination, chloramination, and copper-ion addition provides excellent results for control of both slime and dreissenids. Of these, chloramination was preferred because chloramines were more stable and lasted longer than free chlorine, and it provided good protection against biofilm (slime) growth in conveyance systems. In addition, chloramines are not as reactive as chlorine with organic material in water and thereby produced fewer DBPs and provided a lower oxidant demand.

Copper ion generation was considered a viable control option because it has been used at power generating facilities and seems to be one of the emerging treatments in municipal water systems. It produces no DBPs and is effective against biofouling by both slime and mussels (O'Neill 1993, Yaroslavtseva and Sergeeva 2007). Yet, to protect against harming other aquatic life, the copper ion must be confined to a closed system. Some considerations of copper ion treatment include (1) copper ions are able to penetrate cell walls of small organisms and disrupt cell functions; (2) dreissenids are sensitive to copper ions, and dosages of 5–10 µg/L are usually sufficient for mussel control; (3) copper ions are easily produced through electrolysis; (4) toxic properties of ionic copper inhibit both slime growth and veliger settlement; (5) advantages

Table 14.4 Environmental Conditions Considered to Be Variously Suited (High, Moderate, Low) for Zebra Mussel Growth. Conditions Are Generally Similar for Quagga Mussels, with Slight Differences. Also Given Are Environmental Conditions and Suitability for Mussel Growth in a Lake That Provides Source Water for a Water Supply System

Environmental Parameter	Source	High: Favorable for Optimal Growth	Moderate Growth Potential	Little Growth Potential — For Veligers	Little Growth Potential — For Adult Mussels	Lake's Observed Water Quality	Suitability Evaluation
Alkalinity (as mg $CaCO_3$/L)	Hincks and Mackie (1997)	>60	40–60	17–40	<17	61–88	High to moderate growth potential
	Mackie and Claudi (2010)	100–280	**55–100**	30–55	<30		
Calcium (mg/L)	Whittier et al. (2008)	>28	**20–28**	12–20	<12	18–24	Moderate growth potential
	Mackie and Claudi (2010)	>30	**15–30**	8–15	<8		
Conductivity (µS/cm)	Sorba and Williamson (1997)	>83	37–82	22–36	<22	122–180	High growth potential
	Mackie and Claudi (2010)	>110	60–110	30–60	<30		
pH	Koziowski et al. (2002); Doll (1997)	**7.4–8.7**	6.8–7.4 or **8.7–9.5**	—	<6.8 or >9.5	7.4–9.3	High to moderate growth potential
	Mackie and Claudi (2010)	8.2–8.8	7.8–8.2 or 8.8–9.0	7.0–7.8 or 9.0–9.5	<7.0 or >9.5		
Temperature (°C)	Koziowski et al. (2002)	**15–31**	31–32	—	<10 or >32°C	1.5–26	High to poor growth potential
	Mackie and Claudi (2010)	20–26	10–20	26–32	<10 or >32°C		
Total Hardness (mg $CaCO_3$/L)	Sorba and Williamson (1997)	>90	**45–90**	25–45	<25	65–92	Moderate growth potential
	Mackie and Claudi (2010)	100–280	**55–100**	30–35	<30		

Bold type indicates water quality conditions pertinent to the source water lake in Example 4 of text.

Table 14.5 Relative Suitability of Potential Chemical Treatments Considered for Control of Biofilm and Dreissenid Mussels. Also Given Is the Tendency of Each Treatment to Form Undesirable Byproducts. DBP, Disinfection Byproduct; HAA, Haloacetic Acid; THM, Trihalomethanes

Chemical Treatment	DBP Formation	Biofilm (Slime) Control	Mussel Control
Chlorination	High THMs and HAAs	Very good	Excellent
Chloramine	Low THMs and HAAs	Excellent	Very good
Chlorine dioxide	High chlorite and chlorate	Good	Good
Ozone	No THMs and HAAs	Poor	Poor in conveyance system and reservoirs
Permanganate	No THMs and HAAs	Fair	Poor
Copper ion	No THMs and HAAs	Good	Excellent

include absence of chemical handling and DBP formation; and (6) while copper ion addition may increase concentrations entering water treatment plant systems, water treatment processes can effectively remove copper; for instance, some copper ions may be removed by iron coagulation at the plant. Some negative aspects of copper ion treatment in the water conveyance system include (1) difficulties in regulation and verification of copper doses needed to control mussels; (2) difficulties in maintenance and installation of a point-of-application; and (3) effectiveness of the treatment for slime control may be compromised in such a long conveyance system.

Permanganate treatment was considered a poor treatment option for this system. Although permanganate reduces oxidant demand, controls taste and odors, helps oxidation of iron and manganese, and does not form DBP, it is less effective than chlorine at equal doses, toxic to aquatic life, and notorious for pink water formation (Table 14.1).

Summary of Protection: A Water Conveyance System

Seasonal and/or intermittent chlorination, chloramination, and copper ion addition were selected as potential mussel control measures for this intake structure, reservoir, and associated water supply system.

ACKNOWLEDGMENTS

The material presented in this document is derived from project experience of the authors where various control measures were applied to solve practical problems for the management of infrastructure facility. The authors would like to acknowledge review comments from various individuals, which have helped improve the content substantially. Some of the reviewers/contributors include Ted Davis, Stephanie Harrison, and Paul Wobma of CH2M HILL. Assistance of Dana Rippon, Phil Ryan, Hannah Wilner, and Tim Yamada of CH2M HILL and Kara Lamb of Reclamation during manuscript preparation is highly appreciated. Information provided by Allen Skaja of Reclamation and Charles Fisher of Fujifilm Smart Surfaces, LLC, is highly appreciated. We gratefully acknowledge the support provided by CH2M HILL during preparation of this document. Authors greatly appreciate review comments provided by Thomas F. Nalepa and Donald F. Schloesser.

REFERENCES

Alvarez, M., J. Rose, and B. Bellamy. 2010. Treating algal toxins using oxidation, adsorption, and membrane technologies, Water Research Foundation, Report 2839.

Bierman, V. J. Jr., D. M. Dolan, R. Kasprzyk, and J. L. Clark. 1984. Retrospective analysis of the response of Saginaw Bay, Lake Huron, to reductions in phosphorus loadings. *Environ. Sci. Tech.* 18: 23–31.

Bierman, V. J. Jr., J. Kaur, J. V. DePinto, T. J. Feist, and D. W. Dilks. 2005. Modeling the role of zebra mussels in the proliferation of blue green algae in Saginaw Bay, Lake Huron. *J. Great Lakes Res.* 31: 32–51.

Boelman, S. F., F. M. Neilson, E. A. Dardeau, Jr., and T. Cross. 1997. *Zebra Mussel (Dreissena polymorpha) Control Handbook for Facility Operators*. Misc. Paper EL-97-1. U.S. Army Corps of Engineers Waterways Experiment Station, Vicksburg, MS.

Brady, T. J., J. E. Van Benschoten, and J. N. Jensen. 1996. Chlorination effectiveness for zebra and quagga mussels. *J. Am. Water Works Assoc.* 88: 107–110.

Canale, R. P. and S. C. Chapra. 2002. Modeling zebra mussel impacts on water quality of Seneca River, New York. *J. Environ. Eng.* 128: 1158–1168.

California Department of Fish and Game (CDFG). 2011. Infested reservoirs in California—A summary of the quagga and zebra mussel infestation. http://dfg.ca.gov/invasives/quaggamussel/ (accessed May 15, 2013).

California Science Advisory Panel, 2007. California's response to the zebra/quagga mussel invasion. May 2007.

Carmichael, W. W. 1996. Toxic microcystis and the environment. In *Toxic Microcystis*, M. F. Watanabe, K. H. Harada, W. W. Carmichael, and H. Fujiki eds., pp. 1–11. Boca Raton, FL: CRC Press.

Chakraborti, R. K., J. Kaur, and J. V. DePinto. 2002. Analysis of factors affecting zebra mussel (*Dreissena polymorpha*) growth in Saginaw Bay: A GIS-based modeling approach. *J. Great Lakes Res.* 28: 396–410.

CH2M HILL. 2007. Preliminary assessment of the vulnerability of state water project facilities to potential infestation by quagga mussels. A report prepared for the California Department of Water Resources, Sacramento, California.

CH2M HILL. 2009. Quagga mussel monitoring and control plan for Irvine Lake, California. A report developed for Irvine Ranch Water District and Serrano Water District.

Claudi, R. and G. L. Mackie. 1994. *Practical Manual for Zebra Mussel Monitoring and Control*. Boca Raton, FL: CRC Press.

Claudi, R. and G. L. Mackie. 2000. *The Practical Manual for Zebra Mussel Monitoring and Control*. Boca Raton, FL: CRC Press.

Cohen, A. N. 2007. Potential distribution of zebra mussels (*Dreissena ploymorpha*) and quagga mussels (*Dreissena bugensis*) in California. Phase 1 report. A report prepared for the California Department of Fish and Game.

DeGirolamo, D. J., J. N. Jensen, and J. E. Van Benschoten. 1991. *Inactivation of Adult Zebra Mussels by Chlorine*. American Water Works Association Annual Conference, June 1991, Philadelphia, PA.

De Leon, R. 2008. The silent invasion: Finding solutions to minimize the impacts of invasive quagga mussels on water rates, water infrastructure and the environment. Testimony before the US house of representatives committee on natural resources subcommittee on water and power.

Doll, B. 1997. *Zebra Mussel Colonization: North Carolina's Risks*. Sea Grant North Carolina, University of North Carolina, Raleigh, NC (UNC SG-97-01).

Effler, S. W., S. R. Boone, C. A. Siegfrid, L. Walrath, and S. L. Ashby. 1997. Mobilization of ammonia and phosphorus by zebra mussels (*Dreissena polymorpha*) in the Seneca River, New York. In *Zebra Mussels and Aquatic Nuisance Species*, ed., F.M. D'Itri, pp. 187–207. Boca Raton, FL: CRC Press.

Fahnenstiel, G. L., T. B. Bridgemen, G. A. Lang, M. J. McCormick, and T. F. Nalepa. 1995b. Phytoplankton productivity in Saginaw Bay, Lake Huron: Effects of zebra mussel (*Dreissena polymorpha*) colonization. *J. Great Lakes Res.* 21: 465–475.

Fahnenstiel, G. L., G. A. Lang, T. F. Nalepa, and T. H. Johengen. 1995a. Effects of zebra mussel (*Dreissena polymorpha*) colonization on water quality parameters in Saginaw Bay, Lake Huron. *J. Great Lakes Res.* 21: 435–448.

Fernald, R. T. and B. T. Watson. 2013. Eradication of zebra mussels (*Dreissena polymorpha*) from Millbrook Quarry, Virginia: Rapid response in the real world. In *Quagga and Zebra Mussels: Biology, Impacts and Control*. 2nd Edn., T. F. Nalepa and D. W. Schloesser, eds., pp. 195–213. Boca Raton, FL: CRC Press.

Garton, D. W., R. McMahon, and A. M. Stoeckmann. 2013. Limiting environmental factors and competitive interactions between zebra and quagga mussels in North America. In *Quagga and Zebra Mussels: Biology, Impacts and Control*. 2nd Edn., T. F. Nalepa and D. W. Schloesser, eds., pp. 383–403. Boca Raton, FL: CRC Press.

Great Lakes Water Quality Board. 1989. Report on Great Lakes water quality: Appendix A—Progress in developing and implementing remedial action plans for areas of concern in the Great Lakes Basin. International Joint Commission, Windsor, Ontario.

Heath, R. T., G. L. Fahnenstiel, W. S. Gardner, J. F. Cavaletto, and S. J. Hwang. 1995. Ecosystem-level effects of zebra mussels (*Dreissena polymorpha*): An enclosure experiment in Saginaw Bay, Lake Huron. *J. Great Lakes Res.* 21: 501–516.

Heimowitz, P. and S. Phillips. 2006. Rapid response plan for zebra mussels in the Columbia River Basin: A comprehensive multi-agency strategy to expeditiously guide rapid response activities. Draft report, September 1, 2006.

Hincks, S. S. and G. L. Mackie. 1997. Effects of pH, calcium, alkalinity, hardness and chlorophyll on the survival, growth and reproduction success of zebra mussel (*Dreissena polymorpha*) in Ontario Lakes. *Can. J. Fish. Aquat. Sci.* 54: 2049–2057.

Hitzfeld, B. C., S. J. Hogar, and D. R. Dietrich. 2000. Cyanobacterial toxins: Removal during drinking water treatment and human risk assessment. *Environ. Health Perspect.* 108: 113–122.

Holland, R. E. 1993. Changes in planktonic diatoms and water transparency in Hatchery Bay, Bass Island area, western Lake Erie since the establishment of zebra mussel. *J. Great Lakes Res.* 19: 617–624.

Horgan, M. J. and E. L. Mills. 1997. Clearance rates and filtering activity of zebra mussel (*Dreissena polymorpha*): Implications for freshwater lakes. *Can. J. Fish. Aquat. Sci.* 54: 249–255.

Hosler, D. M. 2011. Early detection of *Dreissenid* Species: Zebra/quagga mussels in water systems. *Aquat. Invasions* 6: 217–222.

James, W. F., J. W. Barko, and H. L. Eakin. 1997. Nutrient regeneration by the zebra mussel (*Dreissena polymorpha*). *J. Freshwat. Ecol.* 12: 209–216.

Jenner, H. A. 1990. Biomonitoring in chlorination antifouling procedure to achieve discharge concentrations as low as possible. In *International Macrofouling Symposium: Symposium Notebook*. pp. 9–11, Palo Alto, CA: Electric Power Research Institute.

Johengen, T. H., T. F. Nalepa, G. L. Fahnenstiel, and G. Goudy. 1995. Nutrient changes in Saginaw Bay, Lake Huron after the establishment of the zebra mussel, *Dreissena polymorpha*. *J. Great Lakes Res.* 21: 449–464.

Johnson, P. D. and R. F. McMahon. 1998. Effects of temperature and chronic hypoxia on survivorship of the zebra mussel (*Driessena polymorpha*) and asian clam (*Corbicula fluminea*). *Can. J. Fish. Aquat. Sci.* 55: 1564–1572.

Kaur, J., J. V. DePinto, V. J. Bierman, Jr., and T. J. Feist. 2000. The effect of zebra mussels on cycling and potential bioavailability of PCBs: Case study of Saginaw Bay. Final Report. EPA contract no. GL985600-01.

Lavrentyev, P. J., W. S. Gardner, J. F. Cavaletto, and J. R. Beaver. 1995. Effects of the zebra mussel (*Dreissena polymorpha*) on protozoa and phytoplankton in Saginaw Bay, Lake Huron. *J. Great Lakes Res.* 21: 545–557.

Lowe, R. L. and R. W. Pillsbury. 1995. Shifts in benthic algal community structure and function following the appearance of zebra mussels (*Dreissena polymorpha*) in Saginaw Bay, Lake Huron. *J. Great Lakes Res.* 21: 558–566.

Mackie, G. L. and R. Claudi. 2010. *Monitoring and Control of Macrofouling Mollusks in Fresh Water Systems*. Boca Raton, FL: CRC Press.

MacIsaac, H. J. 1996. Potential abiotic and biotic impacts of zebra mussels on the inland waters of North America. *Am. Zool.* 36: 287–299.

MacIsaac, H. J., W. G. Sprules, O. E. Johannsson, and J. H. Leach. 1992. Filtering impacts of larval and sessile zebra mussels (*Dreissena polymorpha*) in Western Lake Erie. *Oecologia* 92: 30–39.

MacIsaac, H. J., W. G. Sprules, and J. H. Leach. 1991. Ingestion of small-bodied zooplankton by zebra mussels (*Dreissena polymorpha*): Can cannibalism on larvae influence population dynamics? *Can. J. Fish. Aquat. Sci.* 48: 2051–2060.

Makarewicz, J. C. 1993. Phytoplankton biomass and species composition in Lake Erie, 1970 to 1987. *J. Great Lakes Res.* 19: 258–274.

Malloy, D. P. and D. A. Mayer. 2007. Overview of a novel green technology: Biological control of zebra and quagga mussels with *Pseudomonas flourescens*. New York State Museum. http://www.aquaticnuisance.org/wordpress/wp-content/uploads/2009/01/Dreissena-Novel-Green-Technology-for-Dreissena-Control-4-Malley.pdf (accessed May 15, 2003).

Marsden, J. E. and D. M. Lansky. 2000. Substrate selection by settling zebra mussels, *Dreissena polymorpha*, relative to material, texture, orientation, and sunlight. *Can J. Zool.* 78: 787–793.

McMahon, R. F., M. A. Mathews, T. A. Ussery, R. Chase, and M. Clarke. 1995. Studies of heat tolerance of zebra mussels: Effects of temperature acclimation and chronic exposure to lethal temperatures. U.S. Army Corps of Engineers, Technical Report EL-95-9.

McMahon, R. F., T. A. Ussery, and M. Clarke. 1993. Use of emersion as a zebra mussel control method. Contract Report EL-93-1. U.S. Army Engineer Waterways Experiment Station, Vicksburg, MS. NTIS No. AD A267 665.

Mellina, E., J. B. Rasmussen, and E. L. Mills. 1995. Impact of zebra mussel (*Dreissena polymorpha*) on phosphorus cycling and chlorophyll in lakes. *Can. J. Fish. Aquat. Sci.* 52: 2553–2573.

Miller, A. H. and A. Ignacio. 1994. An approach to identify potential zebra mussel colonization in large water bodies using the best available data and a Geographic Information System. In *Proceedings 4th International Zebra Mussel Conference*, 425–431. Madison, WI.

Nalepa, T. F., W. S. Gardner, and J. M. Malczyk. 1991. Phosphorus cycling by mussels (Unionidea: Bivalvia) in Lake St. Clair. *Hydrobiologia* 219: 239–250.

Nalepa, T. F., J. A. Wojcik, D. L. Fanslow, and G. A. Lang. 1995a. Initial colonization of the zebra mussel (*Dreissena polymorpha*) in Saginaw Bay, Lake Huron: Population recruitment, density, and size structure. *J. Great Lakes Res.* 21: 417–434.

Nalepa, T. F. and G. L. Fahnenstiel. 1995b. *Dreissena polymorpha* in Saginaw Bay, Lake Huron ecosystem: Overview and perspective. *J. Great Lakes Res.* 21: 411–416.

Nalepa, T. F., G. L. Fahnenstiel, and T. H. Johengen. 1999. Impacts of zebra mussels (*Dreissena polymorpha*) on water quality: A case study in Saginaw Bay, Lake Huron. In *NonIndigenous Freshwater Organisms: Vectors, Biology, and Impacts*, R. Claudi and J. H. Leach, eds., pp. 255–271. Boca Raton, FL: CRC Press.

Nalepa, T. F., D. L. Fanslow, and S. A. Pothoven. 2010. Recent changes in density, biomass, recruitment, size structure, and nutritional state of *Dreissena* populations in southern Lake Michigan. *J. Great Lakes Res.* 36: 5–19.

Nicholls, K. H. and G. J. Hopkins. 1993. Recent changes in Lake Erie (North shore) phytoplankton: Cumulative impacts of phosphorus loading reductions and the zebra mussel introduction. *J. Great Lakes Res.* 19: 637–647.

O'Neill, C. R. 1993. Control of zebra mussels in residential water systems. Sea Grant: Coastal Resources Fact Sheet (http://library.marist.edu/diglib/EnvSci/archives/alienspe/oneillcontrol/o'neill%20-%20control%20of%20zebra%20mussels%20in%20residential%20water%20systems.html) (accessed November 20, 2011).

Pillsbury, R. W., R. L. Lowe, Y. D. Pan, and J. L. Greenwood. 2002. Changes in the benthic algal community and nutrient limitation in Saginaw Bay, Lake Huron, during the invasion of the zebra mussel (*Dreissena polymorpha*). *J. N. Am. Benthol. Soc.* 21: 238–252.

Prescott, T. H., R. Claudi, and K. L. Prescott. 2013. Impact of dreissenid mussels on the infrastructure of dams and hydro electric power plants. In *Quagga and Zebra Mussels: Biology, Impacts, and Control*, 2nd edn. T.F. Nalepa and D.W. Schloesser, eds. pp. 244–257. Boca Raton, FL: CRC Press.

Quigley, M. A., W. S. Gardner, and W. M. Gordon. 1993. Metabolism of the zebra mussel (*Dreissena polymorpha*) in Lake St. Clair of the Great Lakes. In *Zebra Mussels: Biology, Impacts and Control*, T. F. Nalepa and D. W. Schloesser, eds., pp. 295–306. Boca Raton, FL: CRC Press.

Ramcharan, C. W., D. K. Padilla, and S. I. Dodson. 1992. Models to predict potential occurrence and density of the zebra mussel, *Dreissena polymorpha*. *Can. J. Fish. Aquat. Sci.* 49: 2611–2620.

Reeders, H. H., A. bij de Vaate, and F. J. Slim. 1989. The filtration rate of *Dreissena Polymorpha* (Bivalvia) in three Dutch lakes with reference to biological water quality management. *Freshwat. Biol.* 22: 33–141.

Shevtsova, L. V., G. A. Zhdanova, V. A. Movchan, and A. B. Primak. 1986. Experimental interrelationship between *Dreissena* and planktic invertebrates. *Hydrobiologia* 22: 36–39.

Skubinna, J. P., T. G. Coon, and T. R. Batterson. 1995. Increased abundance and depth of submerged macrophytes in response to decreased turbidity in Saginaw Bay, Lake Huron. *J. Great Lakes Res.* 21: 476–488.

Sorba, E. A. and D. A. Williamson. 1997. Zebra mussel colonization potential in Manitoba, Canada. Water Quality Management Section, Manitoba Environment Report No. 97-07.

Texas Parks and Wildlife Department—Inland Fisheries Division. 2010. Zebra mussels in Texas: Assessment of relative risks to fishery resources, recommendations for action, and expectations for the future. http://texasinvasives.org/resources/publications/TPWD_.ZebraMussels_in_Texas. pdf (accessed May 15, 2013).

Texas Parks and Wildlife Department, Trinity River Basin in Texas (TPWD). 2012. Zebra mussels found in Lake Ray Roberts, http://www.tpwd.state.tx.us/newsmedia/releases/?req=20120718a (accessed May 15, 2013).

Thomas, C., C. Goemans, and C. Bond. 2010. *The Costs and Benefits of Preventative Management for Zebra and Quagga Mussels*. Fort Collins, CO: The Water Center of Colorado State University, September/October 10–15.

Turner, K., W. H. Wong, S. L. Gerstenberger, and J. M. Miller. 2011. Interagency Monitoring Action Plan (I-MAP) for Quagga Mussels in Lake Mead, Nevada-Arizona, USA. *Aquat. Invasions* 6: 195–204.

U.S. Bureau of Reclamation (USBR). 2011. http://www.usbr.gov/mussels/history/factsheets/currentresearchfactsheet.pdf. (accessed November 20, 2011).

U.S. Bureau of Reclamation (USBR). 2012. Coatings for mussel control—Three years of laboratory and field testing. Technical Memorandum No. MERL-2012-11. March, 2012.

U.S. Geological Survey (USGS). 2011. http://nas.er.usgs.gov/taxgroup/mollusks/zebramussel. (accessed November 20, 2011).

Van Benschoten, J. E., J. N. Jensen, T. J. Brady, D. P. Lewis, J. Sferrazza, and E. F. Neuhauser. 1993a. Response of zebra mussel veligers to chemical oxidants. *Water Res.* 27: 575–582.

Van Benschoten, J. E., J. N. Jensen, D. K. Harrington, and D. J. DeGirolamo. 1995. Zebra mussel mortality with chlorine. *J. Am. Water Works Assoc.* 87: 101–108.

Van Benschoten, J. E., J. N. Jensen, D. Lewis, and T. J. Brady. 1993b. Chemical oxidants for controlling zebra mussels (*Dreissena polymorpha*): A synthesis of recent laboratory and field studies. In *Zebra Mussels: Biology, Impacts and Control*, T.F. Nalepa, and D. W. Schloesser, eds., pp. 599–619. Boca Raton, FL: CRC Press.

Vanderploeg, H. A., J. R. Liebig, W. W. Carmichael, M. A. Agy, T. H. Johengen, G. L. Fahnenstiel, and T. F. Nalepa. 2001. Zebra mussel (*Dreissena polymorpha*) selective filtration promoted toxic *Microcystis* blooms in Saginaw Bay (Lake Huron) and Lake Erie. *Can. J. Fish. Aquat. Sci.* 58: 1208–1221.

Vanderploeg, H. A, T. F. Nalepa, D. J. Jude, E. L. Mills, K. T. Holeck, J. R. Liebig, I. A. Grigorovich, and H. Ojaveer. 2002. Dispersal and emerging ecological impacts of Ponto-Caspian species in the Laurentian Great Lakes. *Can. J. Fish. Aquat. Sci.* 59: 1209–1228.

Vanderploeg, H. A. and J. R. Strickler. 2013. Video Clips 6: Behavior of zebra mussels exposed to *Microcystis* colonies from natural seston and laboratory cultures. In *Quagga and Zebra Mussels: Biology, Impacts, and Control*, 2nd Edn., T.F. Nalepa and D.W. Schoesser, eds., pp. 757–758. Boca Raton, FL: CRC Press.

Whittier, T. R., P. L. Ringold, A. T. Herlihy, and S. M. Peatson. 2008. A calcium-based invasion risk assessment for zebra mussels and quagga mussels. *Front. Ecol. Environ.* 6: 180–184.

Wells, S. and M. Sytsma. 2009. A review of the use of coatings to mitigate biofouling in freshwater. Prepared for the Bonneville Power Administration and the Pacific Marine Fisheries Commission. http://www.tujifilmusa.com/products/speciality-chemicals/smart-surfaces/power-generation/pdf/index/portland-university.pdf (accessed May 15, 2013).

Wolfe, R. L. 1990. Ultraviolet disinfection of potable water. *Environ. Sci. Technol.* 24: 768–773.

Wong, W. H. and S. L. Gerstenberger. 2011. Quagga mussels in the Western United States: Monitoring and management. *Aquat. Invasions* 6: 125–129.

Wong, W. H., S. L. Gerstenberger, J. M. Miller, C. J. Palmer, and B. Moore. 2011. A standardized design for quagga mussel monitoring in Lake Mead, Nevada-Arizona. *Aquat. Invasions* 6: 205–215.

Wong, W. H., G. C. Holdren, T. Tietjen, S. L. 2013. Gerstenberger, B. Moore, K. Turner, and D. Wilson. Effects of invasive quagga mussels (*Dreissena rostriformis bugensis*) on chlorophyll and water clarity in Lakes Mead and Havasu of the lower Colorado River Basin, 2007–2009. In *Quagga and Zebra Mussels: Biology, Impacts and Control*, 2nd Edn., T. F. Nalepa and D. W. Schloesser, eds., pp. 495–509. Boca Raton, FL: CRC Press.

Yaroslavtseva, L. M. and E. P. Sergeeva. 2007. Effect of temperature on the early development of the pacific mussel *Mytilus trossulus* (Bivalvia: Mytilidae) in sea water polluted by copper ions. *Russian J. Marine Biol.* 33: 375–380.

Zequanox. 2012. http://www.marronebioinnovations.com/lightnay/site/wp-content/uploads/2013/05/Zequanox_Golf_trifold_May 2013_web.pdf (accessed May 15, 2013).

CHAPTER 15

Impact of Dreissenid Mussels on the Infrastructure of Dams and Hydroelectric Power Plants

Thomas H. Prescott, Renata Claudi, and Katherine L. Prescott

CONTENTS

Abstract	244
Introduction	244
Dams	244
External Features of Dams	245
Floating Structures	245
Gates and Weirs	245
Isolation Gates	245
Adjustable Weirs	246
Spillways	246
Dam Outlet Works	246
Fish-Handling Facilities	246
Navigation Locks	246
Internal Structures of Dams	247
Conduits, Valves, and Vents	248
Seepage Drains	248
Sumps	248
Uplift Drains	248
Fire Protection	248
Construction Materials	249
Hydroelectric Power Plants	249
Intake and Discharge Structures of Hydroelectric Power Plants	249
Floating Structures	249
Intakes	249
Penstocks	250
Secondary Intakes for Raw and Domestic Water	250
Trashracks and Trashrack Cleaning Systems	251
Discharge Area	251
Internal Structures of Hydroelectric Power Plants	251
Raw Water/Service Water/Cooling Water	251
Heat Exchangers	251
Strainers and Filters	252
Compressed Air Systems	253
Pumps, Turbines, and Generators	253
Piping	254

Valves 254
Transformer Cooling 254
Heating, Ventilation, and Air-Conditioning Systems 254
Domestic Water 255
Level, Flow, and Pressure Sensing Equipment 255
Drainage and Dewatering 255
Fire Protection Water 256
Summary 256
References 257

ABSTRACT

This chapter describes structural components of dams and hydroelectric plants that are vulnerable to fouling by dreissenid mussels (zebra mussel *Dreissena polymorpha* and quagga mussel *Dreissena rostriformis bugensis*). Various external and internal components of these structures are at varied levels of risk to fouling based on their geometry and materials of construction. In general, the larger and more easily accessible the component on a dam or hydroelectric power plant, the lower the problem of fouling and the easier it is to clean and maintain if fouling occurs. The smaller and more complex the component, the harder it may be to mitigate fouling problems. Facilities, systems, and components vulnerable to dreissenid mussel fouling have been identified through facility vulnerability assessments performed in North America, South America, and Europe and are described in detail in this chapter. The same fouling risks posed by dreissenid mussels are also associated with the golden mussel (*Limnoperna fortunei*) in South America.

INTRODUCTION

Dams and hydroelectric power plants would not be thought of as prime targets for impairment by biofouling organisms as limited in size as dreissenid mussels. Large penstocks with diameters of several meters and wide spillways have large flows with immense hydraulic forces. However, these facilities have necessary design features and systems that may make them vulnerable to effects of colonization by mussels. Dreissenid mussels, and the golden mussel in South America, are highly adaptable and tend to affect facilities in different ways than native mollusks. These nuisance species continue to surprise even seasoned experts who have worked with operating facilities and dealt with biofouling issues.

Each facility is unique with diverse arrangements and designs for common functions. Therefore, the risks and solutions to biofouling may be equally diverse. This chapter describes components of dams and hydroelectric power plants that may be vulnerable to problems caused by infestation of zebra mussels (*Dreissena polymorpha*), quagga mussels (*Dreissena rostriformis bugensis*), and golden mussels (*Limnoperna fortunei*). Information presented is based on our field experiences performing vulnerability assessments at a number of dams and hydroelectric facilities in Europe, North America, and South America. Biofouling in some of the vulnerable areas described will not threaten operability but result only in a minor nuisance to maintenance personnel in the form of increased inspection and cleaning. However, biofouling in other vulnerable areas may present serious problems to plant operations and safety. Regardless of ways in which mussels affect a facility, costs to maintain the facility will rise to some extent once mussels become established and become abundant.

DAMS

Dams are often looked on as massive, sturdy objects that back the flow of water (Figure 15.1), but they are much more complex than external appearances suggest. Dams have a variety of construction materials. They may be built of natural rock/earth materials, concrete, or a combination of natural materials and concrete requiring engineered systems to maintain dam safety and performance as designed. They also have a variety of purposes. Dams may be constructed

Figure 15.1 Aerial view of Pueblo Dam on the Arkansas River near Pueblo, Colorado, United States. This is an earth-filled dam with concrete buttresses that was constructed to deliver water for municipal and industrial use. Dreissenid mussels were first found at the dam in 2008 (Benson 2013). (Courtesy of U.S. Bureau of Reclamation, Washington, DC.)

for water resource management (e.g., flood control, recreation, irrigation) or for production of hydroelectric power. Regardless of purpose, dams have structures that allow water to pass through dam boundaries. Resource-management dams have flow control structures, referred to as outlet works, to pass water downstream for such as agriculture, fish hatcheries, natural habitat maintenance, and human use. Dams on navigable waters may have locks for boat traffic or passages for migratory fish. The wide variety of uses for dam-associated water means that all dams will have intakes, conduits, flow control devices, emergency safety structures, and safety monitoring systems to protect against failure.

Despite many unique characteristics of dams in existence, there are common components that perform similar functions. The following is a review of some of the common problem areas that may be subject to mussel fouling. Not all dams will have all structures we describe, and some may have features we have not yet encountered.

Figure 15.2 (See color insert.) Surface of an isolation gate that has been colonized by mussels. (Courtesy of U.S. Bureau of Reclamation, Washington, DC.)

EXTERNAL FEATURES OF DAMS

Floating Structures

Dams commonly have floating surface barriers in areas near water intakes or, in some cases, along the dam face. These barriers prevent ingress of floating debris and prevent boaters from approaching the water intake. At some dams, floats are used to suspend nets and to prevent fish from entering water intakes. Unfortunately, these floating barriers are subject to mussel attachment that can cause them to sink. Mesh filaments of suspended nets are particularly susceptible to mussel fouling since these locations provide good attachment places and provide access to a constant supply of food in passing water. Mussel build-up reduces water flow and places additional strain on the net. Also, dead mussel shells, which slough off the net, enter intakes and may plug strainers and filters.

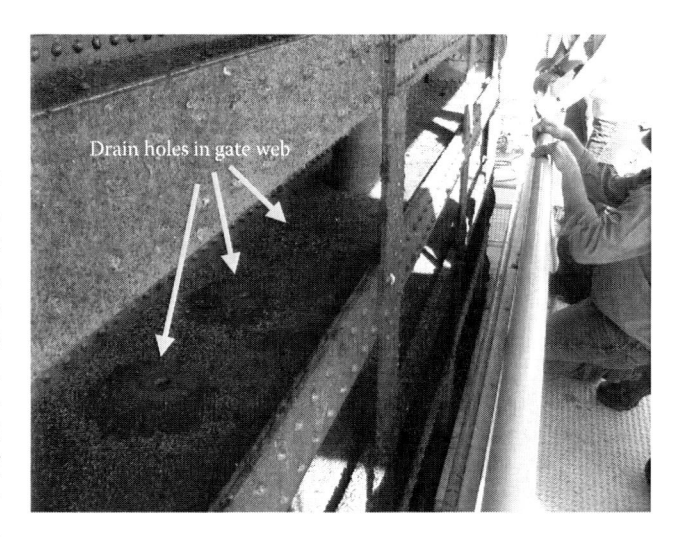

Figure 15.3 (See color insert.) Gate webbing on recently cleaned gate showing small drain holes at risk of being plugged by mussels. (Courtesy of U.S. Bureau of Reclamation, Washington, DC.)

Gates and Weirs

Structures that rely on movable gates to regulate water flow, such as spillways, can be colonized by mussels (Figure 15.2). Mussels attach to outsides and insides of submerged gates, and thereby increase overall weight of a gate placing extra strain on the lifting mechanism. Gates are often constructed of I-beam or C-channel framing material, and webs of the beam or channel are frequently laid horizontally forming a trough that traps water. Small drains are placed in the web to drain water. These and other drains associated with submerged gates can become clogged with mussels and increase weight of gates due to trapped water (Figure 15.3). If design parameters of lifting equipment cannot cope with much additional weight, gates could become inoperable. Improper sealing is also a concern if seals around gate perimeters are colonized by mussels. Mussels trapped between sealing surfaces can increase leakage. Crushed shell fragments are sharp and can score elastomeric seals, reducing seal life.

Isolation Gates

Isolation and emergency gates are often suspended above an intake opening. When needed to control water flow, gates are guided into intake openings by metal slots installed in intake walls. Fouling of gate guide slots could interfere with proper insertion of gates. Gate guides are usually made of carbon steel and formed as C-channel shapes. Mussel fouling in channels could make it difficult to properly seal the gate and, in some cases, cause seal damage.

Adjustable Weirs

Outlets of dams where water is required to flow into channels at a known rate or constant elevation may use adjustable weirs for flow control. At flow velocities less than 1.5 m/s, mussels may accumulate in weir areas and alter flow rates. Weirs may have to be inspected and cleaned regularly of mussels to ensure calibration is maintained.

Spillways

Dam spillways are used infrequently as most dam structures have outlet structures to regulate flows. Spillway problems associated with mussel fouling are unlikely. However, water overflow through spillways often occurs during periods of spring runoff. If mussel veligers are present in raw water, mussels can colonize transverse spillway drains. Over many years, mussel growth could cause drains to become plugged and hold water; a flooded drain could freeze in northern latitudes and crack spillway concrete.

Dam Outlet Works

Most dams have one or more outlet works structures (Figure 15.4). Besides a main outlet, there could be auxiliary outlets and possibly an outlet to maintain minimum discharge flows for recreation or fish habitats in downstream areas (Figure 15.5). Intakes to outlet tunnels are generally fully submerged and covered with metal trashracks (Figure 15.6). Trashracks are the most vulnerable part of these structures because they usually remain submerged and are difficult or impossible to remove for cleaning. If fouled by mussels, manual cleaning by divers will be necessary.

Fish-Handling Facilities

Bypass structures allow fish to migrate past the dam in both upstream and downstream directions (Figure 15.7). For some dams, such as those on the Columbia River, Washington, where great numbers of pacific salmon migrate, bypass systems may be very elaborate with many vulnerable components. Comprehensive fish-handling facilities may include diversion screens, large pumps to direct flow that guides fish into the bypass channel, and conduits and grates to provide even flow distribution in channels. Screens, grates, and nets can become fouled with mussels thereby blocking flow (Figure 15.8). Cooling water to pump bearings, shaft seals, and pump motors can also be blocked resulting in overheated components.

In some dams, the entire fish bypass structure may consist of a simple fish ladder that is dewatered for some time each year. These structures are not particularly vulnerable to infestation as they can usually be readily cleaned of mussel accumulation during normal outage periods.

Navigation Locks

Navigation locks allow vessels to pass a dam both upstream and downstream. Water enters and leaves locks via large concrete conduits. The large size of these conduits allows them to tolerate significant mussel infestation before flows are noticeably reduced. Flow velocities would also be expected to be large enough to prevent mussel settlement, or if mussels do settle and accumulate in multiple layers, they will eventually slough off due to the flushing action of lock operation. In general, it is unlikely that operation of locks would be threatened by mussel settlement even if multiple layers form.

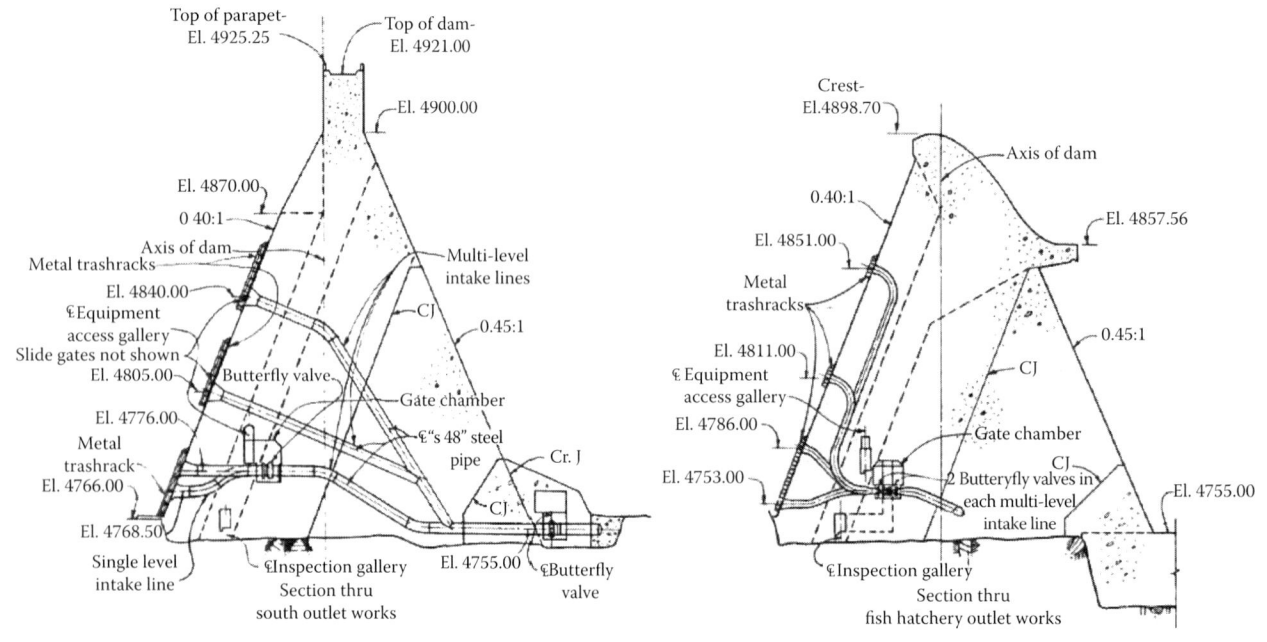

Figure 15.4 Cross sections through a dam showing various outlet works. (Courtesy of U.S. Bureau of Reclamation, Washington, DC.)

Figure 15.5 Large outlet tunnel on the downstream side of a dam.

Figure 15.6 Trashracks at an inlet to a dam discharge tunnel.

Figure 15.7 Example of a fish ladder that can provide large areas for mussel colonization.

Figure 15.8 Fish diversion net fouled with dreissenid mussels.

Accumulation of mussel shells in tracks of lock gates could allow some water leakage to occur. However, if lock gates are cycled frequently, flushing of gate tracks will take place as part of the normal operation of the lock. Motors used to operate gates are usually very powerful and their performance should not be jeopardized by mussel accumulation. In some cases, lock systems incorporate pumps to scavenge water or to accelerate lock cycling periods. These pumps are generally small enough to have air-cooled motors and bearings. Large pumps with water-cooled motors and bearings would be vulnerable to overheating if infested with mussels from the cooling water. Pump seals will have reduced life from overheating and abrasion by mussel shells. Small-pore filters installed on inlets for the seal-flush water are a typical way to protect seals.

Locks also have level gauges as part of gate operations. If these gauges are in contact with raw water, mussel accumulation could adversely affect their accuracy and response time.

INTERNAL STRUCTURES OF DAMS

Other than conduits associated with outlet works, there is little piping within a dam structure that is exposed to raw water. There are usually no components requiring service water, and instrumentation is typically used only for pressure measurements. Lines for pressure instrumentation would not have any flow and are therefore unlikely to be suitable for mussel infestation given the lack of food and low oxygen. Level gauges that use floats could become fouled; therefore, floats should be inspected and cleaned periodically. In some cases, level gauges have pipes that lead to instruments. These pipes can become fouled if level fluctuations are frequent enough so that water flow is maintained and hence suitable for mussels.

Conduits, Valves, and Vents

Outlet works conduits (sometimes called tunnels) have control-valve chambers that divide conduits into upstream and downstream portions. The downstream portion is generally easy to isolate for manual cleaning by closing valves in the control chamber for a period of time to inspect and clean the downstream portion of the conduit. The upstream portion, which starts after the intake grates described earlier, is usually not easy to isolate. If significant mussel infestation occurs in this portion of the conduit, there could be some flow reduction due to increased hydraulic friction caused by mussel accumulation on conduit walls. If conduits are required to pass a predetermined maximum flow for dam safety or operational efficiency, then the degree of mussel accumulation that can be tolerated needs to be assessed and the conduit cleaned as required.

Control-valve chambers frequently have two valves in series, commonly called a pressure gate and a guard gate. Space between the two valves is vented so that it can be filled by separate fill lines. In general, fill lines, wetted portion of vent lines, and air release valves may be fouled by mussels. Of particular concern would be plugging of vent lines as this will slow down filling and draining processes and may cause unsafe negative pressures in associated piping. The condition of vent lines and/or proper operation of vent lines should be verified before draining or filling intervalve chambers and any associated piping.

Accumulations of mussels on conduit walls, valves, and valve seats may interfere with valve operation. In general, these valves are large with powerful operators so they are not likely to be affected by mussels. However, we have encountered smaller outlet works where valve operators are marginally powered. Mussel accumulation could exacerbate operational problems for these marginal valve operators. Monitoring the position of flow valves for a particular flow rate and tracking this position over time can be one way to indicate that conduits have become fouled.

Depending on the shape of the inlet to the pressurized portion of the outlet works conduit, there may be a vent at the conduit inlet to ensure fully developed flow occurs at curved portions of the inlet (this may occur if the transition from inlet to conduit curves more than 90°). Plugging of this vent would increase risk of cavitation and flow reduction in the conduit. Unfortunately, such vent lines are usually located near intake structures and are difficult to clean.

If present, level-gauging systems usually require small-diameter, raw water lines that connect to sensor equipment. These could become plugged and impair performance or response time of the gauges. If float-type measuring systems are used, mussel accumulation on floats may generate a level reading error. If mussels are allowed to fully impact such structures, dams may not be able to control water levels as required.

Seepage Drains

All dams have drains with associated tubes that collect and pass water that seeps through dam seals and percolates through dam walls. Mussel veligers may be able to travel with normal dam seepage into drain tubes where they could settle and grow. Occurrences of such infestations are likely to be rare, but they have been documented at dam facilities and personally observed by the authors. In the unlikely event that a sufficient number of mussels should accumulate to restrict drain flows, reduced drainage should be detected during routine inspections. A plugged or partially blocked drain tube could result in some gallery flooding. Mussels could be removed with a cable-operated pipe cleaner through the access cap opening, or alternatively, mussels could be killed by temporarily closing the drain and introducing a biocide or hot water and then removing empty shells. For earth-filled dams, it is particularly important to keep drain lines free of mussels to prevent any pressure buildup in the dam fill.

Sumps

Seepage through dam walls commonly drains to a sump. Drain piping and sumps are often overlooked as being at risk from mussel fouling, but they can easily become infested. Typically, problems associated with infested drain piping and sumps do not threaten facility operability, but they can be an operational nuisance. Drain lines are often embedded in concrete and not easy to access. Furthermore, drain lines may be long with bends that make cleaning with mechanical pipe cleaners a difficult and costly task. Accumulation of mussels on sump floats could cause unreliable level detection and require increased cleaning efforts in sump areas. Open drains and sumps can often be good places to look for the presence of mussels and other potentially fouling organisms such as clams and snails.

Uplift Drains

Dams also incorporate foundation-uplift drain pipes. Water in these pipes is generally expected to be groundwater seepage and not likely to transport mussels. In addition, reservoir seepage from dams making its way to uplift drains passes through the base material in the dam, which will normally suffocate any mussels. Nevertheless, seepage paths may develop; therefore, it is important that raw water cannot reach uplift drains and that monitoring will pick up unexpected changes in behavior of an uplift monitoring site.

Fire Protection

Fire protection in dams is usually provided by chemical extinguishers. In some cases, protection is provided by raw water-based systems; risks of mussel fouling in these systems are discussed in the section on hydroelectric plants.

Construction Materials

Typical materials used in dam construction such as concrete, carbon steel, and stainless steel are all good substrates for mussel attachment. Less common materials such as aluminum, wood, and plastics also serve as good substrates for attachment. Mussels attach with byssal threads and the site of attachment can cause corrosion pitting on steel and aluminum. When mussels are removed, byssal threads are often left attached, so thorough cleaning is recommended.

Some materials, such as copper and tin, are toxic to mussels so typically these materials remain mussel-free. Tin is not often used as a construction material, but copper piping is very common. Mussels can occasionally be found on copper piping, screens, and grates. This typically occurs in areas of low flow where a biofilm has coated the copper. Mussels attach to the film, not the copper. If these structures are kept clean of biofilm buildup by a high-flow regime, mussels are exposed to the copper ions and do not attach.

HYDROELECTRIC POWER PLANTS

When considering effects of mussels on hydroelectric power plants, size of a plant does not matter; small, one-unit, single-digit megawatt plants may be equally as vulnerable as large, multi-unit plants such as the iconic Hoover Dam (Figure 15.9). However, different design configurations such as base-load plants, peaking plants with spinning reserve, and combined pump/generating units will have some differences in areas affected and hence degree of risk. Like dams, all power plants have structures to pass water (Figure 15.10). As water flow is the primary means of converting energy into electricity, it must flow with high efficiency, and water-moving equipment must operate with high reliability to make a plant productive.

Unlike dams, a small portion of the entire water flow through hydroelectric power plants is diverted and used inside the plant for cooling equipment, fire protection, and station services such as washing floors or equipment, lavatories, domestic water, and even landscaping irrigation. These uses necessitate a large amount of piping and equipment that is at risk of impairment by mussels.

Figure 15.9 View of Hoover Dam on the Colorado River, Arizona/Nevada, United States. The structure serves as a hydroelectric facility, and the upstream reservoir, Lake Mead, serves as a recreational resource and municipal source of water. Dreissenid mussels were first found in Lake Mead in 2007 (Benson 2013). (Courtesy of U.S. Bureau of Reclamation, Washington, DC.)

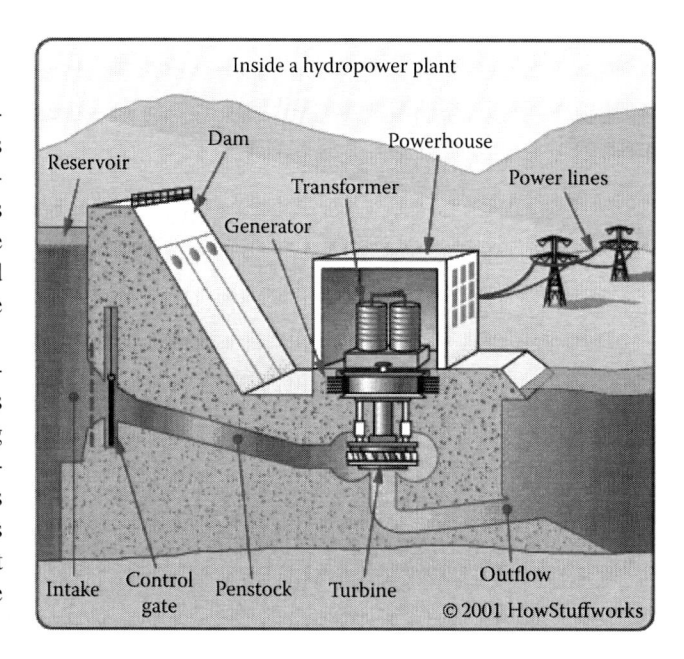

Figure 15.10 Main components of a typical hydroelectric plant.

INTAKE AND DISCHARGE STRUCTURES OF HYDROELECTRIC POWER PLANTS

Floating Structures

Hydroelectric power plants take in water at a high rate, and therefore currents around intakes are far greater than at a dam and would consequently be more dangerous. It is common to have two floating surface barriers near intakes: one barrier is used to capture floating debris, and the other acts as a high-visibility barrier to keep recreational boaters at a safe distance. Mussel attachment can cause barriers to sink, allowing debris and fish to pass into the intake or incur great risks to boaters.

Intakes

Intake systems of hydroelectric power plants provide the means by which water enters a facility. Unlike dams with their deeply-submerged intakes for outlet works, intakes of hydroelectric power plants are usually closer to the water surface. Intake systems encompass large

Figure 15.11 (See color insert.) Photo collage of dreissenid mussels on forebay walls (a and d) and trashracks (b and c) of hydroelectric power plants. (Courtesy of Ontario Power Generation, Toronto, Ontario, Canada.)

static structures such as a concrete-lined forebay, intake pipes, towers, and trashracks. Some structures have moving components like gates, valves, and trashrack cleaning equipment. Concrete surfaces of forebays can tolerate large numbers of mussels, but the main problem caused by infestation of concrete surfaces is sloughing of mussel adults or empty shells that subsequently can enter a plant. For intakes, the greatest vulnerability to mussel infestation occurs on trashracks (Figure 15.11). Plugging of spaces between rack bars reduces water flow through condensers that cool the turbines. Removal of mussels from trashracks may require a unit outage, and the lost power represents a major financial burden if the outage occurs outside planned maintenance cycles.

Penstocks

Penstocks are specialized conduits located immediately downstream of intake structures (Figure 15.12; Hoover Dam penstock). They channel incoming water through a vertical drop toward the turbines. At full operating capacity, flow velocities in penstocks are generally too high for mussels to settle, but during short outages, settlement is possible if a penstock is not dewatered. Once flow is re-established, settled mussels can be firmly attached and difficult to dislodge. Infestation of penstocks by mussels would result in increased hydraulic roughness, which would translate into reduced power production.

Penstocks frequently have air vents to assist in draining and filling the penstock during an outage. On pump/generation plants, vents may be large and similar in shape to a surge tower. Wetted portions of vent pipes can be locations for mussel settlement, as there is typically low velocity flow in these pipes. Adequate air-vent operation should be confirmed whenever penstocks are dewatered.

Secondary Intakes for Raw and Domestic Water

In some hydroelectric plants, domestic water and sometimes raw water are taken from separate intakes on the dam face. These secondary intakes are usually protected by fixed grates. Secondary intake lines tend to be encased in concrete, have small diameters, and have low flow velocities relative to penstock flows. Both fixed grates and pipelines are particularly vulnerable to mussel infestation. At several locations we visited, intake grates had to be removed or cleaned underwater by divers. Intake lines had to be cleaned by hydro-blasting or by addition of chemicals. These cleaning activities represent a costly maintenance addition to power plant operations.

Figure 15.12 Inside one of four main penstocks at Hoover Dam.

Trashracks and Trashrack Cleaning Systems

Most trashracks are highly susceptible to mussel infestation (Figure 15.11). Accessibility of trashracks for cleaning is important and will affect control strategies and associated protection requirements. Trashracks are ideal settlement areas for mussels as water flow around trashracks is generally slow enough to allow mussel settlement but great enough to provide a continuous source of food. Cleaning of submerged trashracks is usually done manually by divers with scraping tools or with pressurized water. Removed mussel debris is flushed downstream when slide gates are opened and the penstock is operating. Alternatively, a hydraulic vacuum can be used together with the scraping process to remove this material. Typically, permits are required for disposal of removed debris.

Given their susceptibility to mussel infestations, trashracks require periodic inspection. If mussels become established in the vicinity of a plant, then inspection frequency should be increased to a quarterly basis until experience is gained with the rate at which mussels accumulate on trashrack structures. Should cleaning become frequent and constitute a maintenance nuisance, consideration could be given to removing the grates and painting them with an anti-fouling paint or foul-release paint. Foul-release paints have performed well even in areas of low flow such as quiescent portions of pump wells.

Automatic trashrack cleaning systems typically use a series of tines connected to a traveling beam to collect the debris. A bar usually rides on the leading edge of trashracks. Traveling beams may be modified to clean spaces between the racks, thus preventing complete closure by mussel in shell overgrowth. If a trashrack cleaning system includes wash nozzles, hoses and nozzles are at risk of being plugged by mussels. In addition to trashracks, intake grates or screens are fixed and require either periodic or continuous cleaning.

Discharge Area

Water that exits from the turbines passes through a draft tube and then into a receiving water body (Figure 15.10). Flow through a discharge area is fast and turbulent; thus, conditions in a discharge area are generally unsuitable for high mussel infestations. However, mussel shell debris may accumulate in the cavity behind grates leading to drain lines. If grates are present in the discharge area of turbines, such as in draft-tube drains, high flow rates and turbulence should prevent mussel settlement on these structures.

INTERNAL STRUCTURES OF HYDROELECTRIC POWER PLANTS

Raw Water/Service Water/Cooling Water

All hydroelectric power plants require water for cooling equipment or for flushing purposes. Water taken from dam reservoirs for use inside a power plant is given many names. Raw water, service water, cooling water, and unit cooling water are terms used, and each may have its own specific meaning. Raw water is considered the most general term and is used here to mean water used for any functions inside a power plant. This water represents a very small percentage of the total flow that passes through a hydroelectric power plant (approximately 0.4%). Although this amount is small, it is important that this water is in good condition (no debris, low level of suspended solids) to protect equipment. After raw water is used for cooling, it is returned to a main water flow.

Raw water is typically taken off a penstock, but may also be taken off a dam face via a separate pipe covered by a grate. On high-head dams, small streams of high-pressure raw water from the penstock may be used to power an eductor pump to draw in raw water from a tail race. It is normal that raw water enters at several points, usually one entry for each turbine unit or penstock. After entering a plant, raw water is fed to a common header. This arrangement complicates the treatment process for protecting plant equipment from mussels. Raw water may be used to cool generator air coolers, heating and ventilating systems, air compressors, pump/turbine seals, pump/turbine guide bearings, generator/motor guides, thrust bearings, transformers, and units used in fire protection (Figure 15.13).

Heat Exchangers

There are three main types of heat exchangers used in a hydroelectric power plant that use raw water. Shell/tube-type heat exchangers usually have raw water on the shell side (as in oil coolers), fin-type heat exchangers have raw

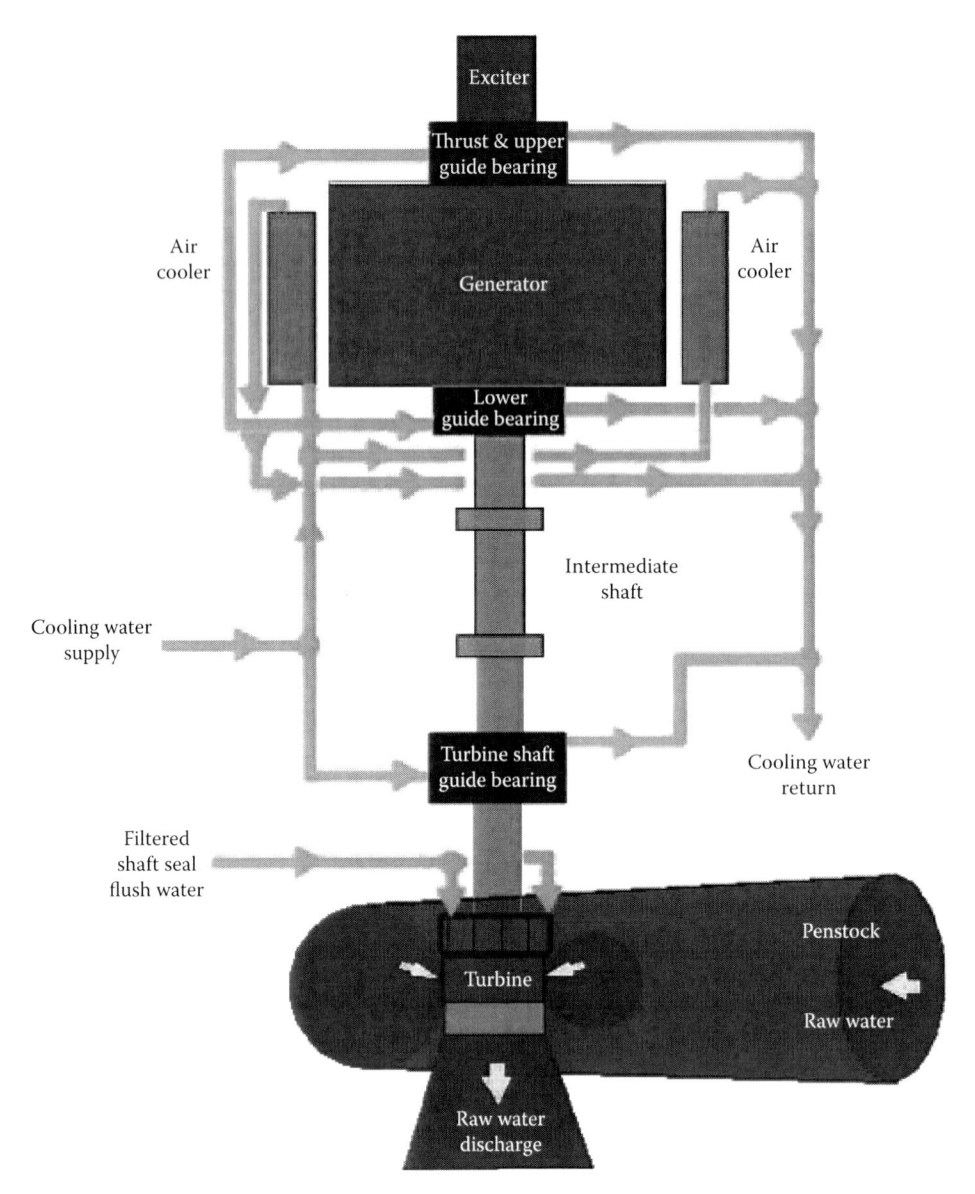

Figure 15.13 Typical flow of raw water used to cool a turbine and generator unit in a hydroelectric power plant.

water on the tube side as in air coolers, and plate-type heat exchangers have raw water throughout the plate. All these heat exchanger types are vulnerable to mussel infestation to a greater or lesser degree. In shell/tube-type heat exchangers, mussels accumulate most frequently in tube inlet and outlet manifolds of the air-to-water heat exchangers and in shell cavities (Figure 15.14). These areas tend to have lower flow velocities, and thus are attractive settlement areas for live mussels. Manifolds and shell side cavities are also natural sedimentation or catchment areas for dead shells or shell fragments that are moved along the piping by flow velocity. Shell material can gradually accumulate in these areas, ultimately blocking tubes and causing poor performance of coolers. Plate-type heat exchangers are particularly vulnerable to plugging by any shell debris due to the narrow gap between heat exchanger plates. The plates act as strainers, retaining most of the debris in the raw water.

As a general guideline, if temperatures of raw cooling water are below 29°C, water flow is less than 1.5 m/s, and tubes are non-copper, then heat exchanger tubes can become settlement areas for mussels. If temperatures of raw cooling water rise above 29°C on the discharge side of heat exchangers, then mussels will not settle in downstream piping. Mussels that have settled when cooling water discharge temperature is below 29°C will die if temperatures rise above this tolerance threshold for several hours.

Strainers and Filters

Strainers are generally the first component of raw water systems. Strainer baskets typically have opening sizes between 3 and 5 mm that will intercept most, but not all, mussels that may be carried into the system or have detached from upstream locations. In raw water systems, typical

Figure 15.14 (See color insert.) Manifold of a heat exchanger fouled with dreissenid mussels.

Figure 15.15 (See color insert.) Downstream side of a self-cleaning filter fouled with dreissenid mussels.

strainers are duplex or self-cleaning. Duplex strainers require more frequent cleaning and self-cleaning strainers backwash more often. Strainer baskets can be filled with dead shells and colonized on both sides by live mussels (Figure 15.15). Although duplex strainers require periodic cleaning with frequency dependent on the infestation level of a raw water source, even self-cleaning strainers have risks. If there are periodic system shutdowns, mussels can attach on an upstream side of a strainer plugging the mesh, or even attach on the downstream side of a strainer, creating a source of mussel shells that may affect downstream components. Depending on the particular design of the strainer, veligers that settle downstream may not be cleaned by the self-cleaning operation and will eventually form adults that can move downstream and impact other components. Strainers are limited in the sizes of particles they can remove. Small juveniles (<3 mm) and settling-age veligers will pass through strainers and colonize downstream piping and components. After a period of time, mussels in the piping will accumulate, grow, die, and slough from pipe walls plugging downstream components such as heat exchanger tubes.

Given limitations of strainer baskets, small-mesh filters are employed to protect systems from mussels. For example, cartridge-type filters of 25 µm are commonly used to protect seals of pumps and turbines. Small-pore, self-cleaning filters are generally an effective and environmentally benign way to keep small mussels out of water systems. However, small-mesh filters can decrease water flow and require frequent maintenance. As raw water remains in a plant for a short time, usually less than 1 h, veligers will pass through the plant before they reach a settling size. Typically, filters of 100 µm nominal mesh size are effective at removing ready-to-settle pediveligers (200–350 µm). If sand filters are used as a means to remove small debris from raw water, it is generally assumed that they will remove ready-to-settle pediveligers.

Compressed Air Systems

Compressed air can be an essential system to operate valve actuators and maintain water-level suppression for rapid-start turbines (i.e., spinning reserve). Compressed air is often used to pressurize oil systems that control turbine inlet vanes, and these systems commonly rely on cooling water for intercoolers. Plugging of intercooler water lines will impair compressor performance. More rarely, compressor motors may have water-cooled bearings or even water-cooled motors. These requirements for cooling water make bearings and motors associated with air compressors vulnerable to mussel infestation and present an operational risk for power plants.

Pumps, Turbines, and Generators

Turbines have guide bearings that are usually oil lubricated with water-cooled oil coolers and also have shaft seals that require some water passage for cooling. Cooling and flush lines of shaft seals can be infested with mussels, resulting in increased temperatures and seal damage. Generators also have both guide bearings and thrust bearings that need water for cooling, and these lines would be susceptible to mussel infestation as well.

Mussel shells are abrasive and could increase the wear rate of shaft-seal sleeve or seal faces on mechanical face seals. Mussels have been known to colonize cavities of seals, such as lantern rings, reducing cooling performance and requiring shutdown of the turbine to clean the cavities. Veligers are small enough to pass through gaps between shafts and seals. Pipe drains from seals could be at risk of mussel fouling, thereby causing flooding problems in the turbine room.

All turbines or turbine/pump units have coolers to cool air around generators/motors. Air coolers transmit heat to raw water, and this heat load is usually the largest use of raw water in a plant. Cooling water can be constrained by mussel infestation that would threaten overall performance of generators/motors.

Piping

Raw water is drawn into and distributed within hydroelectric power plants, and this water will contain free-swimming veligers from nearby populations. Ready-to-settle pediveligers can attach to the piping provided flow velocity is less than 1.5 m/s. As noted previously, carbon steel, stainless steel, aluminum, and concrete piping all have a high risk of settlement, whereas copper piping or alloys with high copper content have a much lower risk. As service water progresses through various pipe branches, branch piping gets smaller. It is these smaller pipes that are at greatest risk of rapid plugging. Most piping of concern is the small-diameter (typically 200 mm or less) piping where water velocities are less than 1.5 m/s. Both mussel settlement and sloughing of mussels from upstream piping can block smaller pipes. Embedded small-diameter piping is of particular concern as in most cases the piping may not be accessible for conventional cleanout.

Valves

Development of dense mussel colonies can place proper operation of valves at risk, particularly valves that do not operate frequently (Figure 15.16). Trapped mussel shells may prevent valves from fully closing, causing leakage past the valve seat. Valves are often placed in series pairs with the inter-chamber cavity having a vent line or pressure-balancing line. Either or both of these lines could be disabled by mussel plugging. Air release and vent valves are of particular interest as these valves usually have a float that can be compromised by mussel fouling. If air-release valve floats do not seal, leakage will occur. If a vent valve is plugged, slow filling and draining may cause pipe walls to collapse.

Transformer Cooling

Older style transformers are often cooled with water from cooling water systems. Transformer coolers may be located remotely from the turbine building and have a separate cooling-water line that needs to be protected. If fouling occurs (Figure 15.17), cooling capacity may be compromised and there is a possibility of transformer damage. Typically, most current style transformers are air cooled that eliminates risk of any impairment from mussels. If older style, water-cooled transformers are near the end of their useful life, transformer replacement may be the most cost-effective way to protect these systems.

Heating, Ventilation, and Air-Conditioning Systems

Some heating, ventilation, and air-conditioning (HVAC) systems use raw water to chill air or to serve as a coolant. Raw water used in these systems may be strained, but it is usually untreated. Piping and cooling units are likely settlement areas for mussels, especially coolers where flow velocities are low in inlet and outlet headers. Coolers may need to be cleaned by chemical or thermal flushing methods if cooler performance deteriorates due to fouling by mussels. HVAC systems are normally associated with

Figure 15.16 (See color insert.) Pressure-regulating valve discharge fouled with dreissenid mussels. (Courtesy of Ontario Power Generation, Toronto, Ontario, Canada.)

Figure 15.17 (See color insert.) Cooling water pipe of a transformer fouled with dreissenid mussels.

personal comfort; therefore, impairment of a system is not usually a high priority as an operational concern. However, sometimes plants contain rooms with essential control equipment, usually electrical in nature, that generate heat and require uninterrupted cooling. These areas need to be examined to make sure any raw water used is protected from impairment by mussels.

Warm discharge water, typically from large heat loads such as the generator air coolers, may be used as a heat source for heat-pump systems. As the need for heating is most likely outside the normal mussel spawning season (typically May to November in North America and Europe), pipes delivering warm discharge water to heat pumps are at low risk of infestation by mussels.

Domestic Water

Many hydroelectric plants are located in areas where municipal water or well water is not available. These plants would have to generate their own water for potable use in sinks, showers, eyewash stations, etc. Sometimes domestic water may be used by equipment that can only function properly with clean, filtered water. Domestic water may have a separate intake from the raw water intake, or it may be taken directly as a branch line from the raw water piping. Depending on actual configuration of a particular domestic water system, the area with greatest potential for mussel fouling is usually within intake pipes up to the point where treatment processes begin. Any check valves or backflow prevention valves used to isolate domestic water from raw water are also subject to fouling. Fouling of these valves can cause leakage and backflow through check valves. Once raw water reaches a point where various water treatment chemicals (such as flocculants or acid) are added, incoming veligers would be eliminated. This will protect all downstream structures of domestic water systems.

Level, Flow, and Pressure Sensing Equipment

The performance and health of all systems is monitored and adjusted by using available instrumentation and control (I&C) systems. Level gauges, sight glasses, flowmeters, pressure transmitters, and various types of control valves are all examples of equipment in I&C systems. Typically, equipment and piping are small relative to the process systems with which they are associated, but their small size makes them particularly vulnerable to mussels (Figure 15.18). Instrumentation equipment may be in contact with raw water, but may not have any flow (such as a pressure tap). Tap lines leading from pipes to pressure transducers may not be at risk if there is no oxygen in lines due to no flow. Entrances to pressure taps then become areas of concern. If tap lines are plugged, instruments could become unreliable.

Figure 15.18 (See color insert.) Thrust-bearing sight glass with live dreissenid mussels.

Equipment used to sense water levels usually includes a tap line, stilling well, and float. Tap lines can become infested with mussels that would impair accuracy of gauges, especially if there are frequent or sudden changes in water flows and water levels. For locations where flow rates and water levels change very slowly, there may not be a reduction in accuracy as a plugged pipe would still allow water to percolate to stilling wells. Should mussels become established in a hydroelectric power plant and it is not practical to clean various instruments that are in contact with water, use of noncontact instrumentation such as ultrasonic or infrared level probes and flowmeters might be considered.

Drainage and Dewatering

When considering mussel infestation in a power plant, drainage systems are often overlooked. Drainage water is easily mistaken as waste and therefore considered unsuitable for mussel settlement and growth. However, this is not the case as mussels are often found in drain channels and drain lines. Drainage water is considered to have served its purpose and not needed to keep a plant operating; hence, drains tend to get less attention. Some drainage lines that have even a trickle of water flow can be suitable for mussel growth. Plugging of these lines is unlikely to affect operability of a plant, but it may be a maintenance nuisance to clean embedded drainage pipes. In addition, periodic large drainage flows through blocked or partially blocked drain lines will result in flooding of floors in associated galleries.

General drainage from plant equipment is directed to the plant sumps that have float switches. These switches should be monitored for mussel attachment since mussels can weigh down floats and cause level errors. Drain lines on penstocks, scroll cases, and draft tubes are probably used infrequently so any residual water between uses of drain lines will become low in oxygen and kill any mussels that settled during drain processes.

During dewatering of long pipes, any air vents should be checked for proper operation in the event mussels have managed to settle in wetted portions of vent lines. If it is impractical to check air vents either prior to or during a draining operation, the collapse pressure of pipe walls should be determined or checked.

Fire Protection Water

If fire protection systems draw water directly from raw water sources, they are as vulnerable to mussel fouling as other systems (Figure 15.19). Fire protection systems are designed to be filled with water and then maintained in a static pressurized state. In reality, hose stations are often a convenient source of water for washing equipment or flushing debris from spills on floors. In addition, fire systems tend to be tested on a regular basis, sometimes monthly. Regardless of purpose, fresh raw water replaces the volume of water used. This replacement water is a main point of entry for mussels into a system. Strainers would prevent mussels from entering firewater branch lines, but some local fire codes do not allow strainers in the fire protection systems.

If time between system uses is long, veligers should not be able to survive as water will become low in dissolved oxygen. Oxygen concentrations drop below the survival threshold of 3 mg/L within 3 weeks. However, should mussels begin to settle in a plant, rates of oxygen decline should be measured as rates can be variable depending on particular water conditions. If oxygen decline rates are low, then the fire protection system is at risk of mussel colonization. Any mussels that slough off colonized portions of fire-water piping could be delivered to the spray nozzles, causing impaired performance of nozzles.

Frequently, fire protection systems have a separate pump to supply water for testing and firefighting. Such a pump is submerged in raw water for long periods of time. The external structures may become heavily infested between test cycles and become a source of shell debris when a pump is started.

Figure 15.19 (See color insert.) Fire-pump bell housing fouled with dreissenid mussels.

SUMMARY

Overall, we have learned from our many facility inspections that each facility is unique and requires its own assessment to properly quantify effects of mussels. Even facilities

Table 15.1 Relative Vulnerability (Low, Medium, High) of Various Areas, Components, and Systems of Dams and Hydroelectric Power Plants Being Fouled by Mussels

Area/Component/System	Relative Vulnerability
Dams	
External structures	
Floating structures	Medium
Gates and weirs	Low
Gates on spillways	Low
Isolation gates	Low
Adjustable weirs	Low
Spillways	Low
Dam outlet works	High
Fish-handling facilities	High
Navigation locks	Low
Internal structures	
Conduits, valves, and vents	Medium
Seepage drains	Low
Uplift drains	Low
Sumps	High
Fire protection	Low
Construction materials	
Concrete, steel	High
Copper	Low
Hydroelectric power plants	
Intake and discharge structures	
Floating structures	Medium
Intakes	High
Penstocks	Low
Secondary intakes for raw and domestic water	High
Fixed grates, trashracks, and trashrack cleaning systems	High
Discharge area	Low
Internal structures	
Raw water/service water/cooling water	Medium
Heat exchangers	High
Strainers and filters	High
Compressed air systems	Low
Pumps, turbines, and generators	Medium
Piping	High
Valves	Medium
Transformer cooling	Low
HVAC	Medium
Domestic water	Medium
Level, flow, and pressure sensing equipment	High
Drainage and dewatering	Medium
Fire protection	High

that are structurally identical may be constructed in different environments where water quality and temperature make the intensity of mussel infestation vastly different between identical facilities. Operational practices may differ between facilities such that treatment options for one may be inappropriate for another. Therefore, vulnerability assessments should include a team of evaluators that includes individuals knowledgeable in the biology and behavior of mussels as well as team members knowledgeable about plant design and its operating regime.

A summary of typical relative vulnerability of various areas, components, and systems of dams and hydroelectric power plants to fouling by mussels is given in Table 15.1. This assessment of vulnerability categorizes the risk of mussel infestation on a particular component relative to other areas/components in a facility. The next step is to add a supplementary column that makes an assessment of risks to a plant if that particular component/area is impaired. Of course, a category chosen for each component in each column may differ depending on the facility design. Completing a table such as this will assist facilities in developing a path forward to prepare for and respond to a mussel infestation.

Depending on facility design, mussels typically increase relative operating costs of a dam or hydroelectric facility, and capital costs can increase significantly if control equipment needs to be installed. When practical, preventing or delaying the arrival of mussels in a water body where a facility is located can reduce or defer costs. In the unfortunate event mussels become established, facilities will require some control measures.

Most commonly, dams and hydroelectric power plants deal with mussel infestation of external structures by manual cleaning. Trashracks are usually the most critical structure. If racks can be removed, then cleaning is easier. Racks that cannot be removed require the forebay to be isolated and drained or the use of divers to scrape the racks underwater. Internal components can be protected from infestation by

- Filters and strainers to keep out shells combined with ultraviolet light to prevent settlement of veligers
- pH adjustment of raw water used in cooling systems
- Periodic draining and manual cleaning of cooling water systems
- Periodic isolation of the system and flushing with hot water
- Periodic application of an approved oxidant usually at the end of the mussel breeding season
- Periodic application of approved molluscicides

REFERENCES

Benson, A. J. 2013. Chronological history of zebra and quagga mussels (Dreissenidae) in North America, 1988–2010. In *Quagga and Zebra Mussels*: *Biology, Impacts, and Control*, 2nd Edn., T.F. Nalepa and D.W. Schloesser, eds., pp. 9–31. Boca Raton, FL: CRC Press.

CHAPTER 16

Managing Expansion of Dreissenids within Traditional Parameters

The Story of Quagga Mussels in Lake Mead National Recreation Area

Valerie Hickey

CONTENTS

Abstract 259
Introduction 260
Lake Mead National Recreation Area 260
Quagga Mussel Crisis 261
Evolution of a Management Response 262
Final Thoughts: Zoning versus Other Management Options 263
References 264

ABSTRACT

Protected areas such as national parks are the cornerstones of conservation. Within these parks, invasive species are second only to habitat destruction as the cause of biodiversity loss and ecological damage. So in January 2007 when a park volunteer discovered the invasive quagga mussel (*Dreissena rostriformis bugensis*) in Lake Mead, which is part of the Lake Mead Recreation Area, the National Park Service (NPS) unit responsible for managing the recreation area was prepared for just such a discovery. It had an early detection program in place and had a clear operational mandate to do something about an invasion if it occurred. Upon finding mussels, the NPS quickly moved to access funding and the best available science to implement a response. After debating four options, the NPS focused on education and enforcing existing laws to prevent mussel spread. Despite this reasonable preparation and response, quagga mussels spread throughout the park, and quagga colonies soon began to appear throughout the southwestern United States.

Given that efforts to stem the invasion were reasonable and rapid, why did they fail? While it is not known if any management strategy would have prevented or stopped the spread of mussels, the NPS was clearly constrained by an underlying commitment to manage the park by "zones." This use of zoning as a management strategy concentrated visitation in the recreation area onto two lakes (Lakes Mead and Mohave), which in turn opened a pathway for invasion, guided management decisions, and likely contributed to the rapid spread of quagga mussels.

Traditional policy efforts in conservation focus on what to conserve and where. However, very little information exists as to what policies actually work in the field. Failure to evaluate how policies translate into actions has led to the adoption of dogma that can be wrong and to an unwavering faith in the connection between a policy directive (e.g., ecotourism, bioprospecting, zoning) and a conservation outcome. Zoning is a common practice in park units worldwide. It reduces traditional stressors on other parts of a park by concentrating the human footprint in one area and reconciles competing mandates to protect nature and allow recreation. However, it also concentrates pathways for invasion by nonindigenous species. The risk of invasion by these species requires that the traditional management approach by zoning be examined and possibly be replaced by a more dynamic approach, particularly in susceptible parks like the Lake Mead National Recreation Area.

INTRODUCTION

The spread of invasive species is a growing global problem. Worldwide, estimated economic damage from invasive species totals more than US $1.4 trillion annually—5% of the global economy—with impacts across a wide range of sectors including agriculture, forestry, transportation, and power generation (Pimentel et al. 2001). In the United States alone, these species cost taxpayers in the United States hundreds of billions of dollars annually in environmental degradation, lost agricultural productivity, and increased health problems (Vitousek et al. 1996, Mack et al. 2000, Sala et al. 2000, Meyerson and Mooney 2007). Invasive species are second only to habitat destruction in depleting biodiversity and damaging protected areas (e.g., Reaser et al. 2003).

Current global stressors such as climate change only exacerbate the expansion of invasive species because the severity of extreme weather events facilitates movement of these species. For example, silver and bighead carps (*Hypophthalmichthys molitrix* and *Hypophthalmichthys nobilis*, respectively) were introduced as biocontrols to maintain aquaculture and wastewater treatment facilities. Having escaped into the Mississippi River after major floods in the early-1990s, they now threaten the ecological integrity of the Great Lakes (USGS 2005, Sea Grant Pennsylvania 2007). A shift in climatic variables favors species with larger bioclimatic ranges. Together, these changes are increasing the vulnerability of certain ecosystems, especially freshwater habitats, to debilitating invasions (SCBD 2009).

Freshwater has a unique social and economic role in everyday life. Its physical properties as a natural but fugitive resource add challenges to its management. Many management systems that deal with difficulties associated with freshwater are based on precedent and tradition. Today, in an era of invasives, as complicated by a changing climate, the politics of water management often collides with environmental science. Institutions that currently manage freshwater as a resource must evolve and adopt evidence-based approaches to address invasives as well as administer allocations and reconcile different uses. This may require inventing options that start not from the status quo of management dogma but from somewhere new. The lessons learned from management responses to the western invasion of quagga mussels (*Dreissena rostriformis bugensis*) at Lake Mead National Recreation Area (henceforth LAME or *the park*) provide a study in just how difficult it is to apply innovative responses when traditional management dogma is steadfastly followed, even in a park replete with political will and resources, and despite facing an invasive that is known to inflict enormous economic damages.

LAKE MEAD NATIONAL RECREATION AREA

Under an agreement with the Bureau of Reclamation, following construction of Hoover Dam in 1936, the park became the nation's first national recreation area managed by the National Park Service (NPS) in October 1964 (USG 1964). Running along the Colorado River in Arizona and Nevada, the park comprises 605,274 ha and has over 1,528 km of shoreline (Figure 16.1), contains important dams and reservoirs, and hosts several endangered and threatened species. As a result of these valuable assets, two other federal agencies, in addition to the NPS, have some level of management jurisdiction over the lakes (the Bureau of Reclamation and Fish and Wildlife Service). In addition, several state-level agencies from Arizona and Nevada also have some management authority, including the Southern Nevada Water Authority, Arizona Game and Fish Department, and Nevada Department of Wildlife.

The park is located at the confluence of three major desert ecosystems (the Mojave, the Great Basin, and the Sonoran) and receives annual rainfall of less than 15 cm per

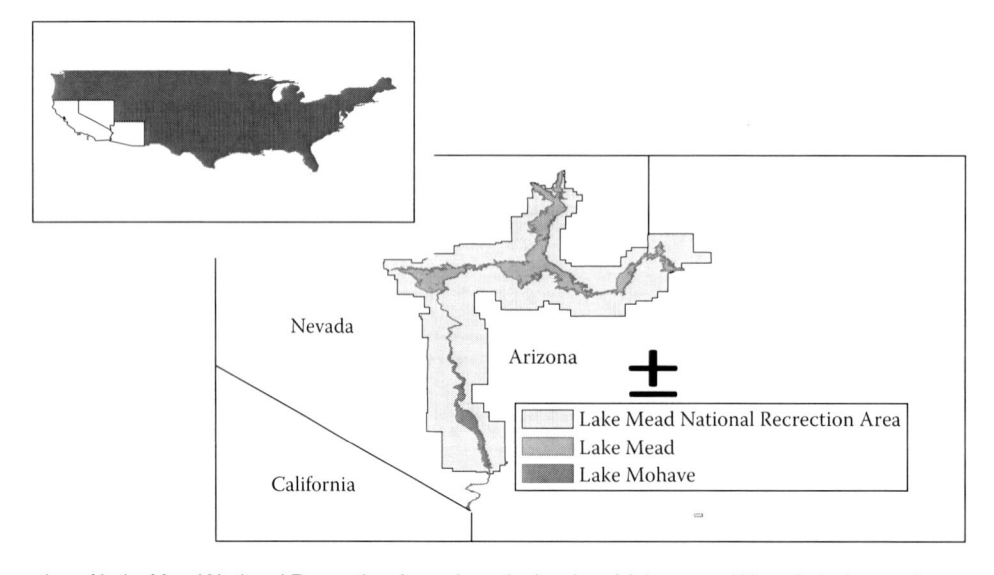

Figure 16.1 Location of Lake Mead National Recreation Area along the border of Arizona and Nevada in the southwestern United States.

year (LAME 2006). Despite these arid conditions, an abundance of plant and animal life exists, each uniquely adapted to the area's conditions. The park is home to over 900 species of flora and 508 species of fauna. Of these, 24 are either federally endangered or threatened (LAME 2001). Adjacent to the rapidly expanding Las Vegas metropolitan area and within a day's drive from Los Angeles and Phoenix, the park hosted almost 8,000,000 visitors in 2005 (NPS 2006). In addition, over one million visitors enter the park every year to tour the Hoover Dam that lies within its boundaries (Bureau of Reclaimation 2013).

The mission statement for the park as defined by NPS reads "to provide diverse inland water recreation opportunities in a spectacular desert setting for present and future generations" (LAME 1999). As a result, NPS management focuses on protecting the park's natural resources while providing recreational opportunities. It does this by adopting the traditional practice of park zoning, whereby visitors are funneled to recreate around several developed areas (marinas) on the park's two lakes, while the rest of the park, including sections of freshwater, is left largely wild, or at least free of a heavy recreational footprint.

QUAGGA MUSSEL CRISIS

Researchers and park staff had long been concerned about Lake Mead as a major potential source of invasive species infestations (Bossenbroek et al. 2007). They were particularly worried about *Dreissena* species (zebra and quagga mussels) that could quickly spread to other water bodies in the western United States (Gerstenberger et al. 2003). Potential pathways for spread from water bodies in the park were extensive, including natural pathways inherent to the Colorado River, as well as assisted overland pathways resulting from the many watercraft coming from across the United States (Figure 16.2). These watercraft occupy Lakes Mead and Mohave in winter and summer and then go to other water bodies. As a result, NPS established an early detection and monitoring program at the park in the early 2000s. This program consisted of three parts: (1) NPS staff placed artificial substrates (modified Portland samplers; Marsden 1992) at each of nine marina sites on both lakes to detect adult mussels. These substrates were placed in shaded water at depths of 1.5, 6.0, and 9 m [NPS 2007]. In addition, NPS staff and volunteers intermittently sent water samples to Portland State University to test for veligers [NPS 2007]); (2) NPS staff, concessionaires, and boating associations received training on identification and management of invasive species (trainers taught participants how to identify dreissenids and to inspect and decontaminate boats); and (3) NPS staff targeted high-risk boats for inspection and decontamination; these high-risk boats came from dreissenid-infested water bodies in the east. Over the 5 years prior to January 2007, park staff intercepted 54 high-risk boats entering LAME. Of those inspected, six vessels had invasive mussels. Park staff decontaminated them prior to launch (LAME 2007).

Despite these precautions, on January 6, 2007, a park volunteer discovered quagga mussels in Lake Mead (LAME 2007). At this time, this discovery extended the distribution of quagga mussels in North America about 1,600 km farther to the west than any other known colony at the time (Figure 16.3; also see Benson 2013). Although we may never know how the initial Lake Mead infestation occurred, park staff presumed that mussels were attached to a watercraft that had been trailered to Lake Mead from an infested lake in the Great Lakes region (LAME 2007). By January 20, 2007, quagga mussels appeared downstream of Lake Mead in Lake Mohave (LAME 2007).

The discovery of quagga mussels in Lake Mead and their rapid spread throughout the Colorado River basin has prompted a number of retrospective questions: Did management agencies lack a mandate to deal with the mussels? Did management agencies have the resources to make political will a reality? Did agencies have access to the best science on how to manage invasive species? Did the rapid spread of mussels occur because the park was unprepared or unwilling to respond, or were mussels simply discovered too late to prevent further spread?

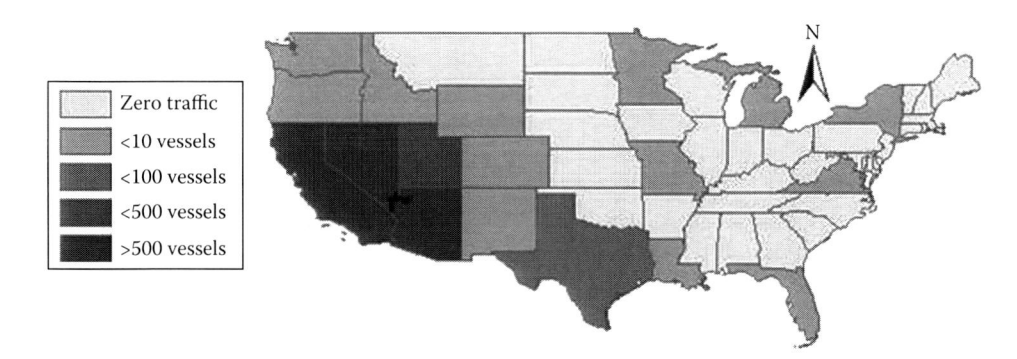

Figure 16.2 Number of watercraft coming from outside the boundaries of the given U.S. state and entering Lake Mead National Recreation Area.

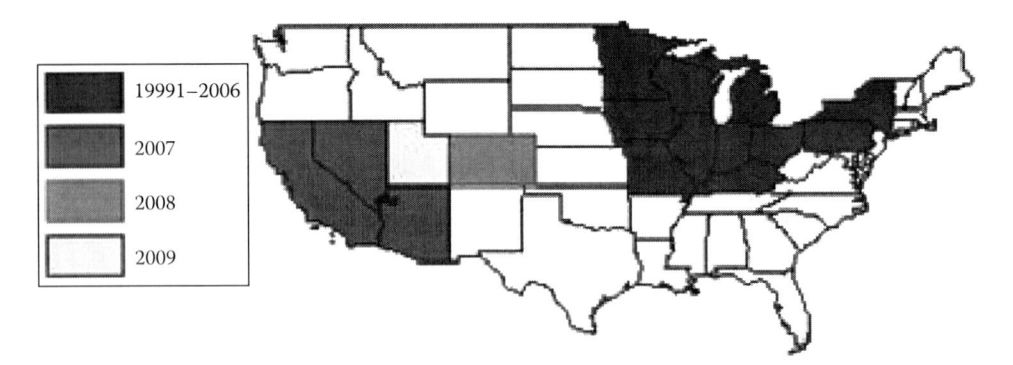

Figure 16.3 U.S. states where quagga mussels were present in 1991–2006, 2007, 2008, and 2009. Note that before quagga mussels were discovered in Lake Mead National Recreation Area in 2007, the presence of quagga mussels was limited to states within and near the Great Lakes region.

EVOLUTION OF A MANAGEMENT RESPONSE

Immediately after quagga mussels were discovered in the park, NPS staff recognized the importance of a rapid response. In January 2007, they quickly established an Interagency Coordination Core Team that was comprised of representatives from each agency with jurisdiction or interest in the lakes. The team grew from the NPS commitment to managing invasive species as outlined in their Management Policies (NPS 2006). The team quickly developed an Initial Response Plan (NPS 2007) to deal with the mussels. This Response Plan was well funded and validated by leading experts. In addition to the appropriated federal funding for LAME that covered fixed costs, NPS secured additional resources from the Southern Nevada Public Lands Management Act (SNPLMA) (USG 1998), which provided funding to better manage southern Nevada's natural capital. NPS also accessed the best available science on invasive mussels. Following the dreissenid discovery, NPS quickly established a scientific advisory team under the leadership of the NPS Biological Resources Division. This team comprised experts in invasive mussels, limnologists, fisheries biologists, and hydrologists familiar with Lakes Mead and Mohave (LAME 2007).

The Initial Response Plan was the culmination of thinking through four potential options. The first option was to do nothing since scientific evidence suggested that little could be done to stop mussels from spreading (LAME 2007). However, this option violated the NPS mandate to conserve the park's resources in a natural state. This option also ignored political pressures on NPS to do "something" (Boxall 2007). The second option, and arguably the most effective way to reduce spread, involved closing the park to all watercraft. NPS staff discarded this option because of the open-access nature of the park, with dozens of unofficial launching sites around the lake. Such an option simply would have been impossible to police. More importantly, the recreation mandate of NPS would not allow such an option, and the NPS commitment to zoning as a way to reconcile protection and recreation meant that closing the lakes effectively closed recreational opportunities. A third option was to prevent recreational watercraft from traveling between different parts of Lake Mead. Staff proffered this option as an attempt to slow colonization of upstream, uninfested sections of the lake. The option called for the so-called gunboat curtains in three areas. These curtains relied on law-enforcement staff to prevent watercraft from moving from one part of the lake to another, despite the ability of trailered watercraft to travel unimpeded by land to different launch sites. By rough estimates, hundreds of vessels passed each "curtain" each day (Graefe and Holland 1997). However, chases of unlawful watercraft movements by enforcement personnel would of themselves provide a pathway for the mussels to colonize upstream. Also, this option for spread prevention also relied on the presumption that mussels would not readily spread downstream from Lake Mead. This presumption was based on studies that showed dreissenid veligers were sensitive to high water turbulence (Rehmann et al. 2003). As part of this assumption, NPS believed at the time that adult mussels found downstream of Hoover Dam were transported there aboard a watercraft rather than being the offspring of upstream adult mussels. The general assumption was that the extremely high turbulence in the intake pipes of Hoover Dam would lead to high veliger mortality, thus providing a natural barrier to spread. Further, it could mean that downstream populations, if eradicated, would have no natural recruitment from upstream populations. This assumption ultimately proved to be untrue (O'Neill 2008) and, along with the impracticality of "gunboat curtains," led to the abandonment of this option.

In coordination with a suite of other agencies with management authority over Lake Mead, NPS ultimately chose a fourth option to prevent the spread of quagga mussels. This option, labeled the Dedicated Invasive Mussel Education and Enforcement (DIME) Program, consisted of four components: education, enforcement, inspection, and impact monitoring. The education component focused on both fixed and dynamic messaging and consisted of a signage system throughout the park that emphasized the ecological dangers of invasive mussels and the threat that mussels poised to private property and existing infrastructure. The signs also made boaters aware of misdemeanor penalties for transporting mussels. Seasonal rangers were employed with SNPLMA funding to staff priority launch ramps and provide in-person messages to boaters as they exited the lakes. These rangers made contact with an average of 1,500 boaters every week and asked them to "clean, drain, and dry" their watercraft prior to launching at another water body (NPS 2007). The enforcement component consisted of law-enforcement rangers that staffed priority launch ramps to create conditions necessary to mandate boat cleaning prior to leaving the park. In Arizona, it was already unlawful to transport mussels (Arizona 2009). Under the Lacey Act, it was also unlawful to transport invasive species like quagga mussels on federal lands (USG 1900). The park worked with local law-enforcement officials to make the transportation of mussels across political boundaries a Class III misdemeanor in Nevada (Nevada 2007). These laws complemented voluntary compliance. The inspection and decontamination component was based on the conclusion of the science advisory team that only watercraft in the lake for more than 3 days posed any real threat of harboring mussels. As a result, the park instituted a program that required all watercraft moored on its lakes to undergo inspection when leaving the park, and watercraft found with mussels must be decontaminated. The mooring contract for each watercraft codified this arrangement (LAME 2007). In 2007, park staff inspected 129 watercrafts under this program and more than one-half (75) were found to have mussels attached. In the first 8 months of 2008, of the 67 watercrafts inspected, almost two-thirds (41) were found to harbor mussels (Hickey 2010). The impact monitoring component was based on a partnership (entitled the Water 2025 Initiative) in which NPS, together with USGS, BOR, University of Nevada Las Vegas, Bureau of Land Management, NDOW, USFWS, Southern Nevada Water Authority, and the Clean Water Coalition, worked to develop water and resource inventories for Lake Mead. This component included a comprehensive monitoring program for invasive species that went beyond passive samplers to test for presence/absence. Importantly, it included a focus on the identification and assessment of biotic indicators of environmental health and how these indicators may be impacted by quagga mussels.

FINAL THOUGHTS: ZONING VERSUS OTHER MANAGEMENT OPTIONS

NPS proceeded through a series of management actions in a reasonable and rapid fashion in efforts to prevent and cope with the introduction of quagga mussels in the park. Nonetheless, implementation of these actions ultimately failed to prevent the introduction and diminish the spread. In hindsight, perhaps one of the key elements that contributed to this failure was the decision to not act beyond traditional zoning as a management strategy, despite a highly favorable political environment. Throughout the mussel crisis, NPS continued to manage visitation by funneling visitors toward the park's two lakes and away from its fragile desert ecosystem. Visitor services remained concentrated in these developed areas. To be sure, the extreme heat and unwelcoming desert landscape served as impediments to altering this management strategy. However, continued use of this management approach was inconsistent with preinvasion efforts to prevent introduction of the mussels. It also led to NPS staff discarding three potential responses to the introduction, including one that had the best chance to hinder mussel spread, that is, closing the lakes to recreation and shifting the focus of recreation toward the desert. At the very least, restricted watercraft use of the lakes and the encouragement of recreational activities in the desert prior to mussel introduction would have diminished the risk of introduction.

Zoning is a politically acceptable, commonplace management strategy to reconcile simultaneous demands for strict conservation and open access in a single landscape. It focuses traditional environmental stressors in certain areas, leaving other areas less disturbed as a result. However, this strategy ignores the interconnectedness of landscapes, particularly in freshwater ecosystems, at a time when the introduction and spread of invasives is of particular concern. Concentration of human "footprints" in the same area over time only serves to promote pathways that increase risks of invasive introductions and creates disturbed areas that increase potential for invasives to gain a foothold. The consequent failure to consider every evidence-based management option in the face of intransigent zoning leaves wild places more vulnerable. Static zoning must make way for dynamic management that could take account of the spread of invasives, particularly in an era of climate change that only adds urgency to the need for shifting conservation strategies. Enhancing the resilience of protected areas demands the adoption of shifting conservation strategies that, like strategies to conserve agricultural landscapes, would allow for natural regeneration after a set period even in those areas heavily used for recreational activities. Though some areas must always remain inviolate, even management of these areas may need to be adjusted over time in response to outside forces, as, for instance, climate change. Other areas should be regularly cycled through greater and lesser recreation footprinting, which allows areas a period to recover. Intermediate actions, such as the introduction of invasive breaks that could physically separate recreation zones from their wilder, less-used counterparts, could bring some much-needed relief from the vectors associated with invasive introduction. At present, it is not known if management strategies other than zoning would have prevented the introduction of quagga mussels into Lake Mead, but agencies and regulatory agencies are encouraged to evaluate philosophies under which park lands are managed, particularly in parks recognized as sensitive and vulnerable to invasives like the Lake Mead Recreation Area.

REFERENCES

Arizona (Government of Arizona). 2009. Revised Statutes, Game and Fish Laws and Rules, R12-4-406. Phoenix, AZ.

Benson, A.J. 2013. Chronological history of zebra and quagga mussels (Dreissenidae) in North America, 1988–2010. In *Quagga and Zebra Mussels: Biology, Impacts, and Control*, 2nd Edn., T.F. Nalepa and D.W. Schloesser, eds., pp. 9–31. Boca Raton, FL: CRC Press.

Bossenbroek, J. M., L. E. Johnson, B. Peters, and D. M. Lodge. 2007. Forecasting the expansion of zebra mussels in the United States. *Conserv. Biol.* 21: 800–810.

Boxall, B. 2007. Quagga mussels in Lake Mead spur high alert. *Los Angeles Times*, January 21, 2007. Available at http://www.post-gazette.com/pg/07021/755227-84.stm

Bureau of Reclaimation. 2013. Hoover tour information. https://www.usbrigov/lc/hooverdam/service/index.html (accessed July 1, 2013).

Gerstenberger, S., S. Powell, and M. McCoy. 2003. The 100th Meridian Initiative in Nevada: Assessing the potential movement of the zebra mussel to the Lake Mead National Recreation Area, Nevada, USA. U.S. Fish & Wildlife Service Report. http://100thmeridian.org/Documents/zebra%20Mussels%20Nevada.pdf

Graefe, A. R. and J. Holland. 1997. *An Analysis of Recreational Use and Associated Impacts at Lake Mead National Recreation Area*. A Social and Environmental Perspective. NPS LAME, Boulder City, NV.

Hickey, V. 2010. The Quagga mussel crisis at Lake Mead National Recreation area, Nevada (USA). *Conserv. Biol.* 24: 931–37.

LAME (Lake Mead National Recreation Area). 1999. *Business Plan*. LAME, Boulder City, NV. Available at http://www.nps.gov/lame/parkmgmt/upload/ExecSum.pdf

LAME (Lake Mead National Recreation Area). 2001. *Strategic Plan 2001–2005*. LAME, Boulder City, NV. Available at http://www.nps.gov/lame/parkmgmt/upload/ACFB12.doc

LAME (Lake Mead National Recreation Area). 2006. Draft business plan. Unpublished.

LAME (Lake Mead National Recreation Area). 2007. *Lake Mead National Recreation Area Draft Quagga Mussel Initial Response Plan*. LAME, Boulder City, NV.

Mack, R. N., D. Simberloff, W. M. Lonsdale et al. 2000. Biotic invasions: Causes, epidemiology, global consequences, and control. *Ecol. Appl.* 10: 689–710.

Marsden, J. E. 1992. Standard protocols for monitoring and sampling zebra mussels. Illinois Natural History Survey Biological Notes 138: 1–39.

Meyerson, L. A. and H. A. Mooney. 2007. Invasive alien species in an era of globalization. *Front. Ecol. Environ.* 5: 199–208.

Nevada (Government of Nevada). 2007. Administrative Code, Chapter 503: Hunting, fishing and trapping; miscellaneous protective measures.

NPS (National Park Service). 2006. Lake Mead National Recreation Area traffic package. Traffic Data Report. NPS, Washington, DC.

NPS (National Park Service). 2007. *Quagga/Zebra Mussel Infestation Prevention and Response Planning Guide*. Natural Resources Program Center, Fort Collins, CO. Available at http://www.nature.nps.gov/biology/Quagga/QuaggaPlanningGuide_ext.pdf

O'Neill, C. R. 2008. Oversight hearing: The silent invasion: Finding solutions to minimize the impacts of invasive quagga mussels on water rates, water infrastructure and the environment. Committee on Natural Resources Water and Power Subcommittee June 24, 2008, Available at http://www.usbr.gov/lc/region/programs/quagga/testimony/cornell.pdf

Pimentel, D., S. McNair, J. Janecka et al. 2001. Economic and environmental threats of alien plant, animal, and microbe invasions. *Agric. Ecosyst. Environ.* 84: 1–20.

Reaser, J. K., A. Gutierrez, and L. Meyerson. 2003. Biological invasions: Does the cost outweigh the benefits? *BioScience* 53: 598–600.

Rehmann, C. R., J. A. Stoeckel, and D. W. Schneider. 2003. Effect of turbulence on the mortality of zebra mussel veligers. *Can. J. Zool.* 81: 1063–1069.

Sala, O. E., F. S. Chapin, J. J. Armesto et al. 2000. Global biodiversity scenarios for the year 2100. *Science* 287: 1770–1774.

SCBD (Secretariat of the Convention on Biological Diversity). 2009. Connecting biodiversity and climate change mitigation and adaptation: Key messages from the Report of the Second Ad Hoc Technical Expert Group on biodiversity and climate change. SCBD, Montreal, Canada.

Sea Grant Pennsylvania, 2007. Asian Carp. Available from http://www.pserie.psu.edu/seagrant/publications/fs/asian-corp2007.pdf (accessed July 1, 2013).

USG (United Stated Federal Government). 1964. To provide an adequate basis for administration of the Lake Mead National Recreation Area, Arizona and Nevada, and for other purposes. Public Law 88-639.

USG (United Stated Federal Government). 1998. Southern Nevada Public Lands Management Act of 1998. Public Law 105-263.

USGS (U.S. Geological Survey). 2005. Foreign non-indigenous carps and minnows (Cyprinidae) in the United States: A Guide to their identification, distribution and biology (Scientific Investigations Report). Department of Interior, Washington, DC.

Vitousek, P. M., C. M. Dantonio, L. L. Loope, and R. Westbrooks. 1996. Biological invasions as global environmental change. *Am. Sci.* 84: 468–478.

CHAPTER 17

Dreissenid Mussels as Sentinel Biomonitors for Human and Zoonotic Pathogens

David Bruce Conn, Frances E. Lucy, and Thaddeus K. Graczyk

CONTENTS

Abstract 265
Introduction 265
Human Pathogens and Diseases 266
Giardiasis 266
Cryptosporidiosis 267
Microsporidioses 267
Development of Sentinel Models 267
Implementation of the Sentinel Model 268
References 271

ABSTRACT

After becoming well-established in newly-colonized waterways, invasive bivalves can prove beneficial as biomonitoring sentinels for human and animal pathogens. Because invasive bivalves often occur in large numbers, filter large volumes of water, and retain filtered particulate material in their systems, they can accumulate suspended microscopic waterborne pathogens. Our experimental work over 15 years has resulted in the development of both *Corbicula fluminea* and *Dreissena* spp. as sentinel biomonitors in diverse waterways in the Laurentian Great Lakes/St. Lawrence River system of North America, and in the Shannon River system in Ireland. Employing sensitive and accurate molecular techniques for diagnosis, we have used these invasive bivalves as sentinels to detect the presence of the eukaryotic pathogens *Cryptosporidium*, *Giardia*, *Cyclospora*, *Enterocytozoon*, and *Encephalitozoon* in laboratory and field settings. Together, these pathogens cause widespread and serious illness and death among both human and animal hosts. By coupling the use of these and other filter-feeding invertebrate sentinels with regularly-deployed navigational aids, we propose that this approach can be applied to develop cost-effective public health biomonitoring programs in navigable waterways throughout the world.

INTRODUCTION

Invasive species often cause considerable changes to their new environments, with most of these changes generally regarded as being undesirable. However, although invasive species are widely considered as the second greatest threat to biodiversity (Simberloff 2000), some human benefits may accrue at locations they invade. Newly established invasives may provide a new exploitable resource, especially when they develop large populations, become ecological keystone species, and are widespread and easily collected within the newly invaded area. Zebra mussels (*Dreissena polymorpha*) and quagga mussels (*Dreissena rostriformis bugensis*) certainly fit this description in many of the aquatic environments they invade since they often become the most dominant and widespread benthic organisms (Dermott and Kerec 1997). These mussels act as dominant suspension-feeding planktivores and turn over large volumes of the water column (Horgan and Mills 1997, Karatayev et al. 1997, 2002, Lucy et al. 2005).

Scientists concerned with public health, agricultural animal health, and wildlife diseases realize that aquatic systems are essential epidemiological reservoirs for pathogens (Zintl et al. 2006). Consequently, they constantly seek effective ways to monitor waterways for the introduction, accumulation, and spread of waterborne pathogens in order to source track potential outbreaks in human or animal populations (Nichols 2008). This is especially important in predicting outbreaks of zoonotic pathogens (i.e., pathogens that can infect both humans and animals) with the ability to build up in one species before infecting another. However, establishing monitoring programs across broad geographic areas, and over long time periods, can be logistically difficult, expensive, and difficult to standardize across the entire spatial and temporal range where monitoring is desirable.

One approach to addressing these problems is the employment of sentinel organisms for biomonitoring. Dreissenid mussels have several characteristics that make them especially well-suited to serve as sentinels for the monitoring of pathogens. They are filter feeders and process large volumes of water each day; they are small, easily collected, and easily processed; and they occur in large numbers and typically in dense concentrations so that many mussels can be collected and pooled for larger sample sizes when necessary. Also, they spread to different catchments, thereby expanding their suitability (Lucy et al. 2008, Benson 2013). Although dreissenids generally harbor their own endosymbionts (Conn and Conn 1995, Conn et al. 1996, 2008, Molloy et al. 1997, Mastitsky et al. 2008), their filter-feeding behavior also results in uptake and retention of parasites and pathogens of other animals, including humans and other vertebrates (Graczyk and Conn 2008). These parasites and pathogens do not truly infect the mussels, but they may be harbored within mussel tissues, thus making them available for identification and quantification as environmental contaminants (Graczyk et al. 2001, 2002).

The purpose of this chapter is to outline the primary waterborne microbial pathogens that are being developed as models for surveillance technologies employing dreissenid mussels and other bivalves as biological sentinel organisms, and to review the published literature on the subject. The sequence of model development beginning with laboratory efficacy tests and extending to pilot field tests is described. Finally, recommendations are made regarding the future development and implementation of these surveillance methods.

HUMAN PATHOGENS AND DISEASES

Waterborne pathogens are a leading cause of human morbidity and mortality worldwide (Dowd et al. 1998, Karanis et al. 2007). Further, many microbial pathogens associated with sediments may become suspended in water columns of lakes and streams, spread by water currents, and ultimately contaminate water available for direct human consumption (Lucy et al. 2008, 2010), contaminate irrigation water that can indirectly affect human foods (Ortega and Sanchez 2010), or contaminate recreational water that can be accidentally ingested by people engaged in swimming or other water sports (Graczyk et al. 2007, Lucy et al. 2008). Furthermore, a large number of these pathogens have zoonotic potential, capable of infecting domesticated livestock and wildlife (Nichols 2008). These pathogens can thus spread from contaminated soil, water, and sediment to cycle between animals and humans by a variety of pathways (Zintl et al. 2006, Conn et al. 2007). The transmission stages of many common pathogens are shed by the millions in feces of infected people or animals and, because they are resistant to environmental stressors, they become ubiquitous in the environment (Wolfe 1992, Karanis et al. 2007). Human pathogens may enter waterways via end products of wastewater treatment, namely, in final effluent and biosolids (Cheng et al. 2009, 2011). These pathogens are extremely diverse, ranging from enteroviruses and enteric bacteria to eggs of parasitic worms (i.e., helminths). Many of the most widespread and problematic pathogens are parasitic protists, which are intermediate in size between bacteria and worms (Graczyk and Conn 2008, Conn 2010a). Thus, these protists have received the greatest amount of attention as candidates to be monitored with sentinel species such as dreissenid mussels and other bivalves. The primary diseases produced by these genetically complex eukaryotic protists vary both epidemiologically and clinically. The most extensively studied of these as candidates for monitoring by sentinel bivalves are the causative agents of the diseases giardiasis, cryptosporidiosis, and various microsporidioses.

GIARDIASIS

The best known among these three protist disease groups is giardiasis, which occurs in waterways around the world and causes diarrhea that can be severe, long lasting, and often difficult to treat (Wolfe 1992). Giardiasis is caused by flagellates of the genus *Giardia*, variously assigned to the species *Giardia lamblia*, *Giardia intestinalis*, and *Giardia duodenalis*, although there is some degree of controversy regarding the taxonomy and actual phylogenetic relationships among various species and strains (Thompson 2004). Different strains apparently have different host specificities, but in general the disease is regarded as zoonotic, occurring in a broad range of wild and domestic animals. Within the host, the actively feeding trophozoite stage adheres to the mucosal lining of the duodenum and interferes with nutrient absorption and other vital intestinal functions. Following asexual reproduction, many trophozoites transform into cyst stages, which possess thick walls that are resistant to environmental stresses, and thus are able to spend considerable time outside a host during transmission to other hosts. Cysts can be filtered and retained by dreissenid mussels and other bivalves, where they may remain viable in tissues and internal cavities for days or weeks, and thus cysts can serve as a diagnostic tool for monitoring (Graczyk et al. 1999).

CRYPTOSPORIDIOSIS

While giardiasis is better known, cryptosporidiosis is a waterborne diarrheal disease that probably poses a greater threat to human and animal health. The disease is caused primarily by the protist *Cryptosporidium parvum*, which is a species that is broadly zoonotic. It occurs at especially high prevalence in cattle populations, causing severe diarrheal scours in calves, and infects humans both intermittently and in large outbreaks (Chalmers and Giles 2010). The largest outbreak in North America occurred in the Lake Michigan watershed near Milwaukee, Wisconsin, in 1993, when approximately 400,000 human cases, 100 human deaths, and an estimated $96.2 million in economic losses occurred from this one outbreak (Corso et al. 2003). Besides *C. parvum*, several other *Cryptosporidium* species also cause diarrhea in humans, domestic animals, and wildlife, and the genomic complexity of the genus is under continual revision (Fayer 2010). *Cryptosporidium* spp. are apicomplexan relatives of the parasites that cause malaria, but are directly transmissible rather than requiring a mosquito vector. They invade the intestinal epithelial cells of their vertebrate hosts where they reproduce by both asexual and sexual processes, destroy host cells, and produce tiny resistant oocysts that pass in feces to contaminate water supplies (Fayer 2008). These oocysts are much smaller and more resistant than cysts of *Giardia* and thus persist longer and travel farther in the environment. They also can be filtered and retained for extended periods by freshwater bivalves (Graczyk et al. 1998a).

MICROSPORIDIOSES

There are many diseases, collectively referred to as emerging infectious diseases, which have been known to occur in humans only within the last few years or decades, usually in immunocompromised patients. Although these diseases may be recognized as recent, they may have been present in undiagnosed form for centuries. Among the most diverse groups of emerging eukaryotic microbial diseases are those that fall within the Microsporidia, which are possibly closer phylogenetically to fungi than to protists (Vossbrinck and Debrunner-Vossbrinck 2005). Microsporidians are obligate intracellular parasites that typically infect a wide range of both vertebrate and invertebrate hosts, and thus are usually regarded as zoonotic when humans become infected (Mathis et al. 2005). Different species occur opportunistically in different sites of the human host, thus causing widely varying forms of microsporidiosis. Perhaps the most widespread of the human-virulent microsporidia is *Enterocytozoon bieneusi*, which primarily affects the intestinal mucosa of a variety of vertebrates. *E. intestinalis* similarly predominates in the intestinal tract but is restricted to mammals. In contrast, *E. hellem* is more likely to cause systemic infections and is primarily a parasite of birds. These and many others that are less known and presumably less prevalent, such as the mammal-infecting *Encephalitozoon cuniculi*, may also cause disseminated infections and frequently affect the respiratory tract. Epidemiologically, most appear to undergo waterborne transmission, entering the human or animal host as a spore, which is even smaller and more resistant than the *Cryptosporidium* oocyst and which also can be filtered and retained by bivalves (Slodkowicz-Kowalska et al. 2006).

DEVELOPMENT OF SENTINEL MODELS

Because of their broad distribution and abundance in large navigable waterways that are associated with vast agricultural lands and major freshwater ports, dreissenids have emerged as the best and most studied sentinels for biomonitoring waterborne-eukaryotic pathogens (Graczyk et al. 2003, Graczyk and Conn 2008, Lucy et al. 2010). However, our own work with the sentinel model began with another important invasive bivalve, *Corbicula fluminea*, which prefers warmer waters and thus has a more south temperate distribution than dreissenids (McMahon 1999). Furthermore, unlike epibenthic dreissenids, *C. fluminea* is an infaunal clam that burrows into the sediment and often lives in smaller streams and ponds that do not support dreissenid populations (McMahon 1999). Moreover, *C. fluminea* is better suited for experimental laboratory purposes than are dreissenids because it is easily kept alive in simple small-volume laboratory aquaria, allowing quantitative exposure to precise numbers of pathogens.

Our initial experiments with *C. fluminea* involved in vitro studies of hemocytes, which demonstrated that hemocytes of this species were able to take up *C. parvum* oocysts efficiently by phagocytosis and maintain them intracellularly in an intact, viable, and infectious state (Graczyk et al. 1997b). Each single clam retained an average of 1.84×10^6 oocysts per mL of hemolymph. A similar study found that when hemocytes of *C. fluminea* were exposed in vitro to *G. duodenalis* cysts, the hemocytes retained 1.6×10^6 cysts per mL of hemolymph (Graczyk et al. 1997a). By demonstrating that *C. fluminea* could effectively concentrate and retain, in viable condition, these pathogens from the environment, both studies set the stage for in vivo tests with *C. fluminea* and pathogens in controlled laboratory exposures.

In the next stage of trials, live *C. fluminea* were exposed in laboratory aquaria to carefully measured concentrations of pathogen transmission stages. It was demonstrated that *C. fluminea* in laboratory aquaria completely cleared the water of experimentally-introduced *C. parvum* oocysts within 24 h and, in addition, retained the majority of oocysts in viable form within their tissues or voided viable oocysts in their feces (Graczyk et al. 1998a). Hence, the pathogens were filtered from the water column and concentrated in live clams or in sediments contaminated by clam feces. A follow-up study demonstrated that *C. fluminea* in aquaria similarly cleared *G. duodenalis* cysts from the water column, and either retained them within their tissues or voided them into the sediment via feces (Graczyk et al. 1999). Another parallel study in laboratory aquaria showed that *C. fluminea* filtered and retained oocysts of another emerging diarrheal protistan pathogen, *Cyclospora cayetanensis* (Graczyk et al. 1998b). Since most oocysts and cysts of these three pathogens were retained within clam tissues, it was reasonable to propose that examining tissues from wild field-collected bivalves could serve as an effective biomonitoring tool. However, testing this in actual field situations was still required.

IMPLEMENTATION OF THE SENTINEL MODEL

As shown, laboratory studies determined that freshwater bivalves efficiently removed human and zoonotic pathogens from the water and retained them for long periods in a viable state. The next step was to determine the extent to which bivalves were naturally contaminated in bodies of water where both they and the pathogens were known to occur. In an initial study, zebra mussels (*D. polymorpha*) were examined for *C. parvum* in the St. Lawrence River, a part of the Laurentian Great Lakes in North America. At a single wastewater discharge site, mussels were found to contain an average of approximately 440 *C. parvum* oocysts (Graczyk et al. 2001). Subsequent studies verified that *D. polymorpha* and *C. fluminea* were similar in their efficacy as biomonitor sentinels for *C. parvum* (Graczyk et al. 2002, 2003, 2006), thus demonstrating that the laboratory model of *C. fluminea* as a sentinel could be compared with the field model of *D. polymorpha* as a sentinel, even though the two live in different types of aquatic environments and *D. polymorpha* is not as easily maintained in the laboratory.

The St. Lawrence River/Great Lakes system was chosen as a pilot site largely because the watershed has been widely and densely colonized by dreissenids and partly because it is a major navigable system of great geographic extent and includes both extensive agricultural lands and many major cities that might serve as sources of pathogen dissemination. Thus, methods that work in this system are likely to be adaptable to other major navigable waterways globally, where most of the world's human populations reside. Because such major navigable waterways are invariably supplied with navigational buoys and other navaids of known location, it seemed that use of these navaids as monitoring platforms would be efficient and effective (Figure 17.1), though the exact nature of such tools would be variable (Figure 17.2).

Figure 17.1 Navigational buoys (navaids) from Lake Ontario densely fouled by dreissenid mussels. The "K" designation on the buoy signs indicates that these were deployed in Kingston Harbor area in the eastern basin of the lake, and the numeral indicates precise location. Variability in mussel density among sites is apparent. These mussels were highly contaminated with waterborne zoonotic pathogens.

Figure 17.2 Design variations in navigational buoys (navaids) from the St. Lawrence River. The number on each buoy corresponds to a specific location where the buoy was deployed. Dreissenid mussels found on buoys in the river were lightly contaminated with waterborne zoonotic pathogens.

At the end of the 2008 navigation season, dreissenid mussels were removed from navaids that had been pulled from the water at locations along the north shore of Lake Ontario and in the upper St. Lawrence River that extended from Toronto to just upstream from Clayton, Ontario (Figure 17.3). Preliminary results indicated that the use of dreissenids attached to navaids does provide consistent data on waterborne pathogens of three major categories (Figure 17.4). These results have not been fully analyzed in terms of identifying point sources of pathogens or patterns of pathogen dissemination, but further work along these lines is underway. Nevertheless, preliminary data do indicate strongly that this model can be adapted for other navigable waterways in diverse geographical locations.

Following initiation of field trials in the Lake Ontario/St. Lawrence River system, the model of *D. polymorpha* was also tested effectively in the River Shannon, Ireland, where the range of pathogens tested was expanded to include *Giardia*, *Cryptosporidium*, and multiple human-virulent microsporidia species (Graczyk et al. 2004). In a later survey of 11 widespread sites in Ireland that utilized *D. polymorpha* to detect *Giardia*, *Cryptosporidium*, and three human-virulent microsporidia (*E. bieneusi*, *E. hellem*, and *E. intestinalis*), at least one pathogen species was detected at each sampling site (Lucy et al. 2008). The sites represented areas with different water quality pressures (primarily sewage outfalls and agriculture) and land uses. One of the sites, a coastal bay area with raw urban sewer discharge, was 100% positive for all analyzed pathogen species. A further study of pathogens in *D. polymorpha* in the same region also suggested long-term water contamination by human pathogens and consequent risk factors to human health (Lucy et al. 2010). In this study, *Cryptosporidium* and *Giardia* were found in *D. polymorpha* at each of three drinking-water plants examined, while *E. hellem* and *E. intestinalis* were found at some sites. Concentrations of pathogens ranged as high as 240 per mussel for *Cryptosporidium*, 112 for *Giardia*, 224 for *E. hellem*, and 32 for *E. intestinalis*. In this study, pathogen concentrations were not related to the size of mussels in the samples (Lucy et al. 2010); this is important because it means that populations with different age classes and size structures because of a range of factors (i.e., time of invasion, food and space availability) can provide statistically comparative results in terms of assessing pathogen concentrations in contaminated waters.

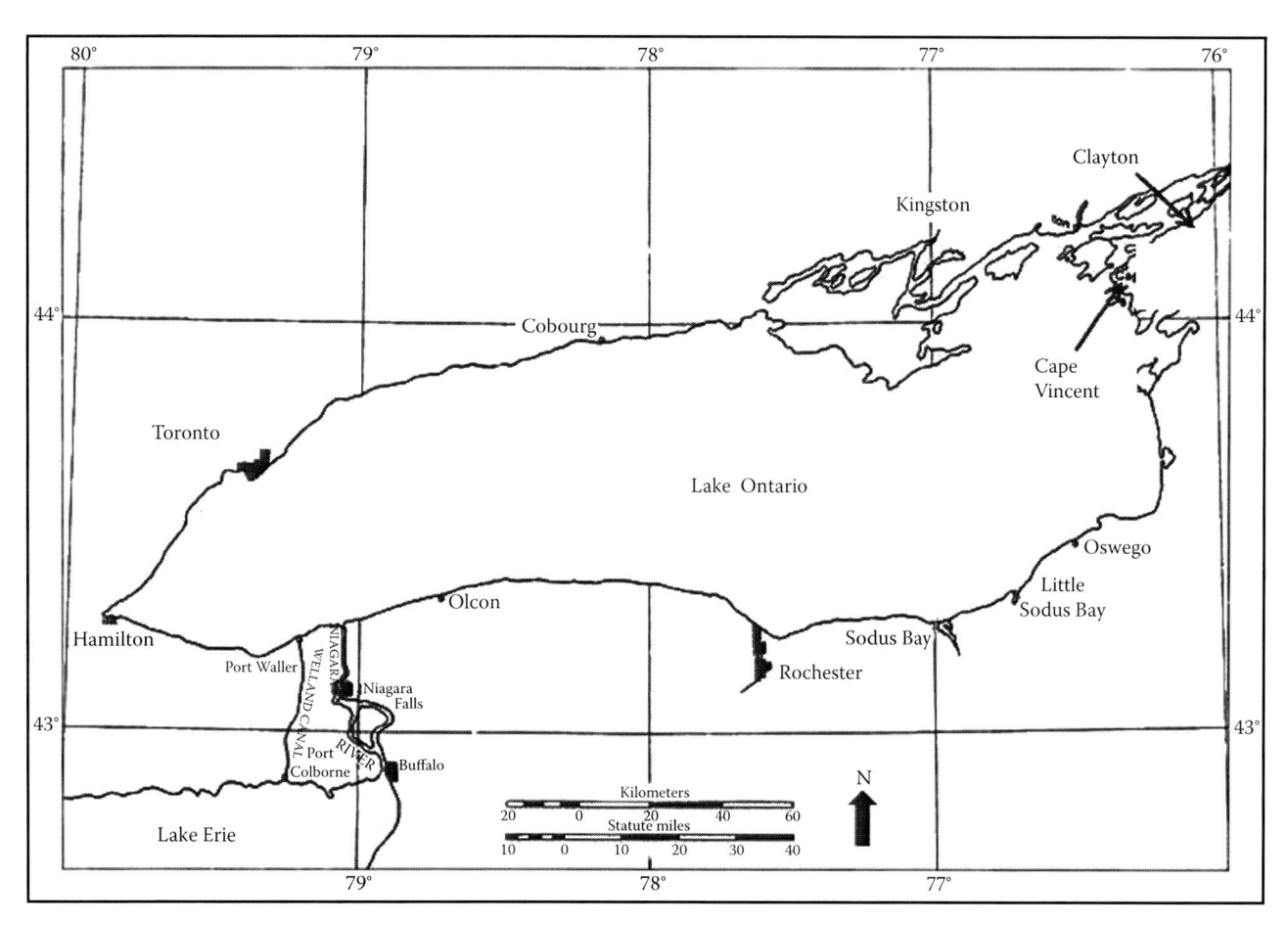

Figure 17.3 Location of major cities/municipalities along the northern shoreline of Lake Ontario and along the St. Lawrence River near where navigational buoys were removed and attached dreissenids analyzed for pathogens. The St. Lawrence River is located in the far northeastern portion of Lake Ontario and flows east.

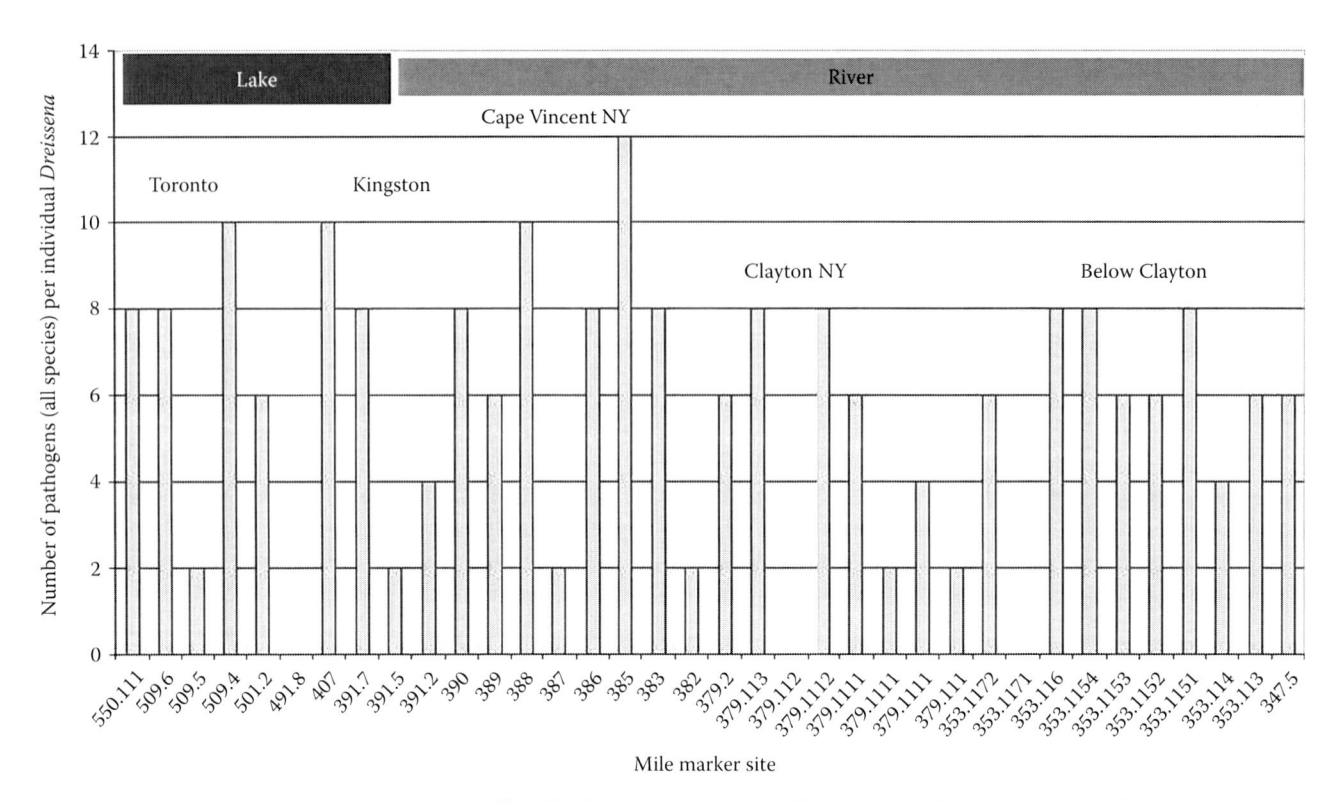

Figure 17.4 Number of waterborne pathogens (*Giardia*, *Cryptosporidium*, and *Enterocytozoon*) found in dreissenid mussels attached to navigational buoys in Lake Ontario and the St. Lawrence River. Each bar represents the mean number of combined pathogens (i.e., all species) per individual mussel on a single buoy of specific location along the Lake Ontario/St. Lawrence navigation. No bar indicates pathogens were not found in dreissenids at that location. Mile markers represent buoy locations located between near Toronto (west) and the upper St. Lawrence River (east). Mile markers are arranged from west (left, higher values) to east (right, lower values). Approximate locations of lake and upper river, as well as principal municipalities, are shown. All data were collected at the end of the 2008 navigational season (early-January 2009 collection date).

Direct comparison of results in the Lake Ontario/ St. Lawrence system and the Shannon River system is difficult since these two waterways are very different in size and extent, and collection protocols differed as a function of available resources. However, results from both waterways show that the use of dreissenids as sentinels for microbial pathogens is effective, and thus can be adapted to other, geographically-diverse water systems. One limitation in using dreissenids is that they prefer north-temperate waters with limited turbidity, and thus could not serve as sentinels in other climates or under different water conditions. Thus, in other climates and in waterways with conditions unsuitable for sustaining dreissenids, other filter-feeding sentinel species should be identified and tested.

In all of these field studies, precise identification of the pathogen species, along with potential viability status, was determined using multiplexed fluorescence in situ hybridization (FISH) and immuofluorescent antibody techniques, as described in detail by Graczyk et al. (2004). These methods allow for easy and inexpensive preservation with ethanol in the field and yet provide accuracy in detection, identification, and determination of viability status. All of these methods are potentially amenable to other waterborne pathogens (Graczyk and Conn 2008).

This sentinel approach can potentially be modified for use in long-term monitoring programs over large geographical ranges in major navigable waterways. For instance, dreissenid mussels attached to navigational-aid platforms (Conn and Conn 2004) can be routinely collected and monitored for pathogens. Such platforms provide precise and replicable sampling sites (Conn 2010a,b; Figures 17.1 and 17.2). Our preliminary work in measuring pathogens in dreissenids attached to navigational buoys in the Lake Ontario and the St. Lawrence River indicates that the entry of these human pathogens from a number of diffuse and point source pathways can be potentially identified and mapped (Figure 17.4). With increased rainfall patterns, we can expect a corresponding increase in entry of pathogens into aquatic systems as a result of greater land runoff.

The use of dreissenid mussels to monitor pathogens provides a useful tool for disease surveillance and risk assessment in waters used for drinking and recreation. Our results indicated that such biomonitoring can assist in source tracking and identification of relevant anthropogenic pressures such as agricultural waste, spread of human biosolids, and inefficient wastewater treatment systems (Lucy et al. 2008, Lucy 2009). This gives water managers the tools and necessary justification to address impacts in catchments and to target public health issues. Examples of outcomes would be upgrades of water treatment plants, prediction of potential outbreaks such as occurred with *Cryptosporidium* in Milwaukee during 1993 (Corso et al. 2003), and public information systems for recreational use of waters near resorts. With increased growth of the human population, more intensive livestock cultivation, and increased rainfall and land runoff because of climate change, the biomonitoring of pathogens will become more urgent in the future.

REFERENCES

Benson, A.J. 2013. Chronological history of zebra and quagga mussels (Dreissenidae) in North America, 1988–2010. In *Quagga and Zebra Mussels: Biology, Impacts, and Control*, 2nd Edn., T.F. Nalepa and D.W. Schloesser, eds., pp. 9–31. Boca Raton, FL: CRC Press.

Chalmers, R. M. and M. Giles. 2010. Zoonotic cryptosporidiosis in the UK—Challenges for control. *J. Appl. Microbiol.* 109: 1487–1497.

Cheng, H.-W. A., F. E. Lucy, T. K. Graczyk, M. A. Broaders, L. Tamang, and M. Connolly. 2009. Fate of *Cryptosporidium parvum* and *Cryptosporidium hominis* oocysts and *Giardia duodenalis* cysts during secondary wastewater treatments. *Parasitol. Res.* 105: 689–696.

Cheng, H.-W. A., F. E. Lucy, T. K. Graczyk, M. A. Broaders, L. Tamang, and M. Connolly. 2011. Municipal wastewater treatment plants as removal systems and environmental sources of human-virulent microsporidian spores. *Parasitol. Res.* 109: 595–603.

Conn, D. B. 2010a. Biomonitoring for zoonotic protists in terrestrial and aquatic environments using insects and bivalve molluscs as sentinel organisms. *Bull. Czech Soc. Parasitol.* 18: 7–8.

Conn, D. B. 2010b. Navigational buoys (navaids) as monitoring platforms. In *Monitoring and Control of Macrofouling Mollusks in Fresh Water Systems*, 2nd Edn., G. Mackie and R. Claudi, eds., pp. 246–248. Boca Raton, FL: CRC Press.

Conn, D. B. and D. A. Conn. 1995. Experimental infestation of zebra mussels, *Dreissena polymorpha* (Mollusca: Bivalvia), by metacercariae of *Echinoparyphium* sp. (Platyhelminthes: Trematoda). *J. Parasitol.* 81: 304–305.

Conn, D. B. and D. A. Conn. 2004. Review and prospective of a long-term monitoring program for assessing the invasion, establishment, and population trends of dreissenid mussels in the Upper St. Lawrence River and Eastern Lake Ontario, 1990–2003. *Aquat. Invaders* 15: 1–7.

Conn, D. B., A. Ricciardi, M. N. Babapulle, K. A. Klein, and D. A. Rosen. 1996. *Chaetogaster limnaei* (Annelida: Oligochaeta) as a parasite of the zebra mussel *Dreissena polymorpha*, and quagga mussel *Dreissena bugensis* (Mollusca: Bivalvia). *Parasitol. Res.* 82: 1–7.

Conn, D. B., S. E. Simpson, D. Minchin, and F. E. Lucy. 2008. Occurrence of *Conchophthirus acuminatus* (Protista: Ciliophora) in *Dreissena polymorpha* (Mollusca: Bivalvia) along the River Shannon, Ireland. *Biol. Invasions* 10: 149–156.

Conn, D. B., J. Weaver, L. Tamang, and T. K. Graczyk. 2007. Synanthropic flies as vectors of *Cryptosporidium* and *Giardia* among livestock and wildlife in a multispecies agricultural complex. *Vector-Borne Zoonotic Dis.* 7: 643–652.

Corso, P. S., M. H. Kramer, K. A. Blair, D. G. Addiss, J. P. Davis, and A. C. Haddix. 2003 Cost of illness in the 1993 waterborne *Cryptosporidium* outbreak, Milwaukee, Wisconsin. *Emerg. Infect. Diseases* 9: 426–431.

Dermott, R. and D. Kerec. 1997. Changes to the deepwater benthos of eastern Lake Erie since the invasion of *Dreissena*, and its ecological implications. *Can. J. Fish. Aquat. Sci.* 54: 922–930.

Dowd, S. E., C. P. Gerba, and I. L. Pepper. 1998. Confirmation of the human-pathogenic microsporidia *Enterocytozoon bieneusi*, *Encephalitozoon intestinalis* and *Vittaforma corneae* in water. *Appl. Environ. Microbiol.* 64: 3332–3335.

Fayer, R. 2008. General biology. In *Cryptosporidium and Cryptosporidiosis*, R. Fayer and L. Xiao (eds.). pp. 1–43. Boca Raton, FL: CRC Press.

Fayer, R. 2010. Taxonomy and species delimitation in *Cryptosporidium*. *Exp. Parasitol.* 124: 90–97.

Graczyk, T. K. and D. B. Conn. 2008. Molecular markers and sentinel organisms for environmental monitoring. *Parasite* 15: 458–462.

Graczyk, T. K., D. B. Conn, and R. Fayer et al. 2002. Asian freshwater clams (*Corbicula fluminea*) and zebra mussels (*Dreissena polymorpha*) as biological indicators of contamination with human waterborne parasites. *Aquat. Invaders* 13: 10–16.

Graczyk, T. K., D. B. Conn, and F. Lucy et al. 2004. Human waterborne parasites in zebra mussels (*Dreissena polymorpha*) from the Shannon River drainage, Ireland. *Parasitol. Res.* 90: 385–391.

Graczyk, T. K., D. B. Conn, D. J. Marcogliese, H. Graczyk, and Y. deLafontaine. 2003. Accumulation of human waterborne parasites by zebra mussels (*Dreissena polymorpha*) and Asian freshwater clams (*Corbicula fluminea*). *Parasitol. Res.* 89: 107–112.

Graczyk, T. K., M. R. Cranfield, and D. B. Conn. 1997a. In vitro phagocytosis of waterborne *Giardia duodenalis* cysts by hemocytes of the Asian freshwater clam (*Corbicula fluminea*). *Parasitol. Res.* 83: 743–745.

Graczyk, T. K., R. Fayer, M. R. Cranfield, and D. B. Conn. 1997b. In vitro interactions of the Asian freshwater clam (*Corbicula fluminea*) hemocytes and *Cryptosporidium parvum* oocysts. *Appl. Environ. Microbiol.* 63: 2910–2912.

Graczyk, T. K., R. Fayer, M. R. Cranfield, and D. B. Conn. 1998a. Recovery of waterborne *Cryptosporidium parvum* oocysts by freshwater benthic clams (*Corbicula fluminea*). *Appl. Environ. Microbiol.* 64: 427–430.

Graczyk, T. K., R. Fayer, D. B. Conn, and E. J. Lewis. 1999. Evaluation of the recovery of waterborne *Giardia* cysts by freshwater clams and cyst detection in clam tissue. *Parasitol. Res.* 85: 30–34.

Graczyk, H., T. K. Graczyk, D. B. Conn, D. J. Marcogliese, and Y. de Lafontaine. 2006. Bioaccumulation of human waterborne parasites by *Dreissena polymorpha* and *Corbicula fluminea*. *Aquat. Invaders* 17: 12–16.

Graczyk, T. K., D. J. Marcogliese, Y. De Lafontaine, A. J. Da Silva, B. Mhagami-Ruwende, and N. Pieniazek. 2001. *Cryptosporidium parvum* oocysts in zebra mussels (*Dreissena polymorpha*): Evidence from the St. Lawrence River. *Parasitol. Res.* 87: 231–234.

Graczyk, T. K., Y. R. Ortega, and D. B. Conn. 1998b. Recovery of waterborne oocysts of *Cyclospora cayetanensis* by Asian freshwater clams (*Corbicula fluminea*). *Am. J. Trop. Med. Hyg.* 59: 928–932.

Graczyk, T. K., D. Sunderland, L. Tamang, F. E. Lucy, and P. N. Breysse. 2007. Bather density and levels of *Cryptosporidium*, *Giardia*, and pathogenic microsporidian spores in recreational bathing water. *Parasitol. Res.* 101: 1729–1731.

Horgan, M. J. and E. L. Mills. 1997. Clearance rate and filtering activity of zebra mussels (*Dreissena polymorpha*): Implications for freshwater lakes. *Can. J. Fish. Aquat. Sci.* 54: 249–255.

Karanis, P., C. Kourenti, and H. Smith. 2007. Waterborne transmission of protozoan parasites: A worldwide review of outbreaks and lessons learned. *J. Water Health* 5: 1–38.

Karatayev, A. Y., L. E. Burlakova, and D. K. Padilla. 1997. The effects of *Dreissena polymorpha* invasion on aquatic communities in Eastern Europe. *J. Shelfish Res.* 16: 187–203.

Karatayev, A. Y., L. E. Burlakova, and D. K. Padilla. 2002. Impacts of zebra mussels on aquatic communities and their role as ecosystem engineers. In *Invasive Aquatic Species of Europe: Distribution, Impacts and Management*, E. Leppakowski, S. Gollasch, and S. Olenin, eds., pp. 433–447. Dordrecht, The Netherlands: Kluwer Academic Publishers.

Lucy, F. E. 2009. Zebra mussels: Review of ecology and impacts since invasion in Ireland. In *Monitoring and Control of Macrofouling Molluscs in Freshwater Systems*, G. Mackie and R. Claudi, eds., pp. 386–389. Boca Raton, FL: CRC Press.

Lucy, F. E., M. Connolly, T. K. Graczyk, L. Tamang, M. R. Sullivan, and S. E. Mastitsky. 2010. Zebra mussels (*Dreissena polymorpha*) are effective sentinels of water quality irrespective of their size. *Aquat. Invasions* 5: 49–57.

Lucy, F. E., T. K. Graczyk, L. Tamang, A. Miraflor, and D. Minchin. 2008. Biomonitoring of surface and coastal water for *Cryptosporidium*, *Giardia*, and human-virulent microsporidia using molluscan shellfish. *Parasitol. Res.* 103: 1369–1375.

Lucy, F., M. Sullivan, and D. Minchin. 2005. Nutrient levels and the zebra mussel population in Lough Key. ERTDI Report Series No. 34. EPA, Wexford, Ireland. www.epa.ie

Mastitsky, S. E., F. E. Lucy, and V. G. Gagarin. 2008. First record of endosymbionts in *Dreissena polymorpha* from Sweden. *Aquat. Invasions* 3: 83–86.

Mathis, A., R. Weber, and P. Deplazes. 2005. Zoonotic potential of the Microsporidia. *Clin. Microbiol. Rev.* 18: 423–445.

McMahon, R. F. 1999. Invasive characteristics of the freshwater bivalve *Corbicula fluminea*. In *Nonindigenous Freshwater Organisms: Vectors, Biology, and Impacts*, R. Claudi and J. Leach, eds., pp. 315–343. Boca Raton, FL: CRC Press.

Molloy, D. P., A. Y. Karatayev, L. E. Burlakova, D. P. Kurandina, and F. Laruelle. 1997. Natural enemies of zebra mussels: Predators, parasites, and ecological competitors. *Rev. Fish. Sci.* 5: 27–97.

Nichols, G. 2008. Epidemiology. In *Cryptosporidium and Cryptosporidiosis*, R. Fayer and L. Xiao, eds., pp. 79–118. Boca Raton, FL: CRC Press.

Ortega, Y. R. and R. Sanchez. 2010. Update on *Cyclospora cayetanensis*, a food-borne and waterborne parasite. *Clin. Microbiol. Rev.* 23: 218–234.

Simberloff, D. 2000. Nonindigenous species: A global threat to biodiversity and stability. In *Nature and Human Society: The Quest for a Sustainable World*, P. Raven and T. Williams, eds., pp. 325–336. Washington, DC: National Academies Press.

Slodkowicz-Kowalska, A., T. K. Graczyk, and L. Tamang et al. 2006. Microsporidia species known to infect humans are present in aquatic birds; implications for transmission via water? *Appl. Environ. Microbiol.* 72: 4540–4544.

Thompson, R. C. 2004. The zoonotic significance and molecular epidemiology of *Giardia* and giardiasis. *Vet. Parasitol.* 126: 15–35.

Vossbrinck, C. R., and B. A. Debrunner-Vossbrinck. 2005. Molecular phylogeny of the Microsporidia: Ecological, ultrastructural and taxonomic considerations. *Folia Parasitol.* 52: 131–142.

Wolfe, M. S. 1992. Giardiasis. *Clin. Microbiol. Rev.* 5: 93–100.

Zintl, A., G. Mulcahy, and T. deWaal et al. 2006. An Irish perspective on *Cryptosporidium*. *IVJ* 59: 442–447.

CHAPTER 18

Contaminant Concentrations in Dreissenid Mussels from the Laurentian Great Lakes
A Summary of Trends from the Mussel Watch Program

Kimani L. Kimbrough, W. Edward Johnson, Annie P. Jacob, and Gunnar G. Lauenstein

CONTENTS

Abstract 273
Introduction 273
Methods 275
    Sites and Sample Collection 275
    Statistical Analysis 276
Results and Discussion 277
Conclusion 281
References 282

ABSTRACT

Bivalves are widely used as sentinel organisms to monitor environmental quality. As part of a nationwide program, bivalves have been used to monitor nearly 150 contaminants in U.S. coastal waters since 1986 and in the Laurentian Great Lakes since 1992. In the Great Lakes, organic contaminants and trace elements in dreissenid mussels (zebra and quagga mussels) are monitored at nearshore sites across the basin on a biennial basis. With nearly two decades of monitoring completed, data reveal a continuing presence of both organic contaminants and trace elements, albeit at predominantly low concentrations at a majority of the sites. The majority of sites showed no trend for organic contaminants (DDT, dieldrin, PAH, and PCBs), mainly because concentrations have reached baseline levels. When a trend was present, it was decreasing, likely because these legacy organic contaminants have been banned for decades. Overall, highest levels of many organic contaminants were found in Lake Michigan. Trends in all trace elements (arsenic, cadmium, lead, and mercury) also showed decreasing or no trend. Arsenic and cadmium had the greatest number of sites with decreasing trends. Trace elements occur naturally in the environment, but elevated levels at some sites may be due to anthropogenic activities. Lessons learned from decades of monitoring bivalves and future directions for enhanced monitoring efforts in the Great Lakes region are discussed.

INTRODUCTION

Bivalve mollusks have long been recognized as environmental sentinels and are used in environmental monitoring programs worldwide. Bivalves are ideal candidates for monitoring because they tolerate and bioaccumulate contaminants, have limited ability to metabolize accumulated contaminants, have a sessile nature, are cosmopolitan in distribution, and are easy to collect and analyze (Farrington 1983). Widespread use of bivalves for contaminant assessment is based on evidence that tissue concentrations reflect contaminant levels in the water and in suspended particles filtered for food in both laboratory (Cunningham and Tripp 1975, Fisher and Teyssie 1986,

Pruell et al. 1987) and field (Roesijadi et al. 1984, Martin 1985, Capuzzo et al. 1989, Sericano 1993) experiments.

In the United States, the first nationwide use of bivalves for chemical contaminant monitoring was initiated by the U.S. Environmental Protection Agency (EPA), which monitored marine species of mussels (*Mytilus*) and oysters (*Ostrea* and *Crassostrea*) from 15 coastal states between 1965 and 1972 (Butler 1973, Goldberg 1975). This program was followed by a second EPA-funded effort that monitored approximately 100 coastal sites nationwide from 1976 to 1978 (Farrington et al. 1983, Goldberg et al. 1983). In 1986, the National Oceanic and Atmospheric Administration National Centers for Coastal Ocean Science Mussel Watch Program (MWP) re-initiated bivalve monitoring at 145 sites (including most of the original EPA sites) in coastal and estuarine regions of the United States.

The MWP expanded to include as many as 300 sites distributed along the entire U.S. coastline, including the Great Lakes, Alaska, Hawaii, and Puerto Rico. Because one single bivalve species is not common to all coastal regions, a variety of species are collected to gain a national perspective. A target species is identified for each site based on abundance and ease of collection. Mussels (*Mytilus* species) are collected from the North Atlantic (Maine to Delaware) and Pacific coasts, while oysters are collected from Delaware Bay southward and along the Gulf Coast (*Crassostrea virginica*), from the Hawaiian coast (*Ostrea sandvicensis*), and from the Puerto Rican coast (*Crassostrea rhizophorae*). The smooth-edged jewel box mollusk (*Chama sinuosa*) is collected from the Florida Keys. Bivalve collection sites are selected to represent coastal areas useful for constructing national and regional assessments of status and temporal trends of chemical contamination.

As part of MWP, a suite of nearly 150 chemical contaminants are measured in bivalve tissue on an annual basis and in sediments approximately every decade. The contaminant component of the monitoring program is complemented by other measurements such as bivalve histopathology and gonadal index. The program monitors the health of bivalve populations by quantifying the prevalence of nearly 70 diseases and parasites found in bivalves and participates in specimen banking, allowing for retrospective analysis of contaminants and biological indicators. Results generated by MWP provide a suite of monitoring data that can be used to assess the environmental impacts of metals, legacy and emerging organic contaminants (surveillance monitoring), the effectiveness of pollution prevention legislation and remediation effectiveness (compliance monitoring), and environmental impacts of catastrophic environmental disasters (impact monitoring). For example, MWP data served as a baseline dataset for evaluating the impact of natural and man-made environmental disasters such as Hurricanes Katrina and Rita, the attack on the World Trade Center, and several oil spills (Lauenstein and Kimbrough 2007, Johnson et al. 2008, Kimbrough et al. 2010).

In the Great Lakes, monitoring efforts were initiated in 1992 (Robertson and Lauenstein 1998) after zebra mussels (*Dreissena polymorpha*) and quagga mussels (*Dreissena rostriformis bugensis*) became established in the late-1980s (Mills et al. 1993, Carlton 2008). Dreissenid mussels are useful environmental bioindicators because of their high filtration rates and an ability to rapidly bioaccumulate both organic and inorganic contaminants (Reeders and bij de Vaate 1992, de Kock and Bowmer 1993, Gossiaux et al. 1998, Berny et al. 2003, Bervoets et al. 2005). The use of these efficient filter feeders in water-quality monitoring is historic and widespread. Field studies on the toxicology and contaminant uptake of dreissenids have been conducted in The Netherlands since 1976 (Reeders and bij de Vaate 1992, Stäb et al. 1995). Dreissenids have been routinely used as contaminant biomonitors in The Netherlands (Kraak et al. 1994) and in the Rhine and Meuse Rivers (Kraak et al. 1991, Mersch and Johansson 1993). In addition, their use as a biological early warning system is also well established (Borcherding 2006). Recently, use of multiple biomarkers to assess cellular and subcellular impacts on dreissenids has also been investigated (de Lafontaine et al. 2000, Osman et al. 2007, Riva et al. 2008, Parolini et al. 2010).

Dreissenid mussels play a significant role in contaminant cycling and biomagnification of contaminants through the food chain (Bruner et al. 1994a,b). Further, large populations of dreissenids have been implicated in slowing the transport of material offshore and in enhancing the nearshore recycling of contaminants through the deposition of feces and pseudofeces (Marvin et al. 2000, Cho et al. 2004, Hecky et al. 2004). The International Joint Commission (IJC 2011) emphasized the need for a greater understanding of the role of dreissenids in physical and biological processes in the nearshore zone, in the management of nearshore impacts, in subsequent offshore impacts, and in the health of the Great Lakes as a whole. To these ends, monitoring of chemical, biological, and toxicological indicators in dreissenid mussels provides a means to further assess temporal trends, biological uptake, and ultimate fate of many contaminants.

This chapter summarizes MWP data for selected chemical contaminants based on long-term monitoring records dating back to 1992. Our discussion includes insights into the lessons learned with regard to the spatial and temporal robustness of the program to assess chemical contaminant status and trends. From the nearly 150 contaminants monitored by the MWP, we selected 4 trace elements and 4 organic contaminant groups, including legacy contaminants with restricted use, and contaminants with ongoing source loadings. We share the pros and cons of the MWP monitoring approach and propose enhancements that address regional needs.

METHODS

Sites and Sample Collection

Contaminant monitoring in the Great Lakes was initiated at 12 sites in the inaugural biennial year (1992–1993) of the MWP. There are 23 core monitoring sites in the Great Lakes that have enough data to perform trend analyses and are the focus of this chapter (Figure 18.1). All MWP sites (Figure 18.1, Table 18.1) lie within watersheds that represent a variety of land uses and occur within the shallow, nearshore zones (<5 m). Additional MWP sites have been added more recently to expand spatial coverage and aid in the assessment of remediation effectiveness efforts (Figure 18.1).

Dreissenid mussels were collected at each of these sites every other year; specifically, sites downstream of and including Stony Point (western Lake Erie) were collected in odd years, and sites upstream of and including the Black River Canal (southern Lake Huron) were collected in even years (Figure 18.1). Since bivalve size, reproductive state, lipid content, and other factors can change seasonally and may affect tissue concentration of chemicals (Kwan et al. 2003, Richman and Somers 2005), these variables were minimized by collecting bivalves of the same size range (15–35 mm) at the same time of year (±3 weeks). In the Great Lakes, mussels were collected in mid-August to mid-September each year; in coastal marine regions, bivalves were collected between November and March.

The specific species of dreissenids (zebra or quagga) was not differentiated for chemical analyses. Mussels were collected by hand, by epibenthic dredge, or by SCUBA-free diving. Individuals were rinsed with water from the collection site to remove mud and debris, put in sealed plastic bags, placed on ice, and then shipped to analytical laboratories within 3 days of collection. Detailed protocols for sample collection and handling are given elsewhere (Lauenstein and Cantillo 1993a–d, 1998, Lauenstein et al. 1997, McDonald et al. 2006).

The MWP uses a performance-based quality assurance process to ensure data quality. This effort has been in operation since 1985 and is designed to document sampling protocols, analytical procedures, and laboratory performance. Analytical laboratories used by the MWP are required to participate in exercises with assistance from the National Institute of Standards and Technology and the National Research Council of Canada to ensure data are comparable in accuracy and precision (Schantz et al. 2000, Willie 2000). Procedures for sample preparation, extraction, and analysis are provided in Technical Memoranda (Kimbrough and Lauenstein 2007, Kimbrough et al. 2007) and available online (http://ccma.nos.noaa.gov/about/coast/nsandt/musselmethods.aspx).

In this chapter, we report the status and trends of four metals (arsenic, cadmium, mercury, and lead) and relevant compounds from four organic compound groups (PCBs, DDT and its metabolites, dieldrin, and PAHs). PCB refers

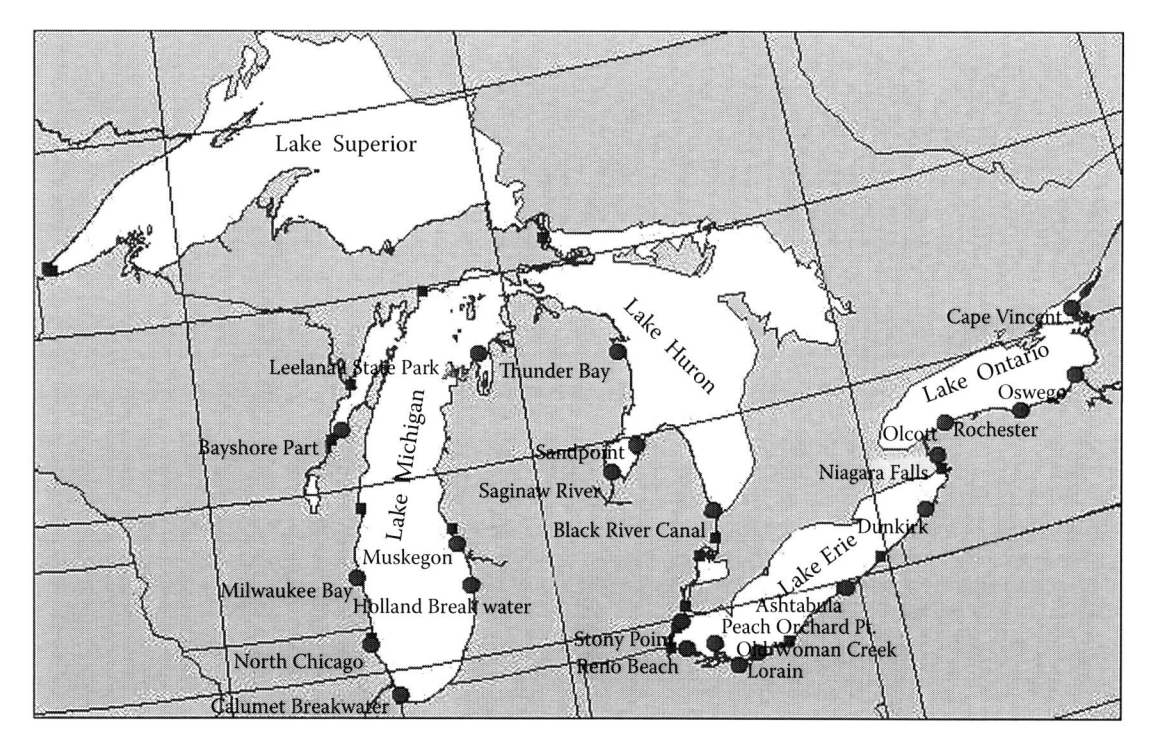

Figure 18.1 General location of core sites of the MWP (●) in the Great Lakes with 6 or more years of data for which trends are presented. Also shown are recently established sites (■) for which data are not examined in this chapter.

Table 18.1 Location of Sites in the Great Lakes Where *Dreissenid* Mussels Were Collected for Analysis of Trace Metals and Organic Compounds as Part of the MWP, 1992–2009

Water Body	Location	State	Latitude	Longitude
Lake Erie	Stony Point	MI	41.9587	−83.2330
Lake Erie	Dunkirk	NY	42.5292	−79.2777
Lake Erie	Reno Beach	OH	41.6745	−83.2262
Lake Erie	Peach Orchard Pt.	OH	41.6597	−82.8250
Lake Erie	Old Woman Creek	OH	41.3850	−82.5187
Lake Erie	Lorain	OH	41.4612	−82.2070
Lake Erie	Ashtabula	OH	41.9247	−80.7183
Lake Huron	Thunder Bay	MI	44.9222	−83.4135
Lake Huron	Black River Canal	MI	43.0443	−82.4387
Lake Michigan	North Chicago	IL	42.3047	−87.8273
Lake Michigan	Calumet Breakwater	IN	41.7272	−87.4950
Lake Michigan	Muskegon	MI	43.2282	−86.3469
Lake Michigan	Holland Breakwater	MI	42.7732	−86.2150
Lake Michigan	Milwaukee Bay	WI	43.0322	−87.8952
Lake Ontario	Olcott	NY	43.3553	−78.6867
Lake Ontario	Rochester	NY	43.2578	−77.4953
Lake Ontario	Oswego	NY	43.4528	−76.5508
Lake Ontario	Cape Vincent	NY	44.1442	−76.3247
Niagara River	Niagara Falls	NY	43.0468	−78.8920
Saginaw Bay	Saginaw River	MI	43.6735	−83.8367
Saginaw Bay	Sandpoint	MI	43.9098	−83.4002
Traverse Bay	Leelanau State Park	MI	45.2057	−85.5368
Green Bay	Bayshore Park	WI	44.6370	−87.8082

Table 18.2 Trace Metals and Organic Compounds Measured in Dreissenid Mussels (Tissue) as Part of the MWP in the Great Lakes, 1992–2009

Category	Contaminant
Trace metals	Arsenic, cadmium, lead, mercury
Organic compounds	
Dieldrin	Aldrin/dieldrin
DDT	2,4′-DDD, 2,4′-DDE, 2,4′-DDT, 4,4′-DDD, 4,4′-DDE, 4,4′-DDT
PAH	Benzo[e]pyrene
PCB	PCB8/5, PCB18, PCB28, PCB44, PCB52, PCB66, PCB101/90, PCB105, PCB118, PCB128, PCB138, PCB153/132/168, PCB170/190, PCB180, PCB187, PCB195/208, PCB206, PCB209

to the sum of 18 PCB congeners, DDT refers to the sum of 6 compounds, dieldrin refers to the sum of 2 compounds, and the representative PAH was benzo[e]pyrene (Table 18.2).

Statistical Analysis

Nonparametric statistical methods were used to perform trend analysis because the tissue concentration data were not normally distributed. Temporal trends in chemical contamination were determined using the Spearman Rank Correlation, which is free of assumptions about normally distributed data. Only sites with 6 or more years of data were used in trend analysis, and sites showing decreasing, increasing, and no trend were reported. Statistical significance was determined when |Rho| > 0.7 and p-value < 0.05. Trend analysis is applied at the site level, and the number of sites with increasing, decreasing, and no trend was calculated for the Great Lakes and for the coastal marine region.

The status summary of the contaminants based on the most recent dataset was determined using cluster analysis. This analysis clusters contaminant concentrations into three significantly different groups such that values contained within a group are more like each other than any other value of a different group. The three categories derived from the clusters were not representative measurements that have exceeded any regulatory thresholds; rather, they denoted concentrations that were significantly higher than the preceding category. There were no regulatory benchmark concentrations for tissue contaminants in dreissenid mussels from the Great Lakes.

Prior to initiation of MWP monitoring in the Great Lakes, use of legacy contaminants was severely restricted and/or banned. The initial rapid decline in legacy contaminant concentrations was easy to detect and, in general, followed first-order kinetics (De Vault et al. 1996). However, trend detection was more difficult to assess over time as contaminant concentrations approached an asymptote. Hence, we calculated a baseline contaminant concentration based on cluster analysis of data from 1992 to 2007. The 90th percentile of the lowest cluster was defined as the baseline measurement.

There were several ways to define such a baseline. One approach was to use analytical results from "clean" or "pristine" sites to define baseline levels. The major drawback of this approach is that pristine sites were difficult to identify, especially in systems impacted with nonpoint source pollution. It can be argued that unlike metals, the baseline concentration for any organic contaminant of human origin should have a concentration of zero (0) as they do not occur naturally in the environment. The advantage of our method of baseline calculation is that it utilized data from multiple sites and a variety of land-use types that characterize a geographic region over several years.

RESULTS AND DISCUSSION

In our results, we provide overall trends, spatial extent, and baseline calculations for both trace element and legacy organic contaminants. Primary objectives were to assess whether concentrations in mussels are increasing, decreasing, or not changing, and identify sites that exceed baseline contamination levels. These objectives are important precursors to informed management decisions and assessment of remedial and legislative actions.

Baseline concentrations of trace elements (arsenic, cadmium, lead, and mercury) in dreissenids from sites in the Great Lakes are presented in Table 18.3. Across all sites, concentrations of examined trace elements in 2008–2009 were predominantly below baseline concentrations (Table 18.3; Figure 18.2). Ranges in concentrations (µg/g dry weight) in 2008/2009 across sites were 3.4–7.5 for arsenic, 0.4–11 for cadmium, 0.4–5.3 for lead, and 0.016–0.086 for mercury. These ranges were consistent with previously reported concentrations in Europe and the Great Lakes. For example, mean arsenic concentrations reported in dreissenid tissue from the St. Lawrence River ranged from 2.49 to 8.23 µg/g dry weight (Kwan et al. 2003), and in Belgium concentrations

Table 18.3 Concentrations of Trace Metals (µg/g Dry Weight) Found in Dreissenid Tissue at Various Sites in the Great Lakes. Estimated Baseline Values Are Presented Adjacent to the Name of the Particular Trace Element. Concentrations for 2008/2009 (S), and Mean Concentrations for Years 1994–2007 (M) and Associated Standard Error (SE) for this later period Are Presented to Bring Perspective to the Recent Measurements. Bolded Values Are Concentrations That Are above Estimated Baseline Estimates

Lake	Location	State	Arsenic 5.6 S	M	SE	Cadmium 4.1 S	M	SE	Lead 3.5 S	M	SE	Mercury 0.054 S	M	SE
Green Bay	Bayshore Park	WI	3.4	4.5	0.35	0.4	0.8	0.07	2.6	1.3	0.15	**0.080**	**0.087**	0.01
Lake Michigan	Milwaukee Bay	WI	4.0	5.6	0.94	1.2	2.2	0.64	2.6	2.3	0.58	0.034	0.044	0.01
Lake Michigan	North Chicago	IL	4.0	**7.7**	0.86	2.3	**5.6**	0.35		**12**	0.35	0.016	**0.078**	0.01
Lake Michigan	Calumet Breakwater	IN	4.5	**7.1**	0.58	2.4	**7.3**	0.54	3.3	2.9	0.19	0.042	**0.069**	0.01
Lake Michigan	Holland Breakwater	MI	4.2	5.6	0.74	2.2	3.7	0.31	1.5	1.0	0.85	0.043	0.053	0.00
Lake Michigan	Muskegon	MI	4.2	5.5	0.50	2.2	3.9	0.47	0.8	0.9	0.30	0.029	0.047	0.01
Traverse Bay	Leelanau State Park	MI	**5.9**	**6.2**	0.32	**4.8**	**7.3**	0.69	0.4	0.6	0.79	**0.058**	**0.097**	0.00
Lake Huron	Thunder Bay	MI	5.2	5.0	0.53	**11**	**10**	1.03	0.9	0.5	0.48	**0.059**	**0.081**	0.01
Saginaw Bay	Saginaw River	MI	5.1	**6.5**	0.57	0.6	1.5	1.23	2.2	2.6	0.14	0.031	0.050	0.01
Saginaw Bay	Sandpoint	MI	4.7	5.2	0.28	1.6	2.4	0.58	3.2	1.4	0.08	**0.056**	0.051	0.01
Lake Huron	Black River Canal	MI	**7.1**	**7.0**	0.99	**6.4**	**9.7**	1.06	1.5	1.5	0.46	**0.056**	**0.081**	0.01
Lake Erie	Stony Point	MI	3.9	**6.2**	0.44	2.1	**6.0**	0.34	**5.3**	**4.4**	0.14	**0.086**	**0.074**	0.00
Lake Erie	Reno Beach	OH	5.0	**6.8**	0.29	1.8	**5.7**	0.41	**4.0**	**3.6**	0.38	0.029	0.038	0.00
Lake Erie	Peach Orchard Pt.	OH	5.1	**7.1**	0.35	2.6	**4.5**	0.21	2.7	3.1	0.15	0.044	**0.061**	0.00
Lake Erie	Old Woman Creek	OH	4.2	**5.8**	0.38	2.1	3.3	0.52	1.2	2.8	2.01	0.034	0.030	0.03
Lake Erie	Lorain	OH	3.9	**8.5**	0.89	2.7	**4.4**	0.70	1.6	2.6	0.32	0.031	0.037	0.01
Lake Erie	Ashtabula	OH	**6.1**	**9.0**	1.24	3.5	**4.7**	0.59	1.4	3.5	0.36	0.027	0.046	0.00
Lake Erie	Dunkirk	NY	**6.6**	**10.1**	0.60	4.0	**4.9**	0.28	2.0	1.9	0.40	0.034	0.038	0.01
Niagara River	Niagara Falls	NY	3.6	**7.2**	0.80	1.2	2.4	0.65	2.8	3.5	0.40	0.037	0.029	0.01
Lake Ontario	Olcott	NY	**6.2**	**9.9**	1.15	3.6	3.8	0.79	0.9	2.1	0.61	0.041	0.054	0.01
Lake Ontario	Rochester	NY	**6.4**	**8.8**	1.03	2.1	3.4	0.39	2.0	1.9	0.68	0.040	0.052	0.01
Lake Ontario	Oswego	NY	3.5	**7.3**	1.10	1.1	2.0	0.19	0.5	1.1	0.19	0.028	**0.062**	0.01
Lake Ontario	Cape Vincent	NY	**7.5**	**9.4**	1.31	1.9	3.4	1.39	0.7	1.1	0.09	0.038	0.048	0.01

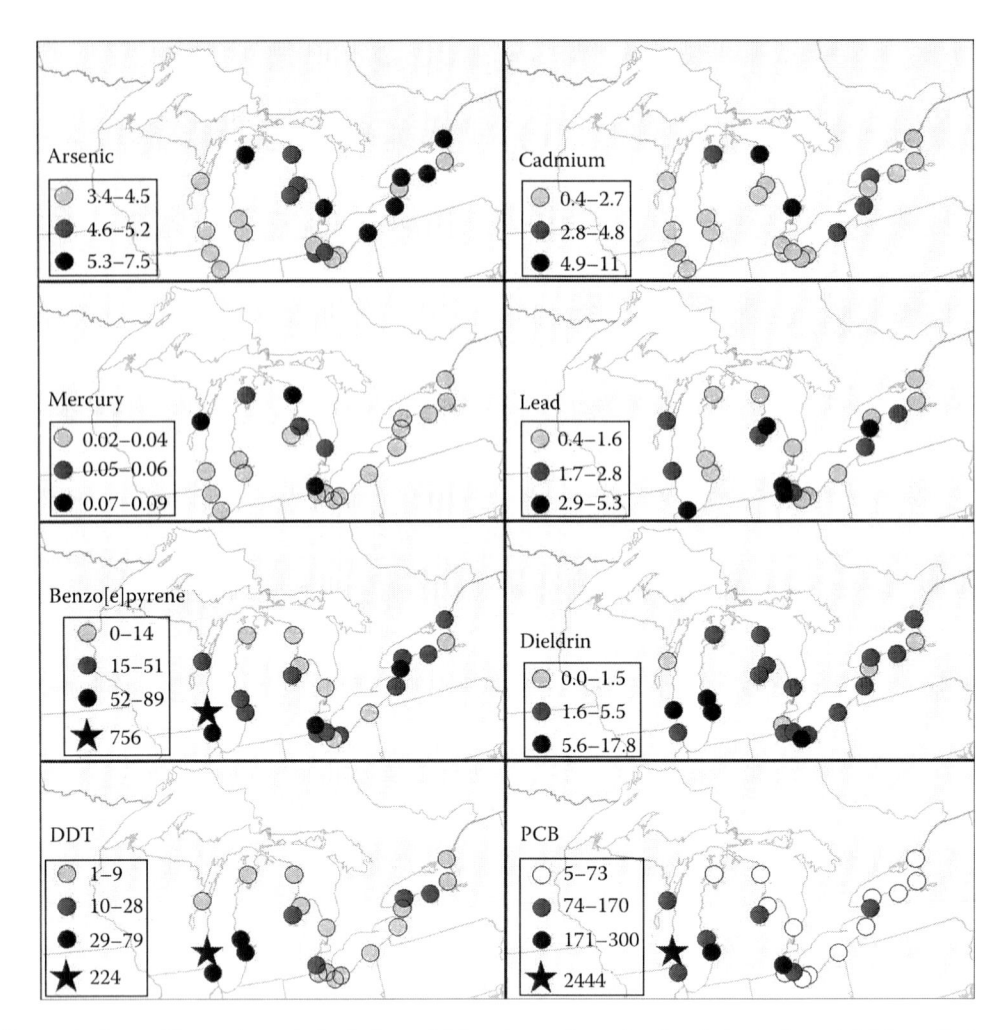

Figure 18.2 Concentrations of trace elements (µg/g dry weight) and organic contaminants (ng/g dry weight) found in tissue of dreissenid mussels at sampling sites in the Great Lakes, 2008–2009.

ranged from 1.59 to 9.84 µg/g dry weight (Bervoets et al. 2004). In the study of dreissenids from a site located near the head of the St. Lawrence River in Lake Ontario, Johns and Timmerman (1998) reported a range for cadmium of 3.4–12.0 µg/g dry weight over a 4-year period.

Arsenic and cadmium had the highest number of sites with decreasing trends in the Great Lakes, and arsenic displayed a trend similar to coastal marine regions of the United States (Figure 18.3). However, trends for cadmium levels were decreasing at more sites in the Great Lakes than many other coastal marine regions in the United States (Kimbrough et al. 2008). Natural levels of arsenic in soil and groundwater in parts of the Great Lakes basin are elevated when compared to the United States as a whole (Thomas 2007), and anthropogenic sources in the region are widespread and include coal, wood, and waste combustion, pesticide use and manufacturing, and metal mining and smelting (ATSDR 2007).

For both lead and mercury, the majority of sites showed no discernible trend (Figure 18.3), and most concentrations in 2008–2009 were below the calculated baselines (Table 18.3). Richman and Somers (2010) reported generally higher concentrations of metals in dreissenid mussels in 1995 compared to 2003. Our results for dreissenids indicated mercury concentrations were similar throughout all the lakes with no outliers or conspicuously high values (Figure 18.2; Table 18.3). This finding is consistent with mercury levels in fish (Carlson and Swackhamer 2006). Notably, many of the sites with elevated concentrations of metals did not display concomitant decreasing trends. However, concentrations of both lead and mercury in Lake Ontario surficial sediments declined from 1968 to 1998 (Marvin et al. 2002). Lower emissions of alkylated lead from automobiles were attributed to the phaseout of leaded gasoline in 1995 (Marvin et al. 2004).

The trend analysis of trace elements could be confounded by many factors including the difference in bioaccumulation rates between quagga and zebra mussels. For example, Johns and Timmerman (1998) reported that bioaccumulation patterns of cadmium, zinc, and copper are different between quagga mussels and zebra mussels. Over the time period of our collections, quagga mussels had been

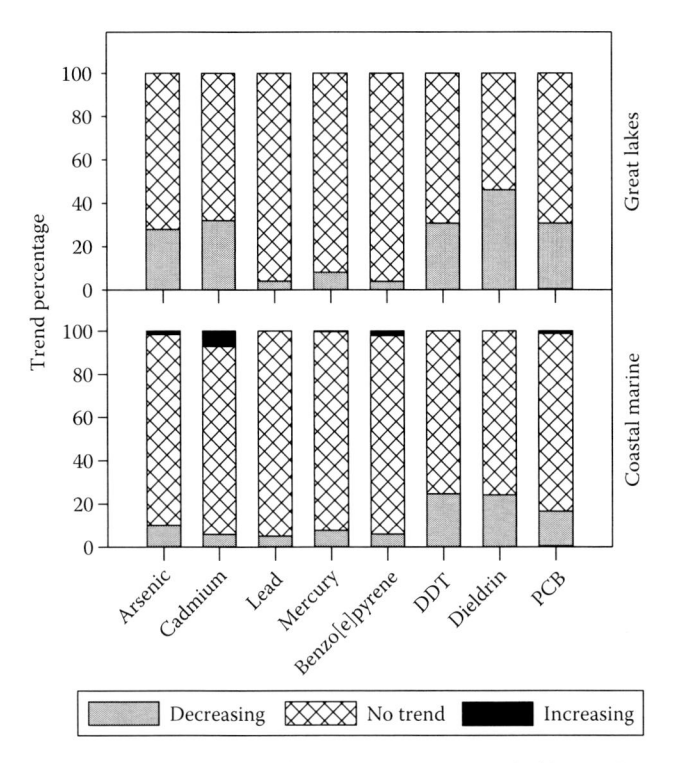

Figure 18.3 Percentage of sites with various trends (decreasing, no trend, increasing) in the Great Lakes relative to trends found in coastal marine regions of the United States. All data collected as part of the MWP. Trends in the Great Lakes are based on data collected between 1992 and 2009, whereas trends in coastal marine regions are based on data collected between 1986 and 2007. Trends were determined using the Spearman Rank Correlation.

gradually replacing zebra mussels as the dominant species in most of the Great Lakes (Mills et al. 1999, Patterson et al. 2005, Nalepa et al. 2010). However, the MWP does not differentiate between species for sample collection.

Metals occur naturally; as a result, elevated environmental concentrations can be linked to both natural and anthropogenic sources (Roesijadi et al. 1984). Natural sources can dictate the existence of a site-specific minimum concentration for each trace element. The measured concentration above this natural baseline can be attributed to human activity; however, distinguishing between sources (natural vs. anthropogenic) is beyond the scope of the MWP. As a surveillance monitoring program, our monitoring reveals sites with elevated concentrations relative to other measurements from the region, but attributing the elevated concentration and increasing trend entirely to anthropogenic sources must be done with caution. Nevertheless, our data provide an early warning signal and help identify sites that may require further investigation.

Environmental monitoring data from across North America documented the initial rapid decline in legacy contaminants in the early years following their restriction and ultimate ban. Some of the strongest evidence came from a suite of studies with undisturbed dated sediment cores collected from freshwater lakes (Swackhamer and Armstrong 1986, Swackhamer et al. 1988, Eisenreich et al. 1989). The MWP in the Great Lakes did not begin until 1992, years after the most significant declining trends occurred for legacy contaminants like DDT, dieldrin, and even PCBs. Therefore, concentrations of organic contaminants generally showed no discernible trend or showed trends that were decreasing (Figures 18.3 and 18.4). The one exception, with regard to the restriction of organic contaminant use, is benzo[e]pyrene, which along with other PAHs has both natural and industrial sources to the environment including forest fires, coal-fired power plants, automobile exhaust, and local releases of oil. Dieldrin was the compound that had the highest number of sites with a decreasing trend, and benzo[e]pyrene had the least number of sites with a decreasing trend (Figure 18.3).

Concentrations of benzo[e]pyrene, DDT, dieldrin, and PCBs were measured above baseline levels in 2008/2009 at some sites (Table 18.4, Figure 18.3), and many sites with elevated concentrations showed no decreasing trend. This emphasized the relevance of this long-term indicator dataset in identifying contaminated sites that may require remediation.

Overall, the highest levels of many legacy organic contaminants were found in Lake Michigan (Figure 18.2, Table 18.4). This finding is consistent with results from the U.S. Great Lakes Fish Monitoring Program that reported that across all lakes, levels of organic compounds in fish were highest in Lake Michigan (Carlson and Swackhamer 2006). Mussels in Milwaukee Bay, Lake Michigan, had elevated concentrations for all four organic contaminants, and three of these concentrations were the highest recorded for the Great Lakes in 2008/2009. Further, none of the four organic contaminants at this site displayed a decreasing trend. Milwaukee Bay has been designated an Area of Concern (AOC), which is a severely degraded area under the Great Lakes Water Quality Agreement (GLWQA 1987).

A broader look at organic compounds and trace metals across all years and sites highlights a clear distinction between these two contaminant types. All four organic compounds had a lower proportion of concentrations that were above baseline values when compared to trace elements, with the exception of lead (Figure 18.4). For organic compounds, the proportion above baselines, calculated using 1992–2007 data, ranged from 1% to 16%, whereas the proportion for trace elements ranged from 21% to 62%. However, most concentrations of metals were only marginally higher than baseline values, and elevated concentrations may have been due to natural factors. In contrast, concentrations of organic compounds that exceeded baselines tended to be higher in magnitude, indicating remediation may be required to reach baseline levels.

Though legacy organic compounds were generally below calculated baseline values, our data showed that these compounds were ubiquitous and measurable in the environment, possibly due to inherent chemical properties (e.g., long half-lives and hydrophobicity) and various global processes such as weather cycles, migration of animals, and global

Table 18.4 Concentrations of Organic Contaminants (ng/g Dry Weight) Found in Dreissenid Tissue at Various Sites in the Great Lakes. Estimated Baseline Values Are Presented Adjacent to the Name of the Particular Organic Contaminant. Concentrations for 2008/2009 (S), and Mean Concentrations for Years 1994–2007 (M) and Associated Standard Error (SE) for this later period Are Presented to Bring Perspective to the Recent Measurements. Bolded Values Are Concentrations That Are above Estimated Baseline Estimates

Lake	Location	State	Benzo[e]pyrene 67 S	M	SE	DDT 26 S	M	SE	Dieldrin 6.6 S	M	SE	PCB 253 S	M	SE
Green Bay	Bayshore Park	WI	14	13	1.8	5.4	**33**	19	1.1	3.4	1.4	170	**380**	159
Lake Michigan	Milwaukee Bay	WI	**756**	**431**	2.4	**224.0**	**176**	4.8	**13.0**	**17.0**	1.2	**2444**	**1092**	9
Lake Michigan	North Chicago	IL	**74**	40	3.5	**58.0**	**84**	2.9	3.9	**10.0**	0.6	125	151	5
Lake Michigan	Calumet Breakwater	IN	.	33	3.9	.	15	3.0	.	**8.1**	0.9	.	62	6
Lake Michigan	Holland Breakwater	MI	62	20	5.4	4.4	**154**	3.6	2.4	**25.0**	1.6	37	**474**	13
Lake Michigan	Muskegon	MI	32	23	2.9	**79.0**	**99**	4.1	**16.0**	**20.0**	1.1	**300**	185	42
Traverse Bay	Leelanau State Park	MI	50	2	4.6	**70.0**	3	3.7	**18.0**	5.5	1.6	132	15	28
Lake Huron	Thunder Bay	MI	0	2	9.3	1.2	6	4.6	3.5	4.1	1.3	5	19	35
Saginaw Bay	Saginaw River	MI	10	46	2.2	2.7	**63**	2.4	2.0	3.1	1.5	6	**255**	7
Saginaw Bay	Sandpoint	MI	45	3	0.6	13.0	9	2.4	4.6	2.5	1.5	108	34	8
Lake Huron	Black River Canal	MI	11	7	2.2	1.4	8	2.7	2.3	5.5	1.4	6	23	10
Lake Erie	Stony Point	MI	**89**	**108**	2.3	**28.0**	**69**	48.0	0.0	5.2	4.9	**291**	**543**	93
Lake Erie	Reno Beach	OH	23	33	65.1	2.9	18	23.0	3.2	**8.0**	4.1	30	135	174
Lake Erie	Peach Orchard Pt.	OH	34	29	4.8	9.4	27	21.0	3.0	6.2	4.1	145	**335**	34
Lake Erie	Old Woman Creek	OH	12	21	7.4	7.5	16	18.0	5.6	**7.3**	2.7	32	63	25
Lake Erie	Lorain	OH	41	27	1.4	7.4	12	3.5	4.3	5.2	1.5	47	58	12
Lake Erie	Ashtabula	OH	13	19	5.4	4.6	12	5.8	3.0	4.5	0.9	63	92	12
Lake Erie	Dunkirk	NY	22	21	1.6	5.4	10	1.4	2.7	3.7	0.4	35	50	7
Niagara River	Niagara Falls	NY	**89**	**101**	5.2	5.3	18	4.7	0.0	1.9	0.8	163	**557**	11
Lake Ontario	Olcott	NY	31	32	14.6	18.0	23	5.2	4.3	5.4	0.7	76	93	76
Lake Ontario	Rochester	NY	51	31	8.3	15.0	20	15.0	3.1	4.8	1.0	51	55	67
Lake Ontario	Oswego	NY	4	6	0.7	4.2	10	2.5	1.5	3.6	0.5	13	41	8
Lake Ontario	Cape Vincent	NY	28	7	0.8	2.8	6	1.2	2.1	4.2	1.1	19	29	5

distillation (Beyer et al. 2000, Ma and Cao 2010, Nizzetto et al. 2010). Given this, monitoring of legacy contaminants should be continued, but perhaps at reduced spatial and temporal scales. More importantly, using MWP data in conjunction with data from other biomonitoring programs in the region can provide a better understanding of the trophic transfer and cycling of these persistent, bioaccumulative pollutants in the food web of the Great Lakes.

The continued presence of legacy organic contaminants coupled with the threat of a multitude of new, emerging contaminants necessitates integration of traditional contaminant monitoring with newer approaches to environmental monitoring, specifically effects-based monitoring (Lam and Gray 2003). With regard to effects-based monitoring, the possibilities being pursued by the MWP are the use of both cellular and molecular biomarkers of chemical exposure in dreissenid mussels to compare polluted and unpolluted sites. Among the biomarkers being considered are lysosomal destabilization, lipid peroxidation, glutathione (GSH), and DNA strand breakage. There is a substantial body of literature validating that environmental pollutants cause destabilization of lysosomes (Lowe et al. 1995, Ringwood et al. 1998, 1999, de Lafontaine et al. 2000). Similarly, lipid peroxidation reflects oxidative damage to lipid-rich components such as cell membranes, which can occur from increased free radicals (Viarengo et al. 1990, de Lafontaine et al. 2000). Cellular damage from free radicals can propagate additional cytotoxic products that can damage DNA (Kehrer 1993, de Lafontaine et al. 2000). Oxidative damage to cellular components can be ameliorated when GSH binds to the oxidant; thus elevated levels of GSH are indicative of pollutant exposure. In addition to cellular biomarkers, the use of molecular techniques such as genomics and metabolomics has rapidly expanded in recent years (Robertson 2010, Domeneghetti et al. 2011, Venier et al. 2011) and shows great promise with bivalve mollusks (Jones et al. 2008, Hines et al. 2010). Hence, comparison of gene expressions or metabolic profiles of bivalves such as dreissenids from polluted and unpolluted environments can potentially become a key component of environmental monitoring by providing early warnings of chemical threats before population effects become apparent.

The long-term effort of monitoring chemical contaminants in dreissenids in the Great Lakes is currently being leveraged to aid in the assessment and effectiveness of remedial actions in specific, degraded areas, such as AOCs. Currently, our data provide a broad view of temporal and spatial trends at relatively few and unreplicated sites. Although calculated contaminant trends for the entire basin are robust, trend

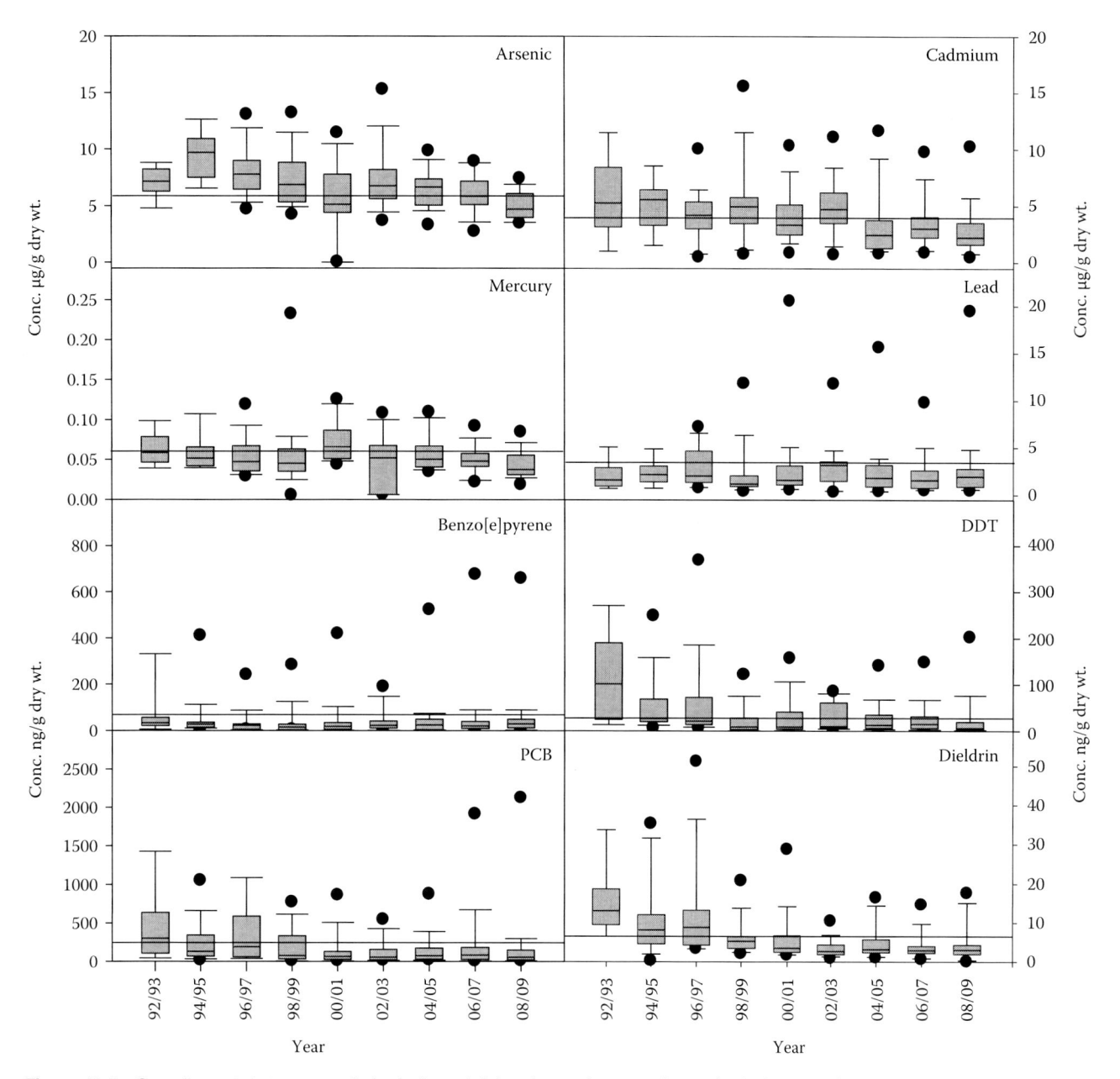

Figure 18.4 Overall trends in trace metals (μg/g dry weight) and organic contaminants (ng/g dry weight) in tissues of dreissenid mussels in the Great Lakes. The continuous horizontal line represents the baseline as defined by the 90th percentile of the lowest of three clusters derived from 1992 to 2007 data. The bottom and top of the box plots represent the 25th and 75th percentiles, respectively, the whiskers represent the 5th and 95th percentiles, the black dots represent outliers that lie beyond the 5th and 95th percentiles, and the line in the middle of the box is the median.

detection at specific locations is weak due to lack of replication across different habitat types and substrates where dreissenids are found. We have added additional sites across the basin in highly impaired areas; this enhancement to the monitoring program allows for a basin-wide comparison of impaired and non-impaired sites. These program enhancements, particularly effects-based monitoring, will address many of the broader ecosystem health concerns and aid resource managers and stakeholders in making informed management decisions.

CONCLUSION

Overall, no trends were observed for most of the contaminants examined in dreissenid tissue, particularly for legacy organic compounds at the site level in the Great Lakes. This was expected because the concentration of legacy organic contaminants at many sites may have reached baseline levels before monitoring began. Although not indicative of the entire basin, dreissenid monitoring sites (nearshore and largely non-impaired) help provide perspective to the extent of contamination. Where contaminant thresholds do not exist, the dreissenid data are informative to identify sites with elevated levels, which in turn will help prioritize management actions and better allocation of funds. Monitoring of contaminant concentrations in dreissenids can support pre- and post-remediation assessments and provide an additional basis for decision making. Monitoring programs can also evolve to meet the newer environmental challenges by strategically adding new approaches and techniques to improve the utility of dreissenid biomonitoring to assess the environmental health of the Great Lakes.

REFERENCES

ATSDR. 2007. Toxicological profile for arsenic. U.S. Department of Health and Human Services, Public Health Service, Agency for Toxic Substances and Disease Registry. Atlanta, GA. 559pp.

Berny, P. J., A. Veniat, and M. Mazallon. 2003. Bioaccumulation of lead, cadmium, and lindane in zebra mussels (*Dreissena polymorpha*) and associated risk for bioconcentration in tufted duck (*Aythya fuligula*). *Bull. Environ. Contam. Toxicol.* 71:90–97.

Bervoets, L., J. Voets, S. G. Chu, A. Covaci, P. Schepens, and R. Blust. 2004. Comparison of accumulation of micropollutants between indigenous and transplanted zebra mussels (*Dreissena polymorpha*). *Environ. Toxicol. Chem.* 23:1973–1983.

Bervoets, L., J. Voets, and A. Covaci et al. 2005. Use of transplanted zebra mussels (*Dreissena polymorpha*) to assess the bioavailability of microcontaminants in Flemish surface waters. *Environ. Sci. Technol.* 39:1492–1505.

Beyer, A., D. Mackay, M. Matthies, F. Wania, and E. Webster. 2000. Assessing long-range transport potential of persistent organic pollutants. *Environ. Sci. Technol.* 34:699–703.

Borcherding, J. 2006. Ten years of practical experience with the *Dreissena*-monitor, a biological early warning system for continuous water quality monitoring. *Hydrobiologia* 556:417–426.

Bruner, K. A., S. W. Fisher, and P. F. Landrum. 1994a. The role of the zebra mussel, *Dreissena polymorpha*, in contaminant cycling. 1. The effect of body-size and lipid-content on the bioconcentration of PCBs and PAHs. *J. Great Lakes Res.* 20:725–734.

Bruner, K. A., S. W. Fisher, and P. F. Landrum. 1994b. The role of the zebra mussel, *Dreissena polymorpha*, in contaminant cycling. 2. Zebra mussel contaminant accumulation from algae and suspended particles, and transfer to the benthic invertebrate, *Gammarus fasciatus*. *J. Great Lakes Res.* 20:735–750.

Butler, P. A. 1973. Organochlorine residues in estuarine mollusks, 1965–72—National pesticide monitoring program. *Pest. Monit. J.* 6:238–362.

Capuzzo, J. M., J. W. Farrington, and P. Rantamaki et al.1989. The relationship between lipid composition and seasonal differences in the distribution of PCBs in *Mytilus edulis* L. *Mar. Environ. Res.* 28:259–264.

Carlson, D. L. and D. L. Swackhamer. 2006. Results from U.S. Great Lakes fish monitoring program and effects of lake processes on bioaccumulative contaminant concentrations. *J. Great Lakes Res.* 32:370–385.

Carlton, J. T. 2008. The zebra mussel *Dreissena polymorpha* found in North America in 1986 and 1987. *J. Great Lakes Res.* 34:770–773.

Cho, Y. C., R. C. Frohnhoefer, and G. Y. Rhee. 2004. Bioconcentration and redeposition of polychlorinated biphenyls by zebra mussels (*Dreissena polymorpha*) in the Hudson River. *Water Res.* 38:769–777.

Cunningham, P. A. and M. R. Tripp. 1975. Factors affecting the accumulation and removal of mercury from tissues of the American oyster *Crassostrea virginica*. *Mar. Biol.* 31:311–319.

de Kock, W. C. and C. T. Bowmer. 1993. Bioaccumulation, biological effects and food chain transfer of contaminants in the zebra mussel *Dreissena polymorpha*. In *Zebra Mussels: Biology, Impacts and Controls*, T.F. Nalepa and D.W. Schloesser, eds., pp. 503–533. Boca Raton, FL: CRC Press.

De Vault, D. S., R. Hesselberg, P. W. Rodger, and T. J. Feist. 1996. Contaminant trends in Lake trout and Walleye from the Laurentian Great Lakes. *J. Great Lakes Res.* 22:884–895.

Domeneghetti, S., C. Manfrin, and L. Varotto et al. 2011. How gene expression profiles disclose vital processes and immune responses in *Mytilus* spp. *Invertebrate Survivor. J.* 8:179–189.

Eisenreich, S. J., P. D. Capel, J. A. Robbins, and R. Bourbonniere. 1989. Accumulation and diagenesis of chlorinated hydrocarbons in lacustrine sediments. *Environ. Sci. Technol.* 23:1116–1126.

Farrington, J. W. 1983. Bivalves as sentinels of coastal chemical pollution: The Mussel (and oyster) *Watch*. *Oceanus* 26:18–29.

Farrington, J. W., E. D. Goldberg, R. W. Risebrough, J. H. Martin, and V. T. Bowen. 1983. US "Mussel Watch" 1976–1978: An overview of the trace metal, DDE, PCB, hydrocarbon, and artificial radionuclide data. *Environ. Sci. Technol.* 17:490–498.

Fisher, N. S. and W. A. Teyssie. 1986. Influence of food composition on biokinetics and tissue distribution of zinc and americium in mussels. *Mar. Ecol. Prog. Ser.* 28:197–207.

GLWQA. 1987. Great Lakes Water Quality Agreement (as amended by Protocol Nov 1987). International Joint Commission United States and Canada. http://www.ijc.org (accessed March 20, 2012).

Goldberg, E. D. 1975. The mussel watch—A first step in global marine monitoring. *Mar. Pollut. Bull.* 6:111–114.

Goldberg, E. D., M. Koide, V. Hodge, A. R. Flegal, and J. Martin. 1983. U.S. Mussel Watch: 1977–1978 results on trace metals and radionuclides. *Estuar. Coast. Shelf Sci.* 16:69–93.

Gossiaux, D. C., P. F. Landrum, and S. W. Fisher. 1998. The assimilation of contaminants from suspended sediment and algae by the zebra mussel, *Dreissena polymorpha*. *Chemosphere* 36:3181–3197.

Hecky, R. E., R. E. H. Smith, and D. R. Barton et al. 2004. The nearshore phosphorus shunt: A consequence of ecosystem engineering by dreissenids in the Laurentian Great Lakes. *Can. J. Fish. Aquat. Sci.* 61:1285–1293.

Hines, A., F. J. Staff, J. Widdows, R. M. Compton, F. Falciani, and M. R. Viant. 2010. Discovery of metabolic signatures for predicting whole organism toxicology. *Toxicol. Sci.* 115:369–378.

International Joint Commission. 2011. 15th biennial report on Great Lakes water quality. International Joint Commission, Canada and the United States. 59pp.

Jones, O. A. H., F. Dondero, A. Viarengo, and J. L. Griffin. 2008. Metabolic profiling of *Mytilus galloprovincialis* and its potential applications for pollution assessment. *Mar. Ecol. Prog. Ser.* 369:169–179.

Johns, C. and B. E. Timmerman. 1998. Total cadmium, copper, and zinc in two dreissenid mussels, *Dreissena polymorpha* and *Dreissena bugensis*, at the outflow of Lake Ontario. *J. Great Lakes Res.* 24:55–64.

Johnson, W. E., K. L. Kimbrough, G. G. Lauenstein, and J. D. Christensen. 2008. Chemical contamination assessment of Gulf of Mexico oysters in response to hurricanes Katrina and Rita. *Environ. Monit. Assess.* 150:211–225.

Kehrer, J. P. 1993. Free radicals as mediators of tissue injury and disease. *Crit. Rev. Toxicol.* 23:21–48.

Kimbrough, K. L., S. Commey, D. A. Apeti, and G. G. Lauenstein. 2010. Chemical contamination assessment of the Hudson Raritan estuary as a result of the attack on the World Trade Center: Analysis of trace elements. *Mar. Pollut. Bull.* 60:2289–2296.

Kimbrough, K. L. and G. G. Lauenstein. 2007. Major and trace element analytical methods of the National Status and Trends Program 2000–2006. NOAA Technical Memorandum NOS NCCOS 29, Silver Spring, MD. 19pp.

Kimbrough, K. L., W. E. Johnson, G. G. Lauenstein, J. D. Christensen, and D. A. Apeti. 2008. As assessment of two decades of contaminant monitoring in the Nation's Coastal Zone. NOAA Technical Memorandum NOS NCCOS 74. 105pp.

Kimbrough, K. L., G. G. Lauenstein, and W. E. Johnson. 2007. Organic contaminant analytical methods of the National Status and Trends Program: Update 2000–2006. NOAA Technical Memorandum NOS NCCOS 30, Silver Spring, MD. 65pp.

Kimbrough, K. L., W. E. Johnson, G. G. Lauenstein, J. D. Christensen, and D. A. Apeti. 2008. An assessment of two decades of contaminant monitoring in the Nation's Coastal Zone. Silver Spring, MD. NOAA Technical Memorandum NOS NCCOS 74, 105 pp.

Kraak, M. H. S., F. Kuipers, H. Schoon, C. J. De Groot, and W. Admiraal. 1994. The filtration rate of the zebra mussel *Dreissena polymorpha* used for water quality assessment in Dutch rivers. *Hydrobiologia* 294:13–16.

Kraak, M. H. S., M. C. T. Scholten, W. H. M Peeters, and W. C. Dekock. 1991. Biomonitoring of heavy-metals in the Western-European Rivers Rhine and Meuse using the fresh-Water mussel *Dreissena polymorpha. Environ. Pollut.* 74:101–114.

Kwan, K. H. M., H. M. Chan, and Y. de Lafontaine. 2003. Metal contamination in zebra mussels (*Dreissena polymorpha*) along the St. Lawrence River. *Environ. Monit. Assess.* 88:193–219.

de Lafontaine, Y., F. Gagné, and C. Blaise et al. 2000. Biomarkers in zebra mussels (*Dreissena polymorpha*) for the assessment and monitoring of water quality of the St Lawrence River (Canada). *Aquat. Toxicol.* 50:51–71.

Lam, P. K. S. and J. S. Gray. 2003. The use of biomarkers in environmental monitoring programmes. *Mar. Pollut. Bull.* 46:182–186.

Lauenstein, G. G. and A. Y. Cantillo. 1993a. Sampling and analytical methods of the National Status and Trends Program National Benthic Surveillance and Mussel Watch Projects 1984–1992: Overview and summary of methods, Volume I NOAA Technical Memorandum NOS ORCA 71, Silver Spring, MD.

Lauenstein, G. G. and A. Y. Cantillo. 1993b. Sampling and analytical methods of the National Status and Trends Program National Benthic Surveillance and Mussel Watch Projects 1984–1992: Comprehensive descriptions of complementary measurements, Volume II NOAA Technical Memorandum NOS ORCA 71, Silver Spring, MD.

Lauenstein, G. G. and A. Y. Cantillo. 1993c. Sampling and analytical methods of the National Status and Trends Program National Benthic Surveillance and Mussel Watch Projects 1984–1992: Comprehensive descriptions of elemental analytical methods, Volume III NOAA Technical Memorandum NOS ORCA 71, Silver Spring, MD.

Lauenstein, G. G. and A. Y. Cantillo. 1993d. Sampling and analytical methods of the National Status and Trends Program National Benthic Surveillance and Mussel Watch Projects 1984–1992: Comprehensive descriptions of trace organic analytical methods, Volume IV NOAA Technical Memorandum NOS ORCA 71, Silver Spring, MD.

Lauenstein, G. G. and A. Y. Cantillo. 1998. Sampling and analytical methods of the National Status and Trends Program Mussel Watch Project—1993–1996 update. NOAA Technical Memorandum NOS ORCA 130, Silver Spring, MD.

Lauenstein, G. G., A. Y. Cantillo, S. Kokkinakis, J. Jobling, and R. Fay. 1997. Mussel Watch project site descriptions, through 1997. NOAA Technical Memorandum NOS ORCA 112, Silver Spring, MD. 354pp.

Lauenstein, G. G. and K. L. Kimbrough. 2007. Chemical contamination of the Hudson-Raritan estuary as a result of the attack on the World Trade Center: Analysis of polycyclic aromatic hydrocarbons and polychlorinated biphenyls in mussels and sediment. *Mar. Pollut. Bull.* 54:284–294.

Lowe, D. M., V. U. Fossato, and M. H. Depledge. 1995. Contaminant-induced lysosomal membrane damage in blood cells of mussels *Mytilus galloprovincialis* form the Venice Lagoon: An in vitro study. *Mar. Ecol. Prog. Ser.* 129:189–196.

Ma, J. M. and Z. H. Cao. 2010. Quantifying the perturbations of persistent organic pollutants induced by climate change. *Environ. Sci. Technol.* 44:8567–8573.

Martin, M. 1985. State Mussel Watch: Toxics surveillance in California. *Mar. Pollut. Bull.* 16:140–146.

Marvin, C. H., M. N. Charlton, and E. J. Reiner et al. 2002. Surficial sediment contamination in Lakes Erie and Ontario: A comparative analysis. *J. Great Lakes Res.* 28:437–450.

Marvin, C. H., E. T. Howell., and E. J. Reiner. 2000. Polychlorinated dioxins and furans in sediments at a site colonized by *Dreissena* in western Lake Ontario, Canada. *Environ. Toxicol. Chem.* 19:344–351.

Marvin, C., S. Painter, D. Williams, V. Richardson, R. Rossmann, and P. VanHoof. 2004. Spatial and temporal trends in surface water and sediment contamination in the Laurentian Great Lakes. *Environ. Pollut.* 129:131–144.

McDonald, S. J., D. S. Frank, J. A. Ramirez, B. Wang, and J. M. Brooks. 2006. Ancillary methods of the National Status and Trends Program: 2000–2006 update. NOAA Technical Memorandum NOS NCCOS 28, Silver Springs, MD. 17pp.

Mersch, J. and L. Johansson. 1993. Transplanted aquatic mosses and freshwater mussels to investigate the trace metal contamination in the Rivers Meurthe and Plaine, France. *Environ. Technol.* 14:1027–1036.

Mills, E. L., J. R. Chrisman, and B. Baldwin et al. 1999. Changes in the dreissenid community in the lower Great Lakes with emphasis on southern Lake Ontario. *J. Great Lakes Res.* 25:187–197.

Mills, E. L., R. M. Dermott, and E. F. Roseman et al. 1993. Colonization, ecology, and population-structure of the quagga mussel (bivalvia, Dreissenidae) in the lower Great Lakes. *Can. J. Fish. Aquat. Sci.* 50:2305–2314.

Nalepa, T. P., D. L. Fanslow, and S. A. Pothoven. 2010. Recent changes in density, biomass, recruitment, size structure, and nutritional state of *Dreissena* populations in southern Lake Michigan. *J. Great Lakes Res.* 36(Suppl. 3):5–19.

Nizzetto, L., M. Macleod, and K. Borgå et al. 2010. Past, present, and future controls on levels of persistent organic pollutants in the global environment. *Environ. Sci. Technol.* 44:6526–6531.

Osman, A. M., H. Van Den Heuvel, and P. C. M. Van Noort. 2007. Differential responses of biomarkers in tissues of a freshwater mussel, *Dreissena polymorpha*, to the exposure of sediment extracts with different levels of contamination. *J. Appl. Toxicol.* 27:51–59.

Parolini, M., A. Binelli, D. Cogni, and A. Provini. 2010. Multibiomarker approach for the evaluation of the cyto-genotoxicity of paracetamol on the zebra mussel (*Dreissena polymorpha*). *Chemosphere* 79:489–498.

Patterson, M. W. R., J. J. H. Ciborowski, and D. R. Barton. 2005. The distribution and abundance of *Dreissena* species (Dreissenidae) in Lake Erie, 2002. *J. Great Lakes Res.* 31(Suppl. 2):223–237.

Pruell, R. J., J. G. Quinn, J. L. Lake, and W. R. Davis. 1987. Availability of PCBs and PAHs to *Mytilus edulis* from artificially resuspended sediments. In *Oceanic Processes in Marine Pollution. Vol. 1. Biological Processes and Wastes in the Ocean*, J.M. Capuzzo, and D.R. Kester, eds., pp. 97–108. Malabar, FL: Krieger.

Reeders, H. H. and A. bij de Vaate. 1992. Bioprocessing of polluted suspended matter from the water column by the zebra mussel (*Dreissena polymorpha Pallas*). *Hydrobiologia* 239:53–63.

Richman, L. A. and K. Somers. 2005. Can we use zebra and quagga mussels for biomonitoring contaminants in the Niagara River? *Water Air Soil Pollut.* 167:155–178.

Richman, L. A. and K. Somers. 2010. Monitoring metal and persistent organic contaminant trends through time using quagga mussels (*Dreissena bugensis*) collected from the Niagara River. *J. Great Lakes Res.* 36:28–36.

Ringwood, A. H., D. E. Conners, and J. Hoguet. 1998. The effects of natural and anthropogenic stressors on lysosomal destabilization in oysters *Crassostrea virginica*. *Mar. Ecol. Prog. Ser.* 166:163–171.

Ringwood, A. H., D. E. Conners, C. J. Keppler, and A. A. DiNovo. 1999. Biomarker studies with juvenile oysters (*Crassostrea virginica*) deployed in situ. *Biomarkers* 4:400–414.

Riva, C., A. Binelli, and A. Provini. 2008. Evaluation of several priority pollutants in zebra mussels (*Dreissena polymorpha*) in the largest Italian subalpine lakes. *Environ. Pollut.* 151:652–662.

Robertson, D. G. 2010. Muscling mussels: Metabolomic evaluation of toxicity. *Toxicol. Sci.* 115:305–306.

Robertson, A. and G. G. Lauenstein. 1998. Distribution of chlorinated organic contaminants in dreissenid mussels along the southern shores of the Great Lakes. *J. Great Lakes Res.* 24:608–619.

Roesijadi, G. J., J. S. Young, A. S. Drum, and J. M. Gurtisen. 1984. Behavior of trace metals in *Mytilus edulis* during a reciprocal transplant field experiment. *Mar. Ecol. Prog. Ser.* 18:155–170.

Schantz, M. M., R. M. Parris, and S. A. Wise. 2000. NIST/NOAA Intercomparison exercise program for organic contaminants in the marine environment: Description and results of 1999 organic intercomparison exercises, NOAA Technical Memorandum NOS NCCOS CCMA 146, Silver Spring, MD.

Sericano, J. L. 1993. The American oyster (*Crassostrea virginica*) as a bioindicator of trace organic contamination. PhD thesis. Texas A&M University, College Station, TX, 242pp.

Stäb, J. A., M. Frenay, I. L. Freriks, U. A. T. Brinkman, and W. P. Cofino. 1995. Survey of nine organotin compounds in The Netherlands using the zebra mussel (*Dreissena polymorpha*) as biomonitor. *Environ. Toxicol. Chem.* 14:2023–2032.

Swackhamer, D. L. and D. E. Armstrong. 1986. Estimation of the atmospheric and non-atmospheric contributions and losses of polychlorinated biphenyls for Lake Michigan on the basis of sediment records of remote lakes. *Environ. Sci. Technol.* 20:879–883.

Swackhamer, D. L., B. D. McVeety, and R. A. Hites. 1988. Deposition and evaporation of polychlorobiphenyl congeners to and from Siskiwit Lake, Isle Royale, Lake Superior. *Environ. Sci. Technol.* 22:664–672.

Thomas, M. A. 2007. The association of arsenic with redox conditions, depth, and ground-water age in the glacial aquifer system of the Northern United States: U.S. Geological Survey Scientific Investigations Report 2007–5036, 26p.

Venier, P., L. Varotto, and U. Rosani et al. 2011. Insights into the innate immunity of the Mediterranean mussel *Mytilus galloprovincialis*. *BMC Genomics* 12:69.

Viarengo, A., L. Canesi, M. Pertica, G. Poli, M. N. Moore, and M. Orunesu. 1990. Heavy metal effects on lipid peroxidation in the tissue of *Mytilus galloprovincialis* Lam. *Comp. Biochem. Physiol.* 97:32–42.

Willie, S. 2000. NOAA National Status and Trends Program thirteenth round intercomparison exercise results for trace metals in marine sediments and biological tissues. NOAA Technical Memorandum NOS NCCOS CCMA 142, Silver Spring, MD.

PART IV

Morphology, Physiology, and Behavior

CHAPTER 19

Morphological Variability of *Dreissena polymorpha* and *Dreissena rostriformis bugensis* (Mollusca: Bivalvia)

Vera Pavlova and Yuri Izyumov

CONTENTS

Abstract	287
Introduction	288
Methods	288
Taxa and Study Sites	288
Shell Form: Traditional and Geometric Morphometric Analysis	289
Pigmentation	290
Results	293
Overall Dimensions and Ratios	293
Morphological Variation: Within a Water Body	293
Morphological Variation: Between Water Bodies	295
Shell Outline Variation	296
Shell Pigmentation	299
Discussion	305
Acknowledgments	312
References	312

ABSTRACT

Features of external shell morphology and pigmentation of zebra mussels (*Dreissena polymorpha*) and quagga mussels (*Dreissena rostriformis bugensis*) from Europe and North America (Lakes Michigan and Erie) were studied. Shell morphology was described and analyzed with traditional and geometric morphometric approaches. Analysis of shell coloration included estimation of degree of pigmentation and descriptions of pigmentation patterns. High variability of morphometric features was found in both mussel species. Most pronounced differences were found between populations from the northern region of the Caspian Sea (water salinity 4‰) and two sites in Lake Michigan (depths of 25 and 45 m, the deepest sites sampled in this study). Zebra mussels from the Caspian Sea were actually a mixture of two subspecies, *D. p. polymorpha* and *D. p. andrusovi*, and some intermediate forms. *D. p. andrusovi* possessed lower and less convex shells, a greater range of pigmentation degree, and a greater diversity of pigmentation pattern than *D. p. polymorpha*. *D. r. bugensis* from the Caspian Sea had the highest and most convex shells of all specimens examined. The shell form and coloration of *D. polymorpha* in Lake Michigan were typical. However, the shell height, width, and degree of pigmentation of *D. r. bugensis* from Lake Michigan (profunda phenotype) were the lowest of all specimens examined. Latitudinal clines in shell morphology and pigmentation in both dreissenid species were observed, and the role of water current velocity, depth, and salinity on geographical gradients was discussed. Generally, *D. r. bugensis* possessed higher variability of shell proportions, outlines, and pigmentation patterns than *D. polymorpha*.

INTRODUCTION

Dreissenid mussels are one of the most studied of all bivalve taxa (Schloesser et al. 1994). Interest in these mussels is a result of their wide geographical distribution, high productivity and abundance, strong impact on ecosystems, and role as biofoulers. Most traditional studies of dreissenids have mainly focused on distributions, population dynamics, and ecological impacts. However, studies of morphological variability are also important to better understand the biology and adaptive capabilities of this taxonomic group.

Studies of the morphology of dreissenid shells extend back more than 200 years. The first description of *Mytilus polymorphus* was made in 1769 (Pallas 1809). We now know this species as *Dreissena polymorpha* (Pallas, 1771). Pallas (1809) described this species from specimens collected in the Yaik (now Ural) and the Volga Rivers and from the northern region of the Caspian Sea. Mussels in rivers (var. *fluviatilis*) were fourfold larger than marine (var. *marina*) ones (Pallas 1809). By the end of the nineteenth century, all known data on the taxonomic diversity of dreissenids were summarized by Andrusov (1897).

The first studies concerned with the effect of environmental factors on dreissenid morphology were conducted on specimens from the Caspian Sea because dreissenids from this area were important food for commercial fishes, and shell morphology was noticeably variable. Spasskiy (1948) reported that shell forms and coloration of *D. polymorpha* (zebra mussel) were related to water salinity. As salinity increased, shells became smaller, lower, and narrower. Karpevich (1955) showed that morphological differences were accompanied by physiological changes as a result of metabolic responses to salt water. A series of papers examined oxygen consumption, thermal tolerance, salinity tolerance, and air-exposure resistance in *D. polymorpha* from a range of water bodies of the Volga River basin (Antonov and Shkorbatov 1983, Shkorbatov et al. 1994, Antonov 1997). Based on these results, it was concluded that *D. polymorpha* can form distinct morphometric and physiological features that can be related to broad and local habitat characteristics. Development of distinct features was evident where environmental conditions were suboptimal (as in Caspian Sea) and also within that part of its range where conditions were optimal.

Studies of shell pigmentation/coloration in *D. polymorpha* showed distinct types that were related to environmental conditions. Consistent differences in pigment extent and frequency of pigmentation patterns led to the discrimination of five population groups defined by geographic region: Aral–Caspian, Ponto–Caspian, Middle Russian, Northeastern, and Baltic (Smirnova et al. 1993, Biochino 1994). Protasov and Gorpinchuk (1997) suggested that dominance of such phenotypes based on shell pigmentation indicated a high adaptive ability under certain conditions. For instance, the ability to adapt to hypoxia was related to shell coloration (Protasov et al. 1997).

To date, there are substantially more studies on morphology of *D. polymorpha* compared to *D. r. bugensis* (quagga mussel), likely because the geographic range of the latter species has been rather limited until recently. From the period when *D. r. bugensis* was first described in 1890 (Andrusov 1890) to the 1940s, this species inhabited a confined area in the lower reaches of the Youzhnyi (South) Bug and the Ingul rivers (Andrusov 1897, Zhuravel 1951). At present, however, the range of *D. r. bugensis* is almost as large as that of *D. polymorpha* (Van der Velde and Platvoet 2007, Imo et al. 2010, Benson 2011a,b), which may indicate that its shell morphology may be as variable as *D. polymorpha*. However, only a few studies describe shell morphology of *D. r. bugensis* (Andrusov 1897, Zhuravel 1951, Dermott and Munawar 1993, Claxton et al. 1998, Lukashev 1999, Protasov 2004, Hubenov and Trichkova 2007, Van der Velde and Platvoet 2007, Peyer et al. 2010). In the majority of studies, only limited amounts of data on shell form and coloration are given, and no generalizations are provided.

After dreissenids invaded the Laurentian Great Lakes of North America in the late 1980s (Hebert et al. 1989, May and Marsden 1992, Carlton 2008), it became important to correctly identify and distinguish between the two species. Various studies were conducted to describe shell morphology and molecular differences (May and Marsden 1992, Pathy and Mackie 1993, Nichols and Black 1994, Ross and Lima 1994, Claxton et al. 1997, 1998). In addition, a previously unknown deepwater phenotype of quagga mussel named "profunda" was discovered in Lake Erie (Dermott and Munawar 1993). Subsequently, this phenotype was found to possess not only morphologic but also physiologic peculiarities, for example, an ability to mature and spawn at lower temperatures compared to the shallow-water phenotype (Claxton and Mackie 1998).

As noted, there have been few studies that document variation in shell morphology of *D. r. bugensis*. Thus, the objective of the present study was to compare and analyze the diversity of shell forms and pigmentation of both *D. polymorpha* and *D. r. bugensis* from different water bodies within their current ranges.

METHODS

Taxa and Study Sites

D. polymorpha in this study was mostly represented by the subspecies *D. polymorpha polymorpha*, but some populations from the northern portion of the Caspian Sea (Malaya Zhemchuzhnaya Banka) consisted of a mixture of *D. p. polymorpha*, individuals that were morphologically close to *D. polymorpha andrusovi* (Logvinenko and Starobogatov 1968) and individuals of an intermediate form. *D. r. bugensis* was represented by the typical epilimnetic phenotype and by a hypolimnetic phenotype (profunda).

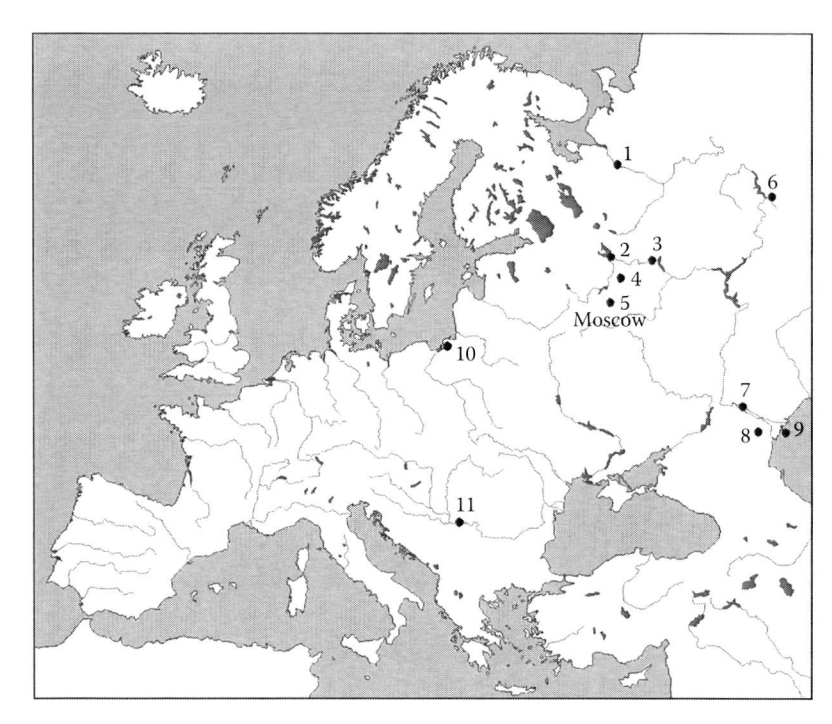

Figure 19.1 Location of water bodies in Europe where *D. polymorpha* and *D. r. bugensis* were collected. 1, Northern Dvina River; 2, Rybinsk Reservoir; 3, Gorky Reservoir; 4, Lake Plescheevo; 5, Moskva River; 6, Kama Reservoir; 7, Volga–Ahtuba floodplain (Ahtuba River [AR], Kalinov channel [KC], Yarovatiy channel [YC], Svyaznye Lakes [SL]); 8, water bodies of Kalmykia Republic of Russian Federation (Lake Sharony [LS], RR-1 supply channel [RR], Chograi Reservoir [CR]); 9, Volga Delta (VD) and north Caspian Sea (NC); 10, Lake Forelevoe; 11, water bodies of Serbian Republic (Perućica Reservoir [PR], Spaića Reservoir [SR], Danube River [DR]).

The profunda phenotype was identified and provided by T.F. Nalepa (GLERL NOAA, Ann Arbor, Michigan).

Dreissenid mussels were collected from water bodies in Europe (Figure 19.1) and North America (Figure 19.2) primarily in autumn 2004–2008. Collections were made on various types of substrate by dredge, bottom trawl, and less frequently by hand (Table 19.1). All sampling locations were in freshwater except in the north Caspian Sea where salinity was 4‰. To determine morphologic variation of shell dimensions within a given water body, we collected dreissenids from 17 sites in the Rybinsk Reservoir (Figure 19.3) and 6 sites in the Gorky Reservoir (Figure 19.4). Sites in these reservoirs were located in stretches and regions that were defined by different physical/chemical conditions. For all water bodies regardless of location, mussels were preserved in ethanol (96%) or formalin (5%) immediately after collection. For purposes of this study, a population was defined as the group of individuals collected at a given site.

Shell Form: Traditional and Geometric Morphometric Analysis

Dreissenid mussels are bivalves that possess shells of mytiloid form. In the evolution of bivalves, this shape developed in epifaunal mussels as a result of the neotenic retention of the byssus through postlarval stages (Stanley 1970). One consequence of byssal retention relative to unspecialized forms was a reduction of anterior parts of the shell, general elongation, and a straightening of the ventral shell margin (Stanley 1970). Such shell form allowed efficient filtration (Yonge 1953) as well as stable attachment to substrates (Yonge and Campbell 1968, Morton 1970).

Traditional morphometric analysis of shell dimensions was performed on 5059 individuals of *D. polymorpha* from 18 different water bodies and on 1968 individuals of *D. r. bugensis* from 10 water bodies (Table 19.1). Shell length (L), height (H), and width (W) of each specimen

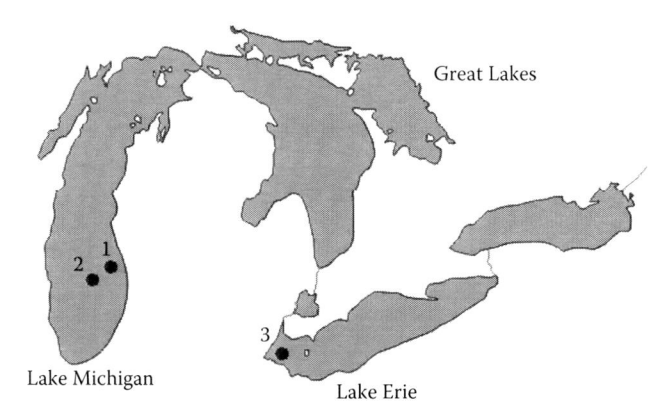

Figure 19.2 Location of sites in the North American Great Lakes where *D. polymorpha* and *D. r. bugensis* were collected. 1, Lake Michigan (25 m depth); 2, Lake Michigan (45 m depth); 3, Lake Erie (3 m depth).

Table 19.1 Country, Water Body, Water Depth, and Substrate Where *Dreissena* Was Collected in Europe and North America

Country	Water Body	Species	Depth (m)	Substrate
Russian Federation	Severnaya (Northern) Dvina River	*D. p. polymorpha*	1	Abandoned fishing net
	Volga River, the Rybinsk Reservoir	*D. p. polymorpha* *D. r. bugensis*	3–20	Submerged logs
	Volga River, the Gorky Reservoir	*D. p. polymorpha* *D. r. bugensis*	5–16	Submerged logs
	Lake Plescheevo	*D. p. polymorpha*	0.5	Unionid mussels
	Moskva River	*D. r. bugensis*	1	Concrete
	The Kama Reservoir	*D. p. polymorpha*	1	Submerged log
	Volga–Ahtuba floodplain (Yarovatiy and Kalinov channels, Svyaznye Lakes)	*D. p. polymorpha* *D. r. bugensis*	0.5–1	Silty sand
	Ahtuba River	*D. p. polymorpha* *D. r. bugensis*	0.5–1	Silty sand
	Volga Delta (Belinskiy Bank)	*D. p. polymorpha* *D. r. bugensis*	3	Sand
	Northern Caspian (Malaya Zhemchuzhnaya Banka)	*D. p. polymorpha* *D. p. andrusovi* *D. r. bugensis*	5	Sand
	Lake Sharony	*D. p. polymorpha* *D. r. bugensis*	0.5	Reed stems
	The Chograi Reservoir	*D. p. polymorpha*	0.5	Unionid mussels
	RR-1 supply channel	*D. p. polymorpha*	0.5	Concrete
	Lake Forelevoe	*D. p. polymorpha*	0.5	Pebble
Serbian Republic	Danube River	*D. p. polymorpha*	5	Silted sand
	The Peručica Reservoir	*D. p. polymorpha*	7–20	Silty sand
	The Spaića Reservoir	*D. p. polymorpha*	5	Silty sand
North America	Lake Erie	*D. p. polymorpha* *D. r. bugensis*	3	Silty sand
	Lake Michigan	*D. p. polymorpha* *D. r. bugensis* profunda	25–45	Silty sand

were measured with a vernier caliper (Figure 19.5). Statistical differences in measured ratios H/L, W/L, and H/W were determined using STATISTICA 8.0 software.

When considering geometric morphometric analysis, it should be noted that general shell outlines are formed by relative positions and angles of the superior, posterior, and ventral shell margins (Figure 19.5). Separate elements of shell outlines and margins are interconnected; that is, if the umbo is curved down, the ventral margin is concave to a greater or lesser extent, and if the umbo is not bent down, the ventral margin is usually convex. Historically, these features of shell margins were used to systematically define species and subspecies of Dreissenidae (Andrusov 1897, Zhadin 1952, Logvinenko and Starobogatov 1968, Babak 1983).

After standard morphometric measurements, shells were cleaned of soft tissues, air-dried, and prepared for geometric analysis. Large individuals of *D. polymorpha* (16.0–24.0 mm length) and *D. r. bugensis* (14.0–22.0 mm length) were selected from each water body and the right shell scanned (scanner Mustek 4800TA Pro II). A total of 486 *D. polymorpha* and 287 *D. r. bugensis* were scanned. Images were processed using SHAPE software (Iwata and Ukai 2002). This software was used to perform elliptic Fourier analysis of shell shapes and principal component (PC) analysis of Fourier descriptors. All other statistical tests were performed with STATISTICA 8.0 software. When ratios of shell dimensions across different water bodies were compared, we first divided samples into four groups based on mean values of shell length: 10.0–13.6 mm (group I), 14.2–18.0 mm (group II), 18.3–21.9 mm (group III), and 22.7–29.2 mm (group IV). Comparisons in ratios between and within water bodies were made only for individuals in the same length group to assure that differences in ratios were not related to different size frequencies of populations.

Pigmentation

Dreissenids have the most variable pigmentation patterns among freshwater mussels. Basically, there are two measures of shell pigmentation in dreissenids: total amount of pigment and pigment pattern. The extent of shell pigmentation can be defined by the amount of melanin present (Starobogatov 1994a). In past studies, the extent of pigmentation was estimated by defining the number and width of stripes on the shells; that is, the amount of shell pigment

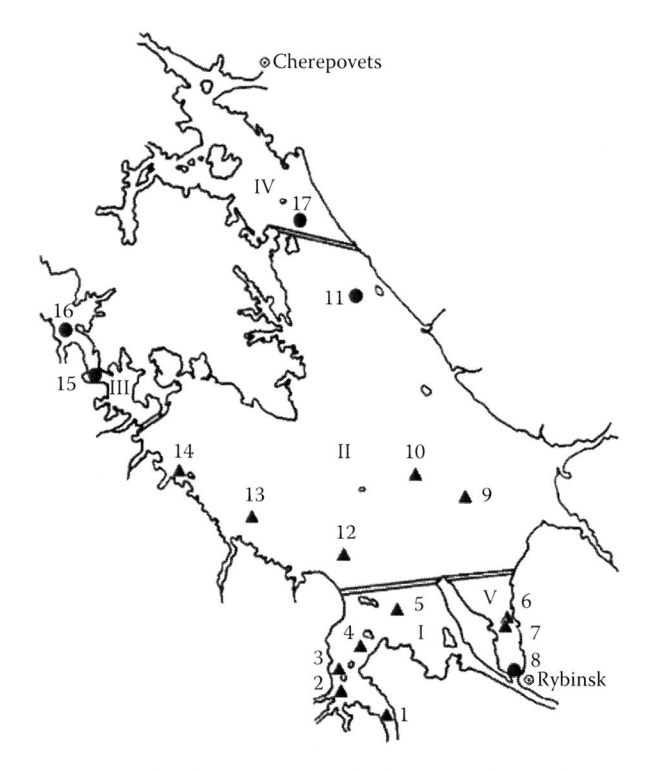

Figure 19.3 Location of sites in Rybinsk Reservoir where *D. polymorpha* and *D. r. bugensis* were collected. Roman numerals denote stretches: I, Volga stretch; II, main stretch; III, Mologa stretch, IV; Sheksna stretch; V, near-dam stretch. Triangles, both *D. polymorpha* and *D. r. bugensis* were collected; circles, only *D. polymorpha* was collected.

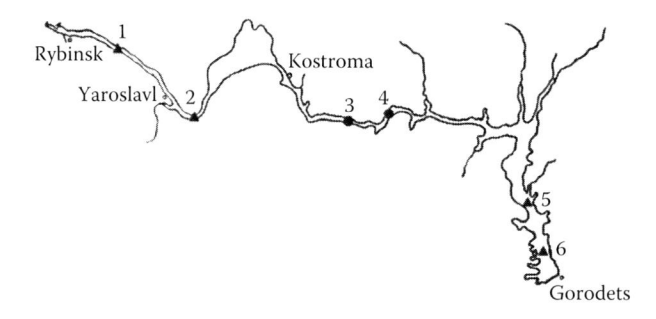

Figure 19.4 Location of sites in Gorky Reservoir where *D. polymorpha* and *D. r. bugensis* were collected. Triangle, both *D. polymorpha* and *D. r. bugensis* were collected; circles, only *D. polymorpha* was collected.

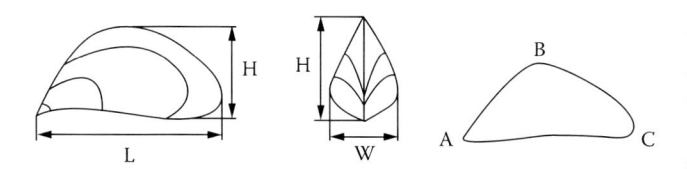

Figure 19.5 Diagrams of shell dimensions and margins for dreissenid mussels: L, length; H, height; W, width; AB, superior margin; BC, posterior margin; and AC, ventral margin. Margins defined according to Babak (From Babak, E.V., *Proc. Palaeontol. Inst. Acad. Sci. USSR*, 204, 1, 1983).

was a function of the number and width of stripes (Biochino 1994, Protasov 1998). We used another method based on colorimetry; pigmentation was determined by measuring the intensity of reflected light using software processing of digital images of shells (Izyumov and Pavlova 2008).

Analysis of pigmentation was possible only for those individuals with an unbroken pigment layer. Therefore, not all samples analyzed for morphological characteristics were acceptable for analysis of pigmentation. In the Rybinsk Reservoir, mussels from only seven sites were analyzed for pigmentation, with *D. polymorpha* analyzed from all seven sites and *D. r. bugensis* analyzed from three sites. In the Gorky Reservoir, mussels from five sites were analyzed, and both species were represented at all five sites. The total amount of pigmentation was determined for mussels from all water bodies given in Table 19.1 except for mussels from Kama Reservoir, Lake Forelevoe, and the Northern Dvina River. Pigmentation pattern was determined for mussels from all water bodies except for mussels from Volga–Ahtuba flood plain and Spaića Reservoir. Overall, about 2700 zebra mussels and 1100 quagga mussels were analyzed for pigmentation.

In the laboratory, preserved specimens were gently cleaned of attached debris. Soft tissues were removed and shells were air-dried at room temperature. Only the right valve of each mussel was analyzed since lateral asymmetry of coloration can occur (Protasov and Gorpinchuk 1997). Shells were placed on a white sheet of paper and photographed with an Olympus 5060 digital camera. Optical axis of the lens was perpendicular to the table. Distance from lens to shells was 20 cm. Photos were taken in a dark room using a built-in flash, and conditions were constant for all samples. Analysis of images was carried out using Image-Pro Plus 3.0 software. For each shell, the mean density of luminosity (DL) was determined. DL corresponds to the quantity of light reflected. A DL of 255 corresponds to full reflection (pure white shell), and a DL of 0 corresponds to complete absorption (pure black shell). Thus, the DL index is inversely proportional to the amount of pigmentation.

Descriptions and analysis of pigmentation patterns on dreissenid shells included a discrimination of simple pattern elements and frequencies of pattern occurrence in populations. This approach to characterize shell pigmentation was first applied by Biochino (1990) for *D. polymorpha*. Biochino (1990) placed every shell pattern in a population into one of five types: dark shell without pattern (OO), light shell without pattern (DD), arcuate pattern (AA), radial pattern (RR), and toothed pattern (CC). The last type was compound, since it included many "microvariations." Later studies described additional characters of shell pigmentation, including background color and elements of pattern that appeared individually or in various combinations (Protasov and Gorpinchuk 1997, Protasov and Sinitsyna 2000). Shell pigmentation patterns of *D. r. bugensis* were first described by Lukashev (1999), and the approach used was similar to that of Protasov and Gorpinchuk (1997).

Pigment patterns on dreissenid shells consist of transversal and longitudinal elements that repeat within the same shell area. Transversal elements can be described by the following terminology (modified from Biochino 1990) (Figure 19.6): A, regular arcuate pattern along the entire shell height; C, zigzag pattern along the entire shell height; AC, intermediate pattern between A and C, with the upper shell half zigzag and the lower shell half with arcs; Ad, arcs with teeth in the middle; and D, light shell without pattern (this type was conditionally included with these other transversal patterns). It should be noted that the AC pattern has two variations: ACs, in which regular arcs occur along nearly the entire shell height with curves only on uppermost part of the shell, and ACi, in which regular arcs occur along nearly the entire shell height with curves only on the lowermost part of the shell. All these patterns were observed in both dreissenid species. In addition, another three variations of the AC pattern were found only in *D. r. bugensis*: ACs-d, ACi-d, and ACs-i-d. These patterns consisted of arcs with teeth in the middle and with curves in the upper part of the shell (s-d), with curves in the lower part (i-d), and with curves in both the upper and lower part (s-i-d).

Longitudinal elements can be described as follows: R indicates a shell with radial lines, Rd indicates a single, light radial line, Ro indicates a single dark radial line, and Rm indicates multiple thin radial lines that alternate between light and dark. F indicates a shell with a border between different patterns along the longitudinal axis (Figure 19.7).

All described pigmentation elements can be combined into various pattern types, and the combination of elements on a shell of a given age was the measured unit used to define pigment variability of a given dreissenid population. Transversal elements form simple types of the same names (Figure 19.6), and longitudinal patterns combine with transversal and/or other longitudinal patterns to form composite types (Figure 19.7).

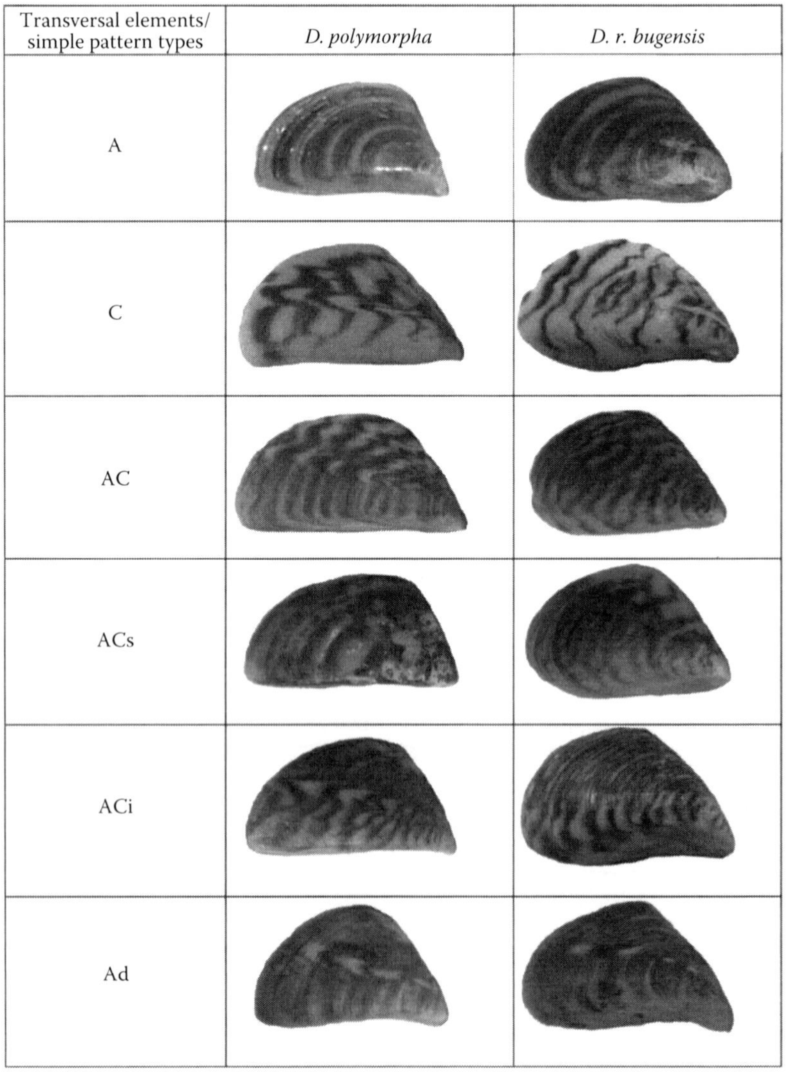

Transversal elements/ simple pattern types	*D. polymorpha*	*D. r. bugensis*
A		
C		
AC		
ACs		
ACi		
Ad		

Figure 19.6 Transversal elements and simple pattern types found in *D. polymorpha* and *D. r. bugensis*.

ACs-d	—	
ACi-d	—	
ACs-i-d	—	
D		

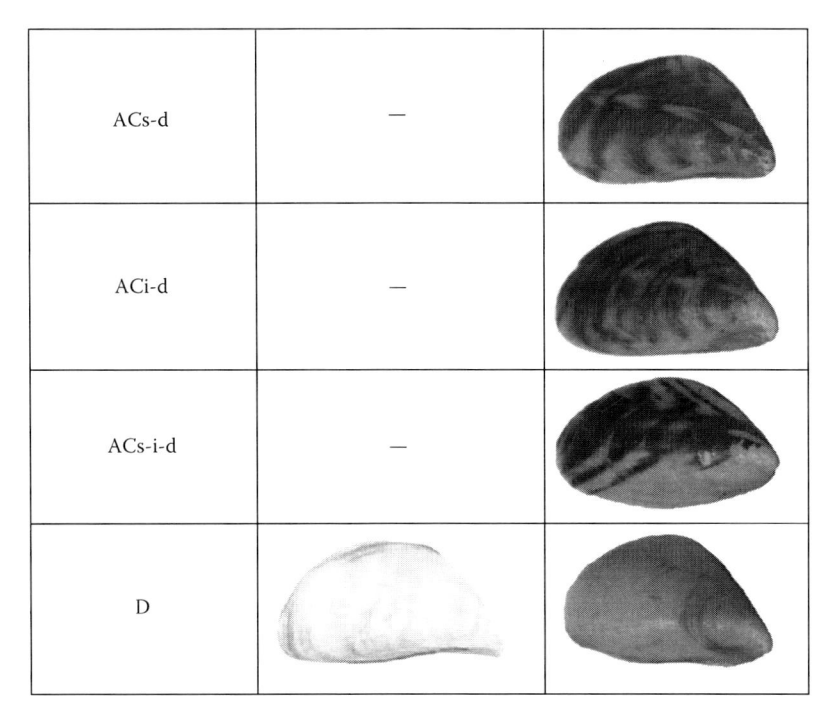

Figure 19.6 (continued)

Our approach to measure shell pigment patterns of dreissenid shells included importance of age-specific variability. Similar to other mussel taxa, dreissenids have structures on their outer shell surface known as "annual rings" that indicate periods of differential growth. Visual analysis showed that shell areas between annual rings usually carried different pigment patterns, that is, a different pattern after a period of growth delay. This phenomenon was noted in other studies (Marsden et al. 1996, Protasov and Gorpinchuk 1997) but has not been examined in detail for dreissenids. In our terminology, the term "age 0+" corresponds to shells of underyearlings and the first-year zones on the shells of older mussels. "Age 1+" corresponds to the zone following the first annual ring of yearling shells and the zone between the first and second annual rings on shells of older mussels. Thus, "age+" does not refer to individuals of this age but to zones of shells secreted at this age (Figure 19.8). A change of patterns after the fourth annual ring was not observed; therefore, patterns were determined only up to age 3+. Pattern types for the different age zones (0+, 1+, 2+, 3+) were determined for each individual. Frequencies of occurrence of pattern types were then determined separately for each age and summarized by age for each population. Similarity between populations was estimated in a pairwise mode according to Zhivotovskiy (1982) as $r_{Zhiv} = \sqrt{p_1 q_1} + \sqrt{p_2 q_2} + \ldots + \sqrt{p_m q_m}$, where r is the similarity between populations and p and q are frequencies of occurrence of pattern types in the first, second, and each subsequent population, respectively. Statistical operations were performed with STATISTICA 8.0 software package.

RESULTS

Overall Dimensions and Ratios

For individuals collected in both Europe and North America, mean shell length and ratios of height/length (H/L), width/length (W/L), and height/width (H/W) for *D. polymorpha* and *D. r. bugensis* are given in Table 19.2. There was a significant difference (p < 0.001) between the two species for all three ratios. Both H/L and H/W were lower and W/L was higher in *D. polymorpha* compared to *D. r. bugensis*. Despite these differences, both species had similar allometry, that is, for both species there was a significant relationship between shell length and each of these ratios (p < 0.001). Typical examples of these relationships for *D. polymorpha* and *D. r. bugensis* are provided for populations in the Perućica Reservoir (Figure 19.9) and in the Ahtuba River (Figure 19.10).

Morphological Variation: Within a Water Body

In the Rybinsk Reservoir, cluster analysis and a subsequent dendrogram of morphologic characteristics (shell ratios) of length groups II and III for *D. polymorpha* indicated individuals from the same stretch of the reservoir clustered together (Figure 19.11). Usually, individuals from the two stretches in the south (I and V) clustered together, as did individuals from the two stretches in the north (III and IV), with minimal overlap between the two geographic regions. Individuals in the main stretch of the reservoir (II) overlapped with individuals in other regions. *D. r. bugensis* was collected at fewer sites than *D. polymorpha* (12 vs. 17) and was only found in stretches I, II, and V. Hence, morphologic differentiation was not as defined as for *D. polymorpha* (Figure 19.12). When the relationship between mean W/L and mean H/L was examined for both species at sites where both species were found, greater variation across sites was apparent for *D. r. bugensis* (Figure 19.13).

In the Gorky Reservoir, samples were collected at only six sites, so cluster analysis was not feasible. However, a scatterplot of the relationship between mean W/L and mean H/L showed a strong spatial difference for *D. polymorpha* (Figure 19.14). Two distinct groupings were apparent: one in the upper reaches of the reservoir (sites 1–3) and the other in the lower reaches (sites 3–5). Individuals in the former group had higher mean values of W/L (range was 0.505–0.523) than individuals in the latter group (range was 0.489–0.493). It is worth noting that individuals from the uppermost site in the Gorky Reservoir (site 1) were distinguished for high values of W/L and also for low values of H/L. As a result, the H/W ratio was the lowest among all sites (0.984) in the Gorky Reservoir. Moreover, values of W/L from sites in the lower reach of the Gorky Reservoir (0.489–0.493) were similar to those of individuals from the Rybinsk Reservoir (0.474–0.498), while values from the sites in the upper reach exceeded those in the Rybinsk Reservoir. We could not compare shell morphometrics of *D. r. bugensis* between sites in the Gorky Reservoir because length–frequency distributions did not overlap. This species was only found at sites in the extreme upper and lower reaches, and the population in the upper reach consisted mostly of large individuals, while the population in the lower reach consisted of mostly small individuals.

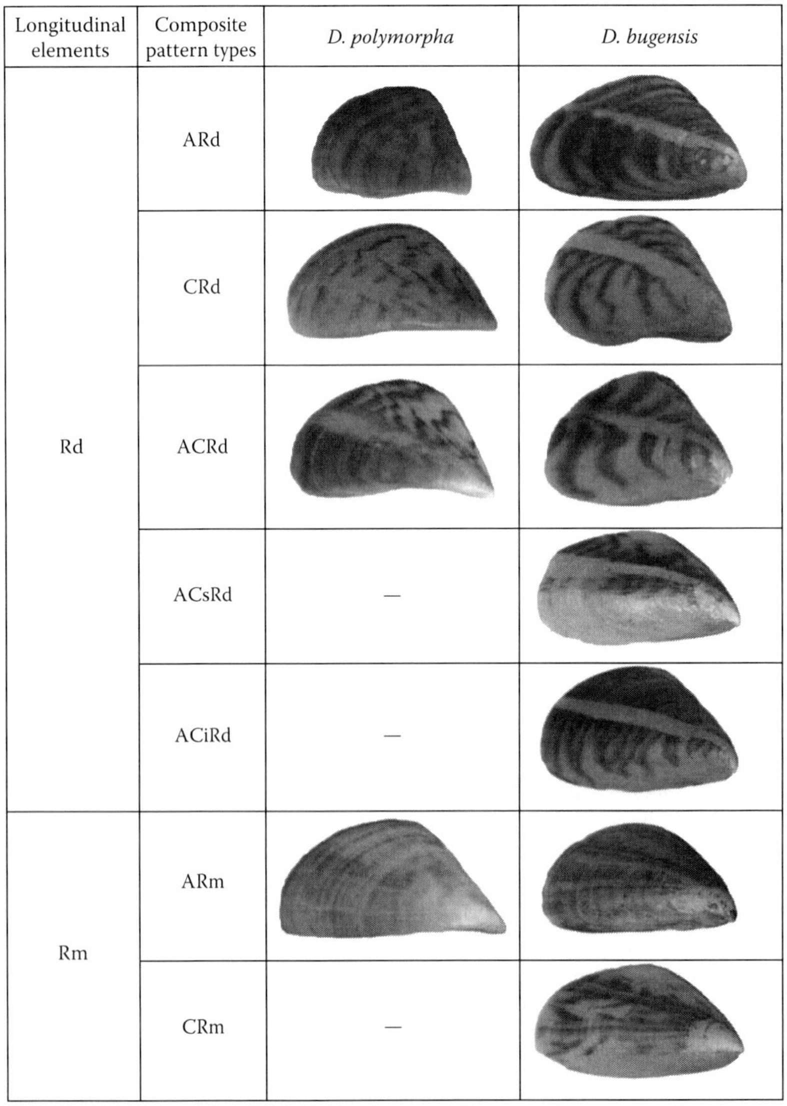

Longitudinal elements	Composite pattern types	*D. polymorpha*	*D. bugensis*
Rd	ARd		
	CRd		
	ACRd		
	ACsRd	—	
	ACiRd	—	
Rm	ARm		
	CRm	—	

Figure 19.7 Longitudinal elements and composite pattern types found in *D. polymorpha* and *D. r. bugensis*.

Morphological Variation: Between Water Bodies

Before we compare shell characteristics of both species across different water bodies, we note that *D. polymorpha* from the north region of the Caspian Sea was difficult to differentiate. With taxonomic guidelines provided by Logvinenko and Starobogatov (1968), we found individuals that fit the description of both *D. p. polymorpha* and *D. p. andrusovi* (Andrusov 1897), along with individuals considered intermediate between the two subspecies. Overall, a total of 254 individuals were collected from the north Caspian Sea, and 10 were identified as *D. p. polymorpha*, 73 as *D. p. andrusovi*, and 171 as intermediate. Shell dimensions of the former two taxa were analyzed separately.

There was a tendency for individuals of *D. p. polymorpha* to have low H/L and W/L ratios in water bodies both in the south (RR-1 supply channel, Chograi Reservoir, north Caspian Sea [*D. p. andrusovi*]) and north (Lake Plescheevo, the Kama Reservoir). Individuals with high values of these ratios were observed mainly in water bodies in the south (Volga–Ahtuba floodplain, north Caspian Sea [*D. p. polymorpha*], Lake Sharony, Danube River) (Figure 19.15). Interestingly, *D. polymorpha* from Lakes Michigan and Erie in the North American Great Lakes had ratios very similar to each other but very different from ratios of east European water bodies. In both lakes, W/L was relatively high but H/L was low compared to *D. polymorpha* from European water bodies.

Longitudinal elements	Composite pattern types	D. polymorpha	D. bugensis
Rm	ACRm	—	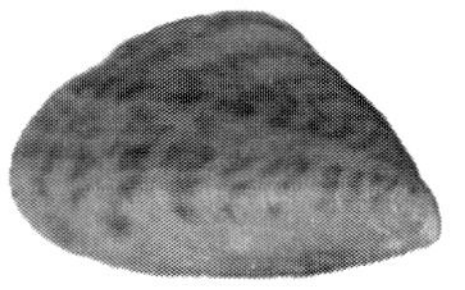
	ACsRm	—	
	ACiRm	—	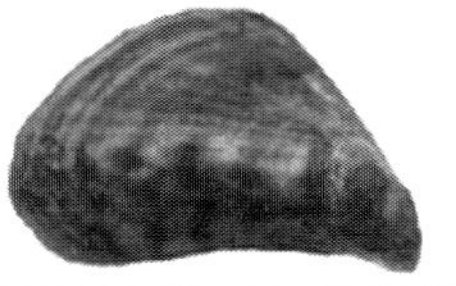
	AdRm	—	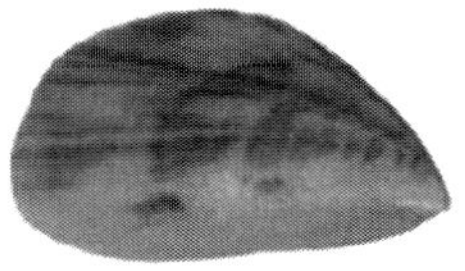
	ACi-dRm	—	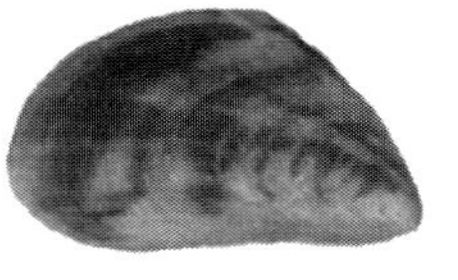
RdRm	ARdRm	—	
	CRdRm	—	

Figure 19.7 (continued)

(continued)

For *D. r. bugensis*, individuals with shell characteristics that were most different from all other water bodies were those from the north Caspian Sea and Lake Michigan (Figure 19.16). Individuals from the North Caspian had relatively high H/L and W/L ratios, whereas those from Lake Michigan had low ratios. Ratios found in these two water bodies were significantly different from all other water bodies (ANOVA, Scheffé test, p < 0.01). Individuals from Lake Michigan were all the "profunda" phenotype as originally described by Dermott and Munawar (1993) from Lake Erie (Table 19.3). Individuals from Lake Erie (3 m depth) were the shallow-water phenotype, and these had shell characteristics that were similar to all other water bodies except Lake Michigan and the north Caspian Sea. Individuals of *D. polymorpha* from deepwater sites in Lake Michigan (25 and 45 m) had shell characteristics that were not significantly different than those of individuals from the shallow site in Lake Erie (ANOVA, Scheffé test, p > 0.05). Thus, a deepwater variant was only evident for *D. r. bugensis*.

For both *D. polymorpha* and *D. r. bugensis*, the W/L ratio was more variable than the H/L ratio (Figure 19.17). As found in the Rybinsk Reservoir (Figure 19.13), variability in these ratios was greater for *D. r. bugensis* than for *D. polymorpha* in all other water bodies.

Shell Outline Variation

An analysis of shell images with SHAPE software indicated that overall shell variability and the specific region of variability were similar for *D. polymorpha* and *D. r. bugensis* (Table 19.4). Based on PC analysis, the greatest source of variability (PC1) was the H/L ratio, and this was confirmed by the strong correlation between this ratio and PC1 for both

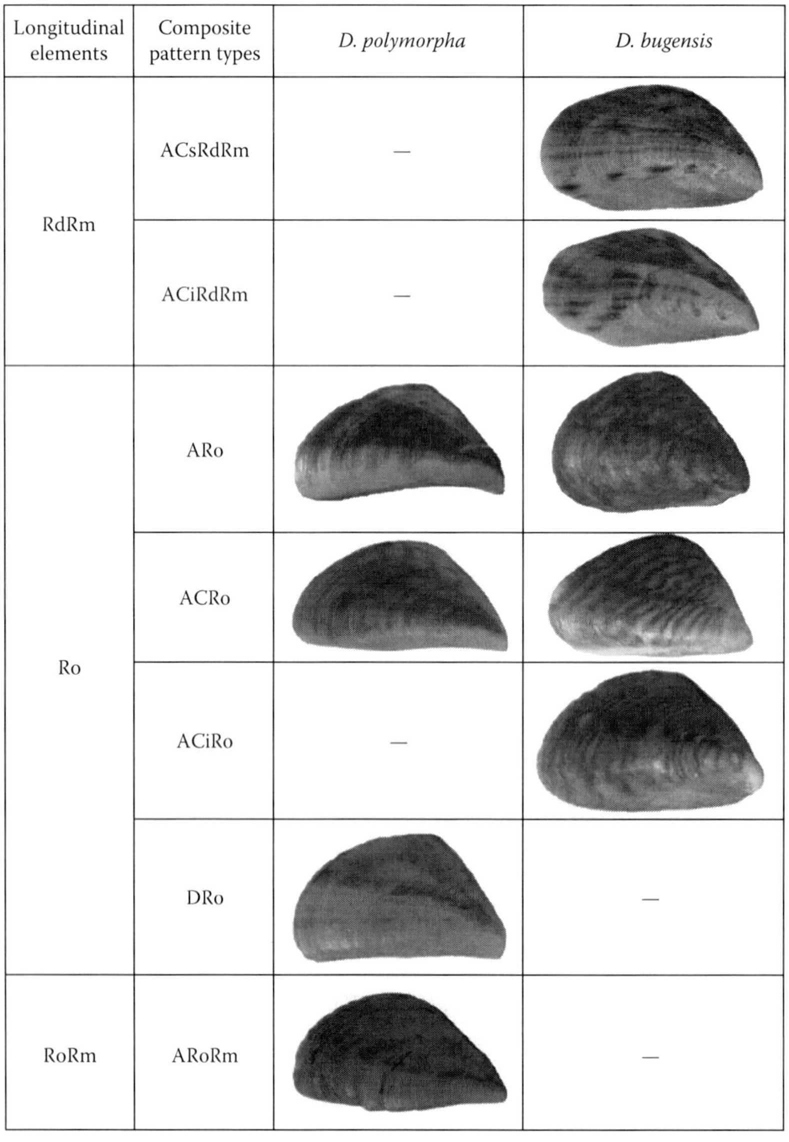

Longitudinal elements	Composite pattern types	*D. polymorpha*	*D. bugensis*
RdRm	ACsRdRm	—	[shell image]
	ACiRdRm	—	[shell image]
Ro	ARo	[shell image]	[shell image]
	ACRo	[shell image]	[shell image]
	ACiRo	—	[shell image]
	DRo	[shell image]	—
RoRm	ARoRm	[shell image]	—

Figure 19.7 (continued)

Longitudinal elements	Composite pattern types	*D. polymorpha*	*D. bugensis*
RdRmRo	ARdRmRo	—	
F	AFD		
	ACFD	—	
	AdFD	—	
	ARdFD	—	
	ARmFD	—	

Figure 19.7 (continued)

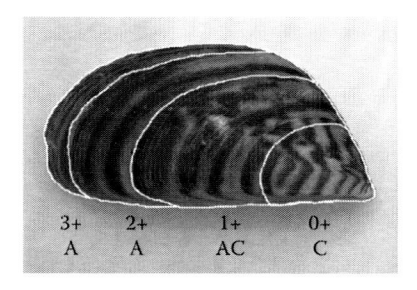

Figure 19.8 Example of a sequence of pattern types on a dreissenid shell (*D. polymorpha*, age 3+ individual). Sequence for this shell is C–AC–A–A.

species (*D. polymorpha*: r = 0.89, p < 0.001; *D. r. bugensis*: r = 0.97, p < 0.01). The second most important source of variability (PC2) was the shape of the ventral margin, followed by the length and shape of the superior margin (PC3). Variability in the superior margin was clearly more apparent for *D. r. bugensis* than for *D. polymorpha*. The fourth and weakest source of variation (PC4) was related to the shape of the connection between the posterior and ventral margins. In general, the diversity of shapes of *D. r. bugensis* was greater than those of *D. polymorpha* for all four first PCs (Figure 19.18).

The extreme diversity of shell shapes of *D. r. bugensis* relative to *D. polymorpha* was striking. Outlines varied from tear-shaped to triangle, and ventral margins varied from strongly convex to strongly concave. A scatterplot of individual scores for the first two PCs over all water bodies showed little overlap between the two species (Figure 19.19). Although some specimens of *D. polymorpha* from the Rybinsk, Gorky, and Perućica Reservoirs and Lake Forelevoe did overlap with some specimens of *D. r. bugensis* from the Volga–Ahtuba floodplain, Volga Delta, and Lake Sharony, variability of shell outlines did not lead to overlap in the same water body. Scatterplots of the four principal scores for each water body indicated some broad geographic patterns for both species (Figures 19.20 and 19.21). PC3 was significantly related to latitude (p < 0.05) for both *D. polymorpha* (R = 0.66) and *D. r. bugensis* (R = 0.84). A rounded superior

Table 19.2 Mean (±SE) Shell Dimensions of *D. polymorpha* and *D. r. bugensis* for All Water Bodies. Ranges Given in Parentheses. L, Shell Length (mm); H, Shell Height; W, Shell Width

Species/Ratio	*D. polymorpha* (N = 5059)	*D. r. bugensis* (N = 1968)
L	17.4 ± 0.08 (5.3–37.4)	19.3 ± 0.14 (5.3–36.5)
H/L	0.519 ± 0.0005 (0.373–0.746)	0.585 ± 0.0009 (0.452–0.767)
W/L	0.493 ± 0.0007 (0.326–0.693)	0.433 ± 0.0014 (0.261–0.714)
H/W	1.064 ± 0.0018 (0.707–1.672)	1.377 ± 0.0043 (0.878–2.031)

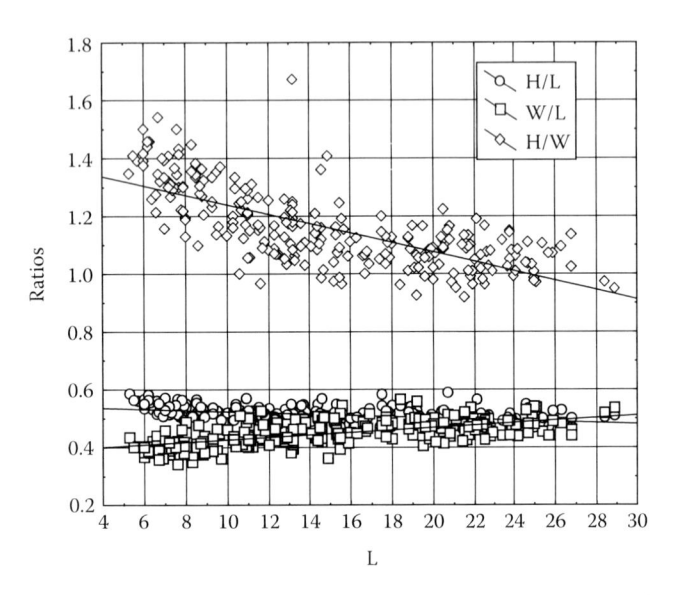

Figure 19.9 Correlations between shell length (L) and height/length (H/L), width/length (W/L), and height/width (H/W) ratios of *D. polymorpha*. Specimens collected from Peručica Reservoir. R-values for the ratios were −0.45, 0.61, and −0.72, respectively. All correlations were significant (p < 0.001).

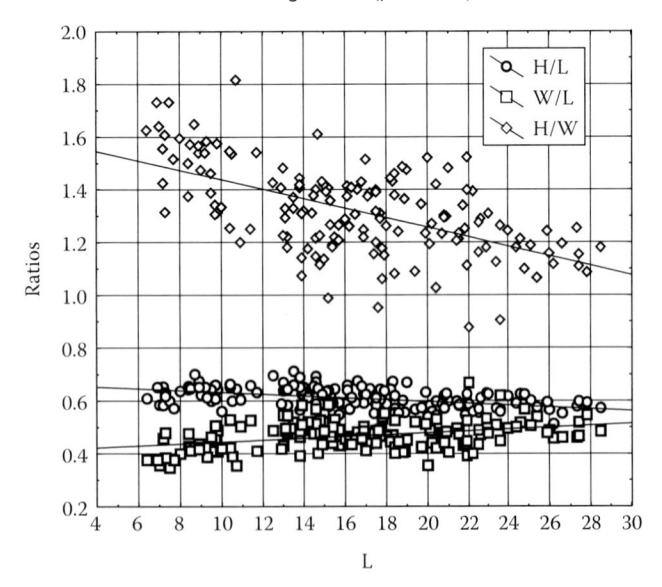

Figure 19.10 Correlations between shell length (L) and height/length (H/L), width/length (W/L), and height/width (H/W) ratios of *D. r. bugensis*. Specimens collected from the Ahtuba River. R-values for the ratios were −0.45, 0.36, and −0.58, respectively. All correlations were significant (p < 0.001).

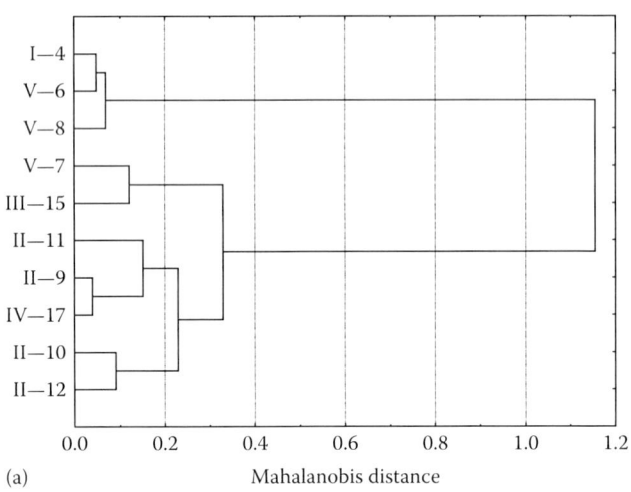

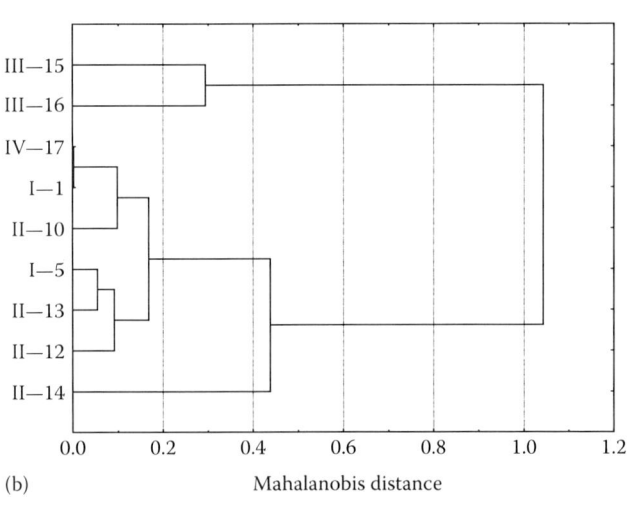

Figure 19.11 Dendrogram showing similarity of shell dimensions for *D. polymorpha* at different sites in Rybinsk Reservoir. Sites and definition of roman numerals are given in Figure 19.3. (a) Length group II; (b) length group III.

margin was typical for the majority of individuals of both species in northern water bodies, whereas a straight superior margin was typical for southern water bodies. Also, for water bodies just within the Volga Basin, there was a significant (p < 0.05) correlation between PC2 and latitude for both species (R = 0.86 for *D. polymorpha* and R = 0.79 for *D. r. bugensis*). As latitude of a water body increased from south to north, ventral margins of shells changed from concave to convex.

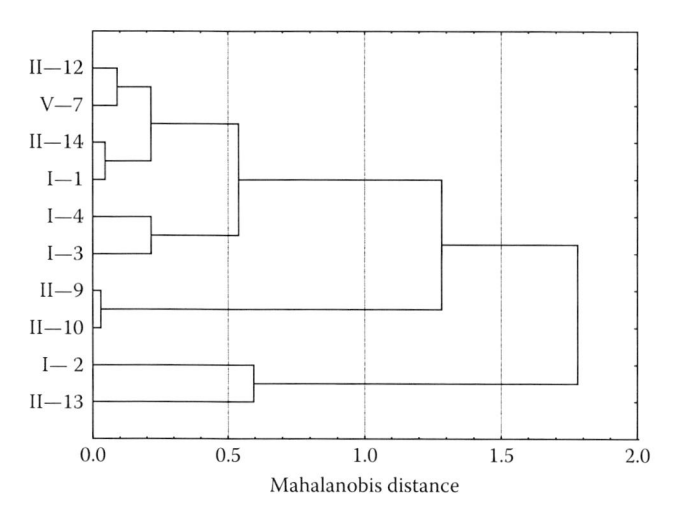

MORPHOLOGICAL VARIABILITY OF *DREISSENA POLYMORPHA*   299

Figure 19.12 Dendrogram showing similarity of shell dimensions for *D. r. bugensis* at different sites in Rybinsk Reservoir. Sites and definition of roman numerals are given in Figure 19.3.

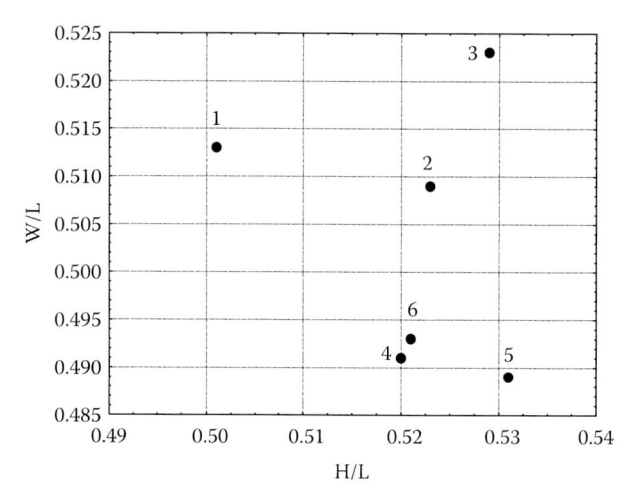

Figure 19.14 Scatterplot of relationship between mean width/length and height/length ratios for *D. polymorpha* (length group II) in Gorky Reservoir. Site designations given as in Figure 19.4.

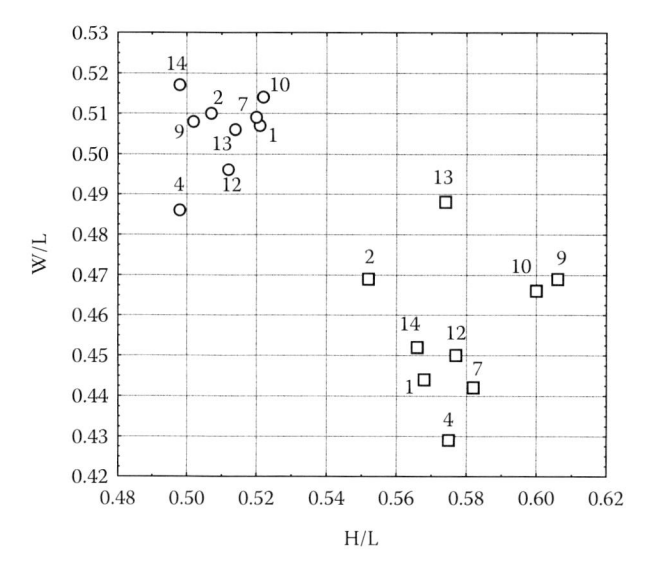

Figure 19.13 Scatterplot of relationship between mean width/length and height/length ratios for *D. polymorpha* and *D. r. bugensis* in Rybinsk Reservoir at sites where the two species were both present. Site designations as given in Figure 19.3. Circles, *D. polymorpha*; squares, *D. r. bugensis*.

Traditional, morphometric measurements combined with shape outlines from geometric morphometric analysis indicated great variability in shell dimensions across the different water bodies for *D. polymorpha* (Figure 19.22) and *D. r. bugensis* (Figure 19.23). Extreme differences in shell characteristics for *D. r. bugensis* in north Caspian Sea and Lake Michigan (profunda phenotype) were evident.

Shell Pigmentation

For all individuals, total amount of pigmentation, as reported in DL units, varied from 86.7 to 226.0 for *D. polymorpha* and from 112.7 to 228.2 for *D. r. bugensis*, and population means varied from 115.9 to 165.8 and from 137.2 to 190.7, respectively (Figure 19.24). Population ranges depended on the presence of light and dark individuals since patterned specimens were present in all populations. Overall, mean values of DL in *D. r. bugensis* were significantly greater than those in *D. polymorpha* in all water bodies (t-test, $p < 0.05$ in all cases). Lowest mean values for both species were found in the Rybinsk Reservoir, and highest mean values were found in the Chograi Reservoir for *D. polymorpha* and in Lake Michigan for *D. r. bugensis*. Across all water bodies, there was a significant correlation between DL and shell length in both *D. polymorpha* ($R = -0.35$ to -0.85; $p < 0.05$) and *D. r. bugensis* ($R = -0.21$ to -0.50; $p < 0.05$). In general, larger and older individuals had a lower DL. A negative correlation was apparent in 83.3% of samples for *D. polymorpha* and in 50% of samples for *D. r. bugensis*. Moreover, a significant correlation ($p < 0.01$) occurred between DL and latitude in both species; DL decreased with increased latitude (Figure 19.25).

A total of 16 pattern types were found for *D. polymorpha* and 35 pattern types were found for *D. r. bugensis* (Tables 19.5 and 19.6). For *D. polymorpha*, primary pattern types found in all populations were C, AC, Ad, and A. The other 12 types (ACs, ACi, D, ARd, CRd, ACRd, ARo, ACRo, DRo, ARm, ARoRm, AFD) were secondary. Among these secondary types, the most common were ACs and ACi, and the least common were AFD, ACRo, and DRo. The four primary types comprised between 73.8% and 100.0% of a population in a given water body and were the only types found in 6 of 26 populations. The greatest diversity of pattern types was observed in *D. p. andrusovi* from the north Caspian Sea (14 types). Primary pattern types for *D. r. bugensis* were A, ARm, and Ad (Table 19.6). Percent frequency of these types varied from 34.1% to 96.6% in individual populations. Other

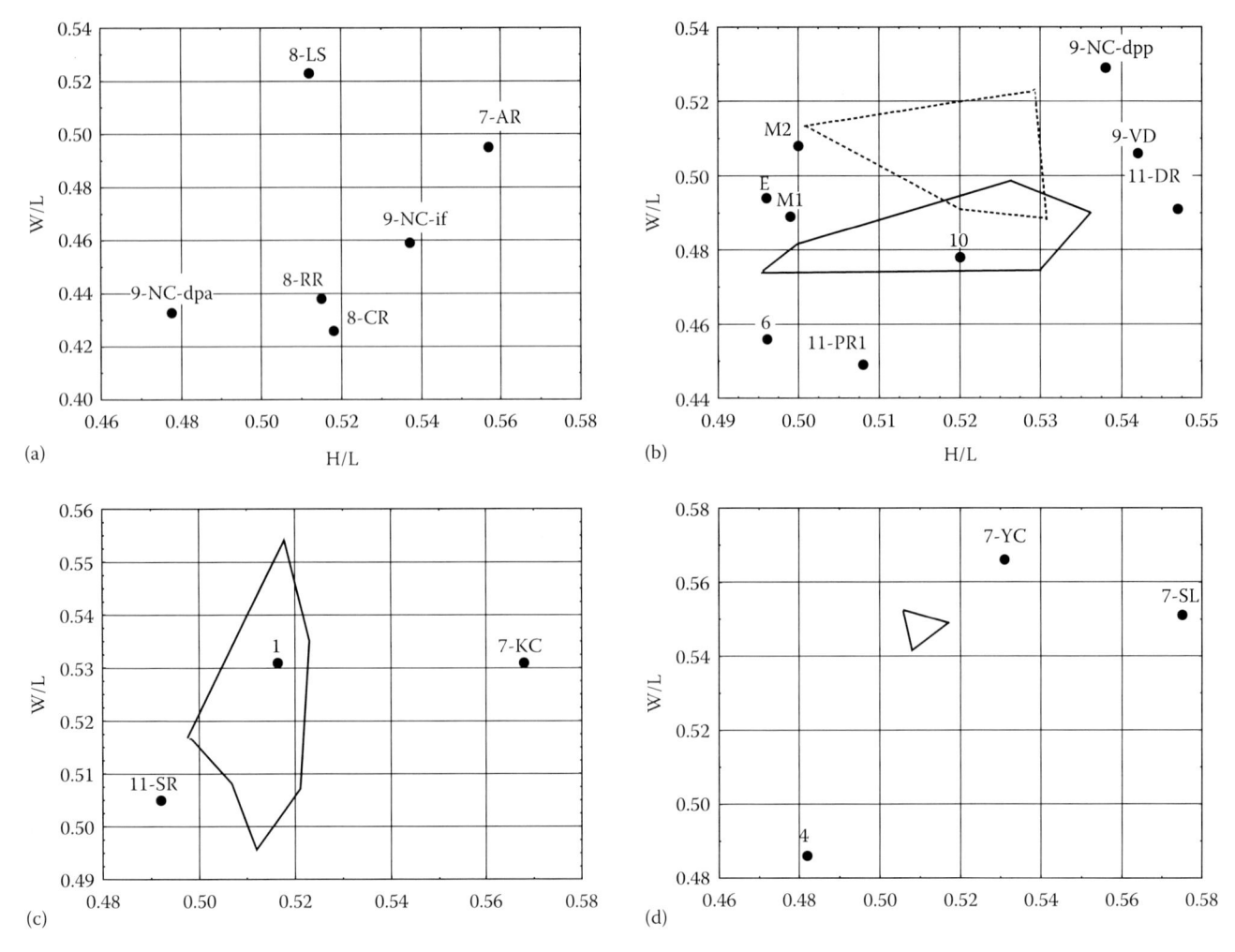

Figure 19.15 Scatterplots of the relationship between mean width/length and height/length ratios for *D. polymorpha* in various water bodies. Designations of European water bodies are given in Figure 19.1; one exception was for the north Caspian Sea in which 9-NC-dpa, *D. p. andrusovi*; 9-NC-dpp, *D. p. polymorpha*; and 9-NC-if, intermediate forms. For sites in the North American Great Lakes, designations are M1, 25 m in Lake Michigan; M2, 45 m in Lake Michigan; and E, 3 m in Lake Erie. Solid line bounds area for populations in Rybinsk Reservoir, and dashed line bounds area for the populations in Gorky Reservoir. (a) Length group I; (b) length group II; (c) length group III; (d) length group IV.

common types were ACi, ARd, and ARdRm, while the least common were ACs-d, ACsRd, CRm, ACi-dRm, CRdRm, ACsRdRm, ACRo, ARdRmRo, AdFD, and ARmFD. Fourteen pattern types were common to both species.

For a given population, a greater number of pattern types occurred in *D. r. bugensis* compared to *D. polymorpha* (12.9 types per population compared to 6.3 types per population). As might be expected given this result, average similarity between populations of *D. polymorpha* was greater than the average similarity between populations of *D. r. bugensis* (r_{Zhiv} of 0.892 for *D. polymorpha* and 0.787 *D. r. bugensis*). Generally, lower values of similarity were found between populations that were most distant from each other, whereas higher values were found between populations situated close to each other. There was a significant correlation (R = 0.53 for *D. polymorpha* and R = 0.29 for *D. r. bugensis*; p < 0.05) between r_{Zhiv} and geographic distance; similarity decreased as distance between populations increased.

There was a relationship between presence of some pattern types and extent of pigmentation (DL). Low pigmentation was associated with elements Rm, Rd, and F. The greatest total frequency of pattern types containing these three elements (47.2%) occurred in *D. r. bugensis* from Lake Michigan; the population in this water body had one of the highest DL values (189.6).

The greatest diversity of pattern types for both species was observed at age 0+ and the lowest diversity at age 3+. For *D. polymorpha*, 15 pattern types were found in individuals at age 0+, 14 at age 1+, 11 at age 2+, and 6 at age 3+. For *D. r. bugensis*, 31, 29, 24, and 10 pattern types were observed for the same ages, respectively. Some pattern types were associated with specific ages. For instance, type C

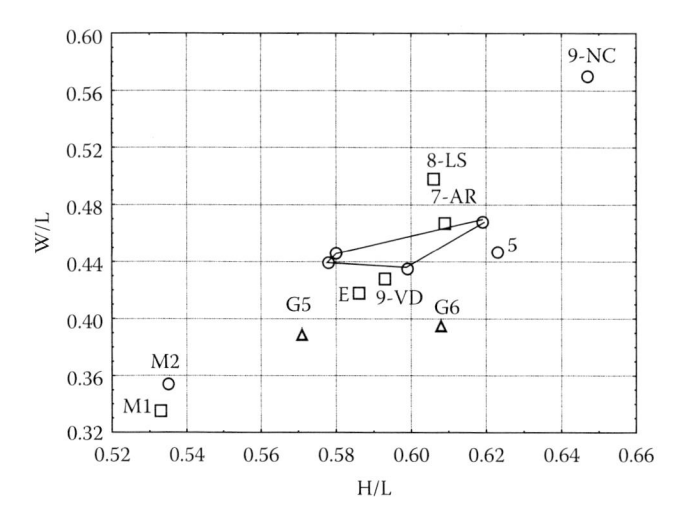

Figure 19.16 Scatterplot of the relationship between mean width/length and height/length ratios for *D. r. bugensis* in various water bodies. Designations of European water bodies are given in Figure 19.1. For sites in Gorky Reservoir, designations are G5 and G6 as defined in Figure 19.4. For sites in North American Great Lakes, designations are as follows: M1, 25 m in Lake Michigan; M2, 45 m in Lake Michigan; E, 3 m in Lake Erie. Solid line bounds area for the populations in Rybinsk Reservoir. Triangles, length group I; squares, length group II; circles, length group III.

Table 19.3 Mean (±SE) Shell Dimensions of *D. r. bugensis* Profunda in Lake Michigan and Lake Erie. Ranges Given in Parentheses. L, Shell Length (mm); H, Shell Height (mm); W, Shell Width (mm). n = 172 for Lake Michigan and n = 15 for Lake Erie

Dimension	Lake Michigan	Lake Erie
L	17.9 ± 0.30 (10.1–28.4)	12.69 ± 0.297
H/L	0.534 ± 0.0022 (0.463–0.607)	0.54 ± 0.008
W/L	0.343 ± 0.0027 (0.261–0.448)	0.33 ± 0.004
H/W	1.569 ± 0.0112 (1.214–2.031)	1.64 ± 0.012

Source: Lake Erie data are taken from Dermott, R. and Munawar, M., *Can. J. Fish. Aquat. Sci.*, 50, 2298, 1993.

was typically found in early ages of *D. polymorpha* for all populations (Figure 19.26). The exception was the population from the Northern Dvina in which type C accounted for 76.3% of pattern types at age 1+. For *D. polymorpha*, frequency of pattern type A increased with age in most populations. Mean frequency of this pattern type did not exceed 13.3% in age 0+ individuals except in the Lake Erie population where it was 40%. Mean frequency of pattern type A in age 3+ individuals was 80%. Frequencies of AC and Ad types were not as strongly related to age as other pattern types. Typical examples of how pattern types changed with age in *D. polymorpha* are given for Rybinsk Reservoir and for Northern Dvina (Figure 19.27). For *D. r. bugensis*, the number of pattern types also decreased with age. Five pattern types were only associated with age 0+ individuals (ARo, ACRo, ACiRo, ARdRmRo, CRdRm), and six

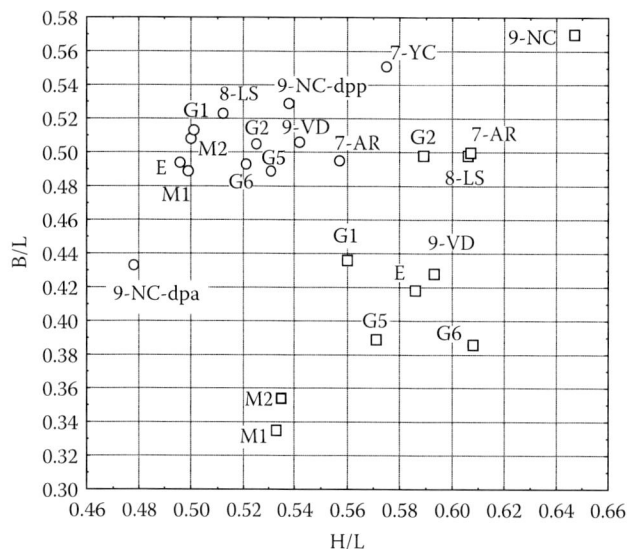

Figure 19.17 Scatterplot of the relationship between mean width/length and height/length ratios for *D. polymorpha* and *D. r. bugensis* in various water bodies (sites where two species presented) excepting those from the Rybinsk Reservoir. Designations of European water bodies are given in Figure 19.1; one exception was for the north Caspian Sea in which 9-NC-dpa, *D. p. andrusovi* and 9-NC-dpp, *D. p. polymorpha*. For sites in the Gorky Reservoir, designations are given in Figure 19.4. For sites in North American Great Lakes, designations are as follows: M1, 25 m in Lake Michigan; M2, 45 m in Lake Michigan; E, 3 m in Lake Erie. Circles, *D. polymorpha*; squares, *D. r. bugensis*.

pattern types only associated with age 0+ and age 1+ individuals (ACs-d, ACRm, AdRm, ACFD, AdFD, ARmFD). Pattern types A and Ad became more prevalent with age (Figure 19.28). Examples of pattern types at different ages for *D. r. bugensis* are given for populations in the Rybinsk Reservoir and in Lake Michigan (Figure 19.29).

Similarity between populations at different ages was lower for *D. r. bugensis* compared to *D. polymorpha*. Values of r_{Zhiv} at ages 0+ to 3+ decreased from 0.736 to 0.675 for *D. r. bugensis* and from 0.856 to 0.730 for *D. polymorpha*. Examination of similarity in pattern types between water bodies revealed some differences. For example, pattern types of zebra mussels in Lake Plescheevo differed considerably from zebra mussels in the Upper Volga River (i.e., Rybinsk and Gorky Reservoirs) even though these water bodies were geographically close. Moreover, *D. polymorpha* in Lake Plescheevo originated from upper Volga River populations (Zhgareva 1992). The population in Lake Plescheevo was distinguished by a very low frequency of pattern type C at age 0+ and a high frequency of type A at ages from 0+ to 2+. Further, the population in the Northern Dvina River was unique in that it exhibited a prevalence of type C at age 0+ and at age 1+ (85.7% and 76.3%, respectively). Most other populations were characterized

Table 19.4 Images of Shell Variability as Defined by Principal Component Analysis (PC). Variability Provided as the Mean ± 2 SD. Also Given Is the Percent Each PC Score Contributed to Total Variability for *D. polymorpha* and *D. r. bugensis*

Principal Component (PC)	*D. polymorpha*	*D. r. bugensis*	Loading (%), *D. polymorpha*/*D. r. bugensis*
PC1			48.3/55.5
PC2			15.3/15.0
PC3			10.1/12.4
PC4			6.0/3.3

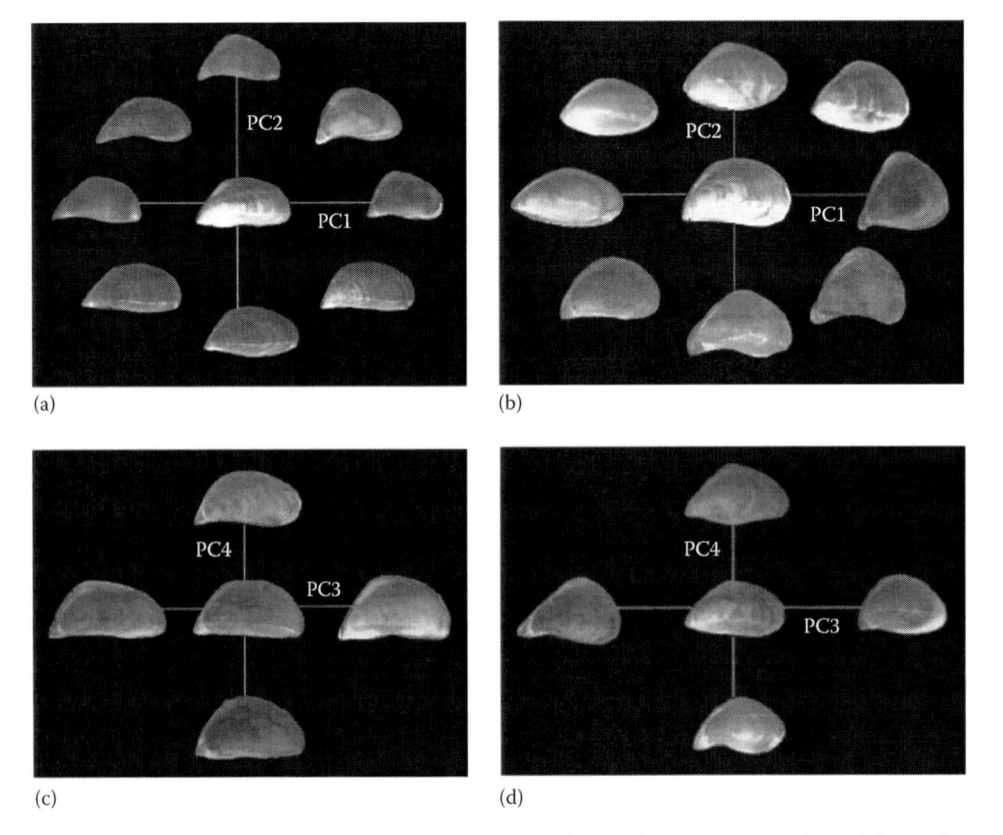

Figure 19.18 Shell images relative to four PCs as defined by PC analysis. (a and c) *D. polymorpha*; (b and d) *D. r. bugensis*.

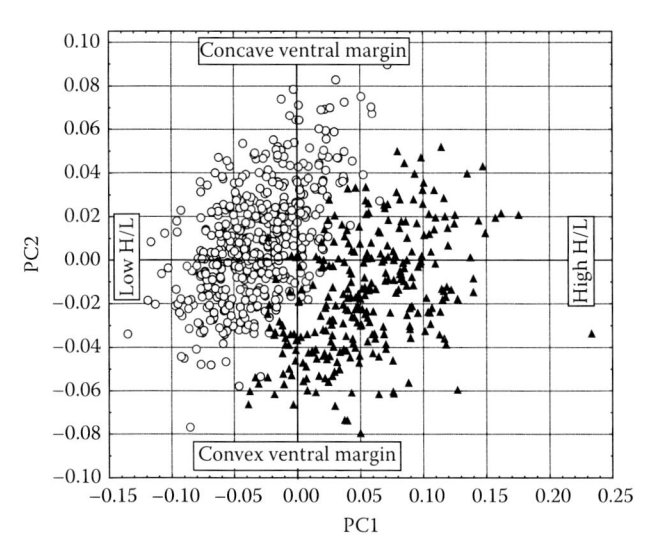

Figure 19.19 Scatterplot of individual scores from PC analysis for *D. polymorpha* (open circle) and *D. r. bugensis* (closed triangle) over all water bodies.

by high frequencies of type A at early ages (north Caspian Sea, RR-1 supply channel, Chograi and Kama Reservoirs, and Lakes Erie and Michigan). In addition, these populations had a high frequency of secondary pattern types (not less than 10% in total). For *D. r. bugensis*, most differences between populations were apparent only in the secondary pattern types of ages 0+ and age 1+. Hence, there was no consistent geographic difference in age-related pattern types in this species.

Since pattern types change during mussel ontogenesis, it was necessary to investigate not only frequencies of pattern types at different ages but also the succession of this change as individuals matured. Analysis of ontogenetic sequences included defining primary and secondary pattern types in different populations. The sequence for individuals age 0+ was represented by only one pattern type (C), age 1+ by two types (C–AC), age 2+ by three types (C–AC–Ad), and age 3+ by four types (C–AC–Ad–A). Types for adjacent ages may be either the same or different (e.g., AC–AC–A–A). Frequency of specimens with either a constant pattern or a pattern that changed varied between populations. When the two species were compared, changes in pattern types with increased age were more frequent in *D. polymorpha* than *D. r. bugensis*. On average, the percentage of individuals with a change in pattern type for *D. polymorpha* and *D. r. bugensis*, respectively, was 83.4% and 70.0% from age 0+ to age 1+, 67.5% and 41.8% from age 1+ to age 2+, and 34.1% and 23.4% from age 2+ to age 3+. Usually during ontogenesis, the pattern changed from zigzag to arched; thus, pattern lines became straighter as age increased. In the sequence of changes of pattern types, this was encoded as C–AC–Ad–A. While some stages in this sequence were absent and others were repeated, the order of types generally changed from zigzag to arched. A reverse order of this sequence was extremely rare and was considered an anomaly. Pattern type A may be considered as a "final" pattern because this pattern did not change after it was observed. In some individuals, this pattern type was observed as early as age 0+.

Aforementioned pattern changes during ontogenesis were typical for *D. polymorpha*. For *D. r. bugensis*, pattern changes were more complex because of the presence of numerous longitudinal elements that tend to be preserved over several ages. This was the reason for a more pronounced consistency of pattern sequences in *D. r. bugensis* than in *D. polymorpha*. Sequences like ARm–ARm–ARm and ARdRm–ARdRm–ARdRm were frequent. The pattern type A was also "final" for *D. r. bugensis*.

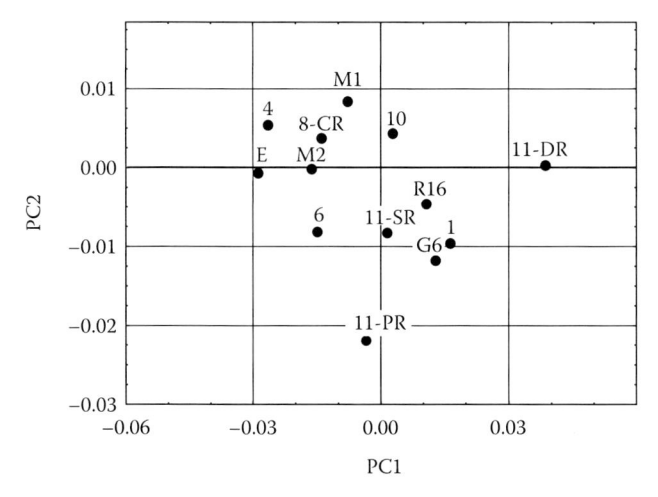

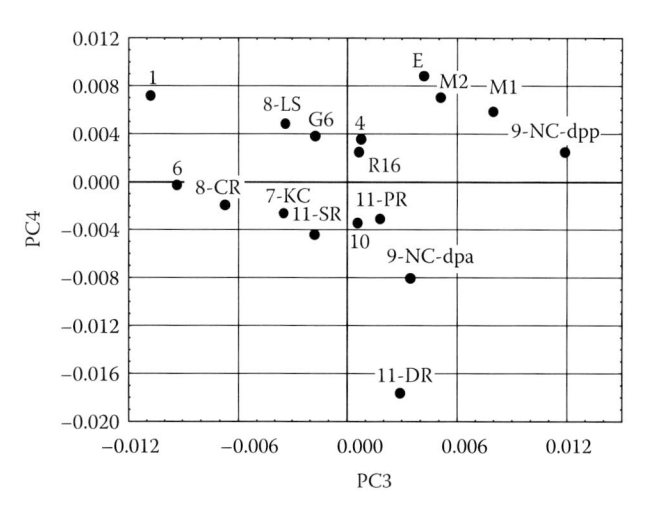

Figure 19.20 Scatterplots of score means from PC analysis for populations of *D. polymorpha* in different water bodies. Designations of European water bodies are given in Figure 19.1; exceptions were for the north Caspian Sea in which 9-NC-dpa, *D. p. andrusovi* and 9-NC-dpp, *D. p. polymorpha* and for sites in the Rybinsk and the Gorky reservoirs, designations are R16 and G6 as defined in Figures 19.3 and 19.4, correspondingly. For sites in North American Great Lakes, designations are M1, 25 m in Lake Michigan; M2, 45 m in Lake Michigan; and E, 3 m in Lake Erie.

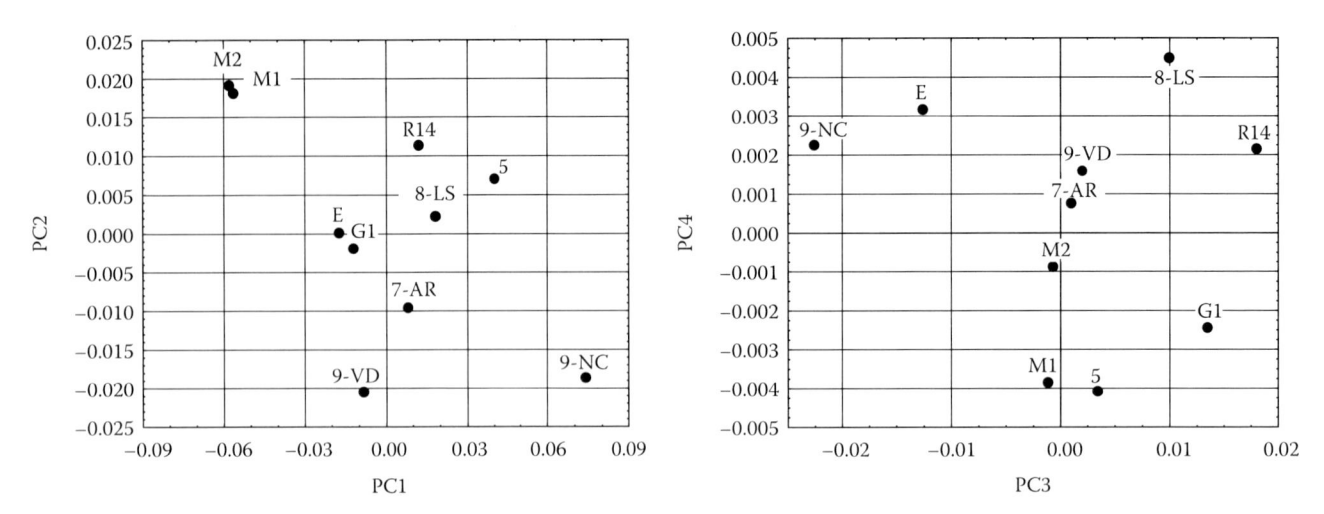

Figure 19.21 Scatterplots of score means from PC analysis for populations of *D. r. bugensis* in various water bodies. Designations of European water bodies are given in Figure 19.1. For sites in the Rybinsk and the Gorky reservoirs, designations are R14 and G1 as defined in Figures 19.3 and 19.4, correspondingly. For sites in North American Great Lakes, designations are M1, 25 m in Lake Michigan; M2, 45 m in Lake Michigan; and E, 3 m in Lake Erie.

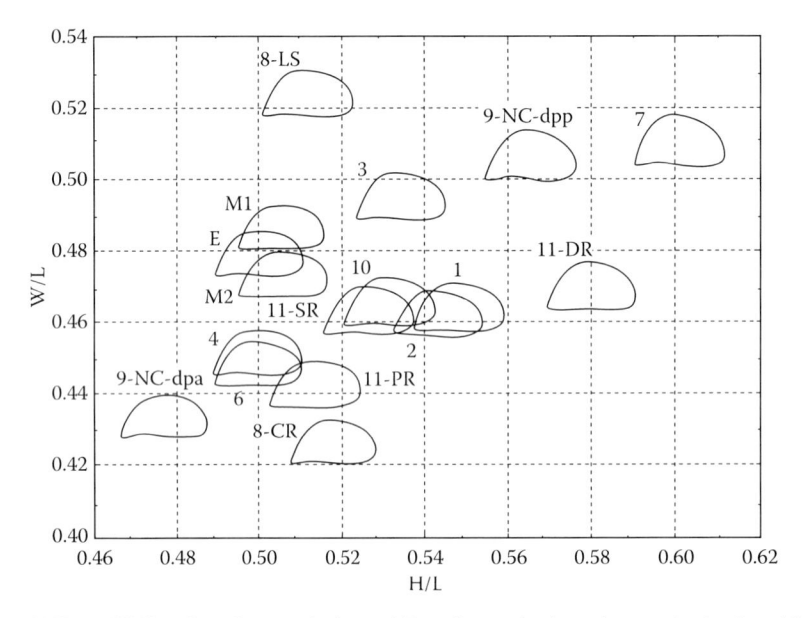

Figure 19.22 Scatterplot of H/L and W/L values for populations of *D. polymorpha* in various water bodies. Mean outlines were defined with SHAPE software. Designations of European water bodies are given in Figure 19.1; one exception was for the north Caspian Sea in which 9-NC-dpa, *D. p. andrusovi* and 9-NC-dpp, *D. p. polymorpha*. For sites in North American Great Lakes, designations are M1, 25 m in Lake Michigan; M2, 45 m in Lake Michigan; and E, 3 m in Lake Erie.

Diversity of pattern types in a population determined the variety of sequences. For example, the greatest variety of sequences (25 at age 1+) was found in a population of *D. p. andrusovi* that also had the greatest number of pattern types. In populations of *D. polymorpha*, the number of sequences varied from 6 to 13, whereas the number of sequences in populations of *D. r. bugensis* varied from 6 to 20. Some sequences in a given population were frequent, while others were rare. A given sequence was mostly determined by the number of secondary rows (Figures 19.30 and 19.31).

While each population possessed a unique set of sequences that repeated, there were some similarities between populations. Populations in the same water body and in water bodies that were connected (such as the Rybinsk and Gorky Reservoirs) differed only slightly in pattern sequences, and differences occurred only in secondary rows. Populations from far-removed water bodies differed to a greater extent, and differences usually occurred in the main sequences. However, in some cases, sequences were very similar even in distant populations as found in the Rybinsk, Gorky and Kama Reservoirs, Lake Forelevoe, and Serbian water bodies (Table 19.7).

Geographic variability in pattern sequences was most apparent from age 0+ to age 1+. For *D. polymorpha*, the sequence C–AC in age 0+ and age 1+ was typical of the

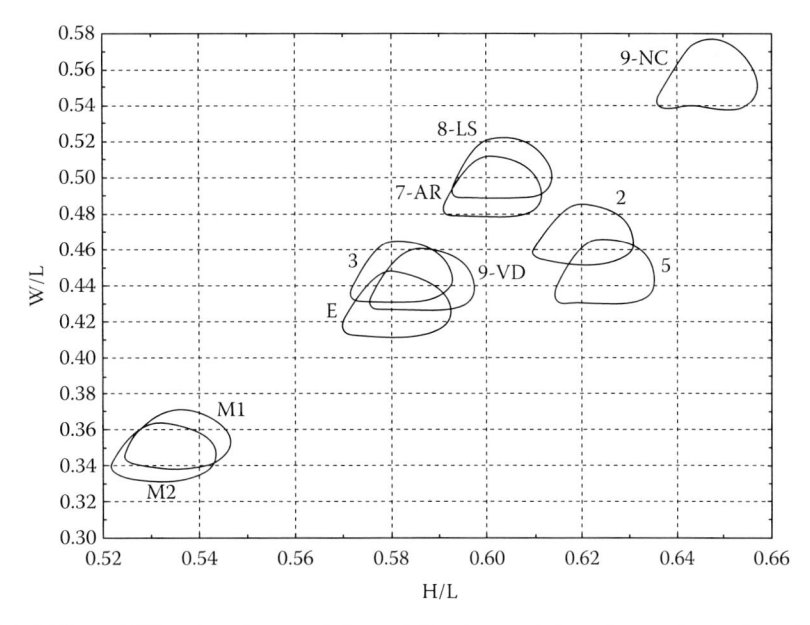

Figure 19.23 Scatterplot of H/L and W/L values for populations of *D. r. bugensis* in various water bodies. Mean outlines were defined with SHAPE software. Designations of European water bodies are given in Figure 19.1. For sites in North American Great Lakes, designations are as follows: M1, 25 m in Lake Michigan; M2, 45 m in Lake Michigan; E, 3 m in Lake Erie.

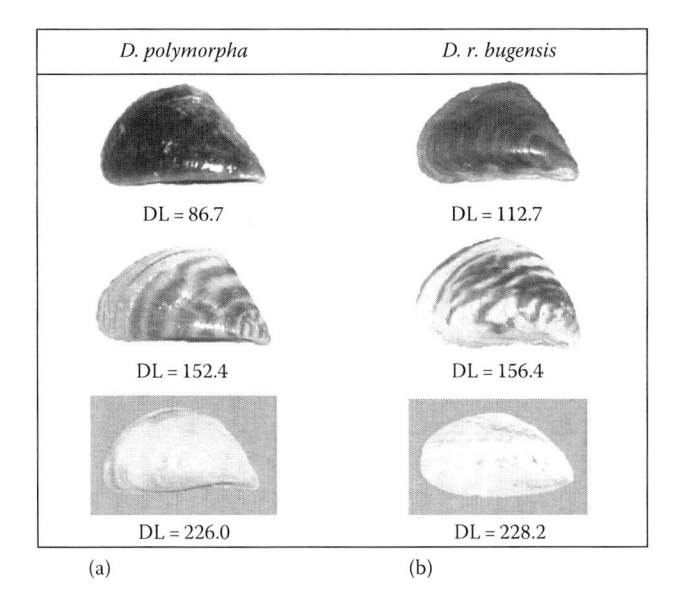

Figure 19.24 Range of pigmentation associated with shells of (a) *D. polymorpha* and (b) *D. r. bugensis* as defined by DL (Density Lum).

sequence found in populations in the northern portion of the Volga Basin, while C–A was typical of populations in the southern portion. The most north of all populations was in the Northern Dvina River, and the typical sequence for this population was C–C. Thus, a rapid shift to straighter pattern lines was observed for south populations (from C to A by age 1+), and the occurrence of straighter pattern lines was gradual from south to north (i.e., AC and Ad pattern types in north populations were still found at ages 1+ and 2+). In the most north population, the trend toward straighter lines was observed only at age 2+ and older. Such geographic tendencies of pattern sequences were not apparent for *D. r. bugensis*.

DISCUSSION

Both *D. polymorpha* and *D. r. bugensis* are epifaunal species that attach to hard substrates with byssal threads. Because of these similar modes of life, the two species have similar morphological features, such as mytiloid shell form, shell allometry, general variability in shell outline, and greater range in width/length than in height/length. Our studies showed that in both species shell width increased more rapidly than shell height as size increased. These results are consistent with earlier studies in various water bodies (Spasskiy 1948, Ovchinnikov 1954, Antonov and Bachurina 1986, Claxton et al. 1998), and similar growth patterns have been reported in other epifaunal bivalves besides *Dreissena* (Seed 1968). Further, in both dreissenid species, the height/length ratio and ventral margin shape were the greatest determinants of shell outlines. These same factors were found to affect shell outlines in other epifaunal bivalves such as *Brachidontes* (Aguirre et al. 2006) and *Mytilus* (Innes and Bates 1999, Krapivka et al. 2007). Apparently, the adaptation of mussels to an epifaunal mode of life with byssal attachment results in alterations of shells in similar ways that probably promote strong attachment to the substrate.

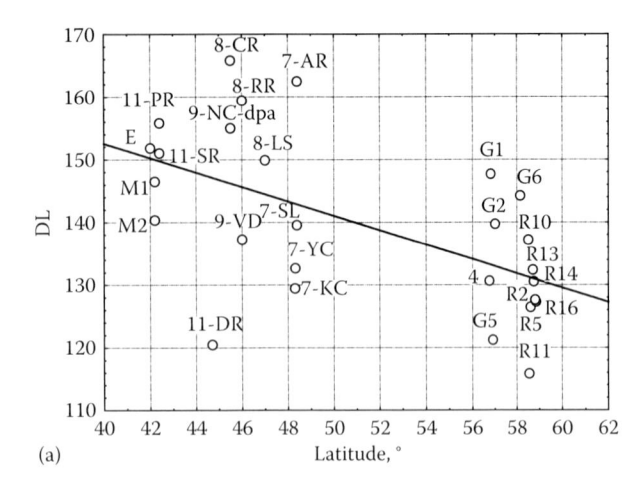

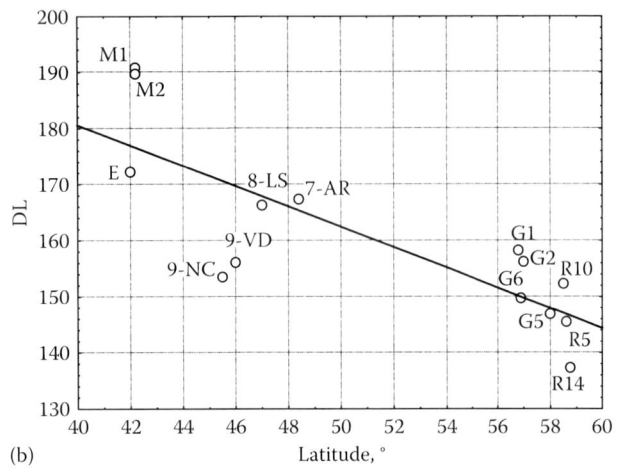

Figure 19.25 Relationship between shell pigmentation (as defined by DL) and latitude in (a) *D. polymorpha* and (b) *D. r. bugensis*. R, −0.60 for *D. polymorpha* (p < 0.05) and R, −0.75 for *D. r. bugensis* (p < 0.05). Designations of European water bodies are given in Figure 19.1. For site in north Caspian Sea, designation is 9-NC-dpa—*D. p. andrusovi*. For sites in the Rybinsk and the Gorky reservoirs, designations are given in Figures 19.3 and 19.4, correspondingly. For sites in the North American Great Lakes, designations are M1, 25 m in Lake Michigan; M2, 45 m in Lake Michigan; and E, 3 m in Lake Erie.

Table 19.5 Pattern Types of *D. polymorpha* Shells. n = 26 Samples from 16 Water Bodies

Pattern Type	Samples with Pattern (% of Total)	Range of Individuals Within a Sample with Pattern Type (%)
C	26 (100.00)	5.5–58.5
AC	26 (100.00)	5.2–50.0
Ad	26 (100.00)	0.5–37.6
A	26 (100.00)	0.5–67.1
ACs	16 (61.54)	0.2–5.6
ACi	13 (50.00)	0.4–5.2
D	8 (30.77)	0.5–9.1
AFD	1 (3.85)	1.3
ARd	4 (15.38)	0.5–2.5
CRd	5 (19.23)	0.3–2.4
ACRd	2 (7.69)	0.5–0.9
ARm	4 (15.38)	0.6–9.4
ACRo	1 (3.85)	0.2
DRo	1 (3.85)	2.1
ARo	3 (11.54)	1.7–1.9
ARoRm	3 (11.54)	0.2–3.5

Observed morphologic differences in *D. polymorpha* and *D. r. bugensis* were mostly related to shell proportions. *D. r. bugensis* has a shell that is higher and less convex than *D. polymorpha*. For *D. r. bugensis*, shell height is prominently greater than the width, while for *D. polymorpha* shell height and width are nearly equal (Table 19.2). Moreover, for *D. r. bugensis* the widest place of the shell in cross section is centrally located, but for *D. polymorpha* it is shifted to the ventral surface. We suggest that these differences reflect a degree of specialization in the two species. *D. polymorpha* is more specialized to an epifaunal mode of life; it possesses a low, wide shell with a flat ventral surface and prominent carina that serves as an adaptation to active hydrodynamic conditions (waves, currents). Shells of *D. r. bugensis* are less specialized to such conditions (lack of carina, greater height).

Lower specialization of *D. r. bugensis* is further confirmed by presence of a profunda phenotype, which can survive unattached on soft substrate and may have at least a partly infaunal mode of life (Dermott and Munawar 1993) that was initially typical for bivalves (Stanley 1970). In general, species with low specialization exhibit greater phenotypic diversity than highly specialized species (Markov and Naimark 1998), and our results are consistent with this finding. *D. r. bugensis* possessed greater variability in shell proportions, shell outlines, and pigmentation patterns than *D. polymorpha*. Overall, *D. r. bugensis* appeared to be more "polymorphic" than *D. polymorpha*. Claxton et al. (1998) came to a similar conclusion based on presence of a deepwater phenotype of *D. r. bugensis* in Lake Erie and lack of a similar phenotype for *D. polymorpha*. Yet it is interesting that molecular–genetic investigations lead to an opposite conclusion. Heterozygosity based on allozyme loci ranged between 0.15 and 0.49 for *D. polymorpha* (Hebert et al. 1989, Boileau and Hebert 1993, Marsden et al. 1995, Andreeva et al. 2001, Soroka 2002, Timmer and Shelley 2005) but ranged between 0.097 and 0.22 for *D. r. bugensis* (May and Marsden 1992, Spidle et al. 1994, Timmer and Shelley 2005). In addition, heterozygosity based on microsatellite DNA showed similar patterns: 0.50–0.94 for *D. polymorpha* (Müller et al. 2002, Astanei et al. 2005, Brown and Stepien 2010) and 0.41–0.88 for *D. r. bugensis* (Wilson et al. 1999, Therriault et al. 2005, Brown and Stepien 2010).

Shell pigmentation in the two species showed a great variety of patterns, and at least some research indicates that the type of pattern may be indicative of phenotypes differentially adapted to environmental conditions. For example, Protasov et al. (1997) found that individuals of *D. polymorpha* with an arched pattern were more resistant to hypoxia than individuals

Table 19.6 Pattern Types of *D. r. bugensis* Shells. n = 16 Samples from 8 Water Bodies

Pattern Type	Samples with Pattern Type (% of Total)	Range of Individuals Within a Sample with Pattern Type (%)
A	16 (100.00)	6.9–83.1
C	12 (75.00)	1.4–5.7
AC	8 (50.00)	0.8–5.7
ACs	6 (37.50)	0.6–1.5
ACi	14 (87.50)	1.6–15.6
Ad	15 (93.75)	3.1–29.2
ACs-d	1 (6.25)	1.2
ACi-d	8 (50.00)	0.8–16.1
ACs-i-d	2 (12.50)	0.6–1.0
D	7 (43.75)	0.6–7.4
ARd	14 (87.50)	1.0–18.3
CRd	7 (43.75)	0.5–3.9
ACRd	6 (37.50)	0.3–2.8
ACsRd	1 (6.25)	2.6
ACiRd	11 (68.75)	0.7–13.8
ARm	16 (100.00)	2.3–27.8
CRm	1 (6.25)	1.3
ACRm	3 (18.75)	0.3–1.1
ACsRm	4 (25.00)	0.6–1.9
ACiRm	8 (50.00)	0.5–4.7
AdRm	2 (12.50)	0.7–1.6
ACi-dRm	1 (6.25)	1.3
ARdRm	13 (81.25)	0.7–15.5
CRdRm	1 (6.25)	1.4
ACsRdRm	1 (6.25)	1.9
ACiRdRm	4 (25.00)	0.5–2.5
ARo	4 (25.00)	0.3–2.9
ACRo	1 (6.25)	1.0
ACiRo	3 (18.75)	0.8–2.9
ARdRmRo	1 (6.25)	0.7
AFD	10 (62.50)	0.6–3.5
ACFD	2 (12.50)	1.0–2.4
AdFD	1 (6.25)	0.6
ARdFD	2 (12.50)	1.9–2.1
ARmFD	1 (6.25)	0.6

with a zigzag pattern. In this study, we found that patterns were age related, so it is unclear if separate patterns, or pattern sequences as a whole, are indicative of specific environmental adaptations as individuals grow. If true, then it may be assumed that ontogenesis results in high adaptability to environmental factors. In this case, ontogenetic sequences reflect a great capacity for dreissenid populations to adapt to conditions at a specific age. As a result, populations possess adaptive potentials to a wide range of conditions, even if conditions change. Further, if this conclusion is correct, even partly, it may help explain apparent high adaptability of both dreissenid species and thus their success as invasive species. If sequences reflect adaptations overall, then a similar conclusion can be reached since sequence types are large in every population.

Formation of certain shell forms can occur as a result of environmental influences, and the variability in shell morphology across the different water bodies in this study can best be explained by such environmental factors as water velocity, water depth, and salinity. The influence of current velocity on dreissenid shell morphology was evident by the comparison of mussels from the Rybinsk and Gorky Reservoirs. The rate of water flow through Rybinsk Reservoir is regulated, and thus bottom currents do not exceed 0.15 m/s, even during spring floods (Butorin 1969). In contrast, water flow rates in Gorky Reservoir are semi-regulated. Flow conditions in the upper reach of this reservoir are similar to riverine with current velocities as high as 0.7 m/s, while velocities in the lower reach are only about 0.04 m/s (Butorin 1969). Overall, average current velocity is higher in Gorky Reservoir than in Rybinsk Reservoir, and *D. polymorpha* from the former reservoir had more convex shells than those from the latter. However, shell widths of individuals from the lower part of Gorky Reservoir were similar to shell widths of individuals from Rybinsk Reservoir. This similarity can most likely be explained by low currents found in the two habitats. High current velocities lead to wider shells in *D. polymorpha* (Ovchinnikov 1954, Karpevich 1955, Antonov 1983, Lajtner et al. 2004). Presumably, such a shell adaptation provides decreased hydrodynamic drag. Wider shells are also typical for other mussels in similar habitats, for example, in tidelands (*Crenomytilus grayanus* [Selin and Vekhova 2003]; *Mytilus galloprovincialis* [Steffani and Branch 2003]; *Mytilus californianus* [Blythe and Lea 2008]). Further, several researchers found that *D. polymorpha* had lower and more convex shells at high water current velocity compared to low velocity (Ovchinnikov 1954, Karpevich 1955, Antonov 1983). Paleontologists note that fossil representatives of the family Dreissenidae from deeper habitats possessed weakly convex and thin shells with a poorly defined carinal bend and byssal groove (Nevesskaya 1965, Babak 1983). Similar findings were recently reported for *D. r. bugensis* collected in the hypolimnion of Lake Erie (Dermott and Munawar 1993, Claxton et al. 1998).

Differences in shell morphology between the deepwater phenotype (profunda) and shallow-water phenotype of *D. r. bugensis* were apparent. The deepwater phenotype (profunda) possessed lower and weakly convex shells with an unpronounced carinal fold and a convex ventral margin. The extent of pigmentation was also lower in the deepwater phenotype than in the shallow-water phenotype. Light coloration of shells of the profunda phenotype is a characteristic feature (Dermott and Munawar 1993) and is evidently a result of an infaunal mode of life and occurrence in deepwater habitats. It is known that infaunal organisms in deepwater habitats often lack pigment. For example, Caspian subspecies of *D. rostriformis* in 20–300 m of water have light, depigmented shells (Logvinenko and Starobogatov 1968). Moreover, individuals with different shell pigmentation also had different growth rates (Yurgens 2006), antioxidant systems (Gostyukhina et al. 2005), processes for protein

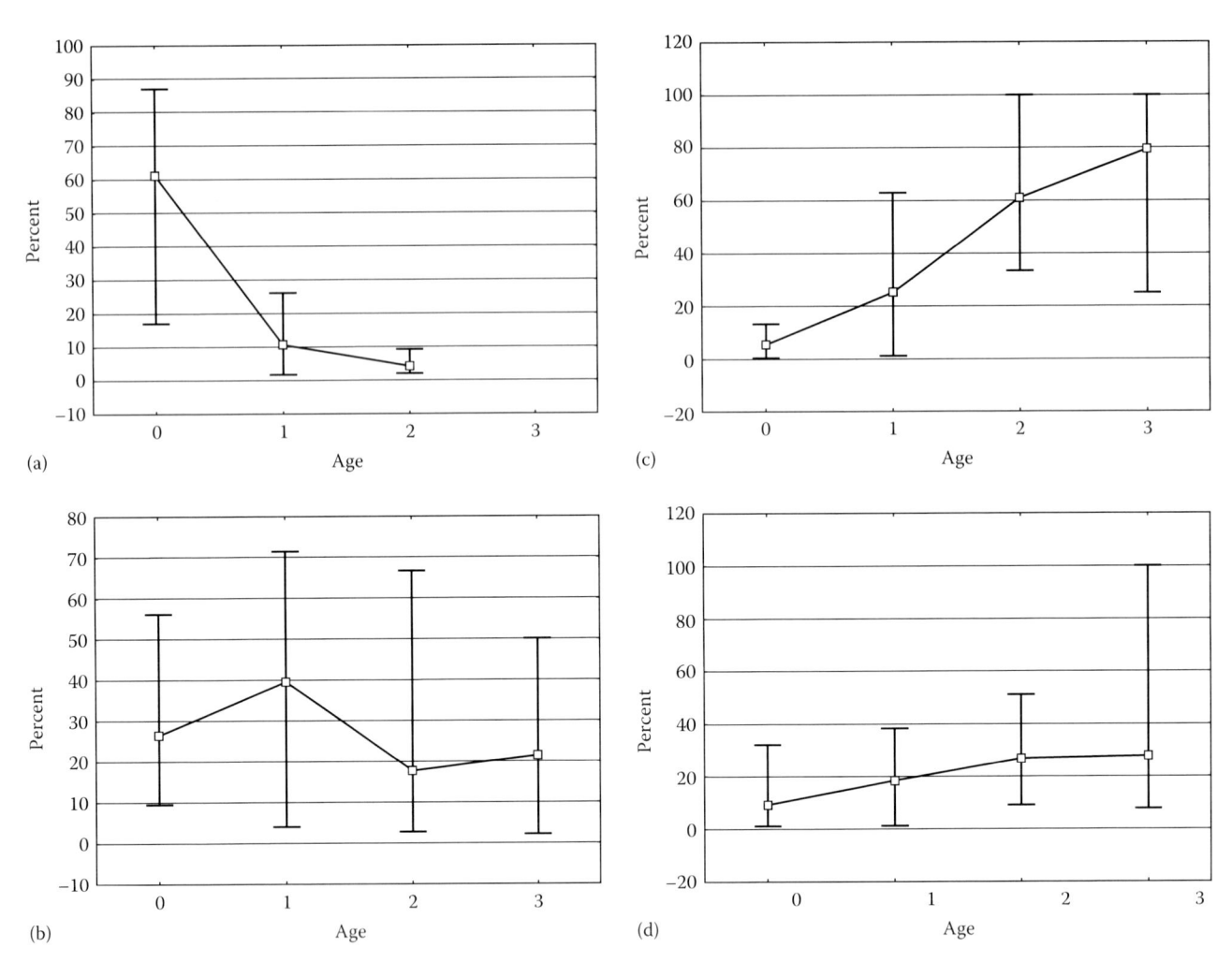

Figure 19.26 Mean (±range) percent frequency of primary pattern types at different ages for *D. polymorpha* across all water bodies. (a) Type C, (b) type AC, (c) type A, (d) type Ad.

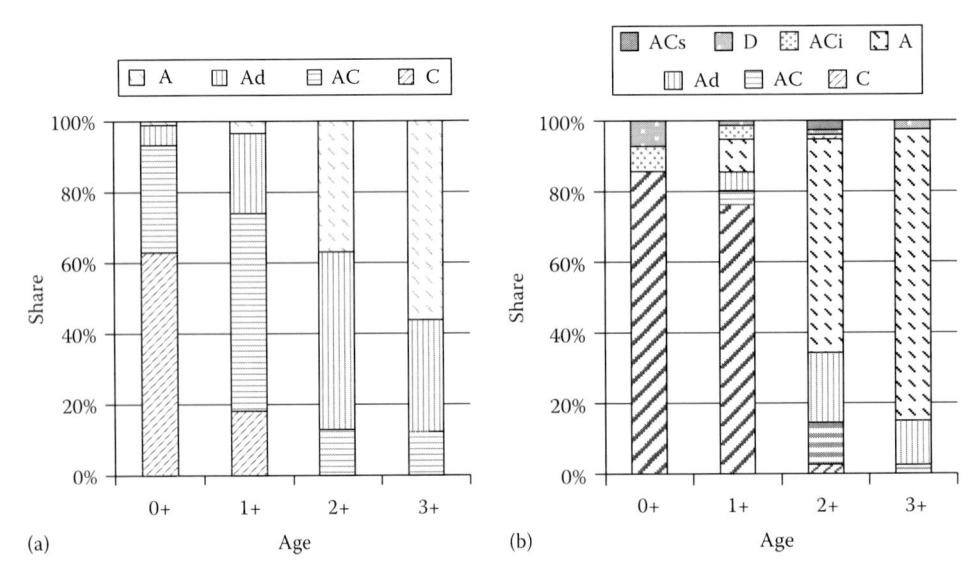

Figure 19.27 Typical examples of how frequency (percent) of pattern type changed with age for *D. polymorpha*. (a) Rybinsk Reservoir, (b) Northern Dvina.

MORPHOLOGICAL VARIABILITY OF *DREISSENA POLYMORPHA*

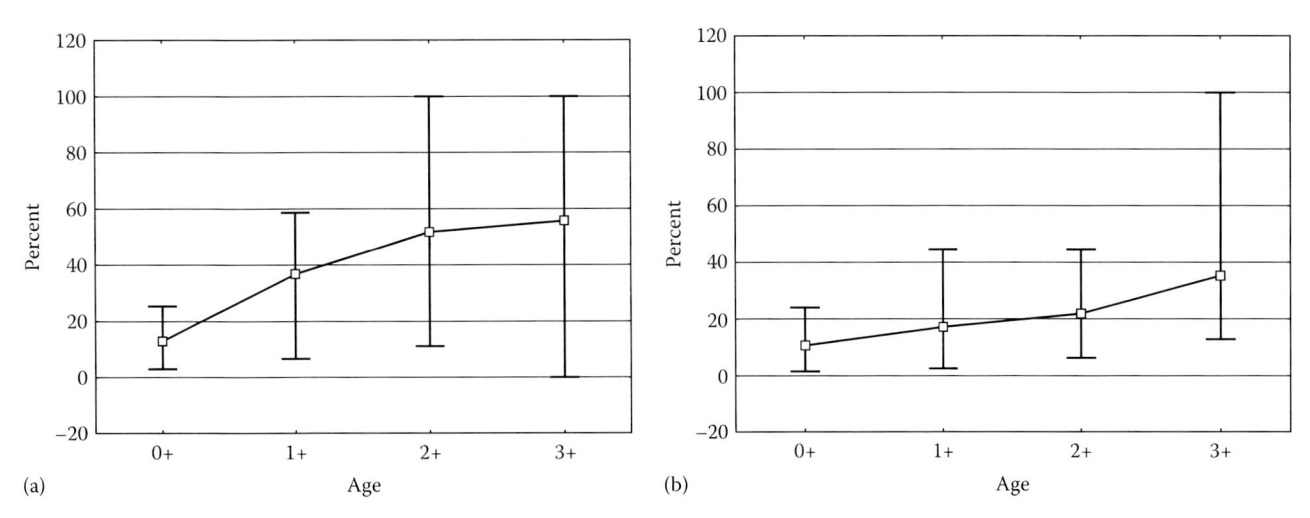

Figure 19.28 Mean (±range) percent frequency of primary pattern types at different ages for *D. r. bugensis* across all water bodies. (a) Type A, (b) type Ad.

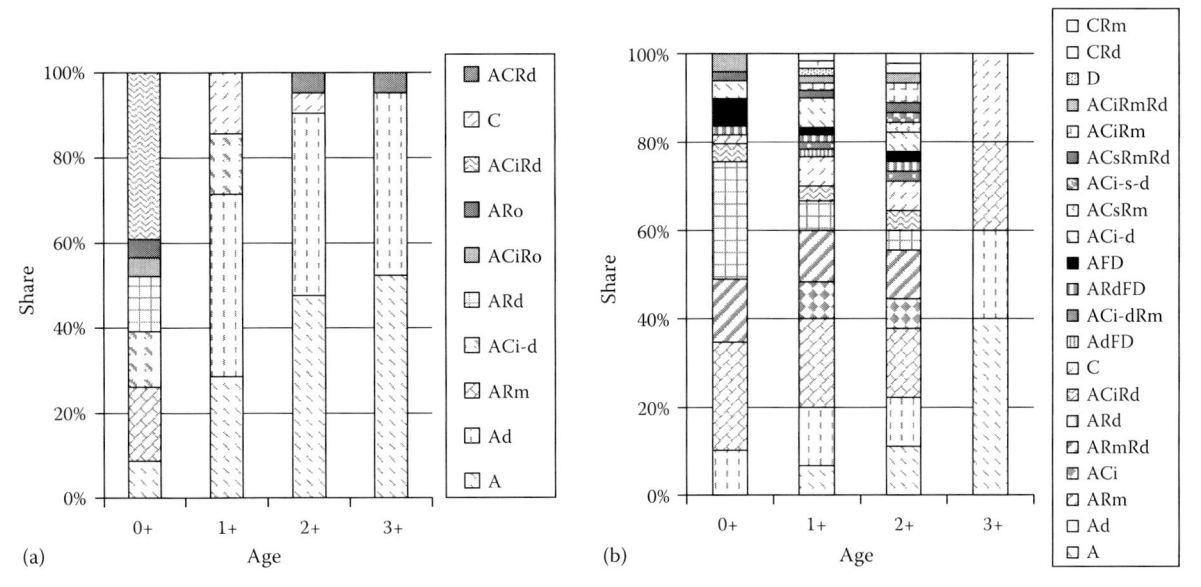

Figure 19.29 Typical examples of how frequency (percent) of pattern type changed with age for *D. r. bugensis*. (a) Rybinsk Reservoir, (b) Lake Michigan.

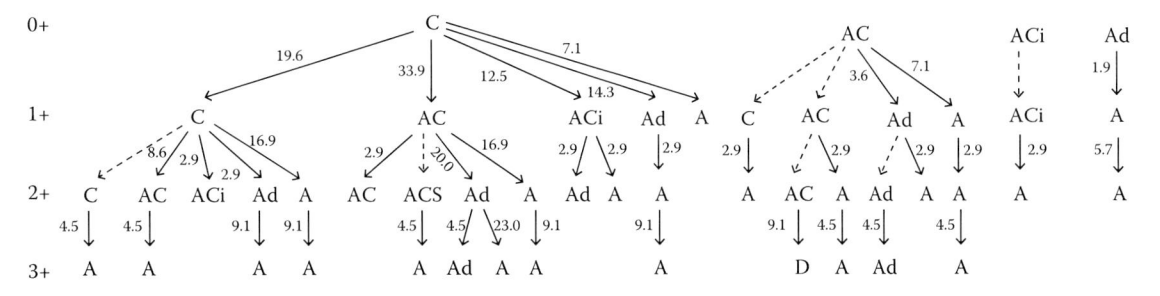

Figure 19.30 Typical ontogenetic scheme of changes in pattern types for *D. polymorpha*. Example given is for the population in Perućica Reservoir. Values given denote percent of individuals with the given sequence at the given age. Since data for ages 1+ to 3+ were aggregated, dotted lines denote pattern change not revealed in mussels of current age but revealed in older ones.

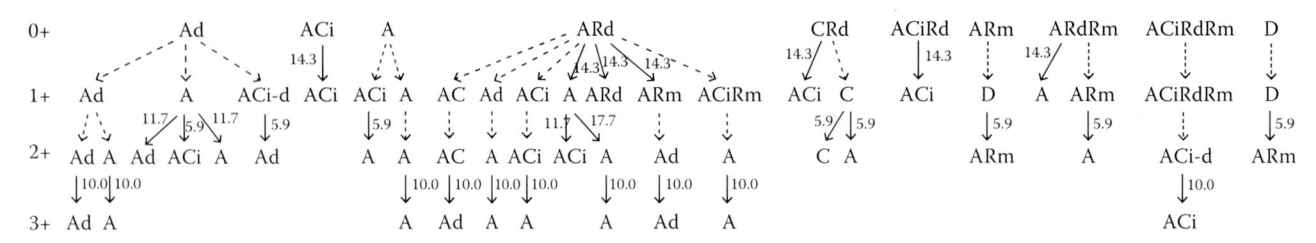

```
0+      Ad        ACi    A        ARd            CRd    ACiRd  ARm    ARdRm  ACiRdRm  D
                  14.3↓           14.3\14.3 14.3 14.3/   ↓14.3  14.3/  ↓      ↓
1+  Ad  A  ACi-d ACi ACi A  AC Ad ACi A ARd ARm ACiRm ACi C ACi D  A  ARm  ACiRdRm  D
      ↙11.7↘11.7 ↓5.9  ↓5.9      ↓  ↓  ↓ 11.↓17.7 ↓    ↓   5.9/↓5.9 ↓5.9 ↓5.9 ↓      ↓5.9
2+ Ad A Ad ACi A Ad   A  A  AC A ACi ACi A  Ad  A   C A         ARm    A  ACi-d ARm
    ↓10.0↓10.0              ↓10.0↓10.0↓10.0↓10.0 ↓10.0 ↓10.0 ↓10.0                      ↓10.0
3+ Ad A              A Ad  A  A      A   Ad     A                              ACi
```

Figure 19.31 Typical ontogenetic scheme of changes in pattern types for *D. r. bugensis*. Example given is for the population in the Volga Delta. Values given denote percent of individuals with the given sequence at the given age. Since data for ages 1+ to 3+ were aggregated, dotted lines denote pattern change not revealed in mussels of current age but revealed in older ones.

Table 19.7 Primary and Secondary Pattern Sequences Found in *D. polymorpha* Populations in Different Water Bodies. Question Marks in the Sequences for North American Great Lakes Indicate Some Uncertainty in Pattern Sequence because of Shell Erosion

Water Body	Primary Sequences	Secondary Sequences
Rybinsk Reservoir	C–AC–Ad	C–C–AC–A,
	C–AC–A–A	AC–AC–Ad–A,
		AC–A–A–A
Gorky Reservoir	C–AC–Ad	AC–A–A,
	C–AC–A–A	C–Ad–A,
		AC–Ad–A
Kama Reservoir	C–AC–Ad	C–Ad–A
	C–AC–A–A	
Lake Forelevoe	C–AC–Ad	AC–Ad–Ad,
	C–AC–A–A	AC–A–A,
		C–AC–AC–Ad
Northern Dvina River	C–C–A–A	C–C–Ad–A
		C–C–Ad–Ad
Lake Sharony	C–AC	AC–Ad
	AC–A	C–A
	AC–AC	
Chograi Reservoir	C–Ad	AC–A
	C–AC	Ad–A
	C–A	
RR-1 channel	C–A	AC–Ad
	C–AC–A	
North Caspian Sea, *D. p. andrusovi*	Ad–A	AC–A,
	D–ARm	C–A,
	C–A	ARd–A,
		DRo–A
		ARo–AC
Serbian water bodies	C–AC	C–Ad
Perućica, Danube	C–AC–Ad–A	
	C–AC–A–A	
North American Great Lakes	?–A–A–A	
	?–Ad–A–A	

synthesis (Scherban and Vyalova 2008), and pearl formation (Brake et al. 2004). In addition, profunda mussels had longer siphons and fewer byssal threads compared to the shallow-water form (Dermott and Munawar 1993). Thus, profunda possesses features typical of endobiont mussels (Stanley 1970, Babak 1983, Dermott and Munawar 1993). Besides these morphologic peculiarities, the deepwater phenotype has physiological adaptations suited to hypolimnetic conditions, for example, maturation and spawning at low temperature (Claxton and Mackie 1998) and effective growth at low food level (Baldwin et al. 2002). Reorganization of metabolic pathways related to the ability to dwell in the hypolimnion may be associated with alteration of pigment synthesis and function of mantle chromatophore cells.

Morphologic features characteristic of a deepwater presence were not found in *D. polymorpha*. Possibly, this species lacks the ability to adapt to hypolimnetic conditions and hence distributions are restricted to shallow water (Zhuravel 1967, Mills et al. 1993). A possible reason for differences in the ability to adapt to deep water in the two species may be a function of their different phylogenetic histories (Orlova et al. 2005). *D. r. bugensis* descended from deepwater ancestors but was later found only in estuaries of the northeastern region of the Black Sea (Babak 1983, Starobogatov 1994b). Presumably, while in these shallow-water habitats, evolution of a deepwater phenotype did not occur even though the genetic potential was present but "hidden" in the genotype (in "silent" genes). This feature was realized only after *D. r. bugensis* colonized water bodies with appropriate conditions, for example, Laurentian Great Lakes in North America (Dermott and Munawar 1993, Claxton et al. 1998, Nalepa et al. 2009) and the Cheboksary Reservoir in Russia (Pavlova 2012). In contrast, *D. polymorpha* evolved in rivers of the northern portion of the Paratethys (Starobogatov 1994b) and inhabited shallow waters initially. This may explain why *D. polymorpha* lacks the genetic ability to adapt to conditions in deep water. To date, there are no known morphological or physiological differences between *D. polymorpha* in shallow and deep waters.

Relative influence of salinity on dreissenid shell morphology was apparent when mussels from the Volga Delta (freshwater) and north Caspian Sea (4‰ salinity) were compared. For *D. polymorpha*, individuals in the north Caspian Sea (especially subspecies *D. p. andrusovi*) had shells that were lower and less convex than shells of individuals from the Volga Delta. However, the opposite was found for *D. r. bugensis*; this species had shells that were higher and more convex in the north Caspian Sea. Karpevich (1955) showed that for *D. polymorpha* the metabolic role of water and bivalent cations declined, while the role of salt and monovalent cations increased as salinity increased. This allowed individuals to maintain homeostasis in water of various salinities. Such a metabolic transformation probably also affects pigment deposition. Our study showed that *D. p. andrusovi* possessed a greater range of pigmentation and a greater diversity of pattern types than *D. p. polymorpha*. There are no comparable metabolic studies of *D. r. bugensis* in salt water. However, experiments conducted by Antonov and Shkorbatov (1990) revealed that *D. r. bugensis* was less tolerant of salt water than *D. polymorpha*. Aleksenko (1992) found that at high levels of salinity in the Dnieper-Bug estuary, *D. polymorpha* was more abundant and attained higher biomass than *D. r. bugensis*. This further suggests that *D. r. bugensis* has a lower adaptation potential to salinity gradients than *D. polymorpha*. Interestingly, adaptation to higher salinity in *D. polymorpha* resulted in formation of small shells, which was opposite to *D. r. bugensis* that formed large shells.

Despite differences that could be attributed to local environmental factors such as water currents and salinity, analysis of shell outlines for both *D. polymorpha* and *D. r. bugensis* showed some broad geographic patterns. Overall, populations in the north had more rounded superior margins than populations in the south. In addition, in the Volga River basin, specimens from populations in the south had shells with a concave ventral margin and specimens from the north had shells with a convex ventral margin. The shape of ventral margins depends on byssal strength; the thicker the byssal bundle the more concave is the margin (Andrusov 1897). This feature is influenced by substrate and habitat hydrodynamics (Bell and Gosline 1997, Carrington 2002, Moeser and Carrington 2006).

Both dreissenid species are epifaunal organisms that evolved in lower reaches of rivers and estuaries (Starobogatov 1994b). In such habitats, water flows are generally high, and thus capacity for strong byssal attachment evolved early. Higher current velocities are apparent in more southern portions of the Volga basin (Volga–Ahtuba floodplain, Volga Delta and north Caspian Sea, and Lake Sharony), and mussels had shells with concave margins. Populations in the north were collected mainly from reservoirs (Rybinsk, Gorky and Kama Reservoirs, and Lake Plescheevo) with low current velocities. In such water bodies, strong attachment would be less important and features such as byssus strength and concave ventral margins were less likely to have been maintained by natural selection. Other studies have found geographic variability in shell outlines of mussel species such as along the western coast of South America where shell height of *Mytilus chilensis* increased from south to north (Krapivka et al. 2007). The opposite occurred for *Brachidontes* along the east coast of Argentina where shell height increased from north to south (Aguirre et al. 2006). In both studies, these spatial gradients were related to changes in salinity concentrations and associated environmental variables (Aguirre et al. 2006, Krapivka et al. 2007).

Colorimetric analysis of shell pigmentation indicated that pigmentation increased from south to north in both *D. polymorpha* and *D. r. bugensis*. A similar trend was found for *D. polymorpha* in an earlier study by Biochino (1990). This geographic tendency may be considered an example of an "inverted" Gloger's rule. According to Gloger's rule, warm-blooded animals in warm, humid climates have brighter coloration than those in cold, dry climates (Gloger 1833). Rapoport (1969) showed that the opposite trend occurred in some insects and mollusks and suggested an "inverted" Gloger's rule for invertebrates. Temperature is one factor that varies strongly with latitude, and temperature can influence many features of organisms. Vernberg (1962) suggested that low temperature favored the production of pigment (melanin), and Mitton (1977) believed that the occurrence of *Mytilus edulis* with white stripes along the west coast of North America was determined by thermal gradients. Mitton (1977) explained that increased frequencies of striped individuals from north to south was a result of greater thermal resistance in striped mussels compared to uniformly dark mussels.

In summary, comparisons of shell morphology in *D. polymorpha* and *D. r. bugensis* revealed several key similarities and differences. Both species had shell features typical of all mussel species with a similar life mode, that is, an epifaunal life habit and byssal attachment. The general shape of shells, properties of allometry, shell outline variability, and a tendency for greater variability in the width index (W/L) than the height index (H/L) were similar in both species. Differences in shell morphology between the two species as related to shell variability and pigmentation were mainly responses to environmental factors such as salinity and water depth. These differences can be explained by the different conditions under which the two species evolved. These same environmental factors likely influenced shell characteristics within a given species at local (individual water bodies) and at wide geographic scales.

ACKNOWLEDGMENTS

I am grateful to all my colleagues who helped me in the mussels sampling (Yuri Gerasimov, Dmitry Pavlov Jr., Igor Stolbunov, Alexey Kasyanov, Irina Voroshilova, Dmitry Vekhov, Elena Nikitenko, DaryaGuseva, Ivan Pozdeev, Thomas Nalepa, Branislav Mićković) and Dmitry Pavlov Sr. for valuable comments and editing of the manuscript. I appreciate the editors' work very much.

REFERENCES

Aguirre, M. L., S. I. Perez, and Y. N. Sirch. 2006. Morphological variability of *Brachidontes* Swainson (Bivalvia, Mytilidae) in the marine Quaternary of Argentina (SW Atlantic). *Palaeogeo. Palaeoclimat. Palaeoecol.* 239:100–125.

Aleksenko, T. L. 1992. Mollusks of the Lower Dnieper-Bug basin and their role in the ecosystem. PhD dissertation, Kiev University, Kiev, Ukraine.

Andreeva, A. M., M. I. Orlova, and Yu. V. Slynko. 2001. Population-genetic analysis of *Dreissena polymorpha* (Pall.) and *Dreissena bugensis* (Andr.) in the Upper Volga reservoirs, the Volga River delta and in the western Gulf of Finland. In *Alien Species in Holarctic, Book of Abstracts of the First U.S.–Russian Symposium*, D. S. Pavlov, ed., pp. 9–11. Borok: Paparin Institute for Biology of Inland Waters.

Andrusov, N. I. 1890. *Dreissena rostriformes* Desch. in the Bug River. *Vestnik Estestvoznaniya*. 6:261–262.

Andrusov, N. I. 1897. Fossil and recent Dreissenidae of Eurasia. *Proc. St. Petersburg Soc. Nat. Dep. Geol. Mineral.* 25:1–683.

Antonov, P. I. 1983. The morphological variability of *Dreissena polymorpha* (Pallas) in the different parts of its distribution range. In *Mollusks: Systematics, Ecology and Dispersion*, I. M. Likharev, ed., pp. 64–67. Leningrad: Nauka.

Antonov, P. I. 1997. Ecological, physiological and morphological features of *Dreissena polymorpha* (Pallas) in the Volga River basin. PhD dissertation, Nizhny Novgorod State University, Nizhny Novgorod, Russia.

Antonov, P. I. and M. Bachurina. 1986. The morphology of *Dreissena polymorpha* in the left-bank tributaries of the Saratov Reservoir. In *Ecological-Physiological and Ecological-Faunistic Aspects of Animal Adaptations*, ed., G. L. Shkorbatov, pp. 82–92. Ivanovo: Ivanovostate University.

Antonov, P. I. and G. L. Shkorbatov. 1983. Ecological and physiological variability of the populations of *Dreissena polymorpha* (Pallas) in the Volga River. In *Species and Its Productivity in the Distribution Range*, V. E. Sokolov, ed., pp. 116–128. Moscow: Nauka.

Antonov, P. I. and G. L. Shkorbatov. 1990. Ecological and physiological features of dreissenid mussels in the lower reaches of the Dnieper River. In *Species in Its Distribution Range: Biology, Ecology and Production of Aquatic Invertebrates*, N. N. Khmeleva, ed., pp. 135–141. Minsk: Navuka i technika.

Astanei, J., E. Gosling, J. Wilson et al. 2005. Genetic variability and phylogeography of the invasive zebra mussel, *Dreissena polymorpha* (Pallas). *Mol. Ecol.* 14:1655–1666.

Babak, E. V. 1983. The Pliocene and Quaternary Dreissenidae of the euxinian basin. *Proc. Palaeontol. Inst. Acad. Sci. USSR* 204:1–204.

Baldwin, B. S., M. S. Mayer, J. Dayton et al. 2002. Comparative growth and feeding in zebra and quagga mussels (*Dreissena polymorpha* and *Dreissena bugensis*): Implications for North American lakes. *Can. J. Fish. Aquat. Sci.* 59:680–694.

Bell, E. C. and J. M. Gosline. 1997. Strategies for life in flow: Tenacity, morphometry, and probability of dislodgment of two *Mytilus* species. *Mar. Ecol. Prog. Ser.* 159:197–208.

Benson, A. J. 2011a. Quagga mussel sightings distribution. Retrieved [11.03.2011] from http://nas.er.usgs.gov/taxgroup/mollusks/zebramussel/quaggamusseldistribution.asp (November 3, 2011)

Benson, A. J. 2011b. Zebra mussel sightings distribution. Retrieved from http://nas.er.usgs.gov/taxgroup/mollusks/zebramussel/zebramusseldistribution.asp (November 3, 2011)

Benson, A. J. 2013. Chronological history of zebra and quagga mussels (Dreossenidae) in North America, 1988–2010. In *Quagga and Zebra Mussels: Biology, Impacts, and Control*, 2nd edn., T. F. Nalepa and D. W. Schloesser, eds., pp. 9–31. Boca Raton: CRC Press.

Biochino, G. I. 1990. Polymorphism and geographic variability of *Dreissena polymorpha* (Pallas). (In *Microevolution of Freshwater Organisms.*) *Proc. Inst. Biol. Inland Waters* 59:143–158.

Biochino, G. I. 1994. Polymorphism and geographic variability of *Dreissena polymorpha* (Pallas). In *Zebra Mussel Dreissena polymorpha: Systematics, Ecology and Practical Meaning*, Ya. I. Starobogatov, ed., pp. 56–66. Moscow, Russia: Nauka.

Blythe, J. N. and D. W. Lea. 2008. Functions of height and width dimensions in the intertidal mussel, *Mytilus californianus*. *J. Shellfish Res.* 27:385–392.

Boileau, M. G. and P. D. N. Hebert. 1993. Genetics of the zebra mussel (*Dreissena polymorpha*) in populations from the Great Lakes and Europe. In *Zebra Mussels: Biology, Impacts, and Control*, T. F. Nalepa and D. Schloesser, eds., pp. 227–238. Boca Raton, FL: CRC Press.

Brake, J., F. Evans, and C. Langdon. 2004. Evidence for genetic control of pigmentation of shell and mantle edge in selected families of Pacific oysters, *Crassostrea gigas*. *Aquaculture* 229:89–98.

Brown, J. E. and C. A. Stepien. 2010. Population genetic history of the dreissenid mussel invasions: Expansion patterns across North America. *Biol. Invasions* 13:3687–3710.

Butorin, N. V. 1969. *Hydrological Processes and Dynamics of Water Masses in the Volga River Reservoirs*. Leningrad, Russia: Nauka.

Carlton, J. T. 2008. The zebra mussel *Dreissena polymorpha* found in North America in 1986 and 1987. *J. Great Lakes Res.* 34:770–773.

Carrington, E. 2002. The ecomechanics of mussel attachment: From molecules to ecosystems *Integr. Comp. Biol.* 42: 846–852.

Claxton, W. T. and G. L. Mackie. 1998. Seasonal and depth variations in gametogenesis and spawning of *Dreissena polymorpha* and *Dreissena bugensis* in eastern Lake Erie. *Can. J. Zool.* 76: 2010–2019.

Claxton, W. T., A. Martel, R. M. Dermott et al. 1997. Discrimination of field-collected juveniles of two introduced dreissenids (*Dreissena polymorpha* and *Dreissena bugensis*) using mitochondrial DNA and shell morphology. *Can. J. Fish. Aquat. Sci.* 54:1280–1288.

Claxton, W. T., A. B. Wilson, G. L. Mackie et al. 1998. A genetic and morphological comparison of shallow- and deep-water populations of the introduced dreissenid bivalve *Dreissena bugensis*. *Can. J. Zool.* 76:1269–1276.

Dermott, R. and M. Munawar. 1993. Invasion of Lake Erie offshore sediments by *Dreissena*, and its ecological implications. *Can. J. Fish. Aquat. Sci.* 50:2298–2304.

Gloger, C. W. L. 1833. *Das Abändern der Vögel durch Einfluss des Klima's*. Breslau, Poland: August Schulz.

Gostyukhina, O. L., A. A. Soldatov, I. V. Golovina et al. 2005. Tissue antioxidant enzymatic complex of the different color morphs of a Black Sea mollusk *Mytilus galloprovincialis* Lam. *Ecol. Giya Morya* 68:42–47.

Hebert, P. D., B. W. Muncaster, and G. L. Mackie. 1989. Ecological genetic studies on *Dreissena polymorpha* (Pallas): A new mollusk in the Great Lakes. *Can. J. Fish. Aquat. Sci.* 46:1587–1591.

Hubenov, Z. and T. Trichkova. 2007. *Dreissena bugensis* (Mollusca: Bivalvia: Dreissenidae)—New invasive species to the Bulgarian malacofauna. *Acta Zool. Bulg.* 59:203–209.

Imo, M., A. Seitz, and J. Johannesen. 2010. Distribution and invasion genetics of the quagga mussel (*Dreissena rostriformis bugensis*) in German rivers. *Aquat. Ecol.* 45:731–740.

Innes, D. J. and J. A. Bates. 1999. Morphological variation of *Mytilus edulis* and *Mytilus trossulus* in eastern Newfoundland. *Mar. Biol.* 133:691–699.

Iwata, H. and Y. Ukai. 2002. SHAPE: A computer program package for quantitative evaluation of biological shapes based on elliptic Fourier descriptors. *J. Hered.* 93:384–385.

Izyumov, Yu. G. and V. V. Pavlova. 2008. The use of colorimetric method for the study the shell coloration variability in *Dreissena polymorpha* (Dreissenidae, Bivalvia). *Zool. Zhur.* 87:1–4.

Karpevich, A. F. 1955. Some data on the morphogenesis in bivalve mollusks. *Zool. Zhur.* 34:46–67.

Krapivka, S., J. E. Toro, A. C. Alcapan et al. 2007. Shell-shape variation along the latitudinal range of the Chilean blue mussel *Mytilus chilensis* (Hupe 1854). *Aquacult. Res.* 38:1770–1777.

Lajtner, J., Z. Marušić, G. I. V. Klobučar et al. 2004. Comparative shell morphology of the zebra mussel, *Dreissena polymorpha* in the Drava River (Croatia). *Biol. Bratislava* 59:595–600.

Logvinenko, B. M. and Ya. I. Starobogatov. 1968. Type Mollusca. In *Atlas of the Invertebrates of Caspian*, Ya. A. Birstein (ed.), pp. 308–385. Moscow, Russia: Pischevaya Promyshlennost.

Lukashev, D. V. 1999. *Dreissena bugensis* Andrusov in the conditions of parts of the Dnieper River with regulated discharge. *Gidrobiol. Zhur.* 35:43–50.

Markov, A. V. and E. B. Naimark. 1998. *Quantitative Regularities of Macroevolution. The Experience of Application of Systemic Approach for the Analysis of Development of Supra-Species Taxa*. Moscow, Russia: Geos.

Marsden, J. E., A. Spidle, and B. May. 1995. Genetic similarity among zebra mussel populations within North America and Europe. *Can J. Fish. Aquat. Sci.* 52:836–847.

Marsden, J. E., A. Spidle, and B. May. 1996. Review of genetic studies of *Dreissena* spp. *Am. Zool.* 36:259–270.

May, B. and J. E. Marsden. 1992. Genetic identification and implications of another invasive species of dreissenid mussel in the Great Lakes. *Can. J. Fish. Aquat. Sci.* 49:1501–1506.

Mills, E. L., R. Dermott, E. F. Roseman et al. 1993. Colonization, ecology, and population structure of the "quagga" mussel (Bivalvia: Dreissenidae) in the Lower Great Lakes. *Can. J. Fish. Aquat. Sci.* 50:2305–2314.

Mitton, J. B. 1977. Shell color and pattern variation in *Mytilus edulis* and its adaptive significance. *Chesapeake Sci.* 18:387–390.

Moeser, G. M. and E. Carrington. 2006. Seasonal variation in mussel byssal thread mechanics. *J. Exp. Biol.* 209:1996–2003.

Morton, B. 1970. The evolution of the heteromyarian condition in the Dreissenacea (Bivalvia). *Palaeonthology* 13:563–572.

Müller, J. C., D. Hidde, and A. Seitz. 2002. Canal construction destroys the barrier between major European invasion lineages of the zebra mussel. *Proc. R. Soc. Lond. B* 269:1139–1142.

Nalepa, T. F., D. L. Fanslow, and G. A. Lang. 2009. Transformation of the offshore benthic community in Lake Michigan: Recent shift from the native amphipod *Diporeia* spp. to the invasive mussel *Dreissena rostriformis bugensis*. *Freshwat. Biol.* 54:466–479.

Nevesskaya, L. A. 1965. *Late-Quaternary Bivalve Mollusks of the Black Sea, Their Systematics and Ecology*. Moscow, Russia: Nauka.

Nichols, S. J. and M. G. Black. 1994. Identification of larvae: The zebra mussel (*Dreissena polymorpha*), quagga mussel (*Dreissena rostriformis bugensis*), and Asian clam (*Corbicula fluminea*). *Can. J. Zool.* 72:406–417.

Orlova, M. I., T. H. W. Therriault, and P. I. Antonov et al. 2005. Invasion ecology of quagga mussels (*Dreissena rostriformis bugensis*): A review of evolutionary and phylogenetic impacts. *Aquat. Ecol.* 39:401–418.

Ovchinnikov, I. F. 1954. *Dreissena polymorpha* in the Rybinsk Reservoir. In *Book of Abstracts of the Third Ecological Conference*, pp. 107–109. Kiev, Ukraine: Kiev State University.

Pallas, P. S. 1809. *The Travel on the Various Provinces of Russian Empire*. Part I. St. Petersburg, Russia.

Pathy, D. A. and G. L. Mackie. 1993. Comparative shell morphology of *Dreissena polymorpha*, *Mytilopsis leucophaeata*, and the "quagga" mussel (Bivalvia: Dreissenidae) in North America. *Can. J. Zool.* 71:1012–1023.

Pavlova, V. 2012. First finding of deepwater *profunda* morph of quagga mussel *Dreissena bugensis* in the European part of its range. *Biol. Invasions* 14:509–514. DOI: 10.1007/s10530-011-0100-1.

Peyer, S. M., J. C. Hermanson, and C. E. Lee. 2010. Developmental plasticity of shell morphology of quagga mussels from shallow- and deep-water habitats of the Great Lakes. *J. Exp. Biol.* 213:2602–2609.

Protasov, A. A. 1998. Shell color intensity as a phenotypic characteristic of populations of *Dreissena polymorpha* (Pallas) (Bivalvia, Mollusca). *Ekologiya* 6:479–482.

Protasov, A. A. 2004. Some features of the phenotypic structure of *Dreissena bugensis* Andr. population in the cooler basin of Chernobyl nuclear power-station. In *Ecological-Functional and Faunistic Aspects of Mollusk Studies and Their Role in Bioindication of Environmental State*, pp. 158–160. Zhitomir, Ukraine: Volin.

Protasov, A. A. and E. V. Gorpinchuk. 1997. The phenotypic structure of *Dreissena polymorpha* populations. *Gidrobiol. Zhur.* 33:21–32.

Protasov, A. A. and O. O. Sinitsyna. 2000. Biotopic variation and phenogeography of *Dreissena polymorpha* (Pallas). *Russ. J. Ecol.* 31:415–421.

Protasov, A. A., O. O. Sinitsyna, E. V. Gorpinchuk et al. 1997. On the adaptivity of the phenotypes in *Dreissena polymorpha* (Pallas). In *Book of Abstracts of Second Conference of Ukrainian Hydroecological Society*, 180pp. Kiev: Institute of Hydrobiology.

Rapoport, E. H. 1969. Gloger's rule and pigmentation of Collembola. *Evolution* 23:622–626.

Ross, T. K. and G. M. Lima. 1994. Measures of allometric growth: The relationship of shell length, shell height, and volume to ash-free dry weight in the zebra mussel, *Dreissena polymorpha* Pallas and the quagga mussel, *Dreissena bugensis* Andrusov. In *Proceedings of the Fourth International Zebra Mussel Conference*, pp. 611–623. Madison, WI.

Scherban, S. A. and O. Yu. Vyalova. 2008. Sexual and phenotypic peculiarities of RNA content in gonads in the Black sea mussels. *Proc. Nat. Acad. Sci. Ukr.* 2:166–170.

Schloesser, D. W., A. bij de Vaate, and A. Zimmerman. 1994. A bibliography of *Dreissena polymorpha* in European and Russian waters: 1964–1993. *J. Shellfish Res.* 13:243–267.

Seed, R. 1968. Factors influencing shell shape in the mussel *Mytilus edulis*. *J. Mar. Biol. Assoc. U.K.* 48:561–584.

Selin, N. I. and E. E. Vekhova. 2003. Morphological adaptations of the mussel *Crenomitilus grayanus* (Bivalvia) to attached life. *Russ. J. Mar. Biol.* 29:230–235.

Shkorbatov, G. L., A. F. Karpevich, and P. I. Antonov. 1994. Ecological physiology. In *Zebra Mussel Dreissena polymorpha: Systematics, Ecology and Practical Meaning*, Ya. I. Starobogatov (ed.), pp. 67–109. Moscow, Russia: Nauka.

Smirnova, N. F., G. I. Biochino, and G. A. Vinogradov. 1993. Some aspects of the zebra mussel, *Dreissena polymorpha* in the former European USSR with morphological comparisons to Lake Erie. In *Zebra Mussels: Biology, Impacts, and Control*, T. F. Nalepa and D. Schloesser, eds., pp. 217–226. Boca Raton, FL: CRC Press.

Soroka, M. 2002. Genetic structure of an invasive bivalve *Dreissena polymorpha* (Pallas) from Poland. I. Geographical and intra-population variation. *Folia Malacol.* 10:175–213.

Spasskiy, N. N. 1948. The variability of *Dreissena polymorpha* Pall. in Northern Caspian and the significance of its varieties as food for vobla roach. *Proc. Volgo-Caspian Sci. Fish. Station* 10:117–128.

Spidle, A. P., J. E. Marsden, and B. May. 1994. Identification of the Great Lakes quagga mussel as *Dreissena bugensis* from the Dnieper River, Ukraine, on the basis of allozyme variation. *Can. J. Fish. Aquat. Sci.* 51:1485–1489.

Stanley, S. M. 1970. *Relation of Shell Form to Life Habits in the Bivalvia (Mollusca)*. Memoir 125. Boulder: The Geological Society of America, Inc.

Starobogatov, Ya. I. 1994a. Morphology. In *Zebra Mussel Dreissena polymorpha: Systematics, Ecology and Practical Meaning*, Ya. I. Starobogatov, ed., pp. 7–17. Moscow, Russia: Nauka.

Starobogatov, Ya. I. 1994b. Systematics and palaeontology. In *Zebra Mussel Dreissena polymorpha: Systematics, Ecology and Practical Meaning*, Ya. I. Starobogatov, ed., pp. 18–46. Moscow, Russia: Nauka.

Steffani, C. N. and G. M. Branch. 2003. Growth rate, condition, and shell shape of *Mytilus galloprovincialis*: Responses to wave exposure. *Mar. Ecol. Prog. Ser.* 246:197–209.

Therriault, T. W., M. I. Orlova, M. F. Docker et al. 2005. Invasion genetics of a freshwater mussel (*Dreissena rostriformis bugensis*) in eastern Europe: High gene flow and multiple introductions. *Heredity* 95:16–23.

Timmer, K. and B. C. L. Shelley. 2005. Population genetic of zebra and quagga mussels in the Finger Lakes: What can genetic analyses tell us about colonization dynamics? *Bull. N. Am. Benthol. Soc.* 23:316–317.

Van der Velde, G. and D. Platvoet. 2007. Quagga mussels *Dreissena rostriformis bugensis* (Andrusov, 1897) in the Main River (Germany). *Aquat. Invasions* 2:261–264.

Vernberg, F. J. 1962. Comparative physiology: Latitudinal effects of physiological properties of animal populations. *Ann. Rev. Physiol.* 24:517–546.

Wilson, A. B., K. A. Naish, and E. G. Boulding. 1999. Multiple dispersal strategies of the invasive quagga mussel (*Dreissena bugensis*) as revealed by microsatellite analyses. *Can. J. Fish. Aquat. Sci.* 56:2248–2261.

Yonge, C. M. 1953. The monomyarian condition in the Lamellibranchia. *Trans. R. Soc. Edinburgh* 62:443–478.

Yonge, C. M. and J. I. Campbell. 1968. On the heteromyarian conditions in the Bivalvia with special reference to *Dreissena polymorpha* and certain Mytilacea. *Trans. R. Soc. Edinburgh* 68:21–43.

Yurgens, E. M. 2006. Ecological characteristic of a mollusk *Macoma balthica* (Linne, 1758) in the southern part of the Baltic Sea. PhD dissertation, Kaliningrad State Polytechnic University, Kaliningrad, Russia.

Zhadin, V. I. 1952. *Mollusks of the Fresh and Brackish Waters of USSR*. Moscow, Russia: Academy of sciences of USSR.

Zhgareva, N. N. 1992. Composition and distribution of the fauna of overgrowths in Lake Plescheevo. In *Factors and Processes of Eutrophication in Lake Plescheevo*, V. P. Semernoy, ed., pp. 95–105. Yaroslavl, Yaroslavl State University.

Zhivotovskiy, L. A. 1982. The indices of population variability of polymorphic features. In *Phenetics of Populations*, A. V. Yablokov, ed., pp. 38–44. Moscow: Nauka.

Zhuravel, P. A. 1951. *Dreissena bugensis* in the Dnieper River and its recent appearance in the Dnieper Reservoir. *Zool. Zhur.* 30:134–136.

Zhuravel, P. A. 1967. Formation of new Caspian fauna nidi in water bodies of different climatic zones of the USSR. *Zool. Zhur.* 46:1152–1162.

CHAPTER 20

Variation in the Quagga Mussel (*Dreissena rostriformis bugensis*) with Emphasis on the Deepwater Morphotype in Lake Michigan

Thomas F. Nalepa, Vera Pavlova, Wai H. Wong, John Janssen, Jeffrey S. Houghton, and Kerrin Mabrey

CONTENTS

Abstract 315
Introduction 316
Study Sites 317
Methods 317
    Siphon Lengths and Shell Form 317
    Shell Dimensions 319
    Density and Biomass in Lake Michigan 320
Results 321
    Relative Length of Inhalant Siphons 321
    Shell Height and Shell Length 322
    Density and Biomass in Southern Lake Michigan 323
Discussion 325
Acknowledgments 328
References 328

ABSTRACT

Two morphological features (relative length of the inhalant siphon to shell length and relative shell height to shell length) of the deepwater morphotype of the quagga mussel (*Dreissena rostriformis bugensis*) were examined among individuals collected at three depths (25, 45, and 93 m) in Lake Michigan. Based on laboratory observations, mean siphon length to shell length ratios were 0.19, 0.44, and 0.20 at the three depths, respectively, and the mean ratio was significantly greater at 45 m than at the other two depths. Mean ratio of in situ deepwater morphs estimated from still-capture frames of underwater video taken at 53–57 m in Lake Michigan was 0.74 but ranged from 0.57 to 0.78 depending on the location. In comparison, the ratio of quagga mussel shallow morphs from western Lake Erie, as observed in the laboratory, was 0.19. We speculate that longer siphons relative to shell lengths at 45 m are an adaptation to greater mussel densities at this depth compared to 25 and 93 m. Siphon lengths of in situ mussels also appeared to be longer relative to shell lengths at locations where mussels were more dense. A longer siphon provides more flexibility in placement and would thus offer a competitive advantage in obtaining food when population densities are high. The mean shell width to shell length ratio of the deepwater morph in Lake Michigan was 0.52, and there was no difference in ratios between the three depths. Water temperatures at the three depths in Lake Michigan were broadly different during seasonal stratification (June to October) and, hence, our results do not agree with previous studies that showed water temperature plays an important role in width to length ratios in shells of quagga mussels. However, some shell plasticity was apparent as ratios decreased with increased shell length regardless of depth. For all individuals, the mean ratio between size categories ranged from 0.54 (shell length 5–10 mm) to 0.48 (shell length > 20 mm), and this difference was significant. The mean ratio for deepwater morphs collected at a 26 m site in a Russian reservoir was 0.54, which was comparable to our mean value, but shell width to shell length ratios for deepwater morphs from the Russian reservoir did not differ significantly among size categories. The mean shell width to shell length ratio for shallow morphs from western Lake Erie was 0.59, which was comparable to values for shallow morphs from the Russian reservoir and Lake Mead (southwestern United States). Similar to the deepwater morph in Lake Michigan, ratios for the shallow morph also decreased with shell size for individuals from western Lake Erie and the Russian reservoir, but did not decrease for individuals from Lake Mead. Overall, the deepwater morphotype in Lake Michigan showed great variation in both siphon length and shell dimensions, which may reflect an ability to readily adapt to various habitat conditions.

INTRODUCTION

After the zebra mussel *Dreissena polymorpha* became established in North America in the late 1980s (Hebert et al. 1989, Carlton 2008), a distinctly different dreissenid was discovered in 1989 in a shallow region of eastern Lake Erie (Mills et al. 1993). May and Marsden (1992) examined specimens of this new dreissenid and found it to be genetically distinct from *D. polymorpha* and gave it the common name "quagga mussel." Based on morphological and genetic characteristics, the quagga mussel was later identified as *Dreissena bugensis* (Rosenberg and Ludyanskyi 1994, Spidle et al. 1994). Later genetic studies showed *D. bugensis* was not a separate species from the Caspian Sea species *D. rostriformis*, but since the former was only found in freshwater and the latter only in mesohaline waters (salinities 3–12 ppt), it was hypothesized that *D. bugensis* was a freshwater subspecies of *D. rostriformis* and hence given the taxonomic designation of *D. rostriformis bugensis* (Therriault et al. 2004).

At about the same time that quagga mussels were discovered in North America, another distinct dreissenid form was found in a relatively deep region of the eastern basin of Lake Erie (>30 m) and termed quagga mussel "profunda" (Dermott and Munawar 1993). Compared to the typical quagga mussel, shells of profunda were elongated and less convex, had rounded ventral edges, and were swollen anteroventrally (Figure 20.1). Physiological differences were also apparent; the profunda variant spawned at a lower temperature than typical quagga mussel, which indicated it was well suited to colder temperatures found in deeper regions (Claxton and Mackie 1998). Despite apparent morphological

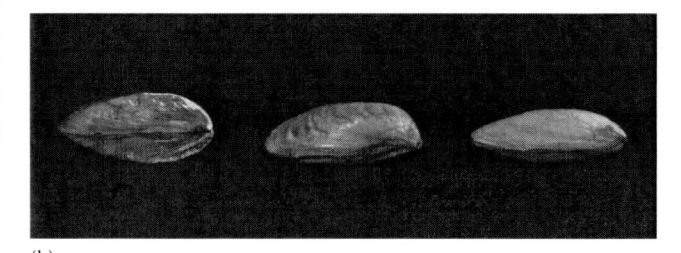

Figure 20.1 Shells of zebra mussel (left), quagga mussel shallow morph (center), and quagga mussel-deepwater morph (right). (a) Dorsal view. (b) Lateral view. Mussels have a shell length of about 24 mm.

and physiological differences, a number of genetic studies found that the profunda variant was not a species separate from the typical quagga mussel (hereafter called shallow morph), but rather a deepwater phenotype (Spidle et al. 1994, Claxton et al. 1998) (hereafter called deepwater morph).

In this chapter we explore variation in morphological features of the deepwater morph, specifically the length of the inhalant siphon and the relative form of the shell. When the deepwater morph was first reported in the eastern basin of Lake Erie, it was noted that the inhalant siphon was elongated and could extend up to 40% of shell length (Dermott and Munawar 1993). This siphon length was greater than found in the shallow morph, and it was suggested that perhaps the greater length was an adaptation to deep, profundal regions. In deep regions, sediments are generally soft, and the longer siphon would allow this morphotype to feed above inorganic, suspended particles usually present at the sediment–water interface of such substrates. In addition, the deepwater morph has a tendency to partially burrow in sediments (Dermott and Munawar 1993), and a longer siphon would be an advantage since it could be extended farther into overlying waters despite the shell being partially buried. While this morphological feature can be an important adaptation to inhabiting deeper regions, variability in the inhalant siphon of the deepwater morph has not been closely examined.

Another feature of the deepwater morph is an elongated shell. Dermott and Munawar (1993) noted that the ratio of shell height to shell length (SH/SL) of deepwater morphs from deeper regions of Lake Erie was significantly lower (by 13%) compared to shallow morphs. Additional measurements of shell morphology over wider areas in Lakes Erie and Ontario confirmed the different shell forms of the two morphs (Claxton et al. 1998). Again, the elongated form along with a more rounded ventral edge of the shell may be an adaptation to inhabiting soft sediments (Peyer et al. 2011). While such a shell form is not efficient for directed movement across substrates, it may allow more effective burrowing in soft sediments (Peyer et al. 2011). Further, the shell form of quagga mussels can be highly plastic. In laboratory experiments, shells of both quagga mussel morphs became more elongated after 2–3 years of exposure to cold temperatures generally found in hypolimnetic regions (6°C–8°C). Hence, consistently cold water temperatures may promote shell adaptations suited for such a habitat.

Recently, the deepwater morph was reported for the first time in eastern Europe (Pavlova 2012). It was found in the deepest region (26 m) of Cheboksary Reservoir, which is located in the middle reaches of the Volga River system, Russia. While the shallow morph was present in shallower regions of this reservoir, the deepwater morph was clearly distinguished by the elongated shell. As found in Lake Erie, the ratio of SH/SL of the deepwater morph in Cheboksary Reservoir was 13% lower than the shallow morph found in the same reservoir. It was also noted that the deepwater morph had an inhalant siphon that was twice as long as the shallow morph (Pavlova 2012).

Given such variability in the genotype, we examined the consistency of the inhalant siphon and shell form of the deepwater morph relative to environmental conditions. We determined the ratio of siphon length to shell length (SP/SL) and the ratio of SH/SL of this morph at sites located at three depths in Lake Michigan: 25, 45, and 93 m. These depths are characterized by widely different temperatures, substrates, and water turbulence. Hence, comparisons across depths should provide some estimate of how responsive these two morphological features are to environmental conditions. For perspective, relative siphon length and shell dimensions were also determined for shallow morph mussels collected from a site in western Lake Erie. Further, shell dimensions were measured in quagga mussels from various depths in Lake Mead, located in the southwestern United States, and in Cheboksary Reservoir, Russia. Quagga mussels became established in Lake Mead in 2007 (Benson 2013). Finally, we provide recent distributions and abundances of the deepwater morph in southern Lake Michigan. These data extend temporal trends given in Nalepa et al. (2010) and provide a context for discussions of depth-related differences in morphological features of this morph.

STUDY SITES

As noted, the three collection sites in Lake Michigan were characterized by greatly different environmental conditions. The 25-m site was located above the thermocline, and the 45- and 93-m sites were located below the thermocline, and hence mussels were subjected to widely different temperature regimes during the stratified period (June to October). Near-bottom temperatures at 25 m reach 20°C during the summer stratified period and can vary widely because of upwellings, longshore currents, and seiches (Nalepa et al. 2010). Bottom temperatures at 45 m are more stable and typically range between 4°C and 6°C during the summer, but can reach 11°C during fall turnover. Bottom temperatures at 93 m generally do not exceed 5°C (Nalepa et al. 2010). Substrates at the 25-, 45-, and 93-m sites were sand, sandy silt, and silt, respectively. The substrate at 25 m is unstable as bottom material is continually resuspended by wave-induced turbulence as evidenced by ripple marks in the sand, and by the general absence of accumulated fine particulates. With increased depth, turbulence is diminished, substrates become more stable, and fine organic material associated with substrates increases as this material settles and accumulates.

METHODS

Siphon Lengths and Shell Form

The three sites were located in southeastern Lake Michigan off Muskegon, Michigan. More specifically, locations were 43°12.00′ N, 86°22.67′ W for the 25-m site (M-25); 43°11.43′ N, 86°25.72′ W for the 45-m site (M-45); and 43°12.00′ N, 86°31.00′ W for the 93-m site (X-2) (Figure 20.2). Specimens of the deepwater morph were collected at each site in September 2011 using a Ponar grab. Only deepwater morphs have been found at these sites (Nalepa et al. 2010). Mussels were transported in coolers and kept at 4°C in lake water until siphon observations were initiated. Lake water was collected at mid-depth of the hypolimnion at site M-45 at the same time mussels were collected. Within 48 h of collection, mussels of different sizes were randomly chosen from each site, superglued onto glass vials, and randomly placed into 10 L aquariums filled with lake water. Vials were placed such that mussel shells were oriented lengthwise to the aquarium wall. When mussels were attached to the vials, care was taken so that glue was placed only on one shell valve so that opening/closing of valves was not impaired. By being glued, each individual remained at the exact same position relative to the aquarium wall during observational periods. Thus, photographs taken from the side would consistently capture the extent of siphon extension relative to shell length. Aquariums were kept in a walk-in cold room at 4°C under lighted conditions. Each day over a 7-day period, mussels were fed a culture of *Cryptomonas* or a commercial mixture designed for

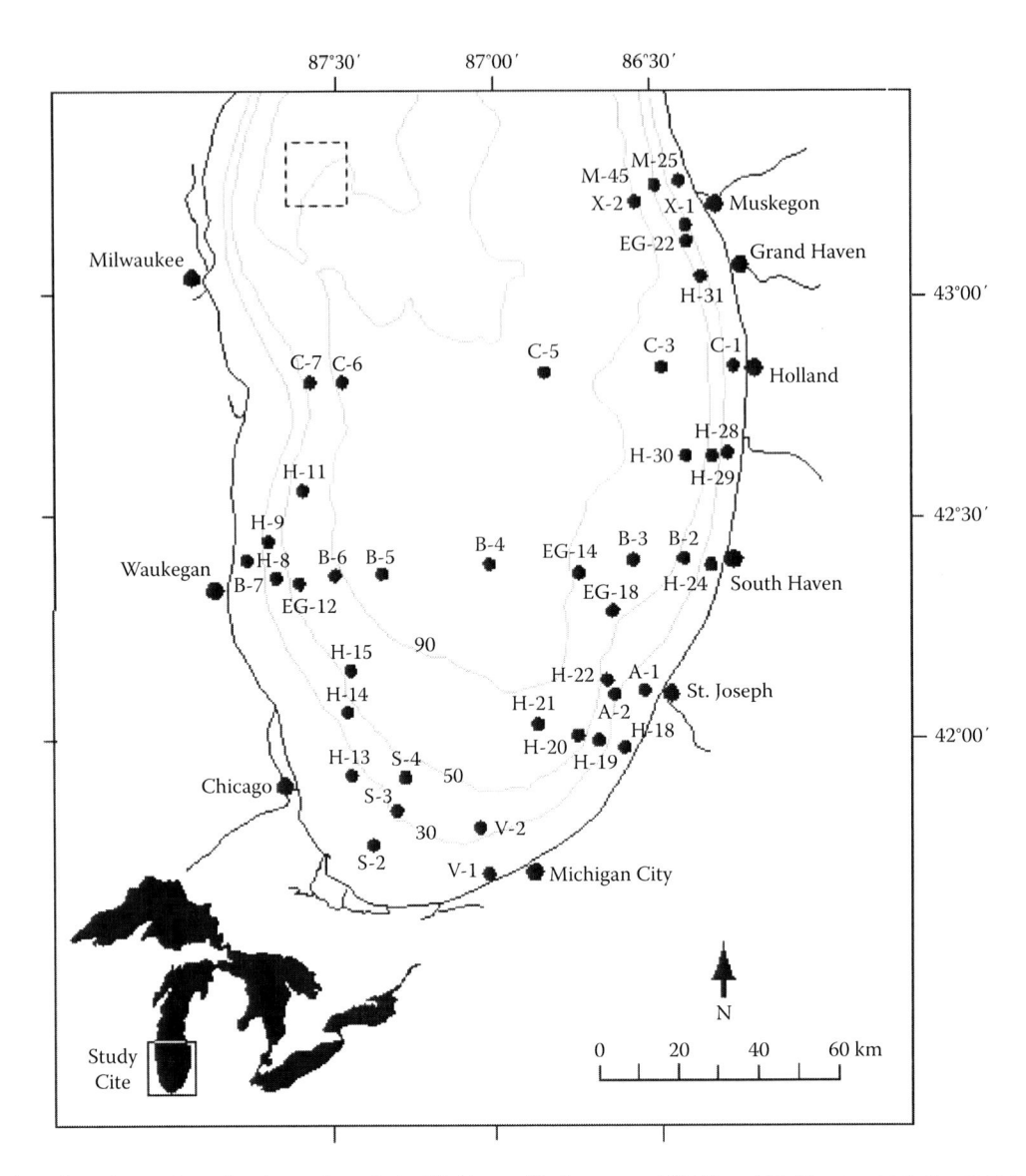

Figure 20.2 Location of sampling sites in southern Lake Michigan. All sites except M-25 and M-45 were part of a long-term monitoring program to assess changes in the benthic community. Quagga mussels for measurements of inhalant siphon lengths and shell dimensions were collected at M-25, M-45, and X-2. The dashed box shows the location of the mid-lake reef where in situ estimates of relative siphon lengths were obtained from videos taken with a ROV. Depth contours are at 30, 50, and 90 m.

shellfish (see Peyer et al. 2011). Photographs were taken 1–4 h after mussels were fed. During the 7-day observation period, a few mussels became unglued and fell to the aquarium bottom. If these mussels displayed extended siphons and were oriented such that both siphon and shell lengths could be accurately photographed, they were included in the data set.

Shallow morphs were collected at a 7-m site in western Lake Erie (41°50.90′ N, 83°13.26′ W) in March 2012. Laboratory procedures followed those for deepwater morphs from Lake Michigan except mussels were kept for about a month until observations were initiated. During this holding period, mussels were kept in an aerated aquarium at 4°C and periodically fed commercial shellfish food.

At the end of the observation period, all photographs were reviewed, and a photograph of each mussel was chosen to best represent maximum siphon extension. A siphon was deemed fully extended if the opening was flared, or if it was the same width as the siphon itself (Figure 20.3). Lengths of siphons and shells were measured to the nearest 0.1 mm using Image-Pro software. Siphon length was measured from a line drawn between the end of the two valves and the midline of the siphon opening (Figure 20.3b). Shell heights and lengths were also measured.

In situ measurements of siphon length and corresponding shell length of deepwater morphs were obtained from still-capture frames from videos taken with a remotely operated vehicle (ROV) at the mid-lake reef in southern Lake Michigan (Figure 20.2, Nalepa et al. 2013). Videos were taken at various locations between 53 and 57 m on Northeast Reef (located about 43°15.0′ N and 87°35.0′ W). Measurements

Figure 20.3 Image of a deepwater morph with inhalant siphon not fully extended (a) and an image of the same mussel with inhalant siphon fully extended (b). Note that the siphon opening is fully open in the latter photo, and hence the mussel was considered to be feeding and the siphon fully extended. This mussel was collected from 45 m in Lake Michigan.

were taken only on those individuals in which both siphon length and corresponding shell length could be readily discerned and/or estimated (see Figure 20.4). Measurements of siphon lengths and shell lengths of the deepwater morph were also obtained on a few individuals in Cheboksary Reservoir, Russia (Figure 20.5). These individuals were collected with a grab, immediately placed in containers with lake water, and photographed when siphons opened and became extended.

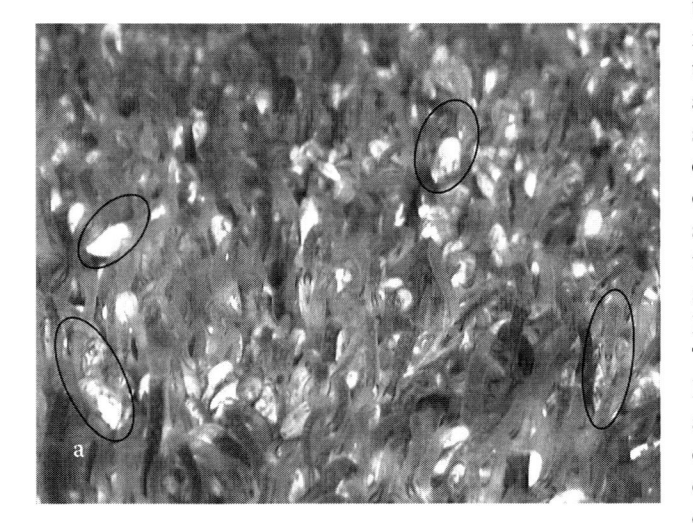

Figure 20.4 (See color insert.) Close-up image of deepwater morphs taken with an ROV on the mid-lake reef (Northeast Reef; see Figure 4a, Houghton et al. 2013) in Lake Michigan. Note the extension of inhalant siphons with openings facing in different directions. The circled individuals show examples of mussels in which relative measurements of siphon length and shell length were obtainable.

Shell Dimensions

Deepwater morphs for measurements of shell dimensions were collected at each of the three sites in Lake Michigan in April, July, and October, 2010, except at the 93-m site where no mussels were collected in October. For each site and date, bottom samples were taken in triplicate with a Ponar grab, contents washed through an elutriation device fitted with a 0.5 mm mesh screen, and retained material preserved in 5% buffered formalin. In the laboratory, retained material was placed in a white enamel pan, and, under a low-power magnifying lamp (1.5×), all dreissenids >5 mm were picked, counted, and placed in vials containing 5% buffered formalin. Individuals <5 mm were only counted. Samples with high numbers of mussels were subsampled by selecting a random quadrant of individuals that were evenly distributed in a gridded tray. Shell dimensions of all individuals >5 mm were obtained by a scanner linked to an IDL software program (Nalepa, unpublished). The scanning program provided shell height and shell length of mussels placed on the scanner screen with an accuracy of ±0.1 mm. The ratio between SH/SL was determined for each mussel and ratios compared between depths. For analysis, mussels were placed into four size categories: 5–10, 10–15, 15–20, and >20 mm, and differences were tested using a two-way ANOVA (size category × depth). Thus, any differences in size-frequency distributions between the three depths would not bias results. Shell dimensions of shallow morphs from western Lake Erie were obtained from individuals used in the siphon-length observations (n = 35).

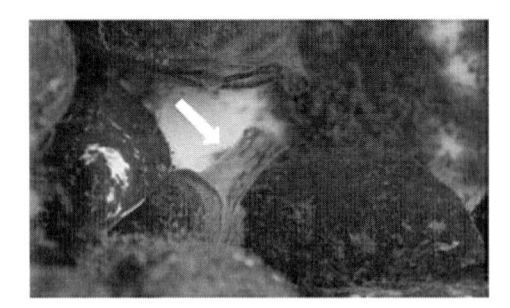

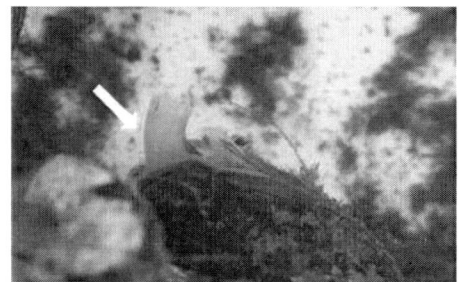

Figure 20.5 Images of the deepwater morph collected from a 26-m site in Cheboksary Reservoir, Russia. Images were taken immediately after collection. Arrows mark the location of inhalant siphons. Note the extension of inhalant siphons relative to shell lengths in individuals in the bottom image.

In addition to quagga mussels from Lake Michigan and western Lake Erie, shell heights and lengths were measured on quagga mussels from Lake Mead, southern United States, and from Cheboksary Reservoir, Russia. For Lake Mead, measurements were obtained on 240 quagga mussels collected from depths of <1, 3, 6, 18, 24, and 30 m using either Image-Pro software or the scanner. Based on having a sinusoidal rather than a straight ventral edge (Figure 20.1), these individuals were considered to be the shallow morph. For Cheboksary Reservoir, shell heights and lengths were obtained for both the shallow morph (n = 454) and the deepwater morph (n = 160). The shallow morph was collected at three sites <10 m, and the deepwater morph was collected at a 26-m site. Shell dimensions were measured with vernier calipers. Pavlova (2012) provided shell dimensions of the two morphs at several of the same sites in Cheboksary Reservoir. However, shell dimensions in this study were given only for individuals that were 12–14 mm in shell length. The present study greatly expands on this previous effort since all sizes of individuals from the sites were measured. For example, the 12–14 mm size range only comprised 10% of all deepwater morphs and 11% of all shallow morphs measured in the present study.

Density and Biomass in Lake Michigan

Trends in density and biomass of dreissenids in the southern basin of Lake Michigan were derived from surveys at 40 sites that were part of a long-term benthic monitoring program (Figure 20.2; Nalepa et al. 1998, 2010). Zebra mussels were first reported in southern Lake Michigan in 1989 (Marsden et al. 1993), and the first post-mussel surveys were conducted in 1992 and 1993. Thereafter, surveys were conducted each year between 1998 and 2010 except in 2009 when no survey occurred. Sites were sampled in late summer/fall, but in some years additional surveys were conducted in spring and midsummer. For these years, mean values were determined. All samples were collected and processed in the laboratory as given above. Biomass at each site (ash-free dry weight of tissue) was estimated from derived size-frequency distributions and length–weight relationships. Mussels for length–weight determinations were collected at select sites located above and below the thermocline. Separate relationships were determined for zebra mussels and quagga mussels. For the three sites in Lake Michigan where siphon lengths and shell dimensions were measured, biomass estimates were derived from length-weight regressions determined at each site in spring, summer, and fall in 2010. Further details of biomass calculations are provided in Nalepa et al. (2009, 2010). For comparisons of long-term trends, the 40 sites were placed into four depth intervals: 16–30 m (n = 11), 31–50 m (n = 12), 51–90 m (n = 11), and >90 m (n = 6). Trends in zebra mussels and quagga mussels between 1992 and 2008 are given in Nalepa et al. (2010). Here, we provide an update of these trends through 2010.

RESULTS

Relative Length of Inhalant Siphons

Of the 45 deepwater morphs and 35 shallow morphs that were photographed over the 7-day observation period, 32 of the former and 19 of the latter were deemed suitable for estimating the ratio of SP/SL. Others were oriented such that a defined ratio could not be discerned, or siphons were deemed never to be fully extended. Mussels usually extended siphons vertically upward within 2 days. For one deepwater morph, the siphon initially extended downward, but over the course of the next 5 days, it progressively moved upward (Figure 20.6). Another deepwater morph that had become dislodged and remained on the aquarium bottom kept its siphon extended downward and appeared to be utilizing food material that had settled to the bottom (Figure 20.7).

Mean SP/SL ratios of deepwater morphs from 25, 45, and 93 m in Lake Michigan were 0.19, 0.44, and 0.20, respectively (Table 20.1). Ratios were significantly different between depths (ANOVA, $F_{[2,29]}$ = 27.66, P < 0.001), with ratios at 45 m significantly greater than ratios at the other two depths (Tukey's HSD, P < 0.001). Shell lengths ranged from 10.9 to 25.4 mm, and lengths were not significantly different between depths (ANOVA, $F_{[2,29]}$ = 1.39, P = 0.27). Although there was some tendency for SP/SL ratios to decrease with shell length at 25 and 93 m (Figure 20.8a and c), the relationship was not significant at any of the depths (linear regression, P ≥ 0.05 for each depth). The mean SP/SL ratio for the shallow morph from western Lake Erie was 0.16. Although lower, this mean ratio was not significantly different (ANOVA, $F_{[2,37]}$ = 2.35, P = 0.11) from ratios for the deepwater morph at 25 and 93 m in Lake Michigan. In contrast to deepwater morphs at the three depths in Lake Michigan,

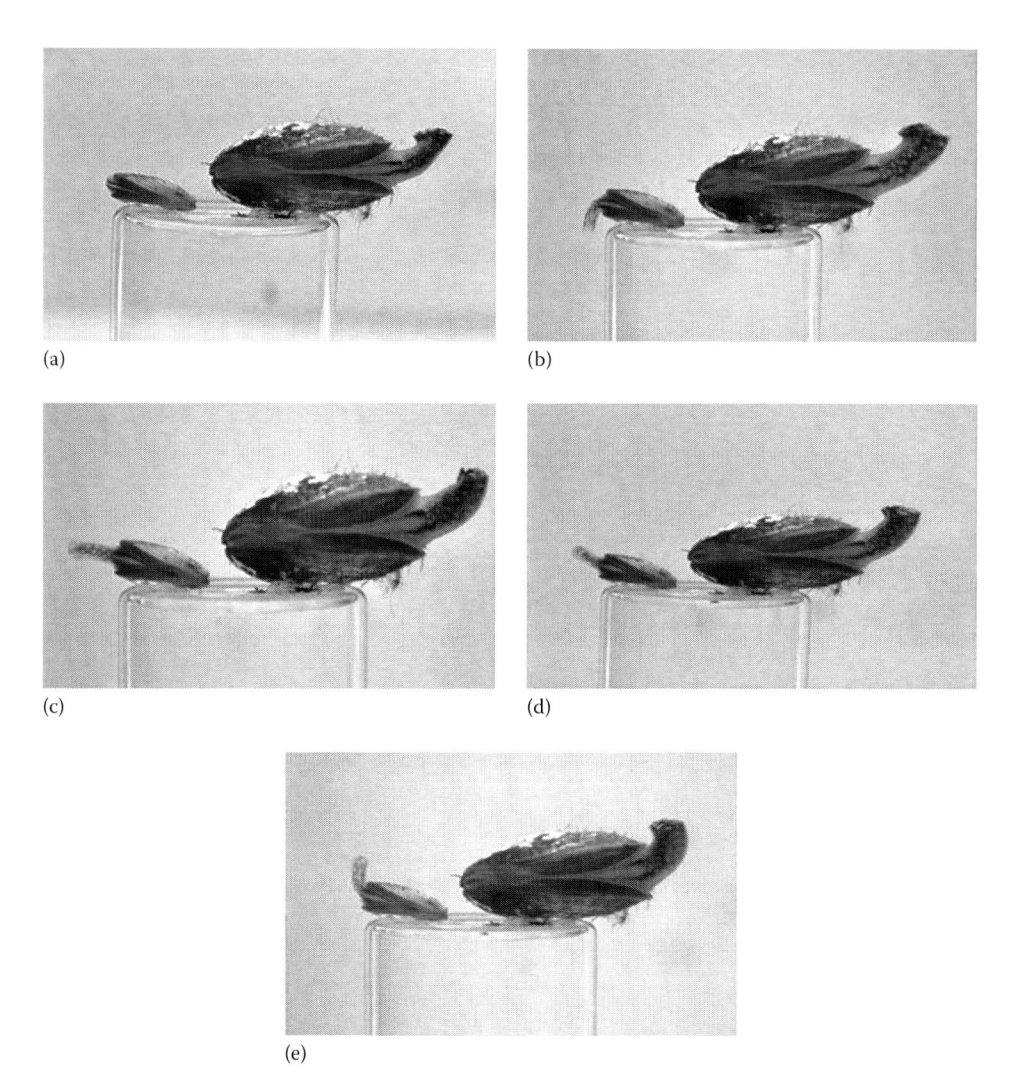

Figure 20.6 Images of two deepwater morphs from the 45 m site at various times over the 7 day observational period: day 1 (a), day 2 (b), day 3(c), day 5 (d), and day 6 (e). The mussel on the left initially extended its inhalant siphon downward, but over time moved the siphon to face upward. In contrast, the mussel on the right kept its siphon extended upward during the entire period. Relative siphon length of the mussel on the left was typical of all mussels from the 45-m site in Lake Michigan.

Figure 20.7 Image of deepwater morph from the 45-m site that had become detached and remained on the aquarium bottom. Note the siphon is faced downward, perhaps because the mussel was feeding on settled detritus.

there was a strong negative relationship between SP/SL ratio and shell length for the shallow morph (linear regression, P > 0.05) (Figure 20.8d). Typical examples of relative siphon lengths for the deepwater morph at the three depths in Lake Michigan and the shallow morph in western Lake Erie are given in Figures 20.6 and 20.9.

Measurements of in situ SP/SL ratios were obtained from two different habitats/substrates at Northeast Reef, southern Lake Michigan. In portions of the video where the ROV stopped and the camera obtained close-up views of extended siphons, mussels were attached to flat bedrock, or to vertical faces of large boulders/rock outcrops (Nalepa et al. 2013). For each habitat/substrate, there were two separate locations where close-ups were obtained, and a total of 16 mussels were identified that had both siphon length and shell length readily visible and measurements obtainable (e.g., see Figure 20.4). As seen in Nalepa et al. (2013), mussel colonies were dense at the habitat locations with flat bedrock, and siphons were closely proximate and extended vertically. Mussels were not as dense on vertical faces of the rock outcrops, and measurements could only be obtained on three mussels. Mean SP/SL ratios obtained from all in situ mussels were 0.74 ± 0.06. Ratios were significantly greater than ratios obtained from laboratory observations of mussels from the 45 m site (t-test, $T_{[1,25]} = 3.95$, $P = 0.001$). While the mean ratio of mussels within dense colonies (0.78 ± 0.10, flat bedrock) was higher than the mean ratio of mussels where densities were not as great (0.57 ± 0.08, rock outcrop), this difference was not significant (t-test, $T_{[1,14]} = 1.50$, $P = 0.16$). SP/SL ratios were obtained for only two deepwater morph individuals from Cheboksary Reservoir (26 m), and ratios were 0.33 and 0.21 (see Figure 20.5).

Shell Height and Shell Length

The relationship between shell height and shell length (SH/SL ratio) was determined for 2956 mussels collected from the three depths in Lake Michigan in 2010. The proportion of mussels in the 5–10, 10–15, 15–20, and >20 mm size categories was 46.7%, 26.8%, 17.6%, and 8.9%, respectively (Table 20.2). The overall mean (±SE) ratio was 0.52 ± 0.01, and there was no significant difference (two-way ANOVA, $F_{[2,2950]} = 0.15$, $P = 0.86$) between depths. However, across all depths there was a significant difference ($P < 0.01$) in ratios between size categories. The ratio declined as mussels increased in size, with overall mean ratios ranging from 0.54 for mussels in the 5–10 mm category to 0.48 for mussels in the >20 mm category (Table 20.2). Further analysis indicated a significant negative relationship (linear regression, $P < 0.001$) between the SH/SL ratio and shell length at each depth (Figure 20.10a through c). For deepwater morphs from the 26-m site in Cheboksary Reservoir, the mean SH/SL ratio was 0.55 ± <0.01, and there was no significant difference (ANOVA, $F_{[3,156]} = 2.47$, $P = 0.06$) between size categories (Table 20.2).

Mean SH/SL ratios of shallow morphs from western Lake Erie, Cheboksary Reservoir, and Lake Mead across all depths were generally similar; mean ratios were 0.59 ± <0.01, 0.60 ± <0.01, and 0.58 ± <0.01, respectively. For the shallow morph from western Lake Erie and Cheboksary

Table 20.1 Mean (±SE) Inhalant Siphon Length to Shell Length Ratio (SP/SL) of the Deepwater Morph Quagga Mussel from 25, 45, and 93 m in Southern Lake Michigan and Mean Ratio of the Shallow Morph Quagga Mussel from 7 m in Western Lake Erie. Also Given Are the Range in Ratios, Mean (±SE) Shell Length (mm) of Mussels from Which Ratios Were Obtained, and the Mean Length (mm) of the Siphon. The Latter Was Determined by Multiplying the Ratio by Actual Shell Length. Numbers of Mussels Are Given in Parentheses

Phenotype/Depth	SP/SL Ratio	Ratio Range	Shell Length	Siphon Length
Deepwater morph				
25 m	0.19 ± 0.02 (11)	0.10–0.27	15.2 ± 1.2	2.9
45 m	0.44 ± 0.04 (11)	0.23–0.60	18.0 ± 1.4	7.9
93 m	0.20 ± 0.03 (10)	0.10–0.40	16.1 ± 1.2	3.2
Shallow morph				
7 m	0.16 ± 0.01 (19)	0.08–0.26	18.2 ± 1.3	2.9

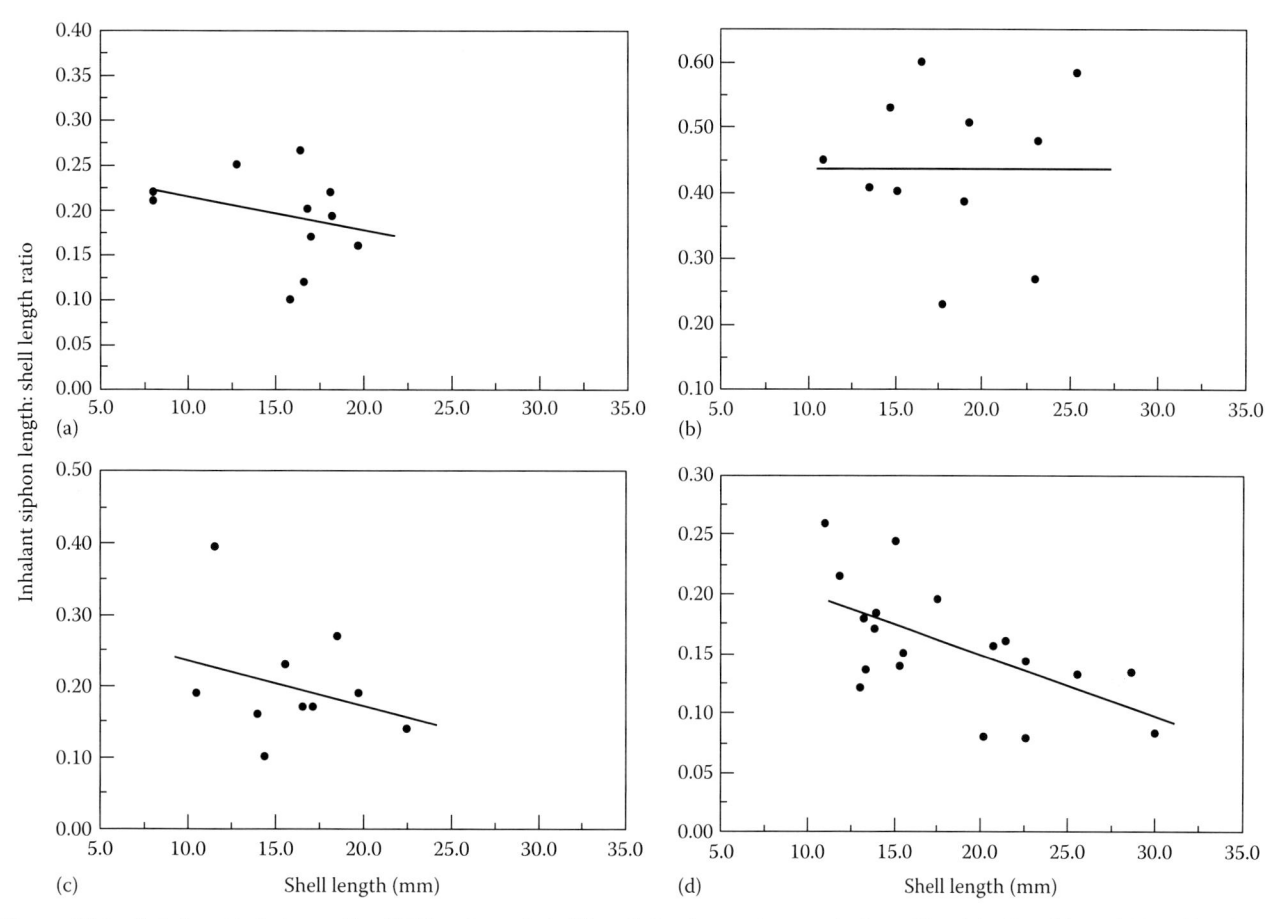

Figure 20.8 Relationship between the SP/SL ratio and shell length for deepwater morphs from 25 m (a), 45 m (b), and 93 m (c) in Lake Michigan and shallow morphs (d) from Lake Erie. Linear regression analysis indicated the relationship was not significant (P>0.05) for deepwater morphs from each of the three depths, but was significant (P < 0.05) for shallow morphs.

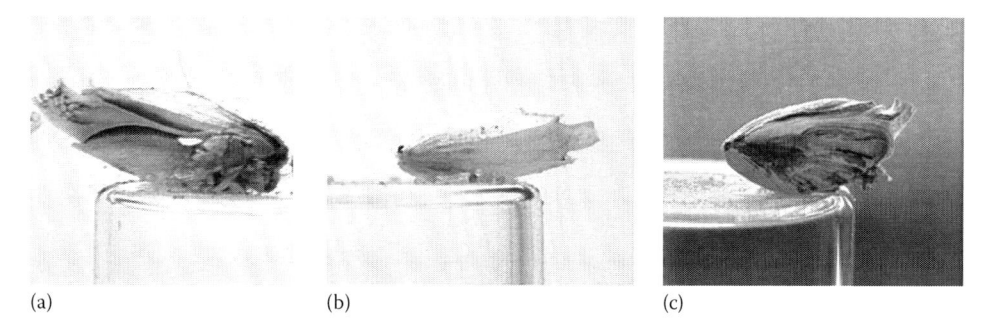

Figure 20.9 (See color insert.) Images of typical siphon lengths of the deepwater morph from the 25-m site (a) and the 93-m site (b) in Lake Michigan and of shallow morph from a 7-m site in western Lake Erie (c). Relative siphon lengths were near the mean siphon lengths of all mussels at these sites. The typical siphon length for mussels at the 45-m site in Lake Michigan is shown in Figure 20.6b through e (mussel on the left).

Reservoir, there was a significant negative relationship (linear regressions, P < 0.01) between SH/SL ratio and shell length (see Figure 20.10d for western Lake Erie). However, for Lake Mead mussels the relationship between the SH/SL ratio and shell length was not significant (linear regression, P > 0.01). Since shallow morphs were collected at six depths in Lake Mead, we also examined differences in the SH/SL ratio between depths. There was a significant difference (ANOVA, $F_{[5,234]}$ = 3.88, P < 0.05) between depths, with a general tendency for lower ratios at deeper depths. Mean ratios for mussels at depths of <1, 3, 6, 18, 24, and 30 m were 0.59, 0.59, 0.59, 0.57, 0.56, and 0.58, respectively.

Density and Biomass in Southern Lake Michigan

Quagga mussels were first found in Lake Michigan in 1997, but were not reported from the southern basin until 2001 (Nalepa et al. 2010). After establishment, densities in this basin

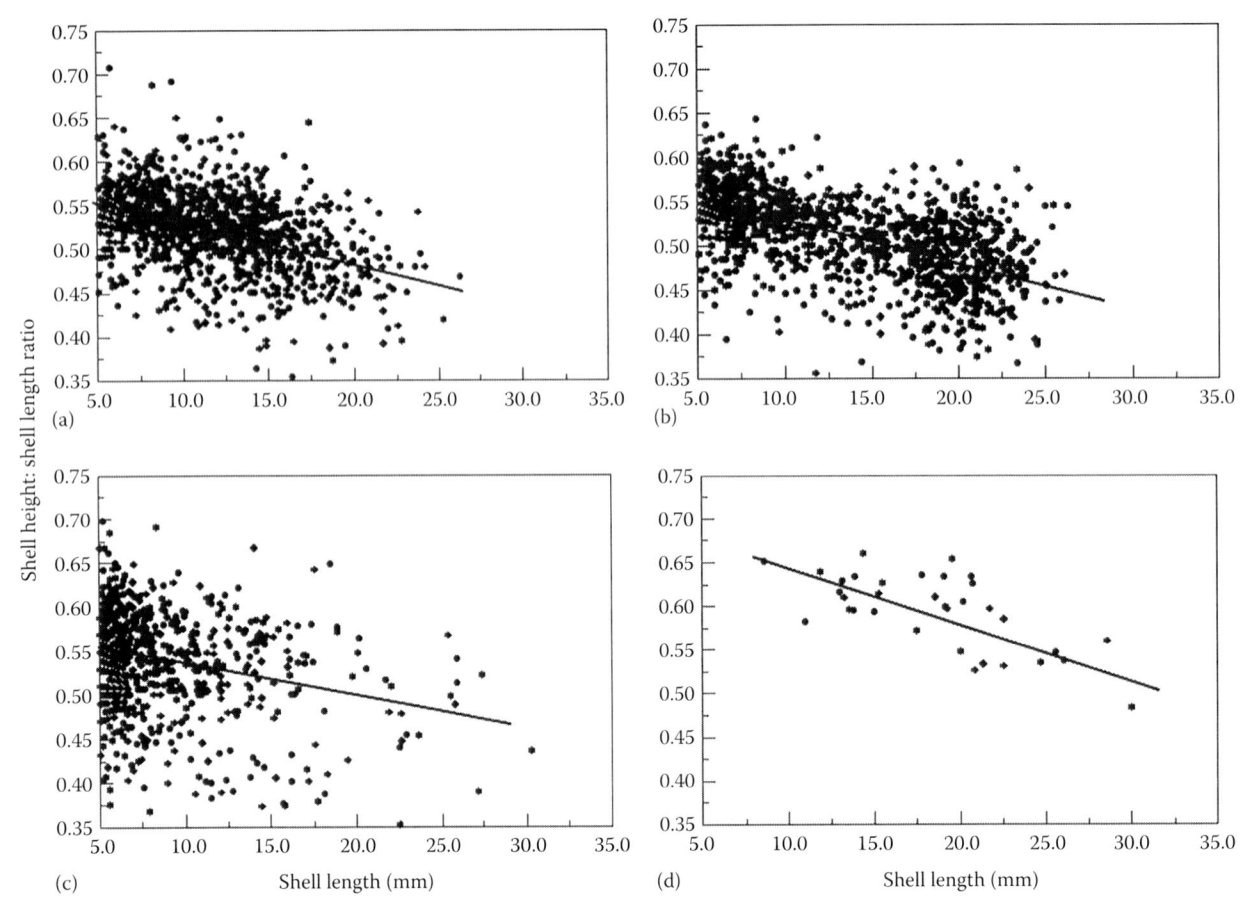

Figure 20.10 Relationship between SH/SL ratio and shell length for deepwater morphs from 25 m (a), 45 m (b), and 93 m (c) in Lake Michigan and shallow morph from western Lake Erie (d). Linear regression analysis indicated the relationship was significant (P > 0.05) at all sites.

Table 20.2 Mean (±SE) Relationship between Shell Height and Shell Length (SH/SL Ratio) of the Deepwater Morph Quagga Mussel from 25, 45, and 93 m in Southern Lake Michigan in Each of Four Size Categories Based on Shell Length: 5–10, 10–15, 15–20, and >20 mm. Values Derived from Mussels Collected in Spring, Summer, and Fall at 25 and 45 m and in Spring and Summer at 93 m. Also Given Are Ratios for the Deepwater Morph from a Site at 26 m in Cheboksary Reservoir, Russia. Numbers of Mussels Are Given in Parentheses

Water body/depth	Shell Length (mm)			
	5–10	10–15	15–20	>20
Lake Michigan				
25 m	0.54 ± 0.04 (451)	0.52 ± 0.04 (444)	0.50 ± 0.04 (192)	0.47 ± 0.04 (38)
45 m	0.54 ± 0.04 (428)	0.51 ± 0.04 (221)	0.49 ± 0.04 (287)	0.48 ± 0.04 (205)
93 m	0.54 ± 0.03 (502)	0.53 ± 0.06 (126)	0.50 ± 0.07 (41)	0.49 ± 0.06 (21)
Cheboksary Reservoir				
26 m	0.55 ± <0.01 (45)	0.54 ± <0.01 (35)	0.55 ± <0.01 (45)	0.56 ± <0.01 (48)

increased sequentially with increased depth; that is, densities initially increased at sites in the 16–30 and 31–50 m depth intervals in 2001–2002, increased at sites in the 51–90 m interval in 2005, and increased at sites in the >90 m interval in 2007 (Figure 20.11). Densities in 2010 were generally consistent with trends observed through 2008 and reported in Nalepa et al. (2010). At sites in the 16–30 m interval, trends remained highly variable, and temporal patterns could not be easily discerned. The population at this interval is dominated by new recruits (<5 mm) (Nalepa et al. 2010), and these small individuals can vary widely depending on the time of settlement and lake conditions (i.e., waves and storms). Trends at the three deeper intervals are more indicative of population patterns in the lake. Densities in 2010 provided evidence that populations at the 31–50 and 51–90 m intervals were no longer increasing, but rather were in the early stages of decline. At 31–50 m, densities

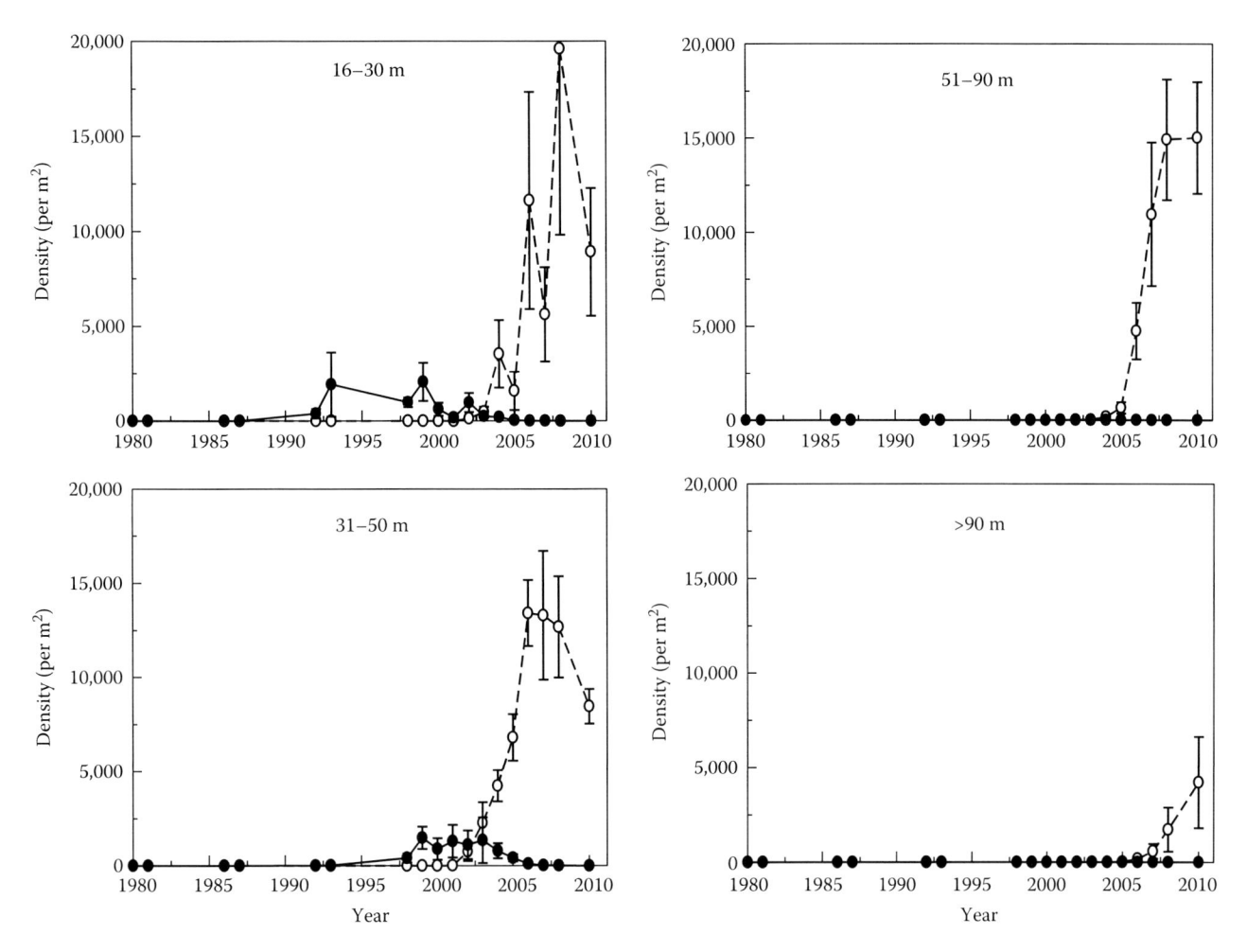

Figure 20.11 Mean (±SE) density (no./m²) of dreissenids at each of four depth intervals in southern Lake Michigan, 1992–2010. Zebra mussels = solid circle/solid line; quagga mussels = open circle/dashed line.

increased to a peak in 2006, remained stable through 2008, and then declined in 2010. Mean density at this depth interval was 13,100/m² in 2006–2008, but only 8,500/m² in 2010. At the 51–90 m interval, densities increased through 2008, but densities in 2010 were similar to 2008, seemingly an indication that the population was no longer increasing, as found at 31–50 m several years earlier. Densities at >90 m were still increasing in 2010. In general, temporal trends in biomass (ash-free dry weight) followed patterns in densities (Figure 20.12). Mean biomass at the 25-, 45-, and 93-m sites where mussels were collected for siphon measurements was 19.3, 52.7, and 7.5 g/m², respectively.

DISCUSSION

Laboratory and in situ observations showed great variation in relative lengths of inhalant siphons of the deepwater morph in Lake Michigan. In the laboratory, mean ratios of SP/SL for mussels from various depths varied from 0.19 to 0.44, and in situ ratios in different habitats ranged from 0.57 to 0.78.

Given the great variation in shell morphology that has been documented for the quagga mussel genotype, it is perhaps not unexpected that relative siphon length would also be highly variable. Certainly one important result of our observations is that the deepwater morph does not consistently present an inhalant siphon that is far longer than the shallow morph as earlier studies would suggest (Dermott and Munawar 1993, Pavlova 2012).

The SP/SL ratio for mussels from 45 m was twice as great as the ratios for mussels from 25 and 93 m. We believe this difference was real and a result of a morphological adaptation to environmental conditions and not simply a variation in maximum siphon extension during laboratory observations. Most bivalves extend their inhalant siphon to the fullest extent possible when feeding (Zwarts et al. 1994), and in our laboratory assessments all individuals deemed not fully open and feeding were excluded from the data set. We are aware of the possibility that siphons of the laboratory mussels could not be fully extended because one valve was glued, and thus a full valve-gap opening could not be achieved. Indeed, this may be the reason why the mean SP/SL ratio for in situ mussels was so much greater than mean ratios

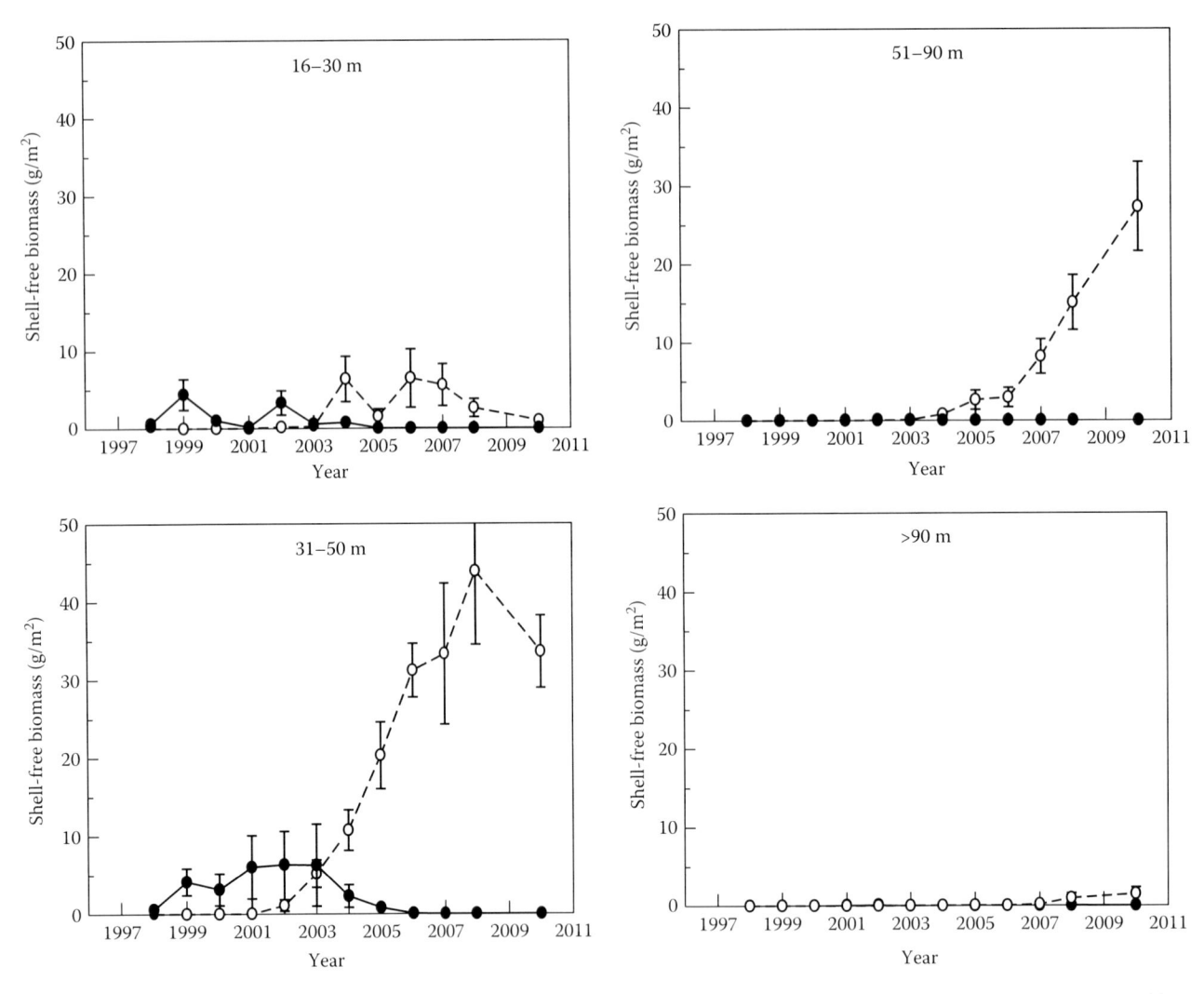

Figure 20.12 Mean (±SE) ash-free dry weight biomass (g/m²) of dreissenids at each of four depth intervals in southern Lake Michigan, 1992–2010. Zebra mussels = solid circle/solid line; quagga mussels = open circle/dashed line.

for the laboratory mussels. Despite this possible impediment all laboratory mussels regardless of depth were treated randomly and exposed to the same experimental conditions.

Given this, what environmental factor(s) may have led to observed longer siphons of mussels at 45 m? In marine bivalves, siphon length is affected by a number of environmental factors, including food availability, current velocities, temperature, and substrate type (Zwarts and Wanink 1989, Levinton 1991, Lin and Hins 1994, Kamerans et al. 1999). All these factors mainly affect burial depth and hence size of siphon needed to extend to the sediment–water interface. Based on the model for marine bivalves, physical conditions that would promote long siphons in the deepwater morph are soft substrates, low water velocities, and stable, cool temperatures. Hence, if anything, we would have expected an increase in siphon length with increased depth at our three collection sites in Lake Michigan. Also, even though the deepwater morph does partially burrow in soft substrates (Dermott and Munawar 1993; see Figure 20.13), it is not clear if such partial burial would affect siphon length.

Figure 20.13 (See color insert.) Image of deepwater morphs that shows individuals partially burrowed in the substrate. The image was taken with a camera attached to an ROV at 40 m depth in Lake Michigan off Muskegon, MI. Note individuals in the relatively open areas in the lower right of the image. The substrate was silty sand.

We suggest that competition for food as promulgated by differences in population density/biomass was the environmental driver of siphon length variation. Notably, mussel biomass at the 45-m site was 2.5 times greater than biomass at the 25-m site and 7 times greater than at the 93-m site. This spatial pattern was consistent with the depth-related pattern observed throughout the southern basin of Lake Michigan; that is, until recent declines, greatest mussel biomass in the southern basin consistently occurred in the 31–50 m interval (Figure 20.12). Even before dreissenids became established, total biomass of benthic macroinvertebrates was greatest at this depth interval in Lake Michigan (Nalepa 1989). At this interval, which is just below the thermocline, near-bottom conditions are more stable than in shallower regions, and organic material produced in the overlying water column and/or transported from more productive shallower regions by offshore currents readily settles and accumulates. Given the far greater density of dreissenids at this depth interval, competition for food is likely intense despite potentially greater amounts of available food compared to other depth intervals, and a longer siphon provides several advantages. A longer siphon enhances attainment of food particles settling from the water column and also provides greater flexibility in avoiding filtering currents of other mussels. Also, mussels in dense colonies become vertically layered as new recruits settle on established individuals, and mussels in lower layers have limited access to food (Tuchman et al. 2004). In such cases, individuals with longer siphons would again have an advantage. The SP/SL ratios of in situ mussels support this contention; mussels in dense colonies had greater ratios compared to mussels that were less densely aggregated. In situ images (Figure 20.13) and a video clip (see Nalepa et al. 2013) of dense colonies of mussels clearly show siphons directed vertically upward but faced in different directions, seemingly to avoid the inhalant currents of other mussels.

Assuming longer siphons at 45 m were an environmental response, specific genetic modes for adaptive expression in siphon length remain unclear, but seemingly a number of mechanisms can apply. Phenotypic responses to environmental conditions can be genetically regulated (1) by regulatory loci that cause other genes to be expressed in particular environments, (2) by differential survival of allelic variants, or (3) by some combination of these two mechanisms (Via et al. 1995). Phenotypic plasticity in soft tissues of mollusks is common (Trussell and Etter 2001), and variations in food availability/food conditions are often the driving mechanism. For instance, in a deposit-feeding bivalve, the gill-to-palp mass ratio in field-collected individuals was correlated to the foraging environment as characterized by the medial grain size of detrital food, and this ratio could be manipulated in laboratory-reared individuals (Drent et al. 2004). Also, larvae of a vermetid gastropod exhibited phenotypic plasticity by developing larger velums (larval feeding structure) under low food conditions compared to high food conditions (Phillips 2011). Interestingly, although not significant, there was an indication that the SP/SL ratio decreased with increased shell length for mussels at 25 and 93 m, but not for mussels at 45 m (Figure 20.8). This suggests some developmental plasticity in relative siphon length based on external drivers that were different at 45 m. As densities of the deepwater morph continue to expand at depths >90 m in Lake Michigan (Figure 20.12; Nalepa, unpublished), further measurements of relative siphon lengths of mussels at these depths should provide a better understanding of the relationship between food availability, densities, and relative siphon length.

Perhaps the most defining morphological difference between the two quagga mussel morphs is the shape of the shell. In the deepwater morph, the shell is elongated relative to the height, and the ratio between height and length is consistently lower compared to the shallow morph. In previous studies, this ratio has been reported to be 0.54 for the deepwater morph and 0.59–0.63 for the shallow morph (Dermott and Munawar 1993, Pavlova 2012). Our mean ratios were generally comparable to these previous studies; the mean ratio for the deepwater morph in Lake Michigan was 0.52, and the mean ratio for the shallow morph in western Lake Erie was 0.59.

Peyer et al. (2010) found that temperature had a strong influence on shell dimensions of quagga mussels during developmental growth. Juveniles (<5 mm) of both deepwater and shallow morphs reared for 2–3 years at water temperatures generally found in deep regions (6°C–8°C) developed shell forms that were more typical of deepwater morphs than those juveniles reared at water temperatures generally found in shallower regions (18°C–20°C). Temperatures at our three depths in Lake Michigan spanned this entire temperature range during the stratified period of 5–6 months (May–October), but were uniformly cold during the unstratified period (November–April) (Nalepa et al. 2010). Given that we did not find any difference in shell dimensions between the three depths for the deepwater morph, perhaps temperatures need to be consistently different (i.e., <8°C vs. >18°C) over long periods as in the experiments of Peyer et al. (2010) to realize developmental differences in shell dimensions. We did, however, find that shell dimensions became more typical of the deepwater morph (lower SH/SL ratio) as shell length increased regardless of depth, which would indicate developmental changes that were not influenced by the range of conditions found at the three depths. Even so, the mean SH/SL ratio of the smallest deepwater morphs (5.0–10.0 mm) was 0.54, which was above the mean ratio of 0.48 found for the largest mussels (>20 mm), but still below the mean ratio of 0.59 for the shallow morph in western Lake Erie.

Mean SH/SL ratios of shallow morphs in western Lake Erie, Cheboksary Reservoir, and Lake Mead were similar (0.58–0.60). Although collection depths in the former two water bodies were limited, collection depths in Lake Mead ranged from <1 to 30 m, and there was a tendency for lower ratios as depth increased. In the future, as the population in Lake Mead increases at deeper depths, further studies to assess shell dimensions over a wider depth range will provide additional insights into effects of temperature on shell morphology. Temperatures in Lake Mead are higher over the entire seasonal period as compared to Lake Michigan, but are still generally lower in the offshore/hypolimnion than in the nearshore/epilimnion (Wong et al. 2013). Thus, shell dimensions of shallow morphs in deeper regions of Lake Mead should be monitored to determine if the SH/SL ratio becomes more similar to the deepwater morph as might be predicted from the results of Peyer et al. (2010).

In southern Lake Michigan, the deepwater morph of the quagga mussel has achieved densities and standing stocks that far exceed those ever achieved by zebra mussels at depths <50 m, and has proliferated at depths >50 m where zebra mussels were rarely found (Figures 20.11 and 20.12). Quagga mussels have several physiological advantages over zebra mussels that allow them to colonize over a wider range of conditions and to replace zebra mussels as the dominant dreissenid where both species occur. Compared to zebra mussels, quagga mussels have higher assimilation rates, particularly at low food levels (Baldwin et al. 2002), and have lower respiration rates (Stoeckmann 2003). Thus, quagga mussels can direct more energy to growth and reproduction. Also, filtration rates of the two species are either similar or greater in quagga mussels depending on season (Diggins 2001). We suggest that, in addition to these physiological advantages, the deepwater morph has a morphological advantage by having longer siphons and the apparent ability to adapt siphon length to habitat conditions. While we did not specifically measure relative siphon length of zebra mussels, lengths are shorter or equal to the shallow morph (see Morton 1993).

Presently, there are wide differences in distributions of the two quagga mussel morphotypes in the Great Lakes. For instance, only shallow morphs and no deepwater morphs are found in Saginaw Bay, a warm, shallow arm of Lake Huron, and mostly deepwater morphs are found at depths >35 m in the main basin of Lake Huron (Nalepa, unpublished). In southern Lake Michigan, all sampled sites were located at depths >16 m, and only the deepwater morph was found (Nalepa et al. 2010). In general, distributions in the Great Lakes seem to indicate depth preferences based on temperature, but Pavlova (2012) suggested that depth-related, hydrostatic pressure may play a role. Whatever the reason, we show that the deepwater morph has great plasticity in morphological traits which may provide an advantage over both the shallow morph and zebra mussels, and additional studies are needed to examine variability in morphological and physiological traits of the deepwater morph. In particular, physiological traits of the two quagga mussel morphs should be better compared. To date, most physiological studies have compared zebra and quagga mussels alone, and frequently the morph of quagga mussel being compared is not even specified.

ACKNOWLEDGMENTS

We thank H. Vanderploeg for providing the *Cryptomonas* used in the siphon length experiments, S. Pothoven for collecting mussels in Lake Michigan, R. Paddock for providing in situ photos, and Abigail Fusario and Carol Stepien for providing useful comments. We also thank the Great Lakes Fishery Trust for its support.

REFERENCES

Baldwin, B. S., M. S. Mayer, and J. Dayton, et al. 2002. Comparative growth and feeding in zebra and quagga mussels (*Dreissena polymorpha* and *Dreissena bugensis*): implications for North American lakes. *Can. J. Fish. Aquat. Sci.* 59: 680–694.

Benson, A. 2013. Chronological history of zebra and quagga mussels (Dreissenidae) in North America, 1988–2010. In *Quagga and Zebra Mussels: Biology, Impacts, and Control*, 2nd Edn., T. F. Nalepa and D. W. Schloesser, eds., 9–31. Boca Raton, FL: CRC Press.

Carlton, J. T. 2008. The zebra mussel *Dreissena polymorpha* found in North America in 1986 and 1987. *J. Great Lakes Res.* 34: 770–773.

Claxton, W. T. and G. L. Mackie. 1998. Seasonal and depth variations in gametogenesis and spawning of *Dreissena polymorpha* and *Dreissena bugensis* in eastern Lake Erie. *Can. J. Fish. Aquat. Sci.* 76: 2010–2019.

Claxton, W. T., A. B. Wilson, G. L. Mackie, and E. G. Boulding. 1998. A genetic and morphological comparison of shallow- and deep-water population of the introduced dreissenid bivalve *Dreissena bugensis*. *Can. J. Zool.* 76: 1269–1276.

Dermott, R. and M. Munawar. 1993. Invasion of Lake Erie offshore sediments by *Dreissena*, and its ecological implications. *Can. J. Fish. Aquat. Sci.* 50: 2298–2304.

Diggins, T. P. 2001. A seasonal comparison of suspended sediment filtration by quagga mussels (*Dreissena bugensis*) and zebra mussels (*Dreissena polymorpha*). *J. Great Lakes Res.* 27: 457–466.

Drent, J., P. C. Luttikhuizen, and T. Piersma. 2004. Morphological dynamics in the foraging apparatus of a deposit feeding marine bivalve: Phenotypic plasticity and heritable effects. *Funct. Ecol.* 18: 349–356.

Hebert, P. D. N., B. W. Muncaster, and G. L. Mackie. 1989. Ecological and genetic studies of *Dreissena polymorpha* (Pallas): A new mollusc in the Great Lakes. *Can. J. Fish. Aquat. Sci.* 46: 1587–1591.

Kamerans, P., H. W. VanderVeer, I. J. Witte, and E. J. Adrians. 1999. Morphological differences in *Macoma balthica* (Bivalvia, Tellinacea), form a Dutch and three southeastern Unites States estuaries. *J. Sea Res.* 41: 213–224.

Levinton, J. S. 1991. Variable feeding behavior of three species of *Macoma* (Bivalvia: Tellinacea) as a response to water flow and sediment transport. *Mar. Biol.* 110: 375–383.

Lin, J. D. and A. H. Hins. 1994. Effects of suspended food availability on the feeding mode and burial depth of the Baltic clam, *Macoma balthica*. *Oikos* 69: 28–36.

Marsden, J. E., N. Trudeau, and T. Keniry. 1993. Zebra mussel study of Lake Michigan. Aquatic Ecology Technical Report 93/14. Illinois Natural History Survey, Champaign, IL.

May, B. and J. E. Marsden. 1992. Genetic identification of Dreissena (Bilvalvia: Dreissenidae), with identification of the quagga mussel as *Dreissena bugensis*. *Can. J. Fish. Aquat. Sci.* 49: 1501–1506.

Mills, E. L., R. M. Dermott, and E. F. Roseman et al. 1993. Colonization, ecology, and population structure of the "quagga" mussel (Bivalvia: Dreissenidae) in the lower Great Lakes. *Can. J. Fish. Aquat. Sci.* 50: 2305–2314.

Morton, B. 1993. The anatomy of *Dreissena polymorpha* and the evolution and success of the Heteromyarian from in the Dreissenoidea. In *Zebra Mussels: Biology, Impacts, and Control*, T. F. Nalepa and D. W. Schloesser, eds., 185–215. Boca Raton, FL: CRC Press.

Nalepa, T. F. 1989. Estimates of macroinvertebrate biomass in Lake Michigan. *J. Great Lakes Res.* 15: 437–443.

Nalepa, T. F., D. J. Hartson, D. L. Fanslow, G. A. Lang, and S. J. Lozano. 1998. Declines in benthic macroinvertebrate populations in southern Lake Michigan, 1980–1993. *Can. J. Fish. Aquat. Sci.* 55: 2402–2413.

Nalepa, T. F., D. L. Fanslow, and G. A. Lang. 2009. Transformation of the offshore benthic community in lake Michigan: Recent shift from the native amphipod *Diporeia* spp. to the invasive mussel *Dreissena rostriformis bugensis*. *Freshwat. Biol.* 54: 466–479.

Nalepa, T. F., D. L. Fanslow, and S. A. Pothoven. 2010. Recent changes in density, biomass, recruitment, size structure, and nutritional state of *Dreissena* populations in southern Lake Michigan. *J. Great Lakes Res.* 36(Suppl. 3): 5–19.

Nalepa, T. F., J. S. Houghton, R. Paddock, and J. Janssen. 2013. Video clip 3: Close-up of inhalant siphons of quagga mussels (*Dreissena rostriformis bugensis*) at two depths in southeastern Lake Michigan. In *Quagga and Zebra Mussels: Biology, Impacts, and Control*, 2nd edn., T. F. Nalepa and D. W. Schloesser, eds., 746–747. Boca Raton, FL: CRC Press.

Pavlova, V. 2012. First finding of deepwater profunda morph of quagga mussel *Dreissena bugensis* in the European part of its range. *Biol. Invasions* 14: 509–514.

Peyer, S. M., J. C. Hermanson, and C. E. Lee. 2010. Developmental plasticity of shell morphology in quagga mussels from shallow- and deep-water habitats in the Great Lakes. *J. Exp. Biol.* 213: 2602–2609.

Peyer, S. M., J. C. Hermanson, and C. E. Lee. 2011. Effects of shell morphology on mechanics of zebra and quagga mussel locomotion. *J. Exp. Biol.* 214: 2226–2236.

Phillips, N. E. 2011. Where are larvae of the vermetid gastropod *Dendropoma maximum* on the continuum of larval nutritional strategies? *Mar. Biol.* 158: 2335–2342.

Rosenberg, G. and M. L. Ludyanskyi. 1994. A review of Russian species concepts of *Dreissena*, with identification of the quagga mussel as *Dreissena bugensis* (Bivalvia: Dreissenidae). *Can. J. Fish. Aquat. Sci.* 60: 126–134.

Spidle, A. P., J. E. Marsden, and B. May. 1994. Identification of the Great Lakes quagga mussel as *Dreissena bugensis* from the Dnieper River, Ukraine, on the basis of allozyme variation. *Can. J. Fish. Aquat. Sci.* 51: 1485–1489.

Therriault, T. W., M. F. Docker, M. I. Orlova, D. D. Heath, and H. J. MacIsaac. 2004. Molecular resolution of the family Dreissenidae (Mollusca: Bivalvia) with emphasis on Ponto-Caspian species, including first report of Mytilopsis leucophaeata in the Black Sea basin. *Mol. Phylogenet. Evol.* 30: 479–489.

Stoeckmann, A. 2003. Physiological energetics of Lake Erie dreissenid mussels: A basis for the displacement of *Dreissena polymorpha* by *Dreissena bugensis*. *Can. J. Fish. Aquat. Sci.* 60: 126–134.

Trussell, G. C. and R. J. Etter. 2001. Integrating genetic and environmental forces that shape the evolution of geographic variation in a marine snail. *Genetica* 112–113: 321–337.

Tuchman, N. C., R. L. Burks, C. A. Call, and J. Smarrelli. 2004. Flow rate and vertical position influence ingestion rates of colonial zebra mussels (*Dreissena polymorpha*). *Freshwat. Biol.* 4: 191–198.

Via, S., R. Gomulkiewicz, G. De Jong, S. M. Schelner, C. D. Schlichting, and P. H. Van Tienderen. 1995. Adaptive phenotypic plasticity: Consensus and controversy. *Tree* 10: 212–217.

Wong, W. H., G. C. Holdren, T. Tietjen, S. Gerstenberger, B. Moore, K. Turner, and D. Wilson. 2013. Effect of invasive quagga mussels (*Dreissena rostriformis bugensis*) on chlorophyll and water clarity in Lakes Mead and Havasu of the lower Colorado River Basin, 2007–2009. In *Quagga and Zebra Mussels: Biology, Impacts, and Control*, 2nd edn., T. F. Nalepa and D. W. Schloesser, eds., 495–508. Boca Raton, FL: CRC Press.

Zwarts, L. and J. Wanink. 1989. Siphon size and burying depth in deposit and suspension feeding bivalves. *Mar. Biol.* 100: 227–240.

Zwarts, L., A. M. Blomert, P. Spaak, and B. DeVries. 1994. Feeding radius, burying depth, and siphon size of *Macoma balthica* and *Scrobicularia plana*. *J. Exp. Mar. Biol. Ecol.* 183: 193–212.

CHAPTER 21

Behavior of Juvenile and Adult Zebra Mussels (*Dreissena polymorpha*)

Jarosław Kobak

CONTENTS

Abstract 331
Introduction 331
Attachment to Substrate 332
Voluntary Detachment 337
Horizontal Movement 337
Vertical Movement 339
Formation of Aggregates 339
Valve Movements 340
Summary 340
Acknowledgments 341
References 341

ABSTRACT

A growing body of evidence shows that the behavior of metamorphosed zebra mussels is important for their survival and distribution. After settlement, mussels live in dense aggregations, byssally attached to substrate. Their attachment strength depends mainly on substrate quality. However, attachment strength is also negatively affected by multiple factors, such as hypoxia, light, and effluents from crushed conspecifics. On the other hand, temperature, water movements, healthy conspecifics, and predator kairomones stimulate adhesion. Zebra mussels often leave their attachment sites, either by choice or because of substrate loss, and crawl in search for another attachment location at a rate of up to 48 cm/h. Mussels avoid light and reduce activity when illumination is unavoidable. They are also less mobile in the presence of crushed conspecifics. Small individuals tend to climb upward, but light and predator kairomones inhibit this behavior. Zebra mussels are able to sense conspecifics and actively seek other individuals, which leads to colony formation. Clumping is further increased by the presence of predators. In general, danger cues (predators, crushed conspecifics, light) seem to affect zebra mussel behavior. When these cues are absent, mussels are active, often relocate, move upward, and keep shell valves open to access suitable conditions of habitat and food. When danger is detected, mussels stop moving, stay near the bottom, aggregate, and close shell valves to increase safety at a cost of increased exposure to less favorable environmental conditions.

INTRODUCTION

A growing body of evidence shows that the behavior of metamorphosed individuals of the zebra mussel *Dreissena polymorpha* is crucial for survival and distribution in the field. Although the distribution of this species is thought to depend mainly on the spread and settlement of planktonic veligers (Lewandowski 1982, 2001), the possibility of active selection of an optimum attachment site by planktonic larvae settling from the water column is very limited because locomotory abilities of larvae are too weak to overcome water movements (Wacker and von Elert 2003). Nevertheless, future survival, growth, and reproductive success of an individual depend to a large extent on its exact position in space. Of importance is whether it occupies the upper or lower surface of an underwater object, inside or outside of a mussel colony, and on or above the bottom. Location strongly affects availability of food, exposure to predators and water movements, and level of intraspecific competition (Tuchman et al. 2004). Therefore, it seems likely that mussels may adjust their position after settlement by crawling over substrates. This has been confirmed by several field observations that showed that newly settled plantigrades were more randomly distributed compared to the highly uneven distribution of older juveniles (Kobak 2004, 2005) and by the ability of juvenile and adult mussels to relocate onto new substrates (e.g., Ackerman et al. 1994, Lauer and Spacie 2004, Czarnecka 2005).

Over a lifespan, mussels can experience a change in substrate, for instance, when individuals settle onto macrophytes that decay in autumn (Lewandowski 2001). Mussels may encounter altered environmental conditions when they attach to unstable or mobile substrates such as vessels, pieces of wood, or anthropogenic waste materials that move to new locations (Johnson and Padilla 1996). Also, mussels actively detach themselves from substrates when local environmental conditions deteriorate (Burks et al. 2002). In such cases, mussels need to find new attachment sites to optimize conditions for further development. Thus, the ability to respond to various environmental cues during movement, site selection, and reattachment could considerably increase growth and survival.

This review summarizes results of studies on zebra mussel behavior in response to abiotic and biotic factors that influence attachment, voluntary detachment, vertical and horizontal locomotion, site selection, colony formation, and valve movements (Table 21.1).

ATTACHMENT TO SUBSTRATE

One prominent feature of adult dreissenids is a bundle of proteinaceous byssal threads used for attachment to hard substrates (Rzepecki and Waite 1993). A mussel uses a groove in its foot as a mold for secretion of a thread, which then stiffens due to quinone tanning (Gosling 2003). Usually, several dozen threads fasten the mussel to the substrate (Clarke and McMahon 1996a,b). Zebra mussels produce byssus even when in poor physiological condition (Clarke 1999) or in the presence of toxins (Rajagopal et al. 2002). Byssal formation is one of the most important types of mussel behavior that may strongly influence survival, growth, and responses to environmental factors.

Attachment strength may be assessed in several ways. Direct measurements have been made with a force gauge (dynamometer) that detects the force needed to detach a mussel (Ackerman et al. 1995, Dormon et al. 1997, Kobak 2006a, 2010, Kobak and Kakareko 2009, Czarnołęski et al. 2010, Kobak et al. 2010). An indirect estimate of attachment strength is the number of produced byssal threads (Clarke and McMahon 1995, 1996a,b, Rajagopal et al. 1996, Clarke 1999, Czarnołęski et al. 2010). Ackerman et al. (1995) used a wall jet apparatus to measure the force needed to flush small mussels off the substrate. Yet another device used to measure attachment strength of young mussels is a rotating disk system; adhesion strength is inferred from the fluid forces generated by rotating a disk-shaped substrate at increased speeds until mussels detach (Ackerman et al. 1992). In general, attachment strength is a function of the number of byssal threads, but thread properties can also affect adhesion. Czarnołęski et al. (2010) noted that adhesion strength per single thread increased significantly with time.

Zebra mussel attachment strength under suitable conditions is usually ca. 1–2 newtons (N) as measured in both long-term (months to years) field studies (Ackerman et al. 1995, 1996, Dormon et al. 1997) and short-time (a week) laboratory experiments (Kobak 2006a, 2010). It should be noted, however, that most individuals tested in these field studies were quagga mussels (*Dreissena rostriformis bugensis*). The maximum value I have ever measured was 7.2 N for a large zebra mussel (>20 mm) that had been attached to a PVC tile for 6 days (Kobak and Kakareko 2009). Values reported for zebra mussels are lower by an order of magnitude than those reported for marine mytilids, which can reach 20–70 N (Bonner and Rockhill 1994, Lachance et al. 2008).

The process of attachment can be divided into two phases. During the first 4–7 days, increase of attachment strength is rapid (Clarke and McMahon 1995, 1996a,b, Rajagopal et al. 1996, Kobak 2006a). In good environmental conditions, a single zebra mussel may produce from 30 (Rajagopal et al. 1996) to 60 (Clarke and McMahon 1996a,b, Clarke 1999) byssal threads during this time. This phase is followed by a stable period with no further changes in attachment strength (Kobak 2006a) and a slower rate of byssal thread formation. The final number of byssal threads produced by a single mussel after 3 weeks has ranged from 70 to 80 (Clarke and McMahon 1995, 1996a,b) to 140 (Clarke 1999).

As expected, larger mussels attach themselves to substrate more strongly than smaller individuals, with an abrupt increase between the size classes of 6–11 and 13–22 mm shell length (Kobak 2006a). The latter size class was about three times more strongly attached than the former, with no significant differences within these two groups (Kobak 2006a).

Table 21.1 Factors Known to Affect Zebra Mussel Behavior

Behavior	Factor	Response Type	Reference
Attachment	Body size	+	Kobak (2006a)
	Substratum material	+ or −	Ackerman et al. (1992, 1995, 1996), Kobak et al. (2002), Kobak (2010)
	Surface free energy	+	Kavouras and Maki (2003)
	Substratum roughness	+	Ackerman et al. (1996)
	Biofilm	+ or −	Kavouras and Maki (2003, 2004), Kobak (2010)
	Light	− up to the second day	Kobak (2001, 2006a)
	Temperature	+≤20°C, −≥30°C	Kobak (2006a)
		+≤15°C, −≥25°C	Rajagopal et al. (1996)
		+up to 31°C	Clarke and McMahon (1996a)
	Water flow	+	Peyer et al. (2009)
		+≤20 cm/s, −≥27 cm/s	Clarke and McMahon (1995)
	Water agitation	−	Clarke and McMahon (1996b)
		+	Rajagopal et al. (1996)
	Hypoxia	−≤15% O_2 at 25°C	Clarke and McMahon (1996b)
	Food quantity	+	Clarke (1999)
	Toxins	−	Kobak et al. (2002), Rajagopal et al. (2002)
	Conspecifics	+	Kobak (2001, 2006a)
	Predators	+	Kobak and Kakareko (2009)
	Crushed conspecifics	−	Czarnołęski et al. (2010)
Detachment	Body size	−	Kobak et al. (2009)
	Light	−	Kobak et al. (2009)
	Substratum inclination	+	Kobak et al. (2009)
	Conspecifics	−	Kobak et al. (2009)
	Hypoxia	+	Mikheev (1964)
Horizontal movement			
Path length	Body size	−	Kobak and Nowacki (2007)
	Light	−	Kobak and Nowacki (2007)
	Crushed conspecifics	−	Toomey et al. (2002), Czarnołęski et al. (2010)
Linear displacement	Light	+	Toomey et al. (2002)
Site selection	Light	−	Kobak (2001), Toomey et al. (2002), Kobak et al. (2009)
	Substratum dark color	+	Kobak (2001)
	Mesh objects	−	Porter and Marsden (2008)
	Solid objects	+	Zhang et al. (1998), Kobak (2003)
	Toxic substrates	−	Kobak et al. (2002)
	Conspecifics	+	Kobak (2003), Kavouras and Maki (2003)
Upward movement	Wastes	+	Burks et al. (2002)
	Hypoxia	+	Mikheev (1964), Burks et al. (2002)
	Gravity	+ (small mussels)	Kobak (2001, 2002, 2006b), Kobak et al. (2009)
		− (large mussels)	
	Light	−	Kobak (2002, 2006b)
	Predators	−	Kobak and Kakareko (2009)
Aggregation	Conspecifics	+	Stańczykowska (1964), Kobak (2003)
	Predators	+	Kobak and Kakareko (2009)
Valve movements			
Opening time	Toxins	−	Borcherding (2006)
	Light	−	Kobak and Nowacki (2007)
Number of events	Toxins	+	Borcherding (2006)
	Light	+	Kobak and Nowacki (2007)
	Temperature	+	Borcherding (2006)

Response types: +, increase or stimulation; −, decrease or inhibition.

Table 21.2 Attachment Strength of Dreissenid Mussels on Selected Substrates

Substrate	Attachment Strength ± SE (N)	Duration	Reference
Limestone rock	1.40 ± 0.40	Long-term[b]	Ackerman et al. (1995)
Wood	0.72 ± 0.04[a]	Long-term[b]	Ackerman et al. (1996)
Concrete (smooth)	0.67 ± 0.05[a]	Long-term[b]	Ackerman et al. (1996)
Concrete (rough)	0.60 ± 0.05[a]	Long-term[b]	Ackerman et al. (1996)
Concrete	3.20 ± 0.22[a]	Long-term[c]	Dormon et al. (1997)
Perspex	0.35 ± 0.03[a]	Long-term[b]	Ackerman et al. (1995)
PVC (smooth)	0.87 ± 0.04[a]	Long-term[b]	Ackerman et al. (1995)
PVC (rough)	0.80 ± 0.10[a]	Long-term[b]	Ackerman et al. (1996)
PVC	3.40 ± 0.29[a]	Long-term[c]	Dormon et al. (1997)
PVC (rough)	0.78 ± 0.06	Short-term[d]	Kobak (2010)
Teflon (smooth)	0.18 ± 0.04[a]	Long-term[b]	Ackerman et al. (1996)
Teflon (rough)	0.80 ± 0.06[a]	Long-term[b]	Ackerman et al. (1996)
Resocart (rough)	1.14 ± 0.09	Short-term[d]	Kobak (2010)
Rubber	0.47 ± 0.05	Short-term[d]	Kobak (2010)
PENATEN cream	0.04 ± 0.01	Short-term[d]	Kobak (2010)
Al (smooth)	0.46 ± 0.05[a]	Long-term[b]	Ackerman et al. (1995)
Al (rough)	0.96 ± 0.09	Short-term[d]	Kobak (2010)
Cu:Ni alloy (70:30)	1.20 ± 0.10[a]	Long-term[b]	Ackerman et al. (1996)
Zinc (rough)	0.16 ± 0.03	Short-term[d]	Kobak (2010)
Mild steel (smooth)	1.60 ± 0.20[a]	Long-term[b]	Ackerman et al. (1996)
Mild steel (rough)	1.30 ± 0.10[a]	Long-term[b]	Ackerman et al. (1996)
Mild steel	1.80 ± 0.46[a]	Long-term[c]	Dormon et al. (1997)
Stainless steel (smooth)	0.67 ± 0.06[a]	Long-term[b]	Ackerman et al. (1995)
Stainless steel (rough)	1.10 ± 0.10[a]	Long-term[b]	Ackerman et al. (1996)
Stainless steel	3.20 ± 0.29[a]	Long-term[c]	Dormon et al. (1997)

N, newtons.
[a] *D. rostriformis bugensis*.
[b] ca. 0.5–1 year in the field, then 1 month in a laboratory, mussel size: 7.3–11.3 mm.
[c] 2 years in the field, then up to 3 months in a laboratory, mussel size: 7–17 mm.
[d] 6 days in a laboratory, mussel size: 12–20 mm.

Strength of mussel attachment is strongly influenced by substrate quality (Table 21.2, Figure 21.1). The most favorable materials for mussel attachment (average attachment strength >1 N) were limestone rock, mild steel (Ackerman et al. 1995, 1996), and resocart plastic (a thermosetting phenol–formaldehyde resin with paper as a filler) (Kobak 2010). Lower attachment strength was noted on Teflon®, Perspex® (Ackerman et al. 1996), and rubber (Kobak 2010) and lowest attachment strength was usually found on toxic materials: zinc, copper, and baby-bottom cream PENATEN® containing zinc oxide (Kobak et al. 2002, Kobak 2010). Low attachment strength was also observed on low free-surface energy materials, usually based on silicone or fluoropolymer resins, which are used as antifouling coatings in hydrotechnical appliances (Race 1992, Kavouras and Maki 2003). Other substrate factors that affect zebra mussel attachment include surface roughness that facilitates mussel adhesion (Ackerman et al. 1996) and bacterial biofilms that facilitate or inhibit adhesion depending on composition (Kavouras and Maki 2003, 2004, Kobak 2010).

Attachment strength of zebra mussels is also affected by light, temperature, water movements, hypoxia, presence/absence of other mussels, and predators (Table 21.1, Figures 21.2 through 21.4). For example, attachment strength is inhibited by light in the first 2 days of settlement (by 42% compared with those exposed in darkness) but not after 6 days (Kobak 2006a; Figure 21.2). Thus, light seems to delay initiation of adhesion, but then mussels exposed to light become as strongly attached as mussels entirely exposed to dark.

Zebra mussel adhesion also depends on temperature. Usually, a gradual increase in attachment strength can be observed up to 15°C–20°C and then a decrease occurs above 20°C–25°C (Rajagopal et al. 1996, Kobak 2006a). However, Clarke and McMahon (1996a) observed a constant increase of byssogenesis up to 31°C, which is close to the lethal temperature for this species (Karatayev et al. 2002). Perhaps, this discrepancy might be related to the elevated upper thermal limits of zebra mussels in North America compared to upper limits of mussels in Europe (McMahon 1996).

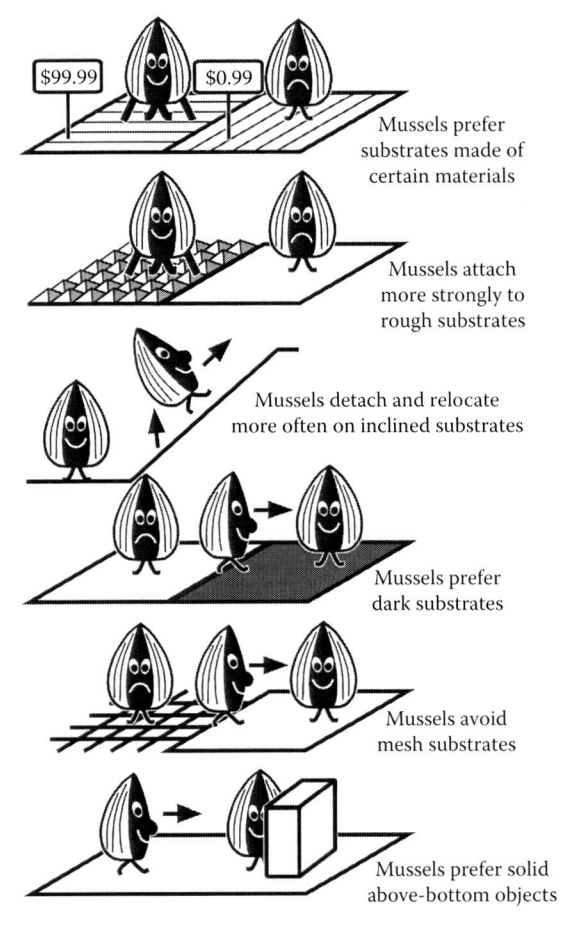

Figure 21.1 Responses of zebra mussels to substrate characteristics.

Results of studies on effects of water movements on mussel attachment are ambiguous. Clarke and McMahon (1995) observed that a water flow up to 20 cm/s stimulated byssogenesis, while at a velocity of 27 cm/s byssal thread production was reduced. A negative effect of water agitation on mussel attachment was also found (Clarke and McMahon 1996b). Conversely, Peyer et al. (2009) observed increases in byssal thread production over a wide range of flow velocities up to 180 cm/s, and Rajagopal et al. (1996) found a positive effect of water agitation on byssogenesis. The two latter studies showed that attachment behavior of zebra mussels might be similar to that of marine *Mytilus* sp., which attaches more strongly when exposed to heavier hydrodynamical forces (Bell and Gosline 1997). The ability of zebra mussels to increase attachment strength with increased flow, and thus survive in turbulent environments, may be particularly beneficial in locations where it co-occurs with the quagga mussel. The quagga mussel is regarded as a superior competitor (Orlova et al. 2005) but has lower byssus production rate and seems to be less adapted to lotic habitats (Peyer et al. 2009).

Zebra mussels were found to reduce byssal production in poor environmental conditions, such as hypoxia (oxygen concentration below 15% saturation at 25°C) (Clarke and McMahon 1996b), food depletion (Clarke 1999), and the presence of toxins, such as chlorine (Rajagopal et al. 2002) or copper ions (Kobak et al. 2002).

Zebra mussels increased reattachment rate and attachment strength in the presence of other mussels by about 40% compared to isolated individuals (Figure 21.3; Kobak 2001, 2006a). These results were evident even when tested mussels had no physical contact with other mussels, which suggests a response to mussel-released chemicals. In another experiment, individual and aggregated mussels were in the same experimental tank, and thus single individuals

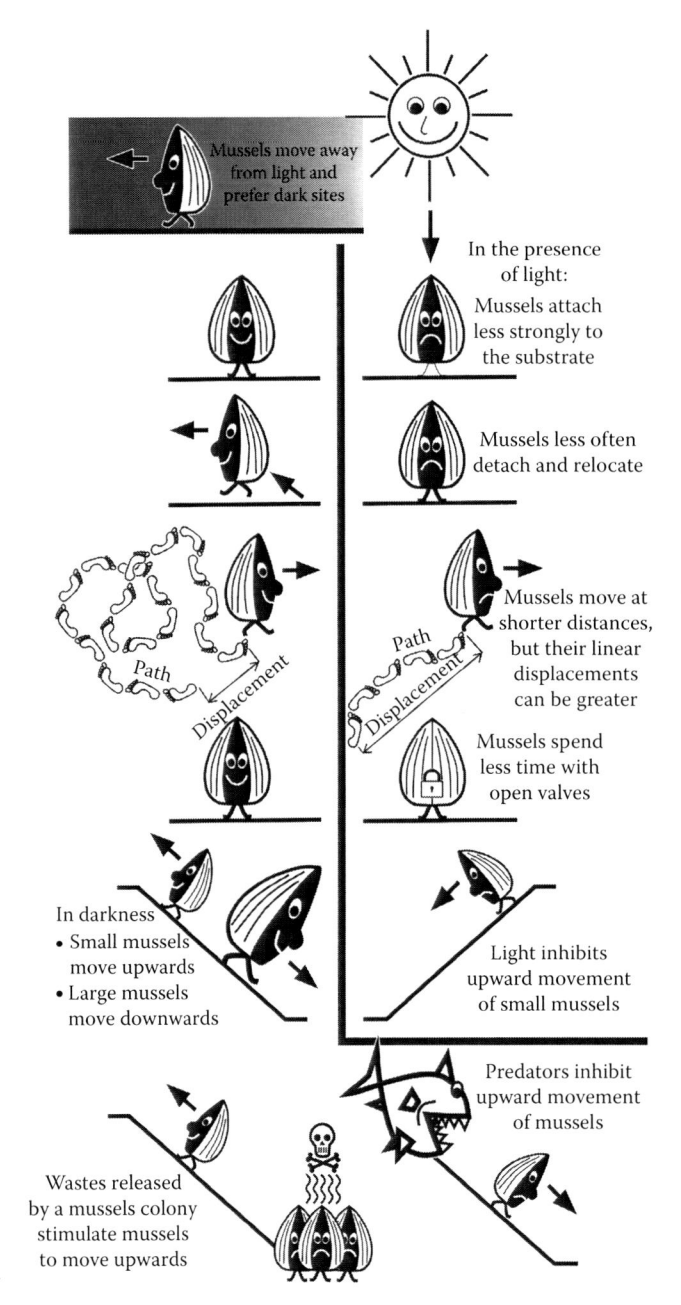

Figure 21.2 Responses of zebra mussels to light and gravity.

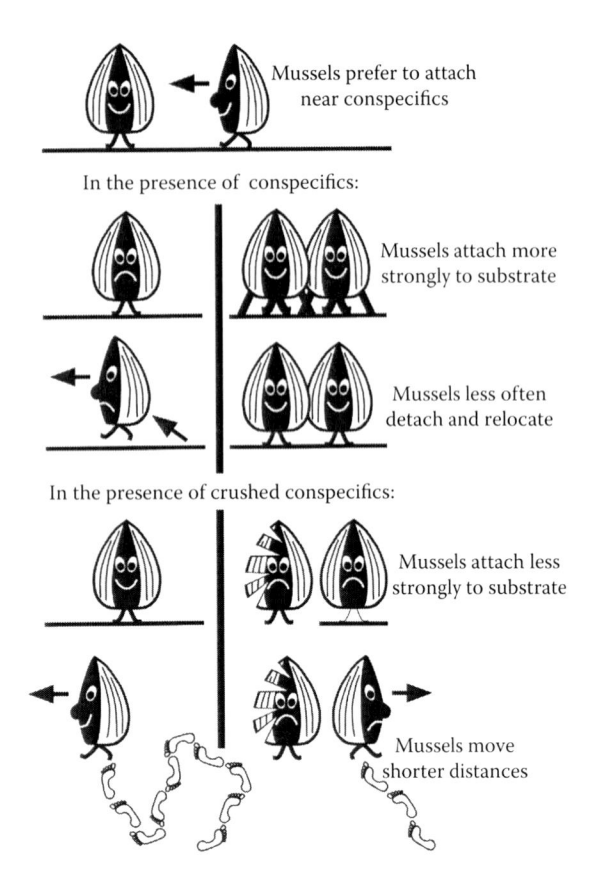

Figure 21.3 Responses of zebra mussels to conspecifics (other mussels).

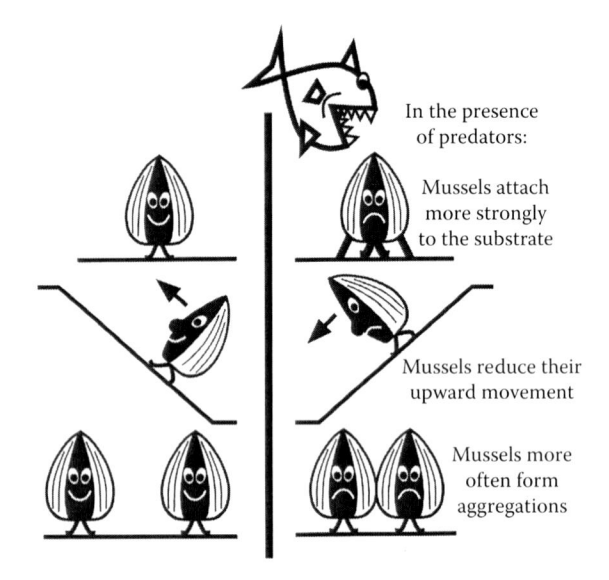

Figure 21.4 Responses of zebra mussels to the presence of predators.

were exposed to any infochemicals (Kobak et al. 2009). Under such conditions, aggregated mussels attached more strongly than single individuals by 65% for small (<10 mm) and by 42% for large mussels (>10 mm). This indicated that physical contact with other mussels also stimulates mussel adhesion. The difference in attachment strength was only observed between single and aggregated mussels; aggregation size (2–19 individuals) did not affect mussel adhesion (Kobak et al. 2009). It should be noted that marine *Mytilus* sp. exhibited an opposite response to conspecifics; that is, aggregated mussels showed reduced attachment strength (Schneider et al. 2005). It was suggested that a reduction in attachment strength led to a more efficient use of energy since the costs of defenses against predators and hydrodynamic forces can be shared among several conspecifics (Schneider et al. 2005).

Firm attachment to substrates can be an efficient antipredator defense. One of the most efficient molluscivores feeding on zebra mussels in Europe is large roach *Rutilus rutilus* (>200 mm total length) (Prejs et al. 1990, Molloy et al. 1997). Foraging roach attempts to detach a mussel for about 2 s and, if it fails, switches to another prey item (Nagelkerke and Sibbing 1996). This suggests that the increase of attachment strength in the presence of predators could positively affect mussel survival. Consistent with this hypothesis, higher attachment strength was shown by small- (<10 mm) and medium-sized (10–17 mm) zebra mussels in the presence of roach (by 103% and 66%, respectively) compared to control mussels not exposed to fish (Kobak and Kakareko 2009; Figure 21.4). The largest mussels (>17 mm) were not affected by the presence of fish (Kobak and Kakareko 2009). This size-dependent response of mussels corroborates the results of the field study by Prejs et al. (1990), who found that large roach prefer mussels up to 17 mm shell length. As roach in the experiments of Kobak and Kakareko (2009) were not allowed to feed on mussels, tested mussels responded to the fish presence rather than to any alarm substances potentially produced by crushed conspecifics. The response of mussels to roach was not immediate but only developed after prolonged exposure to the predators (Kobak et al. 2010). The presence of other potential predators, such as the racer goby *Neogobius gymnotrachelus* and the spiny-cheek crayfish *Orconectes limosus*, which feed on small zebra mussels to some extent (Molloy et al. 1997, Kakareko et al. 2005), and Eurasian perch *Perca fluviatilis*, which do not feed on mollusks (Brylińska 2000), did not affect the attachment strength of mussels (Kobak and Kakareko 2009, Kobak et al. 2010). This suggests that mussels respond specifically to chemicals released by roach (kairomones) rather than to fish exudates in general. Zebra mussels also distinguished between small roach (80–110 mm TL), which were unable to feed on hard-shelled mollusks (Prejs et al. 1990, Brylińska 2000), and large roach (180–250 mm TL) since attachment strength only increased in the presence of the latter (Kobak et al. 2010). The mechanism of this discrimination is unclear, but perhaps it might involve cues specific to mature fish.

Evidence suggests that the stronger a mussel attaches to the substrate the more resistant it is to predators (Green et al. 2008). However, the process of attachment requires a period of increased activity during which a mussel is more vulnerable to detection by predators. Movements of the mussel foot or chemicals released through open valves can be detected by fish. Therefore, in the direct presence of foraging predators, a different strategy to avoid predation may be to reduce byssal production even at the cost of weaker attachment. The attachment strength of mussels exposed to the scent of crushed conspecifics (indicating a direct and near danger) was lower by about 40% compared to control mussels (Czarnołęski et al. 2010; Figure 21.3).

VOLUNTARY DETACHMENT

Mussels may lose their connection with the substrate due to external factors, such as water movements or degeneration of temporary substrates (e.g., submersed plants). They can also actively detach and move to another site. This has been easily observed in the laboratory. When a substrate with attached mussels is placed into an aquarium, many individuals can be found unattached and moving within a few hours.

To test detachment behavior, mussels were allowed to attach to microscope cover slides for a given period of time and then moved to differing conditions of light, substrate inclination, and aggregation (Kobak et al. 2009). When placed in total darkness, 42% of large individuals (>10 mm) and 56% of small mussels (<10 mm) detached and moved to a new site. In lighted conditions, only 10% and 29% of similar sized mussels, respectively, detached (Figure 21.2). Mussels more often detached when on a sloped surface (59% and 73% of large and small individuals, respectively) than on a flat substrate (Figure 21.1). In addition, aggregated mussels detached less frequently (about 50% less) than single mussels (Figure 21.3), but this difference was only found for large mussels. Large mussels were probably responding to the physical contact of other individuals because single mussels in this experiment were exposed to other individuals but did not touch their shells (Kobak et al. 2009).

In all these experiments, there was unavoidable manipulation of the tested mussels that could affect frequency of detachment. Nevertheless, observed rates of detachment suggest that this may be an important process in natural populations. Mussels could benefit from such behavior by colonizing new available substrates and by finding optimum sites after passive relocations (Johnson and Padilla 1996). Another advantage of abandoning an old attachment site would be an escape from deteriorating conditions inside a dense colony (Burks et al. 2002). Mikheev (1964) observed an increased detachment rate of zebra mussels in hypoxic conditions. On the other hand, a mussel that detaches from the substrate faces the energetic costs of producing new byssal threads and temporarily becomes more exposed to various environmental threats, such as predators and excessive water movements.

HORIZONTAL MOVEMENT

Detached mussels crawl over the substrate by use of a wedge-shaped foot to search for new suitable attachment sites. Relocations of juvenile and adult mussels have been observed in the field (see Lesht and Hawley 2013). Lauer and Spacie (2004) noted that patches of substrate that were cleaned of sponges were quickly colonized by adult zebra mussels that had immigrated from nearby sites. Furthermore, Czarnecka (2005) found that zebra mussels colonized artificial substrates deployed in winter at a time when veligers were absent from the water column and settlement had not taken place.

Distances moved by mussels have been measured by placing individuals onto a thin layer of fine sand. The ensuing trail was used to reconstruct the mussel's path (Toomey et al. 2002). It was found that small mussels (<10 mm) traveled a mean distance of 284 mm in 2 h (the maximum distance was 962 mm), while medium (10–20 mm) and large (>20 mm) individuals were less mobile and moved mean distances of only 115 and 47 mm, respectively (Table 21.3).

Table 21.3 Mean and Maximum Distances Moved by Zebra Mussels over the Given Period of Time

Mussel Size (mm)	Mean Distance ± SE; Max (mm)	Duration (h)	Light Conditions	Reference
5–10	284 ± 34; 962	2	Light	Toomey et al. (2002)
5–9	142 ± 17; 381	1	Strobe light	Coons et al. (2004)
6–10	184 ± 31; 404	24	Light (100 lx)	Kobak and Nowacki (2007)
5–9	133 ± 15; 346	1	Darkness	Coons et al. (2004)
6–10	280 ± 17; 826	24	Darkness	Kobak and Nowacki (2007)
6–10	250 ± 17; 770	24	Natural photoperiod	Kobak and Kakareko (2009)
9–10	ca. 300[a]	3	Light	Czarnołęski et al. (2010)
10–20	115 ± 23	2	Light	Toomey et al. (2002)
12–21	59 ± 15; 205	24	Light (100 lx)	Kobak and Nowacki (2007)
12–21	201 ± 16; 703	24	Darkness	Kobak and Nowacki (2007)
12–17	275 ± 15; 803	24	Natural photoperiod	Kobak and Kakareko (2009)
>20	47 ± 4	2	Light	Toomey et al. (2002)

[a] Read from a figure in Czarnoleski et al. (2010).

These results were obtained under constant light conditions. Czarnołęski et al. (2010) reported a mean crawling speed of 200 mm/h for small mussels, which compared to the 142 mm/h reported by Toomey et al. (2002). These results can be compared to those of Kobak and Nowacki (2007) who measured mussel mobility during a 24 h period. In this study, mussels in lighted conditions moved somewhat shorter distances (Table 21.3) than those noted by Toomey et al. (2002) (Table 21.3). This suggests that mussels start to move soon after detachment but stop within a few hours as a longer experimental period did not increase distance traveled.

As found for frequency of detachment, light also affects horizontal movement (Kobak 2001, Kobak and Nowacki 2007, Kobak et al. 2009). Zebra mussels preferred dark over illuminated sites (>80% of individuals) and preferred dark-colored over light-colored substrates (>70% of individuals) (Figure 21.1). Large mussels (>10 mm) were more sensitive to light than small individuals (<10 mm); the former avoided light of an intensity as low as 0.1 lx, while the latter only responded significantly to intensities of 1.2 lx and higher (Kobak and Nowacki 2007). In general, zebra mussels of all sizes responded to light even at very low intensity. Furthermore, mussels could detect a wide range of the visible light spectrum, from 300 to more than 700 nm (Kobak and Nowacki 2007).

Given the photophobicity of zebra mussels, they could be expected to leave unsuitable, illuminated sites and search for more suitable, dark locations. Contrary to these expectations, Kobak and Nowacki (2007) found that strong, uniform light (100 lx) reduced distances moved by mussels. In darkness, small mussels (<10 mm) moved a mean distance of 280 mm, whereas large individuals (>10 mm) moved a mean distance of 201 mm. Light reduced these distances by 34% and 71%, respectively (Table 21.3). A similar tendency, though not statistically significant, was observed at a lower light intensity of 75 lx but not at 12 lx and below. This corresponds to the light-induced decrease of detachment rate described earlier and indicates a general tendency of illuminated mussels to reduce their activity (Figure 21.2). In contrast, several studies found that light had no effect on distance traveled (Toomey et al. 2002, Coons et al. 2004). These studies applied light as a directional stimulus (i.e., from one side) and tested mussels were able to escape from the light source, thus distance traveled was not reduced. Response to illumination was clearly directional as all individuals moved away from light compared to the random movement of control mussels. Furthermore, linear displacement (i.e., length of a straight line between the start and the end of the path) was longer in light than in darkness (Toomey et al. 2002, Coons et al. 2004), which confirms the directional nature of the mussel movement in response to directional light (fewer zigzags and turns).

Thus, zebra mussels are strongly photophobic animals that move away from light when possible (Toomey et al. 2002, Coons et al. 2004) yet reduce locomotory activity when uniform illumination is unavoidable (Kobak and Nowacki 2007). Light may be an indirect indication of potential danger. Lighted regions would signify open water and thus areas devoid of shelters that provide protection against predators and excessive water movements. On the other hand, sheltered sites are usually shadowed, located in crevices, under stones, or among other mussels. Thus, the reduction of activity could be a defense mechanism, preventing the detection by a predator at an exposed site.

Another factor that may affect zebra mussel locomotion is the presence of chemical cues from crushed conspecifics, which may indicate the presence of a foraging predator (Figure 21.3). The presence of crushed conspecifics reduced distances traveled by mussels by 35%–44% (Toomey et al. 2002, Czarnołęski et al. 2010), but neither the duration of movement nor the time spent at the initial position before relocation was affected (Czarnołęski et al. 2010). Thus, decreases in distances traveled must have been a result of a reduction in the rate of movement. Reduction of activity seems to be a general response of mussels to any signs of increased risk, indicated indirectly by the presence of light or directly by alarm substances from crushed conspecifics. Surprisingly, presence of large molluscivorous roach did not cause any significant changes in horizontal movements of mussels (Kobak and Kakareko 2009), though this stimulus did affect other aspects of mussel behavior such as attachment strength, vertical locomotion, and aggregation (see other sections).

Calcium content of water (range 9–204 mg $CaCO_3$/L) had no effect and temperature (range 15°C–27°C) had only a marginal effect on distances traveled (Toomey et al. 2002). Small mussels (<10 mm) responded to an increase in temperature by traveling slightly longer distances, while large mussels (>10 mm) were not affected within this temperature range. It seems likely, however, that lower and higher temperatures than those studied would negatively affect mussel locomotion similar to the influence of temperature on attachment strength as discussed earlier.

Substrate shape and type are other important factors affecting site-selection behavior of moving zebra mussels (Figure 21.1). Mussels avoid mesh structures (Porter and Marsden 2008) and toxic materials such as copper (Kobak et al. 2002) and select solid objects, crevices, corners, and edges (Zhang et al. 1998, Kobak 2003, 2005). A mussel crawling over the substrate is more likely to adhere to an object projecting from the bottom than on a flat surface (Kobak 2003). In addition, mobile mussels stopped movement by 50% (mussels <10 mm) and by 19% (mussels >10 mm) when other live mussels were encountered compared to mussel-shaped stones (Kobak 2003).

Finally, newly-settled, metamorphosed plantigrade and juvenile mussels can actively detach themselves and return to the water column where they drift with currents by means of special byssal threads that increase buoyancy (Martel 1993). This adaptive activity allows juveniles to move to more suitable locations. Well known in marine bivalves as "byssal drifting" (Sigurdsson et al. 1976), this activity in zebra mussels has been documented only in North American populations (Lewandowski 2001).

VERTICAL MOVEMENT

In laboratory cultures, small mussels are frequently observed climbing up aquarium walls to the water surface. They use special, elongated, mucous byssal threads to move on vertical surfaces (Rzepecki and Waite 1993). Upward movement of zebra mussels was also observed in the field as dense clusters of juvenile mussels occurred along the upper edges of vertical tiles (Kobak 2004, 2005). Since the distribution of newly settled plantigrades on the tiles was much more random, it is likely that the observed uneven distribution of juveniles was a result of active upward movement after settlement (Kobak 2004).

Burks et al. (2002) found that 69% of small mussels (<6 mm) in a dense colony migrated upward from the colony base to its surface during a 30-day experiment. Such movement was related to the deterioration of environmental conditions (viz., oxygen and nitrate concentrations) at the colony's base (Figure 21.2). Large mussels (>20 mm) never exhibited such behavior, even though staying at the bottom resulted in increased mortality after a long exposure. Detachment of mussels from colonies and upward migration were also observed under hypoxic conditions (Mikheev 1964).

Subsequent studies have indicated that mussels will move upward even without environmental stimuli. Small mussels (<10 mm) were negatively geotactic in darkness, with 67% climbing up an inclined slope (Kobak 2002, 2006b). On the other hand, large mussels (>10 mm) more often moved downward (73% of individuals) (Kobak 2006b). The minimum angle of inclination at which mussels responded to gravity with either down- or upward movement was 4° (Kobak 2006b).

When moving upward, juvenile mussels may find sites located on a colony surface or on upper surfaces of above-bottom objects. Thus, high concentration of wastes, oxygen deficits, and competition with adults at the base of a colony are avoided (Stańczykowska 1964, Burks et al. 2002, Tuchman et al. 2004). Diggins et al. (2004) suggested that above-bottom objects might constitute refuges for zebra mussels in locations where they live sympatrically with another dreissenid, the quagga mussel *D. rostriformis bugensis*, which is usually a superior competitor on bottom substrates (Orlova et al. 2005). Large mussels are usually less mobile or even totally immobilized by fouling conspecifics (Czarnołęski et al. 2003). Thus, in the inclined surface experiments, large mussels may have selected a downward direction simply because it demanded less effort (Kobak 2006b).

It should be noted that there is a discrepancy between types of behavior described so far; more specifically, small mussels are both negatively geotaxic and phototaxic. Mussels that move upward in the field encounter stronger and stronger illumination and reach surfaces where they are more exposed to adverse environmental factors, such as predator attacks, excessive wave actions, and possible desiccation. Thus, it might be expected that light would modify upward movement of mussels (Figure 21.2). Indeed, inhibition of upward movement by light was observed by Kobak (2002, 2006b), but a complete reversal of movement direction did not occur. Mussels under lighted conditions did not prefer a downward direction but rather moved with similar frequencies in both directions on a slope. Kobak and Nowacki (2007) found that the lowest light intensity inhibiting the upward movement of mussels was 12 lx. Interestingly, this threshold is lower than that inhibiting horizontal movement.

These results are consistent with field observations that mussels live at shallower depths in shadowed locations compared to locations in full sun (Hanson and Mocco 1994, Marsden and Lansky 2000). De Lafontaine et al. (2002) found at a well-illuminated site (more than 1000 lx) in the St. Lawrence River that newly settled mussels avoided the edges of recruitment tiles and preferred shadowed central areas of the tiles. Conversely, I observed the opposite distribution of juveniles in a turbid (<0.1 lx at the depth of substrate exposure) dam reservoir on the lower Vistula River. Newly-settled mussels gathered at the upper edges of the vertical settlement tiles (Kobak 2004, 2005). Probably, such a distribution is only possible when light does not deter the upward movement of mussels. As stated previously, it is likely mussels avoid light because light is an indirect sign of predation risk. Thus, it is not a surprise that small mussels (<10 mm) also responded directly to the presence of their predators, large roach, by reducing upward movement by 44% (Kobak and Kakareko 2009; Figures 21.2 and 21.4).

FORMATION OF AGGREGATES

Zebra mussels usually live in dense clusters or druses that contain from a few to more than 100 individuals (Stańczykowska 1964). One mechanism for the formation of such clusters is the preferential settlement of larvae in the vicinity of adults (Wainman et al. 1996), but clusters can also be a result of the movement of newly-settled recruits (Stańczykowska 1964). Aggregation size increases gradually during several days of mussels crawling over the substrate (Stańczykowska 1964, Kobak et al. 2010). Apart from the selectivity of settling larvae, contagious mussel distribution may have the following causes: (1) lack of suitable hard substrates; (2) preferences of mussels for specific types or parts of substrates (e.g., edges or crevices); and (3) active preferences of mussels for conspecifics.

Stimulation of mussel attachment in the presence of conspecifics (Kobak 2001, Kobak et al. 2009) as described in earlier sections supports the hypothesis that mussels exhibit active preference in substrate selection (Figure 21.3). Stańczykowska (1964) found that the tendency to aggregate was independent of substrate type; experimental mussels aggregated similarly on hard substrates (glass) and on soft, unsuitable materials (sand). On the other hand, mussels in other experiments tended to attach in the vicinity of one another but not directly to conspecific shells (Kobak et al. 2009, 2010, Kobak and Kakareko 2009). Kavouras and Maki (2003) also observed that mussels attached to one another's shell only when the remaining substrate was of poor quality (i.e., devoid of biofilm). These findings suggest that conspecific shells are not always an optimum substrate.

The tendency of mussels to aggregate was affected by the presence of fish (Kobak and Kakareko 2009, Kobak et al. 2010; Figure 21.4). After 6 days of exposure, the percentage of aggregated small mussels (<10 mm) was 36% in empty tanks and 65% in tanks with roach, which is a molluscivore. For medium-sized mussels (10–17 mm), these values were 43% and 66%, respectively, while large mussels (>17 mm) did not respond to the presence of roach (Kobak and Kakareko 2009). Mussels of all sizes did not display increased aggregation rates in the presence of other predators such as the racer goby, spinycheek crayfish, and Eurasian perch (Kobak and Kakareko 2009). However, further studies revealed that aggregation response of mussels to predators was a complex function of exposure time. Mussels exposed for 1 day to large roach (180–250 mm TL), small roach (80–110 mm TL), or perch were clearly more aggregated (small mussels: 46%, medium mussels: 50%) than control mussels not exposed to fish (22% and 26%, respectively) (Kobak et al. 2010). After a longer, 6-day exposure, aggregation levels were greater in the presence of large roach (67% and 77%) than in the control tanks (31% and 47%), but levels in the presence of the other fish were not different from controls. The aggregation level of mussels in the large roach treatment was significantly higher than levels found in all other treatments. This suggests that mussels have two types of responses to fish depending on length of exposure: an initial, non-specific response to all kinds of fish and a later response specific to large roach that occurs after a long, continued exposure to the larger predators.

Aggregated mussels are better protected from predators. Handling costs are greater when predators are forced to detach and separate aggregated mussels compared to handling costs for a single prey. The confusion (the problem of picking a single prey from a group) and dilution effect (the lower probability that a particular individual will be captured) also increase the security of aggregated mussels. On the other hand, mussels in aggregations often have a lower condition index and grow more slowly than individuals that are not aggregated because of the accumulation of wastes, hypoxia, and food depletion (Burks et al. 2002, Tuchman et al. 2004).

VALVE MOVEMENTS

Mussels open their valves to respire and feed. Both these functions are controlled independently by lateral cilia and latero-frontal cirri on gills, respectively (Gosling 2003). Mussel valve movements have been measured by gluing a small magnet to the shell (Borcherding 2006). When a mussel valve opens, the magnet approaches a reed switch that closes an electrical circuit and sends a signal to a computer to collect the data. A similar, though somewhat simplified method, was used by Kobak and Nowacki (2007), who glued two thin wires to both valves. When the valves closed, the wires closed an electrical circuit and this event was recorded by a computer.

Under usual environmental conditions, mussels spend most of their time with valves open and filtering water. A typical stress response results in the reduction of time spent with open valves and an increase in frequency of short valve movements. A decrease of time with open valves can be induced by a wide range of chemical toxicants, for example, ammonium, organic compounds, and heavy metals (Borcherding and Jantz 1997, Borcherding and Wolf 2001, Borcherding 2006), as well as light (Kobak and Nowacki 2007). In darkness, mussel valves were open for 75% of time and moved their valves (by opening or closing them) an average of nine times over a 12 h test period (Kobak and Nowacki 2007). Mussels responded to light of low intensity (3–6 lx) by reducing time spent with open valves (Figure 21.2) by 35% and increasing frequency of valve movements by 79%. Interestingly, mussels were insensitive to stronger light (100 lx) (Kobak and Nowacki 2007). Another stimulus that affects activity of mussel valves is temperature; the frequency of valve movements was positively correlated with temperature (Borcherding 2006).

No direct data are available on the relationship between valve movement patterns and predator effluents, but Naddafi et al. (2007) observed a reduction in filtration rates of mussels in the presence of roach and crayfish. This phenomenon may be related to the overall reduction of mussel activity in the presence of predators and perhaps reflects a decrease in time valves are open, (see review, Czarnoleski and Müller 2013).

SUMMARY

This review shows that behavior of metamorphosed zebra mussels can be regarded as an important factor that determines distribution and survival in the field, apart from the settlement of planktonic larvae. Zebra mussels were found to change attachment sites quite often and to use multiple abiotic and biotic cues (e.g., light, gravity, temperature, conspecifics, predator effluents) to adjust the intensity and direction of their movements. Such mobility may contribute to their fitness in the field, where they can be forced to relocate to new attachment sites. Responses of mussels to light and gravity may help explain differences in field distributions (Marsden and Lansky 2000, De Lafontaine et al. 2002, Kobak 2004, 2005).

There is a growing body of evidence that the behavior of zebra mussels depends to a large extent on direct and indirect signals of danger (light, predator kairomones, crushed conspecific effluents). When these cues are absent from the environment, mussels are more active (in terms of time with open valves and locomotion), and small individuals move upward, thus occupying sites with more suitable environmental conditions (Burks et al. 2002, Tuchman et al. 2004). Conversely, when a danger is detected, mussels prefer to stay near the bottom, more often form aggregations, and become less active. Due to these behavioral changes, mussels occupy safer sites but at a cost of less suitable food and chemical conditions. This can lead to a lower growth rate (Chase and Bailey 1996), weaker physiological condition (Stańczykowska 1964), and even higher mortality (Burks et al. 2002), but at the same time may provide better protection from predators. Thus, the behavior of zebra mussels seems to be based on a compromise between optimal environmental conditions that allow maximum growth and reproduction but greater predation risk, and suboptimal environmental conditions but a diminished predation risk. This situation is similar to that documented for more mobile aquatic animals, such as zooplankton (De Meester et al. 1999) and fish (Gliwicz 2005). Zebra mussel responses to predation pressure are not limited to behavior but also include long-term changes in morphology and life history, as for example, changes in shell thickness or growth patterns (Czarnołęski et al. 2006). More detailed considerations on the evolution of the life-history parameters of zebra mussel populations under variable predation pressure can be found in another chapter of this book (Czarnołęski and Müller 2013).

Many aspects of zebra mussel behavior are still not well known and need further research. For instance, mechanisms for the formation of aggregations remain to be studied. We do not know whether mussels can locate their conspecifics and move directionally to them or, alternatively, whether they just move randomly over the substrate and stop after an accidental contact with another mussel. Factors affecting the voluntary detachment of mussels from the substrate also need further study. Although some factors have been detected, we still do not know how this type of behavior is affected by conspecifics, predators, temperature, food quantity, and other potential stimuli. Moreover, effects of predators on mussel behavior deserve more study, particularly with respect to the actual effectiveness of the antipredator defenses in the presence of molluscivorous fish. Furthermore, interactions between several environmental factors could influence mussel behavior in a manner that is difficult to predict based on separate effects, which opens another field for experimental research.

Studies of mussel behavior are valuable to help explain distribution, survival, and function in the field. Furthermore, these studies can be useful in attempts to utilize zebra mussel as a tool in bioindication or biomanipulation. For instance, zebra mussels are used in the Mussel Monitor Program, which is an early warning system to detect chemical contamination of water by means of automatic observations of mussel valve activity (Borcherding 2006). An increased understanding of behavioral responses of mussels to various environmental stimuli can improve resolution of such systems and reduce the possibility of false alarms (Borcherding 2006). In addition, knowledge of optimal conditions for mussel filtration and site selection can lead to better designs and more efficient substrates when mussels are used as a tool in lake restoration, as has been proposed recently by Stybel et al. (2009) and Dionisio Pires et al. (2010).

ACKNOWLEDGMENTS

I would like to thank the two book editors for linguistic corrections and valuable comments on the earlier draft of this chapter.

REFERENCES

Ackerman, J. D., C. M. Cottrell, C. R. Ethier, D. G. Allen, and J. K. Spelt. 1995. A wall jet to measure the attachment strength of zebra mussel. *Can. J. Fish. Aquat. Sci.* 52:126–135.

Ackerman, J. D., C. M. Cottrell, C. R. Ethier, D. G. Allen, and J. K. Spelt. 1996. Attachment strength of zebra mussels on natural, polymeric, and metallic materials. *J. Environ. Eng. ASCE* 122:141–148.

Ackerman, J. D., C. R. Ethier, D. G. Allen, and J. K. Spelt. 1992. Investigation of zebra mussel adhesion strength using rotating disks. *J. Environ. Eng.* 118:708–724.

Ackerman, J. D., B. Sim, S. J. Nichols, and R. Claudi. 1994. A review of the early life history of zebra mussels (*Dreissena polymorpha*) comparisons with marine bivalves. *Can. J. Zool.* 72:1169–1179.

Bell, E. C. and J. M. Gosline. 1997. Strategies for life in flow: Tenacity, morphometry, and probability of dislodgment of two *Mytilus* species. *Mar. Ecol. Prog. Ser.* 159:197–208.

Bonner, T. P. and R. L. Rockhill. 1994. Ultrastructure of byssus of the zebra mussel (*Dreissena polymorpha*, Mollusca: Bivalvia). *Trans. Am. Micro. Soc.* 113:302–315.

Borcherding, J. 2006. Ten years of practical experience with the *Dreissena*-Monitor, a biological early warning system for continuous water quality monitoring. *Hydrobiologia* 556:417–426.

Borcherding, J. and B. Jantz. 1997. Valve movement response of the mussel *Dreissena polymorpha*—The influence of pH and turbidity on the acute toxicity of pentachlorophenol under laboratory and field conditions. *Ecotoxicology* 6:153–165.

Borcherding, J. and J. Wolf. 2001. The influence of suspended particles on the acute toxicity of 2-chloro-4-nitro-aniline, cadmium, and pentachlorophenol on the valve movement response of the zebra mussel (*Dreissena polymorpha*). *Arch. Environ. Contam. Toxicol.* 40:497–504.

Brylińska, M. 2000. *Ryby słodkowodne Polski.* (*Freshwater Fishes of Poland*). Warsaw, Poland: PWN (in Polish).

Burks, R. L., N. C. Tuchman, C. A. Call, and J. E. Marsden. 2002. Colonial aggregates: Effects of spatial position on zebra mussel responses to vertical gradients in interstitial water quality. *J. N. Am. Benthol. Soc.* 21:64–75.

Chase, M. E. and R. C. Bailey. 1996. Recruitment of *Dreissena polymorpha*: Does the presence and density of conspecifics determine the recruitment density and pattern in a population? *Malacologia* 38:19–31.

Clarke, M. 1999. The effect of food availability on byssogenesis by the zebra mussel (*Dreissena polymorpha* Pallas). *J. Mollusc. Stud.* 65:327–333.

Clarke, M. and R. F. McMahon. 1995. Effects of current velocity on byssal thread production in the zebra mussel (*Dreissena polymorpha*). *Proceedings of the Fifth International Zebra Mussel and Other Aquatic Nuisance Organisms Conference*, pp. 39–45. Toronto, Ontario, Canada.

Clarke, M. and R. F. McMahon. 1996a. Effects of temperature on byssal thread production by the freshwater mussel, *Dreissena polymorpha* (Pallas). *Am. Malacol. Bull.* 13:105–110.

Clarke, M. and R. F. McMahon. 1996b. Effects of hypoxia and low-frequency agitation on byssogenesis in the freshwater mussel *Dreissena polymorpha* (Pallas). *Biol. Bull.* 191:413–420.

Coons, K., D. J. McCabe, and J. E. Marsden. 2004. The effects of strobe lights on zebra mussel settlement and movement patterns. *J. Freshwat. Ecol.* 19:1–8.

Czarnecka, M. 2005. Ekologiczny status epifauny zasiedlającej sztuczne podłoża podwodne (Ecological status of epifauna overgrowing underwater artificial substrata). PhD thesis, Nicolaus Copernicus University, Toruń, Poland (in Polish).

Czarnołęski, M., J. Kozłowski, and P. Kubajak et al. 2006. Cross-habitat differences in crush resistance and growth pattern of zebra mussels (*Dreissena polymorpha*): Effects of calcium availability and predator pressure. *Arch. Hydrobiol.* 165:191–208.

Czarnołęski, M., J. Kozłowski, A. Stańczykowska, and K. Lewandowski. 2003. Optimal resource allocation explains growth curve diversity in zebra mussels. *Evol. Ecol. Res.* 5:571–587.

Czarnołęski, M. and T. Müller. 2013. Anti-predator strategy of zebra mussels: From behavior to life history. In: *Quagga and Zebra Mussels: Biology, Impacts, and Control*. 2nd Edn., T. F. Nalepa and D. W. Schloesser, eds., pp. 345–359. Boca Raton, FL: CRC Press.

Czarnołęski, M., T. Müller, K. Adamus, G. Ogorzelska, and M. Sog. 2010. Injured conspecifics alter mobility and byssus production in zebra mussels *Dreissena polymorpha*. *Fundam. Appl. Limnol. (Arch. Hydrobiol.)* 176:269–278.

De Lafontaine, Y., G. Costan, and F. Delisle. 2002. Testing a new antizebra mussel coating with a multi-plate sampler: Confounding factors and other fuzzy features. *Biofouling* 18:1–12.

De Meester, L., P. Dawidowicz, E. Van Gool, and C. J. Loose. 1999. Ecology and evolution of predator-induced behaviour of zooplankton: Depth selection behaviour and diel vertical migration. In: *The Ecology and Evolution of Inducible Defenses*, R. Tollrian and C. D. Harvel, eds., pp. 160–176. Princeton, NJ: Princeton University Press.

Diggins, T. P., M. Weimer, and K. M. Stewart et al. 2004. Epiphytic refugium: Are two species of invading freshwater bivalves partitioning spatial resources? *Biol. Invasions* 6:83–88.

Dionisio Pires, L. M., B. W. Ibelings, and E. Van Donk. 2010. Zebra mussels as a potential tool in the restoration of eutrophic shallow lakes, dominated by toxic cyanobacteria. In: *The Zebra Mussel in Europe*, G. van der Velde, S. Rajagopal, and A. bij de Vaate, eds., pp. 331–342. Leiden, The Netherlands: Backhuys Publishers.

Dormon, J. M., C. Coish, C. Cottrell, D. G. Allen, and J. K. Spelt. 1997. Modes of byssal failure in forced detachment of zebra mussels. *J. Environ. Eng. ASCE* 123:933–938.

Gliwicz, Z. M. 2005. Food web interactions: Why are they reluctant to be manipulated? Plenary lecture. *Verh. Internat. Verein. Limnol.* 29:73–88.

Gosling, E. 2003. *Bivalve Molluscs: Biology, Ecology and Culture*. Oxford, U.K.: Blackwell Publishing.

Green, N. S., B. A. Hazlett, and S. Prueff-Jones. 2008. Attachment and shell integrity affects the vulnerability of zebra mussels (*Dreissena polymorpha*) to predation. *J. Freshwat. Ecol.* 23:91–99.

Hanson, H. and T. L. Mocco. 1994. *Dreissena* settlement on natural and anthropogenic substrates in the bay of Green Bay (Lake Michigan). *Proceedings of the Fourth International Zebra Mussel Conference*, pp. 409–414. Madison, WI.

Johnson, L. E. and D. K. Padilla. 1996. Geographic spread of exotic species: Ecological lessons and opportunities from the invasion of the zebra mussel, *Dreissena polymorpha*. *Biol. Conserv.* 78:23–33.

Kakareko, T., J. Żbikowski, and J. Żytkowicz. 2005. Diet partitioning in summer of two syntopic neogobiids from two different habitats of the lower Vistula River, Poland. *J. Appl. Ichthyol.* 21:292–295.

Karatayev, A. Y., L. E. Burlakova, and D. K. Padilla. 2002. Impacts of zebra mussels on aquatic communities and their role as ecosystem engineers. In: *Invasive Aquatic Species of Europe: Distribution, Impacts and Management*, E. Leppäkoski, S. Gollasch and S. Olenin, eds., pp. 433–446. Boston, MA: Kluwer Academic Publishers.

Kavouras, J. H. and J. S. Maki. 2003. The effects of natural biofilms on the reattachment of young adult zebra mussels to artificial substrata. *Biofouling* 19:247–256.

Kavouras, J. H. and J. S. Maki. 2004. Inhibition of the reattachment of young adult zebra mussels by single-species biofilms and associated exopolymers. *J. Appl. Microbiol.* 97:1236–1246.

Kobak, J. 2001. Light, gravity and conspecifics as cues to site selection and attachment behaviour of juvenile and adult *Dreissena polymorpha* Pallas, 1771. *J. Mollusc. Stud.* 67:183–189.

Kobak, J. 2002. Impact of light conditions on geotaxis behaviour of juvenile *Dreissena polymorpha*. *Folia Malacol.* 10:77–82.

Kobak, J. 2003. Impact of conspecifics on recruitment and behaviour of *Dreissena polymorpha* (Pallas, 1771). *Folia Malacol.* 11:95–101.

Kobak, J. 2004. Recruitment and small-scale spatial distribution of *Dreissena polymorpha* (Bivalvia) on artificial materials. *Arch. Hydrobiol.* 160:25–44.

Kobak, J. 2005. Recruitment and distribution of *Dreissena polymorpha* (Bivalvia) on substrates of different shape and orientation. *Int. Rev. Hydrobiol.* 90:159–170.

Kobak, J. 2006a. Factors influencing the attachment strength of *Dreissena polymorpha* (Bivalvia). *Biofouling* 22:153–162.

Kobak, J. 2006b. Geotactic behaviour of *Dreissena polymorpha* (Bivalvia). *Malacologia* 48:305–308.

Kobak, J. 2010. Attachment strength of *Dreissena polymorpha* on artificial substrates. In: *The Zebra Mussel in Europe*, G. van der Velde, S. Rajagopal, and A. bij de Vaate, eds., pp. 379–385. Leiden, The Netherlands: Backhuys Publishers.

Kobak, J. and T. Kakareko. 2009. Attachment strength, aggregation and movement of the zebra mussel (*Dreissena polymorpha*, Bivalvia) in the presence of potential predators. *Fundam. Appl. Limnol. (Arch. Hydrobiol.)* 174:193–204.

Kobak, J., T. Kakareko, and M. Poznańska. 2010. Changes in attachment strength and aggregation of zebra mussel, *Dreissena polymorpha* in the presence of potential fish predators of various species and size. *Hydrobiologia* 644:195–206.

Kobak, J., E. Kłosowska-Mikułan, and R. Wiśniewski. 2002. Impact of copper substrate on survival, mobility and attachment strength of adult *Dreissena polymorpha* (Pall.). *Folia Malacol.* 10:91–97.

Kobak, J. and P. Nowacki. 2007. Light-related behaviour of zebra mussel (*Dreissena polymorpha*, Bivalvia). *Fundam. Appl. Limnol. (Arch. Hydrobiol.)* 169:341–352.

Kobak, J., M. Poznańska, and T. Kakareko. 2009. Effect of attachment status and aggregation on behaviour of the zebra mussel, *Dreissena polymorpha*, Bivalvia. *J. Mollusc. Stud.* 75:109–117.

Lachance, A. A., B. Myrand, R. Tremblay, V. Koutitonsky, and E. Carrington. 2008. Biotic and abiotic factors influencing attachment strength of blue mussels *Mytilus edulis* in suspended culture. *Aquat. Biol.* 2:119–129.

Lauer, T. E. and A. Spacie. 2004. Space as a limiting resource in freshwater systems: Competition between zebra mussels (*Dreissena polymorpha*) and freshwater sponges (Porifera). *Hydrobiologia* 517:137–145.

Lesht, B. M. and N. Hawley. 2013. Video Clip 5: Zebra mussel movements on the bottom of Lake Michigan. In: *Quagga and Zebra Mussels: Biology, Impacts, and Control*, 2nd Edn., T. F. Nalepa and D. W. Schloesser, eds., pp. 748–755. Boca Raton, FL: CRC Press.

Lewandowski, K. 1982. The role of early developmental stages in the dynamics of *Dreissena polymorpha* (Pall.) (Bivalvia) populations in lakes. II. Settling of larvae and the dynamics of number of sedentary individuals. *Ekol. Polska* 30:223–286.

Lewandowski, K. 2001. Development of populations of *Dreissena polymorpha* (Pall.) in lakes. *Folia Malacol.* 9:171–213.

Marsden, J. E. and D. M. Lansky. 2000. Substrate selection by settling zebra mussels, *Dreissena polymorpha*, relative to material, texture, orientation, and sunlight. *Can. J. Zool.* 78:787–793.

Martel, A. 1993. Dispersal and recruitment of zebra mussel (*Dreissena polymorpha*) in nearshore area in west-central Lake Erie: The significance of post-metamorphic drifting. *Can. J. Fish. Aquat. Sci.* 50:3–12.

McMahon, R. F. 1996. The physiological ecology of the zebra mussel *Dreissena polymorpha* in North America and Europe. *Am. Zool.* 36:339–363.

Mikheev, V. P. 1964. O skorosti otmiraniya dreisseny w anaerobnych usloviach. [The mortality rate of *Dreissena* in anaerobic conditions]. In: *Biologiya dreisseny i borba s ney*, B. S. Kuzin, ed., pp. 65–68. Leningrad, Russia: Izdatelstvo Nauka, Moskva (in Russian).

Molloy D. P., A. Y. Karatayev, L. E. Burlakova, D. P. Kurandina, and F. Laruelle. 1997. Natural enemies of zebra mussels: Predators, parasites, and ecological competitors. *Rev. Fish. Sci.* 5:27–97.

Naddafi, R., P. Eklöv, and K. Pettersson. 2007. Non-lethal predator effects on the feeding rate and prey selection of the exotic zebra mussel *Dreissena polymorpha*. *Oikos* 116:1289–1298.

Nagelkerke, L. A. J. and F. A. Sibbing. 1996. Efficiency of feeding on zebra mussel (*Dreissena polymorpha*) by common bream (*Abramis brama*), white bream (*Blicca bjoerkna*), and roach (*Rutilus rutilus*): The effects of morphology and behaviour. *Can. J. Fish. Aquat. Sci.* 53:2847–2861.

Orlova, M. I., T. W. Therriault, P. I. Antonov, and G. Kh. Shcherbina. 2005. Invasion ecology of quagga mussels (*Dreissena rostriformis bugensis*): A review of evolutionary and phylogenetic impacts. *Aquat. Ecol.* 39:401–418.

Peyer, S. M., A. J. McCarthy, and C. E. Lee. 2009. Zebra mussels anchor byssal threads faster and tighter than quagga mussels in flow. *J. Exp. Biol.* 212:2027–2036.

Porter, A. E. and J. E. Marsden. 2008. Adult zebra mussels (*Dreissena polymorpha*) avoid attachment to mesh materials. *Northeast. Nat.* 15:589–594.

Prejs, A., K. Lewandowski, and A. Stańczykowska. 1990. Size-selective predation by roach (*Rutilus rutilus*) on zebra mussel (*Dreissena polymorpha*): Field studies. *Oecologia* 83:378–384.

Race, T. 1992. Nontoxic foul-release coatings for zebra mussel control. Zebra Mussel Technical Notes Collection, U.S. Army Engineer Research & Development Center, Vicksburg, MS, ZMR-2-08.

Rajagopal, S., G. Van Der Velde, and H. A. Jenner. 2002. Does status of attachment influence chlorine survival time of zebra mussel, *Dreissena polymorpha* exposed to chlorination? *Environ. Toxicol. Chem.* 21:342–346.

Rajagopal, S., G. Van Der Velde, H. A. Jenner, M. Van Der Gaag, and A. J. Kempers. 1996. Effects of temperature, salinity and agitation on byssus thread formation of zebra mussel *Dreissena polymorpha*. *Neth. J. Aquat. Ecol.* 30:187–195.

Rzepecki, L. M. and J. H. Waite. 1993. The byssus of the zebra mussel, *Dreissena polymorpha*. I: Morphology and in situ protein processing during maturation. *Mol. Mar. Biol. Biotechnol.* 2:255–266.

Schneider, K. R., D. S. Wethey, B. S. T. Helmuth, and T. J. Hilbish. 2005. Implications of movement behaviour on mussel dislodgement: Exogenous selection in a *Mytilus* spp. hybrid zone. *Mar. Biol.* 146:333–343.

Sigurdsson, J. B., C. W. Titman, and P. A. Davies. 1976. The dispersal of young post-larval bivalve molluscs by byssal threads. *Nature* 262:386–387.

Stańczykowska, A. 1964. On the relationship between abundance, aggregations and "condition" of *Dreissena polymorpha* Pall. in 36 Mazurian Lakes. *Ekol. Polska A* 34:653–690.

Stybel, N., C. Fenske, and G. Schernewski. 2009. Mussel cultivation to improve water quality in the Szczecin Lagoon. *J. Coast. Res.* 56:1459–1463.

Toomey, M. B., D. McCabe, and J. E. Marsden. 2002. Factors affecting the movement of adult zebra mussels (*Dreissena polymorpha*). *J. N. Am. Benthol. Soc.* 21:468–475.

Tuchman, N. C., R. L. Burks, C. A. Call, and J. Smarrelli. 2004. Flow rate and vertical position influence ingestion rates of colonial zebra mussels (*Dreissena polymorpha*). *Freshwat. Biol.* 49:191–198.

Wacker, A. and E. von Elert. 2003. Settlement pattern of the zebra mussel, *Dreissena polymorpha*, as a function of depth in Lake Constance. *Arch. Hydrobiol.* 158:289–301.

Wainman, B. C., S. S. Hincks, N. K. Kaushik, and G. L. Mackie. 1996. Biofilm and substrate preference in the dreissenid larvae of Lake Erie. *Can. J. Fish. Aquat. Sci.* 53:134–140.

Zhang, Y., S. E. Stevens, and T. Y. Wong. 1998. Factors affecting rearing of settled zebra mussels in a controlled flow-through system. *Prog. Fish Cult.* 60:231–235.

CHAPTER 22

Antipredator Strategy of Zebra Mussels (*Dreissena polymorpha*)
From Behavior to Life History

Marcin Czarnoleski and Tomasz Müller

CONTENTS

Abstract 345
Introduction 345
Predator-Induced Mortality 346
Predator-Mediated Responses 348
Predation Risk Assessment 351
Adaptive Value of Responses 352
Conclusions and Research Needs 354
Acknowledgments 354
References 355

ABSTRACT

Zebra mussels are subject to spatio-temporal variability in predation pressure, but evidence of their defensive responses is fragmentary and sometimes inconsistent. Here, we integrate experimental and comparative evidence to create a cohesive view of the anti-predator strategy of zebra mussels, with emphasis on the importance of time required for an effective response. In general, zebra mussels use chemosensors to assess predation risks, differentiate between species of predators, tune responses to the type and intensity of alarm cues, and recognize immediacy of predatory attacks. Mussel responsiveness to predators decreases as shells grow in size and become less vulnerable to crushing.

Experiments show that predation risk induces reduced mobility, decreased feeding rates, and increased aggregation. These responses are observable within hours after exposure to predation cues. Over the longer term (days), zebra mussels can increase or decrease strength of byssal attachment in response to cues. Effluents from crushed mussels indicate a risk of immediate predatory attacks. In response, zebra mussels adopt rapid behavioral responses aimed at reduced detection; for instance, they suppress filtering activity to reduce emission of disclosing metabolites. This physiological slowdown may negatively affect byssal production, leading to weaker attachment. If, however, there are only signs of the presence of non-hunting predators and predatory attacks are less likely, zebra mussels either adopt time-demanding allocation responses such as building up stronger byssal attachment or ignore a predator if its actual pressure is low (e.g., non-hunting crayfish, non-molluscivorous fish, or small roach). Comparative evidence suggests that in a lifelong perspective, predator-mediated mortality may stimulate zebra mussels to produce stronger shells to avoid crushing and favor earlier maturation followed by increased reproduction at the expense of growth to maximize expected lifetime offspring production.

INTRODUCTION

Introductions of invasive species to novel environments provide unique opportunities to study ongoing evolutionary and ecological processes. Invasives are exposed to new selection regimes and their successful establishment depends on the strength of this selection and their capacity for adequate responses through rapid adaptation or adaptive phenotypic plasticity. The two-century-long expansion of the Ponto-Caspian zebra mussel *Dreissena polymorpha* (Pall.) throughout Europe and North America has exposed this invader to predatory attacks by various species of birds, fish, and crustaceans in colonized habitats. Apparently, predation has not created an effective barrier against zebra mussels, but the actual role of predator–prey interactions in the range expansion of this bivalve remains unknown. So far, the majority of studies on zebra mussels have overlooked this topic and focused mainly on the ecological and economic impacts of this invader. Typically, physico-chemical conditions have been viewed as the primary factor limiting the abundance of zebra mussel populations, with biotic agents playing a secondary role (Mellina and Rasmussen 1994, Whittier et al. 2008). Hence, there is a paucity of available data on predator–prey interactions of this bivalve. Emerging evidence from experimental studies (e.g., Kobak and Kakareko 2009, Czarnoleski et al. 2010b) demonstrated that zebra mussels adjust their phenotype in a qualitative and quantitative manner relative to predation risk. Such evidence seems to support the conclusion of previous studies that show predation risk can be the reason for differing life histories between zebra mussel populations (reviewed in Czarnoleski et al. 2010a).

In this chapter, we provide an up-to-date review of predator-mediated responses of zebra mussels to gain insight on a capacity of zebra mussels for assessment of predation risks and on the nature of defensive responses. We first characterize predation pressure exerted on zebra mussels by their major predators, focusing on changes through various life stages. Next, we review responses of zebra mussels to waterborne predation cues relative to different timescales. We provide evidence that zebra mussels have the ability to assess predation risks. In order to evaluate the adaptive value of observed responses, we consider the importance of costs and benefits of responses and differences in response time. Finally, we indicate some inconsistencies in reported response patterns and propose how to reconcile them.

PREDATOR-INDUCED MORTALITY

There is evidence showing significant impacts of predation on zebra mussel populations. Cleven and Frenzel (1993) and Werner et al. (2005) reported that zebra mussel abundances were reduced by 90% by diving birds. Interestingly, dense colonies of zebra mussels seem to intensify bird predation as waterfowl can easily collect mussels from the size range that maximizes the energetic effectiveness of dives (Draulans 1982). The potential of fish predation to limit the abundance of zebra mussels has been shown both on theoretical (Eggleton et al. 2004, Madenjian et al. 2010) and empirical grounds (Thorp et al. 1998, Magoulick and Lewis 2002, Bartsch et al. 2005). After introduction, zebra mussels became the most important food type of some North American fish species such as pumpkinseed (*Lepomis gibbosus*) and freshwater drum (*Aplodinotus grunniens*) (Andraso 2005, Watzin et al. 2008). Naddafi et al. (2010) explored factors that determined microscale differences in population density of zebra mussels in a Swedish lake and showed that locations with the highest pressure from roach (*Rutilus rutilus*) had lowest densities of zebra mussels. Although crayfish are regarded important predators within benthic habitats (Nyström et al. 1996), their ability to control zebra mussel populations seems to be limited (Molloy et al. 1997, Perry et al. 2000, Reynolds and Donohue 2001). Martin and Corkum (1994) demonstrated experimentally that consumption of zebra mussels by individual crayfish decreased at higher crayfish densities. These conclusions have been recently supported by the findings of Ermgassen and Aldridge (2011) who studied prospects of using signal crayfish (*Pacifastacus leniusculus*) to curb further spread of zebra mussels. Other crustacean predators such as blue crabs (*Callinectes sapidus*) may temporarily reduce numbers of zebra mussels in isolated populations inhabiting brackish habitats (Molloy et al. 1994, Boles and Lipcius 1997).

Predators feed on all life stages of zebra mussels, and the relative importance of individual predators changes as the mussels grow (Figure 22.1; Molloy et al. 1997, Thorp et al. 1998, Magoulick and Lewis 2002, Eggleton et al. 2004, Czarnoleski et al. 2006). Veligers of zebra mussels are eaten by fish fry and calanoid copepods (Molloy et al. 1997), but predation by copepods decreases once veligers develop a D-shape shell at about 90–100 μm (Liebig and Vanderploeg 1995). The most important predators of sessile forms of zebra mussels are fish (mainly Cyprinidae), diving birds (mainly Anatidae), and crustaceans (mainly those from North America such as spiny-cheek crayfish *Orconectes limosus*) (Molloy et al. 1997). Spiny-cheek crayfish crush zebra mussels up to 12 mm long but usually prefer specimens smaller than 8 mm. Fish, such as roach and carp (*Cyprinus carpio*), and diving birds (*Aythya ferina, Aythya fuligula, Aythya marila, Bucephala clangula, Fulica atra*) eat a wide range of mussel sizes but focus on 8–17 mm long mussels (Piesik 1974, Prejs et al. 1990, De Leeuw and Van Eerden 1992, Hamilton and Ankney 1994, Tucker et al. 1996, Molloy et al. 1997, Czarnoleski et al. 2006). Prejs et al. (1990) reported that small roach (ca. <160 mm) generally do not feed on zebra mussels, and the size of eaten mussels increases as roach become larger.

Most predators tear sessile zebra mussels from hard substrata by breaking their byssal threads and then crush the protective shell to obtain the soft tissue (Czarnoleski

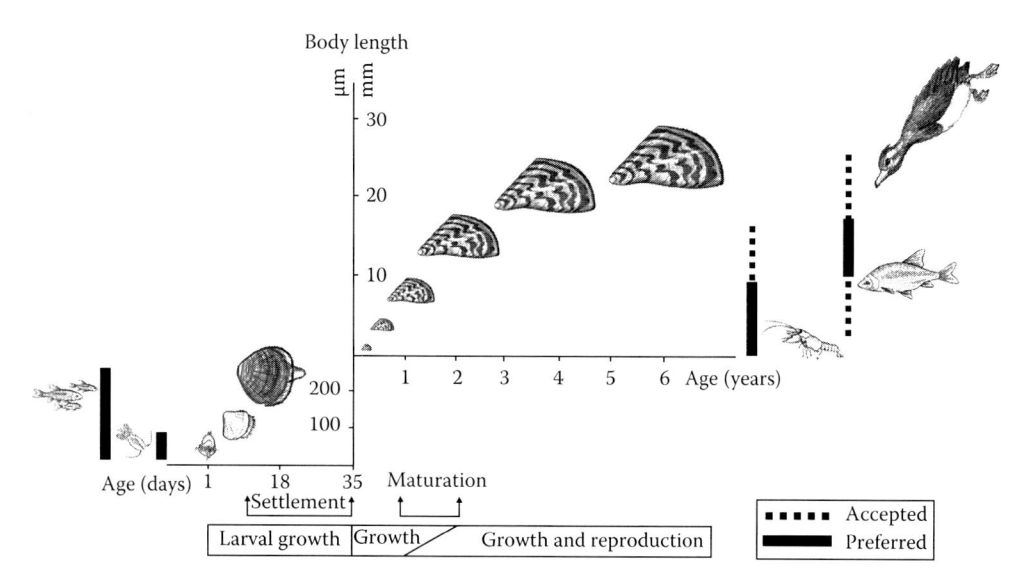

Figure 22.1 Size selectivity of common predators of zebra mussels as reconstructed from the literature. Sizes of life-cycle stages are according to Hopkins and Leach (1993) and Czarnoleski et al. (2003). Size ranges of consumed zebra mussels (accepted and preferred) are indicated by vertical bars in reference to specific predators. (Based on Czarnoleski, M. et al., *Arch. Hydrobiol.*, 165, 191, 2006.)

et al. 2006). Crayfish dislodge individual mussels with their thoracic appendages and crush the sharp edges of mussel shells (Piesik 1974, Perry et al. 1997). Nagelkerke and Sibbing (1996) demonstrated that roach dislodge mussels in a sequence of rapid suction movements. The fish then transfers the prey to the chewing cavity and positions it rostrocaudally lengthwise and dorsoventrally widthwise. A large static force is then applied by the upper pharyngeal teeth to crush shells. In diving birds, the predation mode depends on foraging water depth and prey size. Birds may ingest several mussels directly under water by suction or carry mussel aggregates from the water and dislodge them on land. Prey is swallowed whole and crushed in a muscular gizzard (De Leeuw and Van Eerden 1992).

Assessments of predation pressure should consider the size selectivity and encounter probability of predators. This probability depends on predator energy demands and predator abundance relative to the availability of different prey. In temperate zones where zebra mussels are common, predation pressure varies seasonally. For instance, consumption of zebra mussels by different avian species attained highest levels between autumn and spring when birds flock to migrate or overwinter (bij de Vaate 1991, Wormington and Leach 1992, Molloy et al. 1997). Elevated costs of thermoregulation in endothermic predators such as birds increase food intake in cold temperatures, which should intensify predation pressure on zebra mussels in winter. Indeed, De Leeuw et al. (1999) observed a twofold increase of zebra mussel consumption by *A. fuligula* in winter compared to summer. Czarnoleski et al. (2010a) argued that because waterfowl remain in the winter as long as water is not covered with ice, predation pressure on zebra mussels in temperate zones should be positively linked to the number of days that temperatures exceed 0°C.

Food consumption of ectotherms, such as fish and crayfish, is tightly linked to ambient temperature through a direct temperature dependence of physiological processes. Therefore, predation pressure of ectothermic predators on zebra mussels should be related to ambient temperature and undergo seasonal changes. This is the case in marine planktonic copepods in which predation rates increase with temperature (Hirst and Kiørboe 2002). Liebig and Vanderploeg (1995) argued that the intensity of copepod predation on zebra mussel veligers should intensify with temperature, but this phenomenon has not been directly studied. Eggleton et al. (2004) modeled impacts of predation of several fish species on zebra mussels along a 13° latitudinal gradient in North America. They found that fish from warmer, southern waters have potential to consume twice as many zebra mussels than fish from cooler, northern regions. Besides temperature, other biological and physicochemical characteristics that change seasonally also shape predation pressure. Chybowski (2007) reported that the contribution of animal tissue to the diet of spiny-cheek crayfish, an omnivore, was 45% in spring when plants were not yet developed but only 4% in summer when plant food was abundant. This suggests that the pressure of crayfish on zebra mussels may be lower in summer than in spring, contrary to what can be expected from a strict temperature-dependence model of consumption. According to Watzin et al. (2008), between-year differences in the amount of zebra mussels consumed by freshwater drum (*A. grunniens*) are explained by changes in abundance of crayfish, which is an important component of drum diet.

PREDATOR-MEDIATED RESPONSES

Predators elicit a wide array of different responses in prey, from fast reversible changes in behavior to lifetime modifications of energy allocation. Mollusks under predation pressure reduce feeding rate (Smee and Weissburg 2006), seek refuge (Alexander and Covich 1991), burrow (Griffiths and Richardson 2006), clump (Côté and Jelnikar 1999), increase (Côté 1995) or decrease byssogenesis (Ishida and Iwasaki 2003), delay egg hatching (Miner et al. 2010), increase adductor muscle size (Freeman 2007), increase shell resistance to crushing (Trussell et al. 2003), modify shell shape (DeWitt et al. 2000), accelerate (Cheung et al. 2004) or decelerate growth rates (Nakaoka 2000), and delay maturation (Crowl and Covich 1990).

Zebra mussels respond to predation cues in many different ways. In general, cues from crushed zebra mussels tend to reduce activity of other zebra mussels. For instance, after being placed on sand, detached zebra mussels crawl shorter distances after exposure to effluents from artificially crushed mussels (Toomey et al. 2002). Czarnoleski et al. (2010b) demonstrated that this response was not caused by postponement of crawling activity, by crawling for a shorter period of time, or by an increased tendency to form byssal attachment. It was caused by decreased crawling speed. Distances crawled by zebra mussels became progressively shorter as cue concentrations increased, in concert with a decrease in crawling speed (Figure 22.2a,b). Zebra mussels decreased filtering activity and changed diet composition when exposed to cues released by roach and crayfish that had been fed zebra mussels (Naddafi et al. 2007). Kobak and Kakareko (2009) found that a 6-day-long exposure to cues from non-foraging roach increased the number of zebra mussels that attached in direct physical contact with other mussels, forming aggregates. Further experiments of Kobak et al. (2010) revealed that this tendency for clumping was noticeable already after 24 h of the exposure to predation cues. In all studies that report behavioral responses, the response occurred quickly, within 2 h (Toomey et al. 2002, Naddafi et al. 2007), 3 h (Czarnoleski et al. 2010b), or 1 day (Kobak et al. 2010) after the mussels were exposed to predation cues.

Exposure of zebra mussels to predation cues for several days affects their production of byssal threads, but two different response patterns have been observed. Czarnoleski et al. (2010b) found that detached zebra mussels exposed to effluents from crushed conspecific mussels for 2 and 4 days developed weaker attachment to substratum when compared to mussels in unexposed controls (Figure 22.3a,b). This reduction in attachment strength was caused by a decrease in number of byssal threads with no change in the break resistance of single threads (Czarnoleski et al. 2010b).

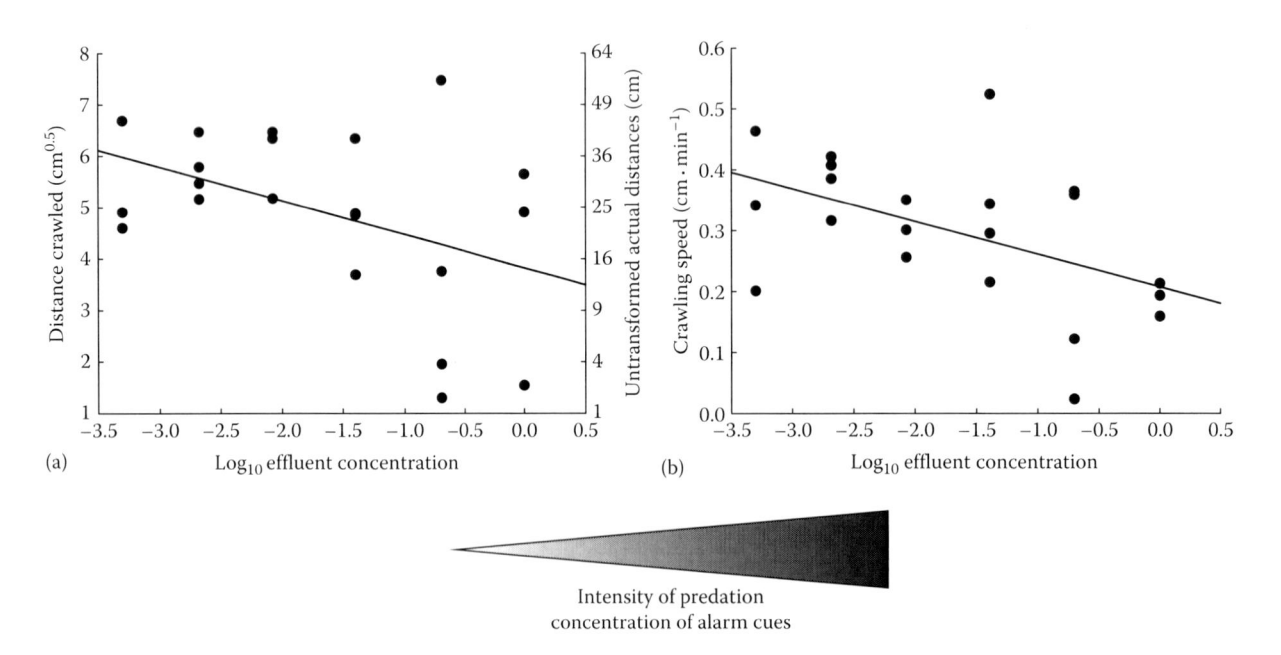

Figure 22.2 Distance (a) and crawling speed (b) of zebra mussels (after 180 min) at different concentrations of effluents containing extracts from crushed zebra mussels. Effluents are a direct indicator of predation risk. Movements measured on sand in aquariums. Each point represents a mean value. Effluent concentrations change geometrically; units represent concentrations relative to that of Toomey et al. (2002). The highest concentration (1) was prepared as given in Toomey et al. (2002); that is, 150 mL of crushed mussels was mixed with 350 mL of water, strained, and 1.3 mL of this strained solution was dissolved in 1 L of water. Concentrations of 1/5, 1/25, 1/125, and 1/625 were obtained by progressively diluting the stock solution with water. A concentration of 0 consisted of just water (no effluent solution). Prior to log transformation, concentration values were coded by adding 0.0005. (Modified from Czarnoleski, M. et al., An evolutionary perspective on the geographic and temporal variability of life histories in European zebra mussels, in *The Zebra Mussel in Europe*, G. Van der Velde, S. Rajagopal, and A. Bij de Vaate, Eds., Backhuys Publishers, Leiden, The Netherlands, pp. 169–182, 2010b.)

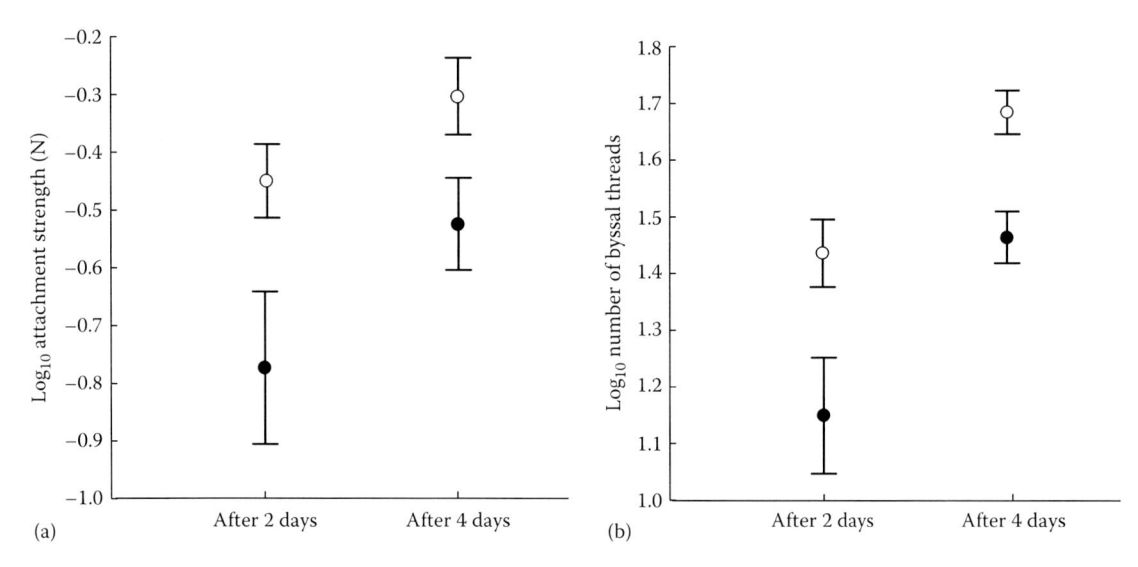

Figure 22.3 Mean (±95 CI) attachment strength (a) and number of byssal threads (b) produced by zebra mussels exposed to effluents from crushed zebra mussels. Measurements were taken after 2 and 4 days of exposure using PLEXIGLAS as a substrate. Open circle, control; solid circle, crushed conspecifics. (Modified from Czarnoleski, M. et al., An evolutionary perspective on the geographic and temporal variability of life histories in European zebra mussels, in *The Zebra Mussel in Europe*, G. Van der Velde, S. Rajagopal, and A. Bij de Vaate, Eds., Backhuys Publishers, Leiden, The Netherlands, pp. 169–182, 2010b.)

Similarly, zebra mussels developed weaker substrate attachment within 3 days when exposed to cues released by other mussels when preyed on by the crayfish *O. limosus* (Czarnoleski et al. 2011). In contrast, both Kobak and Kakareko (2009) and Kobak et al. (2010) showed that cues released by non-foraging roach mediated production of stronger byssal attachment in zebra mussels while other non-foraging fish species and non-foraging crayfish did not affect attachment strength. Results of Kobak et al. (2010) suggested that the development of stronger attachment in response to alarm cues may require considerable amount of time; predator-mediated changes in attachment strength were detectable only after exposure to predation cues for 6 days.

Effects of predation on life history traits are more difficult to define because their detection requires long-term experiments, lasting months or even years in the case of zebra mussels. This explains the scarcity of data. However, there is some indirect evidence from comparative studies that can be used to infer life history responses of zebra mussels to predation pressure (but note that these comparative data cannot fully discern between plastic and genetic sources of phenotypic variability). Czarnoleski et al. (2006) examined mortality rates, individual growth curves, and crush resistance of shells in several European populations as collected from different lakes. In all populations, crush resistance or shell strength was a function of shell length (Figure 22.4). However, they also found that crush resistance of shells was a function of mortality rates (Figure 22.5). Further, populations with higher mortality rates also had a smaller asymptotic body size. This suggested that differences in mortality across populations can be mostly explained by differential predation pressure and that such pressure likely stimulated the production of crush-resistant shells (through increased thickness) and reduced growth rates. Interestingly, these data did not show an energetic trade-off between shell thickness and somatic growth; crush resistance of shells was unrelated to growth curve parameters.

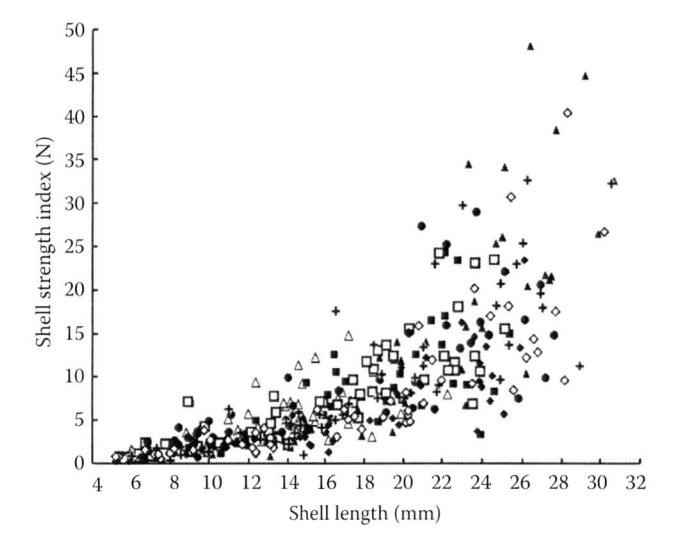

Figure 22.4 Relationship between shell strength and shell length in eight European populations of zebra mussels. Symbols denote population datasets ($n = 50$). Shell strength (crush resistance) was measured by the force needed to puncture the shell. (From Czarnoleski, M. et al., *Arch. Hydrobiol.*, 165, 191, 2006.)

The role of mortality in shaping life history of zebra mussels is further supported by results of large-scale comparative studies of long-established European populations of zebra mussels (Czarnoleski et al. 2003, 2005; reviewed by Czarnoleski et al. 2010a). These studies revealed high

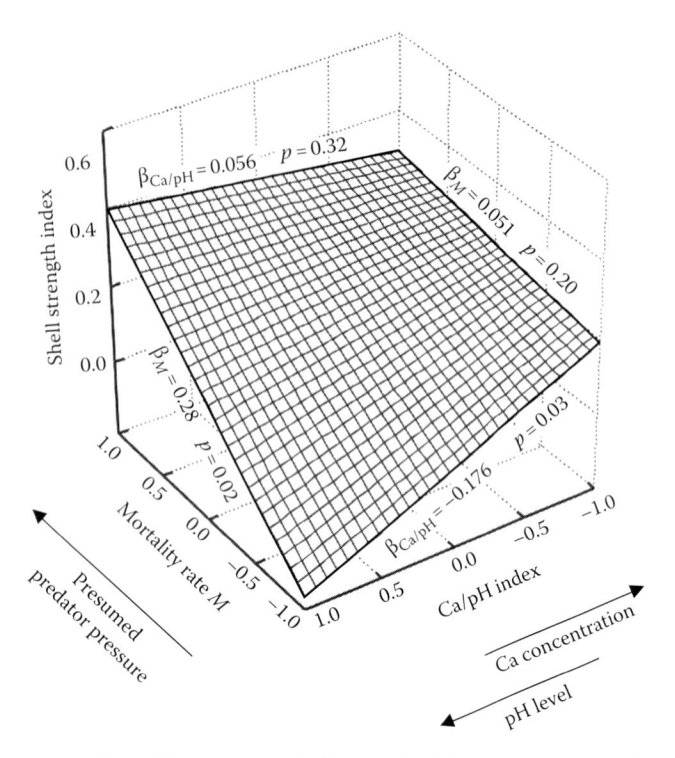

Figure 22.5 The average shell strength of 8 mm zebra mussels from eight European lakes in relation to population mortality rates (M, year^{-1}) and chemical conditions (calcium and pH). Shell strength index is the force (N) needed to puncture a shell. The calcium/pH index is the first principal component derived from a principal component analysis of pH and calcium concentration in lakes. Values of the Ca/pH index correlate positively with pH and negatively with calcium concentration. Partial slopes of shell strength on M (β_M) and shell strength on Ca/pH index ($\beta_{Ca/pH}$) were calculated with a multiple regression model, keeping values of other predictor at ±1 standard deviation. (From Czarnoleski, M. et al., *Arch. Hydrobiol.*, 165, 191, 2006.)

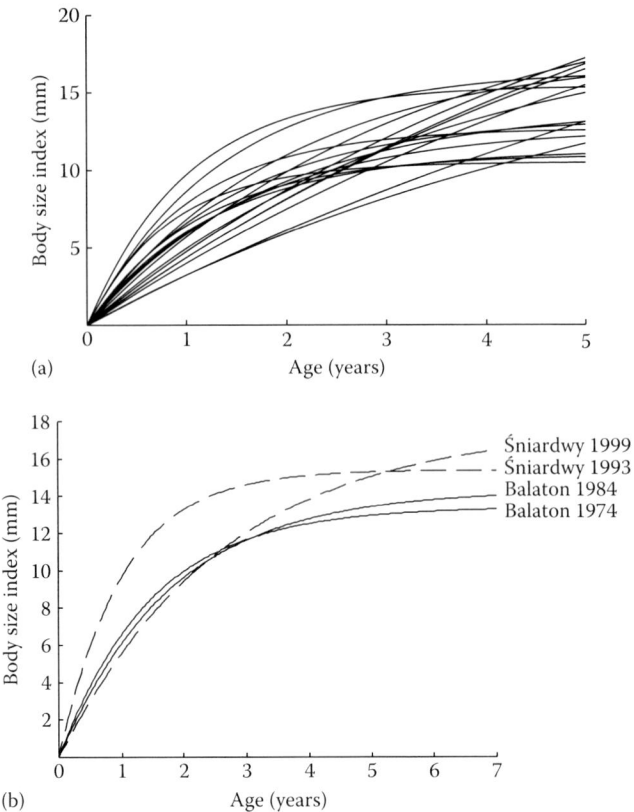

Figure 22.6 Variability of Bertalanffy's growth curves in European populations of zebra mussels. Shown are growth curves in 19 lakes (a). (From Czarnoleski, M. et al., *Evol. Ecol. Res.*, 5, 571, 2003.) An example of the smallest and largest changes of Bertalanffy's growth curves observed in two populations over a given period of time (b). (From Czarnoleski, M. et al., *Evol. Ecol. Res.*, 7, 821, 2005.) The index of body size is expressed in length units and it is the cube root of the product of shell length, width, and height, all measured in mm.

spatiotemporal diversity of growth patterns (Figure 22.6) and demonstrated that a significant portion of this variability could be attributed to differential survivability. Not only did asymptotic body size of zebra mussels correlate negatively with mortality rates across populations in different lakes (Figure 22.7a; Czarnoleski et al. 2003), but it decreased through time in lakes where zebra mussel mortality increased over years (Figure 22.7b; Czarnoleski et al. 2005).

Although comparative studies reveal that mortality may affect the growth pattern of zebra mussels, there have been no direct data to confirm the actual role of predators in this process. The link between abundance of roach and average shell sizes attained by 2-year-old zebra mussels in 17 Polish lakes was examined to test whether differences in risk of roach predation can explain variability in body size between populations (Figure 22.8; Czarnoleski and Müller unpublished data). Roach predation risk was estimated from roach biomass (kg) in routine annual surveys (data courtesy of Dr. A. Wołos, Inland Fisheries Institute, Olsztyn, Poland). For each lake, we calculated the average biomass of roach in a catch from the period 1990–1999 and normalized it to a hectare of lake area. This information was used as a measure of predation level exerted by roach on zebra mussels. Zebra mussels were collected in summer 1999 from the same lakes. Mussel length, width, and height were measured, and mussel age was determined from annual growth rings on shells following methods of Lewandowski (1983). We considered size differences at age 2 when all individuals should be reproductively matured (Czarnoleski et al. 2003). Early maturation and intense reproduction drain resources from growth, which should reduce growth rates. Consequently, in populations that mature earlier, the average size of 2-year-old individuals should be smaller than individuals in populations with later onset of reproduction. Results of

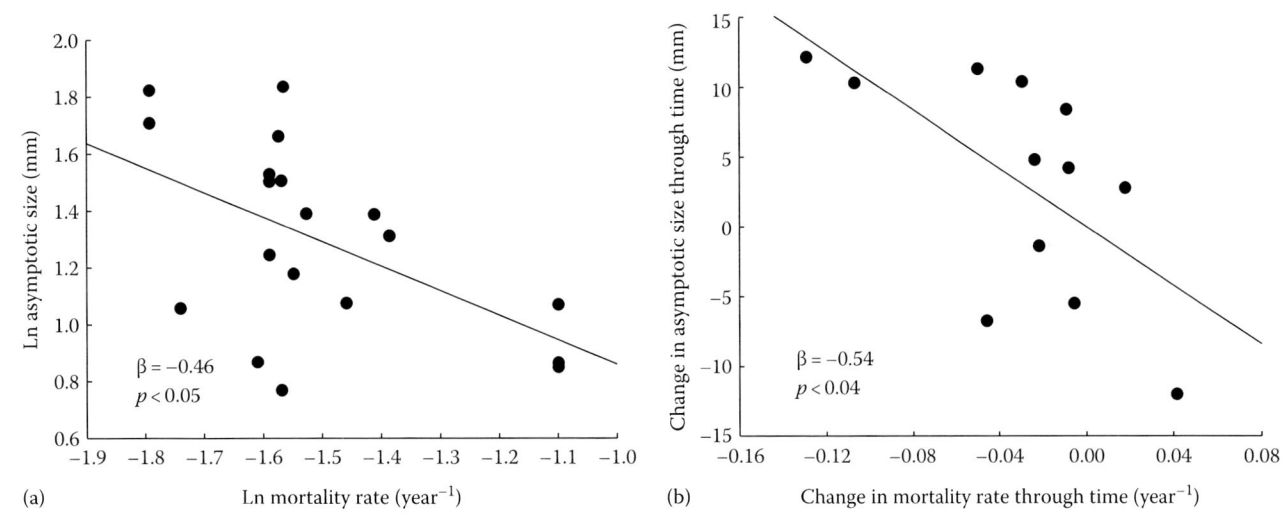

Figure 22.7 Partial regressions showing changes in Bertalanffy's asymptotic size of zebra mussels linked to spatial (a) and temporal (b) changes in population mortality in European lakes. Partial slopes β were calculated with separate multiple regression models with biomass production index as another predictor (data not shown here). Panel (a) shows data from 19 populations. (From Czarnoleski, M. et al., *Evol. Ecol. Res.*, 5, 571, 2003.) Panel (b) shows changes in parameter values observed in 12 populations over a given period of time. (From Czarnoleski, M. et al., *Evol. Ecol. Res.*, 7, 821, 2005.)

our analysis showed that with increased biomass of roach, 2-year-old zebra mussels became smaller (Figure 22.8). This suggests that zebra mussels in natural populations may mature earlier and grow less after maturation under the higher risk of predatory attacks. Certainly, other factors such as production conditions in lakes simultaneously influence the biomass of roach in catches and mussel growth, but we do not have any specific hypothesis to explain the origin of a negative correlation between roach biomass and mussel size driven by such factors.

PREDATION RISK ASSESSMENT

A capacity for accurate assessment of predation threat plays a central role in the evolution of responses to predation, and this evolution is expected if predation risk changes spatiotemporally and antipredator phenotypes incur costs when predators are absent (DeWitt et al. 1998, Tollrian and Harvell 1999, Trussell and Nicklin 2002). Theoretical works on predator–prey interactions commonly assume that organisms possess a capacity for risk assessment (e.g., Lima and Bednekoff 1999), but apparently disadvantageous responses to predation are sometimes observed. For example, Langerhans and DeWitt (2002) described development of defensive traits in snails *Physella virgata* exposed to cues from non-molluscivorous species of fish, whereas Freeman et al. (2009) described nonresponsiveness of blue mussels, *Mytilus edulis*, to a combination of cues from two predators with different hunting modes. Do zebra mussels make chemosensory assessments of predation pressure and respond accordingly?

There is evidence to show that vulnerability to shell crushing may affect the eagerness of zebra mussels to respond to predation cues. Crush resistance increases nonlinearly with mussel size (Figure 22.4; Czarnoleski et al. 2006) and predators of zebra mussels show size preferences (Figure 22.1). Larger zebra mussels that apparently become less prone to predatory attacks show lower responsiveness to alarm cues.

Figure 22.8 Correlation between average biomass of roach (kg/hectare) and the average body size index (mm) of 2-year-old zebra mussels in 17 Polish lakes (Czarnoleski and Müller, unpublished data). Roach biomass was assumed to represent the amount of predation pressure on zebra mussels and was determined from annual fish surveys in the period 1990–1999 in each lake. Zebra mussels were collected in 1999; their body size index is expressed in length units and it is the cube root of the product of shell length, width, and height, all measured in mm.

Toomey et al. (2002) studied zebra mussels from three size groups (5–10, 10–20, >20 mm) and observed that mobility in the smallest size group was most significantly reduced by effluents from crushed mussels. Similarly, Kobak and Kakareko (2009) demonstrated that cues from non-foraging roach stimulated byssal attachment and gregarious behavior only in zebra mussels smaller than 18 mm, matching the size range most vulnerable to predation (Figure 22.1). Comparative data on shell resistance to crushing in European populations of zebra mussels showed that this resistance increased in high-mortality populations but only in mussels smaller than 12 mm, which was just below the upper size threshold of crayfish predation (Czarnoleski et al. 2006). Crush resistance of larger individuals was unrelated to mortality risk. Importantly, the link between mortality level and shell resistance was present only across low-calcium habitats and was absent in high-calcium habitats (Figure 22.5). This suggests that zebra mussels adjust shell production to vulnerability of being crushed by predators. In high-calcium waters, mollusks generally have thick shells that already provide protection from predators, but mollusks in low-calcium waters have thin, less-protective shells, and investments in shell production under predation pressure are evolutionarily justified. Consistent with this view, Rundle et al. (2004) demonstrated that fish exudations elicited production of heavier shells in *Lymnaea stagnalis* snails raised at low calcium concentrations but not in snails grown at high concentrations of calcium. Snails raised without predators in high-calcium water had shells as heavy as snails raised at low calcium concentrations but exposed to predators.

Besides adjusting responses based on predation vulnerability, zebra mussels appear to assess the risk of predator attacks. Kobak and Kakareko (2009) found that zebra mussels increased byssal attachment and gregariousness in response to non-foraging roach, ignoring omnivorous crayfish, goby (predator less common in nature), and non-molluscivorous perch. According to Prejs et al. (1990), roach <160 mm rarely feed on zebra mussels because their pharyngeal teeth cannot efficiently crush zebra mussel shells, which suggests that small roach should not be regarded by zebra mussels as a direct threat. In support of this hypothesis, Kobak et al. (2010) demonstrated that roach smaller than 160 mm did not elicit responses in zebra mussels, while larger roach stimulated attachment of zebra mussels. These findings indicate that zebra mussels can use waterborne cues to discriminate between dangerous predators such as large roach and less dangerous predators such as crayfish, goby, perch, and small roach. Yet, there is another mechanism that might have simultaneously shaped the responsiveness of zebra mussels in these experiments. The amount of alarm cues emitted by predators is positively related to their body mass. Therefore, lower responsiveness of zebra mussels to some predators used by Kobak et al. (2010) may, to some extent, reflect effects of different intensity of alarm cues caused by unequal total biomass of predators in experimental treatments (e.g., the biomass of crayfish and small roach vs. the biomass of large roach).

Chemosensory recognition of predators has been studied extensively in a variety of organisms. Prey have been shown to differentiate predator ability, attack modes, and diet (Sih 1986, Kats and Dill 1998, Pettersson et al. 2000, Reimer and Harms-Ringdahl 2001, Jacobsen and Stabell 2004). While relatively few studies have examined the ability of prey to detect and respond to gradients of predator cues, emerging evidence suggests the occurrence of quantitative risk assessment in protozoans (Wiąckowski and Starońska 1999), insects (Sih 1986, Kesavaraju et al. 2007), crustaceans (Ramcharan et al. 1992, McKelvey and Forward 1995), fish (Brown et al. 2006), and amphibians (Buskirk and Arioli 2002, Schoeppner and Relyea 2008). Concentrations of chemicals released by predators can be a dependable indicator of overall risk of predation (Sih 1986, Kats and Dill 1998). It may help to assess the abundance of predators, predator activity, or predator proximity. There is some evidence to support the view that zebra mussels make a chemo-sensory assessment of predation cue concentrations and fine-tune their responses accordingly. Czarnoleski et al. (2010b) demonstrated a systematic decrease in mobility of zebra mussels with rising concentrations of effluents from crushed zebra mussels (Figure 22.2). Furthermore, comparative evidence showed that growth pattern and shell resistance of zebra mussels gradually changed with mortality level in populations (Figures 22.5, 22.7a, and 22.8; Czarnoleski et al. 2003, 2006). In addition, the magnitude of temporal changes in growth of zebra mussels was proportional to the magnitude of changes in mussel mortality through time (Figure 22.7b; Czarnoleski et al. 2005). If these spatiotemporal patterns reflect phenotypic plasticity rather than microevolution, they indicate that zebra mussels assess the levels of mortality risk and respond accordingly through resource allocation.

ADAPTIVE VALUE OF RESPONSES

Evidence for an adaptive value of predator-mediated responses in zebra mussels is mostly circumstantial. Zebra mussels have a strong tendency to anchor to hard substrata and aggregate with other mussels. This often leads to formation of multilayered age-structured clumps (Stańczykowska and Lewandowski 1993, Strayer and Smith 1996). Soon after mussels are artificially detached, they start to rebuild byssal attachment. Czarnoleski et al. (2010b) observed that 75% of mussels reattached within 1 day of being detached, and 96% reattached within 6 days. During the generation of byssal attachment, resistance to shell removal from the substrate increases asymptotically as a result of increased numbers and strength of threads. When choosing attachment sites, zebra mussels use chemical (Wainman et al. 1996) and physical (Czarnoleski et al. 2004) detection of other mussels and preferentially settle in their proximity (Kobak 2001). Kobak and Kakareko (2009) and Kobak et al. (2010) demonstrated that settling decisions are also mediated by predation cues.

Cues from non-foraging roach stimulated aggregation of zebra mussels within 1 day and led to stronger attachment within 6 days. Thus, predation cues seem to increase resource allocation to byssal production, but before the production of byssal threads begins, threatened mussels choose a site to attach. In general, there is an agreement that firm attachment to substratum and occupation of narrow crevices in substrates and between shells of other mussels reduce the ability of predators to dislodge byssate bivalves (Okamura 1986, Côté 1995). These responses are of protective value to zebra mussels because they need to be dislodged before they are consumed. Green et al. (2008) showed that byssal attachment reduces predation on zebra mussels by crayfish.

Predator-mediated byssus formation might explain why zebra mussels have been observed to reduce mobility on sand in response to effluents from crushed mussels (Toomey et al. 2002). However, Czarnoleski et al. (2010b) demonstrated that this type of alarm cue did not reduce time devoted to crawling but reduced crawling speed and this shortened traveled distances. In fact, zebra mussels tend to produce weaker attachment in response to signals of injured mussels. Such a response has been triggered by effluents from artificially crushed zebra mussels (Czarnoleski et al. 2010b) and by odors released by crayfish foraging on zebra mussels (Czarnoleski et al. 2011). Fitness benefits from decreased attachment strength under predation risk are not clear. Czarnoleski et al. (2010b, 2011) postulated that predator-mediated decreases in mobility and byssal production may result from the suppression in physiology aimed at reduced detection by predators. Predatory fish (Pettersson et al. 2000) and crustaceans (Keller et al. 2001) use chemosensory skills to help locate prey, which suggests that lower emission of metabolites by zebra mussels may decrease their exposure to predation. Injured zebra mussels were observed to stimulate foraging of crayfish under lab (Hazlett 1994) and field (Green et al. 2008) conditions, and Czarnoleski et al. (2011) demonstrated that crayfish detect waterborne odors emitted by live, uninjured zebra mussels. In support of the metabolic-suppression hypothesis, Naddafi et al. (2007) observed reduced feeding by zebra mussels exposed to cues from crayfish and roach that had been fed with zebra mussels. Similar predator-mediated reduction in physiological activity has been observed in other bivalves. For example, the blue mussel *M. edulis* lowered respiration rate and ammonia excretion in the presence of predatory starfish (Reimer et al. 1995), and predatory crustaceans reduced attachment strength of the intertidal mussel *Hormomya mutabilis* (Ishida and Iwasaki 2003) and feeding rates of the hard clam *Mercenaria mercenaria* (Smee and Weissburg 2006). Smee and Weissburg (2006) demonstrated that the predator-driven suppression of feeding reduced mortality caused by predators in natural populations of hard clams.

Taken together, available evidence suggests that zebra mussels adopt alternative antipredator responses. They may either change resource allocation and channel more resources to byssal production to withstand predatory attacks or may suppress physiological activity to reduce detection by predators and sacrifice resistance to dislodgement. Czarnoleski et al. (2011) proposed that fast behavioral suppression of physiological processes may be favored over allocation to byssogenesis when mussels face a sudden danger of immediate attacks. Note that changes in mobility and feeding have been observed within hours after mussels are exposed to alarm cues (Toomey et al. 2002, Naddafi et al. 2007, Czarnoleski et al. 2010b), whereas strengthening of attachment required several days (Kobak and Kakareko 2009). According to Czarnoleski et al. (2011), differences in response time help to explain why byssal attachment is weakened in zebra mussels after being exposed to cues from injured mussels (Czarnoleski et al. 2010b, 2011), while the opposite response or unresponsiveness is observed in mussels confronted with nonhunting predators (Kobak and Kakareko 2009, Kobak et al. 2010). Zebra mussels react by either ignoring a predator (e.g., non-foraging omnivorous crayfish), increasing resource allocation to attachment (e.g., in response to non-foraging roach), or behaviorally suppressing metabolic activity (e.g., in response to active predation of roach and crayfish).

Increased aggregation, enhanced byssal thread production, and metabolic suppression impose apparent fitness costs that may have long-lasting effects on life histories. Mussels in aggregations compete for food and oxygen (Okamura 1986, Czarnoleski et al. 2003). Investments to byssal production drain resources from other functions. According to Lachance et al. (2008), spawning can cause a 32% reduction in byssal attachment of *M. edulis*. Predator-mediated reduction in feeding has been shown to reduce growth rates in *M. mercenaria* clams (Nakaoka 2000). Observations that zebra mussels change quantitative and qualitative responses according to predation cues indicate that costs associated with responses are minimized by the adjustment of responses to the actual risk of predation. There are no good data to assess the role of these costs in shaping life-history strategy of zebra mussels. However, with optimal resource allocation modeling, Czarnoleski et al. (2003) demonstrated that an increased level of aggregation favors earlier maturation and more intense resource allocation to reproduction at the expense of somatic growth. In other words, zebra mussels that would choose to live in denser clumps to reduce predation risk should simultaneously mature earlier and grow slower after maturation. Comparative data to examine this theoretical prediction suggested that such changes in life-history allocation may occur in populations of zebra mussels (Czarnoleski et al. 2003), but experimental works are needed to rigorously test this expectation.

Mollusks have a capacity to alter resource allocation for shell production over the long term and will increase shell thickness when faced with predation pressure (Reimer and Tedengren 1996, Leonard et al. 1999,

Smith and Jennings 2000, Frandsen and Dolmer 2002). Czarnoleski et al. (2006) showed that crush resistance of shells increased with mortality level in zebra mussel populations (Figure 22.5), suggesting that investments in shell strengthening may be beneficial for zebra mussels exposed to predation threats. This response has not been studied experimentally in zebra mussels, but available data on other bivalves suggest that the development of a detectable increase in shell mass may require weeks or months (e.g., Trussell et al. 2003). Such a time requirement suggests that investments in shell production may not be effective if mussels are exposed to sudden increases in predation risks. Another long-term option for prey is a change in resource allocation to growth and reproduction over a lifetime (Stearns 1992, Czarnoleski and Kozłowski 1998). Although this response is not defensive in a strict sense, it may help to maximize fitness under predation pressure equally well as development of direct defenses such as increased attachment or crush resistance. According to an evolutionary paradigm, life forms that pass higher numbers of genetic copies to next generations win the evolutionary competition with less effective reproducers. In organisms such as mollusks and fish whose fertility strongly depends on body size, spending resources on growth can be viewed as a long-term investment in offspring production because the attainment of larger body size increases future fertility (Stearns 1992, Kozłowski 2006). Apparently, organisms should grow as large as physical constraints allow and mature at a size that guarantees the highest reproductive potential. However, natural populations are subject to loss from predation, and starting reproduction earlier in life reduces the risk that death occurs before maturation. It follows that a shorter life span leads to an adaptive strategy of early maturation and the allocation of more resources to reproduction at the expense of growth (Stearns 1992, Czarnoleski and Kozłowski 1998, Kozłowski 2006). Such responses to mortality (e.g., mediated by predation) are consistent with the negative links between an index of roach predation and size of 2-year-old zebra mussels (Figure 22.8) and between mortality of zebra mussels and their asymptotic size (Figure 22.7; Czarnoleski et al. 2003, 2005). Czarnoleski et al. (2010a) provided a detailed discussion on the theory of optimal resource allocation as applied to life-history variability in zebra mussels.

CONCLUSIONS AND RESEARCH NEEDS

Zebra mussels are subject to spatiotemporal differences in predation pressure. In response, zebra mussels appear to assess the risk of predation and adaptively change their behavior and resource allocation. Response to the presence of predators decreases as mussels become less vulnerable to predation. This vulnerability depends on the size and strength of shells. Zebra mussels appear to recognize different species of predators and use the type and intensity of predation cues to assess actual risks of predator attacks. Perceived risk from predators appears to depend on their predatory activity. A mere presence of predators is perceived by zebra mussels as dangerous only if these predators are an important source of mortality in natural conditions (e.g., roach). Responses to lower risk factors (e.g., crayfish) are only triggered if there are signs of active hunting on mussels in the vicinity. Active predation indicates danger of immediate predatory attacks and mussels appear to adjust responses to this threat by considering response times. If there are cues that predators already forage nearby (effluents from crushed conspecifics are present), zebra mussels adopt fast behavioral responses aimed at reduced detection (suppression of disclosing metabolites, but at a cost of weaker attachment). If, however, there are only cues that non-foraging predators are present and immediate predatory attacks are less likely, mussels either adopt more time-consuming allocation responses (building up stronger byssal attachment) or ignore a predator if its pressure is typically not intense. Comparative data suggest that zebra mussels may also follow allocation responses from which benefits can be expected after considerable time (weeks or years). Mussels can channel resources to shell production to increase resistance to crushing, or they can channel resources to increase immediate reproduction at the expense of growth, and thus decrease investments in future reproduction.

Our knowledge of predator–prey interactions in zebra mussels is still very limited, but data published in the last decade provide some understanding of this important issue. The capacity of zebra mussels to recognize predation pressure and to adopt adequate responses suggests that this species is well equipped to colonize environments with a wide range of predation intensities. Yet important questions related to predator–prey interactions remain unresolved. Do zebra mussels respond to the predation of diving birds? How do zebra mussels compromise costs and benefits of short- and long-term responses? How does the effectiveness of different level responses change through zebra mussel ontogeny? When are responses to long-term allocations made? What are the indirect effects of predator-mediated responses of zebra mussels on ecosystems? Future studies are certainly needed to answer these questions and to enhance our understanding of this entire topic.

ACKNOWLEDGMENTS

The work was supported by the Polish Ministry of Scientific Research and Information Technology (grant #N N 304 1176 33) and Jagiellonian University (DS/WBiNoZ/INoŚ/757/10). We thank Justyna Kierat for drawings in Figure 22.1.

REFERENCES

Alexander, J.E. and A.P. Covich. 1991. Predator avoidance by the fresh-water snail *Physella virgata* in response to the crayfish *Procambarus simulans*. *Oecologia* 87: 435–442.

Andraso, G.M. 2005. Summer food habits of pumpkinseeds (*Lepomis gibbosus*) and bluegills (*Lepomis macrochirus*) in Presque Isle Bay, Lake Erie. *J. Great Lakes Res.* 31: 397–404.

Bartsch, M.R., L.A. Bartsch, and S. Gutreuter. 2005. Strong effects of predation by fishes on an invasive macroinvertebrate in a large floodplain river. *J. N. Am. Benthol. Soc.* 24: 168–177.

Bij de Vaate, A. 1991. Distribution and aspects of population dynamics of the zebra mussel, *Dreissena polymorpha* (Pallas 1771), in the Lake Ijsselmeer area (The Netherlands). *Oecologia* 86: 40–50.

Boles, L.C. and R.N. Lipcius. 1997. Potential for population regulation of the zebra mussel by finfish and the blue crab in North American estuaries. *J. Shellfish Res.* 16: 179–186.

Brown, G.E., T. Bongiorno, D.M. DiCapua, I.I. Ivan, and E. Roh. 2006. Effects of group size on the threat-sensitive response to varying concentrations of chemical alarm cues by juvenile convict cichlids. *Can. J. Zool.* 84: 1–8.

Buskirk, V.J. and M. Arioli. 2002. Dosage response of an induced defense: How sensitive are tadpoles to predation risk? *Ecology* 83: 1580–1585.

Cleven, E.J. and P. Frenzel. 1993. Population dynamics and production of *Dreissena polymorpha* (Pallas) in River Seerhein, the outlet of Lake Constance (Obersee). *Arch. Hydrobiol.* 127: 395–407.

Cheung, S.G., S. Lam, Q.F. Gao, K.K. Mak, and P.K.S. Shin. 2004. Induced anti-predator responses of the green mussel, *Perna viridis* (L.), on exposure to the predatory gastropod, *Thais clavigera* Kuster, and the swimming crab, *Thalamita danae* Stimpson. *Mar. Biol.* 144: 675–684.

Chybowski, Ł. 2007. Morphometrics, fecundity, density, and feeding intensity of the spinycheek crayfish, *Orconectes limosus* (Raf.) in natural conditions. *Arch. Pol. Fish.* 15: 175–241.

Côté, I.M. 1995. Effects of predatory crab effluent on byssus production in mussels. *J. Exp. Mar. Biol. Ecol.* 188: 233–241.

Côté, I.M. and E. Jelnikar. 1999. Predator-induced clumping behaviour in mussels (*Mytilus edulis* Linnaeus). *J. Exp. Mar. Biol. Ecol.* 235: 201–211.

Crowl, T.A. and A.P. Covich. 1990. Predator-induced life history shifts in a freshwater snail. *Science* 247: 949–951.

Czarnoleski, M. and J. Kozłowski. 1998. Do Bertalanffy's growth curves result from optimal resource allocation? *Ecol. Lett.* 1: 5–7.

Czarnoleski, M., J. Kozłowski, P. Kubajak, K. Lewandowski, T. Müller, A. Stańczykowska, and K. Surówka. 2006. Cross-habitat differences in crush resistance and growth pattern of zebra mussels (*Dreissena polymorpha* Pallas): Effects of calcium availability and predator pressure. *Arch. Hydrobiol.* 165: 191–208.

Czarnoleski, M., J. Kozłowski, K. Lewandowski, M. Mikołajczyk, T. Müller, and A. Stańczykowska. 2005. Optimal resource allocation explains changes of the zebra mussel growth pattern through time. *Evol. Ecol. Res.* 7: 821–835.

Czarnoleski, M., J. Kozłowski, K. Lewandowski, T. Müller, and A. Stańczykowska. 2010a. An evolutionary perspective on the geographic and temporal variability of life histories in European zebra mussels. In *The Zebra Mussel in Europe*, G. Van der Velde, S. Rajagopal, and A. Bij de Vaate, eds., pp. 169–182. Leiden, The Netherlands: Backhuys Publishers.

Czarnoleski, M., J. Kozłowski, K. Lewandowski, and A. Stańczykowska. 2003. Optimal resource allocation explains growth curve diversity in zebra mussels. *Evol. Ecol. Res.* 5: 571–587.

Czarnoleski, M., L. Michalczyk, and A. Pajdak-Stós. 2004. Substrate preference in settling zebra mussels *Dreissena polymorpha*. *Arch. Hydrobiol.* 159: 263–270.

Czarnoleski, M., T. Müller, K. Adamus, G. Ogorzelska, and M. Sog. 2010b. Injured conspecifics alter mobility and byssus production in zebra mussels *Dreissena polymorpha*. *Fundament. Appl. Limnol.* 176: 269–278.

Czarnoleski, M., T. Müller, J. Kierat, L Gryczkowski, and Ł. Chybowski. 2011. Anchor down or hunker down: An experimental study on zebra mussels' response to predation risk from crayfish. *Anim. Behav.*, 82: 543–548.

De Leeuw, J.J. and M.R. Van Eerden. 1992. Size selection in diving Tufted Ducks *Aythya fuligula* explained by differential handling of small and large mussels *Dreissena polymorpha*. *Ardea* 80: 353–362.

De Leeuw, J., M.R.Van Eerden, and G.H. Visser. 1999. Wintering tufted ducks *Aythya fuligula* diving for zebra mussels *Dreissena polymorpha* balance feeding costs within narrow margins of their energy budget. *J. Avian Biol.* 30: 182–192.

DeWitt, T.J., B.W. Robinson, and D.S. Wilson. 2000. Functional diversity among predators of a freshwater snail imposes an adaptive trade-off for shell morphology. *Evol. Ecol. Res.* 2: 129–148.

DeWitt, T.J., A. Sih, and D.S. Wilson. 1998. Costs and limits of phenotypic plasticity. *Trends Ecol. Evol.* 13: 77–81.

Draulans, D. 1982. Foraging and size selection of mussels by the tufted duck, *Aythya fuligula*. *J. Anim. Ecol.* 51: 943–956.

Eggleton, M.A., L.E. Miranda, L.E., and J.P. Kirk. 2004. Assessing the potential for fish predation to impact zebra mussels (*Dreissena polymorpha*): Insight from bioenergetic models. *Ecol. Freshwat. Fish.* 13: 85–95.

Ermgassen, P.S.E.Z. and D.C. Aldridge. 2011. Predation by the invasive American signal cray fish, *Pacifastacus leniusculus* Dana, on the invasive zebra mussel, *Dreissena polymorpha* Pallas: The potential for control and facilitation. *Hydrobiologia* 658: 303–315.

Frandsen, R.P. and P. Dolmer. 2002. Effects of substrate type on growth and mortality of blue mussels (*Mytilus edulis*) exposed to the predator *Carcinus maenas*. *Mar. Biol.* 141: 253–262.

Freeman, A.S. 2007. Specificity of induced defenses in *Mytilus edulis* and asymmetrical predator deterrence. *Mar. Ecol. Prog. Ser.* 334: 145–153.

Freeman, A.S., J. Meszaros, and J.E. Byers. 2009. Poor phenotypic integration of blue mussel inducible defenses in environments with multiple predators. *Oikos* 118: 758–766.

Green, N.S., B.A. Hazlett, and S. Prueff-Jones. 2008. Attachment and shell integrity affects the vulnerability of zebra mussels (*Dreissena polymorpha*) to predation. *J. Freshwat. Ecol.* 23: 91–99.

Griffiths, C.L. and C.A. Richardson. 2006. Chemically induced predator avoidance behaviour in the burrowing bivalve *Macoma balthica*. *J. Exp. Mar. Biol. Ecol.* 331: 91–98.

Hamilton, D.J. and C.D. Ankney. 1994. Consumption of Zebra Mussels *Dreissena polymorpha* by diving ducks in Lakes Erie and St. Clair. *Wildfowl* 45: 159–166.

Hazlett, B.A. 1994. Crayfish feeding responses to zebra mussel depend on microorganisms and learning. *J. Chem. Ecol.* 20: 2623–2630.

Hirst, A.G. and T. Kiørboe. 2002. Mortality of marine planktonic copepods: Global rates and patterns. *Mar. Ecol. Prog. Ser.* 230: 195–209.

Hopkins, G.J. and J.H. Leach. 1993. A photographic guide to the identification of larval stages of the zebra mussel (*Dreissena polymorpha*). In *Zebra Mussels, Biology, Impacts and Control*, T. F. Nalepa and D. W. Schloesser, eds., pp. 761–772. Boca Raton, FL: CRC Press.

Ishida, S. and K. Iwasaki. 2003. Reduced byssal thread production and movement by the intertidal mussel *Hormomya mutabilis* in response to effluent from predators. *J. Ethol.* 21: 117–122.

Jacobsen, H.P. and O.B. Stabell. 2004. Anti-predator behaviour mediated by chemical cues: The role of conspecific alarm signaling and predator labelling in the avoidance response of a marine gastropod. *Oikos* 104: 43–50.

Kats, L.B. and L.M. Dill. 1998. The scent of death: Chemosensory assessment of predation risk by prey animals. *Ecoscience* 5: 361–394.

Keller, T.A., A.M. Tomba, and P.A. Moore. 2001. Orientation in complex chemical landscapes: Spatial arrangement of chemical sources influences crayfish food-finding efficiency in artificial streams. *Limnol. Oceanogr.* 46: 238–247.

Kesavaraju, B., K. Damal, and S.A. Juliano. 2007. Threat-sensitive behavioral responses to concentrations of water-borne cues from predation. *Ethology* 113: 199–206.

Kobak, J. 2001. Light, gravity and conspecifics as cues to site selection and attachment behaviour of juvenile and adult *Dreissena polymorpha* Pallas, 1771. *J. Moll. Stud.* 67: 183–189.

Kobak, J. and T. Kakareko. 2009. Attachment strength, aggregation and movement of the zebra mussel (*Dreissena polymorpha*, Bivalvia) in the presence of potential predators. *Fundam. Appl. Limnol.* 174: 193–204.

Kobak, J., T. Kakareko, and M. Poznańska. 2010. Changes in attachment strength and aggregation of zebra mussel, *Dreissena polymorpha* in the presence of potential fish predators of various species and size. *Hydrobiologia* 644: 195–206.

Kozłowski, J. 2006. Why life histories are diverse. *Pol. J. Ecol.* 54: 585–605.

Lachance, A.A., B. Myrand, R. Tremblay, V. Koutitonsky, and E. Carrington. 2008. Biotic and abiotic factors influencing attachment strength of blue mussels *Mytilus edulis* in suspended culture. *Aquat. Biol.* 2: 119–129.

Langerhans, R.B. and T.J. DeWitt. 2002. Plasticity constrained: Over-generalized induction cues cause maladaptive phenotypes. *Evol. Ecol. Res.* 4: 857–870.

Leonard, G.H., M.D. Bertness, and P.O. Yund. 1999. Crab predation, waterborne cues, and inducible defences in the blue mussel, *Mytilus edulis*. *Ecology* 80: 1–14.

Lewandowski, K. 1983. Formation of annuli on shells of young *Dreissena polymorpha* (Pall.) *Pol. Arch. Hydrobiol.* 30: 343–351.

Liebig, J.R. and H.A. Vanderploeg. 1995. Vulnerability of *Dreissena polymorpha* larvae to predation by Great Lakes calanoid copepods: The importance of the bivalve shell. *J. Great Lakes Res.* 21: 353–358.

Lima, S.L. and P.A. Bednekoff. 1999. Temporal variation in danger drives anti-predator behavior: The predation risk allocation hypothesis. *Am. Nat.* 153: 649–659.

Madenjian, C.P, S.A. Pothoven, P.J. Schneeberger et al. 2010. Dreissenid mussels are not a "dead end" in Great Lakes food webs. *J. Great Lakes Res.* 36(Suppl. 1): 73–77.

Magoulick, D.D. and L.C. Lewis. 2002. Predation on exotic zebra mussels by native fishes: Effects on predator and prey. *Freshwat. Biol.* 47: 1908–1918.

Martin, G.W. and L.D. Corkum. 1994. Predation of zebra mussels by crayfish. *Can. J. Zool.* 72: 1867–1871.

McKelvey, L.M. and R.B. Forward, Jr. 1995. Activation of brine shrimp nauplii photoresponses involved in diel vertical migration by chemical cues from visual and non-visual planktivores. *J. Plankton Res.* 17: 2191–2206.

Mellina, E. and J.B. Rasmussen. 1994. Patterns in the distribution and abundance of zebra mussel (*Dreissena polymorpha*) in rivers and lakes in relation to substrate and other physiochemical factors. *Can. J. Fish. Aquat. Sci.* 51: 1024–1036.

Miner, B.G., D.A. Donovan, and K.E. Andrews. 2010. Should I stay or should I go: Predator- and conspecific-induced hatching in a marine snail. *Oecologia* 163: 69–78.

Molloy, D.P., A.Y. Karatayev, L.E. Burlakova, D.P. Kurandina, and F. Larulle. 1997. Natural enemies of zebra mussels: Predators, parasites, and ecological competitors. *Rev. Fish. Sci.* 5: 27–97.

Molloy, D.P., J. Powell, and P. Ambrose. 1994. Short-term reduction of adult zebra mussels (*Dreissena polymorpha*) in the Hudson River near Catskill, New York: An effect of juvenile blue crab (*Callinectes sapidus*) predation? *J. Shellfish Res.* 13: 367–371.

Naddafi, R., P. Eklöv, and K. Pettersson. 2007. Non-lethal predator effects on the feeding rate of the exotic zebra mussel *Dreissena polymorpha*. *Oikos* 116: 1289–1298.

Naddafi, R., K. Pettersson, and P. Eklöv. 2010. Predation and physical environment structure the density and population size structure of zebra mussels. *J. N. Am. Benthol. Soc.* 29: 444–453.

Nagelkerke, L.A.J. and F.A. Sibbing. 1996. Efficiency of feeding on zebra mussel (*Dreissena polymorpha*) by common bream (*Abramis brama*), white bream (*Blicca bjoerkna*), and roach (*Rutilus rutilus*): The effects of morphology and behavior. *Can. J. Fish. Aquat. Sci.* 53: 2847–2861.

Nakaoka, M. 2000. Nonlethal effects of predators on prey populations: Predator-mediated change in bivalve growth. *Ecology* 81: 1031–1045.

Nyström, P., C. Brönmark, and W. Graneli. 1996. Patterns in benthic food webs: A role for omnivorous crayfish? *Freshwat. Biol.* 36: 631–646.

Okamura, B. 1986. Group living and the effects of spatial position in aggregations of *Mytilus edulis*. *Oecologia* 69: 341–347.

Perry, W.L., D.M. Lodge, and G.A. Lamberti. 1997. Impact of crayfish predation on exotic zebra mussels and native invertebrates in a lake-outlet stream. *Can. J. Fish. Aquat. Sci.* 54: 120–125.

Perry, W.L., D.M. Lodge, and G.A. Lamberti. 2000. Crayfish (*Orconectes rusticus*) impacts on zebra mussel (*Dreissena polymorpha*) recruitment, other macroinvertebrates and algal biomass in a lake-outlet stream. *Am. Mid. Nat.* 144: 308–316.

Pettersson, L.B., P.A. Nilsson, and C. Brönmark. 2000. Predator recognition and defence strategies in crucian carp, *Carassius carassius*. *Oikos* 88: 200–212.

Prejs, A., I. Lewandowski, and A. Stańczykowska-Piotrowska. 1990. Size-selective predation by the roach (*Rutilis rutilus*) on zebra mussels (*Dreissena polymorpha*): Field studies. *Oecologia* 83: 378–384.

Piesik, Z. 1974. The role of the crayfish *Orconectes limosus* (Raf.) in extinction of *Dreissena polymorpha* (Pall.) subsisting on steel on net. *Pol. Arch. Hydrohiol.* 21: 401–410.

Ramcharan, C.W., S.I. Dodson, and J. Lee. 1992. Predation risk, prey behavior, and feeding rate in *Daphnia pulex*. *Can. J. Fish. Aquat. Sci.* 49: 159–165.

Reimer, O. and S. Harms-Ringdahl. 2001. Predator-inducible changes in blue mussels from the predator-free Baltic Sea. *Mar. Biol.* 139: 959–965.

Reimer, O., B. Olsson, and M. Tedengren. 1995. Growth, physiological rates and behaviour of *Mytilus edulis* exposed to the predator *Asterias rubens*. *Mar. Freshwater Behav. Phys.* 25: 233–244.

Reimer, O. and M. Tedengren. 1996. Phenotypical improvement of morphological defences in the mussel *Mytilus edulis* induced by exposure to the predator *Asterias rubens*. *Oikos* 75: 383–390.

Reynolds, J.D. and R. Donohue. 2001. Crayfish predation experiments on the introduced zebra mussel, *Dreissena polymorpha*, in Ireland, and their potential for biocontrol. *Bull. Fr. Pêche Piscic.* 361: 669–682.

Rundle, S.D., J.I. Spicer, R.A. Coleman, J. Vosper, and J. Soane. 2004. Environmental calcium modifies induced defences in snails. *Proc. R. Soc. Lond. B* 271(Suppl.): S67–S70.

Schoeppner, N.M. and R.A. Relyea. 2008. Detecting small environmental differences: Risk-response curves for predator induced behavior and morphology. *Oecologia* 154: 743–754.

Sih, A. 1986. Anti-predator responses and the perception of danger by mosquito larvae. *Ecology* 67: 434–441.

Smee, D.L. and M.J. Weissburg. 2006. Clamming up: Environmental forces diminish the perceptive ability of bivalve prey. *Ecology* 87: 1587–1598.

Smith, L.D. and J.A. Jennings. 2000. Induced defensive responses by the bivalve *Mytilus edulis* to predators with different attack modes. *Mar. Biol.* 136: 461–469.

Stańczykowska, A. and K. Lewandowski. 1993. Thirty years of studies of *Dreissena polymorpha* ecology in Masurian Lakes of North-eastern Poland. In *Zebra Mussels, Biology, Impacts and Control*, T. F. Nalepa and D. W. Schloesser, eds., pp. 3–37. Boca Raton, FL: CRC Press.

Strayer, D.L. and L.C. Smith. 1996. Relationships between zebra mussels (*Dreissena polymorpha*) and unionid clams during the early stages of the zebra mussel invasion of the Hudson River. *Freshwat. Biol.* 36: 771–779.

Stearns, S.C. 1992. *The Evolution of Life Histories*. Oxford University Press, Oxford, U.K.

Thorp, J.H., M.D. Delong, and A.F. Casper. 1998. In situ experiments on predatory regulation of a bivalve mollusc (*Dreissena polymorpha*) in the Mississippi and Ohio Rivers. *Freshwat. Biol.* 39: 649–661.

Tollrian, R. and C.D. Harvell. 1999. *The Ecology and Evolution of Inducible Defenses*. Princeton, NJ: Princeton University Press.

Toomey, M.B., D. McCabe, and J.E. Marsden. 2002. Factors affecting the movement of adult zebra mussels (*Dreissena polymorpha*). *J. N. Am. Benthol. Soc.* 21: 468–475.

Trussell, G.C., P.J. Ewanchuk, and M.D. Bertness. 2003. Trait-mediated effects in rocky intertidal food chains: Predator risk cues alter prey feeding rates. *Ecology* 84: 629–640.

Trussell, G.C. and M.O. Nicklin. 2002. Cue sensitivity, inducible defense, and trade-offs in a marine snail. *Ecology* 83: 1635–1647.

Tucker, J.K., F.A. Cronin, D.W. Soergel, and C.H. Theiling. 1996. Predation on zebra mussels (*Dreissena polymorpha*) by common carp (*Cyprinus carpio*). *J. Freshwat. Ecol.* 11: 363–372.

Wainman, B.C., S.S. Hincks, N.K. Kaushik, and G.L. Mackie. 1996. Biofilm and substrate preference in the dreissenid larvae of Lake Erie. *Can. J. Fish. Aquat. Sci.* 53: 134–140.

Watzin, M.C., K. Joppe-Mercure, J. Rowder, B. Lancaster, and L. Bronson. 2008. Significant fish predation on zebra mussels *Dreissena polymorpha* in Lake Champlain, U.S.A. *J. Fish Biol.* 73: 1585–1599.

Werner, S., M. Mörtl, H.G. Bauer, and K.O. Rothhaupt. 2005. Strong impact of wintering waterbirds on zebra mussel (*Dreissena polymorpha*) populations at Lake Constance, Germany. *Freshwat. Biol.* 50: 1412–1426.

Whittier, T.R., P.L. Ringold, A.T. Herlihy, and S.M. Pierson. 2008. A calcium-based invasion risk assessment for zebra and quagga mussels (*Dreissena* spp). *Front. Ecol. Environ.* 6: 180–184.

Wiąckowski, K. and Starońska, A. 1999. The effect of predator and prey density on the induced defence of a ciliate. *Funct. Ecol.* 13: 59–65.

Wormington, A. and J.H. Leach. 1992. Concentrations of migrant diving ducks at Point Pelee National Park, Ontario, in response to invasion of Zebra Mussels, *Dreissena polymorpha*. *Can. Field. Nat.* 106: 376–380.

CHAPTER 23

Variation in Predator–Prey Interactions between Round Gobies and Dreissenid Mussels

Christopher J. Houghton and John Janssen

CONTENTS

Abstract 359
Introduction 359
Perspective of the Round Goby as Predator 360
Perspective of Dreissenid as Prey 361
Impacts of the Predator 362
Acknowledgments 365
References 365

ABSTRACT

The predator–prey interaction between the round goby (*Neogobius melanostomus*) and dreissenid mussels (zebra mussel *Dreissena polymorpha* and quagga mussel *Dreissena rostriformis bugensis*) is an important factor determining distributions and abundances of each species in the Laurentian Great Lakes. Round gobies are found throughout the lower Great Lakes, and their spread has probably been facilitated by the fact that this species co-evolved with and readily preys on dreissenids. Reviewed research has shown that round gobies have decreased dreissenid densities at some locations, but there are exceptions, and round goby impacts on dreissenids may not be universal. We propose that a complete understanding of round goby impacts requires knowledge of dreissenid defenses and how dreissenids interact with bio-mechanical adaptations of round gobies. Round goby adaptations include a feeding style in which the fish breaks dreissenid byssal threads through torsion and possession of molariform pharyngeal teeth in larger gobies that aid in crushing mussel shells. Dreissenid defenses include seeking shelter (i.e., rock crevices), aggregating with conspecifics, increasing byssal thread attachment strength, and increasing shell thickness. However, these defenses have mainly been identified relative to other molluscivorous fishes and not the round goby. To illustrate how complex round goby–dreissenid interactions can be, we present results of two field studies. In the first study, we show that where round gobies and quagga mussels have co-existed for about 6 years, densities of dreissenids correlate with rock rugosity as measured by the area of vugs (cavities left by eroded fossils, primarily brachiopods) ($F_{1,27} = 17.8$, $P < 0.001$). In the second study, we present data on the movement of dreissenids from established populations to initially dreissenid-free substrates in the presence of round gobies. Based on study outcomes, we propose that in the presence of high round goby densities, dreissenids strike a balance between lowered risk of predation by occupying sheltered habitats (under rocks and in crevices), and enhanced feeding opportunities (and likely spawning) by occupying more exposed habitats. Because both predator and prey have strategies with costs and benefits that can vary regionally, or even at the level of the microhabitat, we propose that interactions between round gobies and dreissenid mussels will vary at multiple spatial scales.

INTRODUCTION

Interactions between the round goby (*Neogobius melanostomus*) and dreissenids (zebra mussel *Dreissena polymorpha* and quagga mussel *D. rostriformis bugensis*) in the Great Lakes are based on a predator–prey relationship that evolved in the Ponto-Caspian arena. These species invaded the Great Lakes in the 1980s, and they have since spread into a diversity of habitats, ranging from streams to the open coasts, and subsequent predator–prey interactions have been transplanted into new ecological contexts. Dreissenids have dramatically altered energy flow and habitat in the Great Lakes (Nalepa et al. 2010, Vanderploeg et al. 2010) and smaller bodies of water (review by Higgins and Vanderzanden 2010), so trophic links between round gobies and dreissenids are potentially of great importance. Round gobies and dreissenids are now well-established and co-occur in much of the Great Lakes basin.

Most studies of differences between quagga and zebra mussels have focused on physiological differences between the two species and have not closely examined possible differences in behavior or susceptibility to predation (Higgins and Vanderzanden 2010). Because round gobies are mainly restricted to rocky habitat and require cavities among rocks for reproduction (Corkum et al. 2004), pertinent field research on this species has thus far been exclusively conducted on this type of habitat. We herein mostly assume that where quagga mussels have displaced zebra mussels, they fulfill the same role in the food web and also display similar behavior with regard to movement, aggregation, etc. Hence, predator–prey interactions between quagga mussels and round gobies would be similar to zebra mussels and round gobies. However, given different physiological characteristics of the quagga mussel (i.e., adaptations for deep, cold, less productive environments), this assumption may prove not true.

Our review begins with predator and prey "perspectives" considered more or less in isolation. This allowed us to address costs and benefits of alternative predator and prey strategies. We then consider published interactions between round gobies and dreissenids. In addition, we include two studies that demonstrate how round goby-dreissenid interactions are likely to be complex and subject to regional variation. The first study assesses how rock rugosity provides dreissenids shelter from predation, and the second study examines dreissenid migrations from one rock to another. These are simple demonstrations of predator–prey interactions that provide context to the general review. Because of this and because the studies are not much more than simple collections and enumerations, study methods are described with results.

Regional variation in predator–prey dynamics is to be expected because of nuances related to local conditions. For instance, geographical variation related to regional differences in communities is clearly evident in the zonation of blue mussels (*Mytilus* spp.) in rocky, intertidal habitats. *Mytilus* is similar to *Dreissena* in that it is a bivalve that attaches to rocks via byssal threads. Paine (1976) referred to *Mytilus* as a "matrix" taxa because it can often totally cover rocky substrates. Dreissenids also qualify as a matrix taxa. Early studies conducted along the Washington State coast showed that *Mytilus* distributions were strongly affected by predation from sea stars (Paine 1966, Dayton 1971). However, subsequent studies indicated that local flora and fauna strongly affected and even eliminated zonation (Levin and Paine 1974, Paine and Levin 1981, Roughgarden et al. 1988, Paine 2002). Careful consideration of predator–prey interactions in different habitats is therefore necessary when seemingly contradictory results occur.

PERSPECTIVE OF THE ROUND GOBY AS PREDATOR

The round goby is the most molluscivorous of an endemic phylogenetic radiation of Ponto-Caspian gobies of the subfamily Benthophilinae (Neilson and Stepien 2009). Its natural diet includes a diverse assortment of bivalves including dreissenid mussels (Skaskina and Kostyuchenko 1968). However, as with other molluscivorous fishes, gobies have an ontogenetic change of diet as they grow. Numerous studies in North America indicate that, if dreissenids are sufficiently available, round gobies transition from an arthropod-dominated diet to a dreissenid-dominated diet as they increase in size from about 50 to 100 mm TL (Jude and Janssen 1995, Dubs and Corkum 1996, French and Jude 2001, Janssen and Jude 2001, Lederer et al. 2006). However, invasion/establishment of round gobies is not dependent upon the presence of dreissenids. The first well-documented North American river to be invaded by round gobies was the Flint River (Michigan) which, at the time, was not yet invaded by dreissenids (Carman et al. 2006). Other streams lacking dreissenids have also been invaded (Phillips et al. 2003, Pennuto et al. 2010; M. Kornis and U. Wisconsin, Pers. Comm.).

Round gobies have an unusual method for breaking dreissenid byssal threads. Gobies grasp mussels in their mouths and generate torsion through alternating (clockwise–counterclockwise) rotational movements (twisting motion) along their long axis. Djuricich and Janssen (2001) found that the number of twists correlated with zebra mussel length and hypothesized that larger zebra mussels had more byssal threads and thus were more difficult to remove from the substrate. Twisting is apparently a modification of rotational feeding in which a predator spins to tear chunks from prey too large to swallow whole (review by Helfman et al. 2009). Breaking byssal threads that attach a dreissenid to a rock may be analogous to breaking a piece from large prey. Round gobies spin when loosely held by hand, a phenomenon not found in local benthic fishes such as sculpins and darters. Spinning while being grasped also occurs in certain species of the Antarctic family Nototheniidae that are morphologically similar to typical gobies. Members of this family also break off pieces of large prey by rotational feeding (Janssen et al. 1992). Helfman et al. (2009) noted that rotational feeding requires muscular coordination not common in teleost fishes.

Once dreissenid mussels are removed from the substrate, round gobies may either swallow dreissenids whole or crush them before ingestion. Although pharyngeal teeth in most percomorph fishes are sharp, papilliform, and adapted to grasping prey, the crushing of dreissenids by round gobies is facilitated by molariform pharyngeal teeth (Ghedotti et al. 1995). Smaller dreissenids tend to be swallowed whole (Andraso et al. 2011). This may mean that small dreissenids are difficult to crush, implying a scale effect for either the mussel or round goby, or small mussels may not need to be crushed to be swallowed. There is also ontogenetic variation in round goby pharyngeal teeth, with smaller individuals having conical pharyngeal teeth and larger individuals having molariform pharyngeal teeth (Andraso et al., 2011), as is found in certain molluscivorous Cichlidae (Kornfield 1991).

For fish to crush mollusks, specialized and coordinated muscles are required. As yet, coordination of pharyngeal muscles used to crush mussels has not been described for round gobies. In other species, the crushing process varies widely. The molluscivorous pumpkinseed sunfish (*Lepomis gibbosus*), a native North American species, has diverged from congeners by evolving a motor pattern that involves simultaneous contractions of dorsal and ventral pharyngeal muscles. This synchronized contraction generates pressure on shells, whereas alternating their contraction facilitates swallowing (Lauder 1983). There is also variation among drums (Sciaenidae) in pharyngeal processing. For example, the highly molluscivorous black drum (*Pogonias cromis*) produces vice-like contractions of its pharyngeal apparatus, while the generalist red drum (*Sciaenops ocellatus*) produces a shearing action to crush crustacean carapaces (Grubich 2000).

Round gobies generally prey on dreissenids that are within a size range of about 5–7 mm (minimum) to 12–14 mm (maximum) (Ghedotti et al. 1995, Ray and Corkum 1997, Djuricich and Janssen 2001). Maximum size consumed increases with increased round goby length (Andraso et al. 2011, in press). To fish, it is likely that this size range represents a balance between increased energy with mussel size and increased handling time and/or energy required to remove or crush the mussel.

PERSPECTIVE OF DREISSENID AS PREY

Presently, preferences of round gobies for zebra versus quagga mussels are unknown. However, emerging research suggests that there are likely differences in vulnerability and defense strategies of these two mussel species. For dreissenids, there are several likely methods of defense, including attachment strength, shell shape and thickness, and location of attachment (i.e., refugia in crevices, cavities, and among other individuals). Attachment strength and shell morphology may operate by increasing two components of handling time: removal from the substrate, which requires breaking the byssal threads (Djuricich and Janssen 2001), and crushing the shell (Andraso et al. 2011).

Byssal threads are used for substrate attachment by a variety of bivalve species. The best studied are species in rocky intertidal coastal regions such as *Mytilus* spp. Densities and zonation of *Mytilus* are strongly impacted by physical factors, such as dehydration and wave surge (which can be enhanced by impacts of debris such as logs), and by biological factors, such as predator type. The most common predator on *Mytilus* is the sea star, *Pisaster ochraceus*, which opens mussel shells and subsequently everts its stomach to consume mussel tissue (Paine 1966, Dayton 1971). Such predator–prey interactions of intertidal communities may have parallels to round goby and dreissenid interactions.

Because round gobies must break byssal threads to remove dreissenids, the relative strength of attachment is pertinent to the interaction between these two taxa. Zebra mussels synthesize new byssal threads faster than quagga mussels, so in the short term, they are more difficult to remove via tension or shear from their substrate during the early phase of attachment (Peyer et al. 2009). However, after about two to three months, attachment strengths of zebra and quagga mussels are about equal. Although Peyer et al. (2009) addressed resistance to hydrodynamic removal, their results have implications for resistance to removal by predators. Torsion, rather than tension or sheer, is the method used by round gobies to remove dreissenids, so comparative resistance might be different for torsion. Also, differences in shell shape between zebra mussels and quagga mussels, as well as the considerable variation within a species, could impact the ability of round gobies to grasp individual mussels. For example, the carina of the zebra mussel has a sharp edge that might afford a place for the round goby's teeth to attach. However, the carina is typically closely appressed to the substrate, so it may be difficult for a round goby to get its teeth between shell and substrate. The carina of the quagga mussels is more rounded that could be more difficult to grasp because of the lack of a distinct edge or easier to grasp because the shell is not as appressed to a rock substrate.

Most prior work on prey defenses of dreissenids has been conducted with roach (*Rutilus rutilus*), a Eurasian cyprinid that preys heavily on dreissenids (Kobak and Kakareko 2009, Kobak et al. 2010). Evidence suggests that dreissenid defenses involve plasticity in numbers of byssal threads, shell thickness, and tendency to aggregate among conspecifics. These studies showed that small- (<10 mm) and medium-size (10–17 mm) zebra mussels increased attachment strength via changes in byssal thread number in response to the presence of roach and also that zebra mussels aggregated more in the presence of roach. Such aggregation likely makes the mussels more difficult to grasp and remove because the upper and lower jaw must isolate a selected mussel from its neighbors. Smaller mussels also reduced their upward movement and tended to aggregate; both of these responses likely decrease the effectiveness of their suspension feeding. However, Czarnoleski et al. (2010) presented seemingly contrary evidence that, in response to chemical cues from crushed conspecifics, zebra mussels decreased attachment strength and number of byssal threads. Toomey et al. (2002) had contrary results regarding attachment strength in response to crushed conspecifics. These studies should be repeated with round gobies since round gobies, unlike roach, have jaw teeth to grasp their prey.

Because round gobies consume dreissenids mainly in a restricted size range, it is likely that mussels have two periods in their lives when round goby predation would be minimal. The first period is from the veliger stage to about 4 mm. During this period, minimal predation may not be a result of small size per se, but rather the ability to be concealed in crevices and cavities. The second period begins when mussels reach a size of ≥14 mm. The basis for this upper limit is unknown, but is probably because the fish's maximum oral or pharyngeal gape becomes too small to effectively prey on large dreissenids (Andraso et al. 2011). These size-based refugia of dreissenids may contribute to an optimal foraging strategy of round gobies.

IMPACTS OF THE PREDATOR

Field studies and laboratory experiments have shown that predation by round gobies can impact dreissenid populations by decreasing densities (Kuhns and Berg 1999, Barton et al. 2005, Lederer et al. 2006, 2008) and/or by increasing mean mussel size (Djuricich and Janssen 2001, Wilson et al. 2006; Table 23.1). Approaches for field studies fall into several different categories: a spatial approach that compares sites with and without round gobies (Lederer et al. 2006, Wilson et al. 2006), a temporal approach that compares the same sites pre- and post-round goby invasion (Barton et al. 2005), and a combination of both approaches (Lederer et al. 2008, Kipp et al. 2012). The study by Kipp et al. (2012) is of particular interest because it found that the negative effect of goby predation on dreissenids was related to round goby densities and presence of alternative prey. Experimental approaches include the use of colonized artificial substrates with and without barriers to exclude round gobies (Kuhns and Berg 1999, Djuricich and Janssen 2001) or the transfer of rocks from a site without round gobies to a site with round gobies (Lederer et al. 2006).

We propose that rugose rock morphology is a factor that limits vulnerability to predation and may cause regional variation in round goby impacts on dreissenids. Mussels may find refuge from predation under rocks (Djuricich and Janssen 2001) and also in vugs (pits and crevices in rocks caused by differential erosion of component minerals). In field surveys by Lederer et al. (2006, 2008), rocks were collected randomly by hand (snorkeling) in Green Bay, Lake Michigan, along the west side of the Door Peninsula, Wisconsin. This area was chosen because round gobies were actively spreading from the south end of the peninsula to the north end, and the lithology was fairly uniform throughout with most rocks being water-polished flat dolomite "flagstones." In 2003, dreissenids attached to the rocks were a mixture of zebra and quagga mussels (Lederer et al. 2006), but by 2006 dreissenids were mostly all quagga mussels (Lederer et al. 2008). As round gobies spread from south to north between 2003 and 2006, mussel densities in the north decreased (Figure 23.1).

As a follow-up to Lederer et al. (2006, 2008), we observed rocks at several of the same sites and found a scattered number with anomalously high numbers of dreissenids (all quagga mussels at this point) at three sites. Subsequently, we collected rocks having a range of dreissenid numbers for further inspection and found that rocks with high numbers of mussels all contained vugs

Table 23.1 Studies Finding Impacts of Round Gobies on Dreissenid Mussels. Methods of These Studies Varied from Field Manipulations to Observations of Natural Populations with Spatial and/or Temporal Approaches. Results Indicate That Round Gobies Can Impact Size Distributions (Smaller- or Larger-Sized Individuals) and Abundances (Fewer or Greater) of Dreissenid Populations. Kipp et al. (2012) Found Variable Impacts on Dreissenid Populations as Related to Round Goby Density and/or Alternative Prey

Source	Experiment	Spatial	Temporal	Impact
Kuhns and Berg (1999)	X			Fewer
Djuricich and Janssen (2001)	X	X		Larger
Barton et al. (2005)			X	Fewer
Wilson et al. (2006)		X		Larger
Lederer et al. (2006)	X	X		Fewer
Lederer et al. (2008)		X	X	Fewer
Kipp et al. (in press)		X	X	Variable

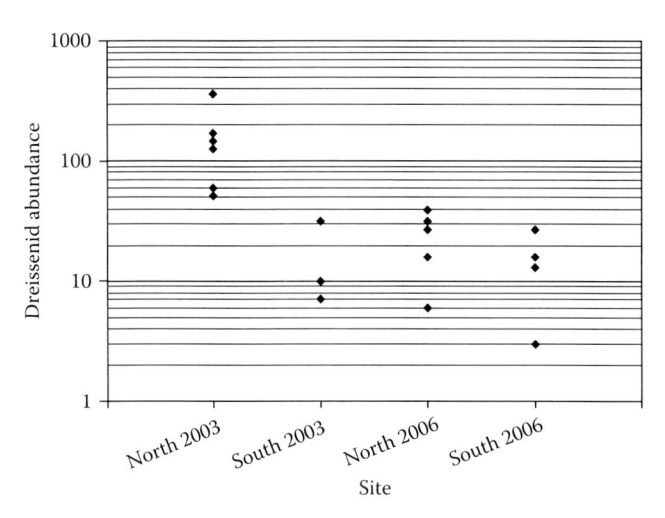

Figure 23.1 Impact of round gobies on dreissenid abundances in the nearshore area off the Door Peninsula in Green Bay, Lake Michigan. In 2003, round gobies were present at sites in the south but not yet found at sites in the north. By 2006, round gobies had fully invaded sites in the north. Dreissenid abundances at north sites in 2003 were significantly higher than abundances at south sites in 2003, at south sites in 2006, and at north sites in 2006. Each point represents an individual site. Dreissenid abundances are given as number per square meter. (From Lederer, A.M. et al., *J. Great Lakes Res.*, 34, 690, 2008.)

Figure 23.2 Typical examples of dolomite rocks with and without vugs. Geologic origin and the number and size of vugs (crevices left in the rock after fossils erode out) found on rocks are important factors affecting the colonization or survival of dreissenid mussels on rocky substrates. The smooth, Burnt Bluff formation dolomite has no vugs and no dreissenids (a), while the Mayville formation rock, with abundant vugs due to fossil pentamerid brachiopod casts and molds, had 16 dreissenids (b).

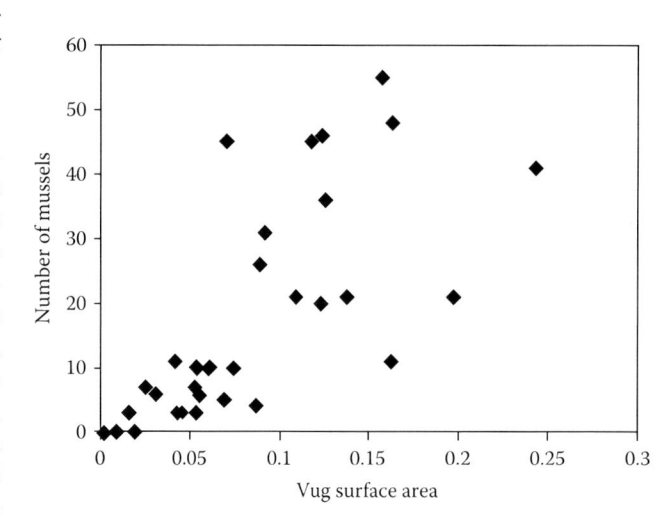

Figure 23.3 Relationship between the number of dreissenids and vug surface area for rocks collected in a nearshore region off the Door Peninsula in Green Bay, Lake Michigan. Each point represents an individual rock.

in which erosion of the calcite shells of fossilized brachiopods (*Pentamerus*) left creases in the dolomite where the shells had been. These rugose rocks were from the Mayville formation, a subunit of the lower Silurian overlain by smoother Burnt Bluff strata. The Mayville formation is exposed at a few sites along the west side of the Door Peninsula (Harris and Waldhuetter 1996; M. Harris, U. Wisconsin-Milwaukee, pers. comm.). We further collected 32 rocks, counted the number of attached mussels, and then measured both the rock and vug surface area from photographs. Because vugs were primarily derived from the same fossil organism, they were of fairly uniform size and shape. Two example rocks are shown in Figure 23.2: one is relatively large and flat, and the other is small but with numerous brachiopod vugs. No quagga mussels were attached to the smooth larger rock, while 16 mussels were attached to vug creases in the smaller rock. An analysis of covariance with site as the group variable and rock area and vug area as covariates indicated no site effect ($F_{1,27} = 2.92$, $P = 0.071$) and no effect of rock surface area ($F_{1,27} = 0.71$, $P = 0.41$), but effect of the vug area was highly significant (Figure 23.3; $F_{1,27} = 17.8$, $P < 0.001$). Evidently either mussels used the vugs as refugia from predation or round gobies did not have access to mussels in the vugs. If the Mayville formation had been more exposed over a wider area, it is entirely possible that results reported by Lederer et al. (2006, 2008) would instead indicate little or no effect of round gobies on mussel densities. The larger implication is that different types of rock fracture and erode in diverse ways, and it is likely that there is great variation in round goby impacts on dreissenid densities based solely on substrate type.

Our second study was concerned with mussel movement from one location to another. The location on a rock surface where a dreissenid will attach likely occurs in at least two stages. The first stage is during veliger settlement and attachment location is generally random. Later stages involve the detachment of byssal threads and movement to a new location to perhaps seek more suitable habitat conditions. Movement of mussels to new locations includes a risk because of exposure to predators or exposure to currents and being moved to unfavorable habitats. However, movement from crevices or rock cavities may be advantageous for feeding and perhaps reproduction. Because of mussel movement, mussel location on a rock surface may not indicate survival of predation by round gobies or other predators, but instead may simply indicate recent migration. However, we expect that mussels large enough to minimize predation risk because they are in a "size refuge" are more likely to move.

We assessed colonization of rocky substrate by mussels well past initial settlement with the use of year-long deployments of bundled-rock incubations at a natural reef (Green Can Reef: lat. N 42°59.3′, long. W 87°49.8′) and a newly constructed artificial reef (lat. N 42°51.3, long. W 87°47.1′) in 2008. Rock incubations consisted of five "cobble" sized crushed dolomite rocks bundled together with steel mesh. Five rock bundles, each with randomly assigned rocks and each bundle randomly assigned to a site, were spread along a "trapline" on the bottom in ~12 m of water at two locations. The artificial reef was completed about two months prior to deployment of the rock bundles, and it had no noticeable dreissenids at the time of bundle deployment. In contrast, rocks at the natural reef were densely covered with dreissenids at the time rock bundles were deployed. Round gobies were first observed near both reefs in 2004 and were abundant when rock bundles were deployed. One sample rock from each bundle (five in total from each reef) was retrieved by SCUBA divers after a year. A Mann–Whitney test showed that sizes of mussels on bundled rocks at the natural reef were significantly greater ($U = 41208.5$, $P < 0.001$) than sizes at the artificial reef. Yet despite this difference, median sizes were similar, 5.5 and 5.0 mm, respectively. This discrepancy occurred because the size range of dreissenids on rock bundles at the natural reef was much greater than the size range at the artificial reef. At the natural reef, 71 of 271 mussels on rock bundles were >10 mm, whereas at the artificial reef only 1 of 179 mussels was >10 mm (Figure 23.4). This finding indicates large mussels at the natural reef migrated from nearby rocks (natural substrate) onto the initially bare bundled rocks. Hence, this seems to verify a possible second stage of settlement that involves mussels large enough to occupy a refuge from round goby predation. Certainly, a second-stage settlement requires further study to understand costs (risks) and benefits to larger mussels that move. Some pertinent variables of such studies would include rock rugosity, predation risk, and potential benefits of increased food availability.

In combination, our two studies indicate complexities in round goby–dreissenid interactions as related to

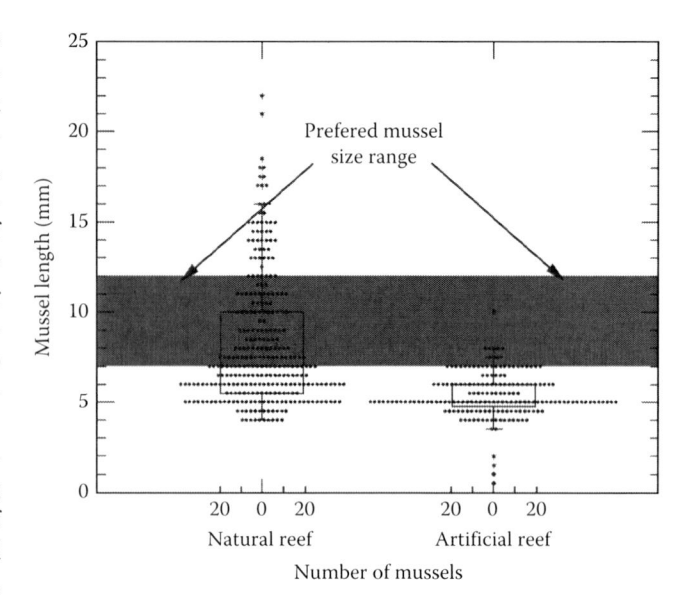

Figure 23.4 Size-frequency distributions of dreissenid mussels attached to bundled rocks that were newly placed on a natural reef (Green Can Reef) and on an artificial reef in Green Bay, Lake Michigan. The rocks were retrieved after one year and the number of attached mussels counted. No large mussels were found on newly placed rocks at the artificial reef, while large mussels colonized newly placed rocks from the surrounding area at the natural reef. The shaded area represents the approximate size range of mussels typically consumed by round gobies.

substrate composition. At one extreme, smooth rocks embedded in bottom substrate lack cavities to shelter small dreissenids from predation. At the other extreme, an abundance of loose rock with vugs provides small dreissenids with abundant shelter. When an individual mussel becomes too large to be eaten, it may colonize more exposed surfaces and gain a location with better access to available food.

We propose that there may be a beneficial comparison between the work of Paine (1966) on rocky intertidal zone community dynamics in which predator–prey organisms mostly coevolved and the combined transplanted plus novel interspecies interactions initiated by the dreissenid–round goby invasions. Dreissenids changed largely barren rocky substrates in the Great Lakes into a less diverse analog of a *Mytilus*-dominated biological matrix (Paine 1976). One of the first experimental demonstrations of the dramatic impacts of predation on a complex community involved *Mytilus* (bivalve prey), *Pisaster* (major sea star predator), as well as lesser predators and competitors for space in intertidal zones (Paine 1966, Dayton 1971). Where they co-occur, *Mytilus* was found almost exclusively in the intertidal but above a lower zone that *Pisaster* occupied. The upper limit for *Pisaster* distribution is determined by desiccation risk. Removal of *Pisaster* allows *Mytilus* to survive lower in the intertidal zone that provides benefits to *Mytilus* because it can suspension feed longer.

Additionally the removal of *Mytilus* by *Pisaster* creates space for other intertidal organisms to attach, so the relative abundance of attached organisms changes in a way that increases species diversity.

The *Mytilus–Pisaster* predator–prey relationship and impact of differential predation on the intertidal community is potentially pertinent to understanding interactions of dreissenids and predators such as round gobies. The first evidence that the relative abundance of quagga mussels and zebra mussels may be differentially impacted by predators was presented by Zhulidov et al. (2006) for the Don River in Russia. In that system, zebra mussels began to displace quagga mussels as the dominant dreissenid, which is opposite of what typically occurs. The authors proposed that this anomaly may be due to a diverse assemblage of mollusk predators (not including round gobies). They cited published and unpublished work (in Russian) that indicated quagga mussels were more easily crushed than zebra mussels. Also, attachment strength is greater in zebra mussels than quagga mussels that may make them less susceptible to predation (Peyer et al. 2009). However, it is also possible that relative abundances of the two dreissenids were impacted by preferential survival by zebra mussels in strong currents. The one site at which Zhulidov et al. (2006) did not find a decline in quagga mussel relative abundance was in a reservoir where currents were minimal.

Another lesson that has emerged since the work of Paine (1966) and Dayton (1971) is that the dynamics they reported are not found at every rocky intertidal zone where *Mytilus* and *Pisaster* co-occur along the North American Pacific coast. In some cases, the *Mytilus* population is limited by veliger recruitment (Roughgarden et al. 1988), and in other cases some *Mytilus* coexist with *Pisaster* because they are too large to be preyed upon (Paine 1976, Paine and Trimble 2004). A dreissenid parallel to limitations of veliger recruitment in *Mytilus* was noted by Wilson et al. (2006). They found that zebra mussels along the north coast of Lake Ontario were slow to become established, perhaps due to a high frequency of cold-water upwellings. Cold upwellings could also impact the metabolism, disrupt spawning, and affect densities of round gobies. In addition, a dreissenid parallel to size refuge effects observed in *Mytilus* would not be surprising because dreissenids do have a "size refuge" where individuals become too large to be effectively preyed upon (Andraso et al. 2011). Paine and Trimble (2004) found that when *Mytilus* too large for *Pisaster* predation were removed by storms, *Mytilus* densities remained low thereafter because *Pisaster* could prey on small, newly-recruited *Mytilus*. Because the introduction and spread of dreissenids preceded that of round gobies in the Great Lakes, there may be areas in which large dreissenids are still abundant despite high densities of round gobies. But, as larger, older dreissenids die, or are removed by other factors, they may be replaced by small individuals that are vulnerable to predation, or perhaps large individuals that move from less exposed areas where they had been sheltered from predation.

The seminal work by Paine (1966) is important because it demonstrated predation as a mechanism that can profoundly impact an entire community and formed the foundation for the study of intertidal community dynamics. We believe that predator–prey interactions between dreissenids and round gobies have the same potential impact on communities in areas where distributions of these two taxa overlap. Round gobies can clearly impact dreissenid populations, but interactions can be complex. Resolving these complexities will help us learn more about communities in the Great Lakes, particularly those found in rocky habitats. Moreover, round gobies and dreissenids are nuisance species that need to be managed. Because of their interactions with each other, management of all of these species will have to be integrated.

ACKNOWLEDGMENTS

We appreciate the insight into Door County geomorphology provided by Mark Harris. Field and laboratory assistance was aided by Amanda Lederer, Chris Malinowski, Yutta Wang, and Liz Boeckmann. Help with fossil and vug identification came from Irene M. Lammers and Margaret Frasier. Greg Andraso provided a helpful review. This work was funded in part by the University of Wisconsin Sea Grant Institute under a grant from the National Sea Grant College Program, National Oceanic and Atmospheric Administration, U.S. Department of Commerce, and from the State of Wisconsin. Federal Grant Number NA06OAR4170011, project number R/AI-1. It was also funded in part by the National Oceanic and Atmospheric Administration, Federal Grant Number NA09NMF4630406.

REFERENCES

Andraso, G. M., M. T. Ganger, and J. Adamczyk. 2011. Size-selective predation by round gobies (*Neogobius melanostomus*) on dreissenid mussels in the field. *J. Great Lakes Res.* 73: 298–304.

Andraso, G. M., J. Cowles, R. Colt, J. Patel, and M. Campbell. 2011. Ontogentic changes in pharyngeal tooth morphology of round gobies (*Neogobius melanostomus*). *J. Great Lakes Res.* 37: 738–743.

Barton, D. R., R. A. Johnson, L. Campbell, J. Petruniak, and M. Patterson. 2005. Effects of Round Gobies (*Neogobius melanostomus*) on Dreissenid mussels and other invertebrates in Eastern Lake Erie, 2002–2004. *J. Great Lakes Res.* 31(Suppl. 2): 252–261.

Carman, S. M., J. Janssen, D. J. Jude, and M. B. Berg. 2006. Diel interactions between prey behaviour and feeding in an invasive fish, the round goby, in a North American river. *Freshwat. Biol.* 51: 742–755.

Corkum, L. D., M. R. Sapota, and K. E. Skora. 2004. The round goby, *Neogobius melanostomus*, a fish invader on both sides of the Atlantic Ocean. *Biol. Invasions* 6: 173–181.

Czarnoleski, M., T. Muller, K. Adamus, G. Ogorzelska, and M. Sog. 2010. Injured conspecifics alter mobility and byssus production in zebra mussels *Dreissena polymorpha*. *Fund. Appl. Limnol.* 176: 269–278.

Dayton, P. K. 1971. Competition, disturbance and community organization: The provision and subsequent utilization of space in a rocky intertidal community. *Ecol. Monogr.* 41: 351–389.

Dubs, D. O. L. and L. D. Corkum. 1996. Behavioral interactions between round gobies (*Neogobius melanostomus*) and mottled sculpins (*Cottus bairdi*). *J. Great Lakes Res.* 22: 838–844.

Djuricich, P. and J. Janssen. 2001. Impact of round goby predation on zebra mussel size distribution at Calumet Harbor, Lake Michigan. *J. Great Lakes Res.* 27: 312–318.

French, J. R. P. and D. J. Jude. 2001. Diets and diet overlap of nonindigenous gobies and small benthic native fishes co-inhabiting the St. Clair River, Michigan. *J. Great Lakes Res.* 27: 300–311.

Ghedotti, M. J., J. C. Smihula, and J. R. Smith. 1995. Zebra mussel predation by round gobies in the laboratory. *J. Great Lakes Res.* 21: 665–669.

Grubich, J. R. 2000. Crushing motor patterns in drum (Teleostei: Sciaenidae): Functional novelties associated with molluscivory. *J. Exp. Biol.* 203: 3161–3176.

Harris, M. T. and A. Waldhuetter. 1996. Silurian of the Great Lakes basin, Part 3. Llandovery strata of the Door Peninsula, Wisconsin. Milwaukee Public Museum. *Contr. Biol. Geol.* 90: 1–162.

Helfman, G. S., B. B. Collette, D. E. Facey, and B. W. Bowen. 2009. *The Diversity of Fishes*. Hoboken, NJ: Blackwell Science.

Higgins, S. N. and M. J. Vanderzanden. 2010. What a difference a species makes: A meta-analysis of dreissenid mussel impacts on freshwater ecosystems. *Ecol. Monogr.* 80: 170–196.

Janssen, J., J. Montgomery, and R. Tien. 1992. Social rotational feeding in *Pagothenia borchgrevinki* (Nototheniidae). *Copeia* 1992: 559–562.

Janssen, J. and D. J. Jude. 2001. Recruitment failure of mottled sculpin *Cottus bairdi* in southern Lake Michigan induced by the newly introduced round goby, *Neogobius melanostomus*. *J. Great Lakes Res.* 27: 319–328.

Jude, D. J. and J. Janssen. 1995. Ecology, distribution, and impact of the newly introduced round and tubenose gobies on the biota of the St. Clair and Detroit Rivers. In *The Lake Huron Ecosystem: Ecology, Fisheries, and Management. Ecovision World Monograph Series, S.P.B.* M. Munawar, T. Edsall, and J. Leach, eds., pp. 447–460. Leiden, The Netherlands: Academic Publishing.

Kipp, R., I. Hebert, M. Lacharite, and A. Ricciardi. 2012. Diet and prey selection by the Eurasian round goby (*Neogobius melanostomus*) in the upper St. Lawrence River. *J. Great Lakes Res.* 38: 78–89.

Kobak, J. and T. Kakareko. 2009. Attachment strength, aggregation and movement of the zebra mussel (*Dreissena polymorpha*, Bivalvia) in the presence of potential predators. *Fund. Appl. Limnol. (Arch. Hydrobiol.)* 174: 193–204.

Kobak J., T. Kakareko, and M. Poznańska. 2010. Changes in attachment strength and aggregation of zebra mussel, *Dreissena polymorpha* in the presence of potential fish predators of various species and size. *Hydrobiologia* 644: 195–206.

Kornfield, I. 1991. Genetics. In *Cichlid Fishes, Behaviour, Ecology and Evolution*. M. H. A. Keenleyside, ed., pp. 103–128. New York: Chapman and Hall.

Kuhns, L. A. and M. B. Berg. 1999. Benthic invertebrate community responses to round goby (*Neogobius melanostomus*) and zebra mussel (*Dreissena polymorpha*) invasion in Southern Lake Michigan. *J. Great Lakes Res.* 25: 910–917.

Lauder, G. V. 1983. Neuromuscular patterns and the origin of trophic specialization in fishes. *Science* 219: 1235–1237.

Lederer, A. M., J. Janssen, T. Reed, and A. Wolf. 2008. Impacts of the introduced round goby (*Apollonia melanostoma*) on dreissenids (*Dreissena polymorpha* and *Dreissena bugensis*) and on macroinvertebrate community between 2003 and 2006 in the littoral zone of Green Bay, Lake Michigan. *J. Great Lakes Res.* 34: 690–697.

Lederer, A., J. Massart, and J. Janssen. 2006. Impact of round gobies (*Neogobius melanostomus*) on dreissenids (*Dreissena polymorpha* and *Dreissena bugensis*) and the associated macroinvertebrate community across invasion front. *J. Great Lakes Res.* 32: 1–10.

Levin, S. A. and R. T. Paine. 1974. Disturbance, patch formation and community structure. *Proc. Nat. Acad. Sci. U S A* 71: 2744–2747.

Nalepa, T. F., D. L. Fanslow, and S. A. Pothoven. 2010. Recent changes in density, biomass, recruitment, size structure, and nutritional state of *Dreissena* populations in southern Lake Michigan. *J. Great Lakes Res.* 36: 5–19.

Neilson, M. E. and C. A. Stepien. 2009. Evolution and phylogeography of the tubenose goby genus *Proterorhinus* (Gobiidae: Teleostei): Evidence for new cryptic species. *Biol. J. Linnean Soc.* 96: 664–684.

Paine, R. T. 1966. Food web complexity and species diversity. *Am. Naturalist* 100: 65–75.

Paine, R. T. 1976. Size-limited predation: An observational and experimental approach with the *Mytilus–Pisaster* interaction. *Ecology* 57: 858–873.

Paine, R. T. 2002. Advances in ecological understanding: By Kuhnian revolution or conceptual evolution? *Ecology* 83: 1553–1559.

Paine, R. T. and S. A. Levin. 1981. Intertidal landscapes: Disturbance and the dynamics of pattern. *Ecol. Monogr.* 51: 145–178.

Paine, R. T. and A. C. Trimble. 2004. Abrupt community change on a rocky shore—Biological mechanisms contributing to the potential formation of an alternative state. *Ecol. Lett.* 7: 441–445.

Peyer, S. M., A. J. McCarthy, and C. E. Lee. 2009. Zebra mussels anchor byssal threads faster and tighter than quagga mussels in flow. *J. Exp. Biol.* 212: 2027–2036.

Pennuto, C. M., P. J. Krkowiak, and C. E Janik. 2010. Seasonal abundance, diet and energy consumption of round gobies (*Neogobius melanostomus*) in Lake Erie tributary streams. *Ecol. Freshwat. Fish.* 19: 206–215.

Phillips, E. C., M. E. Washek, A. W. Hertel, and B. M. Niebel. 2003. The round goby (*Neogobius melanostomus*) in Pennsylvania tributary streams of Lake Erie. *J. Great Lakes Res.* 29: 34–40.

Ray, W. J. and L. D. Corkum. 1997. Predation of zebra mussels by round gobies, *Neogobius melanostomus. Environ. Biol. Fish.* 50: 267–273.

Roughgarden, J., S. Gaines, and H. Possingham. 1988. Recruitment dynamics in complex life cycles. *Science* 241: 1460–1466.

Skaskina, E. P. and V. A. Kostyuchenko. 1968. Food of *Neogobius melanostomus* in the Azov Sea. *Vopr. Iktiol.* 8: 303–311.

Toomey, M. B., D. McCabe, and J. E. Marsden. 2002. Factors affecting the movement of adult zebra mussels (*Dreissena polymorpha*). *J. N. Am. Benthol. Soc.* 21: 468–475.

Vanderploeg, H. A., J. R. Liebig, T. F. Nalepa, G. L. Fahnenstiel, and S. A. Pothoven. 2010. *Dreissena* and the disappearance of the spring phytoplankton bloom in Lake Michigan. *J. Great Lakes Res.* 36(Suppl. 3): 50–59.

Wilson, K. A., E. T. Howell, and D. A. Jackson. 2006. Replacement of zebra mussels by quagga mussels in the Canadian nearshore of Lake Ontario: The importance of substrate, round goby abundance, and upwelling frequency. *J. Great Lakes Res.* 32: 11–28.

Zhulidov, A. V., T. F. Nalepa, A. V. Zhulidov, and T. Yu. Gurtovaya. 2006. Recent trends in relative abundance of two dreissenid species, *Dreissena polymorpha* and *Dreissena bugensis* in the lower Don River. *Arch. Hydrobiol.* 165: 209–220.

CHAPTER 24

Density, Growth, and Reproduction of Zebra Mussels (*Dreissena polymorpha*) in Two Oklahoma Reservoirs

Chad J. Boeckman and Joseph R. Bidwell

CONTENTS

Abstract 369
Introduction 369
Methods 370
    Oologah Lake 370
    Sooner Lake 371
Results 372
    Oologah Lake 372
    Sooner Lake 375
Discussion 377
    Oologah Lake 377
    Sooner Lake 380
Conclusion 381
Acknowledgments 381
References 381

ABSTRACT

Zebra mussels (*Dreissena polymorpha*) were first reported in Oklahoma in 1993 and have since spread to atleast 10 different reservoirs in the state. This study characterizes the population dynamics of zebra mussels in two of these reservoirs. One reservoir had a "natural" temperature regime (Oologah Lake), and the other had a warmwater discharge that significantly altered the thermal profile (Sooner Lake). Zebra mussels were discovered in Oologah Lake in 2003 and monitoring began that year and continued through 2010. In Sooner Lake, zebra mussels were discovered in 2006 and monitoring began in 2007 and also continued through 2010. In both lakes, veliger densities peaked between 500 and 600/L, and peak densities occurred in 2006 in Oologah Lake and in 2010 in Sooner Lake. Seasonal peaks in veliger densities usually occurred in June in both lakes. Densities of adult zebra mussels reached 150,000/m² in the first year of monitoring in each lake study. Maximum growth ranged between 0.1 and 0.14 mm/day, however, in 2009–2010, growth rates moderated slightly. In most years, summer die-offs of older, reproductively-mature *D. polymorpha* were observed. It is hypothesized that die-offs resulted from a combination of poor physiological condition after reproduction and stress induced by water temperatures of 30°C, usually achieved in July and August in Oklahoma reservoirs. Substantial floods in 2007, 2008, and 2009 after one such die-off also contributed to a prolonged recovery of the zebra mussel population in Oologah Lake. At present, it appears temperature may serve as an important control of zebra mussels in Oklahoma reservoirs and, possibly, throughout southern North America.

INTRODUCTION

Zebra mussels (*Dreissena polymorpha*) were first reported in the McClelland–Kerr Navigation System of the Arkansas River, located in the eastern part of

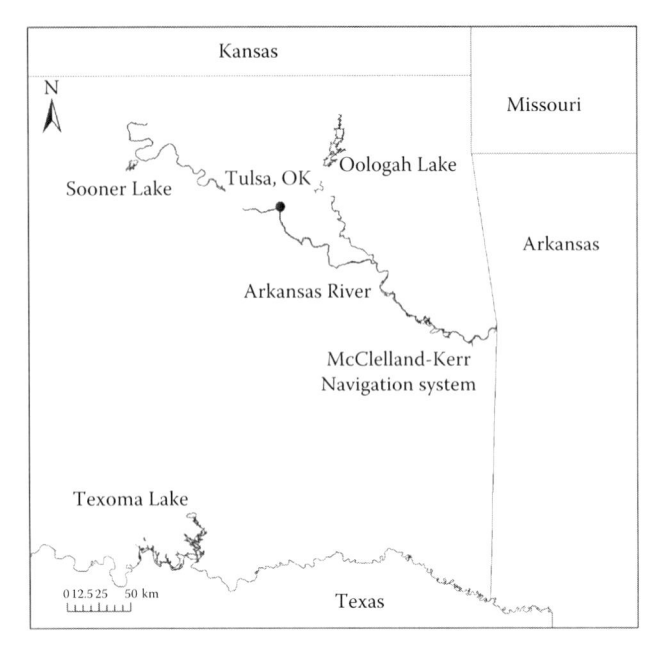

Figure 24.1 Location of Oologah Lake and Sooner Lake in eastern Oklahoma.

Oklahoma, in 1993 (Laney 2010). Mussels were confined to the navigation system for nearly 10 years until 2003 when they were discovered in Oologah Lake (Rogers and Nowata Counties) located northeast of Tulsa, Oklahoma (Figure 24.1). Zebra mussels currently infest at least 10 different reservoirs in Oklahoma, primarily in northeastern portions of the state. However, one reservoir (Lake Texoma) located on the Oklahoma–Texas border has also been colonized. While Oklahoma has a few natural oxbow lakes, most lakes are impounded reservoirs linked to a river. These rivers facilitated the spread of mussels in the state (Havel et al. 2005). For example, mussels clearly spread downstream along the Arkansas River from Kansas into Oklahoma (Bidwell 2010).

Initial predictions of the potential range of zebra mussels in the United States delineated a southern limit near the border of Oklahoma and Texas based on summer water temperatures that were thought to exceed the thermal tolerance of dreissenid mussels (e.g., Strayer 1991). Reservoirs in Oklahoma are warm-monomictic during an average year, with no period of ice cover, and stratification only during the warmest months (July–August). Systems in the northern part of the state have a mean annual temperature range of approximately 5°C–25°C, while those in the southern part of the state have a mean range of 7°C–29°C (OWRB 2007). Reservoirs across the state may have water temperatures that exceed 30°C for several weeks in July and August, which may also be coupled with low-flow conditions. While thermal-tolerance experiments conducted with North American populations indicate zebra mussels are more tolerant of warm temperatures than European populations (e.g., Hernandez et al. 1995, McMahon et al. 1995, McMahon and Ussery 1995), they may still exhibit a negative scope for growth when sustained water temperatures exceed 28°C (Aldridge et al. 1995). As such, there is significant potential for zebra mussels in Oklahoma reservoirs to experience stressful thermal conditions for part of the year. This would be particularly true for systems in southern Oklahoma and into Texas.

Nichols (1996) highlighted the scarcity of data regarding zebra mussel population dynamics in southern North America. Much of what is known about population dynamics in southern latitudes is derived from accounts in the Mississippi River (Allen et al. 1999) where dispersal of zebra mussel veligers and adults from upstream sources is a confounding variable. The present study was initiated as a basic program to monitor the density, reproduction, and growth of zebra mussels in two Oklahoma reservoirs. It represents one of the few long-term studies of zebra mussel population dynamics in southern reservoirs in what has been considered the limit of the southern range for this organism. Data generated from one of these reservoirs stimulated a retrospective consideration of the role temperature may play in influencing zebra mussel populations in this region.

METHODS

Oologah Lake

Oologah Lake is an impoundment on the Verdigris River, completed in 1974 for the purposes of flood control, water supply, and navigation (OWRB 1990). The lake has a storage capacity of 0.7 km^3 at normal pool and drains 11,000 km^2 in northeastern Oklahoma and southeastern Kansas with a maximum depth of approximately 20 m (USACE 2002).

Zebra mussels were first reported in Oologah Lake in 2003. Mussels were most likely brought into the reservoir by recreational boaters or sport fisherman since no source populations upstream of the system were known to occur, and because it is a popular site for sport fishing tournaments. Monitoring of Oologah Lake began in June 2003 at four sites (Spencer Creek, Blue Creek, Redbud Marina, and Hawthorn Bluff) in the south/southeastern part of the lake (Figure 24.2). Lake level was obtained from www.swt-wc.usace.army.mil/OOLOcharts.html accessed on October 26, 2010.

Samples were collected weekly from June 2003 to 2007 and monthly from 2008 to 2010. During each sampling event, temperature (°C), conductivity (mS/cm), dissolved oxygen (mg/L), and pH (standard units) were collected using a Hydro lab Quanta multiparameter probe (Hach Hydromet Corporation, Loveland, CO) at the surface, middle, and bottom of the water column. Secchi-disk depth was also recorded at each site. Mean Secchi-depth data from 2000 to 2002 were provided by the U.S. Army Corps of Engineers, Tulsa District (Tony Clyde, personal communication).

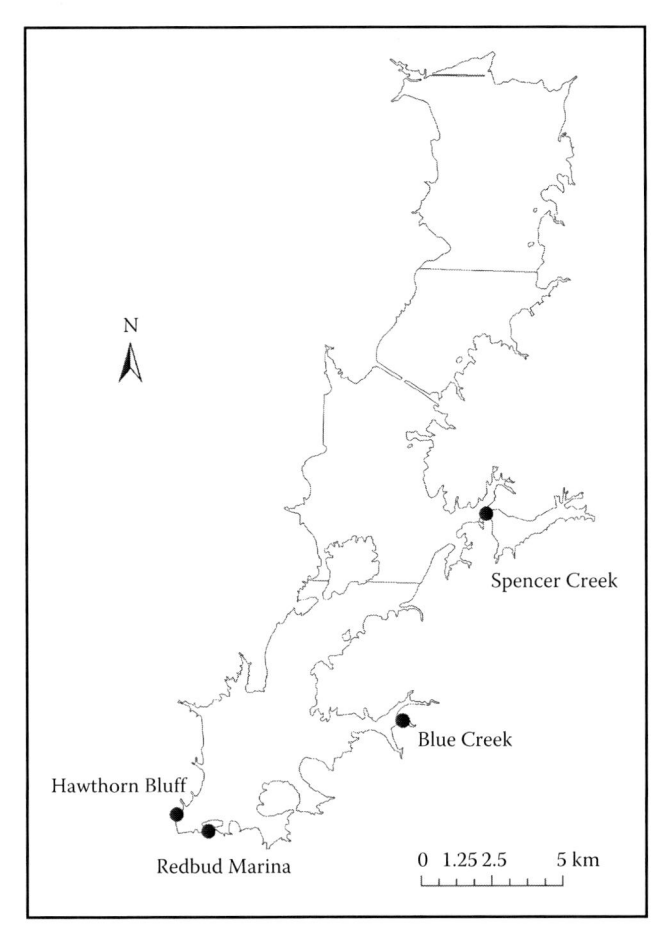

Figure 24.2 Location of four sampling sites in Oologah Lake, Oklahoma.

On each sampling date, four vertical plankton tows were taken from boat docks at each site with a 64 μm mesh Wisconsin-style zooplankton net with a 20 cm aperture. Horizontal tows were obtained at Spencer Creek because this location did not have a boat dock. However, despite differences in tow patterns, similar volumes were sampled at all sites. Redbud Marina was the deepest site, averaging 4 m depth, followed by Hawthorn Bluff and Blue Creek at approximately 3 m, and Spencer Creek at 1.5 m. Each sample was rinsed into individually labeled 125 mL polyethylene bottles and preserved with 70% ethanol. The volume of water sampled was estimated from the equation of the volume of a cylinder ($V = \partial r^2 h$), where r is equal to the radius of the net aperture (0.2 m) and h is equal to the length of the tow.

At Blue Creek, Redbud Marina, and Hawthorn Bluff, densities of settled veligers were obtained from modified glass microscope slide boxes suspended approximately 1 m below the surface. The top and bottom of each box were removed to allow water exchange, and four glass microscope slides were inserted and held in place using plastic zip ties. During each sampling event, slides were removed, placed in 100 mL glass bottles filled with lake water, and transported back to the laboratory for settled veliger enumeration. Veligers collected in plankton tows and on glass slides were enumerated under 12.5× magnification with an Olympus SZX-ILLD100 dissecting microscope (Olympus America, Inc., Center Valley, PA) fitted with cross-polarization filters as described by Johnson (1995).

Densities of adult mussels were obtained from 10 × 20 cm concrete panels suspended at three sampling sites. Five panels were attached to a 1.25 m section of PVC pipe and suspended approximately 1 m under boat docks at Hawthorn Bluff, Redbud Marina, and Blue Creek. Each panel was sampled by laying a wood-framed wire grid (50 total – 2 × 2 cm squares) on its surface and counting the number of zebra mussels in 10 randomly selected grids on each panel. Five grids were counted on the front and five on the back of each panel if mussel densities were high enough to warrant subsampling.

Growth estimates of mussels were obtained from mussels placed in modified polyethylene tackle boxes (10 × 20 cm) with 12 chambers per box. The top and bottom of each box were removed and replaced with a rigid plastic mesh (2 × 2 mm grids) to allow for water exchange. Zebra mussels were collected from boat hulls or the dock itself at Redbud Marina. Individuals of approximately 8–10 mm were isolated by cutting the byssal threads with a scalpel and measured with digital calipers to the nearest 0.01 mm total length. One zebra mussel was placed in each of the 12 chambers per box, and boxes were suspended at several sites and depths for 5–9 weeks. After the growth trial was complete, mussels were remeasured and growth rates were expressed as mm/day.

Sooner Lake

Sooner Lake is an impoundment of Greasy Creek, a tributary of the Arkansas River that was constructed in 1976 to provide cooling water for a coal-fired electricity-generating facility (OWRB 1990). The lake has a capacity of 0.2 km³ at normal pool with an average depth of 8.5 m and a maximum depth of 27 m (Angyal et al. 1987). The power plant on the lake releases heated effluent that is directed from the discharge area to the main body of the lake by a series of dikes that facilitate cooling prior to the water being taken into the plant again via an intake channel. This creates a series of discrete temperature zones in the lake that range from 10°C to 15°C above ambient at the discharge, to ambient temperature in the main lake.

Zebra mussels were first reported in Sooner Lake in 2006. While introduction due to boating/fishing activities is a possibility, the presence of established mussel populations in the Arkansas River above and below Sooner Lake (Bidwell 2010) suggests make-up water taken from the Arkansas River was the most likely source of mussels into the system. Beginning in January 2007, samples were collected

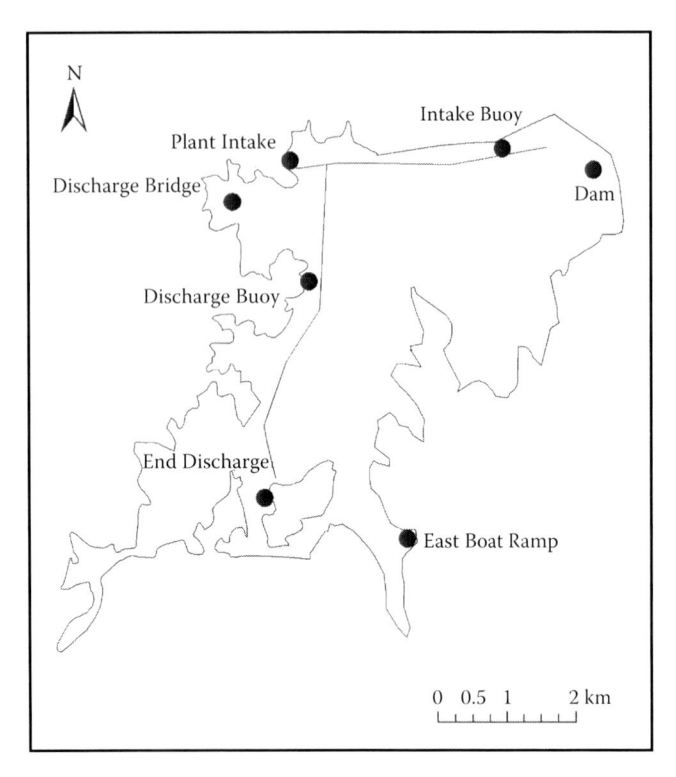

Figure 24.3 Location of six sampling sites in Sooner Lake, Oklahoma. Only temperature was measured at the east boat ramp site.

at six sites within Sooner Lake (Figure 24.3). These sites were selected in part to investigate the potential influence of the thermal effluent on zebra mussel reproduction dynamics. Sites at the discharge buoy, end discharge channel, dam, and intake buoy were accessed by boat, while those at the discharge bridge and plant intake were accessed from the shore. Methods for water chemistry, veliger collection and enumeration, and adult densities were as described for Oologah Lake, with the exception that at the boat-accessed sites, a buoy was deployed with the concrete panel apparatus suspended approximately 1 m below the lake surface, rather than being hung from boat docks. Beginning in July 2008, temperature loggers (HOBO pro v2, Onset Computer Corporation, Pocasset, MA) were suspended below each buoy location (discharge buoy, end discharge, dam, and intake buoy) as well as at the east boat ramp, in association with growth experiments. Loggers were programmed to record temperature every hour.

Growth experiments at Sooner Lake were conducted to investigate the potential effect of the warmwater discharge on zebra mussel growth rates. Experiments were conducted in the same polyethylene tackle boxes as described for Oologah Lake. Zebra mussels used to measure growth were collected from a boat dock outside the zone of the warmwater discharge. Each box was then suspended under the buoys supporting the adult settling panels at approximately 1 m below the surface and remained in the lake for 5–9 weeks with longer deployments occurring in winter months. At the conclusion of the growth trial, each box was collected and the total length of individual mussels was again determined using digital calipers. Growth rates were expressed as the increase in shell length mm/day.

Statistical evaluation of differences in growth rates between sites was accomplished with one-way analysis of variance (ANOVA, $\alpha = 0.05$) followed by Holm–Sidak post hoc tests with appropriate alpha-level corrections. Data that did not meet assumptions of normality or equal variance were natural log-transformed prior to ANOVA. If transformations did not alleviate non-normality or equal variances, an ANOVA on ranks was performed with Dunn's post-hoc procedure (Zar 1999).

RESULTS

Oologah Lake

In Oologah Lake, maximum seasonal temperatures reached just over 30°C in late-August of most years and consistently fell below 5°C in January–February (Figure 24.4). Because of shallow depths, temperature stratification was not observed at any of the monitoring sites. Dissolved oxygen varied widely over a seasonal period, with peaks greater than 10 mg/L observed in late-winter/early-spring, and lows (as low as 2 mg/L) observed in mid- to late-summer. As observed for temperature, vertical stratification of oxygen concentrations was not noted. Water pH was slightly alkaline over the study period with periodic extreme values ranging from 6 to over 9 standard units. Conductivity at Oologah Lake ranged from 0.200 to 0.650 mS/cm. Alkalinity and hardness were indicative of moderately hard water with alkalinity values ranging from 40 to 160 mg/L and hardness from 80 to 240 mg/L as $CaCO_3$.

From 2003 to 2006, mean veliger densities in Oologah Lake increased each year (Figures 24.4 and 24.5). In 2003, veliger density peaked at 30/L in late-September (Redbud Marina) and peak density at a particular site increased in each successive year, from 170/L in 2004 to 480/L in 2006 (Figure 24.5). In 2005 and 2006, veligers were first observed in May when water temperatures ranged between 15°C and 24°C. Peak densities usually occurred in late-May or early-June, associated with water temperatures between 24°C and 28°C. As water temperatures increased in July and August, veliger densities decreased, but a secondary peak in veligers was often observed in September and October when temperatures fell to between 16°C and 26°C (Figures 24.4 and 24.5). This secondary peak in veliger densities was usually much reduced as compared to the spring peak. In 2004, veliger densities did not fit this typical seasonal pattern. Veligers were first observed in early-May and peak densities occurred in July and August. Mean water temperature in July and August at mid-column depth was significantly lower in 2004 than in 2003, 2005, and 2006 (26.8°C vs. 28.5°C, 28.9°C,

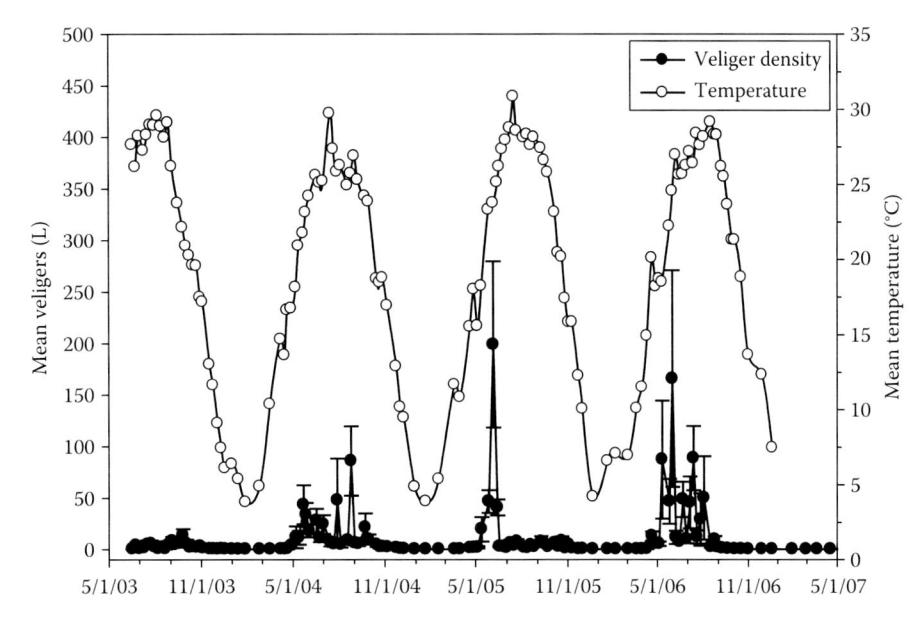

DENSITY, GROWTH, AND REPRODUCTION OF ZEBRA MUSSELS (*DREISSENA POLYMORPHA*)

Figure 24.4 Mean (±SE) veliger densities at four sampling sites in Oologah Lake, Oklahoma, from June 2003 to July 2010. Also, given are mean temperatures at the same sites from June 2003 to April 2007.

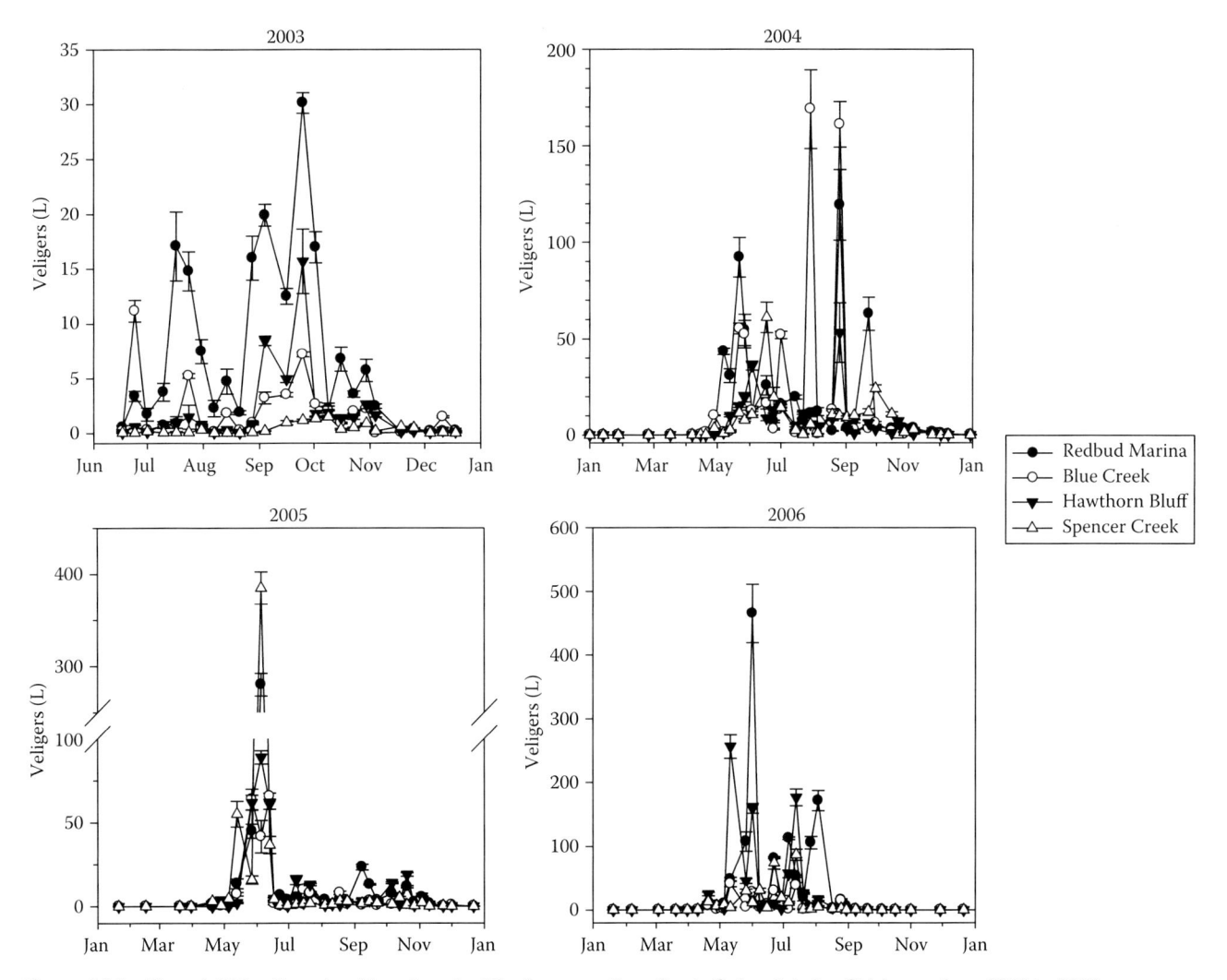

Figure 24.5 Mean (±SE) veliger densities at each of the four sampling sites in Oologah Lake, Oklahoma, from 2003 to 2006.

and 28.2°C, respectively, with $P < 0.005$ for all years). In 2007, veliger densities decreased dramatically to a mean of less than 1/L and remained at his low level through 2010 (Figure 24.4).

Peak densities of settled veligers increased from 200,000/m² in 2003 to near 1.2 million/m² in early-2006 (Figure 24.6). Greatest settling rates were observed 2–3 weeks after maximum densities of veligers were observed in the water column. No settled veligers were detected between 2007 and 2010 (July); this corresponded to the very low densities of veligers found in the water column during this time period.

Densities of zebra mussels on concrete panels also exhibited seasonal trends, with peak densities generally occurring in July shortly after planktonic veligers matured and settled (Figure 24.7). In summer 2004, concrete panels suspended at Redbud Marina were excessively colonized, which made the grid enumeration system (outlined in the Methods section) inadequate to determine density. Those panels were harvested and replaced with clean panels, which were recolonized 6 weeks later.

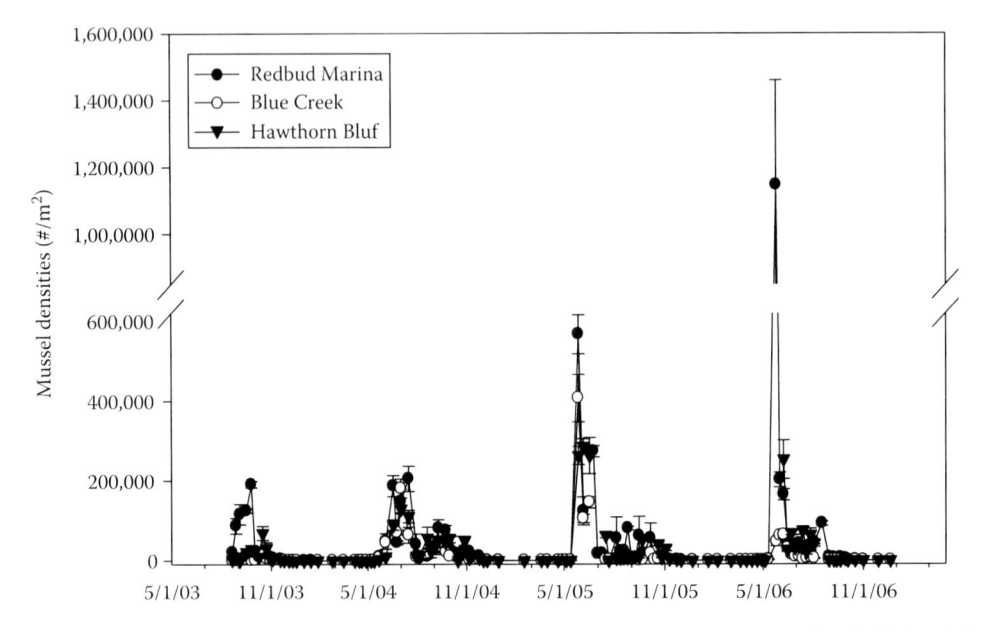

Figure 24.6 Mean (±SE) densities of zebra mussels settled on glass slides suspended at three sites in Oologah Lake, Oklahoma, from 2003 to 2006.

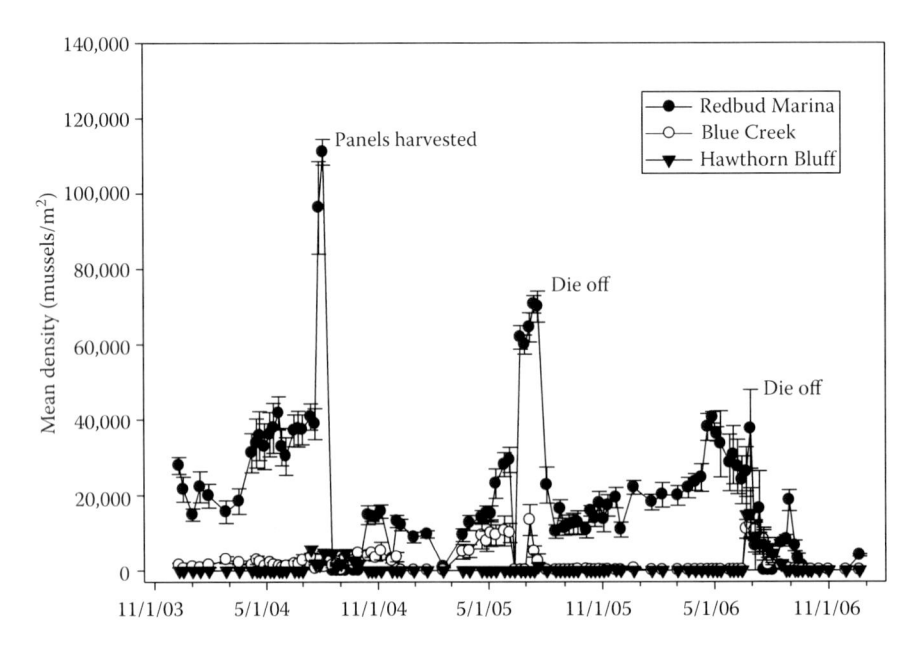

Figure 24.7 Mean (±SE) densities of zebra mussels on concrete panels suspended at three sites in Oologah Lake, Oklahoma, from 2003 to 2006.

Harvested panels were brought into the laboratory where a more accurate estimate of zebra mussel density could be conducted. More thorough examination of the harvested panels revealed a density of 155,000/m^2. In 2005 and 2006, declines in abundance reflected die-offs in late July and August (Figure 24.7). Between 2007 and 2009, no adult mussels were found on concrete panels at any site. However, mussels reappeared at Blue Creek and Redbud Marina in July 2010. Panels suspended at Hawthorn Bluff were repeatedly colonized during the course of the study. Panels at this site were located at the end of a long fetch and were subjected to greater wave action than at the other sites that displaced mussels and lead to artificially low-density estimates for this site.

Initial growth experiments were conducted in Oologah Lake in 2005–2006 with caged mussels. Mussel growth rates approached 0.14 mm/day at Redbud Marina, Blue Creek, and Hawthorn Bluff in June/July 2006 (Figure 24.8). Mean water temperatures during the trial ranged from 26.9°C at Redbud Marina to 27.5°C at Blue Creek. To investigate the influence of depth on growth rates, cages were suspended at 1, 2, and 3 m at Redbud Marina in June 2005 (Figure 24.8). Growth rates approached 0.1 mm/day at 1 and 2 m and were significantly different ($P < 0.001$) from the 0.05 mm/day observed at the 3 m depth. Mean temperatures ranged from 28.9°C at 1 m to 28.2°C at 3 m. A second depth vs. growth trial was initiated at Redbud Marina in September 2005 with cages suspended at 1, 2, and 3 m (Figure 24.8). Growth of zebra mussels was significantly greater ($P < 0.05$) at 2 m than 3 m, but was not different ($P > 0.05$) between 1 and 3 m. Mean temperatures ranged from 22.2°C at 1 m to 22.1°C at 3 m. Finally, to determine winter growth rates, zebra mussels were suspended at 1 m at Redbud Marina in January 2006. Maximum winter growth rates (mean temperature of 6.9°C) approached 0.01 mm/day (Figure 24.8).

From 2000 to 2006, water transparency increased significantly ($P < 0.001$, Table 24.1) in Oologah Lake. Interestingly, a marked decline in water transparency occurred in 2007. While mean Secchi depth values were not significantly different ($P > 0.05$) between 2003–2006 and 2007–2010, mean values declined from 0.99 m in the former period to 0.84 m in the latter, and this decline was consistent with a lake-wide die-off of zebra mussels beginning in 2007. However, substantial flooding occurred in 2007, 2008, and 2009 (Figure 24.9); therefore, the decline in water transparency cannot be strictly caused by the die-off.

Sooner Lake

Based on data from the temperature loggers, sites closest to the warm-water discharge had higher temperatures throughout the year compared to sites farther away. Peak summer temperatures were near 40°C, and winter temperatures varied between 5°C and 10°C at the discharge buoy. Sites located farthest from the warm-water discharge (east boat ramp, dam, and intake buoy) reflected typical Oklahoma reservoir water temperatures of near 30°C in July–August and below 5°C in January–February.

Other water quality parameters were more consistent among sites within Sooner Lake. Dissolved oxygen concentrations ranged between 1 and 14 mg, with highest concentrations recorded during winter and early-spring, and lowest concentrations associated with warmer temperatures in July and August. Sooner Lake stratified during July and August in 2007, 2008, and 2010 although, due to the small size of the lake, length of fetch across the lake, and prevailing winds, two of the three stratification events were quite short in duration. Stratification in 2007 and 2008 lasted about 3 weeks, while the stratification in 2010 lasted for nearly 7 weeks. These stratifications were most evident at the dam location that had a maximum depth of 27 m. During each stratification event, surface dissolved oxygen concentrations of 7 mg/L were depleted to 1.5 mg/L at 15 m below the surface. Water pH ranged from 6.8 to 8.9 standard units during

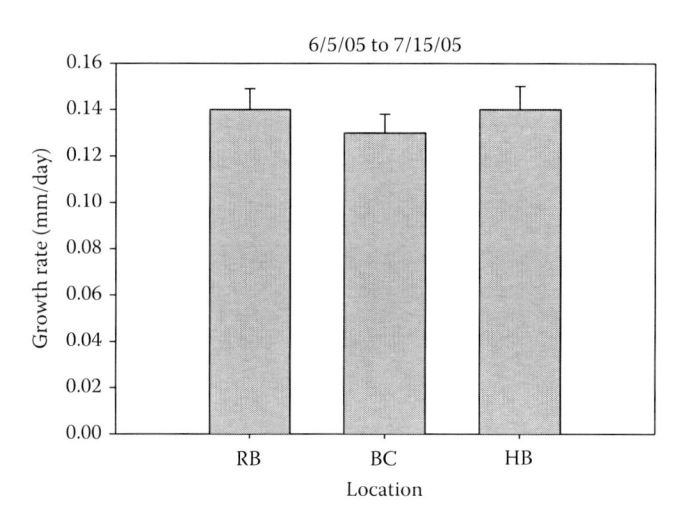

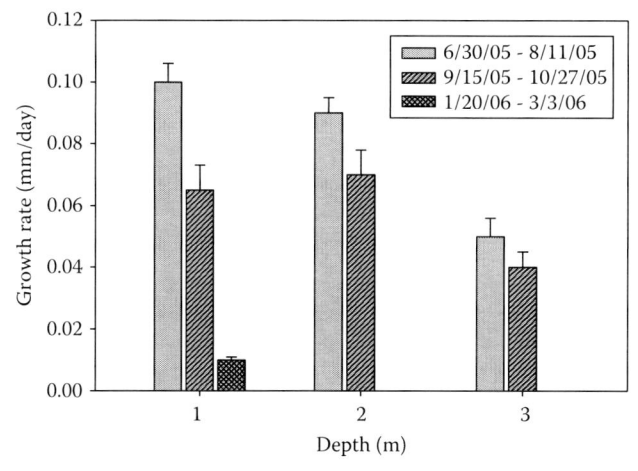

Figure 24.8 Mean (±SE) growth rates of zebra mussels measured from individuals chambered at various locations and depths in Oologah Lake, Oklahoma (top panel): RB, Redbud Marina; BC = Blue Creek, HB = Hawthorn Bluff. Growth rates of mussels suspended in water column at three depths (1, 2, and 3 m) during three different periods in 2005 and 2006 at Redbud Marina (bottom panel).

Table 24.1 Mean (±SD) Secchi-Disk Depth (m) in Oologah Lake during Three Different Periods: Pre-Zebra Mussel Invasion/Low Densities (2000–2002), High Zebra Mussel Densities (2003–2006), and Post-Zebra Mussel Die-Off/Flooding (2007–2010). Differences between Periods Tested with a T-Test

Period	Event	Secchi Depth	Comparison	P-Value
2000–2002	Preinvasion/low densities	0.44 ± 0.17	2000–2002 < 2003–2006	<0.001
2003–2006	High densities	0.99 ± 0.38	2003–2006 = 2007–2010	0.065
2007–2010	Post-die-off/flooding	0.84 ± 0.33	2000–2002 < 2007–2010	<0.001

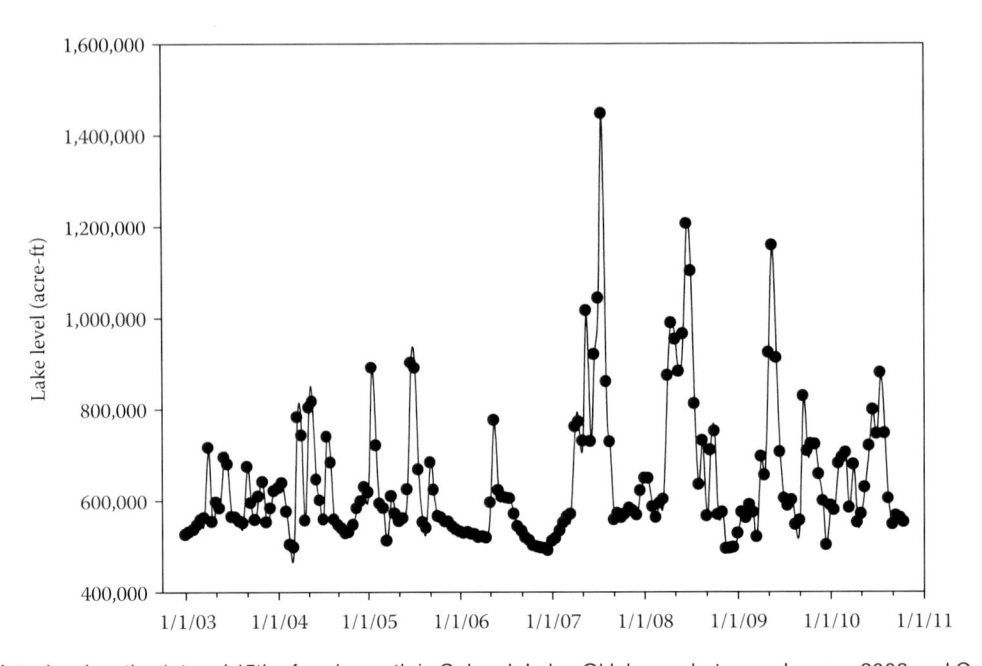

Figure 24.9 Water level on the 1st and 15th of each month in Oologah Lake, Oklahoma, between January 2003 and October 2010.

the course of the study, with conductivity ranging between 1.3 and 2.1 mS/cm. Alkalinity and hardness in Sooner Lake were indicative of hard to very hard water with alkalinity ranging between 100 and 200 mg/L as $CaCO_3$ and hardness between 100 and 240 mg/L as $CaCO_3$.

Zebra mussels were first reported in Sooner Lake in 2006, and peak veliger densities generally increased thereafter from 150/L in 2007 to 580/L in 2008, 350/L in 2009, and 600/L in 2010 (Figure 24.10). The seasonal timing of zebra mussel reproduction at sites away from the heated discharge in Sooner Lake (dam, intake buoy, and plant intake) was generally similar to the seasonal timing in Oologah Lake, with veligers first observed in the pelagic zone in May and peak densities observed in June. However, reproduction occurred earlier at sites nearest the discharge (discharge bridge, discharge buoy, and end discharge sites), with veligers being observed in April and peak densities occurring in May (Figure 24.10).

In contrast to Oologah Lake, densities of settled veligers on glass slides in Sooner Lake decreased each year after first introduction, from a peak of 1.7 million/m^2 in 2007 to a peak of less than 100,000/m^2 in 2010 (Figure 24.11). Peak settling was usually observed in late-June or July, which was 2–3 weeks after peak densities of veligers in the water column. Densities of settled veligers at sites within the discharge channel were usually lower compared to densities at sites located further from the warmwater discharge.

Densities of zebra mussels on concrete panels peaked at 150,000/m^2 in 2007, 50,000/m^2 in 2008, 60,000/m^2 in 2009, and 30,000/m^2 in 2010 (Figure 24.12). At the discharge-buoy location, mean densities never exceeded 10,000/m^2, and were 0/m^2 by July each year; however, at the end of the discharge channel (i.e., end discharge site), densities reached a peak of 60,000/m^2 in 2009 (Figure 24.12). Away from the discharge channel, peak densities recorded in July and August were associated with settlement and growth of young-of-the-year mussels.

Growth experiments conducted in Sooner Lake indicate water temperatures >32°C were lethal to transplanted mussels. Maximum growth rates exceeded 0.1 mm/day in June 2008 and October 2009 at the warmest site (discharge buoy, Figure 24.13). Mean temperatures during these periods were 27.3°C and 25.6°C, respectively. Growth experiments conducted in June–September 2007, July–September 2008, and September 2009 were not successful as mussels died within one week of deployment. Mean temperatures during these unsuccessful mussel growth periods ranged from 32.1°C to 35.3°C. Maximum winter growth rates were 0.05 mm/day in January 2009 when the mean temperature was 16.1°C during the growth period.

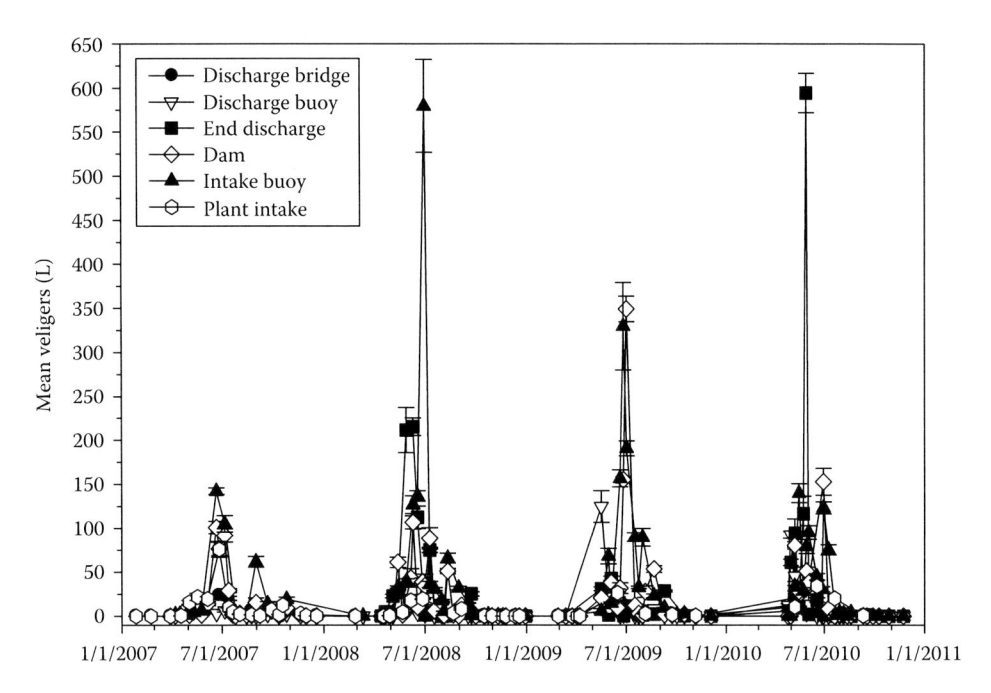

Figure 24.10 Mean (±SE) veliger densities at six sampling sites in Sooner Lake, Oklahoma, from January 2007 to October 2010.

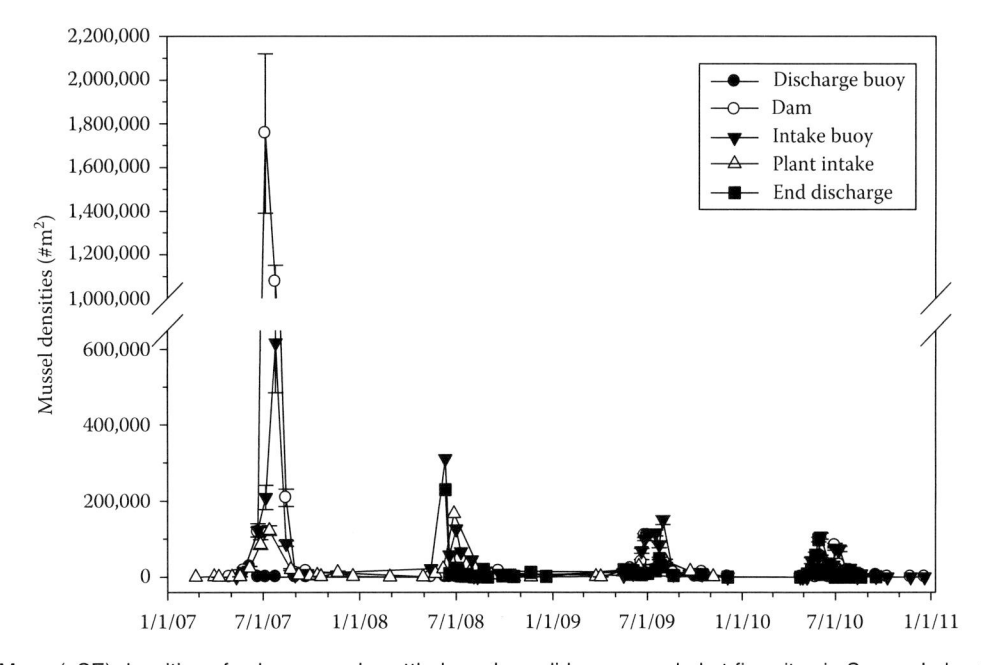

Figure 24.11 Mean (±SE) densities of zebra mussels settled on glass slides suspended at five sites in Sooner Lake, Oklahoma, from February 2007 to September 2010.

Maximum growth rates observed at the intake buoy (site farthest from the warmwater discharge) exceeded 0.11 mm/day in June and September 2007 when mean water temperatures were 19.8°C and 28.0°C, respectively (Figure 24.13). Lowest growth rates of less than 0.04 mm/day occurred in January and May 2009 when mean water temperatures were 10.0°C and 10.4°C, respectively. Mussels at the intake buoy did not survive in September and October 2009 when associated mean water temperatures were 27.3°C and 21.6°C, respectively.

DISCUSSION

Oologah Lake

From 2003 to 2006, the zebra mussel population in Oologah Lake expanded nearly exponentially. Peak veliger densities generally occurred in late May or June, which contrasts to the Great Lakes where peak densities of veligers occurred in July and August during the initial expansion of mussel populations (Fraleigh et al. 1993, Garton and Haag 1993, Smit et al. 1993).

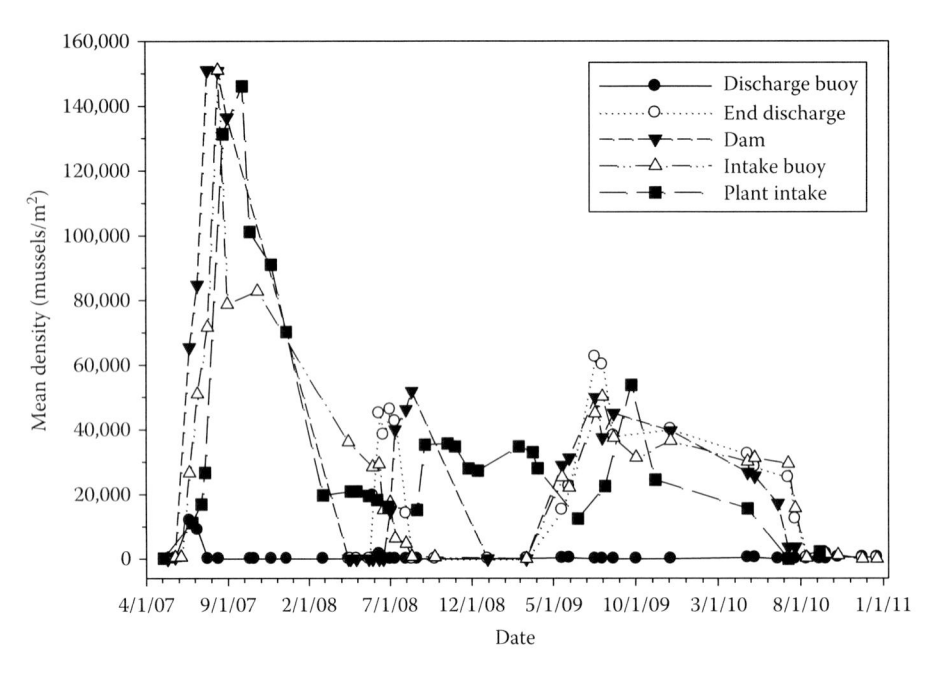

Figure 24.12 Mean densities of zebra mussels on concrete panels suspended at five sites in Sooner Lake, Oklahoma, from April 2007 to September 2010.

In 2004, however, peak veliger densities in Oologah Lake occurred in August when the mean water temperature was 2°C cooler in July and August than in any other year of study. This supports conclusions of McMahon (1996) that water temperatures play an important role in zebra mussel reproduction and population dynamics. Apparently, relatively mild temperatures in 2004 may have provided low enough temperatures to allow mussels to reproduce well into the summer months. Further support for this was provided in 2006 when water temperatures remained below 30°C until August. Veliger densities in 2006 peaked at 480 veligers/L, similar to maximum densities observed in the Great Lakes (Fraleigh et al. 1993, Garton and Haag 1993).

While the specific sequence of events that initiated the 2006 die-off of zebra mussels in Oologah Lake are not explicitly clear, a combination of factors may have played a role. Water temperature in 2006 reached a maximum of 30°C, but this temperature was not greater than maximum temperatures in 2003 and 2005; therefore, it is unlikely that high temperatures alone caused the die-off of mussels in 2006. Unlike other years, 2006 was characterized by low water levels and high zebra mussel densities. Under low-flow conditions, Burks et al. (2002) showed nitrate concentrations to be greater at the base of aggregations of zebra mussels (druses) than at the surface of druses. In addition, they found dissolved oxygen concentrations were lower at the base of druses relative to overlying waters. Similarly, Tuchman et al. (2004) demonstrated zebra mussels on the interior of these druses experience reduced food availability under low-flow conditions. As a result, mussels in the interior of druses in Oologah Lake may have experienced mortality due to low dissolved oxygen, low food resources, and high nitrate concentrations, as well as exposure to lethal temperatures near 30°C. Aldridge et al.

(1995) and McMahon et al. (1995) have shown zebra mussels are sensitive to temperatures above 28°C, with death occurring in a matter of hours when exposed to 30°C. As zebra mussels on the interior of these druses begin to expire, decomposition by-products such as ammonia and nitrite may have increased, particularly under low-flow conditions, and resulted in further declines in water quality within druse interiors. In experiments conducted with *Corbicula* sp., Cherry et al. (2005) and Cooper et al. (2005) showed ammonia spikes within clusters may result in a cascade of changes that increased mortality of individuals in the surrounding environment.

Die-off events of mussels are not unique to reservoirs in Oklahoma, or even to mussel populations located in more temperate climates. Stańczykowska and Lewandowski (1993) noted large-scale and rapid reductions in zebra mussel densities in Mazurian Lakes in northeastern Poland. They noted these events occurred in years after relatively high zebra mussel densities and particularly in eutrophic systems. They also noted some populations later rebounded to previously recorded densities, while in other systems populations were eventually extirpated. Allen et al. (1999) showed that summer die-offs of zebra mussels in the lower Mississippi River were associated with high water temperatures. While the population in Oologah Lake was not extirpated, it remained at very low levels from 2007 through 2010. After near-drought conditions in late 2006, record floods occurred in 2007, 2008, and 2009. These high-water events occurred in June, when peak reproduction occurred in three of the four previous years. While there appears to be little published information available on effects of floods on zebra mussels, these high-water events may have flushed viable veligers out of the reservoir or could have caused veligers settled on structures high in riparian areas to become

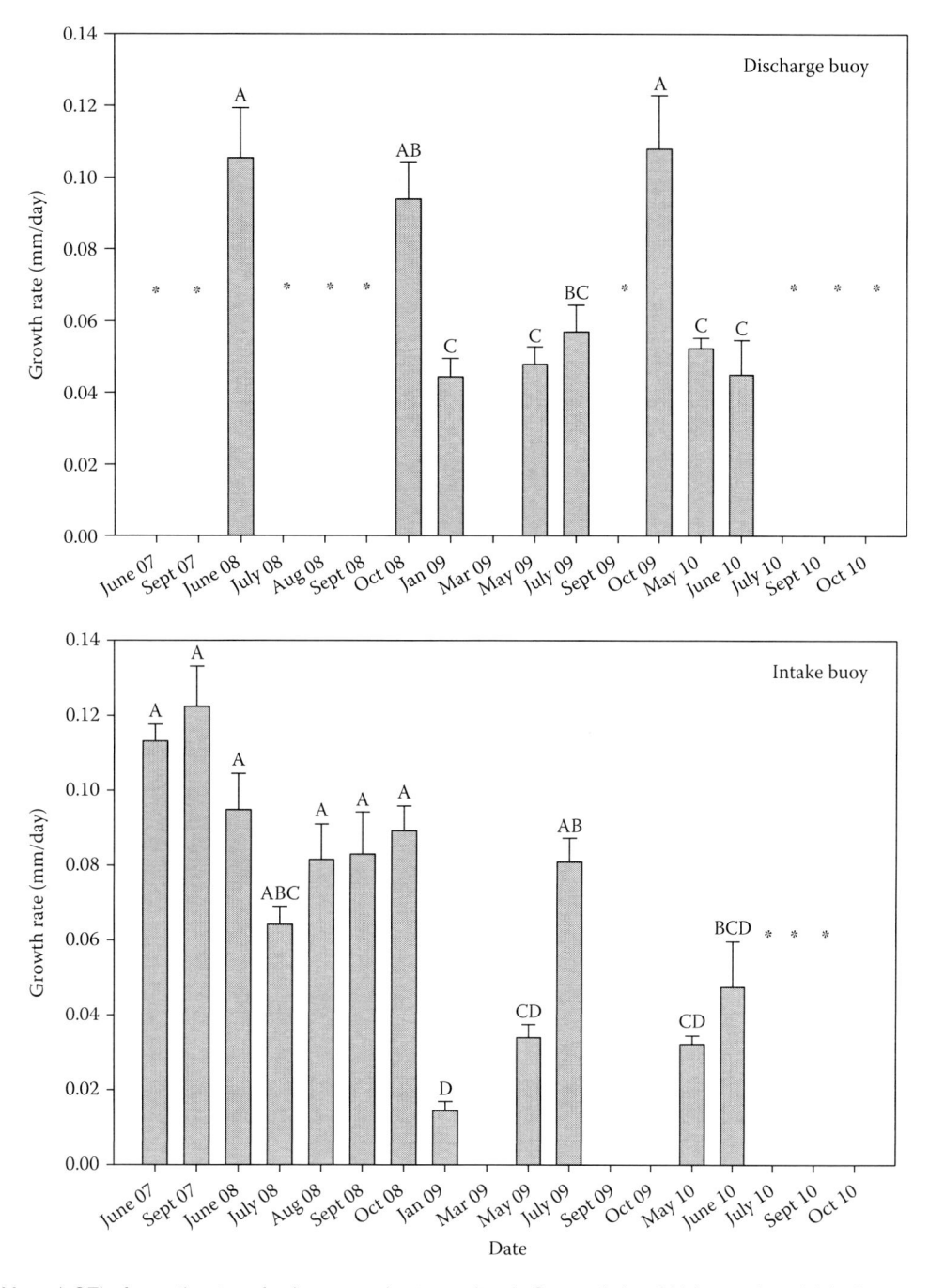

Figure 24.13 Mean (±SE) of growth rates of zebra mussels at two sites in Sooner Lake, Oklahoma. Asterisk indicates death of mussels prior to the end of the growth trial. Bars sharing a common letter are not significantly ($P < 0.05$) different.

exposed to dry conditions once waters receded. Between 2007 and 2009, adults were undetectable and veliger densities were below 1/L until 2010 when adult zebra mussels were once again found at Redbud Marina and Blue Creek and veligers peaked at 1/L. This 3-year-recovery period appears to have resulted from the die-off initiated in 2006 combined with high water levels in peak-reproductive months in 2007, 2008, and 2009. The reduction in water transparency after the die-off in 2006 may have resulted from either flooding events in 2007, 2008, and 2009 or minimal impacts of filtration because of such low densities.

Maximum growth rates of mussels (5–11 mm initial length) in 2005 approached 0.14 mm/day at three of the four sites. These estimates are somewhat greater than previously published growth rates ranging between 0.05 and 0.1 mm/day (Dorgelo 1993, Allen et al. 1999, Karatayev et al. 2006). In addition, zebra mussel growth rates in our study were determined for individual caged mussels that have been shown to decrease growth (Karatayev et al. 2006). Zebra mussels deployed at 3 m depth also exhibited lower growth rates when compared with mussels suspended at 1 and 2 m, which is consistent with data reviewed by Karatayev et al. (2006) who suggest reduced temperature and food availability may explain decreased growth in deeper waters.

While long-term population trends (i.e., survival, growth, and reproduction) of zebra mussels in Oologah Lake are still uncertain, Strayer and Malcom (2006) suggested disturbance may impact population trends of zebra mussels. They state that while regular disturbance tends to stabilize mussel populations, severe and irregular disturbances (such as flooding) may contribute to more variable population trends. Given the inherent variability of reservoir water levels, particularly reservoirs used for flood control, zebra mussel population dynamics may continue on a boom–bust cycle not only in Oologah Lake but also in other reservoirs in the region.

Sooner Lake

Peak veliger densities rapidly increased from 150/L in 2007 to over 500/L in 2008 and 2010. However, these peaks were of short duration, lasting only 1–2 weeks as found in Oologah Lake. Peak densities were similar to those observed in the Great Lakes during the early invasion period (Fraleigh et al. 1993, Garton and Haag 1993). Over a seasonal period, veligers generally first appeared at sites located within the warmwater discharge zone (discharge bridge and discharge buoy). Peak veliger densities usually occurred at the cool-water sites (dam and intake buoy) several weeks after peak densities at warmwater sites. Veliger densities in the discharge zone never exceeded 100/L because adult zebra mussel densities in this area remained low (0–10,000/m^2).

Water temperatures at sites in the discharge zone exceeded 30°C in July of all years, and adult mussels were routinely extirpated from these sites from July through September. After water temperatures declined to below 30°C in the fall, young zebra mussels settled on concrete panels and firm substrates within the discharge zone. Winter water temperatures at the discharge buoy generally ranged between 3°C and 10°C, which allowed young newly settled mussels to grow through much of the winter and reach reproductive maturity by spring. This cycle of repeated extirpations and reintroductions in the discharge zone caused zebra mussel densities to remain low (compared to the cool-water sites) throughout the study period. Cyclical extirpation and reintroduction at warmwater sites was also described by Sinicyna and Zdanowski (2007a,b) at the Konin heated-lakes complex in central Poland.

At sites with ambient temperatures, densities of adult zebra mussels attached to concrete panels peaked near 150,000/m^2 in 2007, and subsequently, the population stabilized near 40,000/m^2 between 2008 and 2010. Isolated die-offs of older zebra mussels were noted at the intake buoy, dam, and end discharge during July and August, associated with water temperatures near 30°C at these locations. These die-offs were restricted to mussels greater than 15 mm in length, while the young-of-the-year mussels were largely able to withstand these high water temperatures. Again, adult densities and juvenile domination during summer are similar to patterns observed in the Konin heated-lakes complex in Poland (Sinicyna and Zdanowski 2007a,b).

Growth rates of zebra mussels varied both seasonally and among sampling sites. At the discharge buoy, maximum growth rates of 0.1 mm/day were achieved in periods that ended in June 2008 and October 2009 with associated mean temperatures of 27.3°C and 25.6°C, respectively. Growth in winter of 2008–2009 was 0.04 mm/day with a mean water temperature of 16°C during the period. During each of the failed attempts to measure growth rates because of poor survival, mean water temperature ranged between 32°C and 35°C. At the intake buoy site, maximum growth rates were slightly greater than in the discharge zone, reaching 0.12 mm/day with a mean temperature of 28°C. However, experiments in 2008 and 2009 resulted in maximum growth rates of only 0.09 mm/day. Karatayev et al. (2006, and references therein) found that growth rates of zebra mussels in lakes varied significantly between years, perhaps brought about by different environmental conditions such as temperature, food availability, and suspended solids.

While isolated die-offs of zebra mussels greater than 15 mm were noted at all locations within Sooner Lake, there was no large-scale die-off similar to what occurred in Oologah Lake in 2006. In Sooner Lake, small-scale die-offs occurred in association with the warmest temperatures in July and August of each year. The largest of these die-offs occurred in 2010, associated with water temperatures that exceeded 30°C at sites farthest from the heated discharge. Also, during the 2010 die-off, the lake stratified because of an uncharacteristic period of relatively low winds and little wind-driven mixing. Dissolved oxygen values in the hypolimnion ranged between 1 and 2 mg/L during the 7 week period in July–August 2010 that may have contributed to the die-off.

As Aldridge et al. (1995) described, given lower metabolic rates of large organisms compared with smaller ones, it seems counterintuitive that large zebra mussels would be more negatively affected by high temperatures than small mussels. However, McMahon (1996) found that large zebra mussels were less tolerant of high temperatures than smaller mussels. This was supported by our observations in Sooner Lake that large mussels were more susceptible to high temperatures compared to small mussels. Although metabolic rates are lower in large mussels as compared to small, energy allocation and energetic demands are quite different.

Sprung (1991, 1993) showed zebra mussels can lose as much as 30% of their body weight after reproduction. Furthermore, Aldridge et al. (1995) demonstrated zebra mussels, 13–17 mm in length, enter negative growth at water temperatures above 28°C because energy lost through metabolic demands exceeded energy gained through food resources. Zebra mussels in these Oklahoma reservoirs reproduce in May or June when lower water temperatures are optimal, but are then exposed to temperatures above their 28°C threshold in July and August. As mature mussels allocate energy toward reproduction rather than somatic growth, stored energy reserves may be inadequate to deal with high water temperatures that generally occurred in July and August in Oklahoma. This hypothesis is further supported by Stoeckmann and Garton (1997, 2001) who found large mussels did not allocate energy to growth and reproduction equally. When exposed to near lethal temperatures, mussels sacrificed somatic growth for reproduction that caused a net negative growth during summer months. Small zebra mussels, on the other hand, allocate no energy for reproduction, but devote most energy to somatic growth and metabolic demands. This may allow them extra energy reserves needed to cope with high water temperatures.

CONCLUSION

Since Oklahoma is positioned near the predicted southern boundary of zebra mussel distributions in North America (Strayer 1991), the present study provided an opportunity to evaluate population dynamics of this organism under environmental conditions that may cause stress during certain times of the year, especially in relation to maximum lethal temperatures. Characterizing zebra mussel population dynamics in Oklahoma reservoirs may allow for a re-evaluation of the southern limit for mussel dispersal. Zebra mussel die-offs occurred in the summer months when temperatures reached and exceeded 30°C. An extended recovery phase of the zebra mussel population in one of the reservoirs (Oologah Lake) appeared to have been brought about by one such die-off, in combination with flood events in three subsequent years. These results indicate that temperature may serve as an important limiting variable for zebra mussel populations in southern North America by affecting adult survival and skewing populations toward juvenile dominance, especially during summer months. With the 2009 discovery of zebra mussels in Lake Texoma on the Oklahoma–Texas border, it is clear zebra mussels will continue to spread to yet warmer habitats, and continued examination of population dynamics under natural conditions will provide further insights into temperature limitations of this species.

ACKNOWLEDGMENTS

The authors would like to thank the City of Tulsa, Oklahoma Gas and Electric Company, U.S. Army Corps of Engineers, and the Ecotoxicology and Water Quality Research Laboratory at Oklahoma State University for providing funding for the various studies outlined here. Everett Laney, U.S. Army Corps of Engineers, Tulsa District, and Jason Goeckler, Kansas Department of Wildlife and Parks, greatly contributed to various zebra mussel-related ideas and projects conducted in the region. Two anonymous reviewers also provided excellent revisions to this chapter.

REFERENCES

Aldridge, D.W., B.S. Payne, and A.C. Miller. 1995. Oxygen consumption, nitrogenous excretion, and filtration rates of *Dreissena polymorpha* at acclimation temperatures between 20 and 32°C. *Can. J. Fish. Aquat. Sci.* 52: 1761–1767.

Allen, Y.C., B.A. Thompson, and C.W. Ramcharan. 1999. Growth and mortality rates of the zebra mussel, *Dreissena polymorpha*, in the Lower Mississippi River. *Can. J. Fish. Aquat. Sci.* 56: 748–759.

Angyal, R., R. Glass, and O.E. Maughan. 1987. The characteristics of the white crappie population of Sooner Lake, Oklahoma. *Proc. Okla. Acad. Sci.* 67: 1–10.

Bidwell, J.R. 2010. Range expansion of the Zebra Mussel, *Dreissena polymorpha*: A review of major dispersal vectors in Europe and North America. In *The Zebra Mussel in Europe*, G. Van der Velde, S. Rajagopal, and A. Bij de Vaate, eds., pp. 69–78. Leiden, The Netherlands: Backhuys Publishers.

Burks, R.L., N.C. Tuchman, C.A. Call, and J.E. Marsden. 2002. Colonial aggregates: Effects of spatial position on zebra mussel responses to vertical gradients in interstitial water quality. *J. N. Am. Benthol. Soc.* 21: 64–75.

Cherry, D.S., J.L. Scheller, N.L. Cooper, and J.R. Bidwell. 2005. Potential effects of Asian clam (*Corbicula fluminea*) die-offs on native freshwater mussels (Unionidae) I: Water-column ammonia levels and ammonia toxicity. *J. N. Am. Benthol. Soc.* 24: 369–380.

Cooper, N.L., J.R. Bidwell, and D.S. Cherry. 2005. Potential effects of Asian clam (*Corbicula fluminea*) die-offs on native freshwater mussels (Unionidae) II: Porewater ammonia. *J. N. Am. Benthol. Soc.* 24: 381–394.

Dorgelo, J. 1993. Growth and population structure of the zebra mussel (*Dreissena polymorpha*) in Dutch Lakes differing in trophic state. In *Zebra Mussels Biology, Impacts, and Control*, T.F. Nalepa and D.W. Schloesser, eds., pp. 79–94. Boca Raton, FL: CRC Press.

Fraleigh, P.C., P.L. Klerks, G. Gubanich, G. Matisoff, and R.C. Stevenson. 1993. Abundance and settling of zebra mussel (*Dreissena polymorpha*) veligers in western and central Lake Erie. In *Zebra Mussels Biology, Impacts, and Control*, T.F. Nalepa and D.W. Schloesser, eds., pp. 129–142. Boca Raton, FL: CRC Press.

Garton, D.W. and W.R. Haag. 1993. Seasonal reproductive cycles and settlement patterns of *Dreissena polymorpha* in western Lake Erie. In *Zebra Mussels Biology, Impacts, and Control*. T.F. Nalepa and D.W. Schloesser, eds., pp. 111–128. Boca Raton, FL: CRC Press.

Havel, J.E., C.E. Lee, and M.J. Vander Zanden. 2005. Do reservoirs facilitate invasions into landscapes? *BioScience* 55: 518–525.

Hernandez, M.R., R.F. McMahon, and T.H. Dietz. 1995. Investigation of geographic variation in the thermal tolerance of zebra mussels, *Dreissena polymorpha*. *Proceedings of the Fifth International Zebra Mussel and Other Aquatic Nuisance Organisms Conference*, pp. 195–209. Toronto, Ontario, Canada.

Johnson, L.E. 1995. Enhanced early detection and enumeration of zebra mussel (*Dreissena* spp.) veligers using cross-polarized light microscopy. *Hydrobiologia* 312: 139–146.

Karatayev, A.Y., L.E. Burlakova, and D.K. Padilla. 2006. Growth rate and longevity of *Dreissena polymorpha* (Pallas): A review and recommendations for future study. *J. Shellfish Res.* 25: 23–32.

Laney, E.E. 2010. Zebra mussel (*Dreissena polymorpha*). Report to United States Corps of Engineers, Tulsa District, Operations Division, Tulsa, OK.

McMahon, R.F. 1996. The physiological ecology of the zebra mussel, *Dreissena polymorpha*, in North American and Europe. *Am. Zool.* 36: 339–363.

McMahon, R.F., M.A. Matthews, T.A. Ussery, R. Chase, and M. Clarke. 1995. Studies of heat tolerance of zebra mussels: Effects of temperature acclimation and chronic exposure to lethal temperatures. United State Army Corps of Engineers Technical Report EL-95-9. Washington, DC.

McMahon, R.F. and T.A. Ussery. 1995. Thermal tolerance of zebra mussels (*Dreissena polymorpha*) relative to rate of temperature increase and acclimation temperature. United States Army Corps of Engineers Technical Report EL-95-10. Washington, DC.

Nichols, S.J. 1996. Variations in the reproductive cycle of *Dreissena polymorpha* in Europe, Russia and North America. *Am. Zool.* 36: 311–325.

OWRB. 1990. *Oklahoma Water Atlas*. Oklahoma Water Resources Board, Oklahoma City, OK.

OWRB. 2007. Report of the Oklahoma Lakes Beneficial Use Monitoring Program. Available online at www.owrb.ok.gov/quality/monitoring/bump.php (accessed December 1, 2010).

Sinicyna, O.O. and B. Zdanowski. 2007a. Development of the zebra mussel, *Dreissena polymorpha* (Pall.), population in a heated lakes ecosystem. I. Changes in population structure. *Arch. Pol. Fish.* 15: 369–385.

Sinicyna, O.O. and B. Zdanowski. 2007b. Development of the zebra mussel, *Dreissena polymorpha* (Pall.), population in a heated lakes ecosystem. II. Life strategy. *Arch. Pol. Fish.* 15: 387–400.

Smit, H., A. bij deVaate, H.H. Reeders, E.H. van Nes, and R. Noordhuis. 1993. Colonization, ecology, and positive aspects of zebra mussels (*Dreissena polymorpha*) in The Netherlands. In *Zebra Mussels Biology, Impacts, and Control*, T.F. Nalepa and D.W. Schloesser, eds., pp. 55–78. Boca Raton, FL: CRC Press.

Sprung, M. 1991. Costs of reproduction: A study on metabolic requirements of the gonads and fecundity of the bivalve *Dreissena polymorpha*. *Malacologia* 33: 63–70.

Sprung, M. 1993. The other life: An account of present knowledge of the larval phase of *Dreissena polymorpha*. In *Zebra Mussels Biology, Impacts, and Control*, T.F. Nalepa and D.W. Schloesser, eds., pp. 39–54. Boca Raton, FL: CRC Press.

Stańczykowska, A. and K. Lewandowski. 1993. Thirty years of studies of *Dreissena polymorpha* ecology in Mazurian Lakes of Northeastern Poland. In *Zebra Mussels Biology, Impacts, and Control*, T.F. Nalepa and D.W. Schloesser, eds., pp. 3–38. Boca Raton, FL: CRC Press.

Stoeckmann, A.M. and D.W. Garton. 1997. A seasonal energy budget for zebra mussels (*Dreissena polymorpha*) in western Lake Erie. *Can. J. Fish. Aquat. Sci* 54: 2743–2751.

Stoeckmann, A.M. and D.W. Garton. 2001. Flexible energy allocation in zebra mussels (*Dreissena polymorpha*) in response to different environmental conditions. *J. N. Am. Benthol. Soc.* 20: 486–500.

Strayer, D.L. 1991. Projected distribution of the zebra mussel, *Dreissena polymorpha* in North America. *Can. J. Fish. Aquat. Sci.* 48: 1389–1395.

Strayer, D.L. and H.M. Malcom. 2006. Long-term demography of a zebra mussel (*Dreissena polymorpha*) population. *Freshwat. Biol.* 51: 117–130.

Tuchman, N.C., R.L. Burks, C.A. Call, and J. Smarrelli. 2004. Flow rate and vertical position influence ingestion rates of colonial zebra mussels (*Dreissena polymorpha*). *Freshwat. Biol.* 49: 191–198.

USACE. 2002. Oologah Lake, Oklahoma watershed study, year 2 interim report of findings: October 2000–September 2001. Report prepared by United States Army Corps of Engineers, Tulsa District, Tulsa, OK.

Zar, J.H. 1999. *Biostatistical Analysis*, 4th Edn. Upper Saddle River, NJ: Prentice Hall.

CHAPTER 25

Limiting Environmental Factors and Competitive Interactions between Zebra and Quagga Mussels in North America

David W. Garton, Robert McMahon, and Ann M. Stoeckmann

CONTENTS

Abstract 383
Introduction 384
Abiotic Factors Influencing Mussel Distributions 384
    Temperature 386
    Calcium 388
    Salinity 389
    pH 390
    Hypoxia 390
    Desiccation and Freezing 391
    Currents and Agitation 392
    Turbidity 392
    Substrate Requirements 393
Biotic Interactions 394
    Diet 394
    Physiological Energetics 395
    Starvation 396
    Longevity 397
Summary 398
References 398

ABSTRACT

Similarities in autecology and niche overlap of the zebra mussel (*Dreissena polymorpha*) and a congener, the quagga mussel (*Dreissena rostriformis bugensis*), predict the potential for strong interspecific competition. During the initial invasion of dreissenids in North America, the zebra mussel dominated; however, subsequently quagga mussels essentially replaced zebra mussels and are now the dominant dreissenid in many benthic habitats. While the range and population densities of quagga mussels increased slowly in North America, this species extended its range faster in the western portions than in the eastern portions of the United States. This chapter reviews studies of abiotic and biotic factors that affect competitive interactions between quagga and zebra mussels.

While differences in tolerance to abiotic factors appear minor, there appear to be appreciable differences in physiological energetics and life history traits between quagga and zebra mussels, which infer that quagga mussels have more "K-strategy" traits (higher competitive advantage but lower invasive potential), whereas zebra mussels have more "r-strategy" traits (greater initial invasive potential but lower competitive ability). Given high niche overlap, complete replacement of one species by the other is expected, but a dynamic coexistence of these species is common. Coexistence can occur when minor differences exist between two species or if those factors responsible for dominance of one species over the other show temporal and/or spatial variation. The complex and dynamic nature of freshwater microhabitats likely contributes to the relatively stable coexistence of these two dreissenid species in many areas of North America.

INTRODUCTION

The capacity of invasive dreissenid mussels, the zebra mussel *Dreissena polymorpha* and a congener, the quagga mussel *Dreissena rostriformis bugensis*, to spread and become abundant was recognized immediately following their initial discovery in North America in the late 1980s (Hebert et al. 1989, Mackie et al. 1991). Primary factors that allow zebra and quagga mussels to dominate in benthic freshwater ecosystems reflect features of dreissenid life history traits such as high genetic variation, short generation times, high fecundity, a dispersive larval stage, and broad tolerance ranges of environmental factors. These traits, coupled with minimal predation, resulted in rapid and explosive population growth of zebra mussels in invaded lakes and rivers of North America in the early 1990s. Following the discovery of the zebra mussel, the confirmation of a second dreissenid mussel, the quagga mussel, in North America generated interest into possible ecological interactions between the two species (Mills et al. 1996).

Initial studies on zebra and quagga mussels in North American suggested that ecological specialization would lead to limited distributional overlap and hence limit interspecific competition (e.g., Dermott et al. 2003). In Lakes Erie and Ontario, quantitative sampling revealed that shallow waters, with extensive hard substrata and typically less than 10 m depth, were dominated by zebra mussels, whereas deeper water habitats characterized by soft/fine sediments were colonized almost exclusively by quagga mussels (Dermott and Munawar 1993, Dermott and Kerec 1997, Diggins et al. 2004). Reasons proposed for ecological separation of these two species included quagga mussels were more sensitive to broader thermal regimes and/or higher-energy environments characteristic of shallow-water habitats, and zebra mussel larvae could not survive the lengthy time period necessary to colonize deepwater habitats and live on soft sediment habitats (e.g., Martel et al. 2001). However, as time progressed it was recognized that zebra mussels in shallow water habitats were being displaced by quagga mussels and that zebra mussels as well as quagga mussels could colonize soft sediments (e.g., Ricciardi and Whoriskey 2004). Thus, ecological interactions between these two species were more intense and complex than initially recognized.

The potential for intense competition between zebra and quagga mussels is a direct consequence of shared evolutionary history and biogeography and nearly identical morphological and autecological traits; that is, niche overlap in these two species is very broad (Orlova et al. 2005). Both are similar in size and appear to have the same dietary and substrate requirements, reproductive and dispersive patterns, and roughly equal vulnerabilities to predation and disease. The potential for quagga mussels to ultimately displace zebra mussels was noted when *D. r. bugensis*, native to the Bug River, invaded the Dnieper River across the Dnieper–Bug River liman (estuary) beginning in the 1950s and in subsequent decades had essentially completely replaced *D. polymorpha* in the lower reaches of the Dnieper River (Mills et al. 1996, Orlova et al. 2004, 2005). However, it remains unclear exactly which specific factor(s), acting alone or in combination, resulted in competitive superiority of quagga mussels.

Multiple factors can lead to competitive dominance by one species over another. In the case of sessile dreissenid mussels, factors include different limits in adult and/or larval stage tolerance to abiotic (physical and chemical) factors (e.g., temperature, dissolved oxygen, pH, turbidity), minimum requirements for dissolved ions (e.g., calcium, hardness), substrate type and stability, and current velocity. Potential biotic factors include measures of fitness, such as life span, fecundity, growth rate, age at maturity, differential survival in response to predation and disease, specific dietary requirements, physiological efficiency, and direct interspecific competition for available substrate. These possible explanations for dominance are not mutually exclusive, are likely to be acting in concert, and may differ in relative importance over temporal and spatial scales.

In this chapter we review studies that investigate abiotic and biotic interactions of zebra and quagga mussels and summarize general traits that may lead to predictions of superiority of one species over the other in regions of sympatry.

ABIOTIC FACTORS INFLUENCING MUSSEL DISTRIBUTIONS

Because zebra and quagga mussels are closely related, they have relatively similar adaptations to abiotic environmental factors such as temperature, hypoxia/anoxia, calcium concentration/hardness, salinity, pH, emersion, current, agitation, turbidity, and substratum (McMahon 1996; Table 25.1). Since *D. polymorpha* has been widespread in western Europe for over 100 years (Mackie and Schloesser 1996, Karatayev et al. 1998, 2007, Minchin et al. 2002, Orlova et al. 2005) and initially dominant and widespread relative to *D. r. bugensis* in North America (Orlova et al. 2005, United States Geological Survey 2010a,b), it has been the far more intensively studied species on both sides of the Atlantic Ocean. Thus, generally far less information exists for the quagga mussel. In addition, there is little comparative information on the biology and ecology of both species

Table 25.1 Comparison of Adaptations to Abiotic Factors by Zebra (*D. polymorpha*) and Quagga Mussels (*D. r. bugensis*)

Physical Factor	*D. polymorpha*	*D. r. bugensis*	Comment	References
Incipient upper thermal limit	29°C–32°C in N. America	28°C in N. America	Can vary based on individual size and nutritional condition, prior temperature experience, and season of the year. Thermal tolerance appears subject to selection by elevated temperatures. Likely to be greater for populations at lower latitudes and in water bodies receiving thermal effluents.	Elderkin and Klerks (2005), Hernandez (1995), Iwanyzki and McCauley (1993), Karateyev et al. (1998), Morse (2009), Spidle et al. (1995)
Incipient lower thermal limit	0°C	0°C	Temperate species that survive over winter in iced over water bodies in the northern portions of their range.	Karateyev et al. (1998), McMahon (1996), McMahon et al. (1995)
Spawning temperature	12°C–24°C	9°C–24°C	Spawning is maximized in *D. polymorpha* $\geq$18°C–20°C that is also likely to be the case in *D. r. bugensis*.	Claxton and Mackie (1998), Karateyev et al. (1998)
Temperature for larval development	12°C–24°C	unknown	The lower limit may be reduced for *D. r. bugensis* but requires experimental confirmation.	Claxton and Mackie (1998), Sprung (1987)
Calcium	<8–12 mg Ca^{2+} l^{-1}	<12 mg Ca^{2+} l^{-1}	Based primarily on presence/absence data in the St. Lawrence River.	Hincks and Mackie (1997), Mellina and Rasmussen (1994), Jones and Ricciardi (2005)
Salinity	<6–12 psu Byssogenesis inhibited at $\geq$4 psu	6–8 psu	Based on laboratory determinations and presence/absence data. Estimates are highly variable among studies.	Kilgour et al. (1994), Orlova et al. (1998), Rajagopal et al. (1996), Strayer and Smith (1993), Walton (1996), Wilcox and Dietz (1998), Wright et al. (1996)
pH	Range = 6.0 to 8.5–9.6 pH 7.4–9.4 for larval development	Unknown	Ca^{2+} and other ion concentrations may impact the tolerated pH range.	Bowman and Bailey (1998), Hincks and Mackie (1997), Sprung (1987)
Hypoxia	P_{O_2} > 2.13 kPa P_{O_2} > 4.3 kPa for larval development Byssogenesis inhibited at $P_{O_2} \leq$ 2.1 kPa	P_{O_2} > 2.13 kPa	Appears to be temperature dependent—tolerance increases at lower temperatures. Likely to be greater in *D. polymorpha*.	Johnson and McMahon (1998), Matthews and McMahon (1999), McMahon and Johnson (unpublished)
Emersion	10.3–27.9 days at 5°C 4.5–12.3 days at 15°C 2.0–5.4 days at 25°C	5–13.5 days at 15°C 3–5 days at 20°C	Emergence tolerance increases with increasing RH.	McMahon et al. (1993), Ussery and McMahon (1995), Ricciardi et al. (1995)
Freezing	>48 h at 0°C to −1.5°C 7.5 h at −3.0°C 4.5 h at −5°C 2.3 h at −7.5°C 2.7 h at −10°C	Unknown	Tissues freeze between −1.8°C and −3.0°C. Freeze tolerance increases in mussel aggregations. Freeze tolerance in *D. r. bugensis* likely to be similar to that of *D. polymorpha* due to similar latitudinal limits but requires study.	McMahon et al. (1993), Paukstis et al. (1996)
Current velocity and agitation	Byssogenesis declines $\geq$27 cycles m s^{-1} and is reduced by agitation $\geq$30 cycles min^{-1}	Filtration inhibited by flows $\geq$ 19 cm s^{-1}	Inhibition or reduction in byssal thread production can lead to mussel dislodgement from the substratum. High current-induced starvation could prevent successful colonization.	Ackerman (1999), Clarke and McMahon (1996a,b), MacIsaac (1996), Rajagopal et al. (1996)
Turbidity	Unknown: likely >100 NTU	Unknown: likely >100 NTU	Further research required to determine incipient upper turbidity limits for both species.	Diggins (2001), Lei et al. (1996), MacIsaac and Rocha (1995), Summers et al. (1996)

(*continued*)

Table 25.1 (continued) Comparison of Adaptations to Abiotic Factors by Zebra (*D. polymorpha*) and Quagga Mussels (*D. r. bugensis*)

Physical Factor	*D. polymorpha*	*D. r. bugensis*	Comment	References
Substratum requirements	Attach to a wide variety of natural and artificial substrata Biofilms stimulate settlement Flows >1–1.5 m s^{-1} prevent settlement	Attach to a wide variety of natural and artificial substrata	Settlement rates are similar on a wide variety of natural substrata including muds and silts. Current negatively impacts settlement rates. Complex or roughened surfaces are preferentially settled. *D. polymorpha* may preferentially settle on submerged macrophytes relative to *D. r. bugensis*. Attachment strength varies among different substrates.	Ackerman et al. (1995), Berkman et al. (1998, 2000), Diggins et al. (2004), Jones and Ricciardi (2005), Karateyev et al. (1998), Kavouras and Maki (2003), Kobak (2001, 2005), Mackie et al. (1991), Marsden and Lansky (2000)
Starvation	100% sample mortality 5°C—945 days (LT$_{50}$—504 days) 15°C—514 days (LT$_{50}$—341 days) 25°C—166 days (LT$_{50}$—118 days)	Unknown	Can survive >75% loss of dry tissue weight. Mortality times likely to be greatly decreased above 25°C. *D. r. bugensis* starvation tolerance likely to be greater than that of *D. polymorpha*.	Chase and McMahon (1995), Chase-Off (1996), Morse (2009), Walz (1978a,b, 1979)

(McMahon 1996, Karatayev et al. 1998, McMahon and Bogan 2001). With establishment of quagga mussels in the southwestern United States and replacement of zebra mussels by quagga mussels in the Great Lakes and St. Lawrence River (United States Geological Survey 2010a,b), undoubtedly more research will be focused on *D. r. bugensis*. However, relatively limited data exist on the abiotic resistance and capacity adaptations of quagga mussels compared to zebra mussels (McMahon 1996, Karatayev et al. 1998, McMahon and Bogan 2001). While both species initially appeared to have similar resistance and capacity adaptations, recent studies suggest that subtle differences do exist that could impact their interspecific competitive success under specific habitat conditions.

Temperature

Temperature tolerance is the one resistance adaptation studied extensively for *D. polymorpha* in both Europe and North America and more recently for *D. r. bugensis* in North America (Karatayev et al. 1998). As both species are adapted to a temperate climate, there is general agreement that their incipient lower thermal limit is 0°C that they tolerate during winter months in the northern portions of their European and North American ranges (McMahon 1996, Karatayev et al. 1998). Unfortunately, use of different assessment procedures has resulted in variable estimates of incipient (i.e., maximum, long-term tolerated limit) upper thermal limits for both species, ranging between 28°C and 36°C (Karatayev et al. 1998). Determination of dreissenid incipient upper thermal limits is influenced by multiple factors. Thus, dreissenid temperature tolerance increases with increased acclimation temperature, decreased latitude, increased nutritional condition, and increased size; tolerance also varies with season of collection (i.e., thermal tolerance is greater in summer-collected individuals even after acclimation to constant temperatures) (Hernandez 1995, Karatayev et al. 1998, Elderkin and Klerks 2005, Morse 2009). Such influences potentially account for the wide variation in published estimates of incipient upper thermal limits for both species.

Zebra mussels collected from the Niagara River, Buffalo, NY, survived exposure to 30°C but experienced 100% mortality (n = 29) after 32.1 days at 31°C (LT$_{50}$ = 13.9 days) and declined exponentially with time to 100% mortality in 1.25 h (LT$_{50}$ = 0.89 h) at 37°C (McMahon and Ussery 1995). Using similar methodology (i.e., individuals exposed to test temperatures ranging from 30°C to 36°C until 100% mortality ensued), an equivalent incipient upper thermal limit of 30°C was estimated for zebra mussels from Lakes Erie and St. Clair acclimated to 25°C (Iwanyzki and McCauley 1993). Using the same methodology, Spidle et al. (1995) found that zebra mussels from Lake Ontario acclimated to 20°C could tolerate 30°C for extended periods. These results suggested that the incipient upper thermal limit of zebra mussels from the Great Lakes region was approximately 30°C. In contrast, Aldridge et al. (1995) maintained zebra mussels from the Niagara River, Buffalo, NY, at 32°C for 43 days on suspensions of dried *Chlorella* sp., suggesting that their incipient upper thermal limit was ≥32°C. More recently, chronic upper thermal limits of zebra mussels from Winfield City Lake, KS, and Hedges Lake, NY, were compared using the methodology of McMahon and Ussery (1995). These two lakes had different thermal regimes, with maximum summer surface water temperatures being 29°C and 24°C at Winfield City Lake and Hedges Lake, respectively. Corresponding incipient upper thermal limits reflected this difference, being 31.7°C for Winfield City Lake and 29.0°C in Hedges Lake. The incipient upper thermal limit for zebra mussels from Winfield City Lake was the highest recorded for this species in North America, apparently a result of natural selection to the elevated thermal regime in the lake (Morse 2009).

Winfield City Lake is a warm, isolated reservoir at the southern edge of the zebra mussel distribution in North America. Similar selection for increased thermal tolerance has been reported for zebra mussel populations in isolated Russian waters warmed above ambient levels by thermal effluents (Karatayev et al. 1998). Selection for thermal-tolerant races of the zebra mussel in warm southwestern U.S. waters could allow this species to invade water bodies at much lower latitudes in North America than previously predicted based on a incipient upper thermal limit of 30°C (Morse 2009).

Thermal tolerance in *D. r. bugensis* has not been as thoroughly investigated as that of *D. polymorpha*. In Europe, the upper thermal limit of quagga mussels from waters receiving a thermal effluent was reported to be 30.5°C (Karatayev et al. 1998). In a comparative study, quagga mussels from Lake Ontario acclimated to 20°C did not survive >14 days at 30°C, while no mortality was recorded among concurrently tested zebra mussels from the same site (Spidle et al. 1995). This suggested that quagga mussels had a lower thermal tolerance than zebra mussels. A low thermal tolerance of quagga relative to zebra mussels was confirmed by tests of acute thermal tolerance (i.e., temperature at death when individuals acclimated to 5°C, 15°C, or 20°C were exposed to temperatures increasing at rates ranging from 0.0167 to 0.2°C min^{-1}). Regardless of acclimation temperature, LT_{50} (i.e., estimated temperature for 50% sample mortality) for quagga mussels was 1.0°C–4.1°C lower than that of zebra mussels, with differences between the two species increasing with decreasing acclimation temperature within the tolerated temperature range (Spidle et al. 1995). When measured using the methodology of McMahon and Ussery (1995), the 28 day incipient upper thermal limit of quagga mussels from the relatively warm waters of Lake Mead (NV and AZ) was 27.2°C, compared to 29.0°C and 31.7°C for zebra mussels from Hedges Lake, NY, and Winfield City Lake, KS, respectively (Morse 2009), which again confirmed a lower thermal tolerance of quagga mussels relative to zebra mussels in North America.

Thus, the evidence suggests that zebra mussels appear to have a 2°C–3°C higher incipient thermal tolerance than quagga mussels in North America, a potential competitive advantage in the warm surface waters of the southern and southwestern United States where ambient summer temperatures reach or exceed 30°C. This is supported by observations that quagga mussels do not compete as well with zebra mussels in waters with elevated temperatures. In the Chernobyl Power Station cooling reservoir on the lower Pripyat River, a tributary of the Dnieper River, Ukraine, the proportion of quagga mussels in the dreissenid population was 32%–63% in all areas except in the cooling pond that received thermal effluent where they comprised only 2% of the dreissenid population (Zhulidov et al. 2010). Two years after power station shut down, proportions of quagga mussels in the dreissenid population in all areas of the cooling reservoir rose to 84%–100% (Zhulidov et al. 2010), which suggested that quagga mussels outcompeted zebra mussels only under natural ambient water temperatures. Thus, the elevated thermal tolerance of zebra mussels could allow them to outcompete quagga mussels in warm water bodies and to colonize water bodies in lower latitudes. In contrast, quagga mussels appear likely to outcompete zebra mussels in colder, north-temperate water bodies.

Thermal tolerances of *D. polymorpha* are much higher when acutely exposed to elevated temperatures. Individual mussels acclimated to 5°C and 32°C maintain normal oxygen consumption rates (V_{O_2}) when exposed to acute (i.e., near instantaneous) temperature changes up to 35°C (Aldridge et al. 1995, Alexander and McMahon 2004). After acclimation to 30°C, the LT_{50} (i.e., estimated temperature for 50% sample mortality) of zebra mussels from the Niagara River, Buffalo, NY, ranged from 35.8°C to 40.1°C when acutely exposed to temperatures increasing at rates of 0.0167–0.2°C min^{-1}, respectively (McMahon and Ussery 1995). Similarly, Spidle et al. (1995) found that the acute LT_{50} of zebra mussels from Lake Ontario acclimated to 20°C ranged from 35.0°C to 37.2°C when exposed to the same rates of temperature increase, while LT_{50} values for similarly acclimated specimens of quagga mussels were lower (35.1°C–36.1°C).

Surface water temperatures display a distinct daily temperature cycle, reaching maximum levels through insolation during daylight hours and minimal levels during dark hours. Zebra mussels from the Niagara River, Buffalo, NY, were exposed to temperatures that simulated daily fluctuations in natural Texas water bodies (range between 25°C–32°C, 27°C–32°C, 26°C–33°C, and 28°C–33°C) and to constant temperatures of 32°C and 33°C. In all cases, survivorship significantly increased in fluctuating temperature regimes relative to constant high-temperature regimes (Table 25.2; Hernandez 1995), which suggested that natural daily water temperature variation may allow both zebra and quagga mussels to survive in water bodies approaching their upper incipient lethal temperatures during daylight hours.

Temperature also impacts dreissenid spawning and larval development. Zebra mussel spawning in Europe and North America is initiated at ≥15°C–18°C. However, low numbers of veligers occur in the plankton at ≥12°C on both continents (McMahon 1996, Karatayev et al. 1998, McMahon and Bogan 2001). Eggs of zebra mussels are incapable of being fertilized at ≤10°C (Sprung 1987). In Lake Erie, epilimnetic populations of zebra and quagga mussels spawned simultaneously after ambient water temperatures reached 18°C–20°C (Claxton and Mackie 1998), suggesting that temperatures required for the initiation of spawning are roughly equivalent in the two species. However, when transplanted to cooler hypolimnetic waters at 23 m depth, quagga mussels initiated spawning at much lower temperatures of only 9°C–11°C, while spawning did not occur in zebra

Table 25.2 Median Lethal Time (Time to 50% Mortality; LT$_{50}$) for *D. polymorpha* (Zebra Mussel) at Fluctuating (12 h/12 h) and Constant Temperature Regimes. Mussels collected from the Niagara River, Buffalo, NY.

Temperature Regime	LT$_{50}$ (h)
Fluctuating 25°C–32°C	185.8
Constant 32°C	67.8
Fluctuating 27°C–32°C	336.6
Constant 32°C	117.6
Fluctuating 26°C–33°C	55.7
Constant 33°C	26.3
Fluctuating 28°C–33°C	40.2
Constant 33°C	34.9

Source: Hernandez, M.R., Thermal response and tolerance in the introduced freshwater bivalve, *Dreissena polymorpha* (zebra mussel). MS thesis, The University of Texas at Arlington, Arlington, TX, 132pp., 1995.

mussels (Claxton and Mackie 1998). These further results suggest that quagga mussels may be able to spawn at much lower temperatures than zebra mussels, allowing them to be more successful in cooler water habitats and colonize more northern latitudes.

Temperature limits for development of fertilized eggs to D-shelled veligers were 12°C–24°C with maximal developmental success at 17.3°C for zebra mussels (Sprung 1987). Similarly, 3-day-old veligers showed 48% and 42% survival after 8 days in artificial pond water at 18°C and 22°C, respectively, but only 6% survival after 7 days at 26°C (Wright et al. 1996). The period of time for zebra mussel veligers to develop and settle progressively decreased with increased temperature from 12°C to 24°C (Sprung 1987). Similarly, veliger growth rate increased and development time decreased exponentially with increased rearing temperature from 15°C to 21°C (Sprung 1995a).

As might be expected, temperature also impacts dreissenid metabolic functions. Zebra mussel oxygen consumption rates (Vo_2) increased up to 30°C (5°C acclimated individuals) and 35°C (15°C and 25°C acclimated individuals) (Alexander and McMahon 2004). Prior acclimation to low temperatures resulted in an increase in Vo_2 relative to individuals acclimated to elevated temperatures (Alexander and McMahon 2004). Increased temperature also reduced the O$_2$ uptake capacity of zebra mussels under hypoxia (Alexander and McMahon 2004), and survival under hypoxia declined exponentially with increased temperature (Johnson and McMahon 1998, Matthews and McMahon 1999).

Filtration rates of zebra mussels increased with increased temperature to 20°C–24°C, above which they progressively declined (Aldridge et al. 1995, Sprung 1995b, Lei et al. 1996). Correspondingly, ingestion rates increased exponentially from 3°C to 24°C (Sprung 1995b) and growth rates increased up to 15°C and progressively declined at higher temperatures (Walz 1978a). Individuals entered negative energy balance at ≥20°C–25°C as maintenance requirements exceeded capacity for energy assimilation even at high levels of food ration (Walz 1978a). Further, increased temperature reduced starvation tolerance in zebra mussels with estimated time for 50% sample mortality (LT$_{50}$) being 117.8, 341.3, and 503.8 days, respectively, at 25°C, 15°C, and 5°C (Chase-Off 1996).

Temperature can also effect dreissenid attachment to substrates and tolerance to salinity. Production of byssal threads used for attachment increased with increased temperature between 5°C and 30°C in zebra mussels (Clarke and McMahon 1996a). In contrast, Rajagopal et al. (1996) recorded maximal byssal thread production at 20°C. During a 6 day experimental period, maximal byssal attachment strength in zebra mussels was attained at 20°C and 25°C and decreased greatly at 30°C, and at 5°C mussels were unable to produce byssal threads (Kobak 2006). A temperature increase from 18°C to 26°C reduced survival of zebra mussel veligers at salinities ranging from 2 to 8 psu (practical salinity units = ‰) (Wright et al. 1996).

Similar studies of the impacts of temperature on the functional biology of quagga mussels have not been conducted. Such studies will be required to more fully understand the influence of temperature on ultimate environmental limits of quagga mussels and competitive interactions with zebra mussels.

Calcium

Minimum calcium requirements for dreissenid mussels are difficult to determine in the laboratory because, like many bivalves, both zebra and quagga mussels can apparently mobilize Ca^{2+} from the shell to their hemolymph under low Ca^{2+} conditions. For example, zebra mussels survived in media depleted of Ca^{2+} for >51 days but did not survive in media containing Ca^{2+} but depleted of any one of four critical ions needed for survival (Mg^{2+}, Na$^+$, Cl$^-$, and K$^+$) (Dietz et al. 1994). For this reason, studies of lower incipient Ca^{2+} limits for dreissenids have been limited primarily to presence/absence data from water bodies of varying Ca^{2+} concentration, and such studies have been much more extensive for zebra than quagga mussels. Zebra mussels were found at 46 of 57 sites in the St. Lawrence River with Ca^{2+} concentrations ranging from 16.2 to 37.6 mg Ca^{2+} L^{-1} (Mellina 1993, Mellina and Rasmussen 1994). They were not found at sites with concentrations ranging from 8.0 to 14.0 mg Ca^{2+} L^{-1}, which indicates a minimal threshold for colonization of 15 mg Ca^{2+} L^{-1}. Based on ability of juvenile mussels to sustain positive shell growth when cultured in water of varying Ca^{2+} concentration from 16 lakes in south central Ontario, Canada, Hincks and Mackie (1997) estimated that the minimum Ca^{2+} concentration for zebra mussels was 8.5 mg Ca^{2+} L^{-1}. There has been only one comparative study of the minimum Ca^{2+} requirements of zebra and quagga mussels (Jones and Ricciardi 2005). This study examined 20 sites in the St. Lawrence River for presence/absence of both species. Zebra mussels occurred at 19 sites with concentrations between 8.0 and 30.0 mg Ca^{2+} L^{-1}, while quagga mussels occurred at 17 sites with concentrations between 12.4 and 30 mg Ca^{2+} L^{-1}. Zebra mussels were not found at a site with a concentration of 7.6 mg Ca^{2+} L^{-1}, while quagga mussels did not occur at three sites with concentrations between 7.6 and 10.0 mg Ca^{2+} L^{-1}. Based on these results, Jones and Ricciardi (2005) suggested that the lower Ca^{2+} limits for zebra and quagga mussels were 8.0 and 12.0 mg Ca^{2+} L^{-1}, respectively. Population density and biomass were positively correlated with Ca^{2+} concentration in both zebra and quagga mussels (Mellina and Rasmussen 1994, Jones and Ricciardi 2005).

While further study of the lower Ca^{2+} limits of quagga mussels is necessary, present data indicate that zebra mussels are more likely to successfully invade lakes of lower Ca^{2+} concentration than quagga mussels (i.e., lower limits are 8 mg Ca^{2+} L^{-1} versus 12 mg Ca^{2+} L^{-1}, respectively). Thus, use of a lower limit of 12 mg Ca^{2+} L^{-1} to predict potential North American distributions for both species (Whittier et al. 2008) may be inappropriate. However, the lower Ca^{2+} concentration limits of both species may be dependent on the synergistic interaction with the concentration of other critical ions including Na^+, Cl^-, K^+, and particularly Mg^{2+}. Indeed, Hallstan et al. (2010) found that Mg^{2+} concentration was a better predictor of the presence of zebra mussels in 30 Swedish lakes than Ca^{2+}. Regardless of differences in Ca^{2+} limits, both species tend to have higher Ca^{2+} requirements than other freshwater bivalves (McMahon 1996, McMahon and Bogan 2001). This is due in part to the high ion permeability of their epithelia (Dietz et al. 1997), which could decrease their capacity to sequester Ca^{2+} from low Ca^{2+} waters compared to other less permeable freshwater bivalves. Poor ability to sequester Ca^{2+} may be the basis for the observation that zebra mussels cannot sustain positive shell growth below 8.5 mg Ca^{2+} L^{-1} (Hincks and Mackie 1997).

Since the majority of data regarding Ca^{2+} concentration limits for both species comes from presence/absence data in the St. Lawrence River (Mellina and Rasmussen 1994, Jones and Ricciardi 2005), further studies of zebra versus quagga mussel distributions relative to Ca^{2+} concentration as well as those of other critical ions (i.e., Na^+, Cl^-, K^+, and particularly Mg^{2+}) should be carried out in other water bodies and drainage systems.

Salinity

Similar to incipient lower limit for calcium, incipient upper limits for salinity have been more intensely examined for *D. polymorpha* than *D. r. bugensis*. Blood solute concentration of 36 mosm (milliosmoles) in zebra mussels is among the lowest recorded for freshwater bivalves (Horohov et al. 1992). This species appears to have relatively little capacity for hyperosmotic regulation; changes in hemolymph ion concentrations in elevated salinities are mostly a passive diffusion process (Dietz and Byrne 2006). Zebra mussels became isosmotic with their medium at 30 mM NaCl and did not survive beyond seven days in media >45 mM NaCl, indicative of salinity tolerance lower than most other freshwater bivalve species (Horohov et al. 1992). Low salinity tolerance in zebra mussels may, in part, result from relatively high epithelial ion permeability resulting in rapid ion uptake in hypertonic media (Dietz et al. 1997) and reduced capacity to increase excretory clearance or filtration rates in elevated salinities (Dietz and Byrne 1997). Survival of zebra mussels in higher salinities also depends on maintenance of Na^+–K^+ balance between hemolymph and the surrounding medium (Dietz et al. 1997).

In a laboratory study, zebra mussels were intolerant of acute transfer to salinities >3.5 psu (practical salinity units = 3.5‰); however, near 100% survival occurred when specimens were stepwise acclimated to salinities as high as 8.8 psu (Wilcox and Dietz 1998). In another laboratory study, lower incipient salinity limits of 4 psu and 4.5 psu were recorded for adult and veliger zebra mussels, respectively (Kilgour et al. 1994), with 1 psu being optimum for adult nutrition condition. At 5°C, 15°C, and 25°C, byssal thread production in zebra mussels was constant up to 3 psu but declined thereafter and ceased at >7 psu (Kilgour et al. 1994). Zebra mussels not acclimated to salinity could be induced to spawn up to 3.5 psu but not at 7 psu, while acclimated mussels spawned up to 7 psu (Fong et al. 1995). Fertilization of spawned gametes occurred up to 3.5 psu (Fong et al. 1995). Patterns of byssal thread production in zebra mussels at different salinities were similar to that for mortality; production progressively declined above 4 psu and essentially ceased at 7 psu (Rajagopal et al. 1996).

Based on a review of presence/absence data for zebra mussels in freshwaters throughout Europe, Strayer and Smith (1993) estimated the seaward limit to be 0.4–2 psu in tidal habitats, depending on degree of short-term salinity fluctuation. In constant salinity nontidal lagoons, estimated tolerance limits rose to 6 psu and reached a maximum of 10–14 psu in sulfate-rich, brackish lakes (Strayer and Smith 1993). Zebra mussel settlement on concrete blocks was observed at sites of varying salinity in the oligohaline portion of the lower Hudson River (New York) (Walton 1996). Settlement and survival occurred at a site where salinity varied between 2 and 6 psu but not at another site where it ranged between 5 and 9 psu.

A few studies have compared salinity tolerance of zebra and quagga mussels. When acutely transferred to elevated salinities, the two species initially exhibited mortality at 8–9 and 4–6 psu, respectively (Orlova et al. 1998). In contrast, when exposed to a stepwise increase in salinity, chronic upper salinity limits of zebra mussels were 11–15 psu, depending on sampling site, while the limit for quagga mussels was 8 psu. The lower salinity limit for both species was 0.007 psu (Orlova et al. 1998). In another study, no significant differences in survivorship were recorded among zebra and quagga mussels when exposed to salinities between 5 and 20 psu (Spidle et al. 1995). A laboratory study by Wright et al. (1996) indicated that embryonic development in both species was significantly inhibited at >2 psu. At 4 psu, no quagga mussel and only a low percentage of zebra mussel embryos attained the trochophore stage of development. At 18°C, early (3-day-old) veliger survival in zebra mussels was equivalent to controls up to 8 psu, while survival relative to controls was reduced at 22°C and 26°C (Wright et al. 1996). At 22°C, the salinity tolerance of early quagga mussel veligers was lower than those of zebra mussels after 10 days at 8 psu. Zebra and quagga mussel pediveligers settled and transformed to juveniles up to 6 psu but not 8 psu (Wright et al. 1996). These comparative studies suggest that veliger and adult quagga mussels may not be as salinity tolerant as zebra mussels of similar life stages. As such, it is possible that zebra mussels may be able to extend further downstream than quagga mussels in the brackish, tidal portions of rivers and brackish inland water bodies at salinities >4 to 8 psu.

pH

Tolerance of pH has been little studied in dreissenid mussels, and what information exists is for *D. polymorpha*. At pH <7.0, uptake rates of Na^+ and Ca^{2+} decreased (Vinogradov et al. 1993) and potentially impaired ability of zebra mussels to osmotically and ionically regulate and maintain shell integrity. Thus, it appears that zebra mussels may be limited to relatively alkaline waters. In a laboratory study, the upper pH limit of zebra mussels was estimated to be 9.3–9.6 (Bowman and Bailey 1998). Based on survivorship in water of varying pH and calcium concentration from 16 lakes in south central Ontario, Canada, Hincks and Mackie (1997) estimated that the incipient pH range of zebra mussels was 6.0–8.5 at Ca^{2+} concentrations <25 mg L^{-1}; low pH tolerance progressively declined as Ca^{2+} concentration declined below 25 mg L^{-1}. Positive shell growth was only recorded in juvenile zebra mussels in lake waters with pH >8.3 regardless of Ca^{2+} concentration (Hincks and Mackie 1997). Successful development from fertilized egg to veliger occurred at pH 7.4–9.4 (Sprung 1987). Thus, the pH tolerance range for zebra mussels appears to be approximately 6.0 to 9.6 at 25 mg Ca^{2+} L^{-1}. However, lower pH limits could increase at <25 mg Ca^{2+} L^{-1}, and upper pH limits decline at >25 mg Ca^{2+} L^{-1} (Hincks and Mackie 1997). To date, pH tolerance of quagga mussels has not been examined in Europe or North America.

Hypoxia

Tolerance to hypoxia (i.e., oxygen concentrations < full air oxygen saturation) has been studied primarily in *D. polymorpha*, and this species is a relatively poor regulator of oxygen uptake rate (V_{O_2}) when oxygen (O_2) concentrations decline (Alexander and McMahon 2004). The percent O_2 regulation value (R) is used to assess the ability of species to regulate V_{O_2} during progressive hypoxia. It is the percentage decline in the capacity of a species to maintain the same level of V_{O_2} from full air oxygen saturation (159.69 mm Hg or 21.29 kPa) to 0 kPa. Thus, R for a perfect O_2 regulator would be 100%, R for a nonregulator would be 50%, and R for a partial regulator would be between 50% and 100% (for further details, see Alexander and McMahon 2004). When exposed to progressive hypoxia, R values for zebra mussels ranged from 51.5% to 81.4% depending on temperature (5°C–25°C) and prior temperature acclimation (5°C, 15°C, and 25°C) (Alexander and McMahon 2004). R values for quagga mussels were 67% and 60% at 5°C and 20°C, respectively (McMahon unpublished data), which suggests that the two species have similar capacities for O_2 regulation of V_{O_2} ranging from nonregulation to moderate regulation dependent on temperature and prior temperature experience.

When zebra mussels were exposed to lethally low levels of oxygen for up to 28 days, both test temperature and prior acclimation temperature effected LT_{50}. Under anoxia, LT_{50} ranged from 2.2 days at 25°C for individuals acclimated at 5°C to 40.3 days at 15°C for individuals also acclimated at 5°C (Matthews and McMahon 1999). In a 28 day exposure to hypoxia varying from 1.05 to 4.24 kPa at 5°C, 15°C, and 25°C, both temperature and O_2 concentration effected zebra mussel survivorship (Johnson and McMahon 1998). Survival increased with increased O_2 concentration and decreased temperature. Relatively high survival occurred at partial pressures of oxygen (P_{O_2}) between 1.05 and 2.12 kPa at 5°C and 15°C and >3.17 kPa at 25°C. Outside this range, LT_{50} values ranged from 120 h at 1.05 kPa and 25°C to 261 h at 1.59 kPa and 5°C. The poor hypoxia tolerances observed for dreissenids are among the lowest recorded for freshwater bivalves, including the highly hypoxia-intolerant Asian clam, *Corbicula fluminea* (Johnson and McMahon 1998).

A minimal P_{O_2} of 4.3 kPa is required for fertilized zebra mussel eggs to successfully develop into veligers (Sprung 1987). Byssal thread production by zebra mussels is inhibited at $P_{O_2} \leq 2.05$ kPa (Clarke and McMahon 1996b). Thus, zebra mussels are highly intolerant of hypoxia, which limits their populations to oxygenated epilimnetic waters in lentic habitats if hypolimnetic waters are hypoxic during summer months.

No comparable data on hypoxia tolerance of *D. r. bugensis* have been published. However, an unpublished investigation of hypoxia tolerance compared survivorship of zebra mussels and quagga mussels under anoxic and hypoxic conditions (i.e., P_{O_2} of 0, 1.06, and 2.13 kPa) at 15°C and 25°C (P. D. Johnson and R. F. McMahon, unpublished data). This study indicated that zebra mussels were significantly more tolerant of hypoxia than quagga mussels under all lethal temperature/P_{O_2} combinations tested. The single exception was 2.13 kPa at 25°C where tolerance levels of the two species were not significantly different (Table 25.3). Over a 30 day test period, survival times of zebra mussels (LT_{50}) were 1.2–3.2 times greater than those of quagga mussels, with relative differences in survival time between the two species being greater at 25°C relative

Table 25.3 Median Lethal Time (Time to 50% Mortality; LT$_{50}$—Kaplan–Meier Survival Analysis) of *D. r. bugensis* (Quagga Mussel) and *D. polymorpha* (Zebra Mussel) at Different Combinations of Temperature and Partial Pressure of Oxygen (Po$_2$). Differences in LT$_{50}$ between the Two Species Were Tested Using the Log Rank Statistic

Variable		LT$_{50}$ (Days)		
Po$_2$ (% Full Air O$_2$ Saturation)	Temperature (°C)	*D. r. bugensis*	*D. polymorpha*	*P*-Value (Log Rank)
0 kPa (0%)	15	18	32	<0.05
0 kPa (0%)	25	5	16	<0.05
1.06 kPa (5%)	15	19	24	<0.05
1.06 kPa (5%)	25	6	5	<0.05
2.13 kPa (10%)	15	26	>30	<0.05
2.13 kPa (10%)	25	15	18	>0.05

Source: Johnson, P.D. and McMahon, R.F., unpublished data

to 15°C at all temperature/Po$_2$ combinations in which >50% mortality was attained to allow computation of LT$_{50}$ values (Table 25.3). These results suggested that zebra mussels are more hypoxia tolerant than quagga mussels. Thus, zebra mussels may be better adapted than quagga mussels to water bodies periodically experiencing moderate hypoxia. The results also suggested that the incipient (30 days) hypoxia limit for both species is Po$_2$ >2.13 kPa, although further research may find it is somewhat higher for zebra mussels. Because both species are relatively intolerant of hypoxia, they are likely to be confined to well-oxygenated habitats and unable to colonize substrates in hypolimnetic regions subject to hypoxia.

Desiccation and Freezing

Dreissenids have a limited capacity to tolerate emersion (i.e., exposure in air) and desiccation. A nearshore, shallow-water population (depth = 1 m) of newly settled and adult zebra mussels in the Włocławek Reservoir, Poland, experienced near 100% mortality when emersed on dry substrate for 1–3 days during summer declines in water levels (Wiśniewski 1992). When artificially emersed on dry sand or dry stone under summer conditions, 100% mortality occurred within 3 days. In contrast, when sand or stone substrates were periodically moistened, mortality was less than 5% (Wiśniewski 1992).

Dreissenid emersion/desiccation tolerance decreased with increased air temperature and decreased relative humidity (RH) (McMahon et al. 1993, Ricciardi et al. 1995). Estimated time to 100% mortality of emersed zebra mussels increased from <5% to >95% RH, being 10.3–27.9, 4.5–12.3, and 2.0–5.4 days, at 5°C, 15°C, and 25°C, respectively (McMahon et al. 1993). At 5°C, 15°C, and 25°C and with RH ranging from <5% to 75%, mortality occurred when internal water (i.e., corporal + extra-corporal mantle cavity water) loss reached 58%–71% of initial values. At lower temperatures (5°C), rates of water loss were highest in the initial stages of air exposure due to valve gaping, presumably for aerial gas exchange. Valve gaping declined at higher temperatures (15°C and 25°C), resulting in water being lost at a more constant rate throughout the exposure period (McMahon et al. 1993). Essentially similar results were reported by Ricciardi et al. (1995) for zebra mussels emersed for a 5 day period. Small mussels (shell length [SL] 10–18 mm) experienced 100% mortality in 10% and 50% RH at 10°C–20°C, while mortalities of 26.7%, 86.7%, and 100% were recorded in 95% RH at 10°C, 20°C, and 30°C, respectively. Large specimens (SL = 21–28 mm) displayed lower mortality rates. At 10°C under 10%, 50%, and 95% RH, mortalities in large individuals were 75%, 0%, and 0%, respectively. At 20°C, 100%, 83.3%, and 52.5%, respective mortalities increased to 100%, 83.3%, and 52.5%, and at 30°C, 100% mortality occurred in all three tested levels of RH (Ricciardi et al. 1995). Similarly, Paukstis et al. (1999) reported that large zebra mussels (SL > 16 mm) after 60 h emersion in 55% RH displayed 0% mortality at 5°C and 71%–75% mortality at 20°C; smaller mussels (SL < 16 mm) displayed greater mortality rates at similar RH and temperatures.

Few studies have compared exposure tolerance of zebra and quagga mussels. At 15°C between <5% and >95% RH, the two species showed no significant differences in percent total water lost with time of exposure or individual size (Ussery and McMahon 1995). In addition, the percent of total water lost just prior to death was not significantly different between the two species, being essentially similar at 45%–60% between <5% and 53% RH. In both species, the percent of total water lost prior to death declined to <30% and <10% at 75% and >95% RH, respectively. Although the percentage of total water loss prior to death was essentially similar for the two species, emersed quagga mussels had higher water loss rates than zebra mussels between 53% and >95% RH, resulting in reduced emersion tolerance times. Thus, while time to 100% sample mortality for quagga mussels was essentially equivalent to that of zebra mussels at 120 h in <5% and 33% RH, it was reduced at 150, 180, and 325 h compared to 310, 225, and 540 h for zebra mussels at 53%, 75%, and >95% RH, respectively

(Ussery and McMahon 1995). Similar results were reported by Ricciardi et al. (1995). They found small quagga mussels (SL = 12–18 mm) emersed at 20°C exhibited 100% mortality after 3 days in 10% and 50% RH and 100% mortality after 5 days in 95% RH. Zebra mussels exhibited 100% mortality after 5 days at 10% and 50% RH but only 86.7% mortality at 95% RH. Thus, both studies indicate that quagga mussels are less emersion tolerant than zebra mussels, which suggests that the zebra mussels may be more successful in shallow-water habitats subject to periodic water level fluctuations. However, both zebra and quagga mussels have relatively poor emersion tolerance compared to other freshwater bivalves, particularly unionids (McMahon 1996, McMahon and Bogan 2001), which suggests that both dreissenid species are likely to be most successful in water bodies experiencing limited annual water level variations.

Capacity to tolerate freezing air temperatures is another aspect of emersion adaptation. Freeze tolerance has only been examined in zebra mussels. Paukstis et al. (1996) found that adult zebra mussels exposed to below freezing temperatures that decline at 1°C 30 min^{-1} acted as exotherms and supercooled to −1.8°C to −3.0°C before tissues froze. Only 2 of 17 individuals survived tissue freezing. McMahon et al. (1993) found that individual adult zebra mussels acclimated at 5°C could survive >48 h at 0°C, but 100% mortality occurred after 15.0 h at −1.5°C, 5.0 h at −3.0°C, 2.5 h at −5°C, 2.0 h at −7.5°C, and 0.7 h at −10°C. In contrast, freeze tolerance increased when adult mussels were clustered (10 individuals), which mimicked aggregated distributions in nature. Tolerance times in clustered mussels increased to >48 h at 0°C and −1.5°C, 7.5 h at −3.0°C, 4.5 h at −5°C, 2.3 h at −7.5°C, and 2.7 h at −10°C. Both studies indicated that zebra mussels are intolerant of emersion at less than −3.0°C and unlikely to survive prolonged exposure to freezing temperatures during winter declines in water levels. Poor freeze tolerance may also be a factor that prevents successful colonization of shallow waters (<1 m depth) in lakes that freeze during winter. To date, freeze tolerance of quagga mussels has not been examined but is likely to be similar to zebra mussels due to their similar latitudinal distributions in Europe and North America.

Currents and Agitation

Population densities of dreissenids appear to be negatively affected by current and wave-induced agitation. On a wave-swept shore of Middle Sister Island in the western basin of Lake Erie (OH), density, biomass, and mean population shell length of zebra mussels increased significantly with depth and distance from shore and increased rock size (MacIsaac 1996). For example, at 7 m and 17 m from shore, most mussels had shell lengths of <16 and <22 mm, respectively, but at ≥27 m from shore mussels with shell lengths >19 mm were dominant, suggesting that larger individuals were more impacted by wave and current disturbance prevalent in nearshore habitats than smaller mussels (MacIsaac 1996). These findings were consistent with observations of a significant reduction in production of byssal threads by adult zebra mussels exposed to current velocities ≥0.27 m s^{-1} in the laboratory (Clarke and McMahon 1996c) and increased production of byssal threads with increased water agitation up to 20 cycles min^{-1} followed by a decline in thread production of 30% and 53% at 30 and 40 cycles min^{-1}, respectively (Clarke and McMahon 1996b, Rajagopal et al. 1996). Current velocity also effects clearance rates in quagga mussels. Rates increased with increased flow rates from 0 to 9 cm s^{-1} but were greatly inhibited at 19 cm s^{-1} (Ackerman 1999). Thus, mussel ingestion rates could also be reduced by elevated flow velocities such as those present on wave-swept shores or in fast-flowing lotic habitats. Further research is required to determine similarities and differences in response to current velocity and agitation in zebra and quagga mussels. However, neither species appears well adapted to habitats with high flow/high agitation conditions such as wave-swept, shallow regions in lakes or rivers with high current velocities. Thus, both species exhibit a preference for settlement in crevices, on undersides of rocks, and on other sites that provide refuge from exposure to high levels of flow and agitation (Czarnoleski et al. 2004, Kobak 2005).

Turbidity

Turbidity can negatively impact metabolic functions of dreissenids. Turbidity ≥ 20 nephelometric turbidity units (NTU) inhibited oxygen consumption rates (V_{O_2}) in zebra mussels acclimated to temperatures of 10°C, 18°C, and 26°C. Decrease in V_{O_2} was maximized at 40%–70% of controls at ≥80 NTU (Alexander et al. 1994). Turbidity also impacts filtration rates. Zebra mussels exposed to suspensions of illite–smectite clay (particle diameter <2 μm) between 25 and 250 mg L^{-1} (i.e., 46–340 NTU) increased frequency of valve closure indicative of irritation but showed little change in proportion of time valves that were held open (MacIsaac and Rocha 1995). Pseudofeces production did not significantly increase with increased turbidity, indicating that mussels progressively decreased water pumping rates as clay particle concentrations increased (MacIsaac and Rocha 1995). Reduction of zebra mussel pumping rates with increased turbidity has been supported by filtration rate studies (Lei et al. 1996). Filtration rates of zebra mussels at 20°C increased up to a microsphere particle concentration of 27 mg L^{-1}. Beyond 27 mg L^{-1}, rates remained constant up to maximal concentration of 64 mg L^{-1} indicative of a corresponding decline in pumping rate. Because filtration rate remained constant at a particle concentration ≥27 mg L^{-1}, particle clearance rates progressively declined with further increase in particle concentration (Lei et al. 1996). Filtration rates of zebra mussels (SL = 14–17 mm) fed with alga *Chlamydomonas eugametos* similarly remained relatively constant (10.6–11.7 mL mussel⁻¹ h⁻¹) over a concentration range of 9,250–92,500 cells mL⁻¹ (Dorgelo and Smeenk 1988). When zebra mussels were fed with alga *Chlamydomonas reinhardtii*, pumping rates declined at ≥16,000 cells mL⁻¹, while ingestion rate remained relatively constant at ≈3.68 × 10⁶ cells h⁻¹ (Sprung and Rose 1988). Both findings (Dorgelo and Smeenk 1988, Sprung and Rose 1988) suggest progressive decline in pumping rate with increased algal cell concentrations beyond a limiting concentration, resulting in near constant algal ingestion rates under progressively elevated algal cell concentrations.

Two studies (Summers et al. 1996, Diggins 2001) compared impacts of increased sediment particle concentration on clearance rates in zebra and quagga mussels. At 8°C, 14°C, and 22°C, individuals of both species collected in different seasons (i.e., winter, spring, summer, and fall) reduced clearance rates from 12 to 1 mL h⁻¹ when exposed to increased concentrations of natural sediment (Diggins 2001). Regardless of season, clearance rates of quagga mussels (15 and 20 mm SL) were significantly greater than those of equivalently sized zebra mussels (Diggins 2001). In another study, oxygen consumption rates (V_{O_2}) in both species decreased in 80 NTU suspensions of bentonite clay at 15°C and 25°C. However, prior acclimation to 80 NTU clay suspensions significantly increased V_{O_2} relative to non-acclimated individuals (Summers et al. 1996). These two studies (Summers et al. 1996, Diggins 2001) suggest that *D. r. bugensis* may be slightly better adapted to elevated turbidities than *D. polymorpha* due to its ability to sustain higher clearance rates (and presumably higher filtration rates) and equivalent V_{O_2}. However, both species appear to have capacity to tolerate and adjust filtration rates and V_{O_2} to suspended sediment concentrations that are greater than naturally found in most North American lotic and lentic habitats. Further research is required to determine the turbidity limits for both species in order to better understand its impact on their invasion success in North America.

Substrate Requirements

Because zebra and quagga mussels are epifaunal bivalves that attach to firm surfaces with proteinaceous byssal threads, available substratum can be a major factor that limits density and biomass (Burlakova et al. 2006). Both species attach to a wide variety of surfaces including natural rock, stone cobble, gravel, course sand, clay, wood, macrophytic aquatic plants, and shells of other live and dead bivalves, including conspecifics. Interestingly, zebra mussels attached to the posterior end of shells of live unionids (*Anodonta cygnea*) sustain higher growth rates than those attached to stones (Hörmann and Maier 2006). Presumably, this occurred because water currents generated by the unionid's siphons increased food availability and O_2 provision to attached mussels.

Zebra and quagga mussels can inhabit mud or silt substrata. In most cases, they initially attach to isolated stones, shells, wood, or other hard-surfaced material exposed on the soft substratum surface and then overgrow the soft substrate as new recruits settle on shells of previously established individuals (Mills et al. 1996, Karatayev et al. 1998, Jones and Ricciardi 2005, Orlova et al. 2005). However, mussels have also been recorded on soft substrata devoid of initial hard attachment sites (Berkman et al. 1998, 2000). Mussel density and biomass tends to rapidly decline with deceased substratum particle size from boulders through cobble, gravel, sand, silt, and mud (Hunter and Bailey 1992, Dermott et al. 2003, Jones and Ricciardi 2005). Hunter and Bailey (1992) reported that zebra mussels found on soft substratum had initially attached to the exposed shells of unionids. Thereafter, newly settled individuals attached to other mussels and formed colonies that extended over the surface of the soft substrate. In western Lake Erie, masses of mixed zebra and quagga mussel colonies were recorded on sediments of mixed silt and clay in which 75%–95% of particles were <63 μm by weight. Even though sediments contained no rocks, shells, or other hard-surfaced objects previously considered necessary for initial settlement of mussels on soft substrata, mussel densities ranged from <50 to 35,200 individuals m⁻² (Berkman et al. 1998, 2000).

Dreissenid mussels can also attach and grow on submerged aquatic macrophytes (Dermott et al. 2003); however, mussels attached to macrophytes are generally younger and considerably smaller than those attached to more permanent substrata since macrophytes generally do not last throughout an entire 3–4 year mussel lifespan (Karatayev et al. 1998). Because of the impermanent nature of macrophytes, dreissenid densities may be less stable in habitats where they are the dominant attachment substratum relative to habitats with a greater variety of substrata (Burlakova et al. 2006). While both zebra and quagga mussels settle and grow on submerged macrophytes, there appear to be differences in their use of macrophytes as an attachment site. For example, at four sites in eastern Lake Erie, proportions of zebra and quagga mussels were determined attached to stems of Eurasian water milfoil (*Myriophyllum spicatum*) and attached to benthic substrata (Diggins et al. 2004). The proportion of zebra mussels relative to quagga mussels was clearly greater on watermill foil stems than on bottom substrata at all four sites (i.e., 99% vs. 50%, 70%–86% vs. 39%, 30% vs. 2%, and 61% vs. 8%, respectively). Preponderance of zebra mussels on water milfoil suggested that (1) quagga mussels could outcompete zebra mussels on typical benthic hard-surfaced substrata, forcing zebra mussels to use macrophytes as an inferior refugium; (2) zebra mussels have a greater preference for macrophytes as a settlement site; or (3) zebra mussels spatially outcompete quagga mussels on macrophytes. In any case, ability of zebra mussels to numerically dominate quagga mussels on submerged macrophytes may allow coexistence with quagga mussels even though the latter may displace them on benthic substrata (Diggins et al. 2004).

Dreissenids display differential preferences for settlement sites. Pediveligers of zebra mussels preferentially settle on substrata with well-developed biofilms and avoid recently submerged surfaces (Kavouras and Maki 2003). They also settle preferentially on complex rather than smooth surfaces (Marsden and Lansky 2000, Czarnoleski et al. 2004) and on filamentous rather than smooth plastic surfaces (Folino-Rorem et al. 2006).

There is conflicting data regarding the impact of surface orientation on zebra mussel settlement. In lentic habitats, settlement rates did differ between horizontal and vertical concrete surfaces (Czarnoleski et al. 2004). In contrast, Marsden and Lansky (2000) found that greater numbers of mussels settled on upper surfaces of horizontally oriented PVC and plexiglass plates, with greater preference for those shaded from sunlight. Substrate selection can also be impacted by current flow. For example, zebra mussel settlement was completely inhibited at velocities >1 to 1.5 m s^{-1} (Mackie et al. 1991). In flows of 0.03–0.27 m s^{-1}, settlement was the same on opposite sides of plastic plates oriented parallel to the flow but significantly lower on the upstream side of vertical plates oriented at right angles to current (Kobak 2005). Zebra mussel settlement increased on the upstream side of concave versus flat or convex glass surfaces oriented vertically into flow but was the same on the downstream surface regardless of plate shape (Kobak 2005). These results suggested that zebra mussel settlement was negatively impacted by increased flow velocities on flat and convex surfaces oriented into currents, while reduced flow rate on similarly oriented concave surfaces supported greater settlement rates (Kobak 2005). Negative impacts of flow on zebra mussel settlement may partially explain their increased settlement on complex, filamentous, or roughened surfaces where laminar flow rates are reduced compared to smooth surfaces.

There have been no comparable studies of impacts of surface orientation, complexity, and current on settlement of quagga mussels. While quagga mussels are replacing zebra mussels in the Great Lakes (Baldwin et al. 2002, Stoeckmann 2003, Wilson et al. 2006), flow inhibition of pediveliger settlement may possibly be a factor in their relatively slow establishment in North American river systems (United States Geological Survey 2010a,b) and the long-term coexistence of quagga mussels with zebra mussels in endemic Ukraine Bug and Dnieper River drainages (Zhulidov et al. 2010).

Juvenile zebra mussels tend to seek dark substrates and shaded areas, which results in populations aggregated in dark, deep places such as the underside of rocks (Kobak 2001). In addition, laboratory experiments determined that illumination inhibited initial (2 days) byssal attachment strength as adult mussels presumably sought shade (Kobak 2006). Similar preference for settlement in shaded areas occurred in the field (Marsden and Lansky 2000). In the same study, settlement rates of zebra mussels varied little on different substrata. Settlement rates were the same on wood, concrete, limestone, steel, fiberglass, and aluminum but were somewhat greater on PVC and black and clear plexiglass compared to glass plates (Marsden and Lansky 2000).

The substrata colonized by dreissenids may also depend on the strength of byssal attachment, particularly in areas where mussels can be dislodged by water currents. Laboratory studies indicated that byssal attachment strength of zebra mussels was greatest on dolomite limestone and became progressively weaker on artificial substrates including PVC, stainless steel, aluminum, and plexiglass, respectively (Ackerman et al. 1995). Byssal attachment strengths of quagga mussels have not been comparatively examined and may be important in their competitive interactions with zebra mussels particularly in lotic, riverine habitats.

BIOTIC INTERACTIONS

Biotic interactions integrate physiological responses to all aspects of the environment, including availability of food, food quality, competitors, predators, and pathogens. All of these interactions ultimately impact fitness (i.e., survival, growth, and reproduction). Physiological responses are density dependent, that is, responses of any dreissenid species will be dependent upon abundance of either or both species, reflecting intra- and inter-specific competition for limited resources. To date, many more studies have examined resistance adaptations to abiotic factors in zebra mussels than in quagga mussels. As one species increases in density, the potential exists for greater exclusion of the other species in areas of sympatry. As the mussel community approaches equilibrium and with limited resources, that species that is more efficient in converting consumed biomass into somatic tissue and gametes will outcompete and replace the less efficient competitor. Hence, studies focusing on the energetic efficiency of these two species are more likely to identify essential features responsible for replacement of zebra by quagga mussels.

Diet

Both species of dreissenid mussels are efficient filter feeders and consume a wide range of suspended material in their diets. Laboratory studies have confirmed that zebra mussels can filter particles well within the size range of bacterioplankton (Frischer et al. 2000), as well a wide range of particle sizes typical of unicellular and filamentous phytoplankton and remarkably even small zooplankton (MacIsaac et al. 1992). Given that dense populations of mussels often occur in invaded lakes, multiple studies have predicted and confirmed that a high percentage (estimates exceeding 50%) of phytoplankton primary production is consumed and diverted to benthic mussel beds (e.g., Madenjian 1995, MacIsaac et al. 1996, Stoeckmann and Garton 1997).

However, quantifying the degree of dietary overlap between zebra and quagga mussels is more challenging as gut contents are difficult to identify for filter-feeding organisms. Stable isotopes of carbon and nitrogen are useful to identify sources of primary production when those sources are isotopically distinct. Macrophytes and phytoplankton are the two principal sources of primary production in freshwater systems, and they have distinct carbon stable isotope ratios ($^{13}C:^{14}C$). In western Lake Erie, stable isotope ratios of zebra and quagga mussels confirm identical diets (Garton et al. 2005). Garton et al. (2005) also found that organic carbon derived from macrophyte detritus suspended in the seston contributed significantly to diets of both zebra and quagga mussels in western Lake Erie but did not contribute to the diet of zebra mussels in nearby, smaller inland lakes. Presumably, wave energy was an important factor contributing to the availability of suspended detritus to filter-feeding mussels, with wave energy greater in western Lake Erie than in inland lakes (Garton et al. 2005). Clearly, when both species co-occur, there is strong and direct competition for food, but it is unknown whether the presence of one species has an inhibitory effect on the other when accessing suspended particulate food.

Physiological Energetics

Energy allocation by an organism follows basic economic principles. Some of the energy acquired from ingested food is used to meet metabolic demands, some is lost in excretion, and some may be unassimilated and lost in feces. When ingested energy exceeds that lost, there is energy available for somatic growth or reproduction. Greater efficiency in converting available energy into either somatic tissue or gametes will favor one species over the other in regions of sympatry as resources become limited and competition is unavoidable. Fundamental studies on the energetics of growth and reproduction are valuable for providing insights into the ecological impact of invasive species as well as competitive interactions between zebra and quagga mussels.

Environmental conditions, including temperature and food, may alter energetic demands or energy available and thereby affect organism survival, growth, and reproduction. Individual species responses may enhance their abilities in particular conditions and give them a competitive edge. For example, at similar food conditions, a species with lower energetic costs would have more surplus energy and thus more energy available for growth and reproduction. Metabolic demands in mussels are influenced by temperature, food conditions, and reproduction (Quigley et al. 1993, Sprung 1995c). These costs are significant for zebra mussels; they use a large proportion of the energy they consume to support metabolic functions (Stoeckmann and Garton 1997). For a zebra mussel population in western Lake Erie, metabolic costs (oxygen consumption and ammonia excretion combined) accounted for about 90% of the seasonal energy budget (May–October) (Stoeckmann and Garton 1997). The effect of food on metabolic demands was not measured in this study, but temperature, which ranged from 15°C to 25°C during the seasonal study period, had a greater affect on oxygen consumption than did reproductive maturity status (Stoeckmann and Garton 1997).

Stoeckmann (2003) compared zebra and quagga mussels and found that the link between metabolic costs and seasonal change in temperature was similar for the two species. For mussels from the western basin of Lake Erie, oxygen consumption of zebra and quagga mussels tracked seasonal water temperatures (15°C–26°C). However, oxygen consumption rates of quagga mussels were never as high as for zebra mussels at any point in the season. Quagga mussels consistently had lower oxygen consumption rates than zebra mussels (15%–26% lower) throughout the summer (Stoeckmann 2003). Higher metabolic costs result in less energy available for growth and reproduction for zebra mussels compared to quagga mussels. Growth and reproduction together averaged only about 5% of the seasonal energy budget for zebra mussels (Stoeckmann and Garton 1997). Because metabolic costs represent such a large proportion of energy demands for zebra mussels, lower oxygen consumption rates imply that quagga mussels have lower metabolic costs than zebra mussels and should have more energy available for growth and reproduction.

Quagga mussels generally have larger shell length and body mass than zebra mussels (Mills et al. 1996, Jarvis et al. 2000, Stoeckmann 2003). Stoeckmann (2003) measured individual shell growth of both species by suspending marked mussels in cages in Lake Erie for 386 days. Shell growth decreased with size for both species, but quagga mussels grew about 21% more than did zebra mussels. Mussel body mass in that location was also measured multiple times during the spring and summer annually from 1998 to 2001 (Stoeckmann 2003). Quagga mussels were always heavier than zebra mussels of the same shell length, and body mass of quagga mussels increased more with shell length compared to zebra mussels (Stoeckmann 2003).

Mussel growth is influenced by both diet quality and quantity (Madon et al. 1998, Stoeckmann and Garton 2001, Baldwin et al. 2002). Zebra mussels had lower body mass, and shell length decreased when fed a poor-quality diet (*Chlorella*) in laboratory experiments (Stoeckmann and Garton 2001). Suspended sediment contributes to reduced food quality, and growth of zebra mussels declined when sediment content of food increased (Madon et al. 1998). In addition, diet may alter food handling and pseudofeces production and thereby negatively affect energetic costs and thus reduce growth. For example, when zebra mussels were fed a toxic cyanobacterium (*Microcystis aeruginosa*), scope for growth decreased, likely due to increased handling costs and pseudofeces production (Juhel et al. 2006). When fed either natural seston or cultured algae (*Chlamydomonas*), quagga mussels had greater growth rates than zebra mussels.

In addition, quagga mussels did not lose body mass at lower concentrations of the cultured algae, whereas zebra mussels did (Baldwin et al. 2002).

Allocation of energy to growth and/or reproduction changes with mussel size. Over a seasonal period, small zebra mussels (<15 mm SL) allocated available energy to body mass (growth), whereas large zebra mussels (>15 mm SL) allocated less energy to body mass and instead allocated greater energy to reproduction (Stoeckmann and Garton 1997). This trend of increased reproduction coupled with decreased growth in body size was consistent for the two species (Stoeckmann 2003). For both species, mass of gametes increased with body mass; however, although quagga mussels were larger than zebra mussels, zebra mussels released greater masses of gametes than did quagga mussels of the same size (Stoeckmann 2003). Smaller body mass coupled with a greater output of gametes indicated that zebra mussels proportionally allocate greater energy and somatic resources to reproduction than quagga mussels (Stoeckmann 2003).

Timing of reproduction for zebra and quagga mussels may differ and has been found to be linked to temperature. Sprung (1987) found that 12°C was the minimum temperature for zebra mussels to produce mature gametes. In contrast, quagga mussels had mature eggs at 4.8°C (Roe and MacIsaac 1997). Further, in a transplant experiment, quagga mussels were capable of spawning at 9°C, whereas zebra mussels were unable to develop mature gametes at the same temperature (Claxton and Mackie 1998). In a sympatric population in western Lake Erie, Stoeckmann (2003) found that only quagga mussels spawned during April (at 9°C). This preceded spawning of zebra mussels at the same location (Haag and Garton 1993).

Basic physiological energy costs must be met before energy is available for growth or reproduction. Sympatric populations of zebra and quagga mussels have subtle but significant differences in energy allocation strategies as well as conversion efficiencies. While zebra mussels divert greater resources to reproduce, quagga mussels delay reproduction in order to increase shell and body mass, suggesting quagga mussels have adopted a K-oriented life history strategy, whereas zebra mussels have adopted more of an r-oriented strategy.

Starvation

Starvation tolerance has only been studied in zebra mussels. In two German lakes with different seston concentrations, newly settled juveniles from the high-seston lake (seston concentration ≈1 mg ash-free dry weight [AFDW] L^{-1}) had growth rates 5.7 times greater than juveniles from the low-seston lake (≈0.3 mg AFDW L^{-1}) (Sprung 1992). Little mortality occurred in juveniles enclosed at 2 m depth in the high-seston lake during their first growing season, whereas 40%–60% mortality occurred in juveniles enclosed in the low-seston lake, presumably as a result of poor nutritional condition. During a 31 day starvation experiment, initial decreases in tissue mass of zebra mussels collected from 2 and 9 m from the Fühlinger See, Germany, occurred mainly in the digestive gland (i.e., digestive diverticula) and were caused primarily by catabolization of lipids (Sprung and Borcherding 1991). Thereafter, other nonlipid energy stores (primarily protein) were utilized for maintenance. Similarly C:N ratios of zebra mussels increased throughout prolonged starvation periods at 5°C (945 days), 15°C (514 days), and 25°C (166 days) indicating preferential catabolism of protein energy stores relative to nonprotein energy stores for maintenance (Chase-Off 1996).

In laboratory starvation experiments, about 25% of digestive gland tubules (i.e., digestive diverticula) disintegrated or degenerated after 10 days in zebra mussels collected from depths of 2 and 9 m in the Fühlinger See, Germany (Bielefeld 1991). Tubule disintegration/degeneration was most pronounced in individuals from 2 m. During starvation, 79% of individuals from 2 m reabsorbed all gametes in their gonads, while complete gamete reabsorption occurred in only 7% of individuals from 9 m. Initially, individuals from 9 m had greater tissue mass and higher levels of lipid reserves than those from 2 m. Thus, individuals from 9 m may have catabolized lipid reserves during starvation, reducing dependency on reabsorption of digestive gland tissue and gametes as an energy source. When fed for 5 days after the starvation period, mussels from both depths showed marked redevelopment of the digestive gland.

Zebra mussels tend to starve (i.e., metabolic rate exceeds energy assimilation rate leading to decline in tissue mass) regardless of food availability at temperatures ≥20°C, with degree of starvation increasing with increasing body size (Walz 1978a, 1979). Thus, zebra mussel populations are likely to experience some degree of starvation when summer water temperatures exceed 20°C. The amount of tissue mass lost likely increases with the length of time that ambient temperatures exceed 20°C. Loss of tissue mass because of high summer temperatures has been reported for populations of zebra mussels in Oklahoma and Kansas (Morse 2009). Indeed, in Lake Ontario both zebra and quagga mussels had tissue dry mass reductions of approximately 60% and 30%, respectively, when water temperatures exceeded 20°C between June and October, 2002 (Casper and Johnson 2010).

Starvation also negatively affects production of byssal threads in zebra mussels. In the laboratory, adult mussels on a low ration of 0.1–0.5 mg C L^{-1} of the alga *Scenedesmus acutus* produced about 90 byssal threads individual^{-1}, while those fed a higher ration of 2.0 mg C L^{-1} of the same alga produced about 130 byssal threads individual^{-1} (Clarke 1999). Unfed adults produced only 60 byssal threads individual^{-1}. These data suggest that starvation can result in a reduction of mussel attachment strength.

During starvation, the rate of tissue loss and times to mortality in zebra mussels are temperature dependent. In laboratory experiments, adult mussels held without food at 25°C exhibited 100% mortality after 166 days (LT_{50} = 118 days and LT_{100} = 143 days) (Chase and McMahon 1995). Under the same conditions, a standard 20 mm individual lost 73.8% of initial dry tissue body mass after 132 days. In contrast, no mortality was observed in adults starved for 229 days at 5°C and 15°C, and a standard 20 mm individual lost 22.9% and 46.9% of body weight, respectively (Chase and McMahon 1995). Shell dry weight mass did not significantly vary during starvation at any test temperature. Walz (1978b) showed that zebra mussels could tolerate a 75% loss of body tissue dry mass after 700 days starvation in water of very low food concentration (i.e., 67 µg C L^{-1}). Similar levels of dry tissue mass loss have been reported among populations of zebra mussels exposed to prolonged summer water temperatures above 25°C in lakes in Kansas and Oklahoma (Morse 2009). When starved at 5°C, 15°C, and 25°C, zebra mussels experienced 100% mortality after 945, 514, and 166 days, respectively, corresponding to LT_{50} values of 504, 341, and 118 days (Chase-Off 1996). Although tolerance of starvation by zebra mussels is extensive at temperatures below 25°C, metabolic demand increases exponentially at higher temperatures, greatly diminishing starvation tolerance (Walz 1978a, Chase-Off 1996, Alexander and McMahon 2004). In response to prolonged starvation, adult zebra mussels reduced Vo_2 by 75% at 25°C, 60% at 15°C, and 20% at 5°C, thereby extending the tolerated starvation period by reducing maintenance metabolic demands on tissue organic energy reserves (Chase-Off 1996). High metabolic rates at temperatures >25°C (Alexander and McMahon 2004) could result in zebra mussel populations with severe negative energy balances during warm summer months in water bodies in the south and southwest United States (Morse 2009).

At present, no corresponding data on starvation tolerance exist for quagga mussels. However, relatively similar rates of Vo_2 for zebra and quagga mussels (Summers et al. 1996, McMahon, unpublished data) suggest similar tolerances of starvation for the two species. Yet, in a mixed species population, reduction of dry tissue weight for quagga mussels between June and October at a site in Lake Ontario was only 50% the reduction found for zebra mussels (Casper and Johnson 2010), suggesting that quagga mussels may be better adapted to surviving periods of negative energy balance than zebra mussels. In a mixed species population in Lake Erie, quagga mussels had Vo_2 up to 50% lower than that of zebra mussels of equivalent tissue mass, as well as higher tissue mass during May–August when ambient water temperatures ranged from 17°C to 26°C (Stoeckmann 2003). This would suggest reduced maintenance costs and/or increased ingestion/assimilation rates in quagga mussels relative to zebra mussels (Stoeckmann 2003).

When cultured at 6°C or 23°C, quagga mussels fed natural seston or *C. reinhardtii* had equivalent survival and generally higher tissue growth rates than zebra mussels (Baldwin et al. 2002). Quagga mussels had an assimilation rate of 81% at low ration levels compared to 63% for zebra mussels, even though equivalent-sized individuals of both species had equivalent clearance rates (Baldwin et al. 2002). Thus, quagga mussels may have lower maintenance requirements and/or increased assimilation efficiencies that could allow them to better survive periods of low food availability that are below maintenance requirements. Such an adaptation could account for the displacement of zebra by quagga mussels in oligotrophic and mesotrophic habitats, such as found in many habitats of the Great Lakes (Stoeckmann 2003). In addition, this adaptation could account for the ability of quagga mussels to inhabit hypolimnetic regions of lakes where food resources are limited compared to epilimnetic regions (Mills et al. 1996). In any case, now that both quagga and zebra mussels have become established in water bodies of the southwestern United States (Morse 2009, McMahon 2011) where surface ambient water temperatures reach or exceed 30°C for prolonged periods during summer months, a comparative study of starvation tolerances in the two species appears highly warranted, particularly at elevated temperatures (>20°C). Differences in ability of the two species to tolerate poor trophic conditions at elevated temperatures could be a major factor in their capacity for invasion and competitive success in oligotrophic water bodies in North America and particularly in the southwest United States.

Longevity

Life spans of zebra and quagga mussels have not been studied rigorously, but estimates range between 4 and 7 years for zebra mussels (McMahon 2002). Environmental conditions that affect survival and therefore longevity have been addressed earlier in this chapter. One biotic factor that affects survival is food quality. In laboratory experiments to test survival at different qualities of food and at different temperatures, survival of zebra mussels was negatively affected by poor-food quality (Stoeckmann and Garton 2001). Mussels altered energy allocation when fed a poor-quality diet and diverted energy to reproduction in lieu of growth and survival. It is not known how quagga mussels would respond to similar conditions. Longevity is also affected by differences in susceptibility to predators, parasites, and disease. Although dreissenids are readily consumed by predators, it is not clear if one species is more vulnerable than the other. Similarly, parasites have been described in dreissenids (Conn et al. 1996), but any species-dependent differences in prevalence or reductions in survival or fitness remain unstudied. However, with its greater size, higher growth rates, and lower energetic costs (Stoeckmann 2003), quagga mussels have better potential to survive adverse conditions than zebra mussels (as displayed in differences to laboratory starvation tolerance). Delayed reproduction, growth to larger size, and enhanced survival imply that quagga mussels have a longer life span than zebra mussels.

SUMMARY

Zebra and quagga mussels both evolved in similar freshwater environments and diverged only recently in geologic time (Orlova et al. 2005), perhaps as recently as 221,000 ± 78,000 years ago (Stepien et al. 2002). Therefore, it is not unexpected that both species possess similar, broad tolerance ranges that overlap for many abiotic factors. Indeed, it can be speculated that, over the broad distribution of dreissenid mussels in North America, these subtle differences will contribute to dominance of one species only at the extreme range of tolerances (see Table 25.1).

Higher thermal tolerances observed for zebra mussels (incipient upper thermal limit 29°C–32°C) compared to quagga mussels (~28°C) provide an advantage in aquatic habitats that are shallower and located where peak summer water temperatures can reach or exceed 30°C for significant periods (e.g., southern United States). This may explain the absence of quagga mussels in surveys of the lower Mississippi River where zebra mussels are well established (United States Geological Survey 2010a,b). Conversely, lower critical temperatures at which quagga mussels can initiate spawning (9°C) compared to zebra mussels (12°C) suggest an advantage in cooler, hypolimnetic waters such as occur widely across the Great Lakes, with quagga mussels present at greater depths than zebra mussels (depth >100 vs. <50 m, respectively, Nalepa et al. 2010).

Other than temperature and salinity, tolerance to various physical and chemical factors has been fairly well studied for zebra mussels but not as well for quagga mussels. Zebra mussels appear to have a lower minimum requirement for calcium ions, greater tolerance of salinity, and greater tolerance to hypoxia, than quagga mussels. Both species appear equally intolerant to aerial exposure and freezing. However, zebra and quagga mussels have less resistance to these potential environmental stressors than native unionid and sphaeriid bivalves (McMahon and Bogan 2001).

While studies of mussel tolerance to abiotic factors are informative for potential competition between the two species, as well as for potential limits to geographic distribution, replacement of zebra mussels by quagga mussels occurs in habitats well within the putative tolerance zones for both species (e.g., warm shallow lotic habitats in western Lake Erie and cooler lentic habitats in the St. Lawrence River). These observations imply that physiological and/or other biotic factors contribute significantly to dominance by quagga mussels in areas of sympatry.

Evidence from comparing physiological energetics of the two species indicates that quagga mussels have evolved a greater efficiency in utilization of available resources (K-selected strategy) compared to zebra mussels (r-selected strategy). The observed replacement of zebra mussels by quagga mussels over the past two decades is consistent with competitive advantages associated with a K-selected species and would be predicted wherever the two species co-occur. Perhaps what is remarkable is that zebra mussels have not been completely replaced in locations colonized by quagga mussels. For instance, the proportion of zebra mussels has remained fairly constant at approximately 20% of the dreissenid community in shallow-water habitats of western Lake Erie.

Species with r-selected strategies can coexist with competitively dominant K-selected species by capitalizing on the dynamic nature of the physical environment, by being in a position to colonize unoccupied habitats, and reproducing before being replaced due to competition. The stable coexistence of quagga and zebra mussels in dynamic, variable environments has been demonstrated using a resource-based model (Krkosek and Lewis 2010). The unpredictable and dynamic features of freshwater systems insure that many mussel communities will be ephemeral, creating short-lived patches of unoccupied habitat. *D. polymorpha*, with higher reproductive effort and shorter generation times, has the capacity to establish itself in these patches prior to *D. r. bugensis*. One prediction from this view of competitive interactions between the two species is that the abundance of zebra mussels relative to quagga mussels should increase in more unstable/dynamic systems, whereas in more stable environments quagga mussels should exclude zebra mussels. Many gaps remain in our knowledge and understanding of the autecology of *D. r. bugensis* as impacts of this species on invaded systems has been overshadowed by the initial explosive growth and expansion of *D. polymorpha*. It is clear that future studies on the impact of dreissenids on North American ecosystems should focus on the now-dominant *D. r. bugensis*.

REFERENCES

Ackerman, J. D. 1999. Effect of velocity on the filter feeding of dreissenid mussels (*Dreissena polymorpha* and *Dreissena bugensis*): Implications for trophic dynamics. *Can. J. Fish. Aquat. Sci.* 56:1551–1561.

Ackerman, J. D., C. M. Cottrell, C. R. Ethier, D. G. Allen, and J. K. Spelt. 1995. A wall-jet to measure the attachment strength of zebra mussels. *Can. J. Fish. Aquat. Sci.* 52:126–135.

Aldridge, D. W., B. S. Payne, and A. C. Miller. 1995. Oxygen consumption, nitrogenous excretion and filtration rates of *Dreissena polymorpha* at acclimation temperatures between 20 and 32°C. *Can. J. Fish. Aquat. Sci.* 52:1761–1767.

Alexander, J. E., Jr. and R. F. McMahon. 2004. Respiratory response to temperature and hypoxia in the zebra mussel *Dreissena polymorpha*. *Comp. Biochem. Physiol.* 137A:425–434.

Alexander, J. E., Jr., J. H. Thorp, and R. D. Fell. 1994. Turbidity and temperature effects in oxygen consumption in the zebra mussel (*Dreissena polymorpha*). *Can. J. Fish. Aquat. Sci.* 51:179–184.

Baldwin, B. S., M. S. Mayer, and J. Dayton. 2002. Comparative growth and feeding in zebra and quagga mussels (*Dreissena polymorpha* and *Dreissena bugensis*): Implications for North American lakes. *Can. J. Fish. Aquat. Sci.* 59:680–694.

Berkman, P. A., D. W. Garton, M. A. Haluch, G. W. Kennedy, and L. R. Febo. 2000. Habitat shift in invading species: Zebra and quagga mussel population characteristics on shallow soft substrates. *Biol. Invasions* 2:1–6.

Berkman, P. A., M. A. Haltuch, E. Tichich et al. 1998. Zebra mussels invade Lake Erie muds. *Nature* 393:27–28.

Bielefeld, U. 1991. Histological observation of gonads and digestive gland in starving *Dreissena polymorpha* (Bivalvia). *Malacologia* 33:31–42.

Bowman, M. F. and R. C. Bailey. 1998. Upper pH tolerance limit of the zebra mussel (*Dreissena polymorpha*). *Can. J. Zool.* 76:2119–2123.

Burlakova, L. E., A. Y. Karatayev, and D. K. Padilla. 2006. Changes in the distribution and abundance of *Dreissena polymorpha* within lakes through time. *Hydrobiologia* 571:133–146.

Casper, A. F. and L. E. Johnson. 2010. Contrasting shell/tissue characteristics of *Dreissena polymorpha* and *Dreissena bugensis* in relation to environmental heterogeneity in the St. Lawrence River. *J. Great Lakes Res.* 36:184–189.

Chase, R. and R. F. McMahon. 1995. Effects of starvation at different temperatures on dry tissue and dry shell weights in the zebra mussel, *Dreissena polymorpha* (Pallas). Contract report EL-95-4, Vicksburg, U.S. Army Corps of Engineers, Water Ways Experiment Station. 11pp.

Chase-Off, R. A. 1996. Effects of prolonged starvation on the zebra mussel, *Dreissena polymorpha* (Pallas). MS thesis, The University of Texas at Arlington, Arlington, TX, 60pp.

Clarke, M. 1999. The effect of food availability on byssogenesis by the zebra mussel (*Dreissena polymorpha* Pallas). *J. Mollus. Stud.* 65:327–333.

Clarke, M. and R. F. McMahon. 1996a. Effects of temperature on byssal thread production by the freshwater mussel, *Dreissena polymorpha* (Pallas). *Am. Malacol. Bull.* 13:105–110.

Clarke, M. and R. F. McMahon. 1996b. Effects of hypoxia and low-frequency agitation on byssogenesis in the freshwater mussel *Dreissena polymorpha* (Pallas). *Biol. Bul.* 191:413–420.

Clarke, M. and R. F. McMahon. 1996c. Effects of current velocity on byssal-thread production in the zebra mussel (*Dreissena polymorpha*). *Can. J. Zool.* 74:63–69.

Claxton, W. T. and G. L. Mackie. 1998. Seasonal and depth variations in gametogenesis and spawning of *Dreissena polymorpha* and *Dreissena bugensis* in eastern Lake Erie. *Can. J. Zool.* 76:2010–2019.

Conn, D. B., A. Ricciardi, and M. Babapulle. 1996. *Chaetogaster limnaei* (Annelida: Oligochaeta) as a parasite of the zebra mussel, *Dreissena polymorpha*, and the quagga mussel *Dreissena bugensis* (Mollusca: Bivalvia). *Parisitol. Res.* 82:1–7.

Czarnoleski, M., L. Michalczyk, and A. Padjak-Stos. 2004. Substrata preference in settling zebra mussels *Dreissena polymorpha*. *Arch. Hydrobiol.* 159:263–270.

Dermott, R., R. Bonnell, S. Carou, J. Dow, and P. Jarvis. 2003. Spatial distribution and population structure of the mussels *Dreissena polymorpha* and *Dreissena bugensis* in the Bay of Quinte, Lake Ontario, 1998 and 2000. Canadian Technical Report of Fisheries and Aquatic Sciences 2479, Fisheries and Oceans Canada, Burlington, Ontario, Canada. 58 pp.

Dermott, R. and D. Kerec. 1997. Changes to the deepwater benthos of eastern Lake Erie since the invasion of *Dreissena*: 1979–1993. *Can. J. Fish. Aquat. Sci.* 54:922–930.

Dermott, R. and M. Munawar. 1993. Invasion of offshore sediments by *Dreissena*, and its ecological implications. *Can. J. Fish. Aquat. Sci.* 50:2298–2304.

Dietz, T. H. and R. A. Byrne. 1997. Effects of salinity on solute clearance from the freshwater bivalve, *Dreissena polymorpha* Pallas. *Exp. Biol.* Online 2: 11. http://www.springerlink.com/content/q48464364m311148/fulltext.pdf (accessed November 24, 2010).

Dietz, T. H. and R. A. Byrne. 2006. Ionic and acid-base consequences of exposure to increased salinity in the zebra mussel, *Dreissena polymorpha*. *Biol. Bull.* 211:66–75.

Dietz, T. H., D. Lessard, H. Silverman, and J. W. Lynn. 1994. Osmoregulation in *Dreissena polymorpha*: The importance of Na, Cl, K, and particularly Mg. *Biol. Bull.* 187:76–83.

Dietz, T. H., S. J. Wilcox, R. A. Byrne, and H. Silverman. 1997. Effects of hyperosmotic challenge on the freshwater bivalve *Dreissena polymorpha*: Importance of K^+. *Can. J. Zool.* 75:697–705.

Diggins, T. P. 2001. A seasonal comparison of suspended sediment filtration by quagga (*Dreissena bugensis*) and zebra (*D. polymorpha*) mussels. *J. Great Lakes Res.* 27:457–466.

Diggins, T. P., M. Weimer, K. M. Stewart, R. E. Baier, A. E. Meyer, R. F. Forsberg, and M. A. Goehle. 2004. Epiphytic refugium: Are two species of invading freshwater bivalves partitioning spatial resources? *Biol. Invasions* 6:83–88.

Dorgelo, J. and J.-W. Smeenk. 1988. Contribution to the ecophysiology of *Dreissena polymorpha* (Pallas) (Mollusca: Bivalvia): Growth, filtration rate and respiration. *Verh. Internat. Verein. Theor. Ang. Limnol.* 23:2202–2208.

Elderkin, C. L. and P. L. Klerks. 2005. Variation in thermal tolerance among three Mississippi River populations of the zebra mussel, *Dreissena polymorpha*. *J. Shelfish Res.* 24:221–226.

Folino-Rorem, N., J. Stoeckel, E. Thorn, and L. Page. 2006. Effects of artificial filamentous substrate on zebra mussel (*Dreissena polymorpha*) settlement. *Biol. Invasions* 8:89–96.

Fong, P. P., K. Kyozuka, J. Duncan, S. Rynkowski, D. Mekasha, and J. L. Ram. 1995. The effect of salinity and temperature on spawning and fertilization in the zebra mussel *Dreissena polymorpha* (Pallas) from North America. *Biol. Bull.* 189:320–329.

Frischer, M. E., S. Nierzwicki-Bauer, R. Parsons, K. Vathanadorn, and K. Waitkus. 2000. Interactions between zebra mussels (*Dreissena polymorpha*) and microbial communities. *Can. J. Fish. Aquat. Sci.* 57:591–599.

Garton, D., C. Payne, and J. Montoya. 2005. Flexible diet and trophic position of dreissenid mussels as inferred from stable isotopes of carbon and nitrogen. *Can. J. Fish. Aquat. Sci.* 62:1119–1129.

Haag, W. R. and D. Garton. 1993. Synchronous spawning in a recently established population of the zebra mussel, *Dreissena polymorpha*, in western Lake Erie, USA. *Hydrobiologia* 234:103–110.

Hallstan, S., U. Grandin, and W. Goedkoop. 2010. Current and modeled distribution of the zebra mussel (*Dreissena polymorpha*) in Sweden. *Biol. Invasions* 12:285–296.

Hebert, P. D. N., B. W. Muncaster, and G. L. Mackie. 1989. Ecological and genetic studies on *Dreissena polymorpha* (Pallas): A new mollusk in the Great Lakes. *Can. J. Fish. Aquat. Sci.* 46:1587–1591.

Hernandez, M. R. 1995. Thermal response and tolerance in the introduced freshwater bivalve, *Dreissena polymorpha* (zebra mussel). MS thesis, The University of Texas at Arlington, Arlington, TX, 132pp.

Hincks, S. S. and G. L. Mackie. 1997. Effects of pH, calcium, alkalinity, hardness and chlorophyll on the survival, growth and reproductive success of zebra mussel (*Dreissena polymorpha*) in Ontario Lakes. *Can. J. Fish. Aquat. Sci.* 54:2049–2057.

Hörmann, L. and G. Maier. 2006. Do zebra mussels grow faster on live unionids than on inanimate substrate? A study with field enclosures. *Int. Rev. Hydrobiol.* 91:113–121.

Horohov, J., H. Silverman, J. W. Lynn, and T. H. Deitz. 1992. Ion transport in the freshwater mussel, *Dreissena polymorpha*. *Biol. Bul.* 183:297–303.

Hunter, R. D. and J. F. Bailey. 1992. *Dreissena polymorpha* (zebra mussel): Colonization of soft substrata and some effects on unionid bivalves. *Nautilus* 106:60–67.

Iwanyzki, S. and R. W. McCauley. 1993. Upper lethal temperatures of adult zebra mussels (*Dreissena polymorpha*). In *Zebra Mussels: Biology, Impacts, and Control*, T. F. Nalepa and D. W. Schoesser, eds., pp. 667–673. Boca Raton, FL: CRC Press.

Jarvis, P., J. Dow, R. Dermott, and R. Bonnell. 2000. Zebra (*Dreissena polymorpha*) and quagga mussel (*Dreissena bugensis*) distribution and density in Lake Erie 1992–1998. *Can. Tech. Rep. Fish. Aquat. Sci.* 2304. Burlington, Ontario, Canada: Department of Fisheries and Oceans.

Johnson, P. D. and R. F. McMahon. 1998. Effects of temperature and chronic hypoxia on survivorship of the zebra mussel (*Dreissena polymorpha*) and Asian clam (*Corbicula fluminea*). *Can. J. Fish. Aquat. Sci.* 55:1564–1572.

Jones, L. A. and A. Ricciardi. 2005. Influence of physicochemical factors on the distribution and biomass of invasive mussels (*Dreissena polymorpha* and *Dreissena bugensis*) in the St. Lawrence River. *Can. J. Fish. Aquat. Sci.* 62:1953–1962.

Juhel, G., J. Davenport, and J. O'Halloran. 2006. Impacts of microcystins on the feeding behavior and energy balance of zebra mussels, *Dreissena polymorpha*: A bioenergetics approach. *Aquat. Toxicol.* 79:391–400.

Karatayev, A. Y., L. E. Burlakova, and D. K. Padilla. 1998. Physical factors that limit the distribution and abundance of *Dreissena polymorpha* (Pall.). *J. Shellfish Res.* 17:1219–1235.

Karatayev, A. Y., D. Boltovskoy, D. K. Padilla, and L. E. Burlakova. 2007. The invasive bivales *Dreissena polymorpha* and *Limropema fortune*: Parallels, contrasts, potential spread and invasion impacts. *J. Shellfish Res.* 26:205–213.

Kavouras, J. H. and J. S. Maki. 2003. Effects of biofilms on zebra mussel postveliger attachment to artificial surfaces. *Invertebr. Zool.* 122:138–151.

Kilgour, B. W., G. L. Mackie, M. A. Baker, and R Kepppel. 1994. Effects of salinity on the condition and survival of zebra mussels (*Dreissena polymorpha*). *Estuaries* 17:385–393.

Kobak, J. 2001. Light, gravity and conspecifics as cues to site selection and attachment behavior of juvenile and adult *Dreissena polymorpha* Pallas, 1771. *J. Mollus. Stud.* 67:183–189.

Kobak, J. 2005. Recruitment and distribution of *Dreissena polymorpha* (Bivalvia) on substrates of different shape and orientation. *Int. Rev. Hydrobiol.* 90:159–170.

Kobak, J. 2006. Factors influencing the attachment strength of *Dreissena polymorpha* (Bivalvia). *Biofouling* 22:141–150.

Krkosek, M. and M. Lewis. 2010. An $R_{(o)}$ theory for source-sink dynamics with applications to *Dreissena* competition. *Theor. Ecol.* 3:25–43.

Lei, J., B. S. Payne, and S. Y. Wang. 1996. Filtration dynamics of the zebra mussel, *Dreissena polymorpha*. *Can. J. Fish. Aquat. Sci.* 53:29–37.

MacIsaac, H. J. 1996. Population structure of an introduced species (*Dreissena polymorpha*) along a wave-swept disturbance gradient. *Oecologia* 105:484–492.

MacIsaac, H. J. and R. Rocha. 1995. Effects of suspended clay on zebra mussel (*Dreissena polymorpha*) faeces and pseudofaeces production. *Arch. Hydrobiol.* 135:53–64.

MacIsaac, H. J., W. G. Sprules, O. E. Johansson, and J. H. Leach. 1992. Filtering impacts of larval and sessile zebra mussels (*Dreissena polymorpha*) in western Lake Erie. *Oecologia* 92:30–39.

Mackie, G. L., W. N. Gibbons, B. W. Muncaster, and I. M. Gray. 1991. *The Zebra Mussel, Dreissena polymorpha: A Synthesis of European Experiences and a Preview for North America*. Toronto, Ontario, Canada: Ontario Ministry of the Environment.

Mackie, G. and D. W. Schloesser. 1996. Comparative biology of zebra mussels in Europe and North America: An overview. *Am. Zool.* 36:244–258.

Madenjian, C. P. 1995. Removal of algae by the zebra mussel (*Dreissena polymorpha*) population in western Lake Erie: A bioenergetics approach. *Can. J. Fish. Aquat. Sci.* 52:381–390.

Madon, S. P., D. W. Schneider, J. A. Stoeckel, and R. E. Sparks. 1998. Effects of inorganic sediment and food concentrations on energetic processes of the zebra mussel, *Dreissena polymorpha*: Implications for growth in turbid rivers. *Can. J. Fish. Aquat. Sci.* 55:401–413.

Marsden, J. E. and D. M. Lansky. 2000. Substrate selection by settling zebra mussels, *Dreissena polymorpha*, relative to texture, orientation, and sunlight. *Can. J. Zool.* 78:787–793.

Martel, A. M., B. Baldwin, and R. Dermott. 2001. Species and epilimnion/hypolimnion-related differences in size at larval settlement and metamorphosis in *Dreissena* (Bivalvia). *Limnol. Oceanogr.* 46:707–713.

Matthews, M. A. and R. F. McMahon. 1999. Effects of temperature and temperature acclimation on survival of zebra mussels (*Dreissena polymorpha*) and Asian clams (*Corbicula fluminea*) under extreme hypoxia. *J. Mollus. Stud.* 65:317–325.

McMahon, R. F. 1996. The physiological ecology of the zebra mussel, *Dreissena polymorpha*, in North America and Europe. *Am. Zool.* 36:239–260.

McMahon, R. F. 2002. Evolutionary and physiological adaptations of aquatic invasive animals: r selection versus resistance. *Can. J. Fish. Aquat. Sci.* 59:1235–1244.

McMahon, R. F. 2011. Quagga mussel (*Dreissena rostriformis bugensis*) population structure during the early invasion of Lakes Mead and Mohave January-March 2007. *Biol. Invasions* 6:131–140.

McMahon, R. F. and A. E. Bogan. 2001. Bivalves. In *Ecology and Classification of North American Freshwater Invertebrates*, 2nd Edn., J. H. Thorp and A. P. Covich, eds., pp. 331–428. New York: Academic Press.

McMahon, R. F. and T. A. Ussery. 1995. Thermal tolerance of zebra mussels (*Dreissena polymorpha*) relative to rate of temperature increase and acclimation temperature. Contract report EL-95-10, Vicksburg, MS: U.S. Army Corps of Engineers.

McMahon, R. F., T. A. Ussery, and M. Clarke. 1993. Use of emersion as a zebra mussel control method. Contract report EL-93-1. Vicksburg, MS: U.S. Army Corps of Engineers.

McMahon, R. F., M. A. Matthews, T. A. Ussery, R. Chase, and M. Clarke. 1995. Studies of heat tolerance of zebra mussels: Effects of temperature acclimation and chronic exposure to lethal temperatures. Contract report EL-95-9. Vicksburg, MS: U.S. Army Corps of Engineers.

Mellina, E. 1993. Patterns in the distribution and abundance of zebra mussels (*Dreissena polymorpha*) in the St. Lawrence River in relation to substrate and other physico-chemical factors. MS thesis, McGill University, Montreal, Quebec, Canada.

Mellina, E. and B. Rasmussen. 1994. Patterns in the distribution and abundance of zebra mussels (*Dreissena polymorpha*) in rivers and lakes in relation to substrate and other physico-chemical factors. *Can. J. Fish. Aquat. Sci.* 51:1024–1036.

Mills, E. L., G. Rosenberg, A. P. Spidle, M. Ludyanskiy, Y. Pligin, and B. May. 1996. A review of the biology and ecology of the quagga mussel (*Dreissena bugensis*), a second species of freshwater dreissenid introduced to North America. *Am. Zool.* 36:271–286.

Minchin, D., F. Lucy, and M. Sullivan. 2002. Zebra mussel: Impacts and spread. In *Invasive Species of Europe: Distribution, Impacts and Management*, E. Leppäkoski, S. Gollasch, and S. Olenin, eds., pp. 135–146. Dordrecht, The Netherlands: Kluwer Academic Publishers.

Morse, J. T. 2009. Thermal tolerance, physiological condition, and population genetics of dreissenid mussels (*Dreissena polymorpha* and *D. rostriformis bugensis*) relative to their invasion of waters in the western United States. PhD dissertation, The University of Texas at Arlington, Arlington, TX, 279pp.

Nalepa, T. F., D. L. Fanslow, and S. A. Pothoven. 2010. Recent changes in density, biomass, recruitment, size structure, and nutritional state of *Dreissena* populations in southern Lake Michigan. *J. Great Lakes Res.* 36(Suppl. 3):5–19.

Orlova, M. I., V. V. Khlebovich, and A. Y. Kpmendantov. 1998. Potential euryhalinity of *Dreissena polymorpha* (Pallas) and *Dreissena bugensis* (Andr.). *Russ. J. Aquat. Ecol.* 7:17–28.

Orlova, M. I., J. R. Muirhead, P. I. Antonov, G. Kh. Shcherbina, Y. I. Starobogatov, G. I. Biochiono, T. W. Therriault, and H. J. McIsaac. 2004. Range expansion of quagga mussels *Dreissena rostriformis bugensis* in the Volga River and Caspian Sea basin. *Aquat. Ecol.* 38:561–573.

Orlova, M. I., T. W. Therriaut, P. I. Antonov, G. Kh. Sccherbina. 2005. Invasion ecology of quagga mussels (*Dreissena rostriformis bugensis*): A review of evolutionary and phylogenetic impacts. *Aquat. Ecol.* 39:401–418.

Paukstis, G. L., F. J. Janzen, and J. K. Tucker. 1996. Response of aerially-exposed zebra mussels to subfreezing temperatures. *J. Freshwat. Ecol.* 11:511–517.

Paukstis, G. L., J. K. Tucker, A. M. Bronikowski, and F. J. Janzen. 1999. Survivorship of aerially exposed zebra mussels (*Dreissena polymorpha*) under laboratory conditions. *J. Freshwat. Ecol.* 14:511–517.

Quigley, M. A., W. S. Gardner, and W. M. Gordon. 1993. Metabolism of the zebra mussel (*Dreissena polymorpha*) in Lake St. Clair of the Great Lakes. In *Zebra Mussels: Biology, Impacts and Control*, T. F. Nalepa and D. W. Schloesser, eds., pp. 295–306. Boca Raton, FL: CRC Press.

Rajagopal, S., G. van der Velde, H. A. Jenner, M. van der Gaag, and A. J. Kempers. 1996. Effects of temperature, salinity, and agitation on byssal thread formation of zebra mussel *Dreissena polymorpha*. *Neth. J. Aquat. Ecol.* 30:187–195.

Ricciardi, A. R., R. Serrouya, and F. G. Whoriskey. 1995. Aerial exposure tolerance of zebra and quagga mussels (Bivalvia: Dreissenidae): Implications for overland dispersal *Can. J. Fish. Aquat. Sci.* 52:470–477.

Ricciardi, A. R. and F. G. Whoriskey. 2004. Exotic species replacement: Shifting dominance of dreissenid mussels in the Soulanges Canal, upper St. Lawrence River, Canada. *J. N. Am. Benthol. Soc.* 23:507–514.

Roe, S. and H. MacIsaac. 1997. Deep-water population structure and reproductive state of quagga mussels (*Dreissena bugensis*) in Lake Erie. *Can. J. Fish. Aquat. Sci.* 54:2428–2433.

Spidle, A. P., E. L. Mills, and B. May. 1995. Limits to tolerance of temperature and salinity in the quagga mussel (*Dreissena bugensis*) and zebra mussel (*Dreissena polymorpha*). *Can. J. Fish. Aquat. Sci.* 52:2018–2119.

Sprung, M. 1987. Ecological requirements of developing *Dreissena polymorpha* eggs. *Arch. Hydrobiol. Suppl.* 1:69–86.

Sprung, M. 1992. Observations on shell growth and mortality of *Dreissena polymorpha* in lakes. *Limnol. Aktuell* 4:19–28.

Sprung, M. 1995a. Physiological energetics of the zebra mussel *Dreissena polymorpha* in lakes. II. Food uptake and gross efficiency. *Hydrobiologia* 304:133–146.

Sprung, M. 1995b. Field and laboratory observations of *Dreissena polymorpha* larvae: Abundance, growth, mortality and food demands. *Arch. Hydrobiol.* 115:537–561.

Sprung, M. 1995c. Physiological energetics of the zebra mussel *Dreissena polymorpha* in lakes. II. Food uptake and gross efficiency. *Hydrobiologia* 304:133–146.

Sprung, M. and J. Borcherding. 1991. Physiological and morphometric changes in *Dreissena polymorpha* (Mollusca: Bivalvia) during a starvation period. *Malacologia* 33:179–191.

Sprung, M. and U. Rose. 1988. Influence of food size and food quality on feeding of the mussel *Dreissena polymorpha*. *Oecologia* 77:526–532.

Stepien, C. A., C. D. Taylor, and K. A. Dabrowska. 2002. Genetic variability and phylogeographical patterns of a nonindigenous species invasion: A comparison of exotic vs. native zebra and quagga mussel populations. *J. Evolut. Biol.* 15:314–328.

Stoeckmann, A. 2003. Physiological energetics of Lake Erie dreissenid mussels: A basis for the displacement of *Dreissena polymorpha* by *Dreissena bugensis*. *Can. J. Fish. Aquat. Sci.* 60:126–134.

Stoeckmann, A. and D. Garton. 1997. A seasonal energy budget for zebra mussels (*Dreissena polymorpha*) in western Lake Erie. *Can. J. Fish. Aquat. Sci.* 54:2743–2751.

Stoeckmann, A. and D. Garton. 2001. Flexible energy allocation in zebra mussels (*Dreissena polymorpha*) in response to different environmental conditions. *J. N. Am. Benthol. Soc.* 20:486–500.

Strayer, D. L. and L. C. Smith. 1993. Distribution of the zebra mussel (*Dreissena polymorpha*) in estuaries and brackish waters. In *Zebra Mussels: Biology, Impacts, and Control*, T. F. Nalepa and D. W. Schoesser, eds., pp. 715–727. Boca Raton, FL: CRC Press.

Summers, R. B., J. H. Thorp, J. E. Alexander, Jr., and R. D. Fell. 1996. Respiratory adjustment of dreissenid mussels (*Dreissena polymorpha* and *Dreissena bugensis*) in response to chronic turbidity. *Can. J. Fish. Aquat. Sci.* 53:1626–1631.

United States Geological Survey. 2010a. NAS—Nonindigenous aquatic species, zebra mussel (*Dreissena polymorpha*). http://nas.er.usgs.gov/queries/factsheet.aspx?speciesid=5 (accessed November 24, 2010).

United States Geological Survey. 2010b. NAS—Nonindigenous aquatic species, quagga mussel (*Dreissena rostriformis bugensis*). http://nas.er.usgs.gov/queries/FactSheet.aspx?speciesID=95 (accessed November 24, 2010).

Ussery, T. A. and R. F. McMahon. 1995. Comparative study of the desiccation resistance of zebra mussels (*Dreissena polymorpha*) and quagga mussels (*Dreissena bugensis*). Contract report EL-95-6, U.S. Army Corps of Engineers, Water Ways Experiment Station, Vicksburg, MS, 18 pp.

Vinogradov, G. A., N. F. Smirnova, V. A. Sokolov, and A. A. Bruznitsky. 1993. Influence of chemical composition on the mollusc *Dreissena polymorpha*. In *Zebra Mussels: Biology, Impacts, and Control*, T. F. Nalepa and D. W. Schoesser, eds., pp. 283–293. Boca Raton, FL: CRC Press.

Walton, W. C. 1996. Occurrence of zebra mussel (*Dreissena polymorpha*) in the oligohaline Hudson River, New York. *Estuaries* 19:612–618.

Walz, N. 1978a. The energy balance of the freshwater mussel *Dreissena polymorpha* Pallas in laboratory experiments and in Lake Constance. III. Growth under standard conditions. *Arch. Hydrobiol. Suppl.* 2:121–141.

Walz, N. 1978b. The energy balance of the freshwater mussel *Dreissena polymorpha* Pallas in laboratory experiments and in Lake Constance. IV. Growth in Lake Constance. *Arch. Hydrobiol. Suppl.* 2:143–1456.

Walz, N. 1979. The energy balance of the freshwater mussel *Dreissena polymorpha* Pallas in laboratory experiments and in Lake Constance. V. Seasonal and nutritional changes in the biochemical composition. *Arch. Hydrobiol. Suppl.* 3/4:235–254.

Whittier, T. R., Ringold, P. L., Herlihy, A. T., and Pierson, S. M. 2008. A calcium-based invasion risk assessment for zebra and quagga mussels (*Dreissena* spp). *Front. Ecol. Environ.* 6:180–184.

Wilcox, S. J. and T. H. Dietz. 1998. Salinity tolerance of the freshwater bivalve *Dreissena polymorpha* (Pallas, 1771) (Bivalvia, Dreissenidae). *Nautilus* 111:143–148.

Wilson, K. A., T. Howell, and D. A. Jackson. 2006. Replacement of zebra mussels by quagga mussels in the Canadian nearshore of Lake Ontario: The importance of substrate, round goby abundance, and upwelling frequency. *J. Great Lakes Res.* 32:11–28.

Wiśniewski, R. 1992. *Dreissena polymorpha* Pallas in the Włocławek Reservoir, its ability to survive during exposure to air. *Proceedings of the Ninth International Malacological Congress*, Edinburgh, U.K., pp. 403–406.

Wright, D. A., E. M. Setzler-Hamilton, J. A. Magee, V. S. Kennedy, and S. P. McIninch. 1996. Effect of salinity and temperature on survival and development of young zebra (*Dreissena polymorpha*) and quagga (*Dreissena bugensis*) mussels. *Estuaries* 19:619–628.

Zhulidov, A. V., A. V. Kozhara, G. H. Scherbina et al. 2010. Invasion history, distribution, and relative abundances of *Dreissena bugensis* in the old world: A synthesis of data. *Biol. Invasions* 12:1923–1940.

CHAPTER 26

Evolutionary, Biogeographic, and Population Genetic Relationships of Dreissenid Mussels, with Revision of Component Taxa

Carol A. Stepien, Igor A. Grigorovich, Meredith A. Gray, Timothy J. Sullivan,
Shane Yerga-Woolwine, and Gokhan Kalayci

CONTENTS

Abstract 403
Introduction 404
    Evolutionary Relationships of the Dreissenidae: *Dreissena*, *Mytilopsis*, and *Congeria* 404
    Biogeography of Dreissenids in Relation to Geologic Events 406
    Background and Evolutionary Questions about *Dreissena* 407
Methods 409
Results and Discussion 421
    Phylogenetic Resolution of the Relationships among *Dreissena* Taxa 421
    Resolution of Potential Subspecies of *D. polymorpha* 422
    Resolution of Putative Subspecies of *D. rostriformis*, Including the Quagga Mussel 425
    Relationships of *D. carinata*, *D.* "*presbensis*," and *D.* "*stankovici*" 427
    Use of Polymerase Chain Reaction to Rapidly Distinguish Zebra and Quagga Mussels 427
    Population Genetic Methods, Including Our Microsatellite Analyses 428
    Genetic Patterns in Eurasia: Native and Invasive Populations 429
    Genetic Diversity of North American Dreissenids 429
    Source Populations of North American Dreissenids 432
    Population Genetics of Dreissenids across North America 433
    Temporal Genetic Changes of North American Dreissenids 436
Acknowledgments 440
References 440

ABSTRACT

The identification of taxa and discernment of evolutionary relationships within the family Dreissenidae have been confounded by morphological plasticity as well as prior lack of a comprehensive DNA sequence data analysis. We thus analyzed the phylogenetic relationships of putative taxa (species and subspecies) in the genus *Dreissena* in relation to its nearest living relatives (*Mytilopsis leucophaeata* and *Congeria kusceri*) using DNA sequence data from the nuclear 28S RNA gene and three mitochondrial genes: cytochrome *c* oxidase subunit I (COI), 16S RNA, and cytochrome (cyt) *b* oxidase. Relationships resolved by maximum likelihood and Bayesian phylogenetic trees are robust and congruent and support division of *Dreissena* into three subgenera: *Dreissena*, *Pontodreissena*, and *Carinodreissena*. The subgenus *Pontodreissena* contains two species: *Dreissena caputlacus* and *Dreissena rostriformis*. Putative subspecies once proposed for *D. rostriformis* lack genetic divergence and likely should no longer be recognized; these include *D. r.* "*bugensis*" (the quagga mussel),

D. r. "*grimmi,*" *D. r.* "*distincta,*" and *D. r.* "*compressa.*" The *Pontodreissena* then comprises the sister group (nearest relative) to a clade comprising the other two subgenera (*Dreissena* and *Carinodreissena*). The subgenus *Carinodreissena* contains the valid taxa *Dreissena carinata* and *Dreissena blanci*; both inhabit ancient lakes in the Balkan Peninsula. We consider the once recognized *Dreissena* "*stankovici*" and *Dreissena* "*presbensis*" to be synonyms of *Dreissena carinata*; DNA and morphological evidence supports this conclusion. The subgenus *Dreissena* includes two species, *Dreissena polymorpha* and *Dreissena anatolica*. *D. anatolica* is endemic to Turkey in lakes north of the Mediterranean, and *D. polymorpha* (the zebra mussel) has been widely introduced throughout much of Eurasia and North America, spreading from its native distribution in the Pontocaspian region.

We additionally analyze population genetic variation for invasive and native populations of the zebra mussel *D. polymorpha* (using 11 nuclear DNA microsatellite loci) and the quagga mussel *D. r.* "*bugensis*" (using 9 microsatellite loci) across North America and Eurasia and compare our results with previous studies that used other markers. Results reveal significant genetic structuring of introduced populations from Eurasia and North America for both species. North American invasions of both species were founded from multiple source populations and a large number of propagules, showing no founder effects and substantial genetic diversity. In contrast, recently colonized quagga mussel populations from the Colorado River and California exhibit some founder effects. Genetic compositions of both species have changed over time at given colonization sites, with some populations adding alleles from adjacent populations, some losing them, and most retaining closest similarity to their original composition. In conclusion, these genetic data comprise a valuable baseline for resolving present and future invasion pathways for dreissenids, as well as interpreting patterns of distributions in their native ecosystems.

INTRODUCTION

Evolutionary Relationships of the Dreissenidae: *Dreissena*, *Mytilopsis*, and *Congeria*

Dreissenidae is the single extant family of the superfamily Dreissenoidea, which contains three living genera: *Mytilopsis* Conrad 1857, the relict cave-dwelling *Congeria* Partsch 1837, and *Dreissena* van Beneden 1835. Two taxa in the *Dreissena* genus, the zebra mussel *Dreissena polymorpha* (Pallas 1771) and the quagga mussel *Dreissena rostriformis* "*bugensis*" (Andrusov 1897), are ecologically significant, nonindigenous invaders of freshwater ecosystems, whereas the dark false mussel *Mytilopsis leucophaeata* (Conrad 1831) has been widely introduced to brackish waters such as the Baltic Sea. Ancestral Dreissenidae are believed to have originated in fresh- and brackish-water basins of the Tethys Ocean in Pangaea during the Triassic period, around the time when the continents of North and South America, Europe, and Africa began to separate (Morton 1993). Dreissenid taxa likely then diversified in the mid- to late-Miocene Epoch during the breakup of the brackish Paratethys Sea (Morton 1993, Harzhauser and Mandic 2010; see Figure 26.1). The Paratethys Sea encompassed the vast region of the present Black, Caspian, and Aral Seas; however, many drainage, elevation, and salinity changes from the late-Miocene to the Pleistocene epochs affected the paleogeographic (and genetic) continuity among these basins (reviewed in Zenkevich 1963).

Dreissenids live mostly on hard substrates in fresh and brackish waters and have undergone considerable adaptive radiation. Their morphological characters include: (1) possession of free-swimming larvae for dispersal; (2) a byssus that attaches the adults and juveniles to the substrate (or to various hosts that may act as dispersal vectors); (3) byssal retractor muscles; and (4) reduced pedal retractor muscles whose primarily function is to attach byssal threads to substrates (the foot is not used for extensive locomotion, which was an ancestral trait (Nuttall 1990, Morton 1993).

The dreissenid genera *Mytilopsis* and *Congeria* are apparent sister taxa (nearest relatives), which in turn comprise the sister group to the genus *Dreissena*, based on analyses of DNA characters (Stepien et al. 2001, 2003). *Mytilopsis* and *Congeria* share the morphological character of an anterior byssal retractor muscle that attaches to the apophysis, which is a small triangular tooth on the anterior shell septum (Morton 1993). In *Dreissena*, the anterior byssal retractor muscle directly attaches to the apical septum and the apophysis is absent (Pathy and Mackie 1993).

The number of species comprising the brackish-water genus *Mytilopsis* is disputed. However, extant members are native to the New World, whereas fossils indicate that it also once was widespread in the Old World where it became extinct (Harzhauser and Mandic 2010). The dark false mussel *M. leucophaeata* later was introduced via shipping back into Europe (van der Velde et al. 1992), and the black-striped mussel *Mytilopsis sallei* (Récluz 1849) invaded Asia (Morton 1993). Notably, in 1835, *M. leucophaeata* colonized the Port of Antwerp, Belgium. Over the twentieth century, its geographical range expanded along the Atlantic Coast of Europe and Great Britain, along the Baltic Sea, and was reported in the Dniester Liman along the coast of the northwestern Black Sea in Ukraine[*] (Grigorovich et al. 2002, van der Velde et al. 2010). It also invaded the Atlantic Coast of Brazil (De Souza et al. 2005). *M. leucophaeata* can tolerate seawater during transport in ship ballast water. This species cannot reproduce at such high salinities; its reproduction also requires warm

---

[*] The occurrence of *M. leucophaeata* in Dniester Liman was not confirmed by subsequent sampling by I. Grigorovich and C. Stepien, suggesting that it failed to establish a self-sustaining population.

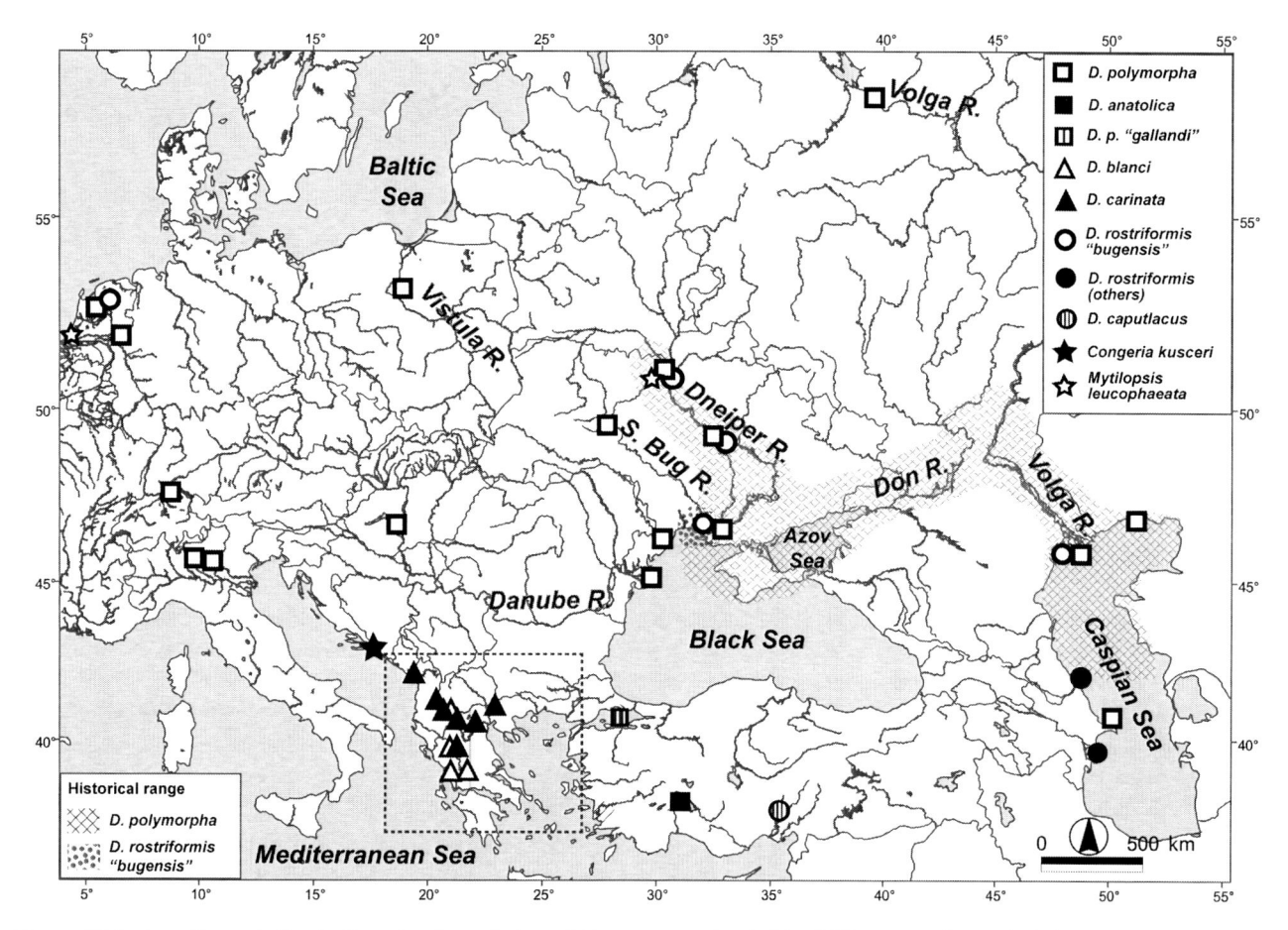

EVOLUTIONARY, BIOGEOGRAPHIC, AND POPULATION GENETIC RELATIONSHIPS OF DREISSENID MUSSELS 405

Figure 26.1 Locations of sites in Eurasia where dreissenid taxa were collected for DNA sequencing and phylogenetic analyses (see Table 26.1). Shaded areas denote original historic ranges of *D. polymorpha* (zebra mussel) and *D. rostriformis* "*bugensis*" (quagga mussel). A close-up view of the area delineated by the dotted square (Balkan region) is given in Figure 26.2.

water temperatures of 15°C–20°C (Schütz 1969). Coolant water discharge by power plants, salination, and global warming are believed to have favored its spread. It often occurs in large densities and causes biofouling problems in brackish-water habitats, similar to problems caused by *Dreissena* in freshwater (van der Velde 2010).

The native range of a related subtropical species, *M. sallei*, extended from the Gulf of Mexico to Colombia; it then invaded Asian ports in the 1960s–1980s and spread widely through India, Taiwan, Japan, Malaysia, and Singapore, where it has caused serious fouling problems (reviewed by Tan and Morton 2006). This spread has been enhanced by its wide temperature (10°C–35°C), salinity (0–50 ppt), and pollution tolerances (reviewed by Galil and Bogi 2009). A genetic study of invasive populations in India, Hong Kong, Taiwan, and Singapore showed high mitochondrial (mt) DNA diversity and implicated multiple and repeated invasions from a number of founding sources (Wong et al. 2011). *M. sallei* then was introduced to the Mediterranean Sea coast of Israel (Galil and Bogi 2009).

The genus *Congeria* contains a wide diversity of extinct Old World coastal species that radiated during the late-Miocene Epoch; all but a single species became extinct during the Messinian salinity crisis about 5 million years ago (Morton 1993). The living relic, *Congeria kusceri* Bole 1962, dates to the Pliocene Epoch and today inhabits a few karst caves in the southern European Dinaric Alps (Morton et al. 1998). *C. kusceri* is distinguished from *Mytilopsis* by its distinct shell form and microstructure, along with its divergent reproductive strategy (Morton 1993, Morton et al. 1998, 2011). Recently, Sket (2011) questioned whether *C. kusceri* actually belongs in the genus *Mytilopsis*; however, Morton maintains that it correctly belongs in that genus (personal communication, 2). *C. kusceri* has a small foot and short thick byssal threads, which indicate that its juveniles and adults are more sedentary than most other dreissenids (Morton et al. 1998). Like other dreissenids, *C. kusceri* occurs in gregarious clusters, but unlike other dreissenids, it is long-lived (reaching 30–40 years) and has low recruitment and low adult mortality (Morton et al. 1998). Thus, there is a fundamental difference in life history traits between the more opportunistic "*r*-selected" species of *Dreissena* and *Mytilopsis* versus the more "*k*-selected" *C. kusceri* (see Stepien et al. 2001).

Stepien et al. (2001) found relatively high mtDNA genetic diversity in *C. kusceri* despite its restricted range and relatively small population sizes, which was attributed to a stable population history in subterranean habitats. Higher genetic diversity of *C. kusceri* compared with lower variability in native populations of *Dreissena polymorpha polymorpha* (zebra mussel) and *D. rostriformis* "*bugensis*" (quagga mussel) indicated that the latter two taxa likely were affected by bottlenecks during population die-offs in ice ages of the Pleistocene Epoch (Stepien et al. 2001).

Biogeography of Dreissenids in Relation to Geologic Events

The fossil record is incomplete, and the appearance of dreissenid taxa in the fossil record likely postdated their evolutionary divergence (see Avise 2004). The mtDNA COI analysis shown in Figure 26.5 conformed to a molecular clock prediction (i.e., the relative divergences among the taxa reflect a relatively steady rate of sequence substitutions over time); this suggests that their evolutionary diversification patterns are relatable to evolutionary time (see Stepien and Kocher 1997, Avise 2004). Geologic history of the dreissenid genus *Mytilopsis* traces back to the ancient Tethys (= Proto-Mediterranean) Sea that once spread over central Europe and western Asia where today's Black, Azov, Caspian, and Aral Seas occur (Starobogatov 1994, Harzhauser and Mandic 2010). Crust movements then portioned the resultant Paratethys Sea into smaller western, central, and eastern basins, whose fauna then evolved independently. After the central Paratethys Sea separated from the ocean, its salinity abruptly declined, which fostered adaptive radiation in *Mytilopsis* (Harzhauser and Mandic 2010).

By the late-Miocene Epoch ~11.6 million years ago (mya), the central Paratethys Sea was replaced by the brackish-water Lake Pannon, at which time ancestral *Congeria* first appeared in the fossil record (Harzhauser and Mandic 2010). Adaptive radiation in this long-lived and physically stable lake produced endemic dreissenid flocks. The earliest representatives of the genus *Dreissena* apparently date to fossil deposits in Lake Pannon ~10.6 mya (Harzhauser and Mandic 2010). By the end of the Miocene Epoch (~6.2 to 5.8 mya), the *Pontodreissena* (*D. rostriformis*/*Dreissena blanci*) lineage had diverged from the common ancestral clade of *Dreissena*/*Carinodreissena* (depicted by the primary bifurcation in the tree of Figure 26.5). Distributional changes of the dreissenid lineages were affected by salinity fluctuations over the course of the early-Pliocene Epoch (Mordukhai-Boltovskoi 1960, Zenkevich 1963). By this time, the Pontic Sea that developed in the area where the today Black, Azov, and Caspian Seas occur was reunited with Lake Pannon, and both ancestral clades of *Dreissena*, including *D. polymorpha* and *D. rostriformis*, penetrated into the Pontocaspian region (Zenkevich 1963). Mountain upheavals (~5.8 to 5 mya) caused the Black Sea and Caspian Sea depressions to separate, although brief periods of contact between them during the late-Pliocene Epoch and the Pleistocene Epoch may have allowed some faunal exchanges (Mordukhai-Boltovskoi 1960). Ancient tectonic lakes then developed in the Balkans between 2 and 4 mya, when *Dreissena carinata* (*Dreissena* "*presbensis*") and *D. blanci* diverged (see Figure 26.2).

The Akchagylian transgression of the Caspian Sea basin (~3.4 mya) temporarily reconnected it with the Black Sea basin through the Manych depression, facilitating faunal interchange between the basins (Zenkevich 1963). Massive immigration of the ancestral Pontocaspian lineages from the Kuyalnik and Chauda Seas to the Apsheron Sea occurred via this link (Mordukhai-Boltovskoi 1960, Zenkevich 1963). This flow of immigrants from the west probably colonized the Baku and Early Khazar Seas (Zenkevich 1963). Formation of the Bosphorus Strait resulted in immigration of the Mediterranean fauna into the ancient Euxinian and the succeeding Uzunlar and Karangat basins (Mordukhai-Boltovskoi 1960, Zenkevich 1963). During the Upper Pleistocene Epoch (~70,000 ya), the link with the Mediterranean Sea was interrupted and Caspian Sea fauna penetrated into the western Girkan Basin through a temporary connection with the Early Khvalyn Sea (Mordukhai-Boltovskoi 1960, Zenkevich 1963). Recent immigration of taxa from the Mediterranean Sea into the Black and Azov Seas resulted from intermittent connections through the Dardanelles and Bosphorus Straits beginning ~9000 to 7000 ya and continuing through the present (Zenkevich 1963).

Throughout the Pleistocene Epoch, salinity of the Black Sea fluctuated. Salinity declined during glacial advances and increased due to saltwater incursions during interglacial periods; in addition, there were two connections and severances from the Mediterranean Sea (Briggs 1974). Most extant genotypes of *Dreissena* species diversified in this epoch, presumably due to these changes in salinity, as well as interglacial range expansions and glacial range contractions. In response to increased salinity, some 159 Pontocaspian taxa occupied isolated freshwater and estuarine habitats, including rivers, limans, and estuaries of the Black, Azov, and Caspian Seas (Mordukhai-Boltovskoi 1960). For example, *D. rostriformis* "*bugensis*" resided in the lower Southern Bug River and Dnieper–Bug Liman, and its Quaternary fossil record appears rich and continuous along the north Black Sea and Azov Sea coasts. Fossil *D. rostriformis* also were described from Pleistocene deposits of the north Black Sea coast area, which were reported as slightly morphologically divergent from modern specimens in the Caspian Sea (Starobogatov 1994). The geological and climatic history of the Paratethys region thus produced the phylogenetic patterns of *Dreissena* diversification seen today.

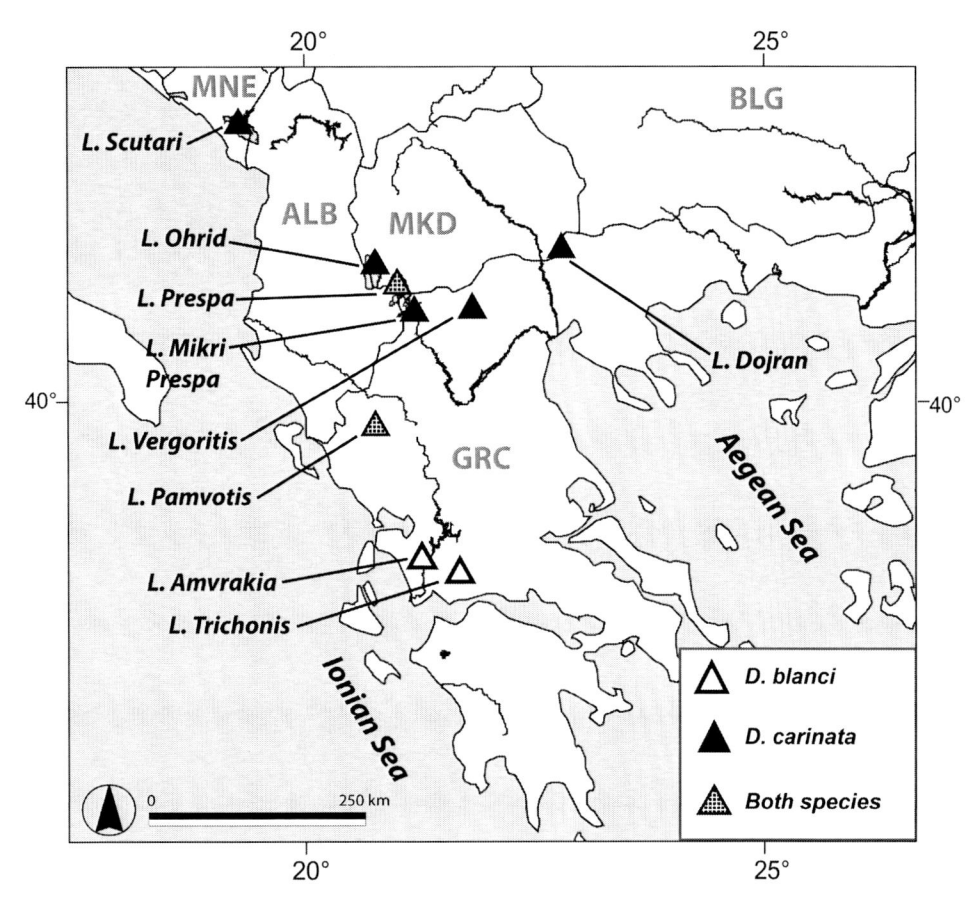

Figure 26.2 Location of sites in the Balkans region of Europe where dreissenid taxa were collected for DNA sequencing and phylogenetic analyses (see Table 26.1).

Background and Evolutionary Questions about *Dreissena*

Traditional taxonomy of dreissenids was assessed by Rosenberg and Ludyanskiy (1994) and revisited by van der Velde et al. (2010). The latter authors summarized some previously published molecular evidence in their report; however, they did not evaluate new or all existing DNA sequence data as is done in this chapter. Dreissenid taxa recognized by van der Velde et al. (2010) included the subgenus *Dreissena* as comprising the extant species *D. polymorpha*, the subgenus *Pontodreissena* containing *D. rostriformis*, and the subgenus *Carinodreissena* with two species—*D. "presbensis"* and *D. blanci*. L'vova and Starobogatov (1982) had recognized *Dreissena "stankovici"* as a full species, which Albrecht et al. (2007) then subsumed into *D. "presbensis."* The two taxa (*D. "presbensis"* and *D. blanci*) then were subsumed into *D. carinata* by Huber (2010; also see World Register of Marine Species, http://www.marinespecies.org), which is evaluated here. We provide a synopsis of the current classification of *Dreissena* (see Table 26.1) and also make recommendations to correct the scientific nomenclature used for *Dreissena* taxa based on our new analyses.

The carinodreissenids—*D. carinata* (*D. "presbensis"*) and *D. blanci*—inhabit ancient Balkan lakes that house high endemic freshwater biodiversity (reviewed by Wilke et al. 2010; see Figure 26.1 and 26.2). The primary ranges of these two taxa differ, with *D. carinata* (*D. "presbensis"*) found primarily in the northern and central Balkan Peninsula, *D. blanci* found in the south, and both found in lakes Pamvotis and Prespa (Albrecht et al. 2007; Figure 26.2). DNA and morphological characters have confirmed their species-level separation (see Albrecht et al. 2007). *D. carinata* (*D. "presbensis"*) has a posterior ventral surface that is convex, whereas *D. blanci* has a surface that is relatively flat; *D. blanci* also has a more extensive keel that extends along most of the shell's length (Albrecht et al. 2007).

Because of their extensive morphological plasticity in various habitats, additional *Dreissena* taxa were hypothesized by some earlier studies, including a variety of putative subspecies for *D. rostriformis* and *D. polymorpha* (see summary by Rosenberg and Ludyanskiy 1994). In the present study, quotation marks are placed around the names of putative taxa that were once proposed, but are not supported by our analyses (Table 26.1). Stepien et al. (2003, 2005) analyzed several putative subspecies of *D. rostriformis* from different depths and substrates in the Caspian Sea,

Table 26.1 Valid Living *Dreissena* Taxa with Taxonomic Authorities, as Supported by DNA Sequence Analyses and Morphological Evidence. Quotations Surrounding a Scientific Name Indicate That the Name Is Invalid and the Taxon Is Not Supported. X = Taxon's Valid Name Is in the Aforementioned Row

Dreissena Subgenus	Authority	*Dreissena* Species	Authority	Subspecies	Authority	Notes on Current Status
Dreissena	van Beneden 1835					Appears to have two valid living species
		D. polymorpha	(Pallas 1771)			Appears to contain a single valid living subspecies
				D. polymorpha polymorpha = zebra mussel	(Pallas 1771)	Valid
				D. p. "gallandi" X	(Locard 1893)	Appears invalid, not divergent from *D. p. polymorpha*
		D. anatolica	Locard 1893			Valid, appears to be supported as a species or subspecies
Pontodreissena	Logvinenko and Starobogatov 1966					Appears to contain two valid species
		D. rostriformis	(Deshayes 1838)	*D. rostriformis*		All subspecies appear invalid
				D. rostriformis "bugensis" = quagga mussel X	Andrusov 1897	Subspecies appears invalid, not separable from Caspian Sea taxa
				D. r. "grimmi" X	Andrusov 1897	Subspecies appears invalid, not separable from *"bugensis"* or others
				D. r. "distincta" X	Andrusov 1897	Subspecies appears invalid
				D. r. "compressa" X	Logvinenko and Starobogatov 1966	Subspecies appears invalid
		D. caputlacus	Shütt 1993			Placement into *Pontodreissena* supported by present study
Carinodreissena	L'vova and Starobogatov 1982					Appears to contain two valid species
		D. carinata	(Dunker 1853)			Valid as a species
		D. "presbensis" X	(Kobelt 1915)			Now *D. carinata* per Huber 2010
		D. "stankovici" X	L'vova and Starobogatov 1982			Now *D. carinata*
		D. blanci	(Westerlund 1890)			Appears to be a distinct taxon in our study from *D. carinata*

Note: Although *D. b. "bugensis"* is used throughout this book, the subspecies name does not merit taxonomic use.

including *Dreissena rostriformis* "grimmi," *Dreissena rostriformis* "distincta," *Dreissena rostriformis* "pontocaspica," and *Dreissena rostriformis* "compressa." Results of this analysis were compared to the quagga mussel *D. r.* "bugensis" from the Black Sea, other areas of Eurasia, and North America and are further evaluated here. The Caspian Sea subspecies once were split on the basis of depth and location, yet their morphological characters overlap extensively (Logvinenko and Starobogatov 1968, Starobogatov 1994). The DNA data of these subspecies are further examined here. A number of *Dreissena* taxa were described from Asia Minor (in Turkey and Iraq), as outlined by van der Velde (2010), whose relationships are further assessed here; these taxa are *D. anatolica*, *D. polymorpha* "gallandi," and *D. caputlacus*. A synopsis of the current classification of *Dreissena*, which includes new results from the present study, is provided in Table 26.1.

METHODS

We surveyed the U.S. National Institutes of Health GenBank database (http://blast.ncbi.nlm.nih.gov) through March 2013 to compile and reevaluate all available DNA sequences from dreissenid taxa encoding the nuclear 28S RNA (ribonucleic acid) gene and three mtDNA genes: cytochrome *c* oxidase subunit I (abbreviated COI), 16S RNA, and cytochrome *b* oxidase (abbreviated cyt *b*) (Tables 26.1 through 26.3). In addition, we sequenced and analyzed new data for representative *Dreissena* taxa from our own collections to provide a more comprehensive analysis. We used primers and modified polymerase chain reaction (PCR) protocols from Stepien et al. (1999, 2001, 2003, 2005) and other sources, including COI, LCO1490 and HCO2198 (Folmer et al. 1994); 16S, 16sarL and 16sbrH (Palumbi 1996); cyt *b*, 151F and 270R (Merritt et al. 1998); Dbucytb and Dpocytb (Brown and Stepien 2010); and 28S, D23F and D6R (Park and O'Foighil 2000), Fbv28S and Du774rc (Hoy et al. 2010), and 5'-CATAGTTCACCATCTTTCGG-3 (from GenBank #FJ455418; Hoy et al. 2010). We compared these sequence data to the reference out-group taxa *M. leucophaeata* and *C. kusceri*, which are the most closely related living relatives of *Dreissena* (see Stepien et al. 1999, 2001, 2003).

Sequence data are biogeographically referenced in Tables 26.1 through 26.3 and Figures 26.1 through 26.4, providing a new and unique resource. Sequences were aligned using CLUSTALX v2.0.12 (Thompson et al. 1997; http://www.clustal.org/), checked manually, and assembled using BIOEDIT v7.0.5 (Hall 1999, 2004; http://www.mbio.ncsu.edu/bioedit/bioedit.html); identical sequences were collapsed to avoid redundancy and confusion (see Table 26.3). Such redundancy has confounded clear interpretation of *Dreissena* data by most other studies, since it was unclear which haplotypes were unique and which were homologous. We eliminated a few sequences that had apparent errors and those that were too short in length, and edited some others (which resulted in combination of equivalent haplotypes). The aim was to include as much sequence data from as many geographic areas, authors, samples, and taxa as possible. Our overall goal thus was to preserve differentiation data, that is, the overall numbers of haplotypes and overall numbers of taxa. Therefore, our data sets contained 538 aligned base pairs (bp) for the mtDNA gene COI, 390 bp for mt 16S RNA, 339 bp for mt cyt *b*, and 615 bp for the nuclear gene 28S RNA.

Our complied data sets for each gene region included new sequences for a variety of taxa that would otherwise have been unavailable, along with sequences from the following published sources: mtDNA COI (Baldwin et al. 1996, Stepien et al. 1999, 2001, Therriault et al. 2004, Gelembiuk et al. 2006, May et al. 2006; Albrecht et al. 2007, Grigorovich et al. 2008, Molloy et al. 2010, Soroka 2010), 16S RNA (Stepien et al. 1999, 2001, Therriault et al. 2004, Molloy et al. 2010), cyt *b* (Stepien et al. 2003, 2005, Brown and Stepien 2010), and nuclear 28S rDNA (Park and O'Foighil 2000, Albrecht et al. 2007, Hoy et al. 2010, Molloy et al. 2010). We submitted all new unique sequences to the NIH GenBank, whose accession numbers are given in Table 26.2. Table 26.2 provides a list of samples, geographic coordinates, genes analyzed for each, sequence haplotypes, and GenBank accession numbers.

Table 26.3 catalogs haplotypes that were determined to be homologous in these sequence regions, including their GenBank designations, geographic locations, and publication sources.

Phylogenetic relationships among dreissenid taxa were analyzed separately for each of the four gene regions using two approaches: maximum likelihood (ML) in PHYML v3.0 (Guindon and Gascuel 2003, Guindon et al. 2005; http://www.atgc-montpellier.fr/phyml/binaries.php) and Bayesian analysis in MrBayes v3.1 (Ronquist and Huelsenbeck 2003; http://mrbayes.csit.fsu.edu/). *M. leucophaeata* and *C. kusceri* were used as out-groups, which together comprise the apparent sister group to *Dreissena* (see Stepien et al. 1999, 2001, 2003; Therriault et al. 2005). MODELTEST v3.7 (Posada and Crandall 1998, Posada 2008; http://darwin.uvigo.es/software/modeltest.html) was employed to determine the simplest best-fit model of evolution for each gene under the Akaike Information Criterion (AIC). The following substitution models were selected: TIM3 + I + G for COI (i.e., Posada 2003 with invariable sites (I) = 0.52 and G-shape parameter = 1.28 of the gamma distribution), TPM2uf + G for 16S rDNA (Kimura 1981 G = 0.13), TVM + G for cyt *b* (Posada 2003, G = 0.65), and TIM3 + G for 28S rDNA (G = 0.16). Bayesian analyses were run for 5 million generations, with the burn-in period determined by plotting log likelihood values at each generation to identify the point in which point stationarity was reached. Nodal support for the ML analyses was determined from 2000 nonparametric bootstrapping pseudoreplications (Felsenstein 1985) and via posterior probability for the Bayesian analyses.

Table 26.2 *Dreissena* Taxa That Have Unique Gene Sequences Used in Our Phylogenetic Analyses (See Trees in Figures 26.5 through 26.8), with Their Geographic Localities, GenBank Accession Numbers, and References

Taxon	Locality	Latitude	Longitude	COI	16S	Cyt *b*	28S	Reference
D. polymorpha polymorpha	St. Lawrence R., NY, USA	44.0806	−76.1944	U47653	—	—	—	Baldwin et al. (1996)
	Eurasia/Great L., N. Amer.	Widespread	Widespread	AF120663	—	—	—	Giribet and Wheeler (2002)
	s. Bug River, UKR	50.2643	30.3328	AF510508 (type 1, F)	—	—	—	Therriault et al. (2004)
	Eurasia/Great L., N. Amer.	Widespread	Widespread	AF510509 (type 2, E)	—	—	—	"
	Eurasia/N. Amer. (Great L./Kansas, USA)	Widespread	Widespread	DQ840121(A)	—	—	—	Gelembiuk et al. (2006)
	s. Volga R. RUS/ s. Ural R., KAZ	44.517/46.1108	48.700/52.4803	DQ840124 (D)	—	—	—	"
	nw. Caspian S. canal, RUS	41.5823	50.4140	DQ840125 (D2)	—	—	—	"
	L. Garda, ITA	45.35	10.31	AM748988–9 (2)	—	—	—	Quaglia et al. (2008)
	L. Konstanz, DEU	47.40	9.10	AM748990	—	—	—	"
	L. Iseo, ITA	45.4411	10.0356	AM748992	—	—	—	"
	L. Como, ITA	46.0156	9.1604	AM748996	—	—	—	"
	L. Garda, ITA	45.35	10.31	AM748997	—	—	—	"
	L. Konstanz, DEU	47.40	9.10	AM749001	—	—	—	"
	w. L. Superior, MN, USA	46.4233	−92.0153	EU484433	—	—	—	Grigorovich et al. (2008)
	"	"	"	EU484435	—	—	—	"
	w. L. Superior, MN, USA/c. Dnieper R., UKR	46.4233/50.2643	−92.0153/30.3328	EU484448	—	—	—	"
	Mohawk R., NY, USA	42.4920	−73.4344	DQ333701	—	—	—	Molloy et al. (2010)
	Eurasia/Great L., N. Amer.	Widespread	Widespread	—	AF038997	—	—	Stepien et al. (1999, 2001, 2003)
	Great L./western USA	Widespread	Widespread	—	DQ333748	—	—	Molloy et al. (2010)/Brown and Stepien (2010)
	s. Volga R., RUS	56.51000	35.3400	—	—	JQ762619 (p-8, ABW5)	—	Stepien et al. (2003, 2005, this study)

	Europe/N. Amer.	Widespread	Widespread	—	—	DQ072117 (p-5, VP5, DpolB)	—	Stepien et al. (2003, 2005)/ Brown and Stepien (2010)
		Danube R., HUN	46.8333	17.3333	—	DQ072120 (p-1, TJ2, Dpol1)	—	"
		w. L. Erie, OH, USA	41.6574	−82.8219	—	DQ072123 (DpolC, VQ10)	—	"
		Great L., N. Amer.	Widespread	Widespread	—	DQ072126 (p-3, QI1, DpoA)	—	"
		L. Wawasee, IN, USA	41.4272	−85.7383	—	GQ988728 (Dpo5, UK5)	—	Brown and Stepien (2010)
		Hudson R., NY, USA	42.2206	−73.8423	—	GQ988729 (Dpo6, AFC18)	—	"
		El Dorado Res., KS, USA	37.8796	−96.8056	—	GQ988730 (Dpo7, AJG8)	—	"
		s. Dnieper R., UKR	46.4089	32.4669	—	GQ988732 (Dpo9, AVI11)	—	"
		Great L., MI, Lake Meade, NV, Amsterdam, NLD	N. Amer./Europe	N. Amer./Europe	—	—	AF131007	Park and O'Foighil (2000)
		c. L. Ontario, NY, USA	43.2059	−77.5304	—	—	FJ455419	Hoy et al. (2010)
		Hedges Lake, NY, USA	43.1076	−73.3832	—	—	FJ455423	"
		s. Dnieper R., UKR	46.4089	32.4669	—	—	JQ700560 (AVI01)	This study
D. p. "gallandi"		L. Buyukcekmece, TUR	41.0167	28.5500	DQ840126	—	—	Gelembiuk et al. (2006)
D. anatolica		L. Beysehir, TUR	37.4619	31.3107	DQ840127	—	—	"
		"	"	"	DQ840129–31 (3)	—	—	"
D. carinata		L. Dojran, GRC	41.2187	22.7793	EF414478	—	—	Albrecht et al. (2007)
		L. Vegoritis, GRC	40.7186	21.7495	EF414479–80 (2)	—	—	"
		L. Ohrid, MKD	41.1051	20.7805	EF414486	—	—	"
		L. Prespa, GRC	40.8203	21.0194	EF414487	—	—	"
		L. Scutari, MNE	42.2352	19.1260	EF414489–90 (2)	—	—	"
		L. Pamvotis, GRC	39.6832	20.8695	EF414491	—	—	"
		L. Vegoritis, L. Dojran, GRC	40.7862	21.8169	—	—	EF414470	"

(continued)

Table 26.2 (continued) *Dreissena* Taxa That Have Unique Gene Sequences Used in Our Phylogenetic Analyses (See Trees in Figures 26.5 through 26.8), with Their Geographic Localities, GenBank Accession Numbers, and References

[b] = both *D. carinata* and *D. "stankovici"*
[a] = *D. "stankovici"*

Taxon	Locality	Latitude	Longitude	COI	16S	Cyt b	28S	Reference
	L. Scutari, MNE/MKD	42.2352	19.126				EF414475[b]	Albrecht et al. (2007)
	L. Ohrid, MKD	41.0236	20.4321	DQ840107–112[a] (6) (N-R)				Gelembiuk et al. (2006)
	L. Prespa, GRC	40.4605	21.0553	DQ840113–20[a] (8)				"
	L. Ohrid, MKD	40.9136	20.7376	EF414484–5[a]				Albrecht et al. (2007)
	"	41.0576	20.8035	DQ333691–92[a] (2)				Molloy et al. (2010)
	"	"	"	DQ333694[a]				"
	"	"	"	DQ333695[a]	DQ333708[a]			"
	"	"	"	DQ333696[a]				"
	"	40.9606	20.7830	DQ333698[a]				"
	"	41.2399	20.7153	DQ333700[a]				"
	L. Ohrid, MKD	41.0200	20.8000		AY302248[a]			Stepien et al. (2003)
	MKD, MNE, GRC	"	"		AF507050[b]			Therriault et al. (2004)
	L. Ohrid, MKD	40.9606	20.7830		DQ333716[a]			Molloy et al. (2010)
	"	41.0576	20.8035		DQ333711[a]			"
	"	"	"		DQ333722–23[a] (2)			"
	"	"	"		DQ333726[a]			"
	"	41.1112	20.7915		DQ333733[a]			"
	"	41.2399	20.7153		DQ333738[a]			"
	"	"	"		DQ333742[a]			"
	"	"	"		DQ333744[a]			"
	L. Ohrid, MKD	41.0200	20.8000			DQ072127[a] (s-1, DsA, ABW3, ADH7; DsC)		Stepien et al. (2003, 2005)
	"	"	"			DQ072128[a] (DsB)		Stepien et al. (2005)
	L. Ohrid, MKD/MNE	40.9136	20.7376				EF414474[a]	Albrecht et al. (2007)
	L. Ohrid, MKD	41.0576	20.8035				DQ333752[a]	Molloy et al. (2010)
	"	40.9606	20.7830				DQ333768[a]	"
	"	41.1112	20.7915				DQ333774[a]	"
	"	41.2399	20.7153				DQ333785[a]	"

	L. Trichonis, GRC	38.5276	21.6561	EF414481–82 (2)	—	DQ333789–90[a] (2) DQ333796[a]	Albrecht et al. (2007)
D. blanci	L. Amvrakia, GRC	38.7679	21.1691	EF414483	—	—	"
	L. Prespa, GRC	40.8203	21.0194	EF414488	—	—	"
	L. Pamvotis, GRC	39.6832	20.8695	EF414492	—	—	"
	L. Trichonis, GRC	38.5276	21.6561	—	—	EF414471–72 (2)	"
D. rostriformis "bugensis"	Eurasia/Great L., N. Amer.	Widespread	Widespread	U47651	—	—	
	c. Dnieper R., UKR/w. L. Erie, OH, USA	50.2643	30.3328	AF479637 (YD08)	—	—	Soroka (2010), This study
	Eurasia/Great L., N. Amer.	Widespread	Widespread	—	AF038996	—	Stepien et al. (1999)
	L. Ontario, NY, USA	43.2880	−77.1414	—	DQ333745	—	Molloy et al. (2010)
	w. L. Erie, OH, USA	41.65740	−82.8219	—	JQ348913 (YD08)	—	This study
	w. L. Erie, OH, USA	41.65740	−82.8219	—	—	DQ072130 (b-1, DbuA, Y18)	Stepien et al. (2003, 2005); Brown and Stepien (2010)
	s. Dnieper R., UKR	46.3800	32.3400	—	—	DQ072133 (Dbu3, AET10)	"
	Eurasia/Great L., N. Amer.	Widespread	Widespread	—	—	DQ072136 (DbuF-G, Dbu5, AEU1)	"
	Eurasia/Great L., N. Amer.	Widespread	Widespread	—	—	DQ072137 (DbuH, AEU2)	"
	n. Volga R., RUS	58.17	37.28	—	—	JQ762620 (b-9, AAW04)	Stepien et al. (2003, 2005, this study)
	e. L. Ontario, NY, USA	44.1318	−76.3390	—	—	JQ762621 (b-10, YL15)	"
	c. Dnieper R., UKR	50.1256	30.8258	—	—	JQ762622 (b-7, ACY06)	"
	s. Dnieper R., UKR	46.3800	32.3400	—	—	JQ762623 (b-6, AAW01)	"
	n. Caspian S., RUS	45.34	47.45	—	—	JQ762624 (b-8, ADF01)	"
	w. L. Erie, OH, USA	41.6574	−82.8219	—	—	JQ762625 (b-1, TS33)	"

(*continued*)

Table 26.2 (continued) *Dreissena* Taxa That Have Unique Gene Sequences Used in Our Phylogenetic Analyses (See Trees in Figures 26.5 Through 26.8), with Their Geographic Localities, GenBank Accession Numbers, and References. Quotation Marks Enclose Former Taxonomic Names That Are Not Supported by Our Study, Which are Regarded as Invalid

Taxon	Locality	Latitude	Longitude	COI	16S	Cyt b	28S	Reference
	w. L. Erie, OH, USA	41.6574	−82.8219	—	—	JQ762626 (b-2, US12)	—	Brown and Stepien (2010), this study
	Eurasia/Great L., USA	"	"	—	—	GQ98873 (Dbu2, AEU6)	—	"
	s. Dnieper R., UKR	46.3800	32.3400	—	—	GQ988738 (Dbu6, AAW2)	—	"
	sw. L. Michigan, MI, USA	42.1163	−86.4937	—	—	GQ988739 (Dbu7, ADV4)	—	"
	s. Volga R., RUS	51.9399	47.3051	—	—	GQ988740 (b-9, Dbu8, ABW4)	—	"
	s. Dnieper R., UKR	46.3800	34.3400	—	—	GQ988741 (Dbu9, ACY8)	—	"
	w. L. Erie, OH, USA	41.6574	−82.8219	—	—	GQ988742 (Dbu10, YD8)	—	"
	s. Dnieper R., UKR	46.3800	32.3400	—	—	GQ988743 (Dbu11, AVH18)	—	"
	w. L. Erie, OH, USA	41.6574	−82.8219	—	—	GQ988744 (Dbu12, ALR2)	—	"
	w. L. Erie, OH, USA	41.6574	−82.8219	—	—	GQ988745 (Dbu13, AGY17)	—	"
	Great Lakes, MI, USA/	Unspecified	Unspecified	—	—	—	AF131008	Park and O'Foighil (2000)
	s. Dnieper R., UKR	46.3800	32.3400	—	—	—	FJ455428	Hoy et al. (2010)
	c. L. Ontario, NY, USA	43.1223	−77.3150	—	—	—	FJ455430	"
	Seneca L., NY, USA	42.4225	−76.5551	—	—	—	FJ455431–2	"
	Dutch Spr. Qry, PA/	40.6262	−75.3704	—	—	—	FJ455435	"
	c. L. Ontario, USA	43.1223	−77.3150	U47650[c]	—	—	—	Baldwin et al. (1996)
D. r. "*bugensis*"/ "*profunda*"	c. L. Ontario, NY, USA	44.0806	−76.1944	—	—	—	—	
D. r. "*bugensis*" "*profunda*"	e. L. Erie, NY, USA	42.2629	−79.5005	—	—	JQ762627 (b-3, WQ3)	JQ700561 (VR16)	Stepien et al. (2003, this study)
	"	"	"	—	—	JQ762628 (b-4, TJ15)	—	"

[c] = both morphotypes

D. r. "grimmi"	c. Caspian S., AZE	40.82	49.82	—	—	JQ762629 (r-4, AAJ03)	—	Stepien et al. (2003, this study)
	"	"	"	—	—	DQ072138 (DrosA)	—	Stepien et al. (2005)
	"	40.20	49.57	—	—	DQ072139 (DrosB)	—	"
	"	40.20	49.57	—	—	DQ072140 (DrosC)	—	"
	c. Caspian S., AZE	42.50	49.00	—	—	—	JQ700564 (ABW07)	"
D. r. "distincta"	s. Caspian S., AZE	39.63	49.77	JQ756297 (AAJ02)	—	JQ762630 (r-2, AAJ02)	—	"
	"	"	"	—	—	JQ762631 (r-3, ABW06)	—	"
D. r. "compressa"	s. Caspian S., AZE	39.63	49.77	JQ756298 (AAW08)	—	JQ762632 (r-5, AAJ04)	—	"
D. r. "distincta"/ D. r. "compressa"	s. Caspian S., AZE	Unspecified	Unspecified	AF510505–7 (3) (1–3, A-C)	—	—	—	Therriault et al. (2004)
D. caputlacus	Seyhan R., TUR	36.5936	35.2241	DQ840099–104 (6)	—	—	—	Gelembiuk et al. (2006)
M. leucophaeata	Hudson R., NY, USA	41.0420	–73.9183	DQ840106	—	—	—	"
				U47649	—	—	—	Baldwin et al. (1996)
	s. Dniester R., UKR	46.7333	33.2666	—	AF507051 (type 1)	—	—	Therriault et al. (2004)
	Europe	Widespread	Widespread	—	AF507052 (type 2)	—	—	"
	s. Miss. R., LA, USA	30.24	–91.11	—	—	JQ762634 (UT05)	—	Stepien et al. (2005, This study)
	Antwerp H., BEL	51.2189	4.4177	—	—	—	EF414468	Albrecht et al. (2007)
C. kusceri	Metković, HRV	43.0523	17.6481	AF325444	AF320601	JQ762633	—	Stepien et al. (2001, This study)
	"	"	"	—	JQ348915 (SR02)	—	—	Stepien et al. (2003, This study)

[a] Originally identified as the former *D. "stankovici"* (now *D. carinata*). Successional unique haplotypes of the same taxon from the same location are designated by dashes, with the total number of haplotypes given in parentheses. Haplotype designation names used by individual authors in their publications are indicated in parentheses under the GenBank number.

Table 26.3 Homologous Gene Sequences to Those Used in Our Phylogenetic Analyses, Identified by Gene, Taxon, GenBank Accession Numbers, Haplotype Identification, Reference, and Locality

Gene	Taxon	GenBank Number in Trees (Haplotype Designation)	Homologous Sequences	Reference	Locality	Latitude	Longitude
COI	*D. polymorpha polymorpha*	AF120663	—	Giribet and Wheeler (2002)	Great Lakes, MI, USA	Unspecified	Unspecified
			AF510510 (type 3, G)	Therriault et al. (2004)	s. Volga R., RUS	48.7043	44.5173
			AF474404	Soroka (2010)	Insko, POL	53.4394	15.5481
		AF510509 (type 2, E)	—	Therriault et al. (2004)	Europe (RUS, UKR)	Widespread	Widespread
			DQ840122 (B)	Gelembiuk et al. (2006)	Eurasia/Great L., USA	Widespread	Widespread
			DQ480123 (C)	"	n. Caspian S. canal, RUS	41.5823	50.4140
			DQ333702	Molloy et al. (2007)	Mohawk R., NY, USA	42.4920	−73.4344
			EF414495	Albrecht et al. (2007)	L. Razim, ROM	44.9458	28.8646
			EU484455	Grigorovich et al. (2008)	w. L. Superior, MN, USA	46.4233	−92.0153
			AF492005	Soroka (2010)	c. Dnieper R., UKR	50.2643	30.3328
		DQ840121 (A)	—	Gelembiuk et al. (2006)	Eurasia/Great L., N Amer.	Widespread	Widespread
			AM749000	Quaglia et al. (2008)	L. Konstanz, DEU	47.40	9.10
			EU484456	Grigorovich et al. (2008)	w. L. Superior, MN, USA	46.4233	−92.0153
			(YJ15)	This study	w. L Erie, OH, USA	41.5176	−81.7069
			(AJG08)	"	El Dorado Res., KS USA	37.8797	−96.8056
		EU484448	—	Grigorovich et al. (2008)	w. L. Superior, MN, USA	46.4233	−92.0153
			AF479636	Soroka (2010)	c. Dnieper R., UKR	50.2643	30.3328
	D. p. "galland"	DQ840126	—	Gelembiuk et al. (2006)	L. Buyukcekmece, TUR	41.0167	28.5500
			EF414493	Albrecht et al. (2007)	L. Buyukcekmece, TUR	41.0167	28.5500
	D. anatolica	DQ084127	—	Gelembiuk et al. (2006)	L. Beysehir, TUR	37.4619	31.3107
			DQ084128	"	"	"	"
	D. carinata	DQ840113	—	Albrecht et al. (2007)	L. Prespa GRC	40.4606	21.0553
			EF414496	Baldwin et al. (1996)	L. Mikri/Prespa GRC	40.6972	21.0383
	D. rostriformis "bugensis"	U47651	—	Claxton et al. (1998)	e. L. Ontario, NY, USA	44.0806	−76.1945
			AF096765	Therriault et al. (2004)	e. Lake Erie, USA	41.5	−81.7
			AF510504 (D)	Gelembiuk et al. (2006)	Eurasia	Widespread	Widespread
			DQ840132	Molloy et al. (2007)	Eurasia/Great L., USA	Widespread	Widespread
			EF080861–62 (2)	Grigorovich et al. (2008)	Hollands Diep, NLD	51.7	4.58389
			EU484436	Soroka (2010)	w. L. Superior, USA	46.4233	−92.0153
		AF495877	—	This study	c. Dnieper R., UKR	50.2643	30.3328
			(AES07)	Soroka (2010)	s. Dnieper R., UKR	46.3800	31.9658
		AF479367	—	"	c. Dnieper R., UKR	50.2643	30.3328
			(YD08)		c. L. Erie, OH, USA	41.4527	−82.1830

	D. r. "*distincta/ compressa*"	AF510505	—	Therriault et al. (2004)	c. Caspian S., AZE	39.63	49.77
	M. leucophaeata	U47649	DQ840133	Gelembiuk et al. (2006)	c. Caspian S., AZE	39.63	49.77
			—	Baldwin et al. (1996)	Hudson R., NY, USA	41.0420	−73.9183
		AF038997	EF414477	Albrecht et al. (2007)	Antwerp H., North S., BEL	51.3597	4.2916
			—	Stepien et al. (1999, 2003, 2005, this study)	Eurasia/Great L., USA	Widespread	Widespread
	D. polymorpha polymorpha		AF507049	Therriault et al. (2004)	Ingul R., UKR/ Gorky Res., RUS	47.685	32.3833
			EF414464	Albrecht et al. (2007)	L. Buyukcekmece, TUR	41.0167	28.5500
			EF414465	"	L. Tressow, GER	53.8484	11.3178
			EF414466	"	L. Razim, ROM	44.9458	28.8646
			DQ333747	Molloy et al. (2010)	Mohawk R., NY, USA	42.4920	−73.4344
		DQ333748		Molloy et al. (2010)	Mohawk R., NY, USA	42.4920	−73.4344
		(YJ15)		This study	c. L. Erie, OH, USA	41.4527	−82.1830
		(AJG08)		"	El Dorado Res., KS, USA	37.8796	−96.8056
	D. carinata/[a] = "*stankovici*," [b] = both	AF507050[b]	—	Therriault et al. (2004)	L. Ohrid, Prespa, MKD	41.02	20.80
			EF414449	Albrecht et al. (2007)	L. Dojran, GRC	41.2187	22.7792
			EF414450	"	L. Vegoritis, GRC	40.7186	21.7495
			EF414451	"	L. Vegoritis, GRC	40.7862	21.8169
			EF414455[a]	"	L. Ohrid, MKD	40.9136	20.7376
			EF414456[a]	"	L. Ohrid, MKD	40.9865	20.7982
			EF414457	"	L. Ohrid, MKD	41.1051	20.7804
			EF414458	"	L. Prespa, GRC	40.8203	21.0193
			EF414460–61	"	L. Scutari, MNE	42.2352	19.1260
			EF414462	"	L. Pamvotis, GRC	39.6832	20.8695
			EF414467	"	L. Mikri, Prespa, GRC	40.6972	21.0383
			DQ333703–06[a] (4)	Molloy et al. (2010)	L. Ohrid, MKD	41.0576	20.8035
			DQ333710[a]	"	"		
			DQ333713–15[a] (3)	"	"	40.9606	20.7830
			DQ333717–21[a] (5)	"	"	"	"
			DQ333724–25[a] (2)	"	"	"	"
			DQ333727[a]	"	"	"	"
			DQ333730–32[a] (3)	"	"	41.1112	20.7915
16S			DQ333734–37[a] (4)	"	"	"	"
			DQ333739[a]	"	"	41.2399	20.7153
			DQ333743[a]	"	"		
		(ABW03)		This study	"	41.02	20.80
							(continued)

Table 26.3 (continued) Homologous Gene Sequences to Those Used in Our Phylogenetic Analyses, Identified by Gene, Taxon, GenBank Accession Numbers, Haplotype Identification, Reference, and Locality

Gene	Taxon	GenBank Number in Trees (Haplotype Designation)	Homologous Sequences	Reference	Locality	Latitude	Longitude
	D. blanci		EF414452	Albrecht et al. (2007)	L. Trichonis, GRC	38.5276	21.6561
			EF414453	"	"	"	"
			EF414454	"	L. Amvrakia, GRC	38.7678	21.1691
			EF414459	"	L. Prespa, GRC	40.8203	21.0193
			EF414463	"	L. Pamvotis, GRC	39.6832	20.8695
	D. rostriformis "bugensis"/c "profunda"/ "grimmi"/"distincta"/ "compressa"	AF038996[c]	—	Stepien et al. (1999)	Eurasia/N. Amer.	Widespread	Widespread
		AY302247		Stepien et al. (2003)	s. Caspian S., AZE	39.63	49.77
		AF507047		Therriault et al. (2004)	s. Dniester R., UKR	46.7333	33.2666
		AF507048			n. Caspian S., RUS	Unspecified	Unspecified
		DQ333745	DQ333746	Molloy et al. (2010)	L. Ontario, NY, USA	43.2880	−77.1414
	D. caputlacus	DQ840102	—	Gelembiuk et al. (2006)	Seyhan R., TUR	36.5936	35.2241
	M. leucophaeata	AF507052	—		s. Dniester R., UKR	46.7333	33.2666
			EF414448	Albrecht et al. (2007)	Antwerp Harbor, BEL	51.3597	4.2916
Cyt b	D. polymorpha polymorpha	DQ072117 (p-1, TJ2)		Stepien et al. (2003, 2005)	w. L. Superior, MN, USA	46.7431	−92.1243
			DQ072118 (p-4, TQ8)	"	L. Dybrzk, POL	53.8412	17.6196
			DQ072119 (p-2, TG9)	"	s. Mississippi R., LA, USA	30.4695	−91.1966
			DQ072121 (DpolF, AFF9)	"	Danube R., HUN	46.8333	17.3333
			DQ072122 (DpolB, YJ4)	"	w. L. Erie, OH, USA	41.6574	−82.8219
			DQ072124 (DpolD, AEZ10)	"	w. L. Superior, MN, USA	46.7431	−92.1243
			DQ072125 (p-2, DpolE, YK3)	"	e. L. Ontario, NY, USA	44.1318	−76.3390
			GQ988725 (Dpo2, YJ4)	Brown and Stepien (2010)	w. L. Erie, OH, USA	41.6574	−82.8219
		DQ072120 (p-1, TJ2)		Stepien et al. (2003, 2005)	Danube R., HUN	46.8333	17.3333
			GQ988727 (Dpoll, TJ2)	Brown and Stepien (2010)	"	"	"
		DQ072123 (DpolC, YJ10)		Stepien et al. (2003, 2005)	w. L. Erie, OH, USA	41.6574	−82.8219
			GQ988726 (Dpo3, YJ10)	Brown and Stepien (2010)	"	"	"
		DQ072126 (p-3, QI1)		Stepien et al. (2003, 2005)	w. L. Erie, OH, USA	41.6574	−82.8219
			GQ988724 (DpoA, VP5)	Brown and Stepien (2010)	w. L. Superior, MN, USA	46.7431	−92.1243
		GQ988730 (DpolT, AJG8)		"	El Dorado Res, KS, USA	37.8796	−96.8056
			GQ988731 (Dpo8)	"	s. Dnieper R., UKR	46.3800	32.3400

D. carinata = *D.* "*presbensis*"/[a] = "*stankovici*"	DQ072127[a] (s-1, ABW3, ADH7; Ds-A)	—	Stepien et al. (2003, 2005)	L. Ohrid, MKD	41.02	20.80
D. rostriformis "*bugensis*"/[c] also includes profunda type	DQ072130[c] (b-1, Y18, TD15, TS33, AAW3, ADH2, ACY7, ACY10, ADH8)	DQ072129[a] (DsC)	Stepien et al. (2003, 2005)	Eurasia/Great L., N. Amer.	Widespread	Widespread
		DQ072131 (DbugH, AFM11)	Stepien et al. (2005)	Kakhovskii C., UKR	45.46	34.77
		DQ072132 (b-7, ACY6)	Stepien et al. (2003, 2005)	c. Dnieper R., UKR	50.2643	30.3328
		DQ072133–4 (b-6, Dbu2)		s. Dnieper R., UKR	46.3800	32.3400
		DQ072135 (b-4, TB11, TJ12, TJ15)		e. L. Ontario, NY, USA	44.1318	−76.3390
		GQ988733 (DbugB, AET10)	Brown and Stepien (2010)	e. L. Erie, OH, USA[b]	42.4333	−79.8333
	DQ072136 (DbugG, AEU1)	—	Stepien et al. (2003, 2005)	Eurasia/Great L., N. Amer.	Widespread	Widespread
		GQ988737 (DbugG, AEU1)	Brown and Stepien (2010)	Eurasia/Great L., USA	Widespread	Widespread
	DQ072137 (DbugF, AEU2)		Stepien et al. (2003, 2005)	s. Volga R., RUS	45.7884	47.8870
		GQ988736 (DbugF, AEU2)	Brown and Stepien (2010)	Eurasia/Great L., USA	Widespread	Widespread
				s. Volga R., RUS	45.7884	47.8870
28S						
D. polymorpha polymorpha	AF131007		Park and O'Foighil (2000)	Great Lakes, MI, USA	Unspecified	Unspecified
		AM779717	Taylor et al. (2007)	Amsterdam, NLD	52.3729	4.8923
		DQ333804–05 (2)	Molloy et al. (2010)	Mohawk R., NY, USA	42.4920	−73.4344
		FJ455425–27 (2)	Hoy et al. (2010)	L. Mead, NV, USA	36.0920	−114.1264
	FJ455419	—		c. L. Ontario, NY, USA	43.2059	−77.5304
		FJ455420–21 (2)	Hoy et al. (2010)			
D. carinata [a] = "*stankovici*"	EF414470	—	Albrecht et al. (2007)	L. Vegoritis, GRC	40.7862	21.8169
		EF414469		L. Dojran, GRC	41.2187	22.7793
	EF414474[a]	—	Albrecht et al. (2007)	L. Ohrid, MKD	40.9136	20.7376
		EF414475		L. Scutari, MNE	42.2352	19.126
		EF414476			42.2368	19.1323
		DQ333749–50[a] (2)	Molloy et al. (2010)	L. Ohrid, MKD	41.0576	20.8035
		DQ333751[a]			"	"
		DQ333753[a]			"	"
		DQ333755–58[a] (4)			"	"

(continued)

Table 26.3 (continued) Homologous Gene Sequences to Those Used in Our Phylogenetic Analyses, Identified by Gene, Taxon, GenBank Accession Numbers, Haplotype Identification, Reference, and Locality

Gene	Taxon	GenBank Number in Trees (Haplotype Designation)	Homologous Sequences	Reference	Locality	Latitude	Longitude
			DQ333759–62[a](4)	"	"	40.9606	20.7830
			DQ333764–67[a](4)	"	"	"	"
			DQ333769[a]	"	"	"	"
			DQ333771[a]	"	"	"	"
			DQ333772–73[a](2)	"	"	41.1112	20.7915
			DQ333775[a]	"	"	"	"
			DQ333777–79[a](3)	"	"	"	"
			DQ333781[a]	"	"	"	"
			DQ333782–84[a](3)	"	"	41.2399	20.7153
			DQ333786–89[a](4)	"	"	"	"
			DQ333791[a]	"	"	"	"
			DQ333793–95[a](3)	"	"	"	"
			DQ333797–98[a](2)	"	"	"	"
			DQ333799–800[a](2)	"	"	41.0576	20.8035
		DQ333752[a]		Molloy et al. (2010)	L. Ohrid, MKD	41.0576	20.8035
			DQ333754[a]	"	"	40.9606	20.7830
			DQ333763[a]	"	"	"	"
			DQ333770[a]	"	L. Ohrid, MKD	"	"
	D. blanci	EF414472		Albrecht et al. (2007)	L. Vegoritis, GRC	40.7186	21.7495
			EF414473	"	L. Trichonis, GRC	38.5276	21.6561
	D. rostriformis "bugensis"	AF131008		Park and O'Foighil (2000)	Great Lakes, MI, USA	Unspecified	Unspecified
			(AES07)	This study	s. Dnieper R., UKR	46.4089	32.4669
		FJ455435		Hoy et al. (2010)	Dutch Spr. Qry, PA, USA	40.6262	−75.3704
			FJ455436	"	"	"	"
			DQ333802–03 (2)	Molloy et al. (2010)	c. L. Ontario, NY, USA	43.28	−77.57
	D. rostriformis "bugensis"[c] = "profunda" (both morphotypes)	JQ700561 (VR16)		This study	e. L. Erie, NY, USA	42.2629	−79.5005
	D. r. "compressa"	JQ700562 (AAJ04)		"	s. Caspian Sea, AZE	39.63	49.77
	D. r. "distincta"	JQ700563 (ABW06)		"	"	"	"

[a] Originally identified as the former *D. "stankovici"* (now synonymized into *D. carinata*).
[b] Haplotype sequence shared by *D. "presbensis"* and *D. "stankovici"* (here as *D. carinata*).
[c] Haplotype shared by *D. rostriformis "bugensis"* shallow water form and "profunda" form. Haplotype designation names used by individual authors in their publications are indicated in parentheses under the GenBank number.

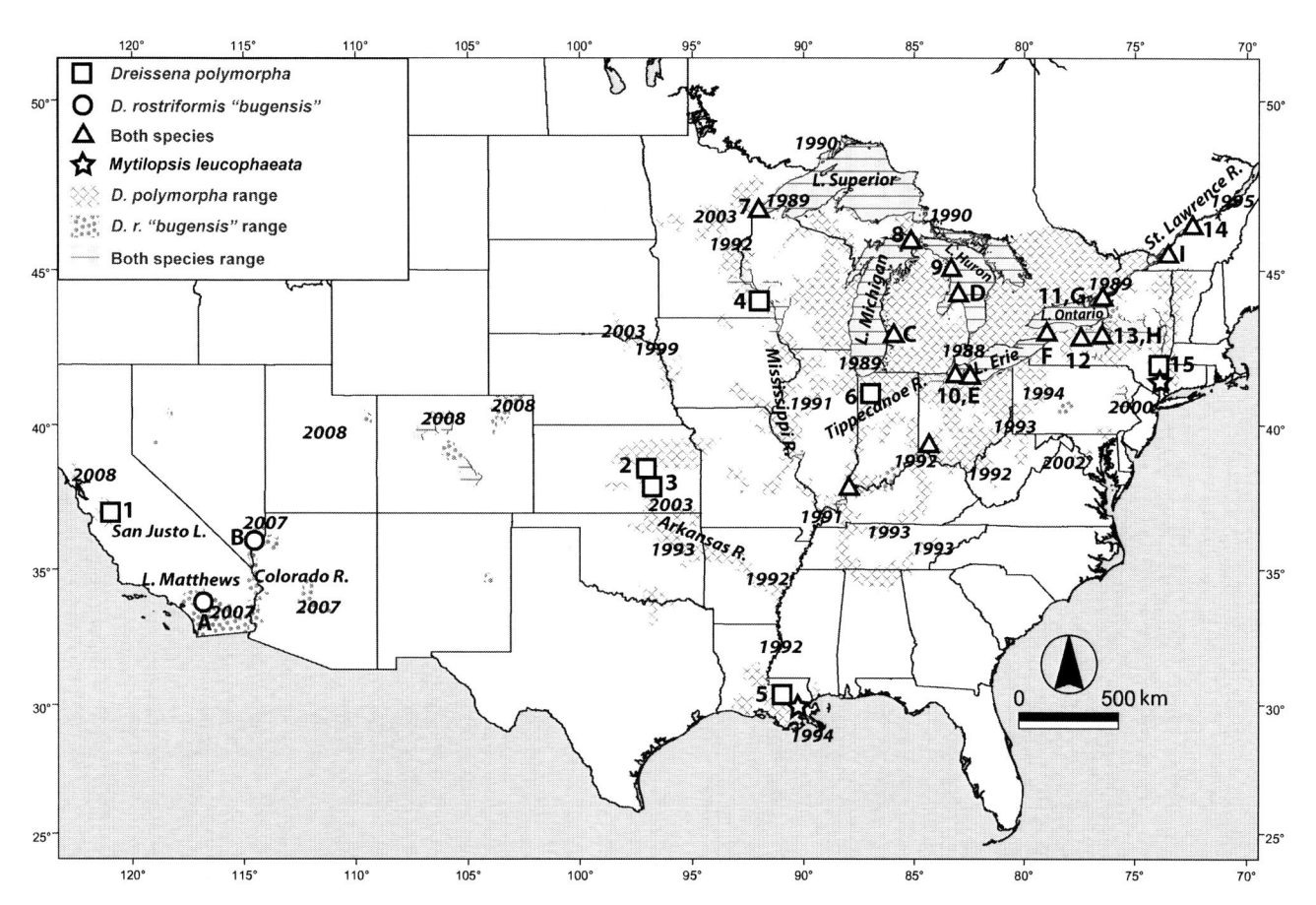

Figure 26.3 Current distribution and dates of first report of zebra mussels and quagga mussels in North America. (Adapted from USGS: Nonindigenous Aquatic Species Website; http://nas.er.usgs.gov/taxgroup/mollusks/zebramussel/) (Benson 2013). Hatched area, zebra mussel; dotted area quagga mussel; striped area, both species. Location of sites where individuals were collected for genetic analysis in this study. (Adapted from Brown, J.E. and Stepien, C.A., *Biol. Invasions*, 12, 3687, 2010.) Zebra mussel, open squares and numbered; quagga mussel, open circles and lettered; both species, open triangles. Also noted are two locations where the dark false mussel *M. leucophaeata* was collected for phylogenetic analyses.

We present tree results from the Bayesian analyses, which are congruent with the trees obtained from ML. The recognition of dreissenid taxa follows the evolutionary species concept (ESC) as described by Wiley and Mayden (2000) and the phylogenetic species concept (PSC) according to Mishler and Theriot (2000). The ESC defines a species as a group of organisms with its own separate evolutionary trajectory over time and space. The PSC defines a species as a monophyletic taxon with diagnosable characters (synapomorphies) and recent genetic coalescence. We additionally employ the operational threshold of Hebert et al. (2004) that a species is distinguished by a degree of variation that is 10 or more times greater than its mean intraspecific variation.

Divergence times among dreissenid COI lineages are evaluated using a penalized likelihood approach (Sanderson 2002) implemented in the program RS v1.71 (Sanderson 2003, 2006: http://loco.biosci.arizona.edu/r8s/). Initially, the complete COI data set was tested for conformance to a molecular clock model, followed by a second analysis with penalized likelihood conducted with an optimal smoothing parameter (= 1.00) determined using cross-validation in RS. Rates of COI nucleotide sequence divergence are related to fossil calibration points based on the divergences between the major clades of dreissenids per Harzhauser and Mandic (2010).

RESULTS AND DISCUSSION

Phylogenetic Resolution of the Relationships among *Dreissena* Taxa

All gene trees show congruent results among *Dreissena* taxa listed in Table 26.1 (Figures 26.5 through 26.8). The mtDNA COI gene has been the focus of most dreissenid research, due to its ease of amplification and popularity as a barcode (e.g., Baldwin et al. 1996, Stepien et al. 1999, 2001, Therriault et al. 2004, Gelembiuk et al. 2006, May et al. 2006, Albrecht et al. 2007, Grigorovich et al. 2008,

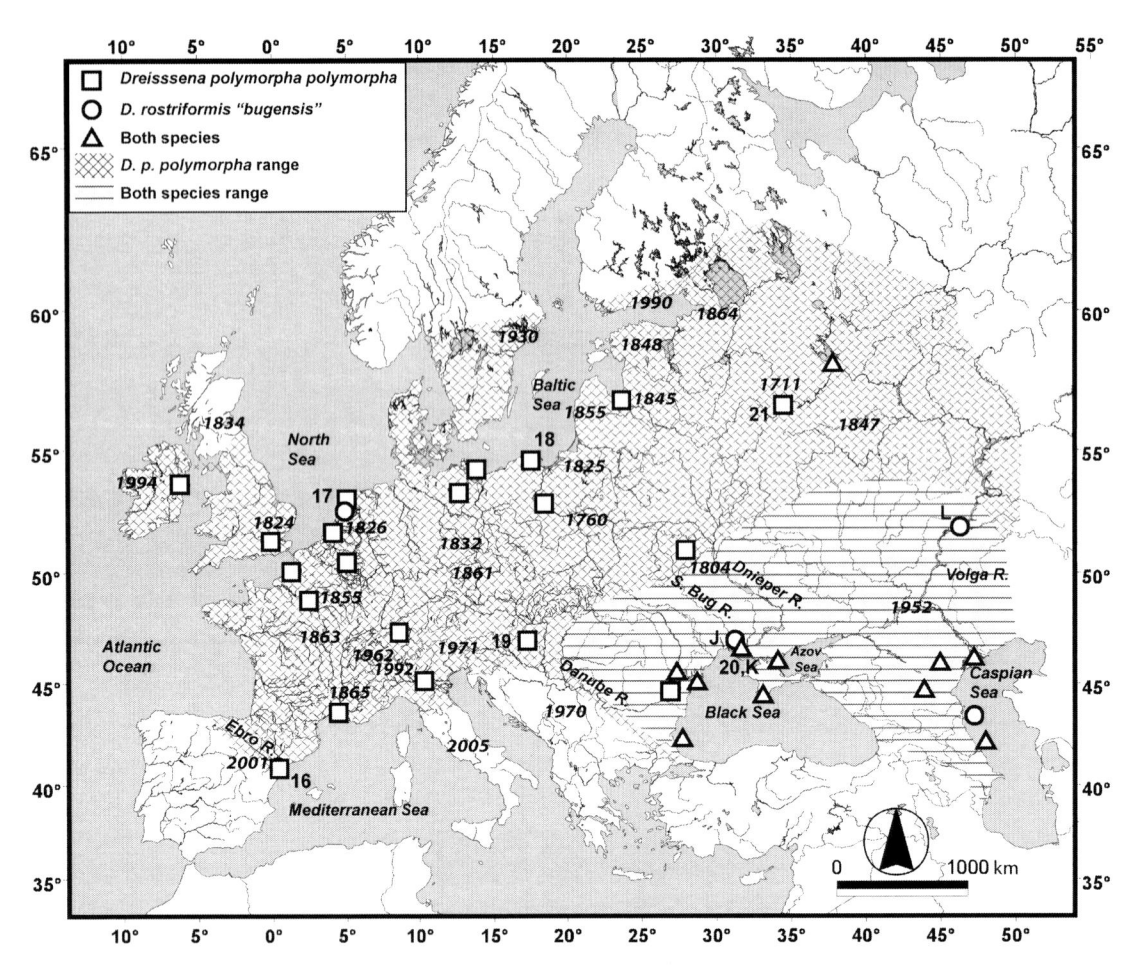

Figure 26.4 Current distribution and date of first report of zebra mussels and quagga mussels in Eurasia. (Adapted from DAISIE; http://www.europe-aliens.org; van der Velde, G. et al., From zebra mussels to quagga mussels: An introduction to the Dreissenidae, In *The Zebra Mussel in Europe*, Van der Velde, G., Rajagopal, S., and Bij de Vaate, A., eds., Backhuys Publishers, Leiden, The Netherlands, pp. 1–10, 2010). Hatched area, zebra mussel distribution; dotted area, quagga mussel distribution; striped area, both species. Location of sites where individuals were collected for genetic analysis in this study. (Adapted from Brown, J.E. and Stepien, C.A., *Biol. Invasions*, 12, 3687, 2010; Feldheim, K.A. et al., *Mol. Ecol. Resour.*, 11, 725, 2011.) Zebra mussel, open squares and numbered; quagga mussel, open circles and lettered; both species, open triangles.

Hoy et al. 2010, Molloy et al. 2010, Soroka 2010). The COI tree (Figure 26.5) depicts three primary *Dreissena* clades, which are the three subgenera—*Pontodreissena*, *Dreissena*, and *Carinodreissena*—and are supported by Bayesian posterior probability values of 0.99%–1.00% and 92%–100% of the ML bootstrap pseudoreplications. The earliest evolutionary division separated the *Pontodreissena* from the common ancestor of the other two subgenera. The subgenera *Dreissena* and *Carinodreissena* are each other's nearest relative (sister group), diverging later from each other (0.99%–1.00%/99%–100%).

The subgenus *Pontodreissena* contains *D. rostriformis* and *D. caputlacus* as clearly definable species (with high support, 0.97%–1.00%/91%–93%). The subgenus *Dreissena* comprises two species, *D. polymorpha* and *Dreissena anatolica* (endemic to Turkish lakes). The subgenus *Carinodreissena* has two taxa: *D. blanci* from Greece (supported by 0.97%/81%) and *D. carinata* in the Balkan Peninsula.

Trees for the other mtDNA genes, 16S RNA (Figure 26.6) and cyt *b* (Figure 26.7), and the nuclear 28S RNA gene (Figure 26.8), are based on fewer sequences and fewer available taxa yet show congruent relationships with the COI tree (Figure 26.5) and each other. Again, the primary bifurcation differentiates the subgenus *Pontodreissena* (*D. rostriformis*) from the clade containing the other two subgenera (*Dreissena* and *Carinodreissena*). All DNA trees thus support all three subgenera. Relationships among primary taxa remain the same in all gene trees, indicating that this phylogeny appears robust and accurate.

Resolution of Potential Subspecies of *D. polymorpha*

All 25 analyzed individuals of the putative subspecies *D. polymorpha* "*gallandi*," which reportedly is endemic to lakes near the Sea of Marmara south of Istanbul (Schütt 1993), possessed a single COI haplotype (DQ840126) that grouped

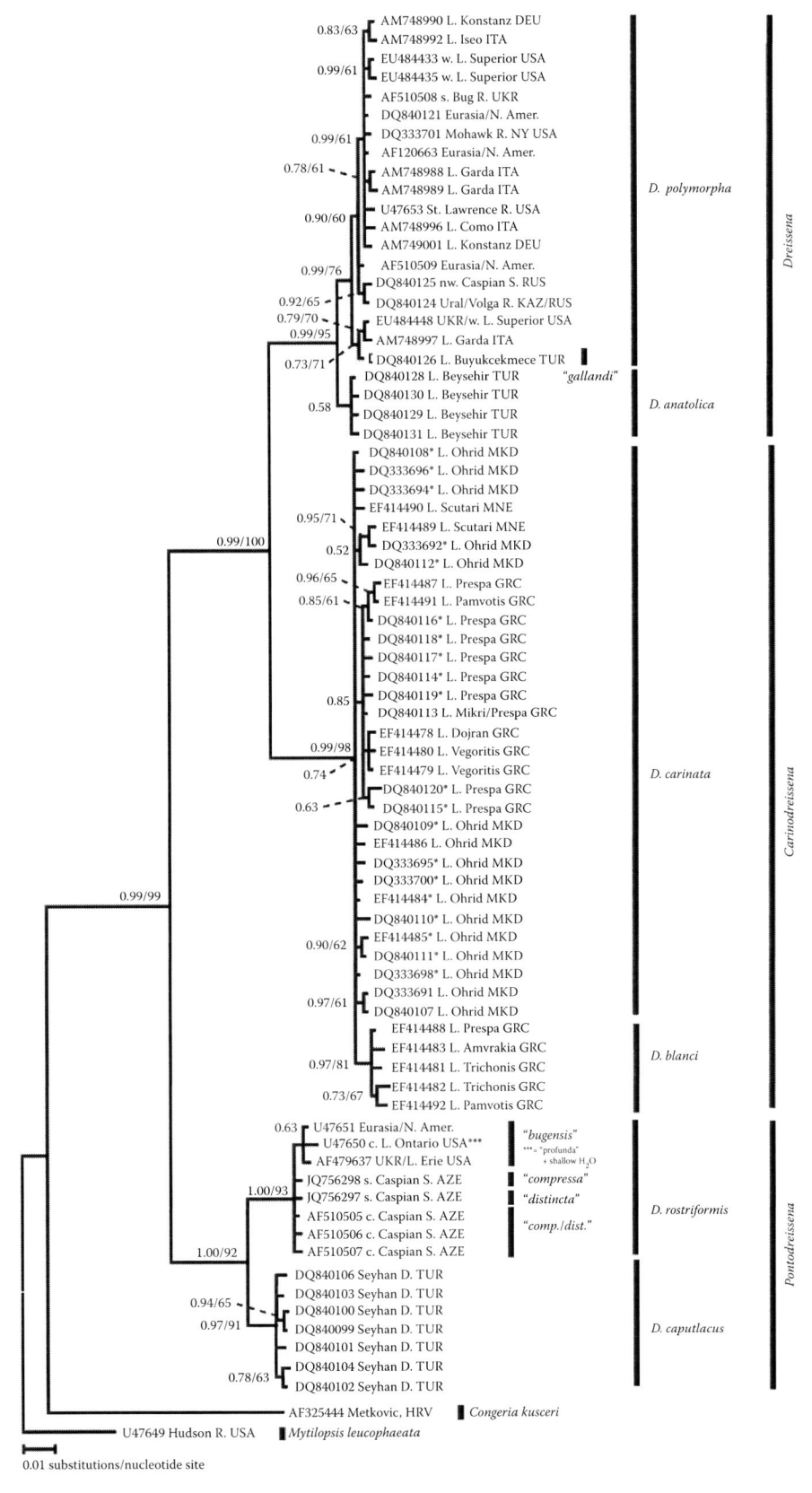

EVOLUTIONARY, BIOGEOGRAPHIC, AND POPULATION GENETIC RELATIONSHIPS OF DREISSENID MUSSELS 423

Figure 26.5 MtDNA COI gene tree of *Dreissena* relationships from Bayesian phylogenetic analysis. Bayesian and ML trees are congruent. GenBank accession sequences (Table 26.1), with geographic location and taxon. Values at nodes, Bayesian posterior probability support/bootstrap percentages from ML tree. *, identified as *D.* "*stankovici*" (now *D. presbensis*); ***, sequence shared by shallow water and "profunda" forms of *D. rostriformis* "*bugensis*."

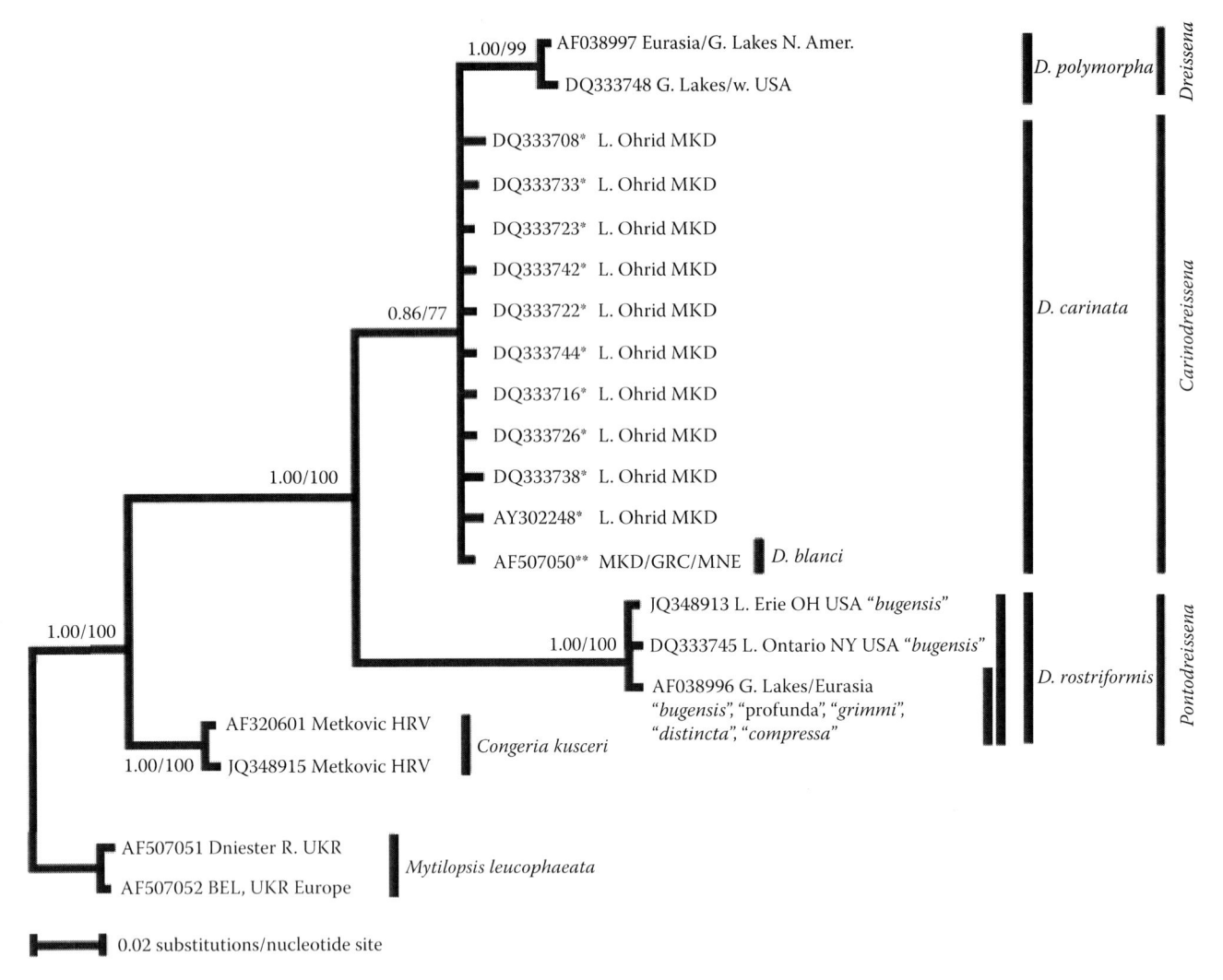

Figure 26.6 MtDNA 16S RNA tree of *Dreissena* phylogenetic relationships derived from Bayesian analysis, which is congruent with the ML tree. GenBank accession sequences (Table 26.1), with geographic location and taxon. Values at nodes are Bayesian posterior probability support/bootstrap percentages from ML tree. *, originally identified as the former *D*. "*stankovici*" (now *D. presbensis*); ***, sequence shared by shallow water and "profunda" forms of *D. rostriformis* "*bugensis*."

with *D. p. polymorpha* haplotypes from Lake Garda, Italy (AM748997), and Lake Superior, United States (EU484448) (Figure 26.5). Our results thus do not support recognition of *D. p.* "*gallandi*." This taxon has not been sequenced for other genes. We recommend that *D. p.* "*gallandi*" should be classified as *D. p. polymorpha*, pending further study with nuclear DNA sequences.

In contrast, four unique COI haplotypes (DQ84028–131) characterized 81 samples of *D. anatolica* from lakes in Turkey (May et al. 2006; Figure 26.5). This clade forms the apparent sister group to a larger clade of 19 haplotypes that encompasses all known sequenced *D. polymorpha* (18 haplotypes) along with the former *D. p.* "*gallandi*" (DQ840126). *D. anatolica* appears to be reciprocally monophyletic and likely comprises a valid taxon that has a separable evolutionary trajectory from *D. polymorpha*. Our results thus lend support to recognizing two species of *Dreissena*: *D. anatolica* and *D. polymorpha* (Figure 26.5). It would be useful to have additional sequences of *D. anatolica*, including the nuclear 28S rDNA gene.

Our analysis resolves 18 unique COI haplotypes for *D. polymorpha* (see Figure 26.5; Table 26.2). The original COI haplotype (U47653) described for zebra mussels from the St. Lawrence River in North America by Baldwin et al. (1996) apparently has not been identified since. Two COI haplotypes of the zebra mussel are widespread in Eurasia and the Great Lakes of North America (AF510590 and DQ9840121), of which the latter has spread westward to Kansas in the United States from the Great Lakes (based on results here). Another haplotype (AF120663) that was identified in the Great Lakes by Giribet and Wheeler (2002) also has been found in the southern Volga River and in Poland by other researchers, but apparently now is relatively rare in North America (Table 26.3). Another haplotype (EU48448) occurs in western Lake Superior as well as in the central Dnieper River in Ukraine. Our analysis

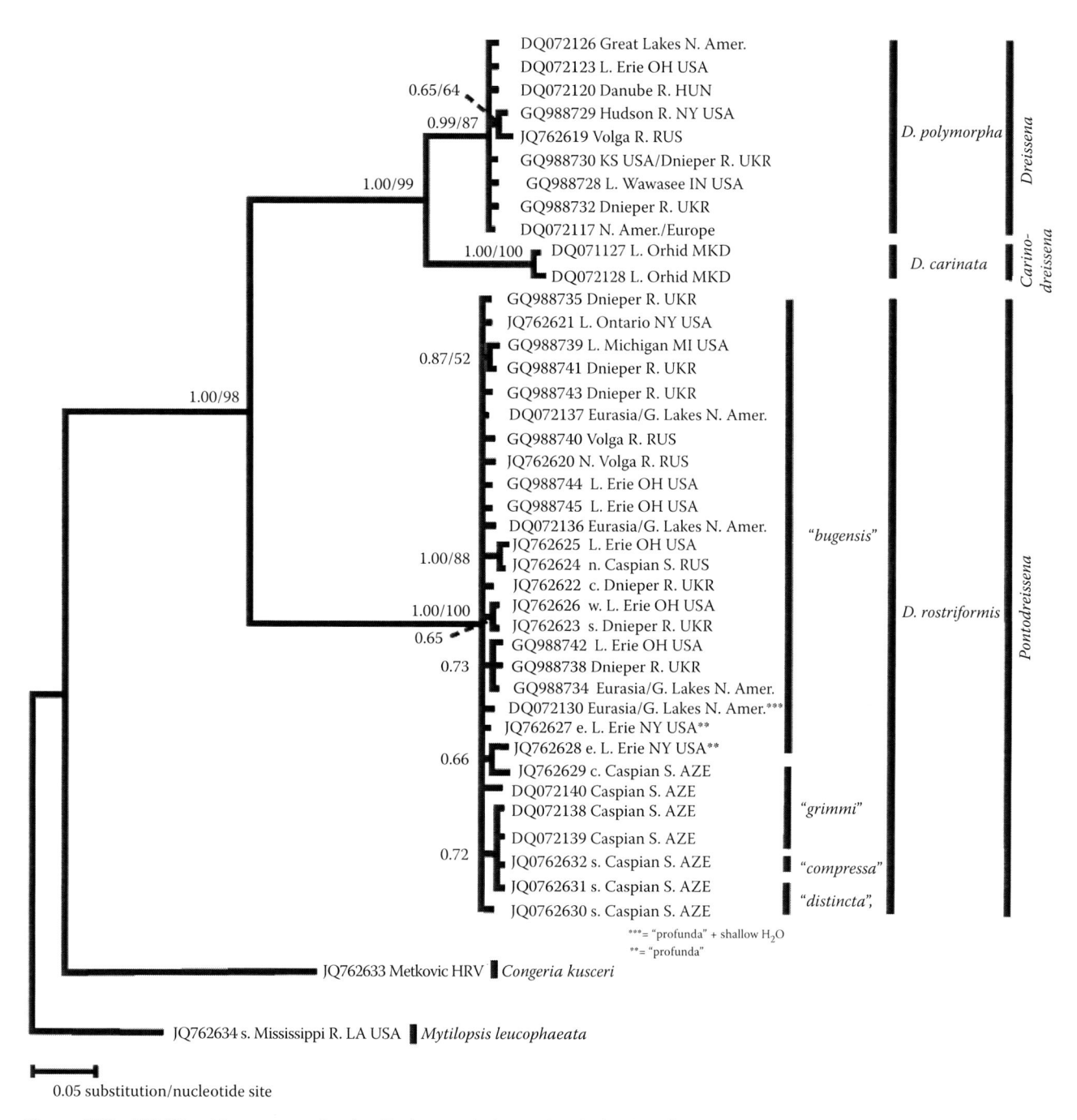

EVOLUTIONARY, BIOGEOGRAPHIC, AND POPULATION GENETIC RELATIONSHIPS OF DREISSENID MUSSELS 425

Figure 26.7 MtDNA cyt *b* gene tree showing *Dreissena* phylogenetic relationships from Bayesian analysis, which is congruent with the ML tree. GenBank accession sequences (Table 26.1), with geographic location and taxon. Values at nodes are Bayesian posterior probability support/bootstrap percentages from ML tree. *, identified as the former *D.* "*stankovici*" (now *D. presbensis*); ***, sequence shared by both the shallow water and "profunda" forms of *D. rostriformis* "*bugensis.*"

identifies additional COI haplotypes; three occur in North America (DQ333701, EU484433 and 35) and 11 are distributed in various Eurasian locations (Tables 26.2 and 26.3). A diversity of nine unique cyt *b* gene zebra mussel haplotypes is shown in Figure 26.7. Together these results reveal a diversity of *D. polymorpha* haplotypes that characterize various zebra mussel populations across its Eurasian and North American ranges.

Resolution of Putative Subspecies of *D. rostriformis*, Including the Quagga Mussel

Several putative subspecies of *D. rostriformis* have been proposed in the literature. The type subspecies *D. r.* "*rostriformis*" is extinct (Rosenberg and Ludyanskiy 1994). Other hypothesized subspecies of *D. rostriformis* have included the quagga mussel *D. r.* "*bugensis*" that was believed to have been

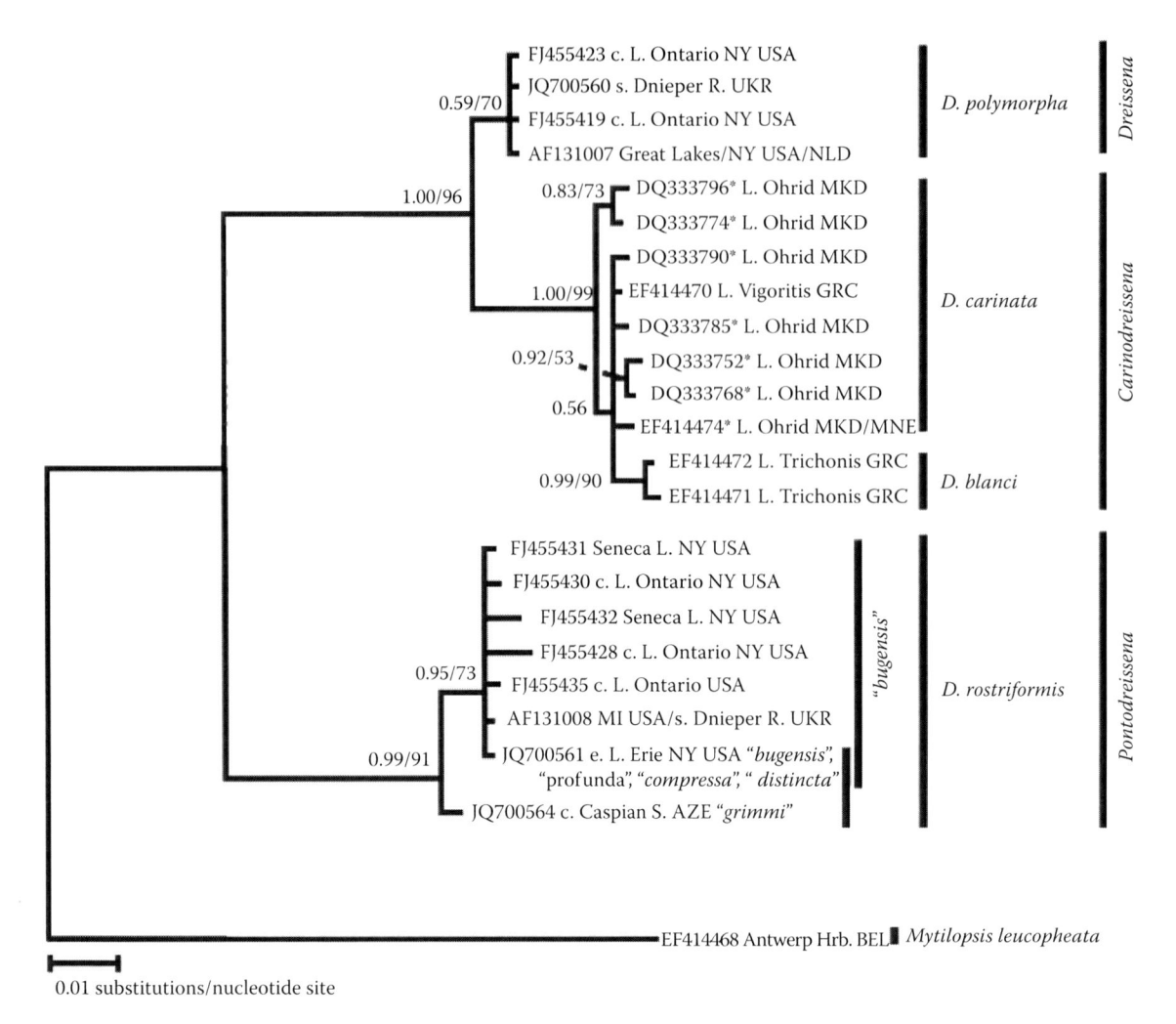

Figure 26.8 Nuclear DNA 28S RNA tree of *Dreissena* phylogenetic relationships from Bayesian analysis, which is congruent with the ML tree. GenBank accession sequences (Table 26.1), with geographic location and taxon. Values at nodes are Bayesian posterior probability support/bootstrap percentages from ML tree. Note that we were unable to obtain congruent sequence from the *C. kusceri* out-group. *, identified as the former *D.* "*stankovici*" (now *D. carinata*); ***, sequence shared by shallow water and "profunda" forms of *D. rostriformis* "*bugensis.*"

endemic to the Southern Bug River estuary in the northern Black Sea region of Ukraine (Figure 26.1A) and the proposed Caspian Sea basin endemics *D. r.* "*grimmi*," *D. r.* "*distincta*," *D. r.* "*compressa*," and *D. r.* "*pontocaspica*" (see Rosenberg and Ludyanskiy 1994, Starobogatov 1994; Figure 26.1). The latter four subspecies were purported to inhabit slightly different ranges and depths in the Caspian Sea, and to vary slightly in shell morphology, however, they overlap in all characters and geographic distributions (Logvinenko and Starobogatov 1968, Starobogatov 1994). Rosenberg and Ludyanskiy (1994) then grouped all proposed Caspian Sea subspecies into *D. r.* "*grimmi*" (thus subsuming *D. r.* "*distincta*," *D. r.* "*pontocaspica*," and *D. r.* "*compressa*"). They stated that *D. r.* "*bugensis*" reaches a larger size, has a more pronounced byssal groove, usually has a broader and more inflated shell, and characteristically is darker in color than individuals from the Caspian Sea. However, Rosenberg and Ludyanskiy (1994) were in doubt as to the taxonomic separation of these subspecies, including *D. r.* "*bugensis*," describing considerable morphological overlap.

We find that COI haplotypes of the quagga mussel *D. r.* "*bugensis*" are only slightly separated from other *D. rostriformis* subspecies (from the Caspian Sea) in the COI tree (Figure 26.5); the quagga mussel lacks Bayesian posterior probability support and has weak bootstrap distinction (63%). We additionally discern five *D. rostriformis* COI haplotypes from the Caspian Sea, which show no distinctiveness. In two of the gene regions tested (mtDNA COI and cyt *b*), haplotype sequences of *D. r.* "*bugensis*" and *D. rostriformis* from the Caspian Sea are not shared (Figures 26.5 and 26.7). For the mtDNA 16S RNA gene, they share an identical sequence (AF038996), including some *D. r.* "*bugensis*" and all other hypothesized subspecies of *D. rostriformis* (Figure 26.6). In the nuclear 28S gene tree, most of the Caspian Sea representatives share a common haplotype with *D. r.* "*bugensis*" from Lake Erie (Figure 26.8). Moreover, the other two genes (COI, cyt *b*) reveal that all putative subspecies of *D. rostriformis* for which sequence data are available (including *D. r.* "*bugensis,*" *D. r.* "*grimmi,*" *D. r.* "*distincta,*" and *D. r.* "*compressa*") are quite closely related (distinguishable at only a single to three nucleotides at most) and are not reciprocally monophyletic (Figures 26.5 and 26.7). Thus, we recommend that all of these putative subspecies be synonymized.

Alternatively, it might be possible to recognize all Caspian types as *D. r.* "*grimmi*" following Rosenberg and Ludyanskiy (1994); however, the lack of clear genetic and morphological distinction is problematic. Side by side, it appears unlikely that taxonomists are able to distinguish between *D. r.* "*bugensis*" and *D. r.* "*grimmi*" or the others. Therriault et al. (2004) also sequenced *D. rostriformis* "*bugensis*" and one or more purported subspecies of *D. rostriformis* (they did not identify the latter to subspecies and provided no morphological characters) and likewise were unable to differentiate them on the basis of the COI or 16S RNA gene sequences. They elected to retain use of the subspecies name of "*bugensis*" since they regarded the two as being geographically separated by salinity tolerances. This practice does not appear to have phylogenetic support or merit, and the salinity tolerances of these taxa have not, to the best of our knowledge and the literature, been empirically tested. Because our study is new, other chapters in this book retained the use of the name "*bugensis,*" which we believe should not be continued in the future. It appears that these groups cannot be reliably distinguished morphologically or genetically.

Dermott and Munawar (1993) described a morphologically distinguishable form of *D. r.* "*bugensis*" they named "profunda" (or alternatively "profundal") from Lake Erie "deeper" water benthic habitats of 10–30 m. The "profunda" form was described as having a broadly rounded, predominately white ventrolateral region and a projecting umbo. The population genetic divergence of the "profunda" form appears unsupported by mt and nuclear DNA sequences, including COI (Baldwin et al. 1996, Claxton et al. 1998), 16S RNA (Stepien et al. 1999; Figure 26.6), cyt *b* sequences (Stepien et al. 2003; Figure 26.7), 28S RNA (Figure 26.8, Table 26.2), or variation at 52 nuclear random amplified polymorphic DNA (RAPD) loci (Stepien et al. 2002). Notably, profunda and shallow water quagga mussel variants share the same sequences for the COI (Claxton et al. 1998), 16S RNA (Table 26.2), cyt *b* (Figure 26.7), and 28S (Figure 26.8) genes. The profunda variant thus likely represents an ecophenotype, reflecting variation in morphology with no overall population genetic differences among the phenotypes. Alternatively, it is possible that it represents the introduction of Caspian Sea region *D. rostriformis* genotypes into the Great Lakes, which then mixed with *D. r.* "*bugensis*" introduced from the Black Sea region. This may merit further investigation, for which we suggest nuclear microsatellite DNA markers, as presented by Brown and Stepien (2010) and Feldheim et al. (2011).

Relationships of *D. carinata*, *D.* "*presbensis,*" and *D.* "*stankovici*"

The species *D. carinata* (formerly *D.* "*presbensis*") contains many haplotypes, which groups together with the formerly recognized *D.* "*stankovici*" (L'vova and Starobogatov 1982) in our phylogenetic trees (Figures 26.5 through 26.8). *D.* "*presbensis*" and *D.* "*stankovici*" were synonymized as *D.* "*presbensis*" by Albrecht et al. (2007); these then were classified as *D. carinata* by Huber (2010), who also included *D. blanci*. We here recognize *D. blanci* as divergent from the *D. carinata* lineage. It may merit recognition as a species or a subspecies, pending further investigation. High genetic diversity of *D. carinata* (*D.* "*presbensis*"; Figures 26.5 through 26.8) indicates that its ancient lake habitats in the Balkan Peninsula (see Figure 26.2) have long been stable environments, including during the Pleistocene glaciations (summarized by Wilke et al. 2010). These lakes contain a large number of other endemic species (Albrecht et al. 2007).

Some individual *D. carinata* (*D.* "*presbensis*") from some locations in the Balkan Peninsula to group together on our COI tree (Figure 26.5), yet there appears to be relatively little genetic differentiation among populations. The geographic range of the formerly recognized *D.* "*stankovici*" is small, encompassing Lake Ohrid (the oldest lake in Europe; Figure 26.2) and possibly other lakes in drainages of the Vardar and Vistritsa Rivers on the border of Albania and Macedonia (L'vova and Starobogatov 1982). Relationships among *D. carinata* (*D.* "*presbensis*") populations from lakes in this region were reviewed by Albrecht et al. (2007) and Wilke et al. (2010). The latter suggested greater genetic distinctiveness of the *D.* "*presbensis*"/*D.* "*stankovici*" clade in Lake Ohrid than was found by Albrecht et al. (2007) and in our analyses (Figures 26.5 through 26.8). Although the genetic diversity of *D. carinata* (*D.* "*presbensis*") is extensive, subspecies are not apparent.

Use of Polymerase Chain Reaction to Rapidly Distinguish Zebra and Quagga Mussels

PCR analysis of DNA sequences can readily and reliably discriminate between zebra and quagga mussels, as well as between other dreissenid species, at all life history stages, including veliger larvae. Sequence differences distinguishing valid dreissenid taxa are apparent in all four of our gene trees (Figures 26.5 through 26.8). Morphological characters (Nichols and Black 1994), however, often are confounded by the plasticity of shell morphology and color pattern in dreissenids. In our experience, most morphologists make some errors in distinguishing dreissenid species, and we have found that most atypical dreissenids, those with highly variable morphologic features across North America, are quagga mussels (*D. rostriformis* "*bugensis*").

Several researchers have developed discrimination tests to separate dreissenid taxa with PCR alone, which may be quicker and less expensive than DNA sequences (but may be less accurate if dependent on presence or absence of the amplicon alone, rather than on the size differences among amplicons). For example, Claxton and Boulding (1998) developed COI primers to amplify dreissenids, followed by restriction digest and gel electrophoresis; their results showed zebra mussels with three bands and quagga mussels with two. Stepien et al. (1999) published a method using PCR-amplified 16S rDNA, followed by restriction digest with two endonucleases, which resulted in species-specific patterns for zebra and quagga mussels, as well as for *M. leucophaeata* and *Corbicula fluminea*. Frisher et al. (2002) developed a technique to distinguish *Dreissena* spp., *M. leucophaeata*, and *C. fluminea* with the nuclear gene 18S rDNA, PCR, a hybridization probe, and a dot blot apparatus. They, however, did not adapt this method to discriminate between zebra and quagga mussels. Another method used a three-PCR primer system for the nuclear 28S rDNA gene, with one primer general to all *Dreissena*, and two others that were specific to either zebra or quagga mussels. However, since amplified products are the same length and cannot be discriminated by size, identification is subject to the vagaries of amplification (Hoy et al. 2010). Also, Hoy et al. (2010) tested their procedure on the unionid *Margaritifera falcata* (western pearl shell mussel) and *C. fluminea*, which did not amplify, but did not test *M. leucophaeata*. Similarly, a PCR test by Ram et al. (2011) used a 16S rDNA primer pair to amplify the zebra mussel (with no product in quagga mussels) and then a COI primer pair to amplify the quagga mussel (and not the zebra mussel). A one-step PCR test that discriminates dreissenid taxa by amplified DNA product length would be an optimal solution that has yet to be realized.

Population Genetic Methods, Including Our Microsatellite Analyses

Despite the ecological importance of dreissenid mussels, relatively little was known of their population genetic patterns until recently (see Brown and Stepien 2010, Feldheim et al. 2011). Notably, most DNA sequence studies focused on the mtDNA COI gene, which amplifies easily, but exhibits relatively modest variability across the population ranges of dreissenid taxa (Tables 26.2 and 26.3 and Figure 26.5). The mtDNA cyt *b* gene shows some population-level resolution (Figure 26.3C), as does the nuclear 28S RNA gene (Figure 26.8), but both also evolve relatively slowly. Since the mtDNA genome is haploid (inherited only from the mother), it has only ¼ of the effective population size of nuclear DNA and is much more sensitive to loss of alleles due to genetic drift from bottlenecks and founder effects (Avise 2004). Founder effects and drift during the Pleistocene glaciations in Europe likely resulted in low numbers of mtDNA haplotypes observed for zebra and quagga mussels (Tables 26.2 and 26.3, Figures 26.5 through 26.8). In contrast, dreissenid taxa in southerly regions that were unglaciated have more mtDNA genetic variability, as evidenced in our phylogenetic analyses with *D. carinata* (*D. "presbensis"*) being a good example (Tables 26.2 and 26.3, Figures 26.5 through 26.8). Faster-evolving nuclear gene regions, such as microsatellites, offer greater potential resolution for discriminating dreissenid population structure, especially in population regions that were glaciated or were founded from them (Brown and Stepien 2010, Feldheim et al. 2011).

Previous studies of nuclear DNA variation of zebra and quagga mussel populations using allozymes (Marsden et al. 1996) and RAPD markers (Stepien et al. 2002) discerned greater genetic variability than found from early mtDNA COI analyses (e.g., Gelembiuk et al. 2006, May et al. 2006; see Tables 26.2 and 26.3). Stepien et al. (2002) described significant genetic divergence of zebra and quagga mussel populations across Europe and North America, including native and introduced regions. Rajagopal et al. (2009) used amplified fragment length polymorphism (AFLP) markers to evaluate origins of the zebra mussel invasion into Spain, but the technique has replication inconsistencies. Studies using nuclear microsatellite loci for dreissenids, until recently, were limited to only a few loci developed by the laboratory of Elizabeth Boulding (University of Guelph, Guelph, Ontario)—six for the quagga mussel (Wilson et al. 1999a,b) and five for the zebra mussel (Naish and Boulding 2001, Astanei et al. 2005). We found that four of those original 11 loci either did not amplify, amplified inconsistently, were not in Hardy–Weinberg equilibrium (HWE), or had null alleles (Brown and Stepien 2010, Feldheim et al. 2011). However, Naish and Boulding (2011) recently corrected errors in their original published primer sequences for zebra mussels. Wilson et al. (1999b) reported genetic divergence among quagga mussel populations in the Great Lakes with the six loci they developed.

Feldheim et al. (2011) developed new assays for several nuclear DNA microsatellites. A total of 386 zebra mussel alleles from 11 loci and 228 quagga mussel alleles from nine loci were recovered, revealing significant genetic diversity and divergence patterns across North America and Eurasia (Tables 26.4 and 26.5; Brown and Stepien 2010, Feldheim et al. 2011). Detailed results of these studies showed that populations of both species generally conformed to HWE expectations following Bonferroni correction, loci were unlinked, and there was little to no evidence of null alleles (based on GENEPOP v. 4.0, http://kimura.univ-montp2.fr/%7Erousset/Genepop.htm, Rousset 2008 and MICRO-CHECKER v2.23, http://www.microchecker.hull.ac.uk, van Oosterhout et al. 2004, 2006). Biological populations (i.e., those with significant divergent genetic compositions) were identified using the *F*-statistic analog θ_{ST} (Weir and Cockerham 1984) and with modified contingency tests (Raymond and Rousset 1995, Goudet et al. 1996); probability values were adjusted via sequential Bonferroni corrections (Rice 1989). Relationships among populations were analyzed with neighbor-joining trees (Saitou and Nei 1987) in PHYLIP v3.69 (Felsenstein 1989; http://evolution.genetics.washington.edu/phylip.html) based on Cavalli-Sforza chord distances (Cavalli-Sforza and Edwards 1967). A 3D factorial correspondence analysis (3D-FCA) (Benzecri 1973) in GENETIX v4.05 (Belkhir et al. 2004; http://www.genetix.univ-montp2.fr/genetix/genetix.htm) explored population divisions without a priori assumptions. Analysis of molecular variance (AMOVA) (Excoffier et al. 1992) in Arlequin v3.5.1.2 (Excoffier and Lishler 2010; http://cmpg.unibe.ch/software/arlequin35/) evaluated hierarchical partitioning of genetic variation (% variance) among geographic groups of populations and sampling years.

The Bayesian analysis programs GENECLASS v2 (Piry et al. 2004; http://www1.montpellier.inra.fr/URLB/) and STRUCTURE v2.3.1 (Pritchard et al. 2000, 2010, Pritchard and Wen 2004; http://pritch.bsd.uchicago.edu/structure_software/release_versions/v2.3.3/html/structure.html) were used to assign individuals to population groups to evaluate possible donor–recipient relationships and expansion pathways. Analyses were conducted with and without prior knowledge of sample identity, and significance was estimated from log likelihood ratios. GENECLASS exclusion tests evaluated sites as likely donors. STRUCTURE assigned individuals to population groups from $K = 1$ (a single population group, i.e., the null hypothesis of panmixia) to $K = N$ (total N of sampling sites); 10 independent runs per K were used with burn-ins of 100,000 replicates and 1,000,000 replicates.

Population genetic bottlenecks usually result in a faster decline of the number of alleles relative to heterozygosity (Luikart et al. 1998a,b). We evaluated population samples for evidence of bottlenecks or rapid range increases with BOTTLENECK v1.2.02 (Cornuet and Luikart 1996, Piry et al. 1999; http://www1.montpellier.inra.fr/URLB/bottleneck/bottleneck.html). In this approach, Wilcoxon tests used a stepwise mutation model to determine whether the number of loci with heterozygosity excess was significantly greater than that expected from equilibrium populations, with 10,000 simulations (Cornuet and Luikart 1996).

Genetic Patterns in Eurasia: Native and Invasive Populations

Previous studies using allozymes and four microsatellite loci discerned significant genetic structure across European populations of zebra mussels, and divergences were greatest in their native distribution area of the northern Black Sea (Müller et al. 2002; see Figures 26.5 through 26.8). Relatively high genetic diversity characterized zebra mussels in the invaded Baltic Sea region, which showed no founder effects (Müller et al. 2001, 2002). Astanei et al. (2005) and Astanei and Gosling (2010) found no significant variation among invasive mussels from Ireland at spatial, temporal, and life history stage scales with five microsatellite loci (including the four used by Müller et al. 2001, 2002). Invasive zebra mussels from Poland likewise showed no differences among various locations, but had high overall polymorphism in allozyme and microsatellite loci (Soroka 2010). Using AFLPs, Rajagopal et al. (2009) found appreciable genetic differences among invasive mussels in northwest Europe, discerning that the Spanish invasion (~2001) appeared to originate in France. Brown and Stepien (2010) and Feldheim et al. (2011) then used 11 microsatellite loci to differentiate among 6 Eurasian populations of zebra mussels, including 4 invasive and 2 native populations (Figures 26.9 through 26.13; Tables 26.4 and 26.5); all had high genetic diversity (average observed heterozygosity (H_O) = 0.70; Table 26.4).

Based on six microsatellite loci, Therriault et al. (2005) found that quagga mussels that invaded the Caspian Sea basin likely entered via multiple pathways from the Black Sea region, showing high gene flow among sites. Using five of those loci, Imo et al. (2010) found no evidence of founder effects in quagga mussel introductions into the Rhine and Main Rivers in Germany, indicating high genetic diversity. The German population traced to at least two independent introductions, attributed to jump dispersal (Imo et al. 2010). Using nine microsatellite loci, Brown and Stepien (2010) and Feldheim et al. (2011) discerned high heterozygosity (average H_O = 0.76) and significant genetic structure, which differentiated two native populations of quagga mussels from the Black Sea, as well as an invasive population in the Volga River (H_O = 0.81; Tables 26.4 and 26.5; Figures 26.9 through 26.13). Results of Therriault et al. (2005) and Brown and Stepien (2010) concurred that high heterozygosity of invasive quagga mussels in the Volga River in the Caspian Sea basin resulted from a large number of founding propagules, likely from multiple source populations (see Table 26.4).

Genetic Diversity of North American Dreissenids

Various genetic studies of zebra and quagga mussel invasions in North America determined diversity levels that appear comparable to their putative Eurasian source populations, suggesting little to no founder effects or bottlenecks. Those genetic studies examined allozymes (Marsden et al. 1995), mtDNA sequences (Stepien et al. 1999, 2003, 2005, May et al. 2006), nuclear RAPD loci (Stepien et al. 2002, 2005), and nuclear microsatellite loci (Brown and Stepien 2010, Feldheim et al. 2011; Table 26.4). Most evidence suggests that North American invasions of both species were founded by large numbers of introduced propagules. This is consistent with the assumption that ballast water discharges by oceanic ships facilitated introductions of large numbers of propagules into the Great Lakes at multiple locations (Kelly et al. 2009), which led to successful establishment (Bax et al. 2003, Drake and Lodge 2004, Holeck et al. 2004).

For various locations in the Great Lakes, our results and those of Brown and Stepien (2010) indicate that overall levels of observed heterozygosities are higher in populations of quagga mussels than in populations of zebra mussels

Table 26.4 Location of Sites Where Populations of Zebra and Quagga Mussels Were Collected for Genetic Studies

	A. Zebra Mussel Sites (Numbered on Figures 26.3 and 26.4 maps)							B. Quagga Mussel Sites (Lettered on Figures 26.3 and 26.4 maps)						
Watershed/Region	Sampling Location	Year Found	N	N_A	H_E	H_O		Sampling Location	Year Found	N	N_A	H_E	H_O	
Western expansion	1. San Justo L., CA, USA	2008	48	114	0.70	0.74		A. L. Matthews, CA, USA	2007	10	39	0.34	0.50
								B. L. Mead, NV, USA	2007	24	56	0.40	0.70
Miss. R. drainage	2. Walnut River, KS, USA	~2003	24	101	0.54	0.71							
	3. Timber Creek, KS, USA	2006	24	104	0.51	0.73							
Upper Miss. R.	4. Lake Pepin, WI, USA	1991	24	94	0.44	0.74							
Lower Miss. R.	5. s. Miss. R., LA, USA	1993	24	126	0.59	0.79							
Ohio R. drainage	6. Tippecanoe R., IN, USA	1994	24	94	0.61	0.69							
Great Lakes													
L. Superior	7. Duluth, MN, USA '95	1989	24	103	0.46	0.59							
	7. Duluth, MN, USA '06		22	81	0.46	0.69							
L. Michigan	8. Mackinac Straits, MI, USA	1989	24	84	0.52	0.66		C. Grand Haven, MI, USA	~1997	24	95	0.46	0.85
L. Huron	9. Alpena, MI, USA	1990	24	74	0.56	0.64		D. Charity Is., MI, USA	~2000	24	110	0.49	0.83
L. Erie	10. Gibraltar Is., OH, USA '02	1998	24	94	0.51	0.76		E. Gibraltar Is., OH, USA '02	~1998	16	49	0.53	0.70
	10. Gibraltar Is., OH, USA '04		23	78	0.54	0.78		E. Gibraltar Is., OH, USA '07		25	90	0.57	0.79
L. Ontario	11. Cape Vincent, NY, USA	1990	24	92	0.56	0.62		F. Olcott, NY, USA	~1990	22	69	0.45	0.72
								G. Cape Vincent, NY, USA	~1990	19	56	0.50	0.64
Erie Canal	12. Clyde, NY, USA	1990	18	75	0.44	0.57							
Oneida Lake	13. Bridgeport, NY, USA	1991	24	116	0.52	0.75		H. Bridgeport, NY, USA	~2008	20	88	0.52	0.77
St. Lawrence R.	14. Becancour, QC, CAN	1989	27	98	0.55	0.71		I. Montreal, QC, CAN	~1992	14	68	0.50	0.75
Hudson R.	15. Catskill, NY, USA '91	1991	22	85	0.49	0.66							
	15. Catskill, NY, USA '03		18	113	0.49	0.66							
Eurasia													
Mediterranean S.	16. Ebro R., ESP	2001	14	65	0.57	0.69							
North S. drainage	17. L. Ijsselmeer, NLD	1827	23	71	0.54	0.67							
Baltic S. drainage	18. Piasnica R., POL	1824	24	78	0.51	0.61							
Black S. drainage	19. Danube R., HUN	1934	24	80	0.50	0.66		J. S. Bug R., UKR	Native	17	57	0.46	0.76
	20. Dnieper R., UKR	Native	30	151	0.62	0.80		K. Dnieper R., UKR	Native	30	75	0.41	0.72
Caspian S. drainage	21. Volga R., RUS	Native	26	107	0.49	0.77		L. Volga R., RUS	1992	24	90	0.51	0.81

Source: Adapted from Brown, J.E. and Stepien, C.A., *Biol. Invasions*, 12, 3687, 2010; Feldheim, K.A. et al., *Mol. Ecol. Resour.*, 11, 725, 2011.
Included are the year of first sighting at the site, sample size (N), number of microsatellite alleles (N_A), and heterozygosity (H_E expected; H_O observed).

EVOLUTIONARY, BIOGEOGRAPHIC, AND POPULATION GENETIC RELATIONSHIPS OF DREISSENID MUSSELS

Table 26.5 Pairwise Tests of Genetic Divergences among Dreissenid Populations from Different Locations as Based on Microsatellite Data

	1	2	3	4	5	6	7'95	7'06	8	9	10'02	10'04	11	12	13	14	15'91
A. Zebra mussel																	
2	0.05+	~															
3	0.05+	0.01	NS	+	+	+	+	+	+	+	+	+	+	+	+	+	+
4	0.08+	0.05+	0.06+	~	+	+	+	+	+	+	+	+	+	+	+	+	+
5	0.08+	0.08+	0.08+	0.05+	+	+	+	+	+	+	+	+	+	+	+	+	+
6	0.08+	0.08+	0.08+	0.10+	0.07+	+	+	+	+	+	+	+	+	+	+	+	+
7'95	0.12+	0.11+	0.11+	0.10+	0.09+	0.05+	+	+	+	+	+	+	+	+	+	+	+
7'06	0.11+	0.09+	0.10+	0.08+	0.09+	0.16+	0.15+	+	+	+	+	+	+	+	+	+	+
8	0.09+	0.13+	0.11+	0.13+	0.10+	0.13+	0.09+	0.11+	+	+	+	+	+	+	+	+	+
9	0.14+	0.18+	0.16+	0.16+	0.15+	0.07+	0.06+	0.15+	0.07+	~	+	+	+	+	+	+	+
10'02	0.16+	0.15+	0.16+	0.13+	0.12+	0.15+	0.18+	0.19+	0.22+	0.21+	~	+	+	+	+	+	+
10'04	0.17+	0.16+	0.16+	0.16+	0.12+	0.17+	0.19+	0.18+	0.22+	0.21+	0.09+	+	+	+	+	+	+
11	0.18+	0.18+	0.18+	0.16+	0.16+	0.16+	0.13+	0.17+	0.22+	0.18+	0.21+	0.23+	~	+	+	+	+
12	0.09+	0.05+	0.05+	0.09+	0.13+	0.16+	0.20+	0.20+	0.16+	0.13+	0.24+	0.26+	0.23+	~	+	+	+
13	0.19+	0.19+	0.18+	0.19+	0.18+	0.16+	0.13+	0.11+	0.13+	0.11+	0.16+	0.16+	0.14+	0.17+	+	+	+
14	0.10+	0.08+	0.08+	0.05+	0.06+	0.06+	0.07+	0.11+	0.16+	0.15+	0.17+	0.18+	0.06+	0.18+	0.08+	~	+
15'91	0.05+	0.03+	0.04+	0.07+	0.08+	0.10+	0.09+	0.16+	0.13+	0.10+	0.21+	0.21+	0.19+	0.05+	0.11+	0.11+	~
15'03	0.15+	0.16+	0.15+	0.13+	0.13+	0.09+	0.10+	0.16+	0.16+	0.10+	0.21+	0.14+	0.10+	0.20+	0.05+	0.09+	0.14+
	0.08+	0.06+	0.05+	0.02+	0.06+	0.11+	0.11+	0.12+	0.16+	0.16+	0.14+						

	A	B	C	D	E'02	E'07	F	G	H	I	J	K					
B. Quagga mussel																	
A. L. Matthews, CA, USA	~	+	+	+	+	+	+	+	+	+	+	+					
B. L. Mead, NV, USA	0.27+	~	+	+	+	+	+	+	+	+	+	+					
C. L. Michigan, MI, USA	0.21+	0.09+	~	+	+	+	+	+	+	+	+	+					
D. L. Huron, MI, USA	0.26+	0.09+	0.01	~	+	+	+	+	+	+	+	+					
E. L. Erie, OH, USA 2002	0.29+	0.17+	0.10+	0.13+		+	+	+	+	+	+	+					
E. L. Erie, OH, USA 2007	0.17+	0.09+	0.04+	0.03+	0.12+		+	+	+	+	+	+					
F. w. L. Ontario, NY, USA	0.21+	0.11+	0.06+	0.08+	0.06+	0.09+		*	+	+	+	+					
G. e. L. Ontario, NY, USA	0.30+	0.18+	0.12+	0.15+	0.09+	0.16+	0.07	~	+	+	+	+					
H. Oneida L., NY, USA	0.18+	0.15+	0.06+	0.05+	0.16+	0.06+	0.11+	0.19+	~	+	+	+					
I. St. Lawrence R., QC, CAN	0.22+	0.15+	0.07+	0.07+	0.19+	0.07+	0.12+	0.20+	0.03+	~	+	+					
J. S. Bug R., UKR	0.12+	0.12+	0.09+	0.10+	0.15+	0.08+	0.09+	0.14+	0.07+	0.10+		+					
K. Dnieper R., UKR	0.25+	0.14+	0.09+	0.09+	0.19+	0.08+	0.13+	0.19+	0.06+	0.06+	0.16+	~					
L. Volga R., RUS	0.21+	0.13+	0.03+	0.03+	0.10+	0.05+	0.06+	0.13+	0.08+	0.09+	0.12+	0.10+					

Source: Adapted from Brown, J.E. and Stepien, C.A., *Biol. Invasions*, 12, 3687, 2010.
F_{ST} analog θ_{ST} below diagonal, contingency tests above diagonal. *$p < 0.05$; +, remains sig. after Bonferroni correction; NS or underline, not sig. Numbers (zebra mussels) and letters (quagga mussels) correspond to locations identified in Figures 26.3 and 26.4 maps.

(average H_O = 0.76 vs. average H_O = 0.68; see Table 26.4). In contrast, populations of zebra mussels outside of the Great Lakes average higher genetic diversity (0.73), whereas quagga mussels that more recently became established in the southwestern United States have lower levels (0.60). Western range expansions of zebra mussels outside the Great Lakes likely had large founder sizes, which may have resulted from continued introductions via commercial and recreational vessels, leading to higher diversity. In contrast, lower genetic diversity in western populations of quagga mussels, especially in California, likely stemmed from a single introduction via the Colorado River aqueduct from Lake Mead. The Lake Mead population appears to have been founded by few individuals, with little subsequent gene influx; this resulted in a founder effect. The latter finding appears to fit the model of low-frequency transport for dreissenid expansions via recreational trailered boats into western North America (Johnson et al. 2001, Bossenbroek et al. 2007).

Source Populations of North American Dreissenids

Population genetic data are fundamental to tracing the source of an invasion(s), determining likely transport pathway(s), and assessing likely vector(s) (Stepien et al. 2005, Rosenthal et al. 2008) and thus may aid targeted management and control efforts (Amsellam et al. 2000, Kang et al. 2007). Correct ecologic comparisons depend on determination of source–colonist relationships. Our studies of microsatellite data indicate that the zebra and quagga mussel introductions to North America (Figure 26.2) were founded from multiple European source locations (Brown and Stepien 2010; Figures 26.3 through 26.4 and 26.9 through 26.13). Genetic evidence suggests that the zebra mussel entered North America in ballast water of ships from the Baltic Sea and northern Europe and the quagga mussel entered in ballast water of ships from northern Black Sea estuaries (Stepien et al. 2003, 2005, Brown and Stepien 2010; see Figures 26.9 through 26.13). These correspond to two of the main invasion pathways identified by Ricciardi and MacIssac (2000) and Grigorovich et al. (2002). Thus, geographic differences in founding sources between the two species suggest that they each arrived in the Great Lakes in different introduction events.

MtDNA cyt *b* sequences, nuclear RAPD analyses, and nuclear DNA microsatellite data suggest that zebra mussel populations in the upper Great Lakes significantly differ from those in the lower Great Lakes (Stepien et al. 2002, 2005, Brown and Stepien 2010; see Figures 26.9 through 26.13); the implication is that theareas had separate founding sources and events. In contrast, an early study with allozyme markers was interpreted as a single source initiating the zebra mussel invasion in the Great Lakes (Marsden et al. 1995). However, a subsequent allozyme study revealed substantial genetic heterogeneity among zebra mussel populations in the Great Lakes and a complex of inland lakes (Lewis et al. 2000).

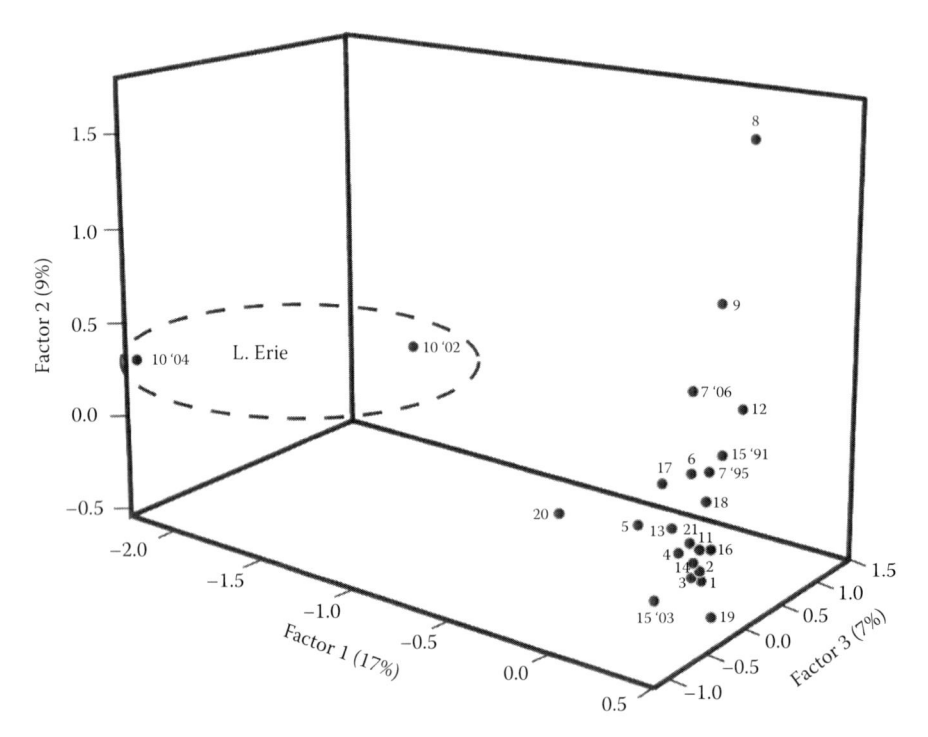

Figure 26.9 A 3D factorial correspondence analysis (GENETIX v4.05; Belkhir et al. 2004) showing population relationships for zebra mussels collected at different sites. Numbers correspond to collection sites as given in Figures 26.3 and 26.4. Years are given for populations that were analyzed for multiple years at the same location. Zebra mussels from Lake Erie were distinct from all mussels in all other locations. ***, sequence shared by both the shallow water and "profunda" forms of *D. rostriformis* "*bugensis*." (Adapted from Brown, J.E. and Stepien, C.A., *Biol. Invasions*, 12, 3687, 2010.)

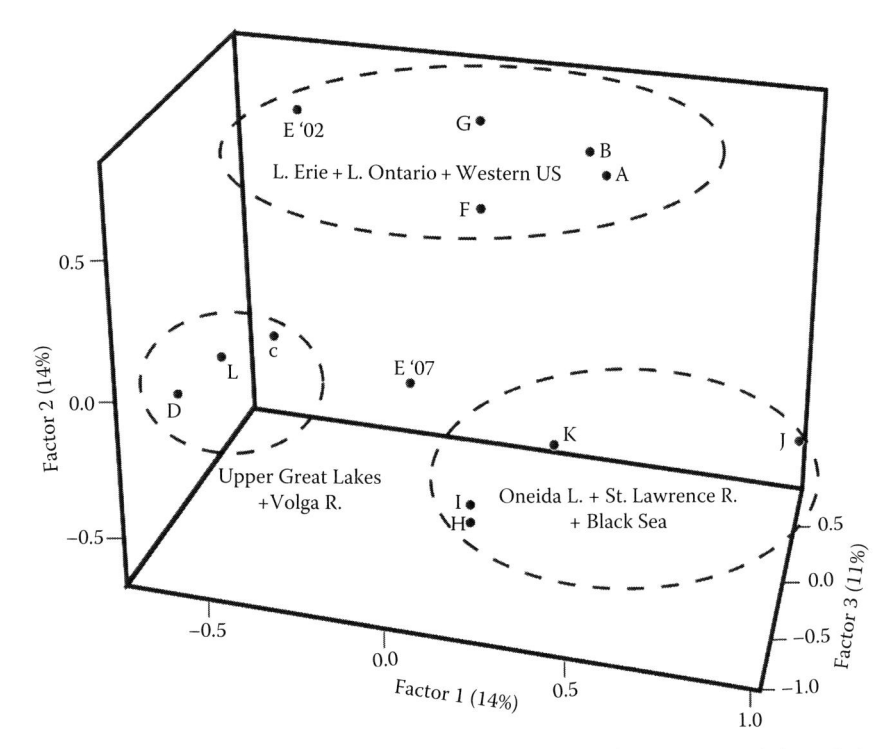

Figure 26.10 A 3D factorial component analysis (GENETIX v4.05; Belkhir et al. 2004) showing population relationships for quagga mussels collected at different sites. Letters correspond to collection sites as given in Figures 26.3 and 26.4. Years are given for populations that were analyzed for multiple years at the same location. Quagga mussel populations clustered into three different groups, with samples from western North America clustering with those from Lake Ontario. (Adapted from Brown, J.E. and Stepien, C.A., *Biol. Invasions*, 12, 3687, 2010.)

An early study of mtDNA COI data was unable to differentiate among populations in North America or most of Europe since their samples only discerned two haplotypes for zebra mussels (with similar frequencies across their sites) and just one for quagga mussels (Gelembiuk et al. 2006, May et al. 2006). However, three types of nuclear DNA data sets have discerned that zebra mussel populations in Lake Erie are distinct from other Great Lakes' locations, which implies separate introduction events (Boileau and Hebert 1993, Stepien et al. 2002, Brown and Stepien 2010; Figures 26.9 through 26.13). Stepien et al. (2002) found that the zebra mussel population in Lake Erie appeared most similar to populations from Poland and The Netherlands (which also had invasive origins). Thus, most studies and the majority of genetic data sets have indicated that the Great Lakes contain multiple and diverse zebra mussel populations, which likely were founded from various European sources.

Brown and Stepien (2010) discerned that the quagga mussel was introduced to North America from at least two founder populations; one colonized Lake Erie and the other apparently colonized most of the other Great Lakes (see Figures 26.9 through 26.13). Thus, in both dreissenid species, populations from Lake Erie appear genetically different from others, suggesting the likelihood of separate introductions from Eurasia. Differentiation of Lake Erie populations appears to coincide with reports by Mills et al. (1996) and Carlton (2008) of both species being in Lake Erie prior to their formal discovery in other regions of the Great Lakes.

Population Genetics of Dreissenids across North America

A common hypothesis is that the spread of an invasive species may lead to high gene flow and homogenize variability across its new range (Viard et al. 2006, Kim et al. 2009). However, Brown and Stepien (2010) disproved this hypothesis for dreissenids because populations of both zebra and quagga mussels show significant genetic divergences across North America (Tables 26.5 through 26.7, Figures 26.9 through 26.13). Such genetic patterns also characterize populations of the Eurasian round goby that invaded the Great Lakes (Dillon and Stepien 2001, Stepien et al. 2005, Stepien and Tumeo 2006, Brown and Stepien 2009). Distinct genetic patterns of the two dreissenid species probably reflect divergent origins from different Eurasian source populations, followed by variable spread to other locations in the Great Lakes and connected rivers. Colonizations of water bodies outside the Great Lakes by dreissenids occurred via trailered boats, as evidenced by genetic patterns from long-distance "jump dispersal" of various founders, with subsequent spread through adjacent rivers and reservoirs (Figures 26.9 through 26.13).

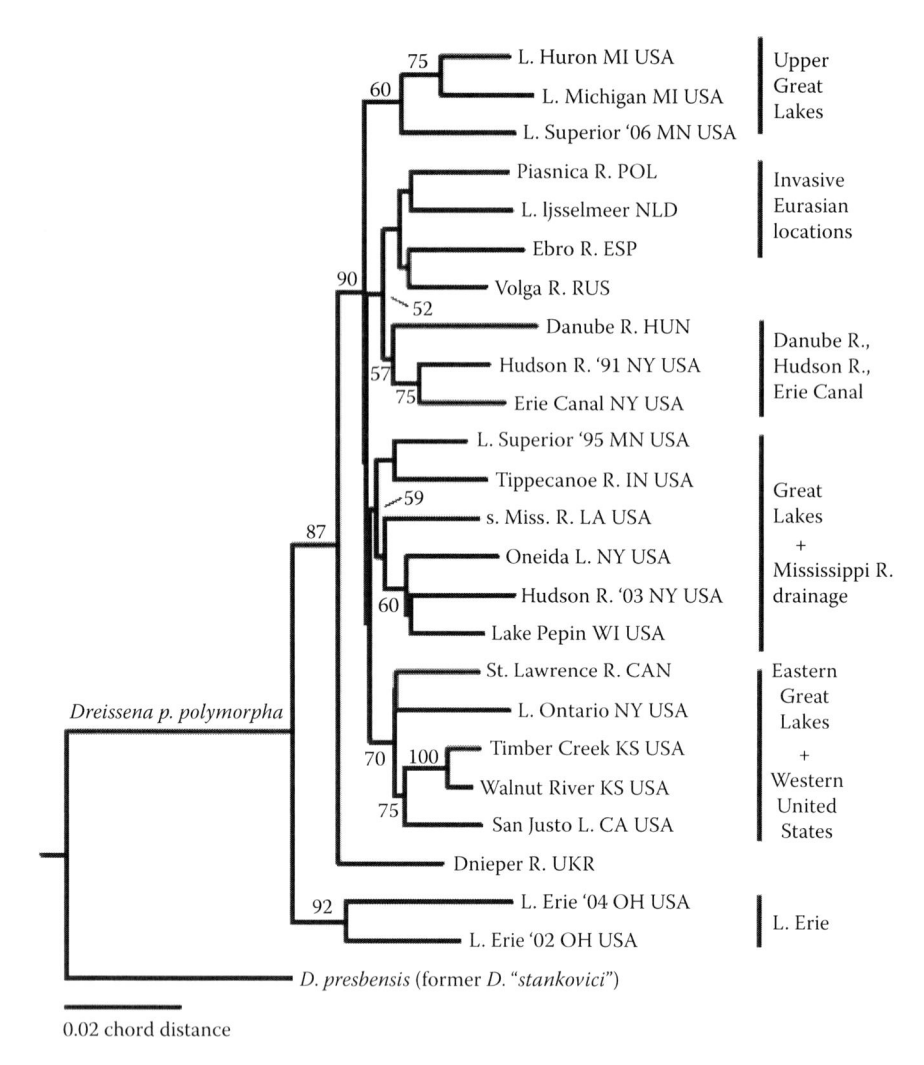

Figure 26.11 Genetic distance tree showing relationships of zebra mussel populations collected from different locations, constructed in PHYLIP v3.6 (Felsenstein 1989) with Cavalli-Sforza chord distances (Cavalli-Sforza and Edwards 1967). Bootstrap percentage values are from 1000 pseudoreplicates. Branch lengths are proportional to genetic divergence. Years are given for samples that were analyzed for multiple years at the same location. (Adapted from Brown, J.E. and Stepien, C.A., *Biol. Invasions*, 12, 3687, 2010.)

Zebra mussel populations sampled from various water bodies in North America differ significantly from each other in pairwise and AMOVA tests (Tables 26.5A and 26.6A), thus revealing appreciable genetic heterogeneity among populations (Brown and Stepien 2010). This finding contrasts with apparent lack of genetic structure in early invasion populations as derived from allozyme makers (Marsden et al. 1996) but is congruent with patterns of heterogeneity discerned with nuclear DNA amplified frequency length polymorphisms (AFLPs; Elderkin et al. 2004), RAPDs (Stepien et al. 2002), and mtDNA sequence variation (Stepien et al. 2003, 2005). Elderkin et al. (2004) attributed this lack of structure with allozyme markers to insufficient polymorphism on a recent time scale and possible influence of selection. For example, Haag and Garton (1995) found significant genetic differences between planktonic larval stages and adult populations of the zebra mussel in Lake Erie, indicating selection at an allozyme locus. Moreover, allozyme markers may be homoplastic, with alleles that genetically differ displaying the same migration patterns (having identical congruent electomorphs; see Stepien and Kocher 1997).

With the exception of populations at two sampling sites in Kansas, all North American populations of zebra mussels significantly diverge, indicating appreciable genetic structure (Tables 26.5A, 26.6A, and 26.7A; Brown and Stepien 2010). For example, zebra mussel populations from the northern and southern portions of the Mississippi River are markedly different (Stepien et al. 2002, Brown and Stepien 2010), corroborating the gradient observed from allozyme data by Elderkin and Klerks (2001) and from nuclear AFLPs (Elderkin et al. 2004). Relationships among populations do not correspond to a genetic isolation pattern based on geographic distance but imply jump dispersal from multiple founding populations (Brown and Stepien 2010).

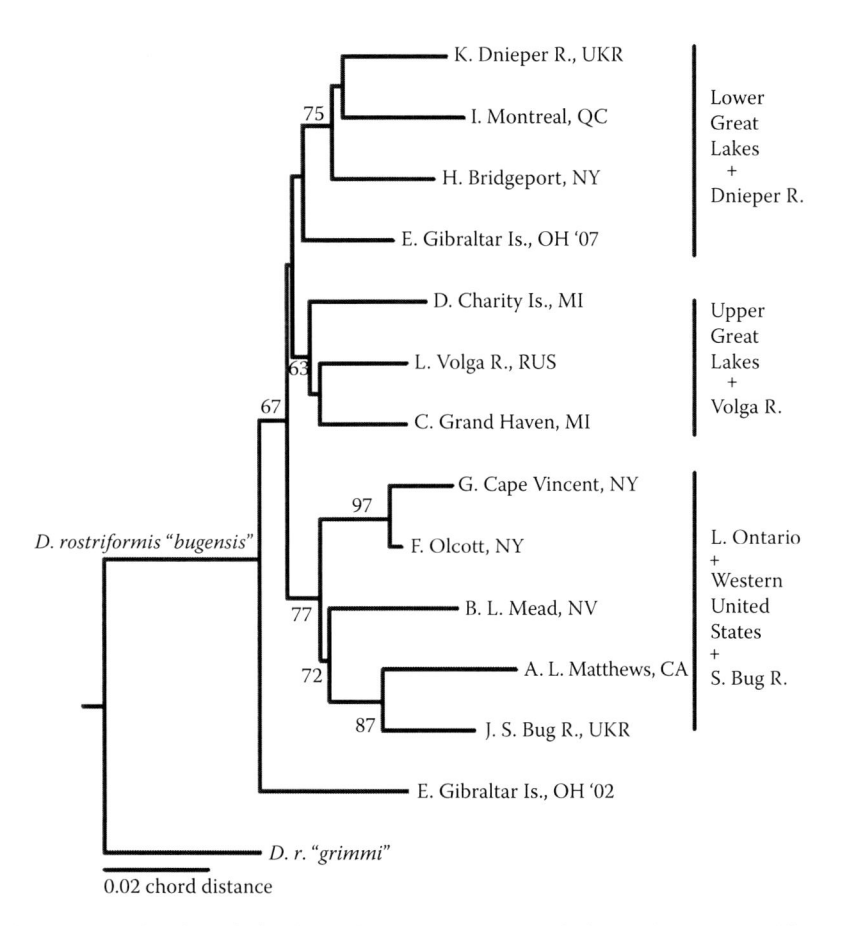

Figure 26.12 Genetic distance tree showing relationships of quagga mussel populations collected from different locations, constructed in PHYLIP v3.6 (Felsenstein 1989) with Cavalli-Sforza chord distances (Cavalli-Sforza and Edwards 1967). Bootstrap percentage values are from 1000 pseudoreplicates. Branch lengths are proportional to genetic divergence. Years are given for samples that were analyzed for multiple years at the same location. (Adapted from Brown, J.E. and Stepien, C.A., *Biol. Invasions*, 12, 3687, 2010.)

AMOVA analyses (Table 26.6) and Bayesian assignment tests (Figure 26.13A) indicate that zebra mussels are differentiated in five primary population groups (among those tested), with additional significant differences occurring among sampling sites within those groups (see Table 26.5; Brown and Stepien 2010). Bayesian assignment analyses indicate that one group (Figure 26.13A) links most of the introduced Eurasian populations (sites 16–19) with the Volga River population (site 21). A second group distinguishes the Lake Erie population (site 10, 2002 and 2004) from the others. The third group links the Dnieper River group (site 20) with those from the Mississippi River drainage (sites 4–6), Oneida Lake (site 13), and the Hudson River (site 15, 2003 only). The fourth group characterizes the upper Great Lakes (sites 7–9) and the earlier Hudson River sample (site 15, 1991). The final group assigns zebra mussels from the western United States (site 1–3) with samples from Lake Ontario (site 11) and the St. Lawrence River (site 14). This indicates that the eastern Great Lakes may have served as the source for this westward expansion.

All pairwise comparisons between quagga mussel populations from the various water bodies significantly differ, which shows appreciable genetic structure across North America and Eurasia (Table 26.5B), as found for zebra mussels (Brown and Stepien 2010). AMOVA analyses indicate that North American quagga mussels appear differentiated among three primary population groups and among sites within those groups (Table 26.6B, $p < 0.0001$). Populations show high genetic self-assignments, also demonstrating significant structure (Table 26.7B). The recovery of three population groups is consistent among the three approaches: 3D-FCA (Figure 26.10), neighbor-joining tree (Figure 26.12), and Bayesian assignment tests (Figure 26.13B). The first group links quagga mussel populations from the upper Great Lakes (sites C–D) with the invasive population in the Volga River (site L). The second group includes populations from Oneida Lake (H), the St. Lawrence River (I), the Dnieper River (K), and some individuals from Lake Erie in 2007 (E). As found for zebra mussels, quagga mussel populations from the western United States (sites A–B) group with those from Lake Ontario (F–G), indicating that Lake Ontario was a likely source of mussel introductions to the Colorado River and southern California. Results do not reflect a genetic

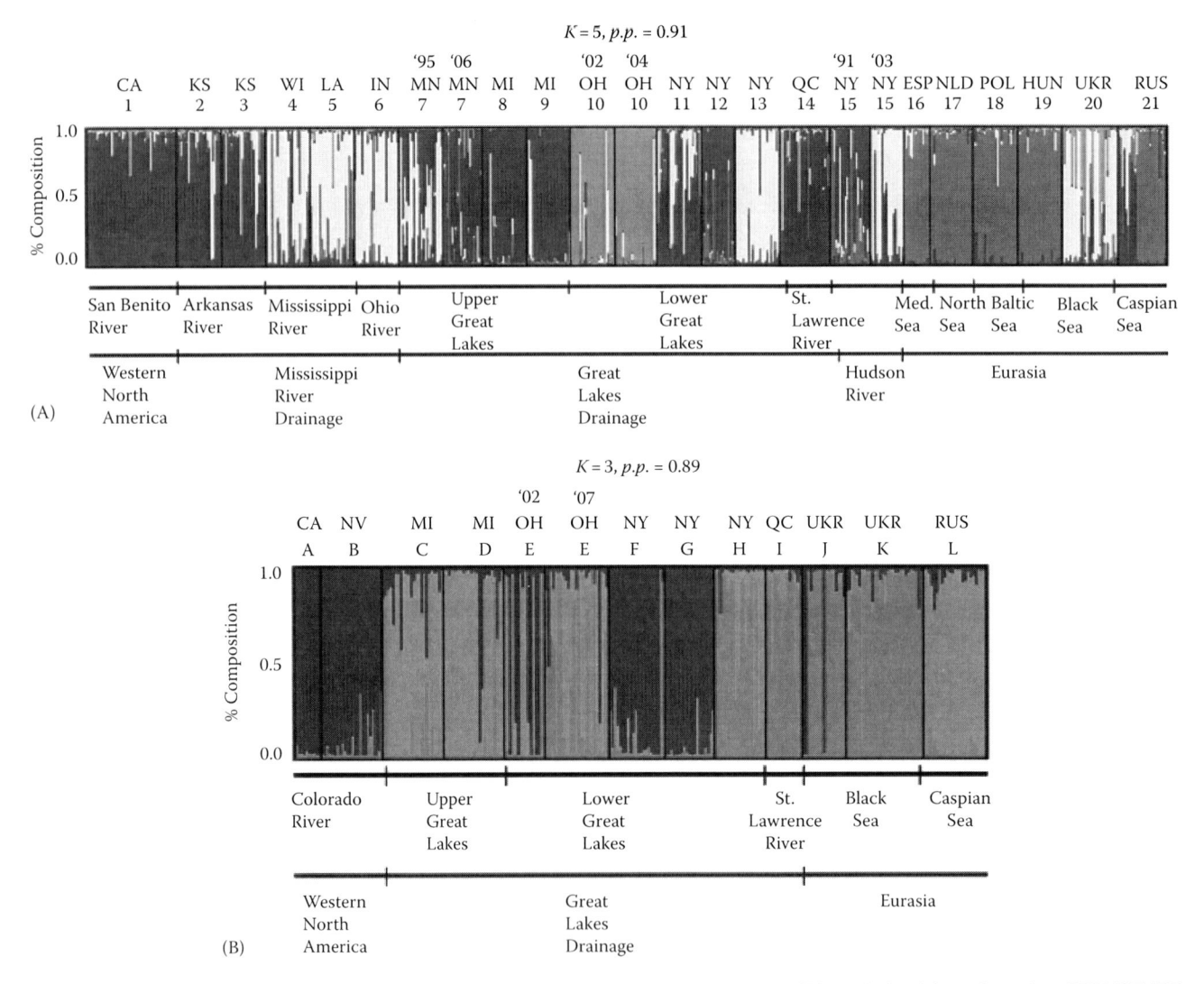

Figure 26.13 Population group structuring for zebra mussels (A) and quagga mussels (B) as derived from Bayesian STRUCTURE analysis (Pritchard et al. 2000; Pritchard and Wen 2004). Analyses discerned five groups (posterior probability = 0.91) for zebra mussels and three groups (posterior probability = 0.89) for quagga mussels. Both species show evidence of multiple introductions. Recent populations established in western North America link to the eastern portions of the Great Lakes. Years are given for samples that are analyzed for multiple years at the same location. (Adapted from Brown, J.E. and Stepien, C.A., *Biol. Invasions*, 12, 3687, 2010.)

isolation with geographic distance correspondence, which is similar to the findings of Wilson et al. (1999b) for quagga mussel populations in the Great Lakes, as well as to our results for zebra mussels (Brown and Stepien 2010).

Bayesian assignment tests reveale that populations of both dreissenid species in the western United States trace to origins from the eastern Great Lakes. Results for both dreissenids thus suggest that overland colonization pathways via trailered boats did not necessarily follow the most proximate connections. This result may have been influenced by the 100th Meridian Initiative (Bossenbroek et al. 2009), which, through education, may have prevented western expansion from boaters leaving proximate sources, but was not as successful for boaters arriving from more distant source localities (Brown and Stepien 2010).

Temporal Genetic Changes of North American Dreissenids

Microsatellite data indicate that genetic compositions of both zebra and quagga mussels have changed over time at some locations in North America (Figure 26.13, Table 26.5). Some populations added alleles from adjacent populations and some lost alleles, but most have retained close similarity to their original genetic composition. Since many genetic signatures remain distinctive and show closest self-assignment, the genotypes that first became established apparently have retained some "genetic population resilience" over time, leading to the divergence patterns among sites. An exception is the sampling site in the Hudson River (site 15 on Figure 26.13A) where genetic composition of

Table 26.6 Results of AMOVA (Excoffier et al. 1992) Showing Partitioning of Genetic Variation within and among Sites Sampled for Zebra and Quagga Mussels

Division of Variation Partitioning Tested	Measure of Variation Partitioning	P-Value
A. Zebra mussel		
Among 5 primary population groups: (1–3, 11, 14); (12, 15–19, 21); (4–7 '95, 13, 15 '03); (10, 20); (7 '06, 8, 9)	Proportion of variation	0.06
	F_{CT}	0.06[a]
Among sampling sites within the groups	Proportion of variation	0.08
	F_{SC}	0.08[a]
Within sampling sites	Proportion of variation	0.86
	F_{ST}	0.14[a]
B. Quagga mussel		
Among 3 primary population groups: (A, B, F, G, J); (C, D, H, I, L); (E, K)	Proportion of variation	0.02
	F_{CT}	0.05[a]
Among sampling sites within the groups	Proportion of variation	0.10
	F_{SC}	0.10[a]
Within sampling sites	Proportion of variation	0.88
	F_{ST}	0.11[a]

Source: Adapted from Brown, J.E. and Stepien, C.A., Biol. Invasions, 12, 3687, 2010.

[a] Significant at $p < 0.0001$. A: Comparisons among five primary zebra mussel population groups: (1–3, 11, 14), (12, 15–19, 21), (4–7 '95, 13, 15 '03), (10, 20), and (7 '06, 8, 9). B: Comparisons among three primary quagga mussel population groups: (A, B, F, G, J), (C, D, H, I, L), and (E, K).

the zebra mussel population changed markedly between 1991 and 2003. This significant temporal change suggests allele contributions from other areas and possible population replacement, as suspected by ecologists (Strayer and Malcolm 2006 and D. Strayer, pers. commun.). Changes in genetic structure of this zebra mussel population may reflect variable population dynamics. Strayer and Malcolm (2006) found that zebra mussel recruitment at this Hudson River site fluctuated greatly in 2- to 4-year cycles, and population abundances differed by an order of magnitude.

Although we did not specifically test for temporal changes in genetic structure in other dreissenid populations, some evidence indicates that this is not uncommon. When Boileau and Hebert (1993) examined the zebra mussel population in Lake Oneida just after it became established, it had much lower heterozygosity than other zebra mussel populations in North America. A decade later, our analysis of the Lake Oneida population showed heterozygosity very similar to zebra mussel populations in the Great Lakes, suggesting that an initial bottleneck faded with time and genetic variation has been augmented by new recruitment. We additionally detect evidence for rapid population expansion and lower heterozygosity in our samples from Lake Superior in 1995, but by 2006, the population had stabilized, and signals of an initial bottleneck were no longer evident. Such signals are ephemeral (Cornuet and Luikart 1996) and would be predicted to fade after a decade, as indicated here.

Temporal comparisons of zebra mussel populations from three locations reveal changes among years as shown in both F_{ST} analog and contingency tests (Table 26.5A). However, populations show strongest self-assignments to their "home" populations (Table 26.7A). Thus, genetic compositions have remained most similar to their initial compositions despite some temporal changes. Temporal comparisons of zebra mussel populations from Lake Erie and Lake Superior show significant net loss of alleles (Table 26.4), whereas the Hudson River population reveals a net gain of alleles. As noted, BOTTLENECK analysis indicates that the Lake Superior population lost the signal of rapid population expansion between 1995 and 2006. However, time pair comparisons for the other two populations do not show demographic signals in BOTTLENECK analysis, suggesting consistency in their reproductive sizes.

Temporal change in genetic composition of quagga mussels is examined only for the population in Lake Erie. There was a significant increase in alleles in 2007 compared to 2002. The genetic composition of the 2007 population more closely resembled populations from the St. Lawrence River and the Black Sea, possibly due to additional introductions into Lake Erie (Figures 26.12 and 26.13B). As found for zebra mussels, genetic composition of the quagga mussel population has remained most similar to its original signature. Thus, although gene flow from other founded populations likely has contributed to overall population

438 QUAGGA AND ZEBRA MUSSELS: BIOLOGY, IMPACTS, AND CONTROL

Table 26.7 Bayesian Population Assignment Tests Using GeneClass2 (Piry et al. 2004) for Zebra Mussel (A) and Quagga Mussel (B) Populations at the Various Sampling Locations. Assignments Are Provided as a Proportion of the Total. When Proportions Do Not Add to 1.0 for a Given Population, it is because Some Individuals Did Not Assign to a Single Population. Most Individuals Assigned to Their Origin Population, and Those That Did Not Usually Assigned to a Nearby Location. Self-Assignment Is in Bold Down Diagonal

Location	1	2	3	4	5	6	7 '95	7 '06	8	9	10 '02	10 '04	11	12	13	14	15 '91	15 '03	16	17	18	19	20	21
A. Zebra mussel																								
1. San Justo L., CA, USA	**0.92**	0	0.02	0	0	0	0	0	0	0	0	0	0	0	0	0	0	0	0	0	0	0	0	0
2. Walnut R., KS, USA	0.08	**0.29**	0.50	0	0	0	0	0	0	0	0	0	0	0	0	0	0	0	0	0	0	0	0	0
3. Timber Crk., KS, USA	0.04	0.65	**0.25**	0	0.04	0	0	0	0	0	0	0	0	0	0	0.04	0	0	0	0	0	0	0	0.04
4. Lake Pepin, WI, USA	0	0	0	**0.46**	0.08	0	0.08	0	0	0	0	0	0	0	0.04	0	0	0.08	0	0	0	0.04	0	0
5. s. Miss. R., LA, USA	0	0	0	0	**0.54**	0	0.13	0	0	0	0	0	0	0	0.04	0	0.04	0	0	0	0	0	0	0
6. Tippecanoe R., IN, USA	0	0	0	0	0	**0.79**	0	0	0	0	0	0	0	0	0	0	0.04	0	0	0	0	0	0.04	0
7. L Superior, MN, USA '95	0	0	0	0	0.04	0.04	**0.54**	0	0	0	0	0	0	0	0.04	0	0.04	0	0	0	0	0	0.04	0.04
7. L Superior, MN, USA '06	0.05	0	0.05	0	0	0	0	**0.68**	0	0	0	0	0	0	0	0	0	0	0	0	0	0	0.05	0
8. L. Michigan, MI, USA	0	0.04	0	0	0	0.04	0	0.04	**0.67**	0	0	0	0	0	0	0	0	0	0	0	0	0	0.04	0
9. L. Huron, MI, USA	0	0	0	0	0	0	0.04	0	0	**0.79**	0	0	0	0	0	0	0	0	0	0	0	0	0.13	0
10. L. Erie, OH, USA '02	0	0	0	0	0	0	0	0	0	0	**0.71**	0	0	0	0	0	0	0	0	0	0	0	0	0
10. L. Erie, OH, USA '04	0	0	0	0	0	0	0	0	0	0	0	**0.74**	0	0	0	0	0	0	0	0	0	0	0.04	0
11. L. Ontario, NY, USA	0.04	0	0	0	0	0	0.04	0	0	0	0	0	**0.58**	0	0	0.04	0	0.04	0	0	0	0	0	0
12. Erie C., NY, USA	0	0	0	0	0	0	0	0	0	0	0	0	0	**0.50**	0	0	0.17	0	0	0	0	0.06	0.06	0
13. L. Oneida, NY, USA	0	0	0	0.04	0.04	0	0	0	0	0	0	0	0	0	**0.56**	0	0	0.04	0	0	0	0	0.04	0
14. St. Lawrence R., QC, CAN	0	0	0	0	0	0	0	0	0	0	0	0	0	0	0	**0.89**	0	0.04	0	0	0	0	0	0
15. Hudson R., NY, USA '91	0	0	0	0	0	0	0.05	0	0	0	0	0	0	0	0	0	**0.73**	0	0.05	0	0	0	0	0
15. Hudson R., NY, USA '03	0.06	0	0	0	0	0	0	0	0	0	0	0	0	0	0	0.06	0	**0.28**	0	0	0	0	0.11	0

Location																					
16. Ebro R., ESP	0	0	0	0	0	0	0	0	0	0	0	0	0	0	0	0.71	0	0	0	0	0.07
17. L. Ijsselmeer, NLD	0.04	0	0	0	0	0	0	0	0	0	0	0	0	0	0	0	**0.70**	0	0	0	0.13
18. Piasnica R., POL	0	0	0	0.04	0	0	0	0	0	0	0	0	0	0	0	0	0.04	**0.58**	0	0.04	0.13
19. Danube R., HUN	0	0	0.04	0	0	0.04	0	0	0	0	0	0	0	0	0	0	0	0	**0.88**	0	0
20. Dnieper R., UKR	0	0	0	0	0.07	0	0	0	0	0	0	0	0	0	0	0	0	0	0	**0.63**	0
21. Volga R., RUS	0.04	0	0	0	0	0	0.04	0	0	0	0	0	0	0	0	0	0	0	0	0	**0.69**

Location	A	B	C	D	E '02	E '07	F	G	H	I	J	K	L
B. Quagga mussel													
A. L. Matthews, CA, USA	**0.20**	0	0	0	0	0	0.10	0	0.10	0	0.40	0	0
B. L. Mead, NV, USA	0	**0.75**	0.04	0.13	0	0.04	0	0	0	0	0	0	0
C. L. Michigan, MI, USA	0	0	**0.38**	0.25	0	0	0	0	0.04	0	0	0	0.04
D. L. Huron, MI, USA	0	0	0.08	**0.46**	0	0	0	0	0	0	0	0	0.08
E. L. Erie, OH, USA 2002	0	0	0	0	**0.69**	0.13	0	0	0	0	0	0	0
E. L. Erie, OH, USA 2007	0	0	0.04	0.08	0	**0.52**	0	0	0.04	0	0	0	0
F. w. L. Ontario, NY, USA	0	0	0.09	0.05	0	0	**0.41**	0.32	0	0.09	0	0	0
G. e. L. Ontario, NY, USA	0	0	0.11	0	0	0	0.47	**0.37**	0	0	0	0	0
H. Oneida L., NY, USA	0	0	0	0.15	0	0	0	0	**0.50**	0.05	0	0	0
I. St. Lawrence R., QC, CAN	0	0	0	0.07	0	0.07	0	0	0.21	**0.14**	0	0	0
J. S. Bug R., UKR	0.12	0.12	0.06	0	0	0	0	0	0.06	0	**0.53**	0.06	0
K. Dnieper R., UKR	0	0	0	0	0	0	0	0	0	0.03	0	**0.57**	0
L. Volga R., RUS	0	0	0	0.08	0	0	0	0	0	0	0	0	**0.63**

Source: Adapted from Brown, J.E. and Stepien, C.A., *Biol. Invasions*, 12, 3687, 2010.

genetic diversity, genetic signatures retain distinctiveness. Genotypes that first became established appear to remain most prevalent and thus exhibit "population genetic resilience" over time. Present populations of both dreissenid mussel species display considerable genetic structure across both North America and Eurasia. Such genetic structure, as shaped by evolutionary factors, provides the basis for genetic adaptations in new habitats.

ACKNOWLEDGMENTS

Data collection was funded by Ohio Sea Grant R/LR-009-PD, an NSF GK-12 fellowship #DGE-0742395 to TJS, and University of Toledo start-up funds to CAS. Samples generously were provided by W. Baldwin, C. Balogh, B. Bodamer, C. Bowen, D. Clapp, N. Claudi, T. Crail, R. Dermott, K. DeVanna, D. Garton, J. Goeckler, J. Gunderson, J. Hageman, S. Hogan, K. Holeck, H. Jenner, D. Jensen, T. Johnson, T. Kakareko, Y. Kvach, E. Laney, Lee, E. Marsden, B. May, C. Mayer, B. Moore, C. Munté i Carrollo, M. Neilson, T. Nalepa, D. Pavlov, J. Postberg, A. Ricciardi, G. Rosenberg, J. Ross, M. Sapota, J. Schooley, B. Sket, A. Spidle, K. Stewart, D. Strayer, M. Thomas, L. Voronina, and M. Watson. Analyses were assisted by J. Brown and Haponski and D. Murphy. We thank P. Uzmann and R. Lohner of the Lake Erie Center for administrative assistance. This is publication 2013-001 from the Lake Erie Research Center, University of Toledo, Oregon, OH.

REFERENCES

Albrecht, C., R. Schultheis, T. Kevrekidis, B. Streit, and T. Wilke. 2007. Invaders or endemics? Molecular phylogenetics, biogeography and systematics of *Dreissena* in the Balkans. *Freshwat. Biol.* 52:1525–1536.

Amsellam, L., J.L. Noyer, T. Le Bourgeois, and M. Hossaert-McKey. 2000. Comparison of genetic diversity in the invasive weed *Rubus alceifolius* Poir. (Rosaceae) in its native range and in areas of introduction, using amplified fragment length polymorphism (AFLP) markers. *Mol. Ecol.* 9:443–455.

Andrusov, N.I. 1897. Fossil and living Dreissenidae of Eurasia. *Proc. St. Petersburg Soc. Nat. Dept. Geol. Mineral.* 25:285–286. [In Russian]

Astanei, I. and E. Gosling. 2010. A microgeographic analysis of genetic variation in *Dreissena polymorpha* in Lough Key, Ireland. In *The Zebra Mussel in Europe*, G. Van der Velde, S. Rajagopal, and A. Bij de Vaate, eds., pp. 127–132. Leiden, The Netherlands: Backhuys Publishers.

Astanei, I., E. Gosling, J. Wilson, and E. Powell. 2005. Genetic variability and phylogeography of the invasive zebra mussel, *Dreissena polymorpha* (Pallas). *Mol. Ecol.* 14:1655–1666.

Avise, J.C. 2004. *Molecular Markers, Natural History, and Evolution*, 2nd Edn. Sunderland, MA: Sinauer Association.

Baldwin, B.S., M. Black, O. Sanjur, R. Gustafson, R.A. Lutz, and R.C. Vrijenhoek. 1996. A diagnostic molecular marker for zebra mussels (*Dreissena polymorpha*) and potentially co-occurring bivalves: Mitochondrial COI. *Mol. Mar. Biol. Biotechnol.* 5:9–14.

Bax, N., A. Williamson, M. Aguero, E. Gonzalez, and W. Geeves. 2003. Marine invasive alien species: A threat to global biodiversity. *Mar. Pollut.* 27:313–323.

Belkhir, K., P. Borsa, L. Chikhi, N. Raufaste, and F. Catch. 2004. GENETIX 4.05, Population genetics software for Windows TM. Université de Montpellier II. Montpellier. http://www.genetix.univ-montp2.fr/genetix/genetix.htm (accessed July 30, 2012).

Benson, A.J. 2013. Chronological history of zebra and quagga mussels (Dreissenidae) in North America, 1988–2010. In *Quagga and Zebra Mussels: Biology, Impacts, and Control*, 2nd Edn., T.F. Nalepa and D.W. Schloesser, eds. pp. 9–31. Boca Raton, FL: CRC Press.

van Beneden, P. 1835. Résultats d'un voyage fait sur le bord de la Méditerranée. *Cr. Hebd. Acad. Sci.* 1:230.

Benzecri, J.P. 1973. *L'analyse des Donnees. Tome II: L'analyse des Correspondances*. Paris, France: Dunod Press.

Boileau, M.G. and P.D.N. Hebert. 1993. Genetics of the zebra mussel (*Dreissena polymorpha*) in populations from the Great Lakes region and Europe. In *Zebra Mussels: Biology, Impacts, and Control*, T.F. Nalepa and D.W. Schloesser, eds., pp. 227–238. Boca Raton, FL: CRC Press.

Bole, J. 1962. *Congeria kusceri* sp. n. (Bivalvia: Dreissenidae). *Biol Vestik* 10:55–61.

Bossenbroek, J.M., D.C. Finnoff, J.F. Shogren, and T.W. Warziniack. 2009. Advances in ecological and economic analyses of invasive species: Dreissenid mussels as a case study. In *Bioeconomics of Invasive Species: Integrating Ecology, Economics, Policy, and Management*, R.P. Keller, D.M. Lodge, M.A. Lewis, and J.F. Shogren, eds., pp. 244–265. New York: Oxford University Press.

Bossenbroek, J.M., L.E. Johnson, B. Peters, and D.M. Lodge. 2007. Forecasting the expansion of zebra mussels in the United States. *Conserv. Biol.* 21:800–810.

Briggs, J.C. 1974. *Marine Zoogeography*. New York: McGraw-Hill, Inc.

Brown, J.E. and C.A. Stepien. 2009. Invasion genetics of the Eurasian round goby in North America: Tracing sources and spread patterns. *Mol. Ecol.* 18:64–79.

Brown, J.E. and C.A. Stepien. 2010. Population genetic history of the dreissenid mussel invasion: Expansion patterns across North America. *Biol. Invasions* 12:3687–3710.

Carlton, J.T. 2008. The zebra mussel *Dreissena polymorpha* found in North America in 1986 and 1987. *J. Great Lakes Res.* 34:770–773.

Cavalli-Sforza, L.L. and A.W.F. Edwards. 1967. Phylogenetic analysis: Models and estimation procedures. *Am. J. Hum. Genet.* 19:526–533.

Claxton, W.T. and E.G. Boulding. 1998. A new molecular technique for identifying field collections of zebra mussel (*Dreissena polymorpha*) and quagga mussel (*Dreissena bugensis*) veliger larvae applied to eastern Lake Erie, Lake Ontario, and Lake Simcoe. *Can. J. Zool.* 76:194–198.

Claxton, W.T., A.B. Wilson, G.L. Mackie, and E.G. Boulding. 1998. A genetic and morphological comparison of shallow- and deep-water populations of the introduced dreissenid bivalve *Dreissena bugensis*. *Can. J. Zool.* 76:1269–1276.

Conrad, T.A. 1831. Description of fifteen new species of recent, and three of fossil shells, chiefly from the coast of the United States. *J. Acad. Natl. Sci. Phila.* 6:256–268.

Conrad, T.A. 1858. Observations of a group of Cretaceous fossil shells, found in Tippah County, Mississippi, with descriptions of fifty-six new species. *J. Acad. Natl. Sci. Phila.* 3:323–336.

Cornuet, J.M. and G. Luikart. 1996. Description and power analysis of two tests for detecting recent population bottlenecks from allele frequency data. *Genetics* 144:2001–2014.

De Souza, J.R.B., C.M.A. Da Rocha, and M.D.P.R. De Lima. 2005. Ocorrencia do bivalve exótico *Mytilopsis leucophaeta* (Conrad) (Mollusca, Bivalvia), no Brazil. *Rev. Brasil Zool.* 22:1204–1206.

Dermott, R. and M. Munawar. 1993. Invasion of Lake Erie offshore sediments by *Dreissena* and its ecological implications, *Can. J. Fish. Aquat. Sci.* 50:2298–2304.

Dillon, A.K. and C.A. Stepien. 2001. Genetic and biogeographic relationships of the invasive round (*Neogobius melanostomus*) and tubenose (*Proterorhinus marmoratus*) gobies in the Great Lakes versus Eurasian populations. *J. Great Lakes Res.* 27:267–280.

Drake, J.M. and D.M. Lodge. 2004. Global hot spots of biological invasions: Evaluating options for ballast-water management. *Proc. R. Soc. Lond. B Biol. Sci.* 271:575–580.

Elderkin, C.L. and P.L. Klerks. 2001. Shifts in allele and genotype frequencies in zebra mussels *Dreissena polymorpha* along the latitudinal gradient formed by the Mississippi River. *J. N. Am. Benthol. Soc.* 20:595–605.

Elderkin, C.L., E.J. Perkins, P.L. Leberg, P.L. Klerks, and R.F. Lance. 2004. Amplified fragment length polymorphism (AFLP) analysis of the genetic structure of the zebra mussel *Dreissena polymorpha* in the Mississippi River. *Freshwat. Biol.* 49:1487–1494.

Excoffier, L. and H.E.L. Lischer. 2010. Arlequin suite ver 3.5: A new series of programs to perform population genetics analyses under Linux and Windows. *Mol. Ecol. Resour.* 10:564–567. http://cmpg.unibe.ch/software/arlequin35/ (accessed July 30, 2012).

Excoffier, L., P.E. Smouse, and J.M. Quattro. 1992. Analysis of molecular variance inferred from metric distances among DNA haplotypes-application to human mitochondrial DNA restriction data. *Genetics* 131:479–491.

Feldheim, K.A., J.E. Brown, D.J. Murphy, and C.A. Stepien. 2011. Microsatellite loci for dreissenid mussels (Mollusca: Bivalvia) and relatives: Markers for assessing invasive and native species. *Mol. Ecol. Resour.* 11:725–732. Doi: 10.1111/j.1755-0998.2011.03012.x

Felsenstein, J. 1985. Confidence limits on phylogenies: An approach using the bootstrap. *Evolution* 39:783–791.

Felsenstein, J. 1989. PHYLIP—Phylogeny inference package (Version 3.2). *Cladistics* 5:164–166. http://evolution.genetics.washington.edu/phylip.html (Version 3.69) (accessed July 30, 2012).

Folmer, O.M., W. Black, R. Hoeh, R. Lutz, and R. Vrijenhoek. 1994. DNA primers for amplification of mitochondrial cytochrome c oxidase subunit I from diverse metazoan invertebrates. *Mol. Mar. Biol. Biotechnol.* 3:294–299.

Frisher, M.E., A.S. Hansen, J.A. Wyllie, J. Wimbush, J. Murray, and S.A. Nierzwicki-Bauer. 2002. Specific amplification of the 18S rRNA gene as a method to detect zebra mussel (*Dreissena polymorpha*) larvae in plankton samples. *Hydrobiologia* 487:33–44.

Galil, B.S. and C. Bogi. 2009. *Mytilopsis sallei* (Mollusca: Bivalvia: Dreissenidae) established on the Mediterranean coast of Israel. *Mar. Biodiv. Rec.* 2:e73.

Gelembiuk, G.W., G.E. May, and C.E. Lee. 2006. Phylogeography and systematics of zebra mussels and related species. *Mol. Ecol.* 15:1033–1050.

Giribet, G. and W.C. Wheeler. 2002. On bivalve phylogeny: A high-level phylogeny of the mollusk class Bivalvia based on a combined analysis of morphology and DNA sequence data. *Invert. Biol.* 121:271–324.

Goudet, J., M. Raymond, T. de Meeus, and F. Rousset. 1996. Testing differentiation in diploid populations. *Genetics* 144:1933–1940.

Grigorovich, I.A., J.R. Kelly, J.A. Darling, and C.W. West. 2008. The quagga mussel invades the Lake Superior basin. *J. Great Lakes Res.* 34:342–350.

Grigorovich, I.A., H.J. MacIsaac, N.V. Shadrin, and E.L. Mills. 2002. Patterns and mechanisms of aquatic invertebrate introductions in the Ponto-Caspian region. *Can. J. Fish. Aquat. Sci.* 59:1189–1208.

Guindon, S. and O. Gascuel. 2003. A simple, fast and accurate algorithm to estimate large phylogenies by maximum likelihood. *Syst. Biol.* 52:696–704. http://atgc.lirmm.fr/phyml/ (Version 2.4.4) (accessed July 30, 2012).

Guindon, S., F. Lethiec, P. Duroux, and O. Gascuel. 2005. PHYML Online—A web server for fast maximum likelihood-based phylogenetic inference. *Nucleic Acids Res.* 33:557–559.

Haag, W.R. and D.W. Garton. 1995. Variation in genotype frequencies during the life history of the bivalve, *Dreissena polymorpha*. *Evolution* 49:1284–1288.

Hall, T.A. 1999. BIOEDIT: A user-friendly biological sequence alignment editor and analysis program for Windows 95/98/NT. *Nucleic Acids Symp. Ser.* 41:95.

Hall, T. 2004. BioEdit v.7.0.0. http://www.mbio.ncsu.edu/BioEdit/ (accessed July 30, 2012).

Harzhauser, M. and O. Mandic. 2010. Neogene dreissenids in central Europe: Evolutionary shifts and diversity changes. In *The Zebra Mussel in Europe*, G. Van der Velde, S. Rajagopal, and A. Bij de Vaate, eds., pp. 11–28. Leiden, The Netherlands: Backhuys Publishers.

Hebert, P.D.N., M.Y. Stoeckle, T.S. Zemlak, and C.M. Francis. 2004. Identification of birds through DNA barcodes. *PLoS Biol.* 2:1657–1663.

Holeck, K.T., E.L. Mills, H.J. MacIsaac, M.R. Dochoda, R.I. Colautti, and A. Ricciardi. 2004. Bridging troubled waters: Biological invasions, transoceanic shipping, and the Laurentian Great Lakes. *BioScience* 54:919–929.

Hoy, M.S., K. Kelly, and R.J. Rodriguez. 2010. Development of a molecular diagnostic system to discriminate *Dreissena polymorpha* (zebra mussel) and *Dreissena bugensis* (quagga mussel). *Mol. Ecol. Res.* 10:190–192.

Huber, M. 2010. *Compendium of Bivalves: A Full-Color Guide to 3,300 of the World's Marine Bivalves. A Status on Bivalvia after 250 Years of Research*. Hackenheim, Germany: Conch Books.

Imo, M., S. Seitz, and J. Johannesen. 2010. Distribution and invasion genetics of the quagga mussel (*Dreissena rostriformis bugensis*) in German rivers. *Aquat. Ecol.* 44:731–740.

Johnson, L.E., A. Ricciardi, and J.T. Carlton. 2001. Overland dispersal of aquatic invasive species: A risk assessment of transient recreational boating. *Ecol. Appl.* 11:1789–1799.

Kang, M., Y.M. Buckley, and A.J. Lowe. 2007. Testing the role of genetic factors across multiple independent invasions of the shrub Scotch broom (*Cytisus scoparius*). *Mol. Ecol.* 16:4662–4673.

Kelly, D.W., G.A. Lamberti, and H.J. MacIsaac. 2009. The Laurentian Great Lakes as a case study of biological invasion. In *Bioeconomics of Invasive Species: Integrating Ecology, Economics, Policy, and Management*, R.P. Keller, D.M. Lodge, M.A. Lewis, and J.F. Shogren, eds., pp. 205–225. New York: Oxford University Press.

Kim, K.S., M.J. Bagley, B.S. Coates, R.L. Hellmich, and T.W. Sappington. 2009. Spatial and temporal genetic analyses show high gene flow among European corn borer (Lepidoptera: Crambidae) populations across the Central US Corn Belt. *Environ. Entomol.* 38:1312–1323.

Kimura, M., 1981. Estimation of evolutionary distances between homologous nucleotide sequences. *Proc. Nat. Acad. Sci.* 78:454–458.

Lewis, K.M., J.L. Feder, and G.A. Lamberti. 2000. Population genetics of the zebra mussel *Dreissena polymorpha* (Pallas): Local allozyme differentiation within midwestern lakes and streams. *Can. J. Fish. Aquat. Sci.* 57:637–643.

Logvinenko, B.M. and Y.I. Starobogatov. 1968. Type Mollusca. In *Atlas of Invertebrates of the Caspian Sea*, Y.A. Birshstein, L.G. Vinogradov, N.N. Kondakov, M.S. Astakhova, and N.N. Romanova, eds., pp. 308–385. Moscow, Russia: Pishchevaya Promyshlennost.

Luikart, G., F.W. Allendorf, J.-M. Cornuet, and W.B. Sherwin. 1998a. Distortion of allele frequency distributions provides a test for recent population bottlenecks. *J. Hered.* 89:238–247.

Luikart, G., W.B. Sherwin, B.M. Steele, and F.W. Allendorf. 1998b. Usefulness of molecular markers for detecting population bottlenecks via monitoring genetic change. *Mol. Ecol.* 7:963–974.

L'vova, A.A. and Y.I. Starobogatov. 1982. A new species of the genus *Dreissena* (Bivalvia, Dreissenida) from Lake Ohrid. *Zool. Zh.* 61:1749–1752.

Marsden, J.E., A. Spidle, and B. May. 1995. Genetic similarity among zebra mussel populations within North America and Europe. *Can. J. Fish. Aquat. Sci.* 52:836–847.

Marsden, J.E., A.P. Spidle, and B. May. 1996. Review of genetic studies of *Dreissena* spp. *Am. Zool.* 36:259–270.

May, G.E., G.W. Gelembiuk, V.E. Panov, M.I. Orlova, and C.E. Lee. 2006. Molecular ecology of zebra mussel invasions. *Mol. Ecol.* 15:1021–1031.

Merritt, T.J.S., L. Shi, M.C. Chase, M.A. Rex, R.J. Etter, and J.M. Quattro. 1998. Universal cytochrome *b* primers facilitate intraspecific studies in molluscan taxa. *Mol. Mar. Biol. Biotechnol.* 7:7–11.

Mills, E.L., G. Rosenberg, A.P. Spidle, M. Ludyanskiy, Y. Pligin, and B. May. 1996. A review of the biology and ecology of the quagga mussel (*Dreissena bugensis*) a second species of freshwater dreissenid introduced to North America. *Am. Zool.* 36:271–286.

Mishler, B.D. and E.C. Theriot. 2000. The phylogenetic species concept (*sensu* Mishler and Theriot): Monophyly, apomorphy, and phylogenetic species concepts. In *Species Concepts and Phylogenetic Theory: A Debate*, Q.D. Wheeler and R. Meier, eds., pp. 44–54. New York: Columbia University Press.

Molloy, D.P., A. Bij de Vaate, T. Wilke, and L. Giamberini. 2007. Discovery of *Dreissena rostriformis bugensis* (Andrusov 1897) in Western Europe. *Biol. Invasions* 9:871–874.

Molloy, D.P., L. Giamberini, L.E. Burlakova, A.Y. Karatayev, J.R. Cryan, S.L. Trajanovski, and S.P. Trajanovska. 2010. Investigation of the endosymbionts of *Dreissena stankovici* with morphological and molecular confirmation of host species. In *The Zebra Mussel in Europe*, G. van der Velde, S. Rajagopal, and A. Bij de Vaate, eds., pp. 227–237. Leiden, The Netherlands: Backhuys Publishers.

Mordukhai-Boltovskoi, F.D. 1960. *Caspian Fauna in the Azov and Black Sea Basins*. Moscow-Leningrad, Russia: USSR AS Press. [In Russian.]

Morton, B. 1993. The anatomy of *Dreissena polymorpha* and the evolution and success of the heteromyarian form in the Dreissenoidea. In *Zebra Mussels: Biology, Impacts, and Control*, T.F. Nalepa and D.W. Schloesser, eds., pp. 185–215. Boca Raton, FL: CRC Press.

Morton, B., F. Belkovrh, and B. Sket. 1998. Biology and anatomy of the "living fossil" *Congeria kusceri* (Bivalvia: Dreissenidae) from subterranean rivers and caves in the Dinaria karst of the former Yugoslavia. *J. Zool. Lond.* 245:147–174.

Müller, J.C., D. Hidde, and A. Seitz. 2002. Canal construction destroys the barrier between major European invasion lineages of the zebra mussel. *Proc. R. Soc. B Biol. Sci.* 290:1139–1142.

Müller, J.C., S. Wöll. U. Fuchs, and A. Seitz. 2001. Genetic interchange of *Dreissena polymorpha* populations across a canal. *Heredity* 86:103–109.

Naish, K.A. and E.G. Boulding. 2001. Trinucleotide microsatellite loci for the zebra mussel *Dreissena polymorpha*, an invasive species in Europe and North America. *Mol. Ecol. Notes* 1:286–288.

Naish, K.A. and E.G. Boulding. 2011. Clarification of the microsatellite loci developed for the zebra mussel. *Mol. Ecol. Resources* 11:223–224.

Nichols, S.J. and M.G. Black. 1994. Identification of larvae: The zebra mussel (*Dreissena polymorpha*), quagga mussel (*Dreissena rostriformis bugensis*), and Asian clam (*Corbicula fluminea*). *Can. J. Zool.* 72:406–417.

Nuttall, C.P. 1990. Review of the Caenozoic heterodont bivalve superfamily Dreissenacea. *Palaeontology* 33:707–737.

Pallas, P.S. 1771. *Reise Durch Verschiedene Provincen des Russischen Reichs*, Vol. 1. St. Petersburg, Russia: Academie der Wissenschaften.

Palumbi, S.R. 1996. Nucleic acids II: The polymerase chain reaction. In *Molecular Systematics*, D.M. Hillis, C. Moritz, and B.K. Mable, eds., pp. 205–248. Sunderland, MA: Sinauer Association.

Park, J.K. and D. O'Foighil. 2000. Sphaeriid and corbiculid clams represent separate heterodont bivalve radiations into freshwater environments. *Mol. Phylogenet. Evol.* 14:75–88.

Partsch, P. 1836. Über die sogenannten versteinerten Ziegenklauen aus dem Plattensee in Ungarn und ein neues urweltliches Geschlecht zweischaliger Conchylien. *Annalen des Wiener Museums für Naturgeschichte* 1:93–102.

Pathy, D.A. and G.L. Mackie. 1993. Comparative shell morphology of *Dreissena polymorpha*, *Mytilopsis leucophaeata*, and the "quagga" mussel (Bivalvia: Dreissenidae) in North America. *Can. J. Fish. Aquat. Sci.* 72:1012–1023.

Piry, S., A. Alapetite, J.-M. Cornuet, D. Paetkau, L. Baudouin, and A. Estoup. 2004. Geneclass2: A software for genetic assignment and first-generation migrant detection. *J. Hered.* 95:536–539. http://www1.montpellier.inra.fr/URLB/ (Version 2.0.g) (accessed March 20, 2012).

Piry, S., G. Luikart, and J.-M. Cornuet. 1999. BOTTLENECK: A computer program for detecting recent reductions in the effective population size using allele frequency data. *J. Hered.* 90:502–503. http://www.ensam.inra.fr/URLB/bottleneck/bottleneck.html (version 1.2.02) (accessed July 30, 2012).

Posada, D. 2003. Using modeltest and PAUP* to select a model of nucleotide substitution. In *Current Protocols in Bioinformatics*, A.D. Baxevanis, D.B. Davison, R.D.M. Page, G.A. Petsko, L.D. Stein, and G.D. Stormo, eds., pp. 6.5.1–6.5.14. Hoboken, NJ: John Wiley & Sons.

Posada, D. 2008. jModelTest: Phylogenetic model averaging. *Mol. Biol. Evol.* 25:1253–1256. http://darwin.uvigo.es/software/jmodeltest.html (Version 0.1.1) (accessed July 30, 2012).

Posada, D. and K.A. Crandall. 1998. MODELTEST: Testing the model of DNA substitution. *Bioinformatics* 14:817–818.

Pritchard, J.K., M. Stephens, and P. Donelly. 2000. Inference of population structure using multilocus genotype data. *Genetics* 155:945–959.

Pritchard, J.K. and W. Wen. 2004. MODELTEST: Testing the model of DNA substitution. *Bioinformatics* 14:817–818.

Pritchard, J.K., W. Wen, and D. Falush. 2010. Documentation for STRUCTURE Software: Version 2.3. Department of Human Genetics University of Chicago, Department of Statistics University of Oxford. http://pritch.bsd.uchicago.edu/structure_software/release_versions/v2.3.3/html/structure.html (Version 2.3.3) (accessed July 30, 2012).

Quaglia, F., L. Lattuada, P. Mantecca, and R. Bacchetta. 2008. Zebra mussels in Italy: Where do they come from? *Biol. Invasions* 10:555–560.

Rajagopal, S., B.J.A. Pollux, J.L. Peters, G. Cremers, S.Y. Moon-van der Staay, T. van Alen, J. Eygensteyn, A. van Hoek, A. Palau, A. Bij de Vaate, and G. van der Velde. 2009. Origin of Spanish invasion by the zebra mussel, *Dreissena polymorpha* (Pallas, 1771) revealed by amplified fragment length polymorphism (AFLP) fingerprinting. *Biol. Invasions* 11:2147–2159.

Ram, J.L., A.S. Karim, P. Acharya, P. Jagtap, S. Purohite, and D.R. Kashian. 2011. Reproduction and genetic detection of veligers in changing *Dreissena* populations in the Great Lakes. *Ecosphere* 2:1–16.

Raymond, M. and F. Rousset. 1995. An exact test for population differentiation. *Evolution* 49:1280–1283.

Récluz, C.A. 1849. Descriptions de quelque nouvelles espéces de coquilles. *Rev. Mag. Zool.* 2:64–71.

Ricciardi, A. and H.J. MacIsaac. 2000. Recent mass invasion of the North American Great Lakes by Ponto-Caspian species. *Trends Ecol. Evol.* 15:62–65.

Rice, W.R. 1989. Analyzing tables of statistical tests. *Evolution* 43:223–225.

Ronquist, F. and J.P. Huelsenbeck. 2003. MrBayes 3: Bayesian phylogenetic inference under mixed models. *Bioinformatics* 19:1572–1574.

Rosenberg, G. and M.L. Ludyanskiy. 1994. A nomenclature review of *Dreissena* (Bivalvia: Dreissenidae), with identification of the quagga mussel as *Dreissena bugensis*. *Can. J. Fish. Aquat. Sci.* 51:1474–1484.

Rosenthal, D.M., A.P. Ramakrishnan, and M.B. Cruzan. 2008. Evidence for multiple sources of invasion and intraspecific hybridization in *Brachypodium sylvaticum* (Hudson) Beauv. in North America. *Mol. Ecol.* 17:4657–4669.

Rousset, F. 2008. GENEPOP'007: A complete re-implementation of the Genepop software for Windows and Linux. *Mol. Ecol. Resour.* 8:103–106. http://genepop.curtin.edu.au/(Version 4.0) (accessed July 30, 2012).

Saitou, N. and M. Nei. 1987. The neighbor-joining method: A new method for reconstructing phylogenetic trees. *Mol. Biol. Evol.* 4:406–425.

Sanderson, M.J. 2002. Estimating absolute rates of molecular evolution and divergence times: A penalized likelihood approach. *Mol. Biol. Evol.* 19:101–109.

Sanderson, M.J. 2003. R8s: Inferring absolute rates of molecular evolution and divergence times in the absence of a molecular clock. *Bioinform. Appl. Notes* 19:301–302.

Sanderson, M.J. 2006. Paloverde: An OpenGL 3-D phylogeny browser. *Bioinformatics* 22:1004–1006.

Schütt, H. 1993. Die Gattung Dreissena im Quartär Anatoliens (Bivalvia: Eulamellibranchiata: Dreissenacea). *Arch. Mollus.* 122:323–333.

Schütz, L. 1969. Ökologische untersuchungen über die benthos fauna im Nordostseekanal. III. Autoecologie der vagilen und hemisessilen arten im bewuchs der pfähle: Makrofauna. *Int. Rev. Ges. Hydrobiol.* 54:553–588.

Sket, B. 2011. Origins of the Dinaris troglobiotic mussel and its correct taxonomical classification. *Congeria* or *Mytilopsis* (Bivalvia: Dreissenidae)? *Acta Biol. Slov.* 54:6–76.

Soroka, M. 2010. Genetic differentiation of *Dreissena polymorpha* from East-European countries. In *The Zebra Mussel in Europe*, G. van der Velde, S. Rajagopal, and A. Bij de Vaate, eds., pp. 133–144. Leiden, The Netherlands: Backhuys Publishers.

Starobogatov, Y.I. 1994. Taxonomy and palaeontology. In *Freshwater Zebra Mussel Dreissena Polymorpha*, J.I. Starobogatov, ed., pp. 47–55. Moscow, Russia: Nauka Press.

Stepien, C.A., J.E. Brown, M.E. Neilson, and M.A. Tumeo. 2005. Genetic diversity of invasive species in the Great Lakes versus their Eurasian source populations: Insights for risk analysis. *Risk Anal.* 25:1043–1060.

Stepien, C.A., A.N. Hubers, and J.L. Skidmore. 1999. Diagnostic genetic markers and evolutionary relationships among invasive dreissenoid and corbiculoid bivalves in North America: Phylogenetic signal from mitochondrial 16S rDNA. *Mol. Phylogenet. Evol.* 13:31–49.

Stepien, C.A. and T.D. Kocher. 1997. Molecules and morphology in studies of fish evolution, In *Molecular Systematics of Fishes*, T.D. Kocher and C.A. Stepien, eds., pp. 1–11, San Diego, CA: Academic Press.

Stepien, C.A., B. Morton, K.A. Dabrowska, R.A. Guarnera, T. Radja, and B. Radja. 2001. Genetic diversity and evolutionary relationships of the troglodytic 'living fossil' *Congeria kusceri* (Bivalvia: Dreissenidae). *Mol. Ecol.* 10:1873–1879.

Stepien, C.A., C.D. Taylor, and K.A. Dabrowska. 2002. Genetic variability and phylogeographic patterns of a nonindigenous species invasion: A comparison of exotic versus native zebra and quagga mussel populations. *J. Evol. Biol.* 15:314–328.

Stepien, C.A., C.D. Taylor, I.A. Grigorovich, S.V. Shirman, R. Wei, A.V. Korniushin, and K.A. Dabrowska. 2003. DNA and systematic analysis of invasive and native dreissenid mussels: Is *Dreissena bugensis* really *D. rostriformis*? *Aquat. Invaders* 14:1–11, 14–18.

Stepien, C.A. and M.A. Tumeo. 2006. Invasion genetics of Ponto-Caspian gobies in the Great Lakes: A "cryptic" species, absence of founder effects, and comparative risk analysis. *Biol. Invasions* 8:61–78.

Strayer, D.L. and H.M. Malcolm. 2006. Long-term demography of a zebra mussel (*Dreissena polymorpha*) population. *Freshwat. Biol.* 51:117–130.

Tan, K.S. and B. Morton. 2006. The invasive Caribbean bivalve *Mytilopsis sallei* (Dreissenidae) introduced into Malaysia and Johur Bahru Singapore. *Raffles Bull. Zool.* 54:429–434.

Taylor, J.D., S.T. Williams, E.A. Glover, and P. Dyal. 2007. A molecular phylogeny of heterodont bivalves (Mollusca: Bivalvia: Heterodonta): New analyses of 18S and 28S rRNA genes. *Zool. Scripta* 36:587–606.

Therriault, T.W., M.F. Docker, M.I. Orlova, D.D. Heath, and H.J. MacIsaac. 2004. Molecular resolution of the family Dreissenidae (Mollusca: Bivalvia) with emphasis on Ponto-Caspian species, including first report of *Mytilopsis leucophaeata* in the Black Sea basin. *Mol. Phylogenet. Evol.* 30:479–489.

Therriault, T.W., M.I. Orlova, M.F. Docker, H.J. MacIsaac, and D.D. Heath. 2005. Invasion genetics of a Freshwater mussel (*Dreissena rostriformis bugensis*) in eastern Europe: High gene flow and multiple introductions. *Heredity* 95:16–23.

Thompson, J.D., T.J. Gibson, F. Plewniak, F. Jeanmougin, and D.G. Higgins. 1997. The CLUSTAL_X windows interface: Flexible strategies for multiple sequence alignment aided by quality analysis tools. *Nucleic Acids Res.* 25:4876–4882.

van der Velde, G., K. Hermus, M. van der Gaag, and H.A. Jenner. 1992. Cadmium, zinc and copper in the body, byssus and shell of the mussels, *Mytilopsis leucophaeta* and *Dreissena polymorpha* in the brackish Noordzeekanaal of The Netherlands. In *The Zebra Mussel Dreissena polymorpha*, D. Neumann and H.A. Jenner, eds., pp. 213–226. New York: Gustav Fischer Verlag Publishers.

van der Velde, G., S. Rajagopal, and A. Bij de Vaate. 2010. From zebra mussels to quagga mussels: An introduction to the Dreissenidae. In *The Zebra Mussel in Europe*, G. Van der Velde, S. Rajagopal, and A. Bij de Vaate, eds., pp. 1–10. Leiden, The Netherlands: Backhuys Publishers.

van Oosterhout, C., W.F. Hutchison, D.P.M. Wills, and P. Shipley. 2004. Micro-checker: Software for identifying and correcting genotyping errors in microsatellite data. *Mol. Ecol. Notes* 4:535–538. http://www.microchecker.hull.ac.uk/ (Version 2.2.3) (accessed July 30, 2012).

van Oosterhout, C., D. Weetman, and W.F. Hutchinson. 2006. Estimation and adjustment of microsatellite null alleles in nonequilibrium populations. *Mol. Ecol. Notes* 6:255–256.

Viard, F., C. Ellien, and L. Dupont. 2006. Dispersal ability and invasion success of *Crepidula fornicata* in a single gulf: Insights from genetic markers and larval-dispersal model. *Helgol. Mar. Res.* 60:144–152.

Weir, B.S. and C.C. Cockerham. 1984. Estimating *F*-statistics for the analysis of population structure. *Evolution* 38:1358–1370.

Wiley, E.O. and R.L. Mayden. 2000. The evolutionary species concept. In *Species Concepts and Phylogenetic Theory: A Debate*, Q.D. Wheeler and R. Meier, eds., pp. 70–89. New York: Columbia University Press.

Wilke, T., R. Schultheiß, C. Albrecht, N. Bornmann, S. Trajanovski, and T. Kevrekidis. 2010. Native *Dreissena* freshwater mussels in the Balkans: In and out of ancient lakes. *Biogeoscience* 7:3051–3065.

Wilson, A.B., E.G. Boulding, and K.A. Naish. 1999a. Characterization of tri- and tetranucleotide microsatellite loci in the invasive mollusk *Dreissena bugensis*. *Mol. Ecol.* 8:692–693.

Wilson, A.B., K.A. Naish, and E.G. Boulding. 1999b. Multiple dispersal strategies of the invasive quagga mussel (*Dreissena bugensis*) as revealed by microsatellite analysis. *Can. J. Fish. Aquat. Sci.* 56:2248–2261.

Wong, Y.T., R. Meier, and K.S. Tan. 2011. High haplotype variability in established Asian populations of the invasive Caribbean bivalve *Mytilopsis sallei* (Dreissenidae). *Biol. Invasions* 13:341–348.

Zenkevich, L.A. 1963. *Biology of the Seas of the USSR*. New York: InterScience Publishers.

CHAPTER 27

Effects of Algal Composition, Seston Stoichiometry, and Feeding Rate on Zebra Mussel (*Dreissena polymorpha*) Nutrient Excretion in Two Laurentian Great Lakes

Thomas H. Johengen, Henry A. Vanderploeg, and James R. Liebig

CONTENTS

Abstract 445
Introduction 446
Methods 447
    Experimental Design 447
    Sampling and Handling 448
    Nutrient Excretion 448
    Variables Affecting Nutrient Excretion 449
        Feeding Rate Variables 449
        Seston Quality and Temperature 449
        Algal Composition 449
    Analytical Procedures 450
Results 450
    Ambient Water Quality Conditions 450
    Seasonal Excretion Rates for Natural Seston 450
    Transplant Experiments 451
    Effects of Temperature and Food Quality and Quantity 453
Discussion 456
    Seasonal Excretion Rates for Natural Seston 456
    Ecosystem Considerations 457
References 458

ABSTRACT

Nutrient excretion rates of zebra mussels from Saginaw Bay, Lake Huron, and Hatchery Bay, western Lake Erie, were compared over a 3-year period. Rates were estimated from nutrient accumulation in bottles of filtered lake water to which mussels were added for short (3–4 h) incubations.

Excretion rates of soluble reactive phosphorus and ammonium were related to gross and net feeding rates measured in companion feeding experiments. Both feeding and excretion rates were related to seston quantity, seston nutrient concentrations, algal composition, and temperature. Excretion rates were significantly different between lakes and were significantly correlated to net feeding (assimilation) rate and algal composition as measured by the biomass percentage of cryptophytes, flagellates, and greens. Excretion rates were also significantly correlated to soluble reactive phosphorus concentrations and the assimilation rate of phosphorus, based on chlorophyll assimilation rate and phosphorus content of the seston. Nitrogen and phosphorus excretion rates showed distinct seasonal patterns and distinct differences between lakes. Notably, nitrogen excretion was most strongly correlated to temperature, while phosphorus excretion was most strongly related to assimilation of chlorophyll and phosphorus. Nitrogen excretion rates were approximately twice as high for mussels from Lake Erie compared to mussels from Saginaw Bay, whereas phosphorus excretion rates were more than 10-times higher for Lake Erie mussels than Saginaw Bay mussels. Higher excretion rates for Lake Erie mussels were associated with the greater availability of phosphorus as reflected by both concentrations and carbon to phosphorus or nitrogen to phosphorus ratios of the seston, as well as a more desirable algal composition as measured by both the percent preferred algae and the lower frequency of cyanophytes that have been shown to be an undesirable food source. Overall, excretion of phosphorus was strongly driven by phosphorus assimilation that was strongly affected by feeding rate, which in turn was driven by algal composition.

It has been broadly reported that mussels highly impact phytoplankton abundance and composition through direct grazing pressure and phytoplankton productivity through recycling of nutrients into the water column. Results of this study reveal, in turn, the strong impact that phytoplankton composition and nutrient availability have on determining mussel feeding rates and nutrient excretion rates. Such ecosystem-specific responses must be considered in any modeling or mass-balance approaches that attempt to quantify the impact of mussel feeding and excretion on phytoplankton and nutrient dynamics.

INTRODUCTION

The invasive Ponto-Caspian mussels, zebra mussel (*Dreissena polymorpha*) and quagga mussel (*Dreissena rostriformis bugensis*), have spread throughout much of Europe and North America and have had profound effects on ecosystems they have invaded, including the Laurentian Great Lakes (e.g., Vanderploeg et al. 2002, Higgins and Vander Zanden 2010). Their filtering impacts have been particularly well studied. Because of high intrinsic filtering rates and high abundances, mussels can greatly reduce algal concentration, increase water clarity (Fahnenstiel et al. 1995, Vanderploeg et al. 2002, 2010), and decrease the chlorophyll (Chl): total phosphorus (TP) ratio (e.g., Higgins et al. 2011). Increases in water clarity can increase the production of benthic algae including nuisance algae such as *Cladophora*, which potentially benefit as well from substrate and nutrients provided by the mussels with which they are often associated (Vanderploeg et al. 2002, Higgins et al 2008, Ozersky et al. 2009). In addition to clearing the water during spring, mussels can promote harmful blooms of the cyanophyte *Microcystis* during summer by selectively rejecting colonial *Microcystis* while ingesting small desirable algae (Vanderploeg et al. 2001, 2009). It is also possible that mussels promote *Microcystis* not only by removing its competitors but also by the nutrients produced during mussel feeding and subsequent excretion (Vanderploeg et al. 2001, 2002).

Thus, nutrient excretion potentially acts synergistically with both mussel feeding impacts on phytoplankton composition and engineering impact of increased light intensity to promote harmful blooms in the water column (*Microcystis*) and nuisance blooms of attached algae on bottom substrates (*Cladophora*).

Mussel filtering, nutrient excretion, and ecosystem engineering are having profound impacts on algae in both nearshore and offshore regions of the Great Lakes (Vanderploeg et al. 2001, 2002, 2009, 2010, Hecky et al. 2004). Dreissenid mussels in the nearshore region can reengineer carbon and phosphorus cycling in the Great Lakes by intercepting phytoplankton and other nutrient-rich particles and retaining these nutrients in the nearshore via increased mussel mass and promoting *Cladophora* growth. This shunt of nutrients to the nearshore potentially reduces available nutrients to the offshore food web (Hecky et al. 2004). Furthermore, high concentrations of quagga mussels now dominating mid-depth regions of the Great Lakes have severely reduced the spring phytoplankton bloom and are possibly acting as a mid-depth sink for nutrients, again reducing nutrients to the offshore food web (Kerfoot et al. 2010, Vanderploeg et al. 2010). In the mid-depth sink, soluble phosphorus excreted by mussels would not be sequestered by *Cladophora* but would be recycled to the water column, a result consistent with lower Chl:TP ratios seen in lakes after invasion by dreissenid mussels. For the lack of a better descriptor, we describe the collective impacts of mussels on nutrient cycling as nutrient-cycling reengineering.

Another way mussels can potentially reengineer nutrient cycling is through the process of homeostatic nutrient excretion (Sterner 1990, Sterner et al. 1992, Vanderploeg et al. 2002). This process is particularly relevant for ammonium and phosphate, which when excreted are immediately available for uptake by algae or cyanobacteria. In this scenario, mussels maintain relatively constant concentrations of nitrogen (N) and phosphorus (P) in their tissues by balancing excretion to match inputs. Since N:P ratios of algae vary depending on nutrient availability, N:P ratios and amounts of nutrients excreted by mussels will vary with algal stoichiometry as well as feeding rate. Theoretically, when seston N:P is low, feeding mussels will excrete P at a high rate, and the N:P ratio of excreted nutrients will be low. When seston N:P is high, mussels will excrete P at a low rate, and the N:P ratio of excreted nutrients will be high. This skewing of N:P ratios has relevance not only to P recycling, but also to algal composition since low N:P ratios lead to cyanobacterial dominance (Smith 1983). Consonant with this theory, Arnott and Vanni (1996) observed that the N:P ratio of soluble N and P excreted by mussels from western Lake Erie was lower than that of the seston ingested, and they argued that this could be one factor responsible for the return of *Microcystis* blooms to Lake Erie subsequent of the zebra mussel invasion.

Despite the importance of N- and P-excretion by mussels to algal dynamics and composition, little work has been done on factors affecting excretion rate and N:P ratios of nutrients excreted. Factors that would need to be considered are concentrations of N and P in seston and mussels, mussel feeding rates, mussel P- and N-excretion, and algal composition since feeding rate is very sensitive to algal composition (Vanderploeg et al. 2001, 2009, 2010). We are not aware of any studies where all these factors were considered. Arnott and Vanni (1996) reported everything but feeding rates, while Conroy et al. (2005) reported excretion rates of mussels kept in cold storage for several days. Further, Naddafi et al. (2008) examined feeding rate of carbon (C), N, and P, but examined only net N and P release because algae present in their experimental incubations removed some portion of the ammonium and phosphate excreted to the water.

In this chapter, we present results of a study to measure the magnitude and seasonal patterns of nitrogen and phosphorus excretion by zebra mussels from two areas in the Great Lakes that differed in trophic status, and to determine what factors contribute to observed trends. The study areas were Saginaw Bay (Lake Huron) and the western basin of Lake Erie—both areas were of particular interest because of the occurrence of *Microcystis* blooms subsequent to invasions by zebra and quagga mussels (Vanderploeg et al. 2001). A major advancement of our study over previous excretion studies is that our studies were conducted in parallel with controlled experiments on feeding behavior (Vanderploeg et al. 2001, 2009) in which seston stoichiometry and alga composition were defined. Finally, we discuss the relative importance of nutrient excretion and food selectivity as mechanisms that could contribute to increases in the occurrence of *Microcystis* blooms.

METHODS

Experimental Design

Nutrient excretion and feeding behavior by zebra mussels were measured under controlled laboratory experiments using natural seston from two sites in Saginaw Bay and one site in western Lake Erie. Filtration and assimilation rates as a function of algal composition were previously reported for mussels from these sites in Vanderploeg et al. (2009). Nutrient excretion rates as reported here were measured using short-term bottle incubations that followed identical collection, handling, and acclimation procedures as in Vanderploeg et al. (2009), and these excretion rates were examined in light of feeding behavior responses. Sites were selected to represent a range of trophic conditions as defined by the quantity and quality of the seston, but yet have similar physical characteristics and well-established zebra mussel populations. In Saginaw Bay, one site was located in the inner bay (SB5) and one site was located in the outer bay (SB19), and in western Lake Erie, one site was located in Hatchery Bay (LE) (Figure 27.1). Site details can be found in Johengen et al. (1995) for Saginaw Bay, and in

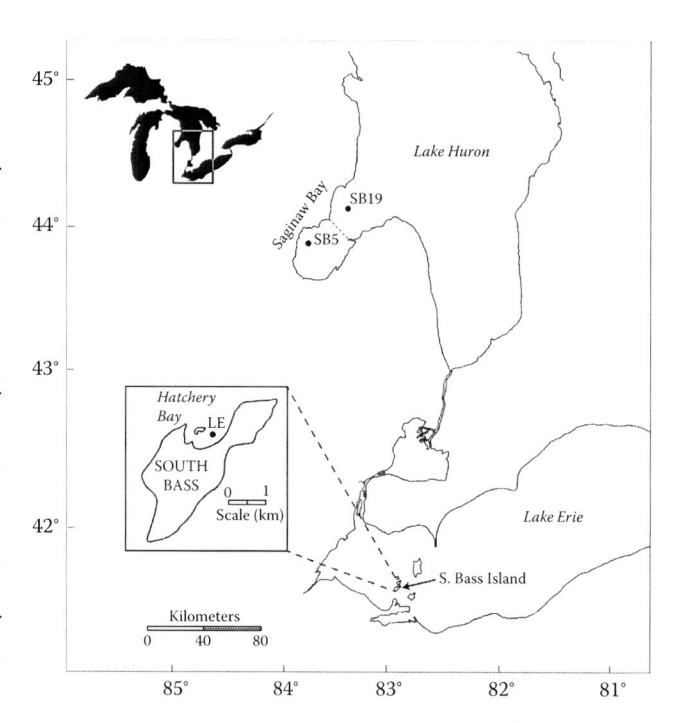

Figure 27.1 Locations where water and zebra mussels were collected in Saginaw Bay, Lake Huron (SB5 and SB19), and in Hatchery Bay, western Lake Erie (LE), 1995–1997.

Holland et al. (1996) for western Lake Erie. Measurements were made monthly between April and November during 1995–1997 at the Saginaw Bay sites, but monthly measurements at the Lake Erie site were not initiated until July 1996. We attempted to directly compare results from multiple sites by conducting the experiments within a few days of each other. In all, nutrient excretion rates were measured 14 times for SB5, 9 times for SB19, and 6 times for LE (Table 27.1).

In addition, two transplant experiments were conducted whereby mussels from Saginaw Bay or Lake Erie were incubated in water from the opposite lake and excretion rates were measured at various time points during these incubations. Transplant experiments were done to confirm that differences in observed excretion rates were a function of the food source and not the mussel populations themselves and to directly quantify the impact of changing seston composition. In September 1996, following measurements of SB5 and LE mussels using ambient seston, LE mussels were acclimated in SB5 water with ambient seston for 38 h and then excretion rates were remeasured after an incubation period of 3–4 h. The SB5 water used for the transplant experiment was collected four days after the original collection at SB5. In June 1997, again following measurements of SB5 and LE mussels using ambient seston, the opposite transplant was performed whereby mussels from SB5 were acclimated in LE water with ambient seston (original collected water) for 118 h and then excretion rates were remeasured after an incubation period of 4 h.

Table 27.1 Ambient Temperature, Chlorophyll, and Particulate and Dissolved Nutrient Concentrations in Water for Each Collection Date and Site in Experiments with Zebra Mussels Conducted during 1995–1997

Date	Temp. (°C)	Site	POC (mg/L)	PON (mg/L)	PP (µg/L)	CHL (µg/L)	NO_3 (mg/L)	NH_4 (µg/L)	SRP (µg/L)
April 24, 1995	7	SB5	0.71	0.08	5.0	2.68	0.50	14.6	0.3
April 26, 1995	6	SB19	0.44	0.03	3.6	1.15	0.33	11.0	0.4
May 22, 1995	16	SB5	0.37	0.05	2.1	0.61	0.48	33.0	0.3
May 24, 1995	12	SB19	0.22	0.02	0.9	0.60	0.41	33.0	0.3
June 19, 1995	20	SB5	0.44	0.07	3.2	1.69	0.34	20.8	0.3
June 21, 1995	20	SB19	0.24	0.03	2.6	0.43	0.35	16.3	0.3
July 10, 1995	20	SB5	2.79	0.36	10.5	16.10	0.03	5.7	0.2
July 12, 1995	20	SB19	1.41	0.18	6.4	7.85	0.16	8.3	0.2
August 14, 1995	25	SB5	3.09	0.30	16.1	9.01	0.04	7.0	0.9
August 16, 1995	25	SB19	2.40	0.32	15.5	9.42	0.26	18.4	0.5
September 18, 1995	18	SB5	3.00	0.34	27.8	6.28	0.03	5.1	1.0
September 26, 1995	15	SB19	0.95	0.14	12.0	6.91	0.08	9.1	0.7
November 7, 1995	5	SB5	1.58	0.25	24.5	8.26	0.04	16.8	1.1
May 14, 1996	9	SB5	0.15	0.03	3.6	1.36	0.54	15.5	0.3
May 16, 1996	9	SB19	0.21	0.03	3.2	0.53	0.31	11.8	0.2
June 12, 1996	17	SB5	0.26	0.04	5.3	0.71	0.52	26.5	0.5
June 26, 1996	18	SB19	0.25	0.04	3.7	0.89	0.41	19.3	0.3
July 18, 1996	25	LE	0.66	0.11	12.9	10.34	0.82	87.8	4.4
July 23, 1996	19	SB19	0.52	0.08	5.5	2.15	0.51	10.1	0.7
July 25, 1996	22	SB5	0.50	0.08	6.0	3.49	0.79	26.4	1.0
August 29, 1996	21	SB5	2.24	0.30	16.5	13.45	0.13	6.7	0.9
September 4, 1996	25	LE	0.50	0.06	14.2	2.31	0.45	15.9	2.2
September 6, 1996	25	LE ZM at SB5[b]	5.90	0.63	14.6	23.5	0.08	4.8	1.1
November 13, 1996[a]	6	SB5	1.37	0.18	10.1	6.01	0.05	6.8	1.0
June 18, 1997	19	SB5	0.41	0.06	7.2	0.60	0.18	27.0	0.5
June 19, 1997	19	LE	0.30	0.03	7.0	0.63	0.36	22.5	2.3
June 23, 1997	19	SB5 ZM at LE[c]	0.42	0.07	12.6	1.61	0.79	42.0	1.8
August 27, 1997	22	LE	0.42	0.05	7.3	1.66	0.74	43.5	10.3
October 16, 1997	16	LE	0.58	0.07	11.7	2.38	0.50	27.7	5.5

[a] POC, PON, and PP data for this date were derived from regressions on measured chlorophyll and average seston stoichiometric ratios for this site.
[b] Transplant experiment in which zebra mussels from LE were fed seston from SB5.
[c] Transplant experiment in which zebra mussels from SB5 were fed seston from LE.

Sampling and Handling

Zebra mussels used for experiments were collected by either scuba divers or with a towed benthic sled. Details of the collection, handling, and acclimation are provided in Vanderploeg et al. (2009). In brief, mussels were brought back to the lab within approximately 4 h of collection, sorted, sized, and cleaned in a temperature-controlled room maintained at the ambient water temperature. After cleaning, mussels were placed in a 30 L aquarium containing ambient-site water and acclimated for 12–18 h before being used in experimental incubations. Water in the aquarium was exchanged with additional site water approximately 2–3 h before animals were placed in experimental containers to ensure that mussels were feeding on seston at concentrations that closely matched ambient levels just prior to the experimental incubations. The acclimation procedure was designed to minimize disturbance effects from prior handling, including removal of any built up waste products, and allowed us to observe that all mussels were filtering normally before the experiments began.

Nutrient Excretion

Nutrient excretion rates were determined by measuring changes in ammonium and soluble reactive phosphorus (SRP) concentrations during incubations of 3–4 h. Incubation bottles contained 500 mL of 0.2 µM filtered site water. Zebra mussels used in experiments ranged in size between 10 and 20 mm, but results were not differentiated by size for this study because experiments indicated that excretion rates differed significantly as a function of size in only 4 of 15 trials (Johengen, unpublished data). Biomass was kept comparable in each bottle by using between two and five mussels depending on the overall size of individuals. For all experiments, two control bottles without any zebra mussels were subjected to identical conditions. There were no significant changes in nutrient concentrations in the control bottles during the course of the incubation for any of the experiments, and these results are not presented further. Excretion rates were determined by subtracting the total mass of nitrogen or phosphorus in each bottle from the 3 or 4 h time point from that determined at time zero, immediately after mussels were added and the bottle well mixed. Changes in volumes and mass removed during sampling were accounted for in all calculations. All nutrient excretion rates were normalized to soft-tissue dry weight. Dry weights were obtained by removing the soft tissue from the shell onto pre-tarred pieces of aluminum foil, drying the tissue at 80°C for 24 h, and then reweighing.

Variables Affecting Nutrient Excretion

Feeding Rate Variables

We expected that nutrient excretion would be related to ingestion and assimilation of C, N, and P. Filtering mussels capture particles and ingest those that are not rejected as pseudofeces. Of the ingested material, some is assimilated and some is egested as feces. Therefore, we can define particle capture rate (gross feeding) and particle assimilation rate (net feeding). The assimilated (net feeding) material is of particular interest because this represents material the mussels can utilize for nutrient storage and excretion. In companion feeding experiments, Vanderploeg et al. (2001, 2009) measured gross clearance rate, net clearance rate, capture rate, and assimilation rate for two size fractions of Chl (>53 and <53 μm) as well as for total Chl, the sum of the two fractions. In essence, these variables are a measure of feeding rate on phytoplankton expressed in terms of Chl concentration. To estimate potential capture and assimilation of C, N, and P, we multiplied Chl capture and assimilation by the ratio of C/Chl, N/Chl, and P/Chl. See details for methodology in Vanderploeg et al. (2001, 2009). We recognize this approach represents an upper bound for assimilation, because not all seston C, N, and P are associated with phytoplankton and could have poor assimilation relative to Chl.

Seston Quality and Temperature

Feeding rate and quality of food for supporting consumer growth depends upon seston quality as determined by algal composition and its C:P or N:P ratio. In addition, nutrients assimilated and excreted depend on these ratios for a given unit of food ingested. To this end, we compared seston nutrient concentration and excretion ratios to the Redfield ratio, which states that the "average" atomic ratio of C:N:P of organisms in the sea is 106:16:1 (Redfield et al. 1963). In turn, net uptake and release of nutrients through biochemical processes tend toward these same ratios (Atkinson and Smith 1983).

Hecky et al. (1993) cautioned that nutrient ratios for lake phytoplankton are much more variable, but concluded that conventional ratios do provide a relative indication of nutrient availability or limitation. Furthermore, Sterner et al. (1993) demonstrated that phytoplankton nutrient content impacted feeding rates in *Daphnia* and thereby regulated *Daphnia* growth and reproduction. Arnott and Vanni (1996) demonstrated that nutrient content of both seston and mussel tissue explained much of the observed variation in their measured mussel excretion rates. Therefore, in addition to examining feeding rate variables as potential drivers of nutrient excretion, we also examined algal composition and nutrient ratios as ultimate drivers of feeding and subsequent nutrient excretion.

It should be noted that when results for the ratios of N- and P-excretion by mussels were examined relative to nutrient availability and seston nutrient concentration, we used a P:N excretion ratio instead of the normal N:P ratio so that we did not lose the eight cases where P-excretion was zero due to a divisor of zero. For comparative purposes, it should be noted that the Redfield mass P:N ratio is 0.139. The loss of eight cases would have had too great an impact on statistical comparisons among sites and among relationships between variables tested.

Temperature is also an obvious variable affecting nutrient excretion through its direct influence on both metabolic processes (Quigley et al. 1993, Jansen et al. 2011) and potential filtration rates (Vanderploeg et al. 2010). We did not directly manipulate temperature in our experiments and all experimental incubations were performed at the ambient temperature observed during sample collection. However, experiments were conducted at ambient temperatures that ranged from 5°C to 25°C, which allowed us to evaluate this environmental variable along with that of seston quality on observed rates of feeding and excretion.

Algal Composition

We also examined the effect of algal composition on measured nutrient excretion rates, but herein present only results for two categories, percent cyanophytes and percent preferred algae (cryptophytes plus flagellates) as based on carbon biomass. Parallel experiments on feeding behavior previously indicated that clearance and assimilation rates were strongly affected by percent contribution of these taxa to phytoplankton composition (Vanderploeg et al. 2009). In addition, feeding experiments coupled with videographic observation of the mussels revealed that high concentrations of the cyanophyte *Microcystis aeruginosa* severely depressed mussel pumping rate and decreased the percent of time mussels spent filtering (Vanderploeg et al. 2001, 2009). Pumping rate was positively correlated with percent of preferred algae, and a modest amount of preferred algae could stimulate the mussels to feed even in the presence of *Microcystis*.

Analytical Procedures

Nutrient concentrations were determined using standard automated colorimetric techniques on a Technicon Auto Analyzer II, as detailed in Davis and Simmons (1979). Nitrate (NO_3) concentrations were determined by the cadmium reduction method, ammonium (NH_4) concentrations were determined by the Bertholet reaction, and SRP concentrations by the molybdate/ascorbic acid method. TP was determined by the same method as SRP after digesting 50 mL of sample with potassium persulfate (1% final concentration) in an autoclave for 30 min (Menzel and Corwin 1965). Particulate phosphorus (PP) was determined by the same method as TP for particles retained on a 0.2 μm nucleopore filter and resuspended in 50 mL of phosphorus-free deionized water. Samples for particulate organic carbon (POC) and particulate organic nitrogen (PON) were processed in triplicate by filtering 100–300 mL of sample through pre-combusted (450°C for 4 h) Whatman GFF filters. Filters were saturated with 1 N HCl to remove inorganic carbonates, dried at 60°C for at least 24 h, and then analyzed on a Perkin Elmer (model 2400) CHN elemental analyzer. Chlorophyll was determined fluorometrically after extracting particles retained on a 47 mm Whatman GFF filter with N,N-Dimethylformamide (Speziale et al. 1984). All mass concentration units for POC, PON, NO_3, NH_4, PP, and SRP are given for the respective elements C, N, and P. Seston C, N, and P ratios were calculated from the POC, PON, and PP data. All excretion and seston ratios were calculated for mass units.

All statistics were performed using SYSTAT (version 11.0). Significant differences in water quality variables among sites were evaluated using the Kolmogorov–Smirnov two-sample test after pooling all observations at each site. Differences in excretion rates for mussels among sites that were measured in pairs a few days apart were examined using a standard t-test. Relationships among excretion rates, water and food quality, and feeding rates were examined with simple correlation. Stepwise linear regression was used to determine if there were interactions with more than one variable.

RESULTS

Ambient Water Quality Conditions

In all, 28 independent sample collections were conducted during this study, including 14 at SB5 and 9 at SB19 in Saginaw Bay and 6 at LE in western Lake Erie. Ambient conditions for temperature, chlorophyll, dissolved nutrients, and particulate nutrients are given in Table 27.1. Means and standard deviations of these variables are given in Table 27.2 to provide a relative comparison in water quality conditions among test sites. Concentrations of dissolved nutrients at LE were significantly higher ($P < 0.05$) than at either SB5 or SB19 throughout the study. Specifically, mean concentrations of NO_3 and NH_4 at the LE site were 1.5 and 1.9 times higher, respectively, than mean concentrations at SB sites, and the mean concentration of SRP was nearly six times greater. Seston nutrient stoichiometry was also significantly different ($P < 0.05$) between the three sites. Mean C:P and N:P ratios for seston at SB sites (see Table 27.2) were considerably higher than Redfield ratios (mass ratio for C:N = 5.7; C:P = 41; N:P = 7.2; Redfield et al., 1963) and well above levels suggested to indicate P-limitation (C:P = 100 and N:P = 9.9; Hecky et al. 1993). In contrast, mean C:P and N:P ratios for seston at LE were near or below Redfield ratios and indicated no P-limitation. C:P for LE seston was 57% and 43% of that for SB5 and SB19, respectively, and N:P for LE seston was 56% and 46% of SB5 and SB19, respectively. Seston C:N ratios were not significantly different ($P > 0.05$) between sites and all tended to be slightly higher than the nominal Redfield ratio.

The greater availability of dissolved nutrients at LE did not result in higher accumulations of phytoplankton biomass. Mean POC concentration at LE was only 36% of that at SB5 and 78% of that at SB19 (Table 27.2). Similarly, mean Chl concentrations at LE were 32% and 65% of that at SB5 and SB19, respectively.

Seasonal Excretion Rates for Natural Seston

Excretion rates of nitrogen and phosphorus exhibited distinctly different seasonal patterns when examined collectively across all three sites (Figure 27.2). Overall,

Table 27.2 Mean (±SE) Temperature, Chlorophyll, and Particulate and Dissolved Nutrient Concentrations for Each Site for All Water Collections in 1995–1997

Site	Temp. (°C)	POC (mg/L)	PON (mg/L)	PP (μg/L)	C:N ratio	C:P ratio	N:P ratio	CHL (μg/L)	NO_3 (mg/L)	NH_4 (μg/L)	SRP (μg/L)
SB5 (n = 14)	16.3 ± 1.9	1.63 ± 0.44	0.20 ± 0.05	11.0 ± 2.1	8.4 ± 0.9	146 ± 28	17.7 ± 3.4	6.70 ± 1.84	0.27 ± 0.07	15.5 ± 2.6	0.7 ± 0.1
SB19 (n = 9)	16.0 ± 2.0	0.74 ± 0.25	0.10 ± 0.03	5.9 ± 1.6	7.0 ± 0.5	109 ± 46	14.4 ± 4.6	3.33 ± 1.21	0.31 ± 0.04	15.3 ± 2.6	0.4 ± 0.1
LE (n = 5)	21.0 ± 1.5	0.58 ± 0.16	0.08 ± 0.02	9.3 ± 1.0	7.9 ± 0.6	63 ± 15	8.1 ± 0.7	2.15 ± 0.82	0.44 ± 0.12	28.3 ± 5.5	3.6 ± 1.5

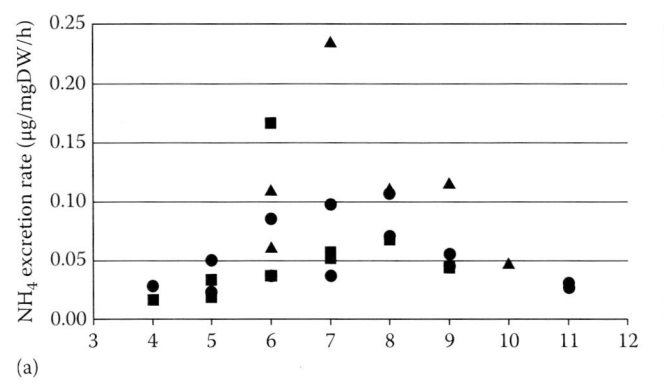

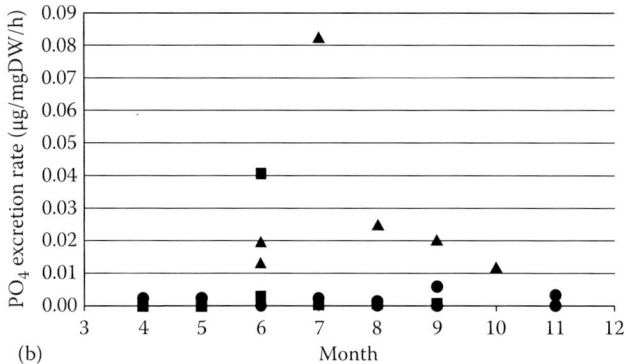

Figure 27.2 Seasonal patterns of ammonium (a) and phosphorus (b) excretion by zebra mussels from Saginaw Bay, Lake Huron (SB5 and SB19), and western Lake Erie (LE) determined from 3 to 4 h incubations in filtered lake water at ambient temperatures. Each point represents a measurement over all 3 years (1995–1997). SB5, circle; SB19, square; LE, triangle.

N-excretion rates ranged between 0.02 and 0.17 µgN/mg/h for mussels from SB sites and between 0.05 and 0.24 µgN/mg/h for mussels from LE. Rates followed a seasonal bell curve similar to temperature and were highest in late summer at all three sites, with notable exceptions at SB19 on June 26, 1996, and at LE on July 18, 1996, (Figure 27.2a) when rates were twice those at other sites at similar times. Overall, mean N-excretion rate for LE mussels was over two times higher than for SB mussels (Table 27.3). This difference in means occurred even when comparing only measurements taken between June to September (0.127 vs. 0.067) when the LE site and the SB sites exhibited similar temperatures. In contrast, P-excretion rates did not follow any consistent seasonal pattern (Figure 27.2b). As seen for N-excretion, P-excretion rates for SB19 mussels on June 26, 1996, and for LE mussels on July 18, 1996, were extremely elevated and almost five times the measured rates at other sites at similar times (Figure 27.2b, Table 27.4). For SB mussels, P-excretion was zero, or below our method of detection (<0.0004 µgP/mg/h), in 8 of 23 experiments (Table 27.4). When P-excretion was measureable, rates ranged from 0.0004 to 0.04 µgP/mg/h. P-excretion rates by LE mussels were significantly higher than SB mussels and ranged from 0.02 to 0.083 µgP/mg/h. Overall, the mean P-excretion rate was similar between the two SB sites when the outlier was omitted, and the mean P-excretion rate at LE was nearly 10 times greater than at the SB sites (Table 27.3).

Feeding and nutrient excretion rates for each experiment are given in Table 27.4. Large differences in gross clearance, net clearance, and net assimilation rates were observed between sites. For comparable months and temperatures (experiments only in June to September), gross and net clearance rates were 4 and 13 times greater, respectively, for LE mussels than for SB mussels. The amount of Chl that was actually assimilated by the mussels averaged 0.001 µgChl/mg/h for SB mussels compared to 0.015 µgChl/mg/h for LE mussels.

Transplant Experiments

In September 1996, after experiments with ambient seston and mussels from SB5 and LE, the same mussels from LE were reacclimated in water from SB5 for 38 h. The quantity and quality of seston differed substantially between the two sites during these experiments. For initial experiments, Chl concentration was six times higher at SB5 than at LE, and seston N:P ratios were 18.3 and 4.3 at the two sites, respectively (Table 27.5). Despite these differences, N-excretion rates were generally similar between sites, with rates for LE mussels only about 7% higher than rates for SB5 mussels (Table 27.5).

Table 27.3 Comparison of Nutrient Excretion Rates for Zebra Mussels from Two Sites in Saginaw Bay, Lake Huron (SB5, SB19), and One Site in Hatchery Bay, Lake Erie (LE), Incubated with Natural Seston. Experiments Were Conducted between April to November, 1995–1997. Reported Values Are the Arithmetic Mean (±SD) for the N Treatments Listed (Redfield Mass P:N = 0.139). Means Are Given with and without (w/o) Outliers on June 26, 1996, and July 18, 1996

Site	N	N-Excretion (µgN/mg/h)	P-Excretion (µgP/mg/h)	P:N Excreted (Mass Ratio)
SB5	14	0.053 ± 0.027	0.002 ± 0.002	0.040 ± 0.043
SB19	9	0.055 ± 0.045	0.005 ± 0.013	0.042 ± 0.079
w/o June 26, 1996	8	0.041 ± 0.018	0.001 ± 0.001	0.017 ± 0.025
LE	6	0.114 ± 0.067	0.029 ± 0.027	0.229 ± 0.065
w/o July 18, 1996	5	0.089 ± 0.032	0.018 ± 0.005	0.205 ± 0.031

Table 27.4 Excretion Rates and Feeding Variables for Each Collection Date and Site for Zebra Mussel Experiments Conducted in 1995–1997. N-Excretion (Nexc) Given in µgN/mg/h, P-Excretion (Pexc) in µgP/mg/h, P:Nexc (Ratio of P-Excretion to N-Excretion, Dimensionless), Gross Clearance Rate (F(chl)) in mL/mg/h, Net Clearance Rate (F(A)) in mL/mg/h, and Assimilation Rate (µgChl/mg/h)

Date	Site	Nexc	Pexc	P:Nexc	F(chl)	F(A)	A(chl)
April 24, 1995	SB5	0.029	0.002	0.080	11.0	2.8	0.004
April 26, 1995	SB19	0.017	0.000	0.000	9.3	7.2	0.006
May 22, 1995	SB5	0.051	0.000	0.000	9.7	2.9	0.001
May 24, 1995	SB19	0.034	0.000	0.000	14.0	11.4	0.005
June 19, 1995	SB5	0.036	0.002	0.047	1.5	0.4	0.000
June 21, 1995	SB19	0.037	0.003	0.072	14.1	3.8	0.001
July 10, 1995	SB5	0.037	0.002	0.062	0.0	0.0	0.000
July 12, 1995	SB19	0.057	0.001	0.007	1.2	0.36	0.003
August 14, 1995	SB5	0.072	0.001	0.019	0.8	0.0	0.0
August 16, 1995	SB19	0.068	0.001	0.012	0.2	0.0	0.0
September 18, 1995	SB5	0.046	0.000	0.000	7.1	1.4	0.007
September 26, 1995	SB19	0.044	0.001	0.014	10.2	0.9	0.002
November 7, 1995	SB5	0.028	0.003	0.116	14.4	1.6	0.010
May 14, 1996	SB5	0.024	0.002	0.092	3.2	0.1	0.000
May 16, 1996	SB19	0.019	0.000	0.000	4.1	1.3	0.000
June 12, 1996	SB5	0.086	0.002	0.027	9.2	7.9	0.006
June 26, 1996	SB19	0.167	0.041	0.243	8.9	5.7	0.004
July 18, 1996	LE	0.236	0.083	0.350	5.6	3.6	0.029
July 23, 1996	SB19	0.052	0.002	0.031	7.1	0.4	0.001
July 25, 1996	SB5	0.098	0.001	0.012	5.9	0.6	0.001
August 29, 1996	SB5	0.107	0.000	0.000	5.4	0.0	0.000
September 4, 1996	LE	0.116	0.020	0.170	18.7	17.1	0.019
September 6, 1996	LE ZM at SB5[a]	0.056	0.006	0.108	0.5	0.0	0.000
November 13, 1996[a]	SB5	0.032	0.000	0.000	7.8	3.7	0.016
June 18, 1997	SB5	0.038	0.000	0.000	4.3	0.0	0.000
June 19, 1997	LE	0.110	0.020	0.177	22.6	14.6	0.006
June 23, 1997	SB5 ZM at LE[b]	0.062	0.013	0.215	21.4	10.1	0.005
August 27, 1997	LE	0.112	0.025	0.222	34.0	22.5	0.017
October 16, 1997	LE	0.047	0.012	0.243	13.9	8.2	0.011

[a] Transplant experiment in which zebra mussels from LE were fed seston from SB5.
[b] Transplant experiment in which zebra mussels from SB5 were fed seston from LE.

Table 27.5 Mean (±SE) Filtering and Excretion Rates for Mussels from Saginaw Bay, Lake Huron (SB5), and Hatchery Bay, Lake Erie (LE), on September 6, 1996. Also Given Are Means for Lake Erie Mussels That Were Acclimated in Saginaw Bay Water for 38 h Before Incubations Were Initiated. Temperature Was 21°C for Saginaw Bay and 25°C for Lake Erie Mussels. N = 4 bottles, with Four Medium-Sized Mussels in Each Bottle. nd, Not Determined

Variable	SB5 Mussels	LE Mussels	LE Mussels in SB5 Water
N-excretion (µgN/mg/h)	0.108 ± 0.009	0.116 ± 0.013	0.056 ± 0.004
P-excretion (µgP/mg/h)	0.000 ± 0.000	0.020 ± 0.003	0.006 ± 0.001
Excreted P:N mass ratio	nd	0.172 ± 0.018	0.106 ± 0.007
Seston N:P mass ratio	18.3	4.3	42.8
Chlorophyll a (µg/L)	13.5	2.3	23.5
Gross clearance rate (mL/mg/h)	5.4 ± 3.6	18.7 ± 3.3	0.5 ± 1.0
Net clearance rate (mL/mg/h)	0.0 ± nd	17.3 ± 3.5	0.0 ± nd
Assimilation rate (µgChl/mg/h)	0.0 ± nd	0.019 ± 0.004	0.0 ± nd

Table 27.6 Mean (±SE) Filtering and Excretion Rates for Mussels from Saginaw Bay, Lake Huron (SB5), and Hatchery Bay, Lake Erie (LE), on June 23, 1997. Also Given Are Means for Saginaw Bay Mussels That Were Acclimated in Lake Erie Water for 118 h before Incubations Were Initiated. All Incubations Were Conducted at 19°C. N = 4 Bottles, with Four Medium-Sized Mussels in Each Bottle. nd, Not Determined

Variable	SB5 Mussels	LE Mussels	SB5 Mussels in LE Water
N-excretion (μgN/mg/h)	0.038 ± 0.002	0.110 ± 0.006	0.062 ± 0.005
P-excretion (μgP/mg/h)	0.000 ± 0.000	0.020 ± 0.001	0.013 ± 0.003
Excreted P:N mass ratio	nd	0.179 ± 0.013	0.208 ± 0.026
Seston N:P mass ratio	8.3	4.3	5.6
Chlorophyll a (μg/L)	0.6	0.6	0.6
Gross clearance rate (mL/mg/h)	4.3 ± 0.6	22.6 ± 0.8	21.4 ± 1.4
Net clearance rate (mL/mg/h)	0.0 ± nd	14.6 ± 1.0	10.1 ± 0.9
Assimilation rate (μgChl/mg/h)	0.0 ± nd	0.006 ± 0.000	0.005 ± 0.000

In contrast, P-excretion rates differed greatly between sites, with P-excretion being 0.020 µgP/mg/h for LE mussels and undetectable for SB5 mussels. Consequently, the P:N excretion ratio also differed greatly between sites, being 0.172 and "zero," respectively. Excretion rates reflected observed feeding behavior. Mussels from LE filtered particles (gross clearance) at a rate 3.5 times faster than the mussels from SB5 (Tables 27.4 and 27.5). Moreover, the net clearance rate of LE mussels was equal to 93% of the gross clearance rate, whereas the net clearance rate of SB5 mussels was zero. Net assimilation of Chl was 0.019 µgChl/mg/h for LE mussels but undetectable for SB5 mussels.

Four days after the original excretion and feeding rate experiments, new water was collected from SB5 for use in the transplant excretion and feeding experiments with LE mussels. There was an extensive *Microcystis* bloom at the time of collection and, consequently, Chl concentrations had increased from 13.5 to 23.5 µg/L, and the seston N:P ratio increased from 18.3 to 42.8 between the two collection dates, August 29 and September 2. The difference in seston composition of the water collected on September 2 would not have affected measured excretion rates for the LE mussels since all prior feeding/acclimation experiments were done with the water initially collected at SB5 on August 29, and incubations for excretion experiments were done in 0.2 µM filtered site water. However, the change in seston composition would have an impact on the feeding results. After LE mussels were acclimated in new SB5 water for 38 h, N- and P-excretion rates declined by 52% and 70%, respectively, which resulted in about a 40% drop in the P:N excretion ratio. Interestingly, this ratio was still considerably more P-rich than the actual food source the mussels were exposed to for the 38 h acclimation period. Gross clearance rates by LE mussels fed SB5 seston were only 3% of those observed under ambient LE seston, and both net clearance rate and net assimilation dropped to zero, similar to the response observed for SB5 mussels.

In June 1997, a second transplant experiment was conducted with SB5 mussels that were acclimated in water from the LE site for 118 h. Chl concentrations at both sites measured only 0.6 µg/L (Table 27.6). The seston N:P ratio at SB5 was only 8.3, which was near the lowest observed throughout the study, but still about twice as high as that for LE seston. For the initial experiments, N-excretion rates by LE mussels fed LE seston were nearly three times greater than for SB5 mussels fed SB seston (Table 27.6). Similarly, P-excretion rates for LE mussels averaged 0.020 µgP/mg/h, whereas rates were again below detection for SB5 mussels. Gross clearance rates by SB5 mussels were only 20% of those for LE mussels and net clearance rate was again zero. The net clearance rate for LE mussels was 14.6 mL/mg/h, which represented 65% of gross clearance rate, and net assimilation of seston was measured at 0.006 µgChl/mg/h (Table 27.6).

After SB5 mussels fed on LE seston, N-excretion rates increased by 63%, and P-excretion rates went from below detection to 0.013 µgP/mg/h. Excretion rates of N and P for the transplanted SB mussels were 56% and 65%, respectively, of rates originally observed for LE mussels. The P:N excretion ratio by transplanted SB5 mussels increased from 0 to 0.208, which was even higher than originally seen for LE mussels (Table 27.6). Gross clearance rates for transplanted mussels increased fivefold from initial rates on SB5 seston and were similar to those observed initially for LE mussels feeding on LE site seston. Net clearance rates of transplanted mussels increased from 0 to 10.1 mL/mg/h and represented approximately 50% of the gross clearance rate on this seston. Net assimilation rate of transplanted SB5 mussels fed LE seston was 0.005 µgChl/mg/h, which was very similar to the original response of LE mussels.

Effects of Temperature and Food Quality and Quantity

Correlation analysis among nutrient excretion rates and ratios, temperature, measures of seston quantity and quality, and feeding behavior was performed on the combined data for all

Table 27.7 Correlation Coefficients for Excretion Rates and Ratios, Ambient Temperature, Measures of Seston Quantity and Quality, and Feeding Behavior for All Experiments Conducted during the Study (Bolded Values Indicate Significant Relationships at P < 0.05)

Variable	N-Excretion	P-Excretion	P:N Excretion	Chl Assimilation Rate
N-excretion	—	**0.88**	**0.66**	**0.39**
P-excretion	**0.88**	—	**0.85**	**0.51**
P:N excretion ratio	**0.66**	**0.85**	—	**0.41**
Temperature	**0.57**	0.34	0.26	−0.19
POC	−0.08	−0.17	−0.17	**−0.54**
Chlorophyll	0.11	0.04	−0.19	**−0.39**
SRP	**0.42**	**0.49**	**0.64**	**0.50**
Seston C:N	−0.28	−0.23	−0.21	−0.08
Seston C:P	−0.27	−0.30	−0.36	**−0.64**
Seston N:P	−0.23	−0.29	**−0.37**	**−0.63**
Gross clearance rate	0.18	0.24	**0.48**	**0.49**
Net clearance rate	0.30	0.33	**0.50**	**0.55**
Chl assimilation rate	**0.39**	**0.51**	**0.41**	—
C assimilation rate	0.19	0.24	0.26	**0.89**
N assimilation rate	0.26	0.31	0.33	**0.90**
P assimilation rate	**0.40**	**0.41**	**0.55**	**0.70**
% Cyanophytes[a]	0.10	−0.10	−0.24	−0.32
% Preferred algae[a]	0.34	**0.36**	**0.55**	**0.36**

[a] Percent composition based on amount of carbon.

sites and results are summarized in Table 27.7. Analysis showed that food quality and feeding rates were important determinants for both NH_4 and SRP excretion. We recognize, however, specific relationships between excretion rates and food quality varied between the SB sites and the LE site as evident in Figures 27.2 through 27.4. Nutrient excretion rates were not correlated to measures of seston quantity such as POC and Chl.

N-excretion was strongly correlated with temperature (Table 27.7), and this relationship was also apparent by the seasonal bell-shaped curve with maximum NH_4 excretion during summer (Figure 27.3a). No significant correlation (P > 0.05) occurred between P-excretion and temperature (Table 27.7); however, the lack of response was much more notable for SB sites than for the LE site (Figure 27.2b), where a temperature effect may have been apparent had a broader range of temperatures been covered during experiments.

Rates and ratios of N- and P-excretion were examined relative to nutrient availability expressed as both dissolved nutrient levels and nutrient content of seston. For these analyses, we used P:N excretion ratio so that we did not lose the eight cases where P-excretion was zero (for comparative purposes it should be noted that the Redfield mass P:N ratio is 0.139). N- and P-excretion were not significantly correlated (P > 0.05) with any of the stoichiometric ratio variables of seston measured (C:N, C:P, or N:P). However, all relationships were negative and indicated a negative response to P-depleted seston (Table 27.7). Despite the lack of a statistical relationship, it was clearly evident that high P-excretion

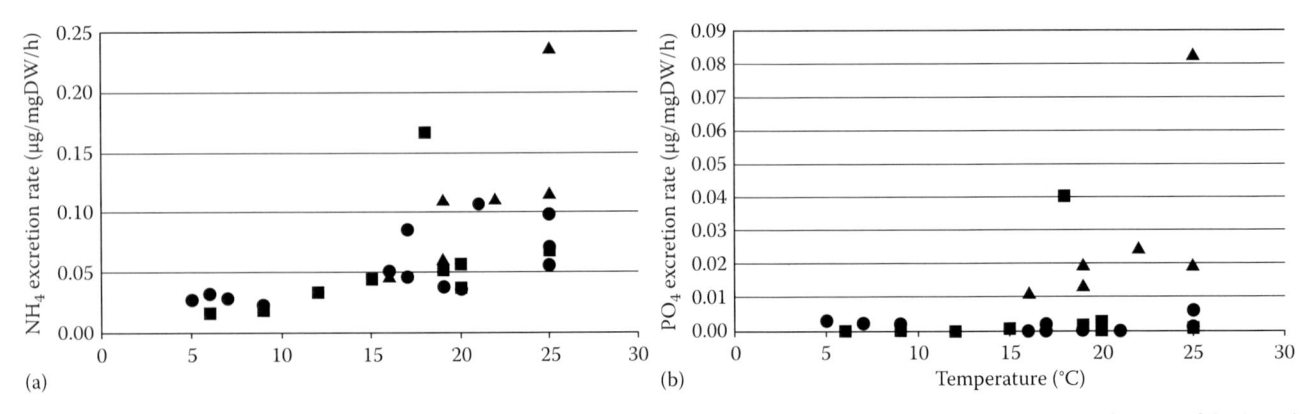

Figure 27.3 Ammonium (a) and phosphorus (b) excretion rates by zebra mussels from Saginaw Bay, Lake Huron (SB5 and SB19), and western Lake Erie (LE) determined from 3 to 4 h incubations in filtered lake water as a function of ambient temperature. Each point represents a measurement over all 3 years (1995–1997). SB5, circle; SB19, square; LE, triangle.

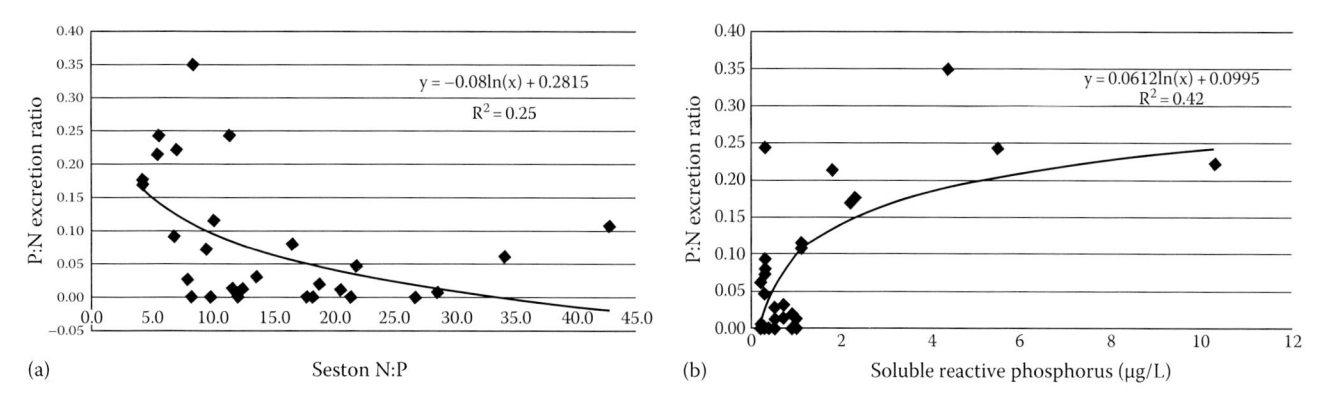

Figure 27.4 Mass ratio of phosphorus and nitrogen excretion rates (P:N) for zebra mussels from Saginaw Bay, Lake Huron, and western Lake Erie as a function of the seston N:P ratio upon which they fed (panel a: $R^2 = 0.25$, $P = 0.006$) and as a function of the SRP concentration at the site during the time of collection (panel b: $R^2 = 0.42$, $P < 0.001$).

rates occurred only at low seston C:P and N:P ratios, and the P:N excretion ratio was significantly ($P < 0.05$) and negatively correlated to seston N:P ratio. Furthermore, both N- and P-excretion rates were positively correlated to SRP concentrations and to the amount of P assimilated from the seston, which again suggests a strong response to the nutritional quality of the seston (Table 27.7). The ratio of P:N excretion was also significantly and negatively correlated to seston N:P ratio and positively correlated to SRP concentration. Relationships between these variables, along with natural log regressions that provided the best fit, are given in Figure 27.4. The importance of seston P-concentration in regulating mussel P-excretion is further evident by the fact that P:N excretion ratios for LE mussels were consistently above the Redfield ratio (more P), and the seston N:P ratios were consistently at or below the Redfield ratio. Seston N:P and excretion P:N ratios for SB mussels were highly variable throughout the study with no consistent relationship, but both ratios generally indicated P depletion relative to Redfield values. On all but two occasions, P:N excretion ratios were below those found in ambient seston indicating a high internal demand for P by SB mussels (Table 27.3). With these two outlier values omitted, mean P:N excretion ratio was lowest at SB19, 2 times higher at SB5 than at SB19, and over 12 times higher at LE (Table 27.3), which closely matched the trend in SRP concentrations between sites.

P-excretion and the P:N ratio of excretion exhibited a significant positive correlation ($P < 0.05$) to percent preferred algae, but correlations to percent cyanophytes were not significant (Table 27.7). N-excretion was also positively correlated to percent preferred algae but only at the 0.07 significance level. While nutrient excretion rates were not significantly correlated to percent cyanophytes for the combined data set, impacts of cyanophytes on clearance and assimilation rates were strongly evident in transplant experiments. For the September 1996 experiment, SB5 seston was comprised of 71% cyanophytes and only 6% preferred algae, compared to 0% cyanophytes and 88% preferred algae for LE seston. When cyanophytes dominated the phytoplankton composition (SB5), we observed reduced gross clearance rates, zero net clearance rates, and zero P-excretion (Table 27.5). Conversely, when preferred algae dominated the phytoplankton composition (LE), we observed high gross and net clearance rates, along with elevated levels of P-excretion. Similarly, for the 1997 transplant experiment, SB5 seston was comprised of 36% cyanophytes and 15% preferred algae compared to 0% cyanophytes and 87% preferred algae for the LE seston. Again when preferred algae comprised a greater percentage of the seston, gross and net clearance rates and nutrient excretion rates were significantly ($P < 0.05$) higher (Table 27.6).

Stepwise multiple linear regressions of all measures of food quantity and quality were used to examine interactions of multiple variables. Results showed that gross and net assimilation rates were significantly correlated with both phytoplankton composition and nutrient stoichiometry of the seston. Multiple linear regression revealed that 55% and 58% of the variance in gross and net assimilation rates, respectively, could be explained by the variables percent preferred algae and seston C:P and N:P ratios ($P < 0.001$). Percentage of preferred algae ranged from 1% to 36% (mean = 11.6%) at SB sites and ranged from 26% to 95% (mean = 68%) at the LE site. In addition to composition differences, P content of LE seston was shown to be over three times more P-rich than SB seston on the basis of both C:P and N:P ratios. Both of these factors presumably contributed to observed differences in assimilation and excretion rates. Conversely, gross clearance rates were significantly ($P < 0.05$) and negatively correlated to percent cyanophytes ($R = -0.57$). During the study, cyanophytes accounted for 0%–75% (mean = 15%) of phytoplankton biomass at SB sites compared to the LE site where none were present on 4 of 6 occasions and only 4%–11% on the other two sampling dates.

N- and P-excretion rates and P:N excretion ratios were significantly ($P < 0.05$) and positively related to the assimilation rate as measured by the amount of Chl removed (Table 27.7). Chl assimilation itself was strongly, negatively correlated with seston C:P and N:P and suggests the importance of P-content in partially regulating observed assimilation rates. P-excretion was also significantly correlated (P < 0.05) to the amount of P assimilated as measured by the stoichiometric relationship between seston Chl and P. Multiple linear regression revealed that 58% of the variation in N-excretion rates could be explained by temperature and assimilation rate (P < 0.001). For P-excretion, 45% of the variation could be explained by these two variables (P < 0.001). Adding percent preferred algae and SRP increased the R^2 slightly to 47%.

DISCUSSION

Seasonal Excretion Rates for Natural Seston

Nutrient excretion rates of zebra mussels varied significantly across seasons and between sites. Seasonal differences were strongly related to temperature; however, differences between sites were more closely related to the nutrient concentration and phytoplankton composition of the seston upon which mussels fed. The most striking observation of this study was the dramatic difference between P-excretion rates of mussels from Saginaw Bay (SB sites) and Lake Erie (LE site), with excretion rates far greater at the Lake Erie site. Differences resulted from higher P content of LE seston and higher assimilation rates of LE mussels. Assimilation rates were significantly correlated with both phytoplankton composition and nutrient stoichiometry. Subsequently, nutrient excretion rates were significantly correlated with assimilation, most notably for assimilation rates based on either chlorophyll or phosphorus compared to assimilation based on carbon or nitrogen. Similar relationships were observed for metabolism and excretion experiments performed on the marine mussel, *Mytilus edulis* (Bayne et al. 1993, Smaal et al. 1997). As summarized by Jansen et al. (2011), metabolism has been shown to be related to both temperature and food; however, interactions between these two environmental parameters make it difficult to directly correlate metabolism to either one of them individually.

Our results indicated that the presence of cyanophytes strongly affected both feeding behavior and excretion rates of zebra mussels, even though there was only a weak correlation between percent cyanophytes and excretion rates. The weak correlation was likely due to the sporadic abundance of cyanophytes. The negative effect of cyanophytes on mussel feeding behavior was clearly apparent in individual experiments, such as the September transplant experiment when LE mussels were exposed to SB5 seston that was predominantly composed of *Microcystis*. Net clearance rates and assimilation rates were basically zero when mussels were exposed to cyanophyte-rich seston, whereas the same mussels filtered normally on cyanophyte-deplete seston (Table 27.5). The lack of feeding and assimilation in turn led to significant declines in nutrients excreted. Lavrentyev et al. (1995) reported a similar effect of *Microcystis* on filtration rates by mussels from Saginaw Bay; grazing rates on microzooplankton and phytoplankton were greatly diminished when cyanophytes comprised a high percentage of the biomass within the seston. In addition, Gardner et al. (1995) anecdotally attributed observed declines in the summer grazing rates of zebra mussels from Saginaw Bay to increased density of *Microcystis*. The overall impact of cyanophytes on mussel feeding behavior is, however, likely to be quite variable. Results obtained in our parallel feeding experiments showed that the impact of *Microcystis* depends on several factors including relative abundance, colony size, and even strain (Vanderploeg et al. 2001, 2009). These studies also indicate that zebra mussels have the ability to feed selectively, so in some instances there may not be an observable negative impact if there is an alternative food available such as cryptophytes. Selective feeding was also observed by Heath et al. (1995) in mesocosm experiments with mussels from Saginaw Bay. Heath et al. (1995) reported that diatoms and chlorophytes were more heavily grazed upon than chrysophytes and cyanophytes. Video observations showed that, on some occasions during Microcystis blooms, mussels were less responsive and spent less time filtering (Vanderploeg et al. 2001, 2009).

Although N-excretion rates, and to a lesser degree P-excretion rates, were correlated to temperature, clearance and assimilation rates had no relation to temperature. This finding suggests that temperature had a greater effect on metabolic processes than on feeding behavior, particularly for N. It also implies that differences in food quality can ultimately drive much of the observed pattern in feeding and excretion by mussels. For example, clearance and assimilation rates were significantly and negatively correlated with seston C:P and N:P ratios. While the effect of nutrient stoichiometry cannot be separated from other potential factors such as composition, these findings are consistent in theory with those of Sterner et al. (1993) who were able to demonstrate that feeding rates, growth, and reproduction in *Daphnia* were all negatively affected by severely P-deficient seston. In our study, much of this relationship was driven by the sharp contrast in conditions between the SB and LE study sites. In shallow, high-energy environments such as Saginaw Bay and western Lake Erie, we might also expect to see a potential effect from inputs of resuspended sediments to the seston. Stoeckmann and Garton (1997) found that assimilation efficiency of Lake Erie mussels was strongly correlated with the percent organic content of seston. This parameter may have reflected both changes in actual phytoplankton composition as well as differences in the amount of inorganic material present due to resuspension. We did not measure total suspended solids concentrations in our study and cannot make a direct assessment of the effects of percent organic concentration; however, our excretion rates were not correlated to POC, Chl, or the C:Chl ratio. Fanslow et al. (1995) found that filtration rates of mussels from Saginaw Bay declined with increased seston concentration; however, they did not observe any statistically significant relationship between filtration rates and seston composition as defined by POC:TSS or Chl:TSS. In our study, Chl assimilation was significantly and negatively correlated to POC and Chl concentrations (Table 27.7). Again, we suspect that most of this relationship was driven by a compositional change of the seston versus direct negative impacts of particle density.

There are a limited number of studies that directly measured nutrient excretion by zebra mussels in the Great Lakes. Calculation of the N-excretion rate for Saginaw Bay mussels, using data from Gardner et al. (1995) and dry weights of our experimental mussels (Johengen, unpublished data), yielded a N-excretion rate of 0.052 μgN/mg/h at a temperature of 19°C. Although this was a net measurement of NH_4 accumulation, the rate is comparable to our rates measured near this temperature. Net P-excretion rates reported in Gardner et al. (1995), as well as those measured using in situ benthic chambers at the same Saginaw Bay sites (Johengen, unpublished data), yielded rates that ranged from below the limit of detection to a high of only 0.002 μgP/mg/h, which are consistent with our results.

Our observed seasonal pattern for N-excretion rates was also similar to previous results of Quigley et al. (1993), in which N-excretion rates of zebra mussels were measured seasonally in Lake St. Clair, Michigan. Conversion of rates given in Quigley et al. (1993) yielded a range for N-excretion of 0.015–0.074 μgN/mg/h over a temperature range of 5.7°C–21°C. Our rates for SB mussels ranged from 0.017 to 0.107 μgN/mg/h over a temperature range from 5°C to 25°C. Our measured N- and P-excretion rates for LE mussels generally agreed with rates for LE mussels measured by Arnott and Vanni (1996) using a similar experimental design. Converting their rates to mass-based units yields mean N- and P-excretion rates of 0.074 μgN/mg/h and 0.032 μgP/mg/h, respectively, for measurements taken monthly between June and September compared to means for our study of 0.114 μgN/mg/h and 0.028 μgP/mg/h. However, removing the outlier data for July 18, 1996, resulted in means of 0.089 μgN/mg/h and 0.18 μgP/mg/h; this value for N-excretion agreed well with Arnott and Vanni (1996), but the value for P-excretion was substantially higher. The average seston N:P ratio for experiments by Arnott and Vanni (1996) was extremely close to that for our study at the mass ratio of 5.9, but Arnott and Vanni (1996) did not provide information about algal composition of actual filtering rates so we cannot make a detailed comparison between studies.

Finally, our rates were within ranges reported by Conroy et al. (2005) for mussels from western Lake Erie despite potential artifacts in their study associated with the use of mussels put in cold storage before experiments began. They found P-excretion rates that ranged from 0.003 to 0.022 μgP/mg/h and N-excretion rates that ranged from 0.082 to 0.365 μgN/mg/h. Also, Conroy et al. (2005) found that excretion rates were significantly different among size classes of mussels, with larger sized mussels excreting at higher rates than smaller mussels. Arnott and Vanni (1996) also found a difference in excretion rates among sizes of mussels; however, we only observed this size effect on certain occasions. P:N excretion ratio also varied among mussel size class in their study and ranged from 0.04 to 0.13. Our P:N excretion ratios for LE experiments were generally higher with a mean of 0.24 for all observations (see Table 27.3).

Ecosystem Considerations

Feeding and excretion by zebra mussels (and more recently quagga mussels) impact many aspects of water quality and ecosystem function. The direct impacts of dreissenid mussel feeding on phytoplankton abundance and composition and water clarity are well documented (e.g., Vanderploeg et al. 2001, 2002, 2009, 2010); however, the indirect impact of nutrient excretion is much less understood. Zebra mussels clearly alter pathways and rates at which particulate nutrients are recycled within the water column. Several articles have suggested that filtering by zebra mussels can actually decouple the existing nutrient–chlorophyll relationship within an ecosystem, leading to higher TP:Chl ratios as phytoplankton are removed and P is excreted (Mellina et al. 1995, Nicholls et al. 1999). Holland et al. (1995) described such a process for Hatchery Bay, Lake Erie, whereby dissolved nutrient concentrations increased, TP levels were unchanged, and diatom abundance declined by nearly 90% after the initial mussel invasion.

It has been suggested that when nutrient excretion by mussels represents a significant nutrient input relative to other pathways there is a potential for mussels to alter phytoplankton composition indirectly by altering the existing ratios of available N and P (Arnott and Vanni, 1996). This suggestion is based on the premise that each algal species has an optimum N:P ratio at which they can be most competitive (Smith, 1982, 1983). Arnott and Vanni (1996) suggested that low N:P excretion ratios by Lake Erie mussels could play a role in promoting cyanophyte (i.e., *Microcystis*) growth through alterations in ratios of available nutrients. Although estimates of N and P turnover from mussel excretion suggest this process is important to nutrient cycling in western Lake Erie (Madenjian 1995, Arnott and Vanni, 1996), it is difficult to prove that low N:P excretion ratios could lead to successional changes in phytoplankton composition. Our results indicate that a low N:P excretion was not a prerequisite in promoting blooms. Specifically, *Microcystis* blooms appeared to be more prevalent in Saginaw Bay during the study period despite mussels in Saginaw Bay having significantly higher N:P excretion ratios than mussels in Lake Erie. Results from our parallel feeding experiments (Vanderploeg et al., 2001, 2009) suggest that the primary mechanism by which zebra mussels may promote *Microcystis* blooms is through their ability to selectively filter other classes of phytoplankton in the presence of *Microcystis* and reject viable

Microcystis as pseudofeces. These findings are also supported by the previous bottle and mesocosm experiments of Heath et al. (1995) and Lavrentyev et al. (1995) who both reported changes in phytoplankton composition after selective zebra mussel grazing.

Seston C:P and N:P ratios in Saginaw Bay typically indicated severe P-limitation in the phytoplankton, and in turn, mussels in Saginaw Bay conserved most of the P that was assimilated and excreted nutrients at N:P ratios that were even higher than that found in the seston. Therefore, at least on short timescales, zebra mussels appear to be a P sink in Saginaw Bay and may actually exacerbate the problem of P-limitation. In western Lake Erie, however, mussels often excreted P in excess of that found within the immediate food source. In this case, nutrient excretion by mussels simply increased the rate at which P turns over in the system and may have potentially promoted higher growth rates for the unfiltered phytoplankton throughout the year. Mellina et al. (1995) computed a P-budget for western Lake Erie and estimated that P-excretion by mussels was equivalent to 1.6% of the average daily phosphorus load. Excretion by zebra mussels could indeed help promote algal blooms in western Lake Erie by increasing the availability of P and potentially yielding greater algal biomass. The low N:P ratio at which nutrients are excreted is similar to the optimum nutrient ratio for *Microcystis* so it may potentially offer a competitive advantage for this species. However, we also saw *Microcystis* blooms occur in Saginaw Bay, which, at times, appeared to be a severely P-limited environment, and N:P excretion ratios tended to be much higher. This contradiction suggests that alternative factors, such as selective filtration, likely play a greater role in promoting and maintaining observed *Microcystis* blooms.

REFERENCES

Arnott, D. L. and M. J. Vanni. 1996. Nitrogen and phosphorus recycling by the zebra mussel (*Dreissena polymorpha*) in the western basin of Lake Erie. *Can. J. Fish. Aquat. Sci.* 53: 646–659.

Atkinson, M. J. and S. V. Smith. 1983. C:N:P ratios of benthic marine plants. *Limnol. Oceanogr.* 28: 568–574.

Bayne, B. L., J. I. P. Iglesias, A. J. S. Hawkins et al. 1993. Feeding behavior of the mussel, *Mytilus edulis*: Responses to variations in quantity and organic content of the seston. *J. Mar. Biol. Assoc. U.K.* 73: 813–829.

Conroy, J. D., W. J. Edwards, R. A. Pontius et al. 2005. Soluble nitrogen and phosphorus excretion of exotic freshwater mussels (*Dreissena* spp.): Potential impacts for nutrient remineralization in western Lake Erie. *Freshwat. Biol.* 50: 1146–1162.

Davis, C. O. and M. S. Simmons. 1979. Water chemistry and phytoplankton field and laboratory procedures, Special Report No. 70. Ann Arbor, MI: Great Lakes Research Division, University of Michigan.

Fahnenstiel, G. L., T. F. Nalepa, and G. A. Lang. 1995. Effects of zebra mussel (*Dreissena polymorpha*) colonization on water quality parameters in Saginaw Bay, Lake Huron. *J. Great Lakes Res.* 21: 435–448.

Fanslow, D. L., T. F. Nalepa, and G. A. Lang. 1995. Filtration rates of the zebra mussel (*Dreissena polymorpha*) on natural seston from Saginaw Bay, Lake Huron. *J. Great Lakes Res.* 21: 489–500.

Gardner, W. S., J. F. Cavaletto, T. H. Johengen, J. R. Johnson, R. T. Heath, and J. B. Cotner, Jr. 1995. Effects of the zebra mussel, *Dreissena polymorpha*, on community nitrogen dynamics in Saginaw Bay, Lake Huron. *J. Great Lakes Res.* 21: 529–544.

Heath, R. T., G. L. Fahnenstiel, W. S. Gardner, J. F. Cavaletto, and S. J. Hwang. 1995. Ecosystem-level effects of zebra mussels (*Dreissena polymorpha*): A mesocosm experiment in Saginaw Bay, Lake Huron. *J. Great Lakes Res.* 21: 501–516.

Hecky, R. E., P. Campbell, and L. L. Hendzel. 1993. The stoichiometry of carbon, nitrogen, and phosphorus in particulate matter of lakes and oceans. *Limnol. Oceanogr.* 38: 709–724.

Hecky, R. E., R. E. H. Smith, D. R. Barton et al. 2004. The near shore phosphorus shunt: A consequence of ecosystem engineering by dreissenids in the Laurentian Great Lakes. *Can. J. Fish. Aquat. Sci.* 61: 1285–1293.

Higgins, S. N., S. Y. Malkin, T. E. Howell et al. 2008. An ecological review of *Cladophora glomerata* (Chlorophyta) in the Laurentian Great Lakes. *J. Phycol.* 44: 839–854.

Higgins, S. N. and M. J. Vander Zanden. 2010. What a difference a species makes: A meta-analysis of dreissenid mussel impacts on freshwater ecosystems. *Ecol. Monogr.* 80: 179–196

Higgins, S. N., M. J. Vander Zanden, L. N. Joppa, and Y. Vadeboncoeur. 2011. The effect of dreissenid invasions on chlorophyll and the chlorophyll: Total phosphorus ratio in north-temperate lakes. *Can. J. Fish. Aquat. Sci.* 68: 319–329

Holland, R. E., T. H. Johengen, and A. M. Beeton. 1996. Trends in nutrient concentrations in Hatchery Bay, Bass Island area, western Lake Erie before and after *Dreissena polymorpha*. *Can. J. Fish. Aquat. Sci.* 52: 1202–1209.

Jansen, H. M., Ø. Strand, T. Strohmeier, C. Krogness, M. Verdegem, and A. Smaal. 2011. Seasonal variability in nutrient regeneration by mussel *Mytilus edulis* rope culture in oligotrophic systems. *Mar. Ecol. Prog. Ser.* 431: 137–149.

Johengen, T. H., T. F. Nalepa, G. L. Fahnenstiel, and G. Goudy. 1995. Nutrient changes in Saginaw Bay, Lake Huron, after the establishment of the zebra mussel (*Dreissena polymorpha*). *J. Great Lakes Res.* 21: 449–464.

Kerfoot, W. C., Y. Foad, S. Green, J. W. Budd, D. J. Schwab, and H. A. Vanderploeg. 2010. Approaching storm: Winter chlorophyll and plankton decline in Lake Michigan. *J. Great Lakes Res.* 36(Suppl. 3): 30–41.

Lavrentyev, P. J., W. S. Garner, J. F. Cavaletto, and J. R. Beaver. 1995. Effects of the zebra mussel (*Dreissena polymorpha* Pallas) on protozoa and phytoplankton from Saginaw Bay, Lake Huron. *J. Great Lakes Res.* 21: 545–557.

Madenjian, C. P. 1995. Removal of algae by the zebra mussel (*Dreissena polymorpha*) population in western Lake Erie—A bioenergetics approach. *Can. J. Fish. Aquat. Sci.* 52: 381–390.

Mellina, E., J. B. Rasmussen, and E. L. Mills. 1995. Impact of zebra mussel (*Dreissena polymorpha*) on phosphorus cycling and chlorophyll in lakes. *Can. J. Fish. Aquat. Sci.* 52: 2553–2573.

Menzel, D. W. and N. Corwin. 1965. The measurement of total phosphorus in seawater based on the liberation of organically bound fractions by persulfate oxidation. *Limnol. Oceanogr.* 10: 280–281.

Naddafi, R., K. Pettersson, and P. Eklov. 2008. Effects of the zebra mussel, an exotic freshwater species, on seston stoichiometry. *Limnol. Oceanogr.* 53: 1973–1987.

Nicholls, K. H., G. J. Hopkins, and S. J. Standke. 1999. Reduced chlorophyll to phosphorus ratios in nearshore Great Lakes water coincide with the establishment of dreissenid mussels. *Can. J. Fish. Aquat. Sci.* 56: 153–161.

Ozersky, T., S. Y. Malkin, D. R. Barton, and R. E. Hecky. 2009. Dreissenid phosphorus excretion can sustain *C. glomerata* growth along a portion of Lake Ontario shoreline. *J. Great Lakes Res.* 35: 321–328

Quigley, M. A., W. S. Gardner, and W. M. Gordon. 1993. Metabolism of the zebra mussel (*Dreissena polymorpha*) in Lake St. Clair of the Great Lakes, 295–306. In *Zebra Mussels: Biology, Impacts, and Control*, T. F. Nalepa and D. W. Schloesser, eds., pp. 295–306. Boca Raton, FL: CRC Press.

Redfield, A. C., B. H. Ketchum, and F. A. Richards. 1963. The influence of organisms on the composition sea-water. In *The Sea*, Vol. 2, M. N. Hill, ed., pp. 26–77. New York: Wiley.

Smaal, A. C., A. Vonck, and M. Baker. 1997. Seasonal variation in physiological energetics of *Mytilus edulis* and *Cerastoderma edule* of different size classes. *J. Mar. Biol. Assoc. U.K.* 77: 817–838.

Smith, V. H. 1982. The nitrogen and phosphorus dependence of algal biomass in lakes: An empirical and theoretical analysis. *Limnol. Oceanogr.* 27: 1101–1112.

Smith, V. H. 1983. Low nitrogen to phosphorus favors dominance by bluegreen algae in lake phytoplankton. *Science* 221: 669–671.

Speziale, B. J., S. P. Schreiner, P. A. Giammatteo, and J. E. Schindler. 1984. Comparison of *N,N*-dimethylformamide, dimethyl sulfoxide, and acetone for extraction of phytoplankton chlorophyll. *Can. J. Fish. Aquat. Sci.* 41: 1519–1522.

Sterner, R. W. 1990. The ratio of nitrogen to phosphorus resupplied by herbivores: Zooplankton and the algal competitive arena. *Am. Nat.* 136: 209–229.

Sterner, R. W., J. J. Elser, and D. O. Hessen. 1992. Stoichiometric relationships among producers, consumers and nutrient cycling in pelagic ecosystems. *Biogeochemistry* 17: 49–67.

Sterner, R. W., D. D. Hagemeier, W. S. Smith, and R. F. Smith. 1993. Phytoplankton nutrient limitation and food quality for *Daphnia*. *Limnol. Oceanogr.* 38: 857–871

Stoeckmann, A. M. and D. W. Garton. 1997. A seasonal energy budget for zebra mussels (*Dreissena polymorpha*) in western Lake Erie. *Can. J. Fish. Aquat. Sci.* 54: 2743–2751.

Vanderploeg, H. A., T. H. Johengen, and J. R. Liebig. 2009. Feedback between zebra mussel selective feeding and algal composition affects mussel condition: Did the regime changer pay a price for its success? *Freshwat. Biol.* 54: 47–63.

Vanderploeg, H. A., J. R. Liebig, W. W. Carmichael et al. 2001. Zebra mussel (*Dreissena polymorpha*) selective filtration promoted toxic *Microcystis* blooms in Saginaw Bay (Lake Huron) and Lake Erie. *Can. J. Fish. Aquat. Sci.* 58: 1208–1221.

Vanderploeg, H. A., J. R. Liebig, T. F. Nalepa, T. G. L. Fahnenstiel, and S. A. Pothoven. 2010. *Dreissena* and the disappearance of the spring phytoplankton bloom in Lake Michigan. *J. Great Lakes Res.* 36(Suppl. 3): 50–59.

Vanderploeg, H. A., T. F. Nalepa, D. J. Jude et al. 2002. Dispersal and emerging ecological impacts of Ponto-Caspian species in the Laurentian Great Lakes. *Can. J. Fish. Aquat. Sci.* 59: 1209–1228.

CHAPTER 28

Chemical Regulation of Dreissenid Reproduction

Donna R. Kashian and Jeffrey L. Ram

CONTENTS

Abstract 461
Introduction 461
Chemical Regulation of Reproductive Behavior 462
Support of the Model 462
    Role of Serotonin 462
    Role of Algal and Gamete-Associated Stimuli 463
    Sperm Attractants 465
    Fertilization and Veliger Development 465
    Seasonal Reproductive Cycle 467
Conclusions 467
Acknowledgments 468
References 468

ABSTRACT

Mechanisms that regulate reproductive behavior of invasive dreissenid mussels, *Dreissena polymorpha* (zebra mussel) and *Dreissena rostriformis bugensis* (quagga mussel), are important to their range expansions, competitive interactions with other species and each other, and vulnerability to control methods. For organisms with external fertilization and planktonic larval forms, such as dreissenid mussels, it is critical that the release of eggs and sperm occurs simultaneously and at a time when habitat conditions and available-food conditions are favorable for larval growth. Although temperature and light are important influences, a primary trigger for synchronous spawning in dreissenids may be provided by environmental chemicals and pheromones. For example, allomones from algae may play a large role in the onset of spawning for zebra and quagga mussels. Algae-associated compounds that stimulate spawning in several invertebrate species are well documented in marine systems; however, little work has been conducted on freshwater species. Dreissenid mussels are excellent freshwater organisms to investigate the potential role of algae-associated compounds on reproduction. Past research established the role of serotonin, a monoamine neurotransmitter, as an internal physiological regulator of spawning in dreissenid mussels and led to the development of sensitive spawning bioassays. Recent bioassay experiments indicated that a freshwater alga, *Chlorella minutissima,* may stimulate spawning in quagga mussels. In the present study, filtrates of *C. minutissima* stimulated spawning in 36% of males compared to 0% for media controls and 14% of females compared to 2% for media controls, supporting the possible role of algal allomones in stimulating spawning in quagga mussels. Here we review the known chemical regulators and algal and other extracts that have been shown to influence spawning in both zebra mussels and quagga mussels and discuss priorities for future research on dreissenid reproduction.

INTRODUCTION

When zebra mussels (*Dreissena polymorpha*) and quagga mussels (*Dreissena rostriformis bugensis*) first arrived in North America in the 1980s, little was known about physiological mechanisms and environmental conditions that controlled reproductive behaviors of these bivalves. Most control methods for bivalves under development in the 1980s were based on non-specific toxic chemicals (e.g., chlorination) directed at controlling adults. Since some organisms were known to have species-specific chemical signals that regulate the coordination of reproductive behaviors between males and females, Ram and Nichols (1993) proposed that increased knowledge of such signals in the zebra mussel might lead to the development of species-specific control methods that avoided the problems of non-specific toxic chemicals. An initial study of effects of serotonin demonstrated this neurochemical could stimulate spawning in male zebra mussels (Ram and Nichols 1990).

Since the work of Ram and Nichols (1993), much has been learned about the mechanisms regulating reproductive behavior of dreissenid mussels, including the role of serotonin in regulating spawning in female zebra mussels (Ram et al. 1993) and the influence of environmental factors such as temperature and food availability (Bacchetta et al. 2010). Stimulation of spawning by serotonin has also been demonstrated in male and female quagga mussels (Miller et al. 1994, Stoeckmann 2003). Although no control method specifically targets dreissenid reproduction, several practical outcomes of this research have resulted, including a patent for a control method based on physiological responses to serotonin (Dietz et al. 1995), a method to induce ripe dreissenid gametes for scientific study (Ram and Fong 1994), and development of methods to detect immature larval dreissenids to monitor presence of mussels in the environment (Ram et al. 2011). Here we review the conceptual and factual basis for a potential relationship between freshwater algae and reproduction in dreissenid mussels and also provide results of experiments that explore this possibility.

CHEMICAL REGULATION OF REPRODUCTIVE BEHAVIOR

For organisms with external fertilization and a planktonic larval form, such as dreissenid mussels, it is critical that spawning of eggs and sperm occur simultaneously and at a time when food is available for larval growth. Although spawning may occur throughout spring and summer, there is often a large pulse or peak in spawning, as evident by a sudden decrease in the presence of ripe oocytes in gonads of adults (Borcherding 1991, Mantecca et al. 2003) and by a corresponding abrupt increase in densities of veligers in plankton (Haag and Garton 1992, Garton and Haag 1993, Ferro et al. 1995, Ram et al. 1996). For example, Bacchetta et al. (2010) report observing the first gamete release from zebra mussels in Lake Como following the spring phytoplankton bloom and then a second spawning event later in the summer following another major increase in algae indicated by peaks in chlorophyll a concentrations. Temperature and light may provide seasonal or diurnal cues for some organisms to spawn, but an additional and primary trigger is

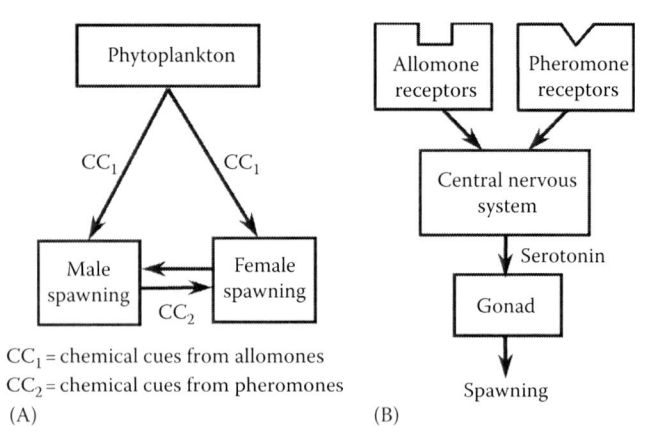

CC_1 = chemical cues from allomones
CC_2 = chemical cues from pheromones
(A) (B)

Figure 28.1 Model of regulation of spawning in dreissenid mussels and other bivalves. (A) External cues and (B) internal regulatory mechanisms.

often provided by environmental chemicals and pheromones. Based on previous literature on bivalve spawning, we propose the response pathway model illustrated in Figure 28.1.

This model indicates the primary stimulus to initiate spawning comes from phytoplankton whose presence would indicate sufficient nutrients may be present for larval growth. To assure synchronized spawning of males and females, chemicals associated with early released gametes would stimulate spawning in the opposite sex, thus providing a feedback between sexes and lead to active near-concurrent mass spawning, which would increase successful fertilizations. Physiological mechanisms that mediate these responses would include receptors for environmental chemicals produced by phytoplankton and gametes and an internal response pathway mediated by serotonin.

After release of gametes, external chemical cues may also influence fertilization and development. For example, Miller et al. (1994) found that fertilization in zebra and quagga mussels was facilitated by the presence of sperm attractants released by eggs. Subsequent to fertilization, embryos develop in the water column for 1–9 weeks, grow in size and function until metamorphosis, and then settle and attach to surfaces with adhesive byssal threads. In some organisms, metamorphosis, settlement, and attachment also appear to be triggered by specific environmental chemicals (Fusetani 1997, Soares et al. 2008). Each life-history stage along this chemical cue/response pathway is a potential stage at which mussels might be controlled by interfering with reproductive or development mechanisms.

SUPPORT OF THE MODEL

Role of Serotonin

Much progress has been made to understand the role of serotonin in dreissenid mussel reproduction. In early experiments on zebra mussels from western Lake Erie conducted in late-summer (August and September, 1990) and mid-spring (early-May, 1991), injection of serotonin initiated spawning in males. However, it was discovered both males and females would spawn in response to serotonin injection in late-spring and mid-summer (Ram et al. 1993). In addition, Ram et al. (1993) found that external application of serotonin by immersion of mussels in serotonin solutions could also elicit spawning in both sexes. Concentrations of serotonin required to elicit spawning by external application was 10-fold higher than needed for injected serotonin. Since zebra mussels are known to have an unusually "leaky" epithelium with high paracellular permeability to solutes (Dietz et al. 1996), it is likely that external application of serotonin may cross body integuments to achieve internal effects.

External application of serotonin rapidly triggers processes in animals that result in spawning and gamete maturation. On average, spawning in males and females begins about 25 and 60 min after serotonin application, respectively, and identical responses and latency can be achieved with as little as 5 min exposure to 1 mM serotonin (Fong et al. 1994c). Typically, >90% of both male and female mussels collected in mid-summer prior to mass spawning can be induced to spawn in response to serotonin (5-HT, Figure 28.2). Response to serotonin consists of not only spawning induction but also maturation of oocytes (Fong et al. 1994c) and activation of sperm motility (Mojares et al. 1995). Whereas oocytes in ripe gonads always have large germinal vesicles, treatment of mussels with serotonin activates the process of germinal vesicle breakdown. Spawned oocytes must have complete vesicle breakdown as a prerequisite for fertilization. The effect of externally applied serotonin is thought to mimic a physiological action of serotonergic neurons on gonads. In support of this hypothesis, serotonin has been shown to activate oocyte maturation and vesicle breakdown in fragments of zebra mussel gonads (Fong et al. 1994c). Furthermore, zebra mussel gonads are innervated by serotonin-containing nerve fibers, as demonstrated by immunohistology and chemical chromatographic studies (Ram et al. 1992).

Numerous studies have used serotonin and other chemicals to characterize serotonin receptors and neural pathway that mediate dreissenid spawning, or to obtain dreissenid gametes for developmental studies (Table 28.1). Experiments on zebra mussels with chemicals known to affect various serotonin receptor subtypes in mammals suggest dreissenid receptors have a unique chemical sensitivity profile (Fong et al. 1993b). For example, of chemicals listed in Table 28.1, the 5-HT$_{1A}$ serotonin receptor agonist 8-hydroxydipropyl-aminotetralin hydrobromide (8-OH-DPAT) is a very effective stimulant of zebra mussel spawning, but the 5-HT$_{1A}$ inhibitor NAN-190 is ineffective at blocking the response to serotonin. Most effective inhibitors of serotonin-induced spawning were cyproheptadine and mianserin, which are 5-HT$_2$ receptor antagonists, and metergoline, which produced stimulation or inhibition, depending on concentration. Other neuroactive chemicals, presented in Table 28.1, that inhibit spawning (dopamine and the tricyclic antidepressants imipramine, desipramine, and clomipramine; Table 28.1), likely act on neural pathways in the central nervous system of mussels (Fong et al. 1994a, Hardege et al. 1997a). The serotonin reuptake inhibitor Prozac enhances sensitivity of zebra mussels to serotonin, which provides further evidence for the role of serotonin in regulating spawning behavior in dreissenid mussels (Fong et al. 2003).

Role of Algal and Gamete-Associated Stimuli

The stimulatory effects of algae and gametes on mussel spawning are summarized in Table 28.1. Algal extracts and filtrates were tested on dreissenid mussels to examine the allomone hypothesis (see Figure 28.1). Marine algal extracts were previously shown to stimulate spawning in male oysters (Miyazaki 1938), several species of sea urchins (Himmelman 1975, Starr et al. 1992), chitons (Himmelman 1975), and marine mussels (Starr et al. 1990). Extracts of a marine alga, *Phaeodactylum tricornutum*, were able to induce spawning in several invertebrate species (Starr et al. 1990), which indicated algal factors that activate spawning responses may not be specific to a given invertebrate species. Therefore, initial experiments of algal effects on dreissenid spawning (Hardege et al. 1997b) tested extracts from *P. tricornutum* and other marine algae on zebra mussels, even though dreissenid mussels are unlikely to encounter such algae in the field.

In our experiments, extracts were derived from cultures of *P. tricornutum*, *Rhodomonas* sp., *Platymonas suecica*, *Dunaliella tertiolecta*, *Oscillatoria* sp., and *Fucus*

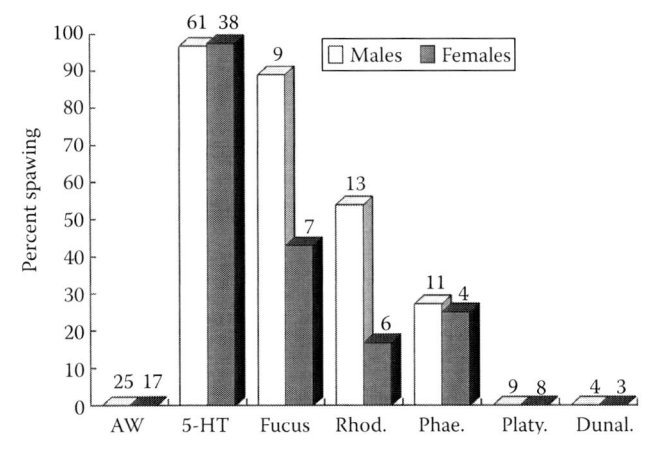

Figure 28.2 Effect of algal extracts on spawning in zebra mussels. Tested solutions were aquarium water (AW = negative control), 5-HT (serotonin = positive control), and extracts of *Fucus*, *Rhodomonas* (Rhod.), *Phaeodactylum* (Phae.), *Platymonas* (Platy.), and *Dunaliella* (Dunal.). Number of animals tested in each category indicated above each bar. (Reproduced from Hardege, J.D. et al., *Exp. Biol. Online*, 2, article 2, http://192.129.24.144/licensed_materials/00898/fpapers/7002001/70020002.htm, 1997b.)

Table 28.1 Effects of Chemicals and Biological Extracts on Dreissenid Mussel Spawning

Chemical	Zebra Mussels Females	Zebra Mussels Males	Quagga Mussels Females	Quagga Mussels Males	References
Serotonin	++	++	++	++	1–11
8-OH-DPAT	++	++			2, 12
NAN-190	NE[a]	NE[a]			12
Metergoline	–	+, –			12, 15
Cyproheptadine	—	—			12, 17
Dopamine	–	–			13
Mianserin	—	—			12, 17
Methiothepin	—	—			14, 15
Imipramine	–	–			16
Desipramine	–	–			16
Clomipramine	–	–			16
Prozac	+	+			17
1-Methyl-5-HT	NE	NE			12
2-Methyl-5-HT	NE	NE			12
Algae and other extracts					
Phaeodactylum tricornutum	+	+			18
Rhodomonas sp.	+	+			18
Platymonas suecice	NE	NE			18
Dunaliella tertiolecta	NE	NE			18
Oscillatoria sp.	NE	NE			18
Fucus vesiculosus	++	+			18
Chlorella minutissima			+	+	19
Sperm	+	+			18
Oocyte	+	+			18

[a] No effect on serotonin-induced spawning; inhibitory to 8-OH-DPAT-induced spawning.
+, treatments activate spawning; –, treatments inhibit spawning; NE, treatments have no effect on spawning; blank, not tested. Double symbols for strongest effects.
References: 1 = Ram et al. (1993); 2 = Vanderploeg et al. (1996); 3 = Stoeckel et al. (2004); 4 = Stoeckmann (2003); 5 = Misamore et al. (2006); 6 = Fallis et al. (2010); 7 = McAnlis et al. (2010); 8 = Pires et al. (2003); 9 = Seaver et al. (2009); 10 = Rehmann et al. (2003); 11 = Kennedy et al. (2006); 12 = Fong et al. (1993b); 13 = Fong et al. (1993a); 14 = Fong et al. (1994b); 15 = Fong et al. (1994a); 16 = Hardege et al. (1997a); 17 = Fong et al. (2003); 18 = Hardege et al. (1997b); 19 = Acharya et al. (2011).

vesiculosus using methods of Hardege et al. (1997b). Extracts were placed in vials with 10 mL of aquarium water (adjusted pH of 7.5) and one adult zebra mussel. All experiments included negative controls (no extracts) and positive controls (serotonin) tested on parallel sets of mussels. Water temperature was kept at 17°C during the experiments, which was high enough to facilitate spawning. As illustrated in Figure 28.2, extracts from *Fucus*, *Rhodomonas*, and *P. tricornutum* caused spawning in 20%–90% of experimental mussels with exact percentage dependent on particular extract and sex of mussel. In contrast, extracts of *Platymonas* and *Dunaliella*, did not induce spawning. *Oscillatoria* extracts also did not cause spawning (data not shown). All non-spawning mussels subsequently spawned when exposed to serotonin, as did nearly 100% of the serotonin-tested positive controls. None of the more than 40 aquarium-water negative controls spawned (subsequently confirmed to be spawnable by testing with serotonin). Thus, algal extracts can stimulate spawning in zebra mussels. However, use of marine algae raised the question: What species of freshwater algae will induce a similar effect on spawning in dreissenid mussels?

Filtrates of cultures of the freshwater alga *Chlorella minutissima* stimulated spawning in quagga mussels (Acharya et al. 2011). Quagga mussels were immersed in individual vials in cell-free filtrates of *C. minutissima* cultures. A total of eight separate trials were conducted on mussels collected from a shallow water site (less than 3 m) in the Detroit River over the period May 20–June 30, 2010. Algal extracts stimulated spawning in 17 of 47 males and 7 of 48 females over the study period. In controls, which consisted of mussels in only dechlorinated water or diluted medium used to grow the algae, 0 of 27 males and 1 of 38 females spawned. This difference was significant ($p < 0.001$, Fisher exact tests) for both males and females. The positive response of one female in the

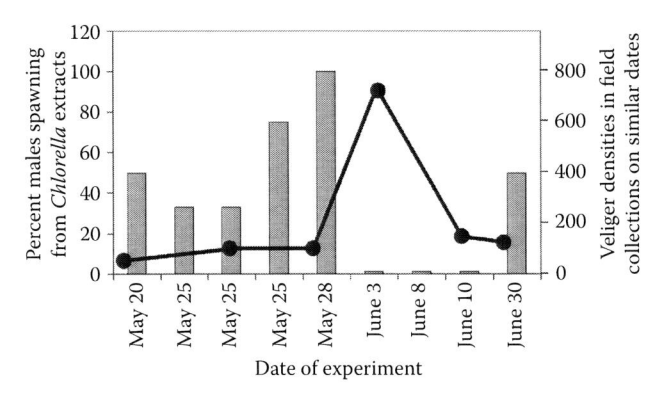

CHEMICAL REGULATION OF DREISSENID REPRODUCTION

Figure 28.3 Percent of male quagga mussels that spawned in response to filtrates of *Chlorella minutissima* cultures tested on various dates (bars), and relative densities of veligers in plankton from field collections (line and solid circles) on similar dates. The three bars for May 25 are for three different groups of mussels tested with different diluents on that date.

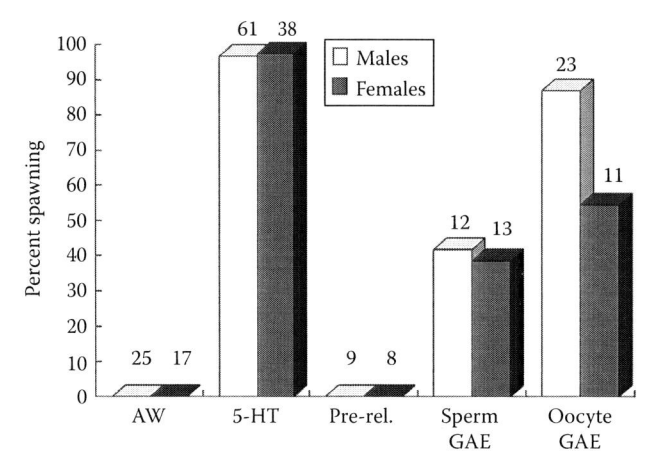

Figure 28.4 Percent of zebra mussels that spawned when exposed to gamete-associated extracts. Experimental treatments were aquarium water (AW; negative control), 5-HT (serotonin; positive control), pre-release water (Pre-rel.), and extracts of sperms and eggs (GAE). See text for description of Pre-rel. and GAE. Number of mussels tested in each treatment indicated above each bar. (Reproduced from Hardege, J.D. et al., *Exp. Biol. Online*, 2, article 2, http://192.129.24.144/licensed_materials/00898/fpapers/7002001/70020002.htm, 1997b.)

controls was very weak (spawning intensity of 1 on a 4-point scale [Ram et al. 1996]). A summary of each experimental trial for male mussels is shown in Figure 28.3. *Chlorella* spp. occur in the Great Lakes (Munawar and Munawar 1982, 1986) and are likely to be encountered by zebra and quagga mussels. Therefore, future experiments should investigate cues associated with *Chlorella* that stimulate spawning in dreissenids, and whether other freshwater algae have a similar effect as outlined in Figure 28.1.

Another feature of our proposed model in Figure 28.1 is that chemicals released with gametes synchronize and stimulate further spawning by providing positive feedback to other ripe but unspawned individuals. To test this, Hardege et al. (1997b) stimulated zebra mussels to spawn with two 5–10 min pulses of serotonin separated by rapid rinses, followed by a third pulse of 15 min. Mussels were placed in serotonin-free water until spawning had occurred. After release of eggs or sperm, water was removed and centrifuged to remove gametes, and the supernatant tested for initiation of spawning. As a further test that serotonin had been effectively removed prior to spawn collection, water removed at the end of the third rinse ("pre-release" water) was also tested. As predicted from the model, gamete extracts from both males and females stimulated spawning more than controls (aquarium water and pre-release water) (Figure 28.4).

Sperm Attractants

Extracts of oocytes of zebra and quagga mussels contain chemicals that serve as sperm attractants for con-specific mussels (i.e., same species) (Miller et al. 1994). Experiments to test inter-species attraction showed extracts of quagga mussel oocytes were less than 5% as attractive to zebra mussel sperm as to quagga mussel sperm, and, conversely, extracts of zebra mussel oocytes were less than 5% as attractive to quagga mussel sperm as to zebra mussel sperm. Sperm attractant mechanisms may prevent high levels of hybridization in areas where both zebra and quagga mussels are prevalent. The potential to form such hybrids has been demonstrated in laboratory fertilizations (Nichols and Black 1994) and in the field (Voroshilova et al. 2010).

Fertilization and Veliger Development

Sperm and oocytes released by serotonin treatment are fully capable of fertilization, which allows observation of post-fertilization development. Released oocytes initiate meiosis after fertilization and undergo first cleavage within 60–70 min. The fertilization process, as observed with scanning and transmission microscopy, involves direct interaction of the sperm acrosomal rod with the oocyte surface (Figure 28.5). Experiments with lectins (chemicals that bind to specific carbohydrate moieties) indicate that the sperm–oocyte interaction is probably mediated in part by carbohydrate moieties on both sperm and oocyte surfaces. Both the acrosomal region of the sperm and the cell surface of unfertilized oocytes can bind various lectins, as demonstrated by the binding of fluorescent lectins, including FITC-concanavalin A, FITC-wheat germ agglutinin, and FITC-fucose binding protein, to both sperm and oocytes (Kyozuka et al. unpublished studies; see also Misamore et al. 2006, Fallis et al. 2010, and McAnlis et al. 2010). Indeed, pretreatment of sperm with concanavalin A caused large reductions in fertilization (from 80% to less than 5%), which was completely prevented in control experiments by co-incubation with its ligand alpha-methyl-D-mannopyranoside (mannoside). Other lectins (wheat germ agglutinin, fucose binding protein,

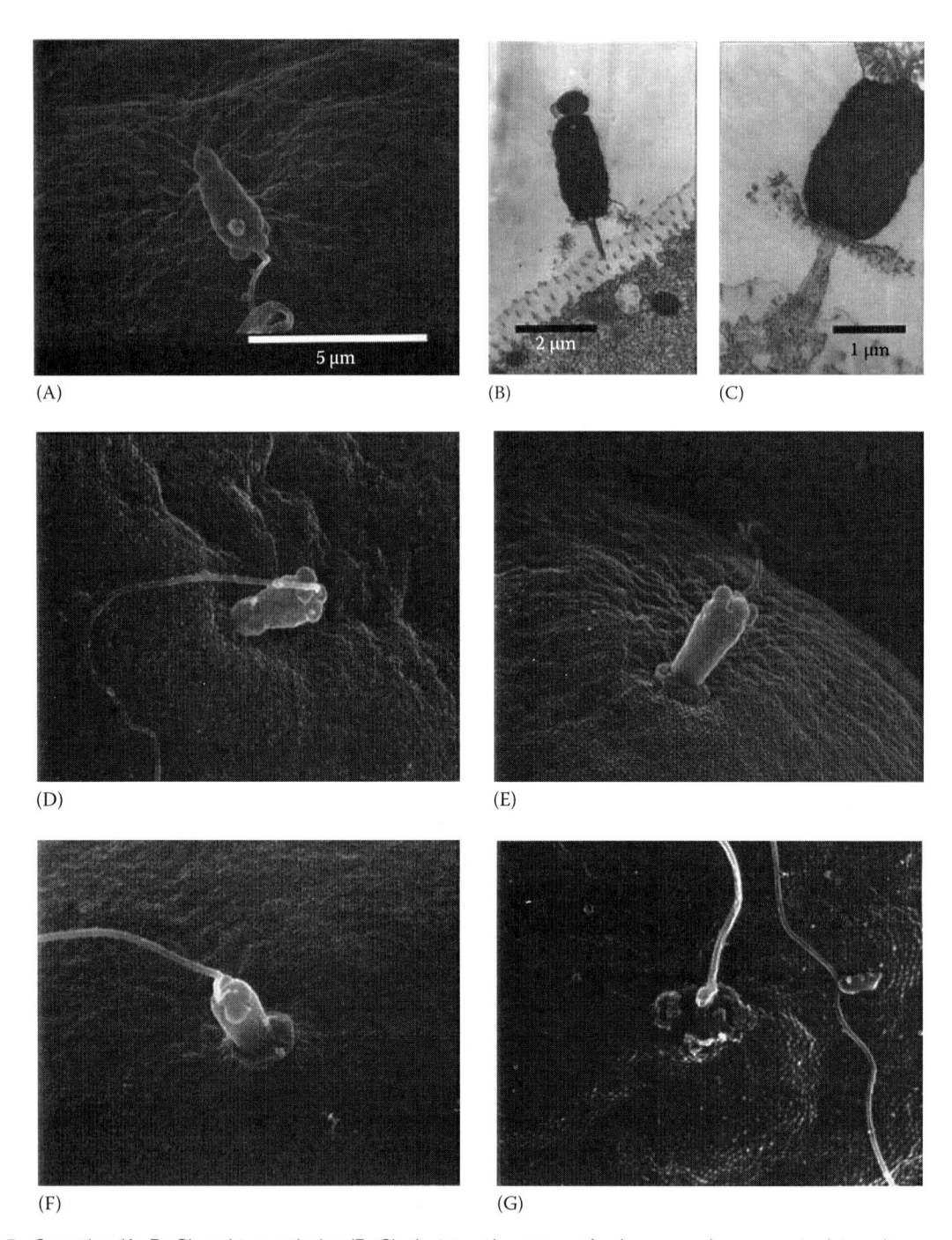

Figure 28.5 Scanning (A, D–G) and transmission (B, C) electron microscopy of zebra mussel sperm entry into zebra mussel oocyte. (A) 30 s after addition of sperm; (B–D) 1 min after addition of sperm; and (E–G) 2, 3, and 5 min after addition of sperm, respectively. Calibration of D–G is the same as in A.

Dolichos biflorus agglutinin) had no effect on fertilization (Kyozuka et al., unpublished data). Similar results (an inhibitory effect of Con A, but no effect of the other lectins) were also obtained with pretreatment of oocytes. Thus, chemical surface signals comprised of mannoside-related carbohydrates may have an important role in sperm–egg interactions in zebra mussels.

With naturally fertilized eggs, embryos develop into viable larvae within 24 h. This sensitive part of the life cycle can be disrupted by various chemicals used to influence spawning. Desipramine, one of the neuroactive chemicals listed in Table 28.1 as having an inhibitory effect on spawning, has been shown to reduce larvae survival even 24 h after fertilization; survival was 70% for controls compared to only 8% for desipramine-treated larvae (Hardege et al. 1997a). Pharmaceuticals found in the environment could have similar effects on early development of dreissenid mussels and other organisms.

Seasonal Reproductive Cycle

Seasonal reproductive cycles of mussels in the field have been measured by responsiveness to serotonin (Ram et al. 2011). Zebra mussels collected biweekly at a site in the Detroit River were tested for serotonin-induced spawning throughout spring and summer in 1994 (Ram et al. 1996). The earliest date spawning could be induced with an average intensity of >2.0 (on a scale of 1–4, where 4 is considered maximal spawning intensity) was early-June, which occurred only after water temperature at the collection site had risen above 15°C. In comparison, in 2010, when quagga mussels had already replaced virtually all zebra mussels at the same site, an average spawning intensity of 2.0 was reached by mid-May even though ambient water temperature at the site was still below 15°C (Ram et al. 2011). Stoeckmann (2003) also observed that quagga mussels in Lake Erie spawned in response to serotonin earlier in the season and at lower ambient water temperatures than zebra mussels from comparable locations.

The positive feedback process proposed in Figure 28.1 would be expected to produce mass-spawning events when mussels are ready to respond to hypothesized phytoplankton initiator(s). In fact, in 2010, veliger densities at the Detroit River site showed a sharp peak in early-June (Figure 28.3), indicative of a mass spawning event in nearby areas. Molecular analyses of veligers in plankton indicated that lower counts of veligers earlier in the season were mostly quagga mussels (Figure 28.6), but the mass-spawning event included a mix of zebra and quagga mussel veligers (Ram et al. 2011). Variation in responses of male quagga mussels to *Chlorella* filtrates

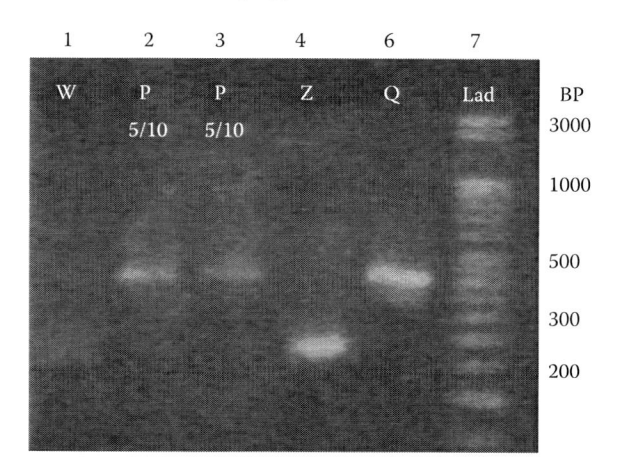

Figure 28.6 Agarose gel analysis of PCR amplification of a DNA extract from (1) Detroit River plankton samples collected on May 10, 2010 (~40 veligers; columns labeled "P 5/10"), (2) an extract of sterile water (column "W," negative control), and (3) extracts of adult zebra and quagga mussels (column "Z" and "Q," respectively; gill tissue, positive controls showing species-specific zebra mussel and quagga mussel PCR products). Also shown is a DNA calibration ladder (column "Lad"). Units of length are given in base pairs (BP). The gel shows that on May 10, 2010, only quagga mussel veligers were detected in the plankton.

in May/June 2010 (and especially the loss of responsiveness around the time of mass spawning in early-June) may reflect these changes that occurred in mussels from the field.

CONCLUSIONS

Since the discovery that male zebra mussels could be stimulated to spawn by serotonin (Ram et al. 1993), accumulated evidence suggests that each component of the model illustrated in Figure 28.1 is governed by maturational events that occur throughout the annual cycle. All components must work in synchrony for successful spawning, fertilization, and development to proceed. According to this model, dreissenid mussels mature in early spring as temperatures rise from winter lows. Initially, as the gonad ripens, mussels are not responsive to serotonin or other activators of spawning. As maturation proceeds during spring, gonads of males become serotonin-responsive first, and gonads of females become serotonin-responsive somewhat later. Only later, as water temperatures increase (albeit to a lower temperature for quagga mussels than for zebra mussels) does a sensory response to algal and gamete factors become evident. We envision ripe animals in the "readiness to respond" state wait, though probably not for very long, for an effective algal trigger of spawning to occur that initiates mass-spawning events evident in dreissenid mussels.

In contrast to this hypothesized "triggered" reproductive behavior of zebra mussels and quagga mussels found at shallow depths, it is important to note that in a recent study (Bacchetta et al. 2010) the reproductive cycle of zebra mussels collected from deep waters (>25 m) did not exhibit a similar pattern of seasonal changes, but instead exhibited no significant response to season. These deepwater mussels live in a more homogeneous environment than their counterparts in shallow water in the photic zone and experience low food levels, low food quality, low temperatures, and no fluctuations in light. Therefore, mussels in deepwater habitats likely depend more on endogenous mechanisms to regulate reproduction. The higher density of quagga mussels than zebra mussels at great depth in Lake Michigan (Nalepa et al. 2009) and elsewhere may be an indication that quagga mussels are better adapted to such "cue-less" environments than zebra mussels.

Priorities for future research especially include identification of the chemical identity of the algal- and gamete-associated factors that stimulate dreissenid mussel spawning. Since such chemical factors have been found for marine bivalves and other broadcast spawners, identification of factors playing a similar role for dreissenid mussels may provide a broader significance for understanding mass-spawning events in general. Comparisons of sensitivities of quagga mussels to zebra mussels and of shallow water animals vs. those that reproduce in the hypolimnion may be helpful in understanding possible differences in reproductive strategies and capabilities of these two species. Another important priority is to understand physiological mechanisms that govern seasonal changes in internal responses to serotonin and external responses to environmental chemicals. Aside from the fundamental significance of understanding what these maturation mechanisms are, a more practical significance is an understanding of spawning in bivalve species that are grown in aquaculture. Serotonin and other chemicals have been used to produce spawning at the most advantageous times for growing progeny. Lack of responsiveness and understanding how to regulate it therefore has important potential economic significance. Finally, knowledge developed in these studies has made the production and study of larval stages of dreissenid mussels more convenient. Further studies of early development stages, metamorphosis, and settlement of dreissenid mussels could still potentially result in novel and potentially species-specific methods for controlling these pest organisms.

ACKNOWLEDGMENTS

The authors wish to thank the funding agencies and the many students and post-docs who have contributed to the work reviewed here. We especially acknowledge the work of Dr. Keiichiro Kyozuka and Dr. Peter Fong who conducted the lectin experiments described here and many other experiments establishing regulators of reproductive behavior in zebra mussels early in this project. In addition to internal Wayne State University funds to the authors and in support of undergraduate student projects, assistance for research on dreissenid mussels in the authors' laboratories has been provided to DK by the National Oceanic and Atmospheric Administration (NOAA) Center for Sponsored Coastal Ocean Research and to JR by NOAA (Michigan Sea Grant), the National Science Foundation, NIH, and NATO. Recent experiments on dreissenid mussel reproduction have been assisted by Hunter Oates, Brittanie Dabney, Payel Acharya, Aos Karim, Sonal Purohit, and Pranav Jagtap. We thank Dr. Steve Salley, Dr. Haiying Tang, and others in the National Biofuels Energy Laboratory for providing the high-density cultures of *Chlorella minutissima* used in the aforementioned experiments.

REFERENCES

Acharya, P., D. Kashian, and J.L. Ram. 2011. Algal regulation of spawning in the freshwater invasive mussel, *Dreissena bugensis*. Abstracts for the American Society of Limnology and Oceanography. http://www.aslo.org/meetings/sanjuan2011/files/asm2011-abs-web.pdf, Waco, TX: Meeting & Marketing Services (accessed May 18, 2003).

Bacchetta, R., P. Mantecca, and G. Vailati. 2010. Reproductive behaviour of zebra mussels living in shallow and deep water in the South Alps lakes. In *The Zebra Mussel in Europe*, G. van der Velde, S. Rajagopal, and A. bij de Vaate, eds., pp. 161–168. Leiden, The Netherlands: Backhuys Publishers.

Borcherding, J. 1991. The annual reproductive cycle of the fresh water mussel *Dreissena polymorpha* Pallas in lakes. *Oecologia* 87:208–218.

Dietz, T.H., J.W. Lynn, and H. Silverman. 1995. Method for controlling bivalves such as zebra mussels, patent number 5417987, filed September 16, 1993. Pages 6 In: http://www.google.com/patents/US5417987, Board of Supervisors of Louisiana State University, Baton Rouge, LA (accessed May 18, 2013.)

Dietz, T.H., S.J. Wilcox, R.A. Byrne, J.W. Lynn, and H. Silverman. 1996. Osmotic and ionic regulation of North American zebra mussels (*Dreissena polymorpha*). *Am. Zool.* 36:364–372.

Fallis, L.C., K.K. Stein, J.W. Lynn, and M.J. Misamore. 2010. Identification and role of carbohydrates on the surface of gametes in the zebra mussel, *Dreissena polymorpha*. *Biol. Bull.* 218:61–74.

Ferro, T.A., H.T. Keppner, and D.J. Adrian. 1995. Annual and seasonal variations in zebra mussel (*Dreissena* spp.) veliger density in the upper Niagara River. In *Proceedings Zebra Mussels and Other Aquatic Nuisance Organisms Conference* 1995, pp. 123–132. Toronto, Ontario, Canada: Professional Edge.

Fong, P.P., J. Duncan, and J.L. Ram. 1994a. Inhibition and sex specific induction of spawning by serotonergic ligands in the zebra mussel *Dreissena polymorpha* (Pallas). *Experientia* 50:506–509.

Fong, P.P., J.D. Hardege, and J.L. Ram. 1994b. Long-lasting, sex-specific inhibition of serotonin-induced spawning by methiothepin in the zebra mussel, *Dreissena polymorpha* (Pallas). *J. Exp. Zool.* 270:314–320.

Fong, P.P., K. Kyozuka, H. Abdelghani, J.D. Hardege, and J.L. Ram. 1994c. In vivo and in vitro induction of germinal vesicle breakdown in a fresh-water bivalve, the zebra mussel *Dreissena polymorpha* (Pallas). *J. Exp. Zool.* 269:467–474.

Fong, P.P., R. Noordhuis, and J.L. Ram. 1993a. Dopamine reduces intensity of serotonin-induced spawning in the zebra mussel *Dreissena polymorpha* (Pallas). *J. Exp. Zool.* 266:79–83.

Fong, P.P., C.M. Philbert, and B.J. Roberts. 2003. Putative serotonin reuptake inhibitor-induced spawning and parturition in freshwater bivalves is inhibited by mammalian 5-HT$_2$ receptor antagonists. *J. Exp. Zool. A-Comp. Exp. Biol.* 298:67–72.

Fong, P.P., D.M. Wall, and J.L. Ram. 1993b. Characterization of serotonin receptors in the regulation of spawning in the zebra mussel *Dreissena polymorpha* (Pallas). *J. Exp. Zool.* 267:475–482.

Fusetani, N. 1997. Marine natural products influencing larval settlement and metamorphosis of benthic invertebrates. *Curr. Org. Chem.* 1:127–152.

Garton, D.W. and W.R. Haag. 1993. Seasonal reproductive cycles and settlement patterns of *Dreissena polymorpha* in western Lake Erie. In *Zebra Mussels: Biology, Impacts, and Control*, T. F. Nalepa and D. W. Schloesser, eds., pp. 111–128. Boca Raton, FL: CRC Press.

Haag, W.R. and D.W. Garton. 1992. Synchronous spawning in a recently established population of the zebra mussel, *Dreissena polymorpha*, in western Lake Erie. USA. *Hydrobiologia* 234:103–110.

Hardege, J.D., J. Duncan, and J.L. Ram. 1997a. Tricyclic antidepressants suppress spawning and fertilization in the zebra mussel, *Dreissena polymorpha*. *Comp. Biochem. Physiol.* 118C:59–64.

Hardege, J.D., J.L. Ram, and M.G. Bentley. 1997b. Activation of spawning in zebra mussels by algae-, cryptomonad-, and gamete-associated factors. *Exp. Biol. Online* 2: Article 2 (http://link.springer.com/article/10.1007%2Fs00898-997-002-y) (accessed May 18, 2003).

Himmelman, J.H. 1975. Phytoplankton as a stimulus for spawning in three marine invertebrates. *J. Exp. Mar. Biol. Ecol.* 20:199–214.

Kennedy, A.J., R.N. Millward, J.A. Steevens, J.W. Lynn, and K.D. Perry. 2006. Relative sensitivity of zebra mussel (*Dreissena polymorpha*) life-stages to two copper sources. *J. Great Lakes Res.* 32:596–606.

Kyozuka, K., P. Fong, and J.L. Ram. 1993. Sperm-egg interactions in *Dreissena polymorpha*: Electron microscopy, lectin-binding, and specific inhibitory effects of lectins on fertilization and development (previously unpublished data).

Mantecca, P., G. Vailati, L. Garibaldi, and R. Bacchetta. 2003. Depth effects on zebra mussel reproduction. *Malacologia* 45:109–120.

McAnlis, K.M.K., J.W. Lynn, and M.J. Misamore. 2010. Lectin labeling of surface carbohydrates on gametes of three bivalves: *Crassostrea virginica*, *Mytilus galloprovincialis*, and *Dreissena bugensis*. *J. Shellfish Res.* 29:193–201.

Miller, R.L., J.J. Mojares, and J.L. Ram. 1994. Species-specific sperm attraction in the zebra mussel, *Dreissena polymorpha*, and the quagga mussel, *Dreissena bugensis*. *Can. J. Zool.* 72:1764–1770.

Misamore, M.J., K.K. Stein, and J.W. Lynn. 2006. Sperm incorporation and pronuclear development during fertilization in the freshwater bivalve *Dreissena polymorpha*. *Mol. Reprod. Dev.* 73:1140–1148.

Miyazaki, I. 1938. On a substance which is contained in green algae and induces spawning action of the male oyster. *Bull. Jpn. Soc. Sci. Fish.* 7:137–138.

Mojares, J.J., J.J. Stachecki, K. Kyozuka, D.R. Armant, and J.L. Ram. 1995. Characterization of zebra mussel (*Dreissena polymorpha*) sperm morphology and their motility prior to and after spawning. *J. Exp. Zool.* 273:257–263.

Munawar, M. and I.F. Munawar. 1982. Phycological studies in Lakes Ontario, Erie, Huron, and Superior. *Can. J. Bot.* 60:1837–1858.

Munawar, M. and I.F. Munawar. 1986. The seasonality of phytoplankton in the North-American Great Lakes, a comparative synthesis. *Hydrobiologia* 138:85–115.

Nalepa, T.F., D.L. Fanslow, and G.A. Lang. 2009. Transformation of the offshore benthic community in Lake Michigan: Recent shift from the native amphipod *Diporeia* spp. to the invasive mussel *Dreissena rostriformis bugensis*. *Freshwat. Biol.* 54:466–479.

Nichols, S.J. and M.G. Black. 1994. Identification of larvae: The zebra mussel (*Dreissena polymorpha*), quagga mussel (*Dreissena rosteriformis bugensis*), and Asian Clam (*Corbicula fluminea*). *Can. J. Zool.* 72:406–417.

Pires, L.M.D., R. Kusserow, and E. Van Donk. 2003. Influence of toxic and non-toxic phytoplankton on feeding and survival of *Dreissena polymorpha* (Pallas) larvae. *Hydrobiologia* 491:193–200.

Ram, J.L., G.W. Crawford, J.U. Walker et al. 1993. Spawning in the zebra mussel (*Dreissena polymorpha*): Activation by internal or external application of serotonin. *J. Exp. Zool.* 265:587–598.

Ram, J.L. and P.P. Fong. 1994. The use of serotonin and related agents for stimulating spawning in zebra mussels. *Dreissena polymorpha Inf. Rev.* 5:10–11.

Ram, J.L., P. Fong, R.P. Croll, S.J. Nichols, and D. Wall. 1992. The zebra mussel (*Dreissena polymorpha*), a new pest in North America: Reproductive mechanisms as possible targets of control strategies. *Invert. Reprod. Dev.* 22:77–86.

Ram, J.L. and Nichols, S.J. 1993. Chemical induction of spawning in the zebra mussel (*Dreissena polymorpha*) and other bivalves. Invited chapter in *Zebra Mussels: Biology, Impact and Control*. eds. D.W. Schloesser and T.F. Nalepa, Lewis Pub., pp. 307–314.

Ram, J.L., P.P. Fong, and D.W. Garton. 1996. Physiological aspects of zebra mussel reproduction: Maturation, spawning and fertilization. *Am. Zool.* 36:326–338.

Ram, J.L., A.S. Karim, P. Archaya et al. 2011. Reproduction and genetic detection of veligers in changing *Dreissena* populations in the Great Lakes. *Ecosphere* 2: Article 3 (http://www.esajournals.org/doi/pdf/10.1890/ES1810-00118.00111).

Ram, J.L. and S.J. Nichols. 1990. Approaches to zebra mussel control through intervention in reproduction. *Proceedings International Zebra Mussel Research Conference* 1990, Columbus, OH, 15pp.

Ram, J.L. and S.J. Nichols. 1993. Chemical induction of spawning in the zebra mussels (*Dreissena polymorpha*) and other bivalves. In *Zebra Mussels: Biology, Impacts and Control*, T.F. Nalepa and D.W. Schloesser, eds., pp. 307–314. Boca Raton, FL: CRC Press.

Rehmann, C.R., J.A. Stoeckel, and D.W. Schneider. 2003. Effect of turbulence on the mortality of zebra mussel veligers. *Can. J. Zool.* 81:1063–1069.

Seaver, R.W., G.W. Ferguson, W.H. Gehrmann, and M.J. Misamore. 2009. Effects of ultraviolet radiation on gametic function during fertilization in zebra mussels (*Dreissena polymorpha*). *J. Shellfish Res.* 28:625–633.

Soares, A.R., B.A.P. da Gama, A.P. da Cunha, V.L. Teixeira, and R.C. Pereira. 2008. Induction of attachment of the mussel *Perna perna* by natural products from the brown seaweed *Stypopodium zonale*. *Mar. Biotech.* 10:158–165.

Starr, M., J.H. Himmelman, and J.C. Therriault. 1990. Direct coupling of marine invertebrate spawning with phytoplankton blooms. *Science* 247:1071–1074.

Starr, M., J.H. Himmelman, and J.C. Therriault. 1992. Isolation and properties of a substance from the diatom *Phaeodactylum tricornutum* which induces spawning in the sea urchin *Strongylocentrotus droebachiensis*. *Mar. Ecol. Prog. Ser.* 79:275–287.

Stoeckel, J.A., D.K. Padilla, D.W. Schneider, and C.R. Rehmann. 2004. Laboratory culture of *Dreissena polymorpha* larvae: Spawning success, adult fecundity, and larval mortality patterns. *Can. J. Zool.* 82:1436–1443.

Stoeckmann, A. 2003. Physiological energetics of Lake Erie dreissenid mussels: A basis for the displacement of *Dreissena polymorpha* by *Dreissena bugensis*. *Can. J. Fish. Aquat. Sci.* 60:126–134.

Vanderploeg, H.A., J.R. Liebig, and A.A. Gluck. 1996. Evaluation of different phytoplankton for supporting development of zebra mussel larvae (*Dreissena polymorpha*): The importance of size and polyunsaturated fatty acid content. *J. Great Lakes Res.* 22:36–45.

Voroshilova, I.S., V.S. Artamonova, A.A. Makhrov, and Y.V. Slyn'ko. 2010. Natural hybridization of two mussel species *Dreissena polymorpha* (Pallas, 1771) and *Dreissena bugensis* (Andrusov, 1897). *Biol. Bull.* 37:542–547.

CHAPTER 29

Role of Fluid Dynamics in Dreissenid Mussel Biology

Josef Daniel Ackerman

CONTENTS

Abstract 471
Introduction 471
Hydrodynamics of Dreissenid Environments 472
    Flow Regimes 472
    Hydrodynamics of Boundary Layers 473
    Hydrodynamics of Benthic Environments of Lakes 473
    Hydrodynamics of Benthic Environments of Rivers 474
Physical Ecology of Dreissenids 474
    External Fertilization 475
    Larval Dispersal 475
    Larval Settlement 477
    Suspension Feeding 478
    Detachment 480
Conclusions 481
Acknowledgments 481
References 481

ABSTRACT

Dreissenid mussels are benthic invertebrates that remain relatively sedentary on the bottom of lakes and rivers except for their early-life history that involves broadcast spawning and a planktotrophic larval stage. This chapter focuses on the interaction of dreissenids with their fluid environment, which is responsible for essential biological processes including the delivery of food and larvae to the benthos as well as the removal of gametes and wastes. The hydrodynamics of lakes and rivers that affect the physical ecology of dreissenids are discussed with specific reference to these transport processes. The physical ecology of important life history functions, which include external fertilization of gametes, dispersal of larvae, settlement of larvae and recruitment of juvenile mussels, suspension feeding, and dislodgement of adult benthic stages, is reviewed. Results indicate that these biological processes are largely driven by hydrodynamics, yet it is also evident that many of the links between hydrodynamics and dreissenid mussels remain unknown and thus provide a challenge for future research.

INTRODUCTION

Like other benthic organisms, dreissenids exist in an environment that is bounded by the water above and by substrates below. Mussels rely on the water column to supply resources and remove wastes, as well as to facilitate fertilization of gametes and to disperse larvae. In short, all aspects of their biology and ecology are influenced and facilitated by the hydrodynamics of water flow. It is relevant then to examine the hydrodynamics of these environments, which is one of the goals of this chapter. The focus will be on periods of the year when most research is conducted (i.e., ice-free, largely stratified periods during late spring and summer) and on typical hydrodynamic conditions.

Unfortunately, most reviews of the physical environments of lakes present the hydrodynamics of the open water, or pelagos, at different vertical depths, but are rarely concerned with the hydrodynamics of the near-bed or benthic environment. This is of lesser concern for rivers, especially small ones, where interactions with the riverbed are of primary importance. There are some fundamental differences that distinguish the pelagos and benthos in terms of the way organisms experience the environment and how they experience water motions. In the pelagic environment, organisms travel with the water (e.g., Lagrangian frame of reference) and, therefore, experience motions relative to their own swimming speed. For dreissenids, this means that larvae (veligers) travel with the current and that larvae may experience the relative safety of large eddy motions except when they experience fluid shear due to passing turbulence. Sessile benthic organisms like postlarval dreissenid mussels are fixed on bottom substrates and experience motion only as water passes by (e.g., Eulerian frame of reference). Resident dreissenids are thus subject to the absolute motions of water, including harmful forces of currents and waves due to storms.

Bottom substrates are special environments both biologically and hydrodynamically because they are influenced primarily by gradients. Gradients in velocity determine the hydrodynamic forces that dreissenids experience—a concern in some shallow depths where such forces can detach mussels from the bed—as well as the exchange processes that occur with overlying waters. These exchanges may include dreissenid eggs and sperm, larvae, food sources, and wastes (i.e., dissolved and particulate scalar quantities). Gradients in scalar quantities such as dissolved gases or food can also affect the speed of exchange processes depending on water flows.

This chapter addresses the role of hydrodynamics in the biology of *Dreissena polymorpha* (zebra mussel) and *Dreissena rostriformis bugensis* (quagga mussel); that is, it focuses on the physical ecology of dreissenids. Key stages in dreissenid life history as well as key biological functions will be addressed. These include the external fertilization of gametes, the dispersal of larvae, the settlement of larvae and recruitment of juvenile mussels, suspension feeding, and dislodgement of adult benthic stages. The chapter begins with a brief review of the hydrodynamics of the environments in which dreissenids are found, with particular attention to flow regimes, the hydrodynamics of boundary layers, and particularly hydrodynamic factors affecting benthic environments of lakes and rivers. It continues with a review of the physical ecology of dreissenid mussels described earlier. Ideas for future research directions are presented at the end of each section.

HYDRODYNAMICS OF DREISSENID ENVIRONMENTS

Flow Regimes

Solar energy, winds, hydraulic gradients, density gradients, and the earth's rotation are responsible for the currents, seiches, waves (both surface and internal), and other motions that characterize inland waters (e.g., Fischer et al. 1979). Together they contribute to the turbulence found in aquatic ecosystems, and turbulence can dictate the rate and scale of physical and biological processes. Turbulence has various definitions, but the simplest is that it concerns water flow where the speed and direction of a water parcel changes rapidly in space and time. The concept of the Reynolds number, $Re = Ul/\nu$, where U is the mean velocity, l is the characteristic length scale, and ν is the kinematic viscosity (Vogel 1994), provides another more familiar definition for turbulence. In this case, turbulence occurs at $Re > 1000$ when inertial forces dominate over viscous ones and flows are chaotic-like in nature with swirling water motions (i.e., with co-occurring eddies of various sizes). Such conditions occur under large spatial scales and moderate water currents, which define most water flows in nature. Conversely, when $Re \ll 1$, viscous forces dominate and flows are reversible and creeping in nature—a state that also defines most water flows but only at small spatial scale and under slow flows such as those experienced by microbes. There is, however, the intermediate case that occurs between $1 < Re < 1000$. This intermediate case defines laminar flow where conditions are more well-defined and somewhat easier to examine conceptually and through equations.

Flow regimes are by definition scale dependent and occur simultaneously at various spatial and temporal scales; that is, Reynolds numbers can be defined for different components simultaneously in the benthos. For example, different values of Re can be calculated to describe the dynamics of particle capture in the gills of a dreissenid mussel, the dynamics of the forces acting on an individual dreissenid attached to a rock, or the dynamics of benthic–pelagic coupling for a reef containing the same individual. Other scales, such as the Kolmogorov length scale, are often used to characterize the scale of the smallest eddies that will dissipate into heat. It has long been held that small plankton, such as dreissenid larvae, do not experience effects of shear and that microbes experience creeping flow at/below this length scale (e.g., Okubo et al. 2001). Some data indicate that this simple characterization might not be accurate as the growth of certain plankton has been found to be reduced even below the Kolmogorov scale (Peters and Marrasé 2000). Regardless, the important issue to consider with regard to the flow regime is the hypothesis or question under investigation. For example, questions of turbulent flow are not relevant to particle sorting within the gills of dreissenids where spatial scales and velocities are small (i.e., small Re). Conversely, considerations of larval swimming speeds are probably not that relevant when examining larval dispersal within a lake basin, which occurs under turbulent flows.

Hydrodynamics of Boundary Layers

Water has a tendency to stick to a boundary (the no-slip condition), which leads to the development of a gradient in velocity (u) with height (z) above the boundary (i.e., $\partial u/\partial z$). Gradients grow with distance downstream (x) and form layers of slower moving water next to the boundary, which is properly defined as a momentum boundary layer (MBL). The vertical extent of the MBL is defined as the point at which the velocity reaches 90% (or often 99%) of the velocity in the free stream away from the boundary. As the MBL grows, it has laminar, transitional, and then turbulent characteristics as can be determined using the local Reynolds number ($Re_x = xU/\nu$). Its thickness (δ_L) has been approximated for some simple geometries like flat plates under laminar flow conditions as

$$\delta_L \approx \frac{5x}{Re_x^{1/2}} \quad (29.1)$$

whereas the MBL thickness (δ_T) under turbulent flow conditions is provided by

$$\delta_T \approx \frac{0.16x}{Re_x^{1/7}} \quad (29.2)$$

A turbulent MBL is comprised of three layers perpendicular to the boundary, a viscous sublayer (VSL) next to the boundary where water forces are primarily viscous, a logarithmic layer where water begins to be dominated by inertial forces, and an outer layer where conditions approach those in the free stream (reviewed in Nishihara and Ackerman 2007). The VSL contains a thin diffusional sublayer (DSL) next to the boundary where diffusion is the primary mode of transport. Velocity within a turbulent MBL can be predicted mathematically using the law of the wall, which describes the idealized logarithmic shape of the velocity profile given by

$$u = \frac{u_*}{\kappa} \ln\left(\frac{z}{z_0}\right) \quad (29.3)$$

where
- u_* is the friction velocity
- κ is the von Karman constant = 0.41
- z_0 is the roughness height (reviewed in Ackerman and Hoover 2001)

Ideal conditions such as flat plates that generate logarithmic velocity gradients exist in nature at times, but natural conditions are characterized mostly by roughness at various spatial scales, including the roughness created by the shells of dreissenid mussels. It is possible to describe different patterns of water flow over idealized two-dimensional roughness in terms of "roughness flow regimes" (Morris 1955). Typically, water will flow over and around (sometimes through if the roughness is porous like a macrophyte bed) any roughness and be caught up in eddies of the turbulent wake that form immediately downstream. This has been termed "isolated roughness flow." As the space between roughness decreases, water in wakes begins to affect the water flowing over the next roughness element downstream leading to "wake interference flow" (Nikora et al. 2001). When the inter-roughness spacing becomes very small, water flowing over the roughness element skims over the spacing and the downstream wakes become isolated resulting in "skimming flow" (reviews in Young 1992, Schindler and Ackerman 2010). These idealized two-dimensional roughness flows are a characterization of what may occur, as rough boundary layer flows may not necessarily follow the law of the wall (Equation 29.3) over large roughness and/ or shallow depth (Lacey and Roy 2008). It is important to recognize that roughness in this context may be defined as an individual dreissenid mussel, a boulder on the bottom, or a large limestone reef. Roughness can include all three types of water flows nested within each other.

Hydrodynamics of Benthic Environments of Lakes

The most conspicuous water motions in lakes are surface waves that occur due to wind stress. Such waves are easily visible as they ultimately encounter lakeshores. Approximately 10% of total wind energy is transferred into the interior of small and medium-sized lakes where it contributes to water column turbulence (Wüest and Lorke 2003). Surface waves also generate orbital motions that extend downward to the bottom in water depths $<\frac{1}{2}\lambda$, where λ is the wavelength. As orbital motions approach the bottom where they become sheared, they change from a circular shape to an elliptical one and, ultimately in shallow regions, to an oscillatory motion at the bottom. Such motion reflects water column turbulence, which is largest near the lake surface and typically dissipates with depth. However, lakes can be stratified thermally with a metalimnion (thermocline) that serves as a cap between the well-mixed epilimnion above and the more-stable hypolimnion below. The metalimnion also serves as the location for the propagation of internal waves, which dissipate ~10% of energy transferred from the lake surface into the lake interior through shear (or Kelvin–Helmholtz) instabilities, and breaking internal waves. The benthic boundary layer, however, dissipates the majority (~90%) of this energy through basin-scale seiches and the breaking of high-frequency internal waves (Wüest and Lorke 2003).

Turbulence in lakes, as measured by energy dissipation (ε), has been shown to comprise three regions that correspond to the locations described earlier. The surface mixed layer, which is exposed to the atmosphere, has the highest level of turbulence ($10^{-8} < \varepsilon < 10^{-7}$ m^2 s^{-3} or W kg^{-1}), which declines linearly with depth to the stratified interior, where the lowest levels of turbulence occur ($10^{-10} < \varepsilon < 10^{-9}$ m^2 s^{-3}). Deeper in the lake, however, turbulence increases as the water depth approaches the benthic boundary layer ($10^{-10} < \varepsilon < 10^{-8}$ m^2 s^{-3}). Even so, the magnitude of ε at the benthic boundary layer is smaller than that of ε at the surface mixed layer (Wüest and Lorke 2003). It is important to recall that the benthic boundary layer exists over all parts of the lakebed, including lake margins. A turbulent characterization of a lake is somewhat different from the more familiar characterization involving thermal stratification. From the former perspective, the lake bottom is not a static, uneventful place. Rather it is the region where substantial physical forces can consequently affect biological activities. Therefore, it would be reasonable to explore the biology of dreissenids in the context of the turbulent realm of lakes which they inhabit.

Hydrodynamics of Benthic Environments of Rivers

Among the distinctions between lakes and rivers, one distinction is that rivers are confined to channels with mostly unidirectional flow driven largely by hydraulic gradients. In addition, the riverbed and consequently the benthic boundary layer are a rather large portion of the channel. One consequence of this is the production of helical secondary flows downstream, which are created by interactions with the riverbed. Another consequence is the presence of river meanders (or braiding depending on the composition of the riverbed), which lead to differences in cross-channel flows. Faster flows occur toward the exterior of the meander where erosion occurs, and slower flows to the interior where deposition occurs. Erosional areas may be sites of scour after high water flow events, whereas depositional areas may become sites of high accumulation of fine sediments under low flows. Such differences would have implications, for example, for the recruitment of and existence of dreissenids in rivers.

As described earlier, turbulence in the benthic boundary layer is strongly affected by material on the riverbed. Isolated roughness, wake interference, and skimming flow regimes represent ways to characterize flows in turbulent environments. There are, however, methods that have been developed to measure both the magnitude and spatial pattern of turbulence along river bottoms, and in some cases, these methods can also be used in lakes. The magnitude of turbulence can be estimated using the aforementioned energy dissipation rate (ε), the eddy diffusivity near the bottom (K_z), or by using shear stresses in the benthic boundary layer. Bed shear stress (τ) is the most frequently used estimate of turbulence, which is taken directly from the law of the wall (Equation 29.3),

$$\tau = \rho u_*^2 \qquad (29.4)$$

where ρ is the density of water. Note that under steady-state conditions, $\varepsilon = U\tau/(\rho H)$, where H is water depth, which indicates that ε in rivers can be much higher than in lakes. The Reynolds shear stress (τ_{Re}) is another estimate of turbulence that is often used, but it suffers from the scale of measurement because it is based on the product of velocity fluctuations in the vertical plane as defined by downstream and vertical directions (u' and w', respectively, where u is the instantaneous velocity and $\bar{u}$ is the mean velocity: $u = \bar{u} + u'$) at height, z, of sampling:

$$\tau_{Re} = \rho \overline{u'w'} \qquad (29.5)$$

The product of u' and w' has been found to provide information on the spatial pattern in the turbulence because it provides information on the patterns of flow conceptually as in the case of an eddy passing a location—first water moves downward, then backward, then upward, and then forward again. Quadrant analysis defines the spatial patterns as: (1) outward interactions of fast-flowing water away from the bottom (quadrant 1; $u'w' > 0$, where $w' > 0$); (2) ejections of slow-flowing water away from the bottom (quadrant 2; $u'w' < 0$, where $w' > 0$); (3) inward interactions of slow-flowing water toward the bottom (quadrant 3; $u'w' > 0$, where $w' < 0$); and (4) sweeps of fast-flowing fluid toward the bottom (quadrant 4; $u'w' < 0$, where $w' < 0$) (Cellino and Lemmin 2004, O'Connor and Hondzo 2008). Sweeps have been found to move particles toward the bottom as well as resuspend particles on the bottom as flow moves horizontally, whereas ejections move particles from the bottom into the water column (Cellino and Lemmin 2004). It seems likely that quadrant analysis may provide information on the distribution of biologically relevant particles such as the gametes and veligers of dreissenids.

PHYSICAL ECOLOGY OF DREISSENIDS

Except for sexual reproduction and larval dispersal, dreissenids have a life history associated with the bottom of lakes and rivers. Hence, mussels remain within the benthic boundary layer for the majority of their existence and face the common problem of obtaining and removing dissolved and suspended material and withstanding the forces that may lead to dislodgement. Some populations in shallow water may be exposed to energetic conditions in which a benthic boundary layer does not form due to water currents but forms due to waves. The following discussion will present a brief review of the various biological and ecological features of dreissenid life history that are linked to hydrodynamics, especially turbulence. Gaps in our current understanding will be identified at the end of each section to stimulate new ways in which to explore and understand the biology of these organisms.

External Fertilization

Dreissenids have conserved life history traits from their marine ancestors (Ackerman et al. 1994), and one of the most prominent traits compared to other freshwater bivalves is external fertilization. This is a subject that has received some interest in marine invertebrates, especially from a physical ecology perspective involving mathematical modeling and field experiments. In this case, local hydrodynamic conditions influence fertilization success when sperm and eggs are broadcast into the water column (Okubo et al. 2001). It has long been believed that sperm, which cannot swim against water currents or turbulence, would become diluted to concentrations below which successful fertilization can occur. The problem of "sperm limitation" has been demonstrated in laboratory and field experiments (reviewed in Levitan 1995) and has been examined using mathematical models. These models suggest that the direction of the sperm cloud is influenced by direction of water currents, whereas the eddy diffusivity in the water column affects sperm spread (Denny and Shibata 1989). Turbulence also has a nonlinear (i.e., dome-shaped) effect on the encounter between eggs and sperm, on their bonding, and on the development of embryos (Mead and Denny 1995, Denny et al. 2002, Riffel and Zimmer 2007). Even though fertilization success in organisms that broadcast spawn relies on turbulent diffusion for the encounter between sperm and eggs, too much turbulence can also dilute gamete concentration and may lead to sperm limitation (Serrão and Havenhand 2009).

It is surprising that there has been little effort devoted to the physical ecology of external fertilization in dreissenids (Table 29.1). Presently, this effort has been limited to several recent studies that examined the affect of sperm concentration on fertilization success in *D. polymorpha* and *D. r. bugensis* in laboratory experiments, flow chambers, and in Lake Erie (Quinn and Ackerman 2011, 2012). Overall, fertilization success was related to sperm concentration, but dreissenids retained some success at low sperm concentrations whereas other benthic species did not. Importantly, fertilization success in both the laboratory and field was related to near-bed turbulence, including turbulence caused by mussel shells. Specifically, higher fertilization success occurred over patches of mussels where there was a higher frequency of ejections of gametes into the water column from the bottom.

A large number of issues remain unknown in our understanding of the physical ecology of external fertilization in dreissenids. For example, what is the fertilization success under oscillatory and other unsteady conditions? Do local hydrodynamic conditions affect sperm–egg encounters as predicted by the aforementioned mathematical models? Are there strategies that dreissenids have evolved to overcome sperm limitation by synchronizing gamete release with physical conditions? (see reviews in Okubo et al. 2001, Serrão and Havenhand 2009).

Larval Dispersal

There have been a number of studies of the life history of dreissenids, including their larval stages (Ackerman et al. 1994, Nichols and Black 1994, Martel et al. 1995, Claxton and Boulding 1998). Unlike native freshwater bivalves, dreissenids have free-swimming planktotrophic larvae similar to their marine ancestors (Ackerman et al. 1994). Marine bivalves and many other benthic organisms have small larvae (<1 mm) with limited swimming ability (less than millimeter per second); that is, larvae cannot swim against ambient currents greater than a few centimeter per second. However, bivalve larvae have behavioral mechanisms (e.g., retract velum and close valves) that allow a change in position in the water column (review in Abelson and Denny 1997, Pineda et al. 2007). Such changes can be very effective because these larvae typically have a Péclet number ($Pe = lU/D$, where D is the molecular diffusivity; the ratio between advection and diffusion) close to 1, and settling velocity (w_s) is typically > swimming velocities, which leads to domination by advection.

Larval dispersal refers to the spread of individuals from the location of spawning to the location of settlement and involves active and passive transport mostly through

Table 29.1 Studies of the Physical Ecology of External Fertilization in Dreissenid Mussels

Species	Type	Technique/Data	Response Measured/Predicted	Reference
D. polymorpha *D. r. bugensis*	Laboratory and model	Sperm dilution/age Dispersion modeling	Fertilization success	Quinn and Ackerman (2012)
D. polymorpha *D. r. bugensis*	Laboratory and field	Flow chamber Field experiments	Fertilization success and turbulence	Quinn and Ackerman (2011)

Table 29.2 Studies of the Physical Ecology of Larval Dispersal in Dreissenid Mussels

Species	Type	Technique/Data	Response Measured/Predicted	Reference
D. polymorpha	Field	River sampling Larval staining	Larval abundance and viability with distance	Horvath and Lamberti (1999)
D. polymorpha	Field	River sampling	Veliger spatial distribution	Barnard et al. (2003)
D. polymorpha	Laboratory	Air-generated turbulence	Veliger mortality	Rehmann et al. (2003)
D. polymorpha	Laboratory	Orbital shaker	Veliger mortality	Horvath and Crane (2010)

hydrodynamics and behavior (Pineda et al. 2007). Many studies model the process of larval dispersal using a number of parameters, as measured or estimated from the environment. Of these, turbulence transport models like the local exchange model use larval w_s and vertical mixing to model the dispersal of settling larvae under turbulent conditions (McNair and Newbold 2001). The settling velocity of the larvae is the most important biological parameter in the model, but it is seldom measured, rather it is often predicted by Stoke's law:

$$w_s = \frac{2r^2 g \left(\rho_{particle} - \rho_{fluid} \right)}{9\mu} \quad (29.6)$$

where
  r is the radius
  g is the acceleration due to gravity
  $\rho_{particle}$ and ρ_{fluid} are the density of the particle and the fluid, respectively
  μ is the dynamic viscosity of the fluid

Unfortunately, Equation 29.6 applies to larvae settling with $Re < 0.5$ in the absence of turbulence and is less effective under higher Re as was evident in a recent examination of w_s of newly metamorphosed unionid mussels whose Re was between 0.2 and 2.1 (Schwalb and Ackerman 2011). Settling becomes more difficult to predict in the presence of turbulence, and there have even been reports that water column turbulence increases the w_s of particles (Ruiz et al. 2004), contrary to the traditional view that turbulence maintains particles in suspension.

Water column turbulence has also been found to affect the biology and ecology of bivalve larvae and other planktonic organisms. For example, low and intermediate levels of turbulence increase feeding and growth rates of plankton, although turbulence affects taxonomic groups differently (Peters and Marrasé 2000). Experimental evidence indicates that organisms that are 10^3 times smaller than the smallest eddies (i.e., millimeter-sized in lakes and oceans given by the Kolmogorov microscale: $\eta_k = (\upsilon^3/\varepsilon)^{1/4}$) are affected by turbulence, even though theoretical constructs predict otherwise (Peters and Marrasé 2000). This was also the case for blue mussel larvae (*Mytilus edulis*) in the laboratory, which responded to turbulence even though they were 2–3× smaller than the η_k (Fuchs and DiBacco 2011). Another complicating factor is that larvae are often associated with discontinuities in the water column including pycnoclines, haloclines, turbidity fronts, tidal bores, and chlorophyll maxima (e.g., Metaxas and Young 1998, Pineda et al. 2007).

There has been considerable interest in the role of currents/turbulence in the dispersal of dreissenid larvae as related to range expansion and as a control mechanism that restricts expansion in industrial settings and rivers (Table 29.2). Although not examined directly, the pattern of initial spread of dreissenids in North America followed a pattern largely consistent with the natural movement of waters in interconnected watersheds and with human-mediated upstream and overland transport (Johnson and Carlton 1996, Wilson et al. 1999). These studies cited similar patterns in Europe with low occurrence of *D. polymorpha* in unconnected lakes. Downstream transport in rivers may be detrimental to dreissenid larvae as both the total number and proportion of live larvae declines downstream from the source (Horvath and Lamberti 1999). Whereas hydrodynamic factors were credited for the decline in live larvae, no evidence of this was provided, and measurements of river hydrodynamics were not made (but see below). A similar explanation was provided in another study for the decline in larval abundances across the freshwater to estuarine transition zones in the St. Lawrence River (Barnard et al. 2003). Limited evidence on distributions of dreissenid larvae in lakes suggests that larvae may concentrate lower in the epilimnion where turbulence may be lower; however, these physical factors were not actually measured (see papers in Nalepa and Schloesser 1993).

The concept that turbulence might negatively affect dreissenid larvae or even cause mortality has, however, been demonstrated in laboratory experiments. Specifically, air bubble–induced turbulence caused mortality of D-shaped veligers and velichoncha at $\varepsilon = 4 \times 10^{-3}$ m^2 s^{-3}, but not of the smaller, younger veligers (Rehmann et al. 2003). Whereas lower levels of turbulence did not lead to mortality, the importance of the ratio of larval size to ε scale was emphasized as it demonstrated that river-induced mortality (Horvath and Lamberti 1999) was plausible based on textbook approximations of turbulence levels. A recent study demonstrated that the damage caused by shaking larvae using an orbital shaker at various speeds (100–400 rpm) increased with exposure time, but no estimate of the hydrodynamic forces involved was attempted (Horvath and Crane 2010). An accurate estimate of the hydrodynamic condition in a laboratory flask is complex and difficult to obtain as ε varies with radius, height above the flask's bottom, and of course with speed; modeling indicates that mean turbulence increased from ε = 4 × 10^{-4} m^2 s^{-3} to 2 × 10^{-3} m^2 s^{-3} between 150 and 200 rpm (Kaku et al. 2006). It should be noted that such high levels of turbulence would not be found in lakes.

There is a great deal to be learned about the role of hydrodynamics in dreissenid larval dispersal, including effects on larval transport. From a practical perspective, settling velocities and swimming speeds of different larval stages should be measured experimentally, and the latter should also be examined in factorial designs with other environmental cues to ascertain ontological changes and the response of larvae to "discontinuities" described earlier. Such information would also provide needed insights for ecosystem modelers but should be verified by field-based studies. Such field studies should, among other things, examine transport of larvae in currents, their response to seiches, internal waves, convection cells, and the benthic boundary layer and examine what effects the interaction of these hydrodynamic factors have on concentrating/dispersing larvae in lakes and rivers. It would also be reasonable to determine the impact of suspension-feeding larvae on lake and river ecosystems under the aforementioned hydrodynamic conditions along with other potential ecological interactions.

Larval Settlement

To date, there has been considerable progress in understanding relationships between hydrodynamics and larval settlement in marine invertebrates (review in Abelson and Denny 1997, Pineda et al. 2007, Marshall et al. 2009). Not surprisingly, a number of hypotheses have been presented to explain patterns of larval settlement, which for marine invertebrates are typically patchy in space and time. Unfortunately, the recruitment of postlarvae is often used synonymously with larval settlement, but this is incorrect as it assumes that pre- and postsettlement are equivalent. Explanations for different patterns in presettlement involve dilution of sperm to levels that limit fertilization (sperm limitation) or variation in fecundity, differential flux of larvae due to hydrodynamics (transport hypothesis or larval supply), differential settlement of larvae due to predation on larvae, larval condition (including older larvae desperate to settle), and larvae with different behavioral strategies. Explanations for postsettlement patterns involve postsettlement mortality and juvenile migration. It is not inconceivable that a number of these explanations occur simultaneously. There is ample evidence from the field to support the larval transport hypothesis, which states that variation in larval settlement among sites is a function of variation in larval flux (e.g., J = larvae m^{-2} s^{-1}), recognizing pre- and postsettlement factors. Flux ($J = UC$) is the product of larval concentration (C) in the water column and the water velocity (U) and provides an integral measure of the number of larvae that are transported past a given location (Gaines and Bertness 1993). Unfortunately, physical conditions of shallow coastal regions, including small-scale hydrodynamics, are never as simple as predictions from the open ocean (Pineda et al. 2007). Similar statements can be made regarding offshore vs. nearshore regions and pelagic vs. benthic regions of lakes. Consequently, care must be taken to examine larvae in space and time, given that there is increasing evidence of restricted larval dispersal due to restricted larval transport and high levels of self-recruitment among other factors, which relate in part to unpredicted physical forcing (Pineda et al. 2007).

Larval settlement in many marine bivalves and also dreissenids can involve two stages, the primary settlement of pediveligers from the water column onto suitable substrates where they undergo metamorphosis and a secondary settlement of recently settled or older juvenile mussels that move to new locations (review in Ackerman et al. 1994). There have been a number of reports of substrate preferences of dreissenids undergoing primary settlement (e.g., Walz 1975, Ackerman et al. 1993b), but few studies have examined secondary settlement of small, juvenile mussels. Martel (1993) recorded the abundance of post-metamorphic juvenile *D. polymorpha* collected in the plankton and on pot-scrubber collectors in the water column. Abundances of those individuals in shallow waters increased during periods of high wave energy, suggesting that resuspension played a role in transport and settlement.

Studies of the influence of hydrodynamics on larval settlement in dreissenids are somewhat limited (Table 29.3). In industrial and experimental settings, settlement of dreissenid larvae, as determined on settling plates/surfaces, was

Table 29.3 Studies of the Physical Ecology of Larval Settlement in Dreissenid Mussels

Species	Type	Technique/Data	Response Measured/Predicted	Reference
D. polymorpha	Field	Scouring pad collectors Plankton samples	Postlarval drift	Martel (1993)
D. polymorpha	Field	Wetland sampling hydrograph	Mussel depth distribution	Bowers and de Szalay (2004)
Dreissena spp.	Laboratory and field	Larval transport near bed	Larval transport and settlement	Quinn and Ackerman (in review)
D. r. bugensis		Scouring pad collectors	Turbulence generated by bottom roughness	

negatively related to free stream velocity (mussel density = $-11 \times 10^3 \log(U) + 2.3 \times 10^3$, $r^2 = 0.99$, $P < 0.001$) (Ackerman and Sim 1994). Specifically, these studies found that settlement of dreissenids declined from 3.5×10^5 m^{-2} at $U = 10^{-3}$ m s^{-1} to zero at $U > 1.5$ m s^{-1}. Settlement in rectangular chambers used in one of the studies increased over the first 15 cm but then declined downstream, a pattern that matched quite nicely with bed shear stress. The bed shear stress (τ) modeled in the experimental chambers using computational fluid dynamic modeling peaked at $\tau = 0.2$ Pa (COMSOL multiphysics; Ackerman unpublished).

Dreissenids have also responded to near-bed turbulence generated by bed roughness caused by their shells in laboratory flow chambers with natural conditions, and in littoral sites in Lake Erie (Quinn and Ackerman in review). In those cases, quadrant analysis revealed that the nature of the near-bed turbulence affected larval settlement. Wake interference flow over mussel patches had high frequencies of turbulent sweeps and ejections and the highest larval settlement, whereas skimming flow over high-density mussel beds led to high rates of larval transport but lower larval settlement. It has been noted that even though dreissenids are capable of settling in many areas, they may not survive due to postsettlement factors (see below). One such factor was fluctuations in water level caused by episodic winds and gravitational seiches. For example, it has been suggested that limited dreissenid survival at shallow depths (<35 cm) in a coastal wetland was due to the exposure of individuals to the atmosphere (Bowers and de Szalay 2004).

The role of hydrodynamics in the settlement of dreissenid larvae has provided some insight into their ecology and control, but there are many aspects of related pre- and postsettlement that remain unknown. In this context, among research needs is an evaluation of the transport hypothesis using combined integrative measures of larval flux and hydrodynamics. This would provide some insight into the more local transport of larvae through the water column and eventual settlement in the benthic region. The response of larvae to near-bed hydrodynamic conditions created by a myriad of roughness elements, including substrates of differing morphometry (e.g., filamentous), should be determined. In addition, it would be reasonable to extend research beyond settlement patterns and associated hydrodynamics to ascertain mechanisms that most influence settlement. Such a study could also examine potential chemical cues and the hydrodynamic dispersion of cues in and around roughness elements, which would provide valuable insights into the behavioral response of settling larvae.

Suspension Feeding

Suspension feeding is the process by which suspended or dissolved material is removed from a water body by organisms and used as source of food (see review in Wildish and Kristmanson 1997). The process of particle removal can rarely be attributed to sieving (i.e., filtering), especially for small particles. However, it is widely recognized that particle capture occurs through hydrosol mechanisms (Riisgård and Larsen 1995). Bivalves are among the most conspicuous of benthic suspension feeders and, to do so, they generate currents to pass the material in suspension across their gills (Riisgård and Larsen 1995). Interest in the effect of hydrodynamics on suspension feeders in bivalves dates to the 1970s and has included a large number of studies, methodologies, and approaches (reviewed in Ackerman 1999, Ackerman and Nishizaki 2004). Among the findings, of utmost importance is the recognition of a dome-shaped response to turbulence; that is, moderate levels of turbulence promote feeding rates and high levels inhibit them (Wildish and Kristmanson 1997). Promotion of the feeding response at moderate turbulence levels is due to the replenishment of material removed from suspension by the bivalves (i.e., a concentration boundary layer; see below) to a point where suspension-feeding rate becomes saturated. The inhibition of feeding at high turbulence levels has been attributed to several factors including: (1) filtration instability, whereby under- or overpressurization of the inhalant and exhalant siphons disrupt the pumping mechanism; (2) behavioral instability, whereby lift and/or drag forces acting on a mussel or its siphon causes it to retract its siphon or close its shell, thereby reducing suspension feeding; and (3) optimal foraging, whereby there is a physiologically optimal particle flux for the processing of food (Wildish and Saulnier 1993).

The nonlinear, dome-shaped suspension response to turbulence has also been observed in *D. r. bugensis* in a laboratory flow chamber (Ackerman 1999; Table 29.4). In that study, the saturation of suspension feeding occurred between 10 and 20 cm s^{-1} (average flow chamber velocity) or more accurately between a flux of 0.68 and 1.35×10^6 cells m^{-2} s^{-1}. The reason for a decline in suspension feeding at the higher flux was most likely due to the behavioral instability mechanism described earlier. Mussels retracted their siphons at 20 cm s^{-1}, which is the Re where the shedding of eddies occurs on circular cylinders (Vogel 1994). This behavioral response to a "pinching" of siphons would, therefore, be responsible for declines in suspension-feeding rates. It is important to note that clearance rates of mussels at the slowest velocity of <1 cm s^{-1} were consistent with published results conducted in the absence of flow (Kryger and Riisgård 1988). It was also noted that clearance rates reported for *D. polymorpha* and *D. r. bugensis* of a given size can vary over two orders of magnitude due to use of unpalatable particles as food, which provides low rates, or due to the hydrodynamics of placing large mussels in small containers, which provides high rates (Ackerman 1999).

Like all benthic suspension feeders, dreissenids depend on water movements to deliver food particles, which can arrive horizontally via advection, arrive from above via particle settling and vertical mixing of the water column, or potentially arrive from below via resuspension of

Table 29.4 Studies of the Physical Ecology of Suspension Feeding in Dreissenid Mussels

Species	Type	Technique/Data	Response Measured/Predicted	Reference
D. polymorpha D. r. bugensis	Laboratory	Flow chamber	Clearance rate	Ackerman (1999)
D. polymorpha	Modeling	Physical and biological measurements	Plankton consumption	MacIsaac et al. (1999)
D. polymorpha	Field and laboratory	Physical and biological measurements Flow chamber	Concentration boundary layer Clearance rate	Ackerman et al. (2001)
D. polymorpha	Laboratory	Flow chamber	Ingestion rate	Tuchman et al. (2004)
D. polymorpha	Modeling	Physical measurements	Plankton biomass	Edwards et al. (2005)
D. polymorpha	Field	Physical and biological measurements	Stratification	Loewen et al. (2007)
D. polymorpha	Modeling	Published data	Plankton biomass	Boegman et al. (2008)

benthic materials. In marine systems, large mussel beds tend to be found in areas of moderate to high water flows (Wildish and Kristmanson 1997) because there are times when suspension feeding can clear the water of algae faster than local hydrodynamics can replenish the algal concentration boundary layer (CBL; Dolmer 2000, Tweddle et al. 2005). The transport of algae in the water overlying the boundary can be described by the advection–diffusion equation given by

$$u\frac{\partial C}{\partial x} + w\frac{\partial C}{\partial z} = \frac{\partial}{\partial z}\left[(D+K)\frac{\partial C}{\partial z}\right] + R_1 \quad (29.7)$$

where R_1 is any reaction that affects C in the CBL (Nishihara and Ackerman 2006, 2007).

There has been considerable interest in examining large-scale impacts of dreissenid suspension feeding on the water column, especially in relation to simple stirred-reactor models as those applied to the western basin of Lake Erie (Table 29.4). Those unrealistic models used population size distribution × size-corrected clearance rates per mussel to estimate that the 7 m deep western basin was being cleared by mussels every 1.3 h. More realistic hydrodynamic models have been applied to the problem (MacIsaac et al. 1999), but they also have limitations if the fundamental physical process of thermal stratification is not considered; thermal stratification is diurnal and episodic in the western basin of Lake Erie (Ackerman et al. 2001, Boegman et al. 2008). An examination of benthic–pelagic coupling by *D. polymorpha* on a reef in the western basin revealed that the impact of suspension feeding was restricted to ~2 m above the bottom (Ackerman et al. 2001). A CBL was detected because stratification had reduced vertical mixing in the water column and because mussels were refiltering bottom water as it flowed over the reef. Besides this localized near-bottom impact, another important finding was that clearance rates determined over the reef were 40% lower than those observed in the laboratory, which was likely due to the more complex nature of natural seston as compared to algal cultures often used in clearance rate studies conducted in the laboratory. These findings have been confirmed in a number of numerical modeling studies. For example, the modeled offshore impact of suspension feeding by dreissenid mussels based on vertical eddy diffusivity measurements was linked to the delivery of algae to the benthic region (Edwards et al. 2005). The importance of weak stratification and wind stress was supported in a more recent modeling study that estimated that dreissenids filtered ~1/2 of the net summer algal growth in the western basin under realistic hydrodynamic conditions as compared to >3/4 under an unrealistic, fully mixed water column condition (Boegman et al. 2008).

Recent changes to the offshore conditions in Lake Michigan have also been attributed to benthic–pelagic coupling by the profunda morph of *D. r. bugensis*, which has rapidly colonized the deeper depths of the lake (Nalepa et al. 2010). Specifically, the large spring diatom bloom has disappeared in the pelagic zone whereas summer plankton productivity levels remain (Fahnenstiel et al. 2010b). Also, large productive basin-scale gyres created by winter storms, which transport nutrient rich waters and sediments offshore, have also declined in productivity since the establishment of large populations of *D. r. bugensis* (Kerfoot et al. 2010). Changes to phytoplankton and zooplankton species composition and alterations of pelagic food webs have also been noted by these studies. The interesting hypothesis presented for the loss of the spring diatom bloom is that isothermal conditions, which prevail in the spring when the water is well mixed, provide mussels access to particles throughout the water column (Fahnenstiel et al. 2010b). Data supporting the ability of dreissenids to clear water at low water temperatures support the concept (Vanderploeg et al. 2010), as do arguments related to nutrient loading presented in the studies (see summary in Fahnenstiel et al. 2010a). Those studies emphasize the need to examine the physical ecology of dreissenids in Lake Michigan, to measure the near- and far-field conditions to which mussels are exposed, and to

Table 29.5 Studies of the Physical Ecology of Detachment in Dreissenid Mussels

Species	Type	Technique/Data	Response Measured/Predicted	Reference
D. polymorpha	Laboratory	Wall jet	Attachment strength	Ackerman et al. (1995)
D. r. bugensis		Mechanical testing	Detachment force	
D. polymorpha	Laboratory	Wall jet	Attachment strength	Ackerman et al. (1996)
D. r. bugensis		Mechanical testing	Detachment force	
D. polymorpha	Laboratory	Flow chamber	Byssal thread production	Clarke and McMahon (1996)
D. polymorpha	Laboratory	Flow chamber	Byssal thread production	Peyer et al. (2009)
D. r. bugensis		Mechanical testing	Byssal thread strength	

determine whether simple stirred-reactor models adequately model/explain recent changes to this aquatic system.

A great number of questions remain unknown with respect to both the suspension feeding and benthic–pelagic coupling of dreissenids, especially related to selective feeding, effects of oscillatory and unsteady water flow, and biogeochemistry of benthic regions including sediments. For example, denitrification was enhanced by the presence of dreissenid feces and pseudofeces in winter periods of low river discharge, but not during other periods of high flow and mixing (Bruesewitz et al. 2006). This is relevant to this discussion as abundances of dreissenid populations are most often reported on a per-unit area (two-dimensional) basis, but populations really occur in multiple layers and thus are three-dimensional. A laboratory flow chamber study revealed a 50% reduction in algal ingestion rate (IR) for each 2 cm depth ($IR = e^{-0.347z}$) within the 7×10^5 mussel m^{-2} aggregation under no flow, but IR increased with flow in the chamber (Tuchman et al. 2004). Interestingly, mussels < 5 mm in shell length had the highest ingestion rates at the maximum water velocity (20 cm s^{-1}), perhaps because they were sheltered from hydrodynamic forces by larger individuals. Research on water residence times and flux of materials and nutrients to/from mussel aggregations and sediments would provide valuable information on the role of dreissenids in the environment.

Detachment

Epibenthic organisms like dreissenids can live in high-energy environments like rocky shores or shores with relative large roughness elements such as boulders, because these features can withstand impacts of breaking waves and storm currents. This subject has received considerable attention in the marine environment especially in rocky, intertidal areas where organism mortality by detachment is an important ecological driver of communities (Denny 1999). Such analysis has revealed that the types of hydrodynamic forces responsible for the dislodgment of benthic organisms include drag ($F_D = 1/2 C_d \rho S U^2$, where C_d is the drag coefficient and S is the projected area), lift ($F_L = 1/2 C_L \rho S U^2$, where C_L is the drag coefficient), shear stress (τ; Equation 29.4), and an acceleration reaction force ($A = \rho Vol C_M du/dt$, where Vol is the volume of the body, C_M is the inertia coefficient, and du/dt is the instantaneous local acceleration of the water past the body) (Denny 1988). It is important to note that relative magnitudes of these forces depend not only on water velocity but also on organism size and shape and on accelerations encountered by the organisms (Koehl 1996).

Dreissenids attach to hard surfaces using byssal threads, which are partly responsible for their success as invasive species (Table 29.5). This single feature—byssal adhesion—allows mussels to colonize an open niche in freshwater systems, as well as cause problems for raw water users (see reviews in Ackerman et al. 1994, Claudi and Mackie 1994). Indeed, byssal adhesion represents the only defense against the hydrodynamic forces discussed earlier. The importance of byssal adhesion has led to the development of techniques to measure adhesive forces (review in Ackerman et al. 1993a, 1995). Attachment strength varies with substrate materials, and resistance to detachment is a function of the byssal thread–substrate interface, the byssal thread, and the byssal thread–pedal retractor muscle interface (Ackerman et al. 1996). Interestingly, byssal thread production relative to ambient hydrodynamic conditions in D. polymorpha was unimodal over the course of 20 days (Clarke and McMahon 1996). The production of byssal threads increased up to a maximum at ~20 cm s^{-1}, but then declined at 27 cm s^{-1}, which was the highest velocity examined. The decline was perhaps due to the inability of mussels to place their foot on the substrate to form/anchor threads at the higher velocities (i.e., a behavioral instability). It is important to note that temporal differences were also noted; rates declined after ~10 days regardless of velocity.

A recent examination of production of byssal threads and attachment strength found significant and important differences between D. polymorpha and D. r. bugensis (Peyer et al. 2009). Specifically, byssal thread production and attachment strength was significantly higher in D. polymorpha than in D. r. bugensis at flow-chamber velocities ranging up to 180 cm s^{-1}. Interestingly, whereas the number of threads produced by D. r. bugensis was relatively constant at each velocity, the number produced by D. polymorpha increased with velocity. The number of threads produced by both species declined over time regardless of velocity as reported earlier. Greater thread production and stronger attachment strength in *D. polymorpha* compared to *D. r. bugensis* were most evident over the course of the first few days and the extent of differences diminished after several months. Differences between the two species may explain the persistence of *D. polymorpha* in high-velocity, shallow-water regions compared to the diminished presence of *D. r bugensis* in these environments (Peyer et al. 2009).

Although a number of studies have provided interesting results on the relationship between hydrodynamic forces and dislodgment of dreissenids, more issues need to be addressed. Specifically, from a mechanistic perspective, the aforementioned influence of organism size and shape on hydrodynamic forces (i.e., C_d, C_L) is not known and will require an ontological comparison. Even more pressing is the need to examine these same factors under acceleration typical of breaking waves (i.e., C_M). Unfortunately, most of our understanding of dreissenid dislodgement is based on steady-state conditions created in flow chambers in the laboratory under relatively low flows. A more sophisticated understanding under unsteady flows is needed along with detailed measurements of the hydrodynamic conditions at spatial and temporal scales appropriate to the individual mussels. Such a mechanistic understanding could lead to the development of predictive models that might address important aspects of dreissenid ecology (e.g., Carrington et al. 2009).

CONCLUSIONS

The ecology of dreissenid mussels is intimately linked to the hydrodynamics of a given water body. For the majority of their life stages, this link is strongest in the benthic rather than the pelagic region, with the exception of the larval stage when dispersal and transport in the pelagic region occurs. Hydrodynamics in benthic regions are more energetic in terms of turbulence then is commonly held, and this is the region in which dreissenids thrive through suspension feeding. Conditions are more energetic in shallow benthic regions (littoral) where mussels can become detached from substrates by hydrodynamic forces. Moreover, this is the region out of which their gametes must pass and become fertilized and into which their larvae must settle. These biological processes are driven by hydrodynamic forces. Although much progress has been made toward understanding these relationships, it is clear that we have only scratched the surface of many of the links between hydrodynamics and the biology/life history of dreissenids. Much remains to be learned about the physical ecology of these ecologically and economically important species.

ACKNOWLEDGMENTS

I would like to thank Anthony Merante for his assistance in the review of the literature, Professor Johny Wüest for valuable discussions, and comments from Tom Nalepa and Don Schloesser. This research was supported by funding from the Natural Sciences and Engineering Research Council of Canada.

REFERENCES

Abelson, A. and M. Denny. 1997. Settlement of marine organisms in flow. *Ann. Rev. Ecol. Syst.* 28: 317–339.

Ackerman, J.D. 1999. The effect of velocity on the filter feeding of zebra mussels (*Dreissena polymorpha* and *D. bugensis*): Implications for trophic dynamics. *Can. J. Fish. Aquat. Sci.* 56: 1551–1561.

Ackerman, J.D., C.M. Cottrell, C.R. Ethier, D.G. Allen, and J.K. Spelt. 1995. A wall jet to measure the attachment strength in zebra mussels. *Can. J. Fish. Aquat. Sci.* 52: 126–135.

Ackerman, J.D., C.M. Cottrell, C.R. Ethier, D.G. Allen, and J.K. Spelt. 1996. Attachment strength of zebra mussels on natural, polymeric and metallic materials. *J. Environ. Eng.* 122: 141–148.

Ackerman, J.D., C.R. Ethier, D.G. Allen, and J.K. Spelt. 1993a. The biomechanics of byssal adhesion in *Dreissena polymorpha*. In *Zebra Mussels: Biology, Impacts, and Control*. T.F. Nalepa and D.W. Schloesser (eds.), pp. 265–282. Boca Raton, FL: CRC Press.

Ackerman, J.D., C.R. Ethier, D.G. Allen, and J.K. Spelt. 1993b. Patterns of recruitment and persistence of zebra mussels on a variety of materials. In *Proceedings, 3rd International Zebra Mussel Conference*, 1993, Toronto, Ontario, Canada. J.L. Tsou and Y.G. Mussalli (eds.), pp. 111–121. Palo Alto, CA: Electric Power Research Institute.

Ackerman, J.D. and T. Hoover. 2001. Measurement of local bed shear stress in streams using a Preston-static tube. *Limnol. Oceanogr.* 46: 2080–2087.

Ackerman, J.D., M.R. Loewen, and P.F. Hamblin. 2001. Benthic-pelagic coupling over a zebra mussel bed in the western basin of Lake Erie. *Limnol. Oceanogr.* 46: 892–904.

Ackerman, J.D. and M.T. Nishizaki. 2004. The effect of velocity on the suspension feeding and growth of the marine mussels *Mytilus trossulus* and *M. californianus*: Implications for niche separation. *J. Mar. Syst.* 49: 195–207.

Ackerman, J.D. and B. Sim. 1994. Fluid dynamic influences on the recruitment of zebra mussels. Poster presentation. *4th International Zebra Mussel Conference*, Madison, WI.

Ackerman, J.D., B. Sim, S.J. Nichols, and R. Claudi. 1994. A review of the early life history of the zebra mussel (*Dreissena polymorpha*): Comparisons with marine bivalves. *Can. J. Zool.* 72: 1169–1179.

Barnard, C., J.J. Frenette, and W.F. Vincent. 2003. Planktonic invaders of the St. Lawrence estuarine transition zone: Environmental factors controlling the distribution of zebra mussel veligers. *Can. J. Fish. Aquat. Sci.* 60: 1245–1257.

Boegman, L., M.R. Loewen, P.F. Hamblin, and D.A. Culver. 2008. Vertical mixing and weak stratification over zebra mussel colonies in western Lake Erie. *Limnol. Oceanogr.* 53: 1093–1110.

Bowers, R. and F.A. de Szalay. 2004. Effects of hydrology on unionids (Unionidae) and zebra mussels (Dreissenidae) in a Lake Erie coastal wetland. *Am. Midl. Nat.* 151: 286–300.

Bruesewitz, D.A., J.L. Tank, M.J. Bernot, W.B. Richardson, and E.A. Strauss. 2006. Seasonal effects of the zebra mussel (*Dreissena polymorpha*) on sediment denitrification rates in Pool 8 of the Upper Mississippi River. *Can. J. Fish. Aquat. Sci.* 63: 957–969.

Carrington, E., G.M. Moeser, J. Dimond, J.J. Mello, and M.L. Boller. 2009. Seasonal disturbance to mussel beds: Field test of a mechanistic model predicting wave dislodgment. *Limnol. Oceanogr.* 54: 978–986.

Cellino, M. and U. Lemmin. 2004. Influence of coherent flow structures on the dynamics of suspended sediment transport in open-channel flow. *J. Hydraul. Eng.* 130: 1077–1088.

Clarke, M. and R.F. McMahon. 1996. Effects of current velocity on byssal-thread production in the zebra mussel (*Dreissena polymorpha*). *Can. J. Zool.* 74: 63–69.

Claudi, R. and G.L. Mackie. 1994. *Practical Manual for Zebra Mussel Monitoring and Control*. Boca Raton, FL: CRC Press.

Claxton, W.T. and E.G. Boulding. 1998. A new molecular technique for identifying field collections of zebra mussel (*Dreissena polymorpha*) and quagga mussel (*Dreissena bugensis*) veliger larvae applied to eastern Lake Erie, Lake Ontario, and Lake Simcoe. *Can. J. Zool.* 76: 194–198.

Denny, M.W. 1988. *Biology and the Mechanics of the Wave-Swept Environment*. Princeton, NJ: Princeton University Press.

Denny, M. 1999. Are there mechanical limits to size in wave-swept organisms? *J. Exp. Biol.* 202: 3463–3467.

Denny, M.W., E.K. Nelson, and K.S. Mead. 2002. Revised estimates of the effects of turbulence on fertilization in the purple sea urchin, *Strongylocentrotus purpuratus*. *Biol. Bull.* 203: 275–277.

Denny, M.W. and M.F. Shibata. 1989. Consequences of surf-zone turbulence for settlement and external fertilization. *Am. Nat.* 134: 859–889.

Dolmer, P. 2000. Algal concentration profiles above mussel beds. *J. Sea Res.* 43: 113–119.

Edwards, W.J., C.R. Rehmann, E. McDonald, and D.A. Culver. 2005. The impact of a benthic filter feeder: Limitations imposed by physical transport of algae to the benthos. *Can. J. Fish. Aquat. Sci.* 62: 205–214.

Fahnenstiel, G., T. Nalepa, S. Pothoven, H. Carrick, and D. Scavia. 2010a. Lake Michigan lower food web: Long-term observations and *Dreissena* impact. *J. Great Lakes Res.* 36(Suppl. 3): 1–4.

Fahnenstiel, G.L., S. Pothoven, T. Nalepa, H. Vanderploeg, D. Klarer, and D. Scavia. 2010b. Recent changes in primary production and phytoplankton in the offshore region of southeastern Lake Michigan. *J. Great Lakes Res.* 36(Suppl. 3): 20–29.

Fischer, H.B., E.J. List, H.C.Y. Koh, J. Imberger, and N.A. Brooks. 1979. *Mixing in Inland and Coastal Waters*. San Diego, CA: Academic Press.

Fuchs, H.L. and C. DiBacco. 2011. Mussel larval responses to turbulence are unaltered by larval age or light conditions. *Limnol. Oceanogr. Fluids Environ.* 1: 120–134.

Gaines, S.D. and M. Bertness. 1993. The dynamics of juvenile dispersal. Why field ecologists must integrate. *Ecology* 74: 2430–2435.

Horvath, T.G. and L. Crane. 2010. Hydrodynamic forces affect larval zebra mussel (*Dreissena polymorpha*) mortality in a laboratory setting. *Aquat. Invasions* 5: 379–385.

Horvath, T.G. and G.A. Lamberti. 1999. Limitation of zebra mussel recruitment in streams by veliger mortality. *Freshwat. Biol.* 42: 69–76.

Johnson, L.E. and J.T. Carlton. 1996. Post-establishment spread in large-scale invasions: Dispersal mechanisms of the zebra mussel *Dreissena polymorpha*. *Ecology* 77: 1686–1690.

Kaku, V.J., M.C. Boufadel, and A.D. Venosa. 2006. Evaluation of mixing energy in laboratory flasks used for dispersant effectiveness testing. *J. Environ. Eng.* 132: 93–101.

Kerfoot, W.C., F. Yousef, S.A. Green, J.W. Budd, D.J. Schwab, and H.A. Vanderploeg. 2010. Approaching storm: Disappearing winter bloom in Lake Michigan. *J. Great Lakes Res.* 36(Suppl. 3): 30–41.

Koehl, M.A.R. 1996. When does morphology matter? *Ann. Rev. Ecol. Syst.* 27: 501–542.

Kryger, R.J. and H.U. Riisgård. 1988. Filtration rate capacities in 6 species of European freshwater bivalves. *Oecologia* 7: 34–38.

Lacey, R.W.J. and A.G. Roy. 2008. The spatial characterization of turbulence around large roughness elements in a gravel-bed river. *Geomorphology* 102: 542–553.

Levitan, D.R. 1995. Ecology of fertilization in free-spawning invertebrates. In *Ecology of Marine Invertebrate Larvae*. R. McEdward (ed.), pp. 123–156. Boca Raton, FL: CRC Press.

Loewen, M.R., J.D. Ackerman, and P.F. Hamblin. 2007. Environmental implications of stratification and turbulent mixing in a shallow lake basin. *Can. J. Fish. Aquat. Sci.* 64: 43–57.

MacIsaac, H.J., O.E. Johannsson, and J. Ye et al. 1999. Filtering effects of introduced bivalve (*Dreissena polymorpha*) in a shallow lake: Application of a hydrodynamic model. *Ecosystems* 2: 338–350.

Marshall, D.J., C. Styan, and C.D. McQuaid. 2009. Larval supply and dispersal. In *Marine Hard Bottom Communities, Ecological Studies*. Vol. 206. M. Wahl (ed.), pp. 165–176. Berlin, Germany: Springer-Verlag.

Martel, A. 1993. Dispersal and recruitment of zebra mussel (*Dreissena polymorpha*) in a nearshore area in west-central Lake Erie: The significance of postmetamorphic drifting. *Can. J. Fish. Aquat. Sci.* 50: 3–12.

Martel, A., T.M. Hynes, and J. Buckland-Nicks. 1995. Prodissoconch morphology, planktonic shell growth, and size at metamorphosis in *Dreissena polymorpha*. *Can. J. Zool.* 73: 1835–1844.

McNair, J.N. and J.D. Newbold. 2001. Turbulent transport of suspended particles and dispersing benthic organisms: The hitting-distance problem for the Local Exchange Model. *J. Theor. Biol.* 209: 351–369.

Mead, K.S. and M.W. Denny. 1995. The effects of hydrodynamic shear stress on fertilization and early development of the purple sea urchin *Strongylocentrotus purpuratus*. *Biol. Bull.* 188: 46–56.

Metaxas, A. and C.M. Young. 1998. Behavior of echinoid larvae around sharp haloclines: Effects of the salinity gradient and dietary conditions. *Mar. Biol.* 131: 443–459.

Morris, H.M. Jr. 1955. Flow in rough conduits. *Trans. ASCE* 120: 373–398.

Nalepa, T.F., D.L. Fanslow, and S.A. Pothoven. 2010. Recent changes in density, biomass, recruitment, size structure, and nutritional state of *Dreissena* populations in southern Lake Michigan. *J. Great Lakes Res.* 36(Suppl. 3): 5–19.

Nalepa, T.F. and D.W. Schloesser. 1993. *Zebra Mussels: Biology, Impacts, and Control*. Boca Raton, FL: CRC Press.

Nichols, S.J. and M.G. Black. 1994. Identification of larvae: The zebra mussel (*Dreissena polymorpha*), quagga mussel (*Dreissena rosteriformis bugensis*), and Asian clam (*Corbicula fluminea*). *Can. J. Zool.* 72: 406–417.

Nikora, V.I., D.G. Goring, I. McEwan, and G. Griffiths. 2001. Spatially-averaged open-channel flow over a rough bed. *J. Hydraul. Eng.* 127: 123–133.

Nishihara, G.N. and J.D. Ackerman. 2006. The effect of hydrodynamics on the mass transfer of dissolved inorganic carbon to the freshwater macrophyte *Vallisneria americana*. *Limnol. Oceanogr.* 51: 2734–2745.

Nishihara, G.N. and J.D. Ackerman. 2007. On the determination of mass transfer in a concentration boundary layer. *Limnol. Oceanogr. Methods* 5: 88–96.

O'Connor, B.L. and M. Hondzo. 2008. Dissolved oxygen transfer to sediments by sweep and eject motions in aquatic environments. *Limnol. Oceanogr.* 53: 566–578.

Okubo, A., J.D. Ackerman, and D.P. Swaney. 2001. Passive diffusion in ecosystems. In *Diffusion and Ecological Problems: New Perspectives*. A. Okubo and S. Levin (eds.), pp. 31–106. New York: Springer-Verlag.

Peters, F. and C. Marrasé. 2000. Effects of turbulence on plankton: An overview of experimental evidence and some theoretical considerations. *Mar. Ecol. Prog. Ser.* 205: 291–306.

Peyer, S.M., A.J. McCarthy, and C.E. Lee. 2009. Zebra mussels anchor byssal threads faster and tighter than quagga mussels in flow. *J. Exp. Biol.* 212: 2027–2036.

Pineda, J., J.A. Hare, and S. Sponaugle. 2007. Larval transport and dispersal in the coastal ocean and consequences for population connectivity. *Oceanography* 20: 22–39.

Quinn, N.P. and J.D. Ackerman. 2011. The effect of near-bed turbulence on sperm dilution and fertilization success of broadcast spawning bivalves. *Limnol. Oceanogr. Fluids Environ.* 1: 176–193.

Quinn, N.P. and J.D. Ackerman. 2012. Biological and ecological mechanisms for overcoming sperm limitation in invasive dreissenid mussels. *Aquat. Sci.* 74: 415–425.

Quinn, N.P. and J.D. Ackerman. In Review. Effects of near-bed turbulence on settlement and resuspension of freshwater dreissenid mussel larvae. *Freshwat. Biol.*

Rehmann, C.R., J.A. Stoeckel, and D.W. Schneider. 2003. Effect of turbulence on the mortality of zebra mussel veligers. *Can. J. Zool.* 81: 1063–1069.

Riffell, J.A. and R.K. Zimmer. 2007. Sex and flow: The consequences of fluid shear for sperm–egg interactions. *J. Exp. Biol.* 210: 3644–3660.

Riisgård, H.U. and P.S. Larsen. 1995. Minireview: Ciliary filter feeding and bio-fluid mechanics—Present understanding and unsolved problems. *Limnol. Oceanogr.* 46: 882–891.

Ruiz, J., D. Macías, and F. Peters. 2004. Turbulence increases the average settling velocity of phytoplankton cells. *Proc. Nat. Acad. Sci.* 101: 17720–17724.

Schindler, R.J. and J.D. Ackerman. 2010. The environmental hydraulics of turbulent boundary layers. In *Advances in Environmental Fluid Mechanics*. D.T. Mihailovic and C. Gualtieri (eds.), pp. 87–126. London, U.K.: World Scientific.

Schwalb, A.N. and J.D. Ackerman. 2011. Settling velocities of juvenile Lampsilini mussels (Mollusca: Unionidae): The influence of behavior. *J. N. Am. Benthol. Soc.* 30: 702–709.

Serrão, E.A. and J. Havenhand. 2009. Fertilization strategies. In *Marine Hard Bottom Communities. Ecological Studies.* Vol. 206. M. Wahl (ed.), pp. 149–164. Berlin, Germany: Springer-Verlag.

Tuchman, N.C., R.L. Burks, C.A. Call, and J. Smarrelli. 2004. Flow rate and vertical position influence ingestion rates of colonial zebra mussels (*Dreissena polymorpha*). *Freshwat. Biol.* 49: 191–198.

Tweddle, J.F., J.H. Simpson, and C.D. Janze. 2005. Physical controls of food supply to benthic filter feeders in the Menai Strait. *Mar. Ecol. Prog. Ser.* 289: 79–88.

Vanderploeg, H.A., J.R. Liebig, T.F. Nalepa, G.L. Fahnenstiel, and S.A. Pothoven. 2010. *Dreissena* and the disappearance of the spring phytoplankton bloom in Lake Michigan. *J. Great Lakes Res.* 36(Suppl. 3): 50–59.

Vogel, S. 1994. *Life in Moving Fluids*. 2nd edn. Princeton, NJ: Princeton University Press.

Walz, N. 1975. The settlement of larvae of *Dreissena polymorpha* on artificial substrates. *Arch. Hydrobiol. Suppl.* 47: 423–431.

Wildish, D. and D. Kristmanson. 1997. *Benthic Suspension Feeders and Flow*. Cambridge, U.K.: Cambridge University Press.

Wildish, D.J. and A.M. Saulnier. 1993. Hydrodynamic control of filtration in *Placopecten magellanicus*. *J. Exp. Mar. Biol. Ecol.* 174: 65–82.

Wilson, A.B., K.A. Naish, and E.G. Boulding. 1999. Multiple dispersal strategies of the invasive quagga mussel (*Dreissena bugensis*) as revealed by microsatellite analysis. *Can. J. Fish. Aquat. Sci.* 56: 2248–2261.

Wüest, A. and A. Lorke. 2003. Small-scale hydrodynamics in lakes. *Ann. Rev. Fluid Mech.* 35: 373–412.

Young, W.J. 1992. Clarification of the criteria used to identify near-bed flow regimes. *Freshwat. Biol.* 28: 383–391.

PART V

Impacts

CHAPTER 30

Meta-Analysis of Dreissenid Effects on Freshwater Ecosystems

Scott N. Higgins

CONTENTS

Abstract 487
Introduction 487
Methods 489
Results 489
    Physical and Chemical Attributes 489
    Pelagic–Profundal Pathway 490
    Benthic–Littoral Pathway 490
Discussion 491
Acknowledgments 493
References 493

ABSTRACT

A meta-analysis of dreissenid effects on the biogeochemistry, flora, and fauna of invaded lake and river ecosystems in North America and Europe was performed. Dreissenid effects on a large number of ecosystem parameters were common, ecologically relevant, and persistent. Dreissenids reduced mean suspended solids concentrations by ~40% in both lake and river habitats and increased mean Secchi depths by 0.9 and 0.6 m in lakes and rivers, respectively. Mean declines in total phosphorus (−18%) were detected in lakes, but not rivers. In rivers, declines in particulate phosphorus were offset by increases in soluble forms (+430%) and total phosphorus remained unchanged. The direction of dreissenid effects on biota was largely dependent on the respective energy pathway. For the pelagic–profundal pathway, mean declines in phytoplankton chlorophyll a (lake −45%, river −78%), zooplankton biomass (lake −47%, river −76%), and biomass of profundal zoobenthos (lake −57%) between pre- and post-invasion periods were detected. For the benthic–littoral pathway, mean increases in periphyton biomass (+170%), macrophyte coverage (+180%), biomass of native zoobenthos (+58%), and total zoobenthic biomass including dreissenid soft tissues (+520%) were detected. Effects of dreissenids on key ecosystem indicators (Secchi depth, total phosphorus, chlorophyll a) were persistent, with no evidence of diminished size of effects within two decades of establishment.

INTRODUCTION

Translocation of species around the globe is common and the cumulative impact is recognized as one of the leading drivers of biodiversity loss and ecological change within terrestrial, marine, and freshwater ecosystems on a global scale (OTA 1993, Simberloff 1996, Sala et al. 2000). For this reason, and because the introduction of nonnative species can provide information on ecological processes (e.g., rapid evolution, dispersal, species interactions) otherwise difficult to obtain, the study of biological invasions has become an important component in the field of ecology (Sax et al. 2007). Among the most notorious ecological invaders in the world (IUCN 2005) is the fresh- and brackish-water mollusk

Dreissena polymorpha, commonly referred to as the zebra mussel. Zebra mussels are relatively small bivalves, reaching 25–35 mm in length, with a high fecundity (>1 million eggs per spawning event) and high dispersal capacity (Ludyanskiy et al. 1993). Native to the Pontocaspian region of Eastern Europe, this species began its range expansion to Western Europe through the opening of transcontinental canal systems during the mid-eighteenth century (Zhadin 1946, Ludyanskiy et al. 1993). Some 150 years later, facilitated by trans-Atlantic commerce, the zebra mussel and a closely related species, the quagga mussel (*Dreissena rostriformis bugensis*), joined a list of >180 nonnative species introduced into the Laurentian Great Lakes of North America. Currently, *D. polymorpha* appears more widely dispersed than *D. rostriformis bugensis*, with reports of established populations across most European nations and within >700 lake and river ecosystems in North America (Benson 2013). While the distribution of the quagga mussel is less well known in Europe (but see bij de Vaate et al. 2013), it has invaded >40 lake and river ecosystems in North America, including many systems in the southwest United States (Benson 2013).

While small in size, high reproductive output and subsequent dispersion within ecosystems can lead to high areal densities (>10,000 m^{-2}) of dreissenid mussels. In some cases, estimates of filtration capacity (FC) suggest that populations can filter an equivalent volume of an entire water body several times in a single day and thus exert considerable influence over particle concentrations, including phytoplankton, in the water column (Higgins and Vander Zanden 2010). In addition, several studies conducted at the ecosystem scale have documented effects on a wide range of physical, chemical, and biological parameters and processes (Karatayev and Burlakova 1995, Nalepa and Fahnenstiel 1995, Idrisi et al. 2001, Strayer 2009).

Quantification of direct and indirect effects of dreissenids on ecosystems is hindered by at least five constraints. First, due to differences in food web constituents (e.g., predators or diseases), dreissenid population densities and their effects may differ substantially between native and invaded ranges. Second, small-scale experiments (e.g., enclosure or bottle experiments) often lack the appropriate spatial and temporal scales, excluding important ecosystem constituents, interactions, and feedbacks (Schindler 1998). Thus, in order to understand and quantify ecological ramifications of dreissenid invasions, studies must include relevant spatial and temporal scales within natural ecosystems in the invaded range. Third, effects of dreissenids may be less (even much less) than natural variability occurring on seasonal or other temporal cycles (e.g., daily, intra-annual). Fourth, it would be relatively easy to attribute a shift in a particular response variable, for example, soluble nutrient concentrations, to dreissenids when in fact it was partly or wholly associated with some other driver (e.g., variations in loading). For this reason, causation cannot be statistically ascertained without controlled studies. Fifth, due to difficulty of containment, it is not anticipated that purposeful introductions of these species for controlled whole ecosystem experiments will be widely permitted for research purposes in systems not already invaded. While these obstacles are problematic, there are several potential approaches to quantify and evaluate effects including long-term monitoring studies, paleo-limnological studies, space-for-time substitutions, and meta-analyses.

Here, I employ a meta-analysis approach to quantify the direction, magnitude, and variance of dreissenid effects to commonly studied ecosystem parameters and processes within naturally invaded lake and river ecosystems and, for comparative purposes, within small-scale experiments. This approach cannot statistically assign causation to changes in ecological parameters between pre- and post-invasion periods. Instead, attributing causation to dreissenids must be accomplished through knowledge of dreissenid ecology and the likelihood that other factors were responsible. One of the benefits of a meta-analysis approach is that it overcomes the idiosyncrasies of individual studies because as the number of studies included in the meta-analysis increases, the likelihood that other factors were responsible for changes in ecosystem attributes is reduced. Another benefit of the approach is the ability to examine variation in the magnitude of dreissenid effects between studies, which provides a better understanding of driving mechanisms and some predictive capacity. No less important, such an approach helps identify gaps in existing knowledge where future studies would be most insightful.

My broad hypothesis is that dreissenid effects on major trophic groups in lakes and rivers are structured according to the framework proposed in Figure 30.1. This framework proposes that the direct effects of dreissenid mussel filtration on water column particles (including phytoplankton) and excretion of particulate and soluble nutrients near the sediment–water interface will alter the flux of energy and nutrients within invaded systems and that such changes will be detectable and ecologically relevant. While fluxes of energy and nutrients are often difficult to measure directly, changes will be evident by the restructuring of biota within the two dominant energy pathways in freshwater ecosystems: the pelagic–profundal pathway and the benthic–littoral pathway (Figure 30.1). The direction of effects on biota will be broadly dependent on the respective energy pathway, with decreases in total biomass or abundance of major trophic groups within the pelagic–profundal energy pathway and increases in major trophic groups within the benthic–littoral pathway (Figure 30.1). The magnitude of effects will be largely dependent on the FC of the mussel population and the degree of hydrodynamic mixing (i.e., access of the mussel population to the water column). Effects on higher trophic levels (i.e., planktivorous, benthivorous, and piscivorous fish) will depend on the magnitude of effects on underlying pathways and their ability to access resources from either pathway.

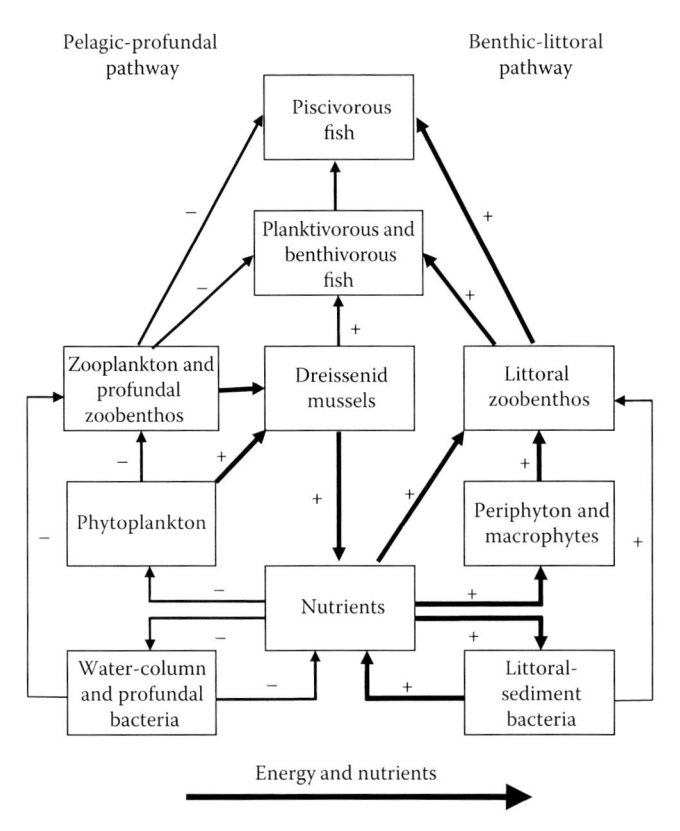

Figure 30.1 Proposed framework for the restructuring of dominant energy and nutrient pathways in freshwater ecosystems after dreissenid invasion. Arrows represent the direction of flow. Bold lines and plus symbols (+) represent an increased flux, and a minus symbol (−) represents a reduced flux.

METHODS

Details of the meta-analysis approach utilized here are described in detail in Higgins and Vander Zanden (2010). In general, the peer-reviewed literature, governmental reports, and on-line databases were searched for studies that contained water quality and food-web information from ecosystems both before and after invasion. Studies included lake and river ecosystems from both North America and Europe. Data from controlled enclosure studies (mesocosms, bottle incubations, etc.) were also included but were treated separately from natural systems. Compiled information from all studies was sorted, assessed for relevance, and recorded in a standardized manner. In order to assess the effect size, an unweighted statistic commonly referred to as the log response ratio (LR) was used. The LR was calculated as $LR = \log_e (X_{+D}/X_{-D})$, where X_{-D} and X_{+D} refer to mean parameter values during the pre- and post-invasion periods, respectively. The significance of effect ($p \leq 0.10$) was tested using a two-tailed t test on the population of LR values using the statistical software R (version 2.7.2). To limit problems associated with non-independence, only one effect size for a given parameter (e.g., Secchi depth) was included for each ecosystem. Such an approach ensured that all ecosystems in the study were equally weighted. If multiple studies on a single ecosystem occurred, effect sizes were calculated using the study with the most robust dataset. Due to a lack of sufficient data on the response of fish populations, a quantitative meta-analysis on these higher trophic levels was not attempted. Instead, physiological and behavioral indicators and the population structure from well-studied ecosystems were qualitatively evaluated against predicted effects. Overall, the analysis included data from 57 lakes, 11 rivers, and 18 enclosure experiments. In some cases, embayments and basins of larger lakes were also included and considered independently from main lake basins. The dataset was continuously updated as new data became available and results here represent an update from previous efforts (Higgins and Vander Zanden 2010, Higgins et al. 2011). Where time series of dreissenid effects were assessed, they were standardized using Z-scores. Z-scores were calculated from each ecosystem independently, using the mean and variance of parameter values during the pre-invasion period. Studies with <3 years of preinvasion data were excluded from the analysis. In all cases mean parameter values were calculated and reported as the geometric mean.

RESULTS

Physical and Chemical Attributes

Following dreissenid invasion, the concentration of suspended solids (SS) declined significantly in lake and river habitats and within enclosure experiments (Figure 30.2). Mean declines in SS were approximately 40% between pre- and post-invasion periods for both lake and river habitats and approximately 60% in enclosure experiments. In response to reductions in SS, water transparency (i.e., Secchi depth) increased significantly in both lake and river habitats. On average, Secchi depth increased by 0.9 m in lake habitats and 0.6 m in river habitats. In lakes, water transparency generally remained above preinvasion values for the two decades where data were available (Figure 30.3). In some cases, particularly in systems without seasonal stratification (mixed), water transparency increased to values 5–15 standard deviations above preinvasion values.

Responses of water-column nutrient concentrations to dreissenid invasion were variable in direction and magnitude (Figure 30.2). Chloride concentration, which was used to interpret potential changes in anthropogenic loadings, displayed no significant changes over the invasion period in lakes, while insufficient data did not allow assessments in rivers and enclosures. Dissolved nitrogen (NO_3 and NH_3) displayed no significant changes in lakes or rivers (NH_3 only) but increased significantly

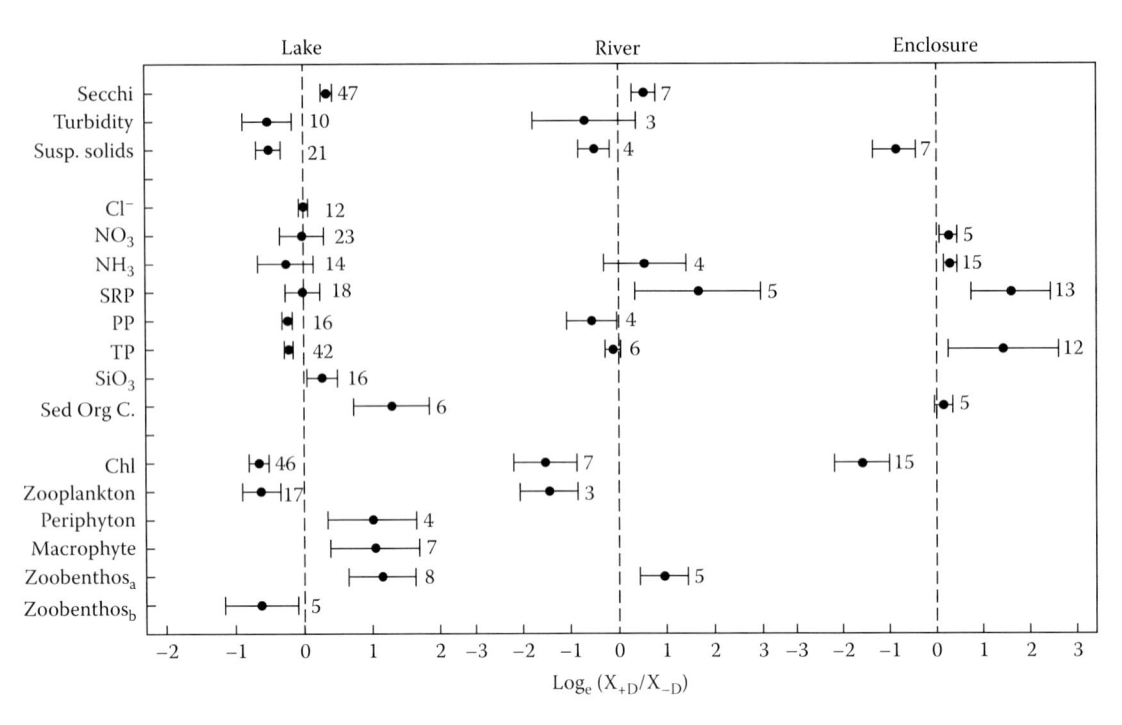

Figure 30.2 The magnitude of dreissenid effects on physical, chemical, and biological parameters in natural lake and river ecosystems and within enclosure or small-scale experiments. Values are the LRs (see methods) of parameters during pre- (X_{-D}) and post-invasion (X_{+D}) periods; error bars represent 90% confidence intervals. Effects were considered significant when the confidence intervals did not include zero (dashed vertical line). Subscripts for zoobenthos refer to the biomass of organisms collected from littoral habitat, a; or profundal habitats, b.

in enclosure experiments (Figure 30.2). In lakes, mean particulate phosphorus (PP) declined by ~21%, mean soluble reactive phosphorus (SRP) did not change significantly, and mean total phosphorus (TP) declined by ~18% (Figure 30.2). The declines in TP were persistent, with post-invasion values remaining within 0–5 standard deviations below preinvasion mean values for the two decades of available data (Figure 30.3). In rivers, while mean PP declined significantly (~43%) over the invasion period, mean SRP increased from 5.7 µg L^{-1} preinvasion to 30.2 µg L^{-1} (i.e., +430%), and water column TP was not significantly affected. The mean concentration of dissolved silica (SiO$_3$), a limiting nutrient for some algal species (e.g., diatoms), increased by ~38% between pre- and post-invasion periods (Figure 30.2).

Pelagic–Profundal Pathway

Significant declines in chlorophyll *a* (chl *a*), the ubiquitous algal pigment used to estimate phytoplankton biomass, were found within lake and river ecosystems and within enclosure experiments (Figure 30.2). Declines in mean chl *a* concentrations between pre- and post-invasion periods were 45%, 78%, and 80% within lakes, rivers, and enclosure experiments, respectively. Chl *a* values during the postinvasion period displayed no evidence of returning to preinvasion levels even two decades after invasion (Figure 30.3) for lakes with and without seasonal stratification.

Total zooplankton biomass declined significantly in both lake and river ecosystems following invasion (Figure 30.2). Declines were similar in magnitude to those of chl *a*, corresponding to a decline in mean zooplankton biomass by 47% and 76% for lake and river ecosystems, respectively. Insufficient data were available to test effects within enclosure experiments. The change in total zooplankton biomass was similar in magnitude and direction to the change in chl *a* (Figure 30.2). The total biomass of native zoobenthos (excluding dreissenid mussel tissues) in the profundal habitats of lakes declined by ~57%. Dreissenid effects on zoobenthos in profundal habitats of rivers were not evaluated due to a lack of sufficient data.

Benthic–Littoral Pathway

Although studies of changes in periphyton were limited, there was a significant increase in mean periphyton biomass (~170% of preinvasion levels) within the littoral zones of lakes (Figure 30.2). Similarly, areal coverage of macrophytes increased by approximately 180% (Figure 30.2). Based on equations that linked maximum depth colonized by macrophytes (i.e., depth of the littoral zone) and water clarity (Higgins and Vander Zanden 2010) and changes to water clarity between pre- and post-invasion periods, the depth of the littoral zone increased by ~1.1 and ~0.6 m across lake and river ecosystems, respectively. Following invasion, the abundance of bacteria increased dramatically (~2000%) in sediments within, or in close proximity to,

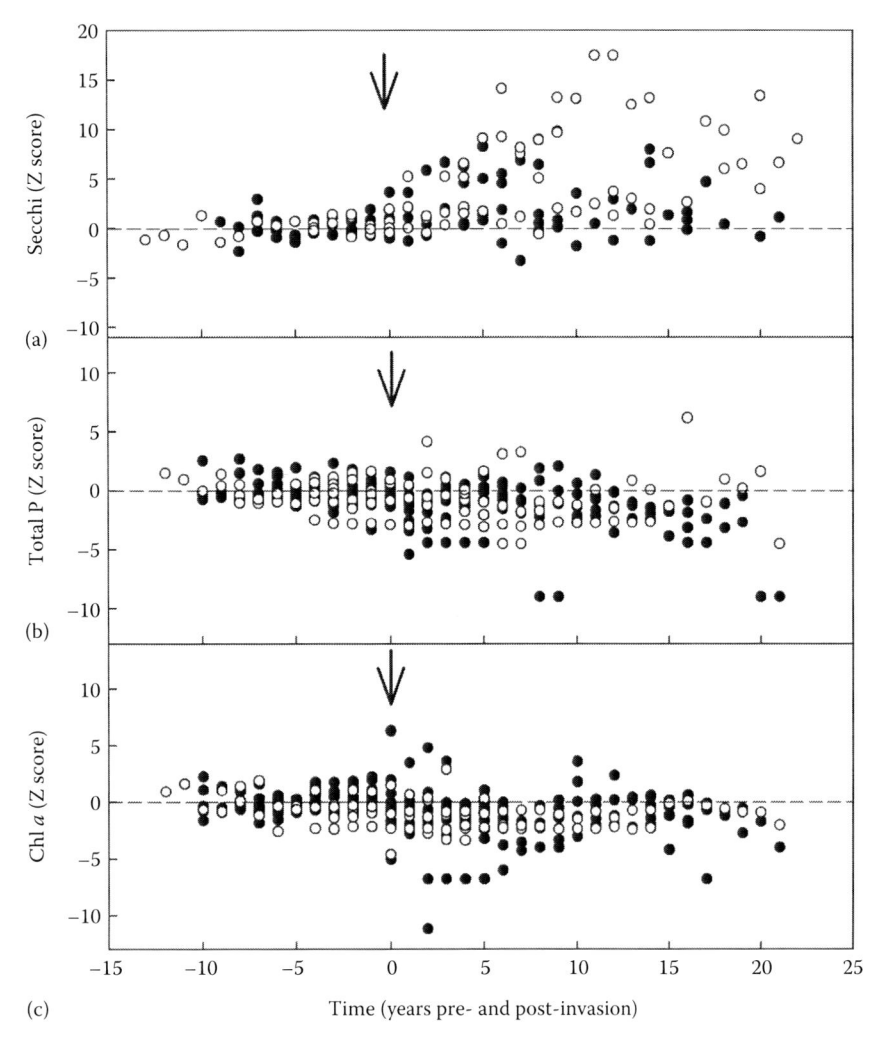

Figure 30.3 Long-term effects of dreissenid mussels on: (a) water clarity (Secchi-depth), (b) total phosphorus, and (c) chlorophyll a (chl a) in lake ecosystems. Z-scores for each lake are based on means and standard deviations of parameter values during the preinvasion period. Closed circles represent lakes that stratify during summer months and open circles represent non-stratified (mixed) lakes. Horizontal dashed lines represent the mean parameter value for each lake during the preinvasion period and arrows represent year of invasion.

dreissenid mussel beds (Higgins and Vander Zanden 2010). Data for changes in bacterial abundance in sediments outside of dreissenid beds or in profundal habitats were not available for assessment. While biomass of native zoobenthos in profundal habitats declined by ~47% following dreissenid invasion, biomass in the littoral increased by ~58% (Figure 30.2). Including dreissenid soft tissues, total zoobenthic biomass in littoral habitats of lakes increased by ~520% of preinvasion values (Figure 30.2), and dreissenid soft tissues accounted for ~80% of total zoobenthic biomass during the post-invasion period (data not shown).

DISCUSSION

During the last decade, improvements to the accessibility of long-term datasets and an ever-growing number of peer-reviewed scientific studies on *D. polymorpha* and *D. rostriformis bugensis* have dramatically improved our understanding of the important role these species play in freshwater ecosystems. Evaluated collectively, these data strongly support the hypothesis that the establishment of these species commonly results in dramatic restructuring of food webs and energy/nutrient fluxes within lake and river ecosystems across their invaded ranges.

Despite considerable variation in the effect sizes for most ecosystem attributes, consistent patterns were apparent. First, significant responses of physical, chemical, and biotic attributes of ecosystems to dreissenid establishment were common and ecologically relevant. Second, evidence from the temporal response of several important ecological indicators (Secchi depth, TP, chl *a*) demonstrated that dreissenid effects were persistent; there was no evidence of diminishing effect sizes within two decades of dreissenid establishment. Third, at the ecosystem scale, the collective response of biota within each trophic level was dependent on the respective energy pathway (Figure 30.1) to which they belonged. Significant declines in total biomass of biota contained within the pelagic–profundal energy pathway (phytoplankton, zooplankton, and profundal zoobenthos) and significant increases in the total biomass of biota contained within the benthic–littoral pathway (sediment bacteria, periphyton, macrophytes, littoral zoobenthos) were a consistent response to dreissenid establishment. Similar effects on phytoplankton and zooplankton biomass were found in an independent analysis using a space-for-time substitution approach to compare 25 stratified lakes with dreissenids to 25 stratified lakes without dreissenids (Kissman et al. 2010). Fourth, the magnitude of effects on components of the pelagic–profundal pathway appeared to be strongly influenced by the collective FC of the dreissenid population. The FC is a function of population density, ecosystem size, hydrodynamics, and factors affecting filtration rates such as temperature, water velocity, turbidity, and food quality (Ackerman et al. 2001, Vanderploeg et al. 2002, Higgins and Vander Zanden 2010). Often, insufficient information on mussel densities and hydrodynamics prevent reliable estimates of FC. In the absence of this information, it is reasonable to suspect that smaller ecosystems such as rivers or shallow lakes, or littoral zones and embayments of large lakes (Hecky et al. 2004), would be more strongly affected by dreissenids than the pelagic zones of larger lake ecosystems. Available data support this prediction. Effects on water clarity, phytoplankton biomass, and zooplankton biomass were largest in rivers, followed by littoral habitats of lakes, then pelagic habitats of lakes (Higgins and Vander Zanden 2010).

Consistent patterns of dreissenid effects were not always apparent between lake and river ecosystems, or between these natural systems and enclosure experiments. An important inconsistency was how dreissenid invasion affected fluxes of the macro-nutrients nitrogen and phosphorus. In both lakes and rivers, PP declined following invasion. However, in lakes, mean SRP did not change following invasion but mean TP declined. These results suggest a significant quantity of phosphorus was directed from the pelagic–profundal pathway to the benthic–littoral pathway or was lost to permanent burial. In contrast, mean SRP in rivers increased following dreissenid invasion, from ~6 μg P L^{-1} preinvasion to ~30 μg P L^{-1} post-invasion, and TP was not affected. These results suggest that, in contrast to the response in lakes, phosphorus in rivers was not as efficiently buried or channeled to the benthic–littoral pathway. Further, the large increase in mean SRP in rivers indicated that benthic and pelagic productivity was not generally limited by phosphorus availability after dreissenid invasion. In lakes, however, phosphorus availability continued to play an important role in controlling phytoplankton and benthic algal productivity after dreissenid invasion (Higgins et al. 2008, 2011). In enclosure experiments, SRP also increased within dreissenid treatments relative to controls, as did soluble nitrogen species (NO_3, NH_3). However, in lakes, there was no evidence to indicate that NO_3 or NH_3 was altered by dreissenids. In rivers, NH_3 generally increased following invasion, but due to high intersystem variability and low sample size, the change was not significant. Further, there was insufficient data to assess effects on NO_3. More data on the response of dissolved and total nitrogen from rivers will be required before statistical relevance can be ascertained. Given the current information, dreissenid effects derived from enclosure experiments tend to overestimate effects on phytoplankton and poorly reflect nutrient cycling dynamics occurring in lake ecosystems.

Not all dreissenid effects are well documented, and further effort will be required to address data gaps before a more complete understanding of the broad ecosystem level changes associated with dreissenid invasions can be achieved. For example, studies on the response of sediment bacteria, periphyton, and zoobenthos in littoral habitats were generally restricted to microhabitats in, or in close proximity to, dreissenid mussel beds. Further, the experimental design of many studies on changes to the zoobenthos did not include periphyton and macrophytes, suggesting that potential effects on algal grazers might be excluded. Effects of dreissenids on biota outside of dreissenid beds were probable, given widespread increases in water clarity and macrophyte coverage that provide habitat and resources. These specific effects need to be quantified before overall effects can be evaluated at the ecosystem scale. Another important omission of this study was a quantitative analysis of how fish communities responded to dreissenid invasion. Given the potential for large changes to their forage base and habitat (e.g., macrophyte coverage), large changes in fish behavior, reproductive capacity, and community structure might be anticipated. Potential effects on fish communities are dependent on the degree to which their forage base is affected and their ability to adjust to such changes. Since the majority of fish species are capable of accessing resources from benthic or pelagic energy pathways (Hecky and Hesslein 1995, Vander Zanden et al. 2006), changes in the behavior and diet of planktivores and deepwater benthivores would be expected in systems where the pelagic–profundal pathway was strongly affected. Further, increases in growth rates and abundance of species that specialize in obtaining resources from benthic–littoral pathways would be expected. Information from systems with long-term datasets qualitatively support these hypotheses. In the Hudson River, for example, increased growth rates (+12%) and abundance (+97%) of fish classified as littoral specialists occurred following dreissenid invasion, while species classified as pelagic specialists declined in terms of growth (−17%) and abundance (−28%) (Strayer et al. 2004). These effects of dreissenids on pelagic specialists were much smaller than reported changes in phytoplankton (approx. −85%) and zooplankton (approx. −70%) biomass (Caraco et al. 1997, Pace et al. 1998) in the Hudson River, suggesting that these fish species may have increased their reliance on benthic–littoral resources. Similar responses of the fish community to dreissenids also occurred in Lake Oneida, New York, and the Bay of Quinte, Lake Ontario (Hoyle et al. 2008, Miehls et al. 2009). Direct evidence of fish populations switching resource pathways has been demonstrated for lake whitefish (*Coregonus clupeaformis*) in Lake Huron (Rennie et al. 2009). Following invasion, diets and isotopic signatures for carbon in lake whitefish shifted from profundal zoobenthos to littoral zoobenthos. Despite the potential for increased reliance on benthic–littoral pathways, however, growth rates of lake whitefish declined in Lake Michigan, Lake Huron, and in the Bay of Quinte, Lake Ontario, following dreissenid invasion (Hoyle et al. 2003, 2008, Pothoven and Madenjian 2008). The degree to which these examples reflect broad patterns occurring in lake and river systems is currently unknown, and more information will be required before a quantitative analysis can be performed using methods described here.

Well-studied indicators of ecosystem condition for lakes (Secchi depth, TP, chl *a*) indicate that the effects of dreissenids are persistent, with no evidence of diminishing effect sizes within two decades following invasion. Since dreissenid effects on pelagic response variables are expected to be related to their FC, this result indicates that dreissenid densities remained sufficiently high over the two decades to maintain effect sizes. Another implication of the importance of mussel FC on the magnitude of effects relates to the potential differences in population densities for zebra mussels and quagga mussels. In the Laurentian Great Lakes, the replacement of zebra mussels by quagga mussels was associated with large increases in total ecosystem densities and filtration capacities since the latter species colonized deeper and colder waters, and softer substrates, to a much greater degree than the former species (Patterson et al. 2005, Vanderploeg et al. 2010, Evans et al. 2011). If this pattern were true for smaller systems across North America and Europe, which are currently dominated by zebra mussels (Benson this issue), any transition toward quagga mussels has the potential to result in greater effect sizes.

Overall, a considerable amount of information is available to assess the ecological effect of dreissenid invasions on lake and river ecosystems. Analyzed collectively, these data indicate that dreissenid invasions commonly redirect energy and nutrients from pelagic–profundal pathways to benthic–littoral pathways. In response, biotic communities associated with these changed pathways are restructured. Typically, total biomass of biota within the pelagic–profundal pathway declines and total biomass of biota within the benthic–littoral pathway increases with the magnitude of these effect sizes dependent on the population density and FC of dreissenids. Significant changes to fish behavior, growth rates, and reproduction following dreissenid invasion are apparent, but the current number of studies is limited. More data are required before a quantitative analysis can be conducted using the approaches described here.

ACKNOWLEDGMENTS

This research was funded by an NSERC postdoctoral fellowship to S. Higgins. Thank you to L. Joppa for data collection, M.J. Vander Zanden for assistance with earlier versions of this manuscript, and M. Rennie for helpful comments. Thank you also to the two anonymous reviewers for their thoughtful and insightful comments and attention to detail.

REFERENCES

Ackerman, J. D., M. R. Loewen, and P. F. Hamblin. 2001. Benthic-Pelagic coupling over a zebra mussel reef in western Lake Erie. *Limnol. Oceanogr.* 46: 892–904.

Benson, A. J. 2013. Chronological history of zebra and quagga mussels (*Dreissenidae*) in North America, 1988–2010. In *Quagga and Zebra Mussels: Biology, Impacts, and Control*, 2nd Edn., T.F. Nalepa and D.W. Schloesser, eds., pp. 9–31. Boca Raton: CRC Press.

bij de Vante, A., G. vander Valde, R. S. E. W. Leuven, and K. C. M. Heilder. 2013. Spread of the quagga mussel, *Dreissena rostriformis bugensis* in western Europe. In *Quagga and Zebra Mussels: Biology, Impacts and Control*, 2nd Edn., T.F. Nalepa and S.W. Schloesser, eds., pp. 83–92. Boca Raton, FL: CRC Press.

Caraco, N. F., J. J. Cole, P. A. Raymond et al. 1997. Zebra mussel invasion in a large, turbid river: Phytoplankton response to increased grazing. *Ecology* 78: 588–602.

Evans, M. A., G. L. Fahnenstiel, and D. Scavia. 2011. Incidental oligotrophication of North American Great Lakes. *Environ. Sci. Technol.* 45: 3297–3303.

Hecky, R. E. and R. H. Hesslein. 1995. Contributions of benthic algae to lake food webs as revealed by stable isotope analysis. *J. N. Am. Benthol. Soc.* 14: 631–653.

Hecky, R. E., R. E. H. Smith, D. R. Barton et al. 2004. The nearshore phosphorus shunt: A consequence of ecosystem engineering by dreissenids in the Laurentian Great Lakes. *Can. J. Fish. Aquat. Sci.* 61: 1285–1293.

Higgins, S. N., S. Y. Malkin, E. T. Howell et al. 2008. An ecological review of *Cladophora glomerata* (Chlorophyta) in the Laurentian Great Lakes. *J. Phycol.* 44: 839–854.

Higgins, S. N. and M. J. Vander Zanden. 2010. What a difference a species makes: A meta-analysis of dreissenid mussel impacts on freshwater ecosystems. *Ecol. Monogr.* 80: 179–196.

Higgins, S. N., M. J. Vander Zanden, L. N. Joppa, and Y. Vadeboncoeur. 2011. The effect of dreissenid invasions on chlorophyll and the chlorophyll:total phosphorus ratio in north-temperate lakes. *Can. J. Fish. Aquat. Sci.* 68: 319–329.

Hoyle, J. A., J. N. Bowlby, and B. J. Morrison. 2008. Lake whitefish and walleye population responses to dreissenid mussel invasion in eastern Lake Ontario. *Aquat. Ecosyst. Health Manage.* 11: 403–411.

Hoyle, J. A., J. M. Casselman, R. Dermott, and T. Schaner. 2003. Resurgence and decline of lake whitefish (*Coregonus clupeaformis*) stocks in eastern Lake Ontario, 1972–1999. In *State of Lake Ontario: Past, Present and Future*, M. Munawar, ed., pp. 149–164. Burlington, VT: World Monograph Series.

Idrisi, N., E. L. Mills, L. G. Rudstam, and D. J. Stewart. 2001. Impact of zebra mussels (*Dreissena polymorpha*) on the pelagic lower trophic levels of Oneida Lake, New York. *Can. J. Fish. Aquat. Sci.* 58: 1430–1441.

IUCN. 2005. 100 of the world's worst invaders. Available from http://www.issg.org/database/species/search.asp?st=100ss [accessed March 3, 2009].

Karatayev, A. Y. and L. E. Burlakova. 1995. The role of *Dreissena* in lake ecosystems. *Russ. J. Ecol.* 26: 207–211.

Kissman, C. E. H., L. B. Knoll, and O. Sarnelle. 2010. Dreissenid mussels (*Dreissena polymorpha* and *Dreissena bugensis*) reduce microzooplankton and macrozooplankton biomass in thermally stratified lakes. *Limnol. Oceanogr.* 55: 1851–1859.

Ludyanskiy, M. L., D. McDonald, and D. Macneill. 1993. Impact of the Zebra Mussel, a bivalve invader—*Dreissena polymorpha* is rapidly colonizing hard surfaces throughout waterways of the United-States and Canada. *Bioscience* 43: 533–544.

Miehls, A. L. J., D. M. Mason, K. A. Frank, A. E. Krause, S. D. Peacor, and W. W. Taylor. 2009. Invasive species impacts on ecosystem structure and function: A comparison of Oneida Lake, New York, USA, before and after zebra mussel invasion. *Ecol. Model.* 220: 3194–3209.

Nalepa, T. F. and G. L. Fahnenstiel. 1995. Preface—*Dreissena polymorpha* in the Saginaw Bay, Lake Huron ecosystem: Overview and perspective. *J. Great Lakes Res.* 21: 411–416.

OTA. 1993. *Harmful Nonindigenous Species in the United States.* U.S. Congress, Washington, DC.

Pace, M. L., S. E. G. Findlay, and D. Fische. 1998. Effects of an invasive bivalve on the zooplankton community of the Hudson River. *Freshwat. Biol.* 39: 103–116.

Patterson, M. W. R., J. J. H. Ciborowski, and D. R. Barton. 2005. The distribution and abundance of *Dreissena* species (Dreissenidae) in Lake Erie, 2002. *J. Great Lakes Res.* 31: 223–237.

Pothoven, S. A. and C. P. Madenjian. 2008. Changes in consumption by alewives and lake whitefish after dreissenid mussel invasions in Lakes Michigan and Huron. *N. Am. J. Fish. Manage.* 28: 308–320.

Rennie, M. D., W. G. Sprules, and T. B. Johnson. 2009. Resource switching in fish following a major food web disruption. *Oecologia* 159: 789–802.

Sala, O. E., F. S. Chapin, J. J. Armesto et al. 2000. Biodiversity—Global biodiversity scenarios for the year 2100. *Science* 287: 1770–1774.

Sax, D. F., J. J. Stachowicz, J. H. Brown et al. 2007. Ecological and evolutionary insights from species invasions. *Trends Ecol. Evol.* 22: 465–471.

Schindler, D. W. 1998. Replication versus realism: The need for ecosystem-scale experiments. *Ecosystems* 1: 323–334.

Simberloff, D. 1996. Impacts of introduced species in the United States. *Consequences* 2: 13–24.

Strayer, D. L. 2009. Twenty years of zebra mussels: Lessons from the mollusk that made headlines. *Front. Ecol. Environ.* 7: 135–141.

Strayer, D. L., K. A. Hattala, and A. W. Kahnle. 2004. Effects of an invasive bivalve (*Dreissena polymorpha*) on fish in the Hudson River estuary. *Can. J. Fish. Aquat. Sci.* 61: 924–941.

Vanderploeg, H. A., J. R. Liebig, T. F. Nalepa, G. L. Fahnenstiel, and S. A. Pothoven. 2010. *Dreissena* and the disappearance of the spring phytoplankton bloom in Lake Michigan. *J. Great Lakes Res.* 36: 50–59.

Vanderploeg, H. A., T. F. Nalepa, D. J. Jude et al. 2002. Dispersal and emerging ecological impacts of Ponto-Caspian species in the Laurentian Great Lakes. *Can. J. Fish. Aquat. Sci.* 59: 1209–1228.

Vander Zanden, M. J., S. Chandra, S. K. Park, Y. Vadeboncoeur, and C. R. Goldman. 2006. Efficiencies of benthic and pelagic trophic pathways in a subalpine lake. *Can. J. Fish. Aquat. Sci.* 63: 2608–2620.

Zhadin, V. I. 1946. Travelling mussel *Dreissena*. *Priroda* 5: 29–37 (In Russian).

CHAPTER 31

Effects of Invasive Quagga Mussels (*Dreissena rostriformis bugensis*) on Chlorophyll and Water Clarity in Lakes Mead and Havasu of the Lower Colorado River Basin, 2007–2009

Wai H. Wong, G. Chris Holdren, Todd Tietjen, Shawn Gerstenberger, Bryan Moore, Kent Turner, and Doyle C. Wilson

CONTENTS

Abstract 495
Introduction 495
Study Site 497
Methods 497
    Statistics 498
Results 499
    Quagga Mussel Density 499
    Chlorophyll and Water Clarity 499
Discussion 500
Acknowledgments 505
References 505

ABSTRACT

Quagga mussels (*Dreissena rostriformis bugensis*) were found in the Boulder Basin of Lake Mead (Nevada–Arizona) in the lower Colorado River Basin on January 6, 2007, and the population continued to grow from 2007 through 2009. This study focused on ecological impacts of quagga mussels in Lake Mead, which is the largest reservoir by water volume in the United States. Annual chlorophyll *a* concentrations decreased significantly during the post-quagga period (2007–2009) compared to the pre-quagga period (2002–2006) in Boulder Basin (−45%) and Virgin Basin (−20%), but not in Las Vegas Bay, Overton Arm, or Gregg Basin. Water clarity increased significantly in Boulder Basin (+13%) in the post-quagga period, but not in other basins in Lake Mead. Although chlorophyll a concentration's and water clarity changed in the post-quagga period, falling water levels and changes in nutrient inputs from wastewater treatment plants are also likely causes for these observations.

INTRODUCTION

Zebra mussels (*Dreissena polymorpha*; Pallas, 1771) and quagga mussels (*Dreissena rostriformis bugensis*; Andrusov 1897) were accidently introduced into the Laurentian Great Lakes in North America in the 1980s (May and Marsden 1992, Mills et al. 1993, Rosenberg and Ludyanskiy 1994, Spidle et al. 1994, Therriault et al. 2004, Carlton 2008). These invasive mussels have created severe ecological, recreational, and economic impacts on many systems because they are biofoulers and efficient ecological engineers (Nalepa and Schloesser 1993, Strayer et al. 1999, Vanderploeg et al. 2002, Ricciardi 2003, Strayer 2009) that cause a variety of changes such as alteration of phytoplankton species composition and nutrient dynamics (Lavrentyev et al. 2000, Wilson and Sarnelle 2002, Conroy et al. 2005, Zhang et al. 2008, Vanderploeg et al. 2010). These mussels affect other organisms by direct colonization or indirect competition for food and/or space (Karatayev et al. 1997,

Strayer 1999, Strayer et al. 2006, Nalepa et al. 2007, 2009) and increase water clarity by removing suspended particles (e.g., phytoplankton, debris, silt, and microzooplankton) in the water column (MacIsaac et al. 1991, 1995, Padilla et al. 1996, Caraco et al. 1997, Thayer et al. 1997, Pace et al. 1998, Strayer et al. 1998, Jack and Thorp 2000, Vanderploeg et al. 2001, 2009, Noonburg et al. 2003, Wilson 2003, Wong et al. 2003, Qualls et al. 2007, Boegman et al. 2008, Kissman et al. 2010). Benthic algae and plants benefit from increased water clarity (light transmittance) and reactive inorganic nutrients (Hecky et al. 2004). Mussels can also alter morphological and physical properties of their habitat areas and affect availability of resources (Karatayev et al. 1997, Vanderploeg et al. 2002, Hecky et al. 2004). Overall, invasive dreissenid mussels are efficient autogenic and allogenic ecosystem engineers that can directly or indirectly change the ecology of a system they invade. The zebra/quagga mussel has become arguably the most serious nonindigenous biofouling pest introduced into North American freshwater systems (LaBounty and Roefer 2007) and one of the most economically and ecologically important pests in the world (Aldridge et al. 2006). In recent years, the spread of dreissenid mussels has continued in North America (LaBounty and Roefer 2007, Benson 2010a,b, McMahon 2011) and Europe (Aldridge et al. 2004, Molloy et al. 2007, Lucy et al. 2009).

On January 6, 2007, quagga mussels were found in the Las Vegas Boat Harbor within the Boulder Basin of Lake Mead, Arizona–Nevada (LaBounty and Roefer 2007, Stokstad 2007), which is the largest reservoir by volume (3.55×10^{10} m^3 at full pool) in the United States and the second largest in terms of surface area (637 km^2) (Holdren and Turner 2010). It is postulated that quagga mussels were introduced into Lake Mead by a boat from the Great Lakes region. This was the first known occurrence of an established dreissenid population in the western United States and the first known North American quagga mussel infestation of a water body not previously infested by zebra mussels. In early 2007, quagga mussels were found primarily in the Boulder Basin; however, by the end of the year, they had rapidly spread throughout the entire lake and downstream to lakes and reservoirs in the Colorado River system (Figure 31.1). The presence of many artificial waterways for drinking water and irrigation along the Colorado River aqueduct exacerbates the spread of quagga mussel veligers. The rate at which quagga mussels spread throughout the southwest was unprecedented (Wong and Gerstenberger 2011). Based on the experience of dreissenids in the Great Lakes region, we anticipated substantial impacts on ecosystems and the economy (Nalepa and Schloesser 1993, MacIsaac 1996, Mills et al. 1996, Pimentel et al. 2005) of Lake Mead and other reservoirs in the Colorado River system (Wong et al. 2011). For instance, an increase in biomass of many benthic organisms would likely occur (Nalepa 2010). Lake Mead provides favorable environmental conditions, such as warm water, high calcium concentrations, rocky substrates, suitable pH, and sufficient dissolved oxygen (LaBounty and Burns 2005, LaBounty and Roefer 2007), for quagga mussel growth, recruitment, and reproduction.

Based on a dataset collected in the Boulder Basin of Lake Mead from 2002 to 2008, there were no basin-wide changes in chlorophyll *a* concentrations or water clarity (Secchi depth transparency) (Wong et al. 2010). However, the analysis was based on only three sampling stations and only 2 years of post-quagga data. The goals of this chapter were to: (1) determine if quagga mussels had an impact on chlorophyll *a* and water clarity in Lake Mead by examining data collected over a longer period (2002–2009) and over a wider area (20 stations located in all basins); and (2) examine if there were any changes in water clarity in Lake Havasu, which is located within the Colorado River system downstream from Lake Mead and also recently colonized by quagga mussels.

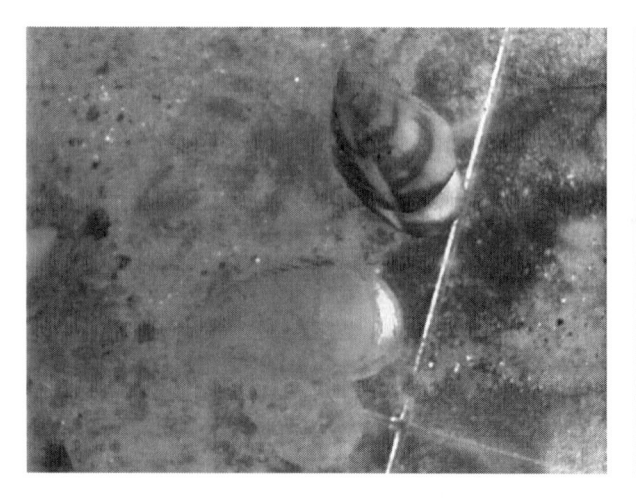

Figure 31.1 (See color insert.) Quagga mussels in Lake Mead. The left image shows a quagga mussel (top individual, 4.2 cm shell length) collected from a shallow area (15 m) near Sentinel Island and a quagga mussel (bottom individual, 4.8 cm in shell length) collected from a deep area (108 m) in the narrows between the Boulder Basin and the Virgin Basin (photo by Wai Hing Wong). The right image shows quagga mussels attached to rocks in a shallow area of Boulder Basin (photo by Bryan Moore).

STUDY SITE

While Lake Mead has traditionally been divided into four basins defined by the Colorado River channel and canyon topography (LaBounty and Burns 2005, Holdren and Turner 2010), to facilitate interpretations of quagga mussel impacts, we divided the lake into five basins: Boulder Basin, Las Vegas Bay, Virgin Basin, Overton Arm, and Gregg Basin (Figure 31.2, Table 31.1). The Colorado River enters Lake Mead at Gregg Basin in the eastern portion of the lake and currently contributes 97% of the lake inflow. Overton Arm receives flows from both the Virgin and Muddy Rivers (1.5% of lake inflow), and these flows combine with the Colorado River flows in the Virgin Basin. Water continues to flow west through the "narrows" and into the Boulder Basin and Las Vegas Bay. Las Vegas Bay is an extension of Boulder Basin and receives water from Las Vegas Wash (1.5% of lake inflow). The wash receives both nonpoint surface and groundwater discharges, runoff from the Las Vegas metropolitan area, and treated effluent from the City of Las Vegas, City of Henderson, and Clark County municipal wastewater treatment facilities (Holdren and Turner 2010). The influx of nutrients from the Las Vegas Wash contributes to higher productivity in Las Vegas Bay and inner Boulder Basin, while the deeper outer portion of this basin is more oligotrophic (LaBounty and Burns 2005). The spatial heterogeneity of these basins can lead to substantial ecological differences relative to quagga mussel impacts.

In recent years, the ecosystem of Lake Mead experienced significant natural and human-caused changes independent of the quagga mussel invasion. For example, lake levels fluctuate yearly, and water surface elevation has been influenced by a severe drought that began in 2000 (Holdren and Turner 2010). A large bloom of the green algae *Pyramiclamys dissecta* occurred in Boulder Basin from early 2001 and persisted through September 2001. The bloom was apparent throughout the entire basin, with chlorophyll *a* peaking at >200 mg/m^3 in the middle and outer portions of the basin (LaBounty and Burns 2005). Algal blooms were apparent downstream in Lake Havasu and as far as a reservoir in San Diego County (about 500 km) and in canals of the Metropolitan Water District of Southern California. Anthropogenic nutrient loading into Las Vegas Wash from wastewater treatment plants has affected chlorophyll concentrations in Las Vegas Bay and Boulder Basin.

METHODS

In the present study, samples were collected for chlorophyll *a* and water clarity measured on a monthly basis from 2002 to 2009 at 20 sites located throughout Lake Mead as part of a lake monitoring program by the Bureau of Reclamation. The monitoring program was initiated to assess potential roles of total phosphorus (TP) loading and water levels that created algal blooms observed in 2001.

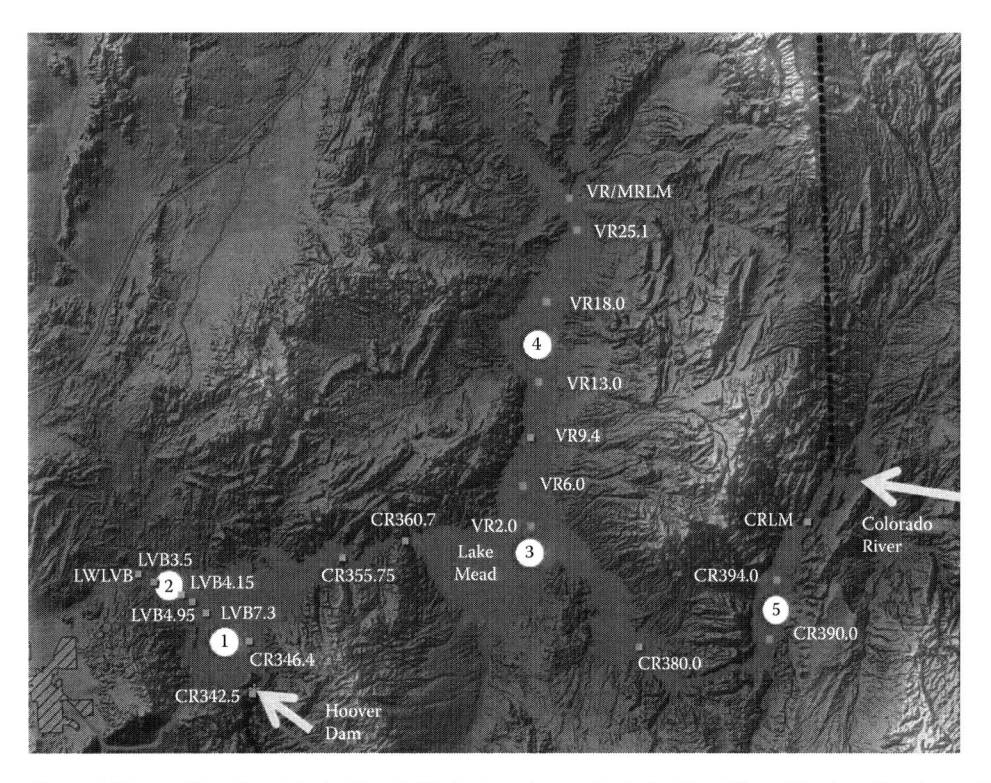

Figure 31.2 Locations of 20 sampling sites in Lake Mead. Circled numbers refer to the five different basins. Exact coordinates are given in Table 31.1.

Table 31.1 Location of 20 Sampling Sites in Each of the Sampled Basins in Lake Mead

Basin	Basin #	Station	Latitude	Longitude
Boulder Basin	1	CR342.5	36° 01' 09.78" N	114° 43' 57.59" W
		CR346.4	36° 03' 41.96" N	114° 44' 20.88" W
		CR355.75	36° 08' 28.97" N	114° 37' 41.66" W
		LVB7.3	36° 05' 29.39" N	114° 47' 14.76" W
Las Vegas Bay	2	LVB3.5	36° 07' 04.92" N	114° 50' 33.42" W
		LVB4.15	36° 07' 00.82" N	114° 49' 49.50" W
		LVB4.95	36° 06' 34.42" N	114° 49' 10.66" W
		LWLVB	36° 07' 04.19" N	114° 50' 56.65" W
Virgin Basin	3	VR2.0	36° 09' 41.01" N	114° 25' 08.83" W
		CR360.7	36° 09' 05.11" N	114° 33' 02.48" W
		CR380.0	36° 02' 48.30" N	114° 16' 24.09" W
Overton Arm	4	VR6.0	36° 12' 50.34" N	114° 25' 03.28" W
		VR9.4	36° 15' 18.17" N	114° 24' 13.20" W
		VR13.0	36° 17' 23.74" N	114° 23' 17.07" W
		VR18.0	36° 21' 36.09" N	114° 23' 15.15" W
		VR25.1	36° 26' 06.90" N	114° 20' 48.07" W
		VR/MRLM	Variable due to the varying lake levels	
Gregg Basin	5	CR390.0	36° 02' 29.59" N	114° 08' 17.24" W
		CR394.0	36° 06' 00.70" N	114° 07' 00.51" W
		CRLM	Variable due to the varying lake levels	

Site locations are given in Table 31.1; three to six sites were located in each of the five major basins. For chlorophyll a, integrated surface water samples (from surface to 5 m) were collected using a pump and flexible hose. Samples were decanted into a sample bottle and, in the laboratory, immediately filtered using a Millipore Corporation manifold and vacuum pump apparatus. Generally, 650 mL of water per sample was filtered through a Whatman GF/C fiberglass filter. Filters were frozen until further analysis. Extraction methods followed standard procedures established by the American Public Health Association (APHA 2005) with acetone as the extracting agent. A spectrophotometer was used to measure absorbance of chlorophyll a, and concentration was calculated based on Method 10200 H (2) described by APHA (2005).

Water clarity was measured with a Secchi disk on each monthly sampling date. All Secchi readings were made using a 1 m long viewscope that was black inside and with a tilted clear plastic base. Use of the viewscope greatly reduced influences of wind (and resulting wave action) and variable light, which influenced readings from time to time (LaBounty 2008). In Lake Havasu, water clarity was measured with a Secchi disk at three locations: Castle Rock, SE; Thompson Bay; and Mid Lake, on a monthly basis from 2000 to 2009 by Arizona Department of Environmental Quality. Sampling years were divided into two periods: pre-quagga (prior to 2007) and post-quagga (2007–2009). Although quagga mussels may have been introduced into Lake Mead in August 2005 (W. B. Wong and B. Moore, unpublished data) or as early as 2003 (McMahon 2011), exponential growth and rapid spread did not occur until 2007 (B. Moore and W. B. Wong). Therefore, impacts, if any, would have been minimal prior to 2007.

Bottom substrates in Lake Mead vary from hard (i.e., rock) to soft (i.e., mud, silt, sand), so a combination of sampling devices/methods was used to estimate mussel densities. On hard substrates, mussels were collected by divers with quadrat frames, while on soft substrates mussels were collected with a Ponar grab (Wong et al. 2011). Samples from hard substrates were placed into a cooler with ice and then frozen at $-20°C$ until mussels were counted, and samples from soft substrates were preserved in 50% ethanol. Samples were not screened, and mussels were counted under a stereomicroscope.

Statistics

T-tests were used to compare annual and seasonal (monthly) differences at each site and basin before (2002–2006) and after (2007–2009) quagga mussels became established in Lake Mead. Differences in annual Secchi depths in Lake Havasu were tested with a t-test to determine if any significant annual changes occurred after 2007. Prior to the t-tests, data were first tested using equality of variances to decide whether the t-test was to be conducted with equal variances or unequal variances. SAS 9.1 (SAS Institute Inc. Cary, NC) was used to perform all statistical analysis. Differences were considered significant at $P \leq 0.05$ (Zar 1996).

RESULTS

Quagga Mussel Density

Adult quagga mussels were first confirmed in the Boulder Basin in early 2007, and by year-end veligers were found throughout the lake (W. H. Wong, unpublished data). Densities on various substrates in the Boulder Basin increased greatly between 2007 and 2009; mean densities increased 14- and 42-fold on rocky (hard) and sand/mud (soft) substrates, respectively, over this 2 year period (Table 31.2). While quantitative estimates of the population on natural substrates in Lake Havasu were not available, at least some data indicated the population had increased in this lake as well (Table 31.2).

Chlorophyll and Water Clarity

Mean annual chlorophyll a and Secchi depths in the pre-quagga (2002–2006) and post-quagga (2007–2009) periods in each of the five basins in Lake Mead are give in Table 31.3. Chlorophyll concentrations were lower in the pre-quagga period at four of the five basins, but differences between periods was only significant (P < 0.05) in the Boulder Basin and Virgin Basin. Overall, mean chlorophyll declined 37.6% in the four basins and ranged from 19.6% (Virgin Basin) to 53.8% (Overton Arm). The only basin where chlorophyll did not decline was Las Vegas Bay where mean chlorophyll actually increased 13.0% in the post-quagga period compared to the pre- quagga period. In the post-quagga period, chlorophyll concentration in Las Vegas Bay was 4.5 times greater than the mean of all the other basins.

Differences in mean Secchi-depth transparency between the pre- and post-quagga periods were not as apparent as for chlorophyll. Secchi-depth transparency increased in two basins and declined in three (Table 31.3). The difference was only significant (P ≤ 0.05) in the Boulder Basin where Secchi-depth transparency increased (+12.7%). For Lake Havasu, the Secchi depth increased by only 6.3% and the difference between 2000–2006 and 2007–2009 was not significant (P > 0.05, Figure 31.3).

When differences between the pre-and post-quagga periods were examined on an monthly basis, significant monthly reductions in chlorophyll concentrations only

Table 31.2 Mean Density of Quagga Mussels on Two Natural Substrate Types in Boulder Basin and Lake Mead (2007 and 2009) and on an Artificial Substrate in Lake Havasu (2008 and 2009) of the Lower Colorado River Basin

Lake	Substrates	Year	Density (mussels/m²)	Standard Deviation	Sites (N)	Increase (%)
Lake Mead[a]	Natural rocky area	2007	624	707	108	1330
	Natural rocky area	2009	8,925	5,994	17	
	Natural sandy/muddy area	2007	80	134	30	4109
	Natural sandy/muddy area	2009	3,350	2,957	3	
Lake Havasu[b]	Artificial snow fence	2008	9,506	8,629	6	120
	Artificial snow fence	2009	20,867	10,777	7	

[a] Wong, W. H. and Moore, B., unpublished data.
[b] Data from Bureau of Land Management.

Table 31.3 Mean (± SD) Chlorophyll a Concentration (mg/m³) and Secchi-Depth Transparency (m) in Different Basins of Lake Mead in the Period before (2002–2006) and after (2007–2009) the Introduction of Quagga Mussels. Differences Were Tested with a Two-Sample t-Test

Variable	Basin	Pre-Quagga Mussel	Post-Quagga Mussel	P-Value
Chlorophyll	Boulder Basin	1.49 ± 0.34	0.82 ± 0.16	0.021
	Las Vegas Bay	7.05 ± 3.10	7.97 ± 2.30	0.676
	Virgin Basin	1.07 ± 0.10	0.86 ± 0.06	0.019
	Overton Arm	7.41 ± 3.77	3.42 ± 1.32	0.136
	Gregg Basin	2.99 ± 1.11	2.03 ± 0.61	0.255
Secchi depth	Boulder Basin	8.48 ± 0.52	9.56 ± 0.72	0.047
	Las Vegas Bay	3.85 ± 0.44	3.28 ± 0.42	0.121
	Virgin Basin	8.92 ± 0.51	8.45 ± 0.28	0.198
	Overton Arm	3.89 ± 1.41	4.81 ± 0.10	0.218
	Gregg Basin	3.06 ± 0.41	2.72 ± 0.41	0.298

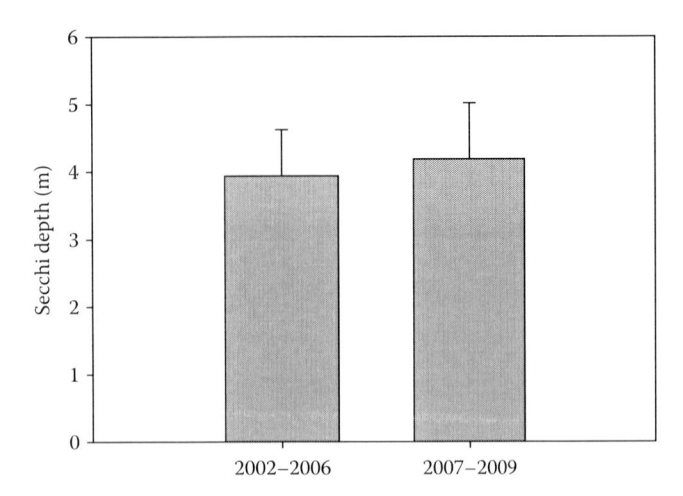

Figure 31.3 Secchi depth transparency (water clarity) in Lake Havasu in the pre-quagga (2000–2006) and post-quagga (2007–2009) periods.

occurred in August and September in the Boulder Basin (P < 0.05), though the concentrations were lower in all other months except February (Figure 31.4A). There were no significant monthly reductions in other basins in any month (Figure 31.4C, E, G, and I). For Secchi depth transparency, a significant increase occurred in July in the Boulder Basin (P < 0.05, Figure 31.4B), and a significant decrease occurred in March in the Gregg Basin (Figure 31.4J), in September in the Las Vegas Bay (Figure 31.4D), and in December in the Virgin Basin (Figure 31.4F).

Monthly chlorophyll *a* data at each of the four sites in the Boulder Basin showed a significant reduction (P < 0.05) in at least one month after quagga mussels became established, especially in the growing seasons from June to September. Sites LVB4.15, CR360.7, and VR9.4 had lower chlorophyll *a* concentrations in January, December, and September (P < 0.05), respectively; while sites LVB3.5 and VR/MRLM had higher chlorophyll *a* concentrations in August and March (P < 0.05), respectively. Mean annual chlorophyll *a* at sites CR342.5, CR355.75, LVB7.3, CR360.7, VR9.4, and VR25.1 showed significantly lower values (P < 0.05) after quagga mussel invaded, while no significant changes were found at the other 14 sites (P > 0.05). Monthly Secchi-depth transparency at sites CR342.5, LVB7.3, and VR13.0 was higher in August, and September (t-test, P < 0.05), respectively, after quagga mussels invaded compared to before, while Secchi-depth transparency at all other stations was either lower or unchanged (P > 0.05).

DISCUSSION

In a previous study, no significant differences in chlorophyll *a* concentrations and water clarity were detected at three sites in the Boulder Basin (CR346.4, LVB 7.3, and LVB4.15) between the pre-quagga period (2002–2006) and the post-quagga period (2007–2008) (Wong et al. 2010). However, with the addition of 2009 data to the analysis, significant reductions of chlorophyll *a* concentrations in the Boulder and Virgin Basins and increased water clarity in the Boulder Basin were apparent. These changes were likely a result of the exponential increase in the quagga mussel population between 2007 and 2009 (Table 31.1). Quagga mussels were first confirmed in the Boulder Basin of Lake Mead in January 2007 (LaBounty and Roefer 2007, Hickey 2010) and gradually spread to upstream basins (i.e., Virgin Basin, Overton Arm, and Gregg Basin) (G. C. Holdren, B. Moore, and W. H. Wong, unpublished data). In this early stage of the invasion, abundances of quagga mussels were higher in the Boulder and Virgin Basins than in upstream basins, and the size of mussels was much greater (W. H. Wong and B. Moore, unpublished data). This is probably the key reason why significant reductions in chlorophyll concentrations were detected in these two basins and not in the others. However, these basins are uniquely different in terms of physiochemical conditions, and differences in mussel filtering rates may also be a contributing factor. Primary production in Lake Mead is primarily driven by phosphorus inputs (Paulson and Baker 1983, LaBounty and Horn 1997, Du 2002, LaBounty and Burns 2005, LaBounty 2008, Holdren and Turner 2010). In Las Vegas Bay where productivity is among the highest in the lake, chlorophyll concentrations have decreased steadily between 1986 and 2009 due to better treated wastewater (i.e., phosphorus reduction) discharged from the Las Vegas Wash (Drury D, personal communication). However, in the period between 2002 and 2009, spanning the pre- and post-quagga mussel invasion (Figure 31.5), there has been no significant change in TP loading to Las Vegas Bay (t-test, P = 0.268). This is probably the reason why no significant decrease in chlorophyll was found in Las Vegas Bay (Figure 31.4C) as primary production in this region is highly affected by TP loads from Las Vegas Wash (LaBounty and Burns 2005, LaBounty 2008).

Dreissenid mussels and most bivalves in general exert top-down pressures on primary production as they are efficient suspension filter feeders (Klerks et al. 1996, Roditi et al. 1996, Baker et al. 1998, Bastviken et al. 1998, Strayer et al. 1999, Diggins 2001, Baldwin et al. 2002, Dionisio Pires et al. 2004). Although the major food resource for dreissenids is phytoplankton, dissolved organic matter (Roditi and Fisher 1999, Baines et al. 2005), bacteria (Silverman et al. 1996a,b), and zooplankton (MacIsaac et al. 1991, 1992, 1995, Pace et al. 1998, Wong et al. 2003, Wong and Levinton 2005) might be utilized as a nutrient source. The high filtration capacity of large dreissenid populations can cause cascading effects on the entire food web. The filtration capacity of the quagga mussel population in Lake Mead was calculated previously based on densities found in 2007 (Wong et al. 2010). It was estimated that the population could filter the water volume of the Boulder Basin in 169 days. Based on the increase in densities in 2009, the population could filter the water volume of Boulder Basin in just 56 days. While filtration capacity was

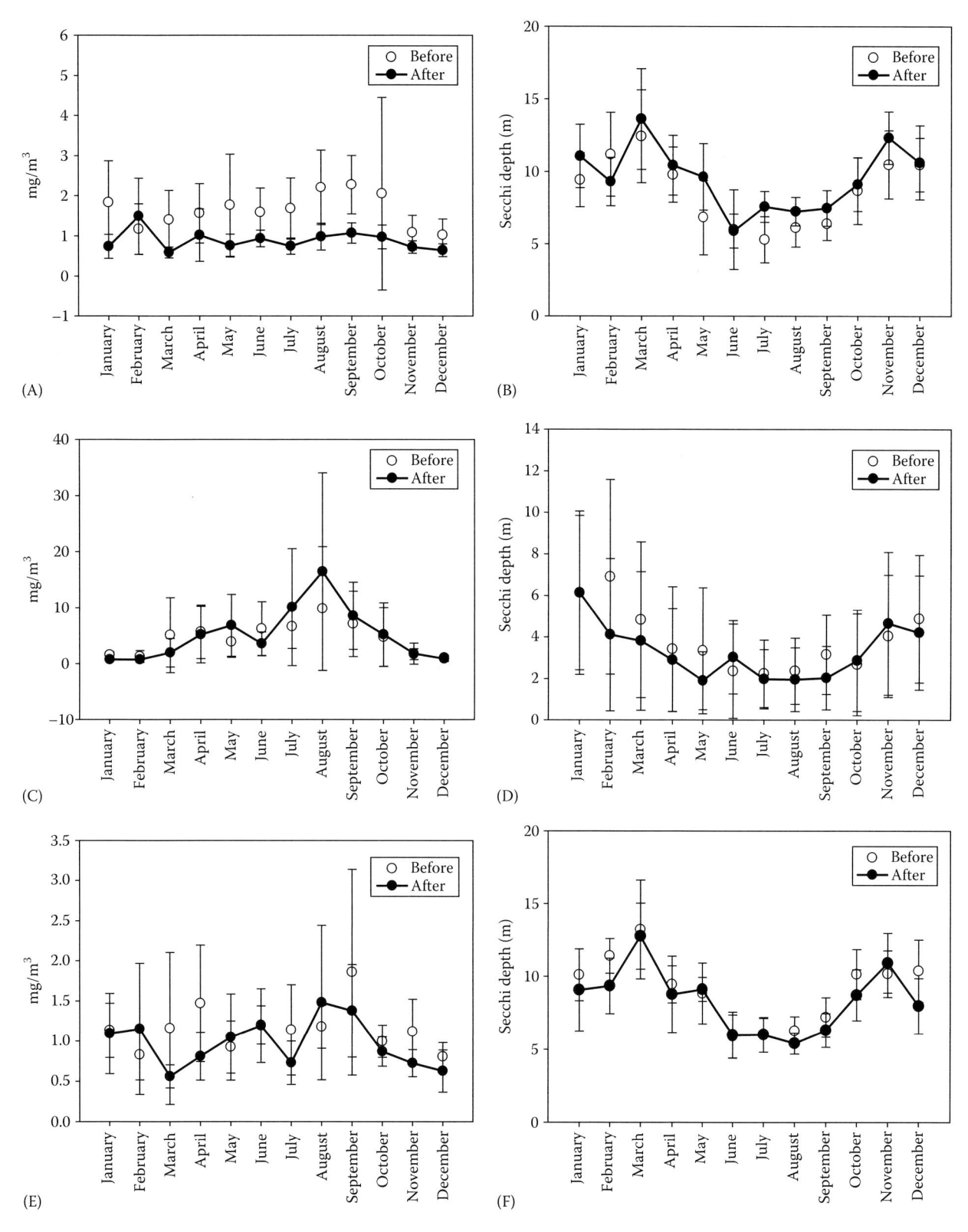

EFFECTS OF INVASIVE QUAGGA MUSSELS ON CHLOROPHYLL AND WATER CLARITY 501

Figure 31.4 Seasonal changes in mean chlorophyll *a* concentration (left) and mean Secchi depth (right) in each of five basins in Lake Mead in the pre-quagga (2002–2006) and post-quagga (2007–2009) periods. (A) Boulder Basin chlorophyll *a*. (B) Boulder Basin water clarity. (C) Las Vegas Bay chlorophyll *a*. (D) Las Vegas Bay water clarity. (E) Virgin Basin chlorophyll *a*. (F) Virgin Basin water clarity.

(continued)

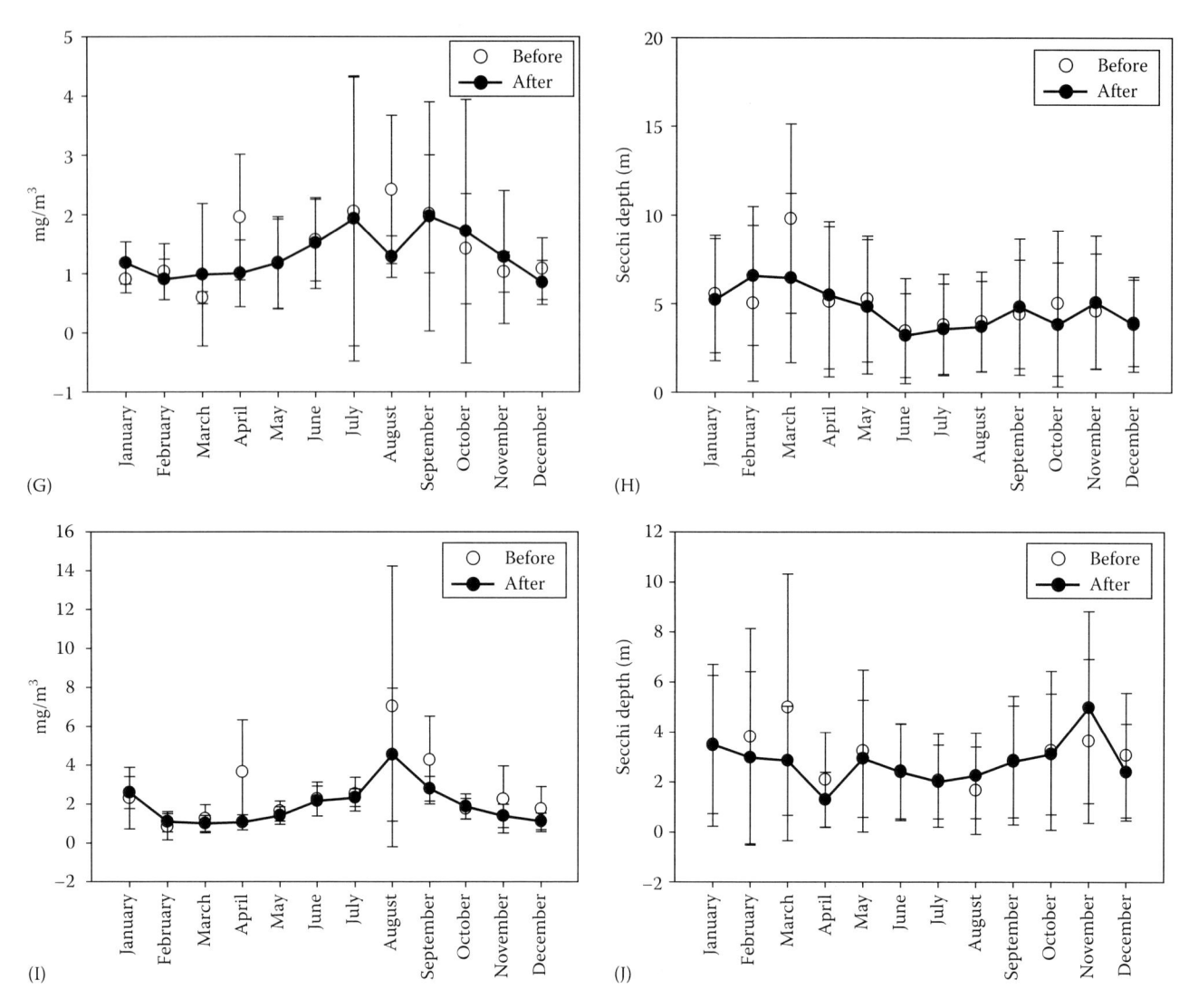

Figure 31.4 (continued) Seasonal changes in mean chlorophyll a concentration (left) and mean Secchi depth (right) in each of five basins in Lake Mead in the pre-quagga (2002–2006) and post-quagga (2007–2009) periods. (G) Overton Arm chlorophyll a. (H) Overton Arm water clarity. (I) Gregg Basin chlorophyll a, and (J) Gregg Basin water clarity.

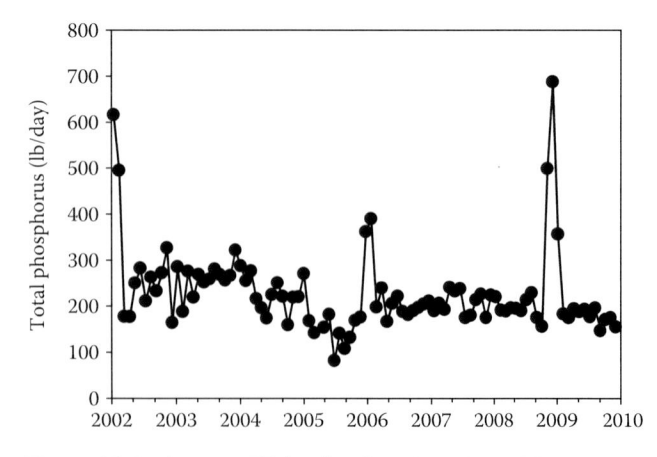

Figure 31.5 Average TP loading from Las Vegas Wash to Las Vegas Bay and Lake Mead from 2002 to 2009. (Data from D. Drury.)

three times greater in 2009 compared to 2007, volume turnover time was still lower than in other systems. For instance, in western Lake Erie and the Hudson River, it only took 1–3 days for the dreissenid populations to filter the entire water body (Bunt et al. 1993, Roditi et al. 1996). However, volume turnover time does not necessarily indicate potential impacts for three reasons. First, mussels may affect the nearshore/epilimnetic region more than the offshore/hypolimnetic region in Lake Mead. Most quagga mussels in the lake were found on hard substrates in the nearshore region (Mueting et al. 2010). More favorable conditions in the nearshore/epilimnetic compared to the offshore/hypolimnetic region (i.e., temperature, oxygen) (LaBounty and Burns 2005) can result in a better grazing environment for quagga mussels. Significant reductions in chlorophyll were found at most sites of the Boulder Basin from June to September (Table 31.3 and Figure 31.3), which corresponds to the period of stratification (LaBounty and Burns 2005). Second, a large portion of the basin has a rocky bottom and a steep slope as Lake Mead was formed when impounded Colorado River water filled deep canyons. In addition, there are several "islands" located in Boulder Basin such as Sentinel Island and Boulder Island. These physical characteristics can minimize boundary-layer effects between stratified water masses. Although such effects may be present in Lake Mead, they may not be as significant as in Lake Erie or in other water bodies where the lake bottom is a large portion of the subsurface area (Ackerman et al. 2001, Zhang et al. 2008). Third, dreissenids in temperate lakes and rivers have limited impacts in the winter when filtering rates are minimal because of low temperatures. In a subtropical system like Lake Mead, dreissenids likely filter at high rates all year due to consistently high water temperature. In Lake Mead, seasonal water temperature range mostly between 11°C and 28°C in both the nearshore/epilimnion and offshore/hypolimnion, and occasionally surface water can reach 30°C.

The relationships between water clarity and chlorophyll concentration and between chlorophyll concentration and TP in the Boulder Basin are strong (LaBounty 2008). Therefore, it is not surprising that water clarity increased with decreased chlorophyll concentration in Boulder Basin. However, water clarity in Virgin Basin did not increase when chlorophyll concentrations decreased, probably because there is no strong relationship between water clarity and chlorophyll concentration in Virgin Basin over the entire 2000–2009 sampling period (Figure 31.6).

In a recent survey of 50 thermally-stratified inland lakes in Michigan with similar nutrient concentrations and morphometries, phytoplankton biomass and water clarity in dreissenid-invaded lakes were 24% lower and 21% greater, respectively, than in non-invaded lakes just three years after the invasion occurred (Kissman et al. 2010). Since zebra mussels were first reported in Lake Erie in 1988, chlorophyll a concentrations decreased by 43% in the western basin and 27% in the west–central basin between 1988 and 1989 (Leach 1993). A 20 year (1983–2002) dataset from the central basin of Lake Erie showed that there was a downward trend of −0.07 µg/L/year in chlorophyll in the period after dreissenids became established but not in the period before dreissenids (Rockwell et al. 2005). In the Hudson River, New York, zebra mussels reduced chlorophyll concentrations by 90% compared to concentrations before mussels (Caraco et al. 2006). In some areas, such as in Green Bay, Lake Michigan, chlorophyll a decreased immediately after dreissenid invasion but increased after several years (Qualls et al. 2007, De Stasio et al. 2008). A recent model of Lake Erie suggested that zebra mussels decreased algal biomass in the period just after invasion, but after several years algal biomass increased. The biomass increase was attributed to an increase in non-diatom inedible algae resulting from the utilization of large amounts of ammonia and phosphate excreted by the dreissenid population (Zhang et al. 2008). In contrast to Lake Mead, where a significant increase of water clarity was found in the deep Boulder Basin within 3 years of dreissenid introduction, the effect of dreissenids on water clarity in the deep eastern basin of Lake Erie was not apparent until quagga mussels had been established for almost 10 years (Barbiero and Tuchman 2004). Even in the shallow western basin of Lake Erie and parts of Green Bay, there were no consistent increases in water clarity after dreissenids invaded (Barbiero and Tuchman 2004, Qualls et al. 2007), and, in the central basin of Lake Erie, water clarity decreased (Burns et al. 2005). These studies attributed the lack of dreissenid impacts on water clarity to large inputs of suspended solids, which may overwhelm impacts of dreissenid filtration, and to sediment resuspension. Studies indicate that because dreissenids do not ingest inorganic particles, they may have only slight impacts on water clarity in systems dominated by such particles (Vanderploeg et al. 2002). In the well-mixed, turbid Hudson River, there was no significant increase in water clarity although there was about 90% reduction in primary production (Strayer et al. 1999). Although there was no significant increase in water clarity in Lake Havasu, aquatic macrophytes have become a nuisance in this reservoir since 2008, which is one year after invasive quagga mussels were found (Figure 31.7). Increased macrophyte growth was likely due to the process of "benthification" (Zhu et al. 2006). As dreissenids increase water clarity, greater amounts of sunlight can reach the bottom that leads to increased macrophyte biomass and species richness. Such changes could be explained by increased light penetration alone (Hecky et al. 2004). Through excretion, dreissenids may also increase nutrients available for many aquatic plants such as *Cladophora* (Hecky et al. 2004). The benthification process increases the importance of benthic primary production over pelagic production in the food web, thereby causing an overall alteration of ecosystem function (Zhu et al. 2006). Unfortunately, there are no consistent historical nutrient and macrophyte information available in Lake Havasu, and a program to monitor macrophytes needs to be established throughout the lower Colorado River Basin.

Some studies show that quagga mussels usually do not reach peak abundance until about 12 years after arrival (Karatayev et al. 2010), but these studies have been conducted in temperate regions such as Europe and Great Lakes. In contrast, Lake Mead is a large subtropical reservoir with thermal and hydrological regimes that are different from water bodies in temperate regions (Holdren and Turner 2010). Therefore, dreissenid population dynamics and potential impacts in Lake Mead and the lower Colorado River Basin are still unclear. For example, it is likely that food could be a potential limiting factor for quagga mussels

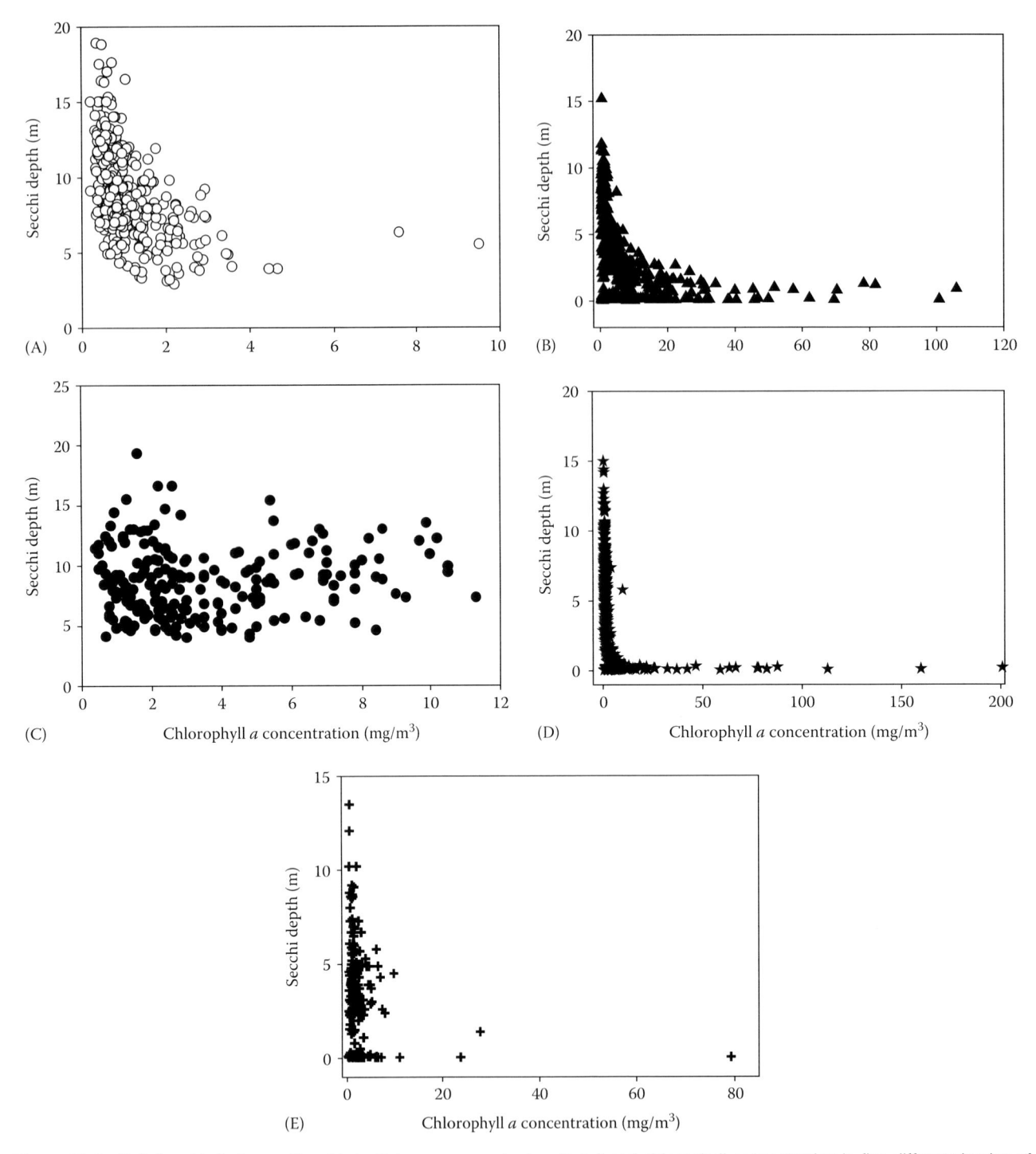

Figure 31.6 Relationship between Secchi-depth transparency (water clarity) and chlorophyll concentration in five different basins of Lake Mead in 2002–2009. Each point represents a monthly value at a given site. (A) Boulder basin, (B) Las Vegas bay, (C) Virgin basin, (D) Overton arm, and (E) Gregg basin.

in Lake Mead as it is a relatively unproductive system, especially in the outer portions of the Boulder and Virgin Basins (Cross et al. 2011). Yet quagga mussels maintain growth at very low chlorophyll a concentration (0.05 mg/m^3) in the Great Lakes (Baldwin et al. 2002). Subsequently, it may not take long for quagga mussels to reach their peak density in Lake Mead.

A long-term standardized monitoring program has been established in Lakes Mead and Mohave. The program tracks changes in mussel size, abundance, and distribution at more than 50 sampling sites. In addition the program monitors changes in water quality parameters (i.e., nutrients, temperature, dissolved oxygen, conductivity, pH, Secchi depth, total organic carbon, fecal coliform bacteria and

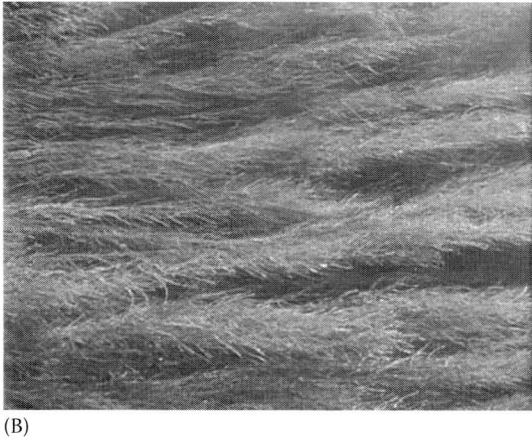

Figure 31.7 Aquatic macrophytes in Lakes Havasu (Image (A): photo by Al Graves) and Mohave (Image (B): photo by David Wong). Lake Mohave is a reservoir in the Colorado River between Lakes Mead and Havasu.

chlorophyll *a*) and biological resources (i.e., phytoplankton biomass and species composition, zooplankton, fish community, and benthic invertebrates) throughout Lake Mead (Turner et al. 2011, Wong et al. 2011). The program was implemented by lake managers and participating agencies in late 2009 and will help lake managers better understand and accurately assess quagga mussel impacts in the lower Colorado River Basin.

ACKNOWLEDGMENTS

Nutrient-loading data from the Las Vegas Wash was provided by Dr. Douglas Drury, and quagga mussel density data in Lake Havasu were collected by U.S. Bureau of Land Management. We thank constructive suggestions from Gerald Hickman and two anonymous reviewers.

REFERENCES

Ackerman, J. D., M. R. Loewen, and P. F. Hamblin. 2001. Benthic-Pelagic coupling over a zebra mussel reef in western Lake Erie. *Limnol. Oceanogr.* 46: 892–904.

Aldridge, D. C., P. Elliott, and G. D. Moggridge. 2004. The recent and rapid spread of the zebra mussel (*Dreissena polymorpha*) in Great Britain. *Biol. Conserv.* 119: 253–261.

Aldridge, D. C., P. Elliott, and G. D. Moggridge. 2006. Microencapsulated BioBullets for the control of biofouling zebra mussels. *Environ. Sci. Technol.* 40: 975–979.

American Public Health Association (APHA). 2005. *Standard Methods for the Examination of Water and Wastewater.* Washington, DC: American Public Health Association.

APHA. 2005. Standard Methods for the Examination of Water and Wastewater American. Public Health Association, Washington, DC.

Baines, S. B., N. S. Fisher, and J. J. Cole. 2005. Uptake of dissolved organic matter (DOM) and its importance to metabolic requirements of the zebra mussel, *Dreissena polymorpha*. *Limnol. Oceanogr.* 50: 36–47.

Baker, S. M., J. S. Levinton, J. P. Kurdziel, and S. E. Shumway. 1998. Selective feeding and biodeposition by zebra mussels and their relation to changes in phytoplankton composition and seston load. *J. Shellfish Res.* 17: 1207–1213.

Baldwin, B. S., M. S. Mayer, J. Dayton et al. 2002. Comparative growth and feeding in zebra and quagga mussels (*Dreissena polymorpha* and *Dreissena bugensis*): Implications for North American lakes. *Can. J. Fish. Aquat. Sci.* 59: 680–694.

Barbiero, R. P. and M. L. Tuchman. 2004. Long-term dreissenid impacts on water clarity in Lake Erie. *J. Great Lakes Res.* 30: 557–565.

Bastviken, D. T. E., N. F. Caraco, and J. J. Cole. 1998. Experimental measurements of zebra mussel (*Dreissena polymorpha*) impacts on phytoplankton community composition. *Freshwat. Biol.* 39: 375–386.

Benson, A. J. 2010a. Quagga mussel sightings distribution. http://nas.er.usgs.gov/taxgroup/mollusks/zebramussel/quaggamusseldistribution.asp. (Accessed on November 10, 2010, 2009).

Benson, A. J. 2010b. Zebra mussel sightings distribution. http://nas.er.usgs.gov/taxgroup/mollusks/zebramussel/zebramusseldistribution.asp. (Accessed on October 1, 2010).

Boegman, L., M. R. Loewen, D. A. Culver, P. F. Hamblin, and M. N. Charlton. 2008. Spatial-dynamic modeling of algal biomass in Lake Erie: Relative impacts of dreissenid mussels and nutrient loads. *J. Environ. Eng.* 134: 456–468.

Bunt, C. M., H. J. MacIsaac, and W. G. Sprules. 1993. Pumping rates and projected filtering impacts of juvenile zebra mussels (*Dreissena polymorpha*) in Western Lake Erie. *Can. J. Fish. Aquat. Sci.* 50: 1017–1022.

Burns, N. M., D. C. Rockwell, P. E. Bertram, D. M. Dolan, and J. J. H. Ciborowski. 2005. Trends in temperature, Secchi depth, and dissolved oxygen depletion rates in the central basin of Lake Erie, 1983–2002. *J. Great Lakes Res.* 31: 35–49.

Caraco, N. F., J. J. Cole, P. A. Raymond et al. 1997. Zebra mussel invasion in a large, turbid river: Phytoplankton response to increased grazing. *Ecology* 78: 588–602.

Caraco, N. F., J. J. Cole, and D. L. Strayer. 2006. Top down control from the bottom: Regulation of eutrophication in a large river by benthic grazing. *Limnol. Oceanogr.* 51: 664–670.

Carlton, J. T. 2008. The zebra mussel *Dreissena polymorpha* found in North America in 1986 and 1987. *J. Great Lakes Res.* 34: 770–773.

Conroy, J. D., W. J. Edwards, R. A. Pontius et al. 2005. Soluble nitrogen and phosphorus excretion of exotic freshwater mussels (*Dreissena* spp.): Potential impacts for nutrient remineralisation in western Lake Erie. *Freshwat. Biol.* 50: 1146–1162.

Cross, C., W. H. Wong, and T. D. Che. 2011. Estimating carrying capacity of quagga mussels (*Dreissena rostriformis bugensis*) in a natural system: A case study of the Boulder Basin of Lake Mead, Nevada-Arizona. *Aquat. Invasions* 6: 141–147.

De Stasio, B. T., M. B. Schrimpf, A. E. Beranek, and W. C. Daniels. 2008. Increased chlorophyll *a*, phytoplankton abundance, and cyanobacteria occurrence following invasion of Green Bay, Lake Michigan by dreissenid mussels. *Aquat. Invasions* 3: 21–27.

Diggins, T. P. 2001. A seasonal comparison of suspended sediment filtration by quagga (*Dreissena bugensis*) and zebra (*D. polymorpha*) mussels. *J. Great Lakes Res.* 27: 457–466.

Dionisio Pires, L. M., R. R. Jonker, E. Van Donk, and H. J. Laanbroek. 2004. Selective grazing by adults and larvae of the zebra mussel (*Dreissena polymorpha*): Application of flow cytometry to natural seston. *Freshwat. Biol.* 49: 116–126.

Du, X. 2002. Algal growth potential and nutrient limitation in the Las Vegas Bay, Lake Mead. Master thesis, University of Nevada Las Vegas, NV.

Hecky, R. E., R. E. H. Smith, and D. R. Barton. 2004. The nearshore phosphorus shunt: A consequence of ecosystem engineering by dreissenids in the Laurentian Great Lakes. *Can. J. Fish. Aquat. Sci.* 61: 1285–1293.

Hickey, V. 2010. The quagga mussel crisis at Lake Mead National Recreation Area, Nevada (U.S.A.). *Conserv. Biol.* 24: 931–937.

Holdren, G. C. and K. Turner. 2010. Characteristics of Lake Mead, Arizona-Nevada. *Lake Reserv. Manage.* 26: 230–239.

Jack, J. D. and J. H. Thorp. 2000. Effects of the benthic suspension feeder *Dreissena polymorpha* on zooplankton in a large river. *Freshwat. Biol.* 44: 569–579.

Karatayev, A. Y., L. E. Burlakova, S. E. Mastitsky, D. K. Padilla, and E. L. Mills. 2010. Invasion paradox: Why *Dreissena rostriformis bugensis*, being less invasive, outcompete *D. polymorpha*? International Conference on Aquatic Invasive Species. San Diego, CA.

Karatayev, A. Y., L. E. Burlakova, and D. K. Padilla. 1997. The effects of *Dreissena polymorpha* (Pallas) invasion on aquatic communities in eastern Europe. *J. Shellfish Res.* 16: 187–203.

Kissman, C. E. H., L. B. Knoll, and O. Sarnelle. 2010. Dreissenid mussels (*Dreissena polymorpha* and *Dreissena bugensis*) reduce microzooplankton and macrozooplankton biomass in thermally stratified lakes. *Limnol. Oceanogr.* 55: 1851–1859.

Klerks, P. L., P. C. Fraleigh, and J. E. Lawniczak. 1996. Effects of zebra mussels (*Dreissena polymorpha*) on seston levels and sediment deposition in western Lake Erie. *Can. J. Fish. Aquat. Sci.* 53: 2284–2291.

LaBounty, J. F. 2008. Secchi transparency of Boulder Basin, Lake Mead, Arizona-Nevada: 1990–2007. *Lake Reserv. Manage.* 24: 207–218.

LaBounty, J. F. and N. M. Burns. 2005. Characterization of boulder basin, Lake Mead, Nevada-Arizona, USA—Based on analysis of 34 limnological parameters. *Lake Reserv. Manage.* 21: 277–307.

LaBounty, J. F. and M. J. Horn. 1997. The influence of drainage from the Las Vegas Valley on the limnology of Boulder Basin, Lake Mead, Arizona-Nevada. *Lake Reserv. Manage.* 13: 95–108.

LaBounty, J. F. and P. Roefer. 2007. Quagga mussels invade Lake Mead. *LakeLine* 27: 17–22.

Lavrentyev, P. J., W. S. Gardner, and L. Yang. 2000. Effects of the zebra mussel on nitrogen dynamics and the microbial community at the sediment-water interface. *Aquat. Microb. Ecol.* 21: 187–194.

Leach, J. H. 1993. Impacts of zebra mussel (*Dreissena polymorpha*) on water quality and fish spawning reefs in Western Lake Erie. In *Zebra Mussels: Biology, Impacts, and Control*, T. F. Nalepa and D. W. Schloesser, eds., pp. 381–397. Boca Raton, FL: CRC Press.

Lucy, F. E., D. Minchin, and R. Boelens. 2009. From lakes to rivers: Downstream larval distribution of *Dreissena polymorpha* in Irish river basins. *Aquat. Invasions* 3: 297–304.

MacIsaac, H. J., W. G. Sprules, and J. H. Leach. 1991. Ingestion of small-bodied zooplankton by zebra mussels (*Dreissena polymorpha*): Can cannibalism on larvae influence population dynamics. *Can. J. Fish. Aquat. Sci.* 48: 2051–2060.

MacIsaac, H. J., W. G. Sprules, O. E. Johannsson, and J. H. Leach. 1992. Filtering impacts of larval and sessile zebra mussels in western Lake Erie. *Oecologia* 92: 30–39.

MacIsaac, H. J., C. J. Lonnee, and J. H. Leach. 1995. Suppression of microzooplankton by zebra mussels: Importance of mussel size. *Freshwat. Biol.* 34: 379–387.

MacIsaac, H. J. 1996. Potential abiotic and biotic impacts of zebra mussels on the inland waters of North America. *Amer. Zool.* 36: 287–299.

May, B. and J. E. Marsden. 1992. Genetic identification and implications of another invasive species of dreissenid mussel in the Great Lakes. *Can. J. Fish. Aquat. Sci.* 49: 1501–1506.

McMahon, R. F. 2011. Quagga mussel (*Dreissena rostriformis bugensis*) population structure during the early invasion of Lakes Mead and Mohave January-March 2007. *Aquat. Invasions* 6: 131–140.

Mills, E. L., R. M. Dermott, E.F. Roseman et al. 1993. Colonization, ecology, and population structure of the quagga mussel (Bivalvia, Dreissenidae) in the lower Great Lakes. *Can. J. Fish. Aquat. Sci.* 50: 2305–2314.

Molloy, D. P., A. bij de Vaate, T. Wilke, and L. Giamberini. 2007. Discovery of *Dreissena rostriformis bugensis* (Andrusov 1897) in western Europe. *Bio. Invasions* 9: 871–874.

Mueting, S. A., S. L. Gerstenberger, and W. H. Wong. 2010. An evaluation of artificial substrates for monitoring the quagga mussel (*Dreissena bugensis*) in Lake Mead, NV-AZ. *Lake Reserv. Manage.* 26: 283–292.

Nalepa, T. F. 2010. An overview of the spread, distribution, and ecological impacts of the quagga mussel, *Dreissena rostriformis bugensis*, with possible implications to the Colorado River System. In *Proceedings of the Colorado River Basin Science and Resource Management Symposium, November 18–20, 2008*, T. S. Melis, J. F. Hamil, G. E. Bennett et al., eds., pp. 113–1221. Scottsdale, AZ: U.S. Geological Survey.

Nalepa, T. F., D. L. Fanslow, S. A. Pothoven, A. J. Foley, and G. A. Lang. 2007. Long-term trends in benthic macroinvertebrate populations in Lake Huron over the past four decades. *J. Great Lakes Res.* 33: 421–436.

Nalepa, T. F., S. A. Pothoven, and D. L. Fanslow. 2009. Recent changes in benthic macroinvertebrate populations in Lake Huron and impact on the diet of lake whitefish (*Coregonus clupeaformis*). *Aquat. Ecosyst. Health Manage.* 12: 2–10.

Nalepa, T. F. and D. W. Schloesser. 1993. *Zebra Mussels: Biology, Impacts, and Control*. Boca Raton, FL: CRC Press.

Noonburg, E. G., B. J. Shuter, and P. A. Abrams. 2003. Indirect effects of zebra mussels (*Dreissena polymorpha*) on the planktonic food web. *Can. J. Fish. Aquat. Sci.* 60: 1353–1368.

Pace, M. L., S. E. G. Findlay, and D. Fischer. 1998. Effects of an invasive bivalve on the zooplankton community of the Hudson River. *Freshwat. Biol.* 39: 103–116.

Padilla, D. K., S. C. Adolph, K. L. Cottingham, and D. W. Schneider. 1996. Predicting the consequences of dreissenid mussels on a pelagic food web. *Ecol. Model.* 85: 129–144.

Paulson, L.J. and J. R. Baker. 1983. *The Limnology in Reservoirs on the Colorado River*. Las Vegas, NV: University of Nevada Las Vegas.

Pimentel, D., R. Zuniga, and D. Morrison. 2005. Update on the environmental and economic costs associated with alien-invasive species in the United States. *Ecol. Econom.* 52: 273–288.

Qualls, T. M., D. M. Dolan, T. Reed, M. E. Zorn, and J. Kennedy. 2007. Analysis of the impacts of the zebra mussel, *Dreissena polymorpha*, on nutrients, water clarity, and the chlorophyll-phosphorus relationship in lower Green Bay. *J. Great Lakes Res.* 33: 617–626.

Ricciardi, A. 2003. Predicting the impacts of an introduced species from its invasion history: An empirical approach applied to zebra mussel invasions. *Freshwat. Biol.* 48: 972–981.

Rockwell, D. C., G. J. Warren, P. E. Bertram, D. K. Salisbury, and N. M. Burns. 2005. The US EPA Lake Erie indicators monitoring program 1983–2002: Trends in phosphorus, silica, and chlorophyll a in the central basin. *J. Great Lakes Res.* 31: 23–34.

Roditi, H. A., N. F. Caraco, J. J. Cole, and D. L. Strayer. 1996. Filtration of Hudson River water by the zebra mussel (*Dreissena polymorpha*). *Estuaries* 19: 824–832.

Roditi, H. A. and N. S. Fisher. 1999. Rates and routes of trace element uptake in zebra mussels. *Limnol. Oceanogr.* 44: 1730–1749.

Rosenberg, G. and M. L. Ludyanskiy. 1994. A nomenclatural review of *Dreissena* (Bivalve, Dreissenidae), with identification of the quagga mussel as *Dreissena bugensis*. *Can. J. Fish. Aquat. Sci.* 51: 1474–1484.

Silverman, H., J. W. Lynn, E. C. Achberger, and T. H. Dietz. 1996a. Gill structure in zebra mussels: Bacterial-sized particle filtration. *Am. Zool.* 36: 373–384.

Silverman, H., J. W. Lynn, and T. H. Dietz. 1996b. Particle capture by the gills of *Dreissena polymorpha*: Structure and function of latero-frontal cirri. *Biol. Bull.* 191: 42–54.

Spidle, A. P., J. E. Marsden, and B. May. 1994. Identification of the Great Lakes quagga mussel as *Dreissena bugensis* from the Dnieper River, Ukraine, on the basis of allozyme variation. *Can. J. Fish. Aquat. Sci.* 51: 1485–1489.

Stokstad, E. 2007. Invasive species—Feared quagga mussel turns up in western United States. *Science* 315: 453–453.

Strayer, D. L. 1999. Effects of alien species on freshwater mollusks in North America. *J. N. Am. Benthol. Soc.* 18: 74–98.

Strayer, D. L. 2009. Twenty years of zebra mussels: Lessons from the mollusk that made headlines. *Front. Ecol. Environ.* 7: 135–141.

Strayer, D. L., N. F. Caraco, J. J. Cole, F. Stuart, and M. L. Pace. 1999. Transformation of freshwater ecosystem by bivalves: A case study of zebra mussels in the Hudson river. *Bioscience* 49: 19–27.

Strayer, D. L., V. T. Eviner, J. M. Jeschke, and M. L. Pace. 2006. Understanding the long-term effects of species invasions. *Trends Ecol. Evol.* 21: 645–651.

Strayer, D. L., L. C. Smith, and D. C. Hunter. 1998. Effects of the zebra mussel (*Dreissena polymorpha*) invasion on the macrobenthos of the freshwater tidal Hudson River. *Can. J. Zool.* 76: 419–425.

Thayer, S. A., R. C. Haas, R. D. Hunter, and R. H. Kushler. 1997. Zebra mussel (*Dreissena polymorpha*) effects on sediment, other zoobenthos, and the diet and growth of adult yellow perch (*Perca flavescens*) in pond enclosures. *Can. J. Fish. Aquat. Sci.* 54: 1903–1915.

Therriault, T. W., M. F. Docker, M. I. Orlova, D. D. Heath, and H. J. MacIsaac. 2004. Molecular resolution of the family Dreissenidae (Mollusca: Bivalvia) with emphasis on Ponto-Caspian species, including first report of Mytilopsis leucophaeata in the Black Sea basin. *Mol. Phylogen. Evol.* 30: 479–489.

Turner, K., W. H. Wong, S. L. Gerstenberger, and J. M. Miller. 2011. Interagency monitoring action plan (I-MAP) for quagga mussels in Lake Mead, Nevada-Arizona, USA. *Aquat. Invasions* 6: 195–204.

Vanderploeg, H. A., T. H. Johengen, and J. R. Liebig. 2009. Feedback between zebra mussel selective feeding and algal composition affects mussel condition: Did the regime changer pay a price for its success? *Freshwat. Biol.* 54: 47–63.

Vanderploeg, H. A., J. R. Liebig, W. W. Carmichael et al. 2001. Zebra mussel (*Dreissena polymorpha*) selective filtration promoted toxic Microcystis blooms in Saginaw Bay (Lake Huron) and Lake Erie. *Can. J. Fish. Aquat. Sci.* 58: 1208–1211.

Vanderploeg, H. A., J. R. Liebig, T. F. Nalepa, G. L. Fahnenstiel, and S. A. Pothoven. 2010. *Dreissena* and the disappearance of the spring phytoplankton bloom in Lake Michigan. *J. Great Lakes Res.* 36: 50–59.

Vanderploeg, H. A., T. F. Nalepa, D. J. Jude et al. 2002. Dispersal and emerging ecological impacts of Ponto-Caspian species in the Laurentian Great Lakes. *Can. J. Fish. Aquat. Sci.* 59: 1209–1228.

Wilson, A. E. 2003. Effects of zebra mussels on phytoplankton and ciliates: A field mesocosm experiment. *J. Plankton Res.* 25: 905–915.

Wilson, A. E. and O. Sarnelle. 2002. Relationship between zebra mussel biomass and total phosphorus in European and North American lakes. *Archiv. Hydrobiol.* 153: 339–351.

Wong, W. H. and S. L. Gerstenberger. 2011. Quagga mussels in the western United States: Monitoring and management. *Aquat. Invasions* 6: 125–129.

Wong, W. H., S. L. Gerstenberger, J. M. Miller, C. J. Palmer, and B. Moore. 2011. A standardized design for quagga mussel monitoring in Lake Mead, Nevada-Arizona. *Aquat. Invasions* 6: 205–215.

Wong, W. H. and J. S. Levinton. 2005. Consumption rates of two rotifer species by zebra mussels *Dreissena polymorpha*. *Mar. Freshwat. Behav. Physiol.* 38: 149–157.

Wong, W. H., J. S. Levinton, B. S. Twining, and N. Fisher. 2003. Assimilation of micro- and mesozooplankton by zebra mussels: A demonstration of the food web link between zooplankton and benthic suspension feeders. *Limnol. Oceanogr.* 48: 308–312.

Wong, W. H., T. Tietjen, S. L. Gerstenberger et al. 2010. Potential ecological consequences of invasion of the quagga mussel (*Dreissena bugensis*) into Lake Mead, Nevada–Arizona, USA. *Lake Reserv. Manage.* 26: 306–315.

Zar, J. H. 1996. *Biostatistical Analysis*. Upper Saddle River, NJ: Prentice Hall.

Zhang, H., D. A. Culver, and L. Boegman. 2008. A two-dimensional ecological model of Lake Erie: Application to estimate dreissenid impacts on large lake plankton populations. *Ecol. Model.* 214: 219–241.

Zhu, B., D. G. Fitzgerald, C. M. Mayer, L. G. Rudstam, and E. L. Mills. 2006. Alteration of ecosystem function by zebra mussels in Oneida Lake: Impacts on submerged macrophytes. *Ecosystems* 9: 1017–1028.

CHAPTER 32

Role of Selective Grazing by Dreissenid Mussels in Promoting Toxic *Microcystis* Blooms and Other Changes in Phytoplankton Composition in the Great Lakes

Henry A. Vanderploeg, Alan E. Wilson, Thomas H. Johengen, Julianne Dyble Bressie, Orlando Sarnelle, James R. Liebig, Sander D. Robinson, and Geoffrey P. Horst

CONTENTS

Abstract 509
Introduction 510
   Toxic Microcystis Blooms Return to Great Lakes 510
   Selective Rejection Paradigm 510
   Possible Interactions with Phosphorus 511
   Overview of Experimental Approach and Objectives 512
Methods 512
Results 513
   Strains Isolated from Inland Lakes 513
   Experiments with Natural Seston 515
   Combined Long-Term/Short-Term Experiments with Natural Seston 515
Discussion 518
   Differences in Grazing Vulnerability of Microcystis among Strains and Lakes 518
   Projecting Mussel Impact on Phytoplankton Community 520
   Microcystis Affects Filtering Intensity 521
   Nutrients and Phytoplankton Growth Rate 521
   Future Directions 521
References 522

ABSTRACT

We investigated the feeding response of zebra and quagga mussels to *Microcystis aeruginosa* strains from culture collection and from natural seston from Saginaw Bay (Lake Huron), western Lake Erie, and enclosures from Gull Lake, an inland lake in Michigan. These experiments were done to evaluate the roles of strain identity, toxin concentration (microcystin), colony size, and environmental phosphorus concentrations as they affect ingestion or selective rejection of *Microcystis* in pseudofeces and potential *Microcystis* bloom promotion through the selective-rejection process.

A combination of traditional feeding experiments with mussels confined in beakers and videotaping of mussel behavior was used. We measured changes in *Microcystis* concentration in the feeding experiments using changes in chlorophyll and the toxin associated with *Microcystis* (microcystin) in small (<53 µm) and large (>53 µm) size fractions. In natural seston, most colonies fell within the large size fraction. Overall, there were complex interactions that could not be simply explained by microcystin concentration, colony size, or environmental P concentration. Experiments with toxic and nontoxic strains from culture collection indicated different reasons for rejection. In one nontoxic strain having colonies in both the small and large fractions, small colonies were ingested, while large colonies were not. In another nontoxic strain, consisting only of large colonies, no colonies were ingested; however, when the colonies were broken apart by sonication, no small colonies or even single cells were ingested. Video observations showed that both of these strains were readily captured and rejected in pseudofeces after a large number were collected. Mussels fed upon the small colonies of a moderately toxic strain, whereas for another less toxic strain, no feeding occurred. When mussels were induced to feed on this latter strain by adding *Cryptomonas*—a favorite food of mussels—to the suspension, one of the mussels showed extreme sensitivity to *Microcystis* by rejecting each colony as they entered the incurrent siphon. Experiments with *Microcystis* having moderate microcystin concentration from both the low P (Saginaw Bay) and high P (Maumee Bay) sites in the Great Lakes were rejected. *Microcystis* from enclosures in Gull Lake was ingested despite having very high microcystin concentrations. Whether the selective-rejection process results in a *Microcystis* bloom depends on both mussel abundance and environmental P concentration as they affect mortality and growth rate of algae competing with *Microcystis*, as well as the composition of different *Microcystis* strains (genetic identities) that can coexist at the same time in the same water body. Questions for future research and research approaches to understand these complex interactions are outlined.

INTRODUCTION

Toxic Microcystis Blooms Return to Great Lakes

Noxious blooms of colonial cyanobacteria such as *Microcystis*, *Anabaena*, and *Aphanizomenon* are well-known symptoms of cultural eutrophication caused by excessive phosphorus (P) loading (Smith 1983, Sommer et al. 1986). Cyanobacterial blooms were common occurrences in Saginaw Bay (Bierman et al. 1984, Stoermer and Theriot 1985) and Lake Erie (Makarewicz 1993) during the 1960s and 1970s. With the control of P loading instituted during the mid-1970s, these blooms diminished. Therefore, it was a surprise that intense toxic *Microcystis* blooms occurred in both Saginaw Bay (Lake Huron) and Lake Erie during the mid- to late-1990s, 3–5 years after zebra mussels became established in these water bodies (Vanderploeg et al. 2001). In addition to the increase in *Microcystis* and its toxins in the Great Lakes, *Microcystis* abundance and its toxins increased in inland lakes in Michigan where zebra mussels became established and total phosphorus (TP) concentrations were <20 µg/L (Raikow et al. 2004, Knoll et al. 2008), suggesting these patterns and underlying mechanisms were not unique to large systems.

Microcystis and other cyanobacterial blooms have serious consequences to ecosystem function and health, to aesthetics, and to wildlife and human health. *Microcystis*, once thought to consist of five or more morphologically distinct species (morphospecies), likely consists of a single species, *M. aeruginosa*, that exhibits considerable intraspecific variation (genetic or strain variation) and phenotypic diversity within strains (Otsuka et al. 2001). A given strain, or genetically distinct population, can exhibit different traits such as colony morphology depending on environment conditions. We will refer to *Microcystis* available from culture collections as strains as these cultures are isolated from single cells or colonies. The genetic identities of some of these cultures, or strains, are described and some are not. *Microcystis* produces a potent diverse class of hepatotoxins (with >75 variants) and other secondary compounds that can poison aquatic organisms and wildlife, pets, and humans that drink the water (e.g., Sivonen and Jones 1999, Wiegand and Pflugmacher 2005). The main exposure pathway to aquatic animals is through ingestion of the cells (Lampert 1982, Nizan et al. 1986, Wiegand and Pflugmacher 2005). Although individual cells are small (~2–6 µm), *Microcystis* is a colonial cyanobacterium that can vary in size from a few cells to over 10,000 cells imbedded in mucilage of varying quantity and viscosity. Larger colonies can exceed 1000 µm. *Microcystis* is a typical dominant of the epilimnion of eutrophic lakes and its main selective advantages are its resistance to grazing and its buoyancy that prevents sedimentation and gives it the capacity to vertically migrate to take advantage of gradients of light and nutrients (Sommer et al. 1986, Reynolds 1997, 2006).

Dominance of *Microcystis* in the plankton can lead to inefficient food webs and shifts in zooplankton community structure. In many systems, zooplankton, including *Dreissena* larvae, can have very low ingestion rates of *Microcystis* colonies, because the colonies are often too large (>50 µm) to be handled by the mouthparts or are too large to fit in the mouth. Small colonies that are within the ingestible size range may be avoided because of toxicity or other taste factors. If ingested, *Microcystis* may lead to lowered survival and growth due for a variety of factors including toxic secondary compounds, such as microcystin, or the poor nutritional quality, for example, low concentration of n-3 polyunsaturated fatty acids (Fulton and Paerl 1987a,b, Vanderploeg et al. 1996, Wiegand and Pflugmacher 2005, Wilson et al. 2006, Ger et al. 2010). Vanderploeg et al. (2009) noted that mussel condition was negatively affected by *Microcystis*; therefore, mussels negatively affected their own populations by promoting *Microcystis* dominance.

Selective Rejection Paradigm

Vanderploeg et al. (1996, 2001) demonstrated that zebra mussels selectively reject *Microcystis* in pseudofeces and, given their high abundance and clearance rates, were likely responsible for promoting *Microcystis* blooms in Saginaw Bay and western Lake Erie. Important to the development of this concept were direct behavioral observations of zebra mussels feeding on *Microcystis* and other seston collected during a *Microcystis* bloom in Lake Erie (Vanderploeg and Strickler 2013; see Sequence 1), shows mussels taking in *Microcystis* and other algae into the incurrent siphon and forcefully expelling pseudofeces containing *Microcystis* every few minutes. Vanderploeg et al. (2001) clearly demonstrated that *Microcystis* colonies (mostly found in the >53 μm screen-size fraction) in Saginaw Bay and Lake Erie were rejected or ingested at a low rate; in addition, both small (<53 μm) and large (>53 μm) colonies of a toxic *Microcystis* strain (LE-3) isolated from Lake Erie were rejected or ingested at a low rate. In contrast, they observed that other toxic (PCC 7820) or nontoxic (CCAP 1450/11) strains of colonial *Microcystis* were readily ingested. This followed observations of Baker et al. (1998) that showed that unicellular toxic (UTEX 2385) and nontoxic strains (UTEX 2386) of *Microcystis* were readily ingested. In addition, zebra mussels cleared the small colonies of *Microcystis* in the Hudson River at 25% of the rate of the preferred phytoplankton, cryptophytes, and unicellular *Microcystis* (Bastviken et al. 1998). Thus, at the outset, there was some controversy as to the importance and relevance of the selective rejection mechanism as the promoter of *Microcystis* blooms. Vanderploeg et al. (2001) reasoned that strain and colony size were major factors. In the case of the Hudson River, Vanderploeg et al. (2001) argued that the strength of selective pressure from mussels is likely reduced, relative to lakes, because of the shorter residence time of *Microcystis* population and lack of grazing selective pressure on *Microcystis* originating upstream of the mussels. Later, two modeling studies—using multiclass-phytoplankton water quality models—corroborated that selective rejection was a necessary condition for the proliferation of *Microcystis* that followed the invasion of mussels in Saginaw Bay (Bierman et al. 2005, Fishman et al. 2009).

Since the time of these early observations, more evidence has been collected that suggests strain identity of *Microcystis* is important to selective rejection, but most of these observations were made with *Microcystis* from culture collections. Dionisio Pires et al. (2005) reported modest clearance (~3 mL/mg dry weight/h; compare with maximum of ~40 mL/mg dry weight/h for a desirable alga; see succeeding text) rates on small (<60 μm) and large (>60 μm) size colonies of toxic (V40) and nontoxic (V131) *Microcystis* strains isolated from Lake Volkerak in The Netherlands. In contrast, Juhel et al. (2006) demonstrated that the unicellular toxic *Microcystis* strain CCAP 1450/10 caused distress to mussels and *Microcystis* was ejected in pseudofeces through the incurrent siphon and the pedal gape.

One problem with *Microcystis* from culture collections is that most of these cultures tend to be axenic and lose their colonial integrity during serial subculture (Otsuka et al. 2001), a possible result of loss of critical co-occurring heterotrophic bacteria that are thought to stimulate mucilage and colony formation (Shen et al. 2011). It is worth noting that the colonial laboratory strains (CCAP 1450/11, LE-3, and PCC 7820) examined by Vanderploeg et al. (2001) and those of Dionisio Pires et al. (2005) did not form particularly large colonies or have a heavy investment of mucilage. In fact, Dionisio Pires et al. (2005) described the colonies of his cultures as cells stuck together without mucilage.

Possible Interactions with Phosphorus

One can hypothesize that rejected *Microcystis* would get a boost from nutrient excretion from mussels that had processed the nutrient competitors of *Microcystis*. Others have argued that mussel nutrient excretion alone can explain the shift to *Microcystis* dominance by excreting more P relative to N and thereby shifting nutrient concentration to lower N:P ratios that favor cyanobacteria (Arnott and Vanni 1996, Bykova et al. 2006, Zhang et al. 2011). However, this advantage would not apply to *Microcystis* since it, unlike other species such as *Anabaena* and *Aphanizomenon*, is not an N-fixer. Although a detailed discussion of nutrient excretion is beyond the scope of this chapter, it is worth considering that the N:P ratio of excreted nutrients varies with seston stoichiometry and algal composition, and *Microcystis* has been reported to increase in mussel-invaded systems with low TP conditions in the seston. As noted by Vanderploeg et al. (2002) and Johengen et al. (2013), N:P ratio by definition would be high in low TP lakes, and the N:P ratio would be further skewed higher by mussels homeostatically retaining P under low TP conditions.

Using a flow cytometer, Dionisio Pires et al. (2005) observed that *Microcystis* consisting of unicells and small colonies (typically <20 μm) were ingested at the same modest-to-moderate rates as other phytoplankton in Lake IJsselmeer, The Netherlands. As a result of observations like these, Pires et al. (2005) advocated use of *Dreissena* as a biomanipulation tool to suppress cyanobacteria and increase water clarity. To increase mussel impact, they recommended the addition of hard substrate to increase zebra mussel populations. Both Lake IJsselmeer and the Hudson River are eutrophic systems having TP concentrations >50 μg/L. The potential role of TP concentration is also underscored by an increase in *Microcystis* in enclosures with zebra mussels fertilized to contain TP ≈ 9 μg/L (Sarnelle et al. 2005) and a decrease when TP = 40 μg/L (Sarnelle et al. 2012). These observations, plus observed increases of *Microcystis* in mussel-invaded lakes with low TP concentrations, have led to the current consensus—or paradigm—that *Dreissena* filtering will lead to *Microcystis* dominance in lakes of low TP. Is the observed lack of *Microcystis* dominance in mussel-invaded lakes with high TP concentrations a result of characteristics of the strains from high TP lakes, or a result of other interactions associated with P concentrations, such as differential growth rates of different algae? Our main focus is the selective rejection mechanism; however, we also consider our results in the context of possible nutrient interactions.

Overview of Experimental Approach and Objectives

To get further insight into the selective rejection mechanism and the role of mussels in promoting *Microcystis* dominance, we used a variety of approaches. First, we did feeding experiments and direct observations of mussels feeding on different laboratory strains of *Microcystis*, which retained their natural colonial form. The strains exhibited a range of toxicity and originated from four inland lakes in Michigan with zebra mussels (Gull Lake, Bear Lake) and without zebra mussels (Gilkey Lake, Hudson Lake). The strains were of known genetic identity and date of isolation (Wilson et al. 2005). These experiments were coupled with video observations to observe the mussel's behavioral response to the different strains.

Second, we examined feeding with natural seston and mussels from outer Maumee Bay, a high TP site in western Lake Erie associated with high TP inflow from the Maumee River, and with natural seston and mussels from inner Saginaw Bay, Lake Huron, which was a low TP site. Experiments were done with both zebra mussels and with quagga mussels, which have now effectively displaced zebra mussels in the Great Lakes. Quagga mussels have an even greater potential for affecting algal composition because they have attained abundances greater than zebra mussels in many nearshore regions and, unlike zebra mussels, are able to colonize deep, offshore regions (Nalepa et al. 2010). Do quagga mussels also selectively reject *Microcystis*?

Lastly, we report on results of a few experiments using water from Gull Lake and results of two enclosure experiments (mesocosm experiments) also in Gull Lake. Gull Lake is considered to be oligotrophic (Knoll et al. 2008) and provides a contrast to eutrophic western Lake Erie and inner Saginaw Bay. The enclosure experiments were designed to explore *Microcystis*–mussel interactions across a trophic gradient of N and P addition to help understand why *Microcystis* is promoted by mussels in some systems but not others (Raikow et al. 2004, Knoll et al. 2008, Sarnelle et al. 2012). Overall, our purpose here is to show a few contrasting results from different lakes or conditions.

After discussing the selective rejection mechanism, we explore the conditions under which *Microcystis* dominance can be promoted in natural systems and the future research that must be done to unravel interactions between mussels and naturally occurring *Microcystis* strains.

METHODS

All experiments, whether with natural seston or laboratory cultures, followed the same protocol: (1) all mussels were typically 14–16 mm in shell length, and (2) mussels were acclimated or re-acclimated to the particular feeding suspension for a minimum of 8 h. The acclimation period was designed to make sure the mussels were familiar with the particular feeding suspension and hence promote feeding and digestive equilibrium with their food prior to measurements. Setup of experiments and calculations of grazing followed standard methods described by Vanderploeg et al. (2001, 2009). In brief, we measured changes in chlorophyll concentration (Chl) in three control beakers without mussels and four experimental beakers with mussels. Samples for Chl were taken from beaker water at the beginning and end of the experiment and from mixed beaker contents at the end of the experiment. This allowed us to calculate both net and gross feeding rate variables; however, we report only net clearance rate (F_A). We typically used five mussels per 2 L beaker, and the duration was typically 2–3 h. Gentle mixing of the beaker water (but not settled material) was provided by bubbling in air through a pipette. We determined feeding on small (<53 μm) and large (>53 μm) size fractions of Chl by collecting seston on a sequential filtering apparatus that collected seston on a 53 μm Nitex screen and seston that passed through the screen to a GF/F filter (Vanderploeg et al. 2001). Summing the material from both fractions allowed us also to do a total mass balance of the constituent Chl collected on the filters. Given algae of moderate to high quality, ~30%–70% of the Chl would be removed with this experimental duration and mussel abundance.

Previous experience with seston from Lake Erie and Saginaw Bay had shown that most *Microcystis* was found in the large size fraction and was often the dominant phytoplankter when present. Clearance rates (mL/h) of mussels were expressed on a per unit ash-free dry weight (AFDW) (mL/mg/h) (Vanderploeg et al. 2001). Water taken from initial control and experimental bottles was preserved with 1% Lugol's solution for later identification of phytoplankton composition (Vanderploeg et al. 2001). Clearance rates were tested to see if they were significantly different ($P < 0.05$) from zero using a two-tailed t-test.

Location for collection of mussels and water for Saginaw Bay experiments with natural seston is the same 3.8 m deep inner bay Site SB5 (43.8953°N, −83.8605°W) we used in earlier studies (Vanderploeg et al. 2001, 2009). For Lake Erie, we collected water and mussels from a 3 m deep station on outer Maumee Bay (41.7417°N, −83.3555°W) because of its high nutrient concentration associated with P loading from the Maumee River.

In experiments evaluating the response to laboratory strains of *Microcystis*, we ran the same experiments with mussels collected at the same time with the cryptophyte, *Cryptomonas ozolini*, or the green alga, *Scenedesmus obliquus*. Clearance rates on these small high-quality algae served as a benchmark for evaluating the clearance rates on *Microcystis* (e.g., Vanderploeg et al. 2001, 2009). Vanderploeg et al. (2001) noted that clearance rates on *Cryptomonas* were somewhat higher than observed by Kryger and Riisgård (1998) for the green alga *Chlorella*; results of the latter study are often used as a benchmark for comparison with other studies. Culture methods for *Microcystis* and algae followed methods described by Vanderploeg et al. (2001, 2009, 2010). Algae were suspended in 0.2 µm filtered lake or river water to create the feeding suspension. In these experiments, we used zebra mussels collected in the Huron River at Argo Dam (near Ann Arbor, Michigan). We generally sought to keep initial Chl at or below 4 µg/L to keep algal concentration low enough to avoid a significant clearance rate drop above the incipient limiting concentration (Vanderploeg et al. 2001, 2009, 2010).

To get further insight on the selective-rejection process with natural seston, we conducted long-term feeding experiments immediately following some of the short-term (2 h) feeding experiments using the same mussels and water used to set up the short-term experiment. Long-term experiments were aimed at obtaining better estimates of low clearance rates for less preferred particles such as *Microcystis*. Long-term experiments, which lasted 16 h with much of the time in the dark, were set up identically to short-term experiment, except beaker contents were mixed every 4 h with a spoon to make sure mussels were always exposed to *Microcystis* during the long duration of the experiment. In these experiments, samples for initial and final Chl concentrations were collected after beaker contents were thoroughly mixed.

In the long-term experiments, we determined feeding rates on *Microcystis*, not only by determining changes in Chl but also by measuring particulate microcystin concentration. Hence, in the latter case, we were measuring feeding on the toxic *Microcystis*. To quantify toxicity and compare it with our results from natural systems, we normalized microcystin concentration to Chl concentration. Microcystin was determined from colonies and cells collected on similar filters in the same filtration system as used for Chl, extracted in 75% MeOH and water, broken apart using a Tekmar probe sonifier, vortexed to remove remaining cellular debris from filters, and centrifuged to segregate fiber filters. The supernatant was analyzed by enzyme-linked immunosorbent assay (ELISA) kits (Envirologix Inc.) read on a Stat Fax 3200 spectrophotometer plate reader (Awareness Technology Inc.) following the Envirologix protocol. For the two experiments with Gull Lake enclosures, we measured zebra mussel feeding on seston from two low-P (no P addition) enclosures having stocked mussel dry-weight densities of 1 g/m^2 and 4 g/m^2 as dry tissue mass (Sarnelle et al. 2005) 1 week after the addition of mussels.

Methods of videotaping mussels followed those of Vanderploeg et al. (2001, 2009). We were particularly concerned about documenting the flow of *Microcystis* into the mussel incurrent siphon and expulsion as pseudofeces (Vanderploeg et al. 2001, 2009) in order to gain further insights into selective rejection mechanisms (further details are given in Vanderploeg and Strickler 2013).

In some experiments, we manipulated the size of the colonies by breaking them apart with a sonic probe. We used this approach with both laboratory cultures and natural colonies of *Microcystis* concentrated with a plankton net (100 µm mesh). The concentrated *Microcystis* was suspended in 80 mL of modified WC media (Vanderploeg et al. 2001), sonicated with several 1 s bursts of a sonic probe, and poured through a 25 µm screen to remove larger colonies not broken apart by sonication. By this method, we were able to produce unicells of *Microcystis* or colonies containing a few cells. The minimum number (typically 20) of bursts necessary to break down the colonies varied with strain and sonic probe used. Microscopic examination of the sonicated cells showed that although the colonies were broken apart, a halo of mucilage was found around the cells. After sonication, the cells were suspended in 0.2 µm filtered lake water. To create a mixture of small and large colonies, sonicated and unsonicated colonies were combined.

RESULTS

Strains Isolated from Inland Lakes

The four strains of *Microcystis* used in the feeding experiments had different toxicity, average colony size, and previous exposure to zebra mussels (Table 32.1), and also came from lakes with average TP concentrations ranging between 8 and 66 µg/L. As noted, strains from Bear Lake (strain: Bear AC) and Gull Lake (strain: Gull 8/23/00) were exposed to zebra mussels, and strains from Gilkey Lake (strain: Gilkey L) and Hudson Lake (strain: Hudson BD) were not exposed to zebra mussels (Table 32.1). Chlorophyll-specific microcystin (MC) concentrations (µg/µg Chl) were highest in Bear AC (0.202) and Gull Lake 8/23/00 (0.190) strains, intermediate in the Gilkey Lake L strain (0.099), and effectively zero in the Hudson BD strain (Table 32.2). The latter strain totally lacked the microcystin gene. The finding of microcystin in the Bear AC strain was a surprise, since it was reported to lack the microcystin gene (Wilson et al. 2005). All strains at the time of our experiments had significant representation in small and large size fractions except for the Hudson BD strain, which had colonies almost exclusively in the >53 µm fraction. The Hudson BD strain also differed from the others in its very heavy investment of mucilage around the cells (Figure 32.1).

Overall, experiments showed an extreme range of mussel clearance rates and behavioral responses that were not necessarily a function of microcystin concentration but rather other properties of the cells, such as mucilage or colony size. Small colonies of the toxic Bear AC strains were cleared at moderate rates—roughly 50% of that observed for the ideal food, *Cryptomonas*; however, large colonies of the Bear AC strain were not cleared at all (Figure 32.2). Both small and large colonies of the toxic Gull 8/23/00 strain were ingested at modest rates (Figure 32.3). Thus, larger colonies of the Gull 8/23/00 strain did not offer protection from grazing by mussels, in contrast

Table 32.1 Characteristics of Colonial *Microcystis* Strains and Lake of Origin of Strains Used in the Zebra Mussel Feeding Experiments with Laboratory Cultures. Strains Were Isolated from the Various Inland Lakes in Michigan by Wilson et al. (2005). Mussels for All Experiments Came from the Huron River, Except That Gull Lake Mussels Were Used for the Experiment with Sonicated Hudson BD Strain of *Microcystis*. TP Is the Total Phosphorus Concentration of the Lake of Origin. Also Listed Is the Control Alga (*Scenedesmus Obliquus* or *Cryptomonas ozolini*) That Was Given to Other Mussels to Serve as a Benchmark of What Feeding Would Be on a Desirable alga. Results of Video Observations on the Mussels Used in the Experiments Are Also Described. Microcystin (MC, μg/L) Was Measured by the ELISA Technique and Normalized to Chlorophyll Concentration (Chl, μg/L). The Gull Lake Strain Was Isolated in August 2000, and All Others in August 2002. The Date Given Is the Date the Experiments Were Conducted; All Experiments Were Run at 20°C

Date	Lake of Origin	Strain/Treatment	TP (μg/L)	MC/Chl Mean	MC/Chl SE	Video Observations
12/20/05	Gull Lake Dreissenids present	Gull 8/23/00 Toxin gene: yes	8	0.190	0.008	No observations
12/20/05		*Scenedesmus*				
1/13/2006	Gilkey Lake Dreissenids absent	Gilkey L Toxin gene: yes	17	0.099	0.015	Mussels exhibited a weak feeding current and stopped feeding. When *Cryptomonas* was added to the *Microcystis* suspension, the mussels resumed feeding and appeared to reject individual colonies of *Microcystis* immediately after they were siphoned.
1/13/2006		*Cryptomonas*				
2/3/2006	Hudson Lake Dreissenids absent	Hudson BD Toxin gene: no	49	0.003	0.0001	Strong filtering current with later forceful (group) rejections of large pseudofeces
2/3/2006		*Cryptomonas*				
3/10/2006	Bear Lake Dreissenids present	Bear AC Toxin gene: no? Mussels in lake	66	0.202	0.014	Strong filtering current and (group) rejections as large pseudofeces
3/10/2006		*Cryptomonas*				
6/22/07	Hudson Lake Dreissenids absent	Hudson BD sonicated Toxin gene: no	49	—	—	No observations

Table 32.2 Conditions for the Short-Term Feeding Experiments in Saginaw Bay (SB) and Lake Erie (LE) during August 2004 and 2005. Temperature (T), Microcystin (MC; μg/L), and Chlorophyll (Chl; μg/L) Were Measured at Time of Experiment. Total Phosphorus (TP) Was Derived from Literature. MC/Chl, Microcystin to Chlorophyll Ratio; ZM, Zebra Mussels; and QM, Quagga Mussel

Date	Treatment/Dominant Alga	T (°C)	TP (μg/L)	MC/Chl Mean	MC/Chl SE
8/24/2004	ZM & SB seston: *Microcystis* in >53 μm fraction	20	12 ± 4[a]	0.355	0.003
8/24/2004	QM & SB seston: *Microcystis* in >53 μm fraction	20	12 ± 4[a]	0.355	0.003
8/31/2005	QM & LE seston: *Aulacoseira* in >53 μm fraction	24	50–90[b]	—	—

[a] Stow (GLERL/NOAA, unpublished data) mean ± SD (N = 14) 2008–2010.
[b] Chaffin et al. (2011), range of 6 sites in western Lake Erie in 2008.

to the Bear AC strain, which did. We do not have detailed size distributions to evaluate whether the Bear AC strain colonies were just too large for ingestion. That clearance rates for the Gull 8/23/00 strain were the same as seen for the control algae *Scenedesmus* should not be interpreted as a very strong feeding rate response because *Scenedesmus* was presented at a very high concentration (Figure 32.3), thus leading to a considerably lower clearance rate (7 mL/mg/h) on this alga compared to higher clearance values seen on the control *Cryptomonas* (ranging between 27–50 mL/mg/h). Net clearance rates observed with the toxic Gilkey L strain and nontoxic Hudson BD strains were zero or not significantly different from zero (Figure 32.2). In the case of the Gilkey L strain, colonies were available

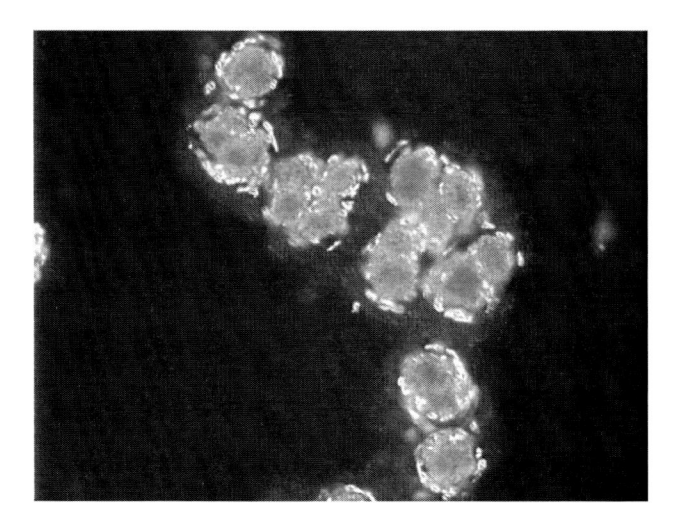

Figure 32.1 The Hudson BD strain of *Microcystis aeruginosa*. The halo around the cells is mucilage.

in both size fractions, whereas in the case of the Hudson BD strain, there were a few small colonies. Note the high negative clearance rate in the <53 μm size fraction is likely a result of particle production (Vanderploeg et al. 1984), that is, the breakdown of larger colonies into colonies falling in the smaller size fraction, which had few colonies to begin with.

Video observations showed different behavioral responses to the strains. Zebra mussels showed strong filtering currents and vigorous rejection of pseudofeces for both Hudson BD and Bear AC strains (Table 32.1; see Sequences 2 and 3 in Vanderploeg and Strickler 2013). This result is particularly interesting for the Hudson BD strain because there was no significant overall net clearance on colonies. In contrast to observations for other strains, mussels showed a strong negative behavioral response to the Gilkey L strain (Table 32.1). Siphons were only partially extended, and feeding currents and expulsion of *Microcystis*-containing pseudofeces were weak (see Sequence 4 in Vanderploeg and Strickler 2013). After *Cryptomonas* was added to the feeding suspension, the mussels exhibited normal feeding behavior (siphons fully extended with a strong feeding current), and it was observed that individual *Microcystis* colonies were rejected immediately after entering the filter chamber (see Sequence 5 in Vanderploeg and Strickler 2013).

To investigate whether the lack of ingestion of the Hudson BD strain was a result of colonies simply being too large for ingestion, we created a mixture of sonicated and unsonicated colonies (Table 32.1; Figure 32.2). Overall, there was no net clearance on either small or large colonies of this nontoxic strain. It is possible that the response was related to some property of the mucilage, since a small halo of mucilage was seen around the sonicated cells and colonies. Also, we cannot rule out other cellular constituents not identified in the present study.

Experiments with Natural Seston

The standard (short-term) clearance-rate experiments conducted in the summers of 2004 and 2005 showed that when *Microcystis* dominated (found in >53 μm fraction) in Saginaw Bay seston, there was no significant feeding by either zebra or quagga mussels; that is, clearance rates for total Chl were not significantly different from zero (Table 32.2; Figure 32.4). The considerable negative clearance rates on the <53 μm fraction suggested breakage of larger colonies into smaller colonies. These experiments represent the only time we were able to collect both quagga and zebra mussels at the same site for comparison when *Microcystis* was dominant. In the experiment of August 31, 2005, with Lake Erie seston, quagga mussels readily fed on the large size fraction dominated by the colonial diatom *Aulacoseira*, demonstrating that other large colonial algae were readily ingested. We do not have details on the upper size range to know if the colonies were really large and beyond some ingestible size range.

Experiments with *Microcystis* colonies from Gull Lake showed that mussels did not feed on either large colonies or small colonies derived from the large colonies by sonication (Table 32.3 and Figure 32.3). In the experiment in which the (unsonicated) colonies collected by net tow were mixed with *Cryptomonas* (August 11, 2005), zebra mussels readily fed on the *Cryptomonas* (present in the <53 μm fraction), but no *Microcystis* colonies (>53 μm fraction) were removed. In both experiments with mixtures of sonicated and unsonicated cells, small colonies of *Microcystis* were removed, but overall, there was no significant ingestion. In the experiment in August, there was a clear large negative clearance rate for the large (>53 μm) size fraction, consistent with particle production most likely in the form of pseudofeces. These results are in marked contrast to the modest feeding seen on both size categories of the Gull 8/23/00 strain. These results are consistent with the much lower microcystin to Chl (MC:Chl) concentration of the Gull 8/23/00 strain (0.190) relative to natural colonies collected in July (0.433) and August (0.777) (Tables 32.1 and 32.3).

Combined Long-Term/Short-Term Experiments with Natural Seston

Both experiments with Saginaw Bay and Lake Erie seston (Table 32.4) clearly showed that no *Microcystis* was ingested in the short- (2–3 h) or long-term (16 h) experimental time periods. There was no change in chlorophyll in either size fraction. The microcystin results show that most microcystin, and by extension toxic *Microcystis*, was found in the >53 μm fraction as expected (Vanderploeg et al. 2001). Microscopic analysis showed that the *Microcystis* colonies were large and would have been present in the larger size fraction. Clearance rates for

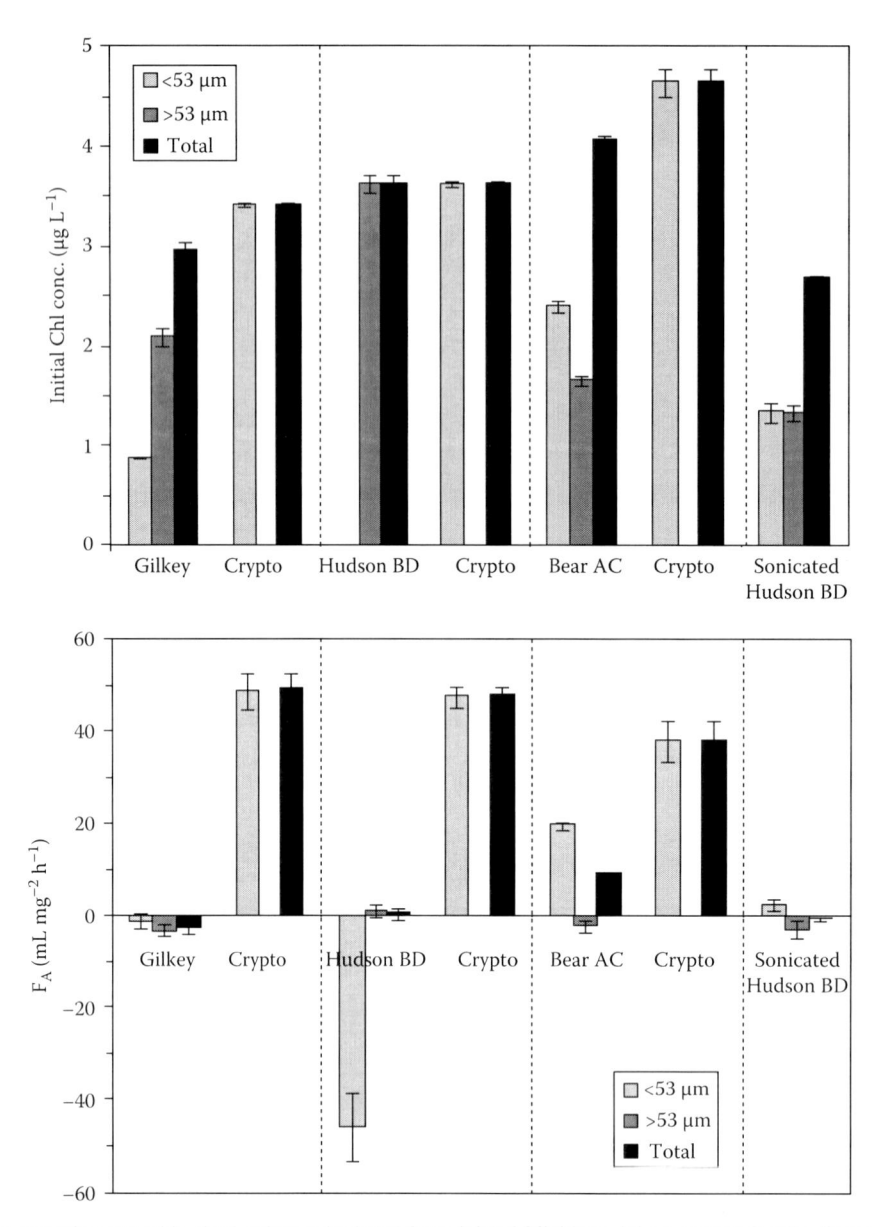

Figure 32.2 Feeding experiments with algal cultures isolated from inland Michigan lakes except for the Gull Lake strain (shown in Figure 32.3). To help put results in context of potential clearance rates, clearance rates on *Cryptomonas*, an ideal food for mussels, from the same collection are compared. Error bars are ±1 SE. Initial chlorophyll concentrations (Chl; upper panel) and net clearance rates (F_A; lower panel) of zebra mussels in feeding experiments designed to determine feeding response to different size fractions (<53 μm, >53 μm) of various *Microcystis* strains. The strains were isolated from various inland lakes in Michigan: Gilkey L strain from Gilkey Lake, Hudson BD strain from Hudson Lake, and Bear AC strain from Hudson Lake. Results for the Gull L strain from Gull Lake are given in Figure 32.3. The Hudson BD strain was sonicated to change the size structure of the colonies. To put clearance rates on the various strains in perspective, clearance rates on *Cryptomonas*, an ideal food, are also given. Error bars are ±1 SE.

microcystin in both these experiments were very low and not significantly different from zero (Table 32.4).

In contrast to results for Saginaw Bay and Lake Erie, results of the enclosure experiments with natural seston in Gull Lake point to modest feeding, including ingestion of *Microcystis* colonies that were codominant with diatoms. MC:Chl ratios were considerably higher in the Gull Lake mesocosms than in the Saginaw Bay and Lake Erie experiments, even though these ratios would be diluted by Chl in co-occurring diatoms, especially in the case of the experiment in mesocosm GL#7, where diatom biomass was greater than *Microcystis* biomass (Vanderploeg, unpublished data). In Saginaw Bay and Lake Erie experiments and in mesocosm GL#13, much of the microcystin and highest MC:Chl ratios were in the >53 μm fraction, whereas microcystin concentrations and

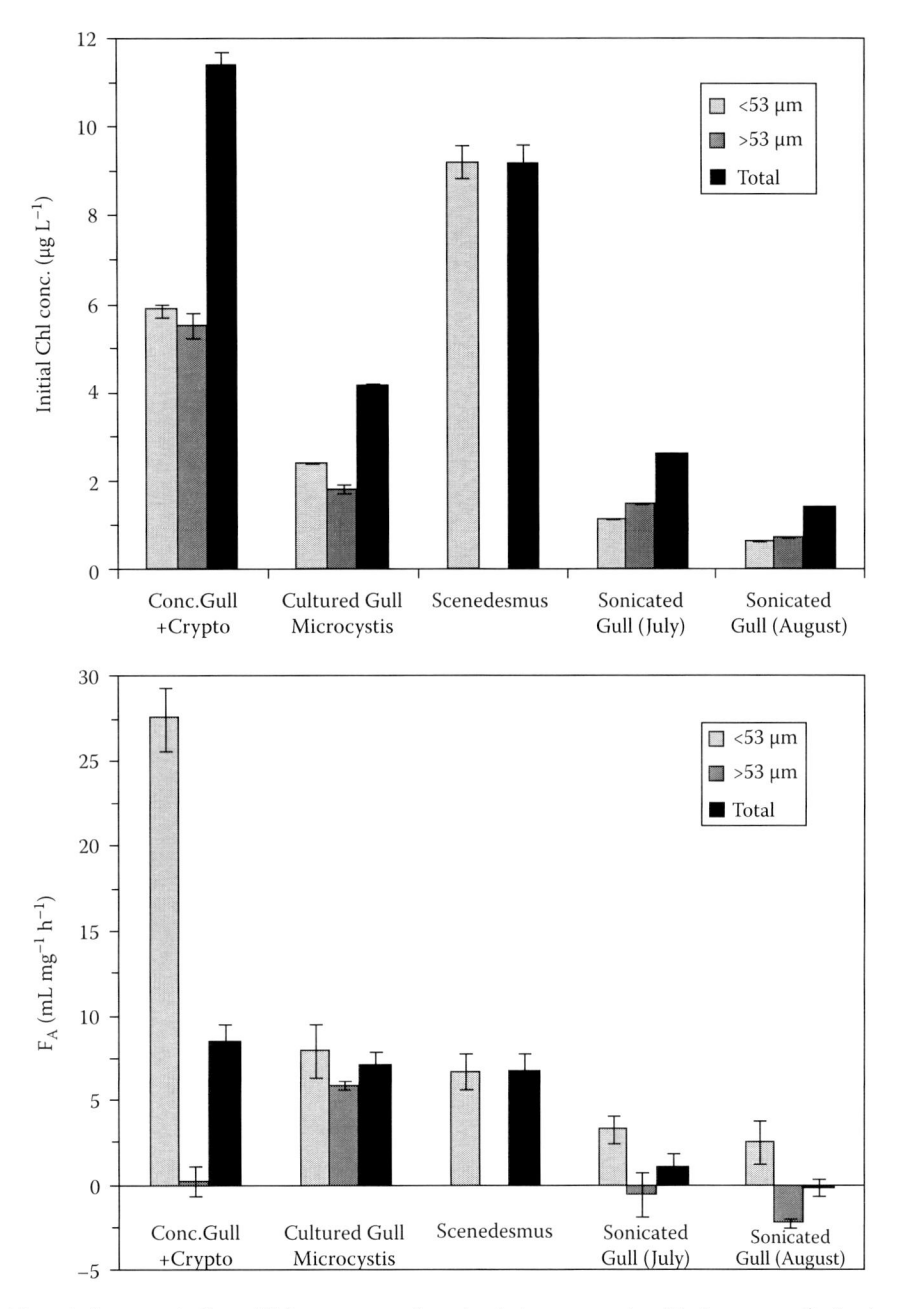

Figure 32.3 Initial chlorophyll concentrations (Chl; upper panel) and net clearance rates (F_A; lower panel) of zebra mussels in feeding experiments designed to determine feeding response to different size fractions (<53 μm, >53 μm) of the Gull 8/23/00 strain from Gull Lake and to different size fractions of *Microcystis* colonies collected in Gull lake with a plankton net. From left to right: *Microcystis* colonies mixed with cultured *Cryptomonas*, the 8/23/00 strain contrasted with results for *Scenedesmus* control culture, and mixtures of sonicated and unsonicated natural colonies from experiments in July and August 2007. Error bars are ± 1 SE. See Table 32.1 for details on the Gull L 8/23/00 strain, and Table 32.3 for details on experiments with naturally occurring Gull Lake *Microcystis* colonies.

M:Chl ratios were more evenly balanced between the two fractions in mesocosm GL#7. In the short-term experiments, we cannot specify whether anything in the >53 μm fraction was ingested because clearance rates had high standard errors (SEs). Moderate clearance rates were seen for the <53 μm fraction and for total Chl overall. Both microcystin and Chl in the long-term treatments were cleared at low, albeit statistically significant rates (2.2–4.7 mL/mg/h). Clearance rates on toxic *Microcystis* (microcystin) in the <53 μm size category were slightly higher than for the >53 μm fraction. Note that we would expect the long-term Chl clearance to be dominated in

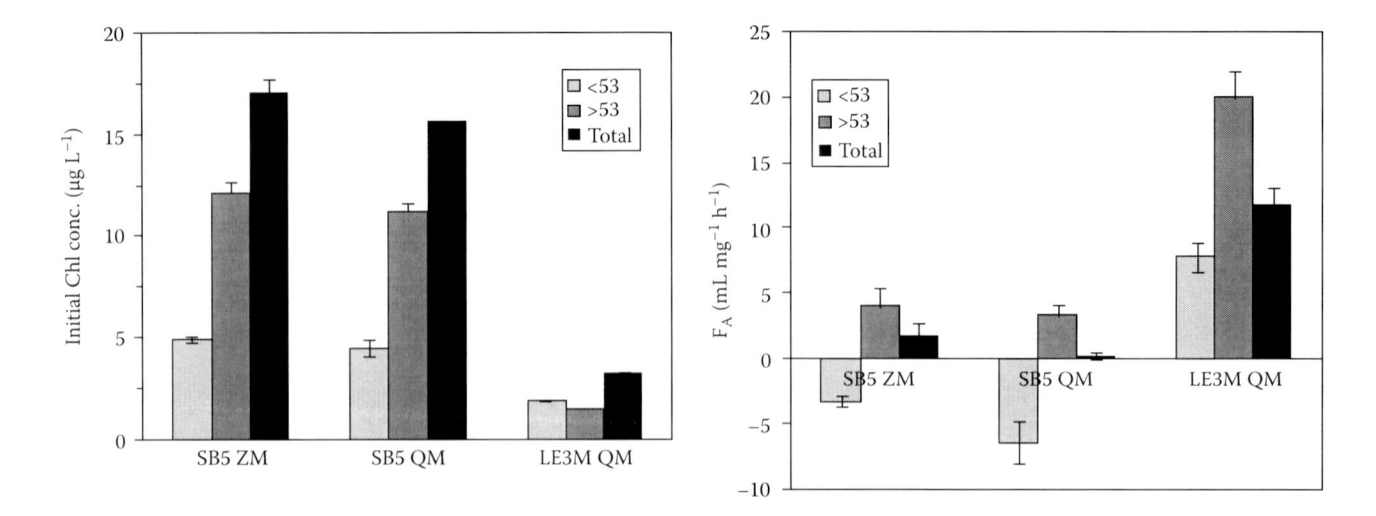

Figure 32.4 Initial chlorophyll concentrations (Chl; upper panel) and net clearance rates (F_A; lower panel) of zebra mussels (ZM) and quagga mussels (QM) in short-term feeding experiments designed to determine feeding response to different size fractions (<53 µm, >53 µm) of seston collected in Saginaw Bay (SB) and western Lake Erie (LE). The >53 µm fraction of the Saginaw Bay seston was dominated by *Microcystis*, and the >53 µm fraction of the Lake Erie seston was dominated by *Aulacoseira*. See Table 32.2 for conditions at the two collection sites. Error bars are ± 1 SE.

Table 32.3 Conditions of Experiments with *Microcystis* Colonies Collected with 100 µm Mesh Plankton Net in Open Water (11 m Depth Contour) of Gull Lake. Treatments as Indicated and Suspended in 0.2 µm Filtered Lake Water. Temperature (T), Microcystin (MC, µg/L), and Chlorophyll (Chl, µg/L) Were Measured at Time of Experiment. MC/Chl = Microcystin to Chlorophyll Ratio. Experiments Conducted with Zebra Mussels

Date	Treatment	T (°C)	MC/Chl Mean	SE
8/11/2005	*Microcystis* colonies mixed with cultured *Cryptomonas*	27	—	—
7/16/2007	Mixture of sonicated and unsonicated colonies	20	0.448	0.021
8/10/2007	Mixture of sonicated and unsonicated colonies	26	0.777	0.088

the long-term experiment by *Microcystis* kinetics because the only Chl left over the longer term would likely come from *Microcystis*.

DISCUSSION

Differences in Grazing Vulnerability of *Microcystis* among Strains and Lakes

Overall, clearance-rate experiments and visual observations reinforce the notion that rejection of *Microcystis* varies with strain as observed with both laboratory cultures and natural seston. We observed extreme variance of zebra mussel response to the four laboratory strains that could not be simply related to microcystin concentration or colony size. Our results reflect observations of extreme variance seen among other strains noted earlier, but our study also showed high variance for cultures that maintained colonial integrity at the time of experiments. Observations with the Gilkey L strain showed complete shutdown of the feeding response when presented with a sole food source, with feeding being resumed only when this *Microcystis* strain was mixed with *Cryptomonas*. This is in contrast to high feeding (enthusiastic filtering) responses to the Bear AC strain, which was ingested, and to the Hudson BD strain that was not ingested whether presented as large colonies or small sonicated colonies. In both Hudson BD and Bear AC strains, there was periodic rejection of larger pseudofeces. The immediate rejection of the Gilkey L colonies as they were captured likely indicated a secondary compound and not microcystin (or at least the microcystin measured by the ELISA method) irritated the mussels. The rejection response was so striking even though microcystin concentrations were only at modest levels.

Table 32.4 Mean (± SE) Clearance Rates (F$_A$) of Zebra Mussels Feeding on Large (>53 µm) and Small (<53 µm) Size Fractions of Seston in Short-Term (2–3 h) and Long-Term (~16 h) Experiments. Seston Was Collected from Inner Saginaw Bay (SB) and Western Lake Erie (LE). Also Given Are Clearance Rates on Seston in the Early Phase (within 1 Week of Setup) of Enclosure Experiments Conducted in Gull Lake. Clearance Rates Based on Changes in Chlorophyll and Microcystin (MC) during the Experimental Period. T = Temperature

Experiment	T (°C)	Size Fraction	Initial Concentration (µg/L) Chl	MC	MC/Chl	F$_A$ (mL/mg/h) Short Term Chl	Long Term Chl	MC
SB; 7/18–19/2006	26	>53 µm	8.19	1.04	0.127	−1.83 ± 0.17	0.19 ± 0.11	−0.28 ± 0.12
Microcystis dominant in >53 µm fraction		<53 µm	7.74	0.03	0.004	0.45 ± 0.57	0.04 ± 0.07	0.26 ± 0.84
		Total	14.81	1.07	0.072	−0.57 ± 0.29	0.11 ± 0.09	−0.27 ± 0.10
LE; 8/29–30/2006	24	>53 µm	16.12	2.97	0.170	0.19 ± 0.11	0.38 ± 0.12	0.11 ± 0.19
Microcystis dominant in >53 µm fraction		<53 µm	4.95	0.30	0.058	−1.20 ± 0.55	−0.11 ± 0.10	−0.48 ± 0.09
		Total	21.08	3.27	0.155	−0.37 ± 0.15	0.26 ± 0.10	0.05 ± 0.17
GL #13 7/10–11/2007	27	>53 µm	0.361	0.194	0.576	7.13 ± 5.81	2.78 ± 0.78	2.82 ± 0.22
4 g/m^2; low P; *Microcystis* and diatoms dominant		<53 µm	1.102	0.072	0.065	6.77 ± 1.38	2.74 ± 0.11	4.72 ± 0.81
		Total	1.463	0.266	0.183	7.28 ± 3.61	2.78 ± 0.20	3.20 ± 0.10
GL #7 7/11–12/2007	27	>53 µm	0.318	0.110	0.351	5.04 ± 5.24	1.64 ± 0.08	2.16 ± 0.24
1 g/m^2; low P		<53 µm	0.917	0.221	0.245	12.02 ± 1.32	1.98 ± 0.26	2.64 ± 0.61
Diatoms and *Microcystis* dominant		Total	1.235	0.331	0.273	9.47 ± 1.62	1.88 ± 0.18	2.44 ± 0.46

Short-term experiments with natural *Microcystis* colonies or with unmodified lake seston also point to mussels selectively rejecting the colonies or shutting down feeding responses. In short-term experiments with Saginaw Bay seston, no significant overall (total) clearance rate was observed for zebra and quagga mussels when *Microcystis* was dominant in the >53 µm size fraction. Likewise, Gull Lake colonies captured in plankton net from the open lake—whether presented to zebra mussels alone or as a combination of sonicated and unsonicated colonies—were not ingested. This was in marked contrast to the Gull 8/23/00 strain that was readily ingested by zebra mussels. The natural Gull Lake colonies had extremely high microcystin levels in contrast to the isolated strain. In the combined long- and short-term experiments, using *Microcystis* from Saginaw Bay and Lake Erie, zebra mussels did not ingest *Microcystis* or any phytoplankton, whether clearance rate was measured from Chl or microcystin content. Experiments in the two enclosures showed that toxic *Microcystis* (measured by microcystin) was cleared at low rates in the long-term experiments. This was also reinforced by observed ingestion of Chl.

There can be multiple *Microcystis* strains in lakes that vary in terms of both colony size and form, and toxicity; these strains can occur at different times or simultaneously in the same lake (e.g., Dionisio Pires et al. 2005, Wilson et al. 2005, van Gremberghe et al. 2009, White et al. 2011). White et al. (2011) used various *Microcystis* strains isolated from Gull Lake, including two isolated on the same date, to study selectivity responses of zebra mussels. Relative to the green alga *Ankistrodesmus*, strain selectivity ranged from one (equal preference relative to *Ankistrodesmus*) to zero.

As noted earlier, the question of whether selective grazing by mussels can promote *Microcystis* dominance in different systems—including those with different TP concentrations—hinges on two issues. First, will the mussels ingest or reject the strains found in a given system? Second, is mussel filtering impact large enough to induce significant mortality on phytoplankton relative to their growth rate? With regard to the first question, strains isolated from both low and high TP lakes can have low or high vulnerability to grazing by mussels. As noted earlier, Dionisio Pires et al. (2005) observed moderate clearance rates on both toxic and nontoxic strains isolated from a high TP lake in The Netherlands. For the strains from Michigan inland lakes, the results were highly variable. Gull 8/23/00 and Gilkey L strains were both from lakes with TP concentrations of <20 µg/L, and clearance rates were high and low, respectively. Bear AC and Hudson BD strains were both from lakes having high (49–66 µg/L) TP concentrations, and clearance rates were high and low, respectively. The strongly rejected LE-3 strain (Brittain et al. 2000) used by Vanderploeg et al. (2001) had a MC:Chl ratio of 0.66 and came from Hatchery Bay of South Bass Island (eastern portion of the western basin of Lake Erie), where TP concentration was 35 µg/L (Holland et al. 1995). As noted earlier, White et al. (2011) noted large variations in zebra mussel grazing across *Microcystis* strains in Gull Lake (TP = 8 µg/L). Also, results of all these experiments show there is no apparent general relationship of grazing response with microcystin concentration.

Despite earlier observations that *Microcystis* in natural seston was readily ingested in high TP systems (e.g., Bastviken et al. 1998, Dionisio Pires et al. 2005), our experiments with seston from Saginaw Bay (TP = 12 ± 4 µg/L) and from western Lake Erie near Maumee Bay (TP = 50–90 µg/L) and previous experiments with seston from western Lake Erie near Hatchery Bay (TP = 35 µg/L) (Vanderploeg et al. 2001) suggest a variable vulnerability of *Microcystis* to mussel grazing that cannot simply be related to TP concentration. Clearance rates in TP-rich western Lake Erie were essentially zero in our experiments and in the experiments of Vanderploeg et al. (2001). Likewise, our experiments with whole and sonicated colonies from the open waters of Gull Lake showed *Microcystis* (derived from plankton net concentrate) was not ingested. These large net-collected colonies presumably survived despite the biomass of mussels in the lake. In this regard, it is worth noting that the concentration of mussels in the lake (~6 g/m^2) was higher than in the enclosures, a necessity because of low survival of mussels stocked at concentrations higher than 4 g/m^2. Interestingly, the combination of short- and long-term feeding experiments did point to an overall low feeding rate on the assemblage of *Microcystis* found in the Gull Lake enclosures. As in the experiments with cultures, no correlation could be made with microcystin concentration.

It is theoretically possible that, in addition to strain effects of *Microcystis*, there can be genetic differences or even phenotypic adaptations among mussel populations to *Microcystis* found in different water bodies. For example, *Daphnia* clones from eutrophic environments were less sensitive to the negative impact of ingestion of a unicellular toxic *Microcystis* strain (Sarnelle and Wilson 2005). Can this same mechanism apply to dreissenid mussels? As mentioned, Dionisio Pires et al. (2005) found that clearance rates of zebra mussels feeding on small colonies of *Microcystis* were similar to rates on other phytoplankton in a eutrophic lake in The Netherlands. Is it possible that zebra mussels in this lake were genetically or phenotypically adapted to the relatively small colonies *Microcystis* found there? Interestingly, although such a phenomenon has not been observed for *Microcystis*–dreissenid interactions, there are reports of population-specific responses by the marine mussel *Mytilus edulis* to the toxic dinoflagellate *Protogonyaulax tamarensis* (Shumway and Cucci 1987). The GT429 strain of this dinoflagellate, which produces a potent neurotoxin, was readily ingested by *M. edulis* from Maine (USA) with no impacts, yet ingestion led to shell closure and death of mussels from New Jersey (USA) and Spain.

Projecting Mussel Impact on Phytoplankton Community

Zebra mussels (and quagga mussels) have clearance rates on a per unit weight basis that are about the same as the filter-feeding zooplankter *Daphnia* (Dionisio Pires et al. 2005). The potential of mussels to affect major changes in water clarity and change phytoplankton composition depends on their ability to develop large populations (Nalepa et al. 2010) relative to other potential grazers (Vanderploeg et al. 2001, 2002, 2009, 2010) and on the ability to feed on a broad size range of particles. In contrast, zooplankton are constrained to feeding on small phytoplankton (typically <30 µm) (e.g., Vanderploeg et al. 1988, Vanderploeg 1994, Dionisio Pires et al. 2005).

The carrying capacity of dreissenids for a given water body can be affected by nutrient loads and calcium concentrations. Wilson and Sarnelle (2002) observed a significant correlation between mussel abundance and TP, but the amount of variation in mussel biomass explained was relatively low (24%). The risk of dreissenid invasion and establishment increases as calcium concentrations increase above a certain threshold (Whittier et al. 2008). With the expansion of energetically efficient quagga mussels, which can utilize soft substrates as well as hard substrates including those found in the hypolimnion of lakes, it is possible that much larger populations can develop, leading to even greater impacts (Nalepa et al. 2010, Vanderploeg et al. 2010). In Lake Michigan, for example, large enough quagga mussel populations developed so as to decimate the spring phytoplankton bloom and leave behind a phytoplankton community with a higher proportion of cyanobacteria than before quagga mussels expanded (Fahnenstiel et al. 2010, Nalepa et al. 2010, Vanderploeg et al. 2010). Note that we would not expect *Microcystis* to be dominant in Lake Michigan because of relatively cool water temperatures in this large, deep lake. The question remains as to how mussel carrying capacity is related to nutrient loading and phytoplankton production, which obviously will dictate mussel impacts.

An important advantage mussels have over zooplankton in affecting phytoplankton communities is their long life span and lower vulnerability to predation. Dreissenids can grow and increase in biomass during spring and winter periods when food quality is good and survive during summer when food quality is poor (e.g., Vanderploeg et al. 2002, 2009). In contrast, *Daphnia* have short life spans and typically occur in spring and summer and then decline as phytoplankton food quality decreases and planktivorous fish populations increase (e.g., Sommer et al. 1986). However, mussels are not immune to predation, particularly in the Great Lakes, where significant predation can occur from round gobies (*Neogobius melanostomus*), which is another invader from the Ponto-Caspian region (Vanderploeg et al. 2002).

Information on temporal changes in mussel biomass in areas of the Great Lakes where thermal (high water temperature) and nutrient characteristics would support *Microcystis* growth during summer is rather limited. However, we know, for example, that abundances of dreissenids in Saginaw Bay are now lower than during the 1990s, when round gobies were not present (Nalepa, unpublished). Round goby abundance in turn is affected by predation from piscivorous fishes (e.g., Madenjian et al. 2011), and the habitat characteristics that promote high removal of gobies is not well defined. Thus, it is difficult to predict what dreissenid biomass will be in the future; likewise, we cannot with certainty predict future impacts.

Microcystis Affects Filtering Intensity

We observed that mussels can selectively reject *Microcystis* and simultaneously feed at high rates on high-quality algae when available. However, it is obvious that once *Microcystis* becomes a major component of the phytoplankton, pumping rates of the mussels decreases—at least for some *Microcystis* strains—as observed in experiments here and those reported by Vanderploeg et al. (2001, 2009). Hence, once *Microcystis* becomes a dominant component of the phytoplankton, mussels are no longer a factor in the promotion of *Microcystis*. This phenomenon has been reported by van Gremberghe et al. (2009) for *Daphnia–Microcystis* interactions. This feeding shutdown may be responsible for the existence of grazing-resistant and grazing-vulnerable *Microcystis* colonies in Gull Lake (White et al. 2011). Thus, at least for some environments, such as Saginaw Bay, promotion of *Microcystis* blooms by dreissenids likely occurs early in the summer season when *Microcystis* first develops. Interestingly, we have seen *Microcystis* blooms quite late in the summer or early fall on Lake Erie (Vanderploeg et al. 2001). This would be consistent with mussels removing palatable algae that would normally start to succeed summer species like *Microcystis*, as waters begin to cool down.

Nutrients and Phytoplankton Growth Rate

For a given *Microcystis* strain having a defined vulnerability to grazing by mussels or zooplankton, nutrient concentrations can potentially affect competitive outcomes between *Microcystis* and other phytoplankton species because grazing impact relative to phytoplankton growth rate will vary with nutrient concentration and mussel density. Thus, unless the fraction of the water column cleared by mussels (FC) is an appreciable fraction of phytoplankton community growth rate, selective rejection of *Microcystis* will not affect the competitive outcome between *Microcystis* and other species (Vanderploeg et al. 2001). Typical growth rates of phytoplankton in nutrient-limited systems range between 0.2 and 0.5/d, with a maximum between 1 and 2/d in systems where nutrients are not limited (Reynolds 1997). Measured growth rates in Saginaw Bay during summer were 0.20–0.25/d. *Microcystis* colony growth is relatively slow (e.g., Fahnenstiel et al. 1995, Reynolds 1997), and large colonies (within and across clones) grow more slowly than small colonies (e.g., Wilson et al. 2010). Wilson et al. (2010) observed considerable variation in growth of *Microcystis* strains under nutrient-saturating conditions, but most values fell between 0.2 and 0.5/d. One can imagine that under high nutrient conditions or during a nutrient-enrichment experiment in an enclosure that competitors of *Microcystis* with high growth rates could overcome effects of selective rejection, unless *Microcystis* rejection rates were high and FC was high enough to overcome the growth rate of the competitors, which could be on the order of 0.5–1/d. Note that other variables such as mixing, temperature, and light can affect the competitive outcome as well. *Microcystis* has the capacity to regulate its position in the water column with gas vacuoles and, as noted earlier, it does best in warm water temperatures and grows better under high light regimes (e.g., Reynolds 1997).

Future Directions

Variable responses of dreissenids to *Microcystis* as observed in the laboratory and field reflect similar responses of zooplankton to *Microcystis* (e.g., van Gremberghe et al. 2009). This variable response means that laboratory experiments with strains available from culture collections, even from the same lake, may sometimes have little meaning for understanding *Microcystis*–grazer interactions in the field, especially when strains lack naturally occurring characteristics such as colony formation. For example, note our contrasting results with the cultures from Gull Lake and colonies captured freshly from the same lake. We observed that the cultures did change over time, and at this writing, none of the colonial *Microcystis* used in the described experiments retained its colonial form. The question may also be asked as to whether the problem of cultures changing over time also applies to secondary compounds that deter grazing. An example is the LE-3 strain, a strain isolated from a bloom in Lake Erie that did not have much colonial integrity, but maintained a chemical deterrent.

Another problem is that many strains of *Microcystis* can co-occur in lakes and this makes generalizations based on a few isolates difficult. Relevant in this regard were observations of White et al. (2011) who showed different responses to strains isolated from Gull Lake, including two isolated on the same day.

It is also possible that *Microcystis*–grazer interactions will not be stable over time as populations adjust to each other. This interplay could include a shift in dominance of particular strains (*Microcystis* and mussels), phenotypic response to changing environmental conditions (including each other), as well as genetic adaptation that can be relatively rapid in the case of *Microcystis* (e.g., Rouco et al. 2011). Understanding important mechanisms associated with dreissenid–*Microcystis* interactions is more urgent than for zooplankton–*Microcystis* interactions since dreissenids are relatively unconstrained by phytoplankton size and can function either to promote harmful cyanobacteria or to control them.

Understanding dreissenid–*Microcystis* interactions in the field has presented difficult experimental challenges. First, there is an enormous amount of work involved in identifying and counting phytoplankton in all the replicates of an experiment, including initial concentrations, concentrations in the water column, and concentrations of settled material of control and experimental beakers (e.g., Vanderploeg et al. 2001). This is especially pertinent to experiments with seston, since taxonomic expertise and estimates of biomass for each of the diversely shaped species are required. *Microcystis* presents the added difficulty of having to estimate the number of cells in colonies of varying sizes and shapes. That is why in many of our experiments, we looked at Chl concentrations, including the larger size fraction, which would account for large colonies. For the same reason we used microcystin—which also can be easily measured—as a surrogate for toxic *Microcystis*. In Gull Lake, microcystin concentration is well correlated with *Microcystis* biomass ($R^2 = 0.87$; Sarnelle et al. 2012). We note that imaging instruments such as flow cytometers (Dionisio Pires et al. 2005) and digital imaging flow cytometers (FlowCAM; Fluid Imaging, Inc) may help automate counting and sizing of phytoplankton and distinguishing them from *Microcystis*. Ultimately, we would like to measure feeding on different strains in nature. In principle, this could be done by examining changes in the genetic material in a feeding experiment.

REFERENCES

Arnott, D. L. and M. J. Vanni. 1996. Nitrogen and phosphorus recycling by the zebra mussel (*Dreissena polymorpha*) in the western basin of Lake Erie. *Can. J. Fish. Aquat. Sci.* 53: 646–659.

Baker, S. M., J. S. Levinton, J. P. Kurdziel, and S. E. Shumway. 1998. Selective feeding and biodeposition by zebra mussels and their relation to changes in phytoplankton composition and seston load. *J. Shellfish Res.* 17: 1207–1213.

Bastviken, D. T. E., N. F. Caraco, and J. J. Cole. 1998. Experimental measurements of zebra mussel (*Dreissena polymorpha*) impacts on phytoplankton community composition. *Freshwat. Biol.* 39: 375–386.

Bierman, V. J., D. M. Dolan, and R. Kasprzyk. 1984. Retrospective analysis of the response of Saginaw Bay, Lake Huron to reductions in phosphorus loadings. *Environ. Sci. Technol.* 18: 23–31.

Bierman, V. J., J. Kaur, J. V. DePinto, T. J. Feist, and D. W. Dilks. 2005. Modeling the role of zebra mussels in the proliferation of blue-green algae in Saginaw Bay, Lake Huron. *J. Great Lakes Res.* 31: 32–55.

Brittain, S. M., J. Wang, L. Babcock-Jackson, W. W. Carmichael, K. L. Rinehart, and D. A. Culver. 2000. Isolation and characterization of microcystins, cyclic hepatotoxins from a Lake Erie strain of *Microcystis aeruginosa. J. Great Lakes Res.* 26: 241–249.

Bykova, O., A. Laursen, V. Bostan, J. Bautista, and L. McCarthy. 2006. Do zebra mussels (*Dreissena polymorpha*) alter lake water chemistry in a way that favours *Microcystis* growth? *Sci. Total Environ.* 371: 362–372.

Dionisio Pires, L. M., B. W. Ibelings, M. Brehm, and E. Van Donk. 2005. Comparing grazing on lake seston by *Dreissena* and *Daphnia*: Lessons for biomanipulation. *Microb. Ecol.* 50: 242–252.

Fahnenstiel, G. L., T. B. Bridgeman, G. A. Lang, M. J. McCormick, and T. F. Nalepa. 1995. Phytoplankton productivity in Saginaw Bay, Lake Huron: Effects of zebra mussel (*Dreissena polymorpha*) colonization. *J. Great Lakes Res.* 21: 465–475.

Fahnenstiel, G., S. Pothoven, H. Vanderploeg, D. Klarer, T. Nalepa, and D. Scavia. 2010. Recent changes in primary production and phytoplankton in the offshore region of southeastern Lake Michigan. *J. Great Lakes Res.* 36: 20–29.

Fishman, D. B., S. A. Adlerstein, H. A. Vanderploeg, G. L. Fahnenstiel, and D. Scavia. 2009. Causes of phytoplankton changes in Saginaw Bay, Lake Huron, during the zebra mussel invasion. *J. Great Lakes Res.* 35: 482–495.

Fulton, R. S. and H. W. Paerl. 1987a. Toxic and inhibitory effects of the blue-green-alga *Microcystis aeruginosa* on herbivorous zooplankton. *J. Plankton Res.* 9: 837–855.

Fulton, R. S. and H. W. Paerl. 1987b. Effects of colonial morphology on zooplankton utilization of algal resources during blue-green-algal (*Microcystis aeruginosa*) blooms. *Limnol. Oceanogr.* 32: 634–644.

Ger, K. A., S. J. Teh, D. V. Baxa, S. Lesmeister, and C. R. Goldman. 2010. The effects of dietary *Microcystis aeruginosa* and microcystin on the copepods of the upper San Francisco Estuary. *Freshwat. Biol.* 55: 1548–1559.

van Gremberghe, I., P. Vanormelingen, K. Van der Gucht et al. 2009. Influence of *Daphnia* infochemicals on functional traits of *Microcystis* strains (Cyanobacteria). *Hydrobiologia* 635: 147–155.

Holland, R. E., T. H. Johengen, and A. M. Beeton. 1995. Trends in nutrient concentrations in Hatchery Bay, Western Lake Erie, before and after *Dreissena polymorpha. Can. J. Fish. Aquat. Sci.* 52: 1202–1209.

Johengen, T. H., H. A. Vanderploeg, and J. R. Liebig. 2013. Effects of algal composition, seston stoichiometry, and feeding rate on zebra mussel (*Dreissena polymorpha*) nutrient excretion in two Laurentian Great Lakes. In *Quagga and Zebra Mussels: Biology, Impacts, and Control.* 2nd Edn., T. F. Nalepa and D. W. Schloesser, eds., pp. 445–459. Boca Raton, FL: CRC Press.

Juhel, G., J. Davenport, J. O'Halloran, S. Culloty, R. Ramsay, K. James, A. Furey, and O. Allis. 2006. Pseudodiarrhea in zebra mussels *Dreissena polymorpha* (Pallas) exposed to microcystins. *J. Exp. Biol.* 209: 810–816.

Knoll, L. B., O. Sarnelle, S. K. Hamilton et al. 2008. Invasive zebra mussels (*Dreissena polymorpha*) increase cyanobacterial toxin concentrations in low-nutrient lakes. *Can. J. Fish. Aquat. Sci.* 65: 448–455.

Kryger, J. and H. U. Riisgard. 1988. Filtration-rate capacities in 6 species of European fresh-water bivalues. *Oecologia.* 77: 34–38.

Lampert, W. 1982. Further studies on the inhibitory effect of the toxic blue-green *microcystis aeruginosa* on the filtering rate of zooplankton. *Archiv. Hydrobiol.* 95: 207–220.

Madenjian, C. P., M. A. Stapanian, L. D. Witzel, D. W. Einhouse, S. A. Pothoven, and H. L. Whitford. 2011. Evidence for predatory control of the invasive round goby. *Biol. Invasions* 13: 987–1002.

Makarewicz, J. C. 1993. Phytoplankton biomass and species composition in Lake Erie, 1970 to 1987. *J. Great Lakes Res.* 19: 258–274.

Nalepa, T. F., D. L. Fanslow, and S. A. Pothoven. 2010. Recent changes in density, biomass, recruitment, size structure, and nutritional state of *Dreissena* populations in southern Lake Michigan. *J. Great Lakes Res.* 36: 5–19.

Nizan, S., C. Dimentman, and M. Shilo. 1986. Acute toxic effects of the cyanobacterium *Microcystis aeruginosa* on *Daphnia magna. Limnol. Oceanogr.* 31: 497–502.

Otsuka, S., S. Suda, S. Shibata, H. Oyaizu, S. Matsumoto, and M. M. Watanabe. 2001. A proposal for the unification of five species of the cyanobacterial genus *Microcystis* Kützing ex Lemmermann 1907 under the Rules of the Bacteriological Code. *Int. J. Syst. Evol. Microbiol.* 51: 873–879.

Pires, L. M. D., B. M. Bontes, E. Van Donk, and B. W. Ibelings. 2005. Grazing on colonial and filamentous, toxic and non-toxic cyanobacteria by the zebra mussel *Dreissena polymorpha*. *J. Plankton Res.* 27: 331–339.

Raikow, D. F., O. Sarnelle, A. E. Wilson, and S. K. Hamilton. 2004. Dominance of the noxious cyanobacterium *Microcystis aeruginosa* in low-nutrient lakes is associated with exotic zebra mussels. *Limnol. Oceanogr.* 49: 482–487.

Reynolds, C. S. 1997. *Vegetation Processes in the Pelagic: A Model for Ecosystem Theory*. Oldendorf/Luhe, Germany: Ecology Institute.

Reynolds, C. S. 2006. *Ecology of Phytoplankton*. Cambridge, U.K.: Cambridge University Press.

Rouco, M., V. Lopez-Rodas, A. Flores-Moya, and E. Costas. 2011. Evolutionary changes in growth rate and toxin production in the cyanobacterium *Microcystis aeruginosa* under a scenario of eutrophication and temperature increase. *Microb. Ecol.* 62: 265–273.

Sarnelle, O. and A. E. Wilson. 2005. Local adaptation of *Daphnia Pulicaria* to toxic cyanobacteria. *Limnol. Oceanogr.* 50: 1565–1570.

Sarnelle, O., J. D. White, G. P. Horst, and S. K. Hamilton. 2012. Phosphorus addition reverses the positive effect of zebra mussels (*Dreissena polymorpha*) on the toxic cyanobacterium, *Microcystis aeruginosa*. *Water Res.* 46: 3471–3478.

Sarnelle, O. and A. E. Wilson. 2005. Local adaptation of Daphnia pulicaria to toxic cyanobacteria, *Limnol. Oceanogr.* 50: 1565–1570.

Sarnelle, O., A. E. Wilson, S. K. Hamilton, L. B. Knoll, and D. F. Raikow. 2005. Complex interactions between the zebra mussel, *Dreissena polymorpha*, and the harmful phytoplankton, *Microcystis aeruginosa*. *Limnol. Oceanogr.* 50: 896–904.

Shen, H., Y. Niu, P. Xie, M. Tao, and X. Yang. 2011. Morphological and physiological changes in *Microcystis aeruginosa* as a result of interactions with heterotrophic bacteria. *Freshwat. Biol.* 56: 1065–1080.

Shumway, S. E. and T. L. Cucci. 1987. The effects of the toxic dinoflagellate Progonyaulax tamarensis on the feeding and behavior of bivalve molluscs. *Aquat. Toxicol.* 10: 9–27.

Sivonen, K. and G. Jones. 1999. Cyanobacterial toxins. In *Toxic Cyanobacteria in Water*. I. Chorus and J. Bartram, eds., pp. 41–111. London, U.K.: E&F Spon.

Smith, V. H. 1983. Low nitrogen to phosphorus ratios favor dominance by blue-green algae in lake phytoplankton. *Science* 221: 669–671.

Sommer, U., Z. M. Gliwicz, W. Lampert, and A. Duncan. 1986. The PEG model of seasonal succession of planktonic events in fresh waters. *Archiv. Hydrobiol.* 106: 433–471.

Stoermer, E. F. and E. Theriot. 1985. Phytoplankton distribution in Saginaw Bay. *J. Great Lakes Res.* 11: 132–142.

Vanderploeg, H. 1994. Zooplankton particle selection and feeding mechanisms. In *The Biology of Particles in Aquatic Systems*. R. S. Wotton, ed., pp. 205–234. Boca Raton, FL: CRC Press.

Vanderploeg, H. A., T. H. Johengen, and J. R. Liebig. 2009. Feedback between zebra mussel selective feeding and algal composition affects mussel condition: Did the regime changer pay a price for its success? *Freshwat. Biol.* 54: 47–63.

Vanderploeg, H. A., J. R. Liebig, and W. W. Carmichael et al. 2001. Zebra mussel (*Dreissena polymorpha*) selective filtration promoted toxic *Microcystis* blooms in Saginaw Bay (Lake Huron) and Lake Erie. *Can. J. Fish. Aquat. Sci.* 58: 1208–1221.

Vanderploeg, H. A., J. R. Liebig, and A. A. Gluck. 1996. Evaluation of different phytoplankton for supporting development of zebra mussel larvae (*Dreissena polymorpha*): The importance of size and polyunsaturated fatty acid content. *J. Great Lakes Res.* 22: 36–45.

Vanderploeg, H. A., J. R. Liebig, T. F. Nalepa, G. L. Fahnenstiel, and S. A. Pothoven. 2010. Dreissena and the disappearance of the spring phytoplankton bloom in Lake Michigan. *J. Great Lakes Res.* 36: 50–59.

Vanderploeg, H. A., T. F. Nalepa, and D. J. Jude et al. 2002. Dispersal and emerging ecological impacts of Ponto-Caspian species in the Laurentian Great Lakes. *Can. J. Fish. Aquat. Sci.* 59: 1209–1228.

Vanderploeg, H. A., G. A. Paffenhöfer, and J. R. Liebig. 1988. *Diaptomus* vs. net phytoplankton: Effects of algal size and morphology on selectivity of a behaviorally flexible, omnivorous copepod. *Bull. Mar. Sci.* 43: 377–394.

Vanderploeg, H. A., D. Scavia, and J. R. Liebig. 1984. Feeding rate of *Diaptomus sicilis* and its relation to selectivity and effective food concentration in algal mixtures and in Lake Michigan. *J. Plankton Res.* 6: 919–941.

Vanderploeg, H. A. and J. R. Strickler. 2013. Video Clip 6: Behavior of zebra mussels (*Dreissena polymorpha*) exposed to *Microcystis* colonies from natural seston and laboratory cultures. In *Quagga and Zebra Mussels: Biology, Impacts, and Control*, 2nd Edn., T. F. Nalepa and D. W. Schloesser, eds., pp. 757. Boca Raton, FL: CRC Press (this volume).

White, J. D., R. B. Kaul, L. B. Knoll, A. E. Wilson, and O. Sarnelle. 2011. Large variation in vulnerability to grazing within a population of the colonial phytoplankter, *Microcystis aeruginosa*. *Limnol. Oceanogr.* 56: 1714–1724.

Whittier, T. R., P. L. Ringold, A. T. Herlihy, and S. M. Pierson. 2008. A calcium-based invasion risk assessment for zebra and quagga mussels (*Dreissena* spp.). *Front. Ecol. Environ.* 6: 180–184.

Wiegand, C. and S. Pflugmacher. 2005. Ecotoxicological effects of selected cyanobacterial secondary metabolites a short review. *Toxicol. Appl. Pharm.* 203: 201–218.

Wilson, A. E., R. B. Kaul, and O. Sarnelle. 2010. Growth rate consequences of coloniality in a harmful phytoplankter. *PLoS One* 5. DOI: 10.1371/journal.pone.0008679.

Wilson, A. E. and O. Sarnelle. 2002. Relationship between zebra mussel biomass and total phosphorus in European and North American lakes. *Archiv. Hydrobiol.* 153: 339–351.

Wilson, A. E., O. Sarnelle, B. A. Neilan, T. P. Salmon, M. M. Gehringer, and M. E. Hay. 2005. Genetic variation of the bloom-forming cyanobacterium *Microcystis aeruginosa* within and among lakes: Implications for harmful algal blooms. *Appl. Environ. Microbiol.* 71: 6126–6133.

Wilson, A. E., O. Sarnelle, and A. R. Tillmanns. 2006. Effects of cyanobacterial toxicity and morphology on the population growth of freshwater zooplankton: Meta-analyses of laboratory experiments. *Limnol. Oceanogr.* 51: 1915–1924.

Zhang, H., D. A. Culver, and L. Boegman. 2011. Dreissenids in Lake Erie: An algal filter or a fertilizer? *Aquat. Invasions* 6: 175–194.

CHAPTER 33

Trends in Phytoplankton, Zooplankton, and Macroinvertebrates in Saginaw Bay Relative to Zebra Mussel (*Dreissena polymorpha*) Colonization
A Generalized Linear Model Approach

Sara Adlerstein, Thomas F. Nalepa, Henry A. Vanderploeg, and Gary L. Fahnenstiel

CONTENTS

Abstract 525
Introduction 526
Methods 526
    Study Site and Data Collections 526
    Data Analysis 527
Results 528
    Zebra Mussels 528
    Phytoplankton 528
    Zooplankton 529
    Benthic Macroinvertebrates 531
Discussion 535
Acknowledgments 541
References 541

ABSTRACT

We quantify temporal and spatial trends in densities of main taxonomic groups of phytoplankton, zooplankton, and benthic macroinvertebrates relative to the colonization of Saginaw Bay, Lake Huron, by zebra mussels *(Dreissena polymorpha)* in 1990–1996. We used data from bay-wide surveys and generalized linear models (GLMs). Mussels were first found in 1991, peaked in 1992, and declined to stable levels in 1993 at stations with hard substrates where they were most abundant. Annual trends in phytoplankton and zooplankton were negatively correlated with zebra mussel trends at different time lags. Most phytoplankton taxa declined within three years of mussel colonization, with disappearance of photosensitive cyanophytes in the first year of colonization. All phytoplankton taxa tended to recover within 5 years except for cyanophytes and chlorophytes.

Diatom annual densities were least affected, but species composition changed. Major zooplankton groups declined after 1 year of mussel colonization. Cyclopoids and cladocerans exhibited lowest densities in 1993, and calanoids and rotifers continued to decline through 1996. Spatial distributions of both plankton groups were fairly homogeneous and remained stable despite changes in densities and patchy mussel distributions. Responses of macroinvertebrate taxa depended on their life history, mobility, and spatial proximity to mussel colonies. Most taxa except for *Gammarus* and *Diporeia* were most abundant in silt and silty-sand substrate in the inner bay where mussel densities were lowest. Oligochaetes declined between 1991 and 1993 but then increased after 1994, chironomids showed no strong patterns but tended to increase after 1992, sphaeriids decreased in the inner bay after 1991, *Gammarus* increased between 1991 and 1993, and *Diporeia* declined after 1991. Our results highlight the importance of prompt analysis of monitoring data collected during the initial-colonization period. Such data can then provide information to optimize survey design during ongoing assessments of zebra mussel impacts.

INTRODUCTION

The zebra mussel (*Dreissena polymorpha*) invaded the Great Lakes in the late-1980s with resultant ecosystem shifts and far-reaching economic impacts. Nevertheless, few areas in the Great Lakes, except for Saginaw Bay, have been monitored to provide comprehensive datasets on environmental quality and biota at different trophic levels during zebra mussel colonization (Nalepa and Fahnenstiel 1995, Nalepa et al. 1996, 2003). Changes in Saginaw Bay during the last several decades illustrate complex responses of Great Lakes aquatic ecosystems to repeated environmental stressors that have included successive invasions of a variety of exotic species of zooplankton, fish, and most importantly the zebra mussel, as well as nutrient enrichment and contaminant loading. More recent studies after 2007 have documented ecosystem impacts of quagga mussels (*Dreissena rostriformis bugensis*) (Fahnenstiel et al. 2010).

To document responses of the Saginaw Bay ecosystem to the establishment of zebra mussels, a water quality and lower food-web monitoring program was conducted from 1990 to 1996 (Nalepa et al. 1996, Johengen et al. 2000). Nalepa et al. (2003) reported that the first large recruitment of zebra mussels occurred in areas with hard substrate in 1991. Mediated by conditions favorable for settlement and survival in Saginaw Bay (Nalepa et al. 2003), zebra mussels rapidly colonized the bay and transformed ecosystem structure through intense filtration activity. A number of studies documented changes in the bay immediately after zebra mussel colonization (1991–1993) (Nalepa and Fahnenstiel 1995). Reported ecosystem changes ranged from alterations of the physical environment to shifts in energy flows of the lower food-web. Physical effects included increased water clarity (Fahnenstiel et al. 1995b) and bottom complexity (Nalepa et al. 1995). Food-web changes included rapid decreases in primary productivity (Fahnenstiel et al. 1995a,b), bacterial communities (Cotner et al. 1995), and zooplankton (Bridgeman et al. 1995), and increases in submerged macrophytes (Skubinna et al. 1995), and benthic algae (Lowe and Pillsbury 1995). Vanderploeg et al. (2002) identified zebra mussels as the cause for the reoccurrence of *Microcystis* blooms in summer. Of the data collected in 1990–1996, only trends in benthic macroinvertebrate communities have been described for the entire period (Nalepa et al. 2003). Macroinvertebrate changes were found to be complex, with increases and decreases varying among taxonomic groups and substrate type.

In this chapter, we analyze temporal and spatial trends in phytoplankton, zooplankton, and benthic macroinvertebrate communities of Saginaw Bay during different stages of invasion and colonization by using a regression approach. We characterize interactive responses of lower food web components by quantifying the timing and relative strengths of observed trends at an ecosystem level. The approach seeks to advance our general knowledge beyond local and immediate responses of ecosystems to disturbances caused by zebra mussels, such as those reported in Nalepa and Fahnenstiel (1995).

METHODS

Study Site and Data Collections

Saginaw Bay is a large (2980 km^2) and productive embayment of Lake Huron (Beeton et al. 1967) that can be divided into a shallow, inner region (mean depth of 5.1 m) and a deeper, outer region (mean depth of 13.7 m) (Nalepa et al. 2003). The Saginaw River constitutes 75% of the total tributary inflow. Data for analyses were from samples collected throughout Saginaw Bay (Figure 33.1) at various time intervals from April to November over the period 1990–1996 (Nalepa et al. 1996, Johengen et al. 2000).

Phytoplankton densities were derived from 336 samples collected at 5 stations (Stations 4, 7, 10, 14, and 16) located in inner Saginaw Bay and 3 stations (Stations 20, 24, and 23) in the outer bay (Figure 33.1). Water samples were collected with a 5L Niskin bottle at a vertical depth of 1–5 m below the water surface. Samples were preserved in Lugol's solution and then filtered onto membrane filters for permanent slide mounts (Fahnenstiel et al. 1998). In the laboratory, phytoplankton were identified to species except for flagellates as described in Vanderploeg et al. (2001).

Zooplankton densities were derived from 604 samples at all the same stations sampled for phytoplankton and at additional stations in 1991 and 1992 (18 total in inner bay and 8 in outer bay). Duplicate samples were collected by vertical tows from 1 m above the bottom to the surface with conical plankton net of 29.5 cm diameter, 113 cm length, and a 63 μm mesh sieve (Bridgeman et al. 1995). Upon collection, samples were narcotized with club soda and preserved in 5% sugar-saturated formalin solution (Haney and Hall 1973). In the laboratory, zooplankton samples other than rotifers were made up to a known volume and subdivided in a Folsom plankton splitter to yield subsamples of 200–300 individuals that were counted and identified. For rotifers, subsamples from the Folsom plankton were split up to a known volume and a Stempel pipette was used to withdraw several 1 mL subsamples and approximately 200 individuals were counted and identified.

Benthic macroinvertebrate densities were from 587 Ponar grab samples collected at 10 stations in spring, summer, and fall and processed as described by Nalepa et al. (2003). Substrate consisted of sand and gravel (Stations 13, 14, and 16), silty sand (Station 11), and silt (Stations 4, 7, and 10) in the inner bay, and silty sand (Stations 20, 23, and 24) in the outer bay. In addition,

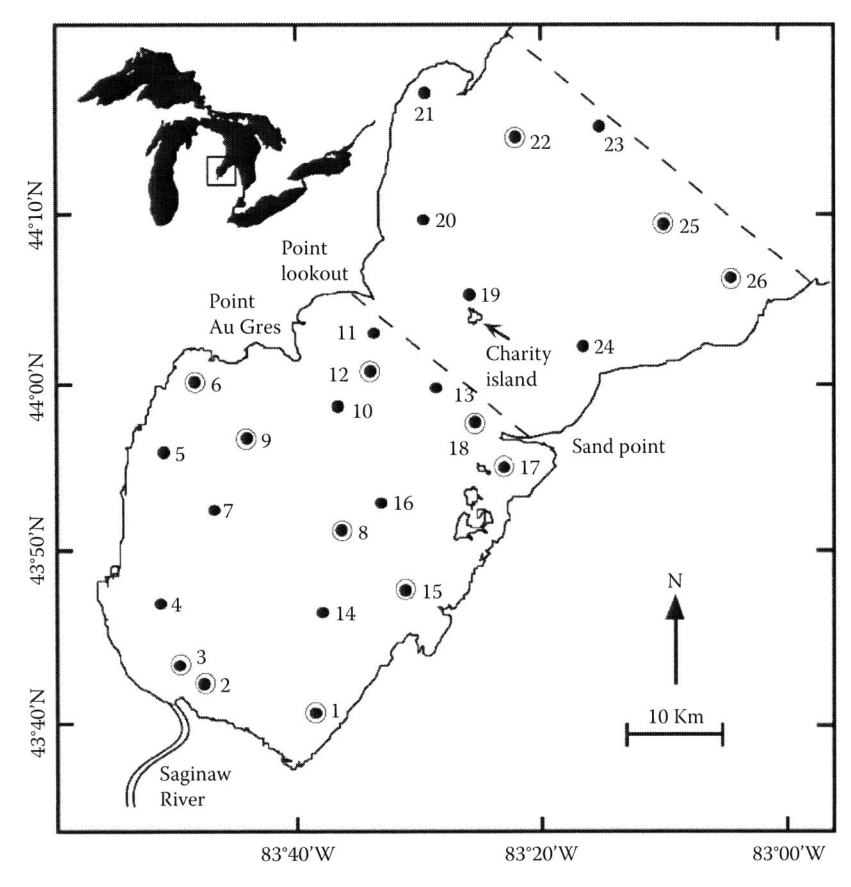

Figure 33.1 Sampling stations for phytoplankton, zooplankton, and benthic macroinvertebrates in Saginaw Bay, 1990–1996. Circled sites were sampled only for zooplankton in 1991 and 1992. Benthic macroinvertebrates were collected at Stations 4, 7, 10, 11, 13, 14, 16, 20, 23, and 24. Additional samples for densities of *D. polymorpha* were collected by SCUBA divers at Stations 5, 6, 13, 14, 15, 16, 19, and 27 (near 26). Dashed line indicates the boundary between inner and outer bays.

during fall, 138 samples for density estimates of zebra mussels were obtained by SCUBA at 5 stations (Stations 5, 6, 15, 19, and 27, which was near Station 26) with hard substrate and 3 stations with sand and gravel (Stations 13, 14, and 16) (Nalepa et al. 2003).

For analysis, we aggregated phytoplankton, zooplankton, and benthic macroinvertebrates into major taxonomic groups. Phytoplankton taxa were placed into seven groups: cyanophytes, bacillariophytes, chlorophytes, chrysophytes, cryptomonads, dinoflagellates, and flagellates. Flagellates comprised unidentified 3–8 μm taxa, while identified taxa were placed within chrysophytes, chlorophytes, etc. Zooplankton taxa were placed into six groups: rotifers, calanoids (including copepodite stages I–V), cyclopoids (including copepodite stages 1–V), cladocerans, nauplii, and dreissenid veligers. Benthic macroinvertebrates were placed into six groups: sphaeriids, chironomids, oligochaetes, *Gammarus* sp., *Diporeia* spp., and zebra mussels.

Data Analysis

Our general approach to investigate responses of phytoplankton, zooplankton, and benthic macroinvertebrates to zebra mussels was to quantify annual, seasonal, and spatial trends in density of individual taxonomic groups. For benthic macroinvertebrates, our approach to trend analysis supplements previous interpretations in Nalepa et al. (2003).

To quantify trends, we implemented generalized linear models (GLMs) (McCullagh and Nelder 1989). GLMs were used to estimate annual abundance trends (year coefficient) and quantify relative importance of spatial and temporal variation (station and month coefficients) and interannual variability (interaction coefficients). GLMs are generalized forms of linear regression models that incorporate covariates and factor variables and provide a powerful statistical framework without normality assumptions for the analysis of skewed data. These models incorporate probability distributions of the exponential family such as normal, binomial, Poisson, and gamma.

We first analyzed temporal and spatial variation of *Dreissena* densities. Annual trends in zebra mussel abundance reported by Nalepa et al. (2003) indicated highest densities in 1991 and 1992 occurred at stations with hard substrate (rock, cobble). However, substrate is patchy within Saginaw Bay and included in the analysis were densities on soft substrates where most other macroinvertebrates were found.

We used the following GLM to analyze densities of phytoplankton, zooplankton, and macroinvertebrates:

$$g(\mu_{yms}) = \alpha + \delta_y + \phi_m + \lambda_s$$

where

- g () is the logarithmic link function relating the response variable and the linear predictor
- μ are the expected densities (phytoplankton cells/mL, zooplankton organisms/m^3, or benthic macroinvertebrates/m^2)
- α is the abundance in a reference year (i.e., for all taxa, it was 1990 prior to zebra mussel colonization, and for zebra mussels it was 1991), month (i.e., April for zooplankton and macroinvertebrates and May for phytoplankton), and station (i.e., Station 5 for SCUBA samples, Station 4 for phytoplankton and macroinvertebrates, and Station 1 for zooplankton)
- δ_y is the factor that represents abundance in year y in relation to the reference year
- ϕ_m is the factor that represents the abundance in month m relative to the reference month
- λ_s is the factor that represents the abundance in station s relative to the reference station

Year represented annual abundance trends, month represented seasonal trends, and station represented spatial variation. Our use of density rather than biomass was of no consequence because preliminary analysis showed no major differences between density and biomass trends. For analysis of zebra mussel densities obtained with SCUBA, month was not included as a predictor because all samples were taken in fall. Interactions among predictors were not included when stations were not sampled in the same months during the study period. Nevertheless, departures from general annual patterns were investigated by GLMs for subsets of data by year. For analysis of macroinvertebrate densities, interactions between year and stations were included to test and quantify responses by substrate.

To identify the most appropriate probability distribution to describe variability in models, we fitted a linear regression of the logarithm of variance as a function of the logarithm of the mean of each response variable with respect to predictor levels. We found slopes close to 2 for all groups and used a gamma distribution accordingly. For each analysis, we performed analysis of deviance and checked model assumptions. Analysis of deviance to test significant variations of each factor was performed by comparison of models with all variables and models excluding tested variables. Tests were performed at 95% confidence level. We ran models with routines available in the S-Plus computing environment (i.e., glm () function) (Becker et al. 1988). Results of analysis of deviance were reported as percentage of variation explained by each model, and also by effects of year, month, and station, to allow comparisons of strength of effects among groups.

RESULTS

Zebra Mussels

Annual mean densities of zebra mussels between 1990 and 1996 as collected with a Ponar grab ranged between 10 and 1,339/m^2, and mean densities collected by SCUBA ranged between 4,123 and 33,886 individuals/m^2 (Table 33.1). Trends in zebra mussels estimated from GLMs varied depending on sampling method. Density indices derived from grab samples were lower at stations with a silt and silty sand substrate (Stations 4, 7, 10, and 11) and in the outer bay (Stations 20, 23, and 24) than at sand and gravel substrates (Stations 13, 14, and 16) in the inner bay (Figure 33.2a). Density indices peaked between 1993 and 1995 in the inner bay (Figure 33.2b) but showed minimal year-to-year differences in the outer bay (Figure 33.2c). Density indices derived from SCUBA-collected samples were similar among stations except for a lower density at Station 13 (Figure 33.2d). For yearly trends, overall densities increased to a peak in 1992 and then declined thereafter in the inner bay (Figure 33.2e), while densities varied without pattern in the outer bay (Figure 33.2f). Spatial and annual variation in zebra mussel densities was significant irrespective of sampling method (probability of F < 0.001), and both models with main effects and interactions explained about 85% of the deviance (Table 33.2).

Phytoplankton

Annual mean total phytoplankton ranged from about 11,600 cells/mL in 1990–1991 to 6,500 cells/mL in 1992 (Table 33.1). The GLM with year, month, and station as predictors explained 53% of the density variation (Table 33.3). Annual variation was significant (probability of F = 0.0002), but seasonal differences contributed more to variation than differences among years or stations. Fitted total densities (indices) decreased starting in 1991 and remained low through 1996 (Figure 33.3), were similar among stations in the inner bay (Stations 4–16) but higher than at stations in the outer bay (Stations 20 and 23) (Figure 33.4), and were highest in August and September (Figure 33.5).

Among phytoplankton groups, cyanophytes were the most abundant, bacillariophytes and flagellates were next in abundance, while chlorophytes, cryptomonads, chrysophytes, and dinoflagellates were generally rare (Table 33.1). Models for each phytoplankton group explained from 34% of the variation for flagellates to 56% for dinoflagellates, with most variation attributed to seasonality (Table 33.3). Differences among years, months, and stations were significant (probability of F < 0.05) for all groups except for differences among stations for dinoflagellates. Annual fitted density for cyanophytes decreased sharply beginning in 1991, whereas most other groups increased between 1990 and 1991 but then eventually decreased; chlorophytes,

Table 33.1 Mean Densities of Phytoplankton (cells/mL), Zooplankton (individuals/m^3), and Benthic Macroinvertebrates (individuals/m^2) at Stations (Figure 33.1) in Saginaw Bay, 1990–1996

Group	Taxa	1990	1991	1992	1993	1994	1995	1996
Dreissena in SCUBA diver samples		—	11,705	33,887	4,124	7,686	4,309	7,452
Dreissena in Ponar grab samples		0	10	1,339	1,073	517	972	559
Phytoplankton	Cyanophytes	8,450	5,021	4,120	2.228	5,450	4,484	3,964
	Bacillariophytes	2,288	4,681	1,579	3,785	2,994	3,370.	2,251
	Chlorophytes	371	919	354	38	95	90	90
	Chrysophytes	17	14	32	9.7	3	3	3
	Cryptomonads	184	250	228	359	92	137	213
	Dinoflagellates	<1	6	<1	<1	<1	<1	<1
	Flagellates	297	708	240	369	233	208	611
	Total	11,608	11,600	6,554	6,788	8,868	8,293	7,133
Zooplankton	Rotifers	241,745	203,653	95,888	22,628	18,303	11,954	6,070
	Calanoids	6,508	4,420	2,155	2,681	1,493	1,682	2,464
	Cyclopoids	13,889	11,123	6,054	2,877	4,152	5,292	8,404
	Cladocerans	74,790	40,696	14,812	10,284	16,136	30,262	32,789
	Nauplii	5,910	11,365	4,246	4,177	3,085	10,591	11,783
	Veligers	4	654	1,736	21,239	5,222	3,644	30,749
	Total	342,846	271,912	124,889	44,771	48,390	63,426	147,259
Macroinvertebrates	Oligochaetes	5,199	3,053	2,808	1,153	974	2,241	2,413
	Chironomids	984	806	966	596	436	1,678	1,543
	Sphaeriids	110	112	80	232	76	46	140
	Gammarus	12	24	15	172	84	86	130
	Diporeia	72	116	24	40	38	6	7
	Total	6,397	4,122	3,895	2,2097	1,612	4,059	4,239

Note: Calanoids and cyclopoids include copepodite stages I–V; sphaeriids are mostly all *Pisidium spp*.

flagellates, and dinoflagellates decreased in 1992, chrysophytes decreased in 1993, and cryptomonads decreased in 1994 (Figure 33.3). Bacillariophytes showed little annual variation; densities decreased in 1992 but then returned to values found in 1990 by 1993. Spatially, fitted densities of bacillariophytes were fairly similar among stations, chrysophytes tended to have higher densities in the outer bay, and other groups tended to have higher densities in the inner bay similar to the spatial pattern of total phytoplankton (Figure 33.4). In general, these spatial patterns remained unchanged over the study period. Seasonally, fitted densities were highest from July to September for most groups, but flagellates, chrysophytes, and cryptomonads peaked in October (Figure 33.5). Seasonal patterns were similar among years for all groups. The exception was in 1990 for cyanophytes. In 1990, densities declined in late summer, while in 1991–1996, densities were at a seasonal maximum in late summer (Figure 33.5).

Zooplankton

Total mean densities ranged between 342,800/m^3 in 1990 and 44,800/m^3 in 1993 (Table 33.1). The model that included year, month, and station as predictors explained 59% of the variation (Table 33.3). Annual differences in total zooplankton were significant (probability of F < 0.001) and more pronounced than differences among months and stations (Table 33.3). Annual fitted densities decreased after 1990 and showed partial recovery after 1993 (Figure 33.6). Densities were fairly similar among stations although they tended to be higher in the inner bay than the outer bay (Figure 33.7). Densities were highest from May to July (Figure 33.8). Both spatial and seasonal patterns remained similar throughout the study period.

Among zooplankton groups, rotifers were most abundant followed by cladocerans and cyclopoid copepods, while calanoid copepods were least abundant (Table 33.1). Dreissenid veligers comprised the greatest fraction of total zooplankton in 1996. The models for zooplankton groups explained from 24% of the variation for calanoids to 55% for veligers, and annual variation accounted for 26%–50% (Table 33.3). Year, month, and station effects were significant for all groups (probability of F < 0.05). Annual fitted density for cladocerans and cyclopoids tended to recover after lowest levels in 1993, fitted density of rotifers and calanoids decreased until 1995, and fitted density of veliger steadily increased (Figure 33.6). Spatially, fitted densities of most groups were similar among stations and tended to be higher in the inner bay than in the outer bay as for total zooplankton. The exception was fitted densities of calanoids that were higher

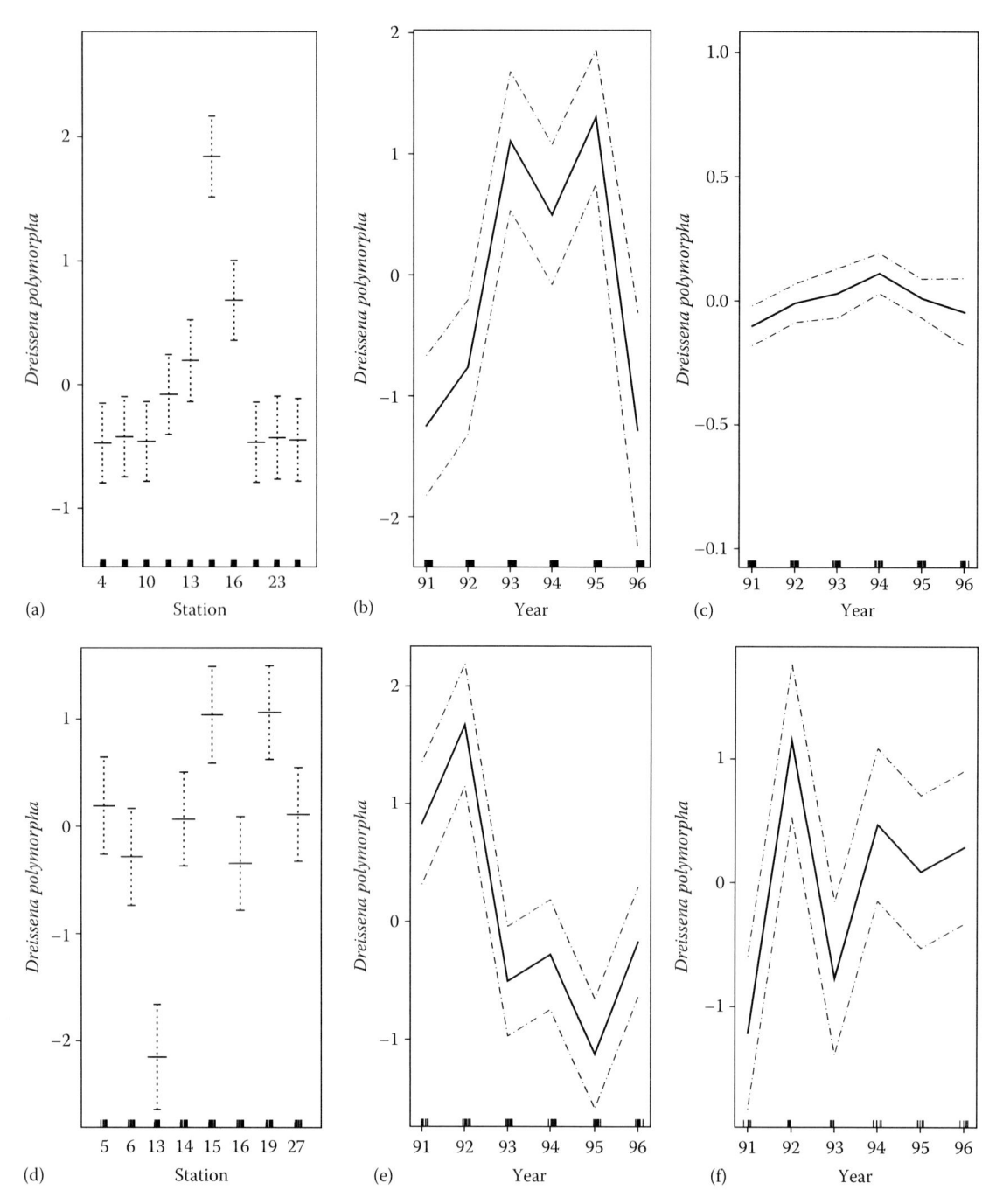

Figure 33.2 Indices of mean densities of *D. polymorpha* as derived from GLMs. Data from samples collected in Saginaw Bay between 1991 and 1996. Top panels (a, b, and c) provide indices derived from samples collected with a Ponar grab, and bottom panels (d, e, and f) provide indices derived from samples collected by SCUBA. (a) Spatial indices (by Stations 4, 7, 10, 11, 13, 14, 16, 20, 23, and 24), (b) annual indices at inner bay stations (Stations 4, 7, 10, 11, 13, 14, and 16), (c) annual indices at outer bay stations (Stations 20, 23, and 24), (d) spatial indices (by Station 5, 6, 13, 14, 15, 16, 19, and 27), (e) annual indices at inner bay stations (Stations 5, 6, 13, 14, 15, and 16), and (f) annual indices at outer bay stations (Stations 19 and 27). Dashed lines give 95% confidence limits. Lines along scales of the x-axes represent the amount of data available for each factor level. Y-axes are scaled so that 0 represents the mean density and scales vary to allow display of overall patterns.

in the outer bay (Figure 33.7). Spatial patterns of zooplankton groups did not change significantly during the study period. For seasonal patterns, fitted densities of calanoids and rotifers remained high between April and July, densities of cladocerans and cyclopoids were high between May and September with a peak in June, and densities of veligers were near zero before June and after September with a peak in July (Figure 33.8). Nauplii generally peaked in May and decreased thereafter. Seasonal fluctuations were fairly similar among years with exceptions in 1990 when nauplii

Table 33.2 Results of Analysis of Deviance for Densities of *Dreissena* in Saginaw Bay as Derived from Samples Collected with a Ponar Grab (1990–1996) and from Samples Collected by SCUBA Divers (1991–1996)

Variable	Residual Degrees of Freedom	Residual Deviance	F Value	Probability of (F)
Ponar Grab				
Null model	585	1003.8		
Year	579	705.9	137.2	0.0000
Station	570	348.5	109.7	0.0000
Month	566	328.9	13.6	1.6e−10
Year/station	512	145.7	9.4	0.0000
SCUBA				
Null model	137	368.4		
Year	132	289.4	36.7	0.0000
Station	125	212.6	25.5	0.0000
Year/station	91	52.7	10.9	0.0000

Table 33.3 Results of Analysis of Deviance from Main Effects Generalized Linear Models (GLM) for Density Variation in Phytoplankton, Zooplankton, and Benthic Macroinvertebrates in Saginaw Bay, 1990–1991. GLM Included as Predictors: Year, Month, and Station. Given Proportions (as a Percentage) Provide the Relative Amount of Variation Explained by the Full GLM, and Relative Amount of the Total Variation Explained by Each Predictor

Taxa	Full GLM%	Year (%)	Month (%)	Station (%)	Year Prob. of F	Month Prob. of F	Station Prob. of F
Total phytoplankton	53	5	72	23	0.000202	<0.000001	<0.000001
Cyanophytes	42	5	83	12	0.000003	<0.000001	0.001991
Bacillariophytes	43	17	64	19	0.000184	<0.000001	0.000002
Chlorophytes	47	56	16	28	<0.000001	0.000341	0.000002
Chrysophytes	37	56	16	28	<0.000001	0.003107	0.000121
Cryptomonads	35	12	16	72	0.000028	0.004823	<0.000001
Dinoflagellates	56	92	5	3	<0.000001	0.047616	0.633845
Flagellates	34	38	15	47	<0.000001	0.000123	<0.000001
Total zooplankton	59	40	27	33	<0.000001	<0.000001	<0.000001
Rotifers	52	50	23	27	<0.000001	<0.000001	<0.000001
Calanoids	24	28	36	36	<0.000001	<0.000001	<0.000001
Cyclopoids	37	29	38	33	<0.000001	<0.000001	0.000001
Cladocerans	45	20	39	41	<0.000001	<0.000001	<0.000001
Nauplii	43	26	62	12	<0.000001	<0.000001	0.002610
Veligers	55	26	60	14	<0.000001	<0.000001	<0.000001
Total macroinvertebrates	56	24	2	74	<0.000001	0.034311	<0.000001
Oligochaetes	57	27	2	72	<0.000001	0.540110	<0.000001
Chironomids	57	43	6	51	<0.000001	0.000036	<0.000001
Gammarus	72	33	6	61	<0.000001	<0.000001	<0.000001
Diporeia (Station 23)	72	74	26	—	0.000910	0.000063	—
Sphaeriids	51	19	2	79	<0.000001	0.208211	<0.000001

peaked for a second time in August and in 1994 when both cyclopoids and nauplii peaked for a second time in August.

Benthic Macroinvertebrates

Annual total densities of benthic macroinvertebrates ranged between 1,600/m^2 in 1994 and 6,400/m^2 in 1990 (Table 33.1). The model explained 56% of the variation, with most variation attributed to differences among stations (Table 33.3). Trends in annual indices as estimated from GLMs were highly variable between station groups defined by substrate type before 1994, but trends became less variable after 1994 (Figure 33.9). In the inner bay before 1994, total densities decreased sharply at stations with silt and silty sand substrates (Figure 33.9a) but showed no apparent trend at stations with sand and gravel substrates (Figure 33.9b). In the outer bay before 1994, total densities were generally stable (Figure 33.9c). Over all years, fitted densities were highest in the inner bay at stations with silt and silty sand substrates (Stations 4–11), intermediate in the outer bay

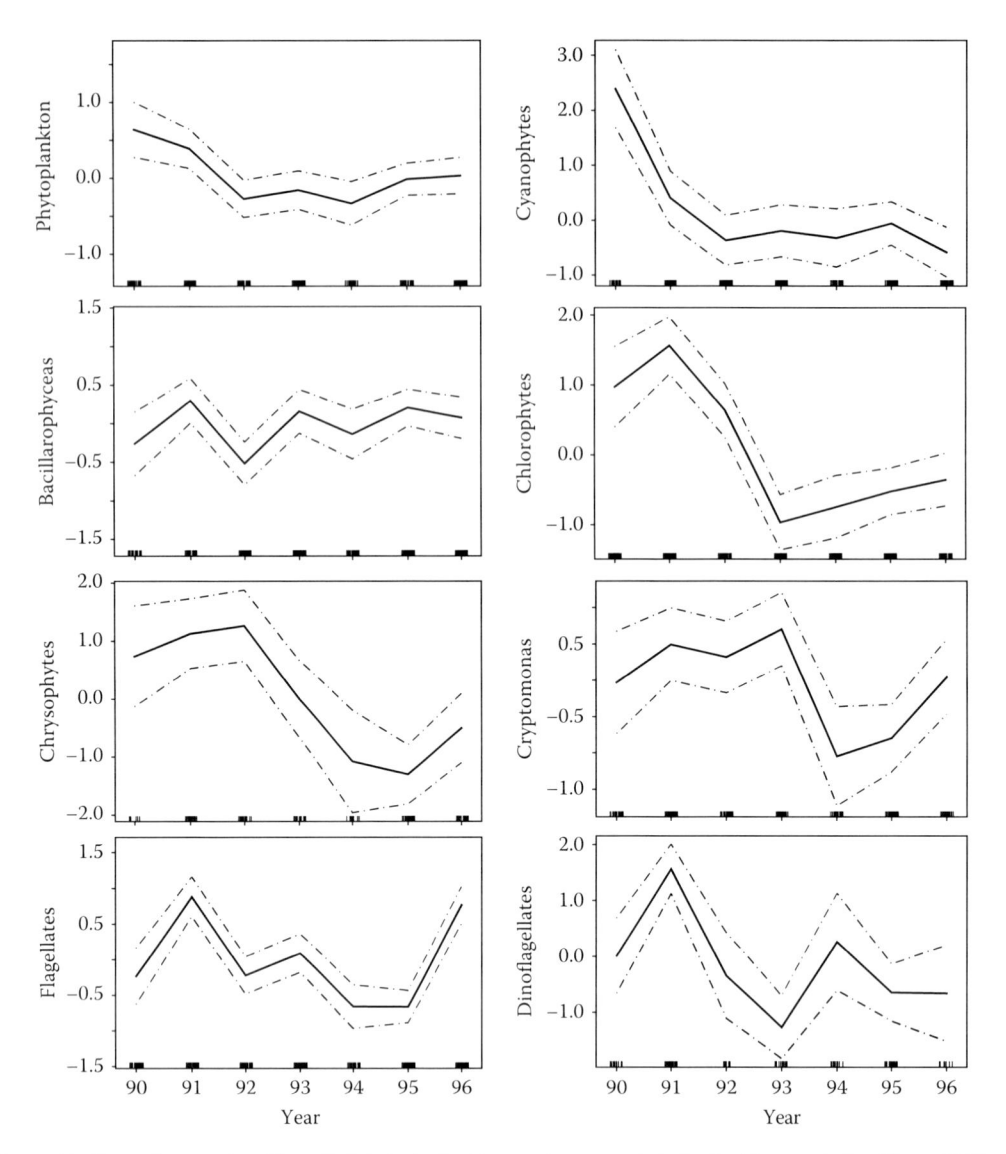

Figure 33.3 Annual indices of mean densities of total phytoplankton and groups of phytoplankton as derived from GLMs. Data collected in Saginaw Bay, 1990–1996. Models included year, station, and month. Dashed lines give 95% confidence limits. Lines along scales of the x-axes represent the amount of data available for each factor level. Y-axes are scaled so that 0 represents the mean density and scales vary to allow display of overall patterns.

(Stations 20 and 24), and lowest in inner bay at stations with sand and gravel substrates (Stations 13–16) (Figure 33.10).

Oligochaetes were the dominant group, followed by chironomids and sphaeriids; *Gammarus* was comparatively rare, and *Diporeia* was found at only one station (Table 33.1). Models explained from 51% of density variation for sphaeriids to 72% for *Gammarus*, with most variation for all groups attributed to differences among stations (Table 33.3). Annual and spatial effects were significant for all groups (probability of F < 0.0009), while monthly variation was significant only for chironomids, *Gammarus*, and *Diporeia* (Table 33.3). Annual trends differed among groups and also within groups as a function of substrate (Figure 33.9). For amphipods, fitted densities of *Gammarus* increased and peaked in 1993, while fitted densities of *Diporeia* steadily declined after 1991. Oligochaetes declined between 1990 and 1994 in the inner bay at stations with silty and sandy silt substrates but showed little change at other stations. After 1994, oligochaetes increased at all stations regardless of substrate. Chironomids showed no annual trends; overall densities were greatest in 1991 and showed greatest annual variation at inner bay stations with silt and silty sand substrates. Sphaeriids in the inner bay decreased at stations with sand and gravel substrates beginning in 1991 and decreased at stations with silt and silty sand substrates beginning in 1993. Sphaeriid annual densities showed little variation at outer bay stations. Spatially, densities of most groups were highest at stations with silt and silty sand substrates in the inner bay, intermediate in the outer bay, and lowest at stations with sand and gravel substrates in the inner bay, similar to total macroinvertebrate densities (Figure 33.10). The two amphipod taxa were

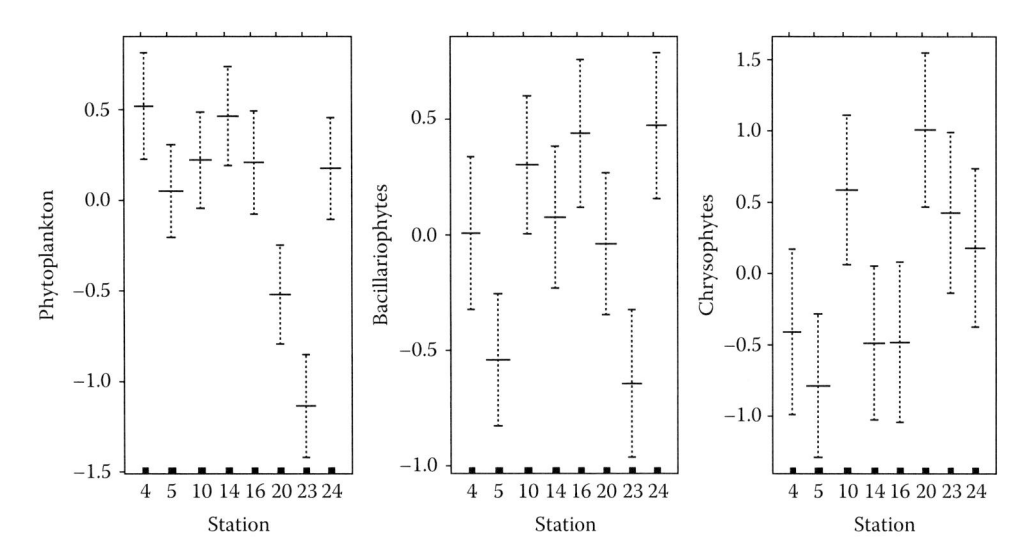

Figure 33.4 Spatial indices of mean densities of total phytoplankton, bacillariophytes, and chrysophytes as derived from GLMs. Data collected in Saginaw Bay, 1990–1996. Indices for cyanophytes, chlorophytes, cryptomonads, dinoflagellates, and flagellates are not shown but were similar to total phytoplankton. Models included year, station, and month. Dashed lines give 95% confidence limits. Lines along scales of the x-axes represent the amount of data available for each factor level. Y-axes are scaled so that 0 represents the mean density and scales vary to allow display of overall patterns.

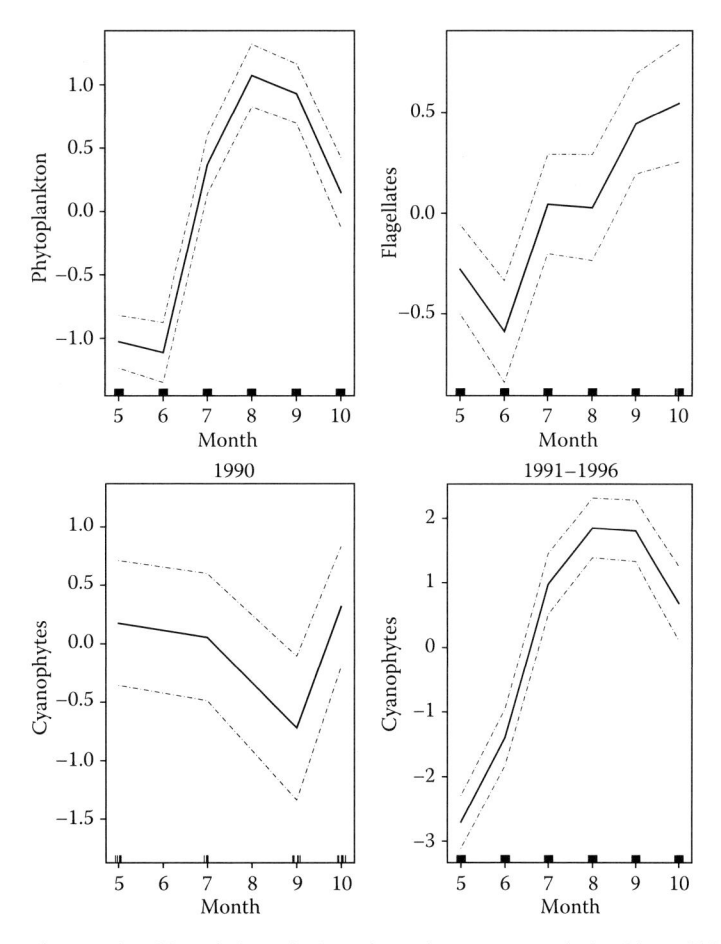

Figure 33.5 Seasonal indices of mean densities of phytoplankton for various years as derived from GLMs. Data collected in Saginaw Bay, 1990–1996. Shown are indices for total phytoplankton 1990–1996, flagellates 1990–1996, cyanophytes 1990 (pre-*Dreissena*), and cyanophytes 1991–1996 (post-*Dreissena*). Indices for bacillariophytes and chlorophytes are not shown but were similar to total phytoplankton, while indices for chrysophytes and cryptomonads were similar to flagellates. Dashed lines give 95% confidence limits. Lines along scales of the x-axes represent the amount of data available for each factor level. Y-axes are scaled so that 0 represents the mean density and scales vary to allow display of overall patterns.

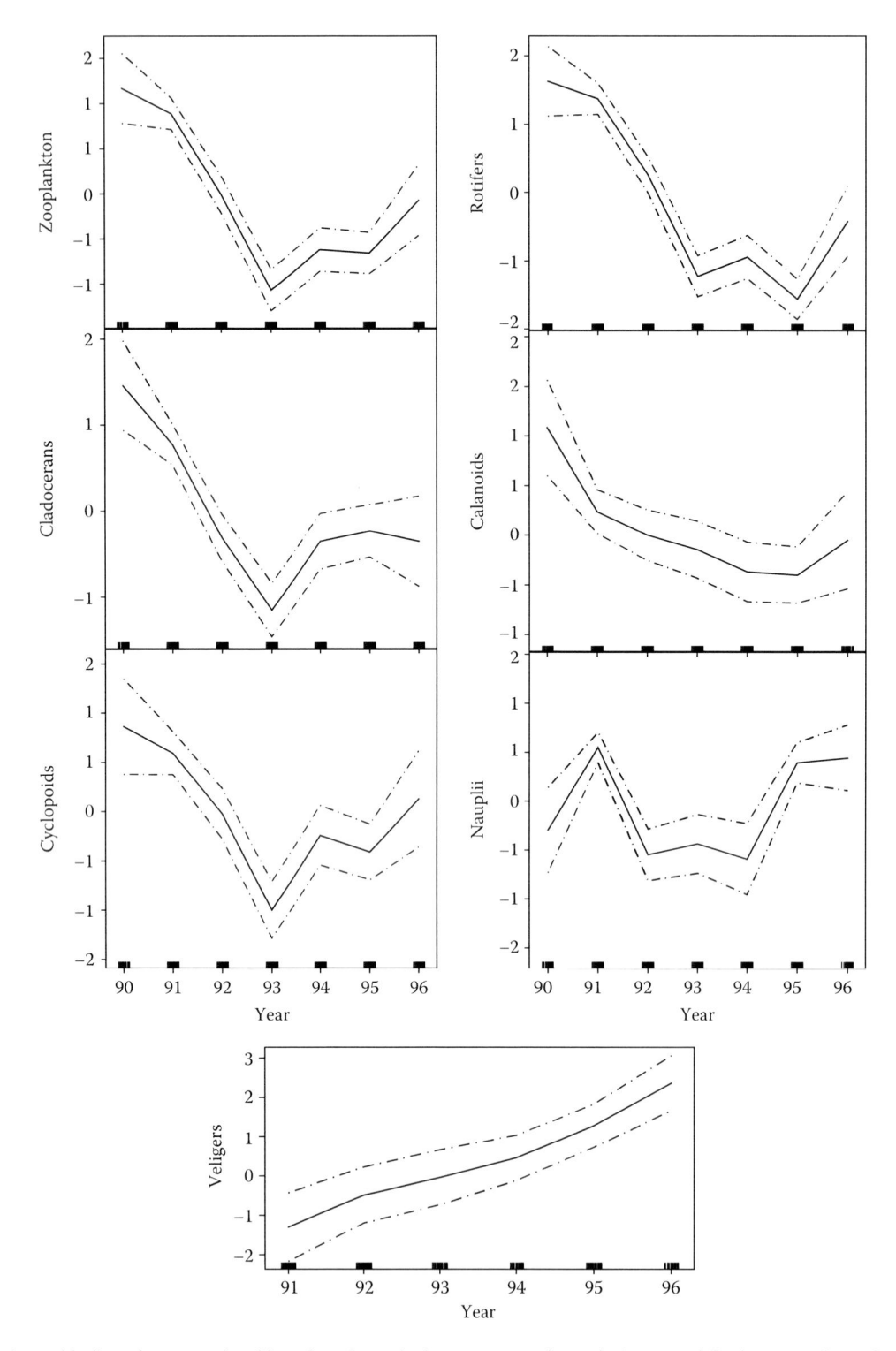

Figure 33.6 Annual indices for mean densities of total zooplankton, groups of zooplankton, and *Dreissena* veligers from GLMs. Data collected in Saginaw Bay, 1990–1996. Models included year, station, and month. Dashed lines give 95% confidence limits. Lines along scales of the x-axes represent the amount of data available for each factor level. Y-axes are scaled so that 0 represents the mean density and scales vary to allow display of overall patterns.

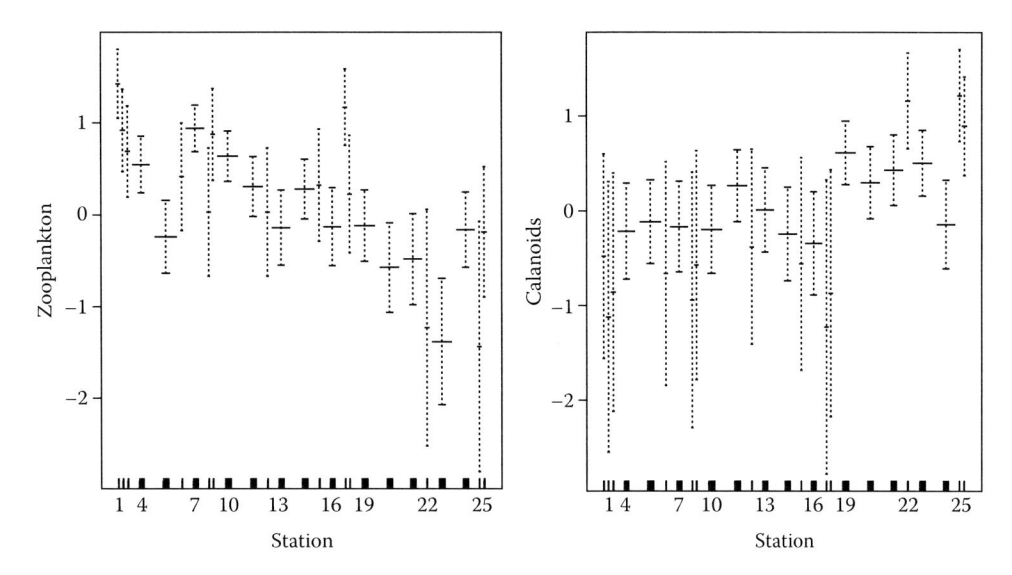

Figure 33.7 Spatial indices for mean densities of total zooplankton, and calanoids as derived from GLMs. Data collected in Saginaw Bay, 1990–1996. Models included year, station, and month. Dashed lines give 95% confidence limits. Lines along scales of the x-axes represent the amount of data available for each factor level. Y-axes are scaled so that 0 represents the mean density and scales vary to allow display of overall patterns.

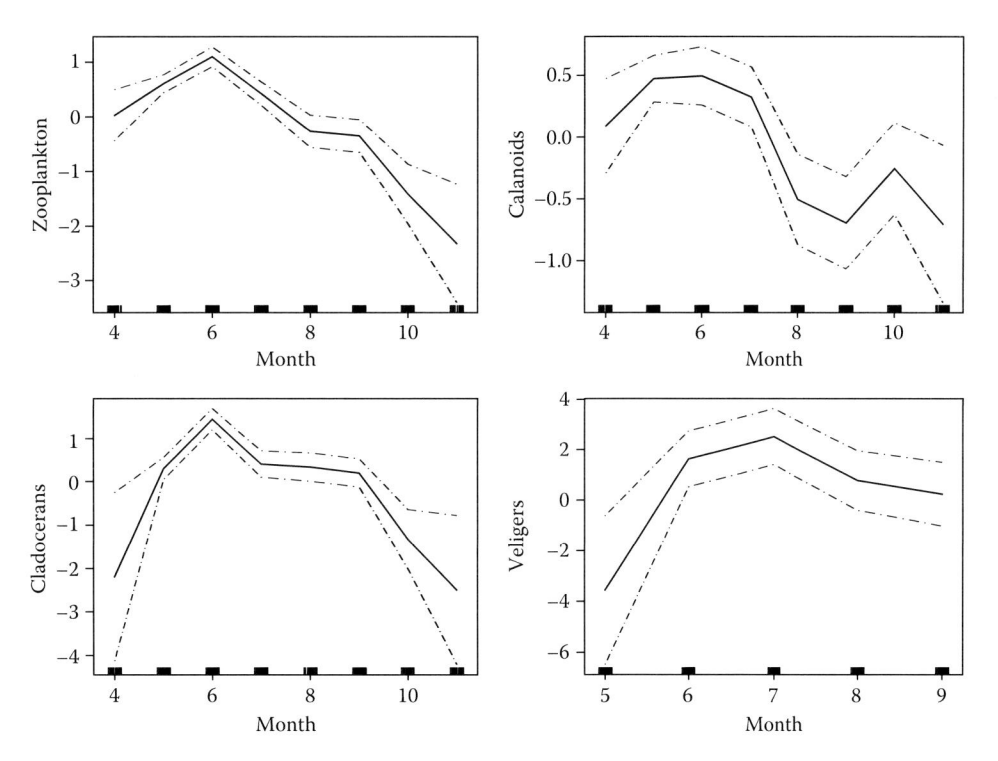

Figure 33.8 Seasonal indices for densities of total zooplankton, calanoids, cladocerans, and veligers as derived from GLMs. Data collected in Saginaw Bay, 1990–1996. Indices for rotifers are not shown but were similar to calanoids, and indices for cyclopoids were similar to cladocerans. Models included year, station, and month. Dashed lines give 95% confidence limits. Lines along scales of the x-axes represent the amount of data available for each factor level. Y-axes are scaled so that 0 represents the mean density and scales vary to allow display of overall patterns.

exceptions to this spatial pattern. Densities of *Gammarus* were greatest at inner bay stations with sand and gravel substrates (Figure 33.10), and *Diporeia* was only collected in the outer bay. Densities of chironomids and *Diporeia* were highest in July and densities of *Gammarus* increased from June to October (Figure 33.11).

DISCUSSION

Changes in abundances of phytoplankton, zooplankton, and benthic macroinvertebrates in Saginaw Bay between 1990 and 1996 confirmed the dramatic and disparate impacts of *Dreissena* introduction to lower foodwebs in freshwater ecosystems. The extent of changes was not surprising because in 1992 and 1993, the zebra mussel population was capable of filtering the entire inner bay in 0.8 and 5 days, respectively (Fanslow et al. 1995). Observed initial trends were generally similar to those documented in other ecosystems (MacIsaac 1996, Karateyev et al. 1997, Strayer et al. 1999, 2006). In particular, trends in Saginaw Bay were consistent with trends in lower food web components in systems with similar success of mussel recruitment (Caraco et al. 1997), characteristics of the water column that affect water transparency (Holland 1993, Leach 1993), structure of resident communities, and strength of interaction pathways

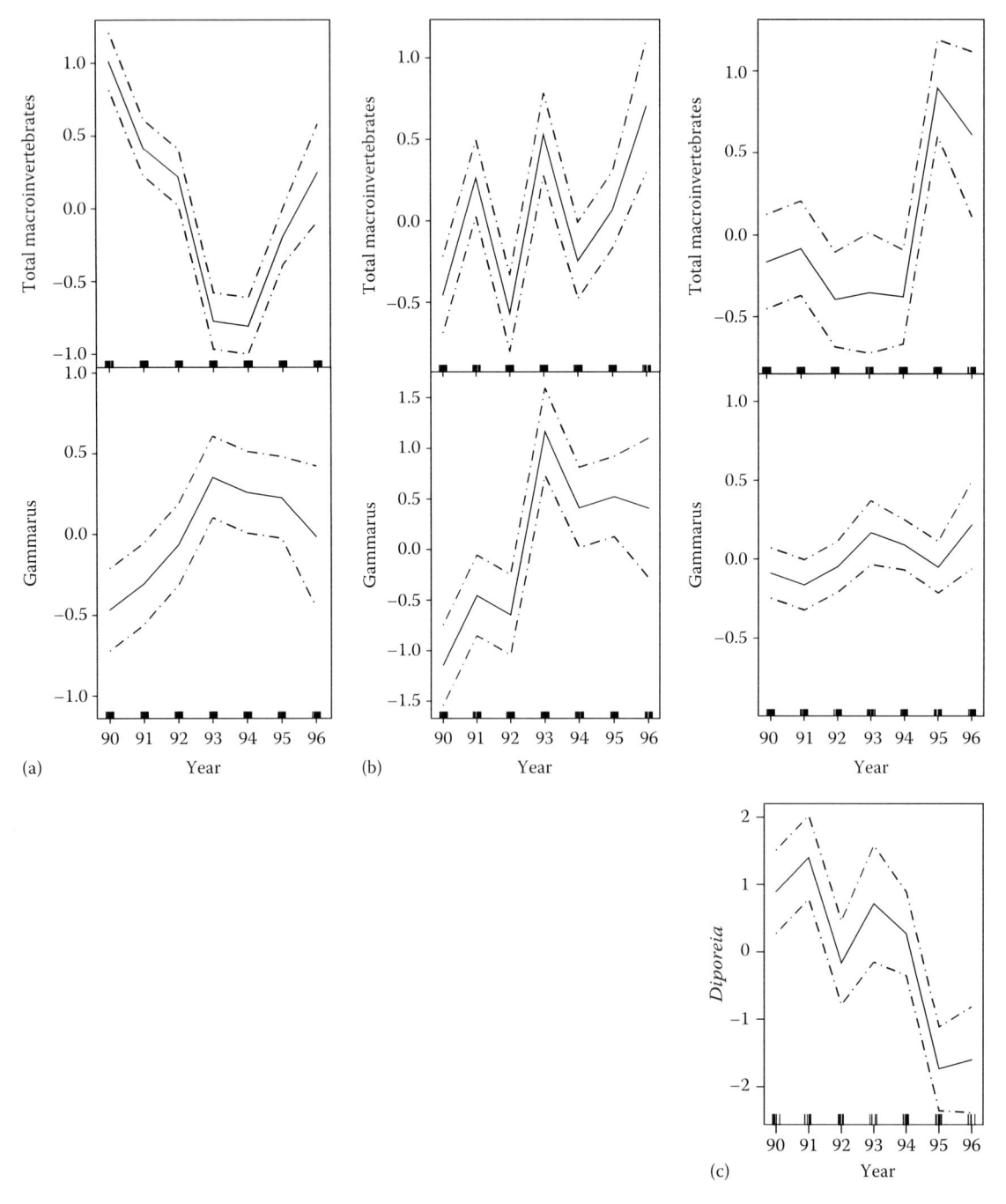

Figure 33.9 Annual indices for mean densities of total benthic macroinvertebrates and macroinvertebrate taxa in inner and outer Saginaw Bay as derived from GLMs. Data collected in Saginaw Bay, 1990–1996 and grouped by substrate type: (a) silt and silty sand in the inner bay (Stations 4, 7, 10, and 11), (b) sand and gravel in the inner bay (Stations 13, 14, and 16), and (c) silty sand in the outer bay (Stations 20, 23, and 24). *Diporeia* was only found in (c). Dashed lines give 95% confidence limits. Lines along scales of the x-axes represent the amount of data available for each factor level. Y-axes are scaled so that 0 represents the mean density and scales vary to allow display of overall patterns.

(continued)

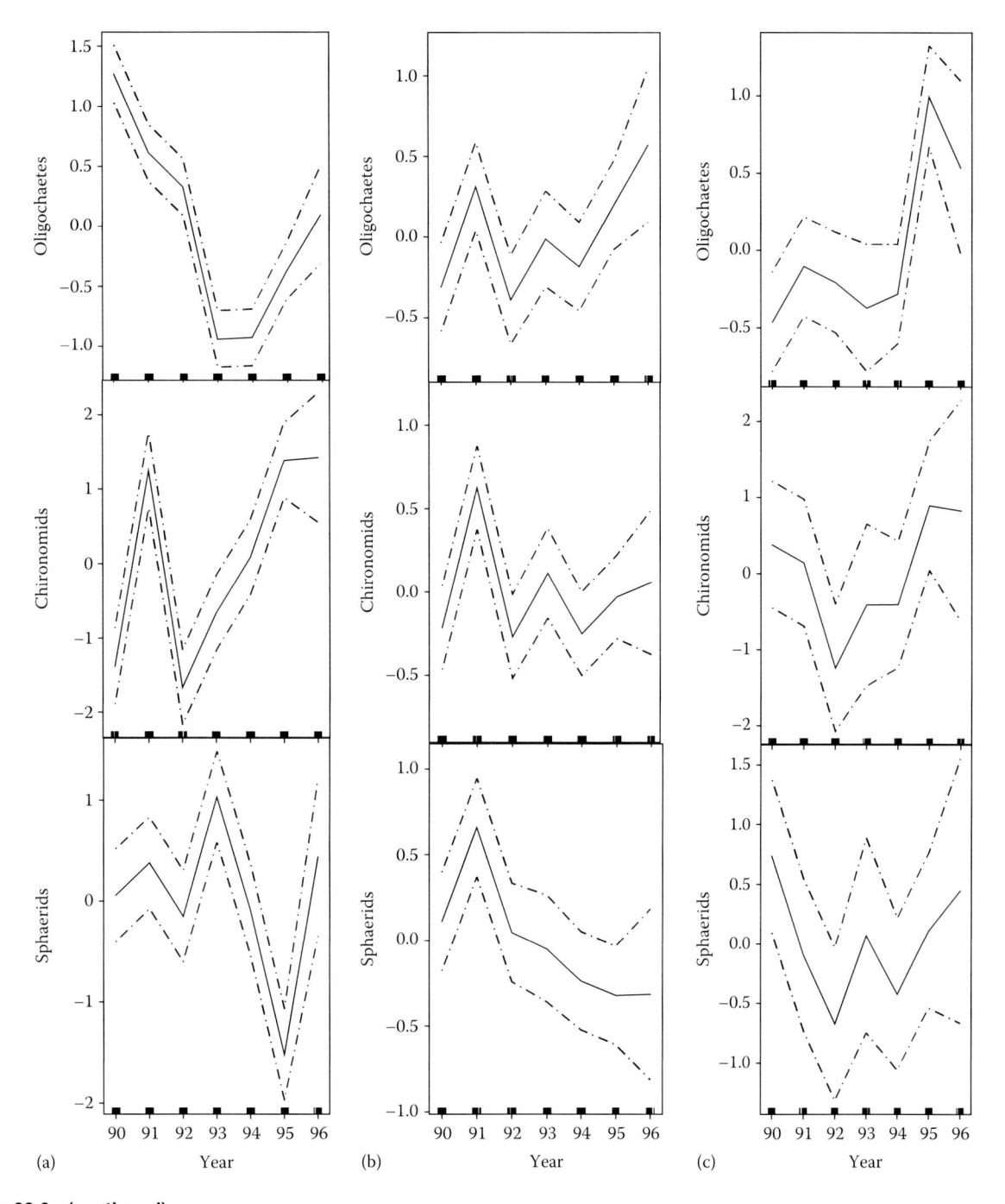

Figure 33.9 (continued)

(Strayer et al. 1999). Changes persisted 5 years after initial colonization. Effects on phytoplankton and zooplankton over the entire monitoring program (1990–1996) were more extensive than effects previously reported for initial and partial portion of the program (1990–1993) (Bridgeman et al. 1995, Fahnenstiel et al. 1995a,b).

Our analysis of annual, seasonal, and spatial trends in phytoplankton, zooplankton, and macroinvertebrate groups relative to trends in zebra mussels showed that most groups were affected. Effects were negative and positive, and correlations had different lag times. *Dreissena* densities increased to a peak in 1992 and declined to stable levels through 1996 based on SCUBA samples (Figure 33.2). We used SCUBA rather than grab samples to interpret annual trends of zebra mussels since based on our field observations, patchy zebra mussel aggregations obstructed closure of the grab and

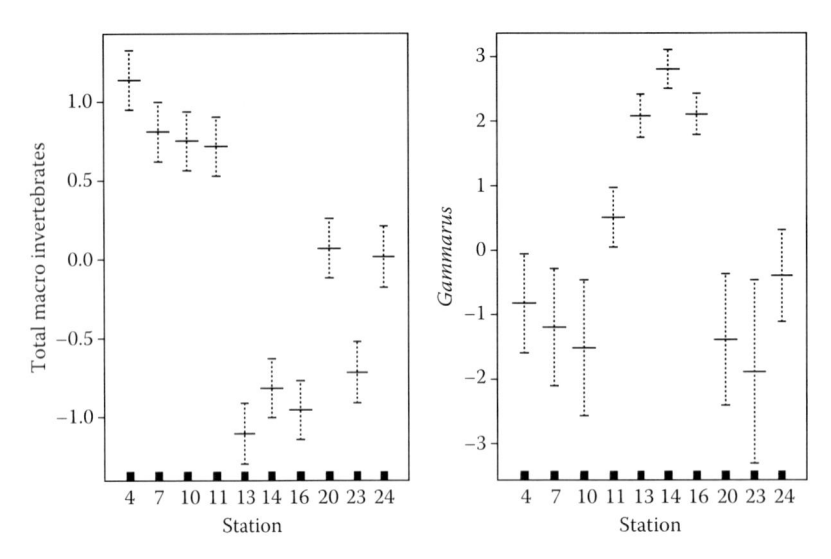

Figure 33.10 Spatial indices for mean densities of total macroinvertebrates and *Gammarus* as derived from GLMs. Data collected in Saginaw Bay, 1990–1996. Indices for oligochaetes, chironomids, and sphaeriids are not shown but were similar to total macroinvertebrates. Stations 4–16 were in the inner bay and substrates were silt and silty sand (Stations 4, 7, 10, and 11) and sand/gravel (Stations 13, 14, and 16). Stations 20–24 were in the outer bay and substrate at both stations was silty sand. Dashed lines give 95% confidence limits. Lines along scales of the x-axes represent the amount of data available for each factor level. Y-axes are scaled so that 0 represents the mean density and scales vary to allow display of overall patterns.

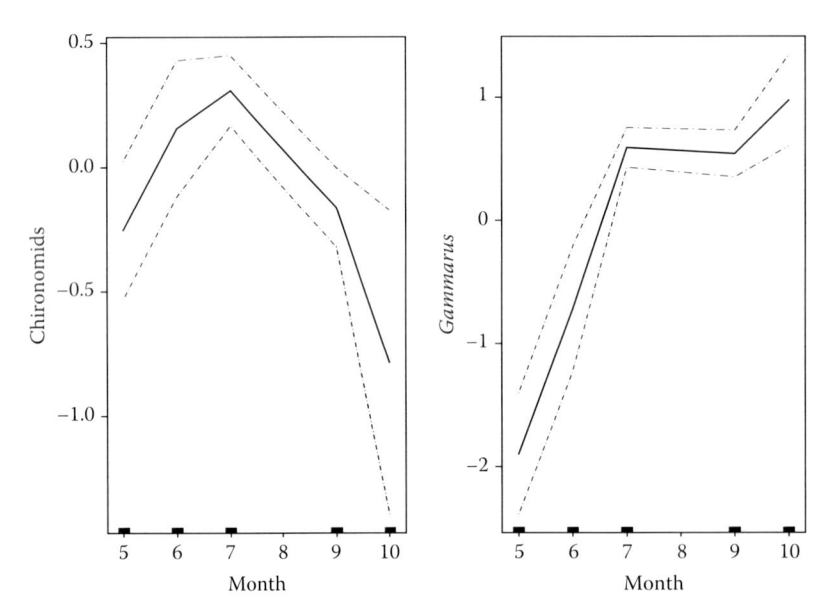

Figure 33.11 Seasonal indices for mean densities of chironomids and *Gammarus* as derived from GLMs. Data collected in Saginaw Bay, 1990–1996. Dashed lines give 95% confidence limits. Lines along scales of the x-axes represent the amount of data available for each factor level. Y-axes are scaled so that 0 represents the mean density and scales vary to allow display of overall patterns.

estimates might be biased. Grab estimates were nevertheless useful to interpret spatial trends. Overall, phytoplankton and zooplankton groups decreased, whereas responses of macroinvertebrate groups varied. Declines in cyanophytes, chlorophytes, flagellates, dinoflagellates, and to a less extent bacillariophytes coincided with the rapid increase in zebra mussels from 1990 to 1992, whereas declines in chrysophytes and cryptomonads lagged two years (Figure 33.3). Further, most groups recovered as zebra mussel densities declined. Most consistent effects occurred in zooplankton as responses were closely correlated with zebra mussel trends; all groups decreased sharply from 1990 to 1992 and tended to recover when zebra mussels declined (Figure 33.6). Density changes in phytoplankton and zooplankton were generally similar throughout the bay even though zebra mussel distributions were patchy. Density changes in benthic macroinvertebrates occurred at the local scale and were a function of substrate. Macroinvertebrate taxa that fluctuated in synchrony with trends in zebra mussels were *Gammarus* and oligochaetes. *Gammarus* increases closely followed *Dreissena* trends, and oligochaetes declines and recoveries mirrored zebra mussel fluctuations in density at stations where they were most abundant and zebra mussels were scarce. Although declines in *Diporeia* correlated well with zebra mussel increases, these taxa did not recover as zebra mussels declined.

Differential impact of zebra mussels among phytoplankton taxonomic groups has been reported by most previous studies in Saginaw Bay, Lake Erie, and other areas (Strayer et al. 1999, Vanderploeg et al. 2001, 2002). In Saginaw Bay, cyanophytes as a group exhibited the most dramatic and persistent decline, a result that generally followed patterns reported elsewhere (e.g., Nicholls and Hopkins 1993, Smith et al. 1998, Strayer et al. 1999, Nicholls et al. 2002). Reported trends vary among individual taxa within this group. For example, while cyanophytes as a group declined, Vanderploeg et al. (2001, 2002) reported summer blooms of *Microcystis* in Saginaw Bay after the zebra mussel invasion. The blooms were attributed to selective rejection of *Microcystis* in pseudofeces. In Green Bay, increases in cyanophytes were reported (Stasio et al. 2008), but little information was provided on overall species composition and abundance of the zebra mussel population. Our interpretation of zebra mussel effects on some groups of phytoplankton entails consideration of habitat requirements of individual species. The persistent decline in cyanophytes in Saginaw Bay was largely a result of disappearance of filamentous colonial *Oscillatoria redekii*, which can be attributed to increased water clarity caused by *Dreissena* filtration, as found by Nicholls et al. (2002) in the Bay of Quinte, Lake Ontario. In Saginaw Bay, water clarity as measured by Secchi depth increased from 0.7 to 1.5 m between fall 1990 and fall 1991 (Fielder et al. 2000) and >2 m over the next 4 years (Nalepa et al. 1996, Johengen et al. 2000). Satellite imagery (Advance Very High Resolution Radiometer) also indicated persistent increases in water clarity after the first large recruitment of *Dreissena* in Saginaw Bay (Budd et al. 2001). *Oscillatoria* (currently *Limnothrix*) are superior competitors under low light intensity conditions and frequently dominate phytoplankton in shallow lakes (mean depth <3 m) with low values of Secchi depth to total depth ratio (<0.3) (Berger 1975, 1989, Romo and Miracle 1993, Rucker et al. 1997). Conversely, *Oscillatoria* are negatively affected by high light intensities (Mur and Schreurs 1995, Havens et al. 1998).

Although the decline in cyanophytes as a group was likely a direct result of zebra mussel colonization, coincident relationships were not as clear for other phytoplankton groups. Most evident was the decline in chlorophytes. Densities in 1990 were already lower than densities found in 1980 when phytoplankton consisted of a mix of bacillariophytes, shade-tolerant cyanophytes, and chlorophytes (Stoermer and Theriot 1985). Chlorophytes were practically eradicated after zebra mussel colonization. Increased abundances of phytoplankton in 1991 (except for cyanophytes) were probably zebra mussel independent because total phosphorous loads were high that year (Figure 33.12) but could also be indirectly linked through facilitation of phytoplankton competitive advantages with cyanophytes and reduction of calanoid grazing.

Reports of *Dreissena* effects on bacillariophytes have varied (Smith et al. 1998, Makarevicz et al. 1999, Idrisi et al. 2001, Nicholls et al. 2002, Barbiero et al. 2006), but declines have been generally explained by mussel filtration activities. Trends in bacillariophytes in Saginaw Bay although significant were minimal after zebra mussel colonization. However, there was a change in species composition of this group, with the most noted change being the replacement of *Cyclotella comensis* by *C. ocellata*. The latter is a faster-growing species (Fahnenstiel unpublished data), and hence, the impact of filtration activities of *Dreissena* may have been moderated by a shift to species with higher specific growth rates, resulting in little change for the group as a whole.

Declines of zooplankton in Saginaw bay a year after zebra mussel colonization are consistent with declines documented for rotifers (Dahl et al. 1995, Beeton and Hageman 2000) and crustaceans (e.g., Karatayev et al. 1997, Pace et al. 1998). Other studies found modest or no declines in some crustacean groups (Dahl et al. 1995, Beeton and Hageman 2000, Mayer et al. 2000, Idrisi et al. 2001). Differences are probably a result of stratification and trophic states of the systems studied. Shallow, well-mixed systems similar to Saginaw Bay are strongly influenced by zebra mussel filtration. At the

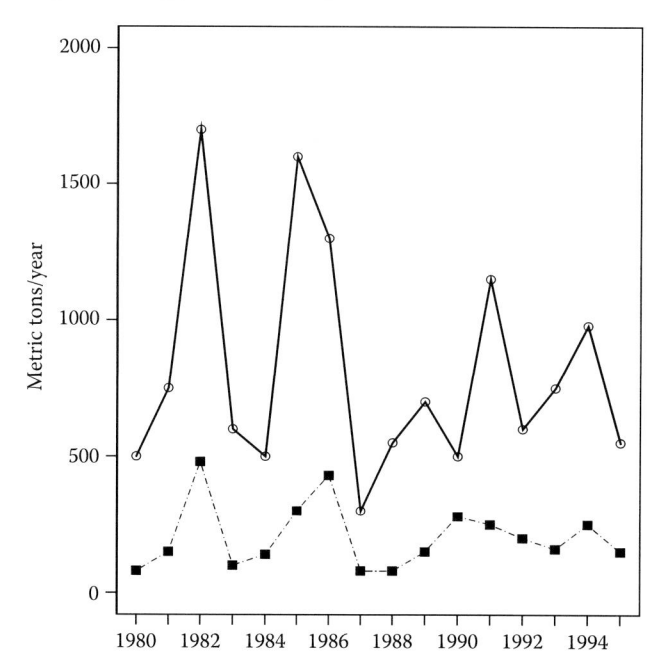

Figure 33.12 Annual phosphorus load (metric tons) into the Saginaw River, 1980 to 1995. Solid line = total phosphorus; metric tons dashed line = dissolved phosphate phosphorus. (Based on Bierman, Jr. V.J. et al., *J. Great Lakes Res.*, 31, 32, 2005.)

same time, Saginaw Bay is phosphorus limited during summer that affects the quality of seston food for consumers (both zooplankton and zebra mussels) and the ability of phytoplankton to grow rapidly relative to grazing pressures. In Saginaw Bay, there was a decline in available food during the early-colonization phase (1991–1993) as evident not only by a decline in total phytoplankton but also by low chlorophyll concentrations throughout the growing season (detailed in Vanderploeg et al. 2001, 2002, 2009). By 1994–1996, low concentrations of chlorophyll occurred in spring, but concentrations in summer/fall increased because of blooms of *Microcystis* that is a low-quality food for zooplankton.

Impacts of zebra mussels on zooplankton were likely mediated through both competition for food and direct predation. All groups were affected despite differences in feeding habits: cladocerans are mostly herbivorous, calanoids are omnivorous, cyclopoids are carnivorous, and rotifers have diverse feeding habits (Vanderploeg 1994, Bundy et al. 2005). Zebra mussels prefer particulate food <50 μm but can filter particles from 0.4 to 450 μm that include microzooplankton (Mikheev 1967, Shevtsova et al. 1986, Cotner et al. 1995, Lavrentyev et al. 1995, MacIsaac et al. 1995). However, maximum retention by zebra mussels is in the 5–35 μm range (Jorgensen et al. 1984, Sprung and Rose 1988). Although many zooplankton groups in this study (or life stages within groups) were within the size range of zebra mussel filtration, rotifers were likely most directly affected by mussel grazing since in general they do not have strong escape responses as do larger copepods and nauplii (e.g., Williamson and Vanderploeg 1988). Further, direct predation and loss of phytoplankton food have been cited as causes for declines in microzooplankton after zebra mussel colonization (MacIsaac et al. 1995, Pace et al. 1998, Vanderploeg et al. 2002). Also, because there is a link between protozoans and crustaceans in the Great Lakes especially for copepods (Carrick et al. 1991, Bundy et al. 2005), crustaceans are likely to be consistently impacted by declines in both phytoplankton and protozoans. Although our analysis did not show reduction in flagellates up to 1994, Lavrentyev et al. (1995) reported filtration by zebra mussels reduced protozoan abundance in Saginaw Bay, and protozoans are prey of some rotifers (Jack and Gilbert 1994), copepods (Burns and Gilbert 1993, Hartmann et al. 1993), and cladocerans (Jack and Gilbert 1994, DeBiase et al. 1990).

Veligers steadily increased in abundance as biomass of the zebra mussel population increased, but the role of veligers in lower food webs remains unknown. Veligers are in the water column for 8–30 days (Sprung 1992) and feed on a variety of small algal species (Vanderploeg et al. 1996). Although veligers are not readily fed upon by most copepods (Liebig and Vanderploeg 1995), they probably are important components of the food web. At times, veliger densities in Saginaw Bay reached maximum densities of $7 \times 10^6/m^3$ in the samples. In Lake Erie, veliger production in 1993 and 1994 ranged from 1%–2% up to 25% of total zooplankton production, which further suggests that they can play a significant role in energy transfer (Johannsson et al. 2000).

Trends among macroinvertebrate groups in response to zebra mussels were more dissimilar than among phytoplankton and zooplankton groups as trends varied relative to times and sites. This was expected as most macroinvertebrate communities consist of taxa with life habits, life histories, and habitat preferences that are much more varied than phytoplankton and zooplankton. Observed declines in oligochaetes in areas with silt substrates in the inner bay that had few mussels were likely mediated by indirect effects of food intake by mussels from other areas. We observed decreased densities of phytoplankton in all areas, and such a decrease would mean a corresponding decrease in organic material settled to the bottom and less food for oligochaetes. Lower food might not be the reason for the decline of *Diporeia* because their diet consists mostly of bacillariophytes (Nalepa et al. 1998, Dermott 2001), a phytoplankton group that experienced little reduction in abundance in Saginaw Bay after zebra mussel colonization. In this particular case, the decline could be related to changes in species composition observed within bacillariophytes or to factors unrelated to food such as disease (Nalepa et al. 2006). Increased densities of *Gammarus* and oligochaetes in areas with high zebra mussel densities in the inner bay were likely due to increased substrate complexity from mussel shells and to increased organic material from mussel biodeposits, as noted by Nalepa et al. (2003) and observed in other systems (Protasov and Afanasyev 1990, Dermott et al. 1993, Griffiths 1993, Stewart and Haynes 1994, Wisenden and Bailey 1995, Botts et al. 1996, Ricciardi et al. 1998). Nevertheless, increased densities of *Gammarus* in areas of low zebra mussel density may reflect movement away from areas with high densities of *Gammarus*. Also noted by Nalepa et al. (2003), overall densities of chironomids in Saginaw Bay did not increase, which contrasts to other studies (Griffiths 1993). However, a very gradual increase of chironomid densities occurred after the initial zebra mussel peak in areas of the inner bay subject to sedimentation (sites with silt substrate at deeper depths), which could reflect benefits from slow accumulation of organic deposition.

Our results have implications for monitoring ecosystem impacts of *Dreissena* in Saginaw Bay and similar systems. The substrate-driven spatial distribution of zebra mussels observed in the present study did not lead to distinct spatial distribution patterns of phytoplankton and zooplankton. Thus, in systems that are well mixed and have populations of dreissenids that have high filtration capacity relative to system volume, dreissenid effects on plankton can be expected to be spatially homogenous. On the other hand, phytoplankton and zooplankton densities displayed important seasonal variations over the 6-year study period. Thus, monitoring dreissenid impacts on plankton communities in systems such as Saginaw Bay can be accomplished with a limited number of stations, but it is essential to obtain broad seasonal coverage. This is particularly important in the initial stages of the colonization period when changes in the pelagic region are most prevalent (i.e., increased water clarity and shifts in phytoplankton composition). In contrast to plankton, broad spatial coverage is important when assessing impacts on benthic macroinvertebrates. Spatial patterns were observed even in areas without zebra mussels, and seasonality was of minor consequence. Thus, monitoring of dreissenid impacts on macroinvertebrates requires extensive coverage of substrate types relative to seasonal coverage. Finally, our results highlight the importance of prompt data analysis collected during initial monitoring stages to help provide information needed to optimize survey design during ongoing assessments of dreissenid impacts.

ACKNOWLEDGMENTS

This study was supported by a grant of the Michigan Sea Grant Program. Valuable editorial comments were provided by Stephen Riley and several anonymous reviewers.

REFERENCES

Barbiero, R. P., D. C. Rockwell, G. J. Warren, and M. Tuchman. 2006. Changes in spring phytoplankton communities and nutrient dynamics in the eastern basin of Lake Erie since the invasion of *Dreissena* spp. *Can. J. Fish. Aquat. Sci.* 63: 1549–1563.

Becker, R. A., J. M. Chambers, and A. R. Wilks. 1988. *The New S Language. A Programming Environment for Data Analysis and Graphics*. Pacific Grove, CA: Wadsworth & Brooks/Cole Advanced Books & Software.

Beeton, A. M. and J. Hageman, Jr. 2000. Changes in zooplankton populations in western Lake Erie after establishment of *Dreissena polymorpha*. *Verh. Int. Verein. Limnol.* 27: 3798–3804.

Beeton, A. M, S. H. Smith, and F. H. Hooper. 1967. Physical limnology of Saginaw Bay. Great Lakes Fisheries Commission Technical Report 12. Ann Arbor, MI: Great Lakes Fishery Commission.

Berger, C. 1975. Occurrence of *Oscillatoria agardhii* GOM. in some shallow eutrophic lakes. *Limnology* 19: 2689–2697.

Berger, C. 1989. In situ primary production, biomass and light regime in the Wolderwijd, the most stable *Oscillatoria agardhii* lake in The Netherlands. *Hydrobiologia* 185: 233–244.

Bierman, Jr. V. J., J. Kaur, J. V. DePinto, T. J. Feist, and D. W. Dilks. 2005. Modeling the role of zebra mussels in the proliferation of blue-green algae in Saginaw Bay, Lake Huron. *J. Great Lakes Res.* 31: 32–55.

Botts, P. S., B. A. Patterson, and D. W. Schloesser. 1996. Zebra mussel effects on benthic invertebrates: Physical or biotic? *J. N. Am. Benthol. Soc.* 15: 179–184.

Bridgeman, T. B., G. L. Fahnenstiel, G. A. Lang, and T. F. Nalepa. 1995. Zooplankton grazing during the zebra mussel (*Dreissena polymorpha*) colonization of Saginaw bay, Lake Huron. *J. Great Lakes Res.* 21: 567–573.

Budd, J., T. D. Drummer, T. F. Nalepa, and G. L. Fahnenstiel. 2001. Remote sensing of biotic effects: Zebra mussels (*Dreissena polymorpha*) influence on water clarity in Saginaw Bay, Lake Huron. *Limnol. Oceanogr.* 46: 213–223.

Bundy, M. H., H. A. Vanderploeg, P. J. Lavrentyev, and A. P. Kovalcik. 2005. The importance of microzooplankton versus phytoplankton to copepod populations during late winter and early spring in Lake Michigan. *Can. J. Fish. Aquat. Sci.* 62: 2371–2385.

Burns, C. W. and J. J. Gilbert. 1993. Predation on ciliates by freshwater calanoid copepods: Rates of predation and relative vulnerabilities of prey. *Freshwat. Biol.* 30: 377–393.

Caraco, N. F., J. J. Cole, P. A. Raymond et al. 1997. Zebra mussel invasion in a large, turbid river: Phytoplankton assemblage response to increase grazing. *Ecology* 78: 588–602.

Carrick, H. J., G. L. Fahnenstiel, E. E. Stoermer, and R. G. Wetzel. 1991. The importance of zooplankton—Protozoan trophic couplings in Lake Michigan. *Limnol. Oceanogr.* 36: 1335–1345.

Cotner, J. B., W. S. Gardner, J. R. Johnson, R. H. Sada, J. F. Cavaletto, and R. T. Heath. 1995. Effects of zebra mussels (*Dreissena polymorpha*) on bacterioplankton: Evidence for size-selective consumption and growth stimulation. *J. Great Lakes Res.* 21: 517–528.

Dahl, J. A., D. M. Graham, R. Dermott, O. E. Johannsson, E. S. Millard, and D. D. Myles. 1995. Lake Erie 1993, western, west central, and eastern basins: Changes in trophic status, and assessment of the abundance, biomass and production of the lower trophic levels. *Can. Tech. Rept. Fish. Aquat. Sci.* 2070, 118p.

DeBiase, A. E., R. W. Sanders, and K. G. Porter. 1990. Relative nutritional value of ciliate protozoa and algae as food for Daphnia. *Microb. Ecol.* 19: 199–210.

Dermott, R. 2001. Sudden disappearance of the amphipod *Diporeia* from eastern Lake Ontario, 1993–1995. *J. Great Lakes Res.* 27: 423–433.

Dermott, R., J. Mitchell, I. Murray, and E. Fear. 1993. Biomass and production of zebra mussels (*Dreissena polymorpha*) in shallow waters of northeastern Lake Erie. In *Zebra Mussels: Biology, Impacts, and Control*, T. F. Nalepa and D. W. Schloesser, eds., pp. 399–413. Boca Raton, FL: CRC Press.

Fahnenstiel, G. L., T. B. Bridgeman, G. A. Lang, M. J. McCormick, and T. F. Nalepa. 1995a. Phytoplankton productivity in Saginaw Bay, Lake Huron: Effects of zebra mussel (*Dreissena polymorpha*) colonization. *J. Great Lakes Res.* 21: 465–475.

Fahnenstiel, G. L., G. A. Lang, T. F. Nalepa, and T. H. Johengen. 1995b. Effects of zebra mussels (*Dreissena polymorpha*) colonization on water quality parameters in Saginaw Bay, Lake Huron. *J. Great Lakes Res.* 21: 435–448.

Fahnenstiel, G. L., A. F. Krause, M. J. McCormick, H. J. Carrick, and C. L. Schelske. 1998. The structure of the planktonic food-web in the St. Lawrence Great Lakes. *J. Great Lakes Res.* 24: 531–554.

Fahnenstiel, G. L., S. Pothoven, H. Vanderploeg, D. Klarer, T. Nalepa, and D. Scavia. 2010. Recent changes in primary production and phytoplankton in the offshore region of southeastern Lake Michigan. *J. Great Lakes Res.* 36(Suppl. 3): 20–29.

Fanslow, D. L., T. F. Nalepa, and G. A. Lang. 1995. Filtration rates of the zebra mussel (*Dreissena polymorpha*) on natural seston from Saginaw Bay, Lake Huron. *J. Great Lakes Res.* 21: 489–500.

Fielder, D. G., J. E. Johnson, J. R. Weber, M. V. Thomas, and R. C. Haas. 2000. Fish population survey of Saginaw Bay, Lake Huron, 1989–1997. Fisheries Research Report 2052, 54p. Lansing: Michigan Department of Natural Resources.

Griffiths, R. W. 1993. Effects of zebra mussels (*Dreissena polymorpha*) on the benthic fauna of Lake St. Clair. In *Zebra Mussels: Biology, Impacts, and Control.*, eds., T. F. Nalepa and D. W. Schloesser, pp. 414–437. Boca Raton, FL: CRC Press.

Haney J. F. and D. J. Hall. 1973. Sugar-coated *Daphnia*: A preservation technique for Cladocera. *Limnol. Oceanogr.* 18: 331–333.

Hartmann, H. J., H. Taleb, L. Aleya, and N. Lair. 1993. Predation on ciliates by the suspension-feeding calanoids copepod *Acanthodiaptomus denticornis*. *Can. J. Fish. Aquat. Sci.* 50: 1382–1393.

Havens, K. E., E. J. Philips, M. F. Cichra, and L. Bai-Lian. 1998. Light availability as a possible regulator of cyanobacteria species composition in a shallow subtropical lake. *Freshwat. Biol.* 39: 547–556.

Holland, R. E. 1993. Changes in planktonic diatoms and water transparency in Hatchery Bay, Bass Island area, Western Lake Erie since the establishment of the zebra mussel. *J. Great Lakes Res.* 19: 617–624.

Idrisi, N., E. L. Mills, L. G. Rudstam, and D. J. Stewart. 2001. Impacts of zebra mussels (*Dreissena polymorpha*) on the pelagic trophic levels of Oneida Lake, New York. *Can. J. Fish. Aquat. Sci.* 58: 1430–1441.

Jack, J. D. and J. J. Gilbert. 1994. Effects of *Daphnia* on microzooplankton communities. *J. Plankton Res.* 16: 1499–1512.

Johannsson, O. E., R. Dermott, D. M. Graham et al. 2000. Benthic and pelagic secondary production in Lake Erie after the invasion of *Dreissena spp.* with implications for fish production. *J. Great Lakes Res.* 26: 31–54.

Johengen, T. H., T. F. Nalepa, G. A. Lang, D. L. Fanslow, H. A. Vanderploeg, and M. A. Agy. 2000. Physical and chemical variables of Saginaw Bay, Lake Huron in 1994–1996. NOAA Technical Memorandum GLERL-115, Ann Arbor, MI.

Jorgensen, C. B., T. Kioboe, and H. U. Riisgord. 1984. Ciliary and mucus-net filter feeding with special reference to fluid mechanical characteristics. *Mar. Ecol. Prog. Ser.* 15: 283–292.

Karateyev, A. Y., L. E. Burlakova, and D. K. Padilla. 1997. The effects of *Dreissena polymorpha* (Pallas) invasion on aquatic communities in eastern Europe. *J. Shellfish Res.* 16: 187–203.

Lavrentyev, P. J., W. S. Gardner, J. F. Cavaletto, and J. R. Beaver. 1995. Effects of the zebra mussel (*Dreissena polymorpha* Pallas) on protozoa and phytoplankton from Saginaw Bay, Lake Huron. *J. Great Lakes Res.* 21: 545–557.

Leach, J. H. 1993. Impacts of the zebra mussel *(Dreissena polymorpha)* on water quality and fish spawning reefs in western Lake Erie. In *Zebra Mussels: Biology, Impacts, and Control*, T. F. Nalepa and D. W. Schloesser, eds., pp. 381–397. Boca Raton, FL: CRC Press.

Liebig, J. R. and H. A. Vanderploeg. 1995. Vulnerability of *Dreissena polymorpha* larvae to predation by Great Lakes copepods: The importance of the bivalve shell. *J. Great Lakes Res.* 21: 353–358.

Lowe, R. L. and R. W. Pillsbury. 1995. Shifts in benthic algal community structure and function following the appearance of zebra mussels (*Dreissena polymorpha*) in Saginaw Bay, Lake Huron. *J. Great Lakes Res.* 21: 558–566.

MacIsaac, H. J. 1996. Potential abiotic and biotic impacts of zebra mussels on the inland waters of North America. *Am. Zool.* 36: 287–299.

MacIsaac, H. J., C. J. Lonnee, and J. H. Leach. 1995. Suppression of microzooplankton by zebra mussels: Importance of mussel size. *Freshwat. Biol.* 34: 379–387.

Makarevicz, J. C., T. W. Lewis, and P. Bertram. 1999. Phytoplankton composition and biomass in the offshore waters of Lake Erie: Pre-and post-*Dreissena* introduction (1983–1993). *J. Great Lakes Res.* 25: 135–148.

Mayer, C. M., A. J. Van De Valk, J. L. Forney, L. G. Rudstam, and E. L. Mills. 2000. Response of yellow perch (*Perca flavescens*) in Oneida Lake New York, to the establishment of zebra mussels (*Dreissena polymorpha*). *Can. J. Fish. Aquat. Sci.* 57: 742–734.

McCullagh, P. and J. A. Nelder. 1989. *Generalized Linear Models*. London, U.K.: Chapman & Hall.

Mikheev, V. P. 1967. Filtration nutrition of the *Dreissena. Trud. Vsesoy. Nauch.-lsslekd. Instit.* 15: 117–129. [Russian with English summary.]

Mur, L. R. and H. Schreurs. 1995. Light as a selective factor in the distribution of phytoplankton species. *Water Sci. Technol.* 32: 25–34.

Nalepa, T. F. and G. L. Fahnenstiel. 1995. *Dreissena polymorpha* in the Saginaw Bay, Lake Huron ecosystem: Overview and perspective. *J. Great Lakes Res.* 21: 411–416.

Nalepa, T. F., G. L. Fahnenstiel, M. J. McCormick et al. 1996. Physical and chemical variables of Saginaw Bay, Lake Huron in 1991–1993. NOAA Technical Memorandum GLERL-91, Ann Arbor, MI.

Nalepa, T. F., D. L. Fanslow, M. B. Lansing, and G. A. Lang. 2003. Trends in the benthic macro-invertebrate community of Saginaw Bay, 1987–1996: Responses to phosphorus abatement and zebra mussel, *Dreissena polymorpha*. *J. Great Lakes Res.* 29: 14–33.

Nalepa, T. F., D. J. Hartson, D. L. Fanslow, G. A. Lang, and S. J. Lozano. 1998. Declines in benthic macroinvertebrate populations in southern Lake Michigan, 1980–1993. *Can. J. Fish. Aquat. Sci.* 55: 2402–2413.

Nalepa, T. F., D. C. Rockwell, and D. W. Schloesser. 2006. Disappearance of the amphipod *Diporeia* spp. in the Great Lakes: Workshop summary, discussion, and recommendations. NOAA Technical Memorandum GLERL-136. Ann Arbor, MI: Great Lakes Environmental Research Laboratory.

Nalepa, T. F., J. M. Wojcik, D. L. Fanslow, and G. A. Lang. 1995. Initial colonization of the zebra mussel (*Dreissena polymorpha*) in Saginaw Bay, Lake Huron: Population recruitment, density, and size structure. *J. Great Lakes Res.* 21: 417–434.

Nicholls, K. H., L. Heintsch, and E. Carney. 2002. Univariate step-trend and multivariate assessments of the apparent effects of P loading reductions and zebra mussels on the phytoplankton of the Bay of Quinte, Lake Ontario. *J. Great Lakes Res.* 28: 15–31.

Nicholls, K. H. and G. J. Hopkins. 1993. Recent changes in Lake Erie (north shore) phytoplankton: Cumulative impacts of phosphorus loading reductions and the zebra mussel introduction. *J. Great Lakes Res.* 19: 637–647.

Pace, M. L., W. E. Findlay, and D. Fisher. 1998. Effects of an invasive bivalve on the zooplankton community of the Hudson River. *Freshwat. Biol.* 39: 103–116.

Protasov, A. A. and S. A. Afanasyev. 1990. Principal types of *Dreissena* communities in periphyton. *Hydrobiol. J.* 26: 15–23.

Ricciardi, A., F. G. Whoriskey, and J. B. Rasmussen. 1998. The role of the zebra mussel (*Dreissena polymorpha*) in structuring macroinvertebrate communities on hard substrate. *Can. J. Fish. Aquat. Sci.* 54: 2596–2608.

Romo, S. and M. R. Miracle. 1993. Long-term periodicity of *Planktothrix agardhii, Pseudoanabaena galeata* and *Geitlerinema* sp. in a shallow hypertrophic lagoon, the Albufera of Valencia (Spain). *Archiv. Hydrobiol.* 126: 469–486.

Rucker, J., C. Wiedner, and P. Zippel. 1997. Factors controlling the dominance of *Planktothrix agardhii* and *Limnothrix redekei* in eutrophic shallow lakes. *Hydrobiologia* 342: 107–115.

Shevtsova, L., G. Zhdanova, B. Mouchan, and A. Primak. 1986. Experimental interrelations between *Dreissena* and planktonic invertebrates. *Hydrobiol. J.* 22: 3639. [English translation of *Gidrobiol. Zhur.* 22: 36–40.]

Skubinna, J. P., T. G. Coon, and T. R. Batterson. 1995. Increased abundance and depth of submersed macrophytes in response to decreased turbidity in Saginaw Bay, Lake Huron. *J. Great Lakes Res.* 21: 476–488.

Smith, T. E., R. J. Stevenson, N. F. Caraco, and J. J. Cole. 1998. Changes in phytoplankton community structure during the zebra mussel (*Dreissena polymorpha*) invasion of the Hudson River (New York*). J. Plankton Res.* 20: 1567–1579.

Sprung, M. 1992. The other life: An account of present knowledge of the larval phase of Dreissena polymorpha. In *Zebra Mussels: Biology, Impacts, and Control*, T. F. Nalepa and D. W. Schloesser, eds., pp. 39–53. Boca Raton, FL: CRC Press.

Sprung, M. and U. Rose. 1988. Influence of body size and food quality on the feeding of the mussel *Dreissena polymorpha. Oecologia* 77: 526–532.

Stasio, B., M. B. Schrimpf, A. E. Beranek, and W. C. Daniels. 2008. Increased Chlorophyll a, phytoplankton abundance, and cyanobacteria occurrence following invasion of Green Bay, Lake Michigan by dreissenid mussels. *Aquat. Invasions* 3: 21–27.

Stewart, T. W. and J. M. Haynes. 1994. Benthic macroinvertebrate communities of southwestern Lake Ontario following invasion of *Dreissena. J. Great Lakes Res.* 20: 479–493.

Strayer, D. L., N. F. Caraco, J. J. Cole, S. Findlay, and M. L. Pace. 1999. Transformation of freshwater ecosystems by bivalves. *BioScience* 49: 19–27.

Strayer, D. L., V. T. Eviner, J. M. Jeschke, and M. L. Pace. 2006. Understanding the long-term effects of species invasions. *Trends Ecol. Evol.* 21: 645–651.

Stoermer, E. F. and E. Theriot. 1985. Phytoplankton distribution in Saginaw Bay. *J. Great Lakes Res.* 11: 132–142.

Vanderploeg, H. A. 1994. Zooplankton particle selection and feeding mechanisms. In *The Biology of Particles in Aquatic Systems*, 2nd Edn., R. S. Wotton, ed., pp. 205–234. Boca Raton, FL: CRC Press.

Vanderploeg, H. A., T. Johengen, and J. Liebig. 2009. Feedback between zebra mussel selective feeding and algal composition affect mussel condition: Did the regime changer pay a price for its success? *Freshwat. Biol.* 54: 47–63.

Vanderploeg, H. A., J. R. Liebig, W. W. Carmichael et al. 2001. Zebra mussel (*Dreissena polymorpha*) selective filtration promoted toxic *Microcystis* blooms in Saginaw Bay (Lake Huron) and Lake Erie. *Can. J. Fish. Aquat. Sci.* 58: 1208–1221.

Vanderploeg, H. A., J. R. Liebig, and A. A. Gluck. 1996. Evaluation of different phytoplankton for supporting development of zebra mussel larvae (*Dreissena polymorpha*): The importance of size and polyunsaturated fatty acid content. *J. Great Lakes Res.* 22: 36–45.

Vanderploeg, H. A., T. F. Nalepa, D. J. Jude et al. 2002. Dispersal and ecological impacts of Ponto-Caspian species in the Great Lakes. *Can. J. Fish. Aquat. Sci.* 29: 1209–1228.

Williamson, C. E. and H. A. Vanderploeg. 1988. Predatory suspension-feeding in *Diaptomus*: Prey defenses and the avoidance of cannibalism. *Bull. Mar. Sci.* 43: 561–572.

Wisenden, P. A. and R. C. Bailey. 1995. Development of macroinvertebrate community structure associated with zebra mussel (*Dreissena polymorpha*) colonization of artificial substrates. *Can. J. Zool.* 73: 1438–1443.

CHAPTER 34

Lake Michigan after Dreissenid Mussel Invasion
Water Quality and Food Web Changes during the Late Winter/Spring Isothermal Period

Steven A. Pothoven and Gary L. Fahnenstiel

CONTENTS

Abstract 545
Introduction 545
Methods 546
Results and Discussion 548
Conclusions 551
References 551

ABSTRACT

This chapter documents water quality and lower foodweb changes in the offshore region of southern Lake Michigan during the late winter/spring isothermal mixing period during three distinct decades: 1983–1989 (pre-dreissenids), 1995–1998 (post-zebra mussel/pre-quagga mussel), and 2007–2010 (post-quagga mussel). Total phosphorus, chlorophyll *a*, and phytoplankton primary productivity did not change or decreased minimally (<22% decrease) between 1983–1989 and 1995–1998 but decreased substantially between 1995–1998 and 2007–2009 (34%, 68%, and 71% decrease, respectively). Secchi depth transparency increased from 6 m in 1985–1989 and 7 m in 1995–1998 to 18 m in 2007–2010. These pronounced changes in water column properties during the isothermal period in 2007–2010 compared to earlier periods were primarily attributed to the filtering activities of the quagga mussel (*Dreissena rostriformis bugensis*) and to a lesser extent to phosphorus load reductions. In contrast to substantial changes in phytoplankton, total zooplankton biomass did not differ between time periods (1986–1988, 1994–1998, and 2007–2009). However, biomass of cyclopoid copepods declined to nearly negligible levels in 2007–2009, perhaps reflecting a more oligotrophic system that favored calanoid copepods. Trends in biomass of calanoid copepods were not readily apparent or possibly increased in 2007–2009. Densities of *Mysis diluviana* declined 60% between 1996–2000 and 2007–2010. These large changes in limnological parameters and the food web have contributed to significant changes in traditional nutrient stoichiometry: Increases in total phosphorus to chlorophyll ratios and decreases in phytoplankton carbon to chlorophyll ratios were noted in 2007–2009 as compared to 1983–1987 and 1995–1998. Our results show that the recent expansion of quagga mussels in Lake Michigan has affected almost all aspects of water column nutrient and foodweb dynamics during the late winter/spring isothermal mixing period. We believe that future actions to manage the Lake Michigan ecosystem must consider these unique impacts of dreissenids and such impacts will affect management actions elsewhere in the Great Lakes.

INTRODUCTION

Lake Michigan has a long history of nonindigenous species introductions that have caused broad ecological changes. Some of these changes resulted from the introduction of fish such as rainbow smelt (*Osmerus mordax*), sea lamprey (*Petromyzon marinus*), and alewife (*Alosa pseudoharengus*) (Wells and McClain 1973, Madenjian et al. 2002). Some changes can also be attributed to nonindigenous invertebrates including *Bythotrephes longimanus* that have altered ecological functions of the lake (Lehman 1987). However, the relatively recent invasion and proliferation of dreissenid mussels may be responsible for some of the greatest ecological changes in the lake to date (Fahnenstiel et al. 2010a). *Dreissena polymorpha* (zebra mussels) was first found in Lake Michigan in 1989, and their population expanded in the 1990s (Nalepa et al. 1998, 2009). *Dreissena rostriformis bugensis* (quagga mussel) was first found in northern Lake Michigan in 1997 and expanded into the main basin of the lake in 2000 (Nalepa et al. 2001). Quagga mussels became abundant in nearshore areas throughout the lake by 2002 and expanded into the offshore region by 2007 where they continue to increase (Nalepa et al. 2010). Zebra mussels were mostly confined to the nearshore region and became rare by 2005 as quagga mussels proliferated (Nalepa et al. 2010).

The recent expansion of quagga mussels into offshore regions of Lake Michigan resulted in a major shift in energy flow and an accumulation of biomass in the benthic region (Nalepa et al. 2009). The ability of dreissenids to filter large amounts of water is linked to changes in energy and nutrient flow throughout the Great Lakes (Nalepa and Fahnenstiel 1995, Hecky et al. 2004). The effects of zebra mussels on phytoplankton populations were studied in the 1990s, and it was clear that zebra mussels had the potential to control phytoplankton abundance and composition, especially in nearshore regions (Holland 1993, Nicholls and Hopkins 1993, Fahnenstiel et al. 1995a,b). Much less is known about the impact of quagga mussels in deep, offshore waters as found in Lake Michigan. Further, assessments of impacts of quagga mussels in Lake Michigan are difficult because nutrient-load reductions initiated in the 1970s have continued into the 1990s and 2000s (Johengen et al. 1994, Mida et al. 2010). These load reductions have possibly altered limnological parameters and the foodweb, masking any changes caused by dreissenids.

This study documents changes in limnological parameters of the offshore region in southern Lake Michigan between 1983 and 2010 during a specific seasonal period, late winter/spring isothermal mixing. Specifically, we examine changes in nutrients, water transparency, phytoplankton chlorophyll and primary production, zooplankton biomass and composition, and abundances of *Mysis diluviana*. The isothermal period is an important component of the seasonal cycle in Lake Michigan as it is a period of disproportionate amounts of primary production (Fahnenstiel and Scavia 1987a). This production subsequently sets the stage for critical secondary production (Gardner et al. 1990). Moreover, the isothermal period is the only period when mussel filtering activities are directly linked to the entire water column (Fahnenstiel et al. 2010a, Vanderploeg et al. 2010). In contrast, during the stratified period (i.e., June–December), epilimnetic plankton are effectively separated from direct filtering by bottom-dwelling dreissenid mussels.

METHODS

Samples were collected at two offshore stations (≥100 m depth; 43°11.99 N, 086°34.19 W (MK) and 43°01.16 N, 086°37.91 W (GH)) in southeastern Lake Michigan (Figure 34.1). All data presented were collected during the late winter/spring isothermal period (surface temperature <5°C), which generally corresponded to March–May. The exception was abundance data for *M. diluviana* that included data from June (surface temperatures <15°C). The two stations were sampled as follows: in the 1980s the GH station was sampled exclusively; in the 1990s the GH station was sampled 8 times and the MK station was sampled 15 times; in 2007–2010 the GH station was sampled 2 times and the MK station was sampled 14 times. Because of the close proximity of these offshore stations and because the measured parameters (chlorophyll, photosynthetic parameters, Secchi depth, and total phosphorus) were similar between stations ($P > 0.05$), these two stations were assumed to represent the offshore region in southeastern Lake Michigan, and data from these stations were combined. For analysis, data were separated into three distinct periods: 1983–1989 (pre-dreissenids), 1995–1998 (post-zebra mussel/pre-quagga mussel), and 2007–2010 (post-quagga mussel expansion). Because earlier data were not available, data for *M. diluviana* were separated into only two periods: 1996–2000 and 2007–2010.

Temperature at depth was measured with an electronic bathythermograph (1983–1989) or Sea-Bird CTD (conductivity, temperature, and depth) equipped with a Sea Tech fluorometer and transmissometer (25 cm beam path) (1995–2010). Secchi depth transparency was measured with a black/white or white 25 cm diameter disk. Underwater light extinction of photosynthetically active irradiation (kPAR) was measured with a LI-COR or Biospherical scalar (4π) light sensor and/or a Biospherical integrating natural fluorometer (INF-3000). Surface incident irradiance was measured with a LI-COR sensor and data logger. During night sampling or when kPAR values were not measured, transmissometer and Secchi disk measurements were converted to kPAR values using empirically determined conversions for each transmissometer or a Lake Michigan empirically determined conversion for Secchi (kPAR = 1.53 (1/Secchi)).

Discrete samples of the water column were taken approximately biweekly with a modified Niskin bottle (Fahnenstiel et al. 2002) and poured into carboys (1 carboy for each depth) from which all water samples were taken. Samples for chlorophyll *a* analysis were filtered onto Whatman GF/F filters, extracted with either 90% acetone (1980s; Strickland and Parsons 1972) or N, N-dimethylformamide (1990s and 2007/2008; Speziale et al. 1984), and analyzed fluorometrically.

Phytoplankton photosynthesis was measured with the clean C-14 technique in a photosynthesis-irradiance incubator (Fahnenstiel et al. 1989, 2000). After incubation, samples

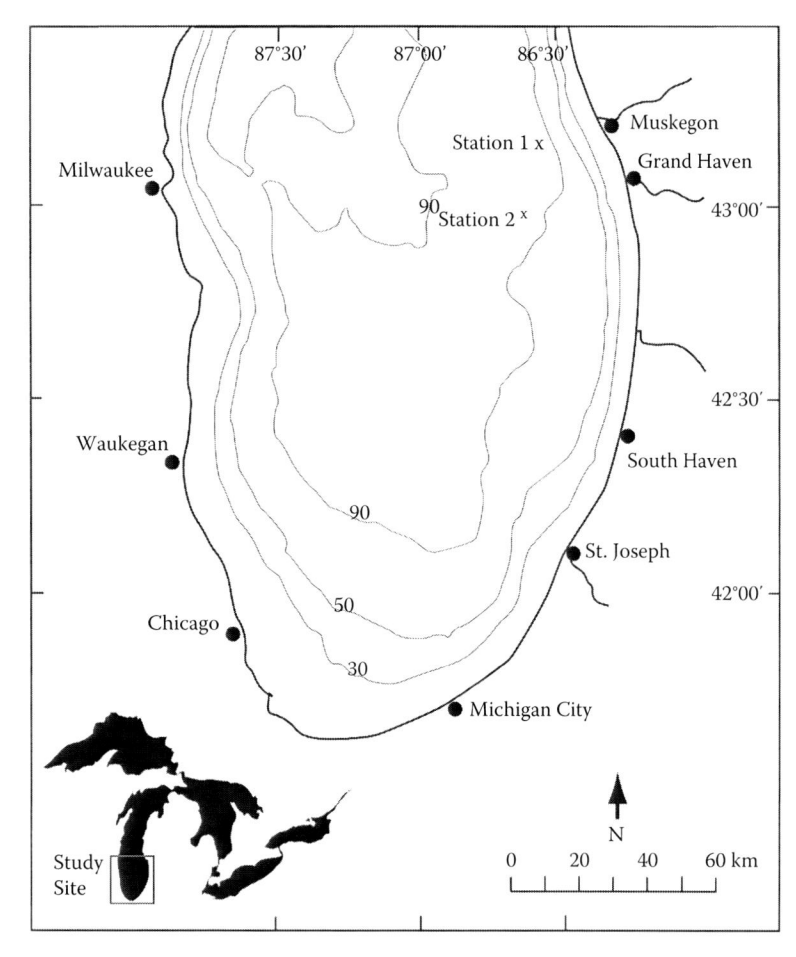

Figure 34.1 Location of two sampling stations in the offshore region of southern Lake Michigan.

were filtered onto 0.45 µm Millipore filters, decontaminated with 0.5 mL of 0.5 N HCL for 4–6 h, placed in scintillation vials with scintillation cocktail, and counted with a liquid scintillation counter. Time-zero blanks were taken and subtracted from all light values. Total carbon dioxide was determined from alkalinity and pH measurements. Photosynthetic rates, normalized to chlorophyll a, were used to construct a photosynthesis-irradiance curve using the methods outlined in Fahnenstiel et al. (1989). Integral daily primary production was determined using the Great Lakes Production Model (Lang and Fahnenstiel 1996), which is based on the model of Fee (1973). Integral production was calculated for a minimum of 4 days preceding and following each sampling day to factor out unusual surface irradiance on the sampling day and to provide a more representative estimate for the sampling period.

Water for phytoplankton counts was preserved in amber bottles with 0.5% Lugol's solution. These samples were then filtered or settled onto microscope slides enumerated under low (10–20× objective) and high (>60× objective) or medium (40–60× objective) magnification. Phytoplankton were identified to the lowest practical taxonomic group. Cell volumes were estimated from average cell dimensions, and these dimensions were converted to volumes using appropriate geometric shapes. Phytoplankton volumes were converted to carbon units with equations from Strathman (1967) for diatoms and Verity et al. (1992) for non-diatoms.

Total phosphorus was measured using standard automatic colorimetric procedures on an autoanalyzer (Davis and Simmons 1979, Laird et al. 1987). Total phosphorus samples were digested in an autoclave after the addition of potassium persulfate (5% final concentration, Menzel and Corwin 1965) and then measured as soluble reactive phosphorus. Particulate phosphorus samples were filtered onto 0.4 µm Nuclepore filters, and these filters were digested and analyzed with procedures used for total phosphorus (Millie et al. 2003). Particulate carbon samples were filtered onto pre-combusted Whatman GF/F filters, and these filters were analyzed on a Model 1110 CHN analyzer (Millie et al. 2003).

Two to four replicate samples for zooplankton and *M. diluviana* were collected approximately monthly with a net towed vertically from 1 to 3 m above bottom to the surface at speeds of approximately 0.5 m/s. Samples for zooplankton were collected during day using a 0.5 m diameter plankton net, (153 µm mesh; 1:5 mouth-to-length ratio), and samples for *M. diluviana* were collected at night using a 1 m diameter plankton net (1000 μm mesh; 1:3 mouth-to-length ratio). Both zooplankton and *M. diluviana* were anesthetized with carbonated water and preserved with sugar-buffered formaldehyde to form a final 3% solution.

To determine zooplankton abundance and composition, aliquots were taken from a known sample volume with a Hensen–Stempel pipette to obtain a minimum of 600 individuals for identification. Length measurements were made on taxa that accounted for over 10% of total abundance and their biomass (dry weight) determined using published length–weight regressions for each taxa (Culver et al. 1985). For zooplankton taxa that comprised less than 10% of the total abundance, biomass was derived from published weights (Hawkins and Evans 1979). Biomass differences between time periods were analyzed for the following zooplankton taxa: nauplii, cyclopoid copepod, *Leptodiaptomus* spp., and *Limnocalanus macrurus*. *Leptodiaptomus* spp. and *L. macrurus* are calanoid copepods that were analyzed individually since they accounted for at least 96% of all calanoids collected and have different feeding strategies; that is, *Leptodiaptomus* spp. are omnivores, and *L. macrurus* is a carnivore (Balcer et al. 1984). Cladocerans were not considered in our analysis since they were essentially absent during the winter/spring period. For *M. diluviana,* all individuals were counted in each sample.

All methods used to determine primary productivity, chlorophyll *a*, phytoplankton, and zooplankton were similar between 1995 and 2009, and thus comparisons between 1995–1998 and 2007–2009 should be robust. Water quality methods were similar from 1983 to 2009. A few changes in zooplankton and phytoplankton methods between 1983 and 1995 were minor and would likely not affect comparisons between periods (1983–1989, 1995–1998, and 2007–2009).

Differences in limnological and biotic parameters between time periods were evaluated with the nonparametric Kruskal–Wallis test. If this test was significant, a nonparametric Tukey-type multiple comparisons test was used to examine pair-wise differences (Zar 1974).

RESULTS AND DISCUSSION

Overall, total phosphorus concentrations were significantly ($P < 0.05$) different between the three periods (Figure 34.2A). While mean concentrations in 1983–1989 and 1995–1998 were not significantly different, mean concentrations in 2007–2009 were significantly lower than in 1995–1998. Mean concentrations in the three successive periods were 6.5, 5.6, and 3.7 μg/L, respectively. The mean in 2007–2009 represented a 34% decrease compared to 1995–1998.

Trends in phytoplankton, as indicated by chlorophyll *a* and primary production, were similar to trends in total phosphorus concentrations, but decreases were more pronounced between the 2007–2009 and 1995–1998 periods (Figure 34.2B and C). Chlorophyll *a* concentrations averaged 3.2 μg/L in 1983–1987, 2.5 μg/L in 1995–1998, and only 0.8 μg/L in 2007–2009 (Figure 34.2B). Chlorophyll concentrations in 2007–2009 were significantly ($P < 0.05$) lower than concentrations in 1983–1987 and 1995–1998, but concentrations in 1983–1987 and 1995–1998 were not significantly different. Phytoplankton areal-integrated primary production rates

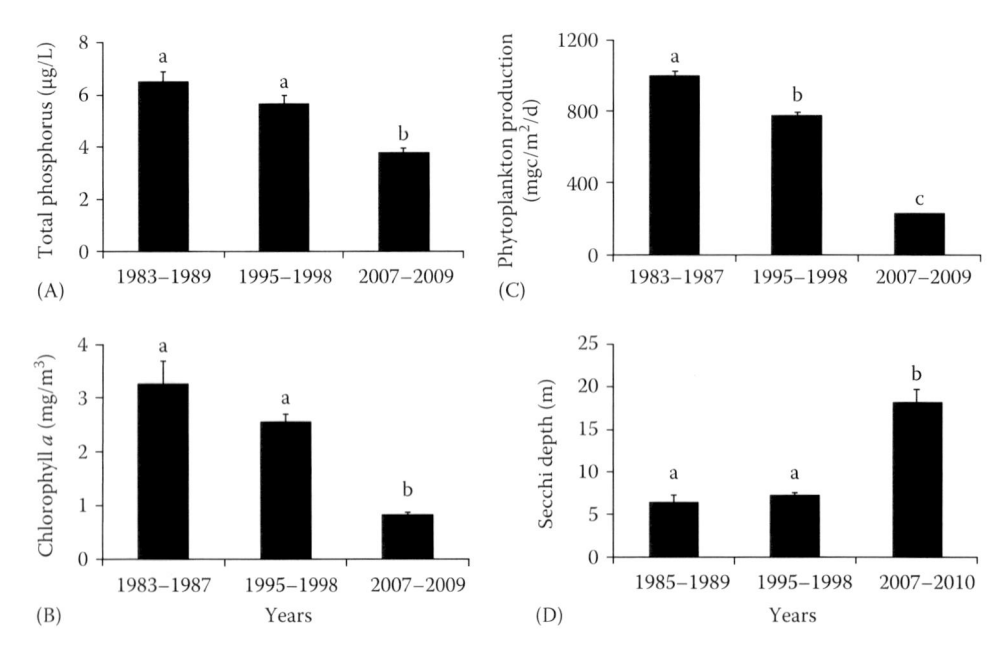

Figure 34.2 Mean (±1 SE) total phosphorus (A), chlorophyll *a* (B), primary production (C), and Secchi depth (D) during the winter/spring isothermal period in offshore southern Lake Michigan. Values represent means for three yearly time periods extending over three decades. Significant differences (nonparametric Tukey-type multiple comparisons, $P < 0.05$) between time periods for each given variable are designated by different letters.

decreased from 993 mg C/m²/day in 1983–1987 to 772 mg C/m²/day in 1995–1998 (P < 0.05; Figure 34.2C). However, areal-integrated primary production was only 222 mg C/m²/day in 2007–2009, which was significantly (P < 0.05) lower than in 1995–1998 and corresponded to a decrease of 71%.

As might be expected from trends in phytoplankton, water transparency did not change significantly (P > 0.05) between 1985–1989 and 1995–1998, but increased significantly (P < 0.05) between 1995–1998 and 2007–2010. Secchi depth transparency averaged 6 m in 1985–1989 and 7 m in 1995–1998, but increased to 18 m in 2007–2010 (Figure 34.2D). Present phytoplankton parameters and water transparency during the late winter/spring mixing period in Lake Michigan are now similar to those found in oligotrophic Lake Superior (Mida et al. 2010, Evans et al. 2011). Although the mean Secchi depth transparency was 18 m in 2007–2010, individual values of 31 m were observed in Lake Michigan in 2010, which suggests that Lake Michigan now has some of the clearest water in the Laurentian Great Lakes.

Changes in total phosphorus, chlorophyll *a*, primary production, and Secchi depth transparency can be explained by two factors: phosphorus load reductions and filtering activities of dreissenid mussels (Fahnenstiel et al. 2010a,b, Mida et al. 2010). Annual phosphorus loads to southern Lake Michigan have gradually decreased from approximately 2800 MT/year in the late 1970s/early 1980s to approximately 1500 MT/year by 2006–2008 (Mida et al. 2010). This gradual decline in phosphorus loading probably was responsible for the small changes in total phosphorus, chlorophyll, primary production, and Secchi depth transparency between 1983–1989 and 1995–1998 and likely a minor contributor to changes between 1995–1998 and 2007–2009. Large decreases in phytoplankton chlorophyll *a* and primary production and increase in Secchi depth transparency between 1995–1998 and 2007–2010 (68%–71%) were most likely attributable to the filtering activities of dreissenid mussels, particularly quagga mussels (Fahnenstiel et al. 2010a,b). Between 1995–1998 and 2007–2009, large populations of quagga mussels became established in Lake Michigan (Nalepa et al. 2009, 2010). In 2007–2008, mussel populations had the capability to filter phytoplankton from the water column at rates that exceeded the phytoplankton growth rate (Vanderploeg et al. 2010) and thus significantly reduce phytoplankton abundance and other suspended particulates. Additional support for the role of quagga mussels in causing these changes was noted in Fahnenstiel et al. (2010b). Such support focused principally on the temporal coherence of environmental changes with mussel populations. For example, observed phytoplankton and water quality changes were restricted mainly to the winter/spring period when the water column is completely mixed and phytoplankton are well distributed throughout the water column and thus available to bottom-dwelling mussels.

In contrast to significant changes in phytoplankton, total zooplankton biomass did not differ among time periods (P > 0.05, Table 34.1). However, there were some differences between time periods when biomass of individual taxa group was examined. Mean cyclopoid biomass was significantly (P < 0.05) lower in 2007–2009 compared to 1986–1988 and 1994–1998 (Table 34.1). Mean biomass was 1.63 mg/m³ in 1994–1998, but only 0.07 mg/m³ in 2007–2009, a decline of 96%. Under oligotrophic conditions and low food availability, cyclopoid copepods are generally outcompeted by calanoid copepods for food resources (Soto and Hurlbert 1991). Calanoid copepod nauplii and adults have lower food thresholds and metabolism than cyclopoid copepods and are able to depress food available to cyclopoid nauplii (i.e., phytoplankton) and adults (i.e., microzooplankton) in low food conditions (Soto and Hurlbert 1991, Adrian 1997, Sommer and Stibor 2002). Also, since adult cyclopoid copepods feed mainly on microzooplankton, the amount of available food may be further reduced since dreissenids are capable of filtering microzooplankton from the water column (Pace et al. 1998). In contrast, biomass of calanoid copepods was stable or, if anything, increased. Although not significantly different from the other two periods, biomass

Table 34.1 Mean Biomass (mg/m³) of Major Zooplankton Taxonomic Groups during Winter/Spring Isothermal Period in the Offshore of Southern Lake Michigan, 1986–1988, 1994–1998, 2007–2009. For Each Zooplankton Group, Significant Differences between Time Periods Are Designated by Different Letters (Nonparametric Tukey-Type Multiple Comparisons, P < 0.05). Standard Error Is Given in Parenthesis

Zooplankton Taxa	1986–1988	1994–1998	2007–2009
Nauplii	0.32 (0.09)	0.60 (0.05)	0.60 (0.14)
Cyclopoid	0.92 (0.22)[a]	1.63 (0.23)[a]	0.07 (0.02)[b]
Leptodiaptomus spp.	11.18 (2.5)[a,c]	7.62 (0.66)[a,b]	12.54 (1.12)[c]
Limnocalanus macrurus	1.39 (0.32)	1.77 (0.32)	2.18 (1.03)
Total	13.82 (2.88)	11.64 (0.73)	15.74 (1.65)

a,b,c denotes where significant differences do or do not exist between pairs of values in each row.

of both *Leptodiaptomus* and *L. macrurus* was greatest in 2007–2009 (Table 34.1). Dreissenid veligers accounted for <1% of total zooplankton biomass in the 1994–1998 and 2007–2009 periods.

M. diluviana densities declined significantly (P < 0.05) from a mean of 139 ± 27/m² in 1996–2000 to 55 ± 9/m² in 2007–2010. Offshore densities of *M. diluviana* in 2007–2010 were similar to those found in the same region of southeastern Lake Michigan in 1985–1989 (Lehman et al. 1990). However, compared to 2007–2010, ecosystem conditions were different in 1985–1989 (e.g., higher fish biomass and primary production in the 1980s) (Pothoven et al. 2010). Pothoven et al. (2010) found that life history characteristics of *M. diluviana* that have been linked to trophic conditions, such as growth, proportion of females with broods, and brood size, did not change between 1996–2000 and 2007–2008. However, results from the same study suggested that in spring, abundances of juvenile *M. diluviana* were significantly related to offshore water-column chlorophyll concentration. Orsi and Mecum (1996) suggested that reduced chlorophyll levels might impact smaller mysids more than larger mysids since the former are more herbivorous.

Besides a decrease in available food, top-down pressures on cyclopoid copepods and *M. diluviana* cannot be entirely discounted as the cause of recent population decreases. In Lake Huron, recent declines in cyclopoid copepods were linked to a combination of decreased primary productivity and increased fish consumption after a major prey item for fish, the benthic amphipod *Diporeia* spp., declined (Barbiero et al. 2009). Similarly in Lake Michigan, despite declines in fish biomass (Bunnell et al. 2009), fish planktivory could still exacerbate declines in cyclopoid copepods and suppress *M. diluviana* because fish predation may now be unevenly distributed among remaining prey. Predation by *Bythotrephes longimanus* and *Cercopagis pengoi* could also have added predatory pressures on cyclopoid copepods; recent abundances of both these predatory cladocerans appeared to be increasing in southern Lake Michigan (Cavaletto et al. 2010).

Large populations of quagga mussels have not only affected plankton and particle composition and abundance in Lake Michigan but almost all aspects of water column foodweb and nutrient dynamics during the winter/spring isothermal mixing period. This is supported by analysis of important nutrient and phytoplankton stoichiometries.

The total phosphorus to chlorophyll ratio (TP/Chl) is an important water quality ratio often used in water quality management (IJC 1988). This ratio increased dramatically in 2007–2009 compared to 1983–1989 and 1995–1998 (P < 0.05, increase 70%; Figure 34.3A). Similar large increases in TP/Chl ratios have been documented in other aquatic systems after establishment of large populations of dreissenid mussels (Mellina et al. 1995, Nicholls et al. 1997). Larger ratios may infer that phytoplankton are more phosphorus replete in the post-dreissenid period, but this may not be the case. The particulate carbon to phosphorus (C/P)

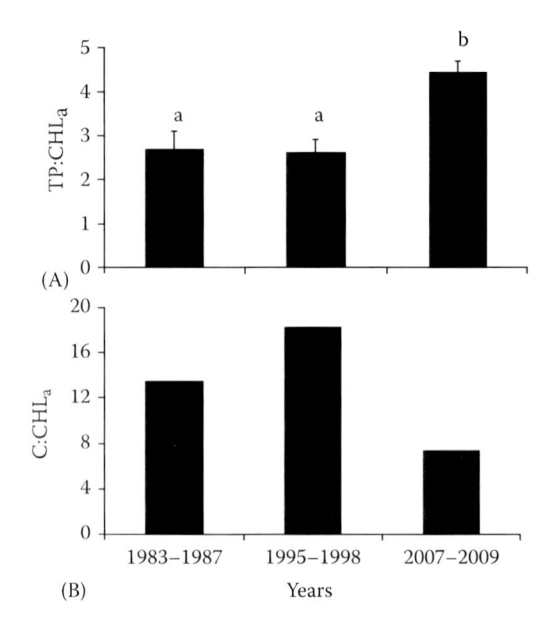

Figure 34.3 Mean (±1 SE) TP/Chl *a* ratio (A) and phytoplankton carbon to chlorophyll *a* ratio (B) during the winter/spring isothermal period in offshore southern Lake Michigan. Values represent means for three yearly time periods extending over three decades. Significant differences (nonparametric Tukey-type multiple comparisons, P < 0.05) between time periods within each given variable are designated by differing letters.

ratio in phytoplankton is a good indicator of phytoplankton phosphorus status (Hecky et al. 1993, Wetzel 2001). In 2007–2009 winter/spring, particulate C/P ratios averaged 214, which indicated a moderate P deficient in phytoplankton (Wetzel 2001). In 1998–1999 C/P ratios averaged only 133, which indicated low P deficiency. Phosphorus deficiency indexes (PDIs) determined from the ratio of the maximum photosynthetic rate (P_{opt}) to the maximum phosphorus uptake (V_{max}) also suggest phosphorus deficiency in 1999–2000 (G. Fahnenstiel, unpublished data). Thus, even though TP/Chl ratios were significantly greater in 2007–2009, C/P ratios and the PDI index suggest that phytoplankton in 2007–2009 were more P deficient than in the 1990s. While impacts of mussel filtering activities on seston in Lake Michigan are well documented (Fahnenstiel et al. 2010a), mussel impacts on the phosphorus cycle are poorly understood.

The most likely explanation for higher TP/Chl ratios in 2007–2009 compared to earlier periods is that non-algal material now accounts for a larger fraction of total phosphorus concentrations and may thus bias historical comparisons of TP/Chl ratios. Even though phytoplankton biomass decreased by over 80% in 2007–2008 compared to 1995–1998 (Fahnenstiel et al. 2010b), crustacean zooplankton biomass did not change between these periods and actually increased slightly (Table 34.1). Thus, crustacean zooplankton made up a larger component of total P in 2007–2009, relative to phytoplankton (i.e., chlorophyll). It should be noted that this same bias does not affect C/P ratios as long as zooplankton have similar C/P ratios as phytoplankton and because zooplankton contribute to both particulate carbon and phosphorus concentrations. In the post-dreissenid period in the Great Lakes, care must be taken when comparing historical stoichiometries as mussels have fundamentally altered food web structure and how carbon and nutrients are cycled.

Another example of how quagga mussels are affecting pelagic food web structure and function in Lake Michigan is seen by examining phytoplankton carbon to chlorophyll ratio, a ratio that is only determined by phytoplankton abundance. The winter/spring phytoplankton carbon to chlorophyll ratios have changed over time. In 2007–2009, this ratio averaged 7, whereas the ratio averaged 13 and 18 in the 1983–1987 and 1995–1998 periods, respectively (Figure 34.3B). A large decrease in this ratio may be explained by a decrease in light availability (Fahnenstiel and Scavia 1987b), but this has clearly not happened as Secchi depth transparency increased in 2007–2009. A more likely explanation for the large decrease is that autotrophic picoplankton are now greater contributors to total phytoplankton carbon. For our ratios, autotrophic picoplankton were not included in phytoplankton carbon as they were not counted with the phytoplankton enumeration techniques, but they were included in chlorophyll values. Phytoplankton carbon concentrations during the winter/spring isothermal period decreased significantly from an average of 45 mg/m^3 in the 1980s and 1990s to an average value of only 6 mg/m^3 in 2007–2009 (Fahnenstiel et al. 2010b). If we assume autotrophic picoplankton biomass in 2007–2009 is similar to 1980s and 1990s values, then autotrophic picoplankton carbon was approximately 5 mg/m^3 in 2007–2009 (Fahnenstiel and Carrick 1992, Fahnenstiel et al. 1998). Thus, in 2007–2009 autotrophic picoplankton biomass was approximately equal to the more traditionally measured phytoplankton (nano and net). However, our assumption of autotrophic carbon in 2007–2009 is probably conservative because actual autotrophic plankton biomass likely increased in 2007–2009. We base this premise on the likely decreased abundance of most picoplankton predators (e.g., rotifers and ciliates) in 2007–2009 due to dreissenid filtering (Pace et al. 1998) and the lack of direct dreissenid filtering on autotrophic picoplankton-sized particles (1 μm; Jorgensen et al. 1984). The exact abundance of picoplankton and their role in food web nutrient and carbon cycling of Lake Michigan remain to be determined in the post-dreissenid period.

CONCLUSIONS

Changes in water quality parameters (total phosphorus, water clarity), phytoplankton abundance and productivity, and zooplankton in the offshore region of Lake Michigan during the late winter/spring isothermal mixing period have been dramatic over the past decade. These changes coincided with the expansion of quagga mussels into the offshore region (Nalepa et al. 2010). Phosphorus concentrations and loads are now well below target levels (Mida et al. 2010), and thus future phosphorus management may need to be reevaluated (Evans et al. 2011). While some changes in water quality and plankton may be a result of nutrient loading reductions, dramatic changes in the period of time after quagga mussel expansion (2007–2009 relative to 1995–1998) combined with limited changes noted in the period before expansion (1983–1998) would suggest that quagga mussels are the primary cause of these changes.

Thus, it is important to note that this oligotrophication of Lake Michigan was not caused primarily by management decisions, but rather by the filtering activities of invasive dreissenid mussels. Because quagga mussel populations in the offshore region are still expanding, whereas populations in the nearshore region seem to be stabilizing (Nalepa et al. 2010), it is difficult to predict the ultimate extent of these changes. In shallower regions of the Great Lakes where zebra mussel populations increased in the early 1990s, present abundances are a fraction of those found during the initial colonization period (Nalepa et al. 2003, Nalepa, unpublished data). If a similar decline occurs in quagga mussel populations in offshore Lake Michigan, it will likely occur over a long time period, mainly because of the slow growth rates and extended life expectancy of quagga mussels at the continuously-cold temperatures found in offshore regions (Nalepa et al. 2010). Thus, it will be difficult to manage this ecological transition caused by mussels even though significant management expectations exist (i.e., fisheries harvest). Given that many of the responses to changing mussel densities are likely during the late mixing/spring isothermal period, we suggest that this time period become a focus of future resource monitoring until dreissenid populations stabilize.

REFERENCES

Adrian, R. 1997. Calanoid-cyclopoid interactions: Evidence from an 11-year field study in a eutrophic lake. *Freshwat. Biol.* 38: 315–325.

Balcer, M. D., N. L. Korda, and S. L. Dodson. 1984. *Zooplankton of the Great Lakes. A Guide to the Identification and Ecology of the Common Crustacean Species*. Madison, WI: The University of Wisconsin Press.

Barbiero, R. P., M. Balcer, D. C. Rockwell, and M. L. Tuchman. 2009. Recent shifts in the crustacean zooplankton community of Lake Huron. *Can. J. Fish. Aquat. Sci.* 66: 816–828.

Bunnell, D. B., C. P. Madenjian, J. D. Holuszko, J. V. Adams, and J. V. P. French. 2009. Expansion of *Dreissena* into offshore waters of Lake Michigan and potential impacts on fish populations. *J. Great Lakes Res.* 35: 74–80.

Cavaletto, J. F., H. A. Vanderploeg, R. Pichlova-Ptacnikova, S. A. Pothoven, J. R. Liebig and G. L. Fahnenstiel. 2010. Temporal and spatial separation allow coexistence of predatory cladocerans: *Leptodora kindtii*, *Bythotrephes longimanus* and *Cercopagis pengoi*, in southeastern Lake Michigan. *J. Great Lakes Res.* 36 (Suppl. 3): 65–73.

Culver, D. A., M. M. Boucerle, D. J. Bean, and J. W. Fletcher. 1985. Biomass of freshwater crustacean zooplankton from length weight regressions. *Can. J. Fish. Aquat. Sci.* 42: 1380–1390.

Davis, C. O. and M. S. Simmons. 1979. Water chemistry and phytoplankton field and laboratory procedures. Special Report 70, Great Lakes Research Division, University of Michigan, Ann Arbor, MI.

Evans, M. A., G. L. Fahnenstiel, and D. Scavia. 2011. Incidental oligotrophication of North American Great Lakes. *Environ. Sci. Technol.* 4: 3297–3303.

Fahnenstiel, G. L., T. B. Bridgeman, G. A. Lang, M. J. McCormick, and T. F. Nalepa. 1995b. Phytoplankton productivity in Saginaw Bay, Lake Huron: Effects of zebra mussel (*Dreissena polymorpha*) colonization. *J. Great Lakes Res.* 21: 465–475.

Fahnenstiel, G. L., C. Beckmann, S. E. Lohrenz, D. F. Millie, O. M. E. Schofield, and M. J. McCormick. 2002. Standard Niskin and Van Dorn bottles inhibit phytoplankton photosynthesis in Lake Michigan. *Verh. Int. Verein. Limnol.* 28: 376–380.

Fahnenstiel, G. L. and H. J. Carrick. 1992. Phototrophic picoplankton in Lakes Huron and Michigan: Abundance, distribution, composition, and contribution to biomass and production. *Can. J. Fish. Aquat. Sci.* 49: 379–388.

Fahnenstiel, G. L., J. F. Chandler, H. J. Carrick, and D. Scavia. 1989. Photosynthetic characteristics of phytoplankton communities in lakes Huron and Michigan: P-I parameters and end-products. *J. Great Lakes Res.* 15: 394–407.

Fahnenstiel, G. L., A. E. Krause, M. J. McCormick, H. J. Carrick, and C. L. Schelske. 1998. The structure of the planktonic food-web in the St. Lawrence Great Lakes. *J. Great Lakes Res.* 24: 531–554.

Fahnenstiel, G. L., G. A. Lang, T. F. Nalepa, and T. H. Johengen. 1995a. Effects of zebra mussels (*Dreissena polymorpha*) colonization on water quality parameters in Saginaw Bay, Lake Huron. *J. Great Lakes Res.* 21: 435–448.

Fahnenstiel, G., T. Nalepa, S. Pothoven, H. Carrick, and D. Scavia. 2010a. Lake Michigan lower food-web: Long-term observations and *Dreissena* impact. *J. Great Lakes Res.* 36 (Suppl. 3): 1–4.

Fahnenstiel, G., S. Pothoven, H. Vanderploeg, D. Klarer, T. Nalepa, and D. Scavia. 2010b. Recent changes in primary production and phytoplankton in the offshore region of southeastern Lake Michigan. *J. Great Lakes Res.* 36 (Suppl. 3): 20–29.

Fahnenstiel, G. L. and D. Scavia. 1987a. Dynamics of Lake Michigan phytoplankton: Primary production and growth. *Can. J. Fish. Aquat. Sci.* 44: 499–508.

Fahnenstiel, G. L. and D. Scavia. 1987b. Dynamics of Lake Michigan phytoplankton: The deep chlorophyll layer. *J. Great Lakes Res.* 13: 285–295.

Fahnenstiel, G. L., R. A. Stone, M. J. McCormick, C. L. Schelske, and S. E. Lohrenz. 2000. Spring isothermal mixing in the Great Lakes: Evidence of nutrient limitation in a suboptimal light environment. *Can. J. Fish. Aquat. Res.* 57: 1901–1910.

Fee, E. J. 1973. A numerical model for determining integral primary production and its application to Lake Michigan. *J. Fish. Res. Bd. Can.* 30: 1447–1468.

Gardner, W. S., M. A. Quigley, G. L. Fahnenstiel, D. Scavia, and W. A. Frez. 1990. *Pontoporeia hoyi*—A direct trophic link between spring diatoms and fish in Lake Michigan. In *Large Lakes: Structural and Functional Properties.*, M. M. Tilzer and C. Serruya, eds., pp. 632–644. New York: Springer.

Hawkins, B. E. and M. E. Evans. 1979 Seasonal cycles of zooplankton biomass in southeastern Lake Michigan. *J. Great Lakes Res.* 5: 256–263.

Hecky, R. E., P. Campbell, and L. L. Hendzel. 1993. The stoichiometry of carbon, nitrogen and phosphorus in particulate matter of lakes and oceans. *Limnol. Oceanogr.* 38: 743–761.

Hecky, R. E., R. E. H. Smith, D. R. Barton et al. 2004. The nearshore phosphorus shunt: A consequence of ecosystem engineering by dreissenids in the Laurentian Great Lakes. *Can. J. Fish. Aquat. Sci.* 61: 1285–1293.

Holland, R. E. 1993. Changes in planktonic diatoms and water transparency in Hatchery Bay, Bass Island Area, western Lake Erie since the establishment of the zebra mussel. *J. Great Lakes Res.* 19: 617–624.

International Joint Commission. 1988. *Great Lakes Water Quality Agreement of 1978 (revised)*. Washington, DC: International Joint Commission.

Johengen, T. H., O. E. Johannsson, G. L. Pernie, and E. S. Millard. 1994. Temporal and seasonal trends in nutrient dynamics and biomass measures in lakes Michigan and Ontario in response to phosphorus control. *Can. J. Fish. Aquat. Sci.* 51: 2570–2578.

Jorgensen, C. B., T. Kiorobe, F. Mohlenberg, and H. U. Riisgard. 1984. Ciliary and mucus-net filter feeding, with special reference to fluid mechanical characteristics. *Mar. Ecol. Prog. Ser.* 15: 283–292.

Laird, G. A., D. Scavia, G. L. Fahnenstiel et al. 1987. Southern Lake Michigan nutrients, temperature, chlorophyll, plankton and water movement during 1983 and 1984. NOAA Technical Memorandum ERL GLER-67. Ann Arbor, MI: Great Lakes Environmental Research Lab.

Lang, G. A. and G. L. Fahnenstiel. 1996. Great Lakes primary production model-methodology and use. NOAA Technical Memorandum ERL GLERL-90. Ann Arbor, MI: Great Lakes Environmental Research Lab.

Lehman, J. T. 1987. Palearctic predator invades North American Great Lakes. *Oecologia* 74: 478–480.

Lehman, J. T., J. A. Bowers, R. W. Gensemer, G. J. Warren, and D. K. Branstrator. 1990. *Mysis relicta* in Lake Michigan: Abundances and relationships with their potential prey, *Daphnia*. *Can. J. Fish. Aquat. Sci.* 47: 977–983.

Madenjian, C. P., G. L. Fahnenstiel, T. J. Johengen et al. 2002. Dynamics of the Lake Michigan food web, 1970–2000. *Can. J. Fish. Aquat. Sci.* 59: 736–753.

Mellina, E., J. B. Rasmussen, and E. L. Mills. 1995. Impact of zebra mussel (*Dreissena polymorpha*) on phosphorus cycling and chlorophyll in lakes. *Can. J. Fish. Aquat. Sci.* 52: 2553–2573.

Menzel, D. W. and N. Corwin. 1965. The measurement of total phosphorus liberated in seawater based on the liberation of organically bound fractions by persulfate oxidation. *Limnol. Oceanogr.* 10: 280–281.

Mida, J. L., D. Scavia, G. L. Fahnenstiel, S. A. Pothoven, H. A. Vandeploeg, and D. M. Dolan. 2010. Long-term and recent changes in southern Lake Michigan water quality with implications for present trophic status. *J. Great Lakes Res.* 36 (Suppl. 3): 42–50.

Millie, D. F., G. L. Fahnenstiel, S. E. Lohrenz, H. J. Carrick, T. H. Johengen, and O. M. E. Schofield. 2003. Physical-biological coupling in southern Lake Michigan: Influence of episodic resuspension on phytoplankton. *Aquat. Ecol.* 37: 393–408.

Nalepa, T. F. and G. L. Fahnenstiel. 1995. *Dreissena polymorpha* in the Saginaw Bay, Lake Huron ecosystem: Overview and perspective. *J. Great Lakes Res.* 21: 411–416.

Nalepa, T. F., D. L. Fanslow, and G. A. Lang. 2009. Transformation of the offshore benthic community in Lake Michigan: Recent shift from native amphipod *Diporeia* spp. to the invasive mussel *Dreissena rostriformis bugensis. Freshwat. Biol.* 54: 466–479.

Nalepa, T. F., D. L. Fanslow, M. B. Lansing, and G. A. Lang. 2003. Trends in the benthic macroinvertebrate community of Saginaw Bay, Lake Huron, 1987–1996: Response to phosphorus abatement and the zebra mussel, *Dreissena polymorpha. J. Great Lakes Res.* 29: 14–33.

Nalepa, T. F., D. L. Fanslow, and S. A. Pothoven. 2010. Recent changes in density, biomass, recruitment, size structure, and nutritional state of *Dreissena* populations in southern Lake Michigan. *J. Great Lakes Res.* 36 (Suppl. 3): 5–19.

Nalepa, T. F., D. J. Hartson, D. L. Fanslow, G. A. Lang, and S. J. Lozano. 1998. Declines in benthic macroinvertebrate populations in southern Lake Michigan, 1980–1993. *Can. J. Fish. Aquat. Sci.* 55: 2402–2413.

Nalepa, T. F., D. W. Schloesser, S. A. Pothoven et al. 2001. First finding of the amphipod *Echinogammarus ischnus* and the mussel *Dreissena bugensis* in Lake Michigan. *J. Great Lakes Res.* 27: 384–391.

Nicholls, K. H., and G. J. Hopkins. 1993. Recent changes in Lake Erie (north shore) phytoplankton: Cumulative impacts of phosphorus loading reductions and the zebra mussel introduction. *J. Great Lakes Res.* 19: 637–647.

Nicholls, K. H., G. J. Hopkins, and S. J. Standke. 1997. *Effects of Zebra Mussels on Chlorophyll, Nitrogen, Phosphorus and Silica in North Shore Waters of Lake Erie*. Toronto, Ontario, Canada: Ontario Ministry of the Environment Energy. ISBN 0-7778-6707-9.

Orsi, J. J. and W. L. Mecum. 1996. Food limitation as the probable cause of a long-term decline in the abundance of *Neomysis mercedis* the opossum shrimp in the Sacramento-San Joaquin estuary. In *San Francisco Bay: The Ecosystem.*, J. T. Hollibaugh, ed., pp. 375–401. San Francisco, CA: Pacific Division American Association for Advanced Science.

Pace, M. L., S. E. G. Findlay, and D. Fischer. 1998. Effects of an invasive bivalve on the zooplankton community of the Hudson River. *Freshwat. Biol.* 39: 103–116.

Pothoven, S. A., G. L. Fahnenstiel, and H. A. Vanderploeg. 2010. Temporal trends in *Mysis relicta* abundance, production, and life-history characteristics in southeastern Lake Michigan. *J. Great Lakes Res.* 36 (Suppl. 3): 60–64.

Sommer, U. and H. Stibor. 2002. Copepods-Cladocera-Tunicata: The role of three major mesozooplankton groups in pelagic food webs. *Ecol. Res.* 17: 161–174.

Soto, D. and S. H. Hurlbert. 1991. Long-term experiments on calanoid-cyclopoid interactions. *Ecol. Monogr.* 61: 245–266.

Speziale, B. J., S. P. Schreiner, P. A. Giammatteo, and J. E. Schindler. 1984. Comparison of N,N—Dimethylformamide, dimethyl sulfoxide, and acetone for extraction of phytoplankton chlorophyll. *Can J. Fish. Aquat. Sci.* 41: 1519–1522.

Strathman, R. R. 1967. Estimating the organic carbon content of phytoplankton from cell volume or plasma volume. *Limnol. Oceanogr.* 11: 411–418.

Strickland, J. D. H. and T. R. Parsons. 1972. *A Practical Handbook of Seawater Analysis*. 2nd Edn. Fisheries Research Board of Canada, Bulletin 167, p. 310.

Vanderploeg, H. A., J. R. Liebig, T. F. Nalepa, G. L. Fahnenstiel, and S. A. Pothoven. 2010. *Dreissena* and the disappearance of the spring phytoplankton bloom in Lake Michigan. *J. Great Lakes Res.* 36 (Suppl. 3): 50–59.

Verity, P. G., C. Y. Robertson, C. R. Tronzo, M. G. Andrews, J. R. Nelson, and M. E. Sieracki. 1992. Relationship between cell volume and the carbon and nitrogen content of marine photosynthetic nanoplankton. *Limnol. Oceanogr.* 37: 1434–1446.

Wells, L. and A. L. McLain. 1973. Lake Michigan: Man's effects on native fish stocks and other biota. Great Lakes Fishery Commission Technical Report 20. Ann Arbor, MI: Great Lakes Fishery Commission.

Wetzel, R. G. 2001. *Limnology: Lakes and River Ecosystems*. San Diego, CA: Academic Press.

Zar, J. H. 1974. *Biostatistical Analysis*. Englewood Cliffs, NJ: Prentice-Hall.

CHAPTER 35

Nutrient Cycling by Dreissenid Mussels
Controlling Factors and Ecosystem Response

Harvey A. Bootsma and Qian Liao

CONTENTS

Abstract 555
Introduction 556
Factors Regulating Nutrient Excretion by Dreissenids 558
    Food Quantity and Quality 558
    Dreissenid Size 560
    Temperature 561
    Other Regulating Factors 561
Ecosystem Impacts of Dreissenid Nutrient Cycling 562
    Influence of Dreissenids on Nutrient Pools and Fluxes 562
    Phytoplankton Response 564
    Role of Hydrodynamics and Lake Morphometry 564
    Benthic Algal Response 566
Management Implications 569
Acknowledgments 570
References 570

ABSTRACT

As a result of their high densities and filtration rates, dreissenid mussels are a major sink for particulate nutrients in lakes and rivers. These consumed nutrients are allocated to one of three fates: egestion as particulate feces and pseudofeces, assimilation into mussel biomass, and excretion in dissolved form. Of these three pathways, nutrient egestion and excretion are usually dominant, with the relative importance of egestion increasing when food supply is high and/or food quality is low. With the exception of several very high values, published mass-specific excretion rates generally range from 1 to 10 μmol gDW^{-1} h^{-1} for nitrogen and 0.1–2 μmol gDW^{-1} h^{-1} for phosphorus. Measured areal N excretion rates range from 155 to 624 μmol m^{-2} h^{-1}, while those for P range from 4.3 to 213 μmol m^{-2} h^{-1}. There is some evidence that P excretion rates for quagga mussels (*Dreissena rostriformis bugensis*) are lower than those for zebra mussels (*Dreissena polymorpha*). Major factors that regulate nutrient excretion include temperature, mussel size, and food quantity/quality, with reproductive stage, dissolved oxygen concentration, and substratum type also playing apparent roles. At the ecosystem scale, the effect of nutrient recycling by dreissenids is modulated by hydrodynamic processes, including vertical and horizontal mixing, which influence both the supply of phytoplankton to dreissenids and the distribution of excreted nutrients. Although the influence of dreissenids on ecosystem-scale nutrient dynamics varies among systems, common ecosystem responses include a decrease in phytoplankton abundance, a decrease in total nutrient concentrations, an increase in the dissolved nutrient to phytoplankton biomass ratio, and an increase in the biomass of benthic macroalgae. In large lakes, the alteration of nutrient dynamics by dreissenids has led to a management conundrum in which maintenance of productive pelagic fish populations and reduction of nearshore nuisance algal blooms are apparently irreconcilable. Biogeochemical models that account for the influence of vertical and horizontal mixing processes on dreissenid grazing and nutrient excretion, as well as the long-term fate of dreissenid biomass and biodeposits, will help to define nutrient management strategies that produce an optimal balance between these two objectives.

INTRODUCTION

Within the Laurentian Great Lakes and many other North American aquatic ecosystems, there are a number of benthic organisms that obtain their food primarily through filter feeding, including clams, bryozoans, sponges, and some insect larvae. These organisms are occasionally present in locally dense aggregations, but their densities at the ecosystem scale are usually relatively low. As a result, prior to the establishment of dreissenid mussels (zebra mussel *Dreissena polymorpha*, quagga mussel *Dreissena rostriformis bugensis*), the trophic niche of benthic filter feeder was one that was largely unoccupied, and the flux of material from the water column to the lake bottom was controlled primarily by the physical processes of settling, resuspension, and advection. Benthic invertebrates were seen primarily as redistributors of material within the benthic region once sedimentation had occurred (Robbins et al. 1977).

After establishment of dreissenid populations in the Great Lakes in the late 1980s, it was almost immediately recognized that dense populations of these filter feeders had the potential to significantly influence plankton abundance and species composition. Initial studies quantified clearance and filtration rates of zebra mussels and used various approaches to extrapolate results to the ecosystem scale (MacIsaac et al. 1992, Bunt et al. 1993, Fanslow et al. 1995). Further studies refined filtration capacities by considering how the relationship between volumetric pumping rate and mass filtration rate is influenced by hydrodynamics and population density (Yu and Culver 1999, Ackerman et al. 2001, Boegman et al. 2008). While these studies led to more moderate estimates of feeding rates and ecosystem impacts, they supported the general contention that filter-feeding dreissenids have the capacity to remove significant quantities of plankton and other particulate material from lakes and rivers.

Measurements of high community filtration rates led to the realization that these organisms also had the potential to play an important role as nutrient cyclers. And so, following initial studies that focused on dreissenid filtering, there have been a number of studies that quantified nutrient excretion. The majority of these studies have measured excretion rates of dreissenids in controlled lab experiments, usually a short time after mussels were removed from their natural environment (Mellina et al. 1995, Arnott and Vanni 1996, Orlova et al. 2004, Conroy et al. 2005, Turner 2010). Others studies have inferred quantitative or qualitative measurements of excretion rates by measuring spatial or temporal changes in natural systems (Johengen et al. 1995, Effler et al. 1997) and in mesocosms (Heath et al. 1995). Very few studies have measured excretion rates for undisturbed populations in the natural environment (Ozersky et al. 2009). In some cases, nutrient excretion rates of dreissenids have been incorporated into numerical ecosystem models, which use values obtained from empirical studies or infer values from mass balance calculations (Schneider 1992, Madenjian 1995, Mellina et al. 1995, Padilla et al. 1996, Canale and Chapra 2002).

A summary of published nitrogen (N) and phosphorus (P) excretion rates for dreissenids and other bivalves is presented in Table 35.1. The table includes recent measurements we have made for *D. r. bugensis* in Lake Michigan (Bootsma et al. unpublished). For dreissenids, the majority of studies included in Table 35.1 apply to *D. polymorpha*, as it was the dominant species in the Great Lakes until recently. However, since the displacement of zebra mussels by quagga mussels in most parts of the Great Lakes (Mills et al. 1999, Stoeckman 2003, Wilson et al. 2006, Nalepa et al. 2010), several studies have measured nutrient excretion of this more recent species. For the genus as a whole, excretion rates of dissolved P normalized to dry tissue weight (excluding shell) range from 0.08 to 3.4 μmol P gDW^{-1} h^{-1}, and excretion rates of dissolved nitrogen range from 1.4 to 26 μmol N gDW^{-1} h^{-1}. The highest P excretion rates were recorded for zebra mussels lying directly on sediment and likely represented both direct mussel excretion and sediment release (Turner 2010). If these measurements are omitted, the highest recorded P excretion rate is about 2 μmol P gDW^{-1} h^{-1}. Data presented in Table 35.1 were obtained under a large variety of experimental conditions and with a variety of measurement methods. Despite this variety, P excretion rates of quagga mussels appear to be lower than those of zebra mussels (mean quagga mussel rate = 0.33 ± 0.18 μmol gDW^{-1} h^{-1}; mean zebra mussel rate = 0.67 ± 0.56 μmol gDW^{-1} h^{-1}; $P < 0.06$). In one of the few studies to directly compare excretion rates of the two species, Conroy et al. (2005) reached this same conclusion. Lower specific excretion rates of the quagga mussel may reflect lower mass-specific clearance rates, lower respiration rates, and/or higher assimilation efficiencies of this species (Baldwin et al. 2002, Stoeckmann 2003). Zebra mussels appear to allocate a greater proportion of food intake to catabolic processes (and hence nutrient excretion), whereas quagga mussels allocate a greater proportion to tissue growth. Conroy et al. (2005) reported that N excretion rates for zebra mussels were greater than those for quagga mussels. This does not appear to be evident in our larger data set, but measurements of N excretion by quagga mussels are fewer in number than those of P excretion (Table 35.1).

Table 35.1 Nutrient Excretion Rates of Dreissenids (*D. polymorpha* and *D. r. bugensis*) and Other Bivalves

Species	Specific P Excretion	Areal P Excretion	Specific N Excretion	Areal N Excretion	Temperature (°C)	Length (mm)	Source	Notes
D. polymorpha	3.22	45.3	13.7	193	17	10–15	Turner 2010	On sediment
D. polymorpha	1.03	14.5	11.0	155	17	10–15		No sediment
D. polymorpha	0.8–2.2	43–121	3.5–7.5	222–624	18–25	<13.5	Arnott and Vanni (1996)	
	0.2–1.3		2.0–8.0			13.5–19		
	0.5–0.95		2.5–9.4			>19		
D. polymorpha		6.8–213					Orlova et al. (2004)	
D. polymorpha	0.14–0.17				17	20	Mellina et al. (1995)	
	0.32–0.37				22	20		
D. polymorpha	0.28	4.3	5.4	163	23	10–15	Conroy et al. (2005)	
D. polymorpha	0.72		26.1		23	20–25		
D. polymorpha	0.181	50.1	1.39	385	20–26		Effler et al. (1997, 2004)	Downstream vs. upstream
D. polymorpha	0.3–0.74	5.24	2.5–10.7		4–19	<14	Naddafi et al. (2008)	
	0.12–0.3		1.0–6.3		4–19	14–20		
	0.11–0.20		0.6–4.8		4–19	>20		
D. polymorpha	0.11	14.1	4.2	530	18	21	James et al. (1997)	
D. polymorpha			1.2		5.7		Quigley et al. (1993)	
			4.8		21			
D. r. bugensis	0.084	2.46			8		Ozersky et al. (2009)	
	0.361	4.15			9			
	0.265	8.11			14			
	0.195	5.47			16			
D. r. bugensis	0.55	41			7.5	17	Bootsma et al., unpublished	In situ measurements
	0.20	10.1			8.9	12.5		
	0.45	24.5			21	13		
	0.66	42.9			20	13.5		
D. r. bugensis	0.22		5.8		23	10–15	Conroy et al. (2005)	
D. r. bugensis	0.31		15.3		23	20–25		
L. radiata	0.042		1.16		10–23		Nalepa et al. (1991)	
L. siliquoidea	0.038–0.093		0.91–1.38		October		Davis et al. (2000)	
P. fasciolaris	0.043		1.16		July		Davis et al. (2000)	
P. fasciolaris	0.023–0.075		0.27–0.42		October		Davis et al. (2000)	
E. dilatata	0.017–0.022		0.36–0.55		October		Davis et al. (2000)	
C. fluminea	1.0				6		Lauritsen and Mozley (1989)	
	18.0				23			
A. wahlbergi	0.48		6.1		25		Kiibus and Kautsky (1996)	
M. edulis		850		320–5500	April–July		Asmus et al. (1990)	
M. edulis		50–429		1000–3700	April–September		Prins and Smaal (1994)	
M. senhousia	1.35		12.2		21	16.7	Magni et al. (2000)	
M. demissus	0.087				6–24		Kuenzler (1961)	
R. philippinarum	3.65		19.2		21	9.4	Magni et al. (2000)	
	1.05		8.6			15.5		
	0.80		4.4			18.9		
	1.45		13.3			23.5		
Various bivalves			~6.0				Table 5 of Magni et al. (2000)	

Source: Bayne, B.L. and Scullard, C. *J. Mar. Biol. Assoc. U.K.* 57, 355, 1977.

Units for specific excretion rates are μmol gDW^{-1} h^{-1}. Units for areal excretion rates are μmol m^{-2} h^{-1}. P excretion refers to the excretion of SRP, except for Mellina et al. (1995) who measured the change in total P after mussels were placed in filtered lake water. Most studies measured N excretion as $NH_3 + NH_4^+$ (see text for exceptions). This likely results in an underestimate of total N excretion because a significant amount of N may be excreted as amino acids.

FACTORS REGULATING NUTRIENT EXCRETION BY DREISSENIDS

Initial studies of dreissenid filtration rates were made with the primary objective of determining ecosystem-scale impacts that these filter feeders would have on the plankton community. Likewise, most studies of dreissenid nutrient excretion rates attempted to place rates into the context of ecosystem-scale nutrient dynamics. To do this, it is necessary to extrapolate rates over time and space, which requires knowledge of how mussel nutrient excretion responds to environmental conditions that are variable in time and space. This knowledge may also facilitate a more realistic inclusion of mussel-mediated nutrient dynamics in ecosystem-scale models. While some existing models account for nutrient excretion (Schneider 1992, Madenjian 1995, Padilla et al. 1996), few direct measurements of dreissenid nutrient excretion were available prior to development of these models, and they therefore relied on measurements made for other bivalves or on assumptions about the fraction of consumed food that is recycled as dissolved nutrients.

Recorded mass-specific N and P excretion rates vary more than an order of magnitude (Table 35.1). A full understanding of the role of dreissenids in ecosystem nutrient dynamics and how this role may be modulated by external drivers, including nutrient loading, weather conditions, and long-term climate change, must account for this variability and factors that control it. Several studies have examined factors that may potentially control dreissenid nutrient excretion rates, either by relating measured rates to environmental conditions (e.g., Arnott and Vanni 1996) or by testing the response of excretion rate to specific factors through direct experimentation. The factor that has received the greatest attention is mussel size, measured either as length or mass (Mellina et al. 1995, Arnott and Vanni 1996, Orlova et al. 2004, Conroy et al. 2005). Surprisingly, temperature effects have received relatively little attention, although its significance has been inferred in some studies (Arnott and Vanni 1996, Ozersky et al. 2009). The following is a summary of these factors and others that are most likely to play a dominant role as regulators of dreissenid nutrient excretion. In a number of cases, the relative importance of a factor is inferred from studies in which filtration rates were measured, because filtration rates and nutrient excretion rates are often correlated.

Food Quantity and Quality

Although several studies have examined the relationship between food supply and dreissenid filtration and clearance rates (Sprung and Rose 1988, Berg et al. 1996), there are virtually no studies that examine the quantitative relationship between food supply and the excretion rates of dissolved nutrients (but see Johengen et al. 2013). A positive correlation between food supply and nutrient excretion might be expected, and the range of recorded excretion rates for different water bodies appears to be explained in part by differences in phytoplankton concentrations (Arnott and Vanni 1996). However, the nature of this relationship will depend on how mussels allocate food resources to various pathways (pseudofeces, feces, tissue growth, dissolved excretion) as a function of food supply. For example, filtration rate may plateau or even decrease above a threshold food concentration (Fanslow et al. 1995), and production of pseudofeces generally increases when food concentration exceeds an incipient limiting threshold (Berg et al. 1996, Madon et al. 1998). Both of these processes will likely result in a nonlinear relationship between nutrient excretion and food supply. The nutrient excretion response to food supply will also differ from the feeding response because, while feeding rate may be dictated by food supply at very low food concentrations, excretion may be driven according to a basal metabolic rate. In fact, nutrient excretion rates may even increase at low food concentrations as a result of mussel emaciation (James et al. 2001). At this time, the paucity of data limits quantitative descriptions of the relationship between food supply and nutrient excretion. A better understanding of this relationship would be valuable to lake managers. Nutrient management strategies are guided by an understanding of the relationship between nutrient loading and algal growth, which is used to set nutrient loading and concentration targets. It now appears that, in many lakes, grazing and nutrient recycling by dreissenids have altered the way in which the algal community responds to nutrient loading, and mussels may be an important nutrient source supporting nuisance algal blooms (Hecky et al. 2004, Malkin et al. 2008, Auer et al. 2010). Any attempts to reduce the supply of nutrients from dreissenids will need to be informed by models that account for the relationship between phytoplankton biomass/composition and mussel nutrient excretion (see section on management implications in the following text).

Food quality can affect nutrient excretion rates by influencing both the amount and fate of ingested food. Various measures have been used to assess food quality. The most common measures are the inorganic–organic mass ratio, phytoplankton species composition, and seston C:N:P ratios. Madon et al. (1998) found that an increase in the inorganic–organic ratio resulted in lower clearance and ingestion rates, more pseudofeces production, and lower assimilation efficiency. Consequences of these responses for nutrient excretion are uncertain. While lower ingestion rates may result in lower nutrient excretion rates, lower assimilation efficiencies imply that a greater proportion of ingested material may be excreted, both in solid (feces) and dissolved forms. Ingestion of inorganic material is of little nutritive value to mussels, but if this inorganic material includes particulate P, this P may be released as dissolved phosphate after passing through the anoxic gut (Hecky et al. 2004).

To our knowledge, no studies have directly examined the potential effect of phytoplankton species on nutrient excretion by dreissenids. However, there is strong evidence that feeding rate is influenced by phytoplankton species composition, and therefore it is likely that there is also a nutrient excretion response. When mussels were fed monocultures of *Chlamydomonas reinhardtii* or *Pandorina morum*, Berg et al. (1996) found that clearance rates and proportions of ingested particles that were ejected as pseudofeces varied, depending on which algal species was being consumed. Both Lavrentyev et al. (1995) and Baker et al. (1998) found that the presence of *Microcystis* affected clearance rates of zebra mussels, although the former authors reported lower rates, while the latter authors reported higher rates. Combined, all these studies suggest that any relationships between food quantity and nutrient excretion rates will likely be moderated by food quality. Incorporation of these relationships into models is difficult given the limited data presently available, and it is unclear which taxonomic level of phytoplankton is mostly strongly linked to feeding selectivity. Results of the earlier studies suggest that at least some effect can be expected at the species level, but further experiments with multispecies phytoplankton assemblages are required to determine how well results of experiments with monocultures apply to natural phytoplankton communities. The contrasting findings of Lavrentyev et al. (1995) and Baker et al. (1998) suggest that dreissenid response to a given species or genus may even vary according to environmental conditions and the physiological status of the algae.

While several studies have measured the N:P ratio of nutrients excreted by dreissenids and other bivalves (Table 35.1), few studies have examined how excretion rates are affected by the nutrient stoichiometry of ingested foods. Yet ecological theory would suggest that nutrient ratios in food supplies will have an influence on nutrient excretion ratios of dreissenids (Elser and Hassett 1994). More specifically, dreissenid nutrient excretion should be regulated by the degree to which C:N:P stoichiometry of dreissenid tissue differs from that of the food source. In one of the few studies to measure the stoichiometry of both dreissenids and their food source, Arnott and Vanni (1996) reported N:P ratios of 7.0–22.3 for phytoplankton and N:P ratios of 16–33 for zebra mussels in Lake Erie, indicating that phytoplankton were N depleted relative to mussel requirements. Theory would predict that these mussels selectively retain N and recycle P. Indeed, the authors reported low N:P excretion ratios ranging from 4.9 to 17.5 and suggested these relatively low ratios could promote the dominance of cyanobacteria. In a similar study with zebra mussels from Lake Erken (Sweden), Naddafi et al. (2008) measured N:P ratios of 16.6–30.8 for seston and N:P ratios of 22.5–40.0 for mussels. Their reported N and P excretion rates (Table 35.1) indicate that N:P excretion ratios varied greatly (~5 to 30). However, N:P excretion ratios were usually similar to or less than N:P ratios for the seston, suggesting selective retention of N and recycling* of P, as would be expected based on the relatively P-rich food supply. At a larger ecosystem scale, Effler et al. (2004) used upstream–downstream comparisons of dissolved N and P concentrations in the Seneca River to derive a N:P excretion ratio of 7.1:1. This is probably a minimum value, as they did not account for possible nitrification of ammonium (NH_4^+) between upstream and downstream sites. The authors did not measure seston nutrient concentrations, but phytoplankton in this river was derived from a hypereutrophic upstream lake and so was probably phosphorus replete. In this case eutrophic conditions appear to promote low N:P excretion ratios that are well below the Redfield ratio of 16:1, which may tend to favor cyanobacteria.

In contrast to the earlier studies, Conroy et al. (2005) reported N and P excretion rates for zebra and quagga mussels in Lake Erie that equate to relatively high N:P excretion ratios (36–52). Conroy et al. (2005) did not report nutrient concentrations of seston, but at the time of sample collection NH_4^+ concentration was 52 µg N L^{-1} and soluble reactive phosphorus (SRP) was undetectable, suggesting that phytoplankton were likely P limited. These results were similar to the findings of Heath et al. (1995) who measured accumulation of dissolved N and P in Saginaw Bay (Lake Huron) mesocosms containing high densities of zebra mussels feeding on P-limited phytoplankton. They observed high dissolved N:P accumulation ratios, ranging from 40:1 to 101:1.

There remains a need for further research on effects of food stoichiometry on nutrient excretion by dreissenids. However, the earlier studies and stoichiometry theory suggest that when N:P ratios of seston are lower than N:P ratios of mussels, N:P ratios of excreted nutrients will be low. Conversely, high N:P ratios of seston will result in selective recycling of nitrogen by dreissenids. As a result, if phytoplankton within a system is N or P limited, nutrient recycling by dreissenids may exacerbate the current condition. Because phytoplankton communities in the Great Lakes tend to be P limited, nutrient recycling by dreissenids will likely reinforce this limitation. Exceptions may occur in more eutrophic systems, such as the western basin of Lake Erie where there is evidence for N limitation at times (Arnott and Vanni 1996).

Although stoichiometric theory and empirical observations suggest that dreissenid nutrient recycling will tend to reinforce the existing nutrient limitation, under certain conditions mussel nutrient recycling may result in a shift between N and P limitation. This is because the optimal N:P ratio for dreissenids may differ from that for phytoplankton. The potential of dreissenids to recycle nutrients and cause a shift between P and N limitation

---

* In this chapter, the terms "excretion" and "recycling" are similar, but "excretion" refers to the release of a dissolved nutrient, whereas "recycling" refers specifically to the conversion of a nutrient from particulate to dissolved form.

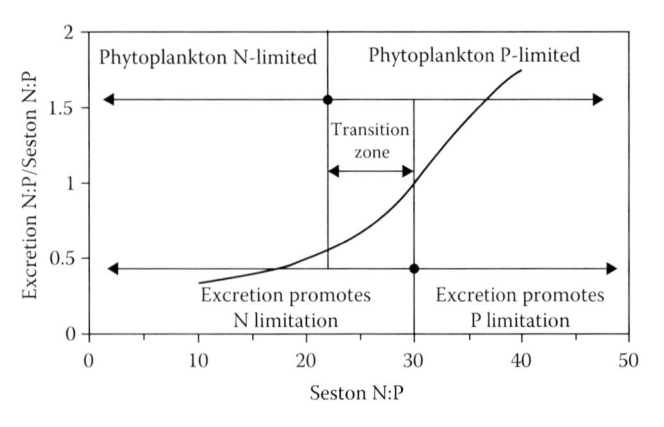

Figure 35.1 Conceptual model illustrating feedback between N:P ratios of seston and N:P ratios of mussel excretion. The model assumes that phytoplankton are P limited when seston N:P ratios are >22 (Healey and Hendzel 1980) and that mussels maintain a constant tissue N:P ratio of 30:1 (an approximate median of measurements made by Arnott and Vanni [1996] and Naddafi et al. [2009]). Excretion N:P was determined as a function of seston N:P, assuming that mussels excrete a constant 25% of ingested N as dissolved N when seston N:P is less than 30 and a constant 25% of ingested P as dissolved P when seston N:P is greater than 30. In the transition zone, where seston N:P ratio is less than that of mussel tissue but greater than the optimal N:P ratio of phytoplankton, recycling by mussels may result in a shift from P limitation to N limitation.

will be greatest when N:P ratios of seston are above the Redfield ratio of 16:1 but lower than N:P ratios of mussels. For example, if seston N:P is 22 and mussel N:P is 30, phytoplankton may be marginally P limited, but dreissenids may selectively retain N and recycle P to meet their stoichiometric requirements, resulting in a low N:P excretion ratio that could shift the phytoplankton community toward N limitation (Figure 35.1). The strength of this interaction will depend in part on the mussels' ability to adjust their tissue N:P ratios. If mussel tissue N:P fluctuates according to changes in food N:P (Naddafi et al. 2009), the ability of mussels to facilitate a switch between P and N limitation, or to reinforce limitation by either nutrient, will be limited. But mussel tissue N:P ratios can differ significantly from those of seston and may be as high as 40 in late summer (Arnott and Vanni 1996, Naddafi et al. 2009), and during these periods mussel excretion can be expected to have a strong influence on the balance between N and P limitation.

Dreissenid Size

Size is a primary determinant of biomass-specific metabolic rates in bivalves, and in general the relationship is inverse (Hamburger et al. 1983). This relationship appears to extend to nutrient excretion (Mellina et al. 1995, Young et al. 1996, Naddafi et al. 2008). In laboratory experiments, Mellina et al. (1995) found that dreissenid shell length explained >70% of the variability in the biomass-specific P excretion rate at a given temperature. Whether this relationship between size and P excretion rate also applies to nitrogen is uncertain. Arnott and Vanni (1996) and Davis et al. (2000) both observed that mussel size explained a significant amount of variability in P excretion but not N excretion. This may reflect different metabolic pathways for the two elements. However, Quigley et al. (1993) provided indirect evidence suggesting that size may have some influence on mass-specific N excretion. They observed a strong inverse relationship between mass-specific oxygen consumption rate and tissue dry weight for zebra mussels collected from Lake St. Clair and, in a separate experiment, they observed a positive correlation between oxygen consumption and N excretion.

Differences in P excretion rates between small and large mussels are large enough that estimates of community excretion rates can include large errors if mussel size distribution is not accounted for. For example, Young et al. (1996) showed that rates of filtration and pseudofeces production in zebra mussels can be up to an order of magnitude greater in large (~25 mm) than in small (~5 mm) individuals. However, mass-specific excretion rates of SRP are much greater for small mussels than for large mussels (Arnott and Vanni 1996, Naddafi et al. 2008). Therefore, it is important to account for both mussel density and size distribution when areal filtration and nutrient excretion rates are calculated. A factor that may alter the observed relationship between size and excretion is the length of time mussels have been deprived of food before measurements are made, which in published studies spans from hours to days. To our knowledge, there are no direct measurements of gut passage time for dreissenids, but Wang and Fisher (1999) report a passage time of less than 3 for marine mussels, and Lauritsen and Mozley (1989) observed a decline in ammonium excretion rate after 2 of food deprivation for the clam, *Corbicula fluminea*. Because gut passage times are likely positively related to mussel size, a food deprivation period of more than a few hours may alter the size–excretion relationship, with excretion by small mussels declining more rapidly than that by large mussels. If mussel size does influence the temporal scales over which nutrient excretion responds to food supply, there will be implications for nutrient dynamics in natural communities. For example, the nearshore zone of large lakes can experience rapid fluctuations in temperature due to upwelling events, and these fluctuations are likely accompanied by large changes in food supply, which may explain lower dreissenid densities in regions of frequent upwelling (Wilson et al. 2006). In areas with populations dominated by small mussels, these fluctuations in food supply will likely lead to rapid changes in nutrient recycling, whereas the nutrient recycling response may be dampened in areas dominated by large mussels.

Temperature

As for most invertebrates, ambient temperature is known to influence metabolic rates of dreissenids (Stoeckmann and Garton 1997) and other bivalves (Kuenzler 1961, Magni et al. 2000). Despite the potentially strong influence of temperature on nutrient excretion rates, few studies have directly quantified this relationship for dreissenids. While temperature is recognized as a factor that may explain observed differences in nutrient excretion rates over time or between systems (Arnott and Vanni 1996, Ozersky et al. 2009), most studies that have extrapolated measured rates to estimate ecosystem-scale or seasonal effects of dreissenid feeding and nutrient excretion have not accounted for temperature differences. Mellina et al. (1995) recorded SRP excretion rates of zebra mussels that were approximately two times greater at 22°C than at 17°C (Table 35.1). At higher temperatures, nutrient excretion may be decoupled from ingestion. For example, rates of ammonium excretion and oxygen consumption of zebra mussels were four times greater at 28°C and 32°C than at 20°C; however, filtration rates were lower at 28°C and 32°C than at 20°C (Aldridge et al. 1995). Over a temperature range that more closely matches that of many temperate lakes, we have observed strong temperature effects, with phosphorus excretion rates increasing more than threefold between 4°C and 20°C (Figure 35.2). Excretion rates measured in situ or measured with mussels that have been exposed to in-lake conditions do not always reveal a relationship to temperature that is predictable (Table 35.1) due to other confounding factors that may change over time and space. However, temperature effects are strong enough that they can often be detected in studies that span one or more seasons (Arnott and Vanni 1996, Naddafi et al. 2008, Ozersky et al. 2009). This, along with the experimental observations cited earlier, indicates that these effects are critical enough such that temperature fluctuations must be accounted for in models of nutrient cycling by dreissenids or when seasonal or annual nutrient excretion rates are estimated from a small number of measurements.

While a number of studies have examined physiological responses of zebra mussels to temperature (Sprung 1995, Alexander et al. 1994), there are few studies that have examined responses of quagga mussels. Quagga mussels are often considered to be better adapted to cold-water conditions compared to zebra mussels. Yet, although quagga mussels appear to be able to reproduce at colder temperatures (Roe and MacIsaac 1997) and may have an upper thermal tolerance lower than that of zebra mussels (Domm et al. 1993, Spidle et al. 1995), the two species appear to exhibit similar clearance rates at both cold and warm temperatures (Baldwin et al. 2002). Therefore, while the recent increase in abundance of quagga mussels and decline of zebra mussels in the Great Lakes may affect nutrient cycling due to changes in absolute numbers of mussels and their spatial distribution, there is no evidence that a switch in dreissenid species will alter the temperature–nutrient excretion relationship.

Other Regulating Factors

Studies based on laboratory experiments and in situ measurements indicate that the factors discussed earlier—food quantity and quality, mussel size, and water temperature—are likely the dominant regulators of nutrient excretion by dreissenids. However, these same studies reveal that not all of the variability in nutrient excretion rates can be explained by these factors alone. In studies that have measured excretion rates of dreissenids at different times of the year, temporal differences in rates were not readily explainable by differences in temperature or food supply. Naddafi et al. (2008) suggested that the positive relationship between mussel size and mass-specific N excretion rate reported by Arnott and Vanni (1996), which is the opposite of findings in other studies, may have been due to nutrient loss during the release of gametes by large mussels. Supportive of this theory is that mussels may have lower P excretion rates and greater N excretion rates during gametogenesis, with a greater proportion of ingested P being allocated to gamete production and N being lost during protein catabolism (Davis et al. 2000). In Lake Erie, Stoeckmann and Garton (1997) found that, although temperature had a strong influence on mussel feeding and metabolism, temperature alone was a poor predictor of seasonal changes in metabolic rate and energy allocation, with other important drivers being reproductive state and food conditions. While they did not specifically address the relationship between these drivers and nutrient excretion, the dependence of excretion on feeding rate and metabolism means that excretion also likely responds to these multiple driving factors.

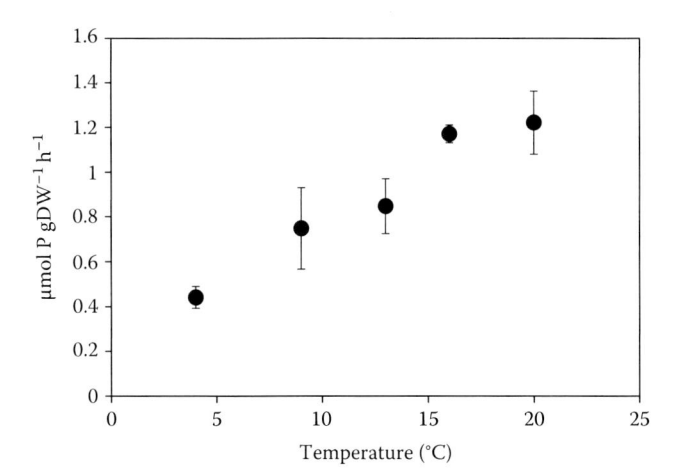

Figure 35.2 Excretion rates of soluble reactive P (μmol P gDW^{-1} h^{-1}) by quagga mussels at five different temperatures. Mussels were collected from Lake Michigan in March 2010. Mussels were fed a monoculture of *Scenedesmus quadricauda* (particulate carbon concentration = 25–33 μmol L^{-1}) and acclimated to experimental temperatures for 3 days prior to measurement of excretion rates. Mussel length was between 16 and 18 mm. Excretion rate was measured as SRP accumulation over 1.5 h after mussels were placed in filtered lake water.

To our knowledge, there have been no direct measurements of the effect of dissolved oxygen concentration on dreissenid nutrient excretion. However, several studies have documented the influence of dissolved oxygen on other metabolic processes. Sprung (1995) noted that filtration rate and ingestion capacity of zebra mussels were reduced at low oxygen concentrations. Similarly, Quigley et al. (1993) observed a significant decline in oxygen consumption rate of zebra mussels when dissolved oxygen concentration decreased. They fitted their data to a Michaelis–Menten model, which indicated that the oxygen consumption rate at 50% dissolved oxygen saturation was ~57% of that at 100% saturation. They also observed a strong correlation between oxygen consumption rate and ammonium excretion rate ($r^2 = 0.76$, $P < 0.03$, using data in their Table 35.1). The implication is that ammonium excretion rate is positively correlated to dissolved oxygen concentration.

In addition to directly affecting mussel metabolism, dissolved oxygen concentrations may influence the fate of nutrients excreted or egested by mussels. There is little data to indicate the direct effect of dreissenids on dissolved oxygen in the benthic boundary layer or in sediment pore water, but it is reasonable to expect that both mussel respiration and bacterial decomposition of mussel biodeposits may lead to hypoxic or anoxic conditions in these environments (Turner 2010). Such conditions would tend to promote nitrogen loss through denitrification while enhancing the release of iron-bound phosphorus, resulting in low N:P recycling ratios. This is supported by experimental observations (Bruesewitz et al. 2006, Bykova et al. 2006, Turner 2010). As Bykova et al. (2006) point out, in lakes where dreissenid nutrient recycling rates are high relative to other internal recycling rates and to external loads, this effect may lead to a state of nitrogen limitation that promotes cyanobacteria dominance. Newell et al. (2005) have postulated a similar relationship between oyster egestion, sediment redox conditions, and N and P recycling in Chesapeake Bay. This link between oxygen dynamics and nutrient recycling will likely be influenced by substrate type, with fine-grained sediments in deep regions more likely to experience hypoxic or anoxic conditions than shallow or rocky substrata.

Most measurements of nutrient excretion by dreissenids have been made using mussels that have been removed from their natural environment and incubated in some sort of enclosure. The unstated assumption is that nutrient excretion in an artificial enclosure is similar to that on the natural bottom substrate, providing that temperature and food supply are similar. However, in the few studies in which dreissenid nutrient excretion has been measured on different substrata, there is evidence that substrate type may have a strong influence, either directly or indirectly. Turner (2010) found that P excretion by zebra mussels in cores with lake sediment was more than three times greater than that in cores without sediment. Because P excretion in the presence of sediment was greater than expected based on food consumption, the elevated excretion rate was attributed to P release from sediment that had passed through mussel guts. Newell et al. (2005) also highlight the important role the substratum and its associated microbial community may play in modifying the effect of bivalves on nutrient recycling. In the study of Turner (2010), the presence of sediment did not appear to promote higher excretion rates of N, and so Turner (2010) suggested that recycling of sediment-bound P by dreissenids could promote low N:P recycling ratios and N limitation of phytoplankton. This raises the question of whether the recent displacement of zebra mussels by quagga mussels in the Great Lakes (Wilson et al. 2006, Nalepa et al. 2010) may further alter nutrient dynamics within these systems, as quagga mussels are more capable of colonizing soft sediments.

ECOSYSTEM IMPACTS OF DREISSENID NUTRIENT CYCLING

Influence of Dreissenids on Nutrient Pools and Fluxes

Dreissenids are major nutrient processors in systems where they are abundant. In Lake Michigan, this is evident by comparing dreissenid P excretion rates to P loads from rivers. Dreissenid dissolved P excretion rates in the lake range from 10.1 to 42.9 μmol P m^{-2} h^{-1} (Table 35.1). Areal imagery indicates that, in the Milwaukee region of the lake, most of the substrate is hard and dreissenids cover ~80% of the bottom in waters shallower than 10 m (Janssen and Bootsma, unpublished). The average distance between the shore and the 10 m isobath in this region is approximately 2 km. Using these values and a median areal excretion rate of 27 μmol P m^{-2} h^{-1}, an approximate recycling rate in the nearshore (0–10 m) zone is 1 mole of P per meter of shoreline per day. In comparison, an approximate loading rate from the Milwaukee River (which receives input from the Menomonee and Kinnickinnic Rivers near its mouth) is 8000 moles of P per day.* Hence, the dreissenid P recycling rate in waters shallower than 10 m over an 8 km stretch of shoreline is similar to the average loading rate via the Milwaukee River. This suggests that for nearshore regions of Lake Michigan that are dominated by rocky substratum and are not in close proximity to a major tributary, mussels are the major source of dissolved P. Arnott and Vanni (1996) reached a similar conclusion for the western basin of Lake Erie, where they estimated that P excreted by zebra mussels exceeded that of all other sources and exceeded external loading by an order of magnitude. Likewise, Ozersky et al. (2009) determined that dissolved P excreted by dreissenids

---

* Based on nutrient measurements made in these rivers by the Milwaukee Metropolitan Sewerage District and river discharge data collected by the U.S. Geological Survey

in an urbanized region of Lake Ontario was greater than external loading. The capacity of dreissenids to be major nutrient recyclers is not surprising considering their ability to ingest large fractions of the primary production in aquatic systems (Stoeckmann and Garton 1997, Naddafi et al. 2008).

Nutrients ingested by dreissenids have one of three fates: assimilation into body mass, excretion in dissolved form, and egestion as feces or pseudofeces. The net ecosystem effect will depend on the relative magnitude of these pathways and how they vary in space and time. Heath et al. (1995) found that nutrient recycling by dreissenids resulted in increases in dissolved P concentrations in mesocosms. A number of other studies suggested that a similar effect occurs in large water bodies (Holland et al. 1995, Makarewicz et al. 2000, Orlova et al. 2004). However, Higgins and Vander Zanden (2010) found that this effect does not appear to apply to lakes in general but does appear to be a common response in rivers. They suggested that the difference in response of rivers versus lakes may be due to less P assimilation by benthic algae in rivers. Another possibility is that excreted P is more diluted in lakes due to greater water volumes relative to bottom areas.

During periods of rapid population growth, as occurred in the decade following the introduction of dreissenids to the Great Lakes, mussel biomass represented a significant nutrient sink. For example, shortly following the establishment of zebra mussels in Saginaw Bay (Lake Huron), the amount of P in mussel mass was equal to approximately one-half the annual P load to the bay, and total phosphorus concentrations decreased by approximately 50% (Johengen et al. 1995). Over the long term, the effect of nutrient assimilation by dreissenids on lake and river nutrient pools will depend in part on the fate of mussel tissue and shell following death. If most nutrients contained in tissues and shells are recycled following death, nutrient pools in the water column may increase once mussel populations have reached steady state. At this point, mussel growth no longer represents a net loss of nutrients from the system (although mussel grazing may still maintain water column total nutrient concentrations at levels lower than if there were no mussels). This dynamic can be seen in the temporal trend of dissolved calcium concentration (using alkalinity as a surrogate) in Lake Erie, which dramatically declined in the decade following the dreissenid invasion but since then has steadily increased toward the pre-dreissenid concentration (Barbiero et al. 2006). In Lake Ontario, calcium concentrations have continued to decline (Barbiero et al. 2006), indicating that calcium dynamics, and probably nutrient dynamics, may take much longer to reach steady state in this lake, perhaps due to its longer hydraulic residence time and continued expansion of its dreissenid population (Watkins et al. 2007).

For dreissenids, few studies have made direct comparisons of nutrients excreted, nutrients allocated to growth, and nutrients egested via feces and pseudofeces. Madenjian (1995) estimated that 22% of the phytoplankton removed by dreissenids from Lake Erie was deposited as pseudofeces. In eutrophic Lake Mikołajskie (Poland), feces/pseudofeces production rates by zebra mussels were several times greater than community assimilation rates (Stańczykowska and Lewandowski 1993). In Lake Erie, mass balance calculations made by Mellina et al. (1995) suggested that most P ingested by zebra mussels was allocated to excretion of dissolved forms, whereas in Lake Oneida biodeposition (feces and pseudofeces) was the dominant P pathway. In contrast, they estimated that P biodeposition in Lake St. Clair was negligible. They attributed these differences to differences in phytoplankton concentrations, arguing that high phytoplankton concentrations in Lake Oneida resulted in ingestion rates that were greater than the maximum assimilation rate, resulting in high rates of biodeposition. This is supported by other studies that have shown that pseudofeces production is related to phytoplankton concentration, phytoplankton species composition, and seston organic content (Berg et al. 1996, Lei et al. 1996, Madon et al. 1998, Schneider et al. 1998).

These observations suggest that, in more productive systems or in systems where a large portion of the seston is nonalgal, pseudofeces production may be a significant nutrient pathway. However, this does not appear to be the case in less productive water bodies. Several studies have reported that pseudofeces production by dreissenids is negligible at seston concentrations typical of those found in the Great Lakes (Lei et al. 1996, Madon et al. 1998). In these systems, nutrients egested via feces are likely a more important nutrient pathway than nutrients ejected via pseudofeces. However, the relative importance of nutrients egested via feces versus dissolved nutrients excreted is uncertain. When zebra mussels were fed algal cultures in the laboratory, Berg et al. (1996) found that absorption efficiencies (proportion of ingested food not egested as feces) were ~80% at low algal densities, indicating that ~20% of ingested food was egested as feces. Others have measured similar assimilation efficiencies when dreissenids were fed high-quality food (Stoeckmann and Garton 1997, Baldwin et al. 2002). In these estimates, "absorption" or "assimilation" generally represents the proportion of ingested food (excluding pseudofeces ejection) that is not egested as feces. This "assimilated" portion is partitioned between growth, reproduction, and metabolic processes, the latter of which produces excretory products. Stoeckmann and Garton (1997) determined that >90% of energy consumed by zebra mussels was allocated to metabolism. Unless the stoichiometry of ingested seston is greatly different from that of mussel tissue, assimilated nutrients will be partitioned in proportions similar to the nutrients ingested. High assimilation efficiencies along with a large fraction of assimilated nutrients being allocated to metabolism suggest that excretion of dissolved nutrients may be greater than fecal egestion of nutrients in systems with low concentrations of high-quality seston. Similar observations have been made for phosphorus excretion and egestion by fish (Brabrand et al. 1990). Further measurements are needed to confirm this suggestion and to determine factors that determine how ingested nutrients are allocated between growth, excretion, and egestion in dreissenids.

The influence of pseudofeces and feces production on ecosystem nutrient dynamics will depend on the ultimate fate of this particulate material. Unlike excreted nutrients, which are readily available to autotrophs, nutrients tied up in pseudofeces and feces represent a loss from the water column. Even if nutrient egestion rates are low relative to excretion rates, feces and pseudofeces formation can significantly alter the total nutrient pool in the water column if this particulate material is retained in the benthos. If this is the case, then the benthic nutrient pool may support other benthic invertebrates (Limén et al. 2005) and fish (Thayer et al. 1997), may be recycled by bacteria and protozoans (Lohner et al. 2007), or may be permanently buried. In the absence of dreissenids, the internal loss of phosphorus in lakes is largely controlled by the passive settling of particulates and subsequent burial at depth. The fact that dreissenids accelerate the sedimentation process is evident since water clarity has increased and phytoplankton concentrations have decreased following dreissenid introductions in many systems (Descy et al. 2003, Hall et al. 2003, Zhu et al. 2007). In addition to accelerating sedimentation, dreissenids may alter the form of sedimented material, packaging it into larger aggregates that are less susceptible to resuspension. Resuspended sediment can make a significant contribution to water column nutrient concentrations (Eadie et al. 2002). If dreissenid production of feces and pseudofeces facilitates retention of particles and associated nutrients in the sediment, it will result in a net loss of nutrients from the water column (Hecky et al. 2004). The significance of this potential nutrient loss mechanism requires further testing and will likely be influenced by hydrodynamics and seston composition. For Lake Erie, Hecky et al. (2004) presented evidence for sediment aggregation by mussels and suggested that this promoted nutrient retention on the lake bottom. A similar mechanism may be acting in Saginaw Bay, where dreissenid filtration appears to have resulted in greater sedimentation rates and reduced inputs of P to Lake Huron (Cha et al. 2011). In contrast, Baker et al. (1998) found that zebra mussels in the Hudson River produced diffuse pseudofeces that were easily resuspended, and mussel filtration did not appear to significantly reduce turbidity. Similarly, Mellina et al. (1995) found that P loss from the water column in Lake Oneida could be accounted for by increased mussel biomass and therefore concluded that virtually all of the feces and pseudofeces were recycled.

Phytoplankton Response

There is some evidence that the excretion of nutrients by dreissenids may promote higher phytoplankton growth rates (Bierman et al. 2005, Higgins and Vander Zanden 2010). However, this effect is generally not strong enough to counter the loss of phytoplankton resulting from mussel filtration. As a result, total phytoplankton biomass either decreases following the establishment of dreissenids (Higgins and Vander Zanden 2010) or the phytoplankton become dominated by a small number of taxa, often cyanobacteria, that avoid grazing and benefit from the nutrients recycled by dreissenids (Bierman et al. 2005, Conroy et al. 2005, Zhang et al. 2008). A positive response in cyanobacteria is especially noticeable in shallower systems with moderate to high dissolved nutrient concentrations. For example, within the Laurentian Great Lakes, blooms of the cyanobacteria *Microcystis* spp. are common in the western basin of Lake Erie (Conroy et al. 2005), in Saginaw Bay (Lake Huron, Bierman et al. 2005), and in the Bay of Quinte (Lake Ontario, Nicholls et al. 2002). However, it appears that dreissenids can also promote cyanobacteria dominance in low-nutrient lakes (Raikow et al. 2004, Knoll et al. 2008). In Lake Michigan, total phytoplankton biomass has decreased following expansion of the dreissenid community, but cyanobacteria concentrations have actually increased (Fahnenstiel et al. 2010). In some cases, this transition to dominance by cyanobacteria has been attributed to a dreissenid-induced decrease in the dissolved N:P ratio (Arnott and Vanni 1996, Conroy et al. 2005). But other mechanisms, such as selective grazing (Vanderploeg et al. 2001, Vanderploeg et al. 2013), are likely also important and may play a larger role in Lake Michigan and other lakes where dissolved N:P ratios are high for most of the year.

The dual effect of filtration and nutrient excretion by dreissenids has altered the relationship between dissolved nutrient concentration and phytoplankton abundance in many lakes, which is frequently reflected in lower chlorophyll–total phosphorus ratios (Mellina et al. 1995, Hall et al. 2003). This "biological oligotrophication" (Holland et al. 1995, Evans et al. 2011) appears to have altered long-standing paradigms describing nutrient and plankton dynamics. Indeed, the ability of dreissenids to so dramatically reduce phytoplankton abundance and increase water clarity in many lakes gives validity to the concept of top-down control (Carpenter et al. 1995). In the Great Lakes, some earlier models of nutrient and plankton dynamics do not account for the effects of dreissenids (Chen et al. 2002, Pauer et al. 2008), but more recent models do account for dreissenids and highlight their potential impact both as a sink for some forms of phytoplankton and as a source of dissolved nutrients (Bierman et al. 2005, Zhang et al. 2008).

Role of Hydrodynamics and Lake Morphometry

As discussed earlier, there are a number of factors that influence filtration and nutrient excretion rates of dreissenids. If relationships between these factors and rates are well quantified and population densities are known, the potential effect of dreissenids on phytoplankton loss rates and nutrient recycling rates can be estimated. However, potential effects will be modulated by hydrodynamics, which influence both the delivery of particulate material to benthic dreissenids and the distribution of dissolved nutrients released by this community. The extent of vertical mixing in the water column will determine the relationship between filtration rate (the volume of water that passes through a mussel per unit time) and clearance rate (the volume of water that a mussel effectively clears of particulate material per unit time). In natural settings, a vertical gradient of particulate material is often observed above mussel beds (Ackerman et al. 2001, Boegman et al. 2008), which indicates that grazing by the mussel community can deplete particulates on time scales that are shorter than time scales over which particulates are replenished by vertical mixing. Under these conditions, clearance rates are lower than filtration rates, and as a result grazing rate is significantly lower than the maximum potential. For example, in situ measurements in Lake Erie indicated that dreissenid clearance rates were as low as 0.14 times filtration rates due to re-filtration (Yu and Culver 1999). Similar results were reported by Edwards et al. (2005) and Boegman et al. (2008).

In deep lakes, three physical properties may limit the effect of dreissenids on phytoplankton populations and nutrient recycling: (1) minimal wind-induced, near-bottom turbulence, which may decrease the rate of phytoplankton being supplied to bottom-dwelling dreissenids; (2) high lake volumes relative to mussel populations, resulting in a small fraction of the lake being filtered per unit time; (3) the presence of a thermocline, which hinders phytoplankton produced in the epilimnion from reaching the dreissenid population for at least part of the year. This third factor also applies to smaller lakes that are deep enough and sheltered enough from wind to establish a thermocline. While these physical properties of large lakes may help minimize the role of mussels in nutrient cycling, in fact a significant portion of total annual phytoplankton growth may occur during the spring bloom before these lakes stratify and the water column is well mixed. Dreissenids can reduce this spring bloom because they have access to the entire water column during this period. In addition, quagga mussels, which are able to maintain relatively high filtration rates at low temperatures (Baldwin et al. 2002), may be able to effectively graze phytoplankton throughout the winter, resulting in lower "seed" populations of phytoplankton during the spring bloom period. The capacity of dreissenids to impact the spring bloom appears to be significant even in deep regions of Lake Michigan, where photosynthesis and phytoplankton abundances are now one-third to one-fifth lower than what they were prior to establishment of dreissenids in these regions (Fahnenstiel et al. 2010). Over this same time span, phytoplankton biomass in summer did not change. Implications of the lost spring bloom for nutrient dynamics are uncertain. Brooks and Edgington (1994) have argued that most recycling of phosphorus from the sediment to the water column occurs during the spring in Lake Michigan, which is the time when uptake of dissolved P by phytoplankton displaces the sediment–water equilibrium with regard to apatite, promoting dissolution of P from the sediment. By depleting the spring phytoplankton population, dreissenids may have effectively turned this mechanism off. This diminished benthic recycling, along with additional nutrients being bound in mussel biomass and biodeposits, may explain the rapid decrease in concentrations of total P found in the water column in the late 1990s (Mida et al. 2010).

Although dreissenid populations in deep regions are isolated from phytoplankton in the epilimnion during the stratified period, populations in shallow, nearshore regions are not. In shallow lakes where a significant proportion of the lake bottom is within the epilimnion, dreissenids can be expected to have a strong influence on phytoplankton and nutrient dynamics throughout the summer-stratified period. In large lakes, the effect of nearshore dreissenid populations on whole-lake plankton and nutrient dynamics will depend on horizontal exchange rates between nearshore and offshore waters. These exchange rates will also determine the degree to which the nearshore dreissenid populations can process particles and nutrients entering directly from rivers, as they will affect the residence time of river plumes in the nearshore zone. Longer nearshore residence times may result in more efficient nutrient retention and a significant reduction of the effective nutrient load to offshore regions, as appears to occur as river water passes through Saginaw Bay into Lake Huron (Cha et al. 2011).

During the stratified period in large lakes, the direction of nearshore currents tends to be parallel to shore within several km of shore, minimizing nearshore–offshore exchange and trapping river effluent close to shore (Rao and Schwab 2007). However, nearshore–offshore exchange can occur as the result of internal seiches, wind-induced upwelling and downwelling, advection by coastal jets, and horizontal mixing resulting from current shear (Rao and Schwab 2007). Because these processes tend to be transient in space and time, they can be difficult to quantify. They are best characterized through combination of approaches including in situ current measurements, satellite imagery, and hydrodynamic modeling (e.g., Rao et al. 2002). Although several studies have considered the role of nearshore–offshore exchange in the transport of pollutants and sediment in the Great Lakes (Shen et al. 1995, Lou et al. 2000), interactions between nearshore dreissenid populations and the offshore plankton community have received little attention. While the expansion of dreissenids into offshore regions has affected plankton abundances in the pelagic waters of even the largest lakes (Fahnenstiel et al. 2010), there remains a need for quantitative data to demonstrate whether nearshore filtration

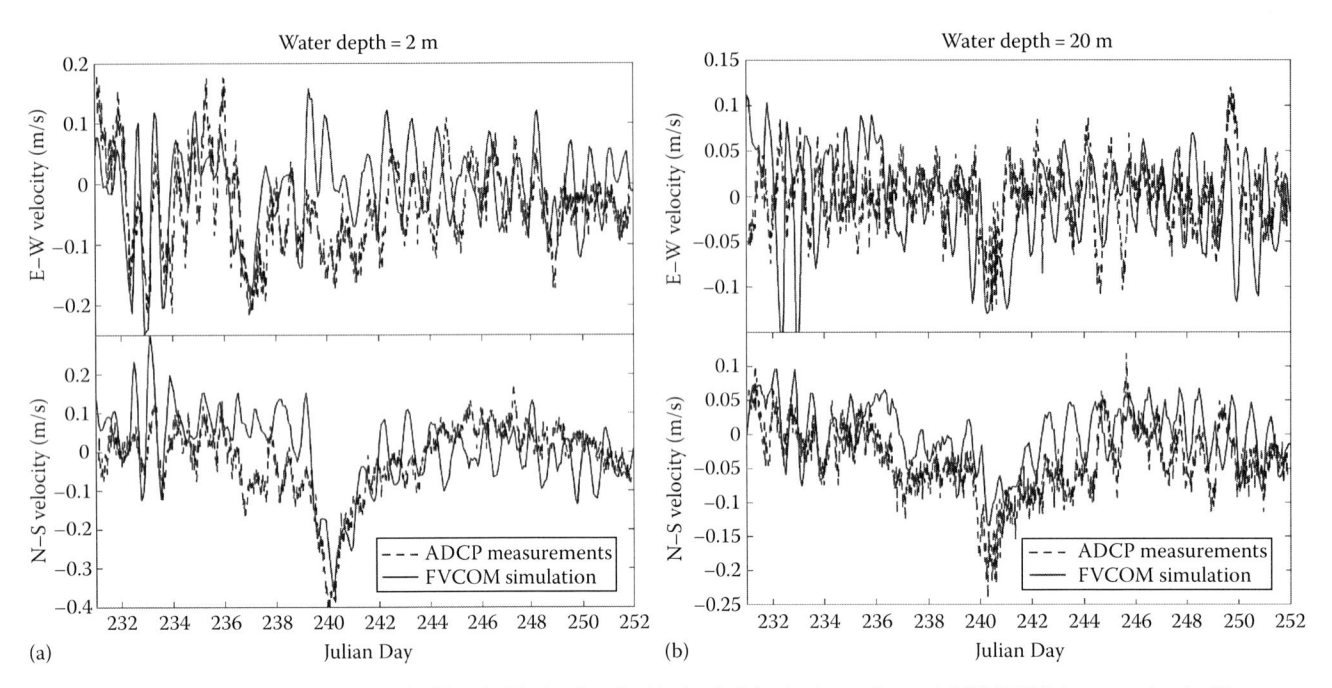

Figure 35.3 Comparison of current velocities (m/s) simulated with the 3-D hydrodynamic model (FVCOM) to current velocities measured with an ADCP at water depths of 2 m (a) and 20 m (b) near the harbor of Milwaukee in summer 2009. Upper panels: east–west current velocity; lower panels: north–south current velocity.

rates and nearshore–offshore exchange rates are sufficient to affect pelagic offshore waters (Barbiero et al. 2011).

To examine the potential impact of dreissenids on the nearshore plankton community and how this impact may be regulated by nearshore–offshore exchange, we adapted and calibrated an unstructured-grid Finite-Volume Coastal Ocean Model (FVCOM; Chen et al. 2003) to simulate the hydrodynamic circulation of Lake Michigan with an average horizontal resolution of 3 km and nearshore resolution of 1.5 km. Good agreement was found between the simulation and nearshore (20 m) ADCP data acquired near Milwaukee Harbor (Figure 35.3). Using simulation results, the horizontal and vertical mixing coefficients were obtained through a large eddy simulation approach (Smagorinsky 1963). Extracted horizontal and vertical mixing coefficients distributed over a cross-lake transect in August are shown in Figure 35.4. Extracted results were used to evaluate the relative importance of mussel filtration and horizontal exchange between nearshore and offshore as factors affecting plankton abundance in the nearshore. A vertical 2-D plane representing a transect starting from Milwaukee Harbor and extending eastward to a water depth of 70 m was selected as being representative of the Lake Michigan nearshore–offshore system. A 2-D turbulent diffusion-driven mass exchange model was then built on this transect. Mussels were assumed to cover the lake bottom with a density of 10,000 m^{-2} at depths of <40 m, and the clearance rate was set constant at 2.7 L mussel^{-1} day^{-1} (based on in situ measurements of particle concentration and particle filtration rate [Liao et al. 2009]). Particle (plankton) concentration (C) was set at 1.0 (C_0) everywhere as the initial condition and was kept constant at the eastern (offshore) boundary. A 30 day simulation was conducted for the month of August when the lake was stratified. Particle concentrations normalized to C_0 after 30 days are shown in Figure 35.5. It is evident that a strong gradient in particle concentration developed between the nearshore and offshore regions at about 6 km eastward of the shoreline (about 30 m water depth). A near-bottom gradient was also noticeable in the nearshore region. This indicates that the filtration capacity of the dreissenid population was significant and the time scale of particulate removal through filtration in the nearshore was shorter than the time scales of nearshore–offshore exchange. From these results, it is reasonable to infer that the impact of nearshore dreissenids on whole-lake nutrient dynamics and energy flow is also regulated by the strength of horizontal exchange.

Benthic Algal Response

While dreissenids have caused declines in phytoplankton abundance in many lakes, they have had the opposite effect on the benthic algal community in the nearshore. In smaller lakes and rivers, dreissenids appear to be responsible for increases in the abundance of macrophytes (Effler et al. 1997, Zhu et al. 2007). In the larger Laurentian Great Lakes, there has been an increase in the abundance of both macrophytes and benthic algae, primarily *Cladophora* sp., coincident with the increase in mussel populations in the nearshore (Skubinna

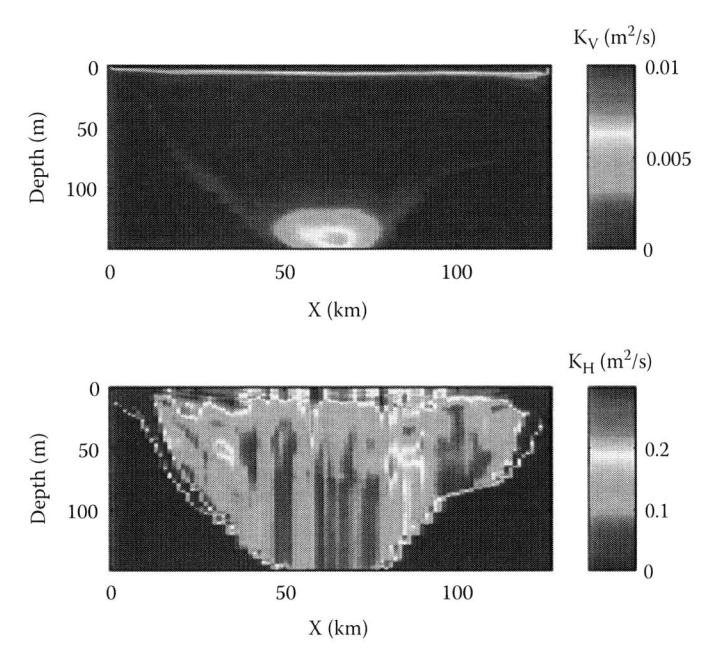

Figure 35.4 (See color insert.) Average monthly vertical (K_V, m²/s, top panel) and horizontal (K_H, m²/s, bottom panel) diffusivity in Lake Michigan as derived from the 3-D hydrodynamic simulation during strongly stratified conditions (August, 2009). The cross section is along a line that extends (in km) between Milwaukee, Wisconsin (left: west shoreline), and Muskegon, Michigan (right: east shoreline).

et al. 1995, Higgins et al. 2008, Auer et al. 2010). During the past decade, *Cladophora* biomass has increased to the nuisance levels observed in the 1960s and 1970s. However, while earlier *Cladophora* problems were generally associated with localized P sources such as river mouths and waste water treatment plants (Canale and Auer 1982), the recent problem appears to be more widespread (Auer et al. 2010). Analysis of historical data suggests that increased water clarity is a primary cause of this resurgence (Malkin et al. 2008, Auer et al. 2010). Increased light flux to the nearshore benthic community as a result of particulate removal by dreissenid filtration has increased the growth rate of *Cladophora* and expanded its depth range. This resurgence of *Cladophora* has occurred despite declines in the concentration of total and dissolved P in most parts of the Great Lakes (Dove 2009, Mida et al. 2010). However, without sufficient nutrients, *Cladophora* would not be able to take advantage of the improved light environment. Hecky et al. (2004) suggested that dreissenids may be a significant nutrient source for *Cladophora*, and measurements made in Lake Ontario indicate that P excretion by dreissenids is more than sufficient to meet the P demand of *Cladophora* in the nearshore region (Ozersky et al. 2009). Studies of other systems have shown that nutrient excretion by bivalves can indeed promote the growth of macroalgae and macrophytes (Peterson and Heck 2001, Pfister 2007, Vaughn et al. 2007). However, while dreissenids have the potential to be a significant nutrient source for *Cladophora* and other benthic algae, there has been no direct quantification of this nutrient pathway.

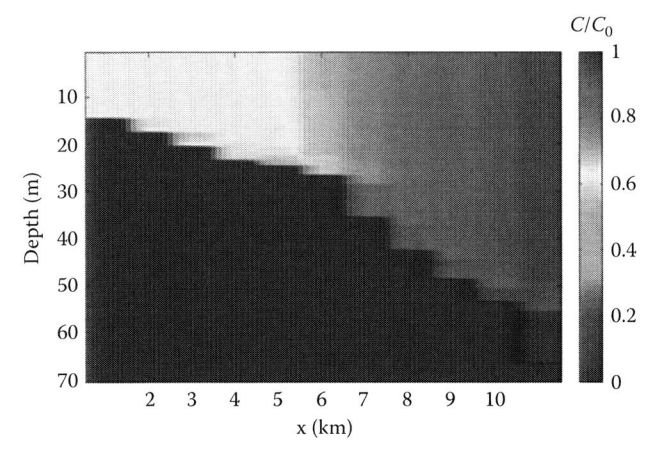

Figure 35.5 (See color insert.) Cross-section distribution of particle concentrations (*C*) normalized to a constant offshore concentration (C_0) after a 30 day simulation period as derived from a 2-D turbulent diffusion-driven mass exchange model. The area was located in Lake Michigan along the western shoreline near Milwaukee, Wisconsin, and extended from the shoreline eastward to a water depth of 70 m, a distance of about 11 km. See text for further details.

Just as vertical mixing regulates the supply of seston from the water column to bottom-dwelling dreissenids and hence the extent by which dreissenid filtration affects phytoplankton abundance, vertical mixing also regulates the fate of dissolved P excreted by dreissenids and the extent *Cladophora* may access this P. The proportion of excreted P that is assimilated by benthic algae will depend on the relative time scales of vertical mixing and P uptake kinetics. Under well-mixed conditions, mussel-derived P will dilute rapidly into the overlying water column, maintaining low concentrations in the benthic boundary layer and minimizing uptake by *Cladophora*. By contrast, under low mixing conditions, dissolved P will accumulate in the benthic boundary layer, allowing for greater uptake by *Cladophora*. The influence of mixing rate can be explored by constructing a simple mass balance model that simulates phosphorus dynamics within the near-bottom layer inhabited by mussels and *Cladophora*. Within this layer, mussel excretion serves as a SRP source, while uptake by *Cladophora* represents a loss. Vertical mixing with the overlying water column represents an input or output, depending on the SRP gradient (positive or negative) between the near-bottom layer and the overlying water column:

$$\frac{dC_1}{dt} = \frac{R - F - U}{z} \quad (37.1)$$

where
C_1 is the SRP concentration in the near-bottom layer
t is the time
R is the areal SRP excretion rate
F is the areal vertical SRP flux rate from the near-bottom layer to the overlying water
U is the areal SRP uptake rate by *Cladophora*
z is the thickness of the near-bottom layer

For the simulations presented here, R was set at 0.17 μg P m^{-2} s^{-1} (an approximate median of the Lake Michigan rates in Table 35.1). F was determined as

$$F = D(C_1 - C_0) \quad (37.2)$$

where
D is the vertical exchange coefficient at the top of the near-bottom layer
C_0 is the SRP concentration in the overlying water, which was set constant at 0.5 μg L^{-1}

Kinetics of SRP uptake by *Cladophora* were determined following the approach of Auer and Canale (1982):

$$U = \hat{U}\frac{P}{K_m + P} \quad (37.3)$$

where
$\hat{U}$ is the maximum phosphorus uptake rate
P is the SRP concentration in the near-bottom layer
K_m is the half-saturation constant for SRP uptake

Auer and Canale (1982) reported $\hat{U}$ values ranging from 0.2 to 1.64 μg P mgDW^{-1} h^{-1}, depending on the P content of *Cladophora*. Here we use an intermediate value of 0.6 μg P mgDW^{-1} h^{-1}. *Cladophora* biomass was set at 100 gDW m^{-2}, which is the approximate median of our measurements over the growing season in Lake Michigan. Like Auer and Canale (1982), we set K_m constant at 125 μg P L^{-1}. The value of z was set at 10 cm, which is the approximate thickness of *Cladophora* beds in Lake Michigan in midsummer.

The model was run for D values ranging between 1 and 50 cm^2 s^{-1}, which encompasses the vertical mixing rates that have been measured over dreissenid beds (Boegman et al. 2008, Liao et al. 2009). For each D value the model was run until steady state was reached, which occurred between ~2 and ~30 min, depending on mixing rate. Model results indicate that, at low mixing rates, P uptake by *Cladophora* may be several times greater than P flux to the water column (Figure 35.6). By contrast, at higher mixing rates, more P was lost to vertical flux than to *Cladophora* uptake. SRP uptake by *Cladophora* was never zero, because even at high mixing rates, when SRP concentration in the near-bottom layer approached the ambient value of 0.5 μg L^{-1}, *Cladophora* was able to assimilate SRP. However, model results indicated that *Cladophora* benefits much more from SRP excretion by dreissenids under slow mixing conditions than under rapid mixing conditions.

These results highlight the potential influence of near-bottom hydrodynamics on the relationship between nutrient excretion by dreissenids and nutrient uptake by *Cladophora*. While dreissenids may provide a large portion of *Cladophora* P demand, there remains a need for quantitative measurements of the nutrient flux between

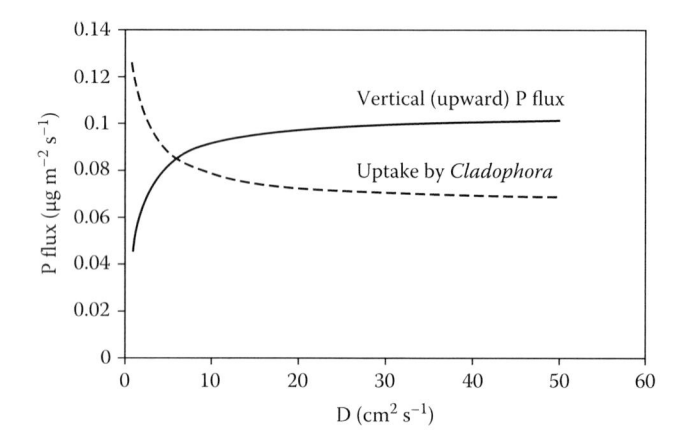

Figure 35.6 Simulated relationship between near-bottom, vertical mixing rates (D, cm^{2}s^{-1}) and the relative fate of dissolved P excreted by dreissenid mussels. The areal flux rate of excreted P (μg m^{-2} s^{-1}) was determined, and the vertical flux rate of P to the water column (solid line) was compared to P uptake by *Cladophora* (dashed line) at different vertical mixing rates. See text for further details.

dreissenids and *Cladophora* and how this flux is modulated by physical processes. Low mixing rates may enhance the ability of *Cladophora* to directly utilize dreissenid-derived nutrients, but such conditions also tend to result in reduced food supply to dreissenids due to increased water column stratification and greater re-filtration of near-bottom water (Boegman et al. 2008). This reduced food supply would likely lead to lower P excretion rates that may offset greater retention of P within *Cladophora* beds. Hence, the net effect of near-bottom mixing on the nutrient relationship between dreissenid and *Cladophora* remains unclear.

MANAGEMENT IMPLICATIONS

Since the arrival of dreissenids in North America over 20 years ago, a large amount of experimental work has greatly improved our understanding of dreissenid physiology and ecology and of the role these bivalves play in the functioning of whole ecosystems. As a result of their high densities, rapid reproduction, high metabolic rates (and hence, filtration rates), and relative resistance to predation, these organisms play a major role in the nutrient dynamics and trophic structure of many rivers and lakes. This role appears to have become even more significant in systems where zebra mussels have been replaced by quagga mussels. The latter species is more widely distributed due to its ability to colonize both hard and soft substrates and to reproduce and grow in cold, hypolimnetic waters (Claxton and Mackie 1998). However, there remains some uncertainty about long-term impacts of dreissenids, especially in large lakes where relatively long residence times of water and nutrients result in long times for these lakes to approach steady state following a major disturbance. In Lake Erie, with a hydraulic residence time of just over two years, it took more than 15 years for alkalinity to approach a return to steady state following the establishment of dreissenids (Barbiero et al. 2006). Depending on the turnover time of the dreissenid nutrient pool, it may take a similar amount of time for nutrient dynamics to return to a steady state. This response time will be even greater in larger lakes (i.e., Lakes Michigan, Huron, and Ontario) where nutrient residence times are longer and where dreissenid populations have continued to expand over the past decade (Wilson et al. 2006, Nalepa et al. 2007, 2009). The immediate effects of dreissenids on plankton abundance and nutrient recycling have been measured in many lakes and rivers, but in large lakes the short-term effects may differ in magnitude or nature from the long-term, ecosystem-scale effects, due to long residence times and lags in biological and biogeochemical feedbacks.

A critical question is whether the decrease in total nutrient concentration that has been observed in many lakes following establishment of dreissenids (Higgins and Vander Zanden 2010) is due to temporary sequestration in shells and tissues of living mussels or to increased burial in sediments as feces, pseudofeces, and dead mussel material. If the former mechanism is dominant, then total nutrient concentrations may partially rebound once populations reach steady state and the decomposition rate of dead mussels equals the production rate of new mussels. But if mussels promote more efficient burial of nutrients, then total nutrient concentrations will remain low, even after populations and nutrient dynamics approach new steady states. Data from European systems, where dreissenids have been established for a much longer period of time, may provide some insight into long-term effects. For example, many North American studies have observed an upward shift in SRP–chlorophyll *a* ratios due to dreissenid filtration, but Mellina et al. (1995) point out that this observation is not as common in European lakes. If this difference is indeed real, there are several mechanisms that may explain it, including higher overall densities of dreissenids in North American lakes (Ramcharan et al. 1992) and higher production–biomass ratios in North American populations (at least those that are still growing exponentially; Mackie and Schloesser 1996).

The ability of dreissenids to reallocate nutrients in aquatic ecosystems has fundamentally changed the structure and function of these systems. This presents a major challenge for researchers and managers. For over half a century, nutrient control has been a focal point for lake management efforts. These efforts have been underpinned by a basic understanding of how critical lake properties, including algal abundance and species composition, dissolved oxygen concentration, and fish production, are linked to nutrient loads and internal nutrient cycles. While the full impact of dreissenids on these relationships continues to be explored, it is obvious that conventional paradigms of nutrient dynamics are being challenged. In the Great Lakes, a particularly challenging conundrum is the simultaneous decrease in pelagic phytoplankton in nearshore and offshore regions and resurgence of nuisance benthic algae, primarily *Cladophora* sp., in the nearshore region. In the 1960s and 1970s, excessive algal growth was a problem in both the benthic and pelagic zones of the nearshore region in the Great Lakes, and both of these zones benefited from phosphorus abatement. This is no longer the case, and it has become more difficult to define a phosphorus loading target that supports pelagic plankton and fish populations while minimizing nuisance algal growth in the nearshore region. Dreissenid filtration has likely altered the relationship between P loading and pelagic phytoplankton concentration, and biogeochemical models will need to be revised to account for this shift. In the nearshore region, increased water clarity resulting from dreissenid filtration has led to increased growth rates of benthic algae, despite the fact that total P concentrations in the water column are at or below target levels

(Auer et al. 2010). This raises the question: is it possible to obtain P loading rates and P concentrations in the water column that will result in acceptably low *Cladophora* biomass? To answer this question, it will be necessary to identify and quantify pathways that link P loading to *Cladophora* growth. In particular, a better understanding of factors that regulate near-bottom dissolved P concentrations in the nearshore zone is required. In addition, models of *Cladophora* dynamics (Higgins et al. 2005, Malkin et al. 2008, Tomlinson et al. 2010) must account for these near-bottom P dynamics. If mussels are able to maintain relatively high concentrations of dissolved P in near-bottom waters where *Cladophora* grows, even when P concentrations in the rest of the water column are low, then high *Cladophora* biomass may be the new "normal" for at least some areas of the Great Lakes (Ozersky et al. 2009). However, large-scale gradients of nutrient concentrations and *Cladophora* biomass in nearshore Lake Michigan (Greb et al. 2005) and apparent localized effects of external P loading on *Cladophora* biomass (Higgins et al. 2012) suggest that acceptable levels of *Cladophora* biomass might be attainable with moderate reductions in P loading. As discussed earlier, such reductions may exacerbate the decline of pelagic plankton populations. Revised biogeochemical models will allow managers to better determine optimal nutrient loads and the time scales over which lakes will respond to changes in nutrient loads and internal cycling by dreissenid mussels.

ACKNOWLEDGMENTS

Previously unpublished data for Lake Michigan presented in this chapter were collected through the research supported by the National Science Foundation (Grant No. OCE 0826477), the Environmental Protection Agency Great Lakes National Program Office (Project GL-00E06901), the Milwaukee Metropolitan Sewerage District, the Wisconsin Coastal Management Program, and the National Oceanic and Atmospheric Administration, Office of Ocean and Coastal Resource Management under the Coastal Zone Management Act, Grant # NA05N0S4191067.

REFERENCES

Ackerman, J. D., M. Loewen, and P. Hamblin. 2001. Benthic-pelagic coupling over a zebra mussel reef in Western Lake Erie. *Limnol. Oceanogr.* 46: 892–904.

Aldridge, D. W., B. S. Payne, and A. C. Miller. 1995. Oxygen consumption, nitrogenous excretion, and filtration rates of *Dreissena polymorpha* at acclimation temperature between 20 and 32°C. *Can. J. Fish. Aquat. Sci.* 52: 1761–1767.

Alexander, J. E. J., J. H. Thorp, and R. D. Fell. 1994. Turbidity and temperature effects on oxygen consumption in the zebra mussel (*Dreissena polymorpha*). *Can. J. Fish. Aquat. Sci.* 51: 179–184.

Arnott, D. L. and M. J. Vanni. 1996. Nitrogen and phosphorus recycling by the zebra mussel (*Dreissena polymorpha*) in the western basin of Lake Erie. *Can. J. Fish. Aquat. Sci.* 53: 646–659.

Asmus, H., R. M. Asmus, and K. Reise. 1990. Exchange processes in an intertidal mussel bed: A Sylt-flume study in the Wadden Sea. *Ber. Biol. Anst. Helgoland.* 6: 1–79.

Auer, M. T. and R. P. Canale. 1982. Ecological studies and mathematical modeling of *Cladophora* in Lake Huron: 2. Phosphorus uptake kinetics. *J. Great Lakes Res.* 8: 84–92.

Auer, M. T., L. M. Tomlinson, S. N. Higgins, S. Y. Malkin, E. T. Howell, and H. A. Bootsma. 2010. Great Lakes *Cladophora* in the 21st century: Same algae, different ecosystem. *J. Great Lakes Res.* 36: 248–255.

Baker, S. M., J. S. Levinton, J. P. Kurdziel, and S. E. Shumway. 1998. Selective feeding and biodeposition by zebra mussels and their relation to changes in phytoplankton composition and seston load. *J. Shellfish Res.* 17: 1207–1213.

Baldwin, B. S., M. S. Mayer, J. Dayton et al. 2002. Comparative growth and feeding in zebra and quagga mussels (*Dreissena polymorpha* and *Dreissena bugensis*): Implications for North American lakes. *Can. J. Fish. Aquat. Sci.* 59: 680–694.

Barbiero, R. P., B. M. Lesht, and G. J. Warren. 2011. Evidence for bottom-up control of recent shifts in the pelagic food web of Lake Huron. *J. Great Lakes Res.* 37: 78–85.

Barbiero, R. P., M. L. Tuchman, and E. S. Millard. 2006. Post-dreissenid increases in transparency during summer stratification in the offshore waters of Lake Ontario: Is a reduction in whiting events the cause? *J. Great Lakes. Res.* 32: 131–141.

Bayne, B. L. and C. Scullard. 1977. Rates of nitrogen excretion by species of *Mytilus* (Bivalvia: Mollusca). *J. Mar. Biol. Assoc. U.K.* 57: 355–369.

Berg, D. J., S. W. Fisher, and P. F. Landrum. 1996. Clearance and processing of algal particles by zebra mussels (*Dreissena polymorpha*). *J. Great Lakes Res.* 22: 779–788.

Bierman, V. J., J. J. Kaur, J. V. DePinto, T. J. Feist, and D. W. Dilks. 2005. Modeling the role of zebra mussels in the proliferation of blue-green algae in Saginaw Bay, Lake Huron. *J. Great Lakes Res.* 31: 32–55.

Boegman, L., M. R. Loewen, P. F. Hamblin, and D. A. Culver. 2008. Vertical mixing and weak stratification over zebra mussel colonies in western Lake Erie. *Limnol. Oceanogr.* 53: 1093–1110.

Brabrand, A., B. Faafeng, and J. Nilssen. 1990. Relative importance of phosphorus supply to phytoplankton production: Fish excretion versus external loading. *Can. J. Fish. Aquat. Sci.* 47: 364–372.

Brooks, A. and D. Edgington. 1994. Biogeochemical control of phosphorus cycling and primary production in Lake Michigan. *Limnol. Oceanogr.* 39: 961–968.

Bruesewitz, D. A., J. L. Tank, M. J. Bernot, W. B. Richardson, and E. A. Strauss. 2006. Seasonal effects of the zebra mussel (*Dreissena polymorpha*) on sediment denitrification rates in Pool 8 of the Upper Mississippi River. *Can. J. Fish. Aquat. Sci.* 63: 957–969.

Bunt, C. M., H. MacIsaac, and G. Sprules. 1993. Pumping Rates and Projected filtering impacts of Juvenile Zebra Mussels (*Dreissena polymorpha*) in Western Lake Erie. *Can J. Fish. Aquat. Sci.* 50: 1017–1022.

Bykova, O., A. Laursen, V. Bostan, J. Bautista, and L. McCarthy. 2006. Do zebra mussels (*Dreissena polymorpha*) alter lake water chemistry in a way that favours *Microcystis* growth? *Sci Total Environ.* 371: 362–372.

Canale, R. P. and M. T. Auer. 1982. Ecological studies and mathematical modeling of *Cladophora* in Lake Huron: 5. Model development and calibration. *J. Great Lakes Res.* 8: 112–125.

Canale, R. P. and S. C. Chapra. 2002. Modeling zebra mussel impacts on water quality of Seneca River, New York. *J. Environ. Eng.* 128: 1158–1168.

Carpenter, S., D. L. Christensen, J. Cole et al. 1995. Biological control of eutrophication in Lakes. *Environ. Sci. Technol.* 29: 784–786.

Cha, Y., C. A. Stow, T. F. Nalepa, and K. H. Reckhow. 2011. Do invasive mussels restrict offshore phosphorus transport in Lake Huron? *Environ. Sci. Technol.* 45: 7226–7231.

Chen, C., H. Liu, and R. C. Beardsley. 2003. An unstructured grid, finite-volume, three-dimensional, primitive equations ocean model: Application to coastal ocean and estuaries. *J. Atmos. Oceanic Technol.* 20: 159–186.

Chen, C., J. Rubao, D. Schwab et al. 2002. A model study of the coupled biological and physical dynamics in Lake Michigan. *Ecol. Model.* 152: 145–168.

Claxton, W. T. and G. L. Mackie. 1998. Seasonal and depth variations in gametogenesis and spawning of *Dreissena polymorpha* and *Dreissena bugensis* in eastern Lake Erie. *Can. J. Fish. Aquat. Sci.* 76: 2010–2019.

Conroy, J. D., W. J. Edwards, R. A. Pontius et al. 2005. Soluble nitrogen and phosphorus excretion of exotic freshwater mussels (*Dreissena* spp.): Potential impacts for nutrient remineralisation in western Lake Erie. *Freshwat. Biol.* 50: 1146–1162.

Davis, W. R., A. D. Christian, and D. J. Berg. 2000. Nitrogen and phosphorus cycling by freshwater mussels in a headwater stream ecosystem. In *Freshwater Mollusk Symposium Proceedings: Part II, Musseling in on Biodiversity*, R. A. Tankersley, D. I. Warmotts, and G. T. Watters et al., eds., pp. 141–151. Ohio Biological Survey Special Publication, Columbus, OH.

Descy, J. P., E. Everbecq, V. Gosselain, L. Viroux, and J. S. Smitz. 2003. Modelling the impact of benthic filter-feeders on the composition and biomass of river plankton. *Freshwat. Biol.* 48: 404–417.

Domm, S., R. W. McCauley, E. Kott, and J. D. Ackerman. 1993. Physiological and taxonomic separation of two dreissenid mussels in the Laurentian Great Lakes. *Can. J. Fish. Aquat. Sci.* 50: 2294–2297.

Dove, A. 2009. Long-term trends in major ions and nutrients in Lake Ontario. *Aquat. Ecosyst. Health Manage.* 12: 281–295.

Eadie, B. J., D. J. Schwab, T. H. Johengen et al. 2002. Particle transport, nutrient cycling, and algal community structure associated with a major winter-spring sediment resuspension event in southern Lake Michigan. *J. Great Lakes Res.* 28: 324–337.

Edwards, W. J., C. R. Rehmann, E. McDonald, and D. A. Culver. 2005. The impact of a benthic filter feeder: Limitations imposed by physical transport of algae to the benthos. *Can. J. Fish. Aquat. Sci.* 62: 205–214.

Effler, S. W., S. R. Boone, C. A. Siegfried, L. Walrath, and S. L. Ashby. 1997. Mobilization of ammonia and phosphorus by zebra mussels (*Dreissena polymorpha*) in the Seneca River, New York. In *Zebra Mussels and Aquatic Nuisance Species*, F. M. Di'Itri, ed., pp. 187–208. Chelsea, MI: Ann Arbor Press.

Effler, S. W., D. A. Matthews, C. M. Brooks-Matthews, M. Perkins, C. A. Siegfried, and J. M. Hassett. 2004. Water quality impacts and indicators of metabolic activity of the zebra mussel invasion of the Seneca River. *J. Am. Water Res. Assoc.* 40: 737–754.

Elser, J. J. and R. P. Hassett. 1994. A stoichiometric analysis of the zooplankton-phytoplankton interaction in marine and freshwater ecosystems. *Nature* 370: 211–213.

Evans, M. A., G. Fahnenstiel, and D. Scavia. 2011. Incidental oligotrophication of North American Great Lakes. *Environ. Sci. Technol.* 45: 3279–3303.

Fahnenstiel, G., S. Pothoven, H. Vanderploeg, D. Klarer, T. Nalepa, and D. Scavia. 2010. Recent changes in primary production and phytoplankton in the offshore region of southeastern Lake Michigan. *J. Great Lakes Res.* 36: 20–29.

Fanslow, D. L., T. F. Nalepa, and G. A. Lang. 1995. Filtration rates of the zebra mussel (*Dreissena polymorpha*) on natural seston from Saginaw Bay, Lake Huron. *J. Great Lakes Res.* 21: 489–500.

Greb, S., P. Garrison, and S. Pfeiffer. 2005. *Cladophora* and water quality of Lake Michigan: A systematic survey of Wisconsin nearshore areas. In *Cladophora Research and Management in the Great Lakes, Special Report 2005-0*, H. A. Bootsma, E. T. Jensen, E. B. Young, and J. A. Berges, eds., pp. 73–80. Milwaukee, WI: Great Lakes WATER Institute, University of Wisconsin-Milwaukee.

Hall, S. R., J. K. Pauliukonis, E. L. Mills et al. 2003. A comparison of total phosphorus, chlorophyll a, and zooplankton in embayment, nearshore, and offshore habitats of Lake Ontario. *J. Great Lakes Res.* 29: 54–69.

Hamburger, K., F. Møhlenberg, A. Randløv, and H. U. Riisgård. 1983. Size, oxygen consumption and growth in the mussel *Mytilus edulis*. *Mar. Biol.* 75: 303–306.

Healey, F. P. and L. L. Hendzel. 1980. Physiological indicators of nutrient deficiency in lake phytoplankton. *Can. J. Fish. Aquat. Sci.* 31: 442–453.

Heath, R., G. Fahnenstiel, W. Gardner, J. Cavaletto, and S.-J. Hwang. 1995. Ecosystem-level effects of zebra mussels (*Dreissena polymorpha*): An enclosure experiment in Saginaw Bay, Lake Huron. *J. Great Lakes Res.* 21: 501–516.

Hecky, R. E., R. Smith, D. R. Barton et al. 2004. The nearshore phosphorus shunt: A consequence of ecosystem engineering by dreissenids in the Laurentian Great Lakes. *Can. J. Fish. Aquat. Sci.* 61: 1285–1293.

Higgins, S. N., R. E. Hecky, and S. J. Guildford. 2005. Modeling the growth, biomass and tissue phosphorus concentration of *Cladophora glomerata* in eastern Lake Erie: Model description and field testing. *J. Great Lakes Res.* 31: 439–455.

Higgins, S. N., S. Y. Malkin, E. T. Howell et al. 2008. An ecological review of *Cladophora glomerata* (Chlorophyta) in the Laurentian Great Lakes. *J. Phycol.* 44: 839–854.

Higgins, S. N., C. M. Pennuto, E. T. Howell, T. Lewis, and J. C. Madarewicz. 2012. Urban influences on *Cladophora* blooms in Lake Ontario. *J. Great Lakes Res.* 38: 116–123.

Higgins, S. N. and M. J. Vander Zanden. 2010. What a difference a species makes: A meta-analysis of dreissenid mussel impacts on freshwater ecosystems. *Ecol. Monogr.* 80: 179–196.

Holland, R. E., T. H. Johengen, and A. M. Beeton. 1995. Trends in nutrient concentrations in Hatchery Bay, western Lake Erie, before and after *Dreissena polymorpha*. *Can. J. Fish. Aquat. Sci.* 52: 1202–1209.

James, W. F., J. W. Barko, and H. L. Eakin. 1997. Nutrient regeneration by the zebra mussel (*Dreissena polymorpha*). *J. Freshwat. Ecol.* 12: 209–216.

James, W. F., J. W. Barko, and H. L. Eakin. 2001. Phosphorus recycling by zebra mussels in relation to density and food resource availability. *Hydrobiologia* 455: 55–60.

Johengen, T., T. Nalepa, G. Fahnenstiel, and G. Goudy. 1995. Nutrient changes in Saginaw Bay, Lake Huron, after the establishment of the zebra mussel (*Dreissena polymorpha*). *J. Great Lakes Res.* 21: 449–464.

Johengen, T. H., H. A. Vanderploeg, and J. R. Liebig. 2013. Effects of algal composition, seston stoichiometry, and feeding rate on zebra mussel (*Dreissena polymorpha*) nutrient excretion in two Laurentian Great Lakes. In *Quagga and Zebra Mussels: Biology, Impacts, and Control*, 2nd Edn., T. F. Nalepa and D. W. Schloesser, eds., pp. 445–459. Boca Raton, FL: CRC Press.

Kiibus, M. and N. Kautsky. 1996. Respiration, nutrient excretion and filtration rate of tropical freshwater mussels and their contribution to production and energy flow in Lake Kariba, Zimbabwe. *Hydrobiologia* 331: 25–32.

Knoll, L. B., O. Sarnelle, S. K. Hamilton et al. 2008. Invasive zebra mussels (*Dreissena polymorpha*) increase cyanobacterial toxin concentrations in low-nutrient lakes. *Can. J. Fish. Aquat. Sci.* 65: 448–455.

Kuenzler, E. J. 1961. Phosphorus budget of a mussel population. *Limnol. Oceanogr.* 6: 400–415.

Lauritsen, D. D. and S. C. Mozley. 1989. Nutrient excretion by the Asiatic clam *Corbicula fluminea*. *J. N. Am. Benthol. Soc.* 8: 134–139.

Lavrentyev, P. J., W. S. Gardner, J. F. Cavaletto, and J. R. Beaver. 1995. Effects of the zebra mussel (*Dreissena polymorpha Pallas*) on protozoa and phytoplankton from Saginaw Bay, Lake Huron. *J. Great Lakes Res.* 21: 545–557.

Lei, J., B. S. Payne, and S. Y. Wang. 1996. Filtration dynamics of the zebra mussel, *Dreissena polymorpha*. *Can. J. Fish. Aquat. Sci.* 53: 29–37.

Liao, Q., H. A. Bootsma, J. Xiao et al. 2009. Development of an in situ underwater particle image velocimetry (UWPIV) system. *Limnol. Oceanogr. Methods*. 7: 169–184.

Limén, H., C. D. A. van Overdijk, and H. J. MacIsaac. 2005. Food partitioning between the amphipods *Echinogammarus ischnus*, *Gammarus fasciatus*, and *Hyalella azteca* as revealed by stable isotopes. *J. Great Lakes Res.* 31: 97–104.

Lohner, R. N., V. Sigler, C. M. Mayer, and C. Balogh. 2007. A comparison of the benthic bacterial communities within and surrounding *Dreissena* clusters in lakes. *Microb. Ecol.* 54: 469–477.

Lou, J., D. J. Swhwab, D. Beletsky, and N. Hawley. 2000. A model of sediment resuspension and transport dynamics in southern Lake Michigan. *J. Geophys. Res.* 105: 6591–6610.

MacIsaac, H. J., W. G. Sprules, O. E. Johannsson, and J. H. Leach. 1992. Filtering impacts of larval and sessile zebra mussels (*Dreissena polymorpha*) in western Lake Erie. *Oecologia* 92: 30–39.

Mackie, G. L. and D. W. Schloesser. 1996. Comparative biology of zebra mussels in Europe and North America: An overview. *Am. Zool.* 36: 244–258.

Madenjian, C. P. 1995. Removal of algae by the zebra mussel (*Dreissena polymorpha*) population in western Lake Erie: A bioenergetics approach. *Can. J. Fish. Aquat. Sci.* 52: 381–390.

Madon, S. P., D. W. Schneider, J. A. Stoeckel, and R. E. Sparks. 1998. Effects of inorganic sediment and food concentrations on energetic processes of the zebra mussel, *Dreissena polymorpha*: Implications for growth in turbid rivers. *Can. J. Fish. Aquat. Sci.* 55: 401–413.

Magni, P., S. Montani, C. Takada, and H. Tsutsumi. 2000. Temporal scaling and relevance of bivalve nutrient excretion on a tidal flat o f the Seto Inland Sea, Japan. *Mar. Ecol. Prog. Ser.* 198: 139–155.

Makarewicz, J. C., P. Bertram, and T. W. Lewis. 2000. Chemistry of the offshore waters of Lake Erie: Pre- and post-*Dreissena* introduction (1983–1993). *J. Great Lakes Res.* 26: 82–93.

Malkin, S. Y., S. J. Guildford, and R. E. Hecky. 2008. Modeling the growth response of *Cladophora* in a Laurentian Great Lake to the exotic invader *Dreissena* and to lake warming. *Limnol. Oceanogr.* 53: 1111–1124.

Mellina, E., J. B. Rasmussen, and E. L. Mills. 1995. Impact of mussel (*Dreissena polymorpha*) on phosphorus cycling and chlorophyll in lakes. *Can. J. Fish. Aquat. Sci.* 52: 2553–2573.

Mida, J. L., D. Scavia, G. L. Fahnenstiel, S. A. Pothoven, H. A. Vanderploeg, and D. M. Dolan. 2010. Long-term and recent changes in southern Lake Michigan water quality with implications for present trophic status. *J. Great Lakes Res.* 36 (Suppl. 3): 42–49.

Mills, E. L., J. R. Chrisman, B. S. Baldwin, R. W. Owens, R. O'Gorman, T. Howell, E. F. Roseman, and M. K. Raths. 1999. Changes in the dreissenid community in the lower Great Lakes with emphasis on southern Lake Ontario. *J. Great Lakes Res.* 25: 187–197.

Naddafi R, P. Eklöv, and K. Pettersson. 2009. Stoichiometric constraints do not limit successful invaders: Zebra mussels in Swedish lakes. *PLoS One* 4: e5345. doi:10.1371/journal.pone.0005345.

Naddafi, R., K. Pettersson, and P. Eklöv. 2008. Effects of the zebra mussel, an exotic freshwater species, on seston stoichiometry. *Limnol. Oceanogr.* 53: 1973–1987.

Nalepa, T. F., D. L. Fanslow, and G. A. Lang. 2009. Transformation of the offshore benthic community in Lake Michigan: Recent shift from the native amphipod *Diporeia* spp. to the invasive mussel *Dreissena rostriformis bugensis*. *Freshwat. Biol.* 54: 466–479.

Nalepa, T. F., D. L. Fanslow, and S. A. Pothoven. 2010. Recent changes in density, biomass, recruitment, size structure, and nutritional state of *Dreissena* populations in southern Lake Michigan. *J. Great Lakes Res.* 36 (Suppl. 3): 5–19.

Nalepa, T. F., D. L. Fanslow, S. A. Pothoven, A. J. Foley, III, and G. A. Lang. 2007. Long-term trends in benthic macroinvertebrate populations in Lake Huron over the past four decades. *J. Great Lakes Res.* 33: 421–436.

Nalepa. T. F., W. S. Gardner, and J. M. Malczyk. 1991. Phosphorus cycling by mussels (Unionidae: Bivalvia) in Lake St. Clair. *Hydrobiologia* 219: 239–250.

Newell, R. I. E., T. R. Fisher, R. R. Holyoke, and J. C. Cornwell. 2005. Influence of eastern oysters on nitrogen and phosphorus regeneration in Chesapeake Bay, USA. In *The Comparative Roles of Suspension Feeders in Ecosystems*. R. Dame and S. Olenin, eds., pp. 93–120. Dordrecht, The Netherlands: Springer.

Nicholls, K. H., L. Heintsch, and E. Carney. 2002. Univariate step-trend and multivariate assessments of the apparent effects of P loading reductions and zebra mussels on the phytoplankton of the Bay of Quinte, Lake Ontario. *J. Great Lakes Res.* 28: 15–31.

Orlova, M., S. Golubkov, L. Kalinina, and N. Ignatieva. 2004. *Dreissena polymorpha* (Bivalvia: Dreissenidae) in the Neva Estuary (eastern Gulf of Finland, Baltic Sea): Is it a biofilter or source for pollution? *Mar. Pollut. Bull.* 49: 196–205.

Ozersky, T., S. Y. Malkin, D. R. Barton, and R. E. Hecky. 2009. Dreissenid phosphorus excretion can sustain C. *glomerata* growth along a portion of Lake Ontario shoreline. *J. Great Lakes Res.* 35: 321–328.

Padilla, D. K., S. C. Adolph, K. L. Cottingham, and D. W. Schneider. 1996. Predicting the consequences of dreissenid mussels on a pelagic food web. *Ecol. Model.* 85: 129–144.

Pauer, J. J., A. M. Anstead, W. Melendez, R. Rossmann, K. W. Taunt, and R. G. Kreis. 2008. The Lake Michigan eutrophication model, LM2-Eutro: Model development and calibration. *Water Environ. Res.* 80: 853–861.

Peterson, B. J. and K. L. Heck Jr. 2001. Positive interactions between suspension-feeding bivalves and seagrass—A facultative mutualism. *Mar. Ecol. Prog. Ser.* 213: 143–155.

Pfister, C. A. 2007. Intertidal invertebrates locally enhance primary production. *Ecology* 88: 1647–1653.

Prins, T. C. and A. C. Smaal. 1994. The role of the blue mussel *Mytilus edulis* in the cycling of nutrients in Oosterschelde estuary (The Netherlands). *Hydrobiologia* 282/283: 413–429.

Quigley, M. A., W. S. Gardner, and W. M. Gordon. 1993. Metabolism of the zebra mussel (*Dreissena polymorpha*) in Lake St. Clair of the Great Lakes. In *Zebra Mussels: Biology, Impacts, and Control*, T. F. Nalepa and D. W. Schloesser, eds., pp. 295–306. Boca Raton, FL: CRC Press.

Raikow, D. R., O. Sarnelle, A. E. Wilson, and S. K. Hamilton. 2004. Dominance of the noxious cyanobacterium Microcystis aeruginosa in low-nutrient lakes is associated with exotic zebra mussels. *Limnol. Oceanogr.* 49: 482–487.

Ramcharan, E. W., D. K. Padilla, and S. I. Dodson. 1992. Models to predict potential occurrence and density of the zebra mussel, *Dreissena polymorpha*. *Can. J. Fish. Aquat. Sci.* 49: 2611–2620.

Rao, Y. R., C. R. Murthy, M. J. McCormick, G. S. Miller, and J. H. Saylor. 2002. Observations of circulation and coastal exchange characteristics in southern Lake Michigan during 2000 winter season. *Geophys. Res. Let.* 29: 1631, doi:1610.1029/2002GL014895.

Rao, Y. R. and D. J. Schwab. 2007. Transport and mixing between the coastal and offshore waters in the Great Lakes: A review. *J. Great Lakes Res.* 33: 202–218.

Robbins, J. A., J. R. Krezoski, and S. C. Mozley. 1977. Radioactivity in sediments of the Great Lakes: Post-depositional redistribution by deposit-feeding organisms. *Earth Planet Sci. Lett.* 36: 325–333.

Roe, S. L. and H. MacIsaac. 1997. Deepwater population structure and reproductive state of quagga mussels (*Dreissena bugensis*) in Lake Erie. *Can J. Fish. Aquat. Sci.* 54: 2428–2433.

Schneider, D. W. 1992. Bioenergetic model of zebra mussel, *Dreissena polymorpha*, growth in the Great Lakes. *Can. J. Fish. Aquat. Sci.* 49: 1406–1416.

Schneider, D. W., S. P. Madon, J. A. Stoeckel, and R. E. Sparks. 1998. Seston quality controls zebra mussel (*Dreissena polymorpha*) energetics in turbid rivers. *Oecologia* 117: 331–341.

Shen, H., I. K. Tsanis, and M. D'Andrea. 1995. A three-dimensional nested hydrodynamic/pollutant transport simulation model for the nearshore areas of Lake Ontario. *J. Great Lakes Res.* 21: 161–177.

Skubinna, J. P., T. G. Coon, and T. R. Batterson. 1995. Increased abundance and depth of submersed macrophytes in response to decreased turbidity in Saginaw Bay, Lake Huron. *J. Great Lakes Res.* 21: 476–488.

Smagorinsky, J. 1963. General circulation experiments with the primitive equations. *Mon. Weather Rev.* 91: 99–164.

Spidle, A. P., E. L. Mills, and B. May. 1995. Limits to tolerance of temperature and salinity in the quagga mussel (*Dreissena bugensis*) and the zebra mussel (*Dreissena polymorpha*). *Can. J. Fish. Aquat. Sci.* 52: 2108–2119.

Sprung, M. 1995. Physiological energetics of the zebra mussel *Dreissena polymorpha* in lakes. II. Food uptake and gross growth efficiency. *Hydrobiologia* 304: 133–146.

Sprung, M. and U. Rose. 1988. Influence of food size and food quality on the feeding of the mussel *Dreissena polymorpha*. *Oecologia* 77: 526–532.

Stańczykowska, A. and K. Lewandowski. 1993. Effect of filtering activity of *Dreissena polymorpha* (Pall.) on the nutrient budget of the littoral of Lake Mikołajskie. *Hydrobiologia* 251: 73–79.

Stoeckmann, A. M. 2003. Physiological energetics of Lake Erie dreissenid mussels: A basis for the displacement of *Dreissena polymorpha* by *Dreissena bugensis*. *Can. J. Fish. Aquat. Sci.* 60: 126–134.

Stoeckmann, A. M. and D. W. Garton. 1997. A seasonal energy budget for zebra mussels (*Dreissena polymorpha*) in western Lake Erie. *Can. J. Fish. Aquat. Sci.* 54: 2743–2751.

Thayer, S. A., R. C. Haas, R. D. Hunter, and R. H. Kushler. 1997. Zebra mussel (*Dreissena polymorpha*) effects on sediment, other zoobenthos, and the diet and growth of adult yellow perch (*Perca flavescens*) in pond enclosures. *Can. J. Fish. Aquat. Sci.* 54: 1903–1915.

Tomlinson, L. M., M. T. Auer, H. A. Bootsma, and E. M. Owens. 2010. The Great Lakes *Cladophora* model: Development, testing, and application to Lake Michigan. *J. Great Lakes Res.* 36: 287–297.

Turner, C. B. 2010. Influence of zebra (*Dreissena polymorpha*) and quagga (*Dreissena rostriformis*) mussel invasions on benthic nutrient and oxygen dynamics. *Can. J. Fish. Aquat. Sci.* 67: 1899–1908.

Vanderploeg, H., J. R. Liebig, W. W. Carmichael et al. 2001. Zebra mussel (*Dreissena polymorpha*) selective filtration promoted toxic Microcystis blooms in Saginaw Bay (Lake Huron) and Lake Erie. *Can. J. Fish. Aquat. Sci.* 58: 1208–1221.

Vanderploeg, H. A., A. E. Wilson, T. H. Johengen et al. 2013. Role of selective grazing by dreissenid mussels in promoting toxic *Microcystis* blooms and other changes in phytoplankton composition in the Great Lakes. In *Quagga and Zebra Mussels: Biology, Impacts, and Control*, 2nd Edn., T. F. Nalepa and D. W. Schloesser, eds., pp. 509–523. Boca Raton, FL: CRC Press.

Vaughn, C. C., D. E. Spooner, and H. S. Galbraith. 2007. Context-dependent species identity effects within a functional group of filter-feeding bivalves. *Ecology* 88: 1654–1662.

Wang, W.-X. and N. S. Fisher. 1999. Assimilation efficiencies of chemical contaminants in aquatic invertebrates: A synthesis. *Environ. Toxicol. Chem.* 18: 2034–2045.

Watkins, J. M., R. Dermott, S. J. Lozano, E. L. Mills, L. G. Rudstram, and J. V. Scharold. 2007. Evidence for remote effects of dreissenid mussels on the amphipod *Diporeia*: Analysis of the Lake Ontario benthic surveys, 1972–2003. *J. Great Lakes Res.* 33: 642–657.

Wilson, K. A., E. T. Howell, and D. A. Jackson. 2006. Replacement of zebra mussels by quagga mussels in the Canadian nearshore of Lake Ontario: The importance of substrate, round goby abundance, and upwelling frequency. *J. Great Lakes Res.* 32: 11–28.

Young, B. L., D. K. Padilla, D. W. Schneider, and S. W. Hewett. 1996. The importance of size-frequency relationships for predicting ecological impact of zebra mussel populations. *Hydrobiologia* 332: 151–158.

Yu, N. and D. A. Culver. 1999. Estimating the effective clearance rate and refiltration by zebra mussels, *Dreissena polymorpha*, in a stratified reservoir. *Freshwat. Biol.* 41: 481–492.

Zhang, H., D. A. Culver, and L. Boegman. 2008. A two-dimensional ecological model of Lake Erie: Application to estimate dreissenid impacts on large lake plankton populations. *Ecol. Model.* 214: 219–241.

Zhu, B., C. M. Mayer, S. A. Heckathorn, and L. G. Rudstam. 2007. Can dreissenid attachment and biodeposition affect submerged macrophyte growth? *J. Aquat. Plant Manage.* 45: 71–76.

CHAPTER 36

Benthification of Freshwater Lakes
Exotic Mussels Turning Ecosystems Upside Down

Christine M. Mayer, Lyubov E. Burlakova, Peter Eklöv, Dean Fitzgerald, Alexander Y. Karatayev, Stuart A. Ludsin, Scott Millard, Edward L. Mills, A. P. Ostapenya, Lars G. Rudstam, Bin Zhu, and Tataina V. Zhukova

CONTENTS

Abstract 575
Introduction 575
Benthification 576
Epidemic of Benthic Stable States 578
Long-Term Data 578
Separating the Relative Importance of Declines in Total Phosphorus and Dreissena 578
Review from Other Systems 580
Upside-Down Ecosystems Past and Future 581
Physical versus Trophic Change 582
Macrophytes versus Benthic Algae 583
Conclusions 583
Acknowledgments 583
References 583

ABSTRACT

Many north-temperate lakes are experiencing a shift in energy production from the open pelagic to the benthic region. This process termed "benthification" is occurring across lakes due to increased water clarity. Benthification alters habitats within aquatic ecosystems by augmenting benthic production and escalating the flow of energy and materials between the pelagic and benthic subsystems. Two anthropogenically driven factors, reduced phosphorus inputs and filter feeding by nonindigenous species (i.e., zebra and quagga mussels, *Dreissena polymorpha* and *Dreissena rostriformis bugensis*, respectively), can both enhance water clarity. However, long-term data from seven lakes in North America and Europe indicate that *dreissenids* are driving benthification more than nutrient reductions. Therefore, ecosystem engineering by these two nonindigenous species is changing the fundamental, physical nature of an entire category of ecosystems.

INTRODUCTION

North-temperate freshwater systems throughout North America and Europe have experienced a number of anthropogenic drivers during the twentieth century. These systems have experienced: (1) excessive phosphorus loading or eutrophication that can promote nuisance levels of phytoplankton and consequently limit water clarity and reduce bottom dissolved oxygen levels (Smith 2003); (2) a subsequent reduction in phosphorus levels via nutrient abatement programs that has been termed oligotrophication (e.g., Sommer et al.

1993, Ludsin et al. 2001); and (3) an accelerated rate of nonindigenous species introductions (e.g., Mills et al. 1994).

Two of these anthropogenically-driven factors, planned reductions in phosphorus inputs and the unplanned introduction and spread of two efficient, invasive filter-feeders, zebra and quagga mussels (*Dreissena polymorpha* and *Dreissena rostriformis bugensis*), are frequently independently cited as the cause of recent increased clarity in north-temperate lakes. Increased water clarity is a potentially important alteration of physical structure that may have implications for a variety of ecosystem-level processes. However, defining the relative importance of *Dreissena* filter feeding and decreased nutrient inputs in promoting water clarity in lakes is complicated by the historical and temporal overlap in these two ecosystem drivers. Nonetheless, understanding which of these two factors, one planned and the other unplanned, is most responsible for returning lakes to a clearer state is crucial to lake and land management practices and to our understanding of ecological processes in aquatic ecosystems.

Indeed, many north-temperate lakes currently have greater water clarity now than during the peak period of eutrophication. Negative consequences of eutrophication have spurred international policies to curb nutrient inputs and thereby ameliorate problems such as nuisance algal blooms. For example, in North America, reductions of nutrient loads in the Great Lakes began after the United States and Canada passed legislation during the early 1970s (e.g., the 1972 Clean Water Act and the Great Lakes Water Quality Agreement) that set target levels for phosphorus inputs. Further, research into the mechanisms underlying eutrophication and possible solutions contributed to the development of important concepts about ecosystem structure and function (Vollenweider 1968, Likens 1972, Shapiro and Wright 1984, Carpenter et al. 1985, McQueen et al. 1986). Phosphorus frequently limits phytoplankton growth in freshwater (Schindler 1977) and is usually implicated as the cause of eutrophication in lakes and rivers. Positive relationships between total phosphorus (TP) and standing crops of phytoplankton have been well documented (e.g., Dillon and Rigler 1974). Therefore, reductions in phosphorus loads via abatement programs likely resulted in lower standing stocks of phytoplankton. However, load reductions and declines in phosphorus levels occurred during the same time period as the introduction of a large number of nonindigenous species (Mills et al. 1994, Holeck et al. 2004), which also affected food web structure and hence productivity (Carpenter et al. 1985, McQueen et al. 1986). Consequently, defining the relative importance of these two potential factors in driving ecosystem-level change is a real challenge.

Dreissenid mussels were introduced into the Great Lakes in 1986 (Carlton 2008) and have since spread to large areas across North America. Dreissenids are also widespread in Europe outside of their native Pontocaspian region. These mussels have been associated with reduced phytoplankton standing stocks and increased water clarity (e.g., Fahnenstiel et al. 1995a,b, Binelli et al. 1997, Higgins 2013). However, their spread coincided with the time period when nutrient loads were decreasing and similar changes in phytoplankton and water clarity were expected. Further, there is likely an interaction between TP and *Dreissena* effects because in addition to reducing phytoplankton standing stocks as measured by chlorophyll *a*, dreissenids modify the relationship between chlorophyll and TP so that chlorophyll levels are lower than would be expected for a given level of phosphorus (Higgins et al. 2011). It is unlikely that there will be an experimental answer to the question of the relative importance of these two anthropogenic drivers of ecosystem change as no intentional, whole-lake-scale studies on *Dreissena* introduction have been conducted. Consequently, the question remains: which of these two anthropogenic drivers of ecosystem change (reductions in phosphorus vs. dreissenid filter feeding) have had a greater impact on water clarity in north-temperate lakes?

In this chapter, we present evidence that supports the theory that *Dreissena*, and not phosphorus reductions, is the more important driver of the observed improvements in water clarity. Further, we argue that changes in water clarity have triggered a suite of connected changes that increase the importance of benthic processes. We term this process "benthification" and propose that it is occurring over a broad geographic range and is having a strong influence on the structure and function of lake ecosystems.

BENTHIFICATION

Benthification is a point process of ecosystem engineering (Jones et al. 1994) wherein increased water clarity (a physical alteration) triggers a predictable suite of modifications to ecosystem structure (e.g., species composition, spatial distribution of primary producers and consumers) and function (e.g., primary production, benthic–pelagic flux) (Figure 36.1). Further, water clarity can affect human aesthetic perception of a lake, with higher water clarity generally being thought of as more desirable. In a turbid, eutrophic lake, large abundances of phytoplankton act much like a forest over-story, limiting light to primary producers in deeper habitats. Increased water clarity affects a lake in a manner similar to cutting down trees in a forest and results in a restructuring of the spatial distribution of primary production, organic material, and energy flow. The spatial extent of potential benthic primary production (algae and macrophytes) will increase because more light reaches a greater proportion of the bottom. In addition to affecting primary producers, increased light penetration will allow visually-feeding

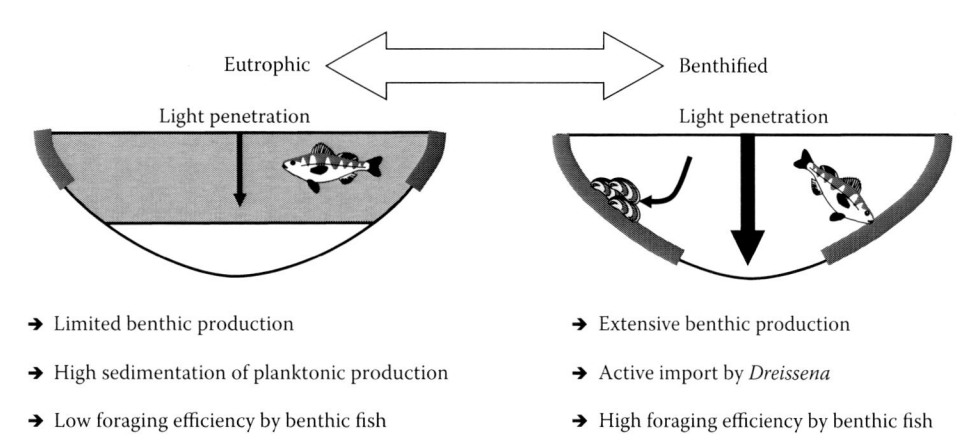

Figure 36.1 Depiction of changes associated with the process of "benthification." An increase in water clarity allows for greater rates and spatial extent of benthic primary production. Transport of material from pelagic to benthic zones changes from passive sedimentation to active importation to areas colonized by *Dreissena*. Transfer of energy and material from the benthic to pelagic zone increases because bottom-foraging fish become more efficient.

fish to forage more efficiently on benthic invertebrates, thereby augmenting the rate of material flux from benthic to pelagic zones. Unlike trophic cascades, in which changes are transmitted to subsequent trophic levels indirectly by a series of linked interactions between organisms, the physical–biotic coupled changes associated with benthification affect organisms directly and simultaneously at multiple trophic levels. As a result, these changes will have intense and rapid system-wide effects.

The degradation of benthic energy pathways (Vadeboncoeur et al. 2003) is a negative consequence of eutrophication that historically has received little attention from limnologists. Phytoplankton blooms, which are promoted in eutrophic waters, shade benthic (bottom associated) primary producers. Benthic primary production is an important, though seldom-measured, component of total ecosystem productivity (Vadeboncoeur et al. 2002), and fish may rely heavily on this energy (Vander Zanden and Vadeboncoeur 2002). Therefore, the consequence of excessive pelagic productivity can be a loss of benthic production and the dwindling of this potentially important energy pathway in aquatic food webs. Consequently, insight into the mechanisms behind the reestablishment of benthic energy fixation and flux is critical to an understanding of total ecosystem function.

Many changes associated with *Dreissena* have been documented in individual systems (Vanderploeg et al. 2002), but there has been little recognition that changes in primary producers, invertebrates, and fish are linked, and such changes constitute a major shift in the function of lake ecosystems across a broad geographic range. In the Great Lakes, direct importation of organic material to the benthic region by the filtration activities of *Dreissena* supports the idea that phosphorus is being redirected to benthic areas by *Dreissena*, as detailed in the "nearshore shunt" hypothesis proposed by Hecky et al. (2004). Benthic primary producers have increased in response to *Dreissena* introduction (Lowe and Pillsbury 1995, Skubinna et al. 1995, Zhu et al. 2006, Cecala et al. 2008, Auer et al. 2010). Also, *Dreissena* beds create a heterogeneous benthic habitat that has had a positive impact on most benthic invertebrates by entrapping dreissenid biodeposits that serve as a food resource and by offering protection from potential predators (e.g., Botts et al. 1996, Ricciardi et al. 1997, Stewart et al. 1998). Benthic invertebrates considered to be grazers have benefited from increased benthic production (Mayer et al. 2002). *Dreissena* provide attachment substrate for invertebrates that require hard substrate (Mayer et al. 2002) as well as for benthic algae such as *Cladophora* (Hecky et al. 2004). Benthic prey consumption by fish has increased after *Dreissena* establishment (Mayer et al. 2000). A review of *Dreissena* effects in Europe (Karatayev et al. 2002) concluded that the role of the benthic community increased dramatically after dreissenid introduction and that the benthic community becomes capable of controlling processes and dynamics in the entire ecosystem. Nevertheless, these changes have generally been interpreted as individual interactions, and there has been little synthetic understanding of the direct linkage of physical changes (ecosystem engineering) to trophic processes and ecosystem function.

The overall effect of benthification, stemming partly from enhanced water clarity and partly from enhanced deposition of organic material, is an increase in benthic primary production and an increase in benthic feeding by fishes. Subsequently, there is an increased flux of organic material associated with the lake's bottom that permeates the off web. Because benthification directly affects organisms, ranging from primary producers to top carnivores, it may in fact initiate both bottom-up and top-down cascading interactions that thus far have been viewed only as trophic cascades without an appreciation for the physically driven mechanism behind the interactions.

EPIDEMIC OF BENTHIC STABLE STATES

The existence of alternative pelagic (phytoplankton) and benthic (algae and macrophytes) stable states has been described for small, shallow lakes (Scheffer et al. 1993) and also suggested for large, shallow lakes such as Lake Erie (Kay and Regier 1999). The concept of benthification builds on the theory of alternate stable states, which suggests that over a range of nutrient levels across systems, an extrinsic disturbance may be required to switch a system from a turbid to clear state. We suggest that dreissenids are indeed providing such a disturbance across a geographically extensive range and are therefore causing an "epidemic" of newly stable benthic states in north-temperate lakes. Further, we suggest that the initial change in water clarity triggers a suite of responses in which multiple trophic levels, not just primary producers, are directly impacted, and therefore the effects may be more pervasive than simple cascading interactions. Lastly, this ecosystem engineering provides a dramatic example of whole-system alteration by a nonindigenous species.

LONG-TERM DATA

To quantify the relative roles of *Dreissena* introduction and nutrient reduction in causing increased water clarity and associated ecosystem change (i.e., benthification), we selected north-temperate lakes in both North America and Europe that had long-term data (>15 years) that spanned both pre- and post-*Dreissena* and pre- and post-phosphorus reductions. We also surveyed available literature to determine the magnitude of water clarity increase that is associated with *Dreissena*. We selected seven water bodies that represented a range of trajectories in phosphorus concentration prior to *Dreissena* introduction, as shown by the relationships of TP to year (Table 36.1). Water bodies that were experiencing significant declines in TP prior to *Dreissena* introduction were Bay of Quinte (two sites, Lake Ontario, Canada), western Lake Erie (United States), and Lake Naroch (Belarus), whereas a marginal decrease was occurring in Oneida Lake (United States). In contrast, TP in Lake Erken (Sweden) was increasing prior to *Dreissena* introduction, and TP in Lakes Myastro and Batorino (Belarus) showed no statistically detectable trend with time. Regardless of the trend prior to *Dreissena* introduction, each water body showed a lower TP level in the post-*Dreissena* period, and the average TP among all the water bodies was significantly lower after *Dreissena* introduction (paired t-test P < 0.0001). The only water body for which the decline in TP may have been temporally connected with *Dreissena* introduction was Lake Erken where an increasing trend in TP reversed around the time of *Dreissena* introduction.

SEPARATING THE RELATIVE IMPORTANCE OF DECLINES IN TOTAL PHOSPHORUS AND DREISSENA

To determine the relative importance of changes in TP levels and *Dreissena* introduction in affecting water clarity, we used an analysis of covariance (ANCOVA) model in which average annual water clarity data (indexed by Secchi depth transparency) from each water body were the response variable. We considered *Dreissena* introduction to occur the first year that populations were reported to be established. *Dreissena* were included as a categorical independent variable and annual TP served as a covariate. All data were $\log_{10}$ transformed to stabilize variance for statistical analyses.

Table 36.1 Regression Parameters for the Relationship between TP ($\log_{10}$ Transformed) and Year for Seven Water Bodies Prior to *Dreissena* Introduction. Effect Considered Significant at P < 0.05. Five of Seven Sites Showed a Significant Decreasing Trend with Time, Whereas One System Showed a Significant Increasing Trend with Time. N, Number of Years Prior to *Dreissena* Introduction. (Data Sources: Bay of Quinte [S. Millard]; Oneida Lake [E. L. Mills and L. G. Rudstam]; Lake Erken [P. Eklöv]; Lakes Batorino, Myastro, Naroch, Belarus [L. E. Burlakova, A. Karatayev, A. P. Ostapenya, T. V. Zhukova]; Western Lake Erie [Ohio Department of Natural Resources and the U.S. Environmental Protection Agency])

Lake	N	Slope	R²	P-Value
Batorino, Belarus	9	−0.01	0.05	0.6450
Erie, United States	17	−0.03	0.31	0.0208
Erken, Sweden	8	0.06	0.96	0.0001
Myastro, Belarus	9	−0.01	0.07	0.5572
Naroch, Belarus	15	−0.03	0.65	0.0003
Oneida, United States	17	−0.01	0.23	0.0532
Bay of Quinte B (Lake Ontario, Canada)	22	−0.02	0.74	0.0001
Bay of Quinte N (Lake Ontario, Canada)	16	−0.02	0.81	0.0001

In no case was there a significant interaction between the *Dreissena* and TP effect, consequently this term was dropped from each model.

We found that average annual Secchi depth transparency increased in all water bodies after *Dreissena* introduction (untransformed mean increase = 0.94 m). The ANCOVAs for each system showed that six of eight *Dreissena* effects were statistically significant, whereas two locations (one Bay of Quinte site and Lake Batorino) showed a statistically significant, positive within-site relationship between TP and annual Secchi depth (Table 36.2). These results indicate that, in a given system, the introduction of *Dreissena* had a greater effect on water clarity than TP declines.

To contrast changes in water clarity across systems, we tested for an increase in water clarity between the pre- and post-*Dreissena* periods using a t-test, paired by site. Water clarity was significantly higher in the post-*Dreissena* period (paired t-test, one tailed, P = 0.0003); moreover, all sites showed some increase (Figure 36.2). In contrast, the relationship between percent change (decrease) in TP and the percent change (increase) in Secchi depth between the pre- and post-*Dreissena* periods was not significant (Figure 36.3; regression, R^2 = 0.05, P = 0.58). This trend confirms that lakes that had large declines in TP did not necessarily experience large increases in clarity and further supports the importance of a prolific grazer, *Dreissena*, in influencing increased water clarity.

Table 36.2 Results of ANCOVA Examining Effects of *Dreissena* Introduction and TP on Water Clarity (Secchi Depth Transparency). *Dreissena* Is a Categorical Effect and TP Is a Covariate. Sums of Squares for *Dreissena* and TP Effects Are Type III (Not Dependent on Model Order). Data Sources Are Given in Table 36.1. Effect Considered Significant at P < 0.05

Lake	Source	DF	SS	F	P
Bay of Quinte, Lake Ontario (station B), Ontario, Canada	Model	2	0.106	18.96	0.000
	Error	27			
	Dreissena	1	0.040	14.42	0.001
	TP	1	0.016	5.83	0.023
Bay of Quinte, Lake Ontario (station N), Ontario, Canada	Model	2	0.324	24.94	0.000
	Error	21	0.098		
	Dreissena	1	0.098	20.91	0.000
	TP	1	0.007	1.60	0.220
Batorino, Belarus	Model	2	0.101	22.67	0.000
	Error	21			
	Dreissena	1	0.005	2.16	0.157
	TP	1	0.025	11.32	0.003
Erie, United States and Canada	Model	2	0.382	34.69	0.000
	Error	25			
	Dreissena	1	0.320	58.10	0.000
	TP	1	0.000	0.01	0.913
Erken, Sweden	Model	2	0.034	6.87	0.006
	Error	19	0.046		
	Dreissena		0.022	8.897	0.008
	TP		0.000	0.005	0.818
Myastro, Belarus	Model	2	0.261	15.46	0.000
	Error	21	0.017		
	Dreissena	1	0.048	2.83	0.107
	TP	1	0.047	2.76	0.112
Naroch, Belarus	Model	2	0.018	0.88	0.000
	Error	21	0.003		
	Dreissena	1	0.119	44.67	0.000
	TP	1	0.004	1.34	0.260
Oneida, NY	Model	2	0.089	10.18	0.001
	Error	24	0.105		
	Dreissena	1	0.045	10.20	0.004
	TP	1	0.000	0.00	0.970

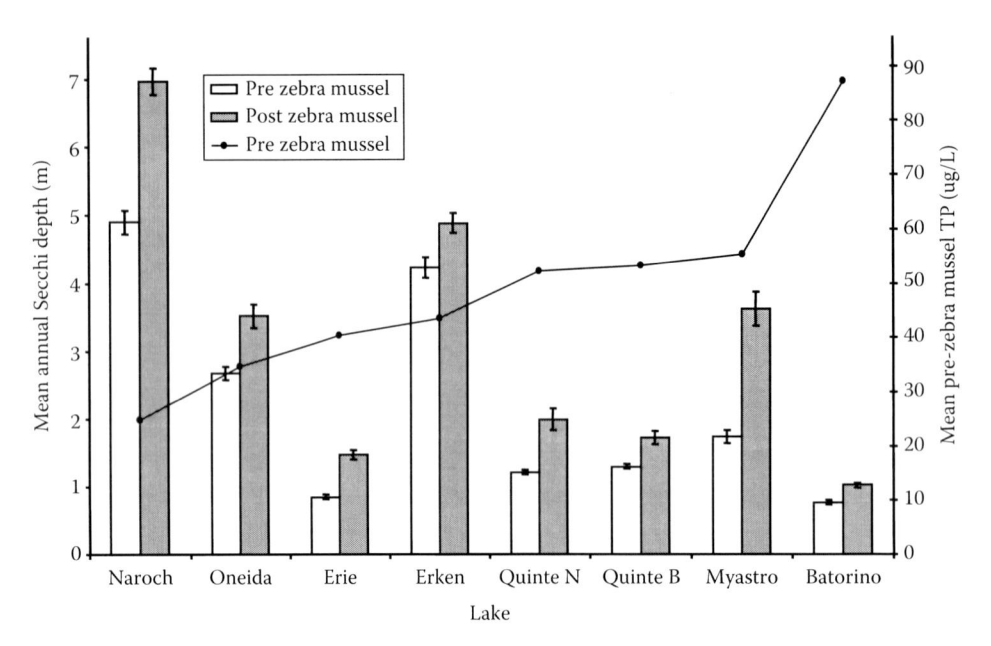

Figure 36.2 Mean (±1 SE) annual Secchi depth transparency (m) in various water bodies in the period before (pre-; open bar) and after (post-; solid bar) establishment of *Dreissena*. Also given is the mean TP (solid line) concentration (μg/L) in each of the water bodies in the period before *Dreissena* establishment. Lakes are arranged in order of increasing TP in the pre-*Dreissena* period.

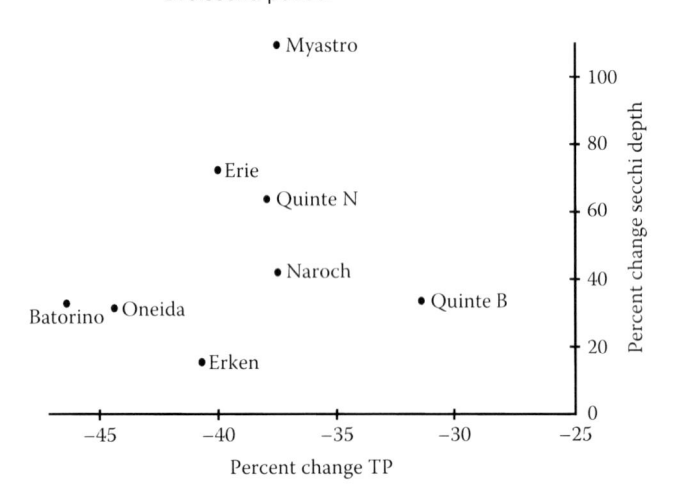

Figure 36.3 Relationship between percent change in TP and Secchi depth transparency (m) in the period before (pre-) and after (post-) establishment of *Dreissena* in various water bodies. The relationship was not statistically significant (regression, P = 0.58).

REVIEW FROM OTHER SYSTEMS

To supplement our analysis, we also searched the published literature for studies that examined water clarity in pre- and post-*Dreissena* periods, but did not necessarily present data collected over the long term. Published papers were obtained by searching the Cambridge Scientific Abstracts database for Biological Sciences and AGRICOLA, using the key words "zebra mussel, *Dreissena*, Dreissenid, water clarity, water quality, water transparency, turbidity, Secchi, and light." We supplemented the published literature with technical reports and data from websites in order to diminish the effects of publication bias. Secchi depth was the most frequently reported index of water clarity; therefore, studies reporting other indices, for example, turbidity (NTU) or attenuation (kPAR), were not used in this analysis. Many of the studies that reported water clarity trends did not report data on nutrient levels and therefore did not help to separate the effects of *Dreissena* from nutrient reduction. However, these studies allowed us to compare the magnitude of change observed in our long-term data sets to that seen in other systems. We found 17 published papers that reported water clarity data during in the pre- and post-*Dreissena* periods; these data were collected in 12 lakes and 1 river (Table 36.3). Different papers examined different time periods or different basins for the same lake (see Lake Erie, Saginaw Bay of Lake Huron, and Oneida Lake). Also, some papers reported trends for specific seasons, whereas some reported means over the entire growing season (Table 36.3). Overall, 33 pre- and post-*Dreissena* values were compared from the 13 water bodies. Of these, 27 showed an increase in post-*Dreissena* water clarity, 3 showed a decrease in water clarity, and 3 showed no change (Table 36.3). The mean change was calculated for each system (lake or river) for which multiple values or studies were available (Figure 36.4). For example, there were 13 measurements reported from different areas of Lake Erie, and the mean of these (0.88 m) was used to represent this system. The overall mean across systems (n = 13) was an increase of 1.02 m (±0.50 SD).

Most of the pre- and post-dreissenid comparisons that did not show an increase in water clarity were in Lake Erie; four of the six instances of no increase in water clarity were in the western or central basins of this lake. The western basin of Lake Erie is a large, shallow system, and physical contributions to turbidity such as wind-driven resuspension

Table 36.3 Comparisons of Secchi Depth Transparency (Meters) in Various Water Bodies in North America and Europe in the Period before and after *Dreissena* Establishment

System	Site Location	Season	Preinvasion Secchi (m)	Post-Invasion Secchi (m)	Difference (m)	Years Compared	Reference
Hargus Lake	Littoral		1.22	2.17	+0.95	93–95	17
Hargus Lake	Pelagic		1.7	2.52	+0.82	93–95	17
Lake Bolduk	Pelagic	Summer	3.70	4.50	+1.20	72 vs. 98	10
Lake Bolshiye Shvakshty	Pelagic	Summer	2.50	3.10	+0.60	71 vs. 98	10
Lake Como			7.5	7.6	+0.10	91–92 vs. 95–96	2
Lake Erie	Central	Spring	4.39	2.84	−1.55	82–04	1
Lake Erie	Central	Summer	6.57	7.12	+0.55	82–04	1
Lake Erie	East	Spring	3.24	7.8	+4.56	82–04	1
Lake Erie	East	Summer	6.11	6.59	+0.48	82–04	1
Lake Erie	Hatchery Bay	Fall	0.99	2.39	+1.41	83–87 vs. 90–95	16
Lake Erie	Hatchery Bay	Spring	0.76	2.14	+1.38	83–87 vs. 90–95	16
Lake Erie	Hatchery Bay	Summer	1.37	3.29	+1.92	83–87 vs. 90–95	16
Lake Erie	Hatchery Bay	Winter	1.28	1.73	+0.45	83–87 vs. 90–95	16
Lake Erie	Hatchery Bay		1.23	2.37	+1.14	84–86 vs. 90–92	6
Lake Erie	West	Spring	2.21	1.31	−0.90	82–04	1
Lake Erie	West	Summer	2.21	2.20	−0.01	82–04	1
Lake Erie	West		1.5	2.74	+1.24	88–89	11
Lake Erie	West–central		3.3	4.05	+0.75	88–89	11
Lake Huron	Saginaw Bay	Spring	1.31	2.42	+1.11	74–80 vs. 91–95	15
Lake Huron	Saginaw Bay	Summer	1.07	1.71	+0.64	74–80 vs. 91–95	15
Lake Huron	Saginaw Bay		1.15	1.79	+0.64	74–80 vs. 91–93	5
Lake Huron	Saginaw Bay		1.26	2.05	+0.79	74–80 vs. 91–93	5
Lake Huron	Saginaw Bay		1.32	2.45	+1.13	74–80 vs. 91–93	5
Lake Lukomskoe	Pelagic	Summer	2.00	4.00	+2.00	65 vs. 80	12
Lake Michigan	Southwest		3.98	5.21	+1.23	81–90 vs. 91–00	8
Lake Myadel	Pelagic	Summer	3.00	4.80	+1.80	80 vs. 98	10
Lake Svir	Pelagic	Summer	1.00	1.77	+0.77	80 vs. 98	10
Lake Volchin	Pelagic	Summer	4.60	3.80	−0.80	80 vs. 98	10
Oneida Lake			3.25	3.88	+0.63	75–91 vs. 92–94	7
Oneida Lake			2.64	3.60	+0.96	88–91 vs. 92–97	9
Oneida Lake			2.56	3.60	+1.04	75–91 vs. 92–97	13
Oneida Lake			2.66	3.18	+0.52	89–91 vs. 92–93	14
Seneca River			0.75	2.1	+1.35	90–91 vs. 93	3 & 4

References are as follows: 1, Barbiero and Tuchman (2004); 2, Binelli et al. (1997); 3, Effler and Siefried (1994); 4, Effler et al. (1996); 5, Fahnenstiel et al. (1995a,b); 6, Holland (1993); 7, Horgan and Mills (1999); 8, http://waterbase.glwi.uwm.edu; 9, Idrisi et al. (2001); 10, Karatayev et al. (2000); 11, Leach (1993); 12, Lyakhnovich et al. (1988); 13, Mayer et al. (2002); 14, Mellina et al. (1995); 15, Nalepa et al. (1999); 16, GLERL/NOAA (1996); 17, Yu and Culver (2000).

of sediment and inputs from the large Maumee River likely added to the variability of water clarity. In a sense, western Lake Erie is much like the Hudson River, where *Dreissena* have lowered standing stocks of phytoplankton (Strayer et al. 1999), but physical forces in this tidal system frequently drive turbidity and water clarity. In western Lake Erie turbidity is driven by both physical and biotic factors. In fact, under some circumstances *Dreissena* themselves may promote late season blooms of toxic algae by selectively filtering competitors (Vanderploeg et al. 2001). In Green Bay, Lake Michigan, Secchi depth transparency did not significantly increase in the post-*Dreissena* period possibly because TP was higher after *Dreissena* than before (Qualls et al. 2007). This suggested that watershed influences can override the grazing effect of *Dreissena*. However, there appears to be an overall trend of increased water clarity in the post-*Dreissena* period for most systems despite the effects of physical disturbance, local eutrophication, and within-lake ecological interactions.

UPSIDE-DOWN ECOSYSTEMS PAST AND FUTURE

The availability of long-term data for water clarity, TP, and *Dreissena* populations for a number of lakes has allowed us to determine the relative importance of two anthropogenic drivers of habitat change. We conclude that,

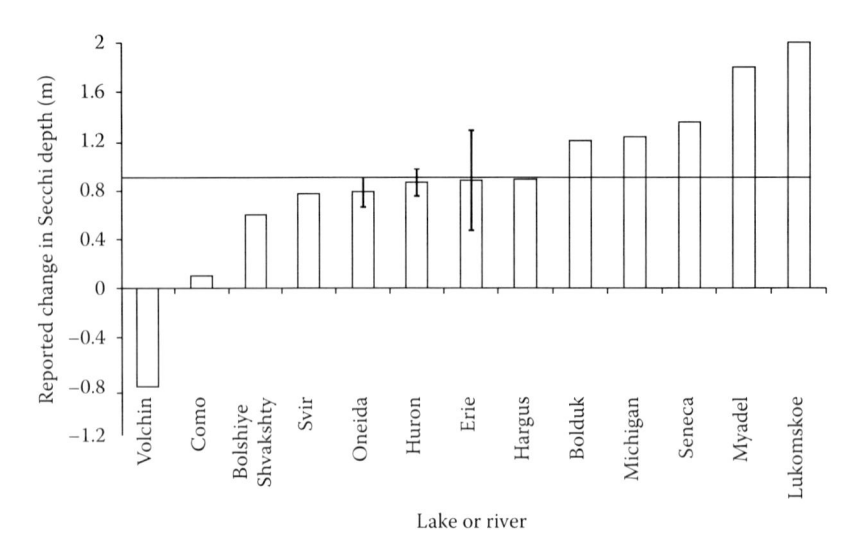

Figure 36.4 Change in mean (±1 SE) Secchi depth transparency (m) in the period before (pre-) and after (post-) establishment of *Dreissena* in various water bodies. Data are derived from published studies that lasted <15 years. Means were calculated when more than one study was available for a given water body. Horizontal line indicates the mean change for all water bodies. See Table 36.3 for data sources.

in the water bodies examined in this study that have over 15 years of data collection, establishment of *Dreissena* rather than phosphorus abatement best explains observed increases in water clarity. In addition, data examined from other published studies with shorter-term data sets support this conclusion. The mean increase in water clarity after *Dreissena* introduction was 0.94 m (as measured by Secchi depth) for lakes with long-term data sets compared to an increase in 1.02 m for lakes with shorter data sets. The shift from a turbid to a clear-water system is not only a change between stable system states (sensu Scheffer et al. 1993); it is a switch between two qualitatively different types of ecosystems, as distinct from each other as forests and grasslands. Paradoxically, benthification, which we attribute primarily to the effects of *Dreissena* filter feeding, may return lakes to a state similar to their prehuman-influenced condition. Prior to anthropogenic eutrophication, many northern temperate lakes probably had water clarity similar to or greater than after *Dreissena* and nutrient abatement. Further, many of these systems likely supported a large biomass of benthic species including gastropods, unionid mussels, and fish. Indeed, benthification may favor benthic fish that were once ecologically and economically important but were in decline due to eutrophication. After *Dreissena* introduction, benthic-feeding littoral fish have shown abundance increases in the Hudson River (Strayer et al. 2004), in Lake Erie (Ludsin et al. 2001), and in many European lakes (Karatayev et al. 2002).

While one consequence of benthification by dreissenids is a directional shift to a more prehuman-influenced state, in reality the process is complex with unexpected feedback loops. *Dreissena* strongly interacts with a variety of native and invasive species in the Great Lakes, and the overall effect promotes macrophytes and shallow-water, benthic invertebrates (DeVanna et al. 2011). Many of the species that are expected to invade the Great Lakes and associated waters in the near future are benthic (Ricciardi and Rasmussen 1998), and many of these species are likely to have strong interactions with *Dreissena*. Thus, changes brought about by *Dreissena* are likely to facilitate continued physical and biotic modifications to lake ecosystems that contribute to uncertainties in predicting eventual outcomes of benthification.

PHYSICAL VERSUS TROPHIC CHANGE

Regardless of directional extent, benthification has strong impacts on ecosystems because habitat change connects directly to a number of trophic levels rather than being transmitted through a series of links. For example, an increase in water clarity and light penetration directly affects the photosynthetic rate of plants and algae, while the same change in light also directly alters the environment for visual predators and prey. The direct, simultaneous connection of habitat change to all trophic levels, from primary producers to top predators, distinguishes this process from both "bottom-up" and "top-down" effects (McQueen et al. 1986, Carpenter et al. 1991), which cascade through a system via a sequence of linkages. However, it is likely that benthification will go on to initiate both top-down and bottom-up trophic cascades. An increase in benthic algae production may favor grazing benthic invertebrates, which may then provide an added resource to benthic-feeding fish, thereby triggering what would be seen as a bottom-up cascade. Alternatively, higher light and lower turbidity may increase efficiency of visual foragers and initiate top-down cascades in either benthic or pelagic zones.

This is not to say that *Dreissena* does not have direct trophic interactions. Grazing on phytoplankton is an obvious trophic connection; however, it is the physical manipulation of the environment by *Dreissena* that is likely to drive the strongest change.

MACROPHYTES VERSUS BENTHIC ALGAE

Macrophytes often play a structural role in lakes and therefore also act as ecosystem engineers. Hence, *Dreissena*-mediated increases in macrophytes can be seen as a "cascading" effect of ecosystem engineering. Filtration of phytoplankton by *Dreissena* and consequent engineering of light levels have indirect ecosystem effects on many types of organisms (algae, invertebrates, and fish) that utilize macrophyte beds as habitat. A further consequence of enhanced macrophyte growth may be an alteration of the quality and distribution of detritus within a lake as macrophytes can provide a major contribution of detrital material to inshore areas (Covich et al. 1999, Schindler and Scheuerell 2002). Ironically, some bottom-dwelling consumers may decline from the benthification process. Organisms that rely on seston settling from the water column, such as the amphipod *Diporeia*, may be negatively affected if benthic input of seston in deep areas is reduced by *Dreissena* (Landrum et al. 2000, Lozano et al. 2001). Other benthic filter feeders may also be negatively affected (Strayer et al. 1998).

Emphasis on shifts between turbid and clear states in lakes has often focused on a dichotomy between phytoplankton and macrophytes, with little emphasis given to the potential importance of increased bottom-associated algae (periphyton and epiphyton). However, many invertebrates readily consume benthic algae, whereas only a small number of specialized taxa directly consume macrophytes (Newman 1991). Therefore, an increase in benthic algae is likely to have direct implications for the spatial dynamics of aquatic food webs.

CONCLUSIONS

We propose that the term benthification is the best descriptor of a process of ecosystem change involving increased water clarity and light penetration, associated primarily with *Dreissena*. This process contemporaneously acts to modify physical habitat and to redirect energy from the pelagic to the benthic habitat. As used here, the term benthification was mostly associated with lakes, but similar processes should occur in any type of ecosystem (marine, freshwater, or terrestrial) where there is a dramatic restructuring of light and energy flow due to physical habitat alteration. We suggest that recognition of the process of benthification will expand our current understanding of energy and trophic pathways leading from lower trophic levels to fish. Insights from research on the process of benthification, along with associated physical and biological changes, may be analogous to how research on the widespread nature and magnitude of the problem of eutrophication led to many novel insights into ecosystem function.

ACKNOWLEDGMENTS

This work was supported in part by the National Oceanic and Atmospheric Administration (NOAA) award NA46RG0090 to the Research Foundation of the State University of New York for New York Sea Grant (project R/CE-20). The U.S. government is authorized to produce and distribute reprints for governmental purposes, notwithstanding any copyright notation that may appear herein. The views expressed herein are those of the authors and do not necessarily reflect the views of NOAA or any of its subagencies. We thank the Ohio Department of Natural Resources, Division of Wildlife, and the U.S. Environmental Protection Agency for providing S. Ludsin with data on western Lake Erie. This is contribution number 289 of the Cornell Biological Field Station and contribution 2013-03 of the University of Toledo's Lake Erie Center.

REFERENCES

Aver, M. T., L. M. Tomlinson, S. N. Higgins, S. Y. Malkin, E. T. Howell, and H. A. Bootsma. 2010. Great Lakes *Cladaphora* in the 21st century: Same algae-different ecosystem. *J. Great Lakes Res.* 36: 248–255.

Barbiero, R. P. and M. L. Tuchman. 2004. Long-term dreissenid impacts on water clarity in Lake Erie. *J. Great Lakes Res.* 30: 557–565.

Binelli, A., A. Provini, and S. Galassi. 1997. Trophic modifications in Lake Como (N. Italy) caused by the zebra mussel (*Dreissena polymorpha*). *Water Air Soil Pollut.* 99: 633–640.

Botts, P. S., B. A. Patterson, and D. W. Schloesser. 1996. Zebra mussel effects on benthic invertebrates: Physical or biotic? *J. N. Am. Benthol. Soc.* 15: 179–184.

Carlton, J. T. 2008. The zebra mussel (*Dreissena polymorpha*) found in North American 1986 and 1987. *J. Great Lakes Res.* 34: 770–773.

Carpenter, S. R., J. F. Kitchell, and J. R. Hodgson. 1985. Cascading trophic interactions and lake productivity. *Bioscience* 35: 634–639.

Carpenter, S. R., T. M. Frost, J. F. Kitchell et al. 1991. Patterns of primary production and herbivory in 25 North American lake ecosystems. In *Comparative Analyses of Ecosystems: Patterns, Mechanisms, and Theories*, J. Cole, S. Findlay, and G. Lovett, eds., pp. 67–96. New York: Springer-Verlag.

Cecala, R. K., C. M. Mayer, E. L. Mills, and K. L. Schulz. 2008. Increased benthic algal primary production in response to zebra mussel (*Dreissena polymorpha*) invasion in Oneida Lake. *J. Integr. Plant Biol.* 50: 1452–1466.

Covich, A. P., M. A. Palmer, and T. A. Crowl. 1999. The role of benthic invertebrate species in freshwater ecosystems. *Bioscience* 49: 119–127.

DeVanna, K. M., B. L. Bodamer, C. G. Wellington, E. Hammer, J. M. Bossenbroek, and C. M. Mayer. 2011. An alternative hypothesis for invasional meltdown: General facilitation by *Dreissena. J. Great Lakes Res.* 37: 632–641.

Dillon, P. J. and F. H. Rigler. 1974. A test of a simple nutrient budget model predicting the phosphorus concentration in lake water. *J. Fish. Res. Bd. Can.* 31: 1771–1778.

Effler, S. W., C. M. Brooks, K. Whitehead et al. 1996. Impact of zebra mussel invasion on river water quality. *Water Environ. Res.* 68: 205–214.

Effler, S. W. and C. Siefried. 1994. Zebra mussel (*Dreissena polymorpha*) populations in the Seneca River, New York: Impact on oxygen resources. *Environ. Sci. Technol.* 28: 2216–2221.

Fahnenstiel, G. L., T. B. Bridgeman, G. A. Lang, M. J. McCormick, and T. F. Nalepa. 1995a. Phytoplankton productivity in Saginaw Bay, Lake Huron: Effects of zebra mussel (*Dreissena polymorpha*) colonization. *J. Great Lakes Res.* 21: 465–475.

Fahnenstiel, G. L., G. A. Lang, T. F. Nalepa, and T. H. Johengen. 1995b. Effects of zebra mussel (*Dreissena polymorpha*) colonization on water quality parameters in Saginaw Bay, Lake Huron. *J. Great Lakes Res.* 21: 435–448.

Hecky, R. E., R. E. H., Smith, D. R. Barton et al. 2004. The near shore phosphorus shunt: A consequence of ecosystem engineering by dreissenids in the Laurentian Great Lakes. *Can. J. Fish. Aquat. Sci.* 61: 1285–1293.

Higgins, S. N. 2013. A meta-analysis of dreissenid effects on freshwater ecosystems. In *Quagga and Zebra Mussels: Biology, Impacts, and Control*, 2nd Edn., T. F. Nalepa and D. W. Schloesser, eds., pp. 487–494. Boca Raton: CRC Press.

Higgins, S. N., M. J. Vander Zanden, L. N. Joppa, and Y. Vadeboncoeur. 2011. The effect of dreissenid invasions on chlorophyll and the chlorophyll: Total phosphorus ratio in north-temperate lakes. *Can. J. Fish. Aquat. Sci.* 68: 319–329.

Holeck, K. T., E. L. Mills, H. J. MacIsaac, M. R. Dochoda, R. I. Colautti, and A. Ricciardi. 2004. Bridging troubled waters: Biological invasions, transoceanic shipping, and the Laurentian Great Lakes. *BioScience* 54: 919–929.

Holland, R. E. 1993. Changes in planktonic diatoms and water transparency in Hatchery Bay, Bass Island area, Western Lake Erie since the establishment of the zebra mussel. *J. Great Lakes Res.* 19: 617–624.

Horgan, M. J. and E. L. Mills. 1999. Zebra mussel filter feeding and food-limited production of Daphnia: Recent changes in lower trophic level dynamics of Oneida Lake, New York, U.S.A. *Hydrobiologia* 411: 79–88.

Idrisi, N., E. L. Mills, L. G. Rudstam, and D. J. Stewart. 2001. Impact of zebra *mussels (Dreissena polymorpha)* on the pelagic lower trophic levels of Oneida Lake, New York. *Can. J. Fish. Aquat. Sci.* 58: 1430–1441.

Jones, C. G., J. H. Lawton, and M. Shachak. 1994. Organisms as ecosystem engineers. *Oikos* 69: 373–386.

Karatayev, A. Y., L. E. Burlakova, D. P. Molloy, and L. K. Volkova. 2000. Endosymbionts of *Dreissena polymorpha* (Pallas) in Belarus. *Int. Rev. Hydrobiol.* 85: 543–559.

Karatayev, A. Y., L. E. Burlakova, and D. K. Padilla. 2002. Impacts of zebra mussels on aquatic communities and their roll as ecosystem engineers. In *Invasive Aquatic Species of Europe. Distribution, Impacts and Management*, E. Leppakoski, S. Gollasch, and S. Olenin, eds., pp. 433–446. Dordrecht, The Netherlands: Kluwer Academic Publishers.

Kay, J. and H. Regier. 1999. An ecosystem approach to Erie's ecology. In *The State of Lake Erie (SOLE)—Past, Present and Future. A Tribute to Drs. Joe Leach & Henry Regier*, M. Munawar, T. Edsall, and I. F. Munawar, eds., pp. 511–533. Dordrecht, The Netherlands: Backhuys Academic Publishers.

Landrum, P. F., D. C. Gossiaux, T. F. Nalepa, and D. L. Fanslow. 2000. Evaluation of Lake Michigan sediment for causes of the disappearance of *Diporeia spp.* in southern Lake Michigan. *J. Great Lakes Res.* 26: 402–407.

Leach, J. H. 1993. Impacts of the zebra mussel (*Dreissena polymorpha*) on water quality and fish spawning reefs in Western Lake Erie. In *Zebra Mussels: Biology, Impacts and Control*, T. F. Nalepa and D. W. Schloesser, eds., pp. 381–397. Boca Raton, FL: CRC Press.

Likens, G. E. 1972. *Nutrients and Eutrophication: The Limiting Nutrient Controversy. Special Symposium*, Volume 1. American Society of Limnology and Oceanography. Lawrence, KS: Allen Press.

Lowe, R. L. and R. W. Pillsbury. 1995. Shifts in benthic algal assemblage structure and function following the appearance of zebra mussels (*Dreissena polymorpha*) in Saginaw Bay, Lake Huron. *J. Great Lakes Res.* 21: 558–566.

Lozano, S. J., J. V. Scharold, and T. F. Nalepa. 2001. Recent declines in benthic macroinvertebrate densities in Lake Ontario. *Can. J. Fish. Aquat. Sci.* 58: 518–529.

Ludsin, S. A., M. W. Kershner, K. A. Blocksom, R. L. Knight, and R. A. Stein. 2001. Life after death in Lake Erie: Nutrient controls drive fish species richness, rehabilitation. *Ecol. Appl.* 11: 731–746.

Lyakhnovich, V. P., A. Y. Karatayev, P. A. Mitrakhovich, L. V. Guryanova, and G. G. Vezhnovets. 1988. Productivity and prospects for utilizing the ecosystem of the Lake Lukoml thermoelectric station cooling reservoir. *Sov. J. Ecol.* 18: 255–259. Translated into English from *Ecologiya*, 1987, 5: 43–48.

Mayer, C. M., J. L. Forney, L. G. Rudstam, A. J. VanDeValk, and E. L. Mills. 2000. The response of yellow perch in Oneida Lake, NY to zebra mussel establishment. *Can. J. Fish. Aquat. Sci.* 57: 742–754.

Mayer, C. M., R. A. Keats, L. G. Rudstam, and E. L. Mills. 2002. Scale-dependent effects of zebra mussels on benthic invertebrates in a large eutrophic lake. *J. N. Am. Benthol. Soc.* 21: 616–633.

McQueen, D. J., J. R. Post, and E. L. Mills. 1986. Trophic relationships in freshwater pelagic systems. *Can. J. Fish. Aquat. Sci.* 43: 1571–1581.

Mellina, E., J. B. Rasmussen, and E. L. Mills. 1995. Impact of zebra mussel (*Dreissena polymorpha*) on phosphorus cycling and chlorophyll in lakes. *Can. J. Fish. Aquat. Sci.* 52: 2553–2573.

Mills, E. L., J. H. Leach, J. T. Carlton, and C. L. Secor. 1994. Exotic species and the integrity of the Great Lakes: Lessons from the past. *Bioscience* 44: 666–676.

Nalepa, T. F., G. L. Fahnenstiel, and T. H. Johengen. 1999. Impacts of the zebra mussel (*Dreissena polymorpha*) on water quality: A case study in Saginaw Bay, Lake Huron. In *Nonindigenous Freshwater Organisms: Vectors, Biology, and Impacts*, R. Claudi and J. H. Leach, eds., pp. 255–271. Boca Raton, FL: CRC Press.

Newman, R. M. 1991. Herbivory and detritivory on freshwater macrophytes by invertebrates: A review. *J. N. Am. Benthol. Soc.* 10: 89–114.

Qualls, T. M., D. M. Dolan, T. Reed, M. E. Zorn, and J. Kennedy. 2007. Analysis of the impacts of the zebra mussel, *Dreissena polymorpha*, on nutrients, water clarity, and the chlorophyll-phosphorus relationship in lower Green Bay. *J. Great Lakes Res.* 33: 617–626.

Ricciardi, A. and J. B. Rasmussen. 1998. Predicting the identity and impact of future biological invaders: A priority for aquatic resource management. *Can. J. Fish. Aquat. Sci.* 55: 1759–1765.

Ricciardi, A., F. G., Whoriskey, and J. B. Rasmussen. 1997. The role of the zebra mussel (*Dreissena polymorpha*) in structuring macroinvertebrate communities on hard substrata. *Can. J. Fish. Aquat. Sci.* 54: 2596–2608.

Scheffer, M., S. H. Hosper, M. L. Meijer, B. Moss, and E. Jeppesen. 1993. Alternative equilibria in shallow lakes. *Trends Ecol. Evol.* 8: 275–279.

Schindler, D. W. 1977. The evolution of phosphorus limitation in lakes. *Science* 195: 260–262.

Schindler, D. E. and M. D. Scheuerell. 2002. Habitat coupling in lake ecosystems. *Oikos* 98: 177–189.

Shapiro J. and D. I. Wright. 1984. Lake Restoration by biomanipulation—Round Lake, Minnesota, the first two years. *Freshwat. Biol.* 14: 371–383.

Skubinna, J. P., T. G. Coon, and T. R. Batterson. 1995. Increased abundance and depth of submersed macrophytes in response to decreased turbidity in Saginaw Bay, Lake Huron. *J. Great Lakes Res.* 21: 476–488.

Smith, V. H. 2003. Eutrophication of freshwater and coastal marine ecosystems: A global problem. *Environ. Sci. Pollut. Res.* 10: 126–139.

Sommer, U., U. Gaedke, and A. Schweizer. 1993. The first decade of oligotrophication of Lake Constance. 2. The response of phytoplankton taxonomic composition. *Oecologia* 93: 276–284.

Stewart, T. W., J. G. Miner, and R. L. Lowe. 1998. Quantifying mechanisms for zebra mussel effects on benthic macroinvertebrates: Organic matter production and shell-generated habitat. *J. N. Am. Benthol. Soc.* 17: 81–94.

Strayer, D. L., N. F. Caraco, J. J. Cole, S. Findlay, and M. L. Pace. 1999. Transformation of freshwater ecosystems by bivalves—A case study of zebra mussels in the Hudson River. *BioScience* 49: 19–27.

Strayer, D. L., K. A. Hattala, and A. W. Kahnle. 2004. Effects of an invasive bivalve (*Dreissena polymorpha*) on fish in the Hudson River estuary. *Can. J. Fish. Aquat. Sci.* 61: 924–941.

Strayer, D. L., L. C. Smith, and D. C. Hunter. 1998. Effects of the zebra mussel (*Dreissena polymorpha*) invasion on the macrobenthos of the freshwater tidal Hudson River. *Can. J. Zool.* 76: 419–425.

Vadeboncoeur, Y., E. Jeppesen, M. J. Vander Zanden, H. H. Schierup, K. Christoffersen, and D. M. Lodge. 2003. From Greenland to green lakes: Cultural eutrophication and the loss of benthic pathways in lakes. *Limnol. Oceanogr.* 48: 1408–1418.

Vadeboncoeur, Y., M. J. Vander Zanden, and D. M. Lodge. 2002. Putting the lake back together: Reintegrating benthic pathways into lake food web models. *BioScience* 52: 44–54.

Vanderploeg, H. A., J. R. Liebig, W. W. Carmichael et al. 2001. Zebra mussel (*Dreissena polymorpha*) selective filtration promoted toxic *Microcystis* blooms in Saginaw Bay (Lake Huron) and Lake Erie. *Can. J. Fish. Aquat. Sci.* 58: 1208–1221.

Vanderploeg, H. A., T. F. Nalepa, D. J. Jude et al. 2002. Dispersal and emerging ecological impacts of Ponto-Caspian species in the Laurentian Great Lakes. *Can. J. Fish. Aquat. Sci.* 59: 1209–1228.

Vander Zanden, M. J. and Y. Vadeboncoeur. 2002. Fishes as integrators of benthic and pelagic food webs in lakes. *Ecology* 83: 2152–2161.

Vollenweider, R. A. 1968. *Scientific Fundamentals of the Eutrophication of Lakes and Flowing Waters, with Particular Reference to Nitrogen and Phosphorus as Factors in Eutrophication*. Paris, France: Organization for Economic Co-operation and Development.

Yu, N. and D. A. Culver. 2000. Can zebra mussels change stratification patterns in a small reservoir? *Hydrobiologia* 431: 175–184.

Zhu, B., D. M. Fitzgerald, C. M. Mayer, L. G. Rudstam, and E. L. Mills. 2006. Alteration of ecosystem function by zebra mussels in Oneida Lake, NY: Impacts on submerged macrophytes. *Ecosystems* 9: 1–12.

CHAPTER 37

Variability of Zebra Mussel (*Dreissena polymorpha*) Impacts in the Shannon River System, Ireland

Dan Minchin and Anastasija Zaiko

CONTENTS

Abstract 587
Introduction 587
Study Area 588
Methods 589
Results 590
    Abundance and Distribution Range of Zebra Mussels 590
    Community Impacts 591
    Habitat Impacts 592
    Ecosystem Impacts 593
Discussion 593
Acknowledgments 595
References 595

ABSTRACT

The biopollution assessment method was used to compare relative impacts of the zebra mussel (*Dreissena polymorpha*) across different regions (assessment units) in the Shannon River system, Ireland. The approach involved defining the zebra mussel abundance and distribution range, as well as the impact of this species on communities, habitats, and ecosystems based on surveys and studies over the period 1997–2007. Zebra mussels were found associated with most habitats ranging from rocky shallows to soft sediments in depths to 37 m. Abundance and biomass were greatest in lakes/reservoirs and lowest in rivers/canals, and impacts were greatest in assessment units having mainly lentic characteristics. Impacts involved total losses of unionids, declines in chlorophyll, and increases in rooted aquatic macrophytes. Variation in the magnitude of impact within each assessment unit depended upon specific features such as pH, available calcium, and most probably turbulence.

INTRODUCTION

The zebra mussel (*Dreissena polymorpha*) was first introduced into Ireland in 1993–1994. The most likely mode of introduction was via used leisure craft imported from Britain (Pollux et al. 2003). Imported craft were found with many live zebra mussels attached to their hulls (Minchin and Moriarty 1998), and associated parasites found in zebra mussels from invaded areas could only have been transmitted with attached mussels and not with the larval stage (Burlakova et al. 2006). The first report of mussels occurred in the Shannon River system (below Lough Derg) (McCarthy et al. 1998), and by 1996 mussels had colonized most lakes and river sections within the navigable area of this system (Minchin et al. 2006b); mussels were carried upstream as attached to the hulls of leissure craft (Minchin et al. 2002, 2003). The main period of boat usage occurs during summer that coincides with the spawning period for mussels when many thousands of individuals could settle on a single vessel (Minchin and

Moriarty 1999). On occasions, such as during boat rallies, there might be intensive mussel spawning events that could enable their establishment on boats and thereby extend their range very rapidly (Minchin and Gollasch 2003). Boats trailered overland also contributed to population expansion both within and outside of the Shannon River system as mussels can survive aerial exposure under damp conditions for several days (Minchin et al. 2002). In addition, downstream dispersal to uncolonized areas almost certainly occurred as veligers from upstream populations (Horvath and Lamberti 1999), and drifting plant materials with attached juveniles are likely to have contributed to spread within lakes (Sullivan et al. 2010).

In some lakes, mussels can form high-density colonies and act as powerful ecosystem engineers that dramatically change habitats, communities, and ecosystems (Botts et al. 1996, Stewart et al. 1998, Karatayev et al. 2002, Zaiko et al. 2009, Vanderploeg et al. 2010). Mussels have the ability to remove energy from the plankton by filtration and deposit it in the form of wastes to the benthos (Karatayev et al. 2007). In Ireland, mussels have greater impacts in lakes than rivers, and abundant populations have had both ecological and socio-economic impacts (Minchin and Moriarty 2002).

In this chapter, we determine environmental impacts of the zebra mussel in different regions (assessment units) of the Shannon River system by using the biopollution assessment method (Olenin et al. 2007). The period of study was 1997–2007, over which time the entire navigable section of the Shannon River system was likely to have been exposed to zebra mussels.

STUDY AREA

The Shannon River is the largest river system in Ireland (Figure 37.1). It drains a surface area of 14,100 km² (17%), has about 220 km of navigable water, and passes through a

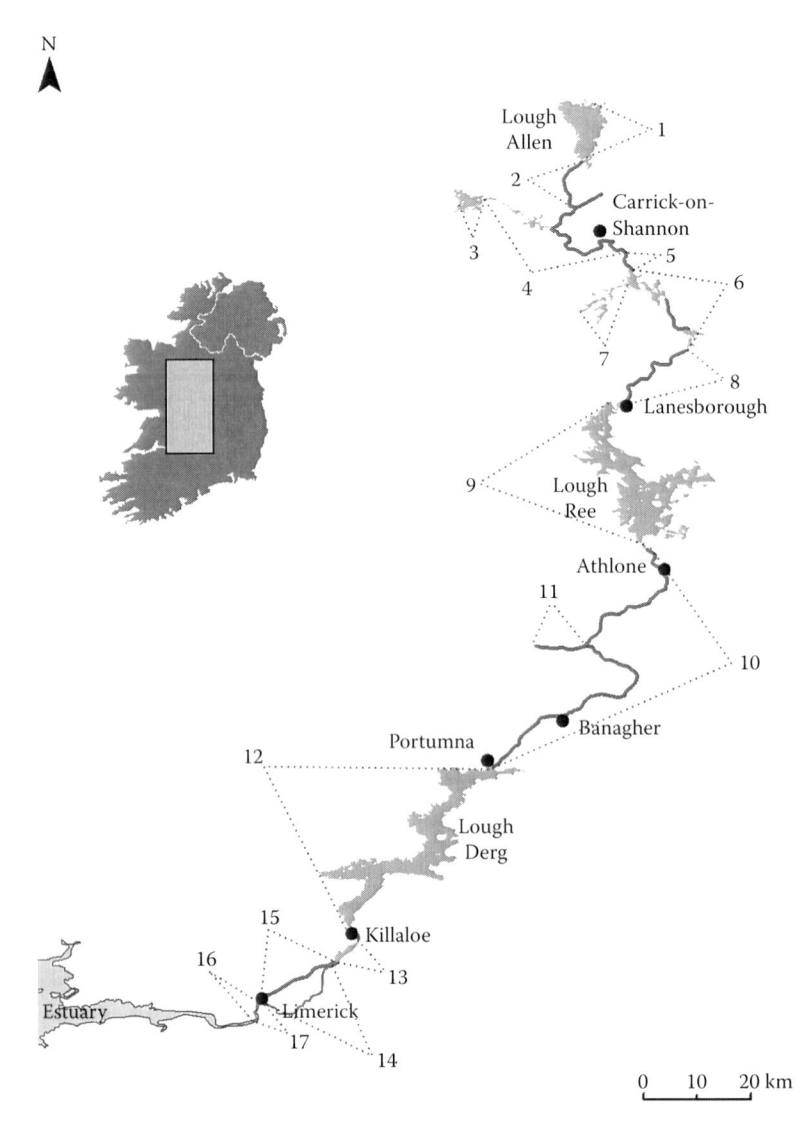

Figure 37.1 Location of defined assessment units on the Shannon River navigation system. See Table 37.1 for detailed descriptions.

chain of lakes interconnected by river sections. Navigation is possible throughout the year although most traffic occurs in summer. There is a decrease in water level of 17 m mediated by nine locks over the ~220 km of length from Lough Allen in the north to the lock at the Ardnacrusha hydroelectric station (Bowman 1998). This last lock lies at the top of the Shannon estuary and has a drop of 28.5 m and is served by a headrace canal 14 km long. The Shannon Navigation is linked to Lough Erne by the Shannon Erne Waterway in the northern portion of Ireland and is also connected to Dublin and to other regions of Ireland by two midland canals: the Grand and the Royal canals.

METHODS

The biopollution assessment method applied in this study refers to "biopollution" as a contraction of "biological pollution" defined by Elliott (2003) and relates to adverse effects of a biological invasion on the environment. It can be defined as any impact of an invasive species that is sufficient to disturb ecological quality by affecting an individual (internal biological pollution by parasites or pathogens), a population (genetic change, i.e., hybridization), a community (structural shift), a habitat (modification of physical–chemical conditions), or an ecosystem (alteration of energy and organic material flow) (Olenin et al. 2010). In the current study, biopollution was considered at the level of community, habitat, and ecosystem. The degree of impact, or the level of biopollution as described by Olenin et al. (2007), was determined by using the assisted means of the free online service BINPAS (Biological Invasion Impact/Biopollution Assessment System) available at http://corpi.ku.lt/databases/binpas/. This computerized system is based on the biopollution assessment method and provides a user-friendly interface to calculate the biopollution level (BPL), accumulates and stores information on abundance and distribution ranges (ADRs) of various alien species in different geographical domains, and also accumulates and stores impacts on communities, habitats, and ecosystem functions (Naršcius et al. 2012).

The biopollution assessment method has proved to be robust enough to quantify an impact in a standard and repeatable way if sufficient information is available to provide a high level of confidence for each result (Olenina et al. 2010, Orendt et al. 2010, Pyšek and Richardson 2010, Zaiko et al. 2011). According to this method, the BPL is delivered for a specified area (termed the assessment unit) and for a selected period of time (the assessment period). In the present study, impacts of the zebra mussel were assessed for 17 different regions of the Shannon Navigation over the period between 1997 and 2007. Included in these assessment areas were entire lakes, river sections, lake and river complexes, and canals (Figure 37.1, Table 37.1). To quantify the BPL, the ADRs of the zebra mussel were first determined within each assessment unit. Mussel ADR was based on annual surveys over the entire assessment period within each defined assessment unit. The magnitude of mussel impacts on community, habitat, and ecosystem function was then evaluated. Specifically, the overall BPL consisted of a combination of mussel ADRs and of mussel impacts on communities (C), invaded habitats (H), and ecosystem functions (E). The ADR was ranked from A (when a zebra mussel population made up only a small part of the macrobenthic community) to E (when a zebra mussel population occurred in high numbers in all available localities). After the ADR was divided into different classes, it was related to the magnitude of impacts. Impacts were divided into five different categories, each having an assigned value from 0 to 4 as follows: 0, no impact; 1, weak impact; 2, moderate impact; 3, strong impact; and 4, massive impact. Thus, impact levels on communities (C0 to C4), habitats (H0 to H4), and ecosystem functions (E0 to E4) were each separately defined. After this was done, the magnitude of bioinvasion impacts was scored at five BPLs ranging from 0 to 4 as previous: 0, no impact; 1, weak impact; 2, moderate impact; 3, strong impact; and 4, massive impact. The BPL is calculated and provided to the user automatically by the BINPAS service, according to the methodology and classifications given in Olenin et al. (2007).

Two levels of confidence—high and moderate—were determined when assessing the BPL for each of the 17 defined assessment units. High confidence level (sensu Olenin et al. 2007) corresponded with an ADR and impact evaluation based on comprehensive studies throughout an entire assessment unit. A medium confidence level corresponded with an ADR and impact evaluation studied in only a part of an assessment unit with results then extrapolated to the entire assessment unit according to expert judgment. All data on performed assessments were stored in BINPAS (see Naršcius et al. 2012) for details on the computerized system.

When performing evaluations, a number of different ecosystem parameters within an assessment unit (obtained prior to and during the study period) were analyzed. A detailed list of the parameters, the methodology applied, and role in a BPL assessment are presented in Table 37.2.

Mussels were sampled by scraping known distances on the vertical surfaces of navigation marks, piles, bridge buttresses, and quays to depths of up to 3.5 m using a WOLF-Garten® hoe with a 15 cm blade and an attached pocket net. Biomass estimates were obtained using an electronic balance for wet weights of all mussels in each sample. Numbers of settled mussels were determined from plates immersed in the summer 1999 in Lough Ree. The plates were retrieved after 6–8 weeks and the number of individuals within 1 cm^2 marked quadrats was counted.

Factorial analysis of variance (ANOVA) (following the Wilks and Shapiro test of normality and Cochran's C test of homogeneity of variances) was used to test for effects of two

Table 37.1 Selected Assessment Units in the Shannon River System, Ireland. See Figure 37.1 for Relative Locations. The Number of Stations Refers to the Number of Sites Sampled for Various Impacts of Zebra Mussels

Assessment Unit No.	Assessment Unit	Water Body Type	Number of Stations	Period of Study	Maximum Depth (m)	References
1	Lough Allen	Lake	3	1998–2006	33	8, 9, 10, 16
2	Lough Allen Canal	Lake and canal	4	1998–2006	12	8, 10
3	Lough Key	Lake	15	1998–2007	24	2, 3, 5, 8, 9, 10, 14, 15
4	Carrick section	Lakes and river	6	1998–2007	16	8, 9, 10, 14
5	Jamestown Canal	Canal	2	1998–2007	2	8, 10
6	Loughs Bofin to Forbes	Lakes and river	9	1998–2007	7	1, 8, 10
7	Fen lakes	Lakes	6	1998–2007	16	8, 10
8	Loughs Forbes to Ree	River	4	1998–2007	12	8, 10
9	Lough Ree	Lake	8	1997–2007	35	8, 9, 10, 15
10	Mid Shannon	River	9	1998–2007	12	5, 8, 9, 10, 14
11	Suck River	River	3	1999–2002	5	8, 10
12	Lough Derg	Lake	29	1997–2007	37	6, 7, 8, 9, 10, 11, 12, 13
13	Parteen Reservoir	Reservoir	2	1997–2004	15	6, 7, 8, 10, 14
14	Lower Shannon River	River	4	2003–2007	2	4, 9
15	Ardnacrusha Headrace	Canal	2	1997–2004	4	4, 8, 10, 14
16	Shannon Upper Estuary	Estuary	2	2007	4	4
17	Limerick Docks	Docks	3	1997–2003	7	8, 9, 10, 12, 13, 14

References are as follows: (1) Atalah et al. (2010), (2) Lucy (2006), (3) Lucy et al. (2005), (4) Lucy et al. (2008), (5) Lucy and Sullivan (2001), (6) McCarthy et al. (1997), (7) McCarthy et al. (1998), (8) Minchin et al. (2002), (9) Minchin et al. (2003), (10) Minchin et al. (2006a), (11) Minchin et al. (2006b), (12) Minchin and Moriarty (1998), (13) Minchin and Moriarty (1999), (14) Pollox et al. (2003), (15) Sullivan et al. (2010), and (16) Timpson and Lucy (1998).

explanatory variables—year of sampling and type of assessment unit (lentic, lake or reservoir, and lotic, river or canal) on zebra mussel densities during the period of peak abundances (2000–2002). Nonparametric Mann–Whitney (Wilcoxon) W test was used to test differences when data sets were unbalanced and did not meet normality assumptions. Such data sets consisted of zebra mussel biomass to abundance ratios in lentic and lotic assessment units and Secchi depth measurements of water transparency and chlorophyll (Chl) concentrations in Lough Derg, before and after zebra mussel establishment.

RESULTS

Abundance and Distribution Range of Zebra Mussels

Mussels varied in abundance throughout the Shannon River system (Table 37.3). The highest ADR class of E was given to 5 of the 17 assessment units. All of these were lakes and had peak biomass values of up to 3–5 kg/m^2. Among lakes with the highest ADR was Lough Ree, where densities of recently-settled mussels on settlement plates were >500,000/m^2 in 1999. The only lake with the lowest ADR class of A was Lough Allen, where some tens of mussels were found only in 1998 and not subsequently. All other lakes had moderate to high mussel densities (ADR of B, C, or D) with lowest classifications generally found where an infested river entered the lake.

In rivers and canals, the ADR did not exceed class C, except where an assessment unit contained small lakes. Densities were generally highest in areas nearest lake discharges and then declined with increased distance downstream especially in areas where there were shallows. Mussels were rarely found in assessment unit 14, which was in the lower portion of the Shannon River (Lucy et al. 2008).

There was a significant difference in mussel densities between lakes and rivers (ANOVA; $F = 408.8$, d.f. = 1, $p < 0.001$), but no significant difference between years 2000 and 2002 when densities were at a peak (ANOVA, $F = 1.6$; d.f. = 2; $p = 0.21$). No significant interaction between sampling year and assessment unit type was found. However, zebra mussel biomass to abundance ratios were significantly greater ($W = 4586.0$, $p < 0.001$) in rivers and channels than in lakes and reservoirs (Figure 37.2), indicating that rivers and canals tended to have larger individuals.

Table 37.2 Methodology Used to Determine the BPL in the Present Study. See "Methods" Section for Applications When Using the Biopollution Assessment Method

Parameter Measured	Habitat	Method Applied	Role in Determination of BPL	References
Zebra mussel abundance and biomass	Vertical surfaces of submerged hydrotechnical constructions, quays, walls, bridge buttresses, navigation poles	Scraping with a WOLF-Garten® hoe (15 cm blade and attached pocket net fixed to a 4 m handle) and measured for numbers and wet weight biomass for a known area	ADR, impact on communities and habitats	http://www.wolf-garten.com/ Minchin et al. (2002, 2006b)
	Soft substrates at varying depths, from near surface to 37 m	Van-Veen grab of 18 cm × 14 cm bite area	ADR, impact on communities and habitats	
	Suspended below floating pontoons	Settlement plates of sanded Perspex of 30 cm × 30 cm	Impact on communities and habitats	Minchin et al. (2006b)
Macrozoobenthos abundance and biomass	All bottom types	Analysis of unpublished data from field surveys, reports, and scientific literature published during 1997–2007	Impact on communities and habitats	Lucy et al.(2005, 2004, 2008), Minchin (1999), Minchin and Moriarty (1999), Minchin et al. (2002, 2003, 2006b)
Zebra mussel larval abundance	Water column	Pumped and filtered larval samples	ADR, impact on communities, ecosystem functioning	Lucy et al. (2005, 2008)
Abundance and fouling of the unionids by the zebra mussels	Soft bottom habitats along the entire navigation	In situ observation of live unionids found in the Van-Veen grab samples and cast empty shells	Impact on communities and habitats	Minchin et al. (2006b)
Incrustation of the macrophytes with zebra mussels	Lake shallows, <4 m depth with *Schoenoplectus lacustris* and *Phragmites australis*	In situ observation and wet weights of mussels attached to single-macrophyte stems	Impact on habitats	Minchin and Boelens (2005), Sullivan et al. (2010)
Windrows of beached zebra mussel shells	Exposed shores of lakes	Height to the crest of the shell windrows was measured to the nearest 5 cm.	Impact on habitats	Lough Derg Science Group (unpublished data)
Water transparency	Water column of lakes	Secchi disk, 30 cm diameter	Impact ecosystem functioning	Bowman (2000), Lucy (2005), Lucy et al. (2008), EPA (unpublished data)
Chl-α concentrations	Water column of the Carrick section, Loughs Bofin to Forbes, Fen lakes, Lough Derg	Hot methanol extraction, without correction for the presence of pheophytin	Impact on ecosystem functioning	Kirk McClure Morton (2001), EPA (unpublished data)
Organic matter content in sediments	Soft sediments colonized by zebra mussels	Loss on ignition	Impact on habitats and ecosystem functioning	Toner et al. (2005), Atalah et al. (2010)

Community Impacts

Following the establishment of zebra mussels, structural changes in the macrozoobenthic communities were found in lakes. Zebra mussels colonized both hard and soft substrates to depths of 37 m. Early in the invasion process, individuals of the widely-distributed unionid, *Anodonta anatina*, became colonized by mussels, and populations were extirpated within about 5 years (Minchin et al. 2006b). Unionids no longer existed in Loughs Derg, Ree, and Key by 2007, but were abundant in all these lakes before zebra mussels became established. Live unionids were still found in many river sections, but at low densities. Because of the loss of the dominant unionid, *A. anatina,* the impact on communities in these large lakes was assessed as C4 (massive) (Table 37.3). In addition, the accumulation of feces and pseudofeces from mussel populations created an energy-rich resource exploited by other macroinvertebrates, including the Ponto-Caspian amphipod *Chelicorophium curvispinum*, as reported elsewhere (Haas et al. 2002). In other assessment units, impacts of zebra mussels ranged from negligible (C0), when no displacement of unionids occurred, to strong (C3) when a former community dominated by unionids remained, but unionid abundance had become severely reduced.

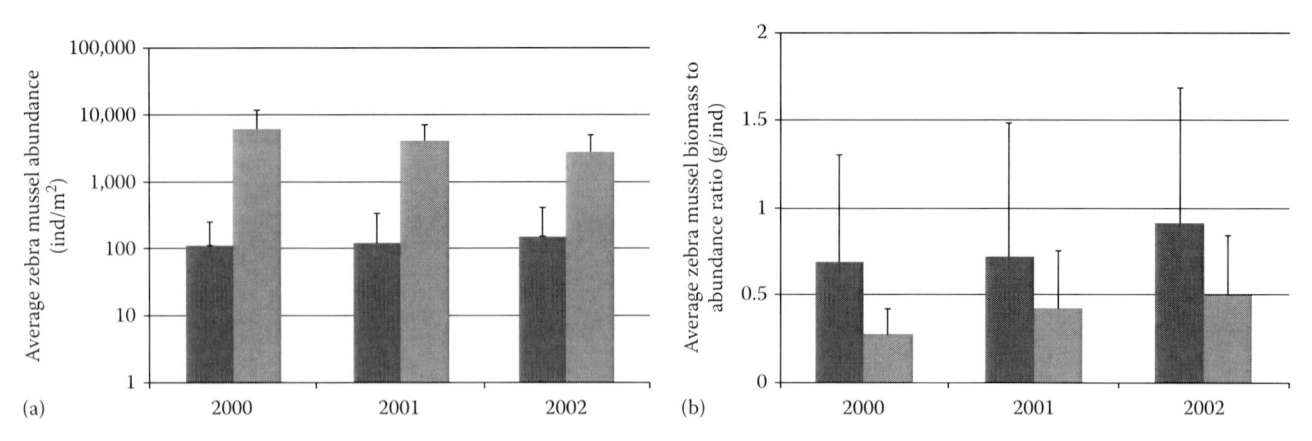

Figure 37.2 Average abundance (a) and average ratio of biomass (wet weight) to abundance (b) for zebra mussels in rivers/canals (dark fill) and lakes/reservoirs (light fill), 2000–2002. The ratio can be defined as wet weight per individual.

Table 37.3 ADR and Impacts on Communities (C), Habitats (H), and Ecosystems (E). Level of Confidence: Bold (High), Unbolded (Moderate). N, No Reliable Information on Zebra Mussel Impacts Was Available (e.g., CN, EN)

Section No.	Assessment Unit	ADR	Communities	Habitats	Ecosystems	Overall BPL
1	Lough Allen	**A**	**C0**	**H0**	**E0**	0
2	Lough Allen Canal	A	C1	H1	EN	1
3	Lough Key	**E**	**C4**	**H4**	**E3**	4
4	Carrick section	C	C2	H2	E2	2
5	Jamestown Canal	B	CN	H1	EN	1
6	Loughs Bofin to Forbes	D	C3	H2	E2	3
7	Fen lakes	D	C3	H2	E2	3
8	Loughs Forbes to Ree	C	CN	H1	EN	2
9	Lough Ree	**E**	**C4**	H4	EN	4
10	Mid Shannon	B	C1	H1	EN	1
11	Suck River	A	C0	H1	EN	1
12	Lough Derg	**E**	**C4**	**H3**	**E3**	4
13	Parteen Reservoir	E	C4	H3	E3	4
14	Lower Shannon River	A	C0	H0	E0	0
15	Ardnacrusha Headrace	C	CN	H1	EN	2
16	Shannon Upper Estuary	C	CN	H2	EN	2
17	Limerick Docks	E	CN	H4	EN	4

Habitat Impacts

In our study, two assessment units were assigned a habitat impact of H0—Lough Allen and the Lower Shannon River (Table 37.3). In both of these units, few zebra mussels were found. In contrast, mussel druses were abundant on hard and soft substrates in Loughs Key, Ree, and Derg, the Parteen Reservoir, and the Limerick Docks, resulting in the highest impact scores (H4). In some of these assessment units, mussels occasionally attached to shells of live gastropods *Lymnaea* spp. and *Viviparus viviparus*, as well as to vacant bivalve shells. During the period of peak abundance, mussels were found in high numbers attached to emergent vegetation (e.g., the reed *Phragmites australis* and the rush *Schoenoplectus lacustris*), and to rhizomes, stems, and leaves of the water lily *Nuphar lutea*. These plants were abundant in shallows of all assessment units (Sullivan et al. 2010). Mussels were also found attached to other submerged plant surfaces such as pondweeds, charophytes, dead branches, exposed roots, and fallen leaves. However, in the years following peak abundance (2000–2002), the number found attached to these surfaces declined (Sullivan. et al. 2010). Mussels that attached to firm surfaces altered substrate texture and created interstitial spaces that entrapped wastes. In addition, empty mussel shells littered surface substrates and altered sediment texture and surface structure; this included shell windrows cast ashore above shallow stony areas in wind-exposed areas of lakes.

Ecosystem Impacts

Despite the brown coloration of water in most lakes caused by runoff from upland and raised bogs, water transparency increased after mussels became abundant (Bowman 2000). In addition, there was a general decline in Chl levels in lakes (Bowman 2000, Lucy 2005, Lucy et al. 2008). In some lakes (assessment units 4, 6, and 7), Chl concentrations declined from 31–53 mg Chl/m^3 in 1998 to 4–7 mg Chl/m^3 in 1999 (Kirk McClure Morton 2001). In Lough Key (assessment unit 3), mussels were sufficiently abundant to turn over the water by filtration within a 10 day period (Lucy et al. 2005). In Lough Derg (assessment unit 12), mussels were first found in 1997 (McCarthy et al. 1998) and had colonized the entire lake by 1998 (Minchin and Moriarty 1998, 1999). Therefore, levels of water transparency and Chl before 1990 and after 2000 reflected pre- and post-invasion conditions. There was a significant increase in water transparency (W = 32259.0, p < 0.001) and decrease in Chl concentrations (W = 39436.0, p < 0.001) following mussel establishment.

Removal of particulates from water and deposition of wastes by mussels has led to benthic enrichment (Toner et al. 2005, Atalah et al. 2010). The combination of nutrient-rich sediments and increased light penetration enabled rooting macrophytes to extend their habitat range to depths of ~3.5 m, and to form extensive beds within some shallow sheltered bays (Minchin and Boelens 2011). The non-native Nuttall's pondweed *Elodea nuttallii* and the water violet *Hottonia palustris* have also become abundant locally (Minchin and Boelens 2005, 2011). Zebra mussels in the Shannon River system are avidly fed upon by coot *Fulica atra* and occasionally by mallard *Anas platyrhynchos* (DM personal observations), and there is evidence that mussels are also fed upon by cyprinids (C. Maguire, personal communication).

DISCUSSION

Impacts of zebra mussels in the Shannon River system were highly variable, ranging from no impact to massive impact. The magnitude of this variation was certainly a function of mussel abundance. Specific features within each assessment unit that most affected abundances were pH, available calcium, and most probably turbulence. Temperatures within assessment units generally ranged from 3°C to 21°C, with extreme ranges of <1°C to 26°C (Lough Derg Science Group, unpublished data). These temperatures lie within the acceptable range for mussel survival (Spidle et al. 1995, Zaiko et al. 2009). No impacts were detected in Lough Allen, an oligotrophic lake, which has low calcium concentrations (maximum 5.8 mg/L) and a low pH (6.6–6.8) (Timpson and Lucy 1998, Bowman 2000). Low pH is a result of run-off from sandstone, silts, and grits, and bog drainage. Although Mikheev (1967) found settled zebra mussels at a pH of 6.6, other authors generally found that pH levels of 7.3–7.4 formed the minimum tolerance level. For instance, Ramcharan et al. (1992) found that mussels placed in two lakes with a pH of 7.3 died within 35 days. Using the model of Hincks and Mackie (1997), which shows predicted mortality in relation to pH and calcium content in water, it would appear that conditions in Lough Allen are generally unsuitable for the colonization of all life-history stages.

In other assessment units, mussel abundances were generally greater in lakes/reservoirs than in rivers/canals (Figure 37.2). Even in small lakes/reservoirs, mussel biomass was ~3–4 kg/m^2, whereas in rivers/canals the biomass was generally <0.5 kg/m^2. In lake assessment units downstream from Lough Allen, acid water becomes buffered by limestone deposits and outcrops, and thus conditions became more suitable for mussels. In river assessment units, mussel abundance varied with size and depth of the river section, with greatest abundances found nearest lake discharges (Lucy et al. 2008). In the study by Lucy et al. (2008), zebra mussels were seldom found downstream of turbulent, shallow streams and rivers. This is a feature also noted by Hovath et al. (1996). Under laboratory conditions veligers are sensitive to turbulent conditions (Horvath and Crane 2010). This could perhaps explain the paucity of mussels found downstream of Parteen Dam that holds the Parteen Reservoir (assessment unit 14). Here veligers would be exposed to many shear forces with a fall of water of ~20 m and then pass through downstream rapids. The upper and middle river sections of the Shannon River system (assessment units 4, 8, and 10) had mussels throughout their length, albeit at lower densities than in assessment units with lakes and reservoirs. In this study, the ADR of zebra mussels generally corresponded with the overall impact score (Figure 37.3). Significant rank correlation between assessed ADR and BPL (Spearman's r = 0.94, p < 0.001) supported our assumption that abundances of zebra mussels might act as a proxy for the BPL score if there is little available data for mussel impacts on communities, habitats, or an ecosystem. This is particularly true for ADR classes D and E. Yet when predicting BPL based on mussel ADR, the level of certainty for each assessment unit should be taken into consideration. For instance, a high level of certainty was set for ADR scores for the assessment units 1, 3, 9, and 12, which reflected the greater number of studies in these units and hence more robust classifications.

Variable impacts of zebra mussels on different elements of an ecosystem are based on high reproductive capacity and rapid growth (Sprung 1991, Lvova et al. 1994, Beisel et al. 2010), high metabolic rate (Shkorbatov et al. 1994), and high filtration rates (Kryger and Riisgård 1988, Ludyanskiy et al. 1993, Mikheev 1994, Daunys et al. 2006, Zaiko and Daunys 2012). The ability of zebra mussels to colonize a wide range of different surfaces, produce extensive shell deposits, attain high densities, and cause extensive ecosystem changes clearly

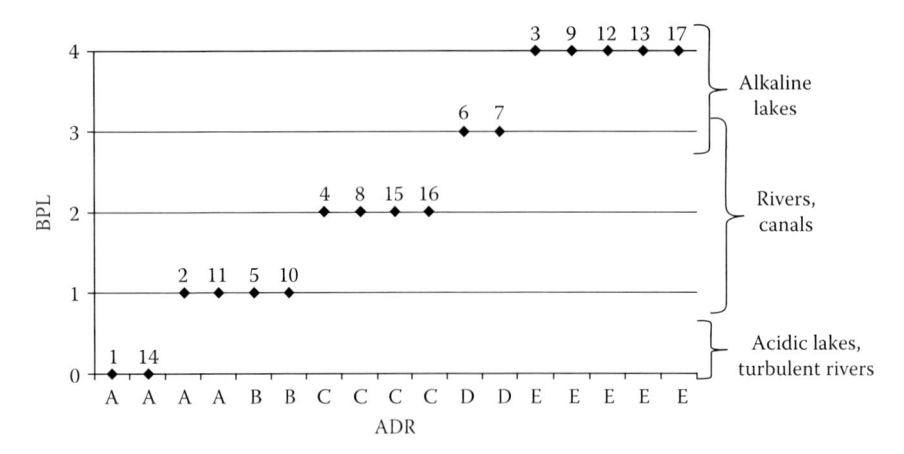

Figure 37.3 Correspondence between assigned ADR and assessed BPL for zebra mussels in the Shannon River. Values above the scored levels refer to assessment units as shown in Figure 37.1.

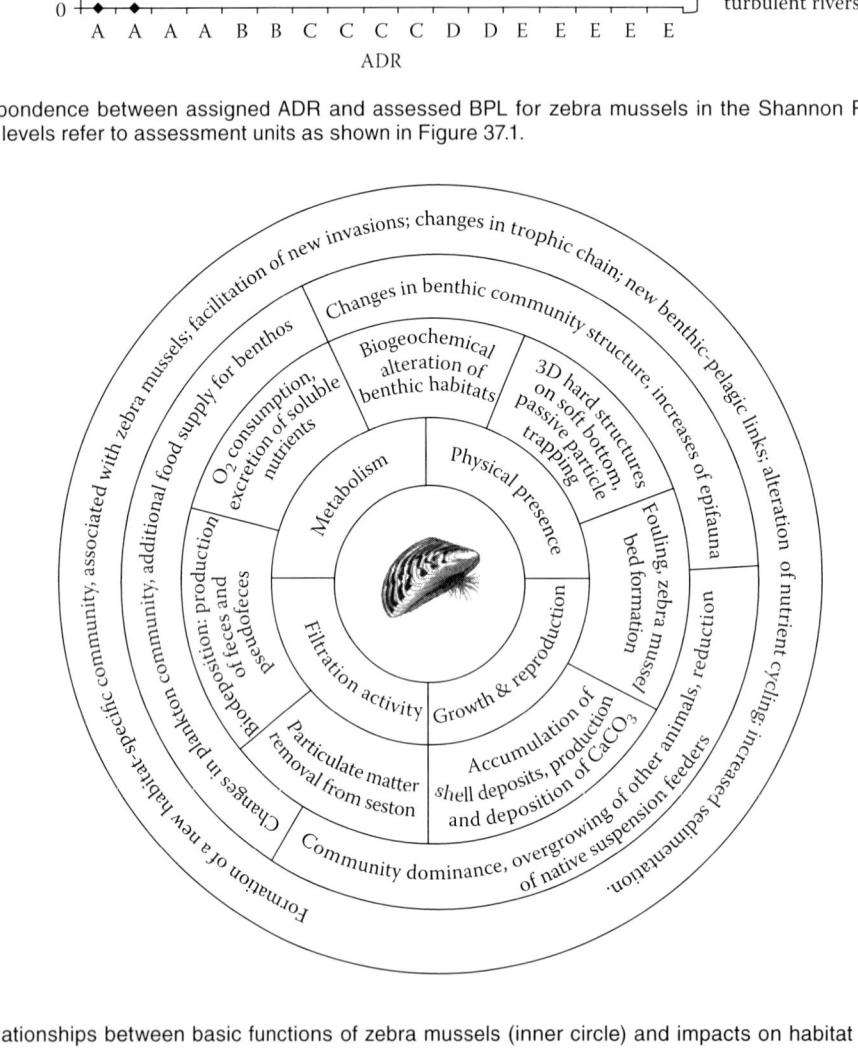

Figure 37.4 Interrelationships between basic functions of zebra mussels (inner circle) and impacts on habitat (second circle from the center), communities (third circle from the center), and ecosystems (outer circle). The cells are ordered correspondingly with impact interactions.

qualifies them as an ecosystem (habitat) engineer (Karatayev et al. 2002 and references therein, Zaiko et al. 2009). A conceptual framework that encompasses the overall effects of zebra mussels at the community, habitat, and ecosystem level is presented in Figure 37.4. Clearly, mussels have had a high impact on the Shannon River system by altering communities, habitats, and ecosystems mainly in lake assessment units by fouling most hard surfaces, forming new surfaces, and colonizing all habitats regardless of depth. This is true even in areas with muddy substrates where zebra mussels are attached to shells, stems, leaves, and exposed rhizomes of aquatic plants.

The five levels of the biopollution assessment method, applied in the current study, do not correspond with the five classes of the ecological status (high, good, moderate, poor, and bad) used in the European Water Framework Directive. This directive addresses assessment of the ecological quality status of water in European rivers, lakes, groundwaters, estuaries, and coasts. Our use of the biopollution assessment method did provide a separate basis for quantifying relative impacts of zebra mussels within assessment units. However, invasive species such as the zebra mussel present problems for managers as impacted areas are unlikely to revert to a good water quality status, as defined by community structure in the European Water Framework Directive, despite progressive management actions. In addition, overall impacts of invasive species can also vary over time independently of any management activities. The arrival and expansion of the zebra mussel is one of several invasive species to have arrived in Ireland in recent years. It is unclear whether the presence of zebra mussels has enabled the colonization and expansion of the Ponto-Caspian amphipod *Chelicorophium curvispinum* (Lucy et al. 2004), or the mysid *Hemimysis anomala* (Minchin and Boelens 2010), both of which have become abundant since zebra mussels arrived. In addition, the recent appearance of the Asian clam *Corbicula fluminea* in the Shannon River system will almost certainly add to future impacts. This bivalve obtains food resources, not only by means of filtration for the water column but also from sediments by means of an extendable foot. Other species likely to arrive in Ireland that will have impacts include the highly predacious, Ponto-Caspian amphipod *Dikerogammarus villosus,* recently having arrived in Britain (MacNeil et al. 2010), and the Asian teleost, *Pseudorasbora parva* (Rosecchi et al. 2006).

ACKNOWLEDGMENTS

We especially thank T. Nalepa, S. Olenin, D. Strayer, and an anonymous referee, for critical comments on drafts. The following agencies supported surveys: the Irish Environmental Protection Agency, which also provided water monitoring data, Marine Institute, Shannon River Basin District Management, Shannon Regional Fisheries Board (now Inland Fisheries Ireland) and Waterways Ireland, and the Lough Derg Science Group. We also thank R. Boelens, D. Dodd, A. Kluttig, F. Lucy, B. Minchin, M. Sullivan, and those who assisted in surveys. Part of this study involved the 6th European Framework Project ALARM. The conclusion of this work was made possible by the 7th EU Framework Project VECTORS, and World Laboratory grant within WFS Planetary Emergencies "Pollution".

REFERENCES

Atalah, J., M. Kelly-Quinn, K. Irvine, and T. P. Crowe. 2010. Impacts of invasion by *Dreissena polymorpha* (Pallas, 1771) on the performance of macroinvertebrate assessment tools for eutrophication pressure in Lakes. *Hydrobiologia* 654: 237–251.

Beisel, J. N., V. Bachmann, and J-Cl, Moreteau. 2010. Growth-at-length model and related life-history traits of *Dreissena polymorpha* in lotic ecosystems. In *The Zebra Mussel in Europe*, G. van der Velde, S. Rajagopal, and A. bij de Vaate, eds., pp. 191–197. Leiden, The Netherlands: Backhuys Publishers.

Botts, P. S., B. A. Patterson, and D. W. Schoesser. 1996. Zebra mussel effects on benthic invertebrates: Physical or biotic? *J. N. Am. Benthol. Soc.* 15: 1778–1788.

Bowman, J. J. 1998. *River Shannon Lake Water Quality Monitoring 1995 to 1997*. Wexford, Ireland: Irish Environmental Protection Agency.

Bowman, J. 2000. *River Shannon, Lake Water Quality Monitoring 1998 to 1999*. Wexford, Ireland: Environmental Protection Agency, Johnstown Castle Estate.

Burlakova, L., D. K. Padilla, A. Y. Karatayev, and D. Minchin. 2006. Endosymbionts of *Dreissena polymorpha* in Ireland: Evidence for the introduction of adult mussels. *J. Mollusc. Stud.* 72: 207–210.

Daunys, D., P. Zemlys, S. Olenin, A. Zaiko, and C. Ferrarin. 2006. Impact of the zebra mussel *Dreissena polymorpha* invasion on the budget of suspended matter in a shallow lagoon ecosystem. *Helgol. Mar. Res.* 60: 113–120.

Elliott, M. 2003. Biological pollutants and biological pollution—An increasing cause for concern. *Mar. Pollut. Bull.* 46: 275–280.

Haas, G., M. Brunke, and B. Streit. 2002. Fast turnover in dominance of exotic species in the Rhine River determines biodiversity and ecosystem function: An affair between amphipods and mussels. In *Invasive Aquatic Species of Europe: Distribution, Impacts and Spread*, E. Leppäkoski, S. Gollasch, and S. Olenin, eds., pp. 426–432. Dordrecht, The Netherlands: Kluwer Press.

Hincks, S. S. and G. L. Mackie. 1997. Effects of pH, calcium, alkalinity, hardness and chlorophyll on the survival, growth and reproductive success of the zebra mussel (*Dreissena polymorpha*) in Ontario lakes. *Can. J. Fish. Aquat. Sci.* 54: 2019–2057.

Horvath, T. G. and L. Crane. 2010. Hydrodynamic forces affect larval zebra mussel (*Dreissena polymorpha*) mortality in a laboratory setting. *Aquat. Invasions* 5: 379–385.

Horvath, T. G. and G. A. Lamberti. 1999. Mortality of zebra mussel, *Dreissena polymorpha*, veligers during downstream transport. *Freshwat. Biol.* 42: 69–76.

Hovarth, T. G., G. A. Lamberti, D. M. Lodge, and W. P. Perry. 1996. Zebra mussel dispersal in lake-stream systems: Source-sink dynamics? *J. N. Am. Benthol. Soc.* 11: 341–349.

Karatayev, A. Y., L. E. Burlakova, and D. Padilla. 2002. Impacts of zebra mussels on aquatic communities and their role as ecosystem engineers. In *Invasive Aquatic Species of Europe: Distribution, Impacts and Spread*, E. Leppäkoski, S. Gollasch, and S. Olenin, eds., pp. 433–466. Dordrecht, The Netherlands: Kluwer Press.

Karatayev, A. Y., D. K. Padilla, D. Minchin, D. Boltovskoy, and L. E. Burlakova. 2007. Changes in global economies and trade: The potential spread of exotic freshwater bivalves. *Biol. Invasions* 9: 161–180.

Kirk McClure Morton. 2001. Lough Ree and Lough Derg catchment monitoring and management system, Final Report. Belfast, Ireland: Kirk, McClure, Morton.

Kryger, J. and H. U. Riisgård. 1988. Filtration rate capacities in 6 species of European freshwater bivalves. *Oecologia* 77: 34–38.

Lucy, F. 2005. The dynamics of zebra mussel (*Dreissena polymorpha*) populations in Lough Key, Co. Roscommon, 1998–2003. PhD dissertation, Institute of Technology, Sligo, Ireland.

Lucy, F. E. 2006. Early life stages of *Dreissena polymorpha* (zebra mussel): The importance of long-term datasets in invasion ecology. *Aquat. Invasions* 1: 171–182.

Lucy, F. E., D. Minchin, and R. Boelens. 2008. From lakes to rivers: Downstream larval distribution of *Dreissena polymorpha* in Irish river basins. *Aquat. Invasions* 3: 297–304.

Lucy, F., D. Minchin, J. M. C. Holmes, and M. Sullivan. 2004. First records of the Ponto-Caspian amphipod *Chelicorophium curvispinum* (Sars, 1895) in Ireland. *Ir. Nat. J.* 27: 461–462.

Lucy, F. and M. Sullivan. 2001. The investigation of an invasive species, the zebra mussel Dreissena polymorpha in Lough Key, Co Rosscommon, 1999. Desktop Study No. 13. Wexford, Ireland: Environmental Protection Agency.

Lucy, F., M. Sullivan, and D. Minchin. 2005. Nutrient levels and the zebra mussel population in Lough Key (2000-MS-5-M1). Synthesis Report. Wexford, Ireland: Environmental Protection Agency.

Ludyanskiy, M. L., D. McDonald, and D. MacNeill. 1993. Impact of the zebra mussel, a bivalve invader. *BioScience* 43: 533–544.

Lvova, A. A., G. E. Makarova, A. F. Alimov, A. Y. Karataev, M. P. Miroshnichenko, V. P. Zakutskii, and M. N. Nekrasova. 1994. Growth and production. In *Freshwater Zebra Mussel Dreissena polymorpha (Pall.) (Bivalvia, Dreissenidae): Systematics, Ecology, Practical Meaning*, J. I. Starobogatov, ed., pp. 156–179. Moscow, Russia: Nauka (in Russian).

MacNeil, C., D. Platvoet, and J. T. A. Dick et al. 2010. The Ponto-Caspian 'killer shrimp', *Dikerogammarus villosus* (Sowinsky, 1894), invades the British Isles. *Aquat. Invasions* 5: 441–445.

McCarthy, T. K., J. Fitzgerald, P. Cullen, D. Doherty, and L. Copley. 1997. Zebra mussels in the River Shannon. A report to the ESB Fisheries Conservation. Clare, Ireland: Ardnacrusha.

McCarthy, T. K., J. Fitzgerald, and W. O'Connor. 1998. The occurrence of the zebra mussel *Dreissena polymorpha* (Pallas. 1771), an introduced biofouling freshwater bivalve in Ireland. *Ir. Nat. J.* 25: 413–415.

Mikheev, V. P. 1967. The nutrition of *Dreissena*. *Trudy Vsesoyuznogo nauchno-isskdovatel'skogo Instituta*, 15: 117–129 (in Russian).

Mikheev, V. P. 1994. Selectivity of zebra mussel feeding. In *Freshwater Zebra Mussel Dreissena polymorpha (Pall.) (Bivalvia, Dreissenidae): Systematics, Ecology, Practical Meaning*, J. I. Starobogatov, ed., pp. 126–129. Moscow, Russia: Nauka (in Russian).

Minchin, D. and R. Boelens. 2005. *Hottonia palustris* L. (water violet) established in Lough Derg, N. Tipperary (H10). *Ir. Nat. J.* 28: 136–137.

Minchin, D. and R. Boelens. 2010. *Hemimysis anomala* is established in the Shannon River Basin District in Ireland. *Aquat. Invasions* 5(Suppl. 1): S71–S78.

Minchin, D. and R. Boelens. 2011. The distribution and expansion of ornamental plants on the Shannon Navigation. *Proc. R. Ir. Acad.*, Dublin 111B: 1–9.

Minchin, D., O. Floerl, D. Savini, and A. Occhipinti-Ambrogi. 2006a. Small craft and the spread of aquatic species. In *The Ecology of Transportation: Managing Mobility for the Environment*, J. Davenport and J. L. Davenport, eds., pp. 99–118. Dordrecht, The Netherlands: Springer.

Minchin, D. and S. Gollasch. 2003. Fouling and ships' hulls: How changing circumstances and spawning events may result in the spread of exotic species. *Biofouling* 19(Suppl.): 111–122.

Minchin, D., F. Lucy, and M. Sullivan. 2002. Zebra mussel: Impacts and spread. In *Invasive Aquatic Species of Europe: Distribution, Impacts and Spread*, E. Leppäkoski, S. Gollasch, and S. Olenin, eds., pp. 135–146. Dordrecht, The Netherlands: Kluwer Press.

Minchin, D., F. Lucy, and M. Sullivan. 2006b. Ireland: A new frontier for the zebra mussel *Dreissena polymorpha*. *Oceanol. Hydrobiol. Stud.* 34: 19–30.

Minchin D., C. Maguire, and R. Rosell. 2003. The zebra mussel (*Dreissena polymorpha*) Pallas invades Ireland: Human mediated vectors and the potential for rapid intranational dispersal. *Proc. R. Ir. Acad.*, Dublin 103B: 23–30.

Minchin, D. and C. Moriarty. 1998. Zebra mussels in Ireland. In *Fisheries Leaflet* No. 177, pp. 1–11. Dublin, Ireland: Marine Institute.

Minchin, D. and C. Moriarty. 1999. Distribution of the zebra mussel *Dreissena polymorpha* (Pallas) in Ireland, 1997. *Ir. Nat. J.* 26: 38–42.

Minchin, D. and C. Moriarty. 2002. Zebra mussels and their impact in Ireland. In *Biological Invaders: The Impact of Exotic Species*, C. Moriarty and D. Murray, eds., pp. 72–78. Dublin, Ireland: Royal Irish Academy.

Narščius, A., S. Olenin, A. Zaiko, and D. Minchin. 2012. Biological invasion impact assessment system: From idea to implementation. *Ecol. Inform.* 7: 46–51.

Olenin, S., D. Minchin, and D. Daunys. 2007. Assessment of biopollution in aquatic ecosystems. *Mar. Pollut. Bull.* 55: 379–394.

Olenin, S., D. Minchin, D. Daunys, and A. Zaiko. 2010. Biological pollution of aquatic ecosystems in Europe. In *Atlas of Biodiversity Risk*, J. Settele, L. Penev, and T. Georgiev et al., eds., pp. 136–137. Sofia, Bulgaria: Pensoft.

Olenina, I., N. Wasmund, S. Hajdu et al. 2010. Assessing impacts of invasive phytoplankton. The Baltic Sea case. *Mar. Pollut. Bull.* 60: 1691–1700.

Orendt, C., C. Schmitt, C. van Liefferinge, G. Wolfram, and E. de Deckere. 2010. Include or exclude? A review on the role and suitability of aquatic invertebrate neozoa as indicators in biological assessment with special respect to fresh and brackish European waters. *Biol. Invasions* 12: 265–283.

Pollux, B., D. Minchin, G. Van der Velde, T. Van Alen, S. Van der Staay, and J. Hackstein. 2003. Zebra mussels in Ireland (*Dreissena polymorpha*), AFLP-fingerprinting and boat traffic both indicate an origin from Britain. *Freshwat. Biol.* 48: 1127–1139.

Pyšek, P. and D. Richardson. 2010. Invasive species, environmental change and management, and health. *Environ. Resour.* 35: 25–55.

Ramcharan, C. W., D. K. Padilla, and S. I. Dodson. 1992. Model to predict potential occurrence and density of the zebra mussel, *Dreissena polymorpha*. *Can. J. Fish. Aquat. Sci.* 49: 2611–2620.

Rosecchi, E., A. J. Crivelli, and G. Catsadorakis. 2006. The establishment and impact of *Pseudorasbora parva*, an exotic fish species introduced into lake Mikri Prespa (north-western Greece). *Aquat. Conserv.* 3: 223–231.

Shkorbatov, G. L., A. F. Karpevich, and P. I. Antonov. 1994. Ecological physiology. In *Freshwater Zebra Mussel Dreissena polymorpha (Pall.) (Bivalvia, Dreissenidae): Systematics, Ecology, Practical Meaning*, J. I. Starobogatov, ed., pp. 67–108. Moscow, Russia: Nauka (in Russian).

Spidle, A. P., E. L. Mills, and B. May. 1995. Limits to tolerance of temperature and salinity in the quagga mussel (*Dreissena bugensis*) and the zebra mussel (*Dreissena polymorpha*). *Can. J. Fish. Aquat. Sci.* 52: 2108–2119.

Sprung, M. 1991. "Costs" of reproduction: A study on metabolic requirements of the gonads and fecundity of the bivalve *Dreissena polymorpha*. *Malacologia* 33: 63–70.

Stewart, T. W., J. G. Miner, and R. L. Lowe. 1998. Quantifying mechanisms for zebra mussel effects on benthic macroinvertebrates: Organic matter production and shell-generated habitat. *J. N. Am. Benthol. Soc.* 17: 81–94.

Sullivan, M., F. Lucy, and D. Minchin. 2010. The association between zebra mussels and aquatic plants in the Shannon River system in Ireland. In *The Zebra Mussel in Europe*, G. van der Velde, S. Rajagopal and A. bij de Vaate, eds., pp. 113–118. Leiden, The Netherlands: Backhuys Publishers.

Timpson, P. and F. Lucy. 1998. *Zebra Mussel Monitoring Programme, Lough Allen*. Sligo, Ireland: School of Science, Institute of Technology.

Toner, P., J. Bowman, K. Clabby, J. Lucey, M. McGarrigle, C. Concannon et al. 2005. *Water Quality in Ireland 2001–2003*. Wexford, Ireland: Environmental Protection Agency.

Vanderploeg, H. A., J. R. Liebig, T. F. Nalepa, G. L. Fahnenstiel, G. L., and S. A. Pothoven. 2010. *Dreissena* and the disappearance of the spring phytoplankton bloom in Lake Michigan. *J. Great Lakes Res.* 36: 50–59.

Zaiko, A. and D. Daunys. 2012. Density effects on the clearance rate of the zebra mussel *Dreissena polymorpha*: Flume study results. *Hydrobiologia* 680: 79–89.

Zaiko, A., D. Daunys, and S. Olenin. 2009. Habitat engineering by the invasive zebra mussel *Dreissena polymorpha* (Pallas) in a boreal coastal lagoon: Impact on biodiversity. *Helgol. Mar. Res.* 63: 85–94.

Zaiko, A., M. Lehtiniemi, A. Narščius, and S. Olenin. 2011. Assessment of bioinvasion impacts on a regional scale: A comparative approach. *Biol. Invasions* 13: 1739–1765.

CHAPTER 38

Impacts of *Dreissena* on Benthic Macroinvertebrate Communities
Predictable Patterns Revealed by Invasion History

Jessica M. Ward and Anthony Ricciardi

CONTENTS

Abstract 599
Introduction 599
Mechanisms of Impacts: Filtration and Substrate Alteration 600
Impacts on Macroinvertebrate Abundance, Diversity, and Composition 601
Impacts on Native Bivalves 603
Predictors of *Dreissena* Impact 604
Temporal Variability of Impact 605
Conclusions and Future Research Directions 605
Acknowledgments 606
References 606

ABSTRACT

Invasive dreissenid mussels (zebra mussel *Dreissena polymorpha* and quagga mussel *Dreissena rostriformis bugensis*) are transforming benthic macroinvertebrate communities in lakes and rivers throughout Europe and North America. Significant changes to these communities are documented typically within a few years to a decade following dreissenid invasion. These changes almost invariably include increased abundance and richness, and reduced evenness, of mussel-associated macroinvertebrate communities. Nearshore communities (excluding *Dreissena*) typically increase by 2–3 times in density and biomass because of increased interstitial habitat from mussel aggregations and nourishment from mussel biodeposits. In contrast, deepwater communities decline by a similar magnitude, likely as a consequence of particulate organic matter being shunted to nearshore areas at the expense of offshore areas by filtration activities of dreissenid populations. Impacts of *Dreissena* on nearshore communities differ among taxa, with disproportionately strong positive effects on mayflies (Ephemeroptera), leeches (Hirudinea), flatworms (Turbellaria), gastropods, and gammarid amphipods. In habitats supporting dense dreissenid populations (>10^3 mussels/m^2), severe declines in native bivalves (Unionidae and Sphaeriidae) commonly occur. The magnitude of these effects varies across space and time with dreissenid population structure and habitat conditions.

INTRODUCTION

Zebra mussels (*Dreissena polymorpha*) and quagga mussels (*Dreissena rostriformis bugensis*) continue to invade and ecologically transform lakes and rivers throughout Europe and North America (Brown and Stepien 2010, Zhulidov et al. 2010). Their impacts on benthic macroinvertebrate communities, in particular, have been the focus of a growing number of studies in recent decades. Collectively, these studies enable quantitative comparisons to determine whether dreissenid impacts are consistent across different ecosystems or, if not, whether their variation follows predictable patterns. A predictive understanding of such impacts is of practical importance to environmental management, given that benthic macroinvertebrates are: (1) valuable indicators of water quality (Washington 1984, Gabriels et al. 2005); (2) functionally important in freshwater ecosystems as mediators of nutrient cycling and energy flow (Covich et al. 1999); and (3) an essential food resource for higher consumers (Vander Zanden and Vadaboncoeur 2002, Schummer et al. 2008). Indeed, changes in the composition and abundance of benthic macroinvertebrates can alter the diet and condition of fish (Owens and Dittman 2003, McNickle et al. 2006). Moreover, accurate models to predict impacts are needed to ensure that observed changes caused by *Dreissena* are not mistakenly attributed to watershed management schemes (e.g., phosphorus abatement) and other environmental stressors, such as contaminants and climate change (MacNeil and Briffa 2009, Atalah et al. 2010).

MECHANISMS OF IMPACTS: FILTRATION AND SUBSTRATE ALTERATION

Dreissena can alter benthic macroinvertebrate communities in lakes and rivers through a variety of mechanisms, but most directly through physical alteration of substrates (Figure 38.1). Like their marine relatives (Suchanek 1986, Lohse 1993), zebra and quagga mussels form dense aggregations (clumped shells of live and dead individuals) by attaching to substrates and to each other's shells with byssal threads; these aggregations provide colonizable surfaces for algae and structurally complex habitat for macroinvertebrates (Ricciardi et al. 1997, Stewart et al. 1998a). Interstitial spaces between mussel shells passively trap sediments and are used by smaller invertebrates as refugia from predation and environmental stressors such as wave action (Gutiérrez et al. 2003, Beekey et al. 2004, Reed et al. 2004). Although common in coastal marine systems where byssally attached molluscs abound (Suchanek 1986), this type of habitat engineering was nonexistent in European and North American freshwater habitats prior to colonization by *Dreissena*.

Dreissenid mussels link offshore planktonic and nearshore benthic communities by filtering suspended particles from the water column and depositing the material as mucus-bound feces and undigested pseudofeces (Karatayev et al. 1997, Gergs et al. 2009). This link has been referred to as the pelagic-to-benthic shunt because it reduces available planktonic resources and increases the organic content of sediments surrounding dreissenid mussel beds (Karatayev et al. 1997, Stewart et al. 1998a, Hecky et al. 2004). Mussel

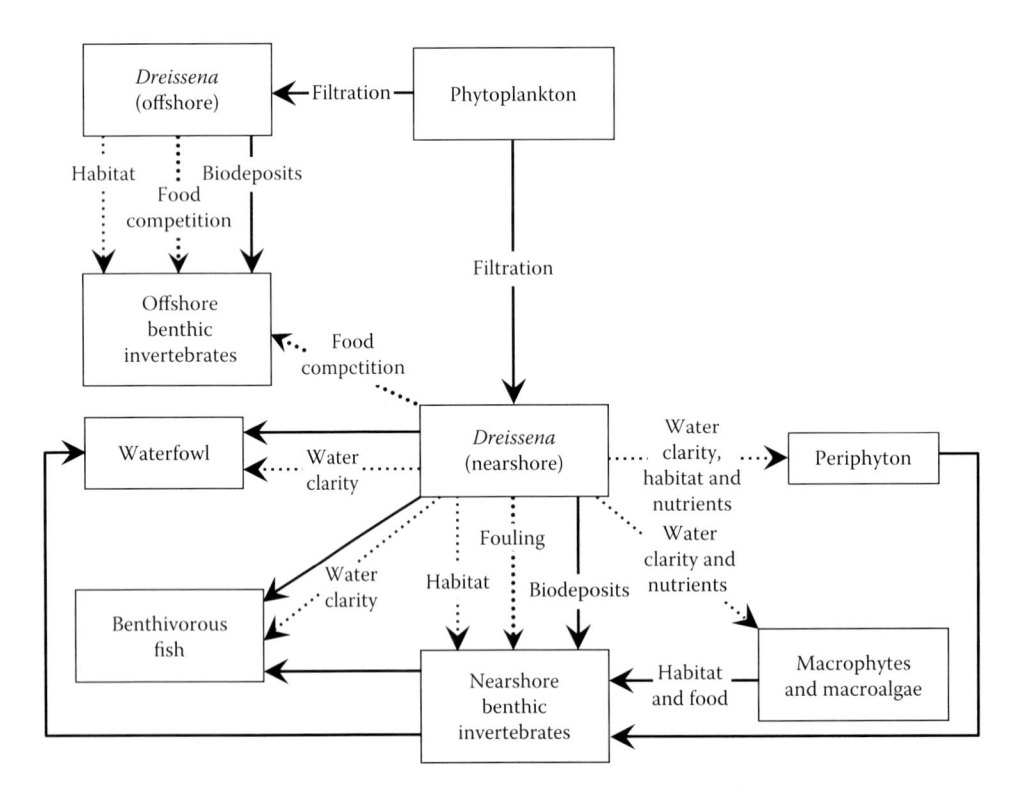

Figure 38.1 Mechanisms of interactions between *Dreissena* and benthic invertebrates through direct food web connections (solid lines), habitat engineering (thick dashed lines), and fouling (thin dashed lines). "Habitat" can be defined as a change in physical conditions (e.g., created by mussel shells or altered sediment composition and chemistry created by mussel biodeposits: feces and pseudofeces). Water clarity potentially improves efficiency of predation by waterfowl and fish on benthic invertebrates. Fouling (dense colonization) of sedentary invertebrates with shells or exoskeletons (e.g., native bivalves and odonate nymphs) impairs metabolism and imposes energetic costs that may reduce survivorship. Offshore invertebrates are affected by competition for seston (from both local and nearshore dreissenid populations) as well as by habitat and biodeposits generated locally.

feces and pseudofeces contain both live and dead phytoplankton, organic detritus, and inorganic particles (Roditi et al. 1997), which can be a source of nourishment to detritivores (Izvekova and Lvova-Katchanova 1972, Stewart et al. 1998a). Growth in gammarid amphipods and chironomid larvae, for example, readily occurs when dreissenid biodeposits are the food source (Izvekova and Lvova-Katchanova 1972, González and Burkart 2004), and both these taxa increase prolifically with dreissenid colonization (Ricciardi et al. 1997, Ward and Ricciardi 2007).

Through their filtration activity, dreissenids increase the depth to which light penetrates in lakes and large rivers, thereby stimulating benthic production even at distances remote from mussel beds (Vanderploeg et al. 2002, Cecala et al. 2008). The resultant enhanced growth of benthic algae and macrophytes can provide food or physical habitat to macroinvertebrates, particularly grazers, causing an expansion of littoral populations (Mayer et al. 2002). Also, at various sites in the lower Great Lakes and St. Lawrence River, there have been observations of increased abundance of freshwater sponges (Early et al. 1996, Ricciardi et al. 1997), whose algal symbionts allow them to grow luxuriantly in enhanced light conditions. On the other hand, blooms of filamentous macroalgae have also been linked to dreissenid filtration activity (Auer et al. 2010) and these blooms have mixed effects on benthic macroinvertebrates (reviewed by Ward and Ricciardi 2010).

Dreissenid colonization has strong negative effects on some benthic taxa (Ward and Ricciardi 2007). High filtration capacity and biomass (Karatayev et al. 1997, Vanderploeg et al. 2002) of dreissenid populations allow them to outcompete native filter feeders for food. Some evidence suggests mussels displace large net-spinning caddisfly larvae (Trichoptera) by fouling their preferred attachment sites on rocky substrates (Ricciardi et al. 1997). Similarly, abundances of large snails (i.e., individuals of comparable or greater size than dreissenids) may be reduced on dreissenid-covered substrates, perhaps because of foraging inefficiency (Wisenden and Bailey 1995, Ricciardi et al. 1997, Ward and Ricciardi 2007). In addition, *Dreissena* attaches to exoskeletons and shells of other macroinvertebrates that have no evolutionary experience with fouling organisms. High densities of attached dreissenids can impair normal activities of snails (Van Appledorn et al. 2007, Van Appledorn and Bach 2007) and, especially, of other bivalves (Baker and Hornbach 1997, 2000; see Impacts On Native Bivalves). Attachment of dreissenids on dragonfly larvae has caused reductions in emergence rates, larval mobility, foraging efficiency, and survivorship (Fincke et al. 2009, Fincke and Tylczak 2011)—the latter possibly resulting from energy loss and increased exposure to predators (Fincke et al. 2009). Attachment to crayfish has also been observed and is believed to impose an energetic cost to individuals (Brazner and Jensen 2000, Ďuriš et al. 2007), but consequences to crayfish fitness have not been studied.

IMPACTS ON MACROINVERTEBRATE ABUNDANCE, DIVERSITY, AND COMPOSITION

Generalizations can be made regarding the impacts of dreissenids on the abundance, diversity, and composition of benthic macroinvertebrate communities. In a meta-analysis of published data (Ward and Ricciardi 2007), we quantified the relative effects of *Dreissena* by dividing the value of a response variable (e.g., macroinvertebrate abundance) after *Dreissena* colonization (X_{+D}) by its value before *Dreissena* colonization (X_{-D}) and then normalized this metric by natural logarithmic (ln) transformation. Typically, dreissenid invasions are associated with increases in the local abundance and taxonomic richness of benthic macroinvertebrates (excluding unionids and dreissenids), whereas community evenness (i.e., Simpson's indices) generally declines following dreissenid invasion (Figure 38.2). Collectively,

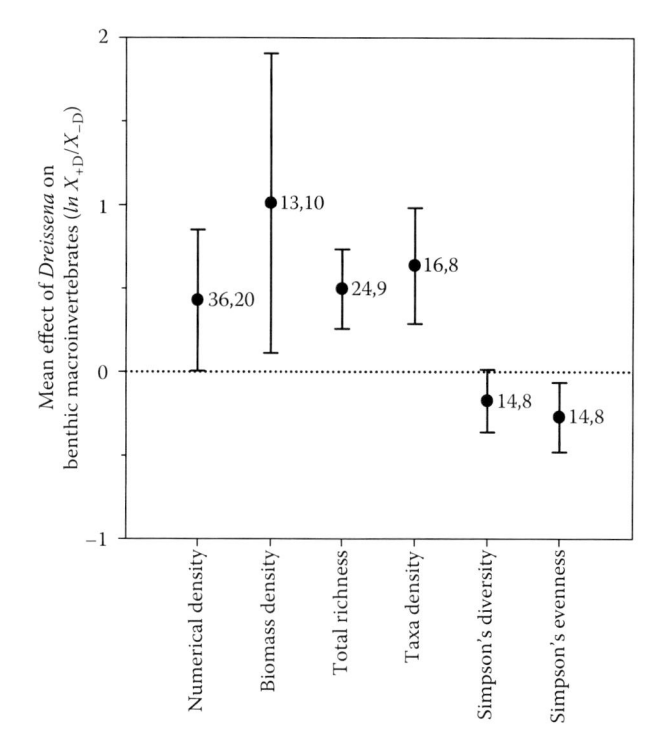

Figure 38.2 Mean impacts of *Dreissena* on benthic macroinvertebrate density (no./m^2), biomass density (g/m^2), total richness (number of taxa), taxa density (no./m^2), and Simpson's diversity and evenness. All calculations exclude Unionidae and *Dreissena*. Relative effects were calculated by dividing the value of a response variable after *Dreissena* colonization (X_{+D}) by its value before *Dreissena* colonization (X_{-D}), followed by ln-transformation (Ward and Ricciardi 2007). Numbers (a,b) shown beside data points are (a) the number of study sites used to calculate each average and (b) the number of studies that served as data sources. Only sites with dreissenid densities >100/m^2 were used as post-colonization data. Positive effects indicate an increase and negative effects indicate a decrease following *Dreissena* invasion. Error bars are 95% confidence intervals. Effect sizes are statistically significant when they do not overlap zero.

these impacts show that dreissenid invasions tend to positively influence benthic macroinvertebrates, but some taxa benefit disproportionately. Experiments and field surveys have demonstrated that communities associated with dense *Dreissena* colonies are dominated by a few abundant taxa such as snails, flatworms, chironomid larvae, and amphipods; these patterns are robust across studies (e.g., Botts et al. 1996, Ricciardi et al. 1997, Stewart et al. 1998a,c, Mörtl and Rothhaupt 2003, Karatayev et al. 2010).

The direction and magnitude of dreissenid effects vary across different substrate types (Figure 38.3). Dreissenid colonization of rocky substrates is almost always accompanied by a contemporaneous increase in densities of other macroinvertebrates. The increase in density may be an order of magnitude higher than the density before dreissenid colonization; this strong positive effect can be attributed to increased interstitial habitat provided by mussel aggregations and, to a lesser extent, by mussel filtration and biodeposition activities (Botts et al. 1996, Ricciardi et al. 1997, Stewart et al. 1998a, Horvath et al. 1999, Mörtl and Rothhaupt 2003). Dreissenid colonization of mixed substrates (including fine sediments) also generally enhances densities of associated macroinvertebrates, but results are much more varied. These patterns reflect differential responses by various functional groups of macroinvertebrates. In general, after dreissenid colonization there is an increase in densities of epifaunal taxa (i.e., those that normally occupy solid surfaces) and a decline in densities of infaunal or burrowing taxa. The latter response reflects the transformation of preferred habitat by dense mussel colonies extending onto soft sediments (Ward and Ricciardi 2007). The inhibition of burrowing macroinvertebrates by byssally attached mussels has also been reported in marine systems where sediments underlying dense mussel beds may become inhospitable for infauna through smothering and anoxia (Creese et al. 1997, Robinson and Griffiths 2002). Similarly, severe oxygen depletion may occur in sediments covered by dreissenid mussels, which can reduce the richness of invertebrates occupying these sediments (Beekey et al. 2004). Despite these negative effects, dreissenid colonization of sandy and silty substrates usually produces an overall increase in richness of local macroinvertebrate communities (Figure 38.3) by offering surfaces for colonization by invertebrates adapted to rocky substrates (Ward and Ricciardi 2007).

After dreissenid invasions of lakes, conspicuous reductions in some nondreissenid benthic macroinvertebrates can occur in the littoral zone (Haynes et al. 1999, Ratti and Barton 2003), but reductions are more pronounced in sublittoral and deepwater regions (Figure 38.4). In the Great Lakes, precipitous declines of the native deposit-feeding amphipod *Diporeia* spp. coincided with the introduction and spread of dreissenids (Nalepa et al. 1998, 2003, Dermott 2001). Prior to its collapse, *Diporeia* was the most abundant benthic macroinvertebrate at deepwater sites, often found at densities of 2000–8000/m² (Owens and Dittman 2003). Declines of *Diporeia* and other macroinvertebrates at offshore sites are thought to be a response to the reduced influx of food particles caused by the enormous filtration capacity of nearshore dreissenid populations (Dermott and Kerec 1997, Nalepa et al. 1998, Dermott 2001, Lozano et al. 2001, Soster et al. 2011).

Effects of *Dreissena* on macroinvertebrate densities are more predictable for some taxonomic groups than for others (Figure 38.5). Dreissenid colonization is frequently associated with increased local densities of leeches (Hirudinea), flatworms (Turbellaria), snails (Gastropoda), and mayfly nymphs (Ephemeroptera). Small snails and other herbivorous taxa (such as heptageniid mayflies) that scrape substrates for food likely

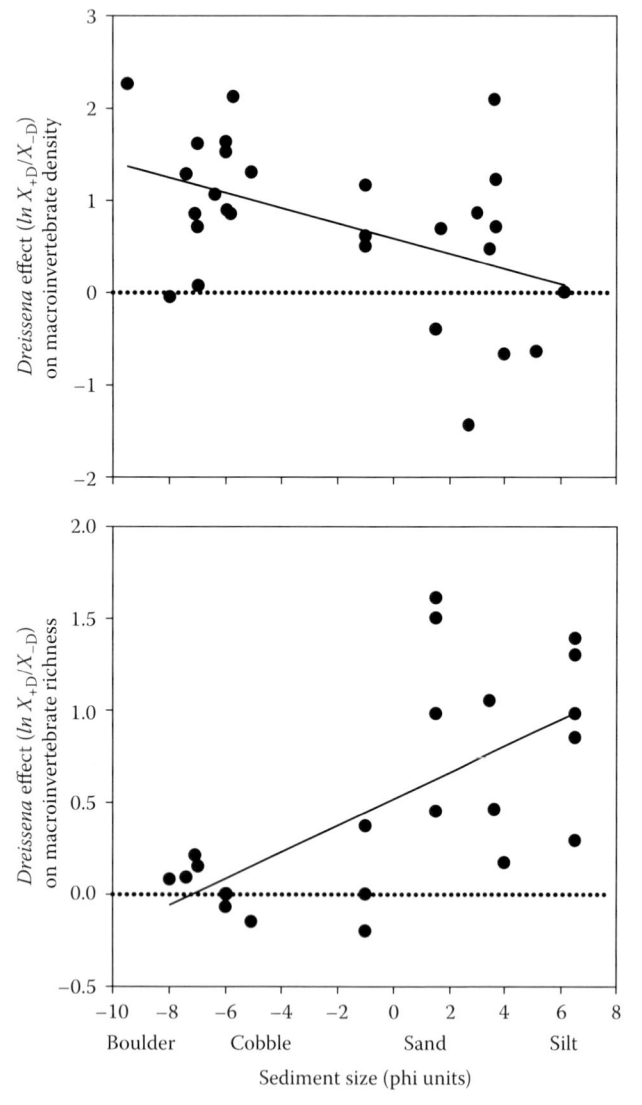

Figure 38.3 Community-level impacts of *Dreissena* as a function of mean size of surrounding substrate (phi units = $-\log_2$ diameter in mm). Effects were calculated as in Figure 38.2. As substrate particle size becomes smaller (increasing phi), the effect of *Dreissena* on density becomes more variable and less positive (a: $r = -0.47$, $P = 0.01$), whereas the effect on total taxonomic richness becomes stronger and more positive (b: $r = 0.66$, $P = 0.0006$). Lines generated by least-squares regression. (After Ward, J.M. and Ricciardi, A., *Divers. Distrib.*, 13, 155, 2007.)

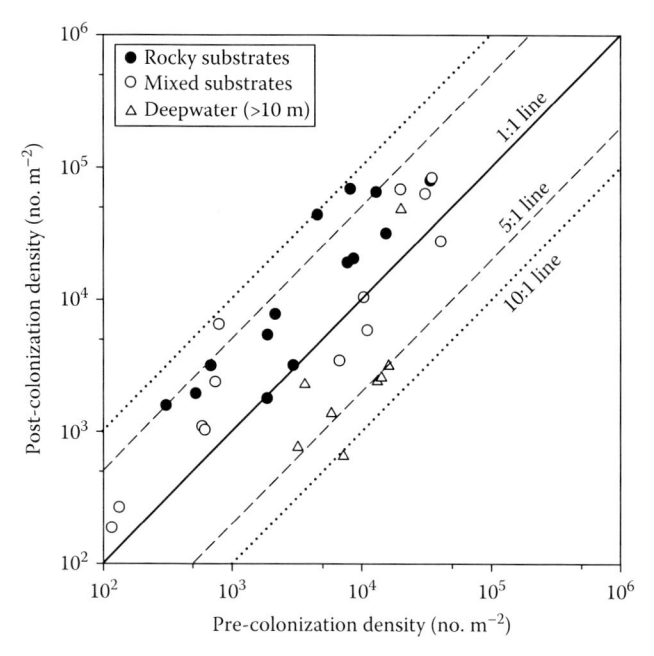

Figure 38.4 Density (no./m²) of benthic macroinvertebrates (excluding *Dreissena* and Unionidae) reported on littoral rocky substrates, littoral mixed substrates, and substrates of deepwater environments in pre- and post-*Dreissena* periods. Each point represents a study site (n = 33 sites in 13 water bodies). Points above the 1:1 line indicate increased density of benthic macroinvertebrates. Data sources: Dusoge (1966), Dermott et al. (1993), Griffiths (1993), Stewart and Haynes (1994), Karatayev and Burlakova (1995), Botts et al. (1996), Dermott and Kerec (1997), Ricciardi et al. (1997), Stewart et al. (1998c), Horvath et al. (1999), Kuhns and Berg (1999), Bially and MacIsaac (2000), Lozano et al. (2001), Strayer and Smith (2001), Mayer et al. (2002), Nalepa et al. (2002), Mörtl and Rauthhaupt (2003), Owens and Dittman (2003), and Beekey et al. (2004).

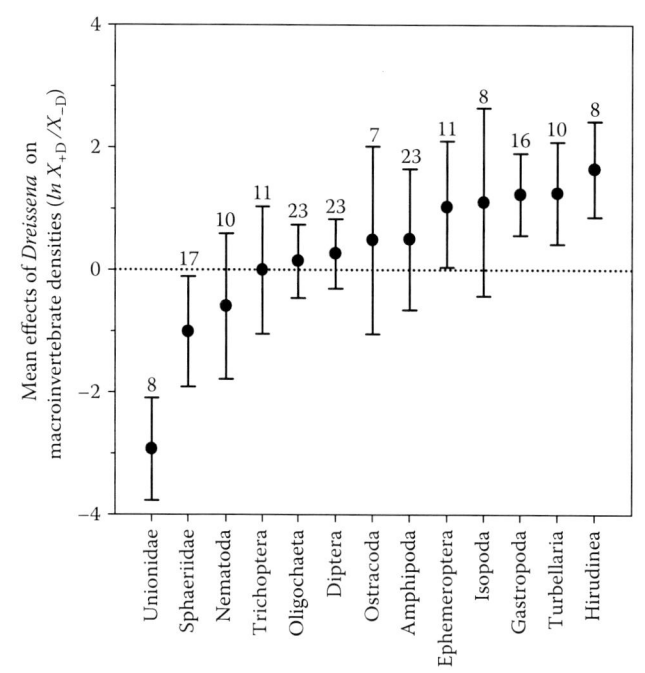

Figure 38.5 Mean effects of *Dreissena* on densities (no./m²) of major taxonomic groups of benthic macroinvertebrates. Effects were calculated as in Figure 38.2. The number of study sites used to calculate each average is given above each data point. Positive values indicate increased density and negative values indicate decreased density in response to *Dreissena*. Error bars are 95% confidence intervals, and effect sizes are statistically significant when they do not overlap zero. Data sources for Unionidae: Ponyi (1992), Nalepa (1994), Ricciardi et al. (1995), Schloesser and Masteller (1999), Martel et al. (2001), Schloesser et al. (2006), Strayer and Malcom (2007), and Maguire et al. (2003). Data sources (20 published studies) for other taxonomic groups are provided in Ward and Ricciardi (2007). (Derived and modified from Ward, J.M. and Ricciardi, A., *Divers. Distrib.*, 13, 155, 2007.)

benefit from increased grazing area and spatial refugia provided by dreissenid shells, although large-bodied snail taxa (e.g., Pleuroceridae) tend to be reduced or displaced (Wisenden et al. 1995, Ricciardi et al. 1997, Ward and Ricciardi 2007). Increased densities of leeches, flatworms, and other predatory species associated with dreissenids likely reflect responses of these taxa to refugia from benthivorous fish and to increased availability of positively affected prey (Ricciardi et al. 1997, Stewart et al. 1998a, Ward and Ricciardi 2007, Gergs and Rothhaupt 2008).

Several broad taxonomic groups are characterized by a mixture of positive and negative responses to dreissenids (Figure 38.5), which reflects variation among species within the group. For instance, families within the group Trichoptera (larval caddisflies) have distinct feeding strategies (Wiggins 1996) and thus may be influenced by dreissenids in diverse ways. Larvae in the Helicopsychidae family, like other small grazers, are positively affected by dreissenid shells (Ricciardi et al. 1997, Horvath et al. 1999). Small filter-feeding caddisflies, such as *Brachycentrus* (Brachycentridae), appear to exploit filtration currents generated by dreissenids, whereas large net-spinning caddisflies in the family Polycentropodidae avoid currents generated by dreissenid filtration or become excluded when dense colonies of *Dreissena* interfere with the construction and placement of their nets (Ricciardi et al. 1997). Similarly, although meta-analysis indicates a nonsignificant, highly variable effect of dreissenids on the density of Amphipoda as a group, more consistent patterns occur at a finer taxonomic resolution (Ward and Ricciardi 2007). Large increases in densities of epibenthic gammarid amphipods (e.g., *Gammarus fasciatus* and *Echinogammarus ischnus*) are correlated with *Dreissena* colonization (Ricciardi 2003), whereas densities of deepwater amphipods (*Diporeia*, as described earlier) typically decline.

IMPACTS ON NATIVE BIVALVES

Dreissena has consistent negative effects on densities of native bivalves (Figure 38.5). Declines in sphaeriid clams (Sphaeriidae) have frequently occurred at local and habitat-wide scales following *Dreissena* invasion and are attributed primarily to effects of competition with *Dreissena* for food (Nalepa et al. 1998, Lauer and McComish 2001, Lozano et al. 2001, Strayer and Smith 2001). Impacts on unionid mussels (Unionidae) are more conspicuous and have received more attention. Unionids had already suffered declines throughout North America in response to habitat degradation during the previous century, but their rate of population loss increased 10-fold, on average, in habitats where dense *Dreissena* populations (>3000/m^2) became established (Ricciardi et al. 1998). Rapid declines of unionid populations were observed following *Dreissena* colonization in the Great Lakes (Nalepa 1994, Schloesser and Nalepa 1994, Schloesser and Masteller 1999, Schloesser et al. 2006), St. Lawrence River (Ricciardi et al. 1996), Hudson River (Strayer and Malcom 2007), Ohio River (Ricciardi et al. 1998, Watters and Flaute 2010), and various smaller water bodies (Harman 2000, Martel et al. 2001). These events have been attributed to both the effects of fouling (overgrowth of unionid shells by attached dreissenids) and, to a lesser extent, competition for food (Parker et al. 1998, Baker and Levinton 2003). Indeed, there is a strong correlation between fouling intensity by dreissenids and mortality of unionids (Figure 38.6). This correlation is supported by experimental studies that show attachment by dreissenids interferes with normal metabolic activities of unionids and weakens their physiological condition (Haag et al. 1993, Baker and Hornbach 1997, 2000, Sousa et al. 2011), whereas the removal of attached dreissenids enhances unionid survival (Hallac and Marsden 2001). Empirical models predict that unionid populations will undergo severe (>90%) declines when individual unionids carry, on average, more than their own weight in attached dreissenids—a condition that generally occurs at mean fouling intensities of >100 dreissenids per unionid and dreissenid population densities on the order of 10^3/m^2 (Ricciardi et al. 1995, 1998). North American water bodies that support such dreissenid densities may experience near total extirpation of unionid populations within 4–8 years following *Dreissena* invasion (Ricciardi et al. 1998). However, unionids may persist at low densities in the presence of *Dreissena* (Strayer and Malcom 2007) or in refugia with locally reduced dreissenid densities within systems invaded by *Dreissena* (Zanatta et al. 2002, Crail et al. 2011).

The European experience has not provided a useful reference for predicting impacts of *Dreissena* on native bivalves in North America. Mass mortalities of unionids have rarely been reported for European water bodies invaded by *Dreissena* (reviewed by Karateyev et al. 1997, Ricciardi et al. 1998), possibly because impacts that might have occurred early in the *Dreissena* invasion (e.g., up to two centuries ago) were undocumented. In addition, *Dreissena* had already expanded into central Europe prior to the last glaciation before subsequently retreating to the Ponto-Caspian basins (Stańczykowska 1977). Thus, unionids in Europe might have already experienced evolutionary pressures to adapt to dominant macrofouling organisms such as *Dreissena* and not have the same ecological sensitivity to fouling as unionids in North America (Ricciardi et al. 1998). Evidence to support this hypothesis may be found in the recent invasion of Irish lakes by *Dreissena*. Unlike continental European assemblages, the fauna of these lakes have no evolutionary experience with *Dreissena*, and their unionids have suffered strong negative impacts resembling those in North America (Maguire et al. 2003).

PREDICTORS OF *DREISSENA* IMPACT

Impacts of an introduced species are expected to be a function of its abundance (Parker et al. 1999). Relationships between impact and abundance have been demonstrated experimentally and statistically for *Dreissena* density and impacts on phytoplankton (Bastviken et al. 1998), nutrient cycling (Mellina et al. 1995), unionid mortality (Ricciardi et al. 1995), and macroinvertebrate abundance on rocky substrates (Ricciardi 2003). Hence, impacts of *Dreissena* are moderated by environmental variables that limit its abundance (e.g., Ramcharan et al. 1992, Mellina and Rasmussen

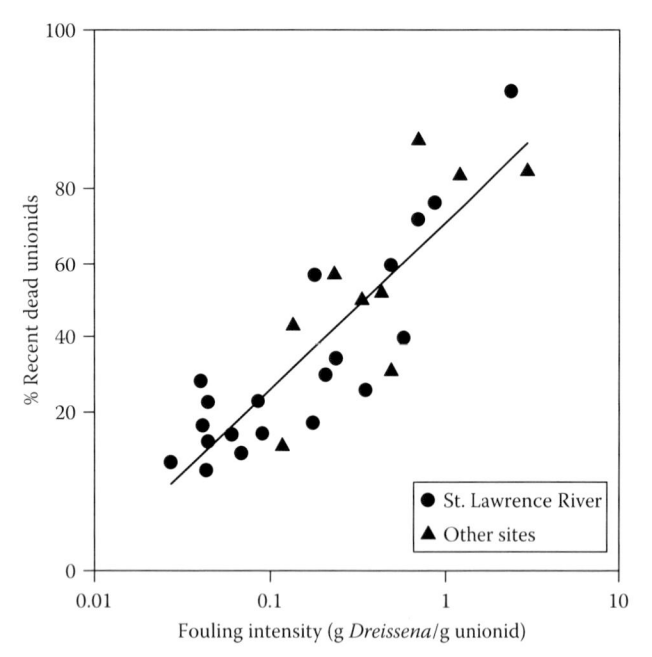

Figure 38.6 Percentage of recent-dead unionids as a function of dreissenid fouling intensity on unionid populations in North America. Fouling intensity defined as the mass of attached dreissenids divided by the mass of unionid host. Data are from the upper St. Lawrence River (20 sites), Detroit River (1 site), western Lake Erie (1 site), Lake St. Clair (1 site), Lake Wawasee (1 site), Richelieu River (4 sites), and Lake Champlain (1 site), referenced in Ricciardi (2003). Line fitted by least-squares linear regression: $\sin^{-1}(y^{0.5}) = 0.48 \log x + 1.0$ ($R^2 = 0.75$, $P < 0.0001$).

1994, Jones and Ricciardi 2005), which implies that the magnitude of impact may be predictable from habitat conditions. In fact, variation in fouling intensity on unionids across sites can be partly explained by differences in water chemistry—specifically, local calcium concentrations (Jokela and Ricciardi 2008). Other important moderators of *Dreissena* impacts on macroinvertebrates include the mean particle size of ambient substrates (Figure 38.3) (Ward and Ricciardi 2007, Jokela and Ricciardi 2008) and the local abundance of molluscivores (Kuhns and Berg 1999). Hydrological variables may also play an important role; there is empirical evidence that the level of fouling on unionids associated with a given *Dreissena* population density is higher in lakes than in rivers (Figure 38.7).

In addition to dreissenid abundances, the extent of dreissenid impacts also depends on the composition of the benthic community prior to invasion. In particular, feeding mode, body size, and substrate preference of individual macroinvertebrate taxa may prove useful to forecast responses to *Dreissena* colonization (Ward and Ricciardi 2007). Impacts are also altered by the presence of other introduced macroinvertebrates (van Overdijk et al. 2003), other habitat engineers (Ward and Ricciardi 2010), and native and exotic predators (Stewart et al. 1998b, Kuhns and Berg 1999, Mörtl et al. 2010). Predation by molluscivorous waterfowl in Lake Constance during winter 1999–2000 and 2001–2002 reduced dreissenid densities by 80% at depths down to 7 m; the resultant loss of *Dreissena*-generated physical habitat and biodeposits was accompanied by reductions in chironomid larvae, mayfly nymphs, and oligochaetes (Mörtl et al. 2010).

TEMPORAL VARIABILITY OF IMPACT

It is increasingly evident that long-term variability in environmental conditions and dreissenid populations (abundance and age structure) can cause dreissenid impacts to change over time (Haynes et al. 1999, Strayer and Malcom 2007, Strayer et al. 2011). In Lake Michigan, the magnitude and spatial extent of impacts on *Diporeia* increased over a decade as dreissenid biomass also increased (Nalepa et al. 2009). In the Hudson River, densities of benthic macroinvertebrates at deepwater sites declined by >80% in the early years of *Dreissena* colonization but began to recover several years later following a long-term decline in adult dreissenid survivorship and aggregate filtration rate (Strayer et al. 2011). Conversely, in southwestern Lake Ontario, increased abundance and diversity of nondreissenid macroinvertebrates observed a few years after *Dreissena* colonization were followed by declines to levels similar to or lower than pre-invasion levels; the development of thick beds of filamentous algae several years after invasion may have been a contributing factor to these declines, particularly for snails that prefer to graze on diatoms (Haynes et al. 1999). Given these disparate outcomes, it is perhaps no surprise that "time since invasion" is not a significant predictor of *Dreissena* impacts on benthic communities (Ward and Ricciardi 2007).

CONCLUSIONS AND FUTURE RESEARCH DIRECTIONS

The invasion history of *Dreissena* in Europe and North America has revealed broad patterns of impacts on benthic communities. Benthic macroinvertebrates in the littoral areas of lakes and rivers typically increase (often doubling) in abundance and taxonomic richness following dreissenid colonization. These increases are accompanied by a decline in community evenness because of disproportionate effects on particular taxonomic and functional groups. For example, *Dreissena* has consistent and strong positive effects on densities of gammarid amphipods, leeches, flatworms, and certain insects, but has strong negative effects on native bivalves. Furthermore, although densities of gastropods generally increase following *Dreissena* colonization, large-bodied snails tend to be displaced from dreissenid-covered substrate; thus, the alteration of hard substrates by *Dreissena* can reduce the mean size of resident gastropods, as has been observed for marine mussels (Griffiths et al. 1992).

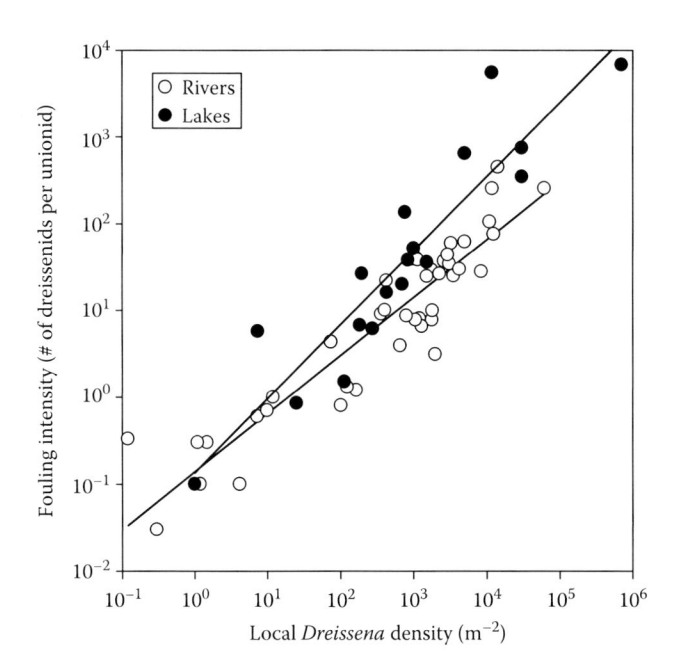

Figure 38.7 Relationship between fouling intensity (the mean number of attached dreissenids per unionid) and local *Dreissena* density at multiple invaded sites in North America and Europe. Lines are fitted by least-squares linear regression. Regression lines for lakes (log $Y = -0.87 + 0.85\ X$, $R^2 = 0.85$, $P < 0.0001$) and rivers (log $Y = -0.85 + 0.67\ X$, $R^2 = 0.86$, $P < 0.0001$) are significantly different (ANCOVA, $P < 0.0001$). (Data are from Ricciardi, A., *Freshwat. Biol.*, 48, 972, 2003.)

The impact of these gastropod shifts on periphyton communities has not been examined. At system-wide scales, macroinvertebrate richness tends to decrease in the presence of *Dreissena*. In littoral regions, the most severe reduction in diversity is the catastrophic decline in populations of native unionid mussels. In deepwater regions, declines in macroinvertebrate abundance and richness (particularly of oligochaete worms, sphaeriid clams, and burrowing amphipods) are hypothesized to be a response to reduced influx of food particles from littoral sites dominated by filter-feeding dreissenids. These and other impacts are typically observed within a few years to a decade of dreissenid colonization.

This review of dreissenid–macroinvertebrate interactions has revealed several research gaps in our understanding of the impacts of *Dreissena* invasions. The most challenging of these from a predictive standpoint is how impacts are altered by habitat conditions. Some impacts are strongly mediated by water chemistry (Jokela and Ricciardi 2008), turbidity (Osterling et al. 2007), or substrate quality (Ward and Ricciardi 2007) and frequently involve complex interactions between dreissenids and other invasive or habitat-engineering species. For example, the round goby (a benthic fish predator) and *Cladophora* (a benthic macroalga) both modify the effects of dreissenids on benthic invertebrate communities (Kuhns and Berg 1999, Lederer et al. 2006, Ward and Ricciardi 2010) and have become common drivers of community-level change in the Great Lakes basin. Such interactions produce synergies that complicate efforts to predict impacts of *Dreissena*, but are rarely studied. There is also a need to investigate how impacts of nearshore dreissenid populations on deepwater benthic communities are moderated by limnological characteristics of lakes (e.g., phytoplankton growth rate, water residence time, water flow patterns). A greater understanding of these interactions could be obtained through the development of models guided by marine studies of the context-dependent ecosystem effects of bivalves (Dame 1996, Prins et al. 1997).

The mechanisms by which byssally attached mussels moderate benthic diversity remain a research focus (e.g., Borthagaray and Carranza 2007, Buschbaum et al. 2009, Ward and Ricciardi 2010). In marine systems, the patch dynamics of mussels affects benthic diversity at multiple spatial and temporal scales (Tsuchiya and Nishihira 1985, Tanaka and Magalhães 2002, Cole 2010). This relationship has hardly been explored for *Dreissena*, although studies have shown that species richness of the macroinvertebrate community occupying dreissenid patches increases with patch area (Bially and MacIsaac 2000), and the topography of dreissenid patches is an important driver of variation in macroinvertebrate richness and evenness over small (<1 m^2) spatial scales (Ward and Ricciardi 2010).

Another research question is whether benthic communities associated with zebra mussels differ in composition and abundance from communities associated with quagga mussels. Recognizing such differences, if they exist, would help predict the community-wide consequences of dreissenid species replacements occurring in North America and Europe (Ricciardi and Whoriskey 2004; Wilson et al. 2006, Zhulidov et al. 2010). Recent independent invasions of the southwestern United States by zebra and quagga mussels (Wittmann et al. 2010) provide opportunities for researchers to compare the effects of these two species in isolation. It would be useful to determine whether byssally attached mussels in general (such as *Dreissena* spp., the golden mussel *Limnoperna fortunei*, and the dark false mussel *Mytilopsis leucophaeata*) have similar impacts on benthic communities; this appears to be the case for *D. polymorpha* and *L. fortunei* (Karatayev et al. 2010). If such effects are concordant, it would suggest universality in the interactions between invasive mussels and benthic macroinvertebrates in freshwater and marine systems.

ACKNOWLEDGMENTS

Funding was provided by the Natural Sciences and Engineering Research Council of Canada in the form of a Canada Graduate Scholarship to JMW and a Discovery Grant to AR.

REFERENCES

Atalah, J., M. Kelly-Quinn, K. Irvine, and T. P. Crowe. 2010. Impacts of invasion by *Dreissena polymorpha* (Pallas, 1771) on the performance of macroinvertebrate assessment tools for eutrophication pressure in lakes. *Hydrobiologia* 654: 237–251.

Auer, M. T., L. M. Tomlinson, S. N. Higgins, S. Y. Malkin, E. T. Howell, and H. A. Bootsma. 2010. Great Lakes *Cladophora* in the 21st century: Same algae-different ecosystem. *J. Great Lakes Res.* 36: 248–255.

Baker, S. M. and D. J. Hornbach. 1997. Acute physiological effects of zebra mussel (*Dreissena polymorpha*) infestation on two unionid mussels, *Actinonaias ligamentina* and *Amblema plicata*. *Can. J. Fish. Aquat. Sci.* 54: 512–519.

Baker, S. M. and D. J. Hornbach. 2000. Physiological status and biochemical composition of a natural population of unionid mussels (*Amblema plicata*) infested by zebra mussels (*Dreissena polymorpha*). *Am. Midl. Nat.* 143: 443–452.

Baker, S. M. and J. S. Levinton. 2003. Selective feeding by three native North American freshwater mussels implies food competition with zebra mussels. *Hydrobiologia* 505: 97–105.

Bastviken, D. T. E., N. F. Caraco, and J. J. Cole. 1998. Experimental measurements of zebra mussel (*Dreissena polymorpha*) impacts on phytoplankton community composition. *Freshwat. Biol.* 39: 375–386.

Beekey, M. A., D. J. McCabe, and J. E. Marsden. 2004. Zebra mussel colonization of soft sediments facilitates invertebrate communities. *Freshwat. Biol.* 49: 535–545.

Bially, A. and H. J. MacIsaac. 2000. Fouling mussels (*Dreissena* spp.) colonize soft sediments in Lake Erie and facilitate benthic invertebrates. *Freshwat. Biol.* 43: 85–97.

Borthagaray, A. I. and A. Carranza. 2007. Mussels as ecosystem engineers: Their contribution to species richness in a rocky littoral community. *Acta Oecol.* 31: 243–250.

Botts, P. S., B. A. Patterson, and D. W. Schloesser. 1996. Zebra mussel effects on benthic invertebrates: Physical or biotic? *J. N. Am. Benthol. Soc.* 15: 179–184.

Brazner, J. C. and D. A. Jensen. 2000. Zebra mussel [*Dreissena polymorpha* (Pallas)] colonization of rusty crayfish [*Orconectes rusticus* (Girard)] in Green Bay, Lake Michigan. *Am. Midl. Nat.* 143: 250–256.

Brown, J. E. and C. A. Stepien. 2010. Population genetic history of the dreissenid mussel invasions: Expansion patterns across North America. *Biol. Invasions* 12: 3687–3710.

Buschbaum, C., S. Dittmann, J.-S. Hong, I.-S. Hwang, M. Strasser, M. Thiel, N. Valdivia, S.-P. Yoon, and K. Reise. 2009. Mytilid mussels: Global habitat engineers in coastal sediments. *Helgol. Mar. Res.* 63: 47–58.

Cecala R. K., C. M. Mayer, K. L. Schulz, and E. L. Mills. 2008. Increased benthic algal primary production in response to the invasive zebra mussel (*Dreissena polymorpha*) in a productive ecosystem, Oneida Lake, New York. *J. Integr. Plant Biol.* 50: 1452–1466.

Cole, V. J. 2010. Alteration of the configuration of bioengineers affects associated taxa. *Mar. Ecol. Prog. Ser.* 416: 127–136.

Covich, A. P., M. A. Palmer, and T.A. Crowl. 1999. The role of benthic invertebrate species in freshwater ecosystems. *BioScience* 49: 119–127.

Crail, T. D., R. A. Krebs, and D. T. Zanatta. 2011. Unionid mussels from nearshore zones of Lake Erie. *J. Great Lakes Res.* 37: 199–202.

Creese, R., S. Hooker, S. De Luca, and Y. Wharton. 1997. Ecology and environmental impact of *Musculista senhousia* (Mollusca: Bivalvia: Mytilidae) in Tamaki Estuary, Auckland, New Zealand. *N Z J. Mar. Freshwat. Res.* 31: 225–236.

Dame, R. F. 1996. *Ecology of Marine Bivalves: An Ecosystem Approach.* Boca Raton, FL: CRC Press.

Dermott, R. 2001. Sudden disappearance of the amphipod *Diporeia* from Eastern Lake Ontario, 1993–1995. *J. Great Lakes Res.* 27: 423–433.

Dermott, R. M. and D. Kerec. 1997. Changes to the deepwater benthos of eastern Lake Erie since the invasion of *Dreissena*: 1979–1993. *Can. J. Fish. Aquat. Sci.* 54: 922–930.

Dermott, R., J. Mitchell, I. Murray, and E. Fear. 1993. Biomass and production of zebra mussels (*Dreissena polymorpha*) in shallow waters of northeastern Lake Erie. In *Zebra Mussels: Biology, Impacts, and Control*, T. F. Nalepa and D. W. Schloesser, eds., pp. 399–413. Boca Raton, FL: CRC Press.

Ďuriš, Z., I. Horká, and A. Petrusek. 2007. Invasive zebra mussel colonisation of invasive crayfish: A case study. *Hydrobiologia* 590: 43–46.

Dusoge, K. 1966. Composition and interrelations between macrofauna living on stones in the littoral of Mikolajski Lake. *Ekol. Pol. – Seria A* 39: 1–8.

Early, T. A., J. T. Kundrat, T. Schorp, and T. Glonek. 1996. Lake Michigan sponge phospholipid variations with habitat: A ^{31}P nuclear magnetic resonance study. *Comp. Biochem. Physiol.* B 114: 77–89.

Fincke, O. M., D. Santiago, S. Hickner, and R. Bienek. 2009. Susceptibility of larval dragonflies to zebra mussel colonization and its effect on larval movement and survivorship. *Hydrobiologia* 624: 71–79.

Fincke, O. M. and L. A. Tylczak. 2011. Effects of zebra mussel attachment on the foraging behaviour of a larval dragonfly, *Macromia illinoiensis*. *Ecol. Entomol.* 36: 760–767.

Gabriels, W., P. L. M. Goethals, and N. De Pauw. 2005. Implications of taxonomic modifications and alien species on biological water quality assessment as exemplified by the Belgian Biotic Index method. *Hydrobiologia* 542: 137–150.

Gergs, R. and K.-O. Rothhaupt. 2008. Effects of zebra mussels on a native amphipod and the invasive *Dikerogammarus villosus*: The influence of biodeposition and structural complexity. *J. N. Amer. Benthol. Soc.* 27: 541–548.

Gergs, R., K. Rinke, and K.-O. Rothhaupt. 2009. Zebra mussels mediate benthic–pelagic coupling by biodeposition and changing detrital stoichiometry. *Freshwat. Biol.* 54: 1379–1391.

González, M. J. and G. A. Burkart. 2004. Effects of food type, habitat, and fish predation on the relative abundance of two amphipod species, *Gammarus fasciatus* and *Echinogammarus ischnus*. *J. Great Lakes Res.* 30: 100–113.

Griffiths, R. W. 1993. Effects of zebra mussels (*Dreissena polymorpha*) on the benthic fauna of Lake St. Clair. In *Zebra Mussels: Biology, Impacts, and Control*, T. F. Nalepa and D. W. Schloesser, eds., pp. 414–437. Boca Raton, FL: CRC Press.

Griffiths, C. L., P. A. R. Hockey, C. van Erkom Schurink, and P. J. Le Roux. 1992. Marine invasive aliens on South African shores: Implications for community structure and trophic functioning. *S. Afr. J. Mar. Sci.* 12: 713–722.

Gutiérrez, J. L., C. G. Jones, D. L. Strayer, and O. O. Iribarne. 2003. Mollusks as ecosystem engineers: The role of shell production in aquatic habitats. *Oikos* 101: 79–90.

Haag, W. R., D. J. Berg, D. W. Garton, and J. L. Farris. 1993. Reduced survival and fitness in native bivalves in response to fouling by the introduced zebra mussel (*Dreissena polymorpha*) in western Lake Erie. *Can. J. Fish. Aquat. Sci.* 50: 13–19.

Hallac, D. E. and J. E. Marsden. 2001. Comparison of conservation strategies for unionids threatened by zebra mussels (*Dreissena polymorpha*): Periodic cleaning vs quarantine and translocation. *J. N. Am. Benthol. Soc.* 20: 200–210.

Harman, W. N. 2000. Diminishing species richness of mollusks in Oneida Lake, New York State, USA. *Nautilus* 114: 120–126.

Haynes, J. M., T. W. Stewart, and G. E. Cook. 1999. Benthic macroinvertebrate communities in southwestern Lake Ontario following invasion of *Dreissena*: Continuing change. *J. Great Lakes Res.* 25: 828–838.

Hecky, R. E., R. E. H. Smith, D. R. Barton et al. 2004. The nearshore phosphorus shunt: A consequence of ecosystem engineering by dreissenids in the Laurentian Great Lakes. *Can. J. Fish. Aquat. Sci.* 61: 1285–1293.

Horvath, T. G., K. M. Martin, and G. A. Lamberti. 1999. Effect of zebra mussels, *Dreissena polymorpha*, on macroinvertebrates in a lake-outlet stream. *Am. Midl. Nat.* 142: 340–347.

Izvekova, E. I. and A. A. Lvova-Katchanova. 1972. Sedimentation of suspended matter by *Dreissena polymorpha* Pallas and its subsequent utilization by Chironomidae larvae. *Pol. Arch. Hydrobiol.* 19: 203–210.

Jokela, A. and A. Ricciardi. 2008. Predicting zebra mussel fouling on native mussels from physico-chemical variables. *Freshwat. Biol.* 53: 1845–1856.

Jones, L. A. and A. Ricciardi. 2005. Influence of physicochemical factors on the distribution and biomass of invasive mussels in the St. Lawrence River. *Can. J. Fish. Aquat. Sci.* 62: 1953–1962.

Karatayev, A. Y. and L. E. Burlakova. 1995. The role of *Dreissena* in lake ecosystems. *Russian J. Ecol.* 26: 207–211.

Karatayev, A. Y., L. E. Burlakova, V. A. Karatayev, and D. Boltovskoy. 2010. *Limnoperna fortunei* versus *Dreissena polymorpha*: Population densities and benthic community impacts of two invasive freshwater bivalves. *J. Shellfish Res.* 29: 975–984.

Karatayev, A. Y., L. E. Burlakova, and D. K. Padilla. 1997. The effects of *Dreissena polymorpha* (Pallas) invasion on aquatic communities in Eastern Europe. *J. Shellfish Res.* 16: 187–203.

Kuhns, L. A. and M. B. Berg. 1999. Benthic invertebrate community responses to round goby (*Neogobius melanostomus*) and zebra mussel (*Dreissena polymorpha*) invasion in southern Lake Michigan. *J. Great Lakes Res.* 25: 910–917.

Lauer, T. E. and T. S. McComish. 2001. Impact of zebra mussels (*Dreissena polymorpha*) on fingernail clams (Sphaeriidae) in extreme southern Lake Michigan. *J. Great Lakes Res.* 27: 230–238.

Lederer, A., J. Massart, and J. Janssen. 2006. Impact of round gobies (*Neogobius melanostomus*) on dreissenids (*Dreissena polymorpha* and *Dreissena bugensis*) and the associated macroinvertebrate community across an invasion front. *J. Great Lakes Res.* 32: 1–10.

Lohse, D. P. 1993. The importance of secondary substratum in a rocky intertidal community. *J. Exp. Mar. Biol. Ecol.* 166: 1–17.

Lozano, S. J., J. V. Scharold, and T. P. Nalepa. 2001. Recent declines in benthic macroinvertebrate densities in Lake Ontario. *Can. J. Fish. Aquat. Sci.* 58: 518–529.

MacNeil, C. and M. Briffa. 2009. Replacement of a native freshwater macroinvertebrate species by an invader: Implications for biological water quality monitoring. *Hydrobiologia* 635: 321–327.

Maguire, C. M., D. Roberts, and R. S. Rosell. 2003. The ecological impacts of a zebra mussel invasion in a large Irish lake, Lough Erne: A typical European experience? *Aquat. Invaders* 14: 10–18.

Martel, A. L., D. A. Pathy, J. B. Madill, C. B. Renaud, S. L. Dean, and S. J. Kerr. 2001. Decline and regional extirpation of freshwater mussels (Unionidae) in a small river system invaded by *Dreissena polymorpha*: The Rideau River, 1993–2000. *Can. J. Zool.* 79: 2181–2191.

Mayer, C. M., R. A. Keats, L. G. Rudstam, and E. L. Mills. 2002. Scale-dependent effects of zebra mussels on benthic invertebrates in a large eutrophic lake. *J. N. Am. Benthol. Soc.* 21: 616–633.

McNickle, G. G., M. D. Rennie, and W. G. Sprules. 2006. Changes in benthic invertebrate communities in South Bay, Lake Huron following invasion by zebra mussels (*Dreissena polymorpha*), and potential effects on lake whitefish (*Coregonus clupeaformis*) diet and growth. *J. Great Lakes Res.* 32: 180–193.

Mellina, E. and J. B. Rasmussen. 1994. Patterns in the distribution and abundance of zebra mussels (*Dreissena polymorpha*) in rivers and lakes in relation to substrate and other physicochemical factors. *Can. J. Fish. Aquat. Sci.* 51: 1024–1036.

Mellina, E., J. B. Rasmussen, and E. L. Mills. 1995. Impact of zebra mussel (*Dreissena polymorpha*) on phosphorus cycling in lakes. *Can. J. Fish. Aquat. Sci.* 52: 2553–2573.

Mörtl, M. and K.-O. Rothhaupt. 2003. Effects of adult *Dreissena polymorpha* on settling juveniles and associated macroinvertebrates. *Int. Rev. Hydrobiol.* 88: 561–569.

Mörtl, M., S. Werner, and K.-O. Rothhaupt. 2010. Effects of predation by wintering water birds on zebra mussels and on associated macroinvertebrates. In *The Zebra Mussel in Europe*, G. van der Velde, S. Rajagopal, and A. bij de Vaate, eds., pp. 239–249. Leiden, The Netherlands: Backhuys Publishers.

Nalepa, T. F. 1994. Decline of native unionids in Lake St. Clair after infestation by the zebra mussel, *Dreissena polymorpha*. *Can. J. Fish. Aquat. Sci.* 51: 2227–2233.

Nalepa, T. F., D. L. Fanslow, and G. A. Lang. 2009. Transformation of the offshore benthic community in Lake Michigan: Recent shift from the native amphipod *Diporeia* spp. to the invasive mussel, *Dreissena rostriformis bugensis*. *Freshwat. Biol.* 54: 466–479.

Nalepa, T. F., D. L. Fanslow, M. B. Lansing, and G. A. Lang. 2003. Trends in the benthic macroinvertebrate community of Saginaw Bay, Lake Huron, 1987 to 1996: Responses to phosphorus abatement and the zebra mussel, *Dreissena polymorpha*. *J. Great Lakes Res.* 29: 14–33.

Nalepa, T. F., D. L. Fanslow, M. B. Lansing et al. 2002. *Abundance, Biomass, and Species Composition of Benthic Macroinvertebrate Populations in Saginaw Bay, Lake Huron, 1987–96*, NOAA Technical Memorandum GLERL-122. Ann Arbor, MI: Great Lakes Environmental Research Laboratory.

Nalepa, T. F., D. J. Hartson, D. L. Fanslow, G. A. Lang, and S. J. Lozano. 1998. Declines in benthic macroinvertebrate populations in southern Lake Michigan, 1980–1993. *Can. J. Fish. Aquat. Sci.* 55: 2402–2413.

Osterling, E. M., E. Bergman, L. A. Greenberg, B. S. Baldwin, and E. L. Mills. 2007. Turbidity-mediated interactions between invasive filter-feeding mussels and native bioturbating mayflies. *Freshwat. Biol.* 52: 1602–1610.

Owens, R. W. and D. E. Dittman. 2003. Shifts in the diets of slimy sculpin (*Cottus cognatus*) and lake whitefish (*Coregonus clupeaformis*) in Lake Ontario following the collapse of the burrowing amphipod *Diporeia*. *Aquat. Ecosyst. Health Manage.* 6: 311–323.

Parker, B. C., M. A. Patterson, and R. J. Neves. 1998. Feeding interactions between native freshwater mussels (Bivalvia: Unionidae) and zebra mussels (*Dreissena polymorpha*) in the Ohio River. *Am. Malacol. Bull.* 14: 173–179.

Parker, I. M., D. Simberloff, W. M. Lonsdale et al. 1999. Impact: Toward a framework for understanding the ecological effects of invaders. *Biol. Invasions* 1: 3–19.

Ponyi, J. E. 1992. The distribution and biomass of Unionidae (Mollusca, Bivalvia), and the production of *Unio tumidus* Retzius in Lake Balaton (Hungary). *Arch. Hydrobiol.* 125: 245–251.

Prins, T. C., A. C. Smaal, and R. F. Dame. 1997. A review of the feedbacks between bivalve grazing and ecosystem processes. *Aquat. Ecol.* 31: 349–359.

Ramcharan, C. W., D. K. Padilla, and S. I. Dodson. 1992. Models to predict potential occurrence and density of the zebra mussel *Dreissena polymorpha*. *Can. J. Fish. Aquat. Sci.* 49: 2611–2620.

Ratti, C. and D. R. Barton. 2003. Decline in the diversity of benthic invertebrates in the wave-zone of eastern Lake Erie, 1974–2001. *J. Great Lakes Res.* 29: 608–615.

Reed, T., S. J. Wielgus, A. K. Barnes, J. J. Schiefelbein, and A. L. Fettes. 2004. Refugia and local controls: Benthic invertebrate dynamics in Lower Green Bay, Lake Michigan following zebra mussel invasion. *J. Great Lakes Res.* 30: 390–396.

Ricciardi, A. 2003. Predicting the impacts of an introduced species from its invasion history: An empirical approach applied to zebra mussel invasions. *Freshwat. Biol.* 48: 972–981.

Ricciardi, A., R. J. Neves, and J. B. Rasmussen. 1998. Impending extinctions of North American freshwater mussels (Unionoida) following the zebra mussel (*Dreissena polymorpha*) invasion. *J. Anim. Ecol.* 67: 613–619.

Ricciardi, A. and F. G. Whoriskey. 2004. Exotic species replacement: Shifting dominance of dreissenid mussels in the Soulanges Canal, upper St. Lawrence River, Canada. *J. N. Am. Benthol. Soc.* 23: 507–514.

Ricciardi, A., F. G. Whoriskey, and J. B. Rasmussen. 1995. Predicting the intensity and impact of *Dreissena* infestation on native unionid bivalves from *Dreissena* field density. *Can. J. Fish. Aquat. Sci.* 52: 1449–1461.

Ricciardi, A., F. G. Whoriskey, and J. B. Rasmussen. 1996. Impact of the *Dreissena* invasion on native unionid bivalves in the upper St. Lawrence River. *Can. J. Fish. Aquat. Sci.* 53: 1434–1444.

Ricciardi, A., F. G. Whoriskey, and J. B. Rasmussen. 1997. The role of the zebra mussel (*Dreissena polymorpha*) in structuring macroinvertebrate communities on hard substrata. *Can. J. Fish. Aquat. Sci.* 54: 2596–2608.

Robinson, T. B. and C. L. Griffiths. 2002. Invasion of Langebaan Lagoon, South Africa, by *Mytilus galloprovincialis* – Effects on natural communities. *Afr. Zool.* 37: 151–158.

Roditi, H. A., D. L. Strayer, and S. E. G. Findlay. 1997. Characteristics of zebra mussel (*Dreissena polymorpha*) biodeposits in a tidal freshwater estuary. *Arch. Hydrobiol.* 140: 207–219.

Schloesser, D. W. and E. C. Masteller. 1999. Mortality of unionid bivalves (Mollusca) associated with dreissenid mussels (*Dreissena polymorpha* and *D. bugensis*) in Presque Isle Bay, Lake Erie. *Northeastern Nat.* 6: 341–352.

Schloesser, D. W., J. L. Metcalfe-Smith, W. P. Kovalak, G. D. Longton, and R. D. Smithee. 2006. Extirpation of freshwater mussels (Bivalvia: Unionidae) following the invasion of dreissenid mussels in an interconnecting river of the Laurentian Great Lakes. *Am. Midl. Nat.* 155: 307–320.

Schloesser, D. W. and T. F. Nalepa. 1994. Dramatic decline of unionid bivalves in offshore waters of western Lake Erie after infestation by the zebra mussel, *Dreissena polymorpha*. *Can. J. Fish. Aquat. Sci.* 51: 2234–2242.

Schummer, M. L., S. A. Petrie, and R. C. Bailey. 2008. Interaction between macroinvertebrate abundance and habitat use by diving ducks during winter on northeastern Lake Ontario. *J. Great Lakes Res.* 34: 54–71.

Soster, F. M., P. L. McCall, and K. A. Herrmann. 2011. Decadal changes in the benthic invertebrate community in western Lake Erie between 1981 and 2004. *J. Great Lakes Res.* 37: 226–237.

Sousa, R., F. Pilotto, and D. C. Aldridge. 2011. Fouling of European freshwater bivalves (Unionidae) by the invasive zebra mussel (*Dreissena polymorpha*). *Freshwat. Biol.* 56: 867–876.

Stańczykowska, A. 1977. Ecology of *Dreissena polymorpha* (Pall.) (Bivalvia) in lakes. *Pol. Arch. Hydrobiol.* 24: 461–530.

Stewart, T. W. and J. M. Haynes. 1994. Benthic macroinvertebrate communities of southwestern Lake Ontario following invasion of *Dreissena*. *J. Great Lakes Res.* 20: 479–493.

Stewart, T. W., J. G. Miner, and R. L. Lowe. 1998a. Quantifying mechanisms for zebra mussel effects on benthic macroinvertebrates: Organic matter production and shell-generated habitat. *J. N. Am. Benthol. Soc.* 17: 81–94.

Stewart, T. W., J. G. Miner, and R. L. Lowe. 1998b. An experimental analysis of crayfish (*Orconectes rusticus*) effects on a *Dreissena*-dominated benthic macroinvertebrate community in western Lake Erie. *Can. J. Fish. Aquat. Sci.* 55: 1043–1050.

Stewart, T. W., J. G. Miner, and R. L. Lowe. 1998c. Macroinvertebrate communities on hard substrates in western Lake Erie: Structuring effects of *Dreissena*. *J. Great Lakes Res.* 24: 868–879.

Strayer, D. L., N. Cid, and H. M. Malcom. 2011. Long-term changes in a population of an invasive bivalve and its effects. *Oecologia* 165: 1063–1072.

Strayer, D. L. and H. M. Malcom. 2007. Effects of zebra mussels (*Dreissena polymorpha*) on native bivalves: The beginning of the end or the end of the beginning? *J. N. Am. Benthol. Soc.* 26: 111–122.

Strayer, D. L. and L. C. Smith. 2001. The zoobenthos of the freshwater tidal Hudson River and its response to the zebra mussel (*Dreissena polymorpha*) invasion. *Arch. Hydrobiol. Suppl.* 139/1: 1–52.

Suchanek, T. H. 1986. Mussels and their rôle in structuring rocky shore communities. In *The Ecology of Rocky Coasts*, eds. P. G. Moore and R. Seed, pp. 70–96. New York: Columbia University Press.

Tanaka, M. O. and C. A. Magalhães. 2002. Edge effects and succession dynamics in *Brachidontes* mussel beds. *Mar. Ecol. Prog. Ser.* 237: 151–158.

Tsuchiya, M. and M. Nishihira. 1985. Islands of *Mytilus* as a habitat for small intertidal animals: Effect of island size on community structure. *Mar. Ecol. Prog. Ser.* 25: 71–81.

Van Appledorn, M. and C. E. Bach. 2007. Effects of zebra mussels (*Dreissena polymorpha*) on mobility of three native mollusk species. *Am. Midl. Nat.* 158: 329–337.

Van Appledorn, M., D. A. Lamb, K. Albalak, and C. E. Bach. 2007. Zebra mussels decrease burrowing ability and growth of a native snail, *Campeloma decisum*. *Hydrobiologia* 57: 441–445.

van Overdijk, C. D. A., I. A. Grigorovich, T. Mabee, W. J. Ray, J. J. H. Ciborowski, and H. J. MacIsaac. 2003. Microhabitat selection by the invasive amphipod *Echinogammarus ischnus* and native *Gammarus fasciatus* in laboratory experiments and in Lake Erie. *Freshwat. Biol.* 48: 567–578.

Vander Zanden, M. J. and Y. Vadaboncoeur. 2002. Fishes as integrators of benthic and pelagic food webs in lakes. *Ecology* 83: 2152–2161.

Vanderploeg H. A., T. F. Nalepa, D. J. Jude et al. 2002. Dispersal and emerging ecological impacts of Ponto-Caspian species in the Laurentian Great Lakes. *Can. J. Fish. Aquat. Sci.* 59: 1209–1228.

Ward, J. M. and A. Ricciardi. 2007. Impacts of *Dreissena* invasions on benthic macroinvertebrate communities: A meta-analysis. *Divers. Distrib.* 13: 155–165.

Ward, J. M. and A. Ricciardi. 2010. Community level effects of co-occurring native and exotic ecosystem engineers. *Freshwat. Biol.* 55: 1803–1817.

Washington, H. G. 1984. Diversity, biotic and similarity indices: A review with special relevance to aquatic ecosystems. *Water Res.* 18: 653–694.

Watters, G. T. and C. J. M. Flaute. 2010. Dams, zebras, and settlements: The historical loss of freshwater mussels in the Ohio River mainstem. *Am. Malacol. Bull.* 28: 1–12.

Wiggins, G. B. 1996. *Larvae of the North American Caddisfly Genera (Trichoptera)*. Toronto, Ontario, Canada: University of Toronto Press.

Wilson, K. A., E. T. Howell, and D. A. Jackson. 2006. Replacement of zebra mussels by quagga mussels in the Canadian nearshore of Lake Ontario: The importance of substrate, round goby abundance, and upwelling frequency. *J. Great Lakes Res.* 32: 11–28.

Wisenden, P. A. and R. C. Bailey. 1995. Development of macroinvertebrate community structure associated with zebra mussel (*Dreissena polymorpha*) colonization of artificial substrates. *Can. J. Zool.* 73: 1438–1443.

Wittmann, M. E., S. Chandra, A. Caires et al. 2010. Early invasion population structure of quagga mussel and associated benthic invertebrate community composition on soft sediment in a large reservoir. *Lake Reserv. Manage.* 26: 316–327.

Zanatta, D. T., G. L. Mackie, J. L. Metcalfe-Smith, and D. A. Woolnough. 2002. A refuge for native freshwater mussels (Bivalvia: Unionidae) from impacts of the exotic zebra mussel (*Dreissena polymorpha*) in Lake St. Clair. *J. Great Lakes Res.* 28: 479–489.

Zhulidov, A. V., A. V. Kozhara, G. H. Scherbina et al. 2010. Invasion history, distribution, and relative abundances of *Dreissena bugensis* in the old world: A synthesis of data. *Biol. Invasions* 12: 1923–1940.

CHAPTER 39

Interactions between Exotic Ecosystem Engineers (*Dreissena* spp.) and Native Burrowing Mayflies (*Hexagenia* spp.) in Soft Sediments of Western Lake Erie

Kristen M. DeVanna, Don W. Schloesser, Jonathan M. Bossenbroek, and Christine M. Mayer

CONTENTS

Abstract 611
Introduction 612
Methods 613
    Analysis of Spatial Association 613
    Laboratory Experiments 614
        Observation Experiments 614
        Habitat-Selection Experiments 614
Results 615
    Analysis of Spatial Association 615
    Laboratory Experiments 615
Discussion 617
    Analysis of Spatial Association 617
    Laboratory Experiments 618
Conclusions and Future Directions 619
Acknowledgments 619
References 620

ABSTRACT

Exotic ecosystem engineers (i.e., *Dreissena polymorpha* and *D. rostriformis bugensis*) have changed soft-bottom habitats of lakes in many ways. The most noticeable change is the presence of hard clusters of mussels found between expanses of otherwise soft sediments. We hypothesized that this shift in available habitat type is likely to affect the distribution of infaunal invertebrates, such as burrowing mayflies (*Hexagenia* spp.). We examined effects of dreissenid clusters on *Hexagenia* presence through analyses of field-measured distributions of mussels and *Hexagenia* throughout western Lake Erie and in laboratory experiments (viewing chamber observation and habitat-selection experiments). In western Lake Erie, distribution analyses indicated that *Dreissena* did not inhibit *Hexagenia* presence and *Hexagenia* were more likely to be present where *Dreissena* were also present. However, there was no spatial cross-correlation between densities of the two species. At sites with no *Dreissena*, *Hexagenia* could achieve very high densities that were unattained at sites with *Dreissena*. In habitat-selection experiments with three habitat types: (1) live *Dreissena* clusters; (2) dead *Dreissena* clusters; and (3) bare sediments, *Hexagenia* strongly selected to colonize below clusters of live dreissenids. In observation experiments, *Hexagenia* also selected to colonize below live *Dreissena* that covered sediments, but this selection was not stronger in the presence of a predator (yellow perch, *Perca flavescens*). Our findings showed that at a small spatial scale in laboratory experiments, *Hexagenia* prefer *Dreissena*-covered sediment, but at a large spatial scale in western Lake Erie, *Hexagenia* do not select for or avoid *Dreissena*. Our results suggested that at a basin-wide scale, *Dreissena* presence does not inhibit *Hexagenia* colonization, but *Dreissena* are not strong determinants of *Hexagenia* distribution and abundance in western Lake Erie.

INTRODUCTION

Zebra mussels (*Dreissena polymorpha*) and quagga mussels (*D. rostriformis bugensis*) have changed the Great Lakes ecosystem in many ways as a result of their effects as ecosystem engineers (Karatayev et al. 2002, Zhu et al. 2006). Ecosystem engineers are organisms that alter physical, chemical, and biotic components of the ecosystem, leading to wide-scale changes that are both direct and indirect (Jones et al. 1994, 1997). Zebra mussels, first discovered in the Great Lakes in 1986 (Carlton 2008), colonize primarily hard substrates. Benthic invertebrates associated with these hard substrates increased as a result of two changes: zebra mussels imported food to the bottom through filter-feeding material from pelagic waters and also increased structural complexity (e.g., Botts et al. 1996, Stewart et al. 1998, González and Downing 1999). The latter change can lead to decreased fish predation on invertebrates within mussel colonies (González and Downing 1999, Mayer et al. 2001, Beekey et al. 2004a).

Quagga mussels were first recorded in the Great Lakes in 1989 (Mills et al. 1993) and, while initially found in deeper, cooler waters, have now replaced some zebra mussel colonies (Mills et al. 1999) and have become the dominant dreissenid species in most areas of the Great Lakes (Stoeckmann 2003). Quagga mussels are capable of inhabiting soft substrates where their colonies fundamentally shift habitat structure because they cover sediments and create a hard, structurally complex substrate. The addition of this structure will likely influence the infaunal (sediment-dwelling) benthic invertebrate community (Dermott and Kerec 1997, Bially and MacIsaac 2000, Freeman et al. 2011). The effects of quagga mussels on native benthic invertebrates are likely to differ from those of zebra mussels because quagga mussels are more likely to be found on soft sediment and will therefore interact more strongly with different guilds of native organisms. In this study, we focused on the effects of dreissenid clusters on *Hexagenia* spp. (*H. limbata* and *H. rigida*) in soft sediments, an infaunal mayfly species important to fish and ecosystem function.

Historically, *Hexagenia* were abundant in many warm, shallow bays and basins of the Great Lakes including western Lake Erie, but populations declined to near extirpation in the 1950s (e.g., Nebeker 1972, Winter et al. 1996, Gerlofsma and Ciborowski 1998). In the early-1990s, abundances began to increase in western Lake Erie and recolonization has now been well-documented (Krieger et al. 1996, Schloesser et al. 2000, Schloesser and Nalepa 2001). The resurgence of *Hexagenia* was temporally coincident with the expansion of dreissenid populations in this portion of the lake (Krieger et al. 1996). *Hexagenia* have been shown to prefer soft sediment colonized by *Dreissena* in small-scale laboratory experiments (DeVanna et al. 2011), similar to the way that epifaunal invertebrates respond to *Dreissena* colonies on hard substrate. However, *Hexagenia* have been shown to select for live *Dreissena* clusters over artificial ones, suggesting that increased substrate structure, and resulting protection from predation, is not the only reason *Hexagenia* select this habitat (DeVanna et al. 2011). Burrowing animals are already protected from predation and, therefore, may respond differently to the threat of predation than invertebrates living on the sediment surface. Although much is known about the effects of dreissenid clusters on epifaunal invertebrates, burrowing infaunal invertebrates like *Hexagenia* may respond very differently to dreissenid clusters due to their presence in the sediment.

Factors influencing spatial distributions of a species occur at multiple biotic and abiotic scales that may be important at one level, but are not always predictive at a different level (Turner et al. 1989, Wiens 1989, Graf et al. 2005). Different levels of scale can be viewed as a hierarchy, from large to small scales, with each level having its own natural cycles and processes structuring it (Senft et al. 1987, Urban et al. 1987). Levels in the hierarchy are not independent of one another, but rather higher-order scales can act to control processes at smaller scales, and smaller scales can drive processes at larger scales (Urban et al. 1987, Peterson 2000). Understanding processes regulating populations at both the local (small) and regional (large) scales has shown to be important due to the connections between hierarchical levels for predicting trout populations in Michigan Rivers (Zorn and Nuhfer 2007), vegetation patterns in North American boreal forests (Peterson 2000), and ecological land classification (Klijn and Udo de Haes 1994). Thus, spatial associations between *Dreissena* and *Hexagenia* may differ depending on what scale observations are made. We hypothesize that *Dreissena* will affect *Hexagenia* at a small spatial scale in a variety of ways including (1) modifying habitat by the addition of shells to soft sediment, (2) providing structural refuge from predation, (3) adding food resources by means of feces and pseudofeces, and (4) increasing flow of well-oxygenated pelagic water to areas close to clusters via filter feeding. Whereas at a larger scale, physical processes, such as sediment type, water currents, and oxygen availability, may be more important in structuring both *Dreissena* and *Hexagenia* distributions than species interactions. Quantifying the spatial relationship between these two taxa at multiple scales may help in understanding what mechanisms are structuring their distributions.

The goal of our study was to examine the spatial association between invasive *Dreissena* and native *Hexagenia* at differing hierarchical levels of scale, from large to small, in soft sediment habitats of western Lake Erie. Firstly, the relationship between *Hexagenia* and *Dreissena* was examined on a large scale by analyzing density relationships of the two taxa at 30 sites sampled over 10 years in the western basin of Lake Erie. Spatial analyses included spatial autocorrelation and cross-correlation across the basin. Secondly, two separate habitat preference experiments were conducted in the laboratory: (1) *observation experiments*, which examined the effect of a predator on the habitat preference of *Hexagenia* when given a choice of bare and *Dreissena*-covered sediments, and (2) *habitat-selection experiments*, which tested whether *Hexagenia* select for *Dreissena*-colonized habitat, artificial *Dreissena* clusters, or bare sediment.

METHODS

Analysis of Spatial Association

We assessed the large-scale spatial association of *Dreissena* and *Hexagenia* at over 30 sites across the western basin of Lake Erie, 1999–2009 (Figure 39.1). Thirty-one sites were sampled in 2000–2002 and 2004–2007, 24 in 1999, 60 in 2003, 19 in 2008, and 14 sites in 2009 for a total of 334 measurements for each taxon. Both taxa were sampled simultaneously with a standard Ponar grab (0.048 m^2 opening; three replicate samples per site). Collection and enumeration methods are in Schloesser et al. (1991).

Our spatial association analyses were designed to determine if there was a relationship between the co-occurrences and densities of *Hexagenia* and *Dreissena*. First, to test the null hypothesis that the presence of *Hexagenia* was independent of the presence of *Dreissena*, we conducted a chi-square test of independence. We also examined the probability of occurrence as determined from the proportion of sites with just *Dreissena*, with just *Hexagenia*, and with both taxa using Bayes' Theorem (McCarthy 2007). Second, we examined the correlation (r) between *Dreissena* and *Hexagenia* densities at each site for all available data, as well as the mean, standard deviation, and coefficient of variation of *Hexagenia* densities when *Dreissena* were present versus absent. We examined the standard deviation of *Hexagenia* densities to understand the dispersion of *Hexagenia* densities from the mean, whereas the coefficient of variation (standard deviation/mean) allowed us to examine the variability of the data relativized to the mean. Third, to examine spatial patterns of each taxon independently, we conducted spatial autocorrelations, as well as cross-correlations between taxa densities for all available data using Moran's correlation coefficient (I). Moran's I is an extension of Pearson's product moment correlation; however, because we assume points close to one another will be more similar, weights are given to each pair of points, with large values given to points close to one another and points further away having smaller weights (Reich et al. 1994, Kalkhan and Stohlgren 2000). When examining the spatial autocorrelation of a species and the cross-correlation between species, values of I range from −1 to +1 with values close to +1 indicate clustering, values close to −1 indicate dispersion, and values near zero suggest randomness

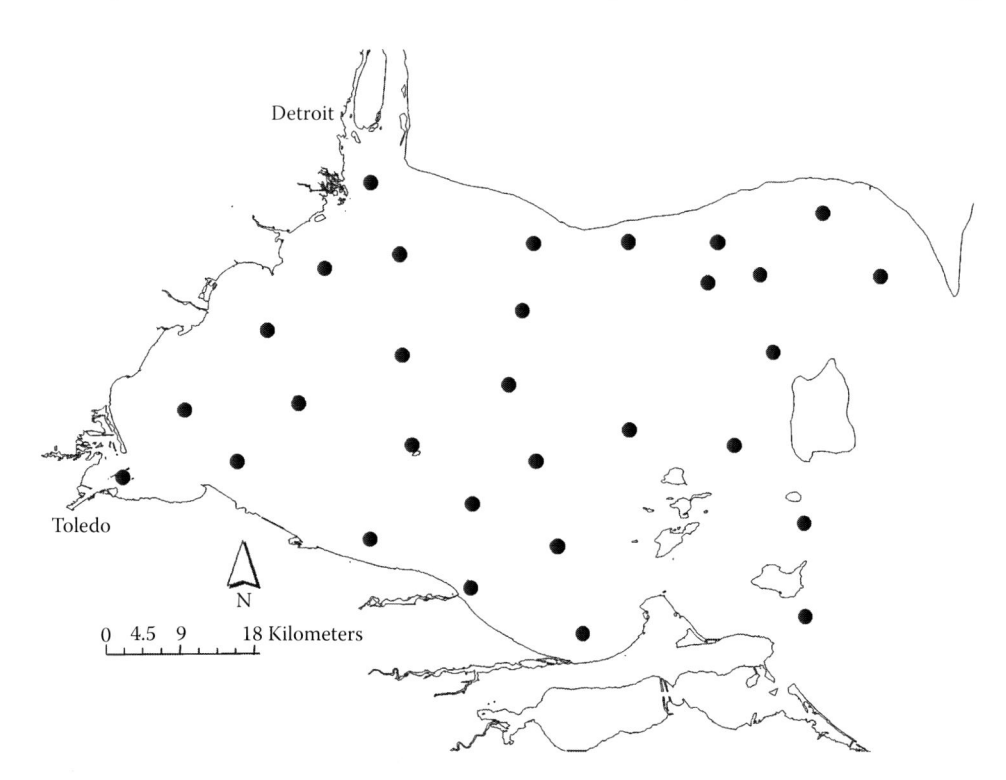

Figure 39.1 Location of 30 sites sampled to examine the distributions and densities of *Hexagenia* and *Dreissena* in the western basin of Lake Erie, 1999–2009.

(Reich et al. 1994, Kalkhan and Stohlgren 2000). The spatial autocorrelation of each taxa and the cross-correlation between taxa (Moran's I) was plotted for the range of distances between points, split into 10 equal distance classes (R, version 2.13.0), with correlations at a distance of zero representing the same site across years.

Laboratory Experiments

To examine small-scale associations between *Hexagenia* and *Dreissena*, both observation experiments and habitat-selection experiments (from DeVanna et al. 2011) were conducted. Sediments (sampled up to 6 cm depth) for the experiments were obtained with a grab sampler at a nearshore site (41.6885 W, 83.4250 N) in western Lake Erie. Sediments were washed through a 1.0 mm mesh sieve to remove both taxa. Sediment composition was silt with a soft texture. *Dreissena* and *Hexagenia* were collected from soft substrates at many sites in western Lake Erie to obtain enough individuals for experiments. Age 1 *Hexagenia* (>10 mm, but without black wing pads) were collected so individuals would be large enough for observation yet be at low risk for emergence during experiments. Both *H. limbata* and *H. rigida* were collected at their natural occurring proportions. Quagga mussels dominated *Dreissena* clusters collected; however, zebra mussels were present in small numbers.

Observation Experiments

Hexagenia behavior in observation experiments was monitored in small (25.4 cm × 1.90 cm × 25.4 cm) chambers that only allowed *Hexagenia* to choose between two habitat types, bare sediment and sediment that was covered by live *Dreissena* clusters (Figure 39.2). Also, *Hexagenia* in both habitats were either exposed (N = 5) or not exposed (n = 5) to a predator (yellow perch, *Perca flavescens*, a common generalist predator). Hence, we tested whether *Hexagenia* selected for habitat type or were distributed randomly and how this selection was affected by the presence of a predator. Prior to experimentation, *Hexagenia* were kept in the laboratory in the same soft sediment as used in the experiment and were able to feed *ad libitum* on organic matter from fresh Lake Erie sediments. Viewing chambers were constructed of acrylic sheets and filled with collected sediments. To establish habitat types (e.g., bare sediment and live *Dreissena*-covered sediment) in chambers, a thin metal sheet divided the chamber into two equal sections that was removed before addition of experimental organisms (e.g., predators and *Hexagenia*). Live *Dreissena* clusters in the chambers were equivalent to a density of 3400 *Dreissena*/m^2 that has been observed in western Lake Erie (Patterson et al. 2005). Organisms were added 24 h after experimental setup; thus, sediments were in place and settled. All treatments were aerated throughout the course of the experiment.

In treatments with fish present, we added a single, age 1 yellow perch 1 h prior to addition of *Hexagenia* nymphs. All chambers, regardless of fish treatment, had a plastic, permeable barrier hung 10 cm from the top of the chamber to allow the fish an area to swim but kept fish 15.4 cm from the sediment and prevented consumption of *Hexagenia*. The barrier did have holes to allow movement of *Hexagenia* through the entire water column. Yellow perch were not allowed to function as predators due to the size of the chambers. After fish acclimated for about an hour, six *Hexagenia* (equivalent density of 1400/m^2; Krieger 1999) were released in the center of the chamber and watched to determine initial habitat selection by nymphs. Initial habitat selection was determined to be the first habitat in which a mayfly began to actively burrow. Observation trials were started one at a time in each of five chambers, and for each chamber, observations lasted 15 min to give *Hexagenia* enough time to choose a habitat and burrow.

Habitat-Selection Experiments

We tested to determine if *Hexagenia* selected for habitats with or without *Dreissena* clusters on soft sediments. We used experimental tanks (circular plastic tubs; 41 cm diameter and 43 cm height) filled with collected sediment and dechlorinated tap water aerated throughout the experiment and separated into three equal "pie-slice"–shaped sections (0.046/m^2) with metal dividers. Three treatment types were then created: (1) bare sediment, (2) live *Dreissena* clusters, and (3) dead *Dreissena* clusters. Live and dead *Dreissena* treatments contained approximately 250 individuals representing a density of 5434/m^2, which has recently been observed in western Lake Erie (Patterson et al. 2005). Dead *Dreissena* clusters were created from empty shells attached together with nontoxic glue. Clusters were glued to 1 g lead weights so they were stable on sediments. Weights were also added to the other two treatments for sake of consistency. Metal dividers were removed after habitat types were established.

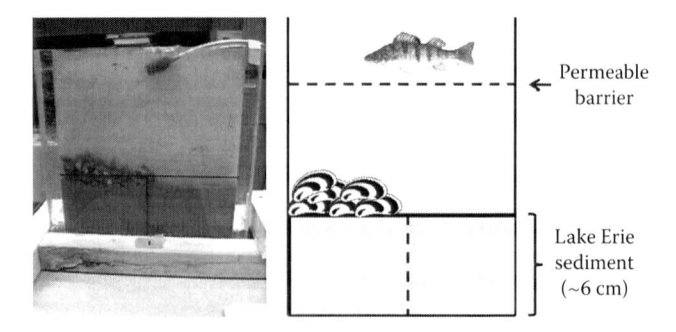

Figure 39.2 Observation chamber constructed of acrylic sheets and filled with sediments from Lake Erie; note two habitat types (bare sediment and live *Dreissena* clusters). In some chambers, fish were added to determine role of the potential predators on *Hexagenia* habitat preferences.

Experiments were conducted using five densities of *Hexagenia* that were within the range of densities found in western Lake Erie (0–2000/m²; Krieger 1999). The number of added individuals and achieved densities per experimental tank were 5 individuals (~100/m²), 9 (~200/m²), 18 (~400/m²), 36 (~800/m²), and 54 (~1200/m²). Each *Hexagenia* density was replicated three times (total N = 15). *Hexagenia* were added to the center of tank at the surface of the water and allowed to select between habitat types. One replicate of each density was run at the same time, and tanks were placed in a straight line in random order. After 48 h, metal dividers were pushed into sediments between habitat types, water was removed, sediments from each habitat were removed and sieved through 250 μm mesh, and *Hexagenia* were counted.

To analyze results of both experiments, the percentage of *Hexagenia* in each habitat type was arcsine square root transformed to help achieve a normal distribution (Zar 1999). Data were analyzed using a split-plot analysis of variance (ANOVA) model (SAS 9.1, α = 0.05), because each experimental unit was split into different habitat types and treatments were applied to different scales (Potvin 2001). In the observation experiment, predators were applied to the whole chamber (main plot factor) but habitat type was applied to only half of the chamber (subplot factor). For the habitat-selection experiment, *Hexagenia* density was applied to the whole mesocosm (main plot factor), whereas each habitat type was applied to only one-third of the experimental unit (subplot factor). When appropriate, split-plot ANOVAs were followed by a Tukey multiple comparison test.

RESULTS

Analysis of Spatial Association

At the basin-wide scale, the presence of *Hexagenia* was related to *Dreissena* presence (chi-square, χ^2 = 7.51, p = 0.006; Table 39.1, Figure 39.3). Of the 334 observations (sites and years), 65% had both *Hexagenia* and *Dreissena* present, 23% had only *Hexagenia*, 6% had only *Dreissena*, and 6% had neither taxa present. The percentage of sites with neither taxa present was very low, and this could be due to the sites being chosen specifically to monitor *Hexagenia*

Table 39.1 Chi-Square Contingency Table Showing Number of Sites (n = 334) in Western Lake Erie Collected between 1999 and 2009 with Both *Hexagenia* and *Dreissena* Present and/or Absent

		Hexagenia		
		Present	Absent	Total
Dreissena	Present	216	21	237
	Absent	78	19	97
	Total	294	40	334

populations. Overall, using Bayes' Theorem (McCarthy 2007), the probability of finding only *Hexagenia* at any site was 0.88, *Hexagenia* at a site with *Dreissena* was 0.91, and the probability of *Hexagenia* at a site without *Dreissena* was 0.80. Although the presence of *Hexagenia* was most likely at sites where *Dreissena* were present, densities were slightly lower than densities at sites where *Dreissena* was absent. *Hexagenia* at sites without *Dreissena* (n = 97) achieved very high densities (>1500/m²); the mean density of nymphs was 384/m² with a high variability (SD = 483.8). However, at sites where *Dreissena* were present (n = 237), the mean density of *Hexagenia* was 270/m², and variability was lower (SD = 341.2) than without *Dreissena* (Figure 39.3), and very high densities of *Hexagenia* (>1000/m²) were unattained. However, once variability was normalized to the mean of the data, no difference existed in the dispersion for sites with and without *Dreissena* (coefficient of variation = 1.26 for both groups). Even though the mean density of *Hexagenia* was lower when *Dreissena* was present, it is within the range rated "excellent" in the Lake Erie Index of Biotic Integrity (Ohio Lake Erie Commission 2004). There was no significant linear correlation between densities of *Hexagenia* and *Dreissena* in western Lake Erie (p = 0.9381, r = −0.0043, Figure 39.3). There was no spatial autocorrelation for either *Hexagenia* or *Dreissena* (Figure 39.4a and b) and no spatial cross-correlation between the two taxa (Figure 39.4c), which means that there was no relationship between the two taxa at distances across the western basin. For *Dreissena* alone, across all distances, the greatest correlation (Moran's I) was 0.06, which is relatively low and suggests no relationship to distance (Figure 41.4a). *Hexagenia* showed a slight correlation at a distance of zero (same site across all years, I = 0.30), but from site to site, no spatial autocorrelation was found (Figure 39.4b).

Laboratory Experiments

Hexagenia selected live *Dreissena* clusters over bare sediment in both types of laboratory experiments (Figures 39.5 and 39.6). In observation experiments, *Hexagenia* selected *Dreissena* clusters over bare sediment (split-plot ANOVA: $F_{1,18}$ = 11.44, p = 0.0017, Figure 39.5). However, there was no significant effect for the presence of a predator (split-plot ANOVA: $F_{1,18}$ = 0.84, p = 0.4408), which indicates *Hexagenia* did not select clusters more often when a predator was present. In habitat-selection experiments, the percentage of *Hexagenia* differed among all habitats (split-plot ANOVA: $F_{2,20}$ = 95.17, p < 0.0001, Tukey: p < 0.05, Figure 39.6). The highest percentage of *Hexagenia* was found in the presence of live *Dreissena* clusters, followed by dead *Dreissena* clusters, and lastly by bare sediment. There was a significant interaction between percentages of *Hexagenia* in each habitat based on *Hexagenia* density (split-plot ANOVA: habitat*density: $F_{8,20}$ = 4.86, p < 0.0001), which indicated the percentage of *Hexagenia* that select each habitat type changed with *Hexagenia* density.

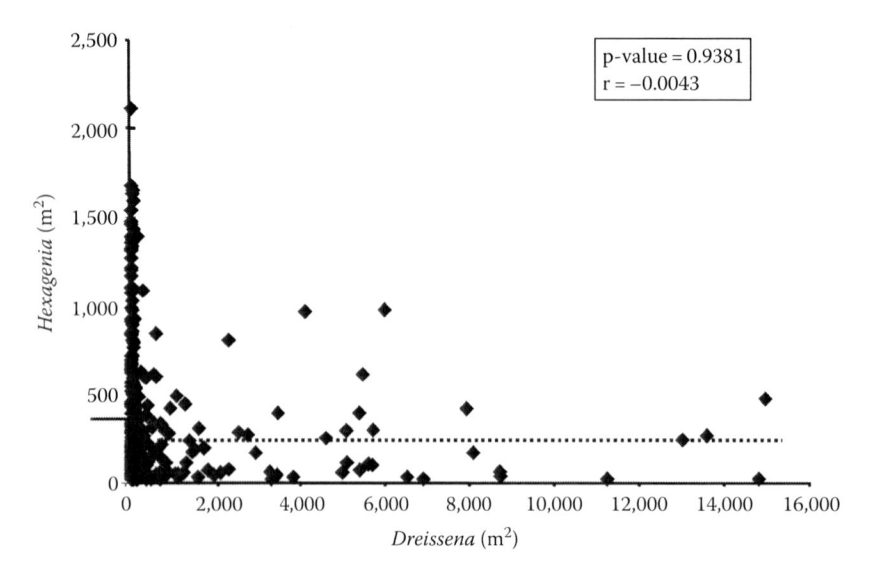

Figure 39.3 Scatter plot of *Hexagenia* and *Dreissena* densities at 30 sites in western Lake Erie 1999–2009. Each point (n = 334) represents the density of *Hexagenia* and *Dreissena* at a site each year. Solid line to the left of the y-axis equals the mean density of *Hexagenia* when *Dreissena* are absent (384/m^2) and dotted line equals the mean density when *Dreissena* are present (270/m^2).

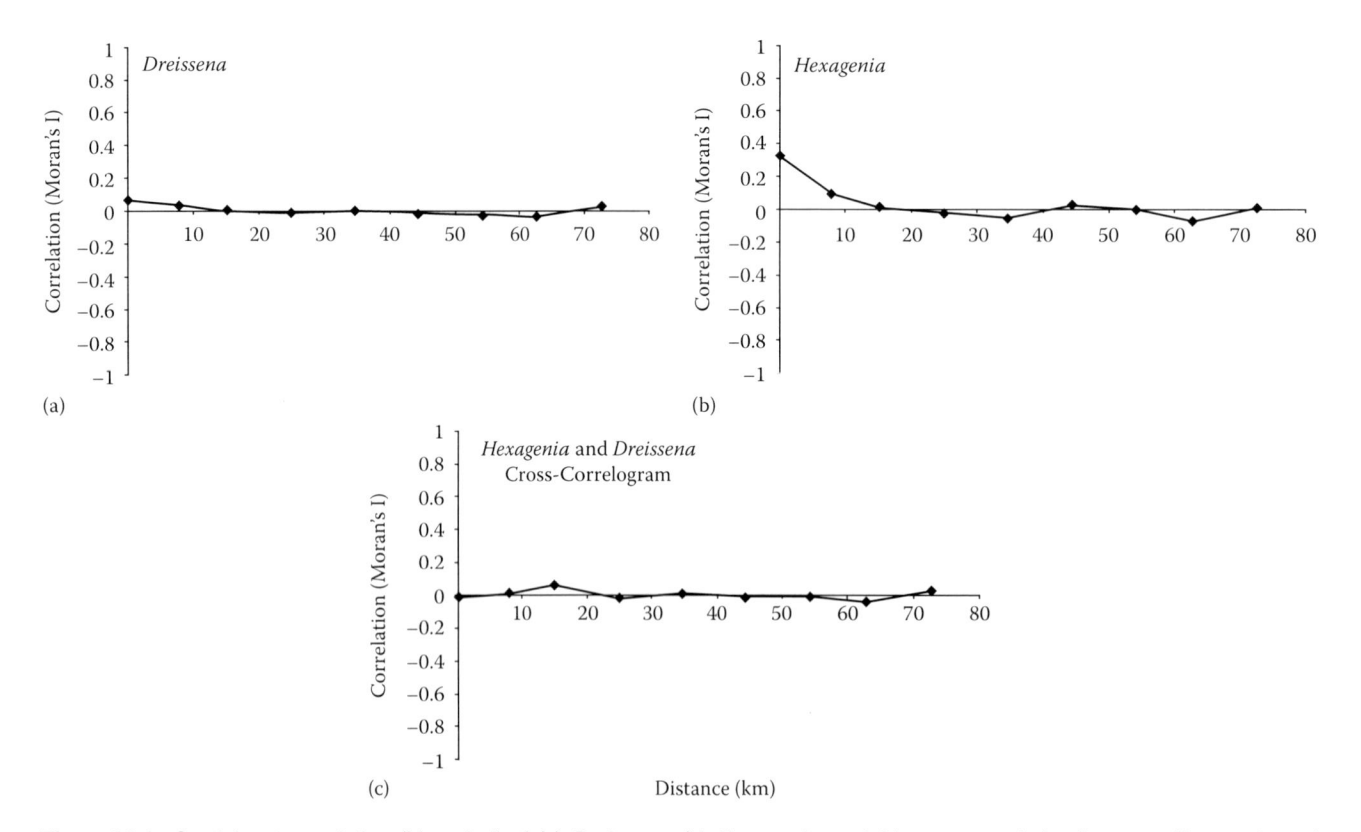

Figure 39.4 Spatial autocorrelation (Moran's I) of (a) *Dreissena*, (b) *Hexagenia*, and (c) cross-correlation between *Hexagenia* and *Dreissena* for 10 distance classes of all sampled sites in western Lake Erie 1999–2009. Correlations at a distance of zero represent the same sampled site across years. Moran's I values range from –1 to +1 where values close to +1 indicated clustering, –1 dispersion, and values near zero indicated randomness.

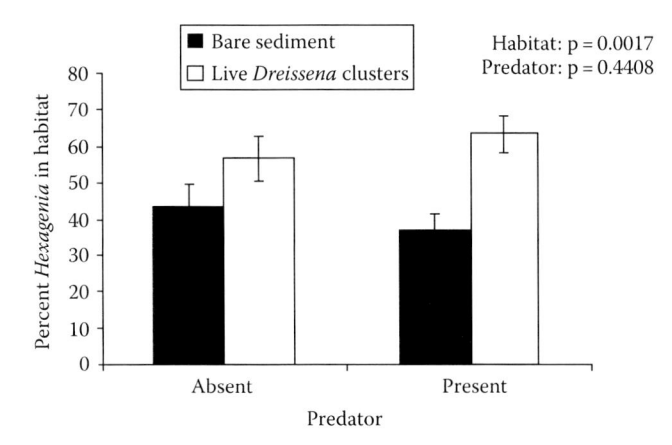

Figure 39.5 Mean percent of total number of *Hexagenia* found in two habitat types (bare sediment and live mussel clusters) in the presence and absence of a fish predator (yellow perch). Bars represent ±1 standard error. Statistics were conducted on arcsine square root transformed values.

DISCUSSION

Analysis of Spatial Association

Our results agree with those of other studies that show *Dreissena* can have great impacts on benthic invertebrates on a small-scale level (e.g., Mayer et al. 2001, Beekey et al. 2004b, Ward and Ricciardi 2007). However, at a large lake scale, many other processes can affect distributions of benthic invertebrates besides presence or absence of *Dreissena*. For our study organism, *Hexagenia*, the distribution and abundance of each life stage may be affected by different factors. For example, eggs of *Hexagenia* are deposited at the surface of the water (Hunt 1951) and so their distribution in the water column is likely to be influenced by large-scale physical processes, such as wind and currents. Once eggs settle out of the water, substrate types and oxygen levels (Gerlofsma and Ciborowski 1999) undoubtedly affect survival. Our laboratory-based, small-scale experiments only examined *Hexagenia* after they hatched, and thus, they were able to move and exhibit habitat selection, which is probably the period of time when small-scale, ecosystem engineer effects of *Dreissena* are important. On the other hand, our large-scale spatial analyses of basin-wide distributions of *Hexagenia* and *Dreissena* incorporated outcomes of many biological and physical processes that can affect different life stages of *Hexagenia*.

Hexagenia and *Dreissena* were found to co-occur at the majority of sites sampled in the western basin of Lake Erie, which suggested *Dreissena* do not inhibit *Hexagenia*. Not only did *Hexagenia* co-occur with *Dreissena*, they were more likely to occur with *Dreissena* than occur without *Dreissena*. Our finding suggests that *Hexagenia*, even at a large scale, are positively associated with the presence of *Dreissena*, which may be due to *Hexagenia* selection for sediment covered with live *Dreissena* clusters as shown in the small-scale experiments. However, associations between *Dreissena* clusters and *Hexagenia* under natural lake conditions are more complex and difficult to interpret than in simple laboratory experiments because we do not know how far individual *Hexagenia* will actively move to select for a habitat type. Physical processes, such as currents, may move planktonic *Dreissena* veligers and *Hexagenia* eggs to similar locations. It is likely that a combination of behavioral selection on a small-scale and physical processes on a large-scale determine the spatial relationship between these two benthic taxa.

Although *Hexagenia* presence was positively associated with *Dreissena*, densities of the two taxa were not correlated. At sites where *Dreissena* were absent, the mean density of *Hexagenia* was high, but a high proportion of these sites had no *Hexagenia*, while a few sites had very high densities (>1500/m^2). Alternatively, at sites where *Dreissena* were present, the mean density of *Hexagenia* was slightly lower. *Hexagenia* densities have previously been shown to be higher at field sites without than with *Dreissena* (Freeman et al. 2011). Therefore, while the presence of *Dreissena* may have a positive influence on *Hexagenia*, dreissenid-induced habitat alterations may serve to limit abundances. *Hexagenia* may not have

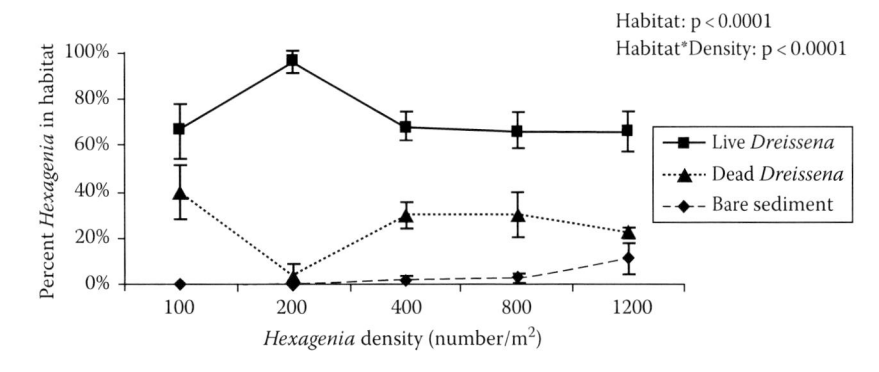

Figure 39.6 Mean percent of total number of *Hexagenia* found in three habitat types (bare sediment and live and dead mussel clusters) at five densities of *Hexagenia* nymphs in laboratory habitat–choice experiments. Bars represent ±1 standard error.

been able to reach high densities (maximum <1000/m^2) when *Dreissena* were present due to low oxygen beneath *Dreissena* (Burks et al. 2002, Beekey et al. 2004b). Also, unlike epifaunal invertebrates that show a positive, linear response to increased densities of *Dreissena* (Mayer et al. 2002), *Hexagenia* in our study did not show such a positive response. These different responses are likely related to differences in habitat preferences of the organisms. As dreissenid density increases, habitat complexity and available surface area for epifaunal invertebrates also increases (e.g., Botts et al. 1996, Stewart et al. 1998, González and Downing 1999); however, for sediment-dwelling invertebrates like *Hexagenia*, favored habitat surface area would not change as dreissenid density increased. *Hexagenia* was more likely to be found in areas with *Dreissena* and, although *Hexagenia* did not occur at densities >1000 nymphs/m^2 when in the presence of *Dreissena*, *Dreissena* presence may decrease *Hexagenia* to population density levels considered to be more "healthy" and sustainable (Ohio Lake Erie Commission 2004).

Densities of *Hexagenia* and *Dreissena* appear to have a high degree of spatial and temporal variability. At the spatial scale of the western basin of Lake Erie, both *Hexagenia* and *Dreissena* densities were distributed independently of distances sampled (Figure 39.4); however, they were both found at the majority of sites. This finding may be a result of the sites sampled. Sites were all well spaced apart, which may inhibit our ability to see spatial autocorrelation at small distances. Also, the distance-independent spatial distributions may be due to both species having a planktonic early life-history stage, that is, *Dreissena* veligers and *Hexagenia* eggs. *Hexagenia* eggs are deposited at a location that is highly variable and dependent on wind speed and direction. Both eggs and veligers act as passive particles carried by water currents until they settle out of the water column (Hannan 1984, Jackson 1986). Although both taxa have planktonic stages, water currents in lakes are highly variable (Beletsky et al. 1999) and, if the two species are not in the water column at the same time or have different settling rates, they may be distributed very differently. *Hexagenia* eggs can sink quickly (1.9 cm/s; Hunt 1951) unlike *Dreissena* veligers, which stay in the water column for 2–4 weeks or longer (Sprung 1989). Once settled from the water column and grown to a developed stage, how far either *Dreissena* or *Hexagenia* can move to select for suitable habitat is not known. We would hypothesize movement over a short distance due to limited mobility of the organisms and susceptibility to predation. For *Hexagenia* across years, there is a weak positive correlation of density at a spatial distance of zero (Figure 39.4b), which indicates a correlation at the same site through time. This suggests that *Hexagenia* densities at a number of sites are consistent from year to year; that is, some sites always have *Hexagenia*, possibly due to favorable sediment conditions, and some sites never have *Hexagenia* due to conditions that are uninhabitable. We present here spatial autocorrelation and cross-correlations across all years. Therefore, temporal variations in data were masked. However, correlograms were run for individual years and yielded similar results. At larger spatial scales, there appear to be many factors influencing the distributions and densities of both *Hexagenia* and *Dreissena*.

Laboratory Experiments

At the small spatial scale examined in laboratory experiments, *Hexagenia* consistently preferred sediments covered by live *Dreissena* clusters compared to bare sediments and sediments covered by dead *Dreissena* clusters in both sets of experiments. Bare soft sediment, thought to be the preferred habitat of *Hexagenia* (e.g., Wang et al. 2001, Bachteram et al. 2005, Chaffin and Kane 2010), was the least selected habitat type in our experiments. However, long-term mesocosm experiments indicate *Hexagenia* survival declined with *Dreissena* (Osterling et al. 2007, Freeman et al. 2011), but *Hexagenia* condition was not affected (Freeman et al. 2011). *Hexagenia* may experience lower survival in the presence of dreissenid mussels because food resources can become limiting in long-term tank experiments. *Hexagenia* reside under and in clusters and may become densely aggregated, leading to high food competition per unit area. Our field data showed that *Hexagenia* were more likely to occur but were slightly less abundant in mussel-dominated habitat. This observation is consistent with both our short-term choice experiments and observational studies. While *Hexagenia* show a behavioral preference for the structured mussel clusters, this habitat may not be beneficial over long time spans (Freeman et al. 2011).

Addition of physical structure in the form of dreissenid clusters was not the only mechanism that affected *Hexagenia* selection because *Hexagenia* preferred live *Dreissena* to dead *Dreissena* clusters. Both live and dead *Dreissena* change the physical structure of available habitat; however, live *Dreissena* also change chemical and biological structure around clusters. For example, live *Dreissena* filter feed, respire, and excrete feces and pseudofeces. Most epifaunal invertebrates in interstitial spaces of mussel clusters located on hard substrates have been shown to occur in equal densities in live and dead *Dreissena* habitats (Botts et al. 1996, González and Downing 1999). Similar to *Hexagenia* in our experiments though, some benthic fauna (tubificid worms and some chironomids and snails) prefer clusters of live mussels over dead clusters (Ricciardi et al. 1997, Stewart et al. 1998). Therefore, the preference of *Hexagenia* for live dreissenid clusters may simply be a response to additional food provided by mussels (Roditi et al. 1997). Another explanation for *Hexagenia* preference for live *Dreissena* clusters may be related to mussel filtration activity. Individual dreissenids filter a relatively high volume of water (between 0.1 and 1 L/h), and the resultant increased flow of oxygenated water near the sediments may benefit *Hexagenia* (Kryger and Riisgard 1988). Although water below *Dreissena* clusters may have lower dissolved oxygen and be diminished in quality (Burks et al. 2002, Beekey et al. 2004b), some of the oxygenated water may be available to *Hexagenia* burrows immediately adjacent to clusters. Therefore, microhabitat alterations in the presence of *Dreissena* clusters may increase selection of this habitat by *Hexagenia*, which is analogous to *Dreissena* effects on other benthic invertebrates.

Sediment covered by live *Dreissena* was the preferred habitat over bare sediment for *Hexagenia* when a predator (yellow perch) was present. However, contrary to expectations, preference for *Dreissena*-covered sediment was not stronger than preference for the same habitat with fish absent (Figure 39.5). We hypothesized that *Hexagenia* would show stronger selection for *Dreissena*-covered sediment when the predator was present, since another genus of mayfly has been shown to change its behavior in the presence of fish, suggesting an ability to detect predators (Kolar and Rahel 1993). The lack of increased selection for *Dreissena* with a predator is consistent with results from the habitat-selection experiment that suggests *Hexagenia* choose live *Dreissena* clusters for reasons other than protection from predation. Although *Hexagenia* did not select for *Dreissena*-covered habitat primarily as protection from predation, *Hexagenia* have been shown to be consumed by fish at lower levels of efficiency when beneath clusters under highly turbid conditions, as compared with levels found at conditions of low turbidity, high light, and no clusters (DeVanna et al. 2011).

CONCLUSIONS AND FUTURE DIRECTIONS

Range expansion of dreissenid mussels onto soft sediments and the observed small-scale habitat selection by burrowing mayflies under and near *Dreissena* clusters may have potential cascading effects to higher trophic levels and overall ecosystem function. Burrowing mayflies of the genus *Hexagenia* are used as a mesotrophic indicator associated with pollution-abatement programs in the Laurentian Great Lakes and other water bodies throughout the world (e.g., Great Lakes [Reynoldson et al. 1989], Mississippi River [Fremling and Johnson 1990], The Netherlands [bij de Vaate et al. 1992]). As a result, it is extremely important that habitat alterations associated with *Dreissena* do not affect the behavior and tolerance of *Hexagenia* to changing oxygen concentrations. The spatial association between *Hexagenia* and *Dreissena* can have dramatic consequences for higher trophic levels. Many scientists were optimistic about the return of *Hexagenia* to western Lake Erie, as *Hexagenia* are an additional food source to many economically important fish species, such as yellow perch (Hayward and Margraf 1987, Schaeffer et al. 2000). It has been shown that consumption of *Hexagenia* by yellow perch decreases in turbid systems when *Hexagenia* are burrowed beneath *Dreissena* clusters (DeVanna et al. 2011). Therefore, given the highly turbid conditions in the western basin where both organisms co-occur, *Hexagenia* may not be available to fish as a food source. As a result, the potential benefit of *Hexagenia* recolonization may be tempered by spatial associations with *Dreissena* clusters.

This study examined the spatial pattern of *Hexagenia* and *Dreissena* at two spatial and temporal scales, and both showed evidence for *Hexagenia* and *Dreissena* co-occurring in western Lake Erie. Future work should examine the processes governing these observed small- and large-scale patterns. On a small scale, it is important to understand what is driving *Hexagenia* to select sediment covered by live *Dreissena* clusters, that is, whether selection is a result of an added food resource, protection from predation, or a combination of both. On a large scale, processes constraining *Hexagenia* densities, such as sediment type, location of adults laying eggs, egg predation by other invertebrates (Plant et al. 2003), and short-term periods of hypoxia (Bridgeman et al. 2006), should be examined using a modeling approach. Connections between these two levels of scale can begin to be explored once the processes constraining and driving spatial relationships between *Hexagenia* and *Dreissena* at each scale are established. Understanding the large- and small-scale processes that interact to determine *Hexagenia* population size and distribution in western Lake Erie will help us better understand *Hexagenia* as an indicator organism and important prey resource for fish, as well as lead to a better understanding of how an invasive ecosystem engineer can have cascading effects throughout the foodweb.

ACKNOWLEDGMENTS

The authors wish to thank P. Armenio, C. Barrett, and K. Doan for their lab and field assistance. We would also like to thank the Mayer and Bossenbroek labs at the University of Toledo for providing constructive comments. We thank the University of Toledo's Department of Environmental Sciences and Lake Erie Center. This research was supported in part by a fellowship from the NSF GK-12 program #0742395 "Graduate Fellows in High School STEM Education: An Environmental Science Learning Community at the Land-Lake Ecosystem Interface," an Ohio Lake Erie Commission Lake Erie Protection Fund Grant to C. Mayer and K. DeVanna, and a Sigma Xi grant-in-aid of research to K. DeVanna. This is contribution number 2013-02 of the University of Toledo, Lake Erie Center. Any use of trade, product, or firm names is for descriptive purposes only and does not imply endorsement by the U.S. Government. This article is Contribution 1707 of the U.S. Geological Survey Great Lakes Science Center.

REFERENCES

Bachteram, A. M., K. A. Mazurek, and J. J. H. Ciborowski. 2005. Sediment suspension by burrowing mayflies (*Hexagenia* spp., Ephemeroptera: Ephemeridae). *J. Great Lakes Res.* 31: 208–222.

Beekey, M. A., D. J. McCabe, and J. E. Marsden. 2004a. Zebra mussels affect benthic predator foraging success and habitat choice on soft sediments. *Oecologia* 141: 164–170.

Beekey, M. A., D. J. McCabe, and J. E. Marsden. 2004b. Zebra mussel colonization of soft sediments facilitates invertebrate communities. *Freshwat. Biol.* 49: 535–545.

Beletsky, D., J. H. Saylor, and D. J. Schwab. 1999. Mean circulation in the Great Lakes. *J. Great Lakes Res.* 25: 78–93.

Bially, A. and H. J. MacIsaac. 2000. Fouling mussels (*Dreissena* spp.) colonize soft sediments in Lake Erie and facilitate benthic invertebrates. *Freshwat. Biol.* 43: 85–97.

bij de Vaate, A., A. Klink, and F. Oosterbroek. 1992. The mayfly, *Ephoron virgo* (Olivier), back in the Dutch parts of the rivers Rhine and Meuse. *Hydrobiol. Bull.* 25: 237–240.

Botts, P. S., B. A. Patterson, and D. W. Schloesser. 1996. Zebra mussel effects on benthic invertebrates: Physical or biotic? *J. N. Am. Benthol. Soc.* 15: 179–184.

Bridgeman, T. B., D. W. Schloesser, and A. E. Krause. 2006. Recruitment of *Hexagenia* mayfly nymphs in western Lake Erie linked to environmental variability, *Ecol. Appl.* 16: 601–611.

Burks, R. L., N. C. Tuchman, C. A. Call, and J. E. Marsden. 2002. Colonial aggregates: Effects of spatial position on zebra mussel responses to vertical gradients in interstitial water quality. *J. N. Am. Benthol. Soc.* 21: 64–75.

Carlton, J. T. 2008. The zebra mussel *Dreissena polymorpha* found in North America in 1986 and 1987. *J. Great Lakes Res.* 34: 770–773.

Chaffin, J. D. and D. D. Kane. 2010. Burrowing mayfly (Ephemeroptera: Ephemeridae: *Hexagenia* spp.) bioturbation and bioirrigation: A source of internal phosphorus loading in Lake Erie. *J. Great Lakes Res.* 36: 57–63.

Dermott, R. and D. Kerec. 1997. Changes in deepwater benthos of eastern Lake Erie since the invasion of *Dreissena*: 1979–93. *Can. J. Fish. Aquat. Sci.* 54: 922–930.

DeVanna, K. D., P. A. Armenio, C. A. Barrett, and C. M. Mayer. 2011. Invasive ecosystem engineers on soft sediment change the habitat preferences of native mayflies and their availability to predators. *Freshwat. Biol.* 56: 2448–2458.

Freeman, K. J., K. A. Krieger, and D. J. Berg. 2011. The effects of dreissenid mussels on the survival and condition of burrowing mayflies (*Hexagenia* spp.) in western Lake Erie. *J. Great Lakes Res.* 37: 426–431.

Fremling, C. R. and D. K. Johnson. 1990. Recurrence of *Hexagenia* mayflies demonstrates improved water quality in Pool 2 and Lake Pepin, Upper Mississippi River. In *Mayflies and Stoneflies*, I. Campbell, ed., pp. 243–248. Dordrecht, The Netherlands: Kluwer Academic Publishers.

Gerlofsma, J. and J. J. H. Ciborowski. 1998. The effects of anoxia on *Hexagenia* mayfly (Ephemeroptera: Ephemeridae) eggs and implications for nymph populations in the western Lake Erie. *Bull. N. Am. Benthol. Soc.* 17: 207.

González, M. J. and A. L. Downing. 1999. Mechanisms underlying amphipod responses to zebra mussel (*Dreissena polymorpha*) invasion and implications for fish-amphipod interactions. *Can. J. Fish. Aquat. Sci.* 56: 679–685.

Graf, R.F., K. Bollmann, W. Suter, and H. Bugmann. 2005. The importance of spatial scale in habitat models: Capercaillie in the Swiss Alps. *Landsc. Ecol.* 20: 703–717.

Hannan, C. A. 1984. Planktonic larvae may act like passive particles in turbulent near-bottom flows. *Limnol. Oceanogr.* 29: 1108–1116.

Hayward, R. S. and F. J. Margraf. 1987. Eutrophication effects on prey size and food available to Yellow Perch in Lake Erie. *Trans. Am. Fish. Soc.* 116: 210–223.

Hunt, B. P. 1951. Reproduction of the burrowing mayfly, *Hexagenia limbata* (Serville), in Michigan. *Florida Entomol.* 34: 59–70.

Jackson, G. A. 1986. Interaction of physical and biological processes in the settlement of planktonic larvae. *Bull. Mar. Sci.* 39: 202–212.

Jones, C. G., J. H. Lawton, and M. Shachak. 1994. Organisms as ecosystem engineers. *Oikos* 69: 373–386.

Jones, C. G., J. H. Lawton, and M. Shachak. 1997. Positive and negative effects of organisms as physical ecosystem engineers. *Ecology* 78: 1946–1957.

Kalkhan, M. A. and T. J. Stohlgren. 2000. Using multi-scale sampling and spatial cross-correlation to investigate patterns of plant species richness. *Environ. Monit. Assess.* 64: 591–605.

Karatayev, A. Y., L. E. Burlakova, and D. K. Padilla. 2002. Impacts of zebra mussels on aquatic communities and their role as ecosystem engineers. In *Invasive Aquatic Species of Europe: Distribution, Impacts and Management*, E. Leppakoski, S. Gollasch, and S. Olenin, eds., pp. 433–446. Dordrecht, The Netherlands: Kluwer Academic Publishers.

Klijn, F. and H. A. Udo de Haes. 1994. A hierarchical approach to ecosystems and its implications for ecological land classification. *Landsc. Ecol.* 9: 89–104.

Kolar, C. S. and F. J. Rahel. 1993. Interaction of a biotic factor (predator presence) and an abiotic factor (low oxygen) as an influence on benthic invertebrate communities. *Oecologia* 95: 210–219.

Krieger, K. A. 1999. Ecosystem change in Western Lake Erie: Cause and effect of burrowing mayfly recolonization. Final Report. Toledo, OH: Ohio Lake Erie Commission.

Krieger, K. A., D. W. Schloesser, B. A. Manny et al. 1996. Recovery of burrowing mayflies (Ephemeroptera: Ephemeridae: *Hexagenia*) in Lake Erie. *J. Great Lakes Res.* 22: 254–263.

Kryger, J. and H. U. Riisgard. 1988. Filtration rate capacities in six species of European fresh water bivalves. *Oecologia* 77: 34–38.

Mayer, C. M., R. A. Keats, L. G. Rudstam, and E. L. Mills. 2002. Scale-dependent effects of zebra mussels on benthic invertebrates in a large eutrophic lake. *J. N. Am. Benthol. Soc.* 21: 616–633.

Mayer, C. M., L. G. Rudstam, E. L. Mills, S. G. Cardiff, and C. A. Bloom. 2001. Zebra mussels (*Dreissena polymorpha*), habitat alteration, and yellow perch (*Perca flavescens*) foraging: System-wide effects and behavioural mechanisms. *Can. J. Fish. Aquat. Sci.* 58: 2459–2467.

McCarthy, M.A. 2007. *Bayesian Methods for Ecology*. Cambridge, MA: Cambridge University Press.

Mills, E. L., J. R. Chrisman, B. Baldwin et al. 1999. Changes in the dreissenid community in the lower Great Lakes with emphasis on southern Lake Ontario. *J. Great Lakes Res.* 25: 187–197.

Mills, E. L., J. H. Leach, J. T. Carlton, and C. L. Secor. 1993. Exotic species in the Great Lakes: A history of biotic crises and anthropogenic introductions. *J. Great Lakes Res.* 19: 1–54.

Nebeker, A.V. 1972. Effect of low oxygen on survival and emergence of aquatic insects. *Trans. Am. Fish. Soc.* 4: 675–679.

Ohio Lake Erie Commission. 2004. State of the Lake Report. Toledo, OH: Ohio Lake Erie Commission.

Osterling, E. M., E. Bergman, L. A. Greenberg, B. S. Baldwin, and E.L. Mills. 2007. Turbidity-mediated interactions between invasive filter-feeding mussels and native bioturbating mayflies. *Freshwat. Biol.* 52: 1602–1610.

Patterson, M. W. R., J. J. H. Ciborowski, and D. R. Barton. 2005. The distribution and abundance of *Dreissena* species (Dreissenidae) in Lake Erie, 2002. *J. Great Lakes Res.* 31(Suppl. 2): 223–237.

Peterson, G. D. 2000. Scaling ecological dynamics: Self-organization, hierarchical structure, and ecological resilience. *Clim. Change* 44: 291–309.

Plant, W., J. J. H. Liborowski, and L. D. Corkum. 2003. Do tube-dwelling midges inhibit the establishment of burrowing mayflies? *J. Great Lakes Res.* 29: 521–528.

Potvin, C. 2001. ANOVA: Experiment layout and analysis. In *Design and Analysis of Ecological Experiments*, S. M. Scheiner and J. Gurevitch, eds., pp. 63–76. New York: Oxford University Press.

Rahbek, C. 2005. The role of spatial scale and the perception of large scale species-richness patterns. *Ecol. Lett.* 8: 224–239.

Reich, R. M., R. L. Czaplewski, and W. A. Bechtold. 1994. Spatial cross-correlation of an undisturbed, natural shortleaf pine stands in northern Georgia. *Environ. Ecol. Stat.* 1: 201–217.

Reynoldson, T. B., D. W. Schloesser, and B. A. Manny. 1989. Development of a benthic invertebrate objective for mesotrophic Great Lakes waters. *J. Great Lakes Res.* 15: 669–686.

Ricciardi, A., F. G. Whoriskey, and J. B. Rasmussen. 1997. The role of the zebra mussel (*Dreissena polymorpha*) in structuring macroinvertebrate communities on hard substrata. *Can. J. Fish. Aquat. Sci.* 54: 2596–2608.

Roditi, H. A., D. L. Strayer, and S. E. G. Findlay. 1997. Characteristics of zebra mussel (*Dreissena polymorpha*) biodeposits in a tidal freshwater estuary. *Arch. Hydrobiol.* 140: 207–219.

Schaeffer, J. S., J. S. Diana, and R. C. Haas. 2000. Effect of long term changes in the benthic community on yellow perch in Saginaw Bay, Lake Huron. *J. Great Lakes Res.* 26: 340–351.

Schloesser, D. W., T. A. Edsall, B. A. Manny, and S. J. Nichols. 1991. Distribution of *Hexagenia* nymphs and visible oil in sediments of the Upper Great Lakes connecting channels. *Hydrobiologia* 219: 345–352.

Schloesser, D. W., K. A. Krieger, J. J. H. Ciborowski, and L. D. Corkum. 2000. Recolonization and possible recovery of burrowing mayflies (Ephemeroptera: Ephemeridae: *Hexagenia* spp.) in Lake Erie of the Laurentian Great Lakes. *J. Aquat. Ecosyst. Stress Recov.* 8: 125–141.

Schloesser, D. W. and T. F. Nalepa. 2001. Changing abundance of *Hexagenia* mayfly nymphs in western Lake Erie of the Laurentian Great Lakes: Impediments to assessment of lake recovery. *Intern. Rev. Hydrobiol.* 86: 87–103.

Senft, R. L., M. B. Coughenour, D. W. Bailey, L. R. Rittenhouse, O. E. Sala, and D. M. Swift. 1987. Large herbivore foraging and ecological hierarchies. *Bioscience* 37: 789–795.

Sprung, M. 1989. Field and laboratory observations of *Dreissena polymorpha* larvae: Abundance, growth, mortality, and food demands. *Arch. Hydrobiol.* 115: 537–561.

Stewart, T. W., J. G. Miner, and R. L. Lowe. 1998. Quantifying mechanisms for zebra mussel effects on benthic macroinvertebrates: Organic matter production and shell-generated habitat. *J. N. Am. Benthol. Soc.* 17: 81–94.

Stoeckmann, A. 2003. Physiological energetics of Lake Erie dreissenid mussels: A basis for displacement of *Dreissena polymorpha* by *Dreissena bugensis*. *Can. J. Fish. Aquat. Sci.* 60: 126–134.

Turner, M. G., R. V. O'Neill, R. H. Gardner, and B. T. Milne. 1989. Effects of changing spatial scale on the analysis of landscape pattern. *Landsc. Ecol.* 3: 153–162.

Urban, D.L., R. V. O'Neill, and H. H. Shugart, Jr. 1987. Landscape ecology: A hierarchical perspective can help scientists understand spatial patterns. *BioScience* 37: 119–127.

Wang, F., A. Tessier, and L. Hare. 2001. Oxygen measurements in the burrows of freshwater insects. *Freshwat. Biol.* 46: 317–327.

Ward, J. M. and A. Ricciardi. 2007. Impacts of *Dreissena* invasions on benthic macroinvertebrate communities: A meta-analysis. *Divers. Distrib.* 13: 155–165.

Wiens, J. A. 1989. Spatial scaling in ecology. *Funct. Ecol.* 3: 385–397.

Winter, A., J. J. H. Ciborowski, and T. B. Reynoldson. 1996. Effects of chronic hypoxia and reduced temperature on survival and growth of burrowing mayflies, *Hexagenia limbata* (Ephemeroptera: Ephemeridae). *Can. J. Fish. Aquat. Sci.* 53: 1565–1571.

Zar, J. H. 1999. *Biostatistical Analysis*, 4th edn. Englewood Cliffs, NJ: Prentice Hall.

Zhu, B., D. M. Fitzgerald, C. M. Mayer, L. G. Rudstam, and E. L. Mills. 2006. Alteration of ecosystem function by zebra mussels in Oneida Lake, NY: Impacts on submerged macrophytes. *Ecosystems* 9: 1017–1028.

Zorn, T. G. and A. J. Nuhfer. 2007. Influences on brown trout and brook trout population dynamics in a Michigan River. *Trans. Am. Fish. Soc.* 136: 691–705.

CHAPTER 40

Zebra Mussel Impacts on Unionids
A Synthesis of Trends in North America and Europe

Frances E. Lucy, Lyubov E. Burlakova, Alexander Y. Karatayev, Sergey E. Mastitsky, and David T. Zanatta

CONTENTS

Abstract 623
Introduction 624
Case Study Sites 624
    Lake St. Clair, United States/Canada 624
    Various Water Bodies, Belarus 625
    Lough Key, Ireland 625
Sampling Protocols 626
Statistical Modeling 626
Results 627
    Case Studies 627
        Lake St. Clair, United States/Canada 627
        Various Water Bodies, Belarus 634
        Lough Key, Ireland 636
Overall Impacts 638
Discussion 640
    General Trends 640
Recommendations 641
Acknowledgments 642
References 642

ABSTRACT

This chapter examines impacts of exotic zebra mussel (*Dreissena polymorpha*) infestation on unionids from water bodies in North America, Belarus, and Ireland over the past two centuries. A variety of methods were used to assess impacts ranging from short-term studies of multiple water bodies to extensive multi-year studies of a single water body. In general, there was a strong positive linear relationship between numbers of zebra mussels per unionid and zebra mussel density. However, datasets indicated that a high percentage of unionids may be infested with zebra mussels even when zebra mussel density in a water body was low. During the first 10 years subsequent to a zebra mussel invasion, there was an overall trend for increased weight of zebra mussels per unionid with increased unionid size, but this trend decreased 10 years after the initial invasion. We discuss possible mechanisms for coexistence of zebra mussels and unionids and make recommendations for management options and unification of research methods. Finally, we identify research priorities that will provide a better understanding of zebra mussel–unionid coexistence and hence aid in the development of unionid survival and management strategies during the initial stages of invasion by dreissenids or other byssate exotic bivalves.

INTRODUCTION

Continued invasion and spread of zebra mussels (*Dreissena polymorpha*) and quagga mussels (*Dreissena rostriformis bugensis*) has threatened survival of native unionid mussels in North America and Europe. Members of the family Unionidae are a major component of freshwater systems, representing the largest and most long-lived (i.e., beyond 100 years) freshwater invertebrate species (Lydeard et al. 2004, Strayer et al. 2004). These bivalves are very important components of aquatic ecosystems, often dominating benthic biomass and production (Negus 1966, Hanson et al. 1989), impacting clarity and quality of water and plankton primary production by removing phytoplankton as well as suspended matter by filtration, affecting nutrient dynamics through excretion and biodeposition of feces and pseudofeces, releasing nutrients from the sediment to the water column, and increasing water and oxygen content in sediments through bioturbation (reviewed in McMahon and Bogan 2001, Vaughn and Hakenkamp 2001, Strayer et al. 2004, Vaughn et al. 2004, 2009, Vaughn and Spooner 2009). Order Unionoida represents the largest freshwater bivalve radiation, with 6 families, 181 genera, and over 800 species distributed across 6 of the 7 continents (Bogan and Roe 2008). Of the six families, Unionidae has the greatest number of species (297) found in North America (Graf and Cummings 2007, Bogan 2008). Unionids are also the most endangered group of freshwater animals, particularly in North America (Bogan 1993, 2008, Stein and Flack 1997, Lydeard et al. 2004, Graf and Cummings 2006). Due to their sensitivity to water and habitat quality, sedentary lifestyle, long life span, complex life cycles with parasitic larvae that require specific host fish, slow growth, and low reproductive rates, over 76% of the North American Unionidae and Margaritiferidae are presumed extinct, threatened, endangered, or of special concern (Williams et al. 1993). In contrast to the high diversity of unionids in North America, only 14 species of the order Unionoida are recognized in Europe (Jaeckel 1967), including 9 species in Belarus. Three species of Unionoida occur in Ireland: *Margaritifera margaritifera* (L., 1758) (pearl mussel), *Anodonta anatina* (L., 1758) (duck mussel), and *Anodonta cygnea* (L., 1758) (swan mussel). Ross (1984, 1988) and Lucey (1995) have noted difficulties in distinguishing between the two common species of *Anodonta* (Kerney 1999), and until molecular phylogenetic investigations are undertaken, there is a distinct possibility that only one true species exists. Therefore, in the present study *A. anatina* and *A. cygnea* will not be distinguished and considered only as *Anodonta*.

The decline of unionids by anthropogenic drivers is well-documented (Bogan 1993, Williams et al. 1993, Richter et al. 1997, Lydeard et al. 2004, Strayer et al. 2004, Strayer and Dudgeon 2010). In addition to impacts due to habitat destruction and loss of water quality, dreissenid mussels have escalated this decline as they settle and attach, often in very large numbers, to the exposed posterior end of the unionid shell,

Table 40.1 Geographic Coordinates of the Water Bodies Studied

Water Body	Latitude	Longitude
USA		
Lake Clark	42°07′13 N	84°19′08 W
Lake St. Clair	42°26′55 N	82°40′35 W
Lake Vineyard	42°05′02 N	84°12′35 W
Belarus		
Lake Batorino	54°50′52 N	26°57′19 E
Lake Bolduk	54°58′27 N	26°25′09 E
Lake Bolshie Shvakshty	54°58′12 N	26°35′09 E
Lake Dolzha	55°27′42 N	26°46′40 E
Lake Drisvyaty	55°36′55 N	27°01′40 E
Lake Lepelskoe	54°54′20 N	28°41′24 E
Lake Malye Shvakshty	54°59′37 N	26°33′02 E
Lake Naroch	54°51′22 N	26°46′57 E
Lake Spory	55°04′11 N	26°47′14 E
Lake Svir	54°48′06 N	26°29′09 E
Lake Volchin	55°00′04 N	26°52′23 E
Lake Volos	55°43′50 N	27°08′18 E
Ireland		
Lough Key	54°00′05 N	08°14′41 W

hampering filter feeding, respiration, locomotion, and reproduction (Mackie 1991, Schloesser et al. 1996, Karatayev et al. 1997, Strayer 1999). Zebra mussels may directly compete with unionids for food (Haag et al. 1993, Strayer and Smith 1996, Caraco et al. 1997), occupy available substrate (Tucker 1994), and induce shell deformities (Lewandowski 1976a, Hunter and Bailey 1992). In most of pre-glaciated Europe, zebra mussels and unionids coexisted (Karatayev et al. 1997), although there is no known zebra mussel fossil record for Ireland. In contrast, unionids in North America have evolved with no adaptive mechanism to deal with epizootic colonization by dreissenids (Haag et al. 1993); consequently, many studies have either forecasted or reported high mortality or extirpation of unionids within the first decade of invasion (Ricciardi et al. 1995, 1998, Nalepa et al. 1996).

In this chapter, we examine three detailed case studies of dreissenid impacts on unionids in different regions of Europe and North America (Table 40.1), analyze data from multiple water bodies across both continents to find general trends, discuss possible mechanisms for coexistence, and make recommendations for future studies.

CASE STUDY SITES

Lake St. Clair, United States/Canada

Lake St. Clair is located in the Laurentian Great Lakes system within central North America. This lake receives water from the St. Clair River through a large delta at the river mouth and drains into Lake Erie via the Detroit River. Other important tributaries include the Clinton River in Michigan and the Sydenham and Thames Rivers in Ontario. Lake St. Clair is heart shaped, with a maximum natural depth of 6.5 m and a surface area of 1115 km². A navigation channel dredged to a depth of 8.3 m bisects the lake approximately along the Canada/U.S. border to accommodate commercial shipping traffic between Lake Erie and Lake Huron (Edsall et al. 1988). The southern and western/northwestern shores of the lake are heavily urbanized, while the northeastern shore consists of a large delta with vast natural marshlands. The eastern shore of the lake is mainly rural farmland and wetland. Leach (1991) described the substrate of Lake St. Clair as muddy sand in the central part and gravel or sand close to shore. Prior to the first report of zebra mussels in 1988 (Hebert et al. 1989), unionids provided the only hard substrate in many areas (Nalepa and Gauvin 1988). At present, most hard substrate in offshore areas consists of unionid shells and druses of live and dead *Dreissena* (Hunter and Bailey 1992, Nalepa et al. 1996).

Various Water Bodies, Belarus

Belarus is a relatively small country located in the geographical center of Europe. Around the turn of the nineteenth century, three interbasin canals were constructed to connect the Black and Baltic Sea basins and thus expand shipping. These canals provided invasion corridors for the introduction of numerous Pontocaspian species from the Black Sea to the Baltic Sea basin (Mordukhai-Boltovskoi 1964, bij de Vaate et al. 2002, Karatayev et al. 2003, 2008b). There are 1040 glacial lakes in Belarus, and these lakes have very different morphology, water chemistry, trophic status, and land use patterns (Karatayev et al. 2005, 2008a). As of 2000, zebra mussels have been found in only 21.2% of the 553 lakes surveyed despite the presence of mussels in this region for almost 200 years (Karatayev et al. 2003, 2010). These lakes, along with several reservoirs, vary in time since zebra mussel colonization (from less than 5 to over 100 years ago) and therefore provide valuable models to study zebra mussel–unionid relationships (Karatayev et al. 2003, 2010) (Table 40.2).

Lough Key, Ireland

Lough Key is a small lake (9 km²) located in the Upper Shannon River basin district in Ireland. Prior to the invasion of zebra mussels in Lough Key, *Anodonta* was described as being "common" in the lake (Ross 1984). Zebra mussels in low densities were discovered in Lough Key in early 1998 (D. Minchin, personal communication), and a rapid expansion occurred in 1999 as densities reached 148,000 m⁻² on stony substrates (Lucy and Sullivan 2001). This small lake

Table 40.2 Unionid Sampling Techniques Employed by Surveys in Lake St. Clair, United States, Water Bodies in Belarus, and in Lough Key, Ireland

Unionid Sampling Technique	Lake St. Clair	Belarus	Lough Key, Ireland
SCUBA: 0.25 m² quadrat, 10 replicates	1986, 1990, 1992, 1994	Lakes Naroch (1990, 1993–1995, 1997), Lukomskoe (1978), Myastro, and Batorino (1993, 1995); reservoirs Drozdy and Chizhovskoe (1995)	
SCUBA: 1 m² quadrat, 20 replicates	1994		
SCUBA: timed searches	1994, 2001		
SCUBA: timed searches and "stake and rope"	2002, 2009		
SCUBA: 0.75 h "loss" survey	2010		
SCUBA: density estimate, no quadrat, not timed			1998, 1999
SCUBA transect, 0.06 m² quadrat, three replicates			2002
Snorkel not timed			2000, 2001, 2003
Snorkel, transect		Lake Lepelskoe (1997); lakes Volchin, Bolshiye Shvakshty, Malye Shvakshty, Bolduk, Dolzha, Myadel, Spory, Svir, and Naroch (1998); Lakes Yuzhny Volos and Drisvyaty (1999)	
Rake			1998
Ekman grab	1994	At depths >2 m in lakes Naroch (1990, 1993–1995), Myastro, and Batorino (1993, 1995); reservoirs Drozdy and Chizhovskoe (1995)	2001
ROV camera			2001

has a maximum depth of 25 m and has one inflow and outflow, the Boyle River. Underlying geology of the lake is composed of sandstone, shale, and limestone that results in calcium-rich waters (Bowman 1998) and therefore highly suitable for zebra mussel colonization. The primary substrate in the lake is mud (>70% of total area), which is present in deeper areas (>3 m) and accounts for between 5% and 100% of the heterogeneous substrate in many littoral areas (Lucy et al. 2005). Other common substrates found in littoral areas are comprised of boulders, cobble, and gravel, and these hard-bottom types provide the main substrate for zebra mussel attachment in the littoral. *Anodonta* and its empty shells were also found to be an important substrate for zebra mussel attachment in the period soon after colonization (Lucy and Sullivan 2001).

SAMPLING PROTOCOLS

A variety of sampling methods have been used in these water bodies (Table 40.2). Methods were developed over a period of time to best document changes in zebra mussel and unionid densities. SCUBA and snorkeling were the most common collection methods used in all three case studies. The depth of sampling sites across all studies varied from 1.0 to 6.5 m (Lake St. Clair), 0.5 to 2.5 m (Belarus), and 1.0 to 5.0 m (Lough Key). Substrate was identified and recorded in situ.

Field surveys in Lake St. Clair were initially performed by SCUBA divers that collected samples in defined quadrats at 29 stations repeatedly sampled across the lake (all in water >2 m) in 1986, 1990, 1992, and 1994 (Nalepa et al. 1996). These same methods were used to survey unionids and zebra mussels at 12 of these stations in 2001 (Hunter and Simons 2004). Additional surveys incorporated timed searches (Nalepa et al. 2001), which were then combined with a semiquantitative "stake and rope" technique (Zanatta et al. 2002, McGoldrick et al. 2009). In 2010, a 0.75 h/diver-snorkel survey was also used in advance of the "stake and rope" technique to sample unionids (Zanatta, this study). In Belarus, unionids were collected while snorkeling within a 1 m wide transect (100 or 500 m, depending on unionid density); however, untimed searches were used at very low densities. Zebra mussels and unionids were collected using replicate 0.25 m^2 quadrats along 8 permanent transects in Lake Naroch, 5 permanent transects in Lake Myastro and Lake Batorino, 6 transects in reservoirs Drozdy and Chizhovskoe, and 14 transects in Lake Lukomskoe (see details in Table 40.2 and Burlakova et al. 2006).

Ekman grabs were also used for sample collection in one Lake St. Clair survey (Gillis and Mackie 1994) and in transect surveys in Belarus and Lough Key. In Belarus, where lake transect surveys were not carried out, zebra mussel density was determined from 10 to 32 Ekman grabs collected from 5 to 16 sites in each lake, depending on lake size. Other methods were also used in Lough Key, namely, raking and filming with an underwater remotely operated vehicle (ROV) camera. Except for the latter method, collected unionids were identified and measured to the nearest millimeter (using calipers or a ruler), and the number of attached zebra mussels counted. In Lake St. Clair and Belarus, unionids were identified to species in most cases. In Lough Key, this was not necessary as only one genus (*Anodonta*) was present. Zebra mussels were removed from the unionids and counted, and in some cases shell lengths measured as mentioned.

In Lake St. Clair, all unionids collected in 2003 (McGoldrick et al. 2009) and 2010 (this study) were returned to the lake. For studies in Belarus, both zebra mussels and some unionids were opened to remove water from mantle cavities and then weighed to the nearest gram after being blotted dry on absorbent paper (wet weight, soft tissue plus shell). For other unionids, weight weights were derived from length–weight relationships for each species. Live unionids were then carefully replaced into the substrate from the water body where they were collected. At Lough Key, weights of both zebra mussels and unionids were determined after they were blotted dry. Unlike studies in Belarus, however, water was not removed from mantle cavities.

STATISTICAL MODELING

Several statistical models were used to describe relationships between the number and total weight of dreissenids per unionid host and the parameters of interest. A *linear regression* model was applied to relate (1) the number of dreissenids per unionid in European and North American water bodies and corresponding densities of dreissenids on other substrates and (2) total weight of dreissenids attached to the unionid and unionid shell length (Lough Key only). A negative binomial distribution-based *generalized additive model* was used to determine whether dreissenid density in Lake St. Clair changed significantly over time (Wood 2006). The model was defined as $Density = \alpha + s(YSI) + \varepsilon$, where α was the average zebra mussel density over all samples, $s(YSI)$ was a smooth function for the year since invasion (calculated as a cubic regression spline), and ε were the model residuals that were assumed to be normally distributed around the mean 0. Data used in this model were repeatedly collected from 12 stations over a span of 11 years (Hunter and Simons 2004). To account for temporal correlation between successive density measurements, an autoregressive component was incorporated into the residual variance of the model (Zuur et al. 2009). We did not incorporate the random effect of sampling station into the model as a preliminary analysis showed it was not necessary.

A *generalized least squares model* was used to examine the relationship between the total weight of attached zebra mussels to the shell length of the corresponding unionid host. In this analysis, we used our original data collected in 15 water bodies: 12 in Belarus, 1 in Ireland, and 2 in the United States. Representatives of seven unionid genera were examined, that is, *Anodonta* (n = 158), *Anodontoides* (n = 4), *Elliptio* (n = 14), *Lampsilis* (n = 8), *Pyganodon* (n = 3), *Unio* (n = 219), and *Villosa* (n = 5). A preliminary exploratory analysis suggested that this dataset could be sufficiently fitted with the following model: *Weight* = $\alpha + \beta_1 L + \beta_2 TSRI + \varepsilon$, where α is the intercept and β_1 and β_2 are the coefficients reflecting the effects of the unionid shell length (*L*) and the time since recognized invasion (*TSRI*) of *Dreissena*, respectively. *TSRI* is considered as a nominal variable that takes a value of 0 if <10 years and 1 if ≥10 years. We used the *TSRI* (= initial detection) rather than the time since actual initial invasion as the latter is rarely known (Burlakova et al. 2006). ε are the residuals, which are approximately normally distributed around the mean 0, and have a variance that was allowed to vary with host length within water body/unionid genus combinations (see Zuur et al. [2009] for more details on this type of parameterization). A preliminary analysis suggested that there was no need to include the random effect of the sampling water body into the model.

Statistical analyses were carried out in the R v2.11.1 computing environment (R Development Core Team 2010) with the help of base R functions as well as an *nlme* package (Pinheiro and Bates 2000). Selection of the optimal model at intermediate steps of analysis was based on the combined use of Akaike's information criterion and analysis of variance (ANOVA) of competing models. Model validation was performed via visual examination of residuals plotted against fitted values and observed values of explanatory variables (Zuur et al. 2009).

RESULTS

Case Studies

Lake St. Clair, United States/Canada

A total of 37 unionid species are known from Lake St. Clair and its tributaries (La Rocque and Oughton 1937, Graf 2002). In 1986, immediately prior to the invasion of the lake by zebra mussels, unionids were abundant throughout the lake with mean densities of 1.9 m^{-2} (Nalepa and Gauvin 1988) (Figure 40.1a). *Lampsilis siliquoidea* was by far the most abundant unionid species. It accounted for 45% of the total unionid community and had a population age structure indicating that the population size was stable (Nalepa and Gauvin 1988). In contrast, the second most abundant unionid in 1986, *Leptodea fragilis*, showed a great deal of variation in recruitment on a yearly basis. Nalepa and Gauvin (1988) noted that while the diversity and community composition of the total unionid community in Lake St. Clair had not changed greatly since the early twentieth century, there were some indications that a few species may have declined.

Lake St. Clair and western Lake Erie were "ground zero" for the invasion of the zebra mussel in North America as these areas were the first colonized on the continent (Hebert et al. 1989, Carlton 2008). Based on size classes present (Hebert et al. 1989, Griffiths et al. 1991) and anecdotal reports (Carlton 2008), it was believed that the zebra mussel had been present since at least 1986. Densities of zebra mussels in Lake St. Clair quickly increased after first detection in 1988 and spread through the lake occurring in two stages: first in the southeast basin (Figure 40.2a) and then into the northwest basin (Figure 40.2b) (Nalepa et al. 1996). In 1990, zebra mussels had a mean density of 1,663 m^{-2}, with a maximum density of 10,389 m^{-2} in the southeast. By 1994, mean density had increased to 3,241 m^{-2}, peaking at 23,037 m^{-2} in the northwest side of the lake (Figure 40.2c). Zebra mussel densities appeared to decline after 1994, with mean densities of 1237 and 1824 m^{-2} recorded in more limited sampling in 1997 and 2001, respectively (Nalepa et al. 2001, Hunter and Simons 2004; Figure 40.3). Trends in population densities in Lake St. Clair varied among sites; densities at some sites increased steadily over time, but densities at other sites gradually or rather abruptly decreased (Figure 40.4). However, the overall trend of increase during the first 8–9 years of colonization was followed by a subsequent decline. This conclusion is supported by the (marginally) significant time smoother term of the generalized additive model fitted to density data from Hunter and Simons (2004) (Figure 40.5).

Following the establishment and spread of zebra mussels, unionid densities and diversity in Lake St. Clair declined massively (Figures 40.1b,c and 40.3). Paralleling spatial patterns of zebra mussel densities, unionids in the southeast part of the lake were affected first, with severe declines between 1986 and 1990 (Nalepa et al. 1996). Unionids in the northwestern side of the lake, with the bulk of its water emanating from the St. Clair River (and Lake Huron), took longer to show impacts by zebra mussels, but massive declines were evident throughout the lake by 1994. Loss of unionid diversity took longer to become evident: 18 species were present in 1986, 17 species in 1990, and 12 species in 1992, and none were found live in 1994.

Additional survey work in 1997 (Nalepa et al. 2001) and 2001 (Hunter and Simons 2004) failed to recover any live unionids from open waters of the lake despite the apparent decline in zebra mussel density. Areas sampled previously by Gillis and Mackie (1994) were exhaustively sampled in 1998 and 1999 but no live unionids were found (Zanatta et al. 2002). Declines in unionid densities in Lake St. Clair mirror those resulting from zebra mussel competition and infestation in the Hudson River (Strayer and Malcom 2007), with mean unionid densities declining to levels two orders of magnitude lower than present pre-*Dreissena*.

After the loss of unionids in the open waters of both Lake St. Clair and western Lake Erie (Schloesser and Nalepa 1994), further surveys documented an apparent refuge for

Figure 40.1 Mean density (m^{-2}) of Unionidae at 29 sampling sites in Lake St. Clair in (a) 1986, (b) 1990, (c) 1992, and (d) 1994. (From Nalepa, T.F. et al., *Great Lakes Res.*, 22, 354, 1996.)

unionids from the effects of zebra mussels in a shallow wetland area in western Lake Erie (Metzger Marsh; Nichols and Wilcox 1997). To determine if similar refuges occurred in Lake St. Clair, shallow bays and wetlands of the St. Clair delta (Figure 40.6) were first examined in 1999 (Zanatta et al. 2002). Surveys of both habitats by Zanatta et al. (2002) and McGoldrick et al. (2009) documented the unionid community (densities, diversity, and health) along with densities of the zebra mussel population (adults and larvae). Virtually all unionids found live in the delta were found in shallow water (<1 m) on sand substrates. The unionid community was diverse with 22 species being present (including several species of conservation concern in Michigan and Ontario, Canada; COSEWIC 2010; Government of Canada 2010; Michigan Department of Natural Resources 2011; Table 40.2); however, densities were very low (0.05 m^{-2}) in comparison to densities in the open lake prior to the zebra mussel invasion (1.9 m^{-2}). Unfortunately, unionid densities in the St. Clair delta in the period before zebra mussels became established are unknown. Open-water sites closest to the delta had low unionid densities in 1986 (<1 m^{-2}, Figure 40.1). Relative frequencies of unionid species in the delta (and higher-level taxonomic groups: Anodontinae, Ambleminae, Lampsilini) were similar to the pre-zebra

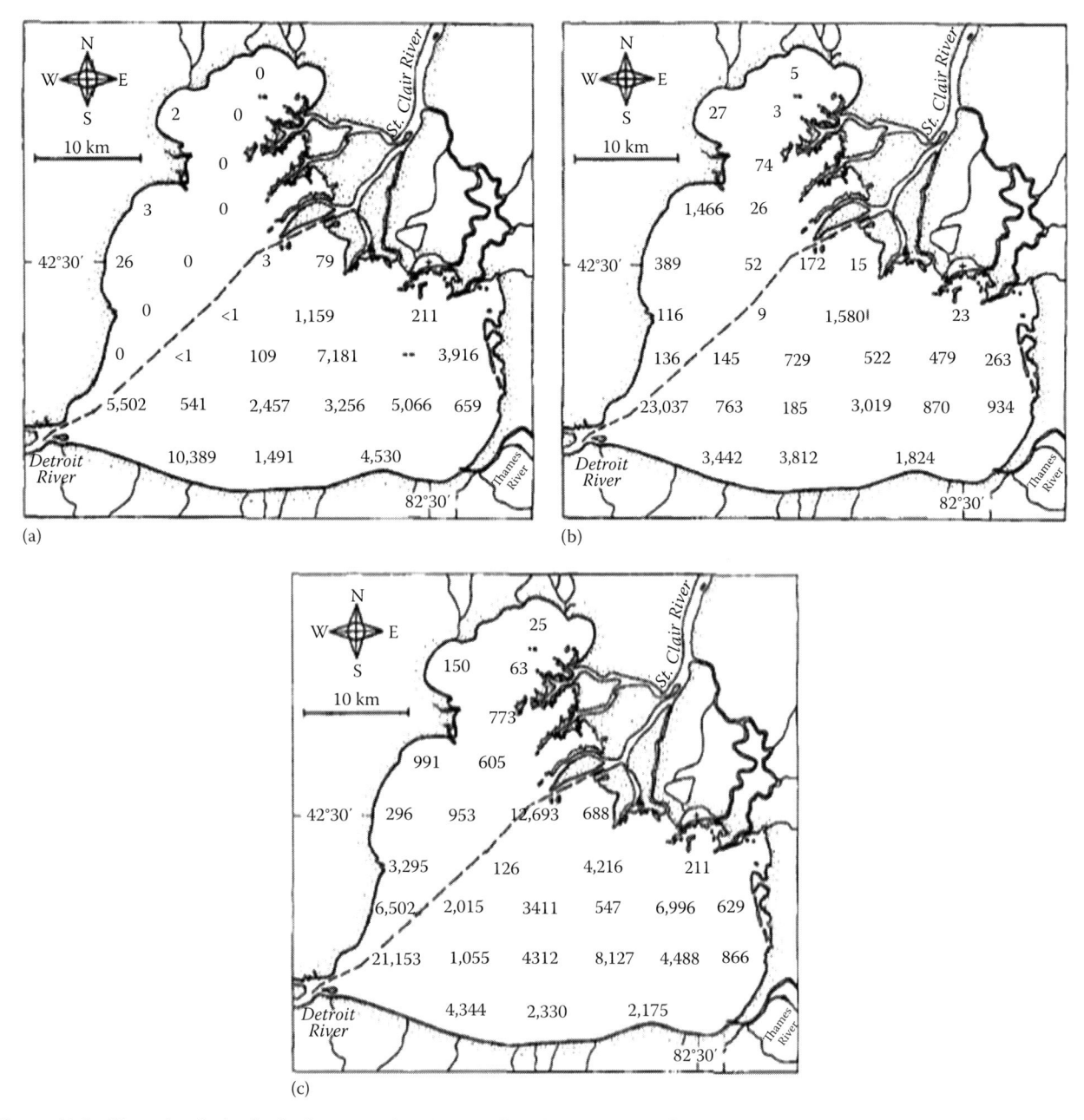

Figure 40.2 Mean density (m^{-2}) of zebra mussels at 29 sampling sites in Lake St. Clair in (a) 1990, (b) 1992, and (c) 1994. (From Nalepa, T.F. et al., *Great Lakes Res.*, 22, 354, 1996.)

mussel community in the open lake. As lampsiline and anodontine mussels were more affected by zebra mussels than other unionid taxa (Schloesser et al. 1998), the dominance of lampsilines in the unionid community of the delta suggests that this community was sheltered from the most severe effects of zebra mussels (McGoldrick et al. 2009). While the delta remains a unionid refugia even though zebra mussels have been abundant in nearby offshore waters for nearly 25 years, the long-term survival of the unionid community in this region remains in question.

In order to assess trends in the unionid community in the delta, the area on the U.S. side of the delta was resurveyed in 2010. Results indicated the unionid community remained in a state of flux. Of the nine sites resurveyed using the same "stake and rope" technique as in 2003 (Zanatta et al. 2002, McGoldrick et al. 2009), unionid densities declined below detectable levels at three sites, declined significantly at two sites (two-tailed *t*-tests, $P < 0.05$), did not change at three sites ($P > 0.05$), and increased at one site ($P < 0.01$) (Figure 40.7). Data used for the *t*-tests were normally distributed (Kolmogorov–Smirnov test) and group variances were not different, thus appropriate for *t*-tests (Zar 1996). Densities at some of the sites sampled with the "stake and rope" method in 2010 but previously sampled with only

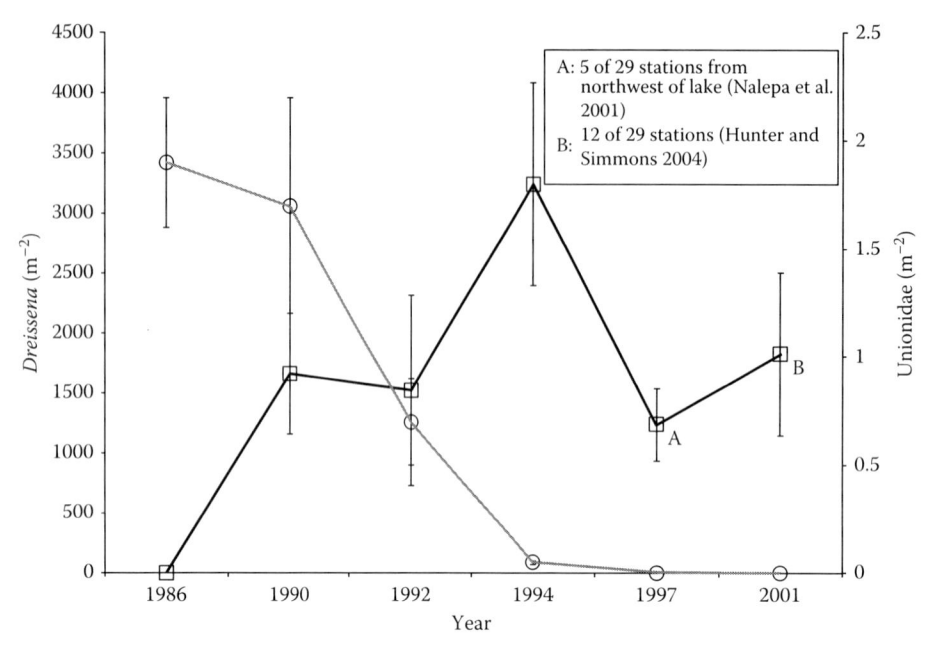

Figure 40.3 Mean (± SE) density (m^{-2}) of *Dreissena* (zebra mussel) and Unionidae in Lake St. Clair (note different scales). Densities in 1986, 1990, 1992, and 1994 were taken from Nalepa et al. (1996), densities in 1997 were taken from Nalepa et al. (2001), and densities in 2001 were taken from Hunter and Simons (2004). *Dreissena* = square/black line; Unionidae = circle/gray line.

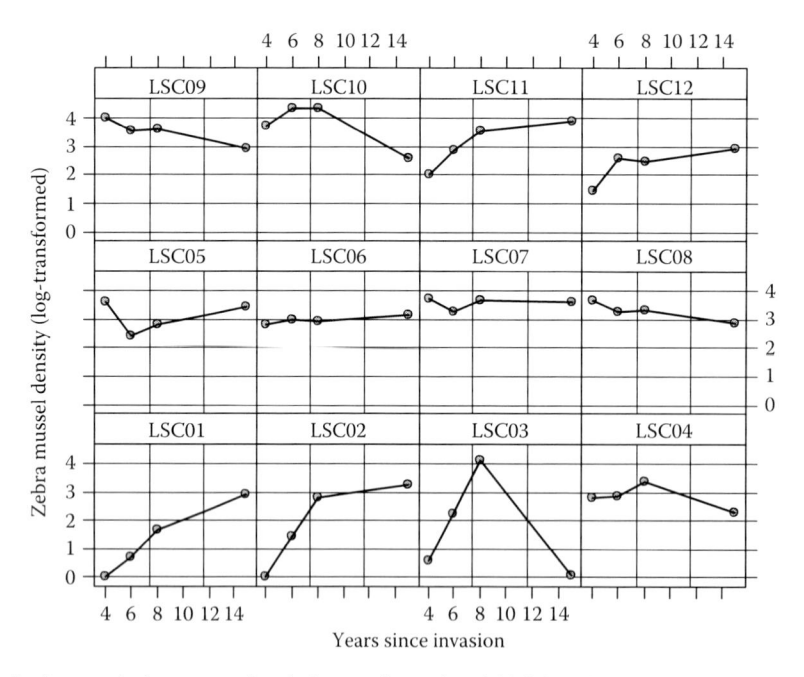

Figure 40.4 Density of *Dreissena* (zebra mussel) relative to time after initial invasion at 12 stations (LSC01 to LSC12) repeatedly sampled in Lake St. Clair. (From Hunter, R.D. and Simons, K.A., *J. Great Lakes Res.*, 30, 528, 2004.) Density was $\log_{10}(x + 1)$-transformed to better depict trends.

timed searches in 2003 were among some of the highest found, especially at sites in Big Muscamoot Bay and Little Muscamoot Bay (Sites 23 and 24; Figure 40.7). Anecdotally, it appeared that densities at some sites that were surveyed using time search only in 2003 increased in 2010. Of note, population densities at sites in the outer bays of the delta (closest to open lake environments, i.e., Sites 9, 12, 13, 17) either declined with individuals still present or declined with individuals no longer found. In contrast, densities at sites further into the bays of the delta were stable or increased between 2003 and 2010. The relative composition of the community remained generally the same over this

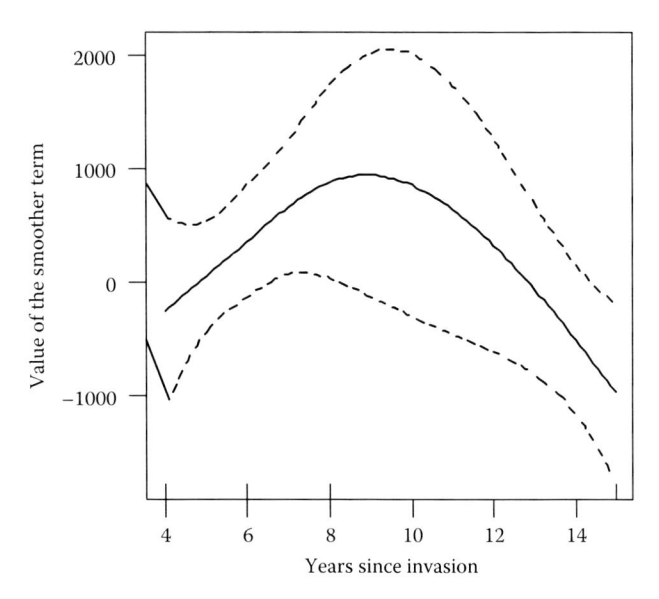

Figure 40.5 Estimated smoother term (s) for the generalized additive model that describes the change in zebra mussel density relative to the number of years after the initial invasion (YSI) at 12 stations in Lake St. Clair. Density = 2838 + s(YSI) (see "Statistical Modeling" for details). Both the intercept and the smoother of this model are significant ($P < 0.001$, z-test, and $P = 0.036$, χ^2-test, respectively). The solid line is the estimated smoother, and the dotted lines are 95% point-wise confidence bands.

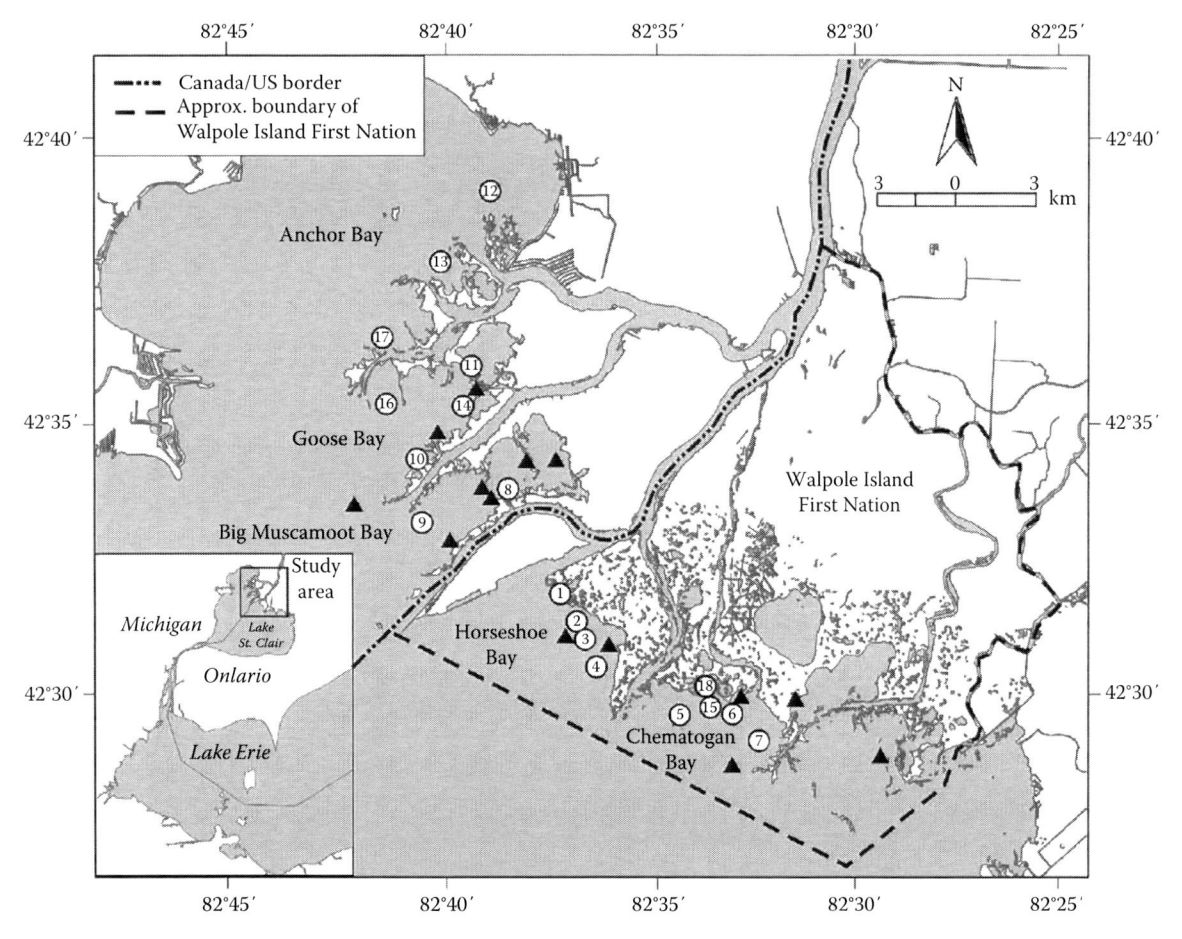

Figure 40.6 Locations of 32 sites surveyed for unionids in the Lake St. Clair delta in 2003, 2005, and 2010. Sites surveyed quantitatively in 2003 and 2005 ($n = 18$) are indicated by circles containing the site number; sites surveyed semiquantitatively (timed search) in 2003/2005 are indicated by black triangles. All sites on the U.S. side were sampled quantitatively in 2010 using the "stake and rope" technique. (From McGoldrick, D.J. et al., *J. Great Lakes Res.*, 35, 137, 2009; Zanatta, D.T. et al., *J. Great Lakes Res.*, 28, 479, 2002.)

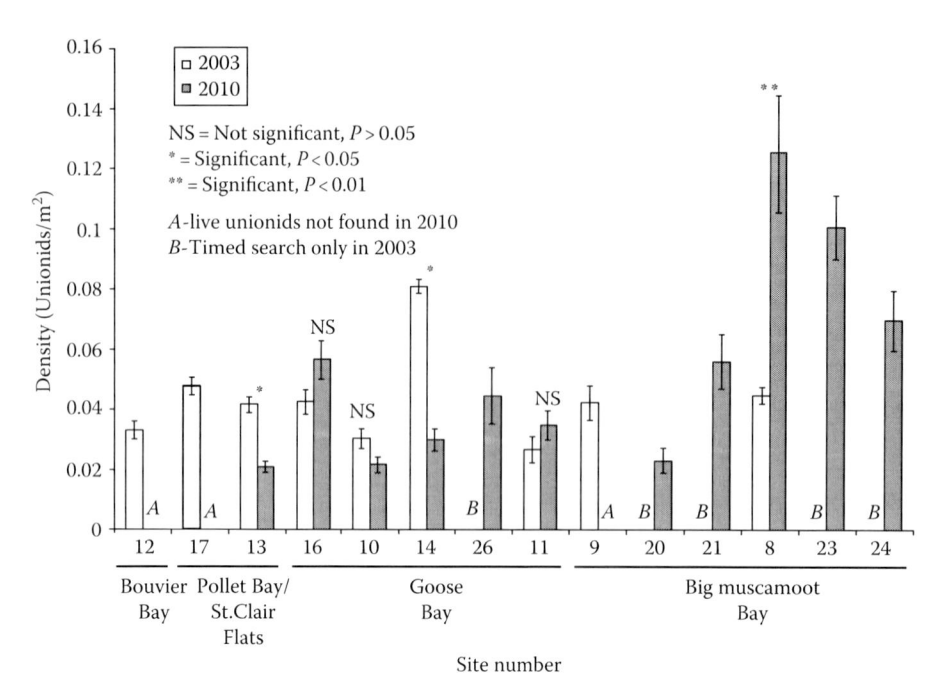

Figure 40.7 Comparison of unionid densities (±SE) in 2003 and 2010 at sites in the St. Clair delta. Densities in both years were determined with the "stake and rope" technique (Zanatta et al. 2002). Differences between years for each site were determined with a two-tailed *t*-test. Asterisk = difference between years was significant; NS = differences between years were not significant.

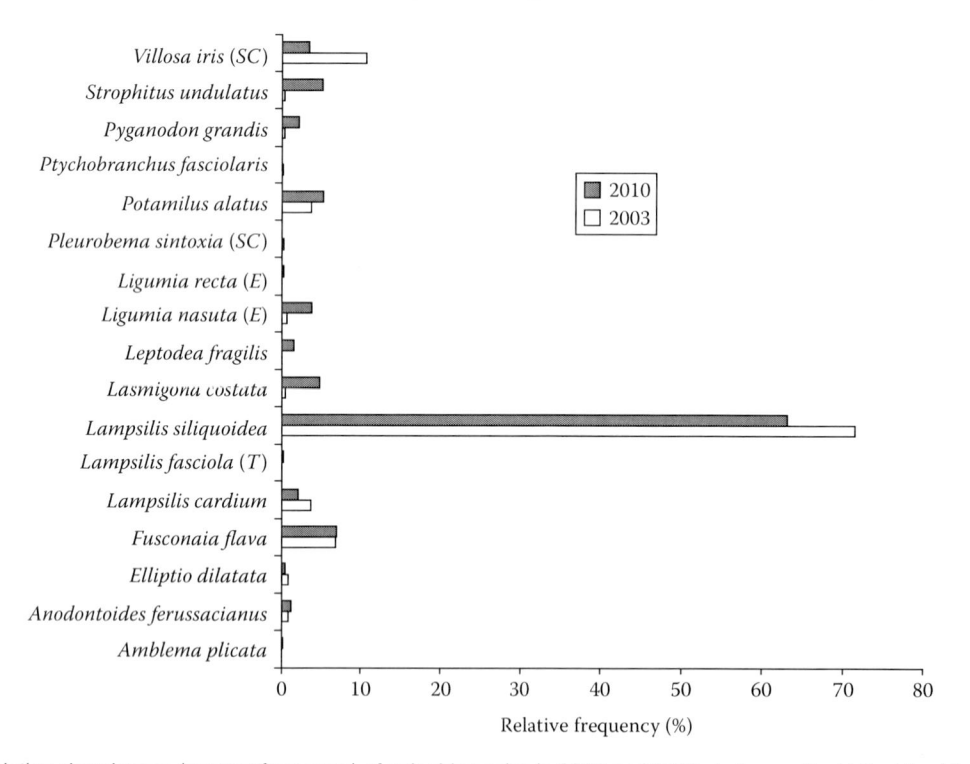

Figure 40.8 Relative abundances (percent frequency) of unionid species in 2003 and 2010 at sites on the U.S. side of the St. Clair delta. Species of conservation concern in Michigan are labeled (*E* = endangered, *T* = threatened, *SC* = special concern).

7-year period. However, there was an increase in frequency of *Ligumia nasuta* (endangered in Michigan and Canada), *Lasmigona costata*, and *Strophitus undulatus* and a decline in frequency of *Villosa iris* (special concern in Michigan, endangered in Canada) (Figure 40.8).

While mechanisms for observed changes in unionid communities in the St. Clair delta remain unclear, there have been significant declines in zebra mussel densities across all sites (Figure 40.9). Sampling in 2010 marked the first year that quagga mussels, *D. rostriformis bugensis*,

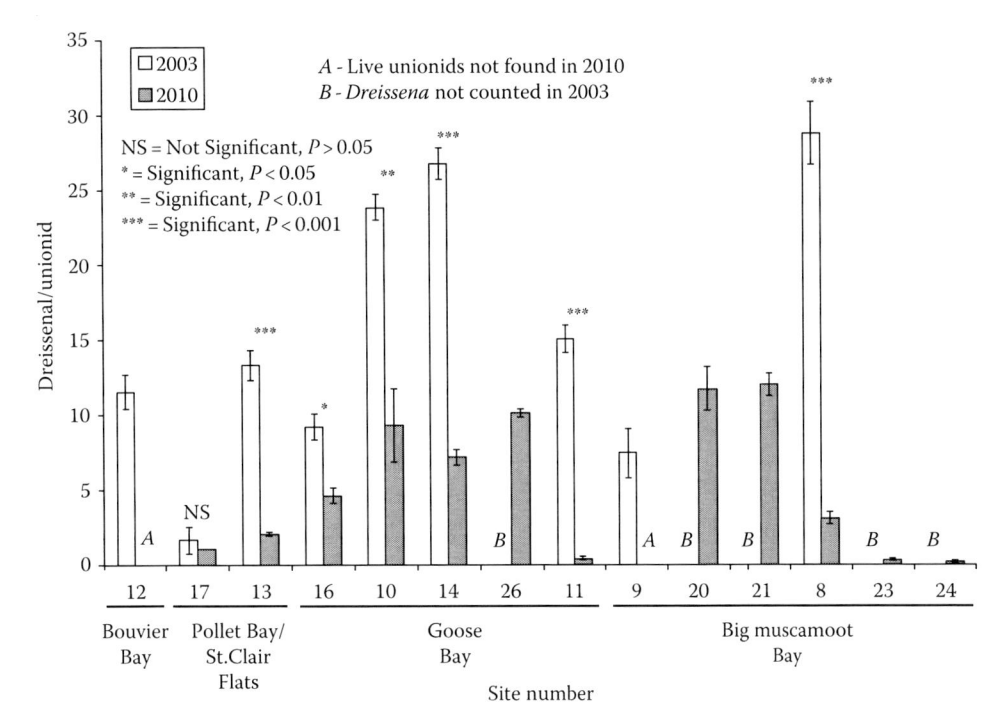

Figure 40.9 Mean number (±SE) of *Dreissena* (zebra mussel) found attached to unionids at sites in the St. Clair River delta in 2003 and 2010. Differences between years at each site were tested using two-tailed *t*-tests. NS = not significant ($P > 0.05$), *$P < 0.05$, **$P < 0.01$, and ***$P < 0.001$.

were detected in the delta. The proportion of quagga mussels to zebra mussels was not quantified, but the former species was very rare. Densities of *Dreissena* (both species), as estimated by the number per unionid, declined significantly (two-tailed *t*-test, $P < 0.05$) between 2003 and 2010 at all sites where unionids were still found alive. Of note, numbers of *Dreissena* per unionid were especially low in the inner bays of the delta, with significant declines ($P < 0.001$) at Sites 11 and 8 between 2003 and 2010, and very low numbers per unionid found at sites not previously documented in 2003 (Sites 23 and 24) (Figure 40.9). As with changes in unionid communities, mechanisms for lower infestation on unionids are unclear.

The unionid community on the U.S. side of the St. Clair delta has persisted since its discovery in 2001 (Zanatta et al. 2002), and findings in 2010 bode well for continued coexistence with *Dreissena*. Although unionid densities at some sites at the outer margins of the delta with relatively high densities in 2003 declined significantly, or declined to zero, densities in 2010 at majority of sites remained high, with unionids having a low number of attached *Dreissena*. Concern remains, however, for many of the rare unionid species (Table 40.3, Figure 40.9).

Several questions remain unanswered with respect to the long persistence of the St. Clair delta as a unionid refuge (many of these questions also apply to other refuges across

Table 40.3 Unionid Species of Greatest Conservation Need (Endangered, Threatened, Special Concern) in Michigan and Ontario That Have Been Found or Formerly Found in Lake St. Clair and/or Tributaries

Alasmidonta marginata	MI	*Quadrula quadrula*[a]	Ontario
Epioblasma t. rangiana[a,b]	MI, Ontario	*Pleurobema sintoxia*[a]	MI, Ontario
Epioblasma triquetra[a,c]	MI, Ontario	*Ptychobranchus fasciolaris*[a]	MI, Ontario
Lampsilis fasciola[a]	MI, Ontario	*Simpsonaias ambigua*[a]	MI, Ontario
Ligumia recta	MI	*Toxolasma parvus*[d]	MI, Ontario
Ligumia nasuta[a]	MI, Ontario	*Toxolasma parvus*[a]	MI, Ontario
Obovaria olivaria[d]	MI, Ontario	*Truncilla truncata*	MI
Obovaria subrotunda[a]	MI, Ontario	*Villosa fabalis*[a,c]	MI, Ontario
Obliquaria reflexa[d]	MI, Ontario	*Villosa iris*[a]	MI, Ontario

[a] Canada, listed by COSEWIC and/or under SARA (Government of Canada 2010).
[b] United States, federally endangered species.
[c] United States, federal candidate species.
[d] Canada, COSEWIC candidate species (COSEWIC 2010).

the lower Great Lakes). Future research on mechanisms for unionid persistence in the delta is critical. Why have zebra mussel populations declined precipitously throughout the delta (and possibly throughout the lake)? Monitoring of zebra mussel veliger densities, settlement rates, and water chemistry (nutrient levels) throughout the delta would provide valuable insights into this question. If the trajectories for *Dreissena* densities in Lake St. Clair continue as modeled (Figure 40.5), ultimately falling below threshold densities that induce unionid mortality (Ricciardi et al. 1995), it may be possible for unionids to recolonize and coexist with dreissenids in the open lake. Also, it is unclear if unionids still exist in the deepwater channels of the delta and the St. Clair River. If unionids still persist in these channels, they may be acting as source populations for unionids in the delta. Anecdotal evidence for this possibility comes from observations that numerous unionid shells were present in recent dredge spoils from channels in the St. Clair River (D. Dortman, Michigan Department Environmental Quality, personal communication). Finally, population genetics research (e.g., Krebs et al. 2003, Krebs 2004, Zanatta et al. 2007, Zanatta and Wilson 2011) can be used to measure genetic variation of unionid species within and among sites, bays, and tributary streams in the St. Clair delta, its tributaries, and other unionid refuges in the Great Lakes. Genetic research will provide valuable information that can be used to interpret source/sink dynamics among sites/bays in and among refuges and tributaries, determine if zebra mussel–induced unionid population crashes caused genetic bottlenecks, determine if gene flow exists among unionid populations in the region, and help predict where unionids are most likely to recolonize. Such studies can assist in prioritizing areas for conservation and give guidelines for managing the unionid communities of the St. Clair delta.

Various Water Bodies, Belarus

To determine the impact of dreissenids on unionid bivalves, 20 water bodies in Belarus were studied from 1990 to 2002. These water bodies had different limnological features and varied in time since zebra mussel colonization and in overall zebra mussel density (Burlakova et al. 2000, Karatayev et al. 2000, 2003, 2005, 2008a, 2010). The longest-established zebra mussel population occurred in Lake Lepelskoe, which was colonized shortly after construction of the Dnieper–Zapadnaya Dvina Canal in 1805 (Karatayev et al. 2008b). This canal was the route through which zebra mussels colonized water bodies in northern Belarus (Burlakova 1998, 1999, Karatayev et al. 2003, 2008b, 2010). The Svisloch River and its reservoirs Chizhovskoe and Drozdy were colonized with zebra mussels in the mid-1980s (Burlakova 1998, 1999). The Braslavskaya Lake system and Lake Lukomskoe were colonized in the late-1960s (Lyakhnovich et al. 1984), and lakes in the Naroch region were colonized in the late-1980s (Karatayev et al. 2003).

The relationship between zebra mussels and unionids was studied in Lake Naroch between 1990 and 2002. The first study in 1990 was conducted during the initial stages of zebra mussel colonization. In that year, the average density of zebra mussels across the whole lake was 7.4 m^{-2} and biomass was 1.5 g m^{-2}. Sixty percent of live unionids ($n = 93$) were colonized by zebra mussels. The average number of mussels per unionid was 9.5, and average biomass was 1.8 g. Unionids were most heavily infested with zebra mussels near a stream that flows into Lake Naroch from Lake Myastro. Lake Myastro was colonized before Lake Naroch, so this connecting stream likely provided a route for zebra mussel invasion into the latter lake (Burlakova et al. 2000). The number of unionids fouled with zebra mussels decreased with increased distance from stream inflow (Pearson $r = -0.82$, $P = 0.023$) (Burlakova et al. 2000).

Between 1990 and 1993, the density of zebra mussels in Lake Naroch increased over 100-fold, and biomass increased 68-fold (Karatayev and Burlakova 1995a, Burlakova et al. 2006). In 1993, all unionids in the littoral zone across the lake were heavily infested with zebra mussels and the majority of unionids were dead. Between 1993 and 1998, only a few live unionids were found during our lake-wide surveys, and in 2002, no live unionids were found.

In contrast, unionids were still abundant in Lake Lepelskoe in 1997, which was ca. 200 years after initial zebra mussel colonization. It was relatively easy for a diver to collect hundreds of unionids within an hour. Although 92% were infested with zebra mussels, the average weight ratio of zebra mussels to unionid was 0.73 ± 0.21 (mean ± SE), lower than that in Lake Naroch (0.91 ± 0.36, average for 1990, 1993–1995, 1998).

Extensive infestation and mass mortality of unionids is typical in the period soon after zebra mussels become established in a given water body and when population growth is rapid (Karatayev et al. 1997, 2002, Burlakova et al. 2000). In contrast to high unionid mortality observed in newly colonized Belarusian water bodies, unionids maintain relatively high densities in lakes where zebra mussels have been found for a long time period. In other cases, unionids appear to be abundant even after the period of initial zebra mussel colonization. Zebra mussels first colonized Lake Lukomskoe in the late 1960s and by 1978 obtained an average density of 758 ± 227 m^{-2} and a biomass of 124 ± 37 g m^{-2} (Karatayev 1983). In 1978, unionid density across the entire lake was 4.7 ± 1.9 m^{-2}, and 75% of the unionids (*Unio pictorum*, *Unio tumidus*, *A. anatina*, and *Anodonta piscinalis*) were fouled with zebra mussels. The maximum number of zebra mussels per unionid was 216, and the average was 30 ± 9. The average ratio of total weight of attached zebra mussels to unionid host weight was 1.20, and the range was 1.04–9.10. In the profundal zone of Lake Lukomskoe, which was dominated by silt, zebra mussels were found only on unionids. On average, about 20% of the total density and biomass of the population of zebra mussels in this lake were comprised of mussels attached to live unionids (Karatayev 1983).

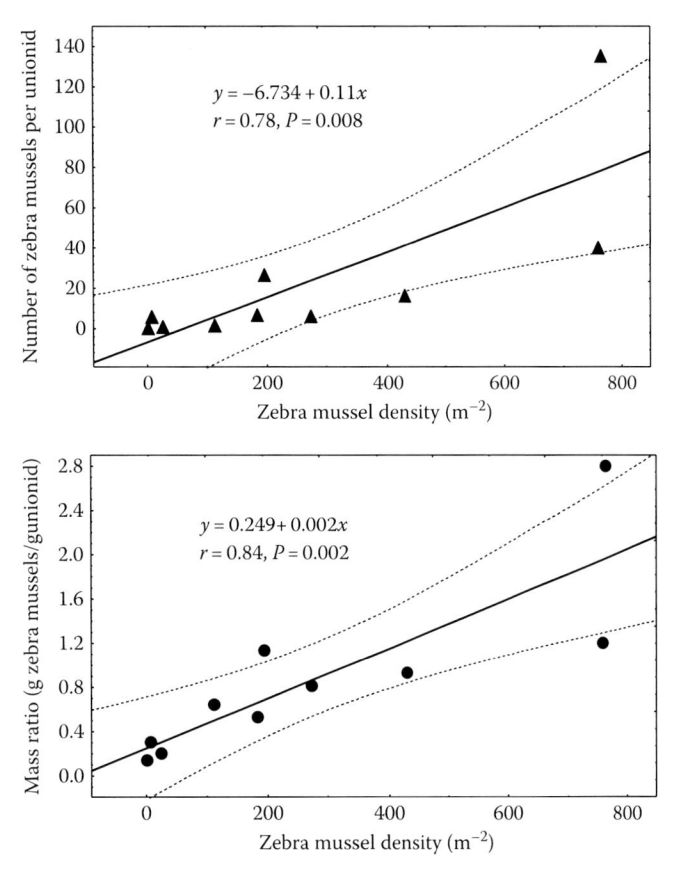

Figure 40.10 The number of attached zebra mussels per unionid relative to zebra mussel density (a) and the mass (weight) ratio of attached zebra mussels and unionid host relative to zebra mussel density (b) for water bodies in Belarus. Density given as number per m², and mass given as total wet weight (g). Each point represents a separate water body. Solid line was derived from the given regression and dotted lines represent 95% confidence limits of the regression line. (Modified from Burlakova, L.E. et al., *Int. Rev. Hydrobiol.*, 85, 529, 2000.)

To determine the relationship between zebra mussel density and unionid infestation, we examined data from Lake Naroch (collected in 1990 and 1993) and eight other lakes in the Naroch region (collected in 1998). We found a direct correlation between zebra mussel density and number of zebra mussels attached to unionids, as well as between zebra mussel density and the ratio of the total wet weight of attached mussels to the total wet weight of the unionid host (Pearson $r = 0.78$, $P = 0.008$, Pearson $r = 0.84$, $P = 0.002$, respectively) (Figure 40.10). From the perspective of the unionid, the weight of attached zebra mussels, or the weight ratio, is probably more important than density (Hebert et al. 1991, Karatayev et al. 1997).

Substrate type may impact the level of *Dreissena* infestation of unionids. In the Svisloch River, where sand and gravel alternate with silt, we found unionids in sand and gravel with high numbers of attached zebra mussels (up to 100 per unionid). In contrast, unionids completely buried in silt were free of zebra mussels, and at several of these silty sites, the density of unionids was around 100 m⁻². The average density of zebra mussels in the Drozdy Reservoir (Svisloch River) was 838 m⁻². All 54 unionids collected from this reservoir had zebra mussels attached, and 15 of the unionids collected were alive. Densities of zebra mussels were much lower (81 m⁻²) in a downstream reservoir, Chizhovskoe, where silt substrates were prevalent. Unionids were abundant in this reservoir, and only 7 of total 107 unionids collected had attached zebra mussels. In Lake Volchin, unionids in sandy substrates were completely infested with zebra mussels, and the number of zebra mussels per unionid was 35.6 ± 8.3. In contrast, in the same lake the number of zebra mussels per unionid in silty substrates was only 13.4 ± 3.3. The difference in number of attached zebra mussels per unionid in the two substrates was significant ($P = 0.049$, ANOVA). The ratio of weight of zebra mussels to weight of host unionid was higher on sand (1.45 ± 0.29) than on silt (0.66 ± 0.17), but the difference was not significant ($P = 0.054$, ANOVA) (Burlakova et al. 2000). Similar results have been obtained in other European water bodies. For instance, in Lake Hallwil, Switzerland, Arter (1989) found that *U. tumidus* was usually buried in silty sediments and rarely overgrown by zebra mussels. However, *A. cygnea* was often only partly buried and was infested more often by zebra mussels relative to *U. tumidus*.

Our studies indicate that unionids may be capable of removing zebra mussels from their shell surface. Often in water bodies where we found unionids heavily infested, we observed some individuals completely free of zebra mussels but had byssal threads attached to their shells. For example, in Lake Bolshiye Shvakshty, 40% of 147 unionids were infested with zebra mussels. Over 70% of these infested unionids had remains of byssal threads on their shells, indicating that more mussels were attached to them before collection. Moreover, 61% of uninfested unionids had zebra mussel byssal threads on their shells. Twenty-three percent of all unionids collected had neither zebra mussels nor byssal threads. Zebra mussel byssal threads were also found on uninfested unionids from other water bodies (e.g., lakes Dolzha, Spory, and Bolduk). In Lake Bolshiye Shvakshty, the average number of zebra mussels on unionids without remains of byssal threads was significantly higher ($P < 0.001$, t-test) than the average number on unionids with remains of byssal threads (7.2 ± 1.3 vs. 1.8 ± 0.2) (Burlakova et al. 2000).

Lough Key, Ireland

The initial introduction of the zebra mussel to Ireland occurred in the lower Shannon River basin in the early 1990s (McCarthy et al. 1998, Minchin and Moriarty 1998). The Shannon River is the largest navigable river system in Ireland, and the most likely vector of introduction was via imported leisure craft. Live zebra mussels were found on hulls of leisure craft that arrived in Ireland from Britain during the period between 1997 and 2001 (Minchin et al. 2002a). Further, zebra mussels in Ireland were genetically most similar to populations from the midlands of Britain (Pollux et al. 2003, Astanei et al. 2005). Once established, zebra mussels spread throughout Ireland via transport by recreational boats (Minchin et al. 2002b). Zebra mussels spread to most lakes in the Shannon River basin district, to heavily fished lakes in western Ireland, and to many other water bodies with moderate to high calcium concentrations (Lucy and Mastitsky, unpublished).

The unionid *Anodonta* was widely distributed in Lough Key (Figure 40.11), and the rest of the Shannon River system at the time zebra mussels arrived in the early 1990s. In April 1998, 23 live *Anodonta* were collected in Lough Key. Of these, 7 were free of attached zebra mussels, while the other 16 had only 1–5 attached mussels that were 3–13 mm in length. These small attached mussels likely settled the previous year. In contrast, *Anodonta* collected in November 1998 ($n = 72$) were more densely infested with mussels (mean = 78 mussels per unionid, range 0–314).

By July 1999, the mean number of zebra mussels found attached to individual *Anodonta* increased to 294 mussels per unionid (range 81–923), which was much higher than found in November 1998. In a limited sample of 12 dead *Anodonta* shells, the maximum number of attached zebra mussels was 1066.

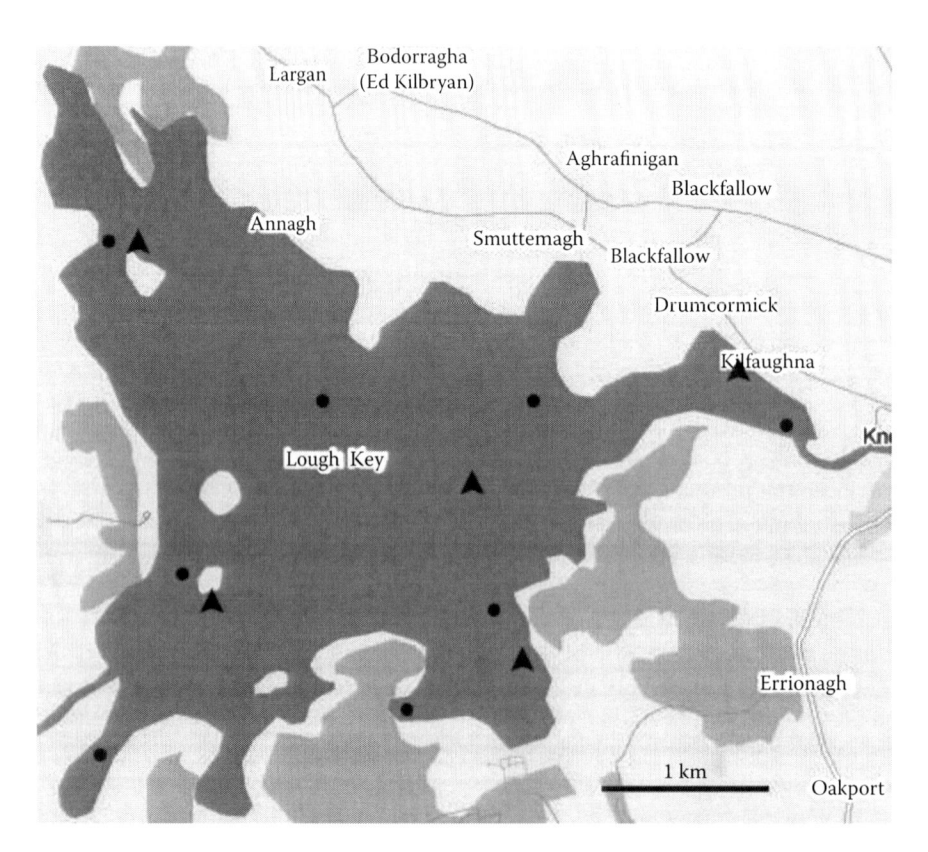

Figure 40.11 Sampling sites in Lough Key, Ireland. Solid circles, snorkel sites; solid triangles, monitoring sites. (From Lucy, F. et al., Nutrient levels and the zebra mussel population in Lough Key, ERTDI Report Series No. 34, Environmental Protection Agency, Wexford, Ireland, 2005; Map: © 2011 Google; Map data © 2011 Tele Atlas.)

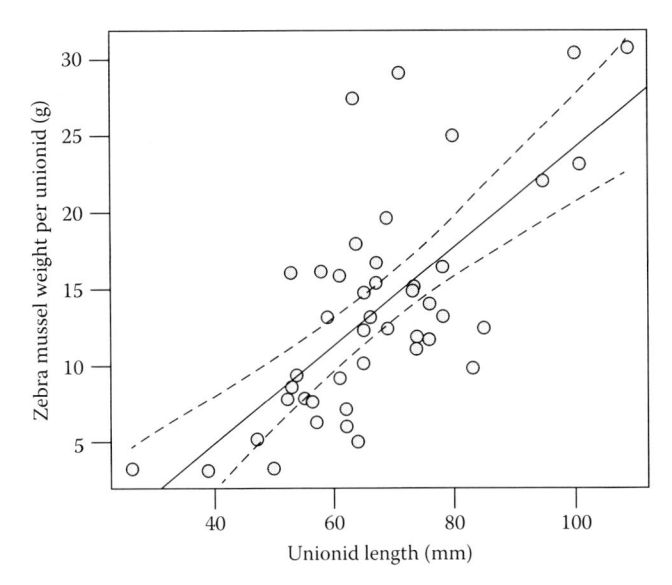

Figure 40.12 Relationship between weight (g) of attached zebra mussels on a unionid and shell length of the unionid (mm) in Lough Key, Ireland. The relationship was defined as *Weight* = −7.80 + 0.32 × *Length*. The model explained 62% of the variance and was highly significant ($P < 0.001$, *F*-test). Dotted lines represent 95% confidence limits of the regression line.

The total number of *Anodonta* collected in July 1999 was 58, and the live/dead ratio was 3.8 to 1 (Lucy and Sullivan 2001). It is not possible to ascertain how many of the dead unionids were attributed to biofouling because no decaying flesh was present to indicate whether shells were fouled prior to or after death, but they were all sampled from extant *Anodonta* beds. Attached zebra mussels were mainly young of the year (<3 mm) or age 1+ (3–19 mm), although age 2+ mussels (≥19 mm) were also found in low numbers. As zebra mussels became further established in Lough Key, it was common by 1999 to find small numbers of age 2+ or older (up to 36 mm) zebra mussels attached to *Anodonta* as well as to stone substrate.

Zebra mussels found on live *Anodonta* occurred mainly on the outside of the posterior end of the shell, but sometimes zebra mussels were found on the inside. The shell length of *Anodonta* collected live ranged from 26 to 85 mm, and the smallest individual collected was heavily infested on one half of its shell. In July 1999, the weight ratio of attached zebra mussels to corresponding live *Anodonta* ranged from 0.24 to 3.48 (mean 0.93).

In November 1999, 36 *Anodonta* were collected from one particular site but only 6 were live. The mean weight ratio of attached zebra mussels to live *Anodonta* was 0.45. Of the 30 dead *Anodonta* collected, 26 were mostly buried in the substrate and infested at the posterior end (4 were half shells only). This indicated that they had been fouled while live and had presumably died during 1999. Zebra mussels were found on 15 of the dead *Anodonta*, with an average of 745 individuals (range 334–1108) per shell and a biomass of 31.6 ± 13.3 g per shell. Collections in 1999 showed a strong positive linear relationship between total weight of attached zebra mussels and unionid shell length ($P < 0.001$, *F*-test) (Figure 40.12). In addition, the weight ratio of attached zebra mussel to unionid weight was significantly negatively associated with unionid shell length ($P < 0.001$, *F*-test).

In 2000, most *Anodonta* were found dead on their sides, with high densities of zebra mussels on the side exposed (outside and often inside) of shells. Zebra mussel colonization often extended from the shell outward onto the substrate. In 2001, a survey with an ROV around lakeshore and island perimeters revealed that the mean density of dead *Anodonta* shells was 2 m^{-2} (range 0–12). Byssal plaques were often noted at the posterior end of dead shells partially covered by substrate (Figure 40.13). By 2003, shells were observed to be primarily sunk in the soft substrate of the lake (Lucy et al. 2005). Mean lengths of dead shells were 8.4 ± 1 cm in 2000, 8.7 ± 1.1 cm in 2001, and 9.2 ± 8.1 cm in 2003, and all were significantly greater (*t*-test, $P < 0.001$) than mean shell

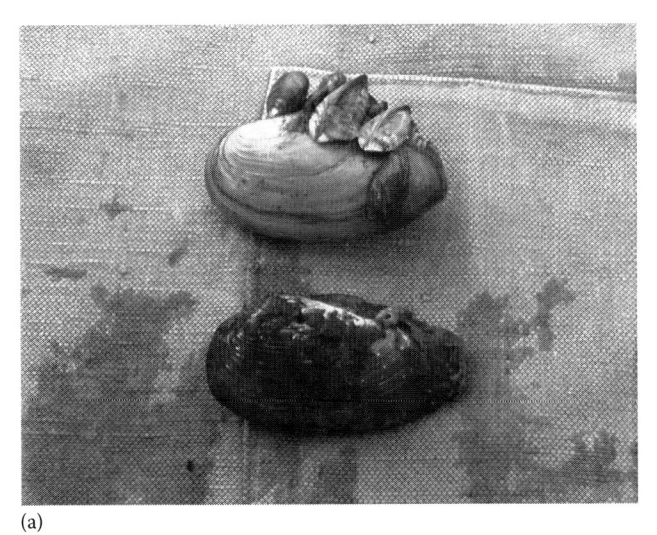

Figure 40.13 Zebra mussels and byssal plaques on unionids collected from (a) Lake Naroch, Belarus, and (b) Lake St. Clair.

lengths of live *Anodonta* collected in 1999. This suggests that, at least in some cases, zebra mussels colonized empty *Anodonta* shells that resulted from individuals that had died of old age/natural causes prior to infestation. No shell deformities were noted on any *Anodonta* (live or dead) during the course of these surveys (1998–2003), indicating mortality was not due to physical damage to unionid shells.

From 2000 to 2003, no live *Anodonta* were recovered or viewed during grab sampling, snorkel, dive, transect, or video surveys. Hence, based on the evidence, the extirpation of *Anodonta* in all sampled areas occurred between November 1999 and August 2000. While it is quite possible that fish may reintroduce *Anodonta* to Lough Key from other parts of the Shannon River basin, without intensive sampling it may take years to detect whether this has happened. In such a case, the rapid extirpation of *Anodonta* following zebra mussel colonization could actually be reversed by a natural reintroduction to the lake. Since zebra mussel populations have subsequently stabilized in Lough Key, it is possible that reintroduction of unionids could result in sustained populations of *Anodonta* in the lake.

OVERALL IMPACTS

Based on these case studies, a high percentage of unionids will be infested even when zebra mussel density in a given water body is low, and infestation grows rapidly with increased zebra mussel density (Figure 40.14). When the average density of zebra mussels in a water body was

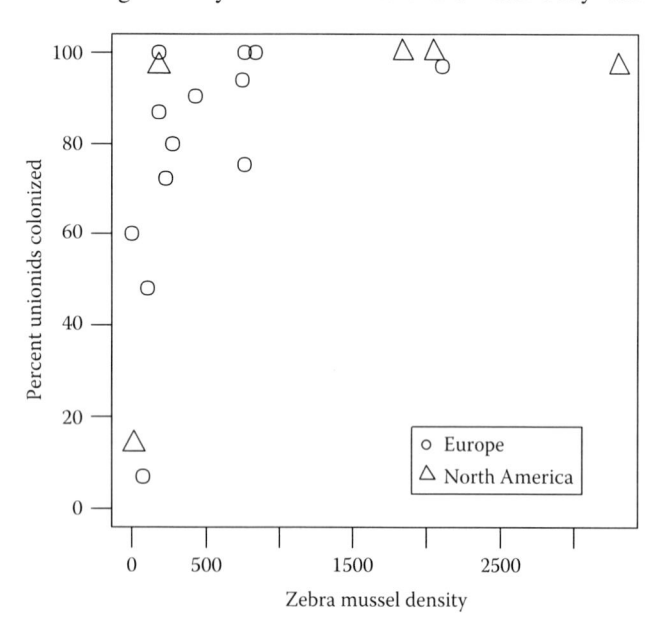

Figure 40.14 Relationship between the percentage of unionids infested with zebra mussels and the average density (No. m^{-2}) of zebra mussels in a water body. European data taken from Karatayev and Burlakova (1995a), Burlakova et al. (2000), and Burlakova and Karatayev (unpublished); North American data taken from Nalepa et al. (1996).

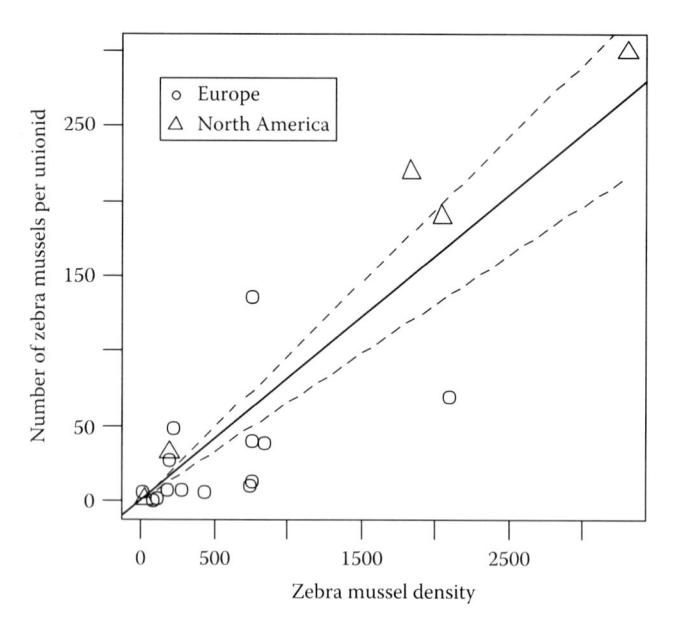

Figure 40.15 Relationship between mean number of zebra mussels per unionid and zebra mussel density (No. m^{-2}) in water bodies in Europe and North America. The relationship was defined as *Number* = 0.081 × *Density*. The model explained 84.2% of the variance and is highly significant ($P < 0.001$, F-test). Dotted lines represent 95% confidence limits of the regression line. European data taken from Karatayev and Burlakova (1995), Burlakova et al. (2000), and Burlakova and Karatayev (unpublished); North American data taken from Nalepa et al. (1996).

ca. >200 m^{-2}, up to 100% of unionids were found to be infested. There was no notable difference in infestation patterns between North American and European data (Figure 40.14). We found a strong positive linear relationship between the number of zebra mussels per unionid host and the zebra mussel density in a given water body (Figure 40.15). Interestingly, most North American observations occurred above the fitted regression line (Figure 40.15), suggesting that this relationship is stronger in North America than in Europe. However, as discussed later, this observation may be an artifact of sampling protocols.

The relationship between the weight of attached zebra mussels and the unionid shell length varied between the 15 international water bodies and also varied with the length of time after zebra mussels invaded (Figure 40.16). Overall, however, datasets were sufficiently fitted by the model: *Weight* = 3.28 + 0.17L − 6.58*TSRI*, where L is the unionid shell length and *TSRI* is the time since recognized invasion of zebra mussel (with value 0 if <10 years, and 1 if ≥10 years). All coefficients of the model were highly significant ($P < 0.001$, t-tests). For *TSRI* < 10 years, there was an overall trend for increase of zebra mussel weight with unionid host's size (Figure 40.17). For *TSRI* ≥ 10 years, such a pronounced trend was not observed (Figure 40.18), indicating that the infestation rate per unionid length declines. In addition, the weight of zebra mussels per unit of unionid length appeared to be significantly lower in water bodies

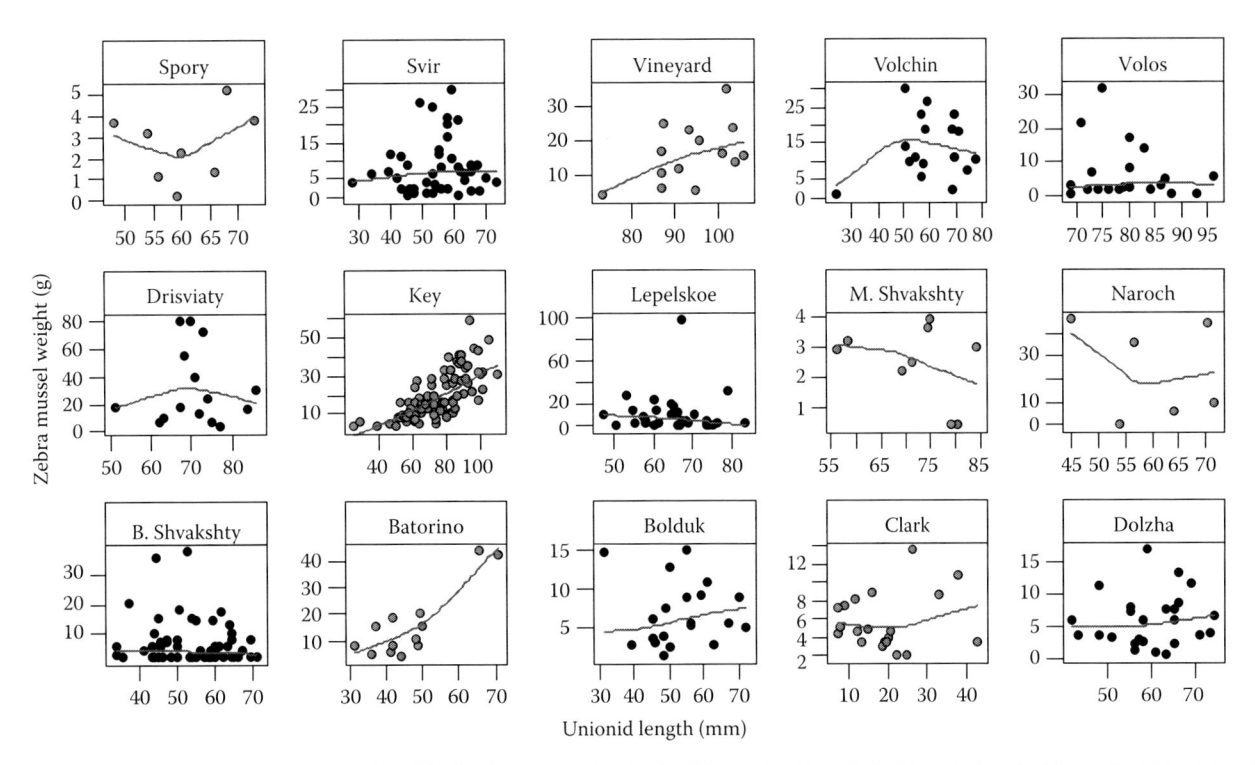

Figure 40.16 Relationship between weight (g) of zebra mussels attached to a unionid and shell length (mm) of the unionid for lakes in Belarus. Lakes were categorized by the *TSRI* of zebra mussel. A smoother was added to each panel to visualize the patterns in data. Gray points represent lakes with a *TSRI* < 10 years, and black points represent lakes with a *TSRI* ≥ 10 years. Data for Lakes Clark, Lepelskoe, Naroch, Volchin, and Vineyard are from Burlakova et al. (2000), data for Lough Key are from Lucy and Sullivan (2001), and data for all other lakes are from Karatayev and Burlakova (unpublished).

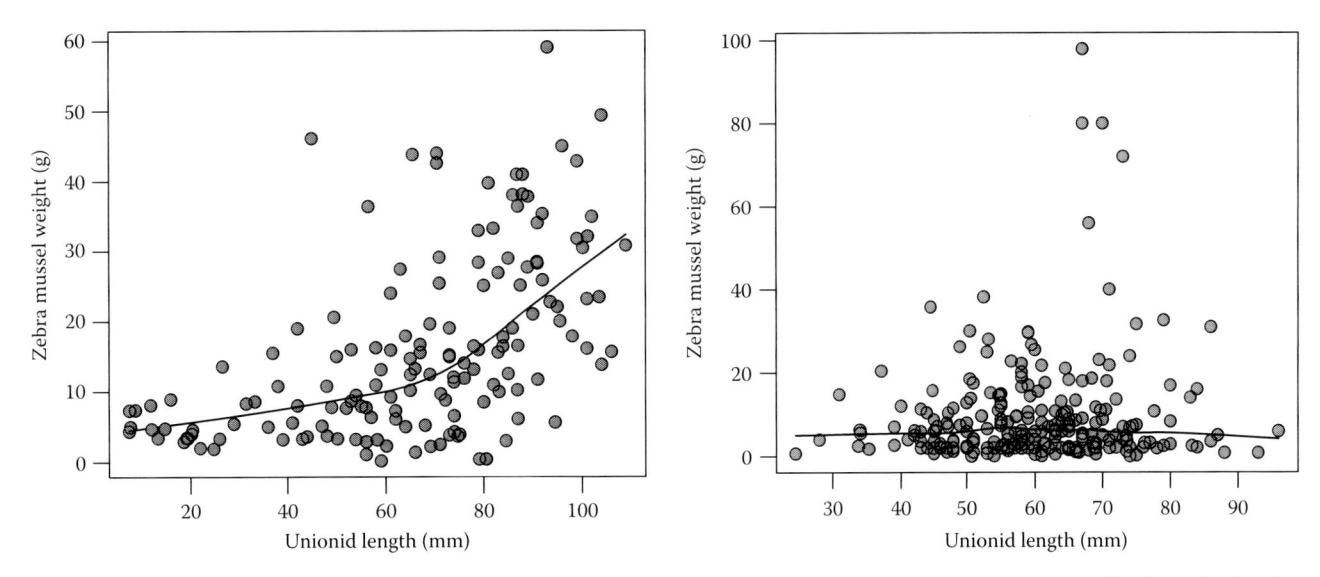

Figure 40.17 Relationship between weight (g) of attached zebra mussels and shell length (mm) of corresponding unionid for lakes with *TSRI* < 10 years (*n* = 162, data pooled for seven lakes). A smoother was added to visualize the pattern.

Figure 40.18 Relationship between weight (g) of attached zebra mussels and shell length (mm) of corresponding unionid for lakes with *TSRI* ≥ 10 years (*n* = 249, data pooled for eight lakes). A smoother was added to visualize the pattern.

colonized for over 10 years (median 0.096 vs. median 0.209; $P < 0.001$, Mann–Whitney test). The negative value of the regression coefficient of *TSRI* suggested that overall zebra mussel weight per unionid weight decreased with time in water bodies colonized by zebra mussels for over 10 years.

As an additional support of this finding, we observed that the ratio of zebra mussel weight to unionid weight was significantly lower when *TSRI* was ≥10 years compared to the ratio when *TSRI* was <10 years ($P < 0.007$, Mann–Whitney test; Figure 40.19). For all water bodies, the median ratio

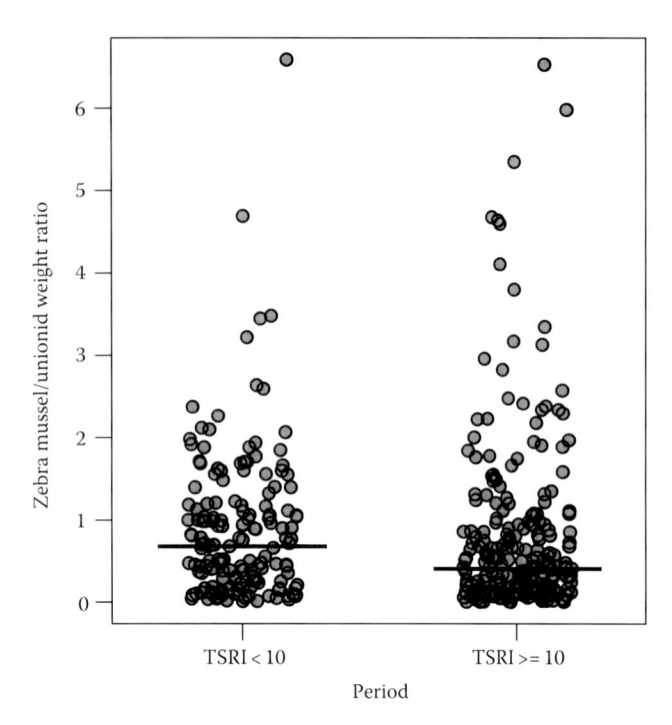

Figure 40.19 The weight ratio of attached zebra mussels to corresponding unionid in seven lakes with *TSRI* < 10 years and in eight lakes with *TSRI* ≥ 10 years (data pooled for all unionid species examined). Horizontal lines are the medians.

was 0.680 for the former time period compared to 0.404 for the latter. In combination, our results suggest that the adverse impact of zebra mussels on unionids decreases about 10 years after zebra mussels invade.

DISCUSSION

General Trends

The population size of an invasive species is one key factor that determines its effects on ecosystems and ecosystem components such as unionids. According to our analysis, impacts of zebra mussels on unionids are high early in an invasion when zebra mussel densities are increasing. Both European and North American data indicated that a high percentage of unionids were infested even when zebra mussel density was low during the early period of expansion. The percentage of unionids infested increased rapidly and reached 100% infestation at mussel densities ca. 200 m^{-2}. In many water bodies, by the time the zebra mussel population reached peak density, the majority of unionids were infested and dead. The dramatic decline of unionids after zebra mussel colonization is well documented both in Europe (Sebestyén 1937, Dussart 1966, Karatayev and Burlakova 1995b, Karatayev et al. 1997, Burlakova et al. 2000) and North America (Haag et al. 1993, Gillis and Mackie 1994, Nalepa 1994, Schloesser and Nalepa 1994, Nalepa et al. 1996, Riccardi et al. 1996, Ricciardi 2003, Strayer and Malcom 2007).

In contrast to the early stages of invasion, effects on unionids seem to decrease as the population of zebra mussels stabilizes or decreases. However, a decline in zebra mussel abundances may not be a general trend. The population may be quite stable for a relatively long period of time after reaching maximum density (Burlakova et al. 2006). Alternatively, the population may cycle (Strayer and Malcom 2006, Strayer et al. 2011) or fluctuate widely (Ramcharan et al. 1992, Stańczykowska and Lewandowski 1993).

The apparent extirpation of unionids from infested water bodies does not happen often (reviewed in Karatayev et al. 1997, Gurevitch and Padilla 2004a,b). Multiple European studies from water bodies colonized with zebra mussels for long periods of time showed that unionids were not extirpated from any of them (Lewandowski 1976b, Karatayev 1983, Miroshnichenko et al. 1984, Miroshnichenko 1987, Ponyi 1992, Burlakova et al. 2000). Although infestation can cause some unionid mortality, unionids not only persevered but also maintained high densities (Karatayev 1983, Miroshnichenko et al. 1984, Miroshnichenko 1987, Karatayev et al. 1997). We found live unionids in water bodies colonized by zebra mussels for >10 years and abundant unionid populations in Lake Lepelskoe where zebra mussels have been present for about 200 years. In the Tsimlyanskoe Reservoir, Russia, unionids (mainly *U. pictorum* and *A. cygnea*) were found to coexist with zebra mussels (Miroshnichenko et al. 1984, Miroshnichenko 1987). The average annual biomass over the entire Tsimlyanskoe Reservoir was 571 g m^{-2} for zebra mussels, 88 g m^{-2} for *U. pictorum*, and 46 g m^{-2} for *A. cygnea* (Miroshnichenko et al. 1984). In Lake Balaton, Ponyi (1992) determined the average density of unionids in 1932 just before the zebra mussel invasion was 3 m^{-2} and was 2 m^2 in 1966–1968. In the Azov Sea, which is within the native range of zebra mussel, one of 13 typical benthic communities included both zebra mussel and Unionidae as subdominant taxa (Vorobiev 1949). Near the mouth of the Don River, Russia, unionids, zebra mussels, and another bivalve *Monodacna* all had high abundances (Vorobiev 1949).

Some unionid communities in North America seem to be recovering after initial declines. Recent studies in the Hudson River found that populations of all native bivalves (unionids and sphaeriids) have stabilized or even recovered (Strayer and Malcom 2007), whereas annual survivorship of adult zebra mussels and aggregate filtration rate of the population declined (Strayer et al. 2011). According to Crail et al. (2011), unionids are extant at several sites outside known refugia in the western basin of Lake Erie, and conditions for unionids in the lake may be improving. Sixteen unionid species were found living in or near Lake Erie, including six sites in the nearshore zone of the lake. Each community consisted of live individuals from two to eight species, with the overall mean density at 0.09 m^{-2} (Crail et al. 2011).

Zebra mussel impacts on unionids may be taxon-specific as based on shell morphology (i.e., thin-shelled Lampsilinae and Anodontinae vs. thick-shelled Ambleminae)

(Lewandowski 1976b, Arter 1989, Haag et al. 1993, Strayer and Smith 1996, Schloesser et al. 1998). It has been shown that unionids "returning" to open-water areas of western Lake Erie were more often thin-shelled, fast-growing species (e.g., *L. fragilis* and *Pyganodon grandis*) (Crail et al. 2011). At least for *L. fragilis*, this is consistent with this species being persistent in a wetland area in western Lake Erie (Nichols and Amberg 1999). Despite the presence of zebra mussels in the wetlands for a number of years, the *L. fragilis* population showed no signs of competition-induced changes in population dynamics, had a limited incidence of recent or past dreissenid infestation (<1%), and displayed successful recruitment on a yearly basis. This seems to contrast with early observations that thin-shelled species were more highly impacted by zebra mussels than thick-shelled species (Schloesser et al. 1996).

Our analysis of multiple water bodies in Europe and North America in different stages of zebra mussel colonization has provided broad insights into unionid–dreissenid interactions. We found a positive relationship between the weight of attached zebra mussels and shell length of the corresponding unionid for water bodies that have been colonized by zebra mussels for less than 10 years. This trend, however, was not observed for water bodies colonized for over 10 years. We also discovered that the ratio of zebra mussel weight to unionid weight was significantly lower in water bodies colonized for over 10 years. These results suggest that adverse impacts of zebra mussels on unionids likely diminish with time.

What are the important features that may allow coexistence of unionids and zebra mussels in a particular habitat? These include factors that inhibit establishment of stable zebra mussel populations and/or allow unionids to escape dreissenid infestation. One of the most frequently reported habitat variables that allow unionid survival is the presence of substrates soft enough for unionids to burrow into and thereby remove attached zebra mussels and/or prevent zebra mussel settlement (Nichols and Wilcox 1997, Nichols and Amberg 1999, Burlakova et al. 2000, Zanatta et al. 2002, Bowers and de Szalay 2004). Factors that are hypothesized to inhibit establishment of stable zebra mussel populations are wave action in shallow water, water level fluctuations, ice scour (Nichols and Wilcox 1997, Burlakova et al. 1998, Bowers and de Szalay 2004, 2005), dense reed beds (Nelson et al. 2009, Sullivan et al. 2010), and remoteness from the source of zebra mussel veligers (Zanatta et al. 2002, McGoldrick et al. 2009). A synergy of various habitat conditions were described from known refuges including large areas of shallow waters (protected bayous) with low flow and warmer temperatures that encourage unionid burrowing (Nichols and Wilcox 1997), hydrological connection of shallow waters to the lake (Bowers and de Szalay 2004, 2005), and fish predation of zebra mussels attached to unionids (Bowers and de Szalay 2007).

Although we found that the percentage of infested unionids increased very quickly with increased zebra mussel density in a water body (Figure 40.14), overall density of zebra mussels is known to depend on substrate availability, morphometry, and trophic status (Stańczykowska and Lewandowski 1993, Lucy et al. 2005, Burlakova et al. 2006). Each of these factors may contribute to low zebra mussel density in some water bodies and thus allow unionids to coexist. The mechanism of coexistence may also vary with water body type: river, lake, reservoir, or interconnecting river and lake systems. Finally, there may be other, yet unidentified, mechanisms that promote long-term coexistence of zebra mussels and unionids.

RECOMMENDATIONS

Our analysis of multiple water bodies in Europe and North America in different stages of zebra mussel colonization suggests that adverse impacts of zebra mussels on unionids are most detrimental during the first stages of invasion but that impacts diminish over time. We found that the negative impact of zebra mussels on unionids may decrease about 10 years after initial zebra mussel colonization. We have some knowledge about habitat parameters that define a potential refuge (see preceding text), and ongoing studies will help to better understand conditions that define a refuge. If refuges can be defined, then they can be located and protected in water bodies that are at the very first stages of invasion or under imminent risk of invasion to promote unionid survival. Some rare unionid species that are intolerant to the conditions in the refuge may be relocated to uninfested waters.

In comparing studies from different countries, we found that different methods of recording the effects of zebra mussels hamper our ability to make generalizations. Therefore, we recommend that a standard set of methods will be a very productive step for future studies on the impact of invasive bivalves (not dreissenids alone) on unionids. For example, we found that most North American observations lay above the fitted regression line presented in Figure 40.15, suggesting that the number of zebra mussels per unionid in North American water bodies is more strongly related to the field density of zebra mussels than in European water bodies. One possible explanation for this may be that North American scientists report all attached mussels regardless of size, while European scientists generally do not include mussels smaller than 1–2 mm (and sometimes <8 mm) in density estimates (reviewed in Karatayev et al. 1997, Burlakova et al. 2000). From the perspective of the unionid, the weight of attached zebra mussels or the weight ratio of attached zebra mussels to unionid host is probably more important than the number found attached (Hebert et al. 1991, Karatayev et al. 1997, Lucy 2005). Therefore, recording the unionid weight, length, and the *Dreissena* weight is very important for further comparisons, especially considering the strong significant relationship between the ratio of zebra mussel weight to unionid weight and zebra mussel density in a water body (Burlakova et al. 2000, Ricciardi 2003). In addition, methods to estimate densities of zebra mussels and unionids need to be standardized or better documented so that rates and frequencies of infestation can be better related to time after initial colonization across different water bodies. Use of comparable methods will allow direct comparisons of studies conducted in different countries and continents and make it possible to improve important generalizations on the impacts of zebra mussels on unionids.

We recommend a number of different studies to better understand zebra mussel–unionid competition and coexistence. Determining where the impacts of zebra mussels would be most detrimental and identifying which species of unionids (especially those that are rare and threatened) would be most vulnerable will assist in the development of global management priorities prior to future invasions of not only dreissenids but also other exotic epifaunal byssate bivalves, such as *Limnoperna fortunei*. Studies to identify the stage of dreissenid colonization when impacts on unionids are diminished (if it occurs) are critical. Additional studies on water bodies that were recently colonized versus water bodies known to host zebra mussels for many years will test our generalizations and aid in the development of management options. To be comparable, these studies have to be designed in parallel and carried out using a unified methodology.

Additional studies are needed to understand differences between impacts of quagga mussels versus zebra mussel on unionids and to understand if the presence of both dreissenid species in a water body will have a stronger impact on unionids than the presence of either species alone. While impacts of zebra mussels on unionids are very well described, we often do not have enough comparable data on the impacts of quagga mussels (Karatayev et al., in review). The zebra mussel has greater rates of byssal thread production and higher attachment strength relative to the quagga mussel (Peyer et al. 2009), which may explain why the zebra mussel is more dominant on unionids when both species are equally prevalent on other substrates (reviewed in Zhulidov et al. 2010, D. Zanatta, personal observation).

Finally, more studies are needed to identify conditions that allow dreissenids and unionids to coexist (e.g., habitat, substrates, hydrodynamics). A number of refuges have been already discovered in Europe and North America, and they provide a unique opportunity to study key habitat attributes and subsequently develop predictive models. Such information provides an opportunity to locate and protect additional unionid refuges and to manage sites to promote unionid colonization, survival, and endangered species recovery. This information is also imperative when creating possible refuges during the initial stage of invasion by dreissenids and other byssate exotic bivalves.

ACKNOWLEDGMENTS

Funding for 2010 sampling on Lake St. Clair was provided by a grant from the Michigan Department of Environmental Quality (DEQ)—Coastal Zone Management Program to DTZ. DTZ acknowledges indispensable field assistance by M. Rowe, J. Sherman, and T. Biber (Central Michigan University); D. Dortman (Michigan DEQ), M. Thomas, K. Koster, and R. Beasley (Michigan DNR); and Dr. R. Gregory (University of Guelph). Special thanks to D. McGoldrick (Environment Canada) for providing the raw Lake St. Clair unionid and *Dreissena* data from 2003. Research Foundation of SUNY provided support for LEB during manuscript preparation. We appreciate the help of the Director of the Naroch Biological Station (Belarusian State University) Dr. T.V. Zhukova and the staff of the station for assistance in the field. We gratefully acknowledge Dmitry Karatayev, as well as Peter A. Mitrakhovich, and Igor Rudakovsky (Belarusian State University) for their technical assistance while sampling in Belarus. We thank David Garton and Ladd E. Johnson for facilitating the study of unionids in lakes Clark and Vineyard in 1996 and Ronald D. Oesch for the identification of North American unionids. FEL acknowledges funding provided by the Irish EPA and the skills provided by Monica Sullivan, Dan Minchin, and Peter Walsh. Thanks to the Institute of Technology, Sligo, particularly to students Ann Skelly and Elaine Ni Chonmhara, and to the support staff in the Department of Environmental Science.

REFERENCES

Arter, H.E. 1989. Effect of eutrophication on species composition and growth of freshwater mussels (Mollusca, Unionidae) in Lake Hallwil (Aargau, Switzerland). *Aquat. Sci.* 51: 87–99.

Astanei, I., E. Gosling, J. Wilson, and E. Powell. 2005. Genetic variability and phylogeography of the invasive zebra mussel, *Dreissena polymorpha* (Pallas). *Mol. Ecol.* 14: 1655–1666.

Bij de Vaate, A., K. Jazdzewski, H.A.M. Ketelaars, S. Gollash, and G. Van der Velde. 2002. Geographical patterns in range extension of Ponto-Caspian macroinvertebrate species in Europe. *Can. J. Fish. Aquat. Sci.* 59: 1159–1174.

Bogan, A.E. 1993. Freshwater bivalve extinctions (Mollusca: Unionoida): A search for causes. *Am. Zool.* 33: 599–609.

Bogan, A.E. 2008. Global diversity of freshwater mussels (Mollusca, Bivalvia) in freshwater. *Hydrobiologia* 595: 139–147.

Bogan, A.E. and K.J. Roe. 2008. Freshwater bivalve (Unioniformes) diversity, systematics, and evolution: Status and future directions. *J. N. Am. Benthol. Soc.* 27: 349–369.

Bowers, R. and F.A. de Szalay. 2004. Effects of hydrology on unionids (Unionidae) and zebra mussels (Dreissenidae) in a Lake Erie coastal wetland. *Am. Midl. Nat.* 151: 286–300.

Bowers, R. and F.A. de Szalay. 2005. Effects of water level fluctuations on zebra mussel distribution in a Lake Erie coastal wetland. *J. Freshwat. Ecol.* 20: 85–92.

Bowers, R. and F.A. de Szalay. 2007. Fish predation of zebra mussels attached to *Quadrula quadrula* (Bivalvia: Unionidae) and benthic molluscs in a Great Lakes coastal wetland. *Wetlands* 27: 203–208.

Bowman, J.J. 1998. *River Shannon Lake Water Quality Monitoring 1995 to 1997*. Wexford, Ireland: Environmental Protection Agency.

Burlakova, L.E. 1998. Ecology of mussel *Dreissena polymorpha* (Pallas) and its role in structure and function of aquatic ecosystems. PhD dissertation. Minsk, Belarus: Belarusian State University.

Burlakova, L.E. 1999. The spread of *Dreissena* in Belarusian lakes. In *International Conference on Aquatic Ecosystems 'The Results and Future of Aquatic Ecology Research,'* A.Y. Karatayev, ed., pp. 30–34. Minsk: BGU Press.

Burlakova, L.E., A.Y. Karatayev, and D.K. Padilla. 2000. The impact of *Dreissena polymorpha* (Pallas) invasion on unionid bivalves. *Int. Rev. Hydrobiol.* 85: 529–541.

Burlakova, L.E., A.Y. Karatayev, and D.K. Padilla. 2006. Changes in the distribution and abundance of *Dreissena polymorpha* within lakes through time. *Hydrobiologia* 571: 133–146.

Caraco, N.F., J.J. Cole, P.A. Raymond, D.L. Strayer, M.L. Pace, S.E.G. Findlay, and D.T. Fischer. 1997. Zebra mussel invasion in a large, turbid river: Phytoplankton response to increased grazing. *Ecology* 78: 588–602.

Carlton, J.T. 2008. The zebra mussel *Dreissena polymorpha* found in North America in 1986 and 1987. *J. Great Lakes Res.* 34: 770–773.

Committee on the Status of Endangered Wildlife in Canada (COSEWIC). 2010. Candidate Wildlife Species, November 1, 2010. http://www.cosewic.gc.ca/eng/sct3/index_e.cfm#9 (Accessed November 3, 2010).

Crail, T.D., R.A. Krebs, and D.T. Zanatta. 2011. Unionid mussels from nearshore zones of Lake Erie. *J. Great Lakes Res.* 37: 199–202.

Dussart, G.B.J. 1966. *Limnologie*. Paris, France: Gauthier-Villars.

Edsall, T.A., B.A. Manny, and C.N. Raphael. 1988. *The St. Clair River and Lake St. Clair, Michigan: An Ecological Profile*. U.S. Fish and Wildlife Service, Biological Report 85(7.3). Ann Arbor, MI: U.S. Fish and Wildlife Service.

Gillis, P.L. and G.L. Mackie. 1994. Impact of the zebra mussel, *Dreissena polymorpha*, on populations of Unionidae (Bivalvia) in Lake St. Clair. *Can. J. Zool.* 72: 1260–1271.

Government of Canada. 2010. Species at risk Public Registry. http://www.sararegistry.gc.ca (accessed August 23, 2010).

Graf, D.L. 2002. Historical biogeography and late glacial origin of the freshwater pearly mussel (Bivalvia: Unionidae) faunas of Lake Erie, North America. Occassional papers on molluscs, The Department of Molluses, Museum of Comparative Zoology, Harward University. 6: 175–211.

Graf, D.L. and K.S. Cummings. 2006. Palaeoheterodont diversity (Mollusca: Trigonioida + Unionoida): What we know and what we wish we knew about freshwater mussel evolution. *Zool. J. Linn. Soc.* 148: 343–394.

Graf, D.L. and K.S. Cummings. 2007. Review of the systematics and global diversity of freshwater mussel species (Bivalvia: Unionoida). *J. Mollusc. Stud.* 73: 291–314.

Griffiths, R.W., D.W. Schloesser, J.H. Leach, and W.P. Kovalak. 1991. Distribution and dispersal of the zebra mussel (*Dreissena polymorpha*) in the Great Lakes Region. *Can. J. Fish. Aquat. Sci.* 48: 1381–1388.

Gurevitch, J. and D.K. Padilla. 2004a. Are invasive species a major cause of extinctions? *Trends Ecol. Evol.* 19: 470–474.

Gurevitch, J. and D.K. Padilla. 2004b. Response to Ricciardi. Assessing species invasions as a cause of extinction. *Trends Ecol. Evol.* 19: 620.

Haag, W.R., D.J. Berg, D.W. Garton, and J.L. Farris.1993. Reduced survival and fitness in native bivalves in response to fouling by the introduced zebra mussel (*Dreissena polymorpha*) in western Lake Erie. *Can. J. Fish. Aquat. Sci.* 50:13–19.

Hanson, J.M., E.E. Prepas, and W.C. Mackay. 1989. Size distribution of the macroinvertebrate community in a freshwater lake. *Can. J. Fish. Aquat. Sci.* 46: 1510–1519.

Hebert, P.D.N., B.W. Muncaster, and G.L. Mackie. 1989. Ecological and genetic studies on *Dreissena polymorpha* (Pallas): A new mollusc in the Great Lakes. *Can. J. Fish. Aquat. Sci.* 46: 1587–1591.

Hebert, P.D., C.C. Wilson, M.H. Murdoch, and R. Lazar. 1991. Demography and ecological impacts of the invading mollusc *Dreissena polymorpha*. *Can. J. Zool.* 69: 405–409.

Hunter, R.D. and J.F. Bailey. 1992. *Dreissena polymorpha* (zebra mussel): Colonization of soft substrata and some effects on unionid bivalves. *Nautilus* 106: 60–67.

Hunter, R.D. and K.A. Simons. 2004. Dreissenids in Lake St. Clair in 2001: Evidence for population regulation. *J. Great Lakes Res.* 30: 528–537.

Jaeckel, S.G.A. 1967. Lamellibranchia. In *Limnofauna Europaea*, ed. J. Illies, pp. 105–108. Stuttgart, Germany: Gustav Fisher Verlag.

Karatayev, A.Y. 1983. Ecology of *Dreissena polymorpha* Pallas in its role in macrobenthos of the cooling waterbody of a thermal power plant. PhD dissertation. Minsk, Belarus: Belarusian State University.

Karatayev, A.Y. and L.E. Burlakova. 1995a. Present and further patterns in *Dreissena polymorpha* (Pallas) population development in the Narochanskaya lakes system. *Vestsi Akademii Navuk Belarusi. Seriya biyalagichnikh navuk* 3: 95–98.

Karatayev, A.Y. and L.E. Burlakova. 1995b. The role of *Dreissena* in lake ecosystems. *Russ. J. Ecol.* 26: 207–211.

Karatayev, A.Y., L.E. Burlakova, and S.I. Dodson. 2005. Community analysis of Belarusian lakes: Relationship of species diversity to morphology, hydrology and land use. *J. Plankton Res.* 27: 1045–1053.

Karatayev, A.Y., L.E. Burlakova, and S.I. Dodson. 2008a. Community analysis of Belarusian lakes: Correlations of species diversity with hydrochemistry. *Hydrobiologia* 605: 99–112.

Karatayev, A.Y., L.E. Burlakova, and D.K. Padilla. 2002. Impacts of zebra mussels on aquatic communities and their role as ecosystem engineers. In *Invasive Aquatic Species of Europe-Distribution, Impacts, Management*. E. Leppakosk, S. Gollasch, and S. Olenin, eds., pp. 433–446. Dordrecht, The Netherlands: Kluwer Academic Publishers.

Karatayev, A.Y., L.E. Burlakova, D.P. Molloy, and L.K. Volkova. 2000. Endosymbionts of *Dreissena polymorpha* (Pallas) in Belarus. *Int. Rev. Hydrobiol.* 85: 543–559.

Karatayev, A.Y., L.E. Burlakova, and D.K. Padilla. 1997. The effects of *Dreissena polymorpha* (Pallas) invasion on aquatic communities in eastern Europe. *J. Shellfish Res.* 16: 187–203.

Karatayev, A.Y., L.E. Burlakova, and D.K. Padilla. 2010. *Dreissena polymorpha* in Belarus: History of spread, population biology, and ecosystem impacts. In *The Zebra Mussel in Europe*, G. Van Der Velde, S. Rajagopal, and A. Bij De Vaate, eds., pp. 101–112. Leiden, The Netherlands: Backhuys Publishers.

Karatayev, A.Y., L.E. Burlakova, D.K. Padilla, and L.E. Johnson. 2003. Patterns of spread of the zebra mussel (*Dreissena polymorpha* (Pallas)): The continuing invasion of Belarussian Lakes. *Biol. Invasions* 5: 213–221.

Karatayev, A.Y., L.E. Burlakova, and D.K. Padilla. 2012. Impacts of zebra mussels on aquatic communities and thier role as ecosystem engineers. In *Invasive Aquatic Species of Europe-Distributor, Impacts and Management*. E. Leppakoski, S. Gollasch, and S. Olenin, eds., pp. 433–436. Dordrecht, The Netherlands: Kluwer Academic Publishers.

Karatayev, A.Y., S.E. Mastitsky, L.E. Burlakova, and S. Olenin. 2008b. Past, current, and future of the central European corridor for aquatic invasions in Belarus. *Biol. Invasions* 10: 215–232.

Kerney, M. 1999. *Atlas of the Land and Freshwater Molluscs of Britain and Ireland*. Essex, U.K.: Harley Books.

Krebs, R.A. 2004. Combining paternally and maternally inherited mitochondrial DNA for analysis of population structure in mussels. *Mol. Ecol.* 13: 1701–1705.

Krebs, R.A., R.N. Vlasceanu, and M.J.S. Tevesz. 2003. An analysis of diversity in freshwater mussels (Bivalvia: Unionidae) of the Cuyahoga and Rocky River Watersheds (Ohio, USA) based on the 16S rRNA Gene. *J. Great Lakes Res.* 29: 307–316.

La Rocque, A. and J. Oughton. 1937. A preliminary account of the Unionidae of Ontario. *Can. J. Res.* 15: 147–155.

Leach, J.H. 1991. Biota of Lake St. Clair: Habitat evaluation and environmental assessment. *Hydrobiologia* 219: 187–202.

Lewandowski, K. 1976a. Long-term changes in the fauna of family Unionidae bivalves in the Mikolajskie Lake. *Ekol. Pol.* 39: 265–272.

Lewandowski, K. 1976b. Unionidae as a substratum for *Dreissena polymorpha*. *Pol. Arch. Hydrobiol.* 23: 409–420.

Lucey, J. 1995. The distribution of *Anodonta cygnea* (L.) and *Anodonta anatina* (L.) (Mollusca: Bivalvia) in southern Irish rivers and streams with records from other areas. *Ir. Nat. J.* 25: 1–40.

Lucy, F. 2005. The dynamics of zebra mussel (*Dreissena polymorpha*) populations in Lough Key, Co. Roscommon, 1998–2003. PhD dissertation. Sligo, Ireland: Institute of Technology.

Lucy, F. and M. Sullivan. 2001. The investigation of an invasive species, the zebra mussel *Dreissena polymorpha* in Lough Key, Co Roscommon. 1999. Desktop study no. 13. Wexford, Ireland: Environmental Protection Agency.

Lucy, F., M. Sullivan, and D. Minchin. 2005. Nutrient levels and the zebra mussel population in Lough Key. ERTDI Report Series No. 34. Wexford, Ireland: Environmental Protection Agency.

Lyakhnovich, V.P., A.Y. Karataev, and G.M. Tischikov. 1984. Distribution of *Dreissena polymorpha* Pallas in Belarus. In *All-Union Conference on Model Species of Aquatic Invertebrates*, pp. 16–20. Vilnius, Lithuania: Viniti Press.

Lydeard, C., R.H. Cowie, W.F. Ponder et al. 2004. The global decline of nonmarine mollusks. *BioScience* 54: 321–330.

Mackie, G.L. 1991. Biology of the exotic zebra mussel, *Dreissena polymorpha*, in relation to native bivalves and its potential impact in Lake St. Clair. *Hydrobiologia* 219: 251–268.

McCarthy, T.K., J. Fitzgerald, and W. O'Connor. 1998. The occurrence of the zebra mussel *Dreissena polymorpha* (Pallas, 1771), an introduced biofouling freshwater bivalve in Ireland. *Ir. Nat. J.* 25: 413–415.

McGoldrick, D.J., J.L. Metcalfe-Smith, M.T. Arts et al. 2009. Characteristics of a refuge for native freshwater mussels (Bivalvia: Unionidae) in Lake St. Clair. *J. Great Lakes Res.* 35: 137–146.

McMahon, R.F. and A.E. Bogan. 2001. Mollusca: Bivalvia. In *Ecology and Classification of North American Freshwater Invertebrates*, 2nd Edn., J.H. Thorp and A.P. Covich, eds., pp. 331–429. San Diego, CA: Academic Press.

Michigan Department of Natural Resources. 2011. Michigan's official list of endangered and threatened species. http://www.michigan.gov/documents/dnr/2007-007_NR_Threatened_Endangered_Species__nonstrike__9–12._274586_7.pdf (accessed May 17, 2011).

Minchin, D., F. Lucy, and M. Sullivan. 2002a. Zebra mussel: Impacts and spread. In *Invasive Aquatic Species of Europe: Distribution, Impacts and Spread*, E. Leppakoski, S. Gollasch, and S. Olenin, eds., pp. 135–146. Dordrecht, The Netherlands: Kluwer Press.

Minchin, D., F. Lucy, and M. Sullivan. 2002b. *Monitoring of Zebra Mussels in the Shannon-Boyle Navigation, Other Navigable Regions and Principal Irish Lakes, 2000 and 2001*. Marine Environment and Health Series, No. 5. Dublin, Ireland: Marine Institute.

Minchin, D. and C. Moriarty. 1998. Zebra mussels in Ireland. Fisheries Leaflet No 177. Dublin, Ireland: Marine Institute.

Miroshnichenko, M.P. 1987. Molluscs of the Tsimlyanskoe reservoir and their role in increasing fish productivity. *Sbornik nauchnykh trudov. Gosudarstvennyj Nauchno-Issledovatel'skij Institut Ozernogo i Rechnogo Rybnogo Khozyajstva* 270: 61–70.

Miroshnichenko, M.P., L.I. Volvich, and B.O. Skabichevskii. 1984. Production and energetic balance of the most abundant molluscs in Tsimlyanskoe reservoir. *Sbornik nauchnykh trudov. Gosudarstvennyj Nauchno-Issledovatel'skij Institut Ozernogo i Rechnogo Rybnogo Khozyajstva* 218: 19–30.

Mordukhai-Boltovskoi, F.D. 1964. Caspian fauna beyond the Caspian Sea. *Int. Rev. Ges. Hydrobiol.* 49: 139–176.

Nalepa, T.F. 1994. Decline of native unionid bivalves in Lake St. Clair after infestation by the zebra mussel, *Dreissena polymorpha*. *Can. J. Fish. Aquat. Sci.* 51: 2227–2232.

Nalepa, T.F. and J.M. Gauvin. 1988. Distribution, abundance, and biomass of freshwater mussels (Bivalvia: Unionidae) in Lake St. Clair. *J. Great Lakes Res.* 14: 411–419.

Nalepa, T.F., D.J. Hartson, D.L. Fanslow, and G.A. Lang. 2001. Recent population changes in freshwater mussels (Bivalvia: Unionidae) and zebra mussels (*Dreissena polymorpha*) in Lake St. Clair, USA. *Am. Malacol. Bull.* 16: 141–145.

Nalepa, T.F., D.J. Hartson, G.W. Gostenik, D.L. Fanslow, and G.A. Lang. 1996. Changes in the freshwater mussel community of Lake St. Clair: From Unionidae to *Dreissena polymorpha* in eight years. *J. Great Lakes Res.* 22: 354–369.

Negus, C.L. 1966. A quantitative study of growth and production of unionid mussels in the River Thames at Reading. *J. Anim. Ecol.* 35: 513–532.

Nelson, K.M., C.R. Ruetz III, and D.G. Uzarski. 2009. Colonization by *Dreissena* of Great Lakes coastal ecosystems: How suitable are wetlands? *Freshwat. Biol.* 54: 2290–2299.

Nichols, S.J. and J. Amberg. 1999. Co-existence of zebra mussels and freshwater unionids: Population dynamics of *Leptodea fragilis* in a coastal wetland infested with zebra mussels. *Can. J. Zool.* 77: 423–432.

Nichols, S.J. and D.A. Wilcox. 1997. Burrowing saves Lake Erie clams. *Nature* 389: 921.

Pinheiro, J.C. and D.M. Bates. 2000. *Mixed-Effects Models in S and S-PLUS*. New York: Springer.

Peyer, S.M., A.J. McCarthy, and C.E. Lee. 2009. Zebra mussels anchor byssal threads faster and tighter than quagga mussels in flow. *J. Exp. Biol.* 212: 2027–2036.

Pollux, B., D. Minchin, G. Van der Velde et al. 2003. Zebra mussels in Ireland (*Dreissena polymorpha*), AFLP-fingerprinting and boat traffic both indicate an origin from Britain. *Freshwat. Biol.* 48: 1127–1139.

Ponyi, J.E. 1992. The distribution and biomass of Unionidae (Mollusca, Bivalvia), and the production of *Unio tumidus* Retzius in Lake Balaton (Hungary). *Arch. Hydrobiol.* 125: 245–251.

R Development Core Team. 2010. *R: A Language and Environment for Statistical Computing*. Vienna, Austria: R Foundation for Statistical Computing. ISBN 3-900051-07-0. http://www.R-project.org.

Ramcharan, C.W., D.K. Padilla, and S.I. Dodson. 1992. Models to predict potential occurrence and density of the zebra mussel, *Dreissena polymorpha*. *Can. J. Fish. Aquat. Sci.* 49: 2611–2620.

Ricciardi, A. 2003. Predicting the impacts of an introduced species from its invasion history: An empirical approach applied to zebra mussel invasions. *Freshwat. Biol.* 48: 972–981.

Ricciardi, A., R.J. Neves, and J.B. Rasmussen. 1998. Impending extinctions of North American freshwater mussel (Unionoida) following the zebra mussel (*Dreissena polymorpha*) invasion. *J. Anim. Ecol.* 67: 613–619.

Ricciardi, A., F.G. Whoriskey, and J.B. Rasmussen. 1995. Predicting the intensity and impact of *Dreissena* infestation on native unionid bivalves from *Dreissena* field density. *Can. J. Fish. Aquat. Sci.* 52: 1449–1461.

Riccardi, A., F.G. Whoriskey, and J.B. Rasmussen. 1996. Impact of the *Dreissena* invasion on native unionid bivalves in the upper St. Lawrence River. *Can. J. Fish. Aquat. Sci.* 53: 1434–1444.

Richter, B.D., D.P. Braun, M.A. Mendelson, and L.L. Master. 1997. Threats to imperiled freshwater fauna. *Conserv. Biol.* 11: 1081–1093.

Ross, E.D. 1984. Studies on the biology of freshwater mussels (Lamellibranchia: Unionacea) in Ireland. MSc dissertation. Galway, Ireland: National University of Ireland.

Ross, E.D. 1988. The reproductive biology of freshwater mussels in Ireland, with observations on their distribution and demography. PhD dissertation. Galway, Ireland: National University of Ireland.

Schloesser, D.W., W.P. Kovalak, G.D. Longton, K.L. Ohnesorg, and R.D. Smithee. 1998. Impact of zebra and quagga mussels (*Dreissena* spp.) on freshwater unionids (Bivalvia: Unionidae) in the Detroit River of the Great Lakes. *Am. Midl. Nat.* 140: 299–313.

Schloesser, D.W. and T.F. Nalepa. 1994. Dramatic decline of unionid bivalves in offshore waters of western Lake Erie after infestation of the zebra mussel, *Dreissena polymorpha*. *Can. J. Fish. Aquat. Sci.* 51: 2234–2242.

Schloesser, D.W., T.F. Nalepa, and G.L. Mackie. 1996. Zebra mussel infestation of unionid bivalves (Unionidae) in North America. *Am. Zool.* 36: 300–301.

Sebestyén, O. 1937. Colonization of two new fauna-elements of Pontus-origin (*Dreissena polymorpha* Pall. and *Corophium curvispinum* G. O. Sars forma devium Wundsch) in Lake Balaton. *Verh. Internat. Verein. Limnol.* 8: 169–181.

Stańczykowska, A. and K. Lewandowski. 1993. Thirty years of studies of *Dreissena polymorpha* ecology in Mazurian Lakes of Northeastern Poland. In *Zebra Mussels: Biology, Impacts, and Control*, eds. T.F. Nalepa and D.W. Schloesser, pp. 3–33. Boca Raton, FL: CRC Press.

Stein, B.A. and S.R. Flack. 1997. *1997 Species Report Card: The State of U.S. Plants and Animals*. Arlington, TX: The Nature Conservancy.

Strayer, D.L. 1999. Effects of alien species on freshwater mollusks in North America. *J. N. Am. Benthol. Soc.* 18: 74–98.

Strayer, D.L., N. Cid, and H.M. Malcom. 2011. Long-term changes in a population of an invasive bivalve and its effects. *Oecologia* 165: 1063–1072.

Strayer, D.L., J.A. Downing, H.R. Haag et al. 2004. Changing perspectives on pearly mussels, North America's most imperiled animals. *BioScience* 54: 429–439.

Strayer, D.L. and D. Dudgeon. 2010. Freshwater biodiversity conservation: Recent progress and future challenges. *J. N. Am. Benthol. Soc.* 29: 344–358.

Strayer, D.L. and H.M. Malcom. 2006. Long-term demography of a zebra mussel (*Dreissena polymorpha*) population. *Freshwat. Biol.* 51: 117–130.

Strayer, D.L. and H.M. Malcom. 2007. Effects of zebra mussels (*Dreissena polymorpha*) on native bivalves: The beginning of the end or the end of the beginning? *J. N. Am. Benthol. Soc.* 26: 111–122.

Strayer, D.L. and L.C. Smith. 1996. Relationships between zebra mussels (*Dreissena polymorpha*) and unionid clams during the early stages of the zebra mussel invasion of the Hudson River. *Freshwat. Biol.* 36: 771–779.

Sullivan, M., F. Lucy, and D. Minchin. 2010. The association between zebra mussels and aquatic plants in the Shannon River system in Ireland. In *The Zebra Mussel in Europe*, G. Van der Velde, S. Rajagopal, and A. Bij De Vaate, eds., pp. 101–112. Leiden, The Netherlands: Backhuys Publishers.

Tucker, J.K. 1994. Colonization of unionid bivalves by the zebra mussel, *Dreissena polymorpha*, in Pool 26 of the Mississippi River. *J. Freshwat. Ecol.* 9: 129–134.

Vaughn, C.C., K.B. Gido, and D.E. Spooner. 2004. Ecosystem processes performed by unionid mussels in stream mesocosms: Species roles and effects of abundance. *Hydrobiologia* 527: 35–47.

Vaughn, C.C. and C.C. Hakenkamp. 2001. The functional role of burrowing bivalves in freshwater ecosystems. *Freshwat. Biol.* 46: 1431–1446.

Vaughn, C.C., S.J. Nichols, and D.E. Spooner. 2009. Community and foodweb ecology of freshwater mussels. *J. N. Am. Benthol. Soc.* 27: 409–423.

Vaughn, C.C. and D.E. Spooner. 2009. Unionid mussels influence macroinvertebrate assemblage structure in streams. *J. N. Am. Benthol. Soc.* 25: 691–700.

Vorobiev, V.P. 1949. *The Benthos of the Azov Sea*. Simferopol, Ukraine: Krymizdat Press.

Williams, J.D., M.L. Warren Jr., K.S. Cummings, J.L. Harris, and R.J. Neves. 1993. Conservation status of freshwater mussels of the United States and Canada. *Fisheries (Bethesda)* 18: 6–22.

Wood, S.N. 2006. *Generalized Additive Models: An Introduction with R*. Boca Raton, FL: Chapman & Hall/CRC.

Zanatta, D.T., S.J. Fraley, and R.W. Murphy. 2007. Population structure and mantle display polymorphisms in the wavy-rayed lampmussel, *Lampsilis fasciola* (Bivalvia: Unionidae). *Can. J. Zool.* 85: 1169–1181.

Zanatta, D.T., G.L. Mackie, J.L. Metcalfe-Smith, and D.A. Woolnough. 2002. A refuge for native freshwater mussels (Bivalvia: Unionidae) from impacts of the exotic zebra mussel (*Dreissena polymorpha*) in Lake St. Clair. *J. Great Lakes Res.* 28: 479–489.

Zanatta, D.T. and C.C. Wilson. 2011. Testing congruency of geographic and genetic population structure for a freshwater mussel (Bivalvia: Unionoida) and its host fish. *Biol. J. Linnean Soc.* 102: 669–685.

Zar, J.H. 1996. *Biostatistical Analysis*, 3rd Edn. Princeton, NJ: Prentice Hall.

Zhulidov, A.V., A.V. Kozhara, G.H. Scherbina et al. 2010. Invasion history, distribution, and relative abundances of *Dreissena bugensis* in the old world: A synthesis of data. *Biol. Invasions* 12: 1923–1940.

Zuur, A.F., E.N. Ieno, N. Walker, A.A. Saveliev, and G.M. Smith. 2009. *Mixed Effects Models and Extensions in Ecology with R*. New York: Springer.

CHAPTER 41

Impacts of Dreissenid Mussels on the Distribution and Abundance of Diving Ducks on Lake St. Clair

David R. Luukkonen, Ernest N. Kafcas, Brendan T. Shirkey, and Scott R. Winterstein

CONTENTS

Abstract 647
Introduction 648
Methods 648
Results 650
    Abundances 650
        Canvasback 650
        Redhead 650
        Scaup 651
    Spatial Distributions 653
        Canvasback 653
        Redhead 653
        Scaup 653
Discussion 653
Summary 659
Acknowledgments 659
References 659

ABSTRACT

Dreissenid mussels (*Dreissena polymorpha* and *Dreissena rostriformis bugensis*) are believed to have impacted use of Lake St. Clair by diving ducks during fall migrations. We characterized fall distributions and abundances of lesser scaup (*Aythya affinis*), greater scaup (*Aythya marila*), canvasback (*Aythya valisineria*), and redhead (*Aythya americana*) ducks in U.S. waters of Lake St. Clair via annual fixed-winged, aerial surveys. We evaluated changes in diving-duck distributions and abundances before (1983–1988) and after (1989–1994, 1995–2000, and 2001–2008) dreissenid mussels became abundant in the lake. Overall, the use of Lake St. Clair during fall migrations increased from 1.1 million use-days before dreissenid establishment to 2.1 million use-days after. Although abundance of diving ducks showed dramatic annual variation, canvasback and scaup use of Lake St. Clair increased after colonization by dreissenid mussels, whereas redhead use of the lake remained relatively unchanged. During this study, continental breeding populations of scaup were in decline and continental populations of canvasbacks were stable, suggesting that Lake St. Clair became a more desirable fall-stopover area that supported proportionately more canvasback and scaup use. Impacts on diving ducks were likely caused by changes in available foods as dreissenid mussels became a new and major prey item for scaup, while canvasbacks likely responded to increased submerged aquatic macrophyte foods associated with greater water clarity attributed to mussel colonization. We observed decreased use of shallow waters (<2 m) and increased use of intermediate water depths (2–6 m) by canvasback and scaup after colonization of the lake by dreissenids. In contrast, redhead use of shallow waters remained high throughout the study. Redhead preference for shallower water may be attributed to their herbivorous food habits and potentially their inability to efficiently forage on submerged aquatic macrophytes made available at intermediate depths. Diving ducks often feed and rest in large groups in fall and are vulnerable to disturbance by boats that complicated interpretation of interactions between dreissenid mussels, food availability, and distributions and abundances of diving ducks on Lake St. Clair.

INTRODUCTION

Lake St. Clair was the early focal point for invasion of zebra mussels (*Dreissena polymorpha*) in North America in the 1980s (Herbert et al. 1989, Carlton 2008), and dreissenid populations have been present in this lake at varying abundances ever since (Nalepa et al. 1996, Hunter and Simons 2004). Lake St. Clair has historically been an important fall concentration area for waterfowl (Bookhout et al. 1989, Prince et al. 1992), especially diving ducks such as lesser scaup (*Aythya affinis*), greater scaup (*Aythya marila*), canvasback (*Aythya valisineria*), and redhead (*Aythya americana*). During fall migration, these ducks rest and acquire foods from Lake St. Clair as one of many stopovers in route from breeding to wintering areas (Bellrose 1976).

Nutrients acquired by diving ducks during migration are needed to sustain flight and for survival and reproduction (Takekawa 1987), and diving ducks began to feed on dreissenids soon after these mussels became established in the Great Lakes (Wormington and Leach 1992, Hamilton and Ankney 1994, Custer and Custer 1996). Although diving ducks that utilize the Great Lakes are generally omnivorous, lesser scaup and greater scaup typically have diets higher in animal foods compared to more herbivorous diets of canvasback and redhead. Lesser and greater scaup have always preyed extensively on mollusks (e.g., snails, fingernail clams), but there was considerable seasonal and geographic variation in their diets prior to invasion of dreissenids (Austin et al. 1998, Ross et al. 2005). After colonization, dreissenids became a dominant prey item for scaup in the lower Great Lakes; however, submerged aquatic macrophytes remained a common food for diving ducks, including scaup on eastern Lake St. Clair (Custer and Custer 1996, Badzinski and Petrie 2006). Both canvasback and redhead are generally more herbivorous than scaup and foraged primarily on submerged aquatic macrophytes (Custer and Custer 1996).

Although researchers documented extensive use of dreissenid mussels by diving ducks, changes in distribution and abundance of diving ducks have not been intensively examined despite availability of long-term monitoring data obtained in the fall. We investigated changes in distribution and abundance of diving ducks from 1983 to 2008 in U.S. waters of Lake St. Clair using data collected by the Michigan Department of Natural Resources. In addition, we used a separate data set, the Mississippi Flyway Coordinated Canvasback Survey, to compare canvasback abundance in Canadian and U.S. waters of Lake St. Clair from 1974 to 2010. We hypothesized that diving ducks increased use of Lake St. Clair after dreissenid introduction. We also hypothesized that responses of diving ducks to dreissenids would be species specific because of varying food habits; scaup would likely respond immediately to dreissenids as a food source, while canvasbacks and redhead would respond to increases in distribution and abundance of submerged aquatic macrophytes resulting from decreased phytoplankton abundance and greater water clarity associated with water filtration by dreissenids (Holland 1993, Leach 1993).

Disturbance by recreational and commercial boaters can influence diving-duck distribution such that birds can be displaced from preferred feeding and resting areas (Martz et al. 1976, Knapton et al. 2000). This yielded another hypothesis that distributions and abundances of diving ducks on Lake St. Clair would shift to areas with limited disturbance (i.e., boating activity).

METHODS

Annual canvasback surveys on U.S. and Canadian waters of Lake St. Clair began in 1974 as part of the Coordinated Canvasback Survey on major migration areas in the Mississippi Flyway (Cordts 2010). These surveys were completed each year between 1974 and 2010, except 1980 when no data were collected (Cordts 2010). Although the target completion date for the Coordinated Canvasback Survey was November 5, unfavorable weather sometimes resulted in surveys completed later in November. In U.S. waters, Coordinated Canvasback Surveys were expanded in 1983 to include all other diving duck species and additional survey dates (October through December). The same survey methodologies were used for Coordinated Canvasback Surveys and expanded diving-duck surveys, and after survey expansion in 1983, a total of 99 surveys were completed.

Expanded diving-duck surveys consisted of three to five flights of the U.S. portion of Lake St. Clair (Figure 41.1), except in 1983 and 2003 when only one flight was completed. Surveys were flown in single-engine, fixed-winged aircraft with a pilot and two observers. Flights did not occur on rainy or windy days so duck distributions were not representative of days with unfavorable weather. Survey data were collected along east to west transects spaced every 3 km, at a flight speed of approximately 150 km/h and at 100 m above water level. Deviations from flight lines occurred when observers needed to better estimate flock size and species composition. Recorded data consisted of locations, size, and species composition of flocks on printed maps of the study area. Greater and lesser scaup cannot be differentiated in aerial surveys so these species were combined and collectively referred to as "scaup." The number and locations of boats were also recorded on study area maps. Fall migrations on Lake St. Clair were defined as October 22–December 2.

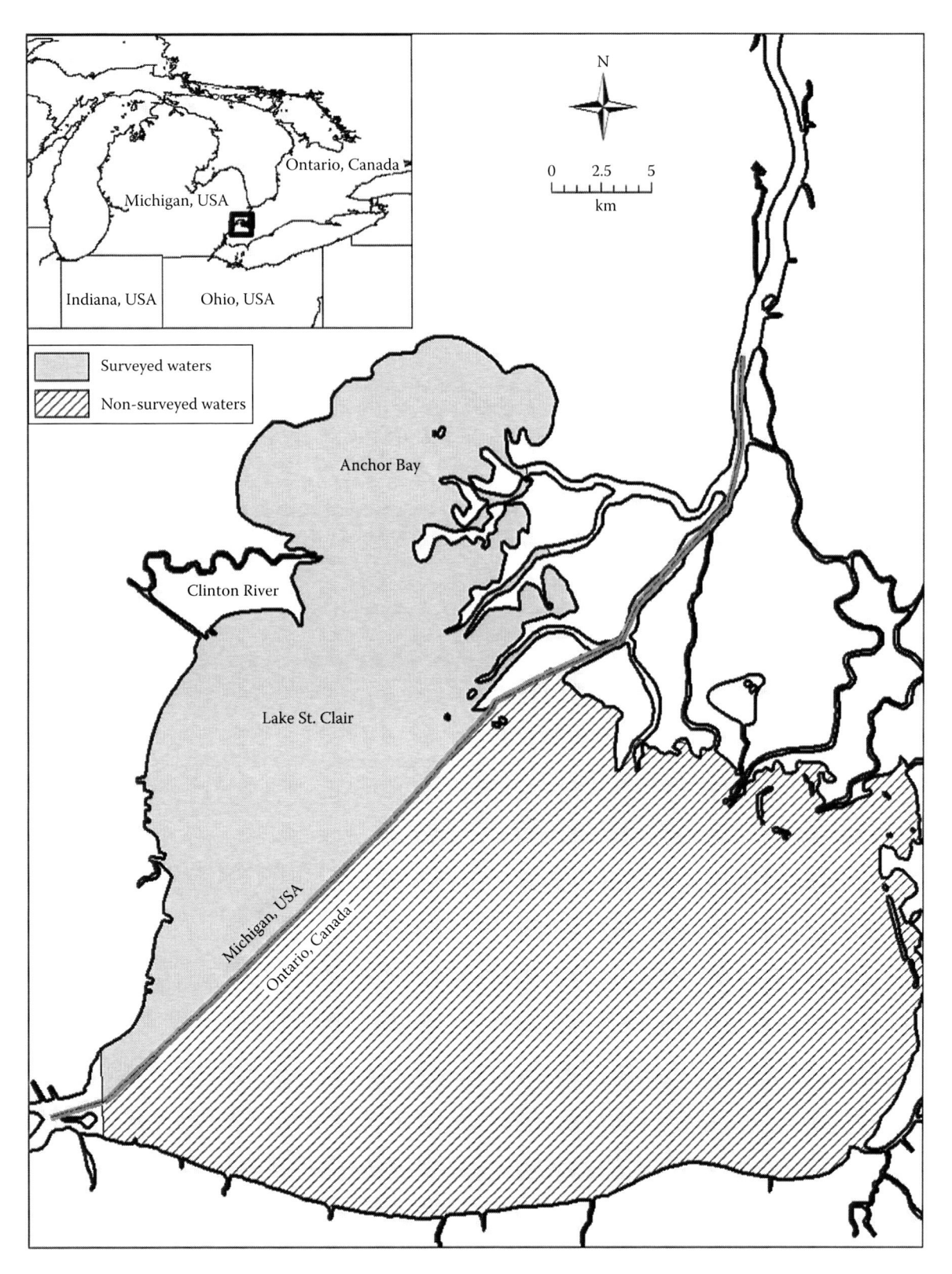

Figure 41.1 Lake St. Clair showing U.S. portion surveyed for diving ducks during fall migrations, 1983–2008.

Study area maps were electronically scanned and georectified using ArcGIS and known coordinates on the study area (ESRI 2006). Records of diving-duck flocks and boats were digitized and displayed using ArcGIS (ESRI 2006). A bathymetric GIS layer (NOAA 2011) was used to associate duck observations with water depths in four categories (<2, 2–4, 4–6, and >6 m) and thereby assess potential temporal changes in diving duck use of these water depths. Water depths of flock locations were adjusted based on deviations from the long-term lake average for the months of October and November (U.S. Army Corps of Engineers 2010). Water depths over the period of study deviated from the long-term average by less than 1 m.

We compared total duck use-days by species for the 42-day fall migration period before (1983–1988) and after (1989–2008) dreissenids became abundant. Use-days were estimated as the product of mean abundances and 42 days; variances for this parameter were determined with methods of Thompson (1992). Temporal changes in diving-duck abundance were modeled with generalized linear models and implemented in SPSS (2009). The number of individuals of each species counted during individual surveys was the response variable and time period was explanatory. The time period was divided into weekly and yearly intervals. For the former, fall migrations were divided into 6 weekly intervals, and for the latter annual surveys were grouped into three 6-year intervals (1983–1988, 1989–1994, and 1995–2000) and one 8-year interval (2001–2008). Dreissenids were first reported in Lake St. Clair in 1988 (Herbert et al. 1989), and the earliest year category (1983–1988) corresponded to the period when dreissenids were not yet abundant in the lake (Nalepa et al. 1996).

Numbers of boats in each survey were included as a covariate in our linear models to account for potential effects of disturbance on diving-duck abundance. Marginal means for effects of week and year in linear models fit to each diving-duck species were estimated with the covariate for boat counts held at the mean over all mean. A geographic information system was used to map distributions of diving ducks in the same year categories used in linear models. Distributions of diving-duck species were modeled with the kernel density estimator routine (ESRI 2006). Kernel density models were fit for each species and year interval and mapped model estimates were displayed in the spatial modeling feature of ArcGIS (ESRI 2006).

RESULTS

Abundances

Canvasback

Mean abundance (± SD) of canvasback (22,165 ± 26,383) between 1988 and 2008 in U.S. waters was highest among the diving ducks studied. Mean annual abundances were highly variable, ranging from a low of 100 canvasbacks in 2003 to a high of 45,300 canvasbacks in 1996 (Figure 41.2a). Canvasback abundances were relatively low before 1989 ($\bar{x} \leq 23,800$) and then increased rapidly with a peak count in U.S. waters in 1991 (maximum = 157,940; Figure 41.2a). After 1991, annual canvasback mean and maximum abundances exhibited dynamic peaks and troughs with an overall downward trend (Figure 41.2a).

The linear model of canvasback abundances in U.S. waters supported effects of season, year, and boat count on abundance (F-values >6.0; p-values <0.01; Figures 41.3 and 41.4). The parameter estimate for the boat effect indicated an expected 648 (±393; 95% confidence interval [CI]) fewer canvasbacks observed for each additional boat counted in the surveys. Mean seasonal abundances of canvasbacks increased after October 22–28, peaked November 19–25, and then decreased but remained relatively high in the last week of fall migration (Figure 41.3). Mean annual canvasback abundances increased after the 1983–1988 period, remained high in the 1989–2000 period, and declined in the 2001–2008 period (Figure 41.4). Estimates of fall duck use-days by canvasbacks in U.S. waters increased from 500,550 use-days per year (±202,445; 95% CI) before 1989 to 1,048,972 use-days per year (±271,342; 95% CI) from 1989 to 2008.

Despite decline in use of U.S. waters during the most recent monitoring period (2001–2008), canvasback abundances increased on Canadian waters of Lake St. Clair during the November Coordinated Canvasback Surveys ($\bar{x}$ = 109,088 canvasbacks; Cordts 2010); an increasing trend in lake-wide (U.S. and Canadian waters combined) abundances occurred after 1988 and abundance peaked in 2009 (maximum = 162,000 canvasbacks; Figure 41.5). While the percentage of canvasbacks observed within U.S. waters varied between 20% and 90% between 1989 and 2001, this percentage declined precipitously to near zero after 2001 (Figure 41.5).

Redhead

Mean (± SD) abundance of redhead (10,238 ± 9,388) was the lowest among diving ducks studied. Although redhead abundance was high in 1983, this was based on a single survey; peak abundance occurred in 1995 (maximum = 34,300 redheads; Figure 41.2b). Variation in mean annual abundances was relatively low between 1984 and 1994 (range 6,333–11,753 redheads; Figure 41.2b). Relatively large variation in annual abundances after 1994 (range 2,483–21,427 redheads) was associated with dramatic fluctuations that included three peak and trough cycles between 1995 and 2008 (Figure 41.2b).

The linear model of redhead abundance supported effects of season and boat usage (F-values >7.0; p-values <0.01; Figure 41.3), but there was weak evidence of abundance differences among year intervals (F = 2.1; p = 0.104;

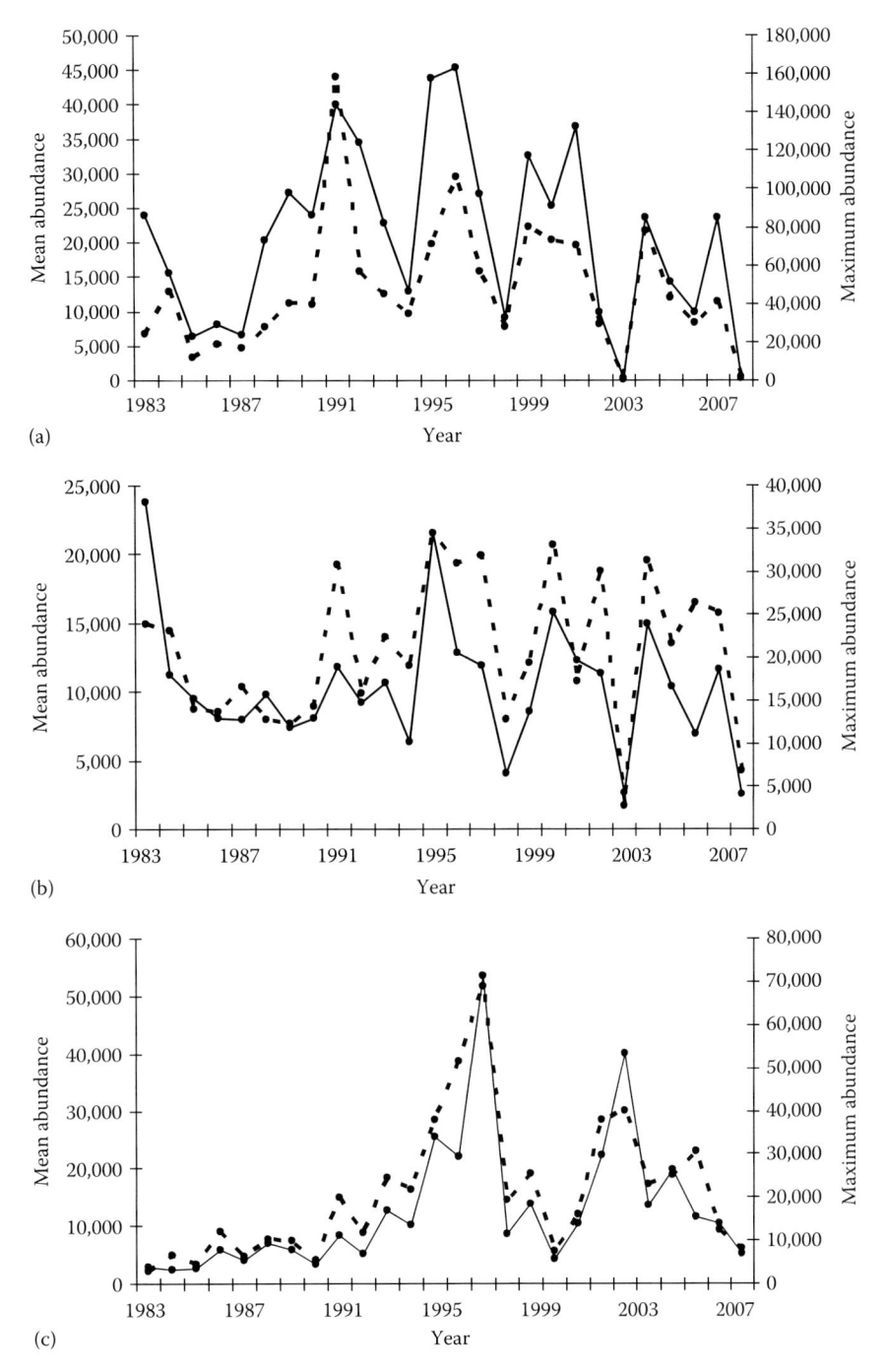

Figure 41.2 Annual mean (solid lines) and maximum (dashed lines) abundances of (a) canvasback, (b) redhead, and (c) scaup observed in aerial surveys of U.S. waters of Lake St. Clair, 1983–2008.

Figure 41.4). The parameter estimate for the boat effect indicated an expected 178 (±122; 95% CI) fewer redheads observed for each additional boat counted during the surveys. Similar to the seasonal pattern observed in canvasback abundance, mean seasonal redhead abundance increased after October 22–28, peaked in November 19–25, and then remained high but decreased in the last week of fall migration (Figure 41.3). Mean annual redhead abundance was relatively stable among year intervals compared to canvasbacks and scaup (Figure 41.4). Estimates of fall duck use-days for redheads in U.S. waters were stable with 413,015 (±113,500; 95% CI) use-days per year before 1989 and 430,575 (±95,371; 95% CI) use-days per year from 1989 to 2008.

Scaup

Mean (± SD) abundance of scaup (12,003 ± 3,208) was intermediate among diving ducks but only slightly greater than

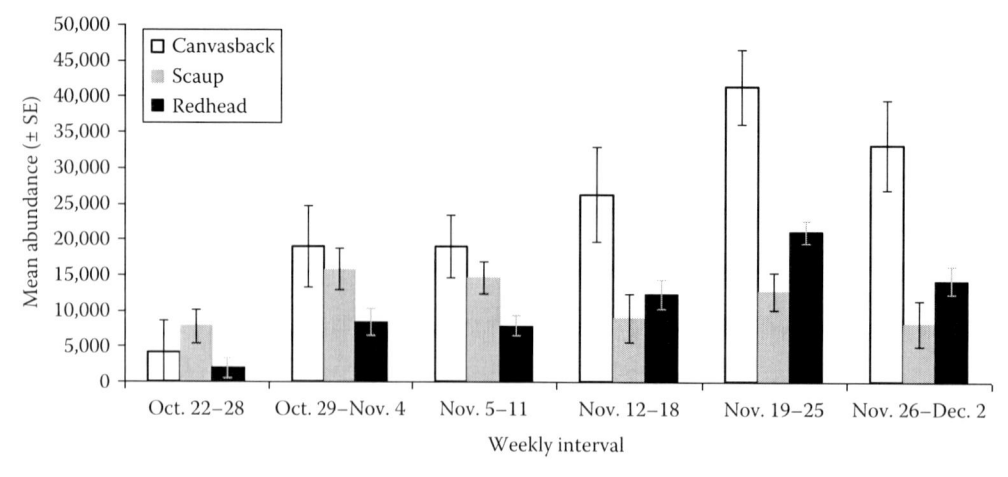

Figure 41.3 Mean (±1 SE) abundance of diving ducks observed in aerial surveys of U.S. waters of Lake St. Clair in 6 weekly intervals, 1983–2008. Estimates are predicted marginal means from species-specific linear models that include week and year intervals with numbers of boats (covariate) held at the mean over all surveys.

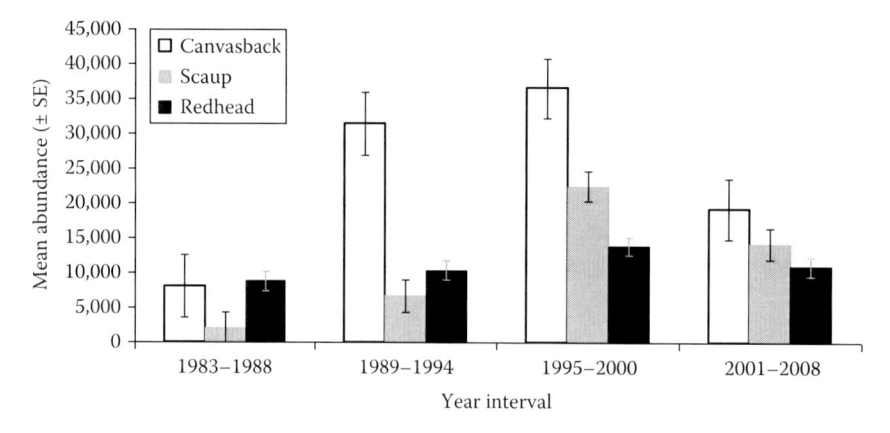

Figure 41.4 Mean (±1 SE) abundance of three diving ducks (canvasback, scaup, and redhead) observed in aerial surveys of U.S. waters of Lake St. Clair in four time periods: 1983–1988, 1989–1994, 1995–2000, and 2001–2008. Dreissenid mussels were not present or had low abundances in 1983–1988. Estimates are predicted marginal means from species-specific linear models that include week and year intervals with numbers of boats (covariate) held at the mean over all surveys.

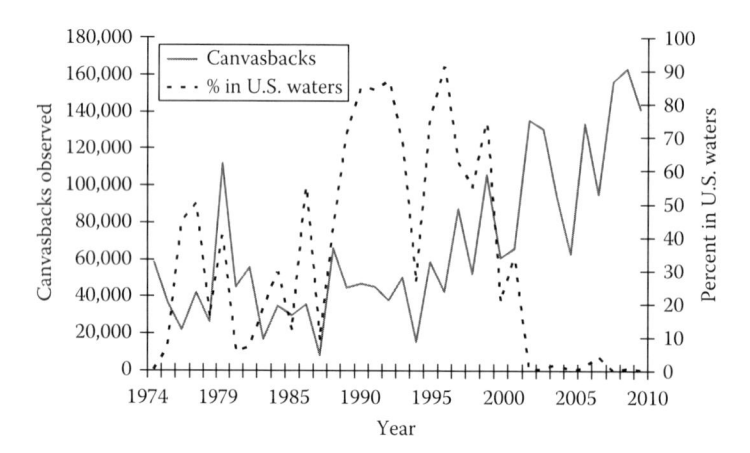

Figure 41.5 Number of canvasbacks observed in aerial surveys of Lake St. Clair (U.S. and Canadian waters) and percent of birds counted in U.S. waters during Mississippi Flyway Coordinated Canvasback Surveys, 1974–2010 (no survey in 1980). (Data from Cordts, S., *Coordinated November Canvasback Inventory*, Minnesota Department of Natural Resources, Bemidji, MN, 2010.)

mean redhead abundance. Mean annual abundances of scaup were variable like other diving ducks (range 2,280–53,450) but exhibited a different temporal trend (Figure 41.2c). Mean abundance was relatively low and stable between 1983 and 1992 (range 2,280–8,110) and then increased rapidly with a peak abundance in 1997 (maximum = 68,600; Figure 41.2c). The peak in 1997 was followed by a decline to a low mean abundance of 4,083 in 2000 that was near mean abundances found in the first 10 survey years. Mean annual scaup abundance between 2001 and 2008 was characterized by another rapid increase with a peak of 40,000 in 2003, followed by a decline and a low of 4,966 in 2008.

The linear model of scaup abundance supported effects of year and boat usage (F-values > 8.0; p-values < 0.01; Figure 41.4), but there was weak evidence of seasonal effects (F = 1.5; p = 0.197; Figure 41.3). The parameter estimate for boat effect indicated an expected 312 (±202; 95% CI) fewer scaup observed for each additional boat counted. Although seasonal abundance of scaup peaked earlier than for other ducks (October 29–November 11), there was less variation in abundance among week intervals compared to other diving duck species (Figure 41.3). Mean annual abundance of scaup increased after 1983–1988, reached a peak in 1995–2000, and then declined in 2001–2008 (Figure 41.4). Estimates of fall duck use-days for scaup in U.S. waters increased from 176,975 (±54,235; 95% CI) use-days per year before 1989 to 599,321(±133,381; 95% CI) use-days per year from 1989 to 2008.

Spatial Distributions

Canvasback

Canvasback distribution in U.S. waters of Lake St. Clair varied throughout the survey period with relatively broader distribution when canvasbacks were most abundant (1989–1994 and 1995–2000; Figure 41.6a through d). Canvasbacks were widely distributed and abundant on the west side of Anchor Bay, north of the Clinton River, in 1983–1988, 1989–1994, and 1995–2000 but were relatively less abundant in this area in 2001–2008 (Figure 41.6a through d). After 1988, canvasbacks were found farther from shore, were relatively more abundant, and were more broadly distributed in west-central Lake St. Clair south of the Clinton River (Figure 41.6a through d). A relatively high percentage (79%) of canvasback observations were in water 2–6 m deep, but temporal changes in distribution resulted in an increased percentage in water 4–6 m deep and a decreased percentage in water <2 m deep (Figure 41.7a).

Redhead

Overall, redhead distribution remained relatively uniform over the survey period (Figure 41.8a through d). In general, abundances were greatest on the west side of Anchor Bay in all yearly time periods. Redheads also used west-central portions of the lake south of the Clinton River and there was a minor shift away from shore after 2000 (Figure 41.8a through d). In general, redheads made greatest use of shallow waters among diving ducks studied with 91% of redhead observations in water <4 m deep (Figure 41.7b). Water-depth use was relatively stable over time, but the greatest percentage (50%) of redheads observed in water <2 m deep was from 1989 to 1994 (Figure 41.7b).

Scaup

Like canvasbacks, scaup became more broadly distributed when abundance increased after 1988 (Figure 41.9a through d). Scaup used Anchor Bay throughout all survey periods and abundance was relatively high in west-central portions of the lake when scaup were most abundant (1995–2000; Figure 41.9c). Over all yearly periods, about 74% of scaup observations were in water 2–6 m deep, and percentages of scaup observations in these depths increased, while the percentage in water <2 m declined (Figure 41.7c). After 2000, scaup persisted in relatively shallow waters of Anchor Bay (Figures 41.7c and 41.9d), resulting in a higher percentage of scaup observations (20%) in water <2 m deep compared to the percentage of canvasback observations (6%).

DISCUSSION

Distributions and abundances of diving ducks in fall migration can vary greatly, and we observed such seasonal and yearly variation on Lake St. Clair. Mean annual abundances of *Aythya* spp. (canvasback, scaup, and redhead ducks combined) in our study area over the entire study period increased from a low of approximately 14,000 birds in the week of October 22–28 to a high of about 75,000 birds in the week of November 19–25. Abundance of canvasbacks and redheads showed distinct seasonal peaks in November, while abundance of scaup was somewhat bimodal, with an earlier peak in abundance as compared to other species, and then a later increase coincident with peak abundance of canvasbacks and redheads. Seasonal changes in scaup abundances were complicated by inability to distinguish between lesser and greater scaup in aerial surveys, and differential migration timing between species could have contributed to the observed seasonal pattern in abundances.

Aggregate abundance of *Aythya* spp. was lowest between 1983 and 1988, increasing to a maximum between 1995 and 2000; abundances decreased in later years of our study but still remained higher than in the earliest survey period. Importance of our study area to diving ducks during fall migration is evident by estimates of duck use-days with about 1.1 million *Aythya* spp. use-days per year before and about 2.1 million use-days per year after colonization of Lake St. Clair by dreissenids. These data also suggest

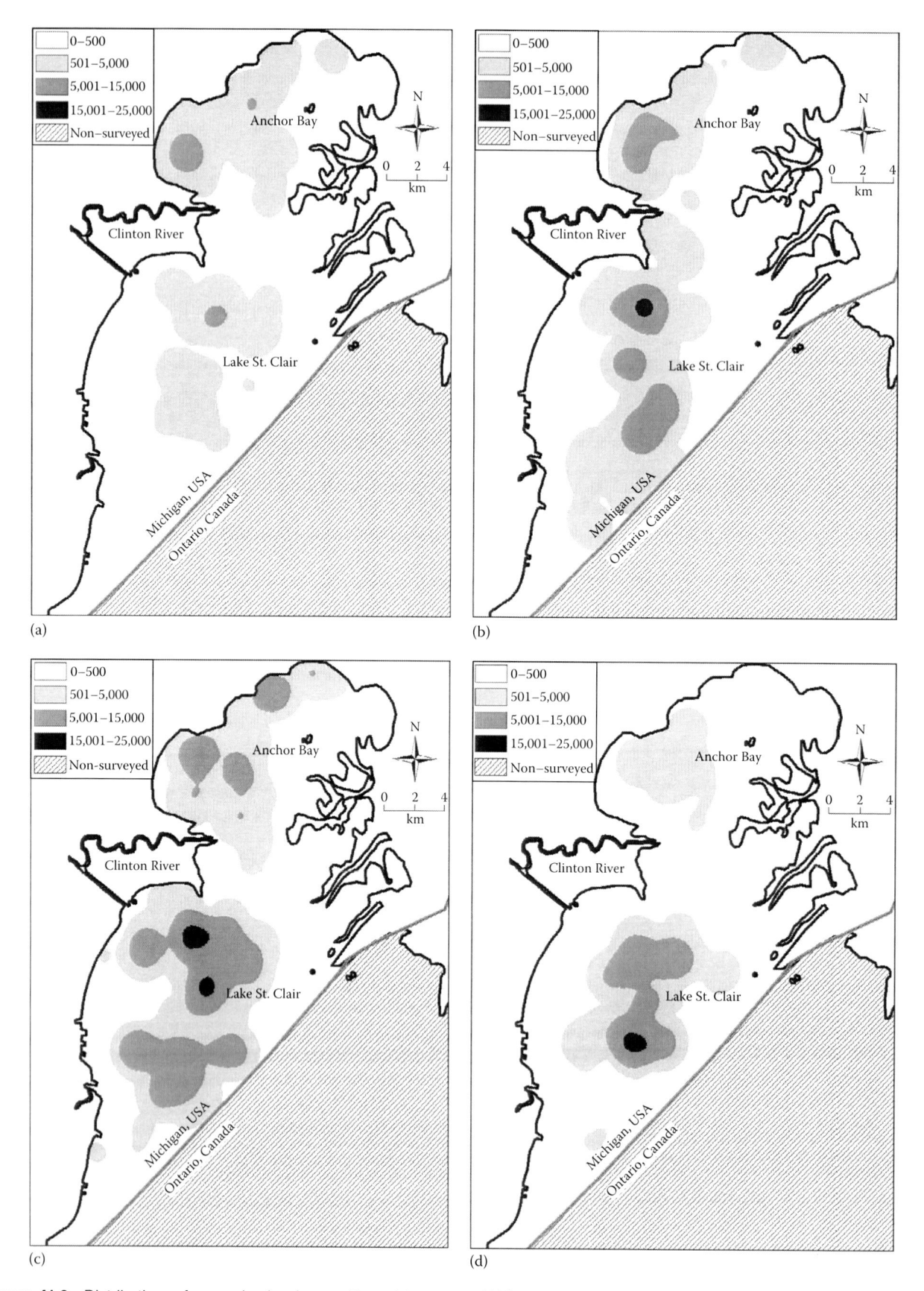

Figure 41.6 Distributions of canvasbacks observed in aerial surveys of U.S. waters of Lake St. Clair in four time periods: (a) 1983–1988, (b) 1989–1994, (c) 1995–2000, and (d) 2001–2008. Abundances (canvasbacks/km^2) interpolated from kernel density models.

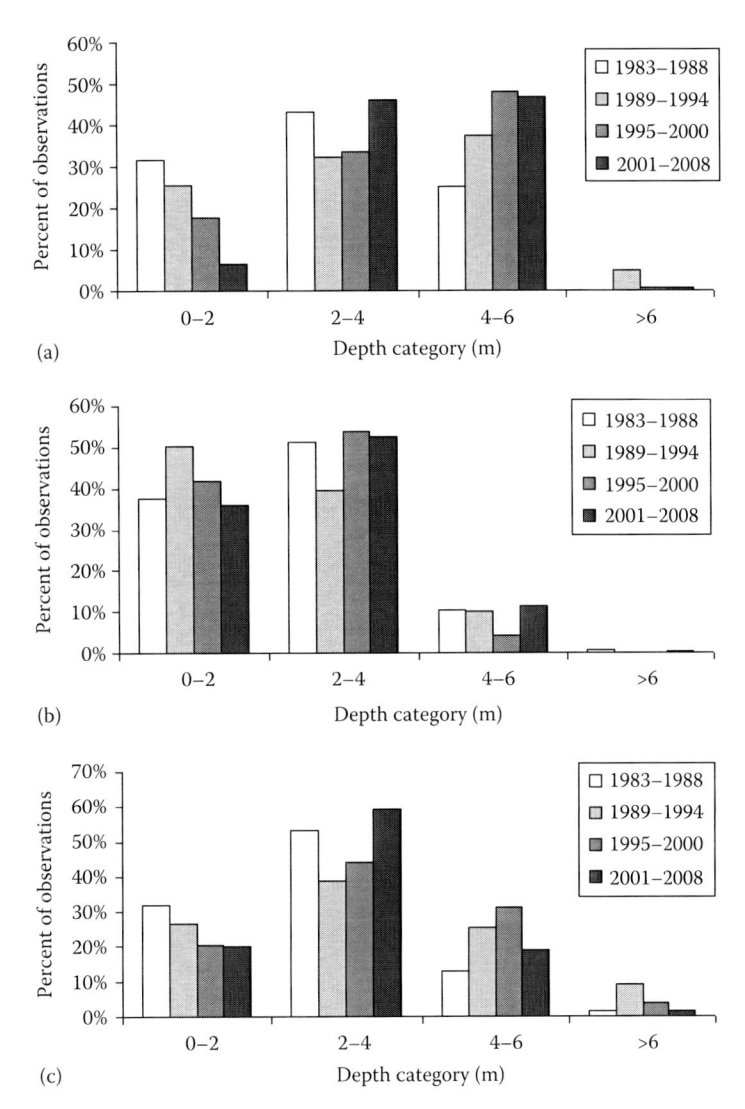

Figure 41.7 Percent of (a) canvasback, (b) redhead, and (c) scaup observed in four water-depth intervals in U.S. waters of Lake St. Clair in 1983–1988, 1989–1994, 1995–2000, and 2001–2008.

relatively large changes in fall abundance and distribution of diving ducks, especially canvasbacks and scaup.

Changes in local abundance of diving ducks in migration areas like Lake St. Clair may be driven by factors operating at multiple spatial scales. At the broadest scale, changes in continental abundance and reproductive success influence abundance of ducks in the fall migration flight. Distributions and reproductive successes of diving ducks at different breeding locales can vary and affect numbers of birds using alternative fall migration routes (Bellrose 1976, Serie et al. 1983).

In major breeding areas of the United States and Canada, breeding canvasback abundance over our study varied from 376,000 to 865,000 birds without a discernible trend (U.S. Fish and Wildlife Service 2010). In the 1995–2000 period, high canvasback abundance on Lake St. Clair corresponded to relatively high breeding abundance of continental canvasbacks ($\bar{x}$ = 736,100). However, the initial increase in canvasback use on Lake St. Clair began in the 1989–1994 time period when canvasback breeding abundance was below the long-term average ($\bar{x}$ = 497,850). Also, counts of the entire lake (U.S. and Canadian waters combined) obtained by the Coordinated Canvasback Survey showed that lakewide use continued to increase through 2010 despite less use in U.S. waters. Thus, canvasback use of Lake St. Clair in fall migration nearly doubled after invasion of dreissenids. Furthermore, increased canvasback use of Lake St. Clair was not closely tied to changes in continental abundance, so changes on Lake St. Clair appeared to have made the lake a more desirable fall migration stopover area that supported proportionately more use.

Abundances of breeding redheads declined throughout the United States and Canada from 1983 to 1988, remained below the long-term average through 1993, increased dramatically until 2001, was near average in 2002–2005, and then rebounded to near-record highs from 2006 to 2008

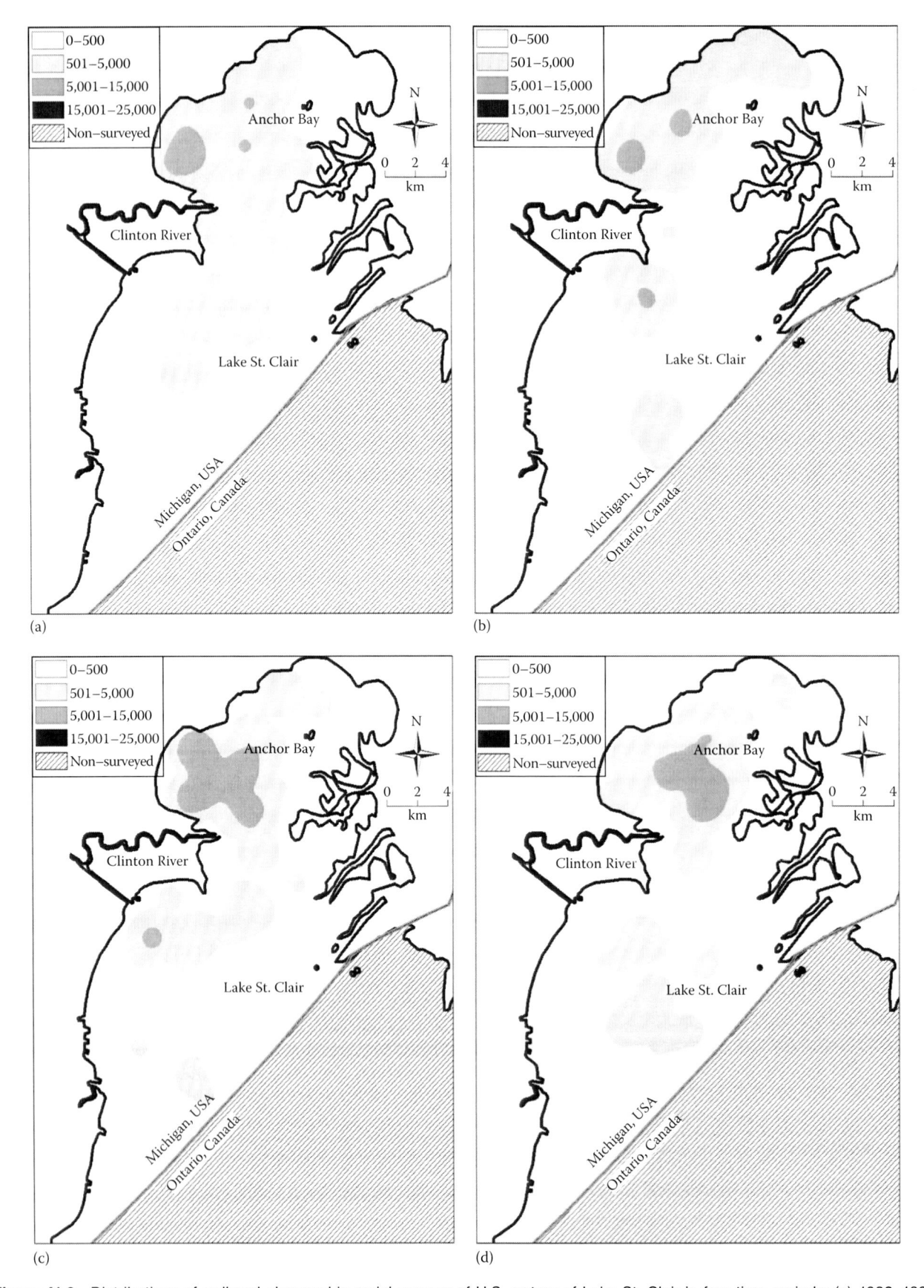

Figure 41.8 Distributions of redhead observed in aerial surveys of U.S. waters of Lake St. Clair in four time periods: (a) 1983–1988, (b) 1989–1994, (c) 1995–2000, and (d) 2001–2008. Abundances (redhead/km^2) interpolated from kernel density models.

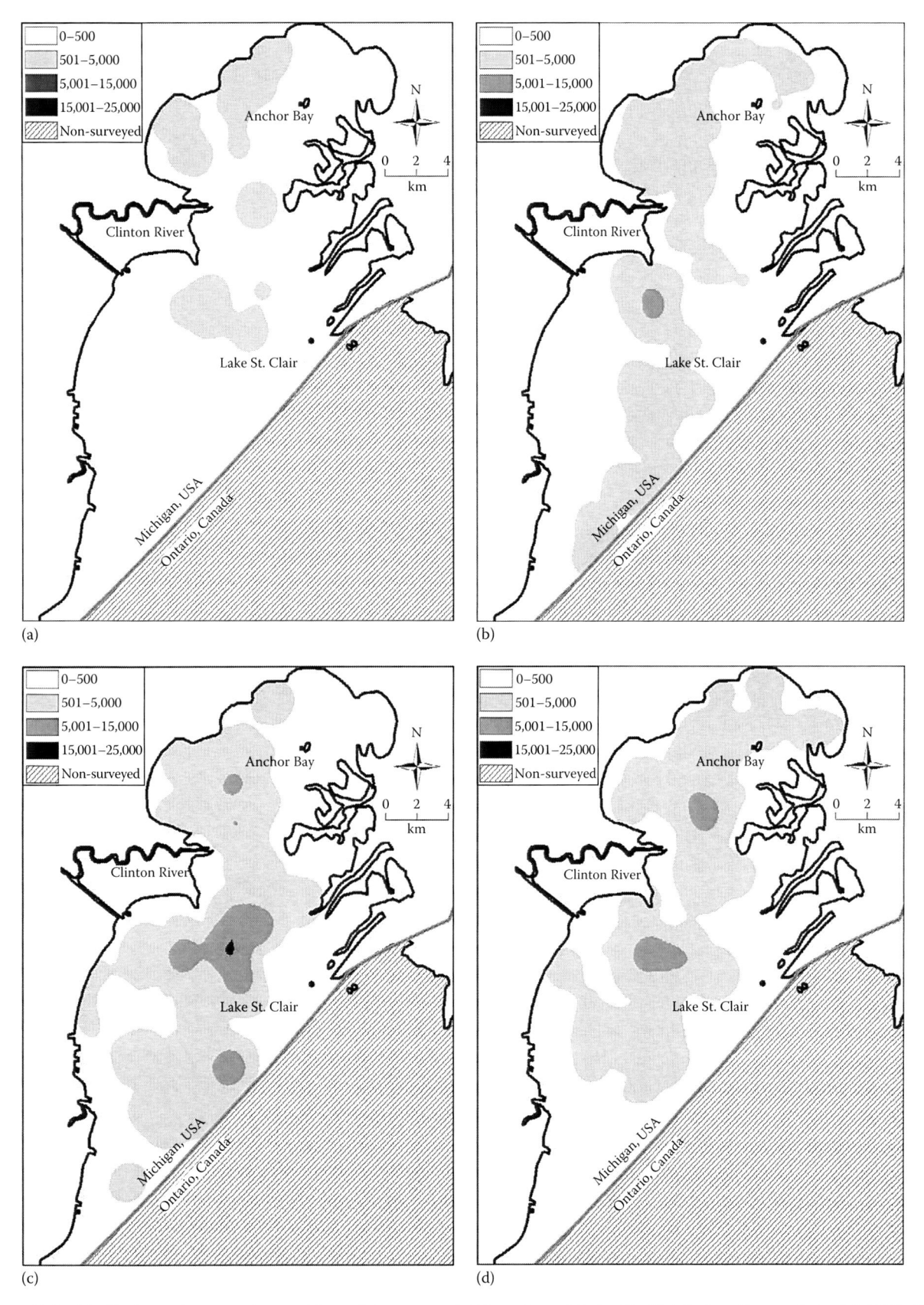

Figure 41.9 Distributions of scaup observed in aerial surveys of U.S. waters of Lake St. Clair in four time periods: (a) 1983–1988, (b) 1989–1994, (c) 1995–2000, and (d) 2001–2008. Abundances (scaup/km^2) interpolated from kernel density models.

(U.S. Fish and Wildlife Service 2010). Although peak redhead abundance in our study area was associated with a period of high continental abundance of breeding redheads (1995–2000), redhead abundance on our study area did not reflect the high continental abundance observed from 2006 to 2008. The cause for low correlation between continental breeding abundance and abundance on Lake St. Clair from 2006 to 2008 is unknown, and dreissenids seem to have had little effect on the distribution and abundance of redheads.

Unlike continental abundances of canvasback and redhead, the continental abundance of scaup (lesser and greater scaup combined) experienced a long-term decline from the mid-1980s through 2008 (U.S. Fish and Wildlife Service 2010). Although we have no data on relative abundance of the two scaup species on Lake St. Clair, the continental decline is typically attributed to reductions in lesser scaup numbers. Increased scaup use of Lake St. Clair in a period when continental abundance declined suggests that proportionately more scaup were using this area compared to pre-dreissenid colonization.

Ultimately, diving duck distributions in fall migration are driven by needs to acquire sufficient nutrients for migration and survival. Our observations of ducks did not contrast specific behaviors (e.g., feeding versus resting) so we could not determine relative importance of different areas of Lake St. Clair as foraging habitats. Also, our surveys were diurnal and diving ducks are known to feed both day and night (Takekawa 1987, Custer et al. 1996). However, we do believe increased use of our study area after colonization by dreissenids reflects a response of canvasbacks and scaup to changes in food availability. Duration of stopovers of canvasbacks in fall migration areas has been inversely related to lipid reserves at time of arrival (Serie and Sharp 1989), and the dramatic increase in canvasback use-days was likely supported by increased food availability.

The role of dreissenids in supporting increased use of our study area by scaup is relatively direct because these mussels became a major prey item for scaup shortly after colonization. Although mussels have not been a primary food item for canvasbacks on Lake St. Clair, it is likely that dreissenids indirectly changed availability of canvasback foods. Dreissenid mussels can increase water clarity (Leach 1993, Custer and Custer 1996), and aquatic macrophytes like wild celery (*Vallisneria americana*) that are important foods for canvasbacks have responded positively to greater water clarity in many areas of Lake St. Clair (Ernest N. Kafcas personal observation). Greater water clarity may have increased available macrophyte foods by expanding macrophyte distributions into deeper waters. Hence, potential foraging areas would increase, and search time needed for ducks to find novel patches of submerged foods would be reduced. Diving ducks were once thought to segregate habitats among species by depth (Siegfried 1976), but more recent studies suggest that foraging is more general and constrained by food abundance and energetic costs of food acquisition (Torrence and Butler 2006). Although there are higher energetic costs for diving ducks to obtain food in deep water, high densities of food can reduce effort and make deepwater foraging a beneficial strategy (Lovvorn 1994). Experimental manipulation of food webs in lacustrine wetlands to increase water clarity has subsequently increased macroinvertebrate and submerged plant foods and ultimately resulted in increased use by diving ducks during migration (Hanson and Butler 1994).

Changes in the ecosystem of Lake St. Clair that promoted increased use by canvasbacks and scaup appeared to have little effect on use by redheads. This might be explained by different feeding habits or food preferences among species. Redheads tend to forage in relatively shallow water and are more herbivorous than other diving ducks (Bellrose 1976). We found that canvasback and scaup decreased use of shallow waters (<2 m) and increased use of intermediate water depths (2–6 m) after colonization of Lake St. Clair by dreissenids; however, redheads remained more common in shallow waters. It may be that redheads could not efficiently use submerged aquatic macrophytes made available by greater water clarity at intermediate depths or that greater water clarity did not promote submerged aquatic macrophyte species that were as attractive to redheads as they were to canvasbacks.

Diving ducks are vulnerable to disturbance from boats (Knapton et al. 2000), and the U.S. portion of Lake St. Clair is among the most heavily used areas of the Great Lakes by recreational boaters (Snider 1999). Fishing and waterfowl hunting have been popular fall recreational activities on Lake St. Clair and diving ducks are often hunted offshore using specialized "layout" boats (Martz et al. 1976). Diving ducks may respond to boating or hunting disturbances by relocating from preferred feeding areas to areas with relatively little disturbance; diving ducks will sometimes return and feed in preferred areas when disturbance is reduced (e.g., nocturnal feeding; Thornburg 1973). Alternatively, refuges with restricted boat use have been established on some migration areas to provide diving ducks undisturbed resting or feeding areas (Kahl 1991). Canadian waterfowl hunting regulations restricted hunters to waters within 300 m of shore. In contrast, hunters in U.S. waters were not restricted to nearshore areas. This difference may have effectively created a refuge for diving ducks in offshore Canadian waters. We hypothesize that increased food availability at appropriate depths reduced constraints on canvasbacks such that disturbance acted to redistribute birds to local food patches where they were relatively undisturbed by boats or hunters. Our abundance models and results of the Coordinated Canvasback Survey suggest that this interaction between food availability and disturbance may have affected distributions of canvasbacks, scaup, and redheads within U.S. waters and redistributed canvasbacks to Canadian waters of Lake St. Clair.

SUMMARY

Our annual surveys indicated that dreissenid mussels impacted fall migration of diving ducks on Lake St. Clair. Although there was high variation in abundance, canvasbacks and scaup increased their use of Lake St. Clair after colonization by dreissenids, but use by redheads remained relatively stable. Dreissenid impacts on abundance and distribution of scaup were direct because dreissenids became a new and abundant food source for this largely carnivorous species of diving duck. Dreissenid impacts on abundance and distribution of canvasbacks were likely a result of increasing water clarity and subsequent increases in submerged aquatic macrophytes that serve as preferred foods for this more herbivorous species. The relative lack of impact of dreissenids on redhead abundance and distribution is not well understood; however, we hypothesize that redheads were not able to efficiently use aquatic macrophytes at intermediate water depths due to foraging constraints. Alternatively, expansion of submerged aquatic macrophytes potentially resulted in increased aquatic macrophyte species that were more preferred by canvasbacks than by redheads.

Although food resources undoubtedly play a large role in the distribution and abundance of diving ducks in fall migration, disturbance also plays a key role on Lake St. Clair and other important stopover areas (Knapton et al. 2000, Kenow et al. 2003). We found negative correlations between boats counted and abundances of all three diving duck taxa. Furthermore, recent data from Coordinated Canvasback Surveys suggest that canvasbacks are being redistributed to the Canadian side of Lake St. Clair where they are relatively free from hunting disturbance in offshore waters. With increases in available food caused largely by dreissenid colonization, diving ducks can likely forage more efficiently, allowing them to spend more time in secluded offshore waters and less time feeding in more productive but also more disturbance-prone shallow portions of the lake. Future diving duck surveys covering the entire lake on the same dates would improve our understanding of the role of disturbance in affecting the distribution of diving ducks during fall migrations.

ACKNOWLEDGMENTS

This study was supported by Federal Aid in Wildlife Restoration Project W-147, Michigan Department of Natural Resources, U.S. Fish and Wildlife Service, Upper Mississippi River and Great Lakes Joint Venture, and Michigan State University. Robert Haas generously demonstrated spatial data digitizing and provided digitized maps of diving duck and boat locations for some years. We thank J. Robison for logistical support, assistance with data collection, and suggestions to improve survey methodologies. G. Soulliere and the editors of this book provided very helpful suggestions on earlier drafts of this chapter.

REFERENCES

Austin, J. E., C. M. Custer, and A. D. Afton. 1998. Lesser scaup (*Aythya affinis*). In *The Birds of North America*. A. Poole and F. Gill, eds., account 338. Philadelphia, PA: Academy of Natural Science, Washington, DC: American Ornithologists' Union.

Badzinski, S. S. and S. A. Petrie. 2006. Diets of lesser and greater scaup during autumn and spring on the lower Great Lakes. *Wildl. Soc. Bull.* 34:664–674.

Bellrose, F. C. 1976. *Ducks, Geese, and Swans of North America*. Harrisburg, PA: Stackpole Books.

Bookhout, T. A., K. E. Bednarik, and R. W. Kroll. 1989. The Great Lakes marshes. In *Habitat Management for Migrating and Wintering Waterfowl in North America*. L. M. Smith, R. L. Pederson, and R. M. Kaminski, eds., pp. 131–156. Lubbock, TX: Texas Tech University Press.

Carlton, J. T. 2008. The zebra mussel *Dreissena polymorpha* found in North America in 1986 and 1987. *J. Great Lakes Res.* 34:770–773.

Cordts, S. 2010. *Coordinated November Canvasback Inventory*. Bemidji, MN: Minnesota Department of Natural Resources.

Custer, C. M. and T. W. Custer. 1996. Food habits of diving ducks in the Great Lakes after the zebra mussel invasion. *J. Field Ornithol.* 67:86–99.

Custer, C. M., T. W. Custer, and D. W. Sparks. 1996. Radio telemetry documents 24-hour feeding activity of wintering lesser scaup. *Wilson Bull.* 108:556–566.

ESRI. 2006. ArcGIS Desktop: Release 9. Redlands, CA: Environmental Systems Research Institute.

Hamilton, D. J. and C. D. Ankney. 1994. Consumption of zebra mussels (*Dreissena polymorpha*) by diving ducks in Lakes Erie and St. Clair. *Wildfowl* 45:159–166.

Hanson, M. A. and M. G. Butler. 1994. Responses to food web manipulation in a shallow waterfowl lake. *Hydrobiologica* 279/280:457–466.

Herbert, P. D. N., B. W. Muncaster, and G. L. Mackie. 1989. Ecological and genetic studies on *Dreissena polymorpha* (Pallas): A new mollusc in the Great Lakes. *Can. J. Fish. Aquat. Sci.* 46:1587–1591.

Hunter, R. D. and K. A. Simons. 2004. Dreissenids in Lake St. Clair in 2001: Evidence for population regulation. *J. Great Lakes Res.* 30:528–537.

Holland, R. E. 1993. Changes in planktonic diatoms and water transparency in Hatchery Bay, Bass Island Area, Western Lake Erie since the establishment of the zebra mussel, *J. Great Lakes Res.* 19:617–624.

Kahl, R. 1991. Boating disturbance of canvasbacks during migration at Lake Poygan, Wisconsin. *Wildl. Soc. Bull.* 19:242–248.

Kenow, K. P., C. E. Korschgen, J. M. Nissen, A. Elfessi, and R. Steinbach. 2003. A voluntary program to curtail boat disturbance to waterfowl during migration. *Waterbirds* 26:77–87.

Knapton, R. W., S. A. Petrie, and G. Herring. 2000. Human disturbance of diving ducks on Long Point Bay, Lake Erie. *Wildl. Soc. Bull.* 28:923–930.

Leach, J. H. 1993. Impacts of the zebra mussel (*Dreissena polymorpha*) on water quality and fish spawning reefs in Western Lake Erie. In *Zebra Mussels: Biology, Impact and Control*. T. F. Nalepa, and D. W. Schloesser, eds., pp. 381–397. Boca Raton, FL: CRC Press.

Lovvorn, J. R. 1994. Biomechanics and foraging profitability: An approach to assessing trophic needs and impacts of diving ducks. *Hydrobiologia* 279/280:223–233.

Martz, G. F., J. Aldrich, and D. Ostyn. 1976. History and future of canvasback populations in Michigan's Great Lakes habitat during fall migration and early winter. Wildlife Division Report 2759. Lansing, MI: Michigan Department of Natural Resources.

Nalepa, T. F., D. J. Hartson, G. W. Gostenik, D. L. Fanslow, and G. A. Lang. 1996. Changes in the freshwater mussel community of Lake St. Clair: From Unionidae to *Dreissena polymorpha* in eight years. *J. Great Lakes Res.* 22:354–369.

NOAA. 2011. National Oceanic and Atmospheric Administration. National Geophysical Data Center. Available online: http://www.ngdc.noaa.gov/mgg/gdas/gd_designagrid.html (accessed January 26, 2011).

Prince, H. H., P. I. Padding, and R. W. Knapton. 1992. Waterfowl use of the Laurentian Great Lakes. *J. Great Lakes Res.* 18:673–699.

Ross, R. K., S. A. Petrie, S. S. Badzinski, and A. Mullie. 2005. Autumn diet of greater scaup, lesser scaup, and long-tailed ducks on eastern Lake Ontario prior to zebra mussel invasion. *Wildl. Soc. Bull.* 33:81–91.

Serie, J. R. and D. E. Sharp. 1989. Body weight and composition dynamics of fall migrating canvasbacks. *J. Wildl. Manage.* 53:431–441.

Serie, J. R., D. L. Trauger, and D. E. Sharp. 1983. Migration and winter distributions of canvasbacks staging on the upper Mississippi River. *J. Wildl. Manage.* 47:741–753.

Siegfried, W. R. 1976. Segregation of feeding behaviour of four diving ducks in southern Manitoba. *Can. J. Zool.* 54:730–736.

Snider, V. W., Jr. 1999. The economic impact of boating in Wayne, St. Clair and Macomb Counties. In *Lake St. Clair: Its Current State and Future Prospects. Conference Summary Report.* Port Huron, MI.

SPSS. 2009. *PASW STATISTICS 18.0 Command Syntax Reference.* Chicago, IL: SPSS Inc.

Takekawa, J. Y. 1987. Energetics of canvasbacks staging on an upper Mississippi River pool during fall migration. PhD dissertation. Ames, IA: Iowa State University.

Thompson, S. K. 1992. *Sampling.* New York: Wiley.

Thornburg, D. D. 1973. Diving duck movements on Keokuk Pool, Mississippi River. *J. Wildl. Manage.* 37:382–389.

Torrence, S. M. and M. G. Butler. 2006. Spatial structure of diving duck (Aythya, Oxyura) guild: How does habitat structure and competition influence diving duck habitat use within northern prairie wetlands? *Can. J. Zool.* 84:1358–1367.

U.S. Army Corps of Engineers. 2010. Detroit district, Great Lakes water levels. Available online: http://www.lre.usace.army.mil/greatlakes/hh/greatlakeswaterlevels/ (accessed January 26, 2011).

U.S. Fish and Wildlife Service. 2010. Waterfowl population status, 2010. Washington, DC: U.S. Department of the Interior.

Wormington, A. and J. H. Leach. 1992. Concentrations of migrant diving ducks at Point Pelee National Park, Ontario, in response to invasion of zebra mussels, *Dreissena polymorpha. Can. Field Nat.* 106:376–380.

CHAPTER 42

Context-Dependent Changes in Lake Whitefish Populations Associated with Dreissenid Invasion

Michael D. Rennie

CONTENTS

Abstract 661
Introduction 662
Diporeia and Dreissenid Mussels 662
Changes in Lake Whitefish Populations with Dreissenids 663
    Great Lakes Populations: Consistencies 663
    Great Lakes Populations: Inconsistencies 670
    Lake Erie 671
    Lake Simcoe 672
Changes in Lake Whitefish Populations without Dreissenids 672
    Lake Superior: *Diporeia* Declines 672
    Lake Nipigon: *Diporeia* Stable 673
Dreissenid Effects via Declines in Resource Abundance 673
Other Evidence for Resource Limitation 674
Alternative Hypotheses/Contributing Factors 675
    Density Dependence 675
    Climate Change 676
    Interspecific Competition 676
    Fishery-Dependent Evolution 676
    Other Invasive Species 676
Conclusion and Recommendations 677
Acknowledgments 677
References 677

ABSTRACT

The manner in which an organism responds to a change in its environment can depend greatly on previous conditions. In this regard, lake whitefish (*Coregonus clupeaformis*) populations from a wide range of aquatic environments have demonstrated a variety of responses to the establishment of dreissenid mussels. A review of the literature indicated that individual growth rates and condition of lake whitefish have typically declined after dreissenid establishment where *Diporeia*—a key prey item of lake whitefish—have also declined in abundance. Temporal declines in lake whitefish growth and condition occurred following dreissenid establishment despite reported increases in lake whitefish consumption rates. A review of lake whitefish populations from noninvaded systems revealed declines in lake whitefish growth and condition as a common response to resource limitation, supporting the hypothesis that typical lake whitefish responses to dreissenid establishment are a function of resource limitation. In contrast, lake whitefish populations from shallow, nutrient-enriched lakes with dreissenids (Lake Erie where *Diporeia* has declined, and Lake Simcoe where *Diporeia* was absent prior to dreissenid establishment) show no evidence of declines in lake whitefish growth and/or condition after dreissenid establishment. Age at maturity was delayed in all but one population of 18 surveyed, regardless of whether dreissenids were established or *Diporeia* had declined in abundance. Body condition of lake whitefish appeared to closely track resource declines in most populations. However, growth declines sometimes appeared to be independent of trends in resource abundance, which suggests effects of other stressors besides dreissenids on lake whitefish growth rates. These stressors may include density dependence, climate warming, and changes in ecosystem community structure that may lead to increased interspecific competition.

INTRODUCTION

Lake whitefish (*Coregonus clupeaformis*) are one of the most economically important species of fish to commercial, recreational, and sustenance fisheries in North America. They are distributed throughout Canada and the northern United States, ranging from the Great Lakes in the south to anadromous populations in the Arctic (Scott and Crossman 1998). In the Great Lakes, lake whitefish accounted for the majority of the commercial catch during the past decade. In 2000, they accounted for 45% (nearly 10,000 metric tons) of the Great Lakes commercial fishery with a dockside value of $40 million (year 2000 Canadian dollars, Kinnunen 2003). Beyond the Great Lakes, lake whitefish are the most heavily exploited species of fish from inland fisheries and second economically only to walleye (Freshwater Fish Marketing Corporation 2010). Lake whitefish are a highly sought after species in recreational winter fisheries on both the Great Lakes and surrounding inland lakes (Evans et al. 1988) and are an important staple fish for many North American First Nations communities (Hopper and Power 1991).

Invasive species have in the past threatened this key economic resource and may be doing so once again. In the Great Lakes, lake whitefish suffered dramatic declines in abundance in the 1950s coincident with the invasion of sea lamprey, *Petromyzon marinus* (Smith and Tibbles 1980), and lamprey control efforts are largely thought to be responsible for the dramatic recovery that followed (Ebener 1997). Sudden and unexpected declines in lake whitefish growth rates and body condition were then observed in the mid-1990s, coincident with the establishment of dreissenid mussels (zebra mussel *Dreissena polymorpha* and quagga mussel *D. rostriformis bugensis*). Simultaneously, abundance of the deepwater amphipod, *Diporeia*—a major prey item for lake whitefish—declined. Following dreissenid establishment in Lake Ontario, declines in *Diporeia* abundance and lake whitefish growth and condition were followed by a substantial and sudden decline in lake whitefish abundance (Hoyle et al. 1999, Dermott 2001). Similar declines in *Diporeia* abundance and lake whitefish growth and condition were subsequently documented following dreissenid establishment in Lake Michigan (Pothoven et al. 2001) and Lake Huron (Dobiesz et al. 2005, Rennie et al. 2009a).

Declines in *Diporeia* abundance and lake whitefish growth and condition following dreissenid establishment have led to hypotheses in the literature that negative impacts of dreissenids on *Diporeia* populations are the cause, under the assumption that *Diporeia* contributed substantially to historical lake whitefish diets (Pothoven et al. 2001). However, these patterns have not been consistent. For example, lake whitefish populations in Lake Erie were relatively stable during dreissenid establishment and a collapse of *Diporeia* (Lumb et al. 2007). Though a number of studies over the past decade have investigated links between dreissenid establishment, *Diporeia* declines, and lake whitefish populations, both within and outside of the Great Lakes basin, these studies have to date not been reviewed and considered together.

This chapter provides a review of existing literature and available unpublished data for the purpose of synthesizing current knowledge of the impacts of dreissenid establishment on North American lake whitefish populations. The potential roles of other factors (climate change, density dependence) that have been shown in the literature as affecting lake whitefish populations are also considered.

DIPOREIA AND DREISSENID MUSSELS

The observed timing between the spread of dreissenids and loss of *Diporeia* among many sites throughout the Great Lakes has been virtually simultaneous (Dermott and Kerec 1997, Nalepa et al. 1998, Dermott 2001). Despite this coordination of events, the exact mechanism behind dreissenid–*Diporeia* interactions remains unknown. While not the focus of the current chapter, some discussion regarding the mechanisms proposed is warranted, given the consequences this interaction is thought to have had for lake whitefish. In a recent survey of expert researchers attending a workshop on Great Lakes *Diporeia* declines (Nalepa et al. 2006a), the top two mechanisms supported by participants were (1) food limitation due to filtration activity by dreissenids and (2) harmful agents affecting *Diporeia*, including yet-to-be-identified metabolic by-products produced by dreissenids that are harmful to *Diporeia*, or pathogenic introductions coincident with or facilitated by dreissenids.

The food-limitation hypothesis seems to be supported by recent work implicating dreissenid filtration in reductions of pelagic primary productivity in Lake Michigan (Fahnenstiel et al. 2010). This is also consistent with the conceptual nearshore phosphorous shunt model (Hecky et al. 2004), which posits a concentration and redirection of productivity from offshore and profundal regions of lakes to the nearshore where dreissenids (particularly zebra mussels) can exist at high densities, and may also act to intercept land-based deposition of nutrients from offshore transport. However, other work examining this hypothesis explicitly in a declining population of *Diporeia* in Lake Michigan found no evidence that *Diporeia* declines were a result of food limitation (Nalepa et al. 2006b). Further, primary productivity does not appear to be driving *Diporeia* declines in Lake Ontario (Watkins et al. 2007).

Research into potential harmful agents that might be affecting *Diporeia* has similarly been inconclusive. Experimental *Diporeia* did not avoid or experience differential mortality when exposed to sediments from locations where the species had previously been extirpated (Nalepa et al. 2006b). Other work found that the exposure of *Diporeia* to sediments where dreissenids were abundant resulted in a minor reduction in *Diporeia* survival (mean survival across all treatments was between 70% and 80%; Dermott et al. 2006). A number of pathogens of *Diporeia* have also been identified (Messick et al. 2004); however, there is currently no clear indication as to how these pathogens might be related to dreissenids or how they may have played a role in the decline of *Diporeia*.

Observations in the New York Finger Lakes further complicate the matter. In these lakes, *Diporeia* abundance appears stable despite dreissenid establishment (largely quagga mussels, Dermott et al. 2006, J. Watkins personal communication). This coexistence in the Finger Lakes suggests an interaction between dreissenids and other stressors may have contributed to different patterns observed in the Great Lakes (Dermott et al. 2006, Rennie et al. 2009a).

Despite the aforementioned difficulties in identifying clear and direct links between dreissenid distributions and *Diporeia* declines, the timing between arrival of dreissenids and loss of *Diporeia* in the Great Lakes makes it difficult to imagine that these changes in benthic communities are totally independent of one another. However, it is clear that more work is needed to better establish the mechanistic relationship between *Diporeia* declines and dreissenids.

CHANGES IN LAKE WHITEFISH POPULATIONS WITH DREISSENIDS

In this chapter, temporal trends in lake whitefish populations that have been exposed to ecosystem changes associated with dreissenids are primarily from the North American Great Lakes (Table 42.1). Where available, additional data were also compiled for lake whitefish populations exposed to dreissenids outside the Great Lakes. In addition, data for uninvaded populations of lake whitefish not subject to dreissenid influences are included to contrast temporal trends with invaded ecosystems.

Great Lakes Populations: Consistencies

Impacts of dreissenids on lake whitefish growth and condition have been nearly ubiquitous following their establishment in the Great Lakes (Table 42.1). Lake whitefish growth rates declined following dreissenid establishment in all but two cases, and all but one population showed declines in condition (Table 42.1). In many cases, the response of lake whitefish to dreissenid establishment has been almost immediate. In Lake Ontario (Hoyle et al. 1999) and in Lake Michigan (Pothoven et al. 2001), declines in lake whitefish growth and condition (and *Diporeia* abundance) were observed to occur within 1–2 years of dreissenid establishment. Of three populations where size-at-age data were available in Lake Huron, one (Southampton) demonstrated a response in growth rates within 1–2 years of reported dreissenid establishment (in 1993; Figure 42.1). Similarly, weight-at-age in northern Lake Michigan stocks appeared to display a change in slope in the early- to mid-1990s (Figure 42.2). For example, lake whitefish growth in Big Bay de Noc began to decline in 1992 shortly after dreissenid establishment and prior to the collapse of *Diporeia* in and around the bay (Nalepa et al. 2006b). While data available to estimate condition from this stock was limited and thus the onset of declines was impossible to pinpoint, relative weight (calculated using Equation 5 in Rennie and Verdon 2008) estimated in 2000–2006 was lower by 30% compared to 1980 values (Figure 42.3). Relative weight in the Southampton stock declined by 5.7% between 1992 and 1994 (Figure 42.3).

Other lake whitefish populations exhibited a lag between dreissenid establishment and declines in growth and condition (Figures 42.1 and 42.2). In these cases, lake whitefish appear to be responding to a decline in *Diporeia* following dreissenid establishment. For instance, dreissenids were established in Georgian Bay, Lake Huron, in 1996 (Rennie 2009), but declines of lake whitefish growth in the Cape Rich stock (located in southern Georgian Bay) were not observed until after 2001 (Figure 42.1). This coincided almost directly with reported declines of *Diporeia* abundance in transects off Cape Rich (Nalepa et al. 2007). Similarly, lake whitefish from Naubinway (northern Lake Michigan) appeared to show major declines in size-at-age in 1997 (Figure 42.2) and marked declines in condition in 1994 and 1997 (Figure 42.3). The timing of these declines coincided with dramatic declines in *Diporeia* abundance in northern Lake Michigan between 1995 and 2000 (Nalepa et al. 2006b) and not with the earlier establishment of dreissenids in the region (Nalepa et al. 2009).

Elsewhere, declines in condition of lake whitefish from near Cheboygan (northwestern Lake Huron) were coincident with dreissenid establishment (Figure 42.3), but declines in growth were delayed until 2000 (Figure 42.2). These responses were similar to those in Georgian Bay (northeastern Lake Huron) where declines in many invertebrate species (including *Diporeia*) occurred in 2000–2003 (Nalepa et al. 2007). In regions of the Great Lakes where the decline of *Diporeia* was less pronounced (e.g., Detour Village, northwestern Lake Huron), declines in lake whitefish growth were similarly delayed (Figure 42.2), and declines in lake whitefish condition were more gradual (Figure 42.3).

Table 42.1 Changes in Measured Parameters of Lake Whitefish Populations during Dreissenid Establishment. Up Arrow (↗) Represents an Increase, Down Arrow (↘) Represents a Decrease, and Equal Sign (=) Represents No Change. Numbers in Parentheses Correspond to References in Table Footnotes.

Location	Growth	Body Condition	Reproductive Investment	Age-at-Maturity	Size-at-Maturity	Energy Density (Fish)	Energy Density (Diet)	$\delta^{13}C$ (Energy Source)	$\delta^{15}N$ (Trophic Position)
Lake Ontario									
Northeastern	↗ (1)	↗ (1)		↗ (2)					↗ (3)
Kingston Basin		↗ (3)						= (3)	
Bay of Quinte		↗ (3)							
Lake Huron									
South Bay	↗ (6)	↗ (6)		↗ (6)	= (28)				
Main Basin	↗ (10)	↗ (11, 27)		↗ (9)	↗ (9)		↗ (7, 8)	↗ (8)	↗ (8)
Northern	↗ (25)	↗ (27)	↗ (12)	↗ (28)	↗ (9); = (28)				
Central	↗ (11)	↗ (11)		↗ (11)					
South									
Lake Huron									
Georgian Bay	↗ (11, 25)			↗ (9, 28)	↗ (9)				

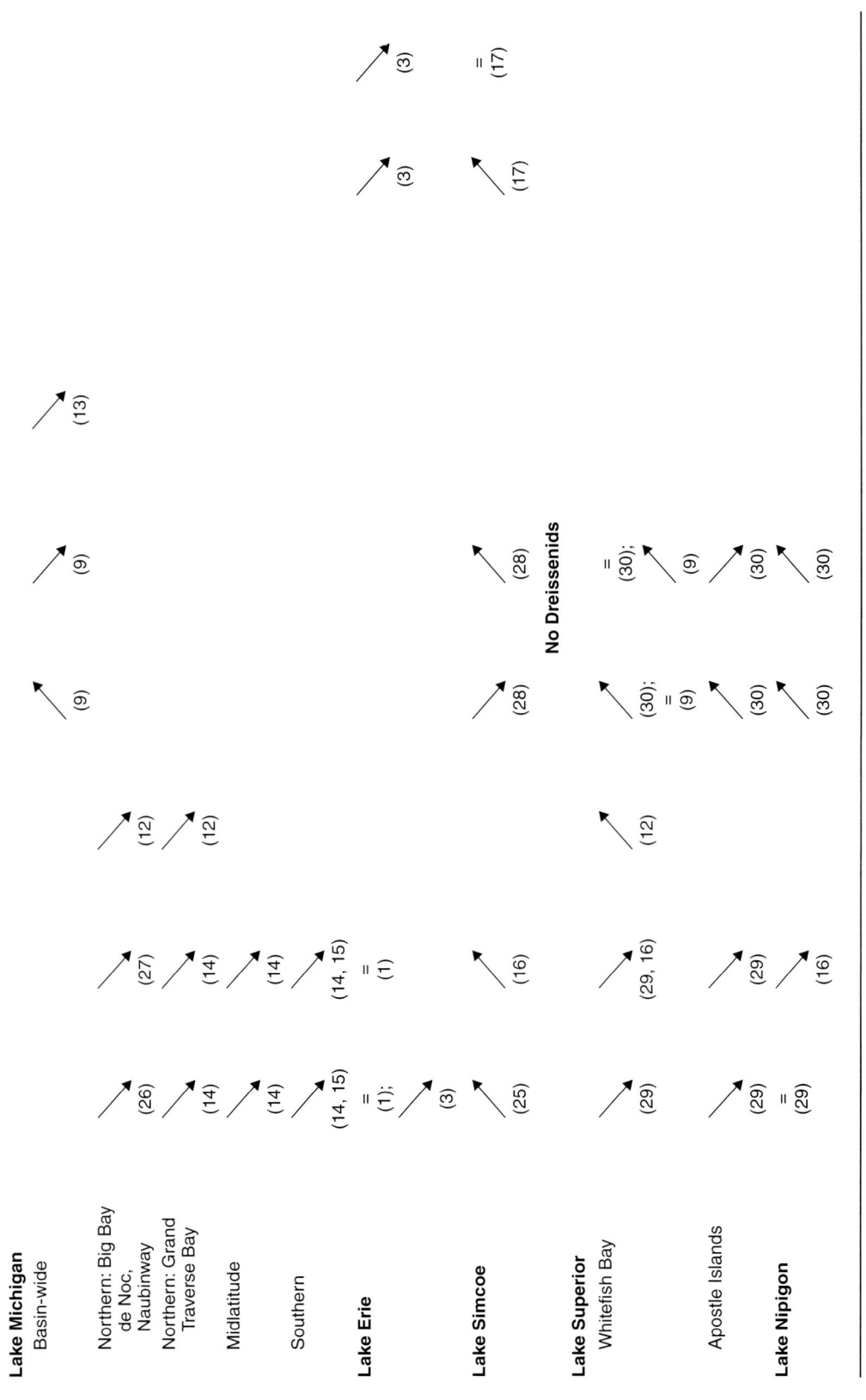

CONTEXT-DEPENDENT CHANGES IN LAKE WHITEFISH POPULATIONS ASSOCIATED WITH DREISSENID INVASION

Lake Michigan
Basin-wide

Northern: Big Bay
de Noc, Naubinway

Northern: Grand
Traverse Bay

Midlatitude

Southern

Lake Erie

Lake Simcoe

Lake Superior
Whitefish Bay

Apostle Islands

Lake Nipigon

[a] Densities based on trawling (active sampling method); all other estimates based on catch from gillnetting (passive sampling method).
1 (Lumb et al. 2007); 2 (Hoyle 2005); 3 (Lumb and Johnson 2012); 4 (Hoyle et al. 1999); 5 (Hoyle et al. 2008); 6 (Rennie et al. 2009a); 7 (McNickle et al. 2006); 8 (Rennie et al. 2009b); 9 (Wang et al. 2008); 10 (Rennie et al. 2012b); 11 (Mohr et al. 2005); 12 (Kratzer et al. 2007); 13 (Pothoven et al. 2006); 14 (DeBruyne et al. 2008); 15 (Pothoven et al. 2001); 16 (Rennie et al. 2010); 17 (M.D. Rennie, D.O. Evans and J.L. LaRose, unpublished data); 18 (Rennie et al. 2012a); 19 (Fernandez et al. 2009); 20 (Riley and Adams 2010); 21 (Pothoven and Madenjian 2008); 22 (Gewurtz et al. 2011); 23 (Gorman et al. 2010); 24 (Nalepa et al. 2005); 25 (Figure 42.1); 26 (Figure 42.2); 27 (Figure 42.3); 28 (Figure 42.5); 29 (Figure 42.6); 30 (Figure 42.7).

Table 42.1 (continued) Changes in Measured Parameters of Lake Whitefish Populations during Dreissenid Establishment. Up Arrow (↗) Represents an Increase, Down Arrow (↘) Represents a Decrease, and Equal Sign (=) Represents No Change. Numbers in Parentheses Correspond to References in Table Footnotes

Location	Depth Distribution	Relative Abundance	Contaminants	Juvenile Growth	Energy Density (Juvenile Diet)	Total Consumption	Feeding Rate	Activity Rate	Conversion Efficiency
				Dreissenid Invaded					
Lake Ontario									
Northeastern Kingston Basin		↗ (5)							
Bay of Quinte			↗ Hg (16)						
Lake Huron									
South Bay	↗ (8)	↗ (6)	= Hg (18)		= (19)		↗ (18)	↗ (18)	↗ (18)
Main Basin	↗ (20)	↗ (11)				= (21)			= (21)
Northern		↗ (11, 12)					↗ (21)		
Central		↗ (20)[a]	= Hg (16)				↗ (21)		
South		↗ (20)[a]					↗ (21)		
Lake Huron									
Georgian Bay			↗ Hg (16)						
Lake Michigan									
Basin-wide									

CONTEXT-DEPENDENT CHANGES IN LAKE WHITEFISH POPULATIONS ASSOCIATED WITH DREISSENID INVASION

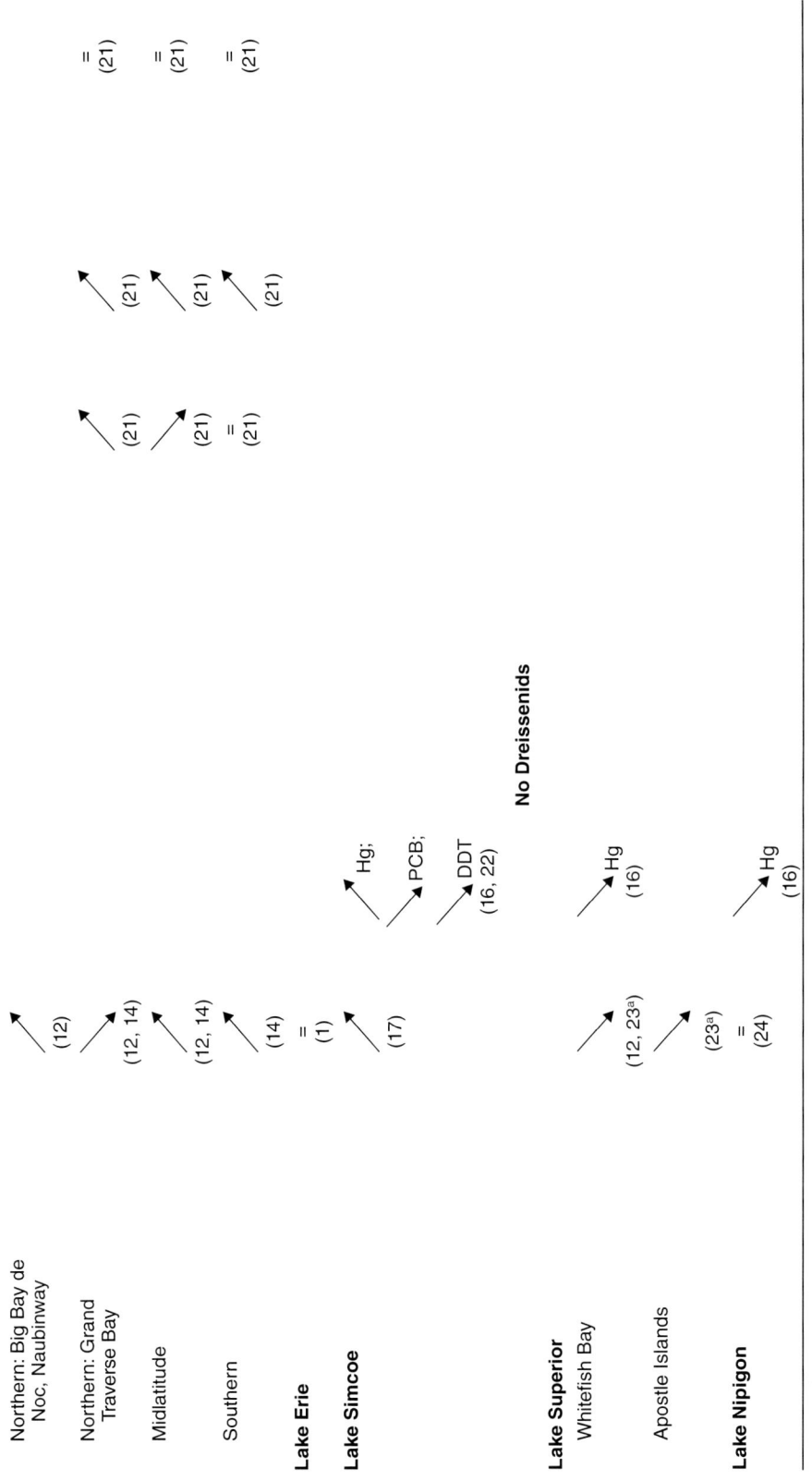

Northern: Big Bay de Noc, Naubinway	↗ (12)			
Northern: Grand Traverse Bay	↗ (12, 14)			
Midlatitude	↗ (12, 14)		↗ (21)	↗ (21)
Southern	↗ (14)		↗ (21)	= (21)
Lake Erie	= (1)		= (21)	= (21)
Lake Simcoe	↗ (17)	↗ Hg; ↗ PCB; ↗ DDT (16, 22)		
Lake Superior Whitefish Bay	↗ (12, 23[a])	↗ Hg (16)		
Apostle Islands	= (24)	↗ Hg (16)	**No Dreissenids**	
Lake Nipigon	= (23[a])			

[a] Densities based on trawling (active sampling method); all other estimates based on catch from gillnetting (passive sampling method).
1 (Lumb et al. 2007); 2 (Hoyle 2005); 3 (Lumb and Johnson 2012); 4 (Hoyle et al. 1999); 5 (Hoyle et al. 2008); 6 (Rennie et al. 2009a); 7 (McNickle et al. 2006); 8 (Rennie et al. 2009b); 9 (Wang et al. 2008); 10 (Rennie et al. 2012b); 11 (Mohr et al. 2005); 12 (Kratzer et al. 2007); 13 (Pothoven et al. 2006); 14 (DeBruyne et al. 2008); 15 (Pothoven et al. 2001); 16 (Rennie et al. 2010); 17 (M.D. Rennie, D.O. Evans and J.L. LaRose, unpublished data); 18 (Rennie et al. 2012a); 19 (Fernandez et al. 2009); 20 (Riley and Adams 2010); 21 (Pothoven and Madenjian 2008); 22 (Gewurtz et al. 2011); 23 (Gorman et al. 2010); 24 (Nalepa et al. 2005); 25 (Figure 42.1); 26 (Figure 42.2); 27 (Figure 42.3); 28 (Figure 42.5); 29 (Figure 42.6); 30 (Figure 42.7).

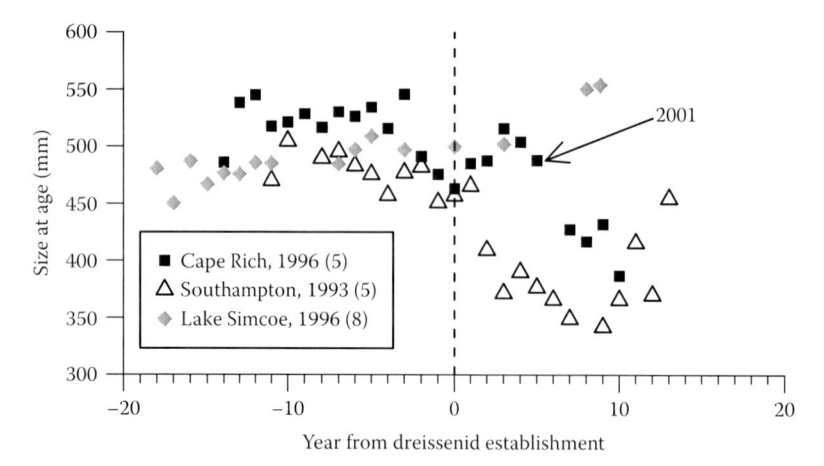

Figure 42.1 Mean size-at-age of lake whitefish in Lake Huron (Cape Rich and Southampton) and in Lake Simcoe in relation to the establishment of dreissenids (year 0 = establishment). Year of dreissenid establishment shown in the legend. (From Rennie, M.D., Influence of invasive species, climate change and population density on life histories and mercury dynamics of *Coregonus* spp., PhD thesis, University of Toronto, Toronto, Ontario, Canada, 2009.) Numbers in parentheses indicate age of lake whitefish examined. Year of major *Diporeia* declines in Cape Rich (2001) is noted. Lake whitefish data taken from Ontario government indexing records.

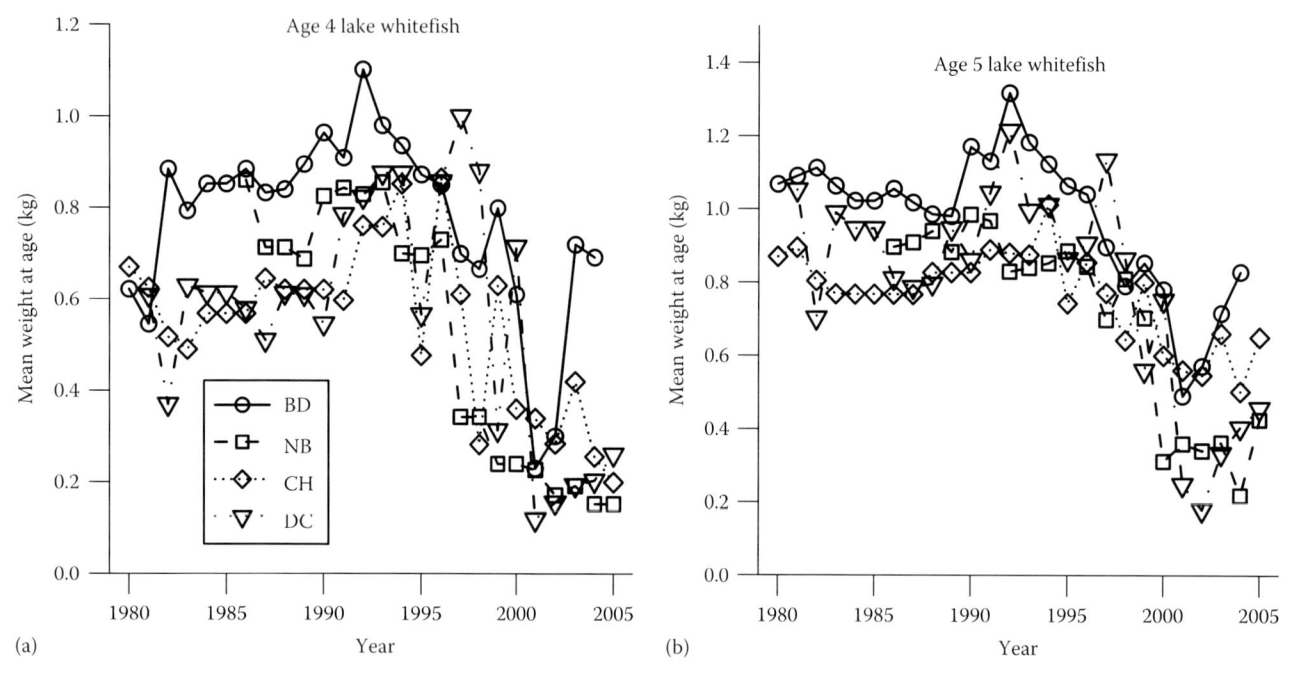

Figure 42.2 Mean weight-at-age for age-4 (panel a) and age-5 (panel b) lake whitefish in northern Lake Michigan (BD = Big Bay de Noc; NB = Naubinway) and northeastern Lake Huron (CH = Cheboygan, DC = Detour-Cedarville). (Reproduced from Rennie, M.D. et al., *Adv. Limnol.*, 63, 455, 2012b. With permission, available at http://www.schweizerbart.de)

Growth, and to some extent condition (Bajer and Hayward 2006), is ultimately related to bioenergetics processes (i.e., the relative rates of energy intake and expenditure) of an organism (Weatherley 1966). Two independent studies have reported an increase in mass-specific consumption rates of lake whitefish following dreissenid establishment (Pothoven and Madenjian 2008, Rennie et al. 2012a). The latter study also estimated higher consumption rates among lake whitefish populations where dreissenids were established compared with populations where dreissenids were not present (Figure 42.4a). Both studies showed that lake whitefish growth rates declined in the presence of dreissenids despite higher rates of food intake, suggesting forage declined in quality. This evidence supports previous work showing a decline in lake whitefish prey quality following dreissenid invasion (McNickle et al. 2006, Rennie et al. 2009b).

While age-at-maturity increased significantly in all but one population exposed to dreissenids, these changes were gradual compared to the abrupt changes in growth and condition more frequently observed (Table 42.1, Figure 42.5a). Further, this pattern was observed over a wide geographic range and in populations without dreissenids (Table 42.1),

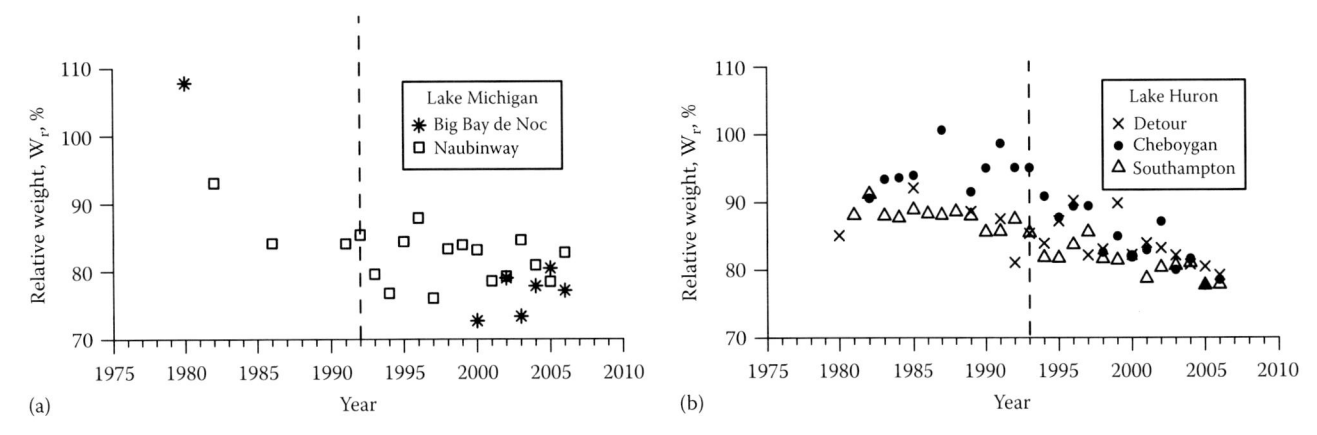

Figure 42.3 Lake whitefish condition (relative weight, W_r, expressed as a percentage of standard weight estimated from Equation 5 in Rennie and Verdon 2008) in northern Lake Michigan (panel a) and Lake Huron (panel b). Dashed line represents year of reported dreissenid establishment in these lakes.

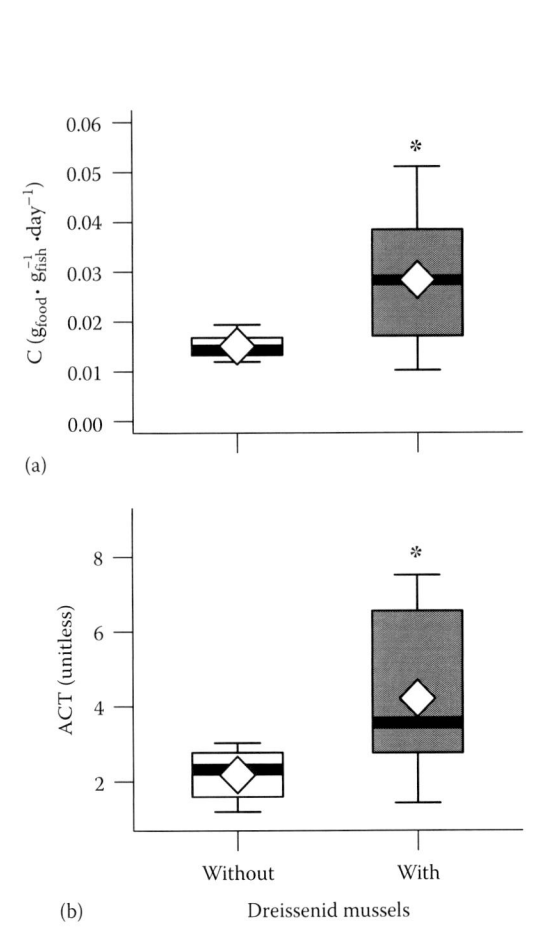

Figure 42.4 Boxplots of relative consumption (C) (panel a) and activity multipliers (ACT, expressed as a multiple of standard metabolism, panel b) obtained from bioenergetics models of lake whitefish populations without (open) and with dreissenids (shaded). Upper and lower bounds of boxes represent the first and third quartiles of the data, whiskers represent 95% confidence intervals, thick bars are medians, and diamonds are mean values. (Reproduced from Rennie, M.D. et al., *Can. J. Fish. Aquat. Sci.*, 69, 41, 2012a. With permission.)

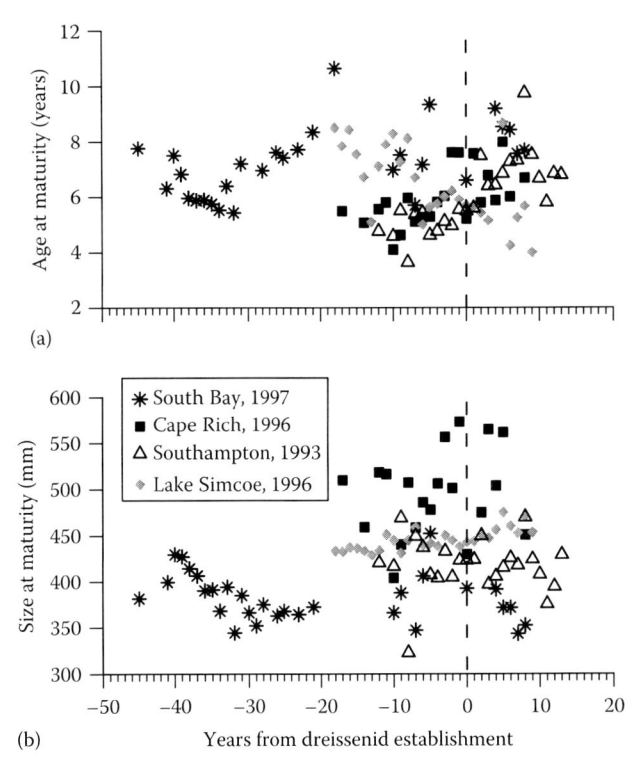

Figure 42.5 Mean age (panel a) and size (panel b) at 50% maturity of lake whitefish populations in various regions of Lake Huron (South Bay, Cape Rich, Southampton) and in Lake Simcoe in relation to the establishment of dreissenids (year 0 = establishment). Dates of dreissenid establishment shown in the legend as reported in Rennie. (From Rennie, M.D., Influence of invasive species, climate change and population density on life histories and mercury dynamics of *Coregonus* spp., PhD thesis, University of Toronto, Toronto, Ontario, Canada, 2008.) Lake Simcoe estimates (from trap nets on spawning shoals that likely underrepresent immature individuals) are estimated as the mean age and size of the youngest and smallest 5% of individuals captured each year, respectively. Linear trends in age-at-maturity are significant and positive for South Bay ($p = 0.08$), Cape Rich ($p = 0.04$), and Southampton ($p < 0.0001$) and negative for Lake Simcoe ($p = 0.0002$). Linear trends in size-at-maturity are significant (positive) for Lake Simcoe only ($p < 0.0001$).

which suggests this variable is likely influenced by factors other than dreissenid establishment. Accompanying this delay in maturity, Kratzer et al. (2007) reported coincident declines in lake whitefish reproductive investment from areas where dreissenids were established, while investment increased at a reference site in Lake Superior where dreissenids were not established.

Though temporal patterns in lake whitefish abundance after dreissenid invasion appear at first glance to lack any consistent pattern, trends do emerge upon closer examination of the data. Observed temporal trends in lake whitefish abundance appear to depend greatly on the collection method employed (Table 42.1). Declines in lake whitefish abundance in central and southern Lake Huron were observed only when data were collected with active sampling methods (bottom trawls, Riley et al. 2008). In contrast, increased abundance was observed in this same region when data were collected with passive methods (e.g., gillnetting, Rennie et al. 2012a).

The different conclusions about trends in relative abundance of lake whitefish between active (decrease) and passive (increase) sampling methods within the same lake (Mohr and Nalepa 2005, Riley et al. 2008) are difficult to reconcile with the consistency in growth and condition declines observed in the same stocks (Mohr et al. 2005, this study). Further, differences exist even when using the same collection methods in a particular region. DeBruyne et al. (2008) reported declines (rather than increases) in lake whitefish relative abundance in northern Lake Michigan from passive sampling gear, but similar sampling from other studies elsewhere in northern Lake Michigan report increases (e.g., Naubinway, Big Bay de Noc, Kratzer et al. 2007). Among sites summarized in Table 42.1 where dreissenids are established, only 2 of 10 regions that employed passive sampling methods report evidence of declines in abundance. In contrast, active sampling programs consistently reported declines.

Reasons for the different conclusions in lake whitefish abundance trends between active and passive sampling methodology remain largely unaddressed, but recent work suggests that the activity (and therefore catchability) of lake whitefish increases in the presence of dreissenids (Figure 42.4b) as a consequence of increased foraging requirements in a depleted-prey field. While it is possible that reported differences in abundance genuinely reflect regional differences among stocks, behavioral changes in fish can also affect encounter rates with sampling gear. Passive collection methods rely on the movements of organisms for capture (Rudstam et al. 1984). As such, a substantial component of catch-per-unit-effort (CPUE) estimates is related to gear encounter, or "catchability" (Spangler and Collins 1992, Biro and Post 2008, Biro and Dingemanse 2009, Rennie et al. 2012a). A hypothesis of increased swimming activity in lake whitefish or increased range dispersal (Rennie et al. 2012b) could manifest itself as increased rates of encounter with passive gear, whereas changes in activity or range distribution would less likely influence catch rates of active sampling gear such as bottom trawls. If this hypothesis is true, changes in CPUE estimated from passive gear may not reflect actual changes in abundance (e.g., Henderson et al. 1983). Other recent work illustrated the influence of factors independent of fish abundance on CPUE estimates of passive collection gear (Deroba and Bence 2009). If catchability rather than population abundance has increased after dreissenid colonization, consequences for sustainable management efforts of lake whitefish fisheries could be substantial because it is CPUE from these passive sampling methods that is frequently used to estimate population abundance and, in turn, set commercial fishing quotas.

Great Lakes Populations: Inconsistencies

Temporal patterns in size-at-maturity of lake whitefish varied widely both within and among populations, and among studies. Size-at-maturity for populations typically either declined or remained unchanged, while it increased in only a minority of cases (Table 42.1, Figure 42.5b). Wang et al. (2008) reported a significant increase in lake whitefish size-at-maturity among cohorts born after 1990, though the increase among these cohorts was on the order of 2.0–2.5 cm, only 3%–4% of typical asymptotic size in this species (50–70 cm; Scott and Crossman 1998). However, increases of this magnitude may relate to delays in age-at-maturation of 1–2 years, based on reported relationships between size- and age-at-maturity for this species (Beauchamp et al. 2004).

Trends in depth-of-capture of lake whitefish following dreissenid colonization were also variable. Commercial fishing habits and a recent analysis of trawling surveys suggest that lake whitefish in the main basin of Lake Huron moved deeper following dreissenid establishment (Mohr et al. 2005, Riley and Adams 2010). In contrast, lake whitefish in South Bay, Lake Huron, exhibited evidence of more shallow distributions following dreissenid establishment (Rennie et al. 2009b). It is possible that fish in the main basin of Lake Huron pursued remnant *Diporeia* populations in deeper waters of the main basin (Nalepa et al. 2007). While a remnant *Diporeia* population did persist in South Bay, it may have been insufficient to keep whitefish offshore compared to increased abundance of invertebrates nearshore in the bay after dreissenid invasion (McNickle et al. 2006).

Changes in lake whitefish growth efficiencies following dreissenid invasion were examined in two studies and the results were not consistent (Table 42.1). In one study, lake whitefish conversion efficiency (or growth efficiency; the proportion of food consumed that is converted into growth) decreased in South Bay lake whitefish following dreissenid invasion and tended to be highest in areas with low *Diporeia* abundance (Rennie et al. 2012a). In another study, conversion efficiencies in Lake Huron and Lake Michigan populations were similar before and after dreissenid invasion (Pothoven and Madenjian 2008). This discrepancy between studies is likely methodological. Pothoven and Madenjian (2008) used a bioenergetics mass-balance model where activity costs are estimated as functions of water temperature and body mass. Therefore, all energetic losses in their model were the result of sub-models and were used with growth rates to estimate consumption. In contrast, Rennie et al. (2012a) estimated consumption from a mercury mass-balance model and used consumption as an input parameter in the bioenergetics model to estimate activity costs expressed as a multiple of standard metabolism (rather than as a sub-model of total metabolism). In the latter case, energetic losses from consumption that are not translated into growth or otherwise accounted for are estimated directly in the mass balance. If lake whitefish activity (associated with increased foraging) increased as a result of dreissenid colonization, the activity sub-model used by Pothoven and Madenjian (2008) in their post-dreissenid colonization models would not reflect this change, resulting in an underestimate of losses to activity. This would then translate into an underestimate of consumption rate and overestimate of growth efficiency in their post-dreissenid bioenergetics models.

Historical studies of isotopic variation and contaminants in whitefish have also been inconsistent between populations. Isotopic signatures of carbon ($\delta^{13}C$) in lake whitefish scales have been observed to increase in at least two populations of lake whitefish where dreissenids colonized (Table 42.1). In South Bay, Lake Huron, this increase was sudden and of substantial magnitude (approx. 4‰ increase), which is consistent with an increased reliance on nearshore carbon following dreissenid establishment. Increases in Lake Simcoe $\delta^{13}C$ were in the same direction as South Bay, but smaller in magnitude (M. D. Rennie, D. O. Evans and J. L. LaRose, unpublished data). Based on a more coarse temporal comparison, $\delta^{13}C$ signatures of scales from Lake Erie and Ontario stocks appear to have declined after dreissenid invasion (Table 42.1). Tissue concentrations of contaminants (primarily mercury) have declined or remained stable regardless of dreissenid establishment, though there is evidence for increased mercury tissue concentrations in lake whitefish from Lake Simcoe (Table 42.1).

Lake Erie

For all regions of the Great Lakes except Lake Erie, the temporal sequence of events was consistent: dreissenids establish, *Diporeia* decline in abundance, and lake whitefish growth and condition decline shortly thereafter. While *Diporeia* was extirpated from Lake Erie after dreissenid establishment (Dermott and Kerec 1997, Barbiero et al. 2011), lake whitefish growth, condition, and abundance generally appear to have been unaffected (Table 42.1). However, a more recent assessment of growth in Lake Erie whitefish suggests slight growth declines (Lumb and Johnson 2012).

If declines in growth and condition of lake whitefish following *Diporeia* collapse are an indication of food limitation (e.g., Pothoven et al. 2001), then changes in diets of lake whitefish might be expected to reflect a reduction of *Diporeia* as a food source (e.g., Rennie et al. 2009b). Historical diets of adult lake whitefish in Lake Erie are unknown, which makes an evaluation of changes in diets impossible. An investigation of isotopic values of scales from archived Lake Erie lake whitefish revealed temporal changes (Lumb and Johnson 2012), but the changes do not explain why Lake Erie lake whitefish responded so differently to dreissenid establishment compared to other stocks in the Great Lakes. Lumb and Johnson (2012) reported a depletion in $\delta^{13}C$ values (approximately 2‰) after dreissenid invasion was observed, which suggests an increased reliance on pelagic or offshore-derived resources. This pattern is smaller in magnitude and opposite in direction compared with another study (Rennie et al. 2009b). A 2‰ depletion is roughly twice as large as would be expected due to atmospheric carbon depletion (Suess 1955, Verburg 2007) and is consistent with contemporary diet data (Lumb et al. 2007) that is dominated by pelagic zooplankton and organisms common in offshore habitats such as sphaeriids and chironomids. Dreissenids were also a major component of lake whitefish diets in Lake Erie (Lumb et al. 2007), and dreissenids are $\delta^{13}C$ depleted relative to other organisms found at similar depths (Rennie et al. 2009b). All of these prey organisms are typically thought to be lower in caloric content than *Diporeia* (Madenjian et al. 2006, Rennie et al. 2011a), and dreissenids likely require a great deal more energy for lake whitefish to process compared to more soft-bodied organisms such as *Diporeia* (Owens and Dittman 2003).

Relative stability of lake whitefish growth and condition in Lake Erie during *Diporeia* declines suggests that: (1) *Diporeia* were never a major component of lake whitefish diets; or (2) the switch to alternative prey of lower caloric value was accompanied by higher consumption rates that were mediated by higher densities and/or rates of production of alternative prey following dreissenid establishment or by increased lake whitefish foraging activity. Increased consumption (Pothoven and Madenjian 2008) and activity rates (Rennie et al. 2012a) have been reported in other lake whitefish populations after dreissenid establishment. The isotopic pattern in lake whitefish in Lake Erie suggests they did not utilize increased abundance and biomass of benthic invertebrates in the nearshore region following dreissenid establishment (Dermott and Kerec 1997). Rather, isotopic patterns suggest that alternative prey would likely consist of offshore zooplankton that are reported to have declined in biomass in the eastern basin but increased in biomass in the central and western basins (Conroy et al. 2005). Lake Erie whitefish undergo seasonal migrations from the eastern basin into the central and western basins (Lumb et al. 2007), and by doing so could gain access to more dense zooplankton communities. Finally, conclusions about changes in lake whitefish growth in Lake Erie may partially depend on how data were analyzed. Previously, temporal trends in growth (size-at-age) in Lake Erie were considered to be more stable than trends observed in Lake Ontario. However, recent work based on growth curves of Lake Erie lake whitefish collected in 1991–2003 suggested growth rates declined relative to those estimated prior to 1986, which was the earliest time period considered in previous assessments (Lumb et al. 2007, Lumb and Johnson 2012).

Lake Simcoe

Like lake whitefish in Lake Erie, the pattern observed in lake whitefish in Lake Simcoe is largely inconsistent with the general pattern observed in the rest of the Great Lakes. Unlike other lake whitefish populations investigated, growth rates, condition, size-at-maturity, and abundance all *increased* in Lake Simcoe after dreissenid invasion, while age-at-maturity *declined*. Also, unlike other lakes discussed in this chapter, *Diporeia* has never been reported in Lake Simcoe (Rawson 1930). Like many lake whitefish populations in the Great Lakes, fish in Lake Simcoe were subject to many simultaneous stressors, including nutrient abatement and urbanization (Evans et al. 1996, Winter et al. 2007). However, differences in response of lake whitefish in Lake Simcoe compared to most in the Great Lakes suggest some additional studies may be warranted.

Total abundance of the benthic invertebrate community in the offshore region of Lake Simcoe declined between 1983 and 2008 (Jimenez et al. 2011). However, biomass over the same time period increased for certain taxa due to a shift in size distributions. Additionally, certain nearshore taxa extended their distribution to deeper waters. Thus, subtle changes in benthic invertebrate communities may be responsible for a sustained lake whitefish population in Lake Simcoe. Ongoing research into ecosystem-level changes in the absence of *Diporeia* may help to better explain trends in Lake Simcoe whitefish populations after dreissenid establishment. In particular, comparisons between lake whitefish populations in Lake Simcoe and Lake Erie may prove to be useful to determine why neither has responded like so many other populations in the Great Lakes to dreissenid establishment.

CHANGES IN LAKE WHITEFISH POPULATIONS WITHOUT DREISSENIDS

Lake Superior: *Diporeia* Declines

Long-term patterns in lake whitefish populations in areas where dreissenids are absent further support the hypothesis that some aspects of lake whitefish life histories are more closely linked to resource abundance (e.g., densities of *Diporeia*), whereas other aspects may be indicative of stressors not yet identified. Declines in *Diporeia* populations in Lake Superior between 1994 and 2000 were observed in regions near Whitefish Bay and the Apostle Islands where dreissenids had not colonized (Scharold et al. 2004). While these declines were not nearly as large or sudden as in the lower Great Lakes where dreissenids were established, lake whitefish from both regions displayed declines in condition (Figure 42.6a) and growth (Figure 42.6b) between the early 1990s and 2000. However, trends over this period may just reflect a general pattern of declines over the longer term (Figure 42.6). Relative and overall abundance of *Diporeia* between the 1970s and early 2000s (Scharold et al. 2004) suggest that other factors besides intraspecific resource limitation influence these more long-term patterns in lake whitefish.

Though data prior to 1990 are sparse, a pattern of gradual increase in age-at-maturity in Lake Superior populations seems evident (Figure 42.7a). This suggests the influence of some unidentified stressor (s) not yet considered. Discrepancies between results presented here for Whitefish Bay, Lake Superior (Table 42.1), and those of Wang et al. (2008) are likely due to differences in methodologies. Although the statistical approach used was identical (Beauchamp et al. 2004), the manner in which data were grouped for estimating age-at-maturity was different. Wang et al. (2008) estimated age- and length-at-maturity for a group of cohorts during two different time periods (before and after 1990; Wang et al. 2008), whereas results

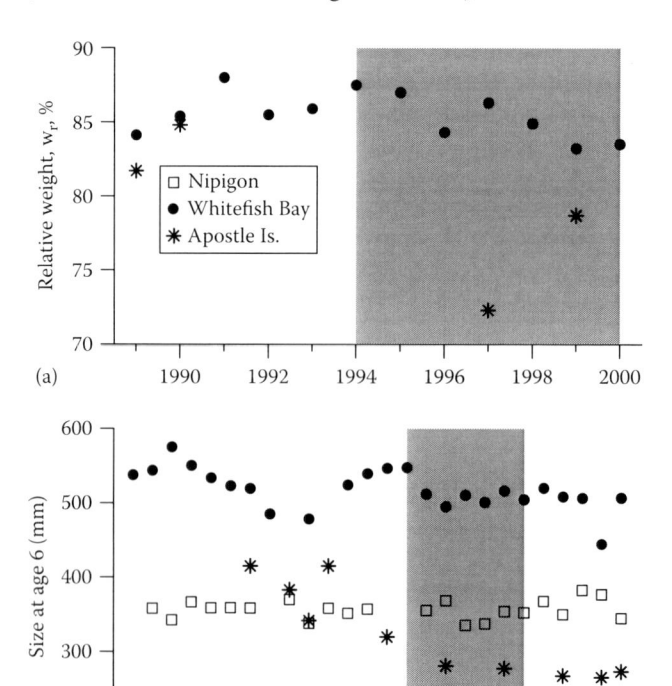

Figure 42.6 Lake whitefish condition (relative weight, W_r, Rennie and Verdon 2008, panel a) and size at age 6 (panel b) from Lake Superior (Whitefish Bay, Apostle Islands) and Lake Nipigon. Gray shaded area represents period of reported *Diporeia* declines in Lake Superior. (From Scharold, J.V. et al., *J. Great Lakes Res.*, 30, 360, 2004.)

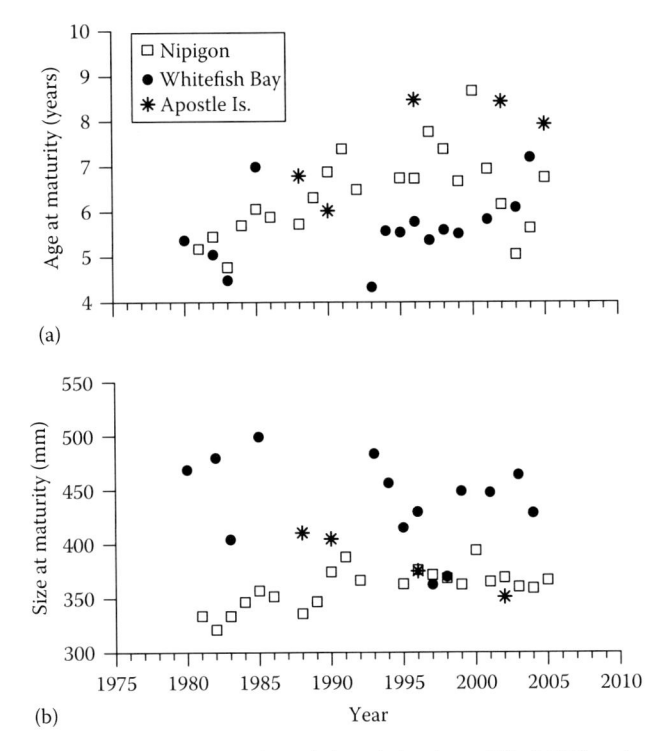

Figure 42.7 Mean age (panel a) and size (panel b) at 50% maturity of lake whitefish populations from Lake Superior (Whitefish Bay, Apostle Islands) and Lake Nipigon as estimated using the method of Beauchamp et al. (From Beauchamp, K.C. et al., *J. Great Lakes Res.*, 30, 451, 2004). Linear trends in age-at-maturity are significant (positive) for Lake Nipigon only (p = 0.04). Linear trends in size-at-maturity are significant (positive) for Lake Nipigon (p = 0.001) and Apostle Islands (negative, p = 0.002).

presented here were based on age-at-maturity estimated for a given year, fitting the statistical model to all cohorts captured within a particular year. Both methods integrate data from many cohorts over time, but the method adopted here permits age-at-maturity estimates for each year in series, for the group of fish spawning in that particular year, rather than limiting estimates for comparison across two time periods only.

Lake Nipigon: *Diporeia* Stable

Like Lake Superior, Lake Nipigon is another large-lake system that has *Diporeia* but few dreissenids; indeed, dreissenids have not been reported from Lake Nipigon. Lake whitefish in Lake Nipigon show no significant detectable trends in growth and no reported changes in abundance (Figure 42.6; Table 42.1); however, recent analysis suggests a decline in condition (Rennie et al. 2010). While historic data are sparse, there is no indication that *Diporeia* in Lake Nipigon has declined. Abundance in the 1920s was about 1000/m^2 (Adamstone 1924), while a recent survey reported that mean amphipod abundance density at 20 m depth was 2300/m^2 (Bentz et al. 2002). Amphipods at this depth were likely *Diporeia*

(McNickle et al. 2006). Lake whitefish length- and age-at-maturity in Lake Nipigon increased, though there was some evidence of earlier maturation in recent years (Figure 42.7).

DREISSENID EFFECTS VIA DECLINES IN RESOURCE ABUNDANCE

The most common and consistent pattern that emerges from this chapter of available information on interactions between dreissenid mussels, lake whitefish populations, and *Diporeia* is that, with one notable exception, lake whitefish growth and condition appear to respond negatively to *Diporeia* declines, even in the absence of dreissenids. Fish condition appears to track temporal declines in *Diporeia* more closely than growth. The correspondence between declines in *Diporeia* abundance and declines in lake whitefish growth and condition suggests the processes that control growth and condition (e.g., food intake, activity) may also be affected by *Diporeia* abundance. Recent work has shown this appears to be true (Rennie et al. 2012a). Rates of lake whitefish consumption, activity, and conversion efficiency were correlated to *Diporeia* abundance (Figure 42.8).

There are some interesting commonalities in the two lakes that were invaded by dreissenids but where growth rates of lake whitefish increased or did not change (Lake Erie and Lake Simcoe). Both lakes are relatively shallow; mean depth of Lake Simcoe is 15 m (Rawson 1930), and mean depth of Lake Erie is 18 m (Rawson 1952). These depths are only 20%–23% of mean depths in the other Great Lakes or in Lake Nipigon (Rawson 1952). Total abundance and biomass of benthos has generally increased in shallow nearshore regions but declined in offshore regions following dreissenid establishment (Higgins and Vander Zanden 2010). Thus, increased food availability in nearshore benthos could mitigate declines in offshore regions, but evidence for such an offset of resources is inconsistent between lakes. In Lake Erie, lake whitefish did not show evidence of increased reliance on nearshore resources (Lumb and Johnson 2012). However, in South Bay (Lake Huron), which is also a relatively shallow system (mean depth 16 m, King et al. 1997), lake whitefish demonstrated an increased reliance on nearshore resources (Rennie et al. 2009b), but growth and condition declined nonetheless (Rennie et al. 2009a). In Lake Champlain, another relatively shallow system, growth and condition of lake whitefish actually appears to have recovered after the establishment of dreissenids (S. Herbst and J. E. Marsden, personal communication). Like Lake Erie, *Diporeia* in Lake Champlain disappeared after dreissenid invasion (Dermott et al. 2006). Unique responses of lake whitefish after dreissenid establishment in these systems (Lake Erie, Lake Champlain, Lake Simcoe) may be associated with differences in nutrient inputs. Although nutrient abatement programs reduced total phosphorus loading

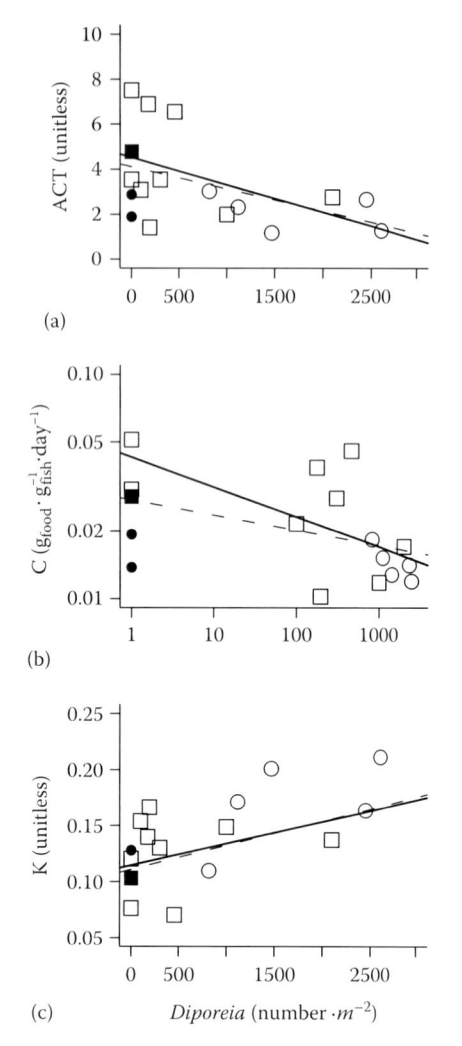

Figure 42.8 Relationships among activity (ACT) (panel a), consumption (C) (panel b), and conversion efficiency (K) (panel c) of lake whitefish with *Diporeia* density. Squares are populations with dreissenids established, and circles are those without. Filled symbols are populations in which *Diporeia* were historically absent or where their absence preceded the appearance of dreissenids. Dashed lines are relationships over all populations; solid lines are relationships excluding filled symbols. (Reproduced from Rennie, M.D. et al., *Can. J. Fish. Aquat. Sci.*, 69, 41, 2012a. With permission.)

in Lake Simcoe (Winter et al. 2007), Lake Erie (Dolan 1993, Dolan and McGunagle 2005), and Lake Champlain (Medalie et al. 2012), current nutrient inputs to all three systems are almost certainly higher compared to South Bay, Lake Huron, where nutrient levels are low and comparable to levels in the main lake basin (Fernandez et al. 2009). Further work that looks in detail at comparisons among populations from lakes of different basin morphology and/or nutrient inputs may provide a clearer explanation for the unique responses of lake whitefish to dreissenid invasion and/or *Diporeia* loss in these lakes versus most others.

OTHER EVIDENCE FOR RESOURCE LIMITATION

This review of observational studies supports the hypothesis that resource limitation is at least a partial explanation for slower growth and reduced condition in lake whitefish following dreissenid establishment. Models provide further evidence to support this hypothesis. Recently, Lumb and Johnson (2012) used bioenergetics models to simulate growth of historic and contemporary lake whitefish populations under reciprocal diet crosses (i.e., historic fish growth based on contemporary diets, and vice versa). They found that simulated growth rates of historic fish were slower than reported growth rates when contemporary diets were applied to their models (holding all other terms constant). Likewise, simulated growth rates of contemporary fish were greater than observed rates when historic diets were used. Additionally, they found that simulated growth rates declined as the proportion of dreissenids in diets increased. This work elegantly illustrated (1) the substantial impact that changes in diet (e.g., loss of *Diporeia*) can have on lake whitefish growth rates and (2) increased proportions of dreissenids in diets that are not also accompanied by increased consumption will result in decreased growth rates. Similarly, Pothoven and Madenjian (2008) estimated contemporary consumption rates of lake whitefish from Lakes Huron and Michigan would need to be up to 122% higher based on contemporary diets in order to achieve growth rates comparable to those observed prior to dreissenid establishment.

Experimental evidence also supports resource limitation as an explanation of declines in lake whitefish growth and condition following dreissenid establishment. Reductions of nutrient inputs to small boreal lakes generated dramatic declines in lake whitefish condition, survival, and abundance (Mills et al. 2002) that were similar to those observed in Lake Ontario after the disappearance of *Diporeia*. Fertilization of an acidified lake caused a pulse in recruitment and survival that resulted in a near order of magnitude increase in lake whitefish abundance (Mills et al. 2002). After nutrient additions were ceased in this lake, lake whitefish growth, condition, survival, and abundance declined dramatically, well beyond pre-manipulation levels (Figure 42.9). Because zooplankton make up a large portion of lake whitefish diets in inland lakes (Carl and McGuiness 2006), it is likely this pattern was the result of food limitation in the system. Zooplankton biomass declined dramatically after cessation of nutrient addition (M. Patterson, personal communication), which likely resulted in intense competition among lake whitefish. In another manipulation, fertilization of a lake resulted in faster growth and improved condition of lake whitefish (Mills 1985). A rapid and dramatic decline in lake whitefish condition (Mills and Chalanchuk 1987) and growth (Mills et al. 1998) in response to cessation of fertilization in the same lake was also documented, though changes in lake whitefish biomass were slower to respond (Mills and Chalanchuk 1987).

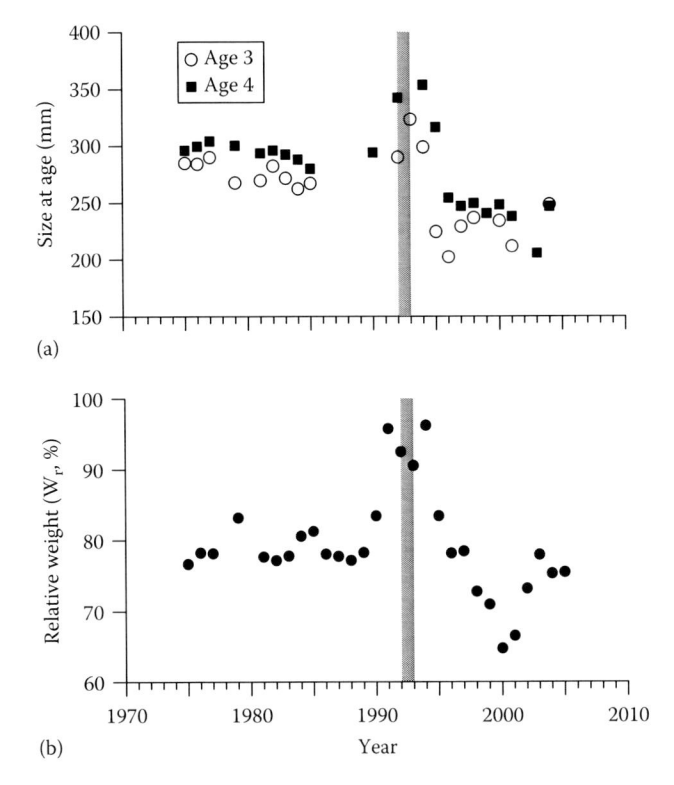

Figure 42.9 Mean size-at-age for age 3 and age 4 lake whitefish only (a) and (b) condition (relative weight, Rennie and Verdon 2008) of all lake whitefish encountered in an experimentally manipulated lake (lake 302 North, Experimental Lakes Area, Ontario). Shaded region indicates period of nutrient (phosphorus) additions.

Both nutrient manipulation experiments suggested that declines in lake whitefish growth, condition, and abundance may be universal responses to reductions in system productivity and/or prey availability. Dreissenids appear to have negatively impacted primary productivity in the Great Lakes (e.g., Fahnenstiel et al. 2010) and hence deposition of primary production in profundal zones would also be reduced. Declines in growth and condition in lake whitefish may therefore reflect a general reduction in profundal resources. This appears to have been the case in South Bay, Lake Huron, where whitefish shifted their foraging habitat and behavior toward shallower regions (Rennie et al. 2009b). Perhaps shallow systems with higher nutrient inputs (e.g., Lakes Simcoe, Erie, and Champlain) may be enriched enough to maintain lake whitefish growth and production, despite the redirection of a substantial proportion of production to nearshore habitats at the expense of offshore habitats.

ALTERNATIVE HYPOTHESES/ CONTRIBUTING FACTORS

In addition to dreissenid colonization/*Diporeia* loss, there are several other stressors that may affect life-history traits of lake whitefish. Age-at-maturity increased in every population except one (Table 42.1). Growth declines in many populations appeared to be more gradual than punctuated declines observed for other life-history traits. In many cases, changes in population abundance appeared to be more gradual (e.g., DeBruyne et al. 2008, Rennie et al. 2009a) than the establishment of dreissenids or loss of *Diporeia* could explain (e.g., Hoyle et al. 1999). A number of possible explanations for these broader, more long-term changes hypothesized in the literature include density dependence, fishery-dependent evolution, climate change, interspecific competition, and other invasive species.

Density Dependence

There is strong evidence for density-dependent growth in lake whitefish (Jensen 1981). In independent studies, experimentally controlled exploitation has resulted in increased lake whitefish growth and recruitment after a proportion of the lake whitefish population was removed, presumably as a result of decreased intraspecific competition (Healey 1975, Healey 1980, Mills and Chalanchuk 1988). Recruitment, juvenile survival, and growth of lake whitefish were all shown to be density dependent in Lake Huron (Henderson et al. 1983). Lake whitefish body condition may also vary with density (Rennie et al. 2009a).

Given the historical tendency for lake whitefish populations to be density dependent, it is perhaps surprising that it is not as clear and apparent among contemporary lake whitefish stocks in the Great Lakes. Abundance in many populations actually declined as growth rates declined (Table 42.1, see Lake Ontario, central and southern Lake Huron, northern Lake Michigan, and southern Lake Superior). Density dependence has been invoked as an explanation for growth declines in central and southern Lake Michigan (DeBruyne et al. 2008). However, the same study also shows declines in growth and condition in northern Lake Michigan where CPUE had also declined. Another study compared temporal changes in lake whitefish growth and density and found that CPUE was less important than other biological (e.g., *Diporeia* abundance) or environmental variables (growing degree days >5°C, epilimnetic volume) in explaining growth declines (Rennie et al. 2009a). Both studies relied on passive sampling gear to estimate CPUE, which is subject to changes in fish activity and catchability as previously noted (Rudstam et al. 1984). Potential behavioral changes in lake whitefish following dreissenid invasion may bias gillnet catch data and influence conclusions regarding density-dependent growth of lake whitefish in the Great Lakes in the period of dreissenid establishment. A study by Stapanian and Kocovsky (2013) also provided evidence of behavioral changes in other fish species in response to dreissenid-induced ecosystem changes in Lake Erie that have potential to influence CPUE and density estimates. Formal studies that examine relationships between annual density estimates from multiple methods (e.g., capture–recapture, active sampling gear) with growth rates and with biologically relevant time lags in response considered may provide a more robust evaluation of density dependence among contemporary Great Lakes stocks.

Climate Change

Authors have speculated about the effects of climate change on lake whitefish populations occurring both in the Great Lakes (Lynch et al. 2010) and in inland lakes (Rennie et al. 2010). The entire Great Lakes region, but especially regions in the upper Great Lakes, has experienced substantial increases in air temperature over the past 30–40 years, particularly since the early 1990s (Jensen et al. 2007, Rennie et al. 2010). Duration of ice cover over the past 30 years has also declined (Jensen et al. 2007), and thermocline depth (and therefore epilimnetic volume) in many lakes appears to have decreased (King et al. 1997, 1999, Snucins and Gunn 2000, Coats et al. 2006, Keller et al. 2006, Keller 2007), as supported by climate model predictions (McCormick 1990, Hondzo and Stefan 1993). Rennie et al. (2010) implicated warmer air temperatures and shallower thermoclines in contributing to reduced rates of primary productivity, which could negatively affect profundal deposition of pelagic algae and abundance of *Diporeia*. Variables associated with climate have been shown to be correlated with *Diporeia* abundance (Rennie et al. 2009a).

In lake whitefish, impacts of warmer air temperatures appear to be stage dependent. In their first year, growth rates of lake whitefish have been shown to be positively related to annual growing degree days (GDD, number of days above a critical temperature) >0°C (Henderson et al. 1983). For older lake whitefish that are capable of behavioral thermoregulation, the opposite was true; that is, third-year growth of fish from south Bay, Lake Huron, was negatively related to annual GDD >5°C but positively related to the percent volume of the epilimnion within the bay (i.e., increasing thermocline depth, Rennie et al. 2009a). Body condition was found to decline in populations in northwest Ontario where climate has warmed dramatically over the past 40 years, whereas body condition was more stable in southern Ontario where climate warming has been less dramatic (Rennie et al. 2010). Taylor et al. (1987) found colder, more severe winters were associated with higher recruitment events and egg survival. Henderson et al. (1983) found that environmental conditions at spawning (November) and hatching (April) did not influence year-class strength, suggesting that at this stage, larval competition for resources can influence juvenile survival rates, independent of environmental conditions (Freeberg et al. 1990).

The sum of evidence suggests that warmer temperatures over the past 40 years would negatively affect lake whitefish. While growth rates in the first year may be more rapid in warmer climates, egg survival and recruitment might be expected to decline (due to milder winters), as would growth rates and body condition of older year classes approaching sexual maturity. This might in turn cause delays in maturity and sizes at maturation. The nearly ubiquitous observation of delayed lake whitefish maturity (Table 42.1) would seem to support the effect of a broad-based stressor such as warming climate. Thus, at least some component of lake whitefish growth declines (and perhaps even declines in population abundance or recruitment) observed in dreissenid-invaded systems might be attributed to regional climate warming rather than impacts of dreissenids on resource abundance.

Interspecific Competition

It is possible that changes in fish communities in coldwater pelagic and profundal regions have influenced lake whitefish. There is evidence of negative interactions between lake whitefish and other deepwater coregonines (Davis and Todd 1998, Carl and McGuiness 2006). In Lake Superior, declines in lake whitefish densities in the mid-1990s were coincident with sharp increases in bloater densities (Bronte et al. 2003). While rainbow smelt (*Osmerus mordax*) have also been suggested to negatively influence lake whitefish (Loftus and Hulsman 1986), densities of rainbow smelt have gradually declined in the upper Great Lakes (Bronte et al. 2003, Riley et al. 2008), as have most other coldwater demersal fishes (Riley et al. 2008). While lake whitefish and round whitefish have considerable dietary overlap, spatiotemporal differences in habitat preferences were concluded to mitigate any substantial competition between these fishes (Macpherson et al. 2010).

Fishery-Dependent Evolution

A number of studies have shown clear relationships between lake whitefish mortality rates and life-history strategies, based on a theoretical foundation of life-history invariants (Jensen 1985, Jensen 1996). Evolutionary models have demonstrated how relationships among life-history traits can lead to fishery-induced evolution through various harvest strategies (Wang and Hook 2009). Typically, evolutionary consequences of fisheries-induced mortality (via selection of the largest individuals following maturity) include more rapid growth and earlier and smaller sizes at maturity as harvest rates increase. Where commercial harvest does appear to have increased in Lakes Michigan and Huron over the past 30 years (Mohr et al. 2005, DeBruyne et al. 2008), patterns of lake whitefish growth and maturity in Great Lakes stocks are opposite those predicted by fishery-dependent evolution (Table 42.1).

Other Invasive Species

It is possible that other invasive species have contributed to patterns observed in lake whitefish after dreissenid establishment. An obvious candidate might be the spiny water flea, *Bythotrephes longimanus*, which invaded the Great Lakes only 5 years before dreissenids (Sprules et al. 1990). *Bythotrephes* can make up a substantial component of lake whitefish diets (Macpherson et al. 2010) and has negatively affected growth rates of other fishes (Parker Stetter et al. 2005). Yet evidence to date suggests that *Bythotrephes* has not significantly impacted lake whitefish. Fernandez et al. (2009) estimated that prey available to larval lake whitefish was similar before and after *Bythotrephes* invaded, despite significant changes in the nearshore zooplankton community. Further, Rennie et al. (2010) found no effect of *Bythotrephes* invasion on condition of lake whitefish populations.

CONCLUSION AND RECOMMENDATIONS

An examination of available data seems to support the hypothesis that effects of dreissenid establishment on lake whitefish have been largely negative in oligotrophic systems but appear to have been less pronounced in more shallow, nutrient-rich systems. Observed declines in lake whitefish growth and condition associated with dreissenid invasion appear to be due to food limitation, consistent with evidence from populations without widespread dreissenid establishment (e.g., Lake Superior), or from populations in experimentally manipulated lakes where a decrease in resources resulted in similar patterns. While more research is needed to help clarify the context dependence of dreissenid establishment on lake whitefish populations and the underlying mechanisms involved, a broader investigation that considers systems both in and outside the Great Lakes basin may ultimately provide the most fruitful approach.

Directed future research may help to fill what appear to be major knowledge gaps in understanding effects of dreissenid establishment on lake whitefish. Quantitative comparisons of contemporary data on larval lake whitefish to historic data prior to dreissenid establishment are lacking. Though Fernandez et al. (2009) estimated little effect of dreissenid establishment on the energy available to larval prey, their conclusions were based on changes observed in the nearshore zooplankton community. With one notable exception (Hoyle et al. 2008), data on changes in lake whitefish recruitment during dreissenid establishment are also sparse. Further investigation into both of these topics would substantially improve our understanding of why differences in abundance patterns exist among stocks where dreissenids have established.

Additionally, focused laboratory studies that investigate assimilation and growth of lake whitefish fed various prey items, including dreissenids, would substantially contribute to our understanding of the role of diet changes (vs. behavioral or environmental changes associated with dreissenid invasion) on lake whitefish life histories. While it is generally assumed that dreissenids are energy poor and difficult to handle relative to other diet items (Owens and Dittman 2003, Pothoven and Madenjian 2008, Rennie et al. 2009b), there have been no laboratory studies to date that attempt to verify this empirically. Indeed, lake whitefish in the Great Lakes are persisting on diets consisting largely of dreissenids and may even exert enough predation pressure so as to influence dreissenid abundance and distribution (Madenjian et al. 2010).

Uncertainty regarding effects of dreissenid establishment on lake whitefish swimming behavior and dispersal—and therefore the potential to affect catchability of passive gear—also deserves investigation (Rennie et al. 2012a). The ability of government agencies to set sustainable harvest quotas is highly dependent on the ability to accurately estimate population abundance. Direct comparisons of active and passive sampling methods between populations in invaded and noninvaded lakes, or among populations known to vary in swimming behavior or resource availability, could provide estimates of the potential magnitude that intraspecific variation in behavior can have on gear encounter and, in turn, population abundance estimates.

Finally, these are not simple systems, and many stressors are likely impacting whitefish populations simultaneously (Rennie et al. 2009a). As such, observational data must be subjected to analytical approaches that consider multiple stressors and partition variance of many potential stressors on response variables of interest (e.g., Deroba and Bence 2009).

ACKNOWLEDGMENTS

Many people contributed data for this chapter: Jake La Rose, Adam Cottrill, and Rick Salmon of the Ontario Ministry of Natural Resources; Mark Ebener of the Chippewa Ottawa Resource Authority; Jason Stockwell and Mark Vinson of the U.S. Geological Survey; and Ken Mills of Fisheries and Oceans Canada (retired). Discussions with Mike Paterson and Scott Higgins (Fisheries and Oceans Canada) were helpful in developing the ideas presented here.

REFERENCES

Adamstone, F. B. 1924. The distribution and economic importance of the bottom fauna of Lake Nipigon with an appendix on the bottom fauna of Lake Ontario. *Univ. Toronto Stud. Biol. Ser.* 24:35–95.

Bajer, P. G. and R. S. Hayward. 2006. A combined multiple-regression and bioenergetics model for simulating fish growth in length and condition. *Trans. Am. Fish. Soc.* 135:695–710.

Barbiero, R. P., K. Schmude, B. M. Lesht, C. M. Riseng, G. L. Warren, and M. L. Tuchman. 2011. Trends in *Diporeia* populations across the Laurentian Great Lakes, 1997–2009. *J. Great Lakes Res.* 37:9–17.

Beauchamp, K. C., N. C. Collins, and B. A. Henderson. 2004. Covariation of growth and maturation of lake whitefish (*Coregonus clupeaformis*). *J. Great Lakes Res.* 30:451–460.

Bentz, J., D. Kwiatkowski, G. Persson, K. Deacon, and V. MacDonald. 2002. Life science values of enhanced management areas and conservation reserves within the Nipigon basin. Geowest Environmental Consultants Ltd., Edmonton, Alberta, Canada.

Biro, P. A. and N. J. Dingemanse. 2009. Sampling bias resulting from animal personality. *Trends Ecol. Evol.* 24:66–67.

Biro, P. A. and J. R. Post. 2008. Rapid depletion of genotypes with fast growth and bold personality traits from harvested fish populations. *Proc. Natl. Acad. Sci.* 105:2919–2922.

Bronte, C. R., M. P. Ebener, and D. R. Schreiner et al. 2003. Fish community change in Lake Superior, 1970–2000. *Can. J. Fish. Aquat. Sci.* 60:1552–1574.

Carl, L. M. and F. McGuiness. 2006. Lake whitefish and lake herring population structure and niche in ten South-central Ontario lakes. *Environ. Biol. Fish.* 75:315–323.

Coats, R., J. Perez-Losada, G. Schladow, R. Richards, and C. Goldman. 2006. The warming of Lake Tahoe. *Climat. Change* 76:121–148.

Conroy, J. D., D. D. Kane, D. M. Dolan, W. J. Edwards, M. N. Charlton, and D. A. Culver. 2005. Temporal trends in Lake Erie plankton biomass: Roles of external phosphorus loading and dreissenid mussels. *J. Great Lakes Res.* 31:89–110.

Davis, B. M. and T. N. Todd. 1998. Competition between larval lake herring (*Coregonus artedi*) and lake whitefish (*Coregonus clupeaformis*) for zooplankton. *Can. J. Fish. Aquat. Sci.* 55:1140–1148.

DeBruyne, R. L., T. L. Galarowicz, R. M. Claramunt, and D. F. Clapp. 2008. Lake whitefish relative abundance, length-at-age, and condition in Lake Michigan indicated by fishery-independent surveys. *J. Great Lakes Res.* 34:235–244.

Dermott, R. 2001. Sudden disappearance of the amphipod *Diporeia* from Eastern Lake Ontario, 1993–1995. *J. Great Lakes Res.* 27:423–433.

Dermott, R., R. Bonnell, and P. Jarvis. 2006. Changes in abundance of deep water amphipod *Diporeia* (Pontoporeiidae) in eastern North American lakes with or without *Dreissena* mussels. Canadian Technical Report of Fisheries and Aquatic Sciences No. 2636.

Dermott, R. and D. Kerec. 1997. Changes to the deepwater benthos of eastern Lake Erie since the invasion of *Dreissena*: 1979–1993. *Can. J. Fish. Aquat. Sci.* 54:922–930.

Deroba, J. J. and J. R. Bence. 2009. Developing model-based indices of lake whitefish abundance using commercial fishery catch and effort data in lakes Huron, Michigan and Superior. *North Am. J. Fish. Manage.* 29:50–63.

Dobiesz, N. E., D. A. McLeish, and R. L. Eshenroder et al. 2005. Ecology of the Lake Huron fish community, 1970–1999. *Can. J. Fish. Aquat. Sci.* 62:1432–1451.

Dolan, D. M. 1993. Point-source loadings of phosphorus to Lake Erie—1986–1990. *J. Great Lakes Res.* 19:212–223.

Dolan, D. M. and K. P. McGunagle. 2005. Lake Erie total phosphorus loading analysis and update: 1996–2002. *J. Great Lakes Res.* 31:11–22.

Ebener, M. P. 1997. Recovery of lake whitefish populations in the Great Lakes—A story of successful management and just plain luck. *Fisheries* 22:18–20.

Evans, D. O., J. J. Houston, and G. N. Meredith. 1988. Status of the Lake Simcoe whitefish, *Coregonus clupeaformis*, in Canada. *Can. Field Nat.* 102:103–113.

Evans, D. O., K. H. Nicholls, Y. C. Allen, and M. J. McMurtry. 1996. Historical land use, phosphorus loading, and loss of fish habitat in Lake Simcoe, Canada. *Can. J. Fish. Aquat. Sci.* 53:194–218.

Fahnenstiel, G. L., S. A. Pothoven, H. A. Vanderploeg, D. M. Klarer, T. F. Nalepa, and D. Scavia. 2010. Recent changes in primary production and phytoplankton in the offshore region of southeastern Lake Michigan. *J. Great Lakes Res.* 36(Suppl. 3):20–29.

Fernandez, R. J., M. D. Rennie, and W. G. Sprules. 2009. Changes in nearshore zooplankton associated with species invasions and potential effects on larval lake whitefish (*Coregonus clupeaformis*). *Int. Rev. Hydrobiol.* 94:226–243.

Freeberg, M. H., W. W. Taylor, and R. W. Brown. 1990. Effect of egg and larval survival on year class strength of lake whitefish in Grand Traverse Bay, Lake Michigan. *Trans. Am. Fish. Soc.* 119:92–100.

Freshwater Fish Marketing Corporation. 2010. 2010 Annual Report. http://www.freshwaterfish.com/system/files/FFMC%20AR%2009-10-EnglishFINAL-July27–10.pdf (accessed 1-1-2011).

Gewurtz, S. B., S. P. Bhavsar, and D. A. Jackson et al. 2011. Trends of legacy and emerging-issue contaminants in Lake Simcoe fishes. *J. Great Lakes Res.* 37(Suppl. 3):148–159.

Gorman, O. T., L. M. Evrard, G. A. Cholwek, J. M. Falck, and M. R. Vinson. 2010. Status and trends of prey fish populations in Lake Superior, 2009. Windsor, Ontario, Canada: Great Lakes Fishery Commission Report.

Healey, M. C. 1975. Dynamics of exploited whitefish populations and their management with special reference to the Northwest Territories. *J. Fish. Res. Board Can.* 32:427–437.

Healey, M. C. 1980. Growth and recruitment in experimentally exploited lake whitefish (*Coregonus clupeaformis*) populations. *Can. J. Fish. Aquat. Sci.* 37:255–267.

Hecky, R. E., H. Smith, D. R. Barton, S. J. Guildford, W. Taylor, M. N. Charlton, and T. Howell. 2004. The nearshore phosphorus shunt: A consequence of ecosystem engineering by dreissenids in the Laurentian Great Lakes. *Can J. Fish Aquat. Sci.* 61:1285–1293.

Henderson, B. A., J. J. Collins, and J. A. Reckahn. 1983. Dynamics of an exploited population of lake whitefish (*Coregonus clupeaformis*) in Lake Huron. *Can. J. Fish. Aquat. Sci.* 40:1556–1567.

Higgins, S. N. and M. J. Vander Zanden. 2010. What a difference a species makes: A meta-analysis of dreissenid mussel impacts on freshwater ecosystems. *Ecol. Monogr.* 80:179–196.

Hondzo, M. and H. G. Stefan. 1993. Regional water temperature characteristics of lakes subjected to climate change. *Clim. Change* 24:187–211.

Hopper, M. and G. Power. 1991. The fisheries of an Ojibway community in northern Ontario. *Arctic* 44:267–274.

Hoyle, J. A. 2005. Status of lake whitefish (*Coregonus clupeaformis*) in Lake Ontario and the response to the disappearance of *Diporeia* spp. In *Proceedings of a Workshop on the Dynamics of Lake Whitefish (Coregonus clupeaformis) and the Amphipod Diporeia spp. in the Great Lakes*. L. C. Mohr and T. F. Nalepa, eds., pp. 47–66. Ann Arbor, MI: Great Lakes Fishery Commission Technical Report 66.

Hoyle, J. A., J. N. Bowlby, and B. J. Morrison. 2008. Lake whitefish and walleye population responses to dreissenid mussel invasion in eastern Lake Ontario. *Aquat. Ecosyst. Health Manage.* 11:403–411.

Hoyle, J. A., T. Schaner, J. M. Casselman, and R. Dermott. 1999. Changes in lake whitefish (Coregonus clupeaformis) stocks in eastern Lake Ontario following Dreissena mussel invasion. *Great Lakes Res. Rev.* 4:5–10.

Jensen, A. L. 1981. Population regulation in lake whitefish, *Coregonus clupeaformis* (Mitchill). *J. Fish Biol.* 19:557–573.

Jensen, A. L. 1985. Relations among net reproductive rate and life-history parameters for lake whitefish (*Coregonus clupeaformis*). *Can. J. Fish. Aquat. Sci.* 42:164–168.

Jensen, A. L. 1996. Beverton and Holt life history invariants result from optimal trade-off of reproduction and survival. *Can. J. Fish. Aquat. Sci.* 53:820–822.

Jensen, O. P., B. J. Benson, and J. J. Magnuson et al. 2007. Spatial analysis of ice phenology trends across the Laurentian Great Lakes region during a recent warming period. *Limnol. Oceanogr.* 52:2013–2026.

Jimenez, A., M. D. Rennie, W. G. Sprules, and J. K. L. La Rose. 2011. Temporal changes in the benthic invertebrate community of Lake Simcoe, 1983–2008. *J. Great Lakes Res.* 37(Suppl. 3):103–112.

Keller, W. 2007. Implications of climate warming for Boreal Shield lakes: A review and synthesis. *Environ. Rev.* 15:99–112.

Keller, W., J. Heneberry, J. Leduc, J. Gunn, and N. Yan. 2006. Variations in epilimnion thickness in small Boreal Shield Lakes: Relationships with transparency, weather and acidification. *Environ. Monit. Assess.* 115:419–431.

King, J. R., B. J. Shuter, and A. P. Zimmerman. 1997. The response of the thermal stratification of South Bay (Lake Huron) to climatic variability. *Can. J. Fish. Aquat. Sci.* 54:1873–1882.

King, J. R., B. J. Shuter, and A. P. Zimmerman. 1999. Signals of climate trends and extreme events in the thermal stratification pattern of multibasin Lake Opeongo, Ontario. *Can. J. Fish. Aquat. Sci.* 56:847–852.

Kinnunen, R. E. 2003. Great Lakes commercial fisheries. Available from Michigan Sea Grant Extension, 710 Chippewa Square, Suite 202, Marquette, MI.

Kratzer, J. F., W. W. Taylor, and M. Turner. 2007. Changes in fecundity and egg lipid content of lake whitefish (*Coregonus clupeaformis*) in the upper Laurentian Great Lakes between 1986–87 and 2003–05. *J. Great Lakes Res.* 33:922–929.

Loftus, D. H. and P. F. Hulsman. 1986. Predation on larval lake whitefish (*Coregonus clupeaformis*) and lake herring (*Coregonus artedii*) by adult rainbow smelt (*Osmerus mordax*). *Can. J. Fish. Aquat. Sci.* 43:812–818.

Lumb, C. E. and T. B. Johnson. 2012. Retrospective growth analysis of lake whitefish (*Coregonus clupeaformis*) in lakes Erie and Ontario, 1954–2003. *Adv. Limnol.* 63:429–454.

Lumb, C. E., T. B. Johnson, H. A. Cook, and J. A. Hoyle. 2007. Comparison of lake whitefish (*Coregonus clupeaformis*) growth, condition, and energy density between lakes Erie and Ontario. *J. Great Lakes Res.* 33:314–325.

Lynch, A. J., W. W. Taylor, and K. D. Smith. 2010. The influence of changing climate on the ecology and management of selected Laurentian Great Lakes fisheries. *J. Fish Biol.* 77:1764–1782.

Macpherson, A., J. A. Holmes, A. M. Muir, and D. L. G. Noakes. 2010. Assessing feeding competition between lake whitefish *Coregonus clupeaformis* and round whitefish *Prosopium cylindraceum*. *Curr. Zool.* 56:109–117.

Madenjian, C. P., D. V. O'Connor, and S. A. Pothoven et al. 2006. Evaluation of a lake whitefish bioenergetics model. *Trans. Am. Fish. Soc.* 135:61–75.

Madenjian, C. P., S. A. Pothoven, and P. J. Schneeberger et al. 2010. Dreissenid mussels are not a "dead end" in Great Lakes food webs. *J. Great Lakes Res.* 36:73–77.

McCormick, M. J. 1990. Potential changes in thermal structure and cycle of Lake Michigan due to global warming. *Trans. Am. Fish. Soc.* 119:183–194.

McNickle, G. G., M. D. Rennie, and W. G. Sprules. 2006. Changes in benthic invertebrate communities of South Bay, Lake Huron following invasion by zebra mussels (*Dreissena polymorpha*), and potential effects on lake whitefish (*Coregonus clupeaformis*) diet and growth. *J. Great Lakes Res.* 32:180–193.

Medalie, L., R. M. Hirsch, and S. A. Ardifield. 2012. Use of flow-normalization to evaluate nutrient concentration and flux changes in Lake Champlain tributaries, 1990–2009. *J. Great Lakes Res.* 38 (Suppl. 1): 58–67.

Messick, G. A., R. M. Overstreet, T. F. Nalepa, and S. Tyler. 2004. Prevalence of parasites in amphipods *Diporeia* spp. from Lakes Michigan and Huron, USA. *Dis. Aquat. Organ.* 59:159–170.

Mills, K. H. 1985. Responses of lake whitefish (*Coregonus clupeaformis*) to fertilization of Lake-226, the Experimental Lakes Area. *Can. J. Fish. Aquat. Sci.* 42:129–138.

Mills, K. H. and S. M. Chalanchuk. 1987. Population dynamics of lake whitefish (*Coregonus clupeaformis*) during and after the fertilization of Lake-226, the Experimental Lakes Area. *Can. J. Fish. Aquat. Sci.* 44:55–63.

Mills, K. H. and S. M. Chalanchuk. 1988. Population dynamics of unexploited lake whitefish (*Coregonus clupeaformis*) in one experimentally fertilized lake and three exploited lakes. *Finn. Fish. Res.* 9:145–153.

Mills, K. H., S. M. Chalanchuk, D. M. Findlay, D. J. Allan, and B. R. McCulloch. 2002. Condition, recruitment and abundance of lake whitefish (*Coregonus clupeaformis*) in a fertilized acid lake. *Arch. Hydrobiol. Spec. Issues Adv. Limnol.* 57:423–433.

Mills, K. H., B. R. McCulloch, S. M. Chalanchuk, D. J. Allan, and M. P. Stainton. 1998. Growth, size, structure and annual survival of lake whitefish (*Coregonus clupeaformis*) during the eutrophication and oligotrophication of lake 226, the Experimental Lakes Area, Canada. *Arch. Hydrobiol. Spec. Issues Adv. Limnol.* 50:151–160.

Mohr, L. C., M. P. Ebener, and T. F. Nalepa. 2005. Status of lake whitefish (*Coregonus clupeaformis*) in Lake Huron. In *Proceedings of a Workshop on the Dynamics of Lake Whitefish (Coregonus clupeaformis) and the Amphipod Diporeia spp. in the Great Lakes*. L. C. Mohr and T. F. Nalepa, eds., pp. 105–126. Great Lakes Fishery Commission Technical Report 66. Ann Arbor, MI: Great Lakes Fishery Commission.

Mohr, L. C. and T. F. Nalepa. 2005. *Proceedings of a Workshop on the Dynamics of Lake Whitefish (Coregonus clupeaformis) and the Amphipods Diporeia spp. in the Great Lakes*. Great Lakes Fishery Commission Technical Report 66. Ann Arbor, MI: Great Lakes Fishery Commission.

Nalepa, T. F., D. L. Fanslow, A. J. Foley III, G. A. Lang, B. J. Eadie, and M. A. Quigley. 2006b. Continued disappearance of the benthic amphipod *Diporeia* spp. in Lake Michigan: Is there evidence for food limitation? *Can. J. Fish. Aquat. Sci.* 63:872–890.

Nalepa, T. F., D. L. Fanslow, and G. A. Lang. 2009. Transformation of the offshore benthic community in Lake Michigan: Recent shift from the native amphipod *Diporeia* spp. to the invasive mussel *Dreissena rostriformis bugensis*. *Freshwat. Biol.* 54:466–479.

Nalepa, T. F., D. L. Fanslow, S. A. Pothoven, A. J. Foley, and G. A. Lang. 2007. Long-term trends in benthic macroinvertebrate populations in Lake Huron over the past four decades. *J. Great Lakes Res.* 33:421–436.

Nalepa, T. F., D. J. Hartson, D. L. Fanslow, G. A. Lang, and S. J. Lozano. 1998. Declines in benthic macroinvertebrate populations in southern Lake Michigan, 1980–1993. *Can. J. Fish. Aquat. Sci.* 55:2402–2413.

Nalepa, T. F., L. C. Mohr, B. A. Henderson, C. P. Madenjian, and P. J. Schneeberger. 2005. Lake whitefish and *Diporeia* spp. in the Great Lakes: An overview. In *Proceedings of a Workshop on the Dynamics of Lake Whitefish (Coregonus clupeaformis) and the Amphipod Diporeia spp. in the Great Lakes*. L. C. Mohr and T. F. Nalepa, eds., pp. 3–20. Ann Arbor, MI: Great Lakes Fishery Commission.

Nalepa, T. F., D. C. Rockwell, and D. W. Schloesser. 2006a. *Disappearance of the Amphipod Diporeia spp. in the Great Lakes: Workshop Summary, Discussions and Recommendations*. NOAA Technical Memorandum GLERL-136. Ann Arbor, MI: Great Lakes Environmental Research Laboratory.

Owens, R. W. and D. E. Dittman. 2003. Shifts in the diets of slimy sculpin (*Cottus cognatus*) and lake whitefish (*Coregonus clupeaformis*) in Lake Ontario following the collapse of the burrowing amphipod *Diporeia*. *Aquat. Ecosyst. Health Manage*. 6:311–323.

Parker Stetter, S. L., L. D. Witzel, L. G. Rudstam, D. W. Einhouse, and E. L. Mills. 2005. Energetic consequences of diet shifts in Lake Erie rainbow smelt (*S*). *Can. J. Fish. Aquat. Sci*. 62:145–152.

Pothoven, S. A. and C. P. Madenjian. 2008. Changes in consumption by alewives and lake whitefish after dreissenid mussel invasions in Lakes Michigan and Huron. *N. Am. J. Fish. Manage*. 28:308–320.

Pothoven, S. A., T. F. Nalepa, C. P. Madenjian, R. R. Rediske, P. J. Schneeberger, and J. X. He. 2006. Energy density of lake whitefish *Coregonus clupeaformis* in Lakes Huron and Michigan. *Environ. Biol. Fish*. 76:151–158.

Pothoven, S. A., T. F. Nalepa, P. J. Schneeberger, and S. B. Brandt. 2001. Changes in diet and body condition of lake whitefish in southern Lake Michigan associated with changes in benthos. *N. Am. J. Fish. Manage*. 21:876–883.

Rawson, D. S. 1930. The bottom fauna of Lake Simcoe and its role in the ecology of the lake. *Univ. Toronto Stud. Biol. Ser*. 40:1–183.

Rawson, D. S. 1952. Mean depth and the fish production of large lakes. *Ecology* 33:513–521.

Rennie, M. D. and R. Verdon. 2008. Development and evaluation of condition indices for the Lake Whitefish. *N. Am. J. Fish. Manage*. 28:1270–1293.

Rennie, M. D. 2009. Influence of invasive species, climate change and population density on life histories and mercury dynamics of *Coregonus* spp. PhD thesis. Toronto, Ontario, Canada: University of Toronto.

Rennie, M. D., W. G. Sprules, and T. B. Johnson. 2009a. Factors affecting the growth and condition of lake whitefish (*Coregonus clupeaformis*). *Can. J. Fish. Aquat. Sci*. 66:2096–3108.

Rennie, M. D., W. G. Sprules, and T. B. Johnson. 2009b. Resource switching in fish following a major food web disruption. *Oecologia* 159:789–802.

Rennie, M. D., W. G. Sprules, and A. Vaillancourt. 2010. Changes in fish condition and mercury vary by region, not *Bythotrephes* invasion: A result of climate change? *Ecography* 33:471–482.

Rennie, M. D., M. P. Ebener, and T. Wagner. 2012b. Can migration mitigate the effects of ecosystem change? Patterns of dispersal, energy acquisition and allocation in Great Lakes lake whitefish (*Coregonus clupeaformis*). *Adv. Limnol*. 63:455–476.

Rennie, M. D., W. G. Sprules, and T. B. Johnson. 2012a. Energy acquisition and allocation patterns of lake whitefish (*Coregonus clupeaformis*) are modified when dreissenids are present. *Can. J. Fish. Aquat. Sci*. 69:41–59.

Riley, S. C., E. F. Roseman, S. J. Nichols, T. P. O'Brien, C. S. Kiley, and J. S. Schaeffer. 2008. Deepwater demersal fish community collapse in Lake Huron. *Trans. Am. Fish. Soc*. 137:1879–1890.

Rudstam, L. G., J. J. Magnuson, and W. M. Tonn. 1984. Size selectivity of passive fishing gear: A correction for encounter probability applied to gill nets. *Can. J. Fish. Aquat. Sci*. 41:1252–1255.

Scharold, J. V., S. J. Lozano, and T. D. Corry. 2004. Status of the amphipod *Diporeia* spp. in Lake Superior, 1994–2000. *J. Great Lakes Res*. 30:360–368.

Scott, W. B. and E. J. Crossman. 1998. *Freshwater Fishes of Canada*. Oakville, Ontario, Canada: Galt House Publications Ltd.

Smith, B. R. and J. J. Tibbles. 1980. Sea lamprey (*Petromyzon marinus*) in lakes Huron, Michigan, and Superior—History of invasion and control, 1936–78. *Can. J. Fish. Aquat. Sci*. 37:1780–1801.

Snucins, E. and J. Gunn. 2000. Interannual variation in the thermal structure of clear and colored lakes. *Limnol. Oceanogr*. 45:1639–1646.

Spangler, G. R. and J. J. Collins. 1992. Lake Huron fish community structure based on gill-net catches corrected for selectivity and encounter probability. *N. Am. J. Fish. Manage*. 12:585–597.

Sprules, W. G., H. P. Riessen, and E. H. Jin. 1990. Dynamics of the *Bythotrephes* invasion of the St. Lawrence Great Lakes. *J. Great Lakes Res*. 16:346–351.

Stapanian, M. A. and P. M. Kocovsky. 2013. Effects of dreissenids on monitoring and management of fisheries. In *Quagga and Zebra Mussels: Biology, Impacts, and Control*, 2nd Edn. T. F. Nalepa and D. W. Schloesser, eds., pp. 681–695. Boca Raton, FL: CRC press.

Suess, H. E. 1955. Radiocarbon concentration in modern wood. *Science* 122:415–417.

Taylor, W. W., M. A. Smale, and M. H. Freeberg. 1987. Biotic and abiotic determinants of lake whitefish (*Coregonus clupeaformis*) recruitment in northeastern Lake Michigan. *Can. J. Fish. Aquat. Sci*. 44:313–323.

Verburg, P. 2007. The need to correct for the Suess effect in the application of d^{13}C in sediment of autotrophic Lake Tanganyika, as a productivity proxy in the Anthropocene. *J. Paleolimnol*. 37:591–602.

Wang, H. Y. and T. O. Hook. 2009. Eco-genetic model to explore fishing-induced ecological and evolutionary effects on growth and maturation schedules. *Evol. Appl*. 2:438–455.

Wang, H. Y., T. O. Hook, M. P. Ebener, L. C. Mohr, and P. J. Schneeberger. 2008. Spatial and temporal variation of maturation schedules of lake whitefish (*Coregonus clupeaformis*) in the Great Lakes. *Can. J. Fish. Aquat. Sci*. 65:2157–2169.

Watkins, J. M., R. Dermott, S. J. Lozano, E. L. Mills, L. G. Rudstam, and J. V. Scharold. 2007. Evidence for remote effects of dreissenid mussels on the amphipod *Diporeia*: Analysis of Lake Ontario benthic surveys, 1972–2003. *J. Great Lakes Res*. 33:642–657.

Weatherley, A. H. 1966. Ecology of fish growth. *Nature* 212:1321.

Winter, J. G., M. C. Eimers, P. J. Dillon, L. D. Scott, W. A. Scheider, and C. C. Willox. 2007. Phosphorus inputs to Lake Simcoe from 1990 to 2003: Declines in tributary loads and observations on lake water quality. *J. Great Lakes Res*. 33:381–396.

(a)   (b)

Figure 2.3 Settling plate array used to monitor seasonal recruitment of juvenile zebra mussels. (a) Plate prior to deployment and (b) plate after 24 weeks deployment at station CHIP in 2003.

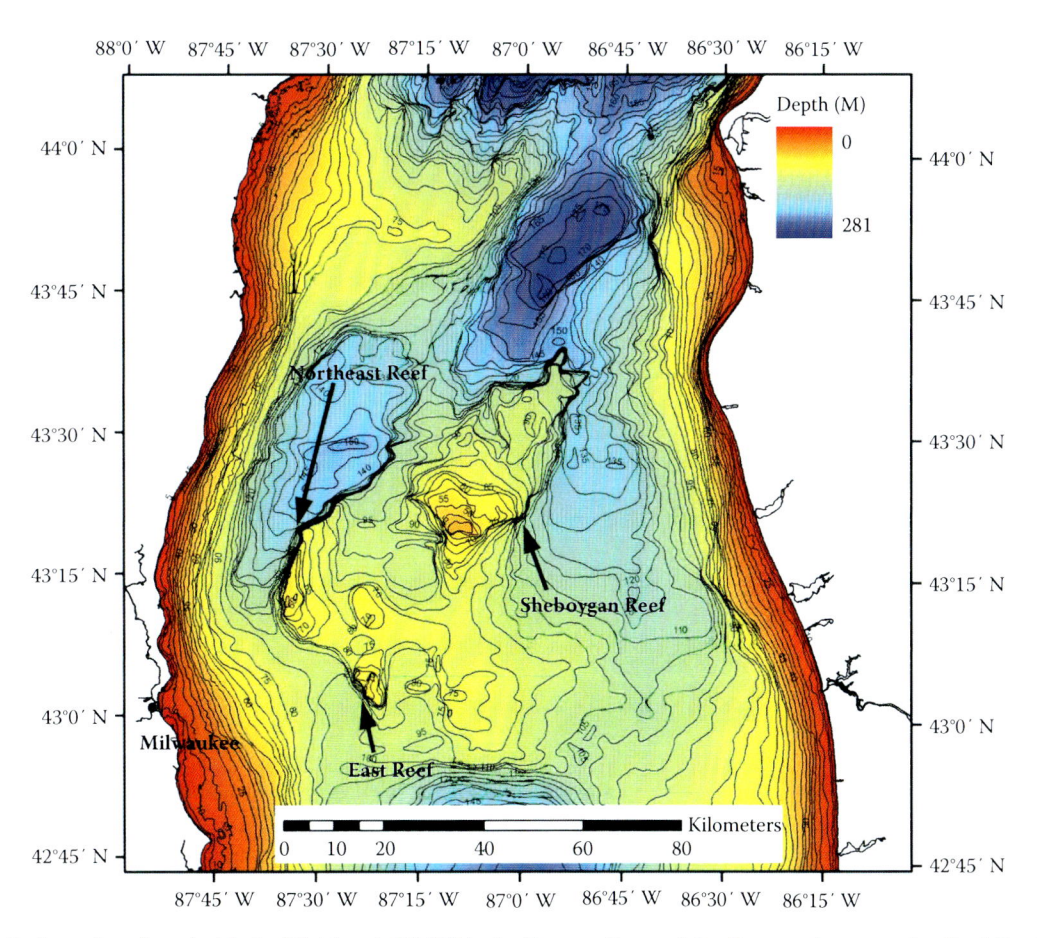

Figure 4.1 Bathymetry of central Lake Michigan's MLRC indicating positions of the three major summits, East Reef, Northeast Reef, and Sheboygan Reef. (Modified from NOAA: Great Lakes Data Rescue Project—Lake Michigan Bathymetry, Area III, 2010.)

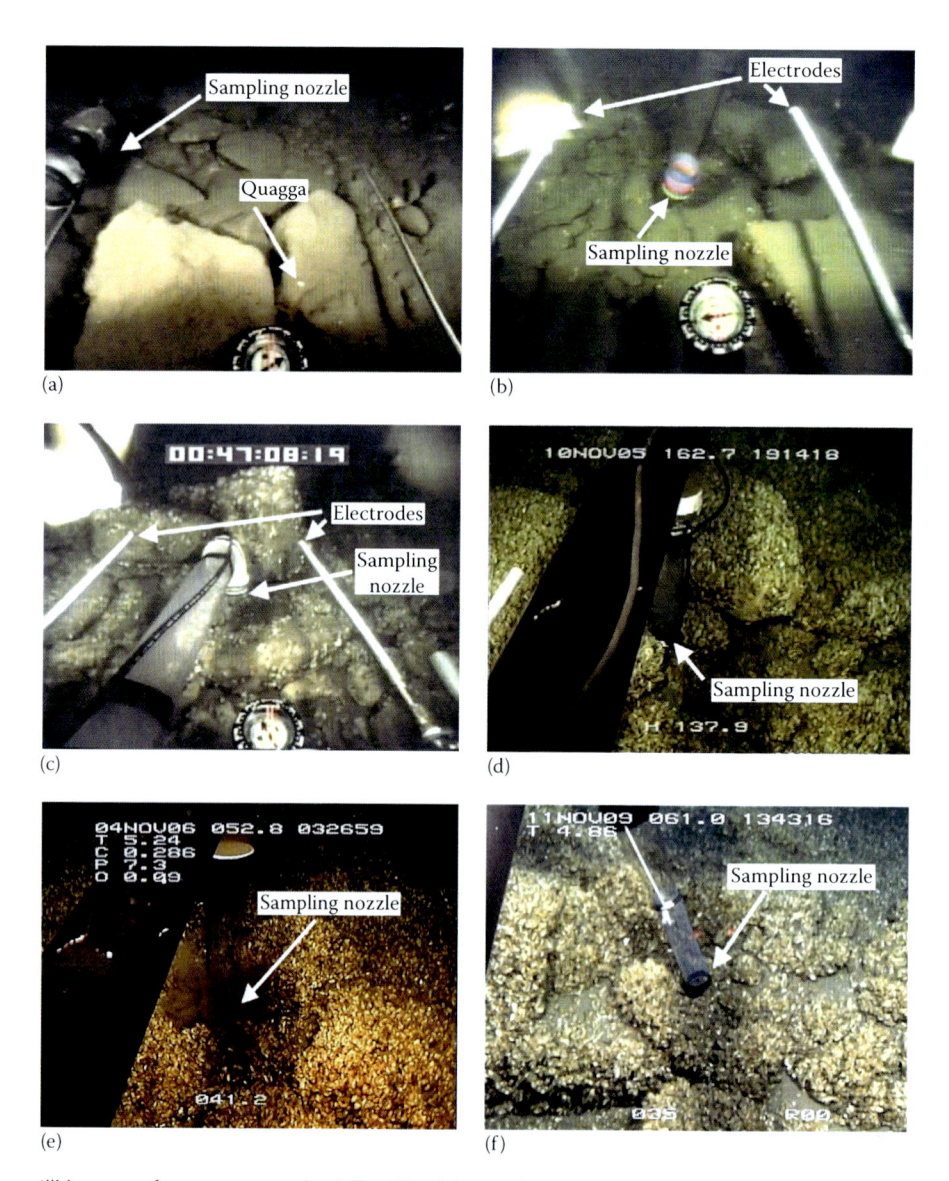

Figure 4.2 Video–still images of quagga mussels at East Reef (approximately 43°01′ N and 87°21′ W). The nozzle (50 mm diameter Plexiglas) and electrodes (35 cm long and 35 cm separation) are labeled for size reference. (a) April 2002; this was our first dive at East Reef, very few quagga mussels were present, and they appeared as scattered white spots on the image. (b) May 2003; quagga mussels were still uncommon. (c) November 2003; quagga mussels were now common and in clusters. (d) November 2005; quagga mussels now covered most rocks, but there are occasional areas of bare rock. (e) November 2006; all cobble and boulder surfaces were now totally covered with quagga mussels. (f) November 2009; quagga mussels covered all boulders and cobbles, but there was less coverage on bedrock underlying cobble and boulders.

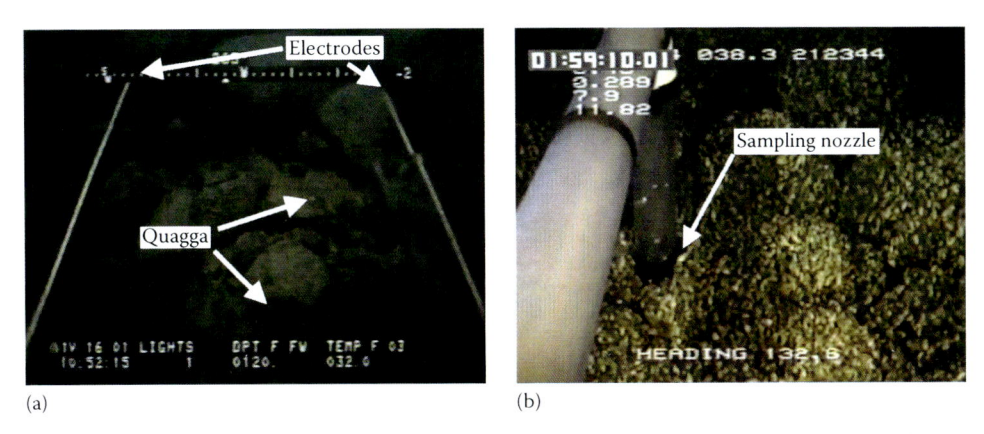

Figure 4.3 Video–still images of quagga mussels at Sheboygan Reef (approximately 43°21′ N and 87°09′ W) at a depth of approximately 40 m. Sheboygan Reef was sampled less frequently than East Reef, but the invasion timing appeared to be similar to that for East Reef. (a) November 2001; only a few quagga mussels were visible. (b) November 2004; quagga mussels covered nearly all rocks.

Figure 4.4 Quagga mussels on smooth bedrock immediately south of the sites in Figure 4.3. (Sheboygan Reef, June 2004, approximately 43° 20.5′ N and 87°09′ W.)

Figure 4.5 Still images of Northeast Reef, November 2010 (approximately 43°15′ N and 87°35′ W). (a) Somewhat lateral view of quagga mussels with their siphons extended. Note that siphons obscured resolution of individual quagga mussels. (b) View of quagga mussels with siphons extended with the camera nearly orthogonal to substrate surface. Note the cluster of *Hydra* (circle), and closer inspection reveals other scattered *Hydra*. (c) Quagga mussels with siphons contracted due to electroshocking. The slimy sculpin (*C. cognatus*) had been stunned by the electroshocker. (d) Quagga mussels with siphons contracted near an electroshocked burbot (*Lota lota*). Away from the electroshocked area, mussel siphons (upper corners) were more extended and the image was hazier.

Figure 4.6 Bycatch of quagga mussels in beam trawl used to collect lake trout fry at East Reef (June 2010). The total catch was about 200 L of live quagga mussels.

Figure 4.7 Quagga mussels with byssal threads attached to coarse sand from East Reef. These mussels were collected via beam trawl.

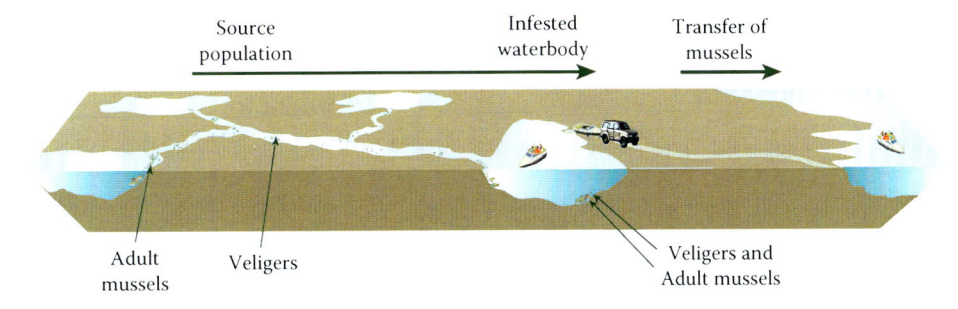

Figure 14.1 Pathways by which dreissenid mussels can be transported between water bodies.

Figure 14.4 Quagga mussels on a carbon-steel debris rack of the water treatment plant in Lake Mead. Pen included for scale.

Figure 14.5 Quagga mussels on the stainless-steel, flare end of the water intake in Lake Mead.

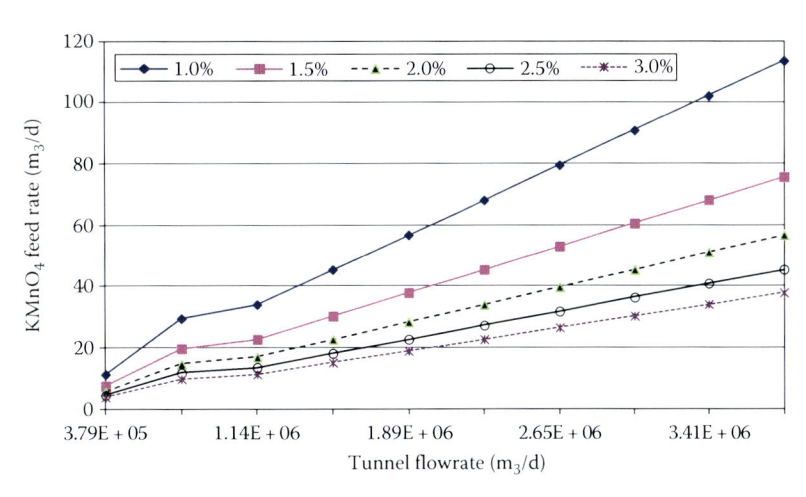

Figure 14.8 Feed rate (m³/day) for various solutions (percent) of potassium permanganate (KMnO$_4$) relative to water flow rate (m³/day) in the intake tunnel at the water treatment plant in Lake Mead. Rates were calculated based on a KMnO$_4$ dose of 0.3 mg/L.

Figure 14.9 Pueblo Dam and Reservoir on the Arkansas River near Pueblo, Colorado. See also Prescott et al. (2013).

Figure 15.2 Surface of an isolation gate that has been colonized by mussels. (Courtesy of U.S. Bureau of Reclamation, Washington, DC.)

Figure 15.3 Gate webbing on recently cleaned gate showing small drain holes at risk of being plugged by mussels. (Courtesy of U.S. Bureau of Reclamation, Washington, DC.)

Figure 15.11 Photo collage of dreissenid mussels on forebay walls (a and d) and trashracks (b and c) of hydroelectric power plants. (Courtesy of Ontario Power Generation, Toronto, Ontario, Canada.)

Figure 15.14 Manifold of a heat exchanger fouled with dreissenid mussels.

Figure 15.15 Downstream side of a self-cleaning filter fouled with dreissenid mussels.

Figure 15.16 Pressure-regulating valve discharge fouled with dreissenid mussels. (Courtesy of Ontario Power Generation, Toronto, Ontario, Canada.)

Figure 15.17 Cooling water pipe of a transformer fouled with dreissenid mussels.

Figure 15.18 Thrust-bearing sight glass with live dreissenid mussels.

Figure 15.19 Fire-pump bell housing fouled with dreissenid mussels.

Figure 20.4 Close-up image of deepwater morphs taken with an ROV on the mid-lake reef (Northeast Reef; see Chapter 6) in Lake Michigan. Note the extension of inhalant siphons with openings facing in different directions. The circled individuals show examples of mussels in which relative measurements of siphon length and shell length were obtainable.

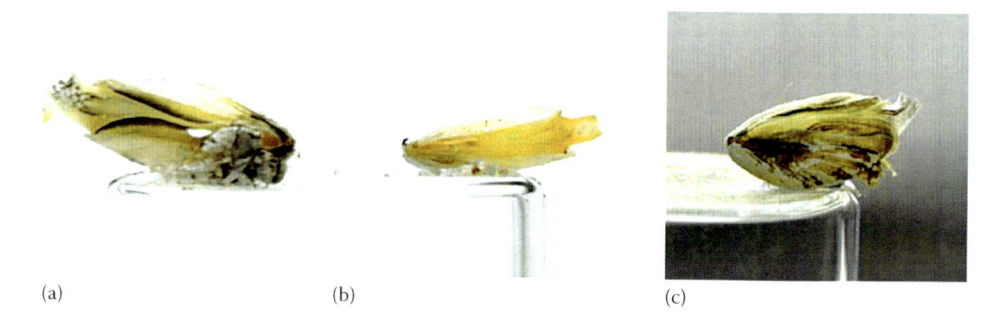

(a) (b) (c)

Figure 20.9 Images of typical siphon lengths of the deepwater morph from the 25 m site (a) and the 100 m site (b) in Lake Michigan and of shallow morph form a 7 m site in western Lake Erie (c). Relative siphon lengths were near the mean siphon lengths of all mussels at these sites. The typical siphon length for mussels at the 45 m site in Lake Michigan is shown in Figure 20.6b through e (mussel on the left).

Figure 20.13 Image of deepwater morphs that shows individuals partially burrowed in the substrate. The image was taken with a camera attached to an ROV at 40 m depth in Lake Michigan off Muskegon, MI. Note individuals in the relatively open areas in the lower right of the image. The substrate was silty sand.

Figure 31.1 Quagga mussels in Lake Mead. The left image shows a quagga mussel (top individual, 4.2 cm shell length) collected from a shallow area (15 m) near Sentinel Island and a quagga mussel (bottom individual, 4.8 cm in shell length) collected from a deep area (108 m) in the narrows between the Boulder Basin and the Virgin Basin (photo by Wai Hing Wong). The right image shows quagga mussels attached to rocks in a shallow area of Boulder Basin (photo by Bryan Moore).

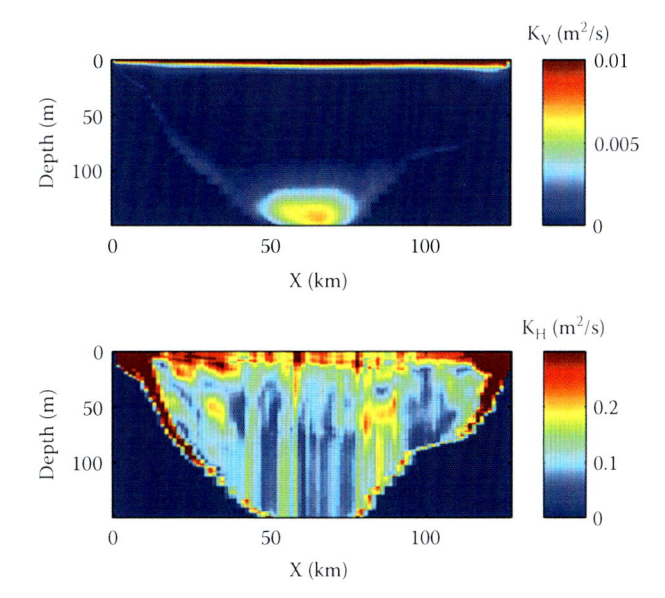

Figure 35.4 Average monthly vertical (K_V, m^2/s, top panel) and horizontal (K_H, m^2/s, bottom panel) diffusivity in Lake Michigan as derived from the 3-D hydrodynamic simulation during strongly stratified conditions (August, 2009). The cross section is along a line that extends (in km) between Milwaukee, Wisconsin (a: west shoreline), and Muskegon, Michigan (b: east shoreline).

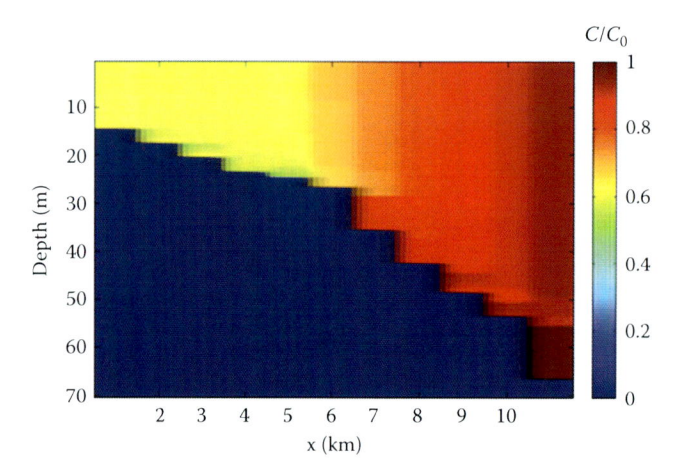

Figure 35.5 Cross-section distribution of particle concentrations (C) normalized to a constant offshore concentration (C_0) after a 30 day simulation period as derived from a 2-D turbulent diffusion-driven mass exchange model. The area was located in Lake Michigan along the western shoreline near Milwaukee, Wisconsin, and extended from the shoreline eastward to a water depth of 70 m, a distance of about 11 km. See text for further details.

(A)

(B)

Figure 44.1 (A) Boat hull heavily colonized by dreissenids (predominantly quagga mussels). This boat was not used during the entire 2010 season, which allowed quagga mussels to attach. (B) Left and right: typical density of dreissenids (predominantly zebra mussels) attached to boats regularly used in Lakes Erie and Ontario in 2010.

Figure 48.3 Image taken in 2007 of the interior of the steamer *Florida* that lies on the bottom at a depth of 61 m in northern Lake Huron. The shipwreck is broken in cross section and hence the interior is open to the water column. Note all surfaces are colonized by quagga mussels. (Courtesy of NOAA Thunder Bay National Marine Sanctuary, Alpena, MI.)

Figure 48.4 Image of the exterior of the steamer *Florida* (see details in earlier text). In the center of the image just below the opening is an iron/wood hatchet uniformly colonized by quagga mussels. (Courtesy of NOAA Thunder Bay National Marine Sanctuary, Alpena, MI.)

Figure 48.5 Image of the exterior surface of the schooner *M.F. Merrick*. This shipwreck was discovered in 2011 on the bottom at a water depth of 91 m in northern Lake Huron. (Courtesy of John Janzen.)

Figure 48.6 Image of the hull interior of the schooner *M.F. Merrick* (see details in earlier text). The hull was nearly intact and the interior confined, yet quagga mussels were present and abundant. (Courtesy of John Janzen.)

CHAPTER 43

Effects of Dreissenids on Monitoring and Management of Fisheries in Western Lake Erie

Martin A. Stapanian and Patrick M. Kocovsky

CONTENTS

Abstract 681
Introduction 682
Methods 682
　Field Collections and Study Area 682
　Data Analysis 683
　　Diel Shift in CPH of Age-0 Yellow Perch 683
　　Effects of Diel Shift on Management of Yellow Perch 684
　　Diel Shifts in Catch Probability Indices of Other Benthic Fishes 684
Results 685
　Diel Shift in CPH of Age-0 Yellow Perch 685
　Effects of Diel Shift on Management of Yellow Perch 686
　Diel Shifts of Catch Probability Indices of Benthic Fishes 687
Discussion 688
Acknowledgments 690
References 691

ABSTRACT

Water clarity increased in nearshore areas of western Lake Erie by the early-1990s mainly as a result of the filtering activities of dreissenid mussels (*Dreissena spp.*), which invaded in the mid-1980s. We hypothesized that increased water clarity would result in greater trawl avoidance and thus reduced ability to capture fish in bottom trawls during daytime compared to nighttime. We examined this hypothesis by summarizing three analyses on fish data collected in western Lake Erie. First, we used a two-tiered modeling approach on the ratio (R) of catch per hour (CPH) of age-0 yellow perch (*Perca flavescens* Mitchill) at night to CPH during daytime in 1961–2005. The best *a priori* and *a posteriori* models indicated a shift to higher CPH at night ($R > 1$) between 1990 and 1991, which corresponded to 3 years after the dreissenid invasion and when water clarity noticeably increased at nearshore sites. Secondly, we examined effects of nighttime sampling on estimates of abundance of age-2 and older yellow perch, which form the basis for recommended allowable harvest (RAH). When data from night sampling were included in models that predict abundance of age-2 yellow perch from indices of abundance of age-0 and age-1 yellow perch, predicted abundance was lower and model precision, as measured by r-squared, was higher compared to models that excluded data collected at night. Furthermore, the use of only CPH data collected at night typically resulted in lower estimates of abundance and more precise models compared to models that included CPH data collected during both daytime and nighttime. Thirdly, we used presence/absence data from paired bottom trawl samples to calculate an index of capture probability (or catchability) to determine if our ability to capture the four most common benthic species in western Lake Erie was affected by dreissenid-caused increased water clarity. Three species of fish (white perch, *Morone americana* Gmelin; yellow perch; and trout-perch, *Percopsis omiscomaycus* Walbaum) had lower mean daytime catchability than nighttime catchability and a positive mean nighttime–daytime difference in catchability after dreissenids became established, which supported the hypothesis of greater trawl avoidance during daytime following establishment of dreissenids. Results for freshwater drum (*Aplodinotus grunniens* Rafinesque) were opposite those of the other three species, which may be a result of behavioral shifts due to freshwater drum feeding on dreissenid mussels. Collectively, these three studies suggest that dreissenids indirectly affected our ability to assess fish populations, which further affects estimates of fish densities and relationships between indices of abundance and true abundance.

INTRODUCTION

Invasive dreissenid mussels (zebra mussels *Dreissena polymorpha* Pallas and quagga mussels *D. rostriformis bugensis* Andrusov) have changed the structure and function of large freshwater ecosystems throughout North America and Europe (Karatayev et al. 1997, 2002). These mussels have had many direct and indirect effects on characteristics of the water column and substrate in the systems they have invaded. Examples include increased water clarity (e.g., Fahnenstiel et al. 1995, Karatayev et al. 1997, Charlton et al. 1999), increased abundances of submerged macrophytes (Zhu et al. 2006), increased biomass of benthic macroinvertebrates (e.g., Botts et al. 1996, Mayer et al. 2000), decreased biomass of phytoplankton (e.g., Caraco et al. 1997), and decreased abundances of small-sized zooplankton (e.g., MacIsaac et al. 1991, MacIsaac 1996, 1999, Pace et al. 1998).

In Lake Erie, zebra mussels were reported as being present in the western basin in 1986 (Charlton et al. 1999), and quagga mussels have been in the lake since 1989 (Mills et al. 1993). By the early-1990s, dreissenid mussels were well established in the western basin and increased water clarity at nearshore sites was well documented (e.g., Holland 1993, Leach 1993, Nichols and Hopkins 1993, Charlton et al. 1999). For example, mean Secchi disk transparency at bottom trawl sites in the western basin during the period 1998–2005 was more than double the mean during 1961–1971 (Stapanian et al. 2009).

Lake Erie has undergone many other changes since the 1960s and early-1970s, when it was eutrophic and highly turbid (Leach 1999). In 1972, the Unites States and Canada signed the Great Lakes Water Quality Agreement (GLWQA) in which the two nations agreed to reduce certain toxic pollutants and loadings of total phosphorus from tributaries and point sources in the Great Lakes. The Federal Water Pollution Control Act (FWPCA) Amendments of 1972 were enacted due to public awareness and concern for controlling water pollution. The FWPCA was amended in 1977 and became the Clean Water Act. Water quality in Lake Erie responded to pollution-abatement programs enacted as a result of these legislations. Total phosphate and chlorophyll in the western basin decreased substantially between 1968–1972 and 1984–1988, and water clarity increased at nearshore sites by the mid-1990s (Charlton et al. 1999).

Changes in water clarity may affect fish behavior, which in turn may affect our ability to collect and hence monitor populations. For example, diel variation in the catch of some fish species has been linked to visibility of survey trawls as a means of escapement (Walsh 1991, Casey and Myers 1998, Stapanian et al. 2009). Fish are more likely to avoid capture when a trawl is more visible, such as during daytime or when light intensity is otherwise greater. Aggregations of yellow perch (*Perca flavescens* Mitchill) have been shown to be positively correlated with light penetration in the water column (Hergenrader and Hasler 1968), and schooling/shoaling behavior has also been linked to greater avoidance (Webb 1980). These behaviors and potentially others (e.g., altered foraging behavior) could result in lower capture probabilities during daytime, which would increase variability in indices and estimates of abundance (Glass and Wardle 1989, Casey and Myers 1998, Kocovsky et al. 2010), and potentially alter the relationship between true abundance of a species and indices used to track trends in abundance (Blanchard et al. 2008).

In this chapter, we summarize our work (Stapanian et al. 2009, Kocovsky et al. 2010, Kocovsky and Stapanian 2011) on changes in the diel catch per effort of fishes at nearshore sites in western Lake Erie before and after the increase in water clarity associated with dreissenid mussels. We reveal that the timing of changes in catch per effort and in an index of capture probability of benthic fishes was associated most strongly with the establishment of dreissenid mussels and subsequent increase in water clarity. We also show how current fisheries monitoring and management programs are affected by these changes.

METHODS

Field Collections and Study Area

All fish for data analysis were collected with a bottom trawl (7.9 m headrope, stretch mesh size of the cod end = 6 mm) at three established sites offshore of East Harbor State Park, Ohio, in western Lake Erie (Figure 43.1) during autumn and summer 1961–2005 (Trometer and Busch 1999, Stapanian et al. 2007, 2009, Kocovsky et al. 2010). Although the area of the study sites is small compared to that of the western basin of Lake Erie, data from these sites have consistently provided precise models that predict abundance of age-2 yellow perch from indices of abundance of age-0 and age-1 yellow perch that are then used to predict

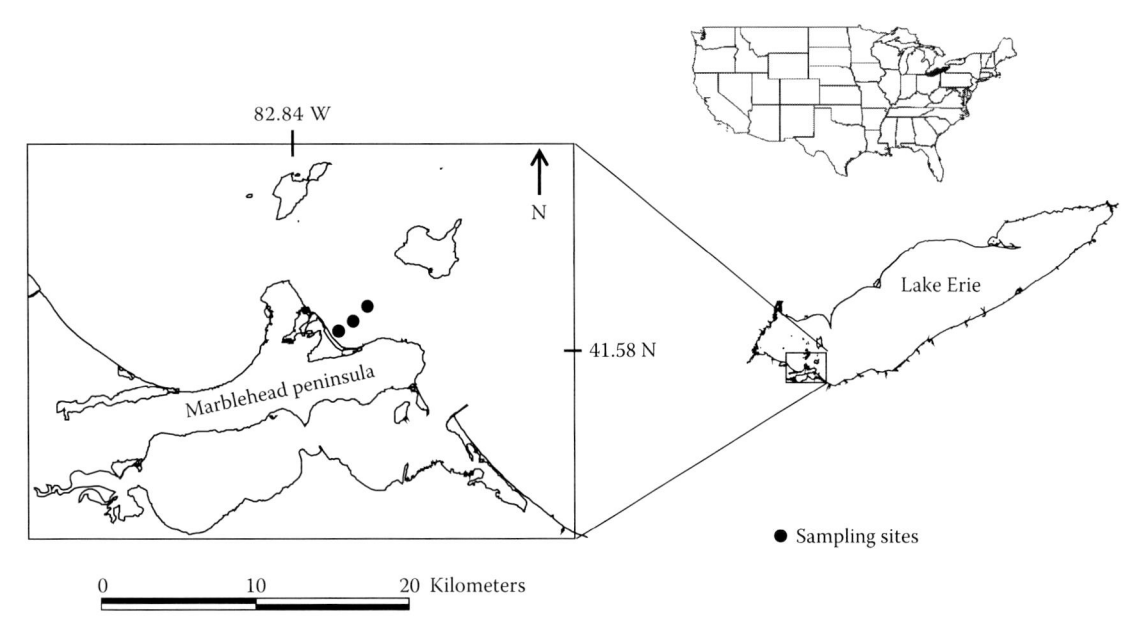

Figure 43.1 Location of study area (inset: Lake Erie) in western Lake Erie near East Harbor State Park, Ohio. Filled circles represent trawling sites.

abundance of age-2 yellow perch for establishing harvest quotas (e.g., YPTG 2007, Kocovsky et al. 2010). The bottom trawl was towed at approximately 3.2 km/h. Duplicate trawl samples were consistently collected at the 3.0, 4.5, and 6.0 m depth contours during morning (beginning at least 30 min after sunrise and ending by 1100), afternoon (1300–1700), and night (beginning at least 30 min after sunset and ending by 2300). Trawl duration was 10 min, with some (<1 % of the total) lasting between 7 and 9.5 min.

Summer sampling was conducted between the first week of August and the first week of September each year. In nearly all years, autumn sampling occurred between the second and fourth weeks of October (Stapanian et al. 2007). Sampling occasionally began earlier (29 September in 2002 and 2003) and ended later (first week of November in 1989, 1994, 1996, 1997, and 2005) owing to inclement weather during prescribed sampling periods. Details of trawling methods, times, and fish collected for data used in our analysis are reported elsewhere (Stapanian et al. 2007, 2009, Kocovsky et al. 2010).

The present study focuses on the most abundant benthic species captured in our bottom trawls during 1961–2008. Yellow perch, white perch (*Morone americana* Gmelin), and freshwater drum (*Aplodinotus grunniens* Rafinesque) caught in trawls were categorized as either age-0, age-1, or age-2 and older based on total length–age keys developed for the western basin of Lake Erie during September and October in each study year (M. Turner, Ohio Department of Natural Resources, personal communication). Trout-perch (*Percopsis omiscomaycus* Walbaum) were categorized as either age-0 or yearling-or-older. For each trawl sample, we calculated the number caught per hour (hereafter: CPH) for the species–age combinations of interest. Mean annual CPH was calculated for morning, afternoon, and night across all depths.

Data Analysis

Diel Shift in CPH of Age-0 Yellow Perch (Stapanian et al. 2009)

We hypothesized that increased water clarity in Lake Erie resulted in decreased CPH of age-0 yellow perch (*P. flavescens*) during daytime compared to during nighttime. We restricted our analysis to age-0 yellow perch because this species is one of the most commercially valuable species in the Great Lakes and because the age-0 size class could be treated as statistically independent across years (Stapanian et al. 2007). To examine the hypothesis that CPH of age-0 yellow perch decreased during daytime compared to nighttime after dreissenids became established, we used two different modeling approaches on trawl data collected during autumn in 1961–2005.

For the first approach, we constructed *a priori* models to test if changes in the ratio of CPH during nighttime to CPH during daytime (R) were associated with the timing of pollution-abatement efforts from the GLWQA or associated with the establishment of dreissenids, with these drivers examined separately and combined. Seven *a priori* models were evaluated with Akaike's information criterion (AIC_c), corrected for sample size (Burnham and Anderson 2002). The first model was a "null" model that represented no change in R over time. Three more models represented whether a change in mean R was associated with passage of water quality legislation (two periods, 1961–1972 and 1973–2005), dreissenids (two periods, 1961–1987 and 1988–2005), and both legislation and dreissenids (three periods, 1961–1972, 1973–1987, and 1988–2005). Three additional models included a 3-year lag before the effects of legislation (1961–1975, 1976–2005), dreissenids

(1961–1990, 1991–2005), or both (1961–1975, 1976–1990, 1991–2005) occurred. The 3-year lag was incorporated to account for the likely scenario that the full effects of water quality legislation or invasion of dreissenids were not immediate. That is, passage of legislation is followed by implementation and environmental response, a sequence that likely takes a few years. Also, dreissenids were first found in western Lake Erie in the late-1980s, and increases in water clarity at offshore sites in the western basin were observed by the mid-1990s (Charlton et al. 1999). The effects of dreissenids on water clarity probably occurred much earlier for nearshore sites like in this study (Holland 1993, Leach 1993, Nichols and Hopkins 1993, Charlton et al. 1999). Large increases in water clarity occurred within 2–3 years following dreissenid invasion of Saginaw Bay, a shallow, well-mixed system similar to the western basin (Fahnenstiel et al. 1995). Finally, a 3-year lag is consistent with large increases in densities of adult and veliger zebra mussels in the western basin of Lake Erie by 1989 (Kovalak et al. 1993, Leach 1993).

In the second approach, we examined all possible combinations of two and three periods during the time series. This *a posteriori* procedure determined temporal transitions to higher R that were best supported by the data, without regard to *a priori* hypotheses. All possible (n = 862) two- and three-period models with a minimum of 2 years per time period were examined to determine if a different temporal transition to higher R than those hypothesized *a priori* was better supported by the data. This approach enabled us to determine how well the analysis using prior knowledge of the system corresponded with all possible scenarios. These *a posteriori* models were evaluated by examining the root mean squared error (RMSE). The best *a posteriori* model was defined as the one with the lowest RMSE.

Effects of Diel Shift on Management of Yellow Perch (Kocovsky et al. 2010)

Following the results of Stapanian et al. (2009), Kocovsky et al. (2010) examined whether a shift toward higher nighttime CPH might have implications for management of harvest of yellow perch. They examined CPH for age-0 and age 1 yellow perch captured at the East Harbor sites during 1991 through 2005 to determine whether sampling during particular diel periods affected management values of CPH used in models to predict abundance of age-2 yellow perch. To build on Stapanian et al. (2009), Kocovsky et al. (2010) determined (1) if sampling during particular diel periods produced more precise models that use indices of abundance of age-0 and age-1 yellow perch to predict abundance of age-2 yellow perch and (2) if predicted abundance varied by diel period from which CPH data were used. Predicted abundance of age-2 yellow perch from such models is the basis for recommended allowable harvest (RAH) used to establish the total allowable catch (TAC) of yellow perch in western Lake Erie.

Kocovsky et al. (2010) calculated four indices of yellow perch abundance (defined as CPH) from our trawl samples: age-0 collected in summer, age-0 collected in autumn, age 1 collected in summer, and age-1 collected in autumn. The modeling procedure used to establish the RAH for yellow perch on Lake Erie (YPTG 2007) was used to assess whether index values calculated from data collected during the three diel periods affected predictions of abundance of age-2 yellow perch in western Lake Erie. Briefly, the RAH is the mean of predicted abundance of age-2 yellow perch generated from regression models that use fishery-independent indices of abundance of age-0 and age-1 yellow perch (such as ours) as the independent variables and a fishery-dependent estimate of abundance of age-2 yellow perch generated primarily from commercial and recreational harvest as the dependent variable. The fishery-dependent abundance of age-2 yellow perch is estimated from a catch-at-age model using Auto Differentiation Model Builder (ADMB) (Otter Research Ltd. 2000). This process for establishing RAH has been in place since 2000. Full details of the modeling procedure are found in Kocovsky et al. (2010) and YPTG (2007).

Kocovsky et al. (2010) then examined the effect of diel period of sampling on the fit of linear regression models that predicted the abundance of age-2 yellow perch from variations of our sampling regime. This was accomplished by comparing predicted age-2 abundance and r^2-values of models using index values from our full sampling regimen (i.e., the full complement of trawl samples from all three diel periods) to those using index values from six reduced sampling regimens, each using data from one or more diel period (Kocovsky et al. 2010). Index values from the full sampling regimen and the reduced sampling regimens (independent variable) were each regressed against age-2 abundance from the ADMB model for each of the four indices using data from 1991 to 2005. Abundance estimates and r^2-values from the full sampling regimen were subtracted from the abundance estimates and r^2-values of regression models using data from reduced sampling regimens for each index to assess differences. A higher r^2-value for a model using an index value calculated from a reduced sampling regimen was interpreted as improved precision for the reduced sampling regimen.

Diel Shifts in Catch Probability Indices of Other Benthic Fishes (Kocovsky and Stapanian 2011)

Besides yellow perch, we were interested in whether capture probability of other benthic species changed after establishment of dreissenids and increase in water clarity. Measuring capture probability for bottom trawls in a large lake is challenging at best. If depletion could be achieved by taking multiple trawl samples in the same area, then capture probability could be estimated directly. During repeated trawl sampling for development of fishing power corrections for bottom trawls in western Lake Erie, Tyson et al. (2006) failed to achieve depletion for any benthic species. Hence, direct estimates of capture probability are not feasible. However, with our replicate trawl samples, we developed an index of capture probability, which we call "catchability." For each pair of trawls at a site, the mean of "successes" (1) and "failures" (0) at capturing a species in the first trawl sample produces a proportion, which represents the catchability (C), for species j in the first trawl sample:

$$C_j = \frac{\sum S_{ij}}{n} \quad (43.1)$$

where

$$S_{ij} = \begin{cases} 1 \text{ if species j was captured in the first} \\ \quad \text{sample of pair i} \\ 0 \text{ if species j was captured in the second sample} \\ \quad \text{of pair i but not the first sample} \end{cases}$$

n is the number of trawl pairs

If a species was not captured in both trawls, it was considered absent and those data were excluded from further analyses. The greater the proportion of successes in capturing a species in the first of a pair of trawls, the greater the catchability of the species.

Daytime and nighttime C_j were calculated for yellow perch, white perch, trout-perch, and freshwater drum, the four most abundant benthic species captured in bottom trawls in western Lake Erie, in the periods 1972–1990 (pre-dreissenid) and 1991–2009 (post-dreissenid). The post-dreissenid period began with the year the full effect of dreissenids was detectable at nearshore sites in western Lake Erie (Holland 1993, Stapanian et al. 2009) and continued through 2009. The same number of years in the pre-dreissenid period was used in the analysis so sample sizes (years) in the two periods would be equal. Furthermore, 1972 was the year the GLWQA was signed, which ultimately reduced total phosphate and chlorophyll in western Lake Erie (Charlton et al. 1999).

Kocovsky and Stapanian (2011) tested for differences in C_j (1) between diel periods within each period and (2) between periods for each diel period with a Kruskal–Wallis test (Noether 1991). The Kruskal–Wallis test was also used to determine if the mean of annual differences between daytime and nighttime catchability varied between the pre-dreissenid and post-dreissenid periods. The nonparametric Kruskal–Wallis test was used because none of the applied transformations of catchabilities achieved the assumptions of normality of residuals and homogeneity of variance required for parametric tests for any species.

RESULTS

Diel Shift in CPH of Age-0 Yellow Perch (Stapanian et al. 2009)

Nighttime CPH of age-0 yellow perch exceeded daytime CPH in only 3 of 12 years (25%) during 1961–1972 (period preceding GLWQA, not lagged) and 5 of 15 years (33%) during 1961–1975 (lagged; Figure 43.2). Higher CPH at night occurred in 10 of 15 years (67%) during 1973–1987 (following GLWQA but preceding dreissenids, not lagged) and 8 of 15 (53%) years during 1976–1990 (lagged). Following establishment of dreissenids, higher night CPH was observed in 14 of 18 years (78%) during 1988–2005 (not lagged) and 14 of 15 years (93%) during 1991–2005 (lagged). During 1991–2005, nighttime CPH was more than double daytime CPH in 10 of 14 years.

The best *a priori* model had two periods, with a break between 1990 and 1991 (Table 43.1). This model accounted for 67% of the weighting of all seven models. The only other *a priori* model that was supported by the data, and accounted for an additional 22% of the weighting, had breaks between 1990 and 1991 and between 1975 and 1976, corresponding to 3 years after the dreissenid invasion and passage of the GLWQA, respectively. Thus, *a priori* models supported a shift toward higher nighttime CPH of age-0 yellow perch 3 years after establishment of dreissenids. Similarly, the best two- and three-period *a posteriori* models both had breaks between 1990 and 1991 (Stapanian et al. 2009). Thus, results supported our hypothesis that age-0 yellow perch exhibited a transition to lower CPH during daytime compared to nighttime, and the timing of the transition coincided with the establishment of dreissenid mussels.

Catches from daytime and nighttime surveys were highly correlated on a log scale that indicated that a day-to-night

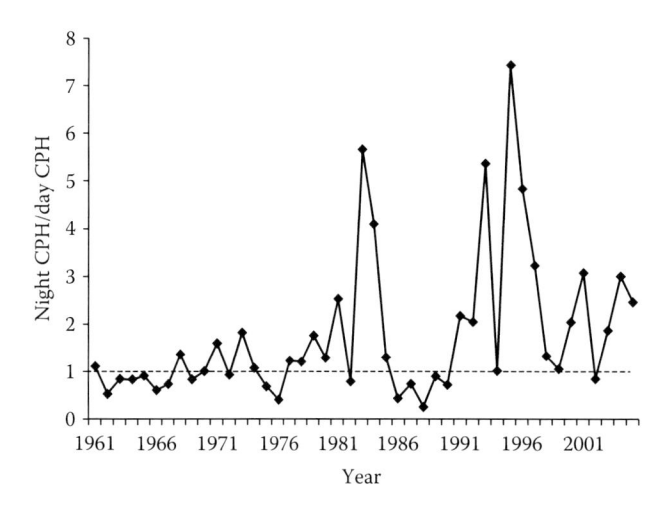

Figure 43.2 Ratio of catch per hour (CPH) of age-0 yellow perch during nighttime to CPH during daytime. Values greater than 1 (dashed line) indicate night CPH was greater. (From Stapanian, M.A. et al., *Freshwat. Biol.*, 54, 1593, 2009. With permission.)

Table 43.1 Summary of *A Priori* Two- and Three-Period ANOVA Models and a One-Period Null Model Predicting the (Natural Log Transformed) Ratio of Nighttime to Daytime Catch per Hour of Age-0 Yellow Perch in Western Lake Erie, 1961–2005 (n = 45). Models Are Described by the Hypothesized Causes of Ratio Changes (Environmental Legislation [CWA] Passed in 1972, the Invasion of Dreissenid Mussels [DI] in 1987, or Both), whether the Timing Was Lagged by 3 Years (Lag), the Dates of the Breaks between Time Periods, and Number of Parameters (p) in the Model. RMSE, Root Mean Squared Error; AIC, Akaike's Information Criterion Statistic; ΔAIC, the Difference in AIC from the Best Fitting Model; AICw, the Relative Importance Weighting of Each Model (Sum = 1)

Model	Break 1	Break 2	p	F	P	RMSE	AIC	ΔAIC	AICw
DI, lag	1990.5	—	2	16.9	<0.001	0.6166	90.2	0.0	0.660
CWA, DI, lag	1975.5	1990.5	3	8.6	0.001	0.6130	91.7	1.5	0.315
DI	1987.5	—	2	5.8	0.02	0.6826	99.3	9.2	0.007
CWA, lag	1975.5	—	2	5.8	0.021	0.6830	99.4	9.2	0.007
CWA, DI	1972.5	1987.5	3	3.7	0.032	0.6702	99.7	9.5	0.006
CWA	1972.5	—	2	5.1	0.029	0.6879	100.0	9.8	0.005
Null	—	—	1			0.7275	103.1	12.9	0.001

Source: From Stapanian, M.A. et al., *Freshwat. Biol.*, 54, 1593, 2009.

Table 43.2 Values of r^2 and Estimated Abundance (Millions) of Age-2 Yellow Perch for Linear Regressions Predicting Abundance of Age-2 Yellow Perch in Western Lake Erie from Indices of Abundance Based on the Number of Age-0 and Age-1 Yellow Perch (Independent Variables) from the Full Sampling Regimen[a] and from Six Reduced Sampling Regimens Using Data from One or More Times of Day

Index		Full Regimen	M	A	N	MA	MN	AN
Summer age 0	r^2	0.57	0.51	0.42	0.65	0.48	0.63	0.57
	Age 2 estimate	2.95	5.30	8.71	−1.30	6.67	0.38	2.59
Summer age 1	r^2	0.83	0.82	0.67	0.76	0.77	0.86	0.80
	Age 2 estimate	17.76	31.04	16.09	9.72	23.59	18.65	11.70
Autumn age 0	r^2	0.60	0.35	0.48	0.73	0.46	0.61	0.66
	Age 2 estimate	1.17	5.41	6.16	1.08	4.24	0.21	1.53
Autumn age 1	r^2	0.69	0.59	0.51	0.72	0.58	0.74	0.67
	Age 2 estimate	3.46	6.58	8.06	1.95	6.54	1.95	3.50

Source: From Kocovsky, P.M. et al., *Fish. Ecol. Manage.*, 17, 10, 2010.
[a] Sampling in daytime (morning, M; afternoon, A) and nighttime (N).

adjustment factor during 1991–2005 might be useful for predicting nighttime CPH from daytime CPH (Stapanian et al. 2009). We selected this yearly range because it corresponded to the last time period in the best models. The adjustment factor was based on the geometric means of the night and day catches. A bias-corrected estimate with 95% confidence interval, as estimated from 1000 bootstrap samples, was 2.32 ± 0.62; that is, nighttime CPH was 2.32 times greater than daytime CPH. Using bootstrapping and cross-validation, Stapanian et al. (2009) found that the use of this correction factor to predict nighttime CPH from daytime CPH reduced RMSE by 25%.

Effects of Diel Shift on Management of Yellow Perch (Kocovsky et al. 2010)

Data collected during morning and afternoon provided less precise estimates of age-2 yellow perch than the full sampling regimen (Table 43.2). For age-0 yellow perch collected in both summer and autumn and for age-1 yellow perch collected in autumn, CPH data from exclusively nighttime sampling provided more precise estimates than CPH data from the full sampling regimen, exclusively morning sampling, exclusively afternoon sampling, and morning and afternoon sampling combined. With the exception of the summer age-1 data, all combinations that included night sampling yielded more precise estimates of age-2 yellow perch. For all indices, CPH data from nighttime sampling alone provided lower estimates of age-2 yellow perch than the full sampling regimen, exclusively morning, and exclusively afternoon.

To determine the effect of using CPH data collected during nighttime on RAH, we examined the ratio of estimated abundance of age-2 yellow perch using our indices (full sampling regimen) to the mean estimated abundance from all

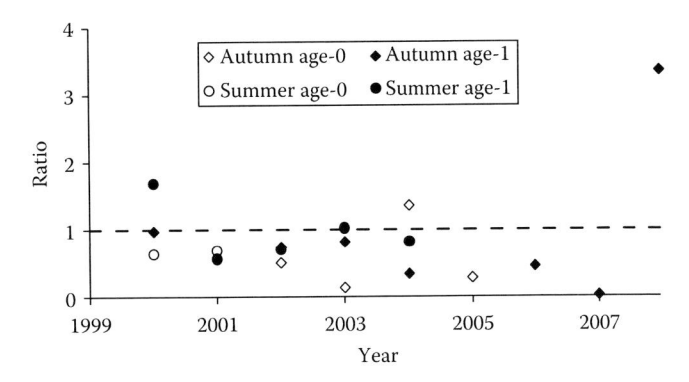

Figure 43.3 Ratio of estimated abundance of age-2 and older yellow perch when data from nighttime bottom trawls are included in indices used in models to estimate abundance of -2 yellow perch to the average estimated abundance of age-2 yellow perch from all models used in estimating recommended allowable harvest of yellow perch from western Lake Erie. Estimates are from linear regression models with index data for age-0 and age-1 yellow perch collected during summer and autumn. Values less than 1 (dashed line) indicate that estimates from indices that included data collected at night were more conservative.

models that were averaged to establish RAH between 2000 and 2008 (data from Kocovsky et al. 2010). Ratios less than 1 indicate that estimates of abundance of age-2 yellow perch from our CPH data, which are the only indices that are generated using data collected at night, produced a more conservative estimate. The ratio was less than 1 for 14 of the 18 (78%) instances in which our indices were included in the suite of models used to establish RAH (Figure 43.3). Therefore, indices that include CPH data collected at night typically provided more conservative estimates of abundance of age-2 yellow perch than indices that included data collected exclusively during daytime. Furthermore, RAH is more conservative whenever indices calculated using data collected at night are included in the suite of models used to establish RAH.

It may seem counterintuitive that models that included nighttime sampling, when CPHs were typically higher than daytime CPHs, produced lower estimates of abundance of age-2 yellow perch (Stapanian et al. 2009, Kocovsky et al. 2010). This was because estimated abundances are based on predictive linear equations that have a slope and y-intercept (YPTG 2007, Kocovsky et al. 2010). An increase in slopes of the predictive linear models calculated from these higher catch rates were offset by a decrease in y-intercepts (Kocovsky et al. 2010). Thus, while greater slopes of regression equations would result from indices that use higher CPH, lower y-intercepts for many regressions that include nighttime data would offset any gains, resulting in lower estimates of abundance.

Diel Shifts in Catch Probability Indices of Benthic Fishes (Kocovsky and Stapanian 2011)

During the pre-dreissenid period, there was no overall difference between daytime and nighttime catchability for three of the four species examined (Table 43.3). Only freshwater drum had lower daytime catchability. Trends in differences between nighttime and daytime mean catchabilities in the pre-dreissenid period varied by species. For white perch and yellow perch, there were no discernible trends within the period (Figure 43.4). For trout-perch, the trend showed a decline from positive in the early-1970s to mostly negative throughout the 1980s, with one high-magnitude positive discrepancy in 1988. The trend for freshwater drum was mostly positive and variable from 1972 through the late-1980s. In the post-dreissenid period, all four species

Table 43.3 Mean Index of Catchability (C) for Four Benthic Fish Species for Periods Pre-Dreissenid (1972–1990) and Post-Dreissenid (1991–2009) at Nearshore Sites (≤6 m Depth) in Western Lake Erie When Sampling in Daytime and Nighttime. "Night–day" Is the Mean of the Annual Differences in the Indices for Nighttime and Daytime; Positive Values Indicate Mean Catchability at Nighttime Was Higher than During daytime. Greater-than and Less-than Symbols Represent Differences in Values between Adjacent Columns (Kruskal–Wallis Test, Single Symbol $P < 0.1$, Double Symbol $P < 0.05$, Triple Symbol $P < 0.01$). Superscript Symbols Indicate Differences in Values between Periods for Daytime, Nighttime, and Nighttime–Daytime (Kruskal–Wallis Tests, * = $P < 0.05$, ** = $P < 0.01$)

	Pre-Dreissenid 1972–1990			Post-Dreissenid 1991–2009		
	Day	Night	Night–Day	Day	Night	Night–Day
Trout-perch	0.826	0.764	−0.061	0.775 <<	0.878	0.102**
Yellow perch	0.952	0.940	−0.012	0.920 <<<	0.981	0.061*
White perch	0.892	0.917	−0.031	0.969 <<	0.984	0.015
Freshwater drum	0.822 <<<	0.974	0.152	0.827	0.861*	0.034*

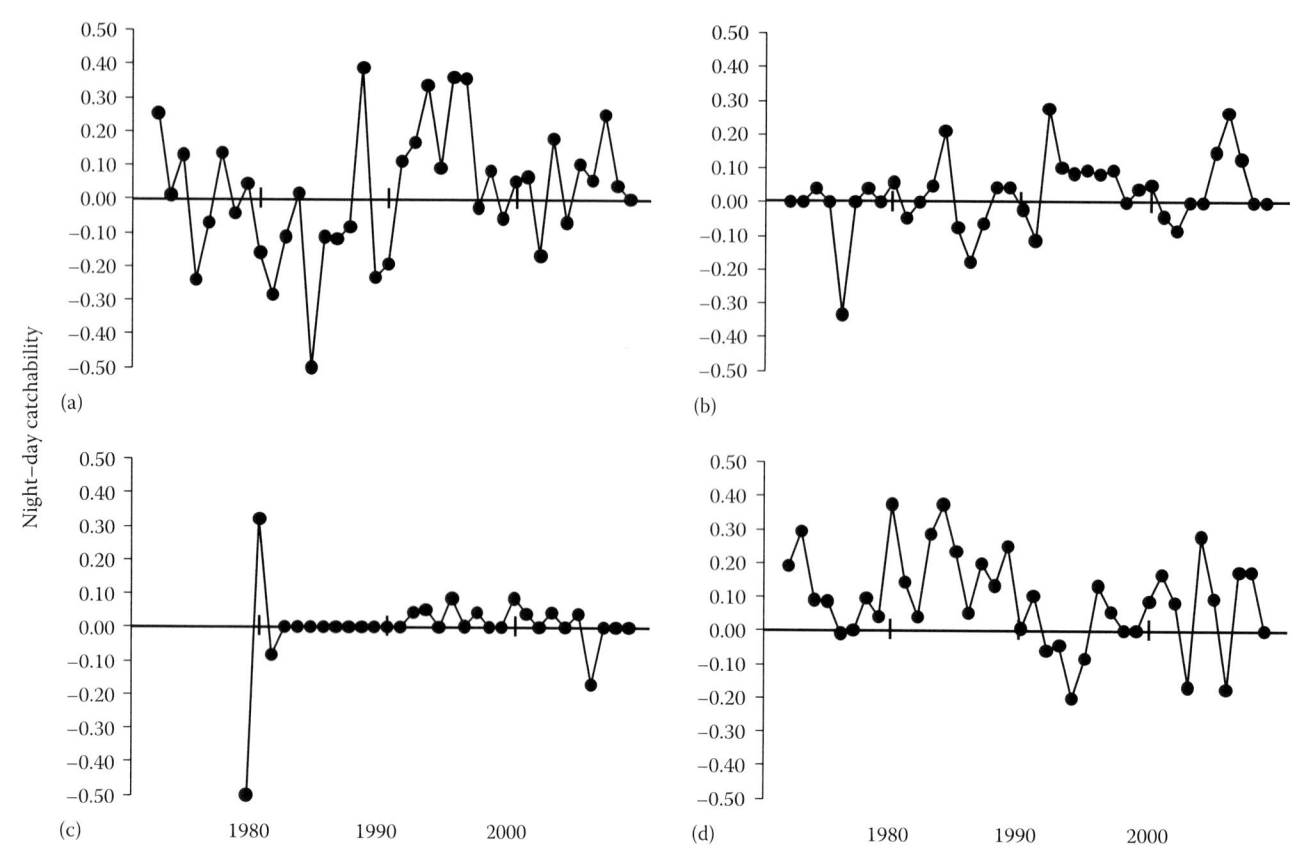

Figure 43.4 Difference between nighttime and daytime catchability of trout-perch (A), yellow perch (B), white perch (C), and freshwater drum (D) captured in bottom trawls in nearshore waters of western Lake Erie, 1972–2009. Positive values indicated nighttime catchability was greater. (From Kocovsky et al. 2011. With permission.)

had mostly positive nighttime–daytime differentials (i.e., nighttime catchability was typically higher, Figure 43.4). Nighttime–daytime differentials for trout-perch decreased from the early 1990s through 2002 but then increased thereafter. The respective differentials for white perch and yellow perch showed no discernible trends within the period. The differential for freshwater drum declined precipitously from 1989 to the low for the time series in 1994 then increased for the next 2 years and remained mostly positive through 2009.

As discussed by Kocovsky and Stapanian (2011), our data analysis generally supported the following conclusions for yellow perch, trout-perch, and white perch: First, daytime catchability was not different than nighttime catchability for all three of these species in the pre-dreissenid period. Second, daytime catchability was significantly lower than nighttime catchability for yellow perch, white perch, and trout-perch in the post-dreissenid period. The opposite response of daytime catchability being higher than nighttime catchability in the post-dreissenid period was not observed. Third, the differential between nighttime and daytime catchability was significantly higher in the post-dreissenid period for trout-perch and yellow perch. Fourth, daytime catchability was inversely related to water clarity for all three species (Kocovsky and Stapanian 2011). Thus, evidence supports a shift in catchability following establishment of dreissenids and that shift in catchability was more strongly influenced by a decrease in daytime catchability. These results are consistent with greater trawl visibility (hence greater fish avoidance) because of increased water clarity from dreissenid filtration, as proposed by Stapanian et al. (2009).

DISCUSSION

Much of the research on effects of dreissenids on aquatic communities has focused on water clarity and quality, nutrient dynamics, and ecology of lower trophic levels, whereas little research has addressed the indirect effects of dreissenids on finned fishes. Analysis of our trawl data in the western basin of Lake Erie provided evidence that the cascading effects of dreissenids can extend to monitoring and management of planktivorous and invertivorous benthic fishes. Our analysis indicated that diel catchability of age-0 yellow perch was greater at nighttime compared to daytime after 1990. The timing of this change corresponded directly to when dreissenid mussels began to have a significant effect on water clarity in nearshore areas of western Lake Erie. Although we did not measure it directly, the most parsimonious explanation for our results was increased trawl avoidance due to clearer water. Further, our analysis showed that the shift in diel catchability for age-0 and age-1 yellow perch had important management implications. Specifically, indices of abundance calculated using CPH data collected at night, when trawl visibility was virtually eliminated as a factor affecting CPH, produced more precise and more conservative estimates of abundance of age-2 yellow perch. Finally, we showed that catchability of the four most abundant benthic species in western Lake Erie was negatively related to water clarity and that a shift toward lower daytime catchability and a greater difference between nighttime and daytime catchability occurred following the establishment of dreissenids. Collectively, these analyses suggested that dreissenids have indirectly affected our ability to monitor and manage fish populations by way of their effect on water clarity. Results also collectively supported the hypothesis that the mechanism for lower fish catches during daytime was increased trawl visibility due to increased water clarity from dreissenid filtration.

The capture of fish in active collecting gears, such as the bottom trawl we used, is the combined result of absolute abundance and spatial distribution of fish, response of fish to sampling gear, selective properties of the gear, and habitat limitations on gear operation (Hayes et al. 1996). The same gear, vessel, and towing speed were used throughout our studies, so potential biases owing to gear selectivity and vessel operations (Tyson et al. 2006) were eliminated. With the exception of the establishment of dreissenids, no changes in bottom substrates occurred during the study period. Absolute annual abundances of the species we examined should not have influenced results because we examined ratios of nighttime CPH to daytime CPH, which adjusts for differences in annual abundance. We found no evidence of changes in diel spatial distribution during the time series for the species we examined that may have occurred because of either vertical or onshore–offshore migration. With the exception of freshwater drum (see succeeding text), we found no evidence that our results were due to changes in distribution and estimated abundance of prey species for these fishes. Therefore, we are not able to reject a change in diel response of fish to the sampling gear as the reason for the shift to greater ratio of nighttime CPH to daytime CPH after 1990. We suggest that increased net avoidance during daytime due to increased water clarity was the most plausible interpretation of the results.

An alternate explanation for changes in catchability is shifts in spatial distributions of fish relative to trawl sites. This may explain the unexpected and opposite response of freshwater drum to the dreissenid invasion. Freshwater drum is one of few species in Lake Erie known to forage actively on dreissenids (French and Love 1995). Dreissenids were present at all trawl sites and were particularly abundant at the 6.1 m site by 1991, and provided a new food source for freshwater drum. This probably resulted in increased daytime foraging activity by freshwater drum, which would explain increased daytime catchability. This possibility is supported by the sudden, high-magnitude, and sustained shift toward higher daytime catchability of freshwater drum during 1992–1995, which was within the period when water clarity increases were greatest in western Lake Erie (Holland 1993) and at our sites (Stapanian et al. 2009). The shift back toward slightly greater daytime catchability of freshwater drum in recent years may be the result of recent decreases in water clarity (Binding et al. 2007, Kocovsky and Stapanian 2011). Thus, unexpected results for freshwater drum may be explained by their unique foraging relationship with dreissenids, which is not shared by any other abundant native fish species.

Small changes or fluctuations in catchability can introduce bias in estimated abundance and trends. Therefore, the shift toward slightly lower catchability, which is an index of capture probability, for benthic fish species following establishment of dreissenids has strong implications for monitoring of trends in fish abundance. For example, Riley and Fausch (1992) demonstrated that low and decreased capture probability resulted in low-bias and imprecise estimates of abundance of trout (Salmonidae) in small streams when using multiple-pass removal electrofishing. Although Riley and Fausch (1992) used different methods in a different system than ours, their results of effects of low capture probability are revealing. In particular, abundance estimates were unbiased only under conditions of highest capture probability (initial capture probability 0.9) and effort (three or four electrofishing passes). Thus, even low-magnitude decreases in capture probability may have introduced bias to abundance estimates and indices of abundance, thus altering the relationship between indices of abundance and the true abundance of the populations being monitored (Blanchard et al. 2008). If this relationship is altered by a decrease in capture probability, then any decrease in an index of abundance may be a result of a decrease in capture probability and not necessarily a decrease in true abundance of the population being monitored.

The day-to-night adjustment factor calculated by Stapanian et al. (2009) for age-0 yellow perch is another example of how management agencies can make better estimates of fish abundance when only daytime data are available. In practice, adjustment factors used by different agencies would account for sampling biases among different vessels and different sampling gears (Tyson et al. 2006) and for changes in diel catchability associated with changes in water clarity over time. If adjustment factors applied to daytime catches are used as a surrogate for nighttime sampling, the ability to detect future changes in diel catchability is potentially lost. Use of an adjustment factor in lieu of sampling at night has the additional liability of adding error to indices and estimates of abundance. Such estimates add additional uncertainty to models that are used to project harvestable surpluses of yellow perch that support recreational and commercial fisheries. Further, trends in water clarity may change over time. In Europe, dreissenids had their greatest effects on water clarity shortly after invasion when population density was greatest (Karatayev et al. 1997). When dreissenid populations declined, water clarity decreased, but water clarity was still greater than preinvasion levels. Recent decreases in water clarity have also been observed in western Lake Erie (Binding et al. 2007). Therefore, we recommend that agencies that do not plan to night sample in annual surveys reassess adjustment factors at regular intervals.

There are a few other studies from North America that report effects of dreissenids on fishes. O'Gorman et al. (2000) reported a shift to deeper water for alewife (*Alosa pseudoharengus* Wilson), rainbow smelt (*Osmerus mordax* Mitchill), and juvenile lake trout (*Salvelinus namaycush* Walbaum) following establishment of dreissenids in Lake Ontario. They did not report Secchi disk depths or other measures of water clarity, but they noted that shifts in fish distributions followed increased water clarity in eastern Lake Erie, which is the major source of water to Lake Ontario. They also did not sample at night, so there was no way to compare daytime and nighttime CPH to assess shifts in catchability or distributions related to level of light.

Rennie et al. (2013) reported evidence of changes in lake whitefish (*Coregonus clupeaformis* Mitchill) behavior following dreissenid invasion. Similar to our results, they suggest that reported increases in catch per unit effort using gillnets may be a reflection of changes in catchability rather than changes in true abundance. Activity rates of lake whitefish estimated from bioenergetics models were significantly higher among populations where dreissenids were established and increased following dreissenid establishment in South Bay, Lake Huron, presumably as a consequence of increased foraging activity. Further, the range of lake whitefish in some Lake Michigan stocks (e.g., Big Bay de Noc) increased dramatically since the establishment of dreissenids (Ebener et al. 2010). Increases in both range and activity would positively influence probability of encounter of gillnets, which is a function of swimming speed and range of fish (Rudstam et al. 1984). Increased capture of lake whitefish because of increased capture probability would create a positive bias in lake whitefish abundance estimates, and therefore a bias in fishing quotas estimated annually by government agencies.

Our results are consistent with studies in the North Atlantic that suggested that the time of day during which sampling takes place affected catchability of fish in bottom trawls (Glass and Wardle 1989, Walsh 1991) and that indices of abundance or recruitment may be affected by time of day the sampling was conducted (Michalsen et al. 1996, Casey and Myers 1998). Nighttime–daytime differences were mostly apparent in Lake Erie only after water clarity increased because of dreissenids. We expect that greater gear visibility in the water column would result in lower capture probability of most fishes during daytime, whether the sampling gear used was active (i.e., gear is pulled through the water, such as bottom- or mid-water trawls) or passive (i.e., gear is stationary and fish swim into the gear, such as gill nets or trap nets). Steinberg (1962) reviewed the effect of visibility of gillnets on catch of European perch (*P. fluviatilis* L.) and roach (*Rutilus rutilus* L.) and commented that the effect of net visibility on catch of fish in gillnets has been known to be "extraordinarily important for some time" and that several predator and prey fishes react to small differences in net visibility. Steinberg (1985) also reported total catch of Atlantic cod (*Gadus morhua* L.) in trammel nets varied with mesh color. Thus, changes in water clarity can be expected to change fish ability to detect and avoid capture in gillnets set during the day.

Similarly, nighttime sampling by electrofishing and trap nets has also been common practice by fisheries management for decades because of the negative effects of daylight on capture probability of fishes using these methods. Our results suggest that night bottom trawling will improve estimates of abundance of benthic fishes in all nearshore areas of the Great Lakes in which dreissenids have become established by reducing net-avoidance behavior. Presently, nearly all bottom trawl sampling in Lake Erie is conducted during daylight hours. Certainly, sampling at night presents additional challenges, but management agencies will have to weigh the additional costs and inconveniences of nighttime sampling against potential benefits to fisheries management.

Potential ramifications of our results are also not limited to Lake Erie. Dreissenid mussels have increased water clarity elsewhere in the Great Lakes (e.g., Fahnenstiel et al. 1995, Vanderploeg et al. 2002), the Hudson River estuary (Caraco et al. 1997 and references therein), Oneida Lake (Zhu et al. 2006), and lakes and streams in eastern Europe (Karatayev et al. 1997). In San Francisco Bay, large decreases in phytoplankton and increases in water clarity were associated with the invasion and establishment of another nonnative bivalve, *Potamocorbula amurensis* (Alpine and Cloern 1992 and references therein). Our research demonstrated that effects of dreissenids extend to fish monitoring and fisheries management and underscored that dreissenids are ecosystem engineers (*sensu* Karatayev et al. 1997) capable of alteration of entire aquatic ecosystems.

ACKNOWLEDGMENTS

We thank the many biologists, technicians, boat crew, and volunteers who served aboard the USGS R/V Musky II during the study period. Constructive reviews were provided by D. Schloesser, T. Nalepa, and two anonymous reviewers. Mention of brand names does not imply endorsement by the U.S. Government. This chapter is Contribution 1732 of the U.S. Geological Survey Great Lakes Science Center.

REFERENCES

Alpine, A.E. and J.E. Cloern. 1992. Trophic interactions and direct physical effects control phytoplankton biomass and production in an estuary. *Limnol. Oceanogr.* 37:946–955.

Binding, C.E., J.H. Jermoe, R.P. Bukata, and W.G. Booty. 2007. Trends in water clarity of the lower Great Lakes from remotely sensed aquatic color. *J. Great Lakes Res.* 33:828–841.

Blanchard, J.L., D.L. Maxwell, and S. Jennings. 2008. Power of monitoring surveys to detect abundance trends in depleted populations: The effects of density dependent habitat use, patchiness, and climate change. *ICES J. Mar. Sci.* 65:111–120.

Botts, P.S., B.A. Patterson, and D.W. Schloesser. 1996. Zebra mussel effects on benthic invertebrates: Physical or biotic? *J. N. Am. Benthol. Soc.* 15:179–184.

Burnham, K.P. and D.R. Anderson. 2002. *Model Selection and Multimodel Inference: A Practical Information—Theoretic Approach*. New York: Springer Verlag.

Caraco, N.F., J.J. Cole, P.A. Raymond, D.L. Stayer, M.L. Pace, S.E.G. Findlay, and D.T. Fischer. 1997. Zebra mussel invasion in a large, turbid river: Phytoplankton response to increased grazing. *Ecology* 78:588–602.

Casey, J.M. and R.A. Myers. 1998. Diel variation in trawl catchability: Is it as clear as day and night? *Can. J. Fish. Aquat. Sci.* 55:2329–2340.

Charlton, M.N., R. LeSage, and J.E. Milne. 1999. Lake Erie in transition: The 1990's. In *State of Lake Erie Past, Present and Future*, M. Munawar, T. Edsall and I. F. Munawar, eds., pp. 97–123. Leiden, The Netherlands: Backhuys Publishers.

Ebener, M.P., T.O. Brenden, G.M. Wright, M.L. Jones, and M. Faisal. 2010. Spatial and temporal distributions of lake whitefish spawning stocks in northern Lakes Michigan and Huron, 2003–2008. *J. Great Lakes Res.* 36:38–51.

Fahnenstiel, G.L., G.A. Lang, T.F. Nalepa, and T.H. Johengen. 1995. Effects of zebra mussel (*Dreissena polymorpha*) colonization on water quality parameters in Saginaw Bay, Lake Huron. *J. Great Lakes Res.* 21:435–448.

French, J.R.P. and J.G. Love. 1995. Size limitation on zebra mussels consumed by freshwater drum may preclude the effectiveness of drum as a biological controller. *J. Freshwat. Ecol.* 10:379–383.

Glass, C.W. and C.S. Wardle. 1989. Comparisons of reactions of fish to a trawl gear at high and low light intensities. *Fish. Res.* 7:249–266.

Hayes, D.B., C.P. Ferreri, and W.W. Taylor. 1996. Active fish capture methods. In *Fisheries Techniques*, 2nd Edn., B. R. Murphy and D. W. Willis, eds., pp. 193–218. Bethesda, MD: American Fisheries Society.

Hebert, P.D.N., B.W. Muncaster, and G.L. Mackie. 1989. Ecological and genetic studies on *Dreissena polymorpha* (Pallas): A new mollusk in the Great Lakes. *Can. J. Fish. Aquat. Sci.* 46:1587–1591.

Hergenrader, G.L. and A.D. Hasler. 1968. Influence of changing seasons on schooling behaviour of yellow perch. *J. Fish. Res. Board Can.* 25:711–716.

Holland, R.E. 1993. Changes in planktonic diatoms and water transparency in Hatchery Bay, Bass Island area, western Lake Erie since the establishment of the zebra mussel. *J. Great Lakes Res.* 19:617–624.

Karatayev, A.Y., L.E. Burlakova, and D.K. Padilla. 1997. The effects of *Dreissena polymorpha* (Pallas) invasion on aquatic communities in eastern Europe. *J. Shellfish Res.* 16:187–203.

Karatayev, A.Y., L.E. Burlakova, and D.K. Padilla. 2002. Impacts of zebra mussels on aquatic communities and their role as ecosystem engineers. In *Invasive Aquatic Species of Europe*, E. Leppäkoski, S. Gollasch, and S. Olenin, eds., pp. 433–446. Dordrecht, The Netherlands: Kluwer Academic Publishers.

Kocovsky, P.M., M.A. Stapanian, and C. Knight. 2010. Evaluating sampling regimens for indices of yellow perch abundance in Lake Erie. *Fish. Ecol. Manage.* 17:10–18.

Kocovsky, P.M. and M.A. Stapanian. 2011. Influence of dreissenid mussels on catchability of benthic fishes in bottom trawls. Submitted to *Trans. Am. Fish. Soc.* 140:1565–1573.

Kovalak, W.P., G.D. Longton, and R.D. Smithee. 1993. Infestation of power plant water systems by zebra mussel (*Dreissena polymorpha* Pallas). In *Zebra Mussels: Biology, Impacts, and Controls*, T. F. Nalepa and D. W. Schloesser, eds., pp. 359–380. Boca Raton, FL: CRC Press.

Leach, J.H. 1993. Impacts of the zebra mussel (*Dreissena polymorpha*) on water quality and fish spawning reefs in western Lake Erie. In *Zebra Mussels: Biology, Impacts, and Control*, T. F. Nalepa and D. W. Schloesser (eds.), pp. 381–397. Boca Raton, FL: CRC Press.

Leach, J.H. 1999. Lake Erie: Passages revisited. In *State of Lake Erie Past, Present and Future*, M. Munawar, T. Edsall, and I. F. Munawar, eds., pp. 5–22. Leiden, The Netherlands: Backhuys Publishers.

Mayer, C.M., A.J. VanDeValk, J.L. Forney, L.G. Rudstam, and E.L. Mills. 2000. Response of yellow perch (*Perca flavescens*) to the establishment of zebra mussels (*Dreissena polymorpha*). *Can. J. Fish. Aquat. Sci.* 57:742–754.

MacIsaac, H.J., W.G. Sprules, and J.H. Leach. 1991. Ingestion of small-bodied zooplankton by zebra mussels (*Dreissena polymorpha*): Can cannibalism on larvae influence population dynamics? *Can. J. Fish. Aquat. Sci.* 48:2051–2060.

MacIsaac, H.J. 1996. Potential abiotic and biotic impacts of zebra mussels on the inland waters of North America. *Am. Zool.* 36:287–299.

MacIsaac, H.J. 1999. Biological invasions of Lake Erie: Past, present and future. In *State of Lake Erie Past, Present and Future*, M. Munawar, T. Edsall, and I. F. Munawar, eds., pp. 305–322. Leiden, The Netherlands: Backhuys Publishers.

Michalsen, K., O.R. Godø, and A. Fernö. 1996. Diel variation in the catchability of gadoids and its influence on the reliability of abundance indices. *ICES J. Mar. Sci.* 53:389–395.

Mills, E. L., R. M. Dermott, and E. F. Roseman et al. 1993. Colonization, ecology, and populations structure of the "quagga" mussel (Bivalvia: Dreissenidae) in the lower Great Lakes. *Can. J. Fish. Aquat. Sci.* 50:2305–2314.

Nichols, K.H. and G.J. Hopkins. 1993. Recent changes in Lake Erie (North shore) phytoplankton: Cumulative impacts of phosphorus loading reductions and the zebra mussel introduction. *J. Great Lakes Res.* 19:637–647.

Noether, G. 1991. *Introduction to Statistics the Non-Parametric Way*. Berlin, Germany: Springer Verlag.

O'Gorman, R., J.H. Elrod, R.W. Owens, C.P. Schneider, T.H. Eckert, and B.F. Lantry. 2000. Shifts in depth distribution of alewives, rainbow smelt, and age-2 lake trout following establishment of dreissenids. *Trans. Am. Fish. Soc.* 129:1096–1106.

Otter Research Ltd. 2000. An introduction to AD Model Builder© version 4.5 for use in nonlinear modeling and statistics. Sidney, British Columbia, Canada: Otter Research Ltd.

Pace, M.L., S.E.G. Findlay, and D. Fischer. 1998. Effects of an invasive bivalve on the zooplankton community of the Hudson River. *Freshwat. Biol.* 39:103–116.

Rennie, M.D. 2013. Context-dependent changes in lake whitefish populations associated with dreissenid invasion. In *Quagga and Zebra Mussels: Biology, Impacts, and Control*, 2 Edn., T.F. Nalepa and D.W. Schloesser, eds., pp. 661–680. Boca Raton, FL: CRC Press.

Riley, S.C. and K.D. Fausch. 1992. Underestimation of trout population size by maximum- likelihood removal estimates in small streams. *N. Am. J. Fish. Manage.* 12:768–776.

Rudstam, L.G., J.J. Magnuson, and W.M. Tonn. 1984. Size selectivity of passive fishing gear: A correction for encounter probability applied to gillnets. *Can. J. Fish. Aquat. Sci.* 41:1252–1255.

Stapanian, M.A., M.T. Bur, and J.V. Adams. 2007. Temporal trends of young-of-year fishes in Lake Erie and comparison of diel sampling periods. *Environ. Monit. Assess.* 129:169–178.

Stapanian, M.A., P.M. Kocovsky, and J.V. Adams. 2009. Change in diel catchability of young- of-year yellow perch in Lake Erie associated with establishment of dreissenid mussels. *Freshwat. Biol.* 54:1593–1604.

Steinberg, R. 1962. Die Fängigkeit von Kiemennetzen für Barsch und Plötze in Abhängigkeit von den Eigenschaften des Netzmaterials, der Netzkonstruktion und der Reaktion der Fische. *Arch. Fischereiwiss.* 12:173–230.

Steinberg, R. 1985. Einfluß der Netzfarbe und Garnstärke auf die Fängigkeit von Dorsch-stellnetzen. *Informationen für die Fischwirtschaft* 32:77–79.

Trometer, E.S. and W.D.N. Busch. 1999. Changes in age-0 fish growth and abundance following the introduction of zebra mussels *Dreissena polymorpha* in the western basin of Lake Erie. *N. Am. J. Fish. Manage.* 19:604–609.

Tyson, J.T., T.B. Johnson, C.T. Knight, and M.T. Bur. 2006. Intercalibration of research survey vessels on Lake Erie. *N. Am. J. Fish. Manage.* 26:559–570.

Vanderploeg, H.A., T.F. Nalepa, and D.J. Jude et al. 2002. Dispersal and emerging ecological impacts of Ponto-Caspian species in the Laurentian Great Lakes. *Can. J. Fish. Aquat. Sci.* 59:1209–1228.

Walsh, S.J. 1991. Diel variation in availability and vulnerability of fish to a survey trawl. *J. Appl. Ichthyol.* 7:147–159.

Webb, P.W. 1980. Does schooling reduce fast-start response latencies in teleosts? *Comp. Biochem. Physiol. Part A* 65:231–234.

Yellow Perch Task Group (YPTG). 2007. Report of the Lake Erie yellow perch task group to the Standing Technical Committee of the Lake Erie Committee. Ann Arbor, Great Lakes Fishery Commission.

Zhu, B., D.G. Fitzgerald, C.M. Mayer, L.G. Rudstam, and E.L. Mills. 2006. Alteration of ecosystem function by zebra mussels in Oneida Lake: Impacts on submerged macrophytes. *Ecosystems* 9:1017–1028.

PART VI

General

CHAPTER **44**

General Overview of Zebra and Quagga Mussels
What We Do and Do Not Know

Alexander Y. Karatayev, Lyubov E. Burlakova, and Dianna K. Padilla

CONTENTS

Abstract 695
Introduction 695
History of Spread 696
Life History 697
Population Biology and Dynamics 698
Competition 698
Vectors of Spread 699
Acknowledgments 700
References 701

ABSTRACT

The zebra mussel, *Dreissena polymorpha*, and quagga mussel, *Dreissena rostriformis bugensis*, are both important invaders in freshwaters of the Northern Hemisphere. These two invaders have similar life habits and life-history characteristics but differ in timing and rates of spread, habitat requirements, growth, and population dynamics. While the zebra mussel is among the best-studied freshwater invertebrates, we do not always have comparable information for the quagga mussel, which limits our ability to predict the spread and ecological impacts of this important freshwater invader. Here we contrast what is known and not known about zebra and quagga mussels and highlight information that is needed, especially for the quagga mussel, if we are to accurately predict its population dynamics and future spread.

INTRODUCTION

Dreissena polymorpha (Pallas 1771) (zebra mussel) and *Dreissena rostriformis bugensis* (Andrusov 1897) (quagga mussel) are both important invasive species in freshwaters of the Northern Hemisphere. These two species have similar life-habits and life-history characteristics and co-occur in their native habitats. The zebra mussel has been an important freshwater invader in Europe for centuries and in North America for over two decades (reviewed in Karatayev et al. 2007, van der Velde et al. 2010a). The quagga mussel did not become an important invader in Europe until the 1940s, and although it was introduced to North America at the same time as the zebra mussel, its rate and pattern of spread have been very different (reviewed in Mills et al. 1996, Orlova et al. 2004, 2005, Karatayev et al. 2007, 2011, van der Velde et al. 2010a). Presently, there is great concern about quagga mussels, as they have recently greatly expanded their range in both Europe and North America (reviewed in Orlova et al. 2004, 2005, Karatayev et al. 2007, van der Velde et al. 2010a). In addition, in many locales within Europe and North America, quagga mussels appear to be displacing zebra mussels as the dominant invader (Zhuravel 1952, 1965, Mills et al. 1996, Orlova et al. 2004, 2005, Ricciardi and Whoriskey 2004, Watkins et al. 2007, Dermott and Dow 2008, Nalepa et al. 2009a,b, 2010, Zhulidov et al. 2010).

Here we review what is known about the life history, biology, and recent spread of both zebra and quagga mussels throughout areas where they have been introduced. We identify similarities and contrast important differences between these two species and, finally, identify gaps in our knowledge that should be priorities for future research.

HISTORY OF SPREAD

Because both zebra and quagga mussels are important invaders, there is a good record of the geographic extent and rate of spread of both species in various different water bodies and countries. Spread in North America has been generally reported at a high level of spatial resolution (Benson 2013), but spread within Europe has been reported with far less detail, particularly prior to 20 years ago.

Zebra mussels were first discovered by Pitter Pallas in 1769 in the backwaters of the Ural River near the Caspian Sea, and the species was subsequently described in 1771 (Zhadin 1946). Until the late-1700s, zebra mussels were found exclusively in their native range of the Pontocaspian basin (Zhadin 1946, Mordukhai-Boltovskoi 1960, Starobogatov and Andreeva 1994). Construction of canals to connect the Dnieper River of the Black Sea basin with rivers of the Baltic Sea basin for commerce led to dramatic increases in distribution (reviewed in Kinzelbach 1992, Starobogatov and Andreeva 1994, Karatayev et al. 2003, 2008, 2010a). After canals were built and international trade increased, zebra mussels were found in the Curonian Lagoon of the Baltic Sea in 1803 (Baltic Sea Alien Species Database 2007) and in London in 1824 (Kerney and Morton 1970). During the first half of the nineteenth century, extensive shipping between major European ports allowed zebra mussels to spread quickly to many countries in Europe (reviewed in Kerney and Morton 1970, Kinzelbach 1992, Starobogatov and Andreeva 1994, Karatayev et al. 2007). In the nineteenth century, the spread of zebra mussels across Europe occurred at an exponential rate for ~70 years (1800–1868), with an average of ~3.9 regions (i.e., countries or geographic provinces within large countries) colonized per decade (Karatayev et al. 2011). After 1867, at the time of the industrial revolution and increased water pollution in Europe, the spread of zebra mussels in Europe essentially stopped (reviewed in Kinzelbach 1992, Karatayev et al. 2007). Over the next 94 years (1869–1962), zebra mussels colonized only two additional regions of Europe (reviewed in Kinzelbach 1992, Karatayev et al. 2011). However, from 1962 to 2009, there was a second period of exponential spread of this invader within Europe, as well as in North America where this species was introduced in the mid-1980s (Carlton 2008). Over this period, the average rate of spread of zebra mussels at the global scale (including both European spread among countries or major regions within countries and spread among North American states) was ~6.6 regions per decade, which is much faster than the original rate of spread across Europe (Karatayev et al. 2011).

The quagga mussel was first described by Andrusov in 1897. In contrast to zebra mussels, quagga mussels have a smaller native range, limited to the Dnieper–Bug Liman (a large coastal lake connected to the Black Sea), the Dnieper River Delta, and lower reaches of the South Bug and Ingulets Rivers (Zhulidov et al. 2010). Although there was extensive ship traffic through canals between areas inhabited by quagga mussels and other regions of eastern and western Europe through the nineteenth and the first half of the twentieth century, quagga mussels did not spread like zebra mussels into western Europe but remained restricted to their native range. It was not until the 1940s that quagga mussels began to spread through the Dnieper River and its tributaries, and by the 1980s they colonized the Don River system (Russia) (reviewed in Zhulidov et al. 2004, Karatayev et al. 2007). By the early-1990s, quagga mussels colonized reservoirs along the Volga River (Antonov and Kozlovsky 2001), and in 2003 quagga mussels were found in the Moscow River within the city of Moscow (Lvova 2004). Finally, in 2006 quagga mussels were found in the Rhine River and The Netherlands (Molloy et al. 2007) and in 2007 in the Maine River, Germany (van der Velde and Platvoet 2007). Quagga mussels were introduced to North America in the late-1980s (Mills et al. 1993).

The initial expansion of quagga mussels across Europe between the 1940s and 1980s was slow, only about 0.03 regions per decade. Since the 1980s, the spread of quagga mussels increased and is now progressing at an exponential rate in both Europe and North America. The rate of spread of quagga mussels from the mid-1980s to the present was 7.4 regions per decade, which is significantly faster than the rate of zebra mussels in the nineteenth century in Europe, but not significantly different than the recent rate of global spread of zebra mussels (6.9 regions per decade). Although both species of *Dreissena* were introduced into North America at about the same time (Mills et al. 1993, Carlton 2008), zebra mussels had colonized 2 times as many states as quagga mussels, almost 8 times more counties, and over 15 times more water bodies by 2008 (Karatayev et al. 2011).

In order to accurately predict the potential spread of any invasive species, it is essential to know what environmental factors limit its distribution or ability to form sustainable populations. While extensive research has been conducted on the biology, physiology, life history, and environmental limits of zebra mussels over the past 100 years (reviewed in Lyakhnovich et al. 1994, Starobogatov 1994, Karatayev et al. 1998, 2007, van der Velde et al. 2010b), few studies have examined similar features of quagga mussels. We do know that zebra mussels are able to tolerate higher salinities than are quagga mussels (6.0‰ vs. 3.5‰, reviewed in Lyakhnovich et al. 1994, Karatayev et al. 1998, 2007), allowing them to invade more estuarine-type habitats. However, both species have very low salinity tolerance. Zebra mussels also have a slightly higher temperature tolerance than quagga mussels

Table 44.1 Life-History Characteristics of Zebra Mussels and Quagga Mussels

Parameter	Zebra Mussel		Quagga Mussel	
Minimal temperature for reproduction (°C)	12°C–15°C	Sprung (1987), Borcherding (1991), Lvova et al. (1994a), Karatayev et al. (1998), Pollux et al. (2010)	5°C–7°C	Roe and MacIsaac (1997), Nalepa et al. (2010)
Duration of spawning period (months)	3–5	Reviewed in Lvova and Makarova (1994)	3	Nalepa et al. (2010)
Fecundity (eggs per reproductive season)	275,000–300,000	Lvova (1977)	N/A	
	≤1,000,000	Sprung (1991)		
	1,700,000	Neumann et al. (1993)		
Typical longevity (years)	4–5	Reviewed in Lvova et al. (1994b) and Karatayev et al. (2006)	N/A	

N/A, no data available.

(33°C vs. 31°C), but the lower temperature limit for both species is 0°C (reviewed in Karatayev et al. 1998, 2007). Quagga mussels appear to be more tolerant of low oxygen conditions than zebra mussels. Shkorbatov et al. (1994) found that all zebra mussels died under anoxic conditions in their experiments at the fourth day of exposure at 20°C, while all quagga mussels survived through the fourth day at the same temperature. Birger et al. (1975) also showed that zebra mussels require higher oxygen concentrations than quagga mussels for survival. The higher tolerance of quagga mussels to lower oxygen conditions relative to zebra mussels is perhaps one reason that quagga mussels can inhabit finer sediments and profundal regions of lakes where zebra mussels are rarely found. These regions can have low oxygen conditions for extended periods of time. Alternatively, Garton et al. (2013) provided data (from unpublished experiments) that showed zebra mussels were more tolerant of hypoxia than quagga mussels. Clearly, more experimental research is needed to determine the effects of low oxygen availability on both the zebra and quagga mussels, including the combined effects of temperature and oxygen conditions.

LIFE HISTORY

Extensive research has been conducted on the biology, physiology, reproduction, and life history of zebra mussels over the past 100 years (reviewed in Starobogatov 1994, van der Velde et al. 2010b). However, quagga mussels have not generally been the subject of similar studies. Often, in the absence of data, the biology and potential impacts of quagga mussels on ecosystems are assumed to be the same as zebra mussels (Keller et al. 2007, Ward and Ricciardi 2007, Higgins and Vander Zanden 2010).

Life histories of both zebra and quagga mussels are unusual among freshwater bivalves in that they have planktotrophic veliger larvae. This type of larval stage allows dispersal through connected waterways and allows these species to spread within a water body in a relatively short period of time after initial introduction. Life history and reproductive characteristics of the zebra mussel have been well studied. For example, fecundity, egg size, larval size, larval development, and larval duration of zebra mussels are fairly well known, but comparable information for quagga mussels is lacking (Table 44.1).

Gonad and oocyte development are different for zebra mussels found in shallow, warm waters versus deep, cold waters of lakes (Walz 1978, Bacchetta et al. 2010). In warmwater regions, gonads are fully developed by spring and spawning occurs in summer. Unspawned eggs may be resorbed in winter and then the cycle repeats (reviewed in Lvova and Makarova 1994). Very few studies have been conducted on zebra mussels in cold, deepwater regions, but the seasonal reproductive pattern appears to be different. In cold deep water, mussels have ripe gonads for extended periods of time and may spawn fewer eggs at once over many months (Bacchetta et al. 2010). There are no similar data following gonad production throughout the year in either shallow or deep waters for quagga mussels, and few studies have examined gonad development in quagga mussels (Nalepa et al. 2010).

Many studies have examined spawning in zebra mussels in shallow waters (Sprung 1987, Borcherding 1991, Lvova et al. 1994a, Karatayev et al. 1998, 2010a, Pollux et al. 2010). In general, these studies have considered the temperature at which spawning occurs, but have not looked at other environmental cues that may drive timing of reproduction, including photoperiod or phytoplankton abundance. Zebra mussels usually spawn when water temperatures reach 12°C–15°C, typically in May to June, and they can continue to spawn until the end of August or September. Quagga mussels have been found to spawn at colder water temperatures in the profundal zones of deep lakes (Roe and MacIsaac 1997, Claxton and Mackie 1998, Nalepa et al. 2010). Although these data are from different lake zones, quagga mussels generally have different temperature thresholds for spawning than zebra mussels.

In a shallow area in Lake Erie, both dreissenid species were found to initiate spawning when the water was 18°C–20°C, whereas in a deeper area of the lake, quagga mussels spawned a week later, when the water temperature reached 9°C (Claxton and Mackie 1998). In Lake Michigan, quagga mussel reproduction was studied at two different water depths: at 25 m, where mean water temperature from April to November was 10.0°C and the maximum annual temperature was 19.7°C, and at 45 m, where mean temperature from April to November was 5.6°C and the maximum annual temperature was 11.2°C (Nalepa et al. 2010). At the 25 m depth, all females had mature oocytes in early-April, but spawning did not occur until September (when water temperatures were at a maximum) and lasted through November. At the 45 m depth, all females had mature oocytes by late-April, and spawning started in early-June when the bottom temperature increased from 2°C to 4.5°C–6.0°C and was complete by early-August (Nalepa et al. 2010). It is obvious that more work is needed to determine spawning cues in quagga mussels, especially those found at different water depths.

If quagga mussels spawn at colder temperatures than zebra mussels, their spawning would be earlier in the spring and may give larval quagga mussels an advantage relative to zebra mussel larvae. Earlier spawning or faster temperature-dependent growth of larvae would allow quagga mussels to metamorphose earlier than zebra mussels and thereby give quagga mussels a competition for limited substrates. To date, there are good data on the length of the larval phase and how larval development changes with water temperature in zebra mussels (Lvova et al. 1994a, Bacchetta et al. 2010), but similar data for quagga mussels are lacking.

Much work has also been done on the typical longevity (life span) of zebra mussels in the field (reviewed in Lvova et al. 1994b, Karatayev et al. 2006; Table 44.1). Again, unfortunately, we do not have comparable data for the quagga mussel. Information about all life phases of the quagga mussel are critically needed if we are to adequately predict their population dynamics as they invade new waters and to understand their potential interactions with zebra mussels.

POPULATION BIOLOGY AND DYNAMICS

Density and population dynamics of zebra mussels can vary widely among lake systems (Ramcharan et al. 1992a, Burlakova et al. 2006). In some lakes, mussel densities increase through time until populations exceed carrying capacity and then densities decrease dramatically as populations collapse in what is considered a "boom and bust" pattern (reviewed in Simberloff and Gibbons 2004). In other lakes, mussel densities will increase and then remain high, or densities will fluctuate through time (Ramcharan et al. 1992b, Parker et al. 1999). Population dynamics of quagga mussels have been less well studied. In part, this may be because quagga mussels generally live much deeper in lakes than zebra mussels and these deep populations are more difficult to sample. As quagga mussels invade more lakes, especially lakes without zebra mussels, more opportunities will be available to examine population trends over time.

For most invasive species, there is a lag time between initial introduction and when the population reaches maximum abundance. This lag time may range from less than a year to decades, depending on the population dynamics of the species and habitats they invade (Kiritani and Yamamura 2003, Simberloff and Gibbons 2004, Daehler 2009). For zebra and quagga mussels, this time lag appears to be quite different. In water bodies where this information is available, the average lag-time between first introduction and when the population became abundant is 2.5 years (SE = 0.2, n = 13) for zebra mussels, and 12.2 years (SE = 1.5, n = 9) for quagga mussels (Karatayev et al. 2011). This shorter lag-time for zebra mussel population growth of zebra mussels may be the key to their invasion success and faster rate of spread compared to quagga mussels.

COMPETITION

In many water bodies where both zebra and quagga mussels co-occur, quagga mussels eventually outcompete zebra mussels (Zhuravel 1952, 1965, Mills et al. 1996, Orlova et al. 2004, 2005, Ricciardi and Whoriskey 2004, Watkins et al. 2007, Dermott and Dow 2008, Nalepa et al. 2010, Zhulidov et al. 2010). Typically when quagga mussels invade a water body already colonized by zebra mussels, they first colonize soft substrates of profundal zones and then spread into littoral zones, where zebra mussels are found. With time, quagga mussels become more abundant than zebra mussels even in littoral zones.

Zebra mussels and quagga mussels were introduced into the Laurentian Great Lakes in the late-1980s. Initially, zebra mussels dominated these lakes. In the mid-1990s, quagga mussels began to increase in abundance in Lakes Erie and Ontario. In 1995, only 37% of all dreissenids in the shallow areas of Lake Ontario were quagga mussels, but by 1998 the proportion of quagga mussels had increased to 59%, by 1999 to 93%, and by 2003 to 99% (Watkins et al. 2007). In Lake Erie, quagga mussels accounted for only 44% of all dreissenids in 1993 (Dermott and Dow 2008). However, by 2002, the relative abundance of quagga mussels had increased to 97% across the entire lake, and zebra mussels were common only in the shallow western basin (Patterson et al. 2005). By 2009, the proportion of quagga mussels reached over 88% in the western basin and over 97% in central and eastern basins (Karatayev et al. in review a). Similar changes in dominance were recently reported in Lake Michigan. Nalepa et al. (2010) sampled zebra and quagga mussel densities in southern Lake Michigan from 1992 to 2008. Zebra mussels were more abundant than quagga mussels until 2002 and dominated dreissenid biomass until 2004. After 2004, quagga mussels greatly increased in density and biomass and are now the most dominant species.

There are water bodies, and certain areas within water bodies, where zebra and quagga mussels coexist and other areas where zebra mussels remain dominant (Ricciardi and Whoriskey 2004, Zhulidov et al. 2004, 2006, 2010, Silaeva and Protasov 2005). For example, in the Soulanges Canal, Canada, quagga mussels are dominant on the bottom and lower portions of the canal walls, while zebra mussels dominate the upper portions (Ricciardi and Whoriskey 2004). In the Kanevskoe Reservoir, Ukraine, zebra mussels dominated in shallow waters up to 1.5 m deep, and quagga mussels dominated below 2 m (Silaeva and Protasov 2005). In the Don River, Russia, both species coexisted for over 25 years, and zebra mussels remain dominant (Zhulidov et al. 2006, 2010). In portions of the Mississippi and Ohio Rivers, quagga mussels remain less than 1% of all dreissenids after 12 years of coexistence (Grigorovich et al. 2008). In addition, both species coexist in their native range in the Dnieper River Delta and Dnieper–Bug Liman, Ukraine (Markovskiy 1954, Moroz and Aleksenko 1983, Zhulidov et al. 2010). All these results suggest that each species may have an advantage under different environmental conditions. Habitat partitioning by these two species is not surprising. During the Miocene–Pliocene period in the Pannon Basin, Central Europe, over 130 species of dreissenids occupied all bottom substrates from the shallow littoral zone to the silty sediments of the profundal zone (Geary et al. 2000). Zebra mussels are likely to be better adapted to the unstable environment of the upper littoral zone where fluctuations in water currents, temperature, and waves are prominent, while quagga mussels may be better adapted to the stable environment found in the deep profundal zone.

A number of hypotheses have been suggested to explain the displacement of zebra mussels by quagga mussels. First, differences in times of reproduction could affect recruitment and population abundance. As noted earlier, quagga mussels may spawn at lower temperatures and earlier in the season than zebra mussels (Roe and MacIsaac 1997, Claxton and Mackie 1998). Second, there may be asymmetrical food competition where quagga mussels are more efficient suspension feeders than zebra mussels (Diggins 2001). Third, quagga mussels appear to have a higher bioenergetic efficiency, resulting in positive growth over a wider range of food concentrations compared to zebra mussels that lose weight under low food levels (Baldwin et al. 2002, Stoeckmann 2003). Thus, zebra mussels seem to be better adapted to high food levels and quagga mussels appear to be better adapted to low food levels (Baldwin et al. 2002).

In laboratory experiments using flow through water from Lake Erie with natural concentrations of seston, quagga mussels had higher survivorship and grew more than zebra mussels under conditions that mimicked the thermal environment of both the profundal and littoral zones of the lake (Karatayev et al. 2010b). In these experiments, neither species grew during the winter. During the growing season at littoral-zone temperatures, the average zebra mussel increased 46% in length, while the average quagga mussel increased 106% in length. At profundal-zone temperatures, the average zebra mussel increased 27% in length, while the average quagga mussel increased 77% in length. For both species, there was a trade-off between survivorship and growth; at littoral-zone temperatures both species had higher growth but lower survival than at profundal-zone temperatures. This result supports the hypothesis that quagga mussels have a greater energetic efficiency than zebra mussels (Mills et al. 1999, Stoeckmann 2003), and this efficiency is not temperature dependent. Because quagga mussels have a higher energetic efficiency regardless of temperature, they are likely to attain larger total population sizes than zebra mussels in a given water body. Larger population sizes will filter larger volumes of water; therefore, quagga mussels may have a greater system-wide effect than zebra mussels, especially in deep lakes and reservoirs with large profundal zones.

VECTORS OF SPREAD

Although both zebra and quagga mussels occurred in waters connected by shipping canals and thus had the same opportunity for spread, zebra mussels began to spread from its native region 150 years before the quagga mussel, suggesting that potential vectors for spread were much less effective for quagga mussels than for zebra mussels. Human-built interbasin canals appear to have been instrumental for the initial spread of the zebra mussel across Europe, but the construction of reservoirs was required for the successful spread of the quagga mussel (reviewed in Karatayev et al. 2007, Therriault and Orlova 2010). In particular, large reservoirs constructed in Ukraine in the second half of the twentieth century appear to have facilitated the spread of quagga mussels, serving as "stepping stones" for invasion.

In the nineteenth and early twentieth centuries, the major mechanism for spread of zebra mussels across Europe was by the attachment of adult mussels to hulls of vessels and rafts transported through shipping canals (Kinzelbach 1992). A greater rate of byssal thread production and higher attachment strength of the zebra mussel relative to the quagga mussel (Peyer et al. 2009) may have contributed to the differential spread of the two species. In addition, habitats in Europe that were initially colonized by zebra mussels included hydrologically unstable, small rivers and canals, which are types of water bodies where quagga mussels are generally not found. In more recent times, increased rates of shipping traffic, ballast water, and overland movement of pleasure boats have probably all contributed to the spread of both species (reviewed in Karatayev et al. 2007).

Recreation boats have been a vector for spread of both species to new water bodies in Europe and especially in North America (Kinzelbach 1992, Padilla et al. 1996,

Figure 44.1 (See color insert.) (A) Boat hull heavily colonized by dreissenids (predominantly quagga mussels). This boat was not used during the entire 2010 season, which allowed quagga mussels to attach. (B) Left and right: typical density of dreissenids (predominantly zebra mussels) attached to boats regularly used in Lakes Erie and Ontario in 2010.

Buchan and Padilla 1999) (Figure 44.1). The ability of both species to be transported by recreation boats depends on a number of factors, including abundance in the source water body, tendency to attach to boats and trailers, and presence of waterweeds that can be transported by trailers, relative attachment strength, and survival during transport. Just as it may have facilitated the spread of zebra mussels relative to quagga mussels through European canals, greater attachment strength may favor the spread of zebra mussels relative to quagga mussels on recreational boats. Although quagga mussels comprised over 98% of dreissenids in eastern Lake Erie (Karatayev et al. in review a) and in Lake Ontario (Pennuto et al. 2012), zebra mussels were often dominant and obtained significantly larger sizes on boats (Karatayev et al. in review b). Moreover, anecdotal observations by marina owners indicate the number of mussels presently found attached to recreational boats is far less than the number found in the 1990s. Therefore, although zebra mussels may be in relatively low abundance, recreational boats are likely to continue to be a vector for spreading both zebra mussels and quagga mussels from Lakes Erie and Ontario.

ACKNOWLEDGMENTS

This study was in part supported by the U.S. EPA grant "The Nearshore and Offshore Lake Erie Nutrient Study" to C. Pennuto, A. Karatayev, A. Pérez-Fuentetaja, L. Burlakova, G. Matisoff, J. Kramer, D. Bade, J. Conroy, and E. Marschall. L.E.B. was supported by the Research Foundation of SUNY.

REFERENCES

Antonov, P. I. and S. V. Kozlovsky. 2001. Spontaneous areal expansion of some Ponto-Caspian species in reservoirs cascades. In *Abstracts of the U.S.–Russia Invasive Species Workshop*, P. G. William, J. S. Gregory, P. B. Ward et al., eds., pp. 16–18. Yaroslavl, Russia: Yaroslavl' State Technical University Press.

Bacchetta, R., P. Mantecca, and G. Vailati. 2010. Reproductive behaviour of zebra mussels living in shallow and deep water in the South Alps lakes. In *The Zebra Mussel in Europe*, G. van der Velde, S. Rajagopal, and A. bij de Vaate, eds., pp. 161–168. Leiden, The Netherlands: Backhuys Publishers.

Baldwin, B. S., M. S. Mayer, and J. Dayton et al. 2002. Comparative growth and feeding in zebra and quagga mussels (*Dreissena polymorpha* and *Dreissena bugensis*): Implications for North American lakes. *Can. J. Fish. Aquat. Sci.* 59:680–694.

Baltic Sea Alien Species Database. 2007. http://www.corpi.ku.lt/nemo/ (accessed November 29, 2010).

Benson, A. J. 2013. Chronological history of zebra and quagga mussels (Dreossenidae) in North America, 1988–2010. In *Quagga and Zebra Mussels: Biology, Impacts, and Control*, 2nd Edn., T.F. Nalepa and D.W. Schloesser, eds., pp. 9–31. Boca Raton: CRC Press.

Birger, T. I., A. Y. Malyarevskaya, and O. M. Arsan. 1975. Physiological aspects of adaptations of molluscs to abiotic and biotic factors, due to blue-green algae. In *Molluscs: Systematic, Evolution and Signification in the Nature. Abstracts of the Fifth Meeting on the Investigation of Molluscs*, I. M. Likharev, ed., pp. 91–94. Leningrad, Russia: Nauka Press (in Russian).

Borcherding, J. 1991. The annual reproductive cycle of the freshwater mussel *Dreissena polymorpha* Pallas in lakes. *Oecologia* 87:208–218.

Buchan, L. A. J. and D. K. Padilla. 1999. Estimating the probability of long-distance overland dispersal of invading aquatic species. *Ecol. Appl.* 9:254–265.

Burlakova, L. E., A. Y. Karatayev, and D. K. Padilla. 2006. Changes in the distribution and abundance of *Dreissena polymorpha* within lakes through time. *Hydrobiologia* 571:133–146.

Carlton, J. T. 2008. The zebra mussel *Dreissena polymorpha* found in North America in 1986 and 1987. *J. Great Lakes Res.* 34:770–773.

Claxton, W. T. and G. L. Mackie. 1998. Seasonal and depth variations in gametogenesis and spawning of *Dreissena polymorpha* and *Dreissena bugensis* in eastern Lake Erie. *Can. J. Zool.* 76:2010–2019.

Daehler, C. C. 2009. Short lag times for invasive tropical plants: Evidence from experimental plantings in Hawaii. *PLoS ONE* 4:e4462.

Dermott, R. and J. Dow. 2008. Changing benthic fauna of Lake Erie between 1993 and 1998. In *Checking the Pulse of Lake Erie*, M. Munawar and R. Heath, eds., pp. 409–438. New Delhi, India: Goodwords Books.

Diggins, T. P. 2001. A seasonal comparison of suspended sediment filtration by quagga (*Dreissena bugensis*) and zebra (*D. polymorpha*) mussels. *J. Great Lakes Res.* 27:457–466.

Garton, D.W., R. McMahon, and A.M. Stoeckman. 2013. Limiting environmental factors and competitive interactions between zebra and quagga mussels in North America. In *Quagga and Zebra Mussels: Biology, Impacts, and Control*, 2nd Edn., T.F. Nalepa and D.W. Schloesser, eds., pp. 383–402. Boca Raton, FL: CRC Press.

Geary, D. H., I. Magyar, and P. Müller. 2000. Ancient Lake Pannon and its endemic molluscan fauna (Central Europe; Mio-Pliocene). *Adv. Ecol. Res.* 31:463–482.

Grigorovich, I. A., T. R. Angradi, and C. A. Stepien. 2008. Occurrence of the quagga mussel (*Dreissena bugensis*) and the zebra mussel (*Dreissena polymorpha*) in the upper Mississippi River system. *J. Freshwat. Ecol.* 23:429–435.

Higgins, S. N. and M. J. Vander Zanden. 2010. What a difference a species makes: A meta-analysis of dreissenid mussel impacts on freshwater ecosystems. *Ecol. Monogr.* 80:179–196.

Karatayev, A. Y., L. E. Burlakova, S. E. Mastitsky, and S. Olenin. 2008. Past, current, and future of the Central European Corridor for aquatic invasions in Belarus. *Biol. Invasions* 10:215–232.

Karatayev, A. Y., L. E. Burlakova, S. E. Mastitsky, D. K. Padilla, and E. L. Mills. 2011. Contrasting rates of spread of two congeners, *Dreissena polymorpha* and *Dreissena rostriformis bugensis*, at different spatial scales. *J. Shellfish Res.* 30: 923–931.

Karatayev, A. Y., L. E. Burlakova, and D. K. Padilla. 1998. Physical factors that limit the distribution and abundance of *Dreissena polymorpha* (Pall.). *J. Shellfish Res.* 17:1219–1235.

Karatayev, A. Y., L. E. Burlakova, and D. K. Padilla. 2006. Growth rate and longevity of *Dreissena polymorpha* (Pallas): A review and recommendations for future study. *J. Shellfish Res.* 25:23–32.

Karatayev, A.Y., L. E. Burlakova, and D. K. Padilla. 2010a. *Dreissena polymorpha* in Belarus: History of spread, population biology, and ecosystem impacts. In *The Zebra Mussel in Europe*, G. van der Velde, S. Rajagopal, and A. bij de Vaate, eds., pp. 101–112. Leiden, The Netherlands: Backhuys Publishers.

Karatayev, A. Y., L. E. Burlakova, D. K. Padilla, and L. E. Johnson. 2003. Patterns of spread of the zebra mussel (*Dreissena polymorpha* (Pallas)): The continuing invasion of Belarusian lakes. *Biol. Invasions* 5:213–221.

Karatayev, A. Y., S. E. Mastitsky, D. K. Padilla, L. E. Burlakova, and M. H. Hajduk. 2010b. Differences in growth and survivorship of zebra and quagga mussels: Size matters. *Hydrobiologia* 668:183–194.

Karatayev, A. Y., D. K. Padilla, D. Minchin, D. Boltovskoy, and L. E. Burlakova. 2007. Changes in global economies and trade: The potential spread of exotic freshwater bivalves. *Biol. Invasions* 9:161–180.

Karatayev, A. Y., L. E. Burlakova, and C. Pennuto et al. Twenty five years of changes in *Dreissena* spp. populations in Lake Erie. *J. Great Lakes Res.* (in review a).

Karatayev, V. A., A. Y. Karatayev, L. E. Burlakova, and D. K. Padilla. Lakeside dominance does not predict the potential for spread of dreissenids. *J. Great Lakes Res.* (in review b).

Keller, R. P., J. M. Drake, and D. M. Lodge. 2007. Fecundity as a basis for risk assessment of nonindigenous freshwater molluscs. *Conserv. Biol.* 21:191–200.

Kerney, M. P. and B. S. Morton. 1970. The distribution of *Dreissena polymorpha* (Pallas) in Britain. *J. Conchol.* 27:97–100.

Kinzelbach, R. 1992. The main features of the phylogeny and dispersal of the zebra mussel *Dreissena polymorpha*. In *The Zebra Mussel Dreissena polymorpha: Ecology, Biological Monitoring and First Applications in the Water Quality Management*, D. Neumann, and H. A. Jenner, eds., pp. 5–17. Stuttgart, Germany: Gustav Fisher.

Kiritani, K. and K. Yamamura. 2003. Exotic insects and their pathways for invasion. In *Invasive Species: Vectors and Management*, G. M. Ruiz and J. T. Carlton, eds., pp. 44–67. Washington, DC: Island Press.

Lithuanian National Invasive Species Database. 2010. http://www.ku.lt/lisd/ (accessed November 29, 2010).

Lvova, A. A. 1977. The ecology of *Dreissena polymorpha* (Pall.) in Uchinskoe Reservoir. PhD dissertation. Moscow, Russia: Moscow State University (in Russian).

Lvova, A. A. 2004. On invasion of *Dreissena bugensis* (Bivalvia, Dreissenidae) in the Ucha Reservoir (Moscow oblast) and the Moscow River. *Zool. Zh.* 83:766–768 (In Russian with English Summary).

Lvova, A. A., A. Y. Karatayev, and G. E. Makarova. 1994a. Planktonic larva. In *Freshwater Zebra Mussel Dreissena polymorpha (Pall.) (Bivalvia, Dreissenidae). Systematics, Ecology, Practical Meaning*, J. I. Starobogatov, ed., pp. 149–155. Moscow, Russia: Nauka Press (in Russian).

Lvova, A. A. and G. E. Makarova. 1994. Reproduction. In *Freshwater Zebra Mussel Dreissena polymorpha (Pall.) (Bivalvia, Dreissenidae). Systematics, Ecology, Practical Meaning*, J. I. Starobogatov, ed., pp. 138–148. Moscow, Russia: Nauka Press (in Russian).

Lvova, A. A., G. E. Makarova, and M. P. Miroshnichenko. 1994b. Growth of the *Dreissena* in different parts of the distribution area. In *Freshwater Zebra Mussel Dreissena polymorpha (Pall.) (Bivalvia, Dreissenidae). Systematics, Ecology, Practical Meaning*, J. I. Starobogatov, ed., pp. 101–119. Moscow, Russia: Nauka Press (in Russian).

Lyakhnovich, V. P., A. Y. Karatayev, S. M. Lyakhov et al. 1994. Habitation conditions. In *Freshwater Zebra Mussel Dreissena polymorpha (Pall.) (Bivalvia, Dreissenidae). Systematics, Ecology, Practical Meaning*, J. I. Starobogatov, ed., pp. 109–119. Moscow, Russia: Nauka Press (in Russian).

Markovskiy, Y. M. 1954. *Fauna of Invertebrates of the Lower Reaches of the Ukrainian Rivers, Living Conditions, and Their Potential Use. Part II. Dnieper-Bug Liman*. Kiev, Ukraine: Academy of Sciences of the Ukrainian SSR Press (in Russian).

Mills, E. L., J. R. Chrisman, B. Baldwin et al. 1999. Changes in the dreissenid community in the lower Great Lakes with emphasis on southern Lake Ontario. *J. Great Lakes Res.* 25:187–197.

Mills, E. L., R. M. Dermott, E. F. Roseman et al. 1993. Colonization, ecology, and population structure of the "quagga" mussel (Bivalvia: Dreissenidae) in the lower Great Lakes. *Can. J. Fish. Aquat. Sci.* 50:2305–2314.

Mills, E. L., G. Rosenberg, A. P. Spidle, M. Ludyansky, and Y. Pligin. 1996. A review of the biology and ecology of the quagga mussel (*Dreissena bugensis*), a second species of freshwater dreissenid introduced to North America. *Am. Zool.* 36:271–286.

Molloy, D. P., A. Bij de Vaate, T. Wilke, and L. Giamberini. 2007. Discovery of *Dreissena rostriformis bugensis* (Andrusov 1897) in Western Europe. *Biol. Invasions* 9:871–874.

Mordukhai-Boltovskoi, F. D. 1960. *Caspian Fauna in the Azov and Black Sea Basins*. Moscow, Russia: Academia Nauk Press (in Russian).

Moroz, T. G. and T. L. Aleksenko. 1983. Benthos of the Dnieper-Bug Liman following the regulation of the Dnieper River run-off. *Gidrobiol. Zh.* 19:33–40 (in Russian with English summary).

Nalepa, T. F., D. L. Fanslow, and G. A. Lang. 2009a. Transformation of the offshore benthic community in Lake Michigan: Recent shift from the native amphipod *Diporeia* spp. to invasive mussel *Dreissena rostriformis bugensis*. *Freshwat. Biol.* 54:466–479.

Nalepa, T. F., D. L. Fanslow, and S. A. Pothoven. 2010. Recent changes in density, biomass, recruitment, size structure, and nutritional state of *Dreissena* populations in southern Lake Michigan. *J. Great Lakes Res.* 36:5–19.

Nalepa, T. F., S. A. Pothoven, and D. L. Fanslow. 2009b. Recent changes in benthic macroinvertebrate populations in Lake Huron and impact on the diet of lake whitefish (*Coregonus clupeaformis*). *Aquat. Ecosyst. Health Manage.* 12:2–10.

Neumann, D., J. Borcherding, and B. Jantz. 1993. Growth and seasonal reproduction of *Dreissena polymorpha* in the Rhine river and adjacent waters. In *Zebra Mussel: Biology, Impacts and Control*, T. F. Nalepa and D. W. Scloesser, eds., pp. 95–110. Boca Raton, FL: CRC Press.

Orlova, M. I., J. R. Muirhead, and P. I. Antonov et al. 2004. Range expansion of quagga mussels *Dreissena rostriformis bugensis* in the Volga River and Caspian Sea basin. *Aquat. Ecol.* 38:561–573.

Orlova, M. I., T. W. Therriault, P. I. Antonov, and G. K. Shcherbina. 2005. Invasion ecology of quagga mussels (*Dreissena rostriformis bugensis*): A review of evolutionary and phylogenetic impacts. *Aquat. Ecol.* 39:401–418.

Padilla, D. K., M. A. Chotkowski, and L. A. J. Buchan. 1996. Predicting the spread of zebra mussels (*Dreissena polymorpha*) to inland waters using boater movement patterns. *Global Ecol. Biogeogr. Lett.* 5:353–359.

Parker, I. M., D. Simberloff, W. M. Lonsdale et al. 1999. Impact: Toward a framework for understanding the ecological effect of invaders. *Biol. Invasions* 1:2–19.

Patterson, M. W. R., J. J. H. Ciborowski, and D. R. Barton. 2005. The distribution and abundance of *Dreissena* species (Dreissenidae) in Lake Erie, 2002. *J. Great Lakes Res.* 31(Suppl. 2):223–237.

Pennuto, C. M., E. T. Howell, T. K. Lewis, and J. C. Makarewicz. 2012. *Dreissena* population status in nearshore Lake Ontario. *J. Great Lakes Res.* 38:161–170.

Peyer, S. M., A. J. McCarthy, and C. E. Lee. 2009. Zebra mussels anchor byssal threads faster and tighter than quagga mussels in flow. *J. Exp. Biol.* 212:2027–2036.

Pollux, J. A., G. van der Velde, and A. bij de Vaate. 2010. A perspective on global spread of *Dreissena polymorpha*: A review on possibilities and limitations. In *The Zebra Mussel in Europe*, G. van der Velde, S. Rajagopal, and A. bij de Vaate, eds., pp. 45–58. Leiden, The Netherlands: Backhuys Publishers.

Ramcharan, C. W., D. K. Padilla, and S. I. Dodson. 1992a. Models to predict potential occurrence and density of the zebra mussel, *Dreissena polymorpha*. *Can. J. Fish. Aquat. Sci.* 49:2611–2620.

Ramcharan, C. W., D. K. Padilla, and S. I. Dodson. 1992b. A multivariate model for predicting population fluctuations of *Dreissena polymorpha* in North American lakes. *Can. J. Fish. Aquat. Sci.* 49:150–158.

Ricciardi, A. and F. Whoriskey. 2004. Exotic species replacement: Shifting dominance of dreissenid mussels in the Soulanges Canal, upper St. Lawrence River, Canada. *J. N. Am. Benthol. Soc.* 23:507–514.

Roe, S. L. and H. J. MacIsaac. 1997. Deepwater population structure and reproductive state of quagga mussels (*Dreissena bugensis*) in Lake Erie. *Can. J. Fish. Aquat. Sci.* 54:2428–2433.

Shkorbatov, G. L., A. F. Karpevich, and P. I. Antonov. 1994. Ecological physiology. In *Freshwater Zebra Mussel Dreissena polymorpha (Pall.) (Bivalvia, Dreissenidae). Systematics, Ecology, Practical Meaning*, J. I. Starobogatov, ed., pp. 67–108. Moscow, Russia: Nauka Press (in Russian).

Silaeva, A. A. and A. A. Protasov. 2005. On co-habitation of alien species in periphyton and benthos. In *Alien Species in Holarctic (Borok-2), Abstracts of the Second International Symposium*, Y. Y. Dgebuadze and Y. V. Slynko, eds., pp. 118–119. Borok, Russia (in Russian).

Simberloff, D. and L. Gibbons. 2004. Now you see them, now you don't!—Population crashes of established introduced species. *Biol. Invasions* 6:161–172.

Sprung, M. 1987. Ecological requirements of developing *Dreissena polymorpha* eggs. *Arch. Hydrobiol. Suppl.* 79:69–86.

Sprung, M. 1991. Costs of reproduction: A study on metabolic requirements of the gonads and fecundity of the bivalve *Dreissena polymorpha*. *Malacologia* 33:63–70.

Starobogatov, J. I. 1994. *Freshwater Zebra Mussel Dreissena polymorpha (Pall.) (Bivalvia, Dreissenidae). Systematics, Ecology, Practical Meaning*. Moscow, Russia: Nauka Press (in Russian).

Starobogatov, J. I. and S. I. Andreeva. 1994. Distribution and history. In *Freshwater Zebra Mussel Dreissena polymorpha (Pall.) (Bivalvia, Dreissenidae). Systematics, Ecology, Practical Meaning*, J. I. Starobogatov, ed., pp. 47–55. Moscow, Russia: Nauka Press (in Russian).

Stoeckmann, A. 2003. Physiological energetics of Lake Erie dreissenid mussels: A basis for the displacement of *Dreissena polymorpha* by *Dreissena bugensis*. *Can. J. Fish. Aquat. Sci.* 60:126–134.

Therriault, T. W. and M. I. Orlova. 2010. Invasion success within the Dreissenidae: Prerequisites, mechanisms and perspectives. In *The Zebra mussel in Europe*, G. van der Velde, S. Rajagopal, and A. bij de Vaate, eds., pp. 59–67. Leiden, The Netherlands: Backhuys Publishers.

van der Velde, G. and D. Platvoet. 2007. Quagga mussels *Dreissena rostriformis bugensis* (Andrusov, 1897) in the Main River (Germany). *Aquat. Invasions* 2:261–264.

van der Velde, G., S. Rajagopal, and A. bij de Vaate. 2010a. From zebra mussels to quagga mussels: An introduction to the Dreissenidae. In *The Zebra Mussel in Europe*, G. van der Velde, S. Rajagopal, and A. bij de Vaate, eds., pp. 1–10. Leiden, The Netherlands: Backhuys Publishers.

van der Velde, G., S. Rajagopal, and A. bij de Vaate. 2010b. *The Zebra Mussel in Europe*. Leiden, The Netherlands: Backhuys Publishers.

Walz, N. W. 1978. The energy balance of the freshwater mussel *Dreissena polymorpha* Pallas in laboratory experiments and in Lake Constance. II Reproduction. *Arch. Hydrobiol. Suppl.* 55:106–119.

Ward, J. M. and A. Ricciardi. 2007. Impacts of *Dreissena* invasions on benthic macroinvertebrate communities: A meta-analysis. *Divers. Distrib.* 13:155–165.

Watkins, J. M., R. Dermott, S. J. Lozano, E. L. Mills, L. G. Rudstam, and J. V. Scharold. 2007. Evidence for remote effects of dreissenids mussels on the amphipod *Diporeia*: Analysis of Lake Ontario benthic surveys, 1997–2003. *J. Great Lakes Res.* 33:642–657.

Zhadin, V. I. 1946. The travelling shellfish *Dreissena*. *Priroda* 5:29–37 (in Russian).

Zhulidov, A. V., A. V. Kozhara, G. H. Scherbina et al. 2010. Invasion history, distribution, and relative abundances of *Dreissena bugensis* in the old world: A synthesis of data. *Biol. Invasions* 12:1923–1940.

Zhulidov, A. V., T. F. Nalepa, A. V. Kozhara, D. A. Zhulidov, and T. Yu. Gurtovaya. 2006. Recent trends in relative abundance of two dreissenid species, *Dreissena polymorpha* and *Dreissena bugensis* in the Lower Don River system, Russia. *Arch. Hydrobiol.* 165:209–220.

Zhulidov, A. V., D. F. Pavlov, T. F. Nalepa, G. H. Scherbina, D. A. Zhulidov, and T. Yu. Gurtovaya. 2004. Relative distributions of *Dreissena bugensis* and *Dreissena polymorpha* in the lower Don River system, Russia. *Int. Rev. Hydrobiol.* 89:326–333.

Zhuravel, P. A. 1952. Liman fauna of the lower Dnieper and predicting its development in the Kakhovskoe Reservoir. *Vestnik Dniepropetr. Inst. Gidrobiol.* 9:77–98 (in Russian).

Zhuravel, P. A. 1965. About broad dispersal and mass abundance of *Dreissena bugensis* across canals and reservoirs of the Ukraine. In *Abstracts of the Second Meeting on the Investigation of Molluscs "Problems of Theoretical and Applied Malacology,"* I. M. Likharev, ed., pp. 63–64. Leningrad, Russia: Nauka Press (in Russian).

CHAPTER 45

Comparative Role of Dreissenids and Other Benthic Invertebrates as Links for Type-E Botulism Transmission in the Great Lakes

Alicia Pérez-Fuentetaja, Mark D. Clapsadl, and W. Theodore Lee

CONTENTS

Abstract 705
Introduction 705
Methods 707
Results 707
Discussion 709
References 710

ABSTRACT

Type-E botulism, a neuroparalytic disease caused by *Clostridium botulinum* type-E and transmitted through diet, has been affecting fish and waterfowl in the Great Lakes region. Dreissenid mussels are suspected to be a key link in the transmission of type-E botulism to upper trophic levels. In this study, we look at *C. botulinum* type-E spore DNA concentrations in benthic organisms common in diets of fish and waterfowl in Lake Erie. We found spore DNA in dreissenid mussels, in dreissenid biodeposits, and in other benthic invertebrates (Chironomidae larvae, Nematoda, Oligochaeta, Ephemeroptera larvae, and Amphipoda). Overall, benthic organisms carried spore DNA in the order of 5.0×10^4 copies/mg at nearshore sites and 3.0×10^3 copies/mg at offshore sites. Dreissenids carried on average 1.0×10^3 copies/mg of spore DNA, but the highest load was found in chironomids with an average of 1.0×10^5 copies/mg. We conclude that because dreissenids are the preferred food of round gobies, they are likely to be a major pathway for the transmission of *C. botulinum* type-E in the Great Lakes. However, other benthic invertebrates can also be implicated in the initiation of a botulism outbreak and act as reservoirs for *C. botulinum* type-E.

INTRODUCTION

Type-E botulism outbreaks were originally reported in the Great Lakes in the 1960s, but it was not until the late-1990s that frequent outbreaks and large fish and waterfowl mortalities were observed in the region (Pérez-Fuentetaja et al. 2006, Sea Grant Michigan 2007, Canadian Cooperative Wildlife Health Center 2008). During the first decade of the 2000s, botulism outbreaks became regular events from spring to autumn in all the Great Lakes except Lake Superior. These outbreaks are caused by an aquatic microorganism, *Clostridium botulinum* type-E, that occurs in sediments. Type-E botulism results from the effect of a potent neurotoxin produced by these bacteria (type-E toxin strain) that causes paralysis and the inability to breathe in vertebrates (Rocke and Bollinger 2007).

Conditions that favor growth, dispersal, and transmission of *C. botulinum* type-E in the Great Lakes are both environmental and biological. *C. botulinum* type-E are cold-tolerant bacteria that have a northern temperate zone distribution with spores germinating optimally at 9°C. The vegetative cells can grow in a wide range of temperatures (6°C–41°C) (Grecz and Arvay 1982, Rocke and

Bollinger 2007), but they thrive at 32.5°C in the presence of anaerobic sediments where decomposition of organic matter occurs. Events that result in warm sediments and favor anaerobic conditions include a warming climate (Sousounis and Glick 2000) with less ice cover in the winter (Assel et al. 2003), and reduced lake levels (Moraska Lafrancois et al. 2010). However, adequate conditions for the proliferation of bacteria are not necessarily sufficient for an outbreak to occur. A number of variables have been identified that can promote disease transmission in aquatic populations. These variables include temperature, the immune response of the host, density of the pathogen, and density of the host population (Riley et al. 2008). Thus, bacterial proliferation and toxin production need to be linked to a benthic foodweb for botulism to occur and spread. Benthic organisms, including but not limited to mollusks, must come in contact with *C. botulinum* type-E and then be consumed by a vertebrate (fish, aquatic salamanders, waterfowl) for the initiation of an outbreak (Pérez-Fuentetaja et al. 2011).

Ponto-Caspian invaders, in particular dreissenid mussels (zebra, *Dreissena polymorpha*, and quagga, *Dreissena rostriformis bugensis*) and one of their natural predators, the round goby (*Neogobius melanostomus*), are suspected to be a vehicle of transmission for type-E botulism in the Great Lakes (Domske and Obert 2001, Robbins 2002, Corkum et al. 2004, Environment DEC 2008). In this view, dreissenid mussels carrying *C. botulinum* type-E would transmit the bacteria, toxin, or spores to round gobies, and gobies would pass them up to higher trophic levels in the foodweb. In 2000, round gobies became very abundant in Lake Erie (Einhouse et al. 2010) and replaced other native fishes in the diets of wildlife in the region. Round gobies are susceptible to the type-E toxin (Steinhart et al. 2004, Yule et al. 2006a,b), but because gobies do not have a swim bladder, they are negatively buoyant and sink after death. As a result, round gobies are not typically found in beach monitoring efforts to evaluate botulism mortality (Domske and Obert 2001, Domske 2004) so we do not know the impact of the disease on their populations.

The invasion and spread of dreissenid mussels have created major changes in the littoral zone of Lake Erie. The sediment in nearshore areas has been modified by an increase in waste products from abundant dreissenid populations. In addition, accumulations of dreissenid shells prevent free-gas exchange between the sediment and overlying water, promoting anoxia. Indirectly, increased water clarity from mussel filtration has resulted in abundant plant growth, and plant decomposition has contributed to oxygen decline in sediments (Vanderploeg et al. 2002, Hecky et al. 2004). The overall intensification of benthic decomposition and anoxia by dreissenids has favored the proliferation of anaerobic sediment bacteria (Fenchel and Findlay 1995), including *C. botulinum* type-E. Given the ubiquitous presence of *C. botulinum* type-E in aquatic sediments (Rocke and Bollinger 2007), benthic organisms such as chironomids and oligochaetes, which feed on sediment bacteria, could become sources of botulism. Therefore, dreissenids would be one of several possible pathways of transmission (Figure 45.1).

In this study, we present data on the depth distribution of *C. botulinum* type-E spores in sediments, dreissenids, and other benthic invertebrates and explore the role that these organisms play in the botulism epizootics that have occurred in Lake Erie.

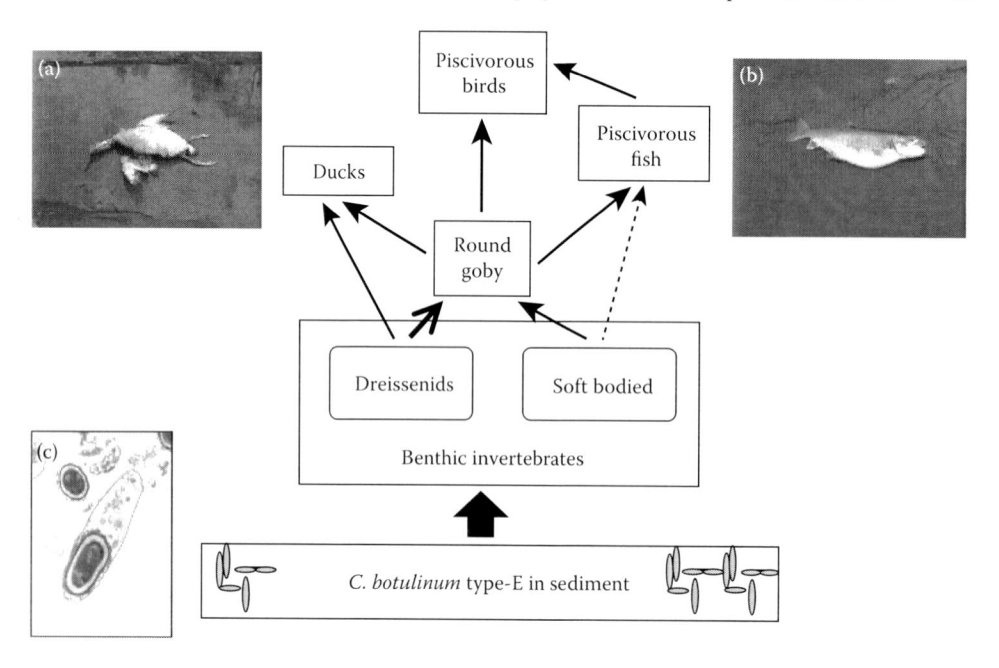

Figure 45.1 Pathways for possible transmission of *C. botulinum* type-E through the foodweb in the Great Lakes. Some victims of botulism in Lake Erie include (a) waterfowl and (b) lake trout. (c) Also shown are dormant spores of *C. botulinum* type-E. (From Stringer, S.C. et al., *Appl. Environ. Microbiol.*, 71, 4998, 2005.) Thick arrows indicate strong connections between organisms and possible transmission of *C. botulinum*. Dotted line indicates an occasional foodweb link.

METHODS

We collected samples from three stations at each of two sites located in eastern Lake Erie near Dunkirk (New York). The two sites were in a nearshore region (mean depth = 6.6 ± 1.4 m) and an offshore region (mean depth = 17.4 ± 1.1 m). Specifically, the former site was southwest of Dunkirk near Van Buren Point, and the latter site was directly north of Dunkirk harbor (Figure 45.2). All stations at these sites were located in areas where large fish and bird mortalities were observed during outbreaks in 2002 (Pérez-Fuentetaja et al. 2006). Stations at each site had mixed substrate of mud/sand/cobble and were 50–80 m apart. In total, about 54 samples were collected nearshore and 58 samples offshore in each of the 2 years.

Sediment, benthic invertebrates, and dreissenids were collected using Ponar grabs and scuba diving. All samples were collected on a weekly basis at each of the sites from June to October in 2002 and from late May to October in 2003, except for the 2002 collection of benthic invertebrates and dreissenids that took place in September. Non-dreissenid groups included Chironomidae, Nematoda, Oligochaeta, Amphipoda, and Ephemeroptera larvae. Collected organisms were alive when sorted, refrigerated, and subsequently frozen at −20°C within 24 h of collection (Pérez-Fuentetaja et al. 2006).

To purify *C. botulinum* type-E genomic DNA from sediment samples, we used the SoilMaster™ DNA extraction kit. DNA purifications were done in duplicate, with 100 mg of starting material each, and were then subjected to PCR and analyzed for the *bont*E gene. The DNA extraction kit used extracted DNA from spores (Pérez-Fuentetaja et al. 2006). We used the *bont*E gene because it was detected in 2001 and 2002 in fish and waterfowl collected during outbreaks of botulism and it has been used in other type-E botulism studies (Kimura et al. 2001, Akbulut et al. 2004, Getchell et al. 2006). To purify *C. botulinum* type-E genomic DNA from benthic invertebrates (dreissenids and non-dreissenids), we used the DNeasy Tissue Kit (Qiagen, Valencia, CA) that also extracts DNA from spores. We included a no-template control (NTC) and diluted plasmid standards on every quantitative polymerase chain reaction (Q-PCR) plate and they were run in triplicate wells. The assay reliably measured 40 *bont*E copies and often below 4 copies per reaction (Getchell et al. 2006). Q-PCR was performed on an Applied Biosystems ABI PRISM model 7700 sequence detector (ABI, Foster City, CA) (refer to Pérez-Fuentetaja et al. [2006] for a discussion on spore DNA). Concentrations of spore DNA were reported as copies/mg with each copy corresponding to one spore.

RESULTS

We found differences between 2002 and 2003 in the number of positive spore DNA samples for *C. botulinum* type-E (Figure 45.3). In 2002, a total of 112 sediment samples including nearshore (54 samples) and offshore (58 samples) were analyzed for spore DNA, and 7 samples tested positive. In 2003, 113 samples of sediment were analyzed (55 samples nearshore and 58 samples offshore) and 3 samples tested positive for spore DNA (Table 45.1). Comparison of the concentrations of spore DNA in sediments (nearshore and offshore sites combined) revealed higher levels in 2002 than in 2003 (t-test, $P < 0.001$).

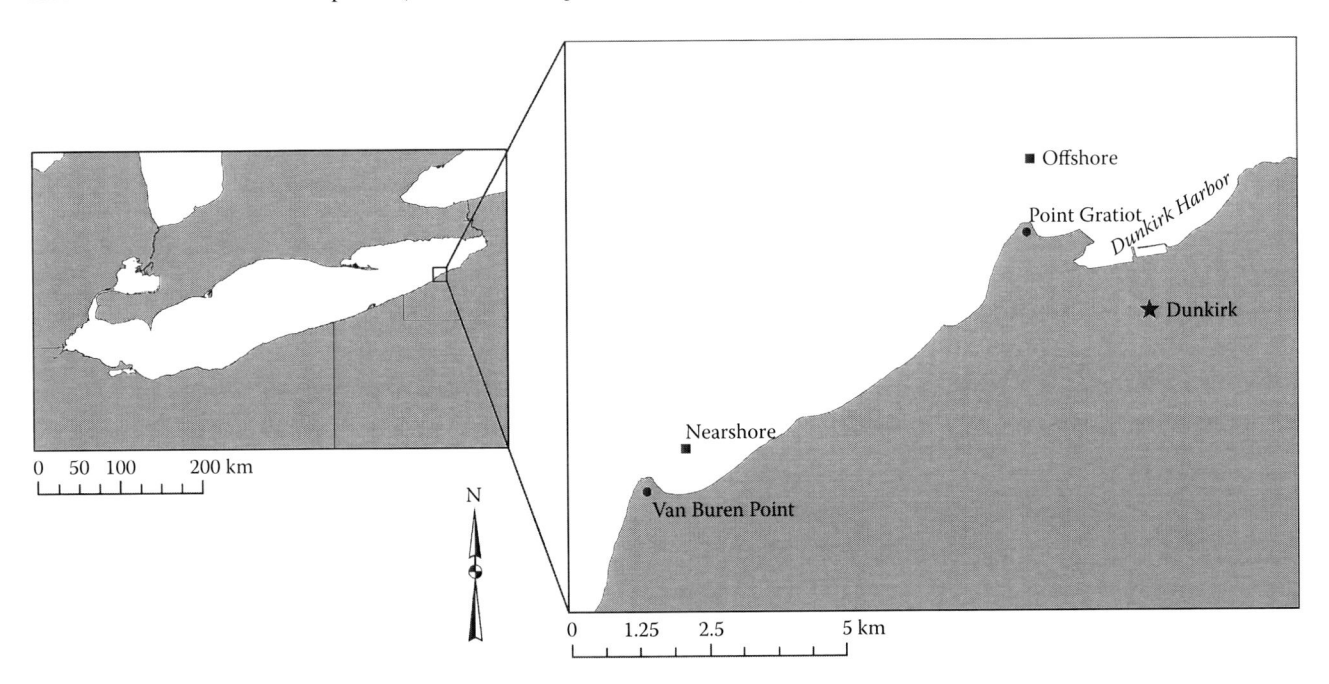

Figure 45.2 Locations of nearshore and offshore sampling sites in Lake Erie for this study. Water depth was on average 7 m for the nearshore site and 18 m for the offshore site. (Courtesy of K. Hastings.)

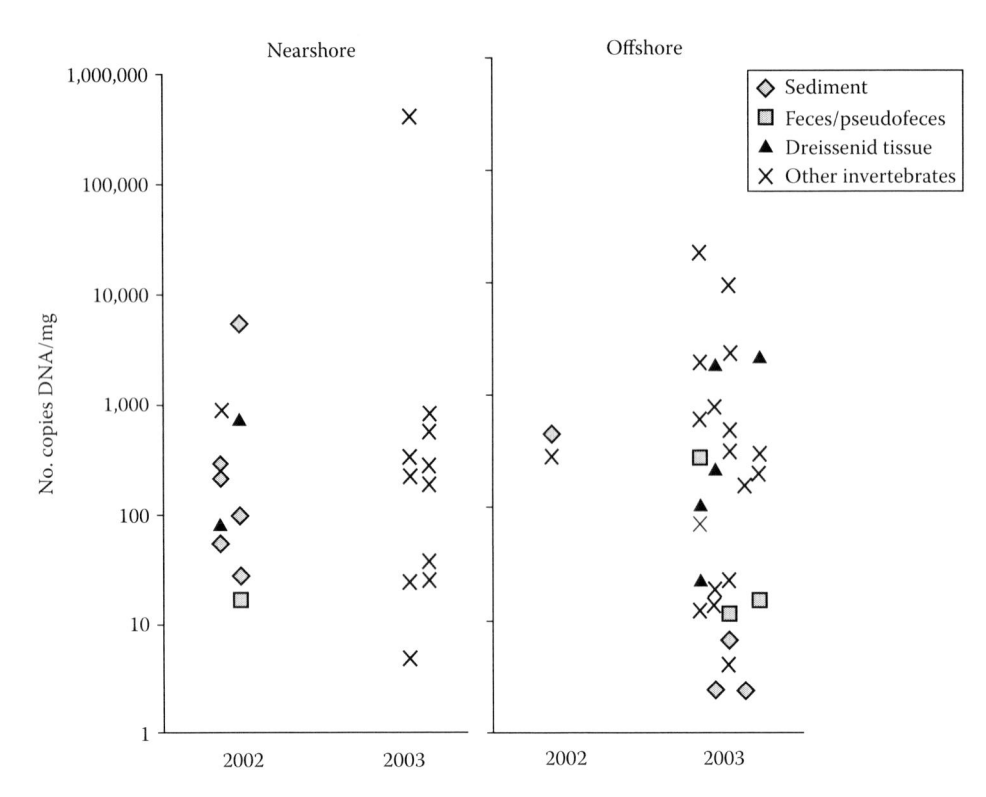

Figure 45.3 Incidence and relative abundance of positive *C. botulinum* type-E spore DNA at the nearshore and offshore sites in 2002 and 2003. Nearshore and offshore samples included sediments, dreissenid tissue, dreissenid biodeposits (feces and pseudofeces), and other benthic invertebrates (chironomid larvae, nematodes, oligochaetes, amphipods, and mayfly larvae). The data presented for 2002 in "other invertebrates" include only one sampling date, while in 2003, samples were collected from May to October.

Table 45.1 Mean (± SD) Concentration (DNA Copies/mg) of *C. botulinum* Type-E Spore DNA (*bontE* Gene) in Samples of Sediment, Dreissenid Feces/Pseudofeces, and "All Benthos" That Tested Positive in 2002 and 2003. Also Given Are the Number of Samples (*N*) and the Number of Those Samples That Tested Positive Each Year. In 2002, Benthos Was Analyzed for Spore DNA on Only One Date. All Other Samples in Both Years Are Collected and Analyzed from Early Summer to Fall. "All Benthos" Includes Dreissenids and Other Benthic Invertebrates

Sample Type	2002			2003		
	Concentration	N	Positive	Concentration	N	Positive
Nearshore site						
Sediment	1032.5 ± 2200.6	54	6		55	0
Feces/pseudofeces	16.8	8	1		0	
All benthos	941.3 ± 993.1	4	3	51,943.3 ± 204,817.7	58	16
Offshore site						
Sediment	456.0	58	1	3.8 ± 2.4	58	3
Feces/pseudofeces	—	0		100.4 ± 151.0	14	3
All benthos	329.3 ± 157.2	3	3	2610.7 ± 7531.3	77	27

In 2002, we found six positive sediment samples with an average (± SD) concentration of 1032.5 ± 2200.6 DNA copies/mg at nearshore stations and only one positive sample with a concentration of 456 DNA copies/mg at offshore stations. In contrast, in 2003, we didn't find any positive sediment samples at nearshore stations and found three positive samples with an average of 3.8 ± 2.4 DNA copies/mg at offshore stations (Table 45.1).

In 2003, benthic invertebrates with spores were found at all sites and at both depths (nearshore, offshore), but no significant differences were found in spore DNA levels among different benthic invertebrate taxa or between organisms from nearshore and offshore stations (ANOVA, $P > 0.05$). However, benthic invertebrates carried a significantly higher level of spore DNA than sediments ($P < 0.001$).

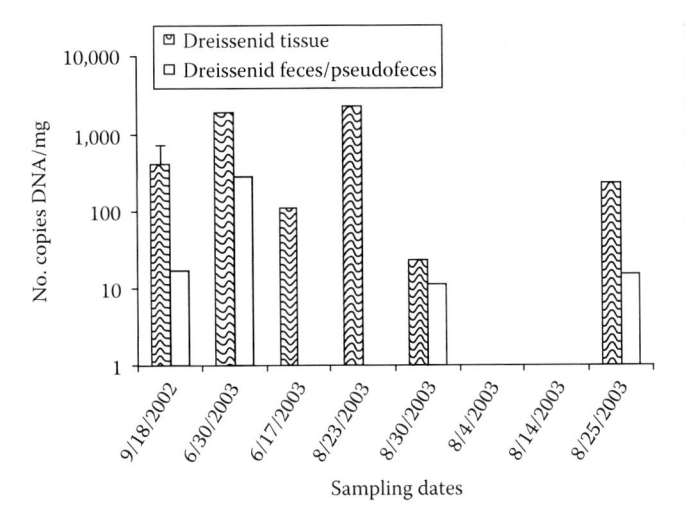

Figure 45.4 Mean (± SE) concentration (# copies DNA/mg) of *C. botulinum* type-E spore DNA in dreissenid tissue and biodeposits on dates when dreissenid mussels were found. In 2002, mussels were found at the nearshore site, and in 2003, mussels were found at the offshore site. On the two dates in 2003 without values, dreissenids were found but tested negative for type-E spore DNA.

Tissue and feces/pseudofeces of dreissenids tested positive for spore DNA mostly during early and late summer (Figure 45.4). Dreissenid tissue consistently contained more spore DNA than surrounding accumulations of sediments and feces/pseudofeces at a given site; the difference between tissue and sediments was significant ($P < 0.01$), but the difference between tissue and feces/pseudofeces was not ($P > 0.05$). Maximum concentrations in tissue reached 1920 DNA copies/mg on May 30, 2003, and 2280 DNA copies/mg on June 23, 2003. Dreissenids carried spore loads (DNA copies per mg tissue) that were similar to other benthic invertebrate taxa except for chironomid larvae at the nearshore site (Figure 45.5). Dreissenid tissue had a mean concentration of 397 ± 317 DNA copies/mg at the nearshore site and 911 ± 1095 DNA copies/mg at the offshore site. In contrast, chironomid larvae had a mean concentration of 102,971 ± 289,724 DNA copies/mg at the nearshore site and 2,322 ± 5,718 DNA copies/mg at the offshore site.

DISCUSSION

The impact of dreissenid mussels and other invasive species on aquatic systems goes beyond ecological alterations when they act as a catalyst for disease/toxin transmission to upper trophic levels (such as fish and waterfowl) (Figure 45.1). Although type-E botulism was not unknown to the Great Lakes region, as noted in the introduction, the intensity and regularity of botulism outbreaks during the last decade is a novel phenomenon. Environmental conditions can play a key role in botulism outbreaks by affecting temperature, which, in turn, triggers changes in sediment chemistry (Rocke and Samuel 1999, Pérez-Fuentetaja et al. 2011). For example, in Lake Erie, outbreaks in 2002 were associated with temperatures near the sediment consistently above 20°C that promoted reducing conditions and a decrease in oxygen near the bottom (Pérez-Fuentetaja et al. 2006). Such conditions would favor spore germination. Yet, even if sediment conditions are favorable for spore germination and proliferation of bacteria, host populations that are trophic links in the foodwebs of aquatic vertebrates have to be present to transmit botulism. Dreissenid mussels are one of those links. The increase in abundance and spread of dreissenid mussels, which can accumulate type-E spores and perhaps even the toxin (Holeck et al. 2004, Getchell and Bowser 2006), has provided a ready food source for another invader, the round goby. Round gobies feed heavily on dreissenid mussels, and in so doing, the two species have coalesced and facilitated the transfer of *C. botulinum* type-E to fish and waterfowl that feed on gobies (Figure 45.1).

In our survey area, large fish and waterfowl mortalities occurred in 2002 due to type-E botulism, while only small sporadic outbreaks occurred in 2003 (Einhouse et al. 2004). In 2002, beach transect surveys along Lake Erie's shoreline in New York yielded 17,301 dead birds (12,616 long-tailed ducks, 2,042 common loons, 839 red-breasted mergansers, 273 horned grebes, 273 ring-billed gulls, and 1,258 others) (Adams et al. 2004). In contrast, in 2003, the beach transect surveys yielded only 3008 dead birds (1969 common loons, 292 ring-billed gulls, 219 long-tailed ducks, 55 red-breasted mergansers, 18 horned grebes, and 455 others) (Adams et al. 2004). Our findings of *C. botulinum* type-E in the benthic region correspond to this yearly difference in wildlife mortalities. Despite a similar effort sampling sediments in 2002 and 2003, more of our samples tested positive and spore concentrations were higher in 2002 when wildlife mortalities

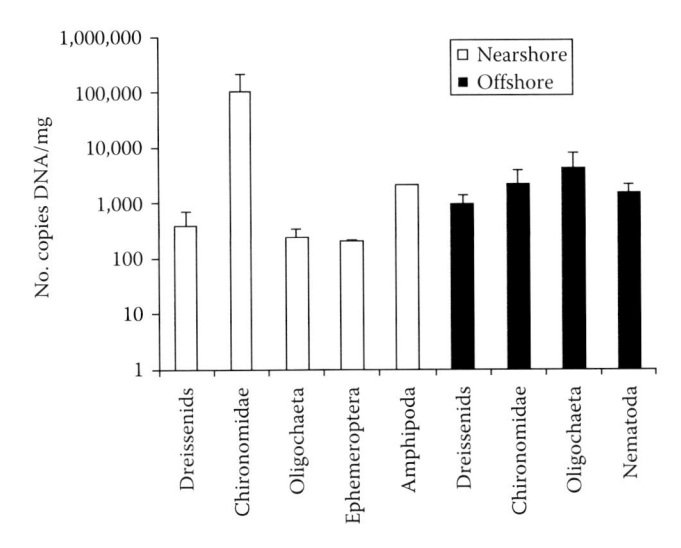

Figure 45.5 Mean (± SE) concentration (# copies DNA/mg) of *C. botulinum* type-E spore DNA in various benthic macroinvertebrate taxa over all sampling dates and sites in 2002 and 2003.

were higher than in 2003. Still, in 2003, benthic invertebrates had very high levels of spore DNA (Figure 45.3) that were above sediment levels. Thus, benthic invertebrates, perhaps through their use of the benthic habitat and feeding activities, accumulate spores. These results point to the importance of benthic invertebrates as reservoirs of *C. botulinum* type-E during years of low botulism occurrence. The high levels of spores found in these organisms probably contribute to the continuity and regularity of the botulism epizootics that have occurred in many of the Great Lakes.

In contrast to other benthic invertebrates that, through bioturbation will oxygenate sediments, dreissenids play multiple environmental roles that could facilitate the initiation and spread of botulism. Accumulations of feces/pseudofeces in mussel druses provide decomposing organic material and anaerobic pockets that create favorable conditions for bacterial germination and production. Mussel druses also function as spore reservoirs that may be disrupted by bioturbation (e.g., amphipods), predators, or water movements. Spores could then be distributed to other parts of the lake or filtered by the mussels and re-deposited around druses. Likewise, accumulations of dead mussel shells contribute to bacterial production by trapping waste materials, thereby facilitating anoxia and increasing spore germination.

We found that burrowing benthic invertebrates carry spores and are also potential links to upper trophic levels in the transmission or initiation of type-E botulism. In particular, chironomid larvae (a common food for young gobies) (Kuhns and Berg 1999, French and Jude 2001) contained the highest levels of spores found in the benthic taxa examined. Chironomids are often found associated with sediments rich in organic matter and low in oxygen levels (Quinlan and Smol 2001), which are habitat to *C. botulinum*, and feed on detritus and bacteria from decomposing material (Thorp and Covich 1991). They are also small-sized organisms, having a larger surface area (body surface and digestive tract) on a per-milligram basis than larger invertebrates like mayflies and dreissenids. For example, the number of copies of spore DNA in this study is reported per milligram of tissue. One milligram of chironomids contains many animals compared to one milligram of dreissenids, which may contain less than one individual. Since spores are either carried on body surfaces (perhaps attached to setae) or found in digestive tracts, a bigger surface area would provide a mechanism for chironomids to carry a high number of spores. Thus, feeding in places where *C. botulinum* bacteria are active or have formed spores combined with a high body surface on a per-milligram basis may explain the high numbers of spores found in chironomids.

Although dreissenids can arguably be a major link in the observed botulism epizootics, other organisms may fulfill this role given the right trophic pathways. Prey selection in this context becomes an important factor. Fish will feed on energetically dense foods (Gerking 1994) and may choose to feed on sessile and readily available dreissenids over smaller burrowing prey such as chironomids and oligochaetes. However, round gobies readily feed upon other benthic organisms besides dreissenids (Lederer et al. 2006, 2008). Thus, amphipods, chironomids, oligochaetes, nematodes, and mayflies could, according to our results, be a prime source for type-E spores to fish. The most affected fish species during the type-E botulism outbreaks in eastern Lake Erie have been benthic feeders such as freshwater drum (*Aplodinotus grunniens*), round gobies (Yule et al. 2006a), and benthic-opportunist smallmouth bass (*Micropterus dolomieu*) (Culligan et al. 2002). These fishes likely encounter and ingest a variety of benthic foods contaminated with *C. botulinum* type-E and are among the most frequent victims of the disease (Domske and Obert 2001).

With climate change, a rise in water temperatures in the Great Lakes could promote the occurrence of botulism epizootics. *C. botulinum* type-E is a plausible candidate to benefit from increased temperatures and pockets of anoxia. As we have seen over the last decade in Lakes Erie, Ontario, Huron, and Michigan, host populations and food web links are in place to transmit *C. botulinum* from sediments to invertebrates and from them to fish and waterfowl when conditions are favorable. The foodweb transfer of *C. botulinum*, however, can follow different pathways. Although dreissenid mussels seem to be rightly positioned to facilitate production and transmission of the bacteria, spores, and toxin, other benthic organisms can transfer *C. botulinum* type-E as well. Dreissenids, however, play a dual role by creating favorable conditions for *C. botulinum* type-E proliferation and by being a preferred food for the abundant round goby. In a more general context, however, benthic invertebrates (dreissenids and otherwise) could be acting as reservoirs for *C. botulinum* type-E during times of unfavorable conditions and low sediment bacterial production. If this is the case, benthic invertebrates may be contributing to the perpetuation of botulism outbreaks in the Great Lakes region by concentrating spores and redistributing them as they move through the sediment.

REFERENCES

Adams, D., K.R. Roblee, and W. Stone. 2004. Botulism caused fish and waterbird mortality in New York waters of Lakes Erie and Ontario, 2003. In *Botulism in Lake Erie Workshop Proceedings*, H.M. Domske, ed., pp. 17–22. Erie, PA: Pennsylvania Sea Grant.

Akbulut, D., K.A. Grant, and J. McLauchlin. 2004. Development and application of real-time PCR assays to detect fragments of the *Clostridium botulinum* types A, B, and E neurotoxin genes for investigation of human foodborne and infant botulism. *Foodborne Pathog. Dis.* 1:247–257.

Assel, R., K. Cronk, and D. Norton. 2003. Recent trends in Laurentian Great Lakes ice cover. *Clim. Change* 57:185–204.

Canadian Cooperative Wildlife Health Center. 2008. Type E botulism in Canada. http://www.glrc.us/documents/botulism/appendixD/canada-summary.pdf (accessed January 3, 2011).

Corkum, L.D., M.R. Sapota, and K.E. Skora. 2004. The round goby, *Neogobius melanostomus*, a fish invader on both sides of the Atlantic Ocean. *Biol. Invasions* 6:173–181.

Culligan, W.J., D.W. Einhouse, J.L. Markham, D.L. Zeller, R.C. Zimmar, and B.J. Beckwith. 2002. *NYS DEC Lake Erie Unit 2001 Annual Report to the Lake Erie Committee.* Albany, NY: New York State Department of Environmental Conservation Division of Fish, Wildlife and Marine Resources.

Domske, H.M. 2004. *Botulism in Lake Erie Workshop Proceedings.* Erie, PA: New York, Ohio, and Pennsylvania Sea Grant.

Domske, H.M. and E.C. Obert. 2001. *Avian Botulism in Lake Erie Workshop Proceedings.* Erie, PA: New York and Pennsylvania Sea Grant.

Einhouse, D.W., J.L. Markham, M.T. Todd, and M.L. Wilkinson. 2010. *NYS DEC Lake Erie 2010 Annual Report to the Lake Erie Committee and the Great Lakes Fishery Commission.* Albany, NY: New York State Department of Environmental Conservation Division of Fish, Wildlife and Marine Resources.

Einhouse, D.W., J.L. Markham, D.L. Zeller, R.C. Zimmar, and B.J. Beckwith. 2004. *NYS DEC Lake Erie Unit 2003 Annual Report to the Lake Erie Committee and the Great Lake Fishery Commission.* Albany, NY: New York State Department of Environmental Conservation Division of Fish, Wildlife and Marine Resources.

Environment DEC. 2008. Special DEC unit to tackle dramatic rise in invasive species. www.dec.ny.gov/environmentdec/41166.html (accessed January 3, 2011).

Fenchel, T. and B.J. Findlay. 1995. *Ecology and Evolution in Anoxic Worlds.* New York: Oxford University Press.

French, J.R.P. and D.J. Jude. 2001. Diets and diet overlap of nonindigenous gobies and small benthic native fishes co-inhabiting the St. Clair River, Michigan. *J. Great Lakes Res.* 27:300–311.

Gerking, S.D. 1994. *Feeding Ecology of Fish.* San Diego, CA: Academic Press.

Getchell, R.G. and P.R. Bowser. 2006. Ecology of type E botulism within dreissenid mussel beds. *Aquat. Invaders* 17:1–8.

Getchell, R.G., W.J. Culligan, M. Kirchgessner, C.A. Sutton, R.N. Casey, and P.R. Bowser. 2006. Quantitative PCR assay used to measure the prevalence of *Clostridium botulinum* type E in fish in the lower Great Lakes. *J. Aquat. Anim. Health* 18:39–50.

Grecz, N. and L.H. Arvay. 1982. Effect of temperature on spore germination and vegetative cell growth of *Clostridium botulinum*. *Appl. Environ. Microbiol.* 43:331–337.

Hecky, R.E., R.E. Smith, D.R. Barton et al. 2004. The nearshore phosphorus shunt: A consequence of ecosystem engineering by dreissenids in the Laurentian Great Lakes. *Can. J. Fish. Aquat. Sci.* 61:1285–1293.

Holeck, K.T., E.L. Mills, H.J. Macisaac, M.R. Dochoda, R.I. Colautti, and A. Ricciardi. 2004. Bridging troubled waters: Biological invasions, transoceanic shipping, and the Laurentian Great Lakes. *BioScience* 54:919–929.

Kimura, B., S. Kawasaki, H. Nakano, and T. Fujii. 2001. Rapid, quantitative PCR monitoring of growth of *Clostridium botulinum* type E in modified-atmosphere-packaged fish. *Appl. Environ. Microbiol.* 67:206–216.

Kuhns, L.A. and M.B. Berg. 1999. Benthic invertebrate community responses to round goby (*Neogobius melanostomus*) and zebra mussel (*Dreissena polymorpha*) invasion in southern Lake Michigan. *J. Great Lakes Res.* 25:910–917.

Lederer, A.A., J. Janssen, and A. Wolf. 2008. Impacts of the introduced round goby (*Apollonia melanostoma*) on dreissenids (*Dreissena polymorpha* and *Dreissena bugensis*) and on macroinvertebrate community between 2003 and 2006 in the littoral zone of Green Bay, Lake Michigan. *J. Great Lakes Res.* 34:690–697.

Lederer, A.A., J. Massart, and J. Janssen. 2006. Impact of round gobies (*Neogobius melanostomus*) on dreissenids (*Dreissena polymorpha* and *Dreissena bugensis*) and the associated macroinvertebrate community across an invasion front. *J. Great Lakes Res.* 32:1–10.

Moraska Lafrancois, B., S.C. Riley, D.S. Blehertand, and A.E. Ballmann. 2010. Links between type E botulism outbreaks, lake levels, and surface water temperatures in Lake Michigan, 1963–2008. *J. Great Lakes Res.* 37:86–91.

Pérez-Fuentetaja, A., M.D. Clapsadl, P.R. Bowser, R.G. Getchell, and W.T. Lee. 2011. *Clostridium botulinum* type E in Lake Erie: Inter-annual differences and role of benthic invertebrates. *J. Great Lakes Res.* 37:238–244.

Pérez-Fuentetaja, A., M.D. Clapsadl, D. Einhouse, P.R. Bowser, R.G. Getchell, and W.T. Lee. 2006. Influence of limnological conditions on *Clostridium botulinum* type E presence in eastern Lake Erie (Great Lakes, USA) sediments. *Hydrobiologia* 563:189–200.

Quinlan, R. and J.P. Smol. 2001. Chironomid-based inference models for estimating end-of-summer hypolimnetic oxygen from south-central Ontario shield lakes. *Freshwat. Biol.* 46:1529–1551.

Riley, S.C., K.R. Munkittrick, A.N. Evans, and C.C. Krueger. 2008. Understanding the ecology of disease in Great Lakes fish populations. *Aquat. Ecosyst. Health Manage.* 11:321–334.

Robbins, J. 2002. Outbreaks of a rare botulism strain stymie scientists. *New York Times*, October 22, 2002, p. F3.

Rocke, T.E. and T.K. Bollinger. 2007. Avian botulism. In *Infectious Diseases of Wild Birds*, N.J. Thomas, D.B. Hunter, and C.T. Atkinson, eds., pp. 377–416. Hoboken, NJ: Wiley-Blackwell.

Rocke, T.E. and M.D. Samuel. 1999. Water and sediment characteristics associated with avian botulism outbreaks in wetlands. *J. Wildl. Manage.* 63:1249–1260.

Sea Grant Michigan. 2007. Type-E botulism confirmed in waterfowl deaths. News Release. http://www.michigan.gov/documents/emergingdiseases/11-16-07-Botulism_215799_7.pdf (accessed January 3, 2011).

Sousounis, P. and P. Glick. 2000. The potential impacts of global warming on the Great Lakes region. In Impacts of Climate Change in the United States. http://www.climatehotmap.org/impacts/greatlakes.html (accessed January 3, 2011).

Steinhart, G.B., R.A. Stein, and E.A. Marschall. 2004. High growth rate of young-of-the-year smallmouth bass in Lake Erie: A result of the round goby invasion? *J. Great Lakes Res.* 30:381–389.

Stringer, S.C., M.D. Webb, S.M. George, C. Pin, and M.W. Peck. 2005. Heterogeneity of times required for germination and outgrowth from single spores of nonproteolytic *Clostridium botulinum*. *Appl. Environ. Microbiol.* 71:4998–5003.

Thorp, J.H. and A.P. Covich. 1991. *Ecology and Classification of North American Freshwater Invertebrates*. San Diego, CA: Academic Press.

Vanderploeg, H.A., T.F. Nalepa, D.J. Jude et al. 2002. Dispersal and emerging ecological impacts of Ponto-Caspian species in the Laurentian Great Lakes. *Can. J. Fish. Aquat. Sci.* 59:1209–1228.

Yule, A.M., I.K. Barker, J.M. Austin, and R.D. Moccia. 2006a. Toxicity of *Clostridium botulinum* type E neurotoxin to Great Lakes fish: Implications for avian botulism. *J. Wildl. Dis.* 42:479–493.

Yule, A.M., V. LePage, J.W. Austin, I.K. Barker, and R.D. Moccia. 2006b. Repeated low-level exposure of the round goby (*Neogobius melanostomus*) to *Clostridium botulinum* type E neurotoxin. *J. Wildl. Dis.* 42:494–500.

CHAPTER **46**

A Comparison of Consumptive Demand of *Diporeia* spp. and *Dreissena* in Lake Michigan Based on Bioenergetics Models

Daniel J. Ryan, Thomas F. Nalepa, Lori N. Ivan, Maria S. Sepúlveda, and Tomas O. Höök

CONTENTS

Abstract 713
Introduction 713
Methods 714
    Model Overview 714
    Consumption Rate 714
    Respiration Rate 715
    Excretion, Egestion, and Specific Dynamic Action Rates 715
    Model of Lake Spatial Structure 715
    Inputs 716
Results 717
Discussion 718
    Model Performance 719
    Ecosystem Consequences 720
Acknowledgments 721
References 721

ABSTRACT

Historically, *Diporeia* spp. were abundant and the dominant benthic consumer in offshore regions of the Great Lakes. Over the past several decades, however, *Diporeia* populations have declined, seemingly in response to *Dreissena* spp. introduction and expansion. As a result, consumption demand of food resources has likely undergone large changes. Bioenergetics models are commonly used to estimate consumption of aquatic organisms from observed growth patterns and environmental conditions. A *Diporeia*-specific bioenergetics model was parameterized from literature values of *Diporeia* growth, water temperature, energy density, and abundances in Lake Michigan. These data were used as input to derive model estimates of consumption. Specifically, the model was used to investigate *Diporeia* consumption of particulate matter in Lake Michigan in 1995, 2005, and 2008, which covered a period of severe declines in *Diporeia* and increased abundances of dreissenids. Comparisons of *Diporeia* consumption before and after their declines indicated that consumption rate decreased by 91%. In contrast, conservatively-estimated consumption rates for dreissenids in Lake Michigan were up to 24 times greater than *Diporeia* consumption rates during 2005, and by 2008, consumption rates of dreissenids in Lake Michigan were more than 330 times greater than consumption by *Diporeia* before their decline in 1995. We hypothesize that higher abundance and consumption by dreissenid mussels, relative to *Diporeia*, altered the function of the Lake Michigan ecosystem.

INTRODUCTION

In the Laurentian Great Lakes, declines of dominant benthic macroinvertebrates in the past two decades (Chironomidae, Sphaeriidae, *Diporeia* spp., and Oligochaeta)

(Nalepa 1987, Nalepa et al. 1998, 2003, 2007, Lozano et al. 2003) have raised concerns about ecosystem function and sustainability of fisheries that depend on benthic macroinvertebrate prey. In particular, benthic amphipods of the genus *Diporeia* spp. (herein, referred to as *Diporeia*) have precipitously decreased, presumably in response to a rapid increase in populations of the invasive mollusks *Dreissena polymorpha* (zebra mussel) and *Dreissena rostriformis bugensis* (quagga mussel) (Dermott et al. 2005a,b, Watkins et al. 2007, Nalepa et al. 2009). While *Diporeia* declines occurred coincident with dreissenid expansion, interactions between these invertebrates are complex and exact mechanisms that caused *Diporeia* to decline remain unknown. A common theory for *Diporeia* declines suggests that competition for food is an important mechanism, since filter-feeding dreissenids can more efficiently graze than deposit-feeding *Diporeia* (Nalepa et al. 1998, 2006a, Dermott 2001). However, some findings are inconsistent with this theory, and a variety of other theories have been proposed to link dreissenids to *Diporeia* declines (Landrum et al. 2000, Messick et al. 2004, Nalepa et al. 2006b, Kiziewicz and Nalepa 2008). Regardless of the exact mechanisms of *Diporeia* declines, it is clear that reduced *Diporeia* populations have led to various ecological consequences in the Great Lakes. Since *Diporeia* is an important prey for fish, it served as an important trophic link between pelagic and benthic nutrients and higher trophic levels (Flint 1986, Gardner et al. 1987, Nalepa et al. 2005a, 2006b). Thus, *Diporeia* declines have been implicated in decreased fish condition, especially for commercially important species such as lake whitefish (*Coregonus clupeaformis*), which historically relied on *Diporeia* as an important energy source (Pothoven et al. 2001, 2006).

Although numerous studies have investigated trophic links between *Diporeia* and fish (Pothoven et al. 2001, Mills et al. 2003, Owens and Dittman 2003, Hondorp et al. 2005, Zimmerman and Krueger 2009), relatively few studies have estimated consumption of particulate matter (Flint 1986, Gardner et al. 1987, Dermott and Corning 1988). With the expansion of dreissenids and the loss of *Diporeia* benthic cycling, the benthic community may now be an energetic sink rather than a pathway to upper trophic levels (Nalepa et al. 2009). While several investigations speculated on the magnitude of this shift in benthic cycling and consumptive demand, actual quantification of this shift has not been determined.

Bioenergetics models describe physiological processes of organisms (Kitchell et al. 1977) and have been extensively used to estimate consumption and growth of various species, including invertebrates (Hanson et al. 1997). We developed a *Diporeia* bioenergetics model by compiling observations from published studies on *Diporeia* physiological processes. Then, we applied this model to estimate consumption by *Diporeia* populations in Lake Michigan during 1994/1995–2005. Finally, we compared our estimates of *Diporeia* consumption with literature-derived estimates of dreissenid consumption.

METHODS

Model Overview

Typical bioenergetics models take a generalized form as proposed by Kitchell et al. (1977):

$$C = G + R + S + U + F \qquad (46.1)$$

where

 C is the consumption rate
 G is the growth rate
 R is the respiration rate
 S is the energy lost to specific dynamic action
 U is the excretion rate
 F is the egestion rate

Species-specific consumption rates are mass and temperature dependent. Species-specific bioenergetics models are generally used to estimate consumption or growth rates from either observed growth or consumption rates, respectively. Model applications also require inputs of water temperature and energy densities of prey and predator (Hanson et al. 1997). Past analyses suggest that consumption and respiration rates strongly influence model predictions, while other rates have less effect (Kitchell et al. 1977, Bartell et al. 1986). We compiled subcomponents of the *Diporeia* bioenergetics model from previously published relationships. To allow for compatibility, we tracked model predictions on a dry mass basis using a wet to dry ratio from Landrum (1988).

Consumption Rate

Two studies have directly measured *Diporeia* consumption rates (C). Dermott and Corning (1988) estimated *Diporeia* consumption rates in Lake Ontario through temporal observation of gut contents. These rates were modeled on a per organism basis, making it difficult to utilize these rates in a bioenergetics model. Lozano et al. (2003) proposed a consumption model for Lake Michigan *Diporeia* on a per mass basis, which could easily be incorporated into a bioenergetics model. Though these models were independently evaluated, consumption rates estimated from these two investigations were similar (Lozano et al. 2003). We selected the model presented by Lozano et al. (2003) to simulate *Diporeia* consumption in our bioenergetics model because of its specificity for Lake Michigan *Diporeia* and ease of use:

$$C = 10^{-1.22} \cdot W \, e^{-0.84} \cdot T^{0.83} \cdot p \qquad (46.2)$$

where

C is the consumption rate, in g (dry) · g (wet mass)$^{-1}$ · day^{-1}
W is the *Diporeia* wet mass, in g
T is the temperature, in °C
p is the proportion of maximum consumption

Respiration Rate

Several studies have examined respiration rates (*R*) of *Diporeia* in the Great Lakes (Johnson and Brinkhurst 1971, Johannsson et al. 1985, Quigley et al. 2002). Johannsson et al. (1985) developed a respiration model for *Diporeia* from Lake Ontario as a function of size and temperature, and a comparative study by Quigley et al. (2002) found that *Diporeia* from Lakes Michigan and Ontario had similar respiration rates. Thus, we used the respiration model proposed by Johannsson et al. (1985) to depict *Diporeia* energetic utilization. An oxycalorific conversion of 3380 cal · g O_2^{-1} was used to convert the amount of oxygen respired to calories respired (Teal 1957):

$$R = e^{-7.19} \cdot W^{-0.219} \cdot e^{0.036T} \qquad (46.3)$$

where

R is the respiration rate, in g O_2 · g (dry mass)$^{-1}$ · day^{-1}
W is the *Diporeia* dry mass, in g
$RQ = 0.036$ is the Q_{10} rate
T is the temperature, in °C

Excretion, Egestion, and Specific Dynamic Action Rates

Excretion rates (*U*) in bioenergetics models are generally modeled as a function of the consumption rate minus the egestion rate (*F*). To our knowledge, excretion rates for *Diporeia* have been measured only as mass-specific amount of ammonia or nitrogen released (Gardner et al. 1987, Quigley et al. 2002). Since studies of *Diporeia* excretion were made independent of consumption rate, interpretation of these excretion rates is difficult. Thus, we assumed that *Diporeia* excretion rates were a constant proportion of energy consumed minus energy egested as in a typical bioenergetics model. We followed the approach of Rudstam (1989), who modeled excretion by *Mysis* using established fish excretion rates while also including molting as a form of excretion (Clutter and Theilack 1971, Toda et al. 1987, Hanson et al. 1997). Since both *Mysis* and *Diporeia* inhabit the same deepwater areas of the Great Lakes, it is likely they have similar excretion rates (Carpenter et al. 1974, Morgan and Beeton 1978). The model takes the form

$$U = 0.18 \cdot (C - F) \qquad (46.4)$$

where

U is the excretion rate, in g (dry) · g (wet mass)$^{-1}$ · day^{-1}
C is the consumption rate
F is the egestion rate

Estimates of *Diporeia* egestion rates are limited, and there is no information on energy lost to specific dynamic action (*S*). Quigley and Vanderploeg (1991) estimated egestion of *Diporeia* by feeding experimental animals a known amount of the alga *Melosira varians* and enumerating *Diporeia* fecal pellet deposition. However, egestion rates were measured independent of *Diporeia* size and thus reflected responses only to *M. varians*. Instead, we used a previously published assimilation efficiency for *Hyalella azteca*, a warm-water deposit-feeding amphipod, to estimate egestion as a constant proportion of *Diporeia* consumption (Hargrave 1970, 1971) (Equation 46.5). In addition, we obtained an estimate of energy needed to assimilate food (specific dynamic action) from the bioenergetics model for *Mysis* developed by Rudstam (1989):

$$F = 0.15 \cdot C \qquad (46.5)$$

where

F is the egestion rate, in g (dry) · g (wet mass)$^{-1}$ · day^{-1}
C is the consumption rate

$$S = 0.15 \cdot (C - F) \qquad (46.6)$$

where

S is the energy lost to specific dynamic action, in g (dry) · g (wet mass)$^{-1}$ · day^{-1}
C is the consumption rate
F is the egestion rate

Model of Lake Spatial Structure

Lake Michigan was divided into 10-minute by 10-minute grid cells (Höök et al. 2004). For each cell, mean depth was calculated by first interpolating bathymetric contour data (Great Lakes Information Network, http://gis.glin.net) via inverse distance weighting (IDW) and then using the zonal statistics toolbox in ArcMap 9.3.1 (ESRI 2009) to calculate cell-specific mean depths. For our analysis, 190 cells were selected with a mean depth ≥30 m (i.e., depths which *Diporeia* were likely to occupy) (Winnell and White 1984, Nalepa 1989, Evans et al. 1990, Auer et al. 2009). First, data from lake-wide surveys in 1994/1995 and 2005 (Nalepa et al. 2008) were used to estimate abundance of *Diporeia* within each cell. Data collected in 1994/1995 (herein, referred to as 1995) were pooled to determine the mean day of year in which *Diporeia* were collected (August and July of 1995 and 2005, respectively). Second, we applied IDW to interpolate abundances of *Diporeia* and, for each year, calculated mean abundances of *Diporeia* in each cell using the zonal statistics toolbox in ArcMap 9.3.1 (ESRI 2009). We explored a variety of interpolation techniques (IDW, kriging, natural neighbor, and spline), which all yielded similar estimates for cell abundances (Ryan 2010). Thus, we used IDW interpolation for parsimony and consistency with previous studies (Nalepa et al. 2008). Next, daily size-specific abundance per cell was approximated by averaging a daily instantaneous mortality rate, Z, from data collected by Nalepa et al. (2006a) (Z = 0.003 day^{-1}). We applied this mortality estimate to extrapolate cell-specific abundances from August and July of 1995 and 2005 to daily abundances from April 1 to November 1 for each year. We multiplied daily total abundance by specific size-class percentages to obtain daily size-specific abundances (T.F. Nalepa, unpublished data).

Bottom water temperatures were obtained from a daily time-step, vertical-thermal model for 1990–2005 (Croley 1989, 1992, Croley and Assel 1994). Thermal-model outputs over 15 years were averaged to calculate a mean depth-specific temperature for each day. Then, daily depth-specific bottom temperature was assigned to each grid cell based on mean depths of individual cells. Modeled bottom-water temperatures varied only slightly among cells, and bottom-water temperatures did not exceed 10°C.

Inputs

In order to use the model for populations of *Diporeia* in Lake Michigan, we relied upon seasonal energy densities of *Diporeia* reported by Gardner et al. (1985) and a single seasonal value for detritus energy density calculated by Dermott (1995). Spatial information for these inputs are lacking; therefore, energy densities were assumed to follow the same temporal patterns throughout the lake.

Reported growth rates for *Diporeia* are highly variable; therefore, we explored three different growth scenarios encompassing observed growth trajectories (Table 46.1). Relatively slow growth rates of *Diporeia* were documented in Batchawana Bay, Lake Superior (Dermott 1995), and in deep water in Lake Michigan (Winnell and White 1984); slow to medium growth rates were observed in Lake Michigan (Lubner 1979) and Lake Ontario (Johannsson et al. 1985); and fast growth rates were observed in shallow waters of Lake Huron (Johnson 1988) and Lake Michigan (Winnell and White 1984). By considering a range of potential growth rates (see Table 46.1), we generated three different scenarios spanning a range of potential *Diporeia* consumption estimates.

Annual *Diporeia* consumption was simulated from April 1 to November 1, and daily consumption rates per cell were modeled over this time and across the three different growth scenarios (slow, medium, and fast). For each growth scenario, consumption was estimated by identifying the consumption rate that matched the final size of *Diporeia* (Table 46.1). That is, for each grid cell and size class, we fitted a *p*-value (proportion of maximum consumption, in 0.001 increments) that led to a predicted final weight that most closely matched the final mass of the growth scenario. To calculate daily size-specific consumption for each growth scenario, we multiplied daily size-specific abundances by their respective size-specific consumption rates. Finally, consumption across all size classes within a cell was summed and multiplied by the area of the cell to obtain total consumption in a grid cell.

We estimated dreissenid consumption (i.e., filtration) for 1995, 2005, and 2008 and qualitatively compared them to *Diporeia* consumption. To estimate lake-wide dreissenid densities in scenarios for 1995 and 2005, we pooled abundance data for *D. polymorpha* and *D. bugensis* within a year (Nalepa et al. 2008), used IDW to interpolate abundance data in cells >30 m in depth, and calculated mean abundance in each cell using the zonal

Table 46.1 Three Different Scenarios for Growth in Length and Mass of *Diporeia* That Were Used to Estimate Consumption in Lake Michigan. Lengths Were Converted to Mass Using the Lake-Wide Length–Mass Relationship Developed by Nalepa et al. (2000)

		Growth Estimates			
	Size Class	Initial Length (mm) April 1	Final Length (mm) October 1	Initial Mass (mg Dry) April 1	Final Mass (mg Dry) October 1
Slow growth	1	1.00	2.00	0.017	0.083
	2	2.00	4.00	0.083	0.397
	3	4.00	6.00	0.397	0.994
	4	6.00	7.00	0.994	1.410
Medium growth	1	1.00	3.00	0.017	0.207
	2	3.00	6.00	0.207	0.994
	4	6.00	7.00	0.994	1.410
Fast growth	1	1.00	6.00	0.017	0.994
	4	6.00	7.00	0.994	1.410

statistics toolbox in ArcMap 9.3.1 (ESRI 2009). Since *D. polymorpha* was the only dreissenid species present in 1995, interpolations were restricted to depths ≤100 m to reflect the life history and spatial distribution of *D. polymorpha* at that time (Mills et al 1993, Nalepa et al. 1998, Stoeckmann 2003). The dreissenid consumption scenario for 2008 was determined from depth-specific dreissenid densities for cells >30 m in mean depth (Nalepa et al. 2009). For all three years, abundances were multiplied by 0.262 g year^{-1}, a bioenergetics estimated growth of *D. polymorpha* in Lake Ontario (Schneider 1992). Under this approach, we assumed that dreissenids have 100% conversion efficiency. While this is clearly an unrealistic assumption, it places a lower bound on consumption estimates. Thus, we multiplied all years by 1.048 g year^{-1} to reflect a more realistic conversion efficiency of 25%. Finally, we scaled all scenarios of dreissenid filtration by 0.58 (i.e., the proportion of the year modeled, April 1 to November 1) such that estimates were comparable with *Diporeia* consumption estimates.

RESULTS

Under all *Diporeia* growth scenarios, estimated total consumption (kg year^{-1}) and mean areal consumption rates (g m^{-2} year^{-1}) were lower in 2005 compared to 1995 (Figure 46.1). For example, under the medium growth scenario, total

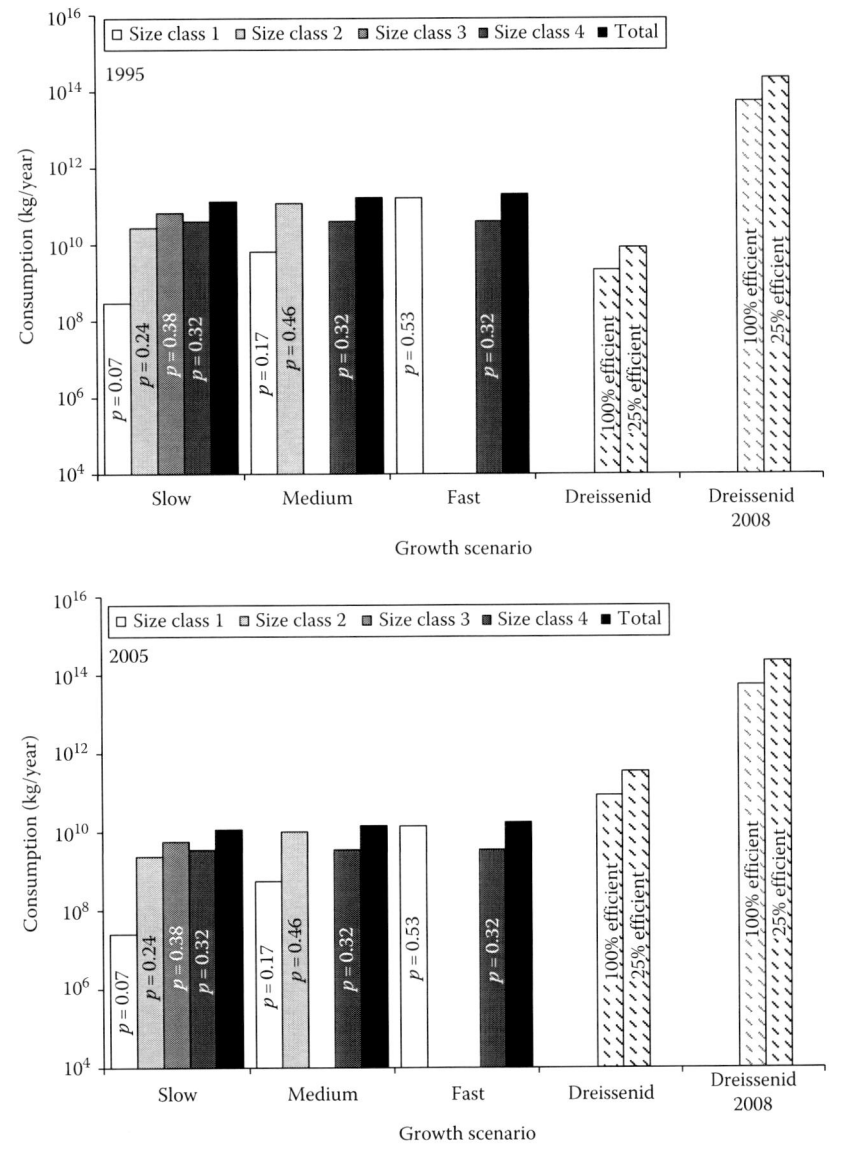

Figure 46.1 Estimated consumption rates and *p*-values by *Diporeia* size classes and growth scenarios for Lake Michigan during 1995 and 2005. Total estimated consumption per growth scenario is present for both years (black bars). For comparison, estimates of dreissenid filtering based on two conservative dreissenid consumption scenarios are presented for 1995, 2005, and 2008. Note the logarithmic scales.

consumption was estimated at 1.7×10^{11} kg year^{-1} in 1995, but this estimate declined 91% by 2005. Greatest spatial consumption rates (>5000 g m^{-2} year^{-1}) for the medium growth scenario were located toward the western shoreline, south of the Door Peninsula, and the majority of cells were characterized by relatively high areal consumption rates (1000–5000 g m^{-2} year^{-1}) in 1995 (Figure 46.2). In contrast, spatial consumption rates for medium growth were substantially reduced in 2005, with the greatest areal consumption rates (500–1000 g m^{-2} year^{-1}) situated in the south central area of the lake, while the majority of cells exhibited low consumption rates (<500 g m^{-2} year^{-1}) (Figure 46.2). Slow and fast growth scenarios exhibited similar trends between years as those observed for the medium scenario.

Consumption estimates for *Diporeia* based upon different growth scenarios differed in *p*-values and total amount of consumption (kg year^{-1}). Total consumption estimates from slow and fast growth scenarios were relatively low (−19%) and high (+23%), respectively, as compared to the medium growth scenario. Interestingly, while the fast growth scenario included fewer size classes of *Diporeia* relative to the slow and medium growth scenarios, faster growth was associated with greater overall consumption (Figures 46.2 and 46.3).

Total dreissenid consumption estimates (kg year^{-1}) for each conversion efficiency scenario were markedly larger in 2005 and 2008 than in 1995. In 1995, dreissenid consumption based on 100% and 25% conversion efficiencies were 2.2×10^6 kg year^{-1} and 8.3×10^6 kg year^{-1}, respectively (Figure 46.1). By 2005, estimates of total dreissenid consumption increased by >40-fold, and by 2008, estimates increased by >25,000-fold.

Comparisons between consumption of *Diporeia* and dreissenids varied temporally. In 1995, total consumption of *Diporeia* based on the medium growth scenario was >95% greater than either dreissenid consumption scenario (Figure 46.1). In 2005, dreissenid consumption estimates for the 100% and 25% scenarios were 6- and 24-fold greater than total *Diporeia* consumption based on the medium growth scenario (Figure 46.1). In 2008, total consumption by dreissenids in Lake Michigan increased over 2005 and was conservatively estimated (i.e., assuming 100% conversion efficiency) to be more than 330-fold greater than consumption by *Diporeia* in the medium growth scenario before their decline in 1995.

DISCUSSION

Application of the *Diporeia* bioenergetics model suggests that total consumption declined by 91% between 1995 and 2005. In contrast, consumption by invasive dreissenid mussels increased >25,000 times between 1995 and 2008.

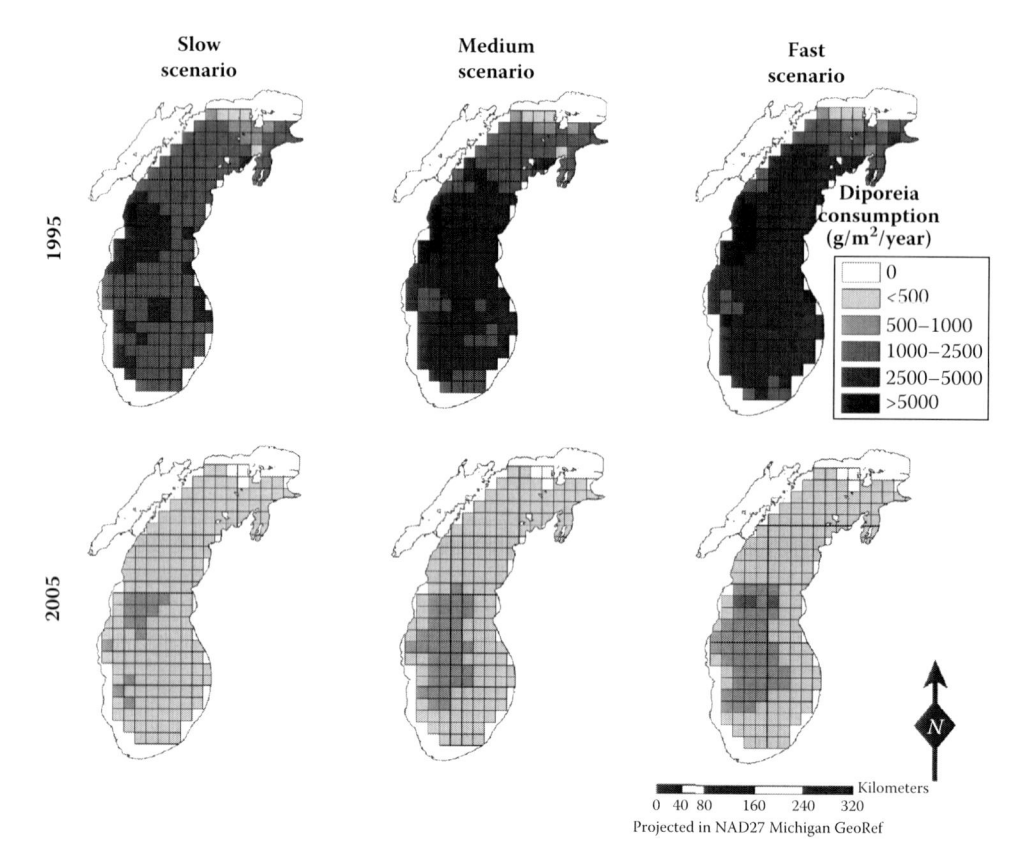

Figure 46.2 Areal *Diporeia* consumption (g · m^{-2} · year^{-1}) for three different growth scenarios in 189, 10 min by 10 min grid cells in Lake Michigan during 1995 and 2005.

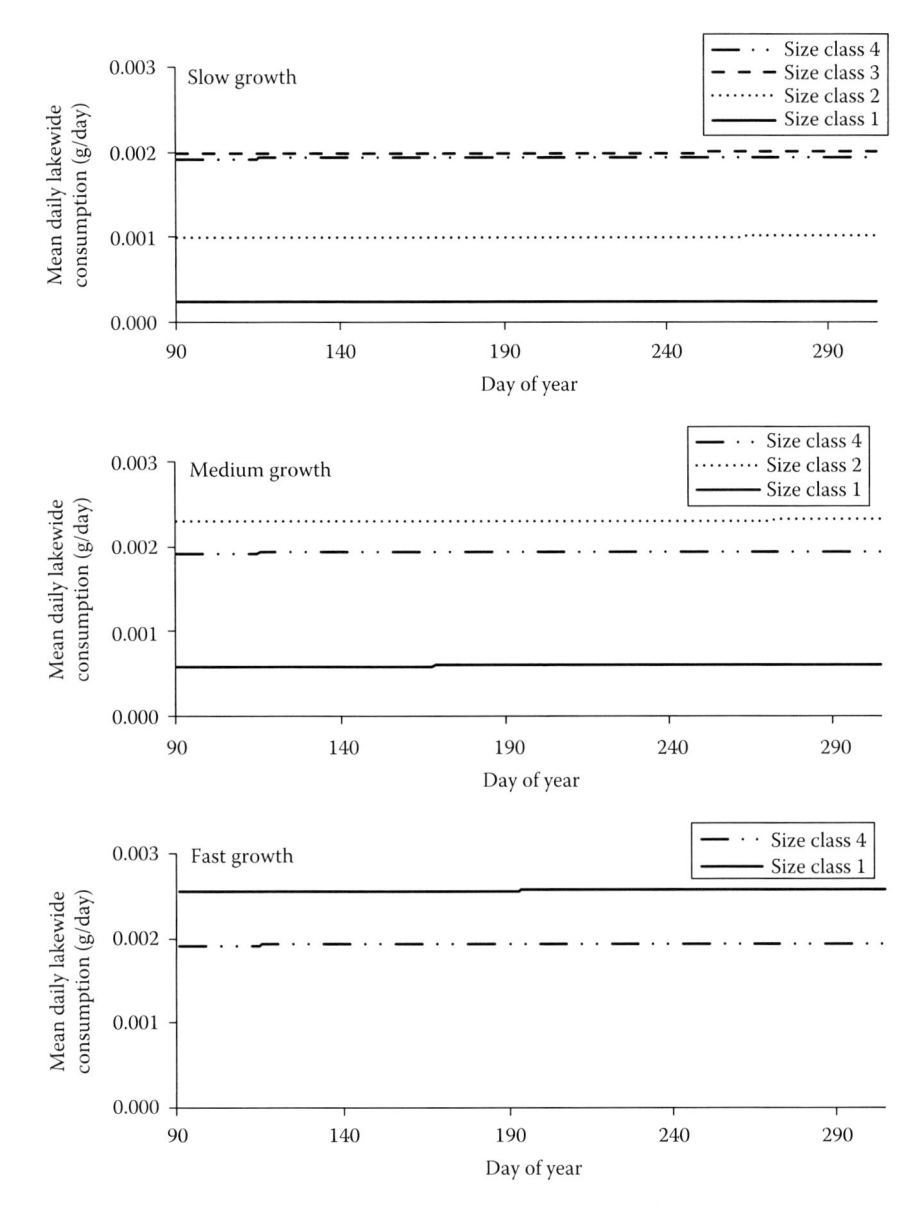

Figure 46.3 Mean daily consumption of *Diporeia* size classes for three different growth scenarios.

Consumption by dreissenids in Lake Michigan in 2008 was conservatively estimated to be 330 times greater than total consumption by *Diporeia* before their decline in 1995. Therefore, consumption by dreissenid mussels not only replaced historic consumption by *Diporeia* but far exceeded previous benthic consumption rates. Comparisons of consumption between dreissenid mussels and *Diporeia*, the historically dominant benthic macroinvertebrate, demonstrate that dreissenid mussels can dramatically alter food web structure and energy pathways of the Lake Michigan ecosystem.

Model Performance

Several bioenergetics models have been corroborated empirically (Kitchell et al. 1977, Kitchell and Breck 1980, Rudstam 1989, Yurista and Schulz 1995), and sensitivity analyses have been used to evaluate parameter influence (Rice et al. 1983, Bartell et al. 1986, Hartman and Brandt 1993). Growth of *Diporeia* in lakes is highly variable, and growth in laboratories is inconsistent, which makes corroboration difficult. Model components were previously developed and evaluated independently, and applications of these models were consistent with these original models. Consumption rates for *Diporeia* estimated by Lozano et al. (2003) were based on laboratory investigations of individuals from Lake Michigan that were ≤6.1 mg (wet weight), which were similar to input weights (sizes) of our model application. In addition, respiration and consumption models for *Diporeia* were developed for temperatures ≤12°C (Johannsson et al. 1985, Lozano et al. 2003), and temperature inputs for our model application never exceeded 10°C. Above this temperature, the model may overestimate consumption and not reflect the life history characteristics of *Diporeia* (Alley 1968, Lubner 1979, Winnell and White 1984). We developed *Diporeia* excretion, egestion, and specific dynamic action rates from several published sources and modeled these rates as constant percentages of consumption. While these published rates have not been as critically developed as consumption and respiration, they likely have less influence on model predictions (Bartell et al. 1986).

Accurate and precise applications of bioenergetics models to estimate consumption are dependent on appropriate input data related to energy densities of food, growth, and temperature. For *Diporeia* in Lake Michigan, spatial and temporal resolutions of these inputs are unknown. For example, it is unknown how energy density of particulate matter varies both spatially and temporally throughout the year. *Diporeia* energy densities were assumed to vary seasonally (Gardner et al. 1985, Nalepa et al. 2006a), but variation of energy density among size classes or between years was not considered. Past studies indicate that lipid levels, and therefore energy densities, have not varied dramatically between years (Gardner et al. 1985, Nalepa et al. 2006a). *Diporeia* size at age is difficult to establish due to the lack of calcified structures, presence of multiple cohorts within sampling areas, and potential variability in growth rates (Green 1968, Lubner 1979, Winnell and White 1984).

A permutation analysis was performed by simply modeling variations in *Diporeia* growth rates, which created upper and lower limits for true *Diporeia* consumption rates in Lake Michigan. As growth scenarios increased from slow to fast, maintenance ratios and overall consumption rates increased, leading to 23% greater consumption by *Diporeia*. It is likely that accurate *Diporeia* growth in Lake Michigan is most similar to the slow and medium growth scenarios than the fast scenario, given *Diporeia* have slow metabolic processes at cold temperatures (Alley 1968, Lubner 1979, Winnell and White 1984, Ryan 2010). In fact, the only field observation of *Diporeia* growth that approached the fast growth scenario was reported by Winnell and White (1984) for populations at relatively warm, shallow depths in Lake Michigan. The fast growth scenario does not accurately reflect *Diporeia* ecology at deepwater (>30 m) locations modeled herein (Winnell and White 1984). Therefore, the fast scenario is most likely an overestimate of *Diporeia* consumption in deepwater areas.

A few studies have examined *Diporeia* consumption in the Great Lakes both directly and indirectly. Dermott and Corning (1988) directly measured *Diporeia* consumption rates in Lake Ontario and found an average consumption rate of 20 g m^{-2} year^{-1} in 1981 and 1982. Reasons for the difference between our consumption estimates and those of Dermott and Corning (1988) are not clear but may be due to how the rates were determined. Dermott and Corning (1988) incorporated intermittent ingestion rates into their calculations, and consequently, their ingestion rate estimates were highly influenced by *Diporeia* gut contents at the time of collection. The rates were also modeled independent of other physiological processes such as respiration, egestion, and excretion rates provided in our model. It is important to note that based on modeled respiration and waste processes, *Diporeia* would need to consume more food than found by Dermott and Corning (1988) in order to sustain positive growth. Conversely, our study estimated the amount of food consumption throughout the year necessary to reach a final size while taking into account respiration, egestion, and excretion rates. Other studies estimated *Diporeia* consumption indirectly through measurement of carbon flow as part of an ecosystem assessment (Flint 1986, Fitzgerald and Gardner 1993). In these assessments, production rates were primarily based on *Diporeia* assimilation efficiency and also imply lower *Diporeia* consumption rates than estimated by our bioenergetics model. If the bioenergetics model presented herein overestimated *Diporeia* consumption, then this would imply that dreissenid-induced shifts in benthic consumption were even greater than our estimates indicate.

Ecosystem Consequences

Regardless of growth assumptions, there were marked differences between amounts of food consumed by *Diporeia* between 1995 and 2005. All growth scenarios exhibited a 91% decline in *Diporeia* consumption between these years, which primarily reflected substantial decreases in *Diporeia* abundances over this 10-year period (Nalepa et al. 2008, 2009). These results offer strong evidence for altered trophic pathways in Lake Michigan. In 1995, particulate organic matter deposited to Lake Michigan sediments had a high likelihood of culminating in *Diporeia* biomass, but this likelihood substantially decreased by 2005 and 2008.

The dreissenid consumption scenarios in 1995 and 2005 allowed for concurrent comparison with *Diporeia* consumption estimates in cells >30 m in depth, whereas the dreissenid consumption scenario in 2008 allowed for comparison between historical *Diporeia* consumption (i.e., in 1995) and consumption of dreissenid after further expansion (i.e., between 2005 and 2008; Nalepa et al. 2009). Dreissenid consumption rates in all scenarios were conservative because shallow-water populations were excluded from our analysis, dreissenid abundances were assumed to be constant throughout each year, and conversion efficiencies were assumed to be relatively high (Nalepa et al. 2008). Nevertheless, consumption estimates for dreissenids were larger than *Diporeia* consumption estimates by 2005. Moreover, dreissenid consumption in 2008 was 330 times greater than *Diporeia* consumption in 1995.

An important pathway for particulate matter in Lake Michigan has shifted from *Diporeia* to dreissenids due to increased dreissenid abundances with effective filtration capacities (Nalepa et al. 2008, 2009). Historically, *Diporeia* incorporated pelagic and benthic detritus into biomass and, in turn, made nutrients and energy available to higher trophic levels (Flint 1986, Fitzgerald and Gardner 1993, Nalepa et al. 2005b). By 2005, particulate matter terminated in a benthic sink associated with dreissenids (Nalepa et al. 2009).

The shift in energy pathway from *Diporeia* to dreissenids has appeared to have a profound effect on several fishes in Lake Michigan. Dreissenid shells pose obstacles for fish and when consumed possess limited nutritional value (Nalepa et al. 2009). Further, dreissenid caloric content (580 cal g^{-1}) (Schneider 1992) is much lower when compared to *Diporeia* (1109 cal g^{-1}) (Pothoven et al. 2001, McNickle et al. 2006). These factors likely contributed to declines in growth, condition, and maturation of many Lake Michigan fishes (Pothoven et al. 2001, 2006, Hondorp et al. 2005, Madenjian et al. 2006, Wang et al. 2008, Bunnell et al. 2009).

In conclusion, while model corroboration is difficult, our *Diporeia* bioenergetics model seems to yield comparative estimates of *Diporeia* consumption. By considering three feasible growth scenarios, we developed a range of *Diporeia* consumption estimates. These estimates suggest that *Diporeia* consumption declined from 1995 to 2005. Moreover, these estimates are orders of magnitude lower than conservative estimates of dreissenid filtering in 2008. Historically, *Diporeia* represented an important trophic link between particulate matter inputs to the benthic zone and higher trophic levels, but by 2005, particulate matter inputs were diverted toward dreissenid biomass. Consequences of this shift in benthic production are only beginning to be understood but likely indicate future decreases in *Diporeia* abundance and fish condition in Lake Michigan.

ACKNOWLEDGMENTS

We would like to thank Brent Lofgren and Timothy Hunter (National Oceanic and Atmospheric Administration, Great Lakes Environmental Research Laboratory) for providing Lake Michigan temperature at depth data, Carolyn Foley (Purdue University) for assisting with ESRI ArcGIS© methods, and Darryl Hondorp for initial discussions about *Diporeia* bioenergetics. Funding for this project was provided by the Great Lakes Fishery Trust (Grant# 2008.886).

REFERENCES

Alley, W.P. 1968. *Ecology of the Burrowing Amphipod, Pontoporeia affinis, in Lake Michigan*. Great Lakes Research Division Special Report No. 36. Ann Arbor, MI: University of Michigan.

Auer, N.A., B.A. Cannon, and M.T. Auer. 2009. Life history, distribution, and production of *Diporeia* near the Keweenaw Peninsula Lake Superior. *J. Great Lakes Res.* 35: 579–590.

Bartell, S.M., J.E. Breck, R.H. Gardner, and A.L. Brenkert. 1986. Individual parameter perturbation and error analysis of fish bioenergetics models. *Can. J. Fish. Aquat. Sci.* 43: 160–168.

Bunnell, D.B., C.P. Madenjian, J.D. Holuszko, J.V. Adams, and J.R.P. French. 2009. Expansion of *Dreissena* into offshore waters of Lake Michigan and potential impacts on fish populations. *J. Great Lakes Res.* 35: 74–80.

Carpenter, G.F., E.L. Mansey, and N.H.F. Watson. 1974. Abundance and life history of *Mysis relicta* in St. Lawrence Great Lakes. *J. Fish. Res. Bd. Can.* 31: 319–325.

Clutter, R.I. and G.H. Theilack. 1971. Ecological efficiency of a pelagic mysid shrimp; estimates from growth, energy budget, and mortality studies. *U.S. Fish. Wildl. Ser. Fish. B.* 69: 93–115.

Croley, T.E. 1989. Verifiable evaporation modeling on the Laurentian Great Lakes. *Water Resour. Res.* 25: 781–792.

Croley, T.E. 1992. Long-term heat storage in the Great Lakes. *Water Resour. Res.* 28: 69–81.

Croley, T.E. and R.A. Assel. 1994. A one-dimensional ice thermodynamics model for the Laurentian Great Lakes. *Water Resour. Res.* 30: 625–639.

Dermott, R. 1995. Production and growth efficiency of two burrowing invertebrates, *Hexagenia limbata* and *Diporeia hoyi*, in Batchawana Bay, Lake Superior. *Can. Data Rep. Fish. Aquat. Sci.* No. 2034.

Dermott, R. 2001. Sudden disappearance of the amphipod *Diporeia* from eastern Lake Ontario, 1993–1995. *J. Great Lakes Res.* 27: 423–433.

Dermott, R., R. Bonnell, and P. Jarvis. 2005a. Population status of the amphipod *Diporeia* in eastern North American lakes with or without *Dreissena*. *Verh. Int. Verein. Limnol.* 29: 880–886.

Dermott, R. and K. Corning. 1988. Seasonal ingestion rates of *Pontoporeia hoyi* (Amphipoda) in Lake Ontario. *Can. J. Fish. Aquat. Sci.* 45: 1886–1895.

Dermott, R., M. Munawar, R. Bonnell et al. 2005b. Preliminary investigations for causes of the disappearance of *Diporeia* spp. from Lake Ontario. In *Proceedings of a Workshop on the Dynamics of Lake Whitefish (Coregonus clupeaformis) and the Amphipod Diporeia spp. in the Great Lakes*, Great Lakes Fishery Commission Technical Report 66. L.C. Mohr and T.F. Nalepa, eds., pp. 203–232. Ann Arbor, MI: Great Lakes Fishery Commission.

ESRI (Environmental Systems Resource Institute). 2009. *ArcMap 9.3.1*. Redlands, CA: ESRI.

Evans, M.S., M.A. Quigley, and J.A. Wojcik. 1990. Comparative ecology of *Pontoporeia hoyi* populations in southern Lake Michigan—The profundal region versus the slope and shelf regions. *J. Great Lakes Res.* 16: 27–40.

Fitzgerald, S.A. and W.S. Gardner. 1993. An algal carbon budget for pelagic-benthic coupling in Lake Michigan. *Limnol. Oceanogr.* 38: 547–560.

Flint, R.W. 1986. Hypothesized carbon flow through the deep water Lake Ontario food web. *J. Great Lakes Res.* 12: 344–354.

Gardner, W.S., T.F. Nalepa, W.A. Frez, E.A. Cichocki, and P.F. Landrum. 1985. Seasonal patterns in lipid content of Lake Michigan macroinvertebrates. *Can. J. Fish. Aquat. Sci.* 42: 1827–1832.

Gardner, W.S., T.F. Nalepa, and J.M. Malczyk. 1987. Nitrogen mineralization and denitrification in Lake Michigan sediments. *Limnol. Oceanogr.* 32: 1226–1238.

Green, R.H. 1968. A summer breeding population of relict amphipod *Pontoporeia affinis* Lindstrom. *Oikos* 19: 191–197.

Hanson, P.C., T.B. Johnson, D.E. Schindler, and J.F. Kitchell. 1997. *Fish Bioenergetics 3.0*. Madison, WI: University of Wisconsin Sea Grant Institute.

Hargrave, B.T. 1970. The effect of a deposit feeding amphipod on metabolism of benthic microflora. *Limnol. Oceanogr.* 15: 21–30.

Hargrave, B.T. 1971. An energy budget for a deposit feeding amphipod. *Limnol. Oceanogr.* 16: 99–103.

Hartman, K.J. and S.B. Brandt. 1993. Systematic sources of bias in a bioenergetics model - Examples for age-0 striped bass. *Trans. Am. Fish. Soc.* 122: 912–926.

Hondorp, D.W., S.A. Pothoven, and S.B. Brandt. 2005. Influence of *Diporeia* density on diet composition, relative abundance, and energy density of planktivorous fishes in southeast Lake Michigan. *Trans. Am. Fish. Soc.* 134: 588–601.

Höök, T.O., E.S. Rutherford, S.J. Brines et al. 2004. Landscape scale measures of steelhead (*Oncorhynchus mykiss*) bioenergetic growth rate potential in Lake Michigan and comparison with angler catch rates. *J. Great Lakes Res.* 30: 545–556.

Johannsson, O.E., R.M. Dermott, R. Feldkamp, and J.E. Moore. 1985. Lake Ontario long term biological monitoring program: Report for 1981 and 1982. *Can. Data Rep. Fish. Aquat. Sci.* No. 1414.

Johnson, M.G. 1988. Production by the amphipod *Pontoporeia hoyi* in South Bay, Lake Huron. *Can. J. Fish. Aquat. Sci.* 45: 617–624.

Johnson, M.G. and R.O. Brinkhurst. 1971. Production of benthic macroinvertebrates of Bay of Quinte and Lake Ontario. *J. Fish. Res. Bd. Can.* 28: 1699–1714.

Kitchell, J.F. and J.E. Breck. 1980. Bioenergetics model and foraging hypothesis for sea lamprey (*Petromyzon marinus*). *Can. J. Fish. Aquat. Sci.* 37: 2159–2168.

Kitchell, J.F., D.J. Stewart, and D. Weininger. 1977. Applications of a bioenergetics model to yellow perch (*Perca flavescens*) and walleye (*Stizostedion vitreum vitreum*). *J. Fish. Res. Bd. Can.* 34: 1922–1935.

Kiziewicz, B. and T.F. Nalepa. 2008. Some fungi and water molds in waters of Lake Michigan with emphasis on those associated with the benthic amphipod *Diporeia* spp. *J. Great Lakes Res.* 34: 774–780.

Landrum, P.F. 1988. Toxicokinetics of organic xenobiotics in the amphipod, *Pontoporeia hoyi*: Role of physiological and environmental variables. *Aquat. Toxicol.* 12: 245–271.

Landrum, P.F., D.C. Gossiaux, T.F. Nalepa, and D.L. Fanslow. 2000. Evaluation of Lake Michigan sediment for causes of the disappearance of *Diporeia* spp. in southern Lake Michigan. *J. Great Lakes Res.* 26: 402–407.

Lozano, S.J., M.L. Gedeon, and P.F. Landrum. 2003. The effects of temperature and organism size on the feeding rate and modeled chemical accumulation in *Diporeia* spp. for Lake Michigan sediments. *J. Great Lakes Res.* 29: 79–88.

Lubner, J.F. 1979. *Population Dynamics and Production of the Relict Amphipod, Pontoporeia hoyi, at Several Lake Michigan Stations*. Milwaukee, WI: The University of Wisconsin-Milwaukee.

Madenjian, C.P., S.A. Pothoven, J.M. Dettmers, and J.D. Holuszko. 2006. Changes in seasonal energy dynamics of alewife (*Alosa pseudoharengus*) in Lake Michigan after invasion of dreissenid mussels. *Can. J. Fish. Aquat. Sci.* 63: 891–902.

McNickle, G.G., M.D. Rennie, and W.G. Sprules. 2006. Changes in benthic invertebrate communities of South Bay, Lake Huron following invasion by zebra mussels (*Dreissena polymorpha*), and potential effects on lake whitefish (*Coregonus clupeaformis*) diet and growth. *J. Great Lakes Res.* 32: 180–193.

Messick, G.A., R.M. Overstreet, T.F. Nalepa, and S. Tyler. 2004. Prevalence of parasites in amphipods *Diporeia* spp. from Lakes Michigan and Huron, USA. *Dis. Aquat. Org.* 59: 159–170.

Mills, E.L., J.M. Casselman, R. Dermott et al. 2003. Lake Ontario: Food web dynamics in a changing ecosystem (1970–2000). *Can. J. Fish. Aquat. Sci.* 60: 471–490.

Mills, E.L., R.M. Dermott, E.F. Roseman et al. 1993. Colonization, ecology and population structure of the 'quagga' mussel (Bivalvia: Dreissenidae) in the lower Great Lakes. *Can. J. Fish. Aquat. Sci.* 50: 2305–2314.

Morgan, M.D. and A.M. Beeton. 1978. Life history and abundance of *Mysis relicta* in Lake Michigan. *J. Fish. Res. Bd. Can.* 35: 1165–1170.

Nalepa, T.F. 1987. Long term changes in the macrobenthos of southern Lake Michigan. *Can. J. Fish. Aquat. Sci.* 44: 515–524.

Nalepa, T.F. 1989. Estimates of macroinvertebrate biomass in Lake Michigan. *J. Great Lakes Res.* 15: 437–443.

Nalepa, T.F., D.L. Fanslow, A.J. Foley III, G.A. Lang, B.J. Eadie, and M.A. Quigley. 2006a. Continued disappearance of the benthic amphipod *Diporeia* spp. in Lake Michigan: Is there evidence for food limitation? *Can. J. Fish. Aquat. Sci.* 63: 872–890.

Nalepa, T.F., D.L. Fanslow, and G.A. Lang. 2009. Transformation of the offshore benthic community in Lake Michigan: Recent shift from the native amphipod *Diporeia* spp. to the invasive mussel *Dreissena rostriformis bugensis*. *Freshwat. Biol.* 54: 466–479.

Nalepa, T.F., D.L. Fanslow, G.A. Lang, D.B. Lamarand, L.G. Cummins, and G.S. Carter. 2008. *Abundances of the Amphipod Diporeia spp. and the Mussels Dreissena polymorpha and Dreissena rostriformis bugensis in Lake Michigan in 1994–1995, 2000, and 2005*, NOAA Technical Report ERL GLERL-144. Ann Arbor, MI: Great Lakes Environmental Research Laboratory.

Nalepa, T.F., D.L. Fanslow, M.B. Lansing, and G.A. Lang. 2003. Trends in the benthic macroinvertebrate community of Saginaw Bay, Lake Huron, 1987 to 1996: Responses to phosphorus abatement and the zebra mussel, *Dreissena polymorpha*. *J. Great Lakes Res.* 29: 14–33.

Nalepa, T.F., D.L. Fanslow, and G. Messick. 2005a. Characteristics and potential causes of declining *Diporeia* spp. populations in southern Lake Michigan and Saginaw Bay, Lake Huron. In *Proceedings of a Workshop on the Dynamics of Lake Whitefish (Coregonus clupeaformis) and the Amphipod Diporeia spp. in the Great Lakes*, Great Lakes Fishery Commission Technical Report. 66. L.C. Mohr and T.F. Nalepa, eds., pp. 157–188. Ann Arbor, MI: Great Lakes Fishery Commission.

Nalepa, T.F., D.L. Fanslow, S.A. Pothoven, A.J. Foley III, and G.A. Lang. 2007. Long term trends in benthic macroinvertebrate populations in Lake Huron over the past four decades. *J. Great Lakes Res.* 33: 421–436.

Nalepa, T.F., D.J. Hartson, J. Buchanan, J.F. Cavaletto, G.A. Lang, and S.J. Lozano. 2000. Spatial variation in density, mean size and physiological condition of the holarctic amphipod *Diporeia* spp. in Lake Michigan. *Freshwat. Biol.* 43: 107–119.

Nalepa, T.F., D.J. Hartson, D.L. Fanslow, G.A. Lang, and S.J. Lozano. 1998. Declines in benthic macroinvertebrate populations in southern Lake Michigan, 1980–1993. *Can. J. Fish. Aquat. Sci.* 55: 2402–2413.

Nalepa, T.F., L.C. Mohr, B.A. Henderson, C.P. Madenjian, and P.J. Schneeberger. 2005b. Lake whitefish and *Diporeia* spp. in the Great Lakes: An overview. In *Proceedings of a Workshop on the Dynamics of Lake Whitefish (Coregonus clupeaformis) and the Amphipod Diporeia spp. in the Great Lakes*, Great Lakes Fishery Commission Technical Report 66. L.C. Mohr and T.F. Nalepa, eds., pp. 3–19. Ann Arbor, MI: Great Lakes Fishery Commission.

Nalepa, T.F., D.C. Rockwell, and D.W. Schloesser. 2006b. *Disappearance of the Amphipod Diporeia spp. in the Great Lakes: Workshop Summary, Discussion, and Recommendations*, NOAA Technical Report ERL GLERL-126. Ann Arbor, MI: Great Lakes Environmental Research Laboratory.

Owens, R.W. and D.E. Dittman. 2003. Shifts in the diets of slimy sculpin (*Cottus cognatus*) and lake whitefish (*Coregonus clupeaformis*) in Lake Ontario following the collapse of the burrowing amphipod *Diporeia*. *Aquat. Ecosyst. Health Manage.* 6: 311–323.

Pothoven, S.A., T.F. Nalepa, C.P. Madenjian, R.R. Rediske, P.J. Schneeberger, and H.X. He. 2006. Energy density of lake whitefish *Coregonus clupeaformis* in Lakes Huron and Michigan. *Environ. Biol. Fishes* 76: 151–158.

Pothoven, S.A., T.F. Nalepa, P.J. Schneeberger, and S.B. Brandt. 2001. Changes in diet and body condition of lake whitefish in southern Lake Michigan associated with changes in benthos. *N. Am. J. Fish. Manage.* 21: 876–883.

Quigley, M.A., P.F. Landrum, W.S. Gardner, C.R. Stubblefield, and W.M. Gordon. 2002. *Respiration, Nitrogen Excretion, and O:N Ratios of the Great Lakes Amphipod Diporeia sp.*, NOAA Technical Report ERL GLERL-120. Ann Arbor, MI: Great Lakes Environmental Research Laboratory.

Quigley, M.A. and H.A. Vanderploeg. 1991. Ingestion of live filamentous diatoms by the Great Lakes amphipod, *Diporeia* sp.: A case study of the limited value of gut contents analysis. *Hydrobiologia* 223: 141–148.

Rice, J.A., J.E. Breck, S.M. Bartell, and J.F. Kitchell. 1983. Evaluating the constraints of temperature, activity and consumption on growth of largemouth bass. *Environ. Biol. Fishes* 9: 263–275.

Rudstam, L.G. 1989. A bioenergetic model for *Mysis* growth and consumption applied to a Baltic population of *Mysis mixta*. *J. Plankton Res.* 11: 971–983.

Ryan, D.J. 2010. Spatial variation in condition and consumptive demand of *Diporeia* spp. in the Great Lakes region. MS thesis. West Lafayette, IN: Purdue University.

Schneider, D.W. 1992. A bioenergetics model of zebra mussel, *Dreissena polymorpha*, growth in the Great Lakes. *Can. J. Fish. Aquat. Sci.* 49: 1406–1416.

Stoeckmann, A. 2003. Physiological energetics of Lake Erie dreissenid mussels: A basis for the displacement of *Dreissena polymorpha* by *Dreissena bugensis*. *Can. J. Fish. Aquat. Sci.* 60: 126–134.

Teal, J.M. 1957. Community metabolism in a temperate cold spring. *Ecol. Monogr.* 27: 283–302.

Toda, H., T. Arima, M. Takahashi, and S. Ichimura. 1987. Physiological evaluation of temperature effect on the growth processes of the mysid, *Neomysis intermedia* Czerniawsky. *J. Plankton Res.* 9: 51–63.

Wang, H.Y., T.O. Hook, M.P. Ebener, L.C. Mohr, and P.J. Schneeberger. 2008. Spatial and temporal variation of maturation schedules of lake whitefish (*Coregonus clupeaformis*) in the Great Lakes. *Can. J. Fish. Aquat. Sci.* 65: 2157–2169.

Watkins, J.M., R. Dermott, S.J. Lozano, E.L. Mills, L.G. Rudstam, and J.V. Scharold. 2007. Evidence for remote effects of dreissenid mussels on the amphipod *Diporeia*: Analysis of Lake Ontario benthic surveys, 1972–2003. *J. Great Lakes Res.* 33: 642–657.

Winnell, M.H. and D.S. White. 1984. Ecology of shallow and deep water populations of *Pontoporeia hoyi* (Smith) (Amphipoda) in Lake Michigan. *Freshwat. Invertebr. Biol.* 3: 118–1318.

Yurista, P.M. and K.L. Schulz. 1995. Bioenergetic analysis of prey consumption by *Bythotrephes cederstroemi* in Lake Michigan. *Can. J. Fish. Aquat. Sci.* 52: 141–150.

Zimmerman, M.S. and C.C. Krueger. 2009. An ecosystem perspective on re-establishing native deepwater fishes in the Laurentian Great Lakes. *N. Am. J. Fish. Manage.* 29: 1352–1371.

CHAPTER 47

Variation in Length–Frequency Distributions of Zebra Mussels (*Dreissena polymorpha*) within and between Three Baltic Sea Subregions*
Szczecin Lagoon, Curonian Lagoon, and Gulf of Finland

Christiane Fenske, Anastasija Zaiko, Adam Woźniczka, Sven Dahlke, and Marina I. Orlova

CONTENTS

Abstract 725
Introduction 726
Material and Methods 727
    Sampling 727
        Statistical Analysis 728
Results 728
    General Environmental Conditions 728
    Length–Frequency Distributions 729
        Szczecin Lagoon 729
        Curonian Lagoon 731
        Gulf of Finland 731
    Comparison of Size Structures across Water Bodies 731
Discussion 732
Acknowledgments 737
References 737

ABSTRACT

We analyzed differences in length–frequency distributions within and between zebra mussel populations in three different subregions of the Baltic Sea (Szczecin Lagoon, Curonian Lagoon, and Gulf of Finland). Populations in all three subregions had length–frequencies that deviated from the normal distribution. Zebra mussel populations in the Gulf of Finland had the highest maximum shell lengths (up to 39 mm), but median shell lengths differed significantly between stations, probably due to different abiotic conditions. Populations in Szczecin and Curonian Lagoons also had length–frequency distributions that were highly variable despite generally similar, soft substrates in both subregions. Differences in distributions between sites were mainly related to variable recruitment and the number of small individuals found in populations. Sampling sites closer to river mouths were distinguished by broader length–frequency distributions and high numbers of small, young-of-the-year mussels that were not evident at some sites located farther away. Greater numbers of small recruits at these sites may be attributed to larvae originating from adult mussels in the river or to consistent, freshwater conditions. Other factors that likely contributed to varying recruitment and survival success in the three subregions were variable water currents, small-scale accumulation of organic debris (fluffy sediment surface layer), oxygen deficiency, high turbidity of water due to dredging, and predation pressure.

* In memory of Robert W. McCauley.

INTRODUCTION

With a surface area of 412,560 km² and a catchment area to surface area ratio of 4:1, the Baltic Sea is one of the largest brackish-water bodies in the world. It developed into its current shape after the last ice age about 7100 years ago (Schiewer 2008). Due to this short time span, the number of macrobenthic species established in the Baltic Sea is rather low. Furthermore, since richness of the macrofauna is lowest at intermediate salinities (Remane 1934), brackish habitats of the Baltic Sea accommodate a lower number of species compared to freshwater or marine habitats.

Recent studies have shown that nonindigenous (or alien) species that are adapted to intermediate salinities have been successfully invading coastal water bodies of the Baltic Sea over the past several decades. Presently, estuaries, coastal lagoons, and inlets of the Baltic Sea generally harbor more nonindigenous species than open-water areas (Paavola et al. 2005, Orlova et al. 2006, Zaiko et al. 2007, 2011). According to Reise et al. (2006), the ratio of nonindigenous to native species in lagoons and estuaries is roughly 1:5, compared to 1:20 for open Baltic Sea coasts, and 1:40 for European marine waters. One of the more widely distributed species in the oligohaline southern and eastern coastal regions of the Baltic Sea is the zebra mussel, *Dreissena polymorpha*. Unlike its status in many other areas, *D. polymorpha* is not an alien species sensu stricto in these regions but rather a postglacial re-immigrant. According to paleontological and geological evidence, *D. polymorpha* existed in the Baltic Sea drainage area during the interglacial period (Meisenheimer 1901, Buynevich et al. 2011) but later became extinct only to become re-established in the early 1800s (Starobogatov and Andreyeva 1994). Currently this species is abundant in lagoons in the south and southeast as well as in eastern inlets of the Baltic Sea.

While northern Baltic Sea coasts (Scandinavian) are mainly rocky and still lifting after glacial retreat, southern coasts are dominated by fine sediments, and are prone to inundation by rising sea levels (Schiewer 2008). Due to predicted sea level rise induced by climate change, thermal expansion of water, and ongoing postglacial sinking of southern Baltic coasts, waters of the Baltic Sea are likely to further inundate coastlines along the southern Baltic. For instance, by 2100 the medium sea level along the German Baltic coast is expected to rise by 20–30 cm (MLUV-MV 2009).

For *D. polymorpha*, higher sea levels and extensions of inner-coastal waters imply an increase in potential settlement area; however, a key factor in future distributions would be a potential change in salinity gradients. Effects of climate change on salinity of the Baltic Sea are still not clear. Some regional scenarios indicate a decrease in average salinity of the Baltic Sea (HELCOM 2007). Alternatively, higher sea levels could also lead to more water exchange with the North Sea and thus increase salinity, which in turn would hamper filtration, respiration, and other physiological processes of *D. polymorpha* (Fenske 2002). Yet even if salinity increased, it is likely that this species would be able to adapt over time and withstand up to 6–7 ppt. So far, *D. polymorpha* does not occur in the open Baltic Sea at salinity >5 ppt, but in experiments with a gradual increase in salinity, dreissenids from the Caspian Sea and Gulf of Finland were able to tolerate salinities even up to 14–15 ppt (Orlova et al. 1998).

Another factor that can affect distributions of *D. polymorpha* in coastal regions of the southern Baltic is nutrient enrichment. Intensive agriculture, high-nutrient runoff, and poor sewage treatment in eastern European countries have caused strong eutrophication in coastal regions of the Baltic Sea (Helcom 2009, 2011). While nutrient input from point sources has decreased as a result of new and updated sewage treatment plants, inputs from diffuse sources (e.g., drained areas, atmospheric deposition) remain high. Eutrophication has led to an excessive increase in suspended organic material, which causes organic enrichment of sediments. In combination with the natural slow water flow in coastal lagoons, this nutrient enrichment results in very soft and unstable sediments. This directly affects recruitment in zebra mussel populations since this species prefers hard substrates. However, by attaching to live and dead shells of other individuals, zebra mussels are able to colonize soft bottoms (Grim 1971). In Lake Erie, Berkman et al. (1998) found that zebra mussels initially colonized sediments without any hard substrate, suggesting that they can bind sediment particles into conglomerates using their byssal threads. Of course, the consistency of the sediment must still be hard enough so mussels do not sink.

In lagoons along coastal regions of the southern Baltic Sea, zebra mussels occur in highest densities on soft-bottom habitats at a depth of 2–4 m. Mussels are found as druses, which are fist-size clumps of mussels of different shell lengths that lie on the soft bottom, or on beds of empty shells. Populations can have strong impacts on ecosystems in these regions (Wolnomiejski and Woźniczka 2008). Large beds of mussels are "islands" of high biodiversity that facilitate the establishment of native and nonnative species in the soft-bottom environment (Zaiko et al. 2007, 2009). Mussel beds increase the benthic surface area available to other invertebrates (Botts et al. 1996, Stewart et al. 1998, Bially and MacIsaac 2000, Karatayev et al. 2002), provide refuges from predators (Gutierrez et al. 2003, Beekey et al. 2004), and increase amounts of organic material in the sediment by depositing feces and pseudofeces (Karatayev and Burlakova 1994, Botts et al. 1996, Stewart et al. 1998). Mussel-induced habitat modifications are primarily a function of abundance, distribution, and size frequencies (size structure) of the population, and such population characteristics can vary strongly within a given habitat (Stańczykowska and Lewandowski 1993, Nalepa et al. 1995, Young et al. 1996, Wolnomiejski and

Woźniczka 2008). Indeed, while growth and size frequencies have been described for different geographical regions, for example, Great Lakes (Chase and Bailey 1999); River Drava, Croatia (Lajtner et al. 2004); and Rivers Moselle and Meuse, France (Beisel et al. 2010), there is often no clear correlation between recruitment, growth, and environmental parameters such as dissolved organic matter, total organic carbon, seston, or turbidity (Jantz and Neumann 1998). In a review of studies from the former Soviet Union, eastern and western Europe, and North America, Karatayev et al. (2006) concluded that differing results for recruitment success, growth, and longevity of zebra mussels may be caused not only by variations in environmental conditions but also by the methods in which these population traits were measured.

There are many factors that influence population development, including both abiotic factors (e.g., temperature, salinity, currents, turbidity, oxygen concentration, suitable settlement substrate) and biotic factors (e.g., food composition and availability, predators, parasites). In order to fully explain variability in size structures of different populations, it would be necessary to have long-term measurements of all these factors—a task that would be difficult to accomplish. A more general approach is to define size structures of populations across habitats that have broadly different environmental conditions. In this chapter, we follow the second approach and describe size–frequency distributions of zebra mussel populations in different habitats along the southern and eastern coasts of the Baltic Sea.

MATERIAL AND METHODS

Sampling

Samples were collected in three locations along the southern and eastern coasts of the Baltic Sea: Szczecin Lagoon, Curonian Lagoon, and the Gulf of Finland (Figure 47.1). Sites in Szczecin Lagoon were located in the western portion of this water body (n = 6; MB1, MB2, MB3, MB5, MB10, MB11), sites in Curonian Lagoon were located in the central portion (n = 3; Preila, Nida, Vente), and sites in the Gulf of Finland were located in the far eastern portion (n = 7; 7(3), 7(5), 8(3), 8(5), F1, PM, RH). Mussel samples in Curonian Lagoon were collected in August and September 2006, and mussels in Szczecin Lagoon and the Gulf of Finland were collected in June and July 2007.

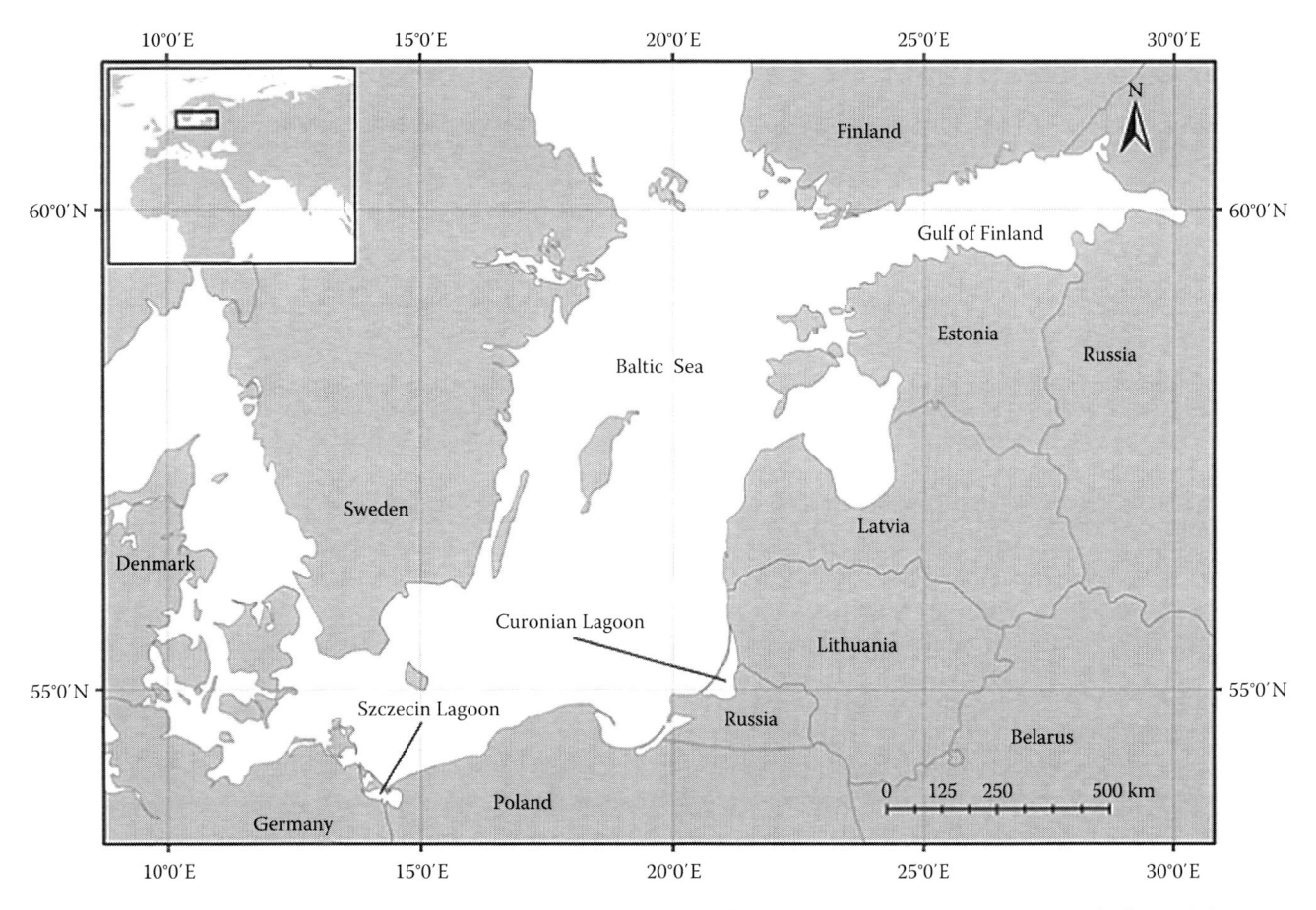

Figure 47.1 General location of the three water bodies where populations of zebra mussels were sampled in this study: Szczecin Lagoon, Curonian Lagoon, and Gulf of Finland.

Both mussels and sediments were collected with a box corer (10 × 10 cm) (Szczecin and Curonian Lagoons) or were hand-collected by SCUBA divers or snorkelers who removed mussels within a square metal frame (0.25 m²) placed on the bottom. Shell lengths of all collected mussels were measured to the nearest mm using digital calipers (Szczecin Lagoon: to hundredth of mm), and length–frequency distributions determined.

Environmental variables (Secchi-depth transparency, salinity, oxygen content, sediment type) were measured at each of the stations at the time mussels were collected. In Szczecin Lagoon, the areal extent of mussel beds and corresponding densities were measured at each of the stations.

Statistical Analysis

Length–frequency distributions of populations at each station were examined for normality, skewness, and kurtosis using standard tests. Skewness characterizes the degree of asymmetry of a distribution around its mean. Positive skewness indicates a distribution with an asymmetric tail extending toward more positive values (mode skewed to the smaller values). Negative skewness indicates a distribution with an asymmetric tail extending toward more negative values (mode skewed to the larger values). Kurtosis characterizes the relative peakedness or flatness of a distribution compared with the normal distribution. Positive kurtosis indicates a relatively peaked distribution and negative kurtosis indicates a relatively flat distribution.

If a length–frequency distribution was not normally distributed, its multimodality was assessed by Hartigan's dip test (H_0 is unimodality; Hartigan 1985, Hartigan and Hartigan 1985), using the dip test package (Maechler and Ringach 2004) of the statistics software R (version 2.13.0). For comparisons of length–frequency distributions between stations, the nonparametric Kruskal–Wallis test was applied followed by a post hoc analysis (multiple pairwise comparisons of medians with Mann–Whitney test, applying the Bonferroni correction for α). The relative similarity of length–frequency distributions between all stations was examined with an MDS plot (multidimensional scaling) using the statistical software NCSS 2001 (Statistical Systems Kaysville, Utah). Finally, the Spearman rank correlation coefficient was calculated to determine if there was a significant relationship between average salinity and maximum mussel shell length over all sites.

RESULTS

General Environmental Conditions

Physical and chemical characteristics of Szczecin Lagoon, Curonian Lagoon, and the Gulf of Finland are shown in Table 47.1. Average salinity content of each of these three water bodies was <2 ppt but can vary from 0 to 8 ppt. Most bottom substrates in Szczecin and Curonian Lagoons consisted

Table 47.1 Key Environmental Variables of the Water Bodies (Szczecin Lagoon, Curonian Lagoon, and Gulf of Finland) where Zebra Mussels Were Collected

Variable	Szczecin Lagoon	Curonian Lagoon	Gulf of Finland Neva River Mouth to Kotlin Island	Gulf of Finland Kotlin Island to Cape Stirsudden
Area (km²)	687	1584	329	1146
Mean depth (m)	3.8	3.8	4.0	12.0
Max. depth (m)	8.5 (10.5)	5.8 (14.0)[b]	20.0	45.0
Mean freshwater inflow (km³/year)	18	23	79	0
Salinity min/max (ppt)	0.3–4.5	0–8	0.06–0.11	0.45–5.0 (near bottom >7.0)
Salinity (ppt) average	1.4	0.1	0.08	1.65
Ice cover (days/year)	59	110	120–130	120–130
Secchi-depth transparency min/max (m)	0.2–3.0 max: 2.2[a]	0.3–2.2	0.1–0.6	0.4–1.2
Secchi-depth transparency average (m)	0.9		0.28	0.84
Substrate type	Mostly silt, sand	Mostly sand, silt	Highly variable, many anthropogenic hard substrates at harbor areas and along storm-surge barrier	Highly variable

Sources: All data from Schiewer, U., Introduction, in *Ecology of Baltic Coastal Waters. Ecological Studies*, ed. U. Schiewer, Vol. 197, pp. 1–21, Springer-Verlag, Berlin, Germany, 2008; Orlova, M., unpublished; Zaiko, A., unpublished, unless otherwise stated.

[a] Data from LUNG (State Agency for Environment, Nature Protection and Geology Mecklenburg-Vorpommern, Germany) are only for the western basin (Kleines Haff) 1975–2001.

[b] Values in parentheses refer to shipping channels or harbor areas.

of fine silt, and even in areas with sandy substrates there was an overlying layer of flocculated organic matter due to high eutrophication. Bottom substrates in the Gulf of Finland were highly variable, and many areas had man-made structures (storm-surge barriers, harbor construction, etc.) that served as settlement areas for zebra mussels. All sites where mussels were collected in the three water bodies were generally well oxygenated. Average Secchi-depth transparency was <1 m in all three water bodies during the sampling period.

Length–Frequency Distributions

Szczecin Lagoon

The size of mussel beds at the six sites in Szczecin Lagoon ranged from 0.06 km² (MB11) to 2.25 km² (MB10), and densities ranged from 1,217/m² (MB 5) to 10,444/m² (MB3), with an average of 3,949/m² (Table 47.2). The size of mussel beds and mussel densities were not correlated (correlation coefficient = −0.0885). Length–frequency distributions at all six sites did not fit the normal distribution (Kolmogorov–Smirnov, p < 0.001 for all sites, Table 47.3). Although sites were not more than a few km apart and apparently had similar abiotic conditions, median shell lengths were significantly different (Kruskal–Wallis, H = 543.211, p < 0.001) when sampled in June/July 2007 (Figure 47.2). Median lengths at two of the sites (MB2, MB3) were significantly lower than at the rest of the stations (Mann–Whitney, p < 0.001). Statistically significant multimodality in zebra mussel size structure was found for most sites (except for MB5) (Table 47.3).

Even when data across all sampling sites in Szczecin Lagoon were pooled (n = 2628), the length–frequency distribution was not normally distributed and there was

Table 47.2 Densities, Median Shell Lengths, and Maximum Shell Lengths of Zebra Mussel Populations at Each Site in Szczecin Lagoon, Curonian Lagoon, and the Gulf of Finland. Also Given Is the Substrate Type, Number of Mussels Measured, and Size of the Mussel Bed. The Latter Was Only Estimated at Stations in Szczecin Lagoon

Water Body/ Station	Substrate	Size of Bed (km²)	Mean Density (No./m²)	Number of Mussels Measured (n)	Median Shell Length (mm)	Maximum Shell Length (mm)
Szczecin Lagoon						
MB1	Soft	0.29	2,759	375	16	26.85
MB2	Soft	0.82	1,404	1480	12	27.39
MB3	Soft	0.33	10,444	187	12	25.01
MB5	Soft	0.48	1,217	156	15	26.26
MB10	Soft	2.25	2,889	104	14	25.50
MB11	Soft	0.06	2,500	325	15	27.67
Curonian Lagoon						
Preila	Soft			69	19	28
Nida	Soft			400	16	28
Vente	Soft			326	6	31
Gulf of Finland						
7(3)	Mixed substrate			61	25	35
7(5)	Mixed substrate			61	28	37
F1	Artificial hard substrate			381	23	39
PM	Artificial hard substrate			408	21	28
8(3)	Mixed substrate			225	14	21
8(5)	Mixed substrate			198	14	26
RH	Mixed substrate			169	17	24

Table 47.3 Values of Various Statistical Tests on Length–Frequency Distributions of Zebra Mussel Populations at Each Site in Szczecin Lagoon. Median Shell Lengths at the Six Sites Were Significantly Different (Kruskal–Wallis, H = 543.211, p < 0.001)

Site	Kolmogorov–Smirnov Test	Skewness	Kurtosis	Dip Test
MB1 n = 376	D = 0.0704; p < 0.001	−0.42 (Z = 1.35; p = 0.18)	−0.38 (Z = −1.8; p = 0.07)	dip = 0.0492; p < 0.001
MB2 n = 1480	D = 0.4524; p < 0.001	0.12 (Z = 0.16; p = 0.9)	0.38 (Z = 2.59; p = 0.01)	dip = 0.0524; p < 0.001
MB3 n = 187	D = 0.4536; p < 0.001	1.05 (Z = 3.6; p < 0.001)	1.72 (Z = 3.12; p < 0.001)	dip = 0.0722; p < 0.001
MB5 n = 156	D = 0.1344; p < 0.001	−0.16 (Z = 0.6; p = 0.6)	−0.52 (Z = −1.68; p = 0.09)	dip = 0.0481; p = 0.02
MB10 n = 104	D = 0.2944; p < 0.001	1.12 (Z = 2.91; p < 0.01)	0.49 (Z = 1.11; p = 0.27)	dip = 0.0962; p < 0.001
MB11 n = 325	D = 0.1713; p < 0.001	0.75 (Z = 3.57; p < 0.001)	−0.31 (Z = −1.24; p = 0.21)	dip = 0.0615; p < 0.001
All stations n = 2625	D = 0.0416; p < 0.001	0.15 (Z = 2.52; p = 0.01)	0.15 (Z = 1.64706; p = 0.0995)	dip = 0.0574; p < 0.85

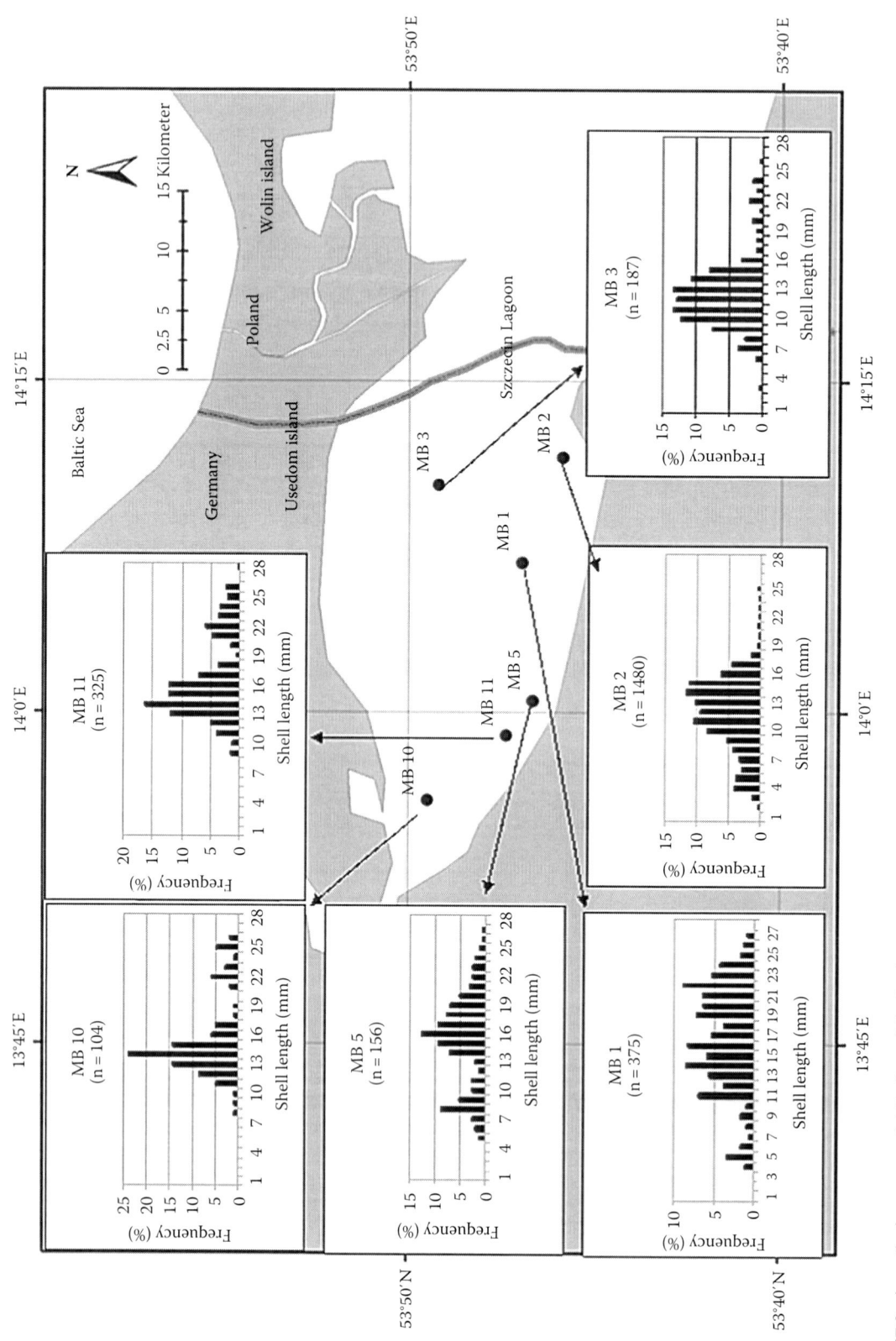

Figure 47.2 Length–frequency distributions (%) of zebra mussel populations at sites in the western portion of Szczecin Lagoon, June/July 2007.

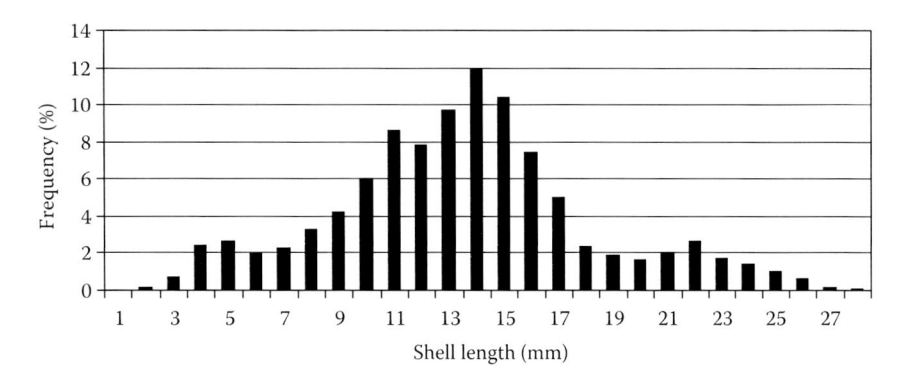

Figure 47.3 Length–frequency distribution (%) of all zebra mussels (n = 2628) collected at the six sites in Szczecin Lagoon, June/July 2007.

Table 47.4 Values of Various Statistical Tests on Length–Frequency Distributions of Zebra Mussel Populations at Each Site in Curonian Lagoon. Median Shell Lengths at the Three Sites Were Significantly Different (Kruskal–Wallis, H = 196.038, p < 0.001)

Site	Kolmogorov–Smirnov Test	Skewness	Kurtosis	Dip Test
Preila	D = 0.02814; p < 0.01	0.06 (Z = 0.16; p = 0.9)	0.27 (Z = 0.66; p = 0.5)	dip = 0.0580; p = 0.1
Nida	D = 0.1213; p < 0.01	−0.37 (Z = 2.09; p = 0.04)	0.16 (Z = 0.73; p = 0.6)	dip = 0.0625; p < 0.0001
Vente	D = 0.5686; p < 0.01	0.97 (Z = 4.41; p < 0.0001)	−0.11 (Z = −0.32; p = 0.7)	dip = 0.0644; p < 0.0001

no statistical evidence for a multimodal distribution (Table 47.3). In the pooled data, shell lengths ranged from 2 to 28 mm, and length at maximum frequency was 14 mm, which probably represented 2-year-old individuals (Figure 47.3).

Curonian Lagoon

Length–frequencies of zebra mussel populations at the three sites in Curonian Lagoon were not normally distributed, and multimodality was significant (p < 0.001) at Nida and Vente (Table 47.4). Mussels at Preila and Nida had significantly larger (Mann–Whitney, p < 0.001) median shell lengths than those at Vente (19, 16, and 6 mm, respectively) (Figure 47.4, Table 47.2). The length–frequency distribution of pooled data for all three sites showed a distinctly bimodal pattern (Table 47.4) with a peak at 4 mm and a greater peak at 17 mm (Figure 47.5).

Gulf of Finland

As found in Szczecin and Curonian Lagoons, length–frequencies of zebra mussel populations at the seven sites in the Gulf of Finland were not normally distributed. The observed maxima indicated a noncontinuous recruitment (Figure 47.6a and b; Table 47.5). Zebra mussel shell lengths varied strongly between the sites (Kruskal–Wallis H = 1039.85; p < 0.001), and median shell lengths were significantly different (Kruskal–Wallis H = 1039.85; p < 0.001). The highest median shell length occurred at Station 7 (7(3), 7(5)) (freshwater conditions, natural mixed substrates, gravel, coarse sand), followed by Station F1 (artificial hard substrates), and Station PM. The lowest median shell length occurred at Station 8 (oligo- to mesohaline conditions) (Table 47.2). The pooled data for all stations (n = 1,503) indicated a multimodal distribution (Table 47.5).

Comparison of Size Structures across Water Bodies

Length–frequency distributions at all sites in the three water bodies were compared using an MDS plot (Figure 47.7). This analysis resulted in site groupings that only partly reflected their water-body locations. One group consisted of sites within the Gulf of Finland, where the highest median shell lengths (21–28 mm) were found. Two other groups consisted of a mixture of sites from all three water bodies and had low and intermediate median shell lengths (6–17 mm). Differences in maximum shell lengths were significant among stations in the Gulf of Finland (Kruskal–Wallis, p < 0.01) but not significant between stations in Szczecin Lagoon or Curonian Lagoon (Kruskal–Wallis, p > 0.05). In general, there was a negative relationship between median shell length and salinity (Figure 47.8; Spearman Rank correlation coefficient −0.59, p = 0.0164), but at the lowest salinities (0.1 ppt in Curonian Lagoon) maximum shell lengths were not as great as found in the other two water bodies. In the Gulf of Finland, a strong salinity gradient was apparent between sites and a strong negative correlation between maximum shell length and average salinity (Spearman Rank correlation coefficient −0.87, p = 0.0103).

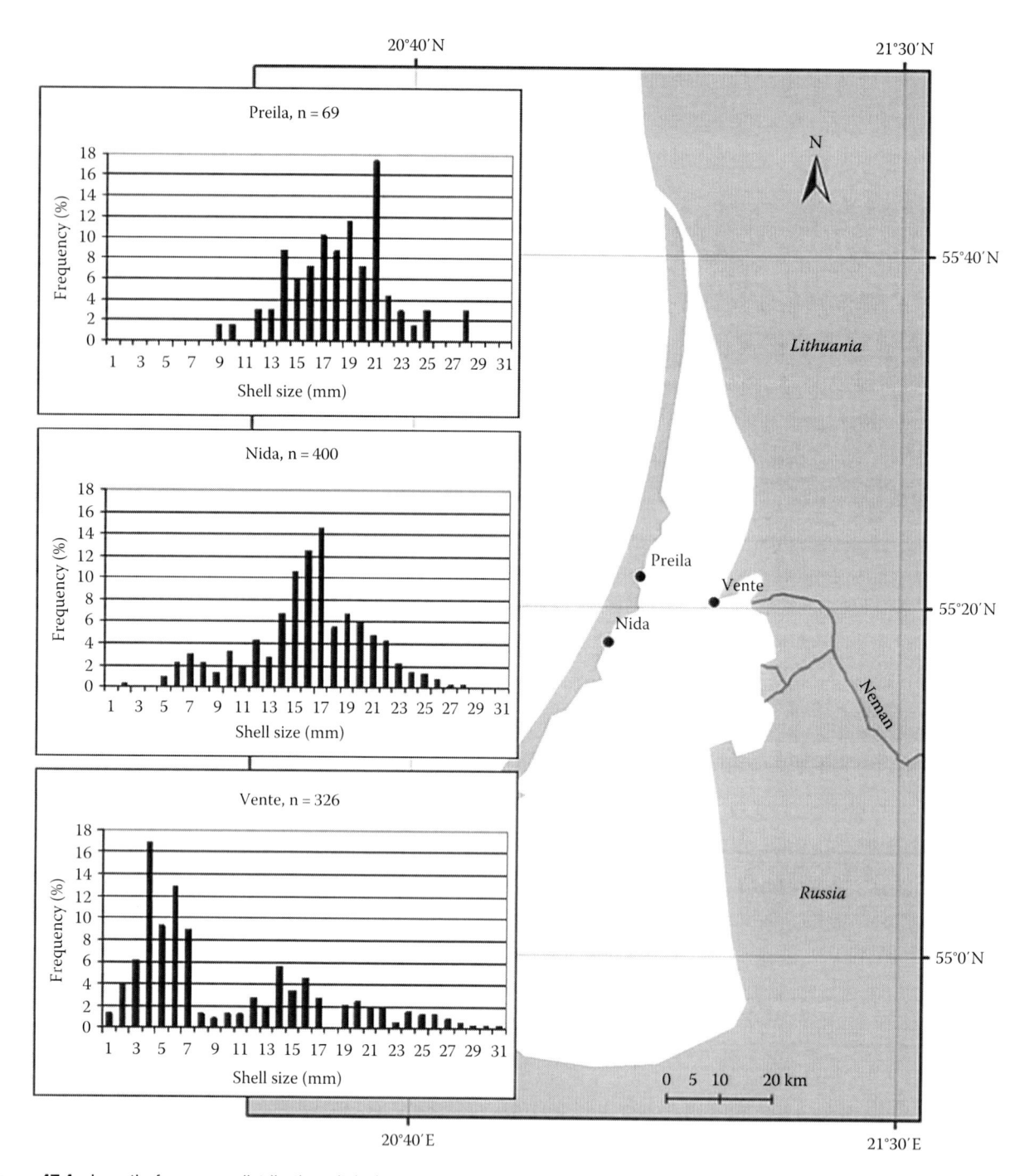

Figure 47.4 Length–frequency distributions (%) of zebra mussel populations at sites in the northern portion of Curonian Lagoon, August/September 2006.

DISCUSSION

Length–frequency distributions of zebra mussel populations are influenced by many factors: water temperature, trophic status, water depth, turbidity, feeding pressure by predators, currents (also determining the type of sediment), and wave action (e.g., Karatayev et al. 2006). These factors affect recruitment and growth and ultimately size of individuals in the population. There were great differences in length–frequency distributions of populations within and between the three water bodies sampled in this study, Szczecin Lagoon, Curonian Lagoon, and the Gulf of Finland, and all distributions at individual sites in the three water bodies were significantly different from the normal distribution.

The multimodality in length–frequency distributions was largely determined by variability in recruitment patterns of the zebra mussel populations. In Curonian Lagoon,

VARIATION IN LENGTH–FREQUENCY DISTRIBUTIONS OF ZEBRA MUSSELS (*DREISSENA POLYMORPHA*)

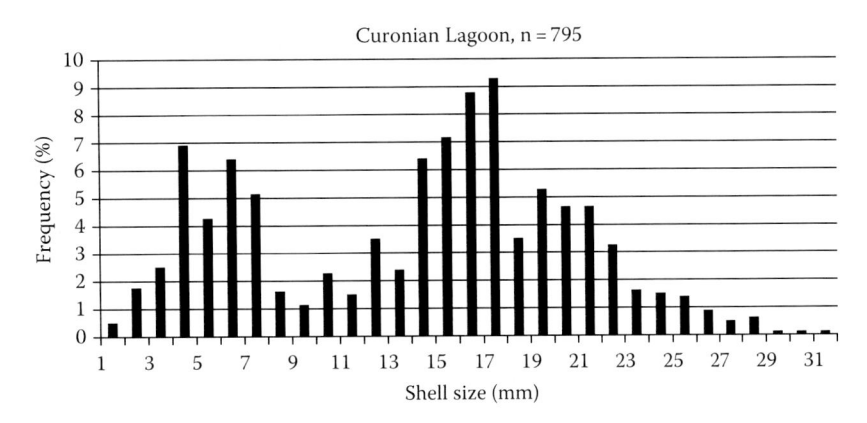

Figure 47.5 Length–frequency distribution (%) of all zebra mussels (n = 795) collected at the three sites in Curonian Lagoon, August/September 2006.

abundances of newly recruited individuals (0+ cohort) were far greater at Vente (peak at 4 mm) than at Nida (peak at 7 mm). The more successful larval settlement at Vente may be due to a bottom substrate consisting of coarser sediments (e.g., fine sand, pebbles, shell deposits) and greater abundances of macrophytes (providing additional substrate for settlement) than at Nida (Daunys and Olenin 1999, Gasiūnaitė et al. 2008). Moreover, Vente was situated closer to the mouth of the river Neman, and this river likely provided food (in the form of suspended matter) and also larvae that were generated from river populations. At Preila, the other site in Curonian Lagoon, small individuals (<10 mm) were not apparent, indicating weak recruitment in the year we sampled (Figure 47.4). Variable recruitment was also apparent in the Szczecin Lagoon (Figure 47.2), with very low numbers of smaller individuals (<10 mm) at two of the six sites

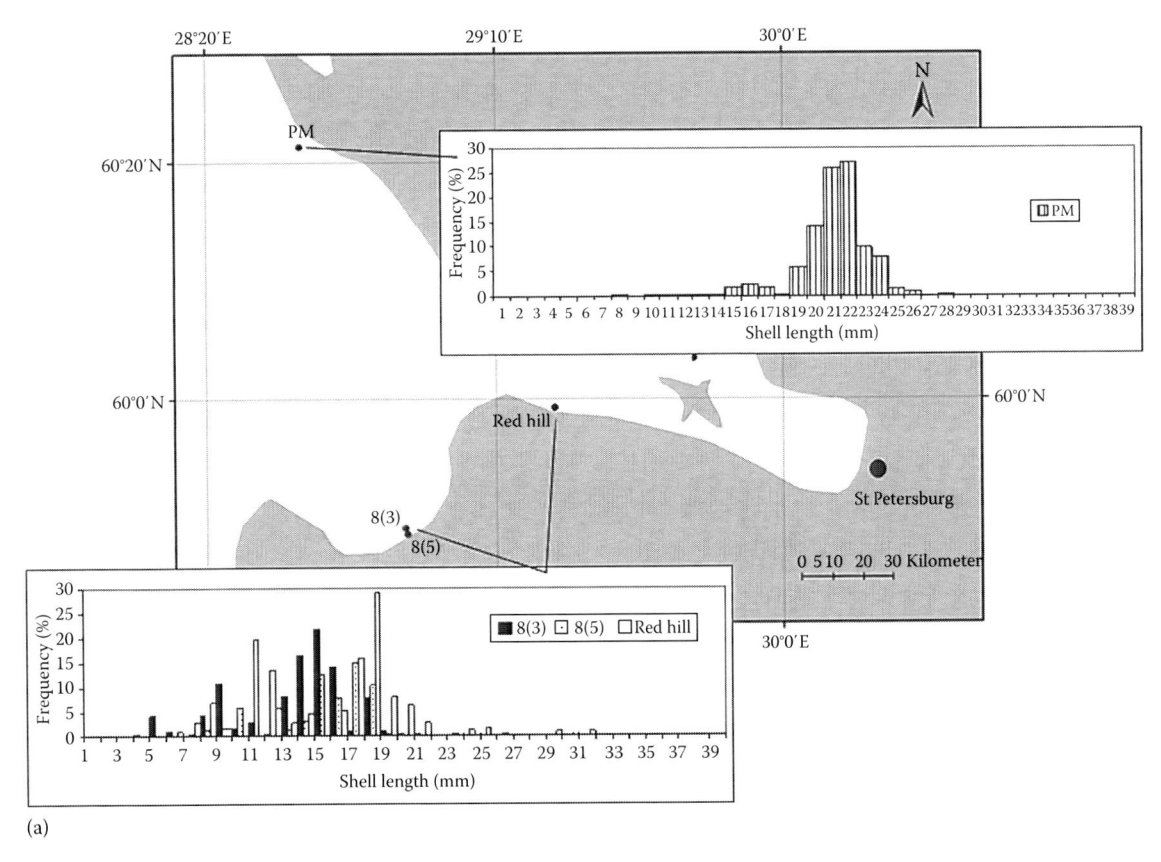

Figure 47.6 (a) Length–frequency distributions (%) of zebra mussel populations at sites in the Gulf of Finland where salinity was 0.4–1.2 ppt (long-term average up to 3.3 ppt, i.e., α-oligohaline, according to Anonymous [1959]). Numbers (n) of mussels measured at sites PM, 8(3), 8(5), and Red Hill (RH) were 408, 225, 198, and 169, respectively.

(continued)

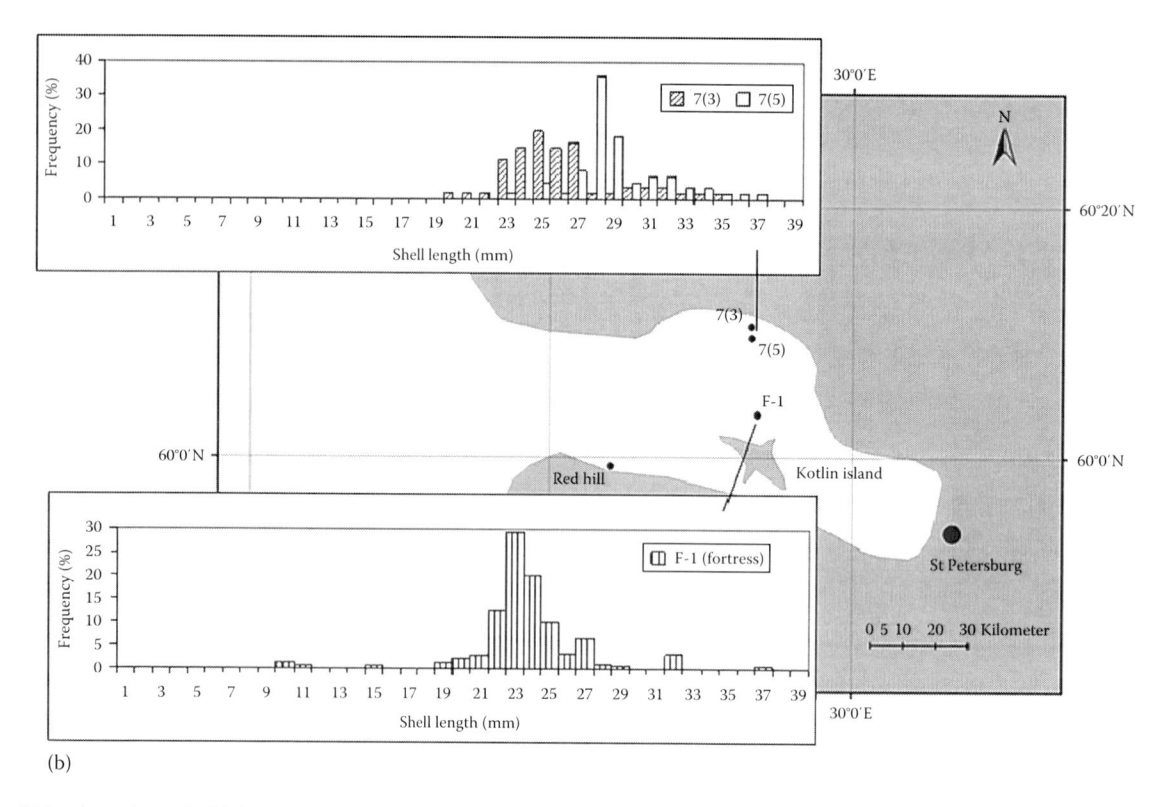

Figure 47.6 (continued) (b) Length–frequency distributions (%) of zebra mussel populations at sites in the Gulf of Finland where salinity was 0.0–0.6 ppt (long-term average 0.8 ppt, i.e., β-oligohaline, according to the Venice System). Numbers (n) of mussels measured at sites 7(3), 7(5), and F1 were 61, 61, and 381.

Table 47.5 Values of Various Statistical Tests on Length–Frequency Distributions of Zebra Mussel Populations at Each Site in the Gulf of Finland. Median Shell Lengths at the Five Sites Were Significantly Different (Kruskal–Wallis, H = 1039.85, p < 0.001). For These Analyses, Data for Stations 7(3) and 7(5) Were Pooled, and Data for Stations 8(3) and 8(5) Were Pooled

Site	Kolmogorov–Smirnov Test	Skewness	Kurtosis	Dip Test
7(3) and 7(5)	D = 0.131; p = 0.03	0.38 (Z = 1.52; p = 0.13)	0.02 (Z = 0.63; p = 0.53)	dip = 0.0615; p < 0.01
F1	D = 0.6232; p < 0.001	16.08 (Z = 17.58; p < 0.001)	294.65 (Z = 14.18; p < 0.001)	dip = 0.0997; p < 0.001
PM	D = 0.842; p < 0.001	−1.65 (Z = 7.09; p < 0.001)	5.93 (Z = 7.52; p < 0.001)	dip = 0.129; p < 0.001
8(3) and 8(5)	D = 0.97; p < 0.001	−0.01 (Z = 0.06; p = 0.96)	0.71 (Z = 2.42; p = 0.02)	dip = 0.0567; p < 0.001
Red Hill (RH)	D = 0.9377; p < 0.001	−0.05 (Z = 0.19; p = 0.85)	1.51 (Z = 2.78; p < 0.01)	dip = 0.0799; p < 0.001
All sites	D = 0.7170; p < 0.001	12.28 (Z = 31.37; p = 0.85)	326.44 (Z = 24.79; p < 0.01)	dip = 0.0531; p < 0.001

(MB10, MB11). Overall, lower numbers of small individuals were found in Szczecin Lagoon than in Curonian Lagoon, which may be explained by the seasonal timing of sample collections. In Curonian Lagoon, samples were collected in August–September, while in Szczecin Lagoon samples were collected in June–July, which was likely before extensive mussel settlement.

In the Gulf of Finland, numbers of small individuals were minimal at all sampling sites. Apparently, recruitment only occurs in warm summers, approximately every 3–5 years (Orlova and Panov 2004), and we assume that recruitment had not occurred in the two previous years since small mussels were only found at one site (8(3), Figure 47.6a). This site was the only one situated outside the area of high turbidity caused by recent dredging activities; thus, besides water temperature, turbidity and food conditions may also play a role in recruitment success. Varying salinity may also have affected larval survival and growth. Salinities of up to 10 ppt can be tolerated by zebra mussels (e.g., in the Caspian and Aral Seas) as long as abrupt fluctuations do not occur, but the maximum salinity for successful spawning and larval development is 6 ppt (McMahon 1996). In general, larvae are more sensitive to unfavorable environmental conditions than adults. Thus, episodic salinity increases in the Gulf of Finland as a result of brackish-water intrusion from the Baltic Sea may have prevented development and settlement of larvae.

The intrusion of brackish water from the Baltic Sea can also affect populations of spawning adults. In the middle area of the Szczecin Lagoon near the German–Polish border,

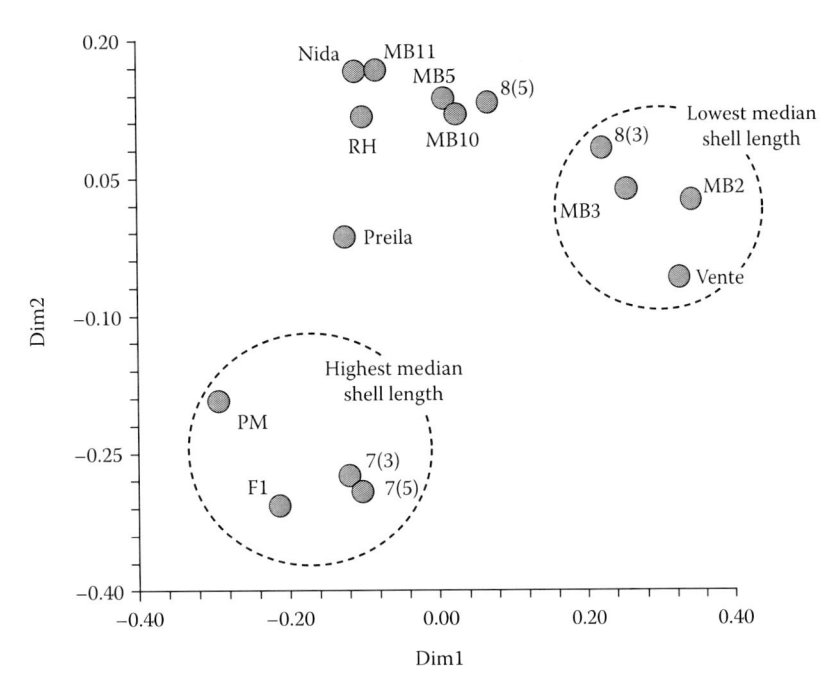

Figure 47.7 MDS plot of the median shell length of zebra mussel populations at all sites sampled in Szczecin Lagoon, Curonian Lagoon, and the Gulf of Finland. Lower left circle: sites with highest median shell length—all sites were in the Gulf of Finland; upper right circle: sites with lowest median shell length—a mix of sites from all three water bodies. The group in the upper left also comprises mussels from all three locations but with intermediate median shell length. Note: the location of MB1 was exactly the same as Nida. Stress value = 0.041, R^2 = 0.852, n = 4925.

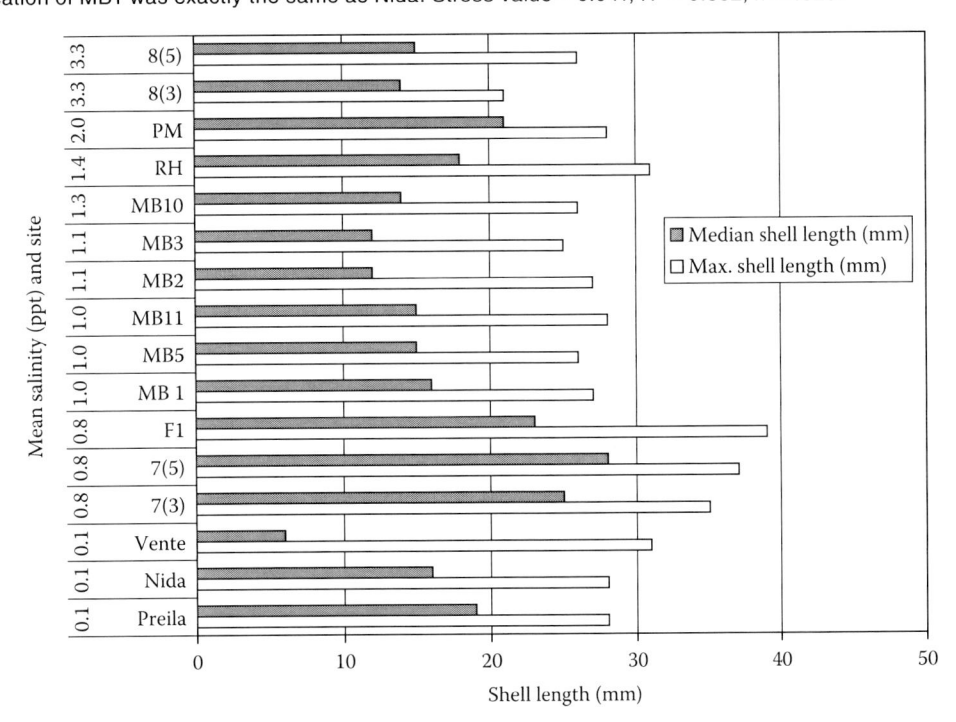

Figure 47.8 Median and maximum shell lengths of zebra mussel populations relative to mean salinity (ppt) at each of the sites sampled in Szczecin Lagoon, Curonian lagoon, and the Gulf of Finland.

varying salinity may have caused the loss of all mussels in a large bed between the late 1990s and 2007 (Fenske et al. 2010). However, differences in salinity alone cannot explain variation in numbers of new recruits between sampling sites in the Szczecin Lagoon. For instance, Stations MB5 and MB11 were located rather close to each other (2.6 km distance), yet length–frequency distributions were highly different. At MB5 smallest individuals were 5 mm and a peak occurred at 8 mm, while at MB11 smallest individuals were 9 mm and a peak occurred at 14 mm. Both sites were relatively far away from inflowing salt water that enters the lagoon from the Baltic Sea. If populations at MB5 or MB11 were affected by wide fluctuations in salinity content, the effect should be apparent at both sites since there is a prevailing ring current in the area of both sites that keeps waters well mixed (Robakiewicz 1993, Radziejewska and Schernewski 2008).

Variations in growth rates can affect length–frequency distributions of zebra mussel populations, but mussels cannot be reliably aged (Karatayev et al. 2006) and thus size at age is not easily determined. However, because of reduced winter growth in temperate climates, length–frequency distributions may provide some estimate of population growth and age structure, at least in the first 2–3 years of a new cohort. Growth slows as mussels get older, making peaks more difficult to distinguish in larger mussels, but there were obviously great differences in numbers of large individuals both within and between the three water bodies. For instance, in Szczecin Lagoon, peaks in number of individuals >20 mm were apparent at Stations MB10 and MB11 but not at MB2 and MB5. In Curonian Lagoon, peaks in large individuals were not apparent at any of the three sites. Comparing the water bodies, greatest maximum lengths were observed in oligohaline to freshwater conditions of the eastern Gulf of Finland where man-made (hard) substrates were present. The maximum length found in this area was 39 mm (reached by a single specimen). In Szczecin Lagoon, where oligohaline conditions prevailed and wide salinity fluctuations were possible, maximum shell length was 28 mm. In the German portion of Szczecin Lagoon, maximum shell lengths of 32–33 mm have been observed but occur rarely (C. Fenske, unpublished). In a recent survey in the Polish portion of Szczecin Lagoon, two empty shells of 35 mm length were found (A. Woźniczka, unpublished). In the central portion of Curonian Lagoon where freshwater conditions prevail, the maximum reported shell length was 31 mm (Table 47.2). Yet salinity alone does not limit growth of zebra mussels. There are several freshwater bodies in Belarus where average shell lengths do not exceed 25 mm, with maxima below 30 mm (Mastitsky 2004). Similarly, Garton and Johnson (2000) reported that zebra mussels rarely obtained shell lengths of >25 mm in a small, freshwater lake in North America.

The length of time zebra mussel larvae spend in the water column before settlement depends mainly on temperature and food availability (Sprung 1993). Food quality, especially content of polyunsaturated fatty acids (PUFA), is relevant to larval growth as well (Wacker and Kraffe 2010). In the initial development stage, larvae still contain a considerable amount of PUFA transferred from adult mussels to their eggs, but for successful settlement they need to obtain some additional nourishment from PUFA-enriched suspended matter. In Szczecin Lagoon, variable food conditions may have led to the great differences in recruitment observed between sites. In this eutrophic lagoon, chlorophyll concentrations were 40–74 μg/L in June/July 2007 (average 61 μg/L, n = 19), and blooms of cyanobacteria (e.g., *Microcystis*) are often observed (Dahlke and Fenske, unpublished). Dense cyanobacteria colonies usually form large gelatinous clumps that are difficult to filter, especially by larvae. Blooms were less prevalent in the area near the mouth of the Oder River, and mussel populations at sites closer to the river mouth (especially MB1 and MB2) had newly settled, small mussels, while populations at sites farther away (MB10 and MB11) lacked small mussels. Alternatively, populations at sites near the Oder River may have been subjected to an influx of larvae from the Oder River, as was suggested for the site near the Neman River in Curonian Lagoon.

As shown in the eastern Gulf of Finland, highly turbid conditions can affect recruitment success. A rapid decrease in mussel abundances occurred in this region in 2005–2007 when turbidity increased as a result of building and dredging activities (Orlova et al. 2008). High turbidity affects populations through deterioration of food quality for both planktonic larvae and settled adults due to increased inorganic particulates in the water. Also, increased sedimentation of fine, inorganic particles may bury hard substrates and make them inaccessible for settling juveniles.

Selective feeding by predators can also affect length–frequency distributions of zebra mussel populations. Especially in winter, birds such as tufted duck (*Aythya fuligula*) and common pochard (*Aythya ferina*) feed on zebra mussels. Other predators include fish as roach (*Rutilus rutilus*), white bream (*Blicca bjoerkna*), vimba (*Vimba vimba*), and invasive round gobies (*Neogobius melanostomus*) (Kublickas 1959 and own observations). Previous investigations noted that coots (*Fulica atra*) and crayfish (*Orconectes limosus*) fed on mussels in Szczecin Lagoon (Piesik 1974, 1983); also, the mitten crab *Eriocheir sinensis* has recently invaded the Szczecin Lagoon and is known to feed on zebra mussels.

Zebra mussel populations in water less than 7 m deep can be highly affected by waterbird predation (Mörtl et al. 2010), and water depths of Curonian and Szczecin Lagoons are predominantly <4 m. However, water transparency in both lagoons is very low, particularly in summer, making it difficult for predators to visually locate mussel prey. Zebra mussels in these two lagoons are often covered by a fluffy layer of organic debris, and most substrates are soft. This makes it difficult for predators to remove individual mussels from druses. Most druses were rather large (e.g., 8 cm diameter) and thus too large for predators to effectively handle or swallow whole. Substrates in the Gulf of Finland were mostly hard, so length–frequency distributions were likely more influenced by predators than distributions in the other two water bodies. Predators usually prefer mussels <10 mm in shell length (Cooley 1991, Nagelkerke and Sibbing 1995, Ray and Corkum 1997) thus predation should not have influenced the maximum shell length of populations in the three water bodies studied.

Finally, the size of the population itself can affect recruitment success, growth, and length–frequency patterns.

Zebra mussel populations are usually not stable, often varying widely over time (Strayer and Malcolm 2006, Stańczykowska et al. 2010). More stable populations tend to live in lakes with a larger surface area, higher levels of calcium, and higher levels of phosphate (Ramcharan et al. 1991). Stańczykowska et al. (2010) described the population in Szczecin Lagoon as being "stable," yet great fluctuations in abundances occur over time (Wolnomiejski and Woźniczka 2008, Fenske et al. 2010). Water bodies with high nutrient levels and considered eutrophic may have conditions that lead to long-term declines in zebra mussel populations or lead to wide variations in abundances from year to year (Stańczykowska et al. 2010). Changes in oxygen content, development of toxic cyanobacteria blooms (e.g., *Microcystis*), and low transparency (as it affects macrophytes) can induce variability in mussel recruitment over both short and long periods of time. However, there may be other factors involved besides trophic conditions in a given system. For instance, in Lake Markermeer (The Netherlands) zebra mussel populations declined unexpectedly in 1992 despite consistently low nutrient levels compared to levels decades before (Noordhuis et al. 2010). The decline in population density (67%) was attributed to climatic conditions and elevated predation pressure (Noordhuis et al. 2010). Since the zebra mussel population has not recovered within 10 years, Noordhuis et al. (2010) described the new conditions as an alternative stable state, following Scheffer (1989).

In summary, zebra mussel populations at sampling sites (even at those located close to each other) had very different size–frequency distributions, with differences mainly a result of variations in the number of small individuals (<10 mm; new recruits). Consistent freshwater conditions and availability of hard substrates provide the best conditions for recruitment success and steady growth. In Curonian and Szczecin Lagoons, recruitment at sites with a soft bottom and varying salinity was limited, as reflected by the lower numbers of small individuals. In contrast, populations at sites located in the vicinity of river mouths had broader size–frequency distributions, including a greater number of small mussels in the 0+cohort. The additional riverine supply of larvae, more stable salinity conditions, and better seston quality (e.g., lower cyanobacteria concentrations) likely contribute to the more successful recruitment at those sites.

ACKNOWLEDGMENTS

In Lithuania, the study was supported by the European Regional Development Fund through the Baltic Sea Region Programme Project. "Sustainable Uses of Baltic Marine Resources (SUBMARINER No. 055)." The study in Germany was supported by the International Bureau of the German Federal Ministry of Education and Research (BMBF), Förderkennzeichen MOE 07/R58. We also acknowledge the support of the University of Greifswald. We thank Thomas Nalepa and Donald Schloesser for constructive comments and Norbert Amelang for help with the maps. This chapter is dedicated to the memory of Robert W. McCauley.

REFERENCES

Anonymous. 1959. Venice system for the classification of marine waters according to salinity. http:/aslo.org/lo/toc/vol_3/issue_3/0346.pdf (accessed May 22, 2013).

Beekey, M.A., D.J. McCabe, and J.E. Marsden. 2004. Zebra mussels affect benthic predator foraging success and habitat choice on soft sediment. *Oecologia* 141: 164–170.

Beisel, J.-N., V. Bachmann, and J.-C. Moreteau. 2010. Growth-at-length model and related life-history traits of *Dreissena polymorpha* in lotic ecosystems. In *The Zebra Mussel in Europe*, G. Van der Velde, S. Rajagopal, and A. Bij de Vaate, eds., pp. 191–197. Leiden, The Netherlands: Backhuys Publishers.

Berkman, P.A., M.A. Haltuch, E. Tichich et al. 1998. Zebra mussels invade Lake Erie muds. *Nature* 393: 27–28.

Bially, A. and H.J. MacIsaac. 2000. Fouling mussels (*Dreissena* spp.) colonize soft sediments in Lake Erie and facilitate benthic invertebrates. *Freshwat. Biol.* 43: 85–97.

Botts, P.S., B.A. Patterson, and D.W. Schloesser. 1996. Zebra mussel effect on benthic invertebrates: Physical or biotic? *J. N. Am. Benthol. Soc.* 15: 179–184.

Buynevich, I.V., A. Damušytė, A. Bitinas, S. Olenin, J. Mažeika, and R. Petrošius. 2011. Pontic–Baltic pathways for invasive aquatic species: Geoarchaeological implications. In *Geology and Geoarchaeology of the Black Sea Region: Beyond the Flood Hypothesis*, eds., I.V. Buynevich, V. Yanko-Hombach, A. Gilbert, and R.E. Martin, pp. 189–196. Geological Society of America Special Paper 473. Boulder Geological Society of America.

Chase, M.I. and R.C. Bailey. 1999. The ecology of the zebra mussel (*Dreissena polymorpha*) in the lower Great Lakes of North America: I. Population dynamics and growth. *J. Great Lakes Res.* 25: 107–121.

Cooley, J.M. 1991. Editorial. Zebra mussels. *J. Great Lakes Res.* 17: 1–2.

Daunys, D. and S. Olenin 1999. Littoral bottom communities of the northern part of the Curonian lagoon. *Ekologija* 2: 19–27 (in Lithuanian with English summary).

Fenske, C. 2002. The ecological importance of mussels, their effect on water quality and their possible use for coastal zone management. In *Baltic Coastal Ecosystems: Structure, Function and Coastal Zone Management*, G. Schernewski and U. Schiewer, eds., pp. 53–64. Berlin, Germany: Springer Verlag.

Fenske, C., S. Dahlke, P. Riel, and A. Woźniczka. 2010. Dynamics of mussel beds in the Szczecin Lagoon. *Coastline Reports* 16: 51–58.

Garton, D.W. and L.E. Johnson. 2000. Variation in growth rates of the zebra mussel, *Dreissena polymorpha*, within Lake Wawasee. *Freshwat. Biol.* 45: 443–451.

Gasiūnaitė, Z.R., D. Daunys, S. Olenin, and A. Razinkovas. 2008. The Curonian Lagoon. In *Ecology of Baltic Coastal Waters. Ecological Studies*, ed., U. Schiewer, Vol. 197, pp. 197–215. Berlin, Germany: Springer-Verlag.

Grim, J. 1971. Tiefenverteilung der Dreikantmuschel *Dreissena polymorpha* (Pallas) im Bodensee. *gwf-Wasser/Abwasser* 112: 437–441.

Gutierrez, J.L., C.G. Jones, D.L. Strayer, and O.O. Iribarne. 2003. Mollusks as ecosystem engineers: The role of shell production in aquatic habitats. *Oikos* 101: 79–90.

Hartigan, P. 1985. Computation of the dip statistic to test for unimodality. Applied Statistics *J. Royal Statist. Soc. Ser.* 34: 320–325.

Hartigan, J. and P. Hartigan. 1985. The dip test of unimodality. *Ann. Stat.* 13: 70–84.

Helcom. 2007. Climate change in the Baltic Sea area. Helcom Thematic Assessment in 2007. *Baltic Sea Environment Proceedings No. 111*. 59p. ISSN 0357 2994. http://www.helcom.fi/stc/files/publications/proceedings/bsepill.pdf (accessed August 31, 2011).

Helcom. 2009. HELCOM Battic Sea Action Plan. http://www.helcom.fi/BSAP/en_GB/intro/ (accessed August 31, 2011).

Helcom. 2011. Helsinki Commission, Baltic Marine Environment Protection Commission. http://www.helcom.fi/projects/jcp/hotspots/en_GB/hotspots/ (accessed February 02, 2011).

In Lithuania, the study was supported by the European Regional Development Fund through the Baltic Sea Region Programme project "Sustainable Uses of Baltic Marine Resources (Submariner No. 055)."

Jantz, B. and D. Neumann. 1998. Growth and reproductive cycle of the zebra mussel in the River Rhine as studied in a river bypass. *Oecologia* 114: 213–225. DOI: 10.1007/s004420050439.

Karatayev, A.Y. and L.E. Burlakova. 1994. Filtration rates. In *Freshwater Zebra Mussel Dreissena polymorpha (Pall.) (Bivalvia, Dreissenidae): Systematics, Ecology, Practical Meaning*, ed. J.I. Starobogatov, pp. 109–120. Moscow, Russia: Nauka (in Russian).

Karatayev, A.Y., L.E. Burlakova, and D.K. Padilla. 2002. Impacts of zebra mussels on aquatic communities and their role as ecosystem engineers. In *Invasive Aquatic Species of Europe - Distribution, Impact and Management*, eds. E. Leppäkoski, S. Gollasch, and S. Olenin, pp. 443–446. Dordrecht, The Netherlands: Kluwer Academic Publishers.

Karatayev, A.Y., L.E. Burlakova, and D.K. Padilla. 2006. Growth rate and longevity of *Dreissena polymorpha* (Pallas): A review and recommendations for future study. *J. Shellfish Res.* 25: 23–32.

Kublickas, A. 1959. Feeding of benthophagous fish in the Curonian lagoon. In *Kursiu Marios*, eds. K. Jankevicius, I. Gasiūnas, A. Gediminas, V. Gudelis, A. Kublickas, and I. Maniukas, pp. 463–521. Vilnius, Lithuania: Lithuanian Academy of Sciences (in Russian).

Lajtner, J., Z. Maruši´, G.I.V. Klobučar, I. Maguire, and R. Erben. 2004. Comparative shell morphology of the zebra mussel, *Dreissena polymorpha* in the Drava River (Croatia). *Biologia* 59/5: 595–600.

LUNG (Landesamt für Umwelt, Naturschutz und Geologie Mecklenburg-Vorpommern, Güstrow, Germany). Data on Water Investigations in Coastal Waters 1975–2001, unpublished.

Maechler, M. and D. Ringach. 2004. The diptest package. Hartigan's dip test statistic for unimodality-corrected code, R package version 0.25–1. http://cran.rproject.org/ (accessed May 22, 2013).

Mastitsky, S.E. 2004. Endosymbionts of bivalve mollusc *Dreissena polymorpha* (Pallas) in waterbodies of Belarus. PhD thesis. Minsk, Belarus: Institute of Zoology of the National Academy of Sciences of Belarus (in Russian).

McMahon, R.F. 1996. The physiological ecology of the zebra mussel, *Dreissena polymorpha*, in North America and Europe. *Am. Zool.* 36: 339–363.

Meisenheimer, J. 1901. Entwicklungsgeschichte von *Dreissena polymorpha* Pall. *Zeitschrift für wissenschaftliche Zoologie* 69: 1–137.

MLUV-MV. 2009. (Ministerium für Landwirtschaft, Umwelt und Verbraucherschutz Mecklenburg-Vorpommern). *Regelwerk Küstenschutz Mecklenburg-Vorpommern*, Schwerin, 201pp.

Mörtl, M., S. Werner, and K.-O. Rothhaupt. 2010. Effects of predation by wintering water birds on zebra mussels and on associated macroinvertebrates. In *The Zebra Mussel in Europe*, G. Van der Velde, S. Rajagopal, and A. Bij de Vaate, eds., pp. 239–250. Leiden, The Netherlands: Backhuys Publishers.

Nagelkerke, L.A.J. and F.A. Sibbing. 1995. Efficiency of feeding on zebra mussel (*Dreissena polymorpha*) by common bream (*Abramis brama*), white bream (*Blicca bjoerkna*) and roach (*Rutilus rutilus*): The effects of morphology and behaviour. *Can. J. Fish. Aquat. Sci.* 53: 2847–2861.

Nalepa, T., J.A. Wojcik, D.L. Fanslow, and G. Lang. 1995. Initial colonization of the zebra mussel (*Dreissena polymorpha*) in Saginaw Bay, Lake Huron: Population recruitment, density, and size structure. *J. Great Lakes Res.* 21: 417–434.

Noordhuis, R., M.R. van Eerden, and M. Roos. 2010. Crash of zebra mussel, transparency and water bird populations in Lake Markermeer. In *The Zebra Mussel in Europe*, G. Van der Velde, S. Rajagopal, and A. Bij de Vaate, eds., pp. 265–278. Leiden, The Netherlands: Backhuys Publishers.

Orlova, M.I., V.V. Khlebovich, and A.Yu. Komendantov. 1998. Potential euryhalinity of *Dreissena polymorpha* (Pallas) and *Dreissena bugensis* (Andr.). *Russ. J. Aquat. Ecol.* 7: 17–28.

Orlova, M.I. and V.E. Panov. 2004. Establishment of the zebra mussel, *Dreissena polymorpha* (Pallas) in the Neva Estuary (Gulf of Finland, Baltic Sea): Distribution, population structure and possible impact on local unionid bivalves. *Develop. Hydrobiol.* 176: 207–217.

Orlova, M.I., D.V. Ryabchuk, and M.A. Siridonov. 2008. Macrozoobenthos. In *Ecosystems of the Neva River Estuary: Biological Diversity and Ecological Problems*, A.F. Alimov and S.M. Golubkov, eds., pp. 272–312. Moscow, Russia: KMK press.

Orlova, M.I., I.V. Telesh, A.E. Antsulevich, A.A. Maximov, and L.F. Litvinchuk. 2006. Effects of non-indigenous species on diversity and community functioning in the eastern Gulf of Finland (Baltic Sea). *Helgo. Mar. Res.* 60: 98–105.

Paavola, M., S. Olenin, and E. Leppäkoski. 2005. Are invasive species most successful in habitats of low native species richness across European brackish water seas? *Estuar. Coast. Shelf Sci.* 64: 738–750.

Piesik, Z. 1974. The role of crayfish *Orconectes limosus* (Raf.) in extinction of *Dreissena polymorpha* subsisting on steelonnet. *Pol. Arch. Hydrobiol.* 21: 401–410.

Piesik, Z. 1983. Biology of *Dreissena polymorpha* (Pall.) settling on stylon nets and the role of this mollusk in eliminating the seston and the nutrients from the water-course. *Pol. Arch. Hydrobiol.* 30: 353–361.

Radziejewska, T. and G. Schernewski. 2008. The Szczecin Lagoon. In *Ecology of Baltic Coastal Waters. Ecological Studies*, U. Schiewer, ed., pp. 115–130. Berlin, Germany: Springer-Verlag.

Ramcharan, C.W., D.K. Padilla, and S.I. Dodson. 1991. A multivariate model for predicting population fluctuations of *Dreissena polymorpha* in North American lakes. *Can. J. Fish. Aquat. Sci.* 49: 150–158.

Ray, W.J. and L.D. Corkum. 1997. Predation of zebra mussels by round gobies, *Neogobius melanostomus*. *Environ. Biol. Fishes* 50: 267–273.

Reise, K., S. Olenin, and D.W. Thieltges. 2006. Are aliens threatening European aquatic coastal ecosystems? *Helgo. Mar. Res.* 60: 77–83.

Remane, A. 1934. Die Brackwasserfauna. *Verh. dt. zoolog. Gesellschaft* 36: 34–74.

Robakiewicz, W. 1993. Hydrodynamic regime of the Szczecin Lagoon and the straits connecting the Lagoon with the Pomeranian Bay. *Bibl. Nauk Hydrotechnol.* Monograph 16. Gdansk, Poland: Polish Academy of Sciences (in Polish).

Scheffer, M. 1989. Alternative stable states in eutrophic shallow freshwater systems: A minimal model. *Hydrobiol. Bull.* 23: 73–83.

Schiewer, U. 2008. Introduction. In *Ecology of Baltic Coastal Waters. Ecological Studies*, U. Schiewer, ed., pp. 1–21. Berlin, Germany: Springer-Verlag.

Sprung, M. 1993. The other life: An account of present knowledge of the larval phase of *Dreissena polymorpha*. In: *Zebra Mussels. Biology, Impacts, and Control*, T.F. Nalepa and D.W. Schloesser, eds., pp. 39–53. Boca Raton, FL: Lewis Publishers.

Stańczykowska, A. and K. Lewandowski. 1993. Thirty years of studies of *Dreissena polymorpha* Ecology in Mazurian Lakes of Northeastern Poland. In *Zebra Mussels. Biology, Impacts, and Control*, eds. T.F. Nalepa and D.W. Schloesser, pp. 3–38. Boca Raton, FL: Lewis Publishers.

Stańczykowska, A., K. Lewandowski, and M. Czarnoleski. 2010. Distribution and densities of *Dreissena polymorpha* in Poland – Past and present. In *The Zebra Mussel in Europe*, G. Van der Velde, S. Rajagopal, and A. Bij de Vaate, eds., pp. 119–126. Leiden, The Netherlands: Backhuys Publishers.

Starobogatov, J.I. and S.I. Andreyeva. 1994. Areal of zebra mussel and its history. In *Freshwater Zebra Mussel Dreissena polymorpha (Pall.) (Bivalvia, Dreissenidae): Systematics, Ecology, Practical Meaning*, J.I. Starobogatov, ed., pp. 47–56. Moscow, Russia: Nauka (in Russian).

Stewart, T.W., J.G. Miner, and R.L. Lowe. 1998. Quantifying mechanisms for zebra mussel effects on benthic macroinvertebrates: Organic matter production and shell-generated habitat. *J. N. Am. Benthol. Soc.* 17: 81–94.

Strayer, D.L. and H.M. Malcolm. 2006. Long-term demography of a zebra mussel (*Dreissena polymorpha*) population. *Freshwat. Biol.* 51: 117–130.

Wacker, A. and E. Kraffe. 2010. Fatty acid nutrition: Its role in the reproduction and growth of zebra mussels. In *The Zebra Mussel in Europe*, G. Van der Velde, S. Rajagopal, and A. Bij de Vaate, eds., pp. 153–160. Leiden, The Netherlands: Backhuys Publishers.

Wolnomiejski, N. and A. Woźniczka. 2008. A drastic reduction in abundance of *Dreissena polymorpha* Pall. in the Skoszewska Cove (Szczecin Lagoon, River Odra estuary): Effects in the population and habitat. *Ecol. Questions* 9: 103–111.

Young, B.L., D.K. Padilla, D.W. Schneider, and S.W. Hewett. 1996. The importance of size-frequency relationships for predicting ecological impact of zebra mussel populations. *Hydrobiologia* 332: 151–158.

Zaiko, A., D. Daunys, and S. Olenin. 2009. Habitat engineering by the invasive zebra mussel *Dreissena polymorpha* (Pallas) in a boreal coastal lagoon: Impact on biodiversity. *Helgo. Mar. Res.* 63: 85–94.

Zaiko, A., M. Lehtiniemi, A. Narščius, and S. Olenin. 2011. Assessment of bioinvasion impacts on a regional scale: A comparative approach. *Biol. Invasions* 13: 1739–1765.

Zaiko, A., S. Olenin, D. Daunys, and T.F. Nalepa. 2007. Vulnerability of benthic habitats to the aquatic invasive species. *Biol. Invasions* 9: 703–714.

CHAPTER 48

A Note on Dreissenid Mussels and Historic Shipwrecks

Russ Green

CONTENT

References 744

Shipwrecks in the Great Lakes and other large, freshwater lakes in North America reflect a rich underwater heritage and possess significant historical, archeological, and recreational value. In the Great Lakes alone, there are an estimated 6000 shipwrecks of various design and construction, including Native American canoes, sailing schooners, grand palace steamers, steel freighters, industrial bulk carriers, and military vessels. Collectively, they provide a record of the Great Lakes' pervasive influence in regional and national history and capture the cultural, personal, environmental, technological, and economic aspects of maritime history. Each of these shipwrecks is unique and non-renewable. Because of continuously-cold temperatures in deep regions of these lakes, most shipwreck sites are extremely well preserved. Also, unlike shipwrecks in marine waters, freshwater shipwrecks until recently have been free of organisms that attach to surfaces and compromise wreck features and integrity.

In the late-1980s, zebra mussels (*Dreissena polymorpha*) and quagga mussels (*Dreissena rostriformis bugensis*) became established in North America, and the propensity of these species to attach to, and proliferate on, hard substrates such as shipwrecks constitutes a significant challenge for archeologists, historians, and preservationists. In a sense, the baseline has shifted in regard to the quality of data that can be obtained (at least easily) from shipwrecks and other submerged archeological sites. Examples of this issue may be provided by recent changes to shipwrecks in and around the Thunder Bay National Marine Sanctuary in northern Lake Huron (Figure 48.1). One of 14 protected areas within the National Oceanic and Atmospheric Administration's National Marine Sanctuary System, the Thunder Bay Sanctuary is home to about 50 shipwrecks dating back to the 1840s; another 40 shipwrecks are located just beyond sanctuary boundaries. While zebra mussels had been present in shallow regions of the sanctuary and found on some wrecks in low numbers since the mid-1990s, it was not until the mid-2000s when quagga mussels became established in deeper waters that the most notable changes occurred. The increase of quagga mussels in and around sanctuary waters was typical of increases found throughout Lake Huron at that time (Nalepa et al. 2007, Nalepa unpublished). One shipwreck just outside the sanctuary is the *Kyle Spangler*, a wooden schooner that sank in 1860 and lies on the bottom at a depth of 55 m. This depth is below the thermocline where water temperatures mostly remain at a constant 4°C–7°C. In 2003, photos were taken of the *Kyle Spangle*'s centerboard winch (Figure 48.2a) and anchor (Figure 48.2b), the latter of which was stowed on the port rail of the vessel. While some mussels were present on both structures, even the smallest features were still easily discerned, providing testimony to the preservation qualities of continuously-cold freshwater. In 2008, photos were again taken of the same two structures after quagga mussels became abundant (Figure 48.2c and d). Both were covered with a thick layer of mussels, and the difficulty of archeological documentation is dramatically apparent. Another typical example of how quagga mussels have obscured features of shipwrecks in the sanctuary is provided by the wooden steamer *Florida*, which lies on the bottom at a depth of 61 m (Figures 48.3 and 48.4). The ship is broken and open in cross section, which allowed mussels easy access to food, and hence every interior surface both wood and metal is heavily colonized (Figure 48.3). Yet even shipwrecks in deeper water and with generally intact hulls tend to be colonized by quagga mussels. In 2011, the schooner *M.F. Merrick* was discovered at a depth of 91 m in an

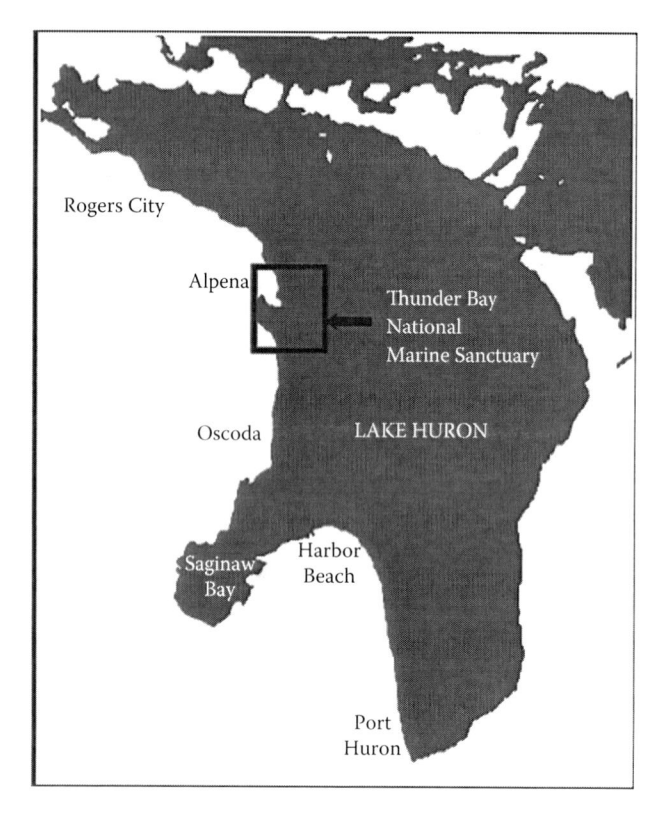

Figure 48.1 Location of Thunder Bay National Marine Sanctuary in northern Lake Huron in the Laurentian Great Lakes.

area near the sanctuary. Structural surfaces of both the exterior and nearly intact, confined interior of this shipwreck were colonized by mussels (Figures 48.5 and 48.6).

In addition to obscuring structural features, dreissenids may affect the structural integrity of many shipwrecks. Wood-constructed vessels of the *Kyle Spangler*'s period are held together by fasteners and fittings made of iron, and research has shown that such fittings undergo increased corrosion when colonized by dreissenids (Watzin et al. 2001). In studies of wooden schooners in Lake Champlain, New York, enhanced pitting was observed on iron fasteners after dreissenid colonization. Indeed, greater iron (Fe^+) concentrations were noted just above mussels on wrecks as compared to parts of the wrecks with no mussels (Watzin et al. 2001). Apparently, accumulated mussel biodeposits (feces and pseudofeces) under and between mussel shells promote bacteria known to cause corrosion. While physical and biological factors that affect corrosion remain unclear, and thus corrosion rates are difficult to predict, it is quite possible that the structural integrity of dreissenid-colonized shipwrecks will be compromised over time.

Ironically, while dreissenids have obscured structures of known shipwrecks, they have increased the ability to view wrecks overall. Because dreissenids are filter feeders and remove particulates from the water, water clarity in lakes where dreissenids have become abundant has increased

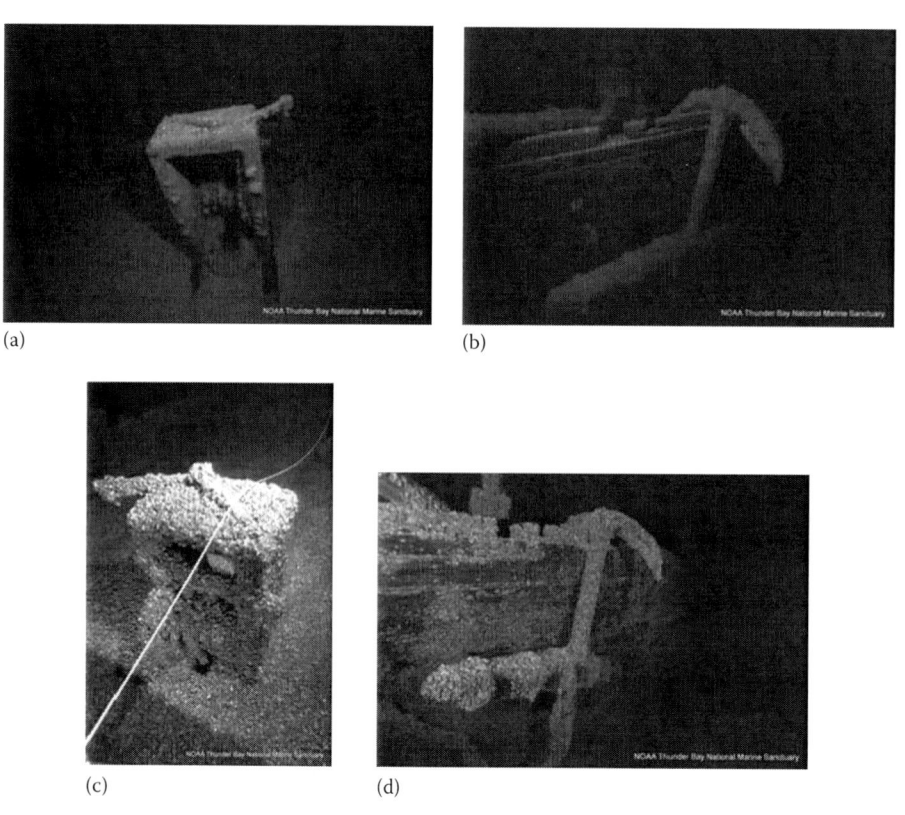

Figure 48.2 Images of a centerboard winch (a) and anchor (b) on the wreck of the wooden schooner *Kyle Spangler* in 2003 before quagga mussels became abundant and of the same winch (c) and anchor (d) in 2008 after quagga mussels became abundant. The schooner sank in 1860 and lies on the bottom at a depth of 55 m in northern Lake Huron. (a and b: Courtesy of Stan Stock; c and d: Courtesy of NOAA Thunder Bay National Marine Sanctuary, Alpena, MI.)

Figure 48.3 (See color insert.) Image taken in 2007 of the interior of the steamer *Florida* that lies on the bottom at a depth of 61 m in northern Lake Huron. The shipwreck is broken in cross section and hence the interior is open to the water column. Note all surfaces are colonized by quagga mussels. (Courtesy of NOAA Thunder Bay National Marine Sanctuary, Alpena, MI.)

Figure 48.5 (See color insert.) Image of the exterior surface of the schooner *M.F. Merrick*. This shipwreck was discovered in 2011 on the bottom at a water depth of 91 m in northern Lake Huron. (Courtesy of John Janzen.)

Figure 48.6 (See color insert.) Image of the hull interior of the schooner *M.F. Merrick* (see details in earlier text). The hull was nearly intact and the interior confined, yet quagga mussels were present and abundant. (Courtesy of John Janzen.)

Figure 48.4 (See color insert.) Image of the exterior of the steamer *Florida* (see details in earlier text). In the center of the image just below the opening is an iron/wood hatchet uniformly colonized by quagga mussels. (Courtesy of NOAA Thunder Bay National Marine Sanctuary, Alpena, MI.)

by nearly 40% (Higgins 2013). In northern Lake Huron, visibility can now be up to 30 m in the spring (personal observation). Greater water clarity has enhanced observations of shipwrecks by divers, boaters, tourists, etc.

Certainly, dreissenids have impacted historic shipwrecks, but yet they are now part of the ecosystem and will likely always be associated with these wrecks. From this viewpoint, shipwrecks can be focal points for ecological research, impact assessments, and studies of dreissenid population trends (see LaValle et al. 1999). In addition, shipwrecks are convenient places to create public awareness and, now more visible, they provide strong images of how invasive species like dreissenids can change general perceptions of underwater habitats and potentially affect lake ecosystems.

REFERENCES

Higgins, S. 2013. A meta-analysis of dreissenid effects on freshwater ecosystems. In *Quagga and Zebra Mussels: Biology, Impacts, and Control*, 2nd Edn., T.F. Nalepa and D.W. Schloesser, eds., pp. 487–494. Boca Raton, FL: CRC Press.

LaValle, P.D., A. Brooks, and V.C. Lakhan. 1999. Zebra mussel wastes and concentrations of heavy metals on shipwrecks in western Lake Erie. *J. Great Lakes Res.* 25: 330–338.

Nalepa, T.F., D.L. Fanslow, S.A. Pothoven, A.J. Foley III, and G.A. Lang. 2007. Long-trends in benthic macroinvertebrate populations in Lake Huron over the past four decades. *J. Great Lakes Res.* 33: 421–436.

Watzin, M.C., A.B. Cohn, and B.P. Emerson. 2001. Zebra mussels, shipwrecks, and the environment. Final Report, 2001. Vergennes, VT: Lake Champlain Maritime Museum.

PART VII

Appendix: Narratives for Video Clips

Video Clip 1: A Visual Documentation of the Eradication of Zebra Mussels (*Dreissena polymorpha*) from Millbrook Quarry, Virginia

Raymond T. Fernald and Brian T. Watson

This video clip on the companion CD contains three sequences that depict distinct phases of our project to eradicate zebra mussels from Millbrook Quarry, Virginia, and is meant to complement our chapter on the same topic (Fernald and Watson 2013). The sequences depict pretreatment zebra mussel distribution, potassium concentrations as measured during the 3 week treatment period, and posttreatment documentation of zebra mussel mortality.

The first video sequence (live pretreatment April 2003; two segments) was recorded by divers with the Virginia Department of Game and Inland Fisheries (VDGIF) during pretreatment assessment of the mussel infestation in April 2003. It presents typical views of zebra mussel distribution on rocks comprising the side slopes and bottom of Millbrook Quarry, from just below the surface to depths of about 21 m. Sparse to dense clusters of zebra mussels were encountered throughout the quarry, both on exposed surfaces and in cracks or fissures within the quarry walls. A number of "attractions" for recreational divers were placed in the quarry (i.e., airplane, cabin cruiser, bus), and sections of these objects were encrusted with zebra mussels from one to three layers deep. Very few dead zebra mussels or empty shells were observed during the field surveys, and the infestation was estimated to have originated 2–3 years prior to it being reported to VDGIF in August 2002. Recreational divers who frequented Millbrook Quarry, in retrospect, confirmed this time period of establishment as being consistent with their first observations of the then-unidentified invasive species. During the population assessment in April 2003, the mussels were most abundant on the undersides of diving platforms suspended approximately 10 m below the surface; abundances on these surfaces were estimated to be approximately 20,000–30,000 mussels per square meter.

Treatment of the quarry with a 12% solution of potassium chloride was initiated on January 31 and ended on February 17, 2006. The second sequence (potassium concentrations February 2006) presents 3D animations of potassium concentrations measured in the water column at various depths along fixed transects after three 5 day periods of daily potassium injections (animations at February 5, 11, and 18). We initially were concerned that, because of differences in densities between the water and potassium solution and because of differences in temperatures through the water column (seasonal thermoclines), the potassium might not rapidly or fully diffuse throughout the water column, potentially leaving clusters of zebra mussels untreated. The long-term effectiveness of potassium was, therefore, a primary consideration in our selection of this chemical as a control treatment, and we had planned to vary the length of the distribution hoses from 1.5 to 15 m to maximize initial distribution of the solution. To our surprise, 24 h after the first week of treatment was completed (all via surface spray or use of hoses <3 m in length), potassium was evenly distributed throughout the water column at approximately 1/3 of our intended target concentration (100 ppm). As noted in the second and third animations, after the second week of treatment (February 11), the potassium was evenly distributed at about 60–70 ppm and, after the third week of treatment (February 18), the potassium concentration was approximately 100 ppm at every monitored point in the water column throughout the quarry.

The third sequence (dead posttreatment April 2006) was recorded by the contractor's robotic submersible camera in April 2006, nearly 2 months after the potassium treatment was completed. The sequence includes three segments extracted from the contractor's documentation of dead zebra mussels throughout the quarry: (1) video of dead zebra mussels on rock substrates, (2) live and apparently unaffected largemouth bass and bluegill, and (3) dead zebra mussels on the underside of one of the dive-training platforms suspended 10 m below the surface. During 2 days of robotic surveys (over 6 h of video was archived and provided to VDGIF), most major dive attractions and training platforms were inspected for evidence of live zebra mussels. No live zebra mussels (or Asian clams) were observed. In addition, the video did not show any evidence of mortality in any organisms other than zebra mussels and Asian clams. In the final segment of this sequence, decayed flesh and dead shells of zebra mussels can be seen falling from the dive platform when disturbed by the water-jet exhaust of the robotic vehicle.

REFERENCE

Fernald, R. T. and B. T. Watson. 2013. Eradication of zebra mussels (*Dreissena polymorpha*) from Millbrook Quarry, Virginia: Rapid response in the real world. In *Quagga and Zebra Mussels: Biology, Impacts, and Control*, 2nd Edn., T. F. Nalepa and D. W. Schloesser, eds., pp. 195–213. Boca Raton, FL: CRC Press.

Video Clip 2: Invasion of Quagga Mussels (*Dreissena rostriformis bugensis*) to the Midlake Reef Complex in Lake Michigan: A Video Montage

Jeffrey S. Houghton, Robert Paddock, and John Janssen

This video on the companian CD documents the invasion of quagga mussels (*Dreissena rostriformis bugensis*) to the Midlake Reef Complex (MLRC) in Lake Michigan and also provides footage of common fauna found on the reef. The MLRC is an ensemble of reefs between the north and south basins of the lake (see Houghton et al. 2013). With summit depths of about 40–50 m, the MLRC lies entirely deeper than the euphotic zone. We compiled representative video taken in 2002, 2003, 2005, 2006, and 2009 with a remotely operated vehicle (ROV; Benthos MiniRover MK-II). Cameras and recording devices were upgraded during the 9 year period, which is apparent from the different ROV configurations seen throughout the time series. While ROV dives and videos during this time span targeted lake trout (*Salvelinus namaycush*) reproduction at the MLRC, they nonetheless provided clear visual documentation of the quagga mussel invasion of the reef complex. The video shows that quagga mussels were nearly nonexistent in 2002, but completely covered cobble substrate of East Reef (approx. 43°01′ N and 87°21′ W) and Sheboygan Reef (approx. 43°21′ N and 87°09′ W) by 2006.

The video begins with footage of mussel colonization at East Reef. In April 2002, mussels were visible as white protrusions on rocks and numbers were sparse (0:07–0:48). By May 2003, mussels were clearly visible as clusters (0:44–1:10), and by November 2003, these clusters nearly completely covered the rocks (1:12–1:40). From May 2004 onward (November 2005 also shown), rocks were completely covered by mussels (1:41–2:38). Note the lake trout that come into view at 1:32 and 2:12.

Colonization of Sheboygan Reef proceeded in the same fashion as East Reef. In November 2001, there were no mussels visible on the rocky substrate (2:42–3:31), but by April 2002, mussels were present and clearly visible as white protrusions on the rocks (3:32–4:00). Mussels completely covered all boulders and cobble of Sheboygan Reef by November 2004 (4:01–4:28).

The last sequence offers a visual compilation of some fauna typically encountered at East Reef besides lake trout and quagga mussels. The sequence provides close-up footage of incurrent siphons of quagga mussels, as well as wider shots of quagga mussel beds (4:39–5:45). *Hydra* are shown attached to the periphery of a rocky overhang (5:48–6:27), as well as nestled among and attached to quagga mussels (6:28–6:50). Slimy sculpin (*Cottus cognatus*) are shown (7:22–8:01), including one that was recently shocked and sampled with the ROV's suction sampler (7:42–8:01). Lastly, both an adult and juvenile burbot (*Lota lota*) are shown at 8:04–8:32 and at 8:33–9:18, respectively.

REFERENCE

Houghton, J. S., R. Paddock, and J. Janssen. 2013. Invasion of quagga mussels (*Dreissena rostriformis bugensis*) to the Mid-Lake Reef Complex in Lake Michigan: A photographic montage. In *Quagga and Zebra Mussels: Biology, Impacts, and Control*. 2nd Edn., T. F. Nalepa and D.W. Schloesser, eds., pp. 65–70. Boca Raton, FL: CRC Press.

Video Clip 3: Close-Up View of Inhalant Siphons of Quagga Mussels (*Dreissena rostriformis bugensis*, Deepwater Morph) on the Midlake Reef Complex in Lake Michigan

Thomas F. Nalepa, Jeffrey S. Houghton, Robert Paddock, and John Janssen

This video clip on the companian CD provides a close-up view of the inhalant siphons of quagga mussels (*Dreissena rostriformis bugensis*, deepwater morph) found on the Midlake Reef Complex (MLRC) in Lake Michigan and is meant to supplement Nalepa et al. (2013). In Nalepa et al. (2013), relative siphon lengths (siphon length/shell length ratio) were determined on mussels observed in the laboratory and estimated from still-frame images of mussels observed in this video clip. The video was obtained with a remotely operated vehicle (ROV; Benthos MiniRover MK-II) in May 2010 at locations between 53 and 57 m on Northeast Reef (located about 43°15.0′ N and 87°35.0′ W), which is a part of the reef complex (see Houghton et al. 2013). The video begins with images of the underside of a rock ledge (0:0–0:23) in which attached *Hydra* can be seen extended downward and moving with the current. It then pans upward to show quagga mussels on the vertical faces of the same ledge (0:44–1:53). As evidenced by unoccupied spaces between individuals, mussel densities at this location were not as great as in other reef locations. While only a few mussels were observed in which the actual length of both the siphon and corresponding shell can be clearly estimated, it is apparent that siphon lengths of all mussels were shorter than siphon lengths seen in later footage. In the next sequence (2:25–4:51), mussels on the horizontal surfaces of another rock outcrop are shown. At this location, mussel densities were high, and individuals tightly occupied all substrate surfaces. Siphons of these mussels were elongated compared to siphons observed in the previous habitat; most siphons were vertically oriented with many faced in different directions. Similarly, mussels on flat bedrock (4:53–5:52) and on the vertical surface of another rock outcrop (6:03–7:37) were densely aggregated, and their siphons were elongated, vertically oriented, and faced in different directions. The suction sampler of the ROV comes into view at 6:11 and 7:33. In the latter sequence, the response of mussels to the approaching sampler can be seen beginning at 7:29 as they rapidly closed their siphons. The two red dots visible in some of the footage, and particularly at 7:14, are parallel lasers situated 10 cm apart. They are typically used as a reference for the ROV operator and can also be used to measure objects during postdive video analysis. Occasional interference (horizontal lines, 7:15) was a result of an electrical shocking event for fish.

While not conclusive, the video supports the hypothesis of Nalepa et al. (2013) that quagga mussels (deepwater morph) in locations where colonies are dense tend to have longer siphons compared to mussels in locations where densities are lower. In dense colonies, there is likely far greater competition for food, and longer siphons provide an advantage by allowing some flexibility in avoiding the feeding currents of other mussels.

REFERENCES

Houghton, J. S., R. Paddock, and J. Janssen. 2013. Invasion of quagga mussels (*Dreissena rostriformis bugensis*) to the Mid-Lake Reef Complex in Lake Michigan: A photographic montage. In *Quagga and Zebra Mussels: Biology, Impacts, and Control*. 2nd Edn., T.F. Nalepa and D.W. Schloesser, eds., pp. 65–70, Boca Raton, FL: CRC Press.

Nalepa, T. F., V. Pavlova, W. H. Wong, J. Janssen, J. S. Houghton, and K. Mabrey. 2013. Morphological variation of the quagga mussel (*Dreissena rostriformis bugensis*), with emphasis on the deepwater phenotype in Lake Michigan. In *Quagga and Zebra Mussels: Biology, Impacts, and Control*. 2nd Edn., T.F. Nalepa and D.W. Schloesser, eds., pp. 315–329. Boca Raton, FL: CRC Press.

Video Clip 4: Visual Documentation of Quagga Mussels (*Dreissena rostriformis bugensis*) at Two Depths in Southeastern Lake Michigan

Russ Miller, Nathan Hawley, and Steven A. Ruberg

This video on the companian CD documents the extent of bottom coverage by quagga mussels (*Dreissena rostriformis bugensis*) at two sites in southeastern Lake Michigan near Muskegon, Michigan, in 2012. The sites were located at about 40 m (43°11.259′ N, 86°25.507′ W) and 110 m (43°13.263′ N, 86°34.9′ W) along a depth-defined transect that is sampled regularly as part of a long-term monitoring program of the lower food web (Fahnenstiel et al. 2010). The video was obtained with cameras mounted on a Model 1000 Outland remotely operated vehicle (ROV) that collected high-definition 1080p video at 30 fps at the 40 m site, and standard-definition video at the 110 m site. Videos at both sites were collected in the vicinity of research moorings that provide physical data (i.e., currents, temperature) as a component to the monitoring program.

In the first sequence, the ROV pans the bottom under natural light at the 40 m site from about 3 m above the bottom. Mussel colonies at this site were dense and areal coverage was about 80%–90%. The research mooring, an acoustic Doppler current profiler, can be seen in the background at 0:51. The substrate at this site is silty sand, and some fine-grained material resuspended by the ROV can be seen at 1:30. Lastly, the ROV rests on the bottom at 1:55, and the extended, vertically oriented inhalant siphons of the mussels can be clearly observed.

In the second sequence beginning at 2:22, the ROV moves just above the bottom at the 110 m site where the substrate is clayey silt. Areal coverage of the mussel population at this site was about 50%–60% and clearly less than at the 40 m site. The ROV stops at the research mooring, and a burbot (*Lota lota*) comes into view at 3:27.

We have begun using videos such as these, along with side-scan sonar imagery, to supplement densities derived from traditional Ponar grab samples to assess changes in mussel populations over the long term in this area of Lake Michigan. Video footage and side-scan sonar provide broad areal coverage and estimates of mussel patchiness, while Ponar grab samples provide actual densities and size–frequency distributions. When towed from the surface, side-scan sonar provides images of mussel beds on the bottom along paths that are 320 m wide. Sonar operating at frequencies of 100 kHz and higher can detect mussels greater than about 1 cm on substrates that are predominantly silt, sand, or gravel (Medwin and Clay 1998). Areal coverage of mussels can be derived from both video and acoustic images using standard image processing techniques. Edge detection and thresholding can be used to estimate areal coverage at specific locations along transects from video imagery, while percentage of coverage can be estimated along the whole transect using side-scan imagery with standard polygon feature digitizing tools contained in graphical information systems (Ruberg, unpublished).

The combination of video, side-scan, and Ponar grab sampling offers the best opportunity to characterize mussel populations. The broad, patchy nature of mussel distributions can be captured by video and side-scan sonar (see distributions at the 110 m site in the video) but not easily estimated with a traditional Ponar grab (area = 0.05 m^2). Yet Ponar grabs are essential for ground truthing the images. For example, while mussels in Ponar samples collected in 2012 have not yet been processed and hence densities are not available, mean densities in samples collected in 2010 at sites near the area shown on the video but at 45 and 93 m were 8,135 and 26,843/m^2, respectively. These values are seemingly inconsistent with the video footage that clearly shows mussels were more abundant at the shallower depth. The reason for this inconsistency is that the deeper depth was dominated by small mussels not seen in the video. The mean proportion of mussels <5 mm was 20.4% at 45 m compared to 89.1% at 93 m (Nalepa, unpublished). However, estimates of biomass, which are inherently based on size structure, were consistent with the video; values at 45 and 93 m were 52.7 and 7.5 g/m^2, respectively (Nalepa et al. 2013). Given the impact of quagga mussels on the foodweb of Lake Michigan (Fahnenstiel et al. 2010), it is essential that all available tools be utilized to characterize mussel populations and document changes over the long term.

REFERENCES

Fahnenstiel, G., T. Nalepa, S. Pohoven, H. Carrick, and D. Scavia. 2010. Lake Michigan lower food web: Long-term observations and *Dreissena* impacts. *J. Great lakes Res.* 36(Suppl. 3): 1–4.

Medwin, H. and C. Clay. 1998. *Fundamentals of Acoustical Oceanography*. New York: Academic Press.

Nalepa, T. F., V. Pavlova, W. H. Wong, J. Janssen, C. S. Houghton, and K. Mabrey. 2013. Morphological variation of the quagga mussel (*Dreissena rostriformis bugensis*), with emphasis on the deepwater phenotype in Lake Michigan. In *Quagga and Zebra Mussels: Biology, Impacts, and Control*. 2nd Edn., T.F. Nalepa and D.W. Schloesser, eds., pp. 315–329. Boca Raton, FL: CRC Press.

Video Clip 5: Zebra Mussel Movements on the Bottom of Lake Michigan

Barry M. Lesht and Nathan Hawley

This movie on the companian CD was created by concatenating a sequence of images made with a modified first-generation, time-lapse digital camera (Kodak DC260) controlled with a timing system of our own design. The camera was located on a bottom-resting tripod in Lake Michigan. Our original purpose for including a time-lapse camera among instruments mounted on the tripod was to document occurrences of sediment transport under stratified conditions in Lake Michigan. Although very little sediment movement occurred during the observation period (July 31–August 26, 2002), the tripod happened to be positioned over preexisting clusters of zebra mussels, and our photographs captured what we believe to be the first long-term in situ observations of zebra mussel movements in the Great Lakes. As such, these observations complement laboratory studies described by Toomey et al. (2002) and descriptions of dreissenid movements given in Chapter 23 by Kobak (2013). Details of the deployment and brief comments on observations follow.

The instrumented tripod, on which we also mounted a current meter, a pressure sensor, two transmissometers, and two temperature sensors, was deployed at 1107 EDT on July 31, 2002, in 37.5 m of water at position 43°12.286′ N, 86°25.620′ W off of Muskegon, Michigan. The bottom substrate at this location was sandy silt. Bottom temperatures were fairly constant throughout the deployment (5°C–10°C) and showed no evidence of the upwelling and downwelling events observed closer to shore. Bottom currents were usually less than 20 cm/s and were oriented along the shore—to the northwest prior to August 6 and to the southeast for the remainder of the deployment. Beam attenuations (which are inversely related to water transparency) were low throughout the deployment.

We mounted the camera on the tripod frame so it was pointing directly downward from a height of 1 m above the bottom. We used an external strobe light (mounted at 0.65 m above the bottom and aimed obliquely across the camera's field of view) keyed to the camera shutter for even illumination. We placed a compass (0.05 m diameter) 0.47 m above the bottom within the field of view to use as a scale reference. The underwater field of view is approximately 0.8 m × 0.6 m. Individual images (available as .jpg files on request to the first author) are separated by 45 min. Dates and times shown in the lower right of the images are Coordinated Universal Time (UTC). The tripod landed on the bottom oriented in such a way that the top of the frame is toward the east.

The photos show a considerable amount of biological activity. The opossum shrimp *Mysis* (small dark elongate shapes) occur frequently, as do benthic fish (slimy sculpin, *Cottus cognatus*). Larger, unidentified fish also appear in several frames. Mussel movement is readily observed as they leave obvious tracks in the sediment; the size (width) of the tracks depends on the size of the individual. Mussel clusters tend to be stable, though occasionally individuals will leave a cluster. For example, an individual leaves a small cluster in the upper right at 02:31 on August 2 and moves to the north (left) and joins a larger central cluster. Also, a larger individual leaves a central cluster at 18:16 on August 15 and moves a considerable distance toward the south (left) before reversing direction and joining a central right cluster. In some instances, small individuals that originally appear to be attached to larger individuals (04:01 on August 19 in the lower right corner) break away and move independently (12:16 on August 2). Although some images show benthic fish immediately adjacent to the mussel clusters (e.g., 19:01 on August 2), it would be difficult to infer that these photos show instances of predation.

Interestingly, much of the movement seen during the observation period involves curved or near circular paths with an apparent (qualitative) tendency toward clockwise movement. Toomey et al. (2002) showed that, under controlled conditions, mussels were negatively phototaxic. The movements seen in our sequence of images, however, occur over fairly short (a few hours) rather than semi-diurnal timescales, and at this water depth, it seems unlikely that directional differences in the external light field would have much influence.

Clearly, mussels observed in this location are not completely sessile creatures. Detailed analysis (yet to be done) of observations such as those shown here may provide further insights into the dynamics of mussel dispersion and colonization. As noted earlier, we are happy to provide the original imagery to colleagues interested in pursuing these questions.

REFERENCES

Kobak, J. 2013. Behavior of juvenile and adult zebra mussels (*Dreissena polymorpha*). In *Quagga and Zebra Mussels: Biology, Impacts, and Control*. 2nd Edn., T.F. Nalepa and D.W. Schloesser, eds., pp. 331–344. Boca Raton, FL: CRC Press.

Toomey, M.B., D. McCabe, and J.E. Marsen. 2002. Factors affecting the movement of adult zebra mussels (*Dreissena polymorpha*). *J. N. Am. Benthol. Soc.* 21:468–475.

Video Clip 6: Behavior of Zebra Mussels Exposed to *Microcystis* Colonies from Natural Seston and Laboratory Cultures

Henry A. Vanderploeg and J. Rudi Strickler

Impacts of grazing by dreissenid mussels or zooplankton on the phytoplankton community are usually determined by traditional feeding experiments in which changes in phytoplankton concentration are measured over time in containers in which mussels or zooplankton are added (Vanderploeg et al. 2001, Dionisio Pires et al. 2005). These observations alone, however, may not provide much insight into feeding mechanisms and may have limited value for the development of general principles. Blooms of the cyanophyte *Microcystis* began to unexpectedly occur in Saginaw Bay, Lake Huron in 1994, which was 3 years after the bay was invaded by zebra mussels (*Dreissena polymorpha*). The blooms were unexpected because phosphorus (P), the usual driver of these blooms, had been decreasing in the bay for many years (Vanderploeg et al. 2001). Traditional feeding experiments with zebra mussels confined in small bottles with Saginaw Bay seston showed that *Microcystis* was removed at a lower rate than other phytoplankton (Lavrentyev et al. 1995); therefore, a link between the blooms and mussels was hypothesized. We had been using a sophisticated video system (Critter Cam; Bundy et al. 1998) to study feeding mechanisms in zooplankton, and when we conducted experiments to examine links between mussel feeding and the blooms of *Microcystis*, we felt further insights could be gained if we videotaped and observed mussel behavior at the same time we did the feeding experiments (Vanderploeg et al. 2001). Methods for making the videotapes were described in Vanderploeg et al. (2001, 2009). Like mussels in companion feeding experiments with *Microcystis* cultures, videotaped mussels received the same overnight acclimation to the suspensions made with these cultures (Vanderploeg et al. 2013). The videotapes were made in an aquarium that received no mixing from external sources; the water motion observed in the video sequences was caused by the filtering activity of the mussels. For point reference, total lengths of the mussels were 14–16 mm.

The five video sequences on the companian CD have only been presented at meetings and conferences in PowerPoint format, which explains why the sequences are so short (2–15 s). Video sequence #1 shows two zebra mussels feeding on natural seston collected during a *Microcystis* bloom in Lake Erie, September 21, 1995 (Vanderploeg et al. 2001). During this 5 s sequence, *Microcystis* colonies flow into the inhalant siphon (siphon on bottom) of the mussel in the foreground and are expelled from the same siphon as pseudofeces, which consisted of *Microcystis* colonies loosely aggregated in mucus. Most of the spherical material in this sequence is *Microcystis*. In this experiment, mussels expelled pseudofeces at a rate of 79 times per hour. Note that in particular loose aggregations, *Microcystis* in the pseudofeces can be readily returned to the water column and continue to grow.

Video sequences #2–5 show results for laboratory strains of *Microcystis* (Wilson et al. 2005) presented at a low concentration (≤4 μg chlorophyll/L; Vanderploeg et al. 2013). Video sequence #2 shows responses of zebra mussels to large, nontoxic colonies (all >53 μm) of the Hudson BD strain of *Microcystis*. In the sequence, mussels with siphons fully extended created strong feeding currents and forcefully expelled large pseudofeces after collecting many colonies. The associated feeding experiments showed that no colonies were assimilated (ingested).Video sequence #3 shows a video sequence from the feeding experiment with the nontoxic Bear AC *Microcystis*, which had an even distribution of small (<53 μm) and large (>53 μm) colonies. The feeding experiment showed moderate feeding on the small colonies with no feeding on large colonies. Again, as with the Hudson BD strain, the Bear AC strain was captured but then rejected; the large, expelled pseudofeces was presumably composed of large colonies of this strain. At first glance, it would be assumed that the large, nontoxic colonies of the Hudson BD and Bear AC strains were rejected because they were too large for ingestion. This indeed is likely the case with Bear AC strain; however, in the case of Hudson BD strain, we sonified the colonies to create a mixture of small and large colonies, and we found that no colonies of any size were ingested (Vanderploeg et al. 2013).

Video sequences #4 and #5 showed there was a very different response to the toxic Gilkey strain of *Microcystis*, which consisted of colonies in both small and large size categories. The feeding experiment showed there was no ingestion of *Microcystis* in either size category. Video observations showed that mussels filtered only intermittently in the presence of the Gilkey L strain. Video sequence #4 shows a mussel weakly expelling pseudofeces with its inhalant siphon only partially extended. Video sequence #5 shows a mussel responding to the same suspension, which was supplemented with *Cryptomonas* at a concentration of 4 μg chlorophyll/L. Note the vigorous rejection response of the mussel as the colony entered the siphon. This mussel was observed to do this many times at the entrance to the siphon. Possibly tentacles lining the siphon or the siphon alone sensed and responded to an irritant associated with the colonies. We noted that in experiments with the Gilkey L and Bear AC strains, the produced pseudofeces were neutrally or positively buoyant. In general, it is thought that pseudofeces sink to the bottom.

Clearly, video observations coupled with the traditional feeding experiments gave valuable insights into the selective-rejection mechanism. Without these observations, we would have been less certain about proposing the mechanism and convincing our colleagues. Seeing is believing.

REFERENCES

Bundy, M. H., T. F. Gross, H. A. Vanderploeg, and J. R. Strickler. 1998. Perception of inert particles by calanoid copepods: Behavioral observations and a numerical model. *J. Plankton Res.* 20:2129–2152.

Dionisio Pires, L. M., B. W. Ibelings, M. Brehm, and E. Van Donk. 2005. Comparing grazing on lake seston by *Dreissena* and *Daphnia*: Lessons for biomanipulation. *Microb. Ecol.* 50:242–252.

Lavrentyev, P. J., W. S. Gardner, J. F. Cavaletto, and J. R. Beaver. 1995. Effects of the zebra mussel (*Dreissena polymorpha* Pallas) on protozoa and phytoplankton from Saginaw Bay, Lake Huron. *J. Great Lakes Res.* 21:545–557.

Vanderploeg, H. A., T. H. Johengen, and J. R. Liebig. 2009. Feedback between zebra mussel selective feeding and algal composition affects mussel condition: Did the regime changer pay a price for its success? *Freshwat. Biol.* 54:47–63.

Vanderploeg, H. A., J. R. Liebig, and W. W. Carmichael et al. 2001. Zebra mussel (*Dreissena polymorpha*) selective filtration promoted toxic *Microcystis* blooms in Saginaw Bay (Lake Huron) and Lake Erie. *Can. J. Fish. Aquat. Sci.* 58:1208–1221.

Vanderploeg, H. A., A. E. Wilson, T. H. Johengen et al. 2013. Role of selective grazing by dreissenid mussels in promoting toxic *Microcystis* blooms and other changes in phytoplankton composition in the Great Lakes. In *Quagga and Zebra Mussels: Biology, Impacts, and Control.* 2nd Edn., T.F. Nalepa and D.W. Schloesser, eds., pp. 509–523. Boca Raton, FL: CRC Press.

Wilson, A. E., O. Sarnelle, B. A. Neilan, T. P. Salmon, M. M. Gehringer, and M. E. Hay. 2005. Genetic variation of the bloom-forming cyanobacterium *Microcystis aeruginosa* within and among lakes: Implications for harmful algal blooms. *Appl. Environ. Microbiol.* 171:6126–6133.

Video Clip 7: Visual Evidence a Native Mussel Population (Unionidae: Bivalvia) in the St. Clair River (Laurentian Great Lakes) Has Survived Despite the Presence of *Dreissena*

Greg Lashbrook* and Kathy Johnson[†]

This video on the companian CD provides visual evidence that native unionid mussels and dreissenids appear to be coexisting in an area in the upper St. Clair River. The unionids were found at a water depth of 2–3 m while we were conducting underwater habitat assessments for a shoreline restoration project near Port Huron, Michigan. The video footage was taken in 2012 and documents six individuals representing five species. Each individual specimen was documented in situ, removed from the substrate, cleaned of attached dreissenids when present, filmed for identification, and replaced in the substrate.

The first unionid shown, *Proptera alata* (pink heel splitter), was deeply embedded in the soft substrate, was covered with a fine layer of sediment, and had attached a cluster of small dreissenids (0:29). The second unionid shown, *Lampsilis siliquoidea* (fat mucket), was found on top of a hard-packed sand and gravel substrate (0:55). This individual had dreissenids covering approximately one-third of its shell. Greater substrate density compared to the previous site was apparent when the unionid was placed back into the substrate. Note the strong current in the river as attached plants wave and debris drift in the background at 0:40–1:00. The third unionid, *Lampsilis cardium* (plain pocketbook), was found on soft substrate less than a meter from the previous individual, indicating the variable nature of the substrate (1:19). This individual was heavily encrusted with dreissenids, which occurred on the top portion of the shell that was exposed above the sediment. The fourth unionid shown (1:39) is *Ligumia recta* (black sand shell), which is a federally-listed species. This individual was found immediately downstream from the previous individuals and was heavily encrusted with dreissenids. The number of dreissenids on the shell would likely impede its movements. The substrate in this area was approximately 6 cm deep and composed of loose, soft sediment. The fifth unionid, *L. siliquoidea* (fat mucket), was found deeply buried in the sediment and was difficult to see because the only visible portion was the inhalant siphon (1:59). The last unionid shown is *Ptychobranchus fasciolaris* (kidney shell), and this individual was found with only a small number of dreissenids attached to the shell (2:29).

Many studies in both North America and Europe have documented the complete loss of unionids soon after dreissenids become established (Ricciardi et al. 1995, Nalepa et al. 1996, Lucy et al. 2013), yet other studies show that unionids and dreissenids can coexist in certain areas (termed "refugia;" Nichols and Wilcox 1997, Zanatta et al. 2002, Crail et al. 2011). Such "refugia" are generally found in shallow water (<2 m) and have bottom substrates that range from firm sand to soft mud. Also, numbers of dreissenids found attached to unionids in these areas are generally lower than the lethal threshold of 100 (Ricciardi et al. 1995). Factors that contribute to the coexistence of unionids and dreissenids in certain areas are poorly understood but could be related to the presence of conditions that are suboptimal for dreissenids yet tolerable for unionids. At any rate, the finding of unionids in this area of the St. Clair River suggests that it is indeed a "refugia," and further studies are needed to document various aspects of unionid populations (density, species composition, number of attached dreissenids, spatial extent of presence) and to characterize associated environmental conditions.

REFERENCES

Crail, T. D., R. A. Krebs, and D. T. Zanatta. 2011. Unionid mussels from nearshore zones of Lake Erie. *J. Great Lakes Res.* 37: 199–202.

Lucy, F. E., L. E. Burlakova, A. Y. Karatayev, S. E. Mastitsky, and D. T. Zanatta. 2013. Zebra mussel impacts on unionids: A synthesis of trends in North America and Europe. In *Quagga and Zebra Mussels: Biology, Impacts, and Control.* 2nd Edn., T.F. Nalepa and D.W. Schloesser, eds., pp. 623–646. Boca Raton, FL: CRC Press.

Nalepa, T. F., D. J. Hartson, G. W. Gostenik, D. L. Fanslow, and G. A. Lang. 1996. Changes in the freshwater community of Lake St. Clair: From Unionidae to *Dreissena polymorpha* in eight years. *J. Great Lakes Res.* 22: 354–369.

Nichols, S. J. and D. A. Wilcox. 1997. Burrowing saves Lake Erie clams. *Nature* 389: 921.

Ricciardi, A., F. G. Whoriskey, and J. B. Rasmussen. 1995. Predicting the intensity and impact of *Dreissena* infestation on native unionid bivalves from *Dreissena* field density. *Can. J. Fish. Aquat. Sci.* 52: 1449–1461.

Zanatta, D. L., G. L. Mackie, J. L. Metcalfe-Smith, and D. A. Woolnough. 2002. A refuge for native freshwater mussels (Bivalvia:Unionidae) from impacts of the exotic zebra mussel (*Dreissena polymorpha*) in Lake St. Clair. *J. Great Lakes Res.* 28: 479–489.

---

* Lead diver and cameraman

[†] Author and editor

Postlude–Synopsis

Thomas F. Nalepa and Don W. Schloesser

As in the First Edition (Nalepa and Schloesser 1993), our strategy in compiling information on dreissenids for this Second Edition was to solicit contributions on specific topics from known experts and to widely announce our intentions to members of the broader scientific community who had experience with dreissenids over the past 20 years. As in any edited book, subject matter is ultimately dependent on those willing to contribute. We are fortunate that such a varied field of experts responded, and this edition contains chapters that address a variety of topics based on both original research and in-depth reviews.

In North America, probably no other freshwater organism has drawn as much attention from such a wide array of researchers, policy-makers, resource managers, and the general public as *Dreissena*. As one indication, the number of published articles on these mussels has increased at a rapid rate since they became established in North America in the late-1980s. Between 1964 and 1993, there was an average of 30 articles published per year on zebra and quagga mussels in the open literature, but between 1994 and 2011 this number increased to 72 per year (Schloesser and Schmuckel 2012). This comparison may be biased by the increase in opportunities for publication in the latter period, but nonetheless, it does indicate continued strong interest in these two species. Simply stated, the reason is that dreissenids affect water resources and ecosystem services and, as such, have caused substantial economic burdens. These burdens are direct, such as those incurred by raw-water users to control biofouling, and also indirect as those incurred by society through diminished recreational opportunities. As noted, we sought to provide a broad spectrum of current and historical information on dreissenid-related topics. While we were successful for the most part, we were not able to include a chapter specifically on economic costs. However, for recent references and broad estimates, we refer readers to Chapters 10 and 14.

As dreissenids continue to spread in North America and, in particular, expand in the west (Chapter 1), both direct and indirect economic costs are sure to increase. To control mussel biofouling, raw-water facilities (municipal water plants, power plants, etc.) in the east and around the Great Lakes region have been retro-fitted and now allocate operating funds on an annual basis for mussel control (e.g., chlorination and mechanical cleaning). Newly-constructed plants have incorporated design features that minimize biofouling or enhance control mechanisms. Facilities in the west face new challenges since many move water over longer distances, or are associated with large dams and reservoirs not common in the east (Chapters 14 and 15).

Disruption of ecosystems and subsequent socioeconomic costs are a function of ecosystem impacts. Impacts of dreissenids on ecosystems and foodwebs are a topic area with most chapters in this book. Several chapters assess these impacts using meta-analysis (Chapters 30 and 38) and provide clear evidence that lakes and river systems invaded by dreissenids have, on average, undergone major physical, chemical, and biological changes. While the magnitude of these changes can be highly variable and a function of dreissenid abundances, it is noteworthy that many of these changes persist even as dreissenid populations stabilize and/or fluctuate around some mean value. Such retrospective analysis has shown that aquatic systems most affected by dreissenids are small, shallower lakes, followed by littoral, nearshore areas of larger lakes, and then by pelagic areas of larger lakes. This paradigm may now have changed since recent evidence suggests that quagga mussel populations in offshore areas of large, deep lakes have increased to such an extent (Chapters 4, 20, and 31) that even pelagic areas of these lakes are now heavily impacted (Chapters 31 and 34). The quagga mussel was only recently reported in North America when our First Edition was published, but this species has since become the dominant dreissenid in the Great Lakes, and it is the dreissenid rapidly spreading in the west (Chapter 1). This species appears particularly well-suited to large lakes and reservoirs as derived from its evolutionary history (Chapters 7 and 26).

This aside, both zebra and quagga mussels impact the structure of ecosystems at the level of individual components (Chapters 40, 41, and 46) and at the level of communities (Chapters 30 and 33). Broad shifts in ecosystem function, as defined by changes in energy and nutrient flow patterns, have been termed benthification (Chapter 36), nearshore nutrient shunt (Chapter 35), and oligotrophication (Chapter 34). While these terms have been previously defined, these cited chapters offer good examples of the mechanisms involved. Most often, economic costs of ecosystem disruption are measured in terms of lost fish production. This book contains chapters on impacts of dreissenids on lake whitefish (Chapter 42), an important commercial species, and on fisheries assessments (Chapter 43).

In retrospect, this might have been better represented with more chapters since links between dreissenids and fish populations are complex, subject to many other variables, and difficult to assess. For a general perspective on fish impacts we refer readers to Johannsson et al. (2000), Strayer et al. (2004), and Riley et al. (2008).

The long-time presence of zebra mussels in Europe, as well as the recent spread of quagga mussels in this geographic region (Chapters 6 and 7), provides an opportunity for researchers in both continents to exchange information and better understand the biology, management, and impacts of these two species. As in our First Edition, European authors have contributed chapters on various topics, including population dynamics and trends (Chapters 8 and 47), behavior (Chapter 21), predation response (Chapter 22), morphology (Chapter 19), and impacts (Chapter 37). A recent book entitled *The Zebra Mussel in Europe* (van der Velde et al. 2010) provides a useful update of dreissenid research in Europe, and supplements many chapters in this book. It also gives some perspective for North American researchers. For example, although two chapters in this Second Edition examine the use of dreissenids as a monitoring tool for contaminants and disease organisms in North America (Chapters 17 and 18), far greater attention is given to this subject in Europe. Also, dreissenids promote blooms of toxic cyanobacteria in North America (Chapter 32), but in Europe dreissenids have been found to graze on cyanobacteria and are promoted as a biocontrol for blooms. Unique to North America is the promotion by dreissenids, of bacteria that produce type-E botulism, with the subsequent movement of this toxin through the foodweb leading to waterfowl mortality (Chapter 45). Also unique in North America is the dramatic loss of the keystone species *Diporeia* (Chapter 46).

In response to all the negative consequences of the dreissenid invasion in North America, the public (Chapter 9), legislators (Chapters 10 and 11), and resource managers (Chapters 12, 13, and 16) are more aware of the general problem of aquatic invasive species, the need to prevent their spread and, most critically, the need to prevent their introduction into North America in the first place.

To be sure, there are nuisance invasive species poised for future introduction into North America, and lessons learned from the dreissenid invasion should not be forgotten. We hope this book serves as a reminder.

REFERENCES

Johannsson, O. E., R. Dermott, and D. M. Graham et al. 2000. Benthic and pelagic secondary production in Lake Erie after the invasion of *Dreissena* spp. with implications for fish production. *J. Great Lakes Res.* 26: 31–54.

Nalepa, T. F. and D. W. Schloesser. 1993. *Zebra Mussels: Biology, Impacts, and Control*. Boca Raton, FL: CRC Press.

Riley, S. C., E. F. Roseman, and S. J. Nichols et al. 2008. Deepwater demersal fish community collapse in Lake Huron. *Trans. Am. Fish. Soc.* 137: 1879–1890.

Schloesser, D. W. and C. Schmuckal. 2012. Bibliography of *Dreissena polymorpha* (zebra mussels) and *Dreissena rostriformis bugensis* (quagga mussels): 1989–2011. *J. Shellfish Res.* 31: 1205–1263.

Strayer, D. L., K. Hattala, and A. Kahnle. 2004. Effects of an invasive bivalve (*Dreissena polymorpha*) on fish populations in the Hudson River estuary. *Can. J. Fish. Aquat. Sci.* 61: 924–941.

van der Velde, G., S. Rajagopal, and A. bij de Vaate. 2010. *The Zebra Mussel in Europe*. Leiden, The Netherlands: Backhuys Publishers.

Index

A

Abiotic factors, mussel distributions
    adaptations, 384–386
    calcium, 388–389
    currents and agitation, 392
    description, 384, 398
    desiccation and freezing, 391–392
    hypoxia, 390–391
    pH, 390
    salinity, 389–390
    substrate requirements, 393–394
    temperature, *see* Temperature, mussel distributions
    turbidity, 392–393
Adaptive value response
    byssal attachment, 352
    crawling and crayfish, 353
    crush resistance, shells, 351, 354
    cues, non-foraging roach, 353
    fitness benefits, 353
    food and oxygen, 353
    mobility and feeding, 353
    mollusks and fish, 354
    mortality and asymptotic size, 351, 354
    resource allocation, 353
Advanced oxidation potential (AOP), 219
Algal composition, nutrient excretion, 449
ANS, *see* Aquatic nuisance species (ANS)
Antifouling coating, 236–237
Antipredator strategy
    adaptive value, responses, 352–354
    description, 346
    invasive species, 345–346
    physico-chemical conditions, 346
    predator-induced mortality, 346–347
    predator-mediated responses, 348–351
    risk assessment, 351–352
AOP, *see* Advanced oxidation potential (AOP)
Aquatic nuisance species (ANS), 129
Aquatic wildlife
    description, 206
    eastern elliptio and lance mussels, 206
    flow and isotope data, 206
    Millbrook Quarry, 206–207
    off-site impacts, 206
    selected treatment option, 206
    TEC, 206
    toxicity data, 207
Asian Clam *(Corbicula fluminea)*, 75
Attachment strength, byssal threads
    formation, 332
    hypoxia, food depletion and toxins, 335
    individual and aggregated mussels, 335–336
    light and gravity, 334, 335
    materials, 334
    measurements, 332
    molluscivores feeding, 336
    newtons (N), 332
    phases, 332
    predators, 336–337
    size class, 332
    temperature, 334
    water movements, 335

B

Behavior, *Dreissena polymorpha*
    abiotic and biotic factors, 332, 333
    aggregate formation
        clusters, 339
        predators, 340
        presence, fish, 340
        stimulation, 339–340
    attachment strength, substrates, *see* Attachment strength, byssal threads
    description, 340–341
    environmental conditions, 332
    horizontal movement, 337–339
    location, 332
    planktonic larvae, 331–332
    valve movements, 340
    vertical movement, 339
    voluntary detachment, 337
Benthic algal response
    areal flux rate, 568
    Cladophora biomass, 567
    Laurentian Great Lakes, 566–567
    nutrient excretion and source, 567, 568
Benthic invertebrates and dreissenids
    bacterial proliferation and toxin production, 706
    botulism epizootics, 710
    C. botulinum, 705–706
    decomposition and anoxia, 706
    DNA extraction kit, 707
    DNA purifications, 707
    environmental roles, 710
    fish and waterfowl mortalities, 709
    macroinvertebrate taxa, 709
    nearshore and offshore sampling sites, 707
    Ponto-Caspian invaders, 706
    Q-PCR, 707
    round gobies, 706
    spore DNA, 707–708
    tissue, 709
    type E botulism, 705
    water temperatures, 710
Benthic macroinvertebrates
    abundance, diversity and composition
        dense colonies, 603
        direction and magnitude, 602
        generalizations, 601
        mussel aggregations, 602
        snails and herbivorous taxa, 602–603
        taxonomic richness, 601
    annual phosphorus load, 531, 539
    annual total densities, 531
    environmental management, 599–600
    filtration and alteration
        colonization, 601
        light penetration, lakes and rivers, 601
        mechanisms, 600
        offshore planktonic, 600
    fitted densities, 531
    Gammarus, 532, 538
    and macroinvertebrate taxa, 531, 536
    native bivalves, 603–604
    oligochaetes, 532
    predictors, dreissena impact, 604–605
    sphaeriid annual densities, 532
    taxonomic richness, 605–606
    temporal variability, 605
Biogeography, dreissenidae
    adaptive radiation, 406
    Akchagylian transgression, Caspian Sea basin, 406
    distributional changes, lineages, 406
    fossil records, 406
    location of sites, Balkans region, 406, 407
    Paratethys Sea, 406
    salinity, glacial advances, 406
Biotic interactions
    diet, 394–395
    longevity, 397
    physiological energetics and responses, 394–396
    starvation, 396–397
Bivalve mollusks, 273–274
Black-striped mussel *Mytilopsis* sp., 199
Boundary Waters Treaty, 187, 188, 194
Buffalo-Syracuse portion, 56

C

California agricultural inspection station, 17
Canadian Governments
    in 1990
        Europeans, 149
        federal agencies, 149
        fisheries coordinator, 147–148
        Inter-Ministerial Coordinating Committee, 148–149
        media reports, 149
        Ministry of Natural Resources, 148
        multiple-user groups, 147
        Ontario Ministry of Environment, 149
        Provincial Zebra Mussel Program, 148
    in 1991
        education program, 155
        initiatives and monitoring results, 155
        invasion-monitoring project, 155–156
        recommendations, 155
        zebra mussels and exotic species, 154
Catch per hour (CPH)
    annual abundance, 689
    daytime, 683
    diel periods, 685
    diel shift, 683
    hypothesis, 683
    linear regression models, 684
    mean annual, 683–684
    nighttime, 684
    "null" model, 683

Chemical regulation, dreissenid reproduction
  algae and gametes
    Chlorella minutissima cultures, 464–465
    pre-release water, 465
    serotonin, 465
    stimulatory effects, spawning, 463, 464
    water temperaure, 464
  description, 461–462
  fertilization and veliger development, 465–466
  primary stimulus, 462
  seasonal reproductive cycle, 467
  serotonin, 462–463
  spawning, eggs and sperm, 462
  species-specific chemical signals, 462
  sperm attractants, 465
Chemical treatment, dreissenid mussels
  advantages and disadvantages, 222
  cationic polymers, 223
  chlorine applications, 222–223
  DBPs, 223
  oxidizing and non-oxidizing chemicals, 222
Chlorine treatment
  bulk tanker storage, 230
  chemical preparation and storage facilities, 231
  CLS, 230, 231
  dosing rates, 230
  gaseous, 230
  injectors, 231
Chronological history, zebra and quagga mussels
  1988, 10, 11
  1989
    Chicago Sanitary and Ship Canal, 12
    fish spawning reefs, 10
    inspection, ozonation system, 12
    Lake Erie near Port Colborne, 12
    locations, new discoveries, 12, 13
    peak mean larval abundance, 10
  1990
    densities, southern and eastern portions, 11
    inland monitoring effort, 12
    Lake Ontario east to Bay of Quinte region, 12
    locations, new discoveries, 12, 13
    navigation buoys, upper St. Lawrence River, 12
    veligers, 11–12
  1991
    expansion, drainage basins, 12
    Illinois River, 13
    locations, new discoveries, 14
    natural lakes, small reservoirs/impoundments and quarries, 14, 15
    small inland lakes, 13
    Susquehanna drainage, 13
  1992, 14, 15
  1993, 14, 16
  1994
    Ausable, Sauble and Trent rivers, 17
    California agricultural inspection station, 17
    electric power stations and waterworks plants, 17
    locations, new discoveries, 17
    natural lakes, small reservoirs/impoundments and quarries, 15, 17
  1995
    Chautauqua Lake, 18
    locations, new discoveries, 18
    from Mississippi River, 18
    in Ontario, 19
    recruitment events, 18
    St. Croix River, 18
    veligers, 18
  1996, 19
  1997, 20
  1998-2001
    Edinboro Lake, 21
    freezing air temperatures, 21
    locations, new discoveries, 20, 21
    Missouri River, 21
    small inland lakes, 21
  2002-2004
    adult mussels, 21–22
    French Creek, 22
    5-hectare quarry, northern Virginia, 22
    locations, new discoveries, 21, 22
    new locations, 22
    small inland lakes, 22
  2005-2006, 22–23
  2007-2010
    Central Arizona Project Canal, 24
    Colorado River system, 24
    Conowingo Dam, 24–25
    country, state/province, water body and year, 25–27
    diver surveys, 23
    genetic evidence, 23
    locations, new discoveries, 23, 24
    mussel-monitoring substrate, Lake Mead, 23
    sampling efforts, 24
  commercial vessels, 27
  description, 10
  federal legislation, 10
  flower shipments, 28
  geographic information system, 10
  industrial water users, 10
  inspection stations, 28
  Lake Mead and Lake Texoma, 28
  mechanism, mussel dispersion, 25
  natural dispersal patterns, 27
  overland movement, recreational boats, 27
  recreational boating and bait buckets, 27
  reproductive strategy of, 28
  upstream locations and disconnected waterways, 25
Coastal Zone Management Act (CZMA), 207
Colorado River system, 24
Competitive negotiation process, 201
Compressed air systems, 253
Congressional Office of Technology Assessment (OTA), 157
Control-valve chambers, 248
Copper ions, 224
Corbicula fluminea, 267–268
Corbicula parvum oocysts, 267, 268
CPH, *see* Catch per hour (CPH)
Cryptosporidiosis, 267
Cyclospora cayetanensis, 268
CZMA, *see* Coastal Zone Management Act (CZMA)

D

Dams, dreissenid mussels
  areas, components and systems, 256, 257
  characteristics, 245
  description, 244
  external features
    adjustable weirs, 246
    fish-handling facilities, 246, 247
    floating structures, 245
    gates and weirs, 245
    isolation gates, 245
    navigation locks, 246–247
    outlet works, 246, 247
    spillways, 246
  internal structures
    conduits, valves and vents, 248
    construction materials, 249
    fire protection, 248
    level gauges, 247
    seepage drains, 248
    sumps, 248
    uplift drains, 248
  pueblo dam aerial view, 244
  resource management, 245
D. carinata, *D. "presbensis"*, and *D. "stankovici"*, 427
Dead zebra mussels, contractor's documentation, 747
Deepwater phenotypes, Lake Michigan
  Cheboksary Reservoir, 317
  cold water temperatures, 317
  deep regions, sediments, 316
  density and biomass
    dreissenids, 320
    length-weight determinations, 320
    southern, 323–325
  depth-related differences, 317
  morphological and genetic characteristics, 316
  physiological differences, 316
  relative length, inhalant siphons, *see* Inhalant siphons
  shell dimensions
    Cheboksary Reservoir, 320
    IDL software program, 319
    measurements, 319
    mussels, 319
  shell height and length
    Lake Mead mussels, 323
    mean SP/SL ratio, deepwater morph quagga mussel, 321, 322
    relationship, 322
  siphon lengths and shell form
    aquariums, 317–318
    images, 318, 319
    location of sampling sites, 317, 318
    26 m site, Cheboksary Reservoir, 319, 320
    ROV, mid-lake reefs, 318, 319

INDEX
765

shallow morphs, 318
vials, 317
study sites, 317
temperatures, substrates and water turbulence, 317
typical shells, 316
3D factorial correspondence analysis, 429, 432
Diporeia spp. and dreissenid mussels
benthic amphipods, 714
bioenergetics models, 714
consumption rate, 714–716
description, 662
ecosystem consequences, 720–721
excretion, egestion and dynamic action rates, 715
food-limitation hypothesis, 662–663
food resources, 713
IDW, 715, 716
Lake Michigan, 718
lake spatial structure, 715–716
lake-wide dreissenid densities, 716
Laurentian Great Lakes, 713–714
mean daily consumption, 718, 719
medium growth scenario, 718
model performance, 719–720
New York Finger Lakes, 663
seasonal energy densities, 716
Dreissenid mussels
aquatic environments and systems, 265, 266
aqueducts and conveyance systems, 217
biological control, 224
blue green algae, 219
breeding canvasback abundance, 655
canvasback, 650
characteristics, 266
chemical costs, 217
chemical treatment, 222–223
construction and coatings materials, 223–224
copper ions, 224
cryptosporidiosis, 267
dams, *see* Dams, dreissenid mussels
description, 217
desiccation, 222
distributions and abundances, diving ducks, 653–654
diving-duck surveys, 648–649
early detection, 221
ecological impacts, 218
ecological keystone species, 265
economic costs, 217
ecosystem, Lake St. Clair, 658
elements and considerations, control plan, 224–225
endosymbionts, 266
environmental requirements
chemical treatments, 220
filtration and metabolic rates, 220
life cycle, natural systems, 219
mussel control approaches, 219
physiological responses and tolerances, 219–220
reservoirs, 220
water temperatures, 220

fixed and moveable infrastructure components, 217
giardiasis, 266–267
and golden mussel, 244
human pathogens and diseases, 266
hydraulic problems, 217
hydroelectric power plants, *see* Hydroelectric power plants
IJC study, *see* International Joint Commission (IJC)
internal components, 257
Kernel density models, 650
Lake St. Clair, 648
microsporidioses, 267
monitoring, 220–221
mussel mitigation programs, *see* Mussel mitigation programs
mussels act, 265–266
nuisance species, 244
nutrients, 648
oxygen deprivation (suffocation), 224
potential cost, 217
quagga mussels, 216–217
recreational and commercial boaters, 648
redhead, 650–651
reservoir drawdown, 222
scaup, 651–652
sentinel models, *see* Sentinel models
spatial distributions, 653
time period, 650
water infrastructure facilities, 221
water resources impacts, 218–219
water treatment plant protection, 225–228
zoonotic pathogens, 266
Dreissenids effects, fisheries
aquatic communities, 688
catchability, 689
CPH, 683–684
day-to-night adjustment factor, 689
diel shift
benthic species, 684–685
yellow perch, 684
dreissenid populations, 690
field collections and study area, 682–683
FWPCA, 682
Lake Erie, 682
nighttime–daytime, 690
phytoplankton, 682
potential ramifications, 690
water clarity, 682
Dreissenids within traditional parameters
climate change, 260
freshwater, 260
invasive species, 260
Lake Mead National Recreation Area, 260–261
management response evolution
DIME program, 262–263
enforcement component, 263
inspection and decontamination component, 263
Interagency Coordination Core Team, 262
mooring contract, 263
NPS, 262
potential options, 262

seasonal rangers, 263
SNPLMA, 262
Quagga Mussel, 261–262
zoning *vs.* management options, 263

E

EA, *see* Environmental assessment (EA)
Ecosystem, dreissenid nutrient cycling
benthic algal response, 566–569
hydrodynamics and lake morphometry, 564–566
management implications, 569–570
nutrient pools and fluxes, 562–564
phytoplankton response, 564
Edinboro Lake, 21
Electric Power Research Institute (EPRI)
environmental and engineering consultants, 147
zebra mussels, 153
Emergency Procurement Solicitation, 200
Endangered Species Act of 1973, 179
Environmental assessment (EA), 202
Environmental compliance and public review
aquatic wildlife, 206–207
CZMA, 207
ESA, 204
FIFRA, 207–208
hydrologic and geochemical setting, 204–206
interagency review and public involvement, 208, 209
Millbrook Quarry zebra mussel eradication, 204, 205
NEPA, 204
state and local environmental requirements, 208
streams, rivers, lakes and surface waters, 207
threatened/Endangered Species, 206
USEPA, 204
VCP, 207
VDACS, 204
VEIA, 204
VEIR, 204
Environmental factors, zebra mussel population expansion
adult densities and sizes, 46, 48
changes, water quality, 46, 49
colonization and growth, 49–50
data analysis, 39
effects of, Lake champlain, 50–51
lake morphometry and spread of
Champlain Canal, 35
Cul-de-sac bays, 35
lake basins, bays and rivers, 34
segment south, Crown Point., 35
water flow and connectivity, 35
location and depth of sites sampled, 35, 37
location of stations, 1994 and 2010, 35, 36
natural and artificial barriers, 34
predators, 34
sampling, *see* Zebra mussel sampling
settled larvae densities, 41, 46–48
temperature and calcium requirements, 34

thermal, chemical and biological patterns
annual mean phytoplankton densities, 41, 43
annual weighted mean zooplankton densities, 41, 42
calcium concentrations, 40
chlorophyll concentration, 40
estimated growing season and calcium concentrations, 39, 40
maximum annual temperatures, 39
phosphorus concentration, 41
phytoplankton densities, 40, 41
secchi-depth transparency, 40
zooplankton densities, 40
upstream and overland expansion, 34
veliger densities, 41, 44–46
water chemistry and plankton sampling, 38–39
EPRI, see Electric Power Research Institute (EPRI)
Eurasian sphaeriids, 75
European Water Framework Directive, 594–595
Evolutionary relationships, Dreissenidae
AIC, 409
Balkans region, location of sites, 407, 409
Bayesian analysis, 409, 421
carinodreissenids, 407
Caspian Sea subspecies, 409
characters, morphological, 404
C. kusceri, mtDNA genetic diversity, 406
COI nucleotide sequence divergence, 421
coolant water discharge, 405
current distribution and first report
Eurasia, 409, 422
North America, 409, 421
dark false mussel *M. leucophaeata,* 404
ESC and PSC, 421
fresh-and brackish-water basins, 404
genus *Congeria,* 405
homologous gene sequences, 409, 416–420
invasive populations, 405
living genera, 404
locations of sites, eurasia, 404, 405
morphological plasticity, habitats, 407
mtDNA genes, 409
Mytilopsis and *Congeria,* 404
Paratethys Sea, 404
penalized likelihood approach, 421
phylogenetic, 409
"r-selected" species, 405
taxa, unique gene sequences, 409–415
traditional taxonomy, 407
valid living taxa, taxonomic authorities, 407, 408
Exotic ecosystem engineers and native burrowing mayflies
Bayes' Theorem, 613
beneath clusters, 619
biotic and abiotic scales, 612
chi-square contingency table, 615
distance-independent spatial distributions, 618
Dreissena, 612–613
habitat-selection experiments, 614–615
habitat types, 617

Hexagenia and *Dreissena* densities, 612, 615–616
laboratory experiments, 614
mussel-dominated habitat, 618
observation experiments, 614
quagga mussels, 612
spatial and temporal variability, 618
spatial association, 613
spatial autocorrelation, 613, 616
Western Lake Erie, 616
zebra mussels, 612

F

False dark mussel *(M. leucophaeata),* 75
Federal Insecticide, Fungicide and Rodenticide Act (FIFRA), 207–208
Federal Water Pollution Control Act (FWPCA), 682
Feeding rate variables, 449
Fertilization and veliger development, 465–466
FIFRA, see Federal Insecticide, Fungicide and Rodenticide Act (FIFRA)
Fire protection
in dams, 248
water, 256
Freshwater ecosystems, dreissenid effects
benthic-littoral pathway, 490–491
containment, 488
densities and filtration capacities, 493
direction, magnitude and variance, 488
ecological processes, 487
energy and nutrients, 488, 489
fish communities, 492–493
food web constituents, 488
Laurentian Great Lakes, 488, 493
methods, 489
nitrogen and phosphorus, 492
patterns, lakes and rivers, 492
pelagic-profundal pathway, 489, 490
physical and chemical attributes, 489–490
phytoplankton and zooplankton biomass, 492
Pontocaspian region, 488
population density, FC, 492
seasonal/temporal cycles, 488
sediment bacteria, periphyton and zoobenthos, 492
small-scale experiments, 488
translocation, species, 487
variation, 491–492
water clarity, 492
Freshwater lakes
anthropogenically-driven factors, 576
benthic stable states epidemic, 578
benthification, 576–577
biological sciences and AGRICOLA, 580
chlorophyll levels, 576
dreissenid mussels, 576
ecological interactions, 581
Lake Erie, 581
macrophytes *vs.* benthic algae, 583
north-temperate freshwater systems, 575
physical *vs.* trophic change, 582–583
pontocaspian region, 576
secchi depth transparency, 580–581

total phosphorus, 576
upside-down ecosystems, 581–582
FWPCA, see Federal Water Pollution Control Act (FWPCA)

G

Gates and weirs, 245
GENECLASS exclusion tests, 429
Generalized linear models (GLMs), 525
Genetic patterns, Eurasia, see Native and invasive populations
Geographical information system (GIS), 220
Giardia duodenalis cysts, 267, 268
Giardiasis, 266–267
GIS, see Geographical information system (GIS)
GLFC, see Great Lakes Fishery Commission (GLFC)
GLMs, see Generalized linear models (GLMs)
Granulated activated carbon (GAC) adsorption, 219
Grazing vulnerability, *microcystis aeruginosa Ankistrodesmus,* 519
feeding experiments, 520
Gull Lake colonies, 519
rejection response, 518
Saginaw Bay seston, 518
TP concentrations, 519
Great Lakes Fishery Commission (GLFC)
binational commission, 179
harmful species, *Gymnocephalus cernuus,* 180
and IJC, 191
productivity, fish stocks, 189
Great Lakes Water Quality Agreement (GLWQA 1987), 279
Great Lakes, zebra mussels
1988, curiosity discovered
beaches and live mussels, 139
biofouling mollusks, 139
Ontario and electric-generating industry, 140
rocks and unionid mussels, 139
"stripped clams", 139
water users and ecology, 139
dispersal-reduction programs, 161
early 1989
funds, resources and agency mandates, 142–143
government agencies and industry newsletters, 143
molluscicides, control biofouling, 144
mussel-control equipment and consulting services, 144
presentations percentage, topic, 143
research and communication needs, 142
workload and effort duplication, 140
workshop participants, 140–142
environmental problems and economic activities, 136
"event sequence", species, 160
government-response programs, 161
IJC and GLFC, 169
1991, implementation
Canadian Governments, 154–156

chlorination, electric-generating
 facilities, 158–159
contaminant cycling, 158
flood of, 159
food limitations, 157–158
presentations, 159–160
research themes, 160
underwater intake pipes, 159
U.S. Governments, 156–157
walleye population, 158
water-supply utilities, 159
invasive species, 136
late 1989
 agricultural pesticides, 146
 beach visitors, 146
 drinking-water-treatment facilities, 144
 economic-cost projection, 147
 electric-generating stations, 145
 environmental and biological
  information, 147
 EPRI and pamphlets, 147
 information and comments, committee
  members, 146
 live organisms, 145
 non-industrial activities, 145
 optimal environment conditions, 144
 population dynamics, 146, 147
 summer secchi-depth measurements, 146
 temporary interruptions, water flow, 145
 unusual events, 146
 veliger abundances, 145
 wide-scale reports, 147
media conferences, 162
Ontario agencies, 163
"policy by tragedy", 168
policy shift, aquatic nuisance species,
 163–164
1990, political responses
 abundances, unionids, 152
 antifouling paints, 154
 aquatic weeds and filamentous
  algae, 151
 binational cooperation, 151
 Canadian governments, 147–149
 cleaning and preventative
  technologies, 152
 control chemicals, 153
 Detroit Edison Company, 152
 habitat characteristics, 152
 industrial and municipal water users,
  152–153
 optimal environmental conditions, 151
 organic contaminants, 154
 presentations, Dutch researchers, 154
 "proof of principal" experiments, 152
 public awareness, veligers, 151
 regional conferences, 153
 U.S. Governments, 149–151
 water-supply industry, 152
presentations, first four North American
 conferences, 136–138
research funding, 164–165
role, science
 changes, Lake St. Clair ecosystem,
  165, 166
 description, 165

fisheries management, 167
model predictions, population
 dynamics, 165
monitoring, 167
structure, Canadian society, 162
unintentional release, veligers, 161
Gunboat curtains, 262

H

HAA, see Haloacetic acids (HAA)
Haloacetic acids (HAA), 223
Heat exchangers, 251–252
Heating, ventilation and air-conditioning
 (HVAC) systems, 254–255
Hoover Dam, 249
Horizontal movement, juvenile and
 adult zebra mussels
 "byssal drifting", 338–339
 calcium content, water, 338
 crushed conspecifics, 336, 338
 detached mussels, 337
 distances, 337–338
 light, detachment, 338
 locomotory activity, 338
 substrate shape and type, 335, 338
Hudson river's bivalve populations
 "body condition", zebra mussels and
  unionids, 73
 conservation biologist, 79
 Dreissena spp., *Corbicula* spp., and
  Limnoperna fortunei, 72
 and ecosystem instability
  aquatic, 78
  decades and long-lasting
   instability, 79
  dramatic changes, 78
  freshwaters, estuaries and coastal
   waters, 78
 exploitation competition, 79
 filtration rate, 73
 invaders, see Invaders, Hudson river's
  bivalve populations
 juvenile unionids, 80
 macrozoobenthos, 72
 native and nonnative, 72
 pea and fingernail clams, 78
 Pearly Mussels (Unionidae), 75–78
 PONAR samples, SCUBA divers, 73
 sphaeriid clams and Rangia cuneata, 73
 study area, freshwater tidal reach, 72
 trophic and engineering roles, 72
 unionid mussels, *Corbicula* and
  Dreissena, 72–73
Human pathogens and diseases, 266
Hydrodynamics
 dreissenids
  benthic environments, 473–474
  boundary layers, 473
  description, 471–472
  flow regimes, 472–473
  gradients, 472
  pelagic environment,
   organisms, 472
  physical ecology, see Physical ecology,
   dreissenids

and lake morphometry
 dreissenid populations, 565
 filtration and nutrient excretion rates,
  564–565
 Lake Michigan, 565
 nearshore plankton community, 566
 phytoplankton populations and nutrient
  recycling, 565
 stratified period, 565
 velocities comparison, 566
Hydroelectric power plants
 areas, components and systems, 256, 257
 components, 249
 description, 249
 discharge area, 251
 floating structures, 249
 Hoover Dam, 249
 intake systems, 249–250
 internal structures
  compressed air systems, 253
  dewatering, 255–256
  domestic water, 255
  drainage, 255
  filters, 253
  fire protection water, 256
  heat exchangers, 251–252
  HVAC systems, 254–255
  level, flow and pressure sensing
   equipment, 255
  piping, 254
  pumps, turbines and generators, 253–254
  raw/service/cooling waters, 251, 252
  strainers, 252–253
  transformer cooling, 254
  valves, 254
 Penstocks, 250, 251
 secondary intakes, raw and domestic
  water, 250
 trashracks and trashrack cleaning
  systems, 251
Hypophthalmichthys molitrix, 260
Hypophthalmichthys nobilis, 260
Hypoxia, 390–391

I

IDW, see Inverse distance weighting (IDW)
IJC, see International Joint Commission (IJC)
Inhalant siphons
 deepwater morphs from 45 m site, 321–322
 mean SP/SL ratio, deepwater morph quagga
  mussel, 321, 322
 SH/SL ratio and shell length, relationship,
  322, 324
 SP/SL ratio and shell length, 321, 323
 in typical lengths, 322, 323
Intake structure protection
 chlorine treatment, 230–231
 final design, 231–232
 materials analysis, 232
 mussel control strategy evolution, 232–233
 MWD, 229
 quagga mussels
  carbonsteel debris rack, 229
  stainless-steel, water intake flare
   end, 229

research, testing, design and
constructing, 233
SNWA, 228–229
system schematic, water intake and pipeline
infrastructure, 230
International Joint Commission (IJC)
awareness and action, 191–192
ballast water treatment, 189, 191
ecosystem/ecotoxicology approach, 190
and GLFC, 189
Great Lakes Water Quality Agreement,
1978, 190–191
industrial wastes, 189
invasive species, 191
"next zebra mussel" invasion prevention,
193–194
physical and biological process, 274
and pollution, 188
regulatory discord
ballast water discharge, 192
government agency and biological
standards, 192
ratification status, 193
re-occurrence, cyanobacterial blooms, 169
sea lamprey, 189
5th Biennial Report, 191
United States and Canada, boundary
waters, 188
Internet age, dreissenid mussels
California, 133
clipping services and "paper" cuttings, 130
Connecticut, 131
Minnesota, 131
Missouri, 131
Montana, 132–133
Nebraska, 131
New Mexico, 132
newspapers and television broadcasting
stations, 131
North Dakota, 131
Oregon, 133
Texas, 132
Utah, 132
Washington, 133
Wyoming, 132
Invaders, Hudson river's bivalve populations
Asian Clam, 75
description, 73
Eurasian sphaeriids, 75
false dark mussel *M. leucophaeata*, 75
quagga mussel, 75
R. cuneata, a brackish-water clam
native, 75
Zebra Mussel, 73–75
Invasion, dreissenid mussels
biological/ecological threat
bioaccumulation, contaminants, 123
explosive growth,
D. polymorpha, 122
issues, 123
Ontario Fisherman magazine, 122
pseudofeces, 123
"cleaner"/"purer", 118
fishing trip, 117
internet age, *see* Internet age, dreissenid
mussels

legislation and politics
Bassmaster piece, 128–129
editorial cartoon, federal funds, 128
media and public attention, 129
Nonindigenous Aquatic Nuisance
Prevention Act, 128
Nonindigenous Species Act, 118
one newspaperman's view, 133–134
perfect storm
1988–1993, 119–122
1994–2011, 129–130
before, 118–119
reeled-in mussels, 118
waterworks and power plants *vs.* zebra
mussels, *see* Waterworks and power
plants *vs.* zebra mussels
Inverse distance weighting (IDW), 715, 716
Isolated roughness flow, 473
Isolation gates, 245

J

Jorka river system, 107, 108

K

Kemmerer/Van Dorn water bottles, 39
Kruskal–Wallis test, 728

L

Lacey Act of 1900, 178–179
Lake Erie
Diporeia, 671–672
relative stability, 671
temporal trends, growth, 671, 672
Lake Mead
basins, 497, 498
bottom substrates, 498
Colorado River, 497
ecosystem, 497
National Recreation Area
arid conditions, 261
Arizona and Nevada in the
Southwestern United States, 260
desert ecosystems and receives annual
rainfall, 260–261
NPS, 260, 261
Pyramiclamys dissecta, 497
Lake Michigan
calanoid copepod nauplii, 549
chlorophyll, 548
description, 545
isothermal period, 546
Leptodiaptomus spp., 548
limnological and biotic parameters, 548
M. diluviana densities, 550
mean biomass, zooplankton, 549
offshore region, 546
PDIs, 550
pelagic food web structure, 551
phytoplankton photosynthesis, 546
quagga mussels, 546
temperature, depth, 546
total phosphorus, 547
TP/Chl ratios, 550–551

water, phytoplankton, 547
water quality parameters, 551
zooplankton, 547
Lake Simcoe, 672
Lake Superior, 672–673
Lake Whitefish (*Coregonus clupeaformis*)
changes, measured parameters, 663–667
climate change, 676
CPUE, 670
density dependence, 675–676
description, 662
Diporeia and Dreissenid mussels, 662–663
dreissenid establishment, 663
effects, dreissenid, 673–674
evidence, resource limitation, 675–676
fishery-dependent evolution, 676
hypotheses/contributing factors, 675
inconsistencies, 670–671
interspecific competition, 676
invasive species, 662, 676–677
Lake Erie, 671–672
Lake Michigan, 663
Lake Nipigon, 673
Lake Simcoe, 672
Lake Superior, 672–673
mean size, 663, 668
sampling methodology, 670
Lampsilis siliquoidea, 627
Larval dispersal
behavioral mechanisms, 475
downstream transport, rivers, 476
marine bivalves, 475
mortality, 476–477
parameters, 476
range expansion, 476
settling, turbulence, 476
spawning, 475–476
velocities and swimming speeds, 477
water column turbulence, 476
Larval settlement
flux, 477
influence, hydrodynamics, 477–478
marine invertebrates, 477
near-bed turbulence, 478
patterns, 477
pediveligers, 477
transport hypothesis, 478
Laurentian Great Lakes, dreissenid mussels
arsenic and cadmium, 278, 279
baseline concentrations, trace elements,
277–278
bivalve mollusks, 273–274
bivalve monitoring, 274
environmental monitoring data, 279
filter feeders, water-quality
monitoring, 274
food chain, 274
GLWQA 1987, 279
IJC 2011, 274
lead and mercury, 278
multiple biomarkers, 274
MWP, *see* Mussel watch program (MWP)
natural *vs.* anthropogenic sources, 279
organic contaminants concentrations,
279, 280
site percentages, 278, 279

INDEX

sites and sample collection
  MWP sites, 275
  trace metals and organic compounds analysis, 275, 276
statistical analysis, 276–277
surveillance monitoring program, 279
trend analysis, trace elements, 278–279, 281
Length–frequency distributions, zebra mussels
  abiotic and biotic factors, 727
  Baltic Sea, 726
  brackish water, 734–735
  Curonian and Szczecin Lagoons, 725
  D. polymorpha, 726
  environmental conditions, 727
  environmental variables, 728
  environmental variables, water bodies, 728
  eutrophic lagoon, 736
  factors, 732
  Finland, 731
  high turbidity, 736
  Kruskal–Wallis test, 728
  Mann–Whitney test, 728
  multimodality, 732
  organic debris, 725
  PUFA, 736
  samples collections, 727
  size structures comparison, water bodies
    northern portion, curonian lagoon, 731–732
    shell lengths, 731
  statistical analysis, 728
  Szczecin lagoon
    curonian Lagoon, 731
    median shell Lengths, 729
    western portion, 730
  Vente, larval settlement, 733
  water bodies, nutrient levels, 737

M

Mann–Whitney test, 728
Masurian Lakeland, *D. polymorpha*
  annual mean densities, 107, 111
  bottom area colonization, 104, 106
  density and biomass, malacofauna, 107
  distributions and population dynamics, 104
  filtration effects, 104
  fish predation, 109
  Great Masurian Lakes region, 104
  improvements, lake water quality, 104, 106
  Jorka river system, 107, 108
  Krutynia River system, 106
  location, water bodies and lakeland regions, 104, 105
  long-term assessments, water quality, 106
  mean annual densities
    Lake Kisajno, 106, 107
    Lake Mikolajskie, 106, 108
    Lake Niegocin, 106, 107
    Lake Śniardwy, 106, 107
  occurrence, 104, 106
  populations, 104
  postveliger settlement, densities, 108
  reduced densities, waterfowl, 109
  seasonal die-off, plants, 109
  sewage treatment plants, 104
  small and shallow Jorka River, 107
  submersed macrophyte taxa, 109
  wind speed/direction and water currents, 108
MBL, *see* Momentum boundary layer (MBL)
McClelland–Kerr Navigation System, 369–370
Meta-analysis approach, dreissenid effects, *see* Freshwater ecosystems, dreissenid effects
Microcystis aeruginosa
  chlorophyll concentration, 512
  colonies, 510
  Cryptomonas, 510
  cryptophyte, 512
  cyanobacterial blooms, 510
  dominance, 510
  filtering intensity, 521
  grazer interactions, 521–522
  grazing vulnerability, 518–520
  Gull Lake, 512
  interactions, phosphorus, 511
  long/short-term experiments
    chlorophyll concentrations, 517
    clearance rates, 515, 519
    microscopic analysis, 515–516
    Saginaw Bay and Lake Erie, 516
  microcystin, 513
  mussel impact, phytoplankton community, 520
  mussels, 510
  natural seston and mussels, 512
  nutrients and phytoplankton growth rate, 521
  polyunsaturated fatty acids, 510
  protocol, natural seston/laboratory cultures, 512
  pseudofeces and potential, 509
  rejection mechanism, 512
  rejection paradigm
    culture collections, 511
    zebra mussels, 510–511
  Saginaw Bay experiments, 512
  seston, Lake Erie and Saginaw Bay, 512
  strains isolate, inland lakes, 513–515
  traditional feeding experiments, 509
  videotaping mussels, 513
Microsatellite analysis, *see* Population genetic methods
Microsporidioses, 267
Midlake Reef Complex (MLRC)
  april 2002, may 2003, 66, 67
  bathymetry, reefs, 65, 66
  beam trawl sampling, 68
  bycatch of, 68, 69
  byssal threads, 68, 69
  cobble and boulder, 68
  colonization, quagga mussel, 65
  distribution and abundance trends, 65
  ecological importance and dynamics, 68, 70
  extended siphons, 67, 69
  hydra, 68
  in Lake Michigan, 749
  lake trout spawning, 66, 70
  lighting, 66
  mussel abundance/biomass, 68
  november 2005, 2006 and 2009, 66, 67
  plexiglass sampling nozzles, 66
  population surveys, 65
  and ROV, 66
  sheboygan reef, 65, 68
  on smooth bedrocks, 67, 68
  water currents, 70
  Wolfvision™ SCB1 XGA, 66
Millbrook quarry zebra mussels eradication
  bathymetry and three-dimensional graphics, 196, 198
  Broad Run, 197, 198
  chlorine, 203
  competitive negotiation process, 201
  copper sulfate ($CuSO_4$), 203–204
  dewater, 203
  diffuser manifold and floating supply line, 208, 209
  The Dive Shop, 196
  emergency procurement solicitation, 200
  environmental compliance and public review, *see* Environmental compliance and public review
  evaluations, 201
  follow-up assessments, 210
  funding, 200
  geochemical and hydrologic investigation, 201
  groundwater, 207
  increasing salinity, 203
  infestation, Virginia, 198
  investigations, 200
  liquid CO_2, 202
  muriate of potash (KCl), 201–202
  negotiation and contractor selection, 202
  "no action", 203
  Occoquan reservoir, Fairfax Water, 197
  pH shift, 203
  potassium concentrations, 210
  regional location, 196, 197
  RFP evaluation panel, 201
  southwest, the northeastern shoreline, 196
  Spectrus CT-1300 (Clamtrol), 202
  storage and delivery tankers, 208, 209
  surrounding roads, rivers and streams, 197, 199
  survey and bathymetric analysis, 196
  treatment methods, 210
  unintentional transport veligers/juvenile mussels, 197–198
  Unionidae, 197
  VCFWRU, 196
  VDGIF, 196
  The Virginia Nonindigenous Aquatic Nuisance Species Act, 199–200
  workgroup, 198–199
MLRC, *see* Midlake Reef Complex (MLRC)
Momentum boundary layer (MBL), 473
Morphological variability, *D. polymorpha* and *D. r. bugensis*
  biology and adaptive capabilities, 288
  environmental factors, 288
  overall dimensions and ratios
    length correlations, 293, 298
    mean shell dimensions, 293, 298

pigmentation patterns, *see* Pigmentation patterns
population groups, geographic region, 288
"profunda", 288
shell outline variation
    H/L and W/L values, 299, 304–305
    PC1, 296–297
    PC analysis, 297, 302
    scatterplot, individual scores, 297, 303
    score means, scatterplots, 297, 303–304
    shell variability, PC analysis, 296, 302
shell pigmentation, *see* Shell pigmentation
taxa and study sites
    country, water body, water depth and substrate, 289, 290
    epilimnetic and hypolimnetic phenotypes, 288
    location, water bodies in Europe, 289
    mean shell dimensions of, 289, 298, 301
    shell variability, PC analysis, 289, 302
traditional and geometric, shell form
    country, water body, water depth and substrate, 289, 290
    neotenic retention, byssus, 289
    pattern types of, 290, 306
    SHAPE software, 290
between water bodies
    characteristics, species, 295
    mean shell dimensions of, 296, 301
    mean width/length and height/length ratios relationship, 295, 300–301
within a water body
    dendrogram, similarity of shell dimensions, 293, 298–299
    mean width/length and height/length ratios relationship, 294, 299
    Rybinsk Reservoir, 294
Multi-barrier control approach
    antifouling coating, 236–237
    chemical treatment plan, 237
    description, 236
    infrequent high doses, 236
    operational actions, 238
    velocity control strategy, 236
    WTP infrastructure, 236
Muriate of potash (KCl)
    concentrations, 202
    floating supply line, 201
    kills mussels, 202
    land disturbance, 202
    treatment, 201
Mussel colonization, East Reef, 749
Mussel mitigation programs
    intake structure protection, *see* Intake structure protection
    multi-barrier control approach, 236–238
    reservoir dam outlet structure, 233–236
    water conveyance system, 238–239
Mussel Watch Program (MWP)
    baseline dataset, 274
    bivalve monitoring, 274
    cellular and molecular biomarkers, 280
    chemical contaminants, 274
    core sites, 275
    performance-based quality assurance process, 275
    trace metals and organic compounds, 275, 276
MWP, *see* Mussel Watch Program (MWP)

N

NANPCA, *see* Nonindigenous Aquatic Nuisance Prevention and Control Act of 1990 (NANPCA)
NANS, *see* Nonindigenous aquatic nuisance species (NANS)
National Environmental Policy Act (NEPA), 203, 204
National Invasive Species Act of 1996 (NISA), 183
National Park Service (NPS)
    adult mussels, 262
    biological resources division, 262
    commitment, managing invasive species, 262
    definition, 261
    detection and monitoring program, 261
    management policies, 261, 262
    monitoring component, 263
    national recreation area managed, 260
    political pressures, 262
    staff
        concessionaires and boating, 261
        discarding, 263
        placed artificial substrates, 261
        targeted high-risk boats, 261
National policy change, zebra mussel
    ANSTF, 181
    catalysts, NANPCA, 180–181
    federal agency programs, 181
    GLFC, 180
    Great Lakes Task Force, 180
    Lake St. Clair in June, 1988, 180
    regional panel concept, 181–182
    states, NANPCA, 181
Native and invasive populations
    3D factorial correspondence analysis, 429, 432
    location of sites, populations, 429, 430
    pairwise tests, divergences, 429, 431
Native bivalves
    dreissena, 603–604
    dreissenid fouling intensity, 604
    littoral mixed substrates, 603
    North American water bodies, 604
    unionid populations, 604
Native Mussel population
    Lampsilis cardium, 759
    Proptera alata, 759
    St. Clair River, 759
Navigation locks, 246–247
NEPA, *see* National Environmental Policy Act (NEPA)
NISA, *see* National Invasive Species Act of 1996 (NISA)
Nonindigenous Aquatic Nuisance Prevention and Control Act of 1990 (NANPCA), 183
Nonindigenous aquatic nuisance species (NANS), 200

North American dreissenids
    genetic diversity
        Lake Mead population, 432
        observed heterozygosities, 429
        putative Eurasian source populations, 429
        Western range expansions, 432
    population genetics of
        AMOVA, genetic variation partitioning, 434, 435, 437
        Bayesian assignment tests, 434, 436, 438–439
        Eurasian source populations, 433
        genetic divergences, pairwise tests, 431, 434
        pairwise and AMOVA tests, 434
    source populations
        3D factorial correspondence analysis, 432–433
        genetic distance tree, 432–435
        population group structuring, 432, 433, 436
    temporal genetic changes
        adaptations, new habitats, 440
        BOTTLENECK analysis, 437
        heterozygosity, Lake Oneida population, 437
        microsatellite data, 436
        populations, St. Lawrence River and Black Sea, 437
        sampling site, Hudson River, 436–437
NPS, *see* National Park Service (NPS)
Nuclear DNA 28S RNA tree, 421, 426
Nutrient cycling, dreissenid mussels
    community filtration rates, 556
    ecosystem impacts, *see* Ecosystem, dreissenid nutrient cycling
    excretion rates, 556–557
    filtration rates, 558
    food quantity and quality, 558–560
    Laurentian Great Lakes, 556
    nitrogen and phosphorus excretion rates, 556
    populations, 556
    regulating factors, 561–562
    size, 560
    temperature effects, 558, 561
Nutrient excretion
    algal composition, 449
    ambient water quality conditions, 450
    analytical procedures, 450
    carbon and phosphorus cycling, 446
    Cladophora growth, 446
    description, 447
    direct and indirect impacts, 457
    dry weights, 449
    experimental design, 447–448
    factors, N:P ratios, 446–447
    feeding rate variables, 449
    filtering rates, 446
    incubation bottles, 448–449
    Microcystis blooms, 457–458
    P-limitation, phytoplankton, 458
    Ponto-Caspian mussels, 446
    sampling and handling, 448

seasonal patterns, *see* Seasonal patterns, nutrient excretion
seston quality and temperature, 449
temperature, food quality and quantity
    Chl assimilation, 455–456
    correlation analysis, 453–454
    C:P and N:P ratios, 454–455
    cyanophytes, 455
    multiple linear regressions, 455
    N-excretion, NH$_4$ excretion, 454
    P-concentration, 455
transplant experiments, 451–453
uptake, cyanobacteria, 446
Nutrient pools and fluxes
    dreissenids, 562
    Lake Erie, 563
    Lake Oneida, 564
    Milwaukee River, 562
    mussel biomass, 563
    sedimented material, 564

O

Oklahoma reservoirs
    McClelland–Kerr Navigation System, 369–370
    Oologah Lake, *see* Oologah Lake
    population dynamics, 370
    Sooner Lake, *see* Sooner Lake
    temperatures, 370
Oologah Lake
    Blue Creek, Redbud Marina and Hawthorn Bluff, 371, 375
    concrete panels, 374–375
    densities, adult mussels, 371
    die-off events, 378–379
    growth estimation, 371
    growth rates, 375, 379–380
    location, sites, 370, 371
    mean veliger densities, 372–374
    mussel populations, 377
    peak densities, settled veligers, 374
    seasonal temperatures, 372
    secchi-disk depth, 370
    vertical plankton, 371
    water temperature, 378
    water transparency, 375, 376
Origin and spread, quagga mussels
    arid and semiarid climatic zones, 95
    comparative size structure
        Kuybyshev Reservoir, 99
        relative size frequency, 99
        relative size structure, 99, 100
        Rybinsk Reservoir, 99
    differentiation between populations, 97
    Dreissenidae and *D.r. bugensis*
        Black Sea basin, 95
        evolutionary origin, 94
        marine environments, 94
        molecular analysis, 94
        Mytilopsis, Congeria and *Dreissena,* 94
    eastern invasion corridor, 97
    genetic diversity, 97
    human activities, 97
    invasion corridors and histories, 94, 95
    liman-relict and Caspian species, 95
    location, corridors, 95, 96
    man-induced, 93
    natural mechanisms, 95
    vs. zebra mussels
        change in relative percentage of, 98
        Dnieper River, 98
        "naturalization" to habitat conditions, 98
        "oligotrophication", 99
        spread of, 97–98
        Volga River system, 98
OTA, *see* Congressional Office of Technology Assessment (OTA)
Oxygen deprivation (suffocation), 224

P

PC, *see* Principal component (PC) analysis
PCR, *see* Polymerase chain reaction (PCR)
PDIs, *see* Phosphorus deficiency indexes (PDIs)
Pea and fingernail clams (Sphaeriidae), 78
Pearly Mussels (Unionidae)
    body condition, *E. complanata,* 77
    mean body condition, 77, 78
    mean densities (+1 SE), 76
    mean riverwide densities, native bivalves, 76
    relationship, mean body condition, 77, 78
    riverwide mean density, juvenile *E. complanata,* 76, 77
    species, 75–76
    trends, mean body condition, 76, 77
Phosphorus deficiency indexes (PDIs), 550
Phylogenetic resolution, *Dreissena* taxa
    DNA trees, 422
    MtDNA COI gene tree, 421, 423
    MtDNA cyt *b* gene tree, 421, 425
    MtDNA 16S RNA tree, 421, 424
    nuclear DNA 28S RNA tree, 421, 426
    subgeneras, 422
    subgenus *Pontodreissena,* 422
Physical ecology, dreissenids
    description, 474–475
    detachment, 480–481
    external fertilization, 475
    larval
        dispersal, 475–477
        settlement, 477–478
    suspension feeding, 478–480
Physiological energetics
    energy allocation, 395
    environmental conditions, 395
    growth and reproduction, 396
    mussel growth, 395–396
    oxygen consumption, 395
    shell length and body mass, 395
    timing, reproduction, 396
Pigmentation patterns
    "annual rings", 293
    colorimetry method, 291
    freshwater mussels, 290
    longitudinal elements and composite pattern types, 292, 294–297
    "microvariations", 291
    soft tissues, 291
    transversal and longitudinal elements, 291–292
    transversal elements and simple pattern types, 292–293
    types sequence, dreissenid shell, 293, 297
Poland, *D. polymorpha,* 110
Polish waters, 103
Polymerase chain reaction (PCR)
    COI primers, 428
    morphological characters, 427
    nuclear 28S rDNA gene, 428
    zebra and quagga mussels, 427
Polyunsaturated fatty acids (PUFA), 736
Pomeranian Lakeland, 110
Ponar grabs, 753
Population genetic methods
    bottlenecks, 429
    faster-evolving nuclear gene regions, 428
    GENECLASS exclusion tests, 429
    genetic diversity and divergence patterns, North America, 428
    mtDNA cyt *b* gene, 428
    native and introduced regions, 428
    Pleistocene glaciations, Europe, 428
Powdered activated carbon (PAC) adsorption, 219
Predation risk assessment
    chemosensory recognition, 352
    concentrations, chemicals, 352
    crush resistance, mussel size, 349, 351–352
    European populations, 352
    low-and high-calcium habitats, 350, 352
    non-foraging roach, 352
    Physella virgata and *Mytilus edulis,* 351
    predator–prey interactions, 351
Predator-induced mortality
    crayfish, 346, 347
    food consumption, ectotherms, 347
    life-cycle stages, 346, 347
    pumpkinseed and freshwater drum, 346
    size selectivity and probability, 347
    spiny-cheek crayfish, 346
Predator-mediated responses
    adaptive value, *see* Adaptive value
    asymptotic size, 350, 351
    attachment strength and byssal threads, 348–349
    distance and crawling speed, 348
    European populations, growth patterns, 349–350
    mollusks, 348
    mortality rates, shells, 349, 350
    non-foraging roach, 349
    roach predation risk, 350–351
    shell strength and length, 349
Principal component (PC) analysis
    correlations with ratios, 297, 298
    Fourier descriptors, 290
    pattern types, dreissenid shell, 297
    source, H/L ratio, 296
Proactive control techniques, 221
Pseudomonas fluorescens, 224
PUFA, *see* Polyunsaturated fatty acids (PUFA)
Putative subspecies, *D. rostriformis*
    Caspian Sea, 426
    COI haplotypes, 426

COI/16S RNA gene sequences, 427
description, 425–426
MtDNA cyt *b* gene tree, 25, 426
MtDNA 16S RNA tree, 424, 426
nuclear DNA 28S RNA tree, 426
population genetic divergence, "profunda", 427

Q

Q-PCR, *see* Quantitative polymerase chain reaction (Q-PCR)
Quagga mussels *(Dreissena rostriformis bugensis)*
    aquatic macrophytes, Havasu and Mohave, 503, 505
    benthic algae and plants, 496
    Boulder and Virgin Basins, 500
    chlorophyll and water clarity, 499–500
    description, 495
    dreissenid mussels, 500
    grazing environment, 502
    in Hudson River, 74, 75
    Lake Havasu, 496
    Lake Mead, *see* Lake Mead
    mean density, 499
    in MLRC, *see* Midlake Reef Complex (MLRC)
    NPS, 261
    potential pathways, 261
    secchi-depth transparency and chlorophyll concentration, 503, 504
    standardized monitoring program, 504–505
    total phosphorus, 497
    T-test, 498
    U.S. states, 261, 262
    water clarity, 498
    watercraft, 261
    zebra mussels, 495
Quantitative polymerase chain reaction (Q-PCR), 707
Quarry treatment, 747

R

Reactive control approaches, 221
Remotely operated vehicle (ROV)
    lake trout spawning, reef complex, 66
    multibeam bathymetry and seabed classification, 68
    mussels observation, 751
    research moorings, 753
    suctioning mussels, 68
Replacement, zebra mussels
    additional evidence, 55
    Buffalo-Syracuse portion, 56
    Buffalo water intake building, 56
    Erie Canal, New York State, 58
    exotic plant and animal species, 59
    locations and substrate types, 56, 57
    locations, 21 sites sampled, 56, 57
    microscopic "veliger" larvae, 55
    multiple sites and times, Erie Canal, 56
    percentage, 58, 59
    physiological/subtle physiochemical factors, 61–62
    Port Colborne, 56
    proportion, 58, 60
    representative length-frequency histograms, 59, 61
    "selective competitive edge", 61
    shallow-warm waters, 61
    study areas, 56
    subsets, 56
    temporal and spatial trends, 59
Representative length–frequency histograms, dreissenids, 56
Request for proposals (RFP), 200
Research and in-depth reviews, dreissenids, 761
Reservoir dam outlet structure and transmission pipeline
    dreissenid mussel control system
        chemical feed systems, 235
        diffuser ring, 235
        dissolved oxygen concentrations, 235–236
        potential impacts, mussel infestation, 235
        protective coatings, 235
    facility, 234–235
    Fry-Ark, 234
    Pueblo dam, 233–234
    SDS, 234
    system components and chemical injection points, 234
    WTP, 234
RFP, *see* Request for proposals (RFP)
Round goby-dreissenid interactions
    approaches, 362
    attachment strength and shell morphology, 361
    brachiopod, dolomite rocks, 362–363
    carina, 361
    chemical cues, 362
    costs and benefits, 360
    covariance, vug area, 363
    densities and zonation, *Mytilus*, 361
    diet, 360
    "flag-stones", 362
    Great Lakes, habitats, 359–360
    Mayville formation, 363
    mussel movement, 363–364
    Mytilus and *Pisaster*, 364–365
    periods, 362
    pharyngeal teeth, 361
    physiological differences, 360
    plasticity, byssal threads, 361
    regional variation, 360
    rock incubations, 364
    size-frequency distributions, 364
    size range, 361
    spread, south to north, 362, 363
    twisting and spinning, 360
    variation, crushing process, 361
ROV, *see* Remotely operated vehicle (ROV)

S

Saint Lawrence Seaway Development Corporation (SLSDC), 193
Seasonal patterns, nutrient excretion
    assimilation rates, 455–456
    calculation, N-excretion rate, 457
    cyanophytes, 456
    description, 450–451
    excretion rates and feeding variables, 451, 452
    filtration rates, 456–457
    potential factors, 456
    Saginaw Bay *vs.* Hatchery Bay, 451
    temperature, 451, 456
Seasonal reproductive cycle, 467
Seepage drains, 248
Sentinel models
    biomonitoring, 270
    coastal bay area, 269
    Cryptosporidium, 269, 271
    development, 267–268
    FISH and immuofluorescent antibody techniques, 270
    Giardia, 269
    Lake Ontario/St. Lawrence River, 269, 270
    navigational-aid platforms, 270
    navigational buoys (navaids), 268
    Shannon River system, 270
    St. Lawrence River/Great Lakes system, 268
    waterborne pathogens, 269, 270
    waterways, 270
Serotonin
    algal effect, 463
    and chemicals, 463, 464
    concentrations, 463
    injection, 462–463
    oocytes, 463
    seasonal reproductive cycles, 467
    stimulation, spawning, 462
    temperature and food availability, 462
    treatment, fertilization, 465
Shannon River system, 270
SHAPE software, 290
Shell pigmentation
    diversity, pattern types, 304
    geographic variability, 304–305
    and latitude, relationship, 299, 306
    mean percent frequency, primary pattern types, 301, 308–309
    ontogenetic sequences, 303
    ontogenetic scheme, pattern types, 304, 310
    pattern frequency changed with age, 301, 308–309
    pattern types, *D. polymorpha* shells, 299, 306–307
    primary and secondary pattern sequences, 304, 311
    range of, 299, 305
    total frequency, pattern types, 300
Shipwrecks and dreissenid mussels
    centerboard winch and anchor, 741, 742
    corrosion rates, 742
    design and construction, 741
    feeders and particulates, water, 742
    invasive species, 743
    marine and freshwaters, 741
    schooner *M.F. Merrick*
        exterior surface, 742, 743
        hull interior, 742, 743

steamer Florida
  exterior, 741, 743
  interior, 741, 743
  thunder bay national Marine sanctuary, 741, 742
  water clarity, 743
  wood-constructed vessels, 742
  wreck, wooden schooner, 741, 742
SLSDC, *see* Saint Lawrence Seaway Development Corporation (SLSDC)
Sooner Lake
  boating/fishing activities, 371
  body weight, reproduction, 380–381
  concrete panels, mean densities, 376, 378
  die-off events, 380
  growth experiments, 372, 376
  growth rates, 377, 379, 380
  location, sampling sites, 371–372
  metabolic rates, 380
  methods, 372
  peak veliger densities, 376, 377, 380
  power plant, 371
  statistical evaluation, growth rates, 372
  water quality parameters, 375–376
  water temperatures, 375, 380
Spectrus CT-1300 (Clamtrol), 202
Sperm attractants, 465
Sphaeriidae, *see* Pea and fingernail clams (Sphaeriidae)
Spillways, 246
Spread of quagga mussels, Western Europe
  advantages, physiological characteristics, 89
  Alps and Pyrenees mountains, 89
  average percentage of, 88
  biomass, 88, 89
  canal–river network, 88
  central, southern and western pathways, 88
  Conrad's false mussel, 84
  description, 84
  dreissenid populations, 89
  filtration activities, zebra mussels, 88
  IJsselmeer and Markermeer Lakes, 88
  lower densities, 89–90
  range expansion
    dispersal pathways and vectors, 86
    Hollandsch Diep, freshwater section, 86
    The Netherlands, 86
    occurrence, 86–87
    populations, Main and Rhine Rivers, 86
    zebra mussel, 84–86
  relative abundance of, 88
  source and vector, 87–88
State and Tribal Wildlife Grant (SWG), 200
St. Lawrence River/Great Lakes system, 268
Strains isolation, inland lakes
  algal cultures, 515–516
  colonial *Microcystis* characteristics, 513, 514
  feeding experiments, 514
  Hudson BD strain, 515
  mussel clearance rates, 513
Subspecies, *D. polymorpha*
  COI haplotypes, 424, 425
  Dreissena taxa, unique gene sequences, 410–415, 424, 425
  homologous gene sequences, 416–420, 424, 425
  MtDNA cyt *b* gene tree, 425
  putative "*gallandi*", 422, 424
Substrate requirements
  byssal attachment strength, 394
  density and biomass tends, 393
  Lake Erie, 393
  lentic habitats, settlement rates, 394
  macrophytes, 393
  pediveligers, 394
  shaded areas, 394
  variety, surfaces, 393
Suspension feeding
  algal ingestion rate, 480
  benthic-pelagic coupling, 479
  bivalves, 478
  denitrification, 480
  food particles, 478–479
  inhibition, 478
  particle removal, 478
  physical ecology, Lake Michigan, 479–480
  phytoplankton and zooplankton species, 479
  "pinching", siphons, 478
  saturation, 478
  stirred-reactor models, 479
  transport, algae, 479
Suwalskie Lakeland
  historic studies, 109
  Lake Wigry, 110
  streams and rivers, *D. polymorpha*, 110
  water bodies and defined lakeland regions, 105, 109
SWG, *see* State and Tribal Wildlife Grant (SWG)

T

Temperature, mussel distributions
  acclimation, Lake Ontario, 387
  byssal thread production, 388
  Chernobyl Power Station, 387
  determination, dreissenid, 386
  development, fertilized eggs, 388
  filtration rates, 388
  fluctuations, 387, 388
  Niagara River, 386
  oxygen consumption rates, 387, 388
  spawning and larval development, 387–388
  starvation, 396–397
  Winfield City Lake, 386–387
  winter months, 386
Texoma Lake, 370
"The Little Dreissenid That Did"
  laws and legislation, 178
  legislative/policy history terminology, 178
  U.S. National policy, 178–179
  NANPCA and NISA, 183–184
  national invasive species policy
    Clinton executive order, 182–183
    federal legislative progress, 183
    NISC, 182
    nonindigenous aquatic nuisance prevention, 182
  U.S. Department of Agriculture, 182–183
  natural aquatic ecosystems, 178
  regulations *vs.* species introductions, 183
  zebra mussel, catalyst for national policy change, 180–182
THM, *see* Trihalomethanes (THM)
Threatened/endangered species, 206
Total phosphorus (TP)
  ANCOVAs, 579
  annual secchi depth, 579
  Dreissena, 578
  post-*dreissena* periods, 579, 580
  water bodies, period, 579, 580
TP, *see* Total phosphorus (TP)
Transplant experiments, nutrient excretion, 451–453
Trashracks and trashrack cleaning systems, 251
Trihalomethanes (THM), 223
Tunnels, 248
Type-E botulism, dreissenids, 762

U

Ultraviolet/hydrogen peroxide (UV/H$_2$O$_2$) oxidation, 219
Unionidae, *see* Pearly Mussels (Unionidae)
Unionids, zebra mussel
  annual biomass, 640
  Anodonta, 637
  anthropogenic drivers, 624
  Belarus, 625
  colonization, 637
  competition and infestation, 627
  densities, 633
  Dreissena weight, 641
  Europe and North America, water bodies, 641
  generalizations, 642
  infestation and mass mortality, 634
  Ireland, 636
  Lake Naroch, 634
  Lake st. Clair, United states/canada, 624–625
  Lampsilis siliquoidea, 627
  Ligumia nasuta, 632
  Lough Key, 625–626
  Muscamoot Bay, 630
  North American and European data, 638
  open-water sites, 628–629
  protocols, 626
  and quagga mussels, 624
  regression line, 638
  relative abundances, 632
  shell deformities, 624
  shell surface, 635
  "stake and rope" method, 630
  statistical models, 626–627
  St. Clair delta, 627–628
  t-tests, 629–630
  veliger densities, 634
  water bodies, 624, 634
  weight and shell length, 639–640

United States Environmental Protection
    Agency (U.S.EPA), 192
USDA, see U.S. Department of Agriculture
    (USDA)
U.S. Department of Agriculture (USDA), 200
U.S. Fish and Wildlife Service (USFWS), 200
USFWS, see U.S. Fish and Wildlife Service
    (USFWS)
U.S. Governments
    in 1990
        environment and public works, 149
        ideas and research themes, 150
        individual workshop participants, 150
        interagency coordinated research
            program, 150
        provisions, NANPCA, 150
        research scientists and natural resource
            managers, 151
    in 1991
        Coast Guard, 156
        control strategies and laboratory
            studies, 157
        Great Lakes Panel, aquatic nuisance
            species, 157
        OTA, 157
        zebra-mussel research, 156–157
    description, 168
U.S. National policy
    1955 Convention, Great Lakes Fisheries, 179
    Endangered Species Act of 1973, 179
    federal and state policies, aquatic
        ecosystems, 178
    independent events and broader-scale
        impacts, 179
    invasive species, 178
    Lacey Act of 1900, 178–179
    orders, executive branch, 179

V

VCP, see Virginia Coastal Resources
    Management Program (VCP)
VDGIF, see Virginia Department of Game and
    Inland Fisheries (VDGIF)
Virginia Coastal Resources Management
    Program (VCP), 207
Virginia Department of Game and Inland
    Fisheries (VDGIF), 747
The Virginia Nonindigenous Aquatic Nuisance
    Species Act, 199–200
Voluntary detachment, 337

W

Wake interference flow, 473
Water chemistry and plankton sampling
    Kemmerer/Van Dorn water bottles, 39
    microscope and Sedgewick-Rafter cells, 39
    secchi-disk depth, 38
    visual transparency, 38
    zooplankton samples, 38
Water conveyance system
    chloramination, 238
    chloramines, 238
    copper ion, 238–239
    environmental conditions, 238, 239
    permanganate treatment, 239
    potential chemical treatments, slime and
        mussels, 238, 239
    protection, 239
Water infrastructure facilities, 221
Water intake structure
    chemical treatment, 227
    components, 226–227
    conceptual toolbox, 226, 227
    gaseous-source chemical feeds, 227
    liquid-source chemicals, 227
    local resource management agencies, 227
    mussel infestation, 226
    mussel-infested pipes, 227
    physical control measurement, 227
Water resources impacts, dreissenid mussels
    boat cleaning training course, 219
    protect non-infested source waters, 219
    surface types, 219
    TOC and DBP, 218
Water treatment plant (WTP) protection
    application feasibility, 226
    comparative analysis, 228
    conceptual toolbox, 226
    cost, 225
    ease of operation, 225
    effectiveness, control measures, 225
    environmental impacts, 225
    health and safety concerns, 226
    mussel monitoring and control plan, 228
    operational flexibility, 225
    record of performance, 225
    regulatory requirements, 225
    source water, 226
    water facility managers, 225
    water intake structure, 226–227
    water treatment plant, 227
Waterworks and power plants vs. zebra mussels
    Bass Islands region, 125
    editorial cartoons, 124, 126
    Electric Power Research Institute, 124
    front page, Monroe Evening News, 124
    graphic issues, 125
    PBS piece, 125
    "pigger", remove mussels, 124, 125
    stemming, 125–128
WHIP, see Wildlife Habitat Incentive
    Program (WHIP)
Wildlife Habitat Incentive Program (WHIP), 200
Wolfvision™ SCB1 XGA, 66

Z

Zebra mussel (Dreissena polymorpha)
    annual abundance trends, 528
    annual mean densities, 528, 530
    behavior, 757
    benthic macroinvertebrate densities,
        526–527
    benthic malacofauna, 110
    cyanophytes and chlorophytes, 525
    diatom annual densities, 525
    distribution, Poland, 111
    Dreissena densities, 527
    ecosystem, Saginaw Bay, 526
    foodwebs, freshwater ecosystems,
        535–536
    Gammarus, 525–526
    GLMs, 525
    Great Lakes, 526
    Hudson BD strain, 757
    in Hudson River
        large filtration rates, 73
        mean body condition, 73, 74
        mean body mass of, 73, 74
        percentage, unionids, 73, 74
        riverwide mean filtration rate, 73, 74
        riverwide mean population
            densities, 73, 74
    invasive species, 110
    Lake Erie, 757
    littoral regions, Masurian Lakes, 111
    movements, Lake Michigan
        biological activity, 755
        instrumented tripod, 755
        observation period, 755
    phytoplankton and zooplankton, 538
    phytoplankton densities, 526
    and quagga mussels
        colder water temperatures, 697–698
        competition, 698–699
        freshwaters, Northern
            Hemisphere, 695
        gonad and oocyte development, 697
        lag-time, 698
        larval stage, 697
        life-history characteristics, 696–697
        oxygen concentrations, 667
        population biology and
            dynamics, 698
        similarities and contrast, 695–696
        temperature tolerance, 696–697
        vectors, spread, 699–700
    Saginaw Bay, 536
    SCUBA, 537
    siphon, 757
    traditional feeding experiments, 758
    variability
        abundance and distribution range,
            590–591
        ADR, 589
        biomass, 592
        biopollution assessment method,
            589–590
        BPL, 594
        Chelicorophium curvispinum, 595
        community impacts, 591
        Corbicula fluminea, 595
        ecosystem engineers, 588
        ecosystem impacts, 593
        environmental impacts, 588
        European Water Framework Directive,
            594–595
        factorial analysis of variance, 589–590
        habitat impacts, 592
        Ireland, 587

INDEX

performing evaluations, 589
Shannon River, 588
zooplankton densities, 526
Zebra mussel sampling
  field methods
    adult, 38
    location of stations, 1994 and 2010, 35, 36
    settled juveniles, 36
    settling plate array, 36, 38
    surface water temperature and secchi-depth transparency, 36
    veligers, 36
  laboratory methods
    analytical procedures, 38
    quadrat samples, 38
    settling plates, 38
Zoning *vs.* management options, 263
Zooplankton
  annual differences, 529
  cladocerans and cyclopoid copepods, 529–530
  generalized linear models, 530–531, 535
  mean densities, 529, 534
  rotifers, 529
  Saginaw bay, 530, 535

Alpena Co. Library
211 N. First Ave.
Alpena, MI 49707